HANDBOOK OF PLASTIC MATERIALS AND TECHNOLOGY

HANDBOOK OF PLASTIC MATERIALS AND TECHNOLOGY

Edited by

Irvin I. Rubin
Robinson Plastics Corporation

A WILEY-INTERSCIENCE PUBLICATION
John Wiley & Sons, Inc.
NEW YORK / CHICHESTER / BRISBANE / TORONTO / SINGAPORE

Library of Congress Cataloging in Publication Data:

Handbook of plastic materials and technology / edited by Irvin
    I. Rubin.
        p.   cm.
    Includes bibliographical references.
    ISBN 0-471-09634-2
    1. Plastics—Handbooks, manuals, etc.   I. Rubin, Irvin I., 1919-

    TP1130.H35   1990
    668.4—dc20   89-48281
                                                            CIP

Printed in the United States of America

10 9 8 7 6 5 4 3

TP 1130
H 35
1990
CHEM

EDITORIAL BOARD

Irvin I. Rubin, Editor
Consultant, businessman, lecturer, Chief Executive Officer of Robinson Plastics Corporation and Irvin I. Rubin Plastics Corporation and co-founder of RLR, Inc., thermoformers. Perhaps most widely recognized for his seminars and as the author of *Injection Molding, Theory and Practice,* the definitive book on the subject. Very active in SPE and a Fellow of the Society, he has authored many articles and chapters in several books on plastics.

Glenn L. Beall, Member of Executive Committee
Designer, consultant, expert witness, lecturer, author, Chief Executive Officer of Glenn Beall/Engineering, Inc. Past President of the Chicago Section of SPE and Past Chairman of the Mold Making & Mold Design Division & Medical Products Divisions of SPE, and is a Fellow of the Society. Past Chairman of the Midwest Section of SPI. Fellow of the British Plastics and Rubber Institute. Member of American Mold Builders Association, the Plastics Academy, the Association of Rotational Molders, the Plastic Pioneers, and is a Trustee of the National Plastics Center and Museum. He holds 35 patents and is Design Editor for *Plastics Design Forum* magazine.

Leonard Berringer, Member of Executive Committee
Editor, author, lecturer. Director of THE Plastics Seminars, best known as Consulting Editor of *Plastics Design Forum, Plastics Machinery & Equipment* and *Plastics Compounding* and former Editor-in-Chief of *Plastics World,* holder of a Jesse Neal Award, given for editorial excellence.

v

CONTRIBUTORS

Brenda A. Bartges, Borg Warner Chemicals Inc., Washington, West Virginia

Ronald R. Bauer, NYCO, Willsboro, New York

Glenn Beall, Consultant, Gurnee, Illinois

Robert J. Beam, PE Technical Service and Development, The Dow Chemical Co., Freeport, Texas

Sidney E. Berger, Consultant, Warwick, Rhode Island

Douglas V. Bibee, PE Technical Service and Development, The Dow Chemical Company, Freeport, Texas

Robert B. Blanchard, Dow Chemical USA, Midland, Michigan

Mort Blumenfeld, Consultant, San Diego, California

P.J. Boeke, Plastics Technical Center, Bartlesville, Oklahoma

William E. Brown, Food Packaging Consultant, Midland, Michigan

Vince Brytus, Ciba-Geigy Co., Hawthorne, New York

Len Buchoff, PCK Elastomerics Inc., Hatboro, Pennsylvania

Michael E. Buck, Textron Specialty Materials, Lowell, Massachusetts

A.E. Campi, Brookpark-Rayalon Inc., Lake City, Pennsylvania

Stuart Caren, Product Director, Unisys CAD/CAM Inc., Boulder, Colorado

Robert F. Cerevenka, Philips Plastics Corp., Phillips, Wisconsin

A.M. Chatterjee, Shell Development, Houston, Texas

James T. Christensen, Crucible Steel, Syracuse, New York

A.W. Coaker, North Olmsted, Ohio

David S. Cordova, Allied Signal Incorporated, Petersburg, Virginia

Paul O. Damm, BASF, Bridgeport, New Jersey

Charles W. Deeley†, Cyro Industries, Orange, Connecticut

P.T. DeLassus, Dow Chemical USA, Midland, Michigan

Barry L. Dickinson, Amoco Performance Products, Inc., Bound Brook, New Jersey

Philip T. Dodge, USI, Quantum Chemicals Div., Rolling Meadows, Illinois

D. Scott Donnelly, Allied-Signal Incorporated, Petersburg, Virginia

Robert E. Duncan, United States Gypsum Division, Chicago, Illinois

Malcom Fenton, Vice President and General Manager, Suzarite Mica Products, Subs Corona Corp., Hunt Valley, Maryland

George Feth, General Electric Co., Selkirk, New York

Harold J. Frey, DuPont, Wilmington, Delaware

K. Furuya, Mitsui & Co., USA, New York, New York

John Gannon, Ciba-Geigy Co., Ardsley, New York

J.W. Gardner, Phillips Chemical Co., Bartlesville, Oklahoma

Michael F. Gardner, Rogers Corp., Manchester, Connecticut

David W. Garrett, Owens-Corning Fiberglas Corp., Granville, Ohio

Saul Gobstein, Sa-Go Associates, Inc., Shaker Heights, Ohio

W. Brandt Goldsworthy, Goldsworthy Engineering, Torrance, California

Leon Grant, TAPPA, Inc., Bow (Concord), New Hampshire

E.A. Greene, Consultant, Hawthorne, California

Rodney J. Groleau, R. J. Groleau Associates, Traverse City, Michigan

Sid Gross, Chief Editor Emeritus, Modern Plastics Magazine, McGraw-Hill, New York, New York

Thomas W. Haas, Director, Cooperative Graduate Engineering, Virginia Commonwealth University, Richmond, Virginia

William J. Hall, Monsanto Company, Springfield, Massachusetts

Stephen Ham, Cashiers Structural Foam Division, Consolidated Metco Inc., Cashiers, North Carolina

Russell D. Hanna, Himont USA, Wilmington, Delaware

Mark Harrington, Cyro Industries, Orange, Connecticut

James E. Harris, Research and Development Specialty Polymers and Composite Division, Union Carbide, Bound Brook, New Jersey

David L. Hartsock, Phillips 66 Co., Bartlesville, Oklahoma

Geoffrey Holden, Shell Development Co., Houston, Texas

John Hull, Vice Chairman, Hull Corporation, Hatboro, Pennsylvania

Denes B. Hunkar, President, Hunkar Laboratories, Inc., Cincinnati, Ohio

Harry S. Katz, Utility Development Corp., Livingston, New Jersey

J.F. Keegan, Occidental Chemical Corp., North Tonowanda, New York

Paul D. Kelley, GE Plastics, Parkersburg, West Virginia

Paul E. Kummer, Cyprus Thompson-Weinman & Co., Montclair, New Jersey

Christopher Irwin, Product Manager, Johnson Controls Inc., Manchester, Michigan

Thomas E. Jacques, Hysol Electronics Chemicals Div., The Dexter Corp., Olean, New York

John A. Jones, Senior Engineer, Borg-Warner Chemicals, Washington, West Virginia

Paul Langston, Textile Fiber Division, DuPont, Wilmington, Delaware

Sidney Levy, Consultant, La Verne, California

William Lewi†, Valiant Plastics, St. Albans, New York

Ruskin Longworth, E. I. du Pont de Nemours & Co., Inc., Wilmington, Delaware

George M. Loucas, Product Engineers, TRW, Mountainside, New Jersey

Sylvio Mainolfi, Branson Ultrasonic Corp., Watertown, Connecticut

Ronald D. Mathis, Phillips 66 Co., Bartlesville, Oklahoma

Paul Matthies, BASF Corp., Wyandotte, Michigan

William K. McConnell, Jr., President, McConnell Co., Fort Worth, Texas

John A. McElman†, Avco Specialty Material, Textron, Lowell, Massachusetts

F.E. McFarlane, Eastman Chemicals Div., Eastman Kodak Co., Kingsport, Tennessee

J. Michael McKinney, PDI, Business Unit of ICI Americas, Edison, New Jersey

Donal McNally, Hoechst Celanese Engineering Resins, Chatham, New Jersey

Virginia B. Messick, The Dow Chemical Company, Freeport, Texas

Raymond W. Meyer, R. W. Meyer & Associates, Tallmadge, Ohio

Melvin A. Mittnick, Textron Specialty Materials, Lowell, Massachusetts

Frank Molesky, Market Division Manager, Salem Industries, Norcross, Georgia

E.L. Moon, Jem Walter Research Corp., St. Petersburg, Florida

James R. Moreland†, Malvern Minerals Co., Hot Springs, Arkansas

Charles T. Moses, Amoco Performance Products Inc., Parma Technical Center, Parma, Ohio

H. Ohi, TPX Export Group, Mitsui Petrochemical Industries, New York, New York

Sigmund Orchon, Consultant, Culver City, California

Tammy J. Pate, PE Technical Service and Development, The Dow Chemical Co., Freeport, Texas

Donald C. Paulson, President, Paulson Seminar Programs, Inc., Southington, Connecticut

C.T. Pillichody, GE Plastics, Parkersburg, West Virginia

Paul I. Prescott, Freeport Kaolin Co., Gordon, Georgia

Leonard Poverno, Grumman Aerospace Corp., Calverton, New York

Joseph A. Radosta, Pfizer Inc., Easton, Pennsylvania

Dewey Rainville, Universal Dynamics, Inc., North Plainfield, New Jersey

Jeff Rainville, Universal Dynamics Inc., Woodbridge, Virginia

Fidel Ramos, Consultant, Newton Highlands, Massachusetts

Christiaan Rauwendal, Manager, Process Research Corp., R&D Raychem, Menlo Park, California

F. Jack Reithel, Engineering Thermoplastics Dept., Dow Chemical USA, Midland, Michigan

Michael C. Restaino, PE Technical Service and Development, The Dow Chemical Co., Freeport, Texas

N.R. Reyburn, Consultant, Erie, Pennsylvania

Lloyd M. Robeson, Principal Research Associate, Air Products & Chemicals, Inc., Allentown, Pennsylvania

Robert C. Rock, Vice President R&D, Plast-O-Meric Inc., Waukegan, Wisconsin

Jordan I. Rotheiser, Jordan Rotheiser Design, Highland Park, Illinois

Irvin I. Rubin, President, Robinson Plastics Corp., Brooklyn, New York

Bruce W. Sands, The PQ Corporation, Lafayette Hill, Pennsylvania

Robert E. Saxtan, NGK Metals Corp., Reading, Pennsylvania

I. Wiliam Serfaty, Ultem Products Section, General Electric Plastics Group, Pittsfield, Massachusetts

Previn L. Shah, The Polymer Corp., Reading, Pennsylvania

Gordon Sharp, Eastman Chemical Products, Kingsport, Tennessee

A.M. Shibley, Consultant, New York, New York

Gerald D. Shook, Consultant, New Albany, Ohio

S.M. Sinker, Plastics Research and Development Ctr., Hoechst Celanese Engineering Plastics Division, Summit, New Jersey

Carleton A. Sperati, DuPont, Parkersburg, West Virginia and Ohio University, Athens, Ohio

F. Melvin Sweeney, Lancaster, Pennsylvania

Gregory M. Swisher, Phillips 66 Co., Bartlesville, Oklahoma

Michael Tapper, Pfizer, Inc., Easton, Pennsylvania

S.C. Temin, Consultant, Walpole, Massachusetts

James L. Throne, Sherwood Technologies, Inc., Naperville, Illinois

W.L. Treptow, Dow Chemical USA, Midland, Michigan

Ivor H. Updegraff, Stamford, Connecticut

Satish K. Wason, Chemical Division, J. M. Huber Corp., Havre de Grace, Maryland

Richard Westdyk, Hoechst Celanese Engineering Resins, Chatham, New Jersey

Mary N. White, The Dow Chemical Co., Freeport, Texas

John V. Wiman, BASF, Jamesburg, New Jersey

Ralph E. Wright, Rogers Corp., Manchester, Connecticut

George E. Zahr, Textile Fiber Division, DuPont Co., Wilmington, Delaware

†Deceased.

FOREWORD

The Society of Plastics Engineers is pleased to sponsor the *Handbook of Plastics Materials and Technology* edited by Irvin I. Rubin, President of Rubin Plastics Corp. in Brooklyn, New York.

After authoring his "best seller," *Injection Molding—Theory and Practice,* Irv Rubin had a driving ambition to develop a comprehensive volume relating materials to processing. This book is that treatise, covering the two subjects in unusual depth. He has selected a galaxy of talented chapter authors, carefully editing their manuscripts to provide a cohesive, free-flowing book.

Mr. Rubin, a Fellow of SPE, has long been involved in advancing the technology of plastics through both the written and spoken word. He has written extensively in the technical press and has instructed in the processing of plastics for over 25 years.

SPE, through its Technical Volumes Committee, has long sponsored books on various aspects of plastics. Its involvement has ranged from identification of needed volumes to recruitment of authors and the professional review and approval of new publications.

Technical competence pervades all SPE activities, not only in the publication of books, but also in other areas such as sponsorship of technical conferences and educational programs. In addition, the Society publishes periodicals, as well as conference proceedings and other publications, all of which are subject to rigorous technical review procedures.

The resource of some 33,000 practicing plastics engineers has made SPE the largest organization of its type worldwide. Further information is available from the Society at 14 Fairfield Drive, Brookfield, Connecticut 06804, U.S.A.

Robert D. Forger
Executive Director
Society of Plastics Engineers

Technical Volumes Committee

Thomas W. Haas, Chairman (1986–1989)
Virginia Commonwealth University

Robert E. Nunn, Chairman (1989–1990)
University of Lowell

PREFACE

This book aims to make you more successful in your business. Plastics, the second largest material used in the United States, impinges everywhere.

Whether you are a seasoned plastics engineer, manufacturer or processor, or simply an interested student or entrepreneur, your need for complete, accurate, up-to-date information about plastics is critical. Now, for the first time, there is a single source written by successful, well-known, plastics professionals that provide in-depth basic information about materials, processing, assembly, and decorating.

The Handbook contains answers to questions most often asked about 119 different plastic topics. From acetal to XT polymer, from blow molding to ultrasonic welding to design, and from simple fillers to exotic reinforcements, each section presents essential information in a convenient, easy-to-find, consistent format.

The Handbook was developed by a prestigious editorial board to ensure that it meets the information needs of people in the plastics industry with high-quality, readable text, comprehensive tables, and handy appendices. The entire text was rewritten by paid editors for easy reading, understanding, and consistency.

The Handbook is divided into three major sections—materials, processes, and technology.

Every material section and *every* processing section is organized into 15 subsections. This allows the user to find information quickly. For example, advantages/disadvantages of materials is always in #7 in every subsection.

The appendices present a wealth of information often used: two kinds of abbreviations, conversion tables from standard to metric and vice versa, and a ranking of materials according to these properties, i.e., density, tensile strength, tensile modulus, flexural modulus, % elongation to break, Izod notched impact, heat-distortion temperature at 264 psi, and dielectric strength.

ACKNOWLEDGEMENTS

The editor is extremely grateful to the following individuals: The Board of Directors who determined the concept of the Handbook, selected the authors, and provided the expert information so necessary in so vast an undertaking. The Executive Committee who helped and guided me in the long task of assembling the articles. Without them the book would not exist; Ms. Deborah Brewer who rewrote most of the articles and Mr. Roland R. MacBride who rewrote the balance; Ms. Loraine McKinney who entered the entire original manuscript into the word processing program.

IRVIN I. RUBIN

Brooklyn, New York

CONTENTS

TECHNOLOGY

HOW TO USE THE HANDBOOK

Both the MATERIAL and PROCESSING sections are arranged in a definite pattern. For example, advantages/disadvantages of materials are always listed as #7 in every section. Below are the section headings. If a previous number (such as number 4 or 6) had not been used, advantages and disadvantages would still be labeled #7.

MATERIALS

1— Chemical name
2— Category
3— History
4— Polymerization
5— Description of properties
6— Applications
7— Advantages and disadvantages
8— Processing techniques
9— Resin forms
10— Specification of properties
11— Processing requirements
12— Processing-sensitive end properties
13— Shrinkage
14— Trade names
15— Information sources

PROCESSING

1— Description and history of process
2— Materials
3— Processing characteristics
4— Advantages and disadvantages
5— Design considerations
6— Processing criteria
7— Machinery
8— Tooling
9— Auxiliary equipment
10— Finishing
11— Decorating
12— Costing
13— Industry practices
14— Industry associations
15— Information sources

1

THE PLASTICS INDUSTRY— AN OVERVIEW

Sid Gross

Chief editor emeritus *Modern Plastics Magazine, McGraw-Hill,*
1221 Sixth Avenue, New York, NY 10020

To get our proper bearings, the first thing we have to accept is that plastics is a material of design and construction. It competes with steel, glass, wood, aluminum, and many other materials (including itself). There is nothing inherently glamorous about it; nothing inherently evil. Like all materials, it stands and falls on how well it does its assigned jobs.

Plastics is the first new material of design in more than 3000 years. During that period, all the other materials had the chance to entrench themselves, to establish the standards, to dictate the tastes of what is proper and what is trash, to take strong hold in the market. None of these traditional materials has been eager to give up any part of its market share. Each has fought (and continues to fight) to hold on to its share, occasionally with campaigns whose purpose was simply to blacken plastics' name.

One result of the assault on plastics by those with vested interest in other materials, or those who are simply misinformed, has been that public opinion about plastics has tended to be strongly negative. Ever since plastics became a factor in the U.S. economy, and to this very day, some people think plastics is inherently bad. Yet the history of plastics consumption has been one of impressive growth. As a matter of fact, a statistician recently determined that, on a volumetric basis, we had left the iron age and entered the plastics age: the character of an "age" being defined by the material most used in manufacture. Plastics, on average, has about 1/7 the specific gravity of steel. That means (theoretically) that for one salad bowl of steel you can produce seven basically identical bowls of high-density polyethylene, which together will also weigh one pound. That's volumetric.

By consensus, the beginning of the plastics industry in the United States dates to the introduction of cellulose nitrate by John Wesley Hyatt in 1870. Hyatt reportedly invented the material as his entry in a contest to find a substitute for ivory in billiard balls, the ivory business having been severely interrupted by the Civil War. The next plastic material was not introduced until about 35 years later. Phenol formaldehyde, or phenolics, was invented in 1907 by Leo Baekeland. It took nearly four more decades and the development of about 18 additional polymers, along with the appropriate processing machinery, before these elements coalesced in 1945 into what we know today as the plastics industry. The event that marks the true beginning of the modern plastics industry occurred at the end of the Second World War with the release to the civilian market of a material that had been a close-kept wartime secret: low density polyethylene. In the 43 years between 1945 and 1988, the U.S. plastics industry grew from a sales volume of about 2.8 billion pounds per year to 56.9 billion pounds (1.3 to 26 billion kg) (see Table 1.1). This impressive growth reflects plastics' pen-

TABLE 1.1   Material Production in the United States, 1988, in 10^6 lb

Plastic	10^6 lb	Plastic	10^6 lb
PE LD	9,865	Packing	13,805
PVC	8,323	Building	11,372
PE HD	8,244	Transportation	2,132
PP	7,304	Elect/electronic	2,068
PS	5,131	Houseware	1,392
Phenolic	3,032	Furniture	1,181
Polyurethane	2,905	Appliance	1,156
Polyester thermoplastic	2,007	Toys	774
Urea and melamine	1,515	Other	20,840
Polyester unsaturated	1,373	Export	2,180
ABS	1,238		
Styrenics, other	1,220		56,900 lbs
Vinyls, other	958		
Acrylic	697		
Nylon	558		
Elastomers thermoplastic	495		
Epoxy	470		
PC	430		
Alkyd	320		
Others	280		
Polyphenylene alloys	180		
Acetal	128		
San	137		
Cellulosics	90		

etration of every industrial market in the United States, plus such nonindustrial markets as agriculture, the military, packaging, mining, and others.

1.1   STRUCTURE OF THE PLASTICS INDUSTRY

The main sectors of the industry are resin suppliers, machinery builders, and processors; with feedstock sources bringing in raw materials at one end, and the markets taking the finished product at the other.

Feedstocks. With negligible exceptions, plastics are derived from crude oil or natural gas through a series of chemical processes. These feeds are described as petrochemicals (chemicals derived from petroleum or petroleum equivalents), and are thus hydrocarbon-based. However, they represent only a small portion of total oil and natural gas production, about 1.5 to 2%. (All petrochemicals combined—and they include many materials other than plastics—take only about 6% of total oil–gas production.) Specific feedstocks needed to make the various plastics are methane, ethylene, propylene, benzene, acetylene, naphthalene, toluene, and xylene. There was concern during the Arab oil embargo of 1973–1974, and again at the start of the Irani–Iraqi war in 1979, that feedstock supply would be curtailed. However, it has since become apparent that supplies will be ample for the foreseeable future. The current lineup of source locations makes even an artificially or politically induced shortage most unlikely.

Resin Supply Sector. Unless it is already in monomer form, feedstock is reacted into monomer. Resin suppliers then polymerize the monomer into polymer, which is the basic plastic raw material. Ethylene monomer becomes polyethylene, styrene monomer becomes polystyrene, vinyl chloride monomer becomes poly(vinyl chloride), etc. These are thermoplastics. Certain classes of plastics (thermosets) are produced by a somewhat different route (addition polymerization) by reacting two different chemicals in the final processing equipment, ie, urea plus melamine.

Polymer (also often called resin) comes in the form of granules, powders, or pellets of different shapes. It is frequently modified by additives that impart fire resistance, lubricity, ultraviolet light protection, impact strength and any number of other performance characteristics. Some plastics need additives just to make them processable. Modified resins are called compounds.

As a round number, there are about 60 distinct and different polymers commercially available today. Through compounding with chemicals, additives, fillers, and reinforcements, this number is swelled into the thousands, from which today's designers can make their choices: They can pick from materials that are transparent or opaque, or anything in between; that float on water or sink; that may be tough enough to resist sulfuric acid or fragile enough to dissolve in water; that may be rigid or flexible; that may be elastic as rubber or hard as rock. In short, plastics come in almost any combination of properties; some compounds are so fine-tuned that they are specified for one single application. Nor are these all the possible combinations and permutations. Through the introduction of blowing agents these resins and compounds can be made into cellular products (foam, sponges) that add an entire new properties dimension. These foams range from the softness of down cushion to the hardness and stiffness of lumber.

Types of Resin. Plastics materials are divided into two groups; thermoplastics and thermosets. Thermoplastics yield finished products that can be remelted and then reshaped into other products over and over again . . . like square ice cubes that can be melted and frozen into round shapes, then melted again and frozen into rectangular form. Thermosets yield products that, once formed, cannot be melted . . . just as dough that has been baked into a cake cannot be made into dough again, or an egg that has been hard-boiled cannot be made liquid again.

The division is not razor sharp. Some materials appear in both forms (polyurethane is one), some start as one and change into the other (eg, cross-linkable PE), and in some compounds the two types are mixed (eg, low-profile polyesters). Still the division is useful.

Thermoplastics in turn are subdivided into commodity and engineering types. Again, the division is not clear-cut, because some resins span both groups (eg, ABS), nor is there definition that precisely distinguishes one from the other. But the distinction holds in a general way, and it is useful. Commodity plastics include polyethylene, polypropylene, polystyrene, poly(vinyl chloride), poly(ethylene terephthalate), and some ABS. The engineering resins are acetals, nylons, polycarbonates, and quite a few more, plus some ABS. Generally, speaking, engineering thermoplastics have better physical properties (heat resistance, impact strength, modulus, etc). As a rule of thumb they also cost at least twice as much per pound as commodity grades.

In a broad sense, commodity plastics find applications where such properties as dimensional stability, stiffness, heat resistance, and toughness are not primary requirements of the end product. This does not make them inferior materials. They have characteristics (price being one) that engineering resins cannot provide. It simply defines their market. Flooring is typically made of vinyls; detergent and bleach bottles are typically made of high density polyethylene, carbonated-beverage bottles are typically made of poly(ethylene terephthalate), bread bags are typically low density polyethylene; backs of TV sets are typically polypropylene, egg cartons are typically polystyrene foam, and on and on. These are all applications where commodity plastics perform perfectly well.

Engineering thermoplastics find markets where product requirements include tight tolerances, higher levels of temperature resistance, and other superior physical properties. Engineering resins go into telephone housings, camera parts (including lenses), under-the-hood automotive components, computer and other electronic assemblies, VCR tape cassettes, mechanical gears of all sorts, business-machine housings, and many more such demanding end uses.

Basic resin suppliers divide pretty much along these same commodity/engineering lines, although again the division is not absolute. With that caveat, commodity resins are generally supplied by companies that are classified as oil companies; engineering thermoplastics are generally produced by so-called chemical companies.

One of the interesting aspects of the resins-supply sector is the constant shift of participants. The economics of the business are such that at various times different companies see opportunities that cannot be ignored, so they get in. As time passes, these opportunities may turn out to be less attractive than they originally seemed, or the basic objectives of the companies change, or better opportunities beckon. At any rate, many leave, frequently to be replaced by other hopefuls. Just to give one example, here is a list of companies making low density polyethylene in 1965; next to it is a list of companies making the same material in 1989:

1965	1989
Allied Chemical	Chevron
Celanese	Dow Chemical
Chemore	Du Pont
Dow Chemical	Eastman
Du Pont	Westlake
Eastman	Exxon
Enjay	Mobil
Grace	Quantum Chemical, USI Div.
Koppers	Rexene
Monsanto	Union Carbide
Phillips	
Rexall	
Shell	
Spencer Chemical	
Union Carbide	
U.S. Industrial Chemical	

Only three companies appear on both lists. Similar lists, though perhaps less dramatic, can be constructed for other of the commodity resins, all of which were commercialized between 1937 and 1957, as follows:

Resin	Year
PVC	1937
Polystyrene	1938
Low density polyethylene	1942
High density polyethylene	1957
Polypropylene	1957

Since 1957, the only new polymer in the volume range of these five has been poly(ethylene terephthalate)s, which have captured the bulk of the carbonated-beverage bottle business. Although new resins are still being discovered and commercialized, those introductions are becoming less frequent, and the tonnages involved are several magnitudes lower than for the volume thermoplastics listed above. Materials innovation has not stopped, however. The advent of truly new polymers has been replaced by new property spectra developed through modification by alloying, blending, compounding . . . and new polymerization technologies that yield an established resin, but with a new set of properties. One of the most significant developments in the plastics industry in recent years was the introduction of technologies for making low density polyethylene by low-pressure polymerization. Linear low density polyethylene is still low density PE, but it has several useful new properties, one of which is "down-gaugeability," which enables the material to be extruded in greatly reduced film thickness without sacrificing performance properties. Other polymerization technology advances have affected most of the other commodity plastics.

The highest level of activity is in the so-called compounding area. There, the resin supplier with a massive capital investment in large reactors is no longer the dominant force. Many smaller companies specialize in compounding. Some have developed proprietary formulations that are the equivalent of fully commercialized basic resin; others compound on a custom basis. And even some of the large basic resin suppliers have started following this trend toward smaller-scale compound production.

A list of all U.S. resin suppliers can be found in the *Modern Plastics Encyclopedia,* a yearly compendium published by McGraw-Hill, and in some other yearbooks.

The New Supply Geography. There are a number of oil-rich regions in the world where new petrochemical complexes have been built, or where they can be built. They are the OPEC nations of the Middle East, Canada, Mexico, and Alaska. Canada has already become an important source of plastics for the United States and elsewhere. OPEC has become a global power, with some of its plants operating in cooperation with U.S. supplier companies. Mexico and Alaska can be expected to become important factors in the years ahead.

The significance of such developments is that the low costs of basic raw materials in these oil-rich regions will give them a competitive edge over traditional suppliers of polyolefins and other resins. This competitive edge is big enough to permit landing of resin almost anyplace in the world at prices that may be below local quotations. This despite the fact that significant transportation cost would be involved. Observers tend to discount the effect of Saudi-produced resin being landed in the U.S., but they can readily picture such resins being shipped to what have been and still are important U.S. export markets.

The ability to produce polyolefin resin practically anywhere in the world with the latest technology and low-cost supply bases will give several countries an important position in the global plastics industry. The United States will not be spared the effects, quite obviously.

The Equipment Sector. No material can prosper without practical ways to process it. The early way for plastics (cellulose nitrate) was the production of "slugs" by a primitive form of compression molding; these slugs were then fabricated into finished products. Refined compression molding technology came into its own after the invention of phenolics.

However, the plastics machinery industry really dates from the introduction of the injection molding machine in the 1920s. Injection molding made possible high-speed plastics production, which in turn provided the manufacturing economics that fueled plastics' growth.

Since those early days, the number of plastics working processes has proliferated. Today we can identify at least 15 distinct types, including injection molding, extrusion, blow molding, thermoforming, compression molding, casting, rotational molding, pultrusion, calendering, filament winding, stamping . . . and more. It is a big business; 1987 dollar value of shipments of major types of machinery in the United States is given in Table 1.2.

Over the years, plastics machinery has become increasingly efficient. As a general rule of thumb, the newest model will have an output perhaps double that of its five-year old predecessor. Machinery has also become more versatile. For example, injection machines have developed into coinjection machines, which can handle more than one material in the same cycle. Thus you can make a two-color cup in one shot. Or you can run offgrade material as the core of a "sandwich" with virgin outer skins.

In the extrusion area, die technology now permits the extrusion of 3-, 5- and 7-layer structures (or more) in a single operation. This permits production of film, sheet, bottles,

TABLE 1.2 Dollars Sales of Machinery, 1987

Type of Equipment	Sales, million $				
	1986	1987	Export	Import	Total
Injection	480.0	257.1	a	230.5	487.6
Structural foam	4.4	3.9	a	0.5	4.4
Extrusion, single-screw	166.0	112.5	a	53.0	165.5
Extrusion, multi-screw	22.0	15.6	a	8.2	23.8
Blow molding	100.0	70.0	a	30.0	100.0
Thermoforming	63.0	60.0	a	10.0	70.0
Reaction injection molding	10.5	6.8	a	4.7	11.5
Compression	16.5	16.2	a	1.2	17.4
Thermoset/transfer injection	6.2	7.1	a		7.1
Expanded bead	9.0	0.5		9.5	10.0
High-intensity mixers	2.5	2.5			2.5
Total	**880.1**	**552.2**	a	**347.6**	**899.8**

aCombined in U.S. sales figure.
Source: industry and *Modern Plastics* estimates.

and other coextrudates that can utilize low-cost substrates for economical production of high-performance film, that can use unapproved or substandard materials by "enveloping" them with accepted films in food-contact applications.

Exciting developments, all. But what is perhaps more important is the emergence of control sophistication for plastics processes that is leading inevitably to the day of the fully-automated production plant. It is in the nature of plastics that the material must be heated to a temperature at which it can be shaped, and then cooled to a temperature at which it can retain its shape. This seems simple enough, but as plastics processors discovered long ago, it really is not. Plastics in the heated or molten state do not behave like conventional materials, and one plastic behaves differently from another. Unless you know what you are doing, peculiar things are liable to happen.

In the early days of plastics there arose a group of operators who seemed to understand the mysteries of the plastics process and who practiced a kind of "black art" . . . which they tended to guard very closely. Today, plastics processing has become more of a science. What made this possible was the introduction of the microprocessor controller with feedback that keeps machine operation conditions within the tolerance limits necessary to produce good parts *consistently*. The application of computers, microelectronics, and video screens makes possible fully automated, unattended operation of many plastics processing machines. Technology has become so advanced that entire processing plants can be run under computer control (from inventory through scheduling, through processing, through invoicing). Full automation has obvious benefits. But perhaps the lasting significance of control systems is that product quality (so-called zero-defect production) can now be guaranteed. This in turn will put to rest the notion that plastics are inferior materials that cannot be trusted for long-term performance. Once that persistent prejudice is buried, we can expect substantial new market penetration by plastics and a continuing high growth rate.

Exciting through electronics, computers and unattended factories may be, the bulk of plastics working machinery is concerned with providing the most effective processing environment. Gear pumps for extruders, cartridge valves for injection machines, vented barrels for hydroscopic materials, robots for parts removal, grinders for inplant recycling, blenders, mixers, hopper-loaders, screen changers, wind-up systems, the whole range of auxiliary equipment—all are indispensable to a functioning processing shop. Nor can we forget the importance of molds and dies. They give the plastic product its final shape, and for many years their production was in the hands of highly skilled toolmakers doing a lot of patient handwork.

But then, in the early 1980s, CAD/CAM entered the scene. CAD/CAM is an acronym for computer assisted design/computer assisted manufacturing. It offers toolmaking a chance at better quality and greater productivity. CAD/CAM lets the designer make drawings on the video screen, modifying them quickly as needed or altering construction details, all without time-consuming reproductions of meticulous ink drawings. Using available software, the experienced designer can locate the gates to assure warp-free moldings, and pinpoint the optimum cavity locations. Other programs help to position cooling channels most effectively and provide just the right dimensions. Then the mold can be "run" through the computer to see if it would actually work. The final design is preserved on tape or on a disk, which can now be used to cut the tool. Time savings and quality improvements are obvious; and toolmaker shortages are no longer a risk.

1.2 THE PROCESSING SECTOR

This is the part of the plastics industry where advances in materials and machinery are combined to create the finished products that the consuming public buys; products that range from the triviality of swizzle sticks to the lifesaving glory of an artificial heart. Basically, there are three types of processor: custom, captive, and proprietary.

Custom Processors. These are operations that in the metal-working field might be known as job shops. They process plastics into components for other manufacturers to use in their products. For example, a manufacturer of refrigerators may retain a custom thermoformer to make his inner door liners. Or a typewriter manufacturer may have his keys made by a custom molder. Custom processors typically have a close relationship with the companies for whom they work. They may be involved (to varying degrees) in the design of the part and of the mold, they may have a voice in material selection, and in general they assume a reasonable level of responsibility for the work they turn out. There is a subgroup in custom processing known as "contract" molders. They have little involvement in the business of their customers. In effect, they just sell machine time.

Captive Processors. These are operations of manufacturers who have acquired plastics processing equipment to make parts they need for the product they manufacture. For example, a car maker may install equipment to mold accelerator pedals, or instrument cluster housings (rather than have a custom molder produce them). A refrigerator manufacturer may acquire a thermoforming machine to produce inner door liners. Generally speaking, manufacturers will install a captive plastics processing operation only when their component requirements are large enough to make it economical. Some manufacturers who are big enough to run their own plastics shops nevertheless will place a portion of their requirements with outside vendors. Reasons: It keeps their own capital investment down, it avoids the problem of internal single-source supply, and it maintains contact with the market and the pricing intelligence it provides. Automobile makers are a good example of this type of operation.

Proprietary Processors. These are operations where the molder makes a product for sale directly to the public under his or her own name. An example: Boonton Molding Co. (no longer in existence) made melamine dinnerware sold to the public under the Boontonware label.

There are no official statistics on the number of plastics processing locations in the United States (or in the world, for that matter). However, consensus places the U.S. number at about 35,000. These plants range in size from two or three machines to hundreds. As to the total number of machines operating in plastics plants in the United States, estimates place it currently at more than 60,000 injection machines and one-third as many extruders. For all other types of processing equipment, the numbers are considerably smaller.

1.3 PLASTICS AND SOCIETY

Since its inception as an industry, plastics has been accused of many societal insults. One of the most recent is that chlorofluorocarbon blowing agents (used in making polyurethane foam) are contributing to the creation of a hole in the earth's ozone layer, permitting the entry of harmful UV rays from the sun. The facts are that chlorofluorocarbons are more widely used in other industries (such as aerosols), and that suppliers are committed by international agreement to a search for effective alternatives to chlorofluorocarbons.

Some of the more persistent indictments of plastics as being environmentally unfriendly are as follows:

- Plastics waste precious energy that could be more suitably employed. It is a frivolous drain on our natural resources.
- Plastics pollute the environment, first in their manufacture and then after their use, because they are not biodegradable.
- Some plastics are carcinogenic, mutagenic, teratogenic, and otherwise injurious to human health.
- Plastics represent a grave fire danger.

Some of these accusations are simply based on ignorance, others have some historic factual basis that is now largely invalid. The facts are these:

Plastics and Energy. Numerous studies have shown that plastics create more energy than they consume in their production. This may sound paradoxical, but a simple energy audit quickly reveals that the energy used in the life cycle of a plastics part (counting raw material, energy for processing, transportation, disposal, and other factors), is generally significantly less than that needed for competing materials. Figures released by the Society of the Plastics Industry (SPI) that plastics products actually save the economy more energy than is used in making them.

Startling as the conclusion may be, this finding says that plastics are in fact a net *contributor* of energy. The public, often just looking at surface data and knowing that plastics come from oil, makes the immediate assumption that other materials, coming from different sources, put no demand on oil. People do not think about the large amounts of energy needed to make aluminum from bauxite, or that used in making glass, and so on. Or that plastics account for less than 2% of crude oil and natural gas consumption.

Plastics and Pollution. It is certainly true that plastics are not naturally biodegradable, but to consider this a detriment is a questionable argument. In fact, it may well be considered an advantage. Of course, plastics left lying around after use do not disappear from view. Such poster-consumer waste as foam cups, detergent bottles and discarded film is a visual annoyance. But properly disposed of in a landfill, it becomes a stabilizing factor at those sites, presenting no danger of leaching into the ground water or nearby bodies of water. Although plastics contribution to the solid-waste stream is only about 1% by weight, its general bulk makes it highly visible and the object of civic concern; recycling has thus become a major objective of the plastics industry.

Systems have been developed that take all the plastics in the waste stream (plus some paper, aluminum, and other nonplastics junk), and reconstitute this mixture into fence posts, bench slats and other relatively low-specification products. Other, higher value projects are under way. The problem of plastics recycling is political and social rather than technological.

Plastics, *per se,* are not dangerous to human health. Their fire hazards are no different than wood, cloth, or other combustibles involved in a fire and when treated with fire retardants can be significantly less.

1.4 OPPORTUNITIES IN PLASTICS

As this section has made clear, the plastics industry is such a diversified economic activity that opportunities to participate in it—either as job holders or as entrepreneurial business operators—are abundant. As the foregoing also has made clear, the prospect for growth in plastics is bright indeed. Taking these two factors together, one can readily agree with the guest's whisper to The Graduate: "Plastics."

In the materials sector, a chemical background is an obvious necessity for any job in the various research and development laboratories maintained by suppliers. There also are substantial opportunities in selling and marketing, where a chemical background is not absolutely essential. Many companies train applicants on the job. A plastics-resin salesperson is usually called a "technical sales representative." And for good reason. The function generally not only involves familiarity with the resin being sold, but also a concern (on a technical level) with customers' needs, consulting on possible causes for processing problems right on the plant floor, and in general being an advisor.

Some chemical engineering work must be done on site; and there are job opportunities in that area as well. All major material suppliers run customer service labs. These consist of a number of processing machines on which the materials being produced by the supplier can be test-run; it is there that molds also are tested. There are job opportunities in running these labs, requiring a background in plastics engineering. Production managers, accountants, financial officers, personnel administrators . . . and all the other normal management functions found in industry are required by material suppliers as well.

The machinery sector primarily requires a cadre of mechanical engineers to design and construct processing equipment, plus associated personnel such as draftsmen and machinists. However, since the machines are going to process plastics, which do not behave like other material, engineers must have a good knowledge of plastics rheology and the resultant behavior of the resin in the machine. The overwhelming shift to computer and microprocessor technology to control machine sequencing and operating conditions suggest electronics as a very advantageous field of knowledge to bring to the job.

The processing sector, be it custom, captive, or proprietary, calls for good grounding in plastics engineering. Personnel must be conversant with the entire materials spectrum and machine functions in order to run the jobs most efficiently. It is still possible for the individual entrepreneur to start a successful business in plastics, but it is getting harder: machinery is more expensive, resin is more expensive, distribution is more involved, and shoddy stuff certainly is not going to sell. Prerequisites of success for product manufacture in the modern plastics industry are a clear vision of what one wishes to contribute, a working knowledge of plastics materials and processes, and reasonable financing. Anyone setting out on a career in plastics, and having no personal contact, would find it useful to get in touch with the Society of Plastics Engineers, 14 Fairfield Dr., Brookfield Center, CT 06805.

2

ACETALS

S. M. Sinker

*Plastics Research & Development Center, Hoechst Celanese Engineering
Plastics Division, 86 Morris Ave., Summit, NJ 07901*

2.1 ACETAL

Polyacetals are thermoplastic polymers of formaldehyde, CH_2O, having the repeat unit
—CH_2—O—, that are available commercially in homopolymer or copolymer form.

2.2 CATEGORY

Acetal homopolymer is a backbone of polyformaldehyde, end-capped as an ester of an
organic acid; that is, polyformaldehyde end-capped with an acetate group. Acetal copolymer
is a backbone of polyformaldehyde with an olefin monomer unit derived from a cyclic ether
or diol/formaldehyde condensation product:

$$-CH_2-CH_2-O-CH_2-O-$$

or:

$$-CH_2-CH_2-CH_2-CH_2-O-CH_2-O-$$

randomly interrupting the chain. The olefin monomer unit also end-caps the chain as an
alcohol:

$$H-O-CH_2-CH_2-O-CH_2-O-$$

or

$$HO-CH_2-CH_2-CH_2-CH_2-O-CH_2-O-$$

The end-capping of the chains (both homopolymer and copolymer) is required to prevent
the irreversible thermal depolymerization of the polymer backbone during melt processing.

2.3 HISTORY

Butlerov[1] first observed polymers of formaldehyde in the mid-1880s. Staudinger[2] did much
of the initial research on linear polymers of formaldehyde and found that thermally stable

high molecular weight polymers were difficult, if not impossible, to produce. The poor thermal stability of these polymers is the result of many different degradation mechanisms. Among the mechanisms are

- Thermal "unzipping" of H—O—CH₂—O—CH₂— end groups to formaldehyde monomer.
- Thermo-oxidative attack on the backbone, causing chain scission to an unstable end group and subsequent depolymerization through "unzipping."
- Acidic attack on the backbone (generally from formic acid formed by the thermo-oxidation of formaldehyde) causing chain scission and "unzipping."
- Thermal chain scission of the CH₂—O— bond [at temperatures above 518°F (270°C)], leading to "unzipping."

In the 1950s, researchers at the Du Pont Co.[3] developed the end-capping technology to prevent thermal unzipping of polymer chains to formaldehyde. Additive packages to prevent thermo-oxidative attack on the polymer backbone and formation of formic acid were developed and a commercial plant went into operation in 1959 to produce Delrin, an acetate end-capped acetal homopolymer.

Shortly afterward, Celanese developed the olefinic copolymer and end-capping[4] technology. Additive packages were also developed to prevent thermo-oxidative attack on this polymer system and a commercial plant went into operation in 1962 to produce Celcon, an acetal copolymer.

Early applications[5] took advantage of acetal's light weight, natural lubricity, solvent resistance, dimensional stability, and strength and stiffness as a metal replacement in gears, wear strips, automotive and appliance instrument housings, pump housings and mechanisms, hose couplings, automotive door handle and window cranks, drain plugs, conveyor belt links, fan blades for electric motors, electric shaver head parts, and aerosol bottle components.

Acetal grades of varying molecular weight are available that enhance moldability or toughness. Lower molecular weight grades exhibit better flow and moldability at a sacrifice in impact resistance. Higher molecular grades exhibit good impact resistance and toughness but are more difficult to mold because they have a higher viscosity at processing temperatures.

Acetals are difficult to modify. The high crystallinity of the base resin makes them incompatible with most polymers, modifiers, reinforcements, and fillers. The standard grades are neat resins and do not contain plasticizers or extenders. To modify the acetals, special compatibilization technology and compounding techniques must be employed. Modification of the base resin is usually done to enhance one or more of the basic properties of the material and usually involves a compromise that diminishes one or more of the other properties (for example, adding elastomeric impact modifiers reduces the strength and stiffness). A wide variety of specialty grades have been developed in both homopolymer and copolymer.

Acetals are considered a specialty resin and domestic use is only a small fraction of total plastic consumption. Since its introduction in 1959, domestic acetal consumption has risen to nearly 100 million pounds in 35 years. In 1960, less than 10 million pounds were produced; in 1970, 46 million pounds, and in 1980, 90 million pounds.[6]

2.4   POLYMERIZATION

A survey of patents[7] indicates that acetal homopolymer is produced from anhydrous formaldehyde that is polymerized to the desired molecular weight in the presence of a catalyst. The polymer is then purified by several washes to remove all traces of catalyst and reaction by-products. Purified polymer is end-capped by acetylating the hydroxyl end-group. End-capped polymer is purified to remove traces of excess acetate end-capping agent, dried, blended with thermal stabilizers and antioxidants, and melt compounded in an extruder to produce the finished pellets.

A similar survey of patents[8] indicates that acetal copolymer is produced from formaldehyde solutions by preparing the trimer intermediate trioxane, which is purified, blended with cylic ether or acetal comonomer, and polymerized to the desired molecular weight in the presence of a catalyst. The polymer is then purified by several washes to remove all traces of the catalyst. Purified polymer is hydrolyzed in caustic aqueous medium to remove unstable end groups and leave the chain-capped with an olefinic alcohol end group. The polymer is then further purified to remove hydrolysis catalyst and liberated formaldehyde, dried, blended with thermal stabilizers and antioxidants, and melt compounded in an extruder to produce finished pellets.

Both processes rely on polymer purification steps involving solvent washing (hydrocarbon, aqueous, or a combination of both). The solvents used in the purification process must be recovered and recycled, typically through distillation. Purification of monomers (anhydrous formaldehyde or trioxane) is also critical. Both the purification of monomers and solvent recovery are energy and capital intensive; typically, the energy and capital costs of these steps far exceeds the raw materials cost. Estimation of the economics of either process is difficult since the volume of solvent per pound of product and the capital costs are regarded as trade secrets. Also, energy prices are currently very volatile. Since raw material cost is significantly less than process energy and capital cost, the finished polymer price is fairly independent of changes in formaldehyde cost.

2.5 DESCRIPTION OF PROPERTIES

Acetals are strong, stiff, tough, and have a low coefficient of friction against metals, acetal, and other plastic resins. The acetals are creep resistant: they exhibit very low levels of cold flow. As a result, they can be used in applications where dimensional stability is important, even when the part is under continuous stress. The creep under load can be predicted and used for design purposes. The strength, stiffness, and toughness are inherent in the material and are also predictable over a wide range of temperatures for very long periods of time. The acetals are also extremely resistant to fatigue under repeated stress loads and their fatigue performance can also be predicted. The predictable nature of strength, toughness, creep, and fatigue over broad temperature ranges for long periods of time make acetals ideal for use in metal replacement. In fact, designers now consider acetal instead of metal for many applications.

Acetals are also extremely resistant to a wide range of solvents and are not hygroscopic. They remain dimensionally stable in harsh environments. Acetals are attacked by strong acids and strong oxidizing agents such as hypochlorite and are not recommended for use in these environments. The homopolymer is also not recommended for use in strong caustic. Copolymers can be used in a pH range of 4–14; homopolymers are limited to a pH range of 4–10.

At 264 psi (1.8 MPa), acetals have a heat-distortion temperature in excess of 230°F (110°C) and can be used in applications up to this temperature intermittently. However, acetals can lose strength and toughness after long-term exposure to hot environments. Homopolymers resist deterioration for up to 1-1/2 years at 180°F (82°C) in air, while the copolymers may be used continuously at temperatures up to 220°F (104°C) in air. The melting point is 338°F (170°C) for the homopolymer and 329°F (165°C) for the copolymer. Thermal conductivity is 1.6 Btu/h/ft^2/°F/in. Coefficient of linear thermal expansion is less than 10^{-4} in./in./°F from -40 to 220°F (-40 to 104°C). Moldings of acetal remain dimensionally stable over the recommended use temperature range. Glass-fiber reinforcement raises the heat-deflection temperature to above 325°F (163°C) at 264 psi (1.8 MPa). The same long-term temperature restrictions and solvent resistance apply to glass-reinforced grades as to unfilled grades.

As shown in the following list, specialty grades are available that are tailored with glass-fiber reinforcement, impact modifiers, wear resistance additives, ultraviolet stabilizers, mineral and glass bead/milled glass fillers, or antistatic/electroconductive additives, or are made electroplatable. In addition, some combinations of the modifiers are available.

2.5.1   Specialty Grades

Low Wear Products. PTFE, silicone fluid, molybdenum disulfide, or other lubricant is added to reduce friction and enhance lubricity for bearing and wear surface applications. Modification is available at several levels.

Toughened/Impact Modified Products. Elastomer modified to improve practical impact and decrease notch sensitivity. Modification is available at several levels.

Ultraviolet/Weathering Stabilized. Special stabilizers are added to protect against deterioration from ultraviolet light (sunshine and so on). Materials are available in a multitude of colors. Specially stabilized black grades are available for out-door weathering applications.

2.5.2   Reinforced Grades

Fiber glass reinforcement is used for exceptional stiffness and creep resistance. Glass-coupled copolymer grades also exhibit higher strength. Modification is available at several levels.

2.5.3   Mineral-filled or Glass Bead/Milled Glass Grades

Mineral or glass bead/milled glass are added as a modification to reduce shrinkage and warpage in molding (such as to enhance dimensional stability). Mineral-reinforced copolymers often exhibit increased impact resistance. Modification is available at several levels.

2.5.4   Antistatic/Electroconductive Grades

Modification made to reduce surface static buildup in electronic and wear applications. Several different modifications are available.

2.5.5   Electroplatable Grades

Modification to allow plating by the standard ABS process after a special etching step.

2.6   APPLICATIONS

Acetals are used in applications where metal, wood, or thermosetting polymers have historically been used. The ease of processing; the ability to predict long-term strength, toughness, creep, and fatigue; the exceptional solvent resistance; the hard, high gloss surface of molded parts; the long-term dimensional stability; and the low friction and good wear resistance are factors that have prompted selection of acetals in a wide variety of such applications.

Acetals have the approval of several agencies: FDA for repeated food contact applications, the NSF and Canadian Standards Association for potable water applications, DFISA for contact with dairy products, and USDA for direct contact with meat and poultry products. They have merited UL ratings for flammability, electrical, and thermal use properties; a general material specification in ASTM (D4181); and the material specification of automotive and industrial corporations. They are listed for use in plumbing applications by several regulatory agencies (IAPMO, BOCA, Southern Standard Building Code, and so on) and are rated by the Plastic Pipe Institute for use in potable water plumbing applications, excluding closed-loop systems or areas where levels of hypochlorite greater than 1 ppm are present.

The following is a general, but not totally inclusive, listing of current application areas:

Industrial. Gears, cams, bearings, chain links, springs, pollution control equipment (noise, air, and waste liquids), valves, and fittings are industrial applications.

Material Handling. Chain links, bearings, cams, conveyor links, valve bodies, hose connectors, pumps (including housings, pistons, valves, and impellers), and wear stops are used in material handling.

Automotive. Fuel level sensors, pump components, and gas caps; cooling fans; trim clips; color coordinated buckle housings, window cranks, shift lever handles, knobs, levers, and visor mounting brackets; instrument cluster gears, bearings, housings, and dials; exterior door pulls, mirror housings and brackets are used in the automotive industry.

Appliances. Appliance uses include refrigerator shelving clips and brackets; gears, bearings, wear strips, and instrument housings in washers and dryers; rollers, spray nozzles, and soap dispensers in dishwashers; bowls, housings, mixing blades, gears, and bearings in counter-top appliances; bodies, tops, and cups in water boilers.

Home Electronics. Key caps, plungers, guides, and base plates in computer keyboards; push buttons, gears, bearings, and springs in telephones; clips, peg boards, and connectors in modular components; tape hubs and guides, cams, gears, reels, and bearings in audio and video tape players.

Plumbing. Water-meter housings, cams, gears, dials, and pressure plates; faucet underbodies, cartridges, stems, packing nuts, and waterways; water softener pumps (housings, pistons, valves, and impellers); filter bodies and valves; pressure regulator valves, bodies, stems, and knobs; and potable water valves are plumbing uses.

Consumer Applications. These include combs, mascara wands and containers, aerosol valves, ski bindings, sprayer pumps and nozzles, small appliance parts (pump components for dental cleaners, gears, cams, bearings, glue applicators, and housings for tools), and toys (shells, frames, wheels, gears, bearings, cams, springs, and connectors).

Hardware. This category includes drapery and venetian blind guides, rollers, bearings, gears, and hangers; furniture casters, slide plates, and locks; tool holders; bearings in adapters (screwdriver for drills).

Irrigation and Agricultural. Pop-up sprinklers (nozzles, arms, gears, housings, and waterways), pumps (housings, impellers, and pistons), metering valves, tractor components (shift lever housings, hydraulic connectors, seed applicators, bearings, and gears) are used in agriculture/irrigation.

2.7 ADVANTAGES/DISADVANTAGES

Acetals are highly crystalline thermoplastic resins that have a unique balance of properties. Using modern design techniques, acetals perform remarkably well in applications where metals were historically used. The ability to predict the long-term strength, toughness, creep, fatigue, chemical resistance, dimensional stability, and wear resistance of acetals in known environments of temperature, pressure, and atmosphere allows engineers the freedom to design parts that area cost-competitive or lower in cost than metals in final form. Acetals are readily injection molded into parts that require no post-processing, such as machining, deburring, riveting, welding, or painting. Acetals can be pigmented to any opaque color.

As noted, acetals are attacked by strong acids and strong oxidizing agents and are not recommended for use in these environments. The homopolymer is not recommended for use in environments where strong caustic is present.

2.8 PROCESSING TECHNIQUES

See the Master Material Outline and the Standard Material Design Chart.

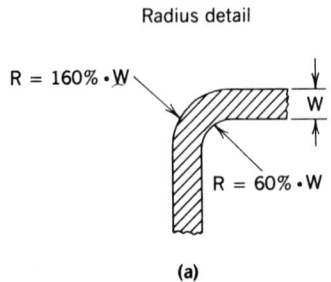

Radius detail

(a)

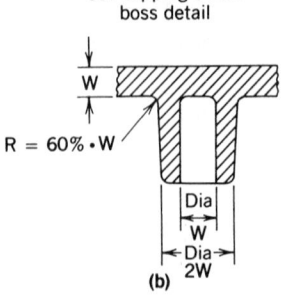

Self tapping screw boss detail

(b)

Solid rib detail

(c)

Hollow rib detail

(d)

Standard design chart for _____ **Acetal** _____

1. Recommended wall thickness/length of flow
   Minimum _____ for _____ distance
   Maximum _____ for _____ distance
   Ideal _____ for _____ distance

2. Allowable wall thickness variation, % of nominal wall _____ 50 _____

3. Radius requirements (scheme **a**)
   Outside: Minimum ___ 160% ___ Maximum _____
   Inside: Minimum ___ 60% ___ Maximum _____

4. Reinforcing ribs (scheme **c** and **d**)
   Maximum thickness ___ 50% of adjacent wall ___
   Maximum height _3/4 in.—ideal 3x wall thickness_
   Sink marks: Yes __X__ No _____

5. Solid pegs and bosses
   Maximum thickness _50% of adjacent wall_
   Maximum weight ___3x wall thickness___
   Sink marks: Yes __X__ No _____

6. Strippable undercuts
   Outside: Yes __X__ No _____
   Inside: Yes __X__ No _____

7. Hollow bosses: (scheme **b**)   Yes __X__ No _____

8. Draft angles
   Outside: Minimum __ $\frac{1}{4}°$ __ Ideal __ 1° __
   Inside: Minimum __ $\frac{1}{2}°$ __ Ideal __ 1° __

9. Molded-in inserts: Yes __X__
                        No _____

10. Size limitations
    Length: Minimum _____ Maximum _____ Process _____
    Width: Minimum _____ Maximum _____ Process _____

11. Tolerances

12. Special process- and materials-related design details
    Blow-up ratio (blow molding) ——————25:1——————
    Core L/D (injection and compression molding) ——————————
    Depth of draw (thermoforming) ——————————
    Other ——————————

Master Material Outline

PROCESSABILITY OF ____Acetal_____

PROCESS	A	B	C	
INJECTION MOLDING	X			
EXTRUSION	X			
THERMOFORMING		X		Low melt strength, narrow window
FOAM MOLDING	X			
DIP, SLUSH MOLDING			X	No commercial solvents if held at melt, poor thermal stability over long resi- dence time
ROTATIONAL MOLDING	X			
POWDER, FLUIDIZED- BED COATING			X	High shrinkage may leave cracked surface
REINFORCED THERMOSET MOLDING				
COMPRESSION/TRANSFER MOLDING				
REACTION INJECTION MOLDING (RIM)				
MECHANICAL FORMING	X			
CASTING				

A = Common processing technique.
B = Technique possible with difficulty.
C = Used only in special circumstances, if ever.

2.9 RESIN FORMS

Acetals are supplied as free-flowing pellets. Standard grades differ in molecular weight, which affects processibility and toughness. (High molecular weight grades are tougher and more impact resistant; lower molecular weight grades are easy to process but lower in toughness and impact resistance.)

2.10 SPECIFICATION OF PROPERTIES

See the Master Outline of Materials Properties.

MATERIAL ___ General-purpose Acetal Copolymer

PROPERTY	TEST METHOD	ENGLISH UNITS	VALUE	METRIC UNITS	VALUE
MECHANICAL DENSITY	D792	lb/ft^3	88	g/cm^3	1.41
TENSILE STRENGTH	D638	psi	8800	MPa	61
TENSILE MODULUS	D638	psi	410,000	MPa	2829
FLEXURAL MODULUS	D790	psi	375,000	MPa	2588
ELONGATION TO BREAK	D638	%	40–75	%	40–75
NOTCHED IZOD AT ROOM TEMP	D256	ft-lb/in.	1.0–1.5	J/m	53–80
HARDNESS	D785		M78–M80		M78–M80
THERMAL DEFLECTION T @ 264 psi	D648	°F	230	°C	110
DEFLECTION T @ 66 psi	D648	°F	316	°C	157
VICAT SOFT-ENING POINT	D1525	°F		°C	
UL TEMP INDEX	UL746B	°F	221	°C	105
UL FLAMMABILITY CODE RATING	UL94		94 HB		94 HB
LINEAR COEFFICIENT THERMAL EXPANSION	D696	in./in./°F	4.7×10^{-5}	mm/mm/°C	8.5×10^{-5}
ENVIRONMENTAL WATER ABSORPTION 24 HOUR	D570	%	0.22	%	0.22
CLARITY	D1003	% TRANS-MISSION	Opaque	% TRANS-MISSION	Opaque
OUTDOOR WEATHERING	D1435	%		%	
FDA APPROVAL			Yes (consult manufacturer)		Yes (consult manufacturer)
CHEMICAL RESISTANCE TO: WEAK ACID	D543		Consult manufacturer		Consult manufacturer
STRONG ACID	D543		Badly attacked		Badly attacked
WEAK ALKALI	D543		Not attacked		Not attacked
STRONG ALKALI	D543		Not attacked		Not attacked
LOW MOLECULAR WEIGHT SOLVENTS	D543		Not attacked		Not attacked
ALCOHOLS	D543		Not attacked		Not attacked
ELECTRICAL DIELECTRIC STRENGTH	D149	V/mil	500	kV/mm	20
DIELECTRIC CONSTANT	D150	10^6 Hertz	3.7		3.7
POWER FACTOR	D150	10^6 Hertz	0.006		0.006
OTHER MELTING POINT	D3418	°F	329	°C	165
GLASS TRANS-ITION TEMP.	D3418	°F		°C	

MATERIAL ___ General-purpose Acetal Homopolymer

PROPERTY	TEST METHOD	ENGLISH UNITS	VALUE	METRIC UNITS	VALUE
MECHANICAL DENSITY	D792	lb/ft^3	86	g/cm^3	1.42
TENSILE STRENGTH	D638	psi	10,000	MPa	69
TENSILE MODULUS	D638	psi	450,000	MPa	3100
FLEXURAL MODULUS	D790	psi	380,000–430,000	MPa	2620–2960
ELONGATION TO BREAK	D638	%	25–75	%	25–75
NOTCHED IZOD AT ROOM TEMP	D256	ft-lb/in.	1.3–2.3	J/m	69–123
HARDNESS	D785		M92–M94		M92–M94
THERMAL DEFLECTION T @ 264 psi	D648	°F	277	°C	136
DEFLECTION T @ 66 psi	D648	°F	342	°C	172
VICAT SOFT- ENING POINT	D1525	°F		°C	
UL TEMP INDEX	UL746B	°F	221	°C	105
UL FLAMMABILITY CODE RATING	UL94		94 HB		94 HB
LINEAR COEFFICIENT THERMAL EXPANSION	D696	in./in./°F	5.8–8.3 × 10^{-5}	mm/mm/°C	10–15 × 10^{-5}
ENVIRONMENTAL WATER ABSORPTION 24 HOUR	D570	%	0.25–0.32	%	0.25–0.32
CLARITY	D1003	% TRANS- MISSION	Opaque	% TRANS- MISSION	Opaque
OUTDOOR WEATHERING	D1435	%		%	
FDA APPROVAL			Yes (consult manufacturer)		Yes (consult manufacturer)
CHEMICAL RESISTANCE TO: WEAK ACID	D543		Consult manufacturer		Consult manufacturer
STRONG ACID	D543		Badly attacked		Badly attacked
WEAK ALKALI	D543		Consult manufacturer		Consult manufacturer
STRONG ALKALI	D543		Consult manufacturer		Consult manufacturer
LOW MOLECULAR WEIGHT SOLVENTS	D543		Not attacked		Not attacked
ALCOHOLS	D543		Not attacked		Not attacked
ELECTRICAL DIELECTRIC STRENGTH	D149	V/mil	500	kV/mm	20
DIELECTRIC CONSTANT	D150	10^6 Hertz	3.7		3.7
POWER FACTOR	D150	10^6 Hertz	0.005		0.005
OTHER MELTING POINT	D3418	°F	347	°C	175
GLASS TRANS- ITION TEMP.	D3418	°F		°C	

MATERIAL General-purpose Acetal Homopolymer, Glass-fiber reinforced

PROPERTY	TEST METHOD	ENGLISH UNITS	VALUE	METRIC UNITS	VALUE
MECHANICAL DENSITY	D792	lb/ft^3	97.2	g/cm^3	1.56
TENSILE STRENGTH	D638	psi	8500	MPa	59
TENSILE MODULUS	D638	psi	900,000	MPa	6200
FLEXURAL MODULUS	D790	psi	730,000	MPa	5030
ELONGATION TO BREAK	D638	%	12	%	12
NOTCHED IZOD AT ROOM TEMP	D256	ft-lb/in.	0.8	J/m	43
HARDNESS	D785		M90		M90
THERMAL DEFLECTION T @ 264 psi	D648	°F	316	°C	158
DEFLECTION T @ 66 psi	D648	°F	345	°C	174
VICAT SOFTENING POINT	D1525	°F		°C	
UL TEMP INDEX	UL746B	°F	221	°C	105
UL FLAMMABILITY CODE RATING	UL94		94 HB		94 HB
LINEAR COEFFICIENT THERMAL EXPANSION	D696	in./in./°F	$2.0\text{–}4.5 \times 10^{-5}$	mm/mm/°C	$3.6\text{–}8.1 \times 10^{-5}$
ENVIRONMENTAL WATER ABSORPTION 24 HOUR	D570	%	0.25	%	0.25
CLARITY	D1003	% TRANSMISSION	Opaque	% TRANSMISSION	Opaque
OUTDOOR WEATHERING	D1435	%		%	
FDA APPROVAL			Yes (consult manufacturer)		Yes (consult manufacturer)
CHEMICAL RESISTANCE TO: WEAK ACID	D543		Consult manufacturer		Consult manufacturer
STRONG ACID	D543		Badly attacked		Badly attacked
WEAK ALKALI	D543		Consult manufacturer		Consult manufacturer
STRONG ALKALI	D543		Consult manufacturer		Consult manufacturer
LOW MOLECULAR WEIGHT SOLVENTS	D543		Not attacked		Not attacked
ALCOHOLS	D543		Not attacked		Not attacked
ELECTRICAL DIELECTRIC STRENGTH	D149	V/mil	490	kV/mm	19
DIELECTRIC CONSTANT	D150	10^6 Hertz	3.9		3.9
POWER FACTOR	D150	10^6 Hertz	0.005		0.005
OTHER MELTING POINT	D3418	°F	347	°C	175
GLASS TRANSITION TEMP.	D3418	°F		°C	

MATERIAL ___ General-purpose Acetal Copolymer, Glass-fiber reinforced

PROPERTY	TEST METHOD	ENGLISH UNITS	VALUE	METRIC UNITS	VALUE
MECHANICAL DENSITY	D792	lb/ft^3	103	g/cm^3	1.65
TENSILE STRENGTH	D638	psi	16,000	MPa	110
TENSILE MODULUS	D638	psi	1,200,000	MPa	8275
FLEXURAL MODULUS	D790	psi	1,050,000	MPa	7240
ELONGATION TO BREAK	D638	%	3	%	3
NOTCHED IZOD AT ROOM TEMP	D256	ft-lb/in.	1.1	J/m	59
HARDNESS	D785		M84		M84
THERMAL DEFLECTION T @ 264 psi	D648	°F	325	°C	163
DEFLECTION T @ 66 psi	D648	°F	331	°C	166
VICAT SOFT-ENING POINT	D1525	°F		°C	
UL TEMP INDEX	UL746B	°F	221	°C	105
UL FLAMMABILITY CODE RATING	UL94		94 HB		94 HB
LINEAR COEFFICIENT THERMAL EXPANSION	D696	in./in./°F	2.2×10^{-5}	mm/mm/°C	4.0×10^{-5}
ENVIRONMENTAL WATER ABSORPTION 24 HOUR	D570	%	0.29	%	0.29
CLARITY	D1003	% TRANS-MISSION	Opaque	% TRANS-MISSION	Opaque
OUTDOOR WEATHERING	D1435	%		%	
FDA APPROVAL			Yes (consult manufacturer)		Yes (consult manufacturer)
CHEMICAL RESISTANCE TO: WEAK ACID	D543		Consult manufacturer		Consult manufacturer
STRONG ACID	D543		Badly attacked		Badly attacked
WEAK ALKALI	D543		Not attacked		Not attacked
STRONG ALKALI	D543		Not attacked		Not attacked
LOW MOLECULAR WEIGHT SOLVENTS	D543		Not attacked		Not attacked
ALCOHOLS	D543		Not attacked		Not attacked
ELECTRICAL DIELECTRIC STRENGTH	D149	V/mil	600	kV/mm	24
DIELECTRIC CONSTANT	D150	10^6 Hertz	3.9		3.9
POWER FACTOR	D150	10^6 Hertz	0.0062		0.0062
OTHER MELTING POINT	D3418	°F	329	°C	165
GLASS TRANS-ITION TEMP.	D3418	°F		°C	

2.11 PROCESSING REQUIREMENTS

Acetals are crystalline materials and hence melt sharply and flow fairly easily. Long flow paths can be used with care. When melt processing, suppliers recommend that no more than 25% regrind be used.

Materials should be stored in a dry area. If excessive moisture is absorbed by the resin pellets or regrind prior to processing, excessive gassing and/or surface appearance problems in finished parts can result. While acetals do not require drying, use of a hot air or desiccant bed dryer at 180°F (82°C) with a three-hour residence time is advantageous, especially if the resin or regrind has been exposed to excessive moisture. Use of processing temperatures in excess of 450°F (232°C) should be avoided since thermal degradation of the polymer can occur. Thermal degradation is usually observable by a high odor level and a discoloration of the resin.

2.12 PROCESSING-SENSITIVE END PROPERTIES

Properties of finished parts are a result of the high crystallinity the materials develop in the molding process. While material temperatures up to 450°F (232°C) are tolerable, acetals should be processed at the lowest temperature consistent with good flow. Homopolymers can be processed at temperatures as low as 400°F (204°C); copolymers at 360°F (182°C). A warm mold temperature of 180°F (82°C) or higher [up to 250°F (121°C)] is desirable for maximizing crystallinity and properties, but mold temperatures as low as 100°F (38°C) are sometimes utilized, especially with thick-walled parts [greater than 1/4 in. (6 mm)]. Thin-walled parts require higher mold temperatures.

2.13 SHRINKAGE

See the Mold Shrinkage MPO.

Mold Shrinkage Characteristics (in./in. or m/m)

MATERIAL _____ Acetal _____

IN THE DIRECTION OF FLOW

THICKNESS

MATERIAL	mm: 1.5	3	6	8	FROM	TO	#
	in: 1/16	1/8	1/4	1/2			
Acetal	0.022	0.022	0.022	0.022			
Glass-reinforced acetal homopolymer	0.013	0.013	0.013	0.013			
GC acetal copolymer	0.004	0.004	0.004	0.004			

	PERPENDICULAR TO THE DIRECTION OF FLOW					
Acetal	0.022	0.022	0.022	0.022		
Glass-reinforced acetal homopolymer	0.021	0.021	0.021	0.021		
GC acetal copolymer	0.018	0.018	0.018	0.018		

NOTE: Mold shrinkage is approximate. It is affected by part design, mold temperature, thickness, injection pressure, packing time, cycle time, orientation, gate design, gate size, gate location, glass content, glass size, and filler content.

Standard Tolerance Chart

PROCESS ___Injection molding___ MATERIAL ___Acetal___

All data given below are approximate; ±

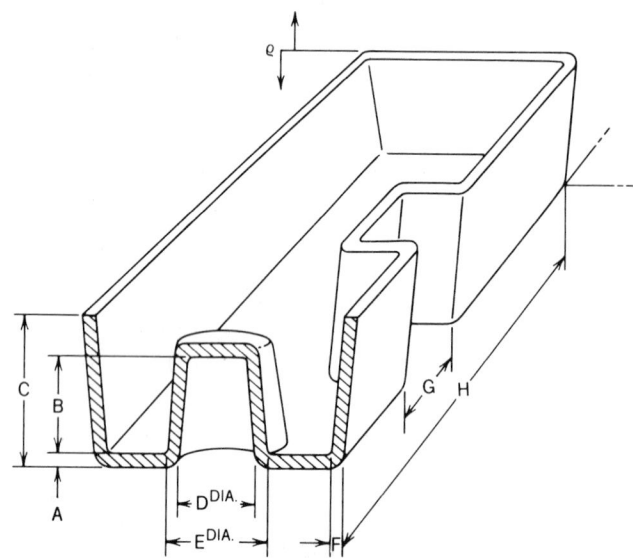

Values are based on a nominal 1/8 in. (3.2 mm) wall thickness and given in plus or minus in./in. and mm/mm.

FINE tolerance — Special care and added cost.
COMMERCIAL tolerance — Normal care required.
COARSE tolerance — Minimal care required, lowest cost.

Tolerance	A	B	C	D	E	F	G	H	
FINE	0.0015	0.0015	0.002	0.0015	0.0015	0.0015	0.002	0.002	in./in.
	0.0015	0.0015	0.002	0.0015	0.0015	0.0015	0.002	0.002	mm/mm
COMMERCIAL	0.002	0.002	0.003	0.002	0.002	0.002	0.003	0.003	in./in.
	0.002	0.002	0.003	0.002	0.002	0.002	0.003	0.003	mm/mm
COARSE	0.004	0.004	0.005	0.004	0.004	0.004	0.005	0.007	in./in.
	0.004	0.004	0.005	0.004	0.004	0.004	0.005	0.007	mm/mm

2.14 TRADE NAMES

Acetals are available under the following trade names from these manufacturers:

Homopolymer: as Delrin, from E.I. Du Pont de Nemours & Co., Plastics Dept., 13020 Du Pont Bldg., Wilmington, DE 19898.

Copolymer: Hoechst Celanese Corp., Engineering Plastics Division, as Celcon, from 26 Main St., Chatham, NJ 07928; and as Ultraform, from Badische Corp., Pureland Industrial Complex, P.O. Box 405, Bridgeport, NJ 08014.

Several custom compounders also provide specialty grades based on the technology described in the text of this section.

2.15 INFORMATION SOURCES

2.15.1 Suppliers

Additional information can be obtained from the suppliers. Some guides that are available are Badische: Ultraform Polyacetal, Product Line, Properties, Processing; Hoechst Celanese: Celcon Acetal Copolymer, Bulletin CE-1A, Properties; Bulletin CE-3A, Injection Molding; Bulletin CE-3B, Extrusion; and Design Manual; E.I. Du Pont de Nemours & Co., Inc.: Delrin Acetal Resin, General Guide to Products and Properties; Delrin Design Handbook, and Delrin Molding Guide.

BIBLIOGRAPHY

1. A.M. Butlerov, *Ann.* **3** 242 (1859).
2. H. Staudinger, *Die Hockmolecularen Organischen Verbindungen,* Springer, Berlin, 1932.
3. U.S. Pat. 2,768,994 (Oct. 30, 1956) R.N. McDonald (to E.I. du Pont de Nemours & Co., Inc.).
4. U.S. Pat. 3,027,352 (March 27, 1962) C. Walling, F. Brown and K. Bortz (to Celanese Corp.).
5. *Mod. Plast.* (Jan. 1961).
6. J. Kroschwity, ed. *Encyclopedia of Polymer Science and Engineering,* Vol. 1, John Wiley & Sons, New York, 1985.
7. Brit. Patent 770,717; U.S. Pat. 2,768,994; U.S. Pat. 2,841,570; and U.S. Pat. 3,172,736.
8. Belg. Pat. 728,306; Fr. Pat. 1,221,148; Belg. Pat. 602,860; and U.S. Pat. 3,156,671.

3

ACRYLONITRILE–BUTADIENE–STYRENE (ABS)

C.T. Pillichody and Paul D. Kelley

GE Plastics, Technology Center, Parkersburg, WV 26101

3.1 INTRODUCTION

ABS plastics are polymerized from the three monomers acrylonitrile, butadiene, and styrene. The family name is based on the first letter of each of these monomers. These plastics consist of a continuous phase of styrene–acrylonitrile (SAN), with a dispersed phase of butadiene-type rubber onto which a similar SAN copolymer has been grafted.

Styrene   Acrylonitrile   Butadiene

3.2 CATEGORY

ABS is not a single material but represents a family of thermoplastic polymers that are hard and tough. ASTM D1788 defines ABS as a plastic containing at least 13% acrylonitrile, 5% butadiene, and 15% styrene.[1] Through variations in composition, molecular weight, and morphology of the rubber phase, the ABS family exhibits a wide range of properties.

3.3 HISTORY

Plastic materials based on acrylonitrile, butadiene, and styrene were introduced in the late 1940s. They were mechanical blends of a copolymer of styrene and acrylonitrile with buna-N (NBR) rubber. A patent for a material of this type was granted in 1948 to L.E. Daly of the U.S. Rubber Co.[2] Suggested applications included golf-ball covers, golf-club heads, protective helmets, screwdriver handles, and telephone switchboard panels. Converting methods suggested for forming these articles were calendering and compression molding. However, the materials were thermoplastic and, therefore, could be reprocessed.

Improvements were made through the early 1950s and a considerable volume was used, particularly in pipe and fittings. However, growth was hindered by poor processibility, poor color stability, and poor low-temperature impact strength, combined with relatively high cost.

In that same decade a breakthrough came with the discovery of methods to graft styrene–acrylonitrile copolymer onto discrete cross-linked particles of polybutadiene.[3] The grafted copolymer served to make the rubber phase compatible with the rigid styrene–acrylonitrile copolymer phase. Through subsequent years, much of the R&D effort in ABS products has been concentrated on creating better elastomeric grafts for use in ABS compositions.

According to data released by the Society of the Plastics Industry, 1987 production capacity for the three U.S. producers of ABS was estimated at 1.175×10^9 lb. Total domestic sales in 1987 were 1.190×10^9 lb, a 10.2% increase over 1986 sales.[4]

Sales of ABS in 1960 were 65×10^6 lb, 1.2% of total U.S. plastics consumption; in 1970, 550×10^6 lb, 3.2% of total U.S. consumption; and in 1980, 1×10^9 lb, 2.9% of total U.S. consumption.

3.4 POLYMERIZATION

Each of the monomers used in ABS can be homopolymerized and the homopolymers are each offered commercially.

Polyacrylonitrile is a rigid, amorphous material with outstanding chemical and solvent resistance. It decomposes before melting and thus is not of importance in articles fabricated by melt processes. However, it can be spun into fibers from a solvent solution. These have been sold under the trade names Orlon (DuPont) and Acrilan (Monsanto).

Polybutadiene is a rubbery material that maintains its flexibility down to extremely low temperatures [below $-76°F\ (-60°C)$]. The presence of double bonds in the polymer permits cross-linking to a thermoset rubber. It is available commercially from many companies, both as a homopolymer and copolymerized as SBR and NBR rubbers. Butadiene contributes a low glass-transition temperature T_g to these elastomers.

Polystyrene is an amorphous thermoplastic that flows easily but is hard and brittle in its unmodified form.

3.4.1 Chemistry

The monomers that constitute ABS are typically polymerized through free-radical reactions. In the ABS manufacturing process, three distinct polymerization reactions take place. First, the elastomeric component, polybutadiene or butadiene copolymer, is produced. Second, styrene and acrylonitrile are copolymerized (grafted) onto this elastomer for compatibility. Third, styrene and acrylonitrile are copolymerized to form the rigid continuous phase.

The elastomeric component is produced separately by emulsion or solution polymerization. This polymer is utilized as a raw material in the ABS polymerization process. The second and third of these polymerizations may take place simultaneously in a single process step. The broad family of different ABS grades results from the many variables that can be modified to achieve a desired property balance. Some of them are listed below.

Rigid phase:
   Ratio of styrene to acrylonitrile
   Molecular weight of the SAN copolymer
   Molecular weight distribution of the SAN copolymer
Elastomeric phase:
   Type and amount of elastomer
   Size of elastomer particles

Size distribution of elastomer particles
Amount of grafting on the elastomer particles
Molecular weight of the grafted SAN copolymer

Additional monomers:
Type and quantity

Polyblends with other polymers:
Type and quantity

Other variables include the use of lubricants, stabilizers, plasticizers, and process aids.

ABS is produced commercially by emulsion, suspension, and mass (bulk) polymerization. ABS materials are also produced commercially as melt-mixed blends of polymers (polyblends), which may have been polymerized by two or more of these processes. Whatever process combinations are used, the final form of the polymer consists of discrete elastomeric particles dispersed in a rigid SAN continuum.

Compositions of commercial ABS materials range from 15 to 30% acrylonitrile, 5 to 30% butadiene, and 45 to 75% styrene.

3.4.2 Other Monomers

The property range of "ABS" can be broadened still further by the use of additional monomers in the polymerization process. In general, these monomers also contain double bonds and copolymerize through reactions similar to the A, B, and S. For example, in some commercial resins, α-methyl styrene or maleic anhydride have been used to increase the heat-deflection temperature. Methyl methacrylate is copolymerized in transparent grades to modify the refractive index.

3.4.3 Compounding

ABS resins are generally melt compounded with such additives as lubricants, stabilizers, and pigments and then formed into pellets. This can be a separate step, utilizing a melt-mixing device like an extruder or Banbury mixer. Recent process innovations sometimes include the processing additives and a melt-pelleting step as the final stage in the polymerization process. Although this has the advantage of reducing costs by eliminating the compounding step, it is not well-suited to producing a variety of colored products, and decreases the flexibility of the process.

3.4.4 Pricing

Early in 1988, a medium-impact grade of ABS listed at $1.00–$1.16/lb.

The complexity of the manufacturing process results in a relatively high production cost for ABS. As with all other plastics, ABS prices vary with order size, reflecting the economics of longer runs and the use of larger equipment. Costs also vary with color, as a result of both pigment costs and the degree of difficulty in maintaining the specific color in production. Some of the major additives (for example, pigments, stabilizers, and halogenated flame retardants) are considerably more expensive and have a significant effect on resin costs, particularly in the specialty grades.

Because of these wide compositional differences within the ABS family of products, prices vary over a significant range.

3.5 PROPERTIES

The degrees of freedom permitted through compositional and structural variations make possible ABS grades with a broad range of strength-to-toughness property balances. The rigid phase provides strength; while toughness is primarily a function of the elastomeric content and the molecular weight of the rigid phase. Since both strength and toughness properties cannot be maximized simultaneously, many ABS grades have been designed to meet specific application or market requirements. This means certain property tradeoffs are necessary in the development of any specific grade.

3.5.1 Impact Resistance

In most applications, impact resistance is the reason for the selection of ABS. Izod impact values of ABS range from 2 to 10 ft-lb/in. (107 to 534 J/m). Typically, ABS grades with higher impact resistance exhibit lower strength, rigidity, hardness, and heat-deflection properties.

The impact resistance of ABS is dependent on the rate of loading and the environmental conditions as well as on the sample preparation used. ABS materials deform in a ductile manner at high strain rates and over a broad temperature range. However, the failure mechanism can shift from ductile to brittle under certain combinations of these two conditions, with the transition dependent on the composition and structure of the ABS grade. Studies have shown the ductility of a high impact grade of ABS to be maintained up to a test speed of approximately 5500 in./min (2.33 m/s) when tested at −20°F (−29°C).[5]

ABS materials are generally less sensitive to notch radii than most other polymers. The critical thickness effects, where significant reductions in impact occur in some materials, does not occur with ABS.

3.5.2 Strength

To obtain increased strength and modulus properties in ABS, some sacrifice in impact and elongation is required. However, additional benefits that are achieved along with higher strength are improved heat distortion, surface hardness, and warp resistance.

The curve shown in Figure 3.1 illustrates the stress–strain behavior of ABS. The proportional limit, which represents the termination of the linear relationship of stress to strain, generally occurs between 0.50 and 0.75% of strain level; it is only slightly influenced by temperature and strain rate. Yield stress is important when potential failure produced by high loads is a major concern. Both yield stress and modulus properties vary directly with changes in the strain rate and inversely with temperature. The yield strain for ABS materials

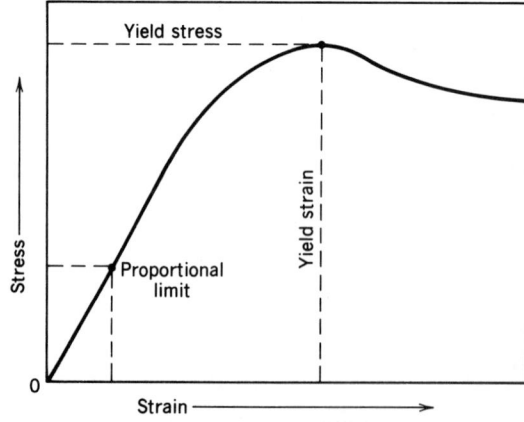

Figure 3.1. Typical stress–strain curve for ABS polymers.

is typically in the 2.5–3.6% range and is characterized by a significant whitening of the material as a result of craze formation and separation of the elastomeric from the rigid phase.

When it is subjected to compressive or flexural loading, ABS will exhibit substantially higher yield strength than that indicated by its tensile properties. However, the response to load at low strains is essentially the same for all three modes of strain. As a result, the tensile, compressive, and flexural tangent modulus values for a specific ABS grade are equivalent.

3.5.3 Creep and Stress Relaxation

Like all thermoplastic polymers, ABS will undergo dimensional changes or force decay when it is subjected to loading for extended periods of time. Commonly referred to as creep and stress relaxation, these phenomena are functions of the viscoelasticity of the polymer. The degree to which either will occur is dependent on the environment, the load or deflection, the length of time, and the characteristics of the ABS grade. Figure 3.2 illustrates the influence of temperature on the creep characteristics of a general-purpose ABS and also shows the apparent creep modulus derived from the creep data. It has been demonstrated that when the initial strain is within the modulus accuracy limit, the creep modulus and relaxation modulus are similar and may be assumed to be the same for design purposes.[6]

Generally, the higher impact grades will exhibit the lowest creep resistance. At elevated temperatures, the highest resistance to creep and stress relaxation is provided by high-heat grades.

3.5.5 Fatigue

When they are tested by the conventional flexural fatigue test at 1700 cycles per minute, ABS materials exhibit a fatigue endurance limit of 0.50 to 0.75% strain. While these figures may be used as a design guideline; actual end-use testing is recommended for material selection.

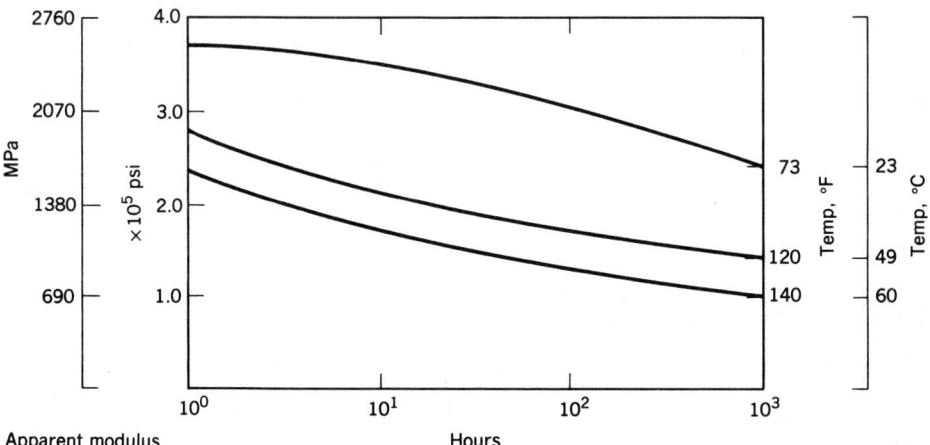

Figure 3.2. Temperature dependence of the apparent creep modulus of a high heat ABS. Data based on flexural creep testing under 1000-psi (6.89-MPa) load.

3.5.6   Heat Deflection

The heat-deflection properties of ABS are also dependent on polymer composition and structure. Higher heat grades contain a lower level of the elastomeric phase and consequently demonstrate lower impact resistance. As with other amorphous polymers, ABS exhibits a relatively flat stress–temperature response. Therefore, heat-deflection temperatures determined according to ASTM D648 at loads of 66 psi (0.45 MPa) and 264 psi (1.8 MPa) differ only 7–14°F (4–8°C). Annealing will increase heat-deflection temperatures an additional 18°F (10°C).

3.5.7   Flammability

ABS grades are available to meet various standards for flammability performance. The general-purpose (nonflame-retardant) grades are generally recognized as 94HB according to Underwriters' Laboratories UL94[7] and are used in applications with a low fire risk. Special flame-retardant (FR) grades have been developed based on alloys with PVC or through the use of halogen–antimony oxide systems. Included are FR grades that meet UL94 VO requirements at a minimum thickness of 0.058 in. (1.4 mm).

3.5.8   Optical Properties

In an unpigmented state, ABS is generally considered to be opaque. However, it is translucent in thin sections. Transparent grades with the property characteristics of ABS are available that are based on copolymers of methyl methacrylate, acrylonitrile, butadiene, and styrene. Transparency results from the addition of the methyl methacrylate in an amount that reduces the refractive index of the rigid phase to match that of the elastomeric phase.[8]

Transparent grades exhibit light transmission of 80–87% at 73°F (23°C) and 0.125 in. (3.2 mm) thickness. Haze ranges from 6–12%; refractive index from 1.51–1.54. Commercial grades have phase-refractive indexes matched for maximum clarity at room temperature. A major change in temperature thus will result in a decrease in light transmission and an increase in haze.

3.5.9   Ultraviolet Resistance

Exposure to ultraviolet (UV) radiation will induce irreversible chemical changes in most organic polymers, including ABS. The extent of degradation depends on the polymer structure, the influence of prior processing conditions, the effects of additives, the exposure conditions, and the presence of protective coatings. UV degradation results in appearance changes characterized by increasing yellowness and an initial embrittlement of the plastic's outer surface.

ABS is used successfully in many products that are subjected to outdoor exposure for brief periods of time. However, for extended outdoor exposure, measures must be taken to minimize property reduction and changes in appearance. Protection is afforded by UV stabilizers, pigments, paints, and film laminates.

3.5.10   Chemical Resistance

The use of acrylonitrile as a comonomer, combined with low water absorption properties, gives ABS materials resistance to attack from a wide range of chemicals. Simple tests[9] show ABS to be resistant to staining from coffee, ketchup, tea, vinegar, and other household items.

In total immersion testing (ASTM D543) involving no external stresses on the specimen during exposure,[1] ABS materials are almost completely resistant to aqueous acids, alkalies, salts, low KB solvents, alcohols, and animal, vegetable, and mineral oils. Concentrated oxidizing acids such as nitric and sulfuric acids will result in disintegration of the material; concentrated phosphoric and hydrochloric acids have little effect.

ABS will undergo stress cracking when it comes in contact with certain chemical agents while the part is under stress. For example, exposure to corn oil under the conditions described for simple immersion testing would have no effect. However, exposure to this same medium under high stress can result in part embrittlement and subsequent failure. It should be recognized that chemical resistance is dependent on the temperature, the media concentration, the exposure time, and the stress present in the application. Consequently, all media that will be in contact with a proposed product should be evaluated under anticipated end-use conditions.

3.5.11 Reinforcement

The addition of glass fillers provides an increase in strength and modulus properties, accompanied by a decrease in overall ductility. Impact resistance can be significantly reduced by glass content as low as 5%, yet remain essentially unaffected by further glass addition. Strength and modulus properties, however, are dependent on glass content, increasing with increasing glass level. Unlike crystalline polymers, the addition glass fillers to ABS does not result in a significant extension of the upper use-temperature range; this is primarily a function of the T_g of the rigid phase.[10] Therefore, although glass reinforcement results in some improvement in modulus and creep properties at high temperatures, it is not of the magnitude expected from the improvement seen in room temperature properties. Glass-filled ABS also has a lower coefficient of thermal expansion and reduced mold shrinkage.

The degree of glass-fiber orientation must be considered in application design. A molded part in which the degree of fiber orientation is very high will exhibit lower strength and stiffness properties perpendicular to the fiber orientation. This may also cause problems in dimensional tolerances, since the coefficient of expansion and mold shrinkage will both be higher perpendicular to the fiber orientation.

3.6 APPLICATIONS

3.6.1 Appliances

Typical applications include injection-molded housings for blenders, mixers, microwave doors, and other kitchen appliances; for floor-care products such as vacuum sweepers and scrubbers; and for other categories of appliances such as hair dryers and power tools. The largest-volume application for ABS in the appliance market is refrigerator door and tank liners. For this particular application, extrusion grades have been developed to meet the requirements of chemical resistance, low-temperature toughness, high surface gloss, and good hot strength for deep-draw thermoforming. A transparent grade is also used for crisper trays.

3.6.2 Automotive

ABS high-heat grades are used for instrument panels, pillar-post moldings, light housings, and other interior trim. These applications require ease of processing combined with functional part distortion temperatures of 176–230°F (80–110°C). In addition to property balances and excellent color consistency, ABS materials were traditionally selected for their high surface gloss. Recently however, consumer preference for the matte finish interiors of the imports have led U.S. automobile designers to specify similar surfaces for ABS parts. As a result, low-gloss ABS grades have been developed to provide the same functional part performance as their high-gloss counterparts.

ABS is also used by the automotive industry because it can easily be decorated. ABS finishes can be smooth or embossed, vinyl covered or wood-grained, painted, vacuum metallized, sputter coated, or even chrome plated. It is estimated that ABS plating grades account for nearly 65% of the plated plastics used in automotive applications. Excellent plate adhesion, low coefficient of expansion, low-temperature ductility, and high heat resistance permit these grades to pass very stringent thermal cycling tests. Typical applications include knobs, light bezels, mirror housing, wheel covers, decorative trim, and grilles.

3.6.3 Building and Construction

Not long ago, the largest single application for ABS was DWV (drain, waste, and vent) pipe and fittings. Two factors leading to a reduction in ABS use were a depressed building industry and a significant cost differential that favored PVC pipe. This differential has since been offset by price increases for PVC and the introduction of ABS foam-core coextrusion technology to provide pipe with lower part density at no sacrifice of required performance properties. Other applications in the building and construction market include roof ventilators, bathtub surrounds, and plated plumbing fixtures.

3.6.4 Business Machines/Consumer Electronics

Application requirements in the business machine and consumer electronics market vary considerably. General-purpose ABS is suitable for applications such as telephones; business machines require a flame-retardant grade. Typical business machine applications include housings, covers, and consoles. The ease with which ABS can be electroplated or vacuum metallized to meet EMI/RFI (electromagnetic or radio frequency interference) shielding requirements is another important advantage.

3.6.5 Other Applications

Low-temperature impact resistance and excellent thermoformability is the basis for selection of ABS in such applications as boat hulls, snowmobile shrouds, and camper components. UV protection for these applications is achieved through the lamination of an acrylic film or a coextruded UV-resistant cap. Other ABS applications include luggage, toys, furniture, and impact modifiers for rigid PVC.

3.7 ADVANTAGES/DISADVANTAGES

The family of ABS products provides a broad range of engineering performance, with a balance of properties at a price that effectively fills a gap between the low-cost commodity plastics and the higher-priced specialty plastics.

Rather than exploiting a unique single property, the utility of ABS lies in its balance of properties. It has excellent chemical resistance and impact resistance with fairly high strength, combined with good processibility and good aesthetics at a reasonable price. For many applications that require higher performance properties than those available in polystyrene, PVC, or polypropylene, a "stepping up" to ABS is successful. Conversely, all of the premium properties of metal or higher priced plastics are not needed in many of their applications, and significant savings can be realized in "stepping down" to ABS.

3.8 PROCESSING TECHNIQUES

ABS materials can be processed by injection molding, extrusion, blow molding, or calendering. However, injection molding and extrusion account for more than 93% of all ABS material usage. Secondary processing operations commonly employed include thermoforming, electroplating, film lamination, ultrasonic welding, and adhesive bonding. ABS materials can also be cold formed, but this technique is not used extensively.

3.8.1   Injection Molding[11]

ABS grades can be injection molded using reciprocating screw or plunger machines; the former is preferred because it provides a more uniform melt and higher available pressure. Processing temperatures range from 350 to 550°F (177 to 288°C), depending on the specific grade. Injection pressures of 10,000–20,000 psi (69–138 MPa) and a clamp pressure of 2–3 t/in. (281–422 kg/cm^2) of projected part surface are usually sufficient. Screws having a compression ratio of 2:1–3:1 and a L/D ratio of 20:1 are recommended.

3.8.2   Extrusion[12]

An extruder having a minimum L/D ratio of 24:1 is recommended to ensure that a uniform melt temperature is delivered to the die. Single- or two-stage screws are acceptable, but the latter is preferred since it also aids in devolatilization. A single-lead, full-flighted, constant-pitch screw with a progressively increasing root diameter and a compression ratio of 2:1–2.5:1 will be suitable for most ABS grades. Straight or coat-hanger type manifold dies can be used, but the coat-hanger type will provide more streamlined material flow and consequently minimize material hangup.

3.8.3   Thermoforming[13]

ABS can be thermoformed over a temperature range of 250–375°F (121–190°C), with the optimum conditions dependent on material grade, part design, draw ratio, sheet thickness, and forming technique. Acceptable techniques include drape forming, plug assist, snapback, pressure forming, or a combination of these. The depth of draw in simple forming is usually limited to the width of the part. However, higher draw ratios are possible if more sophisticated forming techniques are employed or if uniform wall thickness is not critical.

3.8.4   Cold Forming[14]

Some ABS grades can be cold formed using standard metalworking equipment. A reduction in diameter of up to 45% is possible on the first draw, with up to 35% reduction in subsequent redraws. Lubrication is required for successful drawing. An aqueous-type lubricant is preferred for application at the press; a nonaqueous type is usually applied as a precoat. Highly stressed areas may exhibit stress whitening, which can be minimized by reducing the induced stress or masked by white pigment.

3.9   RESIN FORMS

ABS formulations are commercially available as powders (resin with stabilizer but without additives or pigments), beads (resin with stabilizer and lubricants but without pigments), and pellets (resin fully compounded with additives and pigments).

Pellets are available in a variety of grades, formulated to provide higher values for a specific property or group of properties. The available pellet grades include: molding grades, blow-molding grades, extrusion grades, higher flow grades, high-heat grades, transparent grades, foam grades, glass-filled grades, electroplating grades, ABS–PVC alloys, flame-retardant grades, low-gloss grades, ABS–polycarbonate alloys, and ABS–polysulfone alloys.
retardant grades, low-gloss grades, ABS–polycarbonate alloys, and ABS–polysulfone alloys.

3.10   MATERIAL PROPERTIES

Property ranges for the various ABS grades are shown in the accompanying Master Outline of Materials Properties.

MATERIAL ___General-purpose ABS, High Impact___

PROPERTY	TEST METHOD	ENGLISH UNITS	VALUE	METRIC UNITS	VALUE
MECHANICAL DENSITY	D792	lb/ft^3	62.9–64.8	g/cm^3	1.01–1.04
TENSILE STRENGTH	D638	psi	4,800–6,300	MPa	33–43
TENSILE MODULUS	D638	psi	240,000–330,000	MPa	1,650–2,270
FLEXURAL MODULUS	D790	psi	250,000–350,000	MPa	1,720–2,410
ELONGATION TO YIELD	D638	%	2.8–3.5	%	2.8–3.5
NOTCHED IZOD AT ROOM TEMP	D256	ft-lb/in.	6.5–10.0	J/m	347–534
HARDNESS	D785	Rockwell R	80–105		80–105
THERMAL DEFLECTION T @ 264 psi	D648	°F	205–215	°C	96–102
DEFLECTION T @ 66 psi	D648	°F	210–225	°C	99–107
VICAT SOFT-ENING POINT	D1525	°F	196–223	°C	91–106
UL TEMP INDEX	UL746B	°F		°C	60–70
UL FLAMMABILITY CODE RATING	UL94		HB		HB
LINEAR COEFFICIENT THERMAL EXPANSION	D696	in./in./°F	5.3–6.1 ×10^{-5}	mm/mm/°C	9.5–11.0 ×10^{-5}
ENVIRONMENTAL WATER ABSORPTION 24 HOUR	D570	%	0.20–0.45	%	0.20–0.45
CLARITY	D1003	% TRANS-MISSION	Opaque	% TRANS-MISSION	Opaque
OUTDOOR WEATHERING	D1435		Fair		Fair
FDA APPROVABLE			Yes		Yes
CHEMICAL RESISTANCE TO: WEAK ACID	D543				
STRONG ACID	D543				
WEAK ALKALI	D543				
STRONG ALKALI	D543				
LOW MOLECULAR WEIGHT SOLVENTS	D543				
ALCOHOLS	D543				
ELECTRICAL DIELECTRIC STRENGTH	D149	V/mil	400–800	kV/mm	16–31
DIELECTRIC CONSTANT	D150	10^6 Hertz	2.4–3.8		2.4–3.8
POWER FACTOR	D150	10^6 Hertz	0.007–0.015		0.007–0.015
OTHER MELTING POINT	D3418	°F	amorphous	°C	amorphous
GLASS TRANS-ITION TEMP.	D3418	°F	212–239	°C	100–110

Master Outline of Material Properties

MATERIAL ___ General-purpose ABS, Medium Impact

PROPERTY	TEST METHOD	ENGLISH UNITS	VALUE	METRIC UNITS	VALUE
MECHANICAL DENSITY	D792	lb/ft^3	64.2–66.1	g/cm^3	1.03–1.06
TENSILE STRENGTH	D638	psi	4,300–7,500	MPa	30–52
TENSILE MODULUS	D638	psi	300,000–400,000	MPa	2,070–2,760
FLEXURAL MODULUS	D790	psi	320,000–440,000	MPa	2,200–3,030
ELONGATION TO YIELD	D638	%	2.3–3.5	%	2.3–3.5
NOTCHED IZOD AT ROOM TEMP	D256	ft-lb/in.	2.5–6.0	J/m	134–320
HARDNESS	D785	Rockwell R	105–112		105–112
THERMAL DEFLECTION T @ 264 psi	D648	°F	220–220*	°C	93–104*
DEFLECTION T @ 66 psi	D648	°F	215–225*	°C	102–107*
VICAT SOFTENING POINT	D1525	°F	201–225	°C	94–107
UL TEMP INDEX	UL746B	°F		°C	60
UL FLAMMABILITY CODE RATING	UL94		HB		HB
LINEAR COEFFICIENT THERMAL EXPANSION	D696	in./in./°F	3.9–4.8 ×10^{-5}	mm/mm/°C	7.0–8.8 ×10^{-5}
ENVIRONMENTAL WATER ABSORPTION 24 HOUR	D570	%	0.20–0.45	%	0.20–0.45
CLARITY	D1003	% TRANSMISSION	Opaque	% TRANSMISSION	Opaque
OUTDOOR WEATHERING	D1435	%	Fair	%	Fair
FDA APPROVABLE			Yes		Yes
CHEMICAL RESISTANCE TO: WEAK ACID	D543				
STRONG ACID	D543				
WEAK ALKALI	D543				
STRONG ALKALI	D543				
LOW MOLECULAR WEIGHT SOLVENTS	D543				
ALCOHOLS	D543				
ELECTRICAL DIELECTRIC STRENGTH	D149	V/mil	400–800	kV/mm	16–31
DIELECTRIC CONSTANT	D150	10^6 Hertz	2.4–3.8		2.4–3.8
POWER FACTOR	D150	10^6 Hertz	0.007–0.015		0.007–0.015
OTHER MELTING POINT	D3418	°F	amorphous	°C	amorphous
GLASS TRANSITION TEMP.	D3418	°F	221–239	°C	105–115

*Heat-deflection values are for annealed specimens

35

MATERIAL ___ Heat-resistant ABS

PROPERTY	TEST METHOD	ENGLISH UNITS	VALUE	METRIC UNITS	VALUE
MECHANICAL DENSITY	D792	lb/ft^3	65.4–67.3	g/cm^3	1.05–1.08
TENSILE STRENGTH	D638	psi	6,000–7,500	MPa	41–52
TENSILE MODULUS	D638	psi	300,000–380,000	MPa	2,070–2,620
FLEXURAL MODULUS	D790	psi	300,000–400,000	MPa	2,070–2,760
ELONGATION TO YIELD	D638	%	2.8–3.5	%	2.8–3.5
NOTCHED IZOD AT ROOM TEMP	D256	ft-lb/in.	2.0–6.5	J/m	107–347
HARDNESS	D785	Rockwell R	100–111		100–111
THERMAL DEFLECTION T @ 264 psi	D648	°F	220–240	°C	104–116
DEFLECTION T @ 66 psi	D648	°F	230–245	°C	110–118
VICAT SOFT- ENING POINT	D1525	°F	219–244	°C	104–118
UL TEMP INDEX	UL746B	°F		°C	60
UL FLAMMABILITY CODE RATING	UL94		HB		HB
LINEAR COEFFICIENT THERMAL EXPANSION	D696	in./in./°F	3.6–5.1 ×10^{-5}	mm/mm/°C	6.5–9.2 ×10^{-5}
ENVIRONMENTAL WATER ABSORPTION 24 HOUR	D570	%	0.20–0.45	%	0.20–0.45
CLARITY	D1003	% TRANS- MISSION	Opaque	% TRANS- MISSION	Opaque
OUTDOOR WEATHERING	D1435		Fair		Fair
FDA APPROVABLE			Yes		Yes
CHEMICAL RESISTANCE TO: WEAK ACID	D543				
STRONG ACID	D543				
WEAK ALKALI	D543				
STRONG ALKALI	D543				
LOW MOLECULAR WEIGHT SOLVENTS	D543				
ALCOHOLS	D543				
ELECTRICAL DIELECTRIC STRENGTH	D149	V/mil	350–900	kV/mm	14–35
DIELECTRIC CONSTANT	D150	10^6 Hertz	2.4–3.8		2.4–3.8
POWER FACTOR	D150	10^6 Hertz	0.007–0.015		0.007–0.015
OTHER MELTING POINT	D3418	°F	amorphous	°C	amorphous
GLASS TRANS- ITION TEMP.	D3418	°F	230–257	°C	110–125

3.11 PROCESSING REQUIREMENTS

It is just as important to pick the correct processing conditions as the correct ABS grade since both are essential to a successful application.

3.11.1 Drying[15]

ABS plastics are mildly hygroscopic and must be dried before melt processing to avoid surface defects on finished parts. These defects (such as splay, silver streaks, or surface pits) provide evidence of excessive moisture in the material during melt processing. Moisture present during processing does not produce a permanent chemical change (degradation). If it is severe, it may affect ductility; but if the defective part is reground, and the scrap dried and remolded, the original property values will be obtained.

The objective of drying is to bring all the pellets to a moisture level adequate to produce satisfactory parts. In nonvented equipment, these moisture levels are general-purpose molding, 0.1%; molding for electroplating, 0.05%; and extrusion processing, 0.02%.

Generally, these levels can be obtained by subjecting pellets to temperatures of 190°F (88°C) for 2–4 h. Four factors are important to good drying of ABS:

Dry air. A dew point of 0°F (– 18°C) is sufficient.

Temperature. This speeds up the drying time so it can keep up with production rate. Increasing the temperature 20°F (11°C) cuts drying time in half. A low dew point is *not* a substitute for temperature.

Time. The pellets must be hot enough for a period long enough to achieve dryness. Hot residence time is the key: the drying time begins when the pellets are up to temperature.

Air flow. This is an essential factor for good heat transfer. Good air flow is necessary to maximize temperature for pellets in the top of a hopper and gain time at temperature.

Large regrind granules can take several times longer to achieve adequate drying levels than pellets. For better drying, it is important to protect the regrind from picking up moisture and to keep regrind particles as small as practical.

3.11.2 Degradation

Like all thermoplastic materials, ABS resins and alloys are not indestructible and should be processed according to manufacturers' recommendations. This is particularly true for ABS alloys, which must be processed within the limitations of the more sensitive of the alloyed materials. For example, ABS–PVC alloys must be processed within the temperature limits imposed by poly(vinyl chloride).

Signs of material degradation can usually be recognized by color streaking or part discoloration. Severe degradation can also be accompanied by brittleness, specks of degraded polymer or excessive smoke and fumes.

3.11.3 Regrind

Sprues, runners, and trimmings from the same grade and color can be reused. The particle size of regrind should be as small as practical, but fines should be avoided. Rework must be kept clean and free of contamination from other plastics and nonplastic trash. The percentage of regrind that can be satisfactorily blended with virgin material depends almost entirely on the quality (cleanliness and heat history) of the regrind and the end-use requirements. Specific recommendations should be obtained from the manufacturer for the grade of interest.

3.12   PROCESSING-SENSITIVE END PROPERTIES

The properties of an ABS part may depend as much on the processing conditions as on the material grade. Processing at temperatures higher than recommended can result in thermal degradation of the polymer. This, in turn, can result in a part with poor impact performance and, possibly, excessive color shift.

Even when a part is injection molded within the recommended temperature range, its properties may shift significantly as a result of orientation. Studies have shown that certain properties, particularly impact, are very dependent on the processing conditions.[16] Generally, when ABS is molded under conditions of a low melt temperature and a slow fill rate, a high level of orientation will result. The molded part will exhibit a high degree of anisotropy with significant differences in the notched Izod impact and shrinkage values, as measured with and across the direction of major orientation. These conditions also contribute to poor dart impact performance and increased tendency to warp.

Increasing the melt temperature toward the upper limit and increasing the fill rate reduces the anisotropy. Under these conditions, the Izod impact values measured with and across the direction of orientation approach one another, and the dart impact resistance is maximized. The resistance to thermal warpage is also improved, but this will be reduced if excessive packing pressures are used. Tensile and flexural strength are also affected by orientation, but the effect on part performance is usually not significant.

When an ABS plating grade is molded, slow fill rate/high melt temperature conditions should be used. Although these conditions result in a somewhat higher bulk orientation, they also provide for lower skin orientation, which is conducive to improved plate adhesion.

Orientation plays a similar role in influencing the properties of an extruded ABS part. A major factor controlling the level of induced orientation in extrusions is the draw-down ratio. Higher draw-down ratios result in higher levels of orientation.

Die lines, which result from material buildup on the die lips, detract from the part appearance and can affect properties. Sheet or pipe having die lines usually exhibit lower strength properties perpendicular to the direction of extrusion and lower dart impact resistance.

3.12.1   Molding Conditions to Maximize Specific Properties

Molding conditions can be varied to maximize specific properties within the property range of the material being molded. Thus, the molding conditions selected are a very important factor in achieving an acceptable balance of property levels (see Table 3.1).

TABLE 3.1   Injection Molding Conditions to Achieve Specific Part Properties[a]

To	Stock Temperature	Fill Speed
Decrease orientation	High	Fast
Increase gloss	Medium	Fast
Increase warp temperature	High	Fast
Increase dart impact strength	Medium	Fast
Increase plating performance	High	Slow
Increase tensile strength		
With flow	Low	Slow
Cross flow	Medium	Fast

[a]In all cases, use of *hot mold temperature* and *low packing pressure* contributed favorably to the effect shown

3.12.2 Thermoforming

ABS sheet is easy to thermoform over a relatively wide temperature range. The most important parameter in determining the optimum forming temperature is the average temperature of the sheet. Low-temperature forming yields the best hot strength, minimum spot thinning, and, generally, a shorter cycle. Forming at high temperatures results in lower internal stress (orientation) but increases mold shrinkage; and, the material thickness may be less uniform in the part. Because heat enters the sheet through the surface, the surface is always hotter than the core and can be subject to degradation, particularly under conditions of high temperature intensity (fast heating).

3.13 SHRINKAGE

For ABS, the lot-to-lot variation in mold shrinkage is relatively minor. However, the mold shrinkage of a part can vary considerably, depending on mold geometry and molding conditions. For example, parts with thinner walls will exhibit less shrinkage than parts with thick walls, and a higher stock temperature will promote more shrinkage than a lower stock temperature. Conditions that reduce mold shrinkage include packing the mold after fill or constraining the parts by the mold design.

Mold Shrinkage Characteristics (in./in. or mm/mm)

MATERIAL _____ ABS

IN THE DIRECTION OF FLOW

THICKNESS, in./in.

MATERIAL	mm: 1.5	3	6	8	FROM	TO	#
	in.: 1/16	1/8	1/4	1/2			
Medium/high impact		0.006–0.009					
Heat resistant		0.005–0.008					
Flame retardant		0.005–0.008					

PERPENDICULAR TO THE DIRECTION OF FLOW							

NOTE: Mold shrinkage is approximate. It is affected by part design, mold temperature, thickness, injection pressure, packing time, cycle time, orientation, gate design, gate size, gate location, glass content, glass size, and filler content.

The range of values shown in the accompanying shrinkage table attempts to allow for variations encountered in part geometry and molding conditions. These values will be generally satisfactory for determining tooling dimensions. If a part dimension is critical, it is recommended that a more exact value be determined for the specific geometry and molding conditions of interest.

3.14 TRADE NAMES

There are three producers of ABS in the United States. A number of other companies purchase resin from these producers and compound specialty formulations, such as filled, alloyed, and color concentrates.

3.14.1 Producers of ABS

GE Plastics, Pittsfield, Mass., produces Cycolac ABS pellets, Cycovin ABS–PVC alloy pellets, Cycoloy ABS alloys with other polymers, Blendex ABS resin powders, and Marbon ABS resin powders. Dow Chemical Co., Midland, Mich., produces ABS pellets. Monsanto Co., St. Louis, Mo., produces Lustran ABS pellets and Cadon high-heat ABS pellets.

3.14.2 Specialty Formulators and Compounders

Americhem Inc., Cuyahoga Falls, Ohio
Ampacet Corp., Mount Vernon, N.Y.
Bengal Inc., Sepulveda, Calif. (producer of Benstat)
Comalloy International Corp., Brentwood, Tenn.
Coz Chemical Co., Div. of Allied Products, Northbridge, Mass.
Fleet Plastic and Chemical Co., Haverhill, Mass.
General Tire and Rubber Co., Newcomerstown, Ohio (producer of Boltaron)
LNP Corp., Malvern, Pa.
Mobay Chemical Corp., Pittsburgh, Pa. (producer of Bayblend)
H. Muehlstein and Co., Greenwich, Conn.
Occidental Chemical Corp., Pottstown, Pa. (producer of Oxyloy)
PMS Consolidated, Somerset, N.J.
Ponca Pellets Inc., Goddard, Kan.
A. Schulman, Inc., Akron, Ohio (producer of Polyman, Polyflam)
Thermofil, Inc., Brighton, Mich.
Union Carbide Corp., Danbury, Conn. (producer of Mindel)
Uniroyal Inc., Royalite Div., Mishawaka, Ind. (producer of Royalite)
Washington-Penn Plastics, Washington, Pa.
Wilson-Fiberfil International, Evansville, Ind. (producer of Absafil).

BIBLIOGRAPHY

1. *Annual Book of ASTM Standards: Section 8, Plastics;* Vol. I and II, American Society for Testing and Materials, Philadelphia, Pa., 1983.
2. U.S. Pat. 2,435,202 (1948), L.E. Daly (to U.S. Rubber Co.).
3. U.S. Pat. 3,238,275 (filed 1953, issued 1966), W.C. Calvert (to Marbon Chemical Co.).

4. Statistics released by the Society of the Plastics Industry Inc., *Plastics Industry News* **10**(1), (Feb. 1988).

5. R.C. Turner, "Ductile-Brittle Behavior of High-Impact ABS," paper presented at *Automotive Plastics Durability Conference*, Troy, Mich., Dec. 1981.

6. E. Baer, *Engineering Design for Plastics*, Reinhold Publishing Corp., New York, 1964.

7. *Test for Flammability of Plastics Materials for Parts in Devices and Appliances, UL94*, Underwriters Laboratories, Inc., Northbrook, Ill.

8. B.J. Sexton and D.C. Curfman, "Methyl Methacrylate/Acrylonitrile–Butadiene–Styrene Copolymers," in N.M. Bikales, ed., *Encylcopedia of Polymer Science and Technology*, Vol. 1, John Wiley & Sons, Inc., New York, 1976.

9. *Chemical Resistance Considerations for Cyclolac ABS*, Design Tip #6, GE Plastics, Pittsfield, Mass.

10. M.T. Takenori, "Towards an Understanding of the Heat-Distortion Temperature of Thermoplastics," *Polym Eng Sci*, 1104 (Nov. 1979).

11. *Injection Molding with Cycolac, Cycoloy, and Cycovin ABS Resins and Alloys, Tech. Pub. P-403*, GE Plastics, Pittsfield, Mass.

12. *Extrusion Equipment and Procedures for ABS Resin. Technical Publication PB117A*, GE Plastics, Pittsfield, Mass.

13. *Thermoforming Cycolac ABS, Tech. Pub. P406*, GE Plastics, Pittsfield, Mass.

14. J. Frados, *Plastics Engineering Handbook of the Society of the Plastics Industry*, Inc., 4th ed. Van Nostrand Reinhold, New York, 1976.

15. *Cycolac ABS Pellet Drying, Tech. Pub.* SR601A, GE Plastics, Pittsfield, Mass.

16. L.W. Fritch, "How Mold Temperature and Other Variables Affect Falling Dart and Izod Impact," *paper presented at the Annual Technical Conference of the Society of Plastics Engineers*, 1982.

General References

"ABS Resins," in M. Grayson, ed., *Kirk-Othmer Encyclopedia of Chemical Technology*, 3rd ed., Vol. 1, John Wiley & Sons, Inc., New York, 1978, p. 442.

C.H. Basdekis, *ABS Plasstics*, Reinhold Publishing Corp., New York, 1964.

R.D. Beck, *Plastic Product Design*, Van Nostrand Reinhold, New York, 1970.

S. Levy and J.H. DuBois, *Plastics Product Design Engineering Handbook*, Van Nostrand Reinhold Company, New York, 1977.

Modern Plastics Encyclopedia, McGraw-Hill Inc., New York, published annually.

I.I. Rubin, *Injection Molding Theory and Practice*, Wiley Interscience, New York, 1972.

G. Schenkel, *Plastics Extrusion Technology and Theory*, American Elsevier Publishing Co., New York, 1966.

A.V. Tobolsky, *Properties and Structure of Polymers*, John Wiley & Sons, Inc., New York, 1960.

4

ALLYL RESINS (DAP/DAIP)

Ralph E. Wright and Michael F. Gardner

Rogers Corporation, PO Box 550, Mill & Oakland Sts. Manchester, CT 06040

4.1 INTRODUCTION

Allyl resins are available as monomers and as B-staged prepolymers. The monomers are often used as cross-linking agents and flow promoters for other polyesters; the prepolymers are the base resins for molding compounds and prepregs.

The two most widely used allyl resins are diallyl orthophthalate (DAP) and diallyl isophthalate (DAIP), with the following chemical structures:

Diallyl orthophthalate (DAP)

Diallyl isophthalate (DAIP)

4.2 CATEGORY

DAP and DAIP can be converted to rigid thermosetting molding compounds, preimpregnated glass and papers, and coatings or sealants.

4.3 HISTORY

In 1937, Carlton Ellis discovered that unsaturated polyesters would freely copolymerize with monomers containing double-bond unsaturation, yielding rigid thermosetting resins. This was the origin of the family of allyl resins.

In 1958, FMC was issued U.S. Patent 2,832,758 (based on the work of Helberger and J. Thomas) describing a process for producing Dapon allyl prepolymers. The patent claimed a method for preparing DAP prepolymer from DAP monomer using a peroxide catalyst, water, and a "polymerization modifier" such as dimethyl benzyl alcohol. The resulting resin was a white, fluffy powder.

Dapon 35 orthophthalate and Dapon M isophthalate prepolymers (trademarks of the FMC Corp.) became the base resins for a wide range of reinforced molding compounds, decorative laminates, sealants, and coatings. Additionally, diallyl phthalate monomers have come to be utilized in many polyester molding compounds to improve shelf life as well as chemical and weather resistance and to enhance generally their moldability.

In the late 1960s, Osaka Soda of Japan began supplying ortho and isophthalate prepolymers to U.S. compounders. By the spring of 1980, FMC decided to cease production of prepolymers but continue to produce both monomers. In the mid-1980s, Tohto Kasei of Korea became a producer of DAP prepolymer and there are again two suppliers to the marketplace.

U.S. production of DAP/DAIP molding compound in 1960 totaled 2.5×10^6 lb; in 1981, 7×10^6 lb. For prepolymer, 1960 production was 0.8×10^6 lb; in 1980, 3×10^6 lb.

4.4 POLYMERIZATION

As noted, allyl esters based on monobasic and dibasic acids are available as monomers and prepolymers, with the latter serving as base resin for molding compounds and prepregs.

For general-purpose grades, the current (1988) price range is \$2.25–5.00/lb, depending on the volume and type of DAP molding compound.

4.5 DESCRIPTION OF PROPERTIES

Allyl resins retain their electrical insulating properties under severe environmental conditions involving high heat and high humidity. The molded products are dimensionally stable, strong, heat resistant, and chemical resistant.

4.5.1 Mechanical Properties

The dimensional stability of DAP/DAIP compounds is exceptional. Especially noteworthy is its negligible lifetime shrinkage after molding. Abrasion resistance is another important characteristic.

TABLE 4.1 Various Reinforcements on Selected Properties of DAP/DAIP

Reinforcement Resin (DAP) Form Bulk Factor	Dacron Ortho Flake 4.5	Orlon Ortho Granular 3.0	Mineral Ortho Granular 2.3	Mineral Nylon Ortho Granular 2.6	Short Glass Ortho Granular 2.4	Short Glass Ortho Granular 2.3
Specific gravity g/cm^3	1.5	1.40	1.80	1.73	1.80	1.87
Shrinkage (Comp) in./in.	0.010	0.010–0.012	0.003–0.006	0.007–0.010	0.001–0.003	0.0015–0.003
Izod impact, ft·lb/in. (C/T)	4.5	0.65	0.45	0.45	1.0	1.0
Flexural strength, psi	11.500	9000/12,000	11,000/14,000	10,000/13,000	15,000/18,000	16,000/19,000
Flexural modulus, psi	0.64	0.7×10^6	1.6×10^6	1.1×10^6	1.8×10^6	1.8×10^6
Tensile strength, psi	5000	5000	6500	5500	8000	8000
Compressive strength, psi	30,000	29,000	24,000	25,000	28,000	28,000
Water absorption, %	0.20	0.40	0.30	0.40	0.30	0.25
Deflection temperature, °F		290	330	300	500	450
Continuous use temperature, °F	300	225	230	200	375	350
Flammability (IGN/Burn), s	84	87/440	90/373		70/340	110/40
UL rating 1/8 in.				94 HB		94V-O
Dielectric strength						
60 Hz, ST/SS, wet, Vpm	350	375/350	375/350	400/380	375/350	400/375
Dielectric constant						
1 kHz/1 mHz, wet	3.4	3.4/3.1	4.2/3.6	4.3/3.7	4.1/3.9	4.3/4.1
Dissipation factor						
1 kHz/1 mHz, wet	0.012	0.016/0.021	0.013/0.014	0.015/0.019	0.010/0.017	0.010/0.016
Arc resistance, s	125	125	140	135	135	130
MIL-M-14G Type	SDI-30	SDI-5	MDG	MDG	SDG	SDG-F

4.5.2 Thermal Properties

DAP/DAIP compounds are designed for continual use in the range of 350–450°F (177–232°C). DAIP compounds exhibit heat resistance about 90°F (50°C) higher than DAP; accordingly, DAIP compounds are about 75% more costly. Both DAP and DAIP compounds carry a UL thermal listing of 356°F (180°C) and 392°F (200°C), respectively. Fire-retardant grades are rated 94-VO in 1/16-in. (1.6 mm) sections, per UL94.

Various DAP and DAIP molding compounds have been tested for performance in a vapor-phase reflow soldering environment. All materials were capable of withstanding the 410°F (210°C) environment without dimensional change and without significant loss of physical or electrical properties. In addition, both DAP and DAIP performed well in a 487°F (253°C) vapor phase solder environment.

4.5.3 Reinforcements

Reinforcements systems for DAP and DAIP consist of mineral/organic fibers and short glass/long glass fibers. Table 4.1 provides an overview of the effect of various reinforcements on selected properties of DAP.

Numerous specifications exist for reinforced grades, examples of which are given below.

Military specification MIL-M-14, Revision H, will now differentiate between DAP and DAIP compounds and lists the following designations for DAP:

- MDG: Mineral filler, general purpose.
- MDG-F: Mineral filler, general purpose, flame resistant.
- SDG: Short-glass-fiber filler, general purpose.
- SDG-F: Short-glass-fiber filler, flame resistant.
- GDI-30: Long-glass-fiber filler, nominal impact strength, 3.0 ft-lb/in. notch (160 J/m).
- GDI-30F: Long-glass-fiber filler, flame resistant, nominal impact stregth, 3.0 ft-lb/in. notch (160 J/m).

Long Glass Ortho Coarse Granular 3.5	Long Glass Ortho Coarse Granular 3.5	Mineral Iso Granular 2.1	Short Glass Iso Granular 2.5	Short Glass Iso Granular 2.4	Long Glass Iso Coarse Granular 3.5	Long Glass Iso Coarse Granular 3.5
1.74	1.76	1.90	1.75	1.90	1.73	1.76
0.0015–0.004	0.0015–0.004	0.003–0.006	0.001–0.003	0.001–0.003	0.0015–0.004	0.0015–0.004
3.0–7.0	3.0–7.0	0.45	1.0	1.0	3.0–7.0	3.0–7.0
17,000/20,000	18,000/21,000	10,000/13,000	15,000/18,000	14,000/17,000	17,000/20,000	18,000/21,000
1.8×10^6	1.6×10^6	2.0×10^6	1.7×10^6	1.4×10^6	1.4×10^6	1.7×10^6
8500	8500	5500	7500	7500	7500	8500
27,000	27,000	27,000	30,000	26,000	25,000	28,000
0.35	0.35	0.25	0.25	0.25	0.35	0.35
500	500	525	525	525	550	550
400	400	425	425	425	450	450
95/275	110/40	115/222	98/262	119/40	95/339	110/40
	94V-O	94 HB	94 HB	94V-O		
400/350	400/350	400/400	380/360	375/350	400/350	400/350
4.1/4.0	4.2/3.9	4.3/3.8	3.9/3.7	4.1/3.9	3.8/3.6	4.1/3.9
0.011/0.019	0.010/0.018	0.015/0.019	0.011/0.015	0.010/0.013	0.010/0.019	0.010/0.019
135	130	150	150	150	135	135
GDI-30	GDI-30F	MDG	SDG	SDG-F	GDI-30	GDI-30F

- GDI-300: Long-glass-fiber filler, general purpose, nominal impact strength, 30 ft-lb/in. notch (1600 J/m).
- GDI-300F: Long-glass-fiber filler, flame resistant, nominal impact strength, 30 ft-lb/in. notch (1600 J/m).

Military specification MIL-M-14G, under Revision H, will now include military designations for the high heat-resistant DAIP materials. Additional designations include:

- MIG: Mineral filler, heat resistant.
- MIG-F: Mineral filler, flame resistant, heat resistant.
- SIG: Short-glass-fiber filler, heat resistant.
- SIG-F: Short-glass-fiber filler, flame resistant, heat resistant.
- GII-30: Long-glass-fiber filler, heat resistant, nominal impact strength, 3.0 ft-lb/in. notch (160 J/m).
- GII-30F: Long-glass-fiber filler, flame resistant, heat resistant, nominal impact strength, 3.0 ft-lb/in. notch (160 J/m), flame resistant.

DAIP molding compounds can operate in long-term, high-temperature conditions in excess of 490°F (260°C) without substantial loss of physical or electrical properties.

The American Society for Testing and Materials (ASTM) lists DAP and DAIP compounds under D 1636-75A by type, class, and grade as follows:

Type	Class	Grade
I	A Ortho (DAP) B Ortho FR (DAP) C ISO (DAIP) D ISO FR (DAIP)	1. Long glass 2. Medium glass 3. Short glass
II	A Ortho (DAP) B Ortho FR (DAP) C ISO (DAIP) D ISO FR (DAIP)	1. Mineral 2. Mineral/organic fiber
III	A Ortho (DAP)	1. Acrylic fiber 2. Polyester fiber, long 3. Polyester fiber, milled

4.6 APPLICATIONS

The outstanding qualities of DAP/DAIP are used in the very demanding electrical/electronic field where retention of electrical and mechanical properties under extreme conditions is of paramount importance. These applications include connectors, potentiometer housings, switches, relays, circuit breakers, terminal strips, coil bobbins, etc.

4.6.1 Reinforced Laminates

Allyl resins have served as the laminating resin in the manufacture of radomes and aircraft structural members. Here, the key property balance includes heat resistance, retention of flexural modulus, high strength, low dielectric constant, and low dissipation factor, plus excellent weatherability.

When glass reinforcement is added, allylic-based laminates exhibit solvent resistance and high electrical resistivity and retain their excellent dielectric properties over a wide range of temperatures and frequencies. They also display very good creep resistance when exposed to extreme temperatures.

4.6.2 Decorative Laminates

Diallyl phthalate prepolymer has found wide use because of its ability to provide an excellent surface medium for decorative laminate construction. The prepolymer can be utilized in combination with such materials as particleboard, plywoods of various kinds, and hardboard, resulting in laminates with superior surface and decorative qualities.

4.7 ADVANTAGES/DISADVANTAGES

DAP and DAIP have a great advantage over other thermosets, which is their superior retention of insulating properties in severe, hot, humid environments. These materials are capable of all types of molding, displaying excellent heat resistance, superior dimensional stability, and low postmold shrinkage. They are nonoutgassing and noncorrosive to contact pins.

As with other thermosets, DAP and DAIP have a tendency to be brittle because of their high modulus and hard surfaces. Although it has been tried in the past, the use of regrind is not feasible with DAP or DAIP. Slower cure rates have been perceived to be a disadvantage with these compounds. Development work has been ongoing, and new allyl compounds, with much improved cure rates, are available today.

4.8 PROCESSING TECHNIQUES

All grades can be easily molded by all conventional molding methods—compression, transfer, screw transfer, screw injection, and injection-compression molding. The exception is Mil Spec GDI-30F/SDI-30 products, which may require auxiliary feeding equipment for screw transfer or screw injection molding.

DAP/DAIP moldings can be machined, drilled, and tapped by conventional methods.

Start up molding parameters for DAP/DAIP are as follows:

- *Compression, transfer:* preheat temperature, 200–230°F (93–110°C); mold temperature, 325–340°F (166–171°C); Molding pressures, 2,500 psi (17 MPa) for compression, 3,500 psi (24 MPa); for transfer cure rate: 40–45 s for 1/8-in. (3.2-mm) section; 20–30 s for 1/16 in. (1.6-mm) section.
- *Screw injection:* stock temperature, 220–240°F (104–116°C); mold temperature, 320–340°F (160–171°C); screw rpm, 0–50; back pressure, 0–50 psi (0–0.34 MPa).
- *Injection speed:* 3–10 s cure rate, 35–45 s for 1/8-in. (3.2-mm) section; 18–25 s for 1/16-in. (1.6-mm) section.

Master Material Outline

PROCESSABILITY OF ____Allyl Resins_____

PROCESS	A	B	C
INJECTION MOLDING	X		
EXTRUSION			
THERMOFORMING			
FOAM MOLDING			
DIP, SLUSH MOLDING			
ROTATIONAL MOLDING			
POWDER, FLUIDIZED-BED COATING			
REINFORCED THERMOSET MOLDING			
COMPRESSION/TRANSFER MOLDING	X		
REACTION INJECTION MOLDING (RIM)			
MECHANICAL FORMING			
CASTING			

A = Common processing technique.
B = Technique possible with difficulty.
C = Used only in special circumstances, if ever.

Standard design chart for _____**DAP/DAIP**_____

1. Recommended wall thickness/length of flow
   Minimum ____0.010–0.015____ for _____1–2 in._____ distance
   Maximum ____0.015–0.025____ for _____distance
   Ideal   ____0.015–0.025____ for _____1 in._____ distance

2. Allowable wall thickness variation, % of nominal wall _____30%_____

3. Radius requirements
   Outside: Minimum ____1/32 in.____ Maximum _____
   Inside: Minimum ____1/32 in.____ Maximum _____

4. Reinforcing ribs
   Maximum thickness _____1/8 in._____
   Maximum height _____1/4 in._____
   Sink marks: Yes _____ No ___X___ (blisters if undercured)

5. Solid pegs and bosses
   Maximum thickness _____½ in._____
   Maximum weight ____3/4 in.____
   Sink marks: Yes _____ No ___X___ (blisters if undercured)

6. Strippable undercuts
   Outside: Yes _____ No ___X___
   Inside: Yes _____ No ___X___

7. Hollow bosses: Yes ___X___ No _____

8. Draft angles
   Outside: Minimum _____$\frac{1}{4}$°_____ Ideal _____$\frac{1}{2}$°_____
   Inside: Minimum _____$\frac{1}{4}$°_____ Ideal _____$\frac{1}{2}$°_____

9. Molded-in inserts: Yes ___X___
                       No _____

10. Size limitations   1–5 g   5 lb   Transfer
    Length: Minimum ___1–5 g___ Maximum ___3 lb___ Process ___injection___
    Width: Minimum ___1–5 g___ Maximum ___2–3 lb___ Process ___Compression___

11. Tolerances

12. Special process- and materials-related design details
    Blow-up ratio (blow molding) _____
    Core L/D (injection and compression molding) _____ Injection—5:1 unsupported
    Depth of draw (thermoforming) _____ Transfer—8:1 supported
    Other _____ Compression—2.5:1 cold powder
                                      —4:1 preheated preforms

4.9 RESIN FORMS

MDG/SDG/SDG-F/SDI-5 products come in free-flowing granular or diced form and can be readily preformed, preheated, or screw preplasticated in standard equipment. GDI-30/GDI-30F/SDI-30 compounds may require auxiliary equipment. They cannot be easily poured because of their high bulk factor.

4.10 SPECIFICATION OF PROPERTIES

See the Master Outline of Materials Properties.

4.11 PROCESSING REQUIREMENTS

Molds for DAP/DAIP compounds should be well-polished and hard chrome-plated for easy release from the mold and superior molded part appearance. Such mold lubricants as carnauba wax or silicone mold releases can be used when a mold is "broken in," but it is not necessary to continually lubricate mold surfaces.

A normal draft of 1/2° per side on molded surfaces that are parallel with the mold parting line will ensure easy part ejection. All transfer and injection molds must be properly vented at the parting line and, where they are needed, knockout pins should be flat ground to provide more venting, particulary cavity pockets or depressions.

4.12 PROCESSING-SENSITIVE END PROPERTIES

Many processing factors can affect the end-use properties of DAP and DAIP materials.

Key properties, such as percent cure, shrinkage, physical properties, and dimensional

MATERIAL ___ DAP, type MDG (MIL-M-14) — General-purpose, mineral filler, flame resistant

PROPERTY	TEST METHOD	ENGLISH UNITS	VALUE	METRIC UNITS	VALUE
MECHANICAL DENSITY	D792	lb/ft^3	112	g/cm^3	1.80
TENSILE STRENGTH	D638	psi	8000	MPa	55
TENSILE MODULUS	D638	psi	1,600,000	MPa	11,032
FLEXURAL MODULUS	D790	psi	1,700,000	MPa	11,721
ELONGATION TO BREAK	D638	%	3–5	%	3–5
NOTCHED IZOD AT ROOM TEMP	D256	ft-lb/in.	0.60	J/m	32
HARDNESS	D785	Rockwell	E: 65	Rockwell	E: 65
THERMAL DEFLECTION T @ 264 psi	D648	°F	325	°C	163
DEFLECTION T @ 66 psi	D648	°F		°C	
VICAT SOFT-ENING POINT	D1525	°F		°C	
UL TEMP INDEX	UL746B	°F	302	°C	150
UL FLAMMABILITY CODE RATING	UL94		94 VO		94 VO
LINEAR COEFFICIENT THERMAL EXPANSION	D696	in./in./°F	1.1×10^{-5}	mm/mm/°C	2.1×10^{-5}
ENVIRONMENTAL WATER ABSORPTION 24 HOUR	D570	%	0.3	%	0.3
CLARITY	D1003	% TRANS-MISSION	None	% TRANS-MISSION	None
OUTDOOR WEATHERING	D1435	%		%	
FDA APPROVAL					
CHEMICAL RESISTANCE TO: WEAK ACID	D543		No effect		No effect
STRONG ACID	D543		Slight effect		Slight effect
WEAK ALKALI	D543		No effect		No effect
STRONG ALKALI	D543		Slight effect		Slight effect
LOW MOLECULAR WEIGHT SOLVENTS	D543		No effect		No effect
ALCOHOLS	D543		No effect		No effect
ELECTRICAL DIELECTRIC STRENGTH	D149	V/mil	400	kV/mm	15.6
DIELECTRIC CONSTANT	D150	10^6 Hertz	3.7		3.7
POWER FACTOR	D150	10^6 Hertz	0.016		0.016
OTHER MELTING POINT	D3418	°F	None	°C	None
GLASS TRANS-ITION TEMP.	D3418	°F	320–330	°C	160–166

Master Outline of Material Properties

MATERIAL ___ DAP, type SDG-F (MIL-M-14) — Short-glass-fiber filler, flame resistant

PROPERTY	TEST METHOD	ENGLISH UNITS	VALUE	METRIC UNITS	VALUE
MECHANICAL DENSITY	D792	lb/ft³	116	g/cm³	1.87
TENSILE STRENGTH	D638	psi	12,000	MPa	83
TENSILE MODULUS	D638	psi	2,000,000	MPa	13,790
FLEXURAL MODULUS	D790	psi	1,800,000	MPa	12,411
ELONGATION TO BREAK	D638	%	3–5	%	3–5
NOTCHED IZOD AT ROOM TEMP	D256	ft-lb/in.	0.80	J/m	43
HARDNESS	D785		85		85
THERMAL DEFLECTION T @ 264 psi	D648	°F	400	°C	204
DEFLECTION T @ 66 psi	D648	°F		°C	
VICAT SOFT-ENING POINT	D1525	°F		°C	
UL TEMP INDEX	UL746B	°F	356	°C	180
UL FLAMMABILITY CODE RATING	UL94		94 VO		94 VO
LINEAR COEFFICIENT THERMAL EXPANSION	D696	in./in./°F	0.8×10^{-5}	mm/mm/°C	1.5×10^{-5}
ENVIRONMENTAL WATER ABSORPTION 24 HOUR	D570	%	0.25	%	0.25
CLARITY	D1003	% TRANS-MISSION	None	% TRANS-MISSION	None
OUTDOOR WEATHERING	D1435	%		%	
FDA APPROVAL					
CHEMICAL RESISTANCE TO: WEAK ACID	D543		No effect		No effect
STRONG ACID	D543		Slight effect		Slight effect
WEAK ALKALI	D543		Slight effect		Slight effect
STRONG ALKALI	D543		Considerable effect		Considerable effect
LOW MOLECULAR WEIGHT SOLVENTS	D543		No effect		No effect
ALCOHOLS	D543		No effect		No effect
ELECTRICAL DIELECTRIC STRENGTH	D149	V/mil	415+	kV/mm	16.3+
DIELECTRIC CONSTANT	D150	10^6 Hertz	3.5		3.5
POWER FACTOR	D150	10^6 Hertz	0.016		0.016
OTHER MELTING POINT	D3418	°F	None	°C	None
GLASS TRANS-ITION TEMP.	D3418	°F	320–330	°C	160–166

51

MATERIAL ___ DAP, type GDI-30F (MIL-M-14) — Long-glass-fiber filler, flame resistant, nominal impact strength, 3.0 ft-lb/in. notch (160 J/m)

PROPERTY	TEST METHOD	ENGLISH UNITS	VALUE	METRIC UNITS	VALUE
MECHANICAL DENSITY	D792	lb/ft^3	110	g/cm^3	1.76
TENSILE STRENGTH	D638	psi	9000	MPa	62
TENSILE MODULUS	D638	psi	2,000,000	MPa	13,790
FLEXURAL MODULUS	D790	psi	1,700,000	MPa	11,721
ELONGATION TO BREAK	D638	%	3–5	%	3–5
NOTCHED IZOD AT ROOM TEMP	D256	ft-lb/in.	3.0–6.0	J/m	160–320
HARDNESS	D785	Rockwell	E: 85	Rockwell	E: 85
THERMAL DEFLECTION T @ 264 psi	D648	°F	500	°C	260
DEFLECTION T @ 66 psi	D648	°F		°C	
VICAT SOFT-ENING POINT	D1525	°F		°C	
UL TEMP INDEX	UL746B	°F	302	°C	150
UL FLAMMABILITY CODE RATING	UL94		94 VO		94 VO
LINEAR COEFFICIENT THERMAL EXPANSION	D696	in./in./°F	1.2×10^{-5}	mm/mm/°C	2.2×10^{-5}
ENVIRONMENTAL WATER ABSORPTION 24 HOUR	D570	%	0.35	%	0.35
CLARITY	D1003	% TRANS-MISSION	None	% TRANS-MISSION	None
OUTDOOR WEATHERING	D1435	%		%	
FDA APPROVAL					
CHEMICAL RESISTANCE TO: WEAK ACID	D543		No effect		No effect
STRONG ACID	D543		Slight effect		Slight effect
WEAK ALKALI	D543		Slight effect		Slight effect
STRONG ALKALI	D543		Considerable effect		Considerable effect
LOW MOLECULAR WEIGHT SOLVENTS	D543		No effect		No effect
ALCOHOLS	D543		No effect		No effect
ELECTRICAL DIELECTRIC STRENGTH	D149	V/mil	415+	kV/mm	16.3+
DIELECTRIC CONSTANT	D150	10^6 Hertz	3.5		3.5
POWER FACTOR	D150	10^6 Hertz	0.018		0.018
OTHER MELTING POINT	D3418	°F	None	°C	None
GLASS TRANS-ITION TEMP.	D3418	°F	320–330	°C	160–166

stability, are affected by mold temperatures, molding cycles, molding pressures, method of molding, material flow characteristics, and fiber orientation during molding.

4.13   SHRINKAGE

See the Mold Shrinkage Characteristics.

Mold Shrinkage Characteristics (in./in. or m/m)

MATERIAL   ___DAP/DAIP___

IN THE DIRECTION OF FLOW

THICKNESS

MATERIAL	mm: 1.5	3	6	8	FROM	TO	#
	in: 1/16	1/8	1/4	1/2			
Glass filled					0.001	0.003	
Mineral filled					0.003	0.005	

PERPENDICULAR TO THE DIRECTION OF FLOW[a]							

[a]Shrinkage will exceed that which occurs in the direction of flow with cross-sectional thickness and obstructions to flow being the most important factors--which will affect the fiber orientation which, in turn, will greatly affect the amount of shrinkage.

4.14   TRADE NAMES

DAP/DAIP resins are available from Cosmic Plastics, San Fernando, Calif.; from IDI, Indiana; as Durez from Occidental Chemical Corp., Durez Div., North Tonawanda, N.Y.; as Dial from Plaskon Products, Los Angeles, Calif., and as Dapex from Rogers Corp., Manchester, Conn.

BIBLIOGRAPHY

General References

H. Raech Jr., *Allylic Resins and Monomers,* Reinhold Publishing Corp., New York.
R.E. Wright, *Thermoset Molding Manual,* Rogers Corp., Manchester, Conn.

5

CELLULOSICS

Eastman Chemical Products Inc., P.O. Box 431, Kingsport, Tennessee 37662

5.1 INTRODUCTION

Cellulosic plastics are manufactured by chemical modification of cellulose, a naturally occurring polymer available from wood pulp and cotton linters. Cellulose itself is not a thermoplastic material; it does not melt. Cellulose resins include four organic esters: cellulose acetate (CA); cellulose acetate butyrate (CAB), cellulose acetate propionate (CAP), and cellulose acetate triacetate. Because the latter has a very high softening temperature, it is not considered a true thermoplastic material.

When Ⓐ is:	and Ⓑ is:	Plastic
H	H	Cellulose
—C—CH₃ (‖O)	—C—CH₃ (‖O)	Cellulose acetate
—C—CH₃ (‖O)	—C—CH₂—CH₃ (‖O)	Cellulose acetate proprionate
—C—CH₃ (‖O)	—C—CH₂—CH₂—CH₃ (‖O)	Cellulose acetate butyrate
—C—CH₂—CH₃ (‖O)	—C—CH₂—CH₃ (‖O)	Cellulose proprionate
—CH₂—CH₃	—CH₂—CH₃	Ethyl cellulose
—NO₂	—NO₂	Cellulose nitrate

Note: Third hydroxyl unit (OH) is partially esterified in final plastic.

5.2 CATEGORY

Cellulosics are thermoplastic resins.

5.3 HISTORY

The first cellulose derivative used for a commercial synthetic thermoplastic, celluloid patented in 1869, was a cellulose nitrate. Cellulose nitrate for nonthermoplastic usage dates back to 1838, when the product found its way into liquors, explosives, and emulsions for auto safety glass. In 1927, Eastman Kodak Co. commercialized cellulose acetate into photographic film base and later, in 1935, received a patent on the first molding grade CA.

Among the most prominent initial applications were a desk-type staple fastener, a glove compartment door for Chrysler Corp., and later, in 1940, toothbrush handles and lamp parts.

In 1938, Eastman introduced cellulose acetate butyrate and, with the mixed ester cellulosic, a tougher material, more dimensional stability, and better chemical resistance.

After a brief retraction occasioned by stability, cellulose acetate propionate was introduced in 1960. CAP is noted for its excellent processibility, much like butyrate, without the odor associated with CAB.

U.S. production for cellulosic plastics has not changed drastically since 1960 with original existing production equipment still in place. Total production for years 1960, 1970, and 1980 is estimated at 100×10^6 pounds. Production for 1987 was 88.4×10^6 pounds.

5.4 POLYMERIZATION

As noted, cellulosic plastics are manufactured by chemical modification of cellulose, a naturally occurring polymer available from wood pulp and cotton linters. The organic cellulose esters are prepared by reacting chemical cellulose with organic acids and anhydrides, using sulfuric acid as a catalyst in most processes. In the standard synthesis of CA, the reaction proceeds with acetic acid and acetic anhydride to the triester stage. CA is then prepared by hydrolyzing the triester to remove some of its acetyl groups. Plastic-grade cellulose acetate contains 38 to 40% acetyl. The propionate and butyrate esters are made by substituting propionic acid and propionic anhydride or butyric acid and butyric anhydride for some of the acetic acid and acetic anhydride. Plastic grades of CAP contain 39–47% propionyl and 2–9% acetyl; CAB, 26–39% butyryl and 12–15% acetyl.

Chemical cellulose is the main raw material in the production of cellulose plastics, along with odor masking agents for butyrate, stabilizers, and colorants. The base price for each, along with fluctuations in cost, is the basis for current established prices, which range from $1.34 to $1.38 per pound (1988).

5.5 DESCRIPTION OF PROPERTIES

All of the cellulose plastics are characterized as tough materials. Many plastics may be stronger in terms of tensile strength or more rigid in terms of flexural properties, but cellulose plastics offer a combination of desirable properties—easy processibility, transparency, colorability, toughness, and rigidity. The availability of CAB and CAP in special formulations for outdoor use gives additional versatility.

Few plastics offer as wide a range in visual appearance as do the cellulosics. They are available in crystal-like, transparent hues and in an almost unlimited range of transparent, translucent, and opaque colors with variegations, pearlescent, metallic, and specialty effects.

Heat-deflection temperature of 66 psi (0.5 MPa) is average: approximately 160°F (71°C) for acetate, 180°F (82°C) for butyrate, and 200°F (93°C) for propionate.

Although it is not commercially available in a compounded resin, glass-reinforced CAB exhibits significant property improvement in tensile strength, Izod impact, heat-deflection temperature, and hardness.

5.5.1 Selection

A cellulosic plastic is usually selected because of some combination of its properties. All three plastics have excellent colorability.

CAP and CAB are generally preferred to CA when exceptional processing ease is a factor, when weather resistance is needed, when dimensional stability under severe conditions is important, or when some combination of these or other characteristics is desired.

CAP is selected over CAB when greater hardness, tensile strength, and stiffness properties are required.

CA has somewhat different chemical resistance properties than CAB and CAP and is selected in applications where a specific chemical resistance is an advantage. Certain CA formulas and flows also offer greater surface hardness and higher tensile strength than either CAB or CAP.

5.6 APPLICATIONS

Typical CA applications include opticals, handles, sheeting, and toys.

Typical CAB applications are toys, skylights, pens, signs, ski goggles, and automobile trim.

CAP applications encompass toothbrush handles, packaging containers, medical devices, safety glasses, steering wheels, cosmetic parts, and toys.

5.7 ADVANTAGES/DISADVANTAGES

5.7.1 Advantages

- Outstanding combination of physical properties
- Outstanding clarity
- FDA compliance
- ETO and irradiation sterilizability
- Ease of processing
- Chemical resistance
- Dimensional stability

5.7.2 Disadvantages

- Cost per pound, compared to other competitive materials of like physical, chemical, and mechanical properties.

5.8 PROCESSING TECHNIQUES

See the Master Material Outline and the Standard Material Design Chart.

Standard design chart for _____ **Cellulosics** _____

1. Recommended wall thickness/length of flow
   Minimum _____ 0.018 _____ for _____ 3 in. _____ distance
   Maximum _____ none _____ for _____ distance
   Ideal _____ 0.100 _____ for _____ 36 in. _____ distance

2. Allowable wall thickness variation, % of nominal wall _____ 200 _____

3. Radius requirements
   Outside: Minimum _____ 12% _____ Maximum _____ none _____
   Inside: Minimum _____ 12% _____ Maximum _____ none _____

4. Reinforcing ribs
   Maximum thickness _____ 100% w _____
   Maximum height _____ 1.5 w _____
   Sink marks: Yes _____ No ___ X ___

5. Solid pegs and bosses
   Maximum thickness _____ 100% w _____
   Maximum weight _____ 2 times w _____
   Sink marks: Yes _____ No ___ X ___

6. Strippable undercuts
   Outside: Yes ___ X ___ No _____
   Inside: Yes ___ X ___ No _____

7. Hollow bosses: Yes ___ X ___ No _____

8. Draft angles
   Outside: Minimum _____ $\frac{1}{2}°$ _____ Ideal _____ 1° _____
   Inside: Minimum _____ $\frac{1}{2}°$ _____ Ideal _____ 1° _____

9. Molded-in inserts: Yes ___ X ___
   No _____

10. Size limitations   Limited only by size of injection molding machine
    Length: Minimum _____ Maximum _____ Process _____
    Width: Minimum _____ Maximum _____ Process _____

11. Tolerances

12. Special process- and materials-related design details
    Blow-up ratio (blow molding) _____ 4:1 _____
    Core L/D (injection and compression molding) _____ 20:1 _____
    Depth of draw (thermoforming) _____ 3:1 _____
    Other _____

Master Material Outline

PROCESSABILITY OF ___ Cellulosics ___

PROCESS	A	B	C
INJECTION MOLDING	X		
EXTRUSION	X		
THERMOFORMING	X		
FOAM MOLDING			X
DIP, SLUSH MOLDING			
ROTATIONAL MOLDING	X		
POWDER, FLUIDIZED-BED COATING*	X		
REINFORCED THERMOSET MOLDING			
COMPRESSION/TRANSFER MOLDING	X		
REACTION INJECTION MOLDING (RIM)			
MECHANICAL FORMING		X	
CASTING			X

A = Common processing technique.
B = Technique possible with difficulty.
C = Used only in special circumstances, if ever.
*Eastman Plastics does not supply in powder form

5.9 RESIN FORMS

Cellulosics are only available in 1/8 in. (3.2 mm) pellets; no modifications are offered. Because of the uniform size of the pellets, the weight-to-volume ratio is subject to little variation.

The plasticizers required for CA tend to be slightly volatile and slightly water soluble. Plasticizers used in CAB and CAP have higher boiling points, lower vapor pressures at room temperature, and less water solubility than those used in CA; therefore, they remain in the plastic better. Low moisture absorption and good plasticizer retention combine to give CAP and CAB, especially, exceptional permanence and dimensional stability.

5.10 SPECIFICATION OF PROPERTIES

See the Master Outline of Materials Properties.

MATERIAL ___ Cellulose Acetate Butyrate 264MH

PROPERTY	TEST METHOD	ENGLISH UNITS	VALUE	METRIC UNITS	VALUE
MECHANICAL DENSITY	D792	lb/ft^3	74.3	g/cm^3	1.19
TENSILE STRENGTH	D638	psi	5000	MPa	34.5
TENSILE MODULUS	D638	psi	250,000	MPa	1725
FLEXURAL MODULUS	D790	psi	210,000	MPa	1449
ELONGATION TO BREAK	D638	%	50	%	50
NOTCHED IZOD AT ROOM TEMP	D256	ft-lb/in.	3.5	J/m	187
HARDNESS	D785		R–75		R–75
THERMAL DEFLECTION T @ 264 psi	D648	°F	149	°C	65
DEFLECTION T @ 66 psi	D648	°F	162	°C	72
VICAT SOFT-ENING POINT	D1525	°F		°C	
UL TEMP INDEX	UL746B	°F		°C	
UL FLAMMABILITY CODE RATING	UL94		94 HB		94 HB
LINEAR COEFFICIENT THERMAL EXPANSION	D696	in./in./°F	8×10^{-5}	mm/mm/°C	14×10^{-5}
ENVIRONMENTAL WATER ABSORPTION 24 HOUR	D570	%	1.5	%	1.5
CLARITY	D1003	% TRANS-MISSION		% TRANS-MISSION	
OUTDOOR WEATHERING	D1435	%		%	
FDA APPROVAL			Yes		Yes
CHEMICAL RESISTANCE TO: WEAK ACID	D543		Minimally attacked		Minimally attacked
STRONG ACID	D543		Badly attacked		Badly attacked
WEAK ALKALI	D543		Minimally attacked		Minimally attacked
STRONG ALKALI	D543		Badly attacked		Badly attacked
LOW MOLECULAR WEIGHT SOLVENTS[a]	D543				
ALCOHOLS	D543		Minimally attacked		Minimally attacked
ELECTRICAL DIELECTRIC STRENGTH	D149	V/mil	340	kV/mm	13.4
DIELECTRIC CONSTANT	D150	10^6 Hertz	3.6		3.6
POWER FACTOR	D150	10^6 Hertz	0.03		0.03
OTHER MELTING POINT	D3418	°F		°C	
GLASS TRANS-ITION TEMP.	D3418	°F		°C	

[a]Not attacked by aliphatic hydrocarbons; mildly attacked by aromatic hydrocarbons; dissolved by chlorinated hydrocarbons, esters, and ketones; not attacked by diethyl ether

Master Outline of Material Properties

MATERIAL ___Acetate, 036H___

PROPERTY	TEST METHOD	ENGLISH UNITS	VALUE	METRIC UNITS	VALUE
MECHANICAL DENSITY	D792	lb/ft³	79.8	g/cm³	1.28
TENSILE STRENGTH	D638	psi	5800	MPa	40
TENSILE MODULUS	D638	psi	315,000	MPa	2174
FLEXURAL MODULUS	D790	psi	260,000	MPa	1794
ELONGATION TO BREAK	D638	%		%	
NOTCHED IZOD AT ROOM TEMP	D256	ft-lb/in.	3	J/m	160
HARDNESS	D785		R–82		R–82
THERMAL DEFLECTION T @ 264 psi	D648	°F	142	°C	61
DEFLECTION T @ 66 psi	D648	°F	162	°C	72
VICAT SOFT-ENING POINT	D1525	°F		°C	
UL TEMP INDEX	UL746B	°F		°C	
UL FLAMMABILITY CODE RATING	UL94		94 HB		94 HB
LINEAR COEFFICIENT THERMAL EXPANSION	D696	in./in./°F	8×10^{-5}	mm/mm/°C	14×10^{-5}
ENVIRONMENTAL WATER ABSORPTION 24 HOUR	D570	%	2.5	%	2.5
CLARITY	D1003	% TRANS-MISSION	80	% TRANS-MISSION	80
OUTDOOR WEATHERING	D1435	%		%	
FDA APPROVAL			Yes		Yes
CHEMICAL RESISTANCE TO: WEAK ACID	D543		Minimally attacked		Minimally attacked
STRONG ACID	D543		Badly attacked		Badly attacked
WEAK ALKALI	D543		Minimally attacked		Minimally attacked
STRONG ALKALI	D543		Badly attacked		Badly attacked
LOW MOLECULAR WEIGHT SOLVENTS[a]	D543				
ALCOHOLS	D543		Minimally attacked		Minimally attacked
ELECTRICAL DIELECTRIC STRENGTH	D149	V/mil	340	kV/mm	13.4
DIELECTRIC CONSTANT	D150	10^6 Hertz	3.6		3.6
POWER FACTOR	D150	10^6 Hertz	0.03		0.03
OTHER MELTING POINT	D3418	°F		°C	
GLASS TRANS-ITION TEMP.	D3418	°F		°C	

[a]Not attacked by aliphatic hydrocarbons; mildly attacked by aromatic hydrocarbons; dissolved by chlorinated hydrocarbons, esters, and ketones; not attacked by diethyl ether

MATERIAL ___ Cellulose Acetate Propionate 360H2

PROPERTY	TEST METHOD	ENGLISH UNITS	VALUE	METRIC UNITS	VALUE
MECHANICAL DENSITY	D792	lb/ft^3	74.2	g/cm^3	1.19
TENSILE STRENGTH	D638	psi	5075	MPa	35
TENSILE MODULUS	D638	psi	250,000	MPa	1725
FLEXURAL MODULUS	D790	psi	210,000	MPa	1449
ELONGATION TO BREAK	D638	%	60	%	60
NOTCHED IZOD AT ROOM TEMP	D256	ft-lb/in.	7.7	J/m	411
HARDNESS	D785		R–70		R–70
THERMAL DEFLECTION T @ 264 psi	D648	°F	162	°C	72
DEFLECTION T @ 66 psi	D648	°F	176	°C	80
VICAT SOFT-ENING POINT	D1525	°F	212	°C	100
UL TEMP INDEX	UL746B	°F		°C	
UL FLAMMABILITY CODE RATING	UL94		94 HB		94 HB
LINEAR COEFFICIENT THERMAL EXPANSION	D696	in./in./°F	8×10^{-5}	mm/mm/°C	14×10^{-5}
ENVIRONMENTAL WATER ABSORPTION 24 HOUR	D570	%	1.3	%	1.3
CLARITY	D1003	% TRANS-MISSION	80	% TRANS-MISSION	80
OUTDOOR WEATHERING	D1435	%		%	
FDA APPROVAL			Yes		Yes
CHEMICAL RESISTANCE TO: WEAK ACID	D543		Minimally attacked		Minimally attacked
STRONG ACID	D543		Badly attacked		Badly attacked
WEAK ALKALI	D543		Minimally attacked		Minimally attacked
STRONG ALKALI	D543		Badly attacked		Badly attacked
LOW MOLECULAR WEIGHT SOLVENTS*	D543				
ALCOHOLS	D543		Minimally attacked		Minimally attacked
ELECTRICAL DIELECTRIC STRENGTH	D149	V/mil	340	kV/mm	13.4
DIELECTRIC CONSTANT	D150	10^6 Hertz	3.6		3.6
POWER FACTOR	D150	10^6 Hertz	0.03		0.03
OTHER MELTING POINT	D3418	°F		°C	
GLASS TRANS-ITION TEMP.	D3418	°F		°C	

*Not attacked by aliphatic hydrocarbons; mildly attacked by aromatic hydrocarbons; dissolved by chlorinated hydrocarbons, esters, and ketones; not attacked by diethyl ether

5.11   PROCESSING REQUIREMENTS

All cellulosic plastics process easily, although CAB and CAP are easier to mold than CA. Different formulas of each material will process differently, as a result of the type of ester, plasticizer, and amount of additives.

A better molded finish is generally obtainable with CAB and CAP than with CA, but proper drying before molding will often improve the finish of products made from any of these plastics.

CA is not compatible with CAB or CAP and must not be mixed with either of them. CAB can sometimes be mixed with CAP, but their compatibility cannot be assumed in all cases. None of these plastics is compatible with any noncellulosic plastic. Scrap from each material should be kept separate and clean. Reground, clean scrap is used without difficulty in amounts of 20–25%.

5.11.1   Drying

Cellulose ester plastics absorb moisture and nearly always need to be dried before they are processed. CA absorbs moisture faster than CAB or CAP and reaches a high moisture content. The affinity of all three plastics for water increases with temperature, and both the rate of absorption and the equilibrium content increase as atmospheric moisture concentration increases. Adequate drying can help reduce the evolution of fumes or gases during molding. Local exhaust ventilation can also be used as needed to help control the evolution of such fumes.

For most operations, 150°F (66°C) should be considered the minimum operating temperature for drying cellulosic plastics, although the softer flows are sometimes dried at a temperature as low as 130°F (54°C). Temperatures may range up to 185°F (85°C) or even higher when hard flows are dried. In general, it is possible to use temperatures 100–120°F (55–67°C) below the flow temperature of the plastic to be dried, but temperatures above 200°F (93°C) may be detrimental to the plastic.

It is impractical to list specific drying times for all possible combinations of plastic, dryer, and application. In general, however, for an efficient dryer operated at (160°F) (71°C) or higher, and for normal processing operations, the drying times listed in Table 5.1 will be satisfactory.

5.11.2   Gate Design

Restricted Gate.   For molding small articles from cellulosics, it is a good practice to start with a restricted gate approximately 0.015–0.060 in. (0.38–1.52 mm) in diameter and, if needed, increase the diameter in small increments until a satisfactory article is obtained from the mold. Molding temperatures and pressures should be varied after each increase in gate size to determine whether the size is sufficient. A restricted gate can also be used successfully when molding cellulosics in three-plate molds or in molds with tunnel gates.

TABLE 5.1   Typical Drying Times for Cellulose Ester Plastics

Plastic	If Stored Less than One Month	If Stored More than Five Months
Cellulose acetate	2-1/2 h	3-1/2 h
Cellulose butyrate	2 h	3 h
Cellulose proprionate	2 h	3 h

Tunnel Gate. Tunnel or submarine gates can be used only when the runner system is contained in both cavity plates. The runner system is machined along the parting line between both plates, and the tunnel gate is machined into one of the plates below the parting surface. With this type of gating, the molded part and the runner are not usually ejected simultaneously. The molded part is ejected first, shearing the gate. The runner system and gate are then ejected by a separate ejection system.

Fan Gate. Fan gates have a fairly large cross-sectional area for feeding plastic into the mold. They resemble a film die in that the orifice is long and narrow and the flow channel from the runner to the gate is tapered. The degree of taper should be such that the cross-sectional area at any point between the runner and the gate will be the same as it is at the gate.

Fan gates can be used with good results when molding large, heavy parts. A gate 1 × 0.030 × 0.030 in (25.4 × 0.76 × 0.76 mm) (L × W × land) fed by a heavy runner section minimizes the splay effect prevalent on some moldings. Large gates are sometimes difficult to shear from the molded part and often necessitate a finishing operation.

Web Gate. When molding large, flat articles, a web gate may be used successfully to reduce warpage, flow lines, jetting, and strains. This gate allows a long, continuous front of plastic to flow into the cavity, virtually eliminating weld lines in many instances. The gate opens from a runner cut along one edge of the mold cavity. The gate can be as long as the edge of the cavity and should be about 0.012 in (0.31 mm) wide initially. It can be enlarged if necessary. The land between the runner and cavity should be approximately 0.020 in. (0.51 mm).

5.12 PROCESSING-SENSITIVE END PROPERTIES

Cellulosic materials have a wide range of processing conditions. However, clear materials can be overheated if temperatures exceed guidelines for flow used.

Low temperatures in use, both mold and melt, can cause cold temperature strain. This molded-in strain can result in part breakage under given end-use parameters.

5.12.1 Moisture

Moisture causes two types of defects in molded products: roughness of the surface and bubbles in the interior. Surface roughness is the most frequent defect in thin parts; internal bubbles tend to plague thick sections, which may also have rough surfaces. Internal bubbles can also be caused by cooling molten plastic too rapidly, but if bubbles persist even when the molding is cooled slowly, they are probably caused by moisture.

Surface roughness caused by moisture may appear as

- Flow lines and streaks in the gate area (these occur very frequently)
- A scaly surface (frequently seen)
- A blister where a bubble expanded after ejection of the part from the mold (frequently, often in clusters)
- A slight depression or rough, shallow crater when a bubble broke the surface or formed against the mold wall (infrequent; generally occurs singly).

5.12.2 Venting

Air entrapped in a mold cavity causes a void or sink mark to appear on the surface of the molded article, usually at the point most distant from the gate. Charring of the plastic also may occur at the same point because of the heat generated by compression of the entrapped air. For this reason, vents should be provided in appropriate areas around the cavity to permit the rapid escape of air.

Vents should be deep enough to allow for the rapid escape of air but not so deep that they allow molten material to flow into the vent. For cellulosic plastic materials, vent depths should be between 0.001–0.002 in. (0.025–0.051 mm) maximum.

A vent should be located as near as possible to the void, sink mark, or charred material that may appear in a molded article. If it is not properly located, the vent may become plugged with material before the air pocket it is designed to relieve is formed. Proper venting is essential to good mold design and should not be overlooked or added as an afterthought.

5.13 SHRINKAGE

See the Mold Shrinkage Characteristics.

Mold Shrinkage Characteristics (in./in. or mm/mm)

MATERIAL   Cellulosics

IN THE DIRECTION OF FLOW

THICKNESS

MATERIAL	mm: 1.5 in.: 1/16	3 1/8	6 1/4	8 1/2	FROM	TO	#
Acetate, mils/in.	0–4	1–6	4–9	6–12	0.000	0.012	in./in.
Butyrate, mils/in.	0–3	1–5	3–9	5–12	0.000	0.012	in./in.
Propionate, mils/in.	0–3	1–5	3–9	5–12	0.000	0.012	in./in.

PERPENDICULAR TO THE DIRECTION OF FLOW							
Acetate, mils/in.	0–4	1–6	4–9	6–12	0.000	0.012	in./in.
Butyrate, mils/in.	0–3	1–5	3–9	5–12	0.000	0.012	in./in.
Propionate, mils/in.	0–3	1–5	3–9	5–12	0.000	0.012	in./in.

NOTE: Mold shrinkage is approximate. It is affected by part design, mold temperature, thickness, injection pressure, packing time, cycle time, orientation, gate design, gate size, gate location, glass content, glass size, and filler content.

Standard Tolerance Chart

PROCESS _____Injection molding_____ MATERIAL _____Cellulosics_____

All data given below are approximate; ±

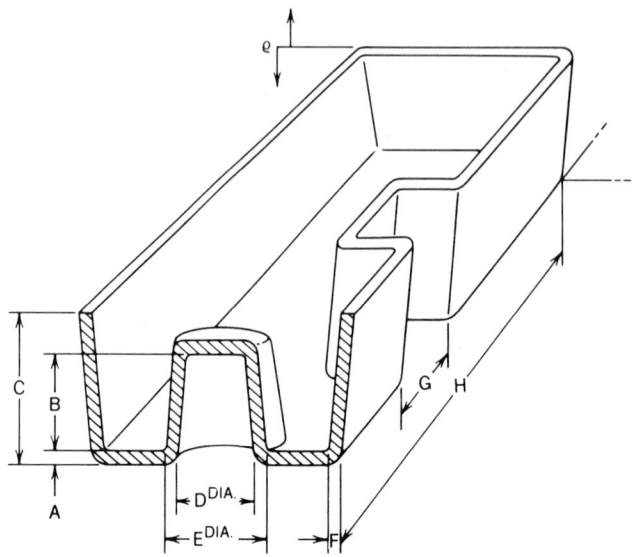

Values are based on a nominal 1/8 in. (3.2 mm) wall thickness and given in plus or minus in./in. and mm/mm.

FINE tolerance – Special care and added cost.
COMMERCIAL tolerance – Normal care required.
COARSE tolerance – Minimal care required, lowest cost.

Tolerance	A	B	C	D	E	F	G	H	
FINE	0.030	0.003	0.003	0.003	0.003	0.020	0.003	0.002	in./in.
	0.030	0.003	0.003	0.003	0.003	0.020	0.003	0.002	mm/mm
COMMERCIAL	0.040	0.004	0.005	0.004	0.005	0.030	0.004	0.003	in./in.
	0.040	0.004	0.005	0.004	0.005	0.030	0.004	0.003	mm/mm
COARSE	0.050	0.006	0.008	0.006	0.007	0.040	0.006	0.005	in./in.
	0.050	0.006	0.008	0.006	0.007	0.040	0.006	0.005	mm/mm

Mold shrinkages are relatively low and uniform for cellulosics. Since the materials are amorphous, shrinkage is virtually the same in the direction of flow as it is across the flow.

Shrinkage subsequent to molding varies somewhat, depending on the formula and flow of the material, and the shape and thickness of the molded article.

5.14 TRADE NAMES

Tenite is the trade name for all Eastman plastics, as in Tenite Cellulose Acetate.

5.15 INFORMATION SOURCES

Considerable information on the properties and processing of cellulosics is available from Eastman Plastics.

6

CHLORINATED POLYETHYLENE (CPE)

Robert R. Blanchard

Dow Chemical USA, Midland, MI 48640

6.1 INTRODUCTION

In the chlorination of polyethylene, a chlorine atom is substituted for a hydrogen atom on the polymer chain, accompanied by the formation of HCl:

| Polyethylene | Chlorine | Chlorinated polyethylene | Hydrochloric acid |

6.2 CATEGORY

Chlorinated polyethylene (CPE) polymers are produced in both elastomeric and thermoplastic forms. In either form, they have extraordinary compatibility with a range of other materials. Readily adaptable to common compounding and curing techniques, CPE polymers can produce end products that are hard and tough, or soft and flexible. Extensibility and recovery of cured and heat-aged elastomers are excellent.

Overall, the properties of CPE depend on the properties of the starting material and on the amount and distribution of the chlorine introduced. Most commercial CPE products contain 27 to 50% chlorine. Because CPE is seldom used alone, the effects on the final properties of the other materials with which CPE is compounded must also be considered.

6.3 HISTORY

The use of forms of halogens (including chlorine) to impart barrier and other properties to long-chain hydrocarbon polymer molecules is a known technology.

The commercialization of polyethylenes in the 1940s demonstrated that the materials had excellent fabrication flexibility along with highly useful electrical properties and a low water-vapor transmission rate. However, they provided less than adequate chemical and ignition-resistance properties for certain applications. European chemical companies conducted early work on the modification of polyethylene with chlorine, and the first commercial CPE polymers were produced by Hoechst, A.G. In the United States in the early 1950s, Allied Chemical and The Dow Chemical Company developed chlorinated polyethylene technology. Allied Chemical commercialized CPE polymers, followed by Dow in the early 1960s. Dow, a basic supplier of ethylene, chlorine, and polyethylenes, purchased the Allied CPE business and product technology in the 1960s. Since then, Dow has been a major global researcher, developer, producer, and marketer of CPE polymers.

The earliest applications of CPE included wire and cable jacketing, impact modification of PVC, hydraulic hose, durable sheeting, and tentative use as a "universal" compatibilizer for scrap and trim thermoplastics.

Global production capacities throughout the 1970s and until 1983 were 85×10^6 pounds; subsequently, capacity grew to 105×10^6 pounds. A further capacity increase of 15×10^6 pounds was announced by The Dow Chemical Company in 1986. Between 80 to 85% of all capacity for CPE is located in the United States. Further capacity increases are likely because of commercial demand for CPE polymers.

6.4 POLYMERIZATION

In most chlorination processes, one molecule of hydrogen chloride is generated for one molecule of chlorine added to the polymer. Considerable costs are incurred in the recovery-utilization of this HCl. The morphology of the polyethylene (PE) polymer used and the accessibility of the PE polymer chain to the chlorination significantly affect both the chlorine distribution and the properties of the CPE polymer produced. The chlorination of polyethylene can be achieved either by solution (solvent), aqueous suspension, or fluid bed processes.

6.4.1 Solution (Solvent) Process

Chlorination of cyrstalline PE begins at the amorphous areas and crystalline surfaces. Carried to an extreme, polymer crystallinity could be destroyed with significant effect on properties. At the same percentage chlorination as PVC (approximately 56%), the Cl must be concentrated in certain regions of the molecules to permit the PE crystals to survive. Such uneven distribution is known as block chlorination. It occurs at low temperatures with the PE particles suspended in a liquid or gaseous medium.

If the PE becomes amorphous because of increasing temperatures, chlorination occurs throughout—but access to some areas may be hindered. The results are not wholly random and are referred to as hindered chlorination. Controlled hindered chlorination does not require the use of solvents. It can be employed to develop products having the desirable properties associated with random chlorination. The chlorination in solution tends to be more uniform.

6.4.2 Aqueous Suspension Process

Chlorination by aqueous suspension process permits excellent temperature control and high reaction rates. However, the low solubility of chlorine in the aqueous phase requires an overpressure of chlorine, and washing and drying the end-product polymer demands much energy and water.

6.4.3 Fluid Bed Process

Anhydrous (particle form) processes make temperature control more difficult but eliminate the need for drying. The HCl produced can be separated in anhydrous, marketable form. Color problems occur more frequently than with aqueous processes, however.

Combinations of these and other polymerization processes are common in the production of CPE. In almost all cases, block chlorination is controlled by reducing the PE to an amorphous or low-crystallinity form so that the CPE is commercially useful.

Depending on the structure of the parent polyethylene and the degree and location of the Cl on the PE chain, CPE end products can be true elastomers or true thermoplastics. Properly chlorinated CPE polymer molecules have essentially saturated linear backbones resulting in significant, commercially valuable property advantages compared to the parent polyethylenes. These advantages include flexibility and toughness, ignition resistance, filler acceptance, and the ability to blend well with a wide variety of polymers.

This and other references to ignition resistance are not intended to reflect hazards under actual fire conditions. Like all organic polymers, CPE materials will burn under the right conditions of heat and oxygen supply, and toxic gases such as carbon monoxide and hydrogen chloride may be released.

6.5 PROPERTIES

Generalization concerning the functional properties of CPE is likely to be misleading. The property behavior discussed below relates to specific commercial polymers whose design in manufacture is tightly controlled, whose performance in complex compounds is maintained or extended, and whose appropriate end-product properties are retained after fabrication. Elastomeric and thermoplastic forms of CPE, and polymers in each category, can provide very different, useful properties.

The combustion characteristics of CPE polymers relate positively to their Cl content, as indicated in Figures 6.1 and 6.2. Not only is CPE an ignition-resistant polyethylene, it is also used in blends to change the ignition characteristics of other polymers favorably. Because of the reluctance of CPE to flow without incident shear, when it burns it usually chars instead of dripping. That can be an advantage, or a deterrent, if CPE is added to a compound utilizing dripping as part of the extinguishing mechanism.

The thermal properites of CPE resins enable them to perform well at most temperatures encountered in the environment. CPE elastomers, which contain essentially no polymer chain unsaturation, can perform well continuously at temperatures of 302°F (150°C) and intermittently at temperatures of 325°F (162°C). Properly compounded wire and cable jacketing and extruded or calendered sheet made of CPE retain flexibility and provide good performance at temperatures below −40°F (−40°C).

The good to excellent chemical resistance properties of CPE polymers earn them much commercial acceptance. They are resistant to most acids, bases, oils, and alcohols; the elastomers can be formulated to provide satisfactory resistance to many solvents.

The general mechanical properties of CPE are perhaps best summarized by describing these polymers as "tough" both in themselves and in their contributions to other polymers. Properly formulated and fabricated, compounds with CPE are commercially noted for their impact resistance, weatherability, abrasion resistance, and resistance to mechanical damage.

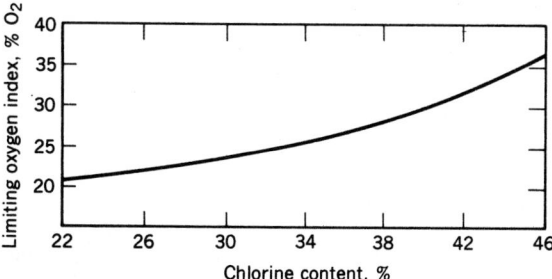

Figure 6.1. Limiting oxygen index vs chlorine content.

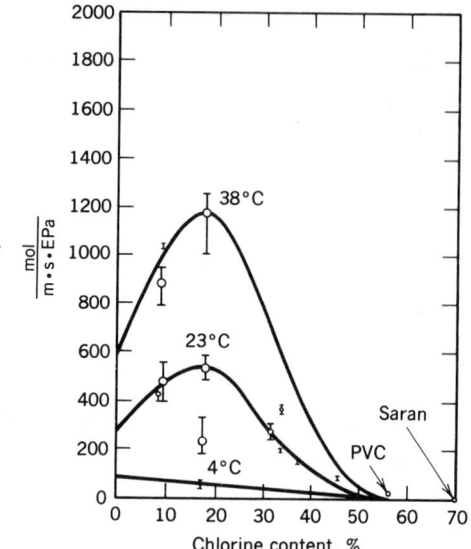

Figure 6.2. Oxygen permeability as a function of chlorine content of chlorinated polyethylene. $\left(\text{To convert } \dfrac{\text{mol}\cdot}{\text{M}\cdot\text{s}\cdot\text{EPa}} \text{ to } \dfrac{\text{cm}^3\cdot\text{mil}}{100\text{ in}^{\cdot2}\cdot\text{d}\cdot\text{atm}}, \text{ multiply by } 0.5.\right)$

Shore A hardnesses of elastomeric CPE compounds can range from 40 to 95; tensile strengths can be as high as 4000 psi (27 MPa); and elongation values can reach 600%.

Compression set resistance of CPE at high temperatures can be superior to that of most other oil-resistant "rubber" on the market. Properly cured CPE elastomer compounds can provide compression set values of 30% [70 h at 302°F (150°C)] compared to values of 50–70% for other polymers.

Ozone and weathering properties of CPE polymers are excellent because of their saturated backbone, which leaves few sites available for ultraviolet light, oxygen, ozone, or other attack. Long-term weathering studies of CPE blended with PVC, in properly formulated thermoplastic compounds, demonstrate excellent impact and color retention.

6.6 APPLICATIONS

Most commercial uses of CPE are high-specification, high-performance applications. The considerable and regular growth experienced by CPE elastomers and thermoplastic resins has occurred in such applications when thorough testing has proven their adequate or advantageous properties, easy fabrication, and cost benefits.

6.6.1 Automotive Hose, Tubing, Wire Jacketing

The combination of high heat resistance, excellent compression set, oil and fuel resistance, electrical properties, abrasion resistance, and effective economics are earning CPE elastomers an increasing share of this performance-demanding industry.

6.6.2 Single-Ply Roofing Membranes

Architects, contractors, and homeowners are using increasing quantities of membranes made with CPE because of the improved profile of ignition resistance, low temperature performance, weatherability, oil and ozone resistance over time, and easy maintenance, repair, and sealability compared to membranes made of competitive elastomers or thermoplastics.

6.6.3 Weatherable Extrusions

A major use for CPE resins is in the impact modification of PVC pipe, sheeting, film, siding, window trim, and similar profile extrusions. Here, CPE can improve impact, retention of physical properties, processability, and economics. Styrenic polymers, such as ABS and SAN, can also gain improved impact and ignition-resistance properties when they are blended with CPE.

6.6.4 Other Applications of CPE

- Industrial hoses—hydraulic, chemical transfer.
- Mining cable jacketing.
- Gaskets, seals, tank linings.
- Wire and cable.
- Molded goods.
- Weather and waterproofing of fabrics.
- Highly filled applications—sound deadening, magnetic strips.
- Compatibility recycled thermoplastics.

6.7 ADVANTAGES/DISADVANTAGES

CPE polymers and the compounds that incorporate them can vary considerably, according to the parent PE, chlorination, blending and compounding, and fabrication. As a result, CPE polymers and compounds can be designed for specific application needs, offering excellent commercial performance. It thus follows that the elastomeric and thermoplastic forms of CPE can be very similar or very dissimilar. This is important to bear in mind in reviewing the following list of advantages and disadvantages of CPE polymers.

6.7.1 Advantages

- Inherently soft and rubbery without addition of extractable or expensive plasticizers; very good weathering and heat resistance; no plasticizer migration (retention of flexibility over time).
- Chemically saturated molecular backbone; better oxidation resistance than common (diene) rubbers; excellent weathering and heat resistance.
- Ignition resistance as a result of Cl content.
- Acceptance of high levels of fillers: barium ferrite for magnets, other fillers for sound deadening; inexpensive filler and as a carrier for additive concentrate.
- Granular form (vs bale form of many other rubbers) permits economical, easy powder blending.
- Conversion from powder to finished product for thermoplastic and elastomeric applications.
- Compatibility with wide range of polymers.
- Vulcanizable by peroxide, radiation, and nonfree-radical processes.
- Oil and chemical resistant.
- Elastomers—good resistance to fuel; excellent resistance to heat, ozone, ignition; excellent compression set resistance.
- Thermoplastics—contribute impact and ignition resistance, weatherability, color retention.

6.7.2 Disadvantages

- High temperature strength is reduced when CPE is used alone in noncross-linked forms.
- Heat stabilizer is required to retard or prevent degradation during high-temperature processing of CPE polymers, as is the case with PVC polymers. Zinc-containing additives should be avoided. Three percent epoxidized soybean oil plus 2% calcium stearate can be used as a heat stabilizer system. Other more effective heat stabilizer systems can be recommended by CPE polymer manufacturers.

6.8 PROCESSING

Conventional molding, extrusion, and rubber compounding equipment and practices are used to fabricate end products of CPE. However, differences in blends and formulations are sufficiently significant that processing generalizations can be misleading. Manufacturers,

Master Material Outline

PROCESSABILITY OF ___Chlorinated Polyethylene (CPE)___

PROCESS	A	B	C
INJECTION MOLDING	X*		X
EXTRUSION	X		
THERMOFORMING		X	
FOAM MOLDING			X
DIP, SLUSH MOLDING		X	
ROTATIONAL MOLDING		X	
POWDER, FLUIDIZED-BED COATING		X	
REINFORCED THERMOSET MOLDING		X	
COMPRESSION/TRANSFER MOLDING	X*		
REACTION INJECTION MOLDING (RIM)		X	
MECHANICAL FORMING			X
CASTING		X	
CALENDERING	X		

A = Common processing technique.
B = Technique possible with difficulty.
C = Used only in special circumstances, if ever.
X* For elastomers.

and their literature, are suggested as reliable resources for processing specifics (see Master Material Outline).

6.9 RESIN FORMS

CPE polymers are produced in soft powder-crumb form. They are commonly marketed in 25-kg bags. In Europe, blends of CPE with PVC also are marketed.

6.10 SPECIFICATION OF PROPERTIES

The typical elastomeric properties developed by given CPE polymers vary according to the choice of curing agent (peroxide or thiadiazole cures, for example) and the influence of all the ingredients used in a compound. Typical thermoplastic resin properties of given CPE resins vary by choice of base polyethylene, by chlorine content, etc. Generally, CPE is used as an impact modifier and impact retention aid in thermoplastics and as an ignition-resistance aid in PVC, ABS, SAN, and other polymers. After fabrication the impact properties of the final resin blend can be improved a factor of 20. CPE elastomers provide outstanding heat, oil, chemical, and weathering resistance; excellent compounding and processing ease; and ignition resistance. Because the specific properties achieved are so dependent on the materials used or compounded with CPE and on the fabrication processes, the tabular data provided should be considered only as general indications of performance expectations. (See Master Outline of Material Properties.)

6.11 PROCESSING REQUIREMENTS

Soft CPE particles can absorb moisture, vapors, or liquids and may fuse quite readily. Thorough dispersion is an important key to successful processing with CPE regardless of the blending or fabrication method chosen: dry blend, Banbury mixing, calendering, extrusion, or injection molding. Good dispersion of CPE with PVC, SAN, or ABS may require temperatures as high as 392°F (200°C) and above.

Dispersion of CPE with ABS, SAN, or polypropylene pellets requires that the resinous portion be fused before the CPE and other ingredients are added. Most CPE polymer powders can pass through a 20-mesh screen. With elastomeric forms, twin-screw extruders are used to process the compound powder blend directly into a finished product. Calendering, extrusion, and molding are all commonly employed, depending on the form and shape desired for the end product.

6.12 PROCESSING-SENSITIVE END PROPERTIES

Most physical properties attained by end products of CPE are influenced by the uniformity of dispersion of the other polymers and fillers used. Care must be used to prevent thermal degradation. The use of a chemical heat stabilizer, and equipment suitable for PVC processing, are usually satisfactory.

6.13 SHRINKAGE

Because CPE resins and elastomers are seldom used in a "pure" state, shrinkage and related design data are more dependent on the other ingredients and polymers used with the CPE in compounds and blends. Tables for such information therefore are not included.

MATERIAL __Chlorinated Polyethylene (CPE)__

PROPERTY	TEST METHOD	ENGLISH UNITS	VALUE	METRIC UNITS	VALUE
MECHANICAL DENSITY	D792	lb/ft^3	72.7–80	g/cm^3	1.08–1.25
TENSILE STRENGTH	D638	psi	1,200–2,500	MPa	8.3–17.2
TENSILE MODULUS	D638	psi	150–500	MPa	1–3.5
FLEXURAL MODULUS	D790	psi		MPa	
ELONGATION TO BREAK	D638	%	200–1000	%	200–1000
NOTCHED IZOD AT ROOM TEMP	D256	ft-lb/in.	No break	J/m	No break
HARDNESS	D785	Shore A	60–90	Shore A	60–90
THERMAL DEFLECTION T @ 264 psi	D648	°F	Flexible	°C	Flexible
DEFLECTION T @ 66 psi	D648	°F	Flexible	°C	Flexible
VICAT SOFT- ENING POINT	D1525	°F	Flexible	°C	Flexible
UL TEMP INDEX	UL746B	°F		°C	
UL FLAMMABILITY CODE RATING	UL94				
LINEAR COEFFICIENT THERMAL EXPANSION	D696	in./in./°F		mm/mm/°C	
ENVIRONMENTAL WATER ABSORPTION 24 HOUR	D570	%		%	
CLARITY	D1003	% TRANS- MISSION		% TRANS- MISSION	
OUTDOOR WEATHERING	D1435	%		%	
FDA APPROVAL					
CHEMICAL RESISTANCE TO: WEAK ACID	D543		Not attacked		Not attacked
STRONG ACID	D543		Not attacked to badly attacked		Not attacked to badly attacked
WEAK ALKALI	D543		Not attacked		Not attacked
STRONG ALKALI	D543		Not attacked to badly attacked		Not attacked to badly attacked
LOW MOLECULAR WEIGHT SOLVENTS	D543		Not attacked to badly attacked		Not attacked to badly attacked
ALCOHOLS	D543		Not attacked to minimally attacked		Not attacked to minimally attacked
ELECTRICAL DIELECTRIC STRENGTH	D149	V/mil	375–500	kV/mm	14.8–19.7
DIELECTRIC CONSTANT	D150	10^6 Hertz	4.3–5.1		4.3–5.1
POWER FACTOR	D150	10^6 Hertz	0.1		0.1
OTHER MELTING POINT	D3418	°F	Most products not crystalline	°C	Most products not crystalline
GLASS TRANS- ITION TEMP.	D3418	°F	−5	°C	−20

6.14 TRADE NAMES

Currently, the following companies are supplying CPE polymers: Tyrin®, The Dow Chemical Company USA; Hostapren, Hoechst AG, Federal Republic of Germany; Kelrinal, DSM (manufactured by Hoechst), Federal Republic of Germany; Ongrovil, Borsodi Vegyi Kombinat, Hungary; Daisolac, Osaka Soda, Japan; Elaslen, Showa Denko, Japan.

BIBLIOGRAPHY

General References

I.A. Abu-Isa, *J. Polym. Sci. Part A-1* **10,** 881 (1972).

I.A. Abu-Isa, *Polym. Eng. Sci.* **15,** 299 (1975).

C.N. Burnell and R.H. Parry, *Appl. Polym. Symp.* **11,** 95 (1969).

Dow CPE Resins: Basics, Product Bulletin, The Dow Chemical Co., Plaquemine, La., 1984, Form No. 305-895-281.

P.J. Canterino, "Ethylene Polymers, Derivatives," in N.M. Bikales *Encyclopedia of Polymer Science and Technology,* 1st ed. Vol. 6, John Wiley & Sons, Inc., New York, pp. 431–454.

D. Jaroszynska, T. Kleps and D. Gdowska-Tutak, *J. Therm. Anal.* **19,** 69 (Jan. 1980).

S. Krozer, *J. Appl. Polym. Sci.* **15,** 1769 (1971).

C.E. Locke and D.R. Paul, *Polym. Eng. Sci.* **13,** 308 (1973).

S. Matsuoka, R.J. Roe, and H.F. Cole, "Dielectric Properties of Polymers," *Proceedings of ACS Symposium,* 161st National Meeting of the ACS, Los Angeles, Calif., 1971.

S. Matsuoka, R.J. Roe, and H.F. Cole, *Polym. Prepr. Am. Chem. Soc. Div. Polym. Chem.* **12,** 192 (Jan. 1972).

B. Schneier, *J. Appl. Polym. Sci.* **16,** 1515 (1972).

UV-Chek AM-595, Ferro Chemical Co., Beford, Ohio.

W.L. Young, *paper presented at the Annual Technical Conference,* Society of Plastics Engineers, Greenwich, Conn., 1978, p. 750.

W.L. Young and R.L. Blanchard in J. Kroschwitz, ed., *Encyclopedia of Polymer Science and Engineering,* Vol. 6, Wiley-Interscience, New York, 1986, pp. 511–513.

W.L. Young and E.D. Serdinsky, *paper presented at the SPE Regional Technical Conference,* Miami, Fla., 1966.

7

ETHYLENE VINYL ACETATE (EVA)

Harold J. Frey

E.I. du Pont de Nemours & Co., Wilmington, Delaware

7.1 INTRODUCTION

Ethylene vinyl acetate (EVA) copolymers are the products of low density polyethylene (LDPE) technology. Their properties are governed by the percentage content of EVA and the melt index. The EVA resins may be represented by:

7.2 CATEGORY

EVA copolymers are thermoplastic materials consisting of an ethylene chain incorporating 5 to 50% vinyl acetate (VA), in general. The VA content controls the resins' crystallinity and flexibility.

7.3 HISTORY

Shortly after the discovery of LDPE, DuPont and ICI began to investigate a series of copolymers of ethylene that could be made by the high-pressure, free-radical process. Vinyl esters and acrylate copolymers were investigated in detail in the early to mid-1940s.

In the period following World War II, most research efforts were directed toward exploring this high-pressure process and the effect of the process variables on the properties of LDPE. It was not until the early 1960s that DuPont and USI (a division of National Distillers) introduced a series of EVA copolymers. Initially, for EVA copolymers with a low VA content were used in film: The presence of 2–7 wt% VA enhanced such desirable film properties as toughness and optical clarity. Higher comonomer content resins (18–40% VA) were later introduced by DuPont as wax-coating tougheners and in hot-melt adhesives.

The initial product lines of DuPont and USI changed very little over subsequent years. Minor elaborations include the development by USI of a range of copolymers with VA contents in the 40–60% range and by DuPont of terpolymers containing low concentrations of acid comonomer.

Production data on EVA resins are not readily available, partly as a matter of definition. There is a general tendency to include EVA copolymers containing less than 5% VA with LDPE figures. With this caveat, the volume of 5% VA content EVA produced in the U.S. in 1970 was estimated at approximately 100×10^6 lb (45,000 metric tons); in 1980, 330×10^6 lb (150,000 metric tons).

7.4 POLYMERIZATION

The free-radical, high-pressure polymerization process is described in the chapter on LDPE. Commercial preparation of EVA copolymer is based on the same process, with the addition of a controlled comonomer stream into the reactor. Both tubular and stirred autoclave units have been used to make EVA copolymers; the autoclave units are preferred for the higher (18% VA) copolymers for better control of such critical variables as molecular weight and comonomer content. A major difference between the conventional low-density process and that employed for EVA is the finishing system. Special equipment is required to handle the soft, low-melting, and frequently tacky EVA copolymer.

These special steps, which can limit the production rate, increase the manufacturing costs for EVA resins—especially those with a higher VA content—and thus affect pricing. VA monomer costs about 30% more than ethylene. Higher VA copolymers thus sell at a substantial premium, both to cover higher raw-material costs and lower polymerization-unit utility.

The early 1987 EVA prices ranged from $0.46/lb for 7.5% VA grades to $0.705/lb for 40% VA grades. Special grades, such as terpolymers and higher-VA grades, are usually sold at an additional premium.

7.5 DESCRIPTION OF PROPERTIES

Incorporation of vinyl acetate in the ethylene polymerization process produces a copolymer with lower crystallinity than conventional ethylene homopolymer. These lower crystallinity resins have lower melting points and heat-seal temperatures, along with reduced stiffness, tensile strength, and hardness. EVAs have improved clarity, low-temperature flexibility, stress-crack resistance, and impact strength but poorer high-temperature properties than LDPE. EVA resins are more permeable to oxygen, water vapor, and carbon dioxide. Chemical resistance is similar to that of LDPE, with somewhat better resistance to oil and grease for EVA resins with a higher VA content. The VA groups contribute to improved adhesion in extrusions or hot-melt-adhesive formulations.

Exposure to temperatures above 401°F (205°C) causes loss of acetic acid, which leaves behind a structure more susceptible to attack by oxygen. This is more noticeable in higher VA resins, in which a combination of temperature and exposure time can produce significant degradation. Upper temperature stability limitations for processing range from 446°F (230°C) for the lower VA grades to 401°F (205°C) for resins with a VA content of 28% or more.

The outdoor stability of natural EVA resins is superior to that of LDPE by virtue of their greater flexibility. Addition of UV stabilizers can extend the outdoor life of clear compounds to three to five years, depending on the degree of exposure. Outdoor life expectancy is also enhanced by the addition of carbon black.

Elvax grades are available with 9–33% VA and a melt index of 0.3–43.

7.6 APPLICATIONS

In addition to specialty applications involving film and adhesives production, EVAs are used in a variety of molding, compounding, and extrusion applications. Some typical end uses include flexible hose and tubing, footwear components, wire-and-cable compounding, toys and athletic goods, bearing pads, color concentrates, extruded gaskets, molded automotive parts (such as energy-absorbing bumper components), molded gaskets for coating plastic lenses, and cap and closure seals.

In footwear applications, EVA resins are used in canvas box toes and flocked or fabric-laminated counters. Foamed and cross-linked EVA is used in athletic or leisure shoe midsoles and in sandals.

Cost range is $0.016–0.024/in³. EVA compares favorably in space-filling characteristics with such other soft thermoplastics as plasticized PVC. EVA has a specific gravity of 0.95 to 1.0, as compared to 1.2 for PVC, which gives it a space-filling advantage per unit weight of 20%.

7.7 ADVANTAGES/DISADVANTAGES

With increasing VA content, EVA resin properties range from those of LDPE to those of highly plasticized PVC. Higher VA resins are soft and flexible, with excellent toughness and good stress-crack resistance. With EVA resins, these are permanent properties; these characteristics do not dissipate with time because of the loss of liquid plasticizer. This is not the case with PVC resins. The density of EVA resins is lower, providing such performance advantages as buoyancy in water.

EVAs also have exceptional low-temperature toughness and flexibility. They are true thermoplastics and can be reprocessed without waste, except for resins that are cross-linked during processing to achieve enhanced elastomeric properties.

7.8 PROCESSING TECHNIQUES

EVA can be processed by all standard plastics processing techniques, including injection and blow molding, thermoforming, and extrusion into sheet and shapes. They accommodate high loadings of fillers, pigments, and carbon blacks. They are compatible with other thermoplastics, and thus are frequently used for impact modification and improvement of stress-crack resistance. This combination of properties makes EVAs highly adaptable vehicles for color concentrates.

EVA resins can be formulated with blowing agents and cross-linking to produce low density foams via compression molding. Such foams are widely used in midsoles for athletic and leisure footwear and in sandals.

Master Material Outline

PROCESSABILITY OF ___ETFE___

PROCESS	A	B	C
INJECTION MOLDING	X		
EXTRUSION	X		
THERMOFORMING	X		
FOAM MOLDING	X		
DIP, SLUSH MOLDING			
ROTATIONAL MOLDING	X		
POWDER, FLUIDIZED-BED COATING		X	
REINFORCED THERMOSET MOLDING	X		
COMPRESSION/TRANSFER MOLDING	X		
REACTION INJECTION MOLDING (RIM)			
MECHANICAL FORMING			
CASTING			

A = Common processing technique.
B = Technique possible with difficulty.
C = Used only in special circumstances, if ever.

Standard design chart for ___Ethylene vinyl acetate___

1. Recommended wall thickness/length of flow
   Minimum ___0.040 in.___ for ___15 in.___ distance
   Maximum ___0.100 in.___ for ___40 in.___ distance
   Ideal _____ for _____ distance

2. Allowable wall thickness variation, % of nominal wall _____ No restriction

3. Radius requirements
   Outside: Minimum ___NA___ Maximum ___NA___
   Inside: Minimum ___NA___ Maximum ___NA___

4. Reinforcing ribs
   Maximum thickness ___None___
   Maximum height ___None___
   Sink marks: Yes _____ No _____ (Depends on cross section)

5. Solid pegs and bosses
   Maximum thickness _____None_____
   Maximum weight _____None_____
   Sink marks: Yes _____ No _____ (Depends on cross section)

6. Strippable undercuts
   Outside: Yes ___X___ No _____
   Inside: Yes ___X___ No _____

7. Hollow bosses: Yes ___X___ No _____

8. Draft angles
   Outside: Minimum ___3° for 12%___ Ideal ___Same___
   Inside: Minimum ___5° for 30%___ Ideal ___Same___

9. Molded-in inserts: Yes ___X___
                       No _____

10. Size limitations
    Length: Minimum ___None___ Maximum ___None___ Process _____
    Width: Minimum ___None___ Maximum ___None___ Process _____

11. Tolerances

12. Special process- and materials-related design details
    Blow-up ratio (blow molding) _____
    Core L/D (injection and compression molding) _____
    Depth of draw (thermoforming) _____
    Other _____

7.9 RESIN FORMS

EVA resins are normally sold in pelletized form, approximately spherical in shape and about 1/8 in. (3 mm) in diameter. A few very soft grades may be coated or may contain an antiblocking agent to provide free-flowing characteristics.

7.10 SPECIFICATION OF PROPERTIES

See the Master Outline of Materials Properties.

7.11 PROCESSING REQUIREMENTS

7.11.1 Injection Molding

To injection mold EVA, the machine hopper throat must be kept cool to prevent caking and bridging. Water cooling of the hopper throat is recommended, although setting the rear zone of the heating cylinder at 250°F (121°C) or below is often sufficient.

Because the feed throat of the cylinder must be kept cool in both ram and screw-type machines, a rising temperature profile from the rear to the front of the cylinder is generally used. This is subject, of course, to the requirements of the cycle and the size of the shot, compared to the capacity of the machine. Fast cycles and large shot sizes may require a more even profile, or, in some cases, a rear zone setting above that of the front zone.

MATERIAL ___ Ethylene Vinyl Acetate

PROPERTY	TEST METHOD	ENGLISH UNITS	VALUE	METRIC UNITS	VALUE
MECHANICAL DENSITY	D792	lb/ft³	58–60	g/cm³	0.93–0.957
TENSILE STRENGTH	D638	psi	900–4,000	MPa	6.2–28
TENSILE MODULUS	D638	psi		MPa	
FLEXURAL MODULUS	D790	psi	1,000–16,000	MPa	6.9–110
ELONGATION TO BREAK	D638	%	650–950 (ultimate)	%	650–950 (ultimate)
NOTCHED IZOD AT ROOM TEMP	D256	ft-lb/in.	No break	J/m	No break
HARDNESS	D785		A: 73–98 D: 24–47		A: 73–98 D: 24–47
THERMAL DEFLECTION T @ 264 psi	D648	°F		°C	
DEFLECTION T @ 66 psi	D648	°F		°C	
VICAT SOFT- ENING POINT	D1525	°F	97–180	°C	36–82
UL TEMP INDEX	UL746B	°F		°C	
UL FLAMMABILITY CODE RATING	UL94				
LINEAR COEFFICIENT THERMAL EXPANSION	D696	in./in./°F		mm/mm/°C	
ENVIRONMENTAL WATER ABSORPTION 24 HOUR	D570	%	0.1	%	0.1
CLARITY	D1003	% TRANS- MISSION	Transparent to translucent	% TRANS- MISSION	Transparent to translucent
OUTDOOR WEATHERING	D1435	%		%	
FDA APPROVAL			Yes		Yes
CHEMICAL RESISTANCE TO: WEAK ACID	D543		Not attacked		Not attacked
STRONG ACID	D543		Attacked by sulfuric acid		Attacked by sulfuric acid
WEAK ALKALI	D543		Not attacked		Not attacked
STRONG ALKALI	D543		Not attacked		Not attacked
LOW MOLECULAR WEIGHT SOLVENTS	D543		Soluble		Soluble
ALCOHOLS	D543		Not attacked		Not attacked
ELECTRICAL DIELECTRIC STRENGTH	D149	V/mil		kV/mm	
DIELECTRIC CONSTANT	D150	10⁶ Hertz			
POWER FACTOR	D150	10⁶ Hertz			
OTHER MELTING POINT	D3418	°F		°C	
GLASS TRANS- ITION TEMP.	D3418	°F		°C	

TABLE 7.1 The Effect of EVA Molding Variables on Finished Part Properties

	Mold Variable						
Property	Melt Temperature	Injection Pressure	Mold Temperature	Screw Forward Time	Fill Speed	Feed Condition	Cycle Time
Minimum shrinkage	Low[a]	Maximum[b]	Low[c]	Long	Fast[d]	Pad	Long[e]
Minimum warpage	Low[a]	Minimum	Medium	Short[f]	Fast[d]	Starve[g]	Long[e]
Maximum dimensional stability	Low[a]	Medium	Medium[c]	Medium	Fast[d]	Starve[g]	Long[e]
Maximum impact strength	High[h]	Medium to high	Low to medium[c]	Medium	Fast[d]	Pad[g]	Medium[e]
Best surface appearance	Medium to high[h]	Maximum[b]	Medium to high[c]	Medium to long	Medium to fast[d]	Pad	Long[e]
Best overall quality	Medium to high[h]	Maximum[b]	Medium[c]	Medium	Medium to fast[d]	Pad	Long[e]

[a]Low side of adequate.
[b]Maximum short of flash.
[c]Low: below 5°C (40°F). Medium: 5–25°C (40–80°F). High: above 25°C (80°F).
[d]As fast as possible, provided surface and venting are adequate.
[e]Greater than minimum.
[f]To prevent packing.
[g]Short of metal-to-metal contact.
[h]Avoid overheating.

Table 7.1 lists recommended cylinder temperatures for the average molding job. These must be modified to suit specific molding requirements. Thin-section, long-flow moldings may require higher melt or mold temperatures. Thicker moldings may need lower melt temperatures for easier ejection and shorter cycles.

Mold coolant temperatures range between 40 and 120°F (5 and 49°C); 60–100°F (15–38°C) is the most common. Above 100°F (38°C), indentation or distortion of the molding often results. On the other hand, coolant temperatures below about 60°F (15°C) may cause flow patterns on the surface of the molding.

EVA resins can be processed on standard screw- and ram-type injection machines. Screws normally used for injection molding of LDPE are satisfactory. Shallow metering-zone screws may cause excessive working of the resin, making melt temperature control more difficult, especially at high output rates.

Because of the relatively low melt temperature used for molding EVA resins and their softness, purging higher melting resins from the heating cylinder is usually more speedily accomplished using a harder resin with a broad molding temperature range, such as high-density polyethylene (HDPE).

Molding pressures for EVA resins are usually low. For most machines, the pressure ranges from 5000 to 15,000 psi (35 to 103 MPa). Higher molding pressures can cause surface roughness and poor gloss.

As for the injection rate, the injection forward speed control should be adjusted to produce a slow ram speed, since rapid filling of the mold cavity usually causes turbulence at the gate. This produces a rough surface on the molded part. Turbulence varies with the type of gate used, but a slow filling speed is generally better than a fast speed. The molding cycle time for EVA resins is equal to or slightly longer than that for LDPE. Because of the flexibility of the resin, some parts may require longer cooling times to obtain satisfactory ejection.

Spray-on lubricants may be needed to aid ejection of some moldings, especially those with little or no draft or taper on the sides. Fluorocarbon lubricant sprays are often used. With moldings that are particularly difficult to eject, aerosol spray waxes have been used with good results.

7.11.2 Extrusion

For extrusion of EVA resins, metering screws of the type commonly used with LDPE are satisfactory. In general, procedures for the extrusion of EVA differ very little from those for LDPE; only a few special precautions are necessary in view of the former's lower melting point. These include a water-cooled hopper feed throat to prevent pellet agglomeration and bridging.

In quenching the extruded product, lower temperatures must be used to provide sufficient stability for the forms. The relative softness of EVA makes extrusions more difficult to cut or trim. Thus it is mandatory that cutting tools be sharp. If the frequency of cutting is high enough to raise the temperature of the tool and cause the polymer to stick, it may be necessary to cool or lubricate the cutters.

7.11.3 Regrind

Because of the rubberlike nature of EVA resins, scrap grinding requires sharp cutter blades, set to minimum clearance of 0.003–0.005 in. (0.075–0.125 mm). The moldings should be cool. Screens used in the grinders should have a minimum opening of 0.250 in. (6.35 mm) and a maximum opening of 0.375 in. (9.525 mm).

Since handling of scrap is difficult, it should be minimized in primary processing by the use of hot-runner molds or minimum-length runners. Care must be taken to ensure that the scrap resin to be reused is not degraded. Heating above the maximum recommended temperature of 425°F (220°C) or excessive residence time in the heating cylinder can seriously affect the physical properties of EVA resins. The residence time in the cylinder at or near the maximum recommended melt temperature should not exceed 20 minutes.

7.12 PROCESSING-SENSITIVE END PROPERTIES

The best overall quality in molded EVA components is achieved at medium-to-high melt temperatures, maximum injection pressures, and medium mold temperatures. Table 7.2 provides a detailed picture of the effect of molding variables on part properties.

TABLE 7.2   Effect of Molding Variables on Finished Part Properties

	Melt Temperature	Injection Pressure	Mold Temperature	Screw Forward Time	Fill Speed	Feed Condition	Cycle Time
Minimum shrinkage	Low[a]	Maximum[b]	Low[c]	Long	Fast[d]	Pad	Long[e]
Minimum warpage	Low[a]	Minimum	Medium	Short[f]	Fast[d]	Starve[g]	Long[e]
Maximum dimensional stability	Low[a]	Medium	Medium[c]	Medium	Fast[d]	Starve[g]	Long[e]
Maximum impact strength	High[h]	Medium to high	Low to medium[c]	Medium	Fast[d]	Pad[g]	Medium[e]
Best surface appearance	Medium to high[h]	Maximum[b]	Medium to high[c]	Medium to long	Medium to fast[d]	Pad	Long[e]
Best overall quality	Medium to high[h]	Maximum[b]	Medium[c]	Medium	Medium to fast[d]	Pad	Long[e]

[a]Low side of adequate.
[b]Maximum short of flash.
[c]Low, below 5°C (40°F). Medium, 5–25°C (40–80°F). High, above 25°C (80°F).
[d]As fast as possible provided surface and venting are adequate.
[e]Greater than minimum.
[f]To prevent packing.
[g]Short of metal-to-metal contact.
[h]Avoid overheating.

7.13 SHRINKAGE

Volume shrinkage varies with the weight percent of the VA in the copolymer. Averages for 10, 20, and 30% VA formulations are given in the Mold Shrinkage MPO.

Mold Shrinkage Characteristics (in./in. or mm/mm)

MATERIAL   Ethylene Vinyl Acetate

IN THE DIRECTION OF FLOW

THICKNESS

MATERIAL	mm: 1.5 / in.: 1/16	3 / 1/8	6 / 1/4	8 / 1/2	FROM	TO	#
10% Vinyl acetate		0.010	0.013				
20% Vinyl acetate		0.003	0.008				
30% Vinyl acetate		0.0013	0.0065				

PERPENDICULAR TO THE DIRECTION OF FLOW							

NOTE: Mold shrinkage is approximate. It is affected by part design, mold temperature, thickness, injection pressure, packing time, cycle time, orientation, gate design, gate size, gate location, glass content, glass size, and filler content.

7.14 TRADE NAMES

The major U.S. trade names and producers of EVA resins are Chemplex, Chemplex Co., Rolling Meadows, Ill.; Elvax, DuPont Co., Wilmington, Del.; Escorene, Exxon Chemical Co., Houston, Tex.; Bakelite, Union Carbide Corp., Polyolefins Div., Danbury, Conn.; Urethane and Vynathene, U.S. Industrial Chemicals Co., Div. of National Distillers & Chemical Corp., New York, N.Y.

BIBLIOGRAPHY

Much of the information on EVA copolymers is available in publications available from the supplier companies. General articles on EVA and data tabulations can be found in the following industry publications:

General References

N.M. Bikales, ed., *Encyclopedia of Polymer Science and Technology,* Vol. 6, Interscience Publishers, New York, 1967.

Modern Plastics Encyclopedia, published annually by Modern Plastics magazine.

Plastics—A Desktop Data Bank, Cordura Publications, 1982.

Plastics Manufacturing Handbook & Buyers' Guide, published annually by *Plastics Technology* magazine.

S.S. Schwartz and S.H. Goodman, *Plastics Materials and Processes,* Van Nostrand Reinhold Co., New York, 1982.

8

FLUOROPOLYMERS, MODIFIED ETHYLENE– TETRAFLUOROETHYLENE COPOLYMER (ETFE)

Carleton A. Sperati

E.I. du Pont de Nemours & Co., Inc., Parkersburg, WV 26104, and Adjunct Professor, Ohio University, Athens, Ohio

8.1 MODIFIED ETHYLENE–TETRAFLUOROETHYLENE COPOLYMER

$$—(\underset{\underset{H}{|}}{\overset{\overset{H}{|}}{C}}—\underset{\underset{H}{|}}{\overset{\overset{H}{|}}{C}}—\underset{\underset{F}{|}}{\overset{\overset{F}{|}}{C}}—\underset{\underset{F}{|}}{\overset{\overset{F}{|}}{C}})_n—$$

8.2 CATEGORY

Modified ethylene–tetrafluoroethylene copolymer (ETFE) is a hard, tough thermoplastic.

8.3 HISTORY

Copolymers of tetrafluoroethylene (TFE) and ethylene were made early in the course of Du Pont's research on fluoropolymers. These materials were not useful, however, until it was discovered that combining new polymerization procedures with a small amount of a modifier, often a termonomer, resulted in polymers with properties suitable for long-term use.[1]

One key requirement, for example, is the ability to withstand exposure at high temperatures without rapid embrittlement. The amount of modifier is very low. The exact amount and nature of the modifier has been kept proprietary by each manufacturer and may be different for different grades within a product line.

Following a period of test marketing with pilot plant resins that began in 1970, modified ETFE fluoropolymer was commercialized by Du Pont in mid-1972. Since then, similar products have been introduced by Ausimont, Daikin, and Hoechst. The most important addition to the modified poly ETFE product line has been the introduction of polymer reinforced with glass fibers.

Official production volume figures are not available. Broad estimates based on general industry information are in 1960, ETFE not yet commercially available; in 1970, pilot plant production started; and in 1980, 1,500,000 lb.

8.4   POLYMERIZATION

Commercial products of modified poly ETFE are made by free-radical initiated addition polymerization, often called radical-chain polymerization. These materials are essentially alternating copolymers.[2] In the molecular formula, they are isomeric with poly(vinylidene fluoride) (PVDF) and can be considered equivalent to a PVDF with a head-to-head, tail-to-tail structure. Many important physical properties of modified ETFE copolymers, however, are very different from and superior to those of available PVDF materials. (These properties do not include the remarkable piezoelectric and pyroelectric characteristics of PVDF.)

The section on PTFE discusses the preparation of tetrafluoroethylene (TFE). The comments in that section on the requirements for safety, high investment to protect against the hazards of TFE, and so on, are equally applicable to modified poly ETFE. Polymerization grade ethylene is readily available by the same processes used for polyethylene.

Prices (1988) for modified ETFE fluoropolymers range from $12.50 to 15.00 per pound, depending on quantity and particular type. Color concentrates are in the range of $20.00 per pound.

8.5   DESCRIPTION OF PROPERTIES

8.5.1   Thermal Properties

Along with a melting point of about 518°F (270°C), modified ETFE fluoropolymers have good thermal properties. As mentioned above, a key factor in development of the modern resins is their ability to withstand long-time exposure to temperatures on the order of 392°F (200°C) with minimal embrittlement. The resins have good performance in various tests for resistance to burning.

8.5.2   Mechanical Properties

Modified ETFE fluoropolymers are strong, tough, and rugged. Their stiffness is similar to PVDF and very much higher than the fluorocarbon polymers PTFE, PFA, and FEP.

8.5.3   Optical Properties

Modified ETFE fluoropolymers do not absorb visible light. At times, however, they may not be water-clear in thick sections as a result of scattering from crystal domains.

8.5.4   Environmental Properties

Modified ETFE fluoropolymers show the excellent resistance to outdoor aging typical of the fluoropolymers. In addition, they have an impressive tolerance for large doses of ionizing radiation. The resin cross-links on radiation, and high temperature performance can be improved by controlled treatment.

8.5.5   Reinforcement

Unlike use of fillers with most fluoropolymers, moldings of glass fiber and modified ETFE fluoropolymers are reinforced, not just filled. The compositions can be molded by conventional means and marked increase is attained in such properties as strength, stiffness, resistance to creep, heat-distortion temperature, and dimensional stability.

8.6   APPLICATIONS

High-performance insulation for wire and cable is a major use for modified ETFE resins. In particular, their resistance to "cut-through" is outstanding. Injection-molding applications also are important. Many water-resistant wristwatch cases, for example, are made from ETFE resins.

8.7 ADVANTAGES/DISADVANTAGES

The main disadvantage of modified ETFE fluoropolymers is their relatively high cost. When value-in-use is considered, however, these materials do jobs for which few other materials are satisfactory.

8.8 PROCESSING TECHNIQUES

Essentially all the common procedures available for thermoplastic polymers can be used with modified poly ETFE. For details, see the Master Material Outline.

Master Material Outline

PROCESSABILITY OF _____ ETFE _____

PROCESS	A	B	C
INJECTION MOLDING	X		
EXTRUSION	X		
THERMOFORMING	X		
FOAM MOLDING	X		
DIP, SLUSH MOLDING			
ROTATIONAL MOLDING	X		
POWDER, FLUIDIZED-BED COATING		X	
REINFORCED THERMOSET MOLDING	X		
COMPRESSION/TRANSFER MOLDING	X		
REACTION INJECTION MOLDING (RIM)			
MECHANICAL FORMING			
CASTING			

A = Common processing technique.
B = Technique possible with difficulty.
C = Used only in special circumstances, if ever.

8.9 RESIN FORMS

Modified poly ETFE is available in molding powder, molding powder containing fillers, and film in a range of thicknesses.

8.10 SPECIFICATION OF PROPERTIES

See the Master Outline of Material Properties.

Master Outline of Material Properties

MATERIAL _____ ETFE

PROPERTY	TEST METHOD	ENGLISH UNITS	VALUE	METRIC UNITS	VALUE
MECHANICAL DENSITY	D792	lb/ft³	106	g/cm³	1.7
TENSILE STRENGTH	D638	psi	6,967	MPa	48
TENSILE MODULUS	D638	psi	120,464	MPa	830
FLEXURAL MODULUS	D790	psi	200,000	MPa	1,378
ELONGATION TO BREAK	D638	%	a	%	a
NOTCHED IZOD AT ROOM TEMP	D256	ft-lb/in.	No break	J/m	No break
HARDNESS	D785		D 70		D 70
THERMAL DEFLECTION T @ 264 psi	D648	°F	165	°C	74
DEFLECTION T @ 66 psi	D648	°F	219	°C	104
VICAT SOFT-ENING POINT	D1525	°F		°C	
UL TEMP INDEX	UL746B	°F		°C	
UL FLAMMABILITY CODE RATING	UL94		VE-0		VE-0
LINEAR COEFFICIENT THERMAL EXPANSION	D696	in./in./°F	9×10^{-5}	mm/mm/°C	16.2×10^{-6}
ENVIRONMENTAL WATER ABSORPTION 24 HOUR	D570	%	<0.02	%	<0.02
CLARITY	D1003	% TRANS-MISSION	b	% TRANS-MISSION	b
OUTDOOR WEATHERING	D1435	%	Excellent	%	Excellent
FDA APPROVAL			No		No
CHEMICAL RESISTANCE TO: WEAK ACID	D543		Not attacked		Not attacked
STRONG ACID	D543		Not attacked		Not attacked
WEAK ALKALI	D543		Not attacked		Not attacked
STRONG ALKALI	D543		Not attacked		Not attacked
LOW MOLECULAR WEIGHT SOLVENTS	D543		Not attacked		Not attacked
ALCOHOLS	D543		Not attacked		Not attacked
ELECTRICAL DIELECTRIC STRENGTH	D149	V/mil	1500	kV/mm	60
DIELECTRIC CONSTANT	D150	10^6 Hertz	2.6		2.6
POWER FACTOR	D150	10^6 Hertz	0.001		0.001
OTHER MELTING POINT	D3418	°F	518	°C	270
GLASS TRANS-ITION TEMP.	D3418	°F	$-184,230^c$	°C	$-120,110^c$

[a]According to ASTM D638, there is no yield point (slope of the stress-strain curve is never zero). The value given is the percentage elongation at which the curve departs from approximate linearity.

[b]Moldings vary from clear to nearly opaque depending on the thickness of the part, the particular grade used, and the conditions of molding.

[c]In its amorphous portion, ETFE fluoropolymer has a loss peak at each of the temperatures listed.[5]

8.11 PROCESSING REQUIREMENTS

Manuals available from Du Pont provide detailed discussion on processing requirements.[3,4]

The main features are that the temperatures required 555–626°F (290–330°C) are higher than those generally used for the common thermoplastics. Special corrosion-resistant alloys are required in the equipment to ensure long-term, trouble-free operation.

8.12 PROCESSING-SENSITIVE END PROPERTIES

Exposure of modified ETFE fluoropolymers to temperatures of 575°F (302°C) for up to one hour or for shorter holdup times to 650°F (343°C) has little or no effect on the color and properties of the fabricated material. Use of excessively high temperatures above these limits can result in thermal degradation with bubble formation and deterioration of mechanical properties.

8.13 SHRINKAGE

Crystallization during cooling results in a relatively high mold shrinkage. See the Mold Shrinkage MPO for sample data.

Mold Shrinkage Characteristics (in./in. or mm/mm)

MATERIAL ___ ETFE ___

IN THE DIRECTION OF FLOW

THICKNESS

MATERIAL	mm: 1.5 / in.: 1/16	3 / 1/8	6 / 1/4	8 / 1/2	FROM	TO	#
		1.5					
		2.0					

	PERPENDICULAR TO THE DIRECTION OF FLOW					
		3.5				
		4.5				

NOTE: Mold shrinkage is approximate. It is affected by part design, mold temperature, thickness, injection pressure, packing time, cycle time, orientation, gate design, gate size, gate location, glass content, glass size, and filler content.

8.14 TRADE NAMES

Modified poly ETFE fluoropolymers are available as Halon ET from Ausimont; as Tefzel fluoropolymer from Du Pont Co., Polymer Products Dept., Wilmington, DE 19898; and as Hostaflon ETFE (TM) from Hoechst (made in Germany).

Daikin makes a modified ETFE copolymer in Japan that is marketed as Neoflox ETFE (TM) in the United States by various distributors.

8.15 INFORMATION SOURCES

The material suppliers are the best source for specific information on modified ETFE fluoropolymers and products made from it. There is no single publication that is comprehensive. The selected list of references, however, provide access to much of the important information in the published literature.

The Fluoropolymers Div. of the Society of the Plastics Industry, Inc. (355 Lexington Ave., New York, NY 10017) is the major trade association concerned with fluoropolymers. The Fluoropolymers Section (D-20.15.12) of ASTM (1916 Race St., Philadelphia, PA 19103) is concerned with standards on fluoropolymers. The main material standard for modified ETFE is ASTM specification D3159.

BIBLIOGRAPHY

1. R.E. Putnam, "Development of Thermoplastic Fluoropolymers" in R.B. Seymour and G.S. Kirk-shenbaum, eds., *High Performance Polymers: Their Origin and Development,* Elvesier, 1986, pp. 279–286.
2. C. Wu, W. Buck and B. Chu, "Light Scattering Characterization of an Alternating Copolymer of Ethylene and Tetrafluoroethylene: 2 Molecular Weight Distributions," *Macromolecules* **20,** 98–103 (1987).
3. *Bulletins E-41337-1 and E-41338-1, Extrusion Guide for Melt Processible Copolymers and its Supplement; Bulletin E-63537-1, Injection Molding Guide for Melt Processible Fluoropolymers,* Du Pont Polymer Products Dept., Wilmington, Del.
4. *Tefzel (TM) Product Bulletins: A-69332, Techniques for the Injection Molding of Tefzel; A-95624, Techniques for Transfer Molding Lined Fittings; E-07229, Product Data,* Du Pont Polymer Products Dept., Wilmington, Del.
5. H.W. Starkweather Jr., "International Motion in an Alternating Copolymer of Tetrafluoroethylene and Ethylene," *J. Polym. Sci. Poly. Phy. Ed.* **11,** 587–593 (1978).

General References

Tefzel (TM) Chemical Use Temperature Guide, Bulletin E-18663-1, Du Pont Polymer Products Dept., Wilmington, Del.

Tefzel (TM) Design Handbook, Bulletin E-31301-1, Du Pont Polymer Products Dept., Wilmington, Del.

R.L. Johnson, "Tetrafluoroethylene Copolymers With Ethylene," in *Encyclopedia of Chemical Technology,* 3rd ed., Vol. 11, John Wiley & Sons, Inc., New York, 1980, pp. 35–41.

R.L. Johnson, "Tetrafluoroethylene Copolymers With Ethylene," in N.M. Bikales, *Encyclopedia of Polymer Science and Technology,* Supp. no. 1, 1976, pp. 269–278.

J.C. Reed and J.R. Perkins, "Tefzel ETFE Fluoropolymer: Temperature Rating and Functional Characterization, *paper presented at 21st International Wire and Cable Symposium,* Atlantic City, N.J., Dec. 6, 1972.

9

FLUOROCARBON POLYMERS, FEP FLUOROCARBON RESIN

Carleton A. Sperati

E.I. du Pont de Nemours & Co., Inc., Parkersburg, WV 26104, and Adjunct Professor, Ohio University, Athens, Ohio

9.1 FEP FLUOROCARBON RESIN

FEP is the acronym for fluorinated ethylene propylene. Its formal name is poly(tetra-fluoroethylene-*co*-hexafluoropropene).

9.2 CATEGORY

FEP resins are hard, tough thermoplastics.

9.3 HISTORY

The fluorocarbon polymers known as FEP fluorocarbon resins were invented in the laboratories of the Du Pont Co.[1] Their invention was initially reported by Sauer[2] in 1951 and by W.T. Miller in 1952.[3] Not until the work of M.I. Bro, B.W. Sandt, and their associates, however,[4-6] were specific polymers developed that combined the melt-flow behavior that would permit processing by (relatively) conventional melt techniques with the desirable physical properties for the final molded or extruded material.

FEP was produced on an experimental basis by Du Pont in the late 1950s and was brought to full commercial status in the early 1960s. Since that time, productive capacity has been increased several times, and additional plants for manufacture of FEP have been built in the Netherlands and Japan.

With the expiration of basic patents on the product and some aspects of the process, competitive materials have come into the U.S. market place from Japan and Germany.

Significant modifications include the availability of the polymer as an aqueous dispersion as well as polymer products where the balance among melt flow, melting point, and mechanical properties has been changed to provide improved toughness and resistance to stress cracking.

There are no official data on the amount of FEP that is made and sold. These estimates are based on a comparison of various published values; 1960, 300,000 lb; 1970, 1,000,000 lb; and 1980, 5,000,000 lb. Depending on the particular type, current (1988) prices for FEP are from \$9.25 to \$10.90 per pound.

9.4 POLYMERIZATION

FEP resins are copolymers of tetrafluoroethylene (TFE), and hexafluoropropylene (HFP). Like PTFE, FEP is made by free-radical initiated polymerization in an aqueous medium. The discussion given on polymerization of PTFE (in the chapter on PTFE) applies equally to FEP. Differences among the commercially available FEP resins stem from differences in the amount of comonomer, the molecular weight, and the control of many other factors in the chemistry of the proprietary processes used.

The basic raw materials for FEP are the same as for PTFE. Since HFP, the comonomer, is made from TFE, the costs of the raw materials for FEP are, inevitably, higher than for PTFE. The same comments on safety, the need for large investment, and so on, which are contained in the chapter on PTFE, apply with equal relevance to FEP resins.

9.5 DESCRIPTION OF PROPERTIES

In most instances, the properties of solid FEP are similar to those of PTFE. In this chapter, the main emphasis will be on the properties that differ most significantly from PTFE.

9.5.1 Thermal Properties

The lower melting point and very much lower melt viscosity of FEP compared to PTFE limit the practical upper use temperature of FEP to 392°F (200°C). On the other hand, the "room temperature" transition of PTFE is not found in FEP; as a result, the dimensional stability of FEP is much less affected by small changes in temperature than PTFE. Properties at very low temperatures are comparable to those of PTFE.

9.5.2 Mechanical Properties

The mechanical properties of FEP at room temperature are similar to those of medium-density polyethylene, ie, relatively soft with high elongation. The properties are retained to a useful extent, however, over the wide range of temperatures mentioned above. FEP is tough and special types are available that have very high resistance to flexing.

9.5.3 Optical Properties

FEP does not absorb electromagnetic radiation in the visible or ultraviolet (UV) range. In contrast to PTFE, its morphology is such that it does not scatter and reflect light; as a result, essentially water clear film can be made. This characteristic has opened attractive opportunities in solar energy applications.

9.5.4 Environmental Properties

Since FEP does not absorb light and thus cannot undergo any photochemical reactions, it has a very high resistance to outdoor weathering under extremely hostile conditions. One set of samples, for example, showed no detectable change after being left on the moon between two of the voyages there. In other work, no changes have been noted after more than 20 years of outdoor exposure.

9.5.5 Electrical Properties

The electrical properties of FEP are also outstanding, although in certain speciality applications its dissipation factor is not quite as low as that of PTFE. The properties include a low dielectric constant than can be decreased even further by specialty foaming techniques, and excellent low-loss factor, high dielectric strength, high surface resistivity, and high resistance to arcing. These properties are maintained over both the wide range of temperatures mentioned above and a very wide range of electrical frequencies. This superior performance leads to a widespread use of FEP in printed circuit boards; so much so, that many of today's electronic products are made possible by the superb electrical properties of FEP.

9.5.6 Chemical Resistance

In its resistance to chemical attack, FEB is similar to PTFE and is an important factor in many applications today. FEP has been dissolved in only a very few, very rare chemicals.

9.5.7 Surface Properties

The —CF_3 group lowers the critical surface tension of FEP to values as low as 17.8 dynes/cm, a value lower than any other available plastic. Film of FEP, therefore, has very useful properties as a release agent in the manufacture of many of the advanced composite materials. The coefficient of friction (μ) of FEP is low, but it does not have the uniquely low values found with PTFE.

9.5.8 Reinforcement

FEP is compounded with chopped glass fiber and many other particulate fillers to enhance certain properties. Resistance to wear in rubbing and sliding applications is the property that is improved the most. Ten to 20 percent (by volume) of such fillers as glass fiber, graphite, bronze, and other materials decreases the wear rate up to five hundred fold.

With most other properties, the added component acts as an inert filler rather than a reinforcing agent. In some instances there is a moderate improvement in resistance to creep, and, with metallic fillers, an increase in thermal conductivity. The major effect of use of fillers, nevertheless, is the improvement in resistance to wear.

9.6 APPLICATIONS

FEP resins are selected for a very wide range of uses that affect every person. The applications fall in the five areas that require one or more of the chemical, electrical, mechanical, thermal, and surface properties. In essentially every instance, a successful application uses at least two of the outstanding properties of this polymer.

In the chemical process industries, products made from FEP are used as gaskets, seals, coatings, or linings in large tanks or process vessels, component parts of valves, piping (solid, lined, or laminated), and a large number of small molded parts for fittings.

Use of FEP as electrical insulation covers the entire wire, cable, and speciality electrical products industry. Its high oxygen index permits FEP to be used in many areas that have stringent standards for combustibility and emission of smoke. To fulfill their function, many high technology products in the electrical industry require the properties and use of FEP. Use of special additives and processing techniques permits control of dielectric constant and loss factor over wide ranges and provides materials that are stable in performance at extremes of temperature and exposure. Many of these serve in classified military applications.

9.7   ADVANTAGES/DISADVANTAGES

The main advantage of FEP is that it is a fluorocarbon polymer that can be processed by essentially conventional injection molding and melt extrusion procedures. In many situations, articles (such as very long lengths of insulated wire) can be made of FEP that are not possible with PTFE, or the articles are less expensive than those made from PTFE.

The main disadvantage of FEP is its lower melting point and its greater sensitivity toward decrease in strength and stiffness properties as temperatures are increased below the melting point. Of course, above its melting point, there is no form stability such as that exhibited by PTFE. The lower critical surface tension of FEP results in slightly better antiwetting properties than found with PTFE. On the other hand, FEP has an appreciably higher coefficient of friction than PTFE. In some applications that require resistance to environmental stress cracking, special forms of FEP that are available must be used to ensure satisfactory performance.

9.8   PROCESSING TECHNIQUES

Essentially all the common procedures available for thermoplastic polymers can be used with FEP, as detailed in the Master Material Outline.

Master Material Outline

PROCESSABILITY OF ____FEP____

PROCESS	A	B	C
INJECTION MOLDING	X		
EXTRUSION	X		
THERMOFORMING			
FOAM MOLDING	X		
DIP, SLUSH MOLDING	X		
ROTATIONAL MOLDING	X		
POWDER, FLUIDIZED-BED COATING	X		
REINFORCED THERMOSET MOLDING			
COMPRESSION/TRANSFER MOLDING		X	
REACTION INJECTION MOLDING (RIM)			
MECHANICAL FORMING			
CASTING			

A = Common processing technique.
B = Technique possible with difficulty.
C = Used only in special circumstances, if ever.

9.9 RESIN FORMS

FEP resins are available in molding powder and aqueous dispersion.

9.10 SPECIFICATION OF PROPERTIES

See the Master Outline of Materials Properties.

9.11 PROCESSING REQUIREMENTS

A detailed discussion of processing requirements is given in a manual from Du Pont.[7] The main features are that the temperatures used (up to 800°F [427°C]) are appreciably higher than those generally used for the common thermoplastics, and thus the equipment must have special corrosion-resistant alloys to ensure long-term, trouble-free operation.

9.12 PROCESSING-SENSITIVE END PROPERTIES

Thermal degradation with bubble formation and deterioration of mechanical properties can result from improper operations; usually, the use of excessively high temperatures.

9.13 SHRINKAGE

Not applicable.

9.14 TRADE NAMES

FEP resins are available under the trade name Teflon FEP Fluorocarbon Resin from Du Pont Co., Polymer Products Dept., Wilmington, Del. 1989, and under the trade name Neoflon FEP from Daikin through various U.S. distributors.

9.15 INFORMATION SOURCES

The material suppliers are the best source for specific information on FEP and products made from it. There is no single publication that is comprehensive. The selected list of references, however, provides access to much of the important information in the published literature.

The Fluoropolymers Div. of the Society of the Plastics Industry, Inc. (355 Lexington Ave., New York, NY 10017) is the major trade association concerned with fluoropolymers. The Fluoropolymers Section (D-20.15.12) of ASTM (1916 Race St., Philadelphia, PA 19103) is concerned with standards on PTFE and related materials.

MATERIAL ___FEP Fluorocarbon Resin___

PROPERTY	TEST METHOD	ENGLISH UNITS	VALUE	METRIC UNITS	VALUE
MECHANICAL DENSITY	D792	lb/ft³	134	g/cm³	2.15
TENSILE STRENGTH	D638	psi	3,048	MPa	21
TENSILE MODULUS	D638	psi	80,000	MPa	551
FLEXURAL MODULUS	D790	psi	95,065	MPa	655
ELONGATION TO BREAK	D638	%	35	%	35
NOTCHED IZOD AT ROOM TEMP	D256	ft-lb/in.	No break	J/m	No break
HARDNESS	D785		D66	D785	D66
THERMAL DEFLECTION T @ 264 psi	D648	°F	129	°C	54
DEFLECTION T @ 66 psi	D648	°F	165	°C	74
VICAT SOFT-ENING POINT	D1525	°F		°C	
UL TEMP INDEX	UL746B	°F		°C	
UL FLAMMABILITY CODE RATING	UL94		VE−0	UL 94	VE−0
LINEAR COEFFICIENT THERMAL EXPANSION	D696	in./in./°F	0.00005	mm/mm/°C	0.0001
ENVIRONMENTAL WATER ABSORPTION 24 HOUR	D570	%	<0.1	%	<0.1
CLARITY	D1003	% TRANS-MISSION		% TRANS-MISSION	
OUTDOOR WEATHERING	D1435	%	Excellent	%	Excellent
FDA APPROVAL			Yes		Yes
CHEMICAL RESISTANCE TO: WEAK ACID	D543		Not attacked		Not attacked
STRONG ACID	D543		Not attacked		Not attacked
WEAK ALKALI	D543		Not attacked		Not attacked
STRONG ALKALI	D543		Not attacked		Not attacked
LOW MOLECULAR WEIGHT SOLVENTS	D543		Not attacked		Not attacked
ALCOHOLS	D543		Not attacked		Not attacked
ELECTRICAL DIELECTRIC STRENGTH	D149	V/mil	1300	kV/mm	50.8
DIELECTRIC CONSTANT	D150	10⁶ Hertz	2.1		2.1
POWER FACTOR	D150	10⁶ Hertz	0.0004		0.0004
OTHER MELTING POINT	D3418	°F	527	°C	275
GLASS TRANS-ITION TEMP.	D3418	°F	−148	°C	−100

BIBLIOGRAPHY

1. R.E. Putnum, "Development of Thermoplastic Fluoropolymers," in R.B. Seymour and G.S. Kirshenbaum, eds., *High Performance Polymers: Their Origin and Development*, Elsevier, 1986, pp. 279–286.
2. U.S. Pat. 2,549,935 (Apr. 24, 1951), J.C. Sauer (to E.I. du Pont de Nemours & Co., Inc.).
3. U.S. Pat. 2,598,283. (May 27, 1952) W.T. Miller (to U.S. Atomic Energy Commission).
4. Novel Perfluorocarbon Polymers, U.S. Pat. 2,946,763 (July 26, 1960), M.I. Bro and B.W. Sandt (to E.I. du Pont de Nemours & Co., Inc.).
5. U.S. Pat. 2,946,763 (June 28, 1960), M.I. Bro (to E.I. du Pont de Nemours & Co., Inc.).
6. U.S. Pat. 2,955,099 (Oct. 4, 1960), R.S. Mallouk and B.W. Sandt (to E.I. du Pont de Nemours & Co., Inc.).
7. *Extrusion Guide for Melt Processible Copolymers and Its Supplement, Bulletins E-41337-1 and E-41338-1,* Du Pont Polymer Products Dept., Parkersburg, W. Va.

General References

R.K. Eby and F.C. Wilson, "Relaxations in Copolymers of Tetrafluoroethylene and Hexafluoropropylene," *J. Appl. Phys.* **33,** 2951–2955 (1962).

S.V. Gangal, "Fluorinated Ethylene-Propylene Copolymers" in M. Grayson, ed., *Encyclopedia of Chemical Technology*, 3rd ed., Vol. 11, Wiley-Interscience, New York, 1980, pp. 24–35.

C.E. Jolly and J.C. Reed, "The Effects of Space Environments on Teflon TFE and FEP Insulation," *paper presented at the 11th Annual Signal Corps Wire and Cable Symposium*, Asbury Park, N.J., Nov. 28–30, 1962.

H.A. Larsen, G.R. Dehoff, and N.W. Todd, "Injection Molding FEP-Fluorocarbon Resin," *Mod. Plast.* **36,** 89–96 (Aug. 1959).

V.J. Leslie and L.S.J. Shipp, "Structure Property Relationships of Fluoropolymers," *IX Int. Symp. of Fluorine Chem.*, Avignon, Sept. 3–7, 1979.

D.I. McCane, "Copolymers with Hexafluoropropylene," in N.M. Bikales, ed., *Encyclopedia of Polymer Science and Technology*, Vol. 13., New York, Wiley, 1970 pp. 654–670.

N.G. McCrum, B.E. Read, and G. Williams, *Anelastic and Dielectric Effects in Polymeric Solids*, John Wiley & Sons, Inc., New York, 1967.

J.C. Reed, E.J. McMahon, and J.R. Perkins, "Effects of High Voltage Stresses on TFE and FEP Fluorocarbon Plastics," *Insulation (Libertyville)* **10,** 35–38 (May 1964).

U.S. Pat. 3,085,083 (Apr. 9, 1963) R.C. Schreyer (to E.I. du Pont de Nemours & Co., Inc.).

H.W. Starkweather Jr., P. Zoller, and G.A. Jones, *J. Polym. Sci. Polym. Phys. Ed.* **22,** 1431–1437 (1984).

J.J. Weeks and co-workers, *Polymer* **22,** 325 (1980).

J.J. Weeks, R.K. Eby, and E.S. Clark, "Disorder in the Crystal Structures of Phases I and II of Copolymers of Tetrafluoroethylene and Hexafluoropropylene," *Polymer* **22,** 1496–1499 (1981).

10

FLUOROPOLYMERS, POLYCHLOROTRIFLUOROETHYLENE (PCTFE)

Carleton A. Sperati

E.I. du Pont de Nemours & Co., Inc., Parkersburg, W.Va. 26104, and Adjunct Professor, Ohio University, Athens, Ohio

10.1 POLYCHLOROTRIFLUOROETHYLENE

$$-(\underset{\underset{F}{|}}{\overset{\overset{F}{|}}{C}}-\underset{\underset{Cl}{|}}{\overset{\overset{F}{|}}{C}})_{n}-$$

10.2 CATEGORY

PCTFE is a hard, tough thermoplastic.

10.3 HISTORY

PCTFE is the oldest of the fluoropolymers. It was made during research at Hoechst in 1934. It was first reported in 1937 in British patent 465,520, from research at I.G. Farbenindustrie.

In the United States, formulation of a useful, high molecular-weight PCTFE polymer came with the invention of W.T. Miller[1] stemming from research he did during World War II for the Manhattan Project.

PCTFE was commercialized by the Kellog Corp. to provide a material that could be used in the extremely hostile environment of the gaseous diffusion process for separating uranium isotopes. When material became available for general use, the trademark Kel-F was selected. The 3M Co. continued to use this trademark after it purchased the business and began marketing PCTFE in 1957.

In the early 1950s the fluoropolymer industry was just starting, and there were great hopes for the future of PCTFE. As the industry developed, however, the superior thermal, electrical, and surface properties of the fluorocarbon polymers (PTFE, FEP, and PFA) and the generally excellent properties combined with the lower costs of the other hard, stiff fluoropolymers (modified poly ETFE and PVDF) resulted in a progressively smaller total market for PCTFE. Today, PCTFE is used for a number of highly specialized applications where its use is required by longstanding specifications or where its unique properties make it the only suitable material.

In the early 1960s, Allied Corp. began to market an excellent barrier film made from PCTFE copolymer with less than 5% vinylidene fluoride. Various processors also extrude PCTFE homopolymer film.

Significant modifications to PCTFE include the introduction of copolymers containing small amounts of vinylidene fluoride. The comonomer improves appreciably the already good resistance of PCTFE to ionizing radiation. In addition, the modified polymers are soluble in conventional solvents, so that they may be used in lacquers, paints, and putties as well as in binders for making plastic bonded propellants and explosives as well as pyrotechnic devices.

A copolymer of ethylene and CTFE also was developed by Allied; it is now part of the Ausimont product line.

Official production volume figures are not available. Based on general industry information, the volume has been estimated to be appreciably less than one million pounds a year, as shown in these broad estimates: 1960, 750,000 lb; 1970, 600,000 lb; 1980, 500,000 lb.

10.4 POLYMERIZATION

Free-radical initiated polymerization has been reported for the preparation of PCTFE in essentially every type of system: bulk, solution, suspension, and emulsion. Many believe that the emulsion system produces the most thermally stable polymer. Moderate temperatures and pressures are required.

A common route to the monomer, chlorotrifluorethylene (CTFE) is dechlorination of the chlorofluorocarbon CFC-113 (1,1,2-trichloro-2,2,1-trifluoroethane). Since preparation of CFC-113 requires the replacement of chlorine by fluorine, the basic raw materials are inherently expensive even though there is a fairly large market for CFC-113 as a solvent. The cost of the CTFE, combined with a relatively low rate in the polymerization reaction and a low volume of production (compared to other fluoropolymers), results in an expensive polymer.

Current (1988) prices for PCTFE molding powders are in the range of $30.00 per pound.

10.5 DESCRIPTION OF PROPERTIES

10.5.1 Thermal Properties

The cryogenic properties of PCTFE are excellent. In high-temperature applications, however, it is quite inferior to the other fluoropolymers (excepting PVDF), because of its relatively low melting point of 412°F (211°C) and also because of its thermally induced crystallization on exposure to temperatures below the melting point, which results in brittleness.

10.5.2 Mechanical Properties

If the thermally induced brittleness is avoided, the mechanical properties of PCTFE are among the best of the fluoropolymers. It has especially good resistance to creep.

10.5.3 Optical Properties

PCTFE does not absorb visible light and can be prepared optically clear in thicknesses up to about 1/8 in. (3.2 mm). The combination of optical clarity, its excellent resistance to most chemicals, and its very good mechanical properties is the basis for many of the high value-in-use applications for PCTFE.

10.5.4 Environmental Properties

PCTFE has the excellent resistance to most harsh environments that is characteristic of the fluoropolymers. It is especially resistant to strong oxidizing agents, liquid oxygen, and the like.

10.5.5 Reinforcement

Incorporation of about 15% glass fiber improves PCTFE's high-temperature properties, increases hardness, but also increases brittleness.

10.6 APPLICATIONS

Typical applications for PCTFE are specialty uses that require its outstanding barrier properties. These uses include high-technology electrical/electronic insulation and parts for medical instrumentation, cryogenic seals, gaskets, and so on.

10.7 ADVANTAGES/DISADVANTAGES

In film form, PCTFE has the lowest moisture permeability of any polymer, it has very good mechanical properties, and it is highly resistant to strong inorganic oxidizing and otherwise corrosive materials. In relatively thin sections [less than 1/8 in. (3.2 mm)], it can be quenched from its molten form to give essentially a water-clear solid that is very useful for such specialties as chemically resistant rotameter tubes and the like.

However, many organic materials attack the polymer, and its relatively limited thermal stability in the molten form requires great care during melt processing in order to maintain the high level of molecular weight required for good mechanical properties in fabricated articles. A special test, zero-strength time (ZST) was developed to permit convenient appraisal of the relative molecular weight of the resin, especially to follow degradation during processing. Lastly, PCTFE is among the most costly of the fluoropolymers.

10.8 PROCESSING TECHNIQUES

PCTFE can be processed by all of the melt procedures used for thermoplastics, including extrusion and compression, injection, and transfer molding. To ensure useful properties in fabricated articles, PCTFE requires melt viscosities higher than those normally used with thermoplastics; as a result, special precautions are required during processing to avoid degradation of the resin.

Master Material Outline

PROCESSABILITY OF __PCTFE__

PROCESS	A	B	C
INJECTION MOLDING	X		
EXTRUSION	X		
THERMOFORMING			
FOAM MOLDING			
DIP, SLUSH MOLDING			
ROTATIONAL MOLDING			
POWDER, FLUIDIZED-BED COATING		X	
REINFORCED THERMOSET MOLDING			
COMPRESSION/TRANSFER MOLDING	X^a		
REACTION INJECTION MOLDING (RIM)			
MECHANICAL FORMING			
CASTING			

A = Common processing technique.
B = Technique possible with difficulty.
C = Used only in special circumstances, if ever.
^{a}Preferred procedure to make parts with the best properties

10.9 RESIN FORMS

PCTFE is available as powder, pellets, pellets containing 15% glass fiber, and dispersions. Low molecular-weight PCTFE is available for use as a special oil or grease. At times, the oil is used to plasticize PCTFE.

Some copolymers with vinylidene fluoride also are produced. The copolymer of ethylene with chlorotrifluorethylene is another member of the fluoropolymer family manufactured and marketed by Ausimont as Halar poly E-CTFE.

10.10 SPECIFICATION OF PROPERTIES

Specific data are given in the Master Outline of Material Properties.

Just as PTFE can be considered a perfluoro linear polyethylene, PCTFE is perfluoro poly(vinyl chloride). Inclusion of the relatively large chlorine on the chain interrupts crys-

MATERIAL __PCTFE__

PROPERTY	TEST METHOD	ENGLISH UNITS	VALUE	METRIC UNITS	VALUE
MECHANICAL DENSITY	D792	lb/ft³	133	g/cm³	2.13
TENSILE STRENGTH	D638	psi	5,725	MPa	40
TENSILE MODULUS	D638	psi	207,000	MPa	1,400
FLEXURAL MODULUS	D790	psi	189,000	MPa	1,250
ELONGATION TO BREAK	D638	%	5	%	5
NOTCHED IZOD AT ROOM TEMP	D256	ft-lb/in.	5	J/m	267
HARDNESS	D785		D 77		D 77
THERMAL DEFLECTION T @ 264 psi	D648	°F	158	°C	70
DEFLECTION T @ 66 psi	D648	°F	266	°C	130
VICAT SOFT-ENING POINT	D1525	°F		°C	
UL TEMP INDEX	UL746B	°F		°C	
UL FLAMMABILITY CODE RATING	UL94		VE–0		VE–0
LINEAR COEFFICIENT THERMAL EXPANSION	D696	in./in./°F	7×10^{-5}	mm/mm/°C	12.6×10^{-5}
ENVIRONMENTAL WATER ABSORPTION 24 HOUR	D570	%	0	%	0
CLARITY	D1003	% TRANS-MISSION	92	% TRANS-MISSION	92
OUTDOOR WEATHERING	D1435	%	Excellent	%	Excellent
FDA APPROVAL			Yes		Yes
CHEMICAL RESISTANCE TO: WEAK ACID	D543		Not attacked		Not attacked
STRONG ACID	D543		Not attacked		Not attacked
WEAK ALKALI	D543		Not attacked		Not attacked
STRONG ALKALI	D543		Not attacked		Not attacked
LOW MOLECULAR WEIGHT SOLVENTS	D543		Minimally attacked		Minimally attacked
ALCOHOLS	D543		Not attacked		Not attacked
ELECTRICAL DIELECTRIC STRENGTH	D149	V/mil	2580	kV/mm	101.6
DIELECTRIC CONSTANT	D150	10^6 Hertz	2.9		2.9
POWER FACTOR	D150	10^6 Hertz	0.03		0.03
OTHER MELTING POINT	D3418	°F	413	°C	212
GLASS TRANS-ITION TEMP.	D3418	°F	131	°C	55

tallization to such an extent that PCTFE often can be molded to give transparent parts in thickness up to 1/8 in., (3.2 mm), as mentioned earlier.

PCTFE is outstanding in its resistance to sunlight, strong inorganic acids, and potent oxidizing agents such as oxygen, ozone, and the fuming, oxidizing acids. It is an excellent barrier material for gases, including water vapor. The combination of excellent cryogenic properties, very low permeability, and resistance to oxidizing agents makes PCTFE a material of choice for liquid oxygen service.

10.11   PROCESSING REQUIREMENTS

PCTFE was developed as a melt processible thermoplastic. However, in order to obtain adequate mechanical properties, a relatively high molecular weight is required that results in melt viscosities that are somewhat higher than those that are usually used with typical thermoplastics. Borderline thermal stability in the melt prevents use of sufficiently high temperatures to permit an appreciable decrease in the melt viscosity during processing. Temperatures during injection molding vary from 490°F (260°C) for the rear cylinder to 662°F (350°C) at the nozzle. Temperature of the melt leaving the nozzle should be in the range of 536–579°F (280–305°C).

For many applications, compression molding at about 599°F (315°C) and 10,000 psi (69 MPa) is the processing technique of choice to ensure the best properties. If injection molding or extrusion must be used, the five modifications outlined below must be considered to ensure good parts:

- Be sure to "pack" the resin in the molten state. This will require pressures of about 80,000 psi (552 MPa) and, of course, it reduces the effective clamping area.
- Maintain a small melt inventory, 10 shots or a maximum of 15 minutes under heat.
- Use precise heat controllers, such as a saturated reactor control system. The molds should be heated in the range of 149–378°F (65–192°C).
- Streamlined flow channels are essential.
- The melt path should be as direct as possible.

10.12   PROCESSING-SENSITIVE END PROPERTIES

Thermal degradation with bubble formation, decrease in ZST values, and deterioration of mechanical properties can result from improper operations, usually use of excessively high temperatures (above 662°F [350°C] at the nozzle).

10.13   SHRINKAGE

The amount of crystallization during cooling affects the mold shrinkage. An allowance of 0.5 to 1.0% for shrinkage is usually satisfactory.

10.14   TRADE NAMES

The only producer of PCTFE in the United States is 3M Co., St. Paul, Minn. which uses the trademark, Kel-F. It is available in copolymer film form as Aclar from Allied Corp., Morristown, N.J. Ausimont manufactures an ethylene–CTFE copolymer that is marketed as Halar fluoropolymer.

PCTFE is available in the United States from foreign producers as Daiflow CTFE from Daikin Kyogo, marketed through various distributors, and as Voltalef from Attochem Inc., Glen Rock, N.J.

10.15 INFORMATION SOURCES

The material suppliers are the best source for specific information on PCTFE and products made from it. There is no single publication that is comprehensive. The selected list of references, however, provides access to much of the important information in the published literature.

The Fluoropolymers Div. of the Society of the Plastics Industry, Inc. (355 Lexington Ave., New York, NY 10017) is the major trade association concerned with fluoropolymers. The Fluoropolymers Section (D-20.15.12) of ASTM (1916 Race St., Philadelphia, PA 19103) is concerned with standards on fluoropolymers. The main material standard for PCTFE is ASTM specification D-1430.

BIBLIOGRAPHY

1. *Kel-F Plastic: Suggested Guidelines for Injection Molding, Suggested Guidelines for Compression Molding Sheets, and Suggested Guidelines for Finishing*, Technical Information Bulletins, 3M Co., St. Paul, Minn.

General References

A.J. Bur and L.A. Wall, ed., *Fluoropolymers*, Wiley-Interscience, New York, 1972, pp. 490–495.

S. Chandrasekaran, "Chlorotrifluoroethylene Homopolymer," in *Encyclopedia of Polymer Science and Engineering*, Vol. 3. John Wiley & Sons, Inc., New York, 1987, pp. 463–480.

Brit. Pat. 465,520 (May 3, 1937) (to I.G. Farbenindustrie).

J.D. Hoffman, "The Specific Heat and Degree of Crystallinity of Polychlorotrifluoroethylene," *J. Am. Chem. Soc.* **74** pp. 1696–1700 (1952).

F.J. Honn, "Chlorotrifluoroethylene Polymers," in *Encyclopedia of Chemical Technology*, 1st ed. Wiley, New York, pp. 691–703.

H.S. Kaufman and M.S. Muthana, *J. Polym. Sci.* **6** p. 251 (1951).

Kel-F Engineering Manual. 3M Co., St. Paul, Minn.

Machine Design, 128–130 (April 19, 1981).

U.S. Pat. 2,564,024 (Aug. 14, 1951) W.T. Miller.

P.C.U.K. Bulletin on Voltalef [PCTFE] Molding Powders.

A.C. West, "Polychlorotrifluoroethylene." in M. Grayson, ed., *Encyclopedia of Chemical Technology* 3rd ed., Vol. 11. Wiley-Interscience, New York, 1980, pp. 49–54.

11

FLUOROCARBON POLYMERS, PFA FLUOROCARBON RESIN

Carleton A. Sperati

E.I. du Pont de Nemours & Co., Inc., Parkersburg, W.Va. 26104, and Adjunct Professor, Ohio University, Athens, Ohio

11.1 PFA FLUOROCARBON RESIN

PFA is the acronym for perfluoro alkoxy when it applies to a fluorocarbon resin.

$$\left[(C-C)_n - (C-C) \right]_m$$

11.2 CATEGORY

PFA resins are hard, tough thermoplastic resins.

11.3 HISTORY

PFA fluorocarbon resins were discovered by J.F. Harris and D.I. McCane[1-3], working with W.F. Gresham and others from Du Pont Fluoropolymers Research at the Experimental Station in Wilmington, Del.[4,5] Pertinent patents were first issued in 1964.

Commercial forms of PFA were designed to meet the need for a melt-processible fluorocarbon polymer that preforms similarly to PTFE rather than to the FEP resins. This is significant in that it means PFA resins can maintain mechanical properties at high temperatures and resist thermal degradation.

Experimental quantities of PFA became available from Du Pont in the early 1970s, and commercial sale began soon after that. An important initial market was the manufacture of injection-molded "wafer baskets" used in automated production of chips for the burgeoning microcomputer industry.

Significant modifications to the original formulation include the availability of PFA resin products with small particle-size powder for powder coating techniques and special grades with superior resistance to environmental stress cracking.

By 1980 production volumes of PFA fluorocarbon resins had reached an estimated 250,000 pounds.

11.4 POLYMERIZATION

PFA resins are copolymers of tetrafluoroethylene, TFE, and a perfluoro vinyl ether of the general formula $CF_2{=}CFOC_nF_{2n+1}$. They are made by free-radical initiated polymerization of the monomers. The discussion on costs, safety, and so on in the chapter on PTFE is equally pertinent to PFA resins.

The major ingredient of PFA resins is the TFE comonomer discussed in the chapter on PTFE. Various syntheses are performed for the perfluoro(vinyl ether)s that might be used for a particular PFA resin. The syntheses are all multistep operations. All the comonomers are made in-house by the manufacturers, and they are expensive.

As of 1988, the price of PFA molding powders ranged from $18.30 to $20.00 per pound, depending on the particular type.

11.5 DESCRIPTION OF PROPERTIES

In most instances the properties of solid PFA are similar to those of PTFE, discussed in that section. Here, the main emphasis will be on the properties of PFA fluorocarbon resins that show significant differences from those of PTFE.

11.5.1 Thermal Properties

Essentially all of the properties of PFA resins below their melting point are very similar or somewhat superior to those of PTFE. PFA's melting point of about 608°F (320°C) is only a few degrees lower than that of PTFE. (The change in properties with increasing temperatures is not a degradation on exposure. It is, rather, related to the reversible softening shown by all plastics as temperature is increased.) The better properties for PFA were unexpected and formed the basis of an upper temperature rating of 500°F (260°C), the same as for PTFE. The "room temperature" transition of PTFE does not occur in PFA. As a result, the latter's dimensional changes are much less sensitive to small changes in temperature than are those of PTFE. Properties at very low temperatures are comparable to those of PTFE.

11.5.2 Mechanical Properties

The mechanical properties of PFA at room temperature are similar to those of medium density polyethylene, ie, relatively soft with high elongation. The properties are retained to a useful extent, however, over the wide range of temperatures mentioned above. PFA is tough and special types are made that have very high resistance to flexing. Although the coefficient of friction of PFA is not as low as PTFE, it is still lower than that of virtually all other plastics.

11.5.3 Optical Properties

PFA does not absorb electromagnetic radiation in the visible or ultraviolet (UV) range. In contrast to PTFE, its morphology is such that it does not scatter and reflect light, so that essentially water clear film can be made. This characteristic has opened attractive opportunities in solar energy applications.

11.5.4 Environmental Properties

Since PFA does not absorb light and thus cannot undergo any photochemical reactions, it has a very high resistance to outdoor weathering under extremely hostile conditions.

11.5.5 Reinforcement

To date, there has been no significant commercial development of glass-fiber filled PFA.

11.5.6 Chemical Resistance

In resistance to chemical attack, PFA is similar to PTFE, and this property is an important factor in many applications today. The comments in the chapter on PTFE are equally pertinent for PFA.

11.5.7 Surface Properties

The critical surface tension of PFA is slightly lower than for PTFE. Therefore, film made of PFA has very useful properties in such applications as bonding film. The coefficient of friction (μ) of PFA is low, but it does not have the uniquely low values found with PTFE.

11.6 APPLICATIONS

PFA resins are used mostly in the electronics and chemical processing industries.

11.7 ADVANTAGES/DISADVANTAGES

The main advantage of PFA resins is that, in contrast to PTFE, they can be used to make complex shapes by injection molding and to make long lengths of insulated wire and other extruded shapes by melt extrusion. The superiority of PFA resins in such applications results from economies gained by using melt processing (injection molding and extrusion) combined with better retention of strength and stiffness properties with increasing temperatures, as compared to PTFE. In spite of their high price per pound, the cost of many fabricated articles made of PFA resins is less than parts with the same utility made from PTFE.

The disadvantage is the high cost that is inherent in these fluorocarbon products. Many useful articles can be made cheaper using PTFE.

In addition to the better retention of physical properties at high temperatures, the main differences between PFA resins and PTFE are the increased transparency (that is, decreased scatter of light) and the very much lower melt viscosity of PFA that permits melt processibility. PFA, however, does not have the stability of form that PTFE exhibits for appreciable periods of time if, inadvertently, temperatures above the melting point are encountered.

11.8 PROCESSING TECHNIQUES

Essentially all the common procedures available for thermoplastic polymers can be used with PFA. For details, see the Master Material Outline.

Master Material Outline

PROCESSABILITY OF ___PFA___

PROCESS	A	B	C
INJECTION MOLDING	X		
EXTRUSION	X		
THERMOFORMING			
FOAM MOLDING	X		
DIP, SLUSH MOLDING	X		
ROTATIONAL MOLDING	X		
POWDER, FLUIDIZED-BED COATING	X		
REINFORCED THERMOSET MOLDING			
COMPRESSION/TRANSFER MOLDING		X	
REACTION INJECTION MOLDING (RIM)			
MECHANICAL FORMING			
CASTING			

A = Common processing technique.
B = Technique possible with difficulty.
C = Used only in special circumstances, if ever.

11.9 RESIN FORMS

PFA resins are available as molding powders and as micropowders for powder coating.

11.10 SPECIFICATION OF PROPERTIES

See the Master Outline of Materials Properties.

11.11 PROCESSING REQUIREMENTS

A detailed discussion of the processing requirements for PFA fluorocarbon resins is given in a manual from Du Pont.[6] The main features are that the temperatures used are higher than those generally used for the common thermoplastics, and thus the equipment must incorporate special corrosion-resistant alloys to ensure long-term, trouble-free operation.

MATERIAL __PFA Fluorocarbon Resin__

PROPERTY	TEST METHOD	ENGLISH UNITS	VALUE	METRIC UNITS	VALUE
MECHANICAL DENSITY	D792	lb/ft³	133	g/cm³	2.14
TENSILE STRENGTH	D638	psi	4,499	MPa	31
TENSILE MODULUS	D638	psi	80,000	MPa	551
FLEXURAL MODULUS	D790	psi	100,145	MPa	690
ELONGATION TO BREAK	D638	%	30	%	30
NOTCHED IZOD AT ROOM TEMP	D256	ft-lb/in.	No break	J/m	No break
HARDNESS	D785		D65	D785	D65
THERMAL DEFLECTION T @ 264 psi	D648	°F	122	°C	50
DEFLECTION T @ 66 psi	D648	°F	162	°C	72
VICAT SOFT-ENING POINT	D1525	°F		°C	
UL TEMP INDEX	UL746B	°F		°C	
UL FLAMMABILITY CODE RATING	UL94		VE−0	UL 94	VE−0
LINEAR COEFFICIENT THERMAL EXPANSION	D696	in./in./°F	0.00006	mm/mm/°C	0.00012
ENVIRONMENTAL WATER ABSORPTION 24 HOUR	D570	%	<0.1	%	<0.1
CLARITY	D1003	% TRANS-MISSION	92	% TRANS-MISSION	92
OUTDOOR WEATHERING	D1435	%	Excellent	%	Excellent
FDA APPROVAL			Yes		Yes
CHEMICAL RESISTANCE TO: WEAK ACID	D543		Not attacked		Not attacked
STRONG ACID	D543		Not attacked		Not attacked
WEAK ALKALI	D543		Not attacked		Not attacked
STRONG ALKALI	D543		Not attacked		Not attacked
LOW MOLECULAR WEIGHT SOLVENTS	D543		Not attacked		Not attacked
ALCOHOLS	D543		Not attacked		Not attacked
ELECTRICAL DIELECTRIC STRENGTH	D149	V/mil	2560	kV/mm	101
DIELECTRIC CONSTANT	D150	10^6 Hertz	2.06		2.06
POWER FACTOR	D150	10^6 Hertz	0.00008		0.00008
OTHER MELTING POINT	D3418	°F	577	°C	304
GLASS TRANS-ITION TEMP.	D3418	°F	<−112	°C	<−80

11.12 PROCESSING-SENSITIVE END PROPERTIES

Since PFA is appreciably more stable thermally than is FEP, there is much less opportunity in the processing of PFA for degradation with bubble formation and for deterioration of mechanical properties that can result from improper operations (usually use of excessively high temperatures).

11.13 SHRINKAGE

See the Mold Shrinkage MPO.

11.14 TRADE NAMES

PFA fluorocarbon resins are available as Teflon PFA Fluorocarbon Resin from Du Pont Co., Wilmington, Del.; as Hostaflon TFA from Hoechst Celanese Corp., Somerville, N.J.; and as Neoflon PFA from various U.S. distributors of Daikin.

11.15 INFORMATION SOURCES

The material suppliers are the best source for specific information on PFA and products made from it. There is no single publication that is comprehensive. The selected list of references, however, provides access to much of the important information in the published literature.

The Fluoropolymers Div. of the Society of the Plastics Industry, Inc. (355 Lexington Ave., New York, NY 10017) is the major trade association concerned with PTFE and related plastics. The Fluoropolymers Section (D-20.15.12) of ASTM (1916 Race St., Philadelphia, PA 19103) is concerned with standards on fluoropolymers.

BIBLIOGRAPHY

1. U.S. Pat. 3,132,123 (May 5, 1964), J.F. Harris, Jr. and D.I. McCane (to E.I. du Pont de Nemours & Co., Inc.).
2. U.S. Pat. 3,159,609 (Dec. 1, 1964), J.F. Harris, Jr. and D.I. McCane (to E.I. du Pont de Nemours & Co., Inc.).
3. U.S. Pat. 3,180,895 (Apr. 27, 1965), J.F. Harris Jr. and D.I. McCane (to E.I. du Pont de Nemours & Co., Inc.).
4. U.S. Pat. 3,450,684 (June 17, 1969), R.A. Darby (to E.I. du Pont de Nemours & Co., Inc.).
5. U.S. Pat. 3,635,926 (Jan. 18, 1972), W.F. Gresham and A.F. Vogelpohl (to E.I. du Pont de Nemours & Co., Inc.).
6. *Extrusion Guide or Melt Processible Copolymers and Its Supplement, Bulletins E-41337-1 and E-41883-1,* Du Pont Polymer Products Dept., Parkersville, W.Va.

General References

Technical Information on PFA, Du Pont Polymer Products Dept., Parkersville, W.Va., July 19, 1985.
Teflon PFA Product Guide, Bulletin E-33272-2, Du Pont Polymer Products Dept., Parkersville, W.Va.
R.L. Johnson, "Tetrafluoroethylene Copolymers With Perfluoroalkoxy Pendant Groups," in M. Grayson, ed., *Encyclopedia of Polymer Science and Technology,* Wiley-Interscience, Supp. No. 1, 3rd ed., 1976.

R.L. Johnson, *Tetrafluoroethylene Copolymers With Perfluorovinyl Ethers,* in M. Grayson, ed., *Encyclopedia of Chemical Technology;* 3rd ed., Vol. 11. Wiley-Interscience, New York, 1980, pp. 42–49.

V.J. Leslie and L.S.J. Shipp, "Structure Property Relationships of Fluoropolymers," *IX Int. Symp. of Fluorine Chem.,* Avignon, Sept. 3–7, 1979.

R.E. Putnam, in R.B. Seymour and G.S. Kirshenbaum, eds. *High-Performance Polymers: Their Origin and Development,* Elvesier, 1986, pp. 279–286.

12

FLUOROCARBON POLYMERS, POLYTETRAFLUOROETHYLENE (PTFE)

Carleton A. Sperati

E.I. du Pont de Nemours & Co., Inc., Parkersburg, WV 26104, and Adjunct Professor, Ohio University, Athens, Ohio

12.1 POLYTETRAFLUOROETHYLENE

$$-(\underset{\underset{F}{|}}{\overset{\overset{F}{|}}{C}}-\underset{\underset{F}{|}}{\overset{\overset{F}{|}}{C}})_n-$$

Polytetrafluoroethylene is commonly known by its acronym, PTFE. It is also referred to as TFE resin or TFE fluorocarbon resin. TFE, the acronym for the monomer (discussed in section 10.5.4, below) is sometimes used misleadingly by itself to mean the polymer, PTFE.

12.2 CATEGORY

PTFE is a hard, tough thermoplastic.

In practice, the exceptionally high melt viscosity of all forms of the resin useful for plastics applications requires processing techniques that are not difficult, but are very different from those used for conventional thermoplastics. In many respects the processing methods used for PTFE are similar to those used in the metals and ceramics industries.

12.3 HISTORY

The period from 1933 until the discovery of PTFE on April 8, 1938, was unusually productive in providing the plastics that are the backbone of this industry today. PTFE was developed from research on new refrigerants within the Du Pont Co. Tetrafluoroethylene, TFE, had been made from one of these materials, difluorochloromethane (CF_2HCl, trade name Freon 22). While he was attempting to run an experiment with TFE, Roy Plunkett found that no gas came out of the cylinder. He had the cylinder cut open and found a waxy-feeling white solid. It was extremely inert, insoluble in all common solvents, and did not appear to melt. Finally it succumbed to simple fabrication techniques, albeit they were novel to normal plastics technology.

Processes for making the monomer and polymer were developed quickly and a pilot plant was built at the Arlington, N.J., facilities of the Du Pont Plastics Dept. The inert nature of this polymer made it a natural candidate for gaskets and other components needed for the Manhattan Project, the secret operation to develop an atomic bomb.

After World War II, the polymer became available for critical industrial uses. A combination of extreme resistance to chemicals and to very high temperatures, superb electrical properties, and unique surface properties indicated a promising future for this new material. Plans for commercial production led to construction of a full-scale plant at the new Du Pont facilities at Parkersburg, W.Va., where operations started in 1950.

The initial PTFE product was a granular powder that could be converted into useful products by procedures similar to those used in powder metallurgy. An aqueous dispersion was soon developed that permitted impregnation and casting of thin films. Coagulation of the dispersion gives a solid "fine powder" or CD (for coagulated dispersion). A unique process termed "lubricated" or "paste" extrusion permitted manufacture of thin-walled insulation on wire, thin-walled tubing, and the like from this form of PTFE.

Other important developments were the introduction of granular powders with very small particle size, which yielded fabricated products with much improved properties, and of agglomerated forms of the finely divided powders, which could be handled much more easily and still retain superior properties in fabricated products.

The production volume of PTFE has never been high by the standards of other plastics. However, its relatively high cost and value in use make it an important part of the plastics business in dollar volume. In 1978, worldwide use of PTFE was estimated to be about $\times 10^7$ lb; its growth is such that by 1988 its world-wide use is estimated to exceed $\times 10^8$ lb.

12.4 POLYMERIZATION

PTFE is made by a free-radical initiated polymerization of the gaseous monomer, tetrafluoroethylene, in an aqueous medium under pressure. The chemistry of this polymerization is such that extremely pure monomer and other ingredients are required in order to achieve the very high molecular weight required for plastics applications. TFE monomer is a very hazardous material having a wide explosive range in air and also having a propensity to disporportionate (rearrange its molecular structure) to form carbon and carbon tetrafluoride with an explosive force similar to that of black powder. As a result, extreme care and very high capital investment are required for the manufacture of the monomer and polymer.

The basic raw materials for TFE are fluorspar (calcium fluoride), sulfuric acid, methane, and chlorine. The methane is chlorinated to form chloroform, and the fluorspar is converted into hydrogen fluoride with the sulfuric acid. Reaction of chloroform with HF produces chlorodifluoromethane, which is pyrolyzed to form TFE. Although the basic raw materials are inexpensive and readily available, the complexity of the processes required, the need for extreme resistance to corrosion, the large amount of by-product hydrogen chloride, and the problems in handling the hazardous TFE result in a polymer that inherently is expensive.

As of 1988, the prices of PTFE powders varied from about $5.60 per pound for the least expensive granular powders to about $8.00 per pound for the most expensive CD products ("fine powders").

12.5 DESCRIPTION OF PROPERTIES

12.5.1 Thermal Properties

The combination of properties exhibited by PTFE represent the extremes available with organic polymers. In a few properties another particular polymer might be superior, but PTFE is unique in offering a combination of properties that is required for many critical applications. Its upper use temperature is given as 500°F (260°C). It is reported to give ductile rather than brittle failures at temperatures just above absolute zero, a useful range of more than 900°F (500°C). In some instances PTFE has been used satisfactorily as a totally enclosed gasket for considerable periods of time at 900°F (500°C), well above the recommended upper use temperature. It is remarkable that any polymer has performed well over this range of more than 1386°F (770°C), as PTFE has done in actual use.

12.5.2 Mechanical Properties

The mechanical properties of PTFE at room temperature are similar to those of medium-density polyethylene—relatively soft with high elongation. However, the properties are retained to a useful extent over the wide range of temperatures mentioned above. PTFE is remarkably tough, showing values in tensile impact strength over 400 ft-lb/in.^{2} (841.2 kJ/m^2), as compared to most other plastics that seldom exceed 100 ft-lb/in.^{2} (210.3 kJ/m^2).

12.5.3 Optical Properties

PTFE does not absorb electromagnetic radiation in the visible or ultraviolet range. Its morphology is such, however, that it has crystalline domains that scatter and reflect light and result in a translucent to opaque appearance, much as the surface of a ground glass plate appears opaque. This property is used in calibrating some spectrometers where blocks of granular PTFE are employed as standards for reflectance in place of magnesium oxide.

12.5.4 Environmental Properties

Since PTFE does not absorb light and thus cannot undergo any photochemical reactions, it has a very high resistance to outdoor weathering. One set of samples, for example, showed no change after 15 years' outdoor aging in Florida, but then blew away in a hurricane, precluding further data from that experiment. In other work, however, no changes have been noted after more than 20 years of outdoor exposure.

12.5.5 Electrical Properties

The electrical properties of PTFE are also outstanding. These include a very low dielectric constant, loss factor as low or lower than any other material, high dielectric strength, high surface resistivity, and high resistance to arcing. These properties are maintained over both the wide range of temperatures mentioned above and a very wide range of electrical frequencies. Many of electronic products today are made possible by the superb electrical properties of PTFE.

12.5.6 Chemical Resistance

Resistance to chemical attack was the first property of PTFE that was used in practice and remains a major factor in current applications. PTFE has been dissolved in only a very few, very rare chemicals. It is, however, attacked by molten alkali metals and by elemental fluorine and pure oxygen at elevated temperatures. It also can react under some conditions with alkaline earth oxides and at elevated temperatures with finely divided aluminum or magnesium. PTFE's chemical resistances is exploited in various flares and other special devices. Sorption of small amounts (for example, 1–2%) of some hydrogen free halo compounds also has been reported.

12.5.7 Surface Properties

The unique surface properties of PTFE also have led to many innovative applications. The critical surface tension of PTFE is so low that it provides an excellent antistick or release surface. The coefficient of friction (μ) of PTFE is also remarkably low. These two properties, incidentally, do not depend on each other. Other fluorocarbon polymers with lower critical surface tension have much higher coefficients of friction. Unlike many materials, the coefficient of friction of PTFE is not constant with increasing load, but decreases as the load increases. Values of (μ) as low as 0.0056 have been measured at loads resulting from a stress of 7111 psi (49 MPa). Surprisingly, PTFE's resistance to wear during rubbing or sliding is poor. As discussed below, however, the great improvement in wear resistance as a result of use of fillers makes it possible to use the low (μ) of PTFE in many practical applications.

12.5.8 Reinforcement

PTFE is compounded with chopped glass fiber and many other particulate fillers to enhance certain properties. The resistance to wear in rubbing and sliding applications is the property that is improved the most. Ten to 20 percent (by volume) of such fillers as glass fiber, graphite, bronze, and other materials decreases the wear rate up to five hundred fold. With most other properties, the added component acts as an inert filler rather than as a reinforcing agent. In some instances, there is a moderate improvement in resistance to creep, and, with metallic fillers, an increase in thermal conductivity. The major effect of use of fillers, nevertheless, is the improvement in resistance to wear.

12.6 APPLICATIONS

PTFE is selected for a very wide range of applications that affect every person. The applications fall in the five areas that require one or more of its chemical, electrical, mechanical, thermal, and surface properties. In essentially every instance, a successful application employs at least two of the outstanding properties of this polymer.

12.6.1 Chemical Process

In the chemical process industries, products made from PTFE include gaskets, seals, coatings or linings in large tanks or process vessels, component parts of valves, piping (solid, lined, or laminated), and a very large number of specialized parts. Most people are familiar with PTFE tape "pipe-dope." Essentially every display in a chemical industry exhibition or show includes one or more items made from PTFE in the products being shown.

12.6.2 Electrical

Use of PTFE in electrical insulation covers the entire wire, cable, and speciality electrical products industry. Its high oxgen index permits PTFE to be used in many areas that have stringent standards for combustibility and emission of smoke. Most high-technology products in the electrical industry require the properties of PTFE in order to function. Use of special additives and processing techniques permits control of dielectric constant and loss factors over wide ranges. These measures result in materials with very stable performance at extremes of temperature and exposure. Many of these serve in classified military applications.

12.6.3 Mechanical

An example of an important mechanical application is the use of filled PTFE in the seal rings in automotive power-steering equipment and automatic transmissions. Similarly, piston rings of filled PTFE permit operation of nonlubricated compressor equipment, resulting in lower power costs or higher capacities for the same machine.

12.6.4 Cookware

The surface and antistick properties of PTFE are employed in many well-known cookware and bakeware products. These same properties make possible coverings on rollers in food processing equipment, xerographic copiers, and saw blades; coatings on snow shovels; and a long list of similar applications. The Alaskan oil pipeline, for example, rests on PTFE-coated steel plates. Most new bridges and tunnels use similar supports to provide easy slip to accommodate dimensional changes that result from thermal expansion.

12.6.5 Medical

Many important new medical products are based on special forms of PTFE, such as high-strength porous products. Outstanding are its uses in cardiovascular grafts, heart patches, the recently approved ligaments for knees, and many others.

12.6.6 Architectural Fabric

An important application for aqueous dispersions of PTFE is the architectural fabric market. This product involves Fiberglas fabric coated with special forms of the aqueous dispersion. The resulting material is used as roofs in a wide variety of buildings, especially where a large area must be enclosed with minimum support. Most notable are the Pontiac "Silver Dome," the airport terminal in Jeddah, in Saudi Arabia (where 105 acres are enclosed), and many college and university stadia or union buildings.

12.7 ADVANTAGES/DISADVANTAGES

The main advantage of products made from PTFE is that they do jobs nothing else will do. The disadvantage is that the cost of the resin is so high that it is not used in many applications for which it is functionally satisfactory but not competitive economically. In addition, preparation of complex shapes out of PTFE requires forming and machining techniques similar to operations with metals. In such situations, PFA or FEP resins, which can be injection molded or extruded by conventional procedures, offer advantages, as explained in the sections on those resins.

12.8 PROCESSING TECHNIQUES

The rheological properties of PTFE are so different from usual thermoplastic materials that the common techniques of melt processing (extrusion or injection molding) are not feasible. A series of techniques have been developed that are unique to the PTFE industry. Broadly, nearly all PTFE is processed by forming the resin to an approximation of the final shape at or near room temperature and then completing the operation by heating (sintering) the material at temperatures above the melting point and cooling to adjust the crystalline content of the final product. Each individual type of operation has its own name in the industry.

Master Material Outline

PROCESSABILITY OF ___PTFE___

PROCESS	A	B	C
INJECTION MOLDING			
EXTRUSION, Ram granular PTFE*	X[a]		
THERMOFORMING	X		
FOAM MOLDING			
DIP, SLUSH MOLDING		X	
ROTATIONAL MOLDING			
POWDER, FLUIDIZED-BED COATING			
REINFORCED THERMOSET MOLDING			
COMPRESSION/TRANSFER MOLDING	X		
REACTION INJECTION MOLDING (RIM)			
MECHANICAL FORMING	X		
CASTING	X		

A = Common processing technique.
B = Technique possible with difficulty.
C = Used only in special circumstances, if ever.
[a]Ram granular PTFE and lubricated (paste) extrusion for coagulant dispersion PTFE.

12.8.1   Granular Powders

Billet Molding and Skiving. The powder is placed in a mold at or slightly above room temperature. The powder is compressed at pressures from 2000–5000 psi (14–34 MPa). After being removed from the mold, the preform is sintered by heating it unconfined in an oven at temperatures in the range of 680–716°F (360–380°C) for times ranging from a few hours to several days. The time–temperature schedule depends on the size and shape of the billet. Billet size varies from less than 2.2 lb (less than a kilogram) up 1540 lb (700 kg), among the largest moldings made of any plastic material. The billets are often in the form of cylinder, which are then mounted on a mandrel. Sheeting is prepared by skiving, much like plywood is cut from large logs. The sheeting is cut in the thickness range of 0.001 in. (0.025 mm) to 0.1 in. (2.5 mm).

Sheet Molding. With sheet molding, the procedures are very similar to those used for billet molding except for the shape and size of the molding. This process is used for sheeting above 0.1 in. (2.5 mm) up to large blocks. The latter are used for a block method skiving

TABLE 12.1. PTFE Molding and Extrusion Materials Based on ASTM-D1457

Type	Comment
Type I	Granular resin used for general-purpose molding and extrusion.
Type III	Resin produced from a coagulated dispersion and normally used with a volatile processing aid. Type III resins are divided into four grades by a series of physical characteristics and each grade is divided into three classes to indicate performance in a test for extrusion pressure.
Type IV	Finely divided resin with an average particle size less than 100 μm.
Type V	A finely divided granular resin typically used in applications requiring improved resistance to creep and stress relaxation in end use.
Type VI	Free-flowing resins, divided into two grades by level of bulk density.
Type VII	Presintered resin that has been treated thermally at or above the melting point of the resin at atmospheric pressure without having been previously molded or preformed.

operation similar to that used traditionally for cellulose nitrate. The molten sheet is sometimes removed from the sintering oven and quench cooled between polished plates under moderate pressure.

Automatic Preforming and Sintering. In this operation, the procedures are similar except for the handling of the powder and the size of the parts. Automatic preforming machines feed the powder, usually Type VI (see Table 12.1), into the mold cavity. The powder is then compressed. The equipment is similar to that used with powder metallurgy. The parts are often transported directly from the preforming press through the sintering oven using automated conveying systems.

Ram Extrusion. With ram extrusion, the powder is fed into a cavity at one end of a heated tube. A reciprocating ram compacts the powder and forces it into the tube. While it is being transported down the length of the tube, the PTFE is melted and coalesced. Continuous lengths of sintered rod or shapes comes out the other end of the extruder. After cooling, the resultant rods, for example, are usually centerless ground and then are used to make a wide variety of parts by means of automatic screw machine procedures. Typically, Type VII powders would be used for small-diameter rods, 0.25 in. (6 mm), and either Type VII or Type VI for large diameters of 1 in. (25 mm) or more.

Screw Extrusion. Unmelted PTFE powders compact so much when they are used in a screw extruder that they are not transported by the screw. However, Type VII powders do not compact in this manner and can be used. Rates of extrusion are very low, and this process is not used to a significant extent in the industry.

12.8.2 "Fine Powders" (CD Resin)

Lubricated Extrusion. This process is often called "paste extrusion" and appears to be unique for these PTFE resins. One of the Type III powders is blended with 15–20% of a "lubricant," which may be any one of a large number of organic liquids. The refined petroleum products usually used include such materials as varnish-makers naphtha, odorless kerosene, or white oil. The blend of powder and lubricant looks much like the powder alone, but the blend is readily compressed and extruded to form thin insulation on wire, thin-walled tubing, ribbon, and many other shapes. These operations are usually carried out at temperatures between room temperature and 212°F (100°C). The procedures are semicontinuous, with the extrudate going from the extruder through a drying operation for removal of the lubricant and then through a sintering operation. The operation must be interrupted occasionally to place a new charge of lubricated powder (or a previously pressed preform) into the cylinder of the extruder.

12.8.3 Aqueous Dispersions

Coating. A convenient way to apply PTFE dispersions to surfaces is by use of aqueous dispersion coatings. In addition to the use of the dispersion as provided by the manufacturer, there are many coating formulations that can be purchased for use in such applications as cookware and bakeware. Coating is done both as dip coating and, with thickened dispersions, as roller coating.

The dispersion should have a concentration of 45–50% PTFE and 6–9% of a wetting agent, based on the amount of PTFE. (Usually a nonionic agent is used. Triton X-100 from Rohm and Haas has been found to be satisfactory; similar materials from other suppliers are also used.) The amount of PTFE deposited on each coat must be restricted to about 3.3 in. (84 mm) to prevent "mud-cracking" when the coating is dried. For drying the coating, infrared lamps or forced-convection ovens at 185–205°F (85–96°C) are usually used. Multiple coats are used to obtain thicker films.

Drying is followed by baking and sintering to remove the wetting agent and then coalesce the PTFE. Temperatures of 500–600°F (260–316°C) are used for the baking and 680–752°F (360–400°C) for the sintering. The time required for each of these steps varies from several seconds to a few minutes, depending on the details of the installation and the thickness of the coating. Good homogeneity is obtained by using multiple dips, with baking and sintering after each.

Casting. Casting is carried out by procedures similar to those used for dip coating. Casting is used to make thin films for use in such applications as heart–lung machines, special electronic equipment, and various speciality applications. Often it is done by allowing the dispersion to flow onto a support surface, usually a polished stainless-steel belt that, will carry the film through the successive steps of coating, drying, baking, sintering, recoating, and, finally, stripping from the belt.

Impregnation. Impregnation of porous materials is used both to make products with the PTFE sintered and also to prepare materials where the PTFE is used in the unsintered state. The former application is illustrated by such materials as the new "architectural fabric"; the latter by such products as the impregnated braided packing made from Teflon fibers.

Since fabric made from glass fibers will withstand the temperatures used to sinter PTFE, there is a large business in impregnating and coating various form of glass fabric with PTFE dispersions. The equipment required is conventional, with feed and take-up rolls, an accumulator, drying tower, and so on. Typical operations involve passing the material to be impregnated through a dip tank followed by passing the coated material through the tower that has separate zones for drying the water, removing the wetting agent by baking, and sintering the composite. Temperatures in these zones are similar to those given above. The comments on coating thickness apply equally to impregnation that is to be followed by sintering.

Co-coagulation. Co-coagulation is a process where the small particle size of the dispersed particles of PTFE provides opportunity to make a wide variety of compositions containing other materials. A dispersion of the second component is mixed with the PTFE dispersion and the components are then coagulated together. After drying, the resulting filled composition can be processed by lubricated extrusion or by conventional molding operations.

12.9 RESIN FORMS

PTFE is available as two major kinds of polymer with various sub-classes within each. These major kinds are the granular and the dispersion-based polymer. ASTM Standard Specification D1457 covers the forms sold as powders and D4441 covers the aqueous dispersions. Table 12.1 lists the types of powders defined in ASTM D1457. ASTM D4441 describes eight types of aqueous dispersion that differ, primarily, in the solids content and the amount of surfactant. The details will not be included here.

As shown in Table 12.1, granular products are available as coarse-ground powder, Type I, used very little at this time. Finely divided powders, Type V, are used for billet molding and exceedingly high-performance applications. Type VI, an agglomerated form of the finely divided type, is used now for general-purpose molding and extrusion applications. Type VII, a presintered powder, gives superior performance in some types of ram or screw extrusion.

Type III PTFE, the CD products ("fine powders"), are used almost entirely in lubricated (often called paste) extrusion. Varius types of "fine powder" are available, optimized for a particular application. The powder that is best for thin insulation on wire, for example, would normally not be used for pipe lining or tape. It is important to contact the material supplier for aid in selecting the best material for a particular application.

Aqueous dispersions are used for impregnating, coating, and preparation of very thin films [down to 0.00025 in. (0.01 mm)].

12.10 SPECIFICATION OF PROPERTIES

See the Master Outline of Materials Properties.

12.11 PROCESSING REQUIREMENTS

Once it has been melted, PTFE undergoes an irreversible change in crystallinity, so scrap cannot be reused for usual processing. There are special uses for scrap, however, that permit most of it to be recovered. For those applications, the scrap is typically ground to a small particle size and then used to make pressure moldings of large billets. At times, virgin resin is mixed with the scrap to reduce somewhat the void content in the final product. Often a processor will add a colorant to the powder to distinguish the product from moldings made from virgin resin. Another outlet for scrap is to use it as a feedstock for preparation of white lubricating powders by a degradation process that uses high energy (ionizing) radiation.

Usually, the powders do not have to be dried. However, if a drum has been stored in a cold location, it is important to allow it to come to equilibrium at room temperature before it is opened and used, in order to prevent condensation of moisture on the cold powder. Liquid water on this form of PTFE will result in defects in moldings or extrusions.

MATERIAL ___Polytetrafluoroethylene (PTFE)___

PROPERTY	TEST METHOD	ENGLISH UNITS	VALUE	METRIC UNITS	VALUE
MECHANICAL DENSITY	D792	lb/ft³	137	g/cm³	2.2
TENSILE STRENGTH	D638	psi	3,000	MPa	20.67
TENSILE MODULUS	D638	psi	60,000	MPa	413
FLEXURAL MODULUS	D790	psi	60,000	MPa	413
ELONGATION TO BREAK	D638	%	40	%	40
NOTCHED IZOD AT ROOM TEMP	D256	ft-lb/in.	No break	J/m	No break
HARDNESS	D785		D55–D70	D785	D55–D70
THERMAL DEFLECTION T @ 264 psi	D648	°F	132	°C	55.6
DEFLECTION T @ 66 psi	D648	°F	250	°C	121
VICAT SOFT- ENING POINT	D1525	°F		°C	
UL TEMP INDEX	UL746B	°F		°C	
UL FLAMMABILITY CODE RATING	UL94		VE–0	UL 94	VE–0
LINEAR COEFFICIENT THERMAL EXPANSION	D696	in./in./°F	5.5	mm/mm/°C	9.9
ENVIRONMENTAL WATER ABSORPTION 24 HOUR	D570	%	<0.05	%	<0.05
CLARITY	D1003	% TRANS- MISSION	Opaque	% TRANS- MISSION	Opaque
OUTDOOR WEATHERING	D1435	%	Superb	%	Superb
FDA APPROVAL			Yes		Yes
CHEMICAL RESISTANCE TO: WEAK ACID	D543		Not attacked		Not attacked
STRONG ACID	D543		Not attacked		Not attacked
WEAK ALKALI	D543		Not attacked		Not attacked
STRONG ALKALI	D543		Not attacked		Not attacked
LOW MOLECULAR WEIGHT SOLVENTS	D543		Not attacked		Not attacked
ALCOHOLS	D543		Not attacked		Not attacked
ELECTRICAL DIELECTRIC STRENGTH	D149	V/mil	645	kV/mm	25.4
DIELECTRIC CONSTANT	D150	10⁶ Hertz	2		2
POWER FACTOR	D150	10⁶ Hertz	<0.0002		<0.0002
OTHER MELTING POINT	D3418	°F	621	°C	327
GLASS TRANS- ITION TEMP.	D3418	°F	−73.3	°C	−100

Overheating PTFE results in thermal degradation, which can be undesirable in some instances. This point is discussed in more detail below.

Type IV granular resin usually has poor powder flow and may pose problems in handling. Type VI resins were developed to provide powders that would feed satisfactorily in automatic preforming equipment.

12.12 PROCESSING-SENSITIVE END PROPERTIES

PTFE is a semicrystalline polymer in its processed form, and many of its mechanical properties are affected by the level of crystallinity. The crystalline content, in turn, is determined by the molecular weight of the resin and the rate of cooling from the melt. In general, the lower the crystallinity, the tougher but less stiff the material. Granular products are often processed into very large moldings and the rate of cooling must be slow to prevent cracking. The molecular weight of the polymer must be kept very high in order to keep the crystallinity as low as possible in the final product. Care must be taken, therefore, to avoid overheating and resulting thermal degradation. Little or no degradation is seen at temperatures at or below 690°F (370°C). Most of the products made by paste extrusion or casting of dispersion are thin and are quenched during processing so that development of crystallinity is minimized.

12.13 SHRINKAGE

The shrinkage that is observed in processing PTFE occurs from the preform to the finished piece. The preform is voidy; the total shrinkage depends on the details of the resin and the conditions used for preforming. In most cases, there is an increase in the total volume. This increase is observed as a large, often 10% or more increase in the direction of preforming, and a smaller, 3 to 5% shrinkage at right angles to the direction of preforming. The type of shrinkage table appropriate for a melt processible thermoplastic is not applicable to processing PTFE.

12.14 TRADE NAMES

PTFE is manufactured within the United States under the following trade names by the following companies:

Halon (TM) polytetrafluoroethylene (TFE) resin, Ausimont, Morristown, N.J. 07960-1838 (virgin and filled granular PTFE)

Teflon TFE Fluorocarbon Resin (TM) Du Pont Co., Polymer Products Dept., Wilmington, Del. 19898 (full line of virgin PTFE resins)

Hostaflon TF, Hoechst Celanese Corp., Somerville, N.J. 08876

Fluon polytetrafluoroethylene (virgin resins) and Tetraloy (filled compositions), ICI Americas, Wilmington, Del. 19897

Fluorocomp (filled PTFE compositions), LNP Corp., Malvern Pa. 19355

PTFE is manufactured outside the United States as:

Polyflon, Daikin, Osaka, Japan
Teflon Tetrafluoroethylene Resin, Du Pont Netherlands
Hostaflon TF, Hoechst Aktiengesellschaft, Frankfort

Fluon polytetrafluoroethylene, Imperial Chemical Industries, Ltd., ICI Plastics Div., Welwyn Garden City, Herts, England

Teflon tetrafluorethylene resin, Mitsui Fluorochemicals, Tokyo, Japan

Algoflon, Montecatini-Edison, Italy

Soreflon, Ugine-Kuhlmann in France

12.15   INFORMATION SOURCES

The material suppliers are the best source for specific information on PTFE and products made from it. There is no single publication that is comprehensive. The selected list of references, however, provides access to much of the important information in the published literature.

The Fluoropolymers Div. of the Society of the Plastics Industry, Inc. (355 Lexington Ave., New York, NY 10017) is the major trade association concerned with PTFE and related plastics. The Fluoropolymers Section (D-20.15.12) of ASTM (1916 Race St., Philadelphia, PA 19103) is concerned with Standards on PTFE and related materials.

BIBLIOGRAPHY

General References

R.C. Doban and co-workers, *The Molecular Weight of Polytetrafluoroethylene,* American Chemical Society, Atlantic City, N.J. Sept. 18, 1956.

R.C. Doban, C.A. Sperati, and B.W. Sandt, "The Physical Properties of "Teflon" Polytetrafluoroethylene," *SPE J.* **11,** 17–21, 24, 30 (1955).

R.K. Eby and K.M. Sinnott, *J. Appl. Phys.* **32,** 1765 (1961).

S.V. Gangal, "Polytetrafluoroethylene," in M. Grayson, ed., *Encyclopedia of Chemical Technology,* 3rd ed., Vol. 11, Wiley-Interscience, New York, 1980, pp. 1–24.

U.S. Pat. 3,953,566 (Apr. 27, 1976), R.W. Gore; U.S. Pat. 3,962,153 (June 8, 1976) and U.S. Pat. 4,187,390 (Feb. 5, 1980) (to W.L. Gore & Associates).

S-F. Lau, H. Suzuki, and B. Wunderlich, *J. Polym. Sci./Polym. Phys. Ed.* **22,** 379–405 (1984).

S-F. Lau, J.P. Wesson, and B. Wunderlich, *Macromolecules* **17,** 1102–1104 (1984).

E.E. Lewis and C.M. Winchester, "Rheology of Lubricated Polytetrafluoroethylene Compositions: Equipment and Operating Variables," *Ind. Eng. Chem.* **45,** 1123–1127 (1953).

J.F. Lontz and W.B. Happoldt Jr., "Teflon Tetrafluoroethylene Resin Dispersion: A New Aqueous Colloidal Dispersion of Polytetrafluoroethylene," *Ind. Eng. Chem.* **44,** 1800–1805 (1952).

J.F. Lontz and co-workers "Teflon Tetrafluoroethylene Resin Dispersion: Extrusion Properties of Lubricated Resin From Coagulated Dispersion," *Ind. Eng. Chem.* **44,** 1805–1810 (1952).

D.I. McCane, "Tetrafluoroethylene Polymers," *Encyclopedia of Polymer Science and Technology,* Vol. 13, Wiley, New York, 1970, pp. 623–670.

R.L. McGee and J.R. Collier, "Solid State Extrusion of Polytetrafluoroethylene Fibers," *Polym. Eng. Sci.* **26,** 239–242 (March 1986).

J.T. O'Rourke, *Design Properties of Filled-TFE Plastics,* Available from LNP Corp., Malvern, Pa.

R.J. Plunkett, "History of Polytetrafluoroethylene: Discovery and Development," Seymour and Kirshenbaum, eds., *History of High Performance Polymers,* Marcel Dekker, New York, 1986.

R.J. Plunkett, "Plastics Hall of Fame," *Mod. Plast.,* 134 (Oct. 1973).

U.S. Pat. 2,230,654 (Feb. 4, 1941), R.J. Plunkett.

M.M. Renfrew and E.E. Lewis, "Polytetrafluoroethylene: Heat-Resistant, Chemically Inert Plastic," *Ind. Eng. Chem.* **38,** 870–877 (1946).

S. Sherratt, "Polytetrafluoroethylene," in *Encyclopedia of Chemical Technology,* 2nd ed., Vol. 9, Wiley, New York, 1966, pp. 805–831.

C.A. Sperati, "Physical Constants of Polytetrafluoroethylene," in E.H. Immergut and Brandrup, eds., *Polymer Handbook,* 2nd ed., sect. V, 1975, pp. 29–40.

C.A. Sperati and H.W. Starkweather Jr., "Fluorine Containing Polymers II: Polytetrafluoroethylene," *Adv. Polym. Sci.* **2,** 465–495 (1961).

C.A. Sperati and J.L. McPherson, *The Effect of Crystallinity and Molecular Weight on Physical Properties of Polytetrafluoroethylene,* American Chemical Society, Atlantic City, N.J. Sept. 18, 1956.

C.A. Sperati, "Polytetrafluoroethylene: History of Its Development and Some Recent Advances," in Seymour and Kirshenbaun, eds. *History of High Performance Polymers,* Marcel Dekker, New York, 1986.

H.W. Starkweather Jr., and co-workers "The Heat of Fusion of Polytetrafluoroethylene," *J. Polym. Sci. Polym. Phys. Ed.* **20,** 751–561 (1982); A. Steininger and P. Stamprecht, *Ram Extrusion of Polytetrafluoroethylene, Kunsttoffe* **60,** 1970, pp. 1–3.

P.E. Thomas and co-workers "Effects of Fabrication on The Properties of "Teflon Resins," *SPE J.* **12,** 89–96 (June 1956).

J.P. Tordella, "An Unusual Mechanism of Extrusion of Polytetrafluoroethylene at High Temperature and Pressure," *Trans. of the Society of Rheology* **7,** 231–239 (1963). See also U.S. Pat. 2,791,806.

13

FLUOROPOLYMERS, POLY(VINYLIDENE FLUORIDE) (PVDF)

Carleton A. Sperati

E.I. du Pont de Nemours & Co., Inc., Parkersburg, WV 26104, and Adjunct Professor, Ohio University, Athens, Ohio

13.1 POLY(VINYLIDENE FLUORIDE)

PVDF is the acronym selected for poly(vinylidene fluoride), although many workers in the field still refer to it as PVF_2. It is also correctly named poly(1,1-difluoroethylene).

$$-(C-C)_n-$$

13.2 CATEGORY

PVDF is a hard, tough thermoplastic fluoropolymer.

13.3 HISTORY

PVDF, patented in 1948, was discovered as part of Du Pont's research on fluoropolymers. The Du Pont patents were licensed originally to Pennsalt Corp. (now Pennwalt Corp.) Starting in 1958, the company has developed this polymer into a major component of the fluoropolymers industry.

PVDF was introduced by Pennwalt in 1960; commercial production was started in 1965. The commercial plant was expanded in 1969, and a new plant was started up in 1985. PVDF manufactured outside the United States is marketed by several companies.

PVDF polymer for use as an architectural finish was introduced in 1965. Other forms of PVDF were introduced in 1983 and 1984. Introduction of copolymers with hexafluoropropylene has further expanded its range of properties and the applications in which it is useful. Discovery of the exceptional piezoelectric and pyroelectric properties of PVDF has stimulated a great deal of basic research on the material and is opening up fields of application never before available to plastics.

Official production volume figures are not available. Following are broad estimates based on general industry information. 1960, PVDF introduced but not yet commercially available; 1970, 2,500,000 lb; 1980, 5,000,000 lb.

13.4   POLYMERIZATION

PVDF is prepared by free-radical initiated polymerization, either in suspension or (usually) in emulsion systems. Changes in the details of the conditions used for polymerization result in some differences in the melting point—309–360°F (154–182°C)—and other important end-use properties.

The basic raw material for PVDF is vinylidene fluoride (VF_2 or VDF). A preferred synthesis is dehydrochlorination of chlorodifluoroethane, CFC-142b. Several other syntheses also are available. Since all of these fluorochemicals require replacement of chlorine with fluorine at some step in the synthesis, they are relatively expensive.

Current (1988) prices for PVDF are about $6.50 per pound. The range of price, based on the volume of purchase and the nature of specialty products, is usually between 5–10%.

13.5   DESCRIPTION OF PROPERTIES

13.5.1   Thermal Properties

PVDF has the lowest melting point of any of the commercial fluoropolymers. As a result, its upper use temperature is limited to about 302°F (150°C), compared to values of 392°F (200°C) for FEP and 490°F (260°C) for PTFE. At temperatures in its useful range, however, PVDF maintains its stiffness and toughness very well.

13.5.2   Mechanical Properties

Toughness along with resistance to creep, fatigue, and "cut-through" are features of PVDF's mechanical properties profile that have stimulated the rapid market growth of PVDF in recent years.

13.5.3   Optical Properties

PVDF does not absorb visible light. However, scattering as a result of its crystalline structure can reduce clarity in thicker sections.

13.5.4   Environmental Properties

PVDF exhibits the excellent resistance to harsh environments characteristic of the fluoropolymers.

13.6   APPLICATIONS

PVDF is widely used in the chemical processing industry in piping systems, valves, tanks (both molded and lined), and other areas where its combination of excellent mechanical properties and superb resistance to most chemicals make it an ideal material for fluid-handling equipment. The high dielectric loss and high dielectric constant of PVDF (8 to 9) both restrict its use in some electrical applications and provide superior performance in others, such as in jackets to increase resistance to cut-through of other insulating materials.

In particular, the piezoelectric and pyroelectric properties of PVDF are opening many new applications with a very high value-in-use. Increasingly important is use of PVDF as the base resin for long-life, exterior coatings on aluminum, steel, masonry, wood, and plastics.

13.7   ADVANTAGES/DISADVANTAGES

PVDF has been used since the early 1960s because of its chemical resistance combined with very good mechanical properties. Its cost, about the lowest of the melt processible fluoropolymers, is an important advantage. In addition, it is now known that PVDF has very unusual and valuable piezoelectric and pyroelectric properties. This performance has stimulated much research with more than one thousand papers being published in the decade from 1972 to 1982.[1]

Master Material Outline

PROCESSABILITY OF ___PVDF_____

PROCESS	A	B	C
INJECTION MOLDING	X		
EXTRUSION	X		
THERMOFORMING	X		
FOAM MOLDING	X		
DIP, SLUSH MOLDING			
ROTATIONAL MOLDING	X		
POWDER, FLUIDIZED-BED COATING	X		
REINFORCED THERMOSET MOLDING			
COMPRESSION/TRANSFER MOLDING	X		
REACTION INJECTION MOLDING (RIM)			
MECHANICAL FORMING			
CASTING			

A = Common processing technique.
B = Technique possible with difficulty.
C = Used only in special circumstances, if ever.

13.8 PROCESSING TECHNIQUES

Essentially all the common procedures available for thermoplastic polymers can be used with PVDF, as detailed in the Master Material Outline.

13.9 RESIN FORMS

PVDF is available as a molding powder, a molding powder containing fillers, and in dispersions containing 44 wt% PVDF in dimethyl phthalate and diisobutyl ketone. Some copolymers are also becoming available for special electrical and other applications.

13.10 SPECIFICATION OF PROPERTIES

See the Master Outline of Materials Properties.

13.11 PROCESSING REQUIREMENTS

A product bulletin from Solvay Polymer Corp.[2] provides specific information about processing of PVDF.

The chief features to be observed in processing PVDF are the temperatures required 365–572°F (185–300°C), which are in the upper range of those generally used for the common thermoplastics. Equipment suitable for use with PVC is usually satisfactory for PVDF for long-term, trouble-free operation. Post-annealing at temperatures just below the melting point is recommended for molded parts where good dimensional stability is required.

13.12 PROCESSING-SENSITIVE END PROPERTIES

Thermal degradation with bubble formation and deterioration of mechanical properties can result from improper operations, usually use of excessively high temperatures [near and above 662°F (350°C)].

13.13 SHRINKAGE

Crystallinization during cooling results in a relatively high mold shrinkage of about 3%.

13.14 TRADE NAMES

Pennwalt Corp., King of Prussia, Pa. is the leading producer and markets a full line of PVDF resins under the trade name, Kynar.

PVDF is also available as Floroflon from Atochem, Inc., Glen Rock, N.J.; as Halon CM from Ausimont, Morristown, N.J.; as KF polymer from Kureha Chemical Industry Co., Ltd., New York, N.Y.; and as Solef from Soltex Polymer Corp., Houston, Tex.

Daikin Industries markets Neoflon VDF through various distributors in the United States.

Liquid Nitrogen Products, Malvern, Pa. supplies PVDF with various fillers and will prepare custom blends.

MATERIAL __PVDF__

PROPERTY	TEST METHOD	ENGLISH UNITS	VALUE	METRIC UNITS	VALUE
MECHANICAL DENSITY	D792	lb/ft³	110	g/cm³	1.77
TENSILE STRENGTH	D638	psi	6,700	MPa	46
TENSILE MODULUS	D638	psi	232,000	MPa	1,600
FLEXURAL MODULUS	D790	psi	218,000	MPa	1,500
ELONGATION TO BREAK	D638	%	15	%	15
NOTCHED IZOD AT ROOM TEMP	D256	ft-lb/in.	6.4	J/m	340
HARDNESS	D785		D 75	D785	D 75
THERMAL DEFLECTION T @ 264 psi	D648	°F	185	°C	85
DEFLECTION T @ 66 psi	D648	°F	284	°C	140
VICAT SOFT- ENING POINT	D1525	°F	284–302	°C	140–150
UL TEMP INDEX	UL746B	°F		°C	
UL FLAMMABILITY CODE RATING	UL94		VE–0	UL 94	VE–0
LINEAR COEFFICIENT THERMAL EXPANSION	D696	in./in./°F	11×10^{-11}	mm/mm/°C	19.8×10^{-11}
ENVIRONMENTAL WATER ABSORPTION 24 HOUR	D570	%	0.05	%	0.05
CLARITY	D1003	% TRANS- MISSION	See note below	% TRANS- MISSION	See note below
OUTDOOR WEATHERING	D1435	%	Excellent	%	Excellent
FDA APPROVAL					
CHEMICAL RESISTANCE TO: WEAK ACID	D543		Not attacked		Not attacked
STRONG ACID	D543		Not attacked		Not attacked
WEAK ALKALI	D543		Not attacked		Not attacked
STRONG ALKALI	D543		Not attacked		Not attacked
LOW MOLECULAR WEIGHT SOLVENTS	D543		Not attacked		Not attacked
ALCOHOLS	D543		Not attacked		Not attacked
ELECTRICAL DIELECTRIC STRENGTH	D149	V/mil	260–1300, depending on thickness	kV/mm	10–51
DIELECTRIC CONSTANT	D150	10^6 Hertz	8–9		8–9
POWER FACTOR	D150	10^6 Hertz	0.018		0.018
OTHER MELTING POINT	D3418	°F	338	°C	170
GLASS TRANS- ITION TEMP.	D3418	°F	–40	°C	–40

Note: Moldings vary from clear to nearly opaque, depending on thickness of the part, the particular resin used, and conditions of molding.

In addition to the U.S. companies listed above, Dynamit Nobel produces Dyflor PVDF and Suddeutsche Kalkstickstoffe-Werke, A.G. makes Vidar PVDF.

13.15 INFORMATION SOURCES

The material suppliers are the best source for specific information on PVDF and products made from it. There is no single publication that is comprehensive. The selected list of references, however, provide access to much of the important information in the published literature.

The Fluoropolymers Div. of the Society of the Plastics Industry, Inc. (355 Lexington Ave., New York, NY 10017) is the major trade association concerned with fluoropolymers. The Fluoropolymers Section (D-20.15.12) of ASTM (1916 Race St., Philadelphia, PA 19103) is concerned with standards on fluoropolymers.

BIBLIOGRAPHY

1. A.J. Lovinger, "Poly(vinylidene fluoride)" in D.C. Bassett, ed., *Developments in Crystalline Polymers 1*, Applied Science Publishers, London, 1982, Chapt. 5.
2. *Solef Solvay PVDF*, Product Bulletin, Solvay & Cie, and Soltex Polymer Corp., Houston, Texas.

General References

W.S. Barnhart and N.T. Hall, "Poly(vinylidene fluoride)" in A. Standen, ed., *Encyclopedia of Chemical Technology*, 2nd ed., Vol. 9, Wiley-Interscience, New York, 1966, pp. 840–847.

M.I. Bro, *J. Appl. Polym. Sci.* **1**, 319–322 (1959).

J.E. Dohany, A.A. Dukert, and S.S. Preston III, "Vinylidene Fluoride Polymers," in N.M. Bikales, ed., *Encyclopedia of Polymer Science and Technology*, Vol. 14, Wiley-Interscience, New York, 1965, pp. 600–610.

J.E. Dohany and L.E. Robb, "Poly(vinylidene fluoride)" in M. Grayson, ed., *Encyclopedia of Chemical Technology*, 3rd ed., Vol. 11, Wiley-Interscience, New York, 1980, pp. 64–74.

Foraflon Polyvinylidene Fluoride, Product Bulletin, PCUK, Produits Chimiques Ugine Kuhlmann (now Atochem, Inc.).

R.L. Johnson, in N.M. Bikales, ed., *Encyclopedia of Polymer Science and Technology*, Suppl. vol. 1, Wiley-Interscience, New York, 1976, pp. 269–278.

KF Polymer, (Polyvinylidene Fluoride), Product Bulletin, Kureha Chemical Co., Ltd.

Kynar Piezo Film, Technical Manual 10-M-11-83-M, Pennwalt Corp.

Kynar Poly(vinylidene fluoride), Product Bulletin, Pennwalt Corp.

W.K. Lee and C.L. Choy. *J. Polym. Sci. Polym. Physics Ed.* **13**, 619–635 (1975).

Mach. Des. 128–130 (April 19, 1981).

H.W. Starkweather, *J. Poly. Sci. Poly. Phy. Ed.* **11**, 587–593 (1973).

Tables of Chemical Resistance of Solef PVDF, Solvay & Cie.

14

IONOMERS

Ruskin Longworth

E.I. du Pont de Nemours & Co., Inc., Wilmington, Delaware
10 Walnut Ridge Road, Greenville, DE 19807.

14.1 IONOMER

$$-(-\underset{\underset{H}{|}}{\overset{\overset{H}{|}}{C}}-\underset{\underset{H}{|}}{\overset{\overset{H}{|}}{C}}-)_n-\underset{\underset{H}{|}}{\overset{\overset{H}{|}}{C}}-\underset{\underset{C=O}{|}}{\overset{\overset{CH_3}{|}}{C}}-(-\underset{\underset{H}{|}}{\overset{\overset{H}{|}}{C}}-\underset{\underset{H}{|}}{\overset{\overset{H}{|}}{C}}-)_m-(-\underset{\underset{H}{|}}{\overset{\overset{H}{|}}{C}}-\underset{\underset{C=O}{|}}{\overset{\overset{CH_3}{|}}{C}}-)-$$

$$O-(Na^+ \text{ or } \tfrac{1}{2}Zn^{++}) \qquad\qquad OH$$

14.2 CATEGORY

Ionomer is a thermoplastic resin.

14.3 HISTORY

Ionomers were discovered by the Du Pont Co. in the early 1960s. The materials discovered at that time were derived from ethylene copolymers in which a copolymer of ethylene with methacrylic acid was neutralized to varying degrees with either sodium or zinc.[1,2] It was soon realized that the acid groups conferred quite distinctive properties on the copolymer. These properties were a consequence of the ionic groups segregating into clusters. Wide-angle x-ray diffraction and electron microscopy revealed that these clusters were between 100Å and 1000Å in size (about 1 to 10 microns).

Thus, whereas ordinary polyethylene consists of crystalline and amorphous hydrocarbon phases, the ionomers additionally contain an ionic phase. All these phases are molecularly interconnected.[2] What is perhaps more revealing than the symbolic formula is a schematic drawing of the crystalline structure of an ionomer, showing the relationship between amorphous hydrocarbon, fringed micellar crystals of polyethylene, and layered ionic clusters (Fig. 14.1).[3]

Since that time, numerous other ionomers have been prepared, with, for example, styrene, butadiene, and fluorocarbons being used to form the backbone. The term ionomer has come to refer to any acid-containing copolymer (carboxylic or sulfonic) with which a small fraction (about 5 wt%) of ionizing material has been combined. The ionomers retain their organic, thermoplastic nature; the amount of combined metal is not so great as to confer water solubility. (It is usually the case that the ionomers will absorb more water than

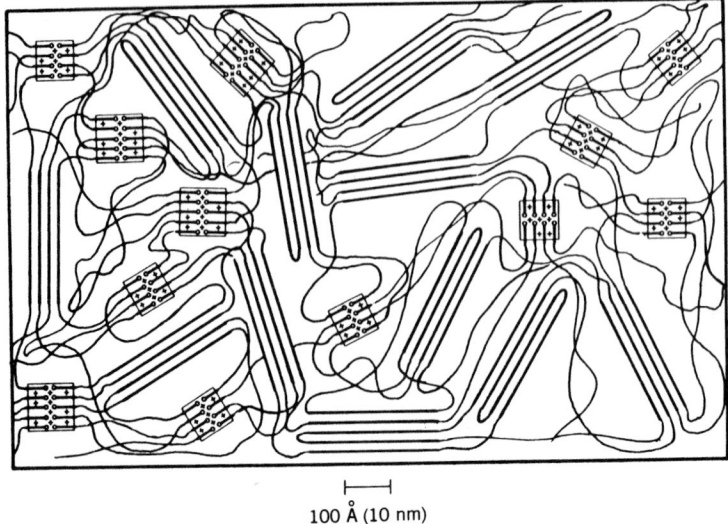

$\vdash\!\!\dashv$
100 Å (10 nm)

Figure 14.1. Schematic structure of ionomer.

the parent copolymer.) However, the only significant commercial production is of the ethylene-based ionomers produced by the Du Pont Co. and marketed under the trade name Surlyn resins. Du Pont also produces a fluorocarbon ionomer under the name Nafian®, however in its final form as a membrane, it cannot really be considered a plastic.

Production exceed of ionomer exceeded 100 million pounds annually in 1983; current (1988) prices range from $1.25 to $1.98 per pound.

14.4 POLYMERIZATION

The Surlyn resins are derived from copolymers of ethylene and methacrylic acid produced by the high pressure process for the manufacture of low density polyethylene. The commercial conversion of the copolymer to the ionomer is a proprietary process.

On a laboratory scale, the preparation of the ionomer from an appropriate acidic copolymer can readily be accomplished either by adding the appropriate base in solution or by combining stoichiometric amounts of base and copolymer in suitable melt-processing equipment, such as a two-roll mill, Banbury mixer, or extruder. Provision must be made to eliminate any water formed during the reaction.

14.5 DESCRIPTION OF PROPERTIES

In the simplest terms, these ionomers resemble LDPE, but it is the differences from polyethylene that determine their applicability and ability to command a higher price. Qualitatively, ionomers may be compared to LDPE as follows:

- Higher tensile strength
- Higher modulus (room temperature)
- Lower modulus [low temperatures, $-4°F$ ($-20°C$)]
- Higher toughness
- Lower softening point
- Much greater clarity, lower haze

- Better abrasion resistance
- Better oil resistance
- Processibility

Recently ionomer compositions have been introduced that incorporate glass fibers. These resemble engineering plastics such as nylon and acetal resins, exhibiting higher modulus than the unmodified ionomers, ie, $(150-325) \times 10^3$ psi $(1034-2241$ MPa) vs $25,000-50,000$ psi $(172-345$ MPa). Compared to engineering plastics, these glass-fiber-reinforced resins have superior low temperature properties [at approximately $-21°$F ($-29°$C), notched Izod impact strength is 3.0 ft.lb/in. (160 J/m)].

14.6 APPLICATIONS

The most important applications for ionomers are in the manufacture of films and sheets, primarily for food packaging of various kinds. Several grades have been approved by the FDA for use in contact with foodstuffs. Additionally, ionomers can be molded by injection molding or extrusion into a wide variety of shapes.

In film applications, the properties that distinguish ionomer films from other varieties of flexible packaging resins are heat sealability, ease of thermoforming, toughness, and excellent optical properties. Also of importance are the resins' resistance to food oils and, in coextrusion with nylon, excellent adhesion. Films coextruded with nylon are widely used in packaging processed meats, such as bacon, frankfurters, etc.

Injection molding is used to make a wide variety of shapes. Toughness and cut-through resistance make for superior golf-ball covers and bowling pin covers. The excellent low-temperature toughness, flexibility, and abrasion resistance make ionomers the preferred choice for ski-boot shells and bumper guards for automobiles (the absence of leachable plasticizers is a key advantage).

Formulations of ionomers are available that give high UV resistance (important for automotive trim rubbing strips) and foamed shapes. The resins can be extruded without difficulty into rods, tubes, pipes, profiles and sheets [sheeting is of 0.010 in. (0.3 mm) and greater thickness].

14.7 ADVANTAGES/DISADVANTAGES

A discussion of the advantages and disadvantages of ionomers is best done with respect to LDPE. Of course, in some applications ionomers will compete successfully against such materials as nylon or polypropylene; in these cases, the glass-filled ionomer resins are of importance.

Generally, ionomers are tougher, stronger, and stiffer than LDPE. The toughness is maintained even at low temperatures. On the other hand, the tear resistance or notch sensitivity is poor. As films, ionomers have outstanding clarity and low haze, but these films have poorer slip and block characteristics than LDPE. Ionomer films adhere well to foils, have excellent hot tack, heat sealability, and abrasion resistance. However, in molding and compounding, ionomers will adhere to metal compounding equipment. Since ionomers have lower freezing points, injection-molding cycles are necessarily longer. Ionomers will absorb moisture and are packaged in hermetically sealed bags and containers; if they are exposed too long to humid conditions, they must be dried before processing. The modulus will change somewhat (decrease) from the dry, as-molded value as a molding absorbs moisture, although this may not be too great to cause a problem. Ionomers formulated with zinc rather than sodium have lower moisture sensitivity, better UV stability, greater adhesion to nylon and polyester, and better low temperature toughness. They generally command a somewhat higher price.

With respect to processability, ionomers have the great advantage over LDPE of being able to be drawn down to very thin films. This fundamental advantage is directly related to the molecular architecture of the resins.

Ionomers will cost more than LDPE.

14.8 PROCESSING TECHNIQUES

Ionomers are readily processed by machinery suitable for thermoplastics. Thus, extrusion into film, sheeting, tubing and pipe, and other profiles can readily be accomplished. Injection molding into solid objects can be accomplished without difficulty.

Master Material Outline

PROCESSABILITY OF ___Ionomer___

PROCESS	A	B	C
INJECTION MOLDING	X		
EXTRUSION	X		
THERMOFORMING	X		
FOAM MOLDING			
DIP, SLUSH MOLDING			
ROTATIONAL MOLDING			
POWDER, FLUIDIZED-BED COATING			
REINFORCED THERMOSET MOLDING			
COMPRESSION/TRANSFER MOLDING			
REACTION INJECTION MOLDING (RIM)			
MECHANICAL FORMING			
CASTING			

A = Common processing technique.
B = Technique possible with difficulty.
C = Used only in special circumstances, if ever.

14.9 RESIN FORMS

Ionomers are normally supplied as free-blowing pellets 1/8 in. (3 mm) in size. This material can be compounded in various ways according to the needs of the application. Thus, ingredients for improved weatherability (UV resistance), colors, flame retardancy, foamed products, or glass reinforcement can all be incorporated.

These ingredients can be added in various ways. Dry blending will be suitable in many cases; often, this can be carried to higher levels than is the case for polyolefins. Where thorough dispersion is needed, such as for UV resistant formulations, then it is desirable to use a mixing screw rather than a polyolefin-type metering screw.

A better method may be to use a concentrate of the particular additive. In this case, the base resin of the concentrate should be an ionomer of lower viscosity; other polyolefins should never be used. A broad range of letdown ratios are used, from 2:1 to 30:1.

Foamed products can be prepared either by incorporating chemical blowing agents or by gas injection of, for instance, a fluorocarbon (Freon). This latter method requires the installation of special equipment. The successful manufacturer of foamed ionomer requires a high melt-viscosity base resin and careful attention to the cooling equipment used, as foamed materials will cool more slowly.

Compounds with higher flex moduli can be made by incorporating glass fiber. The glass fiber will increase the modulus without significantly altering other properties; however, it is better to incorporate the glass in the form of a concentrate prepared by a compounder who is familiar with the necessary surface sizing requirements.

14.10 SPECIFICATION OF PROPERTIES

See the Master Outline of Materials Properties, which shows properties of Surlyn 8528. This resin contains sodium ion and has a melt flow index of 1.3 dg/min.

About twenty other grades of ionomer resin are available. The principal variables are the ion type and content and the melt flow index. Two ion types are available, sodium and zinc. The melt flow indexes vary from 0.7 dg/min up to 14 dg/min.

14.11 PROCESSING REQUIREMENTS

A broad range of melt temperatures is possible for processing ionomer resins; however, temperatures above 464°F (240°C) should be avoided.

14.11.1 Film Extrusion

Broadly, the techniques of manufacture of blown and cast film from LDPE can be applied to the manufacture of film from ionomers. The most important usage of ionomer resins is in the manufacture of blown and cast film. In addition to the standard grade, which is commonly used for these purposes, there are modified grades for improved slip, higher stiffness, and lower heat-seal temperatures (faster heat-seal cycles).

The optical properties of ionomer films are outstanding. Gloss increases and haze decreases as melt temperatures are increased from 360°F (182°C) up to a maximum of about 450°F (232°C). In contrast to LDPE, where fast cooling inhibits the growth of spherulitic crystallinity, ionomers give better optical properties with slow cooling. (In ionomers, although polyethylene crystallinity is present, it is at a much lower level than is typical for

MATERIAL __Ionomer (Surlyn 8528)__

PROPERTY	TEST METHOD	ENGLISH UNITS	VALUE	METRIC UNITS	VALUE
MECHANICAL DENSITY	D792	lb/ft^3	58.6	g/cm^3	0.940
TENSILE STRENGTH	D638	psi	4800	MPa	33
TENSILE MODULUS	D638	psi		MPa	
FLEXURAL MODULUS	D790	psi	31,900	MPa	220
ELONGATION TO BREAK	D638	%	400	%	400
NOTCHED IZOD AT ROOM TEMP	D256	ft-lb/in.	11.4	J/m	608
HARDNESS	D785		Shore D 60	D785	Shore D 60
THERMAL DEFLECTION T @ 264 psi	D648	°F		°C	
DEFLECTION T @ 66 psi	D648	°F	111	°C	44
VICAT SOFT-ENING POINT	D1525	°F	163	°C	73
UL TEMP INDEX	UL746B	°F		°C	
UL FLAMMABILITY CODE RATING	UL94			UL 94	
LINEAR COEFFICIENT THERMAL EXPANSION	D696	in./in./°F	$10-17 \times 10^{-5}$	mm/mm/°C	$18-30 \times 10^{-5}$
ENVIRONMENTAL WATER ABSORPTION 24 HOUR	D570	%		%	
CLARITY	D1003	% TRANS-MISSION	6	% TRANS-MISSION	6
OUTDOOR WEATHERING	D1435	%		%	
FDA APPROVAL			Yes		Yes
CHEMICAL RESISTANCE TO: WEAK ACID	D543		Not attacked		Not attacked
STRONG ACID	D543		Slow attack		Slow attack
WEAK ALKALI	D543		Not attacked		Not attacked
STRONG ALKALI	D543		Not attacked		Not attacked
LOW MOLECULAR WEIGHT SOLVENTS	D543		Slight swell		Slight swell
ALCOHOLS	D543		Slight swell		Slight swell
ELECTRICAL DIELECTRIC STRENGTH	D149	V/mil	1.0	kV/mm	0.04
DIELECTRIC CONSTANT	D150	10^6 Hertz	2.4		2.4
POWER FACTOR	D150	10^6 Hertz	0.003		0.003
OTHER MELTING POINT	D3418	°F	199	°C	93
GLASS TRANS-ITION TEMP.	D3418	°F		°C	

LDPE). For similar reasons, other variables being equal, thicker films have improved opticals because of the slower cooling rate.

With ionomer films, there is the interesting effect that the stiffness will increase by about 50% on aging for up to 10 days after extrusion. This result is obtained with dry film. If moisture is allowed to be absorbed, then the stiffness will decrease and offset the stiffness increase after to aging.

14.11.2 Injection Molding

Ionomers are made from ethylene copolymers and broadly resemble LDPE in their characteristics for injection molding. However, there are important differences to be taken into account.

As mentioned earlier, ionomers have a higher temperature coefficient of viscosity than LDPE. This means that for the same melt index [at 374°F (190°C)] the ionomers will decrease in viscosity twice as rapidly as LDPE. Conversely, if molding is done at the maximum melt temperature, about 480°F (249°C), then there will be a much sharper increase in viscosity on cooling than for LDPE. This will generate relatively more orientation and frozen-in stresses for an ionomer than for LDPE.

Generally, moderate fill rates are preferred for ionomers. Flow orientation is minimized by high melt temperature, lower injection pressure, and shorter injection times. In the case of HDPE, variation in thickness may cause warpage caused by differential shrinkage; with ionomers the warpage results from relaxation of compressive stresses caused by overpacking. Almost uniquely among thermoplastics, ionomers can dilate at high injection pressures and give parts that are oversize with respect to the mold size.

As mentioned above, care should be taken to exclude moisture, although hopper drying is to be avoided lest bridging occurs. While it is recommended to start with a clean machine, purging may be done with other polyolefins (HDPE) while in transition from other thermoplastics (PVC, ABS, acetals, and so on). Purging with ionomers is not recommended. Reground resin can be used provided it is dry; it may be prepared similarly to LDPE. Since wet resin gives off corrosive vapors during processing, the use of stainless steel or corrosion-resistant alloys is recommended (Alloy 306).

Molten ionomers are extremely tacky. Protective covering should be worn.

14.12 PROCESSING-SENSITIVE END PROPERTIES

14.12.1 Moisture Absorption

Ionomers are hygroscopic and will absorb moisture when they are exposed to humid environments. The presence of moisture will alter their processibility; the higher the moisture content, the lower the maximum temperature at which acceptable extrusions can be made. As packaged, the moisture content is less than 800 ppm; the presence of moisture will normally only present a problem when opened containers have been left exposed to the atmosphere. The problem of too high a moisture content will be manifested by surface defects caused by smeared bubbles; at higher levels of moisture there may appear bubbles within the body of the molding. If this is the case, then drying will be necessary. The best way to dry is in trays, preferably in a vacuum oven. If convected-air type ovens are used, the circulating air will have to be dehumidified or moisture may be added. The drying temperature should be no more than about 140°F (60°C). The goal is to lower the moisture content to less than 800 ppm. If the material has been allowed to pick up moisture to, say, 6000 ppm, a drying time of 12 h will be necessary. Hopper drying is not recommended.

Apart from the risk of causing bubbles and foaming during processing, the melt flow index is altered by the presence of moisture and will decrease substantially with increasing

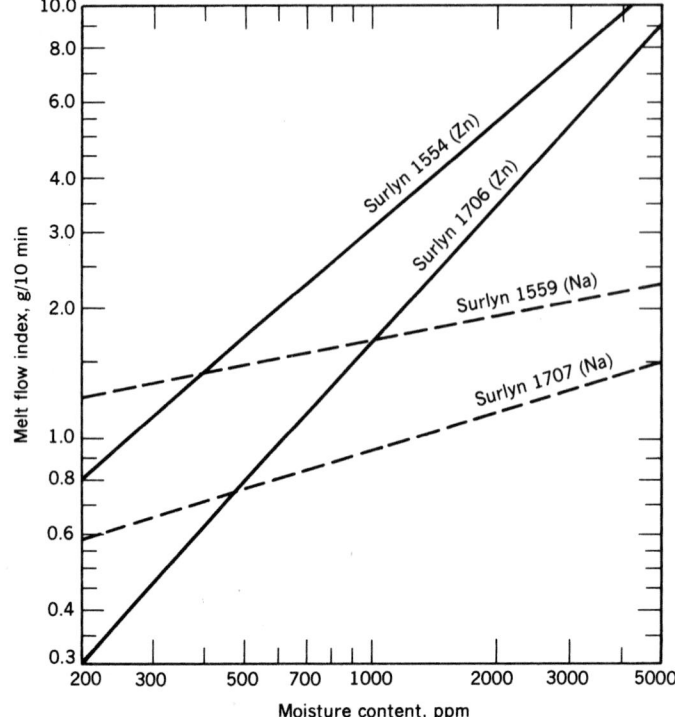

Figure 14.2. Moisture content vs melt flow index for ionomer resins.

water content, as shown in Figure 14.2. Thus, even if the resin is dry at the start of a molding run, it is desirable to purge the feed hopper with nitrogen or dry air. These considerations also apply to bin storage.

14.12.2 Effect of Temperature on the Melt Flow

In contrast to polyethylene, ionomers are more sensitive to temperature changes. This is shown in Figure 14.3, which compares the melt viscosity of Surlyn® 1601 to that of a typical LDPE. Processing temperatures in excess of 540°F (282°C) should be avoided. For the same melt flow index, which is measured at 374°F (190°C), when the temperature is lower, the ionomers will have a higher viscosity than LDPE. This will result in greater horsepower requirements during extrusion operation.

14.13 SHRINKAGE

Because of the lower freezing points typical of ionomers, molds designed for higher melting thermoplastics may have inadequate cooling. A mold surface temperature of 50°F (10°C) is recommended.

Ionomer resins have rather complex shrinkage behavior; and, depending on the injection pressure, there may be a growth of 1.2% or a shrinkage of 0.9%. More specific information is given in Table 14.1.

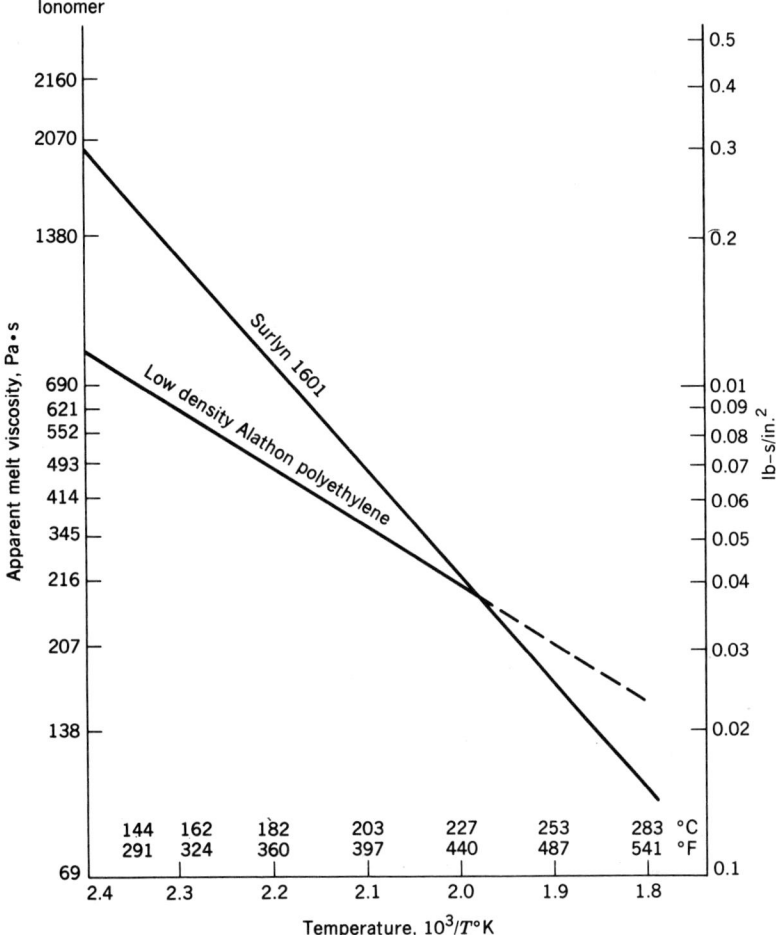

Figure 14.3. Apparent melt viscosity vs temperature. Shear rate = 500 s^{-1}.

14.14 TRADE NAMES

Ionomer resins derived from ethylene copolymers are sold under the trade name Surlyn resins by the Du Pont Co., Polymer Products Dept., Industrial Polymers Division, Wilmington, DE 19898; (302) 772-6025.

14.15 INFORMATION SOURCES

More information is available from Du Pont Co., Polymer Products Dept., Ethylene Polymers Division, Wilmington, DE 19801.

TABLE 14.1. Influence of Molding Conditions and Post-molding Thermal History on Part Shrinkage

Molding Conditions							Part Shrinkage[a]		
Injection Pressure			Mold Surface Temperature		Melt Temperature		72 hrs. After[b] Molding	After 88°C (190°F) Oven Annealing	Total Part Change
MPa	M psi	kg/cm²	°C	°F	°C	°F	mm/mm in./in.[c]	mm/mm in./in.[c]	mm/mm in./in.[c]
68.9	10	703	15	60	250	485	−0.004	−0.008	−0.012
68.9	10	703	38	100	250	485	−0.006	−0.007	−0.013
82.7	12	845	38	100	250	485	−0.005	−0.008	−0.013
89.6	13	915	38	100	250	485	−0.005	−0.007	−0.012
96.5	14	985	15	60	250	485	−0.004	−0.009	−0.012
41.4	6	424	15	60	260	500	−0.006	−0.005	−0.011
55.1	8	565	15	60	260	500	−0.006	−0.006	−0.012
55.1	8	565	49	120	260	500	−0.011	−0.004	−0.015
68.9	10	703	49	120	260	500	−0.014	−0.004	−0.018

[a]Based on 200 mm (8″) long part, 5 mm (0.190″) wall thickness with Surlyn 1554 ionomer resin.
[b]Change in part size dimension immediately after ejection from cavity.
[c]Average of five samples in each condition.

BIBLIOGRAPHY

1. U.S. Pat. 3,264.272, R.W. Rees.
2. R. Longworth and D.J. Vaughan, *Nature* **218,** 85 (1968).
3. E.J. Roche, R.S. Stein, T.P. Russell, and W.J. MacKnight, *J. Polym. Sci. Polym. Phys.* **18,** 1497 (1980).

General References

For a discussion of the physics and chemistry of ionomers in general see:

T.R. Earnest and W.J. MacKnight. *J. Poly. Sci., Macromolecular Reviews* **16,** 41 (1981).

R. Longworth, *Development in Ion-Containing Polymers-I,* Applied Science Publishers, Ltd., Barking, England, 1983, Chapt. 3.

R. Longworth, *Ion-Containing Polymers,* John Wiley & Sons, Inc., New York, 1973, Chapt. 2.

15

LIQUID CRYSTAL POLYMERS (LCP)

15.1 LIQUID CRYSTAL POLYMER

Liquid crystal polymer (LCP) is a generic term attached to a certain family of polymers. The name refers to the tendency of the material to maintain a high degree of crystalline order in the melt phase. LCPs are composed of long, slender, rodlike molecules that orient readily in the direction of material flow. This orientation is maintained as the material solidifies. The degree of order ultimately depends on the chemical structure of the polymer and the conditions under which it was processed.

LCPs can be divided by their thermal performance into three categories. Type I ranks the lowest in thermal performance, with heat-deflection temperature (HDT) values ranging from 180–420°F (82–216°C). Several Type I LCPs are defined in scientific literature; an example of a Type I product is one containing p-hydroxybenzoic acid (PHBA) and poly(ethylene terephthalate) (PET).

p-hydroxybenzoic acid   poly(ethylene terephthalate)
(PHBA)   (PET)

Type II LCPs range in HDT values from 350–465°F (177–241°C). This polymer, sold under the trade name Vectra, is a wholly aromatic copolyester composed of PHBA and p-hydroxynaphthoic acid (PHNA).

p-hydroxybenzoic acid   p-hydroxynaphthoic acid
PHBA   PHNA

Type III LCPs, known by the trade name Xydar, vary in HDT values from 500–670°F (260–354°C). This terpolymer is also a wholly aromatic copolyester containing PHBA, terephthalic acid (TPA), and p,p'-biphenol (ppBP).

p-hydroxybenzoic acid
PHBA

terephthalic acid
TPA

p,p'-biphenol
ppBP

LCPs are characterized by a difference in the morphology of the material at a part's surface from that at its center. At the surface, molecules align in the direction of material flow; in the center molecular orientation is perpendicular to the flow direction, with much higher crystallinity. Between the surface and center is a transitional region where orientation gradually shifts by 90°. The result is a distinct flow-oriented skin layer on the surface and a cross flow-oriented core region in the interior. Testing has shown that much of the strength and stiffness attributed to LCPs is provided by the highly oriented skin layer.[1]

One other distinct characteristic that LCPs share is their inherent anisotropy. The molecular shape and crystalline structure of the polymer cause the mechanical and electrical properties to be anisotropic with flow direction. In general, properties are higher along the direction of material flow. Product designers must consider this when they analyze part performance, as must mold designers when they calculate mold shrinkages. The degree of anisotropy can be reduced by the addition of filler to the material; as a result, most commercially available LCP resin grades will contain some mineral or glass filler.

15.2 CATEGORY

LCPs can best be described as rigid, strong, brittle thermoplastics. They will outperform other thermoplastics in thermal resistance and inflammability. Other characteristics include excellent electrical properties and good chemical resistance.

15.3 HISTORY

The first LCPs were naturally occurring lyotropic biopolymers, such as proteins, viri, and polysaccharides, but these compounds were not recognized as liquid crystalline until the mid-1950s. Crystalline orientation in solution was first observed in 1888 by Reinitzer. In 1949, Onsager theorized the existence of a stable ordered phase in solution based on rigid, rodlike molecules. A lyotropic (nonmelt processible) aromatic polyamide, Kevlar, was introduced in 1965. The first thermotropic (melt-processible) LCP, Ekkcel I2000, was made commercially available in 1972. Patent rights to Ekkcel were purchased by Dart Industries in 1978. There, product development led to a material compounded with mineral particulate being used in Tupperware's Ultra 21 dual ovenable cookware line, beginning in 1984.

Most of the major modifications in LCPs have occurred with the development of filler materials. These fillers are used to tailor the mechanical, thermal, electrical, and processing characteristics of the polymer. The fillers interrupt the material's crystalline structure and reduce the degree of anisotropy present. Commonly used fillers are milled glass, glass fiber, mineral fiber, and mineral particulate.

Total worldwide capacity of LCPs is approximately 22 million pounds per year. At this point about one half is utilized. Indicators and surveys point to the capacity growing substantially in the near future as suppliers gear up for the expected demand.

15.4 POLYMERIZATION

The polymerization of wholly aromatic polyesters is generally carried out through either of two methods.

The Schotten-Bauman reaction utilizes the diacyl chloride and the diphenol of the polymer's aromatic backbone and polymerizes with heat, giving off HCl as the by-product. This reaction can be performed in solution, by interfacial polycondensation, or by melt polycondensation.

The second polymerization method is ester exchange and polycondensation in the melt. Diacid, hydroxy acid, and diacetoxy forms of the aromatic backbone react under heat and a vacuum. The polymer is produced along with acetic acid as the by-product. Both polymerization methods require two steps. The monomers are polymerized to low to medium molecular weight chains in solution or as melt. The polymer is then advanced to a higher molecular weight with heat as a solid.

Since the cost of raw materials represents a significant percentage of the total polymer cost, pricing will fluctuate with raw material prices. Pricing (1988) ranges from $7.35–$30.00/lb, depending on resin grade and quantity.

15.5 DESCRIPTION OF PROPERTIES

15.5.1 Thermal Properties

The most significant benefit LCPs can provide is superior elevated temperature performance, shown by HDT values as high as 671°F (355°C), mechanical strength up to 575°F (302°C), and excellent resistance to long-term creep effects at high temperatures. In addition, LCPs are inherently flame retardant. Most resins carry UL 94 V-0 ratings at thicknesses as low as 0.031 in. (0.79 mm). Other testing has shown Type III LCPs to resist burn-through when exposed to a 2000°F (1095°C) flame for 15 minutes. Rather than burning, the polymer builds an intumescent char on the surface exposed to the flame that protects the layers beneath. The thermal conductivity of LCPs is relatively low, like most plastics, so it serves as a thermal insulator; however, different filler packages can raise the thermal conductivity, if necessary. The thermal expansion/contraction coefficient for LCPs is very low [$5-10 \times 10^{-6}$ in./in./°F ($9-18 \times 10^{-6}$ in./in./°C)] along the direction of flow, and is approximately twice that value when measured perpendicular to flow.

15.5.2 Mechanical Properties

The mechanical properties of LCPs are those of a strong, stiff, brittle material. Tensile and flexural strengths are high [25,000 psi and 30,000 psi (172 and 207 MPa), respectively], particularly when measured in the direction of flow. Tensile and flexural modulus values range from $(2-4) \times 10^6$ psi (13,800–27,600 MPa). Type III compounds have tensile strengths up to 5,000 psi (34 MPa) at 575°F (302°C), making them the only thermoplastic material option at this temperature level. Elongation is low (1–3%), as is Izod impact strength [1–3 ft-lb/in. (53–160 J/m)], typical for a highly crystalline material.

Properties measured perpendicular to material flow typically reach 50–70% of the corresponding property in the flow direction. Applications must be analyzed in both directions to ensure successful part performance.

For Types I and II LCPs, the addition of a reinforcing glass filler will raise the mechanical and thermal properties of the neat resin. Glass reinforcement will improve slightly the mechanical properties of a Type III material but will have no effect on its thermal resistance. Other fillers will reduce the thermal and mechanical properties of Type III compounds and are used to ease processing and reduce anisotropy.

15.5.3 Electrical Properties

The polymer performs as an electrical insulator, with volume resistivity up to 4×10^{15} ohm-cm and dielectric strength as high as 1100 V/mil (43 kV/mm). Arc resistance has been measured as high as 244 seconds. Special filler compositions will allow the specification of LCPs in antistatic and conductive devices. Also of note is the material's high transparency to microwaves, particularly at elevated temperatures, which causes the material to remain cool to the touch over a wide range of temperatures. This feature not only allows the material to perform in microwave cookware applications but also in military tracking devices.

15.5.4 Environmental Properties

LCPs are, by nature, very resistant to chemical attack. They are essentially unaffected by common solvents, household and industrial chemicals, foodstuffs and oils, hydrocarbons, and weak acids and bases (Table 15.1). Their resistance to moisture and humidity, even at high temperatures, is also high. They are somewhat susceptible to attack by very strong acids and bases, such as concentrated H_2SO_4 and 30% NaOH.

Because of their chemical inertness, LCPs comply with the Federal Food, Drug, and Cosmetic Act administered by the FDA and can be used for the commercial processing of food.

15.6 APPLICATIONS

15.6.1 Consumer

By far the largest application area for LCPs to date is cookware, which comprises 80–90% of the LCP consumption today. LCPs are ideal for dual ovenable cookware because of their

TABLE 15.1. Chemical Resistance of 40% Glass-filled LCP

	Weight Change, %	Tensile Strength Change	Tensile Modulus Change	Elongation, %
Chlorobenzene	A	A	A	A
Corn oil (Tested at 218°C)	A	B	A	A
Distilled H_2O	A	A	A	A
Gasoline (unleaded premium)	A	B	A	B
Hydrochloric acid, 37%	A	A	A	A
Isopropyl alcohol	A	A	A	A
Methyl ethyl ketone	A	A	A	A
Sodium hydroxide 20%	C	C	B	A
Trichloroethylene	A	A	A	A

[a]A = Less than 2 standard deviation drop in property values (excellent).
B = Less than 3 standard deviation drop in property values (very good).
C = Less than 4 standard deviation drop in property values (good).
D = 4 or more standard deviation drop in property values (poor).

elevated temperature performance, microwave transparency, stain and chemical resistance to foodstuffs, and hydrolytic stability. These same properties have led to LCP use in microwave oven components, including floor and ceiling parts, stirrer blade assembly components, and turntable assembly parts. The elevated temperature resistance and dielectric properties also permit LCP usage in insulators for small appliance motors.

15.6.2 Machinery/Equipment

The elevated temperature resistance, solvent resistance, dimensional stability, and low-cost manufacturability of LCPs have enabled them to replace metal parts in many industrial applications. One large market is the electronics processing industry, with applications such as circuit board and semiconductor handling devices. These are viable applications because of the material's ability to resist a variety of thermal and environmental processing conditions (wave, vapor phase, and infrared soldering; solvent cleaning; and burn-in testing). LCPs have replaced titanium in these parts because of lower manufacturing cost. The chemical processing industry has cited the elevated temperature chemical and creep resistance of Type III LCPs as the primary reason for their use in tower packing. Thermal and electrical performance of LCPs has led to their selection for electric motors and stators, leading to the design of smaller and more powerful motors.

15.6.3 Electrical/Electronics

The elevated temperature capabilities of LCPs, combined with easy flow and dimensional stability, have stimulated much interest in the electronics market. With the tendency of the electronics industry being toward smaller, more intricate components (and therefore higher temperatures), LCPs are used extensively in sockets, connectors, and chip carriers. The thermal and environmental properties permit the use of LCPs in parts that will see a wide variety of processing conditions, as is often the case for in electronics, where a part must be able to survive the processing cycles of a number of different customers. Additionally, the rapid cycle times of LCP components will reduce the manufacturing costs of components already in plastic as much as 75%. The insulating properties, low thermal expansion/contraction characteristics, and good dimensional stability of LCPs have caused manufacturers to designate the polymers as the material of the future for encapsulating electronic circuitry. The same properties are also responsible for the intense interest in LCPs for printed circuit boards.

15.6.4 Composites

Composite manufacturers point to LCPs as a matrix material for reinforced composite parts. The rigidity and high temperature capability, combined with a high affinity for a variety of fillers, make LCPs an ideal choice for carbon fiber-reinforced composites. This market is expected to mushroom as the technology of composites fabrication advances.

15.7 ADVANTAGES/DISADVANTAGES

15.7.1 Elevated Temperature Properties

The greatest advantage LCPs can offer is elevated temperature performance. LCPs have HDT values as high as 671°F (355°C), higher than any other thermoplastic. Additionally, the retention of tensile strength and modulus at high temperatures (450–575°F [232–302°C]) of Type III LCPs is unsurpassed by any thermoplastic, often making these compounds the only thermoplastic option.

15.7.2 Flammability

LCP resins are inherently flame retardant; They require no additives to achieve UL 94 V-0 and 5V ratings. They also have a limiting oxygen index (LOI) of 45–50, which is among the highest of any engineering thermoplastic. Type III LCPs are the only thermoplastic to pass the FAA's 15-min, 2000°F (1095°C) flame burn-through test, and thus are the only plastic rated as "fireproof" for use on aircraft.

15.7.3 Chemical Resistance

Because of their aromatic structure, LCPs are resistant to most chemical environments. They can be exposed to the chemical agents typically used in industrial and electronics processing without any deleterious effect. Similarly, LCPs will retain their properties when exposed to household chemicals and foodstuffs, and they will not stain easily.

15.7.4 Dimensional Stability

The low thermal expansion/contraction coefficients of LCPs prevent thermal mismatch with mating metal and ceramic components. Elevated temperature warpage of LCPs is also negligible, assuring sound mating during high temperature operation.

15.7.5 Processing

The low viscosity of LCP resins permit the easy filling of very thin-walled, intricate parts. Parts with wall sections as thin as 0.008 in. (0.2 mm) have been successfully molded in Type III LCPs. The shear sensitive nature of the polymer causes it to set up extremely rapidly in the mold, resulting in very short molding cycles. The reduction in cycle time can be as much as 75% over other engineering resins, thus reducing labor costs significantly.

15.7.6 Brittleness

LCPs are characterized by low elongation (1–3%) and low Izod impact values [1–4 ft-lb/in. (53–214 J/m)]. As a result, applications requiring a ductile material, such as those with snap-fit assemblies, should be carefully analyzed. Applications subjected to heavy impact loads should also be avoided.

15.7.7 Weld Lines

LCPs are very shear sensitive, so the melt front will solidify quickly in the mold during processing. This rapid solidification often results in low knit-line strength. Large knit lines should be avoided or placed in low-stress areas whenever possible.

15.7.8 Blistering

When LCP resins are processed at temperatures above recommended levels, polymeric degradation will occur. Parts processed under these conditions will be prone to blistering on exposure to high temperatures or other high energy sources, such as infrared soldering. For this reason the melt temperatures of LCPs must must closely monitored to avoid degradation and subsequent blistering.

15.8 PROCESSING TECHNIQUES

See the Master Material Outline and Standard Material Design Chart.

Master Material Outline

PROCESSABILITY OF ___Liquid crystal polymer___

PROCESS	A	B	C
INJECTION MOLDING	X		
EXTRUSION	a		
THERMOFORMING	b		
FOAM MOLDING	b		
DIP, SLUSH MOLDING			X
ROTATIONAL MOLDING			X
POWDER, FLUIDIZED-BED COATING	X		
REINFORCED THERMOSET MOLDING			
COMPRESSION/TRANSFER MOLDING			X
REACTION INJECTION MOLDING (RIM)			X
MECHANICAL FORMING			X
CASTING			X

A = Common processing technique.
B = Technique possible with difficulty.
C = Used only in special circumstances, if ever.
a To be available.
b Under investigation

Standard design chart for Liquid crystal polymer

1. Recommended wall thickness/length of flow
   Minimum ___0.040 inch___ for ___4 inches___ distance
   Maximum ___NA___ for ___NA___ distance
   Ideal ___0.125 inch___ for ___6 inches___ distance

2. Allowable wall thickness variation, % of nominal wall ___25___

3. Radius requirements
   Outside: Minimum _____ 25% w _____ Maximum _____ none _____
   Inside: Minimum _____ 25% w _____ Maximum _____ none _____

4. Reinforcing ribs
   Maximum thickness _____ 70% w _____
   Maximum height _____ 5 times wall thickness _____
   Sink marks: Yes _____ No __X__

5. Solid pegs and bosses
   Maximum thickness _equal to wall thickness_
   Maximum weight _5 times wall thickness_
   Sink marks: Yes _____ No __X__

6. Strippable undercuts
   Outside: Yes _____ No __X__
   Inside: Yes _____ No __X__

7. Hollow bosses: Yes __X__ No _____

8. Draft angles
   Outside: Minimum _____ 1° _____ Ideal _____ 5° _____
   Inside: Minimum _____ 1° _____ Ideal _____ 5° _____

9. Molded-in inserts: Yes __X__
                        No _____

10. Size limitations
    Length: Minimum __TBD__ Maximum __TBD__ Process __Injection molding__
    Width: Minimum __TBD__ Maximum __TBD__ Process __Injection molding__

11. Tolerances

12. Special process- and materials-related design details
    Blow-up ratio (blow molding) _____
    Core L/D (injection and compression molding) _____
    Depth of draw (thermoforming) _____
    Other _____

15.9 RESIN FORMS

Most LCP resins are produced and sold as filled, pelletized compounds. Fillers vary widely in composition and volume percentage, but most typically are glass, mineral, carbon, or some combination of these. The fillers are used to improve properties, ease processing, and reduce anisotropy. Some unfilled grades are available in pellet and powder form; these generally provide the base resin for specialized compounding or the matrix for reinforced composites.

15.10 SPECIFICATION OF PROPERTIES

See the enclosed Master Outline of Material Properties

MATERIAL _Liquid crystal polymer_

PROPERTY	TEST METHOD	ENGLISH UNITS	VALUE	METRIC UNITS	VALUE
MECHANICAL DENSITY	D792	lb/ft^3	106	g/cm^3	1.70
TENSILE STRENGTH	D638	psi	13,600	MPa	94
TENSILE MODULUS	D638	psi	1,870,000	MPa	12,894
FLEXURAL MODULUS	D790	psi	1,320,000	MPa	9101
ELONGATION TO BREAK	D638	%	1.8	%	1.8
NOTCHED IZOD AT ROOM TEMP	D256	ft-lb/in.	1.6	J/m	85.4
HARDNESS	D785		Rockwell R: 79		Rockwell R: 79
THERMAL DEFLECTION T @ 264 psi	D648	°F	606	°C	319
DEFLECTION T @ 66 psi	D648	°F		°C	
VICAT SOFTENING POINT	D1525	°F	658	°C	348
UL TEMP INDEX	UL746B	°F		°C	
UL FLAMMABILITY CODE RATING	UL94			UL 94	
LINEAR COEFFICIENT THERMAL EXPANSION	D696	in./in./°F	0.83×10^{-5}	mm/mm/°C	1.5×10^{-5}
ENVIRONMENTAL WATER ABSORPTION 24 HOUR	D570	%	<0.1	%	<0.1
CLARITY	D1003	% TRANSMISSION	Opaque	% TRANSMISSION	Opaque
OUTDOOR WEATHERING	D1435	%	NA	%	NA
FDA APPROVAL					
CHEMICAL RESISTANCE TO: WEAK ACID	D543				
STRONG ACID	D543				
WEAK ALKALI	D543				
STRONG ALKALI	D543				
LOW MOLECULAR WEIGHT SOLVENTS	D543				
ALCOHOLS	D543				
ELECTRICAL DIELECTRIC STRENGTH	D149	V/mil	510	kV/mm	20
DIELECTRIC CONSTANT	D150	10^6 Hertz	3.94		3.94
POWER FACTOR	D150	10^6 Hertz	0.031		0.031
OTHER MELTING POINT	D3418	°F	770	°C	410
GLASS TRANSITION TEMP.	D3418	°F		°C	

15.11 PROCESSING REQUIREMENTS

15.11.1 Material Drying

LCPs do not absorb moisture; however, pellets can collect moisture on their surface that can hamper the processing and subsequent performance of the material. For this reason, pellet drying is strongly recommended. Drying pellets for 8 hours at 300°F (149°C) in a hopper drier with a desiccant bed is sufficient for optimum processing.

15.11.2 Melt Profile

The processing melt profile depends on the class (Type I, II, or III) of LCP being molded. The ultimate melt temperature for Type III materials can reach 800°F (422°C) and above. The shear sensitivity and resulting low viscosity of LCPs permits very thin-walled parts to fill easily. Spiral flow testing indicates that the viscosity of the material rises sharply as the melt temperature is lowered. The upper end of the temperature range must be carefully controlled for Type III materials to avoid material degradation. As a result, tight temperature controls are required to ensure top quality parts.

15.11.3 Mold Temperature

The mold temperature does not significantly affect material flow, so thin-walled parts can still be easily filled with LCP in a cold [200°F (93°C)] and lower temperature mold. Despite this, mold temperature can have a great influence on the properties of a part molded in LCP. A low-temperature mold will result in a part with a thicker skin, higher strength and rigidity, and less warpage. High temperature [350–450°F (177–232°C)] molds will improve a part's color uniformity, gloss, and resistance to blistering under elevated temperature exposure. Consequently, the appropriate mold temperature should be chosen according to the application's end-use requirements.

15.11.4 Screw and Barrel Design

LCPs can generally be molded using standard injection molding screws. Best results are obtained with screws that have 2–3:1 compression and 20–24:1 L/D ratios. Because some glass and mineral fillers contained in LCPs are abrasive, screw flights and barrel liners should be hardened for maximum service life. Worn screws and barrels can severely hinder processing by providing insufficient shearing action to melt the polymer; therefore, well-maintained equipment is crucial to successful processing.

Excessive barrel residence time can degrade LCPs and inhibit part performance. Choosing a machine with a 30% or greater shot-to-barrel capacity ratio is recommended to prevent degradation.

15.11.5 Nozzle Design

Straight or reverse-taper nozzles can be used with LCP resins. The best results are achieved with a fully tapered nozzle with a minimum L/D ratio. An independently controlled heater band positioned fully forward to prevent material from freezing in the nozzle is also required.

The nozzle area should include a nonreturn valve; sliding check rings are strongly recommended, and ball checks are discouraged.

15.11.6 Mold Design

LCP molds are constructed of H-13, S-7, and P-20 steels; softer steels and aluminum are generally not recommended for production quality molds because of the abrasive fillers contained in the resins. Mold temperature control can be best maintained by insulating the mold from the mold platens with mica board or some equivalent insulating material. Molds can be heated with either electric cartridge heaters or hot oil lines. Oil heat is preferred for more precise temperature control and the ability to remove as well as add heat to the mold. This is dangerous if done incorrectly as a break in a hot oil line under pressure can cause severe burns and loss of sight.

Standard sprues with full round or trapezoidal runners are suggested for use with LCPs. The sprue, runner, or part should include a cold slug well to capture any material that may have frozen at the nozzle prior to injection. LCPs require no special gating configuration; any gate type commonly used with thermoplastics will work with LCPs. Gate diameters should measure 50% of the part's wall section to eliminate jetting.

15.11.7 Regrind Usage

LCPs can be reground and mixed with virgin material without a noticeable loss in physical properties. The regrind level should not exceed 25% and should be kept constant for best results. Discolored or blistered parts should not be reprocessed.

15.12 PROCESSING SENSITIVE VARIABLES

15.12.1 Mold Temperature vs Properties

LCPs possesses a distinct skin and core, which differ in crystallinity and orientation. The more highly ordered crystalline skin is stronger and more rigid than the internal core; therefore, the properties of a part depend to some degree on the relative thicknesses of its skin and core. A thicker skin results in a stronger, stiffer, more anisotropic part.

The thickness of the skin is affected by the mold temperature. A colder mold surface will produce a thicker skin and correspondingly higher physical properties in a given part.

15.12.2 Mold Temperature vs Appearance

Surface finish is drastically affected by mold temperature. For a high-gloss appearance part, such as a baking/serving casserole dish, the quality of the surface finish is critical to the part's acceptance. These parts must be molded in a hot (350–450°F [177–232°C]) mold in order to produce the required uniformity of color and gloss.

15.13 MOLD SHRINKAGE

See the Mold Shrinkage MPO.

Mold Shrinkage Characteristics (in./in. or mm/mm)

MATERIAL   Liquid crystal polymer

IN THE DIRECTION OF FLOW

THICKNESS

MATERIAL	mm: 1.5 / in.: 1/16	3 / 1/8	6 / 1/4	8 / 1/2	FROM	TO	#
40% mineral-filled		0					
50% glass-filled		0					
30% glass-reinforced		0					

	PERPENDICULAR TO THE DIRECTION OF FLOW					
40% mineral-filled	0.005–0.008					
50% glass-filled	0.005–0.008					
30% glass-reinforced	0.005–0.008					

NOTE: Mold shrinkage is approximate. It is affected by part design, mold temperature, thickness, injection pressure, packing time, cycle time, orientation, gate design, gate size, gate location, glass content, glass size, and filler content.

15.14 TRADE NAMES

LCPs are currently available under the following trade names:
Xydar resins, Amoco Performance Products, Richfield, Conn., and Vectra resins, Hoechst Celanese, Chatham, N.J.

BIBLIOGRAPHY

1. J. Duska, "Liquid Crystal Polymers: How They Process and Why," *Plast. Eng.* (Dec. 1986).

General References

M. Bailey, "Liquid Crystal Polymers," *Modern Plastics Encyclopedia,* McGraw-Hill, New York, Oct. 1986, 64, 10A.

M. Bailey and T. Brasel, "Liquid Crystal Polymers: An Introduction to Properties/Applications," *Proceedings of Regional Technical Conference,* Society of Plastics Engineers, Rochester, N.Y., Nov. 1987.

G. Calundann and M. Jaffee, "Anisotropic Polymers: Their Synthesis and Properties," *Proceedings of the 36th Robert A. Welch Conference on Polymer Research*, 1982.

S.G. Cottis, "Aromatic Polyesters as Engineering Plastics," *Mod. Plast.* **52** (1975).

S.G. Cottis and B.E. Nowak, "Aromatic Copolyester," *Modern Plastics Encyclopedia*, Vol. 10A, McGraw-Hill, New York, Oct. 1976, p. 53.

J. Economy and S.G. Cottis, "Hydroxybenzoic Acid Polymers." in N.M. Bikales, ed., *Encyclopedia of Polymer Science and Technology*, Vol. 15, Wiley-Interscience, New York, 1971, pp. 292–306.

J. Economy and S.G. Cottis, "Linear Aromatic Polyester," *Modern Plastics Encyclopedia*, Vol. 10A, McGraw-Hill, New York, Oct. 1974, p. 48.

J. Economy, B.E. Nowak, and S.G. Cottis. "Ekonol: A High Temperature Aromatic Polyester," *SAMPE J.* (Aug./Sept. 1970).

J. Economy and co-workers, "Synthesis and Structure of the *p*-Hydroxybenzoic Acid Polymer," *J. Polym. Sci.* **14** (1976).

P. Frayer, "High Temperature Liquid Crystalline Composites," *Proceedings of Composites '86, NRCC/ IMRI Symposium*, Nov. 1986.

M. Friedman, "Liquid Crystal Polymers," *Plastic Design Forum* (Jan./Feb. 1986).

F. Jaarsma, J. Chen, and D. Conley, "A New Chemical Resistant Liquid Crystal Polymer," *Proceedings of Plastics Symposium, Managing Corrosion with Plastics*, 1985.

A Tolmie, "Xydar Liquid Crystal Polymers: The Thermoplastic Solution for Innovative Connector Designs," *Proceedings of the 20th Connectors and Interconnections Symposium*, Oct. 1987.

U.S. Pat. 3,637,595 (Jan. 25, 1972) S.G. Cottis, J. Economy, and B.E. Nowak.

U.S. Pat. 3,662,052 (May 9, 1972) B.E. Nowak, J. Economy, and S.G. Cottis.

U.S. Pat. 3,772,250 (Nov. 13, 1973) J. Economy, S.G. Cottis, and B.E. Nowak.

U.S. Pat. 3,816,417 (June 11, 1974) J. Economy, S.G. Cottis, and B.E. Nowak.

U.S. Pat. 3,857,814 (Dec. 31, 1974) J. Economy, B.E. Nowak, and S.G. Cottis.

U.S. Pat. 3,884,876 (May 20, 1975) S.G. Cottis, J. Economy, and A.A. Wosilait.

U.S. Pat. 3,962,314 (Aug. 8, 1976) J. Economy, S.G. Cottis, and B.E. Nowak.

U.S. Pat. 3,974,250 (Aug. 10, 1976) S.G. Cottis, J. Economy, R.S. Storm, and L.C. Wohrer.

U.S. Pat. 3,975,487 (Aug. 10, 1976) S.G. Cottis, J. Economy, and L.C. Wohrer.

U.S. Pat. 3,980,749 (Sept. 14, 1976) S.G. Cottis, J. Economy, and A.A. Wosilait.

16

NYLON

Paul O. Damm and Paul Matthies

BASF Corp., Plastic Materials, 1609 Biddle Avenue, Wyandotte, MI 48192

16.1 NAME

Polyamide is the technical name given to:

Nylon 6; polycaprolactam; —NH(CH$_2$)$_5$CO—

Nylon 6,6; poly(hexamethyleneadipamide); —NH(CH$_2$)$_6$NHCO(CH$_2$)$_4$CO—

Nylon 6,6/6, Nylon 6/6,6; copolyamides of hexamethylenediamine, adipic acid, and caprolactam

Nylon 6,10; poly(hexamethylenesebacamide); —NH(CH$_2$)$_6$NHCO(CH$_2$)$_8$CO—

Nylon 6,12; poly(hexamethylenedodecanamide); —NH(CH$_2$)$_6$NHCO(CH$_2$)$_{10}$CO—

Nylon 11; poly(11-aminoundecanoic acid); —NH(CH$_2$)$_{10}$CO—

Nylon 12; poly(12-aminododecanoic acid); —NH(CH$_2$)$_{11}$CO—

16.2 CATEGORY

All nylons are thermoplastics. Of them, nylon 6,6 possesses the greatest hardness and rigidity and the highest resistance to abrasion and heat deformation.

With respect to toughness, nylon 6,6 is the least tough; nylon 12, the most tough. The order of progression is nylon 6,6, nylon 6,6/6, nylon 6/6,6, nylon 6, nylon 6,10, nylon 11, and nylon 12. The heat distortion and hardness decrease in the same manner.

With respect to water absorption at 73°F (23°C) and 50% relative humidity, nylon 6, nylon 6/6,6, nylon 6,6/6, and nylon 6,6 fall in the range of 3 ± 0.4%. Nylon 6,10 and 6,12 absorb under the same conditions 1.4 ± 0.2%. The water absorption of nylon 11 and nylon 12 is even lower.

16.3 HISTORY[1]

16.3.1 Discovery

Nylon 6. In 1907, J. von Braun at Gottingen University in Germany discovered that the self-condensation of ε-aminocaproic acid led to a caprolactam polymer, probably of low molecular weight, as it appears today. In 1938, P. Schlack at I.G. Farbenindustrie in Berlin polymerized caprolactam to high-molecular-weight nylon 6.

Nylon 6,6. In 1928, W.H. Carothers at Du Pont began systematic studies on condensation polymers, including nylon-type polyamides. In 1935, he first made nylon 6,6 by polycondensation of hexamethylene diamine with adipic acid.

Higher Nylons. In patent applications submitted in 1935, Carothers reported the laboratory preparation of nylon 11 from 11-aminoundecanoic acid and claimed nylon 6,10, to be made from hexamethylene diamine and sebacic acid. The polymerization of ω-dodecalactam to nylon 12 was reported in 1960 by P. Lafont at Rhone-Poulenc in France.

16.3.2 Commercialization

Nylon 6,6, the first polyamide to appear on the market, was introduced in 1938–1939 by Du Pont. The first product sold was a brush bristle. Ladies' stockings followed one year later. Nylon 6,6 molding powder was first offered for sale in the United States in 1941. Early applications were textile machinery gears and electrical coil forms. The I.G. Farbenindustrie in Germany offered nylon 6,6 in limited quantities as early as 1939.

Nylon 6 was commercialized in the period 1929–1940 by I.G. Farben. The first applications included melt-spun bristles and textile filaments; melt-cast tapes used as transmission belts; and injection-molded bushings, bearings, and gears. In the United States production of nylon 6 was taken up in 1955 by Allied Chemical.

Nylon 11 was first made in 1949 on a pilot-plant scale by Organico in France. Full-scale production followed in 1955.

Nylon 12 was introduced in 1966 by Chemische Werke Huls in Germany, in cooperation with Emser Werke in Switzerland.

Nylon 6,12 was commercialized in 1970 by Du Pont for use in such applications as molded insulators and automotive parts.

16.3.3 Significant Modifications

Resin	Year	Producer
Copolyamides 6/66	1939	BASF
Nucleated nylon 6	1959	Bayer, BASF
Glass-fiber-reinforced nylon 6,6	1960	Fiberfil, ICI
Heat stabilized nylon	1954	Du Pont AP 2705227
Nylon 6 by monomer casting	1960	BASF, Bayer, Monsanto
Impact-modified nylon	1965	Du Pont
Flame-retardant nylon	1968	BASF (phosphorus)
Mineral-reinforced nylon	1972	Monsanto
Super tough nylon 6,6	1976	Du Pont
Flexible polyether/polyamide	1981	ATO Chimie
Nylon RIM process	1983	Monsanto

16.3.4 U.S. Sales in Million Pounds Per Year

Year	Total Plastics	Nylon	Total Plastics, %
1960	5550	ca 25	ca 0.45
1970	18,500	101	0.55
1980	35,600	270	0.75

16.4 POLYMERIZATION

16.4.1 General[2]

The thermoplastic nylons are produced by polymerization (or polycondensation) in the molten state. The respective monomers are heated in the presence of various quantities of water at 392–572°F (200–300°C) at normal or elevated pressure depending on the case. The final degree of polymerization is determined by the polycondensation equilibrium; excess water and water formed in the reaction must be evaporated during the process in a controlled manner. At the end, the molten polymer is extruded into strands that are cooled and cut to chips. In some cases, after-condensation in the solid state follows.

16.4.2 Nylon 6[3]

Nylon 6 is formed by ring-opening polymerization (either hydrolytic or anionic) or ε-caprolactam, represented by the formal equation:

$$n\ (CH_2)_5 \quad \longrightarrow \quad -NH\!-\!\left[(CH_2)_5-C\right]_n$$

Raw materials for caprolactam are cyclohexane, toluene, and phenol. The hydrolytic polymerization actually comprises three reactions: hydrolysis of lactam with small amounts of water, polyaddition of lactam molecules to the amino end-groups of the growing chain, and polycondensation. The polymerization leads to an equilibrium with approximately 10% unreacted monomer.

Nylon 6 is manufactured in a three-step, continuous process. First, caprolactam is polymerized at 464–536°F (240–280°C); then, the residual monomer is extracted from the chips with water at 212°F (100°C); lastly, the material is dried at 212–302°F (100–150°C) with nitrogen or under vacuum.

In an alternative process, the hot-water extraction step is replaced by vacuum demonomerization of the polymer melt.

Anionic polymerization of caprolactam is carried out under anhydrous conditions using strong bases as catalysts and acrylating agents as activators. The process is performed below the melting point of the polymer [428°F (220°C)] using so-called monomer casting.

16.4.3 Nylon 12[4]

Nylon 12 is obtained by hydrolytic polymerization of ω-dodecalactam (whose feedstock is butadiene). In contrast to nylon 6, higher polymerization temperatures (about 572°F [300°C]) are required, residual monomer concentration is low, and the extraction step is eliminated.

16.4.4 Nylon 6,6, 6,9, 6,10 and 6,12[5]

Nylon 6,6 is formed by polycondensation of hexamethylene diamine with adipic acid:

$$n \; \underset{\substack{| \\ H}}{\overset{\substack{H \\ |}}{N}}-(CH_2)_6-\underset{\substack{| \\ H}}{\overset{\substack{H \\ |}}{N}} \; + \; n \; HO-\overset{\substack{O \\ ||}}{C}-(CH_2)_4-\overset{\substack{O \\ ||}}{C}-OH \; ---\rightarrow$$

hexamethylene diamine   adipic acid

$$H-\left[\underset{\substack{| \\ H}}{\overset{\substack{H \\ |}}{N}}-(CH_2)C_6-\overset{\substack{H \\ |}}{N}-\overset{\substack{O \\ ||}}{C}-(CH_2)_4-\overset{\substack{O \\ ||}}{C} \right]_n OH \; + \; (2n\text{-}1) \; H_2O$$

Cyclohexane and phenol are the raw materials for adipic acid; adipic acid, butadiene, and acrylonitrile are the raw materials for hexamethylene diamine. The starting material for the manufacture of nylon 6,6 is a 40–60% aqueous solution of hexamethylenediammonium adipate (so-called nylon salt). The polycondensation process involves the following steps: concentrating the nylon salt solution to 75–80%, heating it to reaction temperatures above 392°F (200°C) at autogeneous pressure, evaporating water with a final pressure letdown, and completing the polycondensation at 527–536°F (275–280°C). Discontinuous (autoclave) and continuous processes are used.

Nylon 6,9, 6,10, and 6,12 are obtained in an essentially analogous manner, mostly by a discontinuous process. The dicarboxylic acids required are azelaic acid (from oleic acid), secacic acid (from castor oil), and dodecanedioic acid (from butadiene), respectively.

16.4.5 Nylon 11[6]

Nylon 11 is manufactured by self-condensation of 11-aminoundecanoic acid (whose feedstock is castor oil).

In 1988, the list price of nylon 6 ranged from $1.59 to $1.88 per pound.

16.5 DESCRIPTION OF PROPERTIES

16.5.1 Thermal Properties

The nylon resins are characterized by a sharp melting point at high temperatures:

Nylon	Temperature	
	°F	°C
Nylon 6	428	220
Nylon 66	500	260
Nylon 6,6/6	470	243
Nylon 6,10	419	215

Even at elevated temperatures just short of the melting range, nylon retains its shape because of its partially crystalline structure and the strong cohesive forces exerted between molecules by the hydrogen bonds. Compared to other polymers, mainly amorphous thermoplastics that do have a softening range, the processing temperatures always have to be above the melting point. This melting behavior requires special consideration in the design of hot runner molds.

Nylon is a crystalline thermoplastic with a low thermal coefficient of linear expansion. The reinforced resins, in particular, undergo very little dimensional change when the temperature changes. However, the linear expansion of the glass-reinforced resins depends on the orientation of the glass fiber.

The heat-distortion or heat-deflection temperature (in accordance with ASTM D648) for dry nylon resins is as follows:

	264 psi (1.8 MPa)		66 psi (0.5 MPa)	
	°F	°C	°F	°C
Nylon 6	130–195	54–91	320–356	160–180
Nylon 6,6	212–220	100–104	390	199
Nylon 6,10	150–185	66–85	383	195

Nucleated products show a higher heat-distortion because of their higher crystallinity. This effect is more significant with nylon 6. By reinforcing nylon with glass fiber or minerals, the heat-distortion temperature also increases. Some examples are given as follows:

	264 psi (1.8 MPa)		66 psi (0.5 MPa)	
	°F	°C	°F	°C
Nylon 6, 15% GF	374	190	419	215
25% GF	410	210	428	220
35% GF	419	215	428	220
50% GF	419	215	428	220
25% mineral	392	200	419	215
30% mineral	302	150	392	200
Nylon 6,6, 15% GF	482	250	482	250
25% GF	482	250	482	250
35% GF	482	250	482	250
50% GF	482	250	482	250
20% CF	473	245	482	250
Nylon 6,6/6, 30% GF	437	225	482	250
30% mineral	248	120	410	210
40% mineral	293	145	428	220

The performance of nylon parts exposed to heat depends on the specific thermal properties of the resin concerned, the duration of exposure, the nature of the heat source, and the mechanical load imposed as well as the design of the molding. Consequently, the resistance to heat-deformation displayed by nylon parts cannot be estimated directly from the temperature values obtained in various standardized tests, no matter how valuable they may be as a guide or as a means of comparison.

In order to obtain good resistance to high-temperature aging, it is necessary to stabilize nylon. The performance of the various nylon resins on exposure to sustained high temperatures can be compared by graphic temperature profiles. For a given time, they show the temperature in degrees Farenheit at which the tensile strength drops to 50% of the initial value. The time scale goes from 100 to 20,000 hours. In this regard, it is important to keep in mind that the operating time of a car is in the range of 3000 to 5000 hours.

For a few hours of operation, the maximum service temperature for the various nylon resins is listed below.

Nylon	°F	°C
Nylon 6	355	180
impact modified	320	160
glass-fiber reinforced (15–50%)	390	199
mineral filled (30%)	355	180
Nylon 6,6	390	199
glass-fiber reinforced (15–50%)	465	241
glass-fiber reinforced, FR equipped (25–35%)	430	221
Nylon 6,6/6 mineral filled (30–40%)	355	180
glass-fiber reinforced (30%)	410	210
Nylon 6,10	355	180

16.5.2 Mechanical Properties

Nylon resins are tough, hard thermoplastics. Because of their outstanding properties, they have become indispensable in almost all branches of engineering. The mechanical properties most pertinent to nylon resins are

Mechanical Property	ASTM Test Method
Tensile strength	D638
Elongation	D638
Flexural modulus	D790
Izod impact strength	D256
Hardness, Rockwell	D785

Typical values for the mechanical properties of three general-purpose injection molding grade resins are presented in Table 16.1.

Nylon 6,6 has the greatest hardness and stiffness. It is the best material for parts that must withstand high mechanical loads.

Nylon 6 is tough and hard. It is the best material for parts that must be shock-proof, even at sub-zero temperatures. Nylon 6,10 resin is comparable to nylon 6, but since it absorbs very little moisture, it is more dimensionally stable. It is the best material for precision engineering parts.

TABLE 16.1 Mechanical Properties of Three General-Purpose Injection Molding Grade Resins

Nylon Type	Tensile Strength, psi (MPa)	Elongation, %	Flexural Modulus, psi (MPa)	Izod Impact Strength, ft-lb/in. (J/m)	Rockwell Hardness
6	11,600 (80)	50–100	350,000 (2,410)	1.0 (53)	M86
6/6	11,600 (80)	50	400,000 (2,760)	1.0 (5.3)	M90
6/10	10,100 (70)	70	300,000 (2,070)	1.0 (53)	M75

The most common reinforcing fillers for nylon resins are glass fibers and certain minerals (such as talc). When reinforcing fillers are added to nylon, the resulting toughness decreases with a significant increase in stiffness, impact, tensile strength, and hardness. An example of the effect of reinforcing fillers on a general-purpose nylon 6 is presented in Table 16.2.

Mineral-filled nylons are used to produce parts with enhanced stiffness, good dimensional stability, and a low tendency to warp.

Nylon resins also come with impact modifiers, carbon fibers, glass beads, and flame retardants. All modifiers significantly change one or more of the mechanical proeprties of the nylon resin.

16.5.3 Optical Properties

Injection molded nylon parts have a creamy, off-white, smooth appearance. Most nylon parts are opaque when the thickness is greater than 0.1 in. (2.5 mm) and transparent when the thickness is less than 0.02 in. (0.5 mm). Depending on the molding conditions and mold surface, nylon parts will generally have average to high gloss. Cast film made from nylon 6 resin is used widely because of its clarity and gloss. Typical values are

Optical Property	ASTM Method	Value
Gloss	D2457 (20 deg)	70–100
Haze	D1003	1.0–4.0%

16.5.4 Environmental Properties

Nylons are inherently resistant to lubricants, engine fuels, hydraulic fluids, coolants, refrigerants, paint solvents, cleaners, and aliphatic and aromatic hydrocarbons. They are also resistant to aqueous solutions of many inorganic chemicals (such as salts and alkalies). Nylons are gradually attacked over time by hot, oxygenated, and hot water.

Nylons are attacked by strong acids, phenols, cresols, some heavy metal salt solutions, certain oxidizing agents, and a few chlorinated hydrocarbons at elevated temperatures.

The weatherability (UV resistance) of nylons is relatively poor unless a suitable stabilizer is incorporated. When they are exposed to the outdoors, unstabilized nylons undergo color change and loss of surface gloss and toughness. For maximum performance in outdoor applications, a stabilized nylon with a high carbon black content is recommended.

Nylon resins begin to decompose at temperatures above 570°F (299°C). Combustible gases are formed at temperatures between 840–930°F (449–499°C) that will continue to burn after ignition. The products of decomposition have an odor similar to burnt horn.

TABLE 16.2 Effect of Reinforcing Fillers on the Mechanical Properties of a Nylon 6 Resin

Reinforcement Type	Reinforcement, %	Tensile Strength, psi (MPa)	Elongation, %	Flexural Strength, psi (MPa)	Izod Impact Strength, ft-lb/in. (J/m)
Glass fiber	15	16,700 (115)	3	710,000 (4,900)	1.1 (59)
Glass fiber	30	26,100 (180)	3	1,100,000 (7,590)	2.8 (150)
Glass fiber	50	31,900 (220)	2	1,700,000 (11,700)	3.6 (192)
Mineral	30	12,320 (85)	10	650,000 (4,480)	1.2 (64)

Various test methods have been adopted to evaluate the flame retardancy of plastics, the most famous being UL 94. General-purpose nylon resins are classified as UL 94 V-2. Modified nylon resins run the whole range of the UL 94 test method. Another important test method is the "Oxygen Index," ASTM D2863. General-purpose nylon resins have a % oxygen range between 24–28.

All nylon resins absorb moisture and water to some extent. One consideration that must be addressed by the part designer is the absorption of water-soluble substances (such as dyes, cleaners, and so on) into the nylon part. Their absorption will lead to discoloration of the part and possible carryover of flavors and aromas.

16.6 APPLICATIONS

The most important fields of application for nylon include electrical engineering (that is, power transmission and telecommunications), automotive engineering, mechanical and chemical engineering in general, materials handling, instrumentation, plumbing and sanitary ware, and packaging. Examples in each category are listed below.

Power Transmission
High-voltage insulation
Parts and housings for contact breakers
Cable binders and ducts
Terminal strips and connectors
Plug-and-socket devices
Coil formers household appliances, such as continuous-flow heaters
Solenoid valves
Housings for electric tools
Telecommunications
Circuits and relays
Drive units
Fittings
Contact makers
Cable sheaths
Automotive Engineering
Parts in the engine block and lubricating system, such as timing gears, chain guides, toothed belt-gear covers, oil sumps, oil-filter housings, and cylinder-head covers
Parts for clutches, such as thrust bearing races
Parts and housings for the cooling and ventilation system, such as water-heater tanks for heat exchangers, expansion tanks, hot-water control valves, thermostat casings, and fans and fan casings
Parts in the fuel supply system, such as housings for fuel filters and fuel reservoirs and lines
Parts in transmission systems, gear changes, and tachometer drives, such as bearing races, gearshift lever housings, selector forks, gearshift lever couplings, drive pinions for speedometers, and thrust rings for transmission systems and gas pedal
Parts for the chassis, such as the steering wheel, steering-column bearing, roller bearing races, and clips

Exterior trim parts, such as spoilers, bumpers (single or in combination), rocker panels, radiator grilles, door outer handles, rear-view mirror housings, and decorative wheel hubs and wheel covers

Parts for electrical equipment, such as cable binders, clamps, and connectors; headlamp casings; and fuse boxes

Mechanical and Chemical Engineering in General

Bearings

Gears and transmission systems

Seals

Housings and flanges

Connectors and bolts

Housings for air-pressure gauges

Outer casings for petrol pumps in filling stations

Materials Handling

Idlers and pulleys

Liners and bushings

Transport containers

Conveyor belts and chains

Instrumentation and Precision Engineering

Control disks and cams

Counters

Links and levers

Print wheels for teleprinters mountings and control levers

Sliding parts

Building and Sanitary Ware

Dowels

Fasteners and cable and pipe clamps

Handles, hardware, and fittings

Fans

Furniture, such as chairs

Packaging

Composites with polyolefins (mainly LDPE) for foodstuffs wrapping

Composite film and monofilms for packing medicinal (sterile) articles

Monofilms for sausage skins and baking

16.7 ADVANTAGES/DISADVANTAGES

The advantages of nylons with respect to mechanical properties include mechanical strength, good toughness at low temperatures, and high thermal stability [glass-reinforced heat-stabilized nylon 6,6, for example, can be exposed for short periods of time (hours) to temperatures of up to 465°F (240°C)]. Their resistance to chemicals is excellent and their dielectric properties are very good.

The chief disadvantage among the nylon grades is water absorption. Water absorption rates for nylon 6 and 6,6 fall within the same range; those for nylon 6,10 and 6,12 are significantly lower.

When water is absorbed, a change in properties takes place. The values for tensile strength, tensile modulus, and hardness will decline. On the other hand, the toughness of nylon significantly increases. Unfortunately, the resistance to creep decreases.

Another effect of water absorption is that the volume of the part expands. Assuming that 1% absorbed water is uniformly distributed, there would be an increase in volume of approximately 0.9%; the corresponding average increase in length would be 0.2–0.3%.

The dimensional increase of the glass-reinforced resins (30%) is less than 0.1% in the direction of flow for each 1% of water absorbed. Therefore, these resins retain their dimensions particularly well under conditions of fluctuating humidity. The same applies to mineral-filled resins. When calculating the dimensions of the part, it is essential to consider both the shrinkage and the water absorption.

Because of the fact that nylon absorbs water, it can be colored by immersion in a water solution of dyes. The negative aspect of this property is that nylon parts can be easily stained.

Nylon grades can be molded and extruded without difficulty. Since the melt viscosity of nylon is very low and the nylons have a sharp melting point, some specifics must be considered.

If the temperature in the feeding area of an injection-molding machine, for example, is too high, the material might heat up too quickly, trapping unmelted granules in the screw. If this happens, the screw normally has to be pulled and the material must be removed.

In such a case as when the nozzle is plugged up and the nylon in the cylinder is already molten and under pressure, care must be taken to avoid accidents resulting from the molten nylon when unplugging the nozzle in any of the standard ways (that is, using a torch or a heated nail).

Blow molding of nylons is very difficult. Because of their low melt viscosity, only highly viscous nylon types can be used. They require a lot of attention in order to achieve the right temperature. Better results can be obtained by using additives that strengthen the parison.

Spin welding, vibration welding, and ultrasonic welding can be employed as assembly techniques. It is important to take into account the sharp melting point of nylon and the sudden change from solid to liquid.

16.8   PROCESSING TECHNIQUES

See the Standard Material Design Chart.

All nylon grades are easy flow materials. The flow depends on the viscosity of the type of nylon as well as the processing parameters. Figure 16.1 shows the influence of the stock temperature on the flow, comparing nylon 6, nylon 6,6, and nylon 6,10 as well as nylon 6 and nylon 6,6 reinforced with 35% glass fiber. Curve 6 shows nylon 6,6 reinforced with 35% glass fiber and FR-equipped with red phosphorus.

The injection speed also influences the flow length. Figure 16.2 shows the relation between coil length and injection speed.

The recommended minimum wall thickness of 0.04 in. (1 mm) is not to be understood as a limit. Parts can be molded with wall thickness areas of 0.012–0.02 in. (0.3–0.5 mm) and a size of 0.4 × 0.4 in. (10 × 10 mm).

As a rule of thumb, the wall thickness of parts should be as uniform as possible. The part always should be gated at the area with the highest wall thickness. The wall thickness should decrease slightly towards the end of flow.

For example, it would be no problem to mold a part with a wall thickness of 0.12 in. (3 mm) at the gate area and 0.06 in. (1.5 mm) at the end of the flow. However, the reverse situation would cause problems of voids, sinks, and distortion through different shrinkage.

Because of the notch sensitivity of nylon, the minimum radius should be 0.04 in. (1 mm). The outside minimum, therefore, should be 1 + W.

The maximum radius is not limited; it depends on the article design.

If the rib has a thickness more than 50% of the wall thickness, sinks cannot be avoided. The radius should not be smaller than 0.02 in. (0.5 mm), and preferably 0.04 in. (1 mm).

Master Material Outline

PROCESSABILITY OF ___Nylon 6; 6,6; 6,10; 11; 12_____

PROCESS	A	B	C
INJECTION MOLDING	X		
EXTRUSION	X		
THERMOFORMING			
FOAM MOLDING			X
DIP, SLUSH MOLDING			
ROTATIONAL MOLDING			X
POWDER, FLUIDIZED-BED COATING			
REINFORCED THERMOSET MOLDING			
COMPRESSION/TRANSFER MOLDING			X
REACTION INJECTION MOLDING (RIM)			X
MECHANICAL FORMING	X		
CASTING			X

A = Common processing technique.
B = Technique possible with difficulty.
C = Used only in special circumstances, if ever.

Standard design chart for _____Nylon_____

1. Recommended wall thickness/length of flow, mm
   Minimum ____1____ for ____250____ distance
   Maximum ____5____ for ____1200____ distance
   Ideal ____2____ for ____600____ distance

2. Allowable wall thickness variation, % of nominal wall _____

3. Radius requirements
   Outside: Minimum ____1 + W____ Maximum _____
   Inside: Minimum ____1____ Maximum _____

4. Reinforcing ribs
   Maximum thickness ____0, 5W____
   Maximum height _____
   Sink marks: Yes _____ No ___X___

5. Solid pegs and bosses
   Maximum thickness ____0, 5W____
   Maximum weight _____
   Sink marks: Yes _____ No ___X___

6. Strippable undercuts
   Outside: Yes ___5–6 for unfilled nylon 6, 6/6; 1–2% for glass-filled nylon 6, 6,6___ No _____

   Inside: Yes ___5–6 for unfilled nylon 6, 6/6; 1–2% for glass-filled nylon 6, 6,6___ No _____

7. Hollow bosses: Yes ___X___ No _____

8. Draft angles
   Outside: Minimum ___1°___ Ideal ___2.5°___
   Inside: Minimum ___1.5°___ Ideal ___3°___

9. Molded-in inserts: Yes ___X___
   No _____

10. Size limitations[a]
    Length: Minimum _____ Maximum _____ Process _____
    Width: Minimum _____ Maximum _____ Process _____

11. Tolerances

12. Special process- and materials-related design details
    Blow-up ratio (blow molding) ___1:3___
    Core L/D (injection and compression molding) _____
    Depth of draw (thermoforming) _____
    Other _____

[a] Chairs have been molded from glass-reinforced nylon 6 (35%); bucket seats have been molded from nylon 6. The largest blow-molded nylon 6 (high viscosity) reservoir has a volume of 35 1.

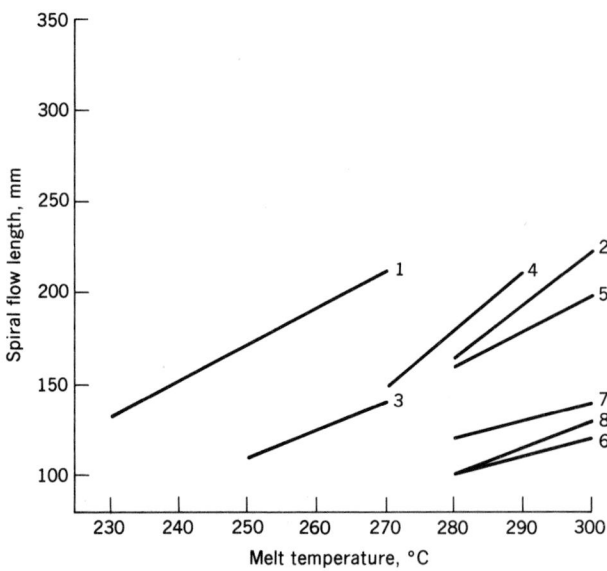

Figure 16.1. Influence of melt temperature on flow capability.

Machine	1300 kN	Curve	Ultramid	Nylon
Screw	30 mm	1	B3S	6, nucleated
Tool	Spiral, 10 mm	2	A3K	6, heat stabilized
	thick	3	S3K	6,10
Cycle time	30 s	4	B3G7	6, 35% glass
Injection Speed	50 mm/s	5	A3G7	6,6, 35% glass
Mold surface temp	176°F (80°C)	6	A3XG7	6,6, 35% glass, + red P
		7	A3X1G7	6,6, 35% glass, flame retardant
		8	KR4601	6,6,6

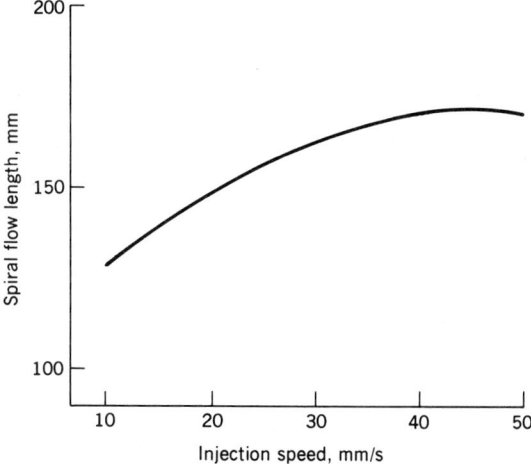

Figure 16.2. Influence of injection speed upon flow

Machine	1300 kN
Screw	30 mm
Tool	Spiral, 10 mm thick
Cycle time	30 s
Injection pressure	1000 bar
Melt temperature	482°F (250°C)
Mold surface temp	176°F (80°C)
Material	Ultramid B3S 6, nucleated

The height of the rib is determined by the mechanical requirements; it is not influenced by sinks. The height depends only on the design.

16.9 RESIN FORMS

Generally, all the nylon resins are supplied in the form of cylindrical or lenticular pellets. For special applications, such as rotational molding or coating, some grades are offered in the form of powder.

Nylon 6 and nylon 6,6 are available as:

- unreinforced
- unreinforced and heat stabilized
- unreinforced and impact modified
- unreinforced and wear friction modified
- unreinforced and flame retardant
- glass-fiber reinforced at 10–50%
- glass-fiber reinforced at 10–50% and heat stabilized
- carbon-fiber reinforced
- impact modified and glass-fiber reinforced
- mineral filled
- glass-fiber reinforced and flame retardant

16.10 SPECIFICATION OF PROPERTIES

See the Master Outlines of Materials Properties.

MATERIAL __General-purpose Nylon 6 for injection molding, heat stabilized__

PROPERTY		TEST METHOD	ENGLISH UNITS	VALUE	METRIC UNITS	VALUE
MECHANICAL DENSITY		D792	lb/ft^3	70.4	g/cm^3	1.13
TENSILE STRENGTH	Dry Moist	D638	psi	11,600 7250	MPa	80 50
TENSILE MODULUS	Dry Moist	D638	psi	435,000 217,500	MPa	3000 1500
FLEXURAL MODULUS	Dry Moist	D790	psi	350,000 140,000	MPa	2414 966
ELONGATION TO BREAK	Dry Moist	D638	%	50–100 200	%	50–100 200
NOTCHED IZOD AT ROOM TEMP	Dry Moist	D256	ft-lb/in.	1.0 1.8	J/m	53 96
HARDNESS		D785		M–86		M–86
THERMAL DEFLECTION T @ 264 psi		D648	°F	131–167	°C	55–75
DEFLECTION T @ 66 psi		D648	°F	>320	°C	>160
VICAT SOFT-ENING POINT		D1525	°F		°C	
UL TEMP INDEX		UL746B	°F		°C	
UL FLAMMABILITY CODE RATING		UL94		94 V–2		94 V–2
LINEAR COEFFICIENT THERMAL EXPANSION		D696	in./in./°F	45 × 10^{-6}	mm/mm/°C	81 × 10^{-6}
ENVIRONMENTAL WATER ABSORPTION AT 73°F, 50% RH, SATURATION			%	3 ± 0.4	%	3 ± 0.4
CLARITY		D1003	% TRANS-MISSION		% TRANS-MISSION	
OUTDOOR WEATHERING		D1435	%		%	
FDA APPROVAL						
CHEMICAL RESISTANCE TO: WEAK ACID		D543		Not resistant		Not resistant
STRONG ACID		D543		Not resistant		Not resistant
WEAK ALKALI		D543		Good resistance		Good resistance
STRONG ALKALI		D543		Good resistance		Good resistance
LOW MOLECULAR WEIGHT SOLVENTS		D543		Good resistance		Good resistance
ALCOHOLS		D543		Good resistance		Good resistance
ELECTRICAL DIELECTRIC STRENGTH	Dry	D149	400 V/mil	2500 1500	16 kV/mm	100 60
DIELECTRIC CONSTANT	Dry Moist	D150	10^6 Hertz	3.5 7.0		3.5 7.0
POWER FACTOR	Dry Moist	Tan 0 D150	10^6 Hertz	0.023 0.3		0.023 0.3
OTHER MELTING POINT		D 2117	°F	428	°C	220
GLASS TRANS-ITION TEMP.		D3418	°F		°C	

Master Outline of Material Properties

MATERIAL ___General-purpose Nylon 6,6___

PROPERTY		TEST METHOD	ENGLISH UNITS	VALUE	METRIC UNITS	VALUE
MECHANICAL DENSITY		D792	lb/ft^3	71.1	g/cm^3	1.14
TENSILE STRENGTH	Dry Moist	D638	psi	13,050 9420	MPa	90 65
TENSILE MODULUS	Dry Moist	D638	psi	493,000 246,500	MPa	3400 1700
FLEXURAL MODULUS	Dry Moist	D790	psi	420,000 320,000	MPa	2896 2206
ELONGATION TO BREAK	Dry Moist	D638	%	20 80	%	20 80
NOTCHED IZOD AT ROOM TEMP	Dry Moist	D256	ft-lb/in.	0.7 1.2	J/m	37 64
HARDNESS		D785		M–90		M–90
THERMAL DEFLECTION T @ 264 psi		D648	°F	220	°C	104
DEFLECTION T @ 66 psi		D648	°F	>392	°C	>200
VICAT SOFT-ENING POINT		D1525	°F		°C	
UL TEMP INDEX		UL746B	°F		°C	
UL FLAMMABILITY CODE RATING		UL94		94 V–2		94 V–2
LINEAR COEFFICIENT THERMAL EXPANSION		D696	in./in./°F	45×10^{-6}	mm/mm/°C	81×10^{-6}
ENVIRONMENTAL WATER ABSORPTION AT 73°F, 50% RH, SATURATION		D570	%	2.8 ± 0.3	%	2.8 ± 0.3
CLARITY		D1003	% TRANS-MISSION		% TRANS-MISSION	
OUTDOOR WEATHERING		D1435	%		%	
FDA APPROVAL						
CHEMICAL RESISTANCE TO: WEAK ACID		D543		Not resistant		Not resistant
STRONG ACID		D543		Not resistant		Not resistant
WEAK ALKALI		D543		Resistant		Resistant
STRONG ALKALI		D543		Resistant		Resistant
LOW MOLECULAR WEIGHT SOLVENTS		D543		Resistant		Resistant
ALCOHOLS		D543		Resistant		Resistant
ELECTRICAL DIELECTRIC STRENGTH	Dry	D149	600 V/mil	3000 2000	24 kV/mm	120 80
DIELECTRIC CONSTANT	Dry Moist	D150	10^6 Hertz	3.2 5.0		3.2 5.0
POWER FACTOR	Dry Moist	Tan^s D150	10^6 Hertz	0.025 0.2		0.025 0.2
OTHER MELTING POINT		D 2117	°F	491	°C	255
GLASS TRANS-ITION TEMP.		D3418	°F		°C	

16.11 PROCESSING CONSIDERATIONS

Under normal processing conditions, melted nylon will not change in viscosity or color. When the melt is exposed to conditions listed below, problems can be expected.

- Resin is too dry or wet
- Poor screw design or worn screw
- Improper temperature settings or faulty instrumentation
- Excessively low or high output rates
- Excessive scrap recycle
- Contamination with other nylon resins or plastics

Nylons are sensitive to moisture during processing. It is important that the processor check with the resin supplier to determine whether the resin is predried or not, and what the shelf-life of the packaged resin is. The water content of the resin should be below 0.1% prior to molding. Each resin supplier should be able to recommend how to dry their particular nylon resin. If the resin comes predried, a hopper dehumidifier (desiccant type) is recommended to keep the resin dry. Overdrying nylon resins can be as detrimental to the part as

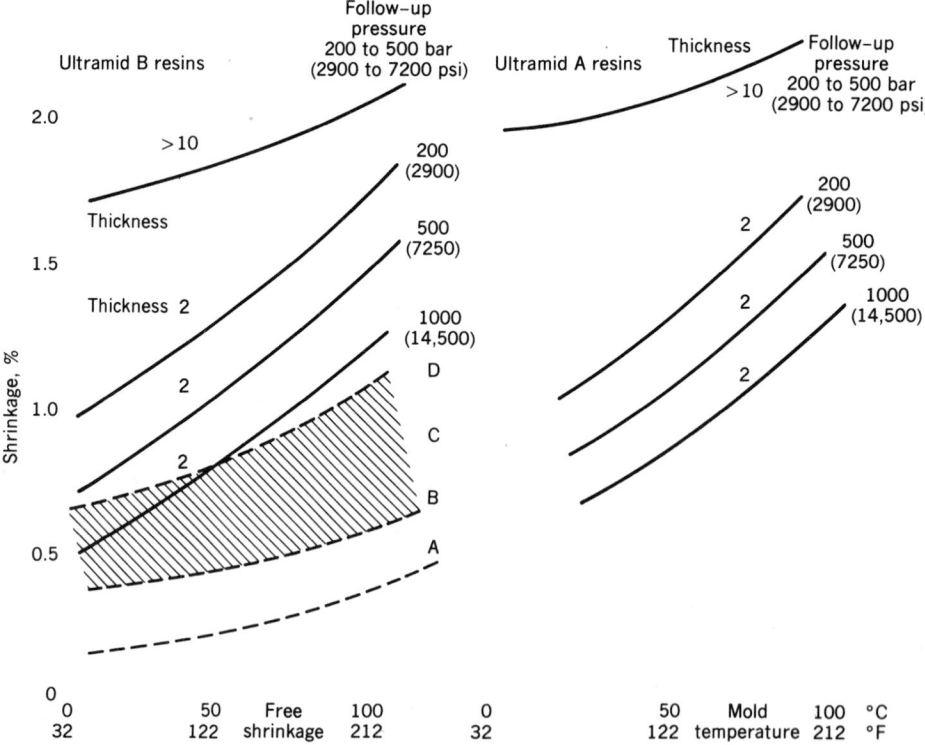

Figure 16.3. Typical shrinkage values for processing nylon 6 and 6,6 grades as a function of mold temperature and pressure. Ultramid B (nylon 6 grades); Ultramid A (nylon 6,6 grades): Ultramid B3G5 (nylon 6 grade, 25% glass): Ultramid B3G7 (nylon 6 grade, 35% glass). For the last two resins curve A is the direction of the flow; curve B has mixed fiber orientation; curve C is the range of values for restricted gates; curve D is at a distance from the sprue and at right angles to the direction of orientation.

processing "wet" resin. If the resin containers have been stored in an unheated warehouse, they should be allowed to stand for 12–24 hours near the processing machine to warm up to room temperature before opening. This will prevent moisture condensation on the resin pellets.

Waste obtained from sprues, rejects, and so on can be ground and recycled, provided that it has not been contaminated (by dirt, oils, or water) or degraded. The proportion that can be added to virgin-prime resin depends on the particular nylon resin in question; normally, it should be less than 10%. If the regrind is not immediately processed its moisture content may exceed the prescribed limit. In such a case it is advised to dry to regrind.

16.12 PROCESSING-SENSITIVE END PROPERTIES

Improper processing of nylons can lead to part distortion (warp), loss of toughness (brittleness), voids and bubbles, loss of gloss or splayed surface, excessive gels in films (fish eyes), and discoloration or poor smell.

16.13 SHRINKAGE

Shrinkage is the difference between the dimensions of the mold and those of the molded part at room temperature. Although primarily a property of the resin, shrinkage also depends on the design, gating, and wall thickness as well as the processing conditions, mold-surface temperature, and holding pressure. The influence of wall thickness, holding pressure, and

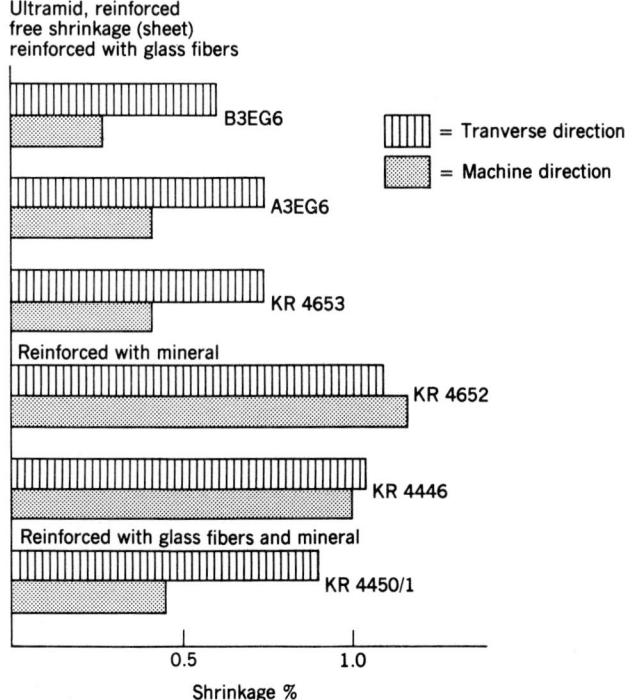

Figure 16.4. Shrinkage of unreinforced nylon sheet at 176°F (80°C) mold temperature, measured after 1 hour at 73°F (23°C).

mold surface temperature is shown for nylon 6 and nylon 6,6 in Figure 16.3, which indicates that the shrinkage increases with the wall thickness. If the wall thickness of a part varies considerably, distortion can result.

The use of some fillers can also influence the shrinkage. How the shrinkage is affected depends on the geometrical shape of the filler. Glass-fiber-reinforced nylons can restrict the shrinkage considerably. The degree of restriction depends on the orientation of the glass fibers. In the flow direction, the restriction is very high. Transverse to the flow direction, the restriction is not as severe. In between these two extremes, it is possible to observe all possible degrees of restriction. The result can be significant distortion problems, depending on the shape of the part and the gate location.

Use of bead-shaped mineral fillers can minimize the difference between the shrinkage in the flow direction and transverse (Fig. 16.4).

16.14 TRADE NAMES

See Table 16.3.

TABLE 16.3 Trade Names and Major Producers of Nylon

Trade Name	Producer	PA-Types
Akulon	AKZO Plastics N.V. NL-6800 Arnhem, Verlperweg 76 Holland	6-, 6,6-, +GF, M
Amilan (Toray)	Toyo Rayon (Toray) Tokyo Japan	6-, 6,6-, 6,10-, 12-, CP(=GF)
Bergamid	Bergmann u. Co. Kunststoffwerk 7560 Gaggenau 2 Federal Republic of Germany	6-, 6,6-, +GF, GS modif. (Regenerate)
Capron	Allied Chemical Corp. Plastics Div. P.O. Box 365 Morristown, NJ	6-, 6-modif., GF
Celanese-Nylon	Celanese Plastics Co. 26 Main Street Chatham, N.J.	6,6-, +GF, +M
Durethan	Bayer AG Leverkusen Federal Republic of Germany	6-, 6,6-, +GF, M, F, CP
Fosta-Nylon	Nycoa, Nylon Corp. of America Manchester, N.H.	6-, CP, modif.
Grilon, Grilamid, Griltex	Emser-Chemie AG Zurich Switzerland	6-, 6,6-, 12-, CP (GF, GS, M)
Kapron	Kunstseidewerk Kapron Klin by Moscow USSR	6-

TABLE 16.3 (*Continued*)

Trade Name	Producer	PA-Types
Latamid	Lati Industria Termo-plastici S.p.A. Italy	6-, 6,6-, 12-, +M and GF
Magnacomp, Nykon, Thermocomp, Thermofil	LNP-Corp. Malvern, Pa.	6- +BA-Ferrit 6,6- +MoS$_2$ 6-, 6,6-, 6,10, 6,12, 6-3-T, 12, +GF, GS, M PTFE, C-Faser
Maranyl	Imperial Chemical Ind. Ltd. Plastics Div. Welwyn Garden City Herts. UK	6-, 6-, 6,10-, CP (GF, GS, M0
Minlon, Zytel	Du Pont de Nemours Wilmington, Del.	6,6- + Fuellstoffe (M)
Nylaglas, Nylode; Plaslube, Xylon	Fiberfil Div., Rexall Chem. Co. Evansville 17, Ind.	6-, 6,6-, 610 (GF)
Nylatron	Polymer Corp. 2120 Fairmont Ave. Reading, Pa.	6,6-, 6-, 6,10- +GF, MoS$_2$
Orgamide	ATO Chemie Tour Aquitaine 92 Paris La Defense France	6-, +GF
Polyloy, Polyamid	Dr. Illing KG Gross-Umstadt Federal Republic of Germany	6-, 6,6-, 11-, 12- (+GF? GS, modif.) partially regenerate
Rilsan	ATO Chemie Tour Aquitaine 92 Paris La Defense France	11-, 12-
Sniamid	Tecnopolimeri S.p.A. Snia-Gruppe Italy	6-, 6,6-, 12- (+GF, GS)
Technyl	Rhone-Poulenc Courbevoie France	6-, 6,6-, 6,10-, CP
Trogamid T	Dynamit Nobel AG, DNAG 5201 Troisdorf Federal Republic of Germany	6-3-T (glass clear and GF)
UBE-Nylon	UBE-Industries Ltd. 7-2, Kasumigaseki 3 Chrome Chiyoda-Ku Tokyo Japan	6-, 12,- CP

TABLE 16.3 (*Continued*)

Trade Name	Producer	PA-Types
Ultramid	BASF AG D-6700 Ludwigshafen/Rhein Federal Republic of Germany (Freeport, Texas)	6-, 6,6-, 6,10, CP
Vestamid	Chemische Werke Huels AG 437 Marl Federal Republic of Germany	12- (GF), 6,12-, PA- Polyether–blockpolymer
Vydyne	Monsanto Plastics & Resins Co. 800 N. Lindbergh Blvd. St. Louis, MO	
Wellamid		

BIBLIOGRAPHY

1. M.I. Kohan, "The History and Development of Nylon-6,6," p. 19; P. Matthies and W.F. Seydl, "The History and Development of Nylon 6," p. 39; G.B. Apgar & M.J. Koskoski, "The History of Development of Nylons 11 and 12," p. 55. R.B. Seymour and G.S. Kirshenbaum, eds., *History of High Performance Polymers,* Elsevier, New York, 1986.

2. M.I. Kohan, ed., *Nylon Plastics,* John Wiley & Sons, Inc. New York, 1973.

3. H.K. Reimschuessel, *J. Polym. Sci. Macromol. Rev.* **12,** 65–139 (1977).

4. W. Griehl and D. Ruestem, *Ind. Eng. Chem.* **62,** 3, 16–22 (1970).

6. D.B. Jacobs and J. Zimmerman, in C.E. Schildknecht, ed., *Polymerization Processes,* Wiley Inter-science New York, 1977, pp. 424–467.

7. M. Genas, *Angew. Chem.* **74,** 535–540 (1962).

17

PHENOLIC RESINS

J.F. Keegan

New Business Development, Occidental Chemical Corp., North Tonawanda, NY 14120

17.1 PHENOLIC

Phenolic resins are, most commonly, the reaction product of phenol and formaldehyde, although substituted phenols and higher aldehydes are used for special applications. A typical repeating monomer unit for a novolak or two-step resin is

hydroxyphenylmethane

For a resole or one-step resin:

trimethyl phenol

17.2 CATEGORY

Phenolics are thermosetting resins, typically hard, stiff (high-modulus), and notch sensitive or brittle.

17.3 HISTORY

Phenolic resins were first investigated by Otto Bayer in the 1880s; he found them to be excellent glues, a function they still serve in many current applications. In 1909, Dr. Leo

Bakeland took advantage of Bayer's work to produce the first commercial phenolic molding compound.

In 1921, Harry H. Dent founded General Plastics, which later became Durez, and developed the first continuous process for making phenolic molding compounds.

Among the first applications for phenolic molding compounds were phonograph records, automotive coil tops, telephone mouthpieces and receivers, and a host of electrical control parts. Phenolic resins were used as coatings for metal cans and castings, wire insulation, paper impregnation, plywood, varnishes, and glues.

As the plastics industry evolved, materials were designed to meet new processing methods. Cold powder automatic presses required "S" granulations or low fines levels. In 1941, with the advent of the electronic preheater, new formulations were called for; by 1950, phenolic television cabinets weighing 40 lb (18 kg) were being produced in two-cavity molds using radio-frequency preheated preforms.

In 1926, with L.E. Shaw's development and patent on transfer molding, longer flow, softer materials were formulated. High impact nodular materials were introduced in the 1940s as means to reduce the bulk factor of these fluffy materials to make them adaptable to automatic molding.

In the early 1960s, the two-stage transfer press using a reciprocating screw was able to plasticize material and place it over a transfer ram for delivery to the mold. Shortly thereafter, the in-line screw injection press for thermosets was developed. It was basically an in-line thermoplastic machine using hot water to control the barrel temperature and employing a screw with no compression zone. These machines eliminated the need to make and preheat preforms. In the years since, injection molding of phenolics has largely replaced transfer molding for noninsert applications, and over 50% of all phenolic molding is now done by the injection method.

In the late 1960s, the screw extruder was developed; it provided a method of plasticizing material to replace preheated preforms for compression and transfer molding. This system was also tied to a rotary press by Alliance Press Co.

The heavy emphasis on injection molding of phenolics has led to the development of innovative molding systems that reduce the scrap-associated sprues and runners.

"Cold manifold" systems, which acted as an extension of the barrel, were developed by Osley & Whitney and Stokes. The plasticized material was maintained in 220–240°F (104–116°C) range, and was fed directly into the part or through tab gates by means of temperature-controlled sprues. Since the material did not set up in the sprues or manifold, it could be run on the next shot.

Live sprue molding incorporated a temperature-controlled sprue bushing which resulted in the saving of the material in the sprue and reduced cycle time.

The runnerless injection-compression system was developed by Durez in 1978. This system eliminated sprues and runners. It combined the advantages of the automatic injection press with the dimensional capabilities of compression molding. In runnerless injection-compression molding the mold is held slightly open, and the material can be injected in 2–3 seconds under very low pressure, as there is no resistance to flow. This corresponds to placing a preheated preform in a compression mold. The mold is then closed by the clamping action of the press, which turns the injection press into a compression press.

The advantages of this system are elimination of sprues and runners, lower compression-type shrinkages, and uniform shrinkage, as there is no flow across the part. Round parts such as automotive thrust washers can be held to tolerance. As the material meets no resistance to flow, the injection time is reduced. Lower pressure is required to close the mold in compression than to keep it closed during injection, so more cavities may be incorporated or a smaller press utilized. Because the mold is open and the material is not subjected to frictional heat, both the stock temperature and the mold temperature may be increased, thus reducing the cure cycle. Another advantage is that the mold is much easier to vent, and staining and burning are eliminated as the mold is open. The reduction in cycle time may reduce capital costs as fewer cavities will give higher production rates, or, again, allow the use of a smaller, less expensive press.

Many thermoset parts are now being handled by robots for finishing and packing. With the proper layout and cycles, one robot can service two presses.

In 1987, phenolic resin sales totaled about 2.8 billion pounds; 64% for plywood and 19% for insulation.

The phenolic molding compound market has held relatively steady: for 1960, 195 million pounds; 1970, 260 million pounds; and 1980, 245 million pounds. Downsizing, the change from electrical to electronic controls, lack of most colors other than black or brown, and inability to regrind scrap have limited growth. On the other hand, the advent of new processing methods that eliminate or greatly reduce scrap and industry needs for higher heat resistance at lower prices have sparked a new growth cycle.

17.4 POLYMERIZATION

Phenolic molding compounds are made from phenolic resin plus fillers, colorants, lubricants, and accelerators, and, in the case of two-stage resins, hexamethylenetetramine (hexa).

Typical molding materials are 40–50% resin by volume and may by either one-stage or two-stage.

17.4.1 Resin Type

The terms "one-step," "one-stage," or "resole" refer to a phenolic resin made by adding more than a one-to-one mol ratio of formaldehyde to phenol in the presence of a basic catalyst. This gives a thermosetting resin that will start to advance or increase in viscosity and molecular weight at room temperature. The phenol:formaldehyde ratio, type of catalyst, and reaction conditions determine the properties of the resin.

Two-stage or two-step resins, or novolaks, are made by reacting less than one mol of formaldehyde with phenol in the presence of an acid catalyst—usually in the 54–84% range. About 90% of all phenolic molding compounds use a two-stage system. A two-step resin is thermoplastic because it does not have enough methylene linkages to set or cure the resin. They are supplied by using hexamethylenetetramine (hexa). Hexa is the reaction product of ammonia and formaldehyde, which breaks down under the heat and pressure of molding and provides methylene groups, —CH$_2$—, which cross-link the resin and transform it into a thermoset. Thus, the term "two-step" refers to the two additions of formaldehyde or methylene groups.

When a two-step material cures it splits off ammonia; when a one-step cures it splits off water. Water is a very large molecule compared to ammonia, which accounts for the differences in dimensional stability between the materials (see Table 17.1).

One-stage materials are used for odor-free applications such as closures, refrigerator controls, hermetically sealed electrical devices, and products that are wet on one side and dry on the other, like pumps and vaporizers. Two-stage materials are used on most other applications. Pure phenolic resins by themselves are not used as molding compounds, as they are brittle, slow curing, high shrink materials. Phenolic resins are excellent glues and are used for laminating wood (64% of the total phenolic resin production is used to make plywood), particleboard, rubber compounding, insulation, shell molding, friction products (brake linings and clutch disks), carbonless copy paper, paper treating, grinding wheels, and a host of other applications, including molding compound.

Phenol is made from cumene, which comes from benzene and propylene. The price of benzene is dependent on oil prices and competition from the upgrading of octane ratings in gasoline. Formaldehyde is made from methanol. Methanol and ammonia are based on natural gas.

TABLE 17.1 Dimensional Stability

Material	Specific Gravity	Impact	Change, %
General purpose phenolic, two-stage	1.36		0.43
one-stage	1.36		0.88
Nylon-filled phenolic, two-stage	1.24		0.35
one-stage	1.24		0.95
Cotton flock phenolic, two-stage	1.40	0.6	0.31
one-stage	1.40	0.6	0.73
Mineral and cotton flock, two-stage	1.64	0.40	0.14
one-stage	1.60	0.40	0.51
Long fibered asbestos, two-stage	1.75	0.70	0.15
Long glass-filled phenolic, one-stage	1.83	15.0	0.09
Short glass fiber-filled phenolic,			
two-stage	1.80	0.50	0.10
one-stage	1.80	0.50	0.18
Mineral-filled DAP	1.65		0.18
Short glass fiber DAP	1.65	0.6	0.06
Long glass fiber DAP	1.65	18.0	0.06
Mineral-filled alkyd	1.90	0.32	0.08
Wood flour urea	1.45		1.84
Melamine, dinnerware grade			2.06
Melamine–phenolic, depending on gravity			0.51–0.98

Thus, pricing of phenolic molding compound depends not only on the base resin but the cost of fillers—glass, sizings, cotton, wood flour; dyes; pigments; freight; and utilities. The hidden costs of complying with environmental and other mandated programs are escalating and significant.

General purpose and heat-resistant materials are 60 cents per pound (bags, truckloads) as of July 1987. Special purpose materials for aerospace applications with silica or graphite fibers may cost from $15–$30 per pound, depending upon filler content and type. Most granular glass-filled materials cost $1.05–$2.00 per pound.

Phenolic piece-part costs have been reduced by material saving, methods of molding, such as live sprue, runnerless injection compression, cold manifold, and hot cone. Thus, the speed of cure and scrap reduction have enabled phenolics to become more cost effective than thermoplastics in selected applications.

17.5 DESCRIPTION OF PROPERTIES

Molded phenolic parts have good thermal and mechanical properties plus retention of properties at elevated temperature. Tensile creep properties (shown in Table 17.2) show that general-purpose phenolic exhibits excellent retention. Replacement of cellulose with glass or mineral gives higher values.

Resistance to creep is important in retention of staked-in inserts in connectors, calibration in circuit breakers, and in automotive transmission, disk brake piston, and pulley applications.

Retention of physical properties at elevated temperature sets phenolics apart from other plastic materials. Some phenolic laminates will exhibit 80,000 psi (552 MPa) flexural strength at room temperatures and 50,000 psi (345 MPa) tested at 55°F (12.8°C) after 500 hours exposure at 500°F (260°C).

Granular glass and mineral-filled heat-resistant types will maintain in excess of 60% of

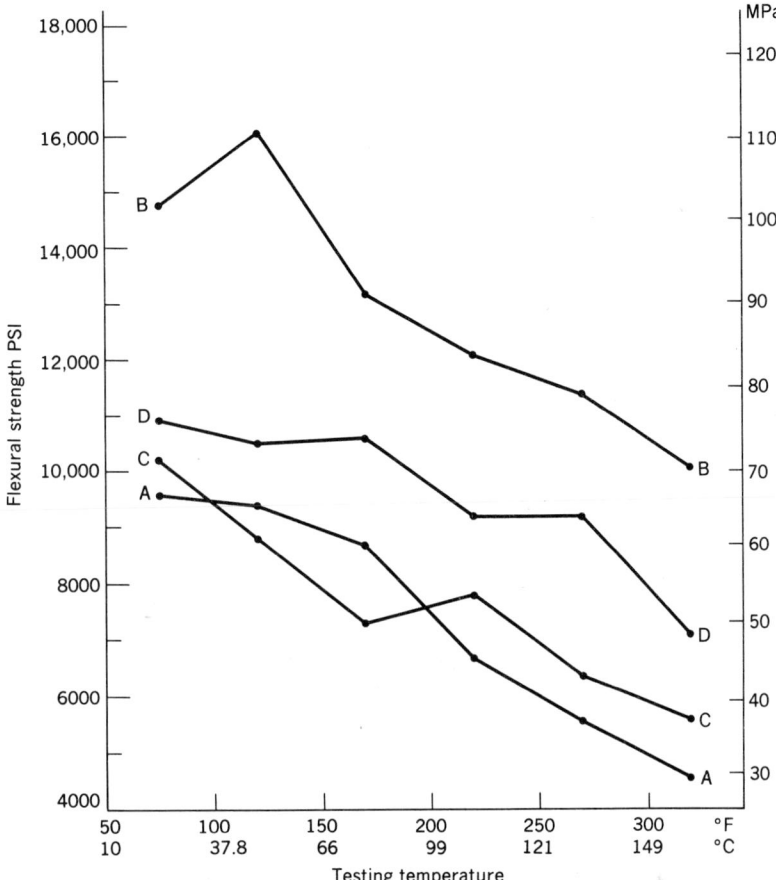

Figure 17.1. Flexural strength as a function of testing temperature. A, Wood flour-filled, general purpose, two-stage phenolics, specific gravity 1.41. B, long glass fiber-filled, one stage phenolic, impact-17.0, specific gravity, 1.83. C, Cotton-flocked, two-stage phenolic, specific gravity, 1.40, CFI-5. D, Long fibered asbestos-filled, two-stage phenolic, specific gravity 1.84, impact 2.2.

their flexural strength and modulus at 320°F (160°C). Cellulose-filled materials in the 1.40 gravity range will retain 50% of their strength at 270°F (132°C) (see Figs. 17.1 and 17.2) (Table 17.2).

Shear strength follows the same pattern as does compressive strength. A high heat-resistant phenolic compound (1.67 specific gravity) was post-baked for 16 hours at 350°F (177°C) and broken with a 1-h soak at the test temperature (Table 17.3). Lineal coefficient of thermal expansion decreases with increasing specific gravity (Table 17.4). Automotive disk brake caliper piston material with a 2.12 specific gravity is 14 mm/mm/°C × 10⁻⁶. Thus, phenolic materials can be picked to match mating metal parts.

Post-baking phenolic promotes further cross-linking and increases the deflection temperature, hardness, flexural, tensile, modulus, and compressive strength. Impact resistance may decline slightly as the material becomes more rigid (Figs. 17.3 and 17.4). Hardness is a function of filler and resin type as well as degree of cure.

Various types of phenolics have been exposed to outdoor weathering for four years (Table 17.5). Phenolic molding compounds retain their properties while losing their surface finish.

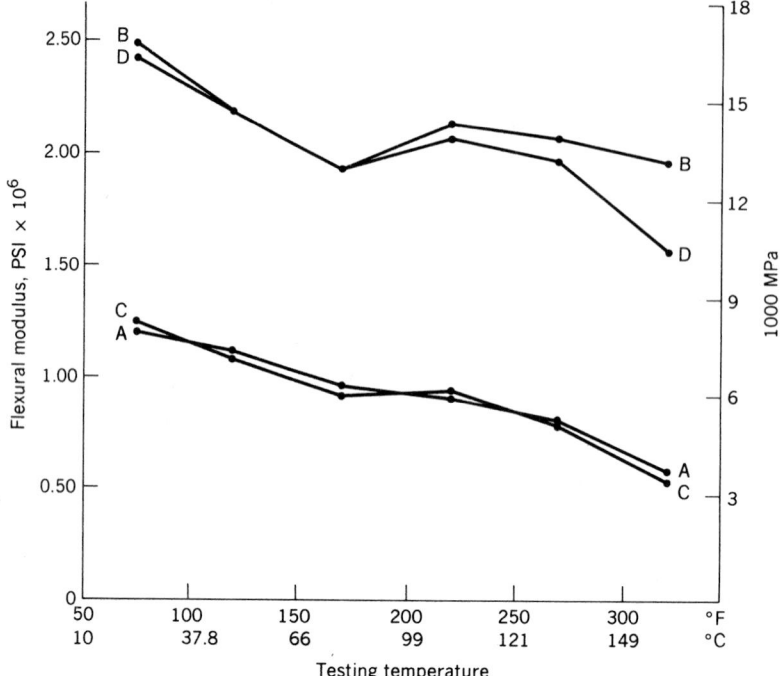

Figure 17.2. Flexural modulus as a function of testing temperature. See Figure 17.1 for explanation of symbols A–D.

TABLE 17.2 Tensile Creep Modulus (Transfer Molded)

	Time	psi	(MPa)	°F	(°C)	Modulus, psi	(MPa)
Cellulose-filled	1000	4000	(27.6)	72	(22.2)	2.2×10^6	(15,172)
general purpose two-	1000	5000	(34.5)	72	(22.2)	2.2×10^6	(15,172)
stage phenolic	1000	1000	(6.9)	250	(121)	2.3×10^5	(1,586)
specific gravity, 1.40	1000	2000	(13.8)	250	(121)	2.5×10^5	(1,724)
Mineral/cellulose	1000	1000	(6.9)	250	(121)	2.4×10^5	(1,655)
medium heat two-stage	1000	2000	(13.8)	250	(121)	3.5×10^5	(2,414)
phenolic, specific gravity,	1000	3000	(20.7)	250	(121)	2.8×10^5	(1,931)
1.48							
Mineral/cellulose high	1000	3000	(20.7)	72	(22.2)	1.6×10^6	(11,034)
heat two-stage phenolic,	1000	4000	(27.6)	72	(22.2)	1.2×10^6	(8,276)
specific gravity, 1.56	1000	500	(3.5)	250	(121)	1.2×10^6	(8,276)
	1000	1000	(6.9)	250	(121)	7×10^5	(4,828)
	1000	2000	(13.8)	250	(121)	4×10^5	(2,759)
Glass/mineral two-stage	1000	3000	(20.7)	250	(121)	1.0×10^6	(6,897)
phenolic, specific gravity,	1000	4000	(27.6)	250	(121)	1.6×10^6	(11,034)
1.75							
Glass/mineral two-	1000	1000	(6.9)	250	(121)	8×10^5	(5,517)
stage phenolic,	1000	2000	(13.8)	250	(121)	9×10^5	(6,207)
specific gravity, 2.10	1000	4000	(27.6)	250	(121)	2×10^6	(13,793)

TABLE 17.3 Compressive Strength vs Temperature

Temperature		Compressive Strength	
°F	(°C)	psi	(MPa)
−40	(−40)	38,740+	(267)
73	(22.8)	32,950	(227)
200	(93)	29,640	(145)
300	(149)	21,410	(148)
400	(204)	30,467	(210)
500	(260)	23,088	(150)
600	(316)	20,480	(141)

TABLE 17.4 Lineal Coefficient of Thermal Expansion[a] vs Specific Gravity

Specific Gravity		Coefficient	
lb/ft³	(g/c)	in./in./°F × 10⁻⁶	(mm/mm/°C × 10⁻⁶)
85.4	(1.37)	23.9	(43.0)
88.6	(1.42)	22.8	(41.0)
97.9	(1.57)	18.9	(34.0)
99.8	(1.60)	17.8	(32.0)
104	(1.67)	14.4	(26.0)
116	(1.86)	13.3	(24.0)
119	(1.90)	10.6	(19.0)
132	(2.12)	7.8	(14.0)

[a]86–140°F (30–60°C range).

Typical applications are powerline quick-disconnect boxes, propellers, and window-tracks. Because they are opaque, they are basically unaffected by ultraviolet light.

Indoor aging of mica-filled electrical grade materials produces little change over six years, as shown in Table 17.6.

Phenolics have excellent flexural fatigue, which is important to designers of power-assist brake systems and transmission components.

Phenolics are used in many different applications where heat resistance, dimensional stability, and low-cost automated production are required.

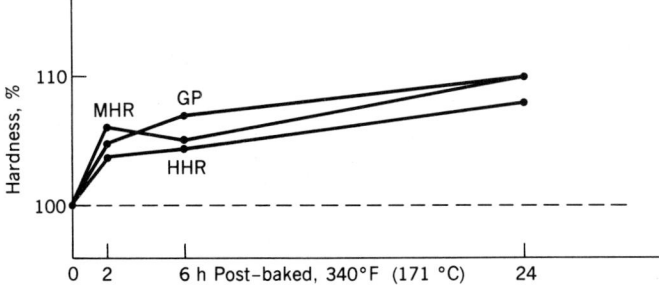

Figure 14.3. Hardness vs curing time. GP = general purpose. MHR = medium heat resistant, HHR = high heat resistant.

TABLE 17.5   Aging Tests

ASTM Test Method	Resin Type	Specific Gravity D792	Dielectric Constant, 1 MHz D150	Power Factor, 1 MHz D150	Insulation Resistance, ohm-cm D257	Impact, ft/lb/in. D256	Flexure, psi D790	Tensile Strength, psi D638	Change Weight, %
Material Type									
Chemical resistant	One-	1.21							
98% resin	stage								
As is			5.11	0.045	2.3×10	0.23	6,300	5,700	
48 mo. shelf			4.39	0.020	Infinity	0.20	10,100	5,900	
48 mo. outdoor			4.55	0.023	3.9×10	0.21	8,200	6,400	−3.0
General Purpose	Two-	1.36							
Woodflour-Filled	stage								
As is			5.14	0.045	5.5×10	0.36	10,800	6,700	
48 mo. shelf			4.96	0.042	1.26×10	0.35	11,900	9,100	
48 mo. outdoor			5.33	0.048	7.75×10	0.32	11,200	7,100	No change
General Purpose	One-	1.36	Not electrical type						
Woodflour-Filled	stage								
As is						0.30	8,300	6,500	
48 mo. shelf						0.30	11,700	8,700	
48 mo. outdoor						0.25	10,100	8,100	−0.15
Medium Impact	Two-	1.38							
Cellulose-Filled	stage								
As is			5.34	0.055	1.3×10	0.56	11,300	7,800	
48 mo. shelf			5.15	0.045	3.48×10	0.50	11,500	7,500	
48 mo. outdoor			5.49	0.056	2.76×10	0.52	10,600	7,500	+0.24
Medium Impact	One-	1.41							
Cellulose Filled	stage								
As is			Not electrical type			0.55	7,800	7,000	
48 mo. shelf						0.47	8,600	7,100	
48 mo. outdoor						0.49	9,900	8,000	−0.27
Heat Resistant	One-	1.53							
Mineral and Cellulose	stage								
As is			5.94	0.078	2.77×10^{12}	0.39	9,000	6,000	
48 mo. shelf			5.32	0.057	9.27×10^{11}	0.30	10,900	7,100	
48 mo. outdoor			5.75	0.052	3.42×10^{10}	0.35	9,200	6,700	−0.80
Heat Resistant	Two-	1.68							
Mineral and Cellulose	stage								
As is			5.37	0.065	6.2×10^{10}	0.51	9,900	5,400	
48 mo. shelf			5.24	0.048	1.42×10^{11}	0.43	10,500	5,300	
48 mo. outdoor	Panel lost								
24 mo. outdoor			5.58	0.070	4.83×10^{9}	0.52	8,300	6,700	
Mineral Filled	Two-	1.75							
	stage								
As is			4.74	0.017	3.9×10^{12}	0.32	7,100	3,500	
48 mo. shelf			4.67	0.019	5.47×10^{12}	0.29	8,600	4,900	
48 mo. outdoor			4.82	0.017	1.55×10^{11}	0.31	9,200	4,200	−0.79
Orlon Filled	DAP	1.32							
As is			3.62	0.017	Infinity	0.60	8,500	5,700	
48 mo. shelf			3.61	0.017	Infinity	0.60	8,800	5,700	
48 mo. outdoor			3.71	0.020	1×10^{14} to Infinity	0.62	7,500	4,800	+0.51

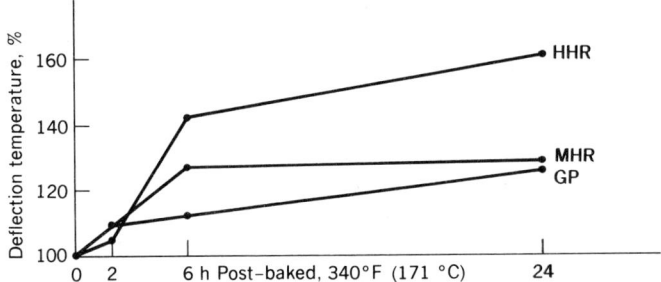

Figure 17.4. Deflection vs curing time. GP = general purpose, MHR = medium heat resistant, HHR = high heat resistant.

Innovative methods of molding and mold design can largely eliminate scrap from runners and sprues, which saves material and shortens cycles.

17.6 APPLICATIONS

Phenolic molding compounds are used in the appliance, automotive, aerospace, closure, electrical/electronic, communications, and heavy electrical markets.

17.6.1 Appliance

Phenolics offer high gloss or cosmetic properties. Different types of mold finishes which impart matte, leatherette, stippled, brushed, and other decorative surfaces are widely used. The heat resistance, dimensional stability, good creep resistance, and electrical properties of phenolic compounds lend themselves well to appliance parts. Broiler end panels, steam irons, coffee-pot bases and handles, knobs, stick handles, bases, and toaster ends all can be phenolic. The Metal Cookware Manufacturers Association has a test for handles (stick handles) intended for pots and pans for in-oven use. This flexural strength test is run on bars conditioned at temperature and tested at conditioning temperature. Parts must retain 50% of the original strength after 1000 hours.

Minimum priced granular phenolic materials are available that will retain 50% of their strength after 1000 hours at 425°F (219°C). Special grades are available that can be chrome plated or decorated wth melamine overlays. Many phenolic parts are painted with epoxy or acrylic finishes using electrostatic, spray, or dip methods.

TABLE 17.6 Indoor Aging vs Electrical Properties of Mica-filled Phenolic[a]

	Mica Low Loss		Mica Arc Resistant	
Time	Power Factor	Dielectric Constant	Power Factor	Dielectric Constant
One day	0.009	4.5	0.025	4.8
One week	0.008	4.7	0.025	4.8
One month	0.008	4.6	0.024	4.8
Six months	0.008	4.5	0.030	4.9
One year	0.008	4.5	0.029	4.9
Two years	0.0082	4.57	0.0296	5.08
Four years	0.0084	4.46	0.0300	5.00
Six years	0.0079	4.65	0.0273	5.01

[a]Mica-filled phenolics run at 1 MCycle aged at room temperature.

Low-cost, high-quality parts are produced automatically using new molding methods (described under History), which improve cycle time and reduce scrap.

17.6.2 Automotive

Automotive applications are basically hidden parts. Phenolics are used in all power-assist brake systems because of their excellent flexural fatigue, heat resistance, ability to hold very close dimensions, and economics of production.

Disk brake caliper pistons of phenolic compounds have replaced steel brake pistons in many car models for lower weight, lower cost, high temperature resistance, and high compressive strength. Low coefficient of thermal conductivity keeps the brake fluid cooler than with metal pistons. Because of their sound dampening abilities, pistons made of phenolic are quieter than those made of metal. As they are unaffected by brake fluid, salt water, grease and oil, they do not corrode in use. They are also very creep resistant and will maintain the close dimensional control required in this harsh environment. Special glass-filled phenolic grades with a low coefficient of thermal expansion are compression molded to produce pistons.

Transmission parts as thrust washers, spacers, and converter reactors are molded from heat-resistant phenolics. Close control of dimension is required as are resistance to hot transmission fluid and high torque loading. Thrust washers and spacers are produced by injection or compression molding.

The Ford converter reactor is produced from glass-reinforced impact-grade phenolic and must withstand tensile and flexural loads as well as high torque loading. The reactor runs in hot, 300°F (149°C) transmission fluid. The low-load side thrust washer is molded as an integral part of the reactor. These close-tolerance parts are produced by pot-type transfer molding.

Phenolic poly "V" or multigroove pulleys are in production using low-gravity impact materials. Very close control of dimensions gives improved belt life. Weight, power, and cost reductions are realized over metal pulleys (Fig. 17.5).

Carburetor spacers are produced from heat-resistant grades and must have good creep resistance as well as thermal insulation properties. They are exposed to gasoline, oil, and under-the-hood environment. These parts are produced by injection molding.

Ignition parts, including distributor caps, coil tops, and rotors, are produced from electrical grade phenolics which retain 20 kV after 1 hour at 180°F (82°C). Speedometer housings, solenoid covers, connectors, and numerous other under-the-hood parts are molded from phenolic heat-resistant and impact materials.

Commutators are molded from glass-filled phenolics for heat resistance and dimensional stability. Adhesion to the copper "exserts" and ability to withstand high revolutions per minute with minimum bar movement is critical to preventing carbon brush wear. Most commutators are pot-type transfer or injection molded. Phenolic commutators are used in windowlifts, starter motors, hand-held appliances, vacuum cleaners, and fractional horsepower motors.

Such electrical parts as circuit breakers, motor control parts, solenoid covers, bus bar supports, wiring device parts, and relays are but a few of the many applications made in phenolics. Designers rely on heat and wear resistance, dimensional stability, UL temperature and flammability properties, and creep resistance. Their low cost per cubic inch and automated processing make phenolic molding compounds very cost effective.

Commutators, coil forms or bobbins, memory frames, and capacitors using glass-filled and specialty low-chloride one-stage resins provide chemical resistance, dimensional stability, and the ability to hold close tolerances in high-volume automated operation.

Phenolic nozzle inserts are used for solid propellant missile applications. Materials may range from cellulose, glass, silica, or carbon-fiber fillers, depending on the individual requirements. Phenolics work well in ablative applications because of the high carbon content of the resin.

Figure 17.5. Multigroove pulley.

Phenolics are used in closures where chemical resistance, odor-free, and nonbleeding requirements are important. Thick sections can be cured much faster than thermoplastics and without sink marks.

Closures may be vacuum plated or sputtered. Most closures are run on cold powder automatic presses.

17.7 ADVANTAGES/DISADVANTAGES

Because they are thermoset, cured phenolics do not melt or soften on exposure to heat. Not only do they have good heat resistance, [phenolics have a generic listing of 300°F (150°C) by Underwriters Laboratories], but they retain a high percentage of their room-temperature properties at elevated temperature. Some phenolics are UL listed at 356°F (180°C).

Phenolics have excellent flexural fatigue properties and good shear strengths. The materials have a wide range of impact properties from 0.30–0.38 ft-lb/in. (16–20 J/m) for general-purpose grades to 17 ft-lb/in. (907 J/m) for grades with 1-in. (25.4 mm) chopped glass rovings. Some glass-filled materials have compressive strengths in excess of 40,000 psi (275 MPa). Granular glass materials are available in the 0.5–0.9 ft-lb/in. (26.7–48 J/m) and nodular materials to 1.5 ft-lb/in. (80 J/m).

Phenolics are generally unaffected by hydrocarbons, brake and transmission fluids, cutting oils, solvents, salt water, soap, and detergent solutions. Certain materials are compounded to have a high degree of chemical resistance. One-stage mineral-filled phenolics are superior to those filled with cellulose, in this respect. Phenolics are attacked by strong mineral acids and bases.

Phenolics are very easy to mold; compression, transfer, and injection-molding techniques are used. Materials are formulated for the molding method to be employed in terms of granulation, plasticity or flow, speed of cure, and lubricant system. Before selecting a material, it is well to check with the raw material supplier to ensure that that particular material will be processed using compatible equipment. Also, shrinkage may vary between methods of molding with the same material and with gate and mold design.

New methods of molding permit cycles on 1/4-in. (6.35 mm) thick sections in excess of 120 shots per hour, and 1/2-in. (12.7 mm) sections above 90 shots per hour with no sink marks. In general, parts over 0.10 in. (2.54 mm) thick can be cured faster than thermoplastics can be frozen.

Extremely complex parts can be molded; size is limited by the capacity of the press. Forty-pound (18 kg) television cabinets were produced in the 1950s and rocket parts have been made up to 732 lb (332 kg).

Processing costs have been reduced by elimination of gates and sprues via runnerless injection compression and runnerless injection. The live sprue concept reduces scrap by keeping the sprue plasticized for molding on the subsequent shot. Thoughtful mold and part design can reduce or eliminate secondary finishing costs.

General-purpose phenolic at 2.7 cents per cubic inch (16.4 cm^3), 300°F (150°C) Underwriters Laboratories listing, deflection temperatures of 330 to 370°F (166 to 188°C), a tensile modulus of 1.2 to 1.4 $\times$ 10^6 psi (8276–9655 MPa), and rapid mold cycles, is an excellent buy. Heat-resistant and impact grades are only slightly more expensive, 2.9 cents and 3.3 cents per cubic inch (16.4 cm^3), respectively. Glass-filled materials run from 7.2 cents to 10.7 cents per cubic inch (16.4 cm^3) (1.75 to 2.15 specific gravity) and are designed for specific applications. Glass-filled materials may be transfer, compression and injection molded.

Phenolics have good electrical properties; 175–250 volts IEC arc-track resistance and ASTM D495 arc resistance above 180 s. Dielectric strength is from 300 to 450 V/mil (11.8–17.7 kV/mm), depending on type.

Colors are limited with phenolics, as the resins darken with exposure to heat and ultraviolet light. Black, shades of brown, greens, and reds are generally commercially available.

Phenolics are rigid, high modulus, notch sensitive materials. Conversely, they have excellent creep resistance, thermal properties, dimensional stability, and corrosion resistance. The threshold damage to radiation is 1 $\times$ 10^7 to 10^8 rads for mineral-filled systems; for unfilled resin, 1 $\times$ 10^6 rads.

Phenolic materials containing organic fillers, such as wood flour, cotton flock, paper flock, or chopped cloth, in that order, show greater moisture absorption in the presence of high humidity. The amount of fungus growth increases as the filler size increases from wood flour through cloth. Mineral-filled materials are generally fungus resistant as are glass-filled materials.

17.8 PROCESSING TECHNIQUES

See Master Material Outline and Standard Material Design Chart.

17.9 RESIN FORMS

Phenolics are available as granular materials 4 to 6 mesh for screw injection. Materials that will be preformed contain 6–10% through 80-mesh fines. Cold power automatic materials are usually less than 0.1% through 80-mesh fines. Impact resistant materials [above 0.5 ft-lb/in. Izod (26.7 J/m)] are available as nodular materials to reduce the bulk factor and may be preformed automatically.

Phenolics may be compounded with almost any filler. Usually a mix of fillers is used to provide a balance of properties. Cellulose as wood flour, paper flock, and cotton flock are used for impact and low specific gravity.

Mica, wollastonite, and glass are used for heat resistance and dimensional stability.

The effect of various fillers is shown in Table 17.7.

Fibrous glass is used to reduce and control shrinkage, and improve physical and electrical properties. Impact will vary depending on the amount and fiber length of the glass used.

In general, as cellulose is removed the water absorption goes down and the heat resistance

PROCESSABILITY OF _____ Phenolics

PROCESS	A	B	C
INJECTION MOLDING	X		
EXTRUSION			X
THERMOFORMING			
FOAM MOLDING			X
DIP, SLUSH MOLDING			
ROTATIONAL MOLDING			
POWDER, FLUIDIZED-BED COATING			
REINFORCED THERMOSET MOLDING	X		
COMPRESSION/TRANSFER MOLDING	X		
REACTION INJECTION MOLDING (RIM)			
MECHANICAL FORMING			
CASTING			X

A = Common processing technique.
B = Technique possible with difficulty.
C = Used only in special circumstances, if ever.

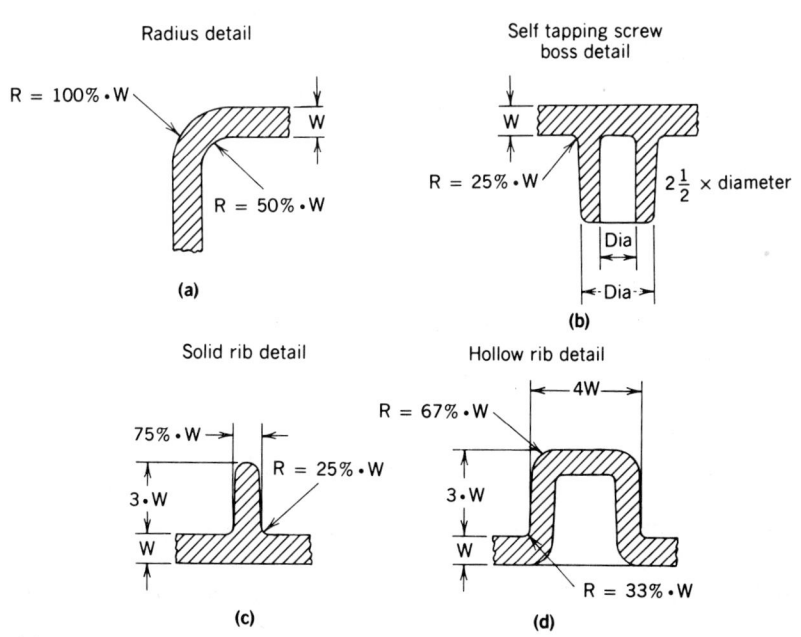

Radius detail

R = 100% • W

W

R = 50% • W

(a)

Self tapping screw boss detail

W

R = 25% • W

$2\frac{1}{2}$ × diameter

Dia

Dia

(b)

Solid rib detail

75% • W

3 • W

W

R = 25% • W

(c)

Hollow rib detail

4W

R = 67% • W

3 • W

W

R = 33% • W

(d)

Injection molded phenolic

Standard design chart for <u>Injection molded phenolic using 152 material and 20 flow</u>
Compression molded phenolic using semipositive mold, 152 heat resistant material, 10 flow, and preheated preforms.

1. Recommended wall thickness (in.)

Minimum	Maximum	Distance from gate	Maximum runner length
0.062	0.125	1–2	10
0.093	0.187	3–4	10
0.187	0.500	3–6	10
0.500	0.750	6–7	12–14

2. Allowable wall thickness variation
   Depends on mold design line-up as it relates to stationary side vs moveable side and to mold-making accuracy

3. Radius requirements (scheme **a**)
   Outside: Minimum ____W____ Maximum ____2 × W (ideal)____
   Inside: Minimum ____50% W____ Maximum ____W____

4. Reinforcing ribs (scheme **c** and **d**)
   Maximum thickness ____75% W____
   Maximum height ____275% W____
   Sink marks: Yes _____ No ___X___

5. Solid pegs and bosses
   Maximum thickness ____0.25–0.33 × diameter____
   Maximum weight _____
   Maximum height ____0.65–0.75 × W____
   Sink marks: Yes _____ No ___X___

6. Strippable undercuts
   Outside: Yes _____ No ___X___
   Inside: Yes ___X___ No _____
   0.093 in radius × 0.010 deep

7. Hollow bosses (scheme **b**): Yes ___X___ No _____

8. Draft angles
   Outside: Minimum ____$\frac{1}{2}°$____ Ideal ____1°____
   Inside: Minimum ____straight wall____ Ideal ____1°____

9. Molded-in inserts: Yes ___X___
   No _____

10. Size limitations
    Length: Minimum _____ Maximum ____10–12 in.____ Process _____
    Width: Minimum _____ Maximum ____3.5–4 lb.____ Process _____

11. Tolerances

12. Special process- and materials-related design details
    Blow-up ratio (blow molding) _____
    Core L/D (injection and compression molding) ____2.5____
    Depth of draw (thermoforming) _____
    Other _____

^{a}Cavity depth: to 1; 1–3 max for 10 material; 3–12 max depth for 12–14 material.

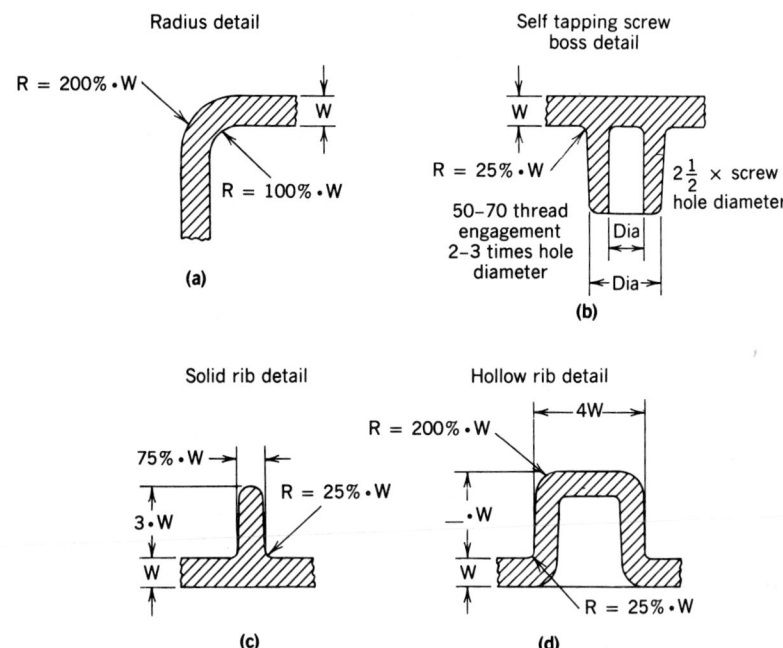

Radius detail

$R = 200\% \cdot W$

W

$R = 100\% \cdot W$

(a)

Self tapping screw
boss detail

W

$R = 25\% \cdot W$

50–70 thread
engagement
2–3 times hole
diameter

Dia

Dia

$2\frac{1}{2}$ × screw
hole diameter

(b)

Solid rib detail

$75\% \cdot W$

$3 \cdot W$

$R = 25\% \cdot W$

W

(c)

Hollow rib detail

4W

$R = 200\% \cdot W$

$\cdot W$

W

$R = 25\% \cdot W$

(d)

Standard design chart for Compression molded phenolic using semipositive mold, 152 heat
resistant material, 10 flow, and preheated preforms.

1. Recommended wall thickness (in.)

Minimum	Maximum	Cavity depth
0.062	0.125	to 1.
0.093	0.187	1–3 max. depth for 10 material
0.187	0.500	3–12 max. depth for 12–14 material

2. Allowable wall thickness variation
Percent depends on concentricity as it relates to core and cavity and semipositive clearance between force and cavity

3. Radius requirements (scheme **a**)
Outside: Minimum ____W____ Maximum ___2 × W (ideal)___
Inside: Minimum ___50% W___ Maximum ____W____

4. Reinforcing ribs (scheme **c** and **d**)
Maximum thickness ____75% W____
Maximum height ____275% W____
Sink marks: Yes _____ No ___X___

5. Solid pegs and bosses
Maximum thickness _0.25–0.33 × diameter_
Maximum weight _____
Maximum height __0.65–0.75 × W__
Sink marks: Yes _____ No ___X___

6. Strippable undercuts
Outside: Yes _____ No ___X___
Inside: Yes ___X___ No _____
0.093 in radius × 0.010 deep

7. Hollow bosses (scheme **b**): Yes ___X___ No _____

8. Draft angles
Outside: Minimum ___$\frac{1}{2}°$___ Ideal ____1°____
Inside: Minimum _straight wall_ Ideal ____1°____

9. Molded-in inserts: Yes ___X___

                             No _____

10. Size limitations
    Length: Minimum _____ Maximum _____ Process _____
    Width: Minimum _____ Maximum _____ Process _____

11. Tolerances

12. Special process- and materials-related design details
    Blow-up ratio (blow molding) _____
    Core L/D (injection and compression molding) _____
    Depth of draw (thermoforming) _____
    Other _____

"Distance from gate	Maximum runner length
1–2	10
3–4	10
3–6	10
6–7	12–14

and modulus, dimensional stability, comparative tracking index, and specific gravity go up. The effect of replacing cellulose with mineral or glass is indicated in Table 17.8.

It should be noted that the replacement of cellulose with glass fibers has a marked effect on dimensional stability. The combination of lower mold shrinkage and lower post-mold shrinkage allows for much closer control of dimensions. These glass-filled materials are the compounds of choice for many automatic transmission and brake piston applications as well as connectors, frames, bases, coil forms, and so on in the electrical/electronic and aerospace industries.

Phenolics are classified as general purpose, impact, heat resistant, closure or nonbleeding, electrical, and special grades.

TABLE 17.7. Effect of Filler on Compression Molded Properties Typical Values

Property	Phenolic Resin	Wood Flour	Cellulose Flock	Cellulose Mineral	Mineral	Mineral Glass	Cellulose Mineral	Mineral	Mineral Glass	Mineral	Mineral Glass
								Alkyd		DAP	
Powder properties	2.2										
Bulk factor	0.60	2.3	2.8	2.4	2.2	2.2	2.4	2.2	2.4	2.3	2.4
Apparent density		0.60	0.50	0.64	0.80	0.80	0.81	1.05	0.90	0.72	0.74
Physical Properties											
Specific gravity, g/cc	1.24	1.36	1.36	1.55	1.80	1.75	1.95	2.2	2.13	1.67	1.78
Shrinkage, %											
Injection		11	11	9	5	4	8	8	6	8	7
Compression	10	7	7	5	3	3	6	6	4	6	5
Tensile × 10^3 psi	5	7	7	8	6	9	6	9	9	7	9
Flexure × 10^3 psi	5	10	10	11	9	14	10	15	15	10	13
Compressive × 10^3 psi	28	31	30	28	25	33	25	35	30	22	28
Tensile modulus × 10^6 psi	0.8	1.1	1.2	1.4	2	2.5	2.5	2.8	3.0	1.2	1.3
Izod impact, ft-lb/in.	0.20	0.28	0.40	0.36	0.32	0.5	0.35	0.36	0.48	0.28	0.50
Deflection temperature, °F	300	340	350	370	375	425	400	450	500	300	450
H_2O absorption, %	0.4	0.50	0.60	0.30	0.20	0.10	0.3	0.1	0.15	0.2	0.2
MIL-M-14G, dimensional											
stability, %	0.78	0.43	0.35	0.52	0.22	0.11	0.43	0.12	0.10	0.18	0.01
Electrical properties											
UL 94 VO at in.		0.240	0.240	0.58	0.021	0.058		0.20	0.20		0.120
ASTM Arc		125	50	150	184	150	181	184	184	130	150
CTI		165	150	220	200	190	600+	600+	650+	600+	650+
Volume resistivity		1	1	1	10	1	100	10	1	100	1000
Dielectric strength,											
V/ml											
S/S		350	350	375	425	400	350	350	375	450	475
S/T		300	300	300	400	325	300	300	300	350	350

TABLE 17.8   Effect on Properties when Cellulose Filler is Replaced with Minerals or Glass

Down	Up
Water absorption	Specific gravity
Impact (depends on formulation)	Modulus
Shrinkage	Electrical
After shrinkage	Dimensional stability
Coefficient of thermal expansion	Heat resistance
Flammability	Deflection temperature
	CTI
	ASTM arc resistance
	Impact (depends on formulation)
	Coefficient of thermal conductivity

Phenolic resins are also used to coat glass and graphite cloth, which is cut on a bias to make tape, and is then wound and autoclaved or pressure bag cured to form solid rocket propulsion motor nozzles.

The treated broad goods can be vacuum bag molded to form large shapes widely used in the aerospace industry.

Phenolic sheet molding compound is available with very fast cures and low pressure requirements [200–300 psi (1.3–2.0 MPa)]. The advantages of phenolic systems are low smoke, NBS smoke chamber less than 40, and very low flame spreads. Their retention of physical properties at elevated temperatures exceeds that of the polyesters. Some phenolic sheet molding compounds can be vacuum bag molded.

Phenolic resins are available as 200-mesh powders, flake, lump, and as liquids in various solvents, including water (liquid one-steps) and dispersions.

17.10 SPECIFICATION OF PROPERTIES

See Master Outline Materials Properties.

17.11 PROCESSING REQUIREMENTS

In general, two-stage phenolics are not adversely affected in storage. If they are stocked at normal conditions under 80°F (26.7°C), the state of polymerization is practically unchanged over two or three years. However, the moisture content may vary considerably, particularly if phenolics are exposed to conditions above 50% relative humidity. This can be corrected by proper drying and preheating before molding.

Single-stage, one-step phenolic molding compounds are considerably more sensitive to aging than two-stage phenolic materials. If stored at normal room temperature, under 80°F (26.7°C), most single-stage materials polymerize slowly. It is strongly recommended that they be used within three months. If longer storage time is required, they should be kept below 40°F (4.4°C).

Phenolics are compression molded from 280–400°F (138–204°C).

Transfer molds normally run 320–380°F (160–193°C), and injection molds from 340–360°F (171–182°C). The type of material, mold design, part geometry, availability of preheat, and press considerations determine the cure and cycle times. In general, molds are run as hot as possible, consistent with part finish and fill considerations, to reduce cure time.

Phenolic materials should be at room temperature before they are processed. In the winter cold, boxed material may not reach room temperature for several days. If it is fed

MATERIAL ___ Phenolic, general-purpose, injection-molded

PROPERTY	TEST METHOD	ENGLISH UNITS	VALUE	METRIC UNITS	VALUE
MECHANICAL DENSITY	D792	lb/ft³	87.3	g/cm³	1.40
TENSILE STRENGTH	D638	psi	7000	MPa	48.3
TENSILE MODULUS	D638	psi	1,200,000	MPa	8,276
FLEXURAL MODULUS	D790	psi		MPa	
ELONGATION TO BREAK	D638	%		%	
NOTCHED IZOD AT ROOM TEMP	D256	ft-lb/in.	0.31	J/m	16.5
HARDNESS	D785		E 85		E 85
THERMAL DEFLECTION T @ 264 psi	D648	°F	302	°C	150
DEFLECTION T @ 66 psi	D648	°F		°C	
VICAT SOFT-ENING POINT	D1525	°F		°C	
UL TEMP INDEX	UL746B	°F	302	°C	150
UL FLAMMABILITY CODE RATING	UL94		V-1 at 0.120 in.		V-1 at 3.1 mm
LINEAR COEFFICIENT THERMAL EXPANSION	D696	in./in./°F	22×10^{-6}	mm/mm/°C	39×10^{-6}
ENVIRONMENTAL WATER ABSORPTION 24 HOUR	D570	%	0.50	%	0.50
CLARITY	D1003	% TRANS-MISSION	Opaque	% TRANS-MISSION	Opaque
OUTDOOR WEATHERING	D1435	%		%	
FDA APPROVAL			Some grades		Some grades
CHEMICAL RESISTANCE TO: WEAK ACID	D543		Minimally attacked		Minimally attacked
STRONG ACID	D543		Attacked		Attacked
WEAK ALKALI	D543		Minimally attacked		Minimally attacked
STRONG ALKALI	D543		Attacked		Attacked
LOW MOLECULAR WEIGHT SOLVENTS	D543		Not attacked		Not attacked
ALCOHOLS	D543		Not attacked		Not attacked
ELECTRICAL DIELECTRIC STRENGTH	D149	V/mil	225	kV/mm	8.87
DIELECTRIC CONSTANT	D150	10^6 Hertz	5.5		5.5
POWER FACTOR	D150	10^6 Hertz	0.06		0.06
OTHER MELTING POINT	D3418	°F		°C	
GLASS TRANS-ITION TEMP.	D3418	°F		°C	

MATERIAL   Phenolic, mineral- and cellulose-filled, heat resistant and electrical grades, compression molded

PROPERTY	TEST METHOD	ENGLISH UNITS	VALUE	METRIC UNITS	VALUE
MECHANICAL DENSITY	D792	lb/ft^3	98.5	g/cm^3	1.58
TENSILE STRENGTH	D638	psi	8,500	MPa	58.6
TENSILE MODULUS	D638	psi	1,500,000	MPa	10,345
FLEXURAL MODULUS	D790	psi		MPa	
ELONGATION TO BREAK	D638	%		%	
NOTCHED IZOD AT ROOM TEMP	D256	ft-lb/in.	0.37	J/m	19.8
HARDNESS	D785		E 82		E 82
THERMAL DEFLECTION T @ 264 psi	D648	°F	360	°C	182
DEFLECTION T @ 66 psi	D648	°F		°C	
VICAT SOFT-ENING POINT	D1525	°F		°C	
UL TEMP INDEX	UL746B	°F	320	°C	160
UL FLAMMABILITY CODE RATING	UL94		V-0 at 0.58 in.		V-0 at 14.8 mm
LINEAR COEFFICIENT THERMAL EXPANSION	D696	in./in./°F		mm/mm/°C	
ENVIRONMENTAL WATER ABSORPTION 24 HOUR	D570	%	0.4	%	0.4
CLARITY	D1003	% TRANS-MISSION	Opaque	% TRANS-MISSION	Opaque
OUTDOOR WEATHERING	D1435	%		%	
FDA APPROVAL			Some grades		Some grades
CHEMICAL RESISTANCE TO: WEAK ACID	D543		Minimally attacked		Minimally attacked
STRONG ACID	D543		Badly attacked		Badly attacked
WEAK ALKALI	D543		Minimally attacked		Minimally attacked
STRONG ALKALI	D543		Badly attacked		
LOW MOLECULAR WEIGHT SOLVENTS	D543		Not attacked		Not attacked
ALCOHOLS	D543		Not attacked		Not attacked
ELECTRICAL DIELECTRIC STRENGTH	D149	V/mil	375	kV/mm	14.8
DIELECTRIC CONSTANT	D150	10^6 Hertz	5		5
POWER FACTOR	D150	10^6 Hertz	0.04		0.04
OTHER MELTING POINT	D3418	°F		°C	
GLASS TRANS-ITION TEMP.	D3418	°F		°C	

MATERIAL __ Phenolic, glass- and mineral-filled, special-purpose (connectors, memory frames, capacitors), injection-moled

PROPERTY	TEST METHOD	ENGLISH UNITS	VALUE	METRIC UNITS	VALUE
MECHANICAL DENSITY	D792	lb/ft³		g/cm³	
TENSILE STRENGTH	D638	psi		MPa	
TENSILE MODULUS	D638	psi		MPa	
FLEXURAL MODULUS	D790	psi		MPa	
ELONGATION TO BREAK	D638	%		%	
NOTCHED IZOD AT ROOM TEMP	D256	ft-lb/in.		J/m	
HARDNESS	D785				
THERMAL DEFLECTION T @ 264 psi	D648	°F		°C	
DEFLECTION T @ 66 psi	D648	°F		°C	
VICAT SOFTENING POINT	D1525	°F		°C	
UL TEMP INDEX	UL746B	°F		°C	
UL FLAMMABILITY CODE RATING	UL94				
LINEAR COEFFICIENT THERMAL EXPANSION	D696	in./in./°F		mm/mm/°C	
ENVIRONMENTAL WATER ABSORPTION 24 HOUR	D570	%	0.10	%	0.10
CLARITY	D1003	% TRANSMISSION	Opaque	% TRANSMISSION	Opaque
OUTDOOR WEATHERING	D1435	%		%	
FDA APPROVAL			Some grades		Some grades
CHEMICAL RESISTANCE TO: WEAK ACID	D543		Minimally attacked		Minimally attacked
STRONG ACID	D543		Attacked		Attacked
WEAK ALKALI	D543		Minimally attacked		Minimally attacked
STRONG ALKALI	D543		Attacked		Attacked
LOW MOLECULAR WEIGHT SOLVENTS	D543		Not attacked		Not attacked
ALCOHOLS	D543		Not attacked		Not attacked
ELECTRICAL DIELECTRIC STRENGTH	D149	V/mil	450	kV/mm	17.7
DIELECTRIC CONSTANT	D150	10⁶ Hertz	5.0		5.0
POWER FACTOR	D150	10⁶ Hertz	0.09		0.09
OTHER MELTING POINT	D3418	°F		°C	
GLASS TRANSITION TEMP.	D3418	°F		°C	

into an injection machine from 1,000-lb (454 kg) boxes, it will give a continuously varying temperature profile. Molding compound cannot be preformed below 55°F (12.8°C).

Flash and scrap parts may be pulverized (200-mesh) and fed back into the material up to 15%. This material is abrasive and reduces the flow of the molding compound. Material with regrind requires an additional UL listing. The use of regrind is limited. Runnerless injection compression and live sprues reduce scrap by eliminating sprues and runner systems.

Granular or nodular phenolic materials may be auger fed or air conveyed directly from container to automatic equipment.

Sprues, gates, and runners should be sized for the type of material being used and will normally be larger than for thermoplastics. Thermosets have a higher filler loading and are much more viscous at normal injection temperatures [240–260°F (116–127°C] than thermoplastics. Also, there is a very rapid chemical reaction taking place during cure, so it is essential to move the material rapidly and at as low pressures as practical.

17.12 PROCESSING-SENSITIVE END PROPERTIES

End properties can be sensitive to processing parameters in that the material must be stored properly to prevent any change in moisture content. Phenolic materials are formulated to a specific moisture content. Loss of moisture reduces the flow or "stiffens" the material and reduces shrinkages. Reduced moisture may also lengthen the cure because of reduced heat transfer. Excessive moisture softens the material and increases the shrinkage, and changes in moisture content can affect the preheating characteristics of the material, thus changing the rate of preheat and flow duration of the material. Excessive moisture will impair electrical properties.

Improper molding conditions may impart flow lines or weld lines, which can also be caused by poor mold design or lack of venting. Low mold temperatures [below 300°F (149°C)] may impair the chemical resistance of one-step materials and thus require an after-bake.

Higher mold temperatures [360–390°F (182–199°C)] will improve physical and thermal properties relative to the standard 340°F (171°C) ASTM molding conditions (Fig. 17.6).

How-to-mold troubleshooting guides are available from material manufacturers.

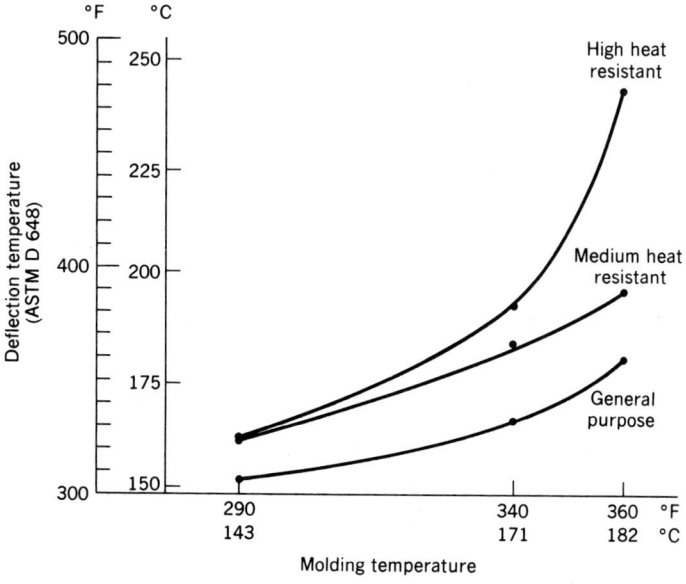

Figure 17.6. Deflection temperature as a function of molding temperature.

17.13 SHRINKAGE

Shrinkage is such a function of mold design that specifications are only a rough guide, at best. Experience has shown that a meeting with the materials supplier, mold maker, molder, and end user before any steel is cut can be most beneficial.

Standard Tolerance Chart

PROCESS ___Transfer molding___   MATERIAL ___Phenolic___

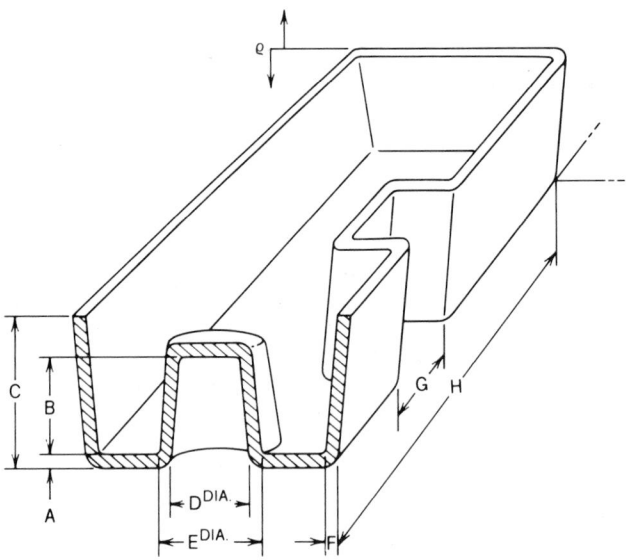

Values are based on a nominal 1/8 in. (3.2 mm) wall thickness and given in plus or minus in./in. and mm/mm.
FINE tolerance – Special care and added cost.
COMMERCIAL tolerance – Normal care required.
COARSE tolerance – Minimal care required, lowest cost.

Tolerance	A	B	C	D	E	F	G	H	
FINE	0.001	0.002	0.003	0.002	0.002		0.003	0.010	in./in.
									mm/mm
COMMERCIAL	0.003	0.004	0.005	0.004	0.004		0.005	0.015	in./in.
									mm/mm
COARSE	0.006	0.010	0.012	0.010	0.010		0.010	0.025	in./in.
									mm/mm

Standard Tolerance Chart

| PROCESS | Compression molding | MATERIAL | Phenolic |

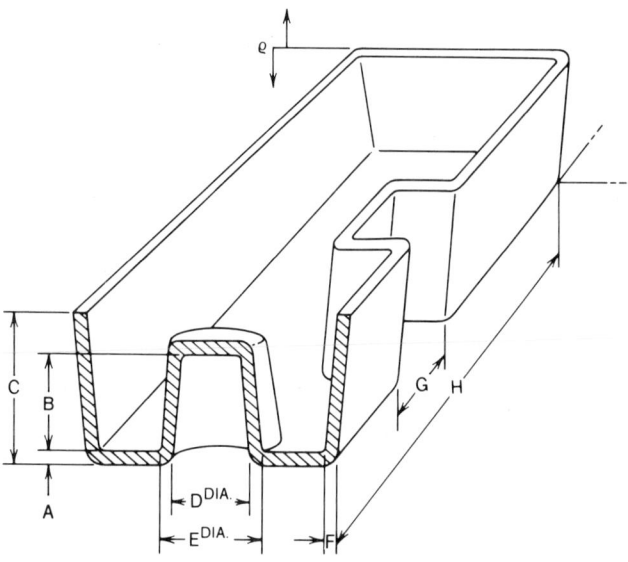

Values are based on a nominal 1/8 in. (3.2 mm) wall thickness and given in plus or minus in./in. and mm/mm.

FINE tolerance – Special care and added cost.

COMMERCIAL tolerance – Normal care required.

COARSE tolerance – Minimal care required, lowest cost.

Tolerance	A	B	C	D	E	F	G	H	
FINE	0.008	0.008	0.008	0.002	0.002		0.003	0.010	in./in.
									mm/mm
COMMERCIAL	0.010	0.010	0.010	0.004	0.004		0.005	0.015	in./in.
									mm/mm
COARSE	0.015	0.015	0.015	0.010	0.010		0.010	0.025	in./in.
									mm/mm

17.14 TRADE NAMES

Phenolic resins are available under the following trade names from the suppliers listed: Fiberite, Fiberite Corp., Div. ICI America, Winona, Minn.; Durez, Durez Div., Occidental Chemical Corp., North Tonawanda, N.Y.; Plaslok, Plaslok Corp., Buffalo, N.Y.; Plenco, Plastics Engineering Co., Sheboygan, Wis.; Resinoid, Resinoid Engineering Corp., Skokie, Ill.; Rogers, Rogers Corp., Manchester, Conn.; Valite, Valite, Div. of Valentine Sugars, Inc., Lockport, La.

BIBLIOGRAPHY

R.L. Brown, *Design and Manufacture of Plastic Parts,* John Wiley & Sons, New York, 1980.

J.H. DuBois and W.I. Pribble, *Plastic Mold Engineering Handbook,* Van Nostrand Reinhold, New York, 1978, pp. 199–343.

J. Frados, *Plastics Engineering Handbook of the SPI,* 4th ed., Van Nostrand Reinhold, New York, 1976.

Military Specification Mil-M-14G, Molding Plastics and Molded Plastic Parts, Thermosetting, Bureau of Ships.

Publications of the SPI Phenolic Molding Div., The Society of the Plastics Industry, Inc., New York.

18

POLY(ARYL SULFONE)

Lloyd M. Robeson
Air Products and Chemicals, Inc., Allentown, PA 18105

Barry L. Dickinson
Amoco Performance Products, Bound Brook, NJ 08805

18.1 STRUCTURE

Poly(aryl sulfone) is a transparent, stable engineering resin. The chemical structure has not been disclosed by the manufacturer.

18.2 CATEGORY

Poly(aryl sulfone) is an amorphous thermoplastic that offers many of the desired property characteristics typical of aromatic polysulfones: high heat-distortion temperature, transparency, good unnotched toughness, excellent hydrolytic stability, creep resistance, environmental stress-crack resistance (relative to other amorphous thermoplastics), injection moldability to close and predictable tolerances, and desirable combustion characteristics (based on laboratory evaluations).

18.3 HISTORY

Poly(aryl sulfone) was introduced in 1983 and is available from Amoco Performance Products under the trade names Radel A-100, A-200, and A-300. It is suitable for high technology applications, including such electrical/electronic parts as printed-circuit boards, snap-fit connectors, and lamp housings. Its overall property profile, combined with its combustion characteristics, suggest its utility in critical areas of the transportation industry. Hydrolytic resistance, toughness, and transparency indicate its suitability for metal or glass replacement in special areas of the processing industry.

18.4 POLYMERIZATION

After a twenty-year development period, Union Carbide created the technology which led to the introduction of poly(aryl sulfone). This technology is summarized in various patents and publications, with key examples noted in the information sources given at the end of this chapter. The resin's polymerization process is considered proprietary.

The truckload price (1989) of poly(aryl sulfone) is $4.40/lb.

18.5 DESCRIPTION OF PROPERTIES

A glass transition of 428°F (220°C) accounts for the high heat-deflection temperature, 400°F (204°C) of poly(aryl sulfone). With a flexural modulus of 400,000 psi (2758 MPa), excellent tensile-impact strength of 160 ft-lb/in.^{2}, and a tensile strength of 12,000 psi (83 MPa), the resin offers the mechanical property balance needed for the demanding performance requirements of emerging high technology markets. Poly(aryl sulfone) exhibits a low temperature relaxation at −148°F (−100°C), which is believed to responsible for its ductility and toughness above that temperature. The property profile from −148°F to −328°F (−100°C to −200°C) is quite constant, since no fundamental transitions occur in that temperature range, and the modulus is reasonably constant over that range. As with other aromatic polysulfones, it is resistant to hydrolysis (eg, water, acids, and bases) over a broad temperature range. Although high temperature polymers generally suffer from processing latitude, poly(aryl sulfone) offers excellent injection molding characteristics in the range of 650–750°F (343–395°C), with a spiral flow of 15 in. (37.5 cm) at 0.80-in. (2-cm) thickness.

18.5.1 Mechanical Properties

Mechanical and other relevant properties of poly(aryl sulfone) at 73.4°F (23°C) are tabulated in Table 18.1. The mechanical properties of poly(aryl sulfone) are functional nearly to its heat-distortion temperature. For design purposes, allowable working stresses in an air environment for short-time exposure can be estimated by dividing the yield stress by an appropriate safety factor. Safety factors of 2 to 3 are commonly employed for engineering thermoplastics.

Temperature		Maximum Working Stress	
°C	°F	Psi	MPa
23	73.4	4000	28
60	140	3400	23
100	212	2700	19
125	257	2250	16
150	302	1900	13

18.5.2 Thermal Properties

With a glass-transition temperature of 428°F (220°C), poly(aryl sulfone) achieves a heat-distortion temperature of 400°F (204°C). Accelerated testing results indicate continuous use potential up to 356°F (180°C). These properties suggest its suitability for high-performance electrical/electronic applications. Indeed, poly(aryl sulfone) will withstand the high temperatures of vapor-phase and wave soldering encountered in electrical connector and printed-circuit-board applications. Its thermal properties are tabulated in Table 18.2.

TABLE 18.1. Typical Mechanical Properties of Radel A-200 Poly(aryl sulfone)

Property	ASTM Method	English	Standard International
		Values and Units	
Tensile modulus	D638	385,000 psi	2,655 MPa
Tensile strength at yield	D638	12,000 psi	82.7 MPa
Tensile elongation yield	D638	6.5%	6.5%
Tensile elongation	D638	40%	40%
Flexural modulus	D790	400,000 psi	2,750 MPa
Flexural strength at 5% strain	D790	16,100 psi	110 MPa
Notched Izod impact strength			
1/8 in. specimen (72°F/ 23°C)	D256	1.6 ft-lb/in.	85 J/m
Tensile impact strength			
1/8 in. specimen (72°F/ 23°C)	D1822	160 ft-lb/in^2.	355 kJ/m^2
Melt flow at 400°C	D1238		30 g/10 min
Density	D1505	85.5 lb/ft^3	1.37 mg/m^3
Mold shrinkage	D955	0.6%	
Rockwell hardness	D785	M85	0.6%
Water absorption (24 h)	D570	0.4%	0.4%
equilibrium	D570	1.85%	1.85%
Refractive index		1.651%	1.651

18.5.3 Electrical Properties

The electrical properties of poly(aryl sulfone) also are listed in Table 15.2. As with mechanical properties, electrical properties will remain reasonably consistent over a broad temperature range. The balance of electrical properties suggests definite utility in emerging electrical/electronic applications.

TABLE 18.2. Thermal and Electrical Property Data For Radel A-200 Poly(aryl sulfone)

Property	ASTM Method	English	Standard International
		Values and Units	
Heat-deflection temperature			
1/4 in. bar, 264 psi	D648	400°F	204°C
Coefficient of linear thermal expansion	D696	2.7 × 10^5 in./in./°F	4.9 × 10^{-5} m/m/°C
Dielectric strength S/T			
0.125 in.	D149	383 v/mil	15,070 V/mm
Arc resistance, 0.125 in.	D495	81 s	81 s
Volume resistivity, 0.125 in.	D257	3.03 × 10^{16} ohm·in.	7.71 × 10^{14} ohm·m
Dielectric constant, 0.125 in.	D150		
60 Hz		3.51	3.51
10^3 Hz		3.50	3.50
10^6 Hz		3.54	3.54
Dissipation factor, 0.125 in.	D150		
60 Hz		0.0017	0.0017
10^3 Hz		0.0022	0.0022
10^6 Hz		0.0056	0.0056

18.5.4 Environmental Properties

Poly(aryl sulfone) is soluble in certain organic solvents, including methylene chloride, dimethylformamide, and dimethyl acetamide. As with other amorphous polymers, it will be subject to environmental stress cracking in organic chemical environments. Union Carbide's comparative data, however, show poly(aryl sulfone) to exhibit better resistance to environmental stress cracking than many other amorphous thermoplastics. The most aggressive organic environments are esters, ketones, and aromatic hydrocarbons; the least aggressive, alcohols and aliphatic hydrocarbons.

Poly(aryl sulfone) exhibits excellent hydrolytic stability. Under severe conditions of stress, elevated temperatures, and acid, base, or salt environments, poly(aryl sulfone) offers outstanding performance. This feature suggests utility in steam autoclaving, processing equipment, and high-temperature fuel cells.

18.5.5 Other Properties

Many of the applications involving high-performance engineering thermoplastics require flammability resistance as defined by various standard laboratory tests. The results for poly(aryl sulfone) listed in Table 18.3.

For improved creep resistance, higher allowable design stresses, and improved environmental stress-crack resistance, glass-fiber filled versions of poly(aryl sulfone) are commercially available. Typical properties for three glass-filled versions (Radel AG-410, AG-420, and AG-430) are tabulated in Table 18.4.

TABLE 18.3. Combustion Characteristics of Radel A-200 Poly(aryl sulfone)[a]

Property	ASTM or Other Test Method	Value
Flammability	UL-94[b]	V-0 (0.023 in.)
Oxygen Index (0.125 in.)	D-2863	33
Autoignition temperature	D-1929	502°C
	NBS flaming smoke chamber (0.063 in. specimen)	
Smoke density		
D_s at 1.5 min.		0
D_s at 4.0 min.		1
D_m at (D_s maximum)		5–15
Gas toxicity ppm		
SO_2 at 1.5–4.0 min		15
SO_2 max		75
CO at 1.5–4.0 min		50
CO max		150

[a]These numerical flame-spread ratings or flammability ratings are not intended to reflect hazards presented by these or any other materials under actual fire conditions.
[b]Test run in Union Carbide laboratory by UL-94 procedure.

TABLE 18.4. Typical Properties of Glass-filled Radel A Resins

Property	ASTM Test Method	Radel		
		AG-410	AG-420	AG-430
Fabrication				
Melt flow, g/10 min	D1238	12.1	8.6	6.3
Mold shrinkage, %	D955	0.5	0.4	0.3
Mechanical				
Tensile strength, psi (MPa)	D638	12,500 (86.3)	15,200 (105)	18,300 (126)
Tensile elongation, %	D638	5.8	3.2	1.9
Tensile modulus, psi (MPa)	D638	555,000 (3,830)	825,000 (5,690)	1,250,000 (8,620)
Flexural strength, psi (MPa)	D790	21,000 (145)	23,500 (162)	26,000 (180)
Flexural modulus, psi (MPa)	D790	590,000 (4,060)	750,000 (5,190)	1,170,000 (8,070)
Notched Izod, ft-lb/in. (J/m)	D256	0.9 (48)	1.1 (58)	1.4 (75)
Tensile impact ft-lb/in.^{2} (kJ/m^2)	D1822	28 (59)	31 (65)	34 (72)
Thermal				
Heat-deflection temperature 264 psi, °F (°C)	D648	410 (209)	415 (213)	415 (213)
Flammability rating UL-94		V-0 at 0.023 in. (V-0 at 0.58 mm)	V-0 at 0.023 in. (V-0 at 0.58 mm)	V-0 at 0.023 in. (V-0 at 0.58 mm)
Electrical				
Dielectric strength, volts/mil (V/mm)	D149	440 (17,300)	440 (17,300)	440 (17,300)
Volume resistivity, ohm·cm	D257	>10^{16}	>10^{16}	>10^{16}
Dielectric constant	D150			
at 60 Hz		3.68	3.84	4.11
at 1 MHz		3.70	3.84	4.13
at 10^6 Hz		3.72	3.88	4.17
Dissipation factor	D150			
at 60 Hz		0.0015	0.0015	0.0019
at 1 MHz		0.0018	0.0018	0.0023
at 10^6 Hz		0.0072	0.0081	0.0094
General				
Density, g/cm^3 (mg/m^3)	D1505	1.43 (1.43)	1.51 (1.51)	1.58 (1.58)

18.6 APPLICATIONS

The high temperature capabilities of poly(aryl sulfone) indicate its utility in advanced composites. Prepregging from a methylene chloride solution is a feasible consolidation route.

Other elevated-temperature applications include under-the-hood automotive parts, lamp housings, oven components, and myriad electrical/electronic parts, such as coil bobbins, printed-circuit boards, breaker housings, and snap-fit connectors.

Excellent hydrolytic stability suggests its application in products that must be steam-autoclaved, and in high-temperature fuel cells, gas-separation or reverse-osmosis membranes, and processing equipment.

The material's overall property profile of high-temperature performance, toughness, hydrolytic stability, and combustion characteristics, combined with the economics of injection moldability, extrusion, or thermoforming, will allow for yet-undefined applications replacing metal, ceramics, and glass.

18.7 ADVANTAGES/DISADVANTAGES

As noted above, poly(aryl sulfone) offers the desirable combination of high heat-distortion temperature, high thermal stability, toughness and ductility, excellent hydrolytic stability (even in acid and base environments), and processing characteristics favorable for injection molding and extrusion on standard equipment. Although its environmental stress-crack resistance is not as good as that of crystalline thermoplastics, it is good relative to other amorphous thermoplastics.

18.8 PROCESSING TECHNIQUES

Injection molding, extrusion, and thermoforming all are suitable methods for processing poly(aryl sulfone).

Prior to melt processing poly(aryl sulfone), drying to less than 0.04 wt% of water will be required to prevent surface streaks or voids in molded or extruded specimens. Minimum drying times in a circulating air oven or hopper drying are 4-1/2 h at 275°F (135°C), 4 h at 300°F (149°C), or 2-1/2 h at 350°F (177°C). Suggested injection molding conditions are given in Table 18.5.

Extrusion conditions will generally be lower in temperature by 36–54°F (20–30°C). Overall, molding and extrusion conditions for this material are quite similar to those for Udel polysulfone P-1700, which is recommended as a purge material to remove lower softening materials prior to running poly(aryl sulfone). Up to 100% regrind is possible for most applications without sacrificing mechanical properties. If color changes need to be minimized, a maximum of 25% regrind is recommended.

Poly(aryl sulfone) can be electroplated using techniques similar to those commercially employed for polysulfone. This is of particular interest in printed-circuit-board applications.

TABLE 18.5. Injection Molding Conditions

Area	Temperature	
	Typical	Range
Barrel, rear	320°C (610°F)	520°F–680°F
center	350°C (660°F)	560°F–730°F
front	350°C (660°F)	570°F–740°F
Stock	365°C (690°F)	635°F–750°F
Mold	140°C (280°F)	260°F–360°F

Master Material Outline

PROCESSABILITY OF ___ Poly(aryl sulfone) _____

PROCESS	A	B	C
INJECTION MOLDING	X		
EXTRUSION	X		
THERMOFORMING	X		
FOAM MOLDING	X		
DIP, SLUSH MOLDING			
ROTATIONAL MOLDING			
POWDER, FLUIDIZED-BED COATING			
REINFORCED THERMOSET MOLDING			
COMPRESSION/TRANSFER MOLDING			X
REACTION INJECTION MOLDING (RIM)			
MECHANICAL FORMING			
CASTING			

A = Common processing technique.
B = Technique possible with difficulty.
C = Used only in special circumstances, if ever.

Standard design chart for ___ Poly(aryl sulfone) ___

1. Recommended wall thickness/length of flow
   Minimum ___ 0.030 in. ___ for ___ 2-3/4 in. ___ distance
   Maximum _____ for _____ distance
   Ideal ___ 0.080 in. ___ for ___ 15 in. ___ distance

2. Allowable wall thickness variation, % of nominal wall ___ 150 ___

3. Radius requirements
   Outside: Minimum _____ Maximum _150%w recommended_
   Inside: Minimum _____ Maximum _50%w recommended_

4. Reinforcing ribs
   Maximum thickness ___ 2/3 ___

Maximum height _____2w_____
Sink marks: Yes _____ No ___X___

5. Solid pegs and bosses
   Maximum thickness _____
   Maximum weight _____
   Sink marks: Yes ___X___ No _____

6. Strippable undercuts
   Outside: Yes ___X___ No _____
   Inside: Yes ___X___ No _____

7. Hollow bosses: Yes ___X___ No _____

8. Draft angles
   Outside: Minimum _____1/2°_____ Ideal _____2°_____
   Inside: Minimum _____1/2°_____ Ideal _____2°_____

9. Molded-in inserts: Yes ___X___
                        No _____

10. Size limitations
    Length: Minimum _____Maximum _____Process _____
    Width: Minimum _____Maximum _____Process _____

11. Tolerances

12. Special process- and materials-related design details
    Blow-up ratio (blow molding) _____
    Core L/D (injection and compression molding) _____
    Depth of draw (thermoforming) _____4 in._____
    Other _____

18.9 RESIN FORMS

Both unmodified and glass fiber-filled versions of Radel A-400 are sold in pellet form.

18.10 SPECIFICATION OF PROPERTIES

See Master Outline of Materials Properties.

18.11 PROCESSING REQUIREMENTS

Poly(aryl sulfone) must be dried to less than 0.04 wt% water to prevent surface streaks or voids in molded or extruded specimens. Overall, molding and extrusion conditions are similar to those of Udel polysulfone P-1700.

Master Outline of Material Properties

MATERIAL ___ Poly(aryl sulfone) RADEL A-200

PROPERTY	TEST METHOD	ENGLISH UNITS	VALUE	METRIC UNITS	VALUE
MECHANICAL DENSITY	D792	lb/ft³	85	g/cm³	1.37
TENSILE STRENGTH	D638	psi	12,000	MPa	82.7
TENSILE MODULUS	D638	psi	385,000	MPa	2,655
FLEXURAL MODULUS	D790	psi	400,000	MPa	2,750
ELONGATION TO BREAK	D638	%	6.5	%	6.5
NOTCHED IZOD AT ROOM TEMP	D256	ft-lb/in.	1.6	J/m	85
HARDNESS	D785				
THERMAL DEFLECTION T @ 264 psi	D648	°F	400	°C	204
DEFLECTION T @ 66 psi	D648	°F		°C	
VICAT SOFT-ENING POINT	D1525	°F		°C	
UL TEMP INDEX	UL746B	°F		°C	
UL FLAMMABILITY CODE RATING	UL94		V-0 (1/16 in.)		V-0 (1/16 in.)
LINEAR COEFFICIENT THERMAL EXPANSION	D696	in./in./°F	0.000 027	mm/mm/°C	0.000 049
ENVIRONMENTAL WATER ABSORPTION 24 HOUR	D570	%	0.4	%	0.4
CLARITY	D1003	% TRANS-MISSION		% TRANS-MISSION	
OUTDOOR WEATHERING	D1435	%		%	
FDA APPROVAL					
CHEMICAL RESISTANCE TO: WEAK ACID	D543		Not attacked		Not attacked
STRONG ACID	D543		Not attacked		Not attacked
WEAK ALKALI	D543		Not attacked		Not attacked
STRONG ALKALI	D543		Not attacked		Not attacked
LOW MOLECULAR WEIGHT SOLVENTS	D543		Soluble, badly attacked by polar or chlorinated solvents		Soluble, badly attacked by polar or chlorinated solvents
ALCOHOLS	D543		Not attacked		Attacked
ELECTRICAL DIELECTRIC STRENGTH	D149	V/mil	383	kV/mm	15.1
DIELECTRIC CONSTANT	D150	10⁶ Hertz	3.54		3.54
POWER FACTOR	D150	10⁶ Hertz	0.0056		0.0056
OTHER MELTING POINT	D3418	°F	Amorphous	°C	Amorphous
GLASS TRANS-ITION TEMP.	D3418	°F	419	°C	215

213

18.12 PROCESSING-SENSITIVE END PROPERTIES

As earlier noted, poly(aryl sulfone) offers excellent molding characteristics over a temperature range of 650–750°F (343–395°C) and does not suffer from this processing latitude, as do high-temperature polymers generally. In secondary finishing operations such as wave and vapor-phase soldering), the material withstands the high temperatures generated in these processes without adverse effect.

Creep resistance, design-stress strength, and environmental stress-crack resistance all are improved with the addition of glass fiber. Drying is required to prevent surface streaks or voids in molded or extruded parts. Color can be affected by the amount of regrind used.

18.13 SHRINKAGE

Typical mold shrinkage of poly(aryl sulfone) as measured by ASTM D955, is 0.6%.

18.14 TRADE NAMES

Amoco Performance Products is the sole manufacturer of poly(aryl sulfone), under the trade names Radel A-100, A-200, and A-300.

BIBLIOGRAPHY

General References

R.N. Johnson and co-workers, *J. Polym. Sci. Part A-1* **5,** 2375 (1967).
Can. Pat. 847,963 (July 28, 1970), R.A. Clendinning and co-workers (to Union Carbide Corp.).
G.T. Kwiatkowski and co-workers, *paper presented at SPE NATEC,* Cleveland, Ohio, Oct. 5–7, 1976.
U.S. Pat. 4,108,837 (Aug. 22, 1978) R.N. Johnson and A.G. Farnham (to Union Carbide Corp.).

19

POLYALLOMER

Gordon Sharp

Eastman Chemical Products, P.O. Box 431 Kingsport, TN 37662

19.1 POLYALLOMER

The various propylene–ethylene polyallomers are unique materials that are quite different in physical properties and crystallinity from blends of polypropylene and polyethylene. They are also distinctly different from copolymers produced from propylene and ethylene by other polymerization processes.

Their insolubility in hexane and heptane evidences the fact that the polyallomer plastics are highly crystalline materials. Infrared spectra confirm that polyallomer chains comprise polymerized segments of each monomer employed. These segments exhibit crystallinities normally associated only with homopolymers of these monomers.

19.2 CATEGORY

Polyallomers are thermoplastic resins.

19.3 HISTORY

Polyallomer was developed in the mid-1960s by staff researchers of Eastman Chemical Products. Its consumption has grown to 15–20 million pounds in 1987.

19.4 POLYMERIZATION

A proprietary polymerization process produces copolymers of 1-olefins that give a degree of crystallinity normally obtained only with homopolymers. The term "polyallomer" was coined to identify the polymers manufactured by this process and to distinguish them from polymer blends previously known as copolymers. The polyallomer materials available today are based on block copolymers of propylene and ethylene.

Current (1988) prices for polyallomer range from 50 to 60 cents per pound.

19.5 DESCRIPTION OF PROPERTIES

Commercially available polyallomer formulations exhibit high or moderately high stiffness and medium, high, and extra-high impact characteristics. Special formulations with added heat endurance or added resistance to ultraviolet radiation, as well as formulations approved for use in contact with food under the regulations of the U.S. Food and Drug Administration, are also available.

The properties of polyallomer film of are especially noteworthy. The dart impact strength of 1 mil (0.025 mm) polyallomer film extruded at optimum conditions averages 55 g when the dart is dropped from a height of 26 in. (660 mm). This compares to 35 and 50 g, respectively, at a drop height of 13 in. (330 mm) for 1 mil (0.025 mm) polypropylene and polyethylene film.

The tear strength of film made from polyallomer is extremely high in the transverse direction; up to three or four times higher than that of either polypropylene or polyethylene film. However, the tear strength is relatively low in the machine direction compared with the other two materials.

19.6   APPLICATIONS

Polyallomer is used in such injection-molded items as fishing tackle boxes, typewriter cases, gas-mask cases, and bowling-ball bags. These lightweight cases can be molded entirely in one piece. Back, front, hinges, handles, and snap clasps can be molded in at the same time in a wide range of colors. Threaded container closures molded of polyallomer have the ability to seal containers without the need for an inner liner.

Shoe lasts molded of polyallomer resist cracking and denting under repeated hammer blows. They withstand temperatures from −40 to 300°F (−40 to 149°C) and can withstand up to 300 pounds of force (1335 N).

Polyallomer sheet with pre-embossed patterns, such as leather graining, can be vacuum formed into luggage shells that retain the leather grain effect, even at the corners where other materials can lose the detail of the pre-embossed finish.

Pre-embossed sheet also is used in fabricating loose-leaf notebooks of all sizes and shapes with integral hinges hot stamped into the sheet.

Antiblushing characteristics and low-temperature toughness are primarily responsible for the success of polyallomer in such applications as extruded automobile windlace and molded cowl panels.

19.7   ADVANTAGES/DISADVANTAGES

Polyallomers combine the most desirable properties of both crystalline polypropylene and high-density polyethylene (HDPE) and can offer impact strengths three to four times that of polypropylene. Some formulations do not exhibit low-temperature brittleness until the temperature drops to about −40°F (−40°C). Resistance to heat distortion is better than that of HDPE but not quite as good as that of polypropylene. Polyallomer has better abrasion resistance than polypropylene and comparable hinge-forming characteristics.

Polyallomers are available in a broad range of fully compounded colors or as color concentrates. Polypropylene color concentrates can be used to color polyallomer, since these two polymers are compatible.

Polyallomers can be processed easily on conventional molding and extruding equipment.

19.8   PROCESSING TECHNIQUES

See the Master Material Outline and the Standard Design Chart.

Master Material Outline

PROCESSABILITY OF _____Polyallomer_____

PROCESS	A	B	C
INJECTION MOLDING	X		
EXTRUSION	X		
THERMOFORMING	X		
FOAM MOLDING	X		
DIP, SLUSH MOLDING			
ROTATIONAL MOLDING		X	
POWDER, FLUIDIZED-BED COATING		X	
REINFORCED THERMOSET MOLDING			X
COMPRESSION/TRANSFER MOLDING	X		
REACTION INJECTION MOLDING (RIM)			X
MECHANICAL FORMING	X		
CASTING			X

A = Common processing technique.
B = Technique possible with difficulty.
C = Used only in special circumstances, if ever.

Standard design chart for _____**Polyallomer**_____

1. Recommended wall thickness/length of flow
   Minimum _____for_____ distance
   Maximum _____for_____ distance
   Ideal _____for_____ distance

2. Allowable wall thickness variation, % of nominal wall _____

3. Radius requirements
   Outside: Minimum _____1/32 in._____ Maximum _____
   Inside: Minimum _____1/32 in._____ Maximum _____

4. Reinforcing ribs
   Maximum thickness _____
   Maximum height _____
   Sink marks: Yes _____ No _____

5. Solid pegs and bosses
   Maximum thickness _____
   Maximum weight _____
   Sink marks: Yes _____ No _____

6. Strippable undercuts
   Outside: Yes ___X___ No _____
   Inside: Yes ___X___ No _____

7. Hollow bosses: Yes ___X___ No _____

8. Draft angles
   Outside: Minimum _____$\frac{1}{4}°$_____ Ideal _____$1°$_____
   Inside: Minimum _____$\frac{1}{4}°$_____ Ideal _____$1°$_____

9. Molded-in inserts: Yes ___X___
                        No _____

10. Size limitations
    Length: Minimum _____ Maximum _____ Process _____
    Width: Minimum _____ Maximum _____ Process _____

11. Tolerances

12. Special process- and materials-related design details
    Blow-up ratio (blow molding) _____
    Core L/D (injection and compression molding) _____
    Depth of draw (thermoforming) _____
    Other _____

19.9   RESIN FORMS

Polyallomers are available in pellet form.

19.10   SPECIFICATION OF PROPERTIES

See the Master Outline of Material Properties.

19.11   PROCESSING REQUIREMENTS/19.12   PROCESSING-SENSITIVE END PROPERTIES

19.11.1   Injection Molding

In general, polyallomer can be injection molded under the same conditions as those used for molding polyproplene of the same flow rate. Large, flat pieces that hold their shape can be easily molded from polyallomer.

Parts molded from polyallomer are comparable to the rubber-modified polypropylene blends in stiffness, hardness, and tensile strength. However, polyallomer is superior to the rubber-modified materials with respect to color, clarity, moldability, dissipation factor, and resistance to blushing when bent or stretched. Parts molded from polyallomer form stronger welds than similar parts molded from rubber-modified polypropylene.

Although the properties of polyallomer are affected less by mold temperature than are those of polypropylene, the mold temperature should be controlled in order to obtain

Master Outline of Material Properties

MATERIAL ___Polyallomer___

PROPERTY	TEST METHOD	ENGLISH UNITS	VALUE	METRIC UNITS	VALUE
MECHANICAL DENSITY	D792	lb/ft^3	55.9–56.1	g/cm^3	0.896–0.900
TENSILE STRENGTH	D638	psi	2800–3800	MPa	19–26
TENSILE MODULUS	D638	psi		MPa	
FLEXURAL MODULUS	D790	psi	90,000–130,000	MPa	621–896
ELONGATION TO BREAK	D638	%		%	
NOTCHED IZOD AT ROOM TEMP	D256	ft-lb/in.	1.6–2.5	J/m	85–133
HARDNESS	D785		41–70		41–70
THERMAL DEFLECTION T @ 264 psi	D648	°F	118–122	°C	48–50
DEFLECTION T @ 66 psi	D648	°F		°C	
VICAT SOFTENING POINT	D1525	°F	253–275	°C	123–135
UL TEMP INDEX	UL746B	°F		°C	
UL FLAMMABILITY CODE RATING	UL94				
LINEAR COEFFICIENT THERMAL EXPANSION	D696	in./in./°F	15–18 × 10^{-5}	mm/mm/°C	27–32 × 10^{-5}
ENVIRONMENTAL WATER ABSORPTION 24 HOUR	D570	%	<0.01	%	<0.01
CLARITY	D1003	% TRANSMISSION		% TRANSMISSION	
OUTDOOR WEATHERING	D1435	%		%	
FDA APPROVAL			Yes		Yes
CHEMICAL RESISTANCE TO: WEAK ACID	D543		[a]		[a]
STRONG ACID	D543				
WEAK ALKALI	D543				
STRONG ALKALI	D543				
LOW MOLECULAR WEIGHT SOLVENTS	D543				
ALCOHOLS	D543				
ELECTRICAL DIELECTRIC STRENGTH	D149	V/mil	800–950	kV/mm	32–37
DIELECTRIC CONSTANT	D150	10^6 Hertz	2.3		2.3
POWER FACTOR	D150	10^6 Hertz	5 × 10^{-4}		5 × 10^{-4}
OTHER MELTING POINT	D3418	°F		°C	
GLASS TRANSITION TEMP.	D3418	°F		°C	

[a] Above 50°F (10°C).

[b] Inert to most aqueous solutions except strong oxidizing acids; resistant to organic solvents and polar substances; swells at room temperature and dissolves at high temperature on exposure to aromatic and chlorinated hydrocarbons.

optimum properties. It is possible, however, to mold a relatively tough part from polyallomer even when the mold cooling is inadequate.

When parts molded from polyallomer are first removed from the mold, they are relatively flexible; with time, they become stiffer. Normally, they reach their ultimate stiffness within 24 hours. This behavior results from the fact that polyallomer remains in a relatively amorphous state for some time after molding and crystallizes on standing. The stiffness of molded articles will increase as the plastic becomes more and more crystalline. It is for this reason that the mold temperature does not appreciably affect the properties of a polyallomer molded article.

Polyallomer can be molded satisfactorily with cylinder temperatures in the range of 450–500°F (232–260°C). If possible, the mold temperature should be kept in the range of 60–70°F (16–21°C). At temperatures in this range, an overall cycle time of around 35 s can be obtained for a part weighing about 6 to 8 oz (170–227 g). The pressure on the plastic during the molding cycle should be in the range of 10,000–20,000 psi (69–138 MPa). The pressure used will depend to some extent on the type of molding machine and the complexity of the article being molded.

19.11.2   Film Extrusion

Flat film can be produced from polyallomer on conventional extrusion and film-handling equipment.

Film with excellent optical and strength properties can be produced from polyallomer, which can be processed over a wide range of extrusion conditions without adverse effects to its properties. Polyallomer film exhibits optical properties comparable to those obtained with polypropylene film, but is much tougher than polypropylene film. Also, better seals are possible, particularly on machines that make side-weld sealed bags. The increased toughness and better sealing characteristics, combined with the high yield per pound of polymer, produces an economical film for packaging applications. Polyallomer yields approximately 31,000 in.^{2} (787,000 mm^2) of 1 mil (0.025 mm) film per pound of polymer.

Changes in melt temperature affect the optical properties of film more than they do any other property. As is the case with all polyolefin films, haze decreases and gloss and transparency increase as the melt temperature rises. Film blocking and coefficient of friction increase slightly as the melt temperature increases. Polyallomer is available with slip and antiblock additives at levels necessary in certain film applications. The other polyallomer film properties are not appreciably affected by changes in melt temperature.

Film properties are not affected appreciably by melt pressure, but melt pressure does affect the output and power requirements of the extrusion machine during the extrusion of film.

Film extruded from polyallomer can be drawn down to a thickness as low as 0.2 mil (0.0051 mm) at a film speed of 300 ft (91 m) per minute.

Polyallomer requires a surface treatment to obtain adhesion of inks or other decorative materials. The same methods of surface treatment as those used for treating other polyolefins can be used successfully for polyallomer. Polyallomer film requires approximately the same level of treatment as that required for polypropylene film.

Table 19.1 gives typical conditions for extruding film from polyallomer by the chill-roll process.

19.11.3   Wire Coating

Polyallomer processes extremely well in wire-coating operations. Its electrical and physical properties make it a very practical wire-coating material. Twenty-six gauge wire can be coated with either solid or cellular polyallomer at rates up to 4000 ft (1219 m) per minute. The coatings produced are smooth and abrasion-resistant.

Polyallomer can be processed on standard equipment used for coating polyethylene, polypropylene, and other thermoplastics on wire. A satisfactory wire coating can be obtained

TABLE 19.1. Typical Extrusion Conditions for Flat Film of Polyallomer

Condition	Value
Cylinder temperatures, °F (°C)	
Zone 1	430 (221)
Zone 2	500 (260)
Zone 3	500 (316)
Zone 4	500 (316)
Gate temperature, °F (°C)	500 (316)
Die temperature, °F (°C)	475 (249)
Melt temperature, °F (°C)	500 (316)
Melt pressure, psi (MPa)	5000 (34)
Screw speed, rpm	43
Chill roll temperature, °F (°C)	
1	50 (10)
2	64 (18)
Film speed, fpm (mpm)	215 (66)
Film thickness, mils	1.0 (0.025)

for polyallomer with the same coating technique as that used for polypropylene. Generally, polyallomer produces a smooth coating at a slightly lower extrusion temperature than does polypropylene.

In some applications, it is necessary for the finished coating to have a high elongation value. This can be accomplished by preheating the wire. Wire-preheat temperatures in the range of 350–400°F (177–204°C) have been found to increase the elongation of polyallomer coatings to a satisfactory value. Polyallomer wire coatings have approximately five times the cut resistance of HDPE coatings.

Table 19.2 shows typical processing conditions used for coating 22-gauge wire with polyallomer.

TABLE 19.2. Typical Wire-Coating Conditions for Polyallomers

Condition	Value	
Die type	Gradually decreasing taper	
Guider tip spacing, in. (mm)	5/16	(8)
Wire size, A.W.G.	22	
Orifice diameter, in. (mm)	0.044	(1.1)
Cylinder temperature, °F	300 (149)350 (177) 350 (177)350 (°C)	
Zone 1	300	(149)
Zone 2	350	(177)
Zone 3	350	(177)
Zone 4	350	(177)
Neck temperature, °F (°C)	350	(177)
Head temperature, °F (°C)	350	(177)
Die temperature, °F (°C)	375	(191)
Melt temperature, °F (°C)	365	(185)
Wire speed, fpm (mpm)	4000	(1210)
Head pressure, psi (MPa)	3000	(21)
Screw speed, rpm	38	
Coated-wire diameter, in. (mm)	0.037	(0.9)
Surface finish of coating	Smooth	
Water bath temperature, °F (°C)	70	(21)
Wire preheat temperature, °F (°C)	350	(177)

19.11.4   Sheet Extrusion

Heavy-gauge sheet with excellent surface finish and thickness uniformity can be extruded from polyallomer on conventional extrusion equipment. A wide range of temperatures on the cylinder, die, and chill rolls can be used to produce good quality sheet. The temperature can be varied as much as 70°F (39°C) without causing any change in the appearance of the product.

With propylene–ethylene polyallomer formulas currently available, gauge control is very good. Sheet 100 mils (2.54 mm) thick can be extruded on calendar rolls with a thickness variation of less than 1 mil across its width. Heavy-gauge sheet can be extruded on a 3-1/4 in. (83 mm) extruder with a 16:1 L/D ratio.

Table 19.3 gives typical processing conditions for extruding heavy-gauge sheet from polyallomer.

19.11.5   Vacuum Forming

Sheeting extruded from polyallomer exhibits vacuum forming characteristics similar to those of both polyethylene and modified polypropylene sheet. Sandwich-type heaters are usually needed to obtain uniform heating. The heating time required is comparable to that used with polypropylene of the same sheet thickness. Sheet made from polyallomer is water-clear in appearance when the forming temperature is reached. If the sheet sags uniformly, a uniform temperature over the entire area of the sheet is indicated. This also indicates that the sheet will stretch uniformly over the forming mold.

Since polyallomer is a crystalline polymer, it requires a great deal of heat to melt the crystallites. When the crystallites melt the sheet becomes very limp, but it has sufficient melt strength to support its own weight in the vacuum-forming operation.

Like other polyolefins, polyallomer is an excellent thermal insulator and, therefore, transmits heat slowly. If a sheet of this material is heated too rapidly, the surface will get too hot while the interior of the sheet is still cold. If the sheet is heated only on one side, heat transfer becomes a significant consideration.

Uniform sheet heating is an absolute necessity for good vacuum forming results. One method of improving heat distribution is to place a heavy screen, eg, 8-mesh stainless-steel wire, between the heaters and the plastic sheet. The heaters heat the screen, which, because

TABLE 19.3.   Typical Conditions for the Extrusion of Heavy-Gauge Sheet

Condition	Value
Cylinder temperatures, °F (°C)	
Zone 1	350 (177)
Zone 2	400 (204)
Zone 3	400 (204)
Gate temperature, °F (°C)	400 (204)
Die temperatures, °F (°C)	
Back	400 (204)
Front	400 (204)
Ends	400 (204)
Upper chill roll temperature, °F (°C)	120 (49)
Lower chill roll temperature, °F (°C)	160 (71)
Calender roll temperature, °F (°C)	180 (82)
Screw speed, rpm	28
Take-off speed, fpm (mpm)	1.5 (0.4)
Sheet thickness, mils	100 (2.54)
Sheet width, in. (mm)	36 (914)

TABLE 19.4. Typical Conditions for Vacuum-Forming
Polyallomer Sheet

Condition	Value
Heater temperature, °F (°C)	
Upper-outer zone	1,100 (587)
Upper-inner zone	700 (371)
Lower entire heater	1,100 (587)
Heating time,	
Upper	60
Lower	40
Cooling time, min (approx)	3

of the rapid heat transfer of metal, becomes heated to a uniform temperature. The screen then acts as the heater and heats the plastic sheet uniformly.

In order to maintain a uniform sheet temperature, drafts should be eliminated around the vacuum-forming machine. Heat loss from the edge of the plastic sheet by radiation and convection can be counteracted by arranging the upper heaters so that the sheet is heated a little more around the outside than it is in the middle. This can be accomplished by installing a band of heaters around the upper heater plate. These heaters are controlled separately from the other heaters in the upper plate. The band would usually be only a few inches (25.4 mm) wide, and its heaters would run approximately 400°F (222°C) hotter than other upper-plate heaters.

Shown in Table 19.4 are the conditions used on a laboratory model vacuum-forming machine to form 0.125 in. (3.2 mm) thick polyallomer sheet.

Polyallomer can be vacuum formed on either a male or female mold with equal success. The selection of the mold type usually depends on the detail required, the depth of draw, and the best location for the thickest and thinnest portions of the formed piece.

19.13 SHRINKAGE

See the Mold Shrinkage MPO.

Mold Shrinkage Characteristics (in./in. or mm/mm)

MATERIAL   Polyallomer

IN THE DIRECTION OF FLOW

THICKNESS

	mm: 1.5	3	6	8	FROM	TO	#
MATERIAL	in.: 1/16	1/8	1/4	1/2			
	0.013	0.015	0.020	0.025			

	PERPENDICULAR TO THE DIRECTION OF FLOW						
	0.013	0.015	0.020	0.025			

NOTE: Mold shrinkage is approximate. It is affected by part design, mold temperature, thickness, injection pressure, packing time, cycle time, orientation, gate design, gate size, gate location, glass content, glass size, and filler content.

19.14 TRADE NAMES

Polyallomer is available under the trade name Tenite from Eastman Chemical Co.

20

POLYAMIDE-IMIDE (PAI)

James L. Throne
Sherwood Technologies, Inc., Naperville, Illinois

20.1 STRUCTURE

20.2 CATEGORY

PAI is a hard, tough thermoplastic.

20.3 HISTORY

PAI was first synthesized as a high-temperature wire enamel in the 1960s. The high performance thermoplastic was an outgrowth of research done by James Stephens, Amoco Chemicals Corporation Research Center, Naperville, Ill. It was introduced as an injection-molding resin in 1976. The first applications were as burn-in electrical pin connectors. By 1980, PAI consumption was approximately 500,000 lb. This article was written in September 1986.

20.4 POLYMERIZATION

PAI is produced by condensation polymerization of trimellitic anhydride (TMA) and a mixture of diamines such as ortho-bisaniline. Acetic acid is the condensation by-product. It can also be produced from trimellitoyl chloride and methylene dianiline. In this case hydrochloric acid and water are evolved. The reaction is frequently carried out in a *n*-methylpyrrolidone (NMP) solution.

Monomers require a very high degree of purity. Thus, raw material costs are very high. Heavy-duty high-temperature processing equipment, the long times required to react PAI fully after the part is molded, and the difficulty in reusing scrap add to the finished cost. The price of finished, unreinforced PAI is about $20/lb. Thirty percent glass-reinforced PAI costs about $15/lb. Thirty percent graphite-reinforced PAI costs about $20/lb.

20.5 DESCRIPTION OF PROPERTIES

PAI has exceptional high-temperature stability. The glass-transition temperature of fully reacted PAI is 525°F (275°C). As shown in Table 20.1, the tensile strength and flexural modulus of unreinforced PAI remain very high at 500°F (260°C).

With 30 wt% glass fiber, the flexural modulus of PAI at 73°F (23°C) is 1.6×10^6 psi (11 GPa) and the tensile strength is 28,000 psi (193 MPa). With 30 wt% graphite fiber, the flexural modulus at 73°F (23°C) is 2.5×10^6 psi (17 GPa) and the tensile strength is 30,000 psi (207 MPa).

The deflection temperature under load (DTUL) at 264 psi (1.82 MPa) for unreinforced PAI is 525°F (275°C). The UL Relative Thermal Index is 428°F (220°C). PAI is inherently flame retardant with an Limiting Oxygen Index value of 43%. PAI is chemically resistant to hydrocarbons and halogenated solvents and to acid and basic solutions below 212°F (100°C). When it is compounded with graphite, molybdenum disulfide, and PTFE powder, a low friction grade has a wear factor of 3×10^{-9} in.³/ft-lb/min at a pressure-velocity level of 50,000 ft-lb/in.²/min. The static and dynamic coefficients of friction at this level are 0.14 and 0.11, respectively.

TABLE 20.1. Properties of PAI 4203L

		Temperature, °F (°C)		
Property		73 (23)	300 (149)	500 (260)
Tensile strength,	psi	27,000	15,000	7,500
D1708	(MPa)	(185)	(100)	(50)
Flex modulus,	psi	660,000	520,000	430,000
D790	(GPa)	(4.55)	(3.59)	(3.00)
Elongation	%	12	17	22

20.6 APPLICATIONS

PAI is used primarily in electrical connectors, bushings, thrust bearings, compressor valve plates and rings, pump and equipment housings, and connecting rods. It is also used in racing engine connecting rods, valves, and piston elements.

PAI is a leading thermoplastic matrix material in continuous-fiber graphite composites. Fully reacted PAI is considered unprocessible. As a result, the PAI polymerization reaction is stopped short of completion to produce injection and extrusion grades. In order to obtain the high-performance properties given above, the molded or extruded product must be solid-stated or postcured in a staged high-temperature oven.

20.7 ADVANTAGES/DISADVANTAGES

PAI provides the designer with a moldable, tough, high-modulus thermoplastic capable of continuous use at temperatures in excess of 392°F (200°C). It can be highly reinforced to provide greater high-temperature strength and outstanding long-term creep resistance. It can be filled to provide very low frictional surfaces, even at very high pressure-velocity conditions. It is inherently flame retardant.

PAI also has some decided shortcomings. It must be processed in a prefinished state, and the molded parts must be postcured to achieve optimum properties. It has a very high melt viscosity and a very high processing temperature. As a result, processing machinery must be high-temperature and robust. Fully reacted PAI is unprocessible and cannot be recycled. PAI is a form of polyamide and thus is quite moisture sensitive. Postcuring of molded parts results in substantial and somewhat nonuniform shrinkage. Part dimensions can be difficult to predict.

Most significantly, PAI is a very expensive polymer, costing typically $20/lb. As a result, only those applications that require its unique blend of properties and performance are fabricated of PAI. Despite its very high price and severe processing restrictions, PAI is a functional substitute for machined metals in aeronautical and aerospace applications. PAI has lower weight and higher strength per unit weight than many high-performance metal alloys, and molding is far less expensive than metal machining and assembly.

20.8 PROCESSING TECHNIQUES

PAI is injection molded in a prefinished state. Its viscosity is highly shear sensitive. At 645°F (340°C), the viscosity is 10^7 poise (1 MPa·s) at a shear rate of 1 s^{-1}, but only 10^5 poise (0.01 MPa·s) at 1000 s^{-1}. As a result, high injection speed is used. A heavy-duty high-temperature injection molding machine is required. A typical 3 zone + nozzle temperature profile for unreinforced PAI is 600 − 625 − 650 + 700°F (315 − 330 − 345 + 370°C). The screw compression ratio should be between 1:1 and 1.5:1. Screw length should be between 18D and 24D. Minimum injection pressure of 20,000 psi (138 MPa) is required and 40,000 psi (276 MPa) recommended. A nitrogen-gas-driven hydraulic accumulator is frequently used to achieve high speed transfer.

PAI continues to react (until it begins to degrade) at elevated temperatures. As a result, the ideal shot size should be about 50% of the machine capacity, with a range of 20 to 80%. The screw should bottom out against the nozzle on each shot. Oil-heat molds to 425°F (220°C) minimize flow freeze-off in thin sections. Gates and runners should be large in diameter and short in length to minimize flow orientation that can result in high residual stresses. PAI is prone to jet into mold cavities at high injection speeds. The high viscosity at low shear rates can cause severe weld-line knitting problems.

Master Material Outline

PROCESSABILITY OF _____ Polyamide–imide

PROCESS	A	B	C
INJECTION MOLDING	X		
EXTRUSION		X	
THERMOFORMING			X
FOAM MOLDING			X
DIP, SLUSH MOLDING			
ROTATIONAL MOLDING			
POWDER, FLUIDIZED-BED COATING			
REINFORCED THERMOSET MOLDING			
COMPRESSION/TRANSFER MOLDING	X		
REACTION INJECTION MOLDING (RIM)			
MECHANICAL FORMING		X	
CASTING			

A = Common processing technique.
B = Technique possible with difficulty.
C = Used only in special circumstances, if ever.

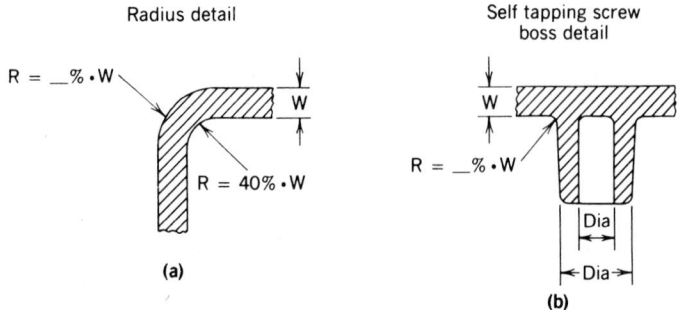

Radius detail

$R = \_\_\% \cdot W$

$R = 40\% \cdot W$

W

(a)

Self tapping screw boss detail

W

$R = \_\_\% \cdot W$

Dia

Dia

(b)

Solid rib detail   Hollow rib detail

(c)   (d)

Standard design chart for __Polyamide–imide (Unreinforced)__

1. Recommended wall thickness/length of flow
   Minimum _____0.020 in._____ for _____0.500 in._____ distance
   Maximum _____None_____ for _____ distance
   Ideal _____ for _____ distance

2. Allowable wall thickness variation, % of nominal wall _____approx. 0.5%_____

3. Radius requirements (scheme **a**)
   Outside: Minimum _____0.015 in._____ Maximum _____none_____
   Inside: Minimum _____0.015 in._____ Maximum _____none_____

4. Reinforcing ribs (scheme **c** and **d**)
   Maximum thickness _____0.5W to 0.8W_____
   Maximum height _____0.5W to 0.8W_____
   Sink marks: Yes __X__ No _____ (Post-curing can remove some measure of sink marks)

5. Solid pegs and bosses
   Maximum thickness _____0.8W_____
   Maximum weight _____0.5W_____
   Sink marks: Yes __X__ No _____ (See note above)

6. Strippable undercuts
   Outside: Yes _____ No __X__
   Inside: Yes _____ No __X__

7. Hollow bosses (scheme **b**): Yes _____ No __X__

8. Draft angles
   Outside: Minimum _____0.5°_____ Ideal _____1°_____
   Inside: Minimum _____0.5°_____ Ideal _____1°_____

9. Molded-in inserts: Yes _____
   No __X__

10. Size limitations
    Length: Minimum __None__ Maximum ____[a]____ Process __injection molding__
    Width: Minimum __None__ Maximum ____[a]____ Process __injection molding__

11. Tolerances
    See attached table

12. Special process- and materials-related design details
    Blow-up ratio (blow molding)_____
    Core L/D (injection and compression molding) _____
    Depth of draw (thermoforming) _____
    Other _____Gating critical to prevent jetting_____

[a]Depends upon the clamp capacity of the injection molding machine. Practically, 20 in. × 20 in. requires 4000T machine.

A twin-screw extruder is recommended for sheet, rod, and profile extrusion. PAI should be extruded at melt temperatures of 700°F (370°C) or more. Very small die land lengths and heavy-duty extruder heads are required. Owing to its very high viscosity at very low shear rates, it is not necessary or desirable to cool PAI on a roll stack. Ambient air cooling followed by water cooling is adequate.

20.9   RESIN FORMS

PAI is available in pellet form for injection molding or extrusion. The reinforced grade contains 0.5 wt% PTFE and 3.0 wt% titanium dioxide. The low-friction grade includes 20 wt% graphite powder (99% < 40 microns), 3 wt% PTFE, and 0.5 wt% molybdenum disulfide. Two fiber reinforced grades have 30 wt% and 40 wt% E-glass fiber for improved stiffness at low relative cost. The high-modulus fiber reinforced grade contains 30 wt% graphite fibers.

Forty-micron (median) reactor powder is available for compression molding and as a matrix in high-performance composites. To minimize residual n-methylpyrrolidone, high temperature [300°F (150°C)] vacuum drying of the reactor powder is recommended.

20.10   SPECIFICATION OF PROPERTIES

The material properties for unreinforced and 30% graphite-reinforced grades of PAI are given in the Master Outline of Material Properties.

20.11   PROCESSING REQUIREMENTS

PAI must be extruded or molded in a prefinished state. It must be thoroughly dried to less than 100 ppm moisture. Normally, drying can be accomplished in 8 hours in a vented, circulating oven with an air temperature of 350°F (175°C) dehumidifed to -40°F (-40°C). Beds should not be more than 2 in. (51 mm) deep.

As-molded PAI has a low glass-transition temperature of about 450°F (230°C), a low modulus and is quite brittle. To achieve high-strength parts, PAI must be solid-stated in a process-controlled temperature-staged vented circulating air oven. The time for solid-stating or postcuring depends on the part thickness, since the reaction by-product, water, must diffuse through the part. Molecular diffusion is strongly dependent on temperature and the diffusion rate is proportional to the square of the part thickness. Very thick parts may take weeks to cure. Care must be taken to minimize the amount of fully reacted PAI on the molded surface, as this inhibits water diffusion from the part interior. This is the major source of out-of-tolerance parts. Postcuring is thought to increase chain length, to fully imidize the chain by closing the ring at the carbonyl, and possibly to cause a small amount of cross-linking. Although the molecular weight of as-molded PAI can be determined by solution viscosity, fully solid-stated PAI has no known solvent. Glass-transition temperature can be used as a measure of the degree of reaction, however.

Fully reacted PAI parts should show a doubling in tensile strength over the as-molded material, as increase in DTUL and glass-transition temperature T_g to 527°F (275°C), an increase in ultimate elongation from 4% to 12–15%, and a ten-fold increase in chemical and wear resistance. About half the total part shrinkage occurs during postcuring. Very high tolerance parts, such as gears and connectors, require substantial care during postcuring to retain dimensions. Scrap production can be substantial in this phase of processing.

MATERIAL __Polyamide–imide (30% wt%, Graphite Reinforced, Fully Reacted)__

PROPERTY	TEST METHOD	ENGLISH UNITS	VALUE	METRIC UNITS	VALUE
MECHANICAL DENSITY	D792	lb/ft^3	88.6	g/cm^3	1.42
TENSILE STRENGTH	D638	psi	30,000	MPa	207
TENSILE MODULUS	D638	psi	2,900,000	MPa	20,000
FLEXURAL MODULUS	D790	psi	2,600,000	MPa	17,900
ELONGATION TO BREAK	D638	%	4	%	4
NOTCHED IZOD AT ROOM TEMP	D256	ft-lb/in.	1.2	J/m	70
HARDNESS	D785		E 94		E 94
THERMAL DEFLECTION T @ 264 psi	D648	°F	525	°C	275
DEFLECTION T @ 66 psi	D648	°F	NA	°C	NA
VICAT SOFT-ENING POINT	D1525	°F	NA	°C	NA
UL TEMP INDEX	UL746B	°F	428	°C	220
UL FLAMMABILITY CODE RATING	UL94		V-0		V-0
LINEAR COEFFICIENT THERMAL EXPANSION	D696	in./in./°F	1.4×10^{-5}	mm/mm/°C	2.5×10^{-5}
ENVIRONMENTAL WATER ABSORPTION 24 HOUR	D570	%	0.8	%	0.8
CLARITY	D1003	% TRANS-MISSION	Opaque	% TRANS-MISSION	Opaque
OUTDOOR WEATHERING	D1435	%	NA	%	NA
FDA APPROVAL			No		No
CHEMICAL RESISTANCE TO: WEAK ACID	D543		Yes		Yes
STRONG ACID	D543		Yes		Yes
WEAK ALKALI	D543		Yes		Yes
STRONG ALKALI	D543		Yes		Yes
LOW MOLECULAR WEIGHT SOLVENTS	D543		Yes to 212°F		Yes to 100°C
ALCOHOLS	D543		Yes		Yes
ELECTRICAL DIELECTRIC STRENGTH	D149	V/mil	NA	kV/mm	NA
DIELECTRIC CONSTANT	D150	10^6 Hertz	NA		NA
POWER FACTOR	D150	10^6 Hertz	NA		NA
OTHER MELTING POINT	D3418	°F		°C	
GLASS TRANS-ITION TEMP.	D3418	°F	527	°C	275

MATERIAL _____ Polyamide–imide (Unreinforced — Fully Reacted Properties)

PROPERTY	TEST METHOD	ENGLISH UNITS	VALUE	METRIC UNITS	VALUE
MECHANICAL DENSITY	D792	lb/ft^3	87.3	g/cm^3	1.40
TENSILE STRENGTH	D638	psi	27,000	MPa	186
TENSILE MODULUS	D638	psi	650,000	MPa	4500
FLEXURAL MODULUS	D790	psi	710,000	MPa	4900
ELONGATION TO BREAK	D638	%	12 (ultimate no yield)	%	12 (ultimate no yield)
NOTCHED IZOD AT ROOM TEMP	D256	ft-lb/in.	3.0	J/m	175
HARDNESS	D785		E 78		E 78
THERMAL DEFLECTION T @ 264 psi	D648	°F	500	°C	260
DEFLECTION T @ 66 psi	D648	°F	NA	°C	NA
VICAT SOFT-ENING POINT	D1525	°F	NA	°C	NA
UL TEMP INDEX	UL746B	°F	428	°C	220
UL FLAMMABILITY CODE RATING	UL94		V-0		V-0
LINEAR COEFFICIENT THERMAL EXPANSION	D696	in./in./°F	2×10^{-5}	mm/mm/°C	3.6×10^{-5}
ENVIRONMENTAL WATER ABSORPTION 24 HOUR	D570	%	1.0	%	1.0
CLARITY	D1003	% TRANS-MISSION	Opaque	% TRANS-MISSION	Opaque
OUTDOOR WEATHERING	D1435	%	NA	%	NA
FDA APPROVAL			No		No
CHEMICAL RESISTANCE TO: WEAK ACID	D543		Yes		Yes
STRONG ACID	D543		Yes		Yes
WEAK ALKALI	D543		Yes		Yes
STRONG ALKALI	D543		Yes		Yes
LOW MOLECULAR WEIGHT SOLVENTS	D543		Yes to 212°F		Yes to 100°C
ALCOHOLS	D543		Yes		Yes
ELECTRICAL DIELECTRIC STRENGTH	D149	V/mil	500	kV/mm	20
DIELECTRIC CONSTANT	D150	10^6 Hertz	NA		NA
POWER FACTOR	D150	10^6 Hertz	NA		NA
OTHER MELTING POINT	D3418	°F		°C	
GLASS TRANS-ITION TEMP.	D3418	°F	527	°C	275

Fully reacted PAI cannot be reprocessed. Molding scrap as powder can be recycled to 30 wt% without a substantial loss in physical properties.

20.12 PROCESSING-SENSITIVE END PROPERTIES

Optimum parts are obtained when PAI is thoroughly dried; when the mold is designed to minimize jetting; when the injection speed, pressure, and temperature profile are optimum; when the machine is sized for the size of the part; and when sufficient time has been allotted for fully reacting the molded parts. For all critical parts, 100% X-ray inspection is recommended.

20.13 SHRINKAGE

See shrinkage tables, Mold Shrinkage Characteristics.

Mold Shrinkage Characteristics (in./in. or mm/mm)

MATERIAL   Polyamide–imide[a]

IN THE DIRECTION OF FLOW

THICKNESS

MATERIAL	mm: 1.5 in.: 1/16	3 1/8	6 1/4	8 1/2	FROM	TO	#
PAI 4203L at 138 MPa, (20,000 psi)					0.008	0.010	
PAI 4203L at 207 MPa, (30,000 psi)					0.006	0.007	in./in.
PAI 4301 30% glass reinforced					0.005	0.007	in./in.

	PERPENDICULAR TO THE DIRECTION OF FLOW						
PAI 4203L at 138 MPa, (20,000 psi)					0.008	0.010	in./in.

[a]Unreinforced; molding and post-curing shrinkage depend upon injection pressure.
NOTE: Mold shrinkage is approximate. It is affected by part design, mold temperature, thickness, injection pressure, packing time, cycle time, orientation, gate design, gate size, gate location, glass content, glass size, and filler content.

Standard Tolerance Chart

PROCESS ___Injection molding___ MATERIAL ___Unreinforced Polyamide-imide___

Values include molding and post-curing shrinkage tolerance

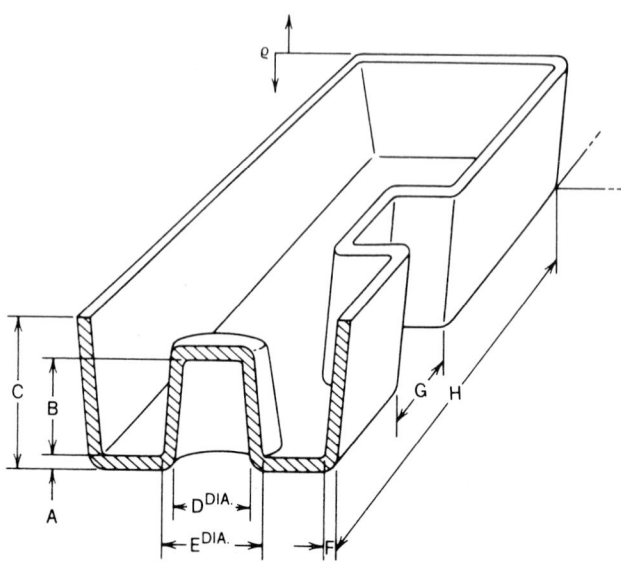

Values are based on a nominal 1/8 in. (3.2 mm) wall thickness and given in plus or minus in./in. and mm/mm.
FINE tolerance – Special care and added cost.
COMMERCIAL tolerance – Normal care required.
COARSE tolerance – Minimal care required, lowest cost.

Tolerance	A	B	C	D	E	F	G	H	
FINE			0.001		0.001		0.001		in./in.
									mm/mm
COMMERCIAL			0.002		0.002		0.002		in./in.
									mm/mm
COARSE			0.003		0.003		0.003		in./in.
									mm/mm

20.14 TRADE NAMES

PAI is sold under the trade name of Torlon by Amoco Performance Products, Atlanta, Ga. In 1985, it was produced by Rho Chemical Company, Joliet, Ill., under exclusive contract to Amoco Chemicals Corporation, Chicago, Ill. It is now produced by Amoco Chemicals Corporation.

BIBLIOGRAPHY

General References

TAT-1, TAT-24a, TAT-44 Amoco Chemicals Technical Bulletins, undated, Chicago, Ill.

C.J. Billerbeck and S.J. Henke, "Torlon Poly(amide–imide)," in J.M. Margolis, ed., *Engineering Thermoplastics: Properties and Applications,* Marcel Dekker, New York, 1985, p. 373.

G.V. Cekis, "Polyamide–imide" in *Modern Plastics Encyclopedia,* Vol. 63, 10A, McGraw-Hill, New York, 1986–1987, p. 37.

A.S. Hay, "Engineering Thermoplastics, Miscellaneous," in M.B. Bever, ed., *Encyclopedia of Materials Science and Engineering,* Vol. 2, Pergamon-MIT, 1986, p. 1556.

21

POLYARYLATE

Lloyd M. Robeson

Air Products & Chemicals Inc., P.O. Box 538, Allentown, PA 18105

21.1 STRUCTURE

21.2 CATEGORY

Polyarylates are a family of aromatic polyesters based on diphenols and aromatic dicarboxylic acids.

12.3 HISTORY

As a result of their processibility, weatherability, high softening temperatures, and excellent physical and mechanical properties, polyarylates have been under intensive industrial investigation since the pioneering efforts in the late 1950s of Conix,[1] Levine and Temin,[2] and Eareckson.[3] They were commercialized in 1974 by Unitika of Japan as U-polymer, in 1978 by Union Carbide as Ardel D-100, in 1979 by Hooker Chemical Co. as Durel, and in 1986 by Du Pont as Arylon.

To date, the polyarylate system that has received the most attention is based on bisphenol A and isophthalic and terephthalic acids. Varying the ratio of isomeric acids in this system leads to both amorphous and crystalline products. In addition, incorporation of more rigid diphenols, such as hydroquinone or 4,4'-biphenol, or *p*-hydroxybenzoic acid into the polymer backbone also results in the generation of crystalline and liquid crystalline materials.

21.4 POLYMERIZATION

Polyarylates can be produced by a variety of processes that, when classified by starting materials, can be divided into two categories: those which utilize acid chlorides and those which do not. The acid chloride processes are usually low-temperature solution processes, whereas the nonacid chloride methods are generally run at high temperatures in the absence of a solvent or as a slurry.

Starting with bisphenol A and isophthaloyl and terephthaloyl chlorides, several routes to polyarylates are feasible. The interfacial process that was first described by Conix[4] and later in detail by Temin[5] is used commercially.[6] This process involves reacting an aqueous solution of the dialkali metal salt of bisphenol A with a solution of the acid chorides in a water insoluble solvent, such as methylene chloride. Phase-transfer catalysts such as quaternary ammonium or phosphonium salts are generally added to the reaction to accelerate the polymerization. The resulting polymer is isolated by phase separation, followed by either evaporation of the organic solvent used or coagulation in a nonsolvent. In addition to the interfacial process, other acid chloride routes to polyarylates have been described by Heck,[7] Hare,[8] and Ueno.[9]

Nonacid chloride processes for the production of polyarylates include the reaction of bisphenol A diacetate with isophthalic and terephthalic acids[1,10–13] and the reaction of bisphenol A with diphenyl iso/terephthalates.[14] Both polymerizations are usually run in the absence of solvent at temperatures exceeding 482°F (250°C). Since acidolysis and transesterification reactions are reversible, continuous removal of acetic acid from the diacetate process and phenol from the diphenyl ester process is essential to obtaining high molecular weight polymer. To facilitate by-product removal at the end of the reaction, when the melt viscosity is extremely high, the final stages of these polymerizations can be run under vacuum or in a vented extruder.[15] Transition metal catalysts[16,17] and heat transfer fluids[18] (eg, Dowtherm A or Therminol 66) have also been used to accelerate both the initial and latter stages of these melt processes.

Polyarylate prices range from $2.40 to $2.80 per pound (1988).

21.5 DESCRIPTION OF PROPERTIES

Polyarylate based on bisphenol A and iso/terephthalic acids (specifically compositions in the range of 25:75 to 75:25 iso/tere ratio) exhibit property profiles typical of a material that can be classified as an engineering polymer. Their high glass-transition temperature [85–88°F (185–190°C)] yields materials with high heat-distortion temperatures [338–356°F (170–180°C)], low creep, and rigidity over a broad temperature range. The impact strength of polyarylate by a number of standard tests places it in a category similar to polycarbonate. Relative to other polymers comprised of only carbon, hydrogen, and oxygen, its flammability resistance (by standard tests) is quite good. Polyarylate's resistance to UV exposure is excellent; in fact, it has been shown to be a UV stabilizer when it is incorporated as an additive into various polymer systems.

The property data reported here pertains specifically to Ardel D-100, a polyarylate commercialized by Union Carbide Corp. (now available from Amoco).

21.5.1 Mechanical Properties

As an amorphous thermoplastic with a glass-transition temperature of 370°F (188°C), polyarylate offers excellent consistency in mechanical properties over a broad temperature range. Over the range of at least −112 to 338°F (−80 to 170°C), polyarylate can be described as rigid, strong, and tough. The impact strength of polyarylate is high and does not exhibit the variation in annealing, notched radius, and sample thickness noted for polycarbonate.[19,20] Annealing of polyarylate of up to 300 h at 266°F (130°C) yields less than a 10% drop in notched toughness [4.3 to 4.0 ft-lb/in. (230–214 J/m) of notch]. After only two hours at 266°F (130°C), polycarbonate goes from an initial value of 16 to 2 ft-lb/in. (854–107 J/m) of notch.

21.5.2 Thermal Properties

The high heat-distortion temperature of 345°F (174°C) combined with high continuous use temperature ratings suggests the utility of polyarylate in electrical/electronic applications. The dynamic mechanical properties of polyarylate are illustrated in Figure 21.1. A significant low temperature relaxation is observed at (-103°F) (-75°C) for polyarylate and is hypothesized as providing the molecular mobility necessary to achieve excellent toughness. The consistency of modulus between this relaxation and the glass-transition temperature indicates minor mechanical property variations over this broad temperature range. The tensile creep of polyarylate (compared with polycarbonate and polysulfone) is illustrated in Figure 21.2.[21] Polyarylate is significantly better in creep resistance than polycarbonate; however, it is not as good as polysulfone (which is well known for its creep resistance).

21.5.3 Electrical Properties

Like its mechanical properties, the electrical properties of polyarylate are expected to be reasonably constant over a broad temperature range. The combination of low flammability (as per UL-94), excellent electrical properties, and a high heat-distortion temperature suggests utility in electrical connectors, switches, electrical appliances and housings, and high-performance electronic applications.

21.5.4 Environmental Properties

The thermal stability of a polyarylate based on bisphenol A and iso-terephthalic acids compared with bisphenol A polycarbonate and bisphenol A polysulfone has been reported

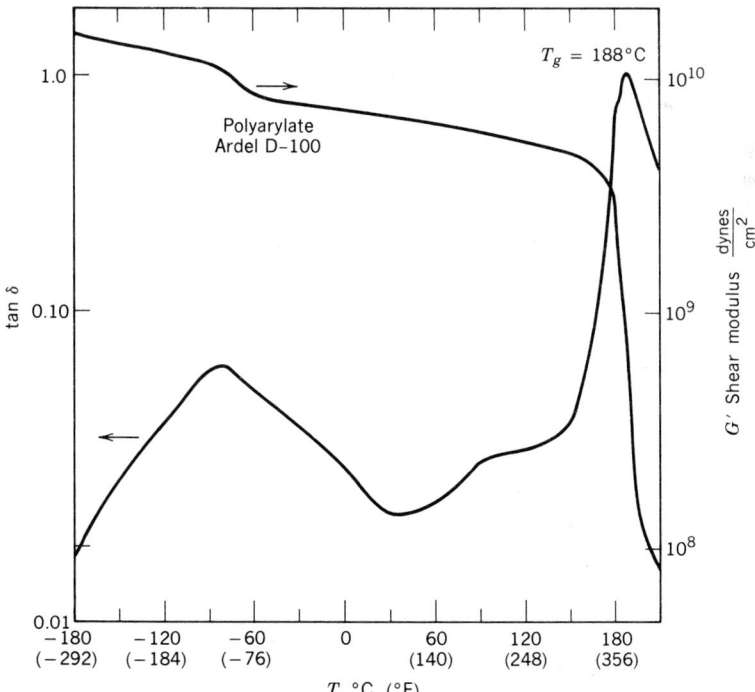

Figure 18.1. Dynamic mechanical results for polyarylate. Courtesy of Amoco Performance Products, Inc.

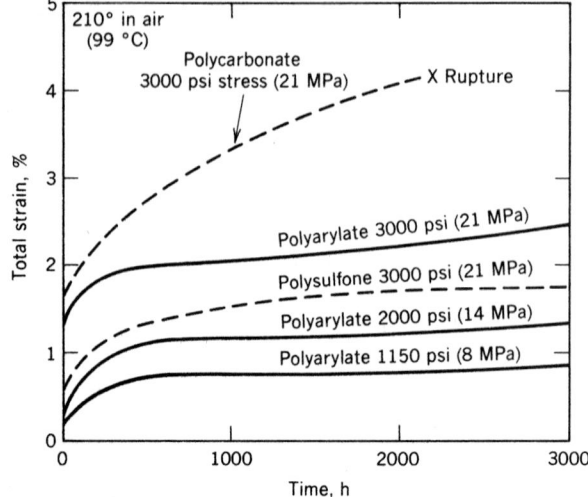

Figure 18.2.   Tensile creep of polyarylate. Courtesy of Amoco Performance Products, Inc.

by Bier.[22] The thermogravimetric analysis test showed the stability of polyarylate to be intermediate between polycarbonate and polysulfone (with polysulfone having the best stability). The continuous-use temperature of polyarylate, as determined by the UL test procedure, is higher than that of most other thermoplastics.

The hydrolytic stability of polyarylate has been only briefly mentioned in the open literature. Bier[22] observed that amorphous polyarylates are slightly more stable with respect to hydrolysis than their amorphous polycarbonate counterparts. Freitag and Reinking[23] noted that a polyarylate based on bisphenol A and iso-terephthalic acids exhibited hydrolysis rates similar to those of polycarbonate. They concluded that polyarylate could be used in water up to 140°F (60°C) for long time exposures. Dickinson[24] indicated definite moisture sensitivity of polyarylate during melt processing, as drying to less than 0.02 wt% water is required before extrusion or injection molding.

The organic chemical resistance of polyarylate is quite similar to that of polycarbonate. Polyarylate exhibits solubility (>10 wt%) in methylene chloride, chloroform, dimethyl-acetamide, and dimethylformamide. Good resistance to aliphatic hydrocarbons, oils, fats, and alcohols is expected. The environmental stress cracking resistance of polyarylate in specific organic media such as aromatic hydrocarbons, esters, ketones, and chlorinated hydrocarbons is limited, and exposure to those environments should be avoided.

The flammability characteristics of polyarylate have been previously published by Domine.[21] Polyarylate (without additives) exhibits a high resistance to ignition and flame spreading. The by-products of combustion of polyarylate (comprised only of carbon, hydrogen, and oxygen) are less toxic than polymers containing halogenated flame retardants.

The aromatic ester structure of polyarylate is analogous to the structure of several UV stabilizers (such as aryl esters). With exposure to UV radiation, the aromatic ester group will rearrange to o-hydroxybenzophenone structures in the main chain.[25] These structures are well known as UV stabilizers. The most extensive study of the UV stability of various polyarylates, including the polyarylate based on bisphenol A and iso-terephthalic acids, was reported by Cohen and co-workers.[25] The photo-Fries rearrangement to the o-hydroxybenzophenone structure was observed with ATR (attenuated total reflectance) infrared spectroscopy. The depth of rearrangement was found to be 0.5 mils (0.1 mm), providing UV protection to the rest of the sample.

The photo-Fries rearrangement is accompanied by a yellowing of the polyarylate surface.

This yellowing will generally be minimal for tinted transparent samples (visual determination); however, it would preclude the use of polyarylate in "water-white" applications. Polyarylate exhibits only limited haze development, even after extended, accelerated UV exposure.

21.5.5 Other Properties

A unique property of polyarylate is its excellent recovery from flexural or elastic extension (up to 12%). This property suggests its utility for clips, springs, snap-fit connectors, and various fasteners.

Polyarylate can be utilized as a permanent, low-cost UV stabilizer particularly suited for addition to polyesters and polycarbonates. Robeson[26] showed that the addition of polyarylate to a number of polymeric materials yields improved UV stability. Additionally, thin films of polyarylate have been found to provide UV protection to a wide variety of substrates.

As previously noted, polyarylates will undergo a surface yellowing on UV exposure. Tinted transparent grades (red, amber, yellow, green, and gray) of polyarylate will generally yield minimal overall color change, as the bulk color will obscure any surface color change. A specific grade of polyarylate (Ardel D-240) has been designed for applications requiring UV exposure.

Glass-fiber-reinforced grades are available for improved mechanical strength and stiffness, lower creep, and greater resistance to environmental stress rupture.

21.6 APPLICATIONS

Polyarylate's flammability ratings, its structure (as noted, it is comprised only of carbon, hydrogen, and oxygen), its high heat-distortion temperature, and its excellent toughness, yield a property balance that is promising for applications in the mass transit interior market, such as aircraft, and in safety equipment housings, fire-protection helmets, and office equipment.

The flexural recovery characteristics of polyarylate combined with its rigidity indicate its suitability for springs, clips, fasteners, snap-lock connectors, high-quality combs, and hinges.

The excellent electrical properties and high heat-distortion temperature, high toughness, and required flammability ratings yield a property balance suitable for electrical/electronic applications such as electrical connectors, snap-lock fit connectors, electrical appliances, relays, and switches.

The UV stability of polyarylate makes it suitable for many exterior applications. These include solar energy collector glazing, traffic lights, automotive lenses, transparent lamp diffusers, exterior glazing, permanent greenhouse glazing, lighting housings, safety reflectors, and so on.

21.7 ADVANTAGES/DISADVANTAGES

The chief advantages of polyarylate are it high heat-distortion temperatures, excellent toughness and weatherability, flame resistance (per UL ratings), transparency, and blending potention (with PET, for example).

Its principal limitations are its hydrolytic stability and resistance to environmental stress rupture.

21.8 PROCESSING TECHNIQUES

Polyarylate can be fabricated using typical thermoplastic processing techniques including extrusion, injection molding, thermoforming, and compression molding.

Master Material Outline

PROCESSABILITY OF ___Polyarylate___

PROCESS	A	B	C
INJECTION MOLDING	X		
EXTRUSION	X		
THERMOFORMING	X		
FOAM MOLDING			
DIP, SLUSH MOLDING			
ROTATIONAL MOLDING			
POWDER, FLUIDIZED-BED COATING			
REINFORCED THERMOSET MOLDING			
COMPRESSION/TRANSFER MOLDING	X		
REACTION INJECTION MOLDING (RIM)			
MECHANICAL FORMING			
CASTING			

A = Common processing technique.
B = Technique possible with difficulty.
C = Used only in special circumstances, if ever.

21.9 RESIN FORMS

Polyarylate is available in pellet form.

21.10 SPECIFICATION OF PROPERTIES

See the Master Outline of Materials Properties.

MATERIAL _____ Polyarylate, Ardel D-100 (TM)

PROPERTY	TEST METHOD	ENGLISH UNITS	VALUE	METRIC UNITS	VALUE
MECHANICAL DENSITY	D792	lb/ft³	75	g/cm³	1.2
TENSILE STRENGTH	D638	psi	9500	MPa	66
TENSILE MODULUS	D638	psi	290,000	MPa	2000
FLEXURAL MODULUS	D790	psi	310,000	MPa	2140
ELONGATION TO BREAK	D638	%	50	%	50
NOTCHED IZOD AT ROOM TEMP	D256	ft-lb/in.	4.2	J/m	224
HARDNESS	D785				
THERMAL DEFLECTION T @ 264 psi	D648	°F	345	°C	174
DEFLECTION T @ 66 psi	D648	°F		°C	
VICAT SOFT-ENING POINT	D1525	°F		°C	
UL TEMP INDEX	UL746B	°F		°C	
UL FLAMMABILITY CODE RATING	UL94		V-0		V-0
LINEAR COEFFICIENT THERMAL EXPANSION	D696	in./in./°F	$2.8–3.4 \times 10^{-5}$	mm/mm/°C	$5.0–6.2 \times 10^{-5}$
ENVIRONMENTAL WATER ABSORPTION 24 HOUR	D570	%		%	
CLARITY	D1003	% TRANS-MISSION		% TRANS-MISSION	
OUTDOOR WEATHERING	D1435	%	surface yellowing on UV exposure	%	surface yellowing on UV exposure
FDA APPROVAL					
CHEMICAL RESISTANCE TO: WEAK ACID	D543				
STRONG ACID	D543				
WEAK ALKALI	D543				
STRONG ALKALI	D543				
LOW MOLECULAR WEIGHT SOLVENTS	D543				
ALCOHOLS	D543		good resistance		good resistance
ELECTRICAL DIELECTRIC STRENGTH	D149	V/mil	400	kV/mm	15.8
DIELECTRIC CONSTANT	D150	10⁶ Hertz	2.62		2.62
POWER FACTOR	D150	10⁶ Hertz	0.01		0.01
OTHER MELTING POINT	D3418	°F		°C	
GLASS TRANS-ITION TEMP.	D3418	°F	370	°C	188

21.11 PROCESSING REQUIREMENTS

Because polyarylate is amorphous, it can be molded to close tolerances and will not exhibit warpage problems. Polyarylate must be well dried prior to processing (less than 0.02 wt% water) to maintain the excellent property balance after fabrication. Typical drying conditions are eight hours at 260°F (127°C) or four hours at 300°F (149°C). Injection molding recommendations include a barrel temperature range of 620–700°F (327–371°C), stock temperatures of 675–735°F (357–390°C), and mold temperatures of 250–300°F (121–149°C). Regrind can be used at a level of up to 25%. Polycarbonate is recommended as a purge material before and after molding.

21.12 PROCESSING-SENSITIVE END PROPERTIES

Polyarylate has a narrow injection-molding range. In addition, it must be kept dry during processing.

21.13 SHRINKAGE

The shrinkage of polyarylate is very similar to that of polysulfone and polycarbonate.

21.14 TRADE NAMES

Currently, polyarylate is produced under the trade name Ardel by Amoco, which purchased the business from Union Carbide; under the name Durel by Hoechst Celanese, which purchased the business from Hooker, and under the name Arylon by Du Pont.

BIBLIOGRAPHY

1. A.J. Conix, *Ind. Chim. Belg.* **22,** 1457 (1957)
2. M. Levine and S.S. Temin, *J. Polym. Sci.* **22,** 179 (1958).
3. W.M. Eareckson, *J. Polym. Sci.* **40,** 399 (1959).
4. U.S. Pat. 3,216,970 (Nov. 9, 1965), A.J. Conix (to Gevaert Photo-Production).
5. S.C. Temin, *Interfacial Synthesis,* Vol. 11, Marcel Dekker, Inc., New York, 1977, pp. 27–63.
6. L.M. Elkin, *High Temperature Polymers,* Supp. A, SRI International, Menlo Park, Calif., 1980, pp. 167–218.
7. U.S. Pat. 3,133,898 (May 19, 1964), M.H. Heck (to Goodyear Tire and Rubber Co.).
8. U.S. Pat. 3,234,168 (Feb. 8, 1966), W.A. Hare (to E. I. du Pont de Nemours & Co., Inc.).
9. U.S. Pat. 3,939,117 (Feb. 17, 1976), K. Ueno (to Sumitomo Chemical Co., Ltd).
10. U.S. Pat. 3,317,464 (May 2, 1976), A.J. Conix (to Gevaert Photo-Production).
11. U.S. Pat 3,329,653 (July 4, 1967), E.M. Beavers, M.J. Hurwitz, and D.M. Fenton (to Rohm & Haas Co.).
12. F.L. Hamb, *J. Polym. Sci.* **10,** 3217 (1977).
13. E.E. Riecke and F.L. Hamb, *J. Polym. Sci.* **15,** 593 (1977).
14. U.S. Pat. 3,395,119 (July 30, 1968), F. Blaschke and W. Ludwig (to Chemische Werke Witten GmBH).
15. Ger. Offen, 2,232,877 (Jan. 17, 1974), K. Eise and co-workers (to Werner & Pfleiderer and Dynamite Nobel A-6).

16. U.S. Pat. 3,972,852 (Aug. 3, 1976), H. Inata, S. Kawase, and T. Shima, (to Teijin Ltd.).

17. U.S. Pat. 3,684,766 (Aug. 15, 1972), W.J. Jackson, Jr., H.F. Kuhfuss, and J.R. Caldwell (to Eastman Kodak Co.).

18. U.S. Pat. 4,067,852 (Jan. 10, 1978), G.W. Calundann (Celanese Corp.)

19. D.C. Prevorsek, Y. Kesten, and B. DeBona, "Polyester Carbonates: A New Class of High Performance Thermoplastics," *Polym. Preprints, Div. Polym. Chem. Am. Chem. Soc.,* **20**(1), 187 (1979).

20. J.T. Ryan, "Impact Properties of Polycarbonate as a Function of Melt Flow Rate, Notch Radius, Thermal History, and Temperature," *34th SPE ANTEC,* Atlantic City, N.J., April 26–29, 1976, p. 205.

21. J.D. Domine, "Polyarylate, A Tough, New Engineering Polyer," *37th SPE ANTEC,* New Orleans, La., May 7–10, 1979, p. 655.

22. G. Bier, "Polyarylates," *Polymer.* **15,** 527 (1974).

23. D. Freitag and K. Reinking, "Aromatic Polyester (APE)—A Transparent Thermoplastic Material with High Thermal Stability," *Kunstoffe,* **71**(1), 46 (1981).

24. B.L. Dickinson, "Polyarylate," *Modern Plastics Encyclopedia,* McGraw-Hill, New York, vol 59(10A), 1982, p. 61.

25. S.M. Cohen, R.H. Young, and A.H. Markhart, "Transparent Ultraviolet-Barrier Coatings," *J. Polym. Sci., Part A-1* **9,** 3263 (1971).

26. U.S. Pat. 4,259,458 (March 31, 1981) L.M. Robeson (to Union Carbide Corp.).

22

POLYBUTYLENE (PB)

A.M. Chatterjee

Senior Research Engineer, Plastics, Shell Development Co., Houston, Texas 77251

22.1 STRUCTURE

The name polybutylene (PB) refers to commercial semicrystalline resins based on high-molecular-weight isotactic poly(1-butene) homopolymer and copolymer. The chemical structure of poly(1-butene) is represented by:

$$\left[\begin{array}{c} \text{C}-\text{H}_3 \\ | \\ \text{C}-\text{H}_2 \\ | \\ -\text{C}-\text{C}- \\ | \quad | \\ \text{H} \quad \text{H}_2 \end{array}\right]_n$$

22.2 CATEGORY

PB resins are thermoplastic. They are polyolefins and exhibit strength and toughness in both rigid and flexible end-product forms.

22.3 HISTORY

Isotactic poly(1-butene) was first synthesized by G. Natta in 1954. Chemische Werke Huels of West Germany started the first commercial production in 1969 and discontinued the operation in 1973. In 1975, Witco Chemical Corp. built a PB plant in Taft, La., which was acquired by Shell Chemical in 1977. Pipe was the first important application of polybutylene and remains to this day as the dominant end-use market for PB.

Shell Chemical is the sole U.S. producer of polybutylene.

22.4 POLYMERIZATION

PB is produced via stereospecific Ziegler-Natta polymerization of 1-butene monomer.[1] Commercial products are based on isotactic high-molecular-weight polymer.

22.5   PROPERTIES

PB resins retain their good mechanical properties at elevated temperatures. Recommended upper and lower use temperatures are approximately 230°F (110°C) and −4°F (−20°C), respectively. Pipe resins, specifically, exhibit excellent flexibility and resistance to creep, environmental stress cracking, chemicals, and wet abrasion.[2] Plumbing code approvals and listings have been received from most states and cities, as well as from major regulatory and standards-making bodies.

Film grades of PB provide good tear and puncture resistance as well as high impact and tensile strength.[2]

Glass reinforcement of polybutylene significantly increases its flexural modulus; it decreases its tensile strength and elongation. For a typical 40 wt% glass-filled pipe-grade PB, tensile properties at break are as follows: tensile strength, 1926 psi (13.3 MPa); elongation 14%; and flexural modulus (1% secant), 258,090 psi (1780 MPa).

22.6   APPLICATIONS

As earlier noted, the main commercial application for PB resins is in pipe.[3,4] PB pipe has received hydrostatic design stress rating of 1000 psi (6.90 MPa) at 73°F (23°C) and 500 psi (3.45 MPa) at 180°F (82°C) from the Plastics Pipe Institute. Applications for small-diameter PB pipe include cold- and hot-water plumbing, the latter including residential and solar plumbing, and underfloor and hydronic heating. Other applications are in well piping, heat-pump piping, fire-sprinkler piping, and specialty hosing. Large-diameter PB pipe finds uses in the transportation of abrasive or corrosive materials at high temperatures in the mining, chemical, and power-generation industries.

Film applications of PB include heavy-duty shipping containers; food, meat and agricultural packaging; compression wraps; hot-fill containers; and industrial sheeting. Other applications include injection molding components, such as fittings for use in conjunction with PB pipe, and compression molding items, such as sheet for abrasion-resistant liner.

22.7   ADVANTAGES AND DISADVANTAGES

The advantages of PB in pipe end uses are its flexibility, toughness, and resistance to creep, environmental stress cracking, wet abrasion, and chemicals. PB films demonstrate high resistance to tear, puncture, and impact, and have high tensile strength.

The limitations of PB are a slower processing rate compared to other polyolefins in the heat state, a polymorphic transformation after melt processing, and low temperature brittleness.

22.8   PROCESSING TECHNIQUES

PB pipe can be fabricated via conventional single-screw extrusion technology using vacuum or pressure sizing for dimension control.[5,6] After it exits from the die, the pipe should be cooled by water. The pipes can be joined via thermal fusion or mechanical fittings of several types.

The blown film process is commonly used for producing PB film[7]; PB film can also be cast on chill rolls. The films are heat-sealable. In addition, PB film can be coextruded or laminated with other films for specific applications. In-line embossing, with good pattern retention, is feasible without the application of heat.

Master Material Outline

PROCESSABILITY OF ___Polybutylene___

PROCESS	A	B	C
INJECTION MOLDING	X		
EXTRUSION	X		
THERMOFORMING			X
FOAM MOLDING		X	
DIP, SLUSH MOLDING			
ROTATIONAL MOLDING	X		
POWDER, FLUIDIZED-BED COATING			X
REINFORCED THERMOSET MOLDING			
COMPRESSION/TRANSFER MOLDING		X	
REACTION INJECTION MOLDING (RIM)			
MECHANICAL FORMING		X	
CASTING		X	
FLAME SPRAYING	X		

A = Common processing technique.
B = Technique possible with difficulty.
C = Used only in special circumstances, if ever.

22.9 RESIN FORMS

PB resins are produced in pellet form. In addition to pipe and film grades, homopolymer and copolymer general-purpose grades are offered.

22.10 SPECIFICATION OF PROPERTIES

Five crystalline forms of poly(1-butene) have been reported.[8] In conventional melt processing, crystallization of the resins initially produces the metasable form II, which transforms to the stable form I over a period of 5 to 7 days at ambient temperature and pressure.[9] During the transformation, density, crystallinity, hardness, rigidity, stiffness, and tensile yield strength all increase to values characteristic of form I. After fabrication and transformation to form I, these resins show crystallinity of 48 to 55% and density of 0.93 to 0.94 g/cm^3.

MATERIAL ___ Polybutylene (Pipe grade, nominal MI — 0.4 g/10 min)

PROPERTY	TEST METHOD	ENGLISH UNITS	VALUE	METRIC UNITS	VALUE
MECHANICAL DENSITY	D792	lb/ft³	58–59	g/cm³	0.93–0.94
TENSILE STRENGTH	D638	psi	4640–5075	MPa	32–35
TENSILE MODULUS	D638	psi	42,050–42,775	MPa	290–295
FLEXURAL MODULUS	D790	psi	54,375–55,100	MPa	375–380
ELONGATION TO BREAK	D638	%	275–300	%	275–300
NOTCHED IZOD AT ROOM TEMP	D256	ft-lb/in.	12–15	J/m	640–800
HARDNESS	D785	Shore D	60	Shore D	60
THERMAL DEFLECTION T @ 264 psi	D648	°F	129–140	°C	54–60
DEFLECTION T @ 66 psi	D648	°F	216–235	°C	102–113
VICAT SOFTENING POINT	D1525	°F	234–237	°C	112–114
UL TEMP INDEX	UL746B	°F		°C	
UL FLAMMABILITY CODE RATING	UL94		HB		HB
LINEAR COEFFICIENT THERMAL EXPANSION	D696	in./in./°F	0.000 073	mm/mm/°C	0.000 013
ENVIRONMENTAL WATER ABSORPTION 24 HOUR	D570	%	<0.03	%	<0.03
CLARITY	D1003	% TRANSMISSION		% TRANSMISSION	
OUTDOOR WEATHERING	D1435	%	Black pigmented grades are weatherable	%	Black pigmented grades are weatherable
FDA APPROVAL			Yes		Yes
CHEMICAL RESISTANCE TO: WEAK ACID	D543		Not attacked		Not attacked
STRONG ACID	D543		Minimal to badly attacked		Minimal to badly attacked
WEAK ALKALI	D543		Not attacked		Not attacked
STRONG ALKALI	D543		Not attacked		Not attacked
LOW MOLECULAR WEIGHT SOLVENTS	D543		Depends on solvent and temperature		Depends on solvent and temperature
ALCOHOLS	D543		Not attacked		Not attacked
ELECTRICAL DIELECTRIC STRENGTH	D149	V/mil		kV/mm	
DIELECTRIC CONSTANT	D150	10^6 Hertz	2.5		2.5
POWER FACTOR	D150	10^6 Hertz	0.000 5		0.000 5
OTHER MELTING POINT	D3418	°F	265–266	°C	120–124
GLASS TRANSITION TEMP.	D3418	°F	–4 to –13	°C	–20 to –25

After fabrication and transformation, the film-grade resins show 46 to 52% crystallinity and 0.90 to 0.92 g/cm^3 density.

Typical properties of PB pipe resins are listed in the Master Outline of Materials Properties.

22.11 PROCESSING REQUIREMENTS

When PB is processed, the amount of regrind should not exceed 30% of the input.

Vacuum or pressure sizing is required for dimensional control in single-screw extrusion of pipe.

22.12 PROCESSING-SENSITIVE END PROPERTIES

Increasing pipe extrusion speed or drawdown reduces the tensile elongation to break, a key property for quality control of PB pipe.[5,6] The effects of blown film processing parameters on mechanical properties can be found elsewhere.[7]

22.13 SHRINKAGE

See Mold Shrinkage Characteristics.

Mold Shrinkage Characteristics (in./in. or mm/mm)

MATERIAL   Polybutylene

IN THE DIRECTION OF FLOW

THICKNESS

MATERIAL	mm: 1.5 in.: 1/16	3 1/8	6 1/4	8 1/2	FROM	TO	#
Aged PB		0.013					

	PERPENDICULAR TO THE DIRECTION OF FLOW					
Aged PB		0.004				

NOTE: Mold shrinkage is approximate. It is affected by part design, mold temperature, thickness, injection pressure, packing time, cycle time, orientation, gate design, gate size, gate location, glass content, glass size, and filler content.

22.14   TRADE NAMES

In the United States, PB resins are made and marketed only by Shell. PB resins are marketed under the trade name of Duraflex®.

BIBLIOGRAPHY

1. I.D. Rubin, *Poly(1-Butene)—Its Preparation and Properties,* Gordon and Breach, New York, 1968.
2. A.M. Chatterjee, "Butene Polymers," *Encyclopedia of Polymer Science and Engineering,* Vol. 2, Wiley Interscience, New York, 1985, p. 590.
3. J.E. Elvers, W. Vermuelen, and N. Spanka, *Plastverarbeiter* **33** 1043 (1982).
4. C.R. Lindegren, *Plast. Eng.* 38 (July 1975).
5. M.P. Schard and O.E. Vera, *Plast. Eng.,* 45 (October 1977).
6. "Processing Polybutylene Pipe," *Technical Bulletin SCC: 544–81,* Shell Chemical Co., Houston, Texas, March 1981.
7. "Processing Shell Polybutylene Film-Grade Resins,"*Technical Bulletin SCC: 391–79,* Shell Chemical Co., Houston, Texas, March 1979.
8. C. Nakafuku and T. Miyaki, *Polymer* **24,** 141 (1983).
9. A.J. Foglia, *Applied Polymer Symposi* **11,** 1 (1969).

General References

A.M. Chatterjee, "Polybutylene," *Modern Plastics Encyclopedia,* McGraw-Hill, New York, 1984–1985, p. 38.

P.W. DeLeeuw, C.R. Lindergren, and R.F. Schimbor. "Polybutylene: Market Status and Outlook," *Chem. Eng. Prog.* 57 (Jan. 1980).

C.R. Lindergren, "Polybutylene: Properties of a Packaging Material," *Polym. Eng. Sci.,* **10,** 163 (1970).

P.W. MacGregor, "Polybutylene: Prospects for the Eighties," *RETEC Technical Papers, Society of Plastics Engineers,* Feb. 1981, p. 71.

C.L. Rohn and H.G. Tinger. "Poly(Butylene), The Engineering Polyolefin," *Proceedings of the 28th ANTEC, Society of Plastics Engineers,* 1970, p. 638.

23

POLYCARBONATE (PC)

General Electric Plastics

1 Plastics Ave., Pittsfield, MA 01201

23.1 POLYCARBONATE

General-purpose polycarbonate is one of the toughest, most versatile engineering polymers. Its chemical structure is

$$\left[\begin{array}{c} \overset{O}{\underset{\|}{C}} - O - \langle\bigcirc\rangle - \overset{\overset{CH_3}{|}}{\underset{\underset{CH_3}{|}}{C}} - \langle\bigcirc\rangle - O \end{array}\right]_n$$

23.2 CATEGORY

Polycarbonate is a thermoplastic resin that spans a range of physical properties at the high end of the performance scale.

23.3 HISTORY

Polycarbonate was introduced commercially 25 years ago. However, like its polystyrene, poly(vinyl chloride), and phenol–formaldehyde predecessors, its gestation period was a long one. Hydroquinone and resorcinol polycarbonates were first reported more than 60 years ago. A second reference to the same polymers was made in 1902. Then, no further activity was recorded until 1956. At that time, General Electric and Farbenfrabriken Bayer independently announced discovery of, as well as plans to produce, a unique class of polycarbonate resins.

The first available commercial quantities of polycarbonate resin found application in two areas: resin cast into film form used in electrical applications and in bases for photographic and graphic arts films, and the resin in pellet form for use in injection molding and in an extrusion compound for a variety of industrial parts.

These applications were evaluated and field-tested by late 1958. By 1959 the resins began moving in substantial quantities. That same year extrusion applications began to surface. Full-scale commercial production started in Germany in 1959. Production at two commercial plants in the United States followed in 1960.

Worldwide annual consumption of polycarbonate has risen from 40×10^6 pounds in 1970 to 218 × 10⁶ pounds in 1980, representing 0.2% and 0.6%, respectively, of all plastics sold, according to industry sources.

23.4 POLYMERIZATION

Polycarbonate is a resin in which groups of dihydric or polyhydric phenols are linked through carbonate groups. A general-purpose polycarbonate is derived from bisphenol A.

For special end uses, formulations combining small amounts of other polyhydric phenols are available. For instance, some meet special industry codes for flame retardance. Other resins have increased melt strength for extrusion and blow molding.

Addition of glass-fiber reinforcement in ranges from 5 to 40% improves creep resistance up to 4000 psi (28 MPa) at temperatures as high as 210°F (99°C) and intermittently up to 10,000 psi (69 MPa). Mold shrinkage decreases, the effect of which is that closer tolerances can be met. Other improvements include up to five times the tensile modulus and more than double the flexural, compressive, and tensile strength of standard polycarbonate.

Moreover, additives or changes in formulation can modify the base resin for greater flame retardance, thermal stability, ductility, ultraviolet (UV) light and color stability, plus easier mold release. Coatings used on polycarbonate sheet improve mar and chemical resistance and weatherability. This is especially critical for glazing, pigmented sign, and clear solar applications.

The properties of foamable polycarbonate are a direct extension of the base polymer. In the foamed state, however, low molded-in stresses improve chemical resistance. It also achieves a high stiffness-to-weight ratio. This occurs because thicker composite wall sections are made possible by using material equivalent to that used in injection molding.

Finally, polycarbonate can be blended with other polymers to enhance specific properties, such as chemical resistance. Recently developed polycarbonate copolymers achieve a thermal performance approaching high-end specialty thermoplastics.

Like most products, the finished polymer price is reflected in the cost of the raw material, how the material is being used, and the volume purchased. Currently (1988), average resin cost ranges from $1.96 per pound for general-purpose polycarbonate to $2.77 per pound for specialty grades.

23.5 DESCRIPTION OF PROPERTIES

Design calculations for polycarbonate resins are no different than those for any other engineering thermoplastic. Such factors as life expectancy, temperature, and stress levels must be considered. Once the end use and environment are defined, standard engineering calculations can fairly accurately predict part performance.

When two or more materials with different coefficients of thermal expansion are used in a common assembly, thermal stresses can become a significant factor in stress analysis. For this reason, thermal stress effects must be considered early in the design of polycarbonate parts incorporating close fits, molded-in inserts, and screw assembly. This hold especially true if the part is assembled at a temperature other than that of the end use.

Polycarbonate resin is generally stable to water, mineral, and organic acids. However, crazing and embrittlement may occur if a part molded from polycarbonate resin is highly stressed and exposed to water or a moist, environment at elevated temperatures. For this reason, a temperature limit of 140–160°F (60–71°C) is recommended.

Also, it is necessary to design to the worst abuse and environmental conditions. Polycarbonate's tensile strength, flexural strength, and flexural modulus decrease steadily as the temperature increases. On the other hand, the effect of temperature on impact strength is the opposite. As temperature drops, the resin becomes stiffer and more brittle. In fact, its notched Izod impact strength drops dramatically near 0°F (-18°C). Still, a notched Izod value of 2 ft-lb/in. (107 J/m) is a very respectable number for many plastics, even at room temperature.

Polycarbonate resins, like most engineering materials, follow Hooke's law of stress and strain. Although stress levels are lower, the shapes of the resin stress/strain curves simulate those of aluminum and steel. Therefore, standard equations can be used to predict the performance of a part.

When working with polycarbonate resins, long-term situations require evaluation of creep phenomena. Creep becomes negligible at a specific stress level, so it can be ignored in long-term, continuous-load applications. The point at which this phenomenon takes effect is called the creep limit for the material. In the case of polycarbonate resin at 73°F (23°C), this limit occurs at about 2000 psi (14 MPa). This is one of the highest creep limits of any unreinforced thermoplastic.

One of polycarbonate resin's outstanding properties is impact strength. Polycarbonate resin also exhibits a property known as "critical thickness." At a specific thickness, the notched Izod impact strength of a test bar will suddenly drop from a high-energy-absorbing ductile failure to a low-energy-absorbing brittle failure. The critical thickness value depends on grade, color, temperature, and notch radius. In general, this will be between 0.160 and 0.200 in (4.0 and 5.0 mm). This agrees favorably with most molding contraints.

In the area of fatigue, calculation of a part's static and dynamic stress levels requires three evaluations: anticipated forces, section modulus, and stress concentration. When cyclic loading is involved, these values should be compared with the fatigue endurance in order to determine life expectancy. This is especially useful in the design of such parts as pressure vessels subject to water hammer as well as cyclically loaded support structures.

Optical properties of polycarbonate resins are excellent. It is for this reason that the product is used extensively for many types of lens and glazing applications. As measured by ASTM D1003, polycarbonate resins have a transmittance rating of 86 to 89% and a haze rating of only 1 to 2%. The refractive index of these resins is 1.586. It should be noted, however, that a scratch-resistant coating is recommended where these products are subject to abrasion.

Environmentally, plastics, like all materials, are susceptible to deterioration through natural weathering. However, polycarbonates receive high marks when it comes to weatherability, particularly with respect to retention of mechanical properties.

Although the polymer system can be degraded by many weathering factors, radiation, temperature, water, contaminants, and normal air constituents, it is ultraviolet (sunlight) and water exposure that impose the primary stresses. While several accelerated aging tests have been developed to approximate the natural effects of weathering, direct-exposure weathering tests in the end-use environments are the best indicator of life expectancy.

It is recommended that UV-stabilized material be used for outdoor applications. The exception is black material, where no further UV stabilization is necessary.

23.6 APPLICATIONS

FDA-sanctioned grades of polycarbonate, long-established in such high-volume food service applications as beverage pitchers, mugs, food processor bowls, and tableware, are also finding growing use in microwave cookware, especially where clarity is required.

In automotive markets, polycarbonate lends higher performance to rear and forward lighting lens applications. Opaque grades are widely used for thin-wall mechanical parts and large exterior parts that are painted to a Class A finish. Polycarbonate offers impact properties that can meet requirements for instrument panels without the need for padding. Coated mar-resistant glazing provides an extra margin of durability in mass transit vehicles.

In construction, polycarbonate is used for door and window components. Hardware applications include drapery fixtures, furniture, and plumbing. The material's inherently nontoxic characteristics provide added consumer safety in fire situations. Polycarbonate sheet glazing products are providing answers to breakage problems, since they conform to U.S. safety glazing regulations and UL Standard 972 burglar-resistance ratings.

Increasingly sophisticated electrical and electronic products, plus demanding UL standards, have expanded polycarbonate's fit in this market. Nonbrominated, flame-retardant grades are particularly suitable because of their processing stability, UL 94 V-0 and 94-5V ratings, oxygen index of 35, low smoke emission, and elimination of corrosive gases. Housings and internal parts of printers, copiers, terminals, and other computer and business equipment, as well as telephone connectors, circuit boards, wiring blocks, and other industrial components, are often molded of polycarbonate.

Polycarbonate sheet manufacturers have developed special products for the solar, agricultural, security, and graphics markets. Coated thin-gauge sheet is growing in popularity in the 30–93 mil (0.030–0.093 in. [0.8–2.4 mm]) range for protective eyewear, appliances, and business machines. Polycarbonate film also offers potential in the metalizing, packaging, and tape markets.

By varying sheet thickness and surface finish, standard glazing grades are a natural fit for sound attenuation and school and industrial glazing.

Evolution of stricter safety standards has prompted the development of highly flame-retardant sheet for aircraft interiors. A widening spectrum of security needs makes laminated polycarbonate sheet a good choice for applications requiring bullet resistance and the ability to withstand forced entrance.

Light transmission and thermal insulation characteristics, coupled with the ability to withstand UV attack, have prompted interest in thin-wall extruded polycarbonate sheet for solar collector and greenhouse glazing applications.

23.7   ADVANTAGES/DISADVANTAGES

The advantages of polycarbonate have been discussed in detail in prior sections.

It should be noted that the material is susceptible to incompatible additives. Although its impact strength is at the high end of the scale, point stress can reduce this property significantly. Reasonable care in blending and structural design will obviate these disadvantages.

23.8   PROCESSING TECHNIQUES

Polycarbonate resin is easily processed by all thermoplastic methods. Although it is most often injection molded or extruded into flat sheets, other options include profile extrusion, structural foam molding, and blow molding.

Secondary operations and finishing methods include solvent and adhesive bonding, painting, printing, hot stamping, ultrasonic welding, and heat staking. It is suggested, however, that solvent-based adhesives, paints, and inks be tested for their compatibility.

Master Material Outline

PROCESSABILITY OF ____Polycarbonate____

PROCESS	A	B	C
INJECTION MOLDING	X		
EXTRUSION	X		
THERMOFORMING	X		
FOAM MOLDING	X		
DIP, SLUSH MOLDING			
ROTATIONAL MOLDING		X	
POWDER, FLUIDIZED-BED COATING			X
REINFORCED THERMOSET MOLDING			
COMPRESSION/TRANSFER MOLDING			
REACTION INJECTION MOLDING (RIM)			
MECHANICAL FORMING	X		
CASTING			X

A = Common processing technique.
B = Technique possible with difficulty.
C = Used only in special circumstances, if ever.

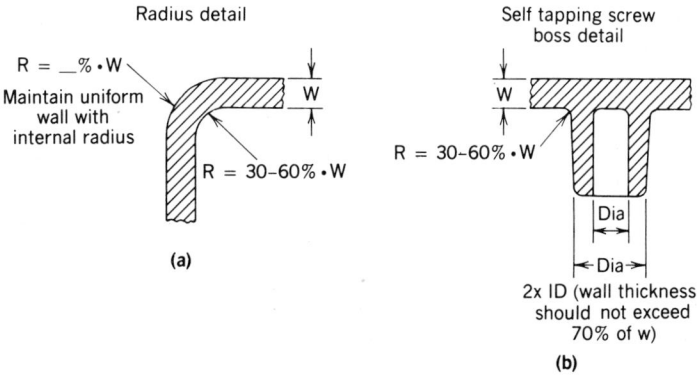

Radius detail

$R = \_\_\% \cdot W$

Maintain uniform wall with internal radius

W

$R = 30\text{--}60\% \cdot W$

(a)

Self tapping screw boss detail

W

$R = 30\text{--}60\% \cdot W$

Dia

Dia

2x ID (wall thickness should not exceed 70% of w)

(b)

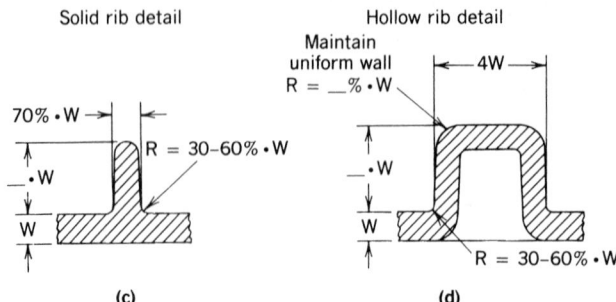

Solid rib detail

Hollow rib detail

(c)

(d)

Standard design chart for ___Polycarbonate___

1. Recommended wall thickness/length of flow
   Minimum ___0.030___ for ___2 in.___ distance
   Maximum ___0.375___ for ___42 in.___ distance
   Ideal ___0.125___ for ___16 in.___ distance

2. Allowable wall thickness variation, % of nominal wall ___5___

3. Radius requirements
   Outside: Minimum ___0.005___ Maximum _____
   Inside: Minimum ___0.020___ Maximum ___60% of wall___

4. Reinforcing ribs
   Maximum thickness ___70%___
   Maximum height _____
   Sink marks: Yes _____ No ___X___

5. Solid pegs and bosses
   Maximum thickness ___70%___
   Maximum weight ___$2\frac{1}{2}$ times wall thickness___
   Sink marks: Yes _____ No ___X___

6. Strippable undercuts
   Outside: Yes _____ No ___X___
   Inside: Yes _____ No ___X___

7. Hollow bosses: Yes _____ No _____

8. Draft angles
   Outside: Minimum ___$\frac{1}{4}°$___ Ideal ___1°___
   Inside: Minimum ___$\frac{1}{4}°$___ Ideal ___1°___

9. Molded-in inserts: Yes _____
   No ___X___

10. Size limitations
    Length: Minimum _____ Maximum _____ Process _____
    Width: Minimum _____ Maximum _____ Process _____

11. Tolerances

12. Special process- and materials-related design details
    Blow-up ratio (blow molding) ___3:1___
    Core L/D (injection and compression molding) _____
    Depth of draw (thermoforming) _____
    Other _____

23.9 RESIN FORMS

Polycarbonate resin is most commonly supplied in pellet form, although both sheet and film forms are available. Pellet packaging normally consists of 50-lb, five-ply kraft bags with a polyethylene coating. Care should be taken in opening the bags to prevent the kraft shreds or other foreign matter from contaminating the pellets.

Several modifications or grades of the polycarbonate resins are available. General-purpose resins, for example, are offered in high, medium, and low melt viscosities. Specialty resins include release grades for close tolerance, minimum draft applications; UV-stabilized grades for outdoor applications; flame retardant products with low smoke generation and toxicity; and grades formulated to meet special FDA requirements.

23.10 SPECIFICATION OF PROPERTIES

See the Master Outline of Materials Properties.

23.11 PROCESSING REQUIREMENTS

Because polycarbonate exhibits hydrolytic sensitivity at the high temperatures needed for processing, it should be dried to less than 0.02% moisture before processing on standard equipment. Tray drying or dehumidifying hopper drying are recommended for resin pellets, while an air-circulating oven operating at 260°F (127°C), plus or minus 5°F (9°C) is a good choice for drying sheet. It is suggested that a minimum of 1 in (25 mm) separation be maintained between the sheets.

Use of vented barrel injection-molding machines may eliminate the need for predrying. However, thorough preliminary testing is suggested, especially when using flame-retardant or highly pigmented grades. Sheet heat-forming techniques requiring lower temperatures (about 315°F [157°C]) may likewise be accomplished without predrying, but these have limited applicability.

When polycarbonate is not degraded or contaminated, it may be reground and remolded without a serious loss of physical proeprties. Blends of regrind to virgin material in proportions up to 20:80 are recommended for applications requiring precise color control or impact strength. Regrind exposed to the atmosphere must be completely redried before processing. Because of the larger particle size, regrind drying should be done at 250°F (121°C) for four to six hours.

Molding cycles for polycarbonate resin are extremely important. The best cycle calls for a quick fill, a hold time just long enough for the gates to freeze, and a brief cooling period.

Stock (and barrel) temperatures are higher for polycarbonates than for most other thermoplastics, ranging from 520 to 620°F (271–327°C). Generally, melt temperature should be 20°F (36°C) above the minimum fill temperature for a specific part. Mold temperatures should range from 160 to 240°F (71–116°C).

Molding pressures usually range from 12,000 psi to 20,000 psi (83–138 J/m), although good parts have been molded up to 30,000 psi (207 J/m).

Thorough purging is essential before or after processing polycarbonate resin on machines used for other plastics. Contamination can cause delamination, black specks, weak spots, and degradation of the polycarbonate resin. The best purging materials are ground acrylic or polycarbonate resin regrind.

MATERIAL ___ High-viscosity Polycarbonate

PROPERTY	TEST METHOD	ENGLISH UNITS	VALUE	METRIC UNITS	VALUE
MECHANICAL DENSITY	D792	lb/ft^3	74.8	g/cm^3	1.2
TENSILE STRENGTH	D638	psi	9000	MPa	62
TENSILE MODULUS	D638	psi	345,000	MPa	2379
FLEXURAL MODULUS	D790	psi	340,000	MPa	2344
ELONGATION TO BREAK	D638	%	110	%	110
NOTCHED IZOD AT ROOM TEMP	D256	ft-lb/in.	2.3 ($^1/_4$ in.)	J/m	123 ($^1/_4$ in.)
HARDNESS	D785		M70		M70
THERMAL DEFLECTION T @ 264 psi	D648	°F	270	°C	132
DEFLECTION T @ 66 psi	D648	°F	280	°C	138
VICAT SOFT-ENING POINT	D1525	°F		°C	
UL TEMP INDEX	UL746B	°F		°C	
UL FLAMMABILITY CODE RATING	UL94				
LINEAR COEFFICIENT THERMAL EXPANSION	D696	in./in./°F	68×10^{-6}	mm/mm/°C	122×10^{-6}
ENVIRONMENTAL WATER ABSORPTION 24 HOUR	D570	%	0.15	%	0.15
CLARITY	D1003	% TRANS-MISSION		% TRANS-MISSION	
OUTDOOR WEATHERING	D1435	%		%	
FDA APPROVAL					
CHEMICAL RESISTANCE TO: WEAK ACID	D543				
STRONG ACID	D543				
WEAK ALKALI	D543				
STRONG ALKALI	D543				
LOW MOLECULAR WEIGHT SOLVENTS	D543				
ALCOHOLS	D543				
ELECTRICAL DIELECTRIC STRENGTH	D149	V/mil	380	kV/mm	15
DIELECTRIC CONSTANT	D150	10^6 Hertz			
POWER FACTOR	D150	10^6 Hertz			
OTHER MELTING POINT	D3418	°F	302	°C	150
GLASS TRANS-ITION TEMP.	D3418	°F		°C	

260

23.12 PROCESSING-SENSITIVE END PROPERTIES

Occasionally, part failures will occur during the development of a new product. These failures are almost always the result of oversight or misinterpretation of the environment, loading, or related factors. Visual examination of the mode of failure often yields information about the reason for failure. For example, a compatibility failure will exhibit crazing throughout the part.

Chemical and environment compatibility are dependent on both stress and temperature. Any material coming in contact with a part, during assembly or end use, should be considered for compatibility—at the actual temperature and load expected during exposure. Chemical attack usually occurs when molded parts come in contact with alkali, alkaline salts, amines, and ozone. There is no dependable substitute for careful testing of prototypes or production parts in typical operating environments.

The best design will not function correctly if the material is significantly degraded during the molding cycle. Excessive heat, moisture, or contamination can cause a breakdown in the molecular structure of polycarbonate resin and cause a reduction in physical properties.

Polycarbonate resin has a very high impact strength, but point stress can sharply reduce this property. In designing impact applications, the designer should give careful attention to sharp corners, notches, and knurls. Severe stress concentration can produce what appears to be a brittle failure, even though the material is ductile.

Walls that are too thick may have sink marks and voids after cooling. Also, wall transitions from thin to thick sections can cause a loss in molding pressure, resulting in sinks and voids.

Polycarbonate is susceptible to incompatible additives and mold release agents. Additives not inherent to polycarbonate resin should not be used.

Like any material, polycarbonate resin can be overstressed to the point of failure if a part is designed improperly. When a part is too complicated for stress analysis, end-use testing may be necessary. This is especially true where high molded-in stresses will contribute to the total stresses in a part.

Even a well-molded part will have some residual molded-in stress that lessens the amount of added stress the part can withstand. For example, improper mold temperature, pressure, or injection can add undesirable stresses. Solvent stress analysis of clear parts can be used to evaluate stress levels.

Finally, whenever a part is subjected to cyclic loading, stress levels should be carefully analyzed to predict part life. Testing of prototypes is the preferred approach.

23.13 SHRINKAGE

The mold shrinkage of polycarbonate resin is very predictable. For unreinforced resins, it is between 0.005–0.007 in./in. (mm/mm). Because polycarbonate resin is noncrystalline, shrinkage is virtually the same in the direction of flow as it is across the flow. Even so, unnecessarily tight tolerances increase part inspection costs and may result in secondary finishing of the part. To avoid this, critical dimensions should show the nominal dimension plus acceptable high and low limits. In the case of unreinforced polycarbonate resin, for example, minimum tolerance should be specified at plus or minus 0.002 in./in. (mm/mm). Tighter limits can be achieved only at a higher cost.

23.14 TRADE NAMES

In the United States, base polycarbonate resins and foamable grades are produced under the trade name Lexan by General Electric Co., Plastics Group, 1 Plastics Ave., Pittsfield, MA 01201; under the trade name Makrolon by the Mobay Chemical Corp., Plastics and

Mold Shrinkage Characteristics (in./in. or mm/mm)

MATERIAL _____Polycarbonate_____

IN THE DIRECTION OF FLOW

THICKNESS

	mm: 1.5	3	6	8	FROM	TO	#
MATERIAL	in.: 1/16	1/8	1/4	1/2			
Unreinforced							0.005– 0.007 in.
10% glass					0 8 in.	8 in. 16 in.	0.002– 0.004 0.0025–
					16 in.	+	0.0045 0.0035– 0.005

PERPENDICULAR TO THE DIRECTION OF FLOW							

NOTE: Mold shrinkage is approximate. It is affected by part design, mold temperature, thickness, injection pressure, packing time, cycle time, orientation, gate design, gate size, gate location, glass content, glass size, and filler content.

Coatings Div., Penn Lincoln Parkway West, Pittsburgh, PA 15205; and under the trade name Calibre by Dow Chemical Co. Dow Center, Midland, MI 48640.

The largest U.S. suppliers of polycarbonate sheet include General Electric Co.; Rohm & Haas Co., Independence Mall West, Philadelphia, PA 19105, under the trade name Tuffak (TM); and Sheffield Plastics Inc., Salisbury Road, P.O. Box 248, Sheffield, MA 01257, under the trade name Poly-glaz (TM).

23.15 INFORMATION SOURCES

Most major manufacturers of polycarbonate resin will provide a great deal of technical assistance, including help in part and mold design, production startup, prototype fabrication and testing, and finished part performance evaluation.

24

POLYETHERIMIDE (PEI)

I. William Serfaty

Ultem Products Section, General Electric Plastics Group, Pittsfield, MA 01201

24.1 STRUCTURE

24.2 CATEGORY

Polyetherimide (PEI) is a high-performance, amorphous thermoplastic based on regular repeating ether and imide linkages. The aromatic imide units provide stiffness, while the ether linkages allow for good melt-flow characteristics and processability. PEI is therefore suitable for applications that require:

- High temperature stability polyetherimide
- High mechanical strength
- Inherent flame resistance with extremely low smoke evolution
- Outstanding electrical properties over a wide frequency and temperature range
- Chemical resistance to aliphatic hydrocarbons, acids, and dilute bases
- UV stability
- Ready processability on conventional equipment

24.3 HISTORY

First introduced to the commercial market in 1982, PEI was the culmination of ten years of laboratory research at General Electric Plastics, under the direction of Joseph Wirth, now head of the technology department at GE Plastics.

24.4 POLYMERIZATION

One of several synthetic routes to polyetherimides of a general structure involves a cyclization reaction to form the imide rings and a displacement reaction to prepare the ether linkage and form the polymer.

The first step of this synthesis is a bis-imide monomer formed by the reaction of nitrophthalic anhydride and a diamine. The second step of polyetherimide synthesis involves the formation of a bisphenol dianion by treatment of a diphenol with two equivalents of base, followed by removal of water. The polymerization step involves displacement of the nitrogroups of the bis-imide by the bisphenol dianion to form the ether linkages of the polymer.

Polymerization is performed under relatively mild conditions in dipolar aprotic solvents or in mixtures of these with toluene. A large number of polyetherimides can be prepared by this synthetic route. The nature of R and Ar, and the position of substitution in the bisimides can all be varied widely in order to achieve molecular structures with desired physical properties.

The 1987 price of PEI is $4.46 per pound for the unreinforced grade and $3.57 per pound for 30% glass reinforced material.

24.5 DESCRIPTION OF PROPERTIES

24.5.1 Mechanical

The mechanical properties of unmodified and three glass-reinforced PEI compounds are shown in Table 24.1. Polyetherimide exhibits a room temperature flexural modulus of 480,000 psi (3309 MPa), outstanding for an unmodified thermoplastic resin. PEI is exceptionally strong with a tensile strength at yield in excess of 15,000 psi (103 MPa) and a flexural strength of 21,000 psi (144 MPa). In load-bearing applications, nonreinforced PEI provides the rigidity frequently associated with other glass-reinforced polymers. Even more impressive is the good retention of these mechanical properties at elevated temperatures. Figures 24.1 and 24.2 depict the tensile and flexural behavior of PEI as a function of temperature. Even at 356°F (180°C) the tensile strength and flexural modulus of PEI remain in excess of 6000 psi (41 MPa) and 300,000 psi (2068 MPa), respectively.

Polyethermide exhibits classical mechanical behavior in accordance with Hookes' law. Under load, it displays a linear stress relationship below the proportional limit. The pro-

TABLE 24.1. Mechanical Properties of Polyetherimide Injection-Molding Resins

Property		Natural	10% glass	20% glass	30% glass
Tensile strength,	psi	15,200	16,500	20,100	24,500
	(MPa)	(105)	(114)	(139)	(169)
Tensile modulus,	psi	430,000	650,000	1,000,000	1,300,000
	(MPa)	(3,000)	(4,483)	(6,897)	(8,966)
Tensile elongation to yield,	%	7–8	5.0		
Tensile elongation to ultimate,	%	60	6	3	3
Flexural strength,	psi	21,000	28,000	31,000	33,000
	(MPa)	(144)	(193)	(207)	(228)
Flexural modulus,	psi	480,000	650,000	900,000	1,200,000
	(MPa)	(3,309)	(4,483)	(6,207)	(8,276)
Notched Izod,	ft-lb/in.	1	1.1	1.6	2
	(J/m)	(53.4)	(58.1)	(85.4)	(106.8)

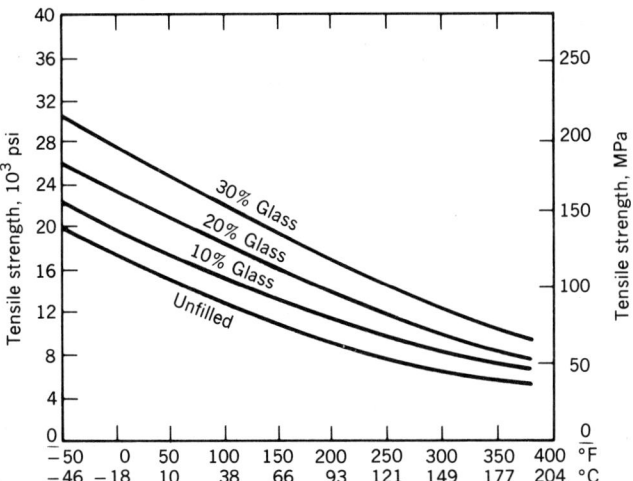

Figure 24.1. Tensile strength vs temperature for polyetherimide.

portional limit for PEI ranges from approximately 8000 psi (55 MPa) for unreinforced resin to 18,000 psi (124 MPa) for 30% glass-reinforced compositions at 73°F (23°C). The stress-strain curves for natural and glass-reinforced PEI are shown in Figure 24.3.

The reinforcing effects of glass fibers in PEI are demonstrated by the substantial increase in tensile yield strength, which increases smoothly from an initial value of 15,000 psi (103 MPa) to as much as 27,000 psi (186 MPa) as glass fiber content is increased to 40%. This results from a combination of the polymer's excellent adhesion to the glass fibers, and the high tensile strength of the fibers themselves. (Micrographs taken of sample fracture surfaces show excellent adhesion of PEI to glass.) Beyond 40 wt% of glass, the tensile strength drops as the maximum packing volume for the fibers is approached and as the extensibility of the composite lowers.

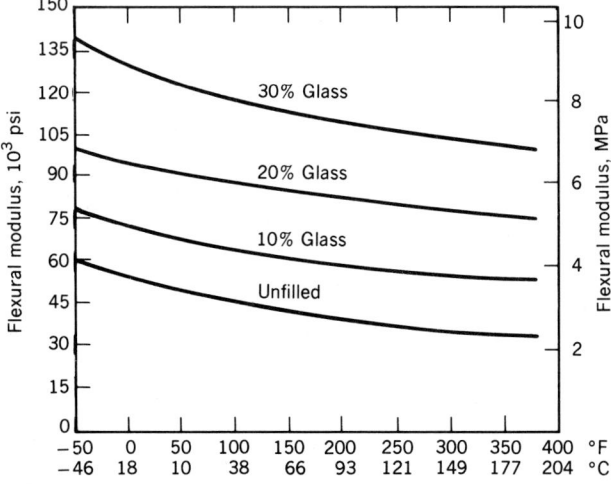

Figure 24.2. Flexural modulus vs temperature for polyetherimide.

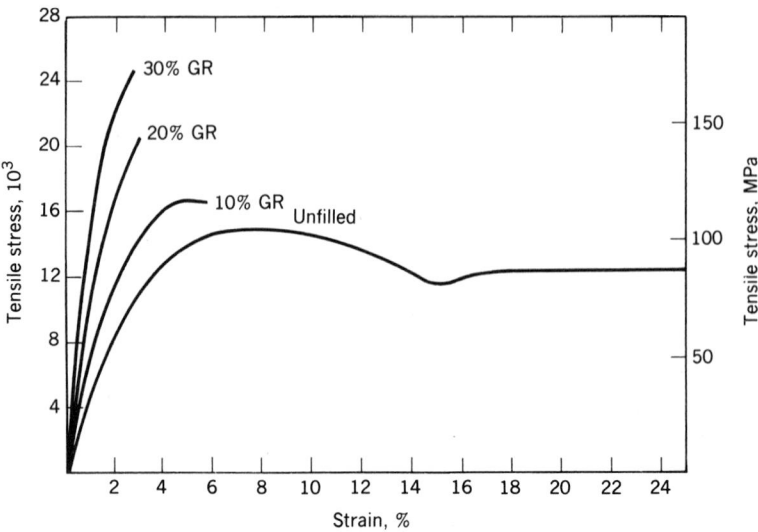

Figure 24.3. Stress–strain curves in tension at 73°F (23°C) for polyetherimide.

24.5.2 Electrical

The resin exhibits an excellent balance of electrical properties which remain stable over a wide range of environmental conditions. Its stable dielectric constant and low disspiation factor make it suitable for use at frequencies in excess of 10^9 Hz, even at elevated temperatures as shown in Figures 24.4 and 24.5.

24.5.3 Thermal

One of the most outstanding properties of polyethermide is its ability to withstand exposure to elevated temperatures. PEI exhibits a glass-transition temperture of 423°F (217°C). The

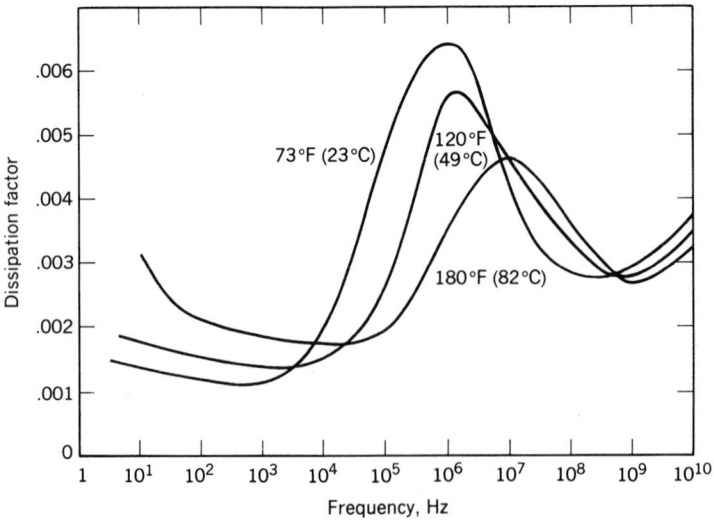

Figure 24.4. Dissipation factor vs frequency at 50% rh for polyetherimide.

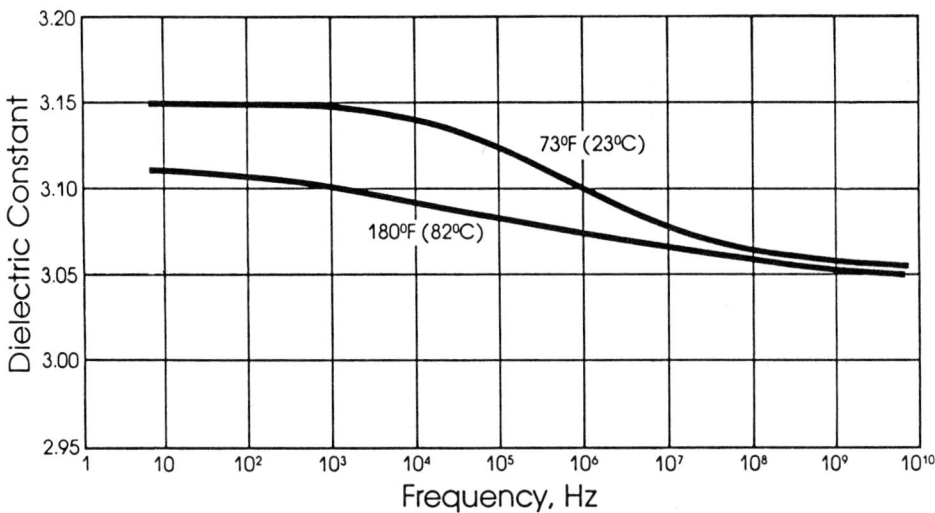

Figure 24.5. Dielectric constant of PEI vs frequency at 50% rh.

high glass-transition temperature T_g of PEI, which provides a heat-deflection temperature under load (DTUL) of 392°F (200°C), accounts for its excellent retention of physical properties at elevated temperatures. The DTUL profile for PEI is given in Figure 24.6. This value provides a useful comparison of material deformation under constant stress conditions at a given temperature. In recognition of its inherent thermal stability, a continuous use temperature listing of 338°F (170°C) with impact has been granted PEI by Underwriters' Laboratories, Inc.

24.5.4 Flammability

Polyetherimide is flame resistant without the incorporation of additives.[1] PEI is listed 94V-0 at 0.016 in. (0.41 mm), according to UL Bulletin 94, and exhibits a limiting oxygen index

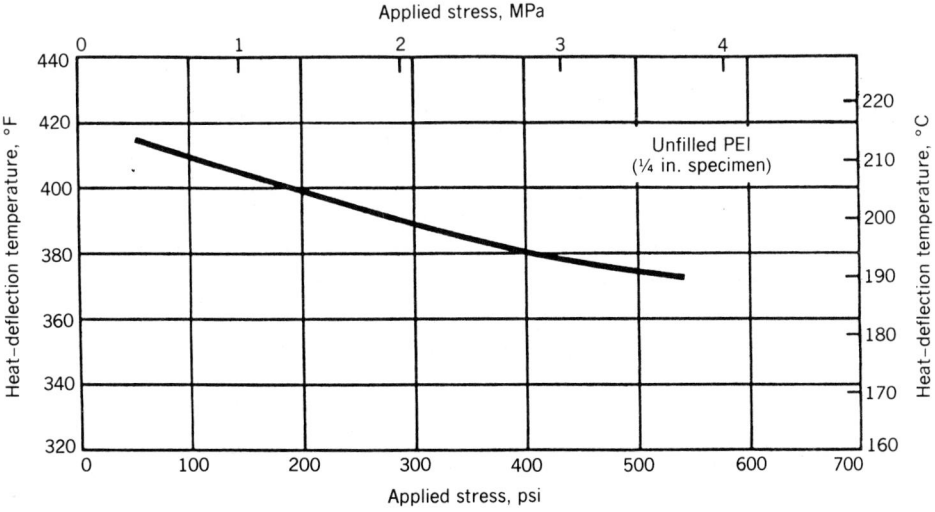

Figure 24.6. Heat-deflection temperature vs applied stress.

TABLE 24.2.   Combustion Characteristics[a]

	Polyetherimide		Polyethersulfone		Polycarbonate	
Condition	FL[b]	SM[c]	FL[b]	SM[c]	FL[b]	SM[c]
D_s 1.5 min	0	0	0	0	13.	0.1
D_s 4.0 min	0.7	0	1.7	0	127.	1.2
D_{max}	31.	0.4	37.	0.6	130.	4.

[a]NBS smoke chamber ASTM E662. Specimen thickness-nominal 0.060 in. (1.52 mm).
[b]FL = Flaming condition.
[c]SM = Smoldering condition.

of 47, among the highest of any thermoplastic. Smoke generation as monitored in the NBS chamber (ASTM E662) is extremely low in comparison with other thermoplastics, as seen in Table 24.2. This combination of low smoke evolution and high oxygen index is unique and rarely seen among engineering thermoplastic materials.

24.5.5 Environmental Resistance

Unlike the majority of amorphous resins, polyetherimide exhibits resistance to a wide range of chemical media. PEI resin is unaffected by most hydrocarbons, alcohols, and fully hal-ogenated solvents. Partially halogenated hydrocarbons such as methylene chloride and chlo-roform will dissolve PEI. Polyetherimide has shown exceptional resistance to mineral acids and tolerates short-term contact with dilute bases.

Hydrolytically stable, PEI retains high tensile strength after 10,000 hours immersion in water at 212°F (100°C), as shown in Figure 24.7. Polyetherimide's physical properties remain

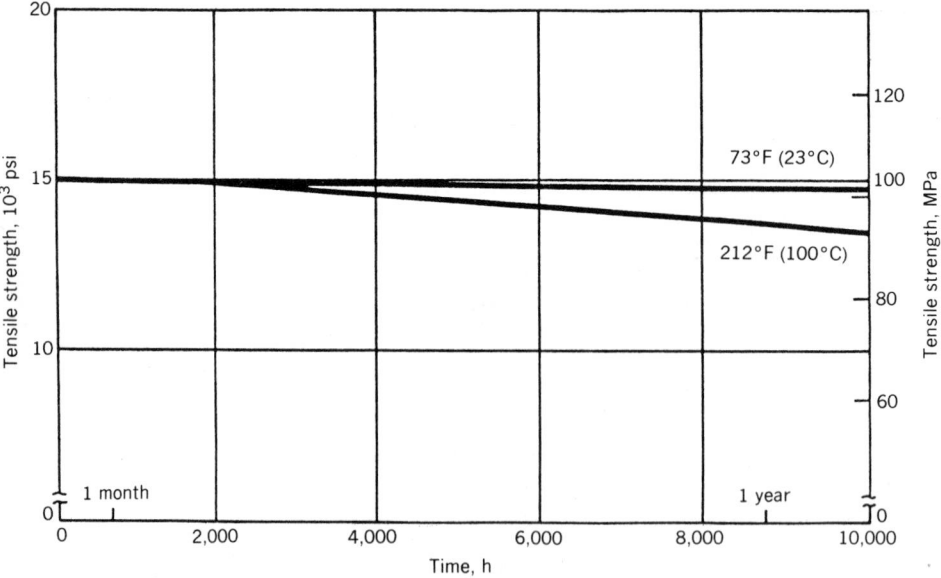

Figure 24.7. Effect of water exposure on tensile strength of polyetherimide.

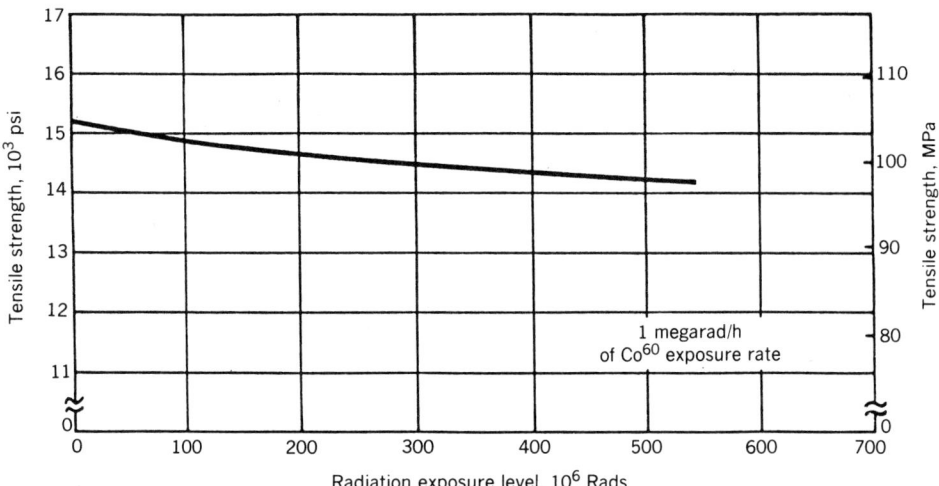

Figure 24.8. Effect of gamma radiation exposure on tensile strength of polyetherimide.

stable following short-term autoclave exposure and drying in vacuum at room temperature. The long-term effects of repeated autoclaving are presently under investigation.

Polyetherimide is resistant to UV radiation without the addition of stabilizers. Exposure to 1000 h of Xenon arc weatherometer irradiation [0.55 W/m^2 irradiance at 340 nm, 145°F (63°C)] has produced a negligible change in the tensile strength of PEI. Quartz-lamp ultraviolet (QUV) test exposure has confirmed the outstanding retention of physical properties exhibited by PEI versus conventional engineering thermoplastics.

PEI also demonstrates excellent resistance to physical property degradation following ionizing radiation exposure. A decrease of less than 5% in tensile strength was observed following 500 megarads of Cobalt 60 exposure at the rate of 1 megarad per hour, as seen in Figure 24.8.

24.6 APPLICATIONS

24.6.1 Automotive

Polyetherimide resin's advantages in this segment are its high strength maintenance at elevated temperatures; good resistance to lubricants, coolants, and fuels; long-term creep resistance; dimensional stability under load; and ductility. Application areas include under-the-hood, electrical, heat-exchange, and mechanical components.

24.6.2 Appliances

Key material features are chemical resistance to foods such as oils, greases, and fats; microwave transparency; heat resistance up to 400°F (204°C); good practical impact strength; flame resistance; and high gloss/colorability.

24.6.3 Electrical and Electronics

Attributes include heat resistance to wave and vapor-phase soldering, intricate design potential, dimensional stability (allowing close point-to-point tolerances), flame resistance, and ductility. Application areas include connectors, switches and controls, integrated circuit carriers and burn-in sockets, and explosion-proof enclosures.

24.6.4 Printed Circuit Boards

PEI's inherent flame resistance, high physical strength, broad chemical resistance, and high temperature stability make it particularly attractive for printed circuit boards.

Key electrical properties for printed circuit applications include PEI's low dissipation factor and stable dielectric constant over a broad frequency and temperature range, as shown in Figures 24.4 and 24.5. The dissipation factor is typically lower than those of comparable thermoplastics and shows less variation with frequency. The dielectric constant of unreinforced PEI is unusually stable, varying only 3% from 10 Hz to 10 GHz at room temperature, making the material particularly suitable for high-frequency applications.

Polyetherimide's high-temperature capabilities allow the material to be processed using most conventional soldering techniques, and also provide a continuous use temperature rating significantly higher than the 266°F (130°C) rating for competitive epoxy/fiberglass based circuitry.

24.6.5 Aerospace

The main strengths of polyetherimide for this market segment are high mechanical strength, flame resistance, low smoke evolution, weight savings versus metals and other heavier materials, chemical resistance, and stain resistance. Applications include lighting, seating, wiring, electrical hardware, and secondary structures. Even engine components comprised of graphite fiber laminates are presently under investigation by various end users.

24.6.6 Other Markets

Fibers represent an emerging market for polyetherimide resin. Applications include protective clothing (for fire fighters, race drivers), dry filtration (fume bags), aircraft fabrics (upholstery, carpets), and drying screens (used in papermaking). Additional market areas include film (dielectric films, insulating tapes, flexible circuitry), and wire insultation.

24.7 ADVANTAGES/DISADVANTAGES

Economic and environmental changes, such as stricter regulations for smoke evolution and flame resistance, energy-efficiency requirements, and electronics miniaturization offer expanded market opportunities for polyetherimide resin.

24.8 PROCESSING TECHNIQUES

See the Master Material Outline and Standard Material Design Chart.

Master Material Outline

PROCESSABILITY OF ___Polyetherimide (PEI)___

PROCESS	A	B	C
INJECTION MOLDING	X		
EXTRUSION	X		
THERMOFORMING	X		
FOAM MOLDING	X		
DIP, SLUSH MOLDING		X	
ROTATIONAL MOLDING			X
POWDER, FLUIDIZED-BED COATING			X
REINFORCED THERMOSET MOLDING			X
COMPRESSION/TRANSFER MOLDING	X		
REACTION INJECTION MOLDING (RIM)			
MECHANICAL FORMING	X		
CASTING	X		

A = Common processing technique.
B = Technique possible with difficulty.
C = Used only in special circumstances, if ever.

Standard design chart for ___Polyetherimide (unreinforced)___

1. Recommended wall thickness/length of flow, mm   (750°F melt, 15,000 psi)
   Minimum ____0.38 mm____ for ____7.62 mm____ distance
   Maximum _____ for _____ distance
   Ideal ____3.16 mm____ for ____364.5 mm____ distance

2. Allowable wall thickness variation, % of nominal wall ____25%____

3. Radius requirements
   Outside: Minimum ____0.015 in. plus wall____ Maximum ____2.0 × wall____
   Inside: Minimum ____0.015 in.____ Maximum ____1.0 × wall____

4. Reinforcing ribs
   Maximum thickness ____0.7 × wall____

Maximum height _____ 3.0 × wall _____
Sink marks: Yes _____ No ___X___

5. Solid pegs and bosses
   Maximum thickness ____ 0.7 × wall ____
   Maximum weight _____
   Sink marks: Yes _____ No ___X___

6. Strippable undercuts
   Outside: Yes _____ No ___X___
   Inside: Yes _____ No ___X___

7. Hollow bosses: Yes ___X___ No _____

8. Draft angles
   Outside: Minimum ____ $\frac{1}{2}°$ ____ Ideal ____ 2° ____
   Inside: Minimum ____ $\frac{1}{2}°$ ____ Ideal ____ 2° ____

9. Molded-in inserts: Yes ___X___
                        No _____

10. Size limitations
    Length: Minimum _____ Maximum _____ Process _____
    Width: Minimum _____ Maximum _____ Process _____

11. Tolerances   ±0.0015 in./in.

12. Special process- and materials-related design details
    Blow-up ratio (blow molding) _____
    Core L/D (injection and compression molding) _____
    Depth of draw (thermoforming) _____
    Other _____

24.9 RESIN FORMS

24.10 SPECIFICATION OF PROPERTIES

See Master Outline Materials Properties—English Units and SI Units

24.11 PROCESSING REQUIREMENTS

Polyetherimide is processed on conventional injection molding, extrusion, or blow molding equipment. It is available in pellet and sheet form. The generally recommended injection molding conditions for PEI are given in the Table 24.3.

24.12 PROCESSING-SENSITIVE END PROPERTIES

24.13 SHRINKAGE

See Mold Shrinkage MPO.

MATERIAL _____ Polyetherimide, 40% glass-reinforced

PROPERTY	TEST METHOD	ENGLISH UNITS	VALUE	METRIC UNITS	VALUE
MECHANICAL DENSITY	D792	lb/ft³	100.4	g/cm³	1.61
TENSILE STRENGTH	D638	psi	27,000	MPa	186
TENSILE MODULUS	D638	psi	1,700,000	MPa	11,724
FLEXURAL MODULUS	D790	psi	1,700,000	MPa	11,724
ELONGATION TO BREAK	D638	%		%	
NOTCHED IZOD AT ROOM TEMP	D256	ft-lb/in.	2.1	J/m	112
HARDNESS	D785		M114		M114
THERMAL DEFLECTION T @ 264 psi	D648	°F	415	°C	213
DEFLECTION T @ 66 psi	D648	°F	420	°C	216
VICAT SOFT-ENING POINT	D1525	°F	454 +	°C	234 +
UL TEMP INDEX	UL746B	°F		°C	
UL FLAMMABILITY CODE RATING	UL94		V-0 at 0.010 in		V-0 at 0.25 mm
LINEAR COEFFICIENT THERMAL EXPANSION	D696	in./in./°F	0.8×10^{-5}	mm/mm/°C	1.4×10^{-5}
ENVIRONMENTAL WATER ABSORPTION 24 HOUR	D570	%	0.13	%	0.13
CLARITY	D1003	% TRANS-MISSION		% TRANS-MISSION	
OUTDOOR WEATHERING	D1435	%		%	
FDA APPROVAL					
CHEMICAL RESISTANCE TO: WEAK ACID	D543		Not attacked		Not attacked
STRONG ACID	D543		Not attacked		Not attacked
WEAK ALKALI	D543				
STRONG ALKALI	D543				
LOW MOLECULAR WEIGHT SOLVENTS	D543				
ALCOHOLS	D543		Not attacked		Not attacked
ELECTRICAL DIELECTRIC STRENGTH	D149	V/mil	610	kV/mm	24
DIELECTRIC CONSTANT	D150	10^6 Hertz			
POWER FACTOR	D150	10^6 Hertz			
OTHER MELTING POINT	D3418	°F		°C	
GLASS TRANS-ITION TEMP.	D3418	°F		°C	

273

MATERIAL _____ Polyetherimide (unreinforced)

PROPERTY	TEST METHOD	ENGLISH UNITS	VALUE	METRIC UNITS	VALUE
MECHANICAL DENSITY	D792	lb/ft^3	79.2	g/cm^3	1.27
TENSILE STRENGTH	D638	psi	15,200	MPa	105
TENSILE MODULUS	D638	psi	430,000	MPa	2966
FLEXURAL MODULUS	D790	psi	480,000	MPa	3310
ELONGATION TO BREAK	D638	%	7–8	%	7–8
NOTCHED IZOD AT ROOM TEMP	D256	ft-lb/in.	1.0	J/m	53
HARDNESS	D785		M109		M109
THERMAL DEFLECTION T @ 264 psi	D648	°F	392	°C	200
DEFLECTION T @ 66 psi	D648	°F	410	°C	210
VICAT SOFT-ENING POINT	D1525	°F	426 +	°C	219 +
UL TEMP INDEX	UL746B	°F	338	°C	170
UL FLAMMABILITY CODE RATING	UL94		V-0 at 0.016 in		V-0 at 0.41 mm
LINEAR COEFFICIENT THERMAL EXPANSION	D696	in./in./°F	3.1×10^{-5}	mm/mm/°C	5.6×10^{-5}
ENVIRONMENTAL WATER ABSORPTION 24 HOUR	D570	%	0.25	%	0.25
CLARITY	D1003	% TRANS-MISSION		% TRANS-MISSION	
OUTDOOR WEATHERING	D1435	%		%	
FDA APPROVAL			Section 21 CFR 177.1595		Section 21 CFR 177.1595
CHEMICAL RESISTANCE TO: WEAK ACID	D543		Not attacked		Not attacked
STRONG ACID	D543		Not attacked		Not attacked
WEAK ALKALI	D543				
STRONG ALKALI	D543				
LOW MOLECULAR WEIGHT SOLVENTS	D543				
ALCOHOLS	D543		Not attacked		Not attacked
ELECTRICAL DIELECTRIC STRENGTH	D149	V/mil	In oil 710 In air 830	kV/mm	In oil 28 In air 33
DIELECTRIC CONSTANT	D150	10^6 Hertz	3.1		3.1
POWER FACTOR	D150	10^6 Hertz	0.006		0.006
OTHER MELTING POINT	D3418	°F		°C	
GLASS TRANS-ITION TEMP.	D3418	°F	423	°C	217

MATERIAL   Polyetherimide (unreinforced)

IN THE DIRECTION OF FLOW

THICKNESS

MATERIAL	mm: 1.5	3	6	8	FROM	TO	#
	in.: 1/16	1/8	1/4	1/2			
Ultem 1000	0.006	0.007	0.008				

	PERPENDICULAR TO THE DIRECTION OF FLOW					
Ultem 1000	0.006	0.007	0.008			

NOTE: Mold shrinkage is approximate. It is affected by part design, mold temperature, thickness, injection pressure, packing time, cycle time, orientation, gate design, gate size, gate location, glass content, glass size, and filler content.

TABLE 24.3. Molding Conditions for PEI

Resin drying	300°F, 4 h, trays (hopper dryer preferred while molding)
Melt temperature	640–800°F (338–427°C)
Melt profile	Nozzle 620–775°F (327–413°C)
	Front zone 610–760°F (321–404°C)
	Center zone 600–740°F (316–393°C)
	Rear zone 590–700°F (310–371°C)
Mold temperature	150 to 350°F water heating is sufficient for most applications. Increased mold temperature will increase flow and improve surface.
Molding pressure	Booster pressure (1st stage) 1000–1800 psig (6.9–12.4 MPa)
	Holding pressure (2nd stage) 800–1500 psig (5.5–10.4 MPa)
	Back pressure 50–200 psig (0.3–1.4 MPa)
Screw design	1.5–4.0:1 Compression ratio: 16 tp 24:1 L/D
Ram speed	Medium to fast
Clamp pressure	3 to 6 t/in.^{2}
Screw speed	50/400 rpm
Purge	HDPE, glass reinforced polycarbonate, or ground cast acrylic.[a] Begin purging at processing temperature and reduce barrel temperature to about 500°F (260°C) while continuing to purge.
Shrink rate	0.005–0.007 in./in.
Mold pressure	Normally not required; if part design causes difficulty in removal, use any standard mold-release agent.

[a]Use only at lower barrel temperatures.

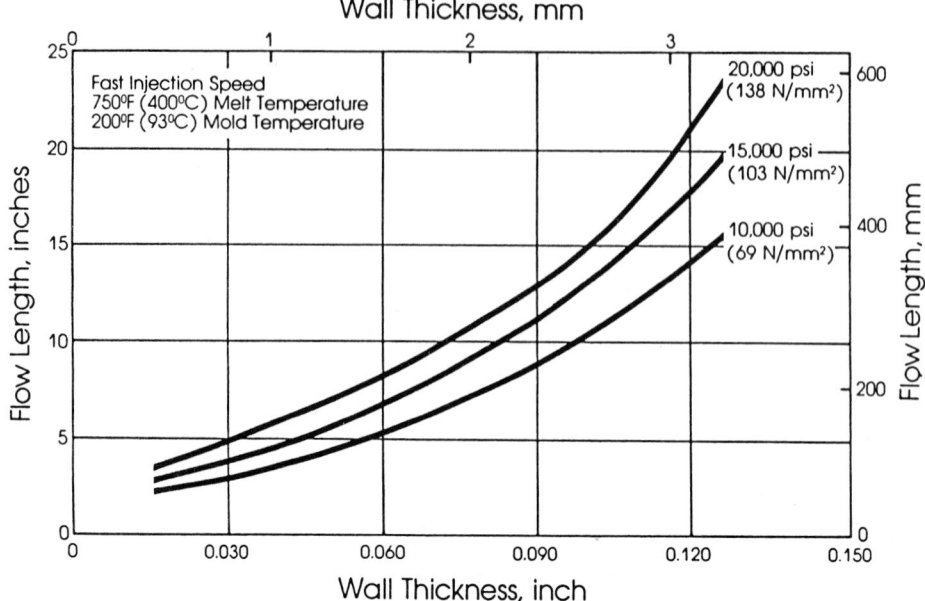

Figure 24.9. Effect of wall thickness on melt flow of PEI.

PEI exhibits a resin viscosity versus shear rate profile comparable to polycarbonate or polysulfone (rather than conventional polyimides), although the processing temperature is higher. In addition, due to its thermal stability, PEI exhibits a much wider processing window than most engineering plastics.

Figure 24.9 depicts the effect of wall thickness and injection pressure on the flow length of PEI injection molding resin. The flow characteristics of polyetherimide permit its use in complex, thin wall parts.

Because of its high melt strength, PEI is readily converted into film, sheet, and profiles via melt extrusion. This melt integrity also allow PEI to be easily thermoformed or blow molded using injection or extrusion techniques.

24.14 TRADE NAMES

Polyetherimide is available commercially from General Electric Company under the Ultem trademark.

BIBLIOGRAPHY

1. D.E. Floryan and G.L. Nelson, *J. Fire Flammability* **11,** 284 (1980).
2. R.O. Johnson and H.S. Burlhis, "Polyetherimide: A New High Performance Thermoplastic Resin," *J. Polym. Sci.,* Special Polymer Symposia edition.
3. W. Serfaty, "PEI—A High Performance Thermoplastic," a paper presented before the Society of Plastics Engineers, 1982.

25

POLYETHERETHERKETONE (PEEK)

Thomas W. Haas

Professor and Director of Graduate Engineering, Virginia Commonwealth University, 900 Park Ave., Richmond, VA 23284-2009

25.1 INTRODUCTION

Polyetheretherketone (PEEK) [T_g 293°F (145°C), T_m 635°F (335°C)] is one of several aromatic polyetherketones. The chemical repeat unit (mer) of PEEK along with that of other ketone polymers of commercial importance.

PEEK

PEKK

PEK

Chain rigidity is derived from the relatively inflexible and immobile phenol and ketone groups. The incorporation of ether links into the polymer backbone renders the chain more flexible, resulting in a decrease in both the glass transition and the melting temperatures relative to polyetherketone (PEK) [T_g 330°F (166°C), T_m 685°F (364°C)] and polyetherketoneketone (PEKK) [T_m 723°F (384°C)].

25.2 CATEGORY

PEEK is a melt-processible, high-performance thermoplastic.[1] It is semicrystalline with a high melting temperature of 635°F (335°C) and a glass-transition temperature (T_g) well above room temperature at 293°F (145°C).[2] At room temperature, PEEK is tough, strong, and rigid. At elevated temperatures, PEEK exhibits good retention of ductility during heat aging with an estimated continuous use temperature of 480°F (250°C), based on the Underwriters Laboratory (UL) Temperature Index.

25.3   HISTORY

PEEK was first prepared in the laboratory in 1977 and then test marketed in 1978 by Imperial Chemical Industries (ICI). ICI Americas subsequently introduced PEEK in North America in 1980.

The earliest applications were wire coating and insulation for high-performance applications in aircraft and computers. PEEK was selected because, in addition to being melt processible, the polymer exhibited excellent cut-through resistance to sharp edges as compared to polytetrafluoroethylene (PTFE) and other fluoropolymers while providing similar thermal aging stability and electrical properties. PEEK also exhibits low smoke emission and flammability, which is important for these applications.

Worldwide production of aromatic ketone polymers is currently estimated to be less than 200,000 pounds per year, with the bulk being PEEK manufactured by ICI.[3] However, a number of suppliers are entering the market, including Du Pont with polyetherketoneketone (PEKK), Amoco with polyetherketone (PEK), and two German (FRG) producers, BASF AG and Hoechst AG, each of which is introducing its own PEK polymers, and firms in Japan. ICI has also recently introduced PEK.[4]

The growth rate is currently more than 100% annually; ICI is anticipating an annual global market of 3 to 5 million pounds by the end of the century.

25.4   POLYMERIZATION

The synthesis of PEEK is the result of the pioneering work of K.W. Johnson and A.G. Farnham,[5] who were the first to prepare aromatic polyethers. Their method involves the formation of ether bonds via the nucleophilic substitution process. It subsequently proved applicable to a number of polysulfones and polyetherketones.

The generalized reaction scheme is

where x = K or Na
     Ar = Aromatic group
     Y = F or Cl
     R = S=O, C=O, N=N, or O=S=O
     Solvent is DMSO or sulfolane

PEEK was made by the reaction of the potassium salt of hydroquinone with difluorobenzophenone in hot tetramethylene sulfone (Sulfolane). However, the resulting polymer was of low molecular weight, owing to crystallization, which caused the growing chains to precipitate from solution and thereby terminate polymerization.[6] It was not until Rose and his associates at ICI[2,7] substituted a high boiling solvent, diphenylsulfone, for Sulfolane at temperatures close to the melting temperature of the polymer that high weight molecular PEEK was made

Current prices for PEEK are from $20 to $32 per pound, which is prohibitive except for premium applications.[8,9] However, the high price is primarily the result of expensive feedstock such as difluorobenzophenone. Process development to reduce costs is currently the objective of intensive ongoing research.

25.5 DESCRIPTION OF PROPERTIES

PEEK is a melt-processible, high-performance thermoplastic.[1] It is semicrystalline with a high melting temperature (T_m) of 635°F (335°C) and a glass-transition temperature (T_g) of 293°F (145°C). The degree of crystallinity can vary from 40% (slow cooling) to essentially amorphous (quenching) but is usually about 35%. When it is crystallized, the morphology from the melt is typically spherulitic. When PEEK is cooled rapidly from 752°F (400°C) in a differential scanning calorimeter (DSC), the overall crystallization rate goes through a maximum at about 446°F (230°C).[9] Postannealing at elevated temperatures of 392–608°F (200–320°C) will increase the thickness of the lamellae and the degree of crystallinity but does not change the size of the spherulites.[10]

X-ray diffraction has shown that the unit cell of PEEK is orthorhombic.[11-16] Reported values of the unit cell dimensions are $a = 0.775$–0.788 nm, $b = 0.586$–0.594 nm, and $c = 0.988$–1.007 nm. The unit cell contains only two thirds of a repeat unit

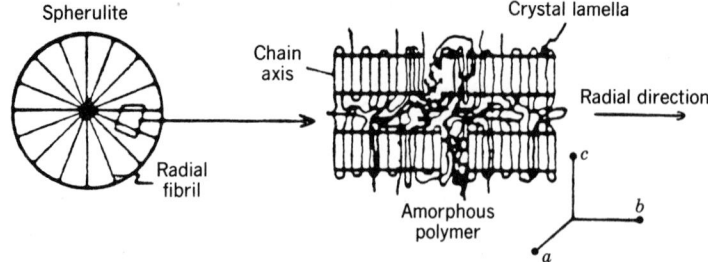

Figure 25.1. Schematic representation of an undeformed spherulite.[17] Courtesy of *Polym. Compos.*

A schematic of a PEEK spherulite is shown in Figure 25.1. In the spherulite, the *b* axis is radial whereas the *a* axis is tangential, *c* is the polymer chain axis which is perpendicular to the fold plane. PEEK forms folded-chain lamellae during crystallization but, because of the rigidity of the chains and the distance between the ether and ketone linkages, the amorphous layers are thick. This gives rise to the polymer's inherently low degree of crystallinity. Thin films of PEEK crystallized from the melt have been shown to form spherulites having essentially cylindrical symmetry.[11]

In addition to the glass transition, PEEK has another, secondary transition. The onset of the beta transition (T_g) is signified by the rotation of ketone linkages at about 284°F (140°C).[18] The gamma transition is very broad and complex, occurring from -202 to 86°F (-130 to 30°C), and is thought to be affected by rotational movement of absorbed water molecules, which form complexes with the phenyl ketone moities.[19]

At room temperature, PEEK is strong and rigid. It exhibits excellent load-bearing properties over long periods of time. As is the case with most semicrystalline polymers, PEEK also has excellent fatigue resistance properties.[1]

The Izod impact strength of neat (unreinforced) PEEK is listed by ICI[20] to be 1.55 ft-lb/in. (83 J/m). Unnotched specimens do not break, an indication of the notch sensitivity

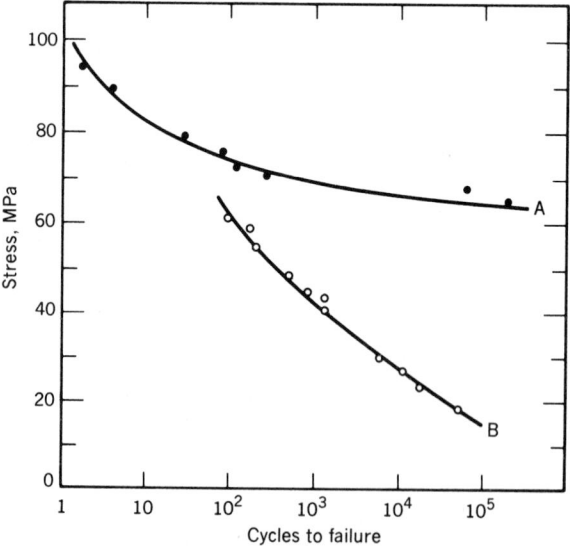

Figure 25.2. Fatigue behavior of unreinforced injection-molded specimens. Tests were done in the transverse direction [73°F (23°C)]. A, Unnotched samples; B, notched samples.[17,21] Courtesy of *Polym. Compos.*

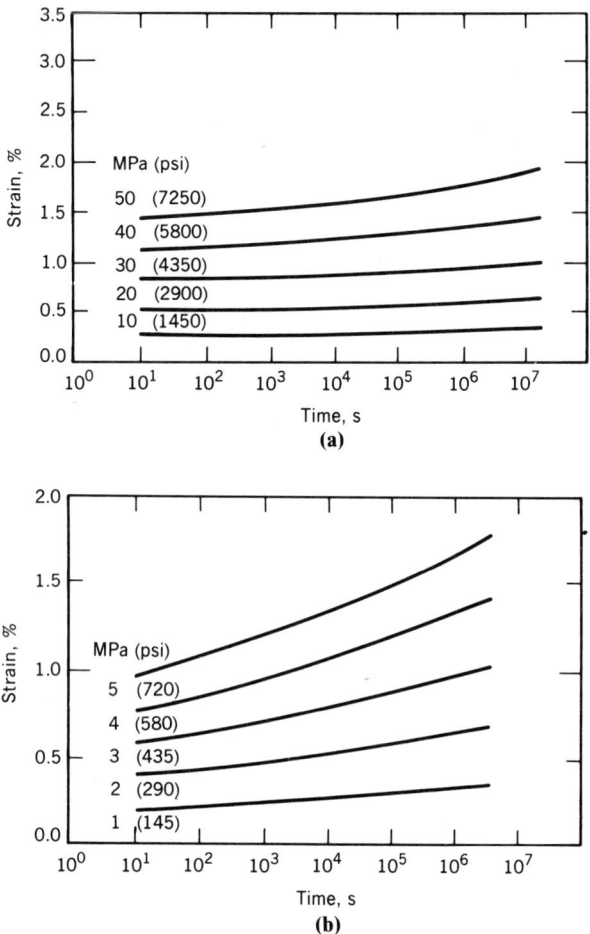

Figure 25.3. Interpolated creep data for general purpose PEEK at **(a)** 73°F (23°C) and **(b)** 350°F (180°C).[20]

of the polymer in this test. Jones and co-workers[21] found that PEEK was also very notch sensitive in fatigue. Data for injection-molded specimens are shown in Figure 25.2.

ICI data[20] from instrumented falling weight impact tests as a function of temperature indicate a maximum impact energy of PEEK at about 104°F (40°C), which is associated with loss mechanisms in the gamma transition. These data also demonstrate that the impact energy drops with decreasing temperature until PEEK will embrittle at about 15°F (−10°C). PEEK is reported to have a brittle to ductile transition at about 5°F (−15°C).[21]

At elevated temperatures, PEEK exhibits good retention of ductility during heat aging with an estimated continuous working temperature of 482°F (250°C).[1,20] The modulus drops rapidly at the T_g of 293°F (145°C), but the semicrystalline structure provides rigidity to give a deflection temperature under load (DTUL) of 320°F (160°C).

The DTUL can be increased to 600°F (315°C) which is close to the melting temperature with the addition of 30% glass-fiber or carbon-fiber reinforcement. The marked increase in DTUL has been associated with the development transcrystallinity at the fiber surface.[22] The addition of reinforcing fibers increases the tensile strength and modulus while greatly reducing the elongation.[17] The effect is more pronounced for carbon fiber as compared to glass.[23] However, the energy to propagate a crack is lower for carbon-reinforced PEEK, resulting in greater impact strength for glass-filled PEEK.[21] Fiber-reinforced PEEK, both

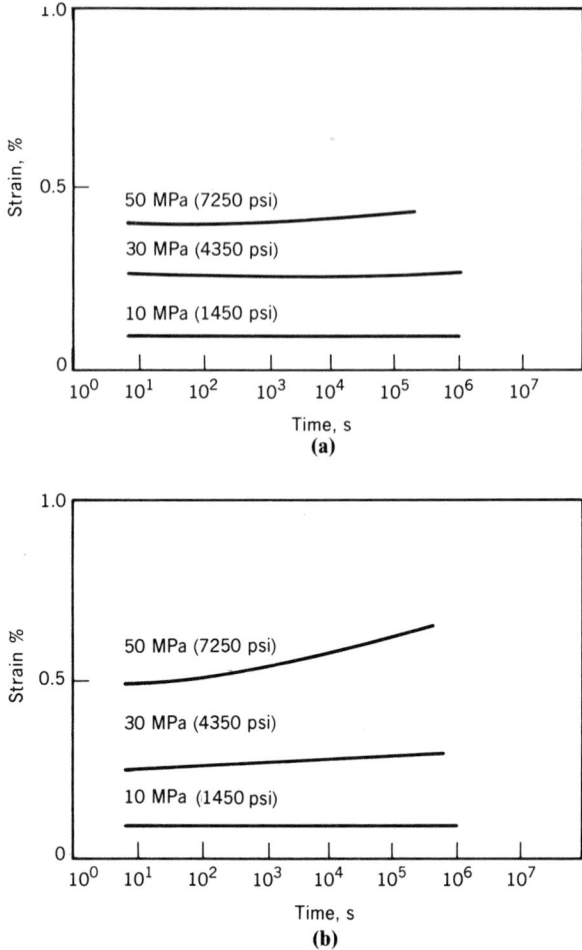

Figure 25.4. Interpolated creep data for 30% glass-fiber reinforced PEEK at **(a)** 73°F (23°C) and **(b)** at 240°F (120°C).[20]

glass and carbon, exhibits brittle fracture behavior because of the domination of the fiber over the matrix polymer.

Continuous carbon-fiber PEEK composites demonstrate different failure/fracture characteristics, as compared to short-fiber composites. Impact tests with quasi-isotropic lay-ups has shown that crack propagation from the initial damage zone is greatly restricted, resulting in a considerable increase in the energy required for total failure.[21] Furthermore, the energy does not fall markedly at low temperatures as is the case for short fiber composites, thereby suggesting that the long fibers dominate the process of crack initiation as well as propagation.

Creep is the deformation with time under constant stress and temperature and is a measure of the viscoelastic response of the polymers. ICI creep data for PEEK are shown in Figures 25.3 and 25.4. These data indicate that the creep strain is strongly dependent on temperature but only slightly on time, confirming the excellent creep resistance of PEEK. Fiber reinforcement reduces creep in all instances.

Because of its semicrystalline morphology, PEEK has enhanced environmental resistance compared to other aromatic thermoplastics. It has excellent chemical resistance to a wide range of organic and inorganic liquids.[20] However, large quantities (15–20%) of methylene chloride have been reported to be absorbed by PEEK films and composite specimens.[24] The only common chemical that will dissolve PEEK is concentrated sulfuric acid. PEEK has

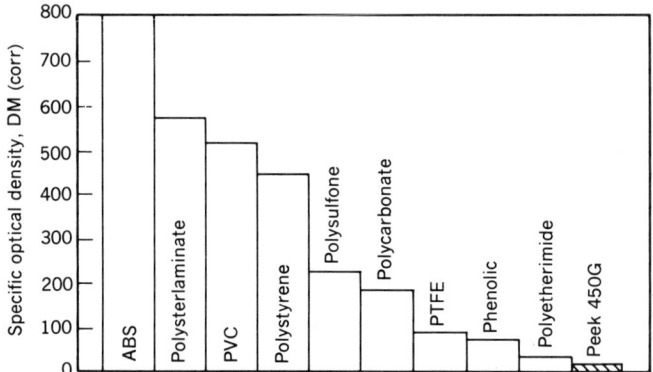

Figure 25.5. Smoke emission on burning of plastics.[21] Test conditions: American National Bureau of Standards smoke chamber, 3.2 mm (0.125 in.) samples, flaming condition. Courtesy of *Polymer.*

exceptional resistance to degradation by hydrolysis, but neat PEEK suffers from the effects of ultraviolet degradation during outdoor weathering. PEEK has excellent electrical insulating properties. It resists significant damage from exposure to gamma radiation at doses less than 1 Grad.[20]

In addition to long-term, high-temperature thermal stability as measured by its UL Temperature Index of 482°F (250°C), PEEK is inherently nonflammable and does not need fire-retardant additives.[20] PEEK has flammability of V-O at 0.08 in. (2 mm) thickness in the UL standard 94 test and a Limiting Oxygen Index as measured by ASTM D2863 of 35% using samples 0.125 in (3.2 mm) thick. In terms of smoke emission as measured by the National Bureau of Standards (NBS) smoke chamber, PEEK gives off less smoke than any other thermoplastic (Fig. 25.5).

25.6 APPLICATIONS

Most commercial uses of PEEK are in premium, high-performance applications, usually involving high temperatures and hostile environments, where a number of its outstanding properties are required.[3,25]

25.6.1 Wire Insulation

Wire coating and insulation are among the earliest applications for PEEK owing to its excellent cut-through resistance, electrical properties, and resistance to thermal aging. PEEK's low smoke emission and flammability are also important in these uses of the polymer, as is its melt processibility.

25.6.2 Electrical Connectors

Connectors and other electrical components can be designed for performance at elevated temperatures and injection molded in PEEK. Fiber-reinforced PEEK will resist distortion during soldering operations.

25.6.3 Fluid Handling

Fluid handling parts such as water meter housings, valve seats, and pump impellers molded in fiber-reinforced PEEK will operate for long periods of time in pressurized water at high temperatures. This is possible because of the hydrolysis resistance of the polymer.

25.6.4 Films

PEEK films are generally comparable in performance with polyimide films but have advantages in certain applications, particularly with regard to fire-safety characteristics. Test panels clad in PEEK film showed the highest resistance to ignition and the lowest smoke emission and toxic products of combustion as compared to panels clad in polyimide, poly(vinyl fluoride), polysulfone, and poly(phenylene sulfide), according to a recent NASA study. As a result, PEEK is the leading contender to replace the poly(vinyl fluoride) sandwich-panel decorative films currently used in the interior of commercial aircraft.

25.6.5 Composites

PEEK/carbon-fiber composites show promise in offering performance characteristics superior to conventional epoxy/carbon-fiber composites for structural aerospace components and other applications where exceptional strength-to-weight properties are required.

25.7 ADVANTAGES/DISADVANTAGES

Because of its high cost, PEEK is currently used only in premium applications. As noted, it retains excellent physical properties at high temperatures and DTULs with 30% fiber reinforcement. However, both neat and reinforced PEEK are notch sensitive, especially in impact and fatigue, and will become brittle at low temperatures. Even at low temperatures PEEK will often retain sufficient impact energy for use at these temperatures.

25.7.1 Advantages

- Tough, strong, and rigid at room temperatures
- Excellent load-bearing properties (creep resistance) over long periods of time
- Excellent fatigue properties
- High deformation temperature under load with 30% fiber reinforcement
- Good retention of ductility during heat aging
- Inherently nonflammable with low smoke emission and a high limiting oxygen index
- Excellent electrical insulating properties
- Excellent chemical resistance to a wide range of organic and inorganic liquids
- Exceptional resistance to degradation by hydrolysis
- Good resistance to gamma radiation

25.7.2 Disadvantages

- Notch sensitive
- Brittle at low temperatures
- Degraded by UV light during outdoor exposure unless painted or used with suitable additives
- High cost

25.8 PROCESSING TECHNIQUES

See the Master Material Outline and the Standard Material Design Chart.

Master Material Outline

PROCESSABILITY OF ___PEEK___

PROCESS	A	B	C
INJECTION MOLDING	X		
EXTRUSION	X		
THERMOFORMING			X
FOAM MOLDING			X
DIP, SLUSH MOLDING			X
ROTATIONAL MOLDING			X
POWDER, FLUIDIZED-BED COATING			X
REINFORCED THERMOSET MOLDING			X
COMPRESSION/TRANSFER MOLDING			X
REACTION INJECTION MOLDING (RIM)			X
MECHANICAL FORMING			X
CASTING			X

A = Common processing technique.
B = Technique possible with difficulty.
C = Used only in special circumstances, if ever.

Standard design chart for ___PEEK___

1. Recommended wall thickness/length of flow, mm
   Minimum ___1.5 mm___ for ___300 mm___ distance
   Maximum ___2.5 mm___ for ___300 mm___ distance
   Ideal ___2 mm___ for ___300 mm___ distance

2. Allowable wall thickness variation, % of nominal wall ___12–15___

3. Radius requirements
   Outside: Minimum ___1.5 mm___ Maximum ___6.0 mm___
   Inside: Minimum ___2.0___ Maximum ___6.0 mm___

4. Reinforcing ribs
   Maximum thickness ___6 mm___

Maximum height _____ 9 mm _____
Sink marks: Yes ___X___ No _____

5. Solid pegs and bosses
Maximum thickness _____ 6 mm _____
Maximum weight _____ 9 mm _____
Sink marks: Yes ___X___ No _____

6. Strippable undercuts
Outside: Yes ___X___ No _____
Inside: Yes ___X___ No _____

7. Hollow bosses: Yes ___X___ No _____

8. Draft angles
Outside: Minimum _____ $\frac{1}{2}°$ _____ Ideal _____ 1° _____
Inside: Minimum _____ $\frac{1}{2}°$ _____ Ideal _____ 1° _____

9. Molded-in inserts: Yes ___X___
No _____

10. Size limitations
Length: Minimum _None_ Maximum _limited by machine size_ Process _injection molding_
Width: Minimum _None_ Maximum _limited by machine size_ Process _injection molding_
*limited by machine size

11. Tolerances Injection molding ±0.2%

12. Special process- and materials-related design details
Blow-up ratio (blow molding) _____
Core L/D (injection and compression molding) _____ 3:1 _____
Depth of draw (thermoforming) _____ 6.0 _____
Other _____

25.9 RESIN FORMS

PEEK is available in pellet form.

25.10 SPECIFICATION OF PROPERTIES

See the Master Outline of Materials Properties.

25.11 PROCESSING REQUIREMENTS

Since PEEK is a thermoplastic, it can be processed to produce precision components using methods typical for these materials.[1] PEEK can be extruded and injection molded, but the process temperatures are higher than those that would be used for polymers with lower melting temperatures. PEEK requires drying prior to use. A minimum of three hours at 300°F (150°C) is recommended. Barrel temperatures in the range of 645–735°F (340–390°C) and mold temperatures of 300–355°F (150–180°C).[26] These temperatures are well within the capabilities of standard processing equipment.

Under processing conditions, the melt viscosity of PEEK is of the order of 100 to 1000 NS/Sm^2, which is generally higher than nylon and polypropylene but comparable to rigid poly(vinyl chloride) and polycarbonate.[26] The melt demonstrates the familiar non-Newtonian behavior, the viscosity decreasing with increasing shear stress or shear rate, which is characteristic of molten polymers.

MATERIAL _____ General-purpose PEEK

PROPERTY	TEST METHOD	ENGLISH UNITS	VALUE	METRIC UNITS	VALUE
MECHANICAL DENSITY	D792	lb/ft³	82.4	g/cm³	1.32
TENSILE STRENGTH	D638	psi	13,300	MPa	92
TENSILE MODULUS	D638	psi	522,100	MPa	3600
FLEXURAL MODULUS	D790	psi	530,800	MPa	3660
ELONGATION TO BREAK	D638	%	50	%	50
NOTCHED IZOD AT ROOM TEMP	D256	ft-lb/in.	1.55	J/m	83
HARDNESS	D785		R126		R126
THERMAL DEFLECTION T @ 264 psi	D648	°F	320	°C	160
DEFLECTION T @ 66 psi	D648	°F		°C	
VICAT SOFT-ENING POINT	D1525	°F		°C	
UL TEMP INDEX	UL746B	°F	480	°C	250
UL FLAMMABILITY CODE RATING	UL94		V-0		V-0
LINEAR COEFFICIENT THERMAL EXPANSION	D696	in./in./°F	26×10^{-6}	mm/mm/°C	47×10^{-6}
ENVIRONMENTAL WATER ABSORPTION 24 HOUR	D570	%	0.5	%	0.5
CLARITY	D1003	% TRANS-MISSION		% TRANS-MISSION	
OUTDOOR WEATHERING	D1435	%	-4	%	-4
FDA APPROVAL					
CHEMICAL RESISTANCE TO: WEAK ACID	D543		Not attacked		Not attacked
STRONG ACID	D543		Attacked		Attacked
WEAK ALKALI	D543		Not attacked		Not attacked
STRONG ALKALI	D543		Not attacked		Not attacked
LOW MOLECULAR WEIGHT SOLVENTS	D543		Not attacked		Not attacked
ALCOHOLS	D543		Not attacked		Not attacked
ELECTRICAL DIELECTRIC STRENGTH	D149	V/mil	480	kV/mm	19
DIELECTRIC CONSTANT	D150	10^6 Hertz	3.2		3.2
POWER FACTOR	D150	10^6 Hertz	0.999		0.999
OTHER MELTING POINT	D3418	°F	635	°C	335
GLASS TRANS-ITION TEMP.	D3418	°F	293	°C	145

MATERIAL ___ PEEK (30% glass-fiber reinforced)

PROPERTY	TEST METHOD	ENGLISH UNITS	VALUE	METRIC UNITS	VALUE
MECHANICAL DENSITY	D792	lb/ft³	93.0	g/cm³	1.49
TENSILE STRENGTH	D638	psi	22,800	MPa	157
TENSILE MODULUS	D638	psi	1,406,800	MPa	9700
FLEXURAL MODULUS	D790	psi	1,495,200	MPa	10,310
ELONGATION TO BREAK	D638	%	2.2	%	2.2
NOTCHED IZOD AT ROOM TEMP	D256	ft-lb/in.	1.8	J/m	96
HARDNESS	D785		R-124		R-124
THERMAL DEFLECTION T @ 264 psi	D648	°F	600	°C	315
DEFLECTION T @ 66 psi	D648	°F		°C	
VICAT SOFT-ENING POINT	D1525	°F		°C	
UL TEMP INDEX	UL746B	°F	480	°C	250
UL FLAMMABILITY CODE RATING	UL94		V-0		V-0
LINEAR COEFFICIENT THERMAL EXPANSION	D696	in./in./°F	12×10^{-6}	mm/mm/°C	22×10^{-6}
ENVIRONMENTAL WATER ABSORPTION 24 HOUR	D570	%	0.06	%	0.06
CLARITY	D1003	% TRANS-MISSION		% TRANS-MISSION	
OUTDOOR WEATHERING	D1435	%		%	
FDA APPROVAL					
CHEMICAL RESISTANCE TO: WEAK ACID	D543		Not attacked		Not attacked
STRONG ACID	D543		Attacked		Attacked
WEAK ALKALI	D543		Not attacked		Not attacked
STRONG ALKALI	D543		Not attacked		Not attacked
LOW MOLECULAR WEIGHT SOLVENTS	D543		Not attacked		Not attacked
ALCOHOLS	D543		Not attacked		Not attacked
ELECTRICAL DIELECTRIC STRENGTH	D149	V/mil		kV/mm	
DIELECTRIC CONSTANT	D150	10⁶ Hertz			
POWER FACTOR	D150	10⁶ Hertz			
OTHER MELTING POINT	D3418	°F	635	°C	335
GLASS TRANS-ITION TEMP.	D3418	°F	293	°C	145

288

MATERIAL ___ PEEK (30% carbon-fiber reinforced)

PROPERTY	TEST METHOD	ENGLISH UNITS	VALUE	METRIC UNITS	VALUE
MECHANICAL DENSITY	D792	lb/ft³	90	g/cm³	1.44
TENSILE STRENGTH	D638	psi	30,200	MPa	208
TENSILE MODULUS	D638	psi	1,885,400	MPa	13,000
FLEXURAL MODULUS	D790	psi	1,885,400	MPa	13,000
ELONGATION TO BREAK	D638	%	1.3	%	1.3
NOTCHED IZOD AT ROOM TEMP	D256	ft-lb/in.	1.6	J/m	85
HARDNESS	D785		R-124		R-124
THERMAL DEFLECTION T @ 264 psi	D648	°F	600	°C	315
DEFLECTION T @ 66 psi	D648	°F	1139	°C	615
VICAT SOFT-ENING POINT	D1525	°F		°C	
UL TEMP INDEX	UL746B	°F	480	°C	250
UL FLAMMABILITY CODE RATING	UL94		V-0		V-0
LINEAR COEFFICIENT THERMAL EXPANSION	D696	in./in./°F	8×10^{-6}	mm/mm/°C	15×10^{-6}
ENVIRONMENTAL WATER ABSORPTION 24 HOUR	D570	%		%	
CLARITY	D1003	% TRANS-MISSION		% TRANS-MISSION	
OUTDOOR WEATHERING	D1435	%		%	
FDA APPROVAL					
CHEMICAL RESISTANCE TO: WEAK ACID	D543		Not attacked		Not attacked
STRONG ACID	D543		Attacked		Attacked
WEAK ALKALI	D543		Not attacked		Not attacked
STRONG ALKALI	D543		Not attacked		Not attacked
LOW MOLECULAR WEIGHT SOLVENTS	D543		Not attacked		Not attacked
ALCOHOLS	D543		Not attacked		Not attacked
ELECTRICAL DIELECTRIC STRENGTH	D149	V/mil		kV/mm	
DIELECTRIC CONSTANT	D150	10^6 Hertz			
POWER FACTOR	D150	10^6 Hertz			
OTHER MELTING POINT	D3418	°F	635	°C	335
GLASS TRANS-ITION TEMP.	D3418	°F	293	°C	145

289

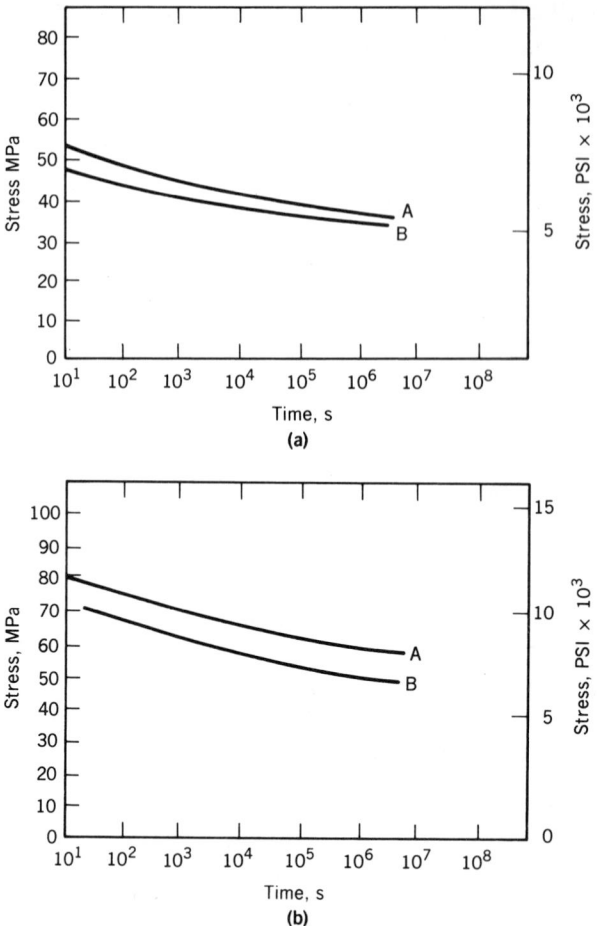

Figure 25.6.   Creep rupture performance of **(a)** 30% glass-fiber reinforced PEEK and **(b)** 30% Carbon-fiber reinforced PEEK at 300°F (150°C). A, 0 to flow, B, 90° to flow.

25.12   PROCESSING-SENSITIVE END PROPERTIES

The crystallization behavior of PEEK is in many ways similar to that of poly(ethylene terephthalate) (PET), except that the melting temperature and glass-transition temperatures of PEEK are about 135°F (75°C) higher.[10] Like PET, PEEK can be quick-cooked from the melt to the glassy state.

Previous thermal history is very influential on the development of the semicrystalline morphology of this material.[17] PEEK may require very slow cooling or postannealing to obtain the optimum crystalline content.[17] Amorphous PEEK would be subject to physical aging at temperatures below but near its T_g. The semicrystalline structure of PEEK is responsible for many of its outstanding properties.

In injection molding, several studies have demonstrated that preferential orientation of the short fibers occurs during mold filling.[21,27] Examination of molded specimens indicates that the fibers in the skin are aligned predominantly in the mold fill direction; transverse fiber orientation exists in a relatively thin central core. This anisotropy will affect the mechanical properties, especially in flexure where surface properties dominate. The effect is demonstrated by ICI data for fiber-reinforced PEEK in creep rupture (Fig. 25.6).

25.13   SHRINKAGE

See the Mold Shrinkage MPO.

Standard Tolerance Chart

| PROCESS | Injection molding | MATERIAL | Peek |

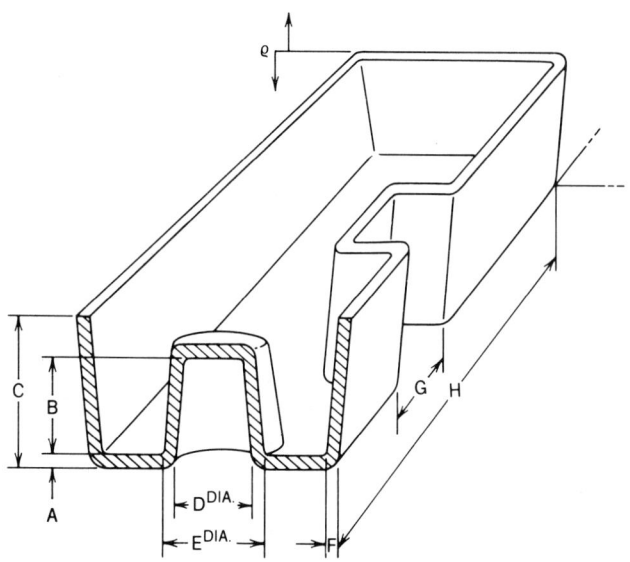

Values are based on a nominal 1/8 in. (3.2 mm) wall thickness and given in plus or minus in./in. and mm/mm.

FINE tolerance – Special care and added cost.

COMMERCIAL tolerance – Normal care required.

COARSE tolerance – Minimal care required, lowest cost.

Tolerance	A	B	C	D	E	F	G	H	
FINE	0.0080	0.0020	0.0017	0.0022	0.0020	0.0080	0.0020	0.0015	in./in.
									mm/mm
COMMERCIAL	0.0240	0.0030	0.0027	0.0037	0.0030	0.0240	0.0030	0.0019	in./in.
									mm/mm
COARSE	0.0416	0.0045	0.0036	0.0053	0.0045	0.0416	0.0045	0.0023	in./in.
									mm/mm

Mold Shrinkage Characteristics (in./in. or mm/mm)

MATERIAL ____ PEEK

IN THE DIRECTION OF FLOW

	THICKNESS				FROM	TO	#
	mm: 1.5	3	6	8			
MATERIAL	in.: 1/16	1/8	1/4	1/2			
General-purpose PEEK	0.010	0.009	0.008	0.008	0.008	0.010	
30% glass-reinforced PEEK	0.003	0.002	0.002	0.002	0.002	0.003	

	PERPENDICULAR TO THE DIRECTION OF FLOW						
General-purpose PEEK	0.013	0.012	0.011	0.011	0.011	0.013	
30% glass-reinforced PEEK	0.008	0.007	0.007	0.007	0.007	0.008	

NOTE: Mold shrinkage is approximate. It is affected by part design, mold temperature, thickness, injection pressure, packing time, cycle time, orientation, gate design, gate size, gate location, glass content, glass size, and filler content.

25.14 TRADE NAMES

PEEK is marketed by ICI under the trademark Victrex ICI also supplies both semicrystalline and amorphous PEEK film under its Stabar label. A continuous-fiber prepreg tape based on carbon fiber impregnated with PEEK is available from ICI and is designated aromatic polymer composite, APC.

BIBLIOGRAPHY

1. O.B. Searle and R.H. Pfeiffer, *Polym. Eng. Sci.* **25** 1474 (1985).
2. T.E. Attwood and co-workers, *Polymer*, **22**, 1096 (1981).
3. A.S. Wood, *Mod. Plast.* **64**, 46 (1987).
4. *Mod. Plast.* **63**, 84 (1986).
5. R.N. Johnson and co-workers, *J. Polym. Sci. A-1*, **5**, 2375 (1967).
6. J.B. Rose, *Am. Chem. Soc., Div. Polym. Chem. Polym.* Prepr. **27**, 40 (1986).
7. U.S. Pat. 4,320,224 (1982) J.B. Rose and P.A. Staniland.
8. Price Lists: Victrex Polyetheretherketone, Natural Unfilled Resins, and Victrex Polyetherether-

ketone, Colored and Reinforced Granular Resins, ICI Advanced Materials Business Group, July 14 and Aug. 4, 1986.

9. D.J. Blundell and B.N. Osborn. *Polymer* **24**, 953 (1983).
10. D.J. Blundell and F. Willmouth, *SAMPE Q.* **17**, 50 (1986).
11. A.J. Lovinger and D.D. Davis, *J. Appl. Physics* **58**, 2843 (1985).
12. S. Kumar, D.P. Anderson and W.W. Adams, *Polymer* **27**, 329 (1985).
13. J.N. Hay and co-workers *Polymer (Commun.)* **25**, 175 (1984).
14. J.M. Chalmers, W.F. Gaskin, and M.W. Mackenzie, *Polym. Bull.* **11**, 443 (1984).
15. P.C. Dawson and D.J. Blundell, *Polymer,* **21**, 577 (1980).
16. D.R. Rueda and co-workers, *Polymer (Commun.)* **24**, 258 (1983).
17. H.X. Nguyen and H. Ishida, *Polym. Compos.* 8, 57 (1987).
18. H.X. Nguyen and H. Ishida, *J. Polym. Sci. Polym. Phys. Ed.* **24**, 1079 (1986).
19. T. Sasuga and M. Hagiwara, *Polymer* **26**, 501 (1985).
20. A Guide to Grades for Injection Molding, Ref. No. VK2, ICI Advanced Materials Business Group.
21. D.P. Jones, D.C. Leach, and D.R. Moore, *Polymer* **26** 1385 (1985).
22. D.A. Luippold, *30th Natl. SAMPE Symp.* **30**, 809 (1985).
23. P.S. Pao, J.E. O'Neil, and C.J. Wolf, *Polym. Mater. Sci. Eng. Proc.* **53**, 677 (1985).
24. E.J. Stober, J.C. Seferis, and J.D. Keenan, *Polymer* **25**, 1845 (1984).
25. *Victrex PEEK—The High Temperature Engineering Thermoplastic,* ICI Advanced Materials Business Group, No. VK1.
26. *Processing,* Ref. No. VK3, ICI Advanced Materials Business Group.
27. S.S. Yau and T.W. Chou, *30th Natl. SAMPE Symp.* **30**, 406 (1985).

General References

H.J. Dillon, "Polyetheretherketone," *Modern Plastics Encyclopedia,* Vol. 52, McGraw-Hill, New York, 1986–1987.

Victrex PEEK Bearing Applications, Ref. No. VKT1, ICI Advanced Materials Business Group.

Victrex PEEK Easy Flow Grades, Ref. No. VKT3, ICI Advanced Materials Business Group.

26

POLYETHERSULFONE (PES)

Thomas W. Haas

Professor and Director of Graduate Engineering, Virginia Commonwealth University, 900 Park Avenue, Richmond, VA 23284-2009

26.1 INTRODUCTION

Polyethersulfone (PES), T_g 446°F (230°C), is one of several aromatic polysulfones. The chemical repeat unit (mer) of PES along with that of other sulfone polymers of commercial importance is

Polyarylsulfone

Polyethersulfone

Polysulfone

Chain rigidity arises from the relatively inflexible and immobile diphenylene sulfone groups. The incorporation of ether links into the polymer backbone introduces flexibility in the polymer, resulting in a decrease in the glass-transition temperature (T_g) that is further augmented by the isopropylidene link. The T_g of PES is intermediate to that of polysulfone T_g 365°F (185°C), and polyarylsulfone, T_g 563°F (295°C).

26.2 CATEGORY

PES is high-temperature, engineering thermoplastic.[1] It is amorphous with a glass-transition well above room temperature at 446°F (230°C).[2] Natural PES is amber/transparent with 65–80% light transmission over the visible range at 20 mil (0.5 mm) thickness. At room temperature, PES is tough, strong, and rigid. At elevated temperatures, PES exhibits good thermal stability during heat aging, with an estimated continuous use temperature of 356°F (180°C), based on the Underwriters Laboratory (UL) Temperature Index.[1]

26.3 HISTORY

PES was introduced by Imperial Chemical Industries (ICI) in 1972. Early uses for this polymer were in such diverse applications as transformer coil formers, oven windows, steam and dry heat sterilizable medical components, and high-intensity lamp housings.

Current U.S. production of aromatic sulfone polymers is estimated at 9 million pounds per year, with about one third being PES.[3] The other U.S. supplier is Amoco, which produces both polysulfone and polyarylsulfone. Current growth of sulfone polymers is about 10–15% per year, well above the average for engineering plastics.[3]

26.4 POLYMERIZATION

The synthesis of PES is the result of the pioneering work of K.W. Johnson and A.G. Farnham[4] who were the first to prepare aromatic polyethers. The method involves the formation of ether bonds via the nucleophilic substitution process, which subsequently proved applicable to a number of polysulfones and polyetherketones.

Poly(arylene ether sulfone)s were prepared by polycondensation of dihalogenodiphenyl sulfones with bisphenoxides in solution in a dipolar aprotic liquid such as dimethyl sulfoxide (DMSO) or tetramethylene sulfone (Sulfolane). Rose and his associates at ICI[5-8] subsequently developed an alternative route by the polycondensation of halogenophenylsulfonyl phenoxides, for example, potassium fluorophenylsulfonyl phenoxide in hot Sulfolane for 24 hours at 392°F (200°C). The use of an excess of phenoxide end-groups to control molecular weight is thought to result in slight branching of the polymer.[7-8]

Current prices for PES are from $4 to $11/lb, depending on the grade.[9] Because of its high cost, PES should not be selected except when other, less expensive engineering thermoplastics are not suitable for the application.

26.5 DESCRIPTION OF PROPERTIES

PES is a high-temperature, engineering thermoplastic. It is amorphous with a T_g well above room temperature at 446°F (230°C). Although PES has a regular structure, it does not crystallize owing to the hindrance of the "awkward" sulfone group.[8,10] In addition to the glass transition, PES has another secondary transition at about −94°F (−70°C).[11]

At room temperature, PES is strong and rigid. It exhibits outstanding creep resistance, as compared to most plastics, but as is typical of glassy amorphous polymers, PES is prone to early failure under fluctuating stresses during fatigue. At room temperature, the impact behavior of PES is similar to nylon, being generally tough but notch sensitive.[12]

The Izod impact strength of neat (unreinforced) PES is listed by ICI to be 1.42 to 2.25 ft-lb/in. (76 to 120 J/m), increasing with increased viscosity or molecular weight.[13] Unnotched specimens do not break, an indication of the notch sensitivity of PES in this test (ASTM D256) Data obtained with the Charpy-type impact test with specimens notched across the line of mold cavity flow indicate that the three most important factors influencing the impact behavior of PES are the notch radius, the amount of moisture absorbed, and the temperature.[2]

ICI impact tests[14] as a function of temperature indicate a maximum in the impact strength of PES specimens with 0.04-in. (1-mm) notch tip radius at about 104°F (40°C). The impact strength at lower temperatures is reduced but often sufficient for use in most applications. Because of the notch sensitive properties of PES, it is recommended practice to avoid designing plastic components with sharp radii. The full potential impact resistance of the material can thereby be utilized.

At elevated temperatures, PES exhibits good retention of ductility during heat aging, with an estimated continuous working temperature of 356°F (180°C), based on the Underwriters Laboratory (UL) Temperature Index.[1]

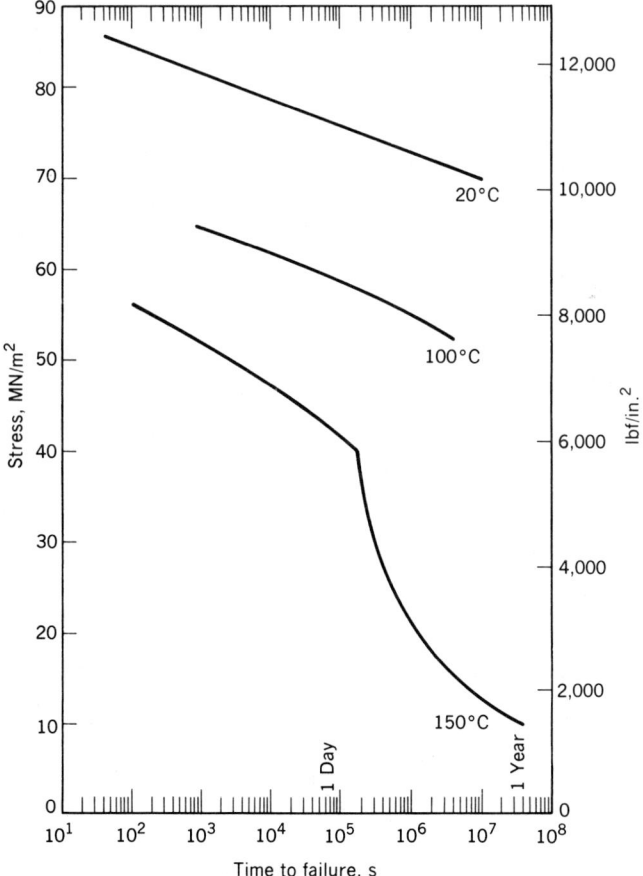

Figure 26.1. Static fatigue of polyethersulfone, unnotched specimens. Stress applied across nominal direction flow.[12] Courtesy of SPE Technical Papers.

The modulus drops rapidly at temperatures approaching the glass-transition temperature, 446°F (230°C). The deflection temperature under load (DTUL) at 264 psi (1.82 MPa) of neat PES is reported to be 397°F (203°C) and can be increased to 421°F (216°C) with 30% glass-fiber reinforcement.[13]

As would be expected of a high temperature polymer, PES retains a significant amount of its room temperature creep resistance at elevated temperatures.[12] Tensile creep rupture data for unreinforced PES at 68°F, 212°F, and 302°F (20°C, 100°C, and 150°C) are shown in Figure 26.1. The failure stress decreases with increasing time and temperature; at 302°F (150°C), it exhibits a ductile/brittle transition at about three weeks under a stress of 5800 psi (40 MPa), where the lifetime curve sweeps downward dramatically.[12]

Tensile creep curves for PES at 302°F (150°C) are shown in Figure 26.2. Also, super-imposed, are the time for the onset of crazing and the brittle failure times from Figure 26.1. The onset of the ductile-brittle transition is strongly dependent on molecular weight, oc-

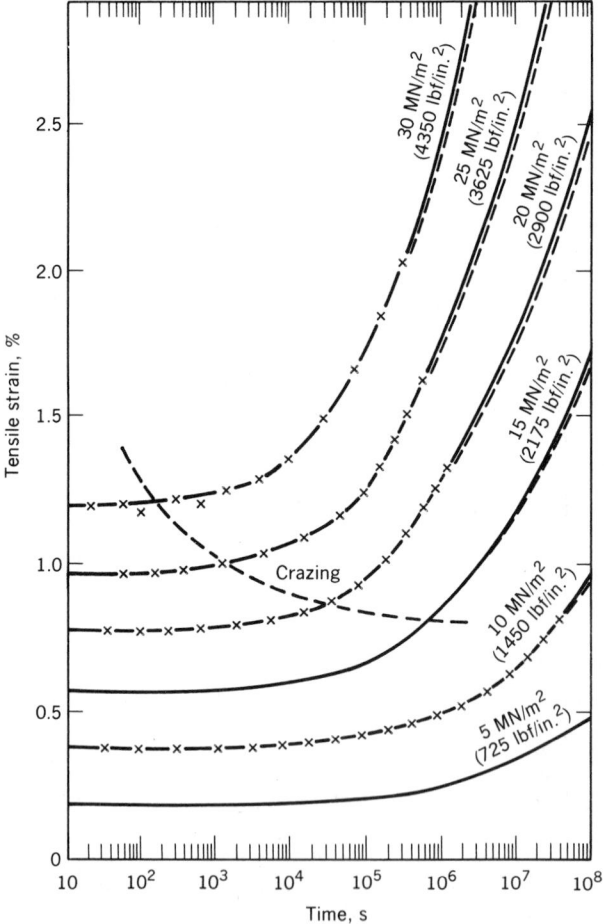

Figure 26.2. Tensile creep of polyethersulfone at 302°F (150°C). Specimens conditioned at test temperature for 12 days prior to creep.[12] —, Computer curves; X, experimental data; ——, notional creep curves. Courtesy of SPE Technical Papers.

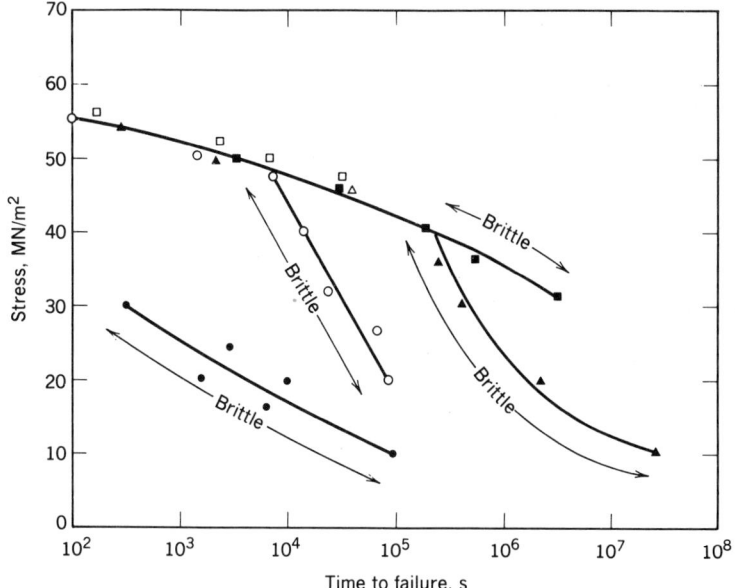

Figure 26.3. Effect of molecular weight on the failure characteristics of polyethersulfone at 302°F (150°C). Courtesy of *Polymer*.

curring at longer times with increased molecular weight (Fig. 26.3). The crazes that form in the high molecular weight polymer are thought to be more resistant to failure.[12]

The creep behavior of PES at 302°F (150°C) is comparable to that of a number of widely used engineering plastics at room temperature. Based on a strain limit in service of 0.8%, PES would be expected to be suitable in many applications at stresses below 1450 lb/ft² (10 MN/m²), depending on the required duration of the load. The incorporation of reinforcing

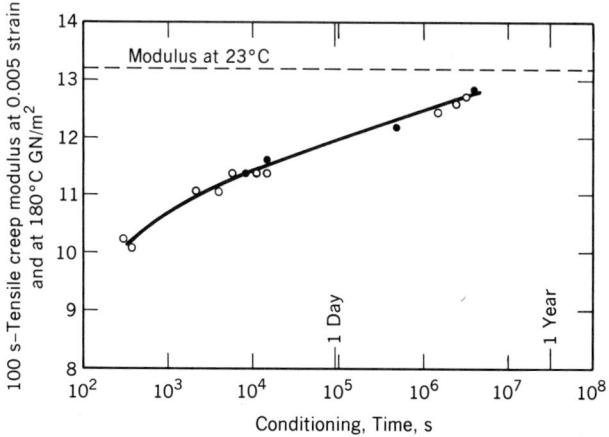

Figure 26.4. Effect of conditioning at 356°F (150°C) of 40% glass-fiber reinforced polyethersulfone.[2] Courtesy of SPE Technical Papers.

fibers into PES reduces creep, making possible the use of 30% glass-reinforced PES at temperatures as high as 356°F (180°C). However, caution should be exercised when the design is intended for long elapsed times at high temperatures where there is significant stress concentration at a notch or inhomogeneity and where stress may fluctuate sharply.[12]

Like all amorphous plastics, PES is subject to physical aging during annealing at temperatures below but near its T_g. In general, when an amorphous polymer undergoes physical aging, the free volume decreases, the modulus increases, and the overall ductility of the material decreases. Physical aging at elevated temperatures increases both the DTUL and the creep resistance of both unfilled and fiber-filled PES.[2] Annealing for a few hours at 356°F (180°C) greatly reduces the time dependence of creep (Fig. 26.4). The combined effect of decreased ductility and increased creep modulus is that the overall time to failure in creep rupture remains insensitive to thermal history.[12]

PES is highly polar and, as expected, is not dissolved by aliphatic hydrocarbons but by organic solvents, including dimethyl sulfoxide, aromatic amines, and nitrobenzene, as well as certain chlorinated hydrocarbons such as dichloromethane, chloroform, and methylene chloride.[14] Neat PES is one of the few transparent plastics not attacked by gasoline or trichlorethylene.[1] Resistance to chemicals is dependent on stress levels, both applied and residual, and on temperature. Chemical stress cracking is markedly reduced with glass-fiber reinforcement.

PES has exceptional resistance to hydrolysis and can withstand long-term exposure to both hot water and super-heated steam. It is little affected by exposure to gamma radiation at doses up to 250 Mrads. During outdoor weathering, however, PES suffers from the effects of ultraviolet degradation.[1]

Although it is polar, the electrical insultation properties of PES are good even at high frequencies.[14] The dielectric constant of dry PES at room temperature remains essentially constant at 3.5 over the frequency range of $60–10^{10}$ Hz and is slightly higher when the polymer absorbs moisture. The loss tangent at 60 Hz remains constant at 0.001 from room temperature to the onset of the T_g at about 392°F (200°C). The volume resistivity decreases from 10^{18} ohm-cm at room temperature to about 10^{13} ohm-cm at 392°F (200°C).

PES is inherently flame resistant without the need for fire retardant additives. Neat PES has a flammability of V-O at 0.020-in. (0.5-mm) thickness in Underwriters' Laboratories Standard 94 test and a Limiting Oxygen Index (ASTM D2863) of 38% using 0.060-in. (1.6-mm) thick samples.[13] In terms of smoke emission as measured by the National Bureau of Standards (NBS) smoke chamber, PES is one of the lowest smoke emitting thermoplastics.

26.6   APPLICATIONS[15]

Most engineering applications of PES are designed to take advantage of its unique combination of properties available at favorable cost. These include toughness, thermal stability, and the ability to withstand loads for long periods of time at elevated temperatures. In some cases, the transparency, resistance to hydrolysis, and low flammability of PES may also be key properties.

26.6.1   Electrical and Electronic

The retention of mechanical and electrical properties, coupled with precision moldability, accounts for PES in use in high-temperature multipin connectors. Relays, terminal blocks, and other electrical components can also be injection molded for performance at elevated temperatures. Printed circuit boards of glass-reinforced PES will resist distortion during high-temperature, solder assembly operations.

26.6.2 Fluid Handling

Components of hot water meters, valves, pumps, and fluid control devices molded in glass-reinforced PES can withstand long-term loads in contact with pressurized water and super-heated steam. This is possible because of the excellent hydrolysis resistance of the material.

26.6.3 Mechanical

Rigidity and creep performance at elevated temperatures makes glass-reinforced PES suitable for many demanding applications as an engineering plastic. Mechanical components can be molded to close tolerances that remain dimensionally stable to the onset of the glass transition. Resistance to motor oils and other automotive fluids makes PES suitable for bearing cages as well as under the hood applications.

26.6.4 Lighting

PES can be electroplated or vacuum metallized to give a high-gloss mirror finish. This makes it suitable for use as high-intensity lamp housings/reflectors because it can withstand the high operating temperatures without distortion.

26.6.5 Films

Cast films of PES offer the same combination of properties as the polymer, enabling use over a wide temperature range. PES films have high transparency, low haze, and compete with films of polyester and polyimide.

26.7 ADVANTAGES/DISADVANTAGES

PES is a high-temperature, engineering thermoplastic that offers a unique combination of properties at favorable costs. It retains its excellent mechanical and electrical properties at high temperatures as demonstrated by its UL Temperature Index of 356°F (180°C) and a DTUL of 421°F (216°C) with 30% glass-fiber reinforcement. However, PES is notch sensitive and prone to early failure during fatigue. Its impact strength is reduced at temperatures below room temperature but often remains sufficient for applications at low temperatures.

26.7.1 Advantages

- Tough, strong, and rigid at room temperature
- Excellent load-bearing properties (creep) over long periods of time
- High deflection temperature under load
- Good retention of ductility during heat aging
- Inherently nonflammable with low smoke emission and a high limiting oxygen index
- Excellent electrical insulating properties
- Exceptional resistance to degradation by hydrolysis
- Good resistance to gamma radiation

26.7.2 Disadvantages

- Notch sensitive
- Poor fatigue resistance
- Reduced impact strength at low temperatures
- Attached by chlorinated hydrocarbons and certain other organic chemicals
- Degraded by UV light during outdoor exposure unless painted or used with suitable additives

26.8 PROCESSING TECHNIQUES

See the Master Material Outline.

Master Material Outline

PROCESSABILITY OF ___PES___

PROCESS	A	B	C
INJECTION MOLDING	X		
EXTRUSION	X		
THERMOFORMING	X		
FOAM MOLDING	X		
DIP, SLUSH MOLDING	X		
ROTATIONAL MOLDING	X		
POWDER, FLUIDIZED-BED COATING	X		
REINFORCED THERMOSET MOLDING			X
COMPRESSION/TRANSFER MOLDING			X
REACTION INJECTION MOLDING (RIM)			X
MECHANICAL FORMING			X
CASTING	X		

A = Common processing technique.
B = Technique possible with difficulty.
C = Used only in special circumstances, if ever.

Radius detail

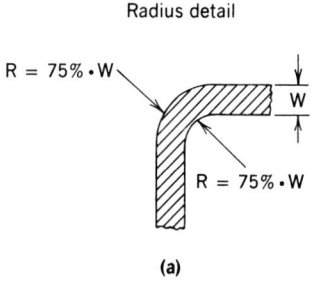

(a)

Self tapping screw
boss detail

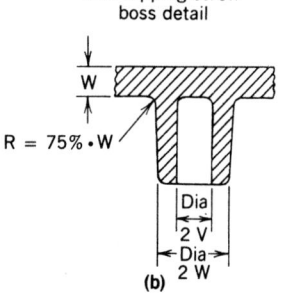

(b)

Solid rib detail

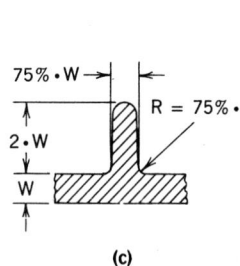

(c)

Hollow rib detail

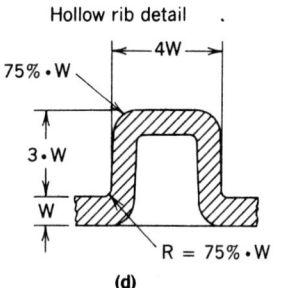

(d)

Standard design chart for _____**PES**_____

1. Recommended wall thickness/length of flow, mm
   Minimum ____1.5 mm____ for ____300 mm____ distance
   Maximum ____3.0 mm____ for ____300 mm____ distance
   Ideal ____2.0 mm____ for ____300 mm____ distance

2. Allowable wall thickness variation, % of nominal wall ____25____

3. Radius requirements (scheme **a**)
   Outside: Minimum ____2.0 mm____ Maximum ____6.0 mm____
   Inside: Minimum ____3.0 mm____ Maximum ____6.0 mm____

4. Reinforcing ribs (scheme **a** and **d**)
   Maximum thickness ____6 mm____
   Maximum height ____12 mm____
   Sink marks: Yes _____ No ____X____

5. Solid pegs and bosses
   Maximum thickness ____6 mm____
   Maximum weight ____12 mm____
   Sink marks: Yes _____ No ____X____

6. Strippable undercuts
   Outside: Yes ____X____ No _____
   Inside: Yes ____X____ No _____

7. Hollow bosses (scheme **b**): Yes ____X____ No _____

8. Draft angles
   Outside: Minimum ____$\frac{1}{2}°$____ Ideal ____1°____
   Inside: Minimum ____$\frac{1}{2}°$____ Ideal ____1°____

9. Molded-in inserts: Yes ___X___

      No _____

10. Size limitations
      Length: Minimum _None_ Maximum _limited by machine size_ Process ___injection molding___
      Width: Minimum _None_ Maximum _limited by machine size_ Process ___injection molding___
      *limited by machine size

11. Tolerances   (molding) ±0.05%

12. Special process- and materials-related design details
      Blow-up ratio (blow molding) _____1.5_____
      Core L/D (injection and compression molding) _____3:1_____
      Depth of draw (thermoforming) _____5.0_____
      Other _____

26.9   RESIN FORMS

PES is available in pellet form.

26.10   SPECIFICATION OF PROPERTIES

See the Master Outline of Material Properties.

26.11   PROCESSING REQUIREMENTS

PES can be processed by methods typically used for thermoplastics. These include injection molding, extrusion, blowmolding, vacuum forming, sintering, and solvent casting.

   PES can be extruded and injection molded in standard equipment but the process temperatures are higher than those that would be used for polymers with lower T_g or melting temperatures. Barrel temperatures of 644–715°F (340–380°C) are recommended with mold temperatures of 284–320°F (140–160°C) in order to minimize residual stress in injection-molded parts.[2,16] PES compounds will absorb small amounts of moisture. Owing to the hydrolytic stability of the polymer, this will not result in degradation but may give rise to bubbles, streaks, and splash marks during mold filling. Therefore, PES should be dried prior to use. A minimum of three hours at 302°F (150°C) is recommended.[16]

   The melt viscosity of PES under processing conditions is of the order of 100 to 5000 NS/ m², which is similar to typical grades of polycarbonate and rigid poly(vinyl chloride) and, therefore, necessitates the use of robust equipment. The melt demonstrates the familiar non-Newtonian behavior, the viscosity decreasing with increasing shear stress or shear rate— a characteristic of molten polymers.

26.12   PROCESSING-SENSITIVE END PROPERTIES

Because of the rigid molecular structure of PES, relaxation times in the melt tend to be long relative to those of more flexible polymers. As a result, PES injection molded parts can be fabricated with a high level of molecular orientation. This occurs unless the parts

MATERIAL _____ General-purpose PES

PROPERTY	TEST METHOD	ENGLISH UNITS	VALUE	METRIC UNITS	VALUE
MECHANICAL DENSITY	D792	lb/ft³	85.5	g/cm³	1.37
TENSILE STRENGTH	D638	psi	12,200	MPa	84
TENSILE MODULUS	D638	psi	354,000	MPa	2440
FLEXURAL MODULUS	D790	psi	377,000	MPa	2600
ELONGATION TO BREAK	D638	%	40–80	%	40–80
NOTCHED IZOD AT ROOM TEMP	D256	ft-lb/in.	1.42–2.25	J/m	76–120
HARDNESS	D785		M-88		M-88
THERMAL DEFLECTION T @ 264 psi	D648	°F	397	°C	203
DEFLECTION T @ 66 psi	D648	°F		°C	
VICAT SOFTENING POINT	D1525	°F	439	°C	226
UL TEMP INDEX	UL746B	°F	356	°C	180
UL FLAMMABILITY CODE RATING	UL94		V-0		V-0
LINEAR COEFFICIENT THERMAL EXPANSION	D696	in./in./°F	3×10^{-5}	mm/mm/°C	5.5×10^{-5}
ENVIRONMENTAL WATER ABSORPTION 24 HOUR	D570	%	0.12–1.7	%	0.12–1.7
CLARITY	D1003	% TRANSMISSION	65–80	% TRANSMISSION	65–80
OUTDOOR WEATHERING	D1435	%	–10	%	–10
FDA APPROVAL			Yes		Yes
CHEMICAL RESISTANCE TO: WEAK ACID	D543		Not attacked		Not attacked
STRONG ACID	D543		Attacked		Attacked
WEAK ALKALI	D543		Not attacked		Not attacked
STRONG ALKALI	D543		Not attacked		Not attacked
LOW MOLECULAR WEIGHT SOLVENTS	D543		Attacked		Attacked
ALCOHOLS	D543		Not attacked		Not attacked
ELECTRICAL DIELECTRIC STRENGTH	D149	V/mil	400	kV/mm	16
DIELECTRIC CONSTANT	D150	10^6 Hertz	3.5		3.5
POWER FACTOR	D150	10^6 Hertz	0.999		0.999
OTHER MELTING POINT	D3418	°F		°C	
GLASS TRANSITION TEMP.	D3418	°F	446	°C	230

MATERIAL ___ PES (30% glass)

PROPERTY	TEST METHOD	ENGLISH UNITS	VALUE	METRIC UNITS	VALUE
MECHANICAL DENSITY	D792	lb/ft^3	99.8	g/cm^3	1.60
TENSILE STRENGTH	D638	psi	20,300	MPa	140
TENSILE MODULUS	D638	psi	1,530,000	MPa	10,550
FLEXURAL MODULUS	D790	psi	1,218,200	MPa	8400
ELONGATION TO BREAK	D638	%	3	%	3
NOTCHED IZOD AT ROOM TEMP	D256	ft-lb/in.	1.59–1.69	J/m	85–90
HARDNESS	D785		M-98		M-98
THERMAL DEFLECTION T @ 264 psi	D648	°F	421	°C	216
DEFLECTION T @ 66 psi	D648	°F		°C	
VICAT SOFT-ENING POINT	D1525	°F		°C	
UL TEMP INDEX	UL746B	°F	356	°C	180
UL FLAMMABILITY CODE RATING	UL94		V-0		V-0
LINEAR COEFFICIENT THERMAL EXPANSION	D696	in./in./°F	1.3×10^{-5}	mm/mm/°C	2.3×10^{-5}
ENVIRONMENTAL WATER ABSORPTION 24 HOUR	D570	%	0.15–0.4	%	0.15–0.4
CLARITY	D1003	% TRANS-MISSION		% TRANS-MISSION	
OUTDOOR WEATHERING	D1435	%	–10	%	–10
FDA APPROVAL			Yes		Yes
CHEMICAL RESISTANCE TO: WEAK ACID	D543		Not attacked		Not attacked
STRONG ACID	D543		Attacked		Attacked
WEAK ALKALI	D543		Not attacked		Not attacked
STRONG ALKALI	D543		Not attacked		Not attacked
LOW MOLECULAR WEIGHT SOLVENTS	D543		Attacked		Attacked
ALCOHOLS	D543		Not attacked		Not attacked
ELECTRICAL DIELECTRIC STRENGTH	D149	V/mil	500	kV/mm	20
DIELECTRIC CONSTANT	D150	10^6 Hertz	3.5		3.5
POWER FACTOR	D150	10^6 Hertz			
OTHER MELTING POINT	D3418	°F		°C	
GLASS TRANS-ITION TEMP.	D3418	°F	446	°C	230

are slow cooled by using high mold temperatures to provide sufficient time for melt stress relaxation during solidification.

The amount of residual stress can be reduced not only by higher mold temperatures, which increase the cycle time, but also by oven annealing. ICI recommends the following procedure for annealing parts molded in PES:[16]

- Dry the molded part for a minimum of three hours at 302°F (150°C)
- Allow the temperature to increase at a rate of 18°F (10°C) per hour until a temperature of 374°F (190°C) is reached for unreinforced PES and 392°F (200°C) for glass-reinforced PES
- Hold at these temperatures for four hours
- Cool at 18°F (10°C) per hour for six hours and then turn off the oven and allow the parts to cool to room temperature.

ICI also recommends a method of estimating the residual strain in neat PES molded parts.[16] The following stress levels can be attributed to moldings that do not exhibit cracking after immersion in one of the test solvents for 100 seconds.

Test solvent	Stress level in the part
Methy ethyl ketone	Very low
Ethyl acetate	Low
Toluene	Medium
Xylene	High

PES parts that exhibit no cracking after immersion in ethyl acetate are considered acceptable for most applications. Moldings that develop cracks in xylene or toluene are not satisfactory for any purpose. In addition to quality control, this test is also recommended for use in the process optimization of injection-molded PES components.

26.13 SHRINKAGE

See Mold Shrinkage MPO.

Mold Shrinkage Characteristics (in./in. or mm/mm)

MATERIAL _____ PES

IN THE DIRECTION OF FLOW

THICKNESS

MATERIAL	mm: 1.5	3	6	8	FROM	TO	#
	in.: 1/16	1/8	1/4	1/2			
General-purpose PES	0.0080	0.0065	0.0060	0.0060	0.0060	0.0080	
30% glass-reinforced PES	0.0040	0.0035	0.0030	0.0030	0.003	0.004	

	PERPENDICULAR TO THE DIRECTION OF FLOW							
General-purpose PES	0.006	0.005	0.004	0.004	0.004	0.006		
30% glass-reinforced PES	0.003	0.002	0.002	0.002	0.002	0.003		

NOTE: Mold shrinkage is approximate. It is affected by part design, mold temperature, thickness, injection pressure, packing time, cycle time, orientation, gate design, gate size, gate location, glass content, glass size, and filler content.

Standard Tolerance Chart

PROCESS _____Injection molding_____ MATERIAL _____PES_____

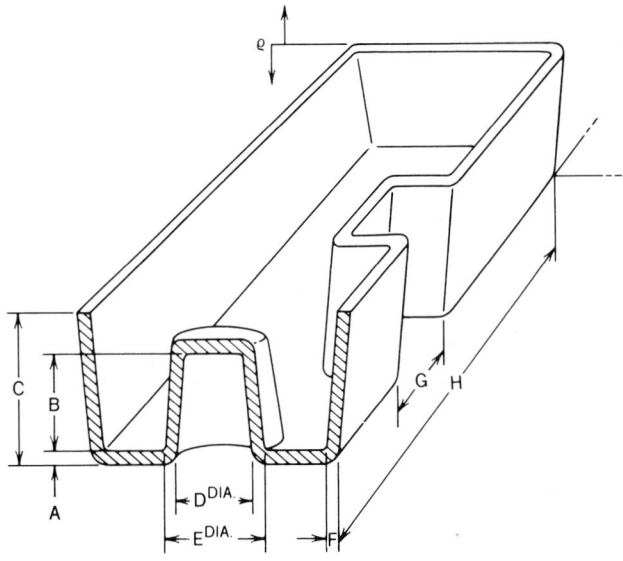

Values are based on a nominal 1/8 in. (3.2 mm) wall thickness and given in plus or minus in./in. and mm/mm.

FINE tolerance – Special care and added cost.
COMMERCIAL tolerance – Normal care required.
COARSE tolerance – Minimal care required, lowest cost.

Tolerance	A	B	C	D	E	F	G	H	
FINE	0.0160	0.0020	0.0016	0.0023	0.0020	0.0160	0.0020	0.0012	in./in.
									mm/mm
COMMERCIAL	0.0320	0.0032	0.0025	0.0040	0.0032	0.0320	0.0032	0.0015	in./in.
									mm/mm
COARSE	0.0520	0.0046	0.0035	0.0056	0.0046	0.0520	0.0046	0.0020	in./in.
									mm/mm

26.14 TRADE NAMES

PES is marketed by ICI under the trade name, Victrex. ICI also supplies cast PES film under its Stabar label. BASF supplies PES in Europe using the trade name, Ultrason E.

BIBLIOGRAPHY

1. O.B. Searle and R.H. Pfeiffer, *Poly. Eng. Sci.* **25,** 474 (1985).
2. D.G. Chasin, J. Feltzin, and S. Cear, *SPE Technical Papers,* **22,** 638 (1976).
3. A.S. Wood, Mod. Plast., **64,** 50 (1987).
4. R.N. Johnson and co-workers, *J. Polym. Sci. A-1* **5,** 2375 (1967).
5. T.E. Attwood and co-workers, *Polymer* **18,** 354 (1977).
6. *Ibid.,* p. 359.
7. *Ibid.,* p. 365.
8. *Ibid.,* p. 369.
9. *Price Lists: Victrex, Polyethersulfone, Natural Unfilled Resins and Victrex, Polyethersulfone, Colored and Reinforced Granular Resins,* ICI Materials Business Group, Aug. 4, 1986.
10. D.C. Leach, F.N. Cogswell, and E. Nield, *Composites* **14,** 251 (1983).
11. G. Makaya, *Polymers* **19,** 601 (1978).
12. K.V. Gotham and S. Turner, *SPE Technical Papers* **19,** 171 (1973).
13. *Properties and Introduction to Processing,* Ref. No. VS2, ICI Advanced Materials Business Group.
14. *Data for Design,* Ref. No. VS4, ICI Advanced Materials Business Group.
15. *Victrex PES—High Temperature Engineering Thermoplastic,* Ref. No. VS1, ICI Advanced Materials Business Group.
16. *Processing,* Ref. No. VS3, ICI Advanced Materials Business Group.

General Reference

O.B. Searle, "Polyethersulfone," *Modern Plastics Encyclopedia,* Vol. 90, McGraw-Hill Co., New York, 1988.

27

POLYETHYLENE, INTRODUCTION

Michael C. Restaino

Polyethylene Technical Service and Development The Dow Chemical Co. Freeport, TX 77541

27.1 INTRODUCTION

Polyethylenes

$$ -(\underset{\underset{H}{|}}{\overset{\overset{H}{|}}{C}} - \underset{\underset{H}{|}}{\overset{\overset{H}{|}}{C}})_n - $$

are major olefin polymers achieving significant growth each year. Their combination of useful properties, easy fabrication, and good economics has earned them global importance as commercial materials.[1]

For the United States alone, polyethylene volumes in billion of pounds were estimated[2] in 1983 at 11.5; 1984, 14.3; 1985, 15.5; and in 1986 at 15.9.

27.2 CATEGORY

Polyethylenes are thermoplastic resins produced by high and low pressure processes using various sophisticated catalyst systems. The result is several families of polymers (low density, linear low-density, and high density), each having very different behavior and performance characteristics. Generally, all polyethylenes possess excellent electrical properties, excellent resistance to water and moisture, and good resistance to organic solvents and chemicals. They are translucent, light weight, tough, and flexible materials.

27.3 HISTORY

The experimentation which led to the discovery of polyethylene stemmed from studies of the effects of high pressures on chemical reactions, which were conducted by the Alkali Div. of ICI in 1932. Concurrently A. Michels at the University of Amsterdam had succeeded in developing high-pressure experimental techniques, including a new pump capable of reaching 3000 atm pressure [44,000 psi (300 MPa)] at temperatures approaching 392°F (200°C). This equipment enabled the researchers at ICI to conduct a series of reactions, including those with ethylene and benzaldehyde, at 338°F (170°C) and 20,000 psi (140 MPa). The resulting reaction caused a white, waxy solid to be deposited on the walls of the test vessel; the solid was identified as a polymer of ethylene.[3]

Further experimentation was attempted in December 1935 using improved equipment. A larger, 80-mL experimental vessel was used that could be supplied with ethylene from a gas intensifier. After the experiment, the vessel was taken apart and 8 g of a white, powdery solid were analyzed: It melted at approximately 239°F (115°C) and had a molecular weight of about 3000.[4] Continued experimentation, early in 1936, led to larger samples of the polymer being produced. It was found to have high electrical resistivity and was capable of being converted into thin, transparent films. Experimentation on both the product and the polymerization process continued. By 1937, a continuously operating laboratory unit was in place, and a small-scale pilot plant was designed.

Product application investigations were favorably aided by chance. A technical staff member at ICI, who previously had worked for the Telegraph Construction Maintenance Co., felt that the new material might be useful as a cable insulator. A joint developmental project was established. Used for submarine telephone cable, the cable and its mechanical properties gave promising results. ICI committed to build a commercial plant, which became operational in the middle of 1939.

Additional early uses for polyethylene, then called "polythene," were accelerated by the needs of World War II. The unique insulating behavior of polyethylene was widely used in telephone cable jacketing, and greatly aided in the development of radar. That the availability of this new insulation material should have coincided with the critical need for such a product was extraordinary. In the development of "radiolocation" (radar), the availability of polyethylene transformed the design, production, installation, and maintenance of airborne radar from a challenge that was nearly insoluble to one that was comfortably manageable.[5] The success of ICI's commercial operations resulted in considerable interest by U.S. companies, who licensed the technology. By 1943, both Du Pont and Union Carbide were operating high-pressure polyethylene production facilities.

Postwar application developments continued in communications and radar but also in such large-scale uses as film extrusion and molding. Blown film extrusion found outlets in the packaging industry where clarity and tear strength were required. Early molding uses in radar components were not economically extended until larger injection presses were developed in the 1950s, particularly in the United States.

27.4 POLYMERIZATION

Continued experimentation and analysis of polyethylene showed it to have a molecular structure composed of both crystalline and amorphous regions. The high pressure (low density, or LDPE) polyethylene obtained by ICI had a rather low molecular weight, a broad molecular weight distribution (MWD), and a density of approximately 57 lb/ft³ (0.920 g/cm³). Until 1940, polyethylene was regarded as a linear long-chain hydrocarbon, but the advent of infrared studies revealed more methyl groups than could be accounted for as terminal end groups. Side-chain branches had to be present to account for the mechanical properties of the polymer. Further work showed that if polymerization techniques were altered to favor the side-chain branching reaction, then the physical properties of the polymer reflected those of a less crystalline, lower density material having highly branched side chains.

During the 1950s, discoveries of certain catalyst systems permitted the low pressure polymerization of ethylene. In 1951, Standard Oil of Indiana patented a process using a supported molybdenum oxide catalyst. In January 1953, Phillips Petroleum filed a patent for ethylene polymerization based on supported chromium trioxide catalysts. Both companies built production facilities and widely licensed their technologies. The polyethylenes made from the Phillips process were straight-chain (high density, or HDPE) homopolymers having a very high density, with melt flow indexes of 0.2–0.5 g/10 minutes.

Another dramatic advance occurred in late 1953. Karl Ziegler, at the Max Planck Institute in Germany, found that triethyl aluminum added to ethylene quite easily, resulting in the formation of higher aluminum alkyl compounds having low molecular weights in the range of 2000–5000. Ziegler sought a cocatalyst that would lead to higher molecular weight polyethylenes. Through continued investigation, Ziegler found that certain transition metal compounds could produce polyethylene in high yield. The most efficient catalysts found were those based on titanium compounds. These could produce polymers having molecular weights in the region of 300,000, and at pressures approaching atmospheric conditions. The HDPE polymers produced by the low-pressure Ziegler process were stiffer [having a density of $58.6 \, \text{lb/ft}^3$ $(0.940 \, \text{g/cm}^3)$] than the high pressure LDPE polyethylene from the ICI process.

Subsequent developments utilizing Ziegler-type catalysts resulted in the production of many new ethylene polymers and copolymers. In 1954, Professor G. Natta of the Italian Chemical Institute announced his discovery of stereoregulated polymers from alpha olefins, such as propylene. The catalysis work of Ziegler and the stereospecific polymerization work by Natta stimulated much research in polymer science.

Ziegler/Natta-type HDPE polyethylenes differed from those produced by Phillips and Standard Oil of Indiana. The latter had higher densities [of 59.7–$60.2 \, \text{lb/ft}^2$ $(0.958$–$0.965 \, \text{g/cm}^3)$] and were linear with very few side-chain branches and a high degree of crystallinity.

One disadvantage of the early Phillips and Ziegler processes was that a separate catalyst removal step was required because the catalyst activity was so low. During the mid-1950s Phillips found that, at high yields, the polymer could be precipitated out of specific solvents and there was no need to remove the catalyst. This discovery led to the development of a continuous slurry process (or particle-form process) carried out in loop reactors.

Phillips catalysts, which are used in continuous reactors, generally provide broader MWDs than do Ziegler/Natta-type catalysts. Early on, low pressure, high density resins produced from the Phillips process were preferred for extrusion applications; resins produced from the Ziegler process found their greatest utility in injection molding. However, refinements in Ziegler catalyst technology have allowed broader MWD resins to be produced, thus expanding their use by the plastics industry.

Phillips catalysts primarily are used in continuous single-stage loop reactors. However, Ziegler-catalyzed systems can employ reactors set up in series in order to produce polymers having broader MWDs. Further, to alter or control resin density, α-olefin comonomers can be introduced into the reaction vessel. These affect the branching distribution and modify the physical properties of the polymer produced.

During the 1960s, Union Carbide developed a low pressure polymerization process capable of producing polyethylene in the gas phase that required no solvents. This process employed a chromium-based catalyst. A commercial plant was in operation by the early 1970s. Catalyst was continuously injected into a vertical reactor as ethylene gas was circulated through the system. Only a small percent of the ethylene was polymerized per pass, but the process had a low overall pressure drop and the energy required to circulate the gas through the system was small. While this gas-phase polymerization process was simple compared to the other processes being used, it did not create much interst among polyethyelene producers. The high density gas-phase process had limited product range capabilities, and changing polymer grades for various applications was difficult.[6] However, the gas-phase process, and the work done with it, led to the next major stage of polyethylene development: The advent in the late 1970s of low density resins produced at low pressures and temperatures.

The low pressure (high density) process plants had been in operation for approximately 20 years and were capable of producing polyethylene copolymers having densities as low as $58.3 \, \text{lb/ft}^3$ $(0.935 \, \text{g/cm}^3)$. However, this was achieved with some difficulty because of agglomeration of resin in the reactor and reduced catalyst efficiencies. Then, in 1977, Union Carbide announced new technology, based on the high density gas-phase process, that permitted the economical production of low density polyethylene under low pressure conditions [less than 100 psi (0.69 MPa)]. The production plant was said to be less expensive than conventional high pressure plants and to have lower energy demands and operating costs.[7] The products produced, using butene-1 as the incorporated comonomer, demonstrated superior mechanical properties and film-drawing tendencies compared to conventional high pressure (low density) products. Films of resins from the new process could be made up to 20–25% thinner but could offer strength characteristics similar to those developed by existing high pressure resins. The process produced granules (powders) of polyethylene directly from the reactor. Union Carbide stated that such resins could be used in unmodified form by fabricators.

During this same time period, The Dow Chemical Co. began producing polyethylene using a proprietary solution process based on Ziegler/Natta-type catalysts. Resins were made at low pressures and densities in a system derived essentially from high-density resin technology. The resulting copolymers, based on the high molecular weight comonomer 1-octene, were introduced and marketed as linear low density polyethylene (LLDPE). Early commercial quantities of the resin were used for film fabrication, especially by the major producers of trash bags. Another difference of the Dow solution products from gas-phase products was that they were produced in standard pellet form, with any needed additives incorporated into the pellet. Therefore, the resins could be handled by fabricators in conventional ways, without the need for equipment modification.

Union Carbide and Dow announced expansion plans for LLDPE resins, however, Du Pont of Canada had been producing LLDPE for over 20 years under the name of Sclair resins. The Du Pont solution process was capable of making polyethylene over a wide range of melt indexes and densities, utilizing 1-butene as the incorporated comonomer. However, Du Pont of Canada did not widely publicize or support its achievement in polymer development.

Overall, LLDPE consists of linear molecules having short side-chain branches. The length of the side chains are largely determined by the comonomer employed during polymerization. The chains can vary from a one carbon atom group (with propylene as the comonomer) to a six carbon group (with octene-1 as the incorporated comonomer). Depending on the resin density required for a particular grade, up to 10 wt% comonomer is employed. All LLDPE resins are characterized as narrow MWD copolymers offering improved mechanical properties over conventional high pressure LDPE homopolymers. The MWD is largely determined by the catalyst; improved polymerization techniques are being researched to permit production of resin grades over a wider range of MWD, so that resins can be tailored more specifically to suit the needs of particular applications. Broader MWD resins are required for wire and cable and pipe applications because of melt strength requirements; narrower MWD resin are preferred for film extrusion and molding applications. The choice of higher α-olefin comonomer, in LLDPEs, has a significant influence on resin and end-product properties. With higher molecular weight comonomers, improvements are noted in impact strength, tear resistance, and environmental stress crack resistance properties. Therefore, physical property improvements often can be realized by incorporating 1-butene over propylene, 1-hexene over butene, and 1-octene over hexene, respectively. Also, the placement of the short-chain branches can enhance resin properties for specific applications. Branching located in the low molecular weight fraction can improve mechanical strength. However, a uniform side-chain branching technique likely will be sought to yield the best overall physical property characteristics.

27.5   DESCRIPTION OF PROPERTIES

Polyethylenes are very versatile thermoplastics earning volume uses in many application areas, particularly in films and injection molding. Each end use requires balanced conditions among variables. The most important of these are melt index, density, molecular weight, MWD, and branching. A proper balance of those property variables is required in determining the best resin choice for a particular application. In order to characterize a particular resin adequately, at least three fundamental properties must be known. These are the melt index, density, and MWD. However, the effects of long-chain branching are very important in understanding the nature of LDPE.[8]

27.6 APPLICATIONS

The characteristics relating to major commercial advantages of LDPE are excellent processibility, excellent optical properties, and flexibility. These properties allow LDPE to be widely used in packaging applications. The presence of long-chain branching in LDPE decreases some mechanical properties, but a relatively broad MWD gives LDPE polymers good melt strength, which makes it particularly suitable for blown film uses.

For HDPE film applications, an excellent degree of mechanical strength can be achieved at thin gauges. The best results often are obtained with HDPE resins exhibiting low melt indexes, high molecular weights, broad MWDs for processing ease, plus linear molecular backbones for a high degree of orientation. Injection molding applications for HDPE require polymers having narrow MWDs to minimize warpage and shrinkage characteristics and to improve impact strength properties. Also, higher melt index resins are needed to improve productivity and to keep overall molding cycles to a minimum.

LLDPEs attempt to develop the best characteristics of both LDPE and HDPE to achieve optimum end-use physical properties. Compared to LDPE, LLDPE has improved thermal properties, a higher stiffness modulus at a comparable density, and much improved environmental stress crack resistance and impact strengths. Further, LLDPE exhibits excellent draw-down capabilities, allowing performance films to be produced at much reduced gauges. In molding applications also, high melt index LLDPE resins are capable of producing end products having reduced wall thicknesses while providing superior physical properties. The markets targeted for LLDPE are thus supported by resin savings and end-product properties intermediate between those of HDPE and LDPE (compared to conventional low pressure, high density and to high pressure, low density polyethylenes).

The future uses of LLDPE are many. Besides effectively competing in traditional low and high density resin markets, specialized blends of LLDPE are being developed and targeted for specific applications. Blends of LDPE and LLDPE in film markets are providing improved physical properties while being produced on unmodified extrusion equipment. Further, continued development work on polymerization catalysts is likely to achieve advances in MWD control, and in the degree and placement of side-chain branching. Also, cost-performance needs are encouraging experimentation with LDPE terpolymers to maximize production performance without the sacrifice of physical properties in the end products.

A recent noteworthy advance in LLDPE technology is provided by the commercialization of extremely low density [56.1–57.1 lb/ft^3 (0.900–0.915 g/cm^3)] polymers that offer greatly improved strength properties. Union Carbide offers them as very low-density polyethylene (VLDPE) resins; Dow offers them as ultra low density polyethylene (ULDPE) resins.

The significance of polyethylene has grown immensely since that first 8 g of a white, powdery solid were scraped off the walls of a reaction vessel approximately 50 years ago. The next 50 years offer technological challenges that will be met by continued, directed process and product research.

BIBLIOGRAPHY

1. J.C. Swallow, *Polyethylene: Technology and Uses of Ethylene Polymers,* Iliffe and Sons, London, 1960.
2. Society of the Plastics Industry, New York, 1986.

3. R.O. Gibson, *The Discovery of Polythene,* Lecture Series No. 1, Royal Institute of Chemistry, 1964.

4. *Ibid.,* p. 3.

5. A. Renfrew, *Trans. Plast. Institute,* London, 1951.

6. E. Lloyd Jones and C.T. Richards, *Polyethylene: Past, Present, and Future* (1983).

7. "New Route to Low Density Polyethylene," *Chem. Eng.* **86,** 26 (1979).

8. D. Romamine, "Synthesis Technology, Molecular Structure and Rheological Behavior of Polyethylene," *Polym. Plast. Technol. Eng.* **19,** (2) (1982).

28

POLYETHYLENE, LOW DENSITY (LDPE)

Douglas V. Bibee

Senior Development Chemist, Polyethylene Technical Service and Development, The Dow Chemical Co., Freeport, TX 77541

28.1 INTRODUCTION

Low density polyethylene (LDPE) is sometimes called high pressure, low density polyethylene (HPLDPE) to differentiate it from low pressure, low density polyethylene (LPLDPE) or linear low density polyethylene (LLDPE). LDPE commonly is made by polymerizing ethylene at high pressure to form polyethylene molecules

$$
\begin{array}{c}
\overset{\displaystyle H}{\underset{\displaystyle H}{C}}{=}\overset{\displaystyle H}{\underset{\displaystyle H}{C}} + \text{High pressure} \longrightarrow -\!(\overset{\displaystyle H}{\underset{\displaystyle H}{C}}\!-\!\overset{\displaystyle H}{\underset{\displaystyle H}{C}})_n\!-
\end{array}
$$

Ethylene gas + High pressure ⟶ Polyethylene

28.2 CATEGORY

The result (in LDPE) is a highly branched long-chain thermoplastic polymer having density of 57.1–57.7 lb/ft^3 (0.915–0.925 g/cm^3) and molecular weight up to 4×10^6. The process also is capable of producing medium density polyethylene (MDPE) up to about 58.3 lb/ft^3 (0.935 g/cm^3). Figure 28.1 illustrates the differences found among LDPE, LLDPE, and HDPE molecules.

28.3 HISTORY

The introductory section on polyethylenes provides historical information on LDPE. U.S. production of high pressure LDPE was

Year	HPLDPE Billion Pounds	% of Total Plastic
1960	1.2	22
1970	3.7	21
1980	5.7	15
1985	4.9	10

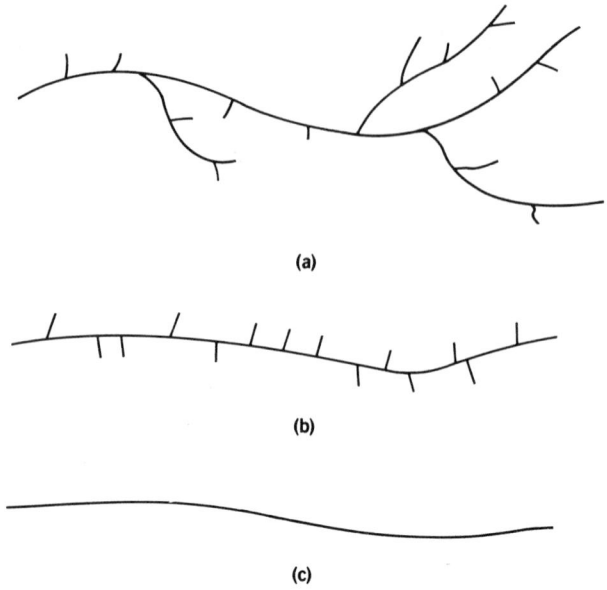

Figure 28.1.   Differences in molecules of **(a)** LDPE, **(b)** LLDPE, and **(c)** HDPE.

28.4   POLYMERIZATION

LDPE is produced by the free-radical polymerization of ethylene at high temperature and high pressure. Temperatures vary from 302–572°F (150–300°C); pressures range from 15,000-50,000 psi (103–345 MPa). The polymerization process involves three basic steps: initiation, propagation, and termination.

Initiation requires an initiator, usually a peroxide, that thermally decomposes into free radicals (eq. 28.1), which react with ethylene (eq. 28.2).

$$\text{Initiator } (R)_2 \text{ ----------------- } 2\ R' \tag{28.1}$$
$$R' + CH_2CH_2 \text{ ---------------- } RCH_2CH_2' \tag{28.2}$$

Propagation occurs as the chain reaction continues (eq. 28.3).

$$RCH_2CH_2' + CH_2CH_2 \text{ ----------- } RCH_2CH_2CH_2CH_2' \tag{28.3}$$

Termination of a growing chain occurs when two free-radical groups combine (eq. 28.4), or when a hydrogen radical transfers from one chain to another (eq. 28.5).

$$RCH_2CH_2' + 'CH_2CH_2R \text{ --------- } RCH_2CH_2CH_2CH_2R \tag{28.4}$$
$$RCH_2CH_2' + 'CH_2CH_2R \text{ --------- } RCH_2CH_3 + RCHCH_2 \tag{28.5}$$

Two commercial methods are used to manufacture LDPE: autoclave and tube. The autoclave process uses a continuous-flow stirred autoclave reactor having an L/D ratio ranging from 2:1 to 20:1. The reactor may be divided by baffles to form a series of well-stirred reaction zones. The autoclave process can produce LDPE resins having a wide range of molecular weight distributions (MWDs).

In the tubular process, the reactor consists of a long tube having L/D ratios greater than 12,000:1. Because there is no mechanical agitation, continuous operation can produce plug flow. Here, the MWD is generally between the extremes achievable by the autoclave.

In both processes, separators downstream from the reactor operate at lower pressures, separating unreacted ehthylene from the polymer. Only 10–30% of the ethylene is converted to polyethylene per pass through the reactor. From the separator, molten polyethylene is extruded through an underwater pelletizer to form pellets. The pellets then are dried and stored in silos until they are loaded into railcars, boxes, or bags.

28.5 DESCRIPTION OF PROPERTIES

The thermal properties of LDPE include a melting range with a peak melting point of 223–234°F (106–112°C). Its relatively low melting point and broad melting range characterize LDPE as a resin that permits fast, forgiving heat-seal operations. The glass-transition temperature T_g of LDPE is well below room temperature, accounting for the polymer's soft, flexible nature. The combination of crystalline and amorphous phases in LDPE make determination of T_g difficult; however, it can be said that there are significant molecular transitions in LDPE at about -4 and $-193°F$ (-20 and $-125°C$).

The molten state mechanical properties of LDPE are affected most by molecular weight and MWD. The average molecular weight is routinely measured by the melt index or gel-permeation chromatography. The melt index is a measure of the molten polymer flow rate at a set of conditions specified in ASTM D1238. The high molecular weight results in a low flow rate and low melt index values, so the molecular weight is inversely proportional to the melt index.

In gel-permeation chromatography, a size exclusion technique is used to measure the range of molecular sizes (lengths) present in a polymer, in order to determine the average molecular weight and MWD. The latter is a graphic representation of the relative amounts of each molecular size present in a whole polymer. Figure 28.2 portrays examples of broad and narrow MWD polymers. Although both have the same average molecular weight, the narrow MWD curve indicates that its distribution of molecular sizes is more uniform than that of the broad MWD curve. Such molten state properties of LDPE as melt strength and drawdown are affected by molecular weight and MWD. Melt strength is an indication of how well the molten polymer can support itself, and drawdown is a measure of how thin the molten polymer can be drawn before breaking. Melt strength is increased with increasing molecular weight and broader MWD, while drawdown is increased with lower molecular weight and narrow MWD.

The solid phase mechanical properties of LDPE are influenced most by molecular weight and density and somewhat by MWD (see Table 28.1). The melt index and density often have opposite effects on properties, necessitating compromises in resin selection.

The optical properties of LDPE are affected by molecular weight and density. High molecular weight molecules produce a rough, low gloss surface; higher density polyethylenes contain more or larger crystalline areas that scatter light and cause a hazy appearance. Fabrication conditions have a significant effect on optics.

The environmental properties of LDPE are subject to thermal and ultraviolet degradation. However, additives are available that can extend outdoor service up to several years.

28.6 APPLICATIONS

LDPE film applications include bread bags, shrink wrap, sandwich bags, and garment bags. Substrates extrusion-coated with LDPE are used for milk cartons and many food packaging applications. Blow molded containers of LDPE are used for milk and chemicals. Injection-

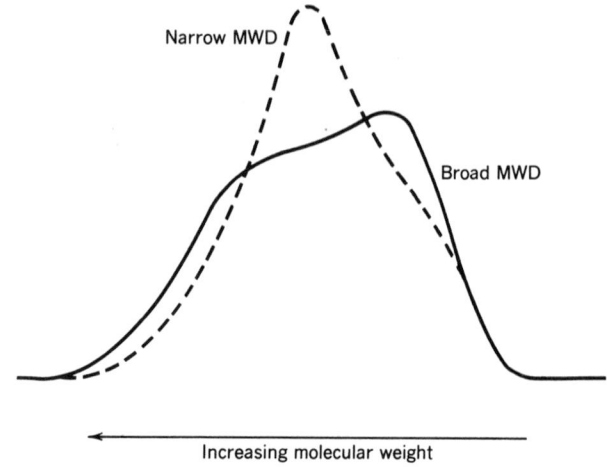

Figure 28.2. Differences in MWD for two LDPE resins of same average molecular weight.

molded items include housewares, can lids, toys, and pails. Other important uses for LDPE include wire and cable jacketing, carpet backing, and foam for life preservers or package cushioning material.

28.7   ADVANTAGES/DISADVANTAGES

LDPE has a good balance of mechanical and optical properties with easy processibility and low cost. It can be fabricated by many different methods for a broad range of applications, making it one of the highest-volume plastics in the world. By comparison, other polymers may excel in a specific property but be restricted to specialty applications by cost, processing limitations, or specific property deficiencies.

LDPE may not be suitable for applications that require extreme stiffness, good barrier properties, outstanding tensile strength, or high temperature resistance.

TABLE 28.1.   Effects of Density, Melt Index, and Molecular Weight Distribution On Solid State Properties of LDPE

Property	As Density Increases	As Melt Index Increases	As MWD Broadens
Dart impact	Decreases	Decreases	Decreases
Tear strength	Decreases	Increases	Decreases
Tensile strength	Increases	Decreases	Negligible
Yield strength	Increases	Negligible	Negligible
Elongation	Decreases	Decreases	Negligible
Stiffness	Increases	Decreases	None
Permeability	Decreases	None	None
Optics	Decreases	Increases	Decreases

28.8 PROCESSING TECHNIQUES

See the Master Material Outline.

Master Material Outline

PROCESSABILITY OF ___LDPE___

PROCESS	A	B	C
INJECTION MOLDING	X		
EXTRUSION	X		
THERMOFORMING	X		
FOAM MOLDING	X		
DIP, SLUSH MOLDING			
ROTATIONAL MOLDING	X		
POWDER, FLUIDIZED-BED COATING			
REINFORCED THERMOSET MOLDING			
COMPRESSION/TRANSFER MOLDING			
REACTION INJECTION MOLDING (RIM)			
MECHANICAL FORMING			
CASTING			

A = Common processing technique.
B = Technique possible with difficulty.
C = Used only in special circumstances, if ever.

Standard design chart for ___LDPE___

1. Recommended wall thickness/length of flow
   Minimum ___0.030___ for ___6___ distance
   Maximum _____ for _____ distance
   Ideal _____ for _____ distance

2. Allowable wall thickness variation, % of nominal wall

3. Radius requirements
   Outside: Minimum ___150% w___ Maximum ___175% w___
   Inside: Minimum ___50% w___ Maximum ___80% w___

4. Reinforcing ribs
   Maximum thickness _____100% w_____
   Maximum height _____
   Sink marks: Yes ____X____ No _____

5. Solid pegs and bosses
   Maximum thickness _____100% w_____
   Maximum weight _____
   Sink marks: Yes ____X____ No _____

6. Strippable undercuts
   Outside: Yes ____X____ No _____
   Inside: Yes ____X____ No _____

7. Hollow bosses: Yes ____X____ No _____

8. Draft angles
   Outside: Minimum _____1°/side_____ Ideal _____2°/side_____
   Inside: Minimum _____1°/side_____ Ideal _____2°/side_____

9. Molded-in inserts: Yes _____
                       No _____

10. Size limitations
    Length: Minimum _____ Maximum _____ Process _____
    Width: Minimum _____ Maximum _____ Process _____

11. Tolerances

12. Special process- and materials-related design details
    Blow-up ratio (blow molding) _____
    Core L/D (injection and compression molding) _____
    Depth of draw (thermoforming) _____
    Other _____

28.9 RESIN FORMS

Most LDPE is sold in pellet form by railcar, 1000-lb and 500-kg cartons, and 50-lb and 25-kg bags. There are many modified versions of the resin that contain additives for slip as well as antiblock, antioxidants, uv stabilizers, antistat tackifiers, and so on. A small amount is sold as powder for special applications.

28.10 SPECIFICATION OF PROPERTIES

See the Master Outline of Materials Properties.

28.11 PROCESSING REQUIREMENTS

In pellet form, LDPE does not require special handling other than protection from contamination. Extrusion equipment at temperatures above 350°F (177°C) should continually purge at a slow rate to prevent thermal degradation of the polymer. The printability of polyethylene end products requires treatment to increase the wetting tension of the surface above that

MATERIAL ___ LDPE

PROPERTY	TEST METHOD	ENGLISH UNITS	VALUE	METRIC UNITS	VALUE
MECHANICAL DENSITY	D792	lb/ft³	57.0–58.3	g/cm³	0.915–0.935
TENSILE STRENGTH	D638	psi	1000–2500	MPa	6.9–17.2
TENSILE MODULUS	D638	psi	20,000–45,000	MPa	138–310
FLEXURAL MODULUS	D790	psi		MPa	
ELONGATION TO BREAK	D638	%	100–700	%	100–700
NOTCHED IZOD AT ROOM TEMP	D256	ft-lb/in.		J/m	
HARDNESS	D785		D45–60		D45–60
THERMAL DEFLECTION T @ 264 psi	D648	°F		°C	
DEFLECTION T @ 66 psi	D648	°F	104–112	°C	41–44
VICAT SOFT-ENING POINT	D1525	°F	194–216	°C	90–102
UL TEMP INDEX	UL746B	°F		°C	
UL FLAMMABILITY CODE RATING	UL94				
LINEAR COEFFICIENT THERMAL EXPANSION	D696	in./in./°F	$100–220 \times 10^{-6}$	mm/mm/°C	$180–396 \times 10^{-6}$
ENVIRONMENTAL WATER ABSORPTION 24 HOUR	D570	%	<0.01	%	<0.01
CLARITY	D1003	% TRANS-MISSION	Translucent	% TRANS-MISSION	Translucent
OUTDOOR WEATHERING	D1435	%		%	
FDA APPROVAL			Yes		Yes
CHEMICAL RESISTANCE TO: WEAK ACID	D543		Not attacked		Not attacked
STRONG ACID	D543		Minimally attacked		Minimally attacked
WEAK ALKALI	D543		Not attacked		Not attacked
STRONG ALKALI	D543		Not attacked		Not attacked
LOW MOLECULAR WEIGHT SOLVENTS	D543		Minimally attacked		Minimally attacked
ALCOHOLS	D543		Not attacked		Not attacked
ELECTRICAL DIELECTRIC STRENGTH	D149	V/mil		kV/mm	
DIELECTRIC CONSTANT	D150	10^6 Hertz	2.2		2.2
POWER FACTOR	D150	10^6 Hertz	0.0003		0.0003
OTHER MELTING POINT	D3418	°F	223–234	°C	106–112
GLASS TRANS-ITION TEMP.	D3418	°F	−195	°C	−126

of the ink to be used. The most common methods for treating polyethylene surfaces are flame treatment and corona treatment. Commercial fabricators routinely recycle LDPE scrap and trim.

28.12 PROCESSING-SENSITIVE END PROPERTIES

Processing conditions affect the molecular orientation and density of LDPE, which, in turn, affect the end-product mechanical and optical properties. Orientation of LDPE commonly results from frozen-in stresses during the fabrication process. Such stresses can be reduced with higher temperatures, slower processing speeds, and larger flow channels. Generally, tear resistance decreases and tensile strength increases in the direction of orientation. The density of LDPE can be affected by processing conditions; the critical parameter is the quench rate, how quickly the molten polymer is cooled to a solid. Fast quench rates produce lower densities, affecting, as noted earlier, the mechanical properties of LDPE film.

28.13 SHRINKAGE

See the Mold Shrinkage MPO.

Mold Shrinkage Characteristics (in./in. or mm/mm)

MATERIAL ____LDPE____

IN THE DIRECTION OF FLOW

THICKNESS

	mm: 1.5	3	6	8	FROM	TO	#
MATERIAL	in.: 1/16	1/8	1/4	1/2			
LDPE					0.015	0.45	

PERPENDICULAR TO THE DIRECTION OF FLOW						

NOTE: Mold shrinkage is approximate. It is affected by part design, mold temperature, thickness, injection pressure, packing time, cycle time, orientation, gate design, gate size, gate location, glass content, glass size, and filler content.

28.14 TRADE NAMES

LDPE is available from Chevron Chemicals Co., Houston, Texas; the Dow Chemical Co., Midland, Mich.; Mobil Chemical Co., Houston, Texas; and Union Carbide Co., Danbury, Conn.

It is available as Alathon from Du Pont Co., Wilmington Del.; as Tenite from Eastman Chemical Products, Kingsport, Tenn. as Rexene from El Paso Products Co., Paramus N.J.; as Escorene from Exxon Chemical Co., Houston, Texas; and as Petrothene from U.S. Industrial Chemicals Co., New York.

29

POLYETHYLENE LINEAR LOW DENSITY (LLDPE)

Tammy J. Pate

Technical Service and Development, Dow Chemical Co., Freeport, TX 77541

29.1 INTRODUCTION LLDP

Linear low density polyethylene can be described as an ethylene/α-olefin copolymer having a linear molecular structure. A general representation for the molecule is

$$
\begin{array}{cccccc}
 & H & H & H & H & \\
 & | & | & | & | & \\
H- & C- & C- & C- & C- & \\
 & | & | & | & | & \\
 & H & H & H & C_{(n-2)}H_{2(n-2)+1} &
\end{array}
$$

where n signifies the number of carbons in the comonomer being used to produce the resin. The comonomers most frequently used commercially are butene, hexene, and octene. LLDPE resins have molecular weights of 10,000–100,000 and varying degrees of crystallinity.

29.2 CATEGORY

LLDPE is a hard, tough, thermoplastic material consisting of a linear backbone with short side-chain branches. The properties of LLDPE in the melt stage and in the finished part are functions of the molecular weight, molecular weight distribution (MWD), and density of the resin. The length and position of the side chains also affect product properties; they are controlled in the production process largely by the comonomer used.

There are many product types: the melt index can range from 0.5–150 g/10 min.; the density from 56.4–58.9 lb/ft³ (0.905–0.945 g/cm³). The resins in the density range of 58.4–58.9 lb/ft³ (0.936–0.945 g/cm³) are often referred to as linear medium density polyethylene resins, while those with densities of 56.4–57.0 lb/ft³ (0.905–0.915 g/cm³) are considered ultra low linear low density polyethylene.

29.3 HISTORY

The introductory section on polyethylenes provides historical information on LLDPE, which has been commercially available for many years. Du Pont Canada developed a process for producing it approximately 20 years ago. Phillips has been producing intermediate-density

LLDPE resins since the early 1960s. However, LLDPE was not a major factor in polyethylene markets until Union Carbide Corp. in 1977 and the Dow Chemical Co. in 1979 announced their commercial activities.

U.S. consumption of LLDPE is increasing rapidly. In 1986, approximately 3.5 billion lb of LLDPE were consumed domestically, representing about 20% of the overall polyethylene consumption. The market share of LLDPE is expected to increase to 35% (6.4 billion lb) by 1990.[1,2]

29.4 POLYMERIZATION

The basic polymerization process requires copolymerizing ethylene and a chosen monomer (α-olefin) using a catalyst. The reactor pressures and temperatures vary depending on the process used. Both the comonomer type and the production process affect the resin's physical properties. Commonly used comonomers are 1-butene, 1-hexene, and 1-octene. Two types of low pressure systems primarily are used to produce LLDPE: the gas-phase fluidized-bed process and the solution processes. LLDPE also can be produced in high pressure, low density polyethylene (LDPE) plants using retrofit technology available (domestically) from Arco and Dow. Table 29.1 lists the North American LLDPE producers, the type of process used, and the announced capacity for each.

The gas-phase fluidized-bed process initially developed by Union Carbide for the production of high density polyethylene (HDPE) has been modified for the production of LLDPE. Gaseous ethylene, hydrogen, a titanium-containing catalyst, and comonomer are continuously fed to a fluidized-bed reactor that operates at a pressure of 304 psi (2.1 MPa) and 176–212°F (80–100°C). Most Union Carbide LLDPE resins currently are produced using 1-butene as the comonomer. The company has recently produced LLDPE grades using higher α-olefins as comonomers in their gas-phase process. Polymer product and gas are intermittently discharged from the reactor; the gas is separated from the polymer. The polymer, in powder form, is then air-veyed to storage or to pelletizers.

Du Pont Canada first commercialized LLDPE resins, producing them by the solution process. Dow currently is the only LLDPE producer using a solution process in the United States. In Dow's proprietary process, polymerization occurs in a well-stirred reactor at temperatures of 302–572°F (150–300°C) and at pressures of 435–725 psi (3–5 MPa). Cold ethylene, solvent, a Zeigler-type catalyst, and comonomer are fed continuously to the re-

TABLE 29.1. LLDPE Producers, United States and Canada

Company	Process	Capacity, million lb (million kg)
Chevron		330 (150)
Dow Chemical, U.S.A.	Solution	849 (386)
Dow Chemical, Canada	Solution	260 (118)
DuPont, Canada	Solution	312 (142)
Eastman	CDF chemie	220 (100)[a]
Enron	Gas phase	249 (113)
Esso, Canada	Gas phase	299 (136)
Exxon	Gas phase	649 (295)
Mobil	Gas phase	374 (170)[a]
Novacor, Canada	Gas phase	598 (272)
Union Carbide, U.S.	Gas phase	1276 (580)
Union Carbide, Canada		229 (104)
USI	Arco	51 (23)

[a]As reported in industry sources.

actor. A wide variety of comonomers can be and are used. The Dowlex family of LLDPE resins from Dow are octene copolymers; Sclair resins from Du Pont Canada are butene copolymers. By recycling the solvent removed from the polymer stream, the heat of reaction is removed from the reaction vessel. The molten polymer exits the reactor and is extruded and pelletized. Additives are added at the feed section of the extruder. The molecular weight of the polymer can be controlled by reactor temperature, catalyst composition, and through use of chain terminators. Polymer density is affected by the amount of comonomer fed to the reactor.

Slurry and stirred-bed processes for LLDPE are in various stages of development and in limited use in the United States. These processes are similar to gas-phase and solution processes in that the reactors operate at low pressures and ethylene, comonomer, and catalyst are combined with some type of agitation.

The cost of producing a pound of LLDPE by the several commercial processes is similar because the raw material, ethylene, is the major material component of the manufacturing cost. Also, process yields are said to be comparable. The least expensive basic process is one where the LLDPE is produced as a powder. The cost of pelletizing the powder has been estimated at 9 cents per lb. The manufacturing cost for producing a pound of LLDPE ranges from $0.26–0.29 per lb, depending on the process, the comonomer used, and the final product form. As of 1988, the range of prices for LLDPE produced in the United States has been $0.38–0.44 per lb for general-purpose butene copolymers and $0.42–0.44 per lb for general-purpose octene copoymers.[3]

29.5 DESCRIPTION OF PROPERTIES

LLDPE is an extremely versatile, low-cost polymer adaptable to many fabrication techniques. Tough, chemically inert, and resistant to solvents, acids, and alkalies, LLDPE also possesses good dielectric characteristics and barrier properties. The resin density has a significant effect on the flexibility, permeability, tensile strength, and chemical and heat resistance of LLDPE. Figure 29.1 lists some of the effects of density on various properties of fabricated LLDPE articles.

LLDPE resins can be pigmented and UV stabilized through conventional means. For-

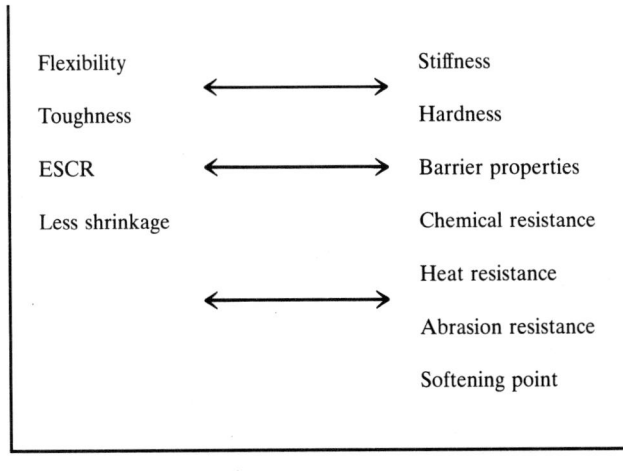

Figure 29.1. The effects of polymer density on fabricated LLDPE articles.

mulations are available for specific coefficient of friction and blocking resistance requirements.

29.6 APPLICATIONS

In the late 1970s, LLDPE was offered to the plastics industry as a new generation polyethylene having outstanding strength properties. LLDPE resins now are used in many application areas including films and coatings, molded articles, and extrusions.

LLDPE films are produced using extrusion blown and cast film equipment and technologies. End-use products made of LLDPE films include garbage bags, stretch cling films, and heavy duty sacks. Such films exhibit exceptional toughness, tear strength, dart impact, and puncture resistance when compared to films of conventional high pressure LDPE. LLDPE films perform well in packaging applications because of excellent heat-seal strength and hot-tack properties. Table 29.2 summarizes the major film applications that account for about 70% of the LLDPE consumed in the United States.

Extrusion coating is a process by which a thin layer of molten polymer is applied to a web substrate (most commonly paper or aluminum foil). The polymer coating often imparts water and grease resistance, heat sealability, and pinhole resistance to the substrate. The polymer must adhere well to the substrate and exhibit good sealing properties. LLDPE materials can perform well in this application but processing may be difficult because of the narrow MWD of the linear polymers. Dow currently markets two LLDPE resins specifically designed for extrusion coating applications.

Molded articles of LLDPE are fabricated by injection and rotational molding. Parts produced by injection molding include trash cans, food containers, lids, pails, and closures. Advantages of LLDPE in these applications include improved environmental stress crack resistance (ESCR), reduced shrinkage and warpage, and excellent low-temperature toughness compared to conventional LDPE materials. Both powder and pellet forms of LLDPE marketed for rotational molding must be ground more finely and more uniformly prior to fabrication. Rotationally molded articles made of LLDPE include shipping containers, storage tanks, and outdoor furniture. LLDPE has good flow and provides rotationally molded parts with excellent ESCR and exceptional strength and toughness.

Extrusions of LLDPE include pipe and telephone and power cable sheathing. LLDPE offers flexibility, good ESCR, and low temperature toughness. Broad MWD LLDPE resins are used for extrusion, making high speed processing posssible and permitting the production of smooth-surfaced cable jacketing and pipe.[4]

29.7 ADVANTAGES/DISADVANTAGES

The advantages and disadvantages of LLDPE materials are dependent on the application for which it is chosen. The resins generally are characterized as narrow MWD polymers.

TABLE 29.2. Major LLDPE Film Applications

Trash bags	Shrink film
Bakery film	Laminating film
Frozen food bags	Packaging film
Heavy duty bags	Agricultural film
Grocery sacks	Construction film
Ice bags	Stretch-cling film
Produce bags	Diaper liners

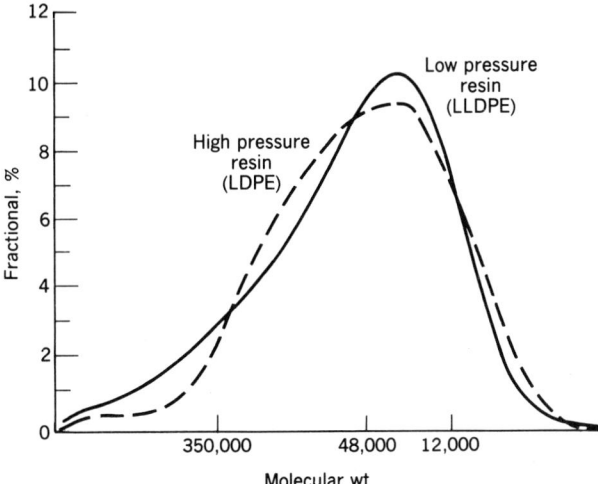

Figure 29.2. Comparison of MWD of high pressure and low pressure LLDPE resins.

Figure 29.2 compares the MWDs of LDPE and LLDPE resin at the same average molecular weight. The LLDPE is considered narrower because the distribution of molecular weights within the polymer is narrower. The MWD along with the comonomer type are the major factors in the polymer's processing and strength properties.

The rheological behavior of a resin is an indicator of its processibility. Typical rheology curves are presented in Figure 29.3. LLDPE is shown to be more viscous than LDPE at a given shear rate; it is also considered less shear-sensitive. The figure illustrates that as shear rate increases, the viscosity of the LLDPE drops more slowly than does that of the two LDPE resins. This indicates that LLDPE is more difficult to process by extrusion.

Many advantages of LLDPE are seen in the finished film, container, or packaged product. Using LLDPE, polyethylene converters can produce a finished product having equivalent and usually better performance than one made of LDPE, using fewer pounds of material. This is possible because of the "excess" strength properties of LLDPE. In molded parts, warpage is often reduced and the higher melting point of LLDPE makes possible the forming of food storage containers that are dishwasher safe. Heat seals using LLDPE are much stronger than those provided by conventional LDPE. This is very advantageous in such packaging applications as hot-fill, form-fill-and-seal, and coextrusions and laminations.

Many of the disadvantages of LLDPE are in processing of the material. As noted earlier, the narrow MWD of the polymer makes it harder to process by extrusion. LLDPE resins typically run hotter, so cooling of the film or part becomes a more critical factor in making products at accepted rates. Equipment manufacturers are constantly upgrading their systems in ways that minimize these disadvantages. There currently are many commercial extrusion systems that enable a converter to process LLDPE resins at rates that are well-accepted in the industry.

29.8 PROCESSING TECHNIQUES

The most common processing techniques used for LLDPE are blown and cast film extrusion, injection molding, rotational molding, and extrusion coating. The processing characteristics of LLDPE are determined by the rheology of the resin and by the processing conditions.

Blown film is produced by melting the polymer in an extruder, which introduces the material to a circular die equipped with some type of air ring for cooling. The bubble then

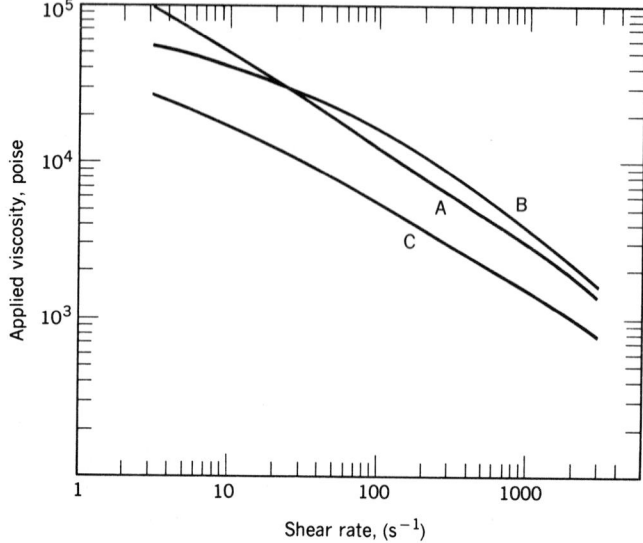

Figure 29.3.   Comparison of rheological behavior.

	Viscosity, P		
Shear rate, s⁻¹	Frac MI LDPE (A)	Octene LLDPE (B)	Linear Grade LDPE (C)
3	98,996	55,570	27,851
7	59,662	44,878	19,852
15	39,598	35,639	13,991
30	26,069	27,785	10,032
75	15,021	18,822	6,336
150	9,768	13,332	4,395
300	6,435	8,910	3,029
750	3,696	4,778	1,816
1,500	2,323	2,785	1,214
3,000	1,419	1,657	792

is pulled from the die through the use of nip rolls, and the slit "flat" film is wound on cores. The chief equipment requirements for producing LLDPE blown film are an extruder with sufficient power and torque, a cooling system allowing acceptable production "rates, and a winding system that can handle the rates required for the production of thin films. The resins used in blown film applications usually have a melt index of 0.8 to 6.0 g/10 minutes.

Typical problems in processing LLDPE blown films are melt fracture and poor bubble stability. Melt fracture can be minimized by increasing processing temperatures, reducing the screw speed, widening the die gap, or lubricating the die. Bubble stability problems are a result of the narrow MWD of LLDPE. Solutions to bubble stability problems include use of dual-lip air rings, stacked air rings, refrigerated air for bubble cooling, internal bubble cooling, and reduction of the melt temperature.[5-17]

The cast film process is similar in that an extruder is employed to melt the polymer; it then transports the melted resin to a coat hanger die. Extruded polymer film is cooled through use of chill rolls or a water bath and is wound onto cores. Again, the rheology of the LLDPE determines its processing characteristics in the extruder. Cast film resins typically have a melt index of 2.0 to 12.0 g/10 min. LLDPE resins are easily produced using cast film because of their excellent draw-down characteristics.

One problem with processing LLDPE by the cast film process is "neck-in." This phenomenon manifests itself by a narrowing of the web of polymer as it is pulled from the dies. This causes thickening of the edges, which then must be trimmed. LLDPE resins are processed at high temperatures, and resin stabilization is very important to avoid gels and polymer degradation. Cast film processes operate at very high speeds and high output rates.

When LLDPE is injection molded, a reciprocating screw is used for melting the polymer and transporting it to the mold cavity. The mixing of color and additives is very important to molded part producers. With injection molding, higher melt index resins (of 5.0 to 100 g/10 minutes) must be used. One concern is the possibility of overshearing the polymer; thus lower back pressures are used in the extruder to minimize such problems.

An extruder is not used in the rotational molding process. Finely ground LLDPE is put into a mold cavity, which is moved to a heated chamber. The polymer melts as the mold is rotated, evenly coating the inside of the mold. Narrow MWD polymers such as LLDPE are easily processed by rotational molding. Polymers with high melt flows comparable to those used in injection molding commonly are rotationally molded.

Extrusion coating is similar to cast film fabrication. A thin film of polymer is drawn down onto a web substrate, such as paper or aluminum foil. Line speeds often are faster than with cast film, so neck-in becomes a more critical factor. The molten polymer must adhere well to the substrate, and the composite structure often must have good heat-sealing properties. LLDPE resins are available that are designed specifically for extrusion coating applications.

Processibility problems are commonly solved through the use of LLDPE/LDPE blends. Blends improve bubble stability and reduce power requirements. However, the resultant end products do not possess the physical strength characteristics of 100% LLDPE products. It is estimated that over 50% of the LLDPE consumed is used in blends.

Coextrusion makes it possible to add, in a multilayer structure, the superior strength properties of LLDPE to the required properties of other polymers, often with economic advantages for the total structure. As a simple example, the puncture resistance of LLDPE can be combined with the stiffness of HDPE. LLDPE has been coextruded with LDPE, HDPE, nylon, EVA, and polypropylene.[18-22] (See the Master Material Outline.)

29.9 RESIN FORMS

LLDPE resins are available in two basic forms, pellet and powder. Both forms are available with slip, antiblock, or antioxidant additives and with various combinations of additives. Specially stabilized LLDPE resins are also available.[23] Particular resins are designed not only for needed properties but also for ease in fabrication (for example, LLDPE resins for special extrusion coating applications).

29.10 SPECIFICATION OF PROPERTIES

See the Master Outline of Materials Properties.

29.11 PROCESSING REQUIREMENTS

Processing of LLDPE involves conveying, blending, and recycling. Conveying systems vary in design, depending on whether the resin is in pellet or in granular form. Many processors add reprocessed resin to the extruder as a filler. Reprocessed resin can be purchased, or scrap or trim material can be extruded and repelletized in house. When reprocessing, it is

Master Material Outline

PROCESSABILITY OF _____LLDPE_____

PROCESS	A	B	C
INJECTION MOLDING	X		
EXTRUSION	X		
THERMOFORMING	X		
FOAM MOLDING	X		
DIP, SLUSH MOLDING			
ROTATIONAL MOLDING	X		
POWDER, FLUIDIZED-BED COATING			
REINFORCED THERMOSET MOLDING			
COMPRESSION/TRANSFER MOLDING			
REACTION INJECTION MOLDING (RIM)			
MECHANICAL FORMING			
CASTING			

A = Common processing technique.
B = Technique possible with difficulty.
C = Used only in special circumstances, if ever.

Standard design chart for _____LLDPE_____

1. Recommended wall thickness/length of flow, mm
   Minimum _____ for _____ distance
   Maximum _____ for _____ distance
   Ideal _____ for _____ distance

2. Allowable wall thickness variation, % of nominal wall ____10%____

3. Radius requirements
   Outside: Minimum ____150% w____ Maximum ____175% w____
   Inside: Minimum ____50% w____   Maximum ____75% w____

4. Reinforcing ribs
   Maximum thickness ____80% w____
   Maximum height ____240% w____
   Sink marks: Yes __X__ No _____

5. Solid pegs and bosses
   Maximum thickness _____ 80% w _____
   Maximum weight _____ 160% w _____
   Sink marks: Yes __X__ No _____

6. Strippable undercuts
   Outside: Yes __X__ No _____
   Inside: Yes __X__ No _____

7. Hollow bosses: Yes __X__ No _____

8. Draft angles
   Outside: Minimum _____ 1°/side _____ Ideal _____ 2°/side _____
   Inside: Minimum _____ 1°/side _____ Ideal _____ 2°/side _____

9. Molded-in inserts: Yes __X__
                  No _____

10. Size limitations
   Length: Minimum _____ Maximum _____ Process _____
   Width: Minimum _____ Maximum _____ Process _____

11. Tolerances

12. Special process- and materials-related design details
   Blow-up ratio (blow molding) _____
   Core L/D (injection and compression molding) _____
   Depth of draw (thermoforming) _____
   Other _____

important that LLDPE resins are at the lowest possible temperature to minimize polymer degradation.

Blending is widely practiced in the industry and can be accomplished by tumble blending dry material or by compounding. Powder and pellets can be blended dry, but a more consistent blend can be obtained through compounding. Color concentrates can be added by tumble-blending with virgin resin resin to produce pigmented films or parts.[24-27]

29.12 PROCESSING-SENSITIVE END PROPERTIES

End properties are affected by processing temperatures, quench rates, and fabrication speeds. These are critical parameters in blown and cast films, molding, extrusion coating, and extrusion. The magnitude of the effects varies, depending on the fabrication processes. Properties usually affected include impact strength, tear properties, tensile behavior, and optical characteristics.

29.14 TRADE NAMES

Manufacturers of LLDPE and the trade names under which their polymers are marketed are Dow Chemical Co., DOWLEX LLDPE; Du Pont, Canada Sclair II; Northern Petrochemical, Norlin; Novacor, Novapol; Union Carbide, G-Resin, Tuflin, VLDPE.

MATERIAL ___ LLDPE

PROPERTY	TEST METHOD	ENGLISH UNITS	VALUE	METRIC UNITS	VALUE
MECHANICAL DENSITY	D792	lb/ft³	56.7–57.6	g/cm³	0.910–0.925
TENSILE STRENGTH	D638	psi	2000–3000	MPa	14–21
TENSILE MODULUS	D638	psi	20,000–27,000	MPa	137–186
FLEXURAL MODULUS	D790	psi	36,000–53,000	MPa	248–365
ELONGATION TO BREAK	D638	%	200–1200	%	200–1200
NOTCHED IZOD AT ROOM TEMP	D256	ft-lb/in.		J/m	
HARDNESS	D785		C73, D45, D47–53		C73, D45, D47–53
THERMAL DEFLECTION T @ 264 psi	D648	°F		°C	
DEFLECTION T @ 66 psi	D648	°F	44–150	°C	6–66
VICAT SOFT-ENING POINT	D1525	°F	176–201	°C	80–94
UL TEMP INDEX	UL746B	°F		°C	
UL FLAMMABILITY CODE RATING	UL94				
LINEAR COEFFICIENT THERMAL EXPANSION	D696	in./in./°F	$89–110 \times 10^{-6}$	mm/mm/°C	$160–198 \times 10^{-6}$
ENVIRONMENTAL WATER ABSORPTION 24 HOUR	D570	%	<0.01	%	<0.01
CLARITY	D1003	% TRANS-MISSION		% TRANS-MISSION	
OUTDOOR WEATHERING	D1435	%		%	
FDA APPROVAL			Yes		Yes
CHEMICAL RESISTANCE TO: WEAK ACID	D543		Not attacked		Not attacked
STRONG ACID	D543		Minimally attacked		Minimally attacked
WEAK ALKALI	D543		Not attacked		Not attacked
STRONG ALKALI	D543		Not attacked		Not attacked
LOW MOLECULAR WEIGHT SOLVENTS	D543		Not soluble		Not soluble
ALCOHOLS	D543		Not soluble		Not soluble
ELECTRICAL DIELECTRIC STRENGTH	D149	V/mil		kV/mm	
DIELECTRIC CONSTANT	D150	10^6 Hertz	2.3		2.3
POWER FACTOR	D150	10^6 Hertz	<0.0005		<0.0005
OTHER MELTING POINT	D3418	°F	257	°C	125
GLASS TRANS-ITION TEMP.	D3418	°F	−202	°C	−130

MATERIAL ___Linear Medium-Density Polyethylene___

PROPERTY	TEST METHOD	ENGLISH UNITS	VALUE	METRIC UNITS	VALUE
MECHANICAL DENSITY	D792	lb/ft^3	57.7–58.6	g/cm^3	0.926–0.940
TENSILE STRENGTH	D638	psi	2000–3500	MPa	14–24
TENSILE MODULUS	D638	psi		MPa	
FLEXURAL MODULUS	D790	psi	36,000–53,000	MPa	248–365
ELONGATION TO BREAK	D638	%	200–1200	%	200–1200
NOTCHED IZOD AT ROOM TEMP	D256	ft-lb/in.		J/m	
HARDNESS	D785		D55, D59		D55, D59
THERMAL DEFLECTION T @ 264 psi	D648	°F		°C	
DEFLECTION T @ 66 psi	D648	°F	44–150	°C	6–66
VICAT SOFT-ENING POINT	D1525	°F	206–237	°C	102–114
UL TEMP INDEX	UL746B	°F		°C	
UL FLAMMABILITY CODE RATING	UL94				
LINEAR COEFFICIENT THERMAL EXPANSION	D696	in./in./°F	83–167 × 10^{-6}	mm/mm/°C	149–300 × 10^{-6}
ENVIRONMENTAL WATER ABSORPTION 24 HOUR	D570	%	<0.01	%	<0.01
CLARITY	D1003	% TRANS-MISSION		% TRANS-MISSION	
OUTDOOR WEATHERING	D1435	%		%	
FDA APPROVAL			Yes		Yes
CHEMICAL RESISTANCE TO: WEAK ACID	D543		Not attacked		Not attacked
STRONG ACID	D543		Minimally attacked		Minimally attacked
WEAK ALKALI	D543		Not attacked		Not attacked
STRONG ALKALI	D543		Not attacked		Not attacked
LOW MOLECULAR WEIGHT SOLVENTS	D543		Not soluble		Not soluble
ALCOHOLS	D543		Not soluble		Not soluble
ELECTRICAL DIELECTRIC STRENGTH	D149	V/mil		kV/mm	
DIELECTRIC CONSTANT	D150	10^6 Hertz	2.3		2.3
POWER FACTOR	D150	10^6 Hertz	<0.0005		<0.0005
OTHER MELTING POINT	D3418	°F	257	°C	125
GLASS TRANS-ITION TEMP.	D3418	°F	−202	°C	−130

BIBLIOGRAPHY

1. R. Martino, "Must LDPE Users Switch to LLDPE?" *Mod. Plast.* (April 1983).
2. B.H. Pickover, "Linear Low-Density Polyethylene: An Overview," *SPE RETEC* (*Proceedings*), Akron, Ohio, Sept. 1982, p. 9.
3. R.S. Pederson, "Linear Low-Density Polyethylene by the Du Pont of Canada Process," *SPE Education Day Technical Symposium,* Montreal, Quebec, March 1979.
4. C. Bird, "An Overview of Applications for LLDPE," *SPE RETEC* (*Proceedings*), Akron, Ohio, Sept. 1982, p. 131.
5. E.W. Veazey and J.E. Suazo, "Linear Low-Density Polyethylene Processability—Where Are We Today?" *TAPPI Paper Synthetics* (*Proceedings*), Chicago, Ill., Sept. 1981.
6. W.A. Fraser, L.S. Scarola, and M. Concha, "Film Extrusion of Low-Pressure LDPE," *TAPPI Paper Synthetics* (*Proceedings*), **64**(4), (April 1981).
7. W.D. Wright and R. Knittel, "Developing Technology for Low-Pressure Film Extrusion," *TAPPI Paper Synthetics Conference,* Sept. 1981.
8. C.I. Chung and R.A. Barr, "Screw Design for LLDPE Blown Film Extrusion," *TAPPI Paper Synthetics Conference,* Sept. 1981.
9. E.W. Veazy, T.L. Barnette, and T.J. Pate, "LLDPE Blown Film Equipment Design," *SPE RETEC* (*Proceedings*), Houston, Texas, Feb. 1984.
10. R.C. Phelps, "The Impact of Linear Low-Density Polyethylene on a Film Extruder," *SPE RETEC* (*Proceedings*), Akron, Ohio, Sept. 1982, p. 61.
11. H. Helmy, "Blown Film Technology for Linear Low-Density Polyethylene and Its Blends," *TAPPI Paper Synthetics* (*Proceedings*), Sept. 1982, p. 93.
12. S.J. Kurtz, L.S. Scarola, and J.C. Miller, "Conversion of LDPE Film Lines for LLDPE Film Extrusion," *SPE ANTEC* (*Proceedings*), May 1982, p. 192.
13. W. Kurzbuch, "Status of Extrusion Equipment Development for LLDPE Blown Film," *SPE RETEC* (*Proceedings*), Akron, Ohio, Sept. 1982, p. 106.
14. R. Knittel, "Developing LLDPE Film Production Technology," *SPE RETEC* (*Proceedings*), Akron, Ohio, Sept. 1982, p. 197.
15. C.J. Rauwendaal, "Analyses of Extrusion Characteristics of LLDPE," *SPE ANTEC* (*Proceedings*), Chicago, Ill., May 1983, p. 151.
16. M. Planeta, "Latest Developments in Blown Film Cooling," *TAPPI Paper Synthetics* (*Proceedings*), Sept. 1983, p. 359.
17. R.D. Krychi, "Recent Changes in Resin Technology Demand a New Approach to Processing Blown Film," *TAPPI Paper Synthetics* (*Proceedings*), Sept. 1983, p. 335.
18. A. Mendelson, "Injection Molding of LLDPE," *SPE RETEC* (*Proceedings*), Akron, Ohio, Sept. 1982, p. 226.
19. R.C. Starr, "Linear Low-Density Polyethylene for Rotational Molding," *SPE RETEC* (*Proceedings*), Akron, Ohio, Sept. 1982, p. 52.
20. M.A. Campbell, "Extrusion Coating of Linear Low-Density Polyethylene," *SPE RETEC* (*Proceedings*), Akron, Ohio, Sept. 1982, p. 39.
21. E.W. Veazy, "The Potential of LLDPE in Coextruded Film," *Pap. Film Foil Converter* (Feb. 1982).
22. J.A. Sneller, "Blowing Multiple-Layer Film Is Less Costly Today; Product Is a Lot Better, Too," *Mod. Plast.* (Jan. 1984).
23. C. Scarry and P.D. Smith, "Stabilization of Linear Low-Density Polyethylene: Effects of Primary and Secondary Antioxidants on Color Development Resulting From Processing," *SPE RETEC* (*Proceedings*), Akron, Ohio, Sept. 1982, p. 83.
24. M.A. Eddleman, "Conveying and Blending of Granular Polyethylene," *SPE RETEC* (*Proceedings*), Akron, Ohio, Sept. 1982, p. 138.
25. P.E. Stout, "Air Conveying of LLDPE," *SPE RETEC* (*Proceedings*), Akron, Ohio, Sept. 1982, p. 166.
26. B. Moller, "Recycling and Blending of LLDPE for Film Extrusion," *SPE RETEC* (*Proceedings*), Sept. 1982, p. 161.
27. J. Nancekivell, "Making the Switch to Linear-Low," *Can. Plast.* (May 1982).

30

POLYETHYLENE, HIGH DENSITY (HDPE)

Robert J. Beam

Data Management Supervisor, Dow Chemical U.S.A., 2040 Willard H. Dow Center, Midland, MI 48674

30.1 INTRODUCTION

For normal molecular weight polyethylenes (melt index >0.5), the density of homopolymer HDPE is fixed at 59.9–60.2 lb/ft^3 (0.960 to 0.965 g/cm^3), depending on the manufacturing process. However, HDPE spans the density range of 58.7–60.3 lb/ft^3 (0.941 to 0.967 g/cm^3) by the use of copolymers that add side-chain branches and thus reduce the density.

The density of HDPE is controlled in the manufacturing process by the amount of comonomer added to the reactor. Typical comonomers used with ethylene in HDPE are propylene, butene, hexene, and octene. As the molecular weight of polyethylene increases, the longer polymer chains do not crystallize as readily and the lower amount of crystallinity further reduces the density of a HDPE homopolymer (melt index <0.5).

30.2 CATEGORY

HDPE is a partially crystalline, partially amorphous thermoplastic material. The degree of crystallinity depends on the molecular weight, the amount of comonomer present, and the heat treatment given. The crystallinity of a given HDPE resin can be varied over a wide range by the rate of cooling from the molten state; slower cooling rates favor crystalline growth. The range of crystallinity for HDPE is normally 50–80%. A density value normally quoted on data sheets for HDPE is determined by a compression molded sheet that has been cooled at the rate of 27°F (15°C) per minute.[1-3] Most commercial fabrication processes cool from the melt at much faster rates; as a result, an article fabricated from HDPE rarely reaches the density quoted on a data sheet. Because the amount of crystallinity in HDPE is variable, HDPE can be considered as an amorphous polymer having a variable amount of crystalline filler.

30.3 HISTORY

The introductory section on polyethylenes provides historical information on HDPE.

30.4 POLYMERIZATION

HDPE is manufactured by a low pressure process; by comparison, low density polyethylene (LDPE) is manufactured by a high pressure process. The pressure used in manufacturing

HDPE is below 2000 psi (14 MPa); in many cases, it is below 1000 psi (7 MPa). [In manufacturing LDPE, pressures commonly exceed 10,000 psi (70 MPa).]

There are three major commercial processes used for polymerization of HDPE: solution, slurry, and gas-phase processes. The catalysts used in the manufacture of HDPE are usually either a transition metal oxide type or a Ziegler/Natta type. It is important to note that the performance of HDPE resins having identical melt indexes, densities, and molecular weight distributions (MWDs) can vary if the resins are produced by different processes. These differences normally are seen only in critical applications having very narrow processing windows. For most applications, HDPE resins selected from more than one supplier will perform adequately, even if the resins are made by different processes.

As mentioned, along with melt index and density, MWD is a distinguishing property of HDPE. As HDPE is polymerized, polymer molecules of many different lengths (molecular weights) are produced. If an HDPE resin has a narrow range of molecule lengths, it is said to have a narrow MWD. Conversely, an HDPE having a broad range of molecule lengths is said to have a broad MWD. The MWD is a plot of molecular weight versus the number or frequency of a given molecular weight. As the MWD of an HDPE is broadened, the processibility and melt strength increase, while impact strength, low-temperature toughness, and warpage resistance decrease. The MWD of HDPE is largely controlled by the type of catalyst used in polymerization and by the type of manufacturing process employed.

30.5 DESCRIPTION OF PROPERTIES

In HDPE, the properties of tensile yield strength, stiffness, creep resistance, impermeability, abrasion resistance, mold shrinkage, and hardness increase with increasing density. On the other hand impact strength, flexibility, and environmental stress crack resistance (ESCR) increase with decreasing density.

As the average molecular weight of HDPE increases, the polymer's molten flow decreases. The standard test for measuring the molten flow of HDPE is called the melt index.[4] Melt index is inversely proprotional to the average molecular weight. The properties of ESCR, impact strength, tensile strength, elongation, melt strength, and die swell improve with decreasing melt index (and with increasing average molecular weight). The properties of processibility, melt drawdown, and optics decrease with decreasing melt index. As the average molecular weight of HDPE increases, there is more shrinkage and warpage present in the molded parts.

These brief comments on melt index and density indicate that it is not possible to maximize all of the properties of HDPE in a single resin. Therefore, compromises are necessary in designing any HDPE resin. For that reason, most manufacturers of HDPE offer many different HDPE resin grades in their product mix. Selection of an HDPE resin for a given application involves careful evaluation of the application requirements, in order to select the HDPE resin that most closely satisfies the most important requirements.

The glass-transition temperature T_g of polyethylene is well below room temperature. This gives polyethylene its more rubbery nature (compared to a polymer such as polystyrene which has a glass-transition point above room temperature). The rubbery nature of HDPE also limits its service temperature compared to that of a polymer having a T_g above room temperature. The T_g for HDPE has been assigned several different values by different measuring techniques and is the subject of much controversey. There are three temperature ranges commonly assigned as the glass-transition point for polyethylene: -207 to $-171°F$ (-133 to $-113°C$), -126 to $-99°F$ (-88 to $-73°C$), and -45 to $9°F$ (-43 to $-13°C$).[5]

Because HDPE is rubbery, its creep modulus is more important than, for example, its flexural modulus in determining the in-service strength of a part fabricated from it. In designing an HDPE part that is intended to bear a load for an extended period of time (greater than one hour), flexural or tensile modulus cannot safely be used to calculate the strength of the fabricated part. Instead, the designer should refer to creep data and select a creep modulus that corresponds to the maximum service time under load for the part. The creep modulus should be employed in strength calculations instead of the flexural and tensile modulus.[6]

Failure to consider creep when parts are designed in HDPE is an invitation to premature part failure. Creep resistance improves in HDPE with increasing density and increasing average molecular weight. Creep resistance can also be improved by the use of such cross-linking techniques as irradiation and chemical cross-linking.

30.6 APPLICATIONS

HDPE is used for many food packaging applications because it provides excellent moisture barrier properties. However, HDPE, like all polyethylenes, is limited to those food packaging applications that do not require an oxygen barrier. In film form, HDPE is used in snack food packages and cereal box liners; in blow-molded bottle form, for milk and some non-carbonated beverage bottles; and in injection-molded tub form, for packaging margarine, whipped toppings, and deli foods.

Because HDPE has good chemical resistance, it is used for packaging many household as well as industrial chemicals. Examples of such injection-molded applications include 5-gal pails of floor cleaner, 1-gal pails of paint, and construction containers of spackling paste. Blow-molded applications include 55-gal HDPE drums of antifreeze.

Although HDPE has good chemical resistance, it is prone to environmental stress cracking (ESC), commonly from such agents as detergents and surfactants. HDPE can still be used to package these items if the designer carefully selects an HDPE with a time-to-ESC-failure longer than the required shelf life of the product. Resistance to ESC for HDPE increases with decreasing density and melt index. Although there are many tests for ESC resistance, there is no substitute for field experience with a given product.

HDPE does not provide good barrier resistance to lower molecular weight hydrocarbon solvents such as kerosene, and it is not recommended for nonvented packages of such solvents. HDPE is used for some noncritical gasoline containers where the loss of gasoline through the wall can be tolerated and the container is vented. It can be used for critical gasoline containers, such as automotive gas tanks, if a barrier is added—by surface sulfonation and fluorination treatments, for example.

General uses of HDPE include injection-molded beverage cases, bread trays, and dunnage trays as well as films (which have wide and growing use in merchandising and grocery sacks). Another interesting HDPE application is extruded sheet that is subsequently thermoformed into such articles as canoes and pickup truck bed liners. HDPE can readily be thermoformed but not as easily as styrenics.

In sum, HDPE is a versatile thermoplastic enjoying many successful applications that maximize its properties.

30.7 ADVANTAGES/DISADVANTAGES

The performance attributes of HDPE have earned this resin family significant commercial uses. Compared to LDPE and LLDPE, HDPE provides greater stiffness and rigidity.

30.7.1 Advantages

- Good moisture barrier properties beneficial in many packaging applications (inner-liner films in paperboard cartons, molded end caps for caulking tubes)
- Good stiffness adequate for some structural applications (beverage crates, tote bins, pallets)
- Load-bearing applications when creep is factored correctly in the part design (load-

bearing capabilities improve with increasing molecular weight; ultra-high molecular weight HDPE affords the greatest load-bearing capabilities)
- Relative chemical inertness (can be used to package some chemicals)
- Good thermal stability over a range of -40 to $600°F$ (-40 to $316°C$).

30.7.2 Disadvantages

- Relatively high gas transmission rates (would not protect a packaged product from oxygen penetration)
- The creep propensity of HDPE prevents it from being considered a true engineering plastic; the load-bearing capability of conventional molecular weight HDPE decreases rapidly with increasing environmental temperature.
- Some chemicals may cause premature failure of HDPE parts because of ESC. Prior to packaging chemicals in containers fabricated of HDPE, testing is recommended as some solvents will penetrate and soften HDPE.
- Higher temperatures may cause degradation of HDPE unless an antioxidant is added to the resin.

30.8   PROCESSING TECHNIQUES

HDPE can be made into film using both the extrusion blown film and cast film processes. It is efficiently extruded into sheet and profiles and is readily injection molded, blow molded, and rotationally molded. When a blowing agent is added, HDPE can be fabricated into foamed products. HDPE can also be compression molded, although it is rare to find a commercial system doing so. HDPE can be forged and molded at temperatures below its melting point, but this is not done commercially. Promising developmental work is under way on the production of fibers from HDPE.

As HDPE is fabricated into film and other articles, orientation-influenced properties become very significant. The crystalline structure of HDPE is oriented during the fabrication process; as a result, films or articles exhibit characteristically anisotropic physical properties.

For example, the impact properties a test specimen cut from an injection-molded part parallel to the line of major flow of plastic into the mold would be significantly different from those of a test specimen cut perpendicular to the major line of flow into the mold. This anisotropic behavior must be considered in HDPE part design. If the fabrication process cannot minimize the anisotropic behavior of fabricated HDPE, then the film or article should be designed to take advantage of the orientation. Fabrication methods commonly used with HDPE are listed in the Master Material Outline.

Most fillers and reinforcing agents can be used with HDPE. The addition of a filler usually improves flexural strength, creep resistance, and hardness, but such gains in physical properties may be offset by reduced impact strength, tensile elongation, and processibility. The addition of a filler can also increase the density of HDPE, and increased density can be a detriment rather than a gain.

Most HDPE resins are sold by the pound by resin producers to fabricators, who, in turn, sell the fabricated articles by the part, not by the pound. The resin producer is concerned with weight, while the seller of fabricated parts is concerned with weight and volume. This relationship of weight and volume can become critical when the density of HDPE (volume per pound) is increased by adding a filler. Any increase in density means that a fabricator must buy (or process) more pounds of filled HDPE to make the same number of parts. This can result in a hidden cost to the fabricator for filled HDPE.

Compounding is another cost factor in filled HDPE systems. Most fillers cannot simply be blended with HDPE and used on normal fabrication equipment. The HDPE and the filler must be compounded by extrusion, intensive mixers, or both, before fabrication, which can add 6 to 12 cents per pound to the cost of filled HDPE. Because of the relatively low

Master Material Outline

PROCESSABILITY OF _____HDPE_____

PROCESS	A	B	C
INJECTION MOLDING	X		
EXTRUSION	X		
THERMOFORMING	X		
FOAM MOLDING	X		
DIP, SLUSH MOLDING			
ROTATIONAL MOLDING	X		
POWDER, FLUIDIZED-BED COATING			
REINFORCED THERMOSET MOLDING			
COMPRESSION/TRANSFER MOLDING			X
REACTION INJECTION MOLDING (RIM)			
MECHANICAL FORMING			X
CASTING			

A = Common processing technique.
B = Technique possible with difficulty.
C = Used only in special circumstances, if ever.

Standard design chart for _____HDPE_____

1. Recommended wall thickness/length of flow, mm
   Minimum _____ for _____ distance
   Maximum _____ for _____ distance
   Ideal   _____ for _____ distance

2. Allowable wall thickness variation, % of nominal wall _____10%_____

3. Radius requirements
   Outside: Minimum ___150% w___ Maximum ___175% w___
   Inside: Minimum ___50% w___   Maximum ___75% w___

4. Reinforcing ribs
   Maximum thickness _____80% w_____
   Maximum height _____240% w_____
   Sink marks: Yes ___X___ No _____

5. Solid pegs and bosses
   Maximum thickness _____80% w_____

Maximum weight _____160% w_____
Sink marks: Yes __X__ No _____

6. Strippable undercuts
   Outside: Yes _____ No __X__
   Inside: Yes __X__ No _____

7. Hollow bosses: Yes __X__ No _____

8. Draft angles
   Outside: Minimum _____1°/side_____ Ideal _____2°/side_____
   Inside: Minimum _____1°/side_____ Ideal _____2°/side_____

9. Molded-in inserts: Yes __X__
                         No _____

10. Size limitations
    Length: Minimum _____ Maximum _____ Process _____
    Width: Minimum _____ Maximum _____ Process _____

11. Tolerances

12. Special process- and materials-related design details
    Blow-up ratio (blow molding) _____1:1 to 3:1_____
    Core L/D (injection and compression molding) _____1:1 to 10:1_____
    Depth of draw (thermoforming) _____1.4 to 1.5:1_____
    Other _____

cost of unfilled HDPE, and the hidden costs of increased density and compounding, it is difficult to add a filler to HDPE and improve economics. This would not be so, of course, if the cost of HDPE exceeded $1 per pound or if the natural density of HDPE were greater than 62 lb/ft³ (1 g/cm³). Some fillers used with HDPE are talc, calcium carbonate, and mica.

The reinforcing material most commonly added to HDPE is glass fiber. (Glass fiber is considered to be a reinforcing material while talc, calcium carbonate, and mica are considered to be fillers because there is more strength gained per pound of glass fiber than per pound of talc, calcium carbonate, or mica.) Glass-fiber-filled HDPE has excellent stiffness and higher temperature service than unreinforced HDPE. The chief drawbacks to using glass fiber in HDPE are increased cost and the abrasiveness of glass to the fabrication equipment. With the cost of surface-treated glass fibers being over $1 per pound, an HDPE so reinforced is much more expensive than unreinforced HDPE.

In deciding whether or not to add a filler or reinforcing agent to HDPE, the designer should consider the following criteria.

- Is there a property requirement in the application that is not met by natural HDPE?
- Can the requirement be met by adding a filler or reinforcing agent?
- Can the application support the additional cost of the filled system?
- Can another polymer (other than HDPE) offer the required properties in an unfilled state as economically as a filled HDPE system?[7]

30.9   RESIN FORMS

HDPE can be purchased from resin suppliers in pellet, granular, or powder form. The most common mode of shipment is by bulk railcar. Bulk truck, 1000-lb and 500-kg cartons, and 50-lb and 25-kg bags shipments also are used. In quantities less than truckload, HDPE is generally sold by brokers or resellers rather than by resin manufacturers.

30.10   SPECIFICATION OF PROPERTIES

See the Master Outline of Materials Properties.

MATERIAL ___ HDPE

PROPERTY	TEST METHOD	ENGLISH UNITS	VALUE	METRIC UNITS	VALUE
MECHANICAL DENSITY	D792	lb/ft^3	58.7–60.3	g/cm^3	0.941–0.967
TENSILE STRENGTH	D638	psi	2700–4400	MPa	18.6–30.3
TENSILE MODULUS	D638	psi		MPa	
FLEXURAL MODULUS	D790	psi	100,000–240,000	MPa	689–1654
ELONGATION TO BREAK	D638	%	100–1000	%	100–1000
NOTCHED IZOD AT ROOM TEMP	D256	ft-lb/in.	0.5–3.0	J/m	27–160
HARDNESS	D785				
THERMAL DEFLECTION T @ 264 psi	D648	°F		°C	
DEFLECTION T @ 66 psi	D648	°F	150–200	°C	65–93
VICAT SOFTENING POINT	D1525	°F	248–266	°C	120–130
UL TEMP INDEX	UL746B	°F		°C	
UL FLAMMABILITY CODE RATING	UL94				
LINEAR COEFFICIENT THERMAL EXPANSION	D696	in./in./°F	60–110 × 10^{-6}	mm/mm/°C	108–198 × 10^{-6}
ENVIRONMENTAL WATER ABSORPTION 24 HOUR	D570	%		%	
CLARITY	D1003	% TRANSMISSION		% TRANSMISSION	
OUTDOOR WEATHERING	D1435	%		%	
FDA APPROVAL			Yes		Yes
CHEMICAL RESISTANCE TO: WEAK ACID	D543		Not attacked		Not attacked
STRONG ACID	D543		Minimally attacked		Minimally attacked
WEAK ALKALI	D543		Not attacked		Not attacked
STRONG ALKALI	D543		Not attacked		Not attacked
LOW MOLECULAR WEIGHT SOLVENTS	D543		Minimally attacked		Minimally attacked
ALCOHOLS	D543		Not attacked		Not attacked
ELECTRICAL DIELECTRIC STRENGTH	D149	V/mil	400–600	kV/mm	15.7–23.6
DIELECTRIC CONSTANT	D150	10^6 Hertz	2.2–3.0		2.2–3.0
POWER FACTOR	D150	10^6 Hertz	0.00005–0.003		0.00005–0.003
OTHER MELTING POINT	D3418	°F		°C	
GLASS TRANSITION TEMP.	D3418	°F		°C	

30.11   PROCESSING REQUIREMENTS

Unfilled, unreinforced HDPE is not hygroscopic and normally does not require drying prior to processing. Filled HDPE will absorb small quantities of moisture and thus may require drying. If the HDPE is stored in outdoor silos, extreme temperature changes can cause moisture to condense on the pellets and thus necessitate drying. HDPE resins should be protected from contamination from dust, fibers, and other polymers as this type of contamination can affect both processing and end-use performance.

Normal processing temperatures for HDPE range from 350–600°F (177–316°C). The thermal stability of HDPE is usually good over the temperature range if the higher-temperature exposure times are kept relatively short; otherwise, the resin can degrade by crosslinking, chain scission, or both. It is normal for a slight change in melt index to occur during processing of HDPE.

Articles fabricated from HDPE can be ground and recycled. The recycled HDPE is normally blended with virgin HDPE before reuse, typically in amounts of 10% or less regrind. Above 10 percent, the physical properties of the fabricated product can be adversely affected. For best results, regrind must be protected from contamination from other polymers, dirt, paper, and so on.

30.12   PROCESSING-SENSITIVE END PROPERTIES

The impact strength, stiffness, and thermal stability of articles fabricated from HDPE are profoundly affected by processing conditions. For example, unidirectional orientation or overpacking a mold can greatly diminish the functional impact strength. If the processing temperatures are kept high and the process has a long resin residence time, the thermal stability of the finished article can be greatly reduced. The cooling rate will affect the stiffness and impact strength of the article: with a slower cooling rate, stiffness increases and impact strength decreases.

30.13   SHRINKAGE

See Mold Shrinkage Characteristics.

Mold Shrinkage Characteristics (in./in. or mm/mm)

MATERIAL _____ HDPE

IN THE DIRECTION OF FLOW

THICKNESS

	mm: 1.5	3	6	8	FROM	TO	#
MATERIAL	in.: 1/16	1/8	1/4	1/2	0.020	0.045	

	PERPENDICULAR TO THE DIRECTION OF FLOW							

NOTE: Mold shrinkage is approximate. It is affected by part design, mold temperature, thickness, injection pressure, packing time, cycle time, orientation, gate design, gate size, gate location, glass content, glass size, and filler content.

30.14 TRADE NAMES

There are many suppliers of HDPE. In brief, HDPE is available as Paxon from Allied Fibers & Plastics Co.; as Hostalen from American Hoechst; as N Resins from the Dow Chemical Co.; as Alathon from Du Pont Co.; as Marlex from Phillips Petroleum Co.; as Fortiflex from Soltex Polymer Corp.; as Petrothene from USI Chemicals; and as Sclair from Du Pont Canada.

BIBLIOGRAPHY

1. *ASTM D1928, Standard Test Method for Preparation of Compression-Molded Polyethylene Test Sheets and Test Specimens,* American Society for Testing and Materials, Philadelphia, Penn.
2. *ASTM D792, Test Method for Specific Gravity and Density of Plastics by Displacement,* American Society for Testing and Materials, Philadelphia, Penn.
3. *ASTM D1248, Standard Specification for Polyethylene Plastics Molding and Extrusion Materials,* American Society for Testing and Materials, Philadelphia, Penn.
4. *ASTM D1238, Standard Test Method for Flow Rates of Thermoplastics by Extrusion Plastometer,* American Society for Testing and Materials, Philadelphia, Penn.
5. G.T. Davis and R.K. Eby, "Glass Transition of Polyethylene: Volume Relaxation," *J. Appl. Physics,* **44** 4274–4281 (Oct. 1973).
6. J. Agranoff, ed., *Modern Plastics Encyclopedia.* McGraw-Hill Inc., New York, 1983, pp. 522–532.
7. H.S. Katz and J.V. Milewski, eds., *Handbook of Fillers and Reinforcements for Plastics.* Van Nostrand Reinhold Co., New York, 1978.

31

POLYETHYLENE, ULTRA-HIGH MOLECULAR WEIGHT (UHMW PE)

Russell D. Hanna

Himont Incorporated, Wilmington, Del.

31.1 INTRODUCTION UHMW PE

Ultra-high molecular weight polyethylene (UHMW PE) is a linear high density polyethylene (HDPE) with a molecular weight in the range of 3 million to 6 million. It is defined by the American Society for Testing & Materials as a minimum relative viscosity of 2.3, using a solution concentration (decahydronaphthalene) of 0.05% at 275°F (135°C).[1] This compares to a molecular weight of 300,000 to 500,000 for extrusion resins, which are sometimes referred to as "HMW high density polyethylene."

$$\left[\begin{array}{cc} \overset{\displaystyle H}{\underset{\displaystyle H}{\overset{|}{\underset{|}{C}}}} & \overset{\displaystyle H}{\underset{\displaystyle H}{\overset{|}{\underset{|}{C}}}} \end{array} \right]_n$$

31.2 CATEGORY

UHMW PE is a thermoplastic with chemical properties similar to those of HDPE. However, its extremely high molecular weight provides it with exceptional impact strength and abrasion resistance as well as special processing characteristics. These unusual properties preclude the use of conventional extrusion and molding techniques.

31.3 HISTORY

The process for manufacturing UHMW PE dates back to the disclosure in 1954 by Professor Ziegler of the Max Planck Institut concerning the use of organo metal catalysts and low pressures to produce linear olefins.[2] It has been produced commercially in Germany and the United States since 1958.

31.4 POLYMERIZATION

Currently, UHMW PE is manufactured under modified Ziegler catalyst systems. Developmental activities to modify the process are being conducted by the polymer producers,

American Hoechst and Himont Incorporated. As of August 1989, prices for UHMW PE ranged from $1.20 to $1.25 per pound.

31.5   DESCRIPTION OF PROPERTIES

UHMW PE has unique combinations of physical and chemical properties, plus other attributes quite similar to those of HDPE. The unique characteristics include

- The highest slurry-abrasion resistance of any thermoplastic.
- Exceptional impact resistance at cryogenic temperatures.
- Low coefficient of friction.
- Self-lubricating properties.
- Outstanding stress-crack resistance.
- High resistance to cyclical fatigue failures.
- Noise and energy attenuation.
- Clearance for use in food processing applications by the Food & Drug Administration and the U.S. Dept. of Agriculture.

The addition of glass (fibers or beads) improves the base resin's stiffness and heat-deflection temperature, with little or no effect on primary properties.

31.6   APPLICATIONS

UHMW PE is chiefly used in sheet form for bulk materials handling. Abrasion resistance, chemical resistance, and light weight make UHMW PE sheet and plate ideal for lining chutes and hoppers in the coal, ore, grain, paper, and chemical processing industries. Food and beverage processing lines, pump components, gaskets, filters, feed screws, guide rails, rollers, gears, and bushings are fabricated from UHMW PE.

Large new application areas recently developed include linings for railcars, self-unloading ships, coal slurry systems, and waste water and sewage treatment systems. In coal preparation plants and coal fired public-utility materials handling systems, UHMW PE liners for chutes, hoppers, bins, and conveyor rollers prevent hang-up, sticking, and bridging of frozen fines. UHMW PE parts are employed on automobile assembly lines to lower energy costs, reduce noise levels, and minimize maintenance costs in conveyor rollers and slide rails.

Forged drive sprockets and idlers, wear plates, and guide shoes fabricated from UHMW PE sheet save weight and eliminate the need for lubrication in farm machinery. Reduction in friction drag extends the life of wood-harvesting chain drag lines and sewage treatment plants. These savings are not only the result of reduced wear on surfaces but also the reduction of horsepower requirements and motor maintenance.

UHMW PE modified with cross-linked polybutadiene and boron carbide is used as neutron shielding in nuclear-driven naval vessels. Self-unloading coal or grain cargo-carrying ships have undergone a design revolution because of UHMW PE liners. The low friction and abrasion resistance of UHMW PE sheet permits minimal slope of the sides of lined holds, thus increasing carrier capacity and ship stability. As a result, unloading and turnaround times for these classes of cargo carriers have been considerably shortened.

It can be anticipated that current development activities on the part of polymer producers, processing equipment suppliers, and UHMW PE fabricators will open new application areas that are not presently served because of high fabrication costs.

31.7 ADVANTAGES/DISADVANTAGES

Key advantages of UHMW PE are implicit in its properties, notably excellent impact strength and abrasion resistance.

The material has few challengers in its combination of toughness, wear resistance, chemical inertness, and lubricity.

The high processing viscosity of UHMW PE makes it difficult to handle the material using conventional techniques.

31.8 PROCESSING TECHNIQUES

The unusual properties of UHMW PE resins are a challenge to processors of other plastics. For all practical purposes, UHMW PE resins have neither melt nor flow characteristics comparable to conventional thermoplastics. The very long-chain molecules preclude use of

Master Material Outline

PROCESSABILITY OF ___Ultra High Molecular Weight Polyethylene 1900___

PROCESS	A	B	C
INJECTION MOLDING			X
EXTRUSION			X
THERMOFORMING		X	
FOAM MOLDING			X
DIP, SLUSH MOLDING			X
ROTATIONAL MOLDING			X
POWDER, FLUIDIZED-BED COATING			X
REINFORCED THERMOSET MOLDING			X
COMPRESSION/TRANSFER MOLDING	X		
REACTION INJECTION MOLDING (RIM)			X
MECHANICAL FORMING		X	
CASTING			X

A = Common processing technique.
B = Technique possible with difficulty.
C = Used only in special circumstances, if ever.

standard extrusion and molding techniques, the result being that UHMW PE fabrication methods have generally been derived from powdered metallurgy and fluorocarbon processing techniques. While UHMW PE displays a crystalline melting point comparable to other HDPE resins [266°F (130°C)], this constitutes a change in appearance from opaque to clear, indicating the disappearance of crystal structures. The extreme high viscosity of this "melt" prevents flow or change in shape, except under high pressures. In compression molding, the particles of resin soften and are pressed together, but microscopic inspection of individual UHMW PE particles will reveal that they are bonded together at their boundary layers.

This melt behavior of UHMW PE has led to investigation of such processing techniques as radio frequency (RF) heating and high-speed fluxing in intensive mixers. Low-pressure forming and sintering processes can be utilized by selecting the appropriate particle size, temperatures, and pressures. Ram extrusion of basic UHMW PE profiles is accomplished by use of long-land, heated dies, which essentially act as "open-end" compression molds.

A process similar to metal stamping is used to form parts from UHMW PE sheet. The sheet is heated to approximately 302°F (150°C) before it is placed in the stamping die. Pressure is maintained for a short time to allow the sheet to cool below its crystallization temperature. This ability to handle a sheet or billet of UHMW PE above the crystalline melting point of the polymer has led to such other specialized fabrication techniques as forging of finished parts. It also permits the sheet to be shaped to fit in lining processes.

Post-fabrication of milled shapes, such as rod, bar, and sheet, can be accomplished by techniques generally used in fabrication of wood or soft metals such as brass. All cutting tools should have wide tooth spacing to provide good chip clearance and minimal heat buildup. Liberal use of coolants will prevent heat softening of the stock.

31.9   RESIN FORMS

UHMW PE is normally supplied to fabricators in the form of fine powder that has appearance of corn meal or sugar. For the most part, these fabricators produce semifinished products, such as plates, billets, rod, bar, tubes, and a variety of profiles using compression molding or ram extrusion techniques.

Although most UHMW PE is used without modification, antioxidants and stabilizers can be added when fabrication techniques or the end uses dictate exposure at elevated temperatures. UHMW PE will accept a variety of fillers and modifiers, including graphite fiber, talc, powdered metal, and glass fiber and beads, which generally are added to improve stiffness and heat-deflection temperatures or decrease deformation under load. A minimal loss in abrasion resistance and impact strength is experienced with most additives. Electrical conductivity and static discharge properties can be obtained through addition of high levels of conductive carbon. Chemical cross-linking will enhance abrasion resistance and reduce deformation under load. Silicone oils and molybdenum disulfide will further reduce its already low coefficient of friction. Incorporation of flame-retardant additives will yield UL 94-V0 characteristics.

31.10   SPECIFICATION OF PROPERTIES

Other properties of UHMW PE are given in the Master Outline of Materials Properties. In all cases, the use of additives to accomplish or enhance a specific property must be evaluated with respect to their effect on the primary properties of toughness and abrasion resistance of UHMW PE.

MATERIAL ___ Ultra-high Molecular Weight Polyethylene (UHMW PE) 1900

PROPERTY	TEST METHOD	ENGLISH UNITS	VALUE	METRIC UNITS	VALUE
MECHANICAL DENSITY	D792	lb/ft³	58.1	g/cm³	0.93
TENSILE STRENGTH	D638	psi	2,900–6,000	MPa	19.9–41.4
TENSILE MODULUS	D638	psi	16,000,000	MPa	110,000
FLEXURAL MODULUS	D790	psi		MPa	
ELONGATION TO BREAK	D638	%	300	%	300
NOTCHED IZOD AT ROOM TEMP	D256	ft-lb/in.	No break	J/m	No break
HARDNESS	D785	Rockwell	R32	Rockwell	R32
THERMAL DEFLECTION T @ 264 psi	D648	°F	115	°C	46
DEFLECTION T @ 66 psi	D648	°F	174	°C	79
VICAT SOFT-ENING POINT	D1525	°F	277	°C	136
UL TEMP INDEX	UL746B	°F		°C	
UL FLAMMABILITY CODE RATING	UL94				
LINEAR COEFFICIENT THERMAL EXPANSION	D696	in./in./°F –22 to 86 86 to 140	$.72 \times 10^{-4}$ 1.1×10^{-4}	mm/mm/°C –30 to 30 30 to 60	1.3×10^{-4} 2.0×10^{-4}
ENVIRONMENTAL WATER ABSORPTION 24 HOUR	D570	%		%	
CLARITY	D1003	% TRANS-MISSION		% TRANS-MISSION	
OUTDOOR WEATHERING	D1435	%		%	
FDA APPROVAL			Yes		Yes
CHEMICAL RESISTANCE TO: WEAK ACID	D543		Not attacked		Not attacked
STRONG ACID	D543		Not attacked		Not attacked
WEAK ALKALI	D543		Not attacked		Not attacked
STRONG ALKALI	D543		Not attacked		Not attacked
LOW MOLECULAR WEIGHT SOLVENTS	D543		Swells slightly		Swells slightly
ALCOHOLS	D543		Not attacked		Not attacked
ELECTRICAL DIELECTRIC STRENGTH	D149	V/mil	450–500	kV/mm	17.7
DIELECTRIC CONSTANT	D150	10^6 Hertz	2.3		2.3
POWER FACTOR	D150	10^6 Hertz	$<5 \times 10^{-3}$		$<5 \times 10^{-3}$
OTHER MELTING POINT	D3418	°F	270	°C	132
GLASS TRANS-ITION TEMP.	D3418	°F		°C	

31.12   PROCESSING-SENSITIVE END PROPERTIES

Extreme pressures encountered in conventional injection molding machines and mold runner and gate systems result in shear-degradation of the polymer. Compression molding of sheet and plates normally requires that trapped air be removed from the polymer particles by cold molding a preform. The preform can then be heated to a fusion temperature of 356–392°F (180–200°C) under reduced pressure. When the preform has reached uniform temperature throughout (dependent on thickness), the cooling cycle is started and pressures increased again to the range of 1500 psi (10 MPa). Appropriately adjusted temperatures/pressures should result in an apparent solid, homogeneous sheet with a specific gravity of approximately 0.935. There is, however, still microscopic porosity because of the discrete particles.

31.13   SHRINKAGE

Not applicable.

31.14   TRADE NAMES

UHMW PE is available under the trade name Hostalen GUR from American Hoechst, Somerville, N.J., and 1900 UHMW from Himont Incorporated, Wilmington, Del.

BIBLIOGRAPHY

1. *ASTM D-4020, Standard Specification for Ultra-High-Molecular-Weight Molding and Extrusion Materials,* American Society for Testing & Materials, Philadelphia, Pa.
2. W.E. Gloor, "Polyethylenes Made by the Ziegler Process," in *Encyclopedia of Chemical Technology,* Vol. 14, 2nd ed., Wiley-Interscience, New York, (1963–1970).

32

POLY(METHYL METHACRYLATE) (PMMA)

Mark Harrington

Technical Service Engineer, Cyro Industries, Orange, CT 06477

32.1 INTRODUCTION

There are many acrylic polymers: acrylonitrile, acrylates, methacrylates, and a host of copolymers. They comprise a wide range of properties and performance characteristics. One of the most important unmodified acrylic materials is poly(methyl methacrylate)(PMMA).

32.2 CATEGORY

At room temperature, PMMA is a hard, stiff, and brittle thermoplastic. Like other amorphous thermoplastics, it softens and loses strength above its glass-transition temperature.

32.3 HISTORY

Acrylic polymers were first reported in 1873. In 1901, Otto Rohm presented his doctoral dissertation on acrylic polymers. Rohm's investigations continued, and by 1927, Rohm & Haas, (former company name of Rohm GmbH Darmstadt, Germany), was producing PMMA for use in lacquers.[1]

Hill and Crawford of ICI were also working with acrylic materials. In 1930, Hill prepared a sample of PMMA, which he described as rigid and glasslike. The raw materials used by Hill were not readily available and were expensive. In 1932, Crawford developed a method of producing the monomer from cheaper materials. Crawford's process is the basis for the process that is used today.[2]

Production of cast sheet and small acrylic articles began in 1933.[3] The first major com-

TABLE 32.1. Use of Cast Materials and Molding Powders[a]

Year	Acrylic Sales Millions of lb	% of Total Plastics Consumption
1960[b]	NA	NA
1964[c]	221	2.4
1970	390	2.1
1980	470	1.5
1986	573	1.2

[a]Ref. 4.
[b]No data published because there were only two producers.
[c]Estimate.

mercial use of PMMA was for aircraft glazing during World War II. Since then, the use of both cast materials and molding powders has increased steadily, as indicated in Table 32.1.

32.4 POLYMERIZATION

32.4.1 General

The production of the monomer, methyl methacrylate, from which PMMA is produced, is still based on the work of Crawford. Acetone, hydrogen cyanide, sulfuric acid, and methanol are combined in various stages and amounts. The methyl methacrylate is then purified by distillation. PMMA is usually produced through a free-radical addition type polymerization using a peroxide or azo initiator. Free-radical polymerization is carried out in bulk, solution, suspension, or emulsion. Temperatures range from 122 to 338°F (50 to 170°C), depending on the process being used and the desired properties of the polymer. Redox polymerization is also possible.

32.4.2 Bulk Polymerization

When cast shapes are made, polymerization usually takes place in two steps. The first step is a partial polymerization before the material is placed in the cell. This is done to thicken the material, so that it will not leak from the cell, and also to minimize the amount of shrinkage during the final cure. The cell is then heated in an autoclave, oven, or water bath for 6 to 12 hours. The exact molecular weight varies greatly from manufacturer to manufacturer, but is generally around 10^6.

Continuous bulk polymerization is a technique used to produce molding and extrusion compounds (MEC). The reaction is brought to about 50% completion with the unconverted methyl methacrylate being recirculated. The molecular weight of the polymers produced runs between 100,000 and 200,000. Consistently high-quality molding and extrusion compounds are produced.

32.4.3 Suspension Polymerization

The high molecular weight of acrylic polymers that are produced by a bulk polymerization which has been brought to completion prevents flow under typical extrusion and injection molding conditions. The first molding and extrusion compounds were produced through a suspension polymerization. In the suspension process, a peroxide catalyst is dissolved in the monomer and dispersed in about two parts water. Suspending agents, emulsifiers, chain-transfer agents, and other additives are also added. The size and shape of the small (50 to 1000 micron) beads can be difficult to control. Polymerization takes place in each dispersed monomer droplet as a localized bulk polymerization. The reaction takes about an hour; a small exotherm is experienced.

32.4.4 Solution Polymerization

PMMA is also produced using solution polymerization techniques. The organic solvent is present in amounts up to 40%. A solvent is used to control the exotherm and to promote flow through the equipment. The reaction is often carried out in several steps; conversion rates approach 100%. The syrups produced are generally put through a vented extruder and then pelletized.

32.4.5 Cost

Molding and extrusion compounds cost about the same in 1987 as they did in 1937, about $0.91 per lb for general-purpose PMMA. Impact-modified acrylic molding compounds cost $0.97 to $1.27 per lb. Cast sheet costs about $1.60 per lb, and extruded sheet costs $1.20 per lb. Methyl methacrylate monomer sells for about $0.52 per lb. The two major raw materials used to make the monomer, acetone and methanol, cost $0.17 to $0.20 per lb and $0.07 to $0.10 per lb, respectively.

32.5 DESCRIPTION OF PROPERTIES

32.5.1 General Properties

As mentioned above, the molecular weight of molding and extrusion compound averages between 100,000 and 200,000. The molecular weight of cast material is much higher (3×10^5 to 6×10^6). This variation accounts for many of the differences between articles made of cast acrylic and those made from molding or extrusion compound. Among different molding compounds, there is also a significant difference in the molecular weight and the comonomer content, and thus certain properties. Properties and processibility can be affected by such things as pellet size and residual monomer or solvent content. Thus, two different grades of PMMA can behave quite differently.

32.5.2 Thermal Properties

ASTM specification D788 divides acrylic molding and extrusion compounds into three groups. The three grades are specified by heat-distortion temperature (HDT) and tensile strength, as shown in Table 32.2.

TABLE 32.2. Grades of Acrylic Molding and Extrusion Compounds, as Specified by ASTM D 788

Property	Grade 5	Grade 6	Grade 8
HDT at 264 psi (1.82 MPa)			
Minimum, °F	156	176	189
(°C)	(69)	(80)	(87)
Maximum, °F	174	187	
(°C)	(79)	(86)	
Tensile strength minimum,			
psi	8000	8000	9000
(MPa)	(55)	(55)	(62)

There are many factors that affect the HDT and tensile strength of a polymer. In practice, the molecular weight, the percent of comonomer, and the amount of certain additives are the factors most used to control material properties. Those materials with a higher molecular weight and a lower percentage of comonomer (ethyl acrylate or methyl acrylate) will, in general, have greater strength and better heat resistance.

PMMA is a polar material; therefore, molecular orientation can have a dramatic effect on properties. The lower the temperature of polymerization, the higher the syndiotacticity of the produced polymer. Polymers that are more syndiotactic will have stronger dipole forces and, therefore, a higher service temperature.

The service temperature of commercial polymers ranges from 150 to 220°F (66 to 104°C). Boiling water will distort most grades of PMMA. Some heat-resistant grades can withstand boiling water.

PMMA loses some of its impact strength at low temperatures. The impact strength remains sufficiently high for the material to be used at temperatures as low as −40°F (−40°C).

Like most plastics, acrylics have a higher coefficient of thermal expansion (CTE) than metals. When they are used with metals, they should not be fixed rigidly to them if wide temperature fluctuations are expected. In glazing applications, it is important to leave adequate room in the frame for thermal expansion. Dimensional changes can also occur when there is a change in humidity.

The glass-transition temperature T_g of unmodified PMMA is about 215°F (108°C) in fact, the transition takes place over a temperature range. Above this temperature the material behaves as a thermoplastic/thermoelastic.

Acrylics are slow burning, flammable materials and can burn to completion if they are not extinguished. The self-ignition temperature is 850°F (454°C). The usual precautions for flammable materials should be followed. Flame-retardant grades of PMMA are available.

Like most plastics, PMMA is a good thermal insulator. This property makes it an excellent glazing material. The thermal conductivity of general-purpose PMMA is less than that of glass.

32.5.3 Physical Properties

PMMA is a polar material and will absorb some moisture. Therefore, its physical, mechanical, and electrical properties depend on temperature, test rate, and humidity. The suitability of an acrylic material is best determined if the material is tested under conditions of use.

Unmodified acrylic is hard and rigid below its glass-transition temperature. It is, therefore, notch sensitive. The impact strength is moderate when the wall thickness is greater than 0.4 in. (10 mm), and transitions in the piece have at least a small radius. The impact strength increases with temperature.

PMMA is tougher than general-purpose polystyrene but not as tough as ABS. It has far more impact strength than glass. PMMA loses some impact strength at low temperatures.

PMMA is often used in applications that require superior optical qualities. PMMA is harder than most thermoplastics, but glass is more scratch resistant. For these applications, PMMA's scratch resistance leaves something to be desired. Scratch resistance can be improved by use of special coatings.

Parts made of acrylic are strong and stiff. The tensile strength at break, compressive strength, and flexural modulus are higher than those of polycarbonate and more than twice those of high-density polyethylene homopolymer.

Unmodified acrylics are brittle. They elongate between 5 and 9% before breaking. General-purpose polystyrene is even more brittle; ABS will elongate more than three times as much as acrylic before breaking.

PMMA is heavier than water but lighter than most metals. It is half as dense as glass. The relative lightness of acrylic sheet makes it useful in building applications. Molding compounds have a bulk density of 40–44 lb/ft³ (0.64–0.70 g/cm³).

Acrylics absorb water, about 1.1% (by weight at equilibrium) at 73°F (23°C) and 70% humidity. The water content will affect electrical properties and dimensions. Short-term mechanical properties are generally unaffected.

32.5.4 Optical Properties

PMMA transmits 92% of daylight. Approximately 4% is reflected at each surface, and absorption is less than 0.5%/in. (25 mm) of thickness. Glass is almost opaque at a thickness of 6 in. (152 mm): PMMA is transparent at six times that thickness. Most PMMA materials have less than 1% haze.

The optical properties of PMMA result in an interesting quality. Rods, tubes, and fibers can be used to "pipe" light. Light can be transmitted through acrylic, around corners, and into restricted areas. There is little loss if the radius of curvature is greater than three times the diameter of the rod or tube (or thickness of sheet). A smooth, scratch-free surface is also required. Wide beams may be transmitted since the critical angle in air is 42°.[5]

Pure PMMA does not absorb much ultraviolet (UV) light. This explains its high degree of light stability. PMMA does absorb small amounts in the 330–400 nm wavelength range. The amount of absorption and, therefore, the degradation rate, increases at wavelengths below 330 nm. The addition of light stabilizers can increase the already excellent light stability of PMMA. The most effective stabilizers are hindered amines and UV absorbers that are efficient in the 290–330 nm wavelength range.

32.5.5 Weathering Characteristics

A substrate can be attacked by the UV light transmitted by acrylic. The life of the substrate can be prolonged by adding stabilizers to the substrate material or by using an acrylic material containing UV absorbers.

Its excellent resistance to sunlight and generally good resistance to chemicals make acrylics the plastic of choice when outdoor weatherability is a concern. Clear acrylic parts and sheet are virtually unaffected by weathering. Acrylic products have been known to last more than 20 years outdoors without any appreciable loss in properties. It is important to note that modification with pigments, impact modifiers, or other additives may reduce PMMA's weatherability.

32.5.6 Chemical Properties

PMMA has good resistance to water, bases, inorganic salt solutions, inorganic acids, most dilute acids, aliphatic hydrocarbons, most household cleaners, fats, oils, and perspiration. It is prone to attack by chlorinated and aromatic hydrocarbons, some dilute acids (hydrocyanic), concentrated sulfuric acid, esters, and ketones.

Resistance to environmental stress cracking (ESCR) depends on the stress the part is under, the material in contact with the PMMA, the temperature, and the time of exposure. Crazing is promoted by solvents and many organic materials that are not solvents, such as aliphatic alcohols. ESCR is improved by annealing. Continuous stresses should not exceed 1500 psi (10 MPa); larger stresses can be sustained for short times.

32.5.7 Electrical Properties

Acrylics exhibit low electrical conductivity, good arc resistance, and high dielectric strength. Because they are polar, acrylics are not as efficient as polyethylene when they are used as an electrical insulator. The polar groups are in the side chain and remain mobile below the T_g; therefore, the dielectric constant and power factor remain relatively high below the T_g.[5]

32.5.8 Effects of Reinforcement

Acrylics are used with reinforcing agents for a few applications. Dentures can be made using an acrylic dough and fillers. Laminates are produced using glass and other fibers, in conjunction with low-molecular-weight acrylics as a binder. The laminates produced are strong, rigid, and lightweight.

32.6 APPLICATIONS

Acrylics find use in many areas and in different forms. In the United States, the largest consumer of acrylic materials is cast sheet (35%). Nearly 26% of acrylic is used for molding or extrusion compounds. (Does not include multipolymers or impact modified materials).

Coatings account for 17% of the usage. The other 21% are specialty grades (coatings, impact modified, etc.).[6]

Typical applications of sheet, molding and extrusion compounds, and rods, tubes, and fibers are given in Table 32.3.

The building and the industrial markets are the largest consumers of acrylic materials. Together, they account for almost half of acrylic consumption.[6]

TABLE 32.3. Typical Applications of PMMA

Sheet	Molding and Extrusion	Rods, Tubes, and Fibers
Glass replacement	Handles and knobs	Medical instruments
Displays, shelving	Light pipes	Light pipes
Signs	Sheet	Sculptures
Enclosures	Taillights	Furniture
Safety glazing	Headlight covers	Electrical parts
Sun screens	Medallions	Lenses
Skylights	Instrument panels	Displays
Greenhouse glazing	Lighting fixtures	Laboratory fixtures
Diffusers	Level indicators	Sight glasses
RV windshields	Displays	
Mirrors	Tissue dispensers	
Aircraft canopies	Street lamp housings	
Plotting boards	Drafting equipment	
Aircraft windows	Refrigerator drawers	
Meter cases	Medical disposables	
Furniture	Dental disposables	
Sanitary ware	Packaging	
Machine guards		
Bug deflectors		

32.7 ADVANTAGES/DISADVANTAGES

32.7.1 Properties

PMMA is most often used because of its superior clarity, transparency, and weatherability; it is rigid, strong, hard, and tough. In general, acrylics have excellent dimensional stability and low mold shrinkage. PMMA is available in a wide range of colors and opacities. There are nearly 100 grades available from which to select many different combinations of properties.

Some of the disadvantages of PMMA are its lack of abrasion resistance (compared with glass), limited stress crack resistance, flammability, vulnerability to attack by some chemicals, limited long-term load-bearing characteristics, a service temperature limited to 200°F (93°C) or lower for a load-bearing part, and notch sensitivity.

32.7.2 Processing

PMMA molding and extrusion compounds process with ease on standard molding (injection and compression), extrusion, and thermoforming equipment. The heat stability is excellent at processing temperatures and, in most applications, regrind can be used at high ratios. In general, acrylics can be injection molded or extruded without drying if a properly designed, vented barrel is used. Mold shrinkage is quite low. There are many grades to choose from, and some suppliers will develop custom grades to meet specific processing and end-use requirements.

Sheet, rods, and tubes are easily machined, welded, and formed. Standard carbide-tipped woodworking tools work well, but special designs are recommended. Acrylics have a wide thermoelastic range, which makes it possible to heat them to forming temperature in ovens.

The viscosity of PMMA at usual processing conditions is higher than that of most polyethylenes, polystyrene, or polypropylene. The equipment must be capable of producing and withstanding high melt pressures [10,000–20,000 psi (69–138 MPa)] in order to maintain reasonable outputs. Viscosity is very sensitive to shear and temperature. Good temperature, speed, and pressure controls are required in order to produce high-quality products consistently. Acrylics are hygroscopic and must be dried in a desiccant-bed type dryer or processed in vented barrels.

Products that are to carry loads or come in contact with certain solvents may require annealing to improve their stress-crack resistance. Materials must be handled carefully in order to prevent scratching.

32.7.3 Costs

General-purpose PMMA has a moderate and stable price. PMMA is about half the price of polycarbonate and twice the price of polystyrene. The material is produced domestically by several companies, some of which also produce acrylic sheet. The consumption of their sheet plants enables them to produce PMMA in large quantities, which helps to keep down the cost of producing the sheet and molding compounds.

There are cheaper transparent plastics, such as polystyrene. Acrylics are used when the additional cost can be justified by their balance of properties.

32.8 PROCESSING TECHNIQUES

See Master Material Outline and the Standard Material Design Chart.

PROCESSABILITY OF ___ Acrylic PMMA

PROCESS	A	B	C
INJECTION MOLDING	X		
EXTRUSION	X		
THERMOFORMING	X		
FOAM MOLDING			
DIP, SLUSH MOLDING	X		
ROTATIONAL MOLDING			X
POWDER, FLUIDIZED-BED COATING			
REINFORCED THERMOSET MOLDING			
COMPRESSION/TRANSFER MOLDING	X		
REACTION INJECTION MOLDING (RIM)			
MECHANICAL FORMING	X		
CASTING	X		

A = Common processing technique.
B = Technique possible with difficulty.
C = Used only in special circumstances, if ever.

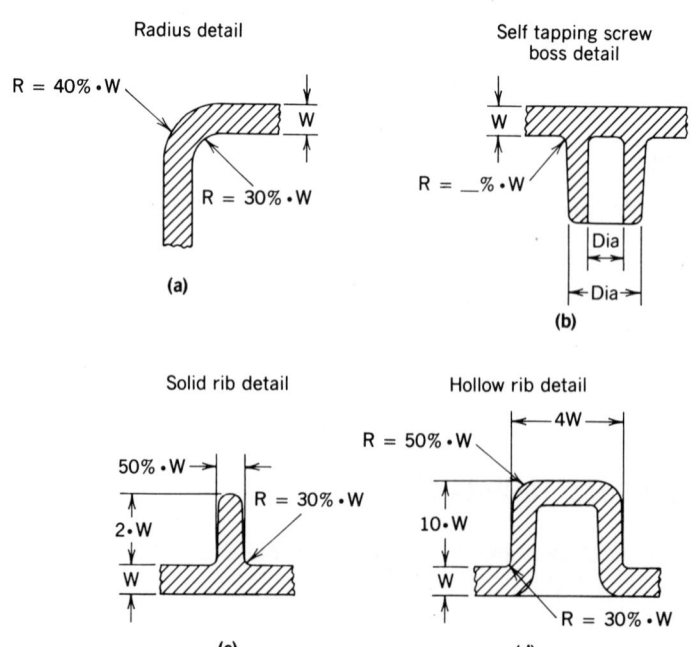

Radius detail

R = 40% • W

W

R = 30% • W

(a)

Self tapping screw boss detail

W

R = ___% • W

Dia

Dia

(b)

Solid rib detail

50% • W

2 • W

W

R = 30% • W

(c)

Hollow rib detail

4W

R = 50% • W

10 • W

W

R = 30% • W

(d)

Standard design chart for ___Acrylic PMMA___

1. Recommended wall thickness/length of flow
   Minimum ___0.035___ for __6 in., 5 in., 2–4 in.__ distance   Grade 5   6   8
   Maximum ___0.75___ for ___40 in.___ distance
   Ideal ___0.10___ for ___6 in.___ distance

2. Allowable wall thickness variation, % of nominal wall ___20___

3. Radius requirements (scheme **a**)
   Outside: Minimum ___30% w___ Maximum ___None___
   Inside: Minimum ___30% w___ Maximum ___None___

4. Reinforcing ribs (scheme **c** and **d**)
   Maximum thickness ___75% w___
   Maximum height ___4 w___
   Sink marks: Yes ___ No ___ depends on cycle

5. Solid pegs and bosses
   Maximum thickness ___2.5 w___
   Maximum weight ___15 w___
   Sink marks: Yes ___ No ___ depends on cycle

6. Strippable undercuts
   Outside: Yes ___ No __X__
   Inside: Yes ___ No __X__

7. Hollow bosses (scheme **b**): Yes ___ No __X__

8. Draft angles
   Outside: Minimum ___$\frac{1}{2}^\circ$___ Ideal ___$1\frac{1}{2}^\circ$___
   Inside: Minimum ___$\frac{1}{2}^\circ$___ Ideal ___$1\frac{1}{2}^\circ$___

9. Molded-in inserts: Yes __X__
   No ___

10. Size limitations
    Length: Minimum _None_ Maximum _None_ Process __Injection molding__
    Width: Minimum _None_ Maximum _None_ Process __Injection molding__

11. Tolerances

12. Special process- and materials-related design details
    Blow-up ratio (blow molding) ___
    Core L/D (injection and compression molding) ___10/1___
    Depth of draw (thermoforming) ___25–100%[a]___
    Other ___

[a]Depends on system, type of sheet, and desired properties

32.9 RESIN FORMS

Molding and extrusion compounds are generally sold as 1/8-in. (3 mm) cylindrical pellets. Fine powders are available for rotational molding. Monomer or prepolymers for casting or coating are supplied as free-flowing liquids or syrups.

Sheet, rods, and tubes are available in many grades and sizes. Cast sheet is made in thicknesses ranging from 0.03 in. to 6.0 in. (0.76 to 153 mm). Extruded, continuously manufactured, and continuously cast sheet is made in thicknesses ranging from 0.06 to 0.6 in. 4 (0.15 to 15 mm). Sheet is available in a wide variety of sizes, textures, and colors. These materials are available in essentially the same grades as the molding and extrusion compounds (high-impact, UV-transmitting or UV absorbing, heat-resistant, or flame-retar-dant) and in a wide rage of colors.

PMMA molding and extrusion compounds are available in uv absorbing or transmitting, high HDT, improved ESCR, high-impact, easy flow, low-contamination level (medical), flame-retardant, improved static dissipation, and abrasion resistant grades as well as in grades that are stable to gamma radiation.

Copolymers and polymer blends containing acrylic comonomers and polymers are be-coming increasingly popular. Blending with PVC improves the impact-resistance, forma-bility, and flame-retardance properties but reduces transparency. Copolymers and polymer blends containing rubber (butyl acrylate, butadiene, or acrylonitrile) offer higher impact strength, better chemical resistance, and reduced notch sensitivity. Alpha methyl styrene copolymers have higher heat resistance. Styrene copolymers have improved melt flow char-acteristics and cost less per pound. Terpolymers, such as MBS (methyl methacrylate–bu-tadiene–styrene), are used as tough transparent plastics or as additives in other plastics, such as PVC. Ethyl and methyl acrylate are often used as copolymers to control the "hard-ness" of the material.

32.10 SPECIFICATION OF PROPERTIES

See the Master Outline of Material Properties.

32.11 PROCESSING TECHNIQUES

32.11.1 General

Viscosity. The melt viscosity of acrylic molding and extrusion compounds is usually higher than that of most polymers, such as polyethylene, polystyrene, or plasticized PVC.

Molding and extruding equipment used to process acrylics must be capable of producing and withstanding high melt pressures. Hardened metals should be used for contact areas.

Like most amorphous materials, acrylics have a small power law index. This indicates that their melt viscosity is very sensitive to the shear rate. Acrylics also have a small con-sistency index, an indication that their viscosity is very sensitive to changes in temperature. This is especially true at temperatures near their glass-transition temperature. A high-quality molding or extrudate cannot be produced consistently if the viscosity of the melt is fluctuating. The temperature and shear rate can be controlled with great accuracy if PID temperature-control and closed-loop process controls are used.

Heat Resistance. The high viscosity of acrylics can lead to overheating of the polymer if high screw speeds are used. The viscous heat generation can be controlled by using lower screw speeds and higher barrel temperatures. Acrylics show little degradation below 482°F (250°C). The degradation rate increases rapidly above 536°F (280°C). Degradation can lead to surface defects, char, and a loss of properties in the finished part.

Drying. PMMA molding and extrusion compounds will absorb moisture. Properly stored PMMA will normally exhibit between 0.3 and 0.5% moisture. Thus, MEC must be dried (or processed in machinery equipped with vented barrels) before processing. On humid days, a desiccant-bed dryer should be used to ensure the material's dryness. Moisture levels should be below 0.03% before extruding and below 0.10% before molding. Regrind need not be dried if it is used immediately. Failure to properly dry the material can result in a slight reduction in transparency, severe surface distortions, or failure of the part.

PMMA should be dried for at least three hours in a desiccant-bed dryer using air with

MATERIAL _____ PMMA (cast sheet)

PROPERTY	TEST METHOD	ENGLISH UNITS	VALUE	METRIC UNITS	VALUE
MECHANICAL DENSITY	D792	lb/ft^3	73–75	g/cm^3	1.17–1.20
TENSILE STRENGTH	D638	psi	8–11 × 10^3	MPa	55–76
TENSILE MODULUS	D638	psi	350,000–450,000	MPa	2,413–3,103
FLEXURAL MODULUS	D790	psi	400,000–500,000	MPa	2,758–3,448
ELONGATION TO BREAK	D638	%		%	
NOTCHED IZOD AT ROOM TEMP	D256	ft-lb/in.	Per inch of notch 0.3–0.6	J/m	Per meter of notch 16.2– 32.4
HARDNESS	D785		Coded scale M80–100		Coded scale M80–100
THERMAL DEFLECTION T @ 264 psi	D648	°F	160–215	°C	71–102
DEFLECTION T @ 66 psi	D648	°F	165–235	°C	74–113
VICAT SOFT-ENING POINT	D1525	°F	203–230	°C	95–110
UL TEMP INDEX	UL746B	°F		°C	
UL FLAMMABILITY CODE RATING	UL94		Coded rating UL94 HB		Coded rating HB
LINEAR COEFFICIENT THERMAL EXPANSION	D696	in./in./°F	28–50 × 10^{-6}	mm/mm/°C	50–90 × 10^{-6}
ENVIRONMENTAL WATER ABSORPTION 24 HOUR	D570	%	0.2–0.4	%	0.2–0.4
CLARITY	D1003	% TRANS-MISSION	91–92	% TRANS-MISSION	91–92
OUTDOOR WEATHERING	D1435	%	1095 days, very slight yellowing	%	1095 days, very slight yellowing
FDA APPROVAL					
CHEMICAL RESISTANCE TO: WEAK ACID	D543		Not attacked		Not attacked
STRONG ACID	D543		Minimally attacked to soluble		Minimally attacked to soluble
WEAK ALKALI	D543		Not attacked		Not attacked
STRONG ALKALI	D543		Not attacked		Not attacked
LOW MOLECULAR WEIGHT SOLVENTS	D543		Soluble to not attacked		Soluble to not attacked
ALCOHOLS	D543		Not attacked to badly attacked		Not attacked to badly attacked
ELECTRICAL DIELECTRIC STRENGTH	D149	V/mil	450–550	kV/mm	18–22
DIELECTRIC CONSTANT	D150	10^6 Hertz	2.8 (Acrylite GP)		2.2–2.8
POWER FACTOR	D150	10^6 Hertz	0.02–0.03		0.02–0.03
OTHER MELTING POINT	D3418	°F	NA	°C	NA
GLASS TRANS-ITION TEMP.	D3418	°F	194–217	°C	90–105

Master Outline of Material Properties

MATERIAL ___ PMMA (general-purpose)

PROPERTY	TEST METHOD	ENGLISH UNITS	VALUE	METRIC UNITS	VALUE
MECHANICAL DENSITY	D792	lb/ft^3	73–74	g/cm^3	1.17–1.19
TENSILE STRENGTH	D638	psi	At break, 7,000– 11,000	MPa	At break, 48–76
TENSILE MODULUS	D638	psi	325,000–470,000	MPa	2,241–3,240
FLEXURAL MODULUS	D790	psi	325,000–460,000	MPa	2,241–3,172
ELONGATION TO BREAK	D638	%		%	
NOTCHED IZOD AT ROOM TEMP	D256	ft-lb/in.	per in. of notch 0.3–0.6	J/m	per m. of notch 16.2–32.4
HARDNESS	D785		Coded scale M68–105		Coded scale M68–105
THERMAL DEFLECTION T @ 264 psi	D648	°F	165–210	°C	74–105
DEFLECTION T @ 66 psi	D648	°F	175–225	°C	79–107
VICAT SOFT-ENING POINT	D1525	°F	180–230	°C	82–110
UL TEMP INDEX	UL746B	°F	122–194	°C	50–90
UL FLAMMABILITY CODE RATING	UL94		Coded rating UL 94 HB		Coded rating HB
LINEAR COEFFICIENT THERMAL EXPANSION	D696	in./in./°F	$28–50 \times 10^{-6}$	mm/mm/°C	$50–90 \times 10^{-6}$
ENVIRONMENTAL WATER ABSORPTION 24 HOUR	D570	%	0.1–0.4	%	0.1–0.4
CLARITY	D1003	% TRANS-MISSION	91–92	% TRANS-MISSION	91–92
OUTDOOR WEATHERING	D1435	%	1095 days, very slight yellowing	%	1095 days, very slight yellowing
FDA APPROVAL			Less than 8% alcohol		Less than 8% alcohol
CHEMICAL RESISTANCE TO: WEAK ACID	D543		Not attacked 0.3–0.7^a		Not attacked 0.3–0.7^a
STRONG ACID	D543		Minimally to soluble 4–100^a		Minimally to soluble 4–100^a
WEAK ALKALI	D543		Not attacked 0.3–0.6^a		Not attacked 0.3–0.6^a
STRONG ALKALI	D543		Not attacked 0.3–0.7^a		Not attacked 0.3–0.7^a
LOW MOLECULAR WEIGHT SOLVENTS	D543		Soluble to not attacked, 0.1–100^a for aromatics and ^{b}		Soluble to not attacked, 0.1–100^a for aromatics and ^{b}
ALCOHOLS	D543		Not attacked to badly 0.4–10^a attached		Not attacked to badly 0.4–10^a attached
ELECTRICAL DIELECTRIC STRENGTH	D149	V/mil	400–500	kV/mm	16–20
DIELECTRIC CONSTANT	D150	10^6 Hertz	Dimensionless 2.2–2.8		Dimensionless 2.2–2.8
POWER FACTOR	D150	10^6 Hertz	Dimensionless 0.3		Dimensionless 0.3
OTHER MELTING POINT	D3418	°F	N/A	°C	N/A
GLASS TRANS-ITION TEMP.	D3418	°F	185–217	°C	85–105

^{a}Percent change in weight, dimensions, or property value.
^{b}Chlorinated hydrocarbons; resistant to aliphatics.

MATERIAL ___ PMMA (impact modified)

PROPERTY	TEST METHOD	ENGLISH UNITS	VALUE	METRIC UNITS	VALUE
MECHANICAL DENSITY	D792	lb/ft³	69–74	g/cm³	1.11–1.18
TENSILE STRENGTH	D638	psi	5000–9000	MPa	55–76
TENSILE MODULUS	D638	psi	200,000–500,000	MPa	2415–3100
FLEXURAL MODULUS	D790	psi	230,000–390,000	MPa	2760–3450
ELONGATION TO BREAK	D638	%	2–40	%	2–40
NOTCHED IZOD AT ROOM TEMP	D256	ft-lb/in.	per in. of notch 0.65–2.5	J/m	per m of notch 35–135
HARDNESS	D785		Coded scale M38–68		Coded scale M38–68
THERMAL DEFLECTION T @ 264 psi	D648	°F	165–210	°C	74–99
DEFLECTION T @ 66 psi	D648	°F	180–215	°C	82–102
VICAT SOFT-ENING POINT	D1525	°F	185–230	°C	85–110
UL TEMP INDEX	UL746B	°F	122	°C	50
UL FLAMMABILITY CODE RATING	UL94		Coded rating UL94 HB		Coded rating HB
LINEAR COEFFICIENT THERMAL EXPANSION	D696	in./in./°F	$27–44 \times 10^{-6}$	mm/mm/°C	$48–80 \times 10^{-6}$
ENVIRONMENTAL WATER ABSORPTION 24 HOUR	D570	%	0.2–0.6	%	0.2–0.6
CLARITY	D1003	% TRANS-MISSION	88–90	% TRANS-MISSION	88–90
OUTDOOR WEATHERING	D1435	%	730 days, slight yellow, embrittlement	%	730 days, slight yellow, embrittlement
FDA APPROVAL			Hot fill less than 8% alcohol		Hot fill less than 8% alcohol
CHEMICAL RESISTANCE TO: WEAK ACID	D543		Not attacked		Not attacked
STRONG ACID	D543		Minimally attacked to soluble		Minimally attacked to soluble
WEAK ALKALI	D543		Not attacked		Not attacked
STRONG ALKALI	D543		Not attacked		Not attacked
LOW MOLECULAR WEIGHT SOLVENTS	D543		Not attacked to soluble		Not attacked to soluble
ALCOHOLS	D543		Not attacked to badly attacked		Not attacked to badly attacked
ELECTRICAL DIELECTRIC STRENGTH	D149	V/mil	380–500	kV/mm	15–20
DIELECTRIC CONSTANT	D150	10⁶ Hertz	Dimensionless 2.7–2.9		Dimensionless 2.7–2.9
POWER FACTOR	D150	10⁶ Hertz	Dimensionless		Dimensionless
OTHER MELTING POINT	D3418	°F	N/A	°C	N/A
GLASS TRANS-ITION TEMP.	D3418	°F	176–212	°C	80–100

a dew point of $-20°F$ ($-29°C$) or below. The temperature of the drying air depends on the grade of acrylic being dried: Grade 8 materials should be dried at $180°F$ ($82°C$); Grade 7 materials at $170°F$ ($77°C$); and Grade 5 materials at $160°F$ ($71°C$).

Regrind and Housekeeping. The processing area should be kept clean and should be climate-controlled, as airborne contaminants will cause surface imperfections. The humidity should be maintained at a reasonably high level to prevent the build up of static charges, which are a safety hazard and attract dust.

PMMA has excellent heat stability under usual processing conditions. First and second generation regrind may be used at any required level without a significant reduction in properties. A slight color shift will be evident as the number of heat histories is increased. The ratio of regrind to virgin resin should be kept constant during a run to minimize color variations and process changes.

32.11.2 Extrusion

PMMA may be processed on single-screw or multiple-screw extruders. Sheets, rods, tubes, and sections are commonly produced. Twin-screw extruders offer improved feeding and degassing characteristics, while single-screw extruders are often preferred because of their great economy.

The high output of extrusion equipment often makes it impractical to dry the material before processing. For this reason, vented extruders are usually employed. A degassing zone of 3 to 6 diameters in length is adequate. High temperatures in the vent zone promote foaming; colder temperatures retard vent flow.

The best results are achieved with longer barrels (30:1 to 40:1 L/D ratios). Longer barrels facilitate melting without overheating the polymer, allowing for improved stability at higher outputs.

Proper screw design is necessary for trouble-free operation. A metering type screw with a gradual compression and constant pitch is preferred. The channel depth in the feed section should be about 16% of the barrel diameter. Compression ratios of 3:1 work well. Gear pumps and static mixers are frequently used to ensure a constant flow of homogeneous material to the die.

The feed throat should encompass at least two flights and be cooled to ensure proper feeding. Cooling is especially important when regrind is processed and when grooved feed sections are used.

Barrels with cast-in channels for water or oil cooling are often required. Without proper cooling, output fluctuations could result from feeding difficulties or temperature-related viscosity variations. Stock temperatures should be between 440 and $480°F$ (227 and $249°C$) for flat sheet and 410 and $450°F$ (210 and $232°C$) for profiles.

The die must be strong and streamlined. The choker bar and the die lips should be adjustable. Extrudate swell is minimized when the land lengths are at least 12 times the thickness of the extrudate. Accurate temperature control is required. Die gaps are set at or slightly above the required thickness.

High-quality polishing rolls are required to produce sheet with a surface quality that approaches that of cast sheet. Rolls should be micro-finished, chrome plated, and hardened. Large-diameter rolls offer improved heat transfer and polishing effectiveness. The speed of take-off equipment should be interlocked with that of the rolls to improve gauge uniformity and operating ease.

Screen packs are often used to filter and homogenize the melt stream. This prevents large particles from damaging the die and improves the quality of the profile. A common mesh arrangement is 20-40-80; finer screens are used in critical applications.

32.11.3 Injection Molding

PMMA can be processed on plunger or screw-type injection molding machines. Screw-type machines have many advantages over the plunger type. This discussion will be limited to processing on screw-type machines.

PMMA's mold shrinkage is similar to that of other thermoplastics. Tooling designed for ABS, impact polystyrene, impact-modified acrylics, polycarbonate, and others is dimensionally compatible with PMMA. Injection pressures are high and molds should be suitably strong.

A cold mold could produce hazy and stressed parts. Mold temperatures of 70–150°F (21–66°C) work well. Warping can be controlled by separately controlling the two halves of the mold, helping thick and thin sections to cool evenly.

The high viscosity of acrylics makes proper vent design and location critical. Vent depths of 0.001 to 0.0015 in. (0.025 to 0.051 mm) work well.

The screw supplies most of the heat required. Barrel temperatures range between 400 and 480°F (204 and 249°C). The higher temperatures are required for large, thin-walled parts molded from stiff grades of PMMA. Back pressure, screw speed, barrel temperatures, and residence time all influence the melt temperature and degree of mixing.

The length of gates, runners, and the sprue should be kept to a minimum; if their diameters are too small, flow resistance could cause filling problems. Hot manifold molds work well with PMMA.

Molded articles have varying degrees of orientation and molded-in stresses. To prevent stress-cracking, parts should be annealed if they will come in contact with solvents or if they will be continuously stressed. Annealing temperatures vary with the grade of material being used, ranging from 140 to 190°F (60 to 88°C). Annealing time must be increased as the wall thickness increases.

32.11.4 Thermoforming

General. The molecular weight of extruded sheet is much lower than that of cast sheet. For this reason, extruded sheet conforms to the shape of the forming tool better than cast sheet; however, cast sheet is thermoelastic over a broader temperature range.

PMMA can be formed by vacuum forming (male or female, with or without a plug assist), drape forming, vacuum snapback forming, line bending, or cold forming (the minimum radius of curvature is 300 times the sheet thickness). When it is heated above the T_g of the polymer, the formed shape will recover about 90% of the deformation.

Both cast and extruded sheet may be heated either in ovens or in place with heating panels. Proper positioning and timing is necessary to ensure even heating. Parts should be cooled to at least 150°F (66°C) before they are removed from the mold.

Cast Sheet. Cast sheet is used for simple shapes, like signs and sanitary ware. Its forming temperatures range from 300 to 360°F (149 to 182°C). Overheating can lead to bubbles. Both cast and extruded material will absorb water if stored improperly; should this occur, the sheet should be dried before it is heated to forming temperature.

Cast sheet can be expected to shrink about 2% upon heating. If the formed part must be dimensionally accurate, the forming tool should be about 1% oversized since the part will shrink further as it goes from the forming temperature to room temperature. Thickness tolerances are larger for cast than for extruded sheet. Variations of up to 20% are not uncommon for cast sheet.

Extruded Sheet. Extruded sheet is used when greater detail or a lower price is required. The extruded sheet should be heated to 290 to 320°F (143 to 160°C) before forming. Extruded sheet will shrink up to 3% in the machine (extrusion) direction and will expand up to 1.5% in the transverse direction. Use of a retaining frame will help minimize these dimensional changes.

32.11.5 Machining

Tools designed for wood or soft metals often work well with acrylic. Tools are also available that have been specifically designed for use with acrylics. Tools should always be sharp, and compatible coolants should be used when thick sections are cut. The sections must be fastened in order to prevent vibration, which could cause cracking. Cast materials are, in general, easier to machine because of their higher molecular weight.

Acrylic parts can be fastened using solvent cements, ultrasonic welding, or mechanical fasteners. Stressed parts should be annealed before solvents are used.

The edges may be finished using sandpaper, scrapers, buffing wheels, or a flame. The surface scratches fairly easily. Protective masking should be left on during as much of the machining process as possible.

32.12 PROCESSING-SENSITIVE END PROPERTIES

32.12.1 General

Most of the properties of acrylic are not affected by normal molding, extrusion, or forming conditions. Acrylics have excellent resistance to the level of heat and shear experienced during processing. Some properties can be affected.

32.12.2 Internal Stresses and Orientation

Many properties can be affected if there are stresses frozen into the part. Internal stresses can reduce the HDT, chemical resistance, ESCR, dimensional stability, and toughness. Highly stressed or oriented areas behave like notches, and this will lower the impact strength. Generally, stress levels are higher in injection-molded parts than in material that has been extruded. The only way to minimize stress in parts is to anneal them after processing. High melt temperatures, slow uniform cooling rates, low and short holding pressures (injection), and small draw downs (extrusion) all help reduce internal stress. Gates and other areas likely to have frozen-in stresses should be located in areas that will not be exposed to solvents or will not support loads.

Parts are stronger in the direction of orientation, while the strength in the transverse direction is decreased. Shrinkage is also increased in the direction of orientation.

32.12.3 Surface Quality and Transparency

The surface quality of a shape reflects the material it contacts during the cooling process. Mold surfaces should be highly polished (except for thermoforming molds) as should the cooling rolls on an extrusion line. When a material is cooled in air, it is best to have good climate control, allow it to cool slowly, provide a large land L/D ratio (extrusion), and maintain a high melt temperature.

A molded part's surface quality is improved by higher mold and melt temperatures. The part should be injected at temperatures just below that at which sinks occur. In extrusion, the rolls should be heated to the temperature at which the ribbon just begins to stick to the rolls.

Bubbles and splay marks are an indication that the material has not been dried or degassed adequately. They can also be caused by overheating the polymer.

Gloss and transparency are affected by processing conditions. In general, the adjustments described above that promote good surface qualities will also reduce haze and increase transparency. The clarity of PMMA is quickly lost when it is contaminated; even very small percentages of other polymers can greatly reduce its transparency. It is also important to keep the processing area clean to minimize the amount of airborne contaminants.

32.12.4 Physical Properties

Physical properties are also reduced by the presence of internal stresses. Notches should be avoided whenever possible, since even small fillet and radii increase a part's strengh tremendously. Overheating PMMA can reduce its molecular weight and physical properties.

Weld lines or knit lines (injection) are weak areas. Higher melt temperatures, high injection rates, and good venting will often eliminate them. Proper gate location can also prevent weld line problems.

When it is overheated, PMMA will begin to depolymerize. This reduces the molecular weight and can produce gases, which can become trapped in the finished part, thereby reducing part strength.

Material that has not been dried or degassed properly will have such surface defects as splay. These surface defects can behave like notches and reduce the strength of the part. These distortions are usually large enough to cause part rejection.

32.12.5 Chemical Properties

Resistance to chemicals is reduced when a PMMA part has internal stresses. Biaxial orientation (thermoforming) can increase chemical resistance. The presence of residual monomer or low boilers will decrease the chemical resistance. Residuals and low boilers can be removed by a properly designed degassing zone.

32.13 SHRINKAGE

The shrinkage tables (Mold Shrinkage MPO) are not exact. Shrinkage depends on many factors, including injection molding conditions, part thickness, mold design, cooling rate and uniformity, flow direction, distance from the gate, and material type.

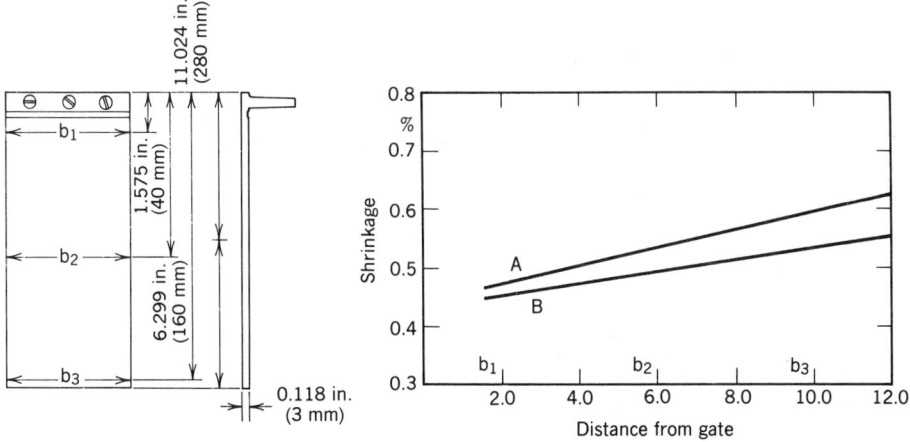

Figure 32.1. Shrinkage transverse to the direction of flow for two grade 8 PMMA materials: A has molecular weight of 150,000 and B a molecular weight of 110,000.

MATERIAL PMMA

IN THE DIRECTION OF FLOW

THICKNESS

MATERIAL	mm: 1.5	3	6	8	FROM	TO	#
	in.: 1/16	1/8	1/4	1/2			
					0.002	0.008	in./in.

	PERPENDICULAR TO THE DIRECTION OF FLOW					
Grade 5 PMMA	0.002	0.004	0.005			
Grade 6 PMMA	0.003	0.004	0.006			
Grade 8 PMMA	0.004	0.005	0.007			

NOTE: Mold shrinkage is approximate. It is affected by part design, mold temperature, thickness, injection pressure, packing time, cycle time, orientation, gate design, gate size, gate location, glass content, glass size, and filler content.

Standard Tolerance Chart

PROCESS Injection molding   MATERIAL PMMA

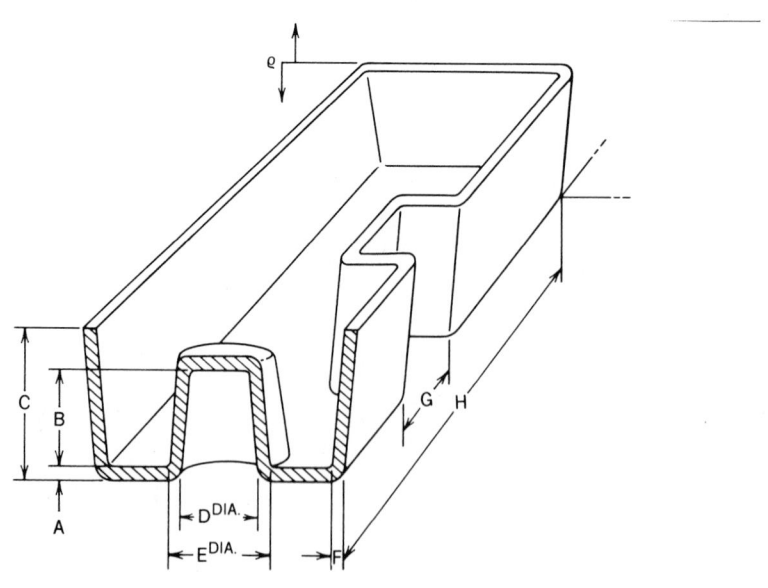

Values are based on a nominal 1/8 in. (3.2 mm) wall thickness and given in plus or minus in./in. and mm/mm.

FINE tolerance – Special care and added cost.

COMMERCIAL tolerance – Normal care required.

COARSE tolerance – Minimal care required, lowest cost.

Tolerance	A	B	C	D	E	F	G	H	
FINE	0.003	0.003	0.003	0.002	0.003	0.003	0.003	0.004	in./in.
									mm/mm
COMMERCIAL	0.005	0.005	0.005	0.003	0.004	0.005	0.004	0.005	in./in.
									mm/mm
COARSE	0.007	0.007	0.008	0.005	0.006	0.007	0.006	0.008	in./in.
									mm/mm

The effects of molecular weight and distance from the gate on shrinkage are illustrated in Figure 32.1. All else being the same, as the distance from the gate increases, lower molecular weight materials shrink less than higher molecular weight materials.

32.14 TRADE NAMES

The trade names and major suppliers for PMMA molding compounds and sheet products are given in Table 32.4.

TABLE 32.4. Suppliers of PMMA

Trade Name	Supplier	Location
Molding Compounds		
CP	Continental Polymers Inc.	Compton, Calif.
Acrylite	CYRO Industries	Mount Arlington, N.J.
Cyrolite		
XT		
Degalan	Degussa Corporation Polymer	Teterboro, N.J.
	Richardson Polymer	Madison, Conn.
Plexiglas	Rohm & Haas	Bristol, Penn.
Sumipex	Sumitomo Chem. Co.	New York, N.Y.
Bakelite	Union Carbide	Danbury, Conn.
	Plaskolite Inc.	Columbus, Ohio
Sheet Products		
	Polycast	Stamford, Conn.
	CYRO Industries	Mount Arlington, N.J.
	Montedison USA Inc.	New York, N.Y.
	Rohm & Haas	Bristol, Penn.
	USS Chemicals	Florence, Ky.
	Conesco Corp.	Berlin, Conn.
	Continental Polymers Inc.	Compton, Calif.
	Koro Corp.	Hudson, Mass.
	Plaskolite Inc.	Columbus, Ohio
	Rotuba Extruders Inc.	Linden, N.J.
	Sumitomo Corp. of America	New York, N.Y.
	Polymer Extruded Products, Inc.	Newark, N.J.

BIBLIOGRAPHY

1. J.A. Brydson, *Plastics Materials,* 3rd ed., Butterworth & Co., London, 1975, p. 328.
2. *Ibid.,* pp. 328–329.
3. R.M. Ogorkiewicz/ICI Ltd., *Engineering Properties of Thermoplastics,* Wiley-Interscience, New York, 1970, p. 215.
4. *Mod. Plast.* **39,** 78 (Dec. 1961); **38** 82, (Jan. 1961); **42,** 97 (Jan. 1965); **48,** 66 (Jan. 1971); **58,** 67 (Jan. 1981); **63,** 60 (Jan. 1986).
5. J.A. Brydson, *Plastics Materials,* 3rd ed., Butterworth & Co., London, 1975, p. 334.
6. *Mod. Plast.* **64,** 56 (Jan. 1987).

General References

J.G. Cook, *Your Guide to Plastics,* Merrow Publishing Co., Ltd., 1964.

Desk Top Data Bank, Plastics, 8th ed., International Plastics Selector, Inc., 1986.

J. Dubois and F. John, *Plastics,* Reinhold Publishing Co., New York, 1967.

R. Gachter and H. Muller, *Plastics Additives,* Hanser Publications, New York, 1985.

C.A. Harper, *Handbook of Plastics and Elastomers,* McGraw-Hill Co., New York, 1975.

A.E. Lever and J. Rhys, *The Properties and Testing of Plastics Materials,* Chemical Publishing Co., Inc., 1962.

Modern Plastics Encyclopedia, 1986/1987, McGraw-Hill, New York, 1986, p. 514.

T. Newman, *Plastics as a Design Form,* Chilton Book Co., New York, 1972.

Plastics Mold Engineering, (SPE Polymer Technology series), Reinhold Publishing Co., New York, 1965.

P.C. Powell, *The Selection and Use of Thermoplastics,* Oxford University Press, London, 1977.

A. Randolph and co-workers, *Plastics Engineering Handbook,* Reinhold Publishing Co., New York, 1960.

I.I. Rubin, *Injection Molding Theory and Practice,* John Wiley & Sons, Inc., New York, 1972.

Technical Data on Plastic Materials, Plastic Materials Manufacturers' Association, 1943.

33

POLYMETHYLPENTENE (PMP)

H. Ohi and K. Furuya

Mitsui Petrochemical Industries, Ltd., New York, New York

33.1 INTRODUCTION

Polymethylpentene (PMP) is a high performance material offering excellent electrical properties and resistance to heat, oil, and chemicals as well as transparency and ease in processing. This properties profile has suited it to such applications as medical instruments, food contact products, industrial materials, electrical appliance and electronic parts, and electrical insulating materials.

propylene   4-methylpentene

polymethylpentene polymer (PMP)

33.2 CATEGORY

Polymethylpentene polymer (PMP) is a stereoregular, crystalline polyolefin.[1]

33.3 HISTORY

PMP was first synthesized by G. Natta and his associates, working in Italy in 1956.

Imperial Chemical Industries (ICI), Ltd., of Britain completed the world's first commercial PMP plant and began to manufacture and sell the resin in 1969. However, because demand was insufficient to fuel growth, ICI discontinued the production of the material. Later, in 1973, Mitsui Petrochemical Industries, Ltd., succeeded ICI by purchasing its entire PMP business, including rights and patents.

Since 1973, demand for PMP has grown steadily at an average annual rate of more than 20%. In 1985, plant capacity doubled from 3000 to 6000 tons per year.

33.4 POLYMERIZATION

PMP is produced from 4-methylpentene-1, a dimer of propylene. The manufacturing process from the raw material to the finished product is short. The first step is dimerization of propylene to produce 4-methylpentene-1 monomer. The monomer is then polymerized. If PMP is produced on a large scale, as is possible within five to 10 years, it can be offered at considerably lower prices. PMP is priced at $2.05–$4.00/lb (1988).

33.5 DESCRIPTION OF PROPERTIES

The properties profile of PMP positions it between such transparent plastics as polycarbonate, polysulfone, polyester, SAN, and acrylic resins and such opaque materials as polyethylene, polypropylene, nylon, and modified poly(phenylene oxide).

For example, PMP is a heat-resistant, transparent material with a white light transmission of 90% or higher and a melting point of 455°F (235°C). Such transparent plastics as polycarbonate, polystyrene, SAN, and acrylic resins are noncrystalline and are not suitable for applications requiring chemical and oil resistance. Such commodity polyolefins as polyethylene and polypropylene are crystalline resins and possess superior processibility, chemical resistance, and electrical properties, but they do not have the heat resistance and transparency of PMP.

To summarize, PMP exhibits transparency and heat resistance, which are the advantages of a noncrystalline resin, as well as the chemical resistance, electrical properties, and moldability, that are the inherent characteristics of a crystalline polyolefin.

PMP is known for several special properties:

- Next to fluorine plastics, the lowest refractive index ($n_D^{25} = 1.463$)
- The lowest specific gravity (52 lb/ft^3 {$d = 0.835$}) of any commercially available plastic material
- The highest dielectric breakdown voltage [1650 V/mil (65 kV/mm)]; for stretched film, 267 V/mil (105 kV/mm)
- Superior flow properties (L/T = 600 or above)
- Good gas permeability. Because of the great difference in its nitrogen and oxygen permeability, PMP is capable of separating nitrogen and oxygen.

33.6 APPLICATIONS

The medical and cosmetic fields were the first to adopt PMP and currently accounts for nearly one third of the demand for PMP. Product applications include hypodermic syringes, blood test cells, laboratory ware, and laboratory animal cages (Fig. 33.1) as well as cosmetic containers.

In the food contact market, PMP has found application in areas that require resistance to heat, oil, and steam; examples include microwave oven trays (a major use), coffee maker parts, automatic dishwasher parts, nursery bottles, popcorn poppers, egg boilers, and whiskey dispensers.

Figure 33.1. Animal cages.

In the industrial materials area, PMP film and resin molds are being used in the process of curing such thermosetting plastics as polyurethane, epoxy, phenolic, melamine, and urea resins. It has found much usage as a release agent, where, thanks to its heat resistance and releasability, PMP is competing with fluoroplastics and triacetate resins. In this field, the primary products are release film for manufacturing printed circuit boards or advanced composite materials, mandrel and sheath for manufacturing high-pressure rubber hoses, and plastic molds for manufacturing contact lenses and LEDs. As an electrical insulating material, PMP offers a low dielectric constant, high dielectric strength, coupled with heat, heat aging and good soldering properties and oil and varnish resistance. It has been specified for use as a covering for heat-resistant wire,[3] power cables,[4] oil well cables, communications cables,[5] and related products.

In the appliance and electronic materials area, PMP is put to use in copying machine parts, acoustic components, coil bobbins, fuse covers, and connectors, among other things. The use of PMP film as a condenser material is being studied to replace polypropylene and polyester films.

Other development work is going on in such application areas as automotive, lighting fixture, and aircraft parts; atomic power-related items; and optical memory and optical fibers, all of which take advantage of the polymer's light weight and optical properties.

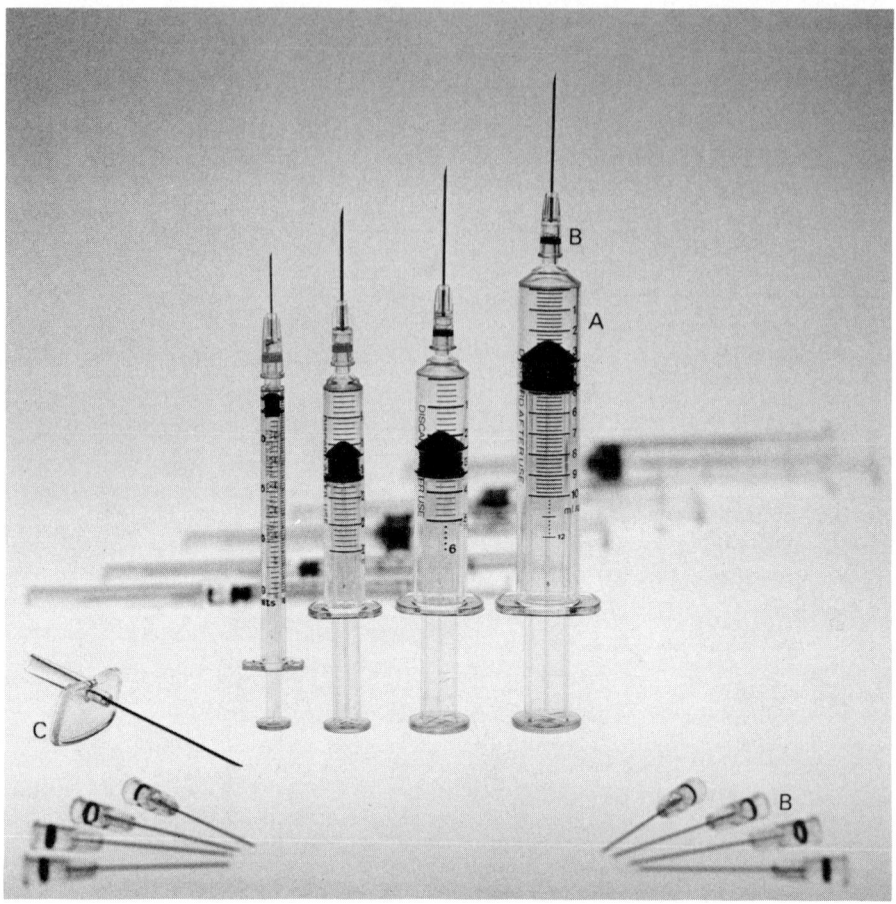

Figure 33.2. Disposable syringes and needle hubs.

33.7 ADVANTAGES/DISADVANTAGES

The chief advantages of PMP are good heat resistance, transparency, chemical resistance, electrical properties, gas permeability, hydrolysis resistance, flowability, releasability, and light weight. In minature moldings, products 0.008–0.010 in. (0.2–0.3 mm) thick are easily molded from PMP.

It exhibits excellent creep characteristics. It can be bonded to metals and other materials through the action of its own shrinkage, without causing toxicity problems from adhesives. Its capability of being joined by forced insertions is advantageous in medical applications, Figure 33.2.

Its limitations are poor rigidity, impact strength at lower temperatures, weatherability, hydrogen solvent resistance, gas barrier properties, melt tension, and adhesion.

33.8 PROCESSING TECHNIQUES

PMP is suited for injection molding and can also be processed by extrusion and blow molding. Its physical properties vary to a great extent, depending on processing conditions.

See the Master Material Outline.

Master Material Outline

PROCESSABILITY OF ___Polymethylpentene___

PROCESS	A	B	C
INJECTION MOLDING	X		
EXTRUSION	X		
THERMOFORMING		X	
FOAM MOLDING			X
DIP, SLUSH MOLDING			
ROTATIONAL MOLDING			X
POWDER, FLUIDIZED-BED COATING			
REINFORCED THERMOSET MOLDING			
COMPRESSION/TRANSFER MOLDING			X
REACTION INJECTION MOLDING (RIM)			
MECHANICAL FORMING			X
CASTING			

A = Common processing technique.
B = Technique possible with difficulty.
C = Used only in special circumstances, if ever.

Standard design chart for ___Injection-molded polymethylpentene___

1. Recommended wall thickness/length of flow, mm
   Minimum ___0.08___ for ___7___ distance
   Maximum ___10___ for ___250___ distance
   Ideal ___2.5___ for ___400___ distance

2. Allowable wall thickness variation, % of nominal wall ___500___

3. Radius requirements
   Outside: Minimum _____ Maximum _____
   Inside: Minimum ___0.5___ Maximum _____

4. Reinforcing ribs
   Maximum thickness ___150% of nominal wall___
   Maximum height _____
   Sink marks: Yes ___X___ No _____

5. Solid pegs and bosses
   Maximum thickness ___12 mm at bottom___

Maximum weight _____

Sink marks: Yes ____X____ No _____

6. Strippable undercuts
   Outside: Yes ____X____ No _____
   Inside: Yes ____X____ No _____

7. Hollow bosses: Yes ____X____ No _____

8. Draft angles
   Outside: Minimum _____0_____ Ideal _____1°_____
   Inside: Minimum _____0_____ Ideal _____2°_____

9. Molded-in inserts: Yes ____X____
                       No _____

10. Size limitations   depends on machine size and thickness; for example, 1,000 mm each for 2.5 mm
                       thickness
    Length: Minimum _____ Maximum _____ Process _____
    Width: Minimum _____ Maximum _____ Process _____

11. Tolerances

12. Special process- and materials-related design details
    Blow-up ratio (blow molding) _____1.5 to 2.5_____
    Core L/D (injection and compression molding) _____30_____
    Depth of draw (thermoforming) _____50%_____
    Other _____

33.9   RESIN FORMS

PMP is available in pellet form.

33.10   SPECIFICATION OF PROPERTIES

See the Master Outline of Materials Properties.

33.11   PROCESSING REQUIREMENTS AND PROCESSING-SENSITIVE END PROPERTIES

PMP exhibits low melt viscosity and high flowability but is readily oriented. To produce good products, it is important to minimize orientation. Products oriented to a small degree can be obtained by molding low-viscosity molten resin at reduced injection speed and high temperature. As in the case of polypropylene, the high-pressure, high-speed molding of PMP causes a higher degree of orientation, resulting in products with low impact strength and high internal strain.

Inherently, PMP is low in melt viscosity and melt tension, which makes it difficult to extrude. In extruding film, round rods, and sheets, it is important to use an extruder with a large heating capacity. But, to obtain greater melt tension, it is very important to extrude PMP at as low a temperature as possible. If PMP is extruded with small preheating capability in the feed section of the extruder, much friction is required to create heat, and this action tends to cause excessive heat buildup.

A number of blow molded PMP products are being produced, including bottles for medical use, cultivating vessels, and laboratory ware. The clarity of the blow molded articles

MATERIAL ___ TPX MX004

PROPERTY	TEST METHOD	ENGLISH UNITS	VALUE	METRIC UNITS	VALUE
MECHANICAL DENSITY	D792	lb/ft^3	52.0	g/cm^3	0.834
TENSILE STRENGTH	D638	psi	28,275	MPa	195.
TENSILE MODULUS	D638	psi	116,000	MPa	800
FLEXURAL MODULUS	D790	psi	108,750	MPa	750
ELONGATION TO BREAK	D638	%		%	
NOTCHED IZOD AT ROOM TEMP	D256	ft-lb/in.		J/m	
HARDNESS	D785		60 (R Scale)		60 (R Scale)
THERMAL DEFLECTION T @ 264 psi	D648	°F	104	°C	40
DEFLECTION T @ 66 psi	D648	°F	185	°C	85
VICAT SOFT-ENING POINT	D1525	°F	320	°C	160
UL TEMP INDEX	UL746B	°F	239	°C	115
UL FLAMMABILITY CODE RATING	UL94		HB		HB
LINEAR COEFFICIENT THERMAL EXPANSION	D696	in./in./°F	0.65×10^{-4}	mm/mm/°C	1.17×10^{-4}
ENVIRONMENTAL WATER ABSORPTION 24 HOUR	D570	%	0.01	%	0.01
CLARITY	D1003	% TRANS-MISSION	92	% TRANS-MISSION	92
OUTDOOR WEATHERING	D1435	%		%	
FDA APPROVAL			Yes		Yes
CHEMICAL RESISTANCE TO: WEAK ACID	D543		Not attacked		Not attacked
STRONG ACID	D543		Not attacked		Not attacked
WEAK ALKALI	D543		Not attacked		Not attacked
STRONG ALKALI	D543		Not attacked		Not attacked
LOW MOLECULAR WEIGHT SOLVENTS	D543		Swells		Swells
ALCOHOLS	D543		Not attacked		Not attacked
ELECTRICAL DIELECTRIC STRENGTH	D149	V/mil	1651	kV/mm	65
DIELECTRIC CONSTANT	D150	10^6 Hertz	2.12		2.12
POWER FACTOR	D150	10^6 Hertz	5×10^{-5}		5×10^{-5}
OTHER MELTING POINT	D3418	°F	464	°C	240
GLASS TRANS-ITION TEMP.	D3418	°F	59	°C	15

depends on the surface condition of the products; it varies according to the temperature of the core and the degree of finish of the mold. At the lip part of the core, the preferable temperature is 536–572°F (280–300°C). The inner and outer die lip finish have better than a mirror finish.

33.13 SHRINKAGE

See the Mold Shrinkage MPO.

Mold Shrinkage Characteristics (in./in. or mm/mm)

MATERIAL   TPX MX004

IN THE DIRECTION OF FLOW

THICKNESS

	mm: 1.5	3	6	8	FROM	TO	#
MATERIAL	in.: 1/16	1/8	1/4	1/2			
	0.014	0.018	0.022	0.023			

	PERPENDICULAR TO THE DIRECTION OF FLOW						
	0.012	0.015	0.018	0.019			

NOTE: Mold shrinkage is approximate. It is affected by part design, mold temperature, thickness, injection pressure, packing time, cycle time, orientation, gate design, gate size, gate location, glass content, glass size, and filler content.

33.14 TRADE NAMES

TPX is the trade name of the resin produced Mitsui Petrochemical Industries Ltd., New York. Crystalor is the trade name of the resin produced by Phillips 66 Co., Bartlesville, Okla.

BIBLIOGRAPHY

1. Ital. Pat. 545,342 (June 30, 1956) (to Montecatini Italy).
2. G. Natta and co-workers, *Chimiea e Industria (Milano)* **38,** 751–765 (1956).
3. Kusui and co-workers, *Dainichi-Nippon Cables Jiho* **64,** 1–9 (April 1979).
4. Jpn. Open Pat. SHO 55-165519 (to Furukawa Electric Co.).
5. Fr. Pat. 1,442,821 (to ICI).

34

POLY(PHENYL SULFONE)

Lloyd M. Robeson

*Union Carbide Corp., Research and Development, Specialty Polymers &
Composites Div., Bound Brook, NJ 08805
Now at Air Products and Chemicals, Inc., Allentown, PA 18195*

34.1–34.4 CATEGORY, HISTORY, AND POLYMERIZATION

Poly(phenyl sulfone), a high temperature thermoplastic from the family of aromatic polysulfones, was commercialized in 1976. The structure and process for poly(phenyl sulfone),
which is known commercially under the trade name Radel R, are covered by a U.S. patent
assigned to Union Carbide (transferred to Amoco).[1-4] The specific structure and synthesis
procedure are considered proprietary information.

34.5 DESCRIPTION OF PROPERTIES

The exact performance balance of poly(phenyl sulfone) includes excellent toughness, excellent hydrolytic stability, excellent creep resistance, high heat-deflection temperature
[399°F (204°C) at 264 psi (1.82 MPa)], good electrical properties, good environmental stress-
crack resistance relative to other amorphous polymers, and low flammability based on
standard laboratory tests.

34.5.1 Mechanical Properties

Poly(phenyl sulfone) exhibits outstanding toughness for a high heat-distortion temperature
polymer. Its toughness is retained over a temperature range of −166°F (−100°C) to the
upper mechanical use range of 392°F (200°C).

After five months at 400°F (204°C), 100% of the tensile strength and 80% of the tensile
impact strength were retained. The high strength, ductility, and toughness are specific attributes that poly(phenyl sulfone) offers over a wide temperature range.

Poly(phenyl sulfone) offers excellent creep resistance even at elevated temperatures.
After 1000 hours in air at 347°F (175°C) and 1000 psi (6.9 MPa) stress, poly(phenyl sulfone)
exhibits less than 1% strain.

34.5.2 Thermal Properties

Poly(phenyl sulfone) is an amorphous polymer (as molded) with a glass-transition temperature T_g of 428°F (220°C). Its dynamic mechanical properties are illustrated in Figure 34.1.
The low temperature relaxation of poly(phenyl sulfone) of −148°F (−100°C) is believed

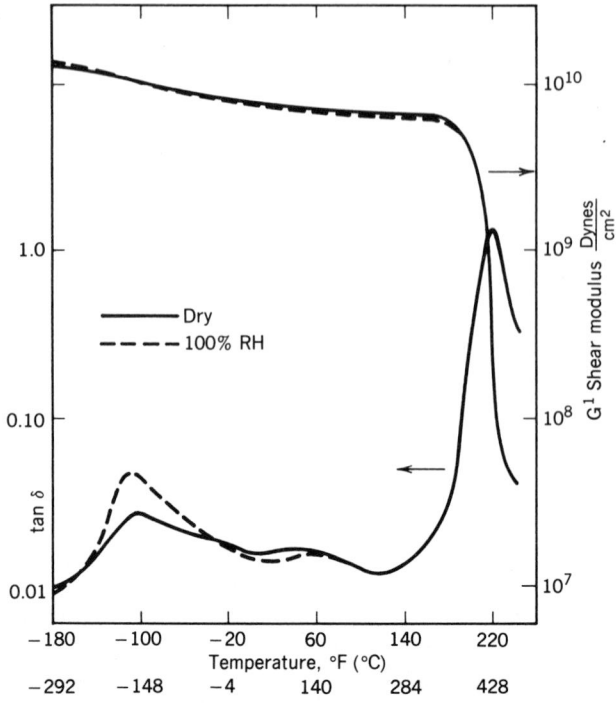

Figure 34.1.   Dynamic mechanical properties of poly(phenyl sulfone).

to be responsible for the resin's excellent toughness above this transition. The constant stiffness (shear modulus) over a broad temperature range suggests minimal property changes over a temperature range not commonly observed with thermoplastics (even those classified as engineering thermoplastics).

As with other aromatic polysulfones, the thermal stability of poly(phenyl sulfone) is excellent. The results of accelerated testing in the Amoco laboratories have projected continuous-use potential of up to 356°F (180°C). Thermogravimetric analysis results show the onset of decomposition to be above 932°F (500°C).

34.5.3   Electrical Properties

As with mechanical properties, the electrical property of poly(phenyl sulfone) data will remain reasonably constant over a broad temperature range. The resin is suitable for emerging high-performance electrical/electronic applications, including printed circuit boards requiring vapor phase or wave soldering exposure conditions.

34.5.4   Environmental Properties

Poly(phenyl sulfone) exhibits excellent hydrolytic stability (for example, to water, acid, base, and salt) even at elevated temperatures. In organic environments, poly(phenyl sulfone) may be subject to environmental stress cracking. The most aggressive environments are ketones, esters, and aromatic hydrocarbons. The least aggressive are aliphatic hydrocarbons and alcohols. Generally, poly(phenyl sulfone) exhibits better environmental stress-cracking resistance than many other amorphous thermoplastics including polysulfone, polycarbonates, and poly(ester carbonates).

34.5.5 Other Properties

The Master Outline of Materials Properties provides fuller data for the above properties, as well as the flammability test results for poly(phenyl sulfone). With respect to its flammability, data from the National Bureau of Standards Flaming Smoke Density tests [for a 1/8-in. (3.2-m) specimen] have demonstrated a low rate of smoke evolution. These data include D_s at 1.5 minutes, 0.28; D_s at 4.0 minutes, 0.37; and D_s maximum (25 min), 35.3. (It is important to note that these numerical ratings are not intended to reflect hazards presented by this or any material under actual fire conditions.)

Poly(phenyl sulfone) can be electroplated using techniques similar to those employed for polysulfone. This material property is highly significant in the manufacture of printed circuit boards.

Glass-fiber reinforcement of poly(phenyl sulfone) will yield improved load-bearing capabilities, lower creep, and improved environmental stress-crack resistance.

34.6 APPLICATIONS

The high-temperature capabilities, combustion characteristics, and excellent toughness of poly(phenyl sulfone) indicate its potential in critical safety-related applications, including safety equipment components, housings, and protective wear (such as helmets) as well as in critical mass-transit interiors (in aircraft, for example). The same property balance, along with excellent electrical properties, suggest its utility in the emerging high-performance electrical/electronic applications, including printed circuit boards and connectors. The resin's excellent hydrolytic stability and transparency suggests its utility in process equipment applications (such as sight glasses and transparent piping). The overall property combination indicates future metal, glass, or ceramic replacement where the economics of thermoplastic processing can be realized.

34.7 ADVANTAGES/DISADVANTAGES

34.7.1 Mechanical Properties

Poly(phenyl sulfone) offers an outstanding combination of toughness for a high T_g polymer. The notched impact is 12 ft-lb/in. (640 J/m). The unnotched toughness is likewise outstanding: greater than 100 ft-lb/in. (5338 J/m), at both 72°F (22°C) and −40°F (−40°C). With the high T_g, creep resistance is excellent, even up to 347°F (175°C) and in the presence of steam.

34.7.2 Thermal Properties

The high T_g of poly(phenyl sulfone) of 428°F (220°C) yields a heat-distortion temperature under load of 399°F (204°C). As a wholly aromatic polysulfone, it also exhibits excellent thermal stability [projected to be 356°F (180°C) continuous-use temperature].

34.7.3 Chemical and Hydrolytic Resistance

Poly(phenyl sulfone) offers excellent hydrolytic stability even in saturated steam and can be repeatedly steam sterilized. Excellent resistance to mineral acids, alkali, and salt solutions is also provided.

Although it has significantly better resistance to environmental stress cracking than polycarbonate or polysulfone, poly(phenyl sulfone) should not be exposed to ketones, esters, and aromatic hydrocarbons.

34.7.4   Processing

Poly(phenyl sulfone) can be processed like polysulfone by such techniques as injection molding and extrusion. In order to fill complex parts, however, high stock and mold temperatures are required. Additionally, poly(phenyl sulfone) must be dried before it is processed to prevent surface streaks or bubble formation in the molded or extruded object.

34.7.5   Dimensional Stability

As an amorphous polymer, poly(phenyl sulfone) exhibits low shrinkage and warpage. The high heat-distortion temperature demonstrates dimensional stability over a broad range of temperatures and applied loads.

34.7.6   Other Properties

The wholly aromatic structure of poly(phenyl sulfone), linked with ether and sulfone groups, allows for desired flame resistance [e.g., V-0 at 0.030-in. (0.762-mm) wall thickness] and low smoke density ratings without the addition of flame-retardant additives.

The electrical properties are consistent over a broad range of temperatures and frequency, as would be expected for a high T_g amorphous polymer with only moderate water sorption.

34.8   PROCESSING TECHNIQUES AND REQUIREMENTS

To eliminate surface streaks or part voids, poly(phenyl sulfone) must be dried to less than 0.04 wt% water prior to melt processing. Minimum drying times of 1-1/2 h at 302°F (150°C) or 3-1/2 h at 275°F (135°C) in a circulating air oven or hopper dryer are required.

Poly(phenyl sulfone) can be injection molded under conditions similar to those used with polysulfone, except 113°F (45°C) higher. Typical injection molding conditions are barrel rear, 680°F (360°C); center, 707°F (375°C); and front, 698°F (370°C); stock, 734°F (390°C); and mold, 302–329°F (150–165°C). Polysulfone is recommended as a purge material. Up to 25% regrind can be used without property changes.

Extrusion at temperatures 86–104°F (30–40°C) lower than injection molding conditions are recommended with casting rolls heated to 247°F (175°C). Thermoforming is possible with predried sheet heated to about 482°F (250°C).

34.9   RESIN FORMS

Poly(phenyl sulfone) is available in pellet form.

34.10   SPECIFICATION OF PROPERTIES

See Master Outline of Materials Properties.

Master Outline of Material Properties

MATERIAL   Poly(phenyl sulfone)

PROPERTY	TEST METHOD	ENGLISH UNITS	VALUE	METRIC UNITS	VALUE
MECHANICAL DENSITY	D792	lb/ft³	80.5	g/cm³	1.29
TENSILE STRENGTH	D638	psi	10,400	MPa	72
TENSILE MODULUS	D638	psi	310,000	MPa	2,140
FLEXURAL MODULUS	D790	psi	330,000	MPa	2,280
ELONGATION TO BREAK	D638	%	60	%	60
NOTCHED IZOD AT ROOM TEMP	D256	ft-lb/in.	12	J/m	640
HARDNESS	D785				
THERMAL DEFLECTION T @ 264 psi	D648	°F	400	°C	204
DEFLECTION T @ 66 psi	D648	°F		°C	
VICAT SOFT- ENING POINT	D1525	°F		°C	
UL TEMP INDEX	UL746B	°F		°C	
UL FLAMMABILITY CODE RATING	UL94		V-0[*]		V-0[*]
LINEAR COEFFICIENT THERMAL EXPANSION	D696	in./in./°F	3.1×10^{-5}	mm/mm/°C	5.5×10^{-5}
ENVIRONMENTAL WATER ABSORPTION 24 HOUR	D570	%		%	
CLARITY	D1003	% TRANS-MISSION		% TRANS-MISSION	
OUTDOOR WEATHERING	D1435	%		%	
FDA APPROVAL					
CHEMICAL RESISTANCE TO: WEAK ACID	D543		Excellent		
STRONG ACID	D543		Good		
WEAK ALKALI	D543		Excellent		
STRONG ALKALI	D543		Excellent		
LOW MOLECULAR WEIGHT SOLVENTS	D543		Fair		
ALCOHOLS	D543		Good		
ELECTRICAL DIELECTRIC STRENGTH	D149	V/mil	371	kV/mm	14.6
DIELECTRIC CONSTANT	D150	10^6 Hertz	3.45		3.45
POWER FACTOR	D150	10^6 Hertz	7.64×10^{-3}		7.64×10^{-3}
OTHER MELTING POINT	D3418	°F		°C	
GLASS TRANS-ITION TEMP.	D3418	°F	428	°C	220

[*] At 0.030 in.; 5V at 0.075 in. (0.75 and 1.90 mm, respectively)

389

34.11 PROCESSING REQUIREMENTS

See Master Material Outline.

Master Material Outline

PROCESSABILITY OF _____Poly(phenyl sulfone)_____

PROCESS	A	B	C
INJECTION MOLDING	X		
EXTRUSION	X		
THERMOFORMING	X		
FOAM MOLDING			X
DIP, SLUSH MOLDING			X
ROTATIONAL MOLDING			X
POWDER, FLUIDIZED-BED COATING			X
REINFORCED THERMOSET MOLDING			X
COMPRESSION/TRANSFER MOLDING		X	
REACTION INJECTION MOLDING (RIM)			X
MECHANICAL FORMING			X
CASTING			X

A = Common processing technique.
B = Technique possible with difficulty.
C = Used only in special circumstances, if ever.

Standard design chart for Poly(phenyl sulfone)

1. Recommended wall thickness/length of flow, mm
   Minimum _____0.05 in._____ for _____1–2 in._____ distance
   Maximum _____ for _____ distance
   Ideal   _____0.12 in._____ for _____12 in._____ distance

2. Allowable wall thickness variation, % of nominal wall _____25%_____ in gradual step

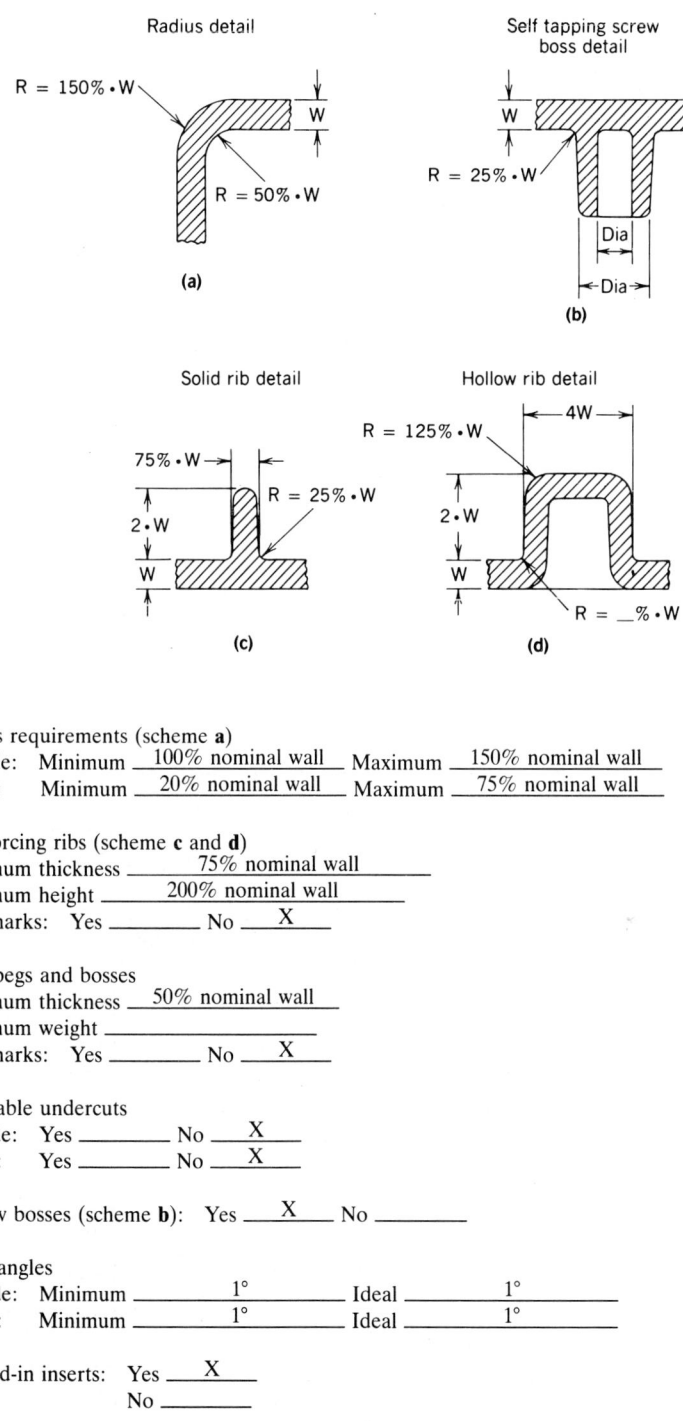

Radius detail

R = 150% • W

W

R = 50% • W

(a)

Self tapping screw
boss detail

W

R = 25% • W

Dia

Dia

(b)

Solid rib detail

75% • W

R = 125% • W

2 • W

R = 25% • W

W

(c)

Hollow rib detail

4W

2 • W

W

R = ___% • W

(d)

3. Radius requirements (scheme **a**)
   Outside: Minimum __100% nominal wall__ Maximum __150% nominal wall__
   Inside: Minimum __20% nominal wall__   Maximum __75% nominal wall__

4. Reinforcing ribs (scheme **c** and **d**)
   Maximum thickness _____75% nominal wall_____
   Maximum height _____200% nominal wall_____
   Sink marks: Yes _____ No ____X____

5. Solid pegs and bosses
   Maximum thickness __50% nominal wall__
   Maximum weight _____
   Sink marks: Yes _____ No ____X____

6. Strippable undercuts
   Outside: Yes _____ No ____X____
   Inside: Yes _____ No ____X____

7. Hollow bosses (scheme **b**): Yes ____X____ No _____

8. Draft angles
   Outside: Minimum _____1°_____ Ideal _____1°_____
   Inside: Minimum _____1°_____ Ideal _____1°_____

9. Molded-in inserts: Yes ____X____
                       No _____

10. Size limitations   None
    Length: Minimum _____ Maximum _____ Process _____
    Width: Minimum _____ Maximum _____ Process _____

11. Tolerances

12. Special process- and materials-related design details
    Blow-up ratio (blow molding) _____
    Core L/D (injection and compression molding) _____ 2.5 _____
    Depth of draw (thermoforming) _____
    Other _____

34.12 PROCESSING-SENSITIVE END PROPERTIES

Poly(phenyl sulfone) must be dried to prevent streaking or bubble formation in injection-molded or extruded parts. Minimum recommended drying times are 2-1/2 h at 302°F (150°C); 3-1/2 h at 275°F (135°C).

34.13 SHRINKAGE

Since poly(phenyl sulfone) is an amorphous polymer, the shrinkage and warpage typical for crystalline polymers is not a problem. Typical mold shrinkage is 0.007 to 0.008 in./in. (0.18 to 0.20 mm/mm).

Mold Shrinkage Characteristics (in./in. or mm/mm)

MATERIAL _____ Poly(phenyl sulfone) _____

IN THE DIRECTION OF FLOW

THICKNESS

MATERIAL	mm: 1.5	3	6	8	FROM	TO	#
	in.: 1/16	1/8	1/4	1/2			
Radel R-5000	0.007	0.007	0.007	0.007			

	PERPENDICULAR TO THE DIRECTION OF FLOW					
Radel R-5000	0.007	0.007	0.007	0.007		

NOTE: Mold shrinkage is approximate. It is affected by part design, mold temperature, thickness, injection pressure, packing time, cycle time, orientation, gate design, gate size, gate location, glass content, glass size, and filler content.

Standard Tolerance Chart

| PROCESS | Injection molding | MATERIAL | Poly(phenyl sulfone) |

Values are based on a nominal 1/8 in. (3.2 mm) wall thickness and given in plus or minus in./in. and mm/mm.
FINE tolerance – Special care and added cost.
COMMERCIAL tolerance – Normal care required.
COARSE tolerance – Minimal care required, lowest cost.

Tolerance	A	B	C	D	E	F	G	H	
FINE	0.002	0.001	0.003	0.001	0.001	0.002	0.001	0.001	in./in.
									mm/mm
COMMERCIAL	0.003	0.003	0.004	0.003	0.003	0.003	0.003	0.003	in./in.
									mm/mm
COARSE	0.005	0.007	0.007	0.005	0.005	0.006	0.006	0.005	in./in.
									mm/mm

34.14 TRADE NAMES

The trade name of poly(phenyl sulfone) is Radel R. It is available in two grades, Radel R-5000 (transparent) and Radel R-5010 (opaque).

BIBLIOGRAPHY

1. R.N. Johnson and co-workers, *J. Polym. Sci. Pt A-1,* **5,** 2375 (1967).
2. Canadian Pat. 847,963 (July 28, 1970), R.A. Clendinning and co-workers (to Union Carbide Corp.).
3. G.T. Kwiatkowski and co-workers, *paper presented at High-Performance Plastics Symposium,* NATEC-SPE, Cleveland, Ohio, Oct. 5–7, 1976.
4. U.S. Pat. 4,108,837 (Aug. 22, 1978), R.N. Johnson and A.G. Farnham (to Union Carbide Corp.).

35

POLY(PHENYLENE ETHER) (PPE)

Brenda A. Bartges

Borg-Warner Chemicals, Inc. Technical Center Washington, WV 26181

35.1 STRUCTURES

Poly(phenylene ether) homopolymer
Poly (2,6-dimethyl-1,4-phenylene ether)
Poly (oxy-(2-6-dimethyl-1,4-phenylene))

Poly(phenylene ether) copolymer

Polyphenylene ether is frequently referred to as poly(phenylene oxide). However, since its functionality is that of an ether, this section will refer to both its homopolymer and copolymer as polyphenylene ethers. Modified poly(phenylene ether) products are used in most commercial applications. These blends contain 20–80% PPE (most commonly, 40–50%) and high impact polystyrene (HIPS).

35.2 CATEGORY

Modified PPE materials are rigid, amorphous engineering thermoplastics. Their unique compatibility with polystyrene (PS), particularly HIPS, results in a wide range of high-temperature, tough, dimensionally stable products.

35.3 HISTORY

In 1959, Hay and co-workers of General Electric Co. first documented that certain 2,6 disubstituted phenols yielded high molecular weight aromatic polyethers upon oxidative polymerization.[1,2]

In June 1966, modified PPE was first introduced as Noryl-1 by General Electric Co.; glass-reinforced blends (Noryl-2 and Noryl-3) entered the marketplace in June 1967. UL94 V-1 recognition (as a self-extinguishing nondrip material) was achieved in early 1968. Glass-fiber-filled PPE products were first produced in March 1969. In September of 1972, the first structural foam product (Noryl FN-215) was introduced; extrusion products (Noryl EN-212 and EN-265) appeared in September 1974. In 1982, Borg-Warner Chemicals introduced its line of modified PPE resins under the trade name Prevex.

Modified PPE production for 1970 was 29 million pounds; for 1980, 132 million pounds; and for 1985, 200 million pounds.[3] While this represents less than 1% of the total plastic consumption for each year, this "specialty" engineering thermoplastic is expanding into new application areas.

35.4 POLYMERIZATION

PPE resin is formed in a liquid-phase stirred reactor at 77–95°F (25–35°C) and atmospheric pressure. Monomer 2,6-xylenol in toluene solvent is mixed with catalyst at a 100:1 monomer:catalyst ratio. The catalyst is a cuprous salt (cuprous chloride or cuprous iodide) and amine complex. The amine portion is a mixture of primary and secondary amines, or straight chain dialkylmonamines (for example, n-dibutyl-amine). Oxygen is sparged into the mixture to generate the exothermic reaction. After the copper residue is removed and the polymer is precipitated with heptane or methanol, it is filtered.

Some 97% of this high molecular weight polymer's theoretical yield is recoverable with this process.[1]

Since the position para to the hydroxy group is the only available active site, the oxidative coupling of 2,6-xylenol produces a linear homopolymer.[4]

A linear copolymer is produced when 5–10% of the monomer is replaced with 2,3,6-trimethyl phenol.[5]

* Unsubstituted monomer reaction permits polymerization at random ortho and para sites, which produces a hard, insoluble, cross-linked material. Large monomer substituents such as ethyl, propyl, or chlorine groups yield a low molecular weight polymer upon reaction.

Unmodified PPE resins are high-temperature polymers with glass-transition temperatures of approximately 410°F (210°C). They are very difficult to process, even at melt temperatures of 575–600°F (302–316°C). In contrast to the high softening point, high melt viscosity of PPE, polystyrene has a low softening point and melt viscosity. PPE and polystyrene are thermodynamically compatible and can be melt-mixed in an extruder in all proportions. The resulting blends exhibit lower viscosity and increased processibility than plain PPE. They display a single glass-transition temperature (between HIPS at 212°F [100°C] and PPE at 410°F [210°C]) at all ratios, a designation of a truly compatible system. This feature permits a blend's viscosity and impact or toughness to be tailored to specific requirements. The addition of HIPS (rubber-modified polystyrene) has been shown to give the PPE blend Izod impact values higher than that of PPE/PS blends or of either component, while retaining complete miscibility.[5] It is these PPE/HIPS blends that are commercially available and will be discussed in further detail.

Typically, the higher the PPE/HIPS ratio, the greater the heat-distortion temperature (HDT) and cost of the product. For example, 80% of the raw material cost is contributed by the phenol, HIPS, and flame-retardant additives (for example, triaryl phosphate esters). It then follows that higher heat-distortion, flame-retardant, or increased modulus materials demand a higher price within the typical range of $1.36–2.36 per pound in 1988 for modified poly(phenylene ether) blends.[3]

High-strength mineral-filled and structural-foam grades are typically priced in the lower portion of this range; glass-reinforced grades fall in the upper portion. The latter provide higher strength advantages at a higher cost because of raw materials and processing. Lower cost mineral-filled grades offer enhanced strength and modulus with good surface properties. The economies of structural-foam grades are further improved with density reduction, by requiring less material to fill a part.

35.5 DESCRIPTION OF PROPERTIES

Modified PPE alloys provide high strength and heat and moisture resistance. Their physical properties are dependent on the additive package used as well as the PPE/HIPS ratio; however, certain generalizations can be made.

Increased PPE content results in blends with improved thermal properties. When the PPE ratio is increased, the material's HDT can range from 180–300°F (82–149°C). Between modified PPE blends and other polymers with lower deflection temperatures under load (DTULs), the room temperature differences in creep rate are small; however, at higher temperatures, PPE/HIPS retains its load-bearing properties better. Not load sensitive, modified PPE blends can outperform nylons and acetals, which suffer dramatic decreases in deflection temperatures under higher stresses. Because of their lower coefficient of thermal expansion, these blends perform well in applications with stringent part tolerances that must withstand temperature fluctuations and stresses.

PPE alloys react much like polycarbonate (PC) and acetal materials to short-term loads. Their tensile yield strength and modulus values range from 6000 to 16,000 psi (41–110 MPa) and 330,000 to 400,000 psi (2350–2760 MPa), respectively, at room temperature. Modified PPE blends are relatively stable and exhibit low creep values under long-term continuous load.

Generally speaking, impact resistance properties are dependent on the ductility and tensile strength of a polymer system. Modified PPE is no exception and has the added benefit of a low degree of notch sensitivity. Although PC has a far greater unnotched Izod value, notched values of some PPE blends approach those of PC at larger thicknesses. They can also outperform ABS, acetal, and unmodified nylons with notched Izod impact values ranging from 2 to 10 ft-lb/in. (107 to 534 J/m).

Modified PPE blends make excellent insulators, thanks to their low loss and dissipation factors over broad temperature and frequency ranges. High humidity has little effect on these properties.

The dimensional stability of modified PPE materials at elevated temperatures and stresses holds environmentally, too. Strong alkali and bases, detergents, and many foodstuffs, as well as hot water, do not alter their properties. The water absorption of modified PPE resins is low, which aids in processing as well as maintaining dimensional accuracy in humid environments. However, chlorinated hydrocarbons and aromatic hydrocarbons will dissolve the blend. This is a function of time of contact and degree of strain. At high levels of molded-in stress, cracking or crazing is apparent with exposure to esters, oils, greases, or alcohols.

Different additive packages are available in PPE/HIPS blends. Although an opaque resin, a wide range of colors is available, including white and light colors. Obviously, these blends cannot be used where light transmission is required. A color shift accompanies ultraviolet (UV) exposure (sunlight or fluorescent); thus, such applications are not recommended unless an appropriate stabilizer package is included. Resins containing carbon black are suitable for outdoor usage, and various colors are offered for fluorescent light exposure. Although UV affects color, it is not detrimental to the properties, as is the case with such polymers as ABS. UL94 classifications up to 5V have been achieved with modified PPE blends. The actual rating is dependent on the particular grade and additive package.

For applications requiring increased modulus and decreased mold shrinkage, glass- and mineral-filled modified PPE resins are available. A three-fold increase in modulus can be produced with loadings as high as 30%. Such fillers are easily worked into the modified PPE material, but ductility and elongation properties suffer and impact decreases.

35.6   APPLICATIONS

Typical application areas for modified PPE materials include communications, computer, and business equipment and housings; appliances; electrical and medical equipment; and automotive components. Television cabinetry; telephone and fiber optic connectors; and computer housings, video display terminals, dust shields, control panels, and printer bases are often made of modified PPE products. Case components and grills of air conditioners; pump housings for dishwashers and clothers washers; and handles, brackets, and motor covers of vacuum cleaners are examples of modified PPE in appliances. The excellent electrical properties of modified PPE materials make them the material of choice in many wall plates, outlet boxes, current-carrying wiring devices, connectors, coil bobbins, and switch applications. Use in automotive instrument panels, wheel covers, and fuse blocks is also common. Metal-plated modified PPE performs well in EMI/RFI shielded enclosures used in the automotive, plumbing, and appliance industries. See Figures 35.1–35.3.

35.7   ADVANTAGES/DISADVANTAGES

Modified PPE materials offer a combination of excellent dimensional stability, heat resistance, and toughness. High DTULs, good creep resistance, and chemical resistance to strong acids, bases, and water are characteristic of these blends. Because of their low moisture absorption, PPE/HIPS resins require only minimal drying prior to processing and have reasonable flow for high temperature materials.

These blends are suitable for thermoforming, extrusion, blow molding, and injection-molding processes, with the latter being most common. Structural-foam materials may be molded on standard injection molding equipment, adding to their cost effectiveness.

Modified PPE blends insulate well, as a result of their inherent low dielectric constant

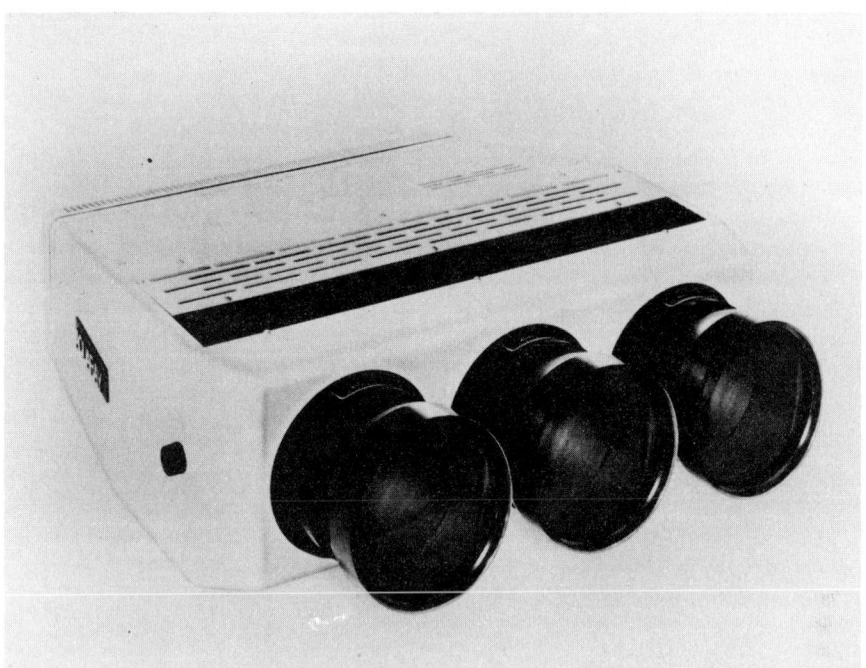

Figure 35.1. Prevex engineering polymer, grade BJA, was selected for the Aquaray-RGB color video projector housing because of its structural and electrical properties, low flammability, price, and availability. Photograph by Borg-Warner Chemicals, Inc.

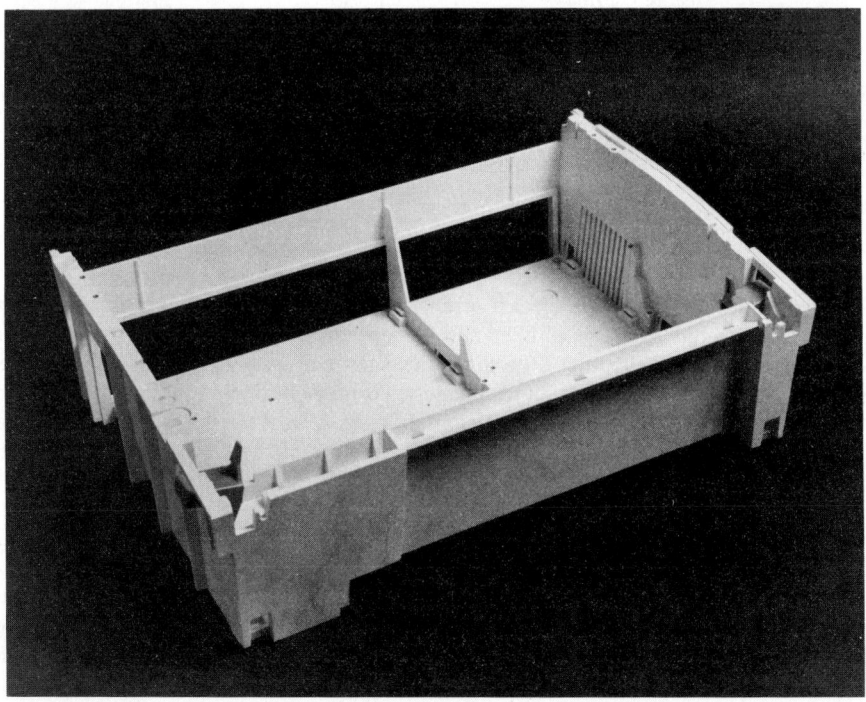

Figure 35.2. Prevex modified PPE polymer is used in this chassis for Telex Computer Products' 187 printer. Photograph by Borg-Warner Chemicals, Inc.

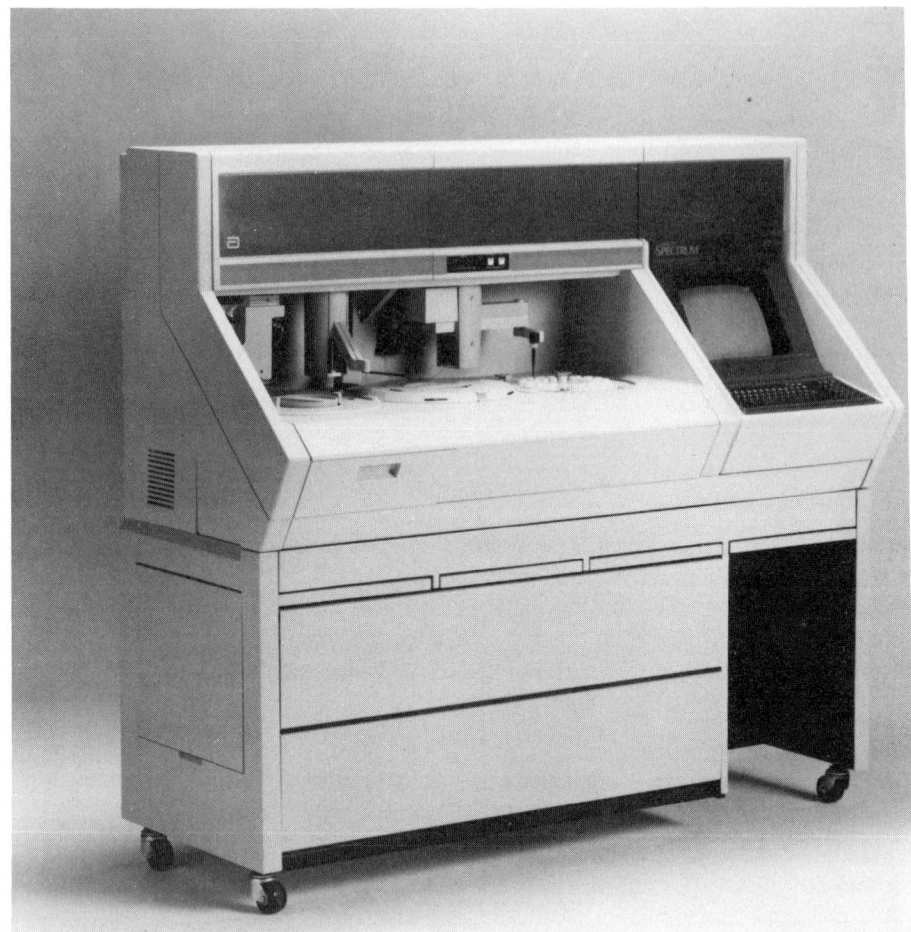

Figure 35.3. The exterior housing panels of the Spectrum Chemistry Analyzer utilize the excellent flow and strength properties of Prevex modified PPE. Photograph by Borg-Warner Chemicals, Inc.

and dissipation factor and high dielectric strength. These materials exhibit good creep resistance and are readily available in filled and flame-retardant grades with UL recognition.

Most disadvantages of modified PPE are processing related: high process temperatures, high viscosity, and higher cost than some polymers. Postfinishing techniques, such as painting or hot stamping, require systems different from those used with polymers such as ABS, but these are, nonetheless, frequently used. The surface gloss of these materials is moderate; if it is not coated, it is usually lower than that of ABS, HIPS, or PC.

The application environment must be taken into account when materials are selected. Although chemically resistant to strong acids and alkali, the hydrocarbons (especially chlorinated hydrocarbons) and aromatic organic solvents will soften and dissolve the blend. Also, color shift accompanying UV exposure complicates color matching. Grades with UV stabilizers are offered by producers for such applications. Color matching can be further limited by the base color of the PPE resin. However, light-colored and white products are available.

35.8   PROCESSING TECHNIQUES

See the Master Material Outline.

Master Material Outline

PROCESSABILITY OF ___Modified PPE Resins___

PROCESS	A	B	C
INJECTION MOLDING	X		
EXTRUSION	X		
THERMOFORMING	X		
FOAM MOLDING	X		
DIP, SLUSH MOLDING			
ROTATIONAL MOLDING			
POWDER, FLUIDIZED-BED COATING			
REINFORCED THERMOSET MOLDING			
COMPRESSION/TRANSFER MOLDING			
REACTION INJECTION MOLDING (RIM)			
MECHANICAL FORMING			
CASTING			

A = Common processing technique.
B = Technique possible with difficulty.
C = Used only in special circumstances, if ever.

Standard design chart for ___Modified PPE resins___

1. Recommended wall thickness/length of flow
   Minimum ___0.04 in.___ for ___8–10 in.___ distance
   Maximum ___0.200 in.___ for ___12–14 in.___ distance
   Ideal ___0.125 in.___ for ___8–10 in.___ distance

2. Allowable wall thickness variation, % of nominal wall ___25%___

3. Radius requirements (scheme **a**)
   Outside: Minimum ___0.015 R___ Maximum ___Dependent on nominal wall thickness___
   Inside: Minimum ___(sharp) 0.015 R___ Maximum ___Dependent on nominal wall thickness___

4. Reinforcing ribs (scheme **c** and **d**)
   Maximum thickness ___50–70% wall thickness___
   Maximum height ___3 times wall thickness___
   Sink marks: Yes ___Slight 70%___ No ___50%___

Radius detail

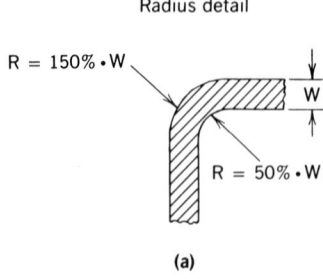

R = 150% · W

W

R = 50% · W

(a)

Self tapping screw
boss detail

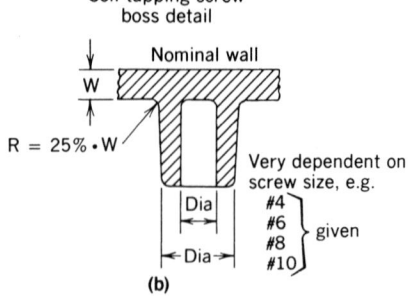

Nominal wall

W

R = 25% · W

Very dependent on
screw size, e.g.

Dia

#4
#6
#8
#10

} given

Dia

(b)

Solid rib detail   Hollow rib detail

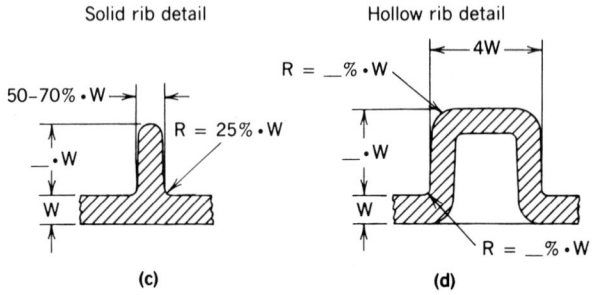

50–70% · W →

_ · W

W

R = 25% · W

(c)

R = _% · W

4W

_ · W

W

R = _% · W

(d)

5. Solid pegs and bosses
   Maximum thickness ____Peg, 50–70% of wall; boss, 50–70% (sink)____
   Maximum weight _____4 times nominal wall thickness_____
   Sink marks: Yes _____ No ___X___ if 70% rule is observed

6. Strippable undercuts
   Outside: Yes _____ No ___X___
   Inside: Yes ___X___ No _____

7. Hollow bosses (scheme **d**): Yes ___X___ No _____

8. Draft angles
   Outside: Minimum _____1°_____ Ideal ___3° minimum___
   Inside: Minimum _____½°_____ Ideal ___1° minimum___

9. Molded-in inserts: Yes ___X___ (for brass, preheating recommended)
                       No _____

10. Size limitations Dependent on machine size (hydraulic pressure)
    Length: Minimum _____ Maximum _____ Process _____
    Width: Minimum _____ Maximum _____ Process _____

11. Tolerances

12. Special process- and materials-related design details
    Blow-up ratio (blow molding) _____20–50%_____
    Core L/D (injection and compression molding) _____20 to 30:1_____
    Depth of draw (thermoforming) _____
    Other _____

35.9 RESIN FORMS

Modified PPE resin is commercially available in pellet form. A variety of HDTs and colored grades are offered. Those with higher HDTs usually contain a higher percentage of PPE resin. A wide range of flame-retardant and UV stabilized grades is also available, as noted above.

Modifications include reinforced high-strength grades, which are glass- or mineral-filled. The modulus and strength of these materials are greater than the unfilled grades; acceptable impact and surface aesthetics are maintained. Structural-foam grades provide improved rigidity and dimensional stability in parts without sink. Foamed products also decrease the weight of a part through density reduction techniques. Extrusion and blow molding grades are also commercially available.

Special unfilled formulations are offered for such postmolding requirements as sputtering or electroplating. These are used for decorative and load-bearing parts in the appliance and automotive industries. Improved chemical resistance to solvents and high heat-distortion properties are achieved with polyblends of PPE and polyamides.

35.10 SPECIFICATION OF PROPERTIES

See the Master Outline of Materials Properties.

35.11 PROCESSING REQUIREMENTS

Because of the high melt strength of modified PPE, extrusion, blow molding, and thermoflowing processes are suitable; however, injection molding is the process most commonly used for modified PPE resins.

Since splay does not affect physical properties in these characteristically low moisture-absorbing materials, drying before molding is optional for nonappearance parts. To eliminate splay, the standard recommended drying times are 2–4 h at 180–220°F (82–104°C), depending on the HDT.

For processing these high-temperature polymers, improved molding machines with L:D ratios of 20:1 and compression ratios of 2:1 to 3:1 are preferred. Typical molding parameters are screw speed, 25–75 rpm; back pressure, 50–100 psi (0.3–0.7 MPa); injection time, 2–15 s; holding pressure 0.5–0.7 times the injection pressure; and mold temperature, 100–150°F (38–66°C). Depending on the PPE:HIPS ratio, processing temperatures will range from 400–600°F (204–316°C).

As the amount of PPE in the blend increases, so does viscosity; the flow decreases and higher processing temperatures are required. The blend exhibits high thermal stability (even FR codes) and withstands high temperatures and residence times. This characteristics permits melt temperature adjustments to fill long-flow or complex tooling. Generally, regrind in virgin pellets is 25% for full performance parts; however, 100 percent regrind is utilized in many instances. Structural-foam grades are available that offer density reductions of 10–20%. The use of chemical blowing agents such as azodicarbonamide and sodium bicarbonate reduces the amount of material required to fill a part. Nominal wall thickness of nonload-bearing foam applications can be 0.150 in., as compared to 0.250 in. for solid parts. To achieve optimum flow and weld-line strength, melt temperatures from 500–580°F (260–304°C) are recommended with a mold temperature of 80–180°F (27–82°C), depending on the rate of production and design tolerances.

Pipe, rod, slab, sheet, and film materials can all be produced using extrusion techniques. Conventional extrusion equipment with barrels (L:D of 24:1 or 30:1), compression ratios

MATERIAL ___ Modified PPE Resins, 30% Glass-Filled

PROPERTY	TEST METHOD	ENGLISH UNITS	VALUE	METRIC UNITS	VALUE
MECHANICAL DENSITY	D792	lb/ft^3	79.2	g/cm^3	1.27
TENSILE STRENGTH	D638	psi	17,000	MPa	117
TENSILE MODULUS	D638	psi	1200	MPa	8
FLEXURAL MODULUS	D790	psi	110,000	MPa	758
ELONGATION TO BREAK	D638	%	3–5	%	3–5
NOTCHED IZOD AT ROOM TEMP	D256	ft-lb/in.	2–3	J/m	107–160
HARDNESS	D785		R115		R115
THERMAL DEFLECTION T @ 264 psi	D648	°F	300	°C	149
DEFLECTION T @ 66 psi	D648	°F	317	°C	158
VICAT SOFT-ENING POINT	D1525	°F		°C	
UL TEMP INDEX	UL746B	°F	194–221	°C	90–105
UL FLAMMABILITY CODE RATING	UL94		94HB		94HB
LINEAR COEFFICIENT THERMAL EXPANSION	D696	in./in./°F	1.4×10^{-5}	mm/mm/°C	2.5×10^{-5}
ENVIRONMENTAL WATER ABSORPTION 24 HOUR	D570	%	0.06	%	0.06
CLARITY	D1003	% TRANS-MISSION		% TRANS-MISSION	
OUTDOOR WEATHERING	D1435	%		%	
FDA APPROVAL					
CHEMICAL RESISTANCE TO: WEAK ACID	D543		Not attacked		Not attacked
STRONG ACID	D543		Not attacked		Not attacked
WEAK ALKALI	D543		Not attacked		Not attacked
STRONG ALKALI	D543		Not attacked		Not attacked
LOW MOLECULAR WEIGHT SOLVENTS	D543				
ALCOHOLS	D543				
ELECTRICAL DIELECTRIC STRENGTH	D149	V/mil	550	kV/mm	22
DIELECTRIC CONSTANT	D150	10^6 Hertz	2.93		2.93
POWER FACTOR	D150	10^6 Hertz	9×10^{-4}		9×10^{-4}
OTHER MELTING POINT	D3418	°F		°C	
GLASS TRANS-ITION TEMP.	D3418	°F		°C	

MATERIAL ___ Modified PPE Resins

PROPERTY	TEST METHOD	ENGLISH UNITS	VALUE	METRIC UNITS	VALUE
MECHANICAL DENSITY	D792	lb/ft³	65.5–68.6	g/cm³	1.05–1.10
TENSILE STRENGTH	D638	psi	6,000–16,000	MPa	41–110
TENSILE MODULUS	D638	psi	34,000 to 40,000	MPa	234–276
FLEXURAL MODULUS	D790	psi	32,000 to 40,000	MPa	221–276
ELONGATION TO BREAK	D638	%	2–3	%	2–3
NOTCHED IZOD AT ROOM TEMP	D256	ft-lb/in.	2–10	J/m	107–534
HARDNESS	D785		R115–120		R115–120
THERMAL DEFLECTION T @ 264 psi	D648	°F	180–300	°C	82–149
DEFLECTION T @ 66 psi	D648	°F	200–280	°C	93–138
VICAT SOFT-ENING POINT	D1525	°F	482–590	°C	250–310
UL TEMP INDEX	UL746B	°F	176–221	°C	80–105
UL FLAMMABILITY CODE RATING	UL94		HB–5V		HB–5V
LINEAR COEFFICIENT THERMAL EXPANSION	D696	in./in./°F	$3.3–7.7 \times 10^{-5}$	mm/mm/°C	$6–14 \times 10^{-5}$
ENVIRONMENTAL WATER ABSORPTION 24 HOUR	D570	%	0.06–0.12	%	0.06–0.12
CLARITY	D1003	% TRANS-MISSION	Opaque	% TRANS-MISSION	Opaque
OUTDOOR WEATHERING	D1435	%		%	
FDA APPROVAL			Yes		Yes
CHEMICAL RESISTANCE TO: WEAK ACID	D543		Not attacked		Not attacked
STRONG ACID	D543		Not attacked		Not attacked
WEAK ALKALI	D543		Not attacked		Not attacked
STRONG ALKALI	D543		Not attacked		Not attacked
LOW MOLECULAR WEIGHT SOLVENTS	D543				
ALCOHOLS	D543				
ELECTRICAL DIELECTRIC STRENGTH	D149	V/mil	400–700	kV/mm	15.7–27.6
DIELECTRIC CONSTANT	D150	10⁶ Hertz	2.7–2.9		2.7–2.9
POWER FACTOR	D150	10⁶ Hertz	0.001–0.0025		0.001–0.0025
OTHER MELTING POINT	D3418	°F		°C	
GLASS TRANS-ITION TEMP.	D3418	°F	212–302	°C	100–150

of 2 to 3:1, moderately deep metering flights, and gradual transitions enable modified PPE materials to be processed very efficiently. A typical material temperature is approximately 550°F (288°C), but this will vary with the rate, design, and material grade selected for the application. To allow for proper draw, shaping, and surface finishing, take-off operations must be adjusted for the blend's high solidification temperature.

Modified PPE blends can be blow-molded using the newer equipment designed for engineering thermoplastics. Generally, low shear screws of 2.0:1 to 2.5:1 compression ratio are recommended, with a L:D ratio of 20:1 or 24:1. The materials should be dried 3–4 h at 180–220°F (82–104°C) before processing. Extruder and accumulator temperatures should be set to achieve a melt temperature between 410–475°F (210–247°C), depending on the actual blend being processed. The surface definition and aesthetics can be modified by adjusting the melt temperature, air pressure, and mold temperature.

Postfinishing techniques common to thermoplastics can be satisfactorily performed on modified PPE parts. However, to obtain good adhesion of these heat-resistant parts, an extra step is necessary. Before painting, it is sometimes recommended that a urethane, alkyd, or acrylic-based primer be applied. When the part is to be electroplated, it is necessary to include a pre-activator step before the catalyst is deposited on the part surface. Sputtering, vacuum metallizing, and hot stamping can also be employed for both decorative and performance purposes. Among the bonding techniques that are practicable are solvent, ultrasonic heat, and vibrational welding; thread-cutting and thread-forming mechanical fasteners and inserts; and such adhesives as acrylic, epoxy, and hot melt.

35.12/13 PROCESSING-SENSITIVE END PROPERTIES AND SHRINKAGE

Molding conditions affect the shrinkage, gloss, and physical property characteristics of modified PPE blends to a limited degree. With these materials such changes are brought about by large molding variations; in general, they are less sensitive to processing conditions than many polymers. Differences are caused by changing the flow of the polymer: this is done by increasing melt temperature or increasing injection pressure. The viscosity decreases 10–15% with a melt temperature increase of 100°F (38°C) or an injection pressure increase of 1000 psi (7 MPa). Use of larger wall sections also improves the flow of the resin through the part (including glass-filled materials).

Modified PPE blends exhibit uniform shrinkage; variations in molding conditions produce uniform changes in mold shrinkage. Mold shrinkage decreases with increasing injection pressure and/or melt temperature, as the polymer chains have sufficient time and energy to realign into less stressful configurations. However, a decrease of only 0.001–0.002 in./in. is seen when the melt temperature is increased 100°F (56°C).

The gloss of modified PPE blends is highly sensitive to molding conditions, and a finished part can be produced to specification by properly adjusting them. The gloss value increases with increasing injection pressure and/or mold temperature. For example, increasing the mold temperature from 125 to 175°F (52 to 80°C) can raise the gloss value from 25° to as high as 60°.

Physical properties are not as sensitive to processing conditions as gloss and are only affected when extreme deviations from the norm occur. Adjustment of injection pressure to prevent flash has no affect on properties, as examination of tensile impact strength and tensile elongation at increasing injection pressure has shown. Moderate changes in mold temperature [±30°F (17°C)] can improve the appearance of parts by minimizing molded-in stresses, but shrinkage and cycle times remain unchanged. If cycle times are short, and shot size and barrel capacity well correlated, modified PPE resins can withstand cylinder temperatures of 600°F (316°C) without degrading. However, this is not recommended as a standard processing procedure.

Mold Shrinkage Characteristics

MATERIAL   Modified PPE

IN THE DIRECTION OF FLOW

THICKNESS

MATERIAL	mm: 1.5 in.: 1/16	3 1/8	6 1/4	8 1/2	FROM	TO	#
Unfilled modified PPE		0.005–0.007	0.006–0.009				
30% glass-filled		0.001	0.003–0.004				
Platable		0.008	0.006				

		PERPENDICULAR TO THE DIRECTION OF FLOW					
Unfilled modified PPE		0.005–0.007	0.006–0.009				
30% glass-filled		0.001	0.003–0.004				
Platable		0.008	0.006				

NOTE: Mold shrinkage is approximate. It is affected by part design, mold temperature, thickness, injection pressure, packing time, cycle time, orientation, gate design, gate size, gate location, glass content, glass size, and filler content.

35.14 TRADE NAMES

Prevex, Borg-Warner Chemicals, Inc., Parkersburg, W.V.; Noryl General Electric Co., Pittsfield, Mass.

BIBLIOGRAPHY

1. U.S. Pat. 3,306,874 (1967) (to General Electric Co.).
2. U.S. Pat. 3,306,875 (1967) (to General Electric Co.).
3. *PPE,* Chemical Systems, Inc., pp. 315–319.
4. U.S. Pat. 4,463,164
5. M. Kramer, *Appl. Polym. Symp.* **13,** 227 (1970).

36

POLY(PHENYLENE OXIDE) (PPO)

George Feth

General Electric Co., Noryl Ave. Selkirk, NY 12158

36.1 INTRODUCTION

PPO is the abbreviated chemical form for polyphenylene oxide resin. The full chemical name of PPO is poly(1,4-(2,6-dimethylphenyl) ether).

36.2 CATEGORY

PPO is a thermoplastic resin. Currently there are more than 30 grades of PPO available for a wide variety of plastics applications.

36.3 HISTORY

PPO was discovered by John Hay in 1956; it was developed commercially by General Electric Co. under the trade name Noryl resin in 1965–1966. General Electric marketing executives realized that the new material would make significant contributions in a wide variety of applications. Areas of early application included business machines, liquid handling, and telecommunications.

36.4 POLYMERIZATION

The chemical formulation is based on the oxidative coupling of substituted phenols and the elimination of a molecule of water. PPO resin is usually alloyed with polystyrene. It acts as a base for a large family of resins with typical heat-distortion temperatures ranging from 180°F (82°C) to more than 300°F (149°C).

Currently (1988), injection-molding grades of PPO are priced at $1.48–1.80 per pound; reinforced grades are priced at $2.14–2.36 per pound.

409

36.5 DESCRIPTION OF PROPERTIES

Because of the large number of resin formulations used to produce filled, unfilled, or flame-retarded versions of PPO, resin properties can vary significantly.

Generally speaking, however, the chemical resistance characteristics of PPO are fairly consistent. PPO resins are noted for their resistance to hyrolysis in acids, bases, salts, and hot water. They may be attacked by certain hydrocarbons, esters, ketones, amines, and halogenated compounds. PPO resins are slightly polar and contain aromatic ring structures.

Although most PPO resins contain polystyrene, normal solvents for styrenic materials such as methyl ethyl ketone (MEK), tetrahydrofuran (THF), or cyclohexanone are not considered solvents of PPO. The resins are soluble in some halogenated solvents such as trichloroethylene (TCE) or chloroform, only partly soluble in dichloromethane, and only slightly soluble in carbon tetrachloride, dichloroethane, and tetrachloroethane. Xylene and toluene are fairly good solvents but are slow to evaporate. Most often, mixtures of tri-chloroethylene and dichloromethane are used as solvents, depending on the speed of evap-oration needed.

The physical, thermal, and environmental properties of PPO resins vary from grade to grade. In selecting a particular grade, it is important to consider end-use temperature, time, load, and chemical environment, as these factors influence the ability to retain physical properties. Such environmental factors as heat and ultraviolet radiation can also have an effect. PPO is available in six standard grades that feature heat resistance, high impact strength, strong tensile and flexural properties at elevated temperatures, and low specific gravities. All standard grades and several specialty grade resins are listed by Underwriters Laboratories for flame retardancy with ratings from HB to V0/5V.

There are four standard grades of glass-filled PPO, two of which are flame retarded. In addition to glass fillers, various fibers, minerals, and particulate fillers are used with PPO resins to create desired properties. Unlike other resins that lose impact strength with the addition of fillers, PPO "high-strength" grades retain excellent Gardner values of 250 in.-lb (28 J) on a 1/8 in. (3.2 mm) thickness and notched Izod values of 3–4 ft-lb/in. (160–214 J/m).

Some filled grades are rated either V-0 or V-1. Glass-filled PPO resins have coefficients of expansion similar to those of aluminum, zinc, or brass.

There are five standard extrusion grades of PPO, ranging from 185 to 265°F (85–130°C) in heat-distortion temperature. Many of these grades are similar to the standard injection grades in physical properties, except for minor quality control specifications.

A foamable grade PPO resin offers excellent dimensional stability, high impact strength, rigidity, UL ratings of V0 and 5V, and a heat-distortion temperature of 205°F (96°C) at 66 psi (0.5 MPa). The resin can be processed with either chemical or nitrogen foaming systems and in high or low pressure injection machines.

36.6 APPLICATIONS

PPO has found a wide range of applications in appliances, electrical, and telecommunications products. For example, one of the fastest growing markets for PPO resins is in such items as computer terminals, printers, modems, typewriter bases, video games, and copiers.

Because of its favorable flammability and high-amp ignition ratings, PPO resins are used extensively in outlet boxes, wire terminal boards, load center boxes, coil bobbins, trans-former cores and cases, house-wired smoke detectors, intrusion alarms, motor covers, and switches. High tensile strengths from 5700 to 17,800 psi (39–123 MPa) and flexural moduli from 250,000 to 1,100,000 psi (1724–7585 MPa) allow dynamic strain rates of up to 8% for use in snap-fit fingers for electrical connectors and enclosure covers. Izod impact strengths of over 2.0 ft.lb/in (107 J/m) at −40°F (−40°C) and from 5.0 to 10.0 ft.lb/in (267–534 J/m) at room temperature can reduce shipping damage and packing costs.

A number of PPO grades are designed primarily for the automotive market. In general, these automotive grades are not flame retardant, but they do possess excellent low-temperature impact resistance, heat-distortion temperatures up to 300°F (149°C) and excellent mechanical strength at a low cost per cubic inch. These automotive grades allow designers flexibility while reducing weight and maintaining strength. The automotive resins have a specific gravity of 66 lb/ft^3 (1.06 g/cm^3) and tensile strengths from 5700 to 10,000 psi (39–69 J/m). Including the standard grades, there are currently more than 19 PPO resins from which to choose for automotive applications.

36.7/.8 ADVANTAGES/DISADVANTAGES AND PROCESSING TECHNIQUES

PPO resins offer the designer a combination of attractive properties including heat resistance, good impact strength, strong tensile and flexural properties at elevated temperatures, and a wide processing range. They also feature low specific gravity and low moisture absorption rates. Thermal aging and outdoor weathering can cause some loss of impact properties, but the tensile and flexural properties change relatively little. On ultraviolet exposure (either simulated sunlight or indoor fluorescent), the resins will yellow or darken somewhat depending on the initial color and the length of exposure. But UV stabilized grades are available that are more stable by a factor of at least three. High-energy radiation such as gamma or x-rays causes little loss of properties, even at high dosages. Molded articles can be sterilized by autoclave, gas, or irradiation, without harm to the material.

Bonding agents such as cyanoacrylate, acrylics, epoxies, silicones, polyurethanes, and polymeric hot melt adhesives are available for PPO resins. However, some formulation variations in adhesives act as aggressive chemical agents rather than bonding agents, causing parts to crack and fail.

36.9 RESIN FORMS

PPO resin is available in powder and pellet form.

36.10 SPECIFICATION OF PROPERTIES

See the Master Outline of Materials Properties.

36.11/.12 PROCESSING REQUIREMENTS AND PROCESSING-SENSITIVE END PROPERTIES

PPO resins can be processed in a number of ways, but the bulk of applications are in injection molding. PPO can also be extruded, and thermoformed, and various processes such as blow molding, compression molding, cold forming, and rotational molding have been used with varying degrees of success.

MATERIAL ___ PPO N190

PROPERTY	TEST METHOD	ENGLISH UNITS	VALUE	METRIC UNITS	VALUE
MECHANICAL DENSITY	D792	lb/ft³	67	g/cm³	1.08
TENSILE STRENGTH	D638	psi	7,000	MPa	48
TENSILE MODULUS	D638	psi	360,000	MPa	2,480
FLEXURAL MODULUS	D790	psi	325,000	MPa	2,200
ELONGATION TO BREAK	D638	%		%	
NOTCHED IZOD AT ROOM TEMP	D256	ft-lb/in.	7	J/m	370
HARDNESS	D785		R 115		R 115
THERMAL DEFLECTION T @ 264 psi	D648	°F	190	°C	88
DEFLECTION T @ 66 psi	D648	°F	205	°C	96
VICAT SOFT-ENING POINT	D1525	°F		°C	
UL TEMP INDEX	UL746B	°F		°C	
UL FLAMMABILITY CODE RATING	UL94		V-0 at 0.058" 5V at 0.120"		V-0 at 1.4 mm 5V at 3 mm
LINEAR COEFFICIENT THERMAL EXPANSION	D696	in./in./°F	0.00004	mm/mm/°C	0.000072
ENVIRONMENTAL WATER ABSORPTION 24 HOUR	D570	%	7	%	7
CLARITY	D1003	% TRANS-MISSION		% TRANS-MISSION	
OUTDOOR WEATHERING	D1435	%		%	
FDA APPROVAL					
CHEMICAL RESISTANCE TO: WEAK ACID	D543		Not attacked		Not attacked
STRONG ACID	D543		Not attacked		Not attacked
WEAK ALKALI	D543		Not attacked		Not attacked
STRONG ALKALI	D543		Not attacked		Not attacked
LOW MOLECULAR WEIGHT SOLVENTS	D543		Attacked		Attacked
ALCOHOLS	D543				
ELECTRICAL DIELECTRIC STRENGTH	D149	V/mil	630	kV/mm	25
DIELECTRIC CONSTANT	D150	10⁶ Hertz	2.78		2.78
POWER FACTOR	D150	10⁶ Hertz			
OTHER MELTING POINT	D3418	°F		°C	
GLASS TRANS-ITION TEMP.	D3418	°F		°C	

412

MATERIAL ___ PPO GFN2 (20% glass-filled)

PROPERTY	TEST METHOD	ENGLISH UNITS	VALUE	METRIC UNITS	VALUE
MECHANICAL DENSITY	D792	lb/ft³	76	g/cm³	1.21
TENSILE STRENGTH	D638	psi	14,500	MPa	100
TENSILE MODULUS	D638	psi	900,000	MPa	6,200
FLEXURAL MODULUS	D790	psi	750,000	MPa	5,100
ELONGATION TO BREAK	D638	%		%	
NOTCHED IZOD AT ROOM TEMP	D256	ft-lb/in.	2.3	J/m	120
HARDNESS	D785		L 106		L 106
THERMAL DEFLECTION T @ 264 psi	D648	°F	290	°C	143
DEFLECTION T @ 66 psi	D648	°F	293	°C	145
VICAT SOFT-ENING POINT	D1525	°F		°C	
UL TEMP INDEX	UL746B	°F		°C	
UL FLAMMABILITY CODE RATING	UL94				
LINEAR COEFFICIENT THERMAL EXPANSION	D696	in./in./°F	0.00002	mm/mm/°C	0.000036
ENVIRONMENTAL WATER ABSORPTION 24 HOUR	D570	%		%	
CLARITY	D1003	% TRANS-MISSION		% TRANS-MISSION	
OUTDOOR WEATHERING	D1435	%		%	
FDA APPROVAL					
CHEMICAL RESISTANCE TO: WEAK ACID	D543		Not attacked		Not attacked
STRONG ACID	D543		Not attacked		Not attacked
WEAK ALKALI	D543		Not attacked		Not attacked
STRONG ALKALI	D543		Not attacked		Not attacked
LOW MOLECULAR WEIGHT SOLVENTS	D543		Attacked		Attacked
ALCOHOLS	D543				
ELECTRICAL DIELECTRIC STRENGTH	D149	V/mil	420	kV/mm	16
DIELECTRIC CONSTANT	D150	10^6 Hertz	2.86		2.86
POWER FACTOR	D150	10^6 Hertz			
OTHER MELTING POINT	D3418	°F		°C	
GLASS TRANS-ITION TEMP.	D3418	°F		°C	

413

36.11.1 Injection Molding

Because of the more uniform heating of the resin and reduced pressure losses for injection-molding applications, injection equipment that uses screws is preferred over ram or combination presses. The length to diameter (L/D) ratios preferred on general-purpose screws are in the range of 20:1 to 24:1; compression ratios of 2:1 to 3:1 are said to be most satisfactory. Vinyl screw tips are not recommended for injection molding applications because they may cause shear degradation and give poor physical properties. Shut-off nozzles or torpedo-type heaters or mixing heads are not recommended either, since they may produce severe localized heating or areas where materials could hang up and become degraded at high molding temperatures. As a general guideline, to estimate the target melt temperature of a particular grade of PPO for injection molding applications, simply add 260°F (144°C) to the heat-distortion temperature. But length of flow, part design, and mold temperature will also affect both the molded part and material flow.

When working with PPO resins, the mold temperature should be kept high to prevent poor knit-line integrity, reduce molded-in stress, and improve surface gloss. Most PPO resins require a minimum mold temperature of 150°F (66°C); for glass-filled and high-heat resins, a 200–250°F (93–121°C) temperature is not uncommon. Clamp tonnage of three to five tons per square inch of projected area and 40 to 80% of the barrel capacity of the machines are recommended. This ensures part dimensional tolerances and avoids long residence times at high temperatures, which can result in degradation of the polymer. Degraded material can cause gases to become entrapped in the melt and produce surface splay or brittle parts. If the shot size and cycle times are not mismatched, barrel temperatures of up to 600°F (316°C) can be used on some grades of PPO without loss of properties. In general, an increase in melt temperature of up to 80°F (44°C) is allowable without loss of properties using the guideline of 260°F (144°C) plus the DTUL of the resin.

Because of their extremely low rate of moisture absorption, PPO resins can be molded without the need for drying. Without drying, however, some surface splay can result from trapped moisture, but this would not affect physical properties. Drying is recommended in injection-molding applications in instances where appearance parts are necessary, or when some kind of secondary finishing is required. Drying times and temperatures vary significantly with the resin grade.

36.11.2 Extrusion

The superior melt strength of PPO resins, coupled with their wide processing ranges and melt stability, can offer extruders fast rates, better sizing, and lighter parts than competitive materials. For extrusion of PPO resins, general-purpose equipment such as single-screw extruders with a minimum L/D ratio of 24:1 is preferred, with stock temperatures running about 180°F (100°C) higher than the DTUL of the resin. Compression ratios of 2.2:1 to 3.5:1 are also recommended, and the metering section of the screw should have 3 to 10 flights. The chance of material stuttering or surging can be reduced by using a long metering section rather than a short one or by adding screens to the screen pack. If a PVC-type screw is being used, then a screen pack using 20-40-60 mesh screens is recommended to create back pressure, increase shear, and reduce surging. Some porosity may be produced in material not dried properly when running sheet, or in pipe profiles more than 0.08 in. (2 mm) thick. The porosity is caused by low levels of moisture and does not affect physical properties, but it may be undesirable for esthetic or design reasons.

Drying temperatures vary from 190 to 240°F (88 to 116°C); drying times range from two to four hours. Air cooling is generally preferred over water cooling because it allows proper sizing and a slow cooldown. Air cooling typically minimizes stress and keeps sections with differing wall thicknesses from warping. Water cooling is effective where wall thicknesses are more uniform, as in extruded pipe, but the bath temperature should be at least 100°F (38°C).

36.11.3 Vacuum/Pressure Forming

Vacuum or pressure forming is easily accomplished with PPO resin sheets, with draw ratios on normal parts being 1:4. Drape forming, using a male mold, allows for deeper draw ratios of up to 2:1, thanks to the excellent melt strength of PPO resins. Deep draws may require the use of metal or asbestos sheets to act as shades in the heating chambers, so that excessive thinning in certain areas does not occur during the draw. Proper temperatures depend on the grade of the resin and run from 380 to 510°F (193 to 266°C). The use of double or sandwich heaters is recommended for sheet thickness of 0.125 in. (3.2 mm) and above. Because of their high heat-distortion temperatures, PPO resins cycle very fast in vacuum forming applications compared to most other resins, and this means less residual stress and faster part removal.

PPO resins can be reground and used with virgin pellets without any loss of physical properties. The recommended use of regrind in injection molding systems is in the range of 25% regrind to 75% virgin pellets. But 100% regrind can be used without loss of physical properties. Under proper conditions, PPO samples have been produced that evidence very little loss of physical properties after seven regrinds.

Regrind can also be used in extrusion applications at levels from 25% up to 100% regrind without loss of properties.

Nearly all finishing and bonding techniques can be used with PPO resins. Such finishing techniques as sputtering or plating are applied using special grades to meet specific requirements. PPO resins can be decorated with most thermal, chemical, or bonding techniques available to the general plastics industry.

36.13 SHRINKAGE

The proper cooling of a part allows for predictable mold shrinkage and proper sizing. PPO shrinkage rates are in the range of 0.005 to 0.007 in./in. (mm/mm).

Mold Shrinkage Characteristics (in./in. or mm/mm)

MATERIAL _____ PPO

IN THE DIRECTION OF FLOW

THICKNESS

MATERIAL	mm: 1.5	3	6	8	FROM	TO	#
	in.: 1/16	1/8	1/4	1/2			
	0.005–7	0.005–7	0.005–7				

	PERPENDICULAR TO THE DIRECTION OF FLOW						

NOTE: Mold shrinkage is approximate. It is affected by part design, mold temperature, thickness, injection pressure, packing time, cycle time, orientation, gate design, gate size, gate location, glass content, glass size, and filler content.

36.14 TRADE NAMES

PPO is a registered trademark of the General Electric Co. PPO resins are sold under G.E.'s Noryl resin trademark.

36.15 INFORMATION SOURCES

Information on PPO resins is available from the manufacturer: General Electric Co., Noryl Ave., Selkirk, NY 12158.

37

POLY(PHENYLENE SULFIDE) (PPS)

J.W. Gardner and P.J. Boeke
Phillips Chemical Co., Bartlesville, Oklahoma

37.1 STRUCTURE

Poly(phenylene sulfide) (PPS) is a polymeric material composed of alternating aromatic rings and sulfur atoms.

$$\left[\bigcirc - S \right]_n$$

37.2 CATEGORY

Poly(phenylene sulfide) is a rigid thermoplastic material with high-performance engineering properties. Among its distinguishing characteristics are high temperature resistance, excellent chemical resistance, inherent flame resistance, and good electrical properties.

37.3 HISTORY

Resins similar to PPS may have been prepared as early as 1888 as by-products in the research of Friedel and Crafts. In 1898, the structure of PPS was first assigned by Grenvesse to the insoluble reaction product of benzene and sulfur in the presence of aluminum chloride. In 1948, MacAllum initiated the development of PPS when he described the synthesis of the resin by the melt reaction of *p*-dichlorobenzene with sodium carbonate and sulfur. The Phillips process for preparing PPS was described by Edmonds and Hill in 1967.[1] This process involves the reaction of *p*-dichlorobenzene with sodium sulfide in the presence of a polar organic solvent. The first commercial plant utilizing this process went onstream in 1973. The 1983 production of PPS powder and filled molding compounds was about 10×10^6 lb.

37.4 POLYMERIZATION

PPS is made commercially by the reaction of *p*-dichlorobenzene with sodium sulfide in a polar organic solvent. The product of this reaction is a low molecular weight polymer insoluble in any known solvent below 392°F (200°C). This product is characterized by a melt flow of 4000–6000 g/10 min. A unique property of PPS is its ability to undergo a slow, controlled curing involving cross-linking and chain extension, which results in a higher molecular weight polymer characterized by increased viscosity (melt flow range of 20–300). PPS has an excellent affinity for glass and mineral fillers; the resulting compounds are especially well-suited to processing by injection molding. Commercially available PPS molding compounds contain both glass fiber and a combination of glass fiber and particulate mineral fillers.

The June 1987 truckload selling price of PPS compounds ranged from $1.45/lb to $3.13/lb.

37.5 DESCRIPTION OF PROPERTIES

37.5.1 Thermal Properties

PPS is a crystalline polymer with a melting point of 545°F (285°C). At this melting transition, the polymer changes from a solid to a low-viscosity melt well-suited to injection molding. PPS compounds have the highest temperature resistance of any of the competitively priced engineering plastics available today. Glass- and mineral-filled versions have heat-deflection temperatures at 264 psi (1.82 MPa) of 428°F (260°C) and UL temperature indexes from 392 to 464°F (200 to 240°C). PPS is also extremely flame resistant with all grades achieving the UL 94 V-0 rating and, in some thicknesses, a 5V rating. Upon thermal degradation at 1472°F (800°C), the polymer releases hydrogen, methane, carbon monoxide, carbon dioxide, and carbonyl sulfide.

37.5.2 Mechanical Properties

PPS is marketed almost exclusively as glass-filled or glass- and mineral-filled compounds tailored to specific injection-molding applications. This is a result of the dramatic improvement in mechanical properties obtained via reinforcement of PPS with glass fiber and minerals. Filled PPS compounds can be classified as very hard and rigid engineering plastics. The flexural modulus of some commercial compounds exceeds 2.4×10^6 psi (17×10^3 MPa). Molding compounds with a tensile strength as high as 9,600,000 psi (66×10^3 MPa) are also available. PPS compounds are characterized by low elongation, ranging from 0.5 to 2%, which results in notched Izod impact values from 0.60 to 1.5 ft-lb/in. (32 to 80 J/m). The surface hardness of PPS compounds is high. PPS reinforced with 40% glass fiber exhibits a Rockwell Hardness (R scale) of 123.

37.5.3 Environmental Properties

PPS is one of the most chemically resistant polymers known. It is inert towards a wide variety of solvents and mineral/organic acids and alkalies. However, members of four classes of compounds—oxidizing agents, strong acids, halogens, and amines—may have some effect on the environmental properties of PPS.

Some coating grades of PPS are approved for food contact in accordance with FDA Regulation 177.2490. In contrast to injection-molding versions, such coating resins made of PPS contain little or no filler. Other kinds of PPS coatings bear NSF approval for transporting potable water.

37.5.4 Electrical Properties

PPS compounds generally possess good insulation properties, as measured by volume resistivity and insulation resistance after extreme exposure to both high humidity and temperature. Low dielectric constants and dissipation factors throughout a broad frequency and temperature range are characteristic of glass-reinforced PPS compound. These properties make PPS particularly well suited for electronic applications where sensitivity and reliability are important.[2,3] In fact, certain PPS compounds have been specially formulated to provide reliable performance under extreme temperature and humidity conditions.[4]

Glass and mineral-reinforced PPS compounds exhibit low arc resistance, as evidenced by various high-voltage, high-current arc tracking tests including the ASTM D495 arc resistance tests. Compounds are available with a high voltage track rate of less than 1 in. (2.5 cm)/min and the comparative tracking index of from 220 to 250 volts. These properties of PPS compounds are widely exploited in sockets, switches, and other higher voltage electrical components.

37.6 APPLICATIONS

37.6.1 Industrial/Mechanical

The high strength-to-weight ratio of PPS compounds, coupled with excellent chemical resistance, dimensional stability, and high temperature resistance, have led to the widespread use of PPS in applications reserved for metals or less durable synthetic materials.[5] Pump housings, pump impellers and vanes, pistons, and valve components are a few examples of applications where PPS compounds are gaining widespread acceptance. In downhole oilfield applications, sucker rod guides made of glass-reinforced PPS offer outstanding chemical resistance and reduced wear over their metal or plastic counterparts. Figure 37.1 shows one such guide made by J.M. Huber Corp. of Borger, Texas. A particularly impressive example of PPS replacing metal is the mist eliminator baffle shown in Figure 37.2. This baffle is produced by the Munters Corp. of Ft. Myers, Fla.; it is used to remove liquid drops from

Figure 37.1. Sucker rod guide molded from glass reinforced PPS (Ryton R-4)

Figure 37.2. Mist eliminator baffle molded from glass reinforced PPS (Ryton R-4)

flowing gases or vapors in such applications as steam turbine separators, evaporators, and absorption towers. PPS is an attractive material for such products because it offers improved chemical resistance over stainless steel at one fifth the weight.

37.6.2 Electrical/Electronics

The insulating properties of certain PPS compounds make them ideally suited for sensitive electronic applications, from connectors to wristwatch components. In fact, connectors represent one of the largest and most important PPS application areas.[6] Some PPS compounds exhibit reliability equivalent or superior to conventionally used thermosets, such as DAP, with one important advantage. PPS compounds offer the sought-after processing advantage of thermoplastics over thermosets; namely, significant cost reduction resulting from greatly reduced cycle times and complete use of runners and other regrind. Figure 37.3 shows several connectors molded in 40% glass-filled PPS. Other specially formulated PPS compounds are also available for connectors.

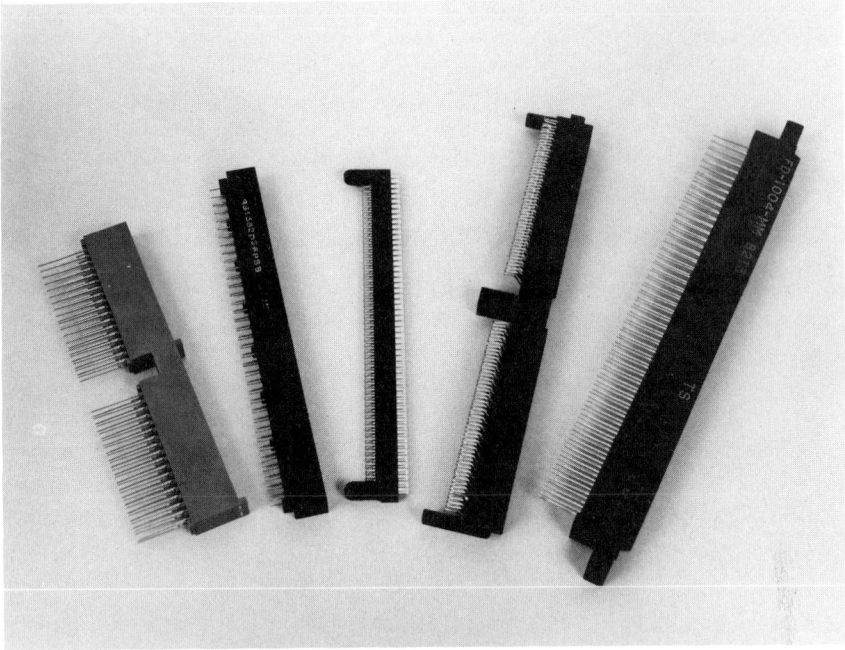

Figure 37.3. Ryton PPS electronic connectors

One recent area of penetration by PPS compounds lies in the encapsulation of electronic circuitry.[7-10] PPS compounds are replacing epoxy thermosets in such encapsulation applications as capacitors and transistors. PPS compounds are also being developed for the encapsulation of sensitive integrated circuits.

Sockets, coil forms, coil bases, and other electrical components require an insulating material with arc resistance, flame retardance, and high temperature resistance. PPS compounds excel in all of these categories. The hostile environment of televisions and electric motors, for example, demands the dimensional stability and temperature resistance offered by PPS compounds.

37.6.3 Automotive

PPS compounds are finding increasing use in under-the-hood automotive applications.[11] Most of these applications require the dimensional stability and high temperature resistance of PPS compounds. EGR valves used in emission control systems, generator elements, sensor devices, light sockets, and cooling system parts are a few examples of successful PPS applications. Figure 37.4 shows an EGR valve produced by Nippondenso, Japan, for Toyota Motor Co. This application requires excellent temperature stability and gas and oil resistance along with high part-to-part molding reproducibility.

37.6.4 Appliances

The combination of good electrical properties coupled with high temperature resistance has led to the use of PPS compounds in many appliance applications, such as handles, housings and bases, hair dryer grilles, protective coatings, and insulators. In addition, certain PPS compounds are designed for microwave cookware applications where PPS offers the added advantage of conventional oven compatibility up to 428°F (220°C).

Figure 37.4.   Automotive exhaust gas recycle valve molded from glass reinforced PPS (Ryton R-4)

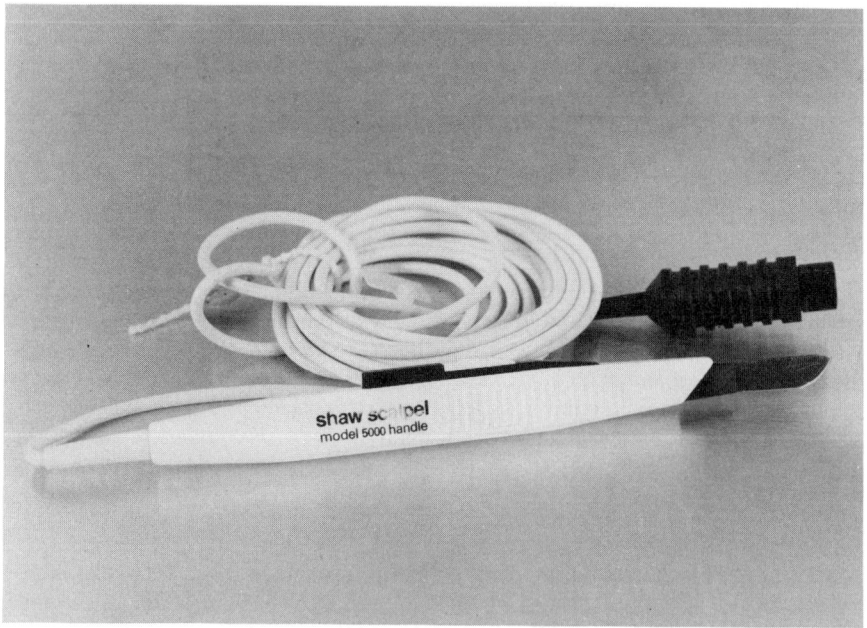

Figure 37.5.   Surgical scalpel with heat insulator molded from PPS molding compound

Figure 37.5 shows another type of application, a specially developed surgical scalpel employing a PPS-insulated handle that allows the blade to heat to temperatures of 428°F (260°C). This is an important feature because the hot blade is used to cauterize tissue as the incision progresses.

37.7 ADVANTAGES/DISADVANTAGES

37.7.1 Processibility

PPS compounds offer the excellent processibility characteristic of a highly crystalline thermoplastic. Compounds are easily injection molded to close tolerances with consistent, reproducible shrinkage. Complete use of regrind is also possible.

These advantages explain why PPS is replacing thermosets in many applications. The generally higher cost of PPS compounds is more than offset by the cost savings realized in production. In addition, the crystalline nature of PPS provides a significant processing advantage over amorphous engineering plastics, which retain a high viscosity even at their recommended molding temperatures.

37.7.2 High-Temperature Properties

PPS compounds are the highest rated competitively priced engineering resins available. The excellent room temperature stiffness and tensile strength of PPS compounds fall off only at elevated temperatures. Even at 392°F (200°C), however, the flexural modulus of 40% glass-filled PPS is equivalent to that of ABS at room temperature.

37.7.3 Flame Resistance

PPS compounds exceed the UL 94 V-0 and 5V requirements without the use of any special flame-retardant additives. PPS compounds have a limiting oxygen index (LOI) ranging from 46 to 53. This is among the highest of any of the engineering plastics. In addition, the fact that PPS compounds achieve this flame resistance with no special additive means that the regrind will exhibit the same flame-resistant performance.

37.7.4 Chemical Resistance

The inert aromatic backbone of PPS lends tremendous chemical resistance to PPS compounds. PPS is generally regarded as second only to PTFE in overall chemical resistance. PPS is unaffected by electronic cleaning solutions and withstands equally well the harsh environments of chemical transport systems and downhole oilfield applications.

37.7.5 Dimensional Stability

The low warpage, long-term stability and part-to-part reproducibility of PPS compounds allow them to perform well in products such as electrical connectors, which require high-temperature dimensional stability and exacting part-to-part reproducibility.

37.7.6 Assembly

One particularly useful advantage of PPS compounds is the ease with which the resulting PPS parts can be assembled. PPS lends itself to mechanical assembly via rivets, bolts, or screws. Semitubular rivets work best, since bolts should anchor into inserts or molded-in threads. Self-threading screws are preferred over the forming type. PPS compounds can be

TABLE 37.1   Equation for Comparing the Costs of PPS Compounds to Thermosets

$$A(B + CD/100) + (E/60 \times F) = \frac{\text{Cost Per Shot}}{G} = \text{Cost Per Part}$$

	DAP	PPS
A = cost per lb of resin, $	2.79	3.13
B = wt of parts produced per shot, lb	0.13	0.13
C = wt of shot, lb	0.26	0.26
D = scrap loss, %	50	0
E = machine rate, $/h	25	25
F = cycle time, min	0.66	0.42
G = number of cavities	8	8

DAP

$$2.79\,(0.13 + 50 \times 0.26/100) + (25/60 \times 0.66)$$
$$0.725 + 0.275 = \$1.00 \text{ per shot}$$
$$1.00/8 = \$0.125 \text{ per part}$$

40% glass-filled PPS

$$3.13\,(0.13 + 0) + (25/60 \times 0.42)$$
$$0.407 + 0.175 = \$0.582 \text{ per shot}$$
$$0.582/8 = \$0.073 \text{ per part}$$

ultrasonically welded quite successfully using a step joint with energy director or a shear joint. There are several adhesives available that bond PPS compounds effectively, provided the assembly has been preceded by an appropriate surface treatment (flame, corona discharge, chromic acid). Finally, some PPS parts have been successfully designed and produced using a snap fit assembly. The baffles shown in Figure 37.2 are assembled using a snap fit with 0.050 in. (12.7 cm) of interference on each snap.

37.7.7   Economic Considerations

PPS compounds are generally classified as being in the higher price range of engineering plastics; so, the economics of PPS might be considered a disadvantage. However, when all factors are considered, the economics of PPS compounds often appear very advantageous. Table 37.1 presents an equation for comparing the cost of PPS compounds to thermosets. Included is an example comparing 40% glass-filled PPS and DAP.

PPS compounds are characterized by elongations from 0.5 to 1.6%. These relatively low elongations result in only moderate impact strengths for PPS compounds. Typical notched Izod impact values from 0.60 to 1.5 ft-lb/in. (32 to 80 J/m) are not as outstanding as other physical properties of PPS. Impact strength is probably the most often cited disadvantage of PPS compounds.

37.8   PROCESSING TECHNIQUES

See Master Material Outline and Standard Material Design Chart.

Master Material Outline

PROCESSABILITY OF _____Poly(phenylene sulfide)_____

PROCESS	A	B	C
INJECTION MOLDING	X		
EXTRUSION			
THERMOFORMING			
FOAM MOLDING			
DIP, SLUSH MOLDING			
ROTATIONAL MOLDING			
POWDER, FLUIDIZED-BED COATING	X		
REINFORCED THERMOSET MOLDING			
COMPRESSION/TRANSFER MOLDING	X		
REACTION INJECTION MOLDING (RIM)			
MECHANICAL FORMING			
CASTING			

A = Common processing technique.
B = Technique possible with difficulty.
C = Used only in special circumstances, if ever.

Standard design chart for _____Polyphenylene Sulfide (PPS)_____

1. Recommended wall thickness/length of flow
   Minimum _____0.030 in._____ for _____5 in._____ distance
   Maximum _____0.125 in._____ for _____20 in._____ distance
   Ideal _____0.080 in._____ for _____16 in._____ distance

2. Allowable wall thickness variation, % of nominal wall _____20_____

3. Radius requirements (scheme **a**)
   Outside: Minimum _____0_____ Maximum _____
   Inside: Minimum _____0_____ Maximum _____60%_____

4. Reinforcing ribs (scheme **c** and **d**)
   Maximum thickness _____75%_____
   Maximum height _____
   Sink marks: Yes _____ No _____

Radius detail

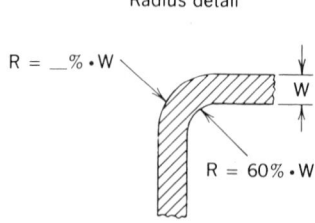

(a)

Self tapping screw
boss detail

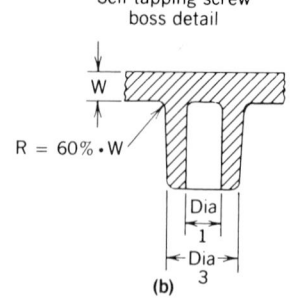

(b)

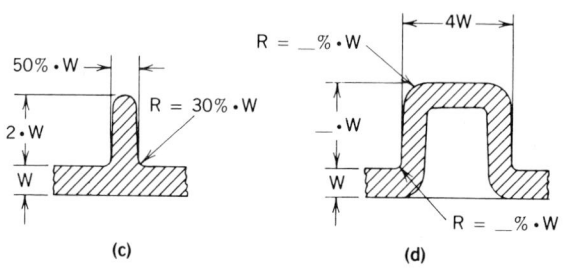

Solid rib detail

(c)

Hollow rib detail

(d)

5. Solid pegs and bosses
   Maximum thickness _____0.125 in._____
   Maximum weight _____0.125 in._____
   Sink marks: Yes _____ No _____

6. Strippable undercuts
   Outside: Yes _____ No _____
   Inside: Yes _____ No _____

7. Hollow bosses (scheme **b**): Yes ___X___ No _____

8. Draft angles
   Outside: Minimum _____0.5_____ Ideal _____1°_____
   Inside: Minimum _____0.5_____ Ideal _____1°_____

9. Molded-in inserts: Yes ___X___
                       No _____

10. Size limitations   Wall thickness 0.125 inch or less
    Length: Minimum _____ Maximum _____ Process _____
    Width: Minimum ___—___ Maximum ___—___ Process _____

11. Tolerances
    ±0.2 mil

12. Special process- and materials-related design details
    Blow-up ratio (blow molding) _____
    Core L/D (injection and compression molding) _____4:1 or less_____
    Depth of draw (thermoforming) _____
    Other _____

37.9 RESIN FORMS

PPS is available as a free-flowing powder for compounding with fillers and additives. Specially formulated coating powders are also available for slurry or fluidized-bed coatings. However, most of the PPS produced is marketed as a pelletized, filled compound. PPS compounds are available with a variety of glass fiber, mineral, and carbon fiber reinforcements. Many are designed for sensitive electrical applications requiring excellent flow and purity. Others are available with filler systems that optimize the material's mechanical strength.

37.10 SPECIFICATION OF PROPERTIES

See Master Outline of Materials Properties.

37.11 PROCESSING REQUIREMENTS

37.11.1 Material Drying

Although PPS itself does not absorb significant amounts of moisture, the glass and mineral fillers used in PPS compounds can be somewhat hygroscopic; therefore, drying is recommended. Generally, a forced air oven or desiccator hopper dryer at 302°F (150°C) for 2–6 hours is sufficient. The time and temperature requirements vary with the particular compound.

37.11.2 Melt Temperature

PPS compounds can be injection molded at melt temperatures from 576 to 675°F (302 to 357°C). The optimum temperature varies, but operation in the recommended range of 599–626°F (315–330°C) ensures good mixing and helps control wear on the machine caused by the filler system. Higher temperatures result in a substantially lower viscosity melt, which can greatly improve filling in very thin-wall parts. The spiral flow of 40% glass-filled PPS nearly doubles when the melt temperature is raised from 550 to 675°F (288 to 357°C).

37.11.3 Mold Temperature

The mold temperature used in molding PPS compounds is an extremely important variable. Mold temperatures of 275–302°F (135-150°C) are necessary to achieve full crystallinity. At temperatures below 199°F (93°C), the molded parts will be amorphous. Amorphous parts characteristically have slightly higher impact, tensile, and flexural strength than do crystalline parts. However, fully crystalline moldings will have a higher flexural modulus and be much more dimensionally stable and heat resistant. If higher use temperatures are a requirement, crystalline parts should be used.

37.11.4 Screw, Barrel and Mold Materials

Glass and mineral fillers are abrasive in PPS compounds, as they are all filled plastic compounds. Therefore, they require the use of the proper materials in the construction of screws, barrel linings, and molds. D-2, D-7, and A-2 are the recommended steels for cores and cavities because they are extremely wear resistant. In addition, D-2 and D-7 are corrosion resistant, which makes them well-suited for tools that are especially difficult to vent adequately.

In certain cases, mold plating can provide added protection for otherwise unsuitable mold materials. Three plating types are resistant to the erosive effects of PPS compounds: elec-

Master Outline of Material Properties

MATERIAL ___ Poly(phenylene sulfide) (PPS) 40% Glass-Filled RYTON-4

PROPERTY	TEST METHOD	ENGLISH UNITS	VALUE	METRIC UNITS	VALUE
MECHANICAL DENSITY	D792	lb/ft³	104	g/cm³	1.67
TENSILE STRENGTH	D638	psi	17,500	MPa	121
TENSILE MODULUS	D638	psi		MPa	
FLEXURAL MODULUS	D790	psi	1,700,000	MPa	11,700
ELONGATION TO BREAK	D638	%	0.9	%	0.9
NOTCHED IZOD AT ROOM TEMP	D256	ft-lb/in.	1.3	J/m	69
HARDNESS	D785		R-123		R-123
THERMAL DEFLECTION T @ 264 psi	D648	°F	500	°C	260
DEFLECTION T @ 66 psi	D648	°F		°C	
VICAT SOFT-ENING POINT	D1525	°F		°C	
UL TEMP INDEX	UL746B	°F	93–104	°C	200–220
UL FLAMMABILITY CODE RATING	UL94		V-0/5-V		V-0/5-V
LINEAR COEFFICIENT THERMAL EXPANSION	D696	in./in./°F	0.000 016	mm/mm/°C	0.000 029
ENVIRONMENTAL WATER ABSORPTION 24 HOUR	D570	%		%	
CLARITY	D1003	% TRANS-MISSION	Opaque	% TRANS-MISSION	Opaque
OUTDOOR WEATHERING	D1435	%		%	
FDA APPROVAL			No		No
CHEMICAL RESISTANCE TO: WEAK ACID	D543		Not attacked		Not attacked
STRONG ACID	D543		40% weight change		40% weight change
WEAK ALKALI	D543		Not attacked		Not attacked
STRONG ALKALI	D543		40% weight change		40% weight change
LOW MOLECULAR WEIGHT SOLVENTS	D543		Not attacked		Not attacked
ALCOHOLS	D543		Not attacked		Not attacked
ELECTRICAL DIELECTRIC STRENGTH	D149	V/mil	450	kV/mm	17.2
DIELECTRIC CONSTANT	D150	10⁶ Hertz	3.8		3.8
POWER FACTOR	D150	10⁶ Hertz	0.001 4		0.001 4
OTHER MELTING POINT	D3418	°F	545	°C	285
GLASS TRANS-ITION TEMP.	D3418	°F	199	°C	93

428

troless nickel plating, slow-deposition chrome plating, and Nye-Carb silicon carbide–nickel plating. The success of any of the above platings is highly dependent on the quality of the application.

Screws and barrels are also subject to the abrasive nature of glass and mineral fillers and should therefore be constructed from the proper abrasion-resistant material. Stellite flighting is recommended to ensure long screw wear. Barrel linings of Xalloy 800 ensure long barrel life.

37.11.5 Shut-Off Nozzles

Certain high flow and specially filled PPS compounds are low enough in viscosity that nozzle drool becomes a problem. Drool can often be reduced or eliminated by drying the material, reducing the stock temperature, or increasing the screw decompression. If these techniques fail, it may be necessary to use a positive shut-off nozzle. Generally, the mechanically actuated needle-type design gives the best results.

37.11.6 Regrind Usage

Runners, sprues, and rejects can all be ground and blended with virgin resin for reprocessing. No effect on part properties is observed as long as the regrind level does not exceed 30–35%.

37.12 PROCESSING-SENSITIVE END PROPERTIES

The processing parameter having the most significant effect on physical properties is mold temperature. Mold temperature dictates the crystallinity of the molded part, which in turn affects the mechanical and physical properties of the final product.

37.12.1 Crystallinity and Mold Temperature

PPS exhibits a glass-transition temperature (T_g) of 190°F (88°C). Above this temperature, the polymer chains possess enough energy to crystallize. However, at temperatures below T_g, the polymer matrix is frozen in a random amorphous state.

Figure 37.6 is a graph of percent crystallinity versus mold temperature. Experience has

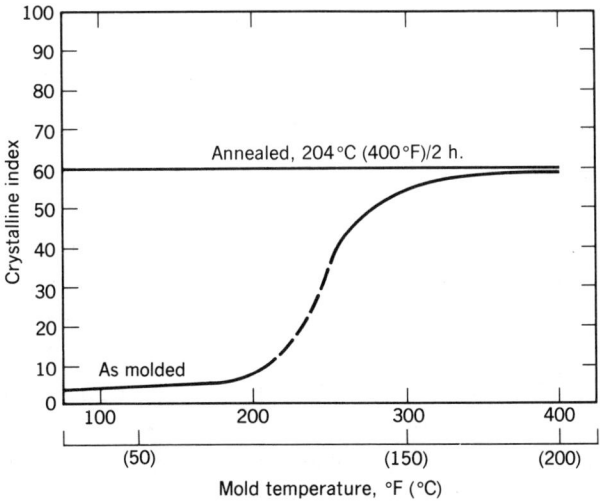

Figure 37.6. Effect of mold temperature upon crystallinity for PPS

TABLE 37.2   40% Glass-Filled PPS

Property	100°F (38°C)	275°F (135°C)
Tensile strength, psi	19,000	17,600
(MPa)	(130)	(121)
Elongation, %	1.2	0.9
Flexural strength, psi	28,000	26,000
(MPa)	(193)	(179)
Flexural modulus 10^3, psi	247	247
(MPa)	(1.7)	(1.7)
Izod impact-notched, ft-lb/in.	1.4	1.3
(J/m)	(74)	(69)
Heat-deflection temperature, 264 psi, °F	446	490
(°C)	(230)	(260)

shown that a mold temperature from 275 to 302°F (135 to 150°C) is adequate to achieve a high level of crystallinity. Mold temperatures from 201 to 250°F (94 to 121°C) generally produce unpredictable levels of crystallinity and are therefore not recommended. Below 201°F (94°C), the crystallinity reaches a minimum and the part is considered amorphous. Regardless of the mold temperature used, any molded part can be brought to maximum crystallinity by a post-mold annealing at about 392°F (200°C) for 2 hours.

37.12.2   Crystallinity and Performance

The crystallinity of the molded PPS part plays a vital role in the final physical and mechanical properties of the part.[12] Table 37.2 presents a side-by-side comparison of 40% glass-filled PPS molded in both a 275°F (135°C) mold and a 100°F (38°C) mold. Generally speaking, the cold-molded amorphous parts possess slightly better room-temperature properties while sacrificing high temperature stability. The hot-molded parts enjoy much better high temperature properties and will have a glossier, smoother surface appearance.

In addition, if cold-molded parts are in service at temperatures above T_g (190°F [88°C]), they will slowly crystallize, thus changing their physical properties during use.

Most PPS compounds are used in applications requiring outstanding physical properties, chemical resistance, and high temperature stability. Therefore, most molders prefer hot molds or post-mold annealing of PPS parts.

37.13   SHRINKAGE

See Mold Shrinkage Characteristics.

Mold Shrinkage Characteristics (in./in. or mm/mm)

MATERIAL _____ Polyphenylene Sulfide (PPS)

IN THE DIRECTION OF FLOW

THICKNESS

MATERIAL	mm: 1.5	3	6	8	FROM	TO	#
	in.: 1/16	1/8	1/4	1/2			
R-4 (in./in.)	0.001	0.002	0.007		0.001	0.003	in.
R-7 (in./in.)	0.001	0.0015	0.004		0.001	0.003	in.
R-10 (in./in.)	0.001	0.0015	0.004		0.001	0.001	in.

	PERPENDICULAR TO THE DIRECTION OF FLOW						
R-4	0.004	0.011	0.025		0.003	0.005	in.
R-7	0.003	0.008	0.015		0.003	0.005	in.
R-10	0.003	0.008	0.015		0.003	0.005	in.

NOTE: Mold shrinkage is approximate. It is affected by part design, mold temperature, thickness, injection pressure, packing time, cycle time, orientation, gate design, gate size, gate location, glass content, glass size, and filler content.

37.14 TRADE NAMES

PPS is available under the following trade names from the manufacturers: RTP 1300 series, Fiberite Corp., Winona, Minn.; Thermocomp OP series, LNP Corp., Malvern, Pa.; and Ryton R series, Phillips Chemical Co., Houston, Tex. PPS is also offered by Hoechst Celanese, Chatham, N.J.

BIBLIOGRAPHY

1. Edmonds and Hill, U.S. Patent.
2. P.G. Kelleher and P. Hubbauer, "Glass Fiber Reinforced Polyphenylene Sulfide for Communication Equipment," *36th ANTEC,* Society of Plastics Engineers, April, 1978.
3. P.J. Boeke and J.S. Dix, "High Performance Polyphenylene Sulfide Interconnection Devices," *13th Annual Connector Symposium Proceedings,* Oct. 1980.
4. P.J. Boeke and J.E. Leland, "PPS Connector Compound With Resistance to High Temperature, High Humidity Environments," *15th Annual Connectors and Interconnection Technology Symposium Proceedings,* Nov. 1982.
5. R.S. Shue and co-workers, "HMW Injection Molding Polyphenylene Sulfide Compound Has More Ductility," *Plast Eng.,* 39 (April 1983).
6. W. Buster, "Understanding PC Connector Materials," *Insulation/Circuits,* (June 1982).
7. H.E. Corlis, "Capacitor Encapsulation Using Reel-Fed Continuous Strip System," *Proceedings of the 15th Electrical/Electronic Conference,* Oct. 1981.
8. S.K. Schuetterle, "Capacitor Encapsulation Using Rotary and Shuttle Thermoplastic Injection Molding," *Proceedings of the 15th Electrical/Electronic Conference,* Oct. 1981.
9. Tsung-yuan Su, "Encapsulation of Ceramic Capacitors With Thermoplastic Materials," *Proceedings of the 32nd Electronics Components Conference,* May 1982.
10. J.S. Dix and J. Demchik, "Encapsulation of Ceramic Capacitors with Polyphenylene Sulfide Molding Compound," *40th ANTEC,* Society of Plastics Engineers, May 1982.
11. J.R. Millers, J.S. Dix, and V.C. Vives. "Resistance of 40 Percent Glass-Reinforced PPS to Automotive Underhood Fluids," *SAE Automotive Plastics Durability Conference Proceedings,* Dec. 1981.
12. D.G. Brady, "The Crystallinity of Polyphenylene Sulfide and Its Effect on Polymer Properties," *Journal of Applied Polymer Science.*

38

POLYPROPYLENE

Russell D. Hanna

Manager, Himont USA Inc., Wilmington, Delaware

38.1 INTRODUCTION

Polypropylene (PP) is a versatile thermoplastic offering a useful balance of heat and chemical resistance, good mechanical and electrical properties, and processing ease.

38.2 CATEGORY

Crystalline polymers of propylene were first described in the literature in 1954 by G. Natta and his associates at the Chimica Industriale del Politechico di Milano.[1] Earlier efforts to initiate propylene polymerization had only resulted in noncrystalline polymers of little or no importance. With the introduction of heterogeneous, stereospecific catalysts discovered by K. Ziegler for the low-pressure polymerization of ethylene,[2] the scene suddenly changed. These reaction products of transition metal compounds with selected organometallic compounds contained active sites for polymerization, such that each new propylene molecule was incorporated in the polymer chain in a regular, geometric manner identical to all preceding methyl groups.[3-5]

Although numerous advances have been made in commercial polymerization processes—multistage continuous reactor chains, propylene solution and vapor phase bulk polymerization, and loop reactors, for example, basic catalyst technology has not changed in 30 years. However, improved techniques for manufacturing catalysts and emphasis on proprietary cocatalysts or adjuvants have increased catalyst efficiency manyfold and optimized crystalline polymer yield.[6]

Three geometric forms of the polymer chain can be obtained (Fig. 38.1). Natta classified them as:

- Isotactic
  All methyl groups aligned on one side of the chain
- Syndiotactic
  Methyl groups alternating
- Atactic
  Methyl groups randomly positioned.[5]

Although both isotactic and syndiotactic forms will crystallize when they are cooled from molten states, commercial injection-molding and extrusion-grade polypropylenes (PP) are generally 94 to 97% isotactic. Fabricated parts are typically 60% crystalline, with a range

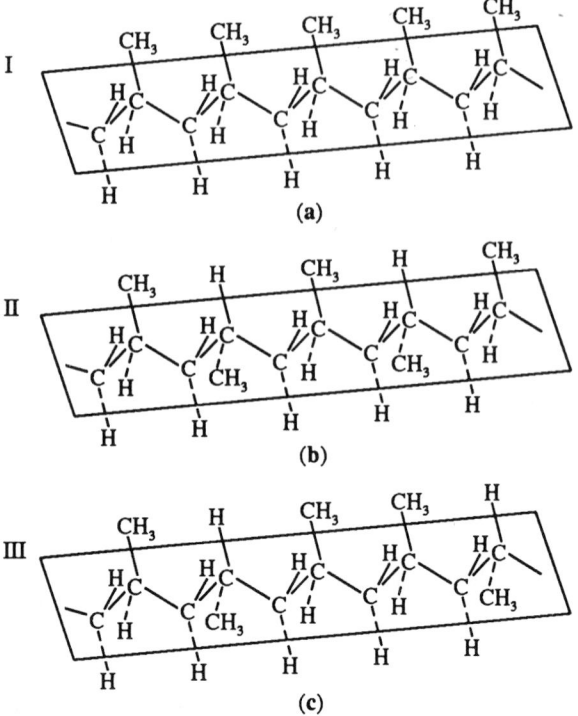

Figure 38.1. Propylene structures. (**a**) Isotactic, (**b**) syndiotactic, (**c**) atactic.

of polyhedral spherulite forms and sizes, depending on the particular mode of crystallization from the melt.[7]

Syndiotactic PP remains a laboratory curiosity. However, atactic PP recovered as residue from solvent or as recycle from slurry diluent distillation is a useful by-product. Although it is not suited to structural plastic uses, major applications have been developed as modifiers in hot-melt adhesives, roofing compounds, and communications cable interstitial-filler gels.[8]

38.3   HISTORY

Working independently from 1953 to 1956, many investigators of organometallic catalysts found it possible to polymerize propylene to the crystalline state. Among these were scientists at Hercules,[9] DuPont,[10] Standard Oil of Indiana (now Amoco),[11,12] Phillips,[13,14] and Montecatini of Italy. In 1971, the U.S. Patent Office awarded priority of invention for crystalline PP to G. Natta at Montecatini.[15-16] This ruling was overturned in 1980, and the basic composition of matter rights were granted to Phillips. The judgment was appealed and finally upheld by the U.S. Supreme Court in March 1983.[17] Several interference actions have been entered with respect to inclusion of certain copolymers in the broad coverage granted Phillips.

Commercial production of crystalline PP was first put on stream in late 1957 by Hercules in the U.S., by Montecatini in Italy, and by Farbewerke Hoechst AG in Germany.[18] It had unique processing characteristics that were not duplicated by any other polymer at the time. PP was the first polymer to establish major applications in all thermoplastic processing categories—injection molding, sheet and profile extrusion, film (both oriented and cast), monofils, and multifilaments. By capitalizing on the basic characteristics of light weight, good stiffness and chemical resistance, a unique ability to form integral (living) hinges, high heat resistance, and toughness when alloyed with elastomers, PP quickly opened new markets in molded luggage, appliance housings, housewares, hospital ware, and automotive components.

Molded applications accounted for two thirds of the estimated 40 million lb sold in the United States in 1960; the remaining third went into monofilaments for seat covers, rope, and twine. U.S. production capacity at that time was estimated to be between 150 and 200 million lb. In addition to Hercules, PP was produced by Avisun (now Amoco), Dow, Eastman, and Humble (now Exxon).[19]

During the 1960s, the number of U.S. producers and their capacity grew rapidly, with sales increasing as much as 35 to 40% in some years. By 1970, nameplate capacities of the eight producers [Amoco, Diamond Shamrock (now Petrofina), Eastman, Exxon, Hercules (now Himont), Montecatini (now USS Chemicals), Phillips, and Shell] were estimated to be just slightly over 1 billion lb.[20] U.S. production was just short of capacity. Some 265 million lb were consumed as multifilament fibers developed for use in carpeting and upholstery fabrics. Automotive applications initiated in the 1960s grew quickly during the 1970s, primarily in battery cases, fan shrouds, and other under-the-hood uses; in fender liners, interior trim, and kick panels. Molded components in washing machines and dishwashers, housewares, luggage, totes, crates, and institutional seating were firmly established during this period, laying a broad base of experience and confidence for growth in the 1980s. Four new producers, Gulf Oil (now part of Amoco), Norchem (now Quantum), Soltex, and El Paso (now Rexene), coming on stream in the 1970s added 1.1 billion lb of new capacity to that of the other eight which, by this time, had grown to an estimated 4 billion-plus lb.[21]

Unfortunately, rosy forecasts of the late 1970s were badly shaken when the recession brought slumps in automotive production and housing starts. Low operating rates and high investment costs stifled incentives for growth, while producers who were late entrants in an overbuilt market kept prices at unprofitable levels. U.S. production reached 4 billion pounds in 1981, then sagged to less than 3.5 billion in 1982.

Since 1982, domestic production of PP has risen at an annual rate of about 13 percent, reaching 5.1 billion lb in 1985. The estimated volume for 1988 is 7.3 billion lb. The severe overcapacity from 1980 through 1982 led to a four-year hiatus in capacity growth. Since late 1986, however, the industry has been operating at or near capacity, and prices have risen sharply to 55–60 cents per lb for general-purpose injection grades (see Table 38.1).

TABLE 38.1. U.S. Producers of Polypropylene, 1989

Supplier	Operating Name-plate Capacity, as of 1/1/89, MT
Himont	1023
Amoco	550
Fina	409
Exxon	323
Aristech	298
Shell	227
Soltex	200
Rexene	193
Phillips	191
Huntsman	148
Quantum Chemical, U.S.I. Div.	114
Eastman	66
Advanced Global	57
Total	3798

There are now approximately 120 PP-producing companies (including joint ventures) in the world, located in North America, Western and Eastern Europe, Africa, Asia, Australia, and South America.

38.4 POLYMERIZATION

One of the older commercial processes for producing PP is solution polymerization. Now generally considered obsolete because of its high costs and limited product range, it is still used by Texas-Eastman owing to its strong position in specialty products. The polymerization is carried out in stirred reactors with feeds of propylene, a hydrocarbon solvent, and Ziegler catalysts. The polymer solution is flash concentrated to remove dissolved propylene and part of the hydrocarbon solvent, then filtered hot to remove catalyst residues. The filtrate is transferred to a stirred tank and cooled to precipitate the isotactic fraction. This insoluble portion is separated by centrifuging, again extracted with hydrocarbon and centrifuged, and washed and dried.[22,23]

The workhorse process for commercial production of PP has been slurry polymerizations in liquid hydrocarbon diluent; for example, hexane or heptane. These are carried out in either stirred batch or continuous reactors. U.S. suppliers producing at least some portions of their PP types via the hydrocarbon slurry process technologies are Amoco, Arco, El Paso, Exxon, Gulf, Himont, Shell, Soltex, and USS Chemicals.[24]

In the slurry process, high-purity (greater than 99.5%) propylene is fed to the reactor containing diluent, as a suspension of solid catalyst particles is metered in.[6] Reaction conditions are commonly in the ranges of 122–176°F (50–80°C) and 5 to 20 atmospheric pressure. Crystalline polymer made under these conditions is insoluble and forms as a finely divided granular solid enveloping the catalyst particles. Monomer addition is continued until the slurry reaches 20 to 40% solids. Residence time varies from minutes to several hours, depending on the catalyst concentration and activity, as well as the specific reaction conditions. Molecular weight during polymerization can be controlled by process temperature and time; however, the addition of hydrogen as the chain-transfer agent is the preferred commercial method because of its low cost.[25]

The polymerization slurry of PP is discharged to a stripping unit where unreacted monomer flashes out for recycling. The catalyst is then deactivated and solubilized by the addition of alcohol. At this point, the bulk of the diluent, solubilized catalysts, and atactic PP are removed by centrifuging. The polymer is purified by steam distillation and/or by water washing with surface-active agents, followed by filtration and centrifuging and then drying. This dried powder is sometimes referred to as "flake" or "fluff," depending on the particle size and bulk density. The dried polymer can be stored, transported, or premixed with stabilizers to be used with or without pelletization.[23] By utilizing state-of-the-art, high-mileage catalyst systems, polymer yields roughly 100 times higher than those obtained by first-generation, Ziegler catalysts are now obtained. This minimizes or eliminates expensive soluble polymer extraction and catalyst residue de-ashing steps required in slurry processes developed for earlier Ziegler catalyst systems.[26]

To cope with relatively poor low-temperature impact strength of homopolymer polypropylene, for a number of years producers have blended various elastomers with homopolymers. Such mixtures are still produced to some extent by extrusion compounding to fulfill specific requirements. (These are now generally referred to as "alloys.") In-situ production of copolymers with ethylene as the comonomer in random or block configurations[27] has, however, been the major mode for producing medium- and high-impact copolymers in recent years. Production of random copolymers can be carried out by adding up to 3.5% comonomer to the propylene feed.[28] Block copolymers are made by sequentially feeding propylene and ethylene to two or more reactors in series. The molecular weight (MW) and molecular weight distribution (MWD) of homopolymer and copolymer fractions are controlled by the use of multiple reactors with appropriate operating conditions and hydrogen concentrations established for each stage.[29]

Propylene is readily polymerized in bulk; that is, in the liquid monomer itself.[30] Arco, El Paso, Phillips, and Shell are practitioners of bulk processing in stirred or loop reactor systems.[31] In either case, liquid propylene (and ethylene if random copolymer is desired) is continuously metered to the polymerization reactor along with a high-activity/high-stereo-specificity catalyst system. Polymerization temperatures are normally in the range of 113–176°F (45–80°C) with pressures sufficient to maintain propylene in the liquid phase [250–500 psi (1.7–3.5 MPa)]. Hydrogen is used for MW control. The polymer slurry, approximately 30 to 50% solids in liquid propylene, is continuously discharged from loop reactors through a series of sequenced valves into a zone maintained at essentially atmospheric pressure and containing terminating agents.[6] In older, low-activity catalyst liquid monomer systems, catalyst residues and some of the atactic PP are removed by either liquid propylene or alcohol/hydrocarbon washing. The concentrate slurry is discharged to a flash chamber for powder formation, drying, and subsequent pelletization with stabilizers. Technologies exist for production of propylene/ethylene block copolymers via bulk polymerization employing stirred or loop reactors; however, details of these are closely held as proprietary information, not yet generally described to the industry.[29]

Modern vapor-phase polymerization is represented in one form by the stirred gas-phase process originally developed by BASF and licensed by Norchem in the U.S.[23] The diluent-free BASF process, as well as Amoco's older and somewhat similar internally-developed system, provides sufficiently high yield of polymer per unit of catalyst so that de-ashing is not required.[32] Although products made in this way contain high levels of titanium and aluminum residues, a unique finishing step during extrusion pelletizing reduces active chlorides to an innocuous level.[23]

BASF process reactors contain a spiral or double-helical agitator to stir the polymer bed. Critical cooling of the bed is maintained by continuous injection of fresh high-purity propylene in a liquid or partly liquified state into the reaction zone. During polymerization, unreacted propylene is removed from the top of the reactor, condensed, and reinjected with the fresh propylene. Evaporation of the unpolymerized propylene absorbs the heat of polymerization and results in intense mixing of the solid polymer particles with the gas phase.[6] Energy costs of the process are economically attractive.

Dramatic improvements in Ziegler-type catalyst technology, which led to the production LLDPE in the late 1970s, have now been extended with process improvements involving granular polymerization products that have the potential to reshape the PP industry. The first such commercial process combines the Spheripol catalyst system jointly developed by Montedison and Mitsui Petrochemical with a Montedison-developed new, simplified bulk (liquid propylene) process technology operating with loop reactors. The Spheripol process is capable of directly producing a relatively large round bead with suitable density to eliminate the need for pelletizing for many applications.[29] Subsequently, Montedison and Hercules, Inc.,[33] formed Himont, Inc., a joint venture worldwide polypropylene company. Himont assumed responsibility for all polypropylene operations and technology of the parent companies on Nov. 1, 1983.

The Himont Spheripol loop reactor process is initiated by injecting specially prepared supported catalyst and cocatalyst into liquid propylene circulated in a relatively simple high L:D ratio loop reactor, followed by monomer removal (Fig. 38.2**a**). The homopolymer so produced can be circulated through successive ethylene and ethylene/propylene gas-phase reactors for insertion of copolymer fractions before final monomer stripping (Fig. 38.2**b**).[29]

In 1977, Union Carbide introduced the Unipol low-pressure gas-phase fluidized-bed process[34,35] for LLDPE, which has recently been adapted to production of PP homopolymers and block copolymers.[36,37] Utilization of the Unipol process was made practical by incorporating Shell Chemicals' high-activity catalyst technology. A simplified flow diagram of the Unipol gas-phase process as widely licensed for production of LLDPE is shown in Figure

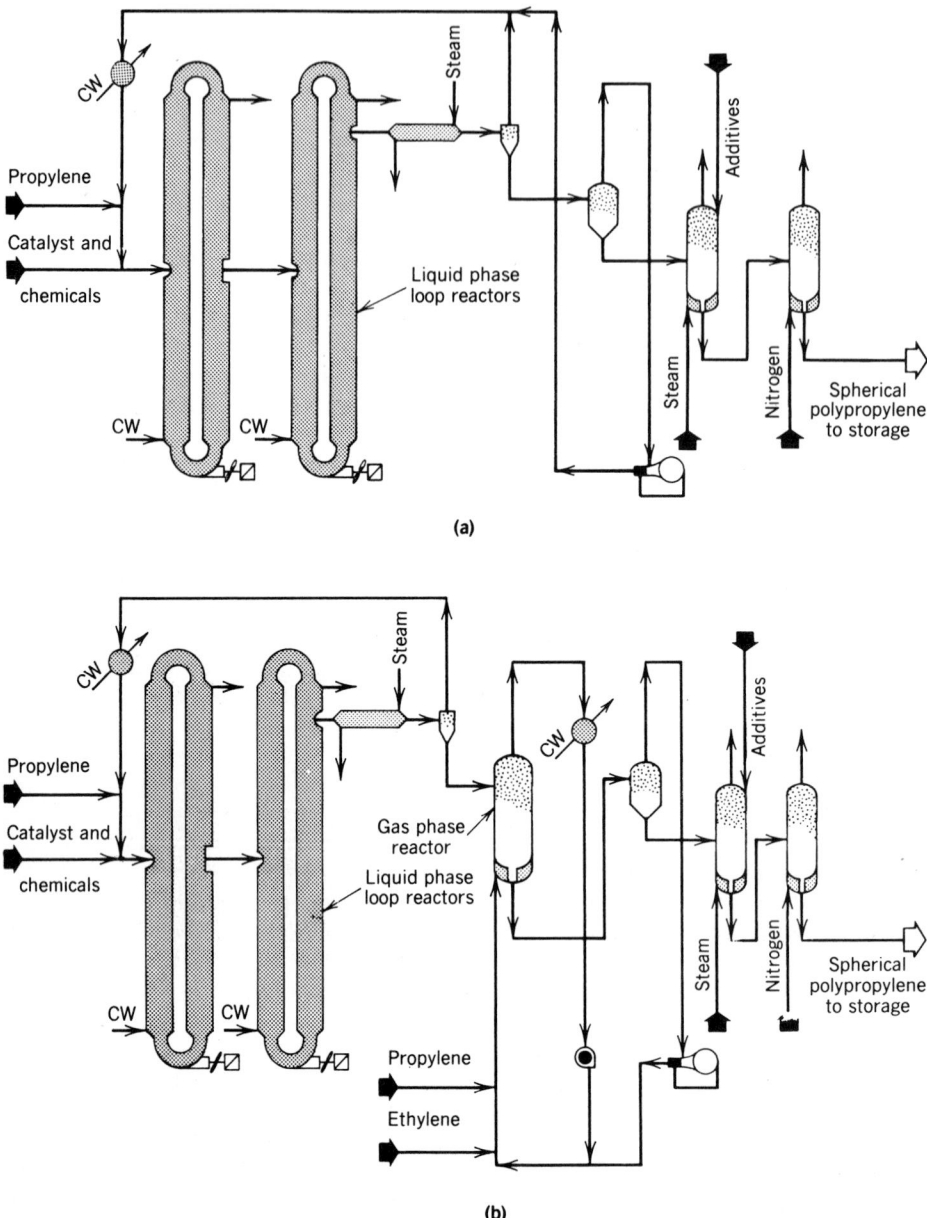

Figure 38.2. (a) Homopolymer Spheripol process. (b) Copolymer Spheripol process. (c) Union Carbide Corporation gas-phase process.

38.2c. Operation of this fluidized-bed reaction system to produce PP is said to be comparable that for production of LLDPE, with the potential for transition from PE to PP and back to PE on relatively short turnarounds.[38]

The Unipol reactor consists of a reaction zone at the bottom and a disengagement zone at the top. A gas-distributor plate is located to ensure distribution of monomer for fluidization. The reaction zone contains olefin particles sufficiently fluidized by the monomer (or monomer-comonomer) gas stream to entrap and distribute the catalyst throughout. Flui-

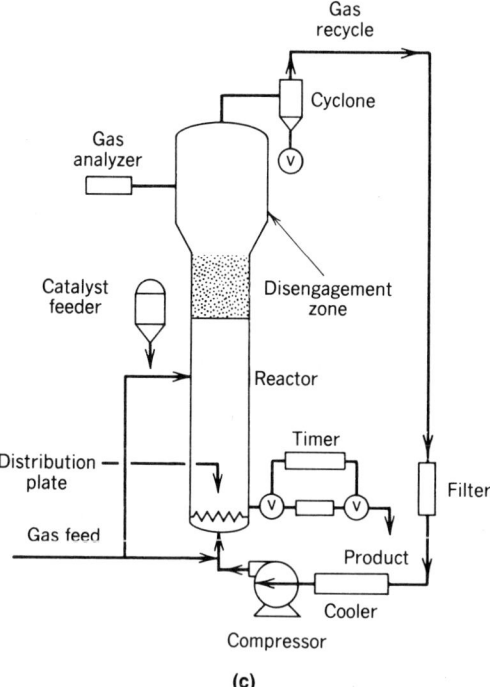

Figure 38.2. *(Continued)*

dization is maintained by the high rate of gas recycle—on the order of 50 times the product discharge/makeup feed. Unreacted monomer is disengaged for recycle while solid particles fall back into the bed for continuous removal through a series of sequenced valves.[39]

The Spheripol and Unipol processes are capable of producing polymer in "crumb" bead or granular forms, with the potential for direct marketing—eliminating any pelletizing finishing operation. Materials handling systems are in hand to store, blend, and transport such products; however, broad-based U.S. market acceptance of unpelletized products remains to be established.

Regardless of the polymerization process used, the propylene homo and copolymer must be stabilized to some degree to prevent oxidative degradation. Otherwise, free-radical chain reaction with the formation of peroxides is autocatalytic and accelerated by heat and ultraviolet or other forms of radiation.[40] The general practice is to incorporate a smally quantity of stabilizer in the polymer prior to its first exposure to the elevated temperatures of a drying operation or long-term storage. (Inert-gas blanketing is also used in some storage/transfer systems.) Additional stabilizers, up to 1%, are blended with the polymer during pelletizing. Selection of additives depends on the end user's fabrication processes, finished product requirements, environmental exposure, and anticipated length of service.[6]

Excellent thermal antioxidants are available to prevent degradation over extended periods of time at temperatures to 248°F (120°C). These are generally hindered phenols and hydroperoxide decomposers or various phosphites; however, most commercial PP compositions contain mixtures of these, since combinations are synergistic.[30,41,42]

In addition to basic stabilization for time/temperature-related effects, other additives are required for special applications. Most common are ultraviolet absorbers or screening agents that capture radicals or decompose the hydroperoxides formed when oxidation takes place. In addition to UV or sunlight resistance, these agents offer excellent protection against

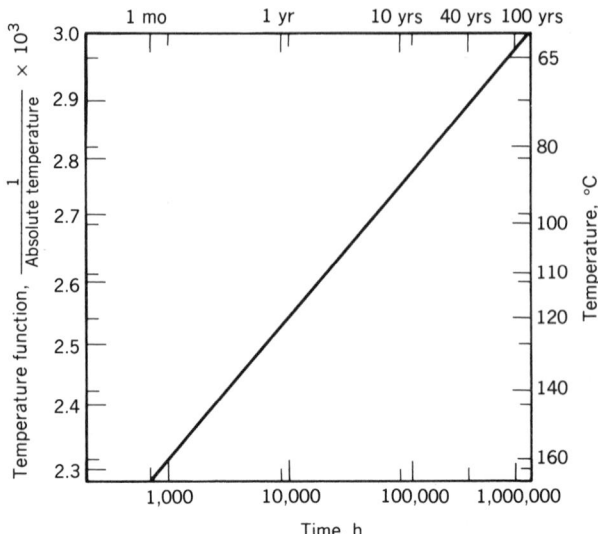

Figure 38.3. Extrapolation of dry heat stability data for general purpose polypropylene.

degradation by gamma or beta energy in radiation sterilization systems. Stabilizers resistant to gas fading are important for textile products, since these are often dried in gas-fired dryers that generate trace amounts of nitrogen oxides.[43] Special stabilizer additives are required for formulations intended for contact with copper wire or molding inserts because a short-term, high-temperature exposure to copper will catalyze thermal degradation of the polymer.[6]

The effectiveness of a stabilizer system in PP can be predicted by extrapolation of accelerated aging test data. Thermal life, that is, the product life that can be expected for a PP article exposed to various temperatures under different conditions, can be estimated by extrapolation of circulating-air oven aging data.[44] For applications involving wet or wet/dry exposures, test specimens are appropriately exposed prior to oven aging. The method of extrapolation used is the Arrhenius plot after the induction period (time to initial failure) at appropriate temperature levels has been determined (Fig. 38.3).[45] Experience has shown that the first sign of chalking can be considered the time to failure, and the retention of such properties as tensile elongation, flexibility, or brittleness closely correlates with this visual end point. In addition to variations in temperature, thermal life testing is influenced by sample thickness, oven air velocity and stabilizer volatility, including its mobility within the polymer.[46-48]

Several major end users of PP have their own accelerated aging tests tailored for specific industry interests (examples include appliances, automotive, wire and cable, fibers, and so on). ASTM D2445, Thermal Oxidative Stability of Propylene Plastics, measures resistance to oxidation at 302°F (150°C); however, such factors as abnormal loss of stabilizer by diffusion and volatilization limit the significance of data obtained in this single, elevated-temperature U-tube test.

Because of the nonpolar nature of polyolefin polymers, all types and grades of PP have exceptionally high resistance to most solvents and chemicals. They are unaffected by aqueous solutions of salts, acids, and alkalies, and only slightly swollen by immersion in oils and hydrocarbons at elevated temperatures. PP is subject to attack by strong oxidizing acids and chorinated solvents.[49] It has excellent resistance to environmental stress cracking as meaured by ASTM D1693.

38.5 DESCRIPTION OF PROPERTIES

"Polypropylene" is not one or even one hundred products. Rather, it is a multidimensional range of products with properties and characteristics interdependent on the

- Type of polymer: homopolymer, random, or block polymer
- Molecular weight and molecular weight distribution
- Morphology and crystalline structure
- Additives
- Fillers and reinforcing materials
- Fabrication techniques

Homopolymers have resistance to deformation at elevated temperatures; high stiffness, tensile strength, surface hardness, and good toughness at ambient temperatures (Table 38.2). Random ethylene–propylene copolymers are characterized by higher melt strengths. They have good clarity and resistance to impact at low temperature, gained at some sacrifice in stiffness, tensile strength and hardness (Table 38.3).

Block copolymers, preferably ethylene, are classed as having medium, high, or extra-high impact resistance with particular respect to subzero temperatures (Table 38.4). Block copolymers consist of a crystalline PP matrix containing segments of EPR-type elastomer particles and/or crystalline PE for energy impact absorption in the rubber phase.[6] The level of the ethylene comonomer between the elastomeric and PE phases, as well as the MW of these segments, has an important bearing on the physical properties of the final block copolymer (Figs. 38.4 and 38.5).

Both MW and MWD of PP polymers are controlled in the reactor by catalyst composition, polymerization process (batch or continuous), and temperatures, pressures, and monomer concentrations. Commercially, the melt flow rate (ASTM D1238, condition L) is used as an expression of MW. As the MW increases, the flow rate decreases. Reactor conditions are established to produce polymer of the particular MW or flow rate desired.

Generally speaking, higher MW leads to better physical properties but poorer processing. Lower MW gives reduced properties but easier processing (Table 38.1).

However, postreactor treatments (generally visbreaking, which refers to chemical deg-

TABLE 38.2. Properties of Typical Polypropylene Homopolymers of Various Flow Rates

Property	Pro-Fax[a] PD-701	Pro-Fax[a] 6323	Pro-Fax[a] 6523	Pro-Fax[a] 6823
Controlled rheology (CR)	Yes	No	Yes	No
Flow rate, D 1238 Cond. L, g/10 min	35	12	4	0.4
Tensile strength at yield D638, MPa	33.2	35.3	35.5	37
Secant modulus of elasticity in bending, D790, MPa	1,470	1,820	1,750	1,655
Izod impact resistance at 23°C, D256, J/m	32	37.3	42.7	161
Izod impact resistance at −18°C, D256, J/m	<16	<16	<16	21
Deflection temperature at 455 kPa Stress, D648, °C	90	102	100	90

[a]Pro-Fax is a registered tradename of Himont Incorporated. Fabrication processing; injection.

TABLE 38.3. Properties of Typical Random Polypropylene Copolymers

Property	Pro-Fax[a] SB-832	Pro-Fax[b] SA-868M
Controlled rheology (CR)	No	No
Flow rate, D1238 Cond. L, g/10 min	2	6
Tensile strength at yield D638, MPa	27.6	
Secant modulus of elasticity in bending, D790, MPa	1,295	1,068
Izod impact resistance at 23°C, D256, J/m	101.4	192
Izod impact resistance at −18°C, D256, J/m	<1,600	<1,335
Deflection temperature at 455 kPa stress, D648, °C	83	75

[a]Blow-molded.
[b]Injection molded.

radation—usually with peroxides) produce lower MW products through polymer chain scission, which also narrows the MWD. Such products are commercially described as "controlled rheology" (CR) grades, with melt flows of 15, 20, 30 and greater. These have impact strengths approximating those of much lower-flow standard resins. A visbroken polymer normally has a better balance of processing and physical properties for most applications than a nonvisbroken polymer of the same average MW. A low MW CR polymer with a narrow MWD gives both good physical and processing properties in most applications.

Several different types of chemical reactions can be used for controlled visbreaking of PP polymers. The most common in commercial practice is the addition of a prodegradant to the polymer before pelletization. The prodegradants used are mainly organic peroxides. Chemical visbreaking technology can be applied to obtain melt flow rates up to 1000. It is also applicable to products containing fillers, reinforcing additives, and most stabilizer/

TABLE 38.4. Properties of Typical Polypropylene Block Copolymers

Property	Pro-Fax[a] SB-751	Pro-Fax[a] 7533	Pro-Fax[a] SB-786	Pro-Fax[a] SB-787	Pro-Fax[b] 7823	Pro-Fax[a] 8523
Controlled rheology (CR)	Yes	No	No	Yes	No	No
Flow rate, D1238 Cond. L, g/10 min	30	4	8	20	0.4	4
Tensile strength at yield D638, MPa	24.5	27.3	27.9	23.4	28	20
Secant modulus of elasticity in bending, D790, MPa	1,050	1,295	1,447	1,089	1,120	1,065
Izod impact resistance at 23°C, D256, J/m	80	134	107	113	534	379
Izod impact resistance at −18°C, D256, J/m	21	32	908[c]	908[c]	53	53
Deflection temperature at 455 kPa stress, D648, °C	75	81	86 113[d]	78 109[d]	78	77

[a]Injection molded.
[b]Extension.
[c]Unnotched specimen.
[d]Annealed specimen.

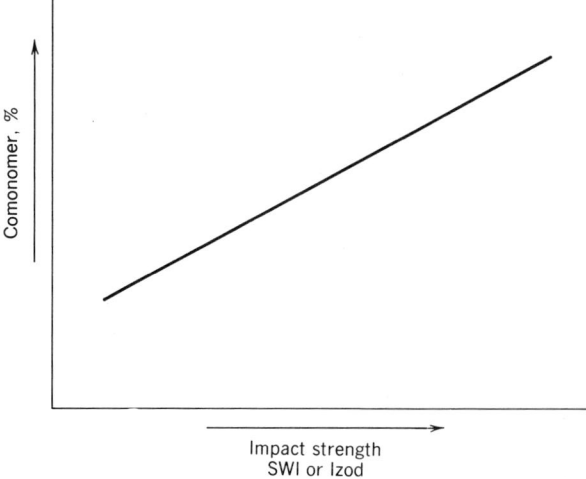

Figure 38.4. Effect of comonomer content.

pigment systems, with minimal or no sacrifice of other properties. CR resins may necessitate some changes in molding process parameters and part and tool design. The advantages of CR resins in film and fiber operations are evidenced in faster line speeds and greater throughput, which may be offset to sone extent by auxiliary or downstream equipment limitations.

New catalyst technologies, in conjunction with state-of-the-art polymerization, have demonstrated capabilities for in-situ production of PP resins comparable in all respects to post-treactor-produced CR resins. These technologies are still highly proprietary; however, licensing arrangements are being offered.

Since PP is essentially a crystalline polymer, its morphology and the nature of its crystalline structure plays a large part in the physical properties of homopolymers, principally with

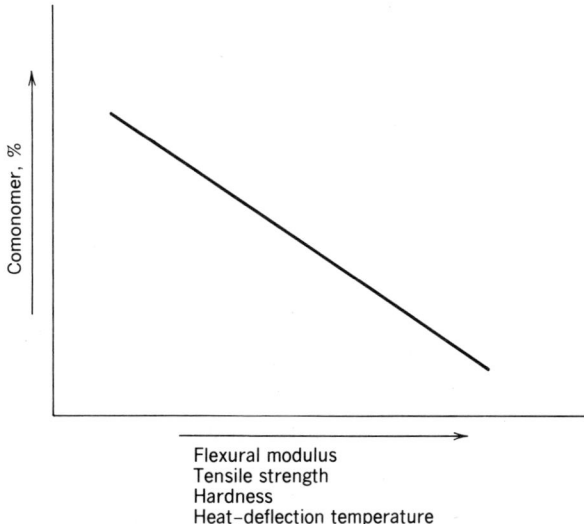

Figure 38.5. Effect of comonomer content.

respect to flexural modulus, surface hardness, and transparency.[50] The rate and manner in which the crystals form as the polymer solidifies from its noncrystalline, molten state affect both physical and processing characteristics. The addition of a small amount (below 0.1%) of a crystalline organic acid or metal salt in the formulation provides crystal nucleation sites which "seed" formation of smaller, more numerous spherulites as the crystal grows. This change in polymer morphology results in a marked increase in crystallization rate, higher crystallization temperature, and improved optical properties. In addition, nucleated products are noted for higher stiffness and strength properties, with better processing characteristics and shorter molding cycles than their nonnucleated counterparts.[51] Some pigments (phthalocyanine blue, in particular) also act as nucleating agents and change the characteristics of the colored compounds accordingly.

Additives that extend the range of properties and uses for PP are typified by flame retardants. Unmodified general-purpose PP products easily qualify as "slow burning" by most industry standards: *Underwriters Laboratories Bulletin 94* (flame class 94HB), ASTM E84, and U.S. Department of Transportation Federal Motor Vehicle Safety Standard (FMVSS) #302. By incorporating chemical substances that interfere with the combustion reaction, reduce flammability of the pyrolysis products, blanket the flame front with noncombustible products, or form solid nondripping ash, products can be made to meet the even more stringent flammability levels of 94 V1 or 94 V0.[42,52]

Foaming agents, both physical and chemical, can be added to PP compounds as dry powders or concentrates in pellet form prior to pelletization, molding, or extrusion.[53] The advantages gained are reduction in density, higher stiffness-to-weight ratio, opacity, and elimination of shrinkage, warpage, and sink marks in molded parts.[54]

Modification with mineral fillers such as glass, talc, mica, and calcium carbonate, or combinations of these, extends the range of properties and uses for PP in numerous ways.[55] The basic concept of reinforcement is utilization of strong, high-modulus fibers or platelets to carry a major part of the load applied to a composite (Table 31.5). Additives that chemically couple the PP matrix to the mineral surface enhance properties at elevated temperatures, particularly tensile, flexural, creep, and dimensional stability characteristics.[55,56] The most commonly used filler is talc, which doubles the stiffness at levels of 40%. Calcium carbonate, which is also widely used, improves the stiffness of homopolymers by

TABLE 38.5. Properties of Typical Filled/Reinforced Propylene Polymers

Properties	Pro-Fax 65F4-4 with 40% Talc	Pro-Fax 65F5-4 with 40% Calcium Carbonate	Pro-Fax GR6523/20 with 20% Glass	Pro-Fax PC-072/20 with 20% Glass Chemically Coupled
Flow rate, D1238 Cond. L, g/10 min	4	4	4	4
Density, D792A-2, g/cm³	1.22	1.22	1.03	1.03
Tensile strength at yield D638, MPa	32.2	26.9	44.8	81.4
Secant modulus of elasticity in bending, D790, MPa	3,450	2,765	3,800	4,180
Izod impact resistance at 23°C, D256, J/m	21	43	60	140
Unnotched impact resistance at 23°C, D256 (Modified), J/m	347	1,286		
Deflection temperature, D648 at 455 kPa, °C	134	120	121	158

only about 50 percent when used at the 40% level, but with a distinct superiority in toughness compared to talc-filled products. Surface-treated mica provides higher stiffness than talc, approaching values obtained with glass-fiber reinforcements.[57] Chemically coupled, 30% glass-filled grades of homopolymer PP are available with flexural moduli over 800,000 psi (5500 MPa) and deflection temperatures at 66 psi (0.5 MPa) of greater than 140°F (60°C).[58]

38.6 APPLICATIONS

U.S. applications for PP approximated 6.5 billion lb in 1987, distributed by the market classes shown in Table 38.6. The largest market area, fibers and filaments, is comprised of several segments with carpeting applications being the largest; these include primary and secondary/woven and nonwoven uses, carpet backing face yarns, indoor/outdoor construc-

TABLE 38.6. Major Markets for Polypropylene

	Million lb.	
Market	1987	1988
Blow molding		
Medical containers	48	53
Consumer		
packaging	66	73
Total blow	*114*	*126*
Extrusion		
Coating	27	28
Fibers and filaments	1750	1904
Film (up to 10 mil)		
Oriented	453	480
Unoriented	107	125
Pipe and conduit	37	39
Sheet (over 10 mil)	82	122
Straws	50	53
Wire and cable	44	46
Other extrusion	14	16
Total extrusion	*2564*	*2813*
Injection molding		
Appliances	141	148
Furniture	60	67
Housewares	183	202
Luggage and cases	9	9
Medical	148	160
Packaging		
Closures	283	305
Containers	238	252
Toys and novelties	32	40
Transportation		
Battery cases	130	133
Other	232	241
Other injection		
molding	191	196
Total injection	*1647*	*1753*
Export	1245	1202
Other[a]	1150	1410
Grand total	*6720*	*7304*

[a]Chiefly resold material and material used for blending.

tions, automotive interior mats and trunk linings, and synthetic turf. PP tying and baler twines are also widely used in industry and agriculture. The good wickability of PP is utilized in such nonwoven applications as disposable diaper and incontinence pad-cover stocks. Other applications include clothing inner liners, wiping cloths, drapes and gowns, tea bags, sleeping bags, and wall coverings. Furniture and automotive[59] upholstery fabrics are produced from both continuous multifilament and staple fibers.

The second largest PP market is film, both oriented (OPP) and cast. Large users of OPP film are packaging for snack foods, bakery products, tobacco products, pet dry foods, candy, gum, cheese wrap, and electrical capacitors. Cast (unoriented) film is used for packaging textile soft goods, cheese, snack foods, and bakery products. Many OPP film applications have grown at the expense of cellophane. Opaque, tenter-oriented films have established positions by displacement of paper, largely as bottle label stocks.[60]

Injection-molded applications are topped by those in transportation, particularly automotive and truck battery cases. PP copolymers have secured 85 to 90% of this market, the result of a drive by automotive manufacturers to reduce weight and cost. In addition to light weight, PP provides outstanding resistance to creep and fatigue, high-temperature rigidity, impact strength, and resistance to corrosion. The ability to mold hinges, undercuts, and other difficult configurations, as well as the availability of a wide variety of filled and reinforced compounds for increased stiffness and strength at temperature extremes, are key factors.

Packaging is the next largest molded product market, specifically closures and containers. Tamperproof, child-resistant, linerless features are important design factors, as are inherent chemical and stress-crack resistance, good color, and high productivity at low cost. Housewares, a closely related category, utilizes random copolymers for refrigerator and shelf-storage containers and lids. Other houseware uses in medium-impact copolymers are hot/cold thermos containers, lunch boxes, coolers, and picnic ware.

PP medical applications such as disposable syringes, hospital trays, and labware are in large part contingent on sterilizability, either autoclaving or radiation. Disposable syringes that are sterilized by radiation require special formulations to avoid discoloration (yellow) or brittleness as consequences of degradation.

Manufacturers of both major and small appliances have long histories of highly successful uses for PP. Washing machines, agitators, tub liners, bleach and detergent dispensing units, valve and control assemblies, drain tubes, and pump housings are some examples. In dishwashers, PP silverware baskets, dispensing units, dish racks, pump housings, and door liners led General Electric to convert from porcelainized metal to a one-piece injection-molded talc-filled-impact copolymer complete inner tub for its top quality models. Applications in refrigerators and freezers have been limited to ice cube trays, inner-door compartments, and breaker strips; however, these are expected to expand to large inner-door and chest liner components as thermoforming products and technologies develop.

Small appliance housings molded from PP are used for coffee makers, hair dryers, vacuum cleaners, can openers, knife sharpeners, room humidifiers and dehumidifiers, floor and ceiling fans, and window air-conditioner units.

38.7 ADVANTAGES/DISADVANTAGES

As "beauty is in the eye of the beholder," so are advantages often in the opinion of the user. Not infrequently, behavior that is usually a disadvantage can turn out to be an advantage. The flammability of unmodified PP, which can be a problem to a maker of electrical equipment, can offer benefits in trash-to-steam waste disposal. A generation of development work has resulted in grades that can overcome any given disadvantage.

Some properties that are usually considered inherent advantages of PP are

- Low specific gravity (density)
- Excellent chemical resistance

- High melting point (relative to volume plastics)
- Good stiffness/toughness balance
- Adaptability to many converting methods
- Great range of special-purpose grades
- Excellent dielectric properties
- Low cost (especially, per unit volume).

Properties usually considered disadvantages of PP are

- Flammability
- Low-temperature brittleness
- Moderate stiffness
- Difficult printing, painting, and gluing
- Low UV resistance
- Reduced extruder output (relative to soft/amorphous resins)
- Haziness
- Low melt strength.

PP grades are available today with brittleness temperatures below $-112°F$ ($-80°C$) of flexural modulus values above 1.5 million psi (10,300 MPa). Deflection temperatures at 264 psi (1.8 MPa) stress in excess of $-247°F$ ($-155°C$) have been demonstrated. PP is routinely glued, painted, and printed. Clearer grades and types with high melt strength are commercially available. Flammability characteristics meeting Underwriters Laboratory standard 94-VO can be obtained in PP. Hindered amine light stabilizers are key to UV-resistant PP grades that withstand 10 years of Florida exposure, in colors other than black. (More than 25 years has been demonstrated for black PP.) Controlled rheology resins and modern extruder/screw designs have largely eliminated extrusion rate problems.

In selecting resins for applications involving serious performance issues, a conversation with potential suppliers is often valuable. A lot of information on properties, service life, chemical resistance, and applications experience is available from the PP manufacturers.

38.8 PROCESSING TECHNIQUES

38.8.1 Injection Molding

An excellent balance of properties and good processing characteristics make PP ideally suited to all types of injection-molding machines. Polymers with a wide range of molding conditions, good flow, and good molten-state welding characteristics are reproducibly and practically molded to close tolerances through careful control of temperatures, ram pressures, and speeds.[61] Polypropylene can be processed in screw-injection, preplasticizing ram-injection, and conventional ram-injection machines. It offers a wide range of molding conditions, good flow, and good welding characteristics. It does not require predrying since no moisture is absorbed.

Start-Up Conditions. The shot size in ounces should be estimated for the particular molded item and compared with the machine's rated capacity. Polyolefin shot sizes have a maximum limit of approximately 75% of the machine's rated capacity in polystyrene. As the shot approaches the maximum limit, the cylinder temperature required to achieve the increased rate of material throughput increases.

The following start-up conditions are suggested for a screw-injection machine:

- Cylinder temperature, 420–470°F (216–243°C)
- Mold temperature, 80–100°F (27–38°C)
- Injection pressure, 60% of full

The molding cycle is an important factor in determining the production conditions, but it is impossible to specify under startup conditions. The molding cycle is dependent on part design, part dimensions, and other factors.

Regardless of the machine being used, the objective should be to attain 500°F (260°C) melt temperature.

Subsequent cylinder temperature increases are not usually made for the purpose of reaching higher melt temperatures; they are made primarily to compensate for greater material throughputs necessitated by higher output rates.

The start-up temperature for the mold is normally room temperature, which usually will ensure filling of the mold. All other variables are started at values that can be easily adjusted to attain optimum operating conditions.

Machine Adjustment to Attain Proper Mold Fill. When machine adjustments are made, it is very important to allow sufficient time for changes to take effect. Temperature adjustments, in particular, will take approximately 15 minutes to show up in the molded part. Experience has shown that many molding problems would have been solved if a more patient evaluation of the effect of variable changes had been undertaken.

Short Shot. A full shot can be achieved by operating with a slight material cushion at the end of the ram stroke. Care should be taken, however, to avoid building up an excessively large material cushion on succeeding strokes. Assuming that sufficient material feed is available, a short shot may still be encountered during startup. If the shot is just slightly short of filling the mold, an increase in injection pressure increments should be made in 20–50 psi (0.13–0.34 MPa) increments until a complete shot is obtained.

If the first shot is substantially short of mold fill, it is necessary to increase the cylinder temperature in 10 to 20° increments. Each incremental change must be fully evaluated before proceeding to further temperature increases or other process changes. Temperatures should not be increased excessively since prolonged exposure of the melt to temperatures above 575°F (302°C) will invite polymer degradation. Final cylinder temperature settings of 560–600°F (293–316°C) are not unusual when operating at fast cycle times.

Sometimes, a short shot is caused by improper venting of the mold. A reduction of clamping pressure might suffice, but if flash occurs at this point, improved venting usually will be necessary before proper molding operation will be feasible. Cavity vents should be held to a maximum of 0.0015 inch (0.04 mm).

Possible causes for a short shot include an undersized cylinder heating capacity; insufficient feed, injection time, or injection pressure; too low a stock temperature; a restriction in flow caused by an undersized nozzle orifice; an unbalanced multiple-cavity mold; foreign material clogging the nozzle or gates; or too low a mold temperature.

Mold Flash. Occasionally, flash may be encountered using the original startup conditions. Assuming that the maximum clamping pressure already has been set, the first step should be to reduce ram pressure in 20–50 psi (0.13–0.34 MPa) increments. This usually will be effective at the startup temperature levels.

At increased temperature levels, it is preferable to reduce the cylinder temperatures first. Reduction of cylinder temperatures in 10–20°F (6–12°C) increments until either the proper fill point or startup temperature levels are reached should suffice. Only if the difficulty persists should the injection pressure be reduced.

Possible causes for mold flash include too low a clamping pressure, too high a stock temperature, foreign material on the mold contact surfaces, or an injection pressure that is too high.

Machine Adjustment to Solve Surface and Part Defects.

Sink Marks. With the problem of sink marks, the best first step is to lower the mold temperature in 10°F (6°C) increments. The low-temperature limit is that point at which an incomplete mold fill is obtained.

If sink marks persist after decreasing the mold temperature to as low as possible, the injection time should be increased. If the mold gate is not solidified when the ram is pulled back, material may be drawn back out of the mold, causing sink marks. However, to avoid excessively long cycle times, the dwell time should be increased only until it is certain that no material is being drawn back.

Another way to eliminate sink marks is to remove the molded parts and quench them in cold water. This technique has proven very helpful when sink marks persist under conditions of gradual cooling.

Only as a last resort should the injection pressure be increased. Although increased injection pressure may solve the problem of sink marks, the packing of molds with high ram pressures can cause serious difficulties as a result of molded-in strains.

Possible causes for sink marks include too high a mold temperature; insufficient dwell time or ram pressure; poor part design, nonuniform sections, or excess wall thickness; an improperly located gate; or too slow a mold cooling.

Flow Marks, Weld Lines, Poor Gloss, Poor Pigment Distribution, Rough Surface. In the screw-injection machine, plasticizing of the material can be improved by varying either the stock temperature, the screw speed, or the back pressure. Increasing the stock temperature will make the melt more fluid and facilitate mixing at a given screw speed. With all other variables constant, better mixing can be obtained by increasing the screw speed. The effect of the higher speed will be more "working" of the melt within the cylinder and a higher temperature through the shearing action of the screw. Increasing the back pressure, with all other variables constant, will retard the flow of the melt in the cylinder, resulting in a longer residence time on the plasticizing section of the screw. This will also result in better mixing and dispersion. Mica specks on the surface are usually caused by excessive moisture, although an excessively high stock temperature or contamination may also contribute.

Possible causes for flow marks, weld lines, poor gloss, poor pigment distribution, or rough surface include too low a stock temperature, filling the mold too quickly, excessive moisture, poor pigment dispersion, a scratched or dirty cavity surface, excess mold lubricant, improper gate location or design, inadequate venting, or too low an injection pressure.

Brittleness. An increase in cylinder temperatures in 10°F (6°C) increments should minimize the possibility of molded-in strains that would contribute to brittleness in the molded part. The temperature must not be increased so high that degradation results.

If the trouble persists, the mold temperature should be increased in 10°F (6°C) increments. This should make filling the mold much easier and thus minimize molded-in strains. Many molders run their molds at 140–160°F (60–71°C) and claim very few brittleness problems. Mold temperatures significantly above this range, however, should be avoided.

Physical tests of the parts produced at various cylinder and mold temperatures should point the way to the optimum operating conditions. Whenever possible, physical tests should be conducted no less than 24 hours after the parts are molded. After 24 hours, the results may differ considerably from those obtained immediately after the part is molded.

Possible causes for brittleness include degraded material from the cylinder; too low a stock temperature; too low a mold temperature; use of improper color concentrates made from another resin; improper design, inadequate radii at corners, notch, or thread design; voids; or contamination.

Warpage. A serious problem encountered with complex-shaped injection-molded parts is warpage. Improper part design is the cause of most warpage problems, and an analysis of the design should be the first step in correcting the problem. Basic tenets of good part design are uniform wall section, radiused corners, and smooth flow of the melt. Since warpage is primarily a result of molded-in strains, the ram-injection pressure should be reduced if the part design cannot be significantly altered. The lower limit is that point at which short shots are obtained.

If warpage persists, the mold temperature of the cavity section—rather than the core—should be varied. Uneven cooling is a frequent cause of molded-in strains. Thus, an even cooling rate throughout the part should be the ultimate objective. Increased booster time to permit faster fill and extra cooling in the mold gate area are advisable.

Increasing the cylinder temperatures should be the last step in solving warpage problems. This can permit molding at reduced injection pressure.

Possible causes for warpage include overpacking in the gate area because of high injection pressure, improperly balanced core and cavity temperatures, molded-in strains caused by low stock temperature or the mold being too cold, too long a flow, insufficient gates, improper design, nonuniform walls, improperly balanced multiple gates, the part being ejected when it is too hot, or inadequate or poor location of the knockout mechanism.

Erratic Quality. When erratic quality is encountered in molded parts, it is usually related to machine performance or the capacity of the machine. Cylinders that are undersized in heating capacity and/or volume will result in a nonuniform melt from shot to shot, thus causing erratic quality. The erratic performance of cycle and temperature controls or poor mold design can also be possible trouble sources.

Possible causes of erratic quality include undersized cylinder volumetric and heating capacity, nonuniform cycle or feed temperature, erratic equipment performance, or unbalanced multiple-cavity layout and runner system.

Cycle-Time Reduction to Maximize Output. To reduce cycle time and thereby maximize output per unit time, the injection time should be decreased in 1-s increments until a time is reached when the ram is held forward 3 to 5 s after ram-forward motion has stopped, or as long as necessary to freeze off the mold gate. The booster time should be held to a minimum. Although a fast fill has been recommended earlier in this section, it is also important to note that a slower fill can correct many molded part defects and may be a preferable technique.

The cure time component should be reduced in 1-second increments. The limitations to reduction of this time component are the appearance and dimensions of the molded part. By a close inspection of the parts produced after each incremental change, the minimum cure time can be determined. The mold should remain open for a minimum period of time to permit the machine to eject or an operator to remove the molded product safely.

The minimum cycle that will maximize productive output can be determined, therefore, by studying each time component in the order that it occurs: reduce the injection time, set booster time, minimize the cure time, and minimize the part removal time.

Procedures for optimizing and troubleshooting molding operations with specific PP types are available through the technical service organizations of most primary suppliers, as are detailed recommendations for part and mold design with emphasis on computer-aided systems.[62]

Living Hinge. The integral "living" hinge has accounted for much growth of PP. It has resulted in many uses offering functionality, durability, appearance, and economy.

Concept. When it is stressed, PP's crystalline structure undergoes a morphological change, resulting in orientation and extremely high tensile strength. The effect of this change is significant—from 5500 psi (38 MPa) in the unoriented state to more than 80,000 psi (550 MPa) after stressing. Furthermore, thin films of PP have virtually unlimited fold endurance, as measured by standard film-testing procedures, and such films are completely resistant to stress cracking. Since this folding stress, or orientation by folding, results in stretching ratios as high as 2 or 3 to 1, PP's high elongation beyond the yield point is ideal for hinge applications.

Design Figure 38.6 is a cross-section sketch of a typical hinge design. This typical design can be modified in a number of ways to suit particular functional requirements. As shown, it could be a simple box with an open lid in the as-molded position. It is important to note the use of radii to improve flow of the melt and reduce notch sensitivity in the area of the hinge. The suggested use of a radiused restriction should also be considered, as this ensures bending at the thinnest point on a straight line along the centerline of the web and provides a controlled fit of the lid to the box. It is also desirable in some cases to include interlocking lips or other features to aid in the development of a straight-line hinge action. Figure 38.6 also indicates that shoulders on the two main bodies can be used to offset any curvature of these parts in the perpendicular plane. It is not possible to develop a web hinge along a curved centerline.

Since there is always tendency for the web to loop, dropping the upper plane of the hinge about 0.005 in. (0.12 mm) is advisable to allow the lid to fit snugly against the box when it is closed. A web-land length of 0.060 in. (1.52 mm) is optimum, with web thicknesses ranging from 0.008–0.15 inch (0.20–3.81 mm) as desired to obtain stiffness of hinge action or to ensure proper mold filling. This design is simply a modification of a flash gate into a cavity. The web will neck down the first time it is flexed.

Mold Design. The best possible location for a concentration of mold temperature control lines is in the area of the hinge. At this point, there will be some buildup of frictional heat, and the mold design necessitates a relatively thin extension of the mold steel in this area. In the design of some parts, it may be necessary to consider other means of controlling a temperature rise in the hinge area. Overheating will result in lamination similar to that which forms around a sprue with a poorly cooled gate or core.

Location of the gate or gates is as important as cooling in preventing lamination or weld

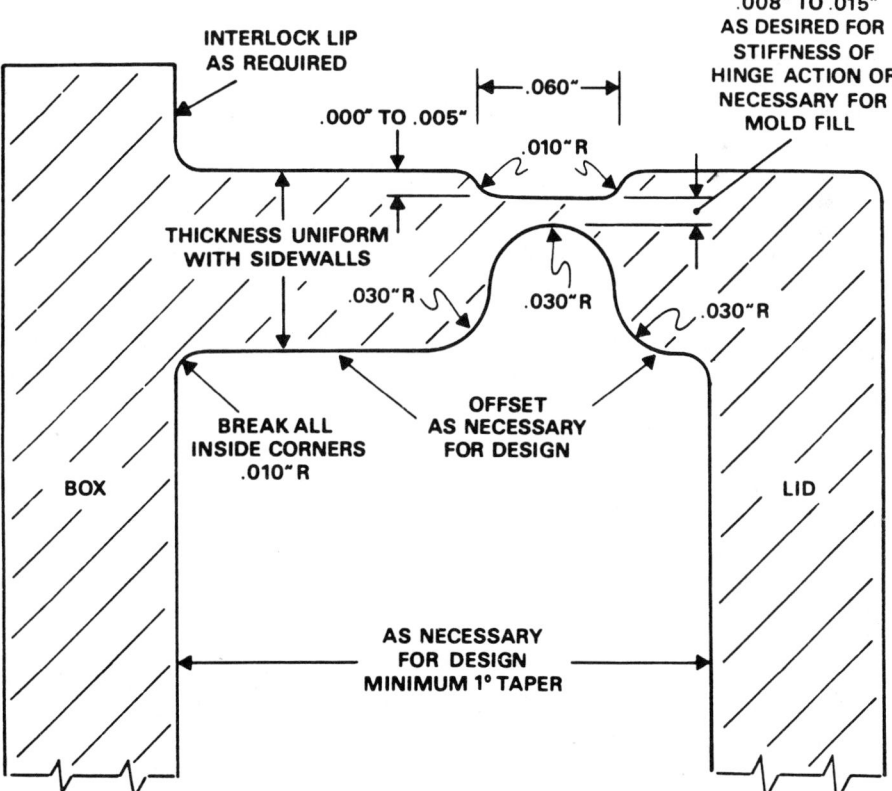

Figure 38.6. Typical polypropylene hinge design.

lines in the hinge area. While it might be expected that the cavity on either side of the hinge restriction should be gated to obtain complete fill, the excellent melt flow of PP at proper molding temperatures makes this unnecessary. In fact, if this is done, it inevitably results in a weld line forming in the hinge web. Such a weld line is weak and will likely fracture when it is flexed. It is commonly thought that, since the use of gating into each cavity is not desirable, the gate or gates into the one cavity should be located as near to the hinge as possible. However, this is not desirable either. Rather, the best placement of the gates is beyond the centerline of the cavity, away from the hinge.

The reason for this gate location becomes clear when the flow within the mold is visualized. If the gate is located near the hinge, the flow will naturally progress equally along the side walls in each direction from the gate until it reaches the hinge. Then it will stop at the hinge restriction until the less restricted sections of the cavity are filled. The continued pressure will break through the frozen section in the hinge to fill the second cavity. This start–stop–start flow in the restriction will inevitably result in lamination and subsequent skin fractures or peeling.

Cold-Formed Hinges. PP hinges can also be successfully cold-formed by a modified stamping process. This technique involves the use of a forming die having a suitable radius for the hinge profile. The forming die is pressed into the PP at a moderate rate, which causes the material to flow from under the die and at right angles to the line of the flex. High-speed stamping operations are not suited to this application, nor does continuous forming with a roller die provide the type of cold flow required to optimum orientation. Cold-formed PP hinge fabrication has made it possible to mold heavier webs and to easily fabricate working prototypes of molded hinge applications.

38.8.2 Fibers

Homopolymer monofilament and multifilament products are produced by conventional extrusion/drawing operations for a variety of uses. Extrusion temperatures generally range from 446–572°F (230–300°C) using special spinnerette dies, followed by quenching and orienting with differential speed-drawing rolls. Drawn fiber can be texturized for better bulk, and set by annealing at the desired draw ratio. Nonwoven fabrics are produced directly from extruded fiber by processes involving air- or stream-jet blowing to tensilize and randomly distribute the hot extrudate on moving screens.[43]

Monofilament is used in cord and rope applications. Tapes produced from fibrillated and slit uniaxially-oriented film are used extensively for baler twine, secondary carpet backing, and woven shipping bags. Upholstery fabrics, craft cord and yarns, filter fabrics, laundry bags, and commercial/institutional carpeting utilize the inherent stain-resistance and good wear properties of PP multifilaments. Nonwoven and spun-bonded fabrics are widely used for medical and personal disposables, carpet backing, geotextiles, and construction fabrics. Staple PP is now a standard in knitting yarn, automotive fabric, carpet, upholstery fabric, woven and nonwoven industrial fabrics, and disposable and durable nonwovens.[63]

38.8.3 Blow Molding

Blow molding of PP is carried out on conventional extrusion-blow and injection-blow systems and on proprietary two-step stretch (orientation) processes.[64] Nonoriented containers made by conventional processes utilize low (1 to 2) melt flow rate homopolymers to the obtain parison melt strengths required. Random copolymers overcome low melt strength limitations at some increase in cost but also improve low-temperature impact resistance and contact clarity. Both homopolymers and copolymers provide excellent resistance to environmental stress cracking (ESC), low moisture vapor transmission (MVT), and applicability in hot-fill applications.

Stretch blow molding, which is generally limited to homopolymers, produces lightweight containers with good clarity and stiffness that can compete with PVC and HDPE for household chemical and detergent uses. Composite blow-molding technologies[41,65] now combine the inherent advantages of PP (low MVT, chemical and ESC resistance, and rigidity) with the oxygen-barrier properties of other polymers, such as PVDC. Containers with these characteristics have opened large markets in food and condiment packaging.

38.8.4 Film

Cast PP film can be made by several different methods including roll-quenched and water-quenched flat processes and water-quenched tubular bubble processes.[66] Composite films that take advantage of the good water-vapor barrier properties of PP and the oxygen-barrier characteristics of PVDC are widely used. Oriented films with high clarity and stiffness are produced by postextrusion stretching, either by tubular or flat film tentering.[60] These films are biaxially oriented to provide balanced strength in both machine and transverse directions. A heat-set step is used for OPP films other than shrink film. Treatments to improve printability, coatings to reduce oxygen permeation, and additives to improve slip, block, and heat sealing are broadly used in all types of PP film.

OPP films made by the tubular process are particularly suited to compete with low MVT coated cellophane in high-speed tobacco packaging machinery. Tenter frame-stretched PP, on the other hand, is better suited for production of heavier films for pressure-sensitive tapes and decorative wrap. Snack-food and bakery product films with special barrier characteristics are produced by coextrusion or lamination processes.[67]

38.8.5 Thermoforming

Relative to other olefinic and styrenic materials, PP is more difficult to thermoform because of its relatively sharp melting point and low melt strength. Machines modified for precise temperature control and management of the increased sag experienced with PP are now available.[68] Both homopolymer and copolymer grades with higher melt strength and processing stability have been introduced, along with a variety of improved processing techniques including plug assist, solid phase pressure forming (SPPF), melt forming, and stamping. Applications for thermoformed PP are essentially thin-wall [15–20 mil (0.04–0.05 mm)] containers for juice and food service, dairy products, and so on. Heavier sheet [30–40 mils (0.7–1 mm)] is used in production of margarine tubs, flower pots, and delicatessen containers using the SPPF process.

A variation of SPPF is the stamping of Azdel (product of PPG Industries) glass-reinforced PP blanks for the automotive and communications industries.[69] Specific applications are seat assemblies, station wagon and van load floors, battery trays, and overhead cable connector covers.

38.9 RESIN FORMS

Polypropylene is available in pellet form.

38.10 SPECIFICATION OF PROPERTIES

See the Master Outline of Materials Properties.

MATERIAL ___Pro-fax 6523 polypropylene___

PROPERTY	TEST METHOD	ENGLISH UNITS	VALUE	METRIC UNITS	VALUE
MECHANICAL DENSITY	D792	lb/ft³	56.4	g/cm³	0.903
TENSILE STRENGTH	D638	psi	5150	MPa	35.5
TENSILE MODULUS	D638	psi	200,000	MPa	1380
FLEXURAL MODULUS	D790	psi	245,000	MPa	1690
ELONGATION TO BREAK	D638	%	Depends on specimen molding, history	%	Depends on specimen molding, history
NOTCHED IZOD AT ROOM TEMP	D256	ft-lb/in.	0.7	J/m	37
HARDNESS	D785		100		100
THERMAL DEFLECTION T @ 264 psi	D648	°F	131	°C	55
DEFLECTION T @ 66 psi	D648	°F	214	°C	101
VICAT SOFTENING POINT	D1525	°F	310	°C	154
UL TEMP INDEX	UL746B	°F	239	°C	115
UL FLAMMABILITY CODE RATING	UL94		HB		HB
LINEAR COEFFICIENT THERMAL EXPANSION	D696	in./in./°F	50×10^{-6}	mm/mm/°C	90×10^{-6}
ENVIRONMENTAL WATER ABSORPTION 24 HOUR	D570	%	<0.03	%	<0.03
CLARITY	D1003	% TRANSMISSION		% TRANSMISSION	
OUTDOOR WEATHERING	D1435	%		%	
FDA APPROVAL			21CFR177.1520(c)1.1		21CFR177.1520(c)1.1
CHEMICAL RESISTANCE TO: WEAK ACID	D543		Excellent		Excellent
STRONG ACID	D543		Varies with acid		Varies with acid
WEAK ALKALI	D543		Excellent		Excellent
STRONG ALKALI	D543		Good		Good
LOW MOLECULAR WEIGHT SOLVENTS	D543		Nonpolar swells; polar excellent		Nonpolar swells; polar excellent
ALCOHOLS	D543		Excellent		Excellent
ELECTRICAL DIELECTRIC STRENGTH	D149	V/mil	600–700	kV/mm	24–28
DIELECTRIC CONSTANT	D150	10^6 Hertz	2.25		2.25
POWER FACTOR	D150	10^6 Hertz	0.0003		0.0003
OTHER MELTING POINT	D3418	°F	327	°C	164
GLASS TRANSITION TEMP.	D3418	°F	-4	°C	-20

454

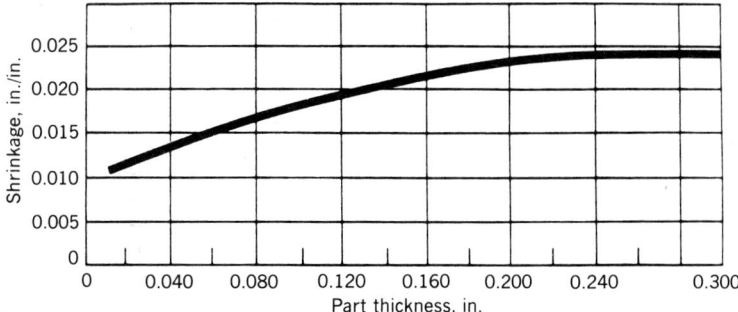

Figure 38.7. Polypropylene shrinkage variation with thickness.

38.13   SHRINKAGE

Shrinkage is the contraction of the injection-molded part that continues for approximately 24 hours after ejection from the machine. It is almost directly related to part thickness; heavy sections experience greater shrinkage than thin areas (Fig. 38.7). Shrinkage provides the fundamental argument in favor of designing molded parts with constant wall thicknesses. A recommended shrinkage shold be figured into the dimensions of an original mold design.

If minor shrinkage problem arises, the mold temperature should be decreased in 10°F (6°C) increments. By running a cold mold, a thick skin and thorough set can be obtained before the part is removed from the mold. If the difficulty persists, the cure time should be increased. This will have the same effect as the running of a colder mold, but it will lengthen the overall cycle time.

Another approach is to lower the cylinder temperature to minimize the necessary cooling. However, reduction of the melt temperature will increase the possibility of molded-in strains.

Only as a last resort should the mold be packed. This will reduce shrinkage, but will often cause molded-in strain or parts sticking in the mold. Excessive shrinkage is one of the more difficult molding problems to solve since the obvious solutions often will run counter to minimizing warpage.

Possible causes of excessive shrinkage include too high a mold temperature, too short a cure time, too high a stock temperature, poor part design, varying wall thickness, or an injection pressure that is too low.

38.14   TRADE NAMES

See Table 38.7.

TABLE 38.7. Trade Names of Polypropylene

U.S. Producers	Trade Name
Advanced Global	
Amoco	Amoco
Aristech	
Eastman	Tenite
Exxon	Escorene
Fina	Fina
Himont	Pro-Fax
Phillips	Marlex
Rexene	
Quantum, U.S.I.	Petrochem
Shell	Shell
Soltex	Fortilene

BIBLIOGRAPHY

1. G. Natta and co-workers, *J. Am. Chem. Soc.* **77,** 1708 (1955).
2. K. Ziegler and co-workers, *Angew Chem.* **67,** 426, 541, 547 (1955).
3. G. Natta, *J. Polym. Sci.* **16,** 143–154 (1955).
4. G. Natta, *Chem. Ind. (London)* **47,** 1520–1530 (1957).
5. G. Natta and P. Corradini, *J. Polym. Sci.* **39,** 29–46 (1959).
6. E. Vandenberg and B. Repka, "Ziegler-Type Polymerizations," *Encyclopedia of Polymer Science and Technology,* Vol. 29, John Wiley & Sons, Inc., New York, 1977, pp. 337–423.
7. G. Oppenlander, *Science,* **159,** (3521), 1311 (1968).
8. B. Maxwell, *J. Polym. Sci. C,* **9,** 43 (1965).
9. A-Fax (R) Product Bulletins, Hercules Incorporated.
10. *Hercules Mixer,* **41,** 1 (Jan. 1958).
11. Brit. Pat. 777, 538 (Nov. 22, 1954).
12. U.S. Pat. 2,692,257 (Apr. 28, 1951). (Hogan/Banks-Phillips).
13. E. Peters and co-workers, *Ind. Eng. Chem.* **49,** 1879 (1957).
14. U.S. Pat. 2,825,721 (Mar. 26, 1956).
15. A. Clark and co-workers, *Ind. Eng. Chem.* **48,** 1152 (1956).
16. *Chem. Week,* **111,** 12 (April 5, 1972).
17. U.S. Pats. 3,715,344, 3,112,300, 3,112,301 (Montecatini).
18. *Chem. Week,* 13–14, (March 23, 1983).
19. *Mod. Plast.,* 37, 39, 156, (Jan. 1958).
20. *Mod. Plast.,* 88, (Jan. 1961).
21. *Mod. Plast.,* 66, (Jan. 1971).
22. *Mod. Plast.,* 73, (Jan. 1981).
23. J. Forsman, *Hydrocarbon Process.* **51,** 130 (Nov. 1972).
24. J. Chriswell, *Chem. Eng. Proc.* 84092 (April 1983).
25. U.S. Pat. 3,051,690 (Vandenberg–Hercules).
26. L. Luciani and co-workers, "New Sterospecific Catalysts for Propylene Polymerization," *37th SPE ANTEC,* May, 1979.
27. J. Short, *Chemtech,* 238 (April, 1981).
28. U.S. 3,200,173 (Schilling-Hercules).
29. R. Tusch, *Polym. Eng. Sci.* **6**(3), 255 (1966).
30. D. Bari, "Polypropylene Technology and Economics" Chemsystems, Inc. (Sept. 1983).
31. G. Crespi and L. Luciani in M. Grayson, ed., *Encyclopedia of Chemical Technology,* 3rd ed., Vol. 16 Wiley-Interscience, New York, pp. 453–469.
32. C. Cipriani and C. Trischman, Jr., *ACS Paper* (March 1981).
33. *Chem. Week,* (Oct. 13, 1982).
34. "Two Hands Are Joined in Polypropylene," *Chem. Week* (May 25, 1983).
35. U.S. Pat. 4,011,382 (Mar. 8, 1977) (Levine and co-workers-Union Carbide).
36. U.S. Pat. 4,003,712 (Jan. 18, 1977) (Miller-Union Carbide).
37. *Plast. Technol.* 89 (Dec. 1983).
38. *Plast. Eng.* 33 (Dec. 1983).
39. *Plast. World,* 6–7, (Dec. 1983).
40. J. Short in M. Grayson, ed., *Encyclopedia of Chemical Technology,* 3rd ed., Vol. 16, Wiley-Interscience, New York, pp. 392–393.
41. L. Reich and co-workers, *Polym. Eng. Sci.* **11**(4), 265 (1971).
42. D. Buchanan, in M. Grayson, ed., *Encyclopedia of Chemical Technology,* 3rd ed., Vol. 16, Wiley-Interscience, New York, pp. 357–385.

43. *ASTM D-3012,* American Society of Testing & Materials, Philadelphia, Penn.

44. Himont USA, Inc, Product Data Bulletin PPD-35C.

45. J. Forsman, *J. Appl. Polym. Sci.* **9,** 2511 (1965).

46. J. Forsman, *SPE J.* **20**(8), 729 (1964).

47. M. Blumberg, *J. Appl. Polym. Sci.* **9,** 3837 (1965).

48. Himont USA, Inc., Product Data Bulletin PPD-32F

49. A. Brockschmidt, "New Polypropylenes: What's Available Today," *Plast. Technol.* 67–70 (Mar. 1982).

50. Kudre and co-workers, *SPE J.* **20** (10), 1113 (1964).

51. H. Beck and co-workers, *J. Appl. Polym. Sci.* **9,** 2131 (1965).

52. D. Scharf, *Modern Plastics Encyclopedia,* Vol. 60, McGraw-Hill Co., New York, 1984, pp. 134–140.

53. R. Heck, *Modern Plastics Encyclopedia,* Vol. 60, McGraw-Hill, New York, 1984, pp. 142–145.

54. *Desk Top Data Bank, Plastics,* Edition 6, Cordura, 1983.

55. L. Cessna, *Hercules Chemist* (58), 7 (April 1969).

56. *Reference Handbook for Pro-fax,* Himont USA Product Bulletin.

57. *Selection Guide to Mineral-filled Pro-fax,* Himont USA Product Bulletin.

58. *Bulletin DFG-14B Optimum Conditions for Injection Molding Pro-fax,* Himont USA Product Bulletin.

59. J. O'Toole, "Design Guide," 391–428, *Modern Plastics Encyclopedia* Vol. 60, McGraw-Hill, New York, 1984.

60. J. Lynch, *Modern Packaging Encyclopedia* 85, (1983).

61. J. Mock, *Plast. Eng.* 20 (Jan. 1984).

62. *Plast. Eng.* 28, (Jan. 1984).

63. G. Smoluk, *Mod. Plast.,* 44–46 (Dec. 1983).

64. *Mod. Plast.,* 66 (Sept. 1983).

65. "Structural Foam PP Truck Battery Cases," *Mod. Plas.,* 40 (Sept. 1983).

66. "Industrial Thermoforming is Ready for Big New Growth," G. Smoluk, *Mod. Plas.,* 58–61 (Aug. 1983).

67. "Surface Modifiers," *Mod. Plas.,* 48–49 (July 1983).

68. R. Wibbens, *Modern Packaging Encyclopedia* McGraw-Hill, New York, 1983, pp. 82–83.

69. Belcher, *Modern Packaging Encyclopedia* McGraw-Hill, New York, 1983, pp. 145–153.

70. A. Brockschmidt, "Additives '83," *Plas. Technol.,* 87–91 (July, 1983).

71. M. Colangelo and co-workers, "Additives '83," *Plast. Technol.,* 103–107 (July 1983).

72. *Plast. World,* 32–33 (Feb. 1983).

73. *Mod. Plast.,* 61 (Jan. 1984).

74. *Pro-fax in Automotive Applications* Himont USA Product Bulletin, July 1982.

75. CEH Marketing Research Report, Polypropylene Resins, *Chemical Economics Handbook* SRI International, Menlo Park, Calif.

76. *World Polyolefin Industry 1982–1983,* Chemsystems, Inc.

39

POLYSTYRENE

William E. Brown

Food Packaging Consultant, Midland, Michigan; formerly senior research manager, Barrier Packaging, The Dow Chemical Co., Midland, Michigan

39.1 INTRODUCTION

Styrene is, along with ethylene, propylene, and vinyl chloride, one of the plastics industry's "big four" building blocks.

Styrene monomer is a clear mobile liquid with the chemical structure

Monomer   Polymer

Polymerization occurs across the $-CH=CH_2$ bond, forming linear (unbranched), noncrystalline molecules hundreds to thousands of units long with corresponding molecular weights of several hundred thousand to millions.

Polymerization of styrene yields a clear, colorless general-purpose polystyrene (GPPS) thermoplastic that is hard and stiff. Rubber (diene and other types) is added to provide the extensibility, toughness, and impact resistance needed in certain applications. These latter styrenic plastics are referred to as "high-impact polystyrene" (HIPS), even when it might be more appropriate to use the term "medium-impact polystyrene" (MIPS; also RMPS for rubber-modified polystyrene) for those grades that develop only modest increases in toughness over the unmodified homopolymer. The physical properties that distinguish nonfiber-reinforced PS from other thermoplastics are summarized as shown in Table 39.1.

39.2 CATEGORY

Polystyrene plastics are available in a variety of types and grades that can be classed as follows:

39.2.1 General-Purpose Polystyrene (GPPS)

The following are nonreinforced styrenic plastics that may contain additives:

459

TABLE 39.1.   Distinguishing Physical Properties
of Polystyrene

Property	Minima	Maxima
Specific gravity	1.04	1.05
Tensile strength, psi	2400	8200
(MPa)	(17)	(56)
Elongation at break, %	1	65
Tensile modulus, psi	240,000	475,000
(MPa)	(1655)	(3275)
Short-time maximum use temperature,		
°F	180	233
(°C)	(82)	(112)
Impact strength,		
ft-lb/in.	0.2	4[a]
(J/m)	(10.7)	(214)[a]

[a]Recent developmental offerings may have impact strengths as high as 7 ft-lb/in. (374 J/m).

- Polystyrene—polymerized styrene with little or no additives and a softening (glass-transition temperature T_g) of 212°F (100°C).
- High-flow and medium-flow grades—These contain flow enhancers that permit the plastic to flow quickly and easily into complex molds and thin sections. These grades soften at temperatures as low as 165°F (74°C) and as high as 230°F (110°C).
- Heat resistant grades—These resins have glass-transition temperatures from 212°F (100°C) to 230°F (110°C). They provide a greater degree of stiffness in the glass temperature region than do other GPPS resins.
- Low residual content grades—These are designed especially for use in food packaging applications. They offer the very lowest taste and odor transfer. Certain recently commercialized resins contain less than 200 ppm monomer and ethylbenzene.

39.2.2   Impact Polystyrenes (HIPS)

HIPS are rubber-modified styrenics that typically contain 1–10% rubber by weight. They exhibit two or more glassy transitions: at least one for the rubber, which is very low; one for the styrenic matrix, which is in the same range as that of the homopolymer; and often one or more for the grafts of rubber with styrene formed during polymerization.

HIPS resins are available in high-flow, high-heat, and high-gloss grades, as are the general-purpose types. Their mechanical properties, particularly impact strength, overlap those of ABS-type polymers.

39.2.3   Specialty Grades—Based on Either GPPS or HIPS

- Structural foam molding grades—These are designed for in-plant addition of blowing agent to permit molding of parts with densities 5–30 percent less than parts of solid PS.
- Foam sheet extrusion grades—These resins are designed to provide optimum rheology for combined extrusion-foaming of sheet, which is later (or in-line) formed into containers and disposable ware.
- Ignition-resistant grades—These contain compounds that modify the burning behavior of PS, either making it more difficult to ignite or slower to burn.

• Fiber or filler-reinforced grades—These are grades having rheological properties permitting them to be efficiently blended with such inorganic materials as glass fibers or grades amenable to admixing with metal powders and so on.

39.2.4 Blends

GPPS or HIPS resins blended with other plastics can contribute to other plastic materials some of the easy processing, quick setting, and stiffness typical of PS. A proprietary blend of PS with polyolefins uses a TPE (thermoplastic elastomer) compatibilizer. Some blends are recognized separately, such as high heat resistant styrene copolymer/PC blends.

39.2.5 Expandable Polystyrene (EPS)

These beads of GPPS contain up to 5–8% of a blowing agent such as pentane. These grades are discussed at greater length in the section on EPS processing.

Most, if not all, manufacturers of PS offer grades that meet U.S. Food and Drug requirements for food and pharmaceutical packaging (see *Code of Federal Regulations, 21CFR 177*).

Often available within these grades are modifications to provide, for example, glossy or matte surface finishes, ease of flow, or release from molds via lubricated particle surfaces. Each manufacturer produces products specifically designed for particular conversion processes and end uses, and should be consulted for recommendations.

39.3 HISTORY

A concise early history of polystyrene presented by Barron[1] is summarized here:

Styrene has been known, at least since 1831, when Bonastre distilled it from a tree resin. Simon polymerized styrene by exposing it to sunlight in 1839. Berthelot, in 1869, discovered the first synthetic method, based on ethylbenzene, for making PS. Matthews in 1911 gave the first suggestions for commercial uses of PS as replacement for traditional materials. High cost and development of fine cracks which could lead to ultimate failure, limited use of PS to Germany. Britain began production of PS in the 1930s and in 1936 commercial production commenced in the U.S.

Warner[2] found evidence to credit Neuman with earlier discovery of styrene as a result of experiments similar to those of Bonastre and described, under the title "Storax," in Nicholson's 1786 publication (*A Dictionary of Practical and Theoretical Chemistry*), a second edition of which was published in 1808. Warner successfully repeated the Neuman and Bonastre experiments with storax (liquid amber resin) from the balsam tree, proving the presence of styrene. Simon retains the distinction of naming styrene. Many other notable early day chemists, their works described in useful detail by Warner, contributed to the extensive analysis and experimentation on which modern styrene chemistry is based.

39.3.1 Initial Uses

The initial uses of PS featured its excellent dielectric and optical properties, although such consumer goods as containers, small appliance parts, toys, and novelties became the most visible applications. Polystyrene has unusually good resistance to conduction of electricity, and so its use in electrical devices spread quickly. Styrene production was diverted during World War II to the making of synthetic rubber, thus supplementing limited natural rubber supplies. The recovery years, 1946–1950, saw rapid growth of styrene from a synthetic rubber precursor to a basic raw material for a new family of plastics.

The easy moldability of polystyrene led to its almost de facto adoption as the standard by which other plastics could be rated in the various molding processes, especially injection molding.

39.3.2   Rubber Modification

Rubber modification, developed and commercialized more or less simultaneously by several manufacturers, emerged in the late 1940s. Rubber was blended on a two-roll mill with GPPS to provide toughness exceeding that of unmodified polystyrene. By 1969, HIPS displaced GPPS as the largest volume grade of styrenic plastic. The early shortcomings of HIPS attributable to oxidation-susceptible rubbers were surmounted with stabilized rubbers, while the tendencies of PS to yellow during exposure to visible light were reduced by use of light stabilizers.

Melt blending has since been supplanted by copolymerization grafting of rubber dispersed in styrene monomer.

39.3.3   Extruded Polystyrene Foam

Extruded polystyrene foam, invented in Sweden in the 1930s, became recognized during the war years as a material especially suited for flotation. It was later developed for thermal insulation and earned rapid growth as a product of The Dow Chemical Company. Typical insulating foams of PS have thermal conductivities of <0.15 to 0.25 Btu-in./h-ft^2-°F (<0.26–0.11 W/m°K), densities in the range <1–2 lb/ft^3, (<0.016–0.032 g/cm^3), and compressive strengths at 10% deflection of 10–40 psi (0.69–28 MPa).

39.3.4   Expandable Polystyrene

Expandable polystyrene (EPS) was developed in the early 1950s by Dow and others and consisted of PS granules made by suspension polymerization and impregnated with a gaseous expanding agent such as pentane. EPS beads are molded into finished shapes by enclosing partially pre-expanded beads in a mold into which steam is introduced. Objects as small as food and drink cups and as large as cooler chests and insulating boards are produced this way. Special forms of expandable beads also are used, after further expansion at point of use, as loose fill cushioning for packaging fragile objects.

39.3.5   Structural Molded Foams

Structural molded foams are high-density foams that are injection molded of PS pellets containing a blowing agent [or by a process by which a gas (nitrogen) is introduced into the polymer melt]. These were brought to market in the 1960s to provide densities in finished objects 5–30% lower than those of solid PS. They have application in furniture, furnishings, containers, and as general wood replacements. Very large parts are readily molded.

39.3.6   Production

Polystyrene was first produced commercially in the United States by The Dow Chemical Company in 1937. U.S. production grew rapidly, to 2 million pounds in 1940, 20 million in 1945, and more than 250 million in 1950.

In 1960, U.S. production totaled 980 million pounds, of which 733 million were molding and extrusion resins (including 120 million for export); 122 million were used in protective paint and paper coatings; and the remaining 125 million represented many of those resins that later became specialties (resins for foam and off-grade resins).

Extraordinary developments in plastics fabrication and processing methods were largely focused on PS. In the period 1957–1961, growth was 13.5%, compounded yearly. At that time, PS accounted for about 17% of the poundage of all plastics produced in the United States. It was exceeded in volume use only by polyethylene, which in 1960 accounted for nearly one fourth of U.S. plastics consumption.

By 1970, the U.S. plastics industry had enlarged production three-fold, to 19,600 million pounds, over its 1960 output. Polystyrene production (including GPPS and HIPS) totaled 1920 million pounds, an increase of 96% over 1960. PS thus accounted for 9.8% of plastics production and more than 10.3% of sales.

Styrene–acrylonitrile (SAN) resins and the terpolymer acrylonitrile–butadiene–styrene (ABS) resins had by then become recognized as separate entities in the statistics of plastics.

During the 1970s, PS consumption grew at a compound annual rate of 6.3%; in 1980, it accounted for about 10% of all plastics sales in the U.S., or 3540 million pounds. By comparison, "other styrenics" reached 770 million pounds of sales in 1980.

In 1987, U.S. sales of PS reached 1.1 billion lb; worldwide sales, 2.3 billion lb. In 1988, The Dow Chemical Company is adding capacity—restarting some 180 million pounds of idle PS capacity and adding 115 million pounds of HIPS capacity.

39.4 POLYMERIZATION

Styrene monomer is prepared by reacting benzene with ethylene to produce ethylbenzene (EB),

Benzene + Ethylene $\longrightarrow$ Ethylbenzene (EB)

followed either by catalytic dehydrogenation to styrene,

Ethylbenzene + Catalyst + Heat $\longrightarrow$ Styrene + Hydrogen

or by oxidation of EB to ethylbenzyl peroxide, followed by reaction with propylene. The resulting methyl benzyl alcohol and propylene oxide are dehydrated to styrene.

Styrene monomer is prevented from polymerizing during shipment and storage by adding to it a small amount of inhibitor and by keeping it cool.

Styrene can be polymerized simply by heating it. However, such mass thermal polymerization results in polystyrene that is too viscous to pump except at temperatures so high as to risk depolymerization. The reaction is also highly exothermic and, in large quantities, is difficult to control. Diluting styrene slows the reaction and, if the diluent is also a solvent for the polymer which forms, pumping is facilitated. EB, a solvent for PS (but not a reactant), is a common diluent in peroxide-catalyzed thermal polymerizations. (Styrene is also a solvent for PS but it is a reactant and thus is used up.) Heat is removed to prevent temperatures from rising above 350°F (177°C).

Polystyrene does not form a second phase in mass solution polymerization. It remains in solution, from which it is later recovered by flashing off EB and unreacted styrene. When impact-resistant grades are desired, polybutadiene and other special rubber compounds are dispersed in the solution.

- At first, batch mass polymerization was used exclusively. Now, continuous polymerization accomplished by pumping diluted monomer through stepped-temperature coil reactors accounts for about two thirds of U.S.-produced PS polymers.
- Most other polystyrene is made via batch suspension polymerization of styrene globules suspended in water. Particles a few thousand microns in diameter are flash dewatered and spray-dried. Water suspension greatly aids in exotherm control.
- Other polymerization methods have been used but continuous mass solution and batch suspension dominate commercial activity.

The reactions are represented by

$$\text{Styrene } + \text{ Heat } + \text{ Catalyst} \rightarrow (\text{Styrene})_n$$

Where n = ca 10,000–20,000.

Close control ensures that the polymer molecules will be the right sizes (lengths) and the right distribution of sizes to yield the desired properties, such as flow during extrusion and molding and mechanical strength in use.

While it is still hot, mass polymer is extruded as strands, chopped (while it is cooling) into pelletlike particles, and air-veyed into bags, packs, railcars, or trucks. When storage is needed at the point of production and use, it is usually in silos.

The forms produced from mass polymerized styrene are most commonly rod-shaped particles 0.10 in. (2.5 mm) or so in diameter by 0.10 in. (2.5 mm) or more in length. Suspension polymer particles are smooth, spherical beads a little smaller than the suspension particles from which they originated.

The densities listed in the Table 39.2 are bulk densities for granules and beads, and material densities for moldings.

Mass solution polymers are best suited for the manufacture of HIPS and crystal GPPS. Suspension polymers are more commonly used for manufacturing expandable bead PS and general product lines having diverse properties. Suspension polymerization offers greater polymer-tailoring versatility than does mass-solution polymerization; the latter, however, offers a cost advantage and requires less downstream cleanup by the manufacturer.

TABLE 39.2. Forms and Densities of Polystyrene

Forms of PS Plastics as Shipped	Typical Sizes		Typical Densities	
	in.	(cm)	lb/ft^3	(g/cm^3)
M&E granules,				
GPPS and HIPS	0.1	(0.25)	40	(0.61)
Foamable beads (EPS)	0.1	(0.25)	40	(0.61)
Moldings and extrusions, solid			66	(1.05)
Structural foam moldings[a]			50	(0.80)
Foamed sheet and moldings			1–6	(0.015–0.09)

[a]30% density reduction from resin at about 71.7 lb/ft^3 (1.15 g/cm^3).

39.4.1 Economics

PS is one of the least expensive plastics available. Large production capacities, efficient manufacturing, and a world market combine to place PS among the most cost-effective of plastics. However, its economics are affected by a variety of competitive forces.

- PS depends on hydrocarbon raw materials and so is subject to the general economics of oil and, to a lesser degree, that of ethylene.
- PS is subject to competition from several other large-volume plastics, notably PE, PP, and PVC, each of which may compete for the same end uses as PS. While substitution can occur on purely economic grounds, a switch in plastic resin often requires costly changes and adjustments in fabrication tooling and processes. Such changes may require a (predetermined) saving in resin cost over a substantial time to finance alterations in tooling and other manufacturing processes.
- Costs of polystyrene are linked to supply/demand balances for the monomer, and hence, energy costs in general.

GPPS and HIPS resins will usually cost about 1.5 to 2 times the market price of monomer and up to about 4 to 5 times that for specialty grades. In 1988, polymer list prices lie in the range of 58 to 65 cents per pound for major GPPS and HIPS grades and 68 to 75 cents per pound for EPS beads. Modified grades, such as ignition-resistant and heat-resistant, are priced at about 70 to 95 cents per pound. Market prices are often lower than list when supplies are plentiful.

The specific gravities of unpigmented GPPS and HIPS are 1.04–1.05, and so the list cost range per cubic inch (1988) is 2.1–3.2 cents.

39.5 DESCRIPTION OF PROPERTIES

GPPS is a hard, crystal-clear, amorphous solid at its common use temperatures. However, PS softens as its glass transition, T_g, which occurs around 212°F (100°C), is approached. Above T_g, PS behaves under stress as a viscous fluid; well above the glass temperature, PS is a readily molded plastic characterized by Newtonian flow at low shear stresses, changing to non-Newtonian flow as shear stress increases. Rubber-modified HIPS grades are opaque and are significantly improved in their toughness and impact resistance.

39.5.1 Thermal Properties

Styrene homopolymer is one of the stiffer plastics, having a flexural modulus upwards of 500,000 psi (3450 MPa) at ambient temperatures. Commercial grades range in this property from a low of 240,000 psi (1655 MPa) for rubber-modified and flame-retarded grades to a high of 475,000 psi (3275 MPa) for GPPS. Polystyrene retains its stiffness to about 36–45°F (20–25°C) below T_g. As the temperature is further raised, PS becomes rubbery and highly extensible, a several hundred percent increase, in fact, at just below the glass temperature. On release of strain, the material either

- Returns to its original dimension, providing the temperature is maintained, or
- Retains its deformed shape if it is quickly quenched but reverts to its preformed shape in obedience to "plastic memory" if it is reheated to the temperature at which it was formed.

Above the glass temperature, PS becomes more and more fluid as the temperature is increased until, at around 284–302°F (140–150°C), it becomes easily moldable by compression, injection, extrusion, transfer, and sheet forming.

The upper use temperature limit (UTL) for a PS article depends on the levels of stress molded into the part and its usage. The limit is defined by the amount of short-term deformation or long-term creep allowed in the part over its life; the corresponding load is derived from the creep modulus of the resin at the temperature of use. For parts free of internal stress and under no external load, the UTL is likely to be only marginally below the glass temperature; that is, 176–194°F (80–90°C). For parts internally stressed or under load, the UTL is lowered in proportion to the stress.

The deflection temperature under flexural load (DTUFL) of unmodified PS approaches the glass (a second order transition) temperature; at 264 psi (1.8 MPa), the DTUFL is 175–226°F (79–108°C). Lower loads raise the deflection temperature. At 66 psi (0.5 MPa), the deflection temperature is increased 9–13°F (5–7°C). Inclusion of glass fibers (for instance, 20 wt%) can increase the deflection temperature 18–45°F (10–15°C) by virtue of decreasing the mobility of the matrix polymer chains.

Another measure of heat resistance of PS is the Vicat temperature (VT), the temperature at which a flat-ended needle of 1-mm area under 1-kg load penetrates the plastic to a depth of 1 mm as the temperature in an immersion bath is raised 50°C per hour. For PS, the VT ranges from 194–226°F (90 to 108°C).

The value of VT, as DTUFL, is dependent in part on the thermal conductivity of the plastic. The temperature rise of the bath, not the specimen, is controlled.

In the absence of aggravation by solvents and stress-crack agents, the upper useful temperature range for PS is, therefore, limited by the glass-temperature range of 165–230°F (74 to 110°C). Maximum continuous use temperature falls from 20–40°F (11–22°C) below the DTUFL. [The glass temperature of polybutadiene rubber, −152°F (−102°C), is well below that of any common use for PS.]

With a thermal conductivity of 0.00012 cal-cm/cm²-s°C at ambient temperature, PS has a low propensity to conduct heat (in comparison to aluminum at 0.5, glass at 0.0017, and cork at <0.0001). This property of PS accounts for 7–10% of the heat conducted by PS foam made up of 3 wt% of PS.

- Thermal conductivity increases linearly with temperature: 0.31 Btu-in./ft²-h-100°F (0.000065 cal-cm/cm²-s-100°C).
- Thermal conductivity increases slightly with pressure; at 16,000 psi (110 MPa) molding pressure and 350°F (177°C), k = 2.05 Btu-in./ft²-h-°F (0.00022 cal-cm/cm²-s-°C), or about 8% more than at zero pressure.

Heating of PS for primary processing into finished parts is mostly facilitated by shearing the plastic, as, for example, in the barrel of an extruder.

On heating, polystyrene expands somewhat more than metals and glass (Table 39.3). When structures, such as containers, incorporate PS with these other materials, the design must take into account the difference in expansion on heating or cooling from the original assembly temperature.

TABLE 39.3. Thermal Expansion Coefficients, Δ1/1-degree

Material	Per °F	Per °C
Polystyrene	0.000028–0.000047	0.00005–0.000085
Bottle glass	0.0000056	0.00001
Steel	0.0000061	0.000011
Aluminum	0.000012	0.000022

39.5.2 Mechanical Properties

Mechanical properties are determined by the inherent nature of PS and, for some properties, by its treatment during fabrication.

- Unmodified and cast into sheets, PS has a tensile strength of 4000 psi (28 MPa), a modulus in tension of 400,000 psi (2760 MPa), and an elongation at break of approximately 1%. It exhibits no yield point. The stress-strain "curve" is, in fact, nearly a straight line to the point of rupture (see Fig. 39.1). During stretching, some crazing (development of apparently fine cracks extending eventually through the specimen as flat planes perpendicular to the direction of stressing) will become visible.
- PS is made considerably more ductile and toughened by including in it, either by copolymerization or blending, rubbery materials such as polybutadiene. The stress/strain curves of modified PS materials show much the same initial behavior of GPPS but, in addition, exhibit a yield point and cold drawing. Rubber may also reduce the stiffness of PS and thus the stress/strain curve will have a lower slope (Fig. 39.1).
- As measured by conventional Izod or Charpy methods, the impact strength of unmodified, unoriented GPPS is susceptible to brittle fracture at levels of 0.25–0.5 ft-lb/in. of notch (13.3–26.7 J/m). HIPS has an Izod impact strength from 0.8 to more than 4 ft-lb/in. of notch (43 to more than 213.5 J/m). It should be noted that HIPS becomes stiffer and less impact resistant at lowered temperatures (Fig. 39.2).

Thermal aging affects PS only slightly. A GPPS specimen 1/8 in. (3.1 mm) thick aged at 122°F (50°C) for 11,000 hours at low continuous stresses maintains 50% or more of its initial dielectric, tensile, and impact strengths when tested by Underwriters' tests. When it is exposed at either 125 or 175°F (52 or 79°C), HIPS declines only 5% in tensile yield strength

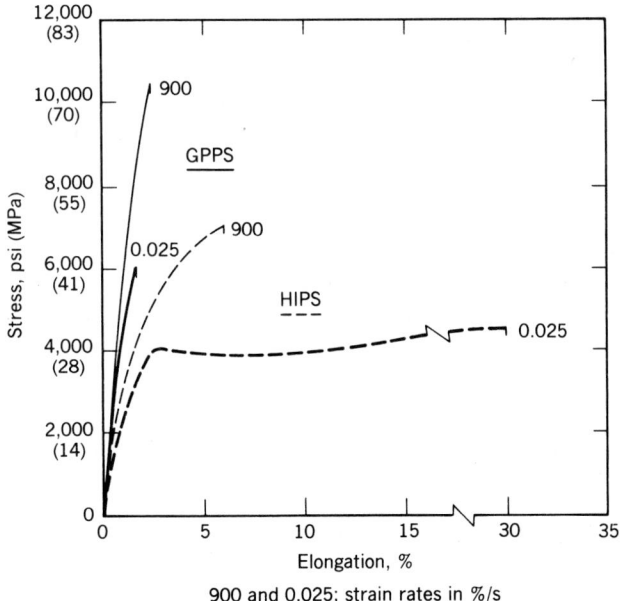

Figure 39.1. Response of polystyrene to different strain rates (%/s). Adapted from ref. 3.

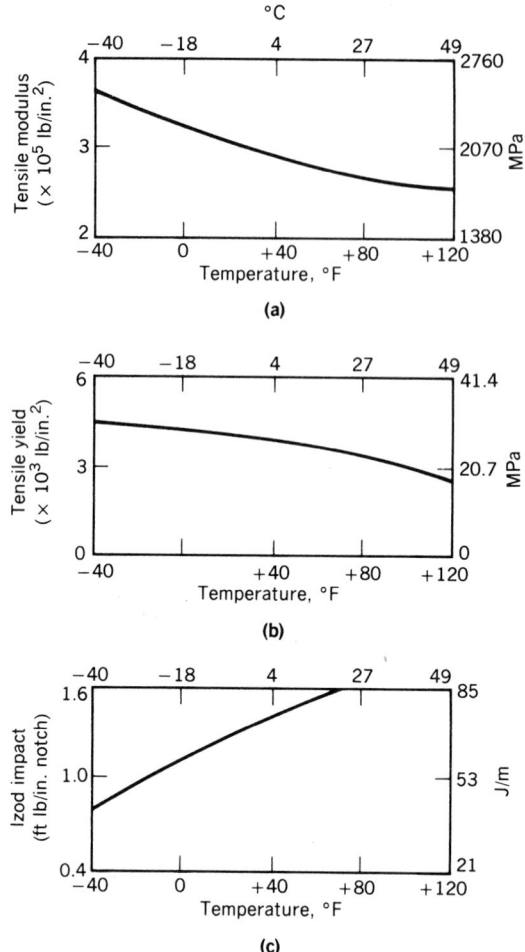

Figure 39.2.   Effects of temperature on (**a**) tensile modulus, (**b**) tensile yield, (**c**) Izod impact of typical HIPS (4).

(tested at room temperature) within 12 weeks, after which no further change occurs over the rest of the 32-week test.

Glass-fiber-reinforced PS is stiffer, stronger, and less extensible than unreinforced polymer. Tensile strengths are 10,000–12,500 psi (69–86 MPa); elongations at break about 1.25–1.5%; and flexural moduli of 900,000–1,200,000 psi (6206–8276 MPa) are reached at common glass-fiber loadings of 20 wt%.

The fatigue strength of PS depends on the stress mode, magnitude of stress, and the frequency of stressing. In typical reversed-bending fatigue tests at 30 cycles per second, GPPS sustains stresses up to about 6000 psi (41 MPa) for 100 cycles.

From 100 to 20,000 cycles, fatigue strength is linked closely to the number of cycles;

$$s = 10,350 - 2175 \log N$$

where s = stress in psi and N = number of reversed bending cycles. When N exceeds 20,000 cycles, s declines with N at much reduced rates (Fig. 39.3).

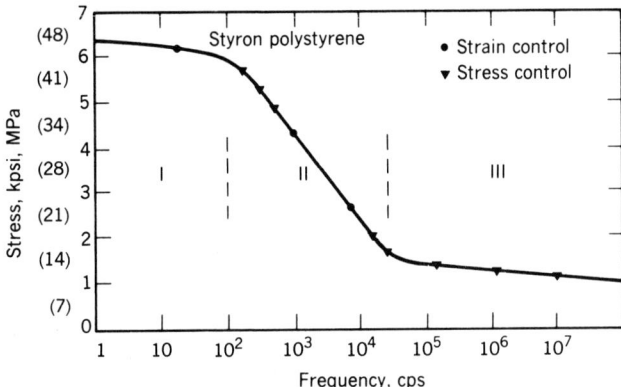

Figure 39.3. Fatigue strength of a typical GPPS part subjected to reversed cyclic stressing (5). Courtesy of Ford Motor Company.

The fatigue strength of typical HIPS at 72°F (22°C) with a 0.25-in. (6.4-mm) radius notch can be represented by the relation, $s = 5000 - 590 \log N$, where N is again the number of reversed-bending cycles and s is the stress. The relation holds over the range of available experimental data; that is, to 2.25 million cycles.

PS resins are not prone to creep much under loads commonly applied in contemporary applications. High loads, those that exceed normal design levels (more than 20–25% of short-time strength), produce noticeable deformation. GPPS compressively stressed at 2000 psi (13.8 MPa), for example, deforms 0.5% in 24 hours at 122°F (50°C); at 4000 psi (27.6 MPa), the deformation is 0.7%.

Stress relaxation of PS occurs whenever strain is sustained, as in the application of a plastic lid to a rigid container such as a glass jar. The corresponding modulus, called the apparent or stress relaxation modulus, declines with time, its initial value corresponding with the "short-time" modulus.

- For GPPS, the stress relaxation modulus [which is used in design formulas (Poisson's ratio for PS is 0.33 to 0.35 at ambient temperatures) for E, the modulus of elasticity] is thus initially (time approximately 0.1 h) the tensile modulus—about 450,000–500,000 psi (3103–3448 MPa) at 73°F (23°C). It declines after a year at 0.16% strain to 200,000 psi (1379 MPa).

- HIPS parallels GPPS in relaxation behavior. A rubber-modified PS with a tensile modulus of 370,000 psi (2450 MPa) at 0.1 h declines in stiffness to 360,000 psi (2480 MPa) at 1 h, 335,000 psi (2310 MPa) at 10 h, 307,000 psi (2115 MPa) at 100 h, 258,000 psi (1780 MPa) at 1000 h, and 200,000 psi (1380 MPa) at 10,000 h under sustained strains of 0.32%. The notch sensitivity of PS depends on the nature of the specific resin and on the geometry of the stress raiser.

- GPPS responds to stress-raising designs and imperfections as a semiductile material under slowly applied stress and as a brittle material under dynamically applied stress. Notch sensitivities thus range from about 0.5 to 1.

- HIPS plastics are usually ductile and, accordingly, exhibit notch sensitivities near zero at ambient temperatures.

- Temperature affects notch sensitivity. A rise in temperature decreases notch sensitivity; a decrease in temperature increases notch sensitivity.

39.5.3 Permanence

PS is resistant to chemical attack by a variety of substances. PS does not react with or absorb such materials or water to any substantial degree. These materials include

- Inorganic acids (except strong oxidizing acids), organic acids, aliphatic amines, bases, inorganic salts
- Foodstuffs, condiments, vegetable oils, beverages
- Pharmaceuticals, soaps, and detergents

PS is attacked by certain classes of materials (listed below). Attack may come in the forms of etching, stress cracking, or solvation, or a combination of forms. These materials include

- Hydrocarbons, aromatic amines, aldehydes, esters, and ketones
- Essential oils and insecticides

In most applications, sufficient data already exist to permit an informed judgment of the ability of PS to resist identifiable chemicals and mixtures. Otherwise, testing is often suggested for substances or mixtures not included in the voluminous collections of data assembled by manufacturers of PS.

39.5.4 Permeability

The relative impermeability of PS to water, water vapor, and many food constituents is the reason for its largest use. Packaging accounts for 30% of U.S. PS consumption, much of it for foods. Moisture, flavor, and atmospheric protection for periods of several weeks is required in many of these applications.

Like other plastics, PS is permeable to gases and moisture. For design at ambient temperatures, the rates shown in Table 39.4 may be used in calculations of mass transfer.

Although PS is not noted as a gas barrier, its use in impeding moisture flow is long-established. Water vapor transmission rates (at 90% relativity) of PS at dairy-case temperatures range from 0.35 to 1 g-mil/100 in.^{2}-day; a PS container 10 mils thick would thus have a permeability of 0.035 to 0.1 g/100 in.^{2}-day at this condition [37.5°F (3°C) and a water-vapor pressure of $0.9 \times 5.7 = 5.13$ mm Hg] or a permeability of 0.0068 to 0.019 g/100 in.^{2}-day-mm Hg.

39.5.5 Optical Properties and Weathering

PS is a clear, transparent, glossy material that is colored or pigmented in many transparent and opaque shades.

TABLE 39.4. Transfer of Gases and Moisture through Polystyrene

Material	Nitrogen[a]	Oxygen[a]	Carbon Dioxide[a]	Water Vapor[b]
Moldings (unoriented)	40–50	300–400	1000–1500	7–10
Oriented sheet (OPS)	50–60	250–350	700–1100	9

[a]Permeability in cm^3-mil/100 in.^{2}-day-atm at 73°F (23°C).

[b]Transmission rate in g-mil/100 in.^{2} at 100°F (38°C), 90% rh for permeability of water vapor, divide by the driving force in mm Hg, which at 100°F (38°C) and 90% rh is 44.28.

- GPPS is available crystal clear and in colors.
- HIPS is nearly always pigmented but may be purchased in natural (which is translucent to opaque, depending on rubber content) and colored in the plant by the molder, thus providing molders with certain economies.

The optical properties of polystyrene earned its earliest applications in lenses, gauges, and windows. Visible light is transmitted by GPPS to the extent of 88 to 91% with a rapid cutoff to less than 10% in the ultraviolet range (250 to 300 millimicrons). Near-infrared transmission is also high, about 70 to 90% at wavelengths up to 3.2 microns, where transmission drops and then rises again above 3.6 microns.

PS is not considered a weather-resistant plastic, although it can be stabilized against the yellowing and embrittling effects of artificial light and sunlight by the addition of ultraviolet (UV) absorbers. Finely divided pigments have increased the useful property retention of HIPS. Satisfactory exposure of two years in Arizona can be obtained with carbon black as pigment. However, longer than seasonal use outdoors is not advised.

39.5.6 Radiation Stability

PS is largely unaffected by sterilizing doses of high-energy radiation absorbed from gamma and electron beam radiation sources; therefore, it is well-suited to medical and food packaging requiring such sterilization methods. Repeated doses at the sterilization levels of 3 to 5 Mrad pose no threat to the properties of GPPS but may lower the toughness of HIPS. Higher doses tend to cross link styrenics.

39.6 APPLICATIONS

Applications for polystyrene exploit its easy and rapid moldability, strength and stiffness, dislike for water, electrical resistivity, and economy. The high visible light transmission of GPPS, coupled with lack of haze, make it valued in applications requiring outstanding optics. One of its earliest applications was in optical lenses.

Styrene plastics are familiar to consumers as containers in delicatessens and restaurants, as cassette cases and bodies, and as foam cups and trays. PS also is used extensively in retail food packaging, appliances, furniture and furnishings, business machines, and communications equipment.

In the early commercial days of polystyrene, much was made of its unusual dielectric behavior—for an organic material—and so it found rapid application in electrical devices. Its low dielectric constant and power factor (accompanied by an insensitivity of these properties to frequencies from household AC levels to billions of cycles per second) and good temperature acceptance [to 176–194°F (80–90°C)], along with a high breakdown voltage, were confirmed by many applications in electrical and communications equipment. Examples include electrical plugs and sockets, switch plates, coil forms, circuit boards, spacers, and housings.

Appliance parts, particularly for refrigerators, gave PS its initial entree into large-volume markets. By the 1950s, these had begun supplementing traditional materials with plastics. GPPS and HIPS were adopted because they offered increased performance at the same or lower costs—especially in the molding of finished parts from plastic granules—compared to the costs for fabrication of metals and glass. The ease of achieving complex shapes with built-in bosses, ribs, and lugs in large-area parts provided designers and molders with important advantages.

Other uses included containers, general housewares, wall tile, extruded wall coverings, toy parts, and disposable ware such as cutlery and dinnerware.

Many of these same uses of PS now have been upgraded with one or more of its many variants, including a growing range of impact (rubber-modified), weather-resistant, high-heat, and light-stabilized grades.

Structural foam is made by injection molding PS that has been impregnated with a blowing agent. Densities of moldings commonly range from 44–60 lb/ft^3 (0.7–1.0 g/cm^3). Large-area and shot size moldings are possible with SF technology.

PS beads containing a blowing agent activated by heating are made using the suspension process (see below) for later use in making such foamed objects as thermal insulation boards, hot drink cups, modular forms, and food trays that have low specific gravities (for example, 0.01 to 0.1).

Modern Plastics recognizes 35 categories of applications for PS in molding and extrusion resins, 7 in extruded foam board and sheet, and 6 in EPS (Table 39.5, which also gives a roughly typical percentage of the total market for GPPS, HIPS, and EPS).

PS is well suited to coextrusion with barrier polymers such as PVDC and EVOH for high-gas and moisture-resistant food and pharmaceutical packages; with foam innerlayers for disposable dishware; and with other plastics for structural, furnishings, furniture, and business machine uses. Coextrusion frees the designer to choose optimum materials and thereby achieve precise surface and bulk properties. Material combinations possible with coextrusion permit best cost-benefit choices for such properties as strength, stiffness, surface condition (such as gloss), stress-cracking resistance, color, and barrier, while meeting economic goals.

PS can be coextruded with most other plastics, including ABS, barrier polymers, blends, other styrenics, PETG, polycarbonates, polyolefins, TPEs, and vinyls. The principal criteria for coextrudeability relate to the rheologies of adjacent layers having flow viscosities and temperatures not sufficiently different to cause instability or upset. Glue layers commonly are required to ensure adequate adhesion between adjacent plastic layers through the subsequent usual forming cycle, which draws the coextruded multilayer sheet into the finished shape. Typical glues are ethylene vinyl acetate copolymers and blends including them.

A few special exceptions to the use of glue layers exist; some research has been directed to eliminating adhesives.

TABLE 39.5. Typical Applications for Polystyrene[a]

Application	%	Application	%
Packaging and disposables,	30	Toys and recreational	7
Rigid packaging		Housewares	6
Closures, lids		Billets from expandable beads	5
Produce baskets		EPS cups and containers	4
Flatware, cutlery		Miscellaneous consumer and industrial	4
Dishes, cups, tumblers		Foam board	3
Oriented film and sheet		Building and construction	2
Dairy containers		Furniture and furnishings	2
Vending and portion cups		All other (except resale and blending)	15
Plates and bowls		Total	100
Foam sheet	13		
Food trays			
Egg cartons			
Single-service ware			
Appliances and consumer electronics	9		
Air conditioners			
Refrigerators and freezers			
Small appliances			
Cassettes, etc			
Radio, TV, stereo cabinets			

[a]*Modern Plastics.*

Styrenic polymers are blended with other plastics to take advantage of the former's stiffness, strength, and quick setting from the easily formed melt state.

Compatibilizers are often necessary to provide bridging between domains of PS and "X." PS is not compatible in its molecular range with most other polymers and otherwise forms weak blends with them.

PS does not sharply change viscosity on melting, as do crystalline polymers, and so it is readily formed over a wide temperature range. Rather than sudden melting to a highly fluid state, PS softens progressively with a rise in temperature; similarly, PS hardens progressively. It develops "green strength" after chilling in the mold or exiting from the die. Moreover, PS releases well from many surfaces after cooling, though release agents (notably silicones) facilitate the process in complex molds, at low draft angles, and for some formulations.

39.7 ADVANTAGES/DISADVANTAGES

39.7.1 Advantages

PS presents the product designer with many useful properties.

First among these is low density—only marginally above that of water, less than 14% that of steel, and 40% that of glass. Combined with a low cost per unit weight, this feature makes PS economical on a volume basis. Among large-volume plastics, only polyolefins are less costly per unit volume than PS. Low density makes handling of PS easier and makes the shipping of it, and products made from it, lower in cost than shipping many other materials.

Clear transparency matching that of glass earns GPPS applications in packaging as bottles, jars, and lids; in such consumer electronics uses as cassette cases; and in disposables—drinking cups, food trays, and deli containers. Transparency allows PS to be used where see-through is important, as in medical and food packaging; in laboratory ware; and in certain electronic uses. PS is easily colored in a wide variety of tones, providing benefits of color matching and coding.

GPPS is stiff; compared to other thermoplastics, it does not deform excessively under load. Creep is not a major concern. HIPS is less stiff than GPPS, but nonetheless has a modulus equivalent to that of polycarbonate and at a lower density. The cost and density-normalized stiffnesses S of the polystyrenics are among the highest of all thermoplastics.

The density and cost-normalized stiffness can be calculated by dividing the modulus E by the density d and by the cost per pound, C:

$$S_{c,d} = \frac{E}{d \times C}$$

S is the cost-normalized specific stiffness in in.-lb/\$; C is the resin cost in \$/lb; and d is the density of resin in lb/in.^{3} Thus, with a modulus of 475,000 psi, a density of 0.0379 lb/in.^{3}, and a price of \$0.60/lb, the $S_{c,d}$ of GPPS is 20,900,000 in.-lb/\$.

Electrical insulation value and high breakdown voltage advantageously supplement other performance requirements. Amorphous melting over a wide temperature range simplifies the molding of PS. It provides the benefits of quick filling with rapid setting, permitting fast cycles, and provides for molding complex shapes with accurate reproduction.

Hydrophobicity of PS obviates the time and cost-consuming drying otherwise necessary in the molding or extruding of some thermoplastics. Also, PS is not subject to swelling in wet environments, so the original dimensions are retained despite such exposure.

Chemical resistance properties of PS allow a range of uses involving exposure to aggressive materials—foods, chemicals, pharmaceuticals, household cleaning agents, and cosmetics. PS can be etched by certain chemicals; this permits placing a write-in space on, e.g., petri dishes.

The Food and Drug Administration has long regulated PS for food contact use and so is quite familiar with its chemistry and its potential interactions with foods and beverages.

39.7.2 Disadvantages

Hot fill and high temperature service may be limited. PS has a glass temperature only slightly above the boiling point of water. Therefore, PS cannot be used above 176–194°F (80–90°C), approximately, because it loses its ability to retain shape. As a result, PS is not used in food retorts [250°F (121°C) at peak] or in sterilizing steam, under auto hoods, or in contact with hot grease (as in a microwave oven). PS solid and foamware are suited for heating water-based foods, however.

Ultraviolet light degrades PS, reducing the molecular weight and producing yellow-colored degradation products. The resulting discoloration and embrittlement preclude outdoor exposure except in uses permitting replacement or in seasonal uses.

PS burns when it is exposed to flame. Ignition-suppressant additives reduce the tendency for PS to continue burning after flame sources are removed. Extra cost and some adverse effects on mechanical properties result.

Hydrocarbons, aromatic amines, insecticides, aldehydes, and ketones attack PS, subjecting it to stress cracking, etching, and softening. The penalty is that certain uses are disallowed and the "universal container" concept is invalidated, which PS otherwise might well satisfy.

The electrical insulating ability of PS, whose surface and volume resistivities rank among the highest of all materials, explains its tendency to attract dust and hold it tenaciously. Dissipation of the surface charge responsible for this is achieved with surface treatments of moisture-attractants such as amines, quaternary ammonium compounds, or anionics. Antistats cost little but represent an inconvenience to the molder.

PS provides a hard, but scratchable, solid surface. For repeated use under conditions of rough surface abuse, PS should be protected with an appropriate coating, such as PVDC or acrylic deposited from a latex, or by a slippery surface coating such as a silicone. Alternatively, designs incorporating ribs and bosses to provide control of contact between articles can be used to advantage.

39.8 PROCESSING TECHNIQUES

Polystyrene is readily processed by all the methods of thermoplastic forming, including injection, extrusion (excepting extrusion blow-molding), sheet forming, compression and transfer molding, and casting (see the Master Material Outline).

The most common methods in use for PS include injection molding and extrusion. The polymer is heated initially by conduction from contact heaters and later, when it is melted, largely by frictional heat generated by shearing the plastic in the injection cylinder or in the extruder barrel.

- Plastic temperatures in these processes range from a low of 300°F (149°C) at the rear section of an extruder to a high of 525°F (274°C) at the molding machine nozzle for resins alone and 350–550°F (176–288°C) for resins reinforced with glass fibers and for mineral-filled resins. Low temperatures favor orientation, desirable in certain parts, while high temperatures favor development of gloss and complete filling of complex molds.

- Pressures in molding are commonly 5000–40,000 psi (34–276 MPa) at the injection nozzle and 1500–4000 psi (10–28 MPa) at the extrusion die. Low-viscosity materials (that is, high-flow resins) require lower pressures and lower temperatures. High-viscosity resins and fiber-reinforced and mineral-filled resins require higher pressures and/or temperatures. Pressures required decrease as the temperatures increase.

Master Material Outline

PROCESSABILITY OF ____Polystyrene____

PROCESS	A	B	C
INJECTION MOLDING	X		
EXTRUSION	X		
THERMOFORMING	X		
FOAM MOLDING	X		
DIP, SLUSH MOLDING			
ROTATIONAL MOLDING			
POWDER, FLUIDIZED-BED COATING			
REINFORCED THERMOSET MOLDING			
COMPRESSION/TRANSFER MOLDING	X		
REACTION INJECTION MOLDING (RIM)			
MECHANICAL FORMING			
CASTING			X

A = Common processing technique.
B = Technique possible with difficulty.
C = Used only in special circumstances, if ever.

Thermoforming is a widely used process for converting PS. Sheet of solid or foamed styrenic is extruded or coextruded and shaped by a variety of forming methods, the most common being by in-line vacuum forming. In that process, a vacuum is drawn between sheet and mold, or pressure is applied to the mold plug while air is drawn out (as in pure vacuum forming) from the space between the sheet and mold.

- Conditions for thermoforming sheet require temperatures several degrees above the glass temperature but low enough to preclude excessive drooping of the sheet as it is thermally conditioned under preheaters and positioned over the molds. Uneven thickness distribution and drag marks may result if the sheet prematurely touches the molds.
- Preforming under pressure is practiced in high-speed manufacturing of containers. Sheet is extruded, cut into squares, brought to the optimum uniform temperature (below the glass temperature but above a temperature at which PS will accept and retain shaping), and then formed under pressure [such as 50,000 psi (345 MPa)] into blanks for further forming between matched metal dies. PS in this "solid phase" can thus be pressed into rounds from squares and given considerable strengthening orientation. Blanks are then formed by any of the conventional thermoforming methods.

Blow-molding is used to produce bottles from injection-molded parisons of PS. Precise neck finish, thread, and body dimensional control result from injection molding parisons. Such precision is carried over to blownware.

Most blow-molding and casting have virtually disappeared from use in shaping PS in favor of the much faster and precise techniques mentioned above. Transfer molding, an early variant of compression molding, consists of heating a mass of plastic to malleable state and transferring it to a press containing matched metal dies wherein it is given form. Casting comprises pouring a molten mass into an open form in which it cools to a single-faced shape. Secondary fabrication processes commonly used with PS parts include:

- Assembly using a variety of fasteners, such as screws
- Machining
- High-speed printing
- Painting, vacuum metallizing, and hot stamping
- Coating by roller, brush, and spray for adhesion and gas barrier
- Electromagnetic and radio frequency interference (EMI and RFI) shielding by coating
- Bonding with adhesives
- Spin welding and ultrasonic welding

The most commonly used part assembly procedures employ snap fits and screws. Foam parts can be assembled with staples.

PS is easily machined with standard woodworking and metalworking tools, but parts can be designed to minimize machining and other secondary methods. Cutting speeds are adjusted to prevent softening and melting of the polymer.

Printing is done at exceptional speeds that keep up with primary processing while providing sharply defined multicolor graphics on a variety of shapes. Preheating the plastic surface with brief exposure to a gas flame (a fraction of a second) both speeds the process and ensures high printing ink adhesion.

Tight bonds achieved with barrier and adhesive coatings of PS parts and films are secured with primers of, for example, styrene–butadiene latexes.

With the rapid growth in electronic equipment, EMI and RFI shielding has become important. PS is shielded by coating molded solid and structural foam parts with paints made conductive by metals such as copper, zinc, or silver, or with graphite. Effectiveness requirements are contained in the United States FCC Docket 20780, Part 15.

Adhesive bonding of PS to itself and other plastics is common practice using either water or solvent-based adhesives. Manufacturers of adhesives and of PS can provide recommended formulations for most tasks. In general, water-based systems are preferred to limit emissions subject to clean air laws.

Spin welding relies on the frictional heat developed quickly between PS and other plastics (or another PS part) when the two are brought together with a spinning/rubbing motion. Practically, this means spinning an article, such as a cup, against a fitting article for a very brief time to melt polymer and so provide the bonding material. Spinning must be stopped abruptly lest undesirable deformation of the parts result. The method is clean, fast, and requires inexpensive fixtures and tools.

The ideal acoustic properties of PS (especially unfilled PS) make it a good candidate for joining by ultrasonic energy; strong, smooth joints result. Glass-fiber-reinforced PS also welds ultrasonically at glass contents up to 30%.

39.9 RESIN FORMS

Forms produced from mass polymerized styrene are most commonly rod-shaped particles a 0.1 in. (2.5 mm) or so in diameter by the same dimension in length. Suspension polymer particles are smooth spherical beads a little smaller than the suspension particles from which they came.

PS pellets are commonly colored in-plant so natural resins account for close to 100 percent of the marketed product. However, on order, manufacturers will provide natural resins preblended with mold release agents, flame retardants agents, and similar specialty additives. Glass-filled PS pellets are available from several sources.

39.10 SPECIFICATION OF PROPERTIES

See the Master Outline of Materials Properties.

- GPPS, the general-purpose PS group, includes unmodified and all heat-resistant grades of crystal PS. However, no copolymers of styrene with other monomers are included.
- HIPS covers all grades of rubber (impact) modified PS.
- GFR-PS (glass-fiber-reinforced PS) includes only GPPS grades that contain 20 percent by weight of chopped strand glass fibers, generally 0.35 mil in diameter (8.89 microns).

PS is used mostly as received, with the common addition of in-plant coloring. Thus, properties are listed for the resins and not for compounds with other materials. Unpigmented resins are assumed; visible light transmission values given are therefore of the crystal material. For HIPS, the light transmission of natural material (hazy to opaque) is of little interest, since HIPS is almost never used in natural color. Visible light transmission is considered not applicable.

HIPS can be purchased in formulations containing ignition suppressants that alter its burning behavior. These grades have properties similar to those of regular resins with the exception of higher densities. They are included within the HIPS property table and noted in particular in the reference to Underwriters' listing.

GFR-PS is GPPS reinforced with glass fibers of the types E or S, usually.

These tables cover the range of commercial materials generally available. It should be noted that the PS industry is large and diversified, with capabilities to make an extensive range of products and product grades. If a need for design performance arises outside the range of properties listed, but feasibly within possibilities for a styrenic plastic, it should be brought to a manufacturer's attention. It is also important to note that several other styrenic copolymers are available: styrene–maleic anhydride (SMA), styrene–acrylonitrile (SAN), styrene–butadiene (K-Resin), and styrene methyl methacrylate (here called polymethyl methacrylate). Each of these is covered in another individual section.

39.11 PROCESSING REQUIREMENTS

PS is readily melt-processed into finished parts by all the methods common to thermoplastics.

Both GPPS and HIPS grades are tolerant of thermal processing and require no special precautions in molding and extrusion. Styrenics can be used as received with no drying, will feed smoothly through air-vey and weigh feed delivery systems, are stable in the melt, develop handleability ("green strength") quickly after shaping by any process, reprocess easily, and present no unusual hazards to a well-run plastics fabrication plant.

- PS rarely requires drying. HIPS water absorption after full immersion does not exceed 0.07% by weight; GPPS absorbs as little as 0.01%. High humidity may condense moisture on pellet surfaces, with resultant bubbles or splay marks in extruded sheet. However, drying, for two hours at 160–180°F (70–82°C) in an air circulating oven or with a hopper dryer, is usually sufficient to avoid formation of blemishes in products.

MATERIAL _____ General-purpose Polystyrene

PROPERTY	TEST METHOD	ENGLISH UNITS	VALUE	METRIC UNITS	VALUE
MECHANICAL DENSITY	D792	lb/ft^3	64.8–65.4	g/cm^3	1.04–1.05
TENSILE STRENGTH	D638	psi	4700–8200	MPa	32.4–56.5
TENSILE MODULUS	D638	psi	450,000–475,000	MPa	3103–3276
FLEXURAL MODULUS	D790	psi	450,000–500,000	MPa	3103–3448
ELONGATION TO BREAK	D638	%	1.2–3.6	%	1.2–3.6
NOTCHED IZOD AT ROOM TEMP	D256	ft-lb/in.	0.25–0.45	J/m	13.3–24.0
HARDNESS	D785		M60–M84		M60–M84
THERMAL DEFLECTION T @ 264 psi	D648	°F	169–226	°C	76–108
DEFLECTION T @ 66 psi	D648	°F	180–233	°C	82–112
VICAT SOFT- ENING POINT	D1525	°F	194–227	°C	90–108
UL TEMP INDEX	UL746B	°F	122	°C	50
UL FLAMMABILITY CODE RATING	UL94		NA to V-0[*]		NA to V-0[*]
LINEAR COEFFICIENT THERMAL EXPANSION	D696	in./in./°F	$2.8–4.7 \times 10^{-5}$	mm/mm/°C	$5–8.5 \times 10^{-5}$
ENVIRONMENTAL WATER ABSORPTION 24 HOUR	D570	%	0.01–0.03	%	0.01–0.03
CLARITY	D1003	% TRANS- MISSION	88–91	% TRANS- MISSION	88–91
OUTDOOR WEATHERING	D1435	%		%	
FDA APPROVAL			Regulated: 21 CFR 177		Regulated: 21 CFR 177
CHEMICAL RESISTANCE TO: WEAK ACID	D543				
STRONG ACID	D543				
WEAK ALKALI	D543				
STRONG ALKALI	D543				
LOW MOLECULAR WEIGHT SOLVENTS	D543				
ALCOHOLS	D543				
ELECTRICAL DIELECTRIC STRENGTH	D149	V/mil	500	kV/mm	19.7
DIELECTRIC CONSTANT	D150	10^6 Hertz	2.54		2.54
POWER FACTOR	D150	10^6 Hertz	0.0002		0.0002
OTHER MELTING POINT	D3418	°F		°C	
GLASS TRANS- ITION TEMP.	D3418	°F	165–230	°C	74–110

[*]For PS containing ignition-suppressant compounds.

MATERIAL _____ PS 20% glass-filled

PROPERTY	TEST METHOD	ENGLISH UNITS	VALUE	METRIC UNITS	VALUE
MECHANICAL DENSITY	D792	lb/ft^3	74.2–74.8	g/cm^3	1.19–1.20
TENSILE STRENGTH	D638	psi	10,000–12,000	MPa	68.9–82.7
TENSILE MODULUS	D638	psi	900,000–1,200,00	MPa	6200–8268
FLEXURAL MODULUS	D790	psi	950,000–1,100,00	MPa	6546–7580
ELONGATION TO BREAK	D638	%	1.5	%	1.5
NOTCHED IZOD AT ROOM TEMP	D256	ft-lb/in.	0.9–2.5	J/m	48.1–134
HARDNESS	D785		M80–M95		M80–M95
THERMAL DEFLECTION T @ 264 psi	D648	°F	200–220	°C	93–104
DEFLECTION T @ 66 psi	D648	°F	220–230	°C	104–110
VICAT SOFT-ENING POINT	D1525	°F		°C	
UL TEMP INDEX	UL746B	°F		°C	
UL FLAMMABILITY CODE RATING	UL94				
LINEAR COEFFICIENT THERMAL EXPANSION	D696	in./in./°F	$1.5–2.4 \times 10^{-5}$	mm/mm/°C	$2.7–4.3 \times 10^{-5}$
ENVIRONMENTAL WATER ABSORPTION 24 HOUR	D570	%	0.01–0.07	%	0.01–0.07
CLARITY	D1003	% TRANS-MISSION	NA	% TRANS-MISSION	NA
OUTDOOR WEATHERING	D1435	%		%	
FDA APPROVAL			Matrix resins regulated in CFR 177		Matrix resins regulated in CFR 177
CHEMICAL RESISTANCE TO: WEAK ACID	D543				
STRONG ACID	D543				
WEAK ALKALI	D543				
STRONG ALKALI	D543				
LOW MOLECULAR WEIGHT SOLVENTS	D543				
ALCOHOLS	D543				
ELECTRICAL DIELECTRIC STRENGTH	D149	V/mil	425	kV/mm	16.7
DIELECTRIC CONSTANT	D150	10^6 Hertz	2.7		2.7
POWER FACTOR	D150	10^6 Hertz	0.0004 (dry)		0.0004 (dry)
OTHER MELTING POINT	D3418	°F		°C	
GLASS TRANS-ITION TEMP.	D3418	°F	239	°C	115

479

Master Outline of Material Properties

MATERIAL ___ High-impact Polystyrene

PROPERTY	TEST METHOD	ENGLISH UNITS	VALUE	METRIC UNITS	VALUE
MECHANICAL DENSITY	D792	lb/ft^3	65.4	g/cm^3	1.05
TENSILE STRENGTH	D638	psi	2325–6000	MPa	16.0–41.3
TENSILE MODULUS	D638	psi	240,000–370,000	MPa	1653–2549
FLEXURAL MODULUS	D790	psi	260,000–390,000	MPa	1791–2687
ELONGATION TO BREAK	D638	%	1.0–1.5	%	1.0–1.5
NOTCHED IZOD AT ROOM TEMP	D256	ft-lb/in.	0.9–4.1	J/m	48.1–219
HARDNESS	D785		L50–L82 M63–M88		L50–L82 M63–M88
THERMAL DEFLECTION T @ 264 psi	D648	°F	156–205	°C	69–96
DEFLECTION T @ 66 psi	D648	°F	163–185	°C	73–85
VICAT SOFT- ENING POINT	D1525	°F	199–220	°C	93–104
UL TEMP INDEX	UL746B	°F	122	°C	50
UL FLAMMABILITY CODE RATING	UL94		NA to V-0[a]		NA to V-0[a]
LINEAR COEFFICIENT THERMAL EXPANSION	D696	in./in./°F	1.9–5.6 × 10^{-5}	mm/mm/°C	3.4–10.1 × 10^{-5}
ENVIRONMENTAL WATER ABSORPTION 24 HOUR	D570	%	0.05–0.07	%	0.05–0.07
CLARITY	D1003	% TRANS- MISSION	NA	% TRANS- MISSION	NA
OUTDOOR WEATHERING	D1435	%		%	
FDA APPROVAL			Regulated: CFR 177		Regulated: CFR 177
CHEMICAL RESISTANCE TO: WEAK ACID	D543				
STRONG ACID	D543				
WEAK ALKALI	D543				
STRONG ALKALI	D543				
LOW MOLECULAR WEIGHT SOLVENTS	D543				
ALCOHOLS	D543				
ELECTRICAL DIELECTRIC STRENGTH	D149	V/mil	300–500	kV/mm	11.8–19.7
DIELECTRIC CONSTANT	D150	10^6 Hertz	2.50–2.52		2.50–2.52
POWER FACTOR	D150	10^6 Hertz	0.0004–0.0008		0.0004–0.0008
OTHER MELTING POINT	D3418	°F		°C	
GLASS TRANS- ITION TEMP.	D3418	°F	199–221	°C	93–105

[a] For HIPS containing ignition-suppressant compounds.

- Weigh feeding is commonly used; pellet sizes are accurately controlled and provide molders with precise sizes and bulk densities.
- GPPS and HIPS have excellent thermal stability and so can be used in a wide range of molding and extrusion equipment. In molding, shot size exceeding 75% of the rated machine capacity minimizes the possibility of thermal degradation.

Regrind handles well at ratios to virgin resin up to 30% in molding and 50% in extrusion resins.

- PS parts made from virgin/regrind blends cannot normally be distinguished from all-virgin parts.
- Use of regrind in food contact applications may not be allowed; it is necessary to consult resin suppliers for the status of applicable Food & Drug law compliance requirements.

PS grades are compatible with one another, but PS is not, with few exceptions, usually compatible with other plastics. Poorly mixed, weak, and hazy parts may result from such blends. Thorough purging of equipment with PS should precede the running of GPPS and HIPS after fabrication of other materials.

39.12 PROCESSING-SENSITIVE END PROPERTIES

The largest effect of fabrication on the properties of PS is obtained through orientation of the molecules into roughly parallel-aligned arrays. PS can be oriented by stressing in the rubbery state, followed by quickly quenching to freeze the orientation into place. The result is an increase in strength in the alignment direction and a sacrifice in strength in the cross direction. There is no accompanying change in stiffness.

Multiaxial orientation strengthens PS in all directions that are so oriented. Biaxial stretching (orientation in both the x and y directions) of sheet from the extrusion die increases strength in both the x and y directions. Orientation can be balanced—strength equal along the perpendiculars—or unbalanced by controlling the orientation (Fig. 39.4). Stiffness, as reflected in the modulus values, is unaffected.

Table 39.6 reviews the mechanical properties of GPPS and HIPS as they are affected by orientation.

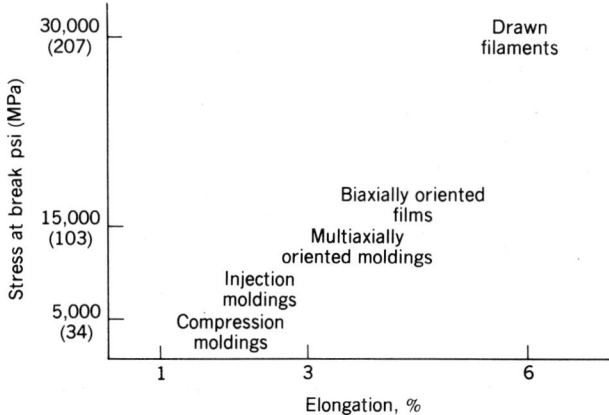

Figure 39.4. Effects of orientation on the strength of polystyrene. Yield for filaments and film; break otherwise. Note: oriented films and filaments typically elongate from 5 to as high as 50% before rupturing, the magnitude increasing with degree of orientation.

TABLE 39.6. Effects of Orientation on Mechanical Properties of Polystyrene

Orientation	GPPS	HIPS
Unoriented	Stiff, brittle; weak if cast	Tough, somewhat flexible
Mono-oriented[a]	Stiff; strong in direction of orientation, weak in cross direction	Flexible; strong in orientation direction, weak in cross direction
Bi-oriented[b]	Stiff; tough, resistant to shattering	Flexible; tougher in all directions, resists shattering

[a]Oriented mainly in one direction. In the case of injection molding, this would mean in the directions of flow in the mold, especially those which favor molecular alignments (as along narrow or thin sections). Fibers extruded of PS are made tough by stretching them as they exit the die.

[b]Oriented in both the longitudinal and lateral directions; for example, extruded sheet can be drawn out of the die at a linear speed greater than its entry speed, and stretched crosswise to match or exceed its width at the die. The resultant sheet is thinned, and its molecules are aligned in both the x and y directions, with some aligned at angles between. Such sheet is stronger and tougher, not stiffer, than unoriented sheet. In cases of great orientation (for example, >3:1 draws, GPPS sheet has been shown able to take nailing without shattering.

There are continuous gradations of properties other than those reviewed in Table 39.6. They are commonly occasioned by both the amount of any rubber added and its effect and by the amount of orientation introduced and its effect.

The temperature of the melt has important effects on the impact resistance of HIPS, as shown in Figure 39.5.

- The higher the injection melt temperature, the lower the Izod and the higher the falling missile impact strength. In both these effects, orientation, its probability declining with temperature increase, is responsible.
- Compression molding, which also reflects the behavior of low orientation molding, as well as extrusion [as exemplified in Figure 32.5 by the 0.5 × 0.5 in. (12.7 × 12.7 mm) cross-section injection molded bar], shows no change in either Izod or falling missile impact strength test as the temperature of 0.5 × 0.125 in. (12.7 × 3.2 mm) test bars is increased.

The gloss of rubber-modified grades of PS is enhanced by use of high mold temperatures. Conversely, low temperatures favor matte finishes. PS grades have inherent glossiness, which can be controlled within the limits set by the chemistry and rheology of each resin.

Color and surface uniformity may be affected by excessive plastic temperatures, with yellowing and silver streaking occurring as a result. Further, high temperatures combined with high pressures may produce flashing at mold parting lines. High pressures alone may promote sticking in the mold, which is both a production nuisance, a hazard, and a source of damaged parts.

The heat resistance of injection moldings is sacrificed when mold temperatures are too low (that is, low enough to produce too-rapid setting of the plastic when it contacts the mold surfaces). This temperature may be determined by experiment. It usually lies in the range below 140°F (60°C).

39.13 SHRINKAGE

Shrinkage is the difference in dimension between a plastic part and the cavity in which it was shaped. Mold shrinkage is the difference noted after temperature equilibrium is reached, nominally 0.1 h after shaping. Age shrinkage is the dimensional change that occurs after mold shrinkage. Total shrinkage is simply the sum of mold and age shrinkage.

PS resins shrink very little after melt processing and are among the most shape and size-retaining thermoplastic materials.

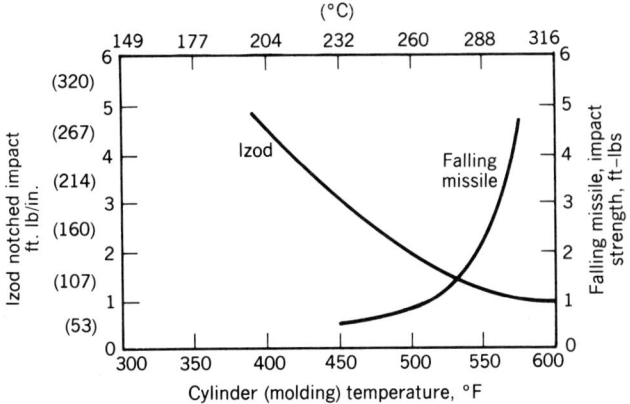

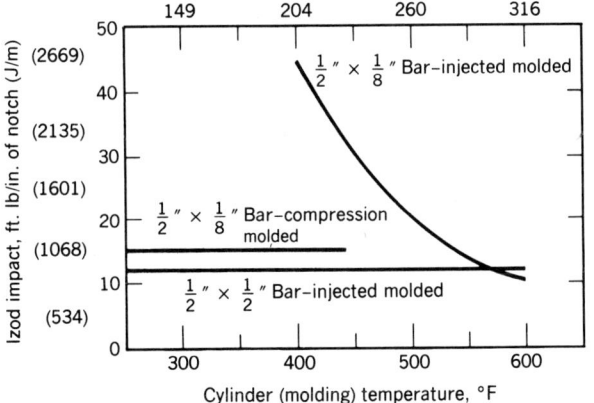

Figure 39.5. Effect of molding cylinder temperature on the impact strength of typical HIPS plastics. Data from The Dow Chemical Company.

The total shrinkage of PS (GPPS and HIPS) ranges from a low in compression moldings of 0.003 in./in. to a high in formed sheet of 0.01 between the time of shaping from the melt to the time at which shrinkage essentially ceases—about 1,000 to 10,000 h at ambient conditions.

- Compression moldings of GPPS and HIPS typically shrink 0.003 to 0.004 in./in. over 1000 to 2000 h, after which dimensional equilibrium is reached. Stress-free compression moldings are isotropic and therefore shrink equally in all directions.
- Extruded sheet shrinks more in the extrusion direction than in the transverse direction; more when highly drawn down than when drawn little; and more when thick than when thin. The least shrinkage that can be expected in extruded PS sheet is in the thinnest sheet at the lowest draw ratio and in the cross direction to extrusion; under these conditions, the numerical values of shrinkage approach those noted for compression moldings.
- Since thermoforming is performed at temperatures well below the glass temperature, thermoformed parts are more prone to shrinkage than parts formed in melt processes.

HIPS parts exhibit mold shrinkage of 0.005–0.008 in./in. after forming from sheet in molds at 100–140°F (38–60°C), and a further 0.002–0.003 in./in. on aging. Virtually no further shrinkage results after one to two months.

- Injection moldings of PS shrink 0.003 to 0.005 in./in. (0.3 to 0.5%) within 0.1 h of molding, and age shrink to a stable shape a further 0.002 in./in. over a year. Increasing pressure above that required to just fill the mold minimizes shrinkage. This is the most influential variable in restricting shrinkage of injection moldings.

Following are general guidelines for the design of injection-molded PS parts.

- Flow length to wall thickness ratio, typically 150. Higher values are achievable among the high-flow grades of PS; lower values apply when molding glass-fiber-reinforced and filled PS.
- Wall thickness variation should be minimized. Total variations exceeding a ratio of 2:1 of thickest to thinnest sections are potential sources for shrinkage distortion in the finished part.
- Abrupt thickness changes should be avoided; local variations approaching a ratio of 0.5 of thickness change signal potential for stress concentration.
- Radii should be no smaller than 0.25 W for corners of moldings (W = wall thickness); 0.005 in. (0.13 mm) for edges (holes, rims, and so on); and 0.005 in. (0.13 mm) for junctions (ribs, bosses, pegs, and so on).
- Holes, whether molded or machined, should be separated by at least one hole diameter.
- Ribs should not exceed the following dimensions relative to the thickness W of the section to which they are attached: 0.8 W for thickness and 2.4 W for height.
- Solid pegs and bosses should stand no higher than twice their diameter measured over base radii.
- Strippable undercuts are not possible in PS.
- Gussets are used to reinforce rims; fillets are used to support pegs and bosses.
- Hollow bosses are commonly designed into PS parts; HIPS is better adapted than GPPS to use of these landing devices for securing screws.
- Draft angle requirements are minimum required in any case, 0.5 degrees, solid pegs and bosses, recommended minimum 1.0 degrees; maximum required in any case, 2.0 degrees.
- Inserts (brass or steel) can be molded into HIPS, but are not normally recommended in GPPS.

If inserts in GPPS are necessary, they should be heated before being placed in the mold to reduce the difference in shrinkage between the plastic and metal, which otherwise leads to cracking and may terminate in fracture and failure.

- Blowup ratios for injection blow-molding of PS typically fall in the range of 3 to 4; the value in a specific case is determined by the rheology of the resin and the design of the parison and part.
- In thermoforming, the normal maximum draw ratio is 2–2.5. With special design and tooling, ratios as high as 4 have been achieved.

39.14 TRADE NAMES

Manufacturers of PS often use codes to designate various grades, but, today, only Amoco, Arco, and Dow apply trademarks (Amoco polystyrene, Dylark polystyrene, and Styron polystyrene, respectively).

Polystyrene manufacturers include Amoco Chemicals, Chicago, Ill.; Arco Chemical, Philadelphia, Penn.; Asoma Polymers, Oxford, Maine; Chevron Chemical, Houston, Texas; Dart Industries (captive manufacturer), Phoenix, Ariz.; The Dow Chemical Company, Midland, Mich.; Fina Oil & Chemical, Dallas, Texas; Huntsman Chemical, Salt Lake City, Utah; Kama Corp., Hazelton, Penn.; Mobil Polymers, Edison, N.J.; and Polysar Inc., Leominster, Mass.

BIBLIOGRAPHY

1. H. Barron, *Modern Plastics*, Chapman and Hall, Ltd., London, 1949, pp. 460–84.
2. R.H. Boundy and R.F. Boyer, eds. *Styrene: Its Polymers, Copolymers and Derivatives,* American Chemical Society Monograph Series, Reinhold Publishing, New York, 1952.
3. S. Strella, *ASTM Bulletin*, 59–60 (Feb. 1959); W.E. Brown, *Design and Engineering Criteria for Plastics,* Publication 1004, Building Research Institute, 1963.
4. *Styron High Impact Polystyrene,* Bulletin, The Dow Chemical Company, Midland, Mich., 1986.
5. *Styron Polystyrene Advance Performance Resins for the Injection Molding Market,* The Dow Chemical Company, Midland, Mich., 1986

General References

W.E. Brown, "Structural Design Criteria for Plastic Materials," *Technical Report No. 93;* reprinted from *Western Plastics* (Feb. 1963).

W.E. Brown, J.D. Striebel, and D.C. Fuccella, "Direct Process Glass Reinforcement of Thermoplastics," *Society of Automotive Engineers Paper No. 680059,* Jan. 1968.

R. Boyer, ed., "Styrene Polymers," *Encyclopedia of Polymer Science and Technology*, Vol. 13, John Wiley & Sons, New York, 1970, pp. 128–447.

E. Baer, ed., *Engineering Design for Plastics, SPE Polymer Science and Engineering Series,* Reinhold Publishing, New York, 1964.

R.B. Seymour, ed., *History of Polymer Science & Technology,* Marcel Dekker, New York, 1982.

J. Frados, ed., *Plastics Engineering Handbook, 4th edition.* Van Nostrand Reinhold, New York, 1976.

M.D. Baijal, ed., *Plastics Polymer Science and Technology,* Wiley-Interscience, New York, 1982.

S.S. Schwartz and S.H. Goodman, *Plastics Materials and Processes,* Van Nostrand Reinhold, New York, 1982.

Trouble Shooting Injection Molding Technology, Bulletin, The Dow Chemical Co., Midland, Mich., 1979.

M. Bakker, ed., *The Wiley Encyclopedia of Packaging Technology,* Wiley-Interscience, New York, 1986.

40

POLYSULFONE (PSO)

James E. Harris

Union Carbide Corp., Bound Brook, New Jersey

40.1 INTRODUCTION

Polysulfone has the following chemical repeat unit:

40.2 CATEGORY

Polysulfone is an amorphous thermoplastic that can be processed by extrusion, injection molding, blow molding, compression molding, and thermoforming.

40.3 HISTORY

Polysulfone was first commercialized by Union Carbide in 1965 as Bakelite Polysulfone. It is now known as Udel polysulfone; the Specialty Polymers and Composites Div. of Union Carbide Corp. is the sole producer.

40.4 POLYMERIZATION

Polysulfone is produced by the nucleophilic aromatic substitution reaction between the disodium salt of 2,2-bis(4-hydroxyphenyl) propane and 4,4'-dichlorodiphenylsulfone. The polymerization is carried out in a polar aprotic solvent.[1,2]

40.5 DESCRIPTION OF PROPERTIES

Polysulfone is a light yellow, transparent thermoplastic with a glass-transition temperature T_g of 374°F (190°C). In addition, it has excellent electrical properties and resistance to acids and bases. The general properties are outlined in the Master Material Outline and in Table 40.1.

40.5.1 Thermal Properties

Polysulfone has a heat-deflection temperature at 264 psi (1.82 MPa) of 345°F (174°C), which is a practical upper limit to its use in continuous load-bearing applications. In steam, the limit is somewhat lower, about 284°F (140°C) because of the plasticizing effect of moisture. Figure 40.1 shows the recommended working stresses in air and steam–water as a function of temperature. Thermal aging studies have shown polysulfone to exhibit no loss in tensile strength or heat-deflection temperature even after two years in air at 302°F (150°C). As with most thermoplastics, annealing reduces impact strength because of volume relaxation. After an initial reduction of some 30 to 40%, tensile impact strength remains relatively constant over a period of 1.5 years.[3]

40.5.2 Mechanical Properties

Polysulfone is a tough, ductile polymer, as indicated by its high elongation to break of greater than 50% and tensile impact strength of 420 ft-lb/in.[2] (883 kJ/m[2]). However, like many thermoplastics, it is notch sensitive, as indicated by a comparison of the notched Izod value given in the Master Material Outline to the corresponding unnotched value of greater than 60 ft-lb/in. (3200 J/m). This factor requires attention during part design; sharp radii and abrupt changes in cross section should be avoided. With these precautions, polysulfone retains its good impact properties to about −148°F (−100°C).

40.5.3 Environmental Properties

Polysulfone exhibits superb hydrolytic stability; that is, resistance to chemical degradation in the presence of water. The effect of moisture on its load-bearing capability was mentioned

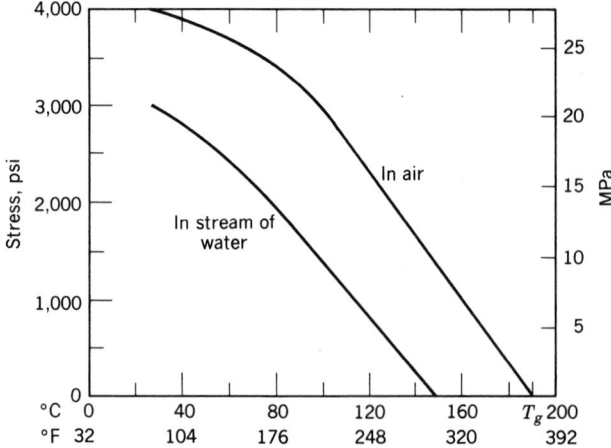

Figure 40.1.   Recommended maximum working stress for polysulfone in air and water or steam as a function of temperature. Source Union Carbide Corp.

TABLE 40.1. Compatibility of Polysulfone with Various Chemical Environments[a]

Inorganic Chemicals	(22°C) 73°F	(60°C) 140°F	(85°C) 185°F	Organic Chemicals	(22°C) 73°F	(60°C) 140°F	(85°C) 185°F
Acids				Acetic acid, 50%	R	R	R
Chromic acid, 60%	NR	*	*	Acetic acid, glacial	LR	*	*
Hydrobromic acid, 20%	R	R	R	Acetic anhydride	NR	*	*
Hydrochloric acid, 20%	R	R	R	Acetone, 5%	R	*	*
Hydrochloric acid, 37%	R	R	LR	Acetone, 100%	NR	*	*
Hydrofluoric acid, 50%	LR	*	*	Acetonitrile, 100%	NR	*	*
Nitric acid, 40%	R	LR	*	Benzene, 100%	NR	*	*
Phosphoric acid, 85%	R	R	R	Carbon tetrachloride, 100%	R	*	*
Sulfuric acid, 40%	R	R	R	Cellosolve solvent, 100%	R	*	*
Sulfuric acid, 85%	R	R	R	Chlorobenzene, 100%	NR	*	*
Sulfuric acid, 95%	NR	*	*	Chloroform, 100%	NR	*	*
				Citric acid, 40%	R	*	*
Bases				Crude oil, Texas, 100%	R	*	*
Ammonia, 29%	R	*	*	Cyclohexane, 100%	R	*	*
Potassium hydroxide, 35%	R	R	*	Cyclohexanone, 100%	NR	*	*
Sodium hydroxide, 50%	R	R	R	Diethyl ether, 100%	NR	*	*
				Dioctyl phthalate, 100%	R	*	*
Other				Ethanol, 100%	R	*	*
Ammonium persulfate, 40%	R	R	*	Ethanolamine, 100%	R	*	*
Calcium chloride, sat.	R	R	R	Ethyl acetate, 100%	NR	*	*
Calcium hypochlorite	R	R	R	Ethylene diamine, 92%	LR	*	*
Cupric chloride, sat.	R	R	R	Ethylene glycol, 100%	R	*	*
Ferrous sulfate, sat.	R	R	R	Formaldehyde, 100%	R	*	*
Hydrogen peroxide, 100%	R	*	*	"Freon" 11, 100%	LR	*	*
Potassium nitrate, sat.	R	R	R	Gasoline, 100%	LR	*	*
Sodium hypochlorite, 17%	R	R	R	Glycerine, 100%	R	*	*
				n-Heptane, 100%	R	*	*
				Isopropanol, 100%	LR	*	*
				Lactic acid, 60%	R	R	R
				MEK, 100%	NR	*	*
				Methylene chloride, 100%	NR	*	*
				1,1,1-Trichloroethane	LR	*	*
				Trichloroethylene, 100%	NR	*	*

[a]R, recommended; NR, not recommended; *, no data; LR, limited recommendation (many applications possible depending on stress level).
NOTE: Detailed chemical compatibility information is available for ALL materials from their manufacturers.

above. The low water absorption of polysulfone results in the ability to maintain close dimensional tolerances. The change in linear dimensions resulting from immersion in water at 212°F (100°C) (equilibrium) is 0.11%.

In general, polysulfone exhibits excellent resistance to aqueous mineral acid, alkali, and salt solutions. Its resistance to detergents and hydrocarbon oils is good even at elevated temperatures under moderate levels of stress. In polar organic solvents such as ketones, chlorinated hydrocarbons, and aromatic hydrocarbons, polysulfone will swell, dissolve or stress crack. Table 40.1 provides a general indication of the applicability of polysulfone insofar as its contact with a wide variety of aggressive materials. Of course, compatibility should be tested under conditions that approximate the intended use as closely as possible because of its strong dependence on time, stress, and temperature.

Like many aromatic polymers, polysulfone exhibits poor resistance to ultraviolet light. Exposure for long periods of time of the neat resin to sunlight is not recommended. This is not a problem for filled or pigmented resin because degradation is limited to the depth on the material's surface to which the UV wavelengths can penetrate. For example, carbon black has been shown to improve the UV stability of polysulfone to the point where it has been used successfully as the absorber plate in solar collectors.

40.5.4   Electrical Properties

The electrical properties of polysulfone deserve special mention. In particular, its low dielectric coefficient and dissipation factor over a wide range of temperatures have permitted such applications as a printed wiring board substrate and microwave cookware. The values at 10^9 Hz are 3.0 and 0.004, respectively, for dielectric and constant dissipation factor at 73°F (24°C) and 50% relative humidity. Polysulfone meets the electrical requirements of MIL-M-14F and MIL-P-19833B, except for the arc resistance of the silicone and melamine types in MIL-M-14F.

40.5.5   Effect of Glass-Fiber Reinforcement

The properties of 10, 20, and 30% glass-fiber-reinforced polysulfone are given in Table 40.2. As with all amorphous thermoplastics, the addition of short glass fiber improves the modulus and strength, reduces ductility and has little effect on the heat-deflection temperature.

40.6   APPLICATIONS

Polysulfone is used extensively in several market areas. The combination of its high heat-deflection temperature, good electrical properties, and excellent hydrolytic stability make it a unique thermoplastic material. Also, it is tough, amorphous, transparent, and can be molded to tight tolerances. These are the keys to its penetration into most of its markets.

One major market for polysulfone is in food-service applications where it competes with

TABLE 40.2.   Typical Properties of Injection Molded Udel GF-110, 120, 130 Resins

Property	ASTM Test Method	GF-110 (10% glass)	GF-120 (20% glass)	GF-130 (30% glass)
Mechanical				
Tensile strength, psi (MPa)	D638	11,300 (78.0)	14,000 (96.6)	15,600 (107.6)
Tensile elongation, %	D638	4	3	2
Tensile modulus, psi (MPa)	D638	530,000 (3,655)	750,000 (5,172)	1,070,000 (7,379)
Flexural strength, psi (MPa)	D790	18,500 (127.6)	21,500 (148.3)	22,400 (154.5)
Flexural modulus, psi (MPa)	D790	550,000 (3,793)	800,000 (5,517)	1,100,000 (7,586)
Notched Izod, ft-lb/in. (J/m)	D256	1.2 (64)	1.3 (69)	1.4 (75)
Tensile impact, ft-lb/in.^{2} (kJ/m^2)	D1822	48 (1,0008)	54 (1,134)	52 (1,092)
Thermal				
Heat-deflection temperature, 264 psi, °F (°C)	D648	354 (179)	356 (180)	358 (181)
Flammability rating, UL-94		V-0 at 0.062 in. (1.57 mm)	V-0 at 0.062 in. (1.57 mm)	V-0 at 0.062 in. (1.57 mm)
Electrical				
Dielectric strength, V/mil (kV/mm)	D149	475 (19)	475 (19)	475 (19)
Volume resistivity, ohm-cm	D257	$>10^{16}$	$>10^{16}$	$>10^{16}$
Dielectric constant	D150			
at 60 Hz		3.3	3.4	3.5
at 1 MHz		3.4	3.5	3.7
Dissipation factor	D150			
at 60 Hz		0.001	0.001	0.001
at 1 MHz		0.005	0.005	0.004
General				
Density, lb/ft^3 (g/cc)	D1505	83 (1.33)	87.3 (1.40)	92.9 (1.49)

stainless steel and glass. Its hydrolytic stability permits repeated exposure to automatic dishwasher cycles, and its chemical resistance is sufficient to stand up to almost any kind of food. Demanding applications such as exposure to hot bacon grease require careful attention to part design and processing conditions for low molded-in stresses. Further, it complies with the Food and Drug Administration food additive regulation 21 CFR 177.2500 and the Dairy and Food Industry Supply Association 3-A sanitary standards for multiple-use plastic materials utilized for product-contact surfaces for dairy equipment (Serial No. 2000). The advantages over stainless steel are its transparency, light weight, and resistance to denting, pitting, and corrosion. Its excellent high-frequency electrical properties also permit service in microwave ovens, where it is lighter and tougher than glass and resists chipping and breaking under both thermal and mechanical shock. Besides microwave cookware, it is typically selected for coffee and other beverage dispensers.

Polysulfone also finds use as a printed wiring-board substrate with performance approaching PTFE-glass at a substantially lower cost. A typical application would be in high-frequency test equipment.

Since its commercialization, use of polysulfone in medical applications has continued to grow. Important considerations are its ability to withstand repeated sterilization with steam, dry heat, and ethylene oxide. Typical applications are in contact-lens cases and surgical-instrument trays.

Because of its excellent resistance to caustic, mineral acids, and dissolved chlorine, polysulfone offers an attractive alternative to metals in certain fluid-handling applications. Further, its higher heat-deflection temperature offers an advantage over PVC, CPVC, and polypropylene. Polysulfone is available as pipe consisting of extruded tubing with a filament-wound glass fiber overwrap. Pump housings have also been constructed of polysulfone.

Finally, a unique and growing application of polysulfone is in membrane separation systems. Bundles of polysulfone hollow fiber membranes[4-6] are being employed primarily to separate hydrogen gas from various process streams. Other commercial systems employ polysulfone membranes in reverse osmosis and ultrafiltration applications.[7,8]

40.7 ADVANTAGES/DISADVANTAGES

As mentioned in the previous section, the major advantages of polysulfone compared with other engineering thermoplastics are its hydrolytic stability, high heat-distortion temperature, excellent thermal stability, and good electrical properties. Disadvantages are poor resistance to ultraviolet light and, in some cases, environmental stress-crack resistance to polar organic solvents. Its light yellow hue and UV stability exclude the use of polysulfone in applications that require both water-white transparency and weatherability. Polysulfone is priced as a specialty resin to be used in those demanding applications where performance is a requisite. The price of the molded article on a cost-per-volume basis should be viewed as the determining factor in materials selection. Polysulfone performs in applications where, for instance, polycarbonate is unsatisfactory.

Polysulfone's HDT of 345°F (174°C) gives it a distinct advantage over other amorphous resins such as polycarbonate [270°F (132°C)] and PPO/PS blends [266–302°F (130–150°C)]. Its hydrolytic stability is significantly better than that exhibited by polycarbonates and even other polyaryl ethers.[9] Table 40.3 compares the change in physical properties of polycarbonate, polysulfone, polyethersulfone, and polyetherimide as a function of time in boiling water. After 3,000 hours, the physical properties of polysulfone are barely affected while those of polycarbonate have all but vanished. Even polyethersulfone and polyetherimide exhibit considerable loss of tensile impact strength after 1,560 hours. The good hydrolytic stability of polysulfone combined with its thermal stability to above 752°F (400°C) permits the use of regrind (discussed below), which reduces the actual cost of resin per part produced.

TABLE 40.3. Mechanical Property (Immersion in Water at 205°F (96°C)[a]

	Initial	264, h	720, h	1560, h	2020, h	2524, h	3000, h
Polycarbonate							
Tensile strength,[b] (MPa)	74	67	69	66	38	13.2	9.0
Notched Izod impact strength,[c] (J/m)	929	132		73			
Tensile impact strength,[d] (kJ/m^2)	549			15			e
Polysulfone							
Tensile strength,[b] (MPa)	72	75	78	81	81	81	82
Notched Izod impact strength,[c] (J/m)	70	83		57			
Tensile impact strength,[d] (kJ/m^2)	281			229			254
Poly(ether sulfone)							
Tensile strength,[b] (MPa)	80	86	83	88	91	93	91
Notched Izod impact strength,[c] (J/m)	88	87		63			
Tensile impact strength,[d] (kJ/m^2)	316			101			f
Polyetherimide							
Tensile strength,[b] (MPa)	112	107	107	109	108	105	94
Notched Izod impact strength,[c] (J/m)	64	61		61			
Tensile impact strength,[d] (kJ/m^2)	128			58			41

[a]With permission from Ref 9.
[b]Tensile strength determined as per ASTM D638.
[c]Notched Izod impact strength determined as per ASTM D256.
[d]Tensile impact strength determined as per ASTM D1822.
[e]Samples were too brittle to test.
[f]No samples available for testing.

40.8 PROCESSING TECHNIQUES

Polysulfone is commonly processed by injection molding, extrusion, compression molding and thermoforming.

40.9 RESIN FORMS

Injection-molding-grade polysulfone is available in pellet and powder form. A higher molecular weight pellet grade is available for extrusion. Flame-retardant, mineral-filled, and glass-reinforced versions are also commercially available. Polymer blends based on polysulfone can be purchased for use in food service, plumbing, and electronics applications.

40.10 SPECIFICATION OF PROPERTIES

See the Master Material Outline. A more complete treatment of the properties of polysulfone can be found in Reference 10.

40.11 PROCESSING REQUIREMENTS

Polysulfone can be easily processed by most modern injection-molding equipment with stock temperatures between 626–752°F (330–400°C). Polysulfone is extremely stable and has been successfully processed at temperatures above 797°F (425°C); however, above 752°F (400°C) in barrels with long residence times, parts may exhibit black streaks caused by thermal

Master Material Outline

PROCESSABILITY OF ___Polysulfone___

PROCESS	A	B	C
INJECTION MOLDING	X		
EXTRUSION	X		
THERMOFORMING			
FOAM MOLDING			X
DIP, SLUSH MOLDING			X
ROTATIONAL MOLDING	X		
POWDER, FLUIDIZED-BED COATING			X
REINFORCED THERMOSET MOLDING	NA		
COMPRESSION/TRANSFER MOLDING	X		
REACTION INJECTION MOLDING (RIM)	NA		
MECHANICAL FORMING	X		
CASTING	X		

A = Common processing technique.
B = Technique possible with difficulty.
C = Used only in special circumstances, if ever.

Standard design chart for ___Polysulfone (PSO)___

1. Recommended wall thickness/length of flow
   Minimum ___0.030___ for ___1 in.___ distance
   Maximum _____ for _____ distance
   Ideal ___0.080___ for ___12 in.___ distance

2. Allowable wall thickness variation, % of nominal wall ___25___ in a "gradual" step

3. Radius requirements
   Outside: Minimum ___125% of nominal wall___ Maximum ___175% of nominal wall___
   Inside: Minimum ___25% of nominal wall___ Maximum ___75% of nominal wall___

4. Reinforcing ribs
   Maximum thickness ___75% of nominal wall at base to avoid sink marks___
   Maximum height ___200% of nominal wall___
   Sink marks: Yes _____ No _____

5. Solid pegs and bosses
   Maximum thickness ___50% nominal wall to prevent sink marks___

Maximum weight _____
Sink marks: Yes_____   No _____

6. Strippable undercuts
   Outside: Yes ___X___ No _____
   Inside: Yes ___X___ No _____

7. Hollow bosses: Yes ___X___ No _____

8. Draft angles
   Outside: Minimum _____1_____ Ideal _____1_____
   Inside: Minimum _____1_____ Ideal _____1_____

9. Molded-in inserts: Yes ___X___ must be preheated to mold temperature
   No _____

10. Size limitations   None
    Length: Minimum _____ Maximum _____ Process _____
    Width: Minimum _____ Maximum _____ Process _____

11. Tolerances

12. Special process- and materials-related design details
    Blow-up ratio (blow molding) _____
    Core L/D (injection and compression molding) _____2.5_____
    Depth of draw (thermoforming) _____
    Other _____

degradation. Mold temperatures range between 140–320°F (60–160°C), with 212°F (100°C) being typical. Although molding wet will not degrade properties, moisture will cause foaming and splay in finished parts. For this reason, it is recommended that the moisture content be reduced to below 0.05% before processing. Pellets dried in trays to a depth of 1–2 in. (25–51 mm) at 275°F (135°C) for 4 hours or 325°F (163°C) for 2–3 hours in a circulating air oven usually yield satisfactory parts. Drying time may have to be increased in humid weather unless the oven is equipped with a dehumidifying unit. Regrind may be used in reasonable quantities provided it is kept dry. Remolding does not degrade polysulfone although it may cause darkening. Polysulfone molded, reground, and remolded four times darkened slightly but did not change in physical properties.

40.12   PROCESSING-SENSITIVE END PROPERTIES

Processing parameters can affect properties, specifically environmental stress-craze resistance, of the molded parts through the development of molded-in stresses. Standard methods for reducing molded-in stresses, such as increasing mold temperatures or reducing injection speed, can be used to improve stress-craze resistance.

40.13   SHRINKAGE

As with other thermoplastics, efficient and economical injection molding of polysulfone begins with proper mold design. Sprues should be short and relatively thick. Runners should be full round or trapezoidal and, in general, larger in diameter than those used for free-flowing materials such as polystyrene or polypropylene. Gates should be located so that the resin flows from thick to thin sections and should be as large as possible. Polysulfone has a pronounced tendency to jet through small gates when they are not properly located. If possible, small gates should be positioned so that the melt will strike a wall or pin as soon

MATERIAL _____ Polysulfone (PSO)

PROPERTY	TEST METHOD	ENGLISH UNITS	VALUE	METRIC UNITS	VALUE
MECHANICAL DENSITY	D792	lb/ft³	77.3	g/cm³	1.24
TENSILE STRENGTH	D638	psi	10,200	MPa	70.3
TENSILE MODULUS	D638	psi	360,000	MPa	2,482
FLEXURAL MODULUS	D790	psi	390,000	MPa	2,689
ELONGATION TO BREAK	D638	%	50–100	%	50–100
NOTCHED IZOD AT ROOM TEMP	D256	ft-lb/in.	1.3	J/m	69
HARDNESS	D785		M-69		R-120
THERMAL DEFLECTION T @ 264 psi	D648	°F	345	°C	174
DEFLECTION T @ 66 psi	D648	°F	357	°C	181
VICAT SOFTENING POINT	D1525	°F	370	°C	188
UL TEMP INDEX	UL746B	°F	320	°C	160
UL FLAMMABILITY CODE RATING	UL94		94V-2 at 0.125 in		94V-2 at 3 mm
LINEAR COEFFICIENT THERMAL EXPANSION	D696	in./in./°F	3.1×10^{-6}	mm/mm/°C	5.6×10^{-5}
ENVIRONMENTAL WATER ABSORPTION 24 HOUR	D570	%	0.3	%	0.3
CLARITY	D1003	% TRANSMISSION	Transparent 5% haze	% TRANSMISSION	Transparent 5% haze
OUTDOOR WEATHERING	D1435	%		%	
FDA APPROVAL			Yes		Yes
CHEMICAL RESISTANCE TO: WEAK ACID	D543		No effect at 200°F		No effect at 93°C
STRONG ACID	D543		No effect at 100°F		No effect at 38°C
WEAK ALKALI	D543		No effect at 200°F		No effect at 93°C
STRONG ALKALI	D543		No effect at 100°F		No effect at 38°C
LOW MOLECULAR WEIGHT SOLVENTS	D543		Attacked under stress		Attacked under stress
ALCOHOLS	D543		No effect		No effect
ELECTRICAL DIELECTRIC STRENGTH	D149	V/mil	425	kV/mm	16.7
DIELECTRIC CONSTANT	D150	10^6 Hertz	3.10		3.10
POWER FACTOR	D150	10^6 Hertz	0.005		0.005
OTHER MELTING POINT	D3418	°F	Not applicable	°C	Not applicable
GLASS TRANSITION TEMP.	D3418	°F	364	°C	185

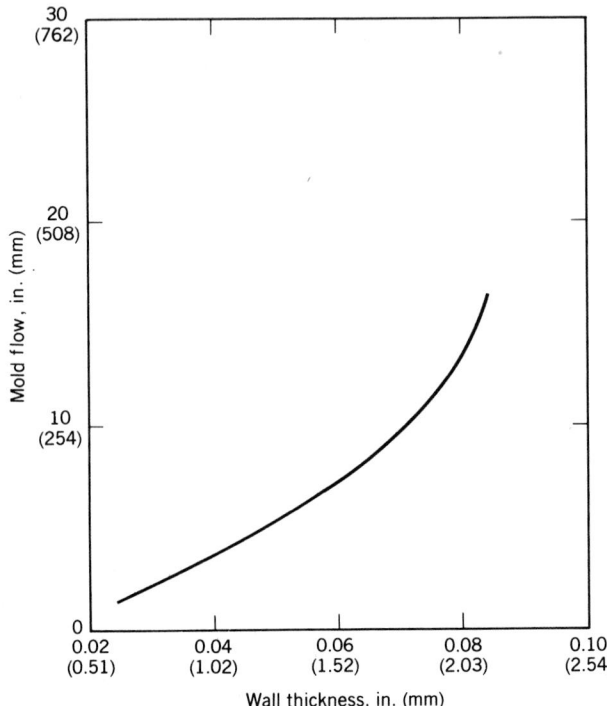

Figure 40.2. Mold flow vs thickness for polysulfone under typical molding conditions. Udel Polysulfone P-1700. Stock temperature, 730°F (387°C); mold temperature, 200°F (93°C); injection pressure, 23,400 psi (161 MPa).

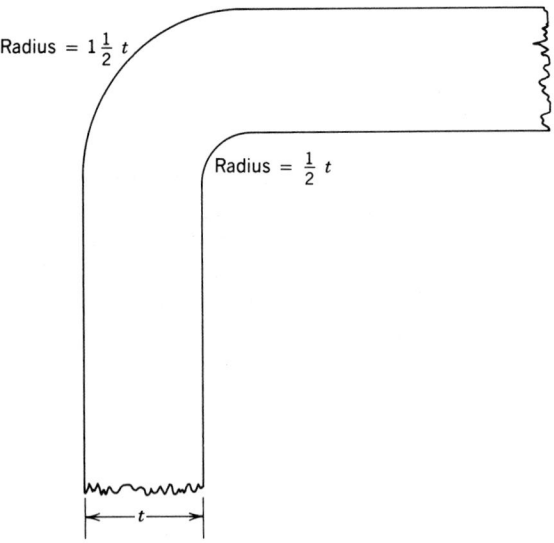

Figure 40.3. Recommended fillets and radii for corner design. Source Union Carbide Corp.

Ribbed cross-section

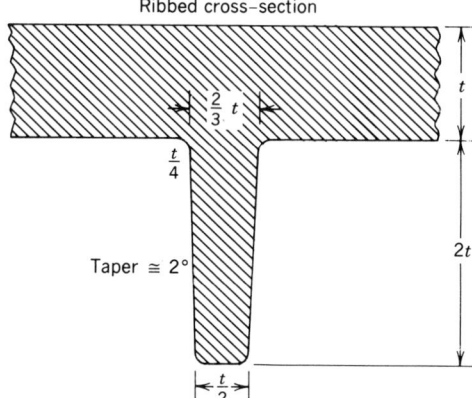

Figure 40.4. Recommended rib construction when designing in polysulfone. Source Union Carbide Corp.

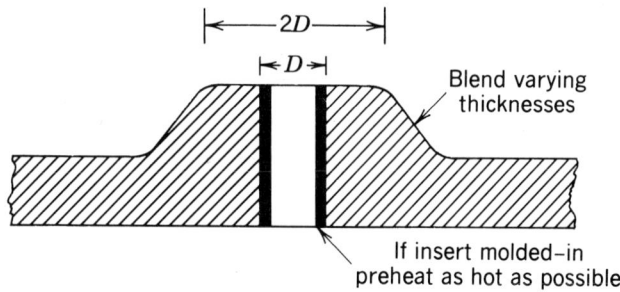

Figure 40.5. Recommended construction when designing bosses and/or molded in inserts. Source Union Carbide Corp.

as it enters the cavity. Polysulfone does not flow easily in thin sections. Therefore, it is best to use a parallel runner and multiple gates to fill a long, thin part. Mold flow versus thickness for typical molding conditions is shown in Figure 40.2. A shrinkage factor of 0.007 mm/mm should be used when molds are designed for polysulfone. Further, the same factor can be used for dimensions parallel to and normal to the direction of flow.

To overcome the notch sensitivity of polysulfone suggestions on radiusing and rib design are given in Figures 40.3–40.5. As previously mentioned, sharp corners and abrupt changes in thickness should be avoided.

40.14 TRADE NAMES

The sole supplier of polysulfone is Union Carbide Corporation, Old Ridgebury Road, Danbury, Conn., under the trade names of Udel and Mindel.

Mold Shrinkage Characteristics (in./in. or mm/mm)

MATERIAL _____ Polysulfone (PSO)

IN THE DIRECTION OF FLOW

THICKNESS

	mm: 1.5	3	6	8	FROM	TO	#
MATERIAL	in.: 1/16	1/8	1/4	1/2			
1700, 1710, 1720 3500	0.007	0.007	0.007	0.007			
GF-210, GF-110	0.004	0.004	0.004	0.004			
GF-205	0.005	0.005	0.005	0.005			
GF-120	0.003	0.003	0.003	0.003			
GF-130	0.002	0.002	0.002	0.002			

	PERPENDICULAR TO THE DIRECTION OF FLOW						
Shrinkages same as above							

NOTE: Mold shrinkage is approximate. It is affected by part design, mold temperature, thickness, injection pressure, packing time, cycle time, orientation, gate design, gate size, gate location, glass content, glass size, and filler content.

Standard Tolerance Chart

PROCESS _____ Injection molding _____ MATERIAL _____ Polysulfone

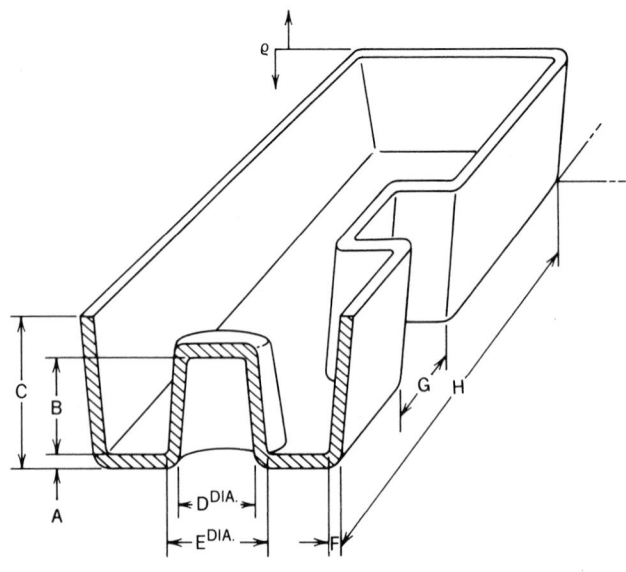

Values are based on a nominal 1/8 in. (3.2 mm) wall thickness and given in plus or minus in./in. and mm/mm.

FINE tolerance – Special care and added cost.

COMMERCIAL tolerance – Normal care required.

COARSE tolerance – Minimal care required, lowest cost.

Tolerance	A	B	C	D	E	F	G	H	
FINE	0.002	0.001	0.003	0.001	0.001	0.002	0.001	0.001	in./in.
									mm/mm
COMMERCIAL	0.003	0.003	0.004	0.003	0.003	0.003	0.003	0.003	in./in.
									mm/mm
COARSE	0.005	0.007	0.007	0.005	0.005	0.006	0.005	0.005	in./in.
									mm/mm

BIBLIOGRAPHY

1. N.J. Ballintyn, in M. Grayson, ed., *Encyclopedia of Chemical Technology,* 3rd ed., Vol. 18, John Wiley & Sons, Inc., New York, 1982 pp. 832–848.
2. R.N. Johnson and A.G. Farnham, *J. Polym. Sci.* **5,** 2415–2427 (1967).
3. T.E. Bugel, *SPE J.* **24**(3), 53–55 (1968).
4. U.S. Pat. 4,230,463 (1980), J.M.S. Henis and M.K. Tripodi.
5. J. Shirley, D. Borzik, and D. Byrnes, *Chem. Process.,* 30–31 (Jan. 1982).
6. D.L. MacLean and T.E. Graham, *Chem. Process.,* 50–55 (Oct. 1980).
7. I. Jitsuhara and S. Kimura, *J. Chem. Eng. Jpn.* **16,** 389–392 (1983).
8. R.J. Peterson, J.E. Cadotte, and J.M. Buettner, *NTIS Report No. PB83-191775,* Springfield, Va.
9. L.M. Robeson and S.T. Crisafulli, *J. Appl. Polym. Sci.* **28,** 2925–2936 (1983).
10. *Polysulfone Design Engineering Handbook,* Union Carbide Corp., Danbury, Conn., 1979.

General Reference

L.M. Robeson, A.G. Farnham, and J.E. McGrath, *Appl. Polym. Sym.* **26,** 373–385 (1975).

41

POLYURETHANE THERMOPLASTIC (TPU)

Sigmund Orchon

Consultant, Culver City, CA 90230

41.1 STRUCTURE

The repeating monomer residue for thermoplastic polyurethanes (TPUs) is[1-7]

$$-O-(-\overset{\overset{O}{\|}}{C}-\overset{\overset{H}{|}}{N}-R-\overset{\overset{H}{|}}{N}-\overset{\overset{O}{\|}}{C}-O-R'-O-\overset{\overset{O}{\|}}{C}-\overset{\overset{H}{|}}{N}-R-\overset{\overset{H}{|}}{N}-\overset{\overset{O}{\|}}{C}-O-R''-O-)_n$$

41.2 CATEGORY

As the name suggests, TPUs are thermoplastic resins.

41.3 HISTORY[8-9]

The TPUs were discovered in the 1950s by C.S. Schollenberger and his associates working in the research laboratories of B.F. Goodrich. They were first commercialized by B.F. Goodrich in 1961. Today, they are offered by Goodrich and several other resin producers.[10-12]

The first most important applications for TPUs were extruded wire and cable jackets and high-performance hose jackets. The introduction of variety of new diisocyanates and/or compounds containing new Tschugaeff-Zerewitinoff hydrogen in the 1960s and 1970s significantly expanded the fields of application of these outstanding materials.[13-15]

After their introduction to the U.S. market in 1961, TPU sales reached 1,000,000 lb in 1963, 22,000,000 lb in 1970, and 33,000,000 lb in 1980.

41.4 POLYMERIZATION

Most TPUs on the market today are synthesized using at least three different types of monomers: a polymeric compound containing a Tschugaeff-Zerewitinoff hydrogen, a diisocyanate, and a chain extender.[16]

(a) Macroglycol   HO—R—OH
(b) Diisocyanate   ONC—R'—NCO
(c) Glycol chain extender   HO—R"—OH

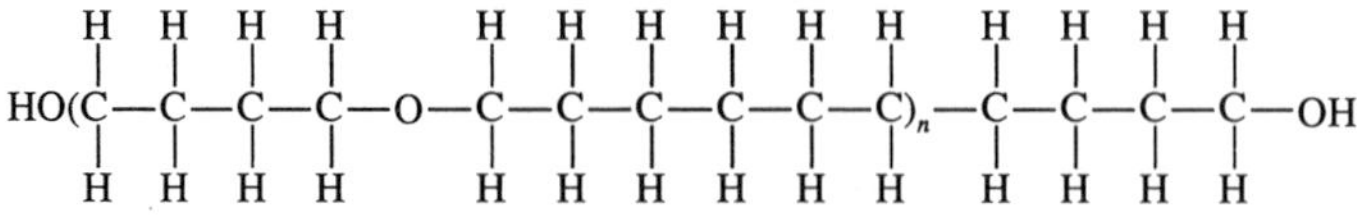

$$HO(\overset{\displaystyle H}{\underset{\displaystyle H}{C}}-\overset{\displaystyle H}{\underset{\displaystyle H}{C}}-\overset{\displaystyle H}{\underset{\displaystyle H}{C}}-\overset{\displaystyle H}{\underset{\displaystyle H}{C}}-O-\overset{\displaystyle H}{\underset{\displaystyle H}{C}}-\overset{\displaystyle H}{\underset{\displaystyle H}{C}}-\overset{\displaystyle H}{\underset{\displaystyle H}{C}}-\overset{\displaystyle H}{\underset{\displaystyle H}{C}}-\overset{\displaystyle H}{\underset{\displaystyle H}{C}}-\overset{\displaystyle H}{\underset{\displaystyle H}{C}}\,)_n-\overset{\displaystyle H}{\underset{\displaystyle H}{C}}-\overset{\displaystyle H}{\underset{\displaystyle H}{C}}-\overset{\displaystyle H}{\underset{\displaystyle H}{C}}-\overset{\displaystyle H}{\underset{\displaystyle H}{C}}-OH$$

Example of (a) Polyester glycol (average molecular weight 1000)

$$HO(\overset{\displaystyle H}{\underset{\displaystyle H}{C}}-\overset{\displaystyle H}{\underset{\displaystyle H}{C}}-\overset{\displaystyle H}{\underset{\displaystyle H}{C}}-\overset{\displaystyle H}{\underset{\displaystyle H}{C}}-O)_n-\overset{\displaystyle H}{\underset{\displaystyle H}{C}}-\overset{\displaystyle H}{\underset{\displaystyle H}{C}}-\overset{\displaystyle H}{\underset{\displaystyle H}{C}}-\overset{\displaystyle H}{\underset{\displaystyle H}{C}}-OH$$

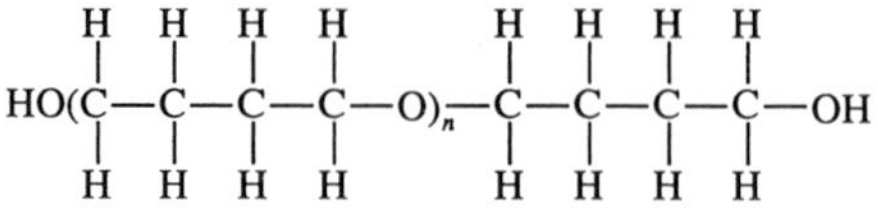

Example of (b) Polyester glycol (average molecular weight 1000)

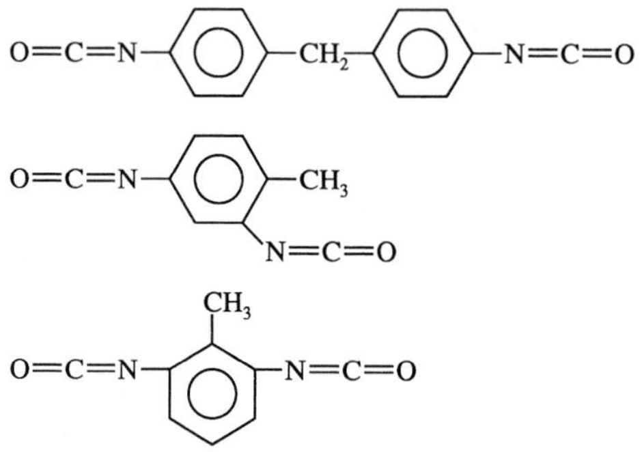

O=C=N—⟨benzene⟩—CH₂—⟨benzene⟩—N=C=O   MDI

O=C=N—⟨benzene ring with —CH₃ and —N=C=O⟩   2,4-TDI

O=C=N—⟨benzene ring with CH₃ and —N=C=O⟩   2,6-TDI

Examples of (c)

Figure 41.1. Structures of raw materials used in thermoplastic polyurethanes

Figure 41.1 shows some of the compounds used in synthesis of TPUs.

Polyurethanes are synthesized by a step-growth polymerization process, in which the chain length of the polymer increases steadily as the reaction progresses. The three basic raw materials listed above react in the following manner:[17]

2 OCN—R—NCO + HO—R'—OH + HO—R"—OH----- ⟶

  (Diisocyanate)   (Polyglycol)   (Chain extender)

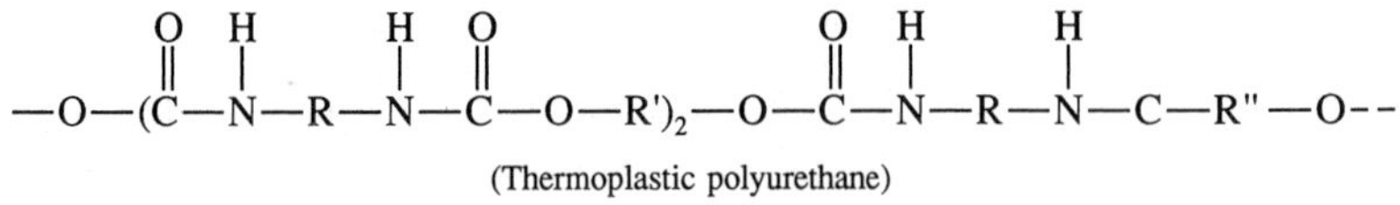

$$-O-(\overset{O}{\overset{\|}{C}}-\overset{H}{\overset{|}{N}}-R-\overset{H}{\overset{|}{N}}-\overset{O}{\overset{\|}{C}}-O-R')_2-O-\overset{O}{\overset{\|}{C}}-\overset{H}{\overset{|}{N}}-R-\overset{H}{\overset{|}{N}}-\overset{H}{\overset{|}{C}}-R"-O---$$

(Thermoplastic polyurethane)

The diisocyanates may be added to the blend of glycol and an extender with the proper catalyst and at optimum temperature conditions; the mixture is then polymerized at random melt polymerization condition.[18-20] The resulting polymer product is in a unique physical state. It has the properties of a thermoset elastomer, without being cross-linked.[21] Strong intramolecular forces, such as hydrogen bonds, van der Waals, Keesom, London, and Debye forces, and intramolecular entanglement of chains, all contribute to the "virtually cross-linked" (VC) state.[22-25] This state, however, depends on temperature.[26] On heating, the action of these forces disappears, permitting the polymer to be processed by standard methods used for thermoplastic systems. On cooling, these forces reappear.

The intramolecular forces of TPUs can be temporarily destroyed by solvation, which enables them to be employed in adhesives and coatings. When the solvents are evaporated, the original properties of the TPUs are restored.

The 1987 price for the pelletized TPUs ranges from $2.13 to 3.82 per pound in truckload quantities.

41.5 DESCRIPTION OF PROPERTIES[16,27-35]

41.5.1 Optical Properties[36]

The optical properties of TPUs are influenced by the ratios of crystalline and amorphous domains in a given polymer system. Because of the force interactions, most TPUs have large crystalline domains; as a result, their transparency increases as the system becomes more amorphous and also as the size of the spherulites decreases. (Therefore, the thermal history of a given polymer will affect its optical properties.) TPUs can be translucent or opaque.

The ease with which the change in crystallinity[37] of TPUs can be achieved provides a latitude for effecting changes in their optical behavior. The crystallinity of the TPUs also affects their electrical properties, notably the dielectric constant.

41.5.2 Environmental Properties[38-40]

The polyester- and polyether-polyols are the polyols most often used. Because of the differences in chemical reactivity, poly(ether–urethanes) are more stable with respect to hydrolysis than poly(ester–urethanes). Ester synthesis is an equilibrium reaction; when exposed to water, it breaks down to the original raw materials of alcohol and acid.

The environmental stability of the TPUs has been studied under both natural and artificial weathering conditions. The dominant mechanism of TPU degradation is the autoxidation process, which is initiated by exposure to the ultraviolet (UV) light. The hydrolytic stability of TPUs depends on their composition, as noted above.

The worst degradation of TPUs occurs with poly(ester–urethanes) as a result of the known reversibility of the ester formation. The reformation of the acid contributes to the self-catalysis of molecular destruction. The various physico-mechanical tests, such as for tensile strength, indicate that the polyester-type polyurethane materials deteriorate substantially more when they are exposed to the humidity of the air than when they are completely immersed in water. This is because during immersion the formed acid is being continuously washed away. There are various ways to improve the stability of polyesters toward hydrolytic decomposition, chiefly by adding different additives or by changing the composition of the esters.[34,40-43]

On the other hand, poly(ether–urethanes) are very resistant to hydrolysis,[9,44-48] since the ether groups do not react with water. Since there are no reactive groups left in the TPUs, they can be extended and their properties improved by addition of other materials. The addition of fillers results in cost savings, improved tear strength and modulus, and increased hardness. These property changes can be observed when about 5–10% of a filler, or combination of fillers, are added to the system. The greatest property improvements can be obtained by the addition of carbon blacks and silicates. Since carbon black absorbs UV rays, it can provide noticeable protection to TPUs from outdoor exposure.[47,49-53]

Various fibers can also be added to TPUs as reinforcing materials, among them asbestos, graphite, and glass. Some aluminum powders can have substantial ultraviolet stabilizing effects over a period of one to two years. Protection against UV effects can also be enhanced by addition of UV stabilizers. Sometimes TPUs require protection against fungus, and the addition of fungicides can prevent destruction of the system.

41.6 APPLICATIONS[16,35,51-61]

Besides wire, cable, and high-performance hose jackets, mentioned above, sheet and film can also be made from TPUs. The variety of TPU applications is tremendous.

Among the most important are extrusion coating of TPUs onto different substrates to improve and/or protect their properties (metal sheets and foils, cotton duck, nylon, animal hide); animal branding with TPU tags; TPU bags used to drop dropping supplies in remote areas [6-ft (1.8-m) long bags have sustained a drop from up to 1000 ft (305-m)]; provision of vapor barriers; manufacture of many household gadgets, gears, cable core end seals, various backing pads for disk assemblies, various pads used in abrasive industry applications, various parts in lawn and orchard sprinklers, and parts made for permanent and temporary artificial hearth devices; specialty mechanical molded goods, such as snowmobile tracks, drives, textile machinery parts, rolls, valve diaphragms and seals, casters and sheets; hot melt adhesives; applications from solutions by standard methods, such as coating, dipping, spraying, brushing, and rolling; and such important application of TPUs as the solvent-adhesives, which, by virtue of rapid bond formation, fit into high-speed production processes.

41.7 ADVANTAGES/DISADVANTAGES

The properties of TPUs[9,61-62] are similar to those of cast polyurethane elastomers of equal hardness, except that TPUs are affected by exposure to elevated temperatures, softening, and loss of their VC (virtually cross-linked) properties. The properties of TPUs are compared to those of other thermoplastic resins in Table 41.1.

The generally recognized advantages of TPUs include:

- Excellent abrasion resistance.
- High tensile strength.

TABLE 41.1. Properties of TPU vs Other Thermoplastics

Property	TPU	Nylon	Vinyl	Acetal	ABS
Hardness	70A-80D	RR111-118	40A-85D	RR118	RR96-118
Resistance to:					
Oxygen	Exc	Exc	Fair	Good	Fair
Ozone	Exc	Exc	Fair	Good	Fair
Heat	Good	Good	Poor	Exc	Fair
Steam	Fair to good	Fair	Fair	Fair	Fair to good
Caustic	Poor to good	Exc	Exc	Poor to fair	Exc
Acid	Poor to good	Fair to exc	Exc	Poor to fair	Fair to exc
Hydrocarbons					
Aromatic	Good	Exc	Fair	Exc	Poor
Aliphatic	Good to exc	Exc	Good	Exc	Exc
Halogenated	Fair to good	Poor	Good	Exc	Poor
Hardness	70a-80D	RR111-118	40A-85D	RR118	RR96-118
Tensile strength, psi	4-9000	7-12000	1.5-9000	8-10000	5-9000
Tear resistance	Exc	Exc	Good	Exc	Good
Elongation %	300-650	25-200	2-450	20-40	10-140
Abrasion resistance	Exc	Exc	Good	Good	Good

TABLE 41.2. Average Parameters for Injection Molding of TPUs

Cylinder temperature, °F (°C)		
Front	335	(169)
Middle	310	(154)
Rear	290	(143)
Nozzle temperature, °F (°C)	345	(174)
Mold temperature, °F (°C)	70–120	(21–49)
Stock temperature, °F (°C)	350–360	(177–182)
Cycle time, s	10–60	10–60

- Excellent tear resistance.
- Excellent load bearing capacity (high compression strength).
- Low compression set.
- Excellent resistance to nonpolar solvents.
- Excellent resistance to fuels and oils.
- Relatively high water vapor transmission.
- Low gas and vapor permeability.
- Good optical properties (ranging from opacity through translucency to transparency).
- Maximum service temperature of 200°F (93°C).
- Minimum service temperature of −70°F (−57°C).
- Very good electrical properties.

41.8 PROCESSING TECHNIQUES

TPUs can be processed by any standard method used for thermoplastic polymers,[9,61,63-64] among them injection, extrusion, compression, blow, and transfer moldings, calendering, extrusion coating, film blow molding, solution, heat sealing, and solvent sealing. The standard TPUs do not require any post-curing. The resins are supplied dry, in the form of pellets or granules.

The average parameters for injection molding of TPUs are given in Table 41.2; for extrusion of TPUs, in Table 41.3; and for calendering of TPUs, in Table 41.4.

Usage of somewhat larger than normal mold gates is advisable for best results. Pin gating should be used only for the smallest shapes, for crosshead and in-line extrusions.

The properties of TPUs are influenced by thermal exposure during and after processing. The thermal exposure affects cross-linking of the resin, produces secondary chemical re-

TABLE 41.3. Average Parameters for Extrusion of TPUs[a]

Barrel temperature, °F (°C)		
Front zone	340	(171)
Middle zone	320	(160)
Rear zone	300	(149)
Die temperature, °F (°C)	340–360	(171–182)
Head temperature, °F (°C)	340–350	(171–177)
Stock temperature, °F (°C)	340–360	(171–182)
Screen pack	20/40/60	20/40/60
Rate lb/h (kg/h)	to 150	(to 68)

[a]2.5 in. (100 m) diameter screen.

TABLE 41.4. Average Parameters for Calendering of TPUs (Inverted L)

Banbury mixer		
Stock temperature at drop, °F (°C)	320–360	(160–182)
Preperation mill		
Stock temperature, °F (°C)	330	(166)
Calender temperature, °F (°C)		
Bottom roll	280	(138)
Middle roll	275	(135)
Top roll	265	(130)
Offset roll	255	(123)

actions, causes changes in crystalline structures as well their shape, and, consequently, influences changes in various properties.

41.9 RESIN FORMS

TPU resins are supplied dry, in the form of pellets or granules. Some aliphatic and cycloaliphatic diisocyanates are used when light stability is required. The stabilization of poly(ester urethane) against autocatalytic hydrolysis is achieved by neutralization of the generated fragments; that is, by addition of poly(carbodiimides) (PCD). The stabilized polymers show pronounced improvements in stress-strain property retention under exposure to humid conditions.

Various manufacturers are adding some other additives to improve their product and/or the performance of the resin during formation of required products; these include lubricants, antioxidants, UV stabilizers, and so on. As with other thermoplastic systems, TPU producers are manufacturing TPU alloys; for example, with ABS, PVC, SAN, and acetals, to name a few. Reinforcement with mineral and glass fillers is also possible.

41.10 SPECIFICATION OF PROPERTIES

See Master Outline of Materials Properties.

41.11 PROCESSING REQUIREMENTS[35,56,65]

The processing of TPUs is affected by their hygroscopicity. Therefore, the control of water absorption during storage is of the utmost importance. To avert problems and disappointments, it is important to understand that the effects of moisture on TPUs are not only physical but also chemical in nature. If the presence of moisture is at all suspected, the resins should be thoroughly dried. While ovens can be used, special hopper-dryers can be more convenient. Storage under a blanket of dry gas will provide a secure barrier against moisture.

The regrind content in virgin TPU will vary, depending on a given resin. In general, 50% of regrind can be used for up to five molding cycles without any adverse effects on TPU properties. If thermal exposure in the extruder or injection molding machine is excessive, the amount of regrind must be adjusted down, to about 25% or less.

MATERIAL ___ Thermoplastic Polyurethanes

PROPERTY	TEST METHOD	ENGLISH UNITS	VALUE	METRIC UNITS	VALUE
MECHANICAL DENSITY	D792	lb/ft³	69.2–79.8	g/cm³	1.11–1.28
TENSILE STRENGTH	D638	psi	1,500–11,000	MPa	10.3–76
TENSILE MODULUS	D638	psi	330–1,450	MPa	2.3–10
FLEXURAL MODULUS	D790	psi	10,000–100,000	MPa	69–690
ELONGATION TO BREAK	D638	%	110–1,000	%	110–1,000
NOTCHED IZOD AT ROOM TEMP	D256	ft-lb/in.		J/m	
HARDNESS	D785		A 50-95 D 46-80		A 50-95 D 46-80
THERMAL DEFLECTION T @ 264 psi	D648	°F	Varies	°C	Varies
DEFLECTION T @ 66 psi	D648	°F		°C	
VICAT SOFT-ENING POINT	D1525	°F		°C	
UL TEMP INDEX	UL746B	°F		°C	
UL FLAMMABILITY CODE RATING	UL94		Medium		Medium
LINEAR COEFFICIENT THERMAL EXPANSION	D696	in./in./°F		mm/mm/°C	
ENVIRONMENTAL WATER ABSORPTION 24 HOUR	D570	%	0.3	%	0.3
CLARITY	D1003	% TRANS-MISSION		% TRANS-MISSION	
OUTDOOR WEATHERING	D1435	%		%	
FDA APPROVAL			Yes		Yes
CHEMICAL RESISTANCE TO: WEAK ACID	D543		Stable		Stable
STRONG ACID	D543		Medium		Medium
WEAK ALKALI	D543		Stable		Stable
STRONG ALKALI	D543		Medium		Medium
LOW MOLECULAR WEIGHT SOLVENTS	D543		No effect to attacked		No effect to attacked
ALCOHOLS	D543		No effect		No effect
ELECTRICAL DIELECTRIC STRENGTH	D149	V/mil		kV/mm	
DIELECTRIC CONSTANT	D150	10^6 Hertz			
POWER FACTOR	D150	10^6 Hertz			
OTHER MELTING POINT	D3418	°F	Varies	°C	Varies
GLASS TRANS-ITION TEMP.	D3418	°F	−60	°C	−51

507

41.14   TRADE NAMES

TPUs are commercially available in the United States from B.F. Goodrich, Cleveland, Ohio, under the trade name Estane and from Mobay Corp., Polyurethane Div., Pittsburgh, Pa., under the trade name Texin.

In the United Kingdom, Elastogran Co., manufactures TPUs under the trade names Caprolan and Elastolan, as does Albis Plastic under the trade name Jectothane. In Germany, the Bayer Corp. manufactures TPUs under the trade name Desmopan.

The polymeric structures of these different TPUs vary. Some companies disclose the basic raw materials; others do not. Therefore, the optimum conditions for processing these polymers should be obtained from the supplier of a particular TPU.

BIBLIOGRAPHY

1. C.S. Schollenberger and co-workers, *Rubber World* **137,** 549–555 (Jan. 1958).
2. C.S. Schollenberger and co-workers, *SPE Transactions* **1,** 31–40 (Jan. 1961).
3. W. Cooper and co-workers, *Ind. Chemist* **36,** 121–126 (1960).
4. W. Neumann and co-workers, *Proceedings of the 4th Rubber Technology Conference*, London, 1962, p. 59.
5. U.S. Pat. 3,193,522 (July 1965), W. Neumann and co-workers.
6. Z.T. Ossefort and co-workers, *Hydrolytic Stability of PUE, 89th Meeting, Rubber Group*, ACS, San Francisco, Calif. May 4, 1966.
7. R.J. Athey, *Rubber Age* **96,** 705 (1965).
8. J.H. Saunders and K.C. Frisch, *Polyurethanes—Chemistry and Technology, High Polymers XVI*, Part I, Wiley-Interscience, New York, 1962.
9. J.H. Saunders and K.C. Frisch, *Polyurethanes—Chemistry and Technology, High Polymers XVI*, Part II, Wiley-Interscience, New York, 1962.
10. F.H. Gahimer and co-workers, "Hydrolytic Stability of PUE and Polyacrylate Elastomers in Humid Environments" *J. Elastoplastic* **1,** 266 (Oct. 1969).
11. G. Magnus and co-workers, *Stability of PUE in Water, Dry Air, and Moist Air Environments*, Rubber Group Chicago Il., ACS, May 3–6, 1966; also, *Rubber Chem. Technol.* **39,** Part 2, 1328–1337 (Sept. 1966).
12. Hydrolysis-Resistant Poly(Ester-Urethanes). U.S. Pat. 3,463,758 (Aug. 26, 1969), F.D. Stewart.
13. W. Dieckmann and F. Breest, *Berichte* **39,** 3052 (1906).
14. G. Magnus, "Polycaprolactone-Based Urethanes," *Rubber Age* **97,** 86–93 (July 1965).
15. *Technical Bulletin #150* Quaker Oats Co., (Aug. 7, 1964).
16. L.F. Fieser and M. Fieser, *Organic Chemistry*, D.C. Heath & Co., Boston, Mass., 1981.
17. K.W. Rausch and co-workers, *J. Elastoplast.* **2,** 144 (April 1970).
18. L. Morbitzer and co-workers, *J. Appl. Polym. Sci.* **16,** 2697 (1972).
19. C.E. Wilkes and co-workers, *J. Macromol. Sci., Phys.* **B7,** 157 (1973).
20. J.H. Saunders, *Rubber Chem. Technol.,* **33,** 1259 (1960).
21. N.V. Seeger and co-workers, *Ind. Eng. Chem.* **45,** 2538 (1953).
22. E.F. Cluff and co-workers, *J. Appl. Polym. Sci.* **3,** 290 (1960).
23. L.C. Kreider and co-workers, U.S. Pat. 2,785,150 (March 1957).
24. D.A. Meyer, "Vulcanization of PUE," in *Vulcanization of Elastomers,* Reinhold, New York, 1964, Chapt. 10.
25. E.E. Gruber and co-workers, *Ind. and Eng. Chem.* **51,** 151 (1959).
26. *Technical Bulletins: Genthane S (GT 53) and Genthan SR (GTOSRI),* Goodyear Corp.
27. M.M. Schwab, *Rubber Age* **92,** 567 (1963).

28. *Technical Bulletins: Adiprene C, A Urethane Rubber and Adiprene CM, A Sulfur-Curable Urethane Rubber,* Du Pont Co., Elastomer Chemicals Dept.

29. D.B. Pattison, U.S. Pat. 2,808,391 (Oct. 1, 1957).

30. W. Kallert, *Kautsch Gummi Kunstst* **19,** 363 (1966).

31. H.S. Kincaid and co-workers, *Polycaprolactone Millable PUE,* Rubber Group, ACS, Montreal, May 1967.

32. P. Wright and co-workers, *Solid PUEs,* Gordon and Breach, New York, 1969.

33. S.V. Urs, *Ind. Eng. Prod. R&D* **1,** (Sept. 1962), p. 199.

34. C.S. Schollenberger and co-workers, *Rubber World* **137,** 549–555 (1958).

35. N.B. Colhup and co-workers, *Introduction to Infrared and Raman Spectroscopy,* Academic Press, New York, 1964.

36. C.S. Schollenberger and co-workers, "Polyurethanes," in O.J. Sweeting, ed., *The Science and Technology of Polymer Films,* Vol. 2, John Wiley & Sons, New York, 1971, Chapt. 12.

37. E.B. Newton and co-workers, *Rubber Chem Technol.* **34,** (1961).

38. S.L. Cooper, *J. Polym. Sci. A1* 1765 (1969).

39. A. Singh and co-workers, *J. Polym. Sci. A1* 2551 (1966).

40. A. Singh, in K.C. Frisch, ed., *Advances in Urethane Science and Technology.* Vol. 1, Technomic Publishing Co., Westport, Conn., 1971, Chapt. 5.

41. C.S. Schollenberger and co-workers, *J. Elastoplast.* **5,** 222 (1973).

42. R. Bonart, *J. Macromol. Sci., Phys. B2,* 115 (March 1968).

43. S.B. Clough and co-workers, *J. Macromol. Sci. Phys.* **B2,** 553 (Dec. 1968).

44. K. Onder and co-workers, *Polymer* **13,** 133 (March 1972).

45. J.H. Saunders, *Rubber Chem. Technol.* **33,** 1259 (1960).

46. U.S. Pat. 3,660,341 (May 1972), K. Dinbergs.

47. D.J. Harmon, B.F. Goodrich R&D Center, Unpublished papers.

48. F.W. Billmeyer Jr., *Textbook of Polymer Chemistry,* Wiley-Interscience, New York, 1967.

49. P.W. Allen, *Techniques of Polymer Characterization.* Academic Press, New York, 1976.

50 E.B. Newton and co-workers, *Rubber Chem. Technol.* **34,** 1 (1961).

51. F. Bueche, *Physical Properties of Polymers.* Wiley-Interscience, New York, 1962.

52. C.S. Schollenberger and co-workers, *J. Elastoplast.* **4,** 4 (1972).

53. R. Adams and co-workers, *J. ACS* **72,** 5154–5157 (1950).

54. L.V. Nevskij and co workers, *Sov. Plast.* **9,** Ce4 (1967).

55. O.G. Tarakanov and co-workers, *J. Polym. Sci. C,* 192–199 (1968).

56. Ger. Pat. Appl. F 15911 IVb/39b, E. Windemuth and co-workers.

57. L.J. Bellamy, *The Infrared Spectra of Complex Molecules,* John Wiley & Sons, New York, 1977.

58. R.G. White, *Handbook of Ind. Infrared Analysis,* Plenum Press, New York, 1964.

59. D.J. David and co-workers, *Analytical Chemistry of Polyurethanes, High Polymers XVI,* Part III, Wiley-Interscience, New York, 1969.

60. M.L. Kaplan and co-workers, *J. Polym. Sci. A1,* 3163–3175 (1970).

61. E.N. Doyle, The Development and Use of PU Products. McGraw-Hill, New York, 1971.

62. H.C. Beachell and co-workers, *J. Appl. Polym. Sci.* 7, 2217–2237 (1963).

63. D.L. Nealy and co-workers, *J. Polym. Sci. A1,* 2063–2070 (1971).

64. M. Day and co-workers, *J. Appl. Polym. Sci.* **16,** 203–215 (1972).

65. L.B. Weisfeld and co-workers, *J. Polym. Sci.* **56,** 455 (1962).

General References

S.L. Cooper and co-workers, *J. Appl. Polym. Sci.* **10,** 1837 (1966).

J.M.G. Cowie, *Polymers: Chemistry and Physics of Modern Materials,* Intertex Books, Aylesbury, England, 1973.

R.A. Dunleavy and co-workers, "PUE: Influence of Structure on High and Low Temperature Properties," *Rubber World* **156,** (June 1967), 53–57.

O.B. Edgar and co-workers, *J. Polym. Sci.* **8,** 1 (1952).

Farbenfabriken Bayer, *Technical Bulletin Urepan E,* An Ester-Based Rubber for Crosslinking With Peroxides (July 1, 1961).

L.L. Harrell Jr., *Macromolecules* **2,** 607 (Nov.–Dec. 1969).

G.W. Miller and co-workers, *J. Appl. Polym. Sci.* **13** (1969) 1277.

M. Morton and co-workers, *Degradation Studies on Condensation Polymers,* U.S. Dept. of Commerce, *Report PB-131,* March 31, 1957, p. 795.

E. Muller and co-workers, *Rubber Chem. Technol.* **26,** 1953, p. 493.

C. Naegli and A. Tyajbi, *Helv. Chim. Acta* **17,** 931 (1934).

Naugatuck Chemical Co., (Uniroyal), Technical Bulletins Vibrathanes 5003 and 5004.

K.A. Piggot and co-workers, *Chem. Eng. Data* **5,** 391 (1960).

Thiokol, *Technical Bulletin: Facts 5,* 1963.

42

POLYURETHANES, THERMOSET

Sigmund Orchon

Consultant, Culver City, CA 92030

42.1 INTRODUCTION

$$-O-(\overset{\overset{O}{\|}}{C}-\overset{\overset{H}{|}}{N}-R)_n \qquad 0 \le n <$$

The polyurethanes[1] are among the most important classes of polymers. The term "polyurethanes" (PU) is used more for convenience than for accuracy, since these polymers are not derived by polymerization of monomeric urethane molecules; neither are they polymers containing primarily urethane residual groups. This explains the presence of $n = 0$ in the chemical equation given above. The polyurethanes do not contain any residual groups of urethane at all; they are called polyurethanes because at one time they did contain residual urethane groups that succumbed to a secondary reaction during the polyurethane process. The list below shows the repeating groups in polyurethanes.

$-(\overset{\overset{H}{\|}}{N}-\overset{\overset{O}{\|\|}}{C}-O-R)-$	Urethanes
$-(\overset{\overset{H}{\|}}{N}-\overset{\overset{O}{\|\|}}{C}-\overset{\overset{H}{\|}}{N}-R')-$	Ureas
$-(R-\overset{\overset{H}{\|}}{N}-\overset{\overset{O}{\|\|}}{C}-\overset{\overset{H}{\|}}{N}-\overset{\overset{O}{\|\|}}{C}-R')-$	Acyl ureas
$-(R-\overset{\overset{H}{\|}}{N}-\overset{\overset{O}{\|\|}}{C}-\overset{\overset{R}{\|}}{N}-\overset{\overset{O}{\|\|}}{C}-R')-$	Biurets
$-(R-\overset{\overset{H}{\|}}{N}-\overset{\overset{O}{\|\|}}{C}-\overset{\overset{R}{\|}}{N}-\overset{\overset{O}{\|\|}}{C}-O-R')-$	Allophanates
$-N\overset{\displaystyle C}{\diagdown}\overset{\displaystyle \|}{}\overset{\displaystyle C}{\diagup}N-$	Phenyl uretidinedione

$$
\begin{array}{c}
O \\
\parallel \\
C \\
\diagup \quad \diagdown \\
-N \qquad N- \\
| \qquad\qquad | \\
O{=}C \qquad C{=}O \\
\diagdown \quad \diagup \\
N \\
|
\end{array}
\qquad \text{Isocyanurates}
$$

42.2 CATEGORY[2–9]

Thermoset polyurethanes are comprised of seven major categories; each will be treated separately in the following sections. These include PU elastomers, foams (flexible, rigid, and RIM), millable PUs, adhesives, coatings, sealants, and fibers.

42.3 HISTORY[10]

In 1937, in an attempt to avoid patent infringement on nylon (which was held by William Carothers of the E.I. du Pont De Nemours & Co., Inc.), Otto Bayer and his associates at I.G. Farbenindustrie in Germany directed their research to the polymerization and development of polyurethanes. Original work of Bayer's resulted in the commercial use of PU for synthetic fibers and bristles, under the trade name Perlon U. At the same time, DuPont researchers in the United States began to investigate isocyanates.

By 1947, very intensive R&D in Germany resulted in limited commercial application of PUs in adhesives, coatings, and foams. In the United States, Lockheed Aircraft Corp. developed the first rigid PU foams. However, U.S. interest in the PUs grew slowly. Only Bayer's development of PU elastomers in 1950 and flexible PU foams in 1952 stimulated the U.S. plastics industry to initiate large-scale commercialization of PUs.

In the years from 1952 through 1954, with Bayer's work in developing polyester-diisocyanate flexible foams in conjunction with machinery suitable for their processing, this major new segment of the plastics industry was launched.

The development of polyetherpolyols in the United States in 1957 changed PU technology. Introduction of various compounds containing Tschugaev-Zerewitinoff hydrogen became the *coup d'eclat* for the PU industry: any compound containing Tschugaev-Zerewitinoff hydrogen reacts with an isocyanate. This discovery gave PU producers a wide selection of raw materials, with the result of a multitude of built-in properties.

$$
\text{OCN-R-NCO} + \text{H----R'---H} \longrightarrow \text{-}\!\left(\!-\overset{\overset{\text{O}}{\parallel}}{\text{C}}-\overset{\overset{\text{H}}{|}}{\text{N}}-\text{R}-\overset{\overset{\text{H}}{|}}{\text{N}}-\overset{\overset{\text{O}}{\parallel}}{\text{C}}-\text{R'}\right)_{\!n}
$$

The isocyanate group (-NCO) can react in different ways, as shown in Figure 42.1.

42.4 PU ELASTOMERS (PUE)[1,10–11]

Depending on certain physical properties, all polymers can be classed into three groups of materials:

- Elastomers are macromolecules characterized by an initial elastic modulus of 10^6–10^7 dynes/cm and by a long range of almost complete and reversible extensibility.

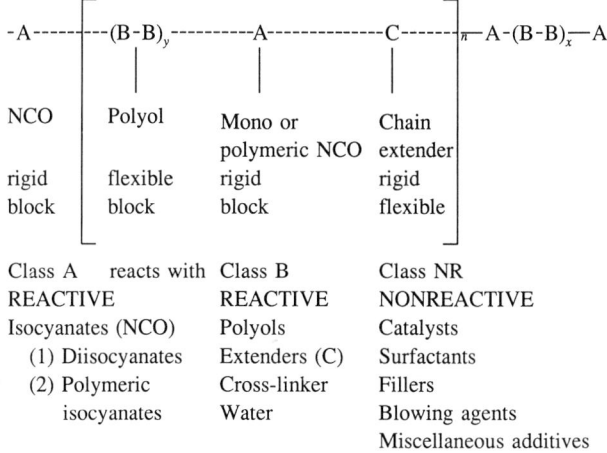

Figure 42.1. Basic structure of thermoset polyurethanes

- Plastics are high polymers with an initial elastic modulus of 10^8–10^9 dynes/cm and a range of deformability of the order of 100–200%, especially under stress. Part of this deformability is reversible, part is permanent.
- Fibers are macromolecules characterized by an initial elastic modulus of 10^{10}–10^{11} dynes/cm and a low range of extensibility—on the order of 10–20%. Part of this deformability is reversible, part shows delayed recovery, and part is permanent.

Second-order transition temperature (T_g), more correctly called glass-transition temperature, is a property that determines whether a PU polymer is an elastomer or a plastic. Any polymer performing above its T_g is an elastomer; any polymer performing below its T_g is a plastic.

PUEs are characterized by the following general properties:

- Performance above its T_g.
- Ability to stretch and retract instantly.
- High modulus and strength when stretched.
- Negligible crystallinity.
- Molar mass large enough for network formation or cross-linking.
- An increase of elastic modulus in response to an increase in temperature.
- Generation of heat when stretched.
- Change in length in response to constant stress and variable temperatures.

To achieve the needed properties for a given PU, more than one polyol and/or Tschugaev-Zerewitinoff-type of hydrogen-containing polymer is used. To achieve thermosetting properties of PUEs, at least one or more tri-functional compound must be used. Many different isocyanate-containing compounds can be used: aromatic isocyanates, such as TDI 80/20, TDI 65/35, MDI (pure), MDI (polymeric), or naphthalene diisocyanate; aliphatic isocyanates, such as hexamethylene diisocyanate (HDI) or dicyclohexylmethane diisocyanates (Mondur W), or isophorondiisocyanate (IPDF). The aliphatic diisocyanates are used when light stability is a must. The liquid PUE process can follow two routes: prepolymer or two-part reaction sequence or one-shot reaction sequence.

Using prepolymer,[10,12,13] PUEs are synthesized prior to casting into the molds or making films. In a one-shot reaction, all the ingredients are mixed at once, producing the final product in situ. Depending on the type of ingredients introduced, reaction temperatures,

Master Outline of Material Properties

MATERIAL ___Thermoset Polyurethane Elastomers___

PROPERTY	TEST METHOD	ENGLISH UNITS	VALUE	METRIC UNITS	VALUE
MECHANICAL DENSITY	D792	lb/ft³	64.2–93.5	g/cm³	1.03–1.5
TENSILE STRENGTH	D638	psi	10,000–15,000	MPa	69–103
TENSILE MODULUS	D638	psi	10,000–100,000	MPa	69–690
FLEXURAL MODULUS	D790	psi	10,000–100,000	MPa	69–690
ELONGATION TO BREAK	D638	%	Above 1,000	%	Above 1,000
NOTCHED IZOD AT ROOM TEMP	D256	ft-lb/in.	25 to flexible	J/m	133 to flexible
HARDNESS	D785	Shore D	10–85 45–80		10–85 45–80
THERMAL DEFLECTION T @ 264 psi	D648	°F	Varies	°C	Varies
DEFLECTION T @ 66 psi	D648	°F		°C	
VICAT SOFT-ENING POINT	D1525	°F		°C	
UL TEMP INDEX	UL746B	°F		°C	
UL FLAMMABILITY CODE RATING	UL94		Medium		Medium
LINEAR COEFFICIENT THERMAL EXPANSION	D696	in./in./°F	$18–36 \times 10^{-5}$	mm/mm/°C	$32–65 \times 10^{-5}$
ENVIRONMENTAL WATER ABSORPTION 24 HOUR	D570	%	0.2–1.5	%	0.2–1.5
CLARITY	D1003	% TRANS-MISSION		% TRANS-MISSION	
OUTDOOR WEATHERING	D1435	%	Varies	%	Varies
FDA APPROVAL			Varies		Varies
CHEMICAL RESISTANCE TO: WEAK ACID	D543		Not attacked		Not attacked
STRONG ACID	D543		Medium		Medium
WEAK ALKALI	D543		Not attacked		Not attacked
STRONG ALKALI	D543		Medium-poor		Medium-poor
LOW MOLECULAR WEIGHT SOLVENTS	D543		Medium-poor		Medium-poor
ALCOHOLS	D543		Not attacked		Not attacked
ELECTRICAL DIELECTRIC STRENGTH	D149	V/mil	200–260	kV/mm	7.9–10.2
DIELECTRIC CONSTANT	D150	10^6 Hertz			
POWER FACTOR	D150	10^6 Hertz			
OTHER MELTING POINT	D3418	°F		°C	
GLASS TRANS-ITION TEMP.	D3418	°F	–76	°C	–60

pressure and/or vacuum applied, extent of time various steps are exposed to certain conditions, and so on, one can obtain one or more reactions, shown in in Figure 42.1, and thereby preplan a product's properties.

42.5 POLYURETHANE FOAMS (PUF)[1,3,10,12–16]

During World War II, natural polymers were not available in Germany. This spurred R&D efforts into the field of rigid PUFs. In the United States, the Air Force conducted research on rigid PUFs. In 1952, flexible PUFs were introduced by Bayer Corp., using polyester polyols at first. They had outstanding properties, but they were too expensive for most applications. In 1957, major changes in PUF technology and marketing were achieved when polyether polyols were introduced. They offered simplified, cheaper, and faster processing. PUFs made from polyether polyols using the prepolymer method had tremendously improved physical and mechanical properties. PUFs tailored to particular specifications were introduced with the introduction of the one-shot process in 1958 by Wyandotte Corp.

The chemistry involved in PUF preparations is, in principle, quite simple. Two reactions are of importance: formation of urethanes and formation of ureas. The first reaction is a prepolymer forming step; the urea formation is the extension step of PUF.[10–13] Typical reactions in thermosetting polyurethanes are given below.

$$RNCO + R'OH \longrightarrow R-\overset{\overset{\textstyle H}{|}}{N}-\overset{\overset{\textstyle O}{\|}}{C}-O-R' \qquad \text{Urethanes}$$

$$RNCO + R'NH \longrightarrow R-\overset{\overset{\textstyle H}{|}}{N}-\overset{\overset{\textstyle O}{\|}}{C}-\overset{\overset{\textstyle H}{|}}{N}-R' \qquad \text{Ureas}$$

$$RNCO + R'COOH \longrightarrow R-\overset{\overset{\textstyle H}{|}}{N}-\overset{\overset{\textstyle O}{\|}}{C}-R' + CO_2 \qquad \text{Amides}$$

$$RNCO + H_2O \longrightarrow R-\overset{\overset{\textstyle H}{|}}{N}-\overset{\overset{\textstyle O}{\|}}{C}-O-H \longrightarrow R-NH_2 + CO_2 \longrightarrow$$

$$R-\overset{\overset{\textstyle H}{|}}{N}-\overset{\overset{\textstyle O}{\|}}{C}-\overset{\overset{\textstyle H}{|}}{N}-R \qquad \text{Urea}$$

$$RNCO + R'-\overset{\overset{\textstyle O}{\|}}{C}-NH_2 \longrightarrow R-\overset{\overset{\textstyle H}{|}}{N}-\overset{\overset{\textstyle O}{\|}}{C}-\overset{\overset{\textstyle H}{|}}{N}-\overset{\overset{\textstyle O}{\|}}{C}-R' \qquad \text{Acyl urea}$$

$$RNCO + R'-\overset{\overset{\textstyle H}{|}}{N}-\overset{\overset{\textstyle O}{\|}}{C}-O-R' \longrightarrow R-\overset{\overset{\textstyle H}{|}}{N}-\overset{\overset{\textstyle O}{\|}}{C}-\overset{\overset{\textstyle R}{|}}{N}-\overset{\overset{\textstyle O}{\|}}{C}-O-R' \qquad \text{Allophanate}$$

$$RCNO + R-\overset{\overset{\textstyle H}{|}}{N}-\overset{\overset{\textstyle O}{\|}}{C}-\overset{\overset{\textstyle H}{|}}{N}-R' \longrightarrow R-\overset{\overset{\textstyle H}{|}}{N}-\overset{\overset{\textstyle O}{\|}}{C}-\overset{\overset{\textstyle R}{|}}{N}-\overset{\overset{\textstyle O}{\|}}{C}-\overset{\overset{\textstyle H}{|}}{N}-R' \qquad \text{Biuret}$$

$$NCO \longrightarrow -N \underset{\underset{\textstyle O}{\|}}{\overset{\overset{\textstyle O}{\|}}{\overset{C}{\diagdown}}}\,\,\overset{\overset{\textstyle O}{\|}}{\underset{\textstyle C}{\diagup}} N- \qquad \begin{array}{l} \text{Phenyl Diisocyanate Dimer} \\ \text{(Phenyl Uretidinedione)} \end{array}$$

$$NCO \longrightarrow$$

Isocyanurates

The two reactions have to work in unison: in order to be successful the polymer growth must be in step with gas formation or evolution to avoid collapse of the PUF. The rates of these reactions are controlled by a catalyst and/or proper combination of types and ratios of catalysts.

The type of Tschugaev-Zerewitinoff hydrogen-containing compound dictates the type of bonding among the polymer chains and determines the degree of cross-linking, which in turn differentiates between flexible and rigid PUF. In flexible PUF, the chains are long and the intramolecular bonds are farther apart, permitting more movement. In rigid PUF, the intramolecular bonds are close together, creating a tight lattice and restricting movement.

The formation of carbon dioxide during the urea reaction creates bubbles, which results in the cellular structure of the PUF.

Other volatile liquids besides water can be used as the blowing agent, especially fluorinated and fluorochlorinated hydrocarbons. These compounds are especially valuable in the production of insulating materials from rigid PUF.

The chemistry of thermoset polyurethane foams (PUF):

$$2 \ R{-}N{=}C{=}O \ + \ HO{-}R' \ + \ H_2O \longrightarrow$$

Isocyanate   +   Polyol   +   Water

$$R{-}N{-}C{-}O{-}R' \longrightarrow R{-}NH_2 + CO_2$$

Carbamic acid

$$R{-}NH_2 + R'{-}N{=}C{=}O \longrightarrow R{-}N{-}C{-}N{-}R'$$

Urea

In addition to the compounds described above, various other ingredients are included in formulations of PUFs to improve stabilization, acceleration, and/or retardation of the reactions, and resistance to ultraviolet light, infrared light, oxygen, and ozone.

42.6 REACTION INJECTION MOLDING (RIM)

RIM processing of PUs involves very reactive ingredients and high-pressure impingements mixer in what is essentially a one-shot process for casting PUEs. The steps involved are described in Figure 42.2. Variations of this basic process include two-part and quasi-prepolymer systems (Fig. 42.3 and 42.4).

Since the isocyanate-hydroxyl (NCO-OH) ratios have to be controlled explicitly, the metering machines must mix the reactant homogeneously at a constant rate. This is accomplished by introduction of all the components into a small chamber through nozzles under high pressure [1500–3000 psi (10 MPa–20 MPa)], whereby they are "triturated" by impingement. Molding pressure is quite low [50 psi (0.35 MPa)]. However, if the part is large,

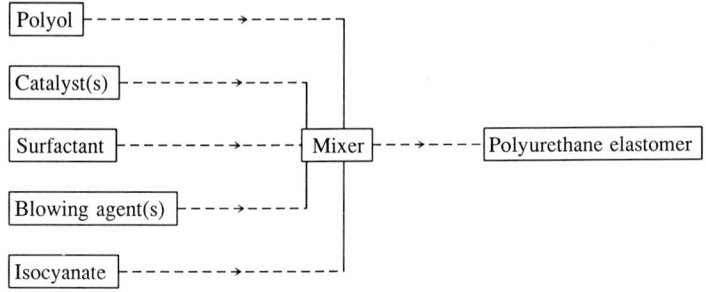

Figure 42.2. RIM one-shot process

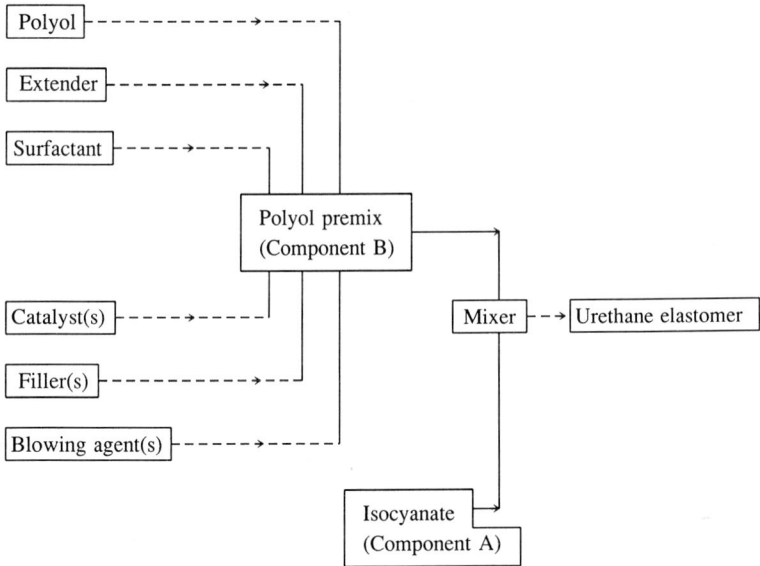

Figure 42.3. RIM two-part process

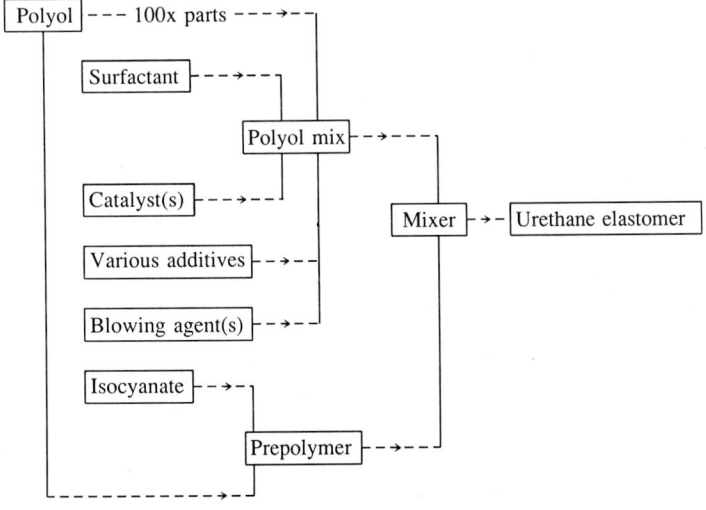

Figure 42.4. RIM quasi-prepolymer process

MATERIAL ___ Rigid Polyurethane Foams

PROPERTY	TEST METHOD	ENGLISH UNITS	VALUE	METRIC UNITS	VALUE
MECHANICAL DENSITY	D792	lb/ft^3	2.3	g/cm^3	0.039
TENSILE STRENGTH	D638	psi	25–75	MPa	0.2–0.52
TENSILE MODULUS	D638	psi		MPa	
FLEXURAL MODULUS	D790	psi	50–85	MPa	0.4–0.6
ELONGATION TO BREAK	D638	%		%	
NOTCHED IZOD AT ROOM TEMP	D256	ft-lb/in.		J/m	
HARDNESS	D785				
THERMAL DEFLECTION T @ 264 psi	D648	°F	7.5–15	°C	0.05–0.1
DEFLECTION T @ 66 psi	D648	°F		°C	
VICAT SOFT-ENING POINT	D1525	°F		°C	
UL TEMP INDEX	UL746B	°F		°C	
UL FLAMMABILITY CODE RATING	UL94				
LINEAR COEFFICIENT THERMAL EXPANSION	D696	in./in./°F	2.7×10^{-4}	mm/mm/°C	4.9×10^{-4}
ENVIRONMENTAL WATER ABSORPTION 24 HOUR	D570	%		%	
CLARITY	D1003	% TRANS-MISSION		% TRANS-MISSION	
OUTDOOR WEATHERING	D1435	%		%	
FDA APPROVAL					
CHEMICAL RESISTANCE TO: WEAK ACID	D543				
STRONG ACID	D543				
WEAK ALKALI	D543				
STRONG ALKALI	D543				
LOW MOLECULAR WEIGHT SOLVENTS	D543				
ALCOHOLS	D543				
ELECTRICAL DIELECTRIC STRENGTH	D149	V/mil		kV/mm	
DIELECTRIC CONSTANT	D150	10^6 Hertz			
POWER FACTOR	D150	10^6 Hertz			
OTHER MELTING POINT	D3418	°F		°C	
GLASS TRANS-ITION TEMP.	D3418	°F		°C	

MATERIAL ___ RIM

PROPERTY	TEST METHOD	ENGLISH UNITS	VALUE	METRIC UNITS	VALUE
MECHANICAL DENSITY	D792	lb/ft³	50–65	g/cm³	0.8–1.04
TENSILE STRENGTH	D638	psi	5,800–6,500	MPa	40–45
TENSILE MODULUS	D638	psi		MPa	
FLEXURAL MODULUS	D790	psi	30–400	MPa	0.20–2.75
ELONGATION TO BREAK	D638	%	10–250	%	10–250
NOTCHED IZOD AT ROOM TEMP	D256	ft-lb/in.		J/m	
HARDNESS	D785 Shore D		45–85		45–85
THERMAL DEFLECTION T @ 264 psi	D648	°F		°C	
DEFLECTION T @ 66 psi	D648	°F		°C	
VICAT SOFT-ENING POINT	D1525	°F		°C	
UL TEMP INDEX	UL746B	°F		°C	
UL FLAMMABILITY CODE RATING	UL94				
LINEAR COEFFICIENT THERMAL EXPANSION	D696	in./in./°F	20–60	mm/mm/°C	36–108
ENVIRONMENTAL WATER ABSORPTION 24 HOUR	D570	%		%	
CLARITY	D1003	% TRANS-MISSION		% TRANS-MISSION	
OUTDOOR WEATHERING	D1435	%		%	
FDA APPROVAL					
CHEMICAL RESISTANCE TO: WEAK ACID	D543				
STRONG ACID	D543				
WEAK ALKALI	D543				
STRONG ALKALI	D543				
LOW MOLECULAR WEIGHT SOLVENTS	D543				
ALCOHOLS	D543				
ELECTRICAL DIELECTRIC STRENGTH	D149	V/mil		kV/mm	
DIELECTRIC CONSTANT	D150	10^6 Hertz			
POWER FACTOR	D150	10^6 Hertz			
OTHER MELTING POINT	D3418	°F		°C	
GLASS TRANS-ITION TEMP.	D3418	°F		°C	

the total force required on the hydraulic presses may be as much as 10–20 tons (9,100–18,200 kg).

In RIM,[1] the molds are part of the process and must be designed accordingly. They must be relatively heavy and include built-in heat controls to regulate the high exotherm developed during the reaction between NCO and OH [about 25 kcal/mol (104kJ/mol)]. Reinforced reaction injection molding (RRIM) is an extension of RIM. In principle, the materials used for RRIM are similar to the ones used for RIM, with the exception of the added reinforcing material—the fiber. The molding process is dissimilar as a result of the abrasiveness and consistency of the system as well as the needed equipment modifications. RIM offers economical advantages where more than 100,000 large parts have to be produced. In certain applications, RRIM PUEs are preferred to metals and elastoplastics, because of their advantageous cost/weight and stiffness ratios and improvements in automatic processing.

42.7 MILLABLE PU ELASTOMERS[1,10]

To produce millable gums, PUEs must be processed in the same fashion as rubbers on two roll mills. Millable PUEs are polymers having a molecular weight of about 15,000–2,500. They are mostly linear, with limited cross-linking. For better stability, the ratio of isocyanate to hydroxyl should equal 1. The reactive groups necessary for curing (vulcanization) consist of —C=C— groups as, for example, in allyl ethers: CH=CH—CH—O—. Sulfur, peroxide, and diiosocyanate cures are employed, with sulfur cures being preferred for better properties, using carbon black for reinforcement.

The millable gums are processed in exactly the same way as standard rubbers—by mastification. The breakdown is a function of both time and temperature [248–302°F (120–150°C)]; when compounded with curative, 230°F (110°C) is sufficient. For plastification of the gum, coumarone–indene resins are sometimes added.

42.8 PU ADHESIVES

Three types of adhesive systems can be produced from PUs: single-component, where the isocyanate group is the main bonding constituent; two-component, where one component contains the prepolymer with isocyanate excess; and thermoplastic hot melts, with, usually, no residual isocyanate groups. Blocked isocyanate groups can be activated when they are exposed to elevated temperature.

The characteristics of PU adhesives include:

- Outstanding reactivity with Tschugaev-Zerewitinoff hydrogen.
- Excellent solubility and permeability of PU molecules.
- Van der Waals, London, Keesom, and Debye forces that produce excellent cohesive bonds to a variety of substrates.
- Hydrogen bonds that can be formulated to produce hard to soft elastomers having excellent vibration-damping properties.
- Possibility of formation of chemical bonds with the surfaces of some substrates, such as metals.
- Ability to withstand cyclic deformations and/or dynamic fatigue.
- Production of many successful joints by reaction with tightly absorbed water on surfaces of some substrates, such as glass.
- Excellent contact angle, permitting almost universal adaptability of PU adhesive to most surfaces.

42.9 PU COATINGS[1,10]

In principle, there is no difference in the basic chemistry of PU in the formulation of adhesives and coatings. Mechanically speaking, in the case of adhesive the PUE film is applied between two substrates (a "sandwich arrangement"). In the case of coatings, the same film is applied on one surface. When applying PUE coatings, one has to consider the exposure of the films to the environment. Proper additives have to be introduced for protection of the coating and for achievement of "cosmetic" features.

42.10 PU SEALANTS[1,10]

PU sealants encompass an exceptional group of elastomers having a wide latitude of deformabilities, cold flow, and elasticities. The PUE sealants are capable of stress-deformation to such an extent that no adhesive bonds with the rigid substrates are destroyed. The properties of PUE sealants have permitted innovative changes in the techniques of the construction industry that were heretofore not possible. Examples include highway and airport expansion joints, architectural construction sealants, clay sewer pipe sealants, marine sealants, aerospace cryogenic sealants, and electrical sealants.

The PU sealants represent a defined group of low-modulus elastomers. The properties of PUE which contribute to the outstanding performance of them as sealants include:

- Excellent elasticity
- High recovery
- Outstanding resistance to abrasion
- Low-temperature flexibility
- Resistance to low concentration of common acids and caustics; excellent resistance to oils.
- Resistance to biodegradation.
- With proper tailoring, resistance to ultraviolet and infrared light, oxygen, ozone, plus fire retardancy.
- For sag control, they can be made thixotropic. (This is important because a sealant's viscosity decreases when it is stirred and resumes its original consistency when it is left undisturbed.)
- Possible manufacture into one-component moisture-cured systems and two-component systems.

42.11 PU ELASTOMERIC FIBERS

Both high fashion and technological developments have permitted the manufacture of high style and comfort in clothing. The more advanced elastomeric fibers, called Spandex, have contributed excellent qualities to fabrics. Spandex is a generic name for PU fibers; it is used to describe as any fiber composed of at least 85% segmented PU. Softness, elongation, and instant recovery when stress-released are the basic features of Spandex fibers; in combination with a high modulus, Spandex surpasses the properties of any elastomeric fiber known.

The unique properties of Spandex include

- Superior oxidation resistance.
- Excellent retention of properties after exposure to ultraviolet light.
- Excellent resistance to dry-cleaning solvents.

- Excellent resistance to body oils.
- Good dyeability.
- Excellent toughness; no need for fiber protection during knitting.
- Finer sizes of fibers possible, permitting sheerer garments.

Spandex fiber is prepared in the same fashion as other fibers—by spinning, a term covering the extrusion of polymer through a small orifice. Spandex fibers are made by either solution or reaction spinning. In solution spinning, the polymer dissolved in a solvent is pumped through orifices of a spinnerette plate and the solvent is diffused out to form the filaments. In reaction spinning, liquid prepolymer, which is extruded through a spinnerette plate, encounters a chain-extending cross-linking component, producing a filament. These two methods can be combined. The two reactions used in these methods are shown below.

42.11.2 Solution Spinning

$$PTMG + 2,4\text{–}TDI \longrightarrow PTMG\text{-dimer}$$

| parts | 92 | 8 | 100 |

$$PTMG\text{-dimer} + MDI \longrightarrow Prepolymer$$

| parts | 100 | 25 | 125 |

$$Prepolymer + DMF + H_2NNH_2{\cdot}H_2O + DMF \longrightarrow Solution$$
20% solids

| parts | 125 | 75 | 2.5 | 425 |

Solution 20% solids + TiO_2 paste + stabilizer = SPINNING PTMG = Polytetramethylene glycol. This formulation is basic for the manufacture of DuPont's version of Spandex called Lycra.

42.11.3 Reaction Spinning

$$Polyester + TiO_2 + Additives + MDI \longrightarrow Prepolymer$$

| parts | 1,000 | | | 250 | 1,250 |

$$Prepolymer + H_2NCH_2CH_2NH_2 \longrightarrow Polymer$$

| parts | 1,250 | excess |

This formulation is basic for the manufacture of Uniroyal's version of Spandex called Vyrene. Many variations are used to obtain different properties in both processes.

42.12 SUPPLIERS OF PU SYSTEMS

Allthane Ltd. Corp., Marina del Rey, Calif.; Armstrong Cork Co., Lancaster, Pa.; Callery Chemical Co., Callery, Pa.; E.I. du Pont de Nemours & Co., Wilmington, Del.; B.F. Goodrich Co., Cleveland, Ohio; Luedor Co., Culver City, Calif.; Mobay Chemical Co., Pittsburgh, Pa.; Polyresins Corp., Sun Valley, Calif.; Reichhold Chemicals Inc., White Plains, N.Y.; and Wyandotte Chemical Corp., Wyandotte, Mich.

BIBLIOGRAPHY

1. P.F. Bruins, *Polyurethane Technology,* Wiley Interscience, New York, 1969.
2. R.G. Arnold and co-workers, *Chem. Revs.* **57,** 47 (1957).
3. O. Bayer. *Angew. Chem. A* **59,** 275 (1947).
4. A. Damusis, K.C. Frisch and co-workers, *Ind. Eng. Chem.* **51,** 1386 (1969).
5. B.A. Dombrow, *Polyurethanes,* Reinhold Publishing Corp., New York, 1957.
6. T. Alfrey and co-workers, *Organic Polymers,* Prentice-Hall, New York & London, 1967.
7. F.W. Bilmeyer, *Textbook of Polymer Science,* John Wiley & Sons, New York, 1962.
8. H.H. Anderson, *J. Am. Chem. Soc.* **66,** 934 (1944); **71,** 1799 (1949); **72,** 193 (1950); **72,** 2091 (1950); **72,** 2761 (1950); and **75,** 1576 (1953).
9. W.J. Bailey and co-workers, *Am. Chem. Soc. Meeting,* April 1959.
10. J.H. Saunders and K.C. Frisch, *Polyurethanes: Chemistry and Technology.* Wiley-Interscience, New York, 1962.
11. J.M. Buist and co-workers, *Advances in Polyurethane Technology,* Maclaren and Sons, London, 1968.
12. *Urethane Abstracts,* Technomic Publications, Westport, Conn., 1972.
13. L.J. Lee, *Rubber Chem. Technol.,* Vol. 153, (3) (July–Aug. 1980).
14. H.F. Staudinger and co-workers, *Berichte* **50,** 1042 (1917).
15. U.S. Pat. 2,834,748 (1958), D.L. Bailey (to Union Carbide).
16. G.E. Molau, *Colloidal and Morphological Behavior of Block and Graft Copolymers,* Plenum Press, New York, 1971.

General References

S.L. Aggarwal, *Block Copolymers,* Plenum Press, New York, 1970.

U.S. Pat. 3,036,020 (1962) and 3,234,153 (1966), J.W. Britain (to Mobay).

Brit. Pat. 1,032,059.

J.M. Buist, *Developments in Polyurethanes, 1,* Applied Science, London, 1978.

Chem. Eng. News, (Oct. 24, 1966).

D.J. David and co-workers, *Analytical Chemistry of Polyurethanes,* Wiley-Interscience, New York, 1969.

P.J. Flory, *Principles of Polymer Chemistry,* Cornell University Press, Ithaca, N.Y., 1953.

P.J. Flory and co-workers, *J. Chem. Phys.* **11,** 521 (1943).

Ger. Offen 2, 1972, pp. 110 and 160.

H. Goldschmidt and co-workers, *Berichte* **22,** 3101 (1889).

A.M. Hermann and co-workers, *J. Macromol. Sci.-Chem., Polym. Lett. Ed.* **9,** 627 (1970).

I.C. Kogon, *J. Org. Chem.* **24,** 83 (1959).

K. Kurita and co-workers, *Makromol. Chem.* **180,** 2331 (1979).

T.S. Mukaiyama and co-workers, *J. Am. Chem. Soc.* **78,** 1946 (1956); **82,** 5359 (1960); and **79,** 73 (1957).

A. Ogino and co-workers, *Toso Gijitsu* **7,** 40 (Sept. 1968).

S. Orchon, "Inducing Thixotropy in Polymer Systems," *Adhesive Age,* 21–24 (July 1972).

A. Rembaum, *J. Elastoplast.* **4,** 280 (1970).

J.H. Saunder, *Rubber Chem. Technol,* **33,** 1293 (1960).

A.M. Schiller, *Soviet Progress in Polyurethanes,* Vol. 1 (1973) and Vol. 2 (1975), Technomic Publications, Westport, Conn.

F.M. Sweeney, *Introduction to RIM,* Technomic Publications, Westport, Conn., 1979.

Ullmann, *Encyclopaedia der Technischen Chemie,* Dritte, voellign neu gestaltete Auflage, Bd. 4, Seite 608–610.

K. Uno and co-workers, *Polym. J.* **6,** 348 (1974).

U.S. Pat. 3,595,839.

U.S. 3,830,785.

S.P.S. Yen and co-workers, *J. Biomed. Res. Symp.* **1,** 83 (1971).

43

FLEXIBLE POLY(VINYL CHLORIDE) (FPVC)

Saul Gobstein

Sa-Go Associates, Inc., Shaker Heights, Ohio

43.1 INTRODUCTION

Flexible poly(vinyl chloride) (PVC) plastic encompasses a wide variety of molding compounds with a broad spectrum of properties, applications, and use in almost all thermoplastic processes. To make this versatile plastic, vinyl chloride polymer is combined with plasticizer, stabilizer, filler, and other additives, depending on the properties desired and the process to be used.

The vinyl chloride polymer may be a homopolymer of repeating units of vinyl chloride monomer (VCM).

$$\left[\begin{array}{cc} \overset{\displaystyle H}{\underset{\displaystyle H}{\overset{|}{\underset{|}{C}}}} & \overset{\displaystyle H}{\underset{\displaystyle Cl}{\overset{|}{\underset{|}{C}}}} \end{array}\right]_n$$

Or, the copolymer, also used in making this plastic, would be repeating units of VCM with other monomers such as vinyl acetate, propylene, or maleate interspersed in a random fashion.

Alloys or polyblends may also be obtained by the incorporation of other polymers such as acrylonitrile, SAN, ABS, SBR, CPE, acrylate, and even some "B" stage thermosetting resins. In addition to these PVC polymers, additives give PVC plastic properties that cannot be achieved any other way.

There is a difference between flexible, plastisol, and rigid PVC technology. For a review of the differences, see the section on rigid PVC.

43.2 CATEGORY

Although at room temperature PVC polymer is horny and brittle[1], heat sensitive, and difficult to process, it can be made into an elastomeric plastic with the addition of plasticizers and stabilizers. Additives help overcome the process limitations of PVC.

525

TABLE 43.1. Additives Used in Flexible PVC Plastics

Type of Additive	Function	Reference
Plasticizers	Impart flexibility	3
	Assist in processing	
Stabilizers	Inhibit degradation	4
Fillers	Reduce cost	5
Lubricants	Improve flow and processing properties	6
Biocides	Inhibit biodeterioration	7
Flame retardants	To delay the initiation and burning	8
Colorants	Impart color	9
Blowing agents	Provide a cellular structure	10
UV absorbers	Protect against UV light	11
Antiblocking and slip agents	Reduce friction	12
Alloying polymers	Special properties	13

43.2.1 Vinyl Additives

The characterization of flexible vinyl products is determined by its additives. When properly dispersed in the PVC polymer matrix, they do not alter its molecular structure but will affect the processing and properties of the polymer.[2]

Table 43.1 presents a list of additives used in flexible PVC. Not all of these additives are used in every formulation. Their selection and amount is determined by the process and end use. In Table 43.1, an additive is considered to be any material added regardless of the amount.

Approximately 60% of all plastic additives are consumed in flexible PVC[14]

43.3 HISTORY

VCM was first prepared by Regnault in 1835. E. Bauman discussed the reaction of vinyl halides and acetylene in a sealed tube in 1872.[15] Although a lot of activity was reported for the next 40 years. The most significant development occured in 1921, when Plausen[16] discovered how to polymerize PVC from dry acetylene which made PVC more than a laboratory curiosity.

Commercial-scale production of PVC plastic started in 1931 in I.G. Farbenindustrie's Bitterfield plant with the development of vinyl paste. Waldo L. Simon, of B.F. Goodrich, made the first major breakthrough by plasticizing PVC.[17] This was the start of commercialization of flexible PVC in the United States.

From this period until the end of World War II, there were many developments in resin, plasticizer, and stabilizer technology. The most significant were the Ried[18] patent on the manufacture of PVC copolymers, the Doolittle[19] patent on the use of lead stabilizers, the Quattlebaum[20] patents on tin stabilizers, and the Gresham[21] patent on the use of DOP (dioctyl phthalate) as a plasticizer. From this two major technologies emerged in the United States: the B.F. Goodrich technology based on the homopolymer and the Union Carbide technology based on the copolymer.

During this same period, process development also took place. Both rubber calenders and extruders were converted to plastic calenders and extruders. Injection molding of vinyls was limited, because the ram injection-molding machine presented too many problems in processing vinyls. These were minimized in the 1950s with the introduction of the reciprocating screw injection molding machine.

In the 1930s, PVC manufactured products included film and sheeting for their waterproofing properties, electrical insulation, and flashlight cases.[22] In the 1940s, molding com-

TABLE 43.2. Flexible PVC Polymer and Plastic Consumption (in billion lb consumed)

Year	Total Polymer	Polymer Used in Flexible	Flexible PVC Plastic
1960	0.904	0.875	1.28
1970	3.05	2.05	2.90
1980	5.76	2.88	4.00

pounds for toys and industrial applications were developed. Since then, there has been a continuous improvement in vinyl plastics. Resins have become more heat stable and additives more effective because of sophisticated techniques.

The growth of flexible PVC plastic and the resin that is used to make the plastic is presented in Table 43.2, which also identifies the difference between resin (polymer) and plastic. For example, in 1984 the average flexible PVC formulation was comprised of 63 parts additives to 100 parts of resin; approximately 39% of a flexible PVC plastic is composed of additives.

In 1970, the PVC market constituted 16.4%[23] of the entire thermoplastic market place. In 1980, the PVC market share increased to 17.8%.[24]

There are two opinions about the first plastisol operation. The choice is between the Bitterfield plant in Germany, and the Union Carbide plant in the United States. In 1946, Union Carbide introduced a new technique for coating.[25] In the description of this technique, the terms plastisol and organosol were mentioned for the first time. At that time, Union Carbide did not realize the potential of this material.

A plastisol is the suspension of a solid in a liquid which does not appreciably dissolve the solid at room temperature but does form a homogeneous mixture at elevated temperatures.

An organosol is a plastisol that contains more than 10 parts of a solvent per 100 parts of resin.

Table 43.3 lists the ingredients, functions, and formulations of plastisols, including references. Plastisols have become an easy-to-use liquid form of a PVC compound converted by heat into a solid product. In Table 43.4, the process and the typical articles produced are presented.

TABLE 43.3. Plastisol Ingredients

Ingredient	Function	Reference
Dispersion resin	A very fine resin, ranging from 0.5 to 2.0 micron, particle that can be homo or copolymer	26
Blending resin	Small particle of resin ranging from 10–150 microns used to decrease cost and viscosity	27
Plasticizer	Suspends small particle Imparts flexibility Assists in processing	28
Stabilizer	Inhibits degradation	29
Filler	Reduces costs	5
Thixotropic agent	Gives pseudoplastic flow	30
Surfactant	Reduces viscosity	31
Blowing agent	Provides a cellular structure	32
Mold release agent	Reduces friction	33
Colorant	Imparts color	34
Solvent and deluent	Reduces viscosity	35
Air release agent	Releases air from plastisol	36
Reactive monomer	Cross-linking agent	37

TABLE 43.4. Plastisols Process and Products[a]

Process	Products
Spread coating	Upholstery transportation
	Landau roof
	Apparel fabrics
	Wall coverings
	Shoes (uppers)
	Boots
	Floor covering
	Carpet backing
	Paper coating
	Roll coating
Molding	
(Rotational)	Syringe bulbs, storage tanks, inflatable balls, dolls, toys, luggage, bicycle seats
(Slush)	Boots, toys, doll faces, head rests, boots for seat belts and gear shifts, auto arm rests, crash pads
(Dip)	Wire racks, plating racks, electrical parts, spark plug covers, tool handles, traffic safety cones, work gloves, fencing
Cast molding	
(Cavity)	Fishing lures, disposal stoppers, toys
(In-place)	Automotive air filters, sewer pipe gaskets, bottle cap liners
(Low-pressure)	Tablecloths, auto mats, dollies, printing plates
Strand coating	Screens (insect protection)
Spray coating	Protective and decorative coatings on furniture, appliances
Cellular coatings	Fabrics, carpet backing, upholstery
Molding	
(Open cell)	Boat bumpers, inner soles, athletic padding
(Closed cell)	Electrical insulation, buoys, life preservers

[a]*Source*: Sa-Go Associates, Inc.

43.4   POLYMERIZATION

For a review of the manufacture of the monomer and polymer, see the section on rigid PVC.

43.5   DESCRIPTION OF PROPERTIES

43.5.1   Physical Properties

The properties of flexible PVC plastics are determined by the type and the amount of additives. Flexible PVC is a formulated plastic; the variations in properties can be significant. For example, specific gravity may vary from a low of 73.5 lb/ft³ (1.18 g/cm³) for a clear compound to a high of 106 lb/ft³ (1.70 g/cm³) for a filled compound.

The molecular weight of the polymer also plays a major role in its physical properties. Low molecular weight polymers are used for injection molding and a high molecular weights for extrusion. However, additives can alter even this practice. The goal is to use the highest molecular weight that can be processed easily.

43.5.2   Thermal Properties

The thermal properties vary in a flexible PVC plastic with the type and amount of plasticizer. The other additives have a minor affect on thermal properties.

Darby[3] says that the addition of plasticizer to a PVC polymer will lower the glass-transition temperature (T_g). The amount by which the T_g is lower is a linear function of the amount and type of plasticizer

The melting point of a flexible PVC plastic will vary with the molecular weight and the type and amount of additives in the formulation. The more solvating the plasticizer, the lower the melting point. Copolymers also have lower melting points.

The thermal conductivity of a flexible PVC plastic will also vary with the type and amount of plasticizer. According to Darby, the thermal conductivity increases to a peak and then decreases as the temperature increases for a particular plasticizer at a particular level. Increasing the level gives greater conductivity at lower temperatures.

Again, depending on the type and amount of plasticizer, the specific heat will vary, increasing with the temperature and leveling off at a particular point.

The thermal expansion of a flexible PVC compound can vary from 7 to 25×10^{-5} in./in. depending on the formulation.

Flexible PVC plastic does not have a heat-distortion temperature because the material is elastomeric.

43.5.3 Mechanical Properties

The addition of plasticizer to a PVC polymer changes the mechanical properties. Ghera[38] demonstrates that the addition of small amounts of plasticizer causes all mechanical properties to change. For example, after adding small amounts of DOP, it was observed that impact goes down until 17 parts are added; then it starts to go back up. Tensile strength will increase until 8 parts per 100 parts resin is added and then decrease, as illustrated in Figure 43.1.

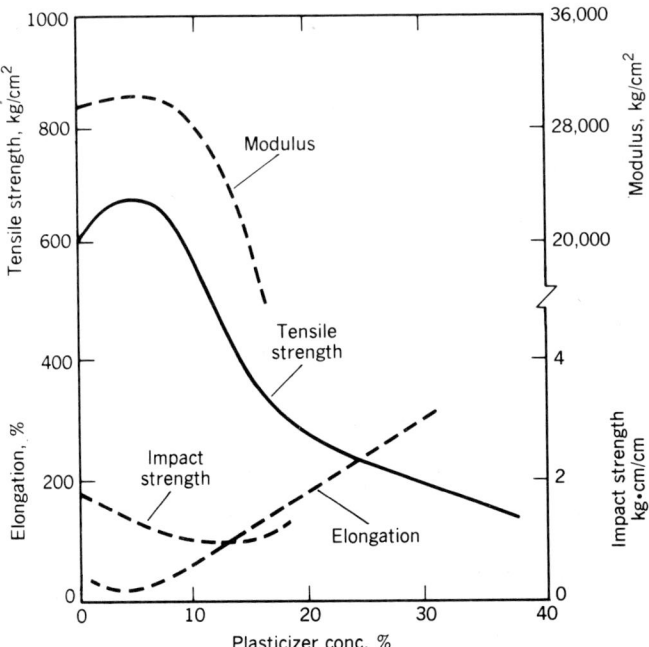

Figure 43.1. Mechanical properties of PVC plasticized with various concentrations of DOP[16]

43.5.4 Optical Properties

The gloss of flexible PVC plastics can be made to vary from a shiny patent leather look to a very dull matte. Compounds vary from clear to translucent to opaque. All types of colors can be incorporated.

43.5.5 Environmental Properties

Flexible PVC is known for its extraordinary chemical resistance. Most dilute acids and bases and many concentrated acids and bases do not attack it. The only solvents to which it is susceptible are ketones, some aromatics, and chlorinated hydrocarbons.

When vinyl compounds are stabilized with lead compounds, impurities in the air that carry sulfur-bearing gases may turn the vinyl black.

43.6 APPLICATIONS

Table 43.5 identifies the major applications and fabrication techniques and the major reasons for their use. Its versatility is demonstrated by the diverse type of applications.

43.7 ADVANTAGES/DISADVANTAGES

The major advantage of flexible PVC plastic is that it is formulated plastic and, therefore, can be made to meet a broad spectrum of requirements. It is the only plastic that can be fabricated by every known plastic process. In addition, the plastic has

- Good chemical resistance
- Good cost performance
- Toughness
- Good outdoor weathering
- Excellent electrical properties
- Ability to be made conductive
- Good surface appearance
- Ease in cleaning
- Ability to be made flame retardant
- Wide color range
- Ability to be made to be glossy
- Ability to be made to be dull.

The disadvantages are

- Heat sensitivity
- Poor resistance to ketones and chlorinated hydrocarbons
- The need to be properly formulated to avoid spewing, blooming, migration, staining, and bleeding problems
- Difficulty in processing

TABLE 43.5. Application and Fabrication Techniques of Flexible PVC Plastics[a]

Applications	Fabrication[b]	Why Used
Building and Construction *(excluding wire, cable)*		
Weatherstrip	2	Weather resistant, maintenance
Swimming pool liners	1	Abrasion and tear resistant, flexible, UV resistant, water resistant
Coated metal components	6	UV resistant, corrosion resistant, maintenance
Flooring (including foam backed)	1,5,6,7	Color, style, wear, maintenance
Upholstery and wall coverings	1,2,7	Color, style, wear
Clothing		
Baby pants	1	Flexible, stain resistant, washable
Outerwear, including foam interliners	1,2,7	Leatherlike style and color, waterproof
Shoes	1,2,3	Grease resistant, long wearing
Rain and footwear	3,8	Clarity, style
Electrical		
Wire and cable, fittings, etc	2,3	Flexible, weather resistant, electrical charac.
Home and Office Furnishings		
Appliances	2,3	Flexible, resilience, color, wear
Garden hose	2	Lightweight, transparent, clean
Shower curtains and closet accessories, tablecloths, etc.	1	Style, color, flexible, translucency
Luggage	1	Water resistant
Agriculture		
Hose	2	Flexible, wear, maintenance
Packaging		
Food wrap and shrink wrap film	2	Oxygen permeable, memory, clarity
Bottle liners	10	Nontoxic, good seal
Recreation		
Balls, dolls, models, inflatables, swim fins, wheels, golf bags, toys	1,2,9	Flexible, color, toughness
Transportation		
Auto mats	1,5	Color, flexible wear
Auto upholstery and seat covers	1,7	Style, flexible, wear, maintenance
Crash pads	1,5,7	Resilience, color, wear
Mud flaps	5	Style, water resistant
Transmission covers	11	Style, wear, flexible
Air filter gaskets	10	Corrosion resistant, grease resis., flexible
Miscellaneous *(typical items)*		
Ensilage covers	1	Weathering, tough
Medical tubing	2	Flexible, transparent, clean
Hammer heads	3	Resilient but with damping action
Steel strip coatings	6	Corrosion resistant
Cather tubes	11	Nontoxic, clarity, clean

[a]*Source*: Sa-Go Associates

[b]
1. Calendering	5. Compression Molding	9. Rotational Molding
2. Extrusion	6. Coating	10. Inplace Molding
3. Injection Molding	7. Laminating	11. Dip Molding
4. Blow Molding	8. Slush Molding	

TABLE 43.6. Effect of Molecular Weight on PVC Properties[a]

Resin	A	B	C	D	E
Intrinsic viscosity	0.50	0.80	0.93	1.03	1.18
Tensile strength, psi	1140	2100	2460	2580	2800
(MPa)	(7.9)	(14.5)	(17)	(17.8)	(19.3)
Ultimate elongation, %	150	280	320	330	360
Tensile stress at 100% elongation, psi	1030	1290	1450	1420	1460
Shore A hardness,					
0 S.	90	91	91	91	92
10 S.	78	80	80	81	82
Tear resistance, lb./in.	167	364	394	386	413
Processing Temperatures					
Calendering, °F	230–260	300–330	310–340	320–350	330–360
(°C)	110–127	149–166	154–171	160–177	166–182
Extrusion, °F		310–340	320–350	330–360	350–380
(°C)		154–171	160–177	166–182	177–193
Molding, °F	260–300	330–370	340–380	350–390	360–400
(°C)	127–149	166–188	171–193	177–199	182–204

[a]Ref. 39.

43.8 PROCESSING TECHNIQUES

Flexible PVC plastics can be processed by more techniques than any other plastic because it contains a relative polar polymer that allows a large range of formulations. The Master Material Outline lists all the process techniques.

The concept, "the higher the molecular weight, the better the physical properties; the higher the molecular weight and the higher are the temperature requirements," is easily demonstrated with flexible PVC. Table 43.6 shows that as the intrinsic viscosity rises (molecular weight) so does the temperature requirement for three process techniques. Also, as the intrinsic viscosity goes up, the physical properties improve.

43.9 RESIN FORMS

Flexible PVC plastic molding compounds are available in fully fluxed pellets, regrind, and powdered dry blend. Fluxed pellets can be cylindrical pellets, face-cut round pellets, or diced in various configurations. Regrind is a random-cut pellet that can vary from fine to large, irregular particles. Dry blend is a free-flowing powder where additives have been dispersed and distributed around the resin particle. By choosing the right type resin that has the right sorptive characteristics, a free-flowing powder is obtained in the blending operation.

43.9.1 Additives

Many of the additives used in flexible and rigid PVC are different. For example, plasticizer is only used in flexible PVC. The stabilizers used in flexible PVC are primarily barium, cadmium, zinc complex liquid, or barium cadmium solid. In rigid PVC, the primary stabilization systems contain tin. There are few or no lubricants used in flexible, while there is a very sophisticated system used in rigid PVC.

Master Material Outline

PROCESSABILITY OF _____ Flexible PVC

PROCESS	A	B	C
INJECTION MOLDING	X		
EXTRUSION	X		
THERMOFORMING	X		
FOAM MOLDING	X		
DIP, SLUSH MOLDING	X		
ROTATIONAL MOLDING	X		
POWDER, FLUIDIZED-BED COATING	X		
REINFORCED THERMOSET MOLDING			X
COMPRESSION/TRANSFER MOLDING			X
REACTION INJECTION MOLDING (RIM)			
MECHANICAL FORMING	X		
CASTING	X		

A = Common processing technique.
B = Technique possible with difficulty.
C = Used only in special circumstances, if ever.
Also, in-place molding, calendering, and blow molding are common techniques; RIM molding is not used at all.

43.9.2 Polyblends

Some flexible PVC compounds incorporate other polymers such as acrylonitrile, styrene–acrylonitrile, or polyurethane. These polyblends usually require less plasticization, and offer better low temperature flexibility and a wider service temperature range.

43.10 SPECIFICATION OF PROPERTIES

The Master Outline of Material Properties presents typical properties of flexible PVC. As shown, a wide variety of properties are available, depending on the type and the amount of additives used.

MATERIAL ___ Flexible PVC

PROPERTY	TEST METHOD	ENGLISH UNITS	VALUE	METRIC UNITS	VALUE
MECHANICAL DENSITY	D792	lb/ft^3	73.5–106	g/cm^3	1.18–1.70
TENSILE STRENGTH	D638	psi	800–3800	MPa	5.5–26.2
TENSILE MODULUS	D638	psi	700–1800	MPa	4.8–12.4
FLEXURAL MODULUS	D790	psi		MPa	
ELONGATION TO BREAK	D638	%	150–450	%	150–450
NOTCHED IZOD AT ROOM TEMP	D256	ft-lb/in.	Varies	J/m	Varies
HARDNESS	D785		Shore A		Shore A
THERMAL DEFLECTION T @ 264 psi	D648	°F		°C	
DEFLECTION T @ 66 psi	D648	°F		°C	
VICAT SOFT-ENING POINT	D1525	°F	Varies	°C	Varies
UL TEMP INDEX	UL746B	°F	Varies	°C	Varies
UL FLAMMABILITY CODE RATING	UL94		Varies		Varies
LINEAR COEFFICIENT THERMAL EXPANSION	D696	in./in./°F	$7-25 \times 10^{-6}$	mm/mm/°C	$12.6-45 \times 10^{-6}$
ENVIRONMENTAL WATER ABSORPTION 24 HOUR	D570	%	Varies	%	Varies
CLARITY	D1003	% TRANS-MISSION	Clear, translu-cent, opaque	% TRANS-MISSION	Clear, translu-cent, opaque
OUTDOOR WEATHERING	D1435	%	Varies	%	Varies
FDA APPROVAL			Possible		Possible
CHEMICAL RESISTANCE TO: WEAK ACID	D543		Not attacked		Not attacked
STRONG ACID	D543		Not attacked		Not attacked
WEAK ALKALI	D543		Not attacked		Not attacked
STRONG ALKALI	D543		Not attacked		Not attacked
LOW MOLECULAR WEIGHT SOLVENTS	D543		Ketones, chlor-inated hydro-carbons attack		Ketones, chlor-inated hydro-carbons attack
ALCOHOLS	D543		Minimally attacked		Minimally attacked
ELECTRICAL DIELECTRIC STRENGTH	D149	V/mil	250–400	kV/mm	9.9–15.8
DIELECTRIC CONSTANT	D150	10^6 Hertz	3.3–4.5		3.3–4.5
POWER FACTOR	D150	10^6 Hertz	0.04–0.14		0.04–0.14
OTHER MELTING POINT	D3418	°F	Varies	°C	Varies
GLASS TRANS-ITION TEMP.	D3418	°F	Varies	°C	Varies

Variability dependent on type and amount of additives.

43.11 PROCESSING REQUIREMENTS[40]

Processing of flexible PVC plastic uses less heat and thus is easier to process than rigid PVC, because of the plasticizer/resin interaction of the flexible PVC. A flexible formula usually includes little or no lubrication, since the plasticizer acts as the lubricant. In compounding the molding powder, it is important to obtain the proper dispersion and distribution of the additives as well as to control the heat and shear, because they will affect the melt behavior in processing.

As in the case of rigid PVC, there must be sufficiently high shear and heat, so that the material reaches the state of fusion. It is only in this state that the cohesive forces are strong enough to bring about the maximum physical properties.

Although there are some flexible PVC applications that can use a dry blend, in most cases a compounded product is preferred. It is in the compounding step that enough shear can be generated to obtain a proper homogeneous product. In some cases, especially when other polymers are included in the formulation, only compounded molding powders can be used.

Throughout this section, the effect of the formulation ingredients has been stressed. Each can have an effect on a particular process. The processor should be aware of the various interactions that may take place and what problems can occur.

Recycling of flexible PVC plastic can be achieved using conventional grinding methods. It is important not to mix rigid and flexible in the recycling process to prevent processing problems.

43.12 PROCESSING-SENSITIVE END PROPERTIES

End property considerations in flexible PVC plastics are usually affected by the ingredients used. In most cases, processing does not affect the properties of flexible PVC as it does rigid PVC. Flexible PVC plastic is a formulated plastic. The formulator has to know what type and how much plasticizer to use (high, medium, or low molecular weight) and what inter-action exists when the various additives are used in order to obtain the desired end properties.

43.13 SHRINKAGE[41]

The design of a flexible PVC plastic product requires an understanding of how the plastic will flow. The viscosity at processing temperatures is not as high as that of rigid PVC plastic and will vary with the amount of plasticizer present. As noted, a variety of flexible PVC products is available, depending on the type and the amount of additives used. Therefore, before a design is begun, the exact product requirements should be known. For example, the variation in shrinkage can be 0.10 to 0.50 in./in. (mm/mm) for a compound without filler and 0.002–0.035 in./in. (mm/mm) for a compound with filler.

43.14 TRADE NAMES

Trade Name	Type	Supplier
ABALYN	Plasticizer	Himont
ABBEY	Conductive vinyl	Abbey Plastics
ABG	Plasticizer	BASF
ABITOL	Plasticizer	Himont
ACE-FLEX	PVC	American Hard Rubber

Trade Name	Type	Supplier
ACE-RIVICLOR	PVC	American Hard Rubber
ACETIN	Plasticizer	Mobay
ACRILAN	Acrylonitrile–vinyl acetate fiber	Monsanto
ACRYLIVIN	PVC/acrylic sheet	General Tire & Rubber
ACRAWAX	Lubricant	Glyco
ADIMOIL	Plasticizer	Mobay
ADIPOL	Plasticizer	FMC
ADMEX	Vinyl plasticizer	Sherex Chemical
ADVASTAB	Vinyl stabilizer	Carstab Corp.
ADVAWAX	Vinyl lubricant	Carstab Corp.
AGILIDE	PVC extrusions	American Agile
AGRIPLAST	Vinyl linings	Agriculture Plastic
AIRCO	VCM and PVC	Air Products & Chemical
ALDO	Plasticizer	Glyco
ALPHA	PVC compounds	Dexter
AMERIFILM	Film and sheet	Ross & Roberts
AMERIPLAST	Film and sheet	Ross & Roberts
AMS	Plasticizer	Amoco
ANVIGL	Rigid PVC extrusions	Anchor Plastics
ANTIBLAZE	Plasticizer	Mobil
APPRETAN	PVAC/VAC/acrylate	Hoechst
AQUAFLEX	PVC sheet	Pantasote
ARMAN	Vinyl film and sheet	Ross & Roberts
ARNAR	Vinyl film and sheet	Ross & Roberts
ASTRA	Flexible PVC extruding tubing	Suflex
ASTROLOGS	PVC profiles	Minter Homes
ATLASTOVON	Plasticized PVC	Atlas Mineral & Chemical
BARR-HYDE	Expanded vinyl	Barr Corp.
BARR-SKIN	Expanded vinyl	Barr Corp.
BENZOFLEX	Plasticizer	Velsicol
BLANDOL	Plasticizer	Inolex
BLAVIN	PVC compounds	Reichold Chemical
BOLTAFLEX	PVC sheet	General Tire & Rubber
BOLTARON	PVC sheet	General Tire & Rubber
BONDALL	Poly(vinyl butyral)	Schenectady Chemicals
BOSILM	Adhesive-coated PVC sheet	USM Chem.
BOUTONNIER	Vinyl asbestos flooring	Amer. Bilbrite Rubber
BRASSLYFE	Vinyl lacquers	Bee Chemical
BRAVEX	Rigid PVC	Apache Industries
BREETH-EES	PVC sheeting	Rand Rubber
BRONCHA	PVC composition	Texileather
BUBBLE-PAK	Saran-coated cushioning	Sealed Air
BUBBLE-WRAP	Saran-coated cushioning	Sealed Air
BUCKRAN	Vinyl film and sheet	Central States Products
BUTACITE	Poly(vinyl acetal)	E.I. du Pont de Nemours & Co., Inc.
BUTVAR	Poly(vinyl butyral)	Monsanto
CAMARGO	Vinyl impregnated fabric	Breneman Harts Horn

Trade Name	Type	Supplier
CAPILAIR	Breathable PVC composite	Uniroyal
CARENA	PVC	Shell Chemical
CARBOWAX	Plasticizer	Velsicol
CASCOREZ	PVC/PVAC	Borden
CASCOVIN	PVC	Borden
CELCA-CET	PVAC emulsions	Celanese
CELL TITE	Rigid vinyl foam	BF Goodrich
CD	Plasticizer	Boron
CERECLOR	Plasticizer	C-I-L (Canada)
CHEMCLAD	Vinyl insulation	Essex Wire
CHEMCON	Conductive vinyl	Chemelec Products
CHEM-FLEX	Vinyls	Eagle Picher
CHEM-O-SET	Cross-linked dispersion	Whittaker
CHEM-O-SOL	Vinyl plastisol	Whittaker
CHEMTRAL	PVC valves	Cabot Piping System
CHLOZEZ	Plasticizer	Dover
CHLOROWAX	Plasticizer	Diamond Shamrock
CHRYSTINE	Vinyls	Chrysler
CIPCO FOAM	Extruded foam	Essex Wire
CIRRUS	Vinyl floor covering	General Tire & Rubber
CITROFLEX	Plasticizer	Morflex Chem.
CLEARFLO	Vinyl tubing	Newage Ind.
CLEARFEAL	PVC film	Borden
CLOPANE	Vinyl film	Clopay
CLOPHEN	Plasticizer	Mobay
CLORAFIN	Plasticizer	Himont
COBON	PVC tube	Cobon Plastics
COBOVIN	Reinforced PVC tube	Cobon Plastics
COLOVIN	PVC film	Borden
COLVELTA	PVC foam	Borden
COMFORTWEAVE	Knitted vinyl	Ford Motor
CORDO	PVC film and foam	Ferro
COVINAX	PVA	Franklin Chemical
COVOL	PVA	CPC International
CPF	Plasticizer	Witco
CP-485	Plasticizer	Occidental
CPH	Plasticizer	C.P. Hall
CREST-FOAM	Vinyl foam	Crest Foam
CRYOFLEX	Plasticizer	Sartomer
CRYSTAL CLEAR	Press-polished vinyl	Jacobson Mfg.
CYCLON	PVC tube	Polykote
CYCLVIN	ABS/PVC alloy	Borg-Warner
DAPCO	PVC tube and sleeve	Davidson Products
DARATAK	PVAc emulsions	W.R. Grace
DAREX	PVAc emulsions	W.R. Grace
DARVAN	Poly(vinylidene cyanide)/ PVAc copolymer	Celanese
DELLATOL	Plasticizer	Mobay
DELTAFLEX	Plasticizer	Reichhold

Trade Name	Type	Supplier
DENFLEX	Vinyl plastisols and primers	Dennis Chemical
DENFOAM	Extruded vinyl	Essex Wire
DIANA FYRBAN	Vinyl coated and impregnated fabrics	Breneman-Hartshorn
DISFLAMOIL	Plasticizer	Mobay
DOMANI	Vinyl flooring	Amer. Biltrite Rubber
DORN	PVC film, sheet, and coated fabric	Gordon-Lacey Chem. Prod.
DOW	Plasticizer	Dow
DURACET	PVA	Franklin
DURAL	Acrylic-modified vinyl	Dexter
DURAN	Vinyl upholstery	Masland Duraleather
DURANYL	Rigid PVC	Mastic
DURASOL	PVC coated materials	Masland Duraleather
DURATEX	Vinyl fabric	General Tire and Rubber
DURELENE	PVC tube	Plastic Warehousing
DUROCEL	Vinyl foam	Mastic
DUROTHERM	PVDC	Carlon Products
DURUG	Glass-reinforced vinyl	Duracote
ECLAT	Vinyl flooring	Amer. Biltrite Rubber
ELASTOFOAM	Vinyl plastisols	Elastomer Chemical
ELASTOLEX	Fabric supported vinyl	General Tire and Rubber
ELMER'S GLUE	PVC/PVAc glues	Borden
ELVACET	PVAc emulsions	E.I. du Pont de Nemours & Co., Inc.
ELVANOL	PVA	E.I. du Pont de Nemours & Co., Inc.
ELVAX	Vinyl	E.I. du Pont de Nemours & Co., Inc.
EMERY	Plasticizer	Emery
ENSOCOTE	PVC lacquer	Uniroyal
ENCOLITE	Vinyl sheet foam	Uniroyal
EPOXOL	Plasticizer	Swift
ESCOFLEX	Plasticizer	Hexagon
ESKAY-LITE	Vinyl film	St. Regis Paper
ESTABEX	Stabilizer	Akzo
ESTYNOX	Plasticizer	Caschem
ETHOSPERSEV	Plasticizer	Glyco
EVERFLEX	PVAc copolymer emulsions	W.R. Grace
EXCELON	PVC tubing	Thermoplastic Processes
EXTRU-FOAM	PVC foam	Extrudyne
EXTRU-GLAZ	PVC sheet	Extrudyne
FABRILITE	Vinyl-coated fabrics	E.I. du Pont de Nemours & Co., Inc.
FASHON	Vinyl film and sheet	General Tire and Rubber
FASSION	Vinyl coating	M.J. Fassler
FASSLUX	Vinyl coating	M.J. Fassler
F-BONDALL	Poly(vinyl formal)/poly (vinyl butyral) wire coat	Viking Wire

Trade Name	Type	Supplier
FEDERAN	Vinyl-coated fabric	Textron
FEDERAN	Vinyl-coated fabric	Airco Plastic
FEDLASTIC	Vinyl-coated fabric	Textron
FEDWELL	Vinyl-coated fabric	Textron
FILMCELLE	Embossed vinyl film	Toscany
FLAMENOL	PVC	General Electric
FLEXAC	PVAc emulsions	Airco Chemicals & Plastics
FLEXBOND	PVAx	Air Prod. & Chem.
FLEXCHLOR	Plasticizer	Pearsall
FLEXILAC	Vinyl resin solutions	Flexible Products
FLEXIPLAST	PVAc	American Hoechst
FLEXMUL	Vinyl emulsions	Flexible Products
FLEX-O-GLAS	Vinyl film	Warp Brothers
FLEXRICIN	Plasticizer	CasChem
FLOFLEX	Vinyl fluid bed powders	Flexible Products
FLOTUBE	Vinyl tubing	Natvar
FOAMEX	Poly(vinyl formal) phenolic	General Electric
FORMATIC	Poly(vinyl formal)	Monsanto
FORMEX	Poly(vinyl acetal)	General Electric
FORMTEX	Poly(vinyl formal)	Essex Wire
FORMVAR	Poly(vinyl formal) resins	Monsanto
FORTUNA	Vinyl flooring	American Biltrite Rubber
FYARASTOR	Plasticizer	Pearsall
GELVA	PVAc resins	Monsanto
GELVATEX	PVAc emulsions	Monsanto
GELVATOL	PVA resins	Monsanto
GEM-FLEX	PVC	Gemloid
GENCASEAL	PVC insulation	General Cable
GENCLOR	PVC	ICI United States
GENOTHERM	PVC film and sheet	American Hoechst
GENO-TUBING	PVC tubing	American Hoechst
GENVAR	Poly(vinyl formal)	General Cable
GEON	PVC, PVC/ poly(vinylidene chloride) copolymers, CPVC and other vinyl materials	B.F. Goodrich Chem.
GLORIA KAYDOL	Plasticizer	Inolex
GLYCOLUBE	Plasticizer	Glyco
GOLD KAST	PVC film	Continental Plastic Co.
GOODALLITE	Vinyl-coated fabric	Goodall-Sanford, Inc.
GRIFFCO	PVA emulsions	Griffin Chemical
GUARD	Vinyl-coated fabric	Columbus Coated Fabrics
HALLITE	PVC tubing	C.P. Hall
HALSTAB	Stabilizers	Halstab
HALCARB	Stabilizers	Halstab
HALLCO	Plasticizer	C.P. Hall
HALLCOMID	Plasticizer	C.P. Hall
HALLOZEZ	Plasticizer	Pearsall

Trade Name	Type	Supplier
HARTEX	Vinyl film and sheet	Harte & Co.
HATCO	Plasticizer	Kalflex
HATCOL	Plasticizer	Kalflex
HB	Plasticizer	Monsanto
HERCOFLEX	Plasticizer	Himont
HERCOLYN	Plasticizer	Himont
HERITAGE	Vinyl film and sheet	Central States Prod. Co.
HEXAPLAS	Plasticizer	C-I-L (Canada)
HI TEMP GEOR	CPVC	B.F. Goodrich
HORCO	Poly(vinyl butyral)	Hodgman Rubber
HOSTOPHAN	PVC emulsions	American Hoechst
HOSTASTAB	Stabilizers	American Hoechst
HOTAVINYL	PVC dispersions	American Hoechst
INDUSOL	Vinyl plastisols and organosols	Industrial Solvents & Chemicals
IRROVIN	Irradiated PVC	ITT Corp.
IRV-O-THIN	Vinyl coating	3 M Co.
INTERSTAB	Stabilizers	Akzo Chemical
JAYON	Vinyl tubing	Johnston Ind. Plastics
KATHARON	PVC sheet and film	Ross & Roberts
KEMESTER	Plasticizer	Humko
KENFLEX	Plasticizer	Kenrich
KENPLAST	Plasticizer	Kenrich
KESSCOFLEX	Plasticizer	Stephan
KEY	Vinyl adhesives	Key Polymer Corp.
KOHINOR	PVC resins and compounds	Pantasote
KODAFLEX	Plasticizer	Eastman
KOILSET	Poly(vinyl butyral) coating	Red Magnet Wire Co.
KORESIN	Vinyl resin on a phenolic base	Akron Chemical
KOROGEL	Plasticized vinyl	B.F. Goodrich Chem.
KOROLAC	Vinyl solutions	B.F. Goodrich Chem.
KOROSEAL	PVC PVF/PVC	B.F. Goodrich Ind. Prod.
KOROWOOD	Vinyl	Koro Corp.
KP	Plasticizer	FMC
KRONITEX	Plasticizer	FMC
KRONOX	Plasticizer	FMC
KRYSTAL	PVC sheet	Allied Chemical
KRYSTALTITE	PVC shrink film	Allied Chemical
KUROFAN	PVC_2 dispersions	BASF Wyandotte
KYDENE	Acrylic/PVC alloy	Rohm & Haas
KYDEX	Acrylic/PVC alloy sheet	Rohm & Haas
LAROFLEX	PVC copolymer	BASF Wyandotte
LASCO PIPE	PVC pipe	Lasco Industries
LAUNDRON	PVC film	Polyval
LECTRO	Stabilizer	Associated Lead
LEMAC	PVAc	Borden Chemical
LEMOL	PVA	Borden Chemical
LINDOL	Plasticizer	Stauffer
LINTREX	Vinyl-coated fabric	Columbus Coated

Trade Name	Type	Supplier
LIQUID ENVELOPE	PVC coatings	Essex Chemical
LUMITE	Saran filaments	Chicopee Mfg.
LUTANOL	Plasticizer	BASF
LUTANOL	Poly(vinyl ether)	BASF Wyandotte
LUTOFAN	PVC emulsion	BASF Wyandotte
LUTONAL	Poly(vinyl ether)	BASF Wyandotte
LUTREX	PVAc	American Hoechst
LUVICAN	Poly(vinyl carbazole)	BASF Wyandotte
LUVITHERM	PVC	BASF Wyandotte
MACCROSAL	Vinyl plastisol	Michigan Chrome & Chem.
MARK	Stabilizer	Argus
MARSHMALLOW	PVC foam	Sommers Plastics
MAYON	Vinyl tubing	Mayon Plastics
MESAMOL	Plasticizer	Mobay
METASAP	Stabilizer	Synthetic Products
METALYN	Plasticizer	Himont
MICRON	PVC coating powder	Michigan Chrome & Chem.
MICROSOL	Vinyl plastisols	Michigan Chrome & Chem.
MICRO-VENT	Porous vinyl film	Canadian Resins & Chem.
MICRO-VENT	PVC/P (VC-Ac) film	Herculite Protective Fab
MIL-O-SOL	PVC plastisols	J.G. Milligan
MIRASTAB	Stabilizer	MRS
MONOPLEX	Plasticizer	C.P. Hall
MONTAGE	Vinyl asbestos flooring	Johns-Manville
MORFLEX	Plasticizer	Morflex
MOVIOL	PVA	American Hoechst
MORVILITH	PVAc	Hoechst, A.G.
MORVIOL	PVA	Hoechst, A.G.
MORVITAL	PVAc	Hoechst, A.G.
MPS	Plasticizer	Occidental
MULTIPRUF	Vinyl-coated materials	Elm Coated Fabrics
NALGON	PVC tubing	Nalge
NATRACHEM	Plasticizer	Harwick
NATVAR	PVC	Natvar
NATROFLEX	Plasticizer	Harwick
NAUGARD	Stabilizer	Uniroyal
NAUGAHYDE	Vinyl-coated fabric	Uniroyal
NAUTAQUILT	PVC sheet	General Tire & Rubber
NAUTOFLEX	Vinyl sheet	General Tire & Rubber
NELCOTE	PVC emulsions	Northeastern Labs.
NEO VAC	PVAc and copolymers	Polyvinyl Chem.
NEVILLAC	Plasticizer	Neville Chemical
NEVILLE	Plasticizer	Neville Chemical
NIACET	Vinyl acetate	Union Carbide
NIEU AMSTERDAM	Vinyl flooring	Amer. Biltrite Rubber
NIOLAN	PVC-coated fabric	Storey Bro. & Co.
NIPEON	PVC resins and copolymers	Nippon Aeon

Trade Name	Type	Supplier
NOB-LOCK	PVC sheet	Americoat
NON TAC	PVC sheet and molding compounds	H.Z. Stauss
NUOPLAZ	Plasticizer	Nuodex
NUOSTAB	Stabilizer	Nuodex
NYFORM	Poly(vinyl formal) coating	Ananconda Wire & Cable
NYGEN-TOLEX	Vinyl-coated nylon fabric	General Tire & Rubber
NYLCLAD	Poly(vinyl formal) with nylon overcoat	Belden Mfg.
OPPANOL	Poly(vinyl isobutyl ether)	BASF Wyandotte
ORIEX	PVC sheet and film	Tenneco Chemicals
PALATINOL	Plasticizer	BASF
PANACHE	PVC flooring	Amer. Biltrite Rubber
PANTASOTE	Vinyl sheet and coated material	Pantasote
PANTEX	Vinyl film	Pantasote
PARACRIL OZO	NBR/PVC Alloy	Uniroyal
PARAPLEX	Plasticizer	C.P. Hall
PARCLOID	PVC and P(VC-Ac) dispers.	Parcloid Chemical
PARICIN	Plasticizer	Caschem
PAROIL	Plasticizer	Dover Chemical
PEEL KOTE	PVC coatings	Polykote
PERBUNAN	PVC/nitrile butadiene	Bayer, A.G.
PEGOSPERSE	Plasticizer and dispersant	Glycol
PERMA-SKIN	Vinyl coatings	Dennis Chemical
PERMAFUSED	Vinyl coating	Anchor Post Products
PEROXIDOL	Plasticizer	Reichhold
PE VE CLAIR	PVC sheet	Brimar
PEXCO	PVC tube	Plastic Extrusion & Eng.
PEXCON	PVC tube	Plastic Ext. & Eng.
PHENOLIT	Phenolic/PVC blend	Soviet Origin
PHOSFLEX	Plasticizer	Stauffer
PICCOLASTIC	Plasticizer	Himont
PICCOVAR	Plasticizer	Himont
PLASTA-GARD	Vinyl	Ashland Chemical
PLASTHALL	Plasticizer	C.P. Hall
PLASTICHLOR	Plasticizer	Harwick
PLASTIFILM	PVC film and sheet	Goss Plastic Film
PLASTIFLEX	PVC compounds	General Fabricators
PLASTIGEN	Plasticizer	BASF
PLASTIMAYD	Vinyl film and sheet	Plastimayd
PLASTOFLEX	Plasticizer	AKZO
PLASTOGLASS	PVC coated fabrics	Storey Bros.
PLASTOLAM	PVC sheet	Storey Bros.
PLASTOLEIN	Plasticizer	Emery
PLASTOLENE	PVC coated fabrics	Storey Bros.
PLASTOMALL	Plasticizer	BASF
PLIADUCT	PVC pipe	Dayco Corp.
PLIOFLEX	PVC	Goodyear Tire & Rubber
PLIOGLAS	Vinyl-coated glass cloth	Flexfirm

Trade Name	Type	Supplier
PLIOHYDE	PVC sheet	Goodyear Tire
PLIOLITE	Vinyl/toluene/butadiene and vinyl/toluene copolymers	Goodyear Tire
PLIOVIC	PVC blending resins and copolymers	Goodyear Tire
PLYAMUL	PVAc adhesives	Reichhold
PLYHYDE	PVC sheet	Plymouth Rubber
PLURACOL	Plasticizer	BASF
PLURONIC	Plasticizer	BASF
POLYCHROMATIC	PVC sheet and film	Ross & Roberts
POLYCIN	Plasticizer	BASF
POLYCLAD	Vinyl	Fortin Laminating
POLYCLAD	Vinyl coating	Carboline
POLYCIZER	Plasticizer	Harwick
POLYDENE	PVC alloy	A. Schulman
POLYPLASTEX	Vinyl sheet and laminates	Polyplastex United
POLYPRENE	Vinyl organosol	Interchemical
POLYSIZER	PVA	Showa High Polymer
POLY-TEX	Acrylic/PVC and PVAc/PVC copolymers	Celanese
POLYVIN	PVC	A. Schulman
PORTPLASTO	PVC extrusions	Brimar
PRIMA	PVC coated fabrics	Dermide
PROFILM	PVC	Protective Lining
PROPIOFAN	Poly(vinyl propionate)	BASF Wyandotte
PVC PEARLS	PVC resin	Escambia Chem.
PV-DUIT	PVC conduit	Carlon Products
PX	Plasticizer	U.S. Chemical
2-PYROL	Plasticizer	GAF
QUELSPRAY	PVC aerosols	Quelcor
QUICKBOND	Polyvinyl butyral coating	Bridgeport Ins. Wire
QUADROI	Plasticizer	BASF
RAYETTE	Composition/vinyl sheet	General Tire & Rubber
REDO	Vinyl-coated fabric	Goodall-Sanford
REET	Vinyl film and sheet	Ross & Roberts
RESIN 18	Processing aid	Amoco
RESOLFLEX	Plasticizer	Cambridge Ind.
RESINITE	PVC tube and shapes	Borden
REISTOFLEX	PVA	Resistoflex
RESPRO	Vinyl sheet, film and coated products	General Tire & Rubber
RESYN	Vinyl copolymer	National Starch
RESYN 3600	Vinylidene chloride copolymerx water disp	National Starch
REXINITE	PVC insulating materials	Borden
REXOBOND	PVC resins	Emkay Chemical
REYLEASE	PVA	Reynolds Metals
REYNOFOAM-V	PVC foams	Reynolds Chem. Prod.
REYNOSOL	Plastisol and organosols	Reynolds Chem.
RICON	Vinyl/butadiene resins	Richardson

Trade Name	Type	Supplier
RIGIDSOL	Plastisol	Watson-Standard
RIGI FLEX	Vinyl extrusion	Crane Plastics
RIGIVIN	PVC	Heil Process Equip.
ROVANA	PVC_2 and copolymers	Dow Badische
RS	Plasticizer	Reichhold
RUCOAM	PVC film and sheet	Hooker Chem. & Plast.
RUCOBLEND	PVC compounds	Hooker Chem. & Plast.
RUCON	PVC resins	Hooker Chem. & Plast.
RUDDFLEX	PVC sheet and film	Rudd Plastic Fabrics
RYERTEX	PVC extrusions	J.T. Ryerson & Son
RYERTEX OMICRON	PVC	J.T. Ryerson & Son
SAFLEX	Poly(vinyl butyral) film	Monsanto
SAIB	Plasticizer	Eastman
SANICIZER	Plasticizer	Monsanto
SARALOY	Vinylidene chloride sheet	Dow Chemical
SARALOY	Vinylidene chloride copolymer	Dow Chemical
SARAN	Vinylidene chloride copolymers, resins, coating film, and microspheres	Dow Chemical
SANTOLITE	Plasticizer	Monsanto
SARANEX	PVC_2/PE Film	Dow Chemical
SARAN-O-LAM	Saran casting, film, and laminates	Printon
SARAN WRAP	Vinylidene chloride copolymer film	Dow Chemical
SARONPAC	PVC_2 film	Dow Chemical
SATINESQUE	Vinyl coated fabrics	Columbus Coated
SATINGIO	Vinyl coated fabrics	Columbus Coated
SEGRANE	PVC foam	Whittaker Corp.
SEILON	Vinyl sheet	Seiberling Rubber
SERFENE	PVC_2 coating	Morton Chemical
SHAWNOL	Organosols and plastisols	Adcote Chemicals
SHUTHANE	PVC/PU blend	Reichhold
SHUVIN	Vinyl	Reichhold
SHUVINITE	PVC/nitrile rubber blend	Reichhold
SHUVIN PERMABOND	PVC molding compounds	Reichhold
SLECK	PVC sheet	Vinyl Plastics
SOA	Plasticizer	Union Carbide
SOBO	PVAc adhesive	Slomons Lab.
SOLVAR	PVAc	Monsanto
SOLVION	PVA film	Polyval
SPARK-L-ITE	PVC_2 monofilament	Chevron Chemical
STAFLEX	Plasticizer	Reichhold
STA-FLOW	PVC–polypropylene copol.	Air Products & Chem.
STARFOL	Plasticizer	Sherex Chemical
STERILKOTE	Vinyl plastisols, organosols, and finishes	Bradley & Vrooman
STIX	Vinyl adhesive	Firestone Plastics
STOKES	PVC	Stokes Molded Prod.
SUPERPIPE	PVC pipe	Vinylplex

Trade Name	Type	Supplier
SURVAC	PVC film	Borden
SWANETTE	Vinyl film and sheet	Harte & Co.
SYLVIN	PVC compound	Sylvania Chemical
SYNPRON	Stabilizers	Synthetic Products
SYNRESYL	PVAc emulsions	Synres Chemical
SYNSKYN	Vinyl coated glass fiber	Polyplastex United
SYNTHINOL	PVC insulation	Cyprus Mines
SYNTHITE	PEs/PVC varnishes	John C. Dolph
SYNTHOVAR	PVC tubing	Varflex
TAM CAST	Plastisols	Tamite Industries
TAM CLAD	Plastisols	Tamite Industries
TAM COTE	Plastisols	Tamite Industries
TAMGUARD	Plastisols	Tamite Industries
TAN-O-LITE	Vinyl coated fabric	Columbia Mills
TAVYNE	Vinyl sheet	O'Sullivan Rubber
TEGMER	Plasticizer	C.P. Hall
TEMPCOR	PVDC Pipe	Colonial Plastics
TENAC	PVAc Resins	Colonial Plastics
TEREZON	Vinyl coated paper	Athol Mfg.
TERSON	PVC coated fabrics	Athol Mfg.
TETRONIC	Plasticizer	BASF
TEXREZ	PVAc solutions	Celanese Coatings & Spec.
THERMASOL	PVC plastisols	Lakeside Plastics
THERMOCHECK	Stabilizer	Ferro
T-LOCK	PVC sheet	Amercoat
TOLEX	Vinyl fabrics	General Tire & Rubber
TOLON	Vinyl sheet	General Tire and Rubber
TOUCHSTONE	PVC sheet and film	Ross and Roberts
TRIACETIN	Plasticizer	Mobay
TRIBASE	Stabilizer	Assoc. Lead
TUFF COTE	PVC insulation	Coleman Wire & Cable
TUFF-HYDE	Laminated vinyl	Barr
TUFF-SKIN	Vinyl foam	Barr
TUFTARP	PVC coated fabric	Grangate & Irwell Rubber
TURBO	PVC tubing	Brand-Rex
TURBOGARD	Cross-linked vinyl	Brand-Rex
TURBOLEX	PVC tubing	Brand-Rex
TURBOTHERM	PVC tubing	Brand-Rex
TURBOTRANS	PVC tubing	Brand-Rex
TURBOTUF	PVC coated glass	Brand-Rex
TURBOZONE	PVC tubing	Brand-Rex
TURON	Vinyl fabric	General Tire & Rubber
TYGOFLEX	Vinyl plastisol	Norton
TYGON	PVC and copolymer tubing	Norton
UNICHEM	PVC compounds	Colorite
UNICHLOR	Plasticizer	Neville
UNICHROME	Vinyl finishes	M&T Chemicals
UNIFLEX	Plasticizer	Union Camp
UNION CARBIDE	PVC resins and compounds	Union Carbide

Trade Name	Type	Supplier
UNIWICK	PVA foam	Unipoint Industries
U.S. DALEX	PVC sheet	Uniroyal
VALTEX	Plastisol-based colors	Interchemical
VALUE	Vinyl sheet	Rand Robbin
VAM	VAc monomer	U.S. Ind. Chemicals
VAN COR	PVC pipe	Colonial Plastics Mfg.
VANSTAY	Stabilizer	Vanderbilt
VARLAN	PVC resins	DSM
VENT-AIR	PVC foam	Barr
VENTURA	PVC foam	Barr
VEREL	Modacrylic fiber copolymers with VAc, VC/and or VC$_2$	Eastman Chemical
VERILON	PVC film	Continental Plastic
VERSILAN	Vinyl upholstery	Landers
VESTOLIT	PVC/PE/PVAc copolymer	Mobay
VICHROME	Vinyl enamels	Interchemical
VICOA	PVC compounds	Vinyl Corp. of America
VICO-VINYL	Compounds	Vinyl Corp. of America
VICRALITE	Vinyl coated products	L.E. Carpenter
VICRATEX	Vinyl coated and impregnated products	L.E. Carpenter
VIKEM	Vinyl spray	Bel-Art Products
VINAC	PVAc emulsion	Air Reduction
VINACRON	Plastisols and organosols	Loes Enterprises
VINALINER	PVC sheet	Goodyear Tire & Rubber
VINALOY	PVC sheet	B.F. Goodrich Chem.
VIN-ALLOY	Vinyl	High Strength Plastics
VINAPEX	PVC film and sheet	Apex Coated Fabrics
VINAPLA-LAC	Vinyl lacquer	Schwartz Chemical
VINASOL	Vinyls	Atlas Synthetics
VIN-CON	Vinyls	Norton
VINE-L	PVC Sheet	Nixon Nitraton Works
VINIDUR	PVC	BASF Wyandotte
VINOL	PVA	Air Reduction
VINOLAST	PVC insulation	American Standard Wire & Co.
VINREX	Vinyl coatings	Rexton Finishes
VIN-SO-LITE	Vinyl foam	Foam King
VINURAN	Styrene-modified PVC	BASF Wyandotte
VINYLFILM	PVC film	Goodyear Tire & Rubber
VINYLOID	PVC coatings	Rowe Products
VINYL-PANE	PVC film	Warp Brothers
VINYLUBE	Processing aid	Glyco
VIPLEX	Plasticizer	Crowley
VIRON	PVC resins	Industrial Vinyls
VIRCOL	Plasticizer	Ruco
VISTA VINYL	PVC film and sheet	Harte & Co.
VISTAFILM	PVC & copolymer	Goodyear Tire & Rubber
VITA TUBE	PVC tube	Goodyear Tire & Rubber

Trade Name	Type	Supplier
VITRONE	PVC sheet, film, tube and profile	Stanley Smith
VOR SEAL	PVC foam	Vorac
VULCO	PVC sheet and shapes	Vulcan Metal
VULT-ACET	PVAc latices	Gen. Latex & Chemical
VXGEN	PVC resins	General Tire & Rubber
VYAN VINYCLAIR	PVC film	Rhodia
VYCANAC	Vinyl film, sheet, molding, and extrusion compounds	Canadian Resins & Chemicals
VYCEL	PVC foam	Goodyear Tire & Rubber
VYDEL	PVC panels & siding	Superior Plastics
VYFLEX	PVC film and sheet	Kaye-Tex Mfg.
VYFLEX	PVC coating powders	Plastic Coatings
VYGEN	PVC resins	General Tire & Rubber
VYLAR	PVC siding	Gotham Industries
VYLENE	Vinyl film and sheet	Elm Coated Fabric
VYLON	PVC pipe	Carlon Products
VYNACLOR	PVC emulsions	Natl. Starch & Chem.
VYNAFOAM	Vinyl plastisol foam	Interchemical
VYNAIR	PVC coated fabrics	I.C.I. (Hyde), Ltd.
VYNALOY	PVC sheet	B.F. Goodrich Chemical
VYNEX	PVC sheet	Nixon-Baldwin Chem.
VYRON	PVC compound	Industrial Vinyls
WALLPOL	PVAc emulsions	Reichhold
WAREFLEX	Vinyl dispersion	Synthetic Prod.
WASCOSEAL	Vinyl moisture barrier	Wasco Products Inc.
WATASEAL	Vinyl film and sheet	Harte & Co.
WATAHYDE	PVC sheet	Harte & Co.
WAUTEX	PVC film	Clopay Corp.
WEBLON	Fiber-reinforced vinyl	Weblon
WICANOL	PVA	Story Chemical
WICASET	PVAc emulsions	Story Chemical
WILFLEX	Vinyl plastisols and organosols	Flexible Products
WILLIAMSBURG	PVC film and sheet	Central States Products
WITE RITE	PVAc adhesive	Passaic Adhesives & Chemicals
WOOD-LOK	Vinyl copolymer adhesive	Natl. Starch & Chemical
WYNENE	Extruded vinyl	Natl. Plastic Products
WYNSOTE	Vinyl-coated fabric	Pantasote
X-LINK	Cross-linking vinyl/acrylic copolymer	Natl. Starch & Chemical
X-PAND	PVC foam	Barr

BIBLIOGRAPHY

1. R.T. Gottesman, "Chemistry and Technology of Polyvinyl Chloride," *Applied Polymer Science,* American Chemical Society, Washington, D.C., 1975, p. 364.
2. L. Mascia, *Role of Additives in Plastics,* Edward Arnold Ltd., London, 1974, p. 1.

3. J.R. Darby and J.K. Sears, "Theory of Solvation and Plasticization," *Encyclopedia of PVC,* Marcel Dekker, New York, 1976, p. 385, 446.

4. L.I. Nass, "Theory of Degradation and Stabilization Mechanism," *Encyclopedia of PVC,* Marcel Dekker, New York, 1976, p. 271.

5. R.P. Braddicks and L.I. Nass, "Fillers," *Encyclopedia of PVC,* Marcel Dekker, New York, 1976, p. 711.

6. E.L. White, "Lubricants," *Encyclopedia of PVC,* Marcel Dekker, New York, 1976, p. 643.

7. S. Gobstein "Miscellaneous Modifying Agents," *Encyclopedia of PVC,* Marcel Dekkar, New York, 1987.

8. I. Touval, "Synergism in Flame Retardants," *Plastic Compounding,* 31 (Sept./Oct. 1982).

9. Saltzmann and R.M. Johnston, *Industrial Color Technology,* American Chemical Society, Washington, D.C., 1971.

10. S. Gobstein, "Miscellaneous Modifying Additives," *Encylopedia of PVC,* Marcel Dekker, New York, 1987.

11. *Encyclopedia of Polymer Science and Technology,* Vol. 14, Wiley-Interscience, New York, 1971, p. 140.

12. A.M. Birks, *Plast. Technol.,* 131 (July 1977).

13. M.F. Day, *High Performance Alloys,* A. Schulman Co. Bulletin.

14. R.M. Berry, *Plastic Additives,* Technomics, Westport, Conn., 1972, p. 63.

15. E. Bauman, "Regarding the Uniting of Vinyl Compounds," *Ann. Der Chemie und Phar.* **163,** 308 (1872).

16. U.S. Pat. 1,425,130 (July 13, 1921), H. Plausen.

17. U.S. Pat. 1,929,453 (Oct. 10, 1933), W.L. Semon (to B.F. Goodrich).

18. U.S. Pat. 1,935,577 (Nov. 14, 1933), E.W. Ried.

19. U.S. Pat. 2,141,126 (Dec. 20, 1938), A.K. Doolittle.

20. U.S. Pat. 2,307,157 (Jan. 5, 1943) W.M. Quattlebaum and C.A. Noffsinger.

21. U.S. Pat. 2,301,867 (Nov. 10, 1942) T.L. Gresham.

22. U.S. Pat. 2,027,961 (Jan. 14, 1936), L.M. Currie.

23. *Mod. Plast.* (Jan. 1961, Jan. 1971, Jan. 1981).

24. *Facts and Figures of the U.S. Plastics Industry, 1982 Ed.,* Society of the Plastics Industry, New York, p. 49.

25. G.M. Powell and R.W. Quarles, "A New Technique in Coatings—Vinylite Resin Dispersions," *Official Digest,* Federation of Paint and Varnish Production Clubs, Dec. 1946.

26. *Dispersion Resin,* B.F. Goodrich Bulletin G-52.

27. E.N. Skiest, "Blending/Dispersion Resin Interaction," *Mod. Plast.* 132 (Oct. 1970).

28. P.R. Graham and J.R. Darby, "Effects of Plasticizer on Plastisol Fusion," *SPE J.,* 91 (Jan. 1961).

29. R.S. Guise, "Effects of Heat Stabilizers on Viscosity of Plastisol Pastes," *Mod. Plast.* (Aug. 1976).

30. H.A. Sarvetnick. *Plastisols & Organisols,* Van Nostrand Reinhold Co., New York, 1972, p. 110.

31. *Ibid.,* p. 111.

32. D.C. Burns and D.C. Dietz, *Trouble Shooting Guide to PVC Dispersion Resins,* Diamond Shamrock Co., p. 9.

33. R.A. Park, "Characteristics of Cellular Vinyls Produced via Chemical Blowing Agents and Mechanical Frothing," *Polymer Plastics Technology Engineering,* Marcel Dekker, New York, 1976, pp. 157–183.

34. R.A. Park, *Chemistry and Technology of Vinyl Resins Used in Coatings,* American Chemical Society, 1975, p. 809.

35. R.A. Park, "Plastisol Color Pigment Selection," *Mod. Plast.,* 64 (March 1975).

36. Ref. 30, p. 17.

37. Ref. 30, p. 113.

38. P. Ghera, "Effects of Small Quantities of Plasticizer in PVC Compounds," *Mod. Plast.,* **36** (2) IV35 (Oct. 1985).

39. General Tire & Rubber Co., 1964, private communication.

40. D.A. Tester, "The Processing of Plasticized PVC," in R.H. Burgess, ed., *Manufacturing and Processing of PVC,* McMillan, New York, 1982, p. 245.

41. B.C. Harris and F.T. Tulley, "Injection Molding of Polyvinyl Chloride," *Encyclopedia of PVC,* Marcel Dekker, New York, 1977, p. 1313.

44

RIGID POLY(VINYL CHLORIDE) (RPVC)

Saul Gobstein

Sa-Go Associates, Inc., Shaker Heights, Ohio

44.1 RPVC

Rigid poly(vinyl chloride) (PVC) is characterized by strength and excellent chemical resistance. The homopolymer has repeating units of vinyl chloride monomer (VCM);

$$\left[\begin{array}{c} H \quad H \\ | \quad\ | \\ -C-C- \\ | \quad\ | \\ H \quad Cl \end{array}\right]_n$$

the copolymer has repeating units of VCM along with another monomer—in various proportions—such as vinyl acetate (the most popular), ethylene, propylene, maleate, and so on.

In Europe, rigid PVC is referred to as unplasticized PVC. In the United States, most papers are published and presented with the term "vinyl" regardless of which material is being discussed.[1]

There are three separate and distinct vinyl technologies: rigid, flexible, and plastisol. The additives used in each technology may be different or may be used for a different purpose. For example, in each technology a lubrication system utilizes different types and amounts of materials. In rigid PVC, the lubrication system may be two parts per hundred parts of resin (PHR) composed of wax, an oxidized polyethylene, and a metallic stearate. In flexible PVC, just 1/4 PHR of stearic acid would be sufficient. And, in plastisols, most applications do not include a lubricant.

All three technologies use PVC resin. However, there are some differences in the use of the resin. Plastisol requires a very small particle size, usually 1 micron; therefore, emulsion-type resin is used for this purpose. In flexibles, a resin particle that would have a very high oil sorption characteristic would be desired so that the plasticizer is absorbed. In rigids, a particle with high bulk density is desired for maximzing throughput.[2]

44.2 CATEGORY

All three technologies are thermoplastic (unless special additives are added for cross linking). Products produced by all three technologies have similar characteristics such as heat sensitivity and the ability to be made clear, opaque, and in a wide range of colors.

TABLE 44.1.   Additives Used in Rigid PVC Plastics

Type of Additive	Function	Reference
Stabilizer	Inhibits degradation	4
Processing aid	Improves hot melt strength, hot elongation, gloss, fusion	5
Impact modifier	Improves impact	5
Lubricant	Improves flow and processing properties	6
Filler	Reduces cost	7
Blowing agent	Provides a cellular structure	8
Colorant	Imparts color	9
UV absorber	Protects against UV light	10
Anti-blocking and slip agents	Reduces friction	11
Reinforcement	Increases strength	12, 13

Flexible PVC is soft; rigid PVC is hard and tough. Rigid PVC finds applications in pipe, conduit, building panels, window frames, house siding, credit cards, profiles, and so on.

Poly(vinyl chloride), horny and brittle at room temperature, is heat sensitive and difficult to process.[3] Since it is polar the polymer readily accepts additives, that overcome basic limitations of the resin, thereby producing a compound with desirable and useful properties.

Additives, fillers, and reinforcements also add valuable properties to the base resin. For example, fiberglass reinforcement can make the plastic tougher and stronger.

Table 44.1 is a list of additives recommended for use in rigid PVC. The general function is also listed as well as a reference for more details.

Between 8 and 12% of additives are used in rigid PVC which play a vital role in its characterization.[14]

44.3   HISTORY

The early development of the technology is covered in the section on flexible PVC. In the 1930s, rigid PVC technology was emphasized in Germany[15] where the materials outstanding properties were demonstrated and where technicians learned how to overcome difficulties in processing this polymer. It is a little known but important historical footnote that PVCs easy processability was exploited by the Germans. During World War II PVC in the form of piping enabled fresh water supply to be restored in the city mere hours after a severe air attack.[16]

Some of the German rigid PVC technology was brought to the United States by the U.S. Army Quartermaster Corp. and the U.S. Bureau of Standards in 1946. However, it was not until the later 1950s when rigid PVC bottle technology was perfected that there was any real commercial interest in the United States. At about that time, the twin screw extruder was introduced in Europe, and came to the United States in the 1960s. This promoted wide use of rigid PVC, especially in the pipe market.[16]

As shown in Table 44.2, PVC has grown by nearly 300% in the years from 1960 to 1980.

44.4   POLYMERIZATION

44.4.1   Manufacture of Vinyl Chloride Monomer

Acetylene, ethylene, chlorine, and oxygen feedstocks are necessary to manufacture vinyl chloride monomer. Feedstock sources are presented in Figure 44.1.

The process by which feedstock is converted to VCM is shown in Figure 44.2. In

TABLE 44.2. Rigid PVC Polymer and Plastics
Consumption, billion pounds[a]

Year	Polymer	Plastic
1960	0.029	0.032
1970	1.00	1.10
1980	2.88	3.20

[a]*Source*: Sa Go Associates.

the early days of VCM production, acetylene was the dominant method; today, only 7% of the VCM is so manufactured. The remainder is produced by oxychlorination (Fig. 44.3).

All of the VCM produced is consumed in the manufacture of PVC resin. In 1976, OSHA[17] and EPA[18] regulations eliminated approximately 3 to 5% of the VCM going into specialty applications such as aerosol propellant hair sprays.[3] These regulations closely monitor free VCM in the workplace and VCM emissions, and must be met by all manufacturers of VCM, polymers, and compounds.

At normal temperature and pressure, VCM is a colorless, dense, pleasantly sweet-smelling gas. It is usually stored in a pressurized liquid state in steel tanks. VCM is soluble in aliphatic and aromatic hydrocarbons and insoluble in water. In the early 1970s, VCM sold for as little as 4.25 cents per pound. In 1989 VCM sold for 23–24 cents per pound.

In the last ten years, VCM prices have varied wildly in response to increased ethylene and chlorine prices and varied demand for VCM. Consequently, many production facilities have been shut down. Since 56.7% of the monomer is chlorine,[19] neither VCM nor PVC prices is dominated by oil and gas prices.

44.4.2 Manufacture of PVC Resin

PVC resin is manufactured by four methods: suspension, emulsion, bulk (or mass), and solution. In the United States, the breakdown by method is 83% suspension, 8% emulsion,

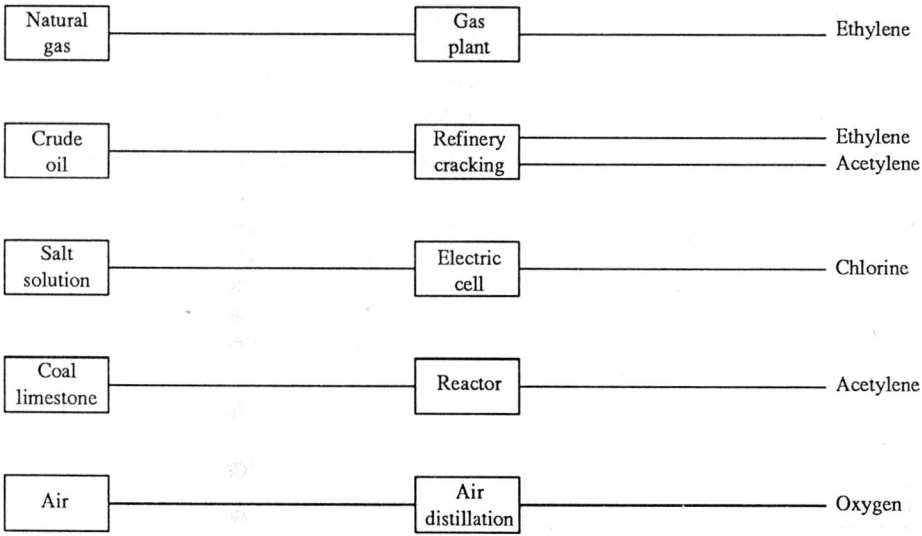

Figure 44.1. Sources of feed stock. Source: Sa-Go Associates, Inc.

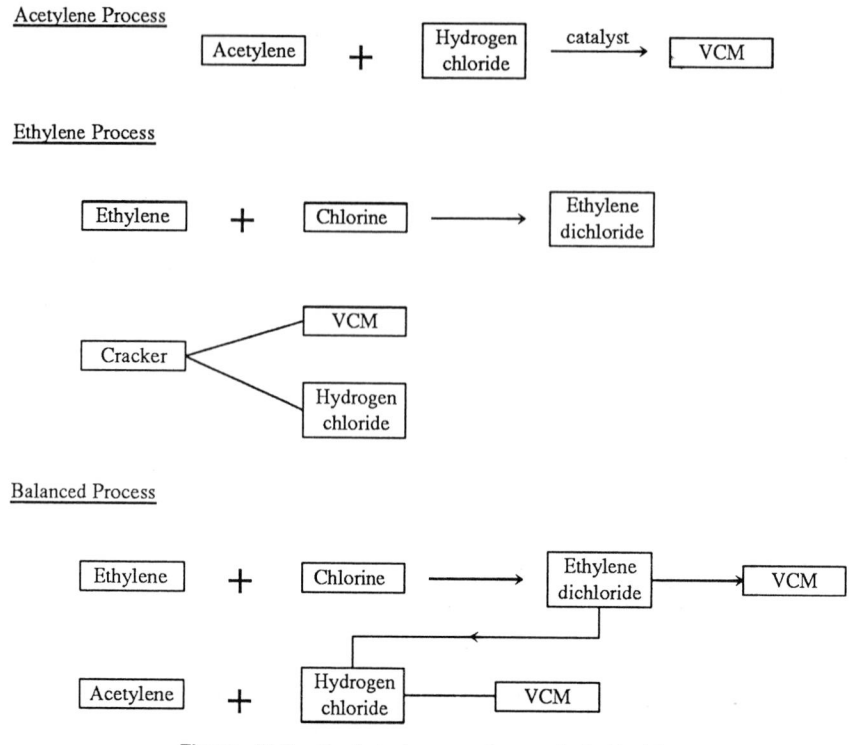

Figure 44.2. Feed stock conversion to vinyl chloride.

7% bulk, and 2% solution.[19] Worldwide, it is estimated that 70% is manufactured by the suspension method, 20% by emulsion, 9% by bulk, and 1% by solution.

44.4.3 Suspension

The suspension polymerization process[20] is sometimes referred to as pearl, bead, or granular processing. PVC so produced has relatively large particles (40 to 200 mesh), low levels of impurities, and resembles, in particle form, shrunken oranges.

In 1984, B.F. Goodrich offered a PVC grade, called Vantage PR,[21] that was manufactured by a new process called microsuspension.[22] It is described as having spherical, uniform particles (20–50 mesh) with high bulk density and improved processing.[23]

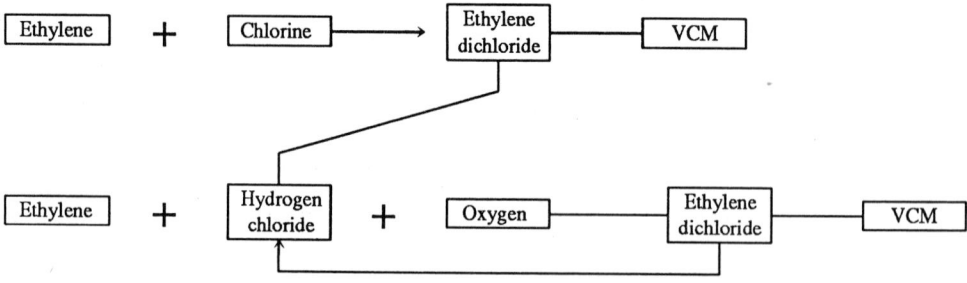

Figure 44.3. Oxychlorination process.

44.4.4 Emulsion

In the United States, the emulsion polymerization process[24] is used to produce plastisols and organasols; in Europe, it is used for production of general-purpose PVCs having a small (1-micron), particle size, a spherical shape, and the highest levels of impurities.

44.4.5 Bulk

Unlike the other processes, bulk polymerization[24] produces particles that are irregular and, in some cases, must be cryogenically ground. The resultant PVC contains little or no impurities.

44.4.6 Solution

The solution polymerization process[20] is used to make highly specialized resins used for metal coatings, record manufacture, powder coatings, and surface coatings that require resistant finishes. The high-quality PVC so produced command a premium price.

44.4.7 Description

PVC is a "head to tail," amorphous polymer[2] with a low degree of crystallinity. It contains 0.4 to 1.1 branches per 100 carbon atom chain. Its glass-transition temperature of about 176°F (80°C) varies with polymerization temperature. The melting point varies with molecular weight, molecular weight distribution, and polymerization temperature.

Commercial PVC resin degrades at moderate processing temperatures. The development of heat stabilizers that inhibit degradation enabled processors to use PVC.

44.5 DESCRIPTION OF PROPERTIES

Typically, a rigid PVC compound contains 90% resin. Additives are included to overcome processing difficulties or to meet end-use requirements. Therefore, there is a constant compromise between processing conditions and physical properties.

For example, a PVC copolymer processes more easily and at a lower temperature than a PVC homopolymer, but the homopolymer has better physical properties and is more heat stable. In deciding between the two, it is necessary to balance the desired physical properties with processing conditions. In the manufacture of phonograph records, a copolymer is preferable since processing is easier, sufficient detail can be achieved in each groove, and impact strength is higher.

The physical properties of PVC are primarily determined by the molecular weight. The higher the molecular weight, the better are the compounds physical properties and the more difficult it is to process. Table 44.3 details resin selection for a particular process.[25]

TABLE 44.3. The Resin of Choice for a Particular Process

Process	K value[a]
Injection molding	55–60
Blow molding	58–68
Extrusion	60–75
Calendaring	65–75

[a] K value is a measure of molecular weight.

There are several ways to indicate molecular weight of a PVC resin, such as inherent viscosity, intrinsic viscosity, specific viscosity,[26] K value (used in Europe), and P value (used in Japan). All three systems and the equivalents are explained in Table 44.4.

Because PVC undergoes states of gellation and fusion, a single rigid PVC compound processed under three different processing conditions can have three completely different physical property profiles. The reason is that a fused product exhibits complete intermolecular mingling and, therefore, optimum physical properties. A product in various stages of gellation does not offer comparable properties. A relatively new theory on microdomain structure carries this further.[27]

44.5.1 Thermal Properties

Commercial grades of PVC resin have a glass-transition temperature (T_g) of approximately 178°F (81°C). The T_g is sometimes referred to as the second order of transition. Above the T_g, the polymer is soft and flexible.[28] It is important to know the T_g with respect to application and service life. Because rigid PVC is a formulated plastic, glass-transition temperature is especially important. The best results are obtained when the resin and additive are blended above the T_g.

The first order of transition, the melting point (T_m), is 347°F (175°C) for rigid PVC. The practical process temperature for rigid PVC plastic will range from 320 to 428°F (160–220°C),[29] depending on the molecular weight of the resin as well as the amount and type of additives in the formulation. To obtain optimum physical properties, the crystalline melting point must be reached; otherwise, the resin will be in a state of gellation and not fusion.

Thermal conductivity of rigid PVC will vary with the ingredients in the formulation. Typical values for rigid PVC, using ASTM C177, range from 3.0 to 5.0 $\times$ 10^{-4} cal/s·cm·°C.

The coefficient of linear thermal expansion for rigid PVC is important, because it is used in outdoor applications where surface temperatures can reach 175°F (80°C). The coefficient of linear expansion is 5.0 to 10.0 $\times$ 10^{-6} in./in./°F (9 to 18 $\times$ 10^{-6} mm/mm/°C), as measured by ASTM D696.

The surface of dark-colored rigid PVC house siding can be distorted by heat. This problem is known as oil-canning.[30] Heat distortion will vary with the molecular weight of the resin and the ingredients in the formulation.

A homegrown heat sag test for rigid PVC panels has been found to be more indicative of the service life of the panel than the ASTM tests.[30]

TABLE 44.4.   Relationship Between Molecular Weight Values

Specific Viscosity ASTM 1243	Specific[a] Viscosity	K Value[b]	P Value[c]
0.21	0.62	50.5	480
0.26	0.77	56	640
0.32	0.92	62	800
0.39	1.10	68.5	960
0.45	1.25	74	1120
0.53	1.43	82	1280
0.60	1.58	89.8	1440

[a]Specific viscosity = 1% in cyclohexonone at 25°C (used in U.S.).
[b]K Value = 0.5 g/100 mL cyclohexonone at 25°C (used in Europe).
[c]P Value = degree of polymerization (used in Japan).

TABLE 44.5. Typical Pipe Properties[a]

	Type I	Type II
Density, lb/ft^3 (g/cm^3)	87.3 (1.40)	84.8 (1.36)
Tensile strength, psi (MPa)	7000 (48)	6000 (41)
Elongation, %	230	280
Tensile modulus, psi (MPa)	430,000 (2,965)	390,000 (2,689)
Hardness, Rockwell R	115	110
Impact, ft-lb/in. (J/m)	0.6 (32)	18.0 (961)
Heat distortion, °F (°C) 264 psi (1.8 MPa)	166 (73)	160 (71)

[a]*Source*: Sa-Go Associates, Inc.

44.5.2 Mechanical Properties

Rigid PVC properties vary according to the types of additives incorporated. For example, the impact strength of a rigid PVC compound can vary from 0.4–22 ft-lb/in. of notch (21–1174 J/m) depending on whether an impact modifier is used and the type and the amount used. Molecular weight has a direct bearing on mechanical properties. The higher the molecular weight, the better the physical properties, the higher the processing temperature requirements, and the more difficult the processing.

Mechanical properties can also be enhanced by glass reinforcement. Some rigid PVC grades contain 10 and 20% glass reinforcement.[12] These compounds are used in special, small-volume applications such as metering valves.

The range of mechanical properties of typical PVC pipe is presented in Table 44.5, along with the variation in Type I and Type II pipe properties.

44.5.3 Optical Properties

Rigid PVC plastic has outstanding optical properties. It can be made into a product that is sparkling clear to one that is opaque. In outdoor applications, it can be formulated to have good to excellent service life.[31]

44.5.4 Environmental Properties

Rigid PVC is widely selected for its excellent acid and alkali resistance. It can be formulated to have good weatherability. Ketones, esters, ethers, and chlorinated hydrocarbons will attack rigid PVC and make it swell or dissolve it.

As noted, environmental agencies such as EPA, OSHA, NIOSH, as well as FDA restrict rigid PVC processes and fabrication.[17,18,32] These restrictions are reviewed in reference 33.

44.6 APPLICATIONS

44.6.1 Pipe and Fittings

Rigid PVC pipe and fittings were first manufactured in Germany during World War II.[34] In subsequent years, the growth of PVC pipe has been phenomenal. In 1983, nearly one third of all PVC resin went into this application—about 2 billion pounds.[35]

PVC's success in pipe and fittings can be attributed to its resistance to most chemicals, imperviousness to attack by bacteria or microorganisms, corrosion resistance (many metal pipes corrode), flexible strength, heavy loads-bearing properties, and ability to overcome

TABLE 44.6.   PVC Pipe Formulation Type I[a]

Component	Typical Single Screw	Typical Multi Screw
Resin (K-68-72)	100.0	100.0
Stabilizer butyl tin	1.5	
Methyl tin		0.3–0.4
Processing aid	1.5	1.0
Filler	3.0	2.5–3.0
Calcium stearate	1.0	0.8
Wax 165	0.6	1.2
Polyethylene		0.2
TiO$_2$	1.0	1.0–1.5
Carbon black	0.2	0.2

[a]*Source*: Sa-Go Associates, Inc.

soil shifting that is not too severe (metal pipe under similar circumstances would break).[36] Tables 44.6 and 44.7 present typical pipe formulations.

44.6.2   House Siding

From 1958 through 1977, the house siding application grew from 0 to more than 175 million pounds per year.[37] Here, the major advantage of PVC is that it does not rust, corrode, or rot and eliminates the need for periodic painting. As noted, dark colors retain the heat of the sun and mud-cracking and warping can occur. This can be overcome by dual extrusion.[38] Table 44.8 presents a typical siding formulation.

44.6.3   Moldings

Injection molding applications for rigid PVC are growing as new resins and compounds improve processibility.[12] Television backs, television cabinets, business machine parts, and appliance parts are examples of PVC moldings. These applications are possible because of development of lower molecular weight PVC, resulting in compounds with lower shear

TABLE 44.7.   PVC Pipe Formulation Type II[a]

	Typical Single Screw	Typical Multi Screw
Resin (K-68-72)	100.0	100.0
Stabilizer (Butyl tin)	2.0	
(Methyl tin)		0.36–0.6
Processing aid	3.0	2.0
Filler	3.0	3.0–10.0
TiO$_2$	2.0	1.0–2.0
Impact modifier (ABS, Acrylic, CPE)	4.0	4.0–6.0
Calcium stearate	1.5	0.5–1.2
Wax 165	0.8	1.2–1.5
Polyethylene	0.1	0.2

[a]*Source*: Sa-Go Associates, Inc.

TABLE 44.8. Typical Siding Recipe (Formulation)[a]

Resin	100
Tin mercaptide	2.0
165 Lubricant	1.0
Oxidized polyethylene	0.25
Calcium stearate	1.00
Processing aid	2.0
Modifier (acrylic)	5–10
TiO$_2$	8–10

[a]Source: Sa-Go Associates, Inc.

sensitivity and improved thermal stability.[39] The significant rigid PVC properties are excellent self-extinguishing characteristics, good stiffness, excellent electricals, low moisture absorption, good surface hardness, good chemical resistance, and lower cost. A typical rigid PVC molding formulation is given in Table 44.9.

44.6.4 Bottles

The market growth of rigid PVC bottles has been negligible since the FDA took PVC off the GRAS list because of its VCM content. However, recent reports indicate that PVC will again be approved. There are already several applications on the market that have FDA approval on an individual basis.[40]

The advantages of rigid PVC in bottles are strength, clarity, resistance to oil and acids (such as salad dressings), and good permeability resistance.

A typical bottle formulation is shown in Table 44.10.

44.6.5 Foams

Cellular rigid PVC products have been popular since the early 1970s. The most popular are extrusion moldings, which have replaced wooden house moldings, that compare favorably with wood in being dent-resistant, less splittable, available in various designer colors, flame-resistant, and easier to make into a complex shape.[41] Other cellular products include pipe, sheet, and so on.

Foamed sheets can be vacuum formed into business machine housings.[42]

TABLE 44.9. Typical Rigid PVC Injection-Molding Compound[a]

Resin (X-55-65)	100	100	100
Acrylic processing aid	1.0–3.0	3.0	
Modifier	0–5.0	3.0	3.0
CaCO$_3$	0–5.0	5.0	
Butyl tin stabilizer	2.0	1.75	2.0
Calcium stearate	0.5		0.4
Wax 165	0.5		
G30			1.7
Polyethylene	0.3	0.2	
CW2		2.0	

[a]The usual procedure is to first pelletize, then injection mold. Injection molding from dry blend has been a problem area.

TABLE 44.10.   Typical Bottle Formulation

PVC (K = 55)	100 parts
Processing aid	2–5 parts
Impact modulus	8–15 parts
Tin mercaptide	2–3 parts
Calcium stearate	0.5
Oxidized polyethylene	0.2
Poly functional Was	0.2

44.6.6   Windows

Rigid vinyl PVC window frames have grown very rapidly, from 1 million in 1981 to an expected 3.5 million units by 1990.[43] The interest in PVC windows has brought about very sophisticated downstream equipment for extruders.[44] The profile, the sealing, and the dual extrusion have all advanced because of this application.

44.6.7   Phonograph Records

Copolymer PVC is the resin of choice in the manufacture of phonograph records. This compound meets the impact property, lubrication, and heat stability requirements and is sufficiently tough that the stylus does not break the fragile walls of the groove. Although there has been little growth in this application in recent years, it nonetheless accounts for 90 million pounds of resin consumption annually.

44.7   ADVANTAGES/DISADVANTAGES

Rigid PVC plastic is a formulated and comounded product with outstanding properties. This versatile product, which offers the possibility of an almost infinite number of compounds, is made from a low-cost feedstock. As noted, 56.7% of the molecule is chlorine. This means that the polymer's availability and pricing are not wholly related to oil or gas feedstocks.

Other advantages of rigid PVC plastics are

- Low cost
- High strength
- Good chemical resistance
- Low water absorption
- High impact (when properly formulated)
- Outstanding pipe characteristics.[45]
- Good outdoor weathering
- Nonburning
- Good stiffness
- Excellent electrical properties
- Good surface appearance.

The disadvantages are

- Processing difficulty caused by polymer stability
- Low heat deflection (low rise temperature)
- Poor creep at elevated temperatures.[46]

44.8 PROCESSING TECHNIQUES

Processing of rigid PVC compounds requires skill and art. The processor has to process the material so that maximum physical properties are obtained under conditions that ensure that degradation does not occur. The various processes for rigid PVC are presented in the Master Material Outline.

Master Material Outline

PROCESSABILITY OF ___Rigid PVC___

PROCESS	A	B	C
INJECTION MOLDING	X		
EXTRUSION	X		
THERMOFORMING	X		
FOAM MOLDING	X		
DIP, SLUSH MOLDING			
ROTATIONAL MOLDING			
POWDER, FLUIDIZED-BED COATING			
REINFORCED THERMOSET MOLDING			
COMPRESSION/TRANSFER MOLDING		X	
REACTION INJECTION MOLDING (RIM)			
MECHANICAL FORMING			
CASTING		X	

A = Common processing technique.
B = Technique possible with difficulty.
C = Used only in special circumstances, if ever.

44.8.1 Injection Molding

Injection molding of rigid PVC requires a special screw—one with a compression of 1.5 to 2.0 (slightly higher for powders)—and a special tip design. The tip is a smear tip and the nozzle should be as small as possible. Additional information can be found in B.F. Goodrich publications and other sources.[47,48]

44.8.2 Extrusion

Extrusion is the process by which the largest poundage of rigid PVC plastic is processed. For rigid PVC, it is the most important processing technique and probably the most controversial. Gomez[49] points out that there are many factors to consider in both screw and die designs.

It is common industry practice to design a die around a particular rigid compound. This is critical because a compound with the same physical properties made by another vendor will not flow the same through the die. Standard extrusion technology cannot be directly applied to the extrusion of these compounds. The first step is to tailor the compound to the process and to the finished product, after which the extruder should follow the special rules for extruding rigid PVC.[50,51]

44.8.3 Blow Molding

Blow molding of rigid PVC began in Europe in the early 1950s. It was not until the late 1960s and early 1970s that blow molding of rigid PVC came into its own.[22,49]

44.8.4 Foamed Rigid PVC

The extrusion of foamed or cellular rigid PVC profiles was developed in the early 1970s and is now being successfully marketed. The product is taking the place of wooden moldings in homes and offices. Foamed pipe and other profiles are still under development.

There are two major areas of foam expansion. Free expansion is where the expansion takes place beyond the die; controlled expansion is where expansion takes place in the die.[22]

44.8.5 Compression Molding

The major application for rigid PVC using this technique is the manufacture of phonograph records.

44.9 RESIN FORMS

Rigid PVC plastic molding compounds are available in fully fluxed pellets, powdered dry blends, compacts, and regrind. The fully fluxed pellets can be cylindrical pellets, face-cut round pellets, or diced pellets. The dry blend is a sandy resin blend with additives that are dispersed and distributed around the resin particle. Compacts are dry blends that are compacted into pellets. Regrind is a random cut pellet of a fluxed material, usually obtained from a reject or a sprue or runner. Care should be taken to separate the fines from the pellets to avoid overheating in processing.

44.9.1 Additives

Additives are incorporated into a rigid PVC plastic to overcome the limitation of the polymer as well as to build in certain properties.

Vinyl Stabilizers. PVC resin is heat sensitive and will degrade if it is exposed to a source of energy such as heat or light. The indication of degradation is a discoloration. It starts as amber, then darkens to deep amber, orange-red brown, and finally black. Degradation leads to chain scission and cross linking; hence a loss in physical properties. Stabilizers are added to the formulation to inhibit the degradation reaction. The typical stabilizers used are

- organo tins.
- mixed metals (barium/tin, barium/cadmium).
- ester tins.

- antimony.
- lead (typical of European system).
- Ca/Zn (typical of FDA system).

Processing Aids.[52] Processing aids are usually polymeric materials that are included in the recipe to overcome processing difficulties. The typical polymers used are acrylic copolymers, styrene, acrylonitrile and styrene, methyl methacrylate, and chlorinated polyethylene.

These materials give the rigid PVC compound homogeneity and better hot melt strength, and bring about peak torque much faster. They also decrease or eliminate plateout.

Impact Modifiers. Impact modifiers are rubbery graft polymers or rubbery alloys that disperse in PVC. Such polymers deform on impact and dissipate the energy of impact. Materials used for this purpose are acrylic copolymers, acrylonitrile–butadiene–styrene, ethylene–vinyl acetate, and chlorinated polyethylene.

Fillers. Fillers are usually incorporated in rigid PVC formulations to reduce cost. In most recipes, 3 PHR is the maximum filler load. More could have a negative effect on viscosity and cause equipment wear. Studies of equipment wear have shown that above 3 PHR, the cost benefits are lost because of equipment wear.

Lubricants[6]**.** Lubricant systems are the most proprietary systems and also the most indispensable in rigid PVC. It is important to obtain a balance between the external and internal lubricants.

Lubricants perform several functions:

- Reduce frictional forces between resin chains (internal).
- Reduce frictional forces between the resin chains and the metal surfaces (external).
- Promote melt flow by reducing melt viscosity.

44.9.2 Chemically Modified Rigid PVCs

Poly(vinyl chloride) can be modified chemically to produce some very desirable results. Chlorinated poly(vinyl chloride) (CPVC), for example, has a service temperature that is 45–65°F (15–30°C) higher than regular PVC. It is used for hot water pipe.

44.9.3 Copolymerization

Copolymers of PVC [primarily poly(vinyl acetate)] are made into molding compounds. They upgrade flow and other properties, compared with conventional PVC.

44.10 SPECIFICATION OF PROPERTIES

The outstanding properties of rigid PVC—nonflammability, chemical resistance, low gas permeability, excellent weatherability, and good impact strength—can be built into the resin at a low cost. The Master Outline of Material Properties identifies key property data.

44.11 PROCESSING REQUIREMENTS

The art of processing rigid PVC is to use a high enough temperature that the compound goes to a state of fusion, which maximizes the physical properties. A temperature that is

MATERIAL ___ Rigid PVC

PROPERTY	TEST METHOD	ENGLISH UNITS	VALUE	METRIC UNITS	VALUE
MECHANICAL DENSITY	D792	lb/ft^3	83.2–98.5	g/cm^3	1.32–1.58
TENSILE STRENGTH	D638	psi	6,000–7,500	MPa	41–52
TENSILE MODULUS	D638	psi		MPa	
FLEXURAL MODULUS	D790	psi	10,000–16,000	MPa	69–110
ELONGATION TO BREAK	D638	%	40–80	%	40–80
NOTCHED IZOD AT ROOM TEMP	D256	ft-lb/in.	0.4–22	J/m	21–1174
HARDNESS	D785		55–85		55–85
THERMAL DEFLECTION T @ 264 psi	D648	°F	150–170	°C	66–77
DEFLECTION T @ 66 psi	D648	°F	145–180	°C	63–82
VICAT SOFTENING POINT	D1525	°F	Varies with formulation	°C	Varies with formulation
UL TEMP INDEX	UL746B	°F	Varies with formulation	°C	Varies with formulation
UL FLAMMABILITY CODE RATING	UL94		V-0		V-0
LINEAR COEFFICIENT THERMAL EXPANSION	D696	in./in./°F	5.0–10.0	mm/mm/°C	9–18
ENVIRONMENTAL WATER ABSORPTION 24 HOUR	D570	%	0.04–4.0	%	0.04–4.0
CLARITY	D1003	% TRANSMISSION	Clear, translucent, transparent, opaque	% TRANSMISSION	Clear, translucent, transparent, opaque
OUTDOOR WEATHERING	D1435	%	Varies with formulation	%	Varies with formulation
FDA APPROVAL			Special formula available		Special formula available
CHEMICAL RESISTANCE TO: WEAK ACID	D543		Not attacked		Not attacked
STRONG ACID	D543		Not to slightly attacked		Not to slightly attacked
WEAK ALKALI	D543		Not attacked		Not attacked
STRONG ALKALI	D543		Not to slightly attacked		Not to slightly attacked
LOW MOLECULAR WEIGHT SOLVENTS	D543				
ALCOHOLS	D543		Resists		Resists
ELECTRICAL DIELECTRIC STRENGTH	D149	V/mil		kV/mm	
DIELECTRIC CONSTANT	D150	10^6 Hertz	32–40		32–40
POWER FACTOR	D150	10^6 Hertz	0.007–0.020		0.007–0.020
OTHER MELTING POINT	D3418	°F	347	°C	175
GLASS TRANSITION TEMP.	D3418	°F	178	°C	81

not high enough will only allow the compound to reach a state of gellation. Although the finished product will have a good appearance, it will not have the maximum required physical properties.

In order to process at a high enough temperature without burning, properly designed tools and balanced lubricated formulations are important. With regard to the latter, one controversy pertains to the order of addition when blending. Some succeed in adding the lubricant early; others require that the lubricant be added last.

If the lubricant is incorporated at a high enough temperature, the equipment will not be coated. By adding the lubricant early, and at a low temperature, one runs the risk of coating the inside of the blender.

Rigid PVC plastic can be ground and reprocessed. Care should be taken to cool the regrind quickly. If rigid PVC is ground by conventional procedures, the regrind can discolor because the plastic retains heat in the container. Since this is a heat-sensitive polymer, degradation can be accelerated.

Another consideration is the fines. It is best to screen them out because they will melt in the barrel faster and thereby may cause pimples or degradation.

44.12 PROCESSING-SENSITIVE END PROPERTIES

End-use properties can be affected by the processing conditions of rigid PVC. If the temperature of the melt does not exceed the crystalline melt point [about 338–410°F (170–210°C)], depending on the molecular weight), then the maximum properties will not develop. An easy test is to place the finished product in acetone for 20 min. If there is no deterioration, then the crystalline melt point was reached.

44.13 SHRINKAGE[2]

The mold shrinkage of rigid PVC is low, from about 0.001 to 0.005 in./in. (mm/mm). The general design parameters for rigid PVC with respect to ribs, bosses, snapfits, venting, radii, draft angles, and so on, are the same as for ABS and PPO.

The most important rule is to build a mold with minimum restrictions to flow. Larger gates, runners, and sprues are recommended to avoid hangup and degradation.

44.14 TRADE NAMES

A list of trade names of vinyl products as well as trade names of raw materials used in vinyl plastics can be found at the end of the section on flexible PVC.

BIBLIOGRAPHY

1. S. Gobstein, "Terminology Dilemma," *Plastics Compounding* 12 (Sept./Oct. 1983).
2. "New B.F. Goodrich PVC Resin," *Plastics World*, 2, 20 (July 1984).
3. R.T. Gottesman, "Chemistry and Technology of Polyvinyl Chloride," *Applied Polymer Science* Chapt. 27, p. 364. ACS, Washington, D.C., 1975.
4. L.T. Nass, "Action and Characteristics of Stabilizers," *Encyclopedia of PVC*, Marcel Dekker, New York, 1976, p. 295.
5. C.F. Ryan and L. Jalbert, "Modifying Resins for Polyvinyl Chloride," *Encyclopedia of PVC*, Marcel Dekker, New York, 1976, p. 602.

6. E.L. White, *Encyclopedia of PVC,* Marcel Dekker, New York, 1976, p. 643.

7. R.P. Braddicks and L.I. Nass, "Fillers," *Encyclopedia of PVC,* Marcel Dekker, New York, 1976, p. 711.

8. H.R. Lasman and J.P. Scull, "Miscellaneous Modifying Additives," *Encyclopedia of PVC,* Marcel Dekker, New York, 1976, p. 801.

9. M. Saltzman and R.M. Johnston, *Industrial Color Technology,* ACS, Washington, D.C. 1971.

10. *Encyclopedia of Polymer Science and Technology,* Vol. 14. Wiley-Interscience, New York, 1971, p. 140.

11. A.M. Birks, *Plastics Technology,* 131 (July 1977).

12. E. Galli, "Improving Performance in Low-Cost Resin." *Plastics Design Forum,* 29 (May 1983).

13. W.V. Titow, "PVC Additives," *Developments in PVC Production and Processing, 1.* Allied Science Publishers, Essex England, 1977, p. 73.

14. L. Mascia, *The Role of Additives in Plastics,* Edward Arnold Ltd., London, 1974, p. 1.

15. M. Kaufman, *The History of Polyvinyl Chloride,* Maclaren & Sons, London, 1969.

16. W.D. Nesbett, *PVC Pipe in Municipal Markets,* Rigid Vinyls RETEC, 1977, p. 115.

17. "Exposure to Vinyl Chloride," *Federal Register.* **39** (194), Part II, (Oct. 4, 1976).

18. "National Emission Standards for Hazardous Air Pollutants—Standard for Vinyl Chloride," *Federal Register,* **41** (205), (October 21, 1976).

19. G.F. Cohan, *Modern Plastics Encyclopedia,* McGraw-Hill, New York, 1981–1982, p. 100.

20. C.W. Johnston, *Encyclopedia of PVC,* Marcel Dekker, New York, 1976, Chapt. 3, p. 76, 91.

21. "New Goodrich PVC Improves Output," *Plastics World,* 10 (July 1984).

22. A. Whelan and J.L. Craft, *Developments in PVC Production and Processing, 1.* Applied Science Publishers, London 1977, pp. 34, 95, 118, 125.

23. L.G. Shaw and co-workers, "Extrusion and Pipe Properties of a Novel Spherical PVC Resin," *ANTEC Proceedings 1984,* p. 820.

24. R.H. Burgess, *Manufacture and Processing of PVC,* MacMillan Publishing Co., New York, 1982, pp. 39, 63.

25. S. Gobstein, *Rigid PVC Technology,* Course Notebook, sponsored by Modern Plastics.

26. *ASTM 1243, ASTM Book of Standards* ASTM, Philadelphia, Pa., 1976, p. 449.

27. J.W. Summers, "The Nature of Poly (Vinyl Chloride) Crystallinity—the Microdomain Structure," *Journal of Vinyl Technology,* **3** (2), 107 (June 1981).

28. R.B. Seymour and C.E. Carraher Jr., *Polymer Chemistry: An Introduction,* Marcel Dekker, New York, 1981, p. 28.

29. W.J. Frissell, "Elucidation of Structure," *Encyclopedia of PVC.* Marcel Dekker, New York, 1976, p. 266.

30. L.I. Gomez, "Testing Rigid PVC Products." *Encyclopedia of PVC,* Marcel Dekker, New York, 1976, p. 1647.

31. L.B. Weisfield, G.A. Thacker, and L.I. Nass, "Photodegradation of Rigid PVC," *SPE Journal* **21,** 649 (1965).

32. "VCM—the Processors Perspective," *SPE RETEC,* Oct. 31, 1974.

33. L.B. Crider and W.C. Holbrook, *Encyclopedia of PVC,* Marcel Dekker, New York, 1976, p. 1725.

34. W.D. Nesbett, "PVC Pipe in Municipal Markets," *Technical Papers, SPE RETEC* 115, (Mar. 1, 1977).

35. "Materials—the Statistical Story," *Modern Plastics,* **60** (1), 70 (Jan. 1983).

36. *Handbook of PVC Pipe,* Unibell Plastic Pipe Assn., 1977.

37. W.P. Hammond, "Markets for Rigid PVC Siding," *Technical Papers, SPE RETEC,* 106 (Mar. 1, 1977).

38. A. Shade, "Compound Requirements for Vinyl Siding," *Journal of Vinyl Technology,* **2,** 64 (June 1979).

39. J.L. Murray and A.J. Dito, "Injection Molding PVC—New Resins Make It Easier," *Plast. Technol.,* 79 (Nov. 1981), p. 79.

40. PET vs. PVC: Processing Economic Key," *Plast. Technol.* **29** (3), 78 (March 1983).

41. M. Batiuk, "Polyvinyl Chloride Extrusions," *Encyclopedia of PVC*, Marcel Dekker, New York, 1976, p. 1308.

42. "Expanded PVC Sheet Covers IBM Processor Requirements," *Plastics Design Forum,* **7**, (5), (Sept. 1982).

43. Extruded PVC Frames the View From a Bay Window, *Plastics Design Forum,* **7** (4), 76 (July 1982).

44. D.E. Neudell, "Profile Downstream Equipment," *Plast. Technol.* **28** (11), 81 (Oct. 1982).

45. "Engineering Design Properties of Unplasticized PVC," *Br. Plast.*, 682 (Dec. 1964).

46. S. Turner, "Creep in Thermoplastics—Unplasticized PVC," *Br. Plast.*, 682 (Dec. 1964).

47. *Injection Molding Guide—Rigid PVC*, B.F. Goodrich Bulletin G-63,

48. F.T. Tulley and B.C. Harris, "Injection Molding of Polyvinyl Chloride," *Encyclopedia of PVC*, Marcel Dekker, New York, 1976, p. 1313.

49. L.I. Gomez, *Engineering With Rigid PVC*, Marcel Dekker, New York, 1984, pp. 153, 246.

50. *Extruding Rigid Geon Vinyls*, B.F. Goodrich Bulletin G-40.

51. *Rigid PVC Extrusion*, Cincinnati Milacron Chemical Bulletin, 1976.

52. M.L. Dannis and F.L. Ramp, "Chemically Modified Polyvinyl Chloride," *Encyclopedia of PVC*, Marcel Dekker, New York 1976, p. 225.

General References

W.M. Coaker, "Calendering Rigid PVC Products," *Engineering With Rigid PVC*, Marcel Dekker, New York, p. 385.

W.H. Doherty, "Rigid Vinyls for Custom Injection Molding," *SPE RETEC*, Oct. 1984.

C.W. Fletcher and Dale Drekman, *Vinyl Technology* **3** (2), 130 (June 1981).

L.T. Nass, "Theory of Degradation and Stabilization Mechanism," *Encyclopedia of PVC*, Marcel Dekker, New York, 1976, p. 271.

J. Lutz, *Tailoring PVC Compounds to the Application—The Choice of Processing Aids and Impact Modifiers*, Rohm & Haas, Briston, Pa.

45

POLY(VINYLIDENE CHLORIDE) COPOLYMERS (PVDC)

P.T. DeLassus and W.L. Treptow

Barrier Resins and Fabrication Laboratory, The Dow Chemical Company, Midland, MI 48674

45.1 INTRODUCTION

Vinylidene chloride, $CH_2{=}CCl_2$, (VDC) can be polymerized to make poly(vinylidene chloride) (PVDC). But since the VDC homopolymer is virtually impossible to fabricate, VDC is not marketed in this form. The commercially available materials commonly known as poly(vinylidene chloride) are actually copolymers with greater than 70 wt % VDC.

Comonomers for VDC include

Vinyl Chloride   Acrylonitrile   Methylacrylate

Butylacrylate

As marketed, the copolymers often contain formulants, plasticizers, and stabilizers, added by the manufacturer. The precise compositions and formulations of the marketed copolymers remain proprietary.

45.2 CATEGORY

VDC copolymers are rich in vinylidene chloride and usually semicrystalline. Common commercial products have 20–50% crystallinity. This results in a thermoplastic polymer that can be melt processed above its melting point (266–356°F [130–180°C]) but below its degradation temperature. Other VDC copolymers are coated as a latex or lacquer onto substrates. VDC copolymers have a glass-transition temperature (T_g) near ambient, which usually results in a stiff and brittle material. Plasticizers may be used to improve the softness and ductility of the polymer.

567

45.3 HISTORY

The discoveries leading to commercialization of VDC copolymers were made by R.M. Wiley and J.H. Reilly of The Dow Chemical Company, Midland, Mich., in 1933 and the years immediately following. Dow began commercialization of PVDC in 1939, under the trademark Saran. Early applications included monolayer films, fibers, molded pipe, and lacquer coatings.

The films and coatings were used to make packages and packaging materials with good oxygen and water vapor barrier properties. The fibers were woven to make articles such as window screens, lawn chair webbing, and curtains for long service in sunny/oxidative environments. The molded piping was used in systems requiring resistance to solvents.

There are no published data on capacity and production of VDC monomer and the various VDC copolymers. However, all these materials are classified as specialty products, and annual production is well below 1% of total annual plastics consumption.

The two U.S. producers of VDC monomer, Dow and PPG Industries, Inc., use some of their production internally and sell the remainder. Among the VDC copolymer producers listed under Trade Names, there are both internal users and merchant sellers.

45.4 POLYMERIZATION

VDC copolymers are prepared commercially by free-radical polymerization in either a suspension or emulsion process. In the suspension process, a single, glass-lined reactor typically is used with a water jacket for temperature control. In a typical suspension process the comonomers, water, a Methocel cellulose ether product, and an oil-soluble, thermally activated, free-radical initiator (lauryl peroxide, for example) are batch-loaded and heated in a sealed reactor to effect polymerization. The polymer precipitates and crystallizes, giving a hard, porous bead.

On the other hand, at the beginning of the emulsion process, the reactor is loaded with the comonomers, water, soap, and water-soluble initiators such as sodium persulfate. Both batch and continuous addition of one or more ingredients have been used in emulsion processing.

VDC copolymers typically sell for $0.75–1.50 per pound, depending on product properties, application, value in end use, and product form (that is, resin, formulated product, or latex). Being specialty materials rather than commodity-type plastics, VDC copolymers are high-value-added products. Prices reflect the high technology necessary for production as well as the high level of research and technical service available to the customer/user from the producer.

45.5 DESCRIPTION OF PROPERTIES

45.5.1 Thermal Properties

VDC homopolymer has a glass-transition temperature of 10°F (-12°C). Copolymers usually have higher T_g. However, for most useful cases, the T_g is less than 86°F (30°C). Plasticizers can suppress the T_g a few degrees.

VDC homopolymer has a crystalline melting temperature, T_m, at about 392°F (200°C). Copolymers have T_m in the range of 266–356°F (130–180°C).

45.5.2 Mechanical Properties

The mechanical properties of any VDC copolymer are functions of the comonomer type and level as well as the degree of crystallinity, orientation, and plasticizer level. Typical, semicrystalline copolymers tend to have high tensile modulus and small elongation at break. The impact strengths are low unless the polymer is reinforced.

45.5.3 Optical Properties

VDC copolymers tend to be clear and to have high gloss. The clarity is preserved even in the presence of crystallinity. A brown hue can occur after long thermal or photo stress.

45.5.4 Environmental Properties

VDC copolymers are generally resistant to solvents and chemicals. Some bases will cause degradation. Although color will develop after long solar exposures, the mechanical properties are not seriously altered by such exposure.

45.5.5 Barrier Properties

VDC copolymers are excellent barriers to mass transport. They are unique in providing a full range of barrier, resisting permeation by oxygen, water vapor, and carbon dioxide as well as flavors and aromas. VDC copolymers are available with room-temperature oxygen permeability as low as 1.2×10^{-14} cm^3(STP)·cm^2/cm^3·s·cm Hg.

45.5.6 Ignition Resistance

In a broad range of experiments, including UL 94 flame spread, limiting oxygen index (LOI), National Bureau of Standards (NBS) smoke chamber, and others, VDC copolymers have shown ignition resistant properties. (Small-scale tests are not intended to reflect performance in actual fire situations).

45.5.7 Solvent Properties

Although VDC copolymers are resistant to solvents, they can be dissolved. Solvents that are useful over broad composition ranges of VDC copolymers include tetrahydrofuran, methyl ethyl ketone, dimethylacetamide, and cyclohexanone.

45.6 APPLICATIONS

45.6.1 Barrier Shrink Wrap

Resin can be extruded into monolayer film with biaxial orientation. This film provides an excellent barrier to oxygen and water vapor. On brief exposure to warm temperatures, the film will shrink to conform to the dimensions of an article such as fresh red meat. Multilayer structures for shrink wrap also are available.

45.6.2 Multilayer Films

Multilayer films including a VDC copolymer layer can be made by either coextrusion or lamination. These structures combine the high barrier of a VDC copolymer layer with the economy and strength of polyethylene, paper, or other layers.

45.6.3 Barrier Coatings

Barrier coatings of a VDC copolymer may be added to articles or films. A latex coating is commonly added to paper or oriented polypropylene or glassine to enhance the barrier properties. A latex coating may be added to a polyester bottle to minimize carbon dioxide transmission. Solvent coatings add a barrier to cellophane, polyester, and other substrates.

45.6.4   Lined Pipe and Molded Parts

Chemical resistance can be added to pipes by the addition of a VDC copolymer liner. The same chemical resistance is useful for such molded parts as gasoline filters, valves, and pipe fittings.

45.6.5   Saran Wrap Plastic Film

This biaxially oriented monolayer film has a formulation package that enhances cling and mechanical properties for household food protection.

45.6.6   Rigid Barrier Containers

VDC copolymers can be used with polypropylene, polystyrene, or polyethylene to make rigid barrier containers. Typically, a multilayer sheet is made by coextrusion. Tie layers adhere the central barrier layer to the structural skins. An additional layer of reprocessed scrap may be included. The sheet can be thermoformed into containers that can protect food for a year or more without refrigeration.

45.7   ADVANTAGES/DISADVANTAGES

45.7.1   Advantages

Perhaps the most important commercial advantage of the VDC copolymers is their usefulness in providing an economical barrier to mass transport of both oxygen and water vapor. Other advantages include high tensile strength, good solvent resistance, and ignition resistance.

45.7.2   Disadvantages

The VDC copolymers are more expensive than most commodity polymers. They have high density and, because of their thermal sensitivity, special care must be taken during extrusion.

45.8   PROCESSING TECHNIQUES

See the Master Material Outline.

45.9   RESIN FORMS

VDC copolymers are marketed in the forms of powders, latexes, and expandable microspheres.

For use as extrusion resins, VDC copolymers produced by either suspension or emulsion processes have historically been available principally as powders.

VDC copolymer powders also are available for use, after dissolution, as lacquer coatings on cellophane and other plastic films.

Latexes of VDC copolymers are available for coating onto polymer films, paper, and other substrates.

Master Material Outline

PROCESSABILITY OF ___PVDC___

PROCESS	A	B	C
INJECTION MOLDING		X	
EXTRUSION	X		
Multilayer THERMOFORMING	X		
FOAM MOLDING			
DIP, SLUSH MOLDING			
ROTATIONAL MOLDING			
POWDER, FLUIDIZED-BED COATING			
REINFORCED THERMOSET MOLDING			
COMPRESSION/TRANSFER MOLDING			
REACTION INJECTION MOLDING (RIM)			
MECHANICAL FORMING	X		
Coextrusion CASTING, film and sheet	X		
LACQUERS COATING	X		
LATEX COATING	X		

A = Common processing technique.
B = Technique possible with difficulty.
C = Used only in special circumstances, if ever.

Expandable microspheres are used as fillers in composite materials.

In whatever form they are marketed, the copolymers incorporate formulants appropriate to the intended use. Most extrusion powders, for example, are formulated with thermal stabilizers and plasticizers.

Reseach on VDC copolymers is wide ranging and vigorous. Numerous recent patents describe process and property improvements with polymer blends or rubber modifications.

45.10 SPECIFICATION OF PROPERTIES

See the Master Outline of Material Properties.

MATERIAL ___ PVDC

PROPERTY	TEST METHOD	ENGLISH UNITS	VALUE	METRIC UNITS	VALUE
MECHANICAL DENSITY	D792	lb/ft^3	106	g/cm^3	1.70
TENSILE STRENGTH	D638	psi	5000	MPa	34.5
TENSILE MODULUS	D638	psi	75,000	MPa	516.8
FLEXURAL MODULUS	D790	psi		MPa	
ELONGATION TO BREAK	D638	%	25	%	25
NOTCHED IZOD AT ROOM TEMP	D256	ft-lb/in.	0.7	J/m	37.4
HARDNESS	D785		60		60
THERMAL DEFLECTION T @ 264 psi	D648	°F	120	°C	49
DEFLECTION T @ 66 psi	D648	°F		°C	
VICAT SOFTENING POINT	D1525	°F	275	°C	135
UL TEMP INDEX	UL746B	°F		°C	
UL FLAMMABILITY CODE RATING	UL94		V-0		V-0
LINEAR COEFFICIENT THERMAL EXPANSION	D696	in./in./°F	19×10^{-5}	mm/mm/°C	34×10^{-5}
ENVIRONMENTAL WATER ABSORPTION 24 HOUR	D570	%	0.1	%	0.1
CLARITY	D1003	% TRANSMISSION	85	% TRANSMISSION	85
OUTDOOR WEATHERING	D1435	%	Moderate discoloration	%	Moderate discoloration
FDA APPROVAL			Yes		Yes
CHEMICAL RESISTANCE TO: WEAK ACID	D543		Not attacked		Not attacked
STRONG ACID	D543		Not attacked		Not attacked
WEAK ALKALI	D543		Minimally to badly attacked		Minimally to badly attacked
STRONG ALKALI	D543		Minimally to badly attacked		Minimally to badly attacked
LOW MOLECULAR WEIGHT SOLVENTS	D543		Not attacked		Not attacked
ALCOHOLS	D543		Not attacked		Not attacked
ELECTRICAL DIELECTRIC STRENGTH	D149	V/mil		kV/mm	
DIELECTRIC CONSTANT	D150	10^6 Hertz			
POWER FACTOR	D150	10^6 Hertz			
OTHER MELTING POINT	D3418	°F	320	°C	160
GLASS TRANSITION TEMP.	D3418	°F	25	°C	-4

*Heat seal temperature, 275°F (135°C).

Barrier properties: Oxygen at 23°C (Oxtran), 0.1 cm^3(STP) ·mil/100-in.^{2}-day-atm; carbon dioxide at 23°C (Permatran C), 0.2 cm^3(STP) ·mil/100-in.^{2}-day-atm; water vapor at 100°F, 90% RH (Permatran w), 0.06 g·mil/100-in.^{2}-day.

572

45.11 PROCESSING REQUIREMENTS

VDC copolymers are among the thermally sensitive plastics (others include PVC and CPE). If they are overheated during processing, they will evolve HCl. Thus, any melt processing of VDC copolymers requires careful control of the melt temperature, a streamlined melt flow, and minimum melt residence time.

Resins formulated for melt processing typically include thermal stabilizers. These may be epoxy compounds or inorganic or organic salts. In some cases organic phosphates or phenolic antioxidants also are used.

A major improvement in melt processing of VDC copolymers was the development by Dow Chemical of coextrusion feedblock technology. By use of the patented feedblock, the VDC copolymer component of a sheet or film structure is completely enveloped by the other materials when coextruded.

Because ferric chloride ($FeCl_3$) acts as a catalyst to initiate rapid degradation of VDC copolymers, materials of construction for melt processing equipment are typically Duranickel (International Nickel Co., Inc.) or Inconel nickel alloys.

In the processing of VDC copolymers, materials feeding is accomplished most commonly by air-veying or by gravity feed.

45.12 PROCESSING-SENSITIVE END PROPERTIES

Orientation during melt fabrication can substantially improve the mechanical properties of VDC copolymers. Uniaxially oriented monofilaments exhibit increased tensile strength but a drop in elongation to rupture. Biaxially oriented films, such as Saran Wrap film, have the expected increase in tensile strength and modulus, with the magnitude of these properties being very dependent on the blown film stretch ratio. Increased amounts of formulants (plasticizers and stabilizers) improve the ductility or overall toughness but reduce tensile strength and modulus.

45.13 SHRINKAGE

Only one datum on shrinkage exits in the public domain. Mold shrinkage measured by ASTM Method D955 is 0.005–0.025 in./in. All other data are the proprietary information of individual molders.

45.14 TRADE NAMES

Saran and Saran Wrap, The Dow Chemical Company, Midland, Mich.; Diofan, BASF, Federal Republic of Germany; Daran, W.R. Grace & Co., Memphis, Tenn.; Haloflex, Imperial Chemical Industries, Ltd., Milbank, London,; Serfene, the Morton Chemical Company, Chicago, Ill.; Polidene, Scott Bader & Co., Ltd., Wollaston, Willingborough, Northamptonshire, UK; Ixan, Solvay, Belgium; and AMSCO RES, Union Oil Company.

Copolymers are also available in Japan from the Asahi Chemical Company and Kureha Chemical Industries.

BIBLIOGRAPHY

General References

R.A. Wessling, *Polyvinylidene Chloride,* Gordon and Breach, New York, 1977.

D.S. Gibbs and R.A. Wessling, "Vinylidene Chloride and Poly(vinylidene chloride)", in M. Grayson, ed., *Kirk-Othmer Encyclopedia of Chemical Technology,* Vol. 23, 3rd ed., Wiley-Interscience, New York, 1983.

SARAN Resins, Monograph 190-289-79, Dow Chemical U.S.A., Midland, Mich., 1979.

R.A. Wessling, D.S. Gibbs, P.T. DeLassus, and B.A. Howell, "Vinylidene Chloride Polymers," in J. Kroschwitz, ed., *Encyclopedia of Polymer Science and Engineering,* 2nd ed., Vol. 17, Wiley-Interscience, New York, 1989.

46

RUBBERY STYRENIC BLOCK COPOLYMERS

Geoffrey Holden

Shell Development Co., Houston, TX 77251

46.1 INTRODUCTION

Styrenic rubber block copolymers combine the properties of thermoplasticity and rubberlike behavior. They are based on block copolymers having a rubber center segment and end segments of polystyrene. Examples are polystyrene-polybutadiene-polystyrene (S-B-S), polystyrene-polyisoprene-polystyrene (S-I-S), and polystyrene-poly(ethylene-*co*-butylene)-polystyrene (S-EB-S).

There are four different monomer structures in these block copolymers, represented as:

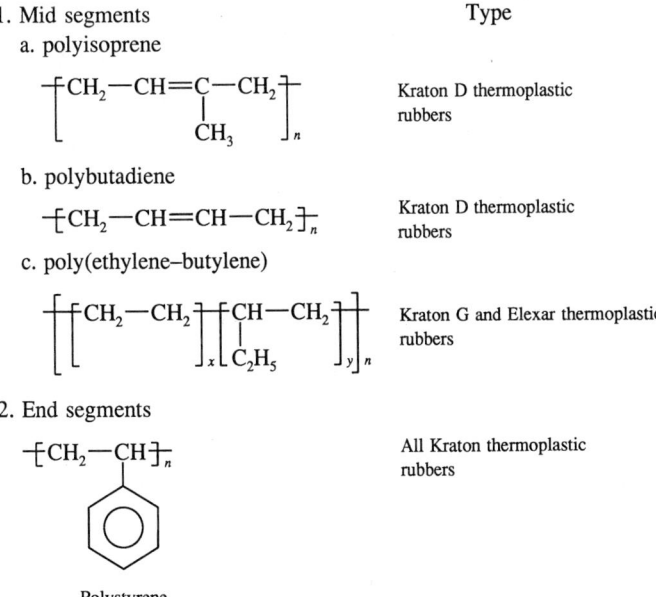

1. Mid segments
   a. polyisoprene   Kraton D thermoplastic rubbers
   b. polybutadiene   Kraton D thermoplastic rubbers
   c. poly(ethylene–butylene)   Kraton G and Elexar thermoplastic rubbers

2. End segments   All Kraton thermoplastic rubbers

Polystyrene

Type

Kraton and Elexar thermoplastic rubbers are registered trade marks of Shell Chemical Company.

TABLE 46.1.   Classification of Styrenic Block Copolymers

| Type | Competitive Materials | |
	Thermosets	Thermoplastics
Rigid	Epoxy, melamine-formaldehyde, SMC	Polypropylene, polystyrene, high density polyethylene
Flexible	Highly vulcanized rubber	Low density polyethylene, ethyl vinyl acetate, plasticized poly(vinyl chloride)
Rubbery	Vulcanized rubber	Kraton and Elexar thermoplastic rubbers

46.2   CATEGORY

The properties of these block copolymers can best be appreciated when they are compared to those other polymers, as in Table 46.1. Polymers are classified by the method used to process them (thermoset or thermoplastic) and their physical properties (hard, flexible, or rubbery). Six classes result. Five have been known for some time; styrenic block copolymers are representative of the sixth.

The structure of these block copolymers gives this unusual combination of properties. The essential feature is that the polystyrene and segment are incompatible with the rubbery center segments and form separate regions or domains. At room temperature, these polystyrene domains are hard and act as physical cross links. That is, they tie the rubber chains together into a network resembling that of conventional vulcanized rubber. At higher temperatures, the domains soften and the material can flow under stress. This behavior is reversible (unlike the chemical cross-linking in conventional vulcanized rubbers), and is the most important characteristic of these block copolymers.

The effects of varying the relative molecular weights and segmental arrangements of these block copolymers have been described in detail.[1] Only two such effects will be mentioned here. First, as the polystyrene content is increased, block copolymers of this type lose their rubbery properties and eventually become hard thermoplastics. Second, block copolymers with structures such as B-S-B or S-B cannot form a network of elastomer chains and so do not have much strength or elasticity.

46.3   HISTORY

Although block copolymers have known since at least 1946[2], the S-B-S and S-I-S block copolymers with the properties of thermoplastic elastomers were first produced on a laboratory scale in 1961.[3] Their novel and useful properties were quickly appreciated, and products based on them were commercially introduced by Shell Chemical Co. in 1965 under the trade name Kraton D thermoplastic rubbers. S-EB-S block copolymers are improved materials, with a saturated elastomer in the center segment; they were commercially introduced in 1972 under the trade name Kraton G and later Elexar. The history of these and similar materials has been reviewed elsewhere in more detail.[4]

Estimated U.S. neat production (that is, the production excluding compounding ingredients) was only 15 million lb/year in 1970; by 1980, it had reached 105 million lb/year (47,000 metric tons). Production is projected to continue to grow at the rate of 16–18%/year.

46.4   POLYMERIZATION

The production of styrenic block copolymers depends on the process of anionic block copolymerization[5] and has been recently reviewed in the literature.[4,6-7] Briefly, an alkyl

TABLE 46.2. Compounding Styrenic Block Polymers, Effects of Components on Compound Properties

Component	Hardness	Processability	Ozone Resistance	Cost	Other
Oil	Decreases	Increases	None	Decreases	Decreases UV resistance
Polystyrene	Increases	Increases	Some increase	Decreases	
Polyethylene	Increases	Variable	Increases	Decreases	Often gives satin finish
Polypropylene	Increases	Variable	Increases	Decreases	Improves high-temperature properties
EVA	Small increase	Variable	Increases	Decreases	
Fillers	Some increase	Variable	None	Decreases	Often improves surface appearance

lithium initiates the polymerization of a monomer (A), which can in turn initiate polymerization of a second monomer (B) to give a block copolymer A-B, and so forth.

Anionic polymerization is limited to three common monomers: styrene, butadiene, and isoprene. Therefore, only two useful A-B-A block copolymers (S-B-S and S-I-S) can be produced directly. In both cases the elastomer segments contains one double bond per molecule of original monomer. These bonds are reactive and limit the stability of the product. To improve this, butadiene monomer may be polymerized in two structural forms, which, on hydrogenation, give essentially a copolymer of ethylene and butylene (EB). This is a fully saturated rubber, and therefore much more stable.

Styrenic block copolymers are never used in the pure form but are compounded with significant amounts of other materials, such as oils, polymers, and fillers. The final block copolymer content is almost always less than 50 wt %; details of the effect of various compounding ingredients are given in Table 46.2, which shows the ingredients used to give rubbery thermoplastics. Modification of existing thermoplastics is another use and compounding with resins, oils, and other materials yield adhesive products.

Current (1987) prices of S-B-S polymers are about 95¢/lb. S-I-S polymers are slightly higher (about $1.05/lb) because of the higher price of isoprene monomer. S-EB-S polymers are more costly (about $1.90/lb) because of the expense of hydrogenation.

46.5 DESCRIPTION OF PROPERTIES

S-I-S polymers are relatively soft and generally reserved for adhesives. S-B-S and S-EB-S polymers are tough, rubbery materials. The pure styrenic block copolymers are blended to give compounded products, most of which contain a rubber processing oil. For this reason, S-B-S and S-EB-S polymers are produced both as pure materials and also as blends with rubber processing oils.

46.5.1 Properties of Compounded Products

Compounds with a wide range of properties are also available. Hardness values range from a very soft Shore A 35 footwear material to a stiff automotive product with Shore D hardness of 55. Similarly, specific gravities range from 0.94 to 1.94, the latter product being intended for sound deadening. Tensile strengths vary from about 1000 to 3000 psi (7 to 20 MPa) and elongations from 300 to 1000%. Altogether, some 25 different grades of compounded products are available, many of them designed for specific end uses.

46.6 APPLICATIONS

46.6.1 Compounded Products

Some of the more important applications for compounded styrenic block copolymers are

- General-purpose molded goods, such as grommets, swim fins, and soap dishes.
- Pharmaceutical and food-contact applications, including milk tubing, catheters, intravenous and blood tubing and bags, syringe bulbs, and rubber tips for spatulas; the absence of vulcanization residues is particularly valuable in these products.
- Footwear (both direct-molded footwear as well as the soles that are adhesively bonded to the upper); the final products show high coefficient of friction.
- Automotive compounds designed to retain their shape when passed through automotive bake ovens; grades with specific end-use properties, such as paintability, flame retardance, and sound deadening, are also available.
- Wire and cable compounds designed to meet various Underwriters Laboratory specifications; some are flame retardant.

46.6.2 Polymer Blending

S-B-S block copolymers, particularly Kraton D1102, can be blended with such thermoplastics as general-purpose polystyrene, high impact polystyrene, polypropylene, high density polyethylene, and low density polyethylene. Two new materials (Kraton D1300 and Kraton G1855) are intended as additives in the production of sheet molding compound.

S-EB-S polymers can be blended with such high-performance engineering thermoplastic resins as poly(butylene terephthalate), poly(ethylene terephthalate), polycarbonate, and nylon.

46.6.3 Adhesives and Sealants

S-I-S block copolymers, particularly Kraton D1107 and D1111, are used to manufacture pressure-sensitive adhesives. In these applications they are combined with tackifying resins compatible with the center polyisoprene segments. S-EB-S block copolymers, such as Kraton G1657, are also used and have the additional advantage of good stability. Other combinations can be used to make assembly adhesives and sealants.[8]

46.7 ADVANTAGES/DISADVANTAGES

46.7.1 Compounded Products

Conventional rubber vulcanate must be mixed with cross-linking agents such as sulfur and cured for about 20 minutes at about 30°F (150°C) to develop useful properties. The scrap produced cannot easily be reused and vulcanization residues can give such problems as taste transfer. In contrast, compounded styrenic block copolymers can be used without further mixing or curing. Cycle times on injection molding machines are short (typically less than two minutes). Scrap is reusable and there are no vulcanization residues.

Because these compounded products are thermoplastic, they cannot be used in applications where resistance to stress at high temperatures is critical; automobile tires, for example. However, in other applications, they can often replace conventional vulcanizates at a considerable savings.

46.7.2 Polymer Blending

The basic advantage of blending these rubbery block copolymers into other thermoplastics is improved impact resistance. It is also sometimes possible to make polymers alloys from otherwise incompatible polymer pairs, such as polypropylene and polycarbonate[9] or poly-ethylene and polystyrene.[10]

46.7.3 Adhesives and Sealants

Traditionally, pressure-sensitive adhesives were applied as solutions. Solvents were then evaporated to give the final product. Such products are costly and introduce problems of air pollution and safety. Hot-melt application is now increasingly used to overcome these drawbacks; products based on styrenic block copolymers S-I-S and S-EB-S are particularly suitable. Even in solution applications, the low molecular weight of styrenic block copolymers allows higher polymer concentrations to be used, so reducing solvent consumption.

46.8 PROCESSING TECHNIQUES

The three most common processing techniques for compounded products and polymer blends are injection molding, blow molding, and extrusion. Thermoforming is practicable but less

Master Material Outline

PROCESSABILITY OF _____ Rubbery Styrenic Block Copolymers _____

PROCESS	A	B	C
INJECTION MOLDING	X		
EXTRUSION	X		
THERMOFORMING		X	
FOAM MOLDING			
DIP, SLUSH MOLDING			
ROTATIONAL MOLDING			
POWDER, FLUIDIZED-BED COATING			
REINFORCED THERMOSET MOLDING			
COMPRESSION/TRANSFER MOLDING			
REACTION INJECTION MOLDING (RIM)			
MECHANICAL FORMING			
CASTING			

A = Common processing technique.
B = Technique possible with difficulty.
C = Used only in special circumstances, if ever.

Standard design chart for __Rubbery Styrenic Block Copolymers__

1. Recommended wall thickness/length of flow, mm
   Minimum _____ for _____ distance
   Maximum _____ for _____ distance
   Ideal   _____ for _____ distance

2. Allowable wall thickness variation, % of nominal wall _____

3. Radius requirements—None, within limits of particle mold design
   Outside: Minimum _____ Maximum _____
   Inside: Minimum _____ Maximum _____

4. Reinforcing ribs
   Maximum thickness __$\frac{1}{2}$ of wall thickness__
   Maximum height __10 times wall thickness__
   Sink marks: Yes __X__ No _____

5. Solid pegs and bosses
   Maximum thickness _____
   Maximum weight _____
   Sink marks: Yes __X__ No _____

6. Strippable undercuts
   Outside: Yes __X__ No _____
   Inside: Yes __X__ No _____

7. Hollow bosses: Yes __X__ No _____
                             yes for rigid materials
                             no for flexible materials

8. Draft angles
   Outside: Minimum __$2\frac{1}{2}°$__ Ideal __$3°$__
   Inside: Minimum __$2\frac{1}{2}°$__ Ideal __$3°$__

9. Molded-in inserts: Yes __X__ must have mechanical application
                                 No _____

10. Size limitations—none, within limits of typical application
    Length: Minimum _____ Maximum _____ Process _____
    Width: Minimum _____ Maximum _____ Process _____

11. Tolerances

12. Special process- and materials-related design details
    Blow-up ratio (blow molding) __$3:1$__
    Core L/D (injection and compression molding) _____
    Depth of draw (thermoforming) _____
    Other _____

common. Injection molded or extruded parts can be foamed by use of hydrazide blowing agents such as the Celogen series available from Uniroyal. Adhesives can be processed either as hot melts or solutions.

46.9   RESIN FORMS

The pure and oiled styrenic block copolymers are available as crumb [irregular particles with an average size of about 0.4 in (1 cm)] powder and pellets. Compounded products are available only as pellets. Most of the compounded products contain mineral filler, usually calcium carbonate or clay.

46.10   SPECIFICATION OF PROPERTIES

Properties of representative pure and oiled styrenic block copolymers are given in Table 46.3 and 46.4. Properties of representative compounded products are given in Table 46.5.

MATERIAL Rubbery Styrenic Block Copolymers (Kraton G and Elexar)

PROPERTY	TEST METHOD	ENGLISH UNITS	VALUE	METRIC UNITS	VALUE
MECHANICAL DENSITY	D792	lb/ft^3	59–115	g/cm^3	0.95–1.84
TENSILE STRENGTH	D638	psi	1,500–2,900	MPa	10–20
TENSILE MODULUS	D638	psi	1,000–2,500	MPa	7–18
FLEXURAL MODULUS	D790	psi	1,500–4,000	MPa	10–28
ELONGATION TO BREAK	D638	%	450–830	%	450–830
NOTCHED IZOD AT ROOM TEMP	D256	ft-lb/in.	No break	J/m	No break
HARDNESS	D785	Shore A	30–95	Shore A	30–95
THERMAL DEFLECTION T @ 264 psi	D648	°F		°C	
DEFLECTION T @ 66 psi	D648	°F		°C	
VICAT SOFT-ENING POINT	D1525	°F		°C	
UL TEMP INDEX	UL746B	°F	221–257	°C	105–125
UL FLAMMABILITY CODE RATING	UL94		HB-V-0		HB-V-0
LINEAR COEFFICIENT THERMAL EXPANSION	D696	in./in./°F	0.000 083–0.000 13	mm/mm/°C	0.000 15–0.000 25
ENVIRONMENTAL WATER ABSORPTION 24 HOUR	D570	%	0.5–2.0	%	0.5–2.0
CLARITY	D1003	% TRANS-MISSION	Opaque	% TRANS-MISSION	Opaque
OUTDOOR WEATHERING	D1435	%	Good	%	Good
FDA APPROVAL			No		No
CHEMICAL RESISTANCE TO: WEAK ACID	D543		Not attacked		Not attacked
STRONG ACID	D543		Minimally attacked		Minimally attacked
WEAK ALKALI	D543		Not attacked		Not attacked
STRONG ALKALI	D543		Not attacked		Not attacked
LOW MOLECULAR WEIGHT SOLVENTS	D543		Badly attacked		Badly attacked
ALCOHOLS	D543		Not attacked		Not attacked
ELECTRICAL DIELECTRIC STRENGTH	D149	V/mil	500–900	kV/mm	20–35
DIELECTRIC CONSTANT	D150	10^6 Hertz	2.0–3.0		2.0–3.0
POWER FACTOR	D150	10^6 Hertz	<0.2–4		<0.2–4
OTHER MELTING POINT	D3418	°F		°C	
GLASS TRANS-ITION TEMP.	D3418	°F		°C	

581

MATERIAL ___ Rubbery Styrenic Block Copolymers (Kraton D)

PROPERTY	TEST METHOD	ENGLISH UNITS	VALUE	METRIC UNITS	VALUE
MECHANICAL DENSITY	D792	lb/ft^3	58–69	g/cm^3	0.93–1.1
TENSILE STRENGTH	D638	psi	500–1,000	MPa	3.5–7
TENSILE MODULUS	D638	psi	400–1,000	MPa	2.5–7
FLEXURAL MODULUS	D790	psi	500–2,000	MPa	3.5–15
ELONGATION TO BREAK	D638	%	300–1,300	%	300–1,300
NOTCHED IZOD AT ROOM TEMP	D256	ft-lb/in.	No break	J/m	No break
HARDNESS	D785	Shore A	40–80		40–80
THERMAL DEFLECTION T @ 264 psi	D648	°F		°C	
DEFLECTION T @ 66 psi	D648	°F		°C	
VICAT SOFT-ENING POINT	D1525	°F		°C	
UL TEMP INDEX	UL746B	°F	140 (max)	°C	60 (max)
UL FLAMMABILITY CODE RATING	UL94		Slow burning		Slow burning
LINEAR COEFFICIENT THERMAL EXPANSION	D696	in./in./°F	0.000 083–0.000 139	mm/mm/°C	0.000 15–0.000 25
ENVIRONMENTAL WATER ABSORPTION 24 HOUR	D570	%	0.2–0.4	%	0.2–0.4
CLARITY	D1003	% TRANS-MISSION	Opaque	% TRANS-MISSION	Opaque
OUTDOOR WEATHERING	D1435	%	Degrades with time	%	Degrades with time
FDA APPROVAL			Some grades		Some grades
CHEMICAL RESISTANCE TO: WEAK ACID	D543		Not attacked		Not attacked
STRONG ACID	D543		Minimally attacked		Minimally attacked
WEAK ALKALI	D543		Not attacked		Not attacked
STRONG ALKALI	D543		Minimally attacked		Minimally attacked
LOW MOLECULAR WEIGHT SOLVENTS	D543		Badly attacked to soluble		Badly attacked to soluble
ALCOHOLS	D543		Minimally attacked		Minimally attacked
ELECTRICAL DIELECTRIC STRENGTH	D149	V/mil	500–900	kV/mm	20–35
DIELECTRIC CONSTANT	D150	10^6 Hertz	2–3		2–3
POWER FACTOR	D150	10^6 Hertz	0.2–4		0.2–4
OTHER MELTING POINT	D3418	°F		°C	
GLASS TRANS-ITION TEMP.	D3418	°F		°C	

TABLE 46.3. Properties of Pure Styrenic Block Copolymers

Property	Kraton D1101 (S-B-S)	Kraton D1102 (S-B-S)	Kraton D1107 (S-I-S)	Kraton G1652 (S-EB-S)
Styrene/rubber ratio	30/70	28/72	14/86	29/71
Hardness, Shore A	71	71	37	75
Tensile strength, MPa	32	32	21	32
(psi)	(4600)	(4600)	(3100)	(4600)
300% modulus, MPa	2.8	2.8	0.7	4.8
(psi)	(400)	(400)	(100)	(700)
Elongation, %	880	880	1300	500
Specific gravity	0.94	0.94	0.92	0.91
Melt flow, condition G, g/10 min	<1	6	9	

These data illustrate the fact that styrenic block copolymers are elastomers, and so must be described by properties relevant to rubbery products (for example, 300% modulus and Shore hardness). Detailed properties of typical grades are given in the Master Outline of Materials Properties, with data expressed in English and SI units.

Pure S-B-S and S-I-S block copolymers are numbered Kraton D11xy (where xy is a two-digit identifier for the individual grades). S-EB-S analogues are numbered Kraton G16xy. Similar oiled products are numbered Kraton D41xy, Kraton D42xy, and Kraton G46xy. Precompounded products for specific end uses are subdivided into several series, depending on their end uses. Kraton D21xy and Kraton G27xy products are intended for medical applications or those involving food contact. Kraton D32xy is a general-purpose series. Kraton D51xy and Kraton D52xy are used to mold shoe soles. The Kraton G7xyz series is intended for automotive applications. The final series, Elexar 8xyz, uses a different trade name to differentiate the products for their specific end use in electrical insulation.

46.11 PROCESSING REQUIREMENTS

46.11.1 Compounded Products

The compounded products are processed in conventional injection molds, extruders, blow molders and thermoformers. As a generalizaton, compounds based on S-B-S (Kraton D)

TABLE 46.4. Properties of Styrenic Block Copolymers Blended with Oils

Property	Kraton D141 (S-B-S)	Kraton D4240 (S-B-S)
Styrene/rubber ratio (pure unoiled polymer)	30/70	44/56
Oil content, wt %	29	46
Hardness, Shore A	47	46
Tensile strength, MPa	19	18
(psi)	(2750)	(1550)
300% modulus, MPa	1.7	1.4
(psi)	(250)	(200)
Elongation, %	1300	1200
Melt flow, condition G (g/10 min)	11	30

TABLE 46.5. Properties of Compounded Styrenic Block Copolymers

Properties	Kraton D2109 (Milk Tubing)	Kraton G2705 (Medical)	Kraton D3202 (General)	Kraton D5119 (Footwear, Direct-molded)	Kraton D5122 (Footwear, Direct-molded)	Kraton D5250 (Footwear, Unit Sole)	Kraton G7150 (Sound Deadening)	Kraton G7720 (Automotive, Soft Parts)	Kraton G7680 (Automotive, Blow Molding)	Elexar 8451 (Wire and Cable)	Elexar 8614 (Wire and Cable, Fire Retardant)
Hardness, Shore A or D	48(A)	55(A)	50(A)	55(A)	65(A)	80(A)	39(D)	60(A)	58(D)	82(A)	48(D)
Tensile strength, MPa	11.1	6.0	5.6	4	6.2	4.8	5.3	5.2	24	14	12
(psi)	(1600)	(1850)	(800)	(580)	(900)	(700)	(750)	(750)	(3500)	(2000)	(1750)
300% modulus, MPa	2.0	2.8	4.1	3.2	6.1	3.2	4.2	2.4	12.4	6.5	7.3
(psi)	(300)	(400)	(600)	(460)	(880)	(450)	(600)	(350)	(1800)	(950)	(1050)
Elongation, %	1200	700	400	380	325	465	450	700	650	650	500
Melt flow, g/10 min (condition E or G)	15(G)		14(E)	27(E)	26(E)	23(E)				22(G)	3(G)
Specific gravity, g/cm³	0.94	0.9	1.0	1.09	1.08	1.10	1.84	1.20	1.03	1.0	1.2
(lb/ft³)	(58.6)	(56.1)	(62.4)	(68.0)	(67.4)	(68.6)	(121)	(74.8)	(64.2)	(62.4)	(74.8)

can be processed under conditions similar to those used for impact polystyrene; compounds based on S-EB-S (Kraton G) can be processed under conditions similar to those used for polypropylene. Blowing agents can be added to produce foamed products. In footwear manufacture, special low-pressure machines are used; in some cases a movable sole plate is required for blown products. Details of recommended processing conditions are given in Table 46.6.

The melt temperature for Kraton D compounds should be kept below 445°F (230°C), and it is not advisable to leave hot material in the barrel of a machine. However, even if decomposition occurs, the products are neither toxic nor corrosive and can be purged from the machine with polystyrene. Kraton G and Elexar rubber compounds are more stable. Most of them should present no problems with degradation, but fire retardant grades can degrade if overheated. Ground scrap from both types of compound is reusable, but grinding of softer products can be difficult. Problems will be reduced if it is realized that rubbery materials must be cut rather than shattered. Thus, grinder blades should be sharp, and clearance between rotating and fixed blades should be minimized.

TABLE 46.6. Typical Processing Conditions for Compounds Based on Styrenic Block Copolymers

Process and Condition	Value S-B-S (Kraton D)	S-EB-S (Kraton G)
Injection Molding		
Feed end, cylinder T, °C	175	210
(°F)	(350)	(410)
Nozzle end cylinder T, °C	195	215
(°F)	(385)	(420)
Nozzle T, °C	200	225
(°F)	(390)	(440)
Mold T, °C	25	30
(°F)	(75)	(90)
Injection pressure, MPa	3.5	3.5
(psi)	(500)	(500)
Injection rate	Slow to moderate	Fast
Back pressure, Mpa	0.5	0.6
(psi)	(80)	(90)
Clamp pressure per unit of projected molding area		
(t/in.^{2})	2	2
(kg/cm^2)	(140)	(140)
Screw speed, rpm	30	40
Cycle time, 1-mm thick,		
(s) (0.04 in.)	20	17
Extrusion		
Screw length:diameter ratio	24:1	24:1
Screw compression ratio	2.5:1	3.0:1
Feed end cylinder T, °C	150	205
(°F)	(300)	(400)
Die end cylinder T, °C	180	225
(°F)	(360)	(435)
Die T, °C	190	235
(°F)	(380)	(455)

46.11.2 Polymer Blending

Blending can often take place on the apparatus used to make the product. Thus, blends of S-B-S and other thermoplastics can be made by co-feeding the two materials into an injection-molding machine set up with a reasonably high back pressure [at least 100 psi (0.7 MPa)]. For extruded products, blending can be done in the extruder barrel, particularly if a mixing screw is used. S-EB-S blends with engineering thermoplastics may require the more intensive mixing of a twin-screw extruder.

46.11.3 Adhesives and Sealants

Adhesives and sealants can be produced on conventional mixing equipment. For hot melt adhesives, either a batch mixer with a high shear zone or a twin screw mixer can be used. Hot melt adhesives based on S-B-S or S-I-S are best produced and stored (in a hot melt tank) under a nitrogen blanket. S-EB-S based analogues are more stable. All these adhesives can be coated onto the substrates at line speeds up to 820 ft/min (250 m/min) using a slot die coater.

46.12 PROCESSING-SENSITIVE END PROPERTIES

The end properties of compounded styrenic block copolymers are sensitive both to processing conditions and processing equipment. They become harder and stiffer as shear is increased. Thus, it is important to produce test samples on equipment and under production conditions used to produce the final part. Misleading results will be obtained if, for example, test or prototype parts are compression molded when the final part is to be injection molded.

46.13 SHRINKAGE

Compared to most thermoplastics, compounded styrenic block copolymers have low mold shrinkage (see Mold Shrinkage). This, combined with the high friction characteristic of rubbers, can give ejection problems with softer products. However, these can usually be avoided if the sides of the mold are tapered. Coating the mold surface with a fluoroplastic helps, as do mold release agents and air ejection and stripper rings. Small-diameter injection pins are not recommended with softer products, since the part will deform rather than eject.

Mold Shrinkage Characteristics

MATERIAL ──── Rubbery Styrenic Block Copolymers

IN THE DIRECTION OF FLOW

THICKNESS

MATERIAL	mm: 1.5	3	6	8	FROM	TO	#
	in.: 1/16	1/8	1/4	1/2			
Kraton D (S-B-S)					0.001	0.005	
Kraton G (S-EB-S) and Elexar					0.005	0.020	

	PERPENDICULAR TO THE DIRECTION OF FLOW						
Kraton D (S-B-S)					0.002	0.010	
Kraton G (S-EB-S) and Elexar					0.010	0.030	

NOTE: Mold shrinkage is approximate. It is affected by part design, mold temperature, thickness, injection pressure, packing time, cycle time, orientation, gate design, gate size, gate location, glass content, glass size, and filler content.

Standard Tolerance Chart

PROCESS	Injection molding	MATERIAL	Rubbery Styrenic Block Copolymers

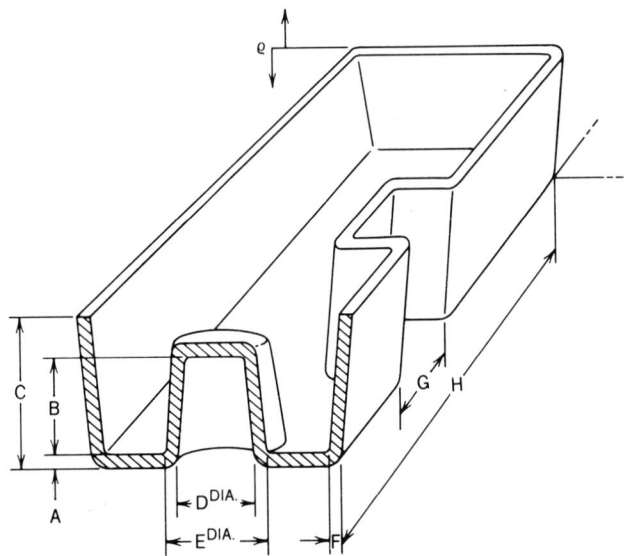

Values are based on a nominal 1/8 in. (3.2 mm) wall thickness and given in plus or minus in./in. and mm/mm.

FINE tolerance – Special care and added cost.
COMMERCIAL tolerance – Normal care required.
COARSE tolerance – Minimal care required, lowest cost.

Tolerance	A	B	C	D	E	F	G	H	
FINE	0.010					0.010			in./in.
	0.250					0.250			mm/mm
COMMERCIAL		0.020	0.020	0.020			0.020	0.020	in./in.
		0.500	0.500	0.500			0.500	0.500	mm/mm
COARSE					0.020				in./in.
					0.500				mm/mm

46.14 TRADE NAMES

In the United States and Canada, the chief trade names are Kraton D, Kraton G, and Elexar (products of Shell Chemical Co.). These and other trade names used in the rest of the world, together with the respective manufacturers, appear below:

- Kraton D (S-B-S and S-1-S), Kraton G (S-EB-S), Elexar (S-EB-S); Shell Chemical; U.S.
- Kraton (S-B-S), Cariflex (S-B-S and S-I-S); Royal Dutch/Shell and associated companies; Federal Republic of Germany and France
- Solprene 400 (S-B-S); Phillips and associated companies; Belgium, Spain, Australia, Mexico
- Europrene Sol T (S-B-S and S-1-S); Enichem; Italy
- Tufprene and Asaprene (S-B-S); Asahi Chemical; Japan
- Stereon (S-B-S); Firestone; U.S.

46.15 INFORMATION SOURCES

References 4, 11, 13, 14, and 15 give an overview of the subject. Others (particularly refs. 12 and 13) provide information on nonstyrenic block copolymers (such as thermoplastic polyurethanes and thermoplastic elastomers not based on block copolymers). References 1, 4, and 13 give a broad description of the theory which explains these properties; reference 12 is a handbook describing the properties of individual grades in detail.

Further details on the properties, processing, and applications of styrenic block copolymers can be obtained from the manufacturers listed above.

BIBLIOGRAPHY

1. G. Holden and co-workers, "Thermoplastic Elastomers," *J. Polym. Sci.* **C-26,** 37 (1969).
2. J.H. Baxendale and co-workers, "Mechanism and Kinetics of Initiation of Polymerization by Stems Containing Hydrogen Peroxide," *Trans. Faraday Soc.* **42,** 155 (1946).
3. U.S. Pat. 3,265,765 (1965) G. Holden and R. Milkovich (to Shell Oil Co.)
4. N.R. Legge and co-workers, "Chemistry and Technology of Block Copolymers," in R.W. Tess and G.W. Poehlein, eds., *Applied Polymer Science,* 2nd ed., American Chemical Society, Washington, D.C., 1985.
5. M. Szwarc and co-workers, "Polymerization Initiated by Electron Transfer to Monomer: A New Method of Formation of Block Polymers," *J. Am. Chem. Soc.* **78,** 2656 (1956).
6. J.E. McGrath, ed., *Anionic Polymerization: Kinetics, Mechanism, and Synthesis,* ACS Symposium Series 166, American Chemical Society, Washington, D.C., 1981.
7. P. Dreyfuss and co-workers, "Elastomeric Block Copolymers," *Rubber Chem. Technol.* **53,** 738 (1980).
8. D.J. St. Clair, "Rubber Styrene Block Copolymers in Adhesive," *Rubber Chem. Technol.* **54,** 208 (1982).
9. G. Holden, "Current Applications of Styrenic Block Copolymer Rubbers." *J. Elastomers Plast.* **14,** 148 (1982).
10. C.R. Lindsey and co-workers, "Mechanical Properties of HDPE-PS-SEBS Blends." *J. Appl. Polym. Sci.* **26** (1981).
11. G. Holden, "Properties and Applications of Elastomeric Block Copolymers," in R.J. Ceresa, ed., *Block and Graft Copolymerization,* John Wiley & Sons, Inc., New York, 1973.

12. B.M. Walker and C.P. Rader, eds., *Handbook of Thermoplastics,* 2nd ed., Van Nostrand Reinhold, New York, 1988.

13. N.R. Legge, G. Holden and H.E. Schroeder, eds., *Thermoplastic Elastomers—A Comprehensive Review,* Hanser, Munich, Federal Republic of Germany, 1987.

14. J.F. Auchter and G. Holden, "Thermoplastic Elastomers," *1983–84 Modern Plastics Encyclopedia,* McGraw-Hill, New York, 1983. Similar articles appear in other editions.

15. A.D. Thorn, "Thermoplastic Elastomers—A Review of Current Information," Rubber and Plastics Research Assn. of Great Britain, Shawbury, Shropshire SY44NR, UK, 1980.

47

STYRENE MALEIC ANHYDRIDE (SMA)

William J. Hall

Technical Leader, Monsanto Co., Springfield, MA

47.1 INTRODUCTION

Styrene maleic anhydride (SMA) thermoplastics are available as copolymers and terpolymer alloys. The copolymer form is the product of mass copolymerization of styrene and maleic anhydride; rubber impact modifier may be incorporated. The terpolymer alloys incorporate acrylonitrile–butadiene–styrene type components. All grades are rated as impact resistant. Following is a representation of the SMA copolymer:

47.2 CATEGORY

Styrene maleic anhydride materials are thermoplastics, characterized in their modified form by hardness and toughness, stiffness, and brittleness.

47.3 HISTORY

Although maleic anhydride was first prepared early in the 19th century, it was not commercially introduced until the early 1930s, when it was made available by National Aniline and Monsanto Co. Its free-radical equimolar copolymerization with styrene monomer was investigated extensively by Alfrey and Lavin[1] in the early 1940s. The resulting copolymers were commercialized in the 1960s by Sinclair Petrochemicals. Today, the principal producers of 1:1 copolymers are Arco Chemical Co. and Monsanto Co.

Important are the high molecular-weight nonequimolar copolymers of styrene maleic anhydride, which were also studied by Alfrey and Lavin. Dylark styrene copolymer by Arco was first used in automotive dashboards in 1972. In 1981, Monsanto introduced its Cadon engineering thermoplastic resins, which are alloy blends of SMA terpolymer and ABS-type terpolymer components. In early 1983, Arco introduced a new series of SMA/polycarbonate alloys under the Arloy trademark.

47.4   POLYMERIZATION

SMA copolymers and terpolymers are usually made by mass copolymerization using a free-radical initiator. There is a strong tendency to form 1:1 equimolar copolymer unless the maleic anhydride (MA) concentration is held at an extremely low level during the entire polymerization sequence. This starve feeding technique is used to make SMA resins containing 5–12% MA.

Impact-modified versions can be produced by copolymerizing these monomers in the presence of polybutadiene. These polymers are sold under the Dylark trademark. However, Monsanto's Cadon resins and Arco's Arloy result from alloying SMA, modified or unmodified, with other thermoplastics.

The price of SMA (1989) is $1.50–2.15/lb in truckload quantities.

47.5   DESCRIPTION OF PROPERTIES

Compared with polystyrene and ABS, SMA polymers and their alloys offer higher heat resistance while maintaining good impact and rigidity. Compared with other engineering thermoplastics such as modified polyphenylene oxide and polycarbonate, SMA polymers offer much softer flow, ease of coloring, and improved chemical resistance.

Table 47.1 provides mechanical property ranges for the three major SMA products offered.

SMA polymers have heat-deflection temperatures ranging from about 204–248°F (96–120°C). For outdoor stability, protective surface coating or a UV-resistant capping layer is required.

Some of the SMA copolymers offered under the Dylark trademark are optically clear and light colored. In the alloy area, only Arloy 2000 is clear. The impact-modified SMA, SMA/ABS alloys, and SMA/BD/PC alloys are opaque and a light straw color.

The chemical resistance of Dylark resembles that of polystyrene; that of Cadon resin resembles ABS. SMA polymers and their alloys are not recommended in strong alkali environments such as strong detergents or some battery electrolytes. Arloy polymers exhibit chemical compatibility characteristics similar to polycarbonate.

TABLE 47.1.   Mechanical Property Range of Styrene Maleic Anhydride

Property	Arloy	Cadon	Dylark
Notched Izod impact, ft-lb/in. (J/m)	1.5–12 (80–640)	2–6 (108–231)	0.4–4 (22–216)
Tensile strength, psi (MPa)	6,500–9,000 (45–63)	5,200–6,000 (36–41)	4,800–7,400 (33–51)
Tensile elongation, (%)	4–80	30–40	1.8–20
Flexural modulus, psi (MPa)	319,000–450,000 (2200–3100)	319,000–348,000 (2200–2400)	319,000–490,000 (2200–3380)
Vicat softning point, 1 kg, °C	131–143	117–133	116–132

Both Cadon and Dylark are frequently reinforced with glass fibers. The affinity of MA to glass is excellent, resulting in good retention of impact strength and providing a substantial improvement in modulus and higher heat resistance.

47.6 APPLICATIONS

The first use of SMA copolymers was in automotive interiors. This is becoming a key application as the rake of the windshield is increased, placing more heat demands on instrument-panel support components. Both glass-reinforced and unreinforced products are finding increased opportunities. The former are used extensively in supporting padded instrument panels. The latter are being used in a variety of interior trim parts and exterior mirror housings.

Other applications for SMA include housewares, small appliances, power tools, steam curlers, audio cassettes, and so on. Flame-retardant V-0 grades are available and are being used in switch covers, modems, and a number of business-machine applications. Outdoor uses include door panels and solar collector components.

47.7 ADVANTAGES/DISADVANTAGES

The three major types of SMA polymers described have a broad range of characteristics. Generally speaking, their greatest advantage is good heat resistance without sacrificing processability. Because of their light color, both the clear and the opaque grades can be colored in a myriad of shades. The processability of Dylark grades, the chemical resistance of Cadon grades, and the extreme toughness of Arloy grades can be used to advantage.

As for disadvantages, none of the SMA family of products is particularly resistant to UV degradation, so some form of exposure protection is advised, as noted. Also, maleic anhydride is chemically reactive even in its copolymer form. It will react with alkalies to form salts and undergo imidization with amine compounds. The polymer will degrade in the 527–581°F (275–305°C) range, releasing some carbon dioxide gas. This will appear as splay marks on injection moldings.

47.8 PROCESSING TECHNIQUES

SMA polymers can be injection molded, extruded, and thermoformed in the conventional manner. Recommended melt temperatures for molding, as given in producers' product bulletins, are as follows: Dylark[2], 400–527°F (204–274°C); Cadon[3], 480–509°F (249–266°C); Arloy 1000[2] series, 530–580°F (277–304°C). Mold temperatures of 129–180°F (49–82°C) are recommended for Dylark and Cadon; 150–210°F (66–99°C) are stated for Arloy.

Arco states that Dylark can be extruded using polystyrene-designed equipment with stock temperatures in the 379–425°F (193–218°C) range. For Arloy, low-shear screws are recommended with melt temperatures in the 500–550°F (260–288°C) range. Monsanto recommends that Cadon resin be extruded and thermoformed using ABS-designed equipment and conditions.[4]

Master Material Outline

PROCESSABILITY OF _____ Styrene Maleic Anhydride _____

PROCESS	A	B	C
INJECTION MOLDING	X		
EXTRUSION	X		
THERMOFORMING	X		
FOAM MOLDING			X
DIP, SLUSH MOLDING			
ROTATIONAL MOLDING			
POWDER, FLUIDIZED-BED COATING			
REINFORCED THERMOSET MOLDING			
COMPRESSION/TRANSFER MOLDING			X
REACTION INJECTION MOLDING (RIM)			
MECHANICAL FORMING			X
CASTING			

A = Common processing technique.
B = Technique possible with difficulty.
C = Used only in special circumstances, if ever.

47.9 RESIN FORMS

The SMA polymers described here are provided as pellets, cubes, and crushed beads. Arco offers a preblended mixture of glass fibers and resin as well as a resin described as being suitable for glass blending in its series of Dylark copolymers. Cadon resin by Monsanto can be blended with glass at the machine and is also available precompounded from custom houses.

47.10 SPECIFICATION OF PROPERTIES

See Master Outline of Material Properties.

MATERIAL ___Styrene Maleic Anhydride___

PROPERTY	TEST METHOD	ENGLISH UNITS	VALUE	METRIC UNITS	VALUE
MECHANICAL DENSITY	D792	lb/ft^3	65.5–71.7	g/cm^3	1.05–1.15
TENSILE STRENGTH	D638	psi	4,790–9,140	MPa	33–63
TENSILE MODULUS	D638	psi	270,000–440,000	MPa	1,860–3,034
FLEXURAL MODULUS	D790	psi	319,000–490,000	MPa	2,200–3,380
ELONGATION TO BREAK	D638	%	Not applicable	%	Not applicable
NOTCHED IZOD AT ROOM TEMP	D256	ft-lb/in.	0.4–11.9	J/m	21.4–635
HARDNESS	D785	Rockwell	L75-108	Rockwell	L75-108
THERMAL DEFLECTION T @ 264 psi	D648	°F	205–259	°C	96–126
DEFLECTION T @ 66 psi	D648	°F	Not applicable	°C	Not applicable
VICAT SOFTENING POINT	D1525	°F	116–143	°C	241–290
UL TEMP INDEX	UL746B	°F	(Cadon) 203	°C	(Cadon) 95
UL FLAMMABILITY CODE RATING	UL94		HB-VO		HB-VO
LINEAR COEFFICIENT THERMAL EXPANSION	D696	in./in./°F	0.000 063–0.000 094	mm/mm/°C	0.000 035–0.000 052
ENVIRONMENTAL WATER ABSORPTION 24 HOUR	D570	%	0.1–0.25	%	0.1–0.25
CLARITY	D1003	% TRANSMISSION	Clear to opaque	% TRANSMISSION	Clear to opaque
OUTDOOR WEATHERING	D1435	%	Not applicable	%	Not applicable
FDA APPROVAL			Dylark copolymers Arloy 2000		Dylark copolymers Arloy 2000
CHEMICAL RESISTANCE TO: WEAK ACID	D543		Not attacked		Not attacked
STRONG ACID	D543		Minimally attacked		Minimally attacked
WEAK ALKALI	D543		Minimally attacked		Minimally attacked
STRONG ALKALI	D543		Badly attacked		Badly attacked
LOW MOLECULAR WEIGHT SOLVENTS	D543		Soluble		Soluble
ALCOHOLS	D543		Minimally attacked		Minimally attacked
ELECTRICAL DIELECTRIC STRENGTH	D149	V/mil	Cadon 420	kV/mm	Cadon 16.4
DIELECTRIC CONSTANT	D150	10^6 Hertz	Cadon 3.5		Cadon 3.5
POWER FACTOR	D150	10^6 Hertz	0.0145 (Dissipation factor)		0.0145 (Dissipation factor)
OTHER MELTING POINT	D3418	°F		°C	
GLASS TRANSITION TEMP.	D3418	°F		°C	

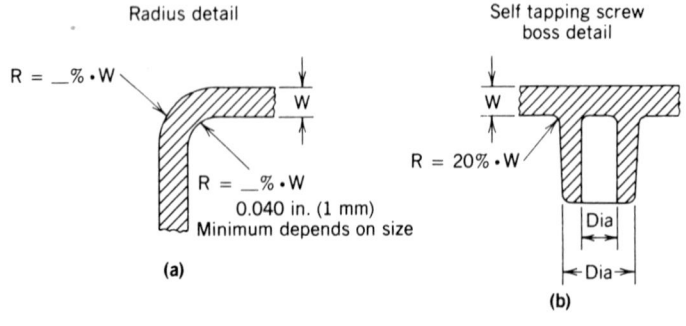

Radius detail

Self tapping screw boss detail

(a)

(b)

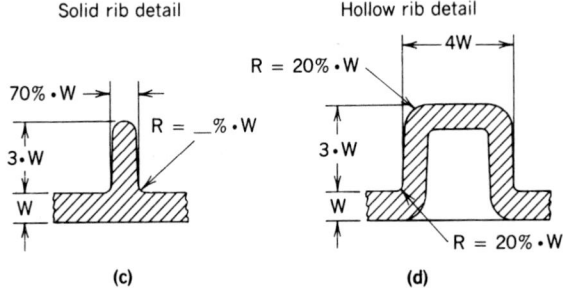

Solid rib detail

Hollow rib detail

(c)

(d)

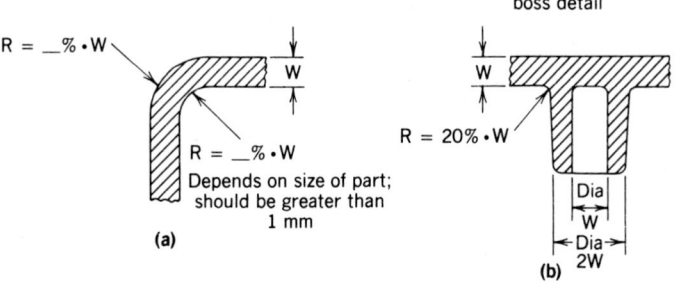

Radius detail

Self tapping screw boss detail

(a)

(b)

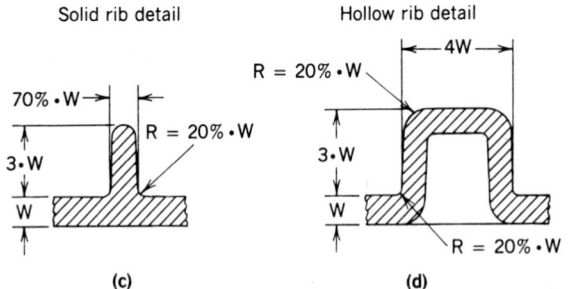

Solid rib detail

Hollow rib detail

(c)

(d)

Standard design chart for <u>**Styrene Maleic Anhydride**</u>

1. Recommended wall thickness/length of flow

Minimum	0.05 in.	for 10 in.	distance
Maximum	0.10 in.	for 10 in.	distance
Ideal		for	distance

2. Allowable wall thickness variation, % of nominal wall _____

3. Radius requirements (scheme **a**)
   Outside: Minimum_____0.04 in_____ Maximum _____
   Inside: Minimum _____0.04 in._____ Maximum _____

4. Reinforcing ribs (scheme **c** and **d**)
   Maximum thickness _____70% of width_____
   Maximum height _____3 × width_____
   Sink marks: Yes _____ No ___X___

5. Solid pegs and bosses
   Maximum thickness _____
   Maximum weight _____
   Sink marks: Yes _____ No ___X___

6. Strippable undercuts
   Outside: Yes _____ No _____
   Inside: Yes _____ No _____

7. Hollow bosses (scheme **b**): Yes ___X___ No _____

8. Draft angles
   Outside: Minimum _____ Ideal _____1°_____
   Inside: Minimum _____ Ideal _____1°_____

9. Molded-in inserts: Yes ___X___
                       No _____

10. Size limitations
    Length: Minimum _____ Maximum _____ Process _____
    Width: Minimum _____ Maximum _____ Process _____

11. Tolerances

12. Special process- and materials-related design details
    Blow-up ratio (blow molding) _____
    Core L/D (injection and compression molding) _____
    Depth of draw (thermoforming) _____
    Other _____

47.11 PROCESSING REQUIREMENTS

Drying is recommended before processing SMA polymers and their alloys. Two to three hours at 194–203°F (90–95°C) with dehumidified air is a good rule of thumb. This is less critical with Dylark than with Cadon, but very important with Arloy because of its polycarbonate component: A 219–230°F (104–110°C) air temperature is recommended.

Regrinds can be used up to 25% if they are of reasonably good quality and clean. SMA/PC alloys should be given special attention in terms of ensuring dry regrind, since moisture present in processing can have deleterious effects on properties. SMA polymers have varying compatibilities and should not be mixed unless recommended by the supplier.

47.12 PROCESSING-SENSITIVE END PROPERTIES

Higher roll-stack temperatures are necessary in order to polish adequately the surface of SMA polymer sheet products. SMA processes at low melt temperatures because of soft melt flow (see section 47.8).

Property and appearance enhancement can be achieved with various finishing techniques, including plating and/or metalizing, painting, and hot stamping.

47.13 SHRINKAGE

SMA polymers are amorphous and exhibit low mold-shrinkage values. These range from 0.005–0.006 in./in. for the series of Dylark resins, 0.007 in./in. for Arloy, and 0.006–0.009 in./in. for the Cadon resins. Inherent mold shrinkage increases with the rubber level, which accounts for the ranges given. The addition of glass fibers has the opposite effect, typically reducing mold shrinkage values by half. Although the values stated here are nominal, the

Standard Tolerance Chart

PROCESS	Injection molding	MATERIAL	Styrene Maleic Anhydride

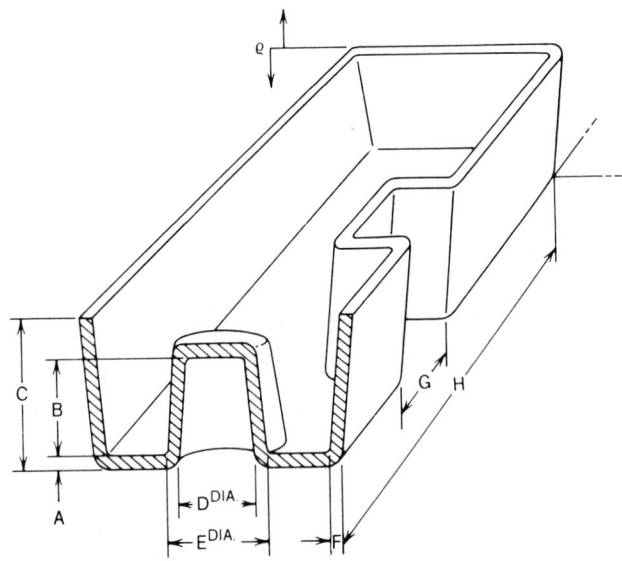

Values are based on a nominal 1/8 in. (3.2 mm) wall thickness and given in plus or minus in./in. and mm/mm.
FINE tolerance – Special care and added cost.
COMMERCIAL tolerance – Normal care required.
COARSE tolerance – Minimal care required, lowest cost.

Tolerance	A	B	C	D	E	F	G	H	
FINE	0.002	0.002	0.0013	0.001	0.001	0.002	0.0013	0.0013	in./in.
									mm/mm
COMMERCIAL	0.003	0.003	0.0022	0.002	0.002	0.002	0.0022	0.0022	in./in.
									mm/mm
COARSE									in./in.
									mm/mm

Mold Shrinkage Characteristics (in./in. or mm/mm)

MATERIAL _____ Styrene Maleic Anhydride _____

IN THE DIRECTION OF FLOW

THICKNESS

	mm: 1.5	3	6	8	FROM	TO	#
MATERIAL	in.: 1/16	1/8	1/4	1/2			
		0.005 to 0.009			0.005	0.009	In./in.

	PERPENDICULAR TO THE DIRECTION OF FLOW						

NOTE: Mold shrinkage is approximate. It is affected by part design, mold temperature, thickness, injection pressure, packing time, cycle time, orientation, gate design, gate size, gate location, glass content, glass size, and filler content.

actual shrinkage of the commercially molded part is affected by melt temperature, injection pressure, mold temperature, mold gating, and mold open time. The cumulative effect of these variables can be more significant than the inherent difference in value within the SMA polymer species.

47.14 TRADE NAMES

SMA is available as Cadon from Monsanto Co., St. Louis, Mo., and as Arloy and Dylark from Arco Chemical Co., Philadelphia, Pa.

BIBLIOGRAPHY

1. T. Alfrey and E. Lavin, "The Copolymerization of Styrene and Maleic Anhydride," *J. Am. Chem. Soc.* **67,** 2044 (1945).
2. Product bulletin, Arco Chemical Co., Philadelphia, Pa.
3. Product bulletin, Monsanto Co., St. Louis, Mo.
4. W.J. Hall and M.A. Cannon, "Thermoforming Cadon Engineering Thermoplastics," *SPE ANTEC* (1982).

General References

A.L. Brant, F.E. Lardi, and A.D. Wambach, "Toughness Retention of Styrenic Copolymer/Polycarbonate Blends After Processing Extremes and Prolonged Aging," *SPE NATEC* (1983).

W.J. Hall and J.O. Campbell, "Chemical Resistance Testing Translated to Molded Part Performance," *SPE NATEC* 1983.

W.J. Hall and co-workers, "New Styrene–Maleic Anhydride Terpolymer Blends," *ACS Symposium Series 229* 49 (1983).

B.C. Trivedi and B.M. Culbertson, *Maleic Anhydride,* Plenum Press, New York, 1982.

48

STYRENE–ACRYLONITRILE COPOLYMER (SAN)

F. Jack Reithel

Engineering Thermoplastics, Technical Service and Development, Dow Plastics, Dow Chemical USA, Midland, MI 48640

48.1 INTRODUCTION

Styrene–acrylonitrile copolymer is a random copolymer of styrene and acrylonitrile having the structure

Styrene + Acrylonitrile = Styrene-acrylonitrile

48.2 CATEGORY

Styrene–acrylonitrile copolymers are stiff, brittle thermoplastics providing significantly improved impact and chemical resistance (compared to general purpose polystyrene) and excellent clarity. Various producers offer grades emphasizing each of those property attributes.

48.3 HISTORY

Copolymerization of the monomers of styrene and acrylonitrile was first discovered and patented by I.G. Farben Industries in 1932.[1] Styrene–acrylonitrile copolymer was first commercialized in 1937 at I.G. Farben Industries. The product was used to produce coatings and lacquers.[2]

Styrene–acrylonitrile copolymer recipes with compositional variations of from 0.05 to 85.0% acrylonitrile were patented variously by General Aniline, 1941[3], Monsanto, 1948[4], and Standard Oil, 1950[5].

Figure 48.1 shows the sales volume and price history of styrene–acrylonitrile copolymers from 1960 through 1985.

601

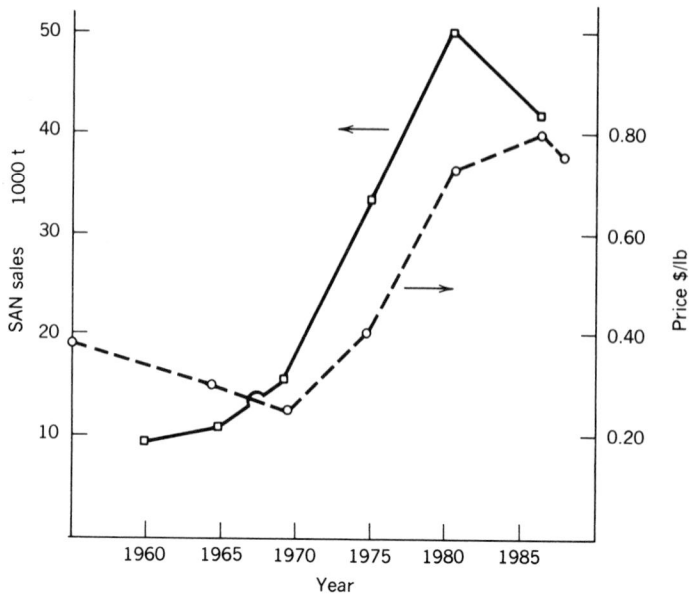

Figure 48.1.   United States Sales volume and price history for SAN.[6,7]

48.4   POLYMERIZATION

Styrene–acrylonitrile copolymer is prepared by reacting styrene and acrylonitrile monomers. Acrylonitrile is a very reactive monomer; its bulk polymerization is therefore hazardous. Styrene is much less reactive. The polymerization of styrene with some amount of initiator, but heated to 212°F (100°C), will take several hours.

The copolymerization of styrene–acrylonitrile also is highly exothermic and can be violent, making control of the reaction temperature difficult. The molecular weight of the SAN copolymer is related to the reaction temperature; lower temperatures result in higher molecular weight products, and higher temperatures result in lower molecular weight products.[8]

In the manufacture of SAN copolymers, it is difficult to make a product that is readily moldable into items that have adequate mechanical properties. Optimum polymer properties are obtained when the styrene–acrylonitrile system is copolymerized at a relatively high reaction temperature, about 266–374°F (130–190°C) in a continuous mass polymerization system.[9]

Three different polymerization techniques are used in the manufacture of SAN resins: emulsion, suspension, and continuous mass processes.

48.4.1   Emulsion Process

In the emulsion copolymerization[10,11] of SAN, both continuous and batch processes are used. Normally the copolymerizing system contains emulsifier, initiator, chain-transfer agent, monomers, and water. The copolymerization is carried out in a temperature range of 158–212°F (70–100°C) usually to a conversion of 97% or higher. Figure 48.2 shows a typical batch emulsion process. A continuous emulsion process has been described[12,13] using two stirred-tank reactors in series followed by a large hold-tank. In a continuous stirred-tank reactor, the theoretical emulsion particle residence time can vary from zero to infinity.[14]

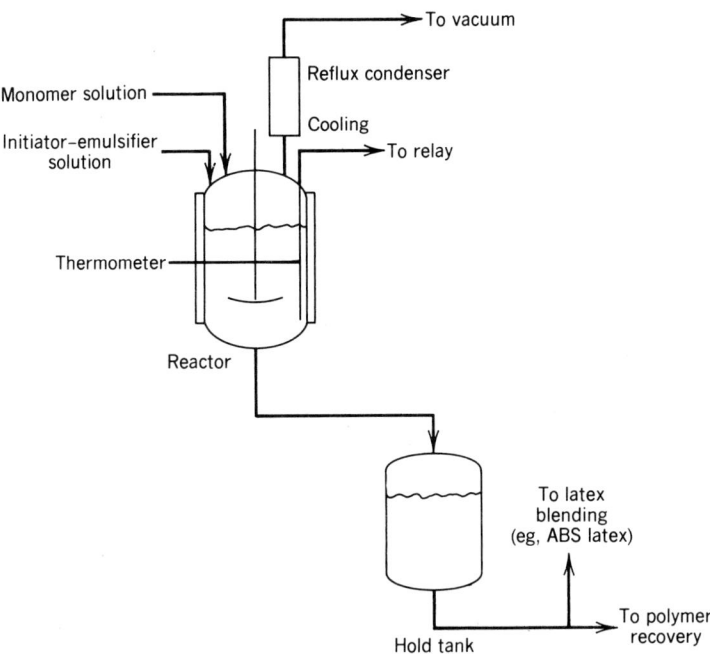

Figure 48.2. Batch emulsion process, SAN.

48.4.2 Suspension Process

In the suspension process, the reaction system contains monomers, chain-transfer agent, initiator, suspending agent, and water. Copolymerization is carried out at temperatures ranging from 140–302°F (60–150°C). Figure 48.3 shows a typical suspension process. Variations of the suspension process have been claimed to produce quality resin having good clarity (low haze level)[15–18] and a good color.[19–23] To avoid compositional drift in nonazeotrope copolymerization, special techniques are employed.[24,25]

48.4.3 Continuous Mass Process

The SAN continuous mass process is conceptually simple, but in practice it becomes complicated. The process can be initiated either thermally or catalytically, and a chain-transfer agent may be used. Copolymerization is carried out between 212–392°F (100–200°C). Solvents can be used to reduce the viscosity, or the copolymerization can be conducted at low conversion levels (40–70%) followed by devolatilization to remove unreacted monomers and solvent. Devolatilization is carried out from 248–536°F (120–280°C) under vacuum lower than 2.9 psi (20 kPa). The devolatilized polymeric melt is fed through a die, stranded, cooled, and pelletized. Because of the high viscosity of the copolymer, sophisticated equipment design is required to process the material to remove the heat of polymerization and the unreacted monomers and solvent, and to maintain a uniform composition. Figure 48.4 illustrates a typical continuous mass process.

48.4.4 Copolymerization Process Comparisons

Because there is no need for emulsifiers, suspending agents, salts, or water in the continuous mass process, it is relatively self contained with minimal waste treatment or environmental

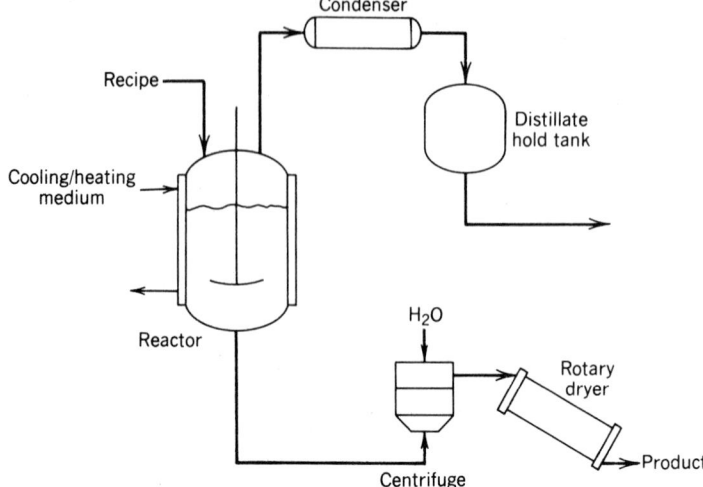

Figure 48.3. Suspension process, SAN.

problems. The other two processes do require additional waste treatment steps. Further, since the SAN copolymers manufactured by the continuous mass process do not contain residual emulsifying or suspending agents, they generally demonstrate superior color and low haze in molded form, and they are preferred for applications requiring a high degree of optical transparency. Both continuous mass and suspension-produced SAN copolymers can be used directly from the reactor for molding applications.

Emulsion SAN copolymers are uniquely suited for manufacture of ABS resins by virtue of the ease of obtaining uniform dispersion of the grafted rubber in an SAN matrix, through blending and coagulation of the SAN and the grafted rubber systems.

The continuous mass process does not use a large quantity of water as do both the

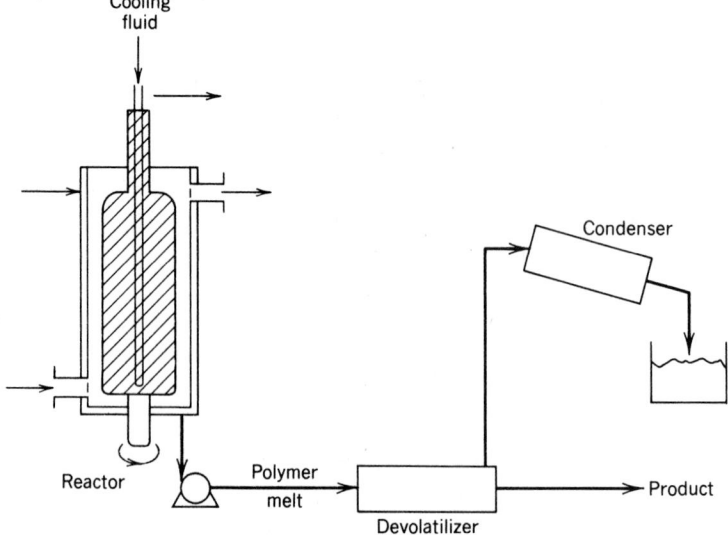

Figure 48.4. Continuous mass process, SAN.

emulsion and suspension processes. It offers very efficient space and time utilization. Because no water is used, it does not require a polymer drying operation. Therefore, it consumes less energy and often is more economical than the other processes. However, because of the residence time, the flow distribution differences, and the backmixing in the continuously stirred tank reactor plus the lateral mixing in the linear flow reactor, it requires greater time to reach steady state conditions. Commonly, some off-grade product is produced during start-up and shut-down procedures. Also, during the transitional period of product change-over when various grades of products are produced in a production line, some indeterminate grade of product is produced that may be undesirable.

Because of the highly viscous nature of the polymeric melt of the continuous mass process, good mixing and adequate heat removal are very difficult to achieve. To overcome these problems, the process operation and the reactor design are complicated. In both emulsion and suspension processes, the polymer particles are dispersed and suspended in a very low viscosity aqueous medium. This makes heat transfer and mixing significantly easier.

48.5 DESCRIPTION OF PROPERTIES

Styrene–acrylonitrile copolymer resins have an excellent balance of physical and chemical properties. Also, these clear, low-cost thermoplastics offer excellent dimensional stability and long-term toughness.

SAN copolymers provide essentially all of the transparency of general-purpose polystyrene (GPPS) but also possess other superior physical properties. They can withstand more impact and can be exposed to a wider range of chemicals without fear of degradation. The SAN copolymers exhibit solvent resistance to liquids such as water, aqueous acids, aqueous alkalies, bleaches, detergents, gasoline, and some chlorinated hydrocarbons. However, they may evidence attack by some aromatic compounds, and they are soluble in ketones.

Compared to GPPS, SAN copolymers have substantially improved resistance to stress in air as well as in other more aggressive environments. Table 48.1 shows the stress that can be sustained without rupture by a 30% acrylonitrile/70% styrene SAN copolymer and a GPPS. The "in air" data shown are the initial sustainable stress psi (MPa). The importance of such stress resistance is magnified when these polymers are placed in an aggressive environment such as milk or butter (common severe stress-cracking agents for polystyrene). As shown in the table, the total stress sustainable by the SAN copolymer drops to 2100 psi (14 MPa) but that of the polystyrene drops to nearly 450 psi (3 MPa).

In practice, the external stress that can be sustained by the working plastic article is reduced from the total stress sustainable by the value of the molded-in internal stress, commonly assumed to be 200–500 psi (1.4–3.5 MPa). In conjunction with their good chemical resistance, their enhanced stress resistance makes SAN copolymers materials of choice where sustained performance in the presence of a wide range of chemical environments is desired.

The stress-crack resistance of SAN copolymers in many environments ranks high along

TABLE 48.1. Critical Stress SAN vs GPPS, % Retention

Polymer	Critical Stress In Air, psi	(MPa)	Critical Stress In Milk, Butter psi	(MPa)	Working Stress Retained, %
30% Acrylonitrile–70% Styrene	3500	(24.1)	2100	(14.5)	60
General purpose polystyrene	1500	(10.3)	450	(3.1)	30

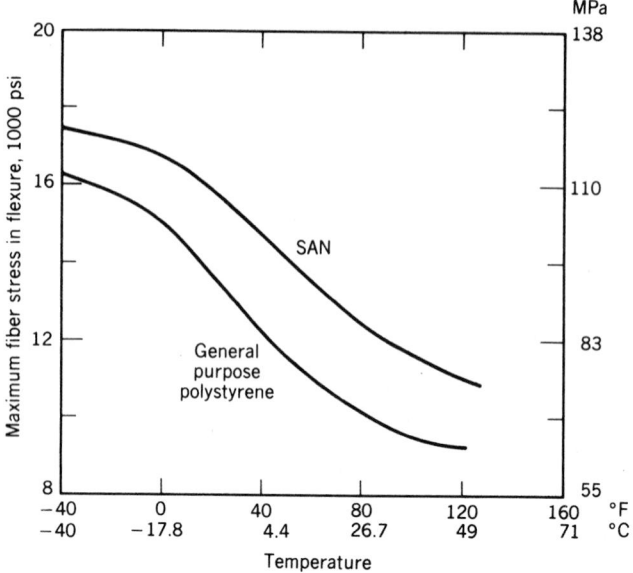

Figure 48.5.   Fiber stress in flexure, SAN vs GPPS.

with excellent tensile and flexural strengths, superior rigidity, abrasion resistance and excellent creep resistance under load.[26,27]

The thermal properties of SAN copolymers, compared to those of general purpose polystyrene, are superior over a broad range of temperatures, as shown in Figures 48.5–48.8. Those data show SAN properties are significantly improved from −40 to 120°F (−40 to 49°C) for fiber stress in flexure, ultimate tensile strength, modulus of elasticity in tension, and tensile yield strength.

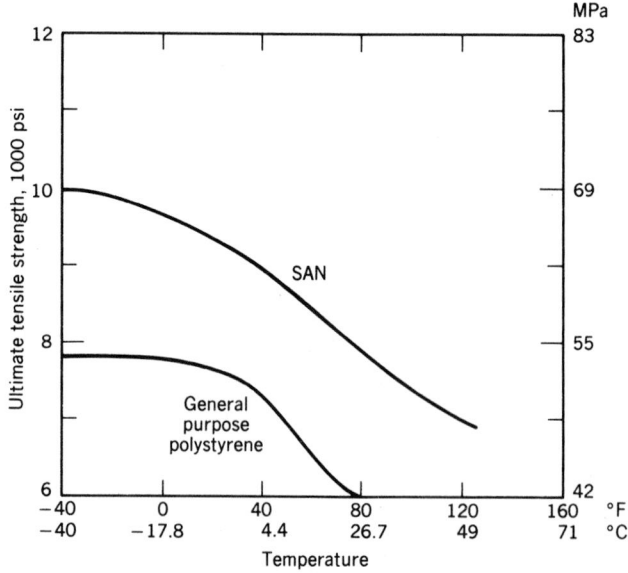

Figure 48.6.   Ultimate tensile strength, SAN vs GPPS.

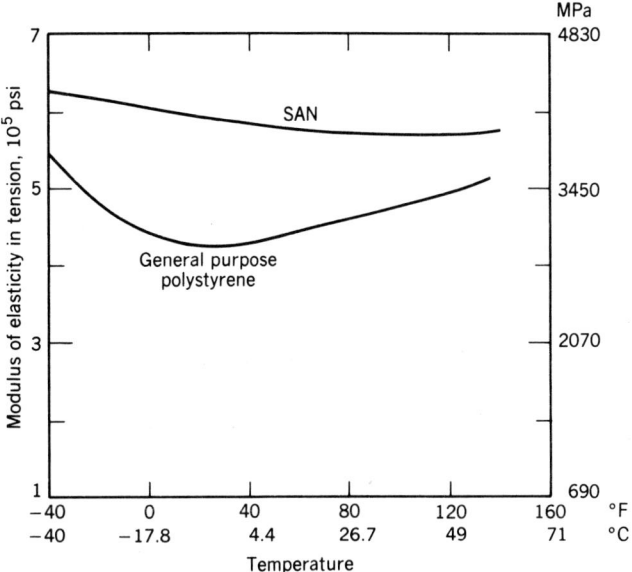

Figure 48.7. Modulus of elasticity in tension SAN vs GPPS.

48.6 APPLICATIONS

The combination of useful properties available from SAN copolymers has fostered both a broad range of applications and solid growth. Properties such as clarity, rigidity, toughness, and heat resistance, coupled with improved chemical resistance, have earned this growth. As stated previously, a SAN copolymer with lower acrylonitrile content will have more

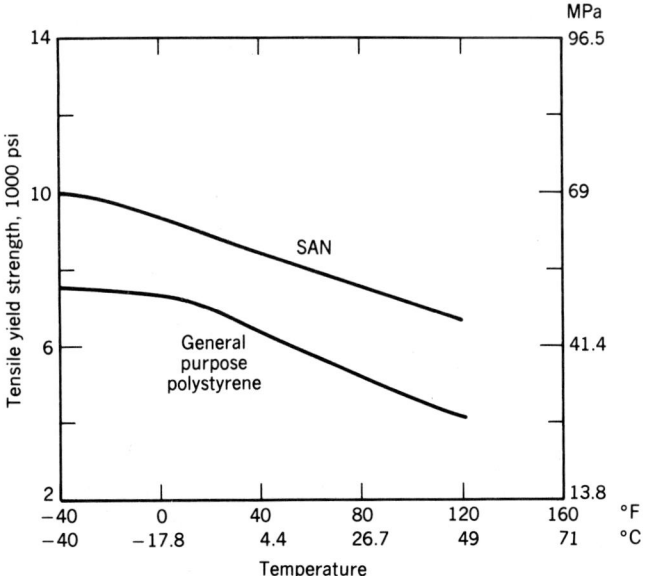

Figure 48.8. Tensile yield strength, SAN vs GPPS.

TABLE 48.2. Typical Styrene–Acrylonitrile (SAN) Applications

Application	Articles or Examples
Electronics	Video cassette windows and reels, telephone parts, terminal boxes
Automotive	Instrument lenses, battery caps, battery cases, instrument-panel support assemblies
Major appliances	Washing machine-drain connectors, air conditioner parts, crisper pans, shelf supports
Small appliances	Blender bowls, juicers, mixers, hair dryers
Housewares	Glasses, tumblers, mugs, bowls, cake covers, brush bristles
Packaging	Display racks, display boxes, lipstick cases, closures, cosmetics containers
Medical	Medical appliances, IV connectors, filter cases
Construction/building	Window panels, storm-door panels, plumbing fixture knobs, toilet seats, disposable cigarette lighters, battery cases, fishing lures, tags, aerosol nozzles, paint jars, filter bowls

moderate chemical resistance and toughness; the copolymers with higher acrylonitrile content will provide greater chemical resistance, heat resistance, and better toughness.

Because of their price[28] and performance capabilities as compared to other clear polymers, SAN copolymers are well accepted as engineering resins. They are now used in electronics applications, small and large appliances, automotive uses, and specialty applications such as disposable cigarette lighters. With the advent of glass-reinforcement, SAN copolymers have replaced a number of metal parts, especially in the automotive industry.[29]

A significant quantity of SAN is consumed in the manufacture of certain proprietary types of ABS because of the ready ability to custom formulate ABS products suitable for specific applications. A list of typical commercial applications for SAN copolymers is seen in Table 48.2

48.7 ADVANTAGES/DISADVANTAGES

SAN copolymers are used in many critical applications because of proper design engineering of the SAN parts, coupled with appropriate part testing. Often there is an economic advantage over other clear polymers that are higher priced or more difficult to mold such as acrylics and polycarbonates.

48.7.1 Advantages

Styrene–acrylonitrile copolymers are optically clear, hard, rigid, and strong products having excellent dimensional stability and good environmental stress-crack resistance. They are readily fabricated using conventional procedures. SAN copolymers also are relatively inexpensive materials.

The thermal properties of SAN copolymers are considerably superior to those of general purpose polystyrene over a wide range of temperatures. Figures 48.5–48.8 compare the thermal properties of a typical SAN copolymer with those of a general purpose polystyrene.

48.7.2 Disadvantages

The primary disadvantage associated with the use of SAN copolymers is an inherently pale straw-yellow color when used in its uncolored state. When this natural color is undesirable for specific applications, a very small amount of blue dye commonly is added to give the final product a very slight bluish color. Some of the current SAN copolymers are essentially color-free and crystal clear in appearance.

As more acrylonitrile is added to the copolymer, the chemical resistance and physical properties improve, but the SAN resin also will display a more distinct yellow color.[27,30,31] Higher level acrylonitrile copolymers also exhibit a greater tendency to increase in yellowness during fabrication.

Because SAN resins are highly polar, they exhibit a slightly hygroscopic tendency, and drying is suggested to ensure adequate performance during fabrication. Excessive moisture present in SAN can cause streaking and splay marks to appear on the surface of the items.

48.8 PROCESSING TECHNIQUES

SAN resins are commercially and satisfactorily fabricated by the processes most commonly used with thermoplastics: injection molding and extrusion (see the Master Material Outline).

Master Material Outline

PROCESSABILITY OF ___ Styrene–Acrylonitrile Copolymer (SAN) ___

PROCESS	A	B	C
INJECTION MOLDING	X		
EXTRUSION	X		
THERMOFORMING		X	
FOAM MOLDING		X	
DIP, SLUSH MOLDING			
ROTATIONAL MOLDING			
POWDER, FLUIDIZED-BED COATING			
REINFORCED THERMOSET MOLDING			
COMPRESSION/TRANSFER MOLDING		X	
REACTION INJECTION MOLDING (RIM)			
MECHANICAL FORMING			
CASTING			

A = Common processing technique.
B = Technique possible with difficulty.
C = Used only in special circumstances, if ever.

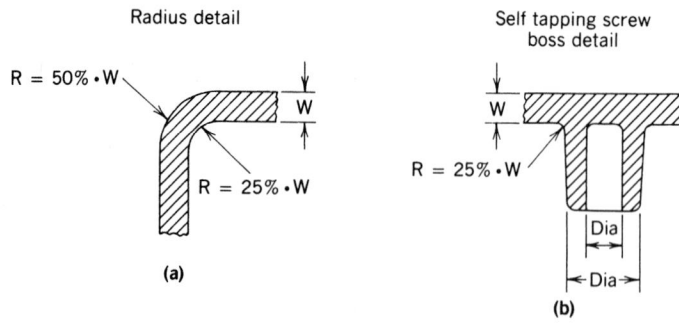

(a)

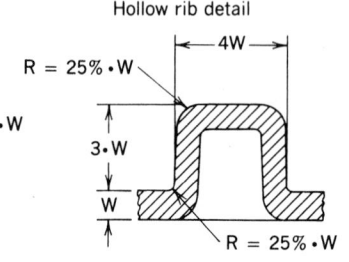

(b)

Solid rib detail   Hollow rib detail

(c)   (d)

Standard design chart for _____ **SAN** _____

1. Recommended wall thickness/length of flow
   Minimum _____ 0.060 in. _____ for _____ 8 _____ distance
   Maximum _____ 0.500 in. _____ for _____ 20 _____ distance
   Maximum _____ 0.120 _____ for _____ 10 _____ distance

2. Allowable wall thickness variation, % of nominal wall _____ 5 _____ No restriction

3. Radius requirements (scheme **a**)
   Outside: Minimum _____ 0.25 × wall T _____ Maximum _____
   Inside: Minimum _____ 0.25 in. × wall T _____ Maximum _____

4. Reinforcing ribs (scheme **c** and **d**)
   Maximum thickness _____ 0.8 × T wall _____
   Maximum height _____ 3× rib thickness _____
   Sink marks: Yes _____ No _____ X _____

5. Solid pegs and bosses
   Maximum thickness _____ 0.8 × T wall _____
   Maximum weight _____ 2× diameter _____
   Sink marks: Yes _____ No _____ X _____

6. Strippable undercuts
   Outside: Yes _____ No _____
   Inside: Yes _____ No _____ X _____

7. Hollow bosses (scheme **b**): Yes _____ X _____ No _____

8. Draft angles
   Outside: Minimum _____ 1° _____ Ideal _____ 3° _____
   Inside: Minimum _____ 1° _____ Ideal _____ 3° _____

9. Molded-in inserts: Yes _____ X _____
                       No _____

10. Size limitations
    Length: Minimum _____ Maximum _____ Process _____
    Width: Minimum _____ Maximum _____ Process _____

11. Tolerances

12. Special process- and materials-related design details
    Blow-up ratio (blow molding) _____4:1_____
    Core L/D (injection and compression molding) ___5:1___
    Depth of draw (thermoforming) _____
    Other _____

48.8.1 Injection Molding

SAN copolymers can be injection molded on the same equipment used for molding of other thermoplastics. A low compression screw design (2.5–3.0:1) is suggested to minimize the potential for frictional burning.[33] Polymer temperatures in the range of 450–500°F (232–260°C) are usually adequate to mold SAN. Mold temperatures in the range of 120–180°F (50–82°C) are generally utilized. Those higher mold temperatures normally yield parts having an improved surface finish and fewer molded-in stresses.

48.8.2 Extrusion

SAN copolymers are readily processed on contemporary extrusion equipment. An extruder with 24:1 length to diameter ratio or greater is recommended, as is a screw compression ratio of 2.5:1. Typical extrusion temperatures utilized for styrene–acrylonitrile copolymers are: 325–425°F (162–218°C), rear; 425–475°F (218–247°C), center; 450–500°F (232–260°C), front and die. The use of cooling water at the hopper zone is suggested.

48.9 RESIN FORMS

Styrene–acrylonitrile copolymers are suppllied in pellet form, normally a 1/8-by-1/8 in. (3 by 3 mm) cylindrical shape. Some very small bead forms are available from some manufacturers, as is a ground pellet form. Pellets containing antistatic agents, UV stabilizers, and internal lubricants are available from some manufacturers. Some pellets treated with a blowing agent are available. These are used in producing low density chemical-resistant styrene–acrylonitrile foam.

48.10 SPECIFICATION OF PROPERTIES

The Master Outline of Materials Properties provides general data on conventional SAN and 30% glass-filled SAN. It is important to note that special copolymers having properties outside the ranges listed may be available. Producers should be consulted.

48.11 PROCESSING REQUIREMENTS

Styrene–acrylonitrile copolymer can be fabricated over a wide range of temperatures without degradation; however, this material may show signs of yellowing when it is processed using very high temperatures or after extended periods of time at temperatures in excess of 500°F (260°C).

SAN copolymer is slightly hygroscopic and absorbs small amounts of moisture. Some drying is suggested before molding or extruding the resins, especially in a warm humid

MATERIAL ___Styrene Acrylonitrile (SAN)___

PROPERTY	TEST METHOD	ENGLISH UNITS	VALUE	METRIC UNITS	VALUE
MECHANICAL DENSITY	D792	lb/ft^3	67	g/cm^3	1.07
TENSILE STRENGTH	D638	psi	8,300–11,000	MPa	57–75
TENSILE MODULUS	D638	psi	500,000–540,000	MPa	3420–3720
FLEXURAL MODULUS	D790	psi	450,000–520,000	MPa	3080–3560
ELONGATION TO BREAK	D638	%	2.0–3.5	%	2.0–3.5
NOTCHED IZOD AT ROOM TEMP	D256	ft-lb/in.	0.25–0.45	J/m	13–24
HARDNESS	D785	Rockwell	M 80–83	Rockwell	M 80–83
THERMAL DEFLECTION T @ 264 psi	D648	°F	210–228	°C	99–109
DEFLECTION T @ 66 psi	D648	°F	218–239	°C	103–115
VICAT SOFTENING POINT	D1525	°F	226–235	°C	108–113
UL TEMP INDEX	UL746B	°F	140	°C	60
UL FLAMMABILITY CODE RATING	UL94		94 HB		94 HB
LINEAR COEFFICIENT THERMAL EXPANSION	D696	in./in./°F	3.7 × 10^{-5}	mm/mm/°C	6.7 × 10^{-5}
ENVIRONMENTAL WATER ABSORPTION 24 HOUR	D570	%	0.05–0.40	%	0.05–0.40
CLARITY	D1003	% TRANSMISSION	87–89	% TRANSMISSION	87–89
OUTDOOR WEATHERING	D1435	%	Yellows	%	Yellows
FDA APPROVAL			No		No
CHEMICAL RESISTANCE TO: WEAK ACID	D543		Not attacked		Not attacked
STRONG ACID	D543		Minimally attacked		Minimally attacked
WEAK ALKALI	D543		Not attacked		Not attacked
STRONG ALKALI	D543		Minimally attacked		Minimally attacked
LOW MOLECULAR WEIGHT SOLVENTS	D543		Minimally attacked		Minimally attacked
ALCOHOLS	D543		Minimally attacked		Minimally attacked
ELECTRICAL DIELECTRIC STRENGTH	D149	V/mil	300–400	kV/mm	12–16
DIELECTRIC CONSTANT	D150	10^6 Hertz	2.6–3.0		2.6–3.0
POWER FACTOR	D150	10^6 Hertz	0.008–0.010		0.008–0.010
OTHER MELTING POINT	D3418	°F		°C	
GLASS TRANSITION TEMP.	D3418	°F	239	°C	115

MATERIAL _____ Styrene Acrylonitrile (SAN) 30% Glass-Filled

PROPERTY	TEST METHOD	ENGLISH UNITS	VALUE	METRIC UNITS	VALUE
MECHANICAL DENSITY	D792	lb/ft^3	76.1	g/cm^3	1.22
TENSILE STRENGTH	D638	psi	20,000	MPa	139
TENSILE MODULUS	D638	psi	1,600,000	MPa	11,000
FLEXURAL MODULUS	D790	psi	1,400,000	MPa	9,600
ELONGATION TO BREAK	D638	%	1.6	%	1.6
NOTCHED IZOD AT ROOM TEMP	D256	ft-lb/in.	1.1	J/m	60
HARDNESS	D785	Rockwell	R 123	Rockwell	R 123
THERMAL DEFLECTION T @ 264 psi	D648	°F	212	°C	100
DEFLECTION T @ 66 psi	D648	°F	225	°C	108
VICAT SOFT-ENING POINT	D1525	°F		°C	
UL TEMP INDEX	UL746B	°F	165	°C	74
UL FLAMMABILITY CODE RATING	UL94		94 HB		94 HB
LINEAR COEFFICIENT THERMAL EXPANSION	D696	in./in./°F	1.06×10^{-5}	mm/mm/°C	1.9×10^{-5}
ENVIRONMENTAL WATER ABSORPTION 24 HOUR	D570	%	0.15	%	0.15
CLARITY	D1003	% TRANS-MISSION	40	% TRANS-MISSION	40
OUTDOOR WEATHERING	D1435	%	Yellows	%	Yellows
FDA APPROVAL			No		No
CHEMICAL RESISTANCE TO: WEAK ACID	D543		Not attacked		Not attacked
STRONG ACID	D543		Minimally attacked		Minimally attacked
WEAK ALKALI	D543		Not attacked		Not attacked
STRONG ALKALI	D543		Minimally attacked		Minimally attacked
LOW MOLECULAR WEIGHT SOLVENTS	D543		Minimally attacked		Minimally attacked
ALCOHOLS	D543		Minimally attacked		Minimally attacked
ELECTRICAL DIELECTRIC STRENGTH	D149	V/mil	500	kV/mm	20
DIELECTRIC CONSTANT	D150	10^6 Hertz	3.6		3.6
POWER FACTOR	D150	10^6 Hertz	0.008		0.008
OTHER MELTING POINT	D3418	°F		°C	
GLASS TRANS-ITION TEMP.	D3418	°F		°C	

613

environment. Drying for 2–4 h at 175–180°F (79–82°C) with a dehumidified forced air system is recommended.[33]

Styrene–acrylonitrile copolymer can be reprocessed after fabrication. Regrind levels of 25% or more can be tolerated if good handling practice is followed and only clean regrind is reprocessed.

The SAN copolymer is compatible with some polymers (ABS, polycarbonate, and PVC), but is incompatible with most others. Because of its incompatibility and the threat of contamination, thorough purging and clean-up should be practiced before fabrication of styrene–acrylonitrile copolymer parts is attempted. Because a large percentage of SAN copolymers is used to produce optically clear parts, the practice of good housekeeping is especially important and is mandatory if high-quality, high-clarity parts are to be produced.

48.12 PROCESSING-SENSITIVE END PROPERTIES

Because they are amorphous polymers, SAN copolymers are highly affected by orientation induced during fabrication. One demonstration of orientation effects is in the manufacture of oriented monofilament for use as brush bristles. In the production of the monofilament, uniaxial orientation levels of 200–300% impart high levels of toughness to an ordinarily brittle product. This allows the final product to maintain its necessary degree of stiffness and chemical resistance, and yet perform well in an application requiring constant flexing.

Orientation effects in injection-molded items are quite different. Thick section or heavy-walled parts can contain high levels of uncontrolled residual biaxial orientation (stress) which will drastically reduce the performance capabilities of the molded part. To combat this biaxial orientation effect, it is common practice to anneal heavy-walled parts after molding in order to reduce the residual orientation levels and to improve the performance capabilities of the molded item. Annealing conditions normally used are 180°F (82°C) in closed air ovens; annealing time is from 4 to 24 h depending on part thickness and geometry. It is extremely important to avoid shock-cooling of parts coming out of an annealing oven; shock-cooling can induce new orientational stresses to the just-annealed part. Annealed parts normally are tested by immersion in a stress-cracking solution[34] for various periods of time. This type of testing is usually a go/no-go test practice evolved by molders in cooperation with their OEMs.

48.13 SHRINKAGE

See Mold Shrinkage and Standard Material Design Chart.

Mold Shrinkage Characteristics (in./in. or mm/mm)

MATERIAL   SAN Copolymer

IN THE DIRECTION OF FLOW

MATERIAL	THICKNESS						
	mm: 1.5	3	6	8	FROM	TO	#
	in.: 1/16	1/8	1/4	1/2			
SAN Copolymer	0.006	0.005	0.005	0.0043			
	0.007	0.007	0.006	0.005			

	PERPENDICULAR TO THE DIRECTION OF FLOW					
SAN Copolymer	0.005	0.004	0.004	0.003		
	0.006	0.006	0.006	0.004		

NOTE: Mold shrinkage is approximate. It is affected by part design, mold temperature, thickness, injection pressure, packing time, cycle time, orientation, gate design, gate size, gate location, glass content, glass size, and filler content.

48.14 TRADE NAMES

Currently, there are three producers of styrene–acrylonitrile copolymers in the United States: Lustran, Monsanto Co., St. Louis, Mo.; SAN, GE/Borg, Parkersburg, W.Va.; and Tyril, The Dow Chemical Company, Midland, Mich.

BIBLIOGRAPHY

1. Brit. Pat. 371,396 (April 14, 1932).
2. Brit. Pat. 459,720 (Jan. 11, 1937).
3. U.S. Pat. 2,228,270 (July 12, 1938) (to General Aniline).
4. U.S. Pat. 2,439,227 (April 6, 1948) (to Monsanto).
5. U.S. Pat. 2,527,162 (Oct. 25, 1950) (to Standard Oil).
6. F.M. Peng, Monsanto, 1975, private communication.
7. Society of the Plastics Industry, Dec. 1986, private communication.
8. R.H. Boundy and R.F. Boyer, eds., *Styrene, Its Polymers, Copolymers, and Derivatives*, ACS Monograph Series, Reinhold Publishing Corp., New York, 1952, pp. 215–289.
9. Brit. Pat. 871,686 (1961).
10. *Polymer Chemistry of Synthetic Elastomers*, Pt. 1, Vol. 23, of *High Polymers*, Wiley-Interscience, New York, 1968, pp. 127–178.
11. *Emulsion Polymerization*, Vol. 9 of *High Polymers*, Wiley-Interscience, New York, 1965.
12. U.S. Pat. 3,547,857 (Dec. 15, 1970), A.G. Murray (Uniroyal Inc.).
13. Brit. Pat. 1,168,760 (Oct. 29, 1969) (to Uniroyal Inc.).
14. *Chemical Reaction Engineering*, John Wiley & Sons, Inc., New York, 1964, Chapt. 9.
15. U.S. Pat. 3,198,775 (Aug. 3, 1965), R.E. Delacretaz and co-workers (to Monsanto).
16. U.S. Pat. 3,258,453 (June 28, 1966), H.K. Chi (to Monsanto).
17. U.S. Pat. 3,681,310 (Aug. 1, 1972), K. Moriyama and co-workers (to Daicel Ltd.).
18. U.S. Pat. 3,243,407 (March 29, 1986) Y.C. Lee (to Monsanto).
19. U.S. Pat. 3,331,810 (July 18, 1967), Y.C. Lee (to Monsanto).
20. U.S. Pat. 3,331,812 (July 18, 1967), Y.C. Lee (to Monsanto).
21. U.S. Pat. 3,287,331 (Nov. 22, 1967), Y.C. Lee and L.P. Paradis (to Monsanto).
22. U.S. Pat. 3,356,664 (Dec. 5, 1967), Y.C. Lee (to Monsanto).
23. U.S. Pat. 3,491,071 (Jan. 20, 1970), R. Lanzo (to Monte Edison).
24. U.S. Pat. 3,738,972 (June 12, 1973), K. Morinyama and T. Osaka (to Daicel Ltd.).
25. Brit. Pat. 1,328,625 (Aug. 30, 1973) (to Daicel Ltd.).
26. H.R. Simonds, ed., *Source Book of New Plastics*, Vol. II, Reinhold Publishing Co., New York, 1961, p. 191.

27. H. McCann, ed., *Modern Plastics Encyclopedia for 1963*, Vol. 40 (1A), Hildreth Press Inc., Bristol, Conn. Sept. 1962.

28. *Ibid.*, Dec. 1986.

29. "Thermoplastics Win New Sales," *Chemical Week* **94**(21), 109 (1964).

30. P. Morgan, ed., *Plastics Progress*, McMillan Co., New York, 1961, p. 42.

31. C.E. Schildenecth, *Vinyl and Related Polymers*, John Wiley & Sons, Inc., New York, 1952, pp. 48–54.

32. The Dow Chemical Company, Midland, Mich., unpublished data.

33. "Tyril SAN", Fabrication Guidelines, The Dow Chemical Company, Midland, Mich., 1986.

49

STYRENE–BUTADIENE (K-RESIN)

Ronald D. Mathis, David L. Hartsock, and Gregory M. Swisher

Phillips 66 Co. Plastics Div., 264 PLB Bartlesville, OK 74004

49.1 STYRENE–BUTADIENE

Styrene–butadiene polymers (hereafter referred to as K-Resin) are block copolymers of styrene and butadiene (about 75:25) featuring polymodal distribution of molecular weight. It should be emphasized that this is only a general description, and there may be multiple blocks of styrene and/or butadiene present in a specific grade of K-Resin polymer.

Following is a molecular diagram representing the copolymer:

49.2 CATEGORY

K-Resin polymers are thermoplastics characterized by high clarity, exceptional toughness, and ease of processibility.

49.3 HISTORY

K-Resin polymers were first synthesized in the mid-1960s by Alonzo G. Kitchen and Frank J. Szalla of Phillips Petroleum Company's Research & Development Division. Phillips commercialized K-Resin polymers with the 1972 start up of a production plant in the United States. The earliest uses were thermoformed packaging applications, followed quickly by injection-molded and blow-molded containers.

Since the introduction of K-Resin polymers, their basic structure has not been modified. Manufacturing techniques, however, are continually being refined to improve the primary benefits of the polymers, especially optical clarity and consistency. A grade suitable for thin-film production was commercialized in 1983 as KR10.

Production has grown steadily since the 1972 commercialization, requiring the construction of a 120×10^6 lb/yr plant that was brought on-stream in 1979. Increasing demand ncessitated further expansion and debottlenecking, which will increase the capacity of this plant to 270×10^6 lb/yr in 1989.

49.4 POLYMERIZATION

K-Resin polymers are produced by a novel solution process involving sequential polymerization of styrene and butadiene in a hydrocarbon solvent at temperatures up to 250°F (121°C). Blocks of polymerized styrene are formed first, after which butadiene is added and subsequently polymerized. Finally, a polyfunctional treating agent is added to form the block copolymer broadly depicted as $(S_m - B_n)_x - Y$. In this equation, S represents polymerized styrene, B is polymerized butadiene, and Y is an atom or group of atoms derived from the polyfunctional treating agent. The subscript x indicates the number and reactivity of the functional groups in the treating agent; the ratio $m:n$ is about $3:1$.

Relative to other transparent plastics, the cost of K-Resin polymers falls between the lower-cost, lower-toughness plastics (such as polystyrene) and the higher-cost, higher-performance plastics (such as polycarbonate).

49.5 PROPERTIES

K-Resin polymers provide a strategic balance of mechanical toughness and optical qualities, including high surface gloss, excellent clarity (90–91% light transmission) and low haze (1–3%). In addition to impact toughness and good elongation, the polymers exhibit good stiffness and heat resistance [200°F (93°C) Vicat softening point]. Furthermore, they permit greater design freedom than most transparent polymers and can be formed easily by conventional processes. K-Resin polymers also duplicate faithfully the mold geometry and surface detail.

Since K-Resin polymers are typically selected for their transparency, grades containing reinforcing agents or fillers are not produced.

49.6 APPLICATIONS

K-Resin polymers are specified for many packaging applications where consumer appeal must be balanced by toughness needed to protect the contents. The low density of the polymers translates into high yields, making unit packaging economical. Furthermore, excellent mold replication allows slim-profile packages to minimize package volume for efficient transport and storage. Unlike most clear styrenic polymers, K-Resin offers long service life in integral living hinges and ball-and-socket hinges. The designer can optimize the filling/closure operations, even for such consumer-reuse packages as hinged boxes.

Consumer appeal is particularly beneficial for the food packaging market: the excellent clarity of K-Resin packaging enhances the display of a wide range of food products. The polymers meet applicable FDA specifications in the Code of Federal Regulations, Title 21, Section 177.1640 (21CFR 177.1640), but their use in packaging fatty foods is limited to certain combinations of polymer, food type, and storage conditions.

Moisture vapor and certain other gases permeate through K-Resin polymers at relatively high rates, which can help gas equalization, resulting in increased product shelf-life in such packaging uses as vegetable wraps. On the other hand, high permeation can limit the shelf life of liquid-containing products unless the resins are coextruded with any of several compatible barrier resins. Finally, coextrusion is required for successful packaging of certain other products because of K-Resin polymers' susceptibility to attack from most common organic solvents, oils and other stress-crack accelerators. Even so, a remarkable variety of food products are still packaged in noncoextruded K-Resin containers with functional and esthetic success.

The medical industry has grown dramatically in the variety and sophistication of its application of K-Resin polymers. Most medical devices and packaging must be clear enough that the nature, amount, or condition of their contents can be determined during use, but also tough enough to resist accidental breakage, which is not only expensive, time-consuming, and hazardous but possibly life-threatening. Beyond those principal features, K-Resin polymers meet the requirements of *U.S. Pharmacopoeia* Class VI-50, are compatible with blood, and exhibit no cytotoxic, mutagenic, or irritancy potential. Finally, the resins can be sterilized by gamma irradiation or ethylene oxide with negligible change in clarity or toughness.

In short, the toughness of K-Resin polymers helps designers to meet safety requirements economically in diverse transparent applications ranging from high-technology chip carriers to novelty items. Following is a brief list of typical applications categorized by processing technique.

- Injection molding: toys, medical ware and diagnostic equipment, office supplies, cassette components, display units, cosmetic containers, storage boxes, tableware, personal-care items, tool handles, overcaps, refrigerator components, hangers, and decorative bulbs.
- Extrusion: disposable cups and deli containers, clam shells, multilayer coextruded and laminated products, blister packages, profiles, tubing, and storage containers.
- Film: shrink wrap, skin packaging, overwraps, stationery supplies, medical packaging, twist wrap, bags, and produce wrap.
- Blow-molding: Bottles for consumables, decorative display containers, medical apparatus, and toys.

49.7 ADVANTAGES/DISADVANTAGES

In the range of transparent plastics, K-Resin polymers fill part of the strategic gap that has long existed between the so-called "commodity" resins and the high-performance engineering polymers. The low side is dominated by polystyrene and the polyolefins, which are either clear or tough but not both. At the other extreme are cellulosics, polycarbonates, and other engineering polymers that are both clear and tough, but comparatively high priced.

In many applications in which it competes with performance resins, K-Resin proves entirely satisfactory. K-Resin grades are more economical than the transparent performance materials with which they compete. This results from a combination of lower cost per pound and lower density, the latter increasing yield of parts per pound. Further, the recommended processing temperatures for K-Resin polymers are comparatively low, which reduces energy consumption, cooling requirements, and cycle times.

Most importantly, many of the high performance materials are very sensitive to moisture and require elaborate drying procedures to achieve full performance or clarity. K-Resin polymers usually require no drying whatever, unless they absorb some surface moisture during high humidity storage.

From the low cost side, the dominant competitor is general-purpose polystyrene, which can be crystal clear but is brittle. Relative to polystyrene, the principal advantage of K-Resin is its enhanced toughness.

When applications require less toughness than K-Resin polymer but more than polystyrene, many fabricators blend general-purpose polystyrene with KR04 grade K-Resin to combine the latter's impact strength with the former's stiffness and surface hardness. The blend ratio may be varied to provide a continuum of property balance and economics. Furthermore, blending may be accomplished conveniently by in-house dry-blending of pellets, so the processor can choose the optimum blend for a specific application or processing requirement. This practice is most common among sheet processors supplying thermoformers of disposable containers. Blends of K-Resin with other compatible resins are used less frequently in other conversion processes.

The tradeoffs of using K-Resin polymers are their relatively low surface hardness, low chemical resistance to polar organic solvents, relatively low stiffness, and poor resistance to direct exposure to sunlight. Low surface hardness and chemical resistance, and stiffness can all be countered with coextrusion techniques; improved resistance to sunlight can be obtained by the addition of ultraviolet stabilizers.

49.8 PROCESSING TECHNIQUES

K-Resins can be processed easily on conventional equipment for sheet and blown or cast-film extrusion, thermoforming, injection molding, and blow-molding. For each of these techniques, K-Resins can accept varying color loads and can be blended with a variety of compatible resins without adverse effects on processibility.

Master Material Outline

PROCESSABILITY OF _____ K-Resin

PROCESS	A	B	C
INJECTION MOLDING	X		
EXTRUSION	X		
THERMOFORMING	X		
FOAM MOLDING		X	
DIP, SLUSH MOLDING			
ROTATIONAL MOLDING			
POWDER, FLUIDIZED-BED COATING			
REINFORCED THERMOSET MOLDING			
COMPRESSION/TRANSFER MOLDING			
REACTION INJECTION MOLDING (RIM)			
MECHANICAL FORMING			
CASTING			

A = Common processing technique.
B = Technique possible with difficulty.
C = Used only in special circumstances, if ever.

Radius detail

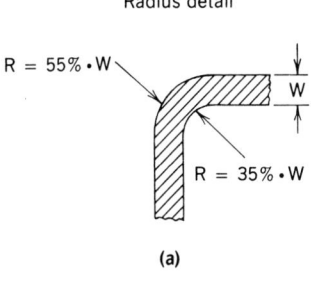

Self tapping screw
boss detail

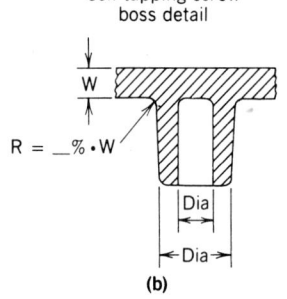

(a)

(b)

Solid rib detail

Hollow rib detail

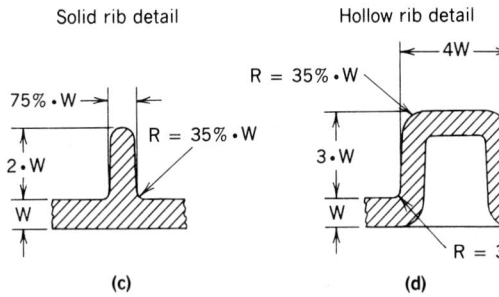

(c)

(d)

Standard design chart for _____K-Resin_____

1. Recommended wall thickness/length of flow
   Minimum _____0.032 in._____ for _____2.5 in._____ distance
   Maximum _____2.50 in._____ for _____48 in._____ distance
   Ideal _____0.090 in._____ for _____26 in._____ distance

2. Allowable wall thickness variation, % of nominal wall _____30_____

3. Radius requirements (scheme **a**)
   Outside: Minimum_____0.000 in._____ Maximum _____none_____
   Inside: Minimum_____0.030 in._____ Maximum _____none_____

4. Reinforcing ribs (scheme **c** and **d**)
   Maximum thickness _____wall thickness_____
   Maximum height _____
   Sink marks: Yes _____ No ___X___

5. Solid pegs and bosses
   Maximum thickness _____2.5 in._____
   Maximum weight _____
   Sink marks: Yes _____ No ___X___

6. Strippable undercuts
   Outside: Yes ___X___ No _____
   Inside: Yes ___X___ No _____

7. Hollow bosses (scheme **b**): Yes ___X___ No _____

8. Draft angles
   Outside: Minimum _____$\frac{1}{2}°$_____ Ideal _____3°_____
   Inside: Minimum _____$\frac{1}{3}°$_____ Ideal _____2°_____

9. Molded-in inserts: Yes ____X____
                              No _____

10. Size limitations
    Length: Minimum __None__ Maximum __None__ Process ___Injection molding___
    Width: Minimum __None__ Maximum __None__ Process ___Injection molding___

11. Tolerances   Shrinkage: 0.003–0.010

12. Special process- and materials-related design details
    Blow-up ratio (blow molding) _____≤3:1_____
    Core L/D (injection and compression molding) _____
    Depth of draw (thermoforming) __Max 4:1 Typical 2:1__
    Other _____

49.9   RESIN FORMS

K-Resin polymers are sold in pellet form.

Various K-Resin polymers are available to fulfill a range of processing and performance needs. KR01 is widely used in injection molding for its good impact strength, stiffness, and warpage resistance.

Other grades are KR03, KR04, KR05, and KR10. Relative to KR01, the KR03 series exhibits improved melt strength and higher impact resistance for applications requiring extra toughness. KR03, KR04, KR05, and KR10 differ only in increasing optical quality as a function of increasing grade number.

KR03 can be extruded but is typically used for injection molding of thicker sections. Intermediate-grade KR04 is usually blended with other transparent resins, such as polystyrene, for extrusion and thermoforming applications. Used as neat polymer, KR05 is normally specified for extrusion, thermoforming, and blow-molding applications requiring greater optical quality. Grade KR10 is selected for extrusion of cast or blown film requiring the maximum optical quality in thin gauges.

49.10   SPECIFICATION OF PROPERTIES

See the Master Outline of Materials Properties.

49.11   PROCESSING REQUIREMENTS

K-Resins polymers are usually specified for their combination of toughness with optical quality. Both features are sensitive to polymer degradation, so the most important processing consideration is accurate control of polymer temperature. To minimize residence time at processing temperatures, the extruder should have a low L/D ratio (preferably less than 30:1). The temperature controllers of the various heater zones should be well maintained and accurately calibrated. The screw must be designed to minimize shear heat input, with a compression ratio less than 3.25:1 and no high-shear mixing sections or barrier flights.

Whether the forming equipment is injection mold or extrusion die, it should also avoid excessive shear from inadequate gate sizes or die gaps. The melt-flow path should be streamlined to eliminate areas where melt could stagnate or hang up long enough to degrade the polymer. Startup and shutdown procedures must be designed to avoid excessive temperatures or soaking times.

MATERIAL ___K-Resin, KR10___

PROPERTY	TEST METHOD	ENGLISH UNITS	VALUE	METRIC UNITS	VALUE
MECHANICAL DENSITY	D792	lb/ft^3	63	g/cm^3	1.01
TENSILE STRENGTH	D638	psi	4350/ 2755[a]	MPa	30/19[a]
TENSILE MODULUS	D638	psi	214,745/ 159,790[a]	MPa	1481/1102
FLEXURAL MODULUS	D790	psi		MPa	
ELONGATION TO BREAK	D638	%	125/200	%	125/200
NOTCHED IZOD AT ROOM TEMP	D256	ft-lb/in.	0.39	J/m	21
HARDNESS	D785		D65		D65
THERMAL DEFLECTION T @ 264 psi	D648	°F	171	°C	77
DEFLECTION T @ 66 psi	D648	°F	185	°C	85
VICAT SOFT-ENING POINT	D1525	°F	200	°C	93
UL TEMP INDEX	UL746B	°F		°C	
UL FLAMMABILITY CODE RATING	UL94				
LINEAR COEFFICIENT THERMAL EXPANSION	D696	in./in./°F	0.66×10^{-4}	mm/mm/°C	1.2×10^{-4}
ENVIRONMENTAL WATER ABSORPTION 24 HOUR	D570	%	0.09	%	0.09
CLARITY	D1003	% TRANS-MISSION	91	% TRANS-MISSION	91
OUTDOOR WEATHERING	D1435	%	3 months	%	3 months
FDA APPROVAL			Nonfatty foods, packaging dry,[a]		Nonfatty foods, packaging dry,[a]
CHEMICAL RESISTANCE TO: WEAK ACID	D543		Minimally attacked		Minimally attacked
STRONG ACID	D543		Badly attacked		Badly attacked
WEAK ALKALI	D543		Minimally attacked		Minimally attacked
STRONG ALKALI	D543		Minimally attacked		Minimally attacked
LOW MOLECULAR WEIGHT SOLVENTS	D543		Not recommended		Not recommended
ALCOHOLS	D543		Minimally attacked		Minimally attacked
ELECTRICAL DIELECTRIC STRENGTH	D149	V/mil	300	kV/mm	12
DIELECTRIC CONSTANT	D150	10^6 Hertz	2.5		2.5
POWER FACTOR	D150	10^6 Hertz			
OTHER MELTING POINT	D3418	°F		°C	
GLASS TRANS-ITION TEMP.	D3418	°F		°C	

[a] Direction of flow/transverse direction.

623

Master Outline of Material Properties

MATERIAL ___K-Resin, KR03, KR04, KR05___

PROPERTY	TEST METHOD	ENGLISH UNITS	VALUE	METRIC UNITS	VALUE
MECHANICAL DENSITY	D792	lb/ft^3	63	g/cm^3	1.01
TENSILE STRENGTH	D638	psi	3770	MPa	26
TENSILE MODULUS	D638	psi	199,810	MPa	1378
FLEXURAL MODULUS	D790	psi	234,755	MPa	1619
ELONGATION TO BREAK	D638	%	80	%	80
NOTCHED IZOD AT ROOM TEMP	D256	ft-lb/in.	0.39	J/m	21
HARDNESS	D785		D65		D65
THERMAL DEFLECTION T @ 264 psi	D648	°F	171	°C	77
DEFLECTION T @ 66 psi	D648	°F	185	°C	85
VICAT SOFT-ENING POINT	D1525	°F	200	°C	93
UL TEMP INDEX	UL746B	°F	122	°C	50
UL FLAMMABILITY CODE RATING	UL94		94HB		94HB
LINEAR COEFFICIENT THERMAL EXPANSION	D696	in./in./°F	0.66×10^{-4}	mm/mm/°C	1.2×10^{-4}
ENVIRONMENTAL WATER ABSORPTION 24 HOUR	D570	%	0.09	%	0.09
CLARITY	D1003	% TRANS-MISSION	90–91	% TRANS-MISSION	90–91
OUTDOOR WEATHERING	D1435	%	3 months	%	3 months
FDA APPROVAL			Nonfatty foods, packaging dry,[a]		Nonfatty foods, packaging dry,[a]
CHEMICAL RESISTANCE TO: WEAK ACID	D543		Minimally attacked		Minimally attacked
STRONG ACID	D543		Badly attacked		Badly attacked
WEAK ALKALI	D543		Minimally attacked		Minimally attacked
STRONG ALKALI	D543		Minimally attacked		Minimally attacked
LOW MOLECULAR WEIGHT SOLVENTS	D543		Not recommended		Not recommended
ALCOHOLS	D543		Minimally attacked		Minimally attacked
ELECTRICAL DIELECTRIC STRENGTH	D149	V/mil	7620	kV/mm	300
DIELECTRIC CONSTANT	D150	10^6 Hertz	2.5		2.5
POWER FACTOR	D150	10^6 Hertz			
OTHER MELTING POINT	D3418	°F		°C	
GLASS TRANS-ITION TEMP.	D3418	°F		°C	

[a] Aqueous, acidic, and alcoholic foods.

624

Regrinding equipment should be well maintained and ventilated to avoid excessive thermal and mechanical degradation during reprocessing. If proper regrinding procedures are employed, the polymer can be routinely reprocessed with negligible loss of mechanical performance or optical quality.

In general, K-Resin polymers do not absorb moisture and thus need not be dried before processing. If the pellets or regrind adsorb surface moisture (as in unprotected storage), the polymers should be dried for one to two hours at 125–140°F (52–60°C). It is very important to avoid excessively high drying temperatures that can soften the resin enough to agglomerate. The resultant blockage can interrupt uniform feeding of the extruder.

Following are specific considerations for each of the major processes for K-Resin.

49.11.1 Injection Molding

To minimize cycle time and obtain optimum parts, the melt should be maintained at the minimum temperature that permits the mold to be filled, usually between 380 and 450°F (193 and 232°C). The melt flow increases as melt temperature increases up to 500°F (260°C), but becomes erratic at higher temperatures.

The relationship between the injection pressure and melt flow is also typical. Surface reproduction is so good, however, that overpacked parts can stick in highly polished molds. Therefore, injection pressure should be minimized, certainly in the second stage, and maintained only as long as necessary.

On the other hand, underpacked parts can show wavy lines or surface ripples, so the parts must indeed be fully packed. If the gates are too small, the injection rate should be slow enough to minimize jetting and excessive shear. If the gauge size is adequate, faster injection rates can improve the weld-line strength, providing the mold is generously vented. Depending on the part, the mold temperature can range from 50 to 150°F (10 to 66°C), with optimum clarity occurring between 80 and 120°F (27 and 49°C). Higher mold temperatures maximize surface gloss and reproduction of mold detail; lower temperatures maximize impact resistance and reduce cycle time.

49.11.2 Blow Molding

In parts blow molded of K-Resin polymers, impact toughness is most heavily dependent on uniform wall distribution and proper pinch-welds. The wall distribution is easily controlled if die tooling is selected to limit the blow-up ratio to below 3:1, allowing for a parison diameter swell of 10–15%. The mold pinch-offs and pinch pocket depths should be similar to those used for high density polyethylene (HDPE).

Achievement of optimum clarity and gloss depends on several processing parameters: a modest melt temperature in the range of 370–385°F (188–196°C), a low blow pressure, and a warm mold temperature of about 75°F (24°C).

49.11.3 Blown Film Extrusion

An attractive, glossy, blown film with good impact strength can be produced if the melt temperature is maintained in the range of 330–395°F (166–202°C). Lower melt temperatures will affect film appearance; higher melt temperatures will contribute to bubble instability. Bubble stability depends on a proper frost-line height, at one to two bubble diameters above the die, with a collapsing frame no more than 4–6 in. (102–152 mm) above that. The film has a high tensile yield and stiffness; as a result, wrinkles cannot be easily pulled out. To minimize wrinkle formation, the bubble must be kept symmetrical by accurate positioning of the die and nip, and careful adjustment of the collapsing frame. The impact strength increases as blow-up and draw-down ratios increase, so tooling must be selected with a generous die gap, preferably 40–50 mils (1–1.3 mm). For several reasons, therefore, most processors prefer to use equipment designed for HDPE rather than LDPE film production. K-Resin polymer can also be cast on a chill roll.

49.11.4 Sheet Extrusion

Optimum clarity and performance are obtained when the stock temperature is maintained at 400–420°F (204–216°C). Degradation can occur above 450°F (232°C). The die should be maintained at a uniform 380–400°F (193–204°C) across its width. The die temperature should not be used to adjust sheet gauge; mechanical die adjustment of the choker bar and/or die lips provides more accurate control. The drawdown ratio should be limited to 1.5:1 to 2:1. The polishing roll nip must be kept level and as close to the die as practical to minimize premature contact of the extrudate with the polishing rolls. At the nip, the first rolls should press firmly against the sheet, but the line speed should be adjusted to minimize the bank of excess melt, which can produce surface imperfections and stresses in the sheet.

For optimum clarity, the roll temperature should be maintained just below the point at which sheet sticks to the rolls, usually about 180°F (82°C).

49.11.5 Thermoforming

Sheet can be thermoformed using any common technique, especially simple male or female drape (with or without plug assist). To maximize clarity, the mold surfaces should be well-maintained and the mold vented generously, but not visibly.

To form most parts, sheet temperature should range from 250 to 300°F (121 to 149°C). The low side of that range yields optimum clarity and the shortest cooling cycles, but warmer sheet can improve uniformity of drawdown and reproduction of part detail. Large masses of unsupported sheet can sag, but sag bands must be positioned so their surface marks do not show on the part. For small, shallow parts, the mold temperature can be as low as 70°F (21°C).

In deep-draw items, especially when a plug assist is used, a mold temperature as high as 120°F (49°C) will facilitate uniform wall distribution. The excellent mold surface and detail reproduction of K-Resin polymer can make part removal difficult, especially from deep-draw male plug assists and highly polished molds. These should be provided with a release coating, a draft angle greater than 3° per side, and a means for blowing air through the vent system to break the natural vacuum for easy ejection.

49.12 PROCESSING-SENSITIVE END PROPERTIES

K-Resin polymers are most sensitive to excessive heat and intense shear, as discussed in the prior section. The resultant polymer degradation can compromise clarity, impact strength, melt flow and/or hinge life. Degradation can also result in elevated gel content or size and mold sticking with part distortion. On the other hand, insufficient processing temperatures can also reduce clarity and impact strength as well as reproduction of mold detail.

49.13 SHRINKAGE

See the Mold Shrinkage MPO for data pertaining to injection-molded K-Resin.

Polymer shrinkage in other forming processes depends heavily on polymer thickness and orientation (indicated by blow-up and draw-down ratios) as well as processing parameters. Specific shrinkage allowances for those processes are available from the resin supplier.

Mold Shrinkage Characteristics (in./in. or mm/mm)

MATERIAL   K-Resin

IN THE DIRECTION OF FLOW

THICKNESS

MATERIAL	mm: 1.5	3	6	8	FROM	TO	#
	in.: 1/16	1/8	1/4	1/2			
KR01	0.006	0.006	0.006		0.004	0.008	in./in.
KR03	0.005	0.005	0.005		0.005	0.009	in./in.

	PERPENDICULAR TO THE DIRECTION OF FLOW						
KR01	0.008				0.004	0.008	in./in.
KR03	0.009				0.005	0.009	in./in.

NOTE: Mold shrinkage is approximate. It is affected by part design, mold temperature, thickness, injection pressure, packing time, cycle time, orientation, gate design, gate size, gate location, glass content, glass size, and filler content.

49.14 TRADE NAMES

K-Resin polymers are produced only by Phillips 66 Company and only in the United States. They are sold only under the name K-Resin.

49.15 INFORMATION SOURCES

Phillips 66 Company provides a comprehensive technical information system on K-Resin polymers. Major titles are as follows.

Technical Information Bulletins (TIB): *TIB 200—Properties & Processing; TIB 201— Sheet Extrusion & Thermoforming; TIB 202—Injection Molding; TIB 203—Blow Molding; TIB 204—Film Extrusion.*

Technical Service Memorandums: *TSM 288 - Food Packageability of K-Resin Polymers; TSM 292—Medical Applications of K-Resin Polymers; TSM 296—Blow Film Extrusion.*

Plastics Technical Center Reports: *PTC 369—K-Resin Polymer Blends With General- Purpose Polystyrene.*

In addition, there are some 40 titles on specific studies.

50

THERMOPLASTIC POLYESTERS POLY(BUTYLENE TEREPHTHALATE) (PBT)

Richard Westdyk and Donal McNally

Hoechst Celanese Engineering Resins, Inc., 26 Main St., Chatham, NJ 07928

50.1 PBT

terephthalic acid   1,4-butanediol

polybutylene terephthalate (PBT)

Most practical processes employ dimethylterephthalate (DMT) instead of terephthalic acid (TA).

50.2 CATEGORY

Thermoplastic polyester (PBT) molding resin is a partially crystalline member of the class of engineering thermoplastics.

50.3 HISTORY

Polyesters of polybasic acids and polyhydric alcohols were first synthesized by Berzelius.[1] In the 1940s, J.R. Whinfield and J.T. Dickson prepared the terephthalates,[2] but commercial interest in PBT did not develop until the late 1960s. The product was taken to market by Celanese, immediately gaining recognition in the form of the IR100 award, an industry honor bestowed on the 100 most significant new products introduced in that year (1970).[3]

50.4 POLYMERIZATION

PBT resins are formed by the polycondensation of 1,4-butanediol with either terephthalic acid or dimethylterephthalate.

In truckload quantities, PBT compounds are priced from about $1.50 to about $2.50 per pound (1989), depending on the formulation.

50.5 DESCRIPTION OF PROPERTIES

PBT molding resins are available in unmodified, glass-reinforced, and flame-retarded grades. They offer high strength and rigidity, low moisture absorption, excellent electrical properties and chemical resistance, rapid molding cycles, and reproducible mold shrinkage. A fast crystallization rate enhances moldability, producing smooth surfaces that can be painted, printed, and ultrasonically welded. UL temperature indexes are relatively high because of a 440°F (227°C) melting point and a 104°F (40°C) glass-transition temperature.

Unreinforced PBT has a heat-deflection temperature of 310°F (154°C) at 66 psi (0.5 MPa). Aside from improving strength and stiffness, glass reinforcement also enhances thermal properties, making possible heat-deflection temperatures to 422°F (217°C).

Water absorption is less than 0.1% after 24 hours. However, PBT is not recommended for extended use in water or aqueous solutions above 125°F (52°C). Besides water, PBT resins are also intrinsically resistant to detergents, weak acids and bases, aliphatic hydrocarbons, fluorinated hydrocarbons, alcohols, ketones, ethylene glycol, carbon tetrachloride, oils, and fats at room temperature. In addition, they exhibit good resistance to motor oil, gasoline, transmission and brake fluids at temperatures to 140°F (60°C).

Unreinforced PBT has a tensile strength of 8000 psi (55 MPa). With 40% glass reinforcement, tensile strength increases to 21,300 psi (147 MPa). Corresponding flexural moduli are 330,000 psi (2280 MPa) and 1,500,000 psi (10,340 MPa). Mineral-filled and mineral/glass-filled grades render intermediate strength and stiffness.

Excellent dimensional stability, principally resulting from low moisture absorption rates, is a distinguishing feature of PBT resins. Notched Izod impact strength ranges from 1.0 ft-lb/in. (53 J/m) for unreinforced grades to 3.5 and 16.0 ft-lb/in. (187 and 854 J/m) respectively, for reinforced and impact-modified unreinforced grades. Glass–reinforced PBT exhibits good resistance to creep at both ambient and elevated temperatures.

The electrical properties of PBT resins are stable over a wide range of temperature and humidity. They are good insulators; volume resistivity is greater than 10^{16} ohm-cm. Arc resistance and dielectric strength are high; the dissipation factor low.

Both unreinforced and reinforced grades of PBT resins are rated UL94 HB. Grades are available to meet UL94 V-0 requirements, and many of them also meet the more restrictive flammability requirements of UL94 5V. Strength, stiffness, and better thermal properties are gained by reinforcing PBT resins with glass fibers. If very low warpage is desired, reinforcement with mineral or mineral/glass fillers is recommended.

Grades with glass contents ranging from 7 to 50 percent are available. Mineral and mineral/glass-filled grades have filler contents between 10 and 40%. The mechanical properties of these grades are between those of unreinforced and glass-reinforced resins.

When specifications call for greater impact strength than that offered by glass-reinforced grades, impact-modified resins are recommended. Special blends can be obtained for applications demanding improved surface appearance and texture.

50.6 APPLICATIONS

PBT resins are used in many types of products: appliances, automobiles, electrical and electronic parts, industrial components, and consumer items. Their resistance to heat and chemicals, plus their wide range of color possibilities, dictate their use in appliances, where they have been found easier to mold and less expensive than thermosetting plastics. Iron and toaster housings, cooker/fryer handles, hair drier nozzles, and food processor blades are all made of PBT.

Strong, light, and weather- and heat-resistant, PBT has been used for both exterior and under-the-hood parts of automobiles. Examples include fender extensions, cowl vents, rear/side louvers and panels, door handles, brake system components, deck lid gear trains, structural brackets, vacuum actuators, air-conditioning valves, distributor caps and rotors, hydraulic transmission parts, ignition coil bobbins, molded-on wire connectors, lamp socket inserts, and rectifier bridges.

The electrical/electronics industry uses PBT for bobbins, connectors, switches, relays, terminal boards, motor brush holders, TV tuners, fuse cases, IC carriers and sockets, and end bells. Tough, dimensionally stable, and having good electrical properties, the material permits automated soldering. It withstands high temperatures and chemical degreasing, and it costs less than epoxies.

In the industrial world, PBT resins are used for pump housings, impellers, valves, brackets, water meter components, tool housings, casings, and replacements for metals in many types of load-bearing parts. Their high strength-to-weight ratio and resistance to corrosion and chemicals make them ideal choices for these and other industrial components.

Because they are light, colorful, and easy to mold with good friction properties, PBT resins are used in drapery hardware, pen barrels, heavy-duty zippers, hair dryers, pocket calculators, and similar consumer products.

50.7 ADVANTAGES/DISADVANTAGES

PBT combines many desirable properties such as high heat-deflection temperatures, high rigidity, broad chemical resistance, low creep, low moisture absorption, excellent electrical properties, and superior dimensional stability at elevated temperatures.

PBT should not be used in hot water applications [above 120–130°F (49–54°C)], which can cause a breakdown in molecular strength and severe degradation. Strong bases and oxidizing acids should likewise be avoided.

50.8 PROCESSING TECHNIQUES

See the Master Material Outline and the Standard Material Design Chart.

50.9 RESIN FORMS

PBT is available in pellet form.

50.10 SPECIFICATION OF PROPERTIES

See the Master Outline of Material Properties.

50.11 PROCESSING REQUIREMENTS

PBT resins are used for a variety of injection molded and extruded parts. The low melt viscosity of PBT at normal operating temperatures allows easy filling of thin-sectioned mold cavities. Melts range between 450 and 500°F (232 and 260°C), mold temperatures between 100 and 200°F (38 and 93°C). Cycle times can be as fast as five seconds for small parts.

Master Material Outline

PROCESSABILITY OF ____PBT____

PROCESS	A	B	C
INJECTION MOLDING	X		
EXTRUSION	X		
THERMOFORMING		X	
FOAM MOLDING	X		
DIP, SLUSH MOLDING			
ROTATIONAL MOLDING			
POWDER, FLUIDIZED-BED COATING			
REINFORCED THERMOSET MOLDING			
COMPRESSION/TRANSFER MOLDING			
REACTION INJECTION MOLDING (RIM)			
MECHANICAL FORMING			
CASTING			

A = Common processing technique.
B = Technique possible with difficulty.
C = Used only in special circumstances, if ever.

Standard design chart for ____PBT____

1. Recommended wall thickness/length of flow
   Minimum ____0.007____ for ____1–2 in.____ distance
   Maximum ____0.125____ for ____7–8 in.____ distance
   Ideal ____0.020–0.060____ for ____3–4 in.____ distance

2. Allowable wall thickness variation, % of nominal wall _____

3. Radius requirements
   Outside: Minimum ____75%____ Maximum ____125%____
   Inside: Minimum ____25%____ Maximum ____75%____

4. Reinforcing ribs
   Maximum thickness ____50%____
   Maximum height ____3____
   Sink marks: Yes _____ No __X (min)__

5. Solid pegs and bosses
   Maximum thickness _____2–3 × wall_____
   Maximum weight ____2 × boss height____
   Sink marks: Yes ___X___ No _____

6. Strippable undercuts
   Outside: Yes _____ No ___X___
   Inside: Yes _____ No ___X___

7. Hollow bosses: Yes ___X___ No _____

8. Draft angles
   Outside: Minimum _____$\frac{1}{2}°$_____ Ideal _____1°_____
   Inside: Minimum _____$\frac{1}{2}°$_____ Ideal _____1°_____

9. Molded-in inserts: Yes ___X___
                        No _____

10. Size limitations
    Length: Minimum ___$\frac{1}{2}$___ Maximum ___6 ft___ Process _Injection molding_
    Width: Minimum _____ Maximum _____ Process _____

11. Tolerances

12. Special process- and materials-related design details
    Blow-up ratio (blow molding) _____
    Core L/D (injection and compression molding) _____
    Depth of draw (thermoforming) _____
    Other _____

Prior to processing, PBT resins should be dried in dehumidifying hopper driers or ovens for four hours at 250°F (121°C) to ensure optimum part performance. The resins can be molded in plunger or screw-type injection machines. They can also be extruded as sheet, film, or profiles, and spun as fiber or monofilament. Reground sprues and runners can be blended with virgin resin for reuse after having been properly dried.

50.11.1 Assembly

PBT is not soluble in most common solvents at room temperature, making the use of solvent adhesives impractical for part assembly. Some epoxy, cyanoacrylate, urethane, and rubber-based adhesives have been found useful, affording reasonable shear strengths but low peel strengths.

Quick, clean part assembly is possible with ultrasonic welding. Little pressure is needed to produce weld strengths that are 90 to 100% of the strength of the base resin. Scarf and interference-type joints have the highest strengths.

A modification of ultrasonic welding permits the staking of metal parts to PBT parts. Cycle times for this procedure are very short.

PBT parts can also be assembled by means of hot bar welding, spin welding, self-tapping screws, intereference and snap fits, and threaded inserts.

50.12 PROCESSING-SENSITIVE END PROPERTIES

To optimize PBT's properties, such as impact strength, it should be dried prior to processing for four hours at 250°F (121°C).

It is important not to overheat the material during processing; melt temperature of 480–520°F (249–271°C) should suffice. Mold temperatures in the range 100°F to 250°F are commonly used with these materials.

MATERIAL ___ PBT, unreinforced

PROPERTY	TEST METHOD	ENGLISH UNITS	VALUE	METRIC UNITS	VALUE
MECHANICAL DENSITY	D792	lb/ft^3	81.7	g/cm^3	1.31
TENSILE STRENGTH	D638	psi	8000	MPa	55
TENSILE MODULUS	D638	psi	370,000	MPa	2551
FLEXURAL MODULUS	D790	psi	330,000	MPa	2275
ELONGATION TO BREAK	D638	%		%	
NOTCHED IZOD AT ROOM TEMP	D256	ft-lb/in.	1.0	J/m	53
HARDNESS	D785		Rm 72		Rm 72
THERMAL DEFLECTION T @ 264 psi	D648	°F	130	°C	54
DEFLECTION T @ 66 psi	D648	°F	310	°C	154
VICAT SOFTENING POINT	D1525	°F		°C	
UL TEMP INDEX	UL746B	°F		°C	
UL FLAMMABILITY CODE RATING	UL94		HB		HB
LINEAR COEFFICIENT THERMAL EXPANSION	D696	in./in./°F	4.1×10^{-5}	mm/mm/°C	7.4×10^{-5}
ENVIRONMENTAL WATER ABSORPTION 24 HOUR	D570	%	0.1	%	0.1
CLARITY	D1003	% TRANSMISSION		% TRANSMISSION	
OUTDOOR WEATHERING	D1435	%	Good resistance	%	Good resistance
FDA APPROVAL			Yes		Yes
CHEMICAL RESISTANCE TO: WEAK ACID	D543		Not attacked		Not attacked
STRONG ACID	D543		Not attacked		Not attacked
WEAK ALKALI	D543		Minimally attacked		Minimally attacked
STRONG ALKALI	D543		Badly attacked		Badly attacked
LOW MOLECULAR WEIGHT SOLVENTS	D543				
ALCOHOLS	D543				
ELECTRICAL DIELECTRIC STRENGTH	D149	V/mil	400	kV/mm	15.8
DIELECTRIC CONSTANT	D150	10^6 Hertz	3.24		3.24
POWER FACTOR	D150	10^6 Hertz	0.019		0.019
OTHER MELTING POINT	D3418	°F	437	°C	225
GLASS TRANSITION TEMP.	D3418	°F	104	°C	40

MATERIAL ___ PBT, 30% glass fiber reinforced

PROPERTY	TEST METHOD	ENGLISH UNITS	VALUE	METRIC UNITS	VALUE
MECHANICAL DENSITY	D792	lb/ft^3	94.8	g/cm^3	1.52
TENSILE STRENGTH	D638	psi	18,000	MPa	124
TENSILE MODULUS	D638	psi	1,400,000	MPa	9650
FLEXURAL MODULUS	D790	psi	1,200,000	MPa	8274
ELONGATION TO BREAK	D638	%	2	%	2
NOTCHED IZOD AT ROOM TEMP	D256	ft-lb/in.	1.5–2.5	J/m	80–133
HARDNESS	D785		Rm 90		Rm 90
THERMAL DEFLECTION T @ 264 psi	D648	°F	410	°C	210
DEFLECTION T @ 66 psi	D648	°F	430	°C	221
VICAT SOFTENING POINT	D1525	°F		°C	
UL TEMP INDEX	UL746B	°F		°C	
UL FLAMMABILITY CODE RATING	UL94		HB		HB
LINEAR COEFFICIENT THERMAL EXPANSION	D696	in./in./°F	1.5×10^{-5}	mm/mm/°C	2.7×10^{-5}
ENVIRONMENTAL WATER ABSORPTION 24 HOUR	D570	%	0.1	%	0.1
CLARITY	D1003	% TRANSMISSION		% TRANSMISSION	
OUTDOOR WEATHERING	D1435	%	Good resistance	%	Good resistance
FDA APPROVAL					
CHEMICAL RESISTANCE TO: WEAK ACID	D543		Not attacked		Not attacked
STRONG ACID	D543		Not attacked		Not attacked
WEAK ALKALI	D543		Minimally attacked		Minimally attacked
STRONG ALKALI	D543		Badly attacked		Badly attacked
LOW MOLECULAR WEIGHT SOLVENTS	D543				
ALCOHOLS	D543				
ELECTRICAL DIELECTRIC STRENGTH	D149	V/mil	525	kV/mm	20.7
DIELECTRIC CONSTANT	D150	10^6 Hertz	3.66		3.66
POWER FACTOR	D150	10^6 Hertz	0.013		0.013
OTHER MELTING POINT	D3418	°F	437	°C	225
GLASS TRANSITION TEMP.	D3418	°F	104	°C	40

635

50.13 SHRINKAGE

See Mold Shrinkage Characteristics.

Mold Shrinkage Characteristics (in./in. or mm/mm)

MATERIAL _____PBT_____

IN THE DIRECTION OF FLOW

THICKNESS

MATERIAL	mm: 1.5 in.: 1/16	3 1/8	6 1/4	8 1/2	FROM	TO	#
Unreinforced PBT		0.017					
Glass-filled PBT		0.002					

PERPENDICULAR TO THE DIRECTION OF FLOW						
Unreinforced PBT		0.023					
Glass-filled PBT		0.006					

NOTE: Mold shrinkage is approximate. It is affected by part design, mold temperature, thickness, injection pressure, packing time, cycle time, orientation, gate design, gate size, gate location, glass content, glass size, and filler content.

Standard Tolerance Chart

PROCESS _____Injection molding_____   MATERIAL _____PBT_____

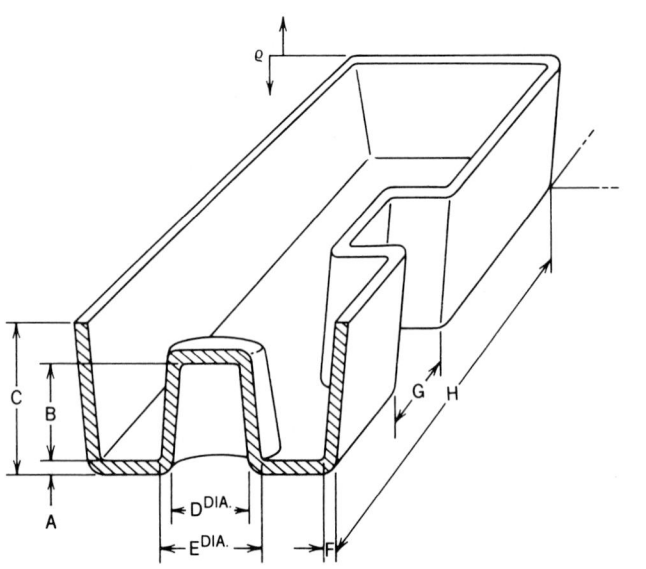

Values are based on a nominal 1/8 in. (3.2 mm) wall thickness and given in plus or minus in./in. and mm/mm.

FINE tolerance – Special care and added cost.

COMMERCIAL tolerance – Normal care required.

COARSE tolerance – Minimal care required, lowest cost.

Tolerance	A	B	C	D	E	F	G	H	
FINE									in./in.
									mm/mm
COMMERCIAL all data ±	0.001	0.001	0.001	0.001	0.001	0.001	0.001	0.001	in./in.
									mm/mm
COARSE									in./in.
									mm/mm

50.14 TRADE NAMES

Currently, PBT is supplied in the U.S. primarily under the trade name Celanex by Hoechst Celanese, and under the trade name Valox by General Electric Plastics.

BIBLIOGRAPHY

1. V.V. Korshak and S.V. Vinogradova, *Polyesters,* Pergamon, Oxford, 1965.
2. U.S. Pat. 2,465,319 (1940), J.R. Whinfield and J.T. Dickson (to E.I. du Pont de Nemours & Co., Inc.
3. Industrial Research Inc., 1970, private communication.

51

THERMOPLASTIC POLYESTERS, POLY(ETHYLENE TEREPHTHALATE) (PET)

F.E. McFarlane

Eastman Chemicals Div., Eastman Kodak Co., Kingsport, TN 37662

51.1 PET

The most widely used thermoplastic polyester is poly(ethylene terephthalate), or PET. The use of PET is growing very rapidly in textiles, packaging, audio and video film, engineering resin applications, and such miscellaneous applications as cable wrap.

Although PET is a homopolymer, the thermoplastic polyester family also includes co-polymers.

dimethyl terephthalate   ethylene glycol

poly(ethylene terephthalate) (PET)   methanol

51.2 CATEGORY

PET is a condensation polymer, which means that the polymerization process involves the elimination of water. Phenolics, polyamides, and polyesters are all examples of products formed by this type of reaction.

51.3 HISTORY

A PET homopolymer is formed by reacting one acid, such as terephthalic acid, with one glycol or diol, such as ethylene glycol. However, when a part of the acid moiety is replaced by another acid, or a part of the diol is replaced by another diol, a copolymer is formed.

The discovery of PET was patented in 1941 by John Rex Whinfield and James Tennant Dickson. The polymer was commercialized in 1953 as a textile fiber and shortly afterward in film form. Biaxially oriented film was introduced by Imperial Chemical Industries Ltd. in the early 1950s.[1]

Initially, PET was considered unsuitable as a thermoplastic molding resin because of the brittleness of thick sections crystallized from the melt. However, in 1966, grades of PET were introduced that were suitable for injection molding and extrusion applications. PET materials now in use are known for their good mechanical, chemical, and electrical properties and for their unusual ability to exist in either an amorphous or a crystalline state. The crystallinity can range from 0 to 60%.

The degree of crystallinity of the polymer will affect several properties, among them chemical resistance, fiber forming ability, thermal stability, and water sensitivity, to name a few.

One of the most significant breakthroughs in modern plastics technology occurred in the late 1960s with the use of PET in carbonated beverage bottles. In this application, stretch blow-molding grades provide the needed toughness and clarity. In the stretch blow-molding process, a parison, or preform, is first injection molded. It is then placed on a mandrel in a quartz heated oven. The preform is heated to just above its glass-transition temperature (T_g) and is stretched to approximately one-and-a-half times its initial length when air is introduced to blow the preform into the contours of the mold.

Later developments include the use of reinforced and unreinforced PET formulations in injection molded and extruded products. These commercial formulations contain such additives as glass and other minerals, flame retardants, impact modifiers, colorants, and stabilizers.

In recent years, copolymers of PET have become increasingly important. One type of copolymer is produced by replacing part of the terephthalic acid in PET with isophthalic acid. In another type, the diol moiety, ethylene glycol, is replaced by 1,4-cyclohexanedimethanol.

PET copolymers are characterized by lower melting points. Some of the copolymers such as PETG® copolyester are completely amorphous or totally noncrystalline. This characteristic allows PET copolymers to be processed at lower temperatures, which minimizes the likelihood of molecular breakdown of the polymer and the resultant loss of properties in the finished part.

51.4 POLYMERIZATION

PET homopolymer is manufactured by the reaction of ethylene glycol with terephthalic acid or dimethyl terephthalate. About 0.35 lb (0.16 kg) of ethylene glycol and 0.87 lb (0.4 kg) of terephthalic acid [or 1.02 lb (0.46 kg) of DMT] are used to make 1 lb (0.45 kg) of PET.

Polymerization takes place in two stages. In the first stage, a prepolymer is formed by heating the reaction mixture and removing the volatile by-products and excess glycol. In the second stage, heat and vacuum are employed to increase the molecular weight with concurrent removal of volatile products, such as water and glycol. Very high molecular weight polymer, that used in bottles or tire cord, eg, is obtained by using a second stage— a solid state polymerization stage—where the polymer is heated to about 54°F (30°C) below the crystalline melting point in a vacuum or in a stream of inert gas.

The molecular weight and polymer composition can be varied in the polymer manufacturing process. Molecular weight is commonly indicated in terms of intrinsic or inherent viscosity (IV). The IV of PET is an indication of the melt viscosity and the resulting physical properties of the end product. By adding one or more comonomers, dramatic changes can be obtained in such properties as the softening point, crystallization rate, glass-transition temperature, flexibility, and toughness.

51.5 DESCRIPTION OF PROPERTIES

PET homopolymer and some of its lower degree of copolymer modifications are known as crystallizable polymers. When held for a period of time in a given temperature range above their T_g of 176°F (80°C), they will crystallize spontaneously. The maximum rate of crystallization occurs in the temperature range of 285–375°F (140–190°C); the melting point is about 489°F (254°C). The crystallization rate can be decreased by chemical modification. Sufficient modification (copolymerization) of PET with an additional glycol (or acid) can result in a polymer that will not crystallize. Such polymers are commonly referred to as having an "amorphous" molecular structure.

PET is known for its clarity and toughness when it is used for the manufacture of oriented film or stretch-blown bottles. It is also a good barrier to gases, such as oxygen and carbon dioxide. PET can be reinforced with glass fibers to produce an injection-molded product with high heat resistance and high stiffness. The resistance of PET to chemicals and solvents is greatly enhanced as the crystallinity of the polymer increases.

51.6 APPLICATIONS

PET finds applications in such diverse end uses as fibers for clothing; films; bottles; food containers; and engineering plastics for precision-molded parts. A wide range of applications is possible because of the excellent balance of properties PET possesses and because the degree of crystallinity and the level of orientation in the finished part can be controlled.

51.7 ADVANTAGES/LIMITATIONS

As noted above, PET is a versatile plastic because of its excellent physical properties and because it can be converted into either amorphous or semicrystalline products. When properly processed, it can be made into oriented and crystallized articles that still possess excellent clarity. Outstanding dimensional stability can be obtained in PET film by controlling orientation and by heat setting during processing. Few other materials offer such a range of processing and property variables. For packaging applications, PET is used because it combines optimum processing, mechanical, and barrier properties.

The barrier properties of PET have been found to be adequate for the retention of carbon dioxide for packaging carbonated beverages. The good oxygen barrier properties of PET allow for its use in many types of food containers. However, PET barrier properties are deficient for packaging beer and certain foods that are highly sensitive to oxygen.

PET homopolymer is injection molded primarily as a crystallized, reinforced material. Applications for glass-reinforced PET include electrical connectors, auto body parts, and other applications where very high temperature environments are not encountered.

PET homopolymer is not generally acceptable for injection molding amorphous (clear) parts because of its propensity to crystallize when used in thick sections and when exposed to temperatures above its T_g. For clear moldings, PET copolymers are more useful. Typical injection molded uses for copolymers include medical devices, toys, and cosmetic jars.

51.8 PROCESSING TECHNIQUES

See the Master Material Outline.

Master Material Outline

PROCESSABILITY OF _____ PET _____

PROCESS	A	B	C
INJECTION MOLDING	X		
EXTRUSION	X		
THERMOFORMING	X		
FOAM MOLDING			
DIP, SLUSH MOLDING			
ROTATIONAL MOLDING			
POWDER, FLUIDIZED-BED COATING			
REINFORCED THERMOSET MOLDING			
COMPRESSION/TRANSFER MOLDING			
REACTION INJECTION MOLDING (RIM)			
MECHANICAL FORMING			
CASTING			

A = Common processing technique.
B = Technique possible with difficulty.
C = Used only in special circumstances, if ever.

Standard design chart for _____ PET (IV = 0.72) _____

1. Recommended wall thickness/length of flow
   Minimum _____ $\frac{1}{16}$ in. _____ for _____ 4 in. _____ distance
   Maximum _____ $\frac{1}{4}$ in. _____ for _____ 24 in. _____ distance
   Ideal _____ $\frac{1}{8}$ _____ for _____ 12 in. _____ distance

2. Allowable wall thickness variation, % of nominal wall _____

3. Radius requirements
   Outside: Minimum _____ Maximum _____
   Inside: Minimum _____ Maximum _____

4. Reinforcing ribs
   Maximum thickness _____
   Maximum height _____
   Sink marks: Yes _____ No _____

5. Solid pegs and bosses
   Maximum thickness _____

Maximum height _____

Sink marks: Yes _____ No _____

6. Strippable undercuts
   Outside: Yes _____ No _____
   Inside: Yes _____ No _____

7. Hollow bosses: Yes _____ No _____

8. Draft angles
   Outside: Minimum _____ Ideal _____
   Inside: Minimum _____ Ideal _____

9. Molded-in inserts: Yes _____
   .   No _____

10. Size limitations
    Length: Minimum _____ Maximum _____ Process _____
    Width: Minimum _____ Maximum _____ Process _____

11. Tolerances

12. Special process- and materials-related design details
    Blow-up ratio (blow molding) _____
    Core L/D (injection and compression molding) _____
    Depth of draw (thermoforming) _____
    Other _____

51.9 RESIN FORMS

Thermoplastic polyester resins are manufactured and supplied generally in 1/16-in. (1.6 mm) to 1/8 in. (3.2 mm) cylindrical or cubed pellets.

51.10 SPECIFICATION OF PROPERTIES

See the Master Outline of Materials Properties.

51.11 PROCESSING REQUIREMENTS

Both homopolymers and copolymers must be dried prior to use. Desiccant dryers designed for thermoplastic polyesters may be used to reduce the moisture content to the desired 0.005% or less. Processing at higher moisture levels will result in a loss of polymer molecular weight and other physical properties.

Acetaldehyde (AA), a product of thermal degradation, is particularly undesirable in beverage-container applications. Acetaldehyde occurs naturally in fruits such as apples; it imparts a fruity off-flavor to cola-type soft drinks. Therefore, when PET bottles are manufactured, it is particularly important to minimize the amount of AA generated by carefully controlling the molding and bottle-blowing conditions.

PET scrap generated in the processing operation can be reground, dried with virgin material and fed to an extruder. Normally, the amount of regrind should not exceed 10%.

Amorphous PETG® copolyester must be dried below its glass-transition temperature of 176°F (80°C). It can be injection molded at 420–435°F (216–224°C).

MATERIAL ___ PETG® Copolymer

PROPERTY	TEST METHOD	ENGLISH UNITS	VALUE	METRIC UNITS	VALUE
MECHANICAL DENSITY	D792	lb/ft^3	79.2	g/cm^3	1.27
TENSILE STRENGTH	D638	psi	7100	MPa	50
TENSILE MODULUS	D638	psi	250,000	MPa	1724
FLEXURAL MODULUS	D790	psi	290,000	MPa	2000
ELONGATION TO BREAK	D638	%	180	%	180
NOTCHED IZOD AT ROOM TEMP	D256	ft-lb/in.	1.7	J/m	90
HARDNESS	D785		R105		R105
THERMAL DEFLECTION T @ 264 psi	D648	°F	145	°C	63
DEFLECTION T @ 66 psi	D648	°F	160	°C	71
VICAT SOFT-ENING POINT	D1525	°F	82	°C	180
UL TEMP INDEX	UL746B	°F		°C	
UL FLAMMABILITY CODE RATING	UL94		HB		HB
LINEAR COEFFICIENT THERMAL EXPANSION	D696	in./in./°F	5.1×10^{-5}	mm/mm/°C	9.1×10^{-5}
ENVIRONMENTAL WATER ABSORPTION 24 HOUR	D570	%	0.5	%	0.5
CLARITY	D1003	% TRANS-MISSION	90	% TRANS-MISSION	90
OUTDOOR WEATHERING	D1435	%	Not suggested	%	Not suggested
FDA APPROVAL			Yes		Yes
CHEMICAL RESISTANCE TO: WEAK ACID	D543		Fair		Fair
STRONG ACID	D543		Poor		Poor
WEAK ALKALI	D543		Fair		Fair
STRONG ALKALI	D543		Poor		Poor
LOW MOLECULAR WEIGHT SOLVENTS	D543		Fair		Fair
ALCOHOLS	D543		Fair		Fair
ELECTRICAL DIELECTRIC STRENGTH	D149	V/mil	1/8"–400 1/16"–560	kV/mm	1/8"–15.7 1/16"–22.1
DIELECTRIC CONSTANT	D150	10^6 Hertz	3.2		3.2
POWER FACTOR	D150	10^6 Hertz			
OTHER MELTING POINT	D3418	°F	None	°C	None
GLASS TRANS-ITION TEMP.	D3418	°F	178	°C	81

MATERIAL ___ PET (0.76 IV Bottle Polymer)

PROPERTY	TEST METHOD	ENGLISH UNITS	VALUE	METRIC UNITS	VALUE
MECHANICAL DENSITY	D792	lb/ft^3	87.3	g/cm^3	1.40
TENSILE STRENGTH	D638	psi	hoop–25,000* axial–10,000*	MPa	hoop–172* axial–69*
TENSILE MODULUS	D638	psi	hoop–620,000* axial–320,000*	MPa	hoop–4275* axial–2206*
FLEXURAL MODULUS	D790	psi		MPa	
ELONGATION TO BREAK	D638	%		%	
NOTCHED IZOD AT ROOM TEMP	D256	ft-lb/in.		J/m	
HARDNESS	D785				
THERMAL DEFLECTION T @ 264 psi	D648	°F		°C	
DEFLECTION T @ 66 psi	D648	°F		°C	
VICAT SOFT-ENING POINT	D1525	°F	174	°C	79
UL TEMP INDEX	UL746B	°F		°C	
UL FLAMMABILITY CODE RATING	UL94				
LINEAR COEFFICIENT THERMAL EXPANSION	D696	in./in./°F		mm/mm/°C	
ENVIRONMENTAL WATER ABSORPTION 24 HOUR	D570	%	<0.1	%	<0.1
CLARITY	D1003	% TRANS-MISSION	88	% TRANS-MISSION	88
OUTDOOR WEATHERING	D1435	%	Not suggested	%	Not suggested
FDA APPROVAL			Yes		Yes
CHEMICAL RESISTANCE TO: WEAK ACID	D543		Good		Good
STRONG ACID	D543		Fair		Fair
WEAK ALKALI	D543		Good		Good
STRONG ALKALI	D543		Fair		Fair
LOW MOLECULAR WEIGHT SOLVENTS	D543		Fair		Fair
ALCOHOLS	D543		Good		Good
ELECTRICAL DIELECTRIC STRENGTH	D149	V/mil	650	kV/mm	26
DIELECTRIC CONSTANT	D150	10^6 Hertz	3.3		3.3
POWER FACTOR	D150	10^6 Hertz	0.03		0.03
OTHER MELTING POINT	D3418	°F	473	°C	245
GLASS TRANS-ITION TEMP.	D3418	°F	176	°C	80

*Properties determined from the sidewalk of a stretch blow-molded bottle.

51.12 PROCESSING-SENSITIVE END PROPERTIES

Properties of polyester parts can be affected to a very significant degree by the processing techniques used to make them. There are three key ways to alter PET properties: control of crystallinity, biaxial orientation, and heat setting. These are not necessarily independent variables. For example, biaxial orientation normally results in some degree of crystallinity in the part produced.

PET film that is chilled after it is extruded will remain amorphous and transparent because its crystallinity is controlled to a very low level by the rapid cooling. If the amorphous film or sheet is vacuum-formed into a tray and cooled quickly, it will remain largely amorphous. By holding the sheet in the mold at 350°F (177°C) for several seconds, the trays will crystallize and become an opaque white. This is an advantage since an amorphous tray would distort badly above its T_g of 176°F (80°C); however, a crystallized tray can be used in an oven at temperatures approaching its crystalline melting point of 485°F (253°C) without distortion. Trays are now in commercial production for use as dual ovenable food trays. Such trays, containing frozen foods, can be taken from storage and placed directly into either a conventional oven or a microwave oven for heating.

The physical properties of PET film and bottles can be changed markedly by the process of biaxial orientation. As crystallizable materials such as PET homopolymer are stretched, their tensile strength and modulus are increased proportionately until strain-induced crystallization begins to occur. This is the process that gives biaxially oriented PET film and two-liter soft-drink bottles their outstanding physical and gas barrier properties. These properties, characterized by high tensile and flexural strength, allow for the packaging of carbonated beverages under pressures up to approximately 60 psig (0.4 MPa).

PET film or bottles made in this manner possess a memory. If the part is reheated beyond its glass-transition temperature, the built-in stress, or memory, will cause it to shrink. To make a product that is acceptable for higher temperature applications, it is necessary to heat set the product by raising its temperature to approximately 400°F (204°C), while restraining it to prevent shrinkage. Heat-setting of PET film is more widely practiced than heat-setting of bottles.

There are many potential PET food-packaging applications in which the food is either added to the container while it is hot or the container is pasteurized after the food is added. Proper heat-setting may permit the use of PET in such applications.

51.13 SHRINKAGE

See the Mold Shrinkage MPO.

Mold Shrinkage Characteristics (in./in. or mm/mm)

MATERIAL _____ PET

IN THE DIRECTION OF FLOW

THICKNESS

MATERIAL	mm: 1.5	3	6	8	FROM	TO	#
	in.: 1/16	1/8	1/4	1/2			
	0.003–0.004	0.003–0.005	0.003–0.006				

PERPENDICULAR TO THE DIRECTION OF FLOW						
0.003–0.004	0.003–0.005	0.003–0.006				

NOTE: Mold shrinkage is approximate. It is affected by part design, mold temperature, thickness, injection pressure, packing time, cycle time, orientation, gate design, gate size, gate location, glass content, glass size, and filler content.

51.14 TRADE NAMES

In the United States, PET is sold as Selar® by DuPont; as Kodapak® and Tenite® by Eastman Chemical Products Inc.; as Cleartuf® by Goodyear; as Hostalen® by Hoechst; and as Melinar® by ICI. The copolymer is sold as Kodar PETG® copolyester by Eastman Chemical Products, Inc.

BIBLIOGRAPHY

1. "Production, Properties, and Applications of the New British Polyethylene Terephthalate Film," *British Plastics*, (Jan. 1953).

52

UREA—MELAMINE

A.E. Campi
Brookpark-Royalon, Inc., Lake City, PA 16423

N.R. Reyburn
Consultant, 4810 Wolf Rd., Erie, PA 16505

52.1 AMINO PLASTICS: MELAMINE AND UREA

Amino plastics (melamine and urea) are thermosetting polymers. They are produced by a polymerization method involving the controlled reaction of formaldehyde and any of various amino-containing compounds. Offered in a variety of physical forms, these resins are widely used in molding, adhesive bonding, and coating applications.

The molecular structures of amino resins are as follows:

Melamine   Urea   Formaldehyde

52.2 CATEGORY

Melamine and urea thermosetting resins are stiff and brittle. Hardness and solvent resistance are among their general attributes.

52.3 HISTORY

Urea was synthesized by Wohler in 1828. It was not until the early 1920s that Kurt Ripper and Fritz Pollac, working in Germany, combined urea with formaldehyde to form a resin. In 1932, as a result of the Toledo Scales Co.'s efforts to produce a white, lightweight housing for scales, urea moldign materials were offered commercially in the United States by American Cyanamid Co. and by Plaskon Co.

Melamine was produced by Liebig in 1834. It was not developed in the United States as a molding material until 1939, through the work of Palmer Griffith and Paul Schroy at American Cyanamid Co.

Originally, these resins were developed for their wide range of colors and little modification has been made in this area. However, to take advantage of urea's arc-tracking resistance, wood-flour-filled brown and black ureas are offered at prices that compete with phenolics for wiring device applications.

Melamine's electrical and heat and hardness resistance properties are sufficiently better than urea's. In spite of its higher cost, melamine has been modified to increase its mechanical strength and dimensional stability. Two approaches have been taken. One, by Plastics Engineering, was to make copolymer melamine–phenol resins with mineral and cellulosic fillers. The other, by Fiberite Corp. and U.S. Polymeric, was to add glass fiber and other fillers to melamine. Fiberite also offers a phenol-modified melamine with chopped cotton filler for military food-service ware.

The development of glaze coatings for in-mold application to melamines has afforded a high-gloss, chinalike surface to melamines, which lengthens product service life by 300%. Ornapres, a Swiss firm, has developed a version of glaze material that affords freedom from coffee stain, which has long been a drawback to acceptance of melamine coffee cups.

The figures given below for total urea and melamine production include both resins and molding compounds.

Year	Production, $\times 10^6$ lb	% of Total Plastics Sales
1960	560	9%
1970	638	3%
1980	1.3	3%

52.4 POLYMERIZATION

When urea or melamine are reacted with formaldehyde, a clear, water-white thermosetting resin is produced.

Formaldehyde is a gas that liquefies at $-2°F$ ($-19°C$); it has a highly irritating odor. Formaldehyde is made from the oxidation of wood alcohol (methanol) in the presence of a catalyst (copper or platinum). $CH_3 - OH \rightarrow H_2 + CH_2O$. About 37% formaldehyde dissolved in water is known as formalin. When it is reacted with urea or melamine, a resinous syrup is produced that, when heated, reacts to form a resin that is very brittle and has very little strength. A 20–25% filler is added to the syrup, which is then processed into molding compound by the addition of a catalyst, curing agent, lubricant, and pigments. The filler can be paper, wood flour, or any other compatible material.

The manufacturing steps in the production of urea or melamine molding compounds are shown in Figure 52.1. As of 1988, the cost of urea molding material is 62 cents/lb for black or brown and 72 cents/lb for white. The cost of melamine molding material (dinnerware grade) is 83 cents/lb.

52.5 PROPERTIES

52.5.1 Thermal Properties

By definition, a thermoset plastic cannot be softened by heat. For example, the urea or melamine handle of a cooking utensil that is exposed to direct flame or an electric insulator that is exposed to an overload temperature will retain the integrity of the functional shape. The product color will darken on exposure to high direct heat. If the duration and severity of the exposure continues, amino-resin molded items will char and ultimately lose their strength, but they will retain their basic shapes.

Since color is important in many applications, suppliers publish continuous-service temperatures based on color stability. For urea, this is 170°F (77°C); for melamine, 210°F (99°C). The UL temperature index (based on maintaining properties other than color) is 212°F (100°C) for urea and 266°F (130°C) for melamine.

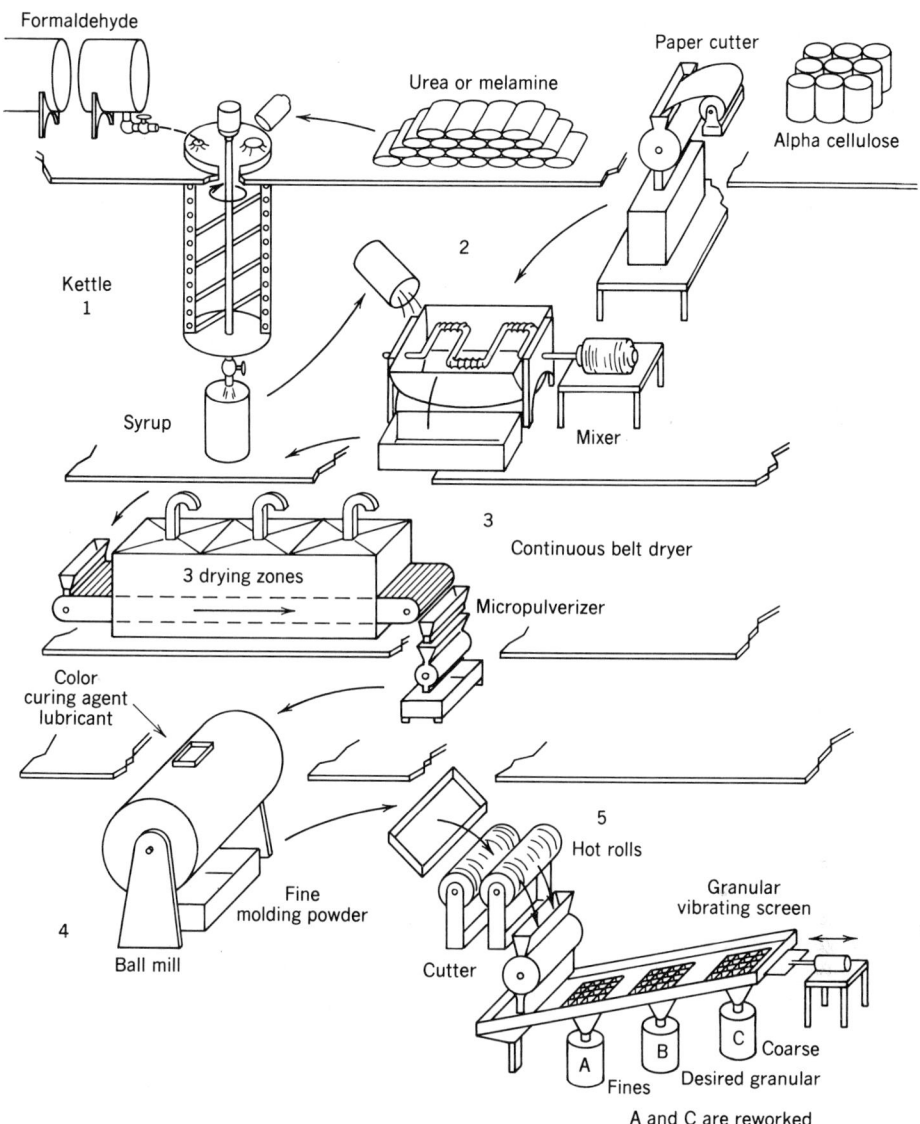

Figure 52.1. Manufacturing steps in the production of melamine or urea molding compounds. *1,* Syrup; *2,* mixing; *3,* drying; *4,* ball milling; *5,* granulation.

A widely used comparative test published for all types of plastic materials is *ASTM D648, Deformation Under Load* test. At 264 psi (1.8 MPa) the value for urea is 266°F (130°C); for melamine, 361°F (183°C). These figures would seem to contradict the definition of thermoset plastics, yet under the condition of the test there is sufficient reduction of the modulus of these materials. At the temperatures indicated, test specimens 1/2 in. (12.7 mm) wide deflect 0.010 in. (0.25 mm) when held on supports 4 in. (102 mm) apart.

52.5.2 Mechanical Properties

Urea and melamine are the hardest of all plastic materials. Urea has a hardness of M110–120 and melamine of M115–125 on the Rockwell scale; each has a very high modulus of elasticity ($1–1.5 \times 10^6$). Although they are brittle at normal room temperatures, these resins withstand low temperatures without significant further embrittlement. Typical notched Izod impact values are 0.25 ft-lb/in. (13.3 J/m), although the addition of glass-fiber fillers to melamine raises the notched Izod to as high as 20 ft-lb/in. (1068 J/m).

52.5.3 Optical Properties

Urea and melamine resins are transparent, but when sufficient filler is added (to overcome the extreme brittleness of the unfilled resin), they become translucent and excellent diffusers of light. The color possibilities of both materials are unlimited.

52.5.4 Environmental Properties

Amino resins are unaffected by solvents. Tests with acetone, gasoline, and detergents do not affect either of these plastics. Dilute acids and alkalies and boiling water (1/2 h) do attack the surface of ureas. Melamines are much more resistant to the acids and alkalies, and easily withstand boiling water. A test for well-cured melamine is 10 min immersion in 1% boiling sulfuric acid solution.

Melamines and ureas do not attract static charge dust, as do many plastic materials.

52.5.5 Reinforcement

Since melamine has an inherently high resistance to arc tracking, compounds are available with various fiber lengths of glass to yield notched Izod impact values of 0.6–20 ft-lb/in. (32–1068 J/m) [compared to 0.35 ft-lb/in. (19 J/m) for alpha cellulose fillers]. Glass also improves dimensional stability. For comparison, short glass-filled melamine shrinks only 0.002–0.004 in. (0.05–0.1 mm), compared to 0.008–0.009 in. (0.20–0.23 mm) for cellulose-filled melamine. With long glass, shrinkage is only 0.0004 in. (0.01 mm).

Glass-filled melamines are sold by Fiberite and U.S. Prolam. Glass-filled ureas are not available.

52.6 APPLICATIONS

In choosing between melamines and ureas, it is important to bear in mind that the materials are very similar. However, higher-cost melamine offers enhancement in physical properties, a hard and chemical resistant surface, and arc resistance.

For wiring devices, closures, small housings, and similar uses, lower-cost ureas are selected because they perform effectively. Electrical-grade brown and black ureas provide better arc resistance than phenolics, with which they are competitively priced in the wiring-device market.

Where the ultimate in resistance to abrasion and repeated exposure to liquids is required, melamine is the appropriate choice. Melamine has been very successful for lightweight bathroom wash basins, and it is well known for its use in dinnerware, cutlery handles, large serving utensils, and bowls. Its resistance to high direct heat makes it the material of choice for trays (cigarette burns will not damage the surface). In this application, the melamine should be at least 0.09 in. (2.3 mm) thick to allow a sufficient heat sink.

Glass-reinforced melamine is used for heavy-duty switch gears, terminal strips, and welding rod holders, where strength, dimensional stability, and arc resistance are essential.

Ureas and melamines are finding increasing use as blast media. The particular hardness of each, lying between walnut shell and sand, makes them valuable, for example, for removal of paint, without damage to the metal beneath the paint.

52.7 ADVANTAGES/DISADVANTAGES

Melamine has the hardest surface of any plastic (it is the resin used in Formica countertops). The tradeoff is its brittleness. To compensate, melamine should be designed with radii at all corners and with sufficient wall thickness to avoid problems.

To maximize melamine's excellent arc-tracking resistance, its brittleness and high shrinkage must be controlled. The addition of glass fillers takes care of both problems, but at a higher cost. If shrinkage reduction is the sole criterion, melamine phenols should be considered.

Processing of melamines and ureas into useful articles has long been assumed to require more expensive molding and additional finishing operations to remove flash. However, formulations of melamine are now available for economical injection molding. To a lesser extent, ureas also can be injection molded. Material suppliers should be consulted for the best molding method and related material for a given application. For large, flat surfaces, compression molding is still probably the better method, since it can minimize flaws.

A basic property of melamine is its ability to accept a molded-in decoration, consisting of a printed sheet of melamine-impregnated paper that is permanently fused to its surface during molding. This ensures all the advantages of a tough, glossy melamine surface in the decorated item.

Another basic property of melamine is that it can be used to impregnate a sheet of paper, which then can be printed and (during molding) fused to a melamine substrate. Thus, a wide variety of extremely durable decorated items can be molded.

52.8 PROCESSING TECHNIQUES

Melamine compounds are most often compression molded, although special grades are adaptable to injection molding. Urea is more easily compression molded than melamine. Like melamine, it can also be injection molded when special formulations are used.

Master Material Outline

PROCESSABILITY OF ____Urea____

PROCESS	A	B	C
INJECTION MOLDING		X	
EXTRUSION			
THERMOFORMING			
FOAM MOLDING			
DIP, SLUSH MOLDING			
ROTATIONAL MOLDING			
POWDER, FLUIDIZED-BED COATING			
REINFORCED THERMOSET MOLDING			
COMPRESSION/TRANSFER MOLDING	X		
REACTION INJECTION MOLDING (RIM)			
MECHANICAL FORMING			
CASTING			

A = Common processing technique.
B = Technique possible with difficulty.
C = Used only in special circumstances, if ever.

Master Material Outline

PROCESSABILITY OF ___Melamine___

PROCESS	A	B	C
INJECTION MOLDING	X[a]		
EXTRUSION			
THERMOFORMING			
FOAM MOLDING			
DIP, SLUSH MOLDING			
ROTATIONAL MOLDING			
POWDER, FLUIDIZED-BED COATING			
REINFORCED THERMOSET MOLDING			
COMPRESSION/TRANSFER MOLDING		X	
REACTION INJECTION MOLDING (RIM)			
MECHANICAL FORMING			
CASTING			

A = Common processing technique.
B = Technique possible with difficulty.
C = Used only in special circumstances, if ever.
[a]Requires special grade

Standard design chart for ___Urea/melamine___

1. Recommended wall thickness/length of flow
   Minimum _____ for _____ distance
   Maximum _____ for _____ distance
   Ideal _____ for _____ distance

2. Allowable wall thickness variation, % of nominal wall ___2___

3. Radius requirements
   Outside: Minimum ___50% W___ Maximum _____
   Inside: Minimum ___25% W___ Maximum _____

4. Reinforcing ribs
   Maximum thickness ___75% W___
   Maximum height ___6 W___
   Sink marks: Yes _____ No ___X___

5. Solid pegs and bosses
   Maximum thickness _____4 W_____
   Maximum weight _____6 W_____
   Sink marks: Yes _____ No ___X___

6. Strippable undercuts
   Outside: Yes _____ No ___X___
   Inside: Yes ___X___ No _____

7. Hollow bosses: Yes ___X___ No _____

8. Draft angles
   Outside: Minimum _____$\frac{1}{2}$°_____ Ideal _____2°_____
   Inside: Minimum _____0_____ Ideal _____1°_____

9. Molded-in inserts: Yes ___X___
                       No _____

10. Size limitations
    Length: Minimum ___none___ Maximum ___none___ Process _compression_
    Width: Minimum ___none___ Maximum ___none___ Process _compression_

11. Tolerances

12. Special process- and materials-related design details
    Blow-up ratio (blow molding) _____
    Core L/D (injection and compression molding) ___L/D 3:1___
    Depth of draw (thermoforming) _____
    Other _____

52.9 RESIN FORMS

Molding compounds are offered in various forms, including:

- Fine powder, 3.5–4 bulk factor—low in cost but difficult to handle; requires special equipment, which usually negates cost savings.
- Granular: 2–2.3 bulk factor—most widely used form.
- Glass filled: 5–10 bulk factor—nodular, flake, and 3/8 to 1-1/2-in. chopped.
- Chopped cotton filled: 10–14 bulk factor—macerated.

Melamine resins for glazing are also available in tablet form (as mentioned, Ornapres offers a stain-proof melamine resin for glazing); melamine resins in dry and in liquid forms are available for foil impregnation and for a wide variety of laminating, adhesive, and bonding applications. Preforms, custom made to the desired size and weight, are offered by Fiberite for high bulk factor materials.

52.10 SPECIFICATION OF PROPERTIES

See the Master Outline of Material Properties.

52.11 PROCESSING REQUIREMENTS

Storage and handling of the raw materials are important considerations. The materials are shipped in drums and should be stored in a cool, dry space. Prolonged storage above 65% relative humidity is not recommended. It is important that hot air from a space heater never

Master Outline of Material Properties

MATERIAL ___Melamine, Alpha-cellulose-filled___

PROPERTY	TEST METHOD	ENGLISH UNITS	VALUE	METRIC UNITS	VALUE
MECHANICAL DENSITY	D792	lb/ft^3	93.5	g/cm^3	1.5
TENSILE STRENGTH	D638	psi	7,000–8,000	MPa	48.2–55.1
TENSILE MODULUS	D638	psi	1,350,000	MPa	9,308
FLEXURAL MODULUS	D790	psi	1,100,000	MPa	7,584
ELONGATION TO BREAK	D638	%	less than 1	%	less than 1
NOTCHED IZOD AT ROOM TEMP	D256	ft-lb/in.		J/m	
HARDNESS	D785	Rockwell	M115-125		M115-125
THERMAL DEFLECTION T @ 264 psi	D648	°F	361	°C	182
DEFLECTION T @ 66 psi	D648	°F		°C	
VICAT SOFT-ENING POINT	D1525	°F		°C	
UL TEMP INDEX	UL746B	°F		°C	
UL FLAMMABILITY CODE RATING	UL94				
LINEAR COEFFICIENT THERMAL EXPANSION	D696	in./in./°F	22.2–25.0 × 10^{-6}	mm/mm/°C	40–45 × 10^{-6}
ENVIRONMENTAL WATER ABSORPTION 24 HOUR	D570	%	0.3–0.5	%	0.3–0.5
CLARITY	D1003	% TRANS-MISSION		% TRANS-MISSION	
OUTDOOR WEATHERING	D1435	%		%	
FDA APPROVAL			DMF 3313		DMF 3313
CHEMICAL RESISTANCE TO: WEAK ACID	D543		Not attacked		Not attacked
STRONG ACID	D543				
WEAK ALKALI	D543				
STRONG ALKALI	D543				
LOW MOLECULAR WEIGHT SOLVENTS	D543		Not attacked		Not attacked
ALCOHOLS	D543		Not attacked		Not attacked
ELECTRICAL DIELECTRIC STRENGTH	D149	V/mil	270–300	kV/mm	10.6–11.8
DIELECTRIC CONSTANT	D150	10^6 Hertz	7.9–8.2		7.9–8.2
POWER FACTOR	D150	10^6 Hertz			
OTHER MELTING POINT	D3418	°F		°C	
GLASS TRANS-ITION TEMP.	D3418	°F		°C	

656

Master Outline of Material Properties

MATERIAL ___Urea, Alpha-cellulose-filled___

PROPERTY	TEST METHOD	ENGLISH UNITS	VALUE	METRIC UNITS	VALUE
MECHANICAL DENSITY	D792	lb/ft³	93.5	g/cm³	1.5
TENSILE STRENGTH	D638	psi	5,500–7,000	MPa	37.9–48.2
TENSILE MODULUS	D638	psi	1,300,000–1,400,000	MPa	8,963–9,653
FLEXURAL MODULUS	D790	psi	1,300,000–1,600,000	MPa	8,963–11,032
ELONGATION TO BREAK	D638	%	Less than 1	%	Less than 1
NOTCHED IZOD AT ROOM TEMP	D256	ft-lb/in.	0.27–0.34	J/m	14.4–18.1
HARDNESS	D785	Rockwell	M110-120		M110-120
THERMAL DEFLECTION T @ 264 psi	D648	°F	266	°C	130
DEFLECTION T @ 66 psi	D648	°F		°C	
VICAT SOFTENING POINT	D1525	°F		°C	
UL TEMP INDEX	UL746B	°F		°C	
UL FLAMMABILITY CODE RATING	UL94				
LINEAR COEFFICIENT THERMAL EXPANSION	D696	in./in./°F	$12.2–20.0 \times 10^{-6}$	mm/mm/°C	$22–36 \times 10^{-6}$
ENVIRONMENTAL WATER ABSORPTION 24 HOUR	D570	%	0.4–0.8	%	0.4–0.8
CLARITY	D1003	% TRANSMISSION		% TRANSMISSION	
OUTDOOR WEATHERING	D1435	%		%	
FDA APPROVAL			DMF 1742		DMF 1742
CHEMICAL RESISTANCE TO: WEAK ACID	D543		Attacked		Attacked
STRONG ACID	D543				
WEAK ALKALI	D543				
STRONG ALKALI	D543		Attacked		Attacked
LOW MOLECULAR WEIGHT SOLVENTS	D543		Not attacked		Not attacked
ALCOHOLS	D543		Not attacked		Not attacked
ELECTRICAL DIELECTRIC STRENGTH	D149	V/mil	330–370	kV/mm	13.0–14.5
DIELECTRIC CONSTANT	D150	10^6 Hertz	6.7–6.9		6.7–6.9
POWER FACTOR	D150	10^6 Hertz			
OTHER MELTING POINT	D3418	°F		°C	
GLASS TRANSITION TEMP.	D3418	°F		°C	

Master Outline of Material Properties

MATERIAL ___Melamine, Glass-filled, Long Glass___

PROPERTY	TEST METHOD	ENGLISH UNITS	VALUE	METRIC UNITS	VALUE
MECHANICAL DENSITY	D792	lb/ft³	121.6	g/cm³	1.95
TENSILE STRENGTH	D638	psi	7,500	MPa	51.7
TENSILE MODULUS	D638	psi		MPa	
FLEXURAL MODULUS	D790	psi		MPa	
ELONGATION TO BREAK	D638	%		%	
NOTCHED IZOD AT ROOM TEMP	D256	ft-lb/in.	3.5–20	J/m	186–1067
HARDNESS	D785				
THERMAL DEFLECTION T @ 264 psi	D648	°F	400	°C	204
DEFLECTION T @ 66 psi	D648	°F		°C	
VICAT SOFT-ENING POINT	D1525	°F		°C	
UL TEMP INDEX	UL746B	°F		°C	
UL FLAMMABILITY CODE RATING	UL94				
LINEAR COEFFICIENT THERMAL EXPANSION	D696	in./in./°F		mm/mm/°C	
ENVIRONMENTAL WATER ABSORPTION 24 HOUR	D570	%	0.1–0.3	%	0.1–0.3
CLARITY	D1003	% TRANS-MISSION		% TRANS-MISSION	
OUTDOOR WEATHERING	D1435	%		%	
FDA APPROVAL					
CHEMICAL RESISTANCE TO: WEAK ACID	D543				
STRONG ACID	D543				
WEAK ALKALI	D543				
STRONG ALKALI	D543				
LOW MOLECULAR WEIGHT SOLVENTS	D543				
ALCOHOLS	D543				
ELECTRICAL DIELECTRIC STRENGTH	D149	V/mil		kV/mm	
DIELECTRIC CONSTANT	D150	10⁶ Hertz			
POWER FACTOR	D150	10⁶ Hertz			
OTHER MELTING POINT	D3418	°F		°C	
GLASS TRANS-ITION TEMP.	D3418	°F		°C	

MATERIAL ___Melamine, Glass-filled, Short Glass___

PROPERTY	TEST METHOD	ENGLISH UNITS	VALUE	METRIC UNITS	VALUE
MECHANICAL DENSITY	D792	lb/ft^3	111.6–114.7	g/cm^3	1.79–1.84
TENSILE STRENGTH	D638	psi	8,000	MPa	55.1
TENSILE MODULUS	D638	psi		MPa	
FLEXURAL MODULUS	D790	psi	2,200,000	MPa	15,169
ELONGATION TO BREAK	D638	%		%	
NOTCHED IZOD AT ROOM TEMP	D256	ft-lb/in.	0.6	J/m	32.0
HARDNESS	D785				
THERMAL DEFLECTION T @ 264 psi	D648	°F	385	°C	196
DEFLECTION T @ 66 psi	D648	°F		°C	
VICAT SOFT-ENING POINT	D1525	°F		°C	
UL TEMP INDEX	UL746B	°F		°C	
UL FLAMMABILITY CODE RATING	UL94				
LINEAR COEFFICIENT THERMAL EXPANSION	D696	in./in./°F	1.1×10^{-5}	mm/mm/°C	2×10^{-5}
ENVIRONMENTAL WATER ABSORPTION 24 HOUR	D570	%	0.15	%	0.15
CLARITY	D1003	% TRANS-MISSION		% TRANS-MISSION	
OUTDOOR WEATHERING	D1435	%		%	
FDA APPROVAL					
CHEMICAL RESISTANCE TO: WEAK ACID	D543				
STRONG ACID	D543				
WEAK ALKALI	D543				
STRONG ALKALI	D543				
LOW MOLECULAR WEIGHT SOLVENTS	D543				
ALCOHOLS	D543				
ELECTRICAL DIELECTRIC STRENGTH	D149	V/mil	130–170	kV/mm	5.1–6.7
DIELECTRIC CONSTANT	D150	10^6 Hertz			
POWER FACTOR	D150	10^6 Hertz			
OTHER MELTING POINT	D3418	°F		°C	
GLASS TRANS-ITION TEMP.	D3418	°F		°C	

be blown at the materials. When a drum is opened, the material it contains is exposed to temperature and humidity, both of which can affect its plasticity. Since the compound is hydroscopic, it should be contained. Of course, the materials will be exposed to ambient air in hoppers and in tote trays around the press, but they should not be left there for any length of time.

Dust collectors should be used to vacuum clean all areas where the material may be exposed, to prevent contamination by dirt or other materials.

Fine powder is difficult to use for preforming, but can be accomplished with a special hopper design and hydraulic machines. Granular resin with a 75–90% pass-through on an 80 mesh screen, and with an apparent density of 41–47 lb/ft³ (0.65–0.75 g/cm³), will perform easily on hydraulic and mechanical machines.

Preform dies should be made of tool steel with a hardness of 54–60 Rockwell C. They should be tapered and polished and chromeplated for good ejection of pills from the die. A guide for clearance of punches and dies when granulated material is used is

Diameter, in. (mm)	Clearance		Comments
	Upper	Lower	
1/4–1/2 (6.4–12.7)	0.003 (0.076)	0.002 (0.051)	For nonround dies, use similar
3/4–1 (19–25.4)	0.005 (0.127)	0.005 (0.127)	clearances for similar dimen-
1–2 (25.4–50.8)	0.010 (0.254)	0.008 (0.203)	sions; clearances may need to
over 2 (over 50.8)	0.012 (0.305)	0.010 (0.254)	be increased by 50–75% for fine powders

Clearances may need to be increased by 50–75% for pine powders.

Resin preheating is done to aid flow, reduce pressure, and eliminate gas, especially on large, thick parts. It also will reduce cure time by 10–25% and improve the surface appearance of molded parts. Preheating can be accomplished using a heated rotary can, infrared lights, or high-frequency heating; the latter is most commonly used. Screw injection is used to preheat material on injection presses. Urea can be preheated to about 180°F (82°C); melamine to 220°F (104°C). Higher molding temperatures are critical, especially when there is a delay after the preheat cycle is completed and before molding begins.

Urea molding can be in flash, semipositive, and positive shut-off compression molds. It is also performed using transfer and occasionally injection molds. The molding temperature range is 270–330°F (132–166°C). The larger the part, the longer the cure and the lower the temperature. Compression molding pressures range from 2,000–8,000 psi (14–55 MPa).

Melamine can be molded in the same molds that are used for urea. Melamine also can be readily molded in transfer and injection molds. Preheating the material 200–240°F (93–116°C) allows the material to flow freely at pressures of 5,000–15,000 psi (34–103 MPa) at mold temperatures of 300–320°F (149–160°C). Compression molding pressures are 1,500–6,000 psi (10–41 MPa) at mold temperatures of 310–340°F (154–171°C).

The mold may be steam or electrically heated. For good control, steam should be dry and saturated (no superheat). The boiler should have a pressure range to 200 psi (1.4 MPa) (gauge). Electric heating should be designed to prevent "hot" spots in the mold. Some fixed and some regulated heaters are recommended.

52.12 PROCESSING-SENSITIVE END PROPERTIES

52.12.1 Compression Molding

The heat and pressure applied during molding causes air, gas, and steam to be released from the material. If they are not removed, the molded part can be blistered or vary widely in its translucency. These released products can be eliminated by a precisely controlled, slow closing rate or a degas or breathe cycle employed just after the mold is closed.

Under high pressures, the mold closing rate can vary 2 to 5 seconds, after which the mold can be opened from 1/16 to 1/2 in. (1.6 to 12.7 mm) for 1 to 3 s or more, to allow the gas to escape. When the mold is opened too far, or when the breathe time is too long, the part will show knit lines. It should be noted that, at this time, the temperature of the compound is increasing at a rapid rate, and the latent catalyst, hexamethylene tetramine, acts to excite the curing agent tetrachlorphthalic acid. Molders have been known to modify the cycle for hours, along with the preheat pressure and temperature, to obtain a good part consistently. A semiautomatic or automatic press must have the needed controls so these adjustments can be duplicated exactly.

When thermoset parts are removed from the hot mold they can be distorted until they cool to below 200°F (93°C). It may be necessary to apply shrink fixtures to the part as soon as it is removed from the mold. The fixture is designed to control the shape or dimensions that could change during the cooling period, which may take 3 to 10 minutes. A fan is sometimes used to decrease the cooling time. When the part has cooled to room temperatures, the critical dimensions should be checked. It is advisable to remember that shrinkage takes place during cooling.

The cure time may vary from 15 seconds to several minutes, depending on mold temperature, type of materials, and part thickness. Preheating will reduce the cure by 20–25% (see Fig. 52.2). The longer the cure, the better the surface gloss.

Inserts can be molded into the piece, but bosses must be at least three times the diameter of the insert to prevent cracking that is caused by shrinkage of the material. Sometimes, inserts are pressed in place before the piece has cooled to room temperature.

Molds should be located over the ram area, but molds extending beyond the ram area are affected by the bending of the platen. (This has been a factor in some molding operations). Most of the molds used with melamine and urea are semipositive shut-off designs. The flash type is usually employed for little-flow, small parts, where aspirin-type pills are used. The semipositive design allows a variable load weight, which is normal in bulk or pill loading, to secure good cutoff. The excess material (flash) can escape and still have enough back pressure to prevent a short. Positive shut-off molds are seldom used, since load weight must be more exact and trapped gas may be a problem.

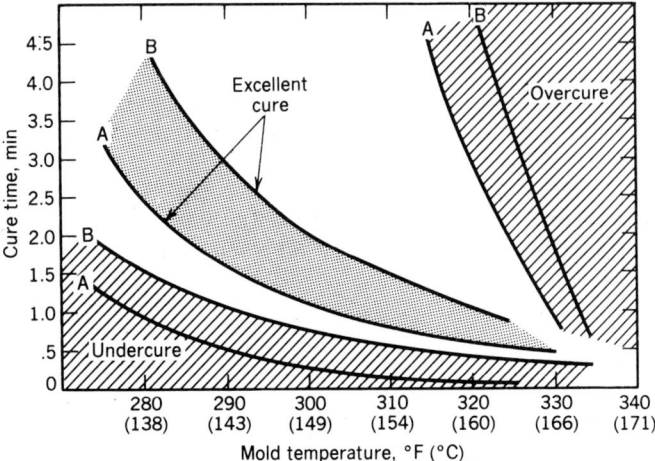

Figure 52.2. Cure rate of urea molding compounds. Section 0.070 in. (1.8 mm). A, fast cure compounds; B, slow cure compounds.

52.12.2 Foil Decoration

Foil decorating is a two-step operation. The piece itself is molded and partially cured; then, the mold is opened and the foil is positioned, after which the mold is reclosed. (Simultaneous molding of foil and substrate causes the foil to tear.) The flash overflow in the mold must be horizontal or angled back at least 20° from the vertical, so that the top half of the mold will lift without disturbing the flash when the mold is opened to place the foil. A reduced cavity temperature of 10–15% is helpful when the molded part remains in the lower half of the mold (so the foil can be added).

Since the foil is quite thin [approximately 0.005 in. (0.13 mm)], relative to the substrate, separate breathing or degassing cycle times are needed for each. The foil may only require a mold opening of 1/8 in. (3.2 mm) or less for 1 or 2 seconds, in order to eliminate blistering. Both melamine and urea can be foil decorated.

Heavy deposits of ink on the foil may cause a poor bond or blistering. The printer may need to reduce the ink. Dusting the foil over the ink with melamine resin "dust" will eliminate the blistering.

52.12.3 Glazing

A very high gloss finish can be obtained on decorated or nondecorated molded parts by adding an extra molding cycle. In this case, a small quantity [a few grams, or enough to provide approximately 0.002 in. (0.051 mm) thickness] of unfilled melamine glazing resin is added after the curing cycle, and the mold is reclosed. The resin flows readily and should just cover the entire surface. Too much resin will result in crazing.

Both decorating and glazing add to the cost of the molded part. Foils may cost from a few cents to 25 cents; glazing material from 2 to 4 cents. The added molding time, plus added scrap losses, have to be included in the cost. Cured melamine, ground and screened to various mesh sizes, is finding increasing use as a blast media for deburring, surface preparation, etc. Its hardness, which falls in the range of sand, walnut shell, and other commonly used blast material, makes it useful for this purpose.

Molders have gone out of business because their losses have been too high. This may be the result of trying to process material not suited to the equipment on hand or of lack of knowledge. Thus, supervisors should be trained to be sensitive to the cost factors. It is essential that man-hours, pounds per 1000, and the percentage of losses be built into the cost. A daily tally should be made to keep these important cost factors under control. Good management relations with the supervisor are a must. The supervisor also has a responsibility to know the job and guide people. The supervisor should check and enforce all safety rules.

In general, there is always peripheral flash that must be removed from thermoset parts, and melamine and urea are no exception. Flash occurs during the period of the molding cycle when the material is flowing: its surface viscosity is very low and it enters mold clearances as small as 0.0002 in. (0.005 mm). Flash is hard and rigid; it can be removed by bulk tumbling, preferably with a pellet blast. Pellet blasts can also be arranged to direct the pellets against conveyorized parts passing continuously through an enclosed tunnel. For large parts, abrasive belts and/or carbide files will eliminate parting lines and restore the original surface luster.

52.13 SHRINKAGE

Urea will shrink 0.006–0.009 in./in. (0.15–0.23 mm/mm) from the cold-mold dimension to the cold piece, and 0.012–0.016 in./in. (0.30–0.41 mm/mm) on aging (Fig. 52.3). Subsequent shrinkage will take place when the piece is stored below a relative humidity of 50–55%; humidities above 55% will cause the part to increase in size. Long-term exposure will cause

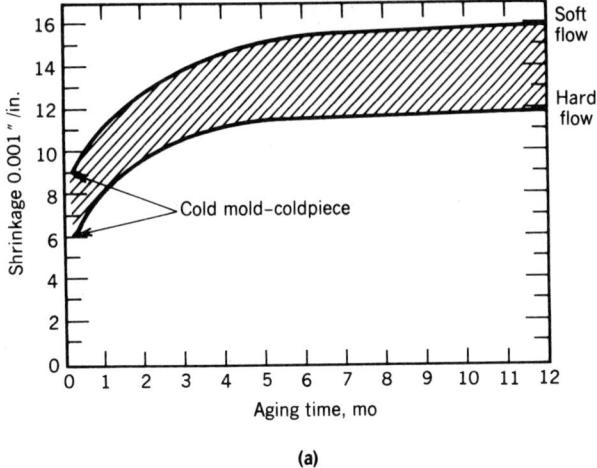

(a)

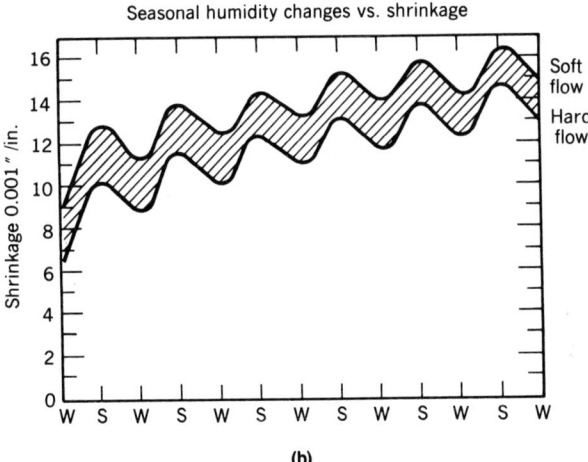

(b)

Figure 52.3. Urea shrinkage behavior. (**a**) Aging at 75°F (24°C) and 30% rh. Note: the longer the cure, the greater the shrinkage. (**b**) Seasonal humidity changes vs shrinkage. W, Winter, indoor low humidity. S, Summer indoor, outdoor humidity. At 50–55% rh very little change.

the part to expand or contract with seasonal humidity changes. Daily wet and dry applications may eventually cause crazing cracks on urea. Thus, under such circumstances, melamine should be used instead.

52.14 MANUFACTURERS

American Cyanamid Co., Plastics Dept., Wayne, NJ 07470.

Fiberite Corp., 501 W. 3rd St., Winona, MN 55987.

Perstorp Compounds, Inc., 238 Nonotuck St., Florence, MA 01060.

Plastics Mfg. Co., 2700 South Westmorland Ave., Dallas, TX 75233.

U.S. Prolam, Inc., P.O. Box 671, Stamford, CT 06904.

Mold Shrinkage Characteristics (in./in. or mm/mm)

MATERIAL _____ Urea and melamine _____

IN THE DIRECTION OF FLOW

THICKNESS

MATERIAL	mm: 1.5 in.: 1/16	3 1/8	6 1/4	8 1/2	FROM	TO	#
Urea		0.008–0.010[a]					
Melamine, alpha-cellulose-filled		0.008–0.010[a]					
Melamine, glass-filled		0.002–0.004					

[a]plus 0.009–0.011 after shrink

PERPENDICULAR TO THE DIRECTION OF FLOW						

NOTE: Mold shrinkage is approximate. It is affected by part design, mold temperature, thickness, injection pressure, packing time, cycle time, orientation, gate design, gate size, gate location, glass content, glass size, and filler content.

52.15   INFORMATION SOURCES

The best sources of further information are data sheets from the manufacturers listed above and the current edition of *Modern Plastics Encyclopedia*.

53

XT POLYMER

Charles W. Deeley

Cyro Industries, 25 Executive Blvd., Orange, CT 06744

53.1 XT POLYMER

XT Polymer is a member of the family variously called "impact acrylics" or "acrylic multipolymers." Details regarding its composition are contained in U.S. Patent 3,354,238.

53.2 CATEGORY

An injection-moldable and extrudable thermoplastic, XT Polymer is a transparent physical blend of a glassy terpolymer and a grafted rubbery polymer. The combination yields a product that is rigid and tough, having several times the notched Izod impact strength of many of the familiar transparent, glassy plastics such as poly(methyl methacrylate), polystyrene (PS), and the methyl methacrylate–styrene and styrene–acrylonitrile copolymers. To accommodate different end-use applications and processing methods, XT Polymer is produced in three grades designated XT-375, XT-X800, and XT-250. The grades XT-375 and XT-X800 are formulated to give a notched Izod impact strength of 1.8–2.0 ft-lb/in. (96–107 J/m). The XT-X800, however, is tailored to have a significantly higher melt flow rate at a given processing temperature and shear stress, making it especially suitable for injection molding where thin sections and/or long flow paths occur. Other property differences are, in general, minor. In the case of XT-250, the level of the rubbery phase is set to give an Izod impact strength of 1.0 ft-lb/in. (53 J/m). Certain other mechanical properties such as stiffness and flexural strength are consequently somewhat greater than for the higher impact grades.

53.3 HISTORY

XT Polymer was introduced in 1963 by a team of researchers at American Cyanamid, which later formed a joint venture company, Cyro Industries, with Rohm GmBH. The latest patent on XT Polymer dates to 1970.

53.4 POLYMERIZATION

The rubbery component of XT Polymer is present as a dispersed discontinuous phase in which the average individual particle has a diameter of about 150 nanometers and in which there occurs agglomerates up to about 1000 nanometers in size. The terpolymer phase has

665

a refractive index (n_{20} = 1.515) which matches closely that of the rubbery component (grafted polybutadiene) so that the resulting blend is transparent.

Current (1988) prices for XT Polymer range from \$1.14 to \$1.22 per pound.

53.5 DESCRIPTION OF PROPERTIES

53.5.1 Optical Properties

The natural color of XT Polymer is a pale transparent yellow. The yellowness results partly from the acrylonitrile component of the terpolymer or "hard" phase and partly from the grafted rubber phase. The majority of XT Polymer production is, however, in a "glass tint" achieved by the addition of small levels of a blue food, drug, and cosmetic dye system. Since the natural yellowness is slight, a wide range of transparent or opaque colors can be realized using appropriate dye or pigment systems. Typical pigment loadings for opaque colors usually reduce the notched Izod impact strength for the XT-375 grade from about 2.0 ft-lb/in. (107 J/m) to about 1.3 ft-lb/in. (69 J/m).

The light transmission of natural XT Polymer determined by ASTM Method D307 is 87%; haze (D-1003) is 9%. The specimens used for these optical measurements are injection molded using tools with a mirror finish surface. Achievement of maximum clarity and surface gloss in a molded part generally requires such a highly polished mold surface and, in addition, a mold metal temperature of about 130–160°F (54–71°C).

An interesting aspect of XT Polymer's transparency is its dependence on temperature. The match of the index of refraction of the hard phase to that of the rubbery phase is closest at room temperature (70°F [21°C]). As the temperature departs from this value, either upward or downward, the match becomes less good since the temperature coefficients of refractive index for the hard and rubbery phase are different. Over the range of most practical importance, −40°F to 125°F (−40 to 52°C) the effect on haze and transparency is small. However, when a molded part is ejected at 160°F (71°C) or a sheet or profile is extruded at a temperature at or above this value, a pronounced haziness is evident. This haziness diminishes to the data sheet value as the plastic cools and the refractive indexes coincide.

In extrusion, achieving a smooth (microscopically) glossy surface on film, sheet, or profiles requires an appreciation of XT Polymer's behavior under processing conditions. In common with many thermoplastics, when an XT Polymer extrudate is allowed to cool freely in air or is quenched in water, it can have a matte or even sharkskinlike surface resulting from the elastic response of the melt. This precursor to "melt fracture" is most evident at lower melt temperatures and higher shear stresses and, as such, can be reduced or eliminated by raising the melt temperature and, if necessary, using a longer die land. In addition to this source of surface irregularity, some of the discrete, spherelike rubber particles of the impact phase tend to protrude from the surface and act as light-scattering centers that result in a haze. In injection molding, the typical high-cavity pressure makes the plastic surface conform to and assume the smoothness of the mold surface on cooling. If the mold surface is highly polished, the plastic molding will be correspondingly smooth and glossy. For extruded film and sheet, a polishing roll stack, judiciously used, can effectively produce a glossy surface on the sheet.

53.5.2 Thermal Properties

The heat-distortion temperature of XT Polymer, as determined by ASTM D-648 using annealed samples and a stress of 264 psi (1.8 MPa) is 186°F (85°C) for XT-375, 180°F (82°C) for XT-X800, and 188°F (87°C) for XT-250.

Maximum continuous service temperature will depend on static and dynamic stress levels and the degree of hostility of the environment as determined, for example, by the presence of potential stress cracking agents.

53.5.3 Mechanical Properties

The room temperature flexural modulus of XT Polymer ranges from 400,000 psi (2758 MPa) for XT-250 to 350,000 psi (2413 MPa) for both XT-375 and XT-X800. Flexural strengths are 13,000 (90 MPa) for XT-250, 11,000 psi (76 MPa) for XT-375, and 10,000 psi (69 MPa) for XT-X800. These values are close to those of other impact acrylics and to that of ABS and impact polystyrene grades having comparable impact strengths.

The rubbery phase of XT Polymer has a glass-transition temperature of about $-110°F$ ($-80°C$). This results in XT Polymer retaining a large fraction of its Izod impact strength at temperatures of $-40°F$ ($-40°C$): 1.2 ft-lb/in. (64 J/m) for XT-375 and 0.7 ft-lb/in. (37 J/m) for both XT-X800 and XT-250.

53.6 APPLICATIONS

53.6.1 Injection-Molded Products

XT Polymer is used in injection-molded medical devices, refrigeration components such as meat, vegetable, and dairy pans, industrial alkaline battery cases, fishing tackle boxes, video cassette shells, and toys.

53.6.2 Extruded and Thermoformed Products

XT Polymer is extruded and/or thermoformed into such products as food tubs and lids, medical packaging, electronic packaging, and engraving stock.

These product areas represent the bulk of XT Polymer usage and reflect applications that require the balance of properties that XT Polymer can provide. The combination of transparency and toughness is common to all these applications.

53.7 ADVANTAGES/DISADVANTAGES

In medical applications, pertinent additional attributes of XT Polymer include:

- ETO and gamma ray sterilizability
- When properly processed, resistance to Freon TF (TM of Du Pont Co.)
- Ultrasonic and solvent bondability
- USP Class VI at 158°F (70°C)
- Approved for food contact use by FDA.

XT-375 Polymer and XT-250 polymer as acrylic multipolymers are regulated for food contact uses under 21 CFR 177.1010. They have been tested at conditions that represent the most severe end-use condition that is recommended-Condition C (hot filled or pasteurized above 150°F (66°C), according to 21 CFR 176.170. They also meet the requirements for all food types except those containing more than 8% alcohol.

Since XT-375 Polymer and XT-250 Polymer contain acrylonitrile in their composition, they are not acceptable for beverage containers and are further regulated by 21 CFR 180.22. This FDA regulation, in addition to listing test conditions, specifies the maximum level of acrylonitrile that may be extracted by a food product and requires that each type of container be evaluated. Experience has showed that virgin XT-375 Polymer or XT-250 Polymer fabricated by conventional extrusion and thermoforming methods will meet the 21 CFR 180.22 requirements for most food products.

For food packaging, refrigerator pans, and fishing tackle boxes, XT Polymer's resistance to cracking by vegetable and mineral oils and attack by plasticizers in "rubber worms" are essential requirements. In specific packaging applications, prudence dictates end-use testing to ensure satisfactory performance.

53.8 PROCESSING TECHNIQUES

See the Master Material Outline.

Master Material Outline

PROCESSABILITY OF ____XT Polymer____

PROCESS	A	B	C
INJECTION MOLDING	X		
EXTRUSION	X		
THERMOFORMING	X		
FOAM MOLDING			
DIP, SLUSH MOLDING			
ROTATIONAL MOLDING			
POWDER, FLUIDIZED-BED COATING			
REINFORCED THERMOSET MOLDING			
COMPRESSION/TRANSFER MOLDING			
REACTION INJECTION MOLDING (RIM)			
MECHANICAL FORMING			
CASTING			

A = Common processing technique.
B = Technique possible with difficulty.
C = Used only in special circumstances, if ever.

53.9 RESIN FORMS

XT Polymer is produced in pellet form and has a bulk density of 40 lb/ft^3 (0.64 g/cm^3).

53.10 SPECIFICATION OF PROPERTIES

See the Master Outline of Materials Properties.

Master Outline of Material Properties

MATERIAL ___ XT Polymer, XT-250

PROPERTY	TEST METHOD	ENGLISH UNITS	VALUE	METRIC UNITS	VALUE
MECHANICAL DENSITY	D792	lb/ft^3	69.8	g/cm^3	1.12
TENSILE STRENGTH	D638	psi	8000	MPa	55
TENSILE MODULUS	D638	psi	430,000	MPa	2965
FLEXURAL MODULUS	D790	psi	400,000	MPa	2758
ELONGATION TO BREAK	D638	%	3.6	%	3.6
NOTCHED IZOD AT ROOM TEMP	D256	ft-lb/in.	1.0	J/m	53
HARDNESS	D785		M68		M68
THERMAL DEFLECTION T @ 264 psi	D648	°F	188	°C	87
DEFLECTION T @ 66 psi	D648	°F		°C	
VICAT SOFT-ENING POINT	D1525	°F	203	°C	95
UL TEMP INDEX	UL746B	°F		°C	
UL FLAMMABILITY CODE RATING	UL94		HB		HB
LINEAR COEFFICIENT THERMAL EXPANSION	D696	in./in./°F	4.4×10^{-5}	mm/mm/°C	7.9×10^{-5}
ENVIRONMENTAL WATER ABSORPTION 24 HOUR	D570	%	0.3	%	0.3
CLARITY	D1003	% TRANS-MISSION	87	% TRANS-MISSION	87
OUTDOOR WEATHERING	D1435	%	Not recommended	%	Not recommended
FDA APPROVAL			Yes		Yes
CHEMICAL RESISTANCE TO: WEAK ACID	D543		Not attacked		Not attacked
STRONG ACID	D543		Minimally to badly attacked		Minimally to badly attacked
WEAK ALKALI	D543		Not attacked		Not attacked
STRONG ALKALI	D543		Not attacked		Not attacked
LOW MOLECULAR WEIGHT SOLVENTS[b]	D543				
ALCOHOLS[c]	D543				
ELECTRICAL DIELECTRIC STRENGTH	D149	V/mil		kV/mm	
DIELECTRIC CONSTANT	D150	10^6 Hertz			
POWER FACTOR	D150	10^6 Hertz			
OTHER MELTING POINT	D3418	°F		°C	
GLASS TRANS-ITION TEMP.	D3418	°F		°C	

[a]Oxygen index (D2863), 19; flammability classification (ASTM D635), burning; burning rate [1/8 in. (3.175 mm)], 1.2 in./minute (30 mm/min).
[b]Soluble in aromatics and chlorinated hydrocarbons; resistant to aliphatics.
[c]Resistant to solutions of about 10% or less concentration.

MATERIAL ___ XT Polymer, XT-X800

PROPERTY	TEST METHOD	ENGLISH UNITS	VALUE	METRIC UNITS	VALUE
MECHANICAL DENSITY	D792	lb/ft^3	69.2	g/cm^3	1.11
TENSILE STRENGTH	D638	psi	7000	MPa	48
TENSILE MODULUS	D638	psi	370,000	MPa	2551
FLEXURAL MODULUS	D790	psi	350,000	MPa	2413
ELONGATION TO BREAK	D638	%	3.6	%	3.6
NOTCHED IZOD AT ROOM TEMP	D256	ft-lb/in	2.0	J/m	107
HARDNESS	D785		M30		M30
THERMAL[a] DEFLECTION T @ 264 psi	D648	°F	186	°C	86
DEFLECTION T @ 66 psi	D648	°F		°C	
VICAT SOFT-ENING POINT	D1525	°F	200	°C	93
UL TEMP INDEX	UL746B	°F		°C	
UL FLAMMABILITY CODE RATING	UL94		HB		HB
LINEAR COEFFICIENT THERMAL EXPANSION	D696	in/in/°F	5.0×10^{-5}	mm/mm/°C	9×10^{-5}
ENVIRONMENTAL WATER ABSORPTION 24 HOUR	D570	%	0.3	%	0.3
CLARITY	D1003	% TRANS-MISSION	87	% TRANS-MISSION	87
OUTDOOR WEATHERING	D1435	%	Not recommended	%	Not recommended
FDA APPROVAL			Yes		Yes
CHEMICAL RESISTANCE TO: WEAK ACID	D543		Not attacked		Not attacked
STRONG ACID	D543		Minimally to badly attacked		Minimally to badly attacked
WEAK ALKALI	D543		Not attacked		Not attacked
STRONG ALKALI	D543		Not attacked		Not attacked
LOW MOLECULAR WEIGHT SOLVENTS[b]	D543				
ALCOHOLS[c]	D543				
ELECTRICAL DIELECTRIC STRENGTH	D149	V/mil		kV/mm	
DIELECTRIC CONSTANT	D150	10^6 HERTZ			
POWER FACTOR	D150	10^6 HERTZ			
OTHER MELTING POINT	D3418	°F		°C	
GLASS TRANS-ITION TEMP.	D3418	°F		°C	

[a] Oxygen index (D2863), 19; flammability classification (ASTM D635), burning; burning rate (1/8 in. [3.175 mm]), 1.2 in. /minute (30 mm/min).
[b] Soluble in aromatics and chlorinated hydrocarbons; resistant to aliphatics.
[c] Resistant to solutions of about 10% or less concentration.

MATERIAL ___ XT Polymer, XT-X800

PROPERTY	TEST METHOD	ENGLISH UNITS	VALUE	METRIC UNITS	VALUE
MECHANICAL DENSITY	D792	lb/ft^3	69.2	g/cm^3	1.11
TENSILE STRENGTH	D638	psi	6300	MPa	43
TENSILE MODULUS	D638	psi	330,000	MPa	2275
FLEXURAL MODULUS	D790	psi	320,000	MPa	2206
ELONGATION TO BREAK	D638	%	3.5	%	3.5
NOTCHED IZOD AT ROOM TEMP	D256	ft-lb/in.	1.8	J/m	96
HARDNESS	D785		M20		M20
THERMAL[a] DEFLECTION T @ 264 psi	D648	°F	180	°C	82
DEFLECTION T @ 66 psi	D648	°F		°C	
VICAT SOFT-ENING POINT	D1525	°F	193	°C	90
UL TEMP INDEX	UL746B	°F		°C	
UL FLAMMABILITY CODE RATING	UL94		HB		HB
LINEAR COEFFICIENT THERMAL EXPANSION	D696	in./in./°F	5.0×10^{-5}	mm/mm/°C	9×10^{-5}
ENVIRONMENTAL WATER ABSORPTION 24 HOUR	D570	%	0.3	%	0.3
CLARITY	D1003	% TRANS-MISSION	87	% TRANS-MISSION	87
OUTDOOR WEATHERING	D1435	%	Not recommended	%	Not recommended
FDA APPROVAL			Yes		Yes
CHEMICAL RESISTANCE TO: WEAK ACID	D543		Not attacked		Not attacked
STRONG ACID	D543		Minimally to badly attacked		Minimally to badly attacked
WEAK ALKALI	D543		Not attacked		Not attacked
STRONG ALKALI	D543		Not attacked		Not attacked
LOW MOLECULAR WEIGHT SOLVENTS[b]	D543				
ALCOHOLS[c]	D543				
ELECTRICAL DIELECTRIC STRENGTH	D149	V/mil		kV/mm	
DIELECTRIC CONSTANT	D150	10^6 Hertz			
POWER FACTOR	D150	10^6 Hertz			
OTHER MELTING POINT	D3418	°F		°C	
GLASS TRANS-ITION TEMP.	D3418	°F		°C	

[a]Oxygen index (D2863), 19; flammability classification (ASTM D635), buming; buming rate [1/8 in. (3.175 mm)], 1.2 in./minute (30 mm/min).
[b]Soluble in aromatics and chlorinated hydrocarbons; resistant to aliphatics.
[c]Resistant to solutions of about 10% or less concentration.

53.11 PROCESSING REQUIREMENTS AND PROCESSING-SENSITIVE END PROPERTIES

XT Polymer is somewhat hygroscopic and therefore usually requires either predrying using an efficient desiccant-type dryer or the use of a vented barrel with an effective degassing zone. For extrusion, the moisture level must be reduced to about 0.03 wt%. Failure to dry adequately results in an extrudate that is hazier than normal, has surface streaks, and, in the extreme, has bubbles throughout the plastic. In addition, material tends to build up at the exit ends of the die lips, which further degrades surface quality.

When extruded sheet or film is to be thermoformed subsequently, the possible reabsorption of moisture prior to forming must be considered. If sufficient reabsorption occurs, some bubbling may be seen when the sheet is heated to forming temperatures (approximately 300–450°F [149–232°C]). Securely wrapping roll stock in polyethylene film ordinarily provides adequate protection against moisture regain.

For injection molding, the drying requirements are less stringent. Moisture levels up to about 0.01% can usually be tolerated. Nonetheless, the presence of surface splay or bubbling is almost always an indication of inadequate drying. Three to four hours residence time with the material at 175–185°F (80–85°C) using an efficient desiccant dryer (-20°F [-30°C] or lower dewpoint with good air flow) ensures adequate drying.

XT Polymer has good thermal stability, so regrind may be used at any required level. As the number of normal heat histories increases, the primary effect will be a color drift with the physical properties little changed. Whenever feasible, a relatively constant ratio of regrind to virgin material is preferred. To avoid compromising the optical properties of XT Polymer, it is essential to avoid cross-contamination with other plastics, even those that may be transparent. For example, the presence of polystyrene at levels as low as a small fraction of 1% will greatly increase the haze in an extrusion or molding. All potential sources of contamination should be considered, such as conveying lines, hoppers, grinders, and airborne particles.

53.11.1 Extrusion

For optimum gloss and transparency of the finished extrudate consistent with melt control, the stock temperature of XT Polymer should be about 480–520°F (249–271°C) for flat sheet. Profile exit temperatures should be in the range of 430–450°F (221–232°C). Within limits, the higher the stock temperature the better will be the gloss and transparency of the final sheet or profile. Extrusion conditions should be such that excessively high localized temperatures and long residence times are avoided. Such a combination of temperature and time tends to cause an increase in melt viscosity. This leads to a gel-like "skin" on the extrudate and results in decreased gloss and transparency. In an extreme case, the color of the plastic is adversely affected and a buildup of gelled material occurs within the die and around its lips.

The general characteristics of a suitable screw are a moderately long, constant-depth feed zone with a gradual transition to a medium-length metering section. The ratio of the feed depth to the meter depth should be in the range of 2.3:1 to 3.5:1. General-purpose screws such as are used for ABS and impact PS have given good results. As with any plastic, the goal is to achieve efficient conveyance of pellets, consistent melting without overheating, and smooth, surge-free delivery of a thermally homogeneous melt to the die.

Since XT Polymer's melt viscosity is a rather sensitive function of temperature, high-quality barrel and die temperature controllers are desirable to minimize fluctuations in machine temperatures. Heater zones with cast-in channels for cooling of the barrel are also desirable. Screws bored for water or oil cooling are useful in certain critical situations. The bore is normally plugged so that cooling extends only into the midpoint of the first transition section of the screw. In this way, the melting of the plastic can be fine-tuned, if necessary.

Robust, streamlined sheeting dies with flexible restrictor bars and lips give excellent results. Precise temperature control with minimal hunting is desirable for maintaining uniform transverse and longitudinal gauge. The die-lip opening is normally set at the desired gauge or several percent higher. Significant drawdown in thickness is possible if the stock temperature is high enough to prevent undesirably high (>20%) orientation.

Screen packs may be desirable for the usual reasons, especially when regrind is used. A pack consisting of 20-40-60 or 20-40-80 meshes are common. Finer screens can be used as needed for critical applications.

An excellent, smooth and glossy surface finish requires the use of high-quality polishing rolls. The rolls should be microfinished, chromed, and hardened to Rockwell C50-60. They should be equipped with accurate and independent temperature and speed controls coupled to rubber pull rolls for best results.

The temperatures of the rolls in a three-roll stack should be found by trial, in order to give the best-appearing flat sheet. The maximum roll temperature is normally limited by sticking of the plastic to the roll.

The larger the roll diameters the better, since polishing effectiveness and heat transfer are improved.

The buildup of a large bank of plastic in the nip of the rolls should be avoided since this can lead to excessive orientation in the sheet. Such orientation can cause brittleness in the across-the-machine direction and difficulties if the sheet is to be thermoformed.

Slitting of XT Polymer sheet can be done using razor-type knives or rotating wheel knives. The razor knife is more likely to cause a slightly raised lip at the edge of the sheet.

Since XT Polymer is a tough impact plastic, the use of heavy-duty grinders is recommended for recovering trim and scrap sheet for reextrusion. Blade clearance should be kept to 0.008 in. (0.2 mm) or less.

Start up is achieved by having the extruder barrel and die preheated to operating temperatures. The screw speed is gradually increased to the desired level while care is taken to avoid excessive motor load current and pressure levels. XT Polymer melt is rather viscous, so normally a short period of purging will suffice to remove any XT Polymer or other plastic that was in the machine at startup. If the material in the extruder has been there for long periods or at excessive temperatures, a longer purge may be needed. It may also be necessary to manually clean the screw, barrel, or die in order to achieve optimum quality sheet. When shutting down, one should run as much material out of the machine as possible and turn all heaters off.

53.11.2 Thermoforming

XT Polymer thermoforms readily using typical commercial equipment. Applications range from thin-wall candy trays to heavy shower stalls. A significant fraction of XT Polymer production is used for thermoformed food tubs and instrument and medical packages. Forming temperatures of the sheet are normally in the range of 300–400°F (149–204°C) with assist plugs at 270–300°F (132–149°C) and molds at 70–130°F (21–54°C).

Orientation in the extruded film or sheet stock should be kept below about 25% to avoid possible brittleness or forming difficulties. Orientation may be determined by immersing for 10 minutes square samples 4 in. (102 mm) on a side in an oil bath kept at 300°F (149°C). The percentage shrinkage in the machine direction after allowing specimens to cool is taken as the orientation. The test samples are placed on a wire rack in the oil bath and covered with a light wire screen to prevent curling. Samples should be taken from various locations across the width of the web, so that uniformity of orientation is confirmed. Significant nonuniformity can lead to localized wrinkling, especially if the forming equipment is such that the sheet is not held along its edges while it is heated and formed.

Optimum trimming results if matched metal dies with zero clearance are used. Steel rule dies should be beveled on both sides and kept sharp. Heated platens and dies can greatly reduce the tendency to generate fines and "angel hair" at the trim line.

53.11.3 Injection Molding

The injection molding of XT Polymer presents no unusual problems. However, the relatively high melt viscosity compared to crystal PS requires that factors contributing resistance to flow be minimized. This implies short sprues and runners and gates with minimum effective diameters of about 0.040 in. (1 mm) and short land lengths. Well-designed runnerless molds are especially desirable, where feasible. When tunnel gating is used an entry angle to the cavity of 30–40° is acceptable. Efficient mold venting is very important when XT Polymer is processed, just as it is with other acrylic and acrylic-based plastics, in order to provide minimum resistance to the egress of vapors generated in the melt. Vent depths of 0.0015–0.002 in. (0.04–0.05 mm) are suggested.

To avoid excessively stressed and consequently potentially brittle regions in XT Polymer moldings, it is important to avoid overpacking. For transparent moldings, such areas are readily apparent when the molding is viewed using white light placed between crossed sheets of polarizing material. A highly colored pattern (usually near the gate or gates) indicates overstressing, which can be eliminated by reduction of one or more factors including cushion size, injection pressure, and hold time under pressure.

53.13 SHRINKAGE

The shrinkage of XT Polymer is in the range of 0.004–0.008 in/in. (mm/mm), which is similar to many other amorphous thermoplastics. Thus, tooling used with ABS, impact PS, poly(methyl methacrylate), polycarbonates, and so on is usually dimensionally compatible with XT Polymer.

53.14 TRADE NAMES

XT Polymer is a registered trade mark of Cyro Industries.

BIBLIOGRAPHY

B.J. Sexton and D.C. Curfman, "Methyl Methacrylate–Acrylonitrile–Butadiene–Styrene Copolymers," in N.M. Bikales, ed., *Encyclopedia of Polymer Science and Technology,* Supplement Vol. 1, John Wiley & Sons, Inc., New York, 1976.

54

ADDITIVES

Harry S. Katz

President, Utility Development Corporation, 112 Naylor Ave., Livingston, NJ 07039

The following is a typical definition of additives: Additives tailor the properties of organic polymers to meet the needs of broad markets or specific end-uses. Their value cannot be overstated. It is no exaggeration to say that without additives the plastics industry would be far smaller and less important than it is today. It certainly would not exert the pervasive and powerful influence on the success of new products, and indeed on the quality of life, that it does.

Defined in the simplest terms, additives are substances incorporated into an organic polymer to alter and improve its basic mechanical, physical, or chemical properties. Additives protect the polymer from the degrading effects of light, heat, and bacteria; change or modify polymer density, lubricity, and thermal processing properties; instill or improve flame retardation or smoke reduction; enhance and protect product color; and provide such special characteristics as biodegradability, electrical conductivity, and improved surface appearance.

Additives can work their "miracles" in loadings as low as 0.1% by weight of the finished compound, as is the case with certain colorants, for example. Or they can be added at levels of 30, 40, or even 50%, as is the case with talc, glass fiber, or other fillers and reinforcements. But whether the load is 0.1% or 50%, additives represent a cost-effective route to better end-use performance.

Because each additive has its own special value to impart, almost all plastics will require the use of more than one of these ingredients to suit the end use. As experience accumulates, and as the concept of niche marketing (supplying compound to meet specific rather than general needs) becomes more widespread, the demand for additives becomes greater.

According to an industry study completed by analyst John Clifford of The Fredonia Group, Cleveland, Ohio, the nearly 10 billion pounds of plastics additives consumed in 1987 is expected to grow 3.4%/year through 1992, to 11.5 billion pounds.[1]

A list of types of additives, and a brief discussion of each, is given below. Many of the definitions and a large part of the information have been derived from recent technical literature.[1-3] Continual updates, and annual reports, on the additives business, are provided by such industry journals as *Modern Plastics, Plastics Engineering, Plastics Technology, Plastics World,* and annual reference books.[4]

54.1 ANTIBLOCKING AGENTS

Antiblocking agents are substances that are added to a polymer to prevent two finished surfaces from adhering. These additives have only partial compatibility with the polymer matrix, thus they exude to the surface to perform their function. Typical examples of antiblocking agents are fatty primary amides, metallic salts of fatty acids, waxes and polysiloxanes. These materials can also be classified with lubricants and slip agents (discussed later).

54.2 ANTIFOGGING AGENTS

These chemicals are usually used in clear films or molded lenses in order to prevent fogging due to condensed moisture.

54.3 ANTIOXIDANTS

An antixoidant is added to a polymer in order to retard or prevent degradation from contact with oxygen during processing or use.

Many polymers are susceptible to degradation by oxygen at room temperature, especially when combined with outdoor exposure. This degradation is accelerated at elevated temperatures. For example, early use of polyethylene film for hothouse applications resulted in film degradation and shredding within weeks after exposure. The problem has been solved by incorporating antioxidants and ultraviolet inhibitors into the film compound, thus providing the capability for long-term outdoor exposure.

Types of antioxidants include hindered amines, secondary amines, hindered phenolics, phosphites, and thioesters.

54.4 ANTISTATIC AGENTS

Plastics have a tendency to generate a high static charge. This can create a problem when a plastic product is grounded, resulting in an electrostatic discharge that can cause fire or explosion. The problem is particularly acute with the enormous recent proliferation of molded and flexible (extruded) plastic electronic circuiting. Antistatic agents prevent the build-up of static charge. This can be accomplished by use of electrically conductive internal antistats (of which there are many) or topical ionic materials such as quaternary ammonium compounds and amines. Topical or surface-applied antistats offer short-term protection; agents compounded in the polymer provide permanent antistatic effects.

54.5 BIOCIDES

Many plastics, including flexible PVC and poly(vinyl alcohol) film, are subject to attack by microorganisms. This can result in degradation of properties, discoloration, or foul odors. Additives that prevent growth of microorganisms are known as biocides, antimicrobials, preservatives, bacetriostats, mildewicides, and fungicides. These chemicals often consist of organosulfur or arsenic compounds. Recent improvements in formulation have widened the use range of biocides to include polyethylene, polypropylene, and thermoplastic elastomers, for such end uses as film, pipe and tubing, and wire-and-cable jacketing.

54.6 BIODEGRADABLE AND PHOTODEGRADABLES

These additives become more important with the growing antagonism against plastics by environmentalists concerned about landfill overload and other problems that ostensibly result from contamination of land and sea by plastics; especially plastic packaging materials. Even though plastics represent a small fraction of wasteload tonnage, a great deal of research is being invested to develop additives that impart rapid biodegradable and/or photodegradable characteristics. It is common knowledge that high concentrations of starch added to blown film is a means for producing a biodegradable product.[5,6] Photodegradable polymers also are used to produce mulch films, which enable farmers to retain moisture on arable land.[6] But at the end of the growing cycle, separation of nondegradable film is costly and difficult.

Therefore, current R & D efforts are focused on developing biodegradable additives that eliminate in-use problems and prevent premature breakdown of polymer molecules.

54.7 BLOWING AGENTS

Blowing (or foaming) agents are additives that decompose by chemical or thermal means, forming a gas that produces a foam product. A typical blowing agent is azodicarbonamide, which decomposes at about 400°F (204°C) to release a gas that is primarily nitrogen. Other chemicals are available that will decompose at different temperature ranges. The direct injection of gases, can also be used to manufacture foamed plastics. The resultant products can range from low density flexible elastomers to rigid structural panels and ornamental furniture components.

There is intensive current global activity to develop practicable alternatives to chlorofluorocarbons −11 and −12, which are the workhorse blowing agents for rigid and flexible polyurethane foam, as well as for polystyrene and polyolefin foams.

Scientific data have linked chlorofluorocarbons (commonly known as CFC, and used in many nonplastics applications as well) to depletion of the earth's ozone layer. Since the ozone layer protects against dangerous ultraviolet radiation from the sun, the potential for a global health hazard has emerged. The problem is so serious that an international agreement known as the Montreal Protocol was issued in 1987, calling for a worldwide effort to limit the use of CFC. In response, the U.S. Environmental Protection Agency issued its final rulings (in August 1988) ordering a freeze on domestic CFC production at 1986 levels, which were slated to become effective in mid-1989. Additional reductions of 20% and 30% were mandated by EPA for 1993 and 1998.

An eventual total phase-out of CFC is the apparent goal of tightening global restrictions. Therefore, a great deal of research effort is being expended by chemical companies. Their common goal is to develop replacements for CFC-11 and CFC-12 that are safe, effective, and affordable.

54.8 BRIGHTENERS

A brightener creates a visual whitening effect by virtue of its fluorescence.

54.9 CATALYSTS

These substances accelerate the cure of a polymer when added in low loads. Other terms are often used as synonyms for "catalyst," such as "accelerator" and "initiator." Organic peroxides are generally used for curing unsaturated polymers, although other chemicals (such as azides) can serve this function as well. There are a wide variety of types of catalysts for epoxy resin systems, including BF_3 complexes, amines and amides.

Initiators are used in coatings and plastics that are cured by means of radiation. Typical initiators for curing ultraviolet curable coatings and plastics are benzoin alkyl ethers or acetophenone derivatives for free-radical systems and aryldiazonium salts for cationic systems. Camphorquinone has been used as a photo initiator for visible-light curable polymers that are used for adhesives and in the dental industry.

54.10 COLORANTS

Colorants are among the most essential of additives, and they are used primarily to improve the appearance of plastics parts. However, colorants provide other benefits, including the use of carbon black in olefins to improve resistance to degradation by sunlight. Also, blue dyes are used in clear vinyls to mask the yellowing appearance that occurs upon aging or exposure to weathering. There is a wide spectrum of colorants (including white, black, primary colors, pearlescents, and phosphorescent pigments) encompassing an infinite number of blends and degrees of gloss. See also the chapter on Colorants.

54.11 COUPLING AGENTS

As the name implies, coupling agents are chemicals that provide an improved bond between a filler or reinforcement and the matrix polymer. A coupling agent can be visualized as a molecule with dual functionality, wherein one part of the molecule will bond to the filler/reinforcement and another will bond to the matrix. Thus, a bonded bridge is formed between two different materials. The coupling agent improves the interface between the filler and matrix, so that the compound's physical properties are optimized and retained during long-term exposure to environmental conditions.

The predominant coupling agents are the silanes.[7] Titanate coupling agents have shown especially remarkable recent improvements in processing characteristics and in synergizing the properties of many composite systems.[8] Many other types of coupling agents are used in composite materials, including zirconates and methacrylato chrome complexes.[9]

54.12 DEFOAMERS

Defoamers are additives for liquid resin systems that reduce or prevent foam formation. Various surfactants, including silicones, fluorochemicals, and acetylene compounds, have been found useful as foam suppressants.

54.13 FILLERS AND REINFORCEMENTS

A number of chapters in this Handbook deal in detail with various fillers and fiber reinforcements. Also, this subject has been well detailed in other recent handbooks.[10,11] Fillers and reinforcements can be designated as additives because they are added to the base polymer to provide improvements. Thus they satisfy the definition that was given at the beginning of this chapter.

Typical fillers are talc, mica, calcium carbonate, calcium sulfate, aluminum trihydrate, synthetic silica, and kaolin.

The early incentive for using fillers in plastics was simply to lower cost since fillers are significantly cheaper than even the lowest-cost plastic resin. But fillers can provide many performance benefits, and that now is often the main reason for using a filler. Among these benefits are reduced shrinkage after molding, reduced coefficient of thermal expansion, increased coefficient of thermal conductivity, increased hardness, and improved abrasion resistance. Fillers can be classified in a number of ways. One way is by shape; fillers are available as microspheres, flakes, short fibers, hollow beads, or irregular particulates.

Reinforcements are used in a polymer to provide a great increase in modulus and strength. Among the most common reinforcements are glass fiber, wollastonite, carbon and graphite fibers, high strength and high modulus polyethylene filaments, and metal filaments.

54.14 FLAME RETARDANTS AND SMOKE SUPPRESSANTS

Materials in this class of additives provide the polymer with resistance to combustion and retard the propagation of flame.[12] Typical fire retardants are antimony oxide, which is used in combination with a halogen compound[13] and aluminum trihydrate.[14] Many other additives are used as flame retardants and smoke suppressants.[15]

Reduction of smoke and toxicity has been and will continue to be a prime focus of research. In no other class of additives does the need to abide by regulations (issued by government agencies and industry groups) play such an important role. Increasingly stringent demands for protection against flammability and the products of incineration (smoke and toxicity) are being made by such authoritative bodies as the Federal Aviation Administration and the National Fire Protection Assn.

All claims for flame retardancy must be accompanied by certification of a material's ability to achieve a specific fire rating (eg, V-O) under tests established by the Underwriters Laboratory.

54.15 HEAT STABILIZERS

Heat stabilizers prevent degradation of the polymer during processing or in elevated-temperature use environments. A polymer that is particularly sensitive to degradation during processing is poly(vinyl chloride) (PVC). It is a serious matter because degradation can lead to corrosion of processing equipment by the hydrochloric acid or gases released when PVC breaks down. Therefore, heat stabilizers for PVC have received a great deal of study and development. Typical products used as heat stabilizers for PVC include barium, cadmium, and zinc stearates. Lead-based and tin stabilizers are also available. Recent increases in the price of cadmium (and its suspected toxicity) have created an incentive toward the development and marketing of barium–zinc stabilizer systems, as well as other mixed stabilizers.

54.16 IMPACT MODIFIERS

An example of an effective impact modifier is the dispersion of an elastomer or rubber in polystyrene to obtain improved impact resistance. Typically, stereoregular polybutadiene elastomers are added to polystyrene at various levels to provide medium and high impact grades. In an analogous manner, reactive liquid rubber polymers have been shown to improve the impact strength of thermoset resin systems.[16,17] The mechanism involved is *in-situ* rubber particle formation during gelation of the thermoset resin. By this means, the failure mechanism can be controlled by the size of individual rubber particles within the resin phase.

54.17 LOW PROFILE ADDITIVES

Many industrial applications require that the molded part have a glossy finish. This is especially true of the automotive industry, where molded plastics must compete both in performance and appearance with high-gloss, painted metal components. Reinforced polyester molded parts, for example, typically have high shrinkage rates during cure, which results in surface defects that include sink marks and microscopic irregularities. Such surface imperfections are unacceptable, and for a long time kept RP out of automotive appearance markets, despite attractive cost-performance benefits. However, additives have been developed that compensate for the shrinkage and provide a dramatic improvement of surface smoothness.

Low profile or low shrink additives are usually thermoplastic materials that are added to the polyester resin.[18] The mechanism whereby these additives improve surface smoothness is not well understood, but a number of theories have been proposed.[19,20]

54.18 LUBRICANTS, SLIP AGENTS, MOLD-RELEASE AGENTS

Additive lubricants improve the movement of a polymeric part against itself or against other materials. This type of additive is often also called a "slip agent." Chemicals of the same types are also used as mold-release agents, and they are designed to prevent a plastic part from sticking to the mold. These chemicals include fatty primary amides, silicones and metallic stearates.

As in the case with antiblocking agents, mold-release agents usually have only partial compatibility with the polymer matrix so that they exude to the surface. Mold-release agents can be external and internal. External products are applied to the surface of the mold or molding equipment, by spraying or other means. Internal products are added to the polymer system, usually at a concentration of less than 1% by weight.

54.19 PLASTICIZERS

A plasticizer is a material incorporated in a polymer to increase its flexibility, processability, elongation or toughness. Addition of plasticizer will usually cause a reduction in melt viscosity, lower the temperature of the second-order transition, or lower the elastic modulus of the solidified polymer.

Plasticizers are the synergistic compounding ingredient in poly(vinyl chloride) and has resulted in the tremendous worldwide growth of the flexible vinyl industry. There are thousands of vinyl plasticizers, including dioctyl phthalate and other phthalates, sebacates, citrates, and phosphate esters.

54.20 THICKENERS

Thickeners increase the viscosity and control the rheology of a polymer system during processing. There are inorganic and organic thickeners. Among the inorganic types are various grades of clay, fumed silicas, and magnesium oxide. The latter is used in sheet molding compounds to control the molding process.[21]

54.21 UV STABILIZERS

Also called light-stabilizers, UV stabilizers are added to a polymer in order to reduce or prevent degradation caused by exposure to sunlight or ultraviolet radiation. Such degradation is a major weakness of polyolefins, but the development of increasingly effective stabilizers, including hindered amines, now enables these plastics to be used for a wide range of outdoor-use products.

54.22 VISCOSITY DEPRESSANTS

By reducing the viscosity of a polymer system, processing characteristics, such as better mold fill and faster cycling, are improved. Some plastics processing methods, such as sheet molding, require high-viscosity compounds; however, in other cases, such as liquid casting systems, it is desirable that the polymer have low viscosity in order to be able to include a high level of filler or short-fiber-reinforcement and still maintain pourability of the mixture.

54.23   WETTING AGENTS

A wetting agent is a substance that reduces the surface tension of a liquid, thereby causing it to spread more readily on a solid surface.

BIBLIOGRAPHY

1. "Annual Additives Update: 10 billion lbs and Growing," *Plast. Eng.* 27 (Aug. 1988).
2. "Plastics Additives: Less Performing Better," *Chem. Eng. News* 35–39, 43–57 (June 13, 1988).
3. "Adding Value with Additives," *Mod. Plast.* 77–131 (Sept. 1988); 85–90 (Oct. 1988).
4. *Additives for Plastics,* DATA Inc., San Diego, Calif.
5. G.J. L. Griffin, *Biodegradable Fillers in Thermoplastics,* in R.D. Deanin and N.R. Schott, *Fillers and Reinforcements for Plastics,* American Chemical Society, Washington, D.C., 1974, p. 159.
6. "Recycling Conference Debates Future of Degradable Plastics," *Mod. Plast.* 174–176 (Sept. 1988).
7. E.P. Plueddemann, *Silane Coupling Agents,* Plenum Press, New York, 1982.
8. S.J. Monte, "Titanates," in H.S. Katz and J.V. Milewski, eds., *Handbook of Fillers for Plastics,* Van Nostrand Reinhold, New York, 1987.
9. H.S. Katz, "Non-silane Coupling Agents," in I. Skeist, ed., *Handbook of Adhesives,* 3rd ed., Van Nostrand Reinhold, New York, 1989.
10. H.S. Katz and J.V. Milewski, eds., *Handbook of Fillers for Plastics,* Van Nostrand Reinhold, New York, 1987.
11. J.V. Milewski and H.S. Katz, eds., *Handbook of Reinforcements for Plastics,* Van Nostrand Reinhold, New York, 1987.
12. A.H. Landrock, *Handbook of Plastics Flammability and Combustion Toxicology,* Noyes Publications, Park Ridge, N.J., 1983.
13. I. Touval, "Antimony Oxide," in Ref. 10.
14. I. Sobolev and E.A. Woychesin, "Aluminum Trihydrate," in Ref. 10.
15. J. Green, H.S. Katz, and J.V. Milewski, "Miscellaneous Flame Retardants and Smoke Suppressants," in Ref. 10.
16. E.H. Rowe and F.J. McGarry, "Improving Damage Resistance of SMC and BMC," *Proceedings of 35th Annual Conference, Reinforced Plastics/Composites Institute,* SPI, Washington, D.C., 1980, paper 18-E.
17. R.S. Drake and A.R. Siebert, "Elastomer Modified Polyesters for Compression and Injection Molded SMC/BMC," *Proceedings of 42nd Annual Conference,* Composites Institute, SPI, Washington, DC, 1987, paper 11-D.
18. L. Kiaee, Y.S. Yang, and L.J. Lee, "Effect of Low Profile Additives on the Curing and Surface Morphology of Sheet Molding Compounds, *Proceedings of 43rd Annual Conference,* Composites Institute, SPI, Washington, D.C., 1988, paper 17-A.
19. L.R. Ross, S.P. Hardebeck, and M.A. Bachman, "Review of Mechanisms of Low-profile Resins, *Proceedings of 43rd Annual Conference,* Composites Institute, SPI, 1988, paper 17-C.
20. K.E. Atkins and co-workers, "Advances in Low-profile Additives for Smoother Surfaces," *Proceedings of 43rd Annual Conference,* Composites Institute, SPI, 1988, paper 17-D.
21. A.H. Horner and R.N. Brill, "Develop Thickening Behavior Conditions to Insure SMC Process Control, *Proceedings of the 39th Annual Conference,* Reinforced Plastics/Composites Institute, SPI, 1984, paper 8-A.

55

FILLERS AND REINFORCEMENTS, INTRODUCTION

Sidney E. Berger

Consultant, 413 Chatham Circle, Warwick, RI 02886

The sections on fillers and reinforcements were prepared under the direction of Sidney E. Berger.

Fillers and reinforcements, the most widely used members of the additive family, have always been in use in some plastics. Their use has grown dramatically in recent years, encouraged by the growing demand for high performance plastics and increasing polymer prices. Usually available as a finely ground powder or fiber, they are added to a polymer to improve properties such as strength, thermal stability, shrinkage and to reduce polymer cost.

ASTM D883 defines a Filler as ". . . a relatively inert material added to a plastic to modify its strength, permanence, working properties, or other qualities, or to lower costs."

Fillers and reinforcements are generally classified into three categories as follows:

Extender. A relatively low-priced material primarily added to reduce composite material costs by occupying space and displacing the more expensive matrix resin. An extender usually has little or no effect on the composite's strength.

Semireinforcement. A moderately priced filler added to displace the matrix resin and promote some improvement in the composite's strength properties.

Reinforcement. A relatively high priced filler primarily added to improve the strength properties of the composite.

Additives are substances that are added to organic polymers to improve their properties and protect them from the degrading effects of light, heat, and bacteria. Additives are also used to change or modify a polymer's density, lubricity, thermal expansion, processing properties, enhance flame retardation, add, enhance and protect color, and other similar characteristics.

The following sections will review the thirteen fillers and reinforcements which are the most widely used. Table 55.1 lists fillers, their specific gravity and particle size.

TABLE 55.1.   Fillers and their Specific Gravity and Particle Size

Filler	Specific Gravity		Particle size, microns, median
	lb/ft^3	g/cm^3	
Alumina trihydrate	151	2.42	<1–100
Calcium carbonate	168	2.7	<1–15
Calcium sulfate	185	2.96	2–10
Carbon fibers	6.2–37	0.1–0.6[a]	3–25[b]
Glass fibers	155–168	2.48–2.70	3–30[c]
Glass microspheres, hollow	9.4–44	0.15–0.70	5–300
Kaolin clay	162–168	2.6–2.7	2–5
Mica	168–193	2.7–3.1	100–325[d]
Processed mineral fibers	175	2.8	200[b]
Silica, microcrystalline	165	2.65	50–75
Silica, synthetic, fused	137	2.2	7–40[d]
,gel	137	2.2	2–20[d]
,precipitated	137	2.2	8–40[d]
Talc	168–175	2.7–2.8	10–60
Wollastonite	181	2.9	50–1250[e]

[a]Bulk density.
[b]Fiber length in mm.
[c]Fiber diameter in microns.
[d]Ultimate particle size in millimicrons.
[e]Typical mesh size by screen analysis.

56

FILLERS, ALUMINA TRIHYDRATE (ATH)

Frank Molesky

Solem Industries, Inc., 4940 Peachtree Industrial Boulevard, Norcross, GA 30071

56.1 CATEGORY

Alumina trihydrate (ATH) is a functional filler, which is primarily employed to impart flame retardance to resin systems. In some resin systems, such as unsaturated polyesters, it is an extender filler replacing more expensive resin in the formulation.

56.2 SOURCE

Major bauxite reserves are found in the United States, South America, Jamaica, Europe, and West Africa. The Bayer process is the most common production technique to convert bauxite ore to ATH.

In Bayer processing, the bauxite ore is crushed. The aluminum hydroxide in the bauxite is then digested with hot caustic soda under pressure and solubilized as sodium aluminate.

The reaction is as follows:

$$2 \; Al(OH)_3 + 2 \; NaOH \longrightarrow 2 \; NaAlO_2 + 4 \; H_2O$$

| Aluminum hydroxide | Sodium hydroxide | Sodium aluminate | Water |

Various insoluble impurities such as iron oxide, silicates, and titania settle out as "red mud." The sodium aluminate solution is then filtered. Under uniform agitation, the solution is seeded with alumina trihydrate to form crystalline aluminum hydroxide. The resulting crystals, which have an average particle diameter of 80 microns, are washed, separated, and dried.

Various techniques can be employed to produce precipitated ATH in various particle sizes, or the ATH can be mechanically ground to a desired particle size.

A second method for producing ATH is known as the "sinter process." Briefly, the "red mud" cited above is blended with limestone and sodium carbonate and then calcined at 1832°F (1000°C) to produce a sinter containing dicalcium silicate and sodium aluminate, which is separated from the insoluble dicalcium silicate. The sodium aluminate is then processed by the Bayer method.

The sinter process produces products that have fewer trace organic contaminants and are generally brighter than the Bayer produced products.

56.3 KEY PROPERTIES

ATH is also known as hydrated alumina or aluminum hydroxide and is chemically designated as $Al(OH)_3$. It has a density of 2.42 g/cm³. The Mohs hardness is 2.5–3.5 and is only moderately abrasive. The refractive index is 1.58, which is very similar to that of many polymer resins. This feature permits the resins to retain their translucency when filled with ATH. ATH is insoluble in water and has a pH of 8–10.

A typical chemical analysis of alumina trihydrate is Al_2O_3 64.9%; SiO_2, 0.02%; Fe_2O_3, 0.01%; Na_2O (total), 0.3%; Na_2O (soluble), 0.02%; and loss on ignition, 34.5%.

There are several suppliers of alumina trihydrate that offer the product in a range of particle sizes from less than 1 to 100 microns. Of course, the finer the particle size, the greater the surface area. Its key properties are its flame retardancy and smoke suppression.

To better understand the flame retardant and smoke suppressant properties of ATH, it is important to study closely the differential thermal analysis and the thermogravimetric analysis graphed in Figure 56.1.

ATH is unreactive to the curing mechanism common to most polymer curing chemistry, such as unsaturated polyester curing, rubber vulcanization, and polyolefinic copolymer cross-linking.

At room temperature, ATH is quite stable. However, between 401–428°F (205–220°C), slow decomposition occurs. Above 428°F (220°C), decomposition becomes rapid as the hydroxyl groups of the ATH begin to decompose endothermically.

The decomposition of the α-alumina trihydrate to transition χ-alumina corresponds to the major endothermic peak at 572°F (330°C). The two smaller endothermic peaks at 446°F (230°C) and 986°F (530°C) relate to a portion of the α-trihydrate decomposing to α-mono-hydrate and the subsequent decomposition of the monohydrate to transition γ-alumina, respectively.

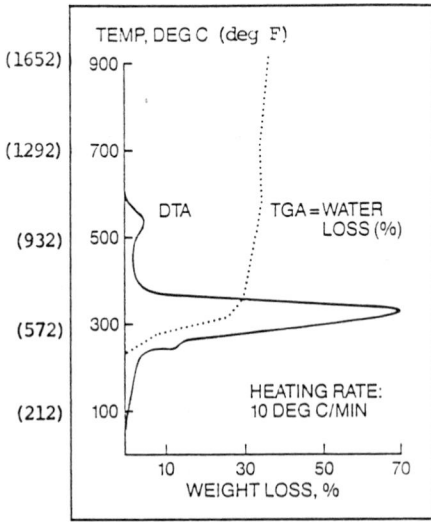

Figure 56.1. DTA/TGA curves for alumina trihydrate. Thermodynamic properties of $Al_2O_3 \cdot 3H_2O$ at 25°C (77°F) are formula weight, 156.0072; heat formation, kcal/mole, −612.5; Gibbs free energy of formulation, kcal/mol, −546.7; entropy, cal/deg-mol, 33.51; heat capacity, cal/deg-mole, 44.49.

The enthalpy of the decomposion (the heat of dehydroxylation) is 280 cal/g. Of the dry powder 34.6 wt% is chemically combined water. The reaction is simply:

$$2 \; Al(OH)_3 \longrightarrow Al_2O_3 + 3 \; H_2O$$

It is endothermic dehydroxylation which is the flame retardance mechanism for burning polymers. ATH retards burning by acting as a heat sink and absorbing a portion of the heat of combustion. The water that is released dilutes the combustion gases and participates in condensed phase reactions. This dilution effect and the endothermic decomposition is credited for the smoke-suppressing properties of alumina trihydrate.

56.4 PROCESSING CHARACTERISTICS

Processing Property	Effect of Filler or Reinforcement
Viscosity	Increases
Melt flow	Decreases
Temperature	Increases
Injection pressure	Increases
Flow in mold	Decreases
Mold shrinkage	Decreases

56.5 APPLICATIONS

There are many application areas for ATH. Most of them make use of ATH for its flame retardance and smoke suppressing properties.

- Fiber reinforced polyesters—Construction-related end uses such as bath tubs, shower stalls, panels, and skylights (both spray-up and lay-up compounds). For this application ATH acts also as a resin extender.
- SMC/BMC laminates—Mostly for electronic equipment as well as appliances and automotive parts.
- Flexible polyurethane foams for seating and mattresses.
- Epoxy resins—Used for encapsulation, potting, and epoxy glass laminates for electrical/electronic uses.
- Styrene–butadiene rubber (SBR) latex foams and adhesives for carpet backing.
- Cross-linked polyethylene, cross-linked ethylene vinyl acetate, and EPDM for wire and cable insulation and jacketing.
- SBR mechanical goods, such as mine belting.
- Coatings, paints, adhesives, and sealants for flame retardance.
- Flexible PVC for wall coverings, upholstery, and wire and cable insulation.

56.6 POLYMERS FILLED

Advantages and limitations of filled polymers are shown in Table 56.1. The formulations containing alumina trihydrate in various polymers shown on pp. 688–691 best describe its use. (Note: All tests, including flame test, were conducted in Solem Industries' laboratories. The results are not intended to reflect hazards of materials under actual fire conditions.)

TABLE 56.1. Filled Polymers, Advantages and Limitations

Polymer	Advantages and Limitations
	Thermosets
Unsaturated polyesters	Flame retardance
	Smoke suppression
	Resin extension hence cost savings
	High loadings tend to adversely affect physical properties
Epoxy	Flame retardance
	Smoke suppression
	High loadings affect physical properties
	Polyolefins
Polyethylene	Flame retardance and smoke suppression
	High loadings need to be used
	Use of surface treatments helps in property retention
	Rubber
SBR latex foams	Flame retardance and smoke suppression
Neoprene foams	High loadings need to be used
EPDM	
SBR	
	Thermoplastics
PVC	Flame retardance and smoke suppression

Unsaturated Polyester

Typical Formulation	A	B
G-P polyester	75	40
ATH (SB-336)		40
1 in. Chopped fiberglass	25	20

Properties		
E-84 tunnel test		
Flame spread	780	70
Smoke development	1000	300

Applications	Grades
Tub/shower units	SB-336
Appliance housings	SB-432
Electrical sheet	431-G
Cultured onyx	Onyx Elite 100

Epoxy

Typical Formulation	Control	A
Liquid epoxy (DGEBA)	100	100
Curing agent (HHDA)	75	75
ATH (SB-136)		120

Properties		
Oxygen index	20.0	27.5

Applications	Grade
Encapsulation	SB-432
Potting compounds	SB-136
Epoxy/glass laminates	SB-336

Cross-Linked Ethylene–Vinyl Acetate

Typical Formulation	Pbw
Ethylene–vinyl acetate	100
Vinyl–tris silane	3
ATH (Micral® 932)	125
Calcium stearate	2
Peroxide	1.7

Properties	
Oxygen index	37

Applications	Grade
Automotive wire	Micral 932
Appliance wire	Micral 932

Urethane

Typical Formulation	Pbw
Polyol (polyether)	100
TDI	33.5
Thermolin 101	30
ATH (SB-632)	120
FR 300 BA	22
Antimony trioxide	8
Other additives	14

Properties	
E162 (radiant panel)	10–25
Smoke density, D_s (4 min)	<200

Applications	Grade
Flexible foam	SB-136
Adhesives	SB-432
Elastomers	SB-632

EPDM

Typical Formulation	Phr
EPDM	100
Escorene LD-400	10
Zinc oxide	5
ATH (Micral 932)	100
Escorez 1102	10
Agerite MA	1.5
Vinyl silane	2.0
Dispersant	1.0
Dicup 40C	8.0

Properties	
Oxygen index	25.0
National Institute of Standards and Technology (NIST) Smoke Density (ASTM E-662)	
flaming	105.0
smoldering	205.0

Applications	Grade
Wire and cable	Micral 932
Roofing membrane	Micral 932
Mine belts	Micral 932

PVC

Typical Formulation	Phr
PVC	100
Plasticizer	40
ATH (Micral 932)	15
Stabilizer	5
Antimony trioxide	3

Properties	
Oxygen index	33.0

Applications	Grade
Communication wire	Micral 932
Wall coverings	Micral 932
Floor tile	SB-136
Conduit	SB-632
Upholstery	SB-632

Polyethylene

Typical Formulation	%
HDPE, including stabilizer package	35
ATH (Micral 932)	65

Properties	
Radiant panel flame spread index	20
National Institute of Standards and Technology Smoke Density (ASTM E-662) D_s (4 min)	10
UL 94	V–0

Applications	Grade
Electrical conduit	Micral 932
Appliance housing	Micral 932

56.7 COMMERCIAL GRADES

ATH is available in a wide range of median particle sizes from less than 1 micron to 100 microns in diameter and in white or near white color. Surface-modified ATH is also commercially available. Common surface modifications include silanes, titanates, and stearates. Table 56.2 is a sampling of ATH suppliers and products and is not intended as a complete list.

TABLE 56.2. Trade Names and Major Grades of Alumina Trihydrate for Extender and Filler Applications

Product	Median Particle Size, Microns	Product	Median Particle Size, Microns
Alcan		C-430	11.0
BACO FRF-5	55–80	Hydral 705	0.5
BACO FRF-10	20–30	Hydral 710-w	1.0
BACO FRF-20	14–20	Lubral 710-w	1.0
BACO FRF-30	11–16	PGA-w	1.0
BACO FRF-40	9–14	Aluchem	
BACO FRF-60	7–9	AC-400 series	6–14
BACO FRF-80	6–7	AC-714	2.2
BACO FRF-85	5–6	AC-722	3.6
BACO Superfine 4	1.2	AC-740	7.0
BACO Superfine 7	0.8	Custom Grinders	
BACO Superfine 11	0.7	Polyfil 115	22–23
Alcoa		Polyfil 113	19–21
C-30	100	Polyfil 110	16–18
C-31	40	Polyfil 203	11–13
C-33	30	Polyfil 204	12–14
C-130		Polyfil 202	9–11
C-230	17.0	Polyfil 301	6.5–8.5
C-231		Polyfil 403	4–5
C-330	7.5	Polyfil 402	3–4
C-331	7.5	Polyfil 401	2.5–3.5
C-333	7.5		

TABLE 56.2. (*Continued*)

Product	Median Particle Size, Microns	Product	Median Particle Size, Microns
Georgia Marble		SB-331	10–12
KC-31	40–60	331-G	10–12
KC-75	20–25	SB-332	10–12
KC-80	16–18	SB-336	15
KC-90	13–15	SB-431	8–10
KC-100	11–13	431-G	8–10
Martinswerk		SB-432	8–10
Martinol OL-104	<1	SB-632	3.0
Martinol OL-107	.5	Micral 932	1.7–2.2
Martinol OL-111		Micral 855	1.7–2.2
Nyco		TAC Industries	
Nycoat	8–10	600	32–34
Solem		753	19–21
SB-30	100	883	17–18
Onyx Elite 100	45	983	15–16
SB-136	20	991	13–14

56.8 COMPOSITE CHARACTERISTICS

Property	Effect of Filler
Tensile strength and modulus	Decreases
Flexural strength	Decreases
Flexural modulus	Increases
Impact strength	Decreases
Retention of properties at elevated temperatures	No effect
Retention of properties under moist conditions	Decreases
Dimensional stability	Increases
Thermal durability (heat aging)	No effect
Thermal conductivity	Increases
Linear coefficient of expansion	Decreases
Heat-deflection temperature	Increases
Surface appearance and texture	No change
Electrical properties	Improved arc tracking and dielectric properties
Chemical resistance properties	No effect

57

FILLERS, CALCIUM CARBONATE

Paul E. Kummer

Cyprus Thompson-Weinman & Co., 92 Greenwood Ave., Monteclair, NJ 07042

57.1 CATEGORY

Calcium carbonate is a widely used extender filler. It is sometimes considered a semireinforcing filler when its use improves a plastic composite's impact resistance, stiffness, and dimensional stability.

57.2 SOURCE

Calcium carbonate is produced by chemical reaction precipitation or, more commonly, by the crushing and subsequent fine grinding of natural calcitic limestone rock.

52.3 KEY PROPERTIES

Calcium carbonate's important properties are good color characteristics because of its high purity. In resin systems, because of its low binder demand (see Table 57.1) it has a high loading capability.

57.4 PROCESSING CHARACTERISTICS

Calcium carbonate has these effects on polymer processing properties shown in Table 57.2.

TABLE 57.1. Key Properties of Calcium Carbonate

Product Ground $CaCO_3$	Supplier	Trade Name	Specific Gravity, g/cm^3	Dry Bulk Density, g/cm^3	Mean Particle Size, μ
Uncoated	Thompson-Weinman	Atomite	2.7	1.1	3.0
Coated	Thompson-Weinman	Kotamite	2.7	1.3	3.0
Uncoated	Thompson-Weinman	Supermite	2.7	0.8	1.0
Coated	Thompson-Weinman	Supercoat	2.7	1.0	1.0
Uncoated	Thompson-Weinman	Snowflake	2.7	1.3	5.5

TABLE 56.2.   Effect of Calcium Carbonate on Polymer Processing

Property	Uncoated Grades	Coated Grades
Viscosity	Increases	Increases
Melt flow	Lowers (volume) index	Lowers index
Compounding	Energy increase required	Less energy increase required
Temperature	No effect	No effect
Injection pressure	Increases	Increases
Flow in mold	Reduces	Reduces less
Mold shrinkage	Reduces	Reduces less

57.5   APPLICATIONS

Current major applications for calcium carbonate as a plastic filler are shown below:

Application	Filler Contribution
Flexible PVC sheeting	Easy processing, good color
Rigid PVC extrusions improvement	Good color, stiffness, impact
SMC and BMC	Low viscosity at high loadings; nonreactive
Injection-molded	Low shrinkage polypropylene

57.6   COMMERCIAL GRADES

Ground mineral grades that range from a mean particle size of approximately 15 microns and a surface area of 0.6 m^2/g to a mean particle size of less than 1 micron with a surface area of 9.0 m^2/g are commercially available today. Many of the finer grades are also available after being coated with stearic acid or calcium stearate. Precipitated grades are generally finer than the finest ground grades; they are usually surface treated or coated for plastic applications.

57.7   COMPOSITES CHARACTERISTICS

Table 57.3 briefly describes the effect that calcium carbonate usually has on plastic composites.

TABLE 56.3.   Effect of Calcium Carbonate on the Properties of Plastic Composites

Property	Effect
Tensile strength	Decreases
Tensile modulus	Increases slightly
Flexural strength	Decreases
Flexural modulus	Increases slightly
Impact strength	Improves
Retention of properties at elevated temperatures	Little effect
Dimensional stability	Improves
Thermal durability	Improves
Thermal conductivity	Slight reduction
Linear coefficient of thermal expansion	Decreases
Heat-deflection temperature	Rises
Surface appearance and texture	Usually improves
Electrical properties	Little effect, if pure
Chemical properties	Possible reduction in resistance to acid

57.8 TRADE NAMES

A number of calcium carbonate producers are listed below, along with the trade names of their best-known products in the plastics industry:

Producer	Trade Name
Genstar Stone Products	Camelwhite
Sylacauga Calcium Products	Micro-White
J.M. Huber	H-White, Micro Fill
Georgia Marble Co.	Gamasperse, CS-11
Pfizer, Inc., MPM Div.	Superpflex 200, Hi Pflix 100
Thompson-Weinman	Snowflake, Kotamite

58

FILLERS, CALCIUM SULFATE

Robert E. Duncan

United States Gypsum Company, Industrial Gypsum Division, 101 S. Wacker Drive, Chicago, IL 60606

58.1 CATEGORY

Calcium sulfate is a common, versatile raw material that is widely distributed in nature[1-4] in both the hydrous and anhydrous forms, which are called gypsum and anhydrite, respectively. Gypsum is industrially important because of the ease with which it can be dehydrated and rehydrated, a property that is especially useful in the building industry.

Although both naturally occurring phases may be used in composites, the anhydrous form, designated $CaSO_4$ II, is currently of greater interest to the plastics industry as an extender filler and semireinforcing filler. This anhydrous material is unique in its physical properties and should not be confused with the common laboratory desiccant, soluble anhydrite or $CaSO_4$ III.

The several different phases of calcium sulfate are shown in Table 58.1. Some of the reactions of calcium sulfate are shown below.

$$1.\ CaSO_4 \cdot 2H_2O \xrightarrow[t>400°C]{} CaSO_4\ II\ +\ 2\ H_2O$$
$$\text{Insoluble anhydrite}$$

$$2.\ CaSO_4 \cdot 2\ H_2O \underset{\text{Liquid } H_2O \quad t<40°C}{\overset{65-140°C}{\rightleftharpoons}} CaSO_4 \cdot 1/2\ H_2O\ +\ 3/2\ H_2O$$
$$\text{Gypsum} \hspace{5cm} \text{Hemihydrate}$$

$$3.\ CaSO_4 \cdot 1/4\ H_2O \underset{\text{Water vapor at room temperature}}{\overset{150-180°C}{\rightleftharpoons}} CaSO_4\ +\ 1/2\ H_2O$$

An understanding of these reactions is important for processors who intend to use the hydrous forms of calcium sulfate in their formulations.

58.2 SOURCE

High-purity gypsum rock is used to manufacture two hydrous products used as extender fillers. The ore is crushed and screened to remove impurities, then ground fine and air classified to produce the final products. The fillers are both nontoxic, have a high purity, and are bright white. The F&P ("food and pharmaceutical") grade fillers conform to the specifications set by the *Food Chemicals Codex* and *United States Pharmacopeia XXI*.[5-6] These products serve many purposes in the food, beverage, and pharmaceutical industries.

TABLE 58.1. Most Common Phases of Calcium Sulfate

	Calcium Sulfate Dihydrate (gypsum)	Calcium Sulfate Hemihydrate	$CaSO_4$ III	$CaSO_4$ II (anhydrite)
Chemical formula	$CaSO_4 \cdot 2H_2O$	$CaSO_4 \cdot \frac{1}{2}H_2O$	$CaSO_4$	$CaSO_4$
Formula wt	172.17	145.15	136.14	136.14
Density, g/cm^3	2.32	α-2.75	α-2.58	2.96
		β-2.63	β-2.48	
Refractive index (average)	1.525	1.56	1.514	1.587
Thermal conductivity M cal/cm·s °C	3.00			11.37
Specific heat, cal/mol/°C,	21.84 + 0.76t^a			14.10 + 0.06t
Solubility at 25°C, g/L	2.1	Unstable	Unstable	2.6
Crystal system	Monoclinic	Hexagonal	Hexagonal	Orthorhombic
Occurrence	Natural and manufactured	Manufactured	Manufactured	Natural and manufactured

$^a t$ = temperature, °C

The anhydrous phase of calcium sulfate, $CaSO_4$ II, is manufactured from the same ore as the hydrous products described above. After the ore is sized, it is processed at a high temperature to effect a change in crystal structure.[4,7] The resultant stable anhydrous material is in the $CaSO_4$ II phase. After further grinding and/or air classification, several $CaSO_4$ II extender fillers are produced.

A fibrous form of calcium sulfate has been produced using a patented hydrothermal technique.[8,9] This filler is available in two different phases, hemihydrate and anhydrous, the latter one existing in the $CaSO_4$ II phase. Both types of the fiber are currently available in commercial quantities and have a nominal aspect ratio of about 30 based on microscopic measurements.

Fibrous calcium sulfate is not durable in biological systems, such as the lungs, but appears to be removed rapidly by natural bodily processes.[10]

58.3 KEY PROPERTIES

See Table 58.2.

58.4 PROCESSING CHARACTERISTICS

Some effects of calcium sulfate fillers on polymer processing are shown in Table 58.3. It should be noted that, in general, the actual processing parameters are not much different from those of the unfilled resin.

58.5 APPLICATIONS

58.5.1 Polyester Resin Systems

$CaSO_4$ II fillers have been used as extender fillers in transformer cores made from halogenated polyester resin. The fillers also perform well in business machine housings, instru-

TABLE 58.2. Physical Properties of Calcium Sulfate Fillers[a]

Product	Trade Name	Specific Gravity	Bulk Density, kg/m³		Particle Size, in μm[b]		OA[c]	pH[d]	Surface Area, m²/g		Brightness[e]	Wt Loss, %, at 225°C	Loss on Ignition, % at 750°C
			Loose	Packed	Top	Mean			Blaine	B.E.T.			
CaSO₄·2H₂O	Terra Alba	2.32	675	1120	100	17	24	7.3	.430	0.9	89	20.33	21.07
CaSO₄,II	Snow White Filler[f]	2.96	705	1300	~75	8	24	11.1	.867	6.3	96	0.27	0.627
	CA-5 Filler	2.96	575	1010	~60	2	22	11.0	1.965	7.2	96	0.48	0.636
	CAS-20-4 Filler	2.96			<20	4		11.0	~.9	2.5	≥96	≤0.20	<0.3
	CAS-20-2 Filler	2.96			<20	2	25	11.0	~1.8	~6.5	≥96	≤0.20	<0.3
	Franklin Fiber Filler, A-30	2.96	270	655	l = 50–100[g]		53	10.4			92	≤0.20	<0.3
CaSO₄·½H₂O	Franklin Fiber Filler, H-30	2.5	235	620	d = 2[g]		59	8.2			92	3–5	4–6

[a]Supplier of all calcium sulfate-based fillers in this table is United States Gypsum Company, Industrial Gypsum Division, 101 S. Wacker Drive, Chicago, Illinois 60606.

[b]Particle size was determined using a Micromeritics, Inc. SediGraph 5000D. The sample size was suspended in A-11, temperature = 30°C, dispersion by ultrasonic bath and magnetic stirrer.

[c]OA = oil absorption and is given in cm³ of oil/100 g of filler.

[d]pH is reported for a 10% slurry after 15 minutes of stirring.

[e]Brightness is reported in percent relative to standard BaSO₄. Measurements were made using an integrating sphere on a Beckman recording UV spectrophotometer over the range of 400–700 nm.

[f]Also available in a food and pharmaceutical grade.

[g]l = length, d = diameter, both in micrometers. The average aspect ratio of Franklin Fiber Filler is ≅30.

699

TABLE 58.3.   Processing Characteristics of Calcium Sulfate Extender and Semireinforcing
Fillers in Polyolefins

Processing Properties	Fillers	Effect
Melt flow	CaSO₄ II extender fillers	Generally observe a reduction in melt flow with these fillers in polyolefins.
	Anhydrous Franklin Fiber Filler	In Himont Pro-Fax® 6523 PM PP. 58% increase at 20% loading. 3% increase at 40% loading compared to virgin resin.
Compounding temperature	CaSO₄ II extender filler	Generally tend to act like a heat sink, requiring a slight increase in compounding/extrusion temperature of 5 to 10°C over that used for processing the unfilled resin. When PP is processed using a Buss Kneader, Model MDK/E 46, a normal temperature profile is 170–220°C.
	Semireinforcing CaSO₄ II fillers	Normally, downstream addition is recommended. Temperature profile is about the same as above.
Injection pressure	CaSO₄ II extender fillers and semireinforcing fillers (Engle 140/40 injection molder.)	The following three zone temperature profile is recommended for Himont Pro-Fax 6523 PM PP: (1) 175°C, (2) 190–205°C, (3) 205°C. Pressures and speeds are in the medium to low ranges; dial reading 1000 psi in 1st stage and 700 psi in 2nd stage (hold).
Mold shrinkage	CaSO₄ II extender fillers	Slightly higher shrinkage at low filler levels. Decreasing shrinkage at higher filler levels.

ment chassis and covers, appliance bases and covers, chemical fittings, parts compression molded from SMC, power mower housings, furniture components, containers, circuit breaker frames, switch gear insulating structures, power tool handles and housings, automotive panels and structures, marine fittings, shower bases, and recreational equipment parts.

58.5.2 Laminate Sheet

In laminate sheet, CaSO₄ II is used at levels of about 50% by weight in electrical and military products. With halogenated polyester resins, CaSO₄ II is not reactive under conditions of high moisture or with polymer degradation products, an important factor when choosing a filler for this application.

In systems that cure at room temperature, such as polyester sprayup, hydrous calcium sulfate fillers provide flame retardancy. This type of filler, CaSO₄·2H₂O, is used in applications such as shower stalls and leisure products. Use of hydrous calcium sulfate has benefitted processors by reducing cost; however, some reformulation work is required in order to use this filler most effectively.

58.5.3 Bulk Molding Compound

When it flows into a mold cavity, $CaSO_4$ II extender filler remains well dispersed and maintains good resin/filler contact. These fillers are available in grades appropriate for food contact and pharmaceutical uses. They offer good resistance to food acids (important in microwave cookware), excellent electrical properties (needed in PC boards), and are unreactive with polydegradation products (halogenated resins, for example).

Extender fillers can extend TiO_2 and act themselves to impart white pigmentation to plastic parts.

58.5.4 PVC Molding Compounds

One particular extender filler is used in cellular vinyl moldings to reduce costs. In addition, because of the softness of $CaSO_4$ II extender fillers, wear on processing equipment is reduced.

58.5.5 PVC Plastisols

Calcium sulfate filler and extender fillers have been found to be very effective in PVC plastisol formulations in which they can replace barytes and provide improved heat aging characteristics. Because of their lower specific gravity, these fillers offer cost savings, compared to barytes. In addition, $CaSO_4$ II fillers are available in grades approved for food contact.

58.6 POLYMERS FILLED

58.6.1 Thermoplastics

PVC[11]. $CaSO_4$ II extender fillers are currently used in cellular vinyl moldings for reduced cost and less wear on processing equipment. Uniform cell structures can be obtained using these fillers.

The finer grades of $CaSO_4$ II extender fillers have good potential in rigid PVC pipe formulations. Some formulations show improved powder flow properties when $CaSO_4$ II is used. This allows increased filler loadings over those possible with other fillers used in PVC pipe and without a concurrent loss of physical properties. The greater softness of $CaSO_4$ II fillers can reduce wear on processing machinery.

Plastisols. In plastisols, replacement of barytes by $CaSo_4$ II helps reduce cost. Food acid and stain resistance are other advantages of $CaSO_4$ II.

The high brightness, purity, low cost, food acid, and stain resistance are advantages for $CaSO_4$ II extender and semireinforcing fillers. The moderate oil absorption could cause a slight increase in viscosity, but some plastisol manufacturers have found that $CaSO_4$ II extender fillers impart quite desirable characteristics to their formulations. Some minor problems, such as blistering, have occasionally been observed some grades of calcium sulfate filler are used in very humid weather. $CaSO_4$ II fillers have become well established in this area.[12]

There is good potential for the use of both extender and semireinforcing types of $CaSO_4$ II fillers in automotive and other applications. Because of their inertness toward acids, these fillers are preferred for applications with halogenated, flame-retardant grade polypropylene. Examples of the properties obtained with PP resin are given in Tables 58.4–58.6. Advantages include softness, brightness, good impact strength, dimensional stability, acid resistance, and smooth surface finish.

HDPE, PPS, PET, PBT, Polyamide-Imide, Liquid Crystal Polymers. Both the extender and semireinforcing types of $CaSO_4$ II fillers are strong contenders for applications in these

TABLE 58.4. Physical Properties of Polypropylene Filled with Snow White Filler (CaSO₄ II)[a]

Property	Filler Level, in phr				
	0	10	20	30	40
Tensile strength at yield, MPa	34.1	34.7	32.6	30.1	29.6
Tensile modulus, GPa	1.10	1.31	1.38	1.31	1.65
Elongation at yield, %	12.9	10.0	9.3	8.4	6.8
Flexural strength, MPa	39.6	42.5	41.4	42.2	
Flexural modulus, GPa	1.17	1.38	1.45	1.93	1.86
Izod impact strength, J/m unnotched	NB	NB	NB	361	383

[a]Resin is Himont Pro-Fax 6523 PM in flaked form.

TABLE 58.5. Physical Properties of Polypropylene Filled with CAS-20-2[a]

Property	Filler Level, in phr				
	0	10	20	30	40
Tensile strength at yield, MPa	35.9	31.9	28.7	26.6	25.2
Tensile modulus, GPa	0.94	0.98	1.10	1.30	1.45
Elongation at yield, %	13.7	10.2	8.2	6.6	4.4
Flexural strength, MPa	40.4	44.6	44.1	44.0	42.5
Flexural modulus, GPa	1.23	1.56	1.76	2.10	2.32
Izod impact strength, J/m unnotched	N.B.	646	555	410	250
Heat-deflection temperature, °C					
0.45 MPa	96	120	122	129	130
1.82 MPa	59	65	66	71	73

[a]Compounding was accomplished by first blending the filler and flaked resin in a twin shell blender. Next the blend was compounded using a Stokes-Windsor twin-screw extruder, stranded, pelletized, and dried. Test specimens were prepared using an Engle 140/40 injection molder equipped with a mold conforming to ASTM requirements for testing plastic parts. Data supplied by the USG Corporation Research Center, 700 North Highway 45, Libertyville, Illinois 60048.

TABLE 58.6. Effect of Anhydrous Fibrous CaSO₄ II on Polypropylene[a]

Property	Filler Level (% by weight)		
	0	20	40
Tensile strength at yield, MPa	33.7	32.9	30.9
Elongation, %	11.0	7.5	5.0
Flexural strength at yield, MPa	42.0	48.8	50.5
Flexural modulus, GPa	1.32	2.55	3.68
Heat deflection, 0.45 MPa, °C	100	123	129
Heat deflection, 1.82 MPa, °C	58	71	80
Impact strength			
Falling wt, J	0.62	0.92	1.35
Izod, notched, J/M	33.3	41.0	52.1
Izod, unnotched, J/M	662	294	200
Melt flow, g/10 min	3.11	4.90	3.20
Specific gravity	0.93	1.01	1.18
Shore D hardness	75	77	80

[a]Himont Pro-Fax 6523 PM, a flaked resin. Compounded resin and Franklin Fiber Filler using a Buss Kneader, Model MDK/E 46, single screw, 46 mm diameter with a 3.5:1 L/D, and 1:1 compression ratio. The cylinder temperature was measured at four positions, screw-190°C, Z1-192°C, Z2-192°C, and cross head-200°C. Vacuum venting was used. The fiber was added at port 2 using an Acrison 1201 feeder. Data supplied by the USG Corporation Research Center, 700 N. Highway 45, Libertyville, IL 60048.

TABLE 58.7. Physical Properties of HDPE Filled with CAS-20-2[a]

Property	Filler Level (% by weight)		
Union Carbide DMDJ-7904-Natural 7-Flaked	0	20	40
Tensile strength at yield, MPa	11.9	12.8	13.4
Elongation at yield, %	20.4	13.2	7.9
Flexural strength at yield, MPa	12.4	14.9	17.4
Flexural modulus, GPa	0.336	0.492	0.810
Izod impact strength, unnotched	NB	NB	NB
Falling weight impact strength, J	NB	16.6	14.7
Melt flow, g/10 min	4.77	4.12	2.44
Dow Chemical Resin #05054	0	20	40
Tensile strength at yield, MPa	22.7	21.8	21.1
Elongation at yield, %	12.2	9.6	4.7
Flexural strength at yield, MPa	25.0	27.3	29.0
Flexural modulus, GPa	0.827	1.23	1.37
Izod impact strength, unnotched	NB	NB	NB
Falling weight impact strength, J	NB	NB	14.2
Heat-deflection temperature, °C			
0.45 MPa	78	86	95
1.82 MPa	48	50	54
Melt flow, g/10 min	2.01	1.49	1.42

[a]Resins were compounded using a Buss Kneader Model MDK/E 46 compounder. The fillers were fed downstream in each case. Test specimen were prepared using an Engel 140/40 injection molder, using three zone temperatures of 205°C and a mold temperature of 35°C. Data supplied by the USG Corporation Research Center, 700 N. Highway 45, Libertyville, Illinois 60048.

polymers. Table 58.7 shows test data obtained for filled HDPE. The results include good surface finish for molded parts, improved impact strength, higher filler loadings, and reduced extruder and mold wear.

Fibrous calcium sulfate semireinforcing filler also has good potential with these resins. Because of its microfibrous nature, it can compliment glass fiber in fiber-glass-reinforced composites and also provide equal or improved physical properties. An improvement in surface appearance can also be achieved. Because compounding and extrusion are high shear processes, downstream addition of the fibrous calcium sulfate filler is recommended. Areas of potential utilization include business machine housings, chemical pump components, and injection-molded printed circuit boards.

58.6.2 Thermosets

Unsaturated Polyesters. $CaSO_4$ II extender fillers have been effective in electrical and electronic applications. The electrical properties obtained are comparable to those where silica (an industry standard) is used, with the added advantage that $CaSo_4$ II is less abrasive than siliceous fillers. Because of the inertness of $CaSO_4$ II fillers and their approval for use in the food and pharmaceutical industries, they show great promise in areas where those tight specifications are necessary, such as microwave cookware, containers for food, and other food contact applications. As previously noted, $CaSO_4$ II fillers maintain good food-acid resistance. The extremely fine grades of $CaSO_4$ II can help processors obtain an excellent surface finish on molded parts. Fibrous grades of calcium sulfate have potential as a partial replacement of fiber glass.

In polyester spray-up, calcium sulfate filler can be used in the barrier coat and in several of the laminate layers to improve physical properties and reduce blistering. An additional benefit is that in some systems calcium sulfate filler helps improve thixotropy.

A potential disadvantage exists if the fibrous fillers are subjected to prolonged high shear mixing, which causes extensive fiber breakdown. Therefore, medium to low shear mixing is recommended in order to realize the full benefit of the fibers.

Epoxies. For calcium sulfate extender fillers, applications exist in potting, casting, and encapsulation compounds where they offer good dielectric properties, excellent compatibility with epoxies, reduced wear on processing equipment, and excellent pigmentability. Surface appearance is improved when the finer fillers are incorporated into the epoxy compound. In low temperature cured systems, the hydrous fillers (see Table 58.2) can be used when improved flame retardancy is required.

The fibrous, semireinforcing calcium sulfate fillers function well as an asbestos replacement in epoxy adhesives and encapsulation compounds. Loading levels of 5 to more than 40% are reportedly being used to provide strength, dimensional stability, and good electrical properties in epoxy composites. The anhydrous variety (see Table 58.2) functions well in high temperature epoxies and in pultrusion applications. Improvements are observed in surface appearance, crack and craze resistance, and fatigue resistance.

Urethane. Calcium sulfate extender fillers work well in urethane sealants where they improve adhesion properties and corrosion resistance.

Urethane RIM. Semireinforcing calcium sulfate filler functions well in urethane RIM because of its relative softness, which means less wear on the impingement mixing head as compared to other fibrous fillers used in RIM formulations—as milled glass, glass flakes, and wollastonite.

58.7 COMMERCIAL GRADES

Anhydrous calcium sulfate fillers for plastics are presently available from only one source. The grades now available are listed in Table 58.2. Fillers for food and pharmaceutical applications are available in the hydrous and anhydrous forms. Table 58.2 lists nine grades (including F&P grades), but more will become available as interest in calcium sulfate based fillers increases and more applications are discovered.

58.8 COMPOSITE CHARACTERISTICS

See Tables 58.4–58.7.

58.9 TRADE NAMES

See Table 58.2.

BIBLIOGRAPHY

1. "Calcium Sulfate and Derived Materials," M. Murat and M. Foucault, eds., *Proceedings of the International RILEM Symposium,* Saint-Revy-les-Chevreuses, France, 1977.
2. M. Grayson, ed., *Kirk-Othmer Encyclopedia of Chemical Technology,* 3rd ed., Vol. 4, John Wiley & Sons, Inc., New York, 1978, pp. 437–448.
3. T.C. Patton, ed., *Pigment Handbook,* Vol. 1. John Wiley & Sons, Inc., New York, 1978, pp. 289–292.

4. *Ullman's Encyclopedia of Industrial Chemistry,* 5th ed., Vol. A4, VCH Publishers, Deerfield Beach, Fl., 1985, pp. 555–584.

5. *Food Chemicals Codex,* 3rd ed., National Academy Press, Washington, D.C., 1981.

6. *United States Pharmacopeia XXI,* 21st rev., *National Formulary,* 16th ed., Official from Jan. 1, 1985, United State Pharmacopeial Convention, Inc., Rockville, MD, 1985, p. 1541.

7. U.S. Pat. 4,080,422 (March 21, 1978), R.E. McCleary.

8. U.S. Pat. 4,152,408 (May 1, 1979), J.G. Winslow.

9. H.S. Kazt and J.V. Milewski, eds., *Handbook of Fillers and Reinforcements for Plastics,* Van Nostrand-Reinhold Co., New York, 1978, p. 431.

10. G.W. Wright, *The Biology of Franklin Fiber Filler,* Publication No. IG-213/12-86, United States Gypsum, Chicago, Ill.

11. S. Walter, *39th SPE ANTEC,* May 1981.

12. R.E. Duncan, K.R. Watkins, and B.D. Jennings III, *37th SPI/RPC Conference, Session 6-A,* 1982, pp. 1–8.

59

CARBON FIBERS

Charles T. Moses

Amoco Performance Products, Inc., Parma Technical Center, 12900 Snow Road, Parma, OH 44130

59.1 CATEGORY

Carbon fiber is primarily employed as a reinforcement to improve the strength properties of plastic composites.

59.2 SOURCE

Carbon fibers used in the reinforcement of plastics are typically produced from three classes of raw materials: rayon-based fibers, acrylic-based fibers and pitch-based fibers. Two types of pitch-based fibers are utilized and the distinction is based on their crystalline properties. Pitches that exhibit liquid crystalline, (anisotropic) behavior are called mesophase pitches; pitches that do not display liquid crystalline properties are called isotropic pitches.

Both mesophase and isotropic pitch fibers are prepared by melt spinning an appropriately selected precursor. The melt spun fiber is then stabilized (thermoset) to prevent it from remelting in the subsequent carbonization process. Carbonization to obtain the desired properties, tensile strength, modulus, conductivity, etc, is conducted by heating the fiber in an inert atmosphere to temperatures above 1000°C (1800°F). Textile products are produced by weaving the carbon tow products.

Acrylic-based carbon fibers are prepared from a range of compositions of polyacrylontrile (PAN) polymer containing a variety of comonomers (such as acrylic acids, acrylates, and sulfonic acids) or even as a homopolymer. The polymeric fiber is spun, oriented by drawing, stabilized, and carbonized to temperatures above 1000°C (1800°F). Textile products are produced by weaving the carbon tow products.

Rayon-based carbon fibers are prepared from specially-formulated cellulosic precursors. Most commonly these rayon materials are woven into a cloth and then carbonized to temperatures above 1000°C (1800°F) to produce a textile carbon fiber product.

59.3 KEY PROPERTIES

For a filament size of 7–10 microns, carbon fiber specific gravity ranges from 1.7–2.2 (105–140 lb/ft³). Table 59.1 identifies supplier and grade name for two major carbon fiber product forms: continuous filament tow and woven fabric textiles. The listing of woven fabric suppliers that follows reflects the current (1988) situation in the U.S. and Europe. As the usage of advanced composites made from carbon fibers continues to expand, this listing will change. Other product forms supplied by some manufacturers include chopped tow and discontinuous

TABLE 59.1.   Suppliers and Trade Names for Continuous Filament and Woven Fabric
Product Forms

Company	Trade Name
Continuous Filament	
Afkim Carbon Fibers	ACIF
Akzo Carbon Fibers Inc.	Fortafil
Amoco Performance Products Inc.	Thornel
Ashai Nippon Carbon Fiber Co. Ltd.	Hi-Carbolon
Ashland Carbon Fibers Division	Carboflex
BASF Structural Materials Inc.	Celion
Courtaulds Grafil Company	Grafil/Apollo
Fiber Materials Inc.	Microfil
Hercules Inc.	Magnamite
Hitco, Materials Div.	Hi-Tex
Mitsubishi Chemical Industries Ltd.	Dialead
Mitsubishi Rayon Co. Ltd.	Pyrofil
Sigri Elektrographit GmbH	Sigrafil
Soficar SA	Filkar/Torayca
Stakpole Fibers Co. Inc.	Panex
Toho Beslon Co. Ltd.	Besfight
Toray Industries Inc.	Torayca
Woven Fabric	
Amoco Performance Products Inc.	Thornel VCK, VCL, and WCA
Avco Specialty Mtls/Textron	Avcarb
Barber-Colman Co.	TWF (triaxial woven fabrics)
Brochier SA	Lyvertex
Burlington Glass Fabrics Co.	(Burlington)
Composite Reinforcements Inc.	Cofab
Courtaulds Advanced Materials	(Courtaulds)
Fabric Developments Inc.	(FDI)
Fiberite Corporation	Weav-rite
Hercules Inc.	Magnamite
Hexcel Corporation	HexStrand
Hitco, Woven Structures Div.	(Hitco)
Mitsubishi Rayon Co. Ltd.	Pyrofil
North American Textiles Inc.	(NATI)
Sigri Elektrographit GmbH	Sigratex
Stackpole Fibers Co. Inc.	Panex
Ten Cate Glas BV	(Ten Cate)
Textile Products Inc.	(TPI)
Textile Technologies Inc.	(TTI)
Toho Beslon Co.	Besfight
Toray Industries Inc.	Torayca

filament (mat) products which are typically granulated for use as a filler in polymeric materials.

59.4   PROCESSING CHARACTERISTICS

In the case of continuous filament or cloth, polymer processing is accomplished through the specialized operation of prepregging, in which the reinforcing material is used as a laminate in a multiply composite. The two principal methods of prepregging are the hot-melt process and the solution process. In hot-melt prepregging, a thermosetting resin is partially reacted, formed into a thin film, and used to impregnate a continuous web of either carbon fiber tows (unidirectional tape) or carbon fiber cloth. In solution prepregging the polymeric matrix (thermosetting or thermoplastic) is dissolved in a solvent to form a solution that is used to impregnate the web of filaments or cloth. Discontinuous product forms (chopped fiber or granulated mat) are typically used as conventional fillers in injection molding or SMC (sheet molding compound) operations.

The following illustrates the relative effect of adding carbon fiber reinforcement as a filler in conventional plastics processing operations. As in any complex technology, these effects are often subject to mitigating factors related to specific equipment or system characteristics.

Processing Property	Effect of Reinforcement
Viscosity	Increases
Melt flow	Decreases
Compounding	Machine dependent
Temperature	Cools faster
Injection pressure	Increases
Flow in mold	Mold dependent
Mold shrinkage	Decreases

59.5 APPLICATIONS

The primary application of carbon fibers is in thermosetting epoxy composites to produce high-strength/high-modulus structural materials. The extraordinary tensile strength and range of modulus values available make carbon fiber–epoxy composites attractive for ultrahigh strength applications typified by military aircraft, rockets, spacecraft, commercial aircraft, and sporting-goods applications. The high electrical conductivity of carbon fiber has also been exploited to produce electromagnetic interference shielding in nonstructural applications. Similarly the unique thermal expansion behavior of carbon fiber composites has lead to its use in composite antennas for satellites. The superior mechanical and thermal performance of carbon fibers has led to their use as friction materials in an amorphous carbon matrix, such as brake disks.

59.6 POLYMERS FILLED

The following summarizes the major families of polymers that have been reinforced with carbon.

Polymer	Advantage	Limitations
THERMOPLASTICS		
ABS	Tough, good shelf	Nylons and polyesters gen-
Acetal	life, reprocessing	erally suffer from creep,
Nylon	capability for all	have relatively low use
Polyamide/imide	thermoplastics	temperature, often poor
Polyarylate		solvent resistance. New
Polyarylsulfone		thermoplastics such as
Polycarbonate		polyarylsulfone, polyary-
Polyester (PBT)		lates and polyamide/im-
Polyetheretherketone (PEEK)		ides do not suffer from
Polyethylene		creep and have excellent
Poly(ethylene sulfide)		solvent resistance. How-
Polypropylene		ever, they do require
Polystyrene		higher forming tempera-
Polysulfone		ture.

Polymer	Advantage	Limitations
THERMOSETS		
Epoxy Polyester Bis maleimide Polyimide	Up to 350°F (177°C) use temperature, high performance composite material	Often brittle and moisture sensitive.

59.7 COMMERCIAL GRADES

Continuous filament carbon fiber is characterized by a variety of tensile strengths, tensile moduli, and filament counts per tow. Typical PAN-based carbon fiber for aerospace applications is produced with a tensile strength from 450,000 psi (3100 MPa) up to 1×10^6 psi (6900 MPa) at a modulus of $33 \times 10^4/43 \times 10^6$ psi (230/290 GPa). Tow sizes of 1000, 3000, 6000, 10,000, and 12,000 filaments are most commonly used. Higher modulus PAN-based fiber at $50 \times 10^4/57 \times 10^6$ psi (340/395 GPa) and ultrahigh modulus PAN-based fibers at $65 \times 10^6/80 \times 10^6$ psi (450/570 GPa) are also available. Filament sizes are typically 4–7 micrometers.

Pitch-based fibers are produced with lower tensile strength than PAN fibers, but with a wider range of moduli—$25 \times 10^6/120 \times 10^6$ psi (160–825 GPa). Tow sizes vary from 500 to 4000 filaments. Filament size is typically 9–10 micrometers.

Continuous filament fibers may be supplied with a variety of finish sizes that are chosen to be compatible with the matrix in which it is used.

Chopped fiber is produced by cutting the continuous filament varieties described above into a variety of lengths [0.12/1 in. (3/25 mm)]. Bulk densities varying from 0.1–0.6 g/cm³ (6–37 lb/ft³) are commercially available.

59.8 COMPOSITE CHARACTERIZATION

The following summarizes the relative effect of adding carbon fibers to polymeric materials. As noted above, these effects are often subject to mitigating factors related to specific equipment or system characteristics.

Property	Effect of Filler on Composite
Tensile strength and modulus	Increases
Flex strength and modulus	Increases
Impact strength	Increases
Retention of property at elevated temperature	Matrix material dominates
Retention of property under moist conditions	Matrix material dominates
Dimensional stability	Increases
Thermal durability	Increases
Thermal conductivity	Increases

59.9 TRADE NAMES

See Table 59.1.

BIBLIOGRAPHY

General References

M. Sittig, ed., *Carbon and Graphite Fibers Manufacture and Applications,* Noyes Data Corp., Park Ridge, N.J., 1980.

J. Delmonte, *Technology of Carbon and Graphite Fiber Composites,* Van Nostrand Reinhold Co., New York, 1981.

G. Lubin, ed., *Handbook of Composites,* Van Nostrand Reinhold Co., New York, 1982.

Carbon & High Performance Fibres Directory 4, Pammac Directories Ltd., Slough, UK, 1988.

60

FILLERS, FIBER GLASS

David W. Garrett

Owens-Corning Fiberglas Corp., Technical Center, 2790 Columbus Rd., Granville, OH 43023

60.1 CATEGORY

Fiber glass is a reinforcement, used primarily to improve composite strength properties.

60.2 SOURCE

Glass fibers are formed by drawing filaments of glass from a furnace containing molten glass through an orifice, and winding the resultant fibers (or strands) on a takeup wheel or collet.

An organic coating, called a size, is applied to the fiber during this forming operation to facilitate processing, maintain fiber integrity, and improve product performance. This treatment generally consists of a silane coupling agent, a film former, and such processing aids as lubricants, antistatic agents, plasticizers, biocides, etc.

The silane coupling agents are added to improve adhesion of the polymer matrix to the glass and retention of composite properties under corrosive conditions. They are generally viewed as offering a direct chemical bond linkage between the glass and the resin, though this is a very simplistic representation of what actually occurs at the interface. (A more detailed review of the bonding mechanisms of these silanes is given in References 1–3.)

The film formers are added for several reasons. They promote polymer compatibility and improve composite strengths; they improve strand integrity and processibility; and they protect the strand from environmental attack.

After treatment, the fibers are dried and further processed (roved, twisted, or woven, for example) as required for the specific end-use application.

60.3 KEY PROPERTIES

See Table 60.1.

Glass fibers are, by far, the most widely used reinforcement for polymers on the market today. This is because they offer a good balance of reinforcing properties (such as strength, modulus, and dimensional and thermal stability) and cost. Mineral fibers and fillers are considerably less expensive, but only offer a fraction of the reinforcing properties. On the other hand, the more exotic fibers (carbon, boron, and aramid, for example) generally are only used in applications where cost is not a significant factor.

Most reinforced products are produced for a glass composition called E-glass. This is a borosilicate glass.[4,5] specifically designed for the reinforcement market. The nomenclature

TABLE 60.1. Properties of Fiber Glass

Product	Supplier[a]	Specific Gravity, g/cm^3	Bulk Density	Particle Size, mμ
E-glass	CSG OCF PPG Manville	2.54	Varies depending on fiber length or package construction	3–30
S-glass	OCF	2.48	0.5 to 1.5 g/cm^3	5–10

[a]CSG, CertainTeed Corp.; OCF, Owens-Corning Fiberglas Corp.; PPG, PPG Industries, Fiber Glass Div.

"E" stems from its originally intended use in electrical applications. E-glass has a tensile strength of 5×10^5 psi (3,450 MPa), a Young's modulus of 10×10^6 psi (72,400 MPa), and an elongation to break of 4.8%.

S-Glass (OCF) and S-2 Glass (OCF) are high strength glasses having roughly the same composition.[4,5] Their tensile strength, Young's modulus, and elongation to break are 65×10^4 psi (4,500 MPa), 12×10^6 psi (85,500 MPa), and 5.7%, respectively. S-Glass is used for critical military and aerospace applications; S-2 Glass for less critical applications. The basic differences in these two glasses are the extremely high level of quality control required for S-Glass plus some variation in sizing. S-Glass also sells for about four times as much as S-2 Glass.

AR-glass (for alkaline-resistant glass) was developed specifically for reinforcing cements and concretes. This is a zirconium-containing glass that has been shown to have more durability than E-glass in strongly alkaline media. Its strength properties are very close to those of E-glass. This material is no longer produced domestically.

There are a few other glass types used in reinforcing polymers. These include C-glass (chemical-resistant glass), A-glass (a soda-lime glass), and R-glass (a cross between E-glass and S-Glass). These glasses are in limited production and are only used in specialty applications.

The strength of a component made with glass fibers increases in proportion to the percentage of fibers added to the polymer. However, this relationship is usually not linear. The orientation of the fibers in a composite determines the direction and level of the ultimate strength and modulus.

The highest strength can be achieved in unidirectional composites (such as pultruded products) along the fiber direction. Properties fall off rapidly, however, in the off-axis directions. As much as 80% reinforcement loadings by weight can be obtained in unidirectional composites.

Bidirectional composites (such as woven roving, filament winding) have some fibers placed at angles to others and positioned in a plane or series of planes (plies). This is the most common form of reinforcement and achieves good, fairly uniform strength in two directions. The strength in the direction perpendicular to the plies, unfortunately, approximates that of the unreinforced polymer. Up to 75% glass loadings by weight can be obtained in bidirectional composites.

Multidirectional or pseudo-isotropic composites will be equally strong in all directions. These composites are very difficult to produce, and almost all reinforced polymers have some strength anisotropy. The industrial process that most closely approximates the multidirectional composite is injection molding. Considerable strength variations related to directionality are still common in these components.

(Numerous reviews and books have been written over the last twenty years on the theoretical and practical properties of composites. Two of the better treatments of this subject are given in References 6 and 7.)

60.4 PROCESSING CHARACTERISTICS

Processing Property	Effect of Reinforcement
Viscosity	Increases
Melt flow	Decreases
Compounding	Depends on polymer system. Usually complicates the process; excess mixing can damage the glass fibers
Temperature	No effect (can act as heat sink)
Injection pressure	Increases
Flow in mold	Decreases
Mold Shrinkage	Decreases

60.5 APPLICATIONS

Glass fibers have been used to reinforce almost every type of commercial polymer. The largest application areas are thermosetting polyesters, epoxies, phenolics, nylons, polypropylene, and thermoplastic polyesters. Fiber glass reinforced plastics are used widely in such market areas as transportation, construction, marine, electrical/electronic, appliance, and aerospace. Some of the advantages are high strength to weight, ease of manufacture, corrosion resistance, dimensional stability under thermal stress, cost, tensile and stiffness improvements, fire retardancy, and good electrical resistivity.

Literally thousands of products in many varied applications are made from glass-fiber reinforced polymers. They include automotive applications such as fender liners, doors, door liners, deck lids, wheels, radiator core supports, transmission supports, front and rear end panels, and under-the-hood components; chemical-resistant tanks and pipes; various equipment used in petroleum, chemical and water–wastewater treatment industries; recreational, commercial and military watercraft; tires; bodies for recreational vehicles; sporting equipment such as tennis rackets, fishing poles, skis, skateboards, protective devices, and vaulting poles; furniture; bathtub, shower and vanity installations; housing and components for business equipment, appliances and power tools; playground and amusement equipment; air-supported building structures; and aerospace and aircraft components.

60.6 POLYMERS FILLED

See Table 60.2.

60.7 COMMERCIAL GRADES

Glass fibers are available in a variety of forms. These include roving, T-30 roving, mats, chopped strands, woven roving, twist and plied yarns, cloth, milled fibers, flakes, papers, and impregnated roving such as tire cord. The sizes on the glass fibers are designed to maximize composite properties and processibility. They are highly proprietary. An end user would need to consult the individual manufacturer to match a specific product to a final part application. The following product forms are available from the manufacturers listed:

Product Form	Manufacturer
Roving	CSG, OCF, PPG
T-30 roving	CSG, OCF, PPG, Manville
Continuous-strand mat	CSG, OCF
Chopped-strand mat	CSG, OCF, PPG
Shingle mat	CSG, OCF, PPG
Yarns	OCF, PPG
Chopped strands	CSG, OCF, PPG, Manville
Milled fibers	OCF
Flake glass	OCF
Woven roving	OCF, CSG, PPG

TABLE 60.2. Properties of Fiber Glass-Filled Polymers

Polymer	Properties	
	Improved	Worsened
Thermoplastics		
Polyethylene	Tensile strength, thermal and dimensional stability, moduli	Impact strength
Polypropylene	Tensile strength, thermal and dimensional stability, moduli	Impact strength
Nylon	Tensile strength thermal and dimensional stability, moduli	Impact strength
Styrenics (polystyrene ABS, SAN, etc)	Tensile strength, thermal and dimensional stability, moduli	Impact strength
Thermoplastic polyesters (PBT, PET)	Tensile strength, thermal and dimensional stability, moduli	Impact strength
Poly(phenylene oxide)	Tensile strength, moduli	Impact strength
Polycarbonate	Tensile strength, moduli, thermal/ dimensional stability	Impact strength
Polysulfone	Tensile strength, moduli low temperature stability	Impact strength
PVC	Tensile strength, moduli	Impact strength
Thermosets		
Polyesters	Tensile strength, moduli, thermal/ dimensional stability, corrosion resistance, impact strength	
Epoxies	Tensile strength, moduli, thermal/ dimensional stability, corrosion resistance, impact strength	
Phenolics	Tensile strength, moduli, thermal/ dimensional stability, impact strength	
Furans	Tensile strength, moduli, thermal/ dimensional stability	
Urethanes	Tensile strength, thermal/ dimensional stability	Impact strength
Melamines	Tensile strength, moduli thermal/ dimensional stability	

TABLE 60.3. Effect of Fiber Glass on Composite Characteristics

Property	Effect of Reinforcement
Tensile strength and modulus	Increases
Flexural strength and modulus	Increases
Impact strength	Generally decreases
Retention of properties at elevated temperatures	Increases
Retention of properties under moist conditions	Depends on polymer system (usually a slight decrease)
Dimensional stability	Increases
Thermal durability	Generally increases
Thermal conductivity	No effect
Linear coefficient of expansion	Decreases (except for PPS)
Heat-deflection temperature	Increases for crystalline thermo-plastics, no effect amorphous systems
Surface appearance and texture	Less smooth (generally)
Electrical properties	Depends on composite
Chemical properties	Depends on composite

60.8 COMPOSITE CHARACTERISTICS

See Table 60.3.

60.9 MANUFACTURERS

Certain Teed Corp., Box 860, Valley Forge, Penn.

Owens-Corning Fiberglas Corp., Fiberglas Tower, Toledo, Ohio.

PPG Industries, Fiber Glass Div., One Gateway Center, Pittsburgh, Penn.

Manville Corp., P.O. Box 5108, Denver, Colorado.

BIBLIOGRAPHY

1. M.R. Rosen, *J. of Coatings Tech,* **50**(644) 70–82 (1978).
2. E.W. Plueddemann, in D.E. Leyden and W.T. Collins, eds., "Chemistry of Silane Coupling Agents," in *Silylated Surfaces,* Gordon Breach, New York, 1980, p. 31.
3. E.W. Plueddemann, *Silane Coupling Agents,* Plenum Press, New York, 1982.
4. F.V. Tooley, ed., *Handbook of Glass Manufacture,* Vol. 1, Ogden, New York, 1953, p. 4.
5. J.G. Mohr and W.P. Rowe, *Fiberglass,* Van Nostrand Reinhold, New York, 1978, p. 20.
6. G.B. McKenna, *Polym. Plast. Technol. Eng.* **5**(1) 23–35 (1975).
7. L.J. Broutman and R.H. Krock, *Modern Composite Materials.* Addision-Wesley, Reading, Mass., 1967.

61

FILLERS, HOLLOW MICROSPHERES

Bruce W. Sands

*The PQ Corporation, 280 Cedar Grove Rd., P.O. Box 258, Lafayette Hill,
PA 19444*

61.1 CATEGORY

Hollow microspheres comprise a functional extender filler.

61.2 SOURCES

Hollow microspheres composed of sodium borosilicate glass are generally produced by spray drying a solution or dispersion and subsequent heat treating and/or acid washing. Products composed of soda lime borosilicate glass are produced by passing glass frit, which contains a blowing agent, through a high temperature furnace.

Ceramic or fly ash spheres are obtained through flotation of coal furnace flue-gas scrubbings. Perlite products are mined and further expanded through additional heating. (These products are multicellular and not true, discrete hollow microspheres. However, they do provide low density, functionality, and resin extension.) Saran microspheres are produced by heating a dispersion of a copolymer of vinylidene chloride and acrylonitrile that contains a volatile organic liquid.

Carbon products are produced by pyrolyzing phenolic hollow spheres or pitch. Phenolic hollow microspheres are produced by spray drying a solution of phenolic resin.

61.3 KEY PROPERTIES

See Table 61.1.

61.4 PROCESSING CHARACTERISTICS

Property	Effect of Filler
Viscosity	Increasing volume loadings increases the viscosity of liquid polymers, but to a considerably lesser degree than increasing volume loadings of irregularly shaped fillers.
Melt flow	Increasing volume loadings will decrease melt flow.

Compounding	Hollow microspheres need to be compounded into a matrix with the lowest shear possible. For high-shear and high-pressure application requirements, product with higher densities, and hence generally higher strength should be used.
Temperature	All hollow microspheres of inorganic glass, ceramic, or perlite compositions will withstand temperatures greater than 662°F (350°C). Some ceramics will withstand up to 2732°F (1500°C). Hollow microspheres of organic compositions are usable at much lower temperatures.
Injection pressure	The hollow microspheres to be used must withstand high pressures or their value will be lost.
Flow in mold	Generally, hollow microspheres will provide better mold flow than will other, irregularly shaped fillers. Surface smoothness of the finished part may be slightly affected by large-particle-size hollow microspheres. Their insulating effect may affect mold flow.
Mold	Hollow microspheres shrinkage may reduce mold shrinkage. Cooling rates will be lower also.

61.5 APPLICATIONS

Application	Major Benefit
Hydrospace (Syntactic foams)	Low density, stiffness
Aerospace (Syntactic foams)	Low density, stiffness
Furniture (Cast polyester, urethane)	Low density, insulative, woodlike, low warp, improved screw and nail holding, high volume loadings, low cost
Plaques (Cast polyester, urethane)	Low density, insulative, woodlike, low warp, high volume loadings, low cost
Marine (Polyester, epoxy)	Low density, high stiffness, improved screw and nail holding, high volume loadings, low cost
Cultured marble (Polyester)	Low density, insulative thermal shock resistant, low cost
Bowling balls (Polyester)	Low density, density (weight) control, high volume loadings, low cost
Automotive body filler (Polyester)	Low density, improved sandability, high volume loadings, low cost, impact resistance, good adhesion
Air filter gasketing (Plastisol)	Low density, low durometer, high volume loadings, low cost
Walk-off mat backing (Plastisol)	Low density, low durometer, low stress whitening, high volume loadings, low cost
Thermoplastics	Low density, low cost

61.6 POLYMERS FILLED

Polymer	Advantages and Limitations
THERMOPLASTICS	
Nylon Polypropylene Polyethylene Other thermoplastics	Low density, better cell uniformity and distribution are advantages. Potential sphere breakage and more careful processing conditions are limitations.

Property	Advantages and Limitations
THERMOSETS	
Polyester	Low density, high-volume loadings insulation, stress release,
Urethane	and low cost are a few advantages. Possible breakage during
Epoxy	processing, minor modification of catalyst levels and/or
Plastisol	temperatures (because of insulative properties) are
	limitations.

61.7 COMMERCIAL GRADES

See Table 61.1.

61.8 COMPOSITE CHARACTERISTICS

Property	Effect of Hollow Microsphere
Density	Decreases
Tensile strength	Decreases
Tensile modulus	Increases or no effect
Flexural strength	Decreases
Flexural modulus	Increases or no effect
Impact strength	Varied results depending on test method
Retention of properties at elevated temperatures	Generally no effect
Retention of properties under moist conditions	May decrease without an appropriate surface treatment
Dimensional stability	Increases or no effect
Thermal stability	No effect
Thermal conductivity	Decreases
Linear coefficient of expansion	Decreases*
Heat-deflection temperature	Generally increases
Surface appearance and texture	Decreases
Electrical properties	Varied effects
Chemical properties	Generally no effect

Organic hollow microspheres will impart a different effect, on retention of properties at elevated temperatures, thermal stability and linear coefficient of expansion.

61.9 TRADE NAMES

Product	Company and Location
Q-CEL Hollow microspheres	The PQ Corporation Valley Forge, Penn.
3M Glass bubbles	3M Company St. Paul, Minn.
Microballoon	Emerson & Cuming Canton, Mass.
Extendospheres	P.A. Industries Chattanooga, Tenn.
Microcel	Glaverbel Fleurus, Belgium
Sil-Cel	Silbrico Corporation Hodgkins, Ill.
Hollow ceramic microspheres (not a trademark)	Fillite USA, Inc., Huntington, W.V.
Carbospheres	Versar Manufacturing Co., Springfield, Va.
Miralite	Pierce & Stevens Chemical Corporation Buffalo, N.Y.
Expancel	KemaNord, Sweden
UCAR Phenolic Microballoons	Union Carbide Corporation Danbury, Conn.

TABLE 61.1. Key Properties of Hollow Microspheres

Microsphere Product	Supplier	Composition	Specific Gravity[a]	Particle Size, μm[b]
Q-CEL 200	PQ Corporation	Sodium borosilicate glass	0.20	75
Q-CEL 300	PQ Corporation	Sodium borosilicate glass	0.24	65
Q-CEL 400[c]	PQ Corporation	Sodium borosilicate glass	0.20	75
Q-CEL 500[c]	PQ Corporation	Sodium borosilicate glass	0.24	68
Q-CEL 600	PQ Corporation	Sodium borosilicate glass	0.43	62
Q-CEL 100	PQ Corporation	Ceamic	0.65	5–300
Q-CEL 110	PQ Corporation	Ceramic	0.65	50–300
Q-CEL 120	PQ Corporation	Ceramic	0.60	5–75
C15/250	3M	Sodalime borosilicate glass	0.15	<177[d]
B23/500	3M	Sodalime borosilicate glass	0.23	<177[d]
B28/750	3M	Sodalime borosilicate glass	0.28	<177[d]
B37/2000	3M	Sodalime borosilicate glass	0.37	<149[e]
B38/4000	3M	Sodalime borosilicate glass	0.38	<177[d]
A16/500	3M	Sodalime borosilicate glass	0.16	<177[d]
A20/1000	3M	Sodalime borosilicate glass	0.20	<177[d]
A32/2500	3M	Sodalime borosilicate glass	0.32	<177[d]
A38/4000	3M	Sodalime borosilicate glass	0.38	<177
D32/4500	3M	Sodalime borosilicate glass	0.32	<74[f]
FTD 202	Emerson & Cuming	Insoluble glass	0.23–0.26	80
FT 102	Emerson & Cuming	Insoluble glass	0.23–0.26	80
SI	Emerson & Cuming	Silica	0.22–0.27	80
R	Emerson & Cuming	Sodium borosilicate glass	0.34–0.38	80
IG-25	Emerson & Cuming	Sodium borosilicate glass	0.23–0.26	80
IG-101	Emerson & Cuming	Sodalime borosilicate glass	0.29–0.30	80
IGD-101	Emerson & Cuming	Sodalime borosilicate glass	0.29–0.30	80
MC-37	Emerson & Cuming	Sodalime borosilicate glass	0.37	40
FAB	Emerson & Cuming	Ceramic	0.67	150
M28	Glaverbel	Sodium borosilicate glass	0.28	5–250
M35	Glaverbel	Sodium borosilicate glass	0.35	5–250
SG	P. A. Industries	Glass	0.70	10–425
CG	P. A. Industries	Glass	0.70	10–200
SF-12	P. A. Industries	Glass	0.70	10–125
SF-14	P. A. Industries	Glass	0.70	10–100
SF-20	P. A. Industries	Glass	0.70	10–75
XOL-200	P. A. Industries	Ceramic	0.22–0.28	10–200
52/7/5	Fillite, U.S.A.	Ceramic	0.70	30–300
100/7	Fillite, U.S.A.	Ceramic	0.70	30–300
200/7	Fillite, U.S.A.	Ceramic	0.00	30–120
PG	Fillite, U.S.A.	Ceramic	0.70	30–250
Sil-32	Silbrico	Perlite	0.18	<250
Sil-42	Silbrico	Perlite	0.26	<150
Type A	Versar	Carbon	0.15	5–150
Type B	Versar	Carbon	0.15	5–100
Type C	Versar	Carbon	0.15	5–50
Type D	Versar	Carbon	0.15	50–150
Miralite 176	Pierce & Stevens	PVDC/AN	<0.04	10–100
Expancel DE	KemaNord	PVDC/AN	0.04	10–100
B50-0840	Union Carbide	Phenolic	0.25–0.35	5–127
B50-0930	Union Carbide	Phenolic	0.21–0.25	5–127

[a]Displacement of air. Liquid displacement will be equal to or greater than.

[b]Average unless otherwise stated.

[c]Surface modified products.

[d]Max of 5 wt% >177.

[e]Max of 3 wt% >149.

[f]Max of 3 wt% >74.

62

FILLERS, KAOLIN

Paul I. Prescott

Freeport Kaolin Co., Gordon, GA 31031

Kaolin is a naturally occurring aluminum silicate also known as kaolinite, kaolin clay, China clay, and hydrous aluminum silicate. Kaolin is a fine particle size, crystalline, clay mineral that occurs as platelets and stacks of thin platelets called booklets. The platelets have a repeating octahedral alumina and tetrahedral silica layer structure bound together by shared oxygen atoms, as shown in Figure 62.1. The molecular weight of kaolin is 258.09. Chemical composition is, Al, 20.9 wt%; Si, 21.8 wt%; H, 1.6 wt%; and O, 55.7 wt%.

$$Al_2O_3 \cdot 2SiO_2 \cdot 2H_2O$$
$$\text{or } Al_2\,Si_2\,H_4\,O_9$$

Since kaolin is a naturally occurring mineral formed through weathering of igneous rock decomposition products, it contains certain impurities. The oxide composition of a purified mid-Georgia kaolin, along with the theoretical oxide composition, is given in Table 62.1.

Kaolin clay occurs throughout the world; however, there are few regions where kaolin exists in sufficient quality and quantity to make mining economically feasible. In the United States, most of the kaolin is mined from a narrow belt extending from mid-Georgia to South Carolina. Mining techniques may depending on topography, but all mining is done in open pits. Once the kaolin has been mined, it is processed in different ways depending on the end use.

The plastics industry utilizes kaolin clay processed in four different ways. Each of the processing techniques and applications in the plastics industry will be discussed in separate sections. The key properties of kaolins sold to the plastics industry are listed in Table 62.2.

62.1 AIR-FLOATED KAOLIN

Crude kaolin ore is removed from the pit and transported to the processing plant where it is shredded to reduce the size of the lumps, dried and pulverized in heated roller mills, and air-classified. The product is then collected in a cyclone or bag collector. The coarse particles rejected by the air classifier are returned to the roller mill to be reground. The grade or quality of the air-floated kaolin product is governed by the particle size of the kaolin, the purity of the ore, and the degree to which coarse particles have been removed. The air-floated kaolins are the least pure of all of the types of kaolin used by the plastics industry. The impurities normally associated with air-floated kaolin are mica, rutile, and quartz as coarse particles and anatase and iron compounds as fine particles.

The key properties of the commercially available air-floated kaolins used in plastics are

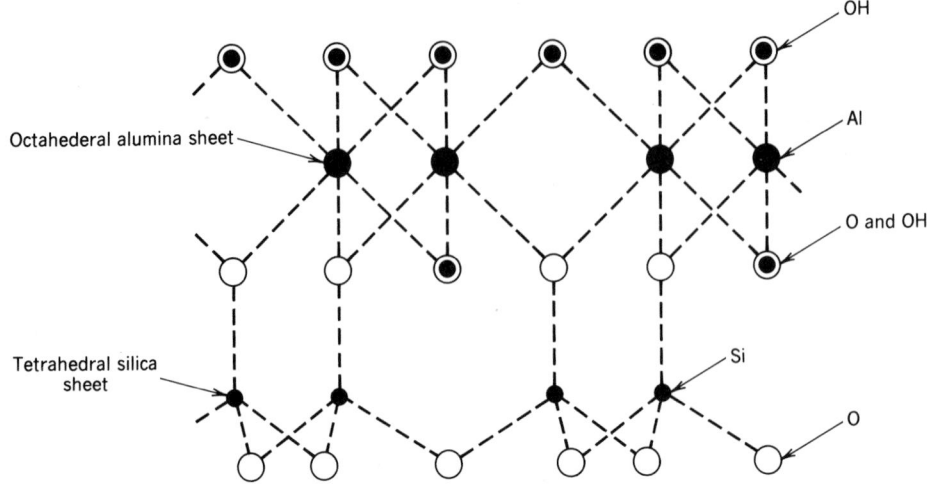

Figure 62.1. Idealized structure for kaolin.

listed in Table 62.3. The fine particle air-floated kaolins tend to exhibit the normal processing characteristics of any fine particle particulate filler, as shown in Table 62.4.

The main applications for air-floated kaolins are as reinforcing or semireinforcing fillers (depending on particle size of the kaolin) in rubber compounds. In plastics, the coarser air-floated kaolins are used as low cost fillers mainly in polyesters, phenolics, and PVC. The advantages and disadvantages of air-floated kaolins are listed in Table 62.5.

The effect of air-floated fillers on filled polymer systems varies with the polymer system and with the filler loading. Table 62.6 lists the general effects of air-floated kaolins in polymer composites.

A list of trademarks and manufacturers of air-floated clays for the plastics industry is given below.

Trademark	Manufacturer
B-80	Thiele Pigment Co.
Dixie	R.T. Vanderbilt Co.
Huber 65A	J.M. Huber Corp.
McNamee	R.T. Vanderbilt Co.
No. 80	Burgess Pigment Co.
Paragaon	J.M. Huber Corp.
Suprex	J.M. Huber Corp.

TABLE 62.1. Oxide Composition, %

Oxide	Theoretical	Beneficiated Mid-Georgia Kaolin Typical Analysis
Al_2O_3	39.50	38.5
SiO_2	46.54	45.4
TiO_2		1.6
Fe_2O_3		0.5
N_2O		0.2
K_2O		0.1
H_2O	13.96	13.7
Total	*100.00*	*100.00*

TABLE 62.2. Key Properties of Kaolin Filler Products

Property	Air-Floated		Water-Washed			Calcined		Surface-Modified	
	Coarse	Fine	Coarse	Medium	Fine	Partial	Full	Calcined	Air-Floated
Specific gravity	2.6	2.6	2.6	2.6	2.6	2.2–2.5	2.6–2.7	2.6	2.6
Particle size, median, μm	1.0–1.8	0.2–0.3	4.0–5.0	0.8–1.0	0.2–0.7	1.7–2.0	0.7–2.0	0.7–2.0	0.3–1.0
pH, 10% solids	4.5–5.5	4.5–5.5	4.2–7.5	4.2–7.5	4.2–7.5	4.5–6.5	4.5–6.5	a	a
BET surface area	10–12	20–24	6–10	14–16	18–22	5–12	5–12	5–12	10–24
Oil absorption	26–30	32–38	30–34	39–41	43–47	45–70	60–90	60–90	26–38

aVaries, some are hydrophobic and cannot be determined.

TABLE 62.3. Key Properties of Air-Floated Kaolins

Suppliers	Trademark	Specific Gravity	Brightness, %	Bulk Density, g/cm^3	Median Particle Size, μm
J.M. Huber	Suprex	2.6	74–76	0.72	0.3
J.M. Huber	Paragon	2.6	74–76	0.72	1.1
J.M. Huber	Huber 65A	2.6	79–81	0.72	1.1
R.T. Vanderbilt	McNamee	2.6	77	0.26–0.56	2.0
R.T. Vanderbilt	Dixie	2.6	70	0.3–0.64	0.2
Thiele	B-80	2.6	80–82	0.48	0.3
Burgess Pigment	No. 80	2.6	80–82	0.48	0.3

TABLE 62.4. Effect of Air-floated Kaolins on Polymer Processing

Processing Property	Effect of Air-Floated Kaolin
Viscosity	Increases
Melt flow	Decreases to no effect[a]
Compounding	No effect
Temperature	Increases to no change[a]
Injection pressure	Increases[b]
Flow in mold	Reduces
Mold shrinkage	Reduces
Mold cycle time	No change

[a]Depends on polymer.
[b]More change as filler loading increases.

TABLE 62.5. Advantages and Disadvantages of Air-floated Kaolins in Plastics

Polymer Type	Advantages	Disadvantages
Thermoplastics	Cost	Color
	Semi-reinforcing increased hardness	Color stability
Thermoset	Cost	Color
		Color stability

TABLE 62.6. Effect of Air-floated Kaolin on Polymer Properties

Property	Effect of Air-floated Kaolin
Tensile strength and modulus	Increases
Flexural strength and modulus	Increases
Impact strength	Decreases
Retention of properties at elevated temperatures	Improves
Retention of properties under moist conditions	No change to slight improvement
Dimensional stability	Improves
Thermal durability	Decreases
Thermal conductivity	Increases
Linear coefficient of expansion	Reduces
Heat-deflection temperature	Moderate improvement
Surface appearance and texture	Improves
Electrical properties	Variable
Chemical properties	Inert

62.2 WATER-WASHED KAOLIN

The crude kaolin ore is suspended in water at the mine or plant using optimum amounts of a deflocculant to form a low-solids, clay/water slurry. The clay, in slurry form, goes through a series of cleaning steps to remove coarse clay, quartz, mica, and so on. The clay slurries from different mine sites are blended to give a uniform feed to the processing plant. At the plant, the blended clay slurry is further purified through a series of screens and cyclones or centrifuges to remove more coarse particles and through a magnetic separator to remove the fine anatase impurities. The cleaned slurries are fractionated in centrifuges into various particle size ranges, leached, and filtered to remove iron staining impurities as well as to

TABLE 62.7. Key Properties of Water-Washed Kaolin Products

Supplier	Trademark	Specific Gravity	Brightness, %	pH	Median Particle Size, μm
Freeport Kaolin	Buca	2.6	86.5–88	6.5–7.5	0.5
Freeport Kaolin	Catalpo	2.6	85.5–87	6.5–7.5	0.8
Freeport Kaolin	Al-Sil-Ate LO	2.6	81–84	6.5–7.5	4.3
J.M. Huber	Huber 80	2.6	85.5–87	6.0–7.0	0.8
J.M. Huber	Huber 35	2.6	82–83.5	4.5–5.5	4.0
Georgia Kaolin	Hydrite PX	2.6	87.90	4.2–5.2	0.7
Georgia Kaolin	Hydrite R	2.6	85–86.5	4.2–5.2	0.8
Georgia Kaolin	Hydrite Flat D	2.6	80–83	4.2–5.2	4.5
Engelhard Corp.	ASP-200	2.6	86.5–88	3.8–4.6	0.6
Engelhard Corp.	ASP-600	2.6	85.5–87	3.8–4.6	0.8
Engelhard Corp.	ASP-400	2.6	80–83	3.8–4.6	4.8
Burgess Pigment	No. 10	2.6	87	4.6	0.5
Burgess Pigment	No. 60	2.6	86	4.6	0.8
Burgess Pigment	No. 40	2.6	81–83	4.6	4.5

increase the slurry solids. The fractionated, purified slurries are then dried and pulverized either as nonredispersed "acid" kaolin or as predispersed kaolin.

The major water-washed kaolin products supplied to the plastics industry are described in Table 62.7. Since the water-washed kaolin products range in median particle size from 0.5 μm to 4.8 μm, the processing characteristics will vary, generally as a function of particle size. The finer the kaolin, the greater the effect on the processing property. Table 62.8 lists the processing properties and the effect kaolin has on properties.

Water-washed kaolins are used mainly in thermoset polyesters where a coarse, low oil absorption, inert filler is reqired. The water-washed kaolins are also used in applications requiring a higher degree of purity and consistency than the air-floated clays can provide. Table 62.9 lists the advantages and limitations of water-washed kaolins as fillers in plastics.

TABLE 62.8. Effect of Water-Washed Kaolin Fillers on Processing Properties

Processing Property	Effect of Water-Washed Filler
Viscosity	Increases
Melt flow	Decreases[a]
Compounding	No effect
Temperature	No effect[a]
Injection pressure	Increases
Flow in mold	Reduces
Mold shrinkage	Reduces

[a]Degree of change depends on polymer.

TABLE 62.9. Advantages and Disadvantages of Water-washed Kaolins in Plastics

Polymer Type	Advantages	Disadvantages
Thermoplastics	Semireinforcing Increased hardness Increased compression strength	Color stability
Thermoset	Chemical resistance Heat sink Reduces cracking and crazing Increased strength	Color Resin demand

TABLE 62.10. Effect of Water-washed Kaolin on Polymer Properties

Property	Effect of Filler
Tensile strength and modulus	Increases
Flexural strength and modulus	Increases
Impact strength	Decrease
Retention of properties at elevated temperatures	Improves
Retention of properties under moist conditions	No effect
Dimensional stability	Improves
Thermal durability	Decreases
Thermal conductivity	Increases
Linear coefficient of expansion	Lowers
Heat-deflection temperature	Moderate improvement
Surface appearance and texture	Improves
Electrical properties	Variable
Chemical properties	Inert

The water-washed kaolins used as fillers affect the properties of polymer composites about the same as air-floated clays. The higher the loading of kaolin and the finer the particle size of the kaolin, the greater the effect on the properties of the component. A list of properties and the effect of water-washed filter clays is shown in Table 62.10. Trademarks and manufacturers of water-washed kaolins for the plastics industry is given below.

Trademark	Manufacturer
Al-Sil-Ate LO	Freeport Kaolin Co.
ASP-200	Engelhard Corp.
ASP-600	Engelhard Corp.
ASP-400	Engelhard Corp.
Buca	Freeport Kaolin Co.
Catalpo	Freeport Kaolin Co.
Huber 35	J.M. Huber Corp.
Hydrite Flat D	Georgia Kaolin Co.
Hydrite PX	Georgia Kaolin Co.
Hydrite R	Georgia Kaolin Co.
No. 10	Burgess Pigment Co.
No. 40	Burgess Pigment Co.
No. 60	Burgess Pigment Co.

62.3 CALCINED KAOLIN

Calcined kaolin pigments or fillers are produced by an extension of the water-washed kaolin process. To produce calcined kaolin, the dried, pulverized, water-washed kaolin is heat treated at 1562 to 1832°F (850 to 1000°C), cooled, pulverized, and air-classified. The calcination process dehydroxylates the alumina layer shown in Figure 62.1.

Calcined kaolins are produced in two grades, partially calcined kaolin, called meta-kaolin, and fully calcined kaolin. The meta-kaolin is used almost exclusively in PVC wire and cable coatings to improve the volume resistivity of the coatings. The fully calcined kaolins are much whiter and, because of the particle configurations, have a higher opacity than non-calcined kaolin. The fully calcined kaolins have excellent dielectric properties in plastics and rubbers. The key properties of the calcined kaolins are shown in Table 62.11.

Calcined kaolins have a higher oil absorption (resin demand) than either air-floated or water-washed kaolins. As such, the calcined kaolins tend to have a more pronounced effect

TABLE 62.11.   Key Properties of Calcined Kaolin

Supplier	Trademark	Specific Gravity	Brightness, %	pH	Median Particle Size, μm
Freeport Kaolin	Whitetex	2.6	90–92	5–6	1.5
	SP-33	2.5	84.5–86	5–6	1.5
	Al-Sil-Ate S	2.7	92–94	5–6	0.75
	No. 100	2.6	91–93	5–6	1.2
J.M. Huber	Huber 40C	2.6	82–86	5–6	1.5
	Huber 70C	2.6	90–92	5–6	1.5
	Huber 80C	2.6	91–93	5–6	1.2
Georgia Kaolin	Glomax LL	2.6	90–92	4.2–5.2	1.8
	Glomax JDF	2.6	90–92	4.2–5.2	0.9
Engelhard Corp.	Satintone 1	2.6	90–92	4.5–5.5	2.0
	Satintone 2	2.5	82–86	4.5–5.5	2.0
	Satintone 5	2.6	90–92	4.5–5.5	1.2
	Satintone Special	2.6	91–93	4.5–5.5	1.2
Burgess Pigment	Iceberg	2.6	90–92	5–6	1.4
	Icecap K	2.6	90–92	5–6	1.0
	30P	2.2	82–84	4–4.5	1.5

TABLE 62.12.   Effect of Calcined Kaolin Fillers on Processing Properties

Processing Property	Effect of Calcined Kaolin
Viscosity	Increases
Melt flow	Decreases
Compounding	No change
Temperature	No change
Injection pressure	Increases
Flow in mold	Decreases
Mold shrinkage	Decreases

on the processing properties of plastics. Changes in the median particle size will have a greater effect on processing properties with calcined kaolins than with the other kaolins. The effects of calcined kaolins on the processing properties of plastics are listed in Table 62.12.

The calcined kaolins have superior color and dielectric properties. The fully calcined kaolins have a whiteness ranging from 90 to 95 percent of a magnesium oxide standard. Color stability is also improved with calcined clays. Calcined kaolins are used as filler reinforcements in engineering thermoplastics where a surface-modifying reagent is compounded in situ. The meta-kaolin exhibits excellent electrical properties in PVC cable insulation compounds and accounts for the major use of calcined kaolin in plastics. The advantages and disadvantages of calcined kaolin are listed in Table 62.13.

The calcined kaolins are higher in oil absorption than the other kaolins and, therefore, tend to have exaggerated effects on filled plastic systems. Strength, hardness, etc, increase while elongation, impact, etc, decrease. The activated surface of calcined kaolins can cause certain plastics to degrade faster at elevated temperatures. A summary of the composite characteristic of calcined kaolin filled plastic is given in Table 62.14.

An alphabetical listing of the commercially available calcined kaolins that are used in plastics is given below.

Trademark	Manufacturer
Al-Sil-Ate S	Freeport Kaolin Co.
Glomax JDF	Georgia Kaolin Co.
Glomax LL	Georgia Kaolin Co.
Huber 40C	J.M. Huber Corp.
Huber 70C	J.M. Huber Corp.
Huber 80C	J.M. Huber Corp.
Iceberg	Burgess Pigment Co.
Icecap K	Burgess Pigment Co.
No. 100	Freeport Kaolin Co.
Satintone 1	Engelhard Corp.
Satintone 2	Engelhard Corp.
Satintone 5	Engelhard Corp.
Satintone Special	Engelhard Corp.
SP-33	Freeport Kaolin Co.
Whitetex	Freeport Kaolin Co.
30 P	Burgess Pigment Co.

62.4 SURFACE-MODIFIED KAOLINS

Air-floated and calcined kaolins are surface-modified to change the plastic composite characteristics. Surface modification, usually with silanes or silicones (but other materials such as organo-titanates can be used), can change the surface energy of the kaolin to make it more compatible with the plastic (resulting in improved impact strength, higher filler loadings, etc). Surface modification can make the kaolin hydrophobic (increased electrical prop-

TABLE 62.13. Advantages and Disadvantages of Calcined Kaolin in Plastics

Polymer	Advantages	Disadvantages
Thermoplastics	Dielectric properties Color Color stability	Oil absorption (platicizer demand)
Thermosets	Dielectric properties	Oil absorption

TABLE 62.14. Effect of Calcined Kaolin on Polymer Properties

Properties	Effect of Calcined Kaolin
Tensile strength and modulus	Increase
Flexural strength and modulus	Increase
Impact strength	Decrease
Retention of properties at elevated temperatures	Decrease to no change
Retention of properties under moist conditions	No effect
Dimensional stability	Increase
Thermal durability	Varies with polymer
Thermal conductivity	Increase
Linear coefficient of expansion	Decrease
Heat-deflection temperature	Increase
Surface appearance and texture	Improves
Electrical properties	Increase
Chemical properties	No charge

TABLE 62.15. Key Properties of Surface-Modified Kaolins

Kaolin Base	Supplier	Trademark	Surface Functionality	pH	Median Particle Size, μm
Calcined	Freeport Kaolin	Translink 37	Vinyl	[a]	1.5
Calcined	Freeport Kaolin	Translink 53	Vinyl	[a]	1.5
Calcined	Freeport Kaolin	Translink 445	Amine	8.5–9.5	1.5
Calcined	J.M. Huber	Nylok 100	Amine	8.5–9.5	1.5
Calcine	J.M. Huber	Polyfil WC	Vinyl	[a]	1.5
Air-floated	J.M. Huber	Nucup 100	Mercapto	4–5	0.3
Air-floated	J.M. Huber	Nulok 321	Amine	7–8	0.3
Calcined	Burgess Pigment	Icecap KE	Vinyl	[a]	0.8
Calcined	Burgess Pigment	2211	Amine	6.5–7.5	1.0
Calcined	Burgess Pigment	5178	Vinyl	[a]	1.0
Air-floated	Thiele Kaolin	RC910	Mercapto	4–5	0.4
Air-floated	Thiele Kaolin	RC990	Mercapto	4–5	0.4
Air-floated	Thiele Kaolin	B2180	Nonfunctional silane	[a]	0.4

[a]Surface modification makes the kaolin hydrophobic.

TABLE 62.16. Effect of Surface-Modified Kaolin on Processing Properties

Processing Property	Effect of Adding Surface-Modified Kaolin	Effect of Surface-modified Kaolin vs Unmodified Kaolin
Viscosity	Increases	Decreases
Melt flow	Slight decrease	Increases
Compounding	No effect	Improves
Temperature	No change	No change
Injection pressure	Increases	Decreases
Flow in mold	Decreases	Increases
Mold shrinkage	Decreases	No change
Degree of dispersion	Good	Improves

[a]At equivalent filler levels.

TABLE 62.17. Advantages and Disadvantages of Surface-modified Kaolin in Plastics

Polymer	Advantages	Disadvantages
Thermoplastics	Increased strength Improved impact Higher loadings Improved moisture resistance	Cost
Thermosets	Increased hydrophobicity Increased strength Improved moisture resistance	Cost

erties, increased moisture resistance etc), and can provide reactive surface groups to enter into cure–crosslinking reactions (increased physical properties). The key properties of the commercially available fillers for plastics are given in Table 62.15.

Surface modification of kaolin not only improves composite properties, but also aids in processing. Since the surface modification can make the kaolin more compatible with the polymer, dispersion of the kaolin in the polymer becomes easier, agglomeration is reduced, and filler loadings can be increased. The processing characteristics of nonsurface modified

TABLE 62.18. Effect of Surface-modified Kaolins on Polymer Properties

Property	Effect
Tensile strength and modulus	Increase
Flexural strength and modulus	Increase
Impact strength	Slight decrease to no change
Retention of properties at elevated temperatures	Increase
Retention of properties under moist conditions	Increase
Dimensional stability	Increase
Thermal durability	Increase
Thermal conductivity	Increase
Linear coefficient of expansion	Decrease
Heat-deflection temperature	Increase
Electrical properties	Increase
Chemical properties	No change to increase

kaolins have been listed in previous sections. Table 62.16 shows not only the effect on processing properties of adding a surface-modified kaolin to the plastic, but also the effect of the surface modification at equivalent filler loadings.

The surface-modified kaolins find their main applications in engineering thermoplastics and in the cable industry as a reinforcing, hydrophobic filler. Table 62.17 lists the advantages and disadvantages of surface-modified kaolin.

The effects of surface-modified kaolins on the filled polymers are listed in Table 62.18. These properties will vary somewhat with the type of polymer and the degree of filling. The commercially available surface-modified kaolins are listed alphabetically below.

Trademark	Manufacturer
2211	Burgess Pigment Co.
5178	Burgess Pigment Co.
B2180	Thiele Kaolin Co.
Icecap KE	Burgess Pigment Co.
Nucup 100	J.M. Huber Corp.
Nulok 321	J.M. Huber Corp.
Nylok 100	J.M. Huber Corp.
Polyfil WC	J.M. Huber Corp.
RC 910	Thiele Kaolin Co.
RC 990	Thiele Kaolin Co.
Translink 37	Freeport Kaolin Co.
Translink 53	Freeport Kaolin Co.
Translink 445	Freeport Kaolin Co.

63

FILLERS, MICA

Malcolm Fenton

Suzorite Mica Products, Subs Corona Corp., Executive Plaza I, Hunt Valley, MD 21031

63.1 CATEGORY

Mica is a reinforcing filler. In some resin systems, it reduces resin costs by as much as a third, but imparts certain selected property improvements reported to surpass those of traditional reinforcements that cost three times as much.

To do this, mica must meet requirements pertaining to size, shape, and, to some degree, composition.

There are two physically occuring categories of mica, "sheet" and "ground." Sheet mica is not used for composites. It is a relatively rare electrical grade, mined and sorted largely by hand. Ground mica, which is used as a reinforcing filler, is a processed flake of uniform size and thickness, generally recovered from pegmatite deposits.

63.2 SOURCE

Mica is part of the mineral class phyllosilicates, which is distinguished by a nearly complete basal cleavage between stacks of two-dimensional silicate crystals and a bond between negatively charged flakes linked by positively charged alkali ions. The layers slide, and thus, mica is "slippery." In making further distinctions between these mica silicates, the various deposits might be classed as muscovite (primarily potassium aluminum silicate), phlogopite (generally magnesium potassium aluminum-ion silicate), biotite (similar to phlogopite but of weaker crystal structure), and rarer compositions that include lepidolite and zinwaidite.

Typical function formulas of the more common micas are Muscovite, $KAl_2(Al\ Si_3O_{10})(OH)_2$; Phlogopite, $KMg_3(Al\ Si_3O_{10})(OH)_2$; and Biotite, $K(MgFe)_3(Al\ Si_3O_{10})(OH)_2$.

Muscovite flake mica used in resin composites occurs very commonly in granitic rock deposits or metamorphosed schists, or gneisses, in North America. These deposits frequently occur concurrently with deposits of feldspar or spogamene in surface outcrop. Phlogopite mica occurs in metamorphosed sediments and is generally associated with pyroxenite. Suzorite mica is a phlogopite that is characterized as having little associated quartz or silica.

Such ore bodies can be mined and harvested by open-pit surface operations. They are shallow in depth: less than 500 ft (152 m). Since mica ore usually occurs with weak horizontal bonds, it lends itself to ripping or recovery by bulldozer or power shovel with minimal blasting.

After the ore has been quarried, it is crushed to a uniform size having individual lumps small enough to feed processing operations uniformly but large enough to minimize retained water from snow or rainfall.

Mica beneficiation is designed to eliminate gangue mineral, reduce particles to fractions of uniform size, and delaminate the flakes into flat, laminar, thin entities. The properties of mica are somewhat in conflict with processing and tend to make it difficult to meet beneficiation objectives.

Ground mica is either wet ground, dry ground, or fluid energy ground. In some cases two processes are employed.

Wet grinding is inefficient and is rarely used for resin composites because of its high cost, limited availability, and narrow range of diameters and aspect ratios.

Dry grinding consists of processes that crush, screen, delaminate, and separate by air or gravity the mica in various stages and sequences suited to the ore used and the desired finished product. The initial stage includes two or three crushing steps, incorporating jaw crushers and hammer mills. These are followed by a closed circuit system of air separators, screens, and mills. In the case of Suzorite mica, special delaminating mills and gangue removal flotation devices are employed.

63.3   KEY PROPERTIES

Muscovites and phlogopites are the only types of mica of importance in resin composites. These types have many similarities as well as some significant differences.

All micas are highly resistant to oxidation, acids or alkalies, fire or heat, electrical energy, water, or organic solvents. Table 63.1 gives properties in which phlogopites and muscovites are similar and dissimilar.

There are differences that are often significant in plastics. Among these are the color, wide variation in decomposition temperature, water content, and stiffness. In general, the phlogopite micas have greater strength and high temperature resistance, however, the muscovites are slightly lighter in color. As a general rule, phlogopite composites tend to be tan or golden brown; muscovites are grey or beige.

Although there are several producers of mica in the United States, relatively few produce grades for plastics composite. They are listed in Table 63.2.

63.4   PROCESSING CHARACTERISTICS

Processing Property	Phlogopite Mica at 40 wt%
Viscosity	Increased
Melt flow	Decreased
Compounding	Easier to compound compared to glass fiber or calcium carbonate
Temperature	Increased
Injection pressure	Slightly higher
Flow in mold	Difficult compared to virgin material
Mold shrinkage	Decreased

63.5   APPLICATIONS

Depending on the individually varying cost/performance relationships, these benefits usually result from incorporating mica in composites:

Compared to all reinforcing fillers, mica provides

- Increased flexural strength, reduced gas permeability.
- Increased stiffness and reduced warp.

TABLE 63.1. Properties of Phlogopites and Muscovites

Property	Phlogopite	Muscovite
Similar		
Specific gravity	2.7–3.1	2.7–3.1
Hardness, mohs	2.5–3.0	2.5–3.0
Thermal conductivity perpendicular to cleavage, cal/cm	16×10^4	16×10^4
Modulus of elasticity, psi	25×10^6	25×10^6
(MPa)	(172,375)	(172,375)
Tensile strength, psi	38–42×10^3	38–42×10^3
(MPa)	(262–290)	(262–290)
Specific heat at 77°F (250°C)	0.20–0.21	0.20–0.21
Dissimilar		
Refractive index	1.57–1.60	1.55–1.57
Maximum temperature with little decomposition, °F (°C)	500 (932)	980 (1796)
Coefficient of thermal expansion perpendicular to cleavage, °C	1–2×10^{-5}	11×10^{-6}
parallel to cleavage, °C	14×10^{-6}	10×10^{-6}
Water of constitution, %	1.0–3.2	4–5
Flexibility	More	Less
Transparency	Less	More

Compared to some reinforcing fillers, mica provides

- Increased heat-distortion temperature.
- Reduced cost and improved creep, and, where useful, such added benefits as improved ultraviolet and infrared resistance, microwave permeability, and heat shielding.

Mica provides little or no advantage over some lower cost fillers with respect to impact resistance, weld-line strength, and, sometimes, color. These drawbacks may be offset by

TABLE 63.2. Producers of Various Grades of Mica

Supplier	Trade Name	Type	Grades Available
J. M. Huber Corp.	Aspra Flex	Muscovite, dry/set ground	4
Unamin Corp.	Mica	Muscovite, dry ground	3
Franklin Mineral	Alsibrone	Muscovite, dry/set ground	4
M.I.C.A.	Mica	Muscovite, dry ground	2
Eagle Quality	Micaflex	Muscovite, dry	2
Suzorite Mica Products[a]	Suzorite	Phlogopite, dry	19

[a]Formerly Martin Marietta Corp.

TABLE 63.3. Physical Properties of Mica-reinforced Polypropylene Polymers

Property	Unfilled[a]	\multicolumn{4}{c}{Suzorite mica Reinforced[a]}
		25% 200S	40% 200S	40% Suzorite 60-S	40% Suzorite 60-NP
Tensile strength, psi	3290	3,110	2,960	3,680	6,160
(MPa)	(23)	(22)	(20)	(25)	(65)
Flexural strength, psi		4,260	5,180	6,540	9360
(MPa)		(29)	(36)	(45)	(65)
Flexural modulus, psi	133,00	317,000	518,000		
(MPa)	(917)	(2,190)	(3,570)		
Tangent flexural modulus, psi				1,040,000	1,200,000
(MPa)				(7,170)	(8,220)
Izod unnotched, ft-lb/in.^{2}					
72°F (22°C)				5.3	4.7
−18°F (0°C)				3.8	3.4

[a]Polypropylene copolymer (melt index 30).

TABLE 63.4. Physical Properties of Mica-reinforced High Density Polyethylene Polymers

	Unfilled[a]	\multicolumn{2}{c}{Suzorite mica Reinforced[a]}	Unfilled[b]	\multicolumn{2}{c}{Suzorite mica Reinforced[b]}
		10% Glass	10% 200NP		20% 200NP	40% 200NP
Tensile strength, psi	3,400	3,790	3,835	3,180	4,290	5,000
(MPa)	(24)	(26)	(27)	(22)	(30)	(35)
Flexural modulus, psi	131,000	207,000	195,000	123,000	330,000	670,000
(MPa)	(903)	(1,430)	(1,350)	(848)	(2,280)	(4,620)
Flexural strength, psi				2,070	3,750	5,850
(MPa)				(14)	(26)	(40)
Izod unnotched						
72°F (22°C) ft-lb/in.^{2}				No break	13.0	6.0
(J/m)					(690)	(320)
−18°F (0°C) ft-lb/in.^{2}				No break	11.0	5.4
(J/m)					(590)	(290)

[a]High molecular weight-high density polyethylene.
[b]High density polyethylene copolymer.

TABLE 63.5. Physical Properties of Mica-reinforced ABS and Calcium Carbonate-filled PVC

	ABS + 5% Long Glass Fiber	ABS + 10% Suzorite[a] mica (92%, 44 Microns)	PVC + 10% CaCO$_3$	PVC + 10% CaCO$_3$ + 33.3% Suzorex mica 200-PO
Tensile strength, psi	7,930	7,480		
(MPa)	(55)	(52)		
Flexural strength, psi	10,300	11,000	7,060	8,950
(MPa)	(71)	(76)	(49)	(62)
Flexural modulus, psi	437,000	619,000	464,000	1,460,000
(MPa)	(3,010)	(4,270)	(3,200)	(10,100)
Notched Izod, ft-lb/in.	5.8	4.1		
(J/m)	(310)	(220)		
Coefficient of thermal expansion[b]				
32–140°F, °F × 10^{-5}			3.69	1.37
0–60°C, (°C × 10^{-5})			(6.65)	(2.47)

[a]Note the equivalant performance of both materials.
[b]The coefficient of expansion of mica-reinforced PVC is the same as aluminum.

combining mica with other fillers, modifying the resin, adding pigment, or using other additives.

Mica has been developed to perform well with polyolefins (chiefly polypropylene, polyethylene, and copolymers), thermoplastic polyesters, nylon, and urethanes. It shows promise with styrenics, PVC, and thermosets. (Additional selected properties are given below and in Tables 63.3–63.10).

Material, Injection Molded	Dielectric Strength	
	V/mil	kV/cm
PBT (unfilled)	360	142
PBT/30 wt% Suzorite mica (60-S)	533	210
PBT/30 wt% Suzorite mica (60-PO)	558	220

Material, Film	Water Vapor Transmission, g.mm/m^2/24 h at 73°F (23°C)
Polystyrene	1.67
Polystyrene + 37% wt% Suzorite mica 200-S	0.901

63.6 POLYMERS FILLED

See Table 63.11.

TABLE 63.6. Physical Properties of Mica-reinforced Nylon 6,6

Property	Unfilled	40% Suzorite mica 325-PO Reinforced
Tensile strength, psi	11,300	12,500
dry (MPa)	(78)	(86)
wet-16 h, 50°C psi	8,460	10,400
(MPa)	(58)	(72)
Flexural strength, psi	12,300	20,000
dry (MPa)	(85)	(138)
wet psi		14,700
(MPa)		(101)
Flexural modulus, psi	443,000	1,360,000
dry (MPa)	(3,060)	(9,380)
wet psi	126,000	840,000
(MPa)	(869)	(5,790)
Izod notched, ft-lb/in.	0.80	0.53
dry (J/m)	(43)	(28)
wet ft-lb/in.		0.93
(J/m)		(50)
Heat deflection, °F	153	401
264 psi (1.82 MPa) (°C)	(67)	(201)

TABLE 63.7. Properties of Polymethylpentene (TPX) Reinforced with Mica

Property	Unfilled	Reinforced with 20% Suzorite mica		
		200HK	200NP	200PO
Tensile strength, psi	3,850	3,870	3,870	3,860
(MPa)	(27)	(27)	(27)	(27)
Flexural strength, psi	3,820	4,540	5,100	4,860
(MPa)	(26)	(31)	(35)	(34)
Flexural modulus, psi	179,000	331,000	402,000	362,000
(MPa)	(1,240)	(2,280)	(2,770)	(2,500)
Izod impact,				
unnotched ft-lb/in.^{2}	2.3	1.6	1.6	1.9
notched ft-lb/in.	0.82	0.80	0.81	0.81
(J/m)	(44)	(43)	(43)	(43)
Gardner impact, in.·lb	4.1		4.1	
(J)	(0.55)		(0.55)	
Heat-deflection temperature, °F	117	124	149	149
at 264 psi (1.82 MPa), (°C)	(47)	(61)	(55)	(55)
Hardness Shore D 73°F (23°C)	75	80	80	80
248°F (120°C)	45	50	47	50
Heat aging,				
302°F (150°C)	Passes 430	96	430	Passes 430

63.7 COMMERCIAL GRADES

The average bulk density for the products given is between 11 and 18 lb/ft^3. The aspect ratio varies between 46 and 85, depending on the grade. There are other grades for commodity uses. They are sometimes tried for composites but generally are not recommended because of the low-purity mica aspect ratio. In the case of Suzorite mica, these are identified as Z grades (for oil-well drilling) and 80SF (a joint cement grade). Other producers have similar identifying grade nomenclatures.

The use of mica as a reinforcing filler will generally provide a reduction in cost, compared to the unfilled resin, on both a weight and a volume basis. Further, this unit cost will frequently be lower than other reinforcements providing similar properties. This latter distinction is important. Use of mica reinforcement is only economically prudent when the

TABLE 63.8. Physical Properties of High Impact Polystyrene Reinforced with Mica

Property	Unfilled	Reinforced with Suzorite mica		
		20% 200S	40% 200S	40% 60S
Tensile strength psi	3,030	2,820	3,090	2,940
(MPa)	(21)	(20)	(21)	(20)
Flexural strength, psi		5,630	5,980	5,970
(MPa)		(39)	(41)	(41)
Flexural modulus, psi	321,000	572,000	1,130,000	1,510,000
(MPa)	(2,210)	(3,950)	(7,800)	(10,400)
Izod impact				
unnotched ft-lb/in.^{2}	18.2	4.38	2.10	1.57

TABLE 63.9. Physical Properties of Thermoplastic Polyesters (PBT and PET) Reinforced with Mica

Property	Unfilled	PBT 40% Suzorite Mica 325 HK	Unfilled	PET 50% Suzorite Mica 60 S
Tensile strength, psi	7,280	9,900	10,400	11,500
(MPa)	(50)	(68)	(72)	(79)
Flexural strength, psi	9,110	16,000	13,100	17,100
(MPa)	(63)	(110)	(90)	(118)
Flexural modulus, psi	347,000	1,680,000	452,000	2,510,000
(MPa)	(2,390)	(11,600)	(3,120)	(17,300)
Izod impact resistance				
Unnotched 73°F (23°C), ft-lb/in.	no break	3.1	7.3	1.9
(J/m)	(no break)	(970)	(390)	(100)
−40°F, °C, ft-lb/in.			6.7	1.9
(J/m)			(360)	(100)
Notched 73°F (23°C), ft-lb/in.			0.52	0.45
(J/m)			(28)	(24)
Notched −40°F, °C, ft-lb/in.			0.52	0.45
(J/m)			(28)	(24)
Heat-deflection temperature, °F	127	336	176	424
at 264 psi (1.82 MPa) (°C)	(53)	(169)	(80)	(218)
Mold shrinkage				
direction of flow, %	2	0.6		
transverse to flow, %	2	1		
Dielectric strength, V/mil	360	533		
(V/mm)	(14.2)	(21.8)		

TABLE 63.10. Properties of Polypropylene Filled with Different Fillers and Mica

Property	Unfilled PP	40% Talc + PP	40% CaCO₃ + PP	30% Glass + PP	40% 200-NP Mica + PP
Tensile strength, psi	4,930	4,270	2,770	6,340	6,190
(MPa)	(34)	(30)	(19)	(44)	(43)
Flexural strength, psi	4,450	6,420	4,720	10,060	9,320
(MPa)	(31)	(44)	(33)	(69)	(64)
Tangent flex. mod, psi	193,000	676,000	421,000	933,000	1,040,000
(MPa)	(1,330)	(4,660)	(2,900)	(6,430)	(7,170)
Izod notched, ft-lb/in.	0.45	0.45	0.75	0.79	0.65
72°F (22°C) (J/m)	(24)	(24)	(40)	(42)	(35)
Heat-deflection					
temperature, °F	136	162	183	257	226
at 264 psi (1.82 MPa), °C	(58)	(72)	(84)	(125)	(108)
Coefficient of thermal					
expansion × 10⁻⁵ °F	7.4	4.0	5.3	2.0	3.3
73–163°F (23–73°C) (°C)	(13.3)	(7.2)	(9.5)	(3.6)	(5.9)
Hardness (D durometer)	68	72	68	69	73
Mold shrinkage					
direction of flow, %	2.0	1.2	1.40	0.8	

TABLE 63.11. Advantages and Limitations of Various Filled Polymers

Polymer	Advantages and Limitations
Thermoplastics	
Polyolefins	Improved flexural strength, dimensional stability, thermal expansion,
Thermoplastic polyesters	heat-distortion temperature, ultraviolet stability, gas and liquid perme-
Nylon	ability.
Polystyrene	Chemical resistance
PVC	Reduced impact strength; tensile strength is usually unchanged.
Polyethersulfone	
Thermosets	
Polyesters	Properties similar to thermoplastics, but increases viscosity and decreases
Phenolics	impact strength. Mica is usually supplemented with glass fiber.
Epoxies	
Reinforced Reaction Injection Molding (RRIM)	
Urethanes	Improves nucleations but limited in some reinforcement properties.

properties sought favor the selection—in other words, for an "engineered" system, but not a general-purpose application.

63.8 COMPOSITE CHARACTERISTICS

See Table 63.12.

TABLE 63.12. Composite Characteristics

Property	Effect of Filler	
	Generally	In Polyolefins
Tensile strength and modulus	Increases	0.3–0.8 times the glass fiber effect
Flexural strength modulus	Significant increase	Double the glass fiber effect
Impact	Decreased	Decreased
Retention of properties at elevated temperature	Improved	Decreased
Retention of properties under moist conditions	Reduces dimensional change with humidity cycling; decreases water absorption	Little effect
Dimensional stability	Much improved: 0.8% vs 2.0% unfilled	Notably reduces warpage, shrinkage
Thermal stability	Increased	Much improved
Thermal durability	Increased	Much improved
Thermal conductivity	Increased	About 4 times that of polypropylene
Linear coefficient of expansion	Decreased	About equal to aluminum
Heat-deflection temperature	Increased	2 times better than unfilled; 1.4 times better than talc filled
Surface appearance and texture	Decreased except for finer grades	Decreased
Electrical properties	Much increased; dielectric strength in PBT increases 40%	Similar
Chemical properties	Somewhat increased acid and solvent resistance	Improved
UV stability	Increased	Increased
Air permeability	Increased	Increased three times in polyethylene

64

FILLERS, PROCESSED MINERAL FIBER

E.L. Moon

Jim Walter Research Corp., 10301 Ninth St. N, St. Petersburg, FL 33742

64.1 CATEGORY

Processed mineral fiber is a fibrous reinforcement (see Fig. 64.1).

64.2 SOURCE

Processed mineral fiber is obtained directly from commodity-grade mineral wool by a patented, mechanical process. Mineral wool is a high-calcium-content aluminosilicate glass produced by remelting the slag waste from smelter operations (predominantly iron) and converting the molten slag into fibers by rotary spinning or blowing processes. To obtain processed mineral fiber, mineral wool is first milled to shorten and disentangle the fibers and then classified to remove nonfibrous material. Thus, the high L/D ratio fibers are separated from the low L/D ratio components.

64.3 KEY PROPERTIES

Processed mineral fiber is supplied by Jim Walter Resources, Inc. and Nyco, Inc. PMF fiber is a registered trademark of Jim Walter Resources, Inc. Key properties are specific gravity, 218; bulk density size, g/cm^3, 0.4; particle size μm: average diameter, 4.5; average length, 200.

64.4 PROCESSING CHARACTERISTICS

Processing Property	Effect of Reinforcement
Viscosity	Increases
Melt flow	Reduces
Compounding	Increases time
Temperature	No effect
Injection pressure	Increases
Flow in mold	Reduces
Mold shrinkage	Reduces

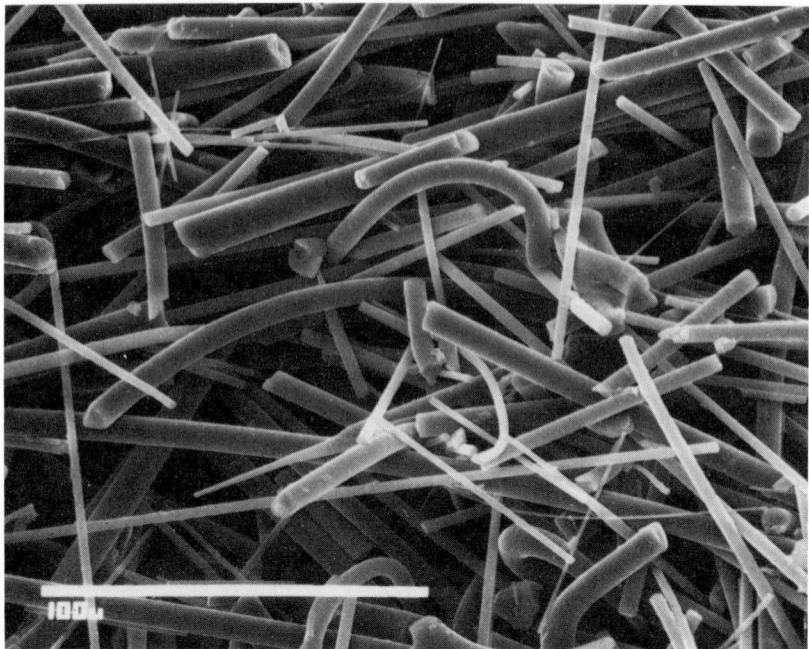

Figure 64.1. Scanning electron microscope image of processed mineral fiber. Magnification 540X. Courtesy Jim Walter Research Corp.

TABLE 64.1. Advantages and Disadvantages of Filled Polymers

Polymer	Advantages and Limitations
Thermoplastic	
Polypropylene	Low polymer adhesion unless chemically coupled, increases modulus, lowers impact strength
Nylon 6, 6; 6,12	Good adhesion with aminosilane coupling agent; improves tensile, flexural, and impact strength; raises heat-deflection temperature
Polyester PBT	Moderate adhesion to polymer matrix, improves tensile and flexural strength, increases impact strength
HIPS	Increases tensile strength, flexural modulus, and heat-deflection temperature
Thermosets	
Phenolic resin molding compound	Reinforcement
Friction materials	Reinforcement, exceptional friction properties, low wear characteristics
Epoxy resin	Flow control, reinforcement, 50–90 parts
Urethane	Low-cost glass fiber, marginal reinforcement, 20–50 parts
RRIM	Partial glass replacement but less effective reinforcement, higher mix viscosity than glass because of greater surface area

TABLE 64.2. Composite Characteristics of Filled Polymers

Property	Effect of Filler
Tensile strength and modulus	Slightly increases
Flexural strength and modulus	Increases
Impact strength	Generally increases
Retention of property at elevated temperatures	Improves
Retention of property under moist conditions	Decreases
Dimensional stability	Increases
Thermal durability	No effect
Thermal conductivity	Increases
Linear coefficient of expansion	Decreases
Heat-deflection temperature	Increases
Surface appearance and texture	May cause mottling
Electrical properties	Reduces
Chemical properties	Mineral acid sensitivity

64.5 APPLICATIONS

Processed mineral fiber is commonly used in phenolic molding compounds, friction materials, mastics, protective coatings, and epoxy and polyester body patching compounds. Processed mineral fiber is a low cost replacement for glass fiber. Common applications include reinforced reaction injection molding (RRIM) systems, flooring felts, nylon, PBT, and elastomeric polymers. The cost of processed mineral fiber is less than one half that of milled and short-chopped glass fiber, but it has many of the properties of E-glass. It is mineral wood, which, by definition, is a man-made vitreous fiber.

64.6 POLYMERS FILLED

See Table 64.1.

64.7 COMMERCIAL GRADES

Name	Characteristics
PMF fiber 204	Average fiber length 200 μm
PMF fiber 204 AX	Aminosilane treatment
PMF fiber 204 CX	Fatty acid treatment

PMF fiber 204 #AX and 204 #CX are developmental, but available in multiton quantities.

64.8 COMPOSITE CHARACTERISTICS

See Table 64.2.

65

FILLERS, SILICA (QUARTZ, FUSED, AND DIATOMACEOUS)

James R. Moreland

Malvern Minerals Co., P.O. Box 1246, 220 Runyon St., Hot Springs, AR 71901

65.1 CATEGORY

The forms and phases of silicon dioxide covered in this chapter perform appropriately at times as an extender filler, a semireinforcing filler, or a reinforcement. The specific function is based on the specific resin system, particle size, surface area, shape, loading, and surface modification.

65.2 QUARTZ

65.2.1 Source

Quartz is better known as crystalline silicas in the applied fields. In plastics, crystalline silica comes from two geologic sources: the quartzite sands or sandstones; and the tripolitic forms, which are viewed as natural microcrystallines. The two types are sometimes referred to as "silica flour." From the standpoint of genesis and geographic distribution, silica is ubiquitous.

The quartzite deposits occur as St. Peter Sandstone or Oriskany quartzite. The St. Peter Sandstone is exposed principally in Northern Illinois. The Oriskany quartzite occurs in a tristate area of Virginia, West Virginia, and Pennsylvania.

The quartzite sandstone operations are huge in scale, requiring transport by rail or rubber-tired trucks capable of carrying more than 100 tons a load. The ore is crushed in primary crushers and then further processed to remove foreign matter. It is subsequently dried, ground in tube mills, and either screened or air-floated into many grades. In production, these raw materials are converted from macro- to microcrystalline.

The tripolitic, or natural microcrystalline quartz silica, is found for the most part in Southern Illinois and in the Novaculite Uplift in West Central Arkansas.

The tripolitic silica in Southern Illinois is mined by the room and pillar method. It is then ground or pulverized and is normally air-floated to many grades of fineness.

The tripolitic silica from West Central Arkansas is strip-mined. The ore is crushed, dried, and mostly air-floated to many grades of fineness. The two tripolis discussed are both natural microforms of quartz. The sandstones are not natural microforms.

65.2.2 Key Properties

See Table 65.1.

TABLE 65.1. Key Properties of Quartz[a]

Product Description	Size Range	Suppliers
Pulverized and air classified	1 millimicron to 44 microns	Ottawa Silica Pennsylvania Glass Sand Unimin
Natural microcrystalline	1 millimicron to 44 microns	Malvern Minerals Tammsco Illinois Minerals

[a]For all types, the specific gravity is 2.65 g/cc; the bulk density, 50–75 lb/ft^3 (42.5 kg/m^3).

65.2.3 Processing Characteristics

Property	Effect of Additive
Viscosity	Increases
Melt flow	Decreases
Compounding	No effect
Temperature	No effect
Injection pressure	Higher
Flow in mold	Lower
Mold shrinkage	Lower

65.2.4 Applications

Microcrystalline types of silica are used in applications requiring hardness, strength, purity, chemical resistance, good flow, electrical insulation, thermal conductivity, heavy loading, dimensional stability, and mar and wear resistance.

Its hardness can be considered a drawback. At 7 on the mohs scale, care must be exercised in its use, as hardness of this nature can cause wear on openings and critical tolerances.

65.2.5 Commercial Grades

Supplier	Quartz Products
Illinois Minerals Company	Imsil Cairo, Ill.
Malvern Minerals Company Hot Springs, Ark.	Novacite Novakup (surface-modified Novacite) Silikup (surface-modified silica other than Novakup)
Tammsco, Inc. Cassopolis, Mich.	Various product types
Pennsylvania Glass Sand	Min-u-Sil SupersilR
Ottawa Silica Co. Ottawa, Ill.	Various product types

65.2.6 Composite Characteristics

See Tables 65.2, 65.3, and 65.4.

65.2.7 Trade Names

See Commercial Grades.

TABLE 65.2. Composite Characteristics of Quartz-filled Silicon Rubbers

Novakup Filled HTV Silicone Rubber Mold Compound				
Compound	1	2	3	4
Stauffer SWS-729	100	100	100	100
Novacite - No Ken-React	100			
Novakup L-207A - TTS 0.5		100		
Novakup - TTOP-12 - 0.5			100	
Novakup - TTM-33S - 0.5				100
Varox	0.8	0.8	0.8	0.8
10-min cure at 300°F				
Tensile		460	500	410
Elongation		380	200	250
Tear, die B		41	42	50
Hardness		60	65	55
15-min cure at 300°F				
Tensile	718	500	530	460
Elongation	178	350	210	280
Tear, die B	65	40	38	37
Hardness	73	65	65	65
Number of parts produced by mold before failure	11	250	195	85
$\frac{1}{4}$ in. section - closed duty system 850–900°F				

TABLE 65.3. Properties of Quartz-filled Thermoplastic Resins

		Nylon 12			Thermoplastic Polyester,[d] 50% Novaculite Silica Filled, 50% Novaculite Silica[e]	
	Unfilled[a]	50% "Novacite L-337"[b] No Silane	50% "Novakup L-337"[c] A-1100	Effect of Aminofunctional Silane, Unfilled	No Silane	Aminosilane[f]
Flexural strength, psi						
Initial	8,400	10,000	11,300	13,500	10,200	14,300
After 16 h in 50°C water	6,200	8,700	9,900	13,000	8,700	13,900
Standard deviation[g]	116	348	120			
Flexural modulus, 10^5 psi						
Initial	1.8	3.1	2.8	3.9	7.8	8.1
After 16 h in 50°C water	0.9	1.8	2.1	3.7	7.9	8.1
Tensile strength, psi						
Initial	5,800	6,000	6,500	7,700	5,300	6,800
After 16 h in 50°C water	5,100	4,600	6,400	7,500	4,900	6,100
Standard deviation[g]	59	113	47			
Ultimate elongation, %	7.5	3.5	24.6	6.5	2.6	2.6
Impact strength, Izod, ft-lb/in. unnotched	31.9	3.6	22.2			
Heat-deflection temperature, °C, 264 psi	52	68	70	63	126	125

[a]HULS Nylon 12 L-1901, granulated natural, L-809-0001-0718.
[b]L-337 Novacite.
[c]L-337 Novakup 1100 1.0.
[d]Celanex J-105 Celanese Plastics.
[e]Novacite 1250 Malvern Minerals Co.
[f]Union Carbide A-1100 applied as a pretreatment on the filler at 0.5 phf.
[g]Based on 5 observations.

TABLE 65.4. Effect of Silane Treatment[a] on the Properties of Quartz-filled Epoxy Resin Castings[b]

Silane Application Method	% Silane		Dielectric Constant						Dissipation Factor						% Weight Gain on 24 h Water Boil	Flexural Modulus, psi × 106	Flexural Strength, psi	Deflection to Failure, in.
			Dry			24 hr Water Boil			Dry			24 h Water Boil						
	Used	Applied	102 Hz	103 Hz	106 Hz	102 Hz	103 Hz	106 Hz	102 Hz	103 Hz	106 Hz	102 Hz	103 Hz	106 Hz				
None	0.000	0.000	4.09	4.01	3.78	5.03	4.67	3.97	0.023	0.015	0.018	0.080	0.036	0.036	1.7	1.00	14.300	0.080
Vapor, low temperature, 1 h	0.300	0.076	4.12	4.04	3.73	5.07	4.90	4.17	0.012	0.014	0.020	0.025	0.026	0.036	1.5	1.12	13.400	0.075
Vapor, low temperature, 20 min.	0.100	0.063	4.12	4.04	3.73	5.09	4.93	4.23	0.013	0.015	0.020	0.026	0.027	0.036	1.5	1.12	11.600	0.065
Dry blend	0.500	0.450	4.10	4.02	3.70	5.21	4.90	4.21	0.012	0.014	0.020	0.087	0.038	0.036	1.6	1.11	18.000	0.120
Methanol slurry	0.300	0.090	4.07	4.00	3.70	5.60	4.87	4.19	0.013	0.017	0.020	0.031	0.028	0.036	1.5	1.04	11.000	0.065
Vapor plus H2O, 1 h	0.540	0.098	4.00	3.92	3.60	5.05	4.84	4.18	0.014	0.016	0.020	0.031	0.027	0.037	0.7	1.10	10.600	0.055
Dry blend, preacidified	0.500	0.430	3.97	3.91	3.60	6.73	4.98	3.95	0.014	0.015	0.020	0.275	0.133	0.037	1.8	1.10	16.700	0.104

[a]Dow Corning Z-6040
[b]Two-part casting formula:

Material	Parts by Weight
A. Epon 828	100
Novacite 1250	125
B. UCC ERL 2793	25

65.3 FUSED SILICA

65.3.1 Source

Fused silica (vitreous) is processed from high-purity quartz sands by fusing in an arc furnace using graphite electrodes. An ingot is created and ground to obtain various products with various grain and particle size characteristics.

Typical chemical analysis of fused silica is SiO_2, 99.6 wt%; Al_2O_3, 0.07 wt%; Fe_2O_3, 0.03 wt%; and Na_2O + K_2O, 0.01 wt%.

65.3.2 Key Properties

The specific gravity of fused silica (vitreous) is 2.18 which is lower than that of quartz (2.65). Its lower specific gravity means increased yields in composites, as compared to ordinary quartz. The significant property of this phase is the low thermal expansion of $0.3 \times 10^{-6}/$°F (0.54×10^{-6}/°C). This characteristic helps reduce overall thermal expansion when desirable.

65.3.3 Processing Characteristics

Property	Effect of Additive
Viscosity	Increases
Melt flow	Decreases
Compounding	No effect
Temperature	No effect
Injection pressure	Higher
Flow in mold	Lower
Mold shrinkage	Lower

65.3.4 Applications

Fused silica (vitreous) is used primarily in thermosetting compounds where it makes for high loading, light weight, low shrinkage, low thermal expansion, and good electrical insulation properties in such resins as silicone, epoxy, and high molecular weight fluorocarbons.

Fused silica is quite expensive; thus its selection is based on high-value requirements.

65.3.5 Commercial Grades

The product Glasgrain is available from Harbison Walker of Calhoun, Ga.

65.3.6 Composite Characteristics

See Table 65.5.

65.3.7 Trade Names

See Commercial Grades.

TABLE 65.5.   Properties of GP 31, 71 Fused Silica Epoxy Systems

Property	Formulation[a]	
	30 p* epoxy 70 p GP3I 5 p hardner	120 p* epoxy 75.2 p GP71 9.6 p hardener
Volume % of powder	52.0%	24.6%
Flexure strength, psi	12,700	12,000
Specific gravity	1.7	1.4
Dielectric constant Measured at 75°F		
Frequency 60	Not determined	4.07
10^3	4.09	3.56
10^6	3.90	3.57
Dissipation factor Measured at 75°F		
Frequency 60	Not determined	0.007
10^3	0.007	0.010
10^6	0.010	0.008

[a]p = parts.

65.4   DIATOMACEOUS SILICA

65.4.1   Source

Diatomite, which is known in industry as diatomaceous silica or earth, is quarried from naturally occurring deposits and transported to primary crushers where it is then crushed, milled, and dried. The product is sold either calcined or in its natural state.

Typical chemical analysis of calcined celite for antiblock: H_2O, 0.1%; ignition loss, 0.2%; SiO_2, 91.9%; Al_2O_3, 3.3%; Fe_2O_3, 1.2%; CaO, 0.5%; MgO, 0.5%; Na_2O, 1.8%; K_2O, 0.3%; and other, less than 0.5%.

65.4.2   Key Properties

In plastics materials, diatomaceous silica imparts good flatting, or gloss reduction, characteristics, and antiblock materials.

Diatomaceous silica is very high in silica content. Uncalcined varieties have surface areas about 15 m^2/g. Oil absorption is as high as 90%. The particles are shaped like skeletal remains and rheologically perform similarly to synthetic amorphous silicas. The silica is microamorphous and opaline silica. See Table 65.6.

TABLE 65.6.   Typical Properties of Three Diatomites[a]

Property	Silver Frost Celite		
	Super Floss	K-5	White Mist
Appearance	Fine, fluffy	White powder	White
Color	White	White	White
325 mesh residue, %	0.1	Trace	Trace
Oil absorption, g/cm	120	115	160

[a]For all types, specific gravity is 2.3; pH, 9–10; refractive index, 1.48; and surface area, 0.7 to 3.5 m^2/g.

65.4.3 Commercial Grades and Trade Names

Supplier	Products
John-Manville Corp. Denver, Colo.	Super Floss Silver Floss K-5 Celite White Mist

66

FILLERS, SYNTHETIC SILICAS (FUMED, PRECIPITATED, AND GEL)

Satish K. Wason

Chemicals Div., J.M. Huber Corp., Havre de Grace, MD 21078

66.1 CATEGORY

Synthetic silicas are inorganic white powders having the chemical formula, $SiO_2 \cdot XH_2O$, where X is the mols of water associated with the silica compound. These materials are also called synthetic amorphous silicon dioxide products. Synthetic silicas are amorphous in nature because they show noncrystalline patterns in x-ray examination.

The three commercially synthetic silicas are

- Fumed (also called pyrogenic or anhydrous silica).
- Precipitated (also called hydrated silica).
- Gel (also called hydrogel, hydrous gel, or hydrated silica).

In each category, several silica grades are available having different particle sizes and surface areas. Figure 66.1 identifies the types of commercially available synthetic silicas. There are two types of silica gels: aerogels and xerogels. There are five types[1] of precipitated silicas, ranging from very high structure (VHS) to very low structure (VLS) types. (The definition of structure is given later in this section.)

Fumed silicas are prepared by the vapor, or thermal, process. Both precipitated and gel type silicas are prepared by the liquid, or wet, process (see Fig. 66.1). Since gels and precipitated silicas are prepared in a liquid medium, these products are also called hydrated silicas.

The surface of synthetic silica contains physically adsorbed water, called moisture or free moisture, and surface hydroxyl groups, called silanol groups. Silica surface can be modified by linking the silanol groups with such coupling agents as silanes and other suitable chemicals. The coupling reaction converts the hydrophilic silica surface to a hydrophobic one. Hydrophobic silica is more compatible and results in improved properties in polymers and plastic systems than its hydrophilic counterpart. The silica refractive index of 1.45 is ideal for producing translucent or transparent filled polymers. Fumed silicas are used as reinforcing filler in silicone rubber and elastomers. Precipitated silicas predominantly, and gels to a lesser degree, are used to reinforce natural and synthetic rubber. (It is estimated that 75% of the precipitated silica produced is employed in reinforcing a variety of rubbers and elastomers.)

Fumed silicas are also used as thixotropic agents in thermoset unsaturated polyester, epoxy resins, and vinyl plastisols.

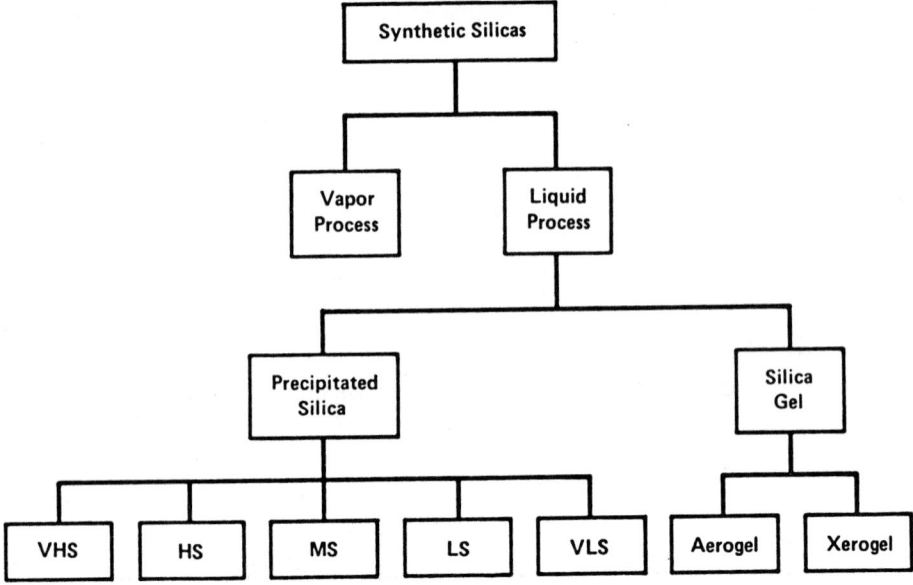

Figure 66.1.   Types of synthetic silicas

Because of their lower cost, precipitated silicas are replacing fumed silicas in many plastic and specialty applications.

66.2   SOURCES

66.2.1   Fumed Silicas

Fumed silica is prepared by the hydrolysis[2-4] of silicon tetrachloride vapor in a flame of hydrogen and oxygen at a temperature of 1832°F (1000°C) or higher.

$$2\ H_2 + O_2 \longrightarrow 2\ H_2O$$

$$SiCl_4 + 2\ H_2O \longrightarrow SiO_2 + 4\ HCl$$

$$2\ H_2 + O_2 + SiCl_4 \longrightarrow SiO_2 + 4\ HCl$$

This reaction produces a silica product having an extremely small particle size and exceptionally high purity. The particle size of fumed silica can be changed by varying the combustion conditions. As a result, fumed silica with surface areas between 50 and 400 m^2/g are readily available—in the United States, from Degussa and Cabot Corp. Degussa markets its products under the Aerosil name; Cabot under the Cab-O-Sil trademark.

66.2.2   Precipitated Silica

Precipitated silica is produced in a liquid medium by the stirred reaction between sodium silicate solution and sulfuric acid or a mixture of CO_2 and HCl and by avoiding the gelation stage during the reaction. The precipitated silica reaction is conducted under alkaline conditions according to the following reaction:

$$Na_2O \cdot X \; SiO_2 + H_2SO_4 \longrightarrow X \; SiO_2 + Na_2SO_4 + H_2O$$
or

$$Na_2O \cdot X \; SiO_2 + CO_2 \longrightarrow X \; SiO_2 + Na_2CO_3$$

$$Na_2CO_3 + HCl \longrightarrow 2 \; NaCl + CO_2 + H_2O$$

In the above equation, X is the molar ratio, SiO_2/Na_2O of the sodium silicate solution. Water glass, a sodium silicate solution of X value 3.3, is generally used in the commercial production of precipitated silicas.

Recently, novel precipitated silica technology has been developed based on the concept of structure control.[5-8] There are five distinct types of controlled-structure precipitated silicas now available in the marketplace (Fig. 66.1): VHS = very high structure, HS = high structure, MS = medium structure, LS = low structure, and VLS = very low structure.

Although the silica structure is related to its particle size and water-pore volume, for convenience the precipitated silica structure definition[1] is based on the oil absorption of silica. VHS type silicas have OA values above 200 cm³/100 g; HS types, between 175–200 cm³/100 g; MS types, between 125–175 cm³/100 g; LS types, between 75–125 cm³/100 g; and VLS types, below 75 cm³/100 g. This relationship is also given in Table 66.1.

66.2.3 Silica Gel

Silica gel is produced by reacting sodium silicate solution with sulfuric acid under the following acidic pH conditions:

$$Na_2O \cdot 3.3 \; SiO_2 + H_2SO_4 \longrightarrow 3.3 \; SiO_2 + Na_2SO_4 + H_2O$$

In the manufacture of silica gel, a hydrasol is first formed and then allowed to age and set to the hydrogel stage. A hydrogel is a precursor material from which two types of silica gels, an aerogel or xerogel, are produced. When a hydrogel is filtered, washed, and dried without the shrinkage of structure,[9,10] it is called an aerogel. When, during processing, a hydrogel results in a shrinkage of structure, it is called a xerogel.

Recently, hydrous gels have been introduced; they are partially dried hydrogels that generally contain more than 20% water.

66.3 KEY PROPERTIES

Synthetic silicas are characterized by various physical and chemical properties such as surface area, particle size, morphology, silanol group density, percent moisture, pH, oil absorption, bulk density, and percent purity.

Because of its high-temperature manufacture, fumed silica exhibits less than 1.5% mois-

TABLE 66.1. Definition of Silica Structure

Silica Structure Level	Oil Absorption, cc/100 g
VHS	Above 200
HS	175–200
MS	125–175
LS	75–125
VHS	Less than 75

ture at 221°F (105°C) and lower silanol group density than the precipitated and silica gels. Precipitated silicas and gels have a moisture content of about 5%. Their higher moisture content limits their use in some plastics and polymer systems.

The key comparative properties of fumed, precipitated, and gel silicas are listed in Table 66.2.

66.3.1 Surface Area

The surface of synthetic silica is determined by the BET nitrogen absorption method.[11] Because of the fineness of their primary particles, all synthetic silicas exhibit a very high surface area. But not all silicas are nonporous in character. Silica gels, for example, exhibit a very high surface area (generally in excess of 300 m^2/g) and most of this surface is attributed to internal porosity. For many reinforcement applications, the internal surface area of silica filler is unavailable for filler/polymer interaction. Therefore, nonporous silicas having a controlled surface area are offered by silica manufacturers for reinforcement applications.

66.3.2 Particle Size

The primary particle diameter of synthetic silica can be calculated from the surface area measurement, assuming spherical morphology for silica particles: Surface area (m^2/g) = $6 \times 10^3/\rho d$, where ρ = 2.2 × the specific gravity of silica in g/cm^3 and d = particle size diameter in nm.

The average primary particle diameter of commercially available synthetic silicas falls in the range of 10–100 nm (0.01–0.1 μm). In real life, primary particles do not exist in silica powder as such, but are further aggregated to a secondary particle structure. During manufacture and shipment, the secondary particles further aggregate to form a tertiary particle structure, called an agglomerated structure. Thus, in silica technology, there are three kinds of particle sizes: primary (also called ultimate), secondary (also called aggregate), and tertiary (also called agglomerate).

When a silica is used in any application, it is essentially purchased in the form of agglomerates. In an application experiencing shear forces, these agglomerates revert back to stable functional aggregates. Based on experience, it is believed that the end-use properties of silicas are dependent on their aggregate size and not on the primary or ultimate particle size (which essentially do not exist). Yet, most silica manufacturers report ultimate particle size data in their literature. Table 66.2 reports the more meaningful aggregate size data along with the ultimate particle size data.

TABLE 66.2. Comparative Properties of Synthetic Silicas

Property	Fumed	Precipitated	Gel
Suface area, m^2/g	50–400	60–300	100–800
Particle size			
Ultimate, nm	7–40	8–40	2–20
Aggregate, μm	0.8–3	2–10	3–10
Bulk density, g/L (Compacted)	10–120	160–200	90–160
5% pH	3.6–4.3	6.5–7.5	4.0–7.5
Silanol groups/nm^2	2–4	8–10	4–8
% Moisture	<1.5	6.0	5.0
Specific gravity	2.2	2.2	2.2
Refractive index	1.46	1.45	1.45
X-ray form	Amorphous	Amorphous	Amorphous
% Silica (ignited basis)	99.8	98.0	98.0

66.3.3 Morphology

The electron micrographs of fumed (Cab-O-Sil), gel (Syloid 244), and precipitated silicas (Zeothix 265) at 69,800 magnification are shown in Figures 66.2–66.4. Examination of electron micrographs reveals that in fumed silica the primary particles are joined together to form a chainlike secondary structure. It is believed that the chainlike structure in fumed silica is responsible for its well known thixotropic behavior in plastic resins.

Examination of silica gel electron micrograph suggests that its primary particles are not visible;[12] instead a large spongelike mass is visible that is supposedly responsible for the gel's very high internal surface area.

In the precipitated silica electron micrograph, the primary particles are connected to form an open grapelike structure. This open structure is believed to be responsible for the rubber reinforcing properties observed with the precipitated silicas.

The electron micrographs of fumed and precipitated silicas have a striking resemblance. Like fumed silicas, precipitated silicas impart thixotropic character to liquid and plastic resins, but in this they are not as efficient as fumed silicas. Since precipitated silicas are

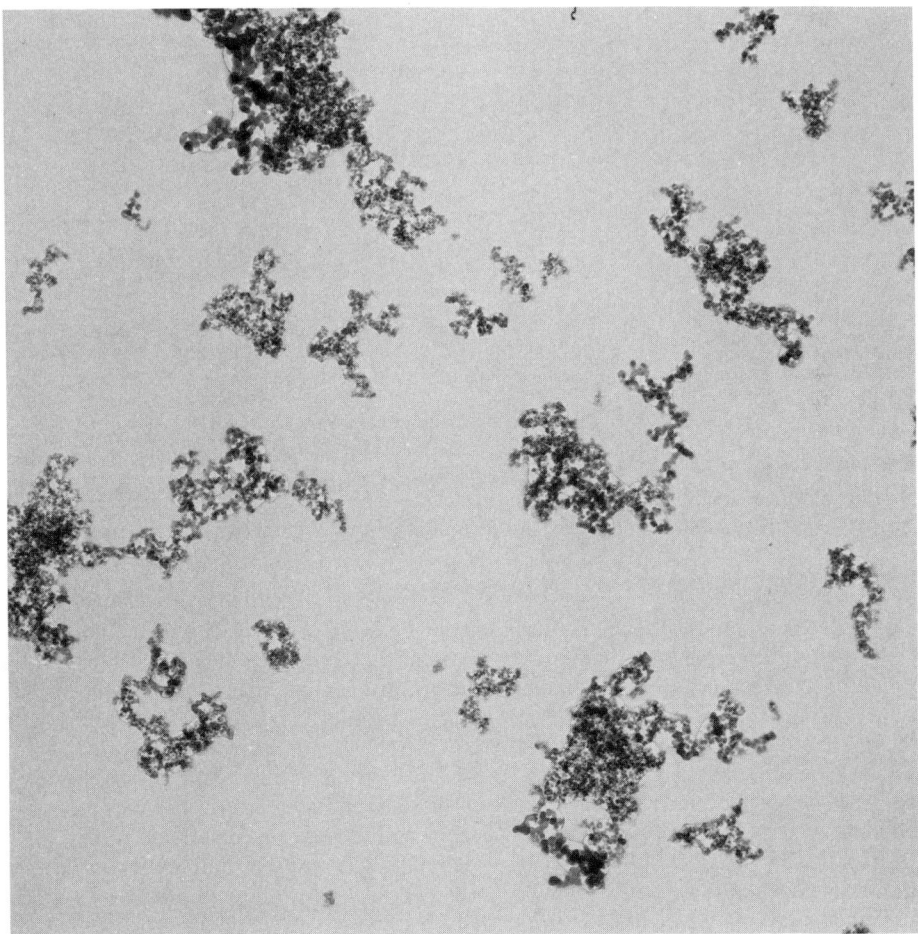

Figure 66.2.   Electron micrograph of fumed silica (Cab-O-Sil)

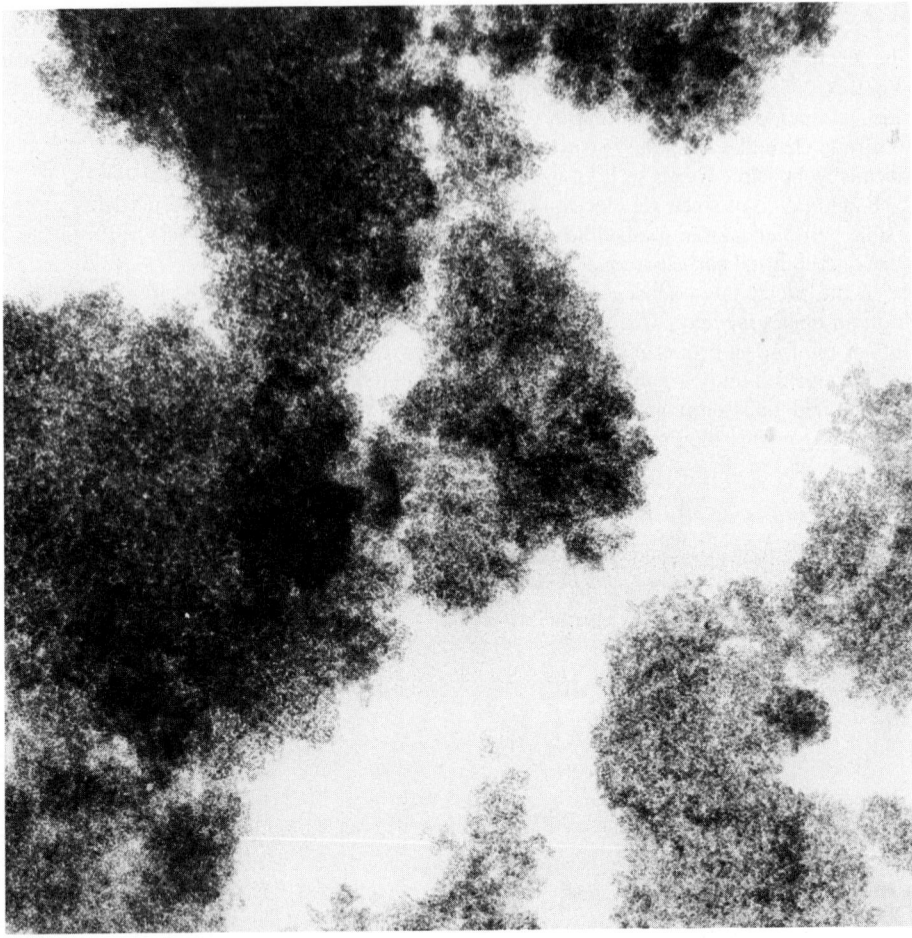

Figure 66.3. Electron micrograph of silica gel (Syloid 244)

significantly lower in price than fumed products, purely on cost basis the performance[13] properties of precipitated silicas are said to be better than fumed silicas.

66.3.4 Silanol Groups

The surface properties of synthetic silicas contain hydroxyl groups called silanol groups.[14,15]

There are three types of silanol groups present on the surface of silica-isolated silanol groups: vicinal (on adjacent silica atoms), silanols, and geminal silanols (two silanol on the same silicon atom). The types of silanol groups on a silicon surface are shown in Figure 66.5.

The silanol groups can be hydrogen bonded with water; in fact, there can be physically adsorbed water present on the silica surface. The silica end-use performance properties can be influenced by the silica silanol group's density and the extent of hydration.

Each ultimate particle of silica can be viewed as consisting of four distinct layers, as shown in Figure 66.6.

The first layer is the inner core; the second layer consists of the silanol groups; the third layer is made up of water molecules hydrogen-bonded to the silanol group; and the fourth layer comprises physically adsorbed water, also called free water.

Figure 66.4. Electron micrograph of precipitated silica (Zeothix 265)

The free water or percent moisture is the water that is released from the silica surface by heating it to 221°F (105°C). The hydrogen-bonded water is removed when silica is heated between 221–392°F (105–200°C). Above 392°F (200°C), the surface silanol groups condense to form siloxane bridges according to the following reaction:

$$2 \text{ SiOH} \longrightarrow \text{Si—O—Si} + \text{H}_2\text{O}$$

Silanol   Siloxane

Precipitated silicas and silica gels exhibit a fully hydroxylated surface and contain between 4–10 silanol groups/nm² surface. Fumed silica contains between 2–4 silanol groups (see Table 66.2).

Because of the difference in silanol group density and a practically anhydrous surface, fumed silicas perform differently than gels and precipitated silicas in many end-use applications.

66.3.5 pH

Among synthetic silicas, silica xerogel and fumed silicas exhibit acidic pH in water. The silica aerogel and precipitated silicas show neutral pH in water.

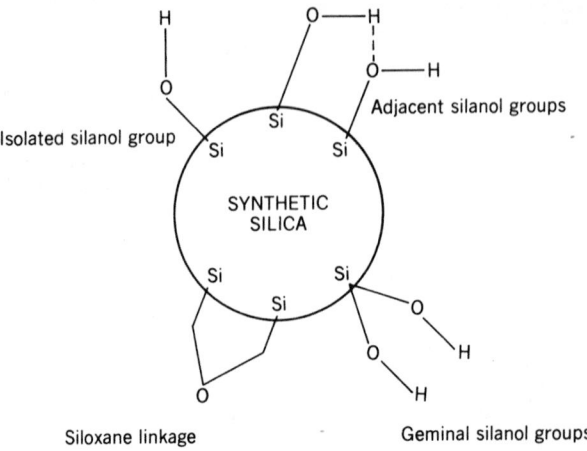

Figure 66.5. Synthetic silica surface hydroxyl groups

66.3.6 Percent Moisture and Loss on Ignition

The percent moisture is the amount of water released by heating silica to 221°F (105°C): the loss on ignition (LOI) is the water that is removed by heating the silica surface to between 221–1832°F (105–1000°C). LOI is also called bound water.

Fumed silicas lose less than 1.5% each of free moisture and LOI water. The silica surface of both gels and precipitated silicas is fully hydrated; when heated, these products release 5% each of free and bound water.

For many plastic and polymer reinforcement applications, it is preferable to choose a silica with minimal free and bound water.

Silica Structure (Segment of Ultimate Particle)	Description	Reaction to Elevated Temperature
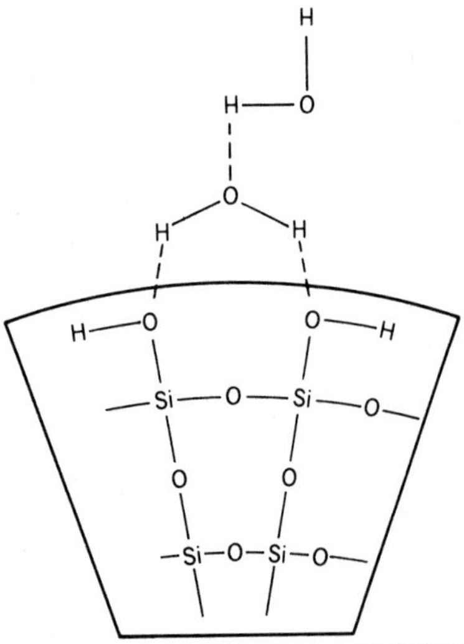	Free water	Released at 221°F (105°C)
	Hydrogen-bonded water	Released at 221° to 392°F (105° to 200°C)
	Silica surface Silanol groups	Converted to siloxane groups at 392°F (200°C) and above $2 \text{ SiOH} \rightarrow \text{Si-O-Si} + H_2O$
	Silica core	Melts at 3092°F (1700°C)

Figure 66.6. Segment of silica structure

66.3.7 Oil Absorption (OA)

The OA of silica is an important property and is predictive of the performance features of silica in many applications. OA is determined by titrating the silica powder with a linseed oil until a stiff, puttylike paste is formed.

Silicas with high OA values are called high structure silicas; silicas with low OA values, low structure silicas.

In plastic polymer and liquid resin systems, high structure silica results in better polymer reinforcement, viscosity building, and other performance properties than the low structure silicas.[16]

66.3.8 Bulk Density

Fumed silicas are very light and fluffy materials. The bulk density of fumed silica is the lowest among the three types of synthetic silicas. It can be a disadvantage in terms of handling and shipping costs.

66.3.9 Percent Density

Among the three types of synthetic silicas, fumed silicas exhibit the highest purity, with 99.8% SiO_2 minimum purity on an ignited basis. The hydrated silicas, precipitated and gels, containing a minimum of 98% silica on an ignited basis.

66.4 PROCESSING CHARACTERISTICS

Processing Property	Effect of Filler
Viscosity	Increases
Melt flow	Decreases
Compounding temperature	Increases
Injection pressure	Increases
Flow in mold	Decreases
Mold shrinkage	Slightly affected

66.5 APPLICATIONS

Fumed silicas are used in three major application areas: as a reinforcing filler, as thixotropic agents, and as a specialty filler.

Precipitated silicas are predominantly used as reinforcing fillers in rubber and plastics. Gel and precipitated silicas are also used as flattening or matting agents in plastics and as an antiblocking agent in film and sheeting.

In recent years, precipitated silicas are being actively used as thixotropic agents in vinyl plastisols, unsaturated polyester laminating resins,[13] gel coat,[17,18] and epoxy resins.[19-21]

66.5.1 Reinforcing Silica

Synthetic silicas are used as reinforcing fillers in silicone rubber.

There are three kinds of reinforcing fillers used in silicone rubber:

- Primary or high reinforcing filler, such as fumed silica,
- Secondary or medium reinforcing filler, such as precipitated silica, and
- Tertiary or low reinforcing filler, such as diatomaceous earth.

TABLE 66.3. Silicone Rubber[a] Reinforcing Properties of Fumed Silica

Silica	PHR	Tensile Strength, psi	Elongation	Tear Die B, ppi	Hardness
Cab-O-Sil	0	700 (4.8 MPa)	500	43	35
Cab-O-Sil	5	925 (6.4 MPa)	500	60	45
Cab-O-Sil	10	1080 (7.4 MPa)	400	80	58
Cab-O-Sil	20	1150 (7.9 MPa)	350	85	68

[a]Dow Corning 432 Base

Synthetic silica is widely used in silicone rubber formulations to increase strength and improve properties of the unfilled gum. The tensile strength of a nonreinforced silicone rubber gum is 49 psi (0.3 MPa). This value can be increased by a factor of about 47 to 2300 psi (16 MPa) by the addition of high-surface-area fumed silica.

Typical properties of fumed silica reinforced silicone rubber are given in Table 66.3, which shows that as the level of fumed silica increases, its tensile strength in silicone rubber increases.

Precipitated silica is used in many silicone rubber formulations. In recent years, high-purity precipitated silica with low electrolyte content has been reported [22] to exhibit excellent reinforcing properties in silicone rubber.

Since precipitated silica is cheaper than fumed silica, it is being used to extend the fumed silica reinforcing properties in many silicone rubber applications.

A low-electrolyte, modified precipitated silica has been evaluated, according to Vondracek and co-workers[22] and Wagner.[23] Precipitated silica does not impart reinforcing properties comparable to fumed silica because of its relatively larger particle size[23] with respect to fumed silica.

Since the refractive index of fumed silicas matches that of silicone rubber, it is possible to produce transparent stocks with the fumed silica.

Precipitated silica is an excellent nonblack reinforcing filler for rubber; it imparts exceptional cut and chip resistance in such applications as off-the-road tires for heavy construction equipment.

At a level of 30–90 phr, precipitated silica is used in high-grade shoe soles and heels to provide stiffness, abrasion resistance, flex resistance, and nonmarking characteristics. Low-grade stocks can be cheapened by decreasing the silica level and increasing the level of inexpensive fillers such as calcium carbonate and clay.

Precipitated silica is used in the composition of ethylene vinyl acetate (EVA) to impart stiffness, improved physicals, and improved weather and age resistance characteristics. Precipitated silica reinforced EVA is used in skate wheels, caster wheels, and automobile bumper strips.

Fetterman[24] studied the properties of precipitated-silica-filled EVA compositions as a function of vinyl acetate content, silica level, and the plasticizer level. He found that materials with a wide range of stiffness and strength were obtained. Fetterman's data are listed in Table 66.4.

66.5.2 Thixotropic Agent

Synthetic silicas are used as viscosity builders, rheology control agents, and thixotropic fillers in plastics and resins.

The rheological properties of synthetic silicas are influenced by the nature of the liquid medium, silica structure, silica concentration, particle size, BET surface area, silanol group density, degree of dispersion, and system pH.

Synthetic silicas are used as thixotropic agents in unsaturated polyester laminating resins

TABLE 66.4. Silica-Reinforced EVA[a]

Vinyl acetate content of EVA, %	28	18	12
Precipitated silica, phr	45	80	80
Dioctyl phthalate, phr	0	15	0
100% Modulus, psi	1,380	1,620	2,600
Tensile strength, psi (MPa)	4,000 (27.6)	3,100 (21.4)	3,710 (25.6)
Elongation, %	450	490	270
Shore D hardness	46	52	66
Tensile strength, psi 158°F (MPa) 70°C	1,880 (12.7)	1,800 (12.4)	2,960 (20.4)
Die C tear, ppi	560	530	610
Flexural modulus, psi (MPa)			
at 73.4°F (23°C)	10,200 (70.3)	28,500 (196.5)	56,000 (386.1)
at −22°F (−30°C)	163,300 (1126.0)	129,700 (894.3)	198,000 (1365.2)

[a]EVA, 100; Silica, as indicated: DOP, as indicated; Silane A-174, 1; stearic acid, 1; zinc stearate, 2; zinc oxide, 2; SR-350, 3; Vulcup 40KE, 4.

and gel coats, epoxy resins, vinyl ester resins, and many specialty resins. The thixotropic behavior of synthetic silica in resins can be further enhanced by using ethylene glycol additive or a suitable coupling agent. The additive functions by forming a bridge between the silica silanol group and the hydroxyl group of the additive, as shown in Figure 66.7.

As a general rule of thumb, the thixotropic behavior of a synthetic silica increases with an increase in BET surface area, a decrease in particle size, and an increase in structure.

Figure 66.8 depicts the rheology properties of four precipitated silicas—Zeothix 177, Zeothix 265, Zeothix 175, and Zeothix 95 in a typical gel coat resin. Zeothix 177 is the smallest in particle size (1.5 μm) and exhibits the best rheological properties. Zeothix 95 is the largest in particle size (2.3 μm) and exhibits the least effective rheological properties in the OCF gel coat resin.

Gel coats are used in fiber-glass-reinforced boats, in tub and shower stalls, and in the transportation industry. Fumed silica is normally used at 1.5–2.0% concentration in gel-coat formulations. Recently, precipitated silica[18] has displaced fumed silica in gel coats at a concentration level of 3–4%.

66.5.3 Specialty Applications

Synthetic silicas are used in small concentrations in many plastic applications to provide special property features. As an example, synthetic silicas (especially silica gels) serve as

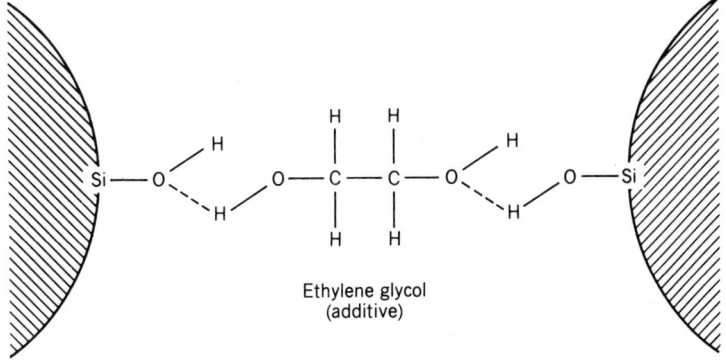

Ethylene glycol
(additive)

Figure 66.7. Bridging of ethylene glycol with silica silanol groups

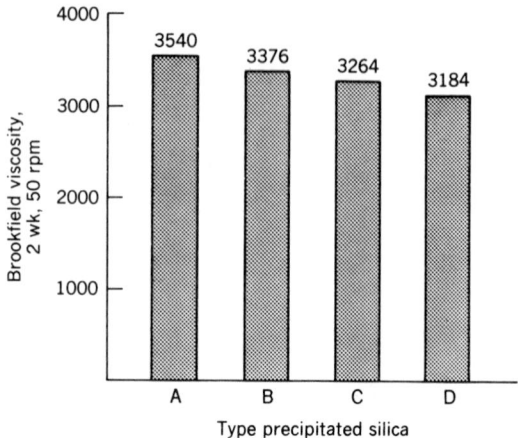

Figure 66.8.   Precipitated silica rheology in gel-coat resin. Zeothix: A, 177; B, 265; C, 175; D, 95.

antiblocking agents in low density polyethylene and polypropylene films. It is believed that the incorporation of synthetic silicas provides roughness to the film surface, which produced the desired antiblocking effect.

Synthetic silicas are used as delustering or matting agents to reduce gloss of films; silica gels and precipitated silicas are widely used as matting agents.

The addition of synthetic silica eliminates the plate-out problems of plasticized PVC by selectively adsorbing the plating ingredients. (Plating out is simply the deposition of certain oily ingredients and colors onto the rolls and hot metal surfaces of the plastic processing equipment.) Synthetic silicas are used at a level of less than 2% to improve the flow properties of PVC master batches, resins, adhesives, and specialty chemicals.

Precipitated silica is used in the ABS pipe formulation to keep the blowing agent in suspension.

Precipitated silica is widely used at the 40–50% level in polyethylene battery separators having high porosity. The separators are prepared[25] by mixing the silica into the PE resin and then extruding it into a film. After extraction with water, a microporous sheet with high puncture resistance and flexibility is produced. Precipitated silica battery separators are used in the production of long-life, maintenance-free storage batteries.

66.6   POLYMERS FILLED

Fumed silicas are used as a reinforcing filler in silicone rubber and as a thixotropic agent in thermosetting resins, such as epoxies and polyesters.

Precipitated silicas are used as reinforcing fillers and property enhancers in rubber and plastics; with respect to the latter, chiefly EVA, PE, and PVC (see Table 66.5).

66.7   COMMERCIAL GRADES

66.7.1   Fumed Silica Commercial Grades

Synthetic silicas with a wide range of physical properties are available from many suppliers. Fumed silicas are available from Cabot and Degussa. Wacker is importing fumed silica into the United States.

The commercial grades of hydrophilic fumed silica available in the United States are listed in Table 66.6. The properties of hydrophobic fumed silica are listed in Table 66.7.

TABLE 66.5. Polymers Filled with Synthetic Silicas

Polymer	Fumed	Gel	Precipitated	Function	Approximate Concentration
	Type Silica				
Silicone rubber	X		X	Reinforcing	40–50 phr
Natural synthetic rubber			X	Reinforcing	40–50 phr
Epoxy	X		X	Thixotrope	2–4%
Laminating resin	X			Thixotrope	0.8–1%
Unsaturated polyester gel coats	X		X	Thixotrope	1.8–4%
LDPE, LP filler/sheeting		X	X	Anti-blocking	0.02–0.2%
PE			X	Porous filler	50%
PVC	X		X	Thixotrope	2–4%
EVA			X	Reinforcing	40%
PVC		X	X	Plate-out	1–2%
Urethane, lacquers			X	Reinforcing, shoe soles	80–100 phr
Plasticized PVC			X	Reinforcing, shoe soles	80–100 phr
HDPE	X			Improved extrusion	1–10%

TABLE 66.6. Properties of Hydrophilic Silicas

Fumed Silica	Source	Surface Area, m²/g	Primary Particle Size, nm	% Moisture	% LOI	4% pH	Bulk Density, (g/L) Uncompacted
Aerosil 130	Degussa	130 ± 25	16	<1.5	1	3.6–4.3	50
Aerosil 150	Degussa	150 ± 15	14	<0.5	1	3.6–4.3	50
Aerosil 200	Degussa	200 ± 25	12	<1.5	1	3.6–4.3	50
Aerosil 300	Degussa	300 ± 30	7	<1.5	2	3.6–4.3	50
Aerosil 380	Degussa	380 ± 30	7	<1.5	2.5	3.6–4.3	50
Aerosil OX50	Degussa	50 ± 15	40	<1.5	1	3.8–4.5	130
Cab-O-Sil M5	Cabot	200 ± 25	14	<1.5	1	3.5–4.2	37
Cab-O-Sil MS-7	Cabot	200 ± 25	14	<1.5	1	3.6–4.2	72
Cab-O-Sil MS-75	Cabot	255 ± 15	11	<1.5	1.5	3.6–4.2	72
Cab-O-Sil HS-5	Cabot	325 ± 25	8	<1.5	2.0	3.6–4.2	37
Cab-O-Sil EH5	Cabot	390 ± 40	7	<1.5	2.5	3.5–4.2	37
Cab-O-Sil PTG	Cabot	200 ± 20	14	<1.5	1	3.8–4.1	37
HDK-S5	Wacker	55 ± 15	40	<1.0	1.5	3.6–4.3	60
HDK-V15	Wacker	150 ± 20	14	<1.0	1.5	3.6–4.3	60
HDK-N20	Wacker	200 ± 30	12	<1.5	1.5	3.6–4.3	60
HDK-T30	Wacker	300 ± 30	7	<1.5	2.0	3.6–4.3	60
HDK-T40	Wacker	400 ± 40	6	<1.5	2.5	3.6–4.3	60

TABLE 66.7. Properties of Hydrophobic Fumed Silicas

Properties	R-972	HDK-H15	HDK-H20	WR50	Tullanox 500
Source	Degussa	Wacker	Wacker	PQ	Tulco
Surface area, m/g	110 ± 20	120 ± 20	170 ± 30	130	225 ± 25
Primary particle size, nm	16	16	11	14	7
% Moisture	0.5	0.6	0.6	1.2	1.5
% LOI[a]	2	2	2	5	4
4% pH[b]	3.5–4.1	4.0–4.8	4.0–4.8	10.4	5.5–6.6
Bulk density, g/L	50	60	60	80	50

[a]Contains approximately 1% chemically bonded carbon.
[b]4% Dispersion in water:methanol (1:1).

TABLE 66.8. Physical Properties of Precipitated Silicas

Product	Average[a] Particle Size (μm)	Surface Area, (m²/g)	OI Absorption[b] (cc/100 g)	Structure Type[c]
	PPG Industries			
HiSil 200 Series (210, 215, 233)		149	200	HS
HiSil EP		60	200	HS
HiSil 404		40	145	MS
HiSil 422		45	147	MS
LoVel 27	3.5	150	200	HS
LoVel 275	5.0	150		HS
LoVel 28	5.5	150	200	HS
LoVel 29	6.5	150	200	HS
LoVel 39A	10.0	150	200	HS
LoVel 66X	3.5	150	200	MS,T[d]
LoVel 70	3.5	150	200	HS,T
HiSil T-600	1.4	150	190	HS
	J. M. Huber Corporation			
Zeothix 177	1.3–1.8	150–200	220–250	VHS
Zeothix 265	1.5–2.0	200–300	200–240	VHS
Zeothix 95	2.0–3.0	120–150	190–230	VHS
Zeosyl 200	4.0–6.0	200–300	190–210	VHS
Zeofree 80	5.0–7.0	120–150	180–200	HS
Zeofree 153	6.0–8.0	100–140	160–170	MS
Zeodent 113	8.0–10.0	150–250	85–100	LS
Zeo 49	8.0–10.0	200–300	85–100	LS
	Degussa/Nasilco			
FK-160		160		VHS
Sipernat 22	80	190	230	VHS
Sipernat 22 S	5	190	230	VHS
FK 500LS		450	300	VHS
Sipernat D17	3	115	125	MS,T
Sipernat 50		450	330	VHS
Sipernat 50 S		450	330	VHS

[a]Secondary particle size.

[b]Oil absorption values reported for PPG production are dibutyl phthalate absorption values. In some cases, the oil absorption values were run by the author to assign structure index.

[c]For structure definition, based on oil absorption, see Table 66.1.

[d]T = Surface treated.

66.7.2 Precipitated Silica Commercial Grades

In the United States, precipitated silicas are manufactured by PPG Industries, North America Silica Co., and J.M. Huber Corp. (see Table 66.8).

66.7.3 Silica Gel Commercial Grades

In the United States, silica gels are predominantly available from W.R. Grace and Glidden, a division of SCM Corp. Commercially available grades of silica gel are listed in Table 66.9.

TABLE 66.9. Physical Properties of Silica Gel

Product	Average[a] Particle Size, μm	Surface Area, m^2/g	Oil Absorption g/100 g	Gel Type[b]
		W. R. Grace (USA)		
Syloid 63	9.0	675	60	H
Syloid 65	4.5	695	75	H
Syloid AL-1	9.0	675	60	H
Syloid 72	4.0	340	220	H
Syloid 74	6.0	340	200	H
Syloid 83	4.0	250	270	H,T
Syloid 161	6.0	340	155	H,T
Syloid 162	7.0	340	155	H,T
Syloid, 169	4.5	250	275	H,T
Syloid 221	6.5	250	275	H
Syloid 234	2.5	250	275	H
Syloid 235	4.0	250	275	H
Syloid 244	3.0	310	280	A
Syloid 245	3.0	250	275	A
Syloid 266	2.0	310	300	A
Syloid 308	4.0	250	200	H,T
Syloid 378	3.0	250	200	H,T
Syloid 620	15.0	320	180	H
Sylox 2	2.0	250	250	A
		SCM (Glidden) USA		
Silcron G-100	3.0	275	270	A
Silcron G-130	3.0	300	220	A,T
Silcron G-300	4.2	300	190	H,T
Silcron G-310	5.0	300	190	H,T
Silcron G-500	5.0	200	175	H,T
Silcron G-510	6.0	225	180	H,T
Silcron G-520	5.0	200	175	H,T
Silcron G-530	6.0	225	180	H,T
Silcron G-550	5.0	200	175	H,T
Silcron G-600	4.7	325	210	H
Silcron G-601	6.5	325	210	H
Silcron G-602	7.1	320	205	H
Silcron G-603	10.1	320	205	H
Silcron G-604	12.0	320	205	H
Silcron G-900	8.5	625	80	H
		Crosfield (Great Britain)		
Gasil 200	6–13	700	90	H
Gasil 23	2–3	300	300	A
Gasil 35	3–6	370	180	H
Gasil 64	6–9	230	170	H
Gasil 544	3–6	250	190	H
Gasil 93	8–12	260	140	H
Gasil 937	6–9	270	150	H

[a]Secondary particle size.
[b]H = hydrogel, A = aerogel, T = treated product.

TABLE 66.10.   Trade Names, Synthetic Silicas

Type Silica	Trade Names	Supplier	Address
Fumed	Cab-O-Sil	Cabot Cab-O-Sil Division	P. O. Box 188 Tuscola, IL 61953
Fumed	Aerosil	Degussa Pigments Division	P. O. Box 2004 Teterboro, NJ 07608
Gel	Syloid	W. R. Grace Davison Chemical Division	P. O. Box 2117 Baltimore, MD 21203
Gel	Silcron	Glidden Division of SCM Corporation	Baltimore, MD 21226
Precipitated	Hi-Sil Lo-Vel	PPG Industries, Inc. Chemical Division	One Gateway Center Pittsburg, PA
Precipitated	Zeothix Zeofree Zeosyl Zeo	J. M. Huber Corporation Chemicals Division	P. O. Box 310 Havre de Grace, MD 21078

66.8   COMPOSITE CHARACTERISTICS

See Tables 66.3 and 66.4.

66.9   TRADE NAMES

See Table 66.10.

BIBLIOGRAPHY

1. S.K. Wason, *J. Soc. Cosmet. Chem.* **29,** 497 (1978).
2. Ger. Pat. 762,723 (1942) (to Degussa).
3. E. Wagner and H. Brunner, *Agnew. Chem.* **72,** 744 (1960).
4. R. Bode, H. Ferch, and H. Fratzscher, *Kaut. Gummi Kunstst.* **20,** 578 (1967).
5. U.S. Pat. 3,693,840 (1975) S.K. Wason, (to J.M. Huber Corp.)
6. U.S. Pat. 4,067,746 (1978) S.K. Wason, (to J.M. Huber Corp.)
7. U.S. Pat. 4,067,746 (1978) S.K. Wason, (to J.M. Huber Corp.)
8. U.S. Pat. 4,132,806 (1979) S.K. Wason, (to J.M. Huber Corp.)
9. R.K. Iler, *The Chemistry of Silica,* John Wiley & Sons, Inc., New York, 1979.
10. D. Barby, "Silicas," in G.D. Parfitt and K.S.W. Sing, eds., *Characterization of Powder Surfaces,* Academic Press, New York, 1976, p. 353.
11. S. Brunauer, P.H. Emmett, and E.J. Teller, *Am. Chem. Soc.* **60,** 309 (1938).
12. F. Kindervater, *Kaut. Gummi Kunstst.* **26,** 7 (1973).
13. S.K. Wason, *Rheology of New Controlled Structure Silicas in Polyester Laminating Resin, 35th Annual Technical Conf.,* RP/C Institute, SPI, Inc., 1980, Sect. 6-E, pp. 1–9.
14. J.A. Hockey, "The Surface Properties of the Silica Powder," *Chem. Ind.,* 57 (1965).
15. M.L. Hair, *Infrared Spectroscopy in Surface Chemistry,* Marcel Dekker, Inc., New York, 1967, Chapt. 4.
16. S.K. Wason, 1970, unpublished data.
17. S.K. Wason and J.W. Maisel, "Thixotropic Properties of New Precipitated Silicas in Gel Coats," *36th Annual Tech. Conf., RP/C Institute,* SPI, Inc., 1981, Sect. 13-A, pp. 1–9.

18. S.K. Wason and J.W. Maisel, Thixotropic Properties of Precipitated Silica in Gel Coats, *Plastic Compounding,* May/June 1981.
19. J.W. Maisel and S.K. Wason, Rheology of Precipitated Silica in Epoxies, *37th Annual Tech. Conf., RP/C Institute,* SPI, Inc. (1982), Sect. 12-E, pp. 1–6.
20. J.W. Maisel and S.K. Wason, Rheology of Precipitated Silica in Epoxies, *Polym. – Plast. Technol. Eng.* **19,** 227–242 (1982).
21. S.K. Wason and J.W. Maisel, Evaluation of New Precipitated Silica Thixotrope in Plastics. *38th Annual Tech. Conf., RP/C, Institute,* SPI, Inc., 1983, Sect. 3F, pp. 1–7.
22. P. Vondracek and M. Schatz, *Kaut. Gummi Kunstst.* **33,** 699 (1980).
23. M.P. Wagner, "Precipitated Silica in Silicone Rubber," *119th Meeting of the Rubber Division of ACS,* Minneapolis, June 2–5, 1981.
24. M.J. Fetterman, *Elast.* and *Plast.* **9,** 226 (1977).
25. U.S. Pat. 3,351,495 (1967), D. Larsen and C. Kehr, (to W.R. Grace).

General References

R.K. Iler, *The Chemistry of Silica,* John Wiley & Sons, New York, 1979.

T.C. Patton, ed., *Pigment Handbook,* Vol. I, John Wiley & Sons, Inc., New York, 1978.

M.P. Wagner in R.B. Seymour, ed., *Additives for Plastics,* Academic Press, Inc., New York, 1978.

67

FILLERS, TALC

Joseph A. Radosta and Michael Tapper

Pfizer Inc., Minerals, Pigments & Metals Div., 640 North 13th Street, Easton, PA 18042-1497

67.1 CATEGORY

Talc, a naturally occurring mineral of hydrated magnesium silicate, covers a wide range of products of diverse composition, shape, purity, and brightness. The talc grades of commercial interest for plastics applications are fine ground products consisting of thin platelets that are white and contain low quantities of impurities. Because of the relatively high platelet aspect ratio of these grades compared to most minerals, they are categorized as a reinforcement and are used primarily to improve composite performance properties.

67.2 SOURCE

Major ore sources of talc in the United States are Montana, California, and Vermont. Depending on the mining location, other mineral contaminants are associated with the talc product; for example, calcite, chlorite, dolomite, magnesite, and tremolite. Methods employed for the production of talcs are dependent on the amounts and characteristics of the associated contaminants.

Talc mining operations can be open pit or underground. Many of the California and Montana mining operations are open pit. Here, the talc seam is exposed through blasting and removal of the overburden and then quarried and sorted according to color and mineralogy. When the ore body is located at greater depths and in mountainous terrain, underground mining operations are usually employed. The mining costs for underground operations tend to be higher than for open pit operations.

For ores that contain low impurity levels, such as those found in Montana, dry processing techniques are usually employed. The crude ore is first reduced in primary jaw crushers or rolls where the large lumps [36 in. (914 mm)] are crushed to a size of 4–6 in. (102–152 mm). The crushed ore is then usually screened, washed, and sorted by hand or electron beam to remove gross impurities. Secondary crushers, such as cone crushers or roller mills, are used to reduce the talc size to at least 0.5 in. (13 mm) and, in many cases as fine as 0.06 in. (1.6 mm).

Particle shape, average size, and size distribution, properties very critical to the reinforcing qualities of the talc in plastics, are essentially determined by the choice of fine grinding method and type of grinding and classification equipment.

Further size reduction is required to produce talcs of commercial value for plastics applications, in which the maximum particle size should not exceed 44 microns and be able to pass through a 325 mesh screen. This fine grinding is achieved in tube mills or Raymond mills.

Higher performance talcs for plastics require extremely fine grinds where the mills employ high-energy input accompanied by air separation. Pebble mills, fluid energy mills (in some cases used with superheated steam), or ultrahigh-speed vertical hammer mills are operated under carefully controlled conditions of pressure, feed rate, and moisture and in a closed loop with air classifiers to obtain products of subsieve size (maximum particle size less than 30 microns).

For ores that contain high levels of impurities deleterious to the use of talc as filler, beneficiation techniques are employed prior to fine grinding. Froth flotation is commonly used for the beneficiation of talc. Using various surfactants, the surface properties of the impurities and/or talc are modified to achieve the separation of one from the other. Subsequent wet grinding and water classification are widely used to achieve the required size distributions of the various products. Where color of the product is a critical parameter, bleaching with suitable bleaching agents can also be employed.

Structurally, talc is made up of layers which consist of a brucite sheet sandwiched between two sheets of silica,[1] as shown in Figure 67.1. The layers are held together by weak bonding forces and are the major reason for the platy shape of particles when they are properly processed in grinding equipment. In Figure 67.2, the thin platelet structure can be readily observed from a scanning electron micrograph of a typical commercial fine ground talc.

67.3 KEY PROPERTIES

Tables 67.1–67.4 specify key properties of commercially available talc used for filling plastics.

67.4 PROCESSING CHARACTERISTICS

Processing of talc-filled compounds is not unusual and can be readily accomplished by any competent molder or extruder of polymers by adjusting machine conditions to accommodate

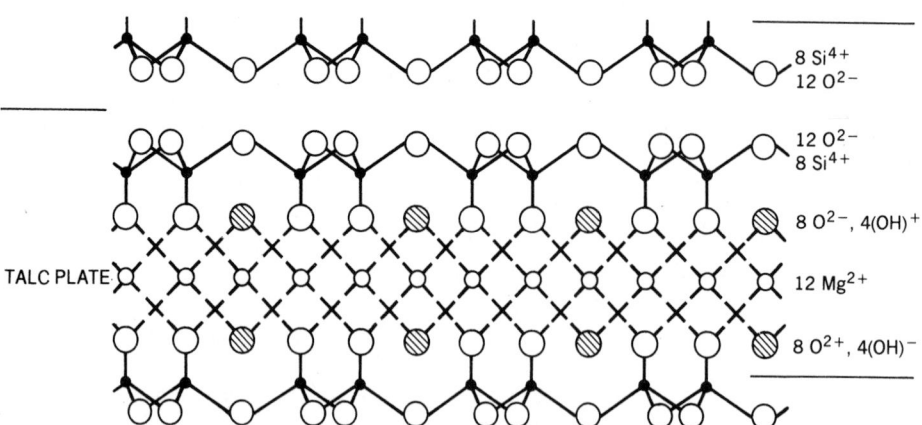

Figure 67.1.   Theoretical structure of talc

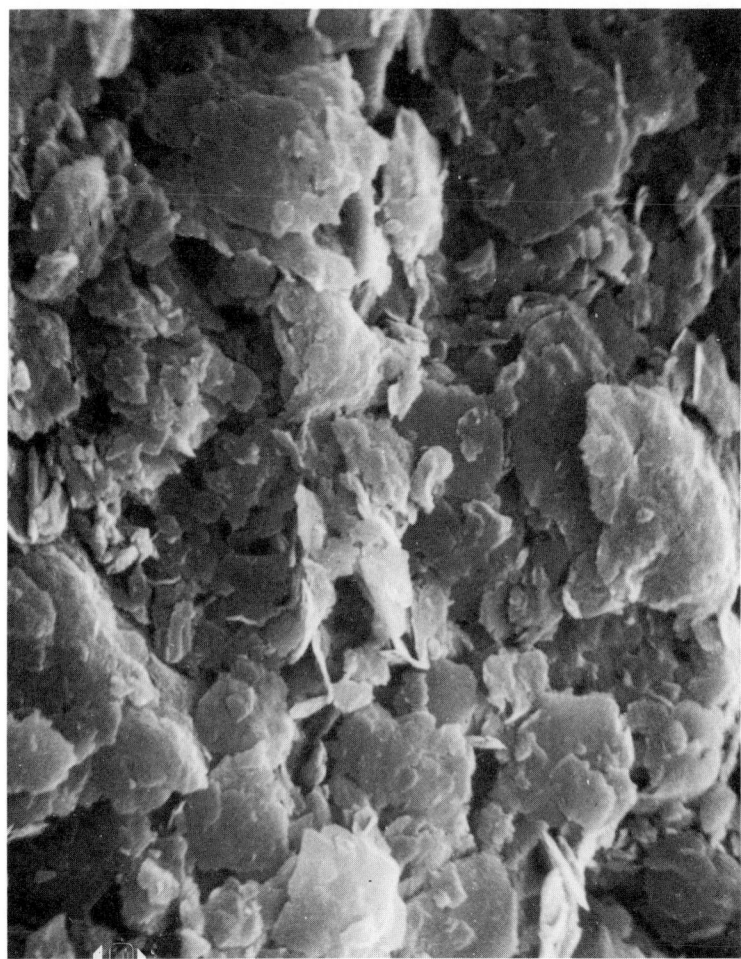

Figure 67.2. Thin platelet structure of talc as seen in scanning electron micrograph

TABLE 67.1. Key Properties of Commercially Available Talc Used for Filling Plastics

Property	Value
Specific gravity	2.7–2.8
Hardness, Moh's scale 1–10	1 (softest)
pH, 5% aqueous solution	9.0–9.5
Average particle size, microns	1.0–10
Hegman grind	0–6^{+}
Bulk density, lb/ft^{3}	10–50
Color	White–gray
% Brightness	85–95
Oil absorption, g/cm^{3}	20–50

TABLE 67.2.   Key Properties and Suppliers of Talc

Supplier	Product Name	Top Particle Size,[a] μ	Maximum +325	Talc Source
Pfizer, MPM Div.	Microtalc MP 12-50	12	Nil	Montana
235 E. 42nd St.	Microtalc MP 15-38	15	Nil	
New York, NY 10017	Microtalc MP 25-38	25	Nil	
(215) 250-7000, X3010	Microtalc MP 40-27	40	Trace	
	Talcron MP 45-26	45	0.5	
Montana Talc Co.	Nicron 660	15	Nil	Montana
28769 Sappington Rd.	Nicron 600	20	Nil	
Three Forks, MT 59752	Nicron 400	40	Nil	
(406) 285-3286	Nicron 325	45	0.5	
R. T. Vanderbilt Co.	Nytal 300	25	0.06	New York
33 Winfield Ave.	Nytal 400	15	Trace	
Norwalk, CT 06855	I.T.X.	NA	2.0	
(203) 853-1400	I.T.5X	NA	0.75	
Cyprus Industrial Minerals	Glacier 325	45	0.3	Montana
Talc Division	Mistron Vapor	12	Nil	
P.O. Box 3419	Stellar 600	15	Nil	Australia
Englewood, CO 80155	Stallar 400	NA	1.0	
(800) 325-0299	Vertal 7	30	1.0	Vermont
	Vertal 700	40	0.3	
	Vertal 300	15	0.03	
Luzenac Inc.	Jet Fill 200	70	3.0	Canada
10175 N. Service Rd. West	Jet Fill 350	50	Trace	
Suite 14	Jet Fill 500	35	Nil	
Oakville, Ontario, Canada	Artic Mist	25	Nil	
L6M 2G2				
(416) 825-3930				

the changes noted below. All effects shown below are proportional to filler-loading levels, higher levels have a greater effect.

Property	Effect
Melt viscosity	Increases
Melt flow	Decreases
Injection pressure	Increases
Flow in mold	Decreases
Mold shrinkage	Decreases
Temperature of melt	Can increase due to shear

TABLE 67.3.   Surface-treated Grades of Talc

Supplier	Trade Name	Type Coating	Major Application
Cyprus Ind. Min.	Mistron Cyprubond	Silane	Elastomers
	Mistron ZSC	Zinc stearate	Elastomers
Pfizer Inc., MPM	MicroPflex 1200	Proprietary	Nylon and Engineering Thermoplastic
	Microtuff 1000	Proprietary	Polyolefins (PP, HDPE)
	Microtuff F	Proprietary	Food contact (PP, HDPE)
	Microbloc	Proprietary	Film antiblock
	Super-Microtuff	Proprietary	Polyolefins (PP, HDPE)

TABLE 67.4. Talc-filled Composite Characteristics

Property	Effect
Tensile strength and modulus	No effect to increases[a]
Flexural strength and modulus	Increases
Impact strength	Decreases[b]
Retention of properties at elevated temperature	Increases
Retention of properties under moist conditions	No effect
Dimensional stability	Increases
Thermal durability	Decreases[c]
Thermal conductivity	Increases
Linear coefficient of expansion	Decreases
Electrical properties	Increases
Chemical properties	No effect to improve
Creep resistance	Increases

[a]Surface treated grades improve tensile strength.
[b]Surface treated grades minimize loss of impact strength.
[c]Additional heat stabilization may be needed at time of compounding.

67.5 APPLICATIONS AND POLYMERS FILLED

Properly selected commercial grades of talc consist of thin platelets where the aspect ratio (ratio of average platelet diameter to thickness) can vary from about 20:1 to 5:1. The greater the aspect ratio of the platelet, the greater is the restraint imposed upon the polymer matrix by the talc and, consequently, the greater its reinforcing properties (stiffness at ambient and elevated temperatures, creep resistance, strength, and dimensional stability). Platy talcs can be used in both thermoplastics and thermosets.

This section discusses both the major application areas for commercial usage of talc and the polymers that are commonly reinforced with talc or that show the most potential for reinforcement. In addition to the polymers described below, talc is being used as a reinforcing filler in various grades of low density polyethylene, various PVC formulations, rubber and polyurethane compounds, and phenolic molding compounds in order to achieve higher strength, stiffness, and dimensional stability.

67.5.1 Polypropylene Homopolymer

The main reason for using talc in polypropylene homopolymer is to obtain values of high stiffness and end-use temperatures that can match those of many unfilled engineering thermoplastics. Additional advantages imparted by the talc are good surface quality of the finished part, lower mold shrinkage, and easier processing. One disadvantage for talc (and any other high surface area mineral filler) in polypropylene is that at high loadings, there could be a significant decrease in thermal stability for a nominally stabilized polypropylene resin. However, this problem can be overcome through the addition of higher levels of a commercial stabilizer/antioxidant package.

The major application area for talc-filled polypropylene homopolymer is automotive parts that require high stiffness and high-temperature creep resistance. These include fluid pump parts, fan blades and shrouds, hot-air blower housings and ducts, air nozzles, battery heat shields, fascia panels, radiator grills, and headlamp housings. Other application areas (which are shared with polypropylene copolymer) are appliance, commercial housing, and electrical.

67.5.2 Polypropylene Copolymer

When impact strength or toughness is a major consideration, particularly at low temperatures, polypropylene copolymer is very often selected as the resin matrix. Here, talc has been used at low loadings (up to 20%) to restore the stiffness and strength to homopolymer levels without significant loss in the desired toughness. There are currently higher performance applications for polypropylene copolymer (competing with tough engineering thermoplastics such as ABS and nylon) that require even greater improvements in stiffness and elevated temperature load-bearing properties while maintaining the desired low-temperature toughness. Toward this end, a special surface-treated grades of talc has been developed for polyolefins that results in a highly improved impact stiffness balance for polypropylene copolymer, Table 67.3.[2]

In the appliance industry, specific applications include refrigerator door liners, washing machine agitators, and housings and parts for dishwasher and washing machine pumps.

67.5.3 High Density Polyethylene (HDPE)

HDPE and its copolymers can be extremely tough at ambient and low temperatures and very often competes with polypropylene copolymer. When reinforced with 20–30% loadings of talc, the excellent toughness of many of the HDPE grades can be maintained along with the increased stiffness. Here again, using a special surface-treated grade of talc gives even greater improvements in the impact strength/stiffness balance.[2] In many cases, HDPE filled with the surface-treated talc can match the stiffness while at the same time dramatically exceed the impact strength of a mica-filled composite.

Many blow molding and thermoforming applications for automotive, industrial, and household parts are under consideration for talc-filled HDPE.

67.5.4 Polystyrene and ABS

Although the present talc usage in styrenics is low, work has been published that shows how polystyrene can be tailored with combinations of talc and the plastic elastomer to provide better overall performance at lower material costs, compared to conventional unfilled polystyrene grades.[3] Talc has been considered for use in polystyrene thermoformed container applications (both foamed and unfoamed) and in ABS pipe and business machine housings.

67.5.6 Nylon and Engineering Thermoplastics

Nylon, thermoplastic polyesters, and other engineering resins are usually selected for their exceptional elevated temperature performance properties of rigidity, strength, toughness, and creep resistance. Many of these resins are reinforced with fiber glass to further increase their strength and stiffness. However, poor impact strength, warpage, and high fiber-glass prices are accompanying disadvantages that can be overcome with the full or partial substitution of talc. In many applications, talc–glass blends are seen as offering a more suitable mix of properties than straight mineral or glass formulations, especially in applications calling for large, stressed parts. In addition to conventional unmodified talc, a special surface-treated grade of talc has been developed for engineering thermoplastics that promises to be a very cost-effective way of replacing all or part of the fiberglass content in reinforced engineering thermoplastics.

67.7 COMMERCIAL GRADES

Major commercial grades are identified in the various tables.

67.14 TRADE NAMES

See the Tables 67.1–67.3.

BIBLIOGRAPHY

1. J.A. Pask and M.F. Warner, "Fundamental Studies of Talc: I, Constitution of Talc," *J. Am. Ceram. Soc.* **37**(3), 118–128 (1954).
2. J.A. Radoata, Improving the Impact/Stiffness Balance of Mineral-Filled Polyolefins With Surface-Modified Talc, *42nd SPE ANTEC,* May 1984.
3. J.A. Radosta, "Improving the Physicals of Impact Polystyrene," *Plast. Eng.* **33,** 28 (Sept. 1977).

68

FILLERS, WOLLASTONITE

Ronald R. Bauer
NYCO, Willsboro, NY 12996-0368

68.1 CATEGORY

Wollastonite is a semireinforcing filler.

68.2 SOURCE

Wollastonite has a relatively short history as an industrial mineral, and an even shorter history as an additive in plastics formulations. The mineral was first noticed as a geological curiosity around 1830 by the famed English geochemist, William Hyde Wollaston, after whom the mineral is named. No commercial use of the mineral was known until 1933, when a deposit in California was mined on a very small scale to supply raw material for the manufacture of mineral wool. No large-scale production took place until 1950, when the Cabot Carbon Co. of Boston purchased mineral rights, production methods, and market data for the wollastonite deposit located in Willsboro, N.Y. The company devised the methods required to beneficiate the ore to a mineralogically and chemically pure form, and performed fundamental market and product development studies.

In 1960, Cabot was instrumental in perfecting the manufacture of what continues to this date to be a unique product. The crystal structure of the wollastonites found in Essex County, N.Y., near the towns of Willsboro and Lewis, lent themselves to production of a high-aspect-ratio now marketed under the trade name NYAD G. The specialized manufacturing techniques (which are of a proprietary nature) permit controlled milling of the wollastonite to a product that has an aspect ratio of 10:1 L/D and a particle size of nominally 85% passing a water-washed 200 mesh screen.

In 1969 the Cabot Carbon Co. sold its interests in the wollastonite operation to Interpace Corp. The new owner proceeded to expand existing sales to the ceramic and coatings industries and promote the uses of wollastonite in a very diverse marketplace. In 1979, Interpace sold the operation to Processed Minerals Inc., which operates the wollastonite business as its NYCO division. Processed Minerals Inc. is a wholly owned subsidiary of Canadian Pacific Enterprises (J.S.).

Property	Value
Chemical formula	$CaSiO_3$
Appearance	White
Particle shape	Acicular
Molecular weight	116

Property	Value
Refractive index	1.63
pH (10% slurry)	9.9
Hardness, Mohs	4.5
Water solubility, mg/100 cc	1.0095
Bulking value, gal/lb	0.0413
Melting point, °F (°C)	2804 (1540)
Transition point, °F (°C)	2128 (1200)
Coefficient of thermal expansion, in./in./°F	3.6×10^{-6}
mm/mm/°C	6.5×10^{-6}

Particle size, microns	Density, lb-ft³	Relative Pressure, mm Hg	Value $\times 10^{-4}$
0.02	10.6	10^{-5}	4.3
0.02–0.07	22.5	10^{-5}	3.2
0.02–0.07	22.5	628	263

Figure 68.1. NYAD G.

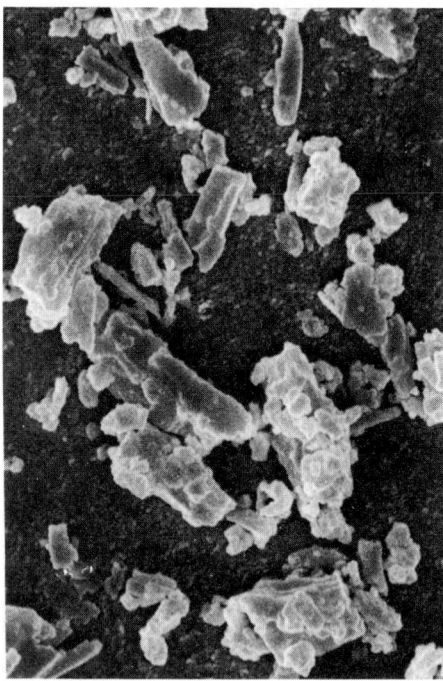

Figure 68.2. NYAD 1250.

68.3 KEY PROPERTIES

Figures 68.1 and 68.2 are typical scanning electron photomicrographs of some wollastonite products supplied by NYCO. Note the relatively high aspect ratio of the wollastonite crystals.

68.4 PROCESSING CHARACTERISTICS

Although is is impossible to assign quantitative values to the many variables encountered in compounding the wide range of today's commercial polymers, it is possible to note some overall effects on the most important aspects of the broad term "rheology." See Table 68.1.

68.5 APPLICATIONS

Wollastonite, and more importantly, surface-modified wollastonite have a broad application as semireinforcing additives in many polymer systems. Some of these applications are listed in Table 68.2.

TABLE 68.1. Effects of Wollastonite on Processing

Properties	Untreated	Surface Treated
Viscosity	Increases	No effect to moderate increase
Melt flow	Decreases	No effect to moderate decrease
Processing temperature	Increase needed	Little or no effect
Injection pressure	No effect to slight increase	No effect to slight decrease
Flow in mold	Moderate reduction	No effect to slight improvement
Mold shrinkage	Reduces	Reduces
Mold cycle time	Reduces	Reduces

TABLE 68.2. Typical Commercial Applications of Wollastonite

Application	Properties Enhanced Through Use
Molded/plated nylon	Promotes good strength and cost performance (vs fiber glass), and chrome platability
Friction products (phenolic resin binder)	Replaces asbestos, improves wet physical properties, good strength retention and heat resistance
Molded epoxy	Coefficient of thermal expansion, good electrical properties
Automotive undercoat	Good corrosion resistance properties
Molded phenolics	Good cost/performance, strength, and electrical properties
Electrical motor commutators, junction boxes	Good electrical properties and processing, cost benefits
Molded rubber	Good physical properties, chemical resistance
Polyester sanitary fixtures	Good surface appearance, strength properties

TABLE 68.3. Properties of Wollastokup Filled Nylon 6 and 6,6 Systems

Property	Nylon 6	Nylon 6 35% filled	Nylon 6 40% filled	Nylon 6,6 40% filled	Nylon 6,6 40% Wollastokup glass-filled	Nylon 6,6 55% Wollastokup glass-filled
Drop weight impact, ft-lb	120	50	30	19		
(J)	(163)	(68)	(41)	(26)		
Izod impact						
Notched, ft-lb	1.1	1.0	.97	.8	1.1	2.0
(J/M)	(59)	(53)	(52)	(43)	(59)	(107)
Unnotched, ft-lb	No break	45	40	33	14	
(J/M)	(No break)	(2,400)	(2,140)	(1760)	(750)	
Flexural stress, psi	16,400	20,000	21,500	20,500	28,000	39,500
(MPa)	(113)	(138)	(148)	(141)	(193)	(272)
Flexural modulus,						
psi × 10^5	3.95	7.4	8.2	8.2	1.4	1.95
(MPa)	(2,720)	(5,100)	(5,700)	(5,700)	(970)	(1345)
Tensile stress, psi	11,800	12,800	13,100	13,500	18,000	25,000
(MPa)	(82)	(88)	(90)	(93)	(124)	(172)
Tensile elongation, %	200	20	15	15	4	
Deflection temperature,						
°C at 264 psi (1.8 MPa)	65	130	130	130	255	260
Specific gravity, lb/ft³	70	90	91	92	92	102
(g/cm³)	(1.13)	(1.45)	(1.46)	(1.48)	(1.48)	(1.63)
Mold shrinkage, w/w	0.013	0.012	0.012	0.012		

TABLE 68.4. Effect of Wollastonite and Wollastokup on Tensile Strength in Brake Linings

Fixed Formula Components[a] (Parts by weight):

Kevlar Pulp	3		Hexamethylene tetramine	2
HRJ-1415 Resin	9		HRJ-2447 (dry wt)	8
Pet Coke	4		HRJ-2448 (dry wt)	8

	NYAD G, 42 parts		G Wollastokup 1108, 42 parts	
Barytes	8	12	8	12
Densimix	8		8	
Fluorspar	8	12	8	12
Tensile Strength				
Room temperature, psi	1125	1160	1630	1960
(MPa)	(7.76)	(7.80)	(11.2)	(13.5)
400 °F, psi	560	540	1020	1155
(MPa)	(3.86)	(3.72)	(7.03)	(7.96)

[a]Resins produced by Schenectady Chemicals Co., Schenectady, N.Y.

68.6 POLYMERS FILLED

One of the key applications of surface-modified wollastonite (10 Wollastokup 10012) is as an additive in nylon compounds (Table 68.3). One function of the Wollastokup in these nylon 6 and nylon 6,6 compounds is to offer metal platability without severely lowering impact strength. By incorporating Wollastokup into the compound and subsequently submitting the molded part to an acid bath, an excellent metal plating anchor pattern is achieved. Additional applications exist where the wollastonite acts as a lower-cost partial replacement for fiber glass in general-purpose molding compounds.

The use of wollastonite as an asbestos replacement is natural because of its high aspect ratio and totally innocuous health effects. One area where this asbestos replacement has taken place is in friction products. The only major shortcoming associated with the use of

TABLE 68.5. Wollastokup in BMC

Compound	Tensile strength, 10^3 psi	Flexural strength, 10^3 psi	Flexural modulus, 10^6 psi	Notched Izod impact, ft-lb/in.
Compression molded				
Standard: all glass	8.16	16.9	2.23	4.9
G Wollastokup, 30%	9.46	15.9	2.19	4.02
10 Wollastokup, 30%	9.05	15.0	2.26	3.42
G Wollastokup, 50%	5.54	10.1	2.3	2.71
10 Wollastokup, 50%	6.31	12.6	2.20	2.68
Screw injected				
Standard: all glass	6.41	11.9	2.3	1.07
G Wollastokup, 30%	5.89	10.60	2.23	1.06
10 Wollastokup, 30%	5.29	8.52	2.18	1.01
G Wollastokup, 50%	5.21	10.60	2.14	1.02
10 Wollastokup, 50%	5.20	9.06	2.2	0.89
Plunger injected				
Standard: all glass	5.07	9.56	2.10	2.38
G Wollastokup, 30%	4.90	11.10	2.04	1.99
10 Wollastokup, 30%	6.91	13.10	2.19	2.18
G Wollastokup, 50%	4.95	11.20	2.18	1.69
10 Wollastokup, 50%	5.38	11.50	2.14	1.67

TABLE 68.6. Commercial Grades of Wollastonite

Mesh Aspect^a	50 Mesh	100 Mesh	160 Mesh	200 Mesh	300 Mesh	325 Mesh	400 Mesh	475 Mesh	1250 Mesh	20:1
NYCO	FW50	Other	G	X		NYAD 325	NYAD 400	NYAD 475	NYAD 1250	NYAD 20:1
Partek				FW200 W-10		FW325 W-20	FW400 W-30	475	1250	
Vanderbilt			Kem-olit	Kem-olit	Kem-olit		Kem-olit			
Wolkem 300				X	Kem-olit		Kem-olit			
Mexican				very coarse				60	100 160 200	

^aAspect ratio is a measurement of the average particle length divided by the average particle width.

786

TABLE 68.7. Effect of Wollastonite on Composites

Property	Effect of Wollastonite	
	Untreated	Treated
Tensile strength and modulus	Slight increase	Large increase
Flexural strength and modulus	Slight increase	Large increase
Impact strength	Decreases	Increases
Retention of properties at elevated temperatures	Increases	Large increase
Retention of properties under moisture	Slight increase	Large increase
Dimensional stability	Slightly improves	Large improvement
Thermal durability	Slight improvement	Large improvement
Heat-deflection temperature	Increases	Large increase
Surface appearance	No effect	Improves
Electrical properties	Improve	Large improvement

wollastonite in place of asbestos is a loss of tensile strength at elevated temperatures. This is overcome through use of other reinforcing additives, generally in very low amounts. Table 68.4 illustrates this phenomenon and shows the 100% improvement in retention of tensile properties gained through the use of G Wollastokup.

Another application where surface-modified wollastonite has shown benefits is in unsaturated polyester bulk molding compound (BMC) formulations (Table 68.5). Surface-modified wollastonite is particularly effective in replacing 25% of the chopped fiber glass in BMC formulations. By using wollastonite, the formulator can achieve a dramatic reduction in raw material costs with little to no sacrifice in physical properties.

The reduction of fiber glass loading must be achieved through use of surface-modified wollastonite for several reasons. First, merely reducing the fiber-glass level by 25% would require incorporation of a larger volume (because of specific gravity differences) of other, nonfunctional fillers such as calcium carbonate and clay. This will destroy properties. Secondly, the wollastonite should be pretreated to permit rapid and uniform dispersion in the paste. Use of nonsilane modification or in situ modification will require higher energy inputs and probable breakdown of the fiber glass fibers.

68.7 COMMERCIAL GRADES

See Table 68.6.

68.8 COMPOSITE CHARACTERISTICS

See Table 68.7.

68.9 TRADE NAMES

NYAD is the trade name of the wollastonite produced by NYCO, Willsboro, N.Y. Other manufacturers are given in the Table 68.6.

69

COLORANTS

J. Michael McKinney

PDI, Business Unit of ICI Americas, 54 Kellog Court, Edison, NJ 08817

69.1 INTRODUCTION

69.1.1 Color and its Measurements

The perception of color is dependent upon the observer's eye, brain, light and the object itself. Portraying a piece of plastic relates to its shape, size, gloss, opacity, surface structure and color.

Color is a personal sensation that is perceived differently by each individual. The stimulus producing the color response is light. One of the essential characteristics of a pigment is that it provides a certain color when illuminated.

Hue, lightness, and saturation are usually sufficient to describe the color of an object. Hue refers to the dominant wavelength of the color such as purple, red, yellow except for black, white, and gray. Lightness is a description of the amount of light reflected by a colored object as compared to a black to white scale. Light colors reflect more light and appear clearer and brighter than dark colors under the same illumination. Saturation simply refers to the strength of a particular color.

The wavelengths of the electromagnetic spectrum from dark red to violet are the ones that give the sensation of color. A beam of white light can be refracted by a prism into its constituent colors and can be recombined to produce white light. However, white light can be produced in addition to all other colors from the three primary colors, red, green, and violet.

The most common color vision defect is when the recognition of red or green is diminished. Of the 8% of the population that suffers from some form of incurable color blindness, 95% are men. Color blindness is carried on the sex link character and is hereditary.

Since "normal" color vision is the average color vision of the population, then there is a range of color vision that is considered good. Two normal people looking at the same object may describe the color slightly different. When using a mechanical instrument to color match the impact is corrected by using a "standard observer" in the instrument.

69.1.2 Light

White light is sunlight reflected back from the sky whose wavelengths fall between 380 and 770 nanometers.

The ratio between the speed of light in a vacuum and its speed in a particular medium is known as the index of refraction of the medium. The greater the index of refraction the greater the degree the light is deflected when leaving or entering that medium. Dispersion is when a light beam of more than one·frequency is split into a corresponding number of beams upon refraction. When light is refracted through a prism the bank of colors that emerge are known as a spectrum. The colors of the spectrum are red, orange, yellow, green, blue, and violet.

Monochromatic light is rarely seen. When certain wavelengths are removed from white light it appears to have a color. So the more continuous the spectrum of light the better the light simulates actual day light, and the more "natural" objects appear.

Because of the inconsistencies in natural daylight it is not advisable to do critical color matching by daylight. Daylight is affected by the seasons, time of day, weather, pollution and other elements in the environment. The best light for colormatching plastics is a well-maintained standardized source of artificial daylight.

Fluorescent lamps are of different wavelengths than daylight and should not be used for color matching. Their only use is to know what the object will look like under fluorescent light.

When all the white light striking an object is reflected the object is white. A black object absorbs all the white light striking it and nothing is reflected back. The absorbed light is converted to heat. When light is differentially absorbed and reflected from an object it changes the spectrum reflected back causing color. For example, a yellow object absorbs all the wavelengths of light except those in the yellow spectrum which it reflects as yellow.

Fluorescence is absorbed radiant energy that is re-emitted at longer wavelengths, usually in the visible region. Optical brighteners are objects that absorb ultraviolet light and emit light in the blue region.

Scattering refers to the redirection of light. Scattered light is redirected through a medium so that a portion of it the light is traveling in all directions. The pigment and/or the plastic itself gives the object its degree of opacity. The scattering of light inside a plastic object which emerges is diffusely transmitted and has the color of the plastic.

An object in which there is no internal scattering yet transmits light is transparent. Unpigmented general purpose styrenics is an example. They appear as if one were looking through a glass. A translucent material is one in which there is a substantial scattering, resulting in a considerable amount of light being reflected. When no light is transmitted because there is so much scattering or because all the light is absorbed, the object is opaque.

Specular reflection occurs when light strikes a smooth surface and reflects it back like a mirror. Specular reflection leads to gloss. Specular reflection is the same color as the light source.

69.1.3 Colorants

Materials that give color and opacity to plastics are chemically characterized as either pigments or dyes. Pigments are finely pulverized natural or synthetic particles which may be of inorganic or organic origin and are insoluble in the matrix in which they are dispersed. Besides providing hue and hiding power, they can enhance chemical properties. The definition includes colorless extenders (ie, calcium carbonate, and silicates). The significant difference between pigments and dyes is that dyes are soluble in the matrix in which they are dispersed, pigments are not.

Pigments containing carbon are organic; those without carbon are inorganic. The pigments used in coloring plastics are produced synthetically. Organic pigments are brighter, cleaner, and richer than their inorganic counterparts. However, they are also more easily affected by the destructive forces of heat, sunlight, chemical attack and migration, and they are more expensive. Performance characteristics, such as chemical and heat resistance, migration tendencies and lightfastness, are contingent upon its chemical composition. Chromophores are specific groups of atoms that determine a pigment's color.

Organic pigments that contain no inorganic constituents are known as toners. Maximum tinctorial strength is achieved when a toner is used. A lake is an organic pigment prepared by combining, extending, or diluting a toner with inorganic base. More than 90% of U.S. production of organic pigments are toners; the rest are lakes.

Inorganic pigments produce opaque dull colors. They have excellent resistance to heat, light, chemical aggression and migration. Typically, inorganic pigments have better exterior weathering durability and have higher densities than organic pigments.

69.2 PIGMENTS AND DYES

69.2.1 Major Organic Pigments

Permanent Red 2B is a mono azo pigment that is widely used in thermoplastics industry because it is inexpensive and has high tinting strength and good bleed resistance. The barium and calcium salts of Red 2B have higher color intensity but inferior lightfastness than the manganese salt derivative. However, the manganese salt accelerates the degradation of ABS, HIS, and polyethylene. The heat stability of this pigment is adequate for many applications.

Lithol Reds are frequently used because they produce sharp strong colors at a low cost in applications where stability is not critical. They are used frequently in PVC.

Quinacridones provide several shades of red that have good heat stability, good light-fastness and bleed resistance in vinyls, polyolefins and polyesters. However, in other systems (such as acrylics, nylons, and polystyrene, where the processing temperature range is high) only the violet and blue shade quinacridones can be used.

Carbozole dioxazine possesses a violet hue. It is used extensively in applications where excellent bleed and heat resistance are required along with good lightfastness. It is such a strong pigment that it is commonly used to redden phthalocyanine blue.

Perylene pigments are diimides of perylene tetracarboxylic acid that provide shades of red or maroon. They provide excellent resistance to chemicals, migration and light. Phthalocyanine green and blue are brilliant colors characterized by being very heat resistant and lightfast. They are the workhorses of the plastic industry. The crystal structure of phthalocyanine determines its color. Phthalocyanine blue has a shade range from red to peacock green, phthalocyanine green has a narrower range from blue to yellow shade. However, the green pigment is more stable than the blue. Shades of the blue pigment are subject to change due to crystallization whereas the green pigment's shades arise from halogenation with chlorine and bromine.

Diarylide yellows and oranges, however, are diazos that allow for many shades and a wide range of properties. Lightfastness, migration, and heat stability are a function of the chemistry involved. Diazo condensation pigments and tetrachorisoindolinones pigments are yellow pigments offering improved performance properties.

PY17 is used extensively in vinyls because of its strength and resistance to migration. HR yellow PY83 is a reddish-yellow that is highly transparent and has fair heat resistance.

Easily dispersed pyrazonone oranges have excellent heat and chemical resistance. They are used widely in polyethylene, polystyrene, vinyls, polyester, phenolics and polypropylene.

Perinone orange, PO43, is an expensive vat pigment that has excellent heat, light and migration resistance is widely used in vinyls.

69.2.2 Inorganic Pigments

Natural inorganic pigments are of mineral origin, mined in the chemical form in which they will be used. Synthetic inorganic pigments are manufactured from inorganic raw materials, making them finer than their natural counterparts. These finer particles are easier to disperse. Chemical resistance is a function of their chemical makeup. For example, iron blue pigments exhibit poor alkali resistance but ultramarine blue is highly stable towards alkalies but sensitive to acids.

Cadmium sulfide and blends of it with cadmium selenide provide a wide array of colors from greenish yellow through orange, red, and maroon. Cadmium sulfide is naturally golden yellow; selenium gives the red shades.

The cadmiums are easily dispersed and provide vivid colors. They are used extensively in thermoplastics because of their excellent resistance to heat and chemical attack. Although cadmium and selenium are considered toxic they are so inert that they can be used in all plastic applications not having direct contact with food.

Blends of cadmium sulfide with cadmium selenide and zinc sulfide (extended with barium sulfate up to 60% by weight) are how lithopone type cadmiums are supplied. This extension with barium sulfate improves dispersibility and economics.

Lead chromate pigments are economical inorganics that yield primrose, lemon yellow, medium yellow, and orange shades. A large range of yellow colors can be produced using these pigments. Although chromates yellow in the presence of atmospheric sulfides and are sensitive to acids and bases, surface treatment of the chromates has produced pigments that are more chemical resistant, heat resistant, and have improved lightfastness.

Another important lead-containing pigment is molybdate orange. This easy-dispersing pigment is often blended with other inorganics and organic pigments. The strong light orange to red colors provided by inexpensive molybdate pigments, coupled with technology that has improved their chemical and heat resistance makes them broadly applicable. Commercially available molybdate orange contains lead chromate, lead molybdate, and lead sulfate.

Cobalt blue is an aluminate that comes in shades from blue to turquoise. Although expensive and weak in tint strength, cobalt aluminate pigments have excellent heat and light stability and chemical resistance. Migration resistance and weatherability are also excellent. Such pigments generally are selected for applications in which the need for performance outweighs cost.

Sodium aluminum sulfosilicates are characterized by permanence, heat resistance, and alkaline stability. By removing sodium during production of this pigment the color shifts from blue towards a pink and violet, which is the commerical color range. They provide a lightly reddish blue pigment that is clean and vivid. However, poor acid resistance and tint strength restricts their use in plastics.

Chromium oxide green is a dull colorant with remarkable resistance to sunlight, heat, acids, and bases. This green pigment is capable of reflecting infrared lighting, making it a colorant of choice for military camouflage.

Chrome green, an altogether different pigment from chromium oxide green, is made by mixing chrome yellow with iron blue. Because this class of pigments is cheap and exhibits good hiding power, it is used in applications where good heat stability and lightfastness are not required. A typical plastic application is in unsaturated polyesters.

Synthetically produced inorganic white titanium dioxide is nontoxic and stable. TiO_2 is widely used as an exteriors colorant because of its outstanding resistance to the destructive forces of sunlight. Two different crystal forms of TiO_2 exist: anatase and rutile. These two species render different chemical and physical properties. Very fine-particle-size rutile TiO_2 (0.2 to 0.35 microns in diameter) is used to provide high strength and blue tone. Superior weathering is a feature of rutile. Rutile also is responsible for giving opacity, brightness, and whiteness to plastics. However, the less abrasive anatase type is an excellent choice where resistance to chalking and ozone is important.

Carbon black is an organic pigment produced by the incomplete combustion of solid, liquid, or gaseous hydrocarbons. There are five classification of carbon blacks; namely, lampblack, channel black, thermal black, furnace black and acetylene black. Each of these blacks finds applications in different markets.

Carbon blacks require lots of energy to achieve a good dispersion, but they are low cost and perform as an excellent heat and light stabilizer. Around 250 shades of blackness or jetness are visible to the naked eye merely by going from the smallest particle size to the coarsest. Usually only 1% carbon black will produce a masstone black part. The major considerations in selecting a carbon black are particle size, structure, surface area, and pH.

69.2.3 Characteristics of Dyes

Dyes produce transparent effects with maximum strength and brilliance. Unlike pigments, dyes cannot protect the resin from ultraviolet-light degradation. However, many dyes are satisfactory in terms of their own lightfastness. When tints of the dyes are made, dye fading is more easily seen.

Dyes are candidates for coloring thermoplastics because of their good to excellent heat stability. Azo dyes are used widely in rigid PVC, styrenics, acrylics, styrene and phenolics because they are readily available, have good fastness qualities, and are economical. Anthraquinone dyes have been adapted as plastics colorants because of superior heat and light stability (even though they cost more). Dyes sensitive to heat and light, such as xanthenes, are used in acrylics, polystyrene, and rigid PVC because of their brilliance and fluorescence.

These three classes of dyes constitute about 60% of the commercial dyes available. The remaining types of interest in the plastic industry are azine, aminoketone, perinone, indigoid, triarylmethane, quinoline, monomethine, indophenol, phthalocyanine, and Fluorescent Brightener 61.

The foregoing information on pigments and dyes is meant to serve as a representation of major types used in plastics. Full details and specifications are beyond the scope of this article, but they can be found in the *Colour Index* (CI), a renowned publication of colorants that provides generic names for identifying dyes and pigments with respect to use, and numbers for identifying them according to chemical composition. This index provides comprehensive data on the performance characteristics, chemical profiles, and use.

69.2.4 Colorant Forms and Functions

There are six basic ways in which color is supplied, each with its own advantages and disadvantages.

1. Dry color, as the name implies, is pigment added to resin for subsequent blending and processing applications.
2. Direct compounding is an inventory-efficient technique by which the customer adds the pigment to the resin inplant, in quantities sufficient for a specific product run.
3. Color concentrates are pellets containing a high level of pigment and are compounded directly into the resin in the molten state. The concentrate is then added to natural resin in predetermined rations dependent on desired color intensity in the finished product.
4. Color pastes are pigments or dyes that are dispersed in a liquid vehicle (such as a plasticizer) compatible with the plastic resin. Pigment loadings of pastes are usually very high.
5. Liquid colors are pigment dispersions in vehicles that possess universality with respect to the resin. These too are of a very high loading and are pumpable.
6. Freeze-dried colors are pigments dispersed in a vehicle that solidifies at room temperature or when chilled. They are usually commercialized in flake or chip form.

Dry color can be a pigment or blends of pigments mixed (usually with a metallic stearate) to facilitate dispersion. Blends are sold in preweighed packets or in bulk quantities in pails. The processor adds dry color to the base resin and tumbles them in drums to thoroughly disperse the color/resin mixture to be fed into the hopper. The advantages of dry color are that large and small quantities are easily accessible at an economical cost. And because the colorant agent is 100% color, storage space is minimal.

Dry colors have some disadvantages. The dust they generate can be a health or cleanup problem. Dust can also contaminate other colors and/or manufacturing equipment. Cosmetic imperfections, such as streaking and specking, can occur when dry color is used; this is a result of dry pigment particles being agglomerated rather than being broken down as small as possible. When pigment particles are used in the agglomerated state their cost effectiveness as colorants is reduced. Finally, the time it takes to purge an injection molder to go from one dry color to another is longer than when color concentrate or dispersed color is used.

Concentrates are melt bends of resin and pigment mixed with a given weight of virgin carrier resin. The advantages of concentrates are ease of handling and color consistency from batch to batch. The negative aspects are that concentrates are relatively expensive and cannot be used with all resins, thereby increasing inventory needs. Also the lead time to fill orders can be lengthy, and the customer is generally required to purchase a minimum quantity of a special color. Metering systems for concentrates involve moderate to high capital expenditure.

Liquid colors provide a great variety of pigmentation possibilities for thermoplastic resins, including engineering grades. A simple yet extremely accurate peristaltic pump has been designed to meter liquid colors to the throat of the plasticating screw. This provides for very critical reproductions in color matching. Liquids require less downtime than any other system when cleaning or purging. A wide array of liquid colors is available and compatible with most thermoplastic resins—therefore reducing inventory and maximizing usable plant space. Once the system has been set up, virtually no supervision is needed.

69.2.5   The Importance of Dispersion

Dispersion, the procedure for producing color concentrates and liquid color, involves three distinct steps. Initial wetting, deagglomeration and air displacement are the prerequisites to disperse a pigment fully and develop it to its maximum potential.

Initial wetting involves blending the vehicle with the pigment. At this stage, the compatibility between pigment and vehicle is readily seen. If insufficient wetting is achieved the pigment may be poorly dispersed no matter how good the following dispersion steps are. Therefore, precise care must be taken in preparation of the premix.

Deagglomeration (size reduction) is the breaking up of clusters of individual pigment particles held together by surface attraction. The right amount of mechanical energy must be applied to isolate the pigment particles so they can be surrounded by vehicle. The mechanical energy required to deagglomerate a pigment is critical, since over-shearing can degrade the dispersion. In order to obtain maximum yield, it is necessary that the dispersion requirement of each pigment be evaluated.

Replacing the air on the pigment surface with vehicle is paramount in developing the color of pigments and is synonymous with intimate wetting. Unless the pigment particles are intimately wetted there is no shearing effect on the pigment aggregates when mechanical energy is dissipated into the liquid dispersion.

Color yield by a pigment dispersion is a function of particle size distribution. Large particles (15 microns) will have rough surfaces with streaking. Small particles (0.3 microns) give maximum intensity accompanied by high transparency. But moderate to small particles result in high hiding power and tinting strength. Measuring the effect of the pigment particles (not particle size) is the method by which to determine the degree of dispersion. The Hegman grind gauge is a common tool used to measure degrees of dispersion. Preparing tints (colors let down with TiO_2) of colored dispersion is another route for determining the degree of dispersion if a control is available.

Master batching is another method of making quantities of a particular color in a particular resin. The larger the batch purchased, the greater the quantity of consistent-color plastic that can be achieved. However, this requires a lot of inventory and space. Usually it is not profitable to fill orders of the colorant in small batches. Special colors must be ordered in large quantities, and they sell for a premium.

69.2.6 Coloring Do's and Dont's

The compatibility of components used in a color system should be known or checked because unexpected problems may occur in production or product use. A liquid color's vehicle must be compatible with the resin. Otherwise the plastic may exude the vehicle. Oil-soluble dyes are not compatible with olefins. They bleed severely. Cadmium pigments are acid sensitive and should not be used with acidic pigments or else the cadmiums will degrade. Phthalocyanine blue causes HDPE to warp. Fluorescent pigments tend to solvate, so certain additives cause these pigments to lose their fluorescence. Red 2B yellows with the addition of zinc stearate. Chrome yellow pigments, which suffer from poor alkali resistance, cannot be used with alkaline extenders such as calcium carbonate.

When color matching in thermoplastic resins such as ABS, SAN, thermoplastic urethane and polycarbonate, it is necessary that these resins be dried before using. Otherwise, residual moisture will cause streaks resembling stress cracks in the finished article. Pigments used in urethane formulations must be dried so that the moisture does not interfere with the cure.

Exposure to outdoor light can dramatically alter color stability. Some pigments get darker, some fade, others change color entirely. The lightfastness of a color is dramatized when it is mixed with a white pigment or when a low-pigment-volume concentration is used.

Migration is the mobilization of some component of a plastic to the surface. There are four subcategories of migration or leaching that are known to occur. *Bleeding* describes diffusion of a colorant from its substrate to an adjacent object. *Crocking* is when the pigment may be removed from the surface of the plastic by rubbing it. After the removal of the color pigment by abrasive rubbing, the surface takes on an appearance of metallic bronze and is known as *bronzing*. Another condition known as *plateout* occurs when an undesirable film forms on the surface of the metal die or calendaring rolls during processing of the plastic.

Heat resistance of a pigment should be greater than that of the processing condition requirements of the plastic. The length of exposure of the pigment to high temperatures is critical when color matching. Pigment burnout occurs when the pigment becomes unstable for the processed time and temperature. Inorganic pigments perform much better in regard to heat resistance than the organics. Most organic pigments are stable for only a few minutes at a maximum of 450°F (232°C), the inorganics generally can be held for over 550°F (288°C) for 5 to 10 minutes.

For decorative effects where opacity is desired, it is important to use enough pigment to produce the desired effect. It is possible to use more pigment than is actually required to achieve opacity. This in turn affects the economics of the finished product. Generally, inorganics alone are sufficient to produce good opacity. But most organic pigments and dyes require opacifiers such as titanium dioxide in order to achieve good hiding power.

Toxicity is an important concern. The U.S. Food and Drug Administration (FDA), under the Federal Food, Drug and Cosmetic Act, makes certain that all the substances in colorants (pigments, vehicles and additives) are acceptable for incorporation into plastics intended for direct food contact. If no prior sanctions exist, the company desiring to use a particular color must apply for and receive approval under a lengthy and complex test protocol.

Occasionally, the ideal colorant for a system is cost-prohibitive and cannot be used. The formulator must then prepare a color match pooling experience and knowledge that will provide the user with all the properties needed to withstand the application's environment. Compromises must be made between seller and user regarding the colorant's performance properties.

69.2.7 Color Measurement and Matching

There are two basic ways of evaluating color: optical and mechanical.

Measuring and describing color can be performed either with a machine or person using a standard as a reference point. In either case catalogs with visual comparison of samples provide useful descriptions of color.

For example, the Munsell book of color contains over 1500 color chips in matte and glossy surfaces. In this systemized book the test swatch and the colored sample in the book (which has a neutral background) are laid side by side and compared in daylight. From this comparison any color can be accurately described by using the Munsell color notation. This notation describes the hue with a number ranging from 1 to 100 or with a letter or letters plus a number. The value scale indicating lightness is characterized by a number where 0 is black and 10 is white. The numbers 0–16 are used to describe the chroma, Munsell's term for saturation. Ten letters make up the hue circle, R, YR, Y, GY, G, BG, B PB, P, and RP which are each divided into 10 parts with some containing decimals such that 10G is the same as OBG. The system of characterizing hue by a number and letter is popularly used in many color notation systems.

The Universal Color Language is another system: it provides a description of colors at a variety of complexity levels. Level I has 13 adjectives describing the hue. Level II increases the accuracy by the addition of 16 more descriptive color adjectives. The Inter-Society Color Council—National Institute of Standards and Technology, which has 267 names for colors, is level III. Level IV contains the 1500 color samples from the Munsell Book of Color. At Level V, a comparison of a colored sample against the nearest color in the Book of Color is required. From the difference in colors from a visual judgment a new value is assigned to the color in question. This allows for roughly 100,000 different colors to be assigned a number. An instrument that measures color differences with mathematical expressions is necessary for Level VI, which allows a few million different colors to be characterized.

Color matching requires that a uniform way of describing color differences or tolerances be postulated. In plastics coloration, allowable tolerances be agreed upon in advance between manufacturer and purchaser. Computer readouts of color matches against a given control allows us to numerically assign a value of just noticeable differences (JND).

David MacAdam has proposed an equation for measuring color differences which is now widely used. Hugh Davidson and Fred Simon have modified this formula so that lightness is incorporated into the effect on color differences. These are called MacAdam units and are used throughout the plastic industry.

The International Commission on Illumination has founded a mathematical treatment of color data so that a common principle can be used in calculating color. The committee has established tristimulus values to describe color, calculated from the transmission values of a sample. The values are designated X, Y, and Z, which correspond to red, green, and blue respectively. Two samples that have the same tristimulus values are considered matches.

There is now a wide variety of instruments that provide an objective way of describing color. And the description of a color can be easily verified from lab to lab.

The spectrophotomer is used in color matching and it is capable of measuring a spectral reflectance curve. The price of these valuable instruments has decreased over the years, now making them affordable to just about any serious color-matching laboratory. Because the spectrophotometers were rather expensive years ago, some laboratories purchased tristimulus colorimeters. Although these colorimeters accurately measure color differences, they cannot measure spectral reflectance curves and subsequent tristimulus values.

Color matching involves a customer sending a sample to a coloring house and specifying how critical the match must be before it will buy the colorant.

The phenomenon of two objects being a color match under a certain light but looking different under another illuminant is called metamerism. If a color match is made using the same pigments and resin, the wavelengths reflected or absorbed are the same regardless of what illuminant is used. In this case, metamerism is eliminated. Preventing a metameric match is a primary consideration in custom color matching. Because one cannot visually see spectral reflectance curves it is necessary to use a spectrophotometer. Whenever the spectrophometric curve of the standard intersects with the curve of the sample a metameric condition exists. The conditions for producing a nonmetameric match require that the same pigments and resin be used. The skilled color matcher can usually determine what pigments are present in the plastic.

Degree of dispersion is important in color matching because the color becomes fully developed only when it is properly dispersed. Maximized color development minimizes the chance of metamerism. A good pigment dispersion supply house can do the job. But color matching must always be performed in the same resin the customer is using. Otherwise the match may appear different and be judged unacceptable.

BIBLIOGRAPHY

General References

G. Liobrowski, *Pigment Handbook* Vols. 1 and 2, 1973.

F. W. Billmeyer, Jr., *Coloring of Plastics* (1979).

T. G. Webber, *Coloring of Plastics* (1979).

G. J. Chamberlin and D. G., Chamberlin, *Colour, its Measurement, Computation and Application,* 1980.

Definitions Committee, of the Federation of Societies for Coatings Technology, *Paint/Coatings Dictionary* (1978).

G. Kuehni, *Computer Colorant Formulation* (1975).

J. Boxall and J. A. Fraunhofer, *Concise Paint Technology* (1977).

T. C. Patton, *Paint Flow and Pigment Dispersion* (1964).

70

REINFORCED PLASTICS, INTRODUCTION

Gerald D. Shook

6528 Cedarbrook Drive, New Albany, OH 43054
The sections on reinforced plastics were prepared with the assistance of Mr. Shook.

In 1988, approximately 2.7 billion lb (1.2 billion kg) of all types of reinforced plastics were shipped including their reinforcements and fillers. The three largest markets were transportation at 0.6 billion lb (0.27 billion kg), construction at 0.5 billion lb (0.23 billion kg), and marine at 0.4 billion lb (0.18 billion kg).

The basic resin matrices for reinforced plastics include alkyd polyesters, epoxies, and vinyl esters. (Their advantages and disadvantages are explored elsewhere in this Handbook). Glass fiber remains the primary reinforcing material, offering a useful combination of performance and economy. However, market demand for assured product integrity under high stress for long periods has created a revolution in high performance reinforcements, including such newer fibers as boron, silicon carbide, carbon/graphite, and polyethylene. There also are new forms in which fibers are placed, such as paper, woven and stitched, and hybrid fabrics.

Many of the high performance reinforcements are used in advanced composites (which constitute only 1–2% of the physical volume but 8% of the dollar volume of reinforced plastics). Their market success is alerting designers to the fact that RP is a strong alternative to many conventional materials. Because reinforced plastics with controlled physical properties can now be molded to shape, eliminating significant assembly, the final product may well be cost-effective even though process and materials may be comparatively expensive. Lightweight strength is an important common asset (see Fig. 70.1).

Advance composites are proving themselves in such critical markets as aerospace, transportation, sporting goods, construction, marine, and pipe. As the costs of boron, carbon/graphite, aramids, and other fibers come down, the technology will spread to broader consumer-goods areas.

A very brief description of different glass-fabric weaves, and an outline of the major reinforced-plastics processes follow.

70.1 FABRIC WEAVES

Plain weave is the oldest and most common style; a basic textile weave wherein one warp thread weaves under, then over one fill thread, etc. This type of fabric is firm and stable, has fair porosity for ease of air removal in laminating, and exhibits balanced strength along its axes.

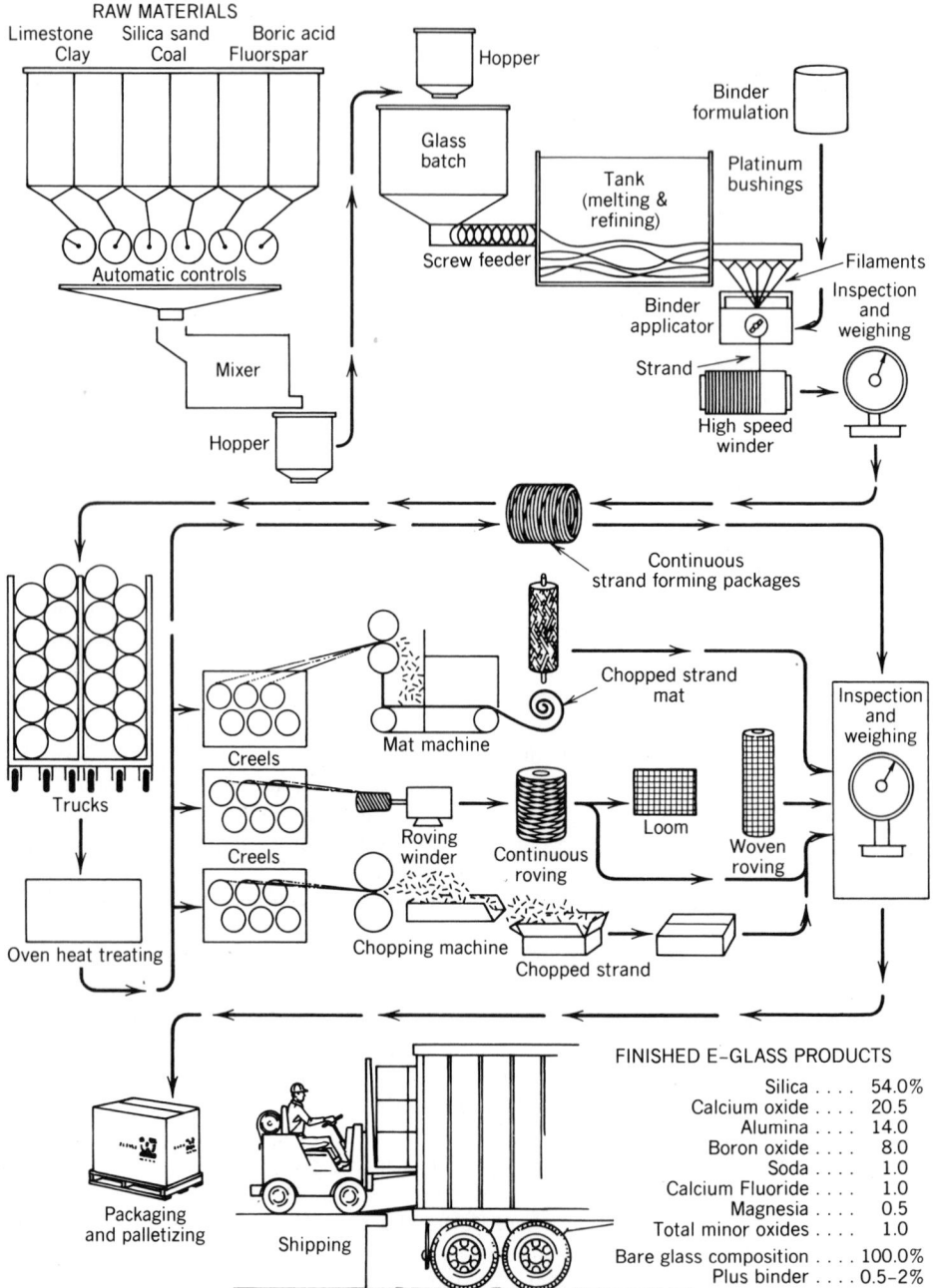

Figure 70.1. The fiber-glass reinforcements manufacturing process.

Basket weave is similar to plain weave, but has two or more yarns weaving as one warp or fill. It is less stable than plain weave, more pliable, and flatter and stronger than equivalent weight or count of plain weave.

Crowfoot satin weave is constructed with one warp and weaving over three and under one fill. It is more pliable than plain or basket weave; it will lay over complex shapes without wrinkles, and it can have more threads/inch in the fabric.

Long-shaft satin weave is constructed with one warp end weaving over four or more and under one fill. It is the most pliable (conforming readily to compound curves); it produces laminates with high strengths; it can be woven with highest density of fill and warp, and is less open than other weaves.

Unidirectional weaves can be adapted to any weave pattern. It is characterized by the gross difference in yarn size between warp and fill. It offers maximum strength in one direction, and imparts high impact resistance to the laminate.

70.2 RP PROCESSES

Hand Lay-up. Hand lay-up also called contact molding is a process wherein the application of resin and reinforcement is done by hand onto a suitable mold surface. The ensuing laminate is allowed to cure in place without further treatment (see Fig. 70.2).

Process steps:

1. Wax mold and/or apply a release agent.
2. Apply gel coat if required.
3. After gel coat cure, apply first coating of catalyzed resin.
4. Apply reinforcement, press into resin and work out the air by brush or roller.
5. Repeat steps 3 and 4, as required.
6. Allow to cure undisturbed.
7. Remove from mold and trim.

Basic tools for hand lay-up include: the mold (almost anything prepared with a parting agent); resin mixing and measuring equipment; rollers, spatulas, blades, scissors, knives; and proper solvent for cleanup.

Advantages of hand lay-up are that it is the simplest RP process, requiring the lowest investment, and having no size restrictions (literally). Design flexibility, acceptance of gel coats, and easy finishing—eliminating sawing or grinding are other advantages. Gel coats are possible.

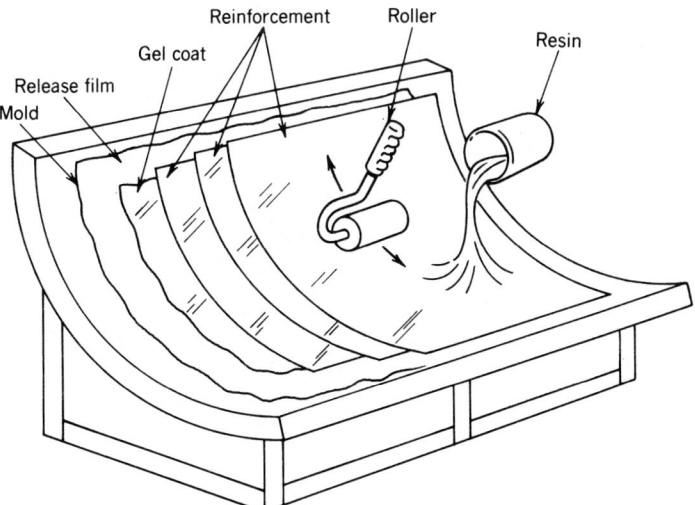

Figure 70.2. Hand lay-up process.

Disadvantages are that scrap is variable and rarely predictable, waste is high (unless carefully monitored), close supervision is required, parts have only one finished surface, and product quality depends on operator skill.

Almost every RP product is first made by this process, including such large items as boats, tanks, flat panels, tooling, and prototypes of all kinds.

Typical physical properties are

Property	Mat	Fabric
Glass, %	20–40	45–55
Tensile strength, 1000 psi (MPa)	10–20 (69–138)	30–50 (207–345)
Flexural strength, 1000 psi (MPa)	20–40 (138–276)	45–75 (360–517)
Tensile modulus, 10^6 psi (MPa)	0.8–1.8 (5500–12,400)	1.5–4.5 (10,000–31,000)

Vacuum Bag Molding. An airtight film is placed over a hand lay-up; air between the bag and the mold is evacuated and excess resin and bubbles of air in the laminate are removed through the combination of atmospheric pressure and hand working (see Fig. 70.3).

Process steps:

1. Make a lay-up as in hand lay-up.
2. Place a bag over the lay-up and seal the edges to the mold.

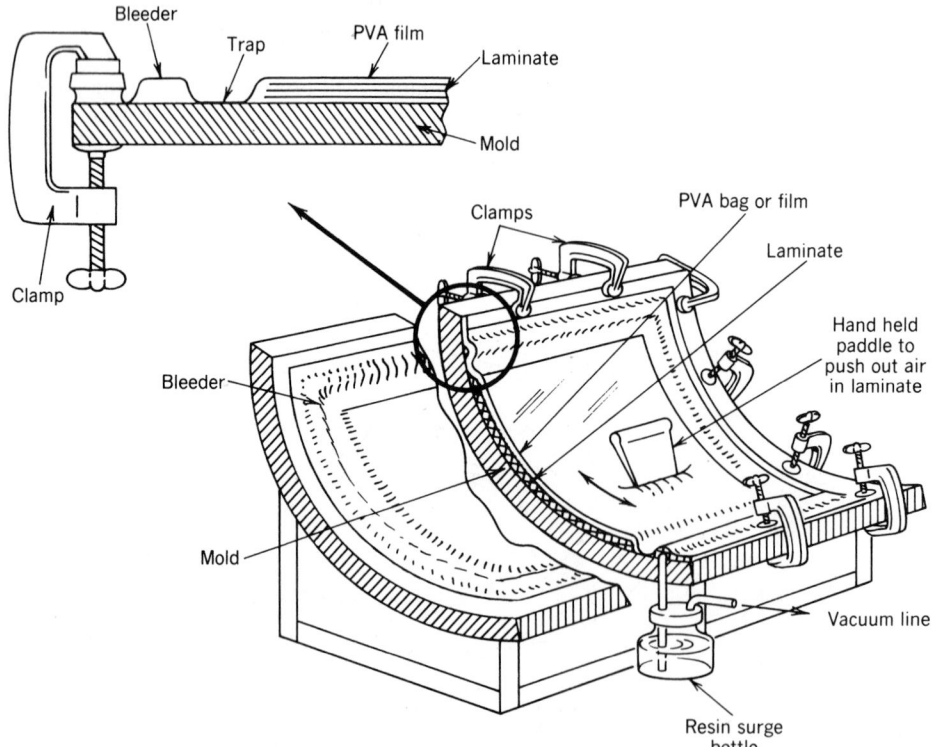

Figure 70.3.   Vacuum bag lay-up process.

3. Draw a vacuum.

4. Hand work the lay-up through the bag to push out excess resin (as well as air in the laminate).

5. Allow to cure.

6. Strip bag from the cured laminate, remove from the mold, and trim.

Basic tools include: suitable mold with large flanges (to permit anchoring the bag to the mold surface); bag clamping or sealing means; special paddles to work the laminate through the bag; vacuum lines, vacuum pump and reservoir, and bleeder material.

Advantages of vacuum bag molding are that it produces a void-free laminate per Mil-P-8013, provides higher glass ratios, retains most of the advantages of contact molding, and makes superior sandwich constructions.

Disadvantages are that it generates much materials waste, requires highly skilled operators, has a slow production rate, and cannot knife trim parts.

Major uses are military and aviation applications where cost is secondary and high physicals and low weight are important.

Physical properties are approximately the same as the higher levels of contact layup.

Pressure Bag Molding. This process wherein the laminate is densified by vacuum under the bag and pressure is applied over the bag to produce parts with higher density and better physicals than vacuum bagging alone (see Fig. 70.4).

Process steps:

1. Make laminates as in contact molding.

2. Place bag over lay-up and seal to the mold.

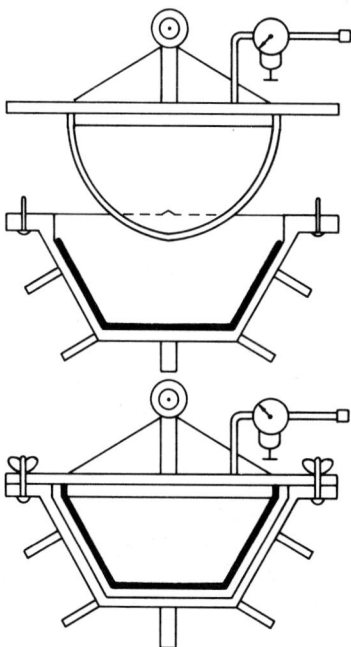

Figure 70.4. Pressure bag molding.

3. Draw vacuum.

4. Handwork the lay-up through bag to push out entrapped air and excess resin.

5. Place pressure plate over the bag and seal to edges of mold.

6. Apply pressure [50 psi (0.3 MPa)] between plate and bag, and allow lay-up to cure.

7. Dismantle, remove pressure and bag and bleeders.

8. Remove part from mold; trim.

Basic tools are a pressure bag (usually neoprene or rubber) shaped to generally match part; a reinforced mold to take internal pressure forces of pressure plate; clamps and clamping ring; and same tools used in vacuum bagging.

Advantages are higher glass ratios, dense void-free molding, ability to make cylindrical shapes; acceptance of undercuts and cores.

Disadvantages are high labor cost, lower production rates, pressure bag is expensive and is limited in use.

Physical properties are

Glass content, % Mat-50%,	Fabric-65%, WR-60%
Tensile strength, 1000 psi (MPa)	40–50 (276–345)
Flexural strength, 1000 psi (MPa)	50–80 (345–552)
Flexural modulus, 10^6 psi (MPa)	2.5–4 (17,200–27,600)

Major uses are the same as for vacuum bag molding.

Autoclave Molding. This is an extension of pressure bag molding; a laminate is densified and cured by application of heat and pressure in a sealed container while under an impermeable membrane or bag to which vacuum has been applied (vacuum bagging in a pressure tank) (see Fig. 70.5).

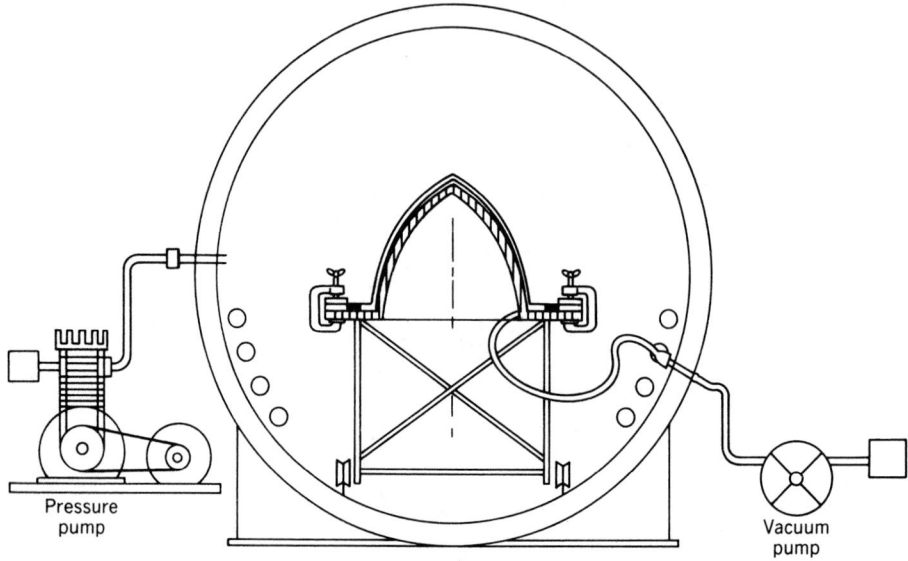

Pressure pump

Vacuum pump

Figure 70.5. Autoclave molding.

Process steps:

1. Make laminate on a mold as in vacuum bagging.
2. Transfer the bagged laminate into the autoclave with the vacuum line intact (usually the line comes from the autoclave and is merely pulled into the tank as the laminate is also entered).
3. Apply heat and pressure and allow to cure.
4. Open autoclave, strip bag, remove part and trim.

Basic tools include an autoclave of suitable size, accessory equipment (pumps, receivers, filters, heat supply, etc), vacuum pump and plumbing, and the same tools as for vacuum bagging.

Advantages are higher laminate density, cores and inserts can be used, many parts can be prepared and simultaneously cured in the autoclave.

Disadvantages are the process is labor-and capital-intensive, it requires very well-trained personnel, and part sizes are governed by the size of the autoclave. Also, molds are more costly because of the extra strengthening required. And, like vacuum bag molding, the process is wasteful of material.

Typical parts are rocket nozzles, nose cones, heat shields, electronic parts, and aircraft parts needing high specific strength. The process also is becoming accepted for less exotic industrial applications.

Physical properties are

Glass content,% Mat, 50%	Fabric, 65%	WR, 60%
Tensile strength, 1000 psi (MPa)	40–55 (276)	379
Flexural strength, 1000 psi (MPa)	50–80 (345)	552
Flexural modulus, 10^6 psi (MPa)	2.5–4.5 (17,200)	31,000

Filament Winding. In this process, continuous strands of reinforcement and other materials are wound on a mandrel, or core of suitable shape, and the materials are carefully positioned to develop required physicals as the mandrel rotates. Resin can be added during or after winding. Curing is accomplished on the mandrel (see Fig. 70.6).

Process steps:

1. A collapsible mandrel is provided for later removal, or the mandrel can become part of the object.
2. Filaments or tapes are wound onto the mandrel under pre-set tension. This reinforcement is usually impregnated with resin in a tank, or is prepregged as it is wound onto the mandrel.
3. Winding patterns are varied as required by design and winding conditions and continue until completed.
4. Laminate and mandrel are placed in an oven and cured while rotating. (For room-temperature cure, no oven is required).
5. After cure and removal of the mandrel, parts and trimmed are finished.

Basic tooling includes: winding machine with appropriate controls; mandrel; curing oven and controls; and resin mix and control equipment.

Advantages: Filament winding uses the lowest-cost form of glass reinforcement. It has high production potential, especially for pipe and tank shells, and yields products with high strength. There is no length limitation for pipes. Large tanks can be made, ie, 60 feet (18 m) in diameter, and it is extremely efficient for making pressure bottles.

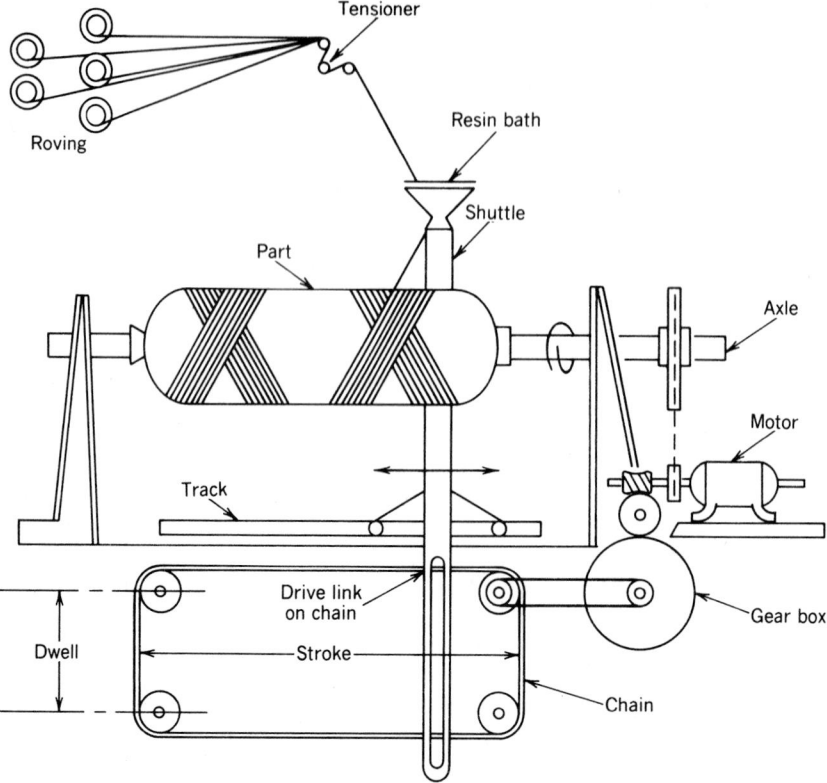

Figure 70.6. Filament winding process.

Disadvantages: Removing of mandrels is difficult. In pressure vessels the holes must be wound in the process and the general production rates are low. Unless laminates are fitted with a bladder porosity occurs, and finally the winding machines are very expensive.

Uses include pressure bottles, pipe tanks, bows, core and fuse tubes, and rifle reinforcements.

Physical properties are

Glass content, %	50–85
Tensile strength, 1000 psi (MPa)	80–250 (552–1720)
Flexural strength, 1000 psi (MPa)	100–200 (690–1380)
Tensile modulus, 10^6 (MPa)	5–7 (34,000–48,000)

Spray-up. Spray-up is a process wherein reinforcement and catalyzed resin are sprayed simultaneously onto a mold surface, densified and allowed to cure (see Fig. 70.7).

Process steps:

1. Prepare mold surface.
2. Test equipment to ensure proper ratios of catalyst and chopped glass.
3. Apply resin and reinforcement in a ventilated area in compliance with OSHA rulings.
4. Roll down to densify and remove air.
5. Allow to cure; and remove and trim.

Basic tools are spray-up equipment, hand tools (rollers, scissors, knives, etc); suitable mold and release.

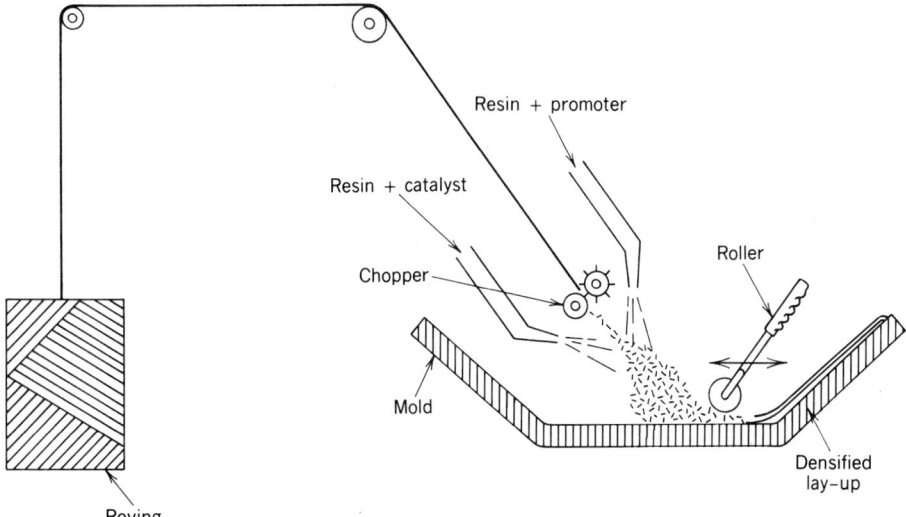

Figure 70.7. Spray-up process.

Advantages of spray-up: Systems are transportable; equipment cost is reasonable ($5000–8000); materials price is low; when automated, high production and part uniformity are possible; no special molds are required; and there are no limitations on part size.

Disadvantages: Tight tolerances are difficult to achieve; part quality is dependent on operator skill; product repeatability is poor; physical properties are low.

Uses are the same as those for hand lay-up, plus automatic spray-up of such large parts as panels, boats, and tanks, etc.

Physical properties are

Glass, %	25–40	
Tensile strength, 1000 psi (MPa)	9–18	62–124
Flexural strength, 1000 psi (MPa)	16–18	160–193
Flexural modulus, 10^6 psi (MPa)	0.8–1.2	5500–8300

Resin Transfer Molding. This is a closed-mold process in which dry reinforcements in the closed mold are injected with a filled catalyzed resin; cure occurs at room temperature (see Fig. 70.8).

Process steps:

1. Force and cavity are waxed, then sprayed with a gel coat.
2. After gel coat cure, dry reinforcements are placed on either force or cavity in a planned manner; molds die closed and clamped.
3. Prepared resin is then injected at 30–50 psi (0.2–0.35 MPa) until resin appears at all vent tubes. The vents are closed, and the part is allowed to cure.
4. Molds are separated and the part removed.
5. Part is trimmed to size and inspected.

Tools include: force and cavity; injection equipment (pump or air pressure pot), handling equipment (clamps, hoists, etc).

Advantages of resin transfer molding are that it permits good surface on both sides (colors can be different front and back) and tight thickness tolerances. Odor is minimal, and products

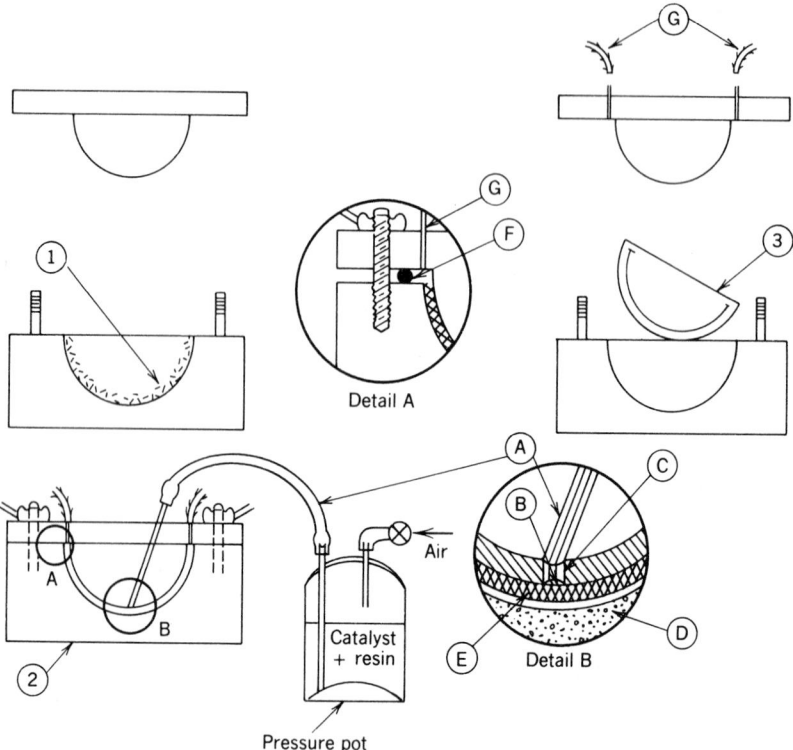

Figure 70.8. Resin transfer molding. 1. Place dry reinforcement. 2. Close mold, clamp, and inject resin. 3. After cure, open mold and remove part-clear vent. A, Probe; B, valve; C, valve body; D, cavity; E. reinforcement; F, seal; G, vent.

very similar to preform press molding can be produced with plastic tooling and room temperature cure, each of which contributes to production speed and parts economy. Other advantages include ease of insert molding, part uniformity nearly equal to that of preform press molding, the ability to make large parts [50–200 ft.² (4.6–18.5 m²)].

Disadvantages are that the process requires well-braced, heavy and costly molds. High glass ratios cannot be achieved, and improperly placed reinforcement can cause dry spots or resin pools. Thus, resin-transfer molding is not recommended for parts that will be highly stressed.

Resin-transfer molding fills the need for press molded parts in small quantities that do not justify metal tools. Physicals of molded products are comparable to those of preformed press molded parts.

Matched Die Molding. This is a mass-production RP processing method. Resin, reinforcement fillers, catalyst, color and internal release all are placed in a mold. When the mold closes, material flows to fill the cavity and is cured in place. Three principal molding materials are sheet molding compound: bulk molding compound or premix; prepreg materials; and preform mat and resin (see Fig. 70.9).

Process steps:

1. Heated, matched molds are set in a press of pre-set capacity, speed of closing, heat, and controls.

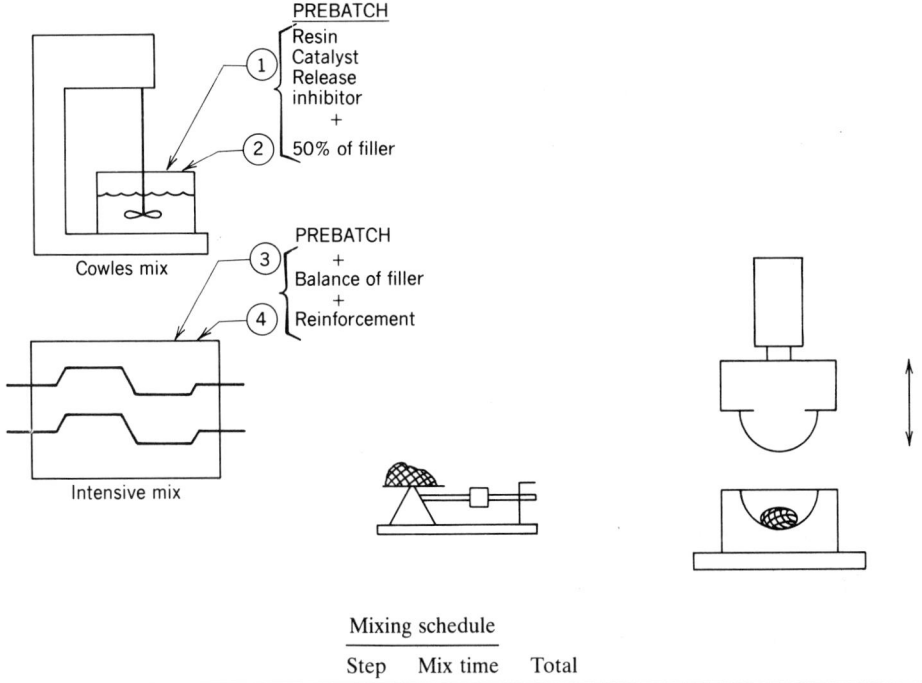

Figure 70.9. Premix molding.

2. Prepared resin-and-reinforcement mix is placed in the mold; the mold is closed.
3. After cure, mold is opened, part is removed, and mold is charged with fresh material for next cycle.

Basic tooling consists of the press; molds with proper attachments, knockouts and heat controls; and equipment for preparing molding materials for charging into the mold. Typical equipment for preparing BMC, SMC, or premix includes mixers, scales, drying oven, preheat oven, and SMC machine. For preform, equipment consists of preform machine with oven, preform screen, chopper/binder, and resin mixer. For prepreg, the essential equipment consists of refrigerator, clicker die machine, electric irons, and clicker dies.

Typical formulations are

BMC, SMC, or Premix-XMC[a]	Preform	Prepeg
Resin, 22%	Resin, 52%	Resin, 24%
Catalyst, 0.15%	Catalyst, 0.5%	Catalyst, 1.2%
Filler, asb., 5%	Filler, 21%	Filler, 12%
Filler, 10–60%	Binder, 1.2%	Color, 1%
Lubricant, 1.15%	Lubricant, 0.1%	Inhibitor, 0.5%
Glass, 10–60%	Glass, 25%	Glass, 61%

[a]A typical Premix formulation: polyester resin, silanc-treated glass fiber, kaolin, calcium carbonate or silica filler, BPO catalyst, zinc stearate lubricant, *t*-butyl inhibitor, and water soluble acrylic binder.

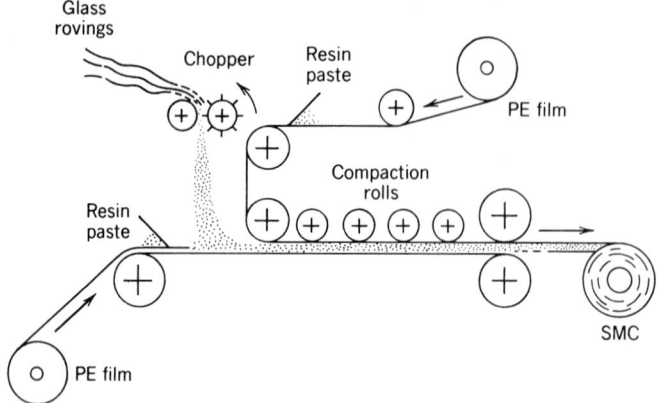

Figure 70.10.   The SMC process.

Sheet Molding Compounds. SMC are essentially the same as premix, except that in SMC the reinforcement is usually formed from chopped fibers much longer than that used in premix. Also, the resin and all its parts are calendered into the fiber reinforcement at a much lower viscosity with eventual high SMC viscosity attained through chemical action in a curing phase (see Fig. 70.10).

Process steps:

1. Resin, modifiers and fillers are prepared in a mixer or extruder, then flowed onto a mat and covered with a release-film carrier.
2. Another layer of mat material and release film are applied; this "sandwich" of mat–resin–mat is conveyed through squeeze and macerating or working rollers to thoroughly blend the resin into the mats.
3. Mat is passed through an oven to hasten the thickening process (or it can be rolled up and allowed to stand).
4. The sheet material is ready for molding shortly after thickening has been accomplished.

Advantages are high production rates, excellent reproductability, requires little operator skill, color throughout the part, good surfaces all over, inserts can be molded in, capable of selective variations in part thickness.

Disadvantages are SMC is not practical unless the quantity can justify the large investment in molds. Equipment and maintenance costs are high, and skilled personnel are required. Finally, lead times are long and there are part size limitations.

Physical properties are

	Preform	Premix	Prepreg
Flexural strength, 1000 psi (MPa)	24–45 (166–360)	6–26 (46–179)	50–80 (345–552)
Tensile strength, (1000 psi (MPa)	9–24 (62–166)	4–10 (28–69)	40–55 (276–379)
Flexural modulus, 10^6 psi (MPa)	1.0–1.8 (6900–12,400)	1.5–2.5 (10,300–17,200)	1.8–1.9 (12,400–13,100)

Prepreg Process. In this method fabric containing a controlled amount of resin is cut and fitted to the male part of a matched die set to make FRP parts. The required plies of prepreg are laid up on a suitable mold and either vacuum bagged or press molded, then cured by the application of heat. Uniformed parts with high specific strength are produced (see Fig. 70.11).

Process steps:

1. Precut sheets of prepreg with release film attached, are brought to the mold area.
2. Prepared mold is then covered with plies of sheet prepreg (with release film removed); each ply is worked with a Teflon paddler and/or heated hand irons to make sure that there is intimate contact between plies.
3. The final layup is then moved into a press and the dies close for final cure. Alternatively, the final lay-up is surrounded with a bleeder strip and double-faced tape and covered with a vacuum bag. Vacuum is applied and the worker ensures that the lay-up is pulled down tight against the mold; the mold assembly under vacuum is put into an oven for final cure.
4. After cure, the mold opens (or the vacuum bag is removed), and the part is trimmed and inspected. Basically the same tools and equipment are used in vacuum bag and autoclave processes.

Advantages of the prepreg process are that is permits precise location of all reinforcement, requires no mixing or calendering, lends itself to complex multi-ply parts, and different weaves and orientation can be used.

Disadvantages include high material cost, high labor/material ratio, high scrap rate, and a high degree of operator skill. Also, prepregs have short shelf life, and usually must be refrigerated. The process is generally limited to aerospace applications because of high cost and low production quantity.

Physical properties are

	Polyester	Epoxy
Tensile strength, 1000 psi (MPa)	50 (345)	47 (324)
Flexural strength, 1000 psi (MPa)	65 (448)	70 (483)
Flexural modulus, 10^6 psi (MPa)	3.2 (22,000)	3.2 (22,000)
Compressive strength, 1000 psi (MPa)	45 (360)	50 (345)

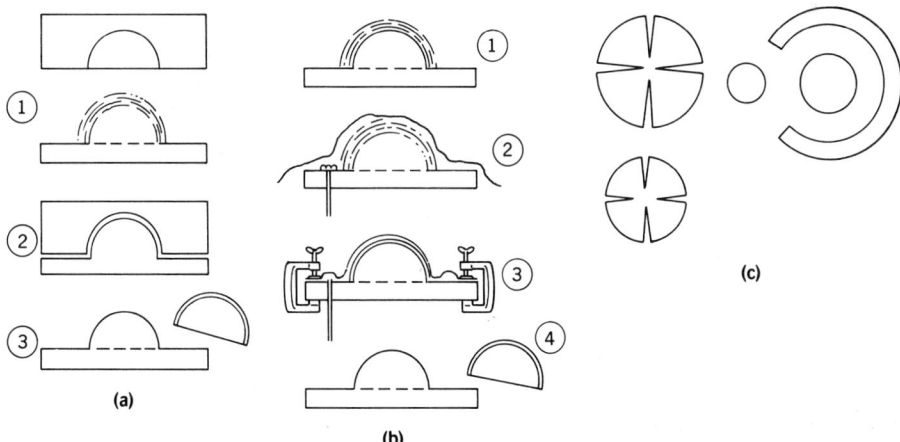

Figure 70.11. Prepeg process. **(a)** Press molded: 1, Prepeg placed on force; 2, molds close for cure; 3, remove part and trim. **(b)** Vacuum bagged: 1, Prepeg put on mold per pattern; 2, place bleeder and bag; 3, clamp and apply vacuum; 4, remove and trim. **(c)** Patterning; 1, Analyze shape of lay up and design patterns to fit on mold without overlaps; 2, cut shape and make dry run; 3, adjust shapes for production; 4, make lay up with resin and/or final prepeg choice.

Prepregs are used in making aerospace radomes, fairings, dive brakes, doors, access panels, communication housings, control surfaces, fan housings, etc.

Preform. A vacuum-energized screen in the shape of the part to be molded is sprayed alternately with chopped fibers and a binder until the screen is filled. When heat set, this material is removed and placed into the mold, saturated with a pre-set amount of filled resin, and allowed to cure (see Fig. 70.12).
Process steps:

1. Using a part-shaped perforated screen placed over the intake of a centrifugal fan, chopped fibers are sprayed alternately with a binder until the screen is filled.
2. Cure is effected by placing the screen and material in a hot-air environment until the binder is cured.
3. Cured preform is removed and placed onto the lower part of the die set.
4. Resin, filler, color, and other ingredients are poured onto the preform in a programmed manner.
5. Mold is closed, allowing the material to flow and cure.
6. Mold is opened; part is removed and trimmed.

Pultrusion. In this process, engineered tapes, fabrics and/or filaments are impregnated with resin, pulled through a shaping die and cured to this shape through addition of catalyzing means. Parts with biaxial strength factors can be produced by pultrusion (see Fig. 70.13).
Process steps:

1. Set up supply of tapes, fabrics, and/or filaments, also set control of tension of these materials.
2. Pull materials into the resin bath.
3. Insert into shaping die and pull partway through.

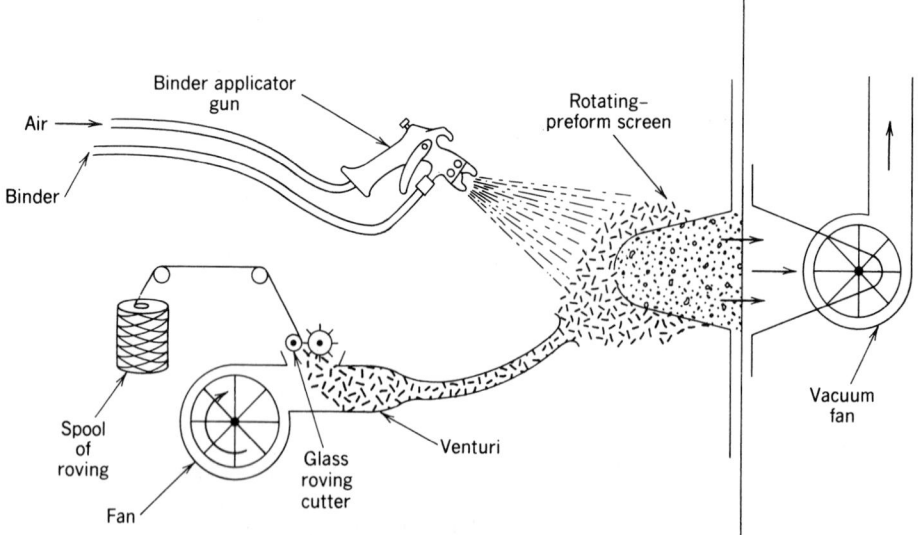

Figure 70.12. Preform manufacturing.

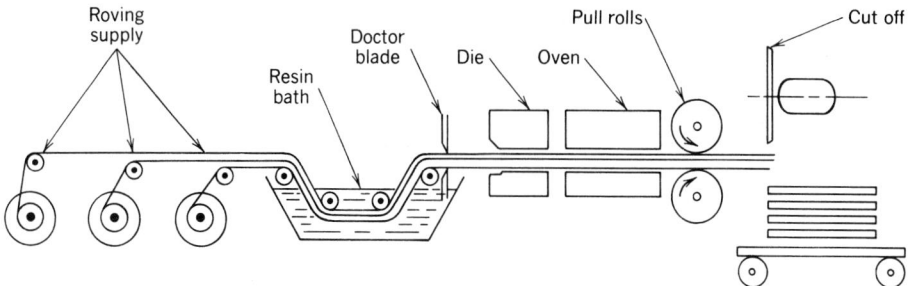

Figure 70.13. Continuous pultrusion process.

4. Apply heat and catalyzing means and then pull this hardened material through the die and into the pulling devices.
5. Adjust heat, pressure, and speed.
6. Set cut-off length and commence production.

Basic tooling includes reinforcement supply frame and tensioning devices, shaping die and heating means, pulling and cut-off device, accessory equipment and small hand tools, resin mixing and quality control.

Advantages of pultrusion include: high production rate of parts with unlimited lengths; very low labor factor, controlled physicals and uniform cross-section, size and shape flexibility.

Disadvantages include high development and equipment costs, difficulty in part-production changeover, high die wear and costly dielectric heating. The process also requires high-level quality control monitoring.

Physical properties are

Glass content, %	up to 80
Tensile strength, 1000 psi (MPa)	40–100 (276–690)
Tensile modulus, 10^6 psi (MPa)	2.5–6 (17,200–41,400)

Pultrusion is used in such applications as slot wedges, pipe, structural shapes for electrical applications, fishing rods, antennas, etc.

Continuous Laminating. This is a process by which laminates are produced with the resin, reinforcement, mold surface, and bagging material (if required) brought together under controlled conditions and rates and allowed to cure on the moving mold surface. The product is then cut into required lengths and packaged for the customer (see Fig. 70.14).

Process steps:

1. Prepare mold surface, or use a firm releasing film.
2. Apply resin while belt (or film) is moving.
3. Apply reinforcement.
4. Densify the composite or laminate through mechanical means (rolls), with or without release film.
5. Apply heat simultaneously and allow to cure.
6. Cut to length and remove film.

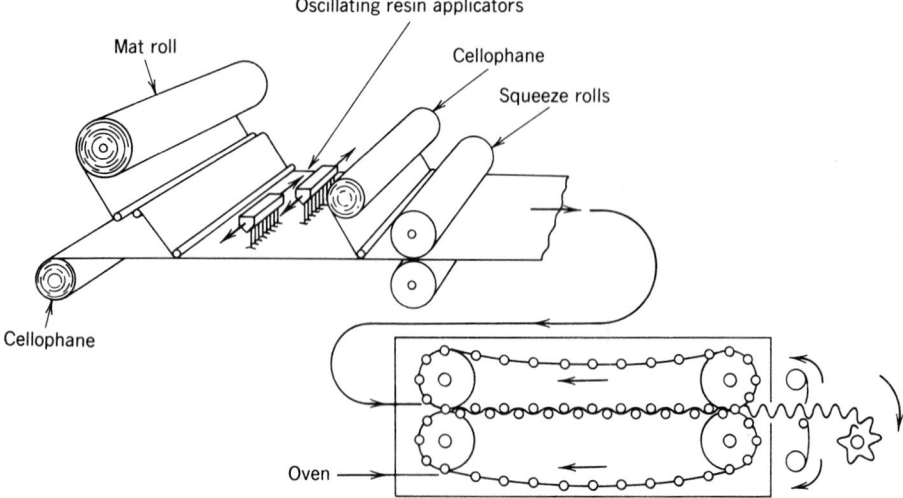

Figure 70.14. Continuous laminating process.

Basic tooling includes the laminating machine (which can be very expensive and occupy large space); mixing and metering gear; reinforcement supply and control; cut-off; stacking; and packing equipment. The foregoing are usually supplied with the basic machine.

Advantages include the ability to produce uniform parts in unlimited lengths and thicknesses, and with good surfaces on both sides. Labor costs are low, and very little scrap is produced once the machine is balanced.

Disadvantages of continuous laminating are high equipment cost with long lead time, the need for a strong sales organization to keep the machine busy (down time is very costly), and difficulty in shape changeover. This usually requires complete machine rebuilding (although certain patented machines can change part shape in an hour or two.)

Physical properties are

Glass, %	15–35	
Tensile strength,		
1000 psi (MPa)	12–18	83–124
Flexural strength,		
1000 psi (MPa)	16–28	110–198
Flexural modulus,		
10^6 psi (MPa)	0.5–1.2	3500–8300

Applications include such relatively flat shapes as glazing, paneling, roofing, tank liners, hoods, and ducts.

Centrifugal Casting. This is a process wherein cylindrical, conical or parabolic shapes [in lengths up to 30 ft (9 m)] can be densified by centrifugal force applied by rotation about the longitudinal axis (see Fig. 70.15).

Process steps:

1. Place reinforcement in mold with release agent.
2. Rotate mold and introduce the resin.
3. Set proper speed for mold size and pressure required.
4. Apply heat and attain cure.
5. Stop rotation and remove the part.

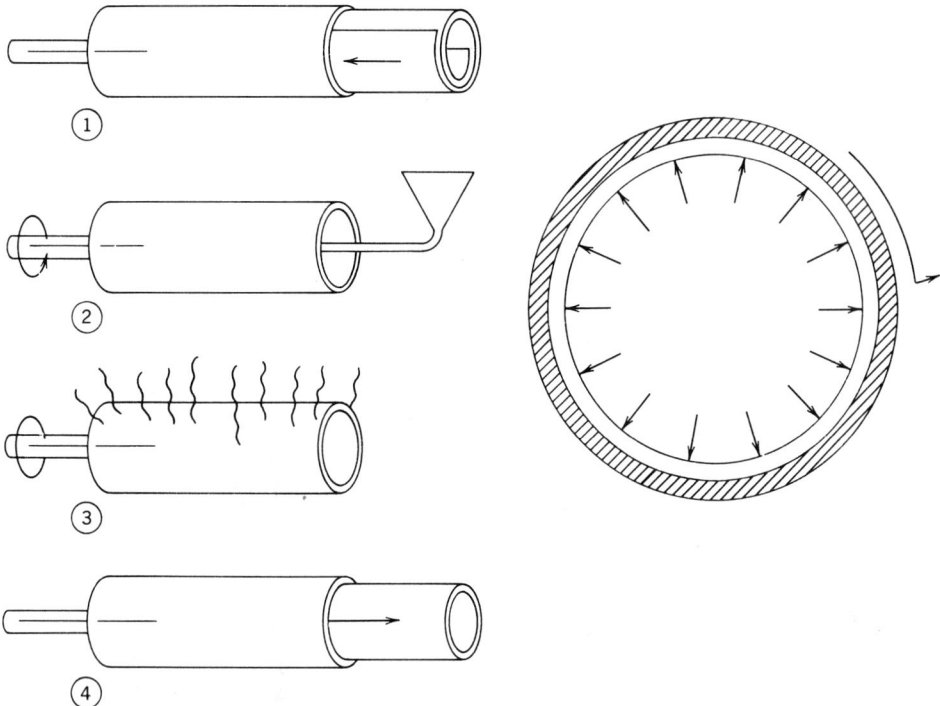

Figure 70.15. Centrifugal molding.

Basic tools include: a dynamically balanced mold; rotating mechanism and oven; and equipment to introduce and form the resin and reinforcement.

Advantages are numerous. Minimum hand tools are required. The process is easy to automate, with excellent uniformity possible. Good surfaces are achieved on both sides. Materials waste is low, and void-free laminates can be produced by minimally skilled operators.

Disadvantages are equipment cost is high and molds must be very carefully balanced. Physical properties of chopped mat or mat, glass content of 40% are

	Mat	Cloth
Tensile strength, 1000 psi (MPa)	10–20 (69–138)	30–50 (207–345)
Flexural strength, 1000 psi (MPa)	20–40 (138–276)	45–75 (360–517)
Flexural modulus, 10^6 psi (MPa)	0.8–1.8 (5500–12,400)	1.5–4 (10,300–27,600)

Uses include such products and shapes as pipes, tanks, hoops, and parabolas.

71

REINFORCED PLASTICS, ALKYD POLYESTERS

Ivor H. Updegraff

407 Den Rd., Stamford, CT 06903

71.1 ALKYD POLYESTERS

Unsaturated polyester alkyd resins are very versatile materials. At room temperature the liquid resins are stable for months or even years but, in a few minutes, can be triggered to cure simply by addition of a peroxide catalyst. Since cure results from an addition reaction, whereby double bonds are converted into single bonds, there is no release of a secondary reaction product.

In contrast to most other plastics that are based on a single prime ingredient, polyesters for reinforced plastic (RP) contain a substantial amount of several components. These may be combined in different ratios; furthermore, each component may be replaced by any of several alternative materials. Consequently, many variations of polyester resins are available. In each type the formulator strives to emphasize the particular properties needed for a specific application. A full line of polyester resins may include as many as fifty or sixty formulations.

71.2 CATEGORY

Polyester alkyd resins for RP are low-molecular-weight linear polymers containing carboxylic ester linkages and carbon/carbon double bonds as recurring units along the polymer chain. The most commonly used type is a combination of maleic anhydride with phthalic anhydride esterified with propylene glycol. Average molecular weight is normally in the range of 1000 to 3000.

To produce a polyester resin for RP, the alkyd is combined with a vinyl monomer, usually styrene. An inhibitor such as hydroquinone is then added to stabilize the mixture. When the resulting polyester resin is applied in the production of RP products, it is catalyzed with a peroxide. This includes polymerization at the double bonds, forming a cross-linked thermoset network.

71.3 HISTORY

It was Carleton Ellis who first recognized the utility of these materials. In the course of his work with unsaturated polyesters made by reacting maleic anhydride with various glycols, Ellis discovered that these unsaturated polyesters could be cured to insoluble solids simply by adding a peroxide catalyst. He applied for a patent on this idea in 1936.[1] Later, he discovered that a more useful product could be made by combining the unsaturated polyester alkyd with a reactive monomer like vinyl acetate or styrene. The viscosity was greatly reduced, which made it easy to add the catalyst and apply the resin to a suitable form or mold. Furthermore, the resulting cure was vigorous and complete. In fact, the polymerization reaction of cure for the mixture was faster than that of either of the components taken separately. A patent for this invention was applied for in 1937 and granted in 1941.[2]

It is evident that Ellis' new product met a definite need in the plastics industry. The annual production of unsaturated polyester resins for reinforced plastics has passed the 1 billion pound level. Polyesters are among the four most important thermoset polymer systems, the other three being phenolics, amino, and epoxy resins. The first important application for polyester resins was in the manufacture of fiber-glass reinforced radomes for military aircraft.

U.S. production of unsaturated polyester resins was approximately 190 million pounds for the year 1960. By 1970 annual production had grown to 570 million pounds. During the next eight years, production increased rapidly, exceeding 1200 million pounds in 1978. Production then declined with the ensuing business recession, falling below the billion pound level in 1980 to 947 million pounds.

Polyester resin production represents approximately 15% of the total production of thermosetting resins.

71.4 POLYMERIZATION

Polyesters for reinforced plastics are generally made by a batch process. The variety of products needed for a complete line of resins favors the batch process since it offers quick and easy changeover. Continuous processes are also in use, but generally only for the large-volume, general-purpose type resins.

Stainless steel is the preferred material of construction for the processing equipment, because it is resistant to the polyester resin and to the materials used in resin manufacture. Since the free-radical polymerization of polyesters is inhibited by ions of iron and copper, these metals should be avoided in processing equipment. Glass-lined reactors are preferred when halogenated starting materials are used.

In general, the glycol is charged to the reaction kettle, followed by phthalic and maleic anhydrides. It is common practice to add 5 to 10% more glycol over the theoretical amount to take care of losses through the condenser as well as side reactions. Air is displaced by inert gas. The first stage of reaction, formation of the half ester, takes place spontaneously at a relatively low temperature. The batch is then heated to complete the ester formation. The rate of flow of inert gas through the batch may be increased to remove the water of reaction. A steam-heated partial condenser is often used to allow the water to be removed while the glycol is returned to the reaction kettle. During the latter stages of esterification, the batch temperature is allowed to rise to 374 to 428°F (190 to 220°C).

Viscosity and acid number tests are run to follow the progress of esterification. When the appropriate requirements are met, the batch is pumped to the cutting kettle. The required amount of styrene monomer may already be present in this kettle, so the polyester alkyd dissolves as rapidly as it is pumped in. Additional inhibitor may be added at this point to avoid any polymerization as the hot alkyd contacts the styrene monomer. Cooling may also be needed to control the batch temperature. The properties of the batch are then determined and adjusted to meet product specifications. Complete production cycles may range from 10 to 20 hours.

This polyester preparation process is often referred to as the fusion process. The reactants are simply fused together and heated until the esterification reaction has progressed far enough to produce the desired product. An alternative method is to use a small amount of xylene or toluene solvent to facilitate removal of the water by azeotropic distillation. Here, the solvent represents only about 8 percent of the kettle charge. It is separated from the water in a decanter and returned to the kettle. When esterification is complete, the condenser is set for straight distillation and the solvent is removed. A final vacuum strip may be applied to obtain more efficient removal of any remaining solvent. Some side reactions may take place during esterification. For example, some branched chain polymer may form by addition of a hydroxyl of the glycol to the maleic or fumaric double bond. It has been estimated that side reactions may consume as much as 10 to 15% of the reactive unsaturation.

The simplest continuous process for the manufacture of unsaturated polyester resins is by the reaction of a mixture of maleic and phthalic anhydrides with propylene oxide. Some glycol must be present to initiate the anhydride/epoxide chain reaction. Since the anhydride/epoxide polymerization reaction proceeds at a relatively low temperature, the maleate double bond is not isomerized to the more active transconfiguration. As a result, a separate heat treatment must be applied after the polymer has been formed to bring about isomerization to the fumarate structure and achieve the desired reactivity with styrene.

Continuous production of unsaturated polyesters is also possible by using a series of stirred reaction vessels.

Unsaturated polyester resins are relatively low in cost, which accounts for their wide usage in the plastics industry. Epoxy resins are somewhat more expensive but are preferred in some applications for low shrinkage, good adhesion, and resistance to attack by chemicals.

In 1988, the list price of a general-purpose type polyester resin made with phthalic anhydride falls in the range of 65–80 cents per pound. A resilient type of resin made with isophthalic acid for greater toughness is somewhat more expensive, 75–83 cents per pound. Other, more complex resin systems such as those making use of bisphenol A to improve chemical resistance or those containing halogenated components to provide fire resistance will also be more expensive.

71.5 DESCRIPTION OF PROPERTIES AND RESIN FORMS

The broad range of properties obtainable with polyester resins makes them suitable for a wide variety of applications. Seven specific types of unsaturated polyester resins that are tailored for particular property requirements are described in the following sections.

71.5.1 General Purpose

This type of polyester is based on a blend of phthalic anhydride with maleic anhydride, esterified with propylene glycol. The mol ratio of phthalic to maleic may range from 2:1 to 1:2. The polyester alkyd is blended with styrene in the range of about two parts of alkyd to one part of styrene. As the name implies, this type of resin is used for a great variety of applications, such as trays, boats, shower stalls, swimming pools, and water tanks.

71.5.2 Flexible

If a straight-chain dibasic acid such as adipic or sebacic is used instead of phthalic anhydride, the resulting unsaturated polyester is much softer and more flexible than the general-purpose type. The use of diethylene or dipropylene glycol in place of propylene glycol also provides flexibility. Such flexible polyesters may be added to the rigid general-purpose type to overcome brittleness and to make the cured product more easily machined. This is an advantage in the manufacture of cast polyester buttons. Flexible resins may also be made by replacing some of the phthalic anhydride with tall oil fatty acids. These monobasic fatty acids will provide flexible groups at the ends of polymer chains.

Flexible resins are often used to make decorative furniture castings and picture frames.

The flexible resin is combined with cellulosic filler such as pecan shell flour and cast in silicone rubber molds. Excellent reproduction of wood carvings may be obtained by using silicone rubber molds made directly from the original carvings.

71.5.3 Resilient

Polyesters of this type fall between the rigid general-purpose and the flexible types. They are intended for use where toughness is needed, such as in bowling balls, safety helmets, guards, gelcoats, aircraft, and automotive parts. It is common practice to use isophthalic acid instead of phthalic anhydride in formulating resilient polyesters. The resin preparation may also be stepwise. First, the isophthalic acid and glycol are combined to give a polyester of low acid number. Maleic anhydride is then added and the esterification is continued. This will produce polyester chains having the reactive unsaturation at the ends or between blocks of glycol/isophthalic polymer, rather than randomly distributed along the polyester chain. Phthalic anhydride does not perform as efficiently as isophthalic acid in this type of esterification because the half ester has a tendency to revert to the anhydride at the higher temperature needed to form the high-molecular-weight polyester.

71.5.4 Low Shrinkage or Low Profile

In ordinary glass-fiber-reinforced polyester moldings, the difference in shrinkage between the resin and the glass fiber results in a pattern of sink marks on the molded surface. Use of low-shrink type polyester can minimize the glass-fiber pattern so that moldings will not have to be sanded smooth before painting. This property is advantageous for automotive and appliance parts. The low-shrink type of polyester resin includes a thermoplastic component such as polystyrene or poly(methyl methacrylate) that is not completely soluble in the system. As curing proceeds, phase changes allow the formation of micro voids that compensate for the normal shrinkage of the polyester resin.

71.5.5 Weather Resistant

This type of polyester is formulated to resist yellowing on exposure to sunlight. In addition to compounding with ultraviolet light absorbers, a part of the styrene may be replaced with methyl methacrylate. It is not practical to replace all of the styrene since the methacrylate does not copolymerize very well with the fumarate double bonds of the unsaturated polyester. Resins of this type are used in gelcoats, outdoor structural panels, and skylights.

71.5.6 Chemical Resistant

Alkali resistance is the principal deficiency of polyester resins. The ester linkages are subject to hydrolysis in the presence of alkalies. Increasing the size of the glycol has the effect of reducing the concentration of ester linkages. Thus, a resin containing "bis glycol" (the reaction product of bisphenol A with propylene oxide), or hydrogenated bisphenol A,

will contain far fewer ester linkages than a corresponding resin of the general-purpose type. These resins are used in the manufacture of chemical processing equipment such as fume hoods, reaction vessels, tanks, and pipes.

71.5.7 Fire Resistant

Ordinary glass-fiber-reinforced polyester moldings and laminates are combustible but have relatively low burning rates. Better resistance to ignition and burning may be achieved by using halogenated dibasic acids in place of phthalic anhydride; for example, tetrachloro-phthalic, tetrabromophthalic, and chlorendic (the addition product to hexachlorocyclopen-tadiene with maleic anhydride) also known as "Het" acid. Dibromoneopentyl glycol has also been used. Further improvement in fire resistance may be achieved with various flame-retardant additives, such as phosphate esters and antimony oxide. Fire resistant polyesters are used in building panels and for certain types of Navy boats.

These seven types include most of the unsaturated polyester resins used in the RP industry. However, some specialty resins are manufactured to meet other specific requirements. For example, the use of triallylcyanurate in place of styrene can greatly improve heat resistance. Diallylphthalate is used in place of styrene in some molding compounds based on unsaturated polyester resins. It is much less volatile than styrene so that the molding compound may be handled as a putty or extruded rope without loss of monomer. Vinyl toluene is also used in this application. Special resins are also made for curing by ultraviolet light. They contain light-sensitive catalytic agents like benzoin or benzoin ethers.

71.6 APPLICATIONS

Polyesters are used in the manufacture of a broad range of products including boats; building panels; structural parts for automobiles, aircraft and appliances; fishing rods; golf club shafts; and so on. The curing reaction is exothermic and often provides adequate heat to ensure satisfactory cure. It is estimated that a normal cure will convert 90% of the reactive double bonds into single bonds. About 80% of the polyester resin made in the United States is used with reinforcement, primarily glass fiber. Applications without reinforcing fibers included buttons, furniture castings, cultured marble, and auto body putty.

71.7 ADVANTAGES/DISADVANTAGES

The major advantages of reinforced polyester systems are

- High strength with light weight: strength to weight ratios are high compared to most metals.
- Low tooling costs: simple forms or molds operating at low pressure are adequate to produce parts.
- Parts consolidation: a single molded part such as a one-piece hood/fender assembly for a truck can replace a number of metal parts and allow great design flexibility. Thus, polyester can offer cost/performance benefits considerable beyond those available with many other materials.

Some limitations must be considered in designing parts made from unsaturated polyesters reinforced with glass fibers. Fatigue resistance of the glass-reinforced polyester may be inadequate for some highly stressed parts. Abrasion resistance and resistance to solvents and chemicals must be considered in some applications. The wicking of moisture along bundles of glass fibers can be troublesome, especially if the surface has been abraded to expose the glass-fiber reinforcement. Special surfacing mats are sometimes used to minimize this problem as well as to improve appearance.

71.8/9 PROCESSING TECHNIQUES

The overall curing behavior of a polyester resin is a delicate balance of the effects of catalyst, inhibitor, and accelerator.

Polyester resins are cured by adding free-radical catalysts to start the chain reaction of polymerization. The free radicals may be derived from peroxides or from other unstable materials much as azo compounds that can be broken into free radicals by heat or irradiation with uv light or other high-energy radiation. Normally, the polyester resin contains an inhibitor, essentially a free radical trap. When the catalyst is added, the inhibitor must first be overcome before the polymerization reaction can start. This induction period allows time for the catalyzed resin to be combined with the reinforcement and placed in position for cure before the polymerization reaction starts. Hydroquinone and some related compounds are good inhibitors, as are some quaternary ammonium halides.

Most peroxide catalysts decompose rather slowly when they are added to the polyester resin. To obtain faster cure, accelerators, also known as promoters, are used to speed up the catalyst decomposition. Essentially, the accelerator is a catalyst for the catalyst. If catalyst and accelerator are mixed together, they can react violently causing fire or explosion. Therefore, it is very important that they be added to the resin separately, making sure that one is completely dissolved the other is added. Many resins are supplied with the promoter already added.

Groups attached to the carbon atoms of the reactive double bond can influence reactivity in two important ways. Steric effects relate to the fact that bulky groups reduce reactivity by shielding the reactive double bond, thereby making it less probable that a second reactive double bond will be in a favorable position to react. Polarity has to do with the tendency of the attached group to attract or donate electrons. Electron-donating groups like methyl, phenyl, and halogen tend to make the double bond electronegative. This is the case for styrene, vinyl toluene, and chlorostyrene. Electron-withdrawing groups like nitrile and carbonyl make the double bond electropositive. This is true for the double bonds in the polyester chain. The opposite polarity of the styrene and polyester alkyd double bonds accounts for the very strong alternating tendency observed in the polymerization reaction for curing unsaturated polyester resins. The greater mobility of the styrene compared to the polyester chain allows it to homopolymerize to some extent. Experience indicates that about two mols of styrene for each alkyd double bond gives optimum results.

Curing behavior is often described in terms of the SPI Gel Test.[3] A quantity of resin is catalyzed and placed in a test tube. A thermocouple probe is immersed in the resin and the assembly is then suspended in a constant temperature water bath maintained at 180°F (82°C). A recorder attached to the thermocouple draws a time/temperature curve as the cure reaction proceeds. This is known as an exotherm curve. The temperature rises slowly as the sample approaches the bath temperature. During this time the inhibitor is being consumed by the free radicals released by the catalyst. Polymer starts to form and the system quickly gels. The temperature then rises rapidly as polymerization accelerates as a result of the "gel effect." It reaches a peak and then falls when the cured plastic cools back to the temperature of the bath. The curing behavior of the system may then be described in terms of the curve by noting the length of the induction period, the time required to reach the peak temperature,

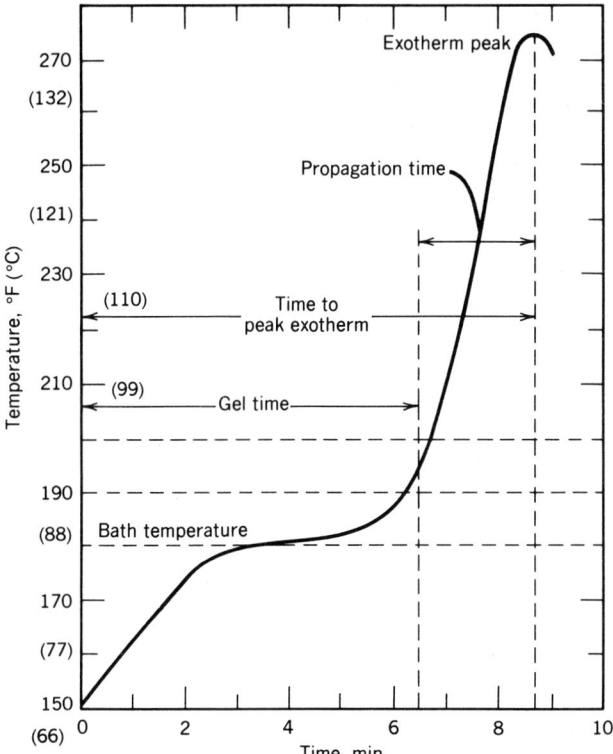

Figure 71.1. Typical polyester resin time vs temperature curve for a 10-g sample in a 19-mm diameter test tube in 180°F (82°C) water bath.

and the peak temperature reached. It is thus possible to make a detailed comparison of the influence of different amounts and kinds of catalysts, inhibitors, and accelerators. A typical exotherm is shown in Figure 71.1.

A wide range of catalyst–accelerator–inhibitor systems are available for use with polyester resins. For example, a general-purpose, hydroquinone-inhibited resin might be cured very rapidly by using an active peroxide catalyst such as methyl ethyl ketone peroxide in combination with an active accelerator such as cobalt naphthenate or octoate. At the other extreme, the same resin might be catalyzed with a much more stable peroxide catalyst such as *t*-butyl perbenzoate and used to make a polyester molding composition that included calcium carbonate filler along with chopped glass fillers. Such a catalyzed premix molding compound might be stable in storage at room temperature for months, yet cure to a firm solid in one minute when it is pressed in a matched metal mold at 284 to 320°F (140 to 160°C).

71.10 SPECIFICATION OF PROPERTIES

Unsaturated polyesters can be formulated to produce cured plastics ranging from hard and brittle to soft and flexible, as indicated in the Master Outline of Materials Properties for typical unreinforced polyester resin castings. A broad range of values is shown for most properties to cover the results obtained with different formulations. However, polyesters are usually combined with reinforcement, generally glass fiber, and this can bring about

Master Outline of Material Properties

MATERIAL _____ Flexible Cast Polyester Resins

PROPERTY	TEST METHOD	ENGLISH UNITS	VALUE	METRIC UNITS	VALUE
MECHANICAL DENSITY	D792	lb/ft³	62.9–74.8	g/cm³	1.01–1.20
TENSILE STRENGTH	D638	psi	500–3000	MPa	3.4–21
TENSILE MODULUS	D638	psi		MPa	
FLEXURAL MODULUS	D790	psi		MPa	
ELONGATION TO BREAK	D638	%	40–310	%	40–310
NOTCHED IZOD AT ROOM TEMP	D256	ft-lb/in.	>7.0	J/m	>370
HARDNESS	D785		84–94 Shore D		84–94 Shore D
THERMAL DEFLECTION T @ 264 psi	D648	°F		°C	
DEFLECTION T @ 66 psi	D648	°F		°C	
VICAT SOFT-ENING POINT	D1525	°F		°C	
UL TEMP INDEX	UL746B	°F		°C	
UL FLAMMABILITY CODE RATING	UL94				
LINEAR COEFFICIENT THERMAL EXPANSION	D696	in./in./°F		mm/mm/°C	
ENVIRONMENTAL WATER ABSORPTION 24 HOUR	D570	%	0.5–2.5	%	0.5–2.5
CLARITY	D1003	% TRANS-MISSION		% TRANS-MISSION	
OUTDOOR WEATHERING	D1435	%		%	
FDA APPROVAL					
CHEMICAL **RESISTANCE** **TO:** WEAK ACID	D543				
STRONG ACID	D543				
WEAK ALKALI	D543				
STRONG ALKALI	D543				
LOW MOLECULAR WEIGHT SOLVENTS	D543				
ALCOHOLS	D543				
ELECTRICAL DIELECTRIC STRENGTH	D149	V/mil	250–400	kV/mm	9.9–15.8
DIELECTRIC CONSTANT	D150	10⁶ Hertz	4.1–5.9		4.1–5.9
POWER FACTOR	D150	10⁶ Hertz	0.023–0.060		0.023–0.060
OTHER MELTING POINT	D3418	°F		°C	
GLASS TRANS-ITION TEMP.	D3418	°F		°C	

Master Outline of Material Properties

MATERIAL ___ Rigid Cast Polyester Resins

PROPERTY	TEST METHOD	ENGLISH UNITS	VALUE	METRIC UNITS	VALUE
MECHANICAL DENSITY	D792	lb/ft^3	68.6–91.0	g/cm^3	1.10–1.46
TENSILE STRENGTH	D638	psi	6,000–13,000	MPa	40–90
TENSILE MODULUS	D638	psi	3,000–6,400	MPa	21–44
FLEXURAL MODULUS	D790	psi	8,700–23,000	MPa	60–160
ELONGATION TO BREAK	D638	%	<5.0	%	<5.0
NOTCHED IZOD AT ROOM TEMP	D256	ft-lb/in.	0.2–0.4	J/m	11–21
HARDNESS	D785		M70-M115		M70-M115
THERMAL DEFLECTION T @ 264 psi	D648	°F		°C	
DEFLECTION T @ 66 psi	D648	°F		°C	
VICAT SOFT-ENING POINT	D1525	°F		°C	
UL TEMP INDEX	UL746B	°F		°C	
UL FLAMMABILITY CODE RATING	UL94				
LINEAR COEFFICIENT THERMAL EXPANSION	D696	in./in./°F		mm/mm/°C	
ENVIRONMENTAL WATER ABSORPTION 24 HOUR	D570	%	0.15–0.60	%	0.15–0.60
CLARITY	D1003	% TRANS-MISSION		% TRANS-MISSION	
OUTDOOR WEATHERING	D1435	%	Yellows	%	Yellows
FDA APPROVAL					
CHEMICAL RESISTANCE TO: WEAK ACID	D543				
STRONG ACID	D543				
WEAK ALKALI	D543				
STRONG ALKALI	D543				
LOW MOLECULAR WEIGHT SOLVENTS	D543				
ALCOHOLS	D543				
ELECTRICAL DIELECTRIC STRENGTH*	D149	V/mil	380–500	kV/mm	14.9–19.7
DIELECTRIC CONSTANT	D150	10^6 Hertz	2.8–4.1		2.8–4.1
POWER FACTOR	D150	10^6 Hertz	0.006–0.026		0.006–0.026
OTHER MELTING POINT	D3418	°F		°C	
GLASS TRANS-ITION TEMP.	D3418	°F		°C	

*Short-time test

dramatic changes in properties. Impact strength, for example, may be increased by as much as 50 times with such reinforcement.

71.11/12 PROCESSING REQUIREMENTS

71.11.1 Hand Lay-Up and Spray-Up

Fabrication of plastic parts with polyester resins is more varied than that of any other type of plastic.

The simplest technique, hand lay-up, merely involves catalyzing the liquid resin and then applying it with layers of fiber-glass cloth or mat to a form or mold using a brush and roller. The resin may contain a thixotropic agent to make it stay in place on a vertical surface. In addition, the resin may contain a wax that is soluble in the liquid resin but less soluble in the cured plastic so that it exudes to the surface as the resin cures. The wax protects the resin from oxygen in the air that would otherwise inhibit cure and leave a tacky surface. Since the catalyzed resin has a limited pot-life, no more should be catalyzed than can be conveniently applied within the pot-life period.

To avoid this inconvenience, it is possible to catalyze the resin as it is being applied. This technique is known as spray-up and is commonly used in the manufacture of boats. The boat mold is first sprayed with a pigmented gelcoat. This will become the exterior surface coating of the completed hull. Before the gelcoat has cured completely, the polyester resin and glass-fiber reinforcement are sprayed up against it, thus making the gelcoat an integral part of the finished molding.

The spray-up type of gun not only sprays catalyzed resin but also feeds fiber-glass roving into a chopper. The pieces of cut roving are then carried with the resin spray to the mold surface. Considerable operator skill is needed to deposit a uniform coating in some areas and to build up the thickness at points of stress. The resin/glass mixture must be rolled to remove trapped air before curing.

Two types of catalyst promoter systems are in common use. The simplest makes use of the promoter, cobalt naphthenate or octoate, in the resin and injects the catalyst, methyl ethyl ketone peroxide, at the spray gun. The other requires two separate resin containers and is known as the two pot system. One is catalyzed with benzoyl peroxide, which is a fairly stable catalyst giving a pot-life of several hours. The other container of resin is promoted with an amine such as dimethylaniline. This solution is also quite stable at room temperature. The spray gun mixes two equal streams to produce a catalyzed and promoted resin that will cure at room temperature. It is still necessary to roll the resin–glass mix to remove trapped air and distribute the resin where needed.

Hand lay-up and spray-up techniques impose no limitation on the size of the object that may be molded.

71.11.2 Preform Molding

It is not uncommon to form the fiber-glass mat reinforcement into the shape of the product to be molded. A small amount of additional binder resin may be used to better hold the preform together. It is then placed in a matched metal mold and the liquid resin is poured over it. Closing the mold forces the resin through the fiberglass mat, eliminating air and excess resin. By using pressure, it is possible to achieve a higher proportion of glass reinforcement; hence greater strength. Furthermore, the glass fibers are mostly oriented parallel to the surface of the molding.

A modification of this technique uses a half mold, or form, closed with a flexible sheet of poly(vinyl alcohol). Pressure is applied either by placing the whole assembly in an autoclave or by drawing a vacuum on the flexible covering or bag.

71.11.3 Centrifugal Casting

Centrifugal casting makes use of centrifugal force to deposit and hold the resin and glass-fiber reinforcement against the inside walls of a spinning mold. The assembly may be placed inside an oven so that the resin may be heated and cured while it is being held against the inside of the spinning mold.

71.11.4 Pultrusion and Filament Winding

Continuous pultrusion and filament winding make use of continuous strands of reinforcement drawn through a bath of catalyzed resin and then through a heated die or wound on a form. It is possible to achieve very high strength by such techniques. Continuous pultrusion can be used to produce structural shapes, golf club shafts, and fishing rods; filament winding, to produce tanks and pressure vessels.

71.11.5 Matched Metal Die Molding

High-volume production items such as automotive and appliance parts are commonly made in matched metal molds.

The polyester molding compound is supplied as a premixed blend of resin, filler, and chopped glass fiber. The premixed molding compound is known as bulk molding compound, or BMC. It is already catalyzed and inhibited against premature curing in storage. The user merely places a weighted quantity of the doughlike molding compound or extruded rope into the hot mold cavity, closes the press, holds the pressure for a minute or two, and then removes the cured molding.

A modification of this technique makes use of a similar type of molding compound in the form of flat sheets. The soft premixed dough of polyester resin and filler is spread on sheets of fiber-glass mat. This sheet of molding compound, or SMC, may then be sandwiched between sheets of polyethylene film and stored until ready for use. Thickening may take place during the first few days of storage, so that when the SMC is unrolled it is a non-tacky sheet that can easily be cut to the desired shape and compression molded into finished parts. The thickening that takes place in storage is not caused by the free radical induced polymerization, but, rather, it is the result of interaction between a thickening agent such as calcium or magnesium oxide and the polyester alkyd. The polyester resin is usually of the low shrinkage/low profile type so that the molded parts will not have to be sanded smooth before painting.

71.11.6 Automatic Injection Molding

The automatic injection molding process is similar to that used for the injection molding of thermoplastics except that, instead of being plasticized in a hot cylinder and then injected into a much cooler mold cavity, the polyester is plasticized in a warm cylinder and then injected into a hot mold cavity where the curing reaction sets the product to a thermoset solid. The great advantage of this process is the high production rate and reduced labor cost. The charge is automatically placed in the mold and the finished part is automatically removed.

Each fabrication method will need specific curing conditions. Hand lay-up, for example, will require about a half hour of time to apply the resin, wet the glass reinforcement with the resin, and make certain that the fiber-glass reinforced polyester film is free of air bubbles.

The catalyst, promoter and inhibitor requirements are best established by a series of trials using small quantities of resin. The test temperature should be the same as that of the fabrication method in question. The two most widely used catalyst/promoter systems are methyl ethyl ketone peroxide (MEKP) with cobalt naphthenate or octoate, and benzoyl peroxide (BPO) with dimethylaniline. MEKP is supplied as a 60% solution in dimethyl phthalate. Benzoyl peroxide is supplied either as a 100% active powder or a 50% active paste. The paste is recommended because it is easier to dissolve in the resin.

A starting concentrating for gel time tests for most resins might be 1% of MEKP solution (60 percent active) with 0.2% of cobalt naphthenate solution (6% cobalt) or 1.5% of BPO with 0.4% of dimethylaniline. After the gel times have been determined for a couple of concentrations, it will be possible to estimate the concentrations required to give the desired rate of cure. To ensure adequate cure, the catalyst concentration should not be less than 1.0% benzoyl peroxide or 0.5% MEKP. A little extra inhibitor such as hydroquinone or 2,5-ditertiarybutyl hydroquinone may be added to extend the gel time, if necessary.

71.13 SHRINKAGE

See Section 71.5.4.

71.14 TRADE NAMES

Unsaturated polyester resins for RP are made by a large number of U.S. companies. Following is a selected list of some of them.

Ashland Oil, Inc., Cook Paint and Varnish Co., Dow Chemical Co., Freeman Chemical Co., ICI Americas, Inc., Koppers Co. Inc., Mobile Oil Corp., Owens-Corning Fiberglas Corp., Plaskon Products, Inc., PPG Industries, Inc., Reichhold Chemicals, Inc. Rohm & Haas Co., Union Carbide Corp., USS Chemicals Div., U.S. Steel Corp.

BIBLIOGRAPHY

1. U.S. Pat. 2,195,362 (March 26, 1940; application made May 21, 1936), C. Ellis (to Ellis-Foster).
2. U.S. Pat. 2,255,313 (Sept. 9, 1941; application made Aug. 6, 1937), C. Ellis (to Ellis-Foster).
3. A.L. Smith, *Sixth Annual Technical Session*, SPI Reinforced Plastics Div., 1951.

General References

H.V. Boenig, *Unsaturated Polyesters: Structure and Properties*, Elsevier Publishing Co., New York, 1964.
J.R. Lawrence, *Polyester Resins*, Van Nostrand Reinhold, New York, 1960.
G. Lubin, *Handbook of Composites*, Van Nostrand Reinhold, New York, 1982, Chapt. 2.
R.M. Nowak and L.C. Rubins, in A. Stander, ed., *Kirk-Othmer Encyclopedia of Chemical Technology*, Vol. 20, 2nd ed., Wiley-Interscience, New York, 1969, p. 791.
R.E. Park and E.E. Parker, in A. Stander, ed., *Kirk-Othmer Encyclopedia of Chemical Technology*, 2nd Supp., 2nd ed., Wiley-Interscience, New York, 1969, p. 902.
E.E. Parker and J.R. Peffer, in C.E. Schildknecht and I. Skeist, eds., *Polymerization Processes*, Wiley-Interscience, New York, 1977, p. 68.

72

REINFORCED PLASTICS, EPOXY RESINS

John Gannon

Ciba Geigy Co., Ardsley, New York

Vince Brytus

Ciba Geigy Co., Hawthorne, New York

72.1 CATEGORY

Epoxies are thermosetting resins. The uncured intermediates are either honey colored liquids or brittle amber solids that become liquid on heating. Enlarged 10 million times, the molecules of resin might resemble short pieces of thread, varying in length from half an inch in some of the liquids to several inches in the solids. When they are cured or hardened, these threads are joined together at the ends and along the sides to form large, cross-linked structures. Each molecule is tied to several others, like the ropes of a fishnet or the filaments of a spider's web, but in an irregular rather than a neat pattern.

These final products are thermoset; they are solids that may become soft on heating but will never liquefy again. The curing of epoxies is as irreversible as the boiling of an egg; just as a 3-min egg differs from one boiled 15 min, the physical properties of an epoxy casting depend on the extent of cure. The greater the density of the cross-links, the better will be the ultimate electrical characterisics and resistance to heat softening or attack by chemicals and water.

In general, the cured materials are characterized by good mechanical strength, high dielectric strength, good dimensional stability, and good heat and chemical resistance. Their most outstanding property is likely to be their excellent adhesion to a wide range of materials. Epoxies are available in many forms, including moldings, castings, adhesives, solders and caulking compounds, laminates, foams, filament wound composites, impregnants, and coatings.

72.2 STRUCTURE

Chemically, the term "epoxy" means a three-membered ring containing one oxygen and two carbon atoms, arranged as follows:

Epoxy resin molecules contain, on the average, more than one epoxy group. The acute angles in the structure strain the interatomic chemical bonds, which makes the three-membered ring highly reactive.

The epoxy content of epoxy resins is referred to as the epoxy (or epoxide) equivalent weight. Epoxy equivalent weight is defined as the weight in grams of the base material that contains one gram equivalent of epoxide. The term "epoxy value," which also is frequently used, indicates the fractional epoxy equivalent weight contained in 100 grams of the resin.

By synthesizing epoxy resins having high epoxy content and high cross-linked density, it is possible to produce materials with improved mechanical, chemical, electrical, and thermal properties. Epoxies are considered cured when nearly all the reactive sites in the resin have been reacted and the system becomes infusible.

72.3   HISTORY

The first correct description of the synthesis of various poly(glycidyl ether)s is found in a patent application filed by Schlack in 1934 (DP 947 632). Schlack did not, however, claim any rights to the new compounds themselves but only to their use in the production of high-molecular polyamines. These were to be used, in the first place, for dressing textiles; no practical application was mentioned.

The Swiss chemist Pierre Castan (of de Trey Bros. in Zurich, Switzerland) was the first to realize that epoxy resins were eminently suited to applications in the field of plastic materials. In 1938, he claimed exclusive rights with regard to plastic materials that cured without evolution of volatile matter and with astonishingly low shrinkage, yielding end products exhibiting outstanding mechanical properties. These were epoxy resins cured with poly(carboxylic acid anhydride)s. A second patent followed in 1943, this time covering the use of basically reactive substances as curing agents. At de Trey Bros., the application of the new resin compositions was limited to the field of dentistry, but Ciba, in agreement with de Trey, began research work on their DP 749912 and EP 518057, the latter being a particularly comprehensive introduction into the plastics industry as a whole. A large number of patents and publications and a wide range of commercial products (adhesives, casting resins, laminating resins, coating resins and tooling resins) have resulted from this work. In July 1945, the first patent application in the field of epoxy adhesives was granted. In the spring of 1946, the first epoxy resin adhesive was offered at the Swiss Industries Fair in Basel.

Work in the field of surface protection can be traced back to about the same period. The first patent was taken out in August 1946.

The development of epoxy resins in the United States took place independently of the pioneer work carried out in Switzerland. It began a few years later and proceeded in a somewhat different direction. The first patent application was that of Greenlee (Devoe and Reynolds), filed in September 1945. Not long afterwards Shell also began work on epoxy resins.

72.4   POLYMERIZATION

Epoxies are cured in two ways: catalysis or polyaddition. In catalytic curing, the epoxy molecules react directly with each other in a reaction started by a catalyst. Such systems are said to be homopolymerized.

Epoxies cured by polyaddition contain hardeners mixed into the resins. These hardeners contain reactive groups that unite with the epoxy groups and become a vital part of the cured material.

The hardeners used for epoxies are Lewis acids and bases, primary and secondary amines, polyamides, polysulfides and mercaptans, organic acids and anhydrides, and phenolic resins.

(There is also a variety of specialized curing agents not included in this classification.) The most important classes of curing agents are the aliphatic and aromatic primary amines, polyamides, anhydrides and phenolic resins.

Aliphatic amine and polyamide hardeners are used with epichlorohydrin bisphenol-A and epoxy-novolac resins. They generally are not used with cycloaliphatic epoxies because of lower reactivity with the latter and the tendency to sacrifice the resin's high-temperature performance. When they are used with standard epoxy and epoxy-novolac resins, these hardeners yield room temperature curing systems that generally have a short pot-life and low heat-deflection temperatures. While flexible epoxy resins are available, certain aliphatic amine hardeners also can be used to achieve flexible systems.

Aromatic amine hardeners are used with both standard epoxies and epoxy-novolacs. They yield polymers with higher heat-deflection temperatures and better chemical resistance than systems based on the aliphatic amines.

Anhydride hardeners are used with all three types of epoxy resins; however, in the case of epoxy-novolacs and cycloaliphatics only certain types of anhydrides are recommended. The anhydride hardeners require an elevated temperature cure but, like the aromatic amines, they generally yield better chemical resistance and better heat distortion temperatures than the aliphatic amines. In addition, some of the anhydrides yield systems with long pot lives and low exotherms. One anhydride hardener in particular, pyromellitic dianhydride (PMDA) can be used with the cycloaliphatics to raise their heat-deflection temperature to about 600°F (316°C). In a similar manner, when methyl nadic anhydride is used with the epoxy–novolacs, heat-deflection temperatures as high as 570°F (299°C) can be achieved.

Table 72.1 summarizes the use of the standard hardeners for epoxy resins.

Phenolic-novolac hardeners are mainly used with the epoxy-novolac resins, although they are sometimes used with the standard epoxies. The phenolic-novolacs require an elevated temperature cure; the cured systems offer good toughness and excellent chemical resistance. They provide a long pot-life and are lower in cost than the aliphatic and aromatic amines. When epoxy-novolacs are cured with phenolic-novolacs, they are tougher than when they are cured with other hardeners.

Other curing and hardening agents also can be used to convert the epoxy resins into usable forms. Amine-terminated polyamide type curing agents are used to achieve moderately flexible and resilient systems. These polyamide hardeners are mainly used with diglycidyl ether of bisphenol-A epoxies. Polyamides yield formulations having a longer pot-life than the aliphatic amines; however, they do cure at ambient temperature. The standard epoxy resins cured with polyamides have outstanding adhesion, but they are restricted to applications where the temperature does not exceed about 250°F (121°C).

Flexibilizing curing agents, such as the polyamides and polysulfides impart flexibility or resiliency to the cured system; however, in general, their use reduces mechanical, electrical, and chemical properties.

Catalytic curing agents, such as tertiary amines and BF_3 monoethylamine, promote epoxy-to-hydroxyl reactions but do not themselves serve as direct cross-linking agents. Certain catalytic curing agents are used in combination with other curing agents to accelerate cures. Some catalytic curing agents have the effect of raising the heat deflection temperature of a system.

72.5 KEY PROPERTIES

Many different types of epoxy resins are available. However, three resins are most prominent:

- Diglycidyl ether of bisphenol-A resins (also called epichlorohydrin–bisphenol-A resins), which are the most widely used epoxies
- Epoxy–novolacs (more accurately, glycidated phenol or o-cresol novolac resins), which are high-temperature materials

TABLE 72.1. Hardeners for Epoxy Resins

	Viscosity, cps, at 25°C, (77°F) or Melting Point, °C (°F)	Recommended Typical Concentration (phr)[a]	Pot Life[b]	Typical Curing Cycles	ASTM Deflection Temperature °C (°F)	Outstanding Characteristics	Applications[c]
Anhydride Hardeners							
Phthalic anhydride (PA)	131°C (268°F)	30–75	3½ h at 110°C to 30 min at 200°C	4 to 24 h at 150°C (302°F)	110–147°C (250–297°F)	Good mechanical strength, excellent resistance to elevated temp; good electrical properties; inexpensive.	CA. E
Methylendomethylene-tetrahydrophthalic anhydride (NMA)	175–225 cps	80–90	At least 24 h at 25°C to 30 min at 100°C	2 h at 140°C (284°F) + 2–192 h at 200°C (392°F)	150–175°C (302–347°F)	Light color, low viscosity and long pot life. Good heat resistance.	A. CA. E. FW. L
Hexahydrophthalic anhydride (HHPA)	35–37°C (95–99°F)	50–100	At least 28 h	2 h at 100°C (212°F) or 2 h at 100°C (212°F) + 2 h at 149°C (300°F)	110–130°C (250–266°F)	Low viscosity mixtures; long pot life; moderate heat resistance; good shock resistance properties.	CA. E. FW
Tetrahydrophthalic anhydride (THPA)	99–101°C (210–214°F)	70–80	2 h at 100°C	2 h at 100°C (212°F) + 4 h at 150°C (302°F)	122°C (252°F)	Low cost; cured properties similar to 907 systems.	A. E. FW
Methyl tetrahydrophthalic anhydride (MTHPA)	50–80 cp	70–80	2 h at 100°C	2 h at 100°C (212°F) + 4 h at 150°C (302°F)	129°C	Low viscosity	A. E. FW
Maleic anhydride (MAZ)	52.5°C (126°F)	(Generally used in conjunction with other anhydrides)				High compressive strength.	CA
Dodecenylsuccinic anhydride (DDSA)	290 cps	130–150	24 h at 60°C	2 h at 121°C (250°F) or 1 h at 100°C (212°F) + 2 h at 204°C (400°F)	60–70°C (140–158°F)	Good electrical properties; low viscosity; long pot life; imparts flexibility.	CA. E. L

Hardener	Viscosity		Pot life	Cure schedule	Heat distortion temp	Properties	Applications
Chlorendic anhydride	240°C (464°F)	100–140	Approx ½ hr at 120°C	24 h at 120°C (248°F)	145°C (293°F)	Excellent mechanical properties at elevated temp; flame retardant; good electrical properties at elevated temp.	A, CA, E, L
Pyromellitic dianhydride	284–286°C (543–547°F)	32	3–4 days	5–20 h at 221°C (430°F)	282°C (540°F)	High physical properties; excellent heat resistance.	A, CA, E, L
Polyamide Hardeners							
Polyamide high mol wt	8,000–12,000 cps at 40°C (104°F)	40	115 min	Gel at 25°C (77°F) + 8 h at 100°C (212°F)	110°C (230°F)	Good electrical properties; excellent adhesives; less sensitive to moisture; low shrinkage.	A, Ca, F, L, T
Polyamide medium mol wt	3,000–6,000 cps	40	75 min	Gel at 25°C (77°F) + 8 hr at 100°C (212°F)	110°C (230°F)	Excellent adhesion to various substrates; outstanding impact resistance; chemical and solvent resistance.	A, CA, L, T
Polyamine Hardeners							
Modified polyamine low mol wt	500–900 cps	35	40 min	14 days at 25°C (77°F) or 16 h at RT + 2 h at 100°C (212°F)	59–89°C (138–192°F)	Cures well in high humidity atmospheric conditions; bonds well to concrete; simple, noncritical mixing ratio.	A, CA, F, T
Modified polyamine low mol wt	300–600 cps	19–25	35 min	7 days at 25°C (77°F) or gel at room temperature + 2 h at 100°C (212°F)	100°C (212°F)	Safety hn; low viscosity; low shrinkage and excellent dimensional stability; good electrical properties and good chemical resistance.	CA, E, F, L, T
Modified liquid aromatic amine medium mol wt	3,000–5,000 cps	20	3 h	2 h at 80°C (176°F) + 4 h at 204°C (399°F) or 12 h at 120°C (248°F) + 4 h at 204°C (399°F)	154–160°C (309–320°F)	Good properties at 150°C; excellent chemical resistance and good electrical properties.	A, CA, E, L

TABLE 72.1. *(Continued)*

	Viscosity, cps. at 25°C, (77°F) or Melting Point, °C (°F)	Recommended Typical Concentration (phr)[a]	Pot Life[b]	Typical Curing Cycles	ASTM Deflection Temperature °C (°F)	Outstanding Characteristics	Applications[c]
Diethylenetriamine (DETA)	Approx 5 cps	8–12	30 min	7 days at 25°C (77°F) 24 h at 65°C (149°F)	80°C (176°F)	Low viscosity at 25°C; good room temperature properties; fast cure.	A, CA, F, L, T
Triethylenetetramine (TETA)	Approx 25 cps	10–13	30 min	7 days at 25°C (77°F) Gel at RT + 2 h at 100°C (212°F)	115–120°C (240–248°F)	Good mechanical and electrical properties; good chemical resistance; low viscosity at 25°C; fast cure.	A, CA, E, F, L
m-Phenylenediamine (*m*PD)	62.6°C (144°F)	13–15	Approx 3½ h	3 h at 154°C (309°F)	165°C (329°F)	High-deflection temperature; good high temperature strength characteristics; excellent chemical resistance.	CA, E, FW, L
Diaminodiphenyl sulfone (DDS)	170–180°C (338–356°F)	36	3 h at 100°C to 75 min at 140°C	24 h at 120°C (248°F) to 24 h at 120°C (248°F) + 4 h at 175°C (347°F)	145–190°C (293–374°F)	High-deflection temperature; excellent thermal stability; outstanding chemical resistance.	A, CA, E, L
BF₃-Monoethylamine (BF₃-MEA)	89°C (192°F)	1–5	At least 6 months	2 h at 105°C (221°F) + 4 h at 150°C (302°F) or 2 h at 105°C (221°F) + 4 h at 200°C (392°F)	125–175°C (257–347°F)	Long pot life at room temperature; good strength and heat resistance; good electrical properties.	A, CA, E, FW, L
Dicyandiamide (Dicy)	208°C (406°F)	4	At least 1 yr	1 h at 176°C (350°F)		Long room temperature storage life; fast high temperature cure.	A, L
Diethylaminopropylamine (DEAPA)	Approx 2 cps	4–8	2–3 h	24 h at 100°C (212°F)	100°C (212°F)	Low viscosity; excellent in adhesives; longer pot life than 950 and 951.	A, CA, F, L, T

Curing agent	Viscosity	phr	Pot life	Cure schedule	Heat distortion temperature	Properties	Applications
Dimethylaminoethanol (DMAE)	Approx 3 cps	6	40 min	Gel at room temperature + 2 h at 80°C (176°F)	100°C (212°F)	Low viscosity; excellent chemical resistance; modifies other amines.	A, CA, F
Diethylaminoethanol (DEAE)	Approx 4 cps	8	Approx 20 h	Gel at room temperature + 2 h at 80°C (176°F)	86°C (187°F)	Low viscosity; excellent chemical resistance; low exotherm.	A, CA, F
N-Aminoethylpiperazine (N-AEP)	Approx 30 cps	13–23	15 min	Gel at room temperature + 8 h at 100°C (212°F)	114°C (238°F)	Low initial viscosity; good dielectric properties at room temperature; high impact strength.	CA, E, L
Piperidine	Approx 10 cps	6	6–15 h	2 h at 150°C (302°F) to 15 h at 150°C (302°F)	100–120°C (212–248°F)	Low exotherm at room temperature; long pot life.	CA
Methylenedianiline (MDA)	77–84°C (171–183°F)	27–30	2½ h at 50°C (122°F)	1½ h at 100°C (212°F) or 24 h at 120°C (248°F) + 4 h at 175°C (347°F)	137°C (279°F)	Outstanding chemical resistance; good high temperature properties.	A, CA, L
Aromatic polyamine	Approx 80°C (176°F)	22–30	Approx 1½ h at 65°C (149°F)	2 h at 80°C (176°F) + 3 h at 150°C (302°F)	108–150°C (226–302°F)	High-deflection temperature; good chemical resistance; good electrical properties at high humidity conditions.	CA, E

[a] phr = parts per hundred parts by weight of resin, depending on resin used and accelerator. Accelerator is usually: BDMA, DMP-10, DMP-30, Argus DB VIII, or Araldite DY 064 0.5 to 8 phr.

[b] At room temperature (23–25°C) unless otherwise indicated.

[c] Applications key: A-Adhesives; CA-Casting; E-Electrical; F-Flooring; FW-Filament winding; L-Laminating; T-Tooling.

• Cycloaliphatic epoxies prepared from cycloolefins and peracetic acid, which are characterized by excellent mechanical properties and good dielectric properties at elevated temperatures, and excellent weatherability.

72.5.1   Diglycidyl Ether of Bisphenol-A Resins

Diglycidyl ether of bisphenol-A epoxy resins are general-purpose materials. Their chief virtue is the broad range of properties they offer. Their strength properties are good, as are their hardness, toughness, and rigidity. Furthermore, these epoxies have very good electrical properties and excellent corrosion resistance, especially to caustic. While not outstanding in heat resistance, standard epoxies are generally useful up to about 300°F (149°C).

The major advantage of the diglycidyl ether of bisphenol-A epoxies is their outstanding versatility in processing. There are numerous epoxy curing agents commercially available for use with liquid and solid epoxy resins. Since the curing agent contributes largely to the performance properties achieved with cured epoxy resins, this large selection of resin-hardener systems means that many specific properties can be designed into the cured epoxy material. Moreover, ambient temperature curing systems (those for which no external heat is needed for curing) are available with a pot-life (the time an epoxy material can be usably worked) ranging from less than a minute to more than a day. Systems based on solid epoxy resins or their solutions, which require thermal curing for maximum performance, also can be designed through appropriate selection of resin and hardener.

Versatility of form is another advantage of the diglycidyl ether of bisphenol-A epoxies. These resins can be used in castings, impregnants, laminates, filament windings, adhesives, foams, moldings, trowelling compounds, and coatings.

72.5.2   Epoxy-Novolacs

Epoxy–novolacs are, basically, either phenol or o-cresol novolac resins whose phenolic hydroxyl groups have been converted to glycidyl ethers. These polyfunctional resins combine the low shrinkage and freedom from volatiles characteristic of epoxies with the high functionality of phenolics. Their high functionality, which ranges from 2 to 5 (as compared with 2 or less for the bisphenol-A epichlorohydrin epoxies), results in epoxy-novolacs that cure as closely knit thermosetting plastics. In the epoxy–novolac molecular structure, the epoxy groups are much closer together than they are in diglycidyl ether of bisphenol-A epoxies.

In addition to greater heat resistance, as compared to standard epoxies, this more compact molecule also resists chemicals better, in particular, some organic solvents that attack the standard epoxy resins.

As expected from their dense cross-linking, the epoxy-novolacs have somewhat less adhesion, toughness, and processing versatility than the diglycidyl ether of bisphenol-A epoxies.

The solid and semisolid epoxy phenol and o-cresol novolacs are handled in two ways: first, their resin viscosity can be reduced by adding standard epoxy resins or epoxy diluents (however, the use of monofunctional diluents reduces the high-temperature performance of the base resin); second, viscosity can be reduced by heating the resins to about 125 to 130°F (52 to 54°C).

72.5.3   Cycloaliphatic Epoxies

Cycloaliphatic epoxies are prepared by the peracetic acid epoxidation of cyclic olefins. Generally, the cycloaliphatics offer several important advantages over conventional epoxies and epoxy-novolacs:

• Easier handling, that is, lower viscosity and longer pot-life resin-hardener systems
• Higher heat-deflection temperatures, often greater than the temperatures required for

curing; for example, a 390°F (199°C) cure to obtain a 480°F (249°C) heat-deflection temperature

- Lower dissipation factors and dielectric constants that are stable as temperature increases, even up to 500°F (260°C)
- Outstanding arc and tracking resistance
- Excellent weatherability.

Similar to the epoxy-novolacs, the cycloaliphatic epoxies outstanding heat resistance results from their compact molecular structure. Furthermore, like the epoxy-novolacs, the cycloaliphatic resins are more brittle than the standard epoxies. The viscosity of the cycloaliphatics ranges from very low viscosity liquids to materials that are solid or semisolid. Cycloaliphatic resins can be cured at room temperature; generally, however, they are cured at elevated temperatures.

The cycloaliphatic epoxies are especially promising for electrical applications. The resins can be used in the form of castings, impregnants, laminates, filament windings, and adhesives.

72.5.4   Other Types

Two other notable types of epoxies are the halogenated resins (usually considered as flame retardant epoxies) and the flexibilizing diepoxides, which are designed to yield cured systems that are flexible and resilient.

Halogenated epoxies are nearly always based on bisphenol-A epichlorohydrin resins. Most of the commercially available epoxies contain bromine, derived from tetrabromo bisphenol-A.

Mechanical, electrical, and chemical properties of these flame retardant materials are quite similar to those of standard epoxy resins. When the bromine content in flame retardant epoxies is about 18 to 20 wt %, glass cloth laminates made from these resins are classified as self-extinguishing, according to the ASTM D 635 flammability test. When bromine concentrations are lower, the cured resins are borderline in their self-extinguishing rating, since they usually exceed the 15-s limit commonly used to classify self-extinguishing materials.

Thermal stability is usually adversely affected by the presence of a halogen in the epoxy resin. This is because the self-extinguishing properties of resins containing chlorine or bromine depend on the release of the halogens by the heat of combustion.

Flexibilizing epoxies are based on epoxidation of long-chain glycols or dimerized fatty acids. Greater elongation, high impact strength, and flexibility are obtained by introducing the long-chain aliphatic or cycloaliphatic segments between the epoxy groups in the molecule. These flexibilizing epoxy resins are generally formulated with bisphenol-A epichlorohydrin resins or epoxy-novolac resins to achieve useful cured systems, since the flexible epoxy resins inherently have low mechanical strengths. Heat-deflection temperatures of the combined resins is lower than those of either the standard or epoxy–novolac resins alone. Furthermore, their solvent and chemical resistance is reduced. Electrical properties at room temperature are essentially unaffected but fall off at elevated temperatures.

72.6   APPLICATIONS

In both properties and variety of end uses, epoxies are one of the most remarkable groups of plastics available. They have the ability to "stick" to many substrate—plastics, metal, glass, and wood. Epoxies have outstanding strength, chemical resistance, water resistance, electrical properties, and heat stability. The main fields of use include adhesives, casting, laminating, surface coatings, surfacing compounds, tooling, electrical potting and encapsulation, body "solders," repair kits, and vinyl stabilizers. The most significant of these are described below.

72.6.1 Adhesives

Adhesives based on epoxy resins are used for bonding such diverse materials as plastics, metal, ceramics, glass, vulcanized rubber, cork, leather, textiles, and so on.

The epoxy-based adhesives range from fluid to viscous, pastelike, and solid resins with corresponding liquid and pastelike hardeners. A few, but not many, contain solvents. When the products are supplied in solid or in powder form, the resin and hardener are already mixed. As a rule, they are applied to preheated surfaces or else to cold surfaces which are then heated. For liquid or paste forms, the hardener is generally added at room temperature immediately before us. The mixture is then applied to the surfaces to be bonded. With adhesives containing a solvent, the latter must be evaporated by predrying after application. In the same way that conventional bonding methods (such as riveting, welding, or soldering) impose special constructional requirements, bonding with epoxy resin-adhesives also requires that the components be designed so that the joints, as far as possible, are strained in the direction of the adhering surfaces (shear).

Heat-curing adhesives are employed for bonding similar materials that have about the same coefficients of thermal expansion. Materials sensitive to heat, ceramics, or other materials with very different coefficients of thermal expansion should be bonded with adhesives that cure at room temperature. Adhesives that are flexible in their cured state are particularly suitable.

Although curing can take place without pressure, the parts to be bonded should nevertheless be so fitted that the thickness of the resulting joint does not exceed 0.004–0.008 in (0.1–0.2 mm); these joint thicknesses not only provide the greatest strengths, but the resin is also kept in place in the joint by its capillary strength. For thicker joints [up to approximately 0.20 in. (5 mm)], pasty adhesives should be selected or else steps must be taken to ensure that the adhesive cannot exude from the joint.

Curing varies according to the adhesive employed and is effected either at room or slightly elevated temperatures, or at elevated temperatures only. The curing time can vary from days to hours or only minutes: the higher the temperature, the shorter the curing cycle. Mechanical strengths of joints made with heat-curing adhesives are generally higher than those bonded with adhesives that set at room temperature. The cured epoxy resin in the joint is infusible, practically insoluble, and resistant to aging. Because of its dielectric properties, it can be employed for certain insulating purposes in the electrical industry.

In many cases it is advantageous to use epoxy resin adhesives with a carrier material, such as fibrous glass cloth. This is advisable, for example, when the coefficient of thermal expansion of the adhesive layer must be brought into line with that of the materials to be bonded, or when the joint must serve at the same time as a positive insulating layer between the parts.

Examples of bonding with epoxy adhesives include, in metal construction, light metal window and door frames, metal letterboxes, sheet metal laminates, light metal photographic tripods, adjustable pipe and hose couplings for fire brigades, milk pails, and other sheet metal containers; in foundries and machine construction, components for durable metal forms, light metal gear housing components, die-cast parts, packing of porous castings and joints, and securing of threads; in the tool and grinding wheel industry, hard metal supports for cutting and measuring tools and centerless grinding machines, and stamps for cutting tools; and in vehicle construction, doors, seats, and luggage carriers, wall linings for railway coaches and buses, ships, and aircraft, supporting constructions and reinforcement, and defrosting equipment for aircraft.

72.6.2 Casting

Epoxy casting resins and their hardeners are supplied either as liquids or solids at room temperature. Curing of the resin-hardener mixture is achieved either at room temperature or by heating, but no pressure need be applied in either case. No perceptible quantity of volatile substances is given off during cure, but the process is accompanied by the development of a varying amount of exothermic heat. Shrinkage during cure is very slight. Cured castings are infusible and isotropic. They are practically insensitive to water, weak acids, and alkalies, to many solvents, and also to atmospheric influences. If adhesion to mold parts is not desired, it must be prevented by means of release agents.

The viscosity and pot-life of the resin-hardener mixtures can be varied to a considerable extent by the choice of resin and hardener, the operating temperature, the addition of fillers, or, in certain cases, by adding accelerators. The viscosity may amount to only a few centipoises; the pot-life can last only minutes or can be extended to days or months.

Before processing, the resins may be extended with various fillers such as quartz flour or sand, porcelain and chalk four, kaolin, talcum, slate flour, certain metallic oxides, glass fibers, or ground graphite and metals. Such additives reduce production costs and, at the same time, alter the properties of the cured resin. These alterations can bring about a reduction of the exothermal reaction, still less shrinkage on curing, a smaller coefficient of linear thermal expansion, greater heat conductivity, increased compressive strength, lower impact flexural strength, and higher specific gravity. The quantity of filler to be added varies according to its type and granular size; it may amount to as much as several times the weight of the resin.

The color of the resin, which is normally straw to yellow, may be varied as desired by the addition of suitable pigments.

Curing of epoxy casting resins occurs at room or at a slightly elevated temperatures (68–212°F (20–100°C)] or at elevated temperatures only [212–392°F (100–200°C)], depending on the components used. The curing time may range from days to hours or only minutes; a higher temperature yields a shorter curing time.

The possibility of processing pure or filled epoxy resins without pressure generally permits the casting of solid pieces in molds, the manufacture of sizable workpieces, and the embedding or bonding of sensitive elements. As solvent-free liquid products, they can also be used in vacuum for nonporous impregnantion of coils or for filling the smallest cavities in metal castings.

Examples of epoxy resin castings include, in the chemical industry, battery casings, centrifugal and membrane pumps, pipes for transport of liquids, pipe fittings such as friction valves, flange couplings, closings, casing and parts of membrane and flap valves, laboratory ware, abrasion-proof underwater slide bearings for centrifugal pumps, and dye baths; in the electrical industry, high-voltage, low-current applications, telephone and high-frequency equipment, x-ray apparatus, electrical measuring instruments, electrical equipment for internal combustion engines, transformers, stators, rotors, commutators, circuit breakers, cable-closing sleeves, and cable end stoppers. In these industries, epoxies are variously used for impregnating cores, transformer coils, capacitors, and so on.

72.6.3 Laminating

Reinforcement materials used singly or in combination with epoxies in the production of laminates include fibrous glass (in the form of woven fabric, mat, roving, and chopped fiber), paper, textile materials, asbestos, mica, metal foils, and metal netting.

Laminates made with epoxy resins are remarkable for their high mechanical and electrical resistance, the values of which are greatly influenced by the reinforcing material. These laminates are practically insensitive to the attacks of humidity, oil, various solvents, weak acids, and alkalies. They retain their properties within a temperature range from −76 to 230°F (−60 to 110°C) and, if resins that are particularly heat-resistant are used, to more than 500°F (260°C).

The following methods may be employed: hand lay-up, vacuum bag, pressure bag, matched-metal molding, casting and impregnating, and preforming.

Epoxy/fibrous-glass laminates have assumed particular significance in the field of laminates inasmuch as epoxy resins adhere extremely well to glass fibers and experience very little shrinkage on curing. The mechanical strength of the laminates is accordingly high, especially in environments where moisture, continuous loads, and vibration are present. The laminates also exhibit high dimensional stability.

Examples of epoxy laminating resins include, in the automotive and aircraft industries, Keller models, checking and assembly fixtures, drill jigs, router jigs, and cutting, wrapping, and welding fixtures, aircraft noses, fuel tanks, floors, and partition walls, bodies, and body components; in plastics processing, molds for low-pressure molding of glass cloth or glass mat laminates with polyester or epoxy resins and for vacuum forming of thermoplastic sheets; in boat construction, all types of boats and coverings for wooden boat hulls; in the electrical industry, pipes and conduits, inserts for switch and quenching chambers, and printed circuits; in the metals industry, models, core boxes, apparatus parts, and casings; and in miscellaneous uses such as fishing rods and helmets.

72.6.4 Surface Coatings

Epoxy resins for surface coatings include three general groups:

- Heat-curing resins, used in acid- and alkali-resistant coatings for drums, cans, and collapsible tubes; punching and deep-drawing coatings for the metal sheet industry; protective coatings for metal foils; enamels for corrosion protection of fittings and machinery; and primers for appliances
- Ester resins of the air-drying and nondrying types, used in manufacturing durable exterior coatings and primers with adhesion and resistance properties; enamels with very good color and gloss retention and excellent resistance to detergents; and in washing machines, refrigerators, and gym floors
- Room-temperature curing resins used in the manufacture of two-component coatings and compounds that cure at room temperature to very hard and highly flexible coatings with excellent chemical resistance. These are used for clear finishes for wood, brass, copper, nickel, and chromium; transparent colored lacquers for aluminum foil; pigmented coatings for machinery, apparatus, containers, and pipes exposed to highly corrosive environments; linings for concrete pipes and tanks; coatings for plastics; protective coatings for the electrical industry; rust-proofing primers; and exterior finishes for water-proofing and decorating masonry, concrete, and cinder block. The curing time of most of these coatings can be cut down to a few minutes by the use of elevated temperatures.

72.6.5 Highway Surfacing

One of the largest potential uses for epoxies is in road and highway building and repair. Many companies are presently conducting extensive development programs with test strips on bridges and roads. From these tests, the superiority of epoxy-based compounds is becoming increasingly evident. The test strips include several different specific uses: sealing membrane, high-friction surfaces, cast traffic reflectors, and adhesive films for repairs, bonding new concrete to old concrete, and reflection strips and traffic bars.

Epoxy systems are rarely used in pure forms for these applications since they are many times greater in adhesive strength and chemical resistance than the conventional construction materials with which they are combined. Instead, the resins are compounded with fillers, pigments, high-friction materials, and aggregate to produce mixtures that fit particular ap-

TABLE 72.2. Commercial Epoxy Resins and Their Characteristics

I. Epoxy Resins Derived From Bisphenol A and Epichlorohydrin

Epoxy eq wt	Softening point or Viscosity, cps at 25°C	Commercial Products	Remarks	Applications
A. Liquid resins				
172–195	500–900 cps	Araldite 506, D.E.R. 334, Epi-Rez 5071, Epon 815, Epotuf 37-130	Lowest viscosity resins diluted by butyl glycidyl ether	Casting, laminating
175–190	5000–10,000 cps	Araldite 6005, Araldite 6008 D.E.R. 330, Epi-Rez 509, Epon 826, Epotuf 37-139	Medium viscosity resins	General purpose
180–200	10,000–16,000 cps	Araldite 6010, D.E.R. 331, Epi-Rez 510, Epon 828, Epotuf 37-140	Standard resins	General purpose
171–181	4,000–5,500 cps	D.E.R. 332, Epi-Rez 50810, Epon 825	Low viscosity diglycidyl ether of bisphenol A	General purpose
190–210	15,000–22,500 cps	Araldite 6020, Epon 830	High viscosity resins	Adhesives, coatings
B. Liquid Resins for Chain Extension				
186–190	2000–7000 cps	Epon 829	Catalyzed 96.5% NV	Intermediate
198	5000 cps	Araldite XU GY 351	Catalyzed 97% NV	Intermediate
199–202	2300–4600 cps	D.E.R. 333	Catalyzed 96.5% NV	Intermediate
C. Solid Resins				
385–500	60–75°C	Araldite 6060, Epi-Rez 519, Epotuf 37-300		Casting, encapsulating
450–550	65–80°C	Araldite 7071, Epi-Rez 520C, Epon 1001, Epotuf 37-301,		Coatings, prepregs
550–725	75–90°C	Araldite 7072, Epi-Rez 522C, Epotuf 37-302, Epon 1002, D.E.R. 664		Coatings, prepregs

TABLE 72.2. *(Continued)*

Epoxy eq wt	Softening point or Viscosity, cps at 25°C	Commercial Products	Remarks	Applications
875–1025	95–105°C	Araldite 6084, D.E.R. 664, Epi-Rez 530C, Epon 1004, Epotuf 37-304		Coatings, laminating
1550–2000	113–123°C	Araldite 7097, D.E.R. 667, Epotuf 37-307		Coatings
2000–2500	125–135°C	Araldite 6097, Epi-Rez 540C Epon 1007		Coatings
3500–5500	135–155°C	D.E.R. 669		Coatings
2400–4000	145–155°C	Araldite 6099, Epi-Rez 550 Epon 1009		Coatings
4000–6000	155–165°C	Epon 1010		Coatings
4000–6000	165–180°C	Epi-Rez 560		Coatings
II. Epoxy Novolac Resins				
172–179	1400–2000 cps at 125°F	Araldite EPN 1139, D.E.N. 431, Epon 152	Phenol novolac	Casting coatings, encapsulation adhesives
176–181	35,000–70,000 cps at 125°F	Araldite EPN 1138, D.E.N. 438, Epon 154 Epotuf 37-170	Phenol novolac	Casting, laminates, moldings, adhesives
230	80°C	Araldite ECN 1280	Cresol novolac	Molding powders, coatings
235	90°C	Araldite ECN 1299	Cresol novolac	Molding powders, coatings
275–285	77–82°C	Epotuf 37-171	Phenol novolac	Molding powders, coatings
295–330	90–95°C	Epi-Rez 529	Phenol novolac	Molding powders, coatings
III. Diglycidyl Ethers of Aliphatic Polyols				
175–205	30–60 cps	D.E.R. 736	Propylene glycol based	Flexibilizer
305–335	55–100 cps	D.E.R. 732	Propylene glycol based	Flexibilizer
400–455	2000–5000 cps	Araldite 508	Modified aromatic based resin	Flexibilizer

IV. Cycloaliphatic Epoxy Resins

70–74	15 cps	ERL 4206	Vinylcyclohexene diepoxide	Reactive diluent
91–102	38–42°C	ER 4205	Bis(2,3-epoxy cyclopentyl) ether	Reactive diluent, reinforced plastics
131–143	350–450 cps	ERL 4221, Araldite CY-179	3,4-Epoxy-cyclohexyl methyl-3,4-epoxy-cyclohexane carboxylate	Filament winding, electrical laminating
205	650 cp	ERL 4299	Bis(3,4-epoxy cyclohexyl methyl adipate)	Castings, filament winding, laminating electrical
161–178	700–900 cps	Araldite CY-184	Hexahydrophthalic acid diglycidyl ester	Castings, electrical

V. Flame-retardant Epoxy Resins

205–225	2,000–5000 cps	14–16% Br	D.E.R. 580	General purpose
223–246	350–540 at 70°C	20% Br	Araldite 8047	Laminating
240–260	500,000–1,000,000 cps	24–26% Br	Epotuf 37-200	General purpose
325–375	45–55°C	45–55% Br	D.E.R. 542	General purpose
350–450	55–65°C	50% Br	Epi-Rez 5163	Laminating
455–520	68–80°C	19–22% Br	D.E.R. 511, Araldite 8011, Epon 1045	Laminating
600–750	90–100°C	42% Br	Epi-Rez 5183	Laminatings, molding powders

plications. The wide formulating latitude of epoxies permits them to be tailored to suit many different end uses and application conditions. All of the compounds developed thus far are two-part systems, but mixing is usually on a simple basis, requiring one-to-one portions or the addition of the contents of one container to that of another.

72.6.6 Floor Surfacing

Compounds based on epoxy resins for floor surfacing are another recent fast-growing innovation in the building field. These materials, based on liquid epoxy resins and curing agents mixed with various concrete aggregate, are supplied as three components that are mixed prior to application. The usual method of applying these materials consists of trowelling a layer about 1/4 in. (6 mm) thick over old flooring or conventional concrete. Potential areas of application include sidewalks, warehouses, chemical plants, swimming pools, silo linings, dairies, and food processing plants.

72.6.7 Other Applications

Other applications include auto body "solders," repair kits, molding compounds, foams, and vinyl stabilizers.

Automotive body solders to replace toxic lead and tin solders adhere well, are light in weight, and can be handled by relatively unskilled workers.

Home repair kits come in a variety of sizes and forms, some with only a few ounces of resin and sufficient hardener for minor repairs, others with several pints of resin, mixing containers, and fibrous glass fillers and cloth for more elaborate work. Kits for repairing auto fenders and bodies are also growing in popularity. Premixed molding compounds are being developed and perfected that can be handled with more ease and reliability than earlier epoxy materials. However, special mold release agents are still a necessity because of the high degree of adhesion of the epoxy resins. Usually, fillers and reinforcements are incorporated into these mixtures.

Foams manufactured from epoxies are sometimes used to encapsulate electrical components for insulation purposes. These foams are very low in density and adhere well to metal parts. The method of manufacture involves the addition of a suitable blowing or foaming agent. Gas is evolved simultaneously with the curing reaction to form a foam structure that conforms to the walls of the container or mold.

Epoxy resins make excellent stabilizers for vinyl-based plastics and coatings. Offering good resistance to both ultraviolet light and high temnperatures, low molecular-weight epoxies function synergistically by acting as hydrogen chloride acceptors.

72.10 COMMERCIAL GRADES

See Table 72.2.

72.14 TRADE NAMES (PRODUCERS)

Araldite, ECN & EPN (Ciba-Geigy); Bakelite, E.R.L. (Union Carbide); D.E.R., D.E.N. (Dow Chemical); Epi-Rez (Hi-Tek); Epon (Shell Chemical); Epotuf (Reichold Chemicals).

73

REINFORCED PLASTICS, BORON AND SILICON CARBIDE CONTINUOUS FILAMENTS

Michael E. Buck, John A. McElman, and Melvin A. Mittnick

Textron Specialty Materials (formerly Avco Specialty Materials), 2 Industrial Ave., Lowell, MA 01851

73.1 CATEGORY

Boron fibers, demonstrated at Texaco Experiments Inc., in 1959, were the first of a family of high-strength, high-modulus, low-density reinforcements developed for advanced aerospace applications.

The initial development of boron was nurtured by the Air Force Materials Laboratory (AFML) now called The Wright R&D Center (WRDC), with the goal of producing higher performance systems through the use of materials with higher specific properties (that is, strength and stiffness divided by density). This Department of Defense interest continued and led to the achievement of routine commercial production of the fibers on a scale where they could be considered a viable structural material by the aerospace industry. As a result, boron–epoxy composites find substantial application in the F-14, F-15, and B-1 aircraft as well as in numerous other aerospace and sporting goods applications.

Although it is best known for its use as a reinforcement in resin–matrix composites,[1,2] boron fiber has also received considerable attention in the field of metal/matrix composites.[3-5] Boron–aluminum was employed for tube-shaped truss members to reinforce the Space Shuttle Orbiter structure and has been investigated as a fan blade material for turbofan jet engines. There are drawbacks, however, in the use of boron in a metal matrix. The rapid reaction of boron fiber with molten aluminum[6] and long-term degradation of the mechanical properties of diffusion-bonded boron–aluminum at temperatures greater than 1000°F (540°C) preclude its use both for high-temperature applications and for potentially more economically feasible fabrication methods such as casting or low-pressure, high-temperature pressing. These drawbacks have led to the development of the silicon carbide (SiC) fiber. SiC fiber will have equivalent, if not better, mechanical properties, offer improved high temperature resistance, and be less expensive than the present boron fiber.

73.2 FIBERS

Both boron and silicon carbide fibers are manufactured by a chemical vapor deposition (CVD) process. In the CVD process, a solid is formed by the decomposition or reduction of one or more gaseous molecules on a heated substrate. Fine-grained materials may be formed at one third or less of the melting point of the material for simple crystal structures and somewhat higher for more complex crystal structures. The crystallite size also depends on the rate of deposition and the surface mobility. Typically, a filament is run through a chamber where the filament is resistance heated, and suitable gases decompose on the heated wire. Although the linear deposition rate is quite high by plating process standards, the contact time required to deposit the necessary coating thickness is in tens of seconds or more. Hence, for any appreciable production rate, many reactors must be run in parallel.

The cost for the substrate and deposition gases and the capital equipment for the reactors and gas handling is quite high. Hence, the filament tends to be rather expensive. However, CVD can be used to deposit almost any type of large-diameter fiber or coating.

73.2.1 Boron Fiber

Boron Filament Production Process. Figure 73.1 shows a schematic diagram of the basic deposition unit. The reactor module consists of a glass deposition tube that is fitted with gas inlet and outlet ports, two mercury-filled electrodes, a tungsten substrate payout system, and a boron filament take-up unit. Tungsten substrate, typically 0.0005 in. (12.5 μm) in diameter, is drawn through the reactor and heated through resistance heating by the DC power supply. Prior to entering the deposition reactor, shown schematically in Figure 73.1, the tungsten substrate is passed through a short "cleaning" stage (not shown in the figure), in which the substrate is heated to candescence in a hydrogen atmosphere in order to remove surface contaminants and residual lubricants that are used in the drawing of tungsten wire. A stoichiometric mixture of boron trichloride and hydrogen is introduced at the top of the

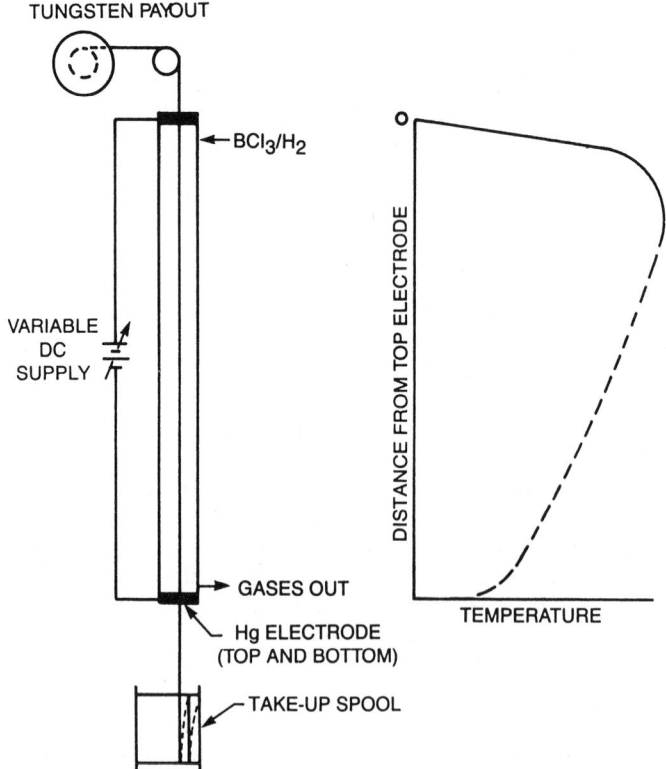

Figure 73.1.   Schematic diagram of boron filament reactor and temperature profile.

reactor. At about 2370°F (1300°C), a mantle of boron is deposited on the tungsten by the reaction

$$BCl_3 + \tfrac{3}{2}\,H_2 \longrightarrow B + 3\,HCl$$

Exhaust gases consisting of HCl, intermediate species, and unreacted H_2 and BCl_3 are removed through the outlet port at the bottom of the reactor and are processed to remove and recycle the unreacted BCl_3. A photograph of about 100 boron filament reactors is shown in Figure 73.2. Standard diameters of the filament exiting from the reactor are 0.0040 in. (100 μm) and 0.0056 in. (140 μm). The filament diameter is changed by varying the drawing rate. Figure 73.3 shows photomicrographs of cross sectional and longitudinal views of a 0.0040-in. (100 μm) boron filament.

Other fiber-forming processes have been attempted. These include the thermal decomposition of diborane and the drawing of fibers from molten boron. The CVD process described above, however, has been found to be the only economical process.

Boron Filament Modifications. Several variations of the standard boron filament were investigated in an effort to improve temperature resistance, wettability (in metal matrices) or mechanical properties, or to decrease cost. These variations involved adding surface coating of B_4C or SiC and replacing the tungsten core with a carbon core.

Each of these modifications were technically successful in that they achieved specific goals, but each had an adverse effect in another area. The B_4C and SiC coatings significantly improved wettability but also increased cost. Because of its larger diameter, the carbon core made it necessary to increase the diameter of the boron filament to attain the same modulus value. For example, a 0.0042 in. (107 μm) boron filament made with the available 0.0013 in. (33 μm) carbon core was required to achieve the same modulus as in 0.0040 in. (100

Figure 73.2. Four banks of boron filament reactors.

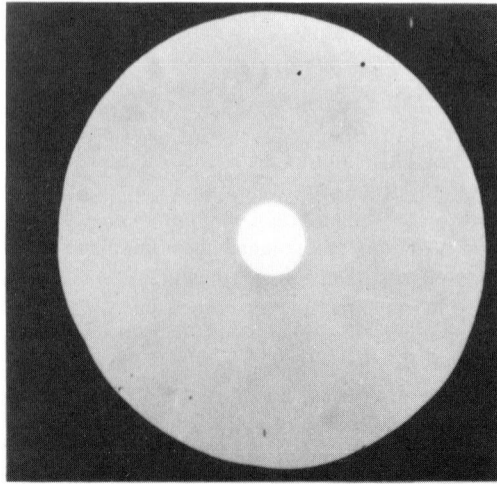

4 MIL FILAMENT CROSS SECTION

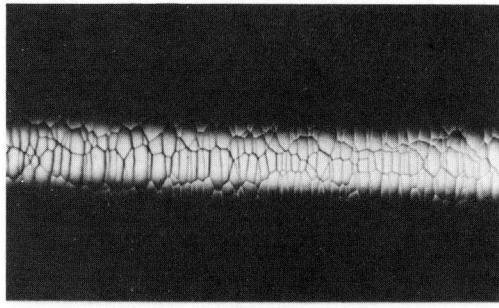

MAGNIFICATION OF BORON FILAMENT SURFACE

Figure 73.3. Cross-section and longitudinal photomicrographs of a 4-mil boron filament.

μm) boron filament with a 0.0005 in. (13 μm) tungsten core. The larger diameter caused problems in then-existing aerospace applications, and work on this fiber was discontinued.

Fiber Properties. Boron filament is unique among composite reinforcement fibers available in production quantities in that it combines superior tensile, compressive, and flexural strengths; high modulus; and light weight in one fiber. Its tensile strength exceeds 5×10^5 psi (3,500 MPa), a compressive strength of about 1×10^6 psi (7,000 MPa), a modulus of 58×10^6 psi (400 GPa), and a density of only 160 lb/ft³ (2.57 g/cm³).

The strength of boron fiber is determined by the statistical distribution of flaws produced during the deposition process. The major flaw types are voids near the tungsten diboride core–boron mantle interface; internal stresses "locked in" during deposition; and surface flaws, principally crystalline or nodular growth. (These flaws are discussed further in reference 1.) Because of this, a histogram of fiber tensile strengths does not follow a normal distribution but, rather, is skewed in having a low-strength tail. The distribution is better described by Weibull statistics. Figure 73.4 provides a histogram of typical boron fiber strengths, showing an average ultimate tensile strength of about 520,000 psi (3,600 MPa) with a coefficient of variation of about 15%. These values have shown steady improvement over the 15 years that these fibers have been in commercial production. Initial values showed average tensile strengths of 450,000 psi (3100 MPa) with 20% coefficients of variation.

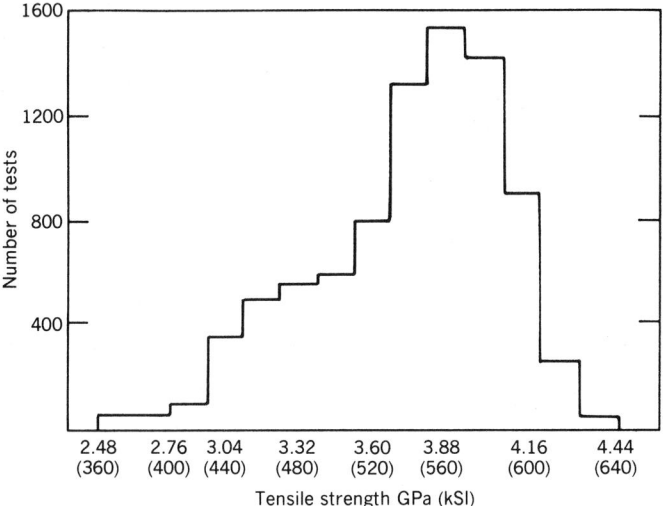

Figure 73.4. Typical histogram of boron filament tensile strengths.

Fiber modulus, on the other hand, is an intrinsic material property and shows very little variation. It is best regarded as a composite property depending on the (reacted) substrate and the volume fraction of pure boron in the fiber. For 4 mil (0.1 mm) boron on tungsten, the value is 58×10^6 psi (4×10^5 MPa). Figure 73.5 compares the strength/weight and stiffness/weight ratios of boron fibers to various materials.

73.2.2 Silicon Carbide Fiber

Silicon Carbide Fiber Production Process. Continuous silicon carbide filament is produced in a tubular glass reactor by CVD, as shown in Figure 73.6. The process occurs in two steps on a carbon monofilament substrate that is resistively heated. During the first step, pyrolytic graphite (PG) approximately 1 μm thick is deposited to smooth the substrate and enhance electrical conductivity. In the second step, the PG-coated substrate is exposed to silane and hydrogen gases. The former decomposes to form beta silicon carbide contin-

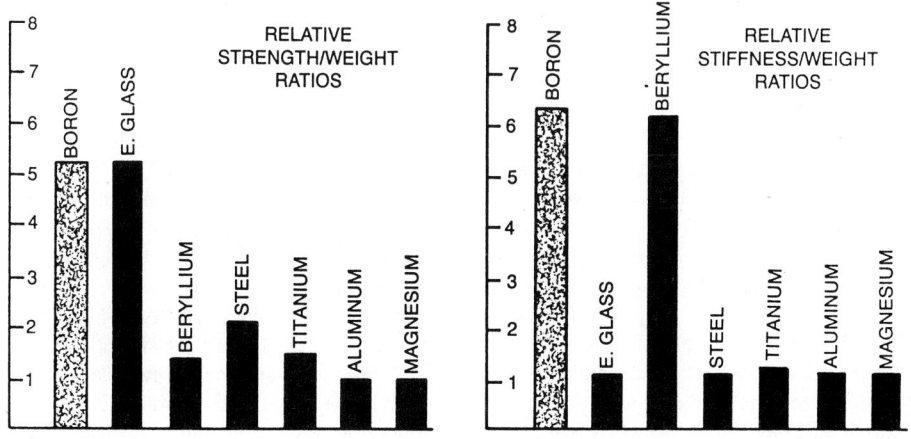

Figure 73.5. Relative strength/weight and stiffness/weight ratios of boron filament.

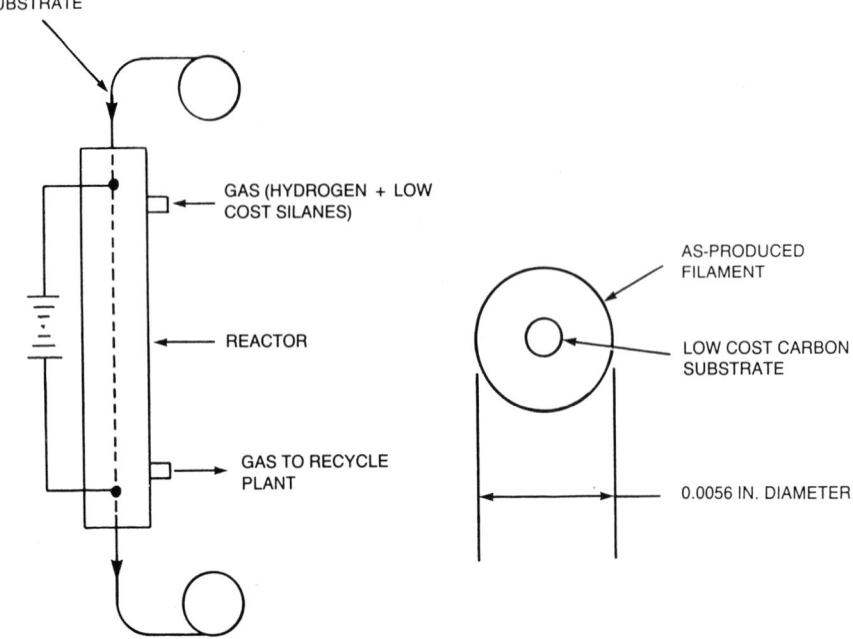

Figure 73.6.   Fabrication of silicon carbide fiber.

uously on the substrate. The mechanical and physical properties of the silicon carbide filament are tensile strength = 500,000 psi (3500 MPa); tensile modulus = 60,000,000 psi (413,000 MPa); density = 187 lb/ft^3 (3.0 g/cm^3); coefficient of thermal expansion (CTE) = 2.7 × 10^{-6}F (1.5 × 10^{-6}C); and diameter = 0.0056 in. (0.140 mm).

Figure 73.7 is a photomicrograph cross section of a fiber surface showing the interior of the fiber and the carbon monofilament substrate. Various grades of fibers are produced, all of which are based on the standard beta silicon carbide deposition process described above where a crystalline structure is grown onto the carbon substrate. The beta SiC is present as such across all of the fiber cross section except for the last few microns at the surface (Fig. 73.8). Here, by altering the gas flows in the bottom of the tubular reactor, the surface composition and structure of the fiber are altered first by an addition of surface strength, then by modification of the silicon-to-carbon ratio to provide improved bonding with the metal.

Processing Considerations and Fiber Properties.   As in any vapor deposition or vapor transport process, temperature control is of the utmost importance in producing CVD SiC fiber. The Textron process calls for a peak deposition temperature of about 2370°F (1300°C).

Temperatures significantly above this cause rapid deposition and subsequent grain growth resulting in a weakening of tensile strength. At temperatures significantly below the optimum, high internal stresses are formed in the fiber that result in a degradation of the metal matrix composite properties upon machining transverse to the fiber.[7]

Substrate quality is also an important consideration in SiC fiber quality. The carbon monofilament substrate, which is melt-spun from coal-tar pitch, has a very smooth surface with an occasional surface anomaly. If it is severe enough, the surface anomaly can result in a localized area of irregular deposition of pyrolytic graphite and silicon carbide (see Figures 73.8 and 73.9), which is a stress-raising region and a strength-limiting flaw in the fiber. The carbon monofilament spinning process is controlled to minimize these local an-

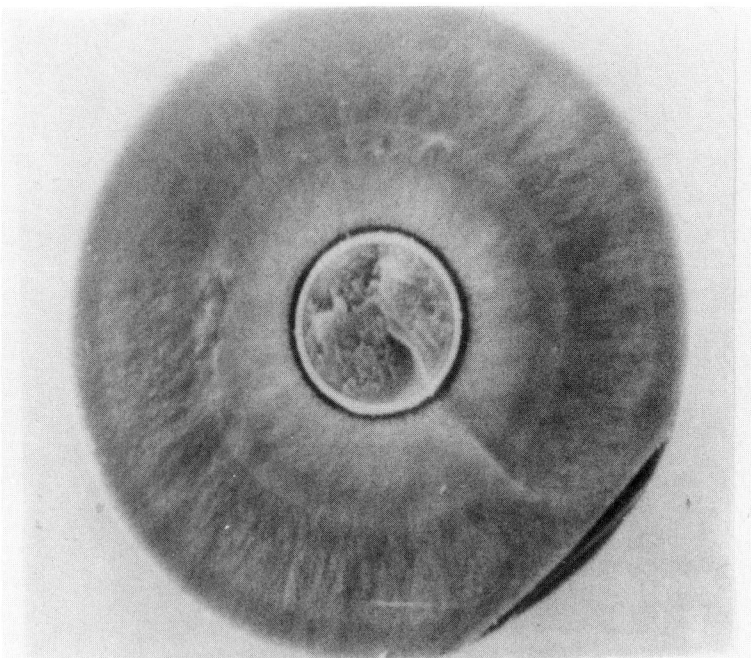

Figure 73.7.   Cross-section of a silicon carbide fiber.

omalies sufficiently to guarantee routine production of high-strength [above 500,000 (3500 MPa)] SiC fiber.

Another strength-limiting flaw that can result from an insufficiently controlled CVD process is the pyrolytic graphite (PG) flaw.[8] This flaw results from irregularities in the PG deposition. Two causes of PG flaws are disruption of the PG layer due to an anomaly in the carbon substrate surface, or mechanical damage to the PG layer prior to the SiC deposition. Examples of both types of flaws are shown in Figures 73.9 and 73.10. PG flaws often cause a localized irregularity in the SiC deposition resulting in a bump on the surface (Fig. 73.10). Poor alignment of the reactor glass can result in mechanical damage to the PG layer by abrasion (Fig. 73.11**a**). A series of PG flaws results in what is called the "string of beads" phenomenon (Fig. 73.11**b**) at the surface of the fiber. The mechanical properties of such a fiber are severely degraded. These flaws are minimized by careful control of the PG deposition parameters, proper reactor alignment, and the minimization of substrate surface anomalies.

The surface region of Textron's SiC fibers are typically carbon rich. This region is important in protecting the fiber from surface damage and subsequent degradation in strength. An improper surface treatment or mishandling of the fiber (such as abrasion) can result in strength-limiting flaws at the surface. Surface flaws can be identified by an optical examination of the fiber fracture face (Fig. 73.12). These flaws are minimized by proper process control and handling of the fiber (minimizing surface abrasion).

Typical mechanical properties of Textron CVD SiC fiber consist of average tensile strengths of 550,000–600,000 psi (3800–4100 MPa) and elastic moduli of $(58–60) \times 10^6$ psi $(4.–4.13 \times 10^5$ MPa). A typical tensile strength histogram (Fig. 73.13) shows an average tensile strength of 580,000 psi (4,000 MPa) with a coefficient of variation of 15%. Figure 73.14 shows the strengths of SiC fibers after exposure at elevated temperatures in argon and oxygen.

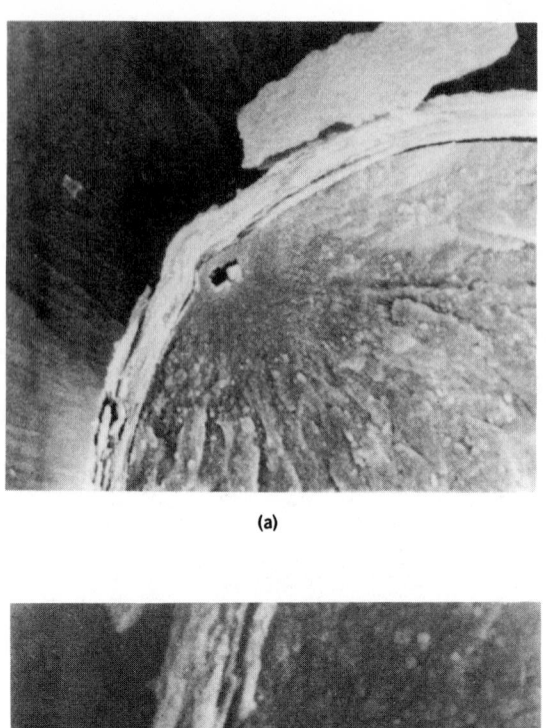

(a)

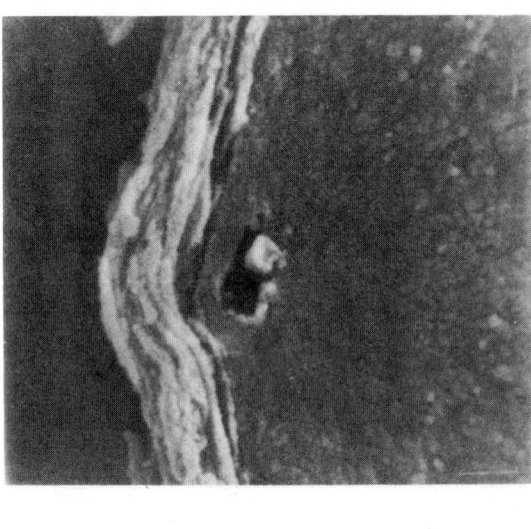

⊢———⊣
1 μm

(b)

Figure 73.8. Strength-limiting flaw at the carbon monofilament/SiC interface caused by a void at the substrate surface (Layered morphology of the pyrolytic graphite is evident in (**b**).

Fiber Variations. The surface region of the SiC fiber must be tailored to the matrix. Shown in Figure 73.15 are schematics of the surface composition of three fiber types. SCS-2 has a 1 μm carbon-rich coating that increases in silicon content as the outer surface is approached. This fiber has been used to a large extent to reinforce aluminum. SCS-6 has a thicker (3 μm) carbon-rich coating in which the silicon content exhibits maxima at the outer surface and 1.5 μm from the outer surface. SCS-6 is primarily used to reinforce titanium.

SCS-8 has been developed as an improvement over SCS-2 to give better mechanical

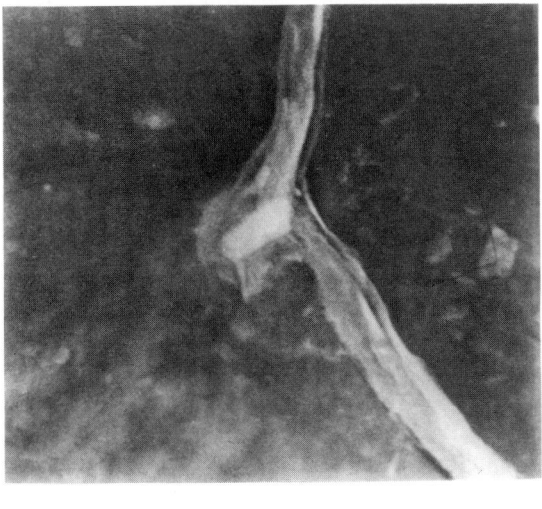

1 μm

Figure 73.9. Irregularity in the PG layer associated with anomaly in substrate surface.

properties in aluminum composites transverse to the fiber direction. The SCS-8 fiber consists of 6 μm of very fine-grained SiC, a carbon-rich region of about 0.5 μm, and a less carbon-rich region of 0.5 μm.

Cost Factors. From an economic standpoint, silicon carbide is potentially less costly than boron for three reasons: the carbon substrate used for silicon carbide is lower in cost than the tungsten used for the boron; raw materials for silicon carbide (chlorosilanes) are less expensive than boron trichloride, the raw material for boron; and deposition rates for silicon carbide are higher than for boron, hence more product can be made per unit of time.

73.3 ORGANIC MATRIX COMPOSITES

73.3.1 Boron-Reinforced Resins

Composite Processing. Most boron filament is sold in the form of continuous boron epoxy preimpregnated tapes, commonly known as prepreg. This section will discuss prepreg manufacture and its use in fabricating composite parts.

Continuous boron epoxy prepreg tapes are typically sold in widths ranging from 0.25 to 6 in. (6.4 to 152 mm). The boron filaments are unidirectionally aligned and occupy about 50% of the composite volume. Typical loadings are upwards of 200 filaments/in. (8 filaments/mm) of width, although this has been and can be varied. The resin content is typically 30–35% by weight.

During the manufacturing process, the tape is backed on one side with a fine layer of style 104 glass scrimcloth with a thickness of 0.0012 in. (0.03 mm). The scrimcloth offers two advantages: it serves to give the tape some lateral integrity during handling and lay-up,

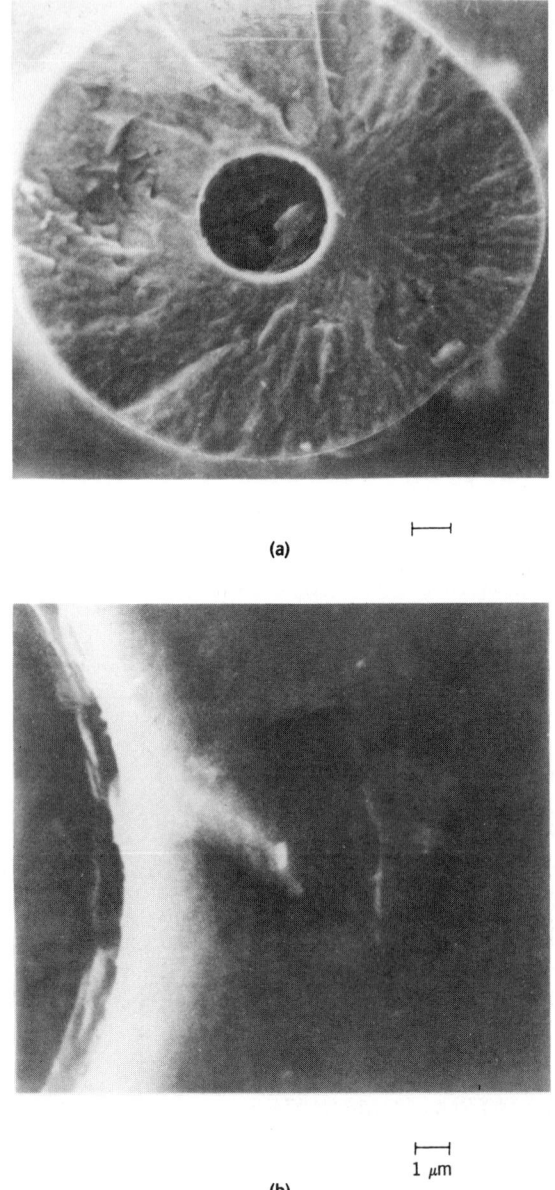

(a)

1 μm

(b)

Figure 73.10.   (a) Surface bump in the SiC fiber; (b) associated with damaged PG layer.

and it helps to maintain filament spacing and collimation during the bonding process. A photograph of the boron epoxy prepreg is shown in Figure 73.16.

There are two production processes commonly used in the manufacture of boron epoxy prepreg tape. The first, which is called the two-step process, begins with the production of 0.25 in. (6.4 mm) ribbon consisting of approximately 50 filaments. Subsequently, tape of greater width, typically 3 or 6 in. (76 or 152 mm) wide, is formed by collimating the narrow ribbons on a separate piece of equipment.[9] In this method, the resin coating of the filaments

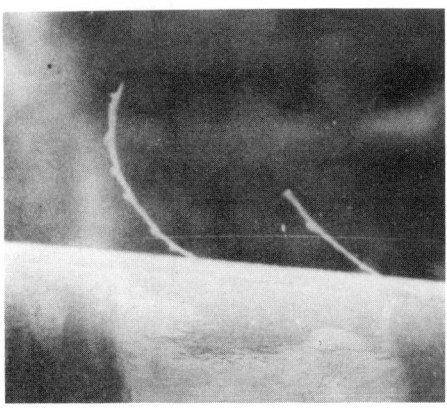

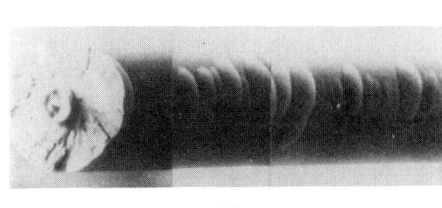

(a)

(b)

Figure 73.11. (a) Damaged pyrolytic graphic layer; (b) associated "string of beads" phenomenon.

is accomplished by means of a hot-melt coating roller directly applying resin to the filaments. A photograph showing the collimating of the narrow ribbons into 6-in. (152-mm) tape is shown in Figure 73.17.

The second, which is called the one-step process, is a proprietary process developed by Textron Specialty Materials. In this process, upwards of 1250 filaments are brought together in a single step and combined with the glass scrimcloth and the resin, forming up to a 6-in.

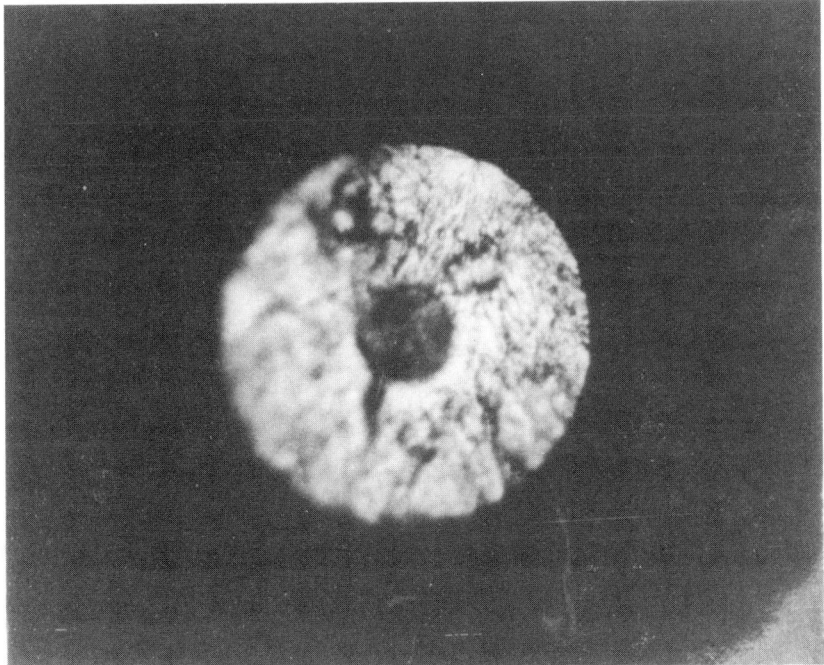

Figure 73.12. Fracture face of SiC fiber showing strength-limiting flaw initiated at the fiber surface (1 o'clock).

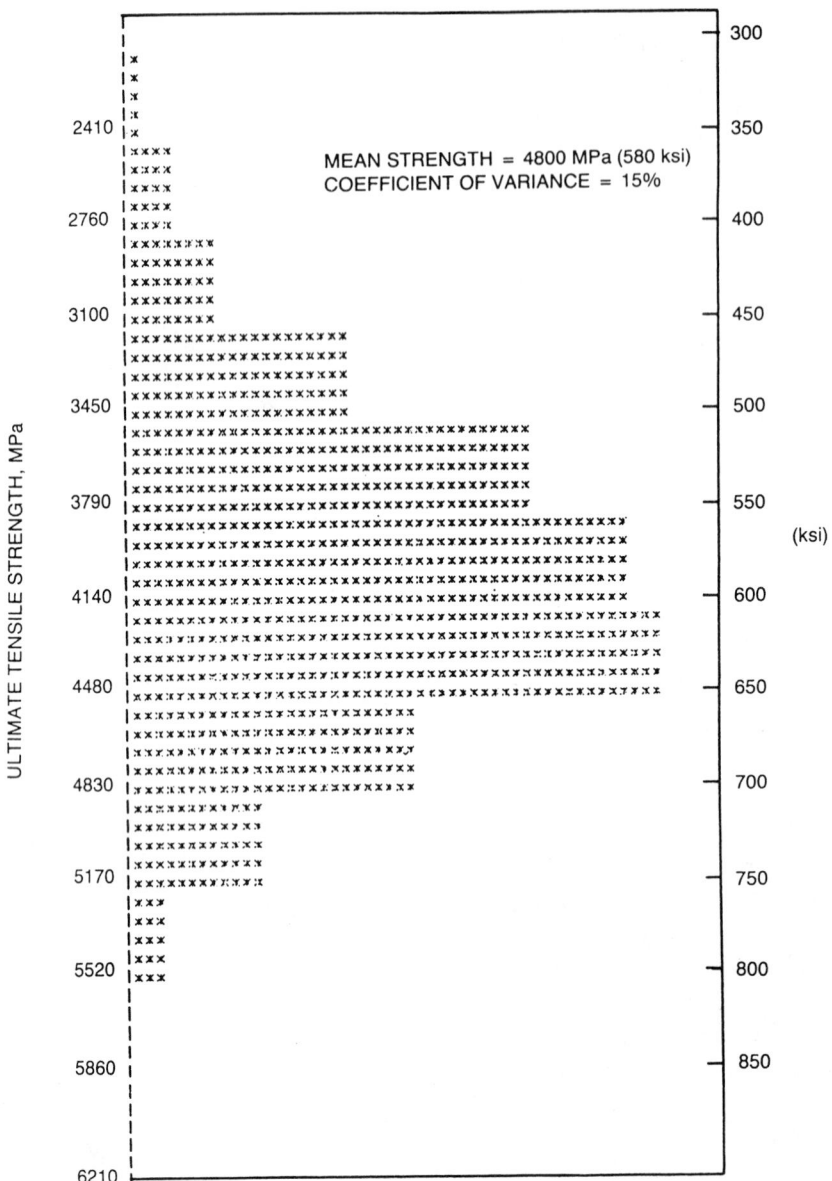

Figure 73.13. Histogram of CVD SiC (SCS-2) fiber tensile strength.

(152-mm) wide tape. In this method, the resin coating of the filaments is accomplished by the transfer of a resin film from a backing paper onto the collimated filaments. The one-step process reduces scrap and labor and is more versatile than the previously described two-step process.

The epoxy resins generally employed cure at either 350°F (177°C), which meets most aerospace specifications, or at 250°F (121°C), which is suitable for most lower-temperature applications such as in sporting goods. Since the curing agent is part of any specific resin formulation, the resin will cure slowly, even at room temperature. At room temperature this will occur over a several-week time span. Therefore, in order to preserve the adhesive

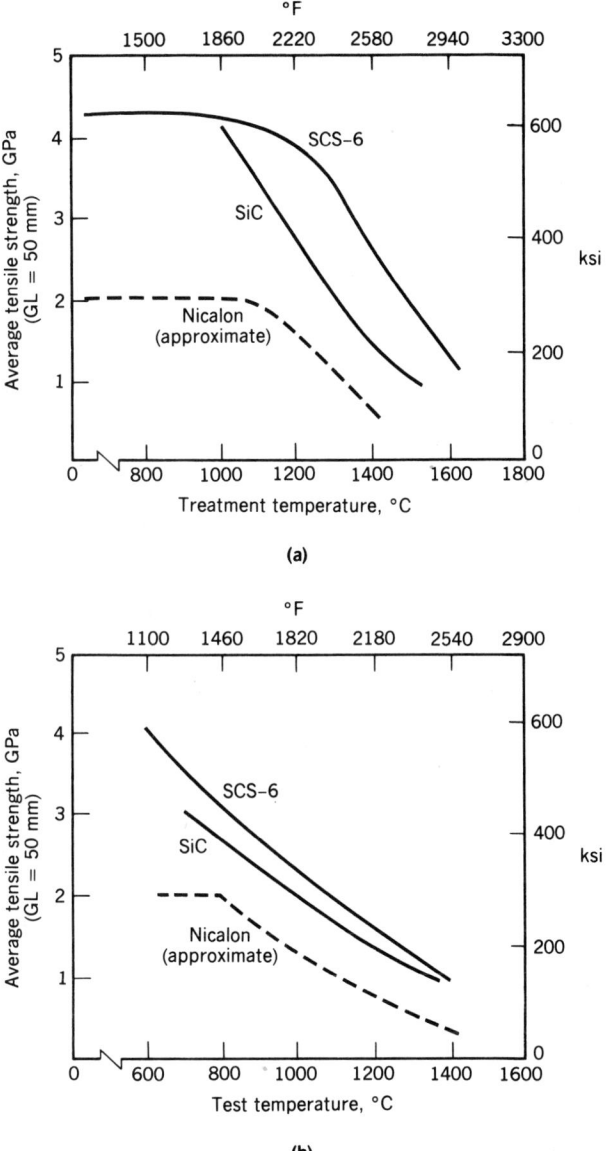

Figure 73.14 (a) Strength at temperature in argon or nitrogen of SiC fibers; (b) strength at temperature in oxygen for SiC fibers. 15 min. exposure.

properties of the boron epoxy prepreg prior to use, the tape must be stored at 0°F (-18°C). The tape has been stored for several years at this temperature without significant loss of adhesion properties.

The prepreg, as manufactured, may then be used in the fabrication of a composite part or assembly. This is accomplished through a multistep process. First, the tape is cut into desired lengths and laid up on a template in a pattern designed to take the best advantage of boron's high mechanical properties. This may be done either manually or by automated means. Once laid up, the part or assembly is encased in a vacuum bag for placement in an autoclave. It is then cured at an elevated temperature and pressure. After removal from

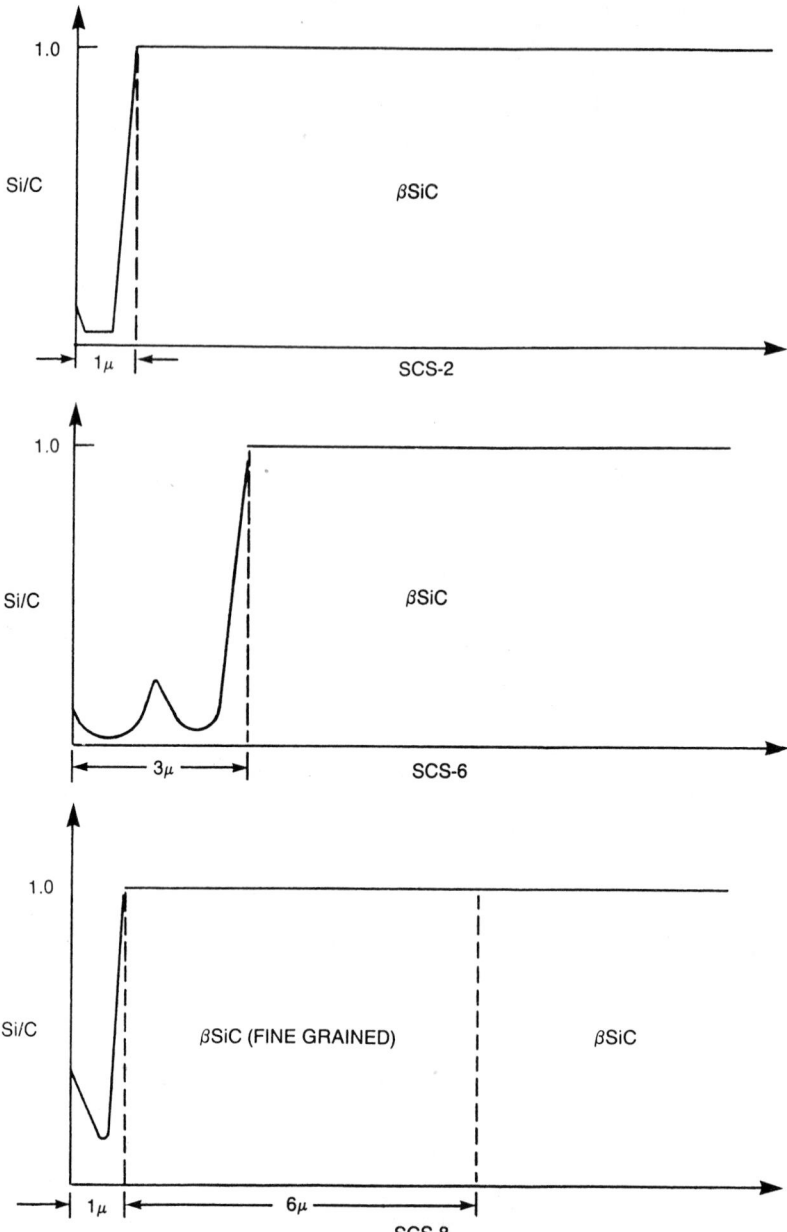

Figure 73.15.   Surface region compositions of Textron's silicon carbide.

the autoclave, any machining or touch-up work may then be done prior to final assembly. Photographs illustrating this procedure as it pertains to the fabrication of the horizontal stabilizer for the F-14 are given in Figures 73.18–73.21.

Although the majority of this work has been done in the reinforcement of epoxy resins, some work has also been done in the reinforcement of high-temperature resins, such as polyimides and poly(aryl sulfones). These types of resins have proven to be viable candidates at service temperatures up to 500–600°F (260–315°C).

Figure 73.16. Boron epoxy prepreg tape.

Figure 73.17. Manufacture of 6-in. (152-mm) wide tape from 1/4-in. (6-mm) wide strips.

Figure 73.18.   Plies of tape are initially laid up and cut in a predesigned pattern. Courtesy of Grumman Aerospace Corp.

Figure 73.19.   Part is encased in a vacuum bag prior to curing in an autoclave at elevated temperature and pressure. Courtesy of Grumman Aerospace Corp.

Figure 73.20. Part after removal from autoclave and finishing. Courtesy of Grumman Aerospace Corp.

Figure 73.21. Aircraft is assembled using completed boron epoxy part. Courtesy of Grumman Aerospace Corp.

Woven "dry" preforms have also been produced. This product consists of collimated arrays of fibers held in either the warp or fill direction by a more flexible interweave material. Interweave fibers have included organic fibers, metallic wires, and ceramic yarns. Since the preforms are woven without a resin, the requirement for refrigerated storage is eliminated. The woven preforms can also be placed in molds for casting or injection molding compositing procedures.

Neither the reinforced high-temperature resin systems nor the woven preforms are currently available on a production basis. There is, however, the capability to manufacture these materials should particular applications arise.

Composite Properties. Boron-reinforced composites exhibit high strengths and stiffnesses in tension, compression, and bending. These properties are dependent on both the intrinsic filament properties and the orientation of the filament layers. Boron epoxy tape lay-up orientations usually consist of either 0°, plus or minus 45°, 90°, or some combination of these. However, boron epoxy lay-up orientations are not limited to these, and practically any desired orientation can be accommodated.

The boron epoxy laminate design data presented in this section are based on a generic 50% v/o boron epoxy composite that consists of 0.004-in. (0.1 mm) diameter fiber embedded in an epoxy resin matrix qualified for continuous 350°F (177°C) service. Generic boron data for 5.6 mil (1.4 mm) diameter fiber composites should be assumed to be the same unless otherwise noted.

The following mechanical test and design data is presented as an introduction to the properties of boron epoxy laminates. For a more complete compilation, the reader should

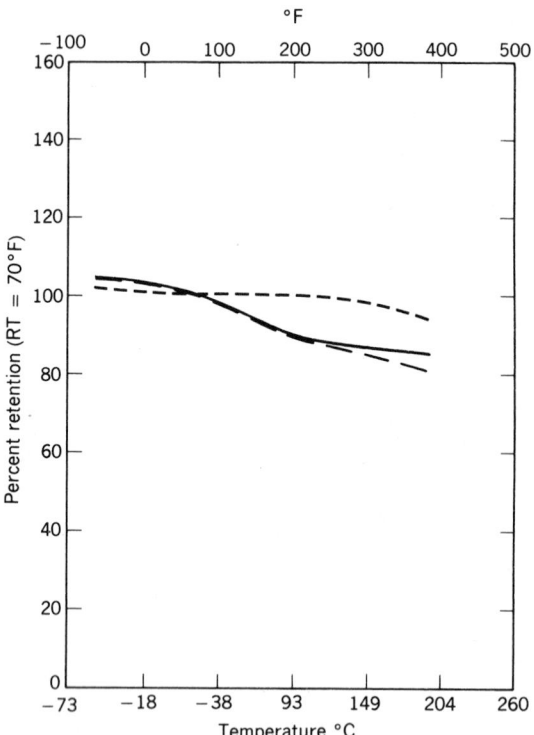

Figure 73.22. Effects of temperature on longitudinal tensile properties.———, strength; ---, modulus; —, elongation.

refer to reference 10, the Advanced Composites Design Guide, which assembles a great deal of information on various aspects of advanced composite design, even though the data are somewhat dated (the guide is more than ten years old, and the properties of boron filament have shown steady improvement in that time).

Figures 73.22 and 73.23 and Table 73.1 give various properties of boron epoxy composites.

Applications. The superior mechanical properties of boron filament reinforced composites have proven attractive to many designers; as a result, boron filaments have been specified for many applications. Currently, primary applications of this material include aerospace and sporting goods.

On the aerospace side, boron composites have been specified in numerous military aircraft and helicopter designs. A few of the larger-volume applications include the F-14 (horizontal stabilizer), the F-15 (horizontal stabilizer, vertical stabilizer, and rudder), and the B-1 bomber (dorsal longeron). In addition to these, there are also various other smaller-volume applications in which boron composites are specified. These include the F-111 (wind doubler),

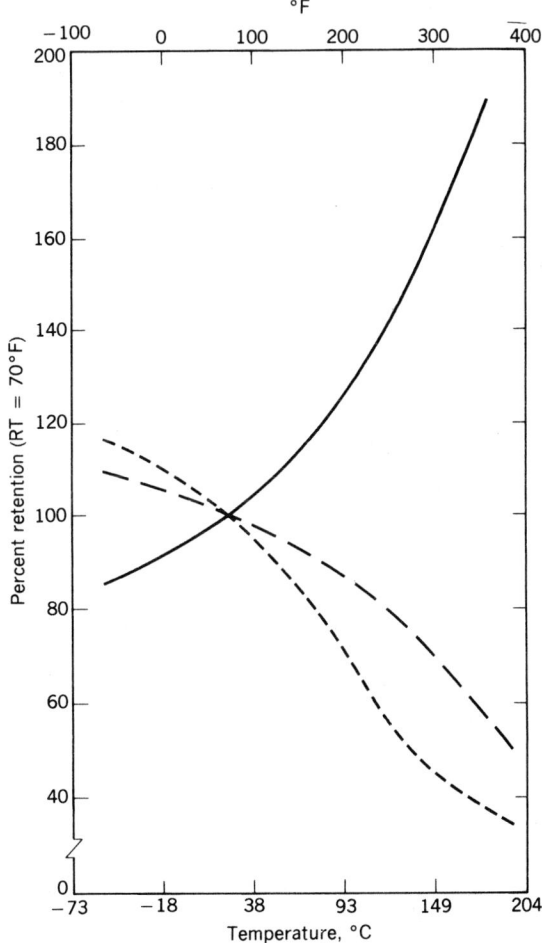

Figure 73.23. Effects of temperature on transverse tensile properties.————, strength; ---, modulus; —, elongation.

TABLE 73.1. Typical Mechanical Properties of (0°) 350°F Cure Boron–Epoxy With 50% Filament by Volume

	Units	RT	177°C (350°F)
Longitudinal tensile strength	MPa (ksi)	1700 (250)	1450 (210)
Longitudinal modulus	GPa (Msi)	210 (30)	210 (30)
Transverse tensile strength	MPa (ksi)	90 (13)	37 (5.4)
Transverse modulus	GPa (Msi)	25 (3.6)	9.7 (1.4)
Longitudinal compressive strength	MPa (ksi)	2600 (375)	950 (140)
Longitudinal modulus	GPa (Msi)	210 (30)	210 (30)
Transverse compressive strength	MPa (ksi)	310 (45)	90 (12.9)
Transverse modulus	GPa (Msi)	24 (3.5)	9.7 (1.4)
Longitudinal flexural strength	MPa (ksi)	2050 (300)	1800 (265)
Longitudinal modulus	GPa (Msi)	190 (28)	170 (25)
Transverse flexural strength	MPa (ksi)	200 (14.6)	75 (10.7)
Transverse modulus	GPa (Msi)	19 (2.7)	11 (1.6)
Interlaminar shear strength	MPa (ksi)	110 (15.6)	48 (6.9)
Longitudinal coefficient of thermal expansion	PPM/°C (PPM/°F)	4.5 (2.5)	4.5 (2.5)
Transverse coefficient of thermal expansion	PPM/°C (PPM/°F)	23.6 (13.1)	36 (20)
Longitudinal Poissons ratio		0.21	0.21
Transverse Poissons ratio		0.019	0.008
Density	g/cm^3 (lb/in.^{3})	2.00 (0.072)	2.00 (0.072)

Mirage 2000 and 4000 (rudder), and the Hawk series and Sea Stallion helicopters (horizontal stabilizer).

On the sporting goods side, boron epoxy composites have been used to improve strength, stiffness, and sensitivity of rackets (tennis, racquetball, squash, and badminton), fishing rods (various kinds), skis (snow and water), and golf club shafts. Boron-reinforced sporting goods have achieved the reputation of being the premier product of the market today. Some of the manufacturers of boron-reinforced sporting goods include Prince, Head, Wilson, Ektelon, Aldila, Honma, Fenwick, and Browning, among other. Figure 73.24 is a photograph showing a sample of these sporting goods applications.

73.3.2 Silicon Carbide Reinforced Resins

The primary purpose for the development of continuous silicon carbide filament was for the reinforcement of metal matrix and ceramic matrix composite structures. Therefore, much of the preliminary design and evaluation of SiC fiber was supported in these areas. It is anticipated that, once SiC fiber is produced on a commercial basis, a greater emphasis will be placed in the area of resin reinforcement, which is currently boron's major area of application. Some preliminary test data developed by the University of Dayton[11] are given in Table 73.2.

73.4 METAL MATRIX COMPOSITES

73.4.1 Boron-Reinforced Metals

Composite Processing. A preform consisting of boron filament and a metal foil is normally used to make a metal matrix composite. The basic process consists of pressing an array of fibers between metal foils. At elevated pressures, the foils deform around the fibers and bond to the fibers and to each other. This preform can then be laid up to form structures

Figure 73.24. Typical boron sporting goods applications.

in the same way prepreg is used. Variations on this process include plasma spraying of metal on the fiber array to make the preform and the step-pressing of continuous boron metal preform tape. In the latter, discrete segments of fiber/foil sandwiches are sequentially diffusion bonded, producing preform continuous in the fiber direction. These preforms are typically fabricated using 0.0056-in. (1.4-mm) diameter fiber. Figure 73.25 demonstrates a typical fabrication scheme to produce boron/aluminum composites from preform.

Boron metal matrix composites typically contain approximately 50% filament by volume; however, v/os ranging from 20 to 60% have been successfully fabricated. The composite can be produced in monolayer and plate form with up to 50 plies; it is typically available in sizes of 3 × 3 ft (915 × 915 mm). Figure 73.26 shows a representative cross section from a six-ply boron/aluminum composite plate.

Even though boron filament's primary composite applications have been to reinforce plastic matrices, a large data base of boron/aluminum metal matrix properties has been developed.[12–15]

TABLE 73.2. Preliminary Test Data on 0° SiC/Epoxy With 50% Filament by Volume

	Units	RT	127°C (260°F)
Tensile strength	MPa (ksi)	1600 (229)	1330 (190)
Tensile modulus	GPa (Msi)	230 (33)	230 (33)
Compressive strength	MPa (ksi)	2280 (326)	1620 (232)
Flexural strength	MPa (ksi)	2200 (314)	2210 (316)
Flexural modulus	GPa (Msi)	224 (32)	210 (30)
Interlaminar shear	MPa (ksi)	105 (15)	63 (9)
Density	g/cm^3 (lb/in.^{3})	2.3 (0.084)	2.3 (0.084)

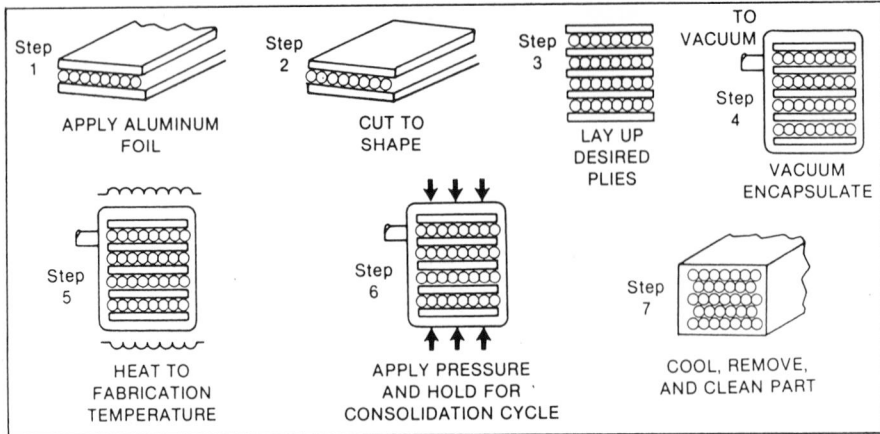

Figure 73.25.   Typical fabrication process for boron/Al composites.

The properties of boron composites depend in a critical way on the lay-up sequence. Properties parallel (longitudinal) to the fiber direction are dominated by fiber properties; those perpendicular (transverse) to the fibers are determined by the matrix material. In a typical application, both ply orientation and number of plies vary across the final part.

Figures 73.27–73.30 and Tables 73.3–73.4 give various properties of boron–aluminum composites.

Applications. The major uses of boron have been in reinforcing resin matrices (typically epoxies) in critical components in military aircraft, such as the F-14, F-15, F-111, B-1, and

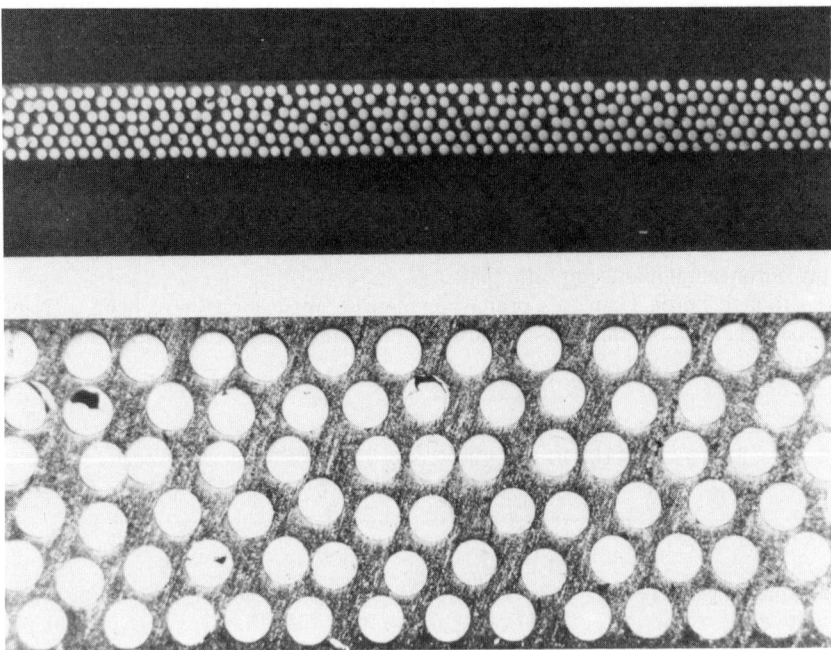

Figure 73.26.   Photomicrograph of cross-section of boron/Al composite top (10× magnification) and bottom (50× magnification).

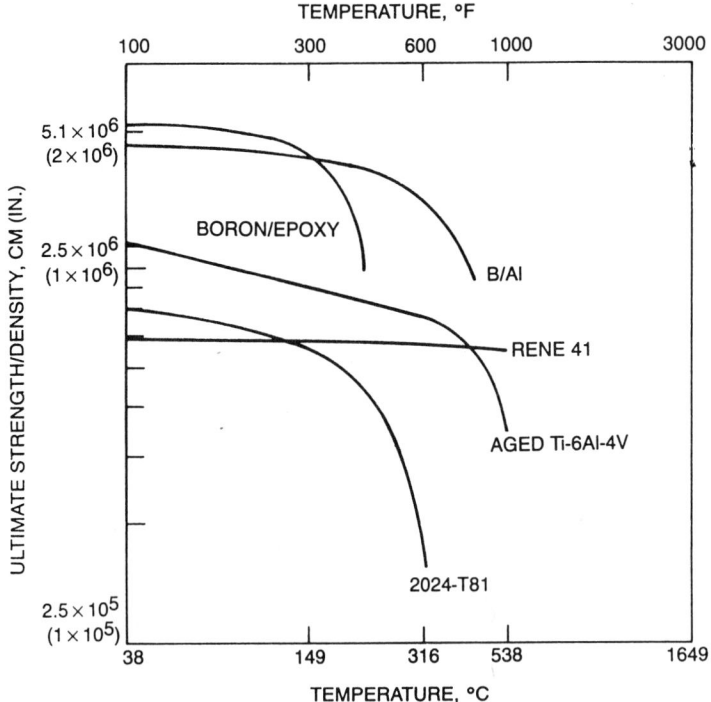

Figure 73.27. Strength of axially reinforced boron composites vs. metals.[3]

several helicopters, and in strengthening and stiffening sporting goods items, such as golf-club shafts, rackets of various kinds, fishing rods, and skis. The operation temperatures seen by these applications are relatively low [less than 350°F (177°C)], so boron filament in resin matrices can readily be used.

Other applications with operating temperature above 350–400°F (177–204°C) have led to the use of metal matrices with boron. Boron metal matrix is used in areas where its

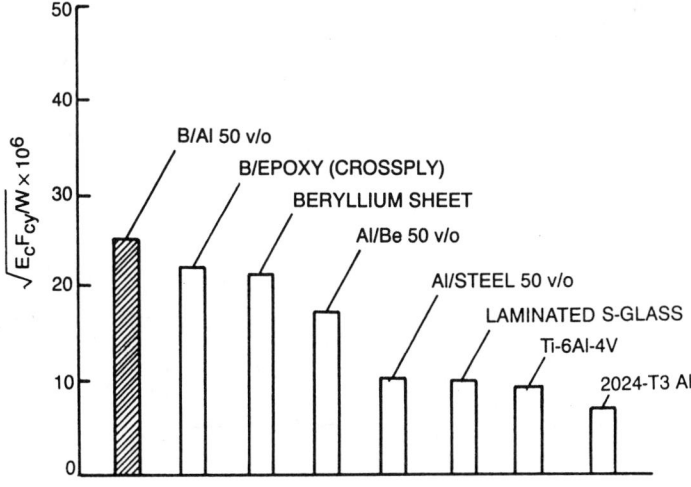

Figure 73.28. Relative crippling efficiency of boron composites vs metals.[3]

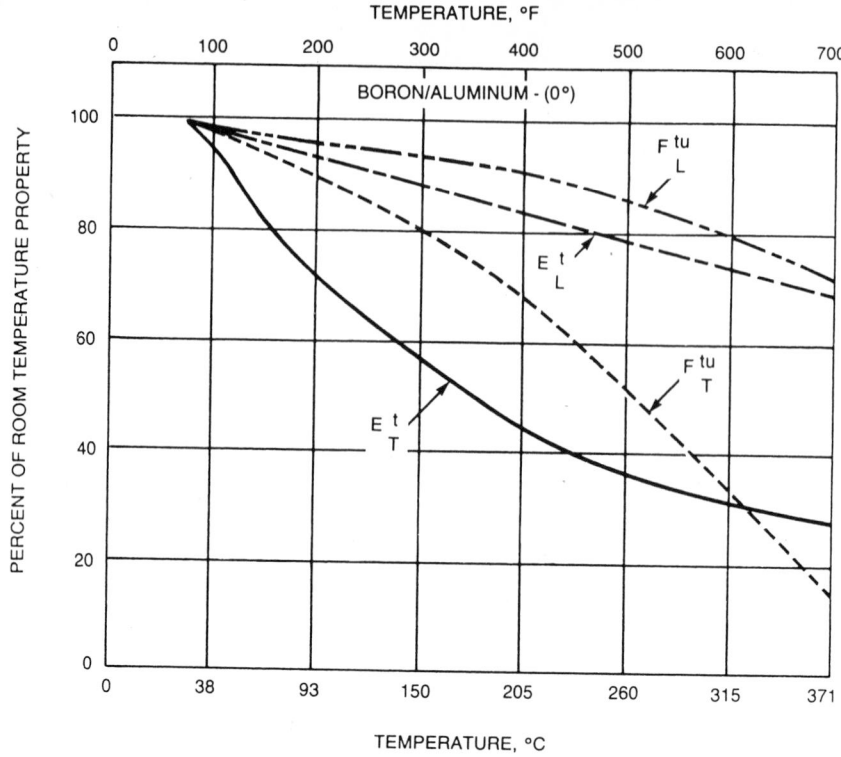

Figure 73.29. Strength and modulus vs temperature of B/Al (0°) in longitudinal and transverse tension.[5]

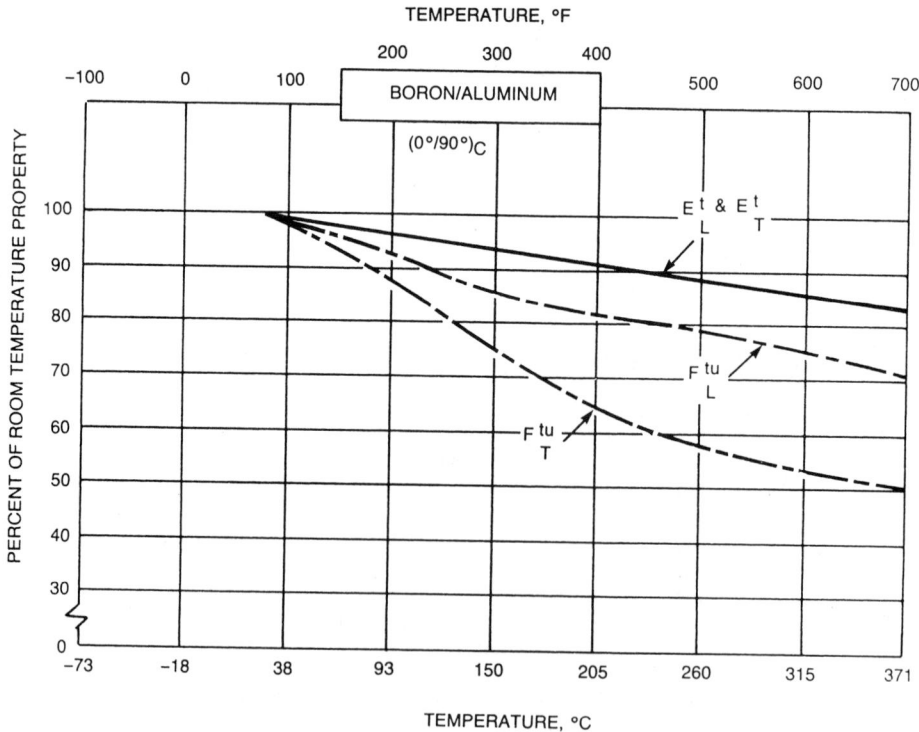

Figure 73.30. Strength and modulus vs temperature of B/Al (0°/90°) cross ply in longitudinal and transverse tension.[5]

TABLE 73.3. Various Properties of Boron–Aluminum Composites (0°)

Design Strengths

Longitudinal tensile ultimate,	MPa (ksi)	F_L^{tu}	1400 (200)
Transverse tensile ultimate,	MPa (ksi)	F_T^{tu}	110 (16.0)
Longitudinal compression ultimate,	MPa (ksi)	F_L^{cu}	3100 (450)
Transverse compression ultimate,	MPa (ksi)	F_T^{cu}	159 (23.0)
In-plane shear ultimate,	MPa (ksi)	F_{LT}^{su}	69 (10.0)
Interlaminar shear ultimate,	MPa (ksi)	F^{isu}	126 (18.3)

Plastic Properties

Longitudinal tension modulus,	GPa (Msi)	E_L^t	235 (34.0)
Transverse tension modulus,	GPa (Msi)	E_T^t	138 (20.0)
Longitudinal compression modulus,	GPa (Msi)	E_L^c	207 (30.0)
Transverse compression modulus,	GPa (Msi)	E_T^c	131 (19.0)
In-plane shear modulus,	GPa (Msi)	G_{LT}	66 (9.5)
Longitudinal Poisson's ratio,		ν_{LT}	0.23
Transverse Poisson's ratio,		ν_{TL}	0.17

Physical Constants

Density	kg/m³ (lb/in.³)	τ	2710 (0.098)
Longitudinal coefficient of thermal expansion	μm/m/°C (μin./in./°F)	α_L	5.8 (3.2)
Transverse coefficient of thermal expansion	μm/m/°C (μin./in./°F)	α_T	19.1 (10.6)

TABLE 73.4. Various Properties of Boron–Aluminum Composites (0°/90°)

Design Strengths

Longitudinal tensile ultimate	MPa (ksi)	F_x^{tu}	483 (70)
Transverse tensile ultimate	MPa (ksi)	F_y^{tu}	483 (70)
Longitudinal compression ultimate	MPa (ksi)	F_x^{cu}	607 (88)
Transverse compression ultimate	MPa (ksi)	F_y^{cu}	607 (88)
In-plane shear ultimate	MPa (ksi)	F_{xy}^{su}	103 (15)
Interlaminar shear ultimate	MPa (ksi)	F^{isu}	96 (10)
Ultimate longitudinal strain	μm/m or μin./in.	ϵ_x^{tu}	6,700
Ultimate transverse strain	μm/m or μin./in.	ϵ_y^{tu}	

Elastic Properties

Longitudinal tension modulus	GPa (Msi)	E_x^t	145 (21)
Transverse tension modulus	GPa (Msi)	E_y^t	145 (21)
Longitudinal compression modulus	GPa (Msi)	E_x^c	145 (21)
Transverse compression modulus	GPa (Msi)	E_y^c	145 (21)
In-plane shear modulus	GPa (Msi)	G_{xy}	
Longtitudinal Poisson's ratio		ν_{xy}	
Transverse Poisson's ratio		ν_{yx}	

Physical Constants

Density	kg/m³ (lb./in.³)	ρ	2710 (0.098)
Longitudinal coefficient of thermal expansion	μm/m/°C (μin./in./°F)	α_x	
Transverse coefficient of thermal expansion	μm/m/°C μin./in./°F)	α_y	

combination of high strength, high stiffness, and light weight are absolutely necessary. Currently, the major application is as a baseline material in the space shuttle. Boron–aluminum tubular struts are used as the frame and rib truss members and the frame stabilizing members in the mid-fuselage section. Photographs and drawings of this are shown in Figures 73.31–73.34.

Boron metal matrix has also been evaluated for uses in jet engine fan blades, aircraft wing skins, I-beam structural support members, landing gear components, bicycle frames and electronic packaging. Boron aluminum cold plates have been shown to greatly improve the solder joint life of IC chips versus aluminum or copper cold plates. Some examples of the metal matrix applications for which boron has been evaluated are given in Figures 73.35–73.37.

73.4.2 Continuous Silicon Carbide Reinforced Metals

Composite Processing. The ability to readily produce acceptable silicon carbide fiber reinforced metals is attributed directly to the ability of the SiC fiber to readily bond to the respective metals and to resist degradation of strength while being subjected to high-temperature processing. In the past, boron fibers have been evaluated in various aluminum alloys and, unless complex solid state (low-temperature/high-pressure) diffusion bonding procedures were adopted, severe degradation of fiber strength has been observed. Likewise in titanium, unless fabrication times are severely curtailed, fiber–matrix interactions produce brittle intermetallic compounds that, again, drastically reduce composite strength.

In contrast, the SCS grade of fibers has surfaces that readily bond to the respective metals without the destructive reactions occurring. The result is the ability to consolidate the aluminum composites using less complicated high-temperature processes such as investment casting and low-pressure (hot) molding. Also, for titanium composites, the SCS-6 filament

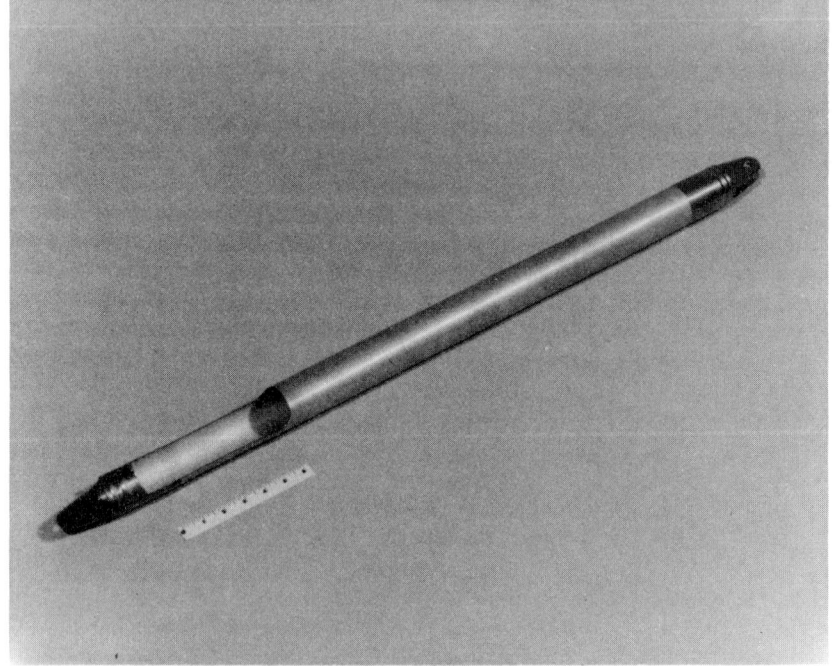

Figure 73.31. B/Al structural tube member used on Space Shuttle.

Figure 73.32. Partially completed Space Shuttle Orbiter, showing B/Al tubes in mid-fuselage structure.[3]

has the ability to withstand long exposure at diffusion bonding temperatures without fiber degradation. As a result, complex shapes with selective composite reinforcement can be fabricated by the innovative superplastic forming/diffusion bonding (SPF/DB) and hot isostatic pressing (HIP) processes. Further details of fabrication techniques are discussed in the following sections; however, the production of intermediary products, such as preforms and fabrics, used in the component fabrication are described first. These are required to simplify the loading of fibers into a mold and to provide correct alignment and spacing of the fibers.

Composite Preforms and Fabrics

- Green tape is an old system consisting of a single layer of fibers that are collimated/spaced side by side across a layer, held together by a resin binder, and supported by a metal foil. This layer constitutes a prepreg (in organic composite terms) that can be sequentially laid up into the mold or tool in required orientations to fabricate laminates. The laminate processing cycle is then controlled so as to remove the resin (by vacuum) as volatilization occurs. Figure 73.38 illustrates the method normally used to wind the fibers onto a foil-covered rotating drum, overspray the fibers with the resin, and cut the layer from the drum to provide a flat sheet of prepreg.
- Plasma sprayed aluminum tape is a more advanced prepreg. It is similar to green tape except that the resin binder is replaced by a plasma sprayed matrix of aluminum. The

Figure 73.33. Location of mid-fuselage structure within Space Shuttle Orbiter vehicle.[3]

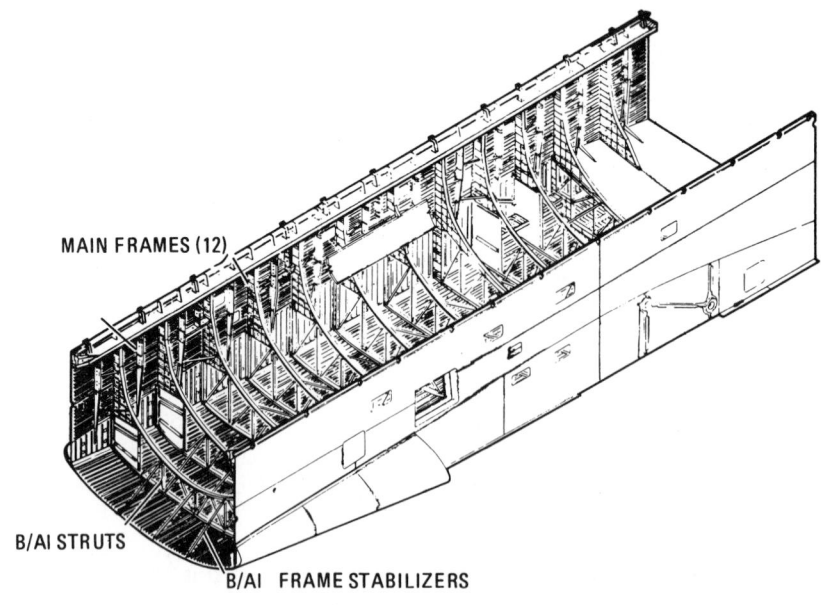

MAIN FRAMES (12)

B/AI STRUTS

B/AI FRAME STABILIZERS

Figure 73.34. Schematic diagram of the mid-fuselage structure of Space Shuttle.[3]

Figure 73.35. B/Al bicycle frame.

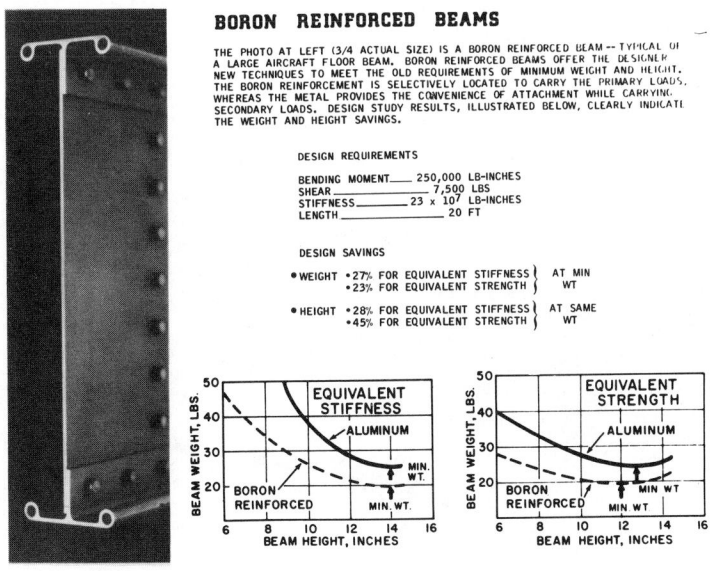

BORON REINFORCED BEAMS

THE PHOTO AT LEFT (3/4 ACTUAL SIZE) IS A BORON REINFORCED BEAM -- TYPICAL OF A LARGE AIRCRAFT FLOOR BEAM. BORON REINFORCED BEAMS OFFER THE DESIGNER NEW TECHNIQUES TO MEET THE OLD REQUIREMENTS OF MINIMUM WEIGHT AND HEIGHT. THE BORON REINFORCEMENT IS SELECTIVELY LOCATED TO CARRY THE PRIMARY LOADS, WHEREAS THE METAL PROVIDES THE CONVENIENCE OF ATTACHMENT WHILE CARRYING SECONDARY LOADS. DESIGN STUDY RESULTS, ILLUSTRATED BELOW, CLEARLY INDICATE THE WEIGHT AND HEIGHT SAVINGS.

DESIGN REQUIREMENTS

BENDING MOMENT_____ 250,000 LB-INCHES
SHEAR_____ 7,500 LBS
STIFFNESS_____ 23 x 10^7 LB-INCHES
LENGTH_____ 20 FT

DESIGN SAVINGS

- WEIGHT •27% FOR EQUIVALENT STIFFNESS } AT MIN
         •23% FOR EQUIVALENT STRENGTH } WT

- HEIGHT •28% FOR EQUIVALENT STIFFNESS } AT SAME
         •45% FOR EQUIVALENT STRENGTH } WT

Figure 73.36. Boron-reinforced aluminum structural member.

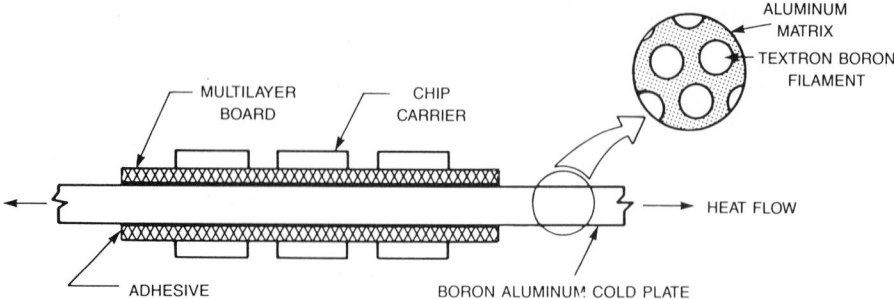

Figure 73.37. Schematic of a conduction cooled chip carrier module using a boron aluminum cold plate.

advantages of this material are the lack of possible contamination from resin residue and faster material processing times, because the hold time required to ensure volatilization and removal of the resin binder is not required. As with the green tape system, the plasma sprayed preforms are laid sequentially into the mold, as required, and pressed to the final shape. Figure 73.39 shows the spraying of the plasma onto the drum of collimated fibers.

• Woven fabric is perhaps the most interesting of the preforms being produced since it is a universal preform concept that is suitable for a number of fabrication processes. The fabric is a uniweave system in which the relatively large-diameter SiC monofilaments are held straight and parallel, collimated at 100–140 filaments per in. (4–6 fil/mm), and held together by a cross weave of a low-density yarn or metallic ribbon. There are

Figure 73.38. Drum winding for SiC—green tape.

Figure 73.39. Plasma spraying of aluminum onto SiC fibers.

now two types of looms that can be specially modified to produce the uniweave fabric (Fig. 73.40). The first is a single arm Rapier type loom capable of producing continuous 60 in. (1524 mm) wide fabric with the SiC filament oriented in the "fill" [60 in. (1524 mm) width] direction. The other is a shuttle type loom in which the SiC monofilaments are oriented in the continuous direction with the lightweight yarn a metal ribbon in the "fill" axis. The shuttle loom can weave fabric up to 6 in. (152 mm) wide. Various types of cross weave materials have been used, such as titanium aluminum and ceramic yarns (see Fig. 73.41).

(a)

(b)

Figure 73.40. (**a**) Rapier loom; (**b**) shuttle loom.

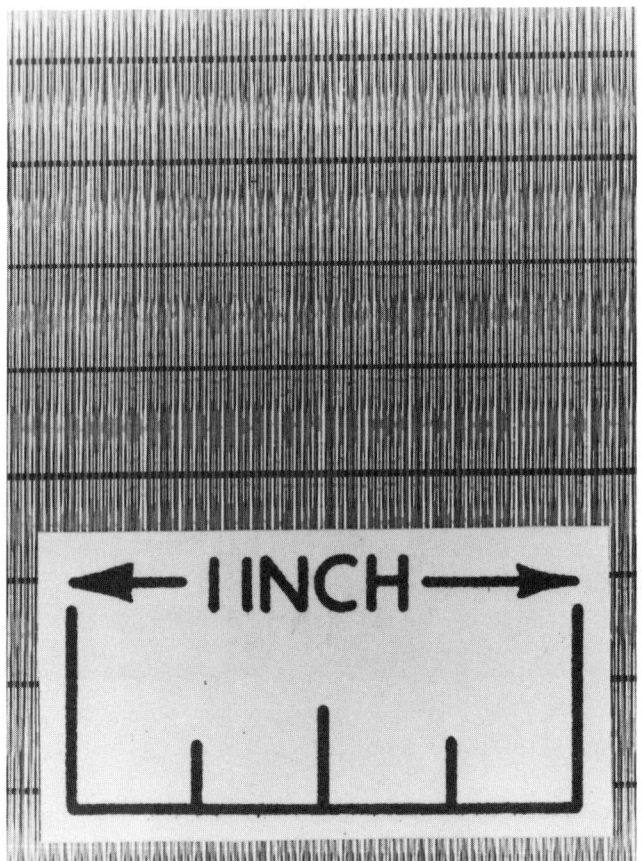

Figure 73.41. SiC—uniweave fabric with an aluminum ribbon cross-weave.

Process Methods. *Investment casting* is a fabrication technique that has been used for many years but is still universally accepted as a very cost-effective method for producing complex shapes. The aerospace business has for some time rejected aluminum castings because of the low strengths that are typically achieved; however, with a material that is now fiber dependent and not predominantly matrix controlled, significant structural improvements have been derived so as to revive the interest in this low-cost procedure. The investment casting technique, sometimes called the "lost wax" process, utilizes a wax replicate of the intended shape to form a porous ceramic shell model. On removal of the wax (by steam heat) from the interior, a cavity is provided for the aluminum. A typical shell mold is shown in Figure 73.42. The mold includes a funnel for gravity pouring with risers and gates to control the flow of the aluminum into the gauge section. Around the neck of the funnel is positioned a seal allowing the body of the mold to be suspended into a vacuum chamber. By a combination of gravity and vacuum (imposed through the porous walls of the shell mold), the total cavity is filled with aluminum.

The SiC fibers are installed in the mold using the fabric described above, either by first placing the fabric into the wax replicate or by simply splitting open the mold and inserting the fabric into the cavity after the wax has been removed. At present, the latter approach is usually used because of contamination and oxidation of the fibers during wax burnout. At some future date, the necessary techniques for including the fiber in the wax, and thereby reducing the processing costs, will probably be developed.

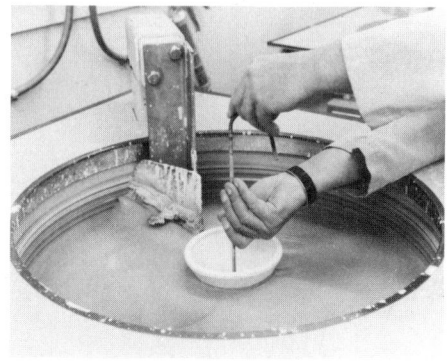

(a)   (b)

(c)

Figure 73.42. SiC/Al shell mold casting.

Hot molding is a term coined by Textron to describe a low-pressure hot-pressing process that is designed to fabricate shaped SiC–aluminum parts at significantly lower cost than the typically diffusion bonding/solid state process. As stated previously, because the SCS-2 fibers can withstand molten aluminum for long periods, the molding temperature can now be raised into the liquid plus solid region of the alloy to ensure aluminum flow and consolidation at low pressure, thereby obviating the requirement for high-pressure die molding equipment.

The best way of describing the hot-molding process is to draw an analogy to the autoclave molding of graphite epoxy where the components are molded in an open-faced tool. The mold in this case is a self-heated slip cast ceramic tool embodying the profile of the finished part. A plasma sprayed aluminum preform is laid into the mold, heated to a near molten aluminum temperature, and pressure consolidated in an autoclave by a "metallic" vacuum bag. The process is schematically presented in Figure 73.43, in which a stiffener is being fabricated in specially prepared molds. The mold can be profiled as required to produce near net shape parts, including tapered thicknesses and section geometry variations. Figure 73.43 shows a "Z" mold tool used for manufacture of 0° to plus or minus 45° SiC/Al stringers for a stiffened panel.

Diffusion bonding of SiC/titanium is accomplished by the hot pressing (diffusion bonding) technology, using fiber preforms (fabric) that are stacked together between titanium foils for consolidation. Two methods are being developed by aircraft and engine manufacturers to manufacture complex shapes. One is based on the HIP technology and uses a steel pressure membrane to consolidate components directly from the fiber/metal preform layer. The other

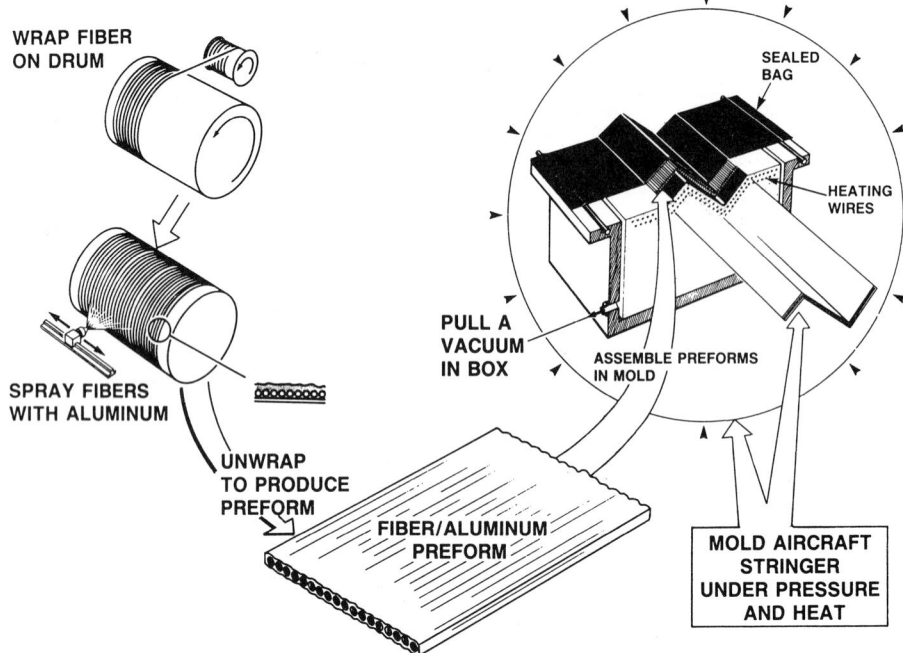

Figure 73.43.   Hot molding of Sic/Al "Zee" stiffeners.

requires the use of previously hot-pressed SCS titanium laminates that are then diffusion bonded to a titanium substructure during subsequent superplastic forming operations. Figure 73.44 illustrates the HIP procedure for a SCS titanium engine drive shaft. This is typical of this fabrication procedure. The fiber preform is placed onto a titanium foil, which is then spirally wrapped, inserted, and diffusion bonded onto the inner surface of a steel tube using a steel pressure membrane. The steel is subsequently thinned down and machined to form

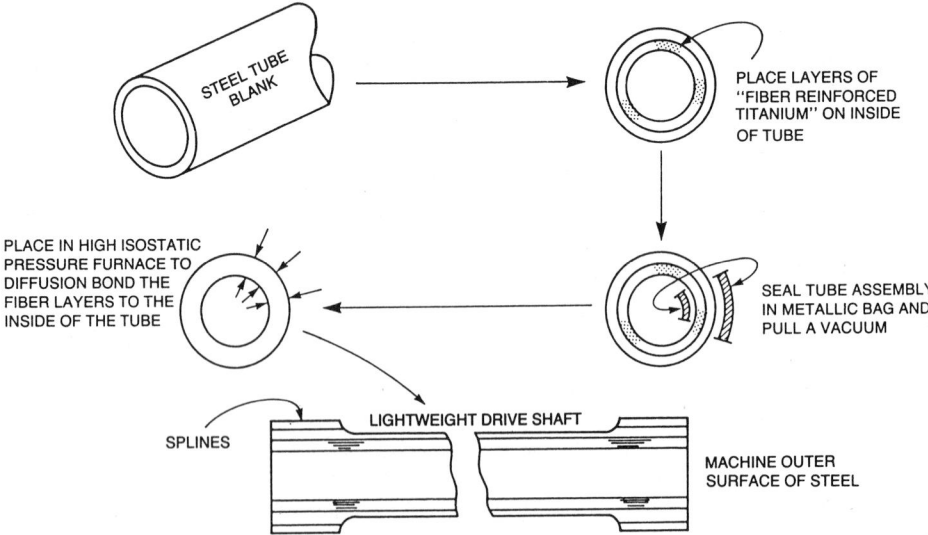

Figure 73.44.   Silicon carbide titanium drive shaft.

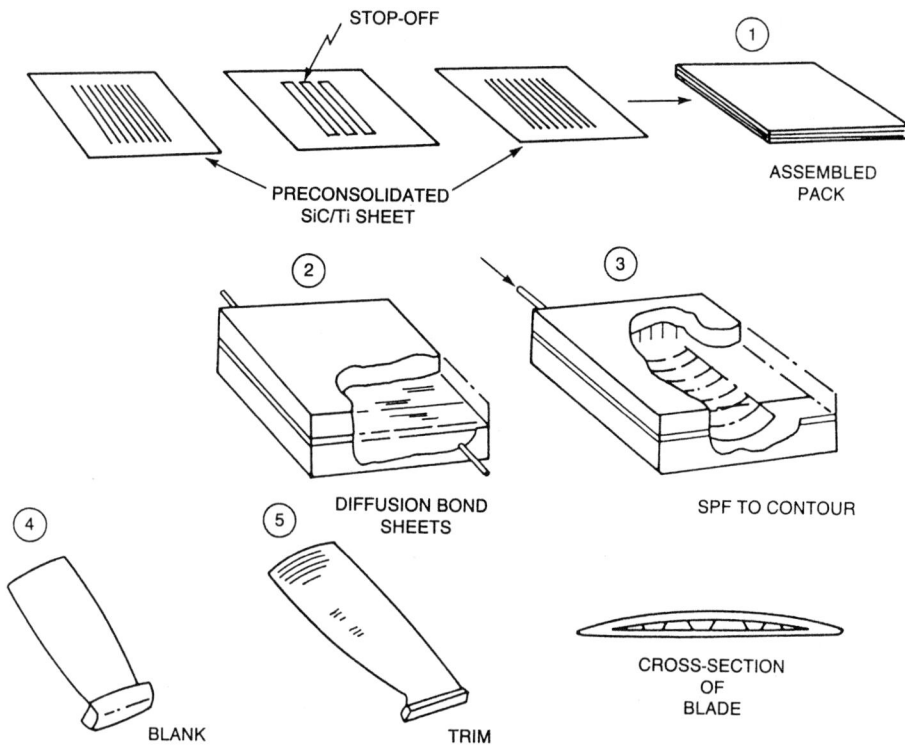

Figure 73.45. Superplastic forming—diffusion bonding of a SiC/Ti blade.

the "spline attachment" at each end. Shafts are also being fabricated for other engine fabricators without the steel sheath. Figure 73.45 illustrates the concept developed for superplastic forming of hollow engine compressor blades. Here the SCS titanium laminates are first diffusion bonded in a press. These are then diffusion bonded to form monolithic titanium sheets, the "stop off" compounds that are selectively positioned to preclude bonding in desired areas. Subsequently, the "stackup" is sealed into a female die. By pressurizing the interior of the stackup, the material is blown into the female die to form the desired shape, stretching the monolithic titanium to form the internal corrugations.

These processes typically require long times at high temperature. In the past, all of the materials used have developed significant matrix-to-fiber interaction that seriously degrades composite strength. However, SCS-6, because of its unique surface characteristics, delays intermetallic diffusion and retains its strength up to seven hours in contact with titanium at 1700°F (925°C).

Composite Properties. Since continuous SiC reinforced metals have been in existence for a relatively short period of time, the property data base has been developed sporadically over this period, depending on funded applications.

SiC/Aluminum. The most mature of the silicon carbide reinforced aluminum consolidation approaches is hot molding, and, therefore, the largest mechanical property data base has been developed using this material. The design data base for hot molded SCS-2/6061 aluminum includes static tension and compression properties, in-plane and interlaminar shear strengths, tension/tension fatigue strengths (SN curves), flexure strength, notched tension data, and fracture toughness data. Most of the data have been developed over a temperature

TABLE 73.5.   Tensile Strength of SCS-2/Aluminum (47% Fiber Volume)

Fiber Orientation	No. of Plies	Tensile Strength MPa	(ksi)	Modulus GPa	(ksi)	Total Strain	Poisson's Ratio	CTE /°C	/°F
0°	6, 8, 12	1462	(212)	204.1	(29.6)	0.89	0.268	6.6	(3.7)
90°	6, 12, 40	86.2	(12.5)	118.0	(17.1)	0.08	0.124	21.3	(11.8)
[0°/90°/0°/90°]$_s$	8	673	(97.6)	136.5	(19.8)	0.90			
[0$_2$°/90°/0°]$_s$	8	1144	(166.0)	180.0	(26.1)	0.92			
[90$_2$°/0°/90°]$_s$	8	341.3	(49.5)	96.5	(14.0)	1.01			
±45°	8, 12, 40	309.5	(44.9)	94.5	(13.7)	10.6	0.395		
[0°/±45°/0°]$_{s+2s}$	8, 16	800.0	(116)	146.2	(21.2)	0.86			
[0°/±45°/90°]$_s$	8	572.3	(83.0)	127.0	(18.4)	1.0			

range of −54°F (−49°C) to 165°F (74°C), with static tension test results up to 900°F (480°C). These data have been summarized in Tables 73.5 through 73.8 and Figures 73.46 and 73.47. As can be seen from these data, the inclusion of a high-performance, continuous silicon carbide fiber in 6061 aluminum yields a very high-strength [200,000 psi (1380 MPa)], high-modulus 30,000,000 psi (207,000 MPa)], anisotropic composite material having a density just slightly greater 178 lb/ft³ (2.85 g/cm³) than baseline aluminum. As in organic matrix composites, cross or angle plying produces a range of properties useful to the designer (Fig. 73.48).

The property data developed for investment cast SCS/aluminum have been limited to static tension and compression. Fiber volume fractions are lower (40% maximum) than the hot-molded laminates (47% typically), because of volumetric constraints in dry loading the shell molds. However, good rule-of-mixture tensile strengths and excellent compression strengths (twice the tensile strength) are being achieved (Table 73.9).

TABLE 73.6.   SCS-2/Aluminum Compression Strength (47% Fiber Volume)

Dir.	Plies	Load N	(lb)	Stress MPa	(ksi)	Modulus GPa	(msi)	Poisson's Ratio
0°	12	36000	(8100)	2647	(383.9)			
		38250	(8600)	2708	(392.7)			
		38700	(8700)	2739	(397.3)			
		40500	(9100)	2878	(417.4)			
		48900	(11000)	3296	(478.0)	212.4	(30.8)	0.241
		53100	(11940)	3689	(535.0)	222.7	(32.3)	
90°	12	4220	(948)	294.4	(42.7)	104.8	(15.2)	
		4380	(985)	300.6	(43.6)	116.5	(16.9)	0.174
		4270	(960)	294.4	(42.7)			
		4230	(950)	292.3	(42.4)	113.1	(16.4)	0.173
		3960	(890)	273.0	(39.6)	115.8	(16.8)	
		3780	(850)	259.2	(37.6)	124.1	(18.0)	
90°	40	13480	(3030)	293.7	(42.6)			
		14610	(3285)	294.4	(42.7)	131.7	(19.1)	0.136
		13280	(2985)	290.0	(42.0)	102.7	(14.9)	
		13430	(3020)	287.5	(41.7)	108.9	(15.8)	
		13520	(3040)	294.4	(42.7)	115.1	(16.7)	
		13680	(3075)	297.2	(43.1)	142.0	(20.6)	0.158

TABLE 73.7. SCS-2/Aluminum Shear Strengths (15° Off-Axis Shear, 47% Fiber Volume)

Test Temperature °C (°F)	Failure Stress MPa	(ksi)	Shear Strength MPa	(ksi)	Shear Modulus MPa	(ksi)
	455.7	(66.1)	113.8	(16.5)	42.5	(6.17)
	452.3	(65.6)	113.1	(16.4)	39.5	(5.73)
RT	479.2	(69.5)	120.0	(17.4)	39.8	(5.77)
	422.6	(61.3)	105.5	(15.3)	40.3	(5.85)
	$\overline{452.5}$	$\overline{(65.6)}$	$\overline{113.1}$	$\overline{(16.4)}$	$\overline{40.5}$	$\overline{(5.88)}$
	437.1	(63.4)	109.6	(15.9)	40.2	(5.83)
75	434.4	(63.0)	108.9	(15.8)	43.2	(6.27)
(165)	424.7	(61.6)	106.2	(15.4)	41.7	(6.05)
	$\overline{432.1}$	$\overline{(62.6)}$	$\overline{108.2}$	$\overline{(15.7)}$	$\overline{41.7}$	$\overline{(6.05)}$
	501.2	(72.7)	125.5	(18.2)	44.5	(6.46)
−55	428.6	(70.0)	120.7	(17.5)	39.6	(5.75)
(−65)	453.0	(65.7)	113.1	(16.4)	39.6	(5.75)
	$\overline{479.0}$	$\overline{(69.4)}$	$\overline{119.8}$	$\overline{(17.3)}$	$\overline{41.3}$	$\overline{(5.98)}$

The use of 6061 aluminum as the matrix material and the capability of the silicon carbide fiber to withstand molten aluminum has made conventional fusion welding (Fig. 73.49) a viable joining technique. Although welded joints would not have continuous fiber across the joint to maintain the very high strengths of the composite, baseline aluminum weld strengths can be obtained. In addition to fusion welding, traditional molten salt bath dip brazing has been demonstrated as an alternative joining method.

An important consideration for emerging materials is corrosion resistance. Testing has been performed on SCS-2/6061 hot-molded material at the David W. Tayler Naval Ship R&D Center[16] under marine atmosphere, ocean splash/spray, alternate tidal immersion,

TABLE 73.8. SCS-2/Aluminum Notched Strength Data (12 Plies of Material, 47% Fiber Volume)

Specimen[a]		Average Gross Stress (RT) MPa	(ksi)	Average Net Stress (RT) MPa	(ksi)	RT Notch Factor	75°C (165°F) Notch Factor	−55°C (−65°F) Notch Factor
DEN	0°	814.8	(118.2)	1269.5	(184.1)	0.92	0.85	0.80
CH (1/16 dia)	0°	1125.9	(163.3)	1163.8	(168.8)	0.84		
CH (1/8 dia)	0°	991.7	(143.8)	1061.6	(154.0)	0.77	0.75	0.72
		898.0[b]	(130.2)	956.1[b]	(138.7)	0.70[b]		
CH (1/4 dia)	0°	842.1	(122.1)	966.4	(140.2)	0.70		
		800.9	(116.2)	911.3[b]	(132.2)	0/66*		
CH (1/8 dia)	0°/±45°	728.1	(105.6)	777.4	(112.8)	0.90		
CH (1/4 dia)	0°/±45°	620.5	(90.0)	710.8	(103.1)	0.83		
CH (1/8 dia)	0°/±45°/90°	437.1	(63.4)	467.5	(67.8)	0.90		
CH (1/4 dia)	0°/±45°/90°	400.6	(58.1)	460.6[b]	(66.8)	0.89		
CH (3/32 dia)	±45°	244.7	(35.5)	256.6	(37.2)	0.85		
CC (1/4)	0°	822.0	(119.2)	944.2	(137.0)	0.68		
CC (1/2)		659.9	(95.7)	886.9	(128.6)	0.64		
		621.6[b]	(90.20)	819.8[b]	(118.9)	0.60[b]		

[a]CH = center hole; CC = center crack (EDM slot); DEN = double edge notch.
[b]40-Ply Material.

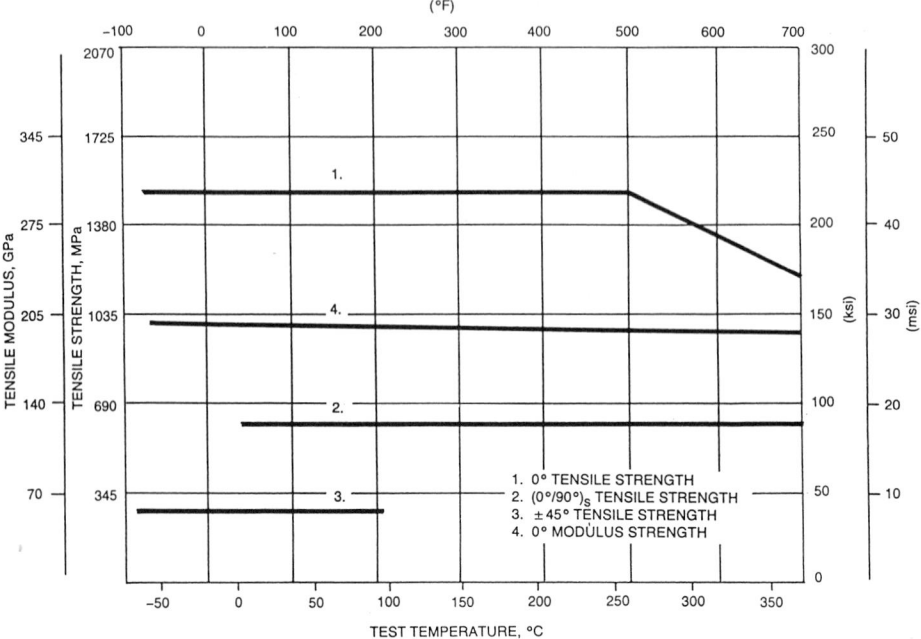

Figure 73.46.   Composite strength vs temperature.

and filtered seawater immersion conditions for periods of 60 to 365 days. The SCS/aluminum material performed well in all tests, exhibiting no more than pitting damage comparable to baseline 6061 aluminum alloy.

SiC/Titanium.   SCS-6/Ti 6-4 composites were originally developed to withstand extended exposure at high temperature.[17] As shown by the data in Table 73.10, composite strengths

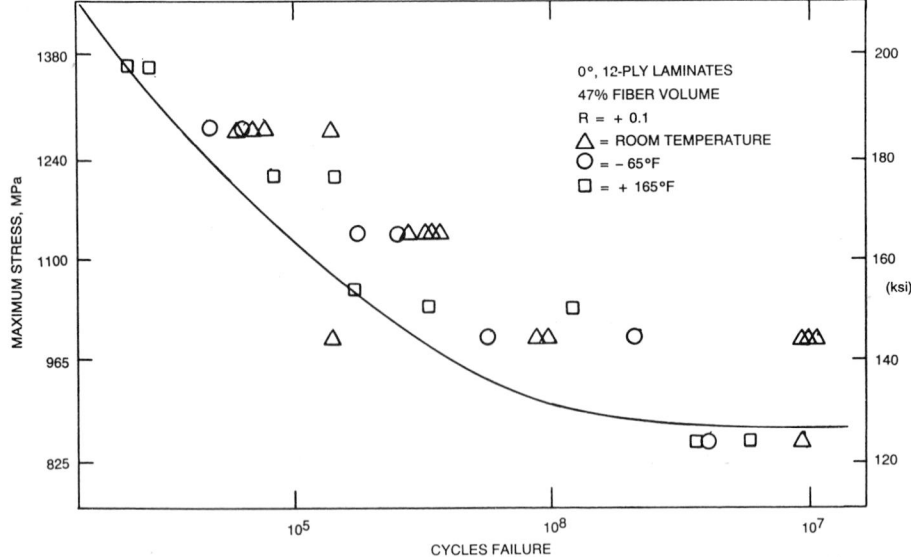

Figure 73.47.   SCS/Al tension/tension fatigue.

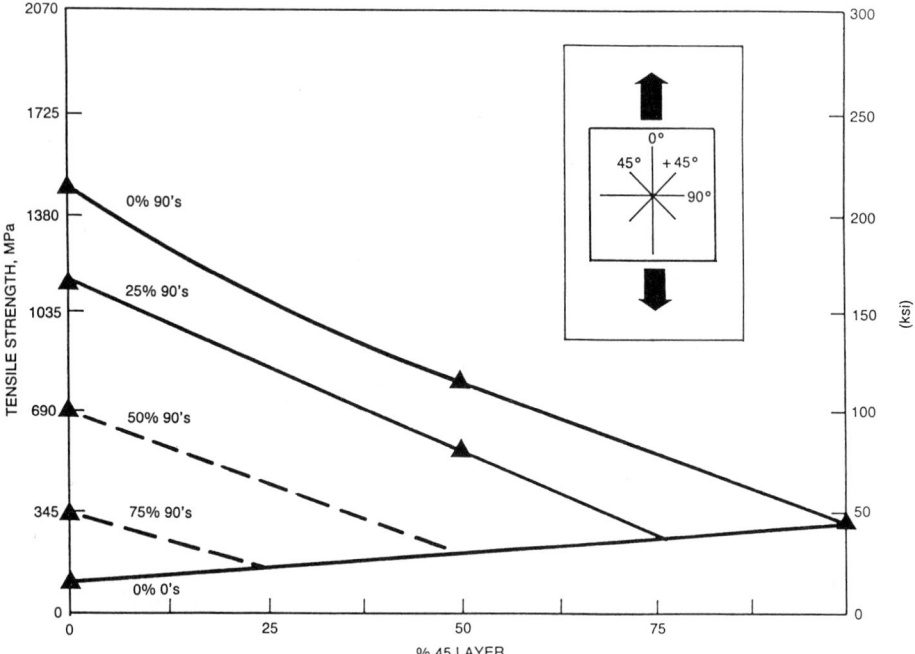

Figure 73.48. Composite tensile strength vs ply orientation.

remained above 2×10^5 psi (1380 MPa) after this extended heat treatment. There has been a successful program to reinforce the beta Ti alloy 15-3-3-3 with SCS fiber, and superior composite properties have been achieved, such as tensile strengths of $(23-28) \times 10^4$ psi (1580–1930 MPa). Fabrication of titanium parts has been accomplished by diffusion bonding and HIP. The latter technique has been particularly successful in the forming of shaped reinforced parts (such as tubes) by the use of woven SiC fabric as a preform. The high-strength, high-modulus properties of SCS-6/Ti represent a major improvement over B_4C-B/Ti composites in which the modulus of the composite is increased relative to the matrix, but the tensile strength is not as high as would be predicted by the rule of mixture. The modulus and strength at elevated temperature are shown in Figures 73.50 and 73.51.

SiC/Mg and SiC/Cu. SCS-2 has been successfully cast in magnesium.[18] The resultant properties are listed in Table 73.11. Under a recent Naval Surface Weapons Center (NSWC) program,[19] development of SiC-reinforced copper has been initiated. At present, about 85% of rule-of-mixture strengths have been achieved at volume fractions of 20–33%. Typical data are presented in Table 73.12.

TABLE 73.9. Cast SCS/Aluminum Data

Fiber Orientation	Fiber Volume	UTS MPa	(ksi)	ROM, %	E_t GPa	(Msi)	ROM, %	UCS MPa	(ksi)	E_c GPa	(Msi)
$0°_3/90°_6/0°_3$	33 v/o	458.5	(66.5)	75	122.0	(17.7)	107	1378.9	(200)	NA	
$90°_3/90°_6/0°_3$	33 v/o	584.0	(84.7)	95	124.8	(18.1)	110	1378.9	(200)	NA	
$0°$	34 v/o	1034.2	(150)	85	172.4	(25)	100	1896.1	(275)	186.2	(27.0)

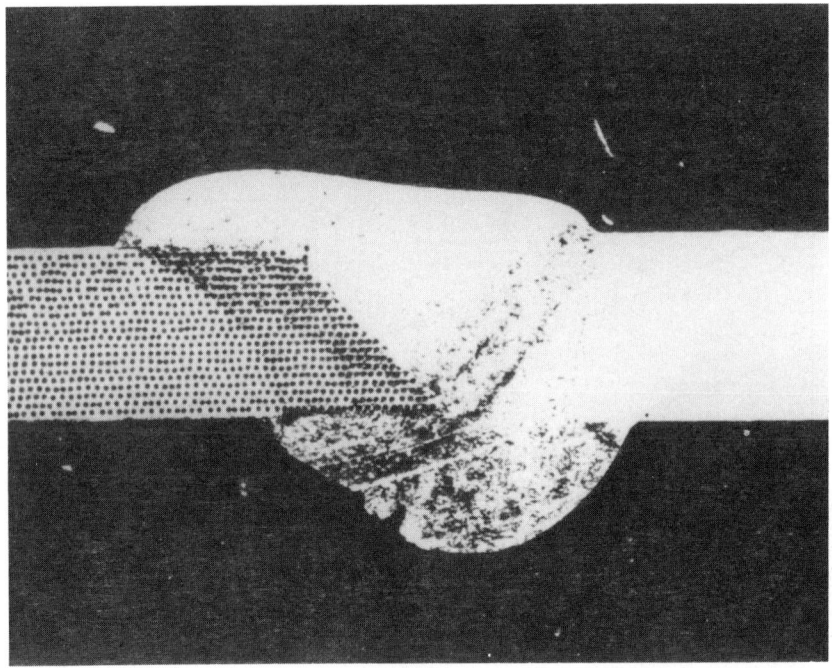

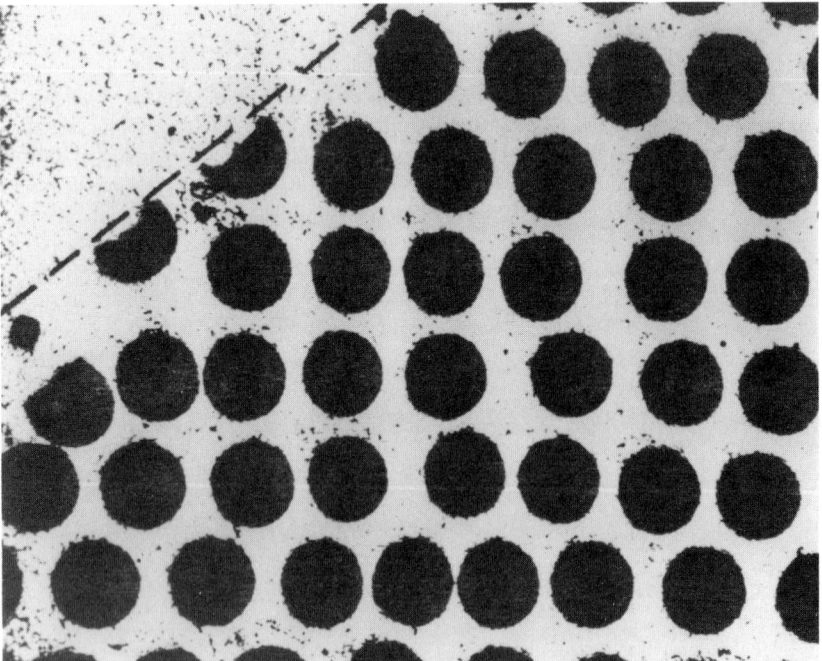

Figure 73.49.   Cross-section of SCS-2/aluminum to 6061 aluminum fusion weld at two magnifications.

TABLE 73.10. SCS-6/Ti Data, Sample Size, 62 Panels

Mechanical Properties of SiC/Ti-6-4 (35 v/o)							
	As Fabricated				After Heating 7 h at 905°C (1660°F)		
	X		SD		X		SD
Ultimate tensile storage, MPa(ksi)	1690	(245)	119.3	(17.3)	1434	(208)	108.9	(15.8)
Elastic modulus, GPa (Msi)	186.2	(27.0)	7.58	(1.1)	190.3	(27.6)	8.3	(1.2)
Strain to failure	0.96		0.091		0.86		0.087	

Mechanical Properties of SiC/Ti-15-3-3-3 (38-41 v/o)							
	As Fabricated				After Heat Treating 16 Hours at 480°C (900°F), 13 Samples		
	X		SD		X		SD
Ultimate tensile strength, MPa (ksi)	1572	(228)	138	(20)	1951	(283)	95.5	(14)
Elastic modulus, GPa (Msi)	197.9	(28.7)	6.21	(0.9)	213.0	(30.9)	4.83	(0.7)
Strain to failure								

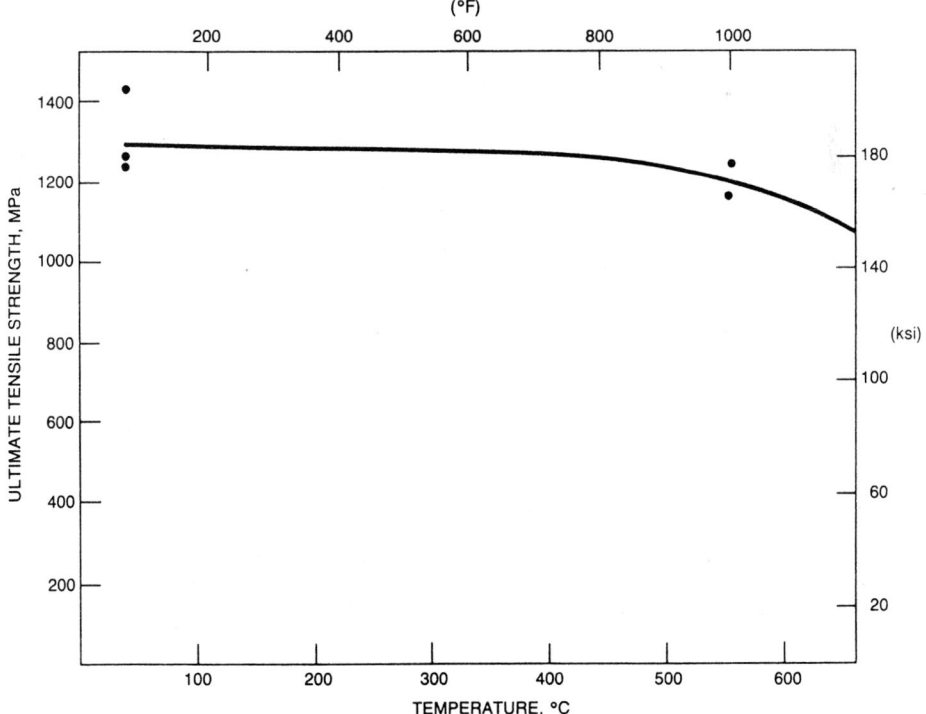

Figure 73.50. Ultimate tensile strength vs temperature for SCS-6/Ti.

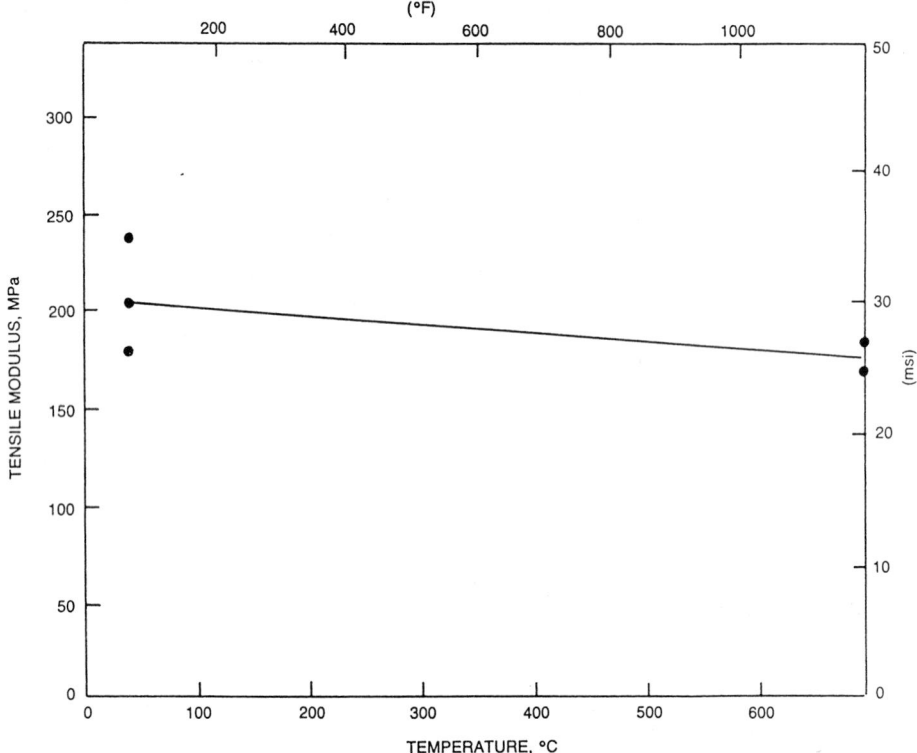

Figure 73.51.   Tensile modulus vs. temperature for SCS-6/Ti.

TABLE 73.11.   SCS/Magnesium Cast Rod Mechanical Data (ZE 41 AT 675°C, 1250°F)

Sample No.	Exposure Time, min	Ultimate Tensile Strength		Strain To Failure, %	Elastic Modulus		Fiber Volume Fraction, %
		MPa	(ksi)	%	GPa	(Msi)	%
VIR (67)	5	1000	(145)	0.83	169.6	(24.6)	34
VIR 69	10	1524	(221)	0.88	209.6	(30.4)	46
VIR 72	10	1331	(193)	0.78	230.3	(33.4)	50
VIR 77	10	1379	(200)	0.95	180.6	(26.2)	37

TABLE 73.12.   SCS/Copper Mechanical Data

Panel	Fiber v/o	Axial Ultimate Tensile Stress		Axial Modulus	
		MPa	(ksi)	GPa	(Msi)
84-014	0.23	690	(100)	172.4	(25.0)
84-153	0.33	965	(140)	202	(29.3)
84-377	0.33	900	(130)	187.5	(27.2)

Applications. The very high specific mechanical properties of silicon carbide reinforced metal matrix composites have generated significant interest within the aerospace industry and, as a result, many research and development programs are now in progress. The principal area of interest is high-performance structures such as aircraft, missiles, and engines. However, as more and more systems are developing sensitivities to performance and transportation weight, other and less sophisticated applications for these newer materials are being considered. The following describes a few of these applications.

- Silicon carbide aluminum wing structural elements are currently being developed. Ten-foot-long "Zee" shaped stiffeners are to be hot-molded and then subsequently rivetted to wing planks for full-scale static and fatigue testing. Figure 73.52 shows sections of a "Zee" stringer, utilizing (0° to plus or minus deg) SCS-2 fibers in a 6061 alloy. Experimental results obtained to date have verified material performance and the design procedures utilized.

- Silicon carbide aluminum bridging elements are being developed for the Army to be used for the lower chord and the king post of a 52-m assault bridge. Future plans call for the development of the top compression tube of the new Tri-Arch bridge being developed by Fort Belvoir.

- Silicon carbide aluminum internally stiffened cylinders are being developed using the previously discussed investment casting process. A way replica is first fabricated that incorporates the total shape of the shell, including internal ring stiffeners and the end fittings. There are the two halves to the split ceramic mold (inner and outer). The fabric containing the SiC fibers is then wound onto the inner shell mold, the two halves of the shell are remated and sealed, and infiltration of the aluminum is accomplished. Figure 73.53 shows the final part.

Figure 73.52. SiC/Al "Zee" stringers.

Figure 73.53. Finished cast stiffened shell.

- Silicon carbide aluminum missile body casings have been fabricated utilizing a unique variation of filament winding. An aluminum motor case is first produced in the conventional manner, this time, however, with significantly less wall thickness than is normally required. The casing is then overwrapped with layers of silicon carbide fibers, in which each layer is sprayed with a plasma of aluminum to build up the matrix thickness. No final consolidation of the 90% dense system is required, because the hydrostatic internal pressure on the circular body imposes no (or very minimal) shear stresses on the matrix. It is hoped that further development of this technique will permit full consolidation of the matrix by vacuum bagging the total section and HIP.

- SiC/titanium drive shafts are being developed and fabricated by the HIP process. These are generally used for the core of an engine, requiring increased specific stiffness to reduce unsupported the length between the bearings and also to increase critical vibratory speed ranges. Figure 73.54 is a shaft blank and a finished shaft. SiC-Ti tubes up to 5 ft (1524 mm) in length have been fabricated and have incorporated into their ends a monolithic load transfer section for ease of welding to the splined or flanged connections.

- Silicon carbide discs for turbine engines are currently under development. Initially, disks were made by winding an SiC-Ti monolayer over a mandrel, followed by hydrostatic consolidation (HIP). The concept now being developed utilizes a "doily" approach

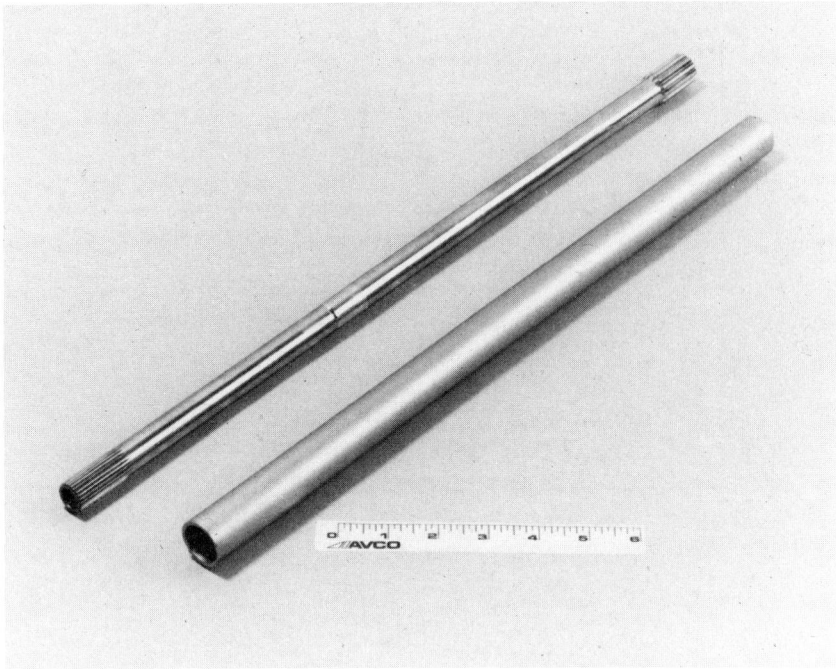

Figure 73.54. SiC/Ti drive shaft.

in which single fibers are hoop wound between titanium metal ribbons to be subsequently pressed together in the axial direction, reducing the breakage of fibers and simplifying the production of tapered cross sections.

- Selectively reinforced silicon carbide titanium hollow fan blades are being developed. A cross section of a prototype blade is shown in Figure 73.55, illustrating the very effective hollow bending section produced.
- Silicon carbide copper materials have been fabricated and tested for high-temperature missile applications. Silicon carbide bronze propellers have been cast (Fig. 73.56) for potential Navy applications where more efficient and quieter propellers are required.

73.5 CERAMIC MATRIX COMPOSITES

73.5.1 Continuous SiC Reinforced Ceramics

High performance ceramic materials offer many outstanding properties for high-temperature structural applications. They have excellent high-temperature strength and corrosion resistance, good thermal shock resistance, low density, and low coefficient of thermal expansion. In addition, starting raw materials are abundant and inexpensive. Despite these qualities, their use in applications such as gas turbine engines has been limited because of their large degree of scatter in strengths and, more importantly, their brittle nature, which often leads to catastrophic failure. Design engineers have been hesitant to work with materials having such low component reliability. Recently, there has been an increasing interest in the tremendous potential offered by fiber reinforcement to toughen ceramics. For example, the fracture toughness of a Corning lithium aluminosilicate glass/ceramic was improved an order of magnitude by the addition of silicon carbide fiber.[20] While a number of different

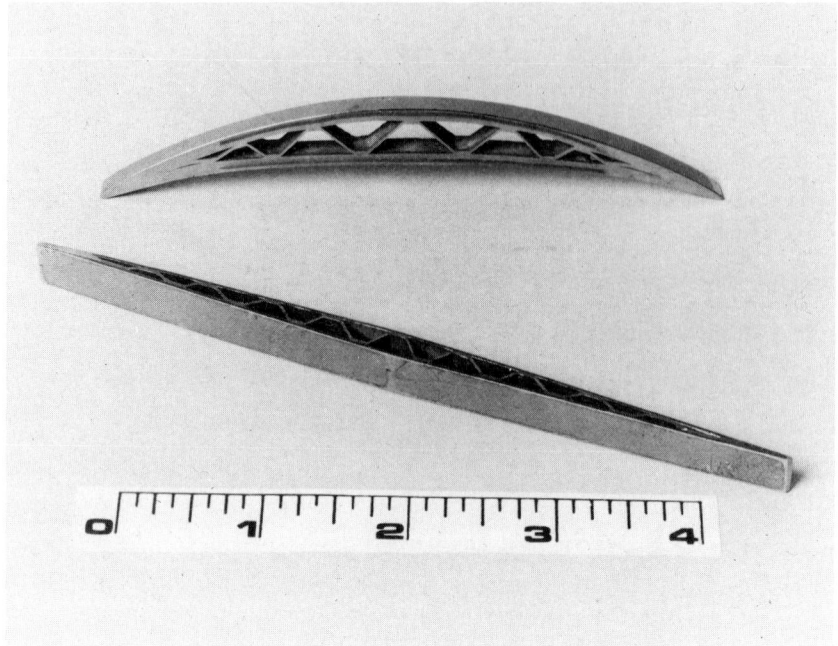

Figure 73.55. Hollow SiC/Ti fan blade section.

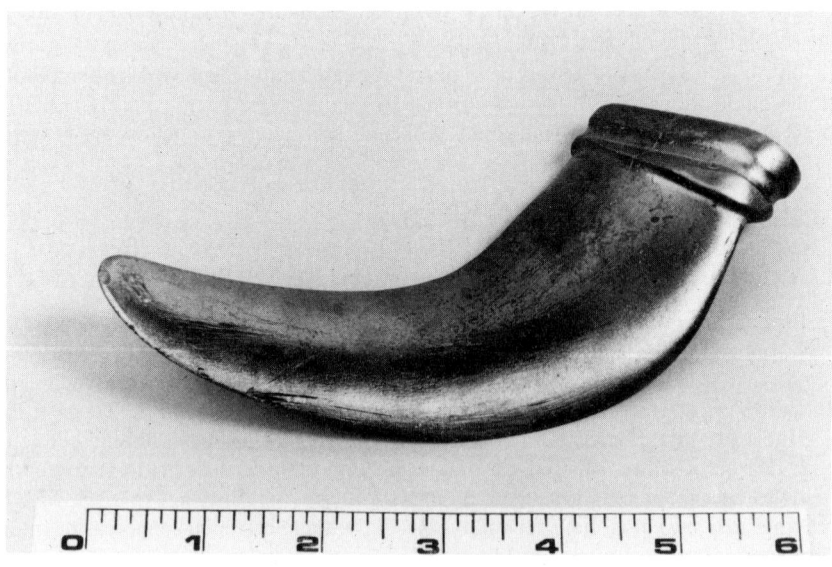

Figure 73.56. SiC bronze propeller.

fiber compositions have been investigated, silicon carbide fibers have produced some of the most promising results.

The two most widely evaluated continuous silicon carbide fibers are Nippon Carbon's Nicalon fiber, marketed by Dow Corning in the U.S., and Textron Specialty Materials' fiber. It should be pointed out that the Nippon fiber was developed by Yajima and co-workers in Japan and is well documented.[21]

73.5.1 Chemistry and Mechanical Properties

To a large extent, the mechanical properties of these fibers, particularly at elevated temperatures, are affected by the chemical composition of the fibers. Textron's fiber is essentially comprised of SiC and carbon, with discrete regions of each. The microstructure of Nippon's fiber is not discrete, consisting of an intimate amorphous combination of Si, C, and O that separates into crystalline β SiC, graphite, and amorphous SiO_2 when it is heat treated above 2192°F (1200°C).

Chemical composition and mechanical properties for both fibers are shown in Table 73.13. Variations in the mechanical properties for both fibers can be partially attributed to low-volume production. As can be seen in Table 73.13, there is typically a range of fiber diameters in processing of Nicalon, which is a major factor in property variations.

Toughening Mechanisms. There are several theories as to what characteristics contribute to the recorded high-fracture toughness in ceramic composites. While understanding of these mechanisms is not complete, several materials factors are considered to be important. The ratio of fiber modulus to matrix modulus should be at least 2:1 to allow a greater load transfer from the matrix to the fiber. Thermal expansion mismatches are also important, as a higher thermal expansion in the fiber produces radial tensile forces around the fiber. When a crack approaches the fiber, it tends to bend around it, thus increasing the toughness by dissipating fracture energy. However, excessive differences in the thermal expansions of the fiber and matrix can cause prestressing of the fiber and can decrease strength and toughness.

Recent work has indicated that chemical interactions at the fiber/matrix interface may

TABLE 73.13. Chemistry and Properties of SiC Fibers

Property	Nicalon	Textron[a]
Composition, wt %[b]		
Si	55.5	70
C	28.4	30
O	14.9	Trace
H	0.13	
Molecular composition, wt %		
SiC	61	99+
SiO_2	28	
Free C	10	Trace
Mechanical properties[c]		
Density, g/cc (lbs/in.[3])	2.5 (0.092)	3.0 (0.110)
Tensile strength, GPa (ksi)	2.4–2.9 (350–425)	3.8–4.1 (550–600)
Tensile modulus, GPa (Msi)	172–200 (25–29)	400–414 (58–60)
Filament diameter, μm (in.)	10–15 (0.0004–0.0006)	143 (0.0056)
Filaments/yarn	500	

[a]Textron - Composition excluding carbon substrate.
[b]Ref. 21.
[c]From suppliers.

have the most profound effect on determining the toughness of ceramic composites. While a good mechanical bond between the fiber and matrix is desirable, a strong chemical bond tends to cause fiber degradation and adversely affects fracture toughness. Fiber coatings that inhibit chemical bonding have proven to be effective in this area. Coating of Nicalon SiC fiber significantly improved toughness over uncoated fibers in a SiO_2 matrix.[22]

Similarly, a ceramic matrix with coated Sumitomo Al_2O_3 fiber had double the strength at considerably higher fracture toughness than a matrix with uncoated fiber.[23]

The significance of this coating work means that engineers considering a ceramic composite materials system need not be totally constrained by the chemical and thermal compatibility of the fiber and matrix. Coating the fiber allows the engineer to choose almost any fiber/matrix combination.

Glasses and Glass/Ceramics. Much of the earliest work in ceramic composites involved both glasses and glass/ceramics. The processing of these systems is easy (dip the fibers in a ceramic slurry and hot press). Relatively low pressure [about 2000 psi (14 MPa)] is needed at hot pressing temperatures to achieve full density, thus lessening the odds of damaging fibers. Glass/ceramics have the added advantage of easy glass processing with subsequent crystallization for higher temperature stability.

Prewo and Brennan evaluated both SiC fibers in a Corning 7740 borosilicate glass.[24] The results are shown in Table 73.14. The higher mechanical properties of Textron's fiber translated into a higher composite strength, stiffness, and toughness than with Nicalon's fiber. However, the larger surface-to-volume ratio of Nicalon was cited as providing better composite microstructure and transverse properties.

The fibers have also been evaluated at low fiber volumes in magnesium aluminosilicate (MAS) glass/ceramic matrices. It was found that strength and toughness values were highly dependent on the heat treatment of the composite. After heat treatment at 1500°F (815°C) for three hours and 2100°F (1150°C) for two hours, they were found to improve significantly.[25] This was attributed to the phase change of the matrix, which put the matrix in compression and produced a weak fiber/matrix bond. As a result, cracks bowed around the fiber and fracture toughness improved. This development work also supported the importance of fiber surfaces in promoting weak bonds and enhancing toughness. Fiber surfaces with the standard treatment of SiC and amorphous carbon effectively stopped cracks, while an untreated surface did not.

The growing interest in glass/ceramic composites has led to new developments in high-temperature matrices. Barium magnesium aluminosilicate (BMAS), developed by Corning Glass, has a usable service temperature of at least 2280°F (1250°C). Both SiC fibers have been evaluated in BMAS matrices. Figure 73.57 shows a plot of strength versus temperature for Nicalon/BMAS composites at 40 vol % in both air and vacuum.

TABLE 73.14.   Properties of SiC Reinforced 7740 Glass[a]

Property	Nicalon	SCS	
Fiber content, vol %	40	35	65
Density, g/cm³ (lb/in.³)	2.4 (0.0087)	2.6 (0.094)	2.9 (0.105)
Axial flexural strength, GPa (ksi)			
70°F	0.29 (42)	0.65 (94)	0.83 (120)
662°F	0.36 (52)		0.93 (135)
112°F	0.52 (75)	0.83 (120)	1.24 (180)
Axial elastic modulus (70°F), GPa (Msi)	117 (17)	186 (27)	290 (42)
Axial fracture toughness, ksi √in.			
70°F	10.5	17.1	
1112°F	6.4	13.0	

[a]Ref. 24.

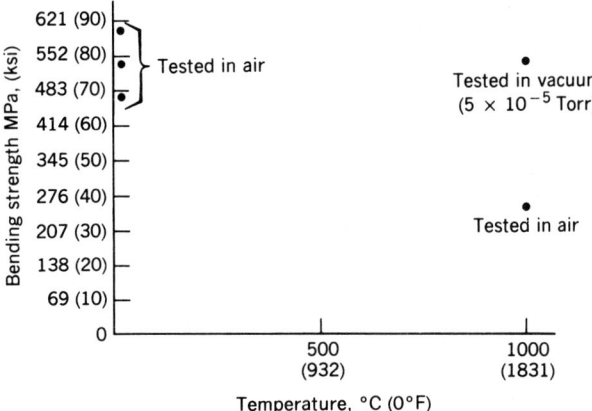

Figure 73.57. Nicalon/bariumosumilite composites bending strength vs temperature.

Once again, there is evidence of high temperature degradation of Nicalon composites in air. Unpublished work on a small scale has been done by Corning Glass with Textron's fiber in BMAS. Preliminary data with 20 vol % show no strength degradation at 1832°F (1000°C) in air. More extensive testing is planned. Additionally, Corning is continuing their development of higher-temperature glass/ceramic matrices, which may result in service temperatures of 2460 to 2550°F (1350 to 1400°C).

Ceramics. Less work has been done in reinforcing ceramics, such as silicon nitride and silicon carbide, largely because of the higher processing temperatures required. Though the data are limited, Textron's fiber has been evaluated in both matrices. Table 73.15 highlights the results of 30 vol % carbon-coated SiC monofilament in a reaction-bonded silicon nitride. Room temperature strength and fracture toughness were increased significantly by adding fibers, and there was only a 20% decrease in strength in testing at 2190°F (1200°C). Equally encouraging was the lack of any apparent fiber degradation from processing, which included exposure of the fiber to 2280°F (1250°C) for 19 hours in flowing N_2 gas.

Battelle Columbus Labs also evaluated Textron's fiber in Si_3N_4. Table 73.16 presents the results of this work. There was evidence of cracks in the fiber, probably caused by processing which involved both high temperature (3180°F [1750°C]) and relatively high pressure [4000 psi (28 MPa)]. Strengths of composites at volume-percentages of 30 and 44% were approximately half that of the unreinforced matrix, clearly demonstrating damaged fiber. Despite this, there was a significant increase in crack growth resistance in the reinforced matrices, indicating that even damaged or broken fibers in place can have a positive effect on toughness.

TABLE 73.15. 30 v/o SCS-0 in Reaction Bonded Si_3N_4[a]

	Monolithic Si_3N_4	Si_3N_4 With 30 v/o SiC
Fracture strength at RT MPa (ksi)	117 (17)	283 (41)
Fracture strength (1200°C, ksi)		220 (32)
Fracture toughness ksi $\sqrt{in.}$	(1.5–2)	7.0–7.2

[a]Ref. 25, 1250°C, 18 h in N_2.

TABLE 73.16.   30 v/o and 44 v/o in Si_3N_4 (Y_2O_3 and Al_2O_3 Sintering Aids, Hot Pressed)[a]

	Si_3N_4	30 v/o SiC[b] in Si_3N_4	44 v/o SiC[b] in Si_3N_4
Modulus of rupture (3 − PT, loading) MPa (ksi)	896 (130)	414 (60)	414 (60)
Crack growth resistance (G_{1C}, J/M^2) (Chevron notch bend test)	82	7,800 8,500	6,200 8,400

[a]1750°C, 4 ksi, N_2.
[b]Evidence of cracks in fiber from processing.

More promising results reinforcing this matrix were obtained by Norton Company, which evaluated both fibers in reaction-bonded silicon nitride. Textron's SCS-6 fiber was more stable under thermal processing conditions than Nicalon, and composite flexural strength was more than twice as high with SCS-6 as the best Nicalon-reinforced composite (Fig. 73.58).

73.6 FUTURE TRENDS AND APPLICATIONS

In addition to boron's high mechanical properties and light weight, boron also offers various other unique properties. Boron is naturally a high neutron absorber, it is exceptionally hard (harder than most ceramics), and it also has unique microwave polarization properties. The combination of these properties with boron's high mechanical properties creates new potential applications in neutron shielding and radiation hardening, cutting and grinding tools, and antenna and radome design.

Boron's exceptional compression properties have resulted in contracts to evaluate boron's potential use in anti-armor projectiles. Work in this area is just beginning, but could lead

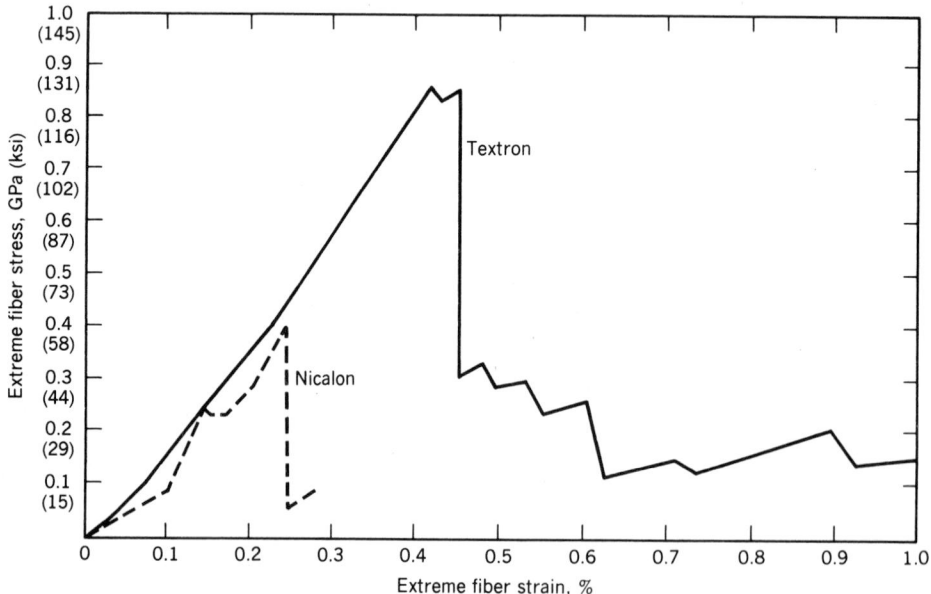

Figure 73.58.   Reaction-bonded Si_3N_4 composites, load vs deflection.

to applications as casings/aeroshells on missiles or projectiles. Additionally, work is ongoing to use boron epoxy doublers to reinforce overfatigued areas on aircraft structures. To date, boron has been used on a number of military aircraft, such as the F-111, C-141, C-130, and Mirage III, among others. Boron epoxy is now being evaluated for use on commercial aircraft, as well.

The SiC fiber is qualified for use in aluminum, magnesium, and titanium. Copper matrix systems are under development. Good results have been obtained using the higher-temperature titanium aluminides as matrix materials. As shown in Figure 73.14, the SCS-6 fiber demonstrates high mechanical properties to above 2500°F (1400°C). It is natural, then, to project systems such as SiC/nickel aluminides/iron aluminide/superalloys, and so on, all of which, on a rule of mixtures basis at least, project very useful properties for engine and hypersonic vehicle applications. Work required in this area includes diffusion barrier coatings and matrix alloy modifications to facilitate high-temperature fabrication processes. Also required is the detailed investigation of any detrimental thermal/mechanical cycling effects that may occur as a result of the mismatch in thermal expansion coefficients between matrix and fiber.

Because of the increasing interest and tremendous potential for advanced composites, new developments in fiber technology are taking place. Dow Corning is being funded by DARPA to develop improved ceramic composites based on an organosilicon polymer-based fiber. The goal of the fiber development phase is to achieve greater mechanical properties than the Nicalon fiber, plus improved high temperature stability in air. Other polymer processes include those developed by Ube, Bayer, and Union Carbide. Textron has developed the capability to weave its fiber with other materials for greatly improved handling of the fiber in composite lay-up. Should any of the polymer-based fibers prove to be successful, Textron could use one as interweave material, yielding a two-dimensional reinforcement with high temperature stability and unique properties in each direction. Other possibilities include three-dimensional reinforcements fabricated by polymer infiltration to produce strong, tough, and very lightweight ceramic composites.

Significant groundwork has been laid to show the improved strength and toughness possible with ceramic composites reinforced with both Nippon's and Textron's fibers. The uniqueness of polymer pyrolysis and CVD-manufactured fibers means that both types of fiber will likely be used. Programs are now needed to demonstrate the efficiency of ceramic composites for specific simple components, followed by programs for more complex components. Once reliability is verified, the excellent potential for ceramic composites in high-temperature structural applications can be realized.

BIBLIOGRAPHY

1. H. DeBolt, "Boron and Other Reinforcing Agents," in G. Lubin, ed., *Handbook of Composites*, Van Nostrand Reinhold Co., New York, 1982, Chapt. 10.

2. V.J. Krukonis, "Boron Filaments," in J.V. Milewski and H.S. Katz, eds. *Handbook of Fillers and Reinforcements for Plastics*, Van Nostrand Reinhold Co., New York, 1977, Chapt. 28.

3. D.L. McKaniels and R. Ravenhall, *Analysis of High-Velocity Ballistic Impact Response of Boron/Aluminum Fan Blades*, NASA TM-83498, 1983.

4. C.T. Salamme and S.A. Yokel, *Design of Impact-Resistant Boron/Aluminum Large Fan Blades*, NASA CR-135417, 1978.

5. J.W. Brantley and R.G. Stabrylla, *Fabrication of J79 Boron/Aluminum Compressor Blades*, NASA CR-159566, 1979.

6. E. Wolff, *Boron Filament, Metal Matrix Composite Materials*, AF33(615)3164.

7. R.J. Suplinskas, *Manufacturing Technology for Silicon Carbide Fiber*, AFWAL-TR-84-4005, 1983.

8. R.J. Suplinskas, *High Strength Boron*, NAS3-22187, 1984.

9. T. Schoenberg and co-workers, *Establishment of a Manufacturing Process for the Production of Boron Epoxy Tape, AFML-TR-73-185*, July 1973.

10. *Advanced Composites Design Guide*, 3rd ed., 2nd rev., Air Force Flight Dynamics Laboratory, Wright-Patterson Air Force Base, Dayton, Ohio, 1976.

11. *AFML F33615-78-C-5172*, University of Dayton, 1978.

12. J.D. Forest, *Boron/Aluminum Tube Constructions for Advanced Vehicle Applications*. General Dynamics, Convair Division, March 1975.

13. C.J. Hilado, ed., *Boron Reinforced Aluminum Systems*, Vol. 6, Materials Technology Series, Technomic Publishing Co., 1974.

14. *Advanced Composites Design Guide*, prepared under *Contract No. F33615-74-C-5075*, by Rockwell International for Wright-Patterson Air Force Base, Sept. 1976.

15. *DoD/NASA Advanced Composites Design Guide*, prepared under *Contract No. F33615-78-C-3203*, by Rockwell International for Wright Patterson Air Force Base, July 1983.

16. D.M. Aylor, *Assessing the Corrosion Resistance of Metal Matrix Composite Materials in Marine Environments*, DTNSRDC/SMME-83/45, 1983.

17. A.J. Kumnick and co-workers, *Filament Modification to Provide Extended High Temperature Consolidation and Fabrication Capability and to Explore Alternative Consolidation Techniques, N00019-82-C-0282*, 1983.

18. J.A. Cornie and Y. Murty, *Evaluation of Silicon Carbide/Magnesium Reinforced Castings, DAAG46-80-C-0076* 1983.

19. J.V. Marzik and A.J. Kumnick, *The Development of SCS/Copper Composit Material, N60921-83-C-0183*, 1984.

20. J.J. Brennan and K.L. Prewo, "Silicon Carbide Fibre Reinforced Glass-Ceramic Matrix Composites Exhibiting High Strength and Toughness," *J. Materials Sci.* **17** (1982).

21. C.H. Andersson and R. Warren, "Silicon Carbide Fibres and Their Potential for Use in Composite Materials, Part 1," *Composites* **15** (Jan. 1984).

22. R. Rice, "Ceramic Fiber Composites with Coated SiC Fibers," *presented at 8th Annual Conference on Composites and Advanced Ceramic Materials*, Cocoa Beach, Fla., Jan. 19–20, 1984.

23. D. Lewis, "Ceramic Fiber Composites with Coated SiC Fibers," *presented at 8th Annual Conference on Composites and Advanced Ceramic Materials*, Cocoa Beach, Fla., Jan. 19–20, 1984.

24. K.M. Prewo and J.J. Brennan, "High Strength Silicon Carbide Fibre-Reinforced Glass-Matrix Composites," *J. Materials Sci.* **15** (1980).

25. M. Tail, *Mechanisms of Ductility, Toughness, and Fracture in High Temperature Materials, AF Contract F33615-81-C-5059*.

74

REINFORCED PLASTICS, GRAPHITE CARBON FIBERS

Leonard Poverno

Grumman Aerospace Corporation, Calveron, NY 11933

74.1 CATEGORY

Graphic carbon fibers are the predominant high-strength, high-modulus reinforcing agent used in the fabrication of high-performance resin/matrix composites.

In general, the term graphite fiber refers to fibers that have been treated above 3092°F (1700°C) and have tensile moduli of elasticity of 5×10^5 psi (3450 MPa) or greater. Carbon fibers are those products that have been processed below 3092°F (1700°C) and consequently exhibit tensile moduli up to 5×10^5 psi (3450 MPa).[1] A further distinction is that the content of carbon fibers is 80 to 95% carbon; of graphite fibers, above 99% carbon. But since the industry has universally adopted the term "graphite," it will be used to describe both product forms in this section.

74.2 HISTORY

Graphic carbon fiber technology is the cutting edge of the reinforced composites industry, albeit comprises a relatively small percentage of the total reinforced plastics (RP) market. However, the most advanced technology and innovative applications are emerging first from this market segment via aerospace applications.

Graphite fibers were first utilized by Thomas Edison in 1880 as part of his incandescent lamp.[2] The lamp filaments were generated by pyrolyzing cellulose (rayon) fibers. When tungsten filaments replaced the graphite in lamps, interest in graphite materials waned until the mid-1950s when Soltes[2] and Abbott[3] created rayon-based graphite fibers via inert atmosphere, 1832°F (1000°C) heat treatments. These products exhibited relatively high tensile strengths of about 4×10^5 psi (2760 MPa) and were designed for rocket/missile ablative component applications.

Further developments by Union Carbide[4] and Bacon[5] resulted in improved fiber strength and stiffness properties by increasing the carbonization temperatures and the preferred crystal basal plane orientation. The theoretical maximum tensile modulus of graphite parallel to the basal planes is 146×10^6 psi (1×10^4 MPa).[6] By 1970, Union Carbide was producing commercially rayon-based carbon fibers with 7.54×10^5 psi (5200 MPa) tensile modulus and 385,000 psi (2650 MPa) tensile strength.

A significant event leading to the development of today's graphite industry was the utilization of polyacrylonitrile (PAN) as a graphite precursor material by Tsunoda[7] in 1960.

Tsunoda cross-linked (stabilized) the PAN in an oxidizing atmosphere at 428°F (220°C) prior to carbonization. Subsequent work led to continued improvement of PAN-based graphite fiber properties by numerous researchers, among them Watt and Johnson of the Royal Aircraft establishment. These developments focused on stretching the PAN precursor to obtain a high degree of molecular orientation of the polymer molecules followed by stabilizing it under tension load, carbonization, and graphitization. PAN-based graphite fibers are currently available with tensile moduli of up to 1.2×10^6 psi (8.28×10^4 MPa) and tensile strengths to 1×10^4 psi (6.9×10^6 MPa).

Pitch was first identified as a graphite precursor by Otani in 1965.[8] These fibers are made by melt spinning a low-cost, isotropic molten pitch petroleum material and then oxidizing the filaments as they are spun. This step is followed by carbonization at 1832°F (1000°C) in an inert atmosphere. Modifications of the initial concept to improve the fiber properties evolved through the 1970s (including mesophase liquid-crystal pitch) until pitch-based graphite fibers with up to 3.75×10^5 psi (2590 MPa) tensile strengths and tensile moduli to 1.2×10^6 psi (8300 MPa) were achievable.

74.3   KEY PROPERTIES AND COMPOSITE CHARACTERISTICS

The excellent properties of graphite are directly attributable to the highly anistropic nature of the graphite crystal.

Graphite fibers are available to the user in a variety of forms: continuous filament for filament winding, braiding, or pultrusion; chopped fiber for injection or compression molding; impregnated woven fabrics and unidirectional tapes for lamination; and paper, felts, or mat for selective reinforcements.

The most common form of graphite sold is the tow, which can be processed secondarily into the other forms. Table 74.1 summarizes relevant properties, such as tensile strength, tensile modulus, and density characteristics, of selected commercially available graphite fibers. These products fall into three general categories, based on their structural characteristics.

The standard grade PAN-based graphite fibers were the first developed and make up the largest part of both the commercial and aerospace markets. These materials have tensile strengths ranging from 4.5×10^5 to 5.5×10^5 psi (3100 to 3800 MPa) and moduli of approximately 340,000 psi (2345 MPa). High-performance aerospace requirements have pushed the industry to develop a family of intermediate-modulus (IM)/high-strain fibers with tensile strengths up to 7×10^5 psi (4800 MPa) with a modulus above 4×10^5 psi (2760 MPa).

The high-strain fibers approach 2% ultimate elongations, with tensile strengths above 800,000 psi (5516 MPa). The last category encompasses the high-modulus fibers, both PAN- and pitch-based. These products have tension moduli ranging from 5×10^5 to 1.2×10^6 psi (3450 to 8280 MPa). Their strain to failures are generally greater than or equal to 1% and, consequently, they are used in high-stiffness/low-strength applications, such as space hardware.

74.4   FIBER MANUFACTURING

The pyrolysis of organic fibers into graphite fibers is a multistage process that begins with structurally acceptable, cost-effective precursors. The three principal graphite precursors are PAN, pitch, and rayon, with PAN as the current predominant product.

Typical PAN precursor manufacturing steps are depicted in Figure 74.1. The starting

TABLE 74.1. Property Summary of Selected Graphite Fibers[a]

Fiber Name	Manufacturer	Precursor	Tensile Strength, ksi (MPa)	Tensile Modulus, msi (GPa)	Ultimate Elongation, %	Density, lb/in.³ (G/cm)
Celion G30-500	BASF	PAN	550 (3792)	34 (234)	1.62	0.064 (1.78)
Celion G30-600	BASF	PAN	630 (4344)	34 (234)	1.85	0.064 (1.78)
Celion G40-600	BASF	PAN	620 (4275)	43.5 (300)	1.43	0.062 (1.73)
Celion G40-700	BASF	PAN	720 (4964)	43.5 (300)	1.66	0.064 (1.77)
Celion G40-800	BASF	PAN	815 (5600)	43.5 (300)	1.9	
Celion GY-70	BASF	PAN	270 (1862)	75 (517)	0.36	0.071 (1.96)
Celion GY-80	BASF	PAN	270 (1862)	80 (572)	0.32	0.071 (1.96)
Magnamite AS-1	Hercules	PAN	450 (3105)	33 (228)	1.32	0.065 (1.80)
Magnamite AS-4	Hercules	PAN	550 (3795)	34 (235)	1.53	0.065 (1.80)
Magnamite AS-6	Hercules	PAN	600 (4140)	35 (242)	1.65	0.066 (1.83)
Magnamite IM-6	Hercules	PAN	635 (4382)	40 (276)	1.50	0.063 (1.73)
Magnamite IM-7	Hercules	PAN	683 (4713)	41 (283)	1.60	0.064 (1.78)
Magnamite HMS4	Hercules	PAN	360 (2484)	49 (338)	0.7	0.065 (1.80)
Magnamite HMU	Hercules	PAN	400 (2760)	55 (380)	0.7	0.067 (1.84)
Grafil XA-S	Hysol/Grafil	PAN	450 (3105)	34 (234)	1.31	0.065 (1.80)
Grafil XA-S High Strain	Hysol/Grafil	PAN	560 (3861)	34 (234)	1.65	0.065 (1.80)
Grafil IM-S	Hysol/Grafil	PAN	450 (3105)	42 (289)	1.07	0.064 (1.78)
Grafil HM-S/6K	Hysol/Grafil	PAN	400 (2760)	54 (373)	0.74	0.067 (1.84)
Thornel T-300	Amoco	PAN	490 (3381)	33 (228)	1.49	0.064 (1.76)
Thornel T-500	Amoco	PAN	565 (3896)	35 (242)	1.61	0.064 (1.76)
Thornel T650/42	Amoco	PAN	720 (4964)	42 (289)	1.71	0.065 (1.79)
Thornel T-40	Amoco	PAN	820 (5634)	41 (283)	2.00	0.065 (1.81)
P-55	Amoco	Pitch	275 (1889)	55 (380)	0.5	0.072 (2.00)
P-75	Amoco	Pitch	300 (2070)	75 (517)	0.4	0.072 (2.00)
P-120	Amoco	Pitch	325 (2242)	120 (827)	0.3	0.079 (2.18)

[a]Data provided by the respective suppliers: BASF, Hercules, Hysol/Grafil and Amoco, in the latest available vendor data sheets.

product is a linear polymer consisting of a carbon hydrogen backbone with polar carbon–nitrogen (nitrile) groups attached (see below).

All commercial production of PAN precursor fiber is based on either dry or wet spinning technology. In both instances, the polymer is dissolved in either an organic or inorganic solvent at a concentration of 5–10% by weight. The fiber is formed when the polymer solution is extruded through spinnerette holes into a hot gas environment (dry spinning) or into a coagulating solvent (wet spinning).[9] The dry spinning results in fiber with a dog bone cross-section, whereas the more popular wet spinning produces a more efficient, round cross-section. The wet spun precursor manufacturing process includes three basic steps: polymerization, spinning, and after-treatments (see Fig. 74.1). In polymerization, the acrylonitrile monomer and other comonomers (methyl acrylate or vinyl acetate) are reacted to form a polyacrylonitrile copolymer. The reactor effluent solution, called "dope," is purified prior to spinning. The unreacted monomers are then removed and the solid contaminants filtered off. The spinning process next extrudes the purified dope through holes in spinnerettes into a coagulation solution. The spun gel fiber then goes through a series of after-treatments such as stretching, oiling, and drying.

In order to produce high-strength, high-modulus graphite fibers from the PAN precursor, it is necessary to induce a preferred molecular orientation parallel to the fiber axis and then stabilize the fiber against the relaxation phenomena and chain scission reactions that result from subsequent carbonization steps.[10,11] Figure 74.2 is a schematic of a typical step-by-step PAN-based graphite manufacturing process, which begins with the aforementioned precursor stabilization followed by carbonization, graphitization, surface treatment, and sizing.

The stabilization or "preoxidation" of the PAN precursor involves heating the fiber in an air oven at 392–572°F (200–300°C) for approximately one hour while controlling the shrinkage/tension of the fiber so that the PAN polymer is converted into an aromatic ladderlike structure that is thermally infusible and inflammable.

The next step in the process is carbonization, which pyrolyzes the stabilized PAN-based fibers until they are transformed into graphite (carbon) fibers. It is during this stage that the high mechanical property levels characteristics of commercially available graphite are developed. The development of these properties is directly related to the formation and orientation of graphitelike fibers or ribbons within each individual fiber (see Fig. 74.3[12]). Carbonization treatment occurs in an inert atmosphere (generally nitrogen) at temperatures greater than 2192°F (1200°C). This step removes hydrogen, oxygen, and nitrogen atoms from the ladder-type polymers whose aromatic rings then collapse into a graphitelike polycrystalline structure. Graphitization is an option in the process that is performed at temperatures above 3272°F (1800°C). The purpose of this step is to improve the tensile modulus of elasticity of the fiber by improving the crystalline structure and preferred orientation of the graphite like crystalites within each individual fiber.

Here again, it should be noted that, technically, the term "graphite fiber" describes fibers that have a carbon content in excess of 99%; the term "carbon fiber" describes fibers that

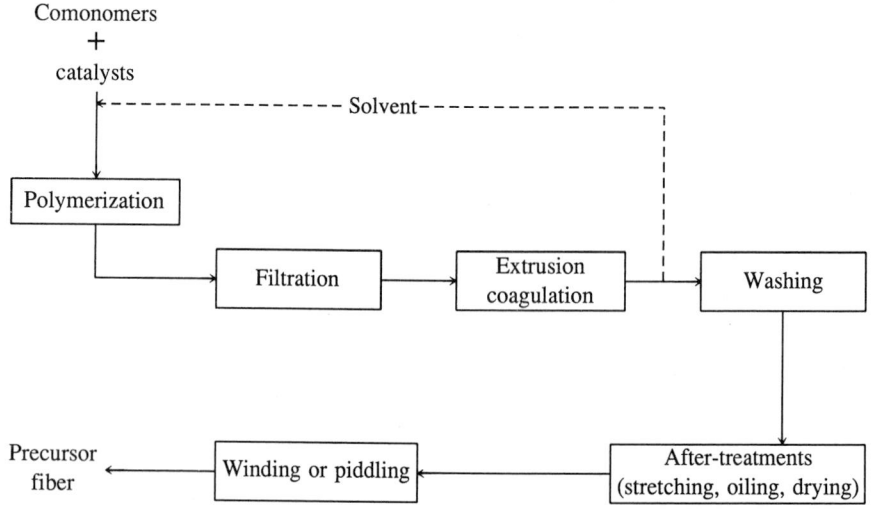

Figure 74.1. Typical PAN precursor process (solution polymerization/wet spinning).

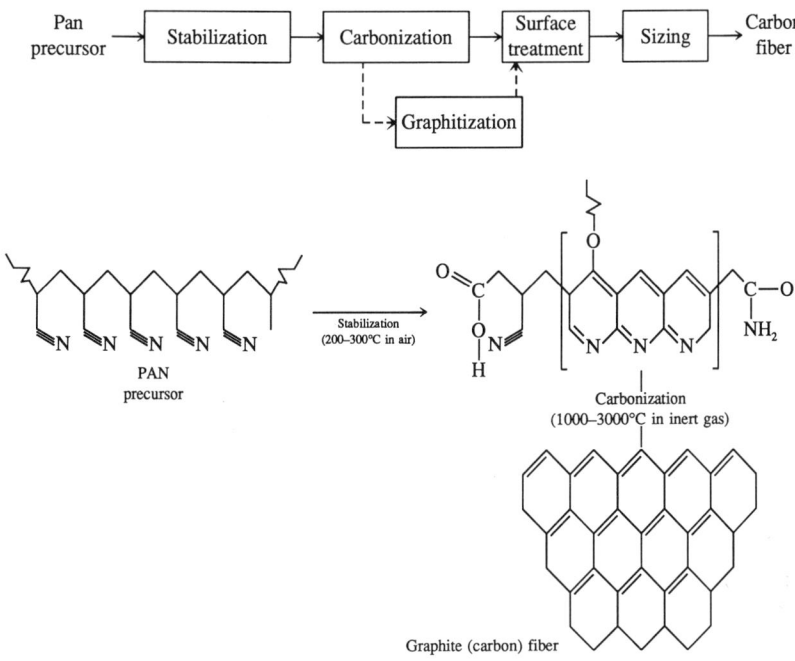

Figure 74.2. Typical carbon fiber process.

have a carbon content of 80–95%. The higher the heat treatment temperature, the higher the carbon content. The more generally used industry term "graphite," has been used throughout this section to describe both.

The final step in the process of producing carbonized or graphitized fiber is surface treatment and sizing prior to bobbin winding the continuous filaments. The surface treatment is an oxidation of the fiber surface to promote wettability and adhesion with the matrix resin

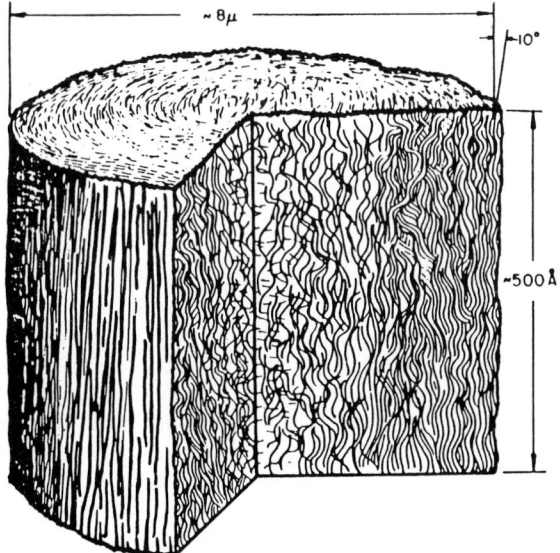

Figure 74.3. Three dimensional structure model of a PAN-based carbon fiber (round cross section) with a tensile modulus of elasticity of 410,000 psi (2830 MPa).[12]

in the composite. The size promotes handleability and wettability of the fiber with the matrix resin. Typical sizing agents are epoxy, poly(vinyl alcohol), polyimide, and water.

Pitch-based graphite fibers are produced by two processes. The first of these processes results in low modulus fibers unless stress graphitization at extremely high temperatures is employed.[13] The precursor for this process is a low softening-point isotropic pitch. The process scheme includes the following steps:

- Melt spin isotropic pitch.
- Thermoset at relatively low temperatures for long periods of time.
- Carbonize in an inert atmosphere at 1832°F (1000°C).
- Stress graphitize at high temperatures 5432°F (3000°C).

The high performance fibers produced in this manner tend to be relatively expensive because of the very long thermosetting times required and the need for a high-temperature stretch procedure.

The more commercially significant processing scheme for making pitch-based fibers is the mesophase process, which includes the following steps:

- Heat treat at 752–842°F (400–450°C) in an inert atmosphere for an extended period of time in order to transform it into a liquid-crystalline (mesophate) state.
- Spin the mesophase pitch into fibers.
- Thermoset the fibers at 572°F (300°C) for 2.5 hours.
- Carbonize the fibers at 1832°F (1000°C).
- Graphitize the fibers at 5432°F (3000°C).

Since long thermosetting times and stress graphitization treatments are not required, the high-performance graphite fibers produced by this process are low in cost.

The process by which rayon precursor is converted to grpahite fibers involves four steps:

- Fiber spinning.
- Stabilization at 752°F (400°C) for long processing times.
- Carbonization at 2372°F (1300°C).
- Stress graphitization at 5432°F (3000°C).

The excessive cost of this processing results from the stress graphitization, which is reflected in the relatively high cost of the rayon-based graphite fibers.

74.5 APPLICATIONS

The major markets for advanced graphite fiber composites are aerospace, marine, automotive, industrial equipment, and recreation. Military aerospace applications will dominate the market for the near future. Military consumption of graphite fiber will rapidly increase in the mid 1990s as new programs, which utilize a very high percentage of composites, move from development to large-scale production. A partial listing of these programs include the ATF, ATA, V22, LHX, and the C-17. If and when SDI applications are moved from R&D to deployment, graphite fiber composites will most likely be utilized. Graphite fiber usage in space applications will also grow as programs, such as the U.S. Space Station and the National Aerospace Plane, approach production.

Nonaerospace military applications are also evolving. Examples of some possible applications currently under development include portable, rapid deployment bridges for the army and propeller shafts for submarines.

Fiber usage in the commercial aerospace sector is also growing, but at a much slower rate than in the military sector. Relatively new commercial planes such as the Boeing 7J7 and the Airbus A320 utilize two to three times the graphite fiber per plane than is used in older commercial models. Innovative new business planes, such as the Beech Starship, whose structural weight is almost entirely composite, will become the standard for the next generation of business jets.

Industrial consumption of graphite fiber should continue to grow as new applications are developed. The biggest industrial market potential, which has thus far defied large-volume use of graphite fiber, is the automotive market. To date, small amounts of graphite are being used in drive shafts for some vans and larger vehicles. Despite this less than successful penetration into the automotive market, many analysts continue to predict that graphite composite automotive usage will rapidly increase as lower-cost fibers become available.

In 1973, a breakthrough led to the use of graphite fiber in golf-club shafts. This was followed by many sporting goods applications such as fishing rods, tennis and racquetball racquets, skis, sail boats, and wind surfers.

In the industrial market, a major and growing use of chopped graphite fibers is as a reinforcement for thermoplastic injection-molding compounds. The inherent advantages include greater strength and stiffness; increased resistance to wear, creep and fatigue resistance; higher electrical conductivity; and improved thermal stability and conductivity.

74.6 TRADE NAMES

See Tables 74.2 and 74.3.[14]

TABLE 74.2. U.S. Production Sources of Carbon Fiber[a]

Carbon/Graphite Fiber Type	Producer	Trade Name	Precursor Source	Production Capacity 1986 × 10³	
				lb	kG
Rayon-base	Polycarbon		AVTEX[b]	300	130
	Amoco	WCA	AVTEX	250	109
	Witco (B.P.)	CCA-4	AVTEX	208	91
				758	330
PAN-base	Hercules	Magnamite		2400	1050
(polyacrylonitrile)[c]	Hysol/Grafil	Grafil		700	305
	Amoco	Thornel-T	Amoco	700	305
	BASF (Celion)	Celion		1000	431
	Stackpole	Panex		250	109
	Avco			150	65
	HITCO (B.P.)	HITEX		200	87
	Great Lakes	Fortafil		450	196
	FMI			4	2
				5854	2250
Pitch-base	Amoco	Thornel-P	Ashland Allied Marathon Mobil	1000	431

[a]Ref. 14.

[b]Some rayon precursor fiber drawn from stockpiled IRC rayon, no longer produced.

[c]As basic acrylonitrile commodity; not as fiber.

TABLE 74.3. Foreign Production Sources of Carbon Fiber[a]

Country/Area	Organization	Trade Name	Precursor Source
United Kingdom	Courtaulds	Grafil	Courtalds
	RK Carbon Fibers Ltd	RK Carbon	Courtalds
France	Serofim	Rigilor	Rhone-Poulenc
	SOFICAR (Toray-Elf)	Filkar	Toray
Federal Republic of Germany	ENKA	TENAX	TOHO
	SIGRI	Sigrafil	Courtaulds
Japan	TOHO	Besfight	TOHO
	Toray	T-300	Toray
	Asahi Nippon	ANC	Asahi
	Mitsubishi Rayon	Pyrofil	M.R.C.
Israel	Afikim	ACIF	Courtalds
Far East	F.P.C.	TAIRYFIL	Courtaulds/M.R.C.
	Shanghai	AS	Domestic

[a]Ref. 14.

BIBLIOGRAPHY

1. G. Lubin and co-workers, *Handbook of Composites,* Van Nostrand Reinhold, New York, 1982, Chapt. 1.
2. U.S. Pat. 3,011,981 (1961), W. Soltes.
3. U.S. Pat. 3,053,755 (1962), W. Abbott.
4. G.E. Cranch, "Unique Properties of Flexible Carbon Fiber", in *Proceedings of the Fifth Conference on Carbon,* Vol. 11. Pergamon Press, New York, 1962, p. 589.
5. R. Bacon, "Growth, Structure and Properties of Graphite Whiskers", *J. Appl. Phys.* **31,** 283 (1960).
6. O.L. Blakslee and co-workers, *J. Appl. Phys.* **41** 3373 (1970).
7. U.S. Pat. 3,286,969 (1966), Y. Tsunoda.
8. S. Otani, *Carbon* **3,** 213 (1965).
9. G.P. Daumit and co-workers, "Latest in Carbon Fibers for Advanced Composites," *Performance Plastics '87 (First International Ryder Conference on Special Performance Plastics and Markets),* Atlanta, Ga., Feb. 11–13, 1987.
10. A.J. Clarke and J.E. Bailey, *Nature* **243,** 146 (1973).
11. J.E. Bailey and A.J. Clarke, *Nature* **234,** 529 (1971).
12. R.J. Diefendorf and E.W. Tokarsky, *The Relationship of Structure to Properties in Graphite Fibers, Part II, AFML-TR-72-133,* 1975.
13. S. Otani and co-workers, *Carbon* **4** 425 (1966).
14. S.L. Channon, *Industrial Base and Qualification of Composite Materials and Structures (An Executive Overview),* Institute of Defense Analysis, March 1984, pp. 28–36.

75

REINFORCED PLASTICS, VINYL ESTERS

Virginia B. Messick and Mary N. White

The Dow Chemical Co., Freeport, Texas

75.1 VINYL ESTER RESINS

Vinyl ester resins are the reaction product of an epoxy resin and a monofunctional ethylenically unsaturated carboxylic acid.

They may be represented by:

$$\left[H_2C = C - \overset{\overset{\displaystyle O}{\|}}{C} - O \right]_{2\text{–}3} R_2$$

where R^1 is H or CH^3 and R^2 is an epoxy resin, usually epoxylated bisphenol A–epichlorohydrin or epoxylated phenol–formaldehyde novolac. Unlike polyester resins, vinyl ester resins have highly reactive terminal unsaturation sites. Polyester resins, which are the reaction product of a difunctional unsaturated carboxylic acid or anhydride such as maleic or phthalic anhydride and a glycol, contain internal unsaturation sites. The toughness of the admittedly stiff and brittle vinyl ester resins actually exceeds that of polyester resins because vinyl ester resins contain an epoxy backbone.

75.2 CATEGORY

Vinyl ester resins are thermosetting polymers. They are stiff, brittle, and tough.

75.3 HISTORY

Vinyl ester resins were patented by Raphael Bowen,[1,2] a dentist who was seeking an acrylic material with improved toughness and bonding to teeth. These resins were first produced commercially in the mid 1960s by The Dow Chemical Co.[3] and Shell Chemical Co.[4] Important early applications of vinyl ester resins include matched metal die molding, filament-wound pipe, and hand-layup fiber-reinforced structures such as vessels, scrubbers, and towers where the toughness and corrosion resistance of vinyl ester resins were used to good advantage.

Since their introduction, vinyl ester resins have been modified to impart properties targeted at specific applications, as follows:

75.3.1 Thermal Resistance

In 1972, Dow Chemical introduced DERAKANE® (a trademark of the Dow Chemical Company) 470 vinyl ester resin,[5] based on phenol formaldehyde epoxy resin. At 36% styrene, DERAKANE 470 has a heat-distortion temperature of 289–1300°F (143–149°C) and excellent corrosion resistance to acid and solvent media. In 1976, Shell Chemical Co. introduced Epocryl DRH-480 vinyl ester resin,[6] based on a bisphenol A–epichlorohydrin epoxy resin. At 40% styrene, Epocryl 480 has a heat-distortion temperature of 250°F (121°C).

75.3.2 Impact Resistance

In 1977 Dow Chemical introduced a rubber-modified vinyl ester resin, XD-8084,[7] with improved impact resistance and adhesion over conventional vinyl ester resins.

75.3.3 Flame Retardance

Beginning in the early 1970s, several flame-retardant vinyl ester resins were made commercially available by Dow Chemical. These include DERAKANE 510-A-40, DERAKANE 510N,[8] and XP-71730. All are based on methacrylates of brominated epoxy resins.

75.3.4 Sheet Molding Compound

Chemically thickenable vinyl ester resins were introduced by Dow Chemical in 1967.[9] DERAKANE 790 has molecular acid functionality, which reacts with bivalent metallic oxides to cause a marked viscosity increase.

75.3.5 Radiation Curing

Radiation curable vinyl ester resins were developed by both Dow Chemical and Shell Chemical in the 1970s by substituting acrylic acid for methacrylic acid in the vinyl ester resin molecule. The acrylate species facilitates rapid curing on exposure to ultraviolet radiation.

No vinyl ester resin was produced in 1960; less than 3 million pounds (1400 metric tons) were produced in 1970 and approximately 14 million pounds (6400 metric tons) were produced in 1980. For those years, vinyl ester resins accounted for 0% and 1% of the total U.S. plastics consumption.

75.4 POLYMERIZATION

Vinyl ester resins are produced via a batch process. An epoxy resin is catalytically reacted with methacrylic acid in the presence of a polymerization inhibitor such as hydroquinone or monomethyl ether of hydroquinone. The esterification reaction is terminated when the desired acid number is reached. A reactive monomer, commonly styrene, is added to the base resin. Additional stabilizing inhibitors may be added at this point.

The choice of the epoxy resin to be used in the synthesis is dictated largely by the properties desired in the final product. A vinyl ester resin targeted for high temperature applications might be based on an epoxylated phenol–formaldehyde novolac; a resin whose end use requires toughness might be based on the diglycidyl ether of bisphenol A that has been advanced to a higher molecular weight with bisphenol A before the esterification reaction with methacrylic acid. In general, vinyl ester resins based on lower molecular weight epoxy resins tend to be more brittle with higher heat-distortion temperatures, while higher molecular weight epoxy resins yield vinyl ester resins that are tougher and more flexible but have lower heat-distortion temperatures.

Currently, corrosion-resistant vinyl ester resins are listed at about $1.05/lb; heat- and corrosion-resistant vinyl ester resins are listed at about $1.20/lb.

75.5 DESCRIPTION OF PROPERTIES

The unique properties of vinyl ester resins are the direct result of their chemical structure, Figure 75.1. The combination of an epoxy backbone with vinyl groups for high reactivity and styrene monomer for low viscosity retains the chemical resistance and strength of the base epoxy resin while improving the reactivity and basic handling characteristics.

75.5.1 Mechanical Properties

Cross-linking in vinyl ester resins is achieved through the terminal vinyl groups. This leaves the entire length of the molecule free to elongate under stress, thus absorbing mechanical and thermal stresses or shocks. The clear casting mechanical properties of various types of vinyl ester resins at 77°F (25°C) are shown in Table 75.1. Of particular interest is the ultimate elongation, 3–17.5%, which is significantly higher than competing resins. The percentage of flexural strength retained after continuous heat aging is given in Table 75.2.

In a glass-reinforced polyester or vinyl ester laminate, the resin will usually crack before the glass breaks. Thus, critical stress (the maximum a laminate can bear without failure) is dependent on strength of the resin and its resistance to cracking.

Critical stress can best be measured by acoustic emission testing, which can identify the critical point by two criteria: the point at which there are 10 events above 70 decibels and the point at which there is a large increase in counts for a small increase in strain. (Acoustic emission counts represent the number of times the transducer signal exceeds a preset threshold as damage occurs in the material being tested. Events are bursts of counts which are separated by a preset time interval.) The results of tests performed on a corrosion-resistant vinyl ester resin and a corrosion-resistant isophthalic polyester resin are shown in Figure 75.2.

It should be observed that this testing was performed in the direction of greatest laminate strength. The critical stress point has been shown to be dependent on both glass type and orientation.

75.5.2 Thermal Properties

Vinyl ester resins show excellent strength retention at temperatures up to their heat-distortion point. Vinyl ester structures are known to tolerate process upsets where heat-distortion temperature is exceeded. This, of course, would be dependent on the actual upset conditions (temperature, length of time, and chemical environment).

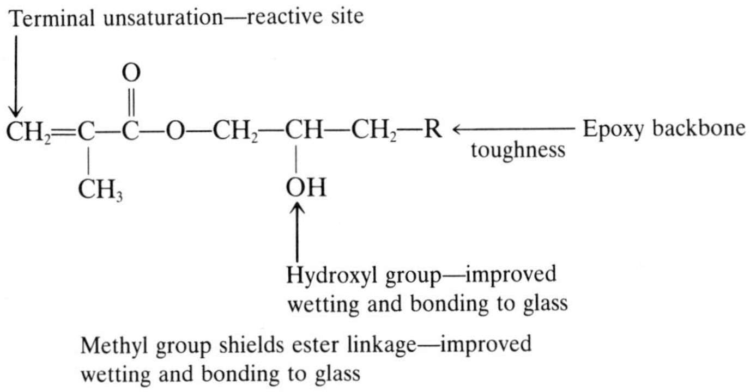

Figure 75.1. Structure imparted characteristics.

TABLE 75.1. Clear Casting Mechanical Properties of Vinyl Ester Resins at 25°C[a]

	Derakane 411-45 General Purpose	Derakane 510-A-40 Brominated	Derakane 470-36 Heat Resistant	XD-8084.05 Impact Resistant
Tensile strength, MPa	78.4	68.9	68.9	64.1
Tensile modulus, MPa	3400	3450	3520	2540
Ultimate elongation, %	5	4	3	7.5
Flexural strength, MPa	117.2	117.2	131.0	108.0
Flexural modulus, MPa	3100	3580	3790	3040
Heat-distortion temperature, °C	100	107	148	77

[a]Data Courtesy of The Dow Chemical Company.

These resins also show a high degree of resistance to thermal aging, Table 75.2. In particular, DERAKANE 470-36 vinyl ester resin based on a phenol–formaldehyde novolac epoxy resin has excellent strength retention at temperatures up to 380°F (193°C). In continuous heat aging tests at that temperature, the resin showed 100% flexural strength retention after six months of exposure. This accounts for its satisfactory service in corrosive environments at elevated temperatures.

75.5.3 Corrosion-Resistance Properties

In vinyl ester resins, the pendant methyl group in the methacrylate structure affords excellent shielding of the ester group, protecting the ester linkage from hydrolysis. Ester hydrolysis is the initial site of attack by acidic or basic media. Vinyl ester resins offer an advantage over polyester resins in that they contain ester groups throughout the molecule, usually alternating with a glycol. The various types of vinyl ester resins can be used in a wide variety of aggressive media, including strong and weak acids, strong and weak bases, organic solvents, and chlorinated organics.

75.6 APPLICATIONS

The largest commercial use of vinyl ester resins is in chemically aggressive environments where corrosion resistance is required. In equipment such as storage tanks, pipe, ducts, scrubbers, and the like, the installed cost plus service life frequently average out to a lower cost per year for vinyl ester resins than for steel. Many case histories of long-term successful

TABLE 75.2. Continuous Heat Aging and % Flexural Strength Retention

Temperatures/ Vinyl Ester Resin	1 Week	2 Weeks	1 Month	2 Months	3 Months	6 Months	9 Months
204°C							
DERAKANE 470-36	102	83	85	92	83	72	62
DERAKANE 510N	101	91	11	F			
DERAKANE 510-A-40	106	20	F				
193°C							
DERAKANE 470-36	76		91		97	100	101
DERAKANE 510N	100		105		90	38	F
DERAKANE 510-A-40	110		107		23	F	F
182°C							
DERAKANE 470-36	90				104	88	88
DERAKANE 510N	99				125	117	139
DERAKANE 510-A-40	98				90	12	F
160°C							
DERAKANE 470-36					84	90	99
160°C							
DERAKANE 510N					111	115	111
DERAKANE 510-A-40					99	111	121

[a]F = Failure; Laminate Construction – V/M/M/WrM/Wr/M; Reprinted from ref. 10. Courtesy of The Dow Chemical Company.

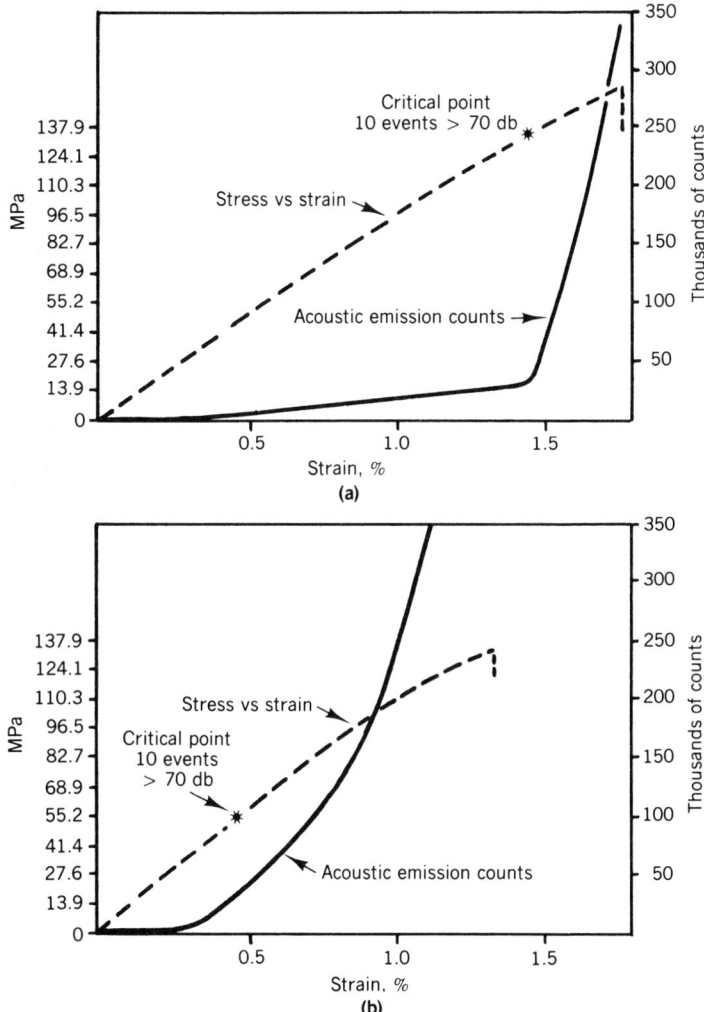

Figure 75.2. Comparative properties of vinyl esters and polyester laminates. **(a)** Derakane 422 vinyl ester resin, 39% glass. **(b)** CR isophthalic polyester resin, 39% glass. Conditions for **(a)** and **(b)**: 1/4-in. laminate VMMWrMWrM; 24-h RT cure; 2-h 100°C post-cure. From ref. 11, vol. VI, Courtesy of the National Association of Corrosion Engineers.

applications of vinyl ester resins in corrosive environments can be found in manufacturers' literature and in preprints of technical conferences.[11]

A growing applications area for vinyl ester resins is for such structures as gratings and reinforced plastic corrugated building panels. High-performance recreational equipment (such as helmets, canoes, and kayaks) is being fabricated from impact-resistant vinyl esters, where the weight saving makes the use of vinyl ester resins very attractive. In the transportation industry, vinyl ester resins have been used to construct reinforced tank trailers. Structural components of automotive equipment have used vinyl ester resin sheet molding compound; it is tough enough to withstand the mechanical stresses associated with this application. Small, lightweight airplanes have also been constructed using vinyl esters.

75.7 ADVANTAGES/DISADVANTAGES

Superior strength and corrosion resistance are key advantages of vinyl ester resins compared to polyester resins. However, vinyl ester resins cost more than polyesters. The price differential between vinyl ester and isophthalic polyester resins is approximately $0.50/lb. Vinyl ester resins are approximately $0.15 to $0.50/lb less expensive than epoxy resins.

Vinyl ester resins have low viscosities, typically 50 to 500 centistokes (0.048 to 0.48 Pa), which aid in processing. The working time (that required for the resin to gel once it has been catalyzed) can vary from several seconds at elevated temperature to several hours.

Vinyl ester resins have a limited shelf life because of the unsaturation present in both the resin molecule and the reactive diluent, usually styrene. Shelf-life typically varies from 3 to 9 months, depending on resin functionality and unsaturation content.

75.8 PROCESSING TECHNIQUES

Vinyl ester resins are processed using a variety of techniques, which are described in the Master Material Outline and the Standard Materials Design Chart.

Master Material Outline

PROCESSABILITY OF _____ Vinyl Ester Resins _____

PROCESS	A	B	C
INJECTION MOLDING			
EXTRUSION			
THERMOFORMING			
FOAM MOLDING			
DIP, SLUSH MOLDING			
ROTATIONAL MOLDING			
POWDER, FLUIDIZED-BED COATING			
REINFORCED THERMOSET MOLDING	X		
COMPRESSION/TRANSFER MOLDING			
REACTION INJECTION MOLDING (RIM)			X
MECHANICAL FORMING			
CASTING		X	

A = Common processing technique.
B = Technique possible with difficulty.
C = Used only in special circumstances, if ever.

Standard design chart for _____ Vinyl Ester Resins _____

1. Recommended wall thickness/length of flow, mm _____ not applicable _____
   Minimum _____ for _____ distance
   Maximum _____ for _____ distance
   Ideal   _____ for _____ distance

2. Allowable wall thickness variation, % of nominal wall <u>Depends on application and specification</u>

3. Radius requirements
   Outside: Minimum _____ 1/4 _____ Maximum _____ none _____
   Inside: Minimum _____ 1/4 _____ Maximum _____ none _____

4. Reinforcing ribs
   Maximum thickness _____ less than 1 in. _____
   Maximum height _____ less than 1 in. _____
   Sink marks: Yes ____ X ____ No _____

5. Solid pegs and bosses
   Maximum thickness _____ less than 1 in. _____
   Maximum weight _____ less than 1 in. _____
   Sink marks: Yes ____ X ____ No _____

6. Strippable undercuts
   Outside: Yes _____ No ____ X ____ (Unless a split mold is used or there is a way to retract the undercut)

   Inside: Yes _____ No ____ X ____

7. Hollow bosses: Yes ____ X ____ No _____

8. Draft angles
   Outside: Minimum _____ Ideal _____ 3° _____
   Inside: Minimum _____ Ideal _____ 3° _____

9. Molded-in inserts: Yes ____ X ____
                      No _____

10. Size limitations _____ none _____
    Length: Minimum _____ Maximum _____ Process _____
    Width: Minimum _____ Maximum _____ Process _____

11. Tolerances ____ Depends on application ____

12. Special process- and materials-related design details
    Blow-up ratio (blow molding) _____
    Core L/D (injection and compression molding) _____
    Depth of draw (thermoforming) _____
    Other _____

75.8.1 Pultrusion and Filament Winding

These are basically wet-fiber continuous or semicontinuous processes. The fibers can be oriented to achieve the desired strength properties. These processes are nearly always heat activated.

75.8.2 Resin Transfer Molding and Continuous Laminating

These are mechanized versions of hand lay-up laminating. In resin transfer molding, the resin is injected onto dry glass. In continuous laminating, layers are applied or wet-out in a sequence of steps, and the layers are then formed and cured as a sheet. Structural panels are formed utilizing this process.

75.8.3 Hand Lay-up or Spray-Up

These processes require minimal equipment yet offer superior flexibility for customized fabrication. They usually employ ambient temperature curing.

75.8.4 SMC, BMC, and Compression Molding

These are high-speed, high-volume manufacturing techniques employing elevated-temperature closed molding. SMC (sheet molding compound) utilizes a thickened resin and chopped glass sheets; BMC (bulk molding compound) utilizes a preformulated resin/filler/chopped glass mixture; matched metal die molding utilizes a thickened resin/filler paste applied to dry glass sheets.

75.9 RESIN FORMS

Vinyl ester resins are supplied as liquids with the resin being dissolved in varying amounts of a reactive monomer, usually styrene. Resin with additives such as promoter or accelerator can be supplied by the manufacturer or distributor. Some formulators supply vinyl ester resins with thixotropic agents or as SMC or BMC.

In laying up vertical surfaces, it is sometimes convenient to have a thixotropic material to avoid runoff of the resin. Table 75.3 shows the effect of two thixotropic additives on several types of vinyl ester resins.

75.10 SPECIFICATION OF PROPERTIES

See Master Outline Materials Properties.

75.11 PROCESSING REQUIREMENTS

Curing of vinyl ester resins is typically done at temperatures ranging from room temperature to 302°F (150°C). A free-radical catalyst or initiator, such as organic peroxides or aliphatic azo compounds, is used alone or in combination with promoters (such as organic cobalt salts) and accelerators (such as dimethyl aniline). The catalyst generates free radicals by thermal decomposition or reaction with promoters or accelerators.

For room-temperature curing, the most commonly used systems are methyl ethyl ketone peroxide and cobalt naphthenate, benzoyl peroxide, and dimethyl aniline, or cumene hydroperoxide with cobalt naphthenate. Gel times are adjusted by the concentrations of the catalyst and promoter; dimethyl aniline is frequently used with methyl ethyl ketone and cumene hydroperoxide systems to decrease gel time.

For elevated-temperature curing, benzoyl peroxide t-butylperbenzoate or 2,5-dimethyl-2,5-di(2-ethylhexanoyl peroxy) hexane are commonly used. Fabricators sometimes use a blend of peroxides which are activated at different temperatures.

It is important to emphasize that care should be taken to avoid directly mixing peroxides and accelerators or promoters. One ingredient should be mixed thoroughly with the resin before adding the other in order to avoid a vigorous decomposition or explosion. Some peroxides may require special handling and storage conditions. Information of this type is readily available from peroxide manufacturers.

Properly cured, vinyl ester resins quickly develop green strength enabling fabricated parts to be removed from the mold without damage. The part is often post-cured at elevated temperatures to develop optimum physical properties. The degree of cure can easily be determined by a hardness measurement. For example, a properly cured Derakane 411-45 vinyl ester resin laminate will have a hardness measurement of 30–35 as measured with a Barber-Colman Hardness Tester, Model 934-1; post-curing the laminate will raise these readings to 35–40.

TABLE 75.3. Effect of Thixotropic Additives[a]

Product	Additive, %	Cab-O-Sil[b] M5 Pigment			Calidria[c] RG-244 Additive		
		Viscosity, Pa·s		Thixotropic Index[d]	Viscosity, Pa·s		Thixotropic Index[d]
		5 rpm	50 rpm		5 rpm	50 rpm	
DERAKANE 411 Vinyl ester resin	0	0.640	0.700	0.914	0.640	0.700	0.91
45% Styrene	1	0.640	0.736	0.869	1.900	1.150	1.65
	2	0.960	0.800	1.20	8.600	3.300	2.60
	3	1.280	0.880	1.45	32.000	8.000	4.00
	4	1.040	0.960	1.08			
	5	1.280	1.280	1.00			
DERAKANE 470	0	0.096	0.083	1.15	0.096	0.083	1.15
45% Styrene	1	0.112	0.096	1.16	0.416	0.213	1.95
	2		0.096		0.416	0.213	1.95
	3	0.176	0.162	1.12	19.100	2.990	6.29
	4						
	5	0.368	0.327	1.12			
DERAKANE 510-A Vinyl ester resin	0				0.320	0.370	0.86
	1				4.160	1.300	3.20
	2				21.500	4.800	4.47
	3				76.00	13.500	5.62

[a]Reprinted from ref. 12. Courtesy of The Dow Chemical Company.
[b]Trademark of Cabot Corporation.
[c]Trademark of Union Carbide Corporation.
[d]Obtained by dividing the viscosity at 5 rpm by that at 50 rpm. The larger values denote a larger thixotropic effect. The Cab-O-Sil pigment was blended in a Cowles dissolver mixer at 1900 rpm. The viscosity was obtained with a Brookfield viscosimeter and a number 4 spindle. All measurements were taken at room temperature.

Master Outline of Material Properties

MATERIAL Vinyl Ester Laminate 1/4" Thick, 40% Glass

PROPERTY	TEST METHOD	ENGLISH UNITS	VALUE	METRIC UNITS	VALUE
MECHANICAL DENSITY	D792	lb/ft^3		g/cm^3	
TENSILE STRENGTH	D638	psi	16,400–31,500	MPa	113.1–217.0
TENSILE MODULUS	D638	psi	$(1.44–1.74) \times 10^6$ 1,440,000–1,740,000	MPa	9,920–11,990
FLEXURAL MODULUS	D790	psi	$(11.03–1.25) \times 10^6$ 1.03–1.25	MPa	7,100–8,610
ELONGATION TO BREAK	D638	%		%	
NOTCHED IZOD AT ROOM TEMP	D256	ft-lb/in		J/m	
HARDNESS	D785				
THERMAL DEFLECTION T @ 264 psi	D648	°F		°C	
DEFLECTION T @ 66 psi	D648	°F		°C	
VICAT SOFT-ENING POINT	D1525	°F		°C	
UL TEMP INDEX	UL746B	°F		°C	
UL FLAMMABILITY CODE RATING	UL94				
LINEAR COEFFICIENT THERMAL EXPANSION	D696	in/in/°F	0.000 017	mm/mm/°C	0.000 03
ENVIRONMENTAL WATER ABSORPTION 24 HOUR	D570	%		%	
CLARITY	D1003	% TRANS-MISSION	Translucent to opaque	% TRANS-MISSION	Translucent to opaque
OUTDOOR WEATHERING	D1435	%	Slight yellowing and chalking	%	Slight yellowing and chalking
FDA APPROVAL			Derakane 411		Derakane 411
CHEMICAL RESISTANCE TO: WEAK ACID	D543		Not attacked (NA)		Not attacked (NA)
STRONG ACID	D543		Minimally to NA		Minimally to NA
WEAK ALKALI	D543		Minimally to NA		Minimally to NA
STRONG ALKALI	D543		Minimally to NA		Minimally to NA
LOW MOLECULAR WEIGHT SOLVENTS	D543		Badly attacked to NA		Badly attacked to NA
ALCOHOLS	D543		Badly attacked to NA		Badly attacked to NA
ELECTRICAL DIELECTRIC STRENGTH	D149	V/mil		kV/mm	
DIELECTRIC CONSTANT	D150	10^6 HERTZ	4.15		4.15
POWER FACTOR	D150	10^6 HERTZ	0.0095		0.0095
OTHER MELTING POINT	D3418	°F		°C	
GLASS TRANS-ITION TEMP.	D3418	°F		°C	

914

MATERIAL __Vinyl Ester Resins, Clear Castings__

PROPERTY	TEST METHOD	ENGLISH UNITS	VALUE	METRIC UNITS	VALUE
MECHANICAL DENSITY	D792	lb/ft³	70–82	g/cm³	1.12–1.32
TENSILE STRENGTH	D638	psi	9,300–12,000	MPa	64.1–82.7
TENSILE MODULUS	D638	psi	$(3.70–5.50) \times 10^5$	MPa	2550–3440
FLEXURAL MODULUS	D790	psi	$(4.4–5.5) \times 10^5$ 440,000–550,000	MPa	3030–3790
ELONGATION TO BREAK	D638	%	3.0–7.5	%	3.0–7.5
NOTCHED IZOD AT ROOM TEMP	D256	ft-lb/in.	0.34–0.44	J/m	18.2–23.5
HARDNESS	D785	Barcol	30–40	Barcol	30–40
THERMAL DEFLECTION T @ 264 psi	D648	°F	170–300	°C	77–149
DEFLECTION T @ 66 psi	D648	°F		°C	
VICAT SOFTENING POINT	D1525	°F		°C	
UL TEMP INDEX	UL746B	°F		°C	
UL FLAMMABILITY CODE RATING	UL94				
LINEAR COEFFICIENT THERMAL EXPANSION	D696	in./in./°F		mm/mm/°C	
ENVIRONMENTAL WATER ABSORPTION 24 HOUR	D570	%	0.1–0.2	%	0.1–0.2
CLARITY	D1003	% TRANSMISSION	Transparent	% TRANSMISSION	Transparent
OUTDOOR WEATHERING	D1435	%	Slight yellowing and chalking	%	Slight yellowing and chalking
FDA APPROVAL			Derakane 411		Derakane 411
CHEMICAL RESISTANCE TO: WEAK ACID	D543				
STRONG ACID	D543				
WEAK ALKALI	D543				
STRONG ALKALI	D543				
LOW MOLECULAR WEIGHT SOLVENTS	D543				
ALCOHOLS	D543				
ELECTRICAL DIELECTRIC STRENGTH	D149	V/mil		kV/mm	
DIELECTRIC CONSTANT	D150	10^6 Hertz			
POWER FACTOR	D150	10^6 Hertz			
OTHER MELTING POINT	D3418	°F		°C	
GLASS TRANSITION TEMP.	D3418	°F	171–300	°C	77–149

915

75.12   PROCESSING-SENSITIVE END PROPERTIES

The strength of fabricated vinyl ester articles is affected by the glass content, glass length, and orientation in SMC, BMC, pultrusion, and filament winding. In hand lay-up fabrication, strength is provided by the use of unidirectional glass and woven roving. Optimum corrosion resistance is obtained when the resin is thoroughly cured.

75.13   SHRINKAGE

Volume shrinkage of unfilled, unreinforced vinyl ester resins varies from 8 to 10 percent. Linear shrinkage will vary with the coefficient of thermal expansion of the particular resin and hence is dependent on the peak exotherm experienced by the part being fabricated. Put another way, the part will lock into shape at peak exotherm and will shrink uniformly in all directions according to the resin's coefficient of thermal expansion as it cools down.

75.14   TRADE NAMES

Vinyl ester resins are available under the following trade names from the following manufacturers:

Hetron, Ashland Chemical Co., Columbus, OH 43216;

Derakane, The Dow Chemical Co., Midland, MI 48076;

CoRezyn, Interplastic Corp., Minneapolis, MN 55413;

Corrolite, Reichhold Chemicals, Inc., White Plains, NY 10603;

and MR, USS Chemicals, Linden, NJ 07036.

Radiation-curable vinyl ester resins are available as Celrad from Celanese Chemical Co., New York, NY 10035 and as Epocryl from Shell Chemical Co., Houston, TX 77001.

Note: The information in this article is presented in good faith, but no warranty is given, nor is freedom from any patent to be inferred.

BIBLIOGRAPHY

1. U.S. Pat. 3,066,112, (Nov. 27, 1962), R.L. Bowen (to the U.S. Government).
2. U.S. Pat. 3,179,623, (April 20, 1965), R.L. Bowen (to the U.S. Government).
3. W.H. Linow and co-workers, *21st SPI Reinforced Plastics/Composites Conference,* 1966, Paper 1-D.
4. C.A. May and H.A. Newey, *20th SPI Reinforced Plastics/Composites Conference,* 1965, Paper 2-D.
5. T.E. Cravens, *27th SPI Reinforced Plastics/Composites Conference,* 1972, Paper 3-B.
6. M.B. Launkitis, *31st SPI Reinforced Plastics/Composites Conference,* 1976, Paper 15-C.
7. K. Hawthorne and co-workers, *32nd SPI Reinforced Plastics/Composites Conference,* 1977, Paper 5-E.
8. L.J. Craigie and co-workers, *34th SPI Reinforced Plastics/Composites Conference,* 1979, Paper 8-B.
9. J.W. Jernigan and co-workers, *22nd SPI Reinforced Plastics/Composites Conference,* 1967, Paper 8-D.
10. *DERAKANE Vinyl Ester Resins from Dow Chemical Resistance Guide,* Form No. 296-320-1182, The Dow Chemical Company, Midland, Mich.

11. "Managing Corrosion with Plastics," NACZE, Publications Dept. PO Box 21830, Houston, TX 77218.

12. *DERAKANE Vinyl Ester Resins for Corrosion Resistance,* Form no. 190-197-74, The Dow Chemical Company, Midland, Mich.

General References

P.F. Bruins, ed., *Unsaturated Polyester Technology,* Gordon and Breach Science Publishers, New York, 1976.

Derakane News, Dow Chemical Co., Midland, Mich.

T.J. Fowler and R.S. Scarpellini "Acoustic Emission Testing of FRP Equipment," *Chemical Engineering,* 145 (Part I) (Oct. 20, 1980) and 293 (Part II) (Nov. 12, 1980).

J. Frados, ed., *Plastics Engineering Handbook of the Society of the Plastics Industry,* Van Nostrand Reinhold Co., New York, 1976.

A. Harper, ed., *Handbook of Plastics and Elastomers,* McGraw-Hill Book Co., New York, 1982.

G. Lubin, ed. *Handbook of Composites,* Van Nostrand Reinhold Co., New York, 1982.

G. Pritchard, ed., *Developments in Reinforced Plastics—I: Resin Matrix Aspects,* Applied Science Publishers Ltd., London, 1980.

S.S. Schwartz, and S.H. Goodman, *Plastics Materials and Processes,* Van Nostrand Reinhold Co., New York, 1982.

76

REINFORCED PLASTICS, EXTENDED-CHAIN POLYETHYLENE FIBERS

David S. Cordova and D. Scott Donnelly

Allied-Signal Incorporated, High Performance Fibers, Technical Center, PO Box 31, Petersburg, VA 23804

76.1 HISTORY

Extended-chain polyethylene (ECPE) fibers are the most recent entrants into the high-performance-fibers field. Spectra ECPE, the first commercially available ECPE fiber and the first in a family of extended chain polymers manufactured by Allied-Signal Incorporated, was introduced in February 1985.

ECPE fibers were developed as a result of fundamental work by researchers in several leading universities. Although the work was supported by industry, the immediate outcome was not foreseen as a commercial entity. The result was the transformation of commodity type polyethylene (PE) into a high performance fiber.

ECPE fibers are arguably the highest modulus and highest strength fibers made. Their availability is significant because it enables polyethylene to be considered among the specialized, high performance materials.

Today, ECPE fibers are being utilized as a reinforcement in areas that, five years ago, were not accessible to any organic fiber. Applications such as ballistic armor, impact shields, and radomes are being developed to take advantage of the fiber's unique properties.

76.2 CATEGORY

ECPE fibers are made from ultrahigh molecular weight polyethylene (UHMPE). Unlike aramids, PE is a flexible molecule that normally crystallizes by folding back on itself. As a consequence, PE fibers made by conventional technology do not possess outstanding physical properties. ECPE fibers, on the other hand, are manufactured by a process that results in most of the molecules being fully extended and oriented in the fiber direction, producing a dramatic increase in physical properties. A simplified analogy of their molecular structure is a bundle of rods, with occasional entangled points that tie the structure together. By comparison, conventional PE is comprised of a number of short-length chain folds that do not contribute to material strength.

The key structural parameters that distinguish ECPE fibers from conventional melt spun materials are illustrated in Figure 76.1. The molecular weight of UHMPE is generally 1 to 5 million; that of conventional PE fibers, 50,000 to several hundred thousand. ECPE fibers

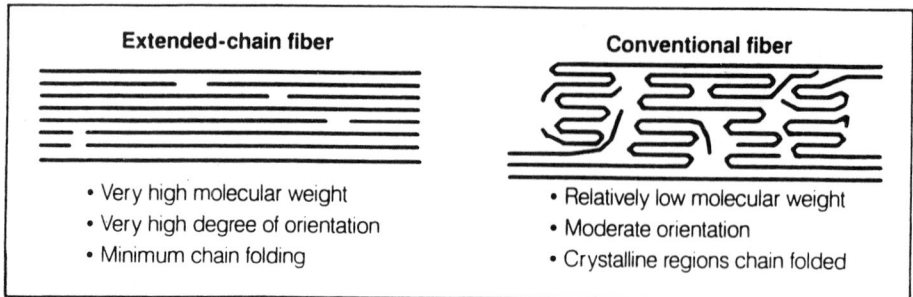

Figure 76.1.   Fiber morphology.

exhibit a very high degree of crystalline orientation (95–99%) and crystalline content (60–85%).

76.3   MANUFACTURING

High modulus PE fibers can be produced by melt extrusion or solid-state extrusion, the latter employing lower molecular weight PE and specialized drawing techniques. These processes lead to a fiber with high modulus but relatively low strength and high creep. Another production method is solution spinning, in which very high molecular weight PE is utilized. With this process modification, a fiber with both high modulus and high strength is produced.

The solution spinning process for a generalized ECPE fiber begins with the dissolution in a suitable solvent of a polymer of approximately 1–5 million molecular weight. The solution serves to disentangle the polymer chains, a key step in achieving an extended chain polymer structure. The solution must be fairly dilute but viscous enough to be spun using conventional melt spinning equipment. The cooling of the extrudate leads to the formation of a fiber that can be continuously dried to remove solvent or later extracted by an appropriate solvent. The fibers are generally postdrawn prior to final packaging.

Unlike most high-performance processes, the solution spinning process is unusually flexible; it can provide an almost infinite number of process and product variations. Fiber strengths of $(3.75–5.60) \times 10^5$ psi (2890–3860 MPa) and tensile moduli of $(15–30) \times 10^6$ psi $[(103–207) \times 10^3$ MPa] have been achieved on a research scale by various companies worldwide. As the solution spinning process is modified, a higher tenacity (stronger) and more thermally stable yarn can be attained. Preliminary evidence (such as increased density, heat of fusion, and x-ray orientation patterns) suggests that the increased strength and stability are caused by higher degrees of molecular orientation.

76.4   KEY PROPERTIES

The comparative strengths of ECPE fibers versus other high performance fibers are summarized in Table 76.1. (Spectra 900, produced by Allied-Signal, will be used to illustrate the general properties of ECPE; by comparison, Spectra 1000 fibers are more stabilized and exhibit a higher strength and modulus.) The tensile properties of ECPE are similar to other high performance fibers; however, because the density of PE is approximately two thirds that of high modulus aramid and half that of high modulus carbon fiber, ECPE fibers possess extraordinarily high specific strengths and specific moduli. The strength of ECPE fiber is at least 35% greater than high modulus aramid or S-glass, and about twice that of conventional

TABLE 76.1. High Performance Fiber Properties

	UHSPE Spectra 1000	Aramid		S-Glass	Graphite HM
		HM	UHM[a]		
Property					
Density, g/cm^3	0.97	1.44	1.47	2.49	1.86
(lb/ft^3)	(60.5)	(89.8)	(91.7)	(155)	(116)
Elongation, %	2.7	2.5	1.5	5.4	0.6
Tensile strength,	435	400	500	665	375
10^3 psi (MPa)	(3000)	(2760)	(3450)	(4590)	(2590)
Specific strength,					
10^6 in.	12.4	7.8	9.5	7.4	5.4
Tensile modulus,	25	19	25	13	57
10^6 psi (10^3 MPa)	(172)	(131)	(172)	(90)	(393)
Specific modulus,					
10^6 in.	714	365	480	140	850

[a]Kevlar 149, epoxy impregnated strand.

high modulus carbon fiber (see Table 76.1). Figure 76.2 compares the specific strength versus specific modulus for currently available fibers.

Polyethylene is known to be a system in which traditional binders and wetting agents are ineffective in improving adhesion levels. However, for ECPE fibers, this characteristic is actually advantageous in specific areas. For instance, ballistic performance is inversely related to the degree of adhesion between the fiber and the resin matrix.

For applications requiring higher levels of adhesion and wetout, the results of extensive research indicate that by submitting ECPE fiber to specific surface treatments, such as corona discharge or plasma treatments, the adhesion of the fiber to various resins is dramatically increased (see Table 76.2).

76.5 APPLICATIONS

The chief application areas being explored and commercialized today for ECPE fibers are divided between traditional fiber applications and high-tech composite applications. The

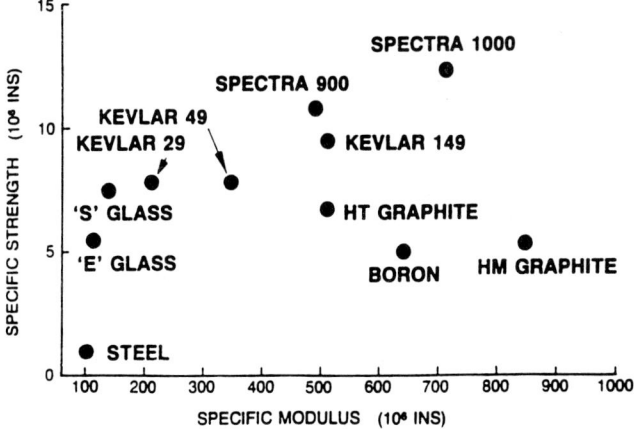

Figure 76.2. Comparative tensile properties of various reinforcing fibers. Kevlar 149 (resin impregnated fibers).

TABLE 76.2. UHSPE Fiber Adhesion Improvements[a]

		Unidirectional			Fabric (Style 903)		
Date	Treatment	SBS, ksi (MPa)	Flex Strength, ksi (MPa)	Flex Modulus, msi (MPa)	SBS, ksi (MPa)	Flex Strength, ksi (MPa)	Flex Modulus, msi (MPa)
10/85[b]	None	1.16 (8)	21.2 (146)	1.2 (8,300)	0.87 (6)	5.7 (39)	0.44 (3,000)
10/86	Corona	2.61 (18)	27.6 (290)	2.6 (18,000)	1.4 (9.7)	10.3 (71)	1.0 (6,900)
10/87	Plasma	4.50 (31)	33.9 (234)	4.5 (31,000)	2.2 (15)	21.0 (145)	2.9 (20,000)

[a]Fiber: Spectra 900; resin: epoxy; fiber loading: 60%.

former include sailcloth, marine ropes, cables, sewing thread, nettings, and protective clothing; the latter, ballistics, impact shields, medical implants, radomes, pressure vessels, boat hulls, sports equipment, and concrete reinforcement.

76.5.1 Sailcloth

As the level of competition in the Americas Cup and other world-class sailing events becomes more competitive, the sail industry has turned to new materials. A winning sailcloth must possess high strength, high modulus, light weight, and minimal distortion during the sailing season. Of the fiber's physical properties, none are more critical than low creep and resistance to seawater and cleaning agents. ECPE fibers (such as Spectra 1000) are well suited for high performance yachting sails, offering, in addition, resistance to seawater and to typical cleaning solutions used in the sailing industry, such as bleach (see Table 76.3).

The creep behavior of ECPE fibers under typical laboratory test loadings of 3–4 g/denier is illustrated in Figure 76.3. These creep levels are substantially below those encountered with conventional PE or the specialized melt-spun high modulus fibers. At this loading, which includes the initial elastic loading component, the creep resistance of certain ECPE grades is comparable to that of a high modulus aramid. (The elastic load component is included in these results on a practical basis since it is an integral part of the sailcloth design.)

TABLE 76.3. Chemical Resistance[a]

	Strength Retention after 6 mo Immersion, %	
Agent	Spectra 900	Aramid
Seawater	100	100
10% Detergent solution	100	100
Hydraulic fluid	100	100
Kerosene	100	100
Gasoline	100	93
Toluene	100	72
Perchlorethylene	100	75
Glacial acetic acid	100	82
1 M Hydrochloric acid	100	40
5 M Sodium hydroxide	100	42
Ammonium hydroxide, 29%	100	70
Hypophosphite solution, 10%	100	79
Clorox	91	0

[a]Immersed in various chemical substances for a period of 6 months, Spectra fibers retained their original strength.

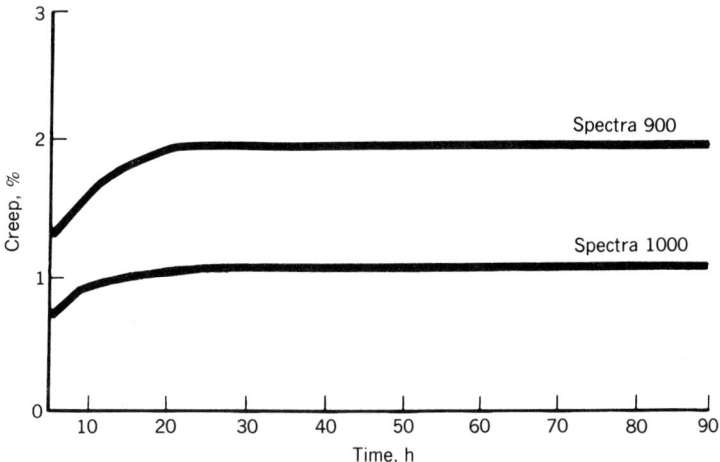

Figure 76.3. Creep resistance of ECPE fibers at 10% loading, room temperature.

76.5.2 Marine Ropes

Its high strength, light weight, low moisture absorption, and excellent abrasion resistance make ECPE a natural candidate for marine rope. Three parameters—diameter, weight per length, and strength—are illustrated in Table 76.4. Since aramid fibers are the accepted standard in the high-performance rope industry, they provide a yardstick by which the ECPE fibers can be measured. The results are given for one grade (Spectra 900) of braid that is 12% smaller, 10% stronger, and 52% lighter than the aramid product.

Two important considerations in marine rope applications are load, cycling, and abrasion resistance. The response of an ECPE fiber rope to load cycling was measured by testing on a sheave device. The rope was repeatedly loaded to 4000 lb (1800 kg) until it broke. In this test, a 12-strand ECPE braid withstood approximately eight times the number of cycles that led to failure in 12-strand aramid braid (Table 76.5). Abrasion resistance was measured by cycling the rope over an oscillating bar. In this test, 0.5-in. (13-mm) diameter ECPE braided rope withstood eight times the abuse of a similar aramid rope (Table 76.5).

76.5.3 Cut-Resistant Gloves and Protective Clothing

Specially toughened and dimensionally stabilized ECPE yarn has been used in a revolutionary new line of cut-resistant products. This technology offers a previously unattainable level of protection from cut and abrasion without sacrificing comfort and launderability. ECPE fibers are being used to produce cut-resistant gloves, arm guards and chaps in such industries as

TABLE 76.4. Comparative Properties of 16-Strand Rope

Property	Spectra 900	Aramid
Diameter, in. (mm)	0.088	0.10
	(2)	(3)
wt/100 ft·lb (kg)	0.153	0.32
	(0.069)	(0.15)
Tensile strength, lb (kg)	1465	1334
	(665)	(605)

TABLE 76.5. Cycle Loading and Wear Tests

Cycle Loading	Spectra 900	Aramid
Cyclic sheave—12 strand braid (10 cycles/min, 4000 lb (1800 kg) tensile load) cycles to break	10,231	1212
Oscillating bar—0.5 in. (13 mm) rope (1.5 cycles/min, 1700 lb (770 kg) tensile load) cycles to break	883	111

meat packing, commercial fishing, and poultry processing and in sheet-metal work, glass cutting, and power tool use. Other applications exploit ECPE fibers' inert chemical nature and cut protection; for instance, in surgical-medical gloves, dental, laboratory testing, and police emergency response over-gloves.

76.5.4   Ballistic Protection

ECPE's high strength and modulus and low specific gravity offer higher ballistic protection at lower density per area than is possible with currently used materials. The significant applications include flexible and rigid armor.

Flexible armor is manufactured by joining multiple layers of fabric into the desired shape. The style of the fabric and number of layers will determine the ballistic resistance that the armor will provide. Typical V50 ballistic limits of plain-weave ECPE fabrics of different denier yarns are plotted as functions of areal density in Figure 76.4. The resultant products include protective vests for military personnel and civilian security forces as well as ballistic blankets that are applied to ceramic and metallic armor as a front spall shield and as a rear spall suppressor. They can also be used to fabricate ballistic protective shelters.

Traditional rigid armor can also be made by utilizing woven ECPE fiber in either thermoset or thermoplastic matrices. These rigid systems exhibit high ballistic protection because

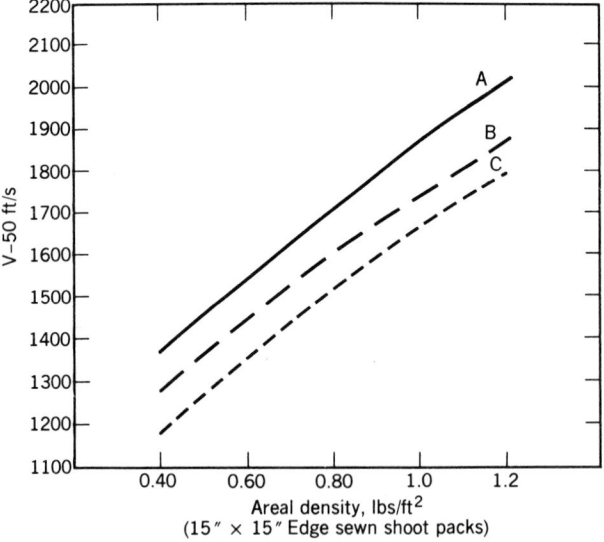

Figure 76.4.   Ballistic performance of ECPE fabrics. Fabric, A, S-185 56x56 PW; B, S-375 51x50 PW; C, S-650 34x34 PW.

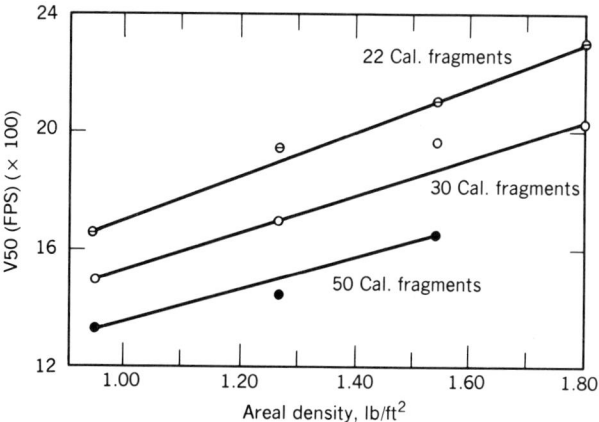

Figure 76.5. ECPE composite ballistic protection vs. 22-, 30-, and 50-caliber fragments.

of the fiber strength and modulus in combination with its low specific gravity, as illustrated in Figure 76.5, which compares V50 values for ECPE fiber and aramid composites against a 22-caliber fragment simulator.

The ECPE fiber ballistic systems can be contoured or formed into armored plates, helicopter seats, army or police helmets, and many other product forms. It is important for these systems to maintain their ballistic protection under a wide range of environmental conditions. In this regard, Figure 76.6 illustrates the superior performance of ECPE fiber armor at temperatures as high as 225°F (107°C). This performance, along with their low

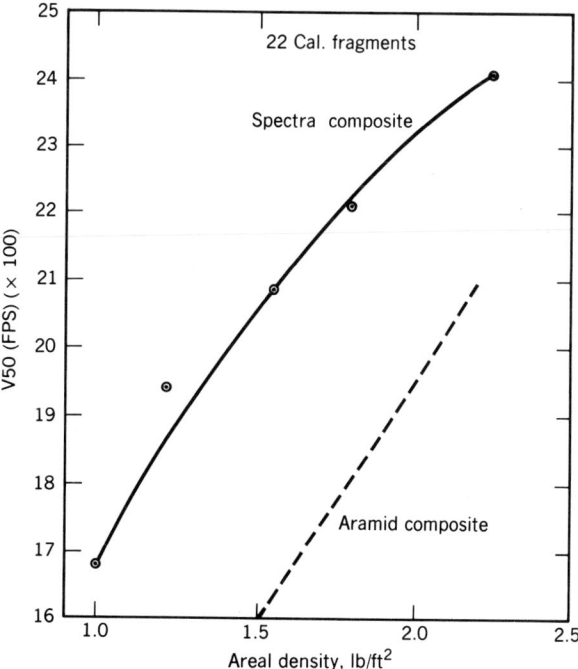

Figure 76.6. Comparative ballistic performance of ECPE and aramid composites.

moisture absorption, chemical inertness, and low weight characteristics, make ECPE fibers extremely well suited for ballistic protection.

76.6 COMPOSITE CHARACTERISTICS

ECPE fibers are recent entrants into the high-performance composites industry, offering high strength and high modulus.

ECPE fibers have been used with a wide variety of resin systems, among them epoxies, polyesters, vinyl esters, silicones, urethanes, and polyethylene. The choice of resin is most often dictated by the end-use application and requirements. The highest mechanical properties are obtained with epoxy and IPN resins; epoxies are used most often by the composites industry, and IPNs are gaining importance in RTM processes. Vinyl ester and urethanes offer the greatest impact and ballistic properties at the expense of mechanical strength. Polyester is intermediate to the two groups, and is most often selected by the radome industry for its electrical properties.

ECPE fibers can be processed essentially the same as aramid, graphite, and glass. Hand layup, matched mold, pressure, and vacuum molding of fabric prepregs are most often used; however, filament winding and pultrusion are also common with continuous filament.

ECPE fibers are available in roving, fabric, continuous mat, and even chopped fiber. Composite applications where high strength (such as in tensile, flexural, or short beam shear) is needed require special fiber to enhance the fiber-to-matrix adhesion. Allied-Signal has developed proprietary treatments for its Spectra fibers to increase the adhesion level and composite properties.

76.6.1 Composite Applications

ECPE fiber reinforced materials are being developed and used widely in ballistics, radar protective domes, aerospace, sports equipment, and industrial applications. Some of these utilize the fiber in hybrid form; that is, in combination with S-2 glass, graphite, aramid, and/or quartz.

Ballistics are currently the dominant market segment. Products include helmets, helicopter seats, automotive and aircraft armor, armor radomes, and other industrial structures.

The radome (radar protective domes) market is also important for ECPE fibers. Because of the excellent electrical properties of polyethylene, ECPE composite systems act as a shield that is virtually transparent to microwave signals, even in high-frequency regions. Hybridization with quartz or glass fiber offers advantages in terms of structure and cost performance.

The major sport equipment applications to date have been canoes, kayaks, and snow and water skis. Numerous other sport applications are under development, among them bicycles, golf clubs, ski poles, and tennis rackets. Further growth is expected in formula race-car bodies.

The industrial market is taking advantage of ECPE fibers in areas where increased strength, impact resistance, noncatastrophic failure, light weight, and corrosion resistance are required. Requirements for corrosion resistance, in particular, have inspired the composites industry to investigate ECPE for applications involving exposure to a wide variety of chemical elements. Previously available high performance fibers cannot function under such adverse conditions.

76.6.2 Composites Properties

Ballistic Protection. As noted, the excellent protection offered by ECPE fabrics can be utilized in hard armor composites. The ballistic protection against 22-, 30-, and 50-caliber

TABLE 76.6. Resistance to Handgun Ammunition of Spectra 1000 and Aramid (Kevlar 29) Composites

Ammunition	No.	Armor System	AD (PSF)	V50 (FPS)
.357 Cal.	1	Spectra/Vinylester 411-45	0.62	1220
158 grain	2	Spectra/Vinylester 411-45	1.12	1443
JSP	3	Kevlar/Polyester	1.15	1281
	4	Spectra/Vinylester 411-45	1.36	1481
	5	Kevlar/Polyester	1.49	1311
9 mm	6	Spectra/Vinylester 411-45	0.62	1082
124 grain	7	Spectra/Latex	0.70	1200
FMJ	8	Spectra/Vinylester 411-45	0.83	1173
	9	Spectra/Latex	1.01	1454
	10	Spectra/Latex	1.23	1594
	11	Kevlar/Polyester	1.28	1241
	12	Kevlar/Polyester	1.46	1372
	13	Spectra/Latex	1.53	1624

threats is summarized in Figure 76.5; Figure 76.6 compares the fragmentation protection of ECPE composites to similar composites reinforced with aramid fibers.

Handgun projectiles present a different type of threat. In comparison to aramid composites, ECPE composites provide reduced weight and increased protection. The resistance to handgun ammunition of ECPE and aramid composites are compared in Table 76.6, which demonstrates the former's lower areal density and/or increased protection.

Impact Resistance. Energy dissipation is an important property of ECPE. The impact strength of fabric composites of ECPE, glass, Kevlar and graphite composites are compared in Table 76.8, which indicates that ECPE composite panels were not penetrated as were the other composite panels under similar conditions.

The performance of ECPE composites under impact loading is shown in Figure 76.7. Toughness gradually increases after each successive impact, working to extend the actual part life.

Drop weight impact tests were performed on honeycomb sandwich composites. The peak forces resisted by the ECPE plates were consistently higher than those of similar aramid

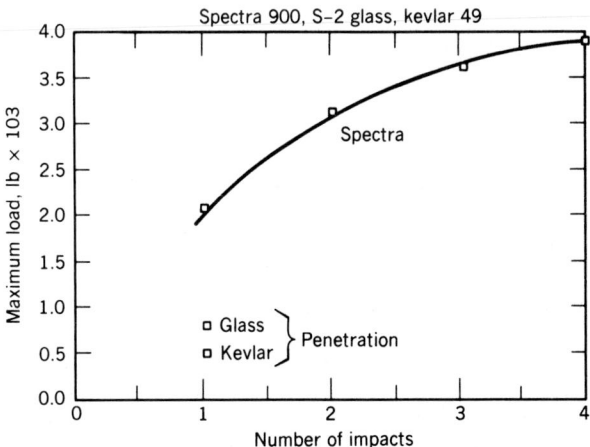

Figure 76.7. Repetitive impact strength of various composites.

TABLE 76.7. Impact Absorption of Sandwich Composites[a]

Skin	No. of Layers	Energy to Peak Force ft-lb. (J)	Total Energy Absorbed ft-lb. (J)
Spectra 900	1	22.4 (30.4)	61.5 (83.4)
Aramid	1	0.7 (0.9)	2.3 (3.2)
Spectra 900	3	33.5 (45.4)	59.8 (81)
Aramid	3	1.5 (2)	10.5 (14.2)

[a]Core: $\frac{1}{2}$-in. honeycomb 3 lb/ft³ (13 mm, 0.05 g/cm³). Resin: Epoxy (Epon 826).

plates (Table 76.7). The peak impact force, total impact energy, and energy absorbed to peak force increase with the increase in face sheet thickness, from 1 to 3 plies.

A practical example is resistance to hailstorm damage. A comparison of the performance ECPE with other reinforcements in a simulated hailstorm test is given in Figure 76.8.

The effects of new surface treatments developed to enhance the fiber/resin interface adhesion on impact performance can be seen in Table 76.8. It should be noted that although the impact properties decrease, the impact resistance of treated ECPE composites is still five times that of glass or aramid, with a significant increase in physical properties.

Electrical Properties. The most desirable attribute for radomes, which are used to protect radar systems, is to be "invisible" or "transparent" to the signal. Because of the low dielectric constant and loss tangent of polyethylene (see Table 76.9), ECPE fiber composite systems are extremely well-suited to fulfill this requirement. The low dielectric constant (2.3–2.5) of ECPE composites has been shown to hold in the high-frequency ranges, even to the millimetric band. The superior electrical properties of ECPE fibers can be utilized in single fiber systems or to improve the properties of glass radomes via hybridization. A dielectric constant of 2.9 has been obtained with an ECPE/glass (25/75) hybrid system.

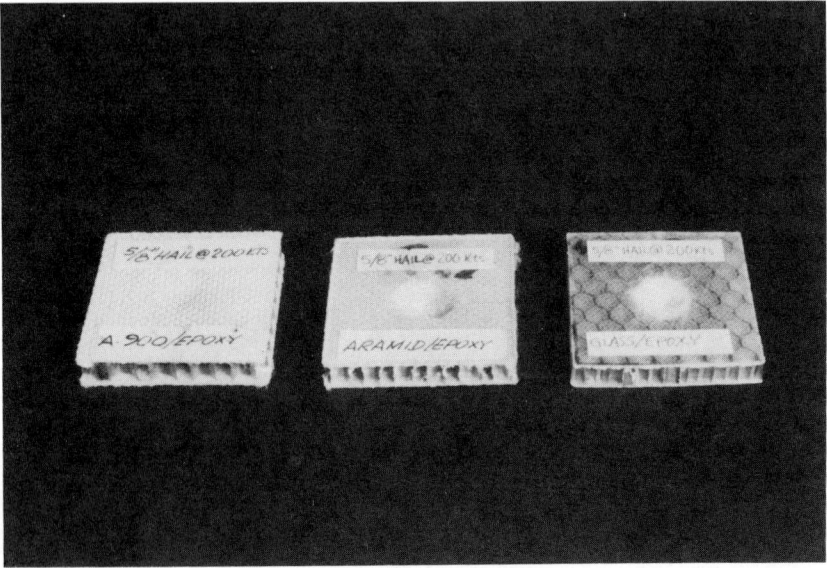

Figure 76.8. Hailstorm test of Type A composite sandwich panels. Courtesy of Norton Company, Ravenna, Ohio.

TABLE 76.8. Instrumented Impact of Fabric Composites[a]

Fiber	Treatment	Max Load lb (kg)	Energy at Max Load ft-lb (J)	Total Energy ft-lb (J)	Observation
Spectra 900	None	1660 (751)	47.4 (64)	54.5 (74)	No penetration
Spectra 900	Plasma	1030 (467)	12.0 (16)	28.0 (38)	Penetration
Kevlar 49	Epoxy compatible	254 (115)	1.3 (1.8)	6.7 (9.1)	Penetration
S-2 Glass	Epoxy compatible	370 (168)	1.8 (2.4)	4.4 (6)	Penetration
HM Graphite	Epoxy compatible	133 (60)	1.2 (1.6)	2.5 (3.4)	Penetration

[a]Resin: Epoxy; Fiber loading: 60%.

TABLE 76.9. Fiber Electrical Properties

Material	Dielectric Constant	Loss Tangent
Spectra	2.0–2.3	0.0002–0.0004
E-Glass	4.5–6.0	0.0060
Aramid	3.85	0.0100
Quartz	3.78	0.0001–0.0002

The advantages of the low-dielectric constant and low-loss tangent is demonstrated by observing the effect of the radome on the transmission ratio. The transmitted signal of a typical ECPE radome matrix is compared with that of a glass radome at various ratios of wall thickness to wavelength in Figure 76.9. The advantages of lower distortion to the signal are even more pronounced in Type A honeycomb sandwich panels (Fig. 76.10).

Other possible electrical applications for ECPE fibers and their reinforced composites are electrical shelters, x-rays tables, optical cables, and other structures where high strength, nonconductive characteristics are needed.

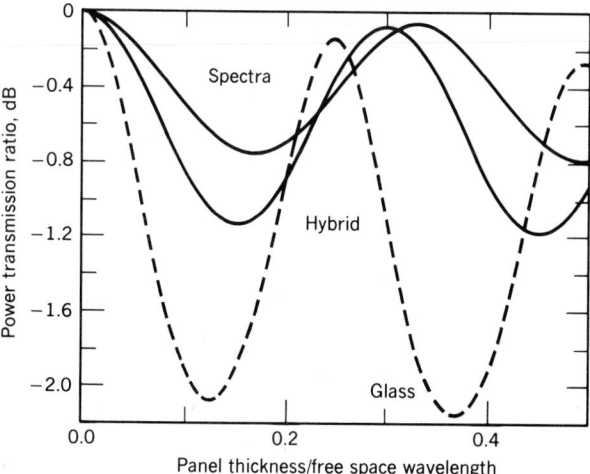

Figure 76.9. Transmission vs relative thickness of flat panels of various composites at 8.5 GHz.

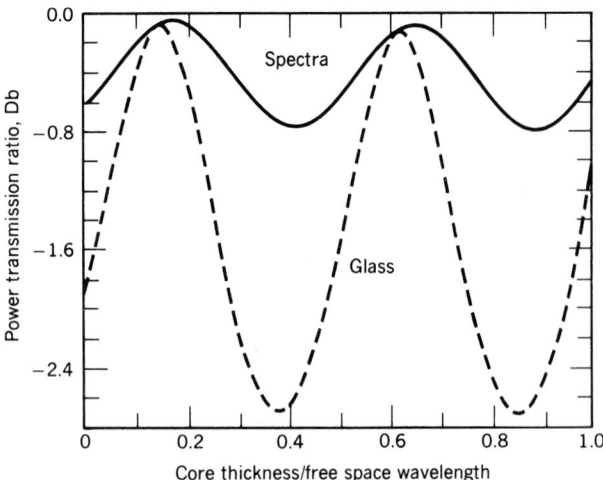

Figure 76.10. Transmission vs relative thickness of type A sandwich radome test panels of various composites at 8.5 GHz.

TABLE 76.10. Properties of Unidirectional Composites (Nontreated fabric)

	Spectra 900	Spectra 1000
Axial tensile strength, 10^3 psi (MPa)	174.0 (1,200)	217.0 (1,500)
Axial tensile modulus, 10^6 psi (MPa)	5.8 (40,000)	9.1 (63,000)
Axial strain to failure, %	3.8	2.6
Major Poisson's ratio	0.32	0.28
Transverse tensile strength, 10^3 psi (MPa)	1.4 (10)	1.5 (10)
Transverse tensile modulus, 10^6 psi (MPa)	0.6 (4,100)	0.2 (1,400)
Axial compressive strength, 10^3 psi (MPa)	15.8 (109)	16.0 (110)
Axial compressive modulus, 10^6 psi (MPa)		3.6 (25,000)
Short beam shear strength, 10^3 psi (MPa)	4.0 (28)	2.5 (17)

Structural Properties. Static test results for two grades of ECPE unidirectional composites are summarized in Table 76.10. All test samples were cut from unidirectional prepregs of corona treated ECPE fiber with Shell Epon 826 epoxy resin and Melamine 5260 cycloaliphatic diamine curing agent. As shown, the strength and modulus of Spectra 1000 are higher than the Spectra 900 composites, thanks to the improved strength of the latter fiber. Further improvements in composite properties can be achieved by applying a plasma surface treatment to the fibers. This treatment increases the interfacial bonding, which translates into even higher composite structural properties.

77

REINFORCED PLASTICS, ARAMID FIBERS

Paul Langston and George E. Zahr

Textile Fiber Division, Du Pont Co., 1007 Market St., Wilmington, DE 19898

77.1 CATEGORY

Aramid fiber is the generic name for aromatic polyamide fibers. As defined by the U.S. Federal Trade Commission, an aramid fiber is a "manufactured fiber in which the fiber-forming substance is a long-chain synthetic polyamide in which at least 85% of the amide linkages are attached directly to two aromatic rings."

77.2 HISTORY, COMMERCIAL GRADES, AND TRADE NAMES

During the past three decades, considerable work has been done on the preparation of fibers from wholly aromatic polymers to obtain high strength, high modulus, and heat resistance. To date, the only commercially available aramid fibers are Du Pont's Nomex and Kevlar 29, 49 and 149; in fact, these trade names are commonly used in lieu of the generic name. In addition to high strength and modulus, Kevlar is corrosion resistant as well as chemically and mechanically stable over a wide range of temperatures. Approximately 22×10^6 lb of aramid fibers were used in 1987. This section focuses primarily on Kevlar and its composites.

77.3 KEY PROPERTIES

77.3.1 Fiber

Kevlar 49 can be used to advantage to obtain composites having lighter weight, greater stiffness, and higher tensile strength than composites incorporating E-glass or S-glass reinforcement. Aramid composites of Kevlar 49 will exhibit slightly lower interlaminar shear and flexural strength, higher flexural and compressive modulus, but less compressive strength than products made with the same volume-fraction of glass reinforcement in the same resin. Weight savings over glass result from the lower specific gravity of aramid fibers, 90.4 lb/in.^{3} (1.45 g/cm^3), versus E-glass, 159.0 lb/in.^{3} (2.55 g/cm^3).

931

Higher stiffnesses reflect a Young's modulus up to 19×10^6 psi (1.31×10^5 MPa) for Kevlar 49 and 27×10^6 psi (1.86×10^5 MPa) for Kevlar 149, compared to 10×10^7 psi (6.9×10^4 MPa) for E-glass and 12×10^6 psi (8.6×10^4 MPa) for S-glass. Aramid fiber-reinforced resin composites are more electrically and thermally insulating than their glass counterparts, more damped to mechanical and sonic vibrations, and transparent to radar and sonar. These high-modulus organic fibers retain the processibility normally associated with conventional textiles, despite their outstanding mechanical properties. This leads to wide versatility in the form of the reinforcement that can be obtained (eg, yarns, rovings, woven and knit goods, felts, and papers). Composites of these fibers have low creep, high impact resistance, very low notch sensitivity, and excellent creep rupture characteristics.

The present cost of high-modulus aramid fibers is higher than E-glass and equivalent to some grades of S-glass on a unit-weight basis. Price differences versus glass are reduced by about half on a unit-volume basis when the lower density of the aramids is taken into account.

Kevlar 49 aramid fibers exhibit a form of material symmetry known as transverse isotropy. This means that the fiber properties in the plane perpendicular to the fiber axis are isotropic, while different properties exist in the direction parallel to the fiber axis. Accordingly, much of the property data provided throughout this section is given for both the axial (fiber direction) and transverse (perpendicular to the fiber) directions.

Because of the geometry of fibers, actual measurement of most of the properties in the transverse plane is impossible. In order to determine these properties, data for Kevlar 49 reinforced unidirectional composites are used in conjunction with analytical expressions relating composite and constituent properties.[1] This treatment yields a consistent set of fiber properties based on measured data.

The data relating to strength are limited to axial properties because there is no way to determine the transverse strength of reinforcing yarns. The axial strength data have been obtained by the resin impregnated strand test (ASTM D2342).

The basic static data shown in Table 77.1 are intended primarily for those investigating the use of Kevlar 49 aramid as a reinforcement in new or unusual matrix materials. Properties such as the transverse modulus of fibers cannot be exploited except when the fibers are utilized in composites. The data allow for the determination of elastic and thermal properties of unidirectional composites through the use of various analytical procedures.[1]

In using these property data, it is important to evaluate the response of the complete

TABLE 77.1. Static Data for Kevlar 49 Fibers

Property	at Room Temperature		at Test Temperatures, 250°F (121°C)	
	U.S. Customary	SI	U.S. Customary	SI
Axial modulus[a]	18×10^6 lb/in.^{2}	124 GPa	15.4×10^6 lb/in.^{2}	114 GPa
Transverse modulus[a]	1×10^6 lb/in.^{2}	6.9 GPa	0.95×10^6 lb/in.^{2}	6.6 GPa
Axial shear modulus[a]	0.4×10^6 lb/in.^{2}	2.8 GPa	0.38×10^6 lb/in.^{2}	2.6 GPa
Axial Poisson ratio[a]	0.36	0.36	0.36	0.36
Transverse shear modulus[a]	0.4×10^6 lb/in.^{2}	2.8 GPa	0.35×10^6 lb/in.^{2}	2.4 GPa
Axial tensile strength[b]	525×10^3 lb/in.^{2}	3.62 GPa	460×10^3 lb/in.^{2}	3.17 GPa
Axial elongation to break[b]	2.9%	2.9%	2.8%	2.8%
Axial thermal expansion[a] coefficient	-2.9×10^{-6} in./in.°F	-5.2×10^{-6} m/m°C	-2.9×10^{-6} in./in.°F	-5.2×10^{-6} m/m°C
Transverse thermal expansion[a] coefficient	23.0×10^{-6} in./in.°F	41.4×10^{-6} m/m°C	23.0×10^{-6} in./in.°F	41.4×10^{-6} m/m°C
Density	0.052 lb/in.^{3}	1.44 g/cm^3	0.052 lb/in.^{3}	1.44 g/cm^3

[a]Determined from unidirectional composite data.
[b]Resin impregnated strand test (ASTM D2343).

TABLE 77.2. Tensile Data Comparative Reinforcing Fiber Data[a]

Property	Kevlar 49 aramid	E-Glass	T-300 Carbon	AS-4 Carbon	S-Glass	Kevlar 149 aramid
Tensile strength,						
10^3 lb/in.^{2}	525	450	470	520	600	500
(MPa)	(3620)	(3100)	(3240)	(3590)	(4140)	(3450)
Tensile modulus,						
10^6 lb/in.^{2}	18.0	10.5	33.5	34.0	12.4	27.0
(GPa)	(124)	(72)	(231)	(234)	(85)	(186)
Elongation to						
break %	2.5	4.3	1.4	1.5	4.8	2.0
Density,						
lb/in.^{3}	0.052	0.092	0.064	0.065	0.090	0.053
(g/cm^3)	(1.44)	(2.55)	(1.77)	(1.80)	(2.49)	(1.47)

[a]Resin Impregnated Strand Test ASTM D2343.

composite system reinforced with Kevlar 49, especially in the investigation of new matrix materials.

Table 77.2 and Figure 77.1 provide information about the tensile modulus and strength of Kevlar 49.

The mechanical properties of fabrics of Kevlar 49 aramid are dependent on the fabric construction. The weave construction parameters for a few widely used weave constructions are given in Table 77.3.

Table 77.4 demonstrates the variation in tensile properties and tear strengths as a function of the fabric weave parameters. The data on tear strength are included for designers needing a bare or coated fabric. As Table 77.4 shows, the tensile strength parameters can be translated into composite strength properties.

Data pertaining to the effects of exposure of bare yarns to various environments are given

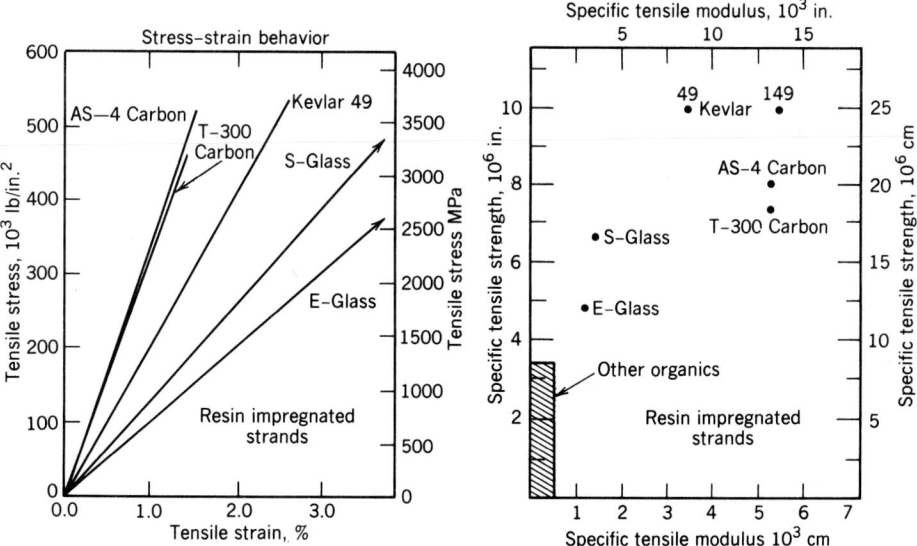

Figure 77.1. Comparison of reinforcing fibers. (**a**) Stress–strain behavior; (**b**) specific strength and modulus.

TABLE 77.3.   Description of Fabrics and Woven Rovings of Kevlar 49[a]

Du Pont Style #	Weave	Basis Weight		Fabric Construction		Yarn Denier	Fabric Thickness	
		oz/yd²	g/m²	ends/in.	ends/cm		10⁻³ in.	mm
				Light Weight				
166*	Plain	0.9	30.6	94 × 94	37 × 37	55	1.5	0.04
199*	Plain	1.8	61.13	60 × 60	24 × 24	55	2	0.05
120	Plain	1.8	61.1	34 × 34	13 × 13	195	4.5	0.11
220	Plain	2.2	74.7	22 × 22	9 × 9	380	4.5	0.11
				Medium Weight				
181	8-Harness Satin	5.0	169.8	50 × 50	20 × 20	380	9	0.23
281	Plain	5.0	169.8	17 × 17	7 × 7	1140	10	0.25
285	Crowfoot	5.0	169.8	17 × 17	7 × 7	1140	10	0.25
328	Plain	6.8	230.9	17 × 17	7 × 7	1420	13	0.33
335	Crowfoot	6.8	230.9	17 × 17	7 × 7	1420	12	0.30
500	Plain	5.0	169.8	13 × 13	5 × 5	1420	11	0.28
				Uni-Directional				
143	Crowfoot	5.6	190.2	100 × 20	39 × 8	380 × 195	10	0.25
243	Crowfoot	6.7	227.5	38 × 18	15 × 7	1140 × 380	13	0.33
				Woven Roving				
1050	4 × 4 Basket	10.5	356.6	28 × 28	11 × 11	1420	18	0.46
1033	8 × 8 Basket	15.0	509.4	40 × 40	16 × 16	1420	26	0.66
1350	4 × 4 Basket	13.5	458.5	26 × 22	10 × 9	2130	25	0.64

[a]Kevlar 49 aramid is often used in fabric reinforced composites. A wide range of fabric styles and weights has been developed. Some of the commonly used varieties are described here. These styles are only available on a special order basis from Du Pont. Custom fabric will be woven to specifications.

in Table 77.5, including the effects of various chemicals on the retained tensile strength and modulus of Kevlar 49 aramid fibers.

77.4 COMPOSITE TYPES

The two basic forms of composite laminate reinforcement are unidirectional and fabric reinforcement.

TABLE 77.4.   Static Room Temperature Strength[a]

Du Pont Style #	Tensile Strength Warp × Fill		Tongue Tear Strength Warp × Fill		Trap Tear Strength Warp × Fill	
	lb/in.	kN/m	lb	N	lb	N
120	250 × 250	44 × 44	60 × 60	267 × 267	22 × 22	98 × 98
181	700 × 700	123 × 123	110 × 110	489 × 489	56 × 56	249 × 249
281	650 × 650	114 × 114	105 × 105	467 × 467	43 × 43	191 × 191
285	650 × 650	114 × 114	[b]	[b]	40 × 40	178 × 178
328	700 × 700	123 × 123	120 × 120	534 × 534	65 × 65	289 × 289
143	1300 × 125	228 × 22	[b]	[b]	15 × 70	67 × 311
243	1500 × 300	263 × 53	[b]	[b]	20 × 100	89 × 445

[a]Property ASTM Test Method; Tensile strength D1117; Tongue tear D2261; Trapezoidal tear D2263.
[b]Construction too loose for testing.

TABLE 77.5. Chemical Resistance Of Bare Yarns of Kevlar 49[a]

Chemical	Retained Tensile Strength, %	Retained Tensile Modulus, %
Acetic acid (99.7% aqueous)	100	99
Formic acid	88	98
Hydrochloric acid (37% aqueous)	100	97
Nitric acid (70% aqueous)	40	95
Sulfuric acid (96% aqueous)	0[b]	
Ammonium hydroxide (28.5% aqueous)	100	98
Potassium hydroxide (50% aqueous)	74	97
Sodium hydroxide (50% aqueous)	90	95
Acetone	100	99
Benzene	100	98
Carbon tetrachloride	100	100
Dimenthylformamide	100	98
Methylene chloride	100	100
Methyl ethyl ketone (MEK)	100	98
Trichloroethylene ("Triclene")	98	99
Chlorothene (1,1,1-Trichloroethane)	100	100
Toluene	100	100
Benzyl alcohol	100	99
Ethyl alcohol	100	98
Methyl alcohol	99	98
Formalin	99	97
Gasoline (Regular)	100	100
Jet fuel (Texaco Abjet K-40)	96	99
Lubricating oil (Skydrol)	100	99
Salt water (5% solution)	100	92
Tap water	100	100

[a]24-h Exposure at room temperature

 Yarns were tested using air-actuated 4-C cord and yarn clamps on an Instron test machine, at 10 in. gauge length with 3 turns per in. twist added, 10% per minute elongation, and at 55% RH and 72°F.

[b]Too weak to test.

77.4.1 Unidirectional Composites

In the unidirectional composite, the fibers are aligned in a single direction, which provides maximum properties in the fiber direction.

When examining the design data for unidirectional composites, the orthotrophy of the composite must be considered. It is typically assumed that properties in the "3" direction are identical to those in the "2" direction. This is because both are perpendicular to the fiber axis.

77.4.2 Fabric Composites

Fabrics woven from Kevlar 49 aramid fiber are often used as composite reinforcement, since fabrics offer biaxial strength and stiffness in a single ply. The composite properties are functions of the fabric weave and the fiber volume fraction. The construction of these fabrics is described in Table 77.6 and Figure 77.2; the basic mechanical properties of various bare fabrics are listed in Figure 77.3. For the fabrics shown, fiber volume fractions range from 50 to 55%, with ply thickness ranging from 5 to 10 mils, depending on fabric construction.

Analytical predictive capabilities for fabric-reinforced composites are not as advanced as those that have developed for unidirectionally composites. The data presented here have all been obtained experimentally.

TABLE 77.6. Fabric Style Description

166, 199	Designed for high strength and stiffness in very thin, ultralightweight laminates.	328	Style 328 is a plain weave using 1420 denier Kevlar; it is used primarily in the boating industry.
120	Designed as the volume equivalent to Style 120 in fiber glass. Kevlar Style 120 is used where a thin laminate with a smooth surface is desired.	335	Style 335 is a 4-harness satin weave version of Style 328. Designed to be used where mold conformability as well as thickness is required.
181	Designed as the volume equivalent to Style 181 in fiber glass. Because Kevlar Style 181 is an 8-harness satin, it conforms well to molds.	500	Initially designed for boats, Kevlar Style 500 is now considered ideal for making tool molds because its open construction results in high resin flow characteristics. It is also made from 1420 denier yarns in a balanced plain weave.
220	When economy is important, Kevlar Style 220 provides minimum thickness and a smooth surface at lower cost than Style 120. Style 220 is made from 380 denier yarn which increases the weight and per ply thickness slightly over Style 120.	143, 243	Unidirectional tape works well when maximum performance is required in one direction.
281, 285	Designed to be a more economical version of Style 181. Both take advantage of a heavier, 1140 denier Kevlar yarn. Style 181 is a plain weave; Style 285 is a crowfoot, or 4-harness satin weave. Style 285 is recommended where a high degree of mold conformability is important.	1050, 1033, 1350	In the boating industry where very heavy fabrics are used, Kevlar is available in three different woven rovings. All three fabrics are basket weave, ranging in weight from 10.5 to 15 oz/yd^2. These fabrics are also effective for cargo liners because of their high resistance to impact in single thickness.

For many applications, fabrics containing more than one fiber type offer promising advantages. Hybrids of carbon and Kevlar 49 aramid yield increased impact resistance over all carbon construction and increased compressive strength over constructions containing only Kevlar 49. Hybrids of Kevlar 49 and glass offer enhanced properties and lower weight than all-glass construction and are less expensive than constructions using Kevlar 49 as the sole reinforcement (Fig. 77.3).

77.4.3 Fabrics and Woven Rovings

Figure 77.4 depicts two basic weave configurations, the plain weave and the satin weave. The plain weave has a one under one over construction, as shown. In the eight-harness satin weave, the pattern is one over and seven under. The crowfoot weave is simply a four harness satin and consists of a one over and three under construction. The basket weave is similar to the plain weave, in which the single yarns are replaced by multiple yarns; a four by four basket weave is characterized by four over and four under construction.

77.4.4 Laminated Composites

Composite materials are typically used in the form of laminates. In this form, multiple plies or laminae of either unidirectionally reinforced or fabric-reinforced or both are laminated to form a plate, shell, beam, or other structural element. The advantage of laminate construction is that the typically low transverse properties of unidirectional plies or low in-plane shear properties of fabric-reinforced plies are augmented by plies at different orientations. The process of selecting the number and orientation of the various plies within a laminate

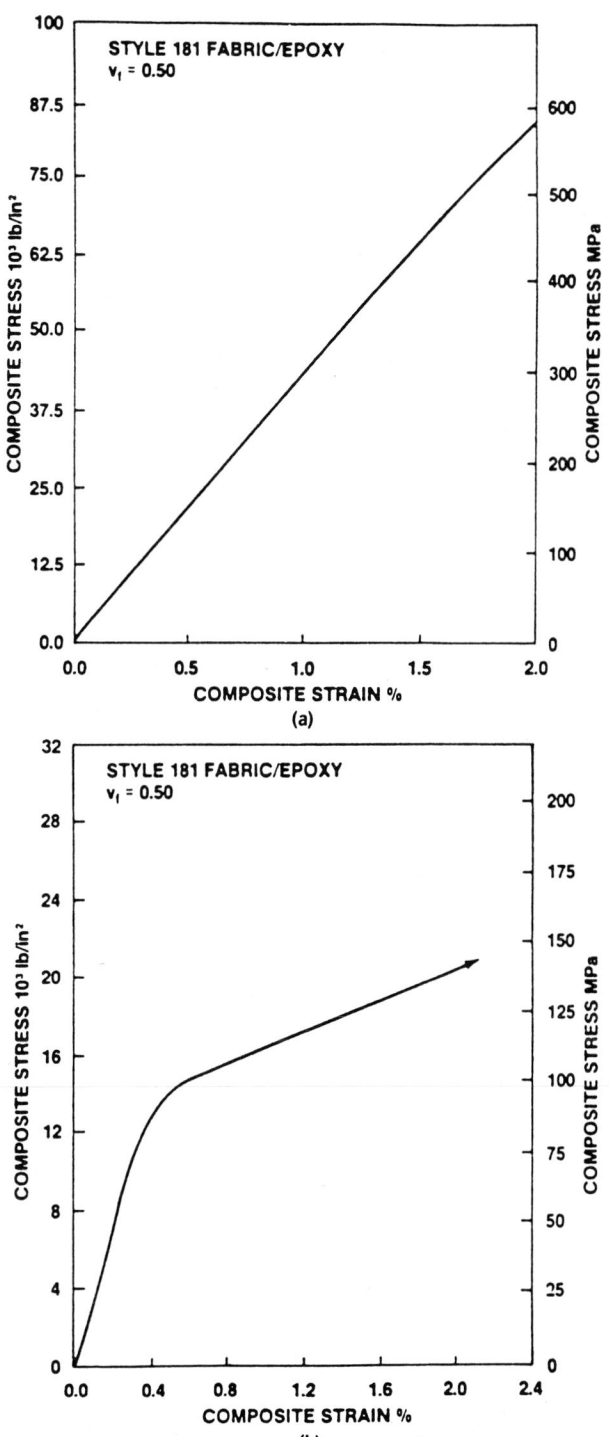

Figure 77.2. Typical fabric composite stress–strain response at room temperature for fabric style 181 of Kevlar 49 reinforced epoxy (warp direction). (**a**) Tension; (**b**) compression.

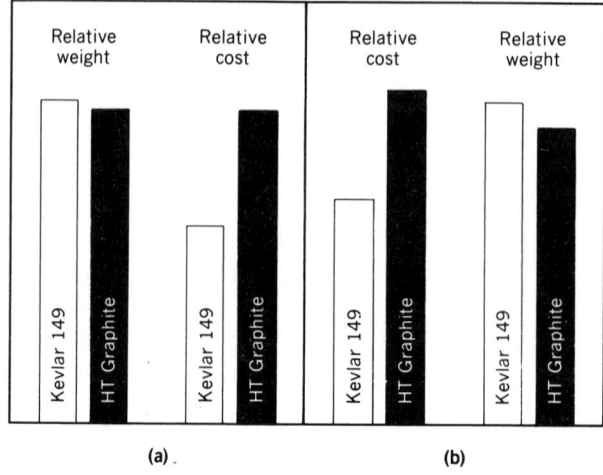

Figure 77.3. Cost comparison between fabric laminates of Kevlar 149 (1140 denier yarn, 4 harness satin weave and graphite (HT Graphite, 3K yarn, plain weave). (**a**) Equal bend stiffness. Bending stiffness of a rectangular section $= S_B = E(bh3/12)$. (**b**) Equal tensile strength.

allows the designer to tailor stiffness, strength, thermal expansion, and other properties to match the loading environment.

77.4.5 Kevlar 149

In early 1987, Du Pont introduced Kevlar 149. Sample property data are given in Table 77.7; its comparative advantages are pointed out in Table 77.8.

- Higher performance (46% modulus increase).
- Lower dielectric properties (65% decrease in moisture regain).

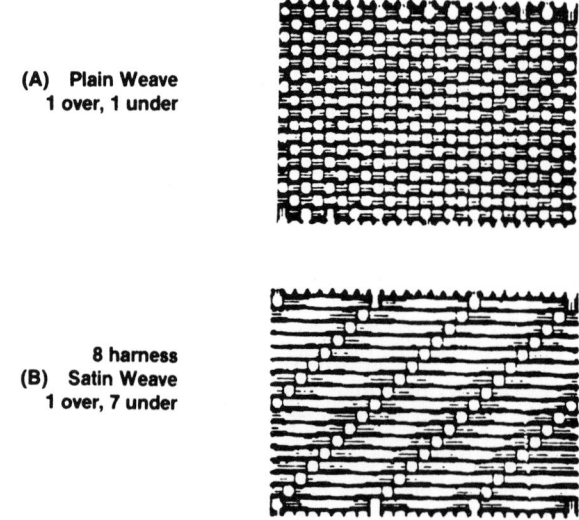

(A) **Plain Weave**
1 over, 1 under

8 harness
(B) **Satin Weave**
1 over, 7 under

Figure 77.4. Basic weave configurations.

TABLE 77.7. Properties of High-Modulus Kevlar 149

Fiber Property[a]	Kevlar 149	Kevlar 49	AS-4 Graphite
Tensile modulus, MPsi	27	19	34
(MPa)	186	131	235
Specific tensile modulus,[b]			
10^8 in	5.1	3.6	5.2
Tensile strength, MPsi	501	525	520
(MPa)	(3455)	(3620)	(3586)
Specific tensile strength,[b]	9.5	10.1	8.0
10^6 in.			
Strain to failure, %	2.0	2.6	1.53
Moisture regain, %	1.5	4.3	0.1%
Density, lb/ft³	92.	90.	112.
g/cm³	1.47	1.44	1.79

[a]Kevlar tensile properties determined by strand test (ASTM D2343-67); modulus measured at 0.5 % strain.
[b]Tensile strength or modulus divided by density.

Advantages vs AS-4 Graphite Fabric

- Reduced acquisition cost (35–45% lower prepreg fabric cost); equal stiffness at equal weight; less complex fastening system required.
- Reduced life-cycle costs [no dissimilar material corrosion problems; system weight reduction resulting from less complex fasteners; greater toughness and vibration damping (Fig. 77.3)]

77.5 APPLICATIONS

Over the past two decades, Kevlar has gained wide acceptance as a fiber reinforcement for composites in many end uses.

The new Boeing 767 makes extensive use of advanced composites, which total 3% of the aircraft structural weight. Figure 77.5, the schematic the exterior of the aircraft, shows numerous composite parts, almost all of which are cored with Nomex. There is also substantial use of Kevlar and hybrids of graphite/Kevlar for lower weight. The possibility of

TABLE 77.8. Comparative Properties—Unidirectional Laminate

Composite Property[a]	Kevlar 149	Kevlar 49
Tensile modulus MPsi	15.4	10.5
(MPa)	(106)	(72)
Tensile strength, MPsi	203.	167.
(MPa)	(1400)	(1152)
Compressive strength, MPsi	38.	40.7
(MPa)	(262)	(281)
Short beam shear, MPsi	5.5	7.2
(MPa)	38.	50.
In-plane shear, MPsi	7.2	7.2
(MPa)	(50)	(50)
Strain to failure, 1%	1.3	1.6

[a]Preliminary data with Fiberite 934 epoxy resin at 350°F (177°C).

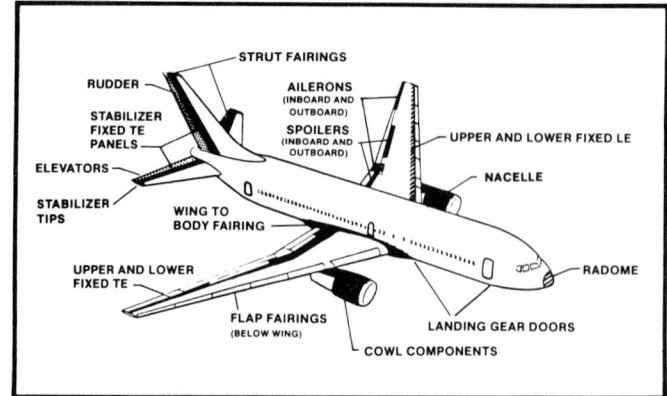

Figure 77.5.   Composite applications in the Boeing 767.

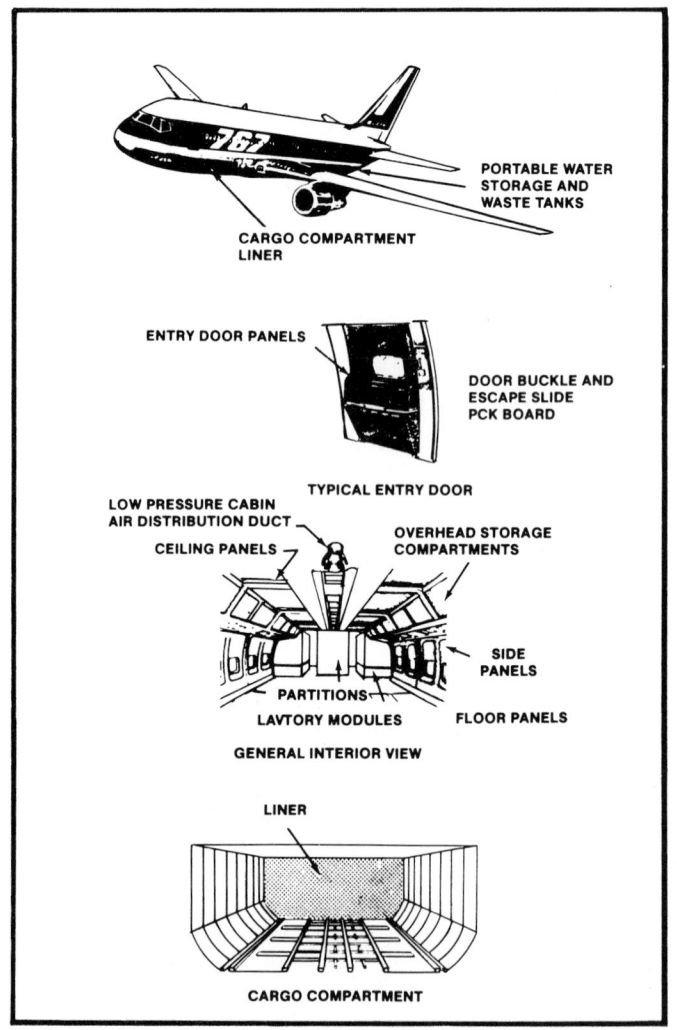

Figure 77.6.   Composite interiors and systems applications in the Boeing 767.

galvanic corrosion between carbon and metal is eliminated by using 100% Kevlar parts where fasteners are used.

Cores of Nomex are used in the floor, ceilings, bulkheads, and side walls in the interior of Boeing's new-generation aircraft (Fig. 77.6). Here, Kevlar can be found in low-pressure cabin air distribution ducts, cargo liners, and entry door panels; advanced composites have enabled Boeing to reduce aircraft weight substantially. Each 767 saves about 2,000 lb through the use of advanced composites; composites faced with Kevlar and hybrids of Kevlar/graphite save more than 500 pounds each.

The Chinook helicopter is as large as many small fixed-wing transports. The commercial version, Model 234, contains 23 wt % of composite materials. The large sponsons on each of the fuel tanks are constructed using composite facings of Kevlar over a honeycomb core of Nomex, with unidirectional carbon fiber reinforcement added selectively for stiffness in critical areas.

Advanced composite construction can do more than reduce weight. Manufacturing costs and maintenance demands can be reduced by changing from a conventional aluminum to a composite structure. A comparison made by Boeing Vertol of an experimental composite helicopter fuselage with a conventional aluminum design showed about a tenfold reduction in part and fastener count. A one-third reduction in weight is projected, together with substantial labor savings in manufacture for the composite structure.

The advanced composite materials data base provided by these and other aircraft has allowed innovative designers such as Leo Windecker and Al Mooney to move ahead in designing aircraft with unprecedented performance. The Avtek 400, Figure 77.7, employs composite skins reinforced with Kevlar over honeycomb of Nomex for more than 70 percent of the airframe weight, producing a six to nine passenger turbo-prop aircraft with one half the weight of a metal aircraft of the same payload capability, three times its fuel efficiency, and half its climb weight.

Other manufacturers of general aviation aircraft have responded to the Avtek challenge by designing and fabricating aircraft that make equivalent use of advanced composite materials.

Figure 77.7. Avtek 400.

The Voyager, the first plane to fly around the world without refueling succeeded because of its "novel and crucial use of polymer composites," according to its designer, Burt Rutan.

BIBLIOGRAPHY

1. Z. Hashin, "Analysis of Properties of Fiber Composites with Anisotropic Constituents," *J. Appl. Mechanics* **46,** 543–550 (1979).

General References

Extensive property and design data on aramid fibers are available from the Du Pont Company, Wilmington, Delaware.

R.E. Alfred, "The Effect of Temperature and Moisture Content on the Flexural Response of Kevlar/Epoxy Laminates," *J. Composite Materials,* **15,** 100, 117 (March 1981).

P.W.R. Beaumont, P.G. Riewald, and C. Zweben, *Methods for Improving the Impact Resistance of Composite Materials, ASTM STP 568, Foreign Object Impact Damage to Composites.*

Characteristics and Use of Kevlar 49 Aramid High-Modulus Organic Fiber, Du Pont Technical Information Bulletin K-5.

C.A.M. de Koning and W.H.M. Dreumel, *Mechanical Jointing in Aramid Fibre Composites: An Experimental Study,* Delft University of Technology, LR-371; DuPont Item #C-94, Jan. 1983.

Fiber Science and Technology, 129 (March 1970).

H.T. Hahn and T.T. Chiao, "Long-Term Behavior of Composite Materials," *Proceedings of the Third International Conference on Composite Materials.*

Z. Hashin and B.W. Rosen, "Fiber Composite Analysis and Design: Composite Materials and Laminates," FAA, in press.

K.E. Hofer and E.M. Olsen, "An Investigation of the Fatigue and Creep Properties of Glass Reinforced Plastics for Primary Aircraft Structures," Final Report under Contract NOW 65-0425-f, NASC, April 1967.

J.C. Norman, *Damage Resistance of High Modulus Aramid Fiber Composites in Aircraft Applications,* DuPont Item #A-15.

Properties of Kevlar 49 Aramid Fabrics and Fabric Reinforced Epoxy and Polyester Composites, DuPont Item #C-28.

P.G. Riewald, *Aramid Fibers in Composite Structures for Transportation Systems,* DuPont Items #A-40.

W.S. Smith, *Environmental Resistance of Kevlar and its Composites—Moisture Absorption, Temperature Resistance, Air Aging, and Hot/Wet Exposures,* DuPont Item #A-34, Sept. 1979.

Vibration Damping of KEVLAR 49 Aramid, Graphite, and Fiberglass Fibers and Composites Reinforced with these Fibers, DuPont Technical Information, Preliminary Information Memo-428.

M.W. Wardle, *Composites of KEVLAR 49, Graphite, and E-glass—Fatigue, Ball-Drop Impact, Fracture Toughness,* DuPont Item #A-36.

C.H. Zweben, *Flexural Strength of Composites of Kevlar,* DuPont Item #A-29.

C. Zweben, *Fracture Toughness of Kevlar 49, E-Glass, and Graphite Composites,* DuPont Item #A-9.

C. Zweben and J.C. Norman, "Kevlar 49/Thornell 300 Hybrid Fabric Composites for Aerospace Applications," *21st National SAMPE Symposium and Exhibit,* April 1976.

C. Zweben and M.W. Wardie, *Flexural Fatigue of Marine Laminates Reinforced with Woven Rovings of Kevlar 49 and E-Glass,* DuPont Item #A-31.

78

REINFORCED PLASTICS, COMPOSITES LAY-UP AND BAGGING

E.A. Greene

4875 W. 134th Place, Hawthorne, CA 90250

78.1 INTRODUCTION

This section is devoted to the little details of lay-up and bagging that have a significant effect on the quality of the finished product.

It is probably reasonable to say that the best source of learning composites fabrication is to work alongside an experienced worker. Such hands-on experience is where the novice can begin to comprehend the handling of materials in the process of lay-up fabrication.

78.1.1 Material Selection

While the available selection of resins and reinforcements is continually expanding, there is an ideal combination of materials and process for most any application. The selection is determined by the requirements of the article to be produced. As an example, the following factors should be considered first:

- Cost limitations
- Quantity
- Quality
- Special property requirements
- Appearance

The materials and process may be selected within these guidelines.

Polyesters and epoxies are the most common resins used for most applications. These two resins may be formulated to any of thousands of characteristics. Polyesters may be catalyzed to cure at room temperature in 25 minutes to 24 hours or heat cure at 150–250°F (66–121°C). Epoxies incorporate a wide range of hardeners that can set the epoxy in five minutes to several hours at room temperature.

Phenolics, silicones and polyimides are special-purpose resins. High temperature thermoplastics such as polysulfone and polyetheretherketone (PEEK) are finding widespread interest because of their special properties and fabrication advantages.

78.1.2 Fiber Reinforcements

The most common choices of reinforcements are fiber glass, aramid (Kevlar 49), and carbon. Other reinforcements, such as boron, are used for special purposes.

Fiber glass is available in the form of mat, roving, or cloth. The rovings can be chopped and sprayed simultaneously with a catalyzed polyester resin directly onto a mold using a special spray-up chopper gun. The same device can be used to spray the chopped fibers onto a perforated form with a diluted binder to make preforms, which are then dried in an oven and stored for future use. Fiber glass cloth can be precut to lay-up patterns for a particular part and stored for later use. This procedure is called kitting; it saves cloth-cutting preparation time and avoids having the molds sit idle while the cloth is being cut. Fiber glass is also available in preimpregnated form. Suppliers can preimpregnate any style of cloth, unidirectional tape, or roving with any type of laminating resin to the volume ratio that is requested by the user.

The advantages of higher priced reinforcements, such as aramid or carbon, is their lighter weight and higher strength. These materials are available as broadgoods or roving and can be supplied in either form as prepregs. Fiber glass or aramid are sometimes combined with carbon fibers to create a hybrid for purposes of vibration damping. The cost may be reduced, depending on the ratio and which material is used. Kevlar 149 will regain moisture at a lower rate than Kevlar 49; both types can contribute to the tensile and strain properties of an aramid/carbon hybrid composite.

78.1.3 Selection of Molds

Proper tooling is essential to successful molding of a fiber-reinforced article. For a contact layup (that is, a room temperature cure with no vacuum bag) almost anything can serve as a tool. However, when a pressurized, elevated temperature cure is required, other factors must be taken into account.

Plastic face plaster tools are ideal for master lay-up models, prototype work or short runs. Steel or aluminum is often used for curing tools and can be quite adequate for single-contour parts. Tools with electroformed nickel surfaces have been serving well for large advanced composite parts. Composite articles with compound contours should be cured on tools with a compatible coefficient of thermal expansion. Here, composite tools become not only more efficient but also more cost effective.

78.2 COMPOSITE TOOL FABRICATION

A master model of wood, ceramic, plaster, plastic face plaster, or metal can serve as a form on which to build a composite tool. Each type of form should be prepared with an adequate interface of release agent.

Gel coats are generally used for fiberglass tool surfaces. A carbon-fiber-filled surface coat or surfacing film is available for carbon/epoxy tools.

To construct a stable composite tool with a minimum of leakage, the lay-up should be debulked or compacted at least every eight plies. This can be accomplished by vacuum bagging each time or, more quickly and economically, by using a permanent silicone rubber sheet mounted on a frame.

Tooling grades of cloth can be utilized for fiber glass tools, but finer weaves should be used at the mold surface. Tools less than 30 in. (76 cm) in length or width should be at least 0.4 in. (10 mm) thick. For larger tools, a thickness of 0.6 in. (15 mm) is recommended.

Aramid tools can also employ tooling grades of materials, but it is important to achieve a balanced layup to avoid warping of the tool. The finer weave at the surface of the tool must be matched on the back side. It is also necessary to construct the backup structure of aramid so as to avoid thermal expansion problems.

Carbon composite tools are also constructed in a balanced manner. The same ply orientation used for the finished article can be used to fabricate the tool. The balanced lay-up is repeated until the desired tool thickness is achieved.

An improved tooling lay-up method being used successfully is a carbon/epoxy tooling prepreg that is cut into small patches and applied to the master model. (Sources are listed later in this section.)

78.2.1 Tool Curing

For curing, a porous release film is positioned over the lay-up and then covered with the proper amount of bleeder plies. A barrier film is placed over the bleeder plies, and the entire master model and lay-up is wrapped with a breather material. The entire assembly is then put into an envelope-type vacuum bag. The latter is used because the master model often leaks, and the autoclave cure is most critical for producing a void-free tool.

Thermocouples should be placed on the face side and the back side of the tool lay-up. As the autoclave is heated, the difference between the two thermocouples should be maintained within 5°F (3°C). The temperature advancement should be stopped at 10°F (6°C) below the gel point of the resin system. The lay-up is then allowed to remain at this temperature for 10 to 20 minutes, depending on the thickness of the lay-up, to allow air and excess resin to flow into the bleeder.

Meanwhile, the autoclave pressure is permitted to climb to 100 psi (0.7 MPa) at a rate of 3 psi (0.02 MPa) per minute. When the pressure reaches 30 psi (0.2 MPa), the vacuum bag is vented to atmosphere. This allows direct autoclave pressure onto the lay-up without additional bleeding of the resin. The temperature can then be elevated to cure the lay-up if the model can withstand higher temperatures. If not, the lay-up can be cured at a lower temperature and then removed from the model and post-cured in an oven. Tooling resins differ in curing requirements. The recommendations on the manufacturer's data sheet should be adhered to, carefully.

78.2.2 Composite Article Fabrication

Of the various processes for producing a fiber glass-reinforced plastic article, probably the most basic is the hand lay-up method, which is a fabrication of fiber glass cloth and polyester resin.

In this process, the mold is treated with a release agent that could be a wax, a parting film, or both. The mold is then covered with a thin coat of resin using a brush. The first layer of dry fiber glass is applied to the mold with care to avoid wrinkles and to achieve proper positioning. Wrinkles can be removed by tugging the edge of the cloth in the warp and fill direction. Any other means will tend to stretch the weave of the cloth, resulting in thinning of fibers in one area and bunching in another, which produces uneven strength and thickness.

The first layer of cloth is then impregnated with resin using a brush or applicator. Subsequent plies are applied in a similar manner. The last ply may be left dry to help bleed out the air and excess resin. A peripheral bleeder/breather, not touching the lay-up, should also be utilized to channel the trapped air to the vacuum line. A vacuum bag of 3-mil (0.08 mm) poly(vinyl alcohol) (PVA) film is used to cover the entire layup.

Small, fairly flat tools may be envelope-bagged. Larger tools are bagged top side only, using a vacuum bag sealant. For high quantity production, a permanent clamping ring can be used.

After the air has been drawn from the bag, the excess resin and trapped air bubbles must be removed. Most commonly, this step involves the use of a squeegee made from polyethylene or PTFE. A lubricant, such as mineral oil, can be applied to the vacuum bag to prevent dragging of the film.

78.3   VACUUM BAGGING

The selection of vacuum bag material is made on the basis of the service temperature. The higher the service temperature requirement, the higher the cost of the material.

PVA film is suitable for temperatures up to 250°F (121°C). Nylon films are available with service temperatures ranging from 350–550°F (177–288°C). Polyimide films can be used safely up to 750°F (399°C).

For fairly flat lay-ups, a silicone rubber sheet can be considered. For large-quantity production of deep contoured parts, a contoured silicone rubber bag could prove to be cost effective because the bags are reusable.

78.3.1   Vacuum Bag Installation

The vacuum bag film should be cut large enough to cover the mold with ample excess material to comfortably avoid bridging. If the film appears to be too small, it should be discarded and replaced with a larger piece. Experience has shown that application of the bag is essentially a one-person operation.

If the mold has raised areas, extra bag material can be provided by using a pig tail of vacuum bag sealant half again as high as the raised part of the mold. The outer edge of the film serves as a guide along the edge of the mold base to keep all of the film over the tool area.

78.4   CURING OPTIONS

Reinforced plastics are cured in the most economical manner that is sufficient for the end use of the article. As an example, decorative fiber glass/polyester parts can be cured at room temperature without pressure. If more strength and less weight is desired, the article can be vacuum bagged and cured by exposure to ultraviolet light or in an oven.

The purpose of designing parts using higher-cost advanced composites is to achieve a higher strength-to-weight ratio and higher service temperatures. Because quality control and reliability are key factors, an autoclave, with its reliable heat and pressure control is used.

78.4.1   Autoclaves

Autoclaves operated at temperatures above 300°F (149°C) should be equipped with an inert atmosphere pressurization system. Simple air pressure can be used at higher temperatures if all of the materials in the autoclave are noncombustible.

A certified operator should load, close, and control the autoclave.

78.4.2   Pressure Bag

The oldest method for curing composites at high pressures is the pressure bag in which a bagged layup is covered with an air chamber. The cover is either bolted to the mold base or is held closed in a press. Air is introduced into the chamber to exert pressure on the bag surface. Vent holes in the mold allow the trapped air in the lay-up to escape.

78.4.3   Rubber Plug

Another, simpler method utilizes a silicone rubber plug. The plug is made by casting silicone rubber into the mold cavity. Allowance for part thickness is made by laying sheet wax in the mold with additional thickness to accommodate the bulk.

The lay-up is made on the rubber plug, which is then placed into the mold cavity. The mold cavity is covered with a retainer that is strong enough to withstand high pressures. No bleeders or bags are required. The rubber plug will provide longer service if a PTFE film is used to separate the rubber from the layup.

The tools can be heated by oven heat or by heating elements embedded in the silicone rubber plug. The oven cure requires close temperature control to permit the rubber to expand before the lay-up cures. Resin systems with a gel point of 280°F (138°C) are best suited for oven cures.

78.5 DEMOLDING

When a cure cycle is completed, the parts should remain under pressure until the temperature is reduced to 120°F (49°C). At this temperature, the bag and bleeder can be removed. The part is now ready for removal from the tool. The object is to remove the part without damaging either the tool or the part. Flat parts can be removed by inserting a soft spatula between the tool and the part. For deeper contoured parts, wedges may be driven between the tool and the part. The wedges and spatulas can be made of nylon, PTFE, or polyethylene. A blast of air can help to break the vacuum between the part and the tool. Several boat-hull manufacturers actually float the hulls out of the molds by forcing water into the mold.

78.6 FINISHING

Even with the molder's best efforts to produce a high-quality composite part, poor finishing technique can spoil it. Delaminations can occur in the trimming or drilling process, which makes the selection of proper machining methods very important.

78.6.1 Material Characteristics

The characteristics of the various composite materials vary widely, making it necessary to use a specialized selection of machining tools.

The most abrasive is boron. With boron composites, diamond or carbide-coated tools survive the longest. The next most abrasive is carbon, which is not nearly as abrasive as boron.

Thickness is a factor in selecting cutting tools. A 1/4-in. (6 mm) wavy set band saw blade with 18 teeth per in. (25 mm) is a good candidate, although a diamond grit blade would dispel any doubts. Drilling can be accomplished with a carbide version of the kind of drill tool that is designed for use with aramid.

Laser or water jet cutting requires expensive equipment. It does, however, offer the advantages of clean cuts and no dust.

The finishing tools that are used for carbon-fiber laminates can also be used for fiber glass, which is less abrasive than carbon.

Aramid fibers are the least abrasive. However, they are very tough, which is why conventional machine tools used for cutting aramid fibers tend to leave a fuzzy finish. Special cutting and drilling tools have been designed to operate in a manner that shears the fibers against the matrix of the laminate. The bottom side of the cut might still have a fiber dangling because there is no more matrix against which to shear. These fibers can be whisked off with the use of a Scotch Brite general-purpose polishing wheel.

High speed-steel, parabolic flute drills with special modified points should be used at 600 rpm. Band saws, using a 1/4″ 24 teeth per in. wavy set blade, should be operated at moderate to low speed to prolong the life of the blade.

The carbide version of the special drills for aramid can be used for carbon-fiber-reinforced

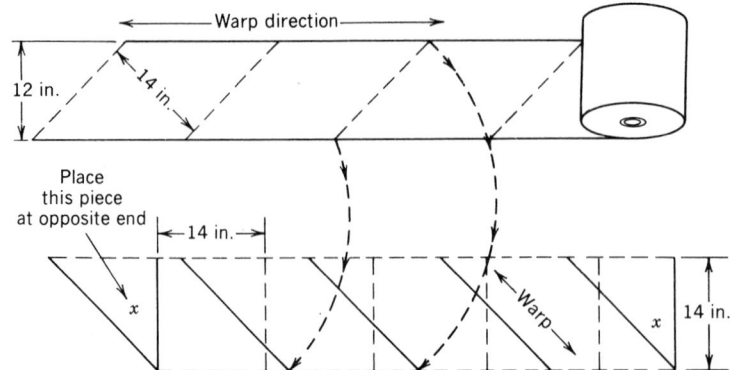

Figure 78.1. Material cutting. Method of cutting 14 in. 45° plies from 12 in. wide unidirectional tape. The number of required cuts from a 12-in. spool may be predetermined by multiplying the number of plies needed by the ply width of the ply and dividing by 17 in.

As an example: If 8 plies are needed and are 24 in. wide, then: 8 × 24 in. = 192 in. ÷ 17 in. = 11.29 or 12 cuts are required. (0.7 of a ply will be left.)

thermoplastics but the revolutions per minute should be reduced to 200 to avoid a temperature buildup.

78.7 MATERIAL CUTTING

See Figure 78.1

78.7.1 Ply Orientation

See Figures 78.2**a** and 78.2**b**.

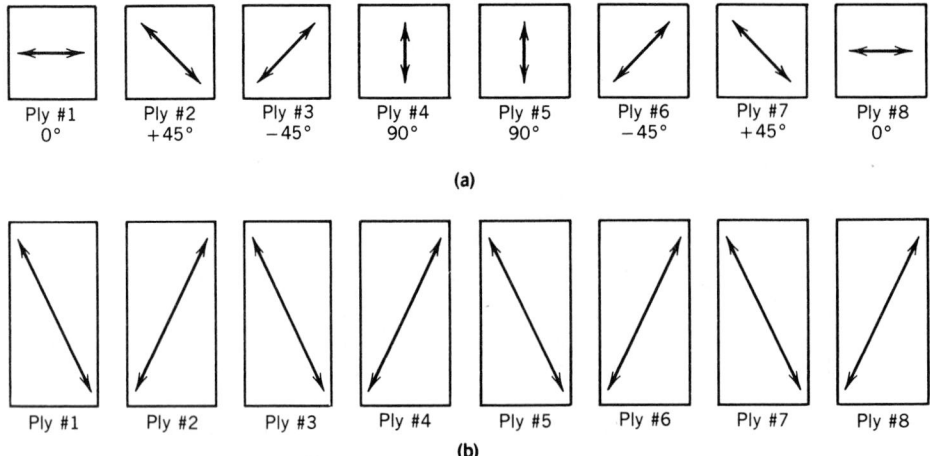

Figure 78.2. Ply orientation: (**a**) Unidirectional carbon-fiber balanced lay-up; (**b**) Fabric or unidirectional aramid orientation plan for tool fabrication.

78.8 PRESSURIZATION OPTIONS

See Figures 78.3–78.7.

78.9 BLEEDER OPTIONS

A vacuum bag and peripheral bleeder only (Fig. 78.8) can be used for a simple fiber glass/
polyester wet lay-up, using a PVA film bag sealed to the tool surface or an envelope bag.
This technique is suitable for room-temperature or ultraviolet curing.

When it is desirable for the bleeder to cover the lay-up (Fig. 78.9), perforated film or
PTFE-coated glass cloth can be used to separate the lay-up and the bleeder.

Another type of bleeding system (Fig. 78.10) is used for autoclave curing of advanced
composite prepreg lay-ups. The dam prevents excessive edge bleeding, and the perforated
film controls the overall flow of resin into the bleeders. The number of bleeders needed can
be determined by test samples prior to part fabrication.

Figure 78.11 shows two specimens containing the same amount of fibers. Specimen B
contains more resin and air, making the laminate thicker. In a tensile test, the gauge readings

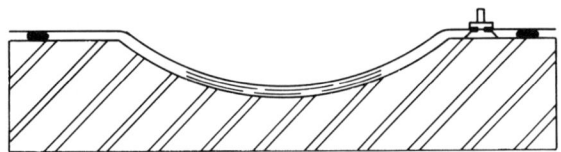

Figure 78.3.   Film bag sealed in place with vacuum bag sealant.

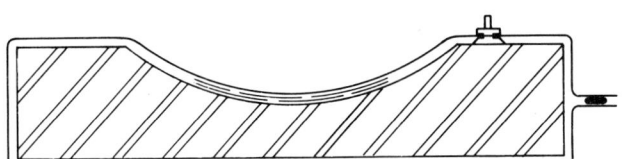

Figure 78.4.   Envelope bag heat-sealed or sealed with vacuum bag sealant.

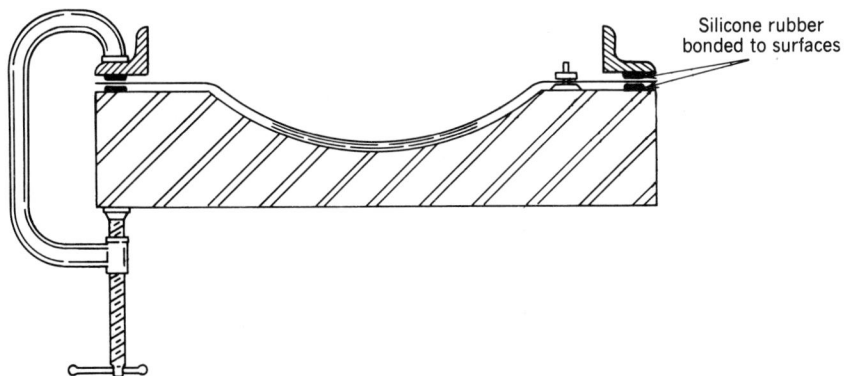

Figure 78.5.   Permanent ring used with C clamps or toggle clamps to reduce bagging time and costs
on longer runs.

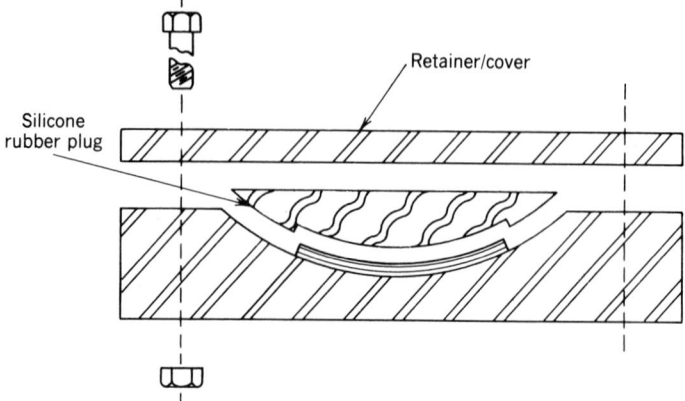

Figure 78.6.   With silicone rubber plug, tool and bolt spacing should be designed to withstand 600 psi (4 MPa).

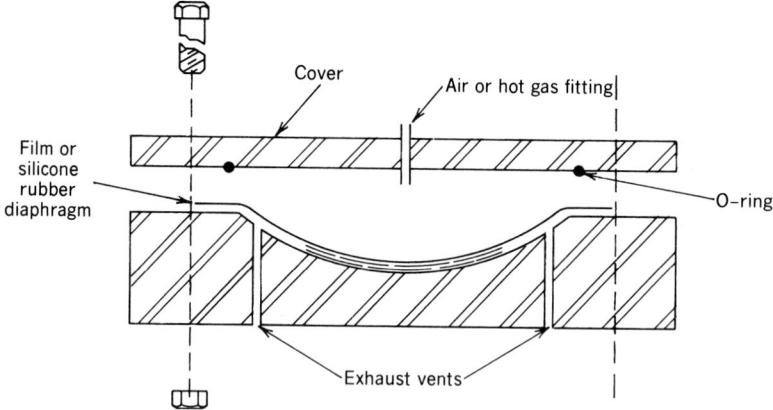

Figure 78.7.   With film or silicone rubber diaphragm, tool and bolt spacing should be designed to withstand 160 psi (1 MPa) or the tool can be closed in a platen press.

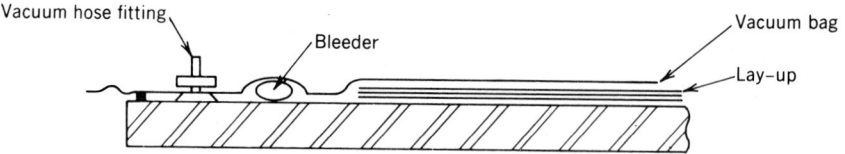

Figure 78.8.   Vacuum bag and peripheral bleeder.

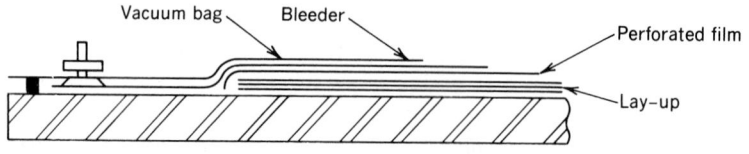

Figure 78.9.   Bleeder covering lay-up.

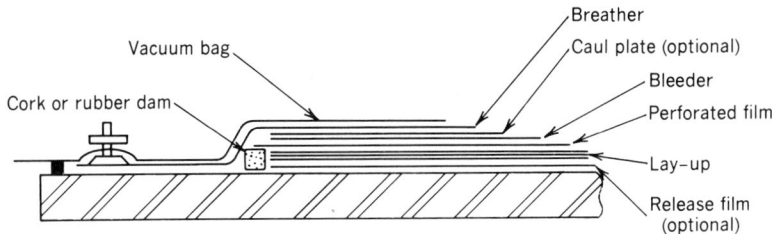

Figure 78.10. Bleeding system for autoclave curing of advanced composite prepreg lay-ups.

may be nearly equal; however, when these readings are calculated with the cross-section area of the specimens, the final strength results would be about one third lower for specimen B. A test of flexural modulus would find specimen B to be much weaker because the layers of fibers do not support each other—the air-laden resin would crumble and offer no resistance.

As a result, specimen A has a higher specific strength because it has the same tensile gauge reading but weighs less. The tensile value will be higher because

$$\frac{\text{Specimen A}}{\text{tensile strength}} = \frac{\text{failure load}}{d \times b, \text{ or } 1 \times 6} \quad \frac{\text{load}}{6}$$

$$\frac{\text{Specimen B}}{\text{tensile strength}} = \frac{\text{failure load}}{d \times b, \text{ or } 1.5 \times 6} \quad \frac{\text{load}}{9} \text{ (1/3 lower)}$$

78.10 MATERIAL AND EQUIPMENT SOURCES

See Table 78.1.

78.11 MANUFACTURERS' ADDRESSES

Airtech International, Inc., 2542 East Del Amo Blvd., P.O. Box 6207, Carson, CA 90749, Phone: (213) 603-9683 Telex: 19-4757

Allied Engineered Plastics, P.O. Box 2332R, Morristown, NJ 07960, Phone: (201) 455-5010

American Cyanamid Company, Engineered Materials Dept., Wayne, NJ 07470

Amoco Performance Products, 39 Old Ridgebury Rd., Dept. F-2311, Danbury, CT 06817

ARO Corporation, One Aro Center, Bryan, OH 43506

Bondline Products, P.O. Box 1473 Norwalk, CA 90651, Phone: (213) 921-1972

Burlington Glass Fabrics, 1345 Ave. of the Americas, New York, NY 10105

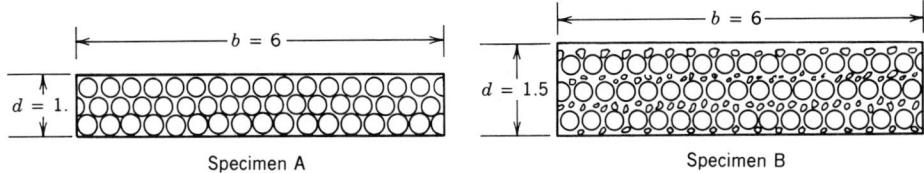

Figure 78.11. Test specimen cross-sections.

TABLE 58.1.   Material and Equipment Sources

Material/Equipment	Supplier
Reinforcements	
Fiber glass	Burlington, Certainteed, Fiberglass Industries, Hexcel, Knytex, Owens Corning, PPG Industries, J.P. Stevens, Thalco
Carbon	American Cyanamid, Amoco Performance Products, Great Lakes Carbon, Hercules
Aramid	E.I. du Pont de Nemours & Co., Inc.
Resins and Adhesives	Alpha, Celanese, Ciba-Geigy, Devcon, Dow, Ferro, Freeman, Furane, Hysol, ICI, Koppers, PPG, Reichhold, Shell, Union Carbide
Carbon Tooling Materials	
Toolmaster (C/E prepeg)	Airtech International
Surface film and fiber-filled resin	Fiber/Resin
ToolRite System	Fiberite
Tooling material	Ren Plastics, Ciba-Geigy
Vacuum Bagging Materials	
Bagging films	Airtech International, Allied Engineered Plastics, Richmond Technologies
Vacuum bag sealants	Airtech International
Bleeders and breathers	Airtech International, Bondline Products, Crawford Fitting Company
Release agents	Airtech International, Hysol Miller-Stephenson, Ram Chemical
Silicone Rubber	Airtech International, CHR, D Aircraft Products, Dow Chemical, General Electric, Mosites Rubber Company
Heating elements	Watlow
Scissors and Shears	
Air powered, Wiss	Airtech International, ARO
Special Tools for Aramid	
Shears	Airtech International, Martin Technologies
Saws, Circular	Simonds
Drills	E.A. Greene & Associates
Nibblers	Bosch, Trumpf
Water jet cutters	Flow Systems, Inc., Hydroforce Systems
Tool Coatings and Anodizing	Tiodize
Automation Equipment	
Robotics	Cincinnati Milicron
Tape laying	Cincinnati Milicron
Broadgoods cutting	GGT
Trimming	Thermwood
Autoclaves	Lipton Industries, Thermal Equipment Corp., United McGill Corp.
Consultants	E.A. Greene & Associates, G.D. Shook

Ciba-Geigy Corporation, 10910 Talbert Ave., Fountain Valley, CA 92708, Phone: (714) 964-2731 Telex: 131411 REPD

Cincinnati Milacron, Cincinnati, OH 45209, Phone: (513) 841-8804

Crawford Fitting Company, 29500 Solon Rd., Solon, OH 44139

D Aircraft Products Inc., 1191 Hawk Circle Anaheim, CA 92807, Phone: (714) 632-8444 Telex: 683 482

DuPont Company, Paul Langston, Room G-15661, Wilmington, DE 19898

E.A. Greene & Associates, 4875 W 134th Place, Hawthorne CA 90250, Phone : (213) 676-0943

Ferro Corporation, 5915 Rodeo Rd., Los Angeles, CA 90016, Phone: (213) 870-7873

Fiber/Resin Corporation, 170 Providencia Ave., Burbank, CA 91503, Phone: (213) 849-4608

Fiberite Corporation, 501 W. Third St., Winona, MN 55987, Phone: (818) 454-3611

Furane Products, 5121 San Fernando Rd., West Los Angeles, CA 90039, Phone: (818) 247-6210

Flow Systems, Inc., 21440 68th Ave Kent, WA 98032, Phone: (206) 872-4900

General Electric Company, Waterford, NY 12188, Phone: (518) 237-3330

G.D. Shook, 6528 Cedarbrook Dr., New Albany, OH 43052, Phone: (614) 855-7796

GGT Inc. Aerospace Division, 24 Industrial Park Road West, Tolland, CT 06084, Phone: (203) 871-8082 Telex: 643 771

Great Lakes Carbon Corp., 320 Old Briarcliff Rd., Briarcliff Manor, NY 10510, Phone: (914) 941-7800 Toll free: (800) 828-6601

Hercules Incorporated, P.O. Box 98, Magna, UT 84044, Phone: (801) 250-5911

Hexcel, 11711 Dublin Blvd., P.O. Box 2312, Dublin, CA 94566-0705, Phone: (415) 828-4200

HydroForce Systems Inc. P.O. Box 5418 Kent, WA 98031, Phone: (206) 735-1888

Hysol P.O., Box 312, Pittsburg, CA 94565, Phone: (415) 687-4201

Lipton Industries, One Lipton Dr., P.O. Box 1159, Pittsfield, MA 01202, Phone: (413) 499-1661 Telex: 95 5317

Martin Technologies, 36270 West 103rd St., DeSoto, KS 66018, Phone: (913) 585-1184 Toll free phone: (800) 255-6314

Miller-Stephenson Chemical Co., Inc., George Washington Hwy., Danbury, CT 06810, Toll free phone: (800) 992-2424

Mosites Rubber Company, Inc., P.O. Box 2115, Fort Worth, TX, Phone: (817) 335-3451

Owens-Corning Fiberglas Corp., Fiberglas Tower, Toledo, OH 43659 Phone: (419) 259-3000

PPG Industries, Inc., 1 Gateway Center, Pittsburgh, PA 15222, Phone: (512) 281-5100

Richmond Technology, P.O. Box 1129, Redlands, CA 92373, Phone: (714) 794-2111

Simonds Cutting Tools, 15619 Blackburn Ave., P.O. Box 505, Norwalk, CA 90650, Phone: (213) 802-2689

J.P. Stevens & Co., Inc., P.O. Box 208, Greenville, SC 29602, Phone: (803) 239-4000

Thermwood Machinery Mfg. Co., Inc., P.O. Box 436, Dale, IN 47523 Phone: (812) 937-4476

Tiodize, 15272 Tiodize Circle, Huntington Beach, CA 92649, Phone: (213) 594-0971

Thermal Equipment Corporation, 1301 West 228th St., Torrance, CA 90501, Phone: (213) 775-6745

United McGill Corporation, One Mission Park, Groveport, OH 43215, Phone: (614) 836-9981

Watlow, 12001 Lackland Rd., St. Louis, MO 63141, Phone: (314) 878-4600

BIBLIOGRAPHY

1. P.R. Young, "Thermoset Matched Die Molding," in G. Lubin, ed., *Handbook of Composites*, Van Nostrand, New York, Chapt. 15.

2. D.L. Denton, "Mechanical Properties Characterization of an SMC-R50 Composite," *34th Annual Conference, SPI RP/CI*, Washington, D.C., Feb. 1979, paper 11-F.

3. B.D. Pratt, "Factors Affecting Arc Resistance of Premix," *Annual Conference SPI RP/CI*, Feb. 1965.

4. Ref. 1, p. 411.

5. R.W. Meyer, *Handbook of Polyester Molding Compounds and Molding Technology*, Chapman & Hall, New York, 1986.

6. *Ibid.*, p. 109.

7. *Ibid.*, p. 77.

8. *Ibid.*, p. 114.

9. *Ibid.*, pp. 118–121.

10. R. Gruenwald and O. Walker, "Fifteen Year's Experience with SMC," *30th Annual Meeting, SPI Reinforced Plastics/Composites Institute*, Washington, D.C., Feb. 4–7, 1975, Sect. 1-A.

11. U.S. Pat. 3,560,294 (Feb. 2, 1971), E.J. Potkanowicz.

12. Ref. 5, p. 184.

13. Ref. 1, p. 406.

14. Ref. 1, p. 444.

General References

G.D. Shook, ed., *Reinforced Plastics for Commercial Composites*, American Society for Metals

G. Lubin, ed., *Handbook of Fiberglass and Advanced Plastics Composites*, Van Nostrand Reinhold, New York.

79

REINFORCED PLASTICS, FILAMENT WINDING

A.M. Shibley

351 West 24th Street, New York, NY 10011

79.1 DESCRIPTION OF PROCESS

Filament winding is a partially automated method for the production of reinforced plastic components. Most filament-wound constructions are cylindrical shapes, although other configurations can be fabricated. Major examples are storage tanks, pipes, and pressure vessels.

The essential step in the process is a machine-controlled operation in which resin-coated reinforcements are wound over a rotating mandrel in predetermined angles and patterns. Angle and pattern control are maintained by coordinating the mandrel rotation with the traverse of the reinforcement payout mechanism. A wound layer completely covers the mandrel; successive layers are added to attain the finished wall thickness.

The reinforcements are in the form of rovings, strands, tapes or monofilaments that are impregnated with a catalyzed resin just prior to being placed on the mandrel (wet winding). Preimpregnated B-staged reinforcements are used in an alternative dry winding variation. The curing stage is normally conducted at an elevated temperature without the application of pressure. The mandrel is removed after the curing cycle is completed. Other operations such as post-curing, machining and the attachment of hardware are performed as required to finish the fabrication.

The controlled placement of the reinforcements is accomplished by either of two basic methods—spiral winding or polar winding (Fig. 79.1). The major axes of motion in spiral winding are a continuous mandrel rotation at a constant speed and a reciprocating traverse of the feed carriage parallel to the mandrel axis. The carriage remains stationary at the end of each traverse for a regulated time interval (dwell) before beginning its return. During this period, the mandrel rotates approximately one-half a revolution (dwell angle). Auxiliary movements frequently employed include a cross-slide of the feed arm perpendicular to the mandrel and a rotation of the feed arm. Both motions facilitate the reinforcement delivery across the mandrel ends.

The relative speeds of the mandrel and feed carriage prescribe the winding angle. The first reciprocating traverse (circuit) locates the reinforcement path. The second circuit does not necessarily place the reinforcement adjacent to the initial path, and several circuits may be required to complete a pattern. The paths then coincide and are advanced one reinforcement band width per circuit. The pattern is called a multicircuit spiral wind. Visually, the reinforcement bands intersect at certain locations (fiber crossovers) and resemble a basket weave. Such crossovers may occur at more than one point, depending on the winding angle and the length of the mandrel, as shown in Figure 79.2. A large winding machine is shown in Figure 79.3.

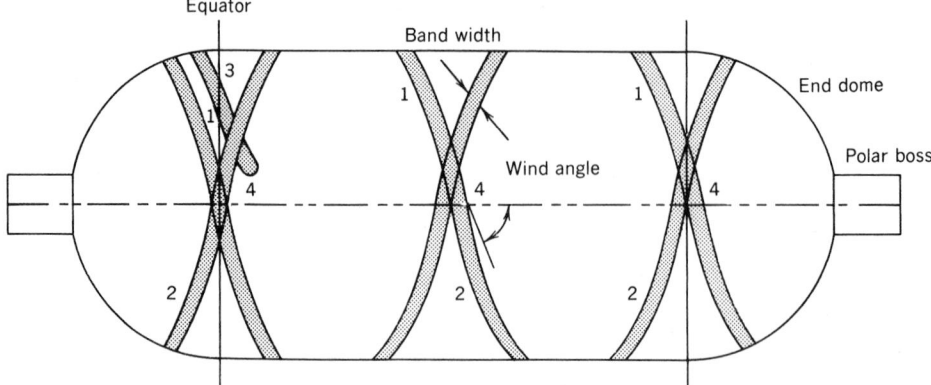

Figure 79.1. Single circuit spiral winding. 1, right hand traverse; 2, left hand traverse; 3, second circuit; 4, crossover points

Patterns are derived by trial and error machine adjustments or are calculated from the geometry.[1,2] A truly accurate pattern completely covers the mandrel in a whole number of repetitive patterns. This arrangement avoids a partial overlapping of layers in attaining total coverage. Slight manipulations of the winding angle, band width, and dwell angle are necessary to establish this pattern. A layer consists of two plies analogous to an axisymmetric laminate with the reinforcements aligned at plus and minus the winding angle. Additional layers are wound at the same or different angles.

Patterns can be reduced to a single circuit repetition if somewhat greater deviations in winding angle, dwell angle, band width, or mandrel length can be tolerated. The path is adjusted to butt against the previous laydown and to advance one band width per circuit. The approximate range of spiral winding is from 15°–80°. A variation, identified as hoop or circumferential winding, is used extensively. The hoop winding angle is close to 90°, and

Figure 79.2.   Large Ogive spiral-wound. Courtesy of McClean-Anderson Inc.

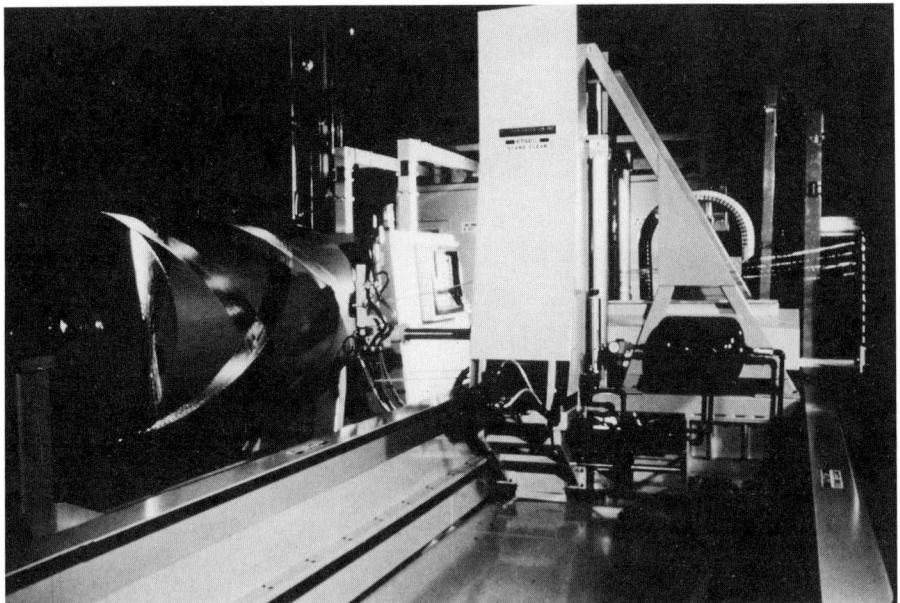

Figure 79.3. Six-axis Ogive polar-wound. Courtesy of McClean-Anderson Inc.

the reinforcement advances one band width per mandrel revolution. Each hoop layer consists of a single ply, usually considered to be at an angle of 90°.

79.1.1 Polar-Winding (Longitudinal Winding)

The reinforcement path in polar winding is the intersection of an inclined plane with a cylinder or a sphere in which the angle of inclination is determined by the polar bosses at each end of the mandrel (Fig. 79.4). Such a path can be generated by the following methods:

- The mandrel remains stationary while the feed arm is rotated about its longitudinal axis
- A feed arm reciprocating traverse is combined with a mandrel rotation per traverse equal to the winding angle
- The feed payout remains stationary while the mandrel, inclined at the winding angle, is tumbled about its longitudinal axis.

In all cases, the feed is advanced one band width per revolution so that the reinforcements lie adjacent to each other without crossovers. A layer consists of a two-ply axisymmetric composite with the reinforcements at plus and minus the winding angle. Polar winding angles normally fall within a 5 to 30° range and usually are reinforced with hoop layers to meet structural requirements.

79.2 MATERIALS

79.2.1 Continuous Reinforcements

Although numerous reinforcing materials have been tried (such as steel and beryllium wire, boron, asbestos, polyester, and polyamide fibers), fiber glass, specifically E-glass roving, is

Figure 79.4. Small Ogive polar-wound. Courtesy of McClean-Anderson Inc.

the major reinforcement used for winding. S-2 glass, graphite, and the Du Pont aramid Kevlar 49 are used to a lesser extent.

Table 79.1 lists typical yields and end counts for E and S-glass winding grade rovings. Winding grades are processed to maintain strand integrity, minimize fuzzing, and reduce catenary (uneven strand lengths or sag within a roving caused by uneven tension).

TABLE 79.1. Fiber Glass Winding Grade Roving Yields, yd/lb

Single Ends		Multiple Ends						
Type 30[a]	Hybon 2079[b]	8	12	15	16	20	30	60
1800	675	462	1125	247	225	675	450	225
1200	250	225	1167			700	467	233
900			1231			738	123	246
675			1250[c]			750[c]		250[c]
450								
225								

[a]Owens-Corning.
[b]PPG Industries.
[c]S-2 Glass.

TABLE 79.2. Graphite/Carbon Fibers

Property	Hercules-Magnamite			Celanese-Celion			Union Carbide-Thornel		
	AS1	AS4	HMS	6000	12000	G50	T-50	T-300	T-500
Filament diameter									
mils	0.315	0.315	0.315	0.280	0.280	0.260	0.256	0.276	0.276
micrometers	8.0	8.0	8.0	7.1	7.1	6.6	6.5	7.0	7.0
Filament count, thousands									
3K		3					3	3	3
6K		6		6			6	6	6
10K	10		10						
12K		12			12	12		12	12
Yield, yd/lb									
3K		2267					2730	2510	2510
6K		1133		1229			1365	1255	1255
10K	619		605						
12K		567			608	584		627	627
Yield, m/g									
3K		4.56					5.50	5.06	5.06
6K			2.28	2.48			2.75	2.53	2.53
10K	1.25		1.22						
12K		1.14			1.23	1.18		1.26	1.26
Density									
lb/in.³	0.065	0.065	0.065	0.064	0.064	0.066	0.065	0.064	0.065
gm/cm³	1.80	1.80	1.83	1.77	1.77	1.83	1.80	1.77	1.80

The more expensive graphite and Kevlar 49 fibers are characterized by high tensile strengths and moduli combined with low densities. They are thus attractive choices for applications based on optimum strength-to-weight ratio and increased rigidity. They are used as the sole reinforcement in these applications or in hybrid constructions with fiber glass. Continuous graphite fibers are marketed as tows, yarns, rovings, and tapes with or without surface treatments or resin compatible sizings. Prepregs with epoxy resins are generally preferred for winding purposes. Examples of graphite fibers are shown in Table 79.2. Kevlar 49 is available as zero twist yarns or rovings at various deniers, as listed in Table 79.3.

79.2.2 Resin Systems

The principal resin systems are formulations of epoxy, polyester, or vinyl ester resins. Polyimides are used in certain high-temperature aerospace applications. The thermoplastic polysulfone is being investigated for potential service in aerospace environments requiring heat resistance, low moisture penetration, and resistance to fuels and solvents.

TABLE 79.3. Kevlar[a] 491 Yarns and Rovings[b]

Denier	Type	Yield, yd/lb	Yield, m/kg	No. Filaments
195	Yarn	22,895	46,155	134
380	Yarn	11,749	23,684	267
1140	Yarn	3,916	7,895	768
1420	Yarn	3,144	6,338	1,000
4560	Roving	980	1,973	3,072
7100	Roving	630	1,268	5,000

[a]Registered trademark, E.I. du Pont de Nemous & Co., Inc.
[b]Ref. 3.

Epoxy winding resins, their curing agents, accelerators, and reactive diluents are essentially the same as their bag molding counterparts used in vacuum bag molding. The diglycidyl ether of bisphenol-A (DGEBA) is the most common type. Epoxy novolacs, resorcinol diglycidyl ethers, and flexible or rubber-modified epoxies are also used. Of the available amine and anhydride curing agents, m-phenylene diamine (MPDA) and nadic methyl anhydride (NMA) are most popular. Accelerators are sometimes added to a resin/curing agent mixture to speed the reaction; conversely, reactions may be retarded by partially reacting the curing agent with a small amount of resin. Reactive diluents such as butyl glycidyl ether, phenyl glycidyl ether, and the diglycidyl ether of 1,4-butanediol are added to reduce the viscosity of the resin system.

The unsaturated polyesters and vinyl esters find extensive use as a result of their lower cost and a balance of physical and chemical properties. They are of particular importance in corrosion-resistant applications. Here again, these resins and their monomers, catalysts, promoters, retarders, and inhibitors are equivalent to the resin systems used for vacuum bag molding and pultrusion.

79.3   PROCESSING CHARACTERISTICS

Initially, filament winding was restricted to cylindrical, conical, and spherical shapes. Components with varying winding angles within a circuit, diameter changes, and other design features were difficult or impractical to fabricate. Recent innovations in winding equipment have removed many of these restraints and permit the fabrication of nonsymmetrical, more complex structures (Fig. 79.4). Production capability has been broadened and now includes such diverse components as helicopter rotor blades, automotive leaf springs, driveshafts, chimney liners, and railroad hopper modular housing. Products are listed in Table 79.4. Figure 79.5 shows a box beam being wound. It is anticipated that these and other markets will be expanded. Currently, however, pipes, storage tanks, and pressure vessels, principally

Figure 79.5. Box beam, Kevlar and carbon fiber. Courtesy of McClean-Anderson Inc.

TABLE 79.4. Some Filament Winding Applications

Component + Resin	Reinforcement + Notes
Rocket motor cases	Aramid, graphite, fiber glass
Epoxy	Hybrids
Frustrum, reentry vehicle	Boron
Phenolic	Development
Railroad hopper car	Fiber glass
Polyester	Demonstration
Tire insert	Fiber glass
Epoxy	Development
Springs, automotive	Fiber glass
Polyester	
Driveshaft	Fiber glass, graphite
Epoxy	Development
Ship hull	
	Feasibility
Wind turbine blades	Fiber glass
Polyester, epoxy	Development
Chimney liners	Fiber glass
Cylinders, gas containing	Fiber glass
	Aluminum, linen
Gas containers, oxygen	Fiber glass, aramid
Epoxy	Life support
Housing modules	Fiber glass
Polyester, epoxy	
Launch tube,	Graphite
Epoxy	Development missile
Bottles, space thruster	Aramid
Epoxy	Aluminum liner
Boosters, space shuttle	Graphite
Epoxy	Development
Cryogenic tanks	Aramid, fiber glass
Epoxy	
Helicopter rotor blades	Fiber glass
Epoxy	Certified
Helicopter rotor	Aramid
Epoxy	Development blades
Fuel tanks, external	Fiber glass, graphite
Epoxy	Development
Helicopter tail section	Aramid, graphite
Epoxy	Development

for corrosion-resistant service, represent the major outlets for the industry. Variations in the manufacture of these latter products are described below.

79.3.1 Pipe

Spiral-wound pipe is produced over a wide range of diameters and pressure ratings for use in oil fields, chemical plants, and other industrial applications. Representative examples are

- 2–3 in. (51–76 mm) diameter high pressure pipe rated at 1000–2000 psi (7–14 MPa).
- 3–4 in. (76–102 mm) diameter pipe at 500–800 psi (3.5–5.5 MPa) and 210°F (99°C) maximum service temperature.

- 18–24 in. (46–61 cm) diameter pipe carrying water at 30–40 psi (0.2–0.3 MPa) and at 80–90°F (27–32°C).

Larger-diameter pipes have also been wound. In some cases, such pipe may be combined with chopped roving or mat. Amine-cured epoxies are generally preferred for high-temperature and high-pressure service. Polyesters and vinyl esters are accepted for less stringent conditions.

Spiral-wound pipes are sometimes lined with thin metal or thermoplastic tubing to prevent fluid weeping or for handling specific corrosive materials. Hoop or spiral winding is also used to reinforce metal or thermoplastic extrusions.

79.3.2 Storage Tanks

Storage tanks are constructed from several material combinations which include high angle (70–80°) spiral or hoop winding with chopped roving, mat, or unidirectional fabrics. The amount of each material depends on the hoop and axial structural requirements. The shells and end domes (heads) for vertical above-ground tanks are usually fabricated separately and are bonded together. The shells consist of spiral winds with mat; the heads are contact molded from mat or chopped roving. Horizontally mounted tanks for above-ground or underground service generally are wound with integral heads and also contain mat or chopped roving. Localized hoop windings are frequently added to increase the rigidity of the tank.

79.3.3 Pressure Vessels

Integrally wound end domes are critical in the performance of the vessel. Their contours depend on whether the winding is to be spiral or polar. The fiber path in spiral-wound end domes is a geodesic tangent to the polar boss; for polar winds, the path is a continuation of the planar intersection. The wall thickness in both cases increases progressively from the equator (cylinder end-dome junction) to the polar boss. Contours and winding patterns are designed to minimize this buildup—a source of discontinuity stresses and unequal strains.

The low angle spiral or polar windings that form the end dome are reinforced by hoop layers over the cylindrical section of the vessel. These hoops are frequently extended beyond the equator as "skirts" for attachment purposes.

High-strength-to-weight pressure vessels, typified by rocket motor cases, have been constructed from S-glass. In later applications, however, graphite or aramid reinforcements were found to be more structurally efficient. Some designs incorporate graphite or aramid longitudinal wraps with S-glass hoops. Epoxy resin systems are used almost exclusively in these applications.

79.4 ADVANTAGES/DISADVANTAGES

A major advantage of filament winding is that reinforcements can be accurately placed in directions that comply with the loading conditions to which the component will be exposed. Maximum structural efficiency can therefore be achieved. Two well-known examples are the 2:1 ratio of hoop to axial reinforcement in pressure vessels and the plus and minus 45° orientation for torsional members.

The high strength reinforcements, i.e., graphite, aramid, S-glass, and E-glass, can be utilized and in their least expensive and most efficient form as rovings or tows.

The amount of reinforcement in the composite can be varied over a wide range, as dictated by strength or economic reasons. The weight percent of fiber glass, for example, may be as high as 80% or as low as 30%.

The winding resins are equivalent to the systems for other processes and can be handled by the same equipment. The excellent physical and chemical properties of these resins is apparent in filament-wound composites. Filament winding can be automated to a high degree. In some applications extremely high production rates are possible. Smooth inner surfaces are the rule. Exterior surfaces rarely require machining.

Large sizes are possible; in-plant windings over 13 ft in diameter by 60 ft in length (4.0 × 18.3 m) are not uncommon. On-site winding methods increase the potential for larger sizes.

On the debit side, the process is restricted to hollow products. Mandrel removal imposes a further limitation on product configurations. The construction and removal of mandrels can be costly and time consuming, particularly when integral end closures are wound. End trimming is necessary for some products. The material removed represents a significant waste. Selection of a winding machine requires careful consideration. The more versatile computerized machine may be too expensive to operate, the less costly, mechanical version may be incapable of handling complex windings.

79.5 DESIGN

79.5.1 Determination of Mechanical Properties

Material properties for the design and analysis of filament-wound structures are obtained by testing of appropriate specimens that closely approximate the intended component. For preliminary design purposes, the necessary mechanical properties can be estimated by use of composite macromechanics and lamination theory. Composite macromechanics utilizes well-known transformation equations to transpose the longitudinal and transverse properties of a single ply into properties at a different set of axes, usually aligned in the loading direction. Lamination theory predicts the properties of an angle-ply laminate from the properties of each ply, the orientation of the ply relative to the loading direction, and the sequence in which the plies are stacked. In addition, the stresses on each ply can be calculated from the stresses imposed on the laminate. The unidirectional properties of the single ply that are used in the calculations are determined experimentally, although they may be estimated by micromechanics based on the properties of the reinforcement and the resin.[4] Table 79.5 lists the unidirectional properties of several reinforcement/resin combinations.

In applying macromechanics to wound structures, the properties of the ply are transformed into axial and hoop components. These values then serve as inputs for specific structural formulae for internal pressure, bending, buckling, torsion, or other loadings. Failures are estimated by maximum stress, maximum strain, or first ply failure theory.

The procedures outlined above involve lengthy calculations that are generally carried out by computer codes.[6] Numerous documents deal with these analytical methods; two of which are listed at the end of this chapter.[7,8] A simplified method for performing these calculations is also presented.[5] This procedure relies on simple equations and graphs depicting angle ply properties versus orientation angles. Calculations can be made with a small calculator rather than a computer.

79.5.2 Wall Thickness

Industrial standards and specifications for wound pipe, tanks, and pressure vessels (listed in Table 79.6) furnish detailed performance and physical requirements for these components. In addition, methods are provided for calculating wall thickness and hoop stress. The American Society for Testing and Materials (ASTM) uses the following basic structural formula relating hoop stress to thickness:

$$S = \frac{P(D - t)}{2t}$$

TABLE 79.5.   Typical Unidirectional Properties at Room Temperature[a]

Property	Reinforcement Resin	E-Glass Epoxy	S-2 Glass Epoxy	S-Glass Epoxy	Graphite Epoxy	Aramid Epoxy
Fiber volume, %		60	60	72	70	54
Density, lb/in.^{3} (g/cm^3)		0.065(1.80)	0.066(1.82)	0.077(2.13)	0.058(1.61)	0.049(1.36)
Modulus, MSI (GPa)						
Longitudinal		5.7(39.3)	6.3(43.4)	8.8(60.6)	26.3(181.3)	12.2(84.1)
Transverse		0.7(4.8)		3.6(24.8)	1.5(10.3)	0.7(4.8)
Shear modulus, ksi (MPa)						
Major		0.30		0.23	0.28	0.32
Minor				0.09	0.01	0.02
Tensile strength, ksi (MPa)						
Longitudinal		160(1103)	180(1242)	187(1289)	218(1503)	172(1186)
Transverse		14(97)		6.7(46)	5.9(41)	1.6(11)
Compressive strength, ksi (MPa)						
Longitudinal		90(621)	110(758)	119(820)	247(1703)	42(290)
Transverse				25.3(174)	35.7(246)	9.4(65)
Flexural strength, ksi (MPa)						
Longitudinal		165(1138)	170(1172)			
In plane shear strength, ksi (MPa)		12.0(82.7)	12.0(82.7)	6.5(44.8)	9.8(67.6)	4.0(27.6)

[a]Ref. 5.

where S = hoop stress, psi (MPa); P = internal pressure, psi (MPa); D = average outside diameter, in. (mm); t = minimum wall thickness, in. (mm).

The hoop stress ratings are determined experimentally, either by long-term pressurization under static conditions, or by cyclic loading of pipe samples.

The maximum allowable stress for tanks (S), usually including a mat layer, is defined as:

$$S = E_t Z$$

where E_t = hoop tensile modulus of the total laminate; Z = allowable strain.

The hoop modulus of the total laminate is calculated by the rule of mixtures, after determining the moduli of the wound layer and mat layer by testing representative coupons:

$$E_t = \frac{E_f t_f}{t} + \frac{E_m t_m}{t}$$

TABLE 79.6.   Specifications for Filament Wound Pipe, Tanks and Pressure Vessels

Document No.[a]	Title
API 5LR-76	Specification for Reinforced Thermosetting Resin Line Pipe (RTRP)
ASME BPV-X	Boiler and Pressure Vessel Code. Section X. Fiber glass Reinforced Plastic Pressure Vessels, 1983
ASTM D2517-81	Reinforced Epoxy Resin Gas Pressure Pipe and Fittings
ASTM D2996-83	Filament-Wound Reinforced Thermosetting Resin Pipe
ASTM D3299-81	Filament-Wound Glass-Fiber-Reinforced Thermoset Resin Chemical-Resistant Tanks
ASTM D4163-82	Reinforced Thermosetting Resin Pressure Pipe

[a]APT = American Petroleum Institute; ASME = American Society of Mechanical Engineers; ASTM = American Society for Testing and Materials.

where E_f = hoop modulus of wound layer, psi (MPa); E_m = modulus of mat layer, psi (MPa); t = total thickness, in. (mm); t_f = thickness of wound layer, in. (mm); and t_m = thickness of mat layer, in. (mm).

The minimum wall thickness for the tank is then given as:

$$t = \frac{PD}{S} = \frac{0.036\ GHD}{2\ E_t Z}$$

where 0.036 = pressure, lb water/in²/in head; G = fluid specific gravity, H = fluid head, in.

The American Society of Mechanical Engineers (ASME) in its code for pressure vessels estimates the thickness of filament-wound cylindrical shells for evaluations based on a netting analysis (only the reinforcements are considered to be load bearing):

$$t = t_h + t_a$$

where $t_h = \dfrac{N_h + t'h}{V_a}$

$$N_h = \frac{PR\ (1 - \tan^2 a)}{t'_h S_h}$$

and

$$t_a = \frac{N_a + t'a}{V_a}$$

$$N_a = \frac{PR}{2\ t'_a S_a \cos^2 a}$$

The symbols are defined as:

a = winding angle
R = inside radius, in.
N_h = required number of hoop layers
N_a = required number of spiral layers
S_h = allowable design stress in hoop reinforcements, psi
S_a = allowable design stress in spiral reinforcements, psi
t = required thickness, in.
t_h = thickness of required hoop layers, in.
t_a = thickness of required spiral layers, in.
t'_h = equivalent thickness of hoop reinforcements per unit width of a two-ply layer, in.
t'_a = equivalent thickness of spiral reinforcements per unit width of a two-ply ($+a$) layer, in.
V_a = volume fraction of spiral reinforcements.

The ASME code specifies the minimum tensile strength of the reinforcements as 90,000 psi (620 MPa) in the hoop direction and 45,000 psi (310 MPa) in the axial direction, determined by hydrostatic burst pressure tests.

79.5.3 Corrosion Resistance

Pipes, tanks, and vessels for service in chemical environments a.e constructed with barriers to prevent corrosive actions and fluid weeping through the wall. ASTM tank standard D3299 defines this barrier as consisting of two layers. The inner surface is a resin-rich layer 0.010 to 0.020 in. (0.25 to 0.50 mm) thick reinforced with a chemical-resistant fiber glass mat or with an organic veil. The mat or veil normally does not exceed 20% by weight. This surface is followed by an interior layer of resin reinforced with chopped strand mat. The combined thickness of the two layers is specified as not less than 0.10 in. (2.5 mm) with a glass content of 27 + 5%. The fittings and tank accessories are similarly protected. For pipes or vessels exposed to corrosive media, thermoplastics or other materials at a minimum thickness of 0.005 in. (0.13 mm) frequently serve as the chemical barrier.

79.6 PROCESSING CRITERIA

Accurate control of the reinforcement tension, resin content, and band width is essential if the winding is to proceed within prescribed design limits. Tension is exerted by guide eyes leading the reinforcement from the package to the resin tank, by drum brakes on the package, by rotary scissor bars in the reinforcement line, and by drag or pressure in the resin tank. Optimum practice is to minimize tension on uncoated reinforcements to prevent abrasion and snarling and to attain maximum pressure at the mandrel after resin impregnation. Tension at this point is in the order of 0.25 to 1.0 lb (1 to 4 N) per end of fiber glass and somewhat higher for dry winding. Guide eyes alone provide sufficient tension for inside-pull roving packages. Braking may be necessary on each roving tube with an outside-pull. Rotary scissor bars facilitate tensioning as the machine traverse is reversed. The reinforcement delivery route should be as direct as possible with a minimum number of guides. In some cases, a multiaxis payout guide is needed for placement accuracy (Figs. 79.6 and 79.7).

The resin viscosity is critical in processing and in control of the resin content. Satisfactory impregnation with epoxy systems requires a viscosity from 300 to 1200 cps (0.3 to 1.0 Pa s). Most epoxies are brought into this range only by the addition of reactive diluents, heating, or both. Polyesters and vinyl esters have viscosities from 250 to 1000 cps (0.3 to 1.2 Pa s) and are easily controlled by the amount of monomer in the system. Regardless of type, sufficient fluidity should be maintained to ensure complete wet-out of the reinforcements and elimination of entrapped air and volatiles during winding and cure. Excessively low viscosities may cause resin migration from the inside layers to the outside as winding tension is applied. Dry inner layers are sources of failure in the cured product. At the other extreme, overly high viscosities lead to fiber fuzzing, increased fiber breakage, uneven resin distribution, air entrapment, and voids in the finished part.

Dry winding removes many of the problems encountered in controlling viscosity but may introduce others. Resin flow is of greater significance with prepregs, and sufficient flow and tack are needed for complete bonding of the layers. Heating the prepreg is indicated in some cases to improve the flow properties.

The uncoated reinforcements are brought together as they enter the resin tank and leave the tank close to the desired band width. Feed bars, located on the feed arm, form the final flat band to be placed on the mandrel. The designs of feed bars are varied and include rings, straight bars, curved bars, bent bars, and combs.[9] The bars are permitted to swivel freely as the traverse is reversed. For greater accuracy in band placement, machines are equipped with a controlled crossfeed and rotating bar.

The band density (ends per inch), the glass yield (yards per pound), the glass content, and the specific gravity of the resin are factors which determine the layer thickness. As an example, a two-ply spiral layer approximately 0.040 in. (1.0 mm) thick can be expected with a band density of 18 ends/in. (1 end/mm) of a 675 yield fiberglass roving at a 60% glass content and a resin specific gravity of 74.8 lb/ft^3 (1.2 g/cm^3).

Winding speeds are variable; the practical limit for commercial windings is about 300 or possibly 350 linear feet per minute (91 or 107 m/min). Lower speeds, 50 to 100 lfpm (15.2 to 30.5 m/min), are typical for more precise windings, particularly with graphite or aramid.

Figure 79.6.   Six-axis payout eye. Courtesy of McClean-Anderson Inc.

In general, winding speed is limited by the ability to maintain proper tension, conform to the winding pattern and ensure uniform resin content and impregnation. Nearly all in-plant windings are cured in conventional hot air ovens. The parts are slowly rotated throughout the cure to prevent sagging. Additional heat is sometimes supplied to large windings by radiant heaters placed inside the mandrel. Time/temperature relations during cure vary for each resin system. Roughly, general purpose epoxy is cured at 250 to 275°F (121 to 135°C), while heat-resistant systems are cured 350 to 375°F (177 to 191°C). Post-cures may be as high as 400°F (204°C). Other epoxies for service below 200°F (93°C) are cured at lower temperatures. Optimum practice for all systems is to gel at a reduced temperature followed by a stage curing in which stepwise increases in temperature, heating rates, and time at each temperature are closely controlled. This procedure precludes excessive resin flow, minimizes exothermic effects, and yields a higher degree of cure.

Polyesters and vinyl esters normally are cured in a 200 to 300°F (93 to 149°C) range. The cure temperatures are determined by the organic peroxide catalysts. Promoters or accelerators can be added to the resin to lower the cure temperature while retarders lower the peak exotherm. Inhibitors prevent premature gellation. It is usually more beneficial to formulate resin systems to reduce peak exotherm and to avoid early gellation, rather than resorting to rapid cure rates.

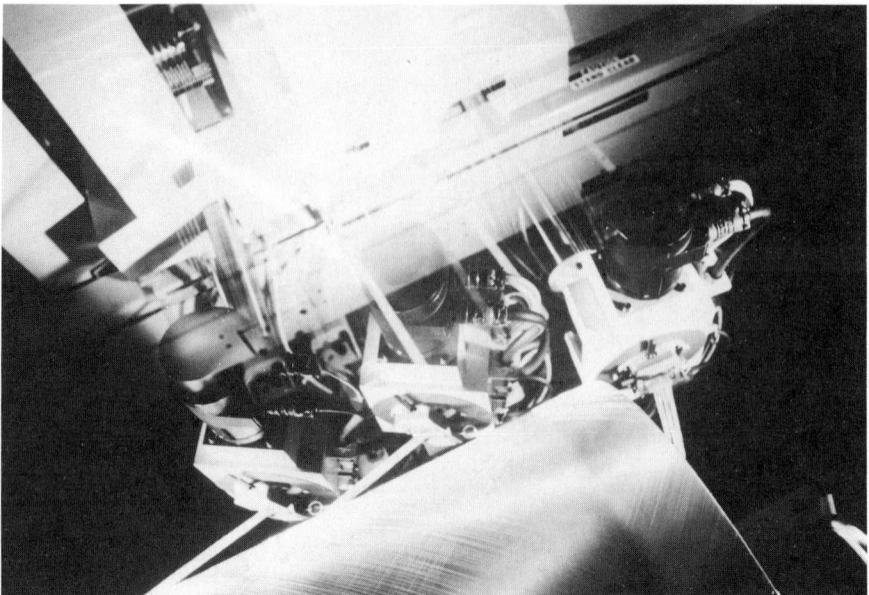

Figure 79.7.   Winding motion. Courtesy of McClean-Anderson Inc.

79.7   MACHINERY

The original concept for a filament winding machine, still in vogue today, is relatively simple and depends on mechanical control of basic motions. These early machines have undergone steady improvement leading to greater flexibility in operation and culminating in computerized control systems. The various machine designs are summarized below.

79.7.1   Basic Spiral Winder

The main components consist of a headstock and adjustable tailstock for mandrel support, a bed with a traversing feed carriage and the drive assembly. A single motor delivers power for mandrel rotation and carriage traverse. The drive assembly contains a series of gear trains and bevel gears that maintain the mandrel and carriage speeds at a fixed ratio. Each change in winding angle, therefore, requires a change in the gear train. The carriage is driven either by a gear chain or a drive screw in a direction parallel to the mandrel axis. The resin tank is usually mounted on the carriage. The carriage on larger machines might also carry a reinforcement creel. The machine is equipped to handle hoop windings and, in some cases, polar windings. Improvements include a controlled dwell, provisions for adding hoop doublers and a crossfeed driven by a separate motor and activated by limit switches.

79.7.2   Planetary Polar Winder

The mandrel is vertically mounted. A rotary arm, tilted at the wind angle, delivers the reinforcements at a constant speed about the mandrel polar shaft. The mandrel is indexed one band width per revolution of the rotary arm. An auxiliary feed system, parallel to the mandrel, provides hoop winds. Dry winding is normal since it is impractical to adapt wet winding to a rotating arm.

79.7.3 Racetrack Polar Winder

The mandrel is supported in a horizontal position. The feed carriage follows a "racetrack" course about the polar shaft of the mandrel, which is rotated at a slow speed or indexed one band width per circuit.

79.7.4 Tumbling Polar Winder

The mandrel is supported at one end only and is placed on a rotating platform. The wind angle is established by tilting the mandrel. The feed remains stationary while the platform is rotated about its central axis. The mandrel is either turned at a slow rate or indexed one band width for each rotation of the platform. The feed traverses for hoop winds as the mandrel is rotated.

79.7.5 Computer Control

Computer control is usually adapted to a basic winder, although it can be used in conjunction with other machine types. This control permits polar as well as spiral winding. Other features are a nonlinear traverse, variable wind angles, the ability to wind irregular shapes, and the elimination of gear changes for each wind angle. The mandrel rotation, carriage traverse, crossfeed, and delivery eye rotation have separate drives. The computer coordinates the dependent variables—carriage, crossfeed, and deliver eye motions—with the independent mandrel rotation.

Control systems are based either on a microprocessor or computerized numerical control.[1,10-12]

79.7.6 Oscillating Mandrel

This microcomputer-controlled machine is designed for the fabrication of tubular products, particularly hybrids. A forward and reverse traverse of the mandrel coupled with mandrel rotation comprise the only motions. The reinforcement delivery system is a stationary ring circumscribing the mandrel. Reinforcements can be placed over the full circumference of the mandrel at the same time so that one traverse yields complete coverage. A second delivery ring is furnished to facilitate hybrid winding.[13]

79.7.7 Ring-Type Winder

This machine utilizes a ring delivery structure that traverses and orbits a stationary mandrel. Changing patterns within a circuit as well as simple spiral patterns can be wound. The machine is equipped with a microcomputer control system that includes a program memory capacity. Roving dispensing mounts and a wet-out system are integral components to the unit.[14]

79.8 TOOLING

79.8.1 Mandrels

Mandrel design and construction are important aspects of the process. Poor designs can result in fiber damage during mandrel removal and deviations from dimensional tolerances. The effects of winding tension and cure shrinkage must be considered. Hoop winds under high tension induce compressive loads that can decrease the mandrel diameter. The mandrel must resist sagging from its own weight as well as buckling loads from external winding forces. As heat is applied during cure, additional loads are imposed. The thermal properties of the mandrel material and the composite shrinkage then become significant factors in the design.

For cylindrical structures with open ends, mandrel design is relatively simple. Either cored or solid steel or aluminum are satisfactory materials. A spider network combined with a central longitudinal shaft is used with large diameter mandrels. In some instances, contoured sections are added at each mandrel end to retain the reinforcement as feed is reversed.

Major problems in the design and construction occur when pressure vessels are wound with integral heads. Concepts that have been tried in coping with this problem include:[15]

- Segmented collapsible metal—A suggested minimum part diameter is 3 ft (0.91 m). Removal may be complicated with small polar openings. Machining costs are high, and a sufficient number of parts is necessary to warrant this construction.
- Low melting alloys—These alloys are high in density and tend to creep under tension. They are limited to small vessels, 1 ft diameter by 1 ft length (0.3 by 0.3 m).
- Eutectic salts—These are a better choice than the alloys and can be used with vessels up to 6 ft (0.6 m) in diameter.
- Soluble plasters—These are easy to contour and are readily washed out.
- Breakout plasters—These are suited for large diameters. Internal support is required; breakout is difficult and can cause damage. Chains are frequently embedded in the plaster to facilitate breakout.
- Sand/PVA—This material dissolves easily in hot water. It requires close control when it is molded into a mandrel and is limited by a low compressive strength.
- Inflatables—These can be used if little or no torque is applied during the winding. They can be combined with an internal supporting structure to increase their useful range.
- Wound spool—This mandrel is constructed by winding a braided cord or filament bundle over a metal core that acts as the polar shaft. The winding is covered with a barrier material to prevent resin penetration. After the wound part is cured, the polar piece is removed and cord or bundle is unwound by an inside-pull. Complex shapes and variable diameters, equivalent to undercuts, are possible.[16]

79.9 AUXILIARY EQUIPMENT

79.9.1 Creels

Creels for holding the reinforcement packages are constructed either for an inside- or an outside-pull. They are centrally located in reference to the winding machine, or mounted on the traversing carriage of spiral winders or on the rotating arm of polar winders.

79.9.2 Tensioning Devices

Adjustable brakes are used to control the tension on rolls with outside-pulls. Spring-loaded compensators may be placed in the feed line to reduce fluctuations in tension.

79.9.3 Impregnation Tanks

There are two basic methods for coating the reinforcements with resin. The strands are passed under a roller located in the tank, or the strands are passed over a metering roller. When necessary, the tank is electrically heated to reduce resin viscosity.

79.9.4 Metering Equipment

Optimum results are obtained when the resin and catalyst are continuously mixed and metered into the tank. Standard equipment is available for this purpose.

79.9.5 Mandrel Extractors

Mechanical or hydraulic extractors are employed to remove the winding from the mandrel. Mechanical types are satisfactory for components up to 12 in. (0.30 m) in diameter. Hydraulic jacks are required for greater diameters.

79.9.6 Curing Ovens

Conventional hot-air circulating ovens are typical for curing. Custom-built ovens are necessary for extremely large windings.

79.10 FINISHING

79.10.1 Installation of Hardware

Machining and bonding procedures for filament-wound structures closely follow the recommended procedures for angle ply and other laminates. Details in regard to tool design, selection of adhesives, strengths of bonded joints, and so on are found in other sources.[17,18] Typical examples of these operations as performed on pipes, tanks, pressure vessels, and containers are described below.

End Trimming. Excess material at the mandrel ends is removed by steel, carbide, or diamond tipped cut-off wheels. Thick-walled or builtup end sections may require lathe cutting, drilling, or grinding as preparation for the placement of attachments or joints.

Above-ground Vertical Storage Tanks. Top and bottom heads are adhesive bonded to the tank wall. Walls are sometimes constructed in two sections that are joined together by adhesive bonded couplings. Surfaces to be bonded are prepared by sanding or grinding. Attachments such as nozzles, vents, manways, and hold down or lifting lugs are bonded to the walls or heads. Compensating reinforcements and corrosion barriers are added whenever cut-outs are made in the walls or heads (see ASTM D3299, in Table 79.6).

Underground Horizontal Storage Tanks. Tanks are usually made in two sections, each with an integral head. The sections are put together by butt or bell and spigot joints that are reinforced by bonded overlayers. Penetrations and lifting lugs are treated in the same manner as vertical tanks.

Pressure Vessels. Polar and closures are usually bolted to metal inserts previously wound into the integral end dome.

Cylindrical Containers. Sections are joined or closed by flat-faced metal flanges. The flanges are either bonded or bolted to the container wall by single or double lap joints.[19]

Joining Pipe Sections. Pipes are joined by a variety of methods including flanges, bell and spigot joints, and couplings in which the fittings are adhesive bonded to the pipe sections. Surfaces to be bonded are prepared by sanding or grinding. Normally, the fittings are installed in the field, although some, such as bell and spigot joints, are bonded to the pipe sections by the manufacturer.

BIBLIOGRAPHY

1. C.D. Hermansen and R.R. Roser, "Filament Winding Machine: Which Type Is Best for Your Application?" *36th Annual Conference, Reinforced Plastics/Composites Institute,* Society of the Plastics Industry, Feb. 16–20, 1981, Sect. 5-A.

 2. G. Lubin, ed. *Handbook of Composites*, Van Nostrand Reinhold Co., New York, 1982, pp. 458–459.

 3. *Kevlar Aramid*, Bulletin K-2, E.I. du Pont de Nemours & Co., Inc. Feb. 1978.

 4. C.C. Chamis and J.H. Sinclair, "Mechanics of Interply Hybrid Composites, Properties, Analysis, and Design," *34th Annual Conference, Reinforced Plastics/Composites Institute*, Society of the Plastics Industry, 1979, Sect. 20-E.

 5. C.C. Chamis, "Prediction of Fiber Mechanical Behavior Made Simple," *35th Annual Conference, Reinforced Plastics/Composites Institute*, Society of the Plastics Industry, 1980, Sect. 12-A.

 6. C.C. Chamis, *Computer Code for the Analysis of Multilayered Fiber Composites—User Manual*, NASA TN D-7013, 1971.

 7. B.D. Agarwal and L.J. Broutman, *Analysis and Performance of Fiber Composites*, John Wiley & Sons, Inc., New York, 1980.

 8. S.W. Tsai and H.T. Hahn, *Introduction to Composite Materials*, Cambridge University Press, New York, 1981.

 9. *Publication No. 5-CR-6516*, Owens-Corning Fiberglas, 1974.

10. T.M. Harper and J.S. Roberts, "Advanced Filament Winding Machines for Large Structures," *39th Annual Conference, Reinforced Plastics/Composites Institute*, Society of the Plastics Industry, 1984, Sect. 9-E.

11. J.F. Kober, "Complex Filament-Wound Composites Utilizing a Microcomputer Controlled Filament Winder," *35th Annual Conference, Reinforced Plastics/Composites Institute*, Society of the Plastics Industry, 1980, Sect. 19-A.

12. J.F. Kober, "Microcomputer-Controlled Filament Winding," *34th Annual Conference, Reinforced Plastics/Composites Institute*, Society of the Plastics Industry, 1979, Sect. 1-F.

13. J. Sabo, "The Filament Winding of Composite Driveshafts," *25th National SAMPE Symposium and Exhibition*, May 6–8, 1980.

14. D.E. Beck, "A Ring-Type Filament Winding Machine With a Dedicated Microcomputer Control System," *36th Annual Conference, Reinforced Plastics/Composites Institute*, Society of the Plastics Industry, 1980, Sect. 5-B.

15. Ref. 2, p. 481.

16. M.A. Yates and S.B. Driscoll, "A Closed Structure Mandrel System for the Filament Winding Process, *37th Annual Conference, Reinforced Plastics/Composites Institute*, Society of the Plastics Industry, 1982, Sect. 4-C.

17. Ref. 2, pp 602–632.

18. U.S. Material Development and Readiness Command, Dar Com P 706-316. *Engineering Design Handbook, Joining of Advanced Composites*, Alexandria, Va., March 1979.

19. D.A. MacNab and S.T. Peters, "Graphite/Epoxy Launch Tube for MX," *SAMPE J.* (Nov./Dec. 1983).

80

REINFORCED PLASTICS, THERMOSET MATCHED-DIE MOLDING

Raymond W. Meyer

R.W. Meyer & Associates, 233 Timothy Dr., Tallmadge, OH 44278

80.1 DESCRIPTION AND HISTORY OF PROCESS

Currently, nearly half of all reinforced plastics products are produced by the thermoset matched-die molding process. This process is used when production requirements are large and/or when precision and reproducibility are key factors. Often, it offers the best quality at the lowest price; the effect of operator skill on part quality is slight.

Thermoset matched die molding can be broadly defined as a molding process in which the loading and closing of the mold cause the molding material to conform to the mold configuration, and in which the cure takes place while the material is confined in the mold.

The mold, or mold set, is usually in two main parts (called male and female, or core and cavity), with one part fitting inside the other with a controlled space between them. Complex parts can require intricate molds having several major elements.

80.1.1 Reinforced Molding Compounds

Reinforced molding compounds include bulk molding compounds [in Europe, called dough molding compounds (DMC)], previously known as "premix," and sheet molding compounds.

Bulk Molding Compounds (BMC). Bulk molding compounds are completely formulated, preimpregnated fiber glass-reinforced plastics molding materials that are either prepared by the molder or are purchased from a compounder. BMC materials can be chemically thickened; if they are, it is important to control the storage environment to minimize the maturation period or reaction time. When thermoplastic additives are included to control surface finish or appearance the material is referred to as low shrink (LS) or low profile (LP) BMC.

Sheet Molding Compounds (SMC). Sheet molding compounds are completely formulated, preimpregnated chemically thickened fiber glass-reinforced plastics molding materials that are supplied to the press in a dry or nontacky sheet form normally sandwiched between two carrier films. The carrier films are removed and discarded just before molding.

80.1.2 History

Present-day molding compounds had a rather ignominious start. This start is generally attributed to a technician at a molder's plant who was assigned the task of finding a use for the mountains of trim scrap left over on compression-molded parts reinforced with tailored mat. This technician cut the mat trimmings into small squares and hand mixed them into a catalyzed polyester for use as a localized reinforcement. The term "gunk" was easily associated with such a mixture; its chief advantage was the ability to mold thick and thin sections in a single part.

It was soon discovered that when the mixture was dried slightly by the addition of an inert filler, the molding material was much easier to handle. Electrical insulators were a natural product for such material, known as "premix," which was a mixture of resin, reinforcement, fillers, and so on, that was not in web form and was usually prepared by the molder just before use.

The advantages of premix compounds were

- The ability to mold thick and thin sections in a part.
- The ability to mold in metal inserts.
- Excellent electrical properties.
- Lower raw materials costs.

Disadvantages of premix compounds were

- Poor surface finish.
- Internal cracks and voids in thick areas.
- A tendency for the resin to flow away from the glass reinforcement on long flow paths.
- Relatively low physical and mechanical properties.
- Difficulty in preparing mold charges.

The addition of various thermoplastics offset the polymeric shrinkage that was responsible for the poor surfaces and also eliminated the cracking in thick sections. These improved molding compounds are the bulk molding compounds, which generally contain from 15 to 20% glass content by weight in lengths of 0.25–0.50 in. (6–13 mm).

The addition of Group II metal oxides and hydroxides or chemical thickeners to increase the compound viscosity helped overcome the problem of the resin flowing away from the glass reinforcement. A further development in this chain of materials evolution was the increase in glass content to 30 wt % and the glass fiber lengths of 1–2 in. (25–51 mm) to improve greatly the general property level and to simplify the manufacturing and handling of the compound by providing it in convenient sheet form. These latter materials are the sheet molding compounds.

Each of the changes described above resulted in a higher viscosity molding compound that produced more back pressure in the mold during the molding process.

80.1.3 Properties

BMC and SMC offer an extraordinary range of desirable properties. The variability of resin, filler, and reinforcement types permits a compound to be tailored to a specific design or performance requirement. In general, BMC and SMC compounds are characterized by:

- High strength, particularly impact strength.
- Excellent electrical performance, especially resistant to arc tracking.
- Heat and flame resistance.
- Dimensional stability.
- Chemical resistance.

- Relatively rapid cure.
- Low mold shrinkage.
- Low molding pressure.
- Variable thickness moldings.
- Low cost.

Physical Properties. The strength and stiffness that distinguish reinforced molding compounds result from the length, content, and quantity of the reinforcing fiber. Other ingredients in the formulation have the primary function of keeping the reinforcing fiber in place and protecting it during processing.

In BMC, glass fiber longer than 1/4 in. (6 mm) produces little, if any, practical improvement in physical properties. Fiber longer than 1/2 in. (12 mm) creates mixing and molding problems.

The strength of BMC is much more responsive to changes in fiber content than to any other factor. While the mechanical properties generally increase with greater fiber content, a point of diminishing returns is reached at about 35%, when mixing, moldability, and surface finish are taken into consideration (Table 80.1).

In BMC and SMC alike, the quantity of reinforcing fiber is critical to the compound's strength. Figure 80.1 shows the effect of glass concentration versus physical properties for a commercially available SMC. Figure 80.2 shows the effect of several glass lengths and glass concentrations on the impact strength of an SMC.

Because of the different manner in which glass fibers are incorporated into the compounds, SMC can utilize wider range of fiber lengths and content than BMC. Despite this, most SMC compounds are made with 1-in. (25-mm) fibers at a concentration of 25–30%. With shorter fibers the properties are essentially the same as those of BMC at a slightly higher compounding cost. Longer fibers [2–3 in. (51–76 mm)] that can be handled on a conventional SMC machine provide a small increase in strength with a disproportionate reduction in moldability (especially in filling bosses and ribs) and a poorer surface finish.

Very long or continuous fibers arranged in a deliberate pattern, parallel to each other, or at a controlled angle provide very high strength in the fiber direction and much lower strength transverse it. A very high strength SMC can be obtained if the mold charge is arranged with the fibers in the direction of the expected loads on the molded part. To compensate for the transverse strength deficiencies, a parallel-fiber SMC can be plied with conventional random-fiber types (Table 80.2). Poor surface finish and moldability are attendant disadvantages, but in areas where surface finish is not a criteria, such SMC parts can replace steel and aluminum parts.

Electrical Properties. An important electrical property is the ability to resist carbon arcing. This, in combination with strength, heat resistance, dimensional stability, moldability, and

TABLE 80.1. Mechanical Properties vs Glass Content[a]

Glass Fiber Content, % 1/4-in. Fiber	Notched Izod Impact, ft-lb/in. (MPa)	Flexural Strength, psi (MPa)	Flexural Modulus psi × 10⁶ (GPa)	Tensile Strength, psi (MPa)
10	3.66 (195)	9,800 (68)	1.53 (10.6)	3,559 (24)
20	6.13 (327)	16,000 (110)	1.64 (11.3)	6,340 (44)
30	7.37 (393)	19,600 (135)	1.66 (11.5)	7,920 (55)
40	10.65 (568)	22,300 (154)	1.72 (11.9)	9,530 (66)
50	11.03 (588)	23,200 (160)	1.59 (11.0)	6,890 (48)

[a]Ref. 1.

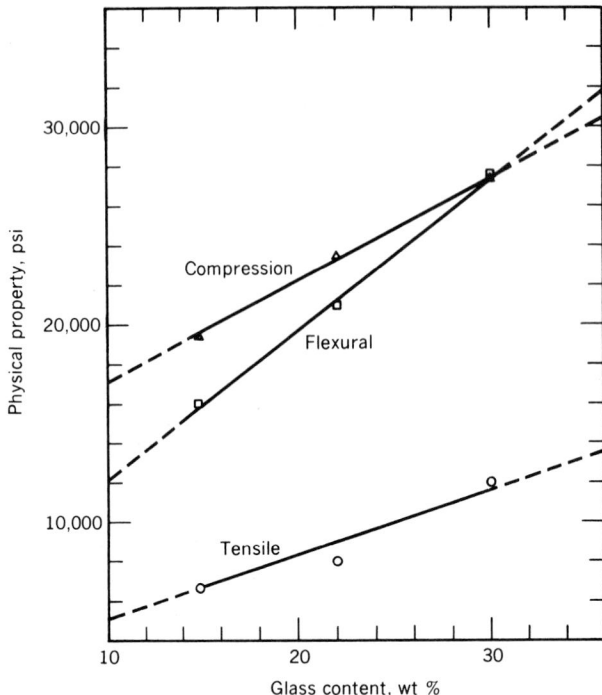

Figure 80.1.   Physical properties vs glass content for Premi-Glas® 7200 SMC.

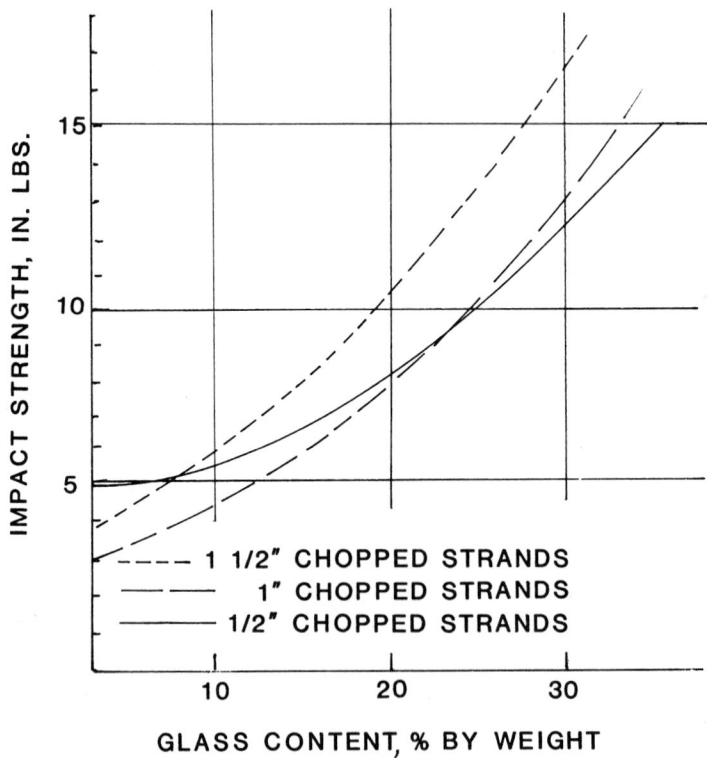

Figure 80.2.   Effect of glass length and concentration on the impact strength of SMC.

TABLE 80.2. Selected Mechanical Properties of Various SMC Constructions[a]

Designation	Fiber Arrangement	Fiber Content, %	Flexural Strength, psi $\times 10^3$ (MPa)	Flexural Modulus, psi $\times 10^6$ (GPa)	Tensile Strength, psi $\times 10^3$ (MPa)	Notched Izod Impact, ft-lb/in. (J · m)
TMC-25[b]	Random, $\frac{1}{2}$	25	17 (117)	1.0 (6.9)	7 (48)	10 (533)
TMC-28[b]	Random	28	27 (186)	1.4 (9.7)	14 (97)	10 (533)
SMC-R15[c]	Random, 1 in.	15	15 (103)	1.4 (9.7)	8 (55)	8 (426)
SMC-R30[c]	Random, 1 in.	30	23 (159)	1.9 (13.1)	11 (76)	
SMC-R40[c]	Random, 1 in.	40	30 (207)	2.0 (13.8)	17 (117)	
SMC-R50[c]	Random, 1 in.	50	37 (255)	2.2 (15.2)	23 (159)	
SMC-R65[c]	Random, 1 in.	65	48 (331)	2.3 (15.9)	30 (207)	
HMC-65[d]	Random	65	59 (407)	2.2 (15.2)	30.5 (210)	20.5 (1092)
SMC-C3OR20[d]	Continuous/Random	30/20	85 (585)	3.3 (22.8)	55 (379)	
XMC-3[d]	Continuous/Random	50/25	125 (862)	5.5 (38)	75 (517)	
SMC-C60	Continuous	60	130 (897)	5.4 (37.3)	81 (559)	
XMC-2[d]	Continuous/Random	75	155 (1069)	5.5 (38)	90 (621)	

[a]Ref. 2.
[b]USS Chemicals.
[c]Owens-Corning Fiberglas Corp.
[d]PPG Industries.

reasonable cost, has resulted in widespread use of BMC and SMC materials in high-voltage applications. Their arc resistance qualities result largely from the inclusion of such inert inorganic filler materials as hydrated alumina, silica, and china clay and from small quantities of fine, powdered polyethylene (about 5 wt %) and nylon fiber. For maximum arc resistance, it is necessary to minimize both the resin and the glass-fiber content, which reduces physical properties.

Table 80.3 shows some relationships among resin, glass fiber, and filler contents with respect to arc resistance. If the arc resistance is satisfactory, it is likely that other electrical properties, such as dielectric strength, dielectric constant, dissipation factor, volume resistivity, and arc quenching will also be satisfactory for most purposes.

Heat Resistance. Heat resistance, short-term hot strength, and flammability tend to be considered together, but they are not necessarily common properties of a given molding compound.

Heat resistance is the long-term resistance to thermal degradation below the flammability temperature. Short-term hot strength (or simply hot strength) has to do with the thermoplasticity of the resin. Heat resistance and hot strength are properties derived largely from the resin, although there is evidence that certain fillers do enhance heat resistance.

Flammability is the measure of burning. In spite of the fact that all organic materials will burn if heat above the ignition temperature continues to be applied, compounds are variously classified as nonburning, self-extinguishing, and fire-retardant, depending on their ease of

TABLE 80.3. Arc Resistance as Affected by Formulation[a]

Glass Fibers, %	Filler, %	Resin, %	Filler	Arc Resistance s[b]
15	64	21	Hydrated alumina	202
9	73	18	Hydrated alumina	240
7.5	75	17.5	Hydrated alumina	264
0	85	15	Hydrated alumina	300

[a]Ref. 3.
[b]ASTM D495.

ignition, their ability to extinguish when heat is removed, and their rate of burning. Because molding compounds are used in building and construction applications, smoke generation and flame spread are other burning characteristics to be considered.

Flame resistance in polyester resins is achieved with halogenated resins (plus antimony trioxide for a synergistic effect), by adding halogen and phosphorus-containing compounds to conventional resins, and by using hydrated alumina as the principal or sole filler in a molding compound. Table 80.4 provides some data on flame resistance of various polyester BMC formulations.

Halogen and phosphorous compounds, as well as the use of chlorendic anhydride (50% or more) as a hardener, will provide a measure of flame resistance to epoxies.

Dimensional Stability. The dimensional stability of BMC and SMC formulated with polyester resin, glass-fiber reinforcement, and china clay, silica, or alumina fillers is unexcelled. Postmold shrinkage is minimal. Water absorption is very low, resulting in only slight changes in weight and dimensional changes that are difficult to detect. The coefficient of thermal expansion is close to or lower than that of aluminum. Continuous exposure to elevated temperature can produce only minimal dimensional changes.

Chemical Resistance. With proper selection of resin and inert filler, the chemical resistance of molding compounds can be excellent. The degree of chemical resistance is determined by the properties of the resin.

Acid- and alkali-resistant polyesters, vinyl esters, furans, and acid-resistant epoxies can be used in BMC. There are limitations on resin selection for SMC, since the resin must be capable of responding to the thickener.

Of the common fillers, china clay and silica offer excellent chemical resistance. Calcium carbonate has good alkali resistance but reacts poorly in acid environments and should not be used in such.

Shrinkage. Glass-reinforced molding compounds shrink very little. A maximum of 0.004 in./in. (0.10 mm/mm) is typical. The glass fibers and inorganic fillers that have very little thermal shrinkage contribute most to low overall mold shrinkage. Figure 80.3 shows the shrinkage trends of various formulations.

TABLE 80.4.   Flame Resistance of Alumina Hydrate vs Halogen–Antimony Trioxide Systems in BMC Formulations

	Mix 1[a]	Mix 2[a]	Mix 3[a]	Mix 4[a]	Mix 5[b]
Formulations					
Plaskon 9520 resin	30	30	None	None	30
Halogenated resin (28% Cl)	None	None	34	34	None
Benzoyl peroxide paste	0.6	0.6	0.6	0.6	0.6
Alumina hydrate (Hydral 710)	53	38	None	15	None
ASP-400 clay	None	15	44	29	53
Antimony trioxide	None	None	5	5	None
¼-in. Fiber glass chopped strands	15	15	15	15	15
Zinc stearate	1.4	1.4	1.4	1.4	1.4
Flame resistance					
By ASTM D635	NB	NB	NB	NB	SE
By Federal method 2023					
Ignition time, s	154	135	95	150	93
Burning time, s	31	60	56	27	234

[a]NB = Nonburning.
[b]SE = Self extinguishing.

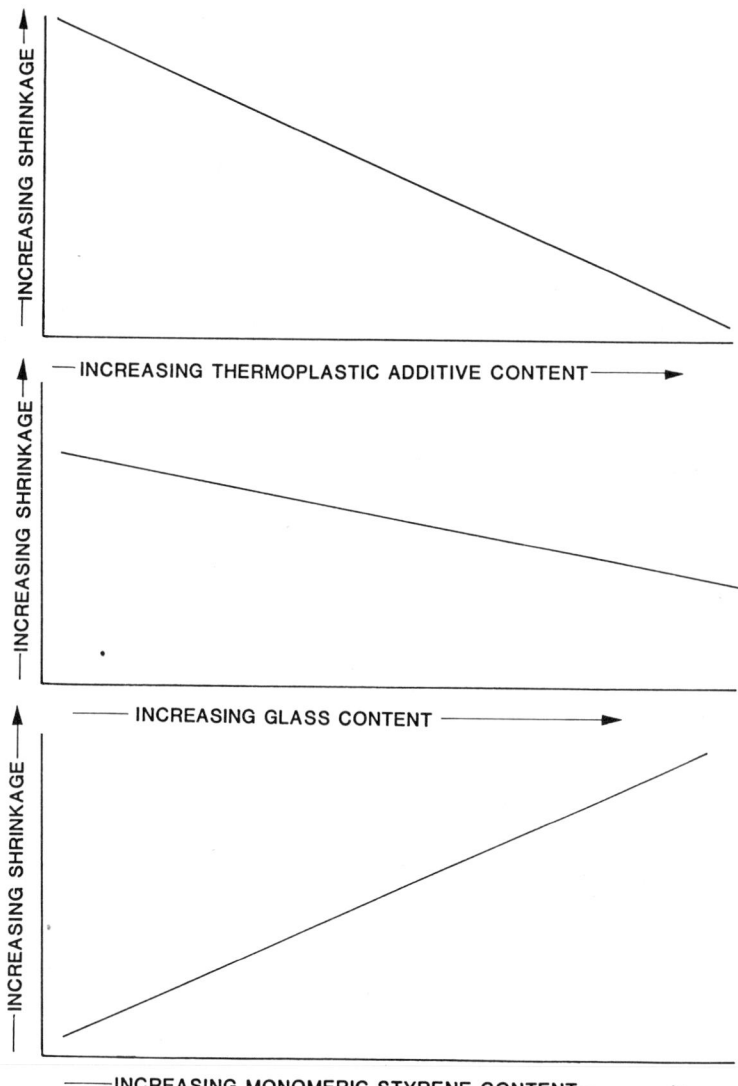

Figure 80.3. General trends of shrinkage with varying formulations.

However, the combination of low-shrink, high-strength fibers with the high cure and thermal shrinkage of conventional polyester resins can create considerable stresses in the resin matrix. The secondary effects of these stresses are surface waviness, warping, surface cracks, and internal voids. The chemical thickening process used in SMC and some BMC materials reduces both the cure shrinkage and orientation of the fibers as the charge flows to fill the mold. To obtain a very good surface and almost complete freedom from distortion, it is also necessary to incorporate a thermoplastic additive in the formulation. These include polyethylene, polystyrene, poly(vinyl acetate), methyl methacrylate, polycaprolactone, and others.

Acrylic monomer limits shrinkage because it does not cross-link with the polyester but homopolymerizes during the exothermic reaction of the polyester resin, leaving foamlike occlusions that apparently resist the polyester polymerization shrinkage. The most obvious advantage of low-shrink resin systems is surface smoothness. An acrylic system provided the first Class A surface for automotive moldings but these have since been matched, and in many cases surpassed, by other thermoplastic systems.

While mechanical properties suffer when these thermoplastic additives are included, the overall mix of mechanical, chemical, and electrical properties is adequate for many applications. Products molded with these thermoplastic additive formulations can be painted with a minimum of surface preparation and/or pigmented for satisfactory molded-in color.

Part Thickness. If proper care is taken in designing the transition from one thickness to another, wide variations in thickness are possible within the same molded part without major molding problems. With conventional molding compounds, sinks or depressions (often with small cracks) will appear opposite stiffening ribs in thin, flat areas unless precautions are taken. The maximum rib thickness should be no more than that of the surface it is stiffening. Radii at the junction are desirable from a structural viewpoint but contribute to the sink problem; two small ribs are better than one thick one. Unsupported edges are better stiffened by turning than by increasing the wall thickness, except where a thicker edge is necessary to resist handling damage.

Other faults that occur with thick and thin sections have much to do with fiber orientation. The fibers tend to align themselves with the direction of flow in thin sections and crosswise in thin to thick sections. This latter condition is the result of the flowing front cascading in the mold (explained in more detail below). In addition to appearance defects, this results in wide and unexpected variations in strength. Chemical thickening and the use of flow control fillers help keep the fibers in their original, heterogeneous relationship.

Molding Pressure. Low molding pressure requirements are a major advantage of BMC and SMC. Indeed, the highest pressure required for BMC and SMC is at the low end of the pressure scale required for typical phenolic, melamine, and urea compounds. No pressure is needed to keep reaction products such as water or residual solvents from gassing and forming blisters or voids. Unthickened BMC can be formulated to mold simple shapes at 200 psi (1.4 MPa). Chemically thickened compounds require higher molding pressures in the range of 500–1500 psi (3.4–10 MPa).

Generally, the higher the viscosity of the thickened compounds, the better the quality of the molding but also the higher the molding pressure required. SMC must be nontacky so that it can be handled at the press. Some compounds can be formulated to be nontacky at 2–12 million centipoise, but most are nontacky at 25–30 million centipoise and up.

Cure Rate. Fast to rapid cures at temperatures below that required for other thermosets are possible with polyester-based compounds. The exothermic nature of polyester curing results in an imbalance of cure rate to thickness; thick sections cure in shorter times than might be expected. Many parts that are nominally thin-walled but have thick ribs or bosses can be removed from the mold at an early stage, and cure will continue to completion in the interior of the thick sections. Standard polyester BMC and SMC require a 45–60 second cure at 280–300°F (138–149°C) for 0.1 in. (2.5 mm) thickness.

80.1.4 Applications

Reinforced molding compounds have been applied successfully in place of concrete, wood, ceramics, sheet metal, die-cast zinc and aluminum, and metal castings; in nearly every such material replacement, they have offered lower cost as well as improved product performance. In addition, with the development of low-shrink and low-profile formulations, reinforced molding compounds are no longer limited to nonappearance industrial products.

Automotive. The first large-volume market for reinforced molding compounds was in sisal-reinforced automotive heater housings and air-conditioning ducts. Nearly every car model on the market had such a product under its hood when the cost of sisal was only a small fraction of the cost of chopped glass fibers. Now that the cost of sisal is approaching that of glass fibers, this application is made in low-glass content BMC, injection-molded thermoplastics reinforced with glass fibers and/or wollastonite, and, to a very limited extent, the original sisal-reinforced BMC.

Today, the largest volume of automotive applications involve SMC and include exterior panels on the Corvette and Fiero, a variety of truck and van parts, spoilers on sports cars, wind deflectors on truck cabs, and grille opening panels and fender extensions on passenger automobiles. Molded-in provisions for mounting headlamps, grille trim, and other accessory components result in a one-piece molding that replaces 15 or more metal stampings and/or die castings. These parts, painted to match the exterior of the car, require a material as smooth as sheet metal that will withstand the operating temperatures of the paint ovens. Low-profile SMC formulations offer surfaces nearly equal to sheet metal, better heat resistance than cost competitive thermoplastics, and lower cost than stamped metal assemblies, zinc die castings or heat-resistant thermoplastics.

Electrical and Electronic. Electrical and electronic molded parts are applications in which reinforced molding compounds have made significant contributions. Electrical switchgear housings, formerly fabricated from flat sheet laminates in a metal framework, are now molded of BMC and SMC, with the accompanying advantages of lower cost, better electrical performance, and considerably decreased dimensions. The elimination of the metal structure and the excellent arc resistance of the reinforced moldings has reduced the clearances needed to prevent arcing under high voltage conditions. Even though one of the earliest and continuing uses of SMC (in Germany) is for low voltage electrical distribution station housings, unpainted outdoor applications have been few, since weathering characteristics of reinforced molding compounds are poor. However, polyurethane coatings have proven to be very effective and have resulted in the use of SMC for such items as medium voltage insulator brackets for above-ground power distribution applications.

Other electrical uses include housings for power tools (electrical drills, sanders, and so on). Precision molding with low-shrink resin systems has changed the concept of tool construction. The stability and strength of reinforced molding compounds permit their use as both the structural housing and the electrical insulation.

One of the more recent electrical applications is in electrical outlet boxes for construction (Fig. 80.4). These parts are made both by compression and by injection molding of thermosetting molding compounds.

Appliances and Business Equipment. Reinforced molding compounds have found many applications in appliances; eg, in air conditioners, where the key properties are corrosion resistance, electrical insulation, mechanical strength, and the ability to mold complex shapes. Air conditioner housings can have molded-in blower scrolls, air ducts, mounting brackets for controls, blower motor, switches, and so on. A substantial measure of thermal and sound insulation is achieved, and no painting is required. Other appliance uses are in food disposers, refrigeration, humidifiers, dishwashers, and laundry equipment.

Reinforced molding compounds have found use in business equipment as housings for computers and terminals, cash registers, typewriter cases, and adding machine housings. One of the first successful such applications was airline computer terminals and displays, which made use of textured surfaces, molded-in color, electrical insulation, lower costs, and lighter weight compared to the previously used metal housings.

Miscellaneous. Shower-stall bases molded of polyester BMC using synthetic fabric reinforcement have been a successful application for many years. Such a molding can weight

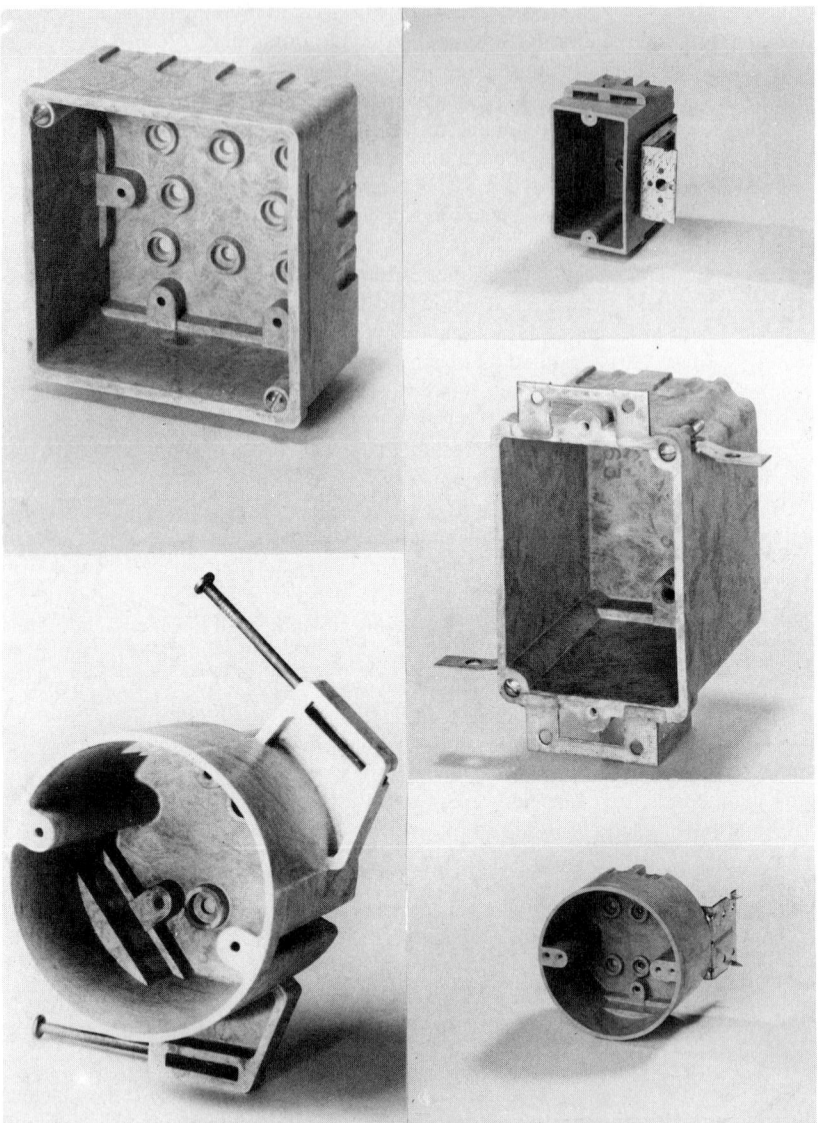

Figure 80.4.   Various premix electrical outlet boxes. Courtesy of Allied Moulded Products.

60 lb (27 kg) and yet be lightweight compared to the cast concrete terrazo it replaces. It offers water and stain resistance, much easier installation, and lower shipping costs. It is also nonporous and easy to clean, and available in many colors at a very competitive price.

One model of sewing machine molded of reinforced molding compounds weighs considerably less than comparable metal models. Molded-in bosses and ribs made the design feasible; the dimensional stability of the moldings made the project successful.

80.2 MATERIALS

A typical molding compound contains resin, low profile additive, inert fillers, internal release agent, colorant, peroxide catalyst, thickener, and glass reinforcement.

80.2.1 Resin

Ideally, the resin should have a viscosity low enough to permit easy mixing but high enough not to liquefy and separate from the other ingredients as the compound flows in the mold. It should cure relatively quickly, possess enough hot strength to permit easy removal of the molded part from the hot mold without damage, and be sufficiently resilient to permit some deformation of the part without cracking.

Most polyester resins for fiber-glass-reinforced-plastics (FRP) compounds are in the 25-poise viscosity range, although resins from 10–2500 poise are being used. Resins having up to 600 poise viscosity can be mixed in conventional equipment without solvents that must later be evaporated. Polyester resins for molding compounds are often classified by:

• Their basic polymer ingredients (for example, orthophthalic, isophthalic, bisphenol, and so on).
• Their cross-linking monomer (styrene, vinyl toluene, DAP, and so on).

Generally, the lowest-cost resin is an orthophthalic-based polymer dissolved in styrene. Isophthalic, bisphenol, and HET anhydrides offer better physical properties, corrosion resistance, and reduced flammability, respectively. Vinyl toluene is less volatile than styrene, and compounds made with it do not suffer from monomer evaporation on exposure to air. DAP is even less volatile and provides better electrical properties.

Epoxy resins offer substantial advantages in strength and chemical resistance, yet their much higher cost, limitations on compound selections, and longer cure rate have restricted their applications in reinforced molding compounds. However, vinyl esters, a close relative built on an epoxy backbone but cross-linked with styrene and peroxide-cured, offers toughness, chemical resistance, and flexibility in compounding, that, in spite of its higher price, makes it a viable product for high-performance applications.

In addition, there are resins especially formulated for chemical thickening purposes and low-shrink devices. Some typical properties resin/monomer systems are shown in Table 80.5.

The resin content in reinforced molding compounds may range from approximately 18–50% by weight, with 30% being a good starting point for most compounds. Where very low absorption fillers, such as calcium carbonate, can be used, a resin content at the low level will produce a moldable compound. Where the special properties of a high absorption filler, such as china clay, are required, the resin requirement will be higher.

80.2.2 Reinforcements

Reinforcements for BMC include chopped glass fibers and various organic fibers. Formerly, asbestos floats and sisal were popular reinforcements for molding compounds, but asbestos is considered a carcinogenic agent and sisal has been dropped in most compounds because of its large price increases in recent years.

Chopped glass fibers are available in lengths of 1/8, 1/4, and 1/2 in. (3, 6, and 12 mm) and in soft and hard varieties. Soft glass fibers (styrene monomer soluble) are less noticeable on the surfaces of molded parts, while hard glass fibers (insoluble styrene binders) usually provide more mechanical properties and better fiber integrity on long flow paths. Glass fibers are used in concentrations of 5–50% by weight in BMC. Less than 5% affords no detectable structural advantages, and more than 35% creates molding problems. Up to 20% provides compounds that can be extruded or compacted for easy handling. Higher glass concentrations makes fluffy, springy compounds that do not hang together well. Compounds high in resin content are very sticky. Sisal fibers can be used up to about 20% by weight in a compound.

A very low-cost molding with excellent water and stain resistance plus good electrical properties can be made with diced nylon tricot fabric. Loadings of up to 15 wt % are used

TABLE 80.5. Molding Compound Resins: Properties of Unfilled Castings[a]

Polymer Monomer	Orthophthalic Styrene	Iso-Styrene	Iso-Vinyl Toluene	Ortho-Styrene	Ortho-DAP	Vinyl Ester Styrene	HET Acid Styrene
Viscosity, P	26	27	26	24	400	5	5
Specific gravity	1.22	1.2	1.09	1.17	1.25	1.04	1.04
Heat distortion, °F (°C)	162 (72)	176 (80)	214 (101)	165 (74)	392 (200)	190 (88)	212 (100)
Flexural strength, psi (MPa)	17,500 (121)	17,600 (121.4)	18,500 (127.6)	23,000 (158.7)	12,900 (89)	20,000 (138)	16,000 (110.4)
Flexural modulus, psi (GPa)	0.56×10^6 (3.9)	0.42×10^6 (2.9)	0.55×10^6 (3.8)	0.50×10^6 (3.1)	0.45×10^6 (3.1)	0.50×10^6 (3.4)	0.5×10^6 (3.45)
Tensile strength, psi (MPa)	8900 (61.4)	9500 (65.6)	7730 (53.3)	13,000 (89.7)	8,000 (55.2)	11,000 (75.9)	12,000 (82.8)
Tensile elongation, %	1.8	6.5	2.9	4.0	1.5	5.2	4.0

[a] Ref. 4.

in such compounds. The fabric makes a light, fluffy mix that is difficult to handle. The mold shrinkage of this compound is quite high, but since there is no differential between the nylon and the polyester resin, the surfaces are remarkably smooth. The color and weave of the nylon fabric are difficult to mask and the mechanical properties are low. The lower molded density permits thicker sections, which sometimes can compensate for the lower physical properties.

Some compounds have been made with carbon and aramid fibers. These materials increase some physical properties, but not in proportion to their much greater basic fiber strength. Their high cost restricts their use to some specific applications.

SMC is commonly reinforced with glass roving chopped on the impregnator into 1 in. (25 mm) lengths, although lengths of 1/2–3 in. (12–75 mm) are used. Commonly, SMC formulations contain about 30 wt % glass fiber. Roving is classified as being hard or soft, in the same manner as chopped glass strands. The hard types chop more easily, wet out poorly, but mold well. The soft types are harder to cut into lengths, do not mold as easily, give a poorer surface finish, and wet out easily. Figure 80.5 shows the effect of both glass roving type and concentration on the impact strength of an SMC.

Originally, SMC was made from soluble binder chopped strand mat using 2 in. (51 mm) fiber lengths. Some SMC is still made this way (mainly in Europe). To increase physical properties, continuous glass rovings, carbon fibers, and aramid fibers have also been included in SMC compositions.

80.2.3 Fillers

The fillers in common use in reinforced molding compounds are silicas and silicates, carbonates, sulfates, and oxides. However, almost any inert material that can be reduced to a

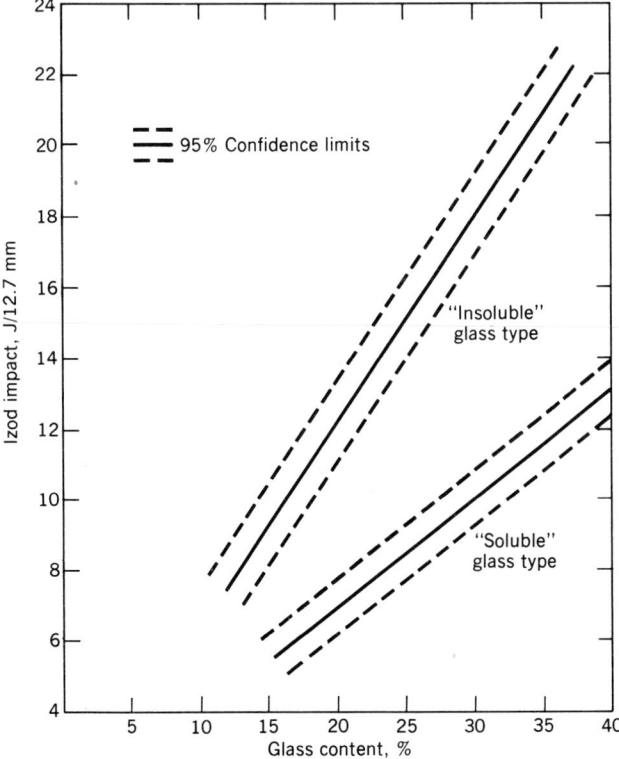

Figure 80.5. Impact strength vs glass content for SMC.[5] Courtesy of Marcel Dekker, Inc.

particle size in the range of 1/2–50 microns can be used. The first group includes talc, china clay, silica (sand), diatomaceous earth, and volcanic ash; the second is composed entirely of various calcium carbonates; the third includes barium sulfate (barytes) and calcium sulfate; and the fourth, hydrated alumina. In these groups are the so-called natural materials that are brought to a useful state by wet and/or dry grinding, and those produced by chemical precipitation. The latter can be of the smallest particle size as well as the most uniform. The specific gravity ranges from 2 for diatomaceous earth, to 4.45 for barytes, with the predominately used materials (china clay and calcium carbonate) in the 2.6–2.7 range.

The bulking fillers are china clays and calcium carbonates. Calcium carbonates have the lowest oil absorption and, consequently, more can be added to a mix. However, their characteristics are such as to cause poor flow. Clay-filled compounds have better flow, and molded properties are better in many respects but not in color. Frequently, a combination of clays and calcium carbonates provides a high filler loading with good flow. The addition of smaller quantities of high oil-absorption talc to predominately calcium carbonate-filled compounds will also improve flow with less effect on color.

Desirable filler properties in a filler for a compound include:

- Low specific gravity.
- Low oil absorption.
- Nonporous.
- Nonabrasive.
- Low cost.
- Readily dispersible without agglomeration.
- Chemical purity and whiteness.
- Wide particle size range (1–15 microns) with mean diameter of 5 microns.

The filler content of a given compound is inversely proportional to the amount of reinforcement needed to satisfy the physical property requirements. However, in formulations for electrical and fire retardancy, the opposite is true. Sufficient filler must be included to achieve the desired properties, and then the maximum amount of reinforcement is added. Usually, fillers can replace the reinforcement with minimal effect on the general moldability of the compound.

Additives that might be considered fillers are also available to achieve a specific modification of the compound. Some, such as antimony trioxide used in conjunction with halogenated resins or waxes, improve flame retardancy; some, such as magnesium oxide, magnesium hydroxide, and calcium oxide, provide chemical thickening; and others, such as powdered polyethylene and polystyrene, reduce shrinkage.

Particle Size. Fillers are commonly classified by particle size in terms of the fineness of sieve through which a given percentage of the product will pass (for example, 99.8% through 325 mesh screen) or by the particle size expressed in microns. For molding compounds, the foregoing example represents a reasonable low limit. The space between the wires of a 325 mesh screen is 44 microns; any substantial quantity of larger particles is unsatisfactory.

Particle Size Distribution. The particle size classification provides information about the maximum particle size but not about the smallest particles or to the sizes in between. In order to obtain a good packing density of fillers in the molding compound, a fairly wide range of particle sizes is required.

Oil Absorption. This property, which is stated as the percent linseed oil required to wet-out a given weight of filler, provides an approximation of the relative amounts that could be used to attain the same viscosity. This information is useful in filler selection.

TABLE 80.6. Melting Points of Several Metallic Stearates[a]

Metallic Stearate	Melting Point, °C
Zinc stearate	120–125
Calcium stearate	150–160
Aluminum distearate	145–160
Magnesium stearate	154
Lithium stearate	200–210

[a]Ref. 6.

The oil absorption is a function of specific surface of the particles. Porous particles have higher oil absorption values than nonporous ones of the same size. The lowest oil absorption permits the highest proportion of filler, and, in most cases, the principal filler in a resin/filler system can have a low oil absorption value.

Thixotrophy. Thixotropy is a phenomena in which the nominal viscosity of a material is markedly reduced when the material is disturbed, returning to the original state when the disturbance ceases. Some high oil-absorption fillers that provide high viscosity also have a substantial thixotropic effect. Mold closing forces usually are sufficiently disturbing that viscous thixotropic filler systems can exhibit the flow properties of lower-viscosity, nonthixotropic systems.

80.2.4 Internal Release Agents

Internal mold-release agents are used in all molding compounds. Zinc stearate is the most common since it has a melting point slightly lower than the common molding temperatures employed. Calcium and aluminum stearates sometimes are used in injection-molding compounds because they are usually molded at higher temperatures. Stearic acid can be used for lower temperature molding. An alkyl phosphate in liquid form, Zelec UN®, is useful in wet molding systems. Table 80.6 lists the melting points of several popular metallic stearates.

Metallic stearates and stearic acid are used at a 1–3 wt % level in compounds. Zelec UN® is effective at 0.5%. Some reduction in physical properties can occur with excessive use of internal mold release agents.

80.2.5 Colorants

Dispersions of pigments in a compatible resin system are widely used in formulating molding compounds. Table 80.7 lists the effects of some pigments on the curing characteristics of

TABLE 80.7. The Effects of Pigments on the Cure Characteristics of Polyester Resins[a]

Pigments	Effect
Cadmium salts	Accelerators
Carbon black	Inhibitor
Iron oxide	Accelerator
Copper salts	Inhibitor/accelerator (Depending on concentration)
Aluminum salts	Accelerators
Titanium dioxide	Slight accelerator
Phthalocyanine green	Slight inhibitor
Organic dyes and pigments	Inhibitors
Kaolin clays	Slight accelerators

[a]Ref. 7.

polyester resins. The effects of the pigment on the polyester resin should be investigated before a decision is made about a pigment in a formulation. Economic considerations can sometimes dictate that lower-cost, dry pigments be used in BMC formulations. The intensive mixing required to make a BMC often permits use of dry pigments.

80.2.6 Curing Agents

Benzoyl peroxide is a good, economical curing agent for BMC that is used shortly after mixing. However, if long-term storage is required, it should not be used because its shelf life is poor. *Tert*-Butyl perbenzoate (TBPB) requires higher molding temperatures but is very stable and permits higher temperature mixing. It is the standard curing agent in BMC/ SMC compounds. More recent developments include peroxyesters and peroxyketals, which give shelf lives equal to TBPB with slightly faster cure times. Combinations of TBPB with more reactive peroxides, such as *tert*-butyl peroctoate (TBPO), can significantly reduce cure times, providing that shelf life is not a factor. Table 80.8 shows the reactions of some typical peroxides and combinations in a standard isophthalic polyester SMC. The platen gel time at 295°F (146°C) is a reasonable indication of the relative mold flow time that can be expected.

80.2.7 Thickeners

Many of the Group II oxides and hydroxides will thicken carboxy terminated unsaturated polyester resins to some extent; however, the most popular thickeners are magnesium oxide, magnesium hydroxide, calcium oxide, calcium hydroxide, or combinations of these materials.

Magnesium oxide does not occur naturally. It is made by calcining magnesium salts. The type used to thicken polyester resins is made be a controlled calcination of magnesium hydroxide, magnesium basic carbonate, or mixtures of these two products to give a porous, spongy, large surface area MgO. The surface area of highly reactive MgO is about 200 m²/ g, based on nitrogen absorption. In general, increases in thickening are directly related to higher surface area.

Besides reacting with polyester resins, MgO also reacts readily with water to form the hydroxide and, with carbon dioxide, the carbonate. In practical terms, this means that the handling and storing of MgO is quite important in the SMC plant. For this reason some early SMC manufacturers arranged to purchase their MgO in small, sealed, moisture-impervious bags of 1–3 lb (0.5–1.4 kg) each, rather than in 50–lb (23 kg) bags. This special handling added considerably to the cost, so several companies offered preweighed MgO in

TABLE 80.8. Press Molding Peroxides: Reaction Data[a,b]

Peroxide	Concentration, Weight, %	Gel Time, s	Exotherm Time, s	Peak Temperature, °F (°C)	Platen Gel Time at 295°F (146°C) s
tert-Butyl perbenzoate	1.0	152	179	329 (165)	27
tert-Butyl perbenzoate	1.4	142	168	330 (166)	25
Peroxyester	1.0	124	152	313 (156)	22
Peroxy ketal	1.0	106	133	305 (152)	18
tert-Butyl perbenzoate	0.8	104	133	307 (153)	19
tert-Butyl peroctoate	0.2				
tert-Butyl perbenzoate	0.7	99	127	310 (154)	15
tert-Butyl peroctoate	0.3	99	127	310 (154)	15
tert-Butyl perbenzoate	0.5	86	112	315 (157)	13
tert-Butyl peroctoate	0.5	86	112	315 (157)	13

[a]Ref. 1. Resin: Typical SMC isophthalic polyester.
[b]Apparatus: Modified hot block tests at 270°F (132°C).

polystyrene bags. The entire bag could be added to the mix where it would dissolve without coming into contact with air in the plant.

An alternative solution is the purchase of commercially available dispersions of reactive MgO in inert polymeric carrier vehicles.

Another thickening system consists of the formation of a three-dimensional polyurethane rubber network distributed throughout the polyester matrix. This system is called the interpenetrating thickening process (ITP) and is a development of ICI Americas Inc. The ITP system involves a finite chemical reaction. When the ingredients are converted, the thickening process is complete, and the viscosity remains constant. Maturation time can be very short; in practical terms, time must be allowed for wet-out of the glass reinforcement, just as in metal oxide systems.

80.2.8 Low-Profile Additives

Most of the older lower cost thermoplastic resins have found use as low-profile additives in FRP systems. Table 80.9 lists some of the thermoplastic resins that have been successfully employed. (Not all thermoplastics that can be used in an FRP system can be used economically.)

Thermoplastics that are soluble in monomeric styrene usually are charged to the formulation as a syrup. (The usual solids content is also given in Table 80.9.) Thermoplastic syrups that do not contain carboxyl termination can be used as carrier materials for the Group II metal thickeners.

Thermoplastic syrups that are compatible with the polyester resin (those that do not separate on standing) usually are mixed with the resin by the vendor and sold as "one pack" or "one component" systems. An example of such a material is poly(vinyl acetate). This eliminates the choice of thermoplastic family and the concentration of additive in the formulation by the molder. Some molders prefer to be able to vary both the type and concentration of thermoplastic additive in order to be able to control the compound shrinkage.

80.2.9 Tougheners

The highly reactive resins used in low-profile BMC and SMC produce a fairly brittle product that sometimes is subject to profuse surface cracking during handling, shipping, and subsequent secondary operations. In an attempt to toughen these compounds to resist cracking, synthetic rubber compounds are added to the formulation. They increase the impact resistance and elongation of the compound, usually with a reduction of flexural strength and stiffness.

B.F. Goodrich Chemical Company offers its Hycar® reactive liquid polymer (RLP) as an elastomeric additive. The recommended dosage is 10 phr (parts additive per hundred parts resin); it is claimed that in calcium carbonate-filled systems a considerable improvement in Gardner impact strength results.

80.2.10 Viscosity Depressants

As inert fillers are added to a polyester resin, the mix viscosity increases rapidly. Certain materials act to depress the viscosity curve as fillers are added; they are called viscosity depressants or viscosity control agents.

Calcium stearate acts as a viscosity depressant when china clays are added to a polyester resin system.

Union Carbide markets a material, Y-9306 Dry Silane concentrate, which acts as a viscosity depressant when alumina trihydrate is added to a polyester resin. A loading of 4% concentrate, based on the alumina trihydrate content, is recommended.

TABLE 80.9. Thermoplastic Additives[a]

Item	Thermoplastic Family	Solids Content, %	Commercial Material Designation	Vendor[b]	Carboxyl Termination	Non-carboxyl Containing	Remarks
1	Acrylic	40	Paraplex P-681	OCF	X		Polymethylacrylate/acrylate copolymer
2		40	Paraplex P-701	OCF	X		
3	Polyethylene	100	Microthene FN-510	USIC		X	
4	Polystyrene	100	M9-C2	Amoco		X	Medium impact grade, pellet form
5		35	LP-80	UCC		X	Medium impact grade
6	Polycaprolactone	100	LPS-60	UCC	X		
7	Vinyl	35	LP-35	UCC	X		
8	Poly(vinyl acetate)	40	LP-40A	UCC	X		
9		40	LP-90	UCC		X	
10	Cellulose acetate butyrate	100	EAB-551-0.08	EC		X	
11		100	EAB-551-0.2	EC		X	
12		100	EAB-551-0.1	EC		X	

[a]Ref. 8.

[b]OCF = Owens-Corning Fiberglas Corp.
USIC = U.S. Industrial Chemicals Co.
Amoco = Amoco Chemicals Corp.
UCC = Union Carbide Corp., Thermosetting Resins and Compounds
EC = Eastman Chemical Products Inc., Chemicals Div.

Union Carbide also markets a product known as Viscosity Reducer-3, or VR-3, which is an effective viscosity control agent for calcium carbonate-filled polyesters.

80.2.11 SMC Carrier Films

Two carrier films are used to transport the SMC through the impregnator. Normally the carrier films are 3–4 in. (76–102 mm) wider than the SMC product to prevent the resin mix from flowing out of the roll before the mix viscosity is controlled by chemical thickening. Untreated extruded polyethylene film of 200 gauge [2 mils (0.05 mm) thick] is the most popular carrier film because of its more favorable cost.

The carrier film should be pigmented in a color that contrasts with the SMC product to make it easier for the press operator, while preparing mold charges, to be certain that all the film has been removed. Nylon film is a very good carrier film for SMC; since it is much tougher than polyethylene, half the thickness generally will do the same job. Nylon also is impervious to styrene evaporation. Some molders are using only one nylon film and folding the edges over to form a satisfactory edge seal. One grade of uncoated cellophane has been tried as a carrier film but it was partially solubilized by the styrene in the SMC resin mix and inhibited the cure of the SMC. When heated rolls are used on the impregnator, the cellophane tends to become brittle. Polyethylene-coated cellophane has worked satisfactorily as an SMC carrier film but it is not competitive with extruded blown polyethylene film.

80.2.12 SMC Outer Wrapping

Polyethylene film is fairly porous and will allow styrene monomer to evaporate. For this reason, it is necessary to securely wrap the completed roll of SMC product with a barrier film, such as Mylar® polyester film, aluminized kraft paper, cellophane, or nylon film, and to tape all joints to prevent evaporation. Mylar® works very well, but it is expensive and transparent, allowing light to penetrate the outer SMC layers. Aluminized kraft paper is a good barrier film, opaque, and fairly inexpensive.

80.3 FORMULATIONS

A formulation is a compound designed to obtain the desired properties in the completed molding, within the limitations of the molding conditions. Compounds may be formulated to obtain a specific strength, stiffness, toughness, electrical insulation, corrosion resistance, fire retardancy, or combinations of two or more of these qualities, but, first of all, they must be moldable.

A reinforced molding compound must remain homogeneous as it flows to the mold extremities. If the resin/filler/reinforcement separate, serious variations in properties will occur throughout the molding, and some of the usefulness of the reinforcement will be lost.

The ideal compound will flow easily and fill the extremities of the mold. These two molding characteristics are seldom easy to combine with the other necessary requirements, and the best formulation generally is a compromise.

Flow properties are essentially a function of the degree to which the resin mix is absorbed by, or adsorbed on, the filler and the reinforcement. Each of the dry ingredients in a formulation has its own particular resin absorption, or drying effect on the resin. For example, in the case of two very common fillers, china clay has more than twice the absorptive power of calcium carbonate. Longer glass fibers are less drying than short ones, and hard types less than soft types. The drier the compound, the lower the plasticity or flow. Some guidance on fillers is available from the linseed oil absorption data published by the paint industry. Table 80.10 lists the oil absorption properties of some commonly used inert fillers. Not only do the fillers vary in resin absorption, but the resins vary in their ability to wet the fillers,

TABLE 80.10.   Oil Absorption of Some Fillers Used in Reinforced Molding Compounds[a]

Filler	Particle Size, Microns	Specific Gravity	Oil Absorption[b]
Antimony trioxide	44	5.70	11
Asbestos	50	2.56	38
Barytes	7.5	4.40	11
Calcium carbonate	2.5	2.71	14–16
Calcium carbonate	5	2.71	9–10
Calcium carbonate	7.5	2.71	5.5–6.5
Calcium carbonate	14	2.71	5–6
China clay	1	2.58	60
China clay	5	2.58	32
Diatomaceous earth	7	2.05	88
Talc	5	2.71	55–59
Alumina trihydrate	12	2.42	30
Alumina trihydrate	1	2.42	60
Feldspar	14	2.60	30
Feldspar	9	2.60	35

[a]Ref. 1.
[b]Grams linseed oil per 100-mL filler.

depending on their viscosity, basic chemical construction, and quantity of monomer. A high viscosity resin will carry reinforcement and filler well, but high viscosity makes mixing more difficult. Combinations of smaller quantities of high absorption fillers, such as china clay or talc, with low absorption ones, such as calcium carbonate, work well to solve both the flow and homogenity problems.

The chemical thickening process can also be used as a flow control device in making BMC. A resin with a low initial viscosity readily permits addition of large quantities of filler or reinforcement. The thickening effect simulates the effect of a high viscosity resin, or the use of high absorption fillers.

80.4   COMPOUNDING

Compounding is the conversion of the raw materials into reinforced molding compounds. Table 80.11 lists typical compound formulations. The order of ingredient addition to the mix is important, and the preferred order is given in Table 80.12, together with the terminology applied to the various phases of a mix.

For BMC, when items 1 through 6 are combined the resultant mix is referred to as a resin mix. Resin mixes should have long shelf lives. Resin mixes for BMC are charged to a double-arm intensive batch mixer or are pumped to a continuous mixer where the filler is added to form a paste. If the compound is to be thickened, the chemical thickener is added at this point. The glass reinforcement is then added on a controlled mix cycle to complete the BMC. Regular unthickened compounds are ready to be molded at this point. For high glass content compounds it generally is desirable to age the compound at least one day to permit the glass reinforcement to more fully wet out in order to develop full physical properties. Thickened mixes must be aged in a temperature-controlled environment to complete the maturation to a predetermined molding viscosity. The length of this maturation is determined by the nature of the polyester resin, the type and quantity of chemical thickener used, and the temperature of the maturation, which is normally maintained in the 90–105°F (32–41°C) range. The maturation period can range from a day to a week. The tendency is to reduce the maturation period so that the quantity of materials in storage is minimized.

TABLE 80.11. Typical Compound Formulations

Ingredient	General Purpose Glass Reinforcement Premix[a]	Arc Resistant Premix[a]	Fire Retardant Premix[a]	Chemically Thickened BMC[a]	Low Profile High Strength SMC[b]	Automotive Grade SMC[b]	Low Profile Injection Grade BMC[b]
Resin	Orthophthalic, styrene, 28%	Isophthalic, vinyl toluene, 18%	Plaskon 9520, 30%	Orthophthalic, styrene, 32.9%	SG-30 resin, 26.2	Selectron 50239, 16.13	Paraplex P19A, 33.45
Catalyst	Benzoyl peroxide, 0.3%	Benzoyl peroxide, 0.2%	Benzoyl peroxide paste (50%), 0.6%	Dicup 40C 0.8%	LP-40A, 4.6	Selectron 5990, 9.94	Microthene
Release agent	Zinc stearate, 1%	Zinc stearate, 1%	Zinc stearate, 1.4%	Zinc stearate, 0.8%			FN-510, 0.97
Colorant	Titanium dioxide, 5%				TBPB, 0.3	Styrene mon., 0.81	TBPB, 0.97
Filler	China clay, 15%	Hydrated alumina, 72%	Hydrated alumina, 53%	Calcium carbonate, 49.3%	Zinc stearate, 0.9	TBPB, 0.4, 0.4	Zelec UN, 1.91
Filler	Calcium carbonate, 35%			MgO, 1.7%	Atomite calcium carbonate, 6.2	Zinc stearate, 0.54, 0.54	Calcium carbonate, 43.16
Filler					MgO dispersion, 1.8	Calcium carbonate, 40.31	Triton GR-7, 0.47
Reinforcement	Hard glass fibers, 15%	Hard glass fibers, 9%	1/4 in. Hard glass fibers, 15%	1/4 in. Hard glass fibers, 14.5%	Glass fibers, 1 in. length, 60	Mg(OH)$_2$, 0.87	Glass fibers, 1/4 in. length, 19.07
Reinforcement						Glass fibers, PPG-518, 60.00	

[a]Ref. 1.
[b]Ref. 6.

TABLE 80.12. Molding Compound Raw Materials Order of Addition and Terminology[a]

<table>
<tr><td colspan="4" align="center">Bulk Molding Compounds</td></tr>
<tr><td>*Item*</td><td>*Ingredient*</td><td></td><td></td></tr>
<tr><td>1.</td><td>Polyester resin</td><td rowspan="9"></td><td rowspan="9"></td></tr>
<tr><td>2.</td><td>Thermoplastic additive</td></tr>
<tr><td>3.</td><td>Styrene monomer (if required)</td></tr>
<tr><td>4.</td><td>Organic peroxide catalyst</td></tr>
<tr><td>5.</td><td>Internal release agent</td></tr>
<tr><td>6.</td><td>Pigments</td></tr>
<tr><td>7.</td><td>Inert fillers</td></tr>
<tr><td>8.</td><td>Chemical thickeners (if required)</td></tr>
<tr><td>9.</td><td>Glass reinforcement</td></tr>
</table>

Resin mix (items 1–6 or 7), BMC

<table>
<tr><td colspan="2" align="center">Sheet Molding Compounds</td></tr>
<tr><td>*Item*</td><td>*Ingredient*</td></tr>
<tr><td>1.</td><td>Polyester resin</td></tr>
<tr><td>2.</td><td>Thermoplastic additive</td></tr>
<tr><td>3.</td><td>Styrene monomer (if required)</td></tr>
<tr><td>4.</td><td>Organic peroxide catalyst</td></tr>
<tr><td>5.</td><td>Internal release agent</td></tr>
<tr><td>6.</td><td>Pigments</td></tr>
<tr><td>7.</td><td>Inert fillers</td></tr>
<tr><td>8.</td><td>Chemical thickeners</td></tr>
<tr><td>9.</td><td>Glass reinforcement</td></tr>
</table>

Resin mix, Paste, SMC

[a]Ref. 9.

For SMC, the order of addition is the same as for BMC except that the inert filler is added directly to the resin mix in the mix tank. Generally a lower quantity of inert filler is used for SMC than for BMC. The inert filler is weighed and gradually charged to the mix tank as the agitator continues to rotate at a moderate speed (900–1200 rpm). Some systems use a hanging weigh hopper, which transfers the filler into the mix tank via an inclined U-plane containing a vibrator. Other systems use a rotary auger to meter the weighed charge to the mix tank. The best results occur when the filler is continuously charged to the tank into the high side of the mix near an edge, and not into the vortex of the mix. (Addition of filler into the vortex carries along a considerable quantity of air.) Many mix tanks contain an angle bolted or welded to the inside sides of the mix tank at two or three places to fold the rotating mix away from the sides, and thereby increase turbulence. On mixes that contain a high filler loading (more than 200 phr), it is generally necessary to raise the revolving blade several times during the filler addition to maintain the proper vortex. Mixing should be continued until the resin mix reaches a predetermined temperature, usually 100°F (38°C). This will require more mixing time during winter months than for summer periods.

For batch-type SMC systems, the chemical thickener should be added to the paste immediately before the paste is transferred to the SMC impregnator. If the mix has been allowed to stand for an hour or more, or if its temperature has decreased, the paste should be remixed until its temperature is 100°F (38°C) before the chemical thickener is added. When the thickener is added, the paste should be mixed until its temperature is 100°F (38°C), or for a minimum of 10 minutes.

80.4.1 Preparing Resin Mixes

A variable-speed, high shear, shingle-shaft mixer (usually called a dissolver or dispersator) is excellent for preparing BMC and SMC resin mixes. Such a mixer is shown in operation in Figure 80.6. For fastest dispersion, the blade diameter should be approximately one third the mix tank diameter. Turbine mixers and propeller-type agitators also are used. Usually the polyester resin and then the thermoplastic additive are weighed into the mix tank. These two ingredients are agitated until the thermoplastic is thoroughly dispersed in the polyester resin. The minor ingredients (styrene monomer, catalyst, release agent, pigments, and any other small-quantity additives) are weighed separately and charged to the mix tank while the blade rotates at a nominal speed (500–900 rpm). This prepared resin mix is then dumped into the double-arm high intensity mixer or pumped to a continuous mixer.

A rotating horizontal shaft mixer containing plows inside a drum (made by Day Mixing Company) is sometimes used to prepare BMC and SMC pastes. On some models the BMC can be completely formulated in this mixer by addition of the glass reinforcing strands directly to the mixer at the end of the mix cycle. Figure 80.7 shows such a mixer.

Table 80.13 is a troubleshooting guide for use in overcoming problems when SMC is being prepared.

80.4.2 Resin Mix Dispensing Systems

One of the first dispensing systems for SMC pastes consisted of a 50–100 gallon (0.2–0.4 m³) hold tank and a Grayco or Alemite air-operated pump to transfer the paste to the doctor

Figure 80.6. Typical dissolver-type mixer preparing a resin mix.

Figure 80.7. Littleford Mixer capable of preparing resin mixes and BMC. Courtesy of Day Mixing Co.

TABLE 80.13.   Troubleshooting SMC Processing Problems[a]

Problem	Description	Possible Cause	Remedy
Film wrinkles	Film wrinkles at doctor box station	Low film tension	Increase film tension
		Misaligned slat expander	Readjust the slat expander
		Insufficient carrier film to expander roll contact	Reposition the slat expander to increase the carrier film contact
		Roll of film out of alignment with film travel	Realign the carrier film roll
		Film drag on the doctor box platens	Clean the plates to remove any contamination
Film tears	At the edges of the doctor boxes	Insufficient clearance between side dams and film	Raise the side dams
		Edge damage on the film roll	Replace the roll of film
	In paste deposit area of the doctor blades	Filler lumps or contamination in the paste	Strain the paste
Streaky paste deposit	Dry lines in paste parallel to film travel	Filler lumps or contamination in the paste	Remix and strain the paste

TABLE 80.13. (*Continued*)

Problem	Description	Possible Cause	Remedy
Dry fibers	Uneven (thick-thin) sections in SMC	Fiber clumps dropping from cutter or framework of cutter	Prevent fiber glass accumulations on ledges and static bars. Install air jets.
		Resin starved and/or glass rich SMC	Adjust resin mix feed. Adjust side dams on roving chopper
		Too high viscosity	Control the humidity in mix room and in SMC room. Check mix ingredients for both proper type and concentration. Verify by assay if necessary. Check thixotropy of SMC soups and pastes. Verify water content of major mix ingredients
		Insufficient number of compaction rolls	Install additional compaction rolls
		Premature chemical thickening	Control and humidity in mix room and SMC room. Check mix ingredients for both proper type and concentration. Verify water content of all major ingredients.
		Eddy currents in the glass deposition area	Enclose the chopper area, remove any fans from the area
		Static accumulations	Install static bars. Ground all framework and rolls
Nonuniform fiber distribution	Thick and thin areas across SMC width	Uneven roving spacing	Restring all rovings
		Insufficient number of rovings	Add additional rovings to creel
		Static accumulation	Install static eliminators, ground framework and rolls
	High fiber content near SMC edges	Roving pattern into chopper too wide	Narrow roving pattern at cutter comb
		Side shields improperly adjusted	Readjust chopper side shields
		Air currents under the chopper	Enclose hopper frame to eliminate air currents
	Low fiber content near SMC edges	Roving pattern too narrow at chopper	Widen roving pattern at cutter comb
		Side shields improperly adjusted	Readjust side shields on chopper
		Uneven roving pattern into chopper	Adjust roving pattern at chopper comb
Long fibers	SMC otherwise appears normal	Chopper operating improperly	Check cutter cot and roll alignment
			Replace cutter blades
			Replace worn cutter cot
			Increase pressure on cutter cot

997

TABLE 80.13. (*Continued*)

Problem	Description	Possible Cause	Remedy
Fiber buildup on cutter frame/ rolls	Fiber clumps on framework, on cutter roll, and/or on cot	Excessive static	Check static eliminators for proper operation, replace damaged pins if necessary
		Abrasion of rovings at creel, guides, tubes, etc	Check for roughness at all roving contact points
		Low humidity in area	Increase humidity
		Roving pull speed too high	Increase number of rovings to chopper and reduce chopper speed
Air between fibers and SMC		Insufficient compaction roll pressure	Increase roll pressure
		Resin-rich SMC edges	Adjust for small amount of dry edge glass
Poor fiber wet-out	Random distribution	Uneven glass blanket	Check for static and chopper condition
		Paste viscosity too high	Reduce paste viscosity
		Insufficient compaction pressure	Increase compaction roll pressures
		Glass content too high for machine capabilities	Increase roll temperatures, change glass to one more easily wet-out
	Localized lane of SMC	Improper spacing of rovings into the chopper	Respace the rovings in the comb
		Nonuniform paste deposit on films	Check doctor blade set-up, check for contamination on back-up plate
		Nonuniform compaction pressure	Check compaction rolls for distortion, check for contamination build-up on compaction rolls
Squeeze-out at winding station		Excessive winding tension	Reduce winding tension
		Paste viscosity too low for glass content being used	Adjust paste viscosity
		Film too narrow for SMC width	Get wider film or reduce SMC width
Telescoping of SMC at wind-up station	Rolls are egg shaped	Excessive winding tension	Reduce winding tension
		Poor alignment of winder with SMC machine	Realign the winder with the machine
Improper weight per unit area	Sheet appears normal and uniform	Glass and/or resin quantities need adjustment	Check glass content, increase glass content (raise cutter speed) and/or paste feed (raise doctor blade) as required
		Incorrect machine calibration	Recalibrate machine settings
		Resin mix ingredients incorrectly proportioned	Check specific gravity of resin mix

[a]Ref. 10.

998

boxes on the SMC impregnator. A tee was placed in the supply line to divert the stream to the two doctor box stations with a ball valve on each side of the tee to control paste flow. This system worked very well on the early SMC pastes, which did not reach a viscosity of over 100,000 centipoise in the first two hours after the chemical thickener was added. Some laboratories still use this system of dispensing. Even with the early mixes, a variation in glass wet-out frequently was observed from start of a batch to the end of the batch of paste.

As resins and chemical thickeners changed to quicker maturation systems, the variation in glass wet-out from start to finish of a batch became intolerable. A considerable amount of development work was devoted to improving paste dispensing systems. The idea of the two-pot system evolved, using an A side (resin side) and a B side (thickener side). The A side contains all the polyester resin, thermoplastic low-profile additive, the catalyst, release agent, and most of the inert filler. The B side consists of an unthickenable carrier resin, the pigment, chemical thickener, and the balance of the inert filler.

One of the first dispensing systems to use the A and B side concept consisted of two connected double-acting cylinders with adjustable pistons mounted on a single shaft. Air-operated solenoid valves were used on the cylinder ports. The driving force for moving the piston was furnished by the pressure of the resin mixes pumped to the cylinders. A limit switch interrupted the piston travel and reversed the travel direction, so that the exit ports became the entrance ports and vice versa. At one installation, two holding tanks were used for the A and B sides and gear pumps were used to move the two resin mixes through the cylinders. Another used pressure pots with a nitrogen charge over the two resin mixes as the moving force. Both molders used air-driven piston pumps to move the resin mixes from the mix house to the holding tanks. While this system did permit continuous operation of the SMC impregnator, it was not without its problems. Even though the seals on the cylinders were made of Teflon, they wore out quickly from the abrasive mix and had to be replaced weekly. The connecting lines were small in diameter, and the solenoid valves tended to stick. Maintenance costs were high and downtime was frequent.

The Koppers Company perfected a dispensing system in its laboratory that used a Moyno pump on the A side and a gear pump on the B side. This was a considerable improvement over previous systems.

After considerable effort and many changes, the Marco Chemical Division of W.R. Grace & Company built a similar system but used differently sized Moyno pumps for both the A and B sides.

Some of the above mentioned systems are now commercially available from Finn & Fram, E.B. Blue Company, and others.

80.4.3 Static Mixers

In using the two-pot system, the A and B mixes must be combined before they are fed to the doctor box stations. This can be accomplished by the use of a static mixer. Such a unit is designed to achieve a homogeneous mix by flowing the two streams through geometric patterns formed by elements in a tubular barrel. Experience has shown that a stainless-steel mixer of 1 in. (25 mm) pipe size, with a minimum of 21 elements (preferably 24 elements) will do a reasonable job with SMC resin mixes. The static mixer should be installed in such a manner that the elements are easily removable for periodic cleaning in methylene chloride. Addition of all the pigment to the B side mix allows the operator to visually judge the degree of mixing at the doctor boxes. A homogeneous mix without streaks is required.

80.4.4 Dynamic Mixers

Although static mixers do a good job on combining the two resin mix sides when the mix lines and elements are cleaned thoroughly on a regular basis, they leave much to be desired for continuous operations 24 hours a day, six days a week. For such operations a dynamic in-line mixer is required. One such mixer that has proven to be satisfactory on SMC impregnators is made by Chemineer Inc.

80.4.5 In-Process Testing of SMC

SMC Batch Process. The minimum in-process tests recommended for keeping an SMC batch process under control include:

- Weight per unit area
- Glass content
- Glass wet-out
- Initial paste viscosity
- Final paste viscosity
- Monitoring of critical process temperatures

SMC Continuous Process. The minimum in-process tests recommended for keeping a continuous or semicontinuous SMC process under control include:

- Weight per unit area
- Glass content
- Glass wet-out
- Initial paste viscosity
- Final paste viscosity
- Process temperature monitoring
- Ratio check of A and B streams

Test Frequency. Weight per unit area should be checked:

- At impregnator startup
- Once each hour thereafter
- After each process adjustment

Glass content should be determined:

- At impregnator startup
- Once per hour thereafter
- After each process adjustment

Glass wet-out should be checked:

- At impregnator startup
- Once per hour thereafter
- Whenever weight per unit area, glass content, or viscosity checks are high
- After a process adjustment or a new lot of glass roving has been added to the line.

Initial paste viscosity should be measured:

- At impregnator startup
- Whenever a new lot of paste is charged to the line.

Final paste viscosity should be measured:

- After 1 hour aging of the paste sample
- After 24 paste aging
- Daily thereafter until all the SMC has been molded

Critical process temperatures should be monitored:

- Continuous recording of all critical process temperatures is recommended.

Ratio on A and B sides should be checked:

- At impregnator startup
- Once per hour thereafter
- Whenever a new batch of A side resin mix or B side thickener is charged to the line

80.4.6 In-Process Testing of BMC

Usually the Brookfield viscosity and SPI cure characteristics of the BMC resin mix are checked on each batch prepared. After the inert filler is added, the mix is usually not checked.

80.4.7 Molding Compound Process Equipment

A variety of mixers are used in the preparation of reinforced molding compounds. Mixers for handling resin mixes are described in other sections.

BMC Batch Mixers. Intensive double-arm batch mixers for preparing premixes and BMCs are made by the Day Mixing Company, by Baker-Perkins, and by several other equipment manufacturers. A variety of arms or blades are available; some of these styles are shown in Figure 80.8. For the chemical and bakery industries these double-arm mixers are supplied

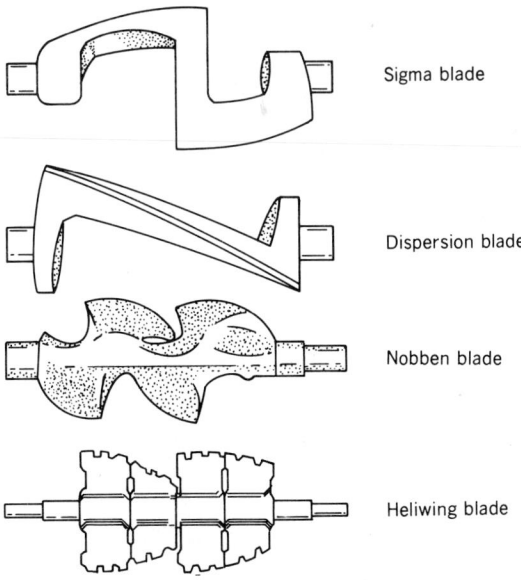

Sigma blade

Dispersion blade

Nobben blade

Heliwing blade

Figure 80.8. Several double-arm mixer blade styles.

with sigma blades and with a blade-to-shell clearance of from 0.010–0.050 in. (0.25–1.27 mm). It has been found that sigma blades entrap resin, which will not mix in with the balance of the mix. Sigma blades are also very difficult to clean. A dispersion type or single-curve blade is much more acceptable. The dispersion blade has sharp edges; the single curve blade (having a 135 or 180° curve) has no sharp edges and is considerably easier on the glass reinforcement during mixing. The two blades rotate toward each other tangentially with blade speed ratios of 3:2. For FRP molding compounds, a blade-to-shell clearance of 1/8 in. (3 mm) is used for the small laboratory mixer with a capacity of 2.25 (8.5 L) and from 1/4–3/8 in. (6–10 mm) clearance on 50- and 100-gal (0.2–0.4-m³) production mixers.

Other Mixers. The Ross/AMK Kneader Extruder made by Charles Ross & Son Company and shown in Figure 80.9 is an unusual mixing device. This unit prepares the BMC in a conventional manner in a double-arm mixer, but instead of dumping the completed mix for additional processing, the mix is extruded into log form from the mixer by an extrusion discharge screw located in the cavity beneath the mixing blades. A pneumatic guillotine cutter can be added to automatically cut the extruded logs to length. The unit combines the efficiency of the double-arm mixer with the convenience of the extrusion screw. Some materials handling problems are thereby eliminated, and the risk of operator injury is minimized since the mix is not dumped. A variety of mixer sizes are available from 1–1000 gal (0.004–3.8 m³) working capacity.

The use of a Ross planetary mixer to prepare premix batches is another unusual application (Fig. 80.10). The reported advantages include a decrease in total mixing time and an improvement in compound strength resulting from less fiber degradation. The low viscosity ingredients are mixed at high speed, and final mixing is done at a lower speed to protect the glass reinforcement. Mixer sizes of from 1-300 gallons (0.004–1.1 m³) are available. No seals come into contact with any mix ingredient, so maintenance problems are minimal compared to other compound mixers. A vacuum can be supplied on some models to remove trapped air from the mix.

The handling of premix and BMC materials is a persistent problem. Densifying the charge by running it through a clay extruder is expedient in compression molding. Before the

Figure 80.9. Ross/AMK kneader-extruder. Courtesy of Charles Ross & Son Co.

Figure 80.10. Ross planetary mixer. Courtesy of Charles Ross & Son Co.

success of BMC injection molding, several different types of loading equipment or stuffers were developed to transfer the bulk compound into the injection machine. One of the most satisfactory stuffers is made by Martin Hydraulics (Fig. 80.11). A hydraulic ram feeds the BMC into the injection molding machine. When the ram is retracted, it is loaded by means of a hopper mounted directly over an opening above the ram. BMC inside the hopper is transferred to the ram opening by means of a close fitting plate that is attached to another hydraulic cylinder. Models are available to fit most injection molding machines.

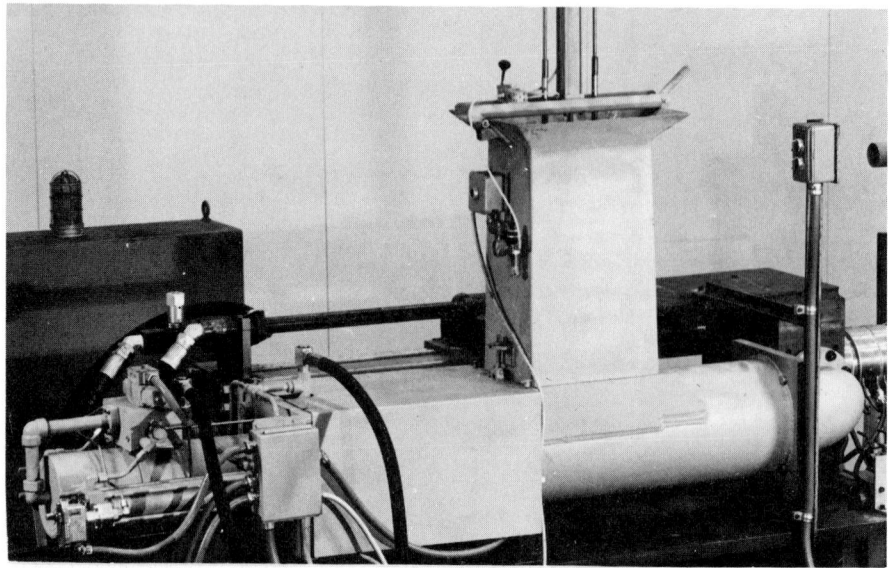

Figure 80.11.   Martin Hydraulics stuffer mounted on a Van Dorn injection-molding machine. Courtesy of Martin Hydraulics Inc.

Clay-Type Extruders.   In order to densify BMC charges for handleability, they are generally run through some type of extruder. Most hydraulic type extruders are designed and built in-house. Clay-type extruders, similar to those used for making bricks, are used for extruding BMC. These extruders have very large-diameter screws that are polished and chrome plated to make them easier to clean. Some have pneumatic cutters built in to cut the extruded logs to length. Figure 80.12 shows a late production model clay-type extruder having a pneumatic cutoff. Figure 80.13 shows an earlier model Bonnot clay-type extruder being used in production.

SMC Impregnators.   Three basic types of SMC impregnators are in general use today: nip rolls used with glass mat reinforcement; belt types used with roving and glass mat reinforcement; and the hollow can or roll type used primarily with roving reinforcements.

Nip Roll Impregnators.   Gruenwald and Walker[11] have described the evolution in West Germany of first-generation SMC impregnators that contained nip rolls for use with glass-strand mat reinforcement. The initial machine consisted of a pair of driven rolls with a narrow gap between them. The carrier film and the reinforcing mats are fed into the nip between the rolls, into which resin mix is pumped. The rolls force the impregnation of the mat while it is sandwiched between polyethylene films and remove some of the trapped air. Dams on each side of the rolls prevent the resin mix from running out. Several plies of mat can be fed into the unit at one time to make a heavier product.

One problem with nip roll impregnators was the tendency for loose glass fibers to accumulate in the resin mix. Those glass fiber agglomerations interfered with impregnation. When they were drawn into the roll nip, they tore the glass mat and sometimes damaged the rolls. These problems were overcome by installation of a wide slit hopper above the rolls to introduce the resin mix between the rolls. Of course, the result was a great restriction on the amount of resin that could be fed to the machine.

As requirements for SMC increased, it became necessary to speed up the material travel

Figure 80.12. Late-model Bonnot BMC extruder. Courtesy of The Bonnot Co.

through the impregnator, and unwet glass areas resulted. Increasing the roll diameters had no appreciable effect. Then, doctor blades were installed over idler rolls to coat the poly- ethylene films with a layer of resin mix. While that change solved the dry glass problem, slippage then was encountered between the main driven rolls and the polyethylene sandwich. The slippage problem was solved by texturizing the rolls.

Belt-Type SMC Impregnators. Figure 80.14 shows a typical belt-type SMC impregnator. After completion of the moving sandwich (after the top film has been applied), working disk rolls force the resin mix into the glass fibers. Some machine designs use widely spaced disks followed by more closely spaced disks. A perforating roll was included on early machines to pierce the top carrier film and allow air to escape. In practice, it was found that the paste quickly seals the perforations and does not allow much air to escape. Most perforating rolls have since been removed from impregnators.

Roll-Type SMC Impregnators. Belt-type machines posed certain problems. First, it was necessary to keep the belt tracking properly. While that technology is readily available, it does add considerably to the cost of the machine. Secondly, the SMC product was worked only on the top side. It was felt that working the SMC sandwich top and bottom would ensure better wet-out.

PPG Industries developed a can or roll-type SMC impregnator that works very well; a patent was issued on this unit.[12] Since the rolls are hollow, it is simple to fit them with rotary joints and to pump heat transfer fluid through the rolls to control their temperature. Some machine designs call for the rolls to be of slightly smaller diameters as the product proceeds downstream, so as to lightly stretch the carrier films. Also, it is simple to place working rolls so that they come in contact with both sides of the sandwich.

Figure 80.13. Early model Bonnot BMC extruder in production.

Figure 80.15 illustrates one roll-type SMC impregnator. Figure 80.16 shows a commercially available roll-type SMC impregnator from I.G. Brenner Company.

80.4.8 Preparation for Molding

Between mixing and molding, some time can elapse during which the material must be protected from various types of degradation. The most common type of degradation is monomer loss by evaporation. Keeping the mix well covered in a bag or box (cellophane, aluminum film, or aluminized kraft paper) will suffice for short periods. If the mix is to be kept for long periods, sealed containers and refrigerated storage areas are necessary, and, of course, formulation plays an important role in controlling this problem. Polyethylene film is not a suitable wrapping material for long storage periods as it will pass styrene fumes, permitting the compound to dry out.

80.5 MOLDING

It now is readily apparent that a satisfactory molded FRP part is achieved only after the part has been adequately designed, a proper mold has been procured, and a suitable molding

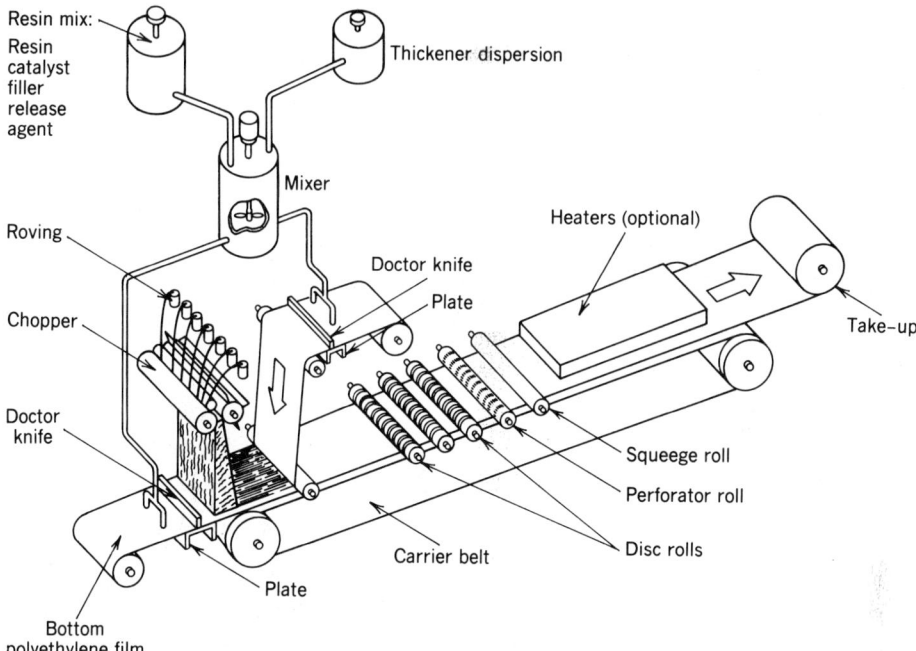

Figure 80.14. Belt-type SMC impregnator.

material has been employed. Insufficient attention to any of these three factors can result in disaster for the molder. In the early days of compound development, the primary emphasis was on the materials since they were new and novel. Attempts to resolve problems generally took the form of materials changes. Some of the headaches incurred by molders were the result of use of poorly designed molds or attempts to mold improperly designed parts.

80.5.1 Compression Molding Process Variables

Table 80.14 lists important compression molding variables that must be controlled in order for satisfactory molding. The first four items are dependent on the press setup and generally are outside the influence of the press operator. The last four items are generally the sole

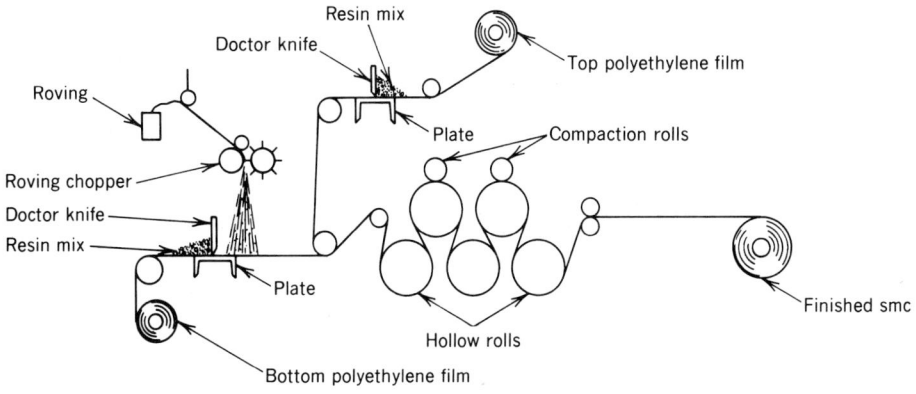

Figure 80.15. Roll-type SMC impregnator.

Figure 80.16.   Roll-type SMC impregnator. Courtesy of I.G. Brenner Co.

responsibility of the press operator. His or her attitude and the degree of reproducibility of these items directly affect the molded part defect rate.

Slow closing of the press is considerably more important in molding BMC and SMC than in mat or preform molding. The objective is to flow the compound in a uniform front at a rate that allows the air in the cavity to escape through the shear edge without being trapped by the advancing fronts. Too slow a closure can allow the material to pregel on the hot mold; too fast a closure will trap air.

Modern presses can close at a rate of 800 in. (20 m) per minute to within 1 in. (25 mm) of the shear edge, at which point they are slowed down to a variable closure rate of 1–20 in. (25–500 mm) per minute. A press closure faster than 4 seconds for the final 1 in. (25 mm) travel might damage the shear edges. A slow close of 5–15 seconds for the final 1 in. (25 mm) travel is generally satisfactory. The time from the first compound contacting the hot mold to full pressure should not exceed 20 seconds.

Molding pressures of 100–300 psi (0.7–2.1 MPa) on the parts projected area are usual for mat and preform parts. Molding pressures of 500–1200 psi (3.4–8.3 MPa) are required

TABLE 80.14.   Important Molding Variables[a]

1. Press slow close
2. Molding pressure
3. Mold temperature
4. Cure cycle
5. Charge size
6. Charge weight
7. Charge shape
8. Charge placement

[a]Ref. 13.

for the stiffer thickened compounds. A pressure of 1000 psi (7 MPa) on the part projected area is quite common. Higher pressures increase sink marks. Lower pressures tend to cause scumming of the mold and porosity.

Mold temperatures depend on the particular resin/catalyst system being used. In order to obtain a reasonable cure cycle, most materials are specified at approximately 300°F (149°C) molding temperature. It is advisable to maintain the female or cavity portion of the mold 10° hotter than the male or core portion. Temperature variations within a hot mold portion should be maintained within ±10°F (6°C). To accomplish this the mold must be properly designed for good heat transfer. Steam traps should be mounted lower than the mold to prevent condensate buildup and must be kept free of scale.

Minimum cure cycles are desirable. The point at which complete cure has been reached is a matter of conjecture. The use of bosses and ribs in BMC and SMC parts further complicates this conjecture. In breaking in a new mold, a safe cure time is customarily employed and then is progressively reduced.

Unfortunately, the selection of charge size, weigh, shape, and placement are a matter of trial and error. Past experience counts considerably in being able to start a job in a minimum of time.

Determination of the proper charge size is more complicated. Generally, the BMC charge is first compacted by extruding to convert it into a convenient log form. The larger the diameter of the log, the less likely is loss of physical properties resulting from fiber degradation. On complicated and large parts, two differently sized extrusions are sometimes used. A simple charge shape is best. Charge shape depends on the geometry of the part and can be adapted to fit the shape of the mold. When multiple extrusions are used, they should not be butted in the charge placement, which might entrap air. Overlapped extrusions are more satisfactory.

Small plugs of BMC should not be stuffed into bosses (a popular practice several years ago). Pyrolytic investigations of such molded sections revealed folded fibers with pockets that prevent the escape of trapped air. It is preferable that the glass fibers flow into such cavities. When multiple charges are necessary, they should be connected, if at all possible. In older presses, or in molds that deflect so that one corner closes before the others, the charge has to be placed on the corner that closes first so that it flows away from this area.

80.5.2 Testing Molding Compounds

Viscosity Increase of Thickened Resin Mixes. The viscosity of retained resin mix samples is checked periodically using a Brookfield recording Rheolog viscosimeter.

Reactivity or Cure. Typically, the reactivity of a resin mix is measured using an Audrey II dielectric analyzer (made by Tetrahedron Associates) in conjunction with an X-Y recorder and a modified flat sheet mold to determine the exact moment of cure and the rate of cure of a fully compounded molding compound.

Spiral Flow. One of the more important parameters of molding compounds is spiral flow length. Duralastic Products Company developed its Hydratec spiral flow mold to measure this property under controlled conditions.

Physical Properties of Molding Compounds. The manner in which physical property data are obtained has a direct bearing on the results. It is not enough to specify the ASTM or other test procedure to be used; the manner in which the test coupons are obtained must also be spelled out. Accordingly, many laboratories use standard ASTM molds for preparing test bars (Fig. 80.17).

When glass fibers are included in a molding compound formulation, the orientation of the glass in the final coupons has a direct bearing on the results obtained. Data should be obtained both in the direction of flow and perpendicular to flow. Flexural strength is more sensitive to glass content than tensile strength.

Figure 80.17 shows the relationship of compression molded coupons to those machined from molded flat sheets.

80.5.3 Molding Compound Flow

One of the important characteristics of BMC and SMC is their unusual ability to flow in the mold as heat and pressure are applied. Since the fiber lengths usually are different in BMC and SMC, a somewhat different orientation as these two materials are molded might be expected. A study was made in a square flat sheet mold using the usual charge shape for each material with the charge placed in the center of the mold. Figure 80.18 shows the flow pattern and fiber orientation obtained by stopping the press platen before it rested on stops and allowing the charge to cure. Molded charges of each material were checked for tensile strength across the width of the panels as molded. In general, lower tensile strength values are given for BMC, with the edges furthest from the charge having the lowest tensile strengths. In the SMC panel, the variation from one location to another across the panel width was not as great.

Other variables are examined in Figures 80.19–80.21.

80.5.4 In-Mold Coating

The technique of applying coatings to FRP molded parts during the molding operation has been common practice for several years. Gel coats have been applied to the surface of the mold in hand layup molding for over 30 years. It is still general practice in the marine industry to apply such coatings to boat hulls for both surface protection and for improved appearance.

The advent of matched metal molding and the use of presses to handle the molds created interest in gel coats that could be applied to the hot mold surfaces and would transfer and adhere to the molded parts. Despite considerable effort, no feasible technique of application was developed. The problems included compounding a coating to apply to a hot mold at 275–300°F (135–149°C) and transfer and adhere to the part and uniformly applying such a coating to the surface of a complex mold cavity.

As a result, the practice of painting the parts to improve their surface appearance was adopted. This required considerable hand sanding to obtain a satisfactory surface, which prevented FRP from being used in large volume markets. With the development of low profile SMC systems, the surface smoothness problem was solved. Subsequently, many automotive applications were attempted.

Porosity in the part surface, which had to be visually located and then hand filled, slowed the manufacturing output for automotive use and by 1975 threatened to obviate automobile part applications. About this time, General Tire & Rubber Company (now GenCorp) developed its in-mold coating process, which is designed to completely eliminate porosity. This is done by raising the mold approximately 5 mils (0.13 mm), injecting a polymerizable coating, and then curing the coating on the part. Although this procedure succeeds in eliminating porosity, it has no effect on other SMC molding defects.

Since the introduction of the two-component isocyanate coating system, several alternative systems have become available. One component coating systems are now available from GenCorp, Sherwin-Williams Company, and PPG Industries.

The original systems available were injected into the mold at rather low pressures [100–200 psi (0.7–1.4 MPa)]. The Sherwin-Williams Company has introduced a high pressure system [5000 psi (34 MPa)] that eliminates the requirement for a programmed force velocity system and thereby permits older presses to be used for in-mold coating. However, the cost

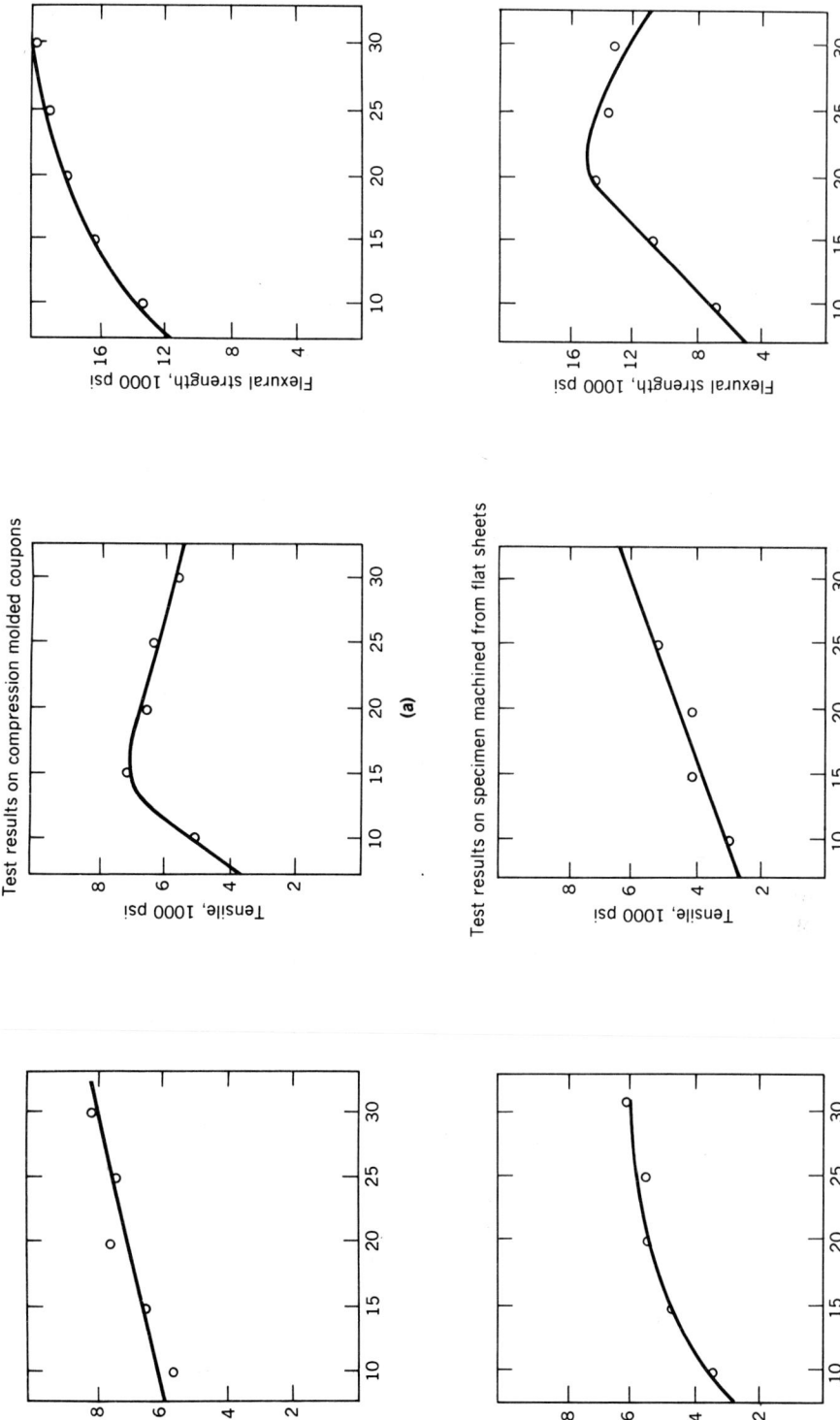

Figure 80.17. A comparison of test results-molded vs machined specimens. (**a**) Test results on compression molded coupons. (**b**) Test results on specimen machined from flat sheets.

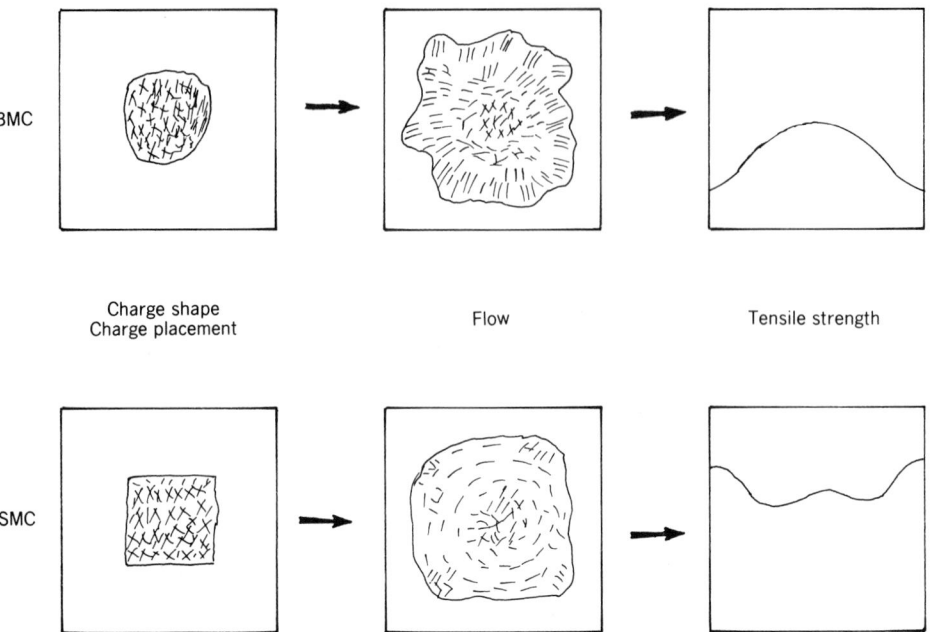

Figure 80.18.   Fiber orientation caused by flow.

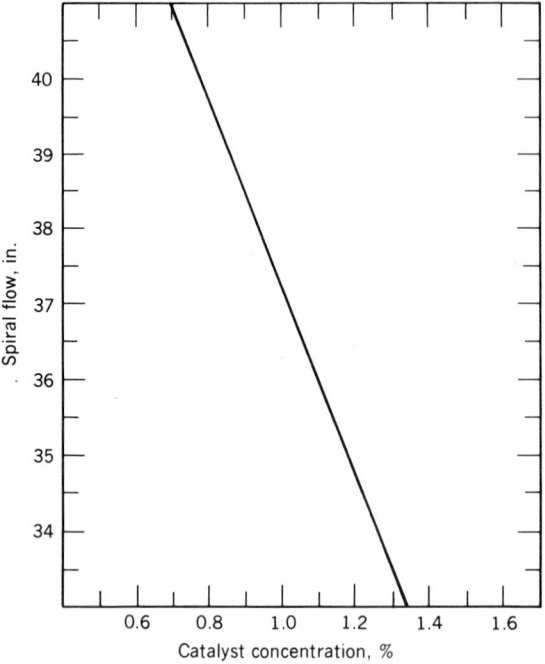

Figure 80.19.   The effect of catalyst concentration on flow properties of BMC. Pressure, 625 psi; cure time, 2 min; mold temperature, 300°F; 30% glass content BMC.

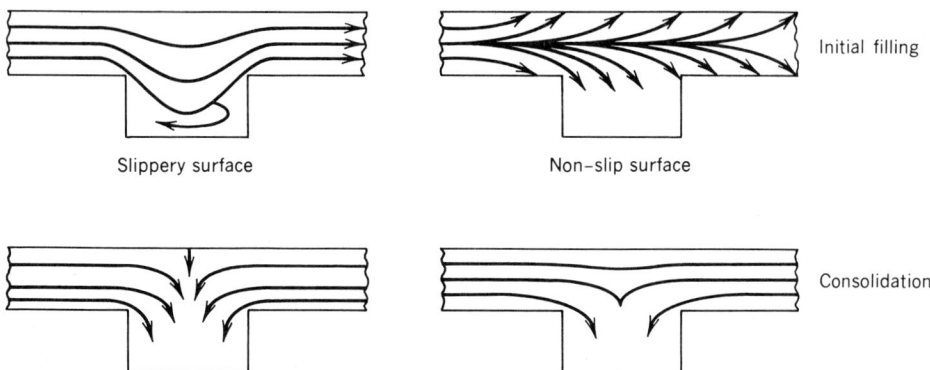

Figure 80.20. Effect of mold surface texture on flow.

of the ejector nozzle in the mold and the dispensing equipment for this high-pressure system is high.

Ferro Corporation has developed a powdered in-mold coating system that is sprayed onto the hot mold and cured before the compound is charged to the mold. Special spray techniques and containment equipment were developed to make this system practical.

Most of the exterior panels on the Corvette sports car are in-mold coated SMC. Powdered in-mold coating is reported to be useful in the production of plumbing articles.

80.6 MATCHED METAL MOLDS

A good matched metal chrome plated steel mold is required for optimum production in molding BMC and SMC materials.

The mold can be designed by the molder or by the moldmaker. Larger molders maintain

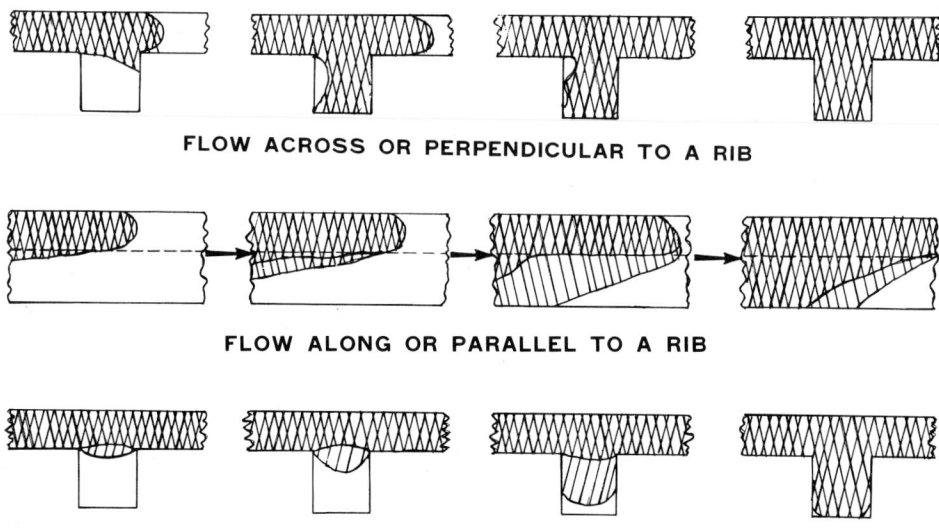

Figure 80.21. Methods of filling ribs during flow.

their own engineering staffs for the design of molds and secondary fixtures, and to follow the progress of the mold through the moldmaker's shop. Smaller molders employ the moldmaker's engineering staff for the development of working mold drawings.

Either way it is necessary for the molder to develop a satisfactory mold specification that will standardize such items as handling bolts, mounting means, heating channels, and so on. When the moldmaker designs the mold, the molder must furnish:

- Part print
- Master model
- Size of press to be used, including mounting hole patterns
- Expected molding pressure
- Type of molding compound to be used
- Molding material shrinkage
- Ejector cylinder line pressure
- Operating temperature ranges for both core and cavity
- Type of steel to be used.

Figure 80.22 shows FRP compression mold nomenclature.

80.6.1 Mold Steel Selection

The anticipated production quantities expected from the mold and/or the product end use dictate the choice of mold steel, as shown in Table 80.15.

When steel forgings are used, a certificate of the steel composition, vacuum degassing, and hardness should be supplied by the steel company to the moldmaker.

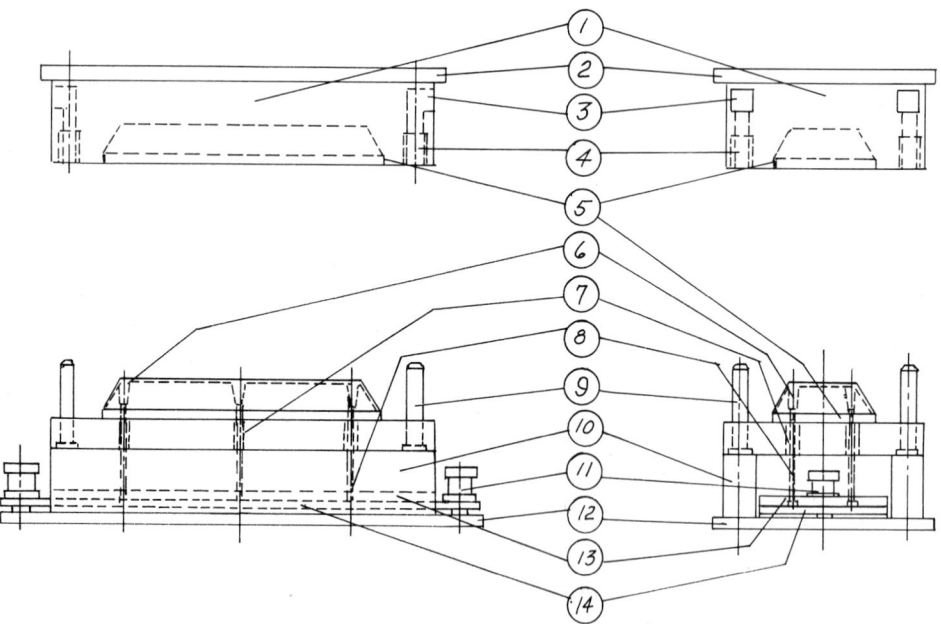

Figure 80.22. FRP compression mold nomenclature. 1, Mold cavity; 2, Cavity clamping plate; 3, Flash clean out opening; 4, Bushing; 5, Mold shear edge; 6, Part boss; 7, Ejector sleeve; 8, Ejector pin; 9, Shoulder type guide pin; 10, Core parallel; 11, Operating cylinder; 12, Core clamping plate; 13, Ejector pin retaining plate; 14, Ejector plate.

TABLE 80.15. Mold Steel Selector

Planning Volumes	Type of Steel[a]	
	Core	Cavity
Production Planning Volumes		
5,000–20,000 parts/yr	AISI-1045 steel.	AISI-1045 steel.
20,000–30,000 parts/yr	AISI-4140 forged steel pre-hardened to Rockwell C of 28–32.	AISI-4140 forged steel pre-hardened to Rockwell C of 28–32.
Over 30,000 parts/yr	AISI-4140 forged steel pre-hardened to Rockwell C of 28–32.	P-20 forged steel prehardened to Rockwell C of 28–32.
100,000 parts or less for mold life	AISI-1045 steel.	AISI-1045 steel.
100,000–200,000 parts for mold life	AISI-4140 forged steel pre-hardened to Rockwell C of 28–32.	AISI-4140 forged steel pre-hardened to Rockwell C of 28–32.
Over 200,000 parts during mold life	AISI-4140 forged steel pre-hardened to Rockwell C of 28–32.	P-20 forged steel prehardened to Rockwell C of 28–32.
Product End Use		
Structural items where surface appearance is not critical such as reinforcing panels, truck front ends, etc, where molded surface quality is of secondary importance.	AISI-1045 steel	AISI-1045 steel
High quality surface appearance decorative items such as grille opening panels, head lamp surrounds, quarter wheel opening covers, etc. where a high degree of polish is required on the outer part of cavity surface.	AISI-4140 forged steel	P-20 forged steel

[a]AISI = American Iron and Steel Institute.

Stress relieving of both the core and cavity is necessary after rough duplicating. The steel company should provide the time/temperature requirements for proper stress relieving.

Where slight surface porosity may not be a serious concern, some molds can be made of Meehanite and other casting steels, with considerable cost savings over machining from steel billets. Cast-in-steam cavities can provide better and quicker heat transfer than drilled steam passages. For medium to smaller molds, prehardened steels (32–35 Rockwell C) are a good choice, as they can be freely machined to a high finish. For small, high-production molds, air hardening tool steels that are easily machined in the annealed state are useful. These can be readily heat treated to R_c 50–55 hardness with a minimum of distortion. The pinch-off or resin seal area should be flame hardened to R_c 50–52.

80.6.2 Mold Design

Generally, mold design is relatively the same as for phenolic compression molds except that lighter sections may be used owing to the lower molding pressures required. When molds are to be heated from platens, better heat transfer can be obtained by using aluminum alloys for pillars, ejection bar spacers, and other supporting parts.

Mold Seals or Shear Edges. Most compression molded parts are made in positive or semipositive molds. In the positive mold, the seal, shear edges (left over from preform molding), or juncture of the male and female mold sections at the edge of the molded part telescope with only sufficient clearance to permit the escape of air and not the molding materials. While this type of mold is very critical as to charge weight, it compresses the material to its maximum density; this is the most satisfactory way to mold parts to meet critical electrical and mechanical requirements. The positive mold is most commonly used with an external land, which, with a slight mold overcharge, permits dense parts to be molded with a minimum of size variation.

In the semipositive mold, the female or cavity section is relieved with a taper so that the excess material can escape as the mold closes. It is then increasingly restricted until the closure is positive over the final short length of travel. This type of seal is less sensitive to overcharging in the range of 2–5%, and it provides parts having uniform size, weight, and density.

There are many variations of these seals. Shear edges for SMC molds which cut off excess compound as the mold opens are flame hardened on these edges to minimize wear.

Injection and transfer molds generally do not telescope. The male and female come together in a horizontal plane. Since the mold halves come together before the mold is charged, there is no need to provide relief for excess material. It is practically impossible to have the two surfaces meet exactly, so there is usually enough space to vent air trapped in the mold. For particularly difficult venting problems, local areas can usually be ground a few thousandths of an inch to provide venting.

Venting of blind pockets can be accomplished by a special ejector pin design, as shown in Figure 80.23. Vent pins can be operated by the ejector plate (and will also act as ejector pins). The stroke must be long enough to expose the reduced pin diameter so that it can be completely cleaned.

Mash-Off Designs. When it becomes necessary to include a mash-off area in a mold, the following designs are suggested:

- Small areas less than 1 in. (25 mm). For mash-off areas smaller than 1 in. (25 mm) diameter, square, or rectangular areas with sides of less than 1 in. (25 mm), a uniform clearance of 0.010 in. (0.25 mm) is recommended, as shown in Figure 80.24A.
- Medium sized areas. For areas larger than 1 in. (25 mm) and smaller than 3 in. (76 mm), an outside area with clearances of 0.010 in. (0.25 mm) with interior clearances of 0.045 in. (1.1 mm) is recommended, as shown in Figure 80.24B.
- Large areas. For areas larger than 3 in. (76 mm), a pinch-off area of 0.010 in. (0.25 mm) with an internal clearance of 0.060 in. (1.5 mm) is recommended, as shown in Figure 80.24C.
- Mash-off areas in a mold must be flame hardened to a R_c 50–55. Metal areas opposite the mash-off areas also must be flame hardened.
- Mash-off areas should be highly polished to prevent material from sticking to the mold.

Heel Blocks. Molds should have heel blocks, made of hardened steel against bronze, suitable to withstand all lateral forces at 1800 psi (12 MPa) molding pressure.

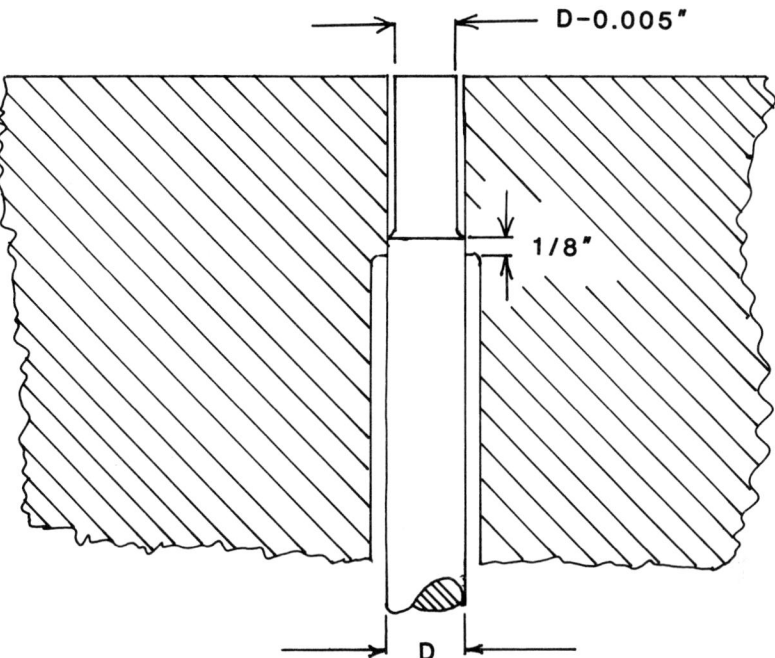

Figure 80.23. Self-cleaning knock-out pin design.

- Heel blocks should be an integral part of the mold (not bolted on).
- Heel blocks should be vertical.
- Heel blocks should engage a minimum of 2 in. (51 mm) before the shear edges engage.
- Heel blocks should have bronze plates on one side.
- Wear plates should have a 1/8 in. (3 mm) minimum chamfer or radius lead in.
- Contact areas opposite the bronze wear plates should be flame hardened to R_c 50–55.
- Grease fittings should be provided on wear plates with facilities to obtain grease from outside the mold if the heel blocks are not readily accessible with a brush.
- Heel blocks must be balanced in area on opposite sides of the mold.
- Wear plates should be bolted to the heel blocks with socket head cap screws and contain dowel pins to prevent them from being sheared off.

Internal Heating Conduits. All steam or oil lines should be connected internally. Exterior looping should be avoided.

- All steam lines should be 3/4 in. (American standard taper pipe threads).
- All inlets and outlets must be permanently identified on the mold.
- Heat line spacing in the core and cavity should be 4 in. (102 mm) center-to-center.
- All pipe plugs should be of the steel hexagon socket type. The threads should be wrapped with Teflon tape before installation.
- A heating pipe pattern should be designed to provide a uniform temperature [within

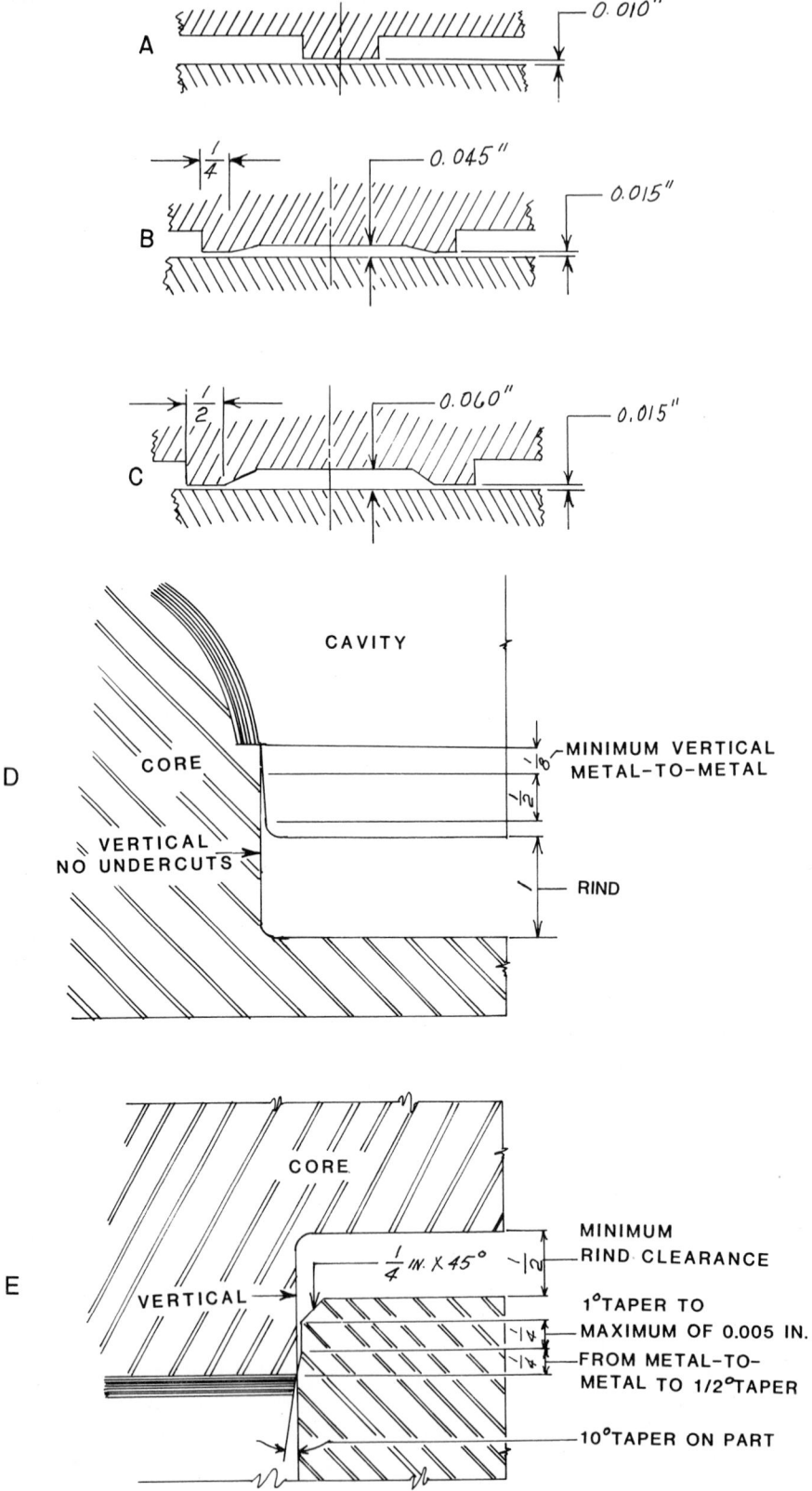

Figure 80.24. Mold seal details

+ 10°F (6°C)] in both halves of the mold with the capability of heating to 320°F (160°C) with 100 psig (0.7 MPa) steam pressure.

• All steam lines should be cleared of all drill chips or other obstructions.

Leader Pins. Leader pins or guide pins should be provided on all molds. At a minimum, the diameter should be 2% of the mold width plus length.

• Guide pins should have a shoulder-type construction to prevent them from being pulled through the core mold half.
• Guide pins should be installed on the core half of the mold; guide bushings on the cavity half of the mold.
• All guide pins should be of identical height and at least 1/4 in. (6 mm) longer than the highest point on the core, with the ends chamfered or tapered for lead-in.
• The length of the guide pin retention in the bushing should be at least 1.5 times the pin diameter.
• Guide pins should be set in from the outside of the mold by at least one pin diameter.
• One pin should be offset to prevent misalignment of the core and cavity.
• A clearance of 0.0005 in. (0.012 mm) per inch (millimeter) diameter should be provided.
• Guide bushings should be bronze plated on steel type with figure eight oil grooves.

Mold Stops. Mold stops should be provided to limit vertical travel during closure. These should be the split type with a maximum thickness of 1 in. (25 mm) or 1/2 in. (13 mm) each mold half. Stops should be made of oil-hardened steel with a R_c 55–60. All stops should be flat and contact each other simultaneously. Stops should be bolted to the mold with 1/4 in. (6 mm) diameter steel socket head cap screws. They should be large enough to withstand the full press tonnage if the press were closed without a charge in place.

Flash Blowout Openings. Openings should be provided for flash removal over the guide bushings at a point beyond the guide pin travel in the bushing. This opening should be at least 1.5 times the pin diameter.

Mold Operating Cylinders. Hydraulic cylinders normally are used to actuate ejector pins and slides in a mold. In specifying hydraulic cylinders for FRP molds:

• All hydraulic cylinders should have rod ends cushioned at both ends, spring-loaded Teflon cups, and Teflon seals.
• Hydraulic cylinders should not be mounted directly to the mold surface. A steel spacer should be used between the cylinder and the mold steel to minimize heat transfer.
• All piston rods should be keyed to prevent unintentional backout.
• Port outlets on cylinders should be positioned to yield minimum maintenance hook up and servicing.
• All ejector cylinders should be installed with linear alignment couplers to reduce rod, seal, and bearing wear and to prevent binding or erratic movement caused by misalignment. A minimum of 1/8 in. (3 mm) clearance must be provided for the alignment coupler in the ejector plate.

Ejector Assembly.

- The core and/or cavity ejector pattern should be agreed upon by the molder and the moldmaker prior to the completion of the mold design.
- All ejector plates should be solid steel, not welded, construction.
- Ejector retainer plates and ejector backup plates should be made of ANSI 1020 hot-rolled steel plate.
- The top and bottom surfaces of all ejector plates should be Blanchard ground to within 0.005 in. (0.13 mm) parallel and flat, prior to any drilling operation.
- Positive return of the ejector plate must be provided for on all molds, with return pins approximately 1 in. (25 mm) in diameter.
- D-M-E type stop pins, or equivalent, should be provided on both sides of the ejector assembly.
- Clearance between ejector pins and ejector sleeves should be a maximum of 0.0005 in. (0.013 mm). To achieve this, fit pins must be selectively picked from stock or plated.
- Ejector assemblies must be provided with a minimum of four ground and hardened guide pins and bushings.
- Ejector plates should be bolted together with socket head capscrews with the head on the clamp plate side of the assembly. Access to the capscrews must be provided for with clearance holes in the clamp plate.

Ejector Pins.

- Use standard D-M-E, or equivalent, ejector pins.
- The smallest diameter pin permissible is 5/16 in. (8 mm). Pin length should not exceed 80 times pin diameter.
- Ejector pins on contoured surfaces must be keyed to prevent rotation and to provide one-way assembly.
- Pin lengths should not be altered more than 1/2 in. (13 mm).
- Head clearances on all moving ejector pins is shown in Figure 80.23.
- Fit areas of ejector pins and ejector sleeves should be 5/8 in. (16 mm).

Ejector Sleeves.

- Use standard D-M-E, or equivalent, ejector sleeves.
- Ejector sleeves and ejector sleeve extensions should of solid, not welded, construction.
- All sleeves and sleeve extensions should be numbered and cross-referenced to their locations in the mold with a like number.
- The ejector sleeve internal edges and surface just below the edge where the fit surface ends must have a 60° included lead angle.

Core Pins.

- Use standard D-M-E, or equivalent, ejector pins.
- The smallest diameter pin permissible is 5/16 in. (8 mm). The pin length should not exceed 80 times pin diameter.
- Pin lengths should not be altered more than 1/2 in. (13 mm).
- Head clearances on all moving core pins are shown in Figure 80.23.
- Fit areas of ejector pins and ejector sleeves should be 5/8 in. (16 mm).

- Stationary core pins are to be held fast with a cover plate.
- All core pins for holes in the line of draw should extend through to the rear clamp plate so they can be removed without affecting the balance of the mold. The head clearance counterbore should be a minimum of 1/2 in. (13 mm) diameter larger than the head diameter.
- All pins should be numbered and cross referenced to their location in the mold with a like number.
- Pin diameter should be 1/16 inch (1.5 mm) minimum larger than pin tip.

80.6.3 Mold Finishing

The very high polish customarily used on thermoplastic molds is not required on FRP molds. The filler and fiber types and content of the formulation appear limit gloss, regardless of the mold finish.

A 600 finish is satisfactory for most FRP molds. This means that the final polishing was done with a 600 grit abrasive. Successively finer abrasive grits are used in approximately 100 grit intervals, starting with a grit sufficiently coarse to remove the machining tool marks. The final polishing should be in the draw or ejection direction to minimize sticking of the part in the mold. The grind area also should be polished to prevent the flash from sticking.

After the mold has been sampled and the part approved to part print, the mold is hard chrome plated from 0.0003–0.001 in. (0.008–0.025 mm) thickness to protect the mold surface from corrosion and minor abrasive damage and to promote part release.

80.6.4 Handling Requirements

- Eye bolt holes must be provided on all four sides of the cavity and core blocks, and located with respect to the block center of gravity.
- Thread diameters of the eye bolts on either mold block should be large enough to lift the entire mold.
- All assemblies that weigh over 200 lb (90 kg) must have their own eye bolt holes for handling during mold maintenance.
- Holes for tie straps for use during shipment and storage must be provided on two sides of the mold. The cavity and core should be blocked open a minimum of 1/2 in. (13 mm) with cold rolled steel plate before the tie strap holes are located.

80.6.5 BMC Mold Specific Design Details

BMC Mold Travel After Shear. Figure 84.24D, and 84.24E show two SMC shear edge designs that have telescoping lengths of 1/2–5/8 in. (13–16 mm). The telescoping shear edge design allows the trapped air to escape before pressure is applied to the molding compound inside the mold cavity. For BMC materials, the depth of this shear edge usually is 1/4 in. (6 mm). A 1/16–1/8 in. (1.5–3 mm) metal-to-metal contact generally is provided to obtain this seal; it is followed by a 1/2-deg taper to the full 1/4 in. (6 mm) depth. A 1/8 in. (3 mm) radius or a 1/4 in. (6 mm) by 45-deg chamfer usually is included to protect the mold.

Loading Wells. As the BMC glass content increases, so does the bulk factor. Normally, no real problems are encountered with BMCs having glass levels of 10–20%. Above 25%, the filler level must be reduced since the bulk factor increases rapidly. Bulk factor is the ratio of the volume occupied by the bulk material to that required for the molded part. Bulk factors usually fall in the range of 2 to 4, but can be as high as 7–8 to 1. For the molding compound to be contained in the cavity of the open compression mold, a loading well must be included. Figure 80.25 shows the design of a mold for a high glass level cup that includes a loading well.

80.6.6 Release and Ejection of Parts from the Mold

Effective part release from the mold begins with the part design. The first item to be decided is in which half of the mold should the part remain as the mold opens. Normally, the part should remain in that portion of the mold that is more easily handled by the operators.

Next, the method to be used for keeping the part in the desired mold portion must be decided. These are

- Incorporate areas in the part without draft.
- Provide shallow undercuts, preferably in large flat areas. These will keep the part in place as the mold opens but will spring out when the part is ejected.
- If all the part surfaces are critical, undercuts in the flash sometimes can keep the part in the desired mold half. The flash may have to be thickened for this method to be effective.
- Provide follower pins (similar to ejector pins) actuated for a short distance as the mold opens to force the part to stay in the desired mold half.

Removal of the hot part from the mold once it is effectively retained is the next consideration. If the part is a simple shape and is retained in the cavity, the cure shrinkage plus the thermal shrinkage as the part cools will often release the part so that it can be picked up by hand (using with suction cups if large flat areas are available). Some parts can be

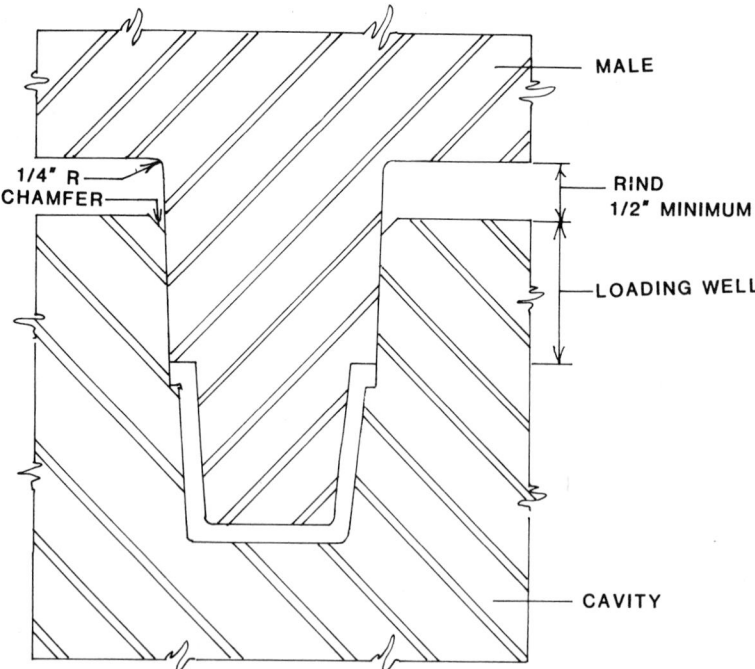

Figure 80.25.   Loading wells in FRP molds for high glass-content compound

removed by blowing air with a hand gun around the part periphery. Air sometimes can be used by incorporating in the mold a spring-loaded poppet valve connected to an air passage. Air pressure will break the vacuum formed during molding and partially eject the part.

The most common and most successful removal method is to incorporate ejector pins into the mold. The greater strength of FRP usually permits fewer pins to be used but its generally lower hot strength requires larger diameter pins. Straight pins with close clearance work well with BMC and SMC, but the hole should be relieved from a point 1.5–2 diameters below the mold surface to provide space for material flowing past the pin.

80.7 MOLDING PRESSES

Several kinds of presses are used to mold reinforced molding compounds. The most common is a four-column upward or downward acting moving platen with a self-contained hydraulic system.

Because of increasing demands for rigidity and parallelism, welded frame presses are becoming more popular. These have four rectangular parallel ways on which the moving platen rides. Adjustable wear plates (gibs) on the moving platen permit very close control of its parallelism. Disadvantages compared to round column presses are that, for the same useful mold space, they are larger and access for mold changes and press loading are restricted.

Some of the more important factors in selecting a press are press capacity, breakaway force, stroke, daylight, operating speeds, controls, and upward or downward acting operation.

80.7.1 Press Capacity

A press intended for BMC or SMC should ideally be able to apply 4000 psi (28 MPa) on the projected area of the part, which provides for a large safety factor for the normal molding range. Press tonnage should be adjustable so that the maximum force is not applied to very small molds.

80.7.2 Breakaway Force

The breakaway or return force normally should be 20–25% of rated capacity, usually attained by the use of auxiliary cylinders.

80.7.3 Stroke and Daylight

Stroke and daylight determine the depth of parts that can be molded on the press.

Daylight should be slightly more than three times the depth of the deepest part to be molded plus an allowance for mold thickness, heating platens, knockout mechanism, and so on. The larger the mold, the thicker the mold plates required to resist distortion. More complicated part designs require more knockout pins over bosses and ribs.

At the minimum, stroke must be at least twice the depth of the deepest part to be molded. Although it can be longer to permit easier handling of shallow parts, it generally is more economical to absorb the extra daylight with bolsters (spacers) under the smaller molds or with ram spacers that fit between the moving platen and the cylinder.

80.7.4 Operating Speeds

The current trend is to increase press operating speeds to minimize cure cycles. Formerly 400–600 in./minute (10–15 m/minute) was satisfactory, recent specifications now call for a range of 800–1200 in./min (20–30 m/min) for closing speeds.

An adjustable slow close cycle (usually engaged with limit switches) is necessary to protect the mold shear edges. The final closing speed should be adjusted to permit trapped air to escape from the mold just before the mold is sealed at the shear edges.

Three press opening speeds also are required; a slow breakaway (to protect the mold shear edges), a rapid return, and a slowdown before the ejector mechanism operates. The first and third usually are about the same speed as the final pressing speed, with the second corresponding to the rapid advance speed.

80.7.5 Controls

Most presses operate semiautomatically. After loading the mold, the operator initiates the cycle with dual pushbuttons. As the platen moves through its stroke, adjustable cams contact limit switches to effect changes in speed from rapid advance to intermediate to final pressing speed. When the dialed final pressing tonnage is reached, a pressure switch starts the cure cycle timer. When the timer runs out, the return phase of the cycle is automatically started with a slow breakaway speed, then a rapid return, followed by a slowdown near the end of the stroke. Most large presses for BMC and SMC now are operated with programmable controllers. These are essentialy if in-mold coating is used.

80.7.6 Upward or Downward Acting Operation

Medium-sized, upward acting presses are considered safer than their downward acting counterparts because the closing mold of the former is generally out of reach of the operator. Upward acting presses of even moderate size require pits to position the platen at a convenient working height. Access must be provided in the form of removable floor panels. The hydraulic pump must be mounted separately, using additional space. Small to moderately sized downward acting presses can be mounted directly to the floor. Figure 80.26 shows a modern SMC press.

80.8 PART DESIGN

The following general principles apply to both BMC and SMC molded parts.

80.8.1 Locate Flash So That It Can Easily Be Removed

On simple parts, the designer can choose whether to locate the mold parting line on the vertical plane or the horizontal plane of the part as it is molded. For optimum production, it is best to locate the parting line so as to make removal of the flash as simple as possible. The glass reinforcing fibers toughen the flash as well as the molding compound.

A dish-type part, for example, can be designed to have either a vertical or horizontal flash. In both cases, a jitterbug sander probably would be used to remove flash. Flash removal is fairly simple and straightforward when a vertical flash is used because the part is supported in a jig with the flash extended upward; the operator simply moves the sander in a horizontal plane over the flash. In other configurations, the part has to be rotated; this is hazardous because the vertical walls near the flash can be gouged by the sander.

80.8.2 Avoid Parting Lines on Show Surfaces

If a rectangular hospital tray made of molded-in color BMC required both top and bottom edges of the tray to be rounded, the mold parting line has to be placed on the horizontal center line of the part. When the flash is sanded off, an unsightly white line is left. This resulted in a large quantity of rejected trays that had to be salvaged by painting, adding

Figure 80.26. Modern SMC press

extra expense. A part redesign allowing the bottom edge sharp, would have permitted a vertical parting line. Then the flash could have been sanded off without interfering with the critically exposed top edge surface, considerably lowering the part cost.

80.8.3 Avoid Blind Pockets

Blind pockets usually occur over deep bosses and some ribs. When air is trapped in those pockets, it is compressed as the compound fills the part, causing voids. In extreme cases diesel burn results. Knockout pins can be placed on bosses and over ribs to act as vents for bleeding trapped air. Figure 80.23 shows the design of a self-cleaning pin.

80.8.4 Provide a Generous Draft on Vertical Walls, Bosses, and Ribs

Since reinforced molding compounds shrink as they cure, it is necessary to provide a draft on all vertical walls, ribs, and bosses in order for the part to be released from the mold. On parts with depths of up to 3 in. (76 mm), a minimum draft of 1° is recommended. On parts 3–6 in. (76–152 mm) deep, a minimum draft of 2 deg is necessary. A 3° draft is necessary for parts deeper than 6 in. (152 mm).

80.8.5 Avoid Molded Through Holes When Maximum Strength Is Required

Whenever reinforced molding compounds are forced to flow around a core pin in a mold, a knit line results that is weaker than the surrounding areas of the part. Cracks frequently occur at that knit line. For maximum strength, it is advisable to drill holes in a secondary operation rather than to mold them in, especially those holes near the outer periphery of the part. If molded-in holes are absolutely required, it is advisable to place the charge over the hole areas and to force the compound to flow away from the holes in all directions, thereby avoiding knit lines in those areas.

80.8.6   Avoid Press-Outs and Mash-Outs Whenever Possible

Inclusion of press-outs and mash-outs in a part design is common practice in preform and mat reinforced parts since the reinforcement is already in place when the mold closes and the thin liquid resin mix still has an opportunity to saturate the reinforcement. However, in BMC and SMC molding, the reinforcement must flow with the resin and filler. A press-out area in a mold creates an obstruction to the advancing fronts. The result is orientation of the glass reinforcement and part weakening in that direction. For this reason, it is advisable not to include press-out areas in BMC and SMC part designs. (An exception is picture-frame parts, which are covered later in this section.) If press-outs or mash-outs cannot be avoided, the compound charge should be placed over the press-out area in the mold to force the flow away from that area.

80.8.7   Avoid Flow From a Thin Section to a Thicker Section

When the moving front in a mold is allowed to flow from a thin section to a thicker section, the back pressure is lost and the moving front begins to cascade.

In one case, problems were encountered with severe porosity on the outer edges of an automobile fender extension. The original part had been made of die-cast zinc. The outer edge contained an indentation to receive a molded rubber gasket. When that part was converted to FRP, it was found that an area of high porosity coincided with the enlarged edge section. The porosity was greatly reduced when the mold was reworked to minimize the transition from the thin to the thicker section.

80.8.8   Molding Picture-Frame Parts Presents Problems

Picture-frame parts present no unusual problems for preform or mat reinforced parts, but they do for BMC and SMC parts. Both these materials must flow in the mold to develop optimum properties. If a mold is charged, as shown in Figure 80.27, a weak knit line will develop. SMC can be charged, as shown in Figure 80.27, but that requires considerable time on the operator's part and precure could result before the mold is fully closed.

One option is to place four narrow strips of continuous strand mat on the mold before the charge is added. The mat protects the compound from the hot mold and reiforces the knit lines to prevent cracks and weakened physical properties.

If large quantities of such parts are to be made of SMC, it generally is less expensive to build a press out in the mold, as shown in Figure 80.27, and charge the mold, as shown in Figure 80.27D. With that procedure, the center section is trimmed in the mold, and the problem with knit lines is eliminated.

80.8.9   Avoid Internal Undercuts in a Part

Internal undercuts on an SMC or BMC part must be avoided, but external undercuts may be molded using slides or split molds. Such slides should be located around the periphery of the part. Slides may be operated mechanically or hydraulically.

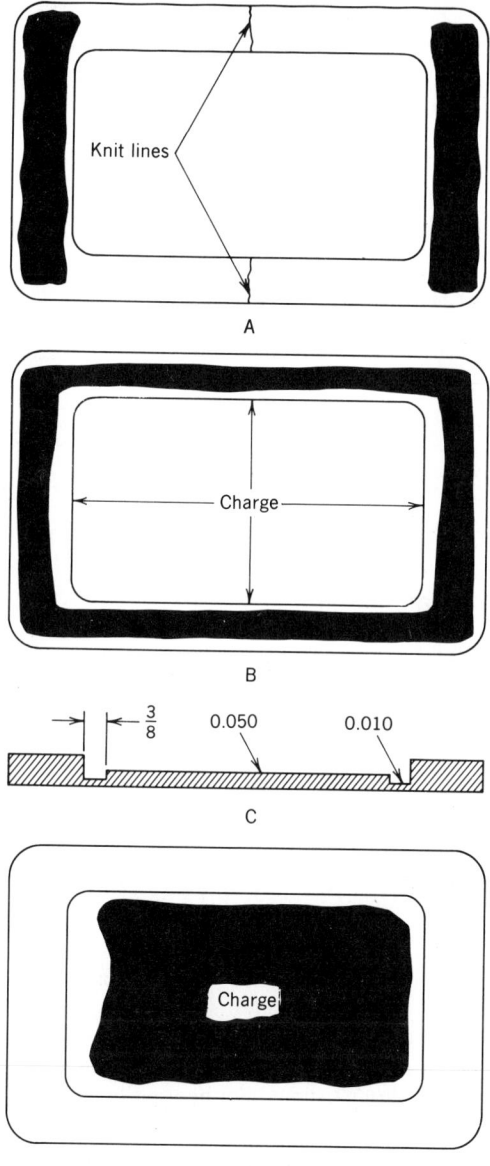

Figure 80.27. Problems presented by picture frame-type parts

80.8.10 Ribs on Nonappearance Parts

Because rib mass affects part cure time, ribs should be used only where they are needed to provide rigidity to a part. Figure 80.28 lists the basic formulae for calculating rib stiffness. Design considerations are not as critical for ribs on nonappearance parts as they are for exterior surfaces. The following design principles apply:

- Ribs should have a minimum taper of 1° per side.
- Lead-in fillets at the rib bases generally should be a minimum of 1/16 in. (1.5 mm).

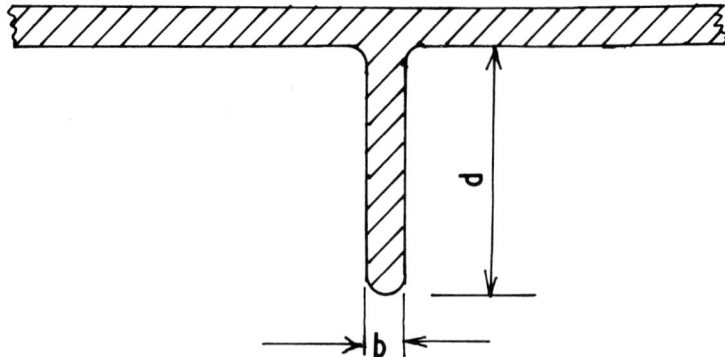

Figure 80.28.   Stiffness calculations for ribs

Ultimate stiffness, $Su = \dfrac{3\,PL}{2\,bd^2}$

Elastic modulus, $Eb = \dfrac{L^3 M}{4bd^3}$

L = length of rib; P = Force in pounds; M = Slope of stress/strain curve.

80.8.11   Ribs on Appearance Parts

Incorporation of ribs into appearance parts is more complex. To start, the proper molding compound must be selected to ensure minimum shrinkage. Even with proper material selection, ribs on appearance surfaces will telescope through unless they are hidden under styling lines or contoured surface areas. Figure 80.29 provides guidlelines and recommendations for designing ribs into appearance parts.

80.8.12   Bosses

Figure 80.30 provides guidelines for bosses on both BMC and SMC parts.

80.8.13   Inserts

Metal inserts can readily be molded into BMC and SMC parts. Inserts should be located perpendicular to the parting lines and secured so that they will not be displaced during compound flow under pressure. Nickel and cadmium plated steel and brass are frequently used. Aluminum has been used on large inserts where minimum weight is important. Unplated copper, brass, and zinc inserts should not be used.

Metal inserts can be either male or female. The portion of the insert to be embedded in the molding compound should be cylindrical in shape and contain undercut grooves or deep knurling, or both, to provide good holding power between the molding compound and the metal insert. Square or hexagonal corners on inserts tend to produce areas of high-stress concentration that can crack under load. Metal inserts should be degreased before use and preferably should be at the same temperature as the mold when they are inserted into it. Otherwise, a cold insert will act as a heat sink and may interfere with the cure of the compound. Proper insert can be easily achieved by storing a supply of metal inserts on the steam platen adjacent to the mold or on corners of the mold. The operator must use gloves or a tool to handle the hot metal inserts.

80.8.14   Corner Details

On box-type parts where corners are involved, it is important to use the largest corner radius possible. Without generous radii, orientation will occur, which can cause the wall of thin

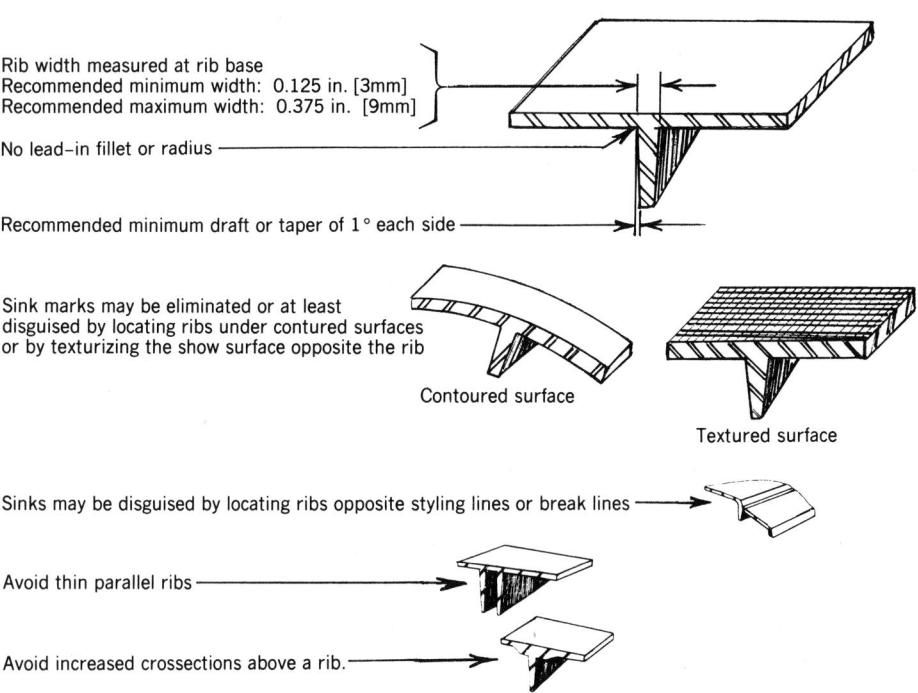

Rib width measured at rib base
Recommended minimum width: 0.125 in. [3mm]
Recommended maximum width: 0.375 in. [9mm]

No lead-in fillet or radius

Recommended minimum draft or taper of 1° each side

Sink marks may be eliminated or at least
disguised by locating ribs under contoured surfaces
or by texturizing the show surface opposite the rib

Contoured surface

Textured surface

Sinks may be disguised by locating ribs opposite styling lines or break lines

Avoid thin parallel ribs

Avoid increased crossections above a rib.

The use of short glass fibers and high glass concentrations are recommended.

Figure 80.29. Guidelines for ribs on appearance parts

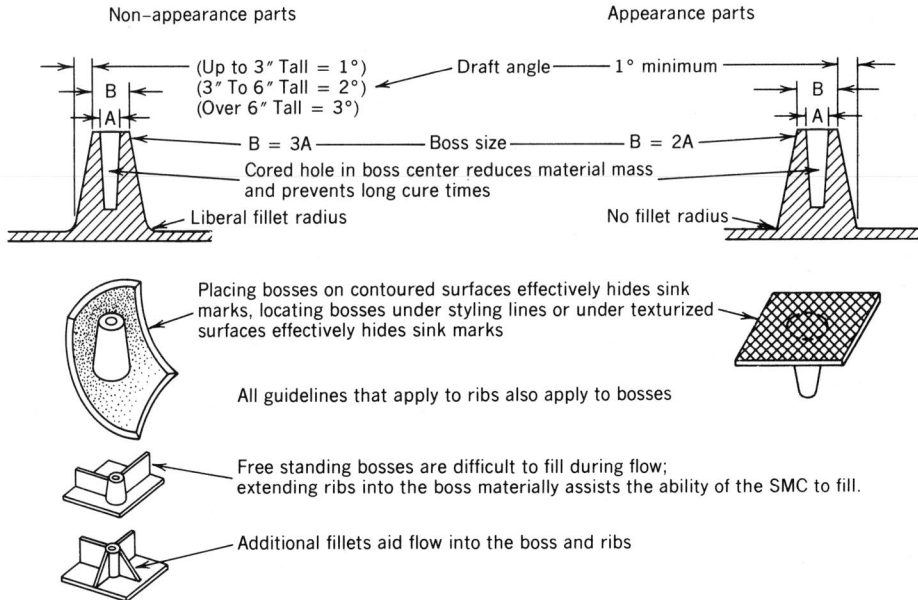

Non-appearance parts

Appearance parts

(Up to 3″ Tall = 1°)
(3″ To 6″ Tall = 2°)
(Over 6″ Tall = 3°)

Draft angle

1° minimum

B = 3A

Boss size

B = 2A

Cored hole in boss center reduces material mass
and prevents long cure times

Liberal fillet radius

No fillet radius

Placing bosses on contoured surfaces effectively hides sink
marks, locating bosses under styling lines or under texturized
surfaces effectively hides sink marks

All guidelines that apply to ribs also apply to bosses

Free standing bosses are difficult to fill during flow;
extending ribs into the boss materially assists the ability of the SMC to fill.

Additional fillets aid flow into the boss and ribs

Figure 80.30. Guidelines for bosses

boxes to warp; particularly important if the boxes have been designed to nest within each other or when lids must mate with the box. Proper charge placement is also important in molding box-type parts. The use of pyramid charges (bottom plies larger than succeeding top plies) helps avoid weak knit lines on box-type parts. Since BMC generally contains shorter glass lengths than SMC, glass orientation caused by flowing around corners is reduced in BMC parts.

80.9 PREFORM MOLDING

Preform molding, also called wet mat or mat die molding, is essentially the same insofar as molding techniques are concerned. It differs in the amount of reinforcement preparation required and in the limited part complexity that can be achieved.

The reinforcement (preform or tailored mat) is made to cover slightly more area than the finished part. The resin mix (polyester resin, fillers, pigments, internal release agent, and catalyst) is applied to the reinforcement, usually after it has been placed on the hot mold. Closing the mold:

- Forces the resin mix to flow and saturate the reinforcement
- Shears off the excess reinforcement at the point of mold engagement (shear edges or seal), at the same time trapping the resin mix until it cures under the influence of heat and pressure.

Although preform molding has been largely superseded by SMC, it remains a useful, though limited, process. When it is design details are not too complex, preform molding can cost 10 percent less than SMC. Uniformity of mechanical properties is excellent because the reinforcing fibers are evenly distributed throughout the part.

Bosses, ribs, abrupt thickness variations, and molded-in inserts are not recommended with preform molding. Also disadvantageous is the high cost strength-to-cost ratio of preform molded products.

Although the number of molders capable of producing preform molded parts is low, there is evidence of a revival of interest in the process. With the addition of low profile additives to the resin mix, the problem of surface smoothness appears to have been overcome. Preform reinforcements also are being used in deep-draw reaction injection molding (RIM).

80.9.1 Properties

As with reinforced molding compounds, the physical properties of preform moldings are largely influenced by the amount and type of reinforcement they contain. Where continuous strand glass mat can be used, it provides the highest strength; however, its use is limited to parts with simple shapes and shallow draws.

The properties of chopped strand mat and preform-reinforced moldings are about 20% lower than those reinforced with continuous strand mat. The practical range of glass content in mat and preform molding is 25–50%, with most falling in the 25–35% range. Table 80.16 identifies some properties of continuous strand mat and preform-reinforced moldings.

Although the physical properties of mats and preform moldings are high compared to those of SMC and BMC, the properties directly influenced by the filler, such as electrical insulation and fire retardancy, usually are inferior. Because the resin mix must flow through and wet the reinforcement as the mold closes, its maximum viscosity and hence its filler content is limited. The different methods of wetting the reinforcement in BMC and SMC processing readily permit as much as 50 percent filler compared to 35% for preform or mat moldings, with 25% glass fiber reinforcement. It is also practical to make BMC and SMC with lower glass contents (less than 25%) and higher filler loadings.

TABLE 80.16. Typical Preform Properties[a]

Resin type	Polyester
Glass content	25%
Specific gravity	1.61
Izod impact strength, ft, lb/in. (J/m)	10 (533)
Flexural strength, psi (MPa)	25,000 (172.5)
Flexural modulus, 10^6 psi (GPa)	1.1 (7.6)
Tensile strength, psi (MPa)	15,000 (103.5)

[a]Ref. 14.

80.9.2 Preform Process

The preform process is a method of collecting chopped glass fibers on a screen in the shape of the item to be molded. A high volume flow of air draws the fibers to the screen and retains them there in a relatively even distribution. A binder, usually in an aqueous emulsion, is sprayed onto the collected fibers and is cured in place to hold the preform to the shape of the screen. The binder content of the finished preform is approximately 5 wt %. The glass is in the form of a roving that is fed to a roving chopper. The chopper cuts the fibers into lengths of 2 in. (51 mm) for most applications. Combinations of several lengths may be used to better conform to the screen contour. For deep-draw, relatively straight-sided moldings, the preforms must be very compact; otherwise, they can be damaged by the shear edge of the cavity as the mold closes. High air velocity and consequently high horsepower is required to make compact preforms. Molded part thickness is limited by the suction of the preform machine; usually, to 1/4 in. (6 mm) maximum. (This requires 3000 cfm (85 m³/minute) and about 6 hp/ft² (48 kW/m²). For greater thicknesses, it is possible to make two separate preforms and nest them, which means two different sizes of preform screens. If the heavier thickness is required only in a localized area of the molding, pieces of glass mat can suffice. Two types of preform machines are in use: the plenum chamber type (Figure 80.31) and the directed fiber type (Figure 80.32).

With the plenum chamber, the glass rovings are fed into a cutter molded on its top. The falling strands are drawn onto the preform screen by suction as the binder is sprayed into the chamber. The perforated screen is usually mounted on a rotating table for better distribution of the deposited glass fibers. A set of timers controls fiber and binder deposition. After sufficient glass has been deposited, the preform is placed in an oven where it is dried and the binder is cured. The preform is then removed from the screen, and the screen is returned to the plenum chamber. The process can be mechanized so that one screen is in the plenum chamber while others are in the oven. Some machine designs use a vertical merry-go-round.

In the directed fiber process, air carries a stream of chopped fibers directed by the operator onto a perforated screen. The air inside the screen is exhausted by a fan, creating a suction which draws the chopped glass to the screen and holds it there. The binder is sprayed on simultaneously with the glass deposition from a spray gun. The operator manipulates the glass/binder stream to suit the contours of the preform screen, building up thicker sections as required. The preform screen usually rotates to ensure that each surface is brought into the operator's range. The roving cutter can be preset to turn off when the proper quantity of glass has been deposited. The rate of deposition depends on the speed of the chopper but is usually maintained at approximately 1 lb/minute (0.4 kg/minute). After the stream of glass fibers is turned off, the operator continues to spray binder to ensure thorough wetting of the preform. The preform with the screen is then placed in an oven to remove water and cure the binder. A parting agent is usually applied to the screen to prevent sticking.

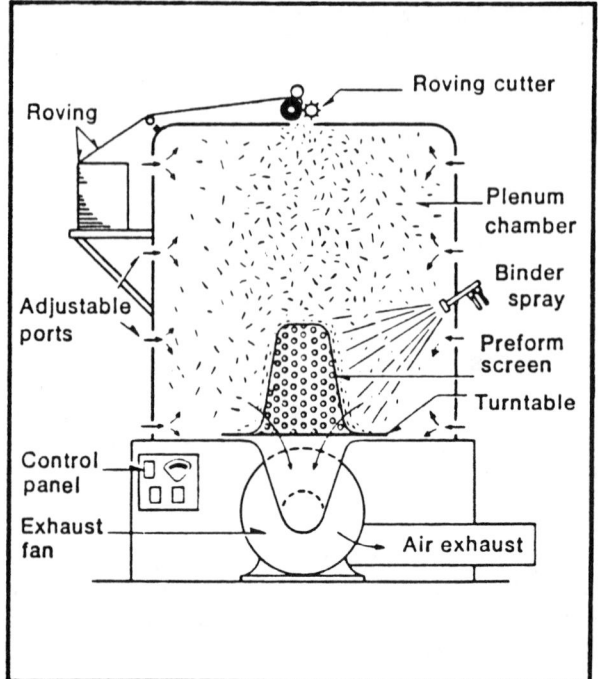

Figure 80.31. Plenum chamber-type preform machine

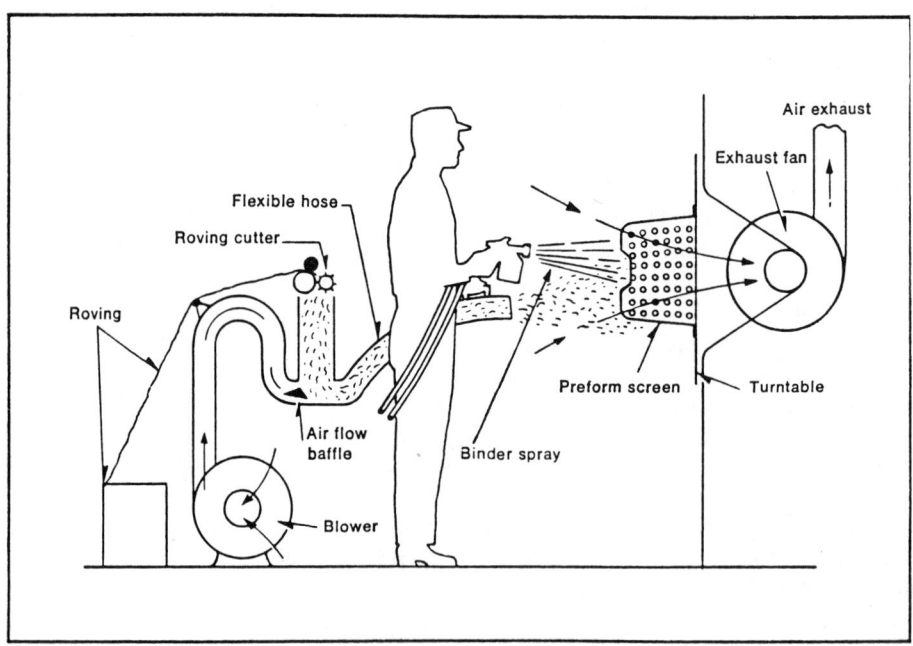

Figure 80.32. Directed fiber-type preform machine

80.9.3 Preform Screens

Preform screens are fabricated from 16–18 gauge mild steel perforated sheet metal with approximately 40% open area. The holes may be 1/8 in. (3 mm) in diameter on approximately 3/16 in. (5 mm) staggered centers. If the part is round or ogive shaped, no problems should be encountered with hole spacing. If the part is rectangular or complex, larger holes may be required at the corners and edges to ensure even deposition of fibers. On complex shapes internal baffling is sometimes used to control air flow.

Teflon emulsions can be baked onto the outside of the preform screen to effect good release of the preform, or a release agent must be periodically sprayed onto the screens. The screens should be cleaned regularly to remove excess binder and release agent. The best cleaning procedure is to soak the screens in a bath of hot alkali solution followed by wire brushing of stubborn areas.

80.9.4 Preform Binders

Preform binders are an important part of the preform process. Most binders are aqueous emulsions of reactive polyester resins that are peroxide catalyzed. Some contain wetting agents and additives to control pH. Emulsions in the range of 5–10% solids are sprayed onto the preform to give a final ignition loss of from 5–7%. In addition to polyester resins, other polymers have been used including melamine–formaldehyde, two-component modified urea–formaldehyde and polyamide resin, and so on.

Solvent binder systems have been used in the past, with some success. Their cost usually is higher than the emulsion systems, and they present fire dangers.

Solid catalyzed polyester resins also have been successfully used as preform binders. They cost considerably more than the emulsion polyesters.

80.9.5 Molding

When chopped strand mat is used as the reinforcement, it is tailored to fit the mold and sometimes held in place by stitching or the use of an adhesive. Continuous strand mat has more ability to conform to the shape of the mold. Unless the part is very complex, continuous strand mat is generally used without tailoring other than cutting to the peripheral shape required. In most cases, the mat material is molded in the cavity (female) side. The bottom portion of the male mold pushes the reinforcement into place with a minimum of disturbance or tearing.

Preforms generally are loaded into the male mold (Figure 80.33), which minimizes disturbance to the preform in most configurations. However, there are some cases (such as parts with a great deal of draft) where loading the preform into the cavity is advantageous.

On small parts, the resin mix can be applied to the preform (never on the mold surface where precure would result). On large parts, it is necessary to spread the mix over a larger area. A charge pattern must be worked out by trial-and-error to arrive at one that will properly fill out the preform. Careful control of the final pressing speed is necessary so that the resin mix flow is not so fast as to disturb the reinforcement, yet fast enough to prevent precure of the resin mix.

Molding conditions for preform molding are similar to those for reinforced molding compounds except that the unit pressure required is substantially less. A maximum molding pressure of 500 psi (3.5 MPa) may be required; 200 psi (1.4 MPa) is usually sufficient. Often, the required pressure is dictated by the shearing force required to cut the reinforcement. In thick parts, this may amount to several hundred pounds per lineal inch.

Molding temperature and cure time are controlled principally by the time required to charge the resin mix and fill out the mold. To prevent precure during this portion of the cycle, it is often necessary to compromise and use a longer cure time.

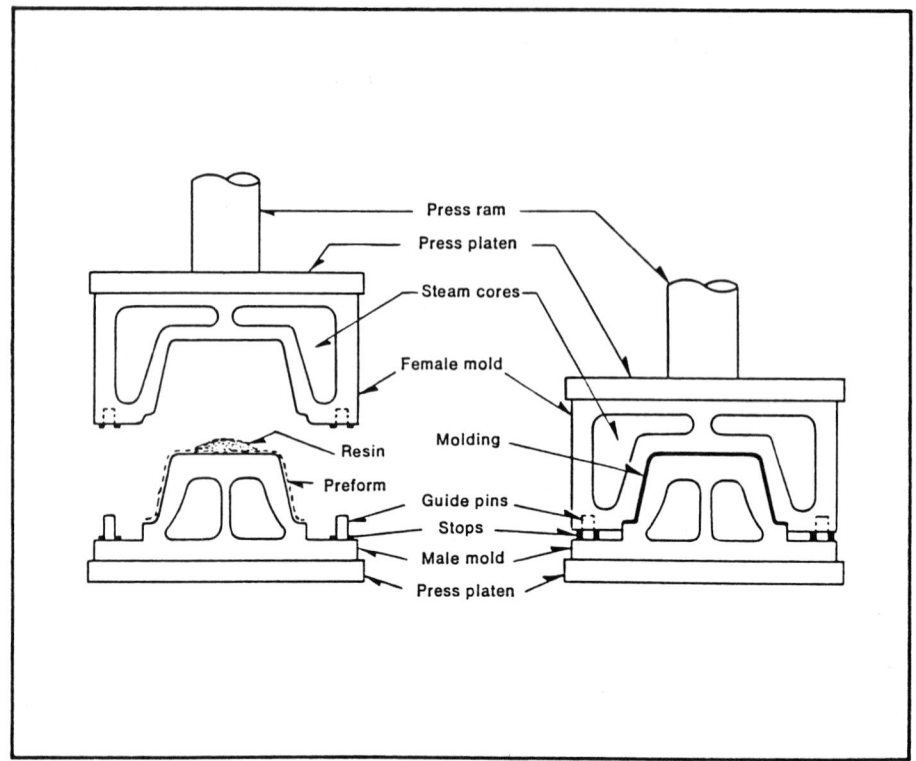

Figure 80.33. Schematic diagram of typical preform molding

Small parts may have an overall molding cycle of three minutes; very large parts can require 15 minutes. Mold temperatures range from 220–300°F (104–149°C), with larger parts favoring the lower temperatures. Usually, internal mold release is added to formulations. External mold releases in the form of waxes, low molecular weight polyethylene, and solutions of superphosphate are applied to the mold surfaces.

Removal of parts from the mold is usually accomplished with the aid of an air blast from a hand-held nozzle or suction cup lifters. Ejector mechanisms employed with reinforced molding compounds are not practical or as necessary as with preform molding. After removal from the mold, most parts must be placed on cooling fixtures to prevent distortion. Since small, molded holes are not feasible in preform molding, provisions for punching, drilling, routing, and so on may be included in the cooling fixture.

80.9.6   Resin Mix Formulations

Most types of polyester resins (isophthalic, orthophthalic, bisphenol, HET anhydride) and vinyl esters can be used in mat and preform molding, but they must be formulated to a relatively low viscosity in the range of 800–1500 centipoise or be able to be reduced in viscosity by additional monomer. The quantity of filler loading is limited by the low viscosity requirement, and the dominant filler must be one that does not separate from the resin as the mix flows in the mold. China clay is commonly used alone or in combination with a low oil absorption calcium carbonate. Aluminum hydrate, effective in BMC and SMC for improved flame retardancy and electrical properties, cannot be used in sufficient loading to have a similar effect in preform and mat formulations. However, when aluminum hydrate is used with a halogenated resin, some reduction in smoke generation and flame spread can be achieved. Some typical formulations are given in Table 80.17.

TABLE 80.17. Typical Formulations for Preform and Mat Molding[a]

Ingredients	Percent of Resin Mix				Percent of total
	1	2	3	4	
Polyester resin	65	49	69	33	
Low profile additive	0	0	0	0	
Catalyst	$\frac{1}{2}$–1	$\frac{1}{2}$–1	$\frac{1}{2}$–1	$\frac{1}{2}$–1	
Release agent	1	1	1	1	60–80
Pigment	5	5	5	5	
China clay	29	20	25	20	
Calcium carbonate	0	25	0	25	
Preforo or mat reinforcement					40–20

[a]Ref. 15.

80.9.7 Mold Construction

Matched metal molds for mat and preform molding differ from molds for BMC and SMC in two ways:

- The need for a shearing action to cut off the excess reinforcement at the edge of the part
- A requirement for close control of the mold space dimensions.

The shear edges on both mold halves must be flame hardened to be able to cut the highly abrasive glass fibers effectively. The fit between the molds at this shear edge must be quite close [0.002–0.004 in. (0.05–0.1 mm)] to ensure a shearing action. This degree of fit is expensive to obtain and difficult to maintain. Even small temperature differences in large molds can change the clearance from too much to interference (which usually results in mold damage). Guide pins must be substantial and close fitting to prevent displacement and damage to the shear edges. Since the reinforcement will not flow to accommodate variations in mold space, both the mold space and reinforcement thickness must be closely controlled to prevent resin-rich and resin-starved areas in the molded part. Variations greater than 0.010 in. (0.25 mm) in a 0.100 in. (2.5 mm) thick part can cause problems. When variations do occur, the only way to obtain a good appearance part is to reduce the reinforcement to accommodate the minimum mold space. This, of course, reduces the strength of the part.

80.9.8 Presses

Except for lower tonnage requiremenmts, press specifications for preform and mat molded parts are similar to those used for BMC and SMC.

80.9.9 Part Design

In general, preform and mat molding are best used for products of relatively constant thickness with generous radii at any change in direction. The practical thickness range is from about 1/16 in. (1.5 mm) to about 1/4 in. (6 mm). Transitions from one thickness to another should be made over a distance of several times the part thickness, but every effort should be made to use a constant thickness. Radii should be about four times the thickness. Molded holes require shear edges in the mold around the periphery. Bosses and ribs should be avoided, if at all possible. When they are required insertion at the time of molding of pieces of BMC or extra pieces of mat can help. Stiffening of unsupported edges is best accomplished by turning (flanging) rather than by thickening. Draft angle on vertical walls should be the maximum allowable but never less than 1° and preferably 2°. Sometimes, one side of a part can be straight if the other has very generous draft and the mold can be tilted to provide molding draft.

80.10   COLD PRESS MOLDING

Cold press molding is a matched die molding process using low pressure and unheated molds, usually constructed of FRP.

The first phase of the process is essentially the same as making a hand layup, sprayup, or rigid thermoplastic product, except that it is done on one set of marched molds that are closed before cure takes place. Principal advantages over the open mold processes are

- Modestly smooth and dimensionally controlled back surface
- Relatively low tooling cost (compared to heated metal molds)
- Low press cost since not much pressure is required [50 psi (0.34 MPa)]
- Rapid toolup time only slightly longer than for open mold processes
- Possibility of gel coats or thermoformed thermoplastic surfaces on one side.

Disadvantages include:

- Long molding cycle, essentially the same as open mold processes
- Short mold life, resulting from high exotherm during resin cure with no way for heat to dissipate from poorly conductive plastic molds. (Molds can be made more elaborate and costly with spray metal faces and cast in cooling coils.)
- Parts must be machine-trimmed after molding.
- Surfaces not gel coated (or otherwise treated) need substantial postmold finishing for a quality finish. ´
- Ribs, bosses, inserts, and complex shapes are not practical.

Materials include those normally suitable for open mold processes:

- Low-viscosity, highly reactive polyester resins
- Chopped and/or continuous strand mats
- Sprayup roving chopped in place
- Open mold gel coats
- Room temperature cure catalyst/accelerator systems.

Applications include any product of relatively simple shape that could be made by open mold processes but needs more precision in its thickness or relatively good finish on both sides, and that will be made in sufficient volume to justify the mold cost.

80.11   RESIN INJECTION MOLDING

Resin injection molding (also referred to as resin transfer molding) is another process for obtaining two relatively good appearance surfaces with better dimensional control than that obtained with open mold processes.

In this process, the reinforcement (preform or tailored mat) is loaded into the mold, the mold is closed, and the resin is injected or transferred into the mold to impregnate the enclosed reinforcement. The process has most of the advantages and disadvantages of the cold press molding process. However, a press is not required. With simple parts injected at low pressure, trunk latches can be strong enough to clamp the molds shut during cure. However, when more complex parts are molded, the mold structure and clamping device can become quite massive. While a press is not required to apply pressure for molding, some device must be used to handle the molds. This could be a low-tonnage press or an electric hoist. When the mold is closed, catalyzed resin is injected through a port at a low point. As it rises in the mold, it pushes the trapped air out, at the edge of the part or from vents. Where the two molds join, the reinforcement is usually partially compressed so that air can pass but flow of the viscous resin is slightly restricted. As with cold press molding, the exotherm of the cure has a tendency to degrade the plastic mold, and more elaborate and costly mold construction can be employed to speed production. Where production volumes justify the additional costs, molds are made of aluminum and contain embedded copper coils that circulate warm water to speed up the cure. For aircraft parts, aluminum molds have been used for years with moderately curing polyester and epoxy resins. These products generally use stain weave glass cloth reinforcement. To impregnate this material effectively, it is necessary to draw a vacuum to evacuate the trapped air and pump the resin in under pressure. After charging the mold, it is brought to curing temperature by the use of build-in heaters or by placing the mold in an oven.

80.12 FOAM RESERVOIR MOLDING

Foam reservoir molding (also known as elastic reservoir molding) consists of making a sandwich of resin impregnated open-cell flexible polyurethane foam between outer layers of glass-fiber mat reinforcement. When the sandwich is placed in a mold and pressure applied, the foam is compressed, forcing the resin into the outer layers. The elastic foam exerts sufficient pressure to force the face layers into contact with the mold surfaces.

When the components are properly proportioned, it is possible to produce a molding having the advantages of sandwich construction; that is, a low-density core with strong faces that provide a favorable stiffness-to-weight ratio.

The required molding pressures are low (less than 100 psi [0.7 MPa]). Applications include relatively large, flat shapes, such as decks, hoods, and vehicle roofs. The advantages include:

- Low-density parts
- High impact strength
- Good bending strength
- Relatively fast molding cycle.

Acknowledgments This section was prepared by updating P. Robert Young's "Thermoset Matched Die Molding," in G. Lubin, ed., *Handbook of Composites* Van Nostrand Reinhold, New York, 1982, Chapter 15, and by including large portions from the author's *Handbook of Polyester Molding Compounds & Molding Technology*, Chapman & Hall, New York, 1986. The author thanks P. Robert Young, Chapman & Hall, Van Nostrand Reinhold for the use of this material.

BIBLIOGRAPHY

1. P.R. Young, "Thermoset Matched Die Molding," in G. Lubin, ed., *Handbook of Composites,* Van Nostrand, New York.

2. D.L. Denton, "Mechanical Properties Characterization of an SMC-R50 Composite," *paper presented at the 34th Annual SPI Conference, RP/CI,* Washington, D.C., Feb. 1979, paper 11-F.

3. B.D. Pratt, "Factors Affecting Arc Resistance of Premix," *paper presented at the annual SPI Conference,* Feb. 1965.

4. Ref. 1, p. 411.

5. R. Burns, *Polyester Molding Compounds,* Marcel Dekker, Inc., New York, 1982.

6. R.W. Meyer, *Handbook of Polyester Molding Compounds and Molding Technology,* Chapman & Hall, New York, 1986.

7. *Ibid.,* p. 109.

8. *Ibid.,* p. 77.

9. *Ibid.,* p. 114.

10. *Ibid.,* pp. 118–121.

11. R. Gruenwald and O. Walker, "Fifteen Year's Experience with SMC," *Sec. 1-A Proceedings SPI 30th Meeting,* Washington, D.C., 1975.

12. U.S. Pat. 3,560,294 (Feb. 2, 1971) W. H. Potkanowicz.

13. Ref. 6, p. 184.

14. Ref. 1, p. 406.

15. Ref. 1, p. 444.

81

REINFORCED PLASTICS, PULTRUSION

W. Brandt Goldsworthy

President, Goldsworthy Engineering, Inc., 23930 Madison Street, Torrance, CA 90505

81.1 DESCRIPTION AND HISTORY OF THE PROCESS

Pultrusion, a coined word, refers to a single-step, continuous reinforced plastics raw materials conversion system similar to extrusion. As materials of construction are pulled through a heated die (hence the term "pultrusion"), polymerization of the resin occurs to form continuously a rigid cured profile corresponding to the die orifice shape. Various combinations of unidirectional and off-axis reinforcing filaments or fabrics may be used. The emerging end product is a constant cross-sectional shape of infinite length. No further processing is required, except to cut the stock to the desired size.

This section will cover fiber reinforcement and resin matrices only, not metal or ceramics.

The process consists, essentially, of those elements depicted in Figure 81.1. Resin-wet reinforcements are drawn into the system through squeeze-out bushings, which remove excess resin, and are optionally preheated by dielectric radio frequencies (5–100 MHz) or microwave frequencies (915–2450 MHz) or, in cases where conductive reinforcements such as graphite are included, by induction preheating. The resin-impregnated reinforcing roving then enters the heated forming/curing die, and the cured stock is pulled out of the die by suitable pulling devices.

History goes back to the late 1940s, when the pultrusion was invented and the process and equipment were developed. By the early 1950s, pultrusion equipment for production of simple power pople lanyard-rod stock existed in several locations. Another early machine for a captive operation was a pultruder built in 1958 specifically for hammer-handle rod stock.

Pultrusion's debut was marked by considerable variety in machine types, including continuous pull and intermittent pull, horizontal product direction and vertical product direction, caterpillar tractor-type pullers and reciprocating clamp pullers. In these early years, all pultruded products had one common characteristic: they were all purely unidirectionally reinforced, and the industry was therefore excluded from markets that required products with off-axis fiber orientation for structural purposes.

The process was revolutionized with the development of in-line infeeding of multidirectional reinforcement, which enabled the production of pultrusions with off-axis structural properties.

In the mid-1960s, experimental efforts focused on pulling fiber-glass roving in conjunction with continuous swirl mats, primarily for production of ladder rail channel used in electrical safety ladders. Although it is a relatively simple channel shape, the fiber construction requires extra longitudinal rovings at strategic points, making it a complex profile for its time. After

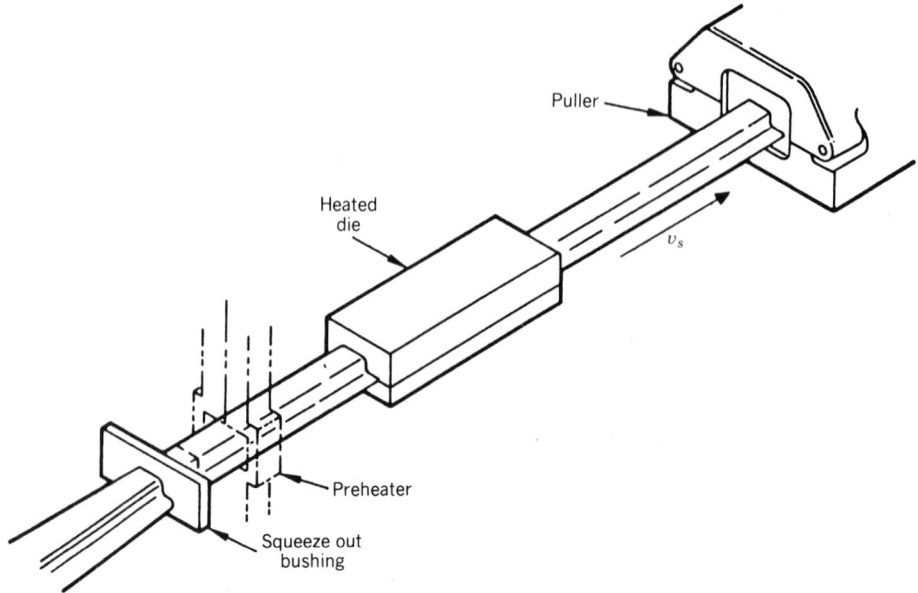

Figure 81.1. Pultrusion process elements.

considerable development effort, the pultruded ladder rail was approved by Western Electric in 1972 for its own company ladder use. Since that time, literally millions of lineal feet have been produced annually by several suppliers.

Further refinements in off-axis capability were achieved in the early 1970s with the development of a high-strength square tube pultrusion application. This led to later round tube manufacturing equipment, in which it was proved that virtually any fiber orientation and laminate schedule could be pultruded to produce stock with structural properties equivalent to traditional metal stock.[1]

As shown in Figure 81.2, this round tubing construction starts with a combination of an inside ply of polyester wall mat brought into the system through a top-side spool, a ply of random glass fiber, and a ply of longitudinal fibers. Because of the thick wall, this initial material package is then taken through an impregnating bushing. After the longitudinals and the first two plies of veil and mat are impregnated, another ply of random glass mat is added, and a ply of pure circumferentials is wrapped around the package by a winding wheel. Another ply of random mat is added, following which a +45 deg ply and a −45 deg ply are added by large winding wheels. Another impregnating bushing wets out this last group of plies. The package then enters the radio frequency (RF) system, where energy is introduced. In this actual case, after this tooling had been designed and built, the customer requested another ply or random fiber mat and a ply of C-glass (chemical glass) on the outside. Therefore, although it is not recommended, these plies were added after the fact downstream of the RF cabinet. The product then proceeds through the pullers in the usual manner.[2]

At the same time, development work was under way to standardize equipment and increase production rates. The first universal pultrusion machine, the Glastruder (which came on the market in 1967) was a complete production system comprising all the necessary machine elements: fiber-glass roving creels for 420 roving packages, universal resin tank for single or multiple profiles, start-up device, profile adjustable reciprocating clamp pullers, and automatic cutoff saw. A dielectric cure station was also incorporated, which was dramatically upgraded two years later by another key innovation in curing heat systems.[3]

Fundamental to the pultrusion process is control of the resin/catalyst chemical reaction

Figure 81.2. Pultrusion machine tooled for round tubing, with in-line feeding of off-axis reinforcements. Courtesy of Goldsworthy Engineering Inc.

under the influence of heat such that the resin gellation point and the peak exotherm point both occur inside the steel die while the curing mass is moving downstream. Otherwise, cure is incomplete, and exotherm occurs outside the die, resulting in profile cracking and uncontrolled surface finish.[4] Throughout the 1950s and 1960s, all pultrusion machines used one of two curing methods: dielectric curing with radio frequency (RF) energy in RF transparent dies, or thermal curing with conductive heat in metallic dies. Both systems had inherent disadvantages. Steel dies were expensive, and thermal heating is considerably slower than dielectric curing. On the other hand, the less expensive RF transparent dies did not lend themselves to complex shapes. They also wore out far more rapidly than steel dies, making tolerance control a constant vigil. These disadvantages were resolved in 1969 with the invention of the so-called "augmented cure" process and equipment, which combined the two curing methods in synergistic fashion.[5] Augmented cure is achieved by installing an RF generator between the wet-out and die sections, which provides instantaneous molecular frictional heat throughout the wet fiber/resin mass prior to die entry.

The preheated mass then enters a steel die, which, in conjunction with augmented cure, is heated by the common methods of electric strip heaters or similar means (primarily to prevent heat loss rather than provide cure heat). The result is that less of the steel die length is used to reach the resin gellation point temperature, and material die residence time is consequently shortened. As a result, this significant invention dramatically increased production speeds, particularly for heavy sections.

It should perhaps be noted here that while no theoretical technical reasons prevent off-axis production using augmented cure, typical off-axis reinforcing mat materials do not perform well with RF heat. Mat resin/binder systems tend to soften under the influence of RF heat, and, because hydraulic shear forces at the die entrance increase exponentially with an increase in speed rate, the mat cannot be run successfully at the higher speeds that are the goal of augmented curing.

81.2 MATERIALS

All types of filamentary reinforcements are pultrudable, rovings, tows, mat, cloth, and any combination thereof. Standard E-glass in one form or other is used in the bulk of commercial products, but high-modulus fibers such as graphite may also be pultruded.

Pultrusion-grade resin matrices are also available in a variety of systems utilizing polyester, epoxy, and vinyl ester resins, with polyester by far the most common. As can be seen by the partial list on Table 81.1, an extensive number of resin systems have been specially formulated for the pultrusion process. This is by no means an exhaustive list, only an indication of the types and variety of pultrusion-grade resin systems available and some of the better-known suppliers.[6] A purchasing catalog such as the *Thomas Register, Modern Plastics Encyclopedia, SPI Buyers Guide, Plastics World Directory,* or the like will reveal a far more extensive number of resins and suppliers.[7-10] Desirable characteristics include high heat distortion, fast cure, and good wet-out.

Processing variables can be manipulated to complement specific resin types. For example, to obtain high-quality composites with epoxies, pressure should be applied simultaneously with gellation, as exotherm occurs prior to the time that gellation occurs, rather than after, as is the case with polyester. Epoxies frequently exhibit higher viscosities than polyester, and thus require longer filament wet-out times. In addition, it is more difficult to remove entrapped air and dissipated volatiles in an epoxy system, although solventless systems are available that are particularly suited to pultrusion. Other main differences are lower polymerization shrinkage percentages and higher interface adhesion forces exhibited during pultrusion with epoxies than with polyesters. However, interface adhesion can be minimized by such techniques as slip sheets, die conditioning (plating and waxing), and internal die releases.[11]

In an apparent no-pressure process like pultrusion, the use of condensation reaction resin systems would seem to be impractical, as the reaction water should create unacceptable porosity. However, development work done in the pultrusion of specific phenolic resins indicates that they are pultrudable without porosity, and this may open a whole new marketing area; that is, building and construction materials where nonflammability, low smoke generation, and low toxicity are important.

Special pultrusion grades of vinyl ester appeared on the market in the late 1970s, and they have been very successful. Although they are not yet widely used, vinly esters have several striking advantages and show great promise for pultrusion in the future. Their primary advantage is that they offer physical properties and heat-distortion temperatures approaching those of medium-grade epoxies but with better performance at better prices than epoxies. Vinyl ester resin system prices generally fall between the prices of polyester and epoxy systems. The processing parameters are essentially the same for vinyl ester as polyester. The former tends to be a low-viscosity, fast-curing system.

Reinforcement type (glass, graphite, boron, and so on) and form (tow, roving, mat, cloth) are also important considerations, which influence such properties as:

- Intrinsic strength and stiffness of the part (adequate axial strength must be present to sustain process pulling forces)
- Thermal conductivity and specific heat (second-order effects, influencing the rate of heat transmitted from die and so on)
- Volume ratio of reinforcement-to-matrix (controls mass effect during curing). If dielectric or inductive preheat is to be used, the electricial properties of the filament also become significant (see Table 81.2).

Dual-compatible (polyester and epoxy) and resin-system-specific rovings are available in single- and multiple-end packages with high yields. Cloth and mat materials tend to slow

TABLE 81.1. Pultrusion Grade Resin Systems

Manufacturer	Resin Type	Trade Name/ Number	Pultrusion Grade/Application
Koppers	Polyester	Dion 8200	Said to solve exotherm cracking problems by relieving large shrinking strains caused by high exotherm.
	Polyester	Dion	Capable of continuous use at temperatures over 300°F. (New development aimed at sucker rod market.)
Ashland	Polyester	EP 34456	"Reactive flexibilizer" blended with standard grade to reduce exotherm/ cracking and increase line speed (for 1-in. dia. rods).
	Brominated polyester	Hetron 613	For flame retardant applications.
	Polyester	Hetron 197A	High chemical resistance, especially in acid environments (but not caustic- and hypochlorite-resistant).
	Vinyl ester	Hetron 922	High chemical resistance; resists caustic and hypochlorite.
	Vinyl ester	Hetron 980	High physical property retention at elevated temperatures. (Sucker rod resin system under development).
Dow	Vinyl ester	Derakane 411-35 and 470-25	Low styrenated, high viscosity (aimed at sucker rod market). Good physical properties and high temperature resistance at low filler loadings.
	Vinyl ester	Derakane 411-35	Low inhibitor version offers higher line speeds.
Reichold	Polyester	Polylite 31-020	Highly reactive isophthalic polyester. Said to greatly increase line speeds in small diameter rods and profiles.
	(Polyesters)	(under development)	(For thicker profiles; very high heat resistance. Aimed at sucker rod market.)
	(Epoxies)	(under development)	(For increased line speed; higher physical properties; accelerated B-stage; good green strength.)
Shell	Epoxy	Epon 9102 and 9302	Increased line speed (up to 3–4 ft/min for 1-in. rod). RF curable (first epoxy with this claim; without RF, line speed approx. 6–10 in./min.). Viscosity and shrinkage similar to polyesters.
	Epoxy	9310	Also recommended for pultrusion.
Alpha Resins Collierville, Tenn.			Resins to improve wetout, cure speed, flame resistance, weatherability, paintability, and surface appearance under development.

TABLE 81.1 *(Continued)*

Manufacturer	Resin Type	Trade Name/ Number	Pultrusion Grade/Application
Freeman Chemicals, Port Washington Wisc.			Resins to increase speed and improve electrical properties under development.
Silmar Div., Sohio, Hawthorne, Calif., & Covington, Ky.			Existing resin systems being adapted for pultrusion.

production, as discussed, relative to off-axis reinforcement; however, recent innovations in nonwoven, knitted reinforcements show great promise as they appear not to be subject to the weaknesses of conventional glass mat or cloth. The main reinforcements are stitched together in straight lines, rather than woven over and under each other; there are thus considerably fewer "kinks" to create stress problems when the laminate is under tension. Stitched reinforcements are now on the market in uniaxial, biaxial, and triaxial form, consisting of glass, glass–graphite, or various hybrids of glass–graphite–Kevlar. This material will give structural pultruded profiles better transverse tensile strength. When they are used in conjunction with continuous strand mat, they will also give the pultruder a choice of either a higher-performance profile or equal performance to that achieved with roving and continuous strand mat in structural profiles—but in a thinner laminate.

Another benefit of this material is the ability to precisely position anywhere within the warp and fill directions the type of reinforcement needed for the application's desired properties. Hybrid reinforcement system designs can now be considered; they function quite specifically in the pultruded laminate. Another exciting development is the introduction of pultrudable polyester fibers, which would fit applications that do not require the rigidity of glass fibers but could benefit from the impact resistance of polyester. Hybrid glass fiber–polyester fiber composites would give the dual advantages of the lower modulus polyester

TABLE 81.2. Electrical and Thermal Properties For Filament and Resin[a]

Material	Electrical		Thermal	
	Loss Tangent	Dielectric Constant	Spec. Heat, cal/°C/g	Thermal Cond. cal/cm/s cm^{10} C$(10)^4$
E-Glass	1–3 5	6.32	0.19–0.25	8–3
Graphite	NA	NA	0.172	2400
Epoxy	20–60	3.2–4.8	0.25	4–32
Polyester	6–30	2.8–4.2	0.30	4
Polyimide	6–15	3.5–3.8[b]	0.25	4–32
Glass-polyester	13–20	3.9–4.2[b]	0.25	10–16
Graphite-epoxy				400 (p)[c]
				15 (t)[d]
Kevlar-49-epoxy	10	3.2		40.3 (p)[c]
	14–45	3.6–4.0		3.4 (t)[d]

[a]Ref. 11.
[b]1 MHz.
[c]p = parallel to fibers.
[d]t = transverse to fibers.

TABLE 81.3. Pultrusion Grade Fiber Reinforcements

Manufacturer	Type and Form	Trade Name/ Number	Comments
Allied	PET roving		Poly(ethylene terephthalate) fiber low modulus, low performance, low cost.
Owens-Corning Fiberglas	Fiber glass roving	E-glass	Standard commercial product.
	Fiber glass roving	S-2 glass 463 and S-2 glass 449	Glass compatible; improved shear performance, low cost/high performance; suitable for military applications.
	Fiber glass roving	425	Conventional roving for general purpose pultrusion. Available in 56 and 113 yields; low catenary, very good dispersion; good processing and properties in all pultrusion applications.
	Fiber glass roving	424, Type 30	Good wetout with little or no strand working; zero catenary; available in 113 and 225 yields; good strand dispersion and tensile properties; recommended for use with polyester resin only. Strand will have broken filaments and/or fuzz if processed under harsh conditions; may require more ends of glass than conventional roving.
PPG	Fiber glass roving	E-glass	Standard commercial products.
Dupont	Organic fiber	Kevlar 49	Light weight, high performance organic fiber; suitable for aerospace/aircraft and military applications; intermediate cost; primarily used in tensile applications and panels.
Hercules	Graphite tow	AS4	Light weight, high performance, intermediate carbon fiber; more expensive than Kevlar or glass; used in stiffness-critical applications.
Union Carbide	Graphite tow	T300	Same as Hercules AS4.

fiber allowing greater flexibility and the higher modulus glass fiber giving the composite enough memory to return to its original straightness.

Some common, currently available reinforcements are listed in Table 81.3. Again, there are so many fine pultrusion reinforcements on the market today, in such a variety of types and forms, many designed for specific use and properties, that no attempt has been made to compile and exhaustive list. Table 81.3 gives an indication, only, of the more common varieties and suppliers; purchasing directories such as those referenced[7-10] will provide a far broader scope of information.

81.3 PROCESSING CHARACTERISTICS

As noted in the preceding discussion, end product properties are dependent on materials used and processing parameters, within the inherent constraints of the materials and process themselves. However, basic characteristics of reinforced plastics/composites (RP/C), in general, which apply across the board to the advantage of pultrusion products, include the following:[12]

- Light weight
- Corrosion resistance

- Dimensional stability
- Electrical insulation
- Resistance to thermal transfer
- Flame resistance
- Parts consolidation
- Relatively low tooling costs
- Composite design capability
- Machinability/ease of fabrication
- Hybrid material options
- Pigmentability/decorative characteristics
- Resiliency
- Electromagnetic transparency
- Moisture/rot resistance
- Low maintenance/cleanability
- Vandal resistance
- Ease of repair

In addition to these general RP/C advantages, the pultrusion process offers a number of important features:[12]

- Infinite length
- Complex shapes
- Smooth finished surfaces
- Variable wall thickness
- Hollow sections
- Unlimited size
- Precise positioning of reinforcements
- Insert/encapsulation material options
- Reproducible quality
- Multiple cavities
- Low labor costs

81.3.1 Parts Produced

Figure 81.3 illustrates the scope and variety of configurations that can be produced by the pultrusion process and equipment. As indicated in specific end-product applications, the resiliency of a pultruded design to absorb some impact or to experience even major deflection and return to its original position has enabled it to be used in lightning rods, highway delineator posts, dunnage bars, sailboat battens, and spring applications, among others. Continuous, infinite-length capability accommodates such products as 37-1/2 ft-(11/25 m) sucker rods, 2.2-km optical fiber core materials, and 10 ft-(3.3 m) third rail coverboards—all of which can be made in relatively inexpensive dies only 3 ft (0.9 m) long.

Since several channel sizes can be pultruded at one time, the manufacturing cost per foot allows increased market penetration, with lower selling prices. Unlimited size capability is possible because pultrusion is essentially a no-pressure process. Continuous stock 18 × 36 in. (45 × 90 cm), consisting of multiple hollow sections with intercostal webs, is in production today for the automotive industry. And development of an aerospace application is in work on 24 × 36 in. (61 × 90 cm) capacity pultrusion equipment.

Figure 81.3. Pultrusion profile samples.

Primary markets and material/product characteristics are listed in Table 81.4. This is by no means a definitive list; the possibilities for new pultrusion development range over the entire broad scope of end products, from grating and gutter applications to structural profiles for exotic lunar and space construction projects. (Test methods and standards established for pultruded products are discussed later in this section.)

Some of the primary pultruders who manufacture these and other parts are listed in Table 81.5. Again, this is by no means a complete list of all pultruders but is representative of the larger and better-known producers in the United States. The buyers' guides provide additional listings.[7–10]

81.4 ADVANTAGES/DISADVANTAGES

The tremendous advantages offered by pultrusion materials and process must of course be weighed against inherent disadvantages and limitations, which are probably best delineated by published standards and specifications.

A joint Society of the Plastics Industry (SPI)/American Society for Testing and Materials (ASTM) task force known as the D20.18.02 Subcommittee issued in 1981 a draft copy of standard recommended practice for classifying visual defects in thermosetting reinforced plastic pultruded products.[13] This draft formed the basis for ASTM's D3918 standard,[14] which is a compilation of terms used in the reinforced plastics pultrusion industry. Included are such common defects as black markings (black smudges that result from excessive die pressure), crazing (multiple fine separation cracks), resin-rich areas (areas that lack sufficient

TABLE 81.4. Primary Markets and Material/Product Characteristics

Products	Design Flexibility	High Strength-to-Weight Ratio	Light Weight	Corrosion Resistance	Dimensional Stability	Resistance to Thermal Transfer	Flame Resistance	Parts Consolidation	Relatively Low Tooling Costs	Composite Design Capability	Machinability/Fabrication	Hybrid Material Options	Pigmentability/Decorative	Resilience	Electromagnetically Transparent	Moisture/Rot Resistant	Low Maintenance/Cleanability	Vandal Resistant	Repairability	Electrical Insulation
Electrical																				
Cable tray/cover sheet	X	X	X	X			X	X			X									X
Bus support panel		X					X													X
Aerial booms	X	X	X	X				X			X		X							X
Third-rail coverboard		X	X				X	X		X							X			X
Ladders		X	X				X		X	X	X		X	X		X	X	X		X
Wind turbine blades		X	X						X	X	X	X				X	X			X
Hot line tools		X	X				X			X	X		X							X
Pole line hardware		X		X			X		X				X		X	X	X			X
Transformer spacer stick		X															X			X
Conduit snakes		X					X								X					X
Motor slot wedges		X		X	X		X			X						X	X			X
Tree-trimming poles		X	X	X			X													X
Light poles		X	X	X			X	X	X		X	X	X	X			X	X	X	X
Conduit		X	X		X		X		X	X	X		X			X	X			X
Fuse tubes	X	X					X		X				X			X				X
Cable TV enclosures		X					X	X	X	X	X					X				X
Insulating channel					X		X	X		X	X		X		X	X		X		X

Characteristics

Corrosion Resistant Market

- Sucker rods
- Filter cell sheets
- Flume trough
- Water distribution tubes
- Vertical media support grids
- Flue-gas scrubber blades
- Exhaust ducts
- Industrial guttering
- Flooring/grating
- Sewage treatment flights
- Weir gates
- Span bridge
- Hog grating
- Hog penning units
- Grape picker sticks
- Feeding troughs
- Pipe supports
- Building panels
- Safety hand rail/ stair supports
- Fabricated girders
- Structural beams
- Truss beams
- Room beams
- Nuts and bolts
- Concrete forms
- Standard structural shapes

Consumer/Recreation

- Fishing rods
- Tool handles
- Umbrella standards
- Atennas
- Sail battens
- Flag poles

TABLE 81.4 *(Continued)*

Characteristics

Products	Design Flexibility	High Strength-to-Weight Ratio	Light Weight	Corrosion Resistance	Dimensional Stability	Resistance to Thermal Transfer	Flame Resistance	Parts Consolidation	Relatively Low Tooling Costs	Composite Design Capability	Machinability/Fabrication	Hybrid Material Options	Pigmentability/Decorative	Resilience	Electromagnetically Transparent	Moisture/Rot Resistant	Low Maintenance/Cleanability	Vandal Resistant	Repairability	Electrical Insulation
Consumer/Recreation																				
Fence posts	X	X	X				X			X	X		X	X		X				X
Drapery rod pulls	X							X			X		X	X						X
Catscan patient support table	X	X	X	X	X					X	X	X			X					
Hockey sticks		X	X				X		X	X	X		X	X		X	X			
Stadium seating		X									X	X	X			X		X		
Golf flags/shafts		X	X							X	X		X	X		X	X	X		
Bows/arrows		X			X					X		X		X		X	X	X		
Snowmobile trackrod		X		X	X					X	X		X	X			X			
Bike flags		X										X	X	X		X				
Ski poles		X	X								X									
Construction																				
Walk-in food cooler insulated panels/shelves, etc	X	X	X	X		X	X	X		X	X	X	X	X		X	X			
Highway delineator posts	X	X		X	X		X	X		X	X		X			X				
Pre-engineered building thermal breaks						X				X	X									
Building panel systems	X	X	X	X		X	X	X		X	X	X	X	X	X	X	X	X	X	X
Railroad crossing arms	X	X	X							X	X		X			X				
Window spacer strips		X	X		X	X	X		X		X		X							
Window frames and sashes	X	X			X	X	X		X		X		X			X	X			

1050

Door jambs	X			X	X	X	X			X	X		X			X				
Mine roof support beams		X	X	X						X	X			X		X				
Highway safety barricades	X		X		X						X		X	X		X	X			
Sign spring		X								X	X	X	X	X		X				
Surveyor poles	X		X		X		X						X	X		X	X			
Street sign blanks	X	X									X		X	X				X		
Transportation/ Material Handling																				
Air conditioner ducts	X	X				X	X	X		X	X		X							
Bus luggage rack	X	X	X				X	X		X	X		X				X			
Outer belt panels	X	X	X	X		X	X	X		X	X		X			X	X	X		
Wire raceway channels	X		X				X				X									X
Hand rails	X	X	X				X				X									
Window latch bars	X	X					X				X									
Window molding	X	X		X	X	X	X			X	X		X			X	X			
Refrigerated truck trailer doors	X	X	X	X	X	X	X	X		X	X			X		X				
Scuff plate (trailer)		X					X			X	X			X			X		X	
Refrigerated trailer structurals		X	X			X	X			X	X									
Truck airfoils	X	X	X								X			X						
Dunnage bars	X	X	X							X	X	X		X						
Bumper beams	X	X	X							X	X	X		X						
Overhead monorail system	X	X	X			X	X			X	X									
Material handling components	X	X	X					X		X	X	X		X					X	

TABLE 81.5.   U.S. Pultrusion Companies[a] (By State)

California	*Oklahoma*
Glastrusions, Inc.	Fibercase
Rancho Dominguez	Sand Springs
	Pennsylvania
Structron, Inc.	Creative Pultrusions
Encinitas	Alum Bank
Glasforms, Inc.	Permali, Inc.
San Jose	Mt. Pleasant
J. Miller Industries	Robroy Industries
Santa Ana	Vebrona
Poly-Trusions, Inc.	Westinghouse Electric
Santa Ana	West Miflin
Illinois	*South Carolina*
Dasco Products	Shakespeare Corp.
Rockford	Columbia
R.D. Werner Co.	Coastal Engineered Products
Franklin Park	Varnville
Bennett Industries	
Peotone	*Tennessee*
	IKG Industries
Massachusetts	Nashville
Chase Shawmut	
Newburyport	*Texas*
	Fiberflex Products
Missouri	Big Spring
A.B. Chance Co.	JBC Enterprises, Inc.
Centralia	Houston
Nevada	*Virginia*
Carsonite International	Morrison Molded Fiber Glass
Carson City	Bristol
Ohio	
Pultrusion Technology	
Twinsburg	
Pultrusions Corp.	
Aurora	
Glastic Company	
Cleveland	

[a]Ref. 1.

reinforcement), and twist (longitudinal progressive rotation). The method and frequency of sampling and acceptance levels of such visual defects should be agreed upon in advance by the purchaser and the seller.

ASTM has also published the following test method standards:[15] D3914, In-Plane Shear Strength of Pultruded Glass-Reinforced Plastic Rod; D3916, Tensile Properties of Pultruded Glass-Fiber-Reinforced Plastic Rod.

These and other RP/C industry standards and specifications, as well as general resource information have been covered in some detail by McDermott, former manager of the SPI RP/C Institute, in a paper presented in 1981.[16]

81.5 DESIGN

Design data covering the typical elements of sections and standard tolerances should be taken into consideration where pultruded stock is to be used.

Morrison Molded Fiber Glass (MMFG) and other pultruders publish design guides to be used by potential customers as well as for their own internal use. The guides also detail factors of safety. When allowable stress is not specified, MMFG states that the factor of safety must be carefully chosen based on type of load and the accuracy of estimated loads on the structure; precision of analysis and stress determination (fiber-glass shapes cannot be stress relieved in areas of stress concentration, such as notches and holes); and deterioration caused by environmental conditions.[17]

Weights for shapes shown in their tables are calculated and are based on their respective densities and nominal dimensions. Two typical elements of sections, hat sections and I-beams are reproduced from the MMFG design guide as Figures 81.4 and 81.5.

In addition, this design guide provides standard tolerances for pultruded round and square bar and tubes, and other structural shapes.

The following ASTM standard ASTM D3917, Dimensional Tolerance of Thermosetting Glass-Reinforced Plastic Pultruded Shapes also gives dimensional tolerance data.[15]

81.6 PROCESS CRITERIA

Processing variables of time, pressure, viscosity, and temperature are expressed graphically in Figure 81.6.[18] Where feasible, augmented cure preheating and multiple zone die heaters can be used to enhance process variable controls and thereby achieve both increased production rates and higher quality product.

81.7 MACHINERY

A typical pultrusion machine is illustrated in Figure 81.7. Glass-fiber roving is dispensed from center-pull packages sitting on bookshelf-type racks such as those illustrated. Glass-

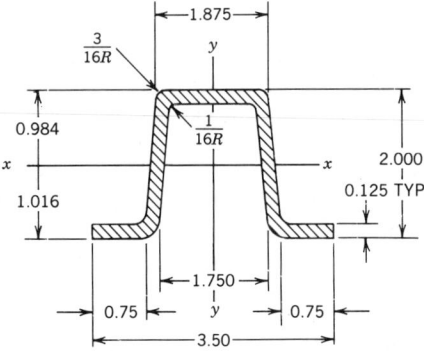

Figure 81.4. Typical elements of sections: hat section. Courtesy of MMFG.

		Properties						
		X-X axis				Y-Y axis		
Wt	A	1	S_1	S_2	r	1	S	r
0.71	0.864	0.493	0.485	0.501	0.744	0.847	0.484	0.975

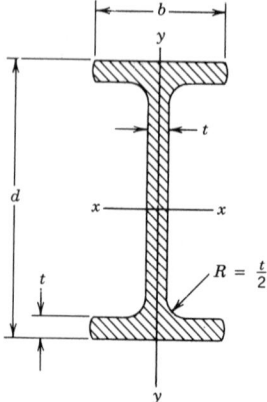

Figure 81.5. Typical elements of sections:-I-beams. Courtesy of MMFG.

| Size | | | | | Axis X-X | | | Axis Y-Y | | | Flange |
d	b	t	Wt	A	I	S	r	I	S	r	Toe Radius
2	1	⅛	0.35	0.468	0.277	0.277	0.77	0.021	0.042	0.210	1/8
3	1½	¼	1.07	1.380	1.750	1.170	1.13	0.143	0.191	0.322	1/4
4	2	¼	1.42	1.888	4.400	2.200	1.54	0.338	0.338	0.425	
6	3	¼	2.09	2.880	15.92	5.32	2.36	1.132	0.755	.628	1/4
	3	⅜	3.08	4.225	22.30	7.43	2.31	1.711	1.14	.640	3/8
8	4	⅜	4.16	5.725	55.45	13.85	3.12	4.032	2.016	.840	3/8
	4	½	5.48	7.510	70.62	17.65	3.08	5.406	2.710	.850	1/2
10	5	⅜	5.27	7.225	111.67	22.33	3.93	7.854	3.140	1.040	3/8
	5	½	6.94	9.510	143.48	28.70	3.90	10.514	4.220	1.06	1/2
12	6	½	8.56	11.510	254.10	42.30	4.70	18.11	6.05	1.26	1/2

fiber mats, supplied in large rolls, may be pulled off simple spindles on the outside tangent. Roving may also be dispensed tangentially where roving twist would constitute a detrimental factor in the end product. Other reinforcements and surfacing materials, such as graphite fibers and nonwoven veils, are generally dispensed tangentially.

Die tooling for simple profiles can be mounted on the die station. For complex parts, special tooling may be stationed just upstream and/or downstream from the wet-out section.

The resin wet-out section shown here consists of a wet-out table; pass-through tank, with perforated polyurethane end plates that provide for roving passage so as to accommodate several streams of material; and resin drip tray underneath the tank, with resin reservoir and pump assembly for recirculating resin into the tank for reuse. These are standard accessory items. This model also has an optional water jacket with a circulating water supply system and a Mokon heater for resin temperature control.

A combined system of roving guidance and squeeze-out bushings is mounted at the die station entrance. Excess resin is funneled back to the drip tray under the resin tank.

The type of die heating system used is dependent on specific requirements. Strip heaters are commonly fastened to the outside faces of the die tooling and connected to the electrical system. In addition, where mandrels are used to produce hollow sections, cartridge heaters may be installed in the mandrel interior, where the mandrel size permits. The model shown here utilizes thermocouple probes for continuous multiple zone die heat monitoring, with corresponding digital readouts for up to 10 channels.

A cooling section about 15 ft (4.6 m) long allows the cured stock to cool before it reaches the saw section and is removed from the machine.

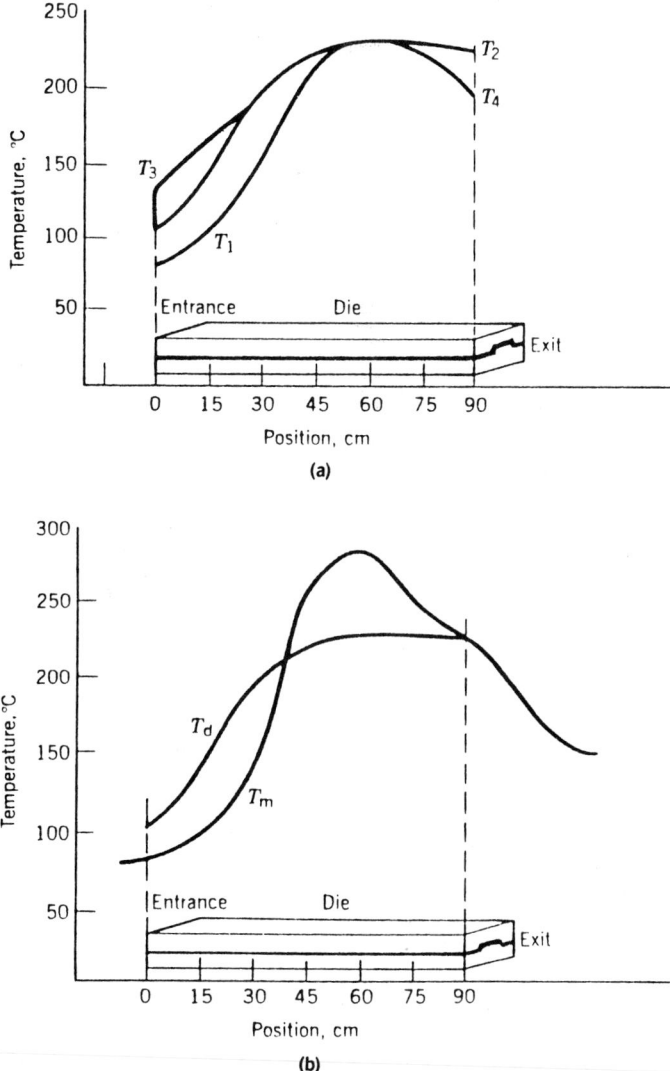

Figure 81.6. (**a**) Die temperature profiles: T_1, start-up; T_2, steady-state; T_3, with preheat; and T_4, cool-down. (**b**) Die (T_d) and material (T_m) temperatures. (**c**) Viscosity and temperature: V, viscosity; T_d, die temperature; and T_m, material temperature. (**d**) Pressure, viscosity, and temperature: P, pressure; V, viscosity; T_d, die temperature; and T_m, material temperature.

The primary mechanical feature of all pultrusion machines is the gripper/puller system. Because of the precision required to hold the cure points within the die while moving at high speeds, the pullers must be accurately controlled.

Two fundamentally different gripper/puller systems are used: the opposed-tread caterpillar tractor-type and the hand over hand reciprocating clamp type.

The caterpillar type employs two constantly rotating, opposed, cleated belts between which the cured profile is pulled. While less costly than reciprocating clamps, the caterpillar type pullers have the distinct disadvantage of requiring, for complex profiles, a relatively large number of specially shaped gripping pads covering both upper and lower tread surfaces. Often, each totals more than 8 lineal ft (2.4 m). An even more significant disadvantage of

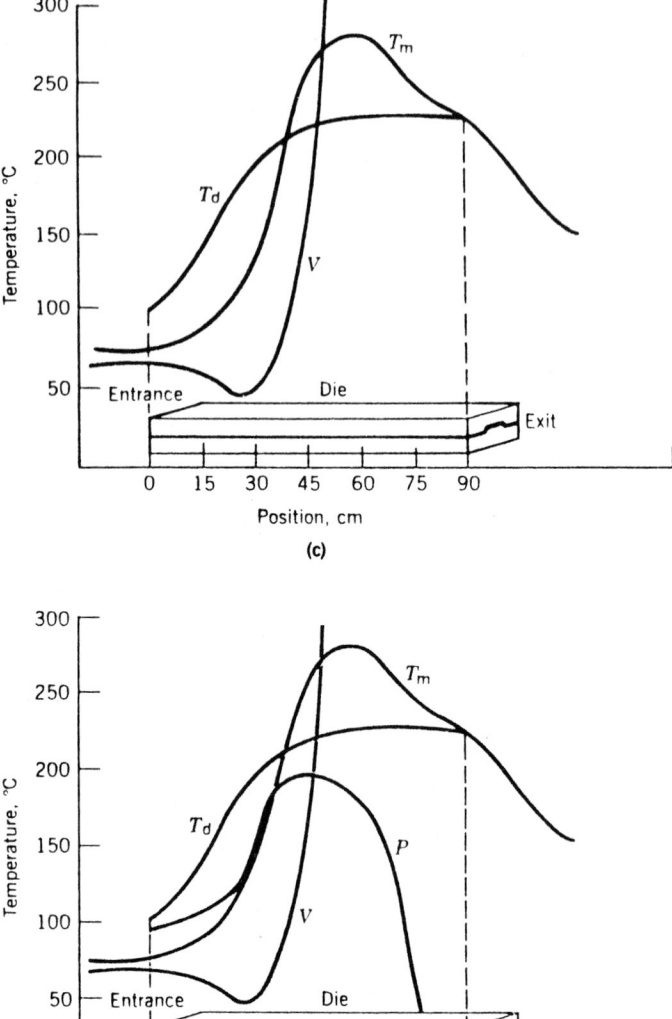

Figure 81.6   (*Continued*)

caterpillar types is evident when it is necessary to pull large profiles and heavy leads of around 10,000 lb (4500 kg). Because the gripping force cannot be isolated from the pulling force, these substantial gripping loads are transferred into the bearings of the rotating pulling mechanism, creating an undesirable condition.

The reciprocating clamp system which is used in the Goldsworthy Engineering Pulmaster model (Fig. 81.7) eliminates both problems inherent in the tractor belt puller. It employs only two pairs of shaped pads approximately 2 ft (0.6 m) long around the profile. Each

Figure 81.7. Pulmaster pultrusion machine Model PM 8-24-0. Courtesy of Goldsworthy Engineering Inc.

reacts against the other hydraulically; with the profile in between, the clamping forces are isolated within the clamp itself.

The final station in a typical pultrusion machine is the cutoff saw, which consists in this model of a diamond-rimmed circular saw blade, wet-cooled and lubricated, synchronized with the movement of the pullers, and automatically actuated by a preset, cut-to-length switch.

Hydraulic servo drive and programmable controls equip this unit with the latest drive/ control technology.

81.8 TOOLING

Since pultrusion's early days when equipment was tooled exclusively for solid profiles, a broad variety of tooling has been developed for product types and sizes over the entire spectrum of commercial, industrial, and military applications.

Continuous convolute folding type tooling for tubular pultrusion is based in part on technology used in the Japanese paper industry for continuously folding materials. Because glass mat materials, which are widely used to bring off-axis reinforcements into the fiber/ matrix bundle, closely resemble paper in their mutual inability to stretch, the technology was directly applicable and was successful from the start (circa 1971). The fundamental principle involved is that the material web can be folded continuously without wrinkling (stretching) if every point in any plane normal to the web width is made to travel the same distance at the same speed.

The hoop strength limitation of convolute folding type tooling can be overcome where necessary by adding in-line hoop winding wheels. Borrowing from filament winding to some extent, the +45° overwinding wheel carrying spools of roving or tow is revolved around the axis of the part being pultruded (Fig. 81.8). By controlling the ratio of the longitudinal throughput speed with the overwinding wheel rotational speed, the filament laydown angle can be varied from nearly 90° circumferential to any lesser angle. Common overwinding angles are 90° for tubular hoop direction loads and +45° for torsional loads.

Theoretically, any tubular wall thickness may be built up using overwinding wheel techniques. Advantages of this technology are significant both in terms of raw material economics and the strength characteristics of the pultruded profile. Roving, the least expensive form of fiber glass, may be used in lieu of more costly cloths and mats. And, most importantly, because the roving filaments are placed accurately in the desired load paths, the laminate specific strength is optimized.

For simple shapes of low mass, upstream tooling feeding unidirectional and off-axis reinforcements through the matrix and directly into segmented steel dies is sufficient. RF augmented cure preheat may be used with this type of tooling setup as long as nonconductive materials are used.

Larger, more complex parts require correspondingly larger, more complex tooling, such as that illustrated in Figure 81.9, which feeds into the 18 × 36 in.(45 × 90 cm) capacity machine shown in Figure 81.10. Utilizing mat spools, roving racks and carding plates, folding tools, and segmented steel mandrels in an open die, the upstream tooling provides the flexibility of adding mat and/or other off-axis reinforcement to longitudinal fiber reinforced plastic stock. This large-capacity unit is currently producing continuous stock consisting of multiple hollow sections with intercostal webs for the automotive industry.

Figure 81.8. Pultrusion machine tooled with convolute folding tools and in-line hoop winding wheels. Courtesy of Goldsworthy Engineering, Inc.

Figure 81.9. Tooling station for 18 × 36 in. (45 × 90 cm) pultrusion machine. Courtesy of Goldsworthy Engineering Inc.

Figure 81.10. 18 × 36 in. (45 × 90 cm) capacity pultrusion machine. Courtesy of Goldsworthy Engineering Inc.

81.9 AUXILIARY EQUIPMENT

A wide variety of auxiliary equipment is available to the pultrusion industry today, some of which is listed below with suggested sources and round order-of-magnitude prices. This is by no means a complete list, but it does give some idea of the scope and price range available.

81.9.1 Tooling

- Winding wheels add 90–45° off-axis roving for optimized high strength pultrusions. Dual, four package per wheel, combined, or opposed rotation wheels are available from Goldsworthy Engineering Inc. (GEI) for $10,000–30,000.
- Folding tooling adds off-axis glass mat. It must be customized for the specific application. One convolute "tulip" for 10-in. (254-mm) wide material is available from GEI for around $2500.
- Mandrels for hollow sections are available from local tool and die grinding shops. Price is dependent on size and profile, ranging from $2000 to $10,000.

81.9.2 Preheat Equipment

- RF generator is used to achieve high speed for massive section. It is available from GEI in the range of $20,000 to $60,000.
- Induction generator is used to initiate preheat for conductive stock. It is available from Lepel Inc. Price depends on the power requirements and ranges from $5000 to $50,000.

81.9.3 Resin Wet-Out System

- With a pass-through resin tank with recirculating pump assembly, the pass-through tank ensures thorough fiber wet-out, and cost savings are effected through return of resin drip to tank. Several streams of material can be accommodated. It is available from GEI in models ranging from $2000 to $13,000.
- Resin temperature control system maintains a constant preset resin temperature, thereby aiding in control of heat and viscosity variables. It is available from Mokon (Buffalo, N.Y.) for about $3500.

81.10/81.11 FINISHING AND DECORATING

A number of in-line coating systems are available that can be applied to provide ultraviolet- and weather-resistant coatings for outdoor applications or simply for improved aesthetics. The system is generally stationed between the puller section and the saw section.

81.12 COSTING

As noted, so many variables exist in pultrusion processing, material type and form, profile complexity and mass, influence of heat, and so on, it has been impossible to develop standard cost formulae for stock pricing. Costing essentially must be done on a case-by-case basis, taking into account specific design factors, cost of raw materials, and so on.

81.13 INDUSTRY PRACTICE

Three promising new directions in pultrusion are worthy of note.[19]

The first is newly developed preform/postform techniques that are enabling pultrusion of products exhibiting combinations of curve and twist previously considered unattainable. Initial applications are in aerospace, but the potential is industry wide. In somewhat related processes, special tooling has been used to bend or deform hot pultrusions as they exit the die.

Second, the advent of a new generation of high-temperature, high-physical-property thermoplastic resins with relatively low melt viscosities holds tremendous potential for thermoplastic matrix pultrusions. This new technology could not only enhance current pultrusion properties and processibility but could also greatly improve toughness, abuse resistance, and postformability of pultruded composites, thereby eliminating the source of many automotive industry and consumer reservations about "plastics." It is even possible to predict that thermoplastics might become the dominate matrix materials some five to 20 years from now. Processing difficulties definitely exist, however, and an entirely new processing technology, from resin formulation to fiber finishing to processing parameters, will need continuing development.

The third trend is the recent development of very powerful, large cross-sectional capacity pultrusion machines with pull load capacities of 10 tons (98 kN) and cross-sectional areas of up to 24 by 36 in. (46 by 92 cm) or more. Further implementation of this machinery is expected to open new markets in large structural profiles.

Although most of the emphasis currently is directed toward developing processing techniques for these high performance polymers, an even larger market in consumer goods and industrial application exists with the use of many of the standard, lower-cost polymers, such as nylon, polycarbonate, polypropylene, and so on. These types of polymers should present considerably fewer processing difficulties than the high performance polymers.

81.14 INDUSTRY ASSOCIATIONS

Prominent plastics industry associations that service pultrusion processors and suppliers include the following:

- Society of the Plastics Industry (SPI)
- Society of Plastics Engineers (SPE)
- Society of Manufacturing Engineers (SME)
- The Composites Group of the Society of Manufacturing Engineers (COGSME)
- American Society of Testing and Materials (ASTM)
- Society for the Advancement of Material and Process Engineers (SAMPE)

BIBLIOGRAPHY

1. G. Ewald, "Pultrusion Machines: Past, Present, and Future", *35th Annual Technical Conference, SPI Reinforced Plastics/Composites Institute Conference,* 1980.
2. W. Brandt Goldsworthy, "Fabrication Techniques," in *Advanced Thermoset Composites,* Van Nostrand Reinhold, New York, 1986.
3. U.S Pat. 3,556,888 (Jan. 19, 1971), W. Brandt Goldsworthy (to Glastrusions Inc.) and U.S. Pat. 3,684,622 (Aug. 15, 1972) (to Glastusions, Inc.).
4. G. Ewald, "Pultrusion and Pulforming," *Modern Plastics Encyclopedia 1985–1986,* McGraw-Hill, New York, 1985.

5. U.S. Pat. 3,674,601 (July 4, 1972), W. Brandt Goldsworthy (to Glastrusions Inc.) and U.S. Pat. 3,793,108 (Feb. 19, 1974) (to Glastrusions, Inc.).

6. M. Colangelo and M. Naitove, "Pultrusion Process Technology: Beyond Infancy, Not Yet Mature," *Plastics Technology*, 49–53 (Aug. 1983).

7. *Thomas Register of American Manufacturers and Thomas Register Catalog File*, 75th ed., 19 vols., Thomas Publishing Co., New York, annual.

8. *Modern Plastics Encylcopedia*, McGraw-Hill, New York, annual.

9. *SPI Membership Directory & Buyers Guide*, Society of the Plastics Industry, New York, annual.

10. *Plastics World Plastics Directory*, Cahners Publishing Co., Newton, Mass., annual.

11. B. Jones, "Pultruding Filamentary Composites—An Experimental and Analytical Determination of Process Parameters," *29th Annual Technical Conference*, SPI Reinforced Plastics/Composites Institute, 1974.

12. W. Brandt Goldsworthy and J. Martin, "Pultruded Composites—A Blueprint for Market Penetration," *40th Annual Conference*, SPI Reinforced Plastics/Composites Institute, 1985.

13. C. Thomas, Draft copy of standard recommended practice for classifying visual defects in thermosetting reinforced plastic pultruded products: Memo to Members of ASTM D20.18.02 from Task Force on Classification of Visual Defects in Reinforced Plastic Pultruded Products, Dec. 21, 1981.

14. "ASTM D3218-80. Standard Definitions of Terms Relating to Reinforced Plastic Pultruded Products," in *1985 Annual Book of ASTM Standards*, Vol. 8.03, American Society for Testing and Materials, Philadelphia, Pa. 1985.

15. *Annual Book of ASTM Standards*, Vol. 8.03. American Society for Testing and Materials, Philadelphia, Pa., 1985.

16. J. McDermott, "Codes, Standards, and Design Data on Reinforced Plastics," *36th Annual Conference*, SPI Reinforced Plastics/Composites Institute, Feb. 1981.

17. *Extren Fiberglass Structural Shapes–Engineering Manual*, Morrison Molded Fiber Glass Company, Briston, Va., 1978.

18. J.E. Sumerak and J.D. Martin, "It's Time We Really Understood Pultrusion Process Variables," *Plastics Technology* (Feb. 1983).

19. D. Beck, "New Processes and Prospects in Pultrusion," *38th Annual Conference*, SPI Reinforced Plastics/Composites Institute, (Feb. 1983).

General References

G. Lubin, ed., *Handbook of Composites*, Van Nostrand, New York, 1982.

R. Meyer, *Pultrusion Technology*, Methuen Publ., New York, 1985.

M.M. Schwartz, *Composite Materials Handbook*, McGraw-Hill, New York, 1984.

Modern Plastics, (monthly journal) and *Modern Plastics Encyclopedia* (annual), McGraw Hill, New York.

Plastics Technology, New York (monthly journal).

Plastics World, Cahners Publishing Co., Newton, Mass. (monthly journal)

SAMPE Quarterly, SAMPE, Covina, Calif. (quarterly journal).

82

BLOW MOLDING

Christopher Irwin

Johnson Controls, Inc., Plastics Technology Group, Manchester, MI

82.1 DESCRIPTION AND HISTORY OF PROCESS

Blow molding is a process for the production of hollow objects in which air or occasionally nitrogen is used to expand a hot preform (or parison) against a female mold cavity. A common feature of all blow molded articles is "re-entrant" curves; for example, a bottle having an opening much smaller than the body. Some items, however, are designed for structure: toy tricycles and wheels utilize the strength of the unit box sections. An important feature for some items is the capability to provide very thin wall sections with relatively low stress. Another characteristic is that the dimensions of part detail are more accurately held on the outside than on the inside, where material thickness can easily vary the shape.

The first attempt to blow mold hollow plastic objects, more than 100 years ago, employed two sheets of cellulose nitrate clamped between two mold halves. Steam was injected between the sheets, which softened the material, sealed the edges, and expanded it against the mold cavity.[1] However, the highly flammable nature of cellulose nitrate limited the usefulness of the technique.

In the early 1930s more suitable materials, cellulose acetate and polystyrene, were developed, which led to the development of some automated equipment by the Plax Corp. and by Owens-Illinois, based on "glass blowing" techniques.[1,2] Unfortunately, the high cost and poor performance of these materials discouraged rapid adoption, as they offered no advantage over glass bottles.

The advent of low density polyethylene (LDPE) in the mid-1940s provided the needed advantage. The "squeezability" of this material gave the plastic bottle a feature glass could not match.[1] The real beginning of blow molding occurred in the late 1950s with the development of high density polyethylene (HDPE) and the availability of commercial blow molding equipment.[1] HDPE solved many of the problems of LDPE, and, most importantly, bottles could now be lightweight and stiffer. Commercial equipment gave more firms the opportunity to start blow molding. Until that time, all blow molding was done by a select few, using proprietary technology.

The acceptance of HDPE for packaging bleach, detergent, household chemicals, and milk in the 1960s and 1970s and the acceptance of poly(ethylene terephthalate) (PET) for packaging carbonated beverages in the 1980s has expanded the blow molding process to unprecedented levels. It is truly one of the high-growth technologies.

Although it is still principally used for bottles, blow molding is being discovered and utilized more and more by designers for industrial part applications. Examples include automotive rear-deck air spoilers, seat backs, toy tricycles and wheels, typewriter cases, surfboards, flexible bellows, fuel tanks, and so forth. Figure 82.1 shows a fuel tank being molded on an accumulator head machine.

Figure 82.1. Blow-molded Volkswagen fuel tank made of high molecular weight polyethylene. Equipment is Krupp-Kautex Model KB250-S120 with 120 mm extruder and 75 kg accumulator head. Photograph courtesy of Krupp-Kautex.

Today, blow molding encompasses two fundamental process approaches based on the method used to create the preform or parison. These are injection blow molding, which uses an injection molded "test tube" shaped preform and extrusion blow molding, which uses an extruded tube parison.

These basic approaches are, in turn, used to create other process variations, namely:

- Stretch blow molding, in which preform shape, resin, and careful processing combine to cause a biaxial orientation of the resin in the bottle.
- Multilayer blow molding, in which two or more layers of different resins are combined in a single preform.
- Multilayer/stretch blow molding, in which one of the resins is also biaxially oriented.

Each process variation has its own family of resins, applications, advantages and disadvantages.

82.2   MATERIALS

The activity in blow molding has led to more and more resins being considered and formulated for blow molding. Although most thermoplastic resins can be blow-molded, in practice only a few are considered. For many resins, a blow-molding opportunity has not evolved. Specific resin characteristics will also often limit its use to only one of the process approaches. A resin could be very easy to process by one approach and very difficult or impossible by another. The following discussion covers the resins most often considered and the processes most often utilized.

- Polyester, PET [poly(ethylene terephthalate)], with careful temperature conditioning of an injection-molded preform, this resin can be biaxially oriented by stretch blow molding. For smaller bottles or parts, the resin is also used in the basic injection blow-molding process. Because PET lacks melt strength, it is not generally suitable for the extrusion blow-molding process; however, it may be coextruded in a multilayer process.

MASTER PROCESS OUTLINE   **PROCESS**   Blow molding

Instructions:
For "ALL" categories use   Y = yes
For "EXCEPT FOR" categories use   N = no
                                         D = difficult

□ **ALL Thermoplastics**
Except for:

Acetal
Acrylonitrile–Butadiene–
Styrene
Cellulosics
Chlorinated Polyethylene
Y Ethylene Vinyl Acetate
Ionomers
Liquid Crystal Polymers
Y Nylons
Poly(aryl sulfone)
Polyallomers
Polyamid-Imid
Polyarylate
Polybutylene
Y Polycarbonate
Polyetherimide
Polyetherketone (PEEK)
Polyethersulfone

Y Acrylonitrile
Y Thermoplastic elastomers
Y Polyethylene
Y HDPE
Y LDPE
Y UHMWPE
   Poly(methyl methacrylate)
   Poly(methyl pentene) (TPX)
   Poly(phenyl sulfone)
   Poly(phenyl ether)
Y Poly(phenylene oxide) (PPO)
   Poly(phenylene sulfide)
Y Polypropylene
Y Polystyrene
   Polysulfone
   Polyurethane Thermoplastic
Y Poly(vinyl chloride)-PVC
   Poly(vinylidine chloride)-Saran
   Rubbery Styrenic Block Polyms
   Styrene Maleic Anhydride
Y Styrene–Acrylonitrile (SAN)
Y Styrene–Butadiene (K Resin)
Y Thermoplastic Polyesters
   XT Polymer

□ **ALL Thermosets**
Except for:

Allyl Resins
Phenolics
Polyurethane Thermoset
RP-Mat. Alkyd Polyesters
RP-Mat. Epoxy
RP-Mat. Vinyl Esters
Silicone
Urea Melamine

□ **ALL Fluorocarbons**
Except for:

Fluorocarbon Polymers-ETFE
Fluorocarbon Polymers-FEP
Fluorocarbon Polymers-PCTFE
Fluorocarbon Polymers-PFA
Fluorocarbon Polymers-PTFE
Fluorocarbon Polymers-PVDF

For basic extrusion blow-molding applications, a noncrystalline glycol modified polyester called PETG has been formulated; however, its noncrystalline molecular structure does not permit biaxial orientation.

- Nylon materials are available for injection and extrusion blow molding. It is also used as the barrier layer of some multilayer structures.
- Poly(phenylene oxide), the styrene-modified version of this material is extrusion blow-molded.
- Polycarbonate resin is both extrusion and injection blow-molded.
- Styrene–butadiene resin is extrusion blow-molded but can also be injection blow-molded.
- Polyethylene (HDPE) is the resin most common used in blow molding in both extrusion and injection blow-molding processes. This inert, low-cost, forgiving resin is perfect for many package applications. LDPE is also easily processed by both methods; although LDPE applications are not as common.
- Ultrahigh molecular weight polyethylene (UHMWPE) is often extrusion blow-molded, particularly in applications where environmental stress-crack resistance is important.
- Rigid poly(vinyl chloride) (PVC) is another common resin routinely used in "continuous" extrusion blow molding. Because of its heat sensitivity, it is not suitable for the "intermittent" extrusion approach, as the rapid parison formation of this approach can add considerable shear heat. Special heat stable compounds are also routinely injection blow-molded, although the resultant cavitation and bottle size are limited. Rigid PVC can also be biaxially oriented by stretch blow molding.

- Polystyrene–general-purpose polystyrene is routinely injection blow-molded. A slight amount of biaxial orientation permits this brittle material to be practical for some applications. Impact grades with higher performance and melt strength, but often lacking clarity, can be formulated for extrusion blow molding.
- Polypropylene–both homopolymer and copolymer grades are routinely injection and extrusion blow-molded. The resin can also be biaxially oriented with stretch blow molding. Polypropylene is often used as the main layer of many multilayer food containers.
- Ethylene vinyl alcohol (EVOH) is a resin that is used as the main barrier layer of many multilayer food containers.
- Acrylonitrile and styrene acrylonitrile resins can be injection blow-molded. Some grades are used on "continuous" extrusion equipment. The material can also be biaxially oriented.
- Thermoplastic elastomers (TPEs) comprise a wide family of resins with grades that can either be injection or extrusion blow-molded.

The extrusion blow-molding process requires the resin to have a capability known as "melt strength." The extruded parison must always hang in the air by its own strength until the mold can capture it for blow molding. Resins that have a relatively "sharp" melting point or that do not have melt strength, as is the case with many injection grades, are difficult to blow mold. What usually occurs is that the hanging parison begins to stretch or draw down from its own weight. The top part of the parison begins to thin, which, in turn, will cause poor material distribution and ultimately poor part performance.

The injection blow-molding process has more resin processing flexibility. Even though the parison is exposed to the air, melt strength is not as crucial because of the support provided by the parison core rod.

82.3 PROCESSING CHARACTERISTICS

Injection blow molding is generally used for small bottles and parts of less than 500 mL in volume. The process is scrap free with extremely accurate part weight control and neck finish detail. On the other hand, part proportions are somewhat limited and handled containers are not practical. Tooling costs are also relatively high.

Extrusion blow molding, the most common process, is used most often for bottles or parts 250 mL in volume or larger. Tanks as large as 5000 liters, weighing 120 kg, have been blow-molded (Fig. 82.2). Tooling cost is lower. Part proportions are not severely limited. Containers with handles and offset necks are commonplace. On the other hand, flash or scrap resin must be trimmed from each part and recycled. Operator skill is more crucial to the control of part weight and quality.

Stretch blow molding is used for bottles between 500 mL and 2 L in size, although some bottles as large as 25 L have been molded with this process. With precise temperature conditioning, stretch blow molding manipulates the molecular structure of certain resins by orienting the molecules biaxially. This orientation enhances the stiffness, impact, and barrier performance of the bottle, permitting a reduction in weight or use of a lower grade of material.[3,4] The technique is limited to extremely simple bottle shapes.

82.4 ADVANTAGES/DISADVANTAGES

The two basic process approaches, injection blow molding and extrusion blow molding, have unique advantages. Equipment designs in specific subcategories of these two approaches approaches also have unique advantages, which are detailed in the machinery section.

Figure 82.2. Blow-molded 3000-L heating oil fuel tank made of high molecular weight polyethylene. Equipment is Krupp-Kautex Model B200 with four 120-mm extruders and one 136-kg accumulator head. Tanks as large as 5000 L have been blow-molded. Courtesy of Krupp-Kautex.

82.4.1 Injection Blow Molding

Injection blow molding is typically used for bottles or parts that are 500 mL or smaller in size. Larger parts can be molded, but small equipment size and reduced cavitation will usually limit the economic feasibility of such parts. On the other hand, for large quantities of small bottles, injection blow molding is usually the lowest cost alternative.

Injection blow molding equipment can process a wider variety of resins. For some resins such as polystyrene, the process is the only alternative. General-purpose polystyrene is often used for low cost pill bottles.

The process is scrap free with no flash to trim or recycle. No pinch-off scar is created where the flash would have been once attached.

The injection-molded neck provides far more accurate neck and finish dimensions than is possible with extrusion blow molding. This permits special shapes with intricate internal and external contours. The finish or open end of the part does not need to be round or flat, which offers an entire realm of design possibilities.

Because the preform is accurately injection molded, part weight and, in turn, bottle volume are far more consistent.

For a blow-molded article, injection blow molding provides outstanding part surface finish or texture. The highly polished injection and blow cavity surfaces ensure the high quality finish.

82.4.2 Extrusion Blow Molding

Extrusion blow molding is typically used for bottles or parts 250 mL or larger in size.

The "continuous" extrusion process is best suited for heat sensitive PVC resins. Many other resins are moldable, provided that melt strength is adequate.

Although dimensional accuracy of the finish is not as good, the extrusion process offers far fewer limitations on part proportions. It permits extreme dimensional ratios; for example, long and narrow, flat and wide, double walled, offset necks, molded-in handles, and odd shapes.

Extrusion blow-mold tooling is often made of aluminum at as much as one third the cost of injection blow-mold tooling. Besides the blow-mold cavities, injection blow molding requires a set of steel preform mold cavities and two, three, or four sets of core rods, depending on the type of machine. The extrusion process is, therefore, ideal for short-run production applications.

The ability to easily adjust parison weight and material distribution make the process appropriate for prototype work. The ability to adjust easily can cause some inconsistency in production.

82.5 DESIGN

Although they are not called "product design guidelines," several references to guidelines will be made throughout this section in the form of machinery and tooling limitations. Examples include parison blow-up ratios, ovality ratios, tooling sizes, etc. Good product design begins with a clear understanding of the machinery and the process. To that, several other general guidelines and characteristics can be added. If these are understood, they can make the difference between a good product and a not-so-good product.

Most blow-molded articles perform better if the designer remembers to radius, slant, and taper all surfaces. Square, flat surfaces with sharp corners do not work very well. The result? Wall thickness varies considerably from side panels to corners. Corners become thin and weak; heavy side panels, thick and distorted. Flat panels never are flat and flat shoulders offer little strength. Likewise, highlight accent lines should be "dull" with a radius of 0.06 in. (1.5 mm) or more. If they are any sharper, the parison will not penetrate, and trapped air marks will result along the edge.

A blow-up ratio of 4:1 for extrusion blow-molded bottles or parts is considered the maximum. The rule applies not only to overall size and shape but also to isolated sections. For example, bottle handle designs that are deeper than they are wide across the mold parting face are difficult to mold and often become thin and weak.

Ribs do not always stiffen. Often, blow-molded ribs create more surface area to be covered, which, in turn, thins the wall. A bellows or accordion effect is created that flexes more easily. If flexing is to occur, then where might the "hinge" points be? The design could be altered to interrupt the hinge action.

It is important for bottle designers to know bottle performance test procedures. The Society of the Plastics Industry has identified several recommended standard practices. The most important ones are

- Vertical compression or top load strength.
- Drop impact resistance.
- Product compatibility and permeability.

- Closure torque.
- Top load stress crack resistance.

Blow-molding process conditions can influence not only bottle dimensions but also bottle volume. Seven conditions have been identified that produce significant changes in bottle volume.

HDPE bottles shrink over time, and about 80–90% of the shrinkage takes place in the first 24 h. Lighter weight bottles are not only bigger inside from less plastic, but they also bulge more. A 4-L bottle, 5 g less in weight, will increase in volume about 12 mL, 5 mL for the difference in weight and 7 mL for bulge. Faster cycle times, lower parison, expanded air pressure, and higher melt and mold temperatures will reduce bottle volume. Storage temperature is very important. After 10 days, significant volume changes can occur in bottles stored at 140°F (60°C) as compared to bottles stored at 68°F (20°C). Table 82.1 lists many other process conditions that can influence the quality of an extrusion blow-molded bottle. Figure 82.3 illustrates basic bottle and finish nomenclature.

82.6 PROCESSING CRITERIA

The process of blow molding is heavily dependent on the machinery used. For this reason, many process details related to a particular approach are also discussed here in relation to the machinery.

82.7 MACHINERY

82.7.1 Injection Blow Molding

In the injection blow-molding process, melted plastic resin is injected into a parison cavity and around a core rod. This "test tube" shaped parison, while still hot and on the core rod, is transferred to the bottle blow-mold cavity. Air is then passed through the core rod, expanding the parison against the cavity, which cools the part.[5,6]

Early injection blow-molding techniques, all two-position methods, were adaptations of standard injection-molding machines fitted with special tooling. The Piotrowski method used a 180° rotating arbor with two sets of core rods and one set of parison and bottle cavities. The Farkas and Moslo methods used an alternating shuttle with basically two sets of core rods, one set of parison cavities, and two sets of bottle cavities. The primary difficulty with these methods is that the injection mold and blow-mold stations must stand idle while the finished parts are removed. This led to the 1961 invention in Italy by Gussoni of the three-position method, which used a horizontal 120° indexing head with split mold parison and bottle cavities and three sets of core rods. The third station was to be used for part removal while the parison and bottle-molding phases were completed simultaneously. The method was not developed immediately because a special machine was required. During the early 1960s, Wheaton made this process commercial in a captive operation. In the late 1960s Jomar and Rainville (now Johnson Controls, Inc.) made machines available for purchase. Their use then began to expand rapidly. This method is now the basis for virtually all injection blow molding today (Figs. 82.4, 82.5, and 82.6).

The three-position method has evolved into four-position or more equipment. Modular equipment design has added a great deal of process flexibility. An added position between bottle stripping and preform injection stations can provide additional time to detect a bottle not stripped or to temperature condition the core rod. An added position between bottle blow mold and stripping stations can provide time to decorate the bottle. An added position between parison injection and bottle blow-mold stations can provide additional time to temperature condition or preblow the parison.

TABLE 82.1. Problems Encountered in Blow Molding

Problem	Probable Causes	Correction
Excess parison stretch	High melt temperature, slow extrusion rate	Reduce stock temperature; high die temperature may be contributing factor; increase extrusion rate
Rough parison surface	Usually caused by low melt temperature; can also be caused by melt fracture or die fracture of material	Reduce extrusion rate; raise parison die temperature gradually; if condition persists, check materials; try higher melt index resin; check die entrance angle; streamline overall die design.
Uneven wall thickness circumferentially in product	Mandrel not centered in die; uneven melt temperature; extrusion rate too high; faulty die design; unsymmetrical product shape	Center die/mandrel; check melt condition (probably too hot); reduce extrusion rate; check die design: shape die to increase thickness; use preblow; use larger parison
Uneven wall thickness lengthwise in product	Parison necking down; irregular product shape	Reduce parison necking by increasing extrusion rate and lowering melt temperature; extreme case requires programmed control of parison extrusion
Bubbles in parison	Moisture causes many small bubbles; trapped air causes larger bubbles	For moisture, dry pellets; for air, increase screw speed; increase screen pack to increase back pressure
Streaks in parison	Contamination from "hang-up" in equipment; melt overheated and degrading material	Clean die head; if condition repeats with clean system, check flow channels for hang-up areas; check for hot spots
Parison forms into a roll	Mandrel too hot, die too cold; parison clings to cooler surface	Increase die temperature; cool mandrel
Poor gloss on product	Poor melt flow	Try resin with higher melt index
Ripples in parison, grooves or lines in product	Contamination in die gap; possibly poor die design; extrusion rate too high	Clean gap; check die design: check for nicks; reduce extrusion rate
Product surface shows pits, fish scales, etc; parison smooth	Trapped air in mold; condensed water on mold surface; insufficient blow pressure	Vent mold, sand blast surface for polyethylene; polish surface for poly(vinyl chloride); increase size of air line and blowing pressure; raise die temperature to reduce condensation; check for water leaks; if condition is localized, check for air leak around the blow pin
Product warpage	Improper cooling	Check uniformity of die cooling; wall thickness variation of product may be excessive, may require redesign or excessive cooling cycle; stress crack potential increases under conditions producing warpage.
Container welds on weld seam	Inadequate weld; melt temperature too low	Raise melt temperature

TABLE 82.1. (*Continued*)

Problem	Probable Causes	Correction
Thin wall streak at parting line	Dies not completely closed; material stays hot at parting line; blow pressure causes stretching	Increase clamp pressure; check to be sure dies do not bounce on contact
Black specks	Material contaminated or hold-up material flaking off melt channel	Check incoming material; clean melt channel
Part sticks in mold	Mold temperature too high, cycle too short	Improve mold cooling, increase cycle
Parison pinch off sticks to product	Parison tail too long	Shorten parison, provide additional pinch-off; relieve area to cool tail
Parison blow-out	Blow-up too rapid; melt temperature too high; thin section in parison due to contamination; pinch-off could be too sharp or too hot; product may have too high a blow-up ratio	Program air blow-up start with low pressure and increase; check parison condition and temperature; check pinch-off; if blow-up ratio is too high, use larger parisons
Excess shrinkage	Poor cooling; melt temperature too high	Increase blow pressure to obtain better cooling contact with mold surfaces; improve cooling in mold; increase cycle; reduce melt temperature
Thin wall at pinch-off	Pinch-off too sharp; inadequate blow; flash pocket too deep	Increase pinch-off land width; reduce relief area to hold and cool pinched off material
Thick wall at pinch-off	Pinch-off clearance too high; pinch-off angle too small	Reduce pinch-off clearance; open pinch angle to about 30°
Undercuts fail to strip	Product overcooled; shrinkage excessive; undercuts too severe	Reduce cycle, relieve undercut; use movable die insert to product undercut

Key suppliers of three- and/or four-station injection blow-molding machines include Johnson Controls, Inc. and Jomar Corp. Costs for machines range from $100,000 to $300,000. For some applications, as many as 25 to 30 million bottles can be produced annually.

Another process somewhat related to injection blow-molding is dip blow molding used for small containers. The process has the advantage of reducing some of the molded-in stress. A core rod is inserted into a cupel, displacing resin and packing it into the neck finish area. As the core rod moves away, a predetermined amount of resin is permitted to remain on the rod. A hot knife cuts the hot plastic, separating the material on the rod from the melt pool in the cupel. The core rod is then rotated to the blow station for blow molding. Hesta is a supplier of this type of machinery.

82.7.2 Extrusion Blow Molding

In the extrusion blow-molding process, melted plastic is extruded as a tube into free air. The tube, also called a parison, is captured by the two halves of the bottle blow mold. A blow pin is inserted, through which air enters and expands and then cools the parison against the cold mold cavity. Unlike injection blow molding, flash is a by-product of the process which must be trimmed and reclaimed. The excess is formed when the parison is pinched together and sealed by the two halves of the mold.

Extrusion blow molding is divided into two basic categories: continuous extrusion and intermittent extrusion. These in turn are divided into other subcategories.

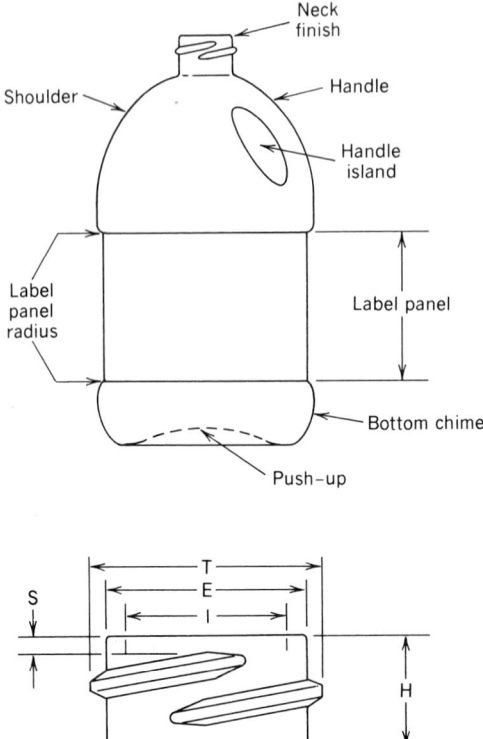

Figure 82.3. Bottle and finish nomenclature.

In the continuous extrusion blow-molding process, the parison is continuously formed at a rate equal to the rate of part molding, cooling, and removal. To avoid interference with the parison formation, the mold clamping mechanism must move quickly to capture the parison and return to the blowing stations where the blow pin will enter. The process has three subcategories: the rising mold method, the rotary method, and the shuttle method.

In the rising mold method (Fig. 82.7), the parison is continuously extruded directly above the mold cavity. When it is at the proper length, the mold rises quickly to capture the parison and returns downward to the blow station. After blowing the bottle, the mold opens, the part is removed, and the process repeats. ADS is a supplier of this type of machinery.

In the rotary method, up to twenty clamping stations are mounted to either a vertical or horizontal wheel (Figs. 82.8 and 82.9). As the wheel rotates past the extruder, simultaneously, a parison is captured, bottles are molded and cooled, and a cooled bottle is removed. The method can provide high production yields; however, a disadvantage is the complexity and setup of the multiple mold clamps. It is usually not suited for short production runs.

Many rotary machines are private designs produced by bottle blow molders, for example Owens-Illinois and Continental. Suppliers of commercial equipment include: Johnson Controls, Inc. and Graham Engineering Co. Machinery systems are often as high as $1.5 million.

In the shuttle method a blowing station is located on one or both sides of the extruder (Figs. 82.10, 82.11). As the parison reaches the proper length, the blow mold and clamp quickly shuttle to a point under the extrusion head, capture and cut the parison, and return to the blowing station. With dual sided machines, the clamps shuttle on an alternating basis. For increased production output, multiple extrusion heads are used.

The continuous extrusion process, although well suited for most resins, is best for

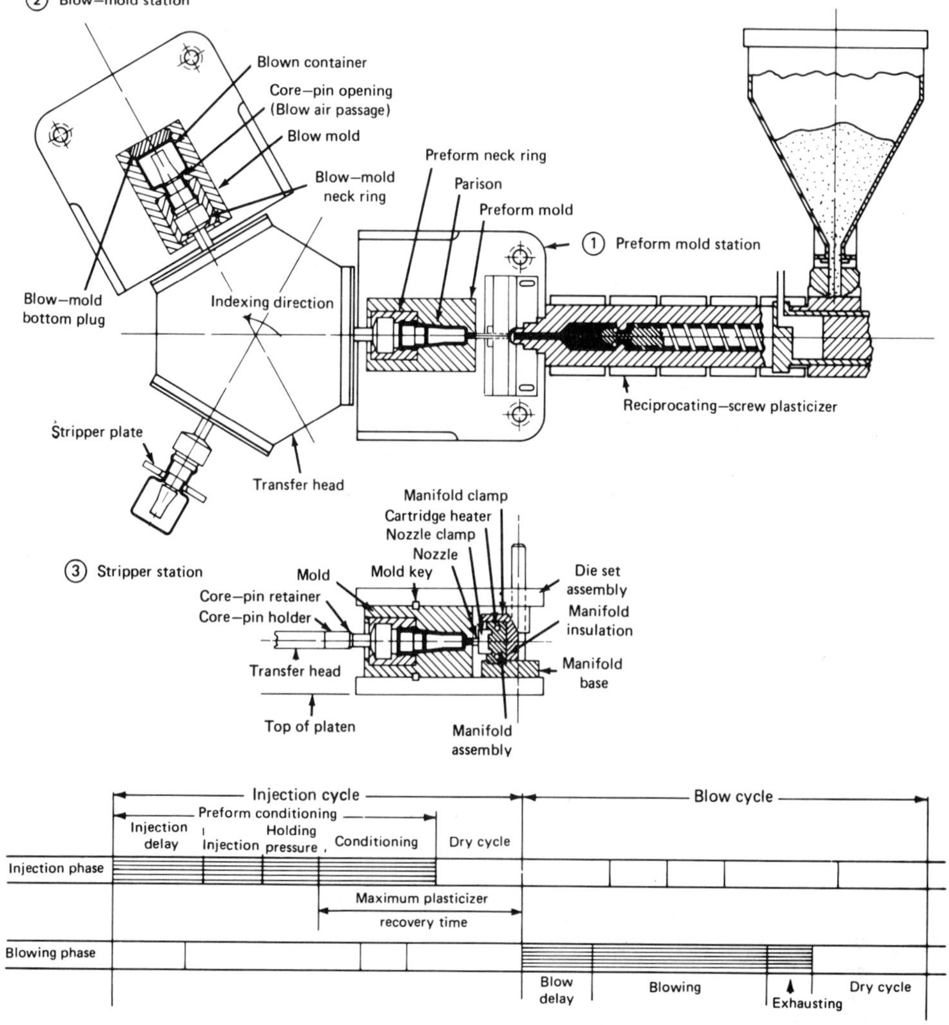

Figure 82.4. Process description of a typical three-station injection blow-molding machine. Courtesy of Johnson, Controls Inc.

poly(vinyl chloride) resins. Poly(vinyl chloride) can rapidly degrade if overheated slightly. The relatively slow uninterrupted flow of material in this process reduces the tendency for "hot spots" to occur which would damage the material. Generally, the process is used for bottles or parts 4.1 or less in volume; however, with ultra high molecular weight polyethylene resins providing superior melt strength, the process could be used for some large industrial parts as well. Several firms supply this type of machinery. Key suppliers include: Johnson Controls, Inc., Bekum, Battenfeld/Fischer, Automa, Hayssen, Hesta, and Kautex. Costs for these machines typically range from $75,000 to $350,000.

Intermittent extrusion is the second basic extrusion blow-molding category. In this process the parison is quickly extruded after the bottle is removed from the mold. The mold clamping mechanism does not need to transfer to a blowing station. Blow molding, cooling, and part removal all take place under the extrusion head (Fig. 82.12), which also allows the clamping system to be more simple and rugged. The stop/start aspect of the extrusion method makes this process more suitable for polyolefin and other materials that are not heat-sensitive. The

Figure 82.5. Uniloy Model 88-3, three-station injection blow-molder. The machine is fitted with core rod cooling for the processing of PET. Courtesy of Johnson Controls, Inc.

Figure 82.6. The indexing table of a three-station injection blow-molder. On the right are the parison preform molds. On the left are the bottle blow molds. In the center is the stripper bar. Mounted to the indexing table are three sets of core rods. Courtesy of Johnson Controls, Inc.

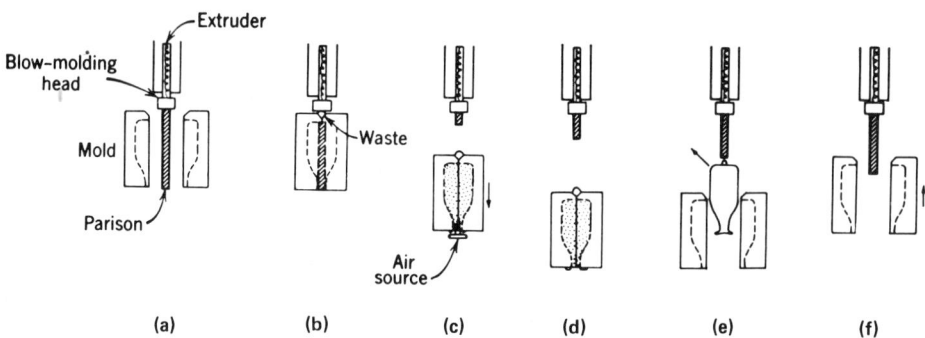

Figure 82.7. Continuous extrusion with rising molds: **(a)** complete extrusion of parison; **(b)** close mold; **(c)** blow; **(d)** cool; **(e)** eject; **(f)** raise mold.

process has three subcategories: the reciprocating screw method, the ram accumulator method, and the accumulator head method.

The reciprocating screw method is normally used for bottles between 250 mL and 10 L in volume. After the parison is extruded, the screw moves backward, accumulating melt in front of its tip. When the previously molded bottle has cooled the mold opens, the bottle is removed and the screw quickly moves forward, pushing plastic melt through the extrusion head and thereby forming the next parison. Up to twelve parisons have been extruded simultaneously. Figure 82.13 shows a reciprocating screw-extrusion blow-molder for a dairy bottle. Major suppliers of this type of machinery include Johnson Controls Inc. and Improved Machinery. Costs range from $120,000 to $400,000.

The ram accumulator method, which is no longer in widespread use, is intended for heavy parts weighing 2 kg or more. Much like the reciprocating screw, the system is used to quickly extrude heavy parisons that might sag or draw down from gravity. The accumulator is a reservoir mounted alongside the extruder. A piston or plunger pushes the melt through the extrusion head. Unlike the reciprocating screw, melt that enters the reservoir first is the

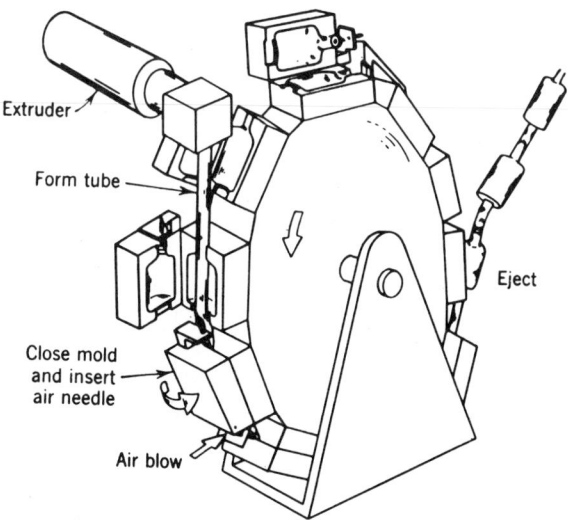

Figure 82.8. Multicavity continuous-tube blow-molding machine. Individual bottles are usually separated from the continuous parison tube at the time of mold opening.

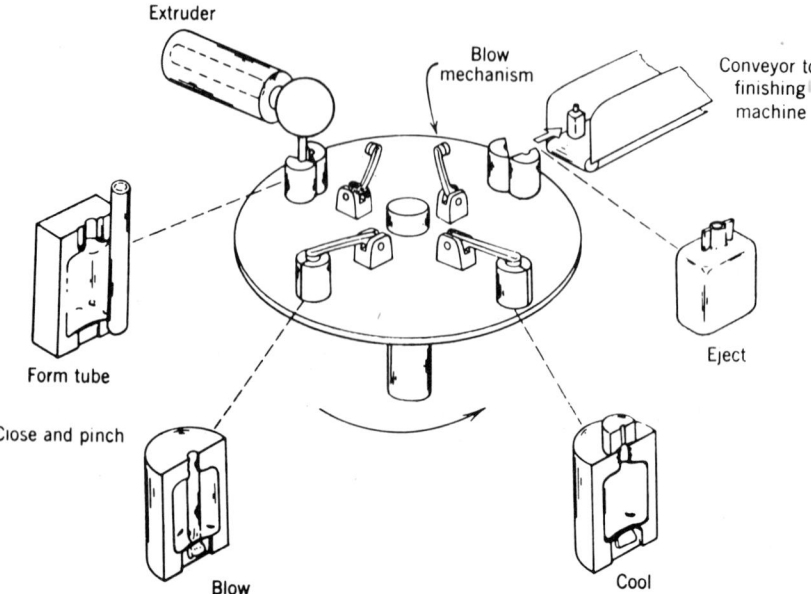

Figure 82.9. Horizontal rotary blow-molding machine. A four-station machine is illustrated; the turntable indexes each mold intermittently.

last to leave. As a result, the melt history of the resin is not uniform. No major suppliers of machinery currently use this approach, though a few minor suppliers still do.

The accumulator head method (Fig. 82.14) has replaced the ram accumulator in its applicability for heavy parts. The reservoir, tubular in shape, is a part of the extrusion head itself. Plastic melt that enters the head first is also first to leave. A tubular plunger quickly extrudes the melt from the head annulus with a low uniform pressure, which helps to reduce the stresses found in other systems. Several firms supply this type of machinery. Key suppliers include Johnson Controls Inc., Bekum, Battenfeld/Fischer, Hayssen, Kautex, APV, and Hartig. Costs range from $100,000 to $1,000,000 or more.

Another process somewhat related to extrusion blow molding is the extrusion/molded neck process (Fig. 82.15). Still used today by Owens-Illinois, this is a proprietary process

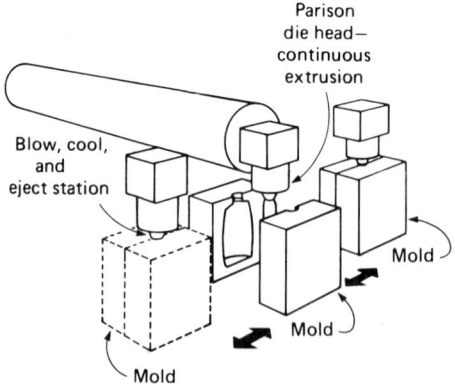

Figure 82.10. Shuttle continuous extrusion blow-molding machine. Molds on this dual-sided system alternately move to capture a parison.

Figure 82.11. Bekum BM-602D double-sided continuous extrusion blow molding machine. Courtesy of Bekum.

that traces its roots to glass-blowing technology. It is unique in that the bottle neck finish is injection molded and the bottle body is extrusion blow-molded. In this process, the two halves of a neck finish cavity or neckring are mounted to an actuating head assembly, which intermeshes with the two halves of the blow-mold cavity. The process cycle begins with the main body mold cavity open and the neckring cavity closed. The actuating head assembly moves downward to contact the extrusion die head. When it is in position, extrusion pressure fills the neck section with plastic melt. After it is held for one to two seconds, the head assembly moves upward while, simultaneously, the parison is extruded. When the head assembly reaches the top of its stroke, the blow-mold cavity closes on the parison. The remaining steps of parison pinch-off, blowing, and part removal all follow conventional techniques. Although the production cycle is somewhat slow, the process offers the quality

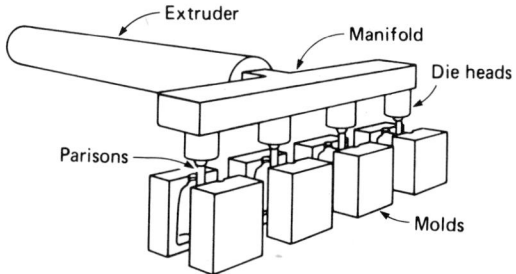

Figure 82.12. Intermittent extrusion blow-molding machine. Parisons are extruded quickly; molding, cooling, and part removal all take place under the extrusion die head.

Figure 82.13. Uniloy 350R2 eight-head intermittent extrusion blow-molder for the manufacture of 2 L (half gallon) handled milk bottles. Production rates of more than 65 bottles per minute have been achieved. Also shown is the flash trimmer. Courtesy of Johnson Controls, Inc.

advantage of an accurately molded neck and the process advantage of a parison held at both ends.

82.7.3 Stretch Blow Molding

In stretch blow molding, four main plastic resins are used: poly(ethylene terephthalate) (PET), poly(vinyl chloride) (PVC), polypropylene, and polyacrylonitrile.[3,6] The stretch blow-molding process is based on the molecular behavior of the resin; a parison or preform is

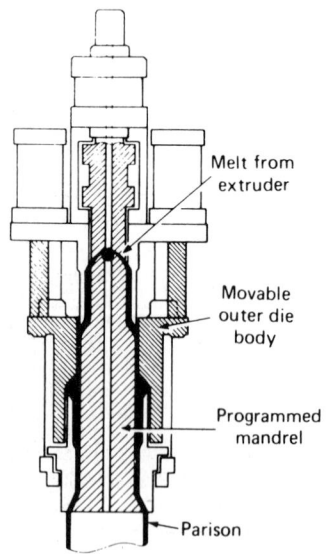

Figure 82.14. Typical accumulator head.

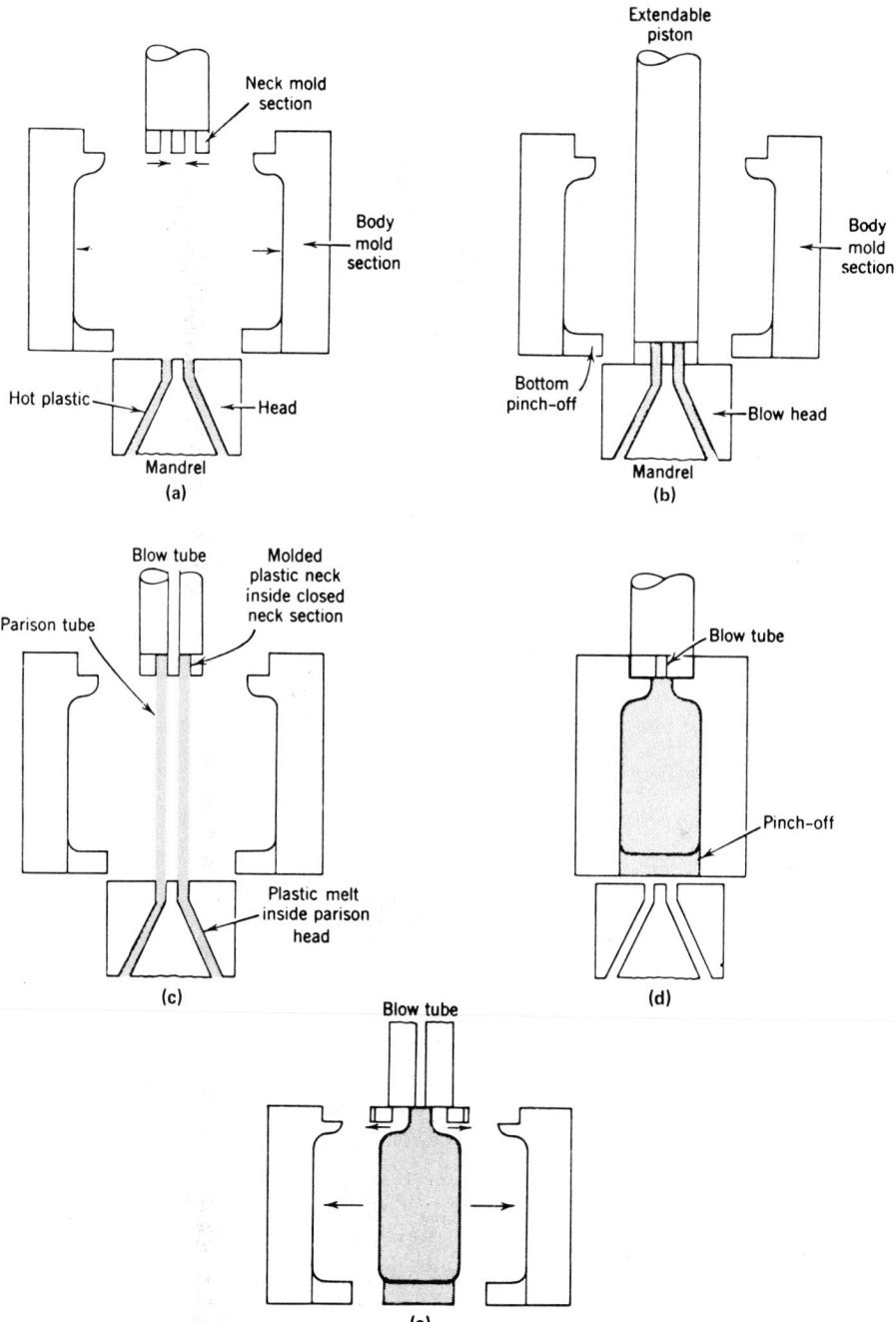

Figure 82.15. Extrusion molded neck blow-molding process: **(a)** body section open, neck section closed, neck section retracted; **(b)** neck section extruded to mate with parison nozzle (plastic fills neck section); **(c)** neck section retracted with parison tube attached; **(d)** body section closed, making pinch-off (parison blow to body sidewalls); **(e)** body molds open, neck molds open, bottle about to be ejected.

temperature conditioned and then rapidly stretched and cooled. For best results with PET the resin molecules must be conditioned, stretched, and oriented at just above the glass-transition temperature. At this point the resin can be moved without the risk of crystallization (Figs. 82.16 and 82.17).

The advantage of the process is an improvement in resin performance, such as bottle impact strength, cold strength, transparency, surface gloss, stiffness, and gas barrier, which permits lighter weight, lower cost bottles, use of lower cost resins with less impact modifiers, or products packaged that normally would not be suitable. The process uses injection molded, extruded, or extrusion blow-molded parisons with either a one-step or two-step approach.

In the one-step approach, the stages of parison production, stretching, and blowing all take place in the same machine. In the two step approach, parison production is done separately from parison stretching and blowing.

The main advantage of the one-step approach is the savings in energy, as the parison is rapidly cooled to the stretch temperature. With the two-step approach, the parison is completely cooled to room temperature and then reheated to the stretch temperature (Fig. 82.17). On the other hand, the two-step approach can be productively more efficient; a minor breakdown in one of the steps does not halt the other. The optimum balance of product design versus preform design versus production output is also easier to achieve with the two-step approach. Limits on parison production, for example, will not force a compromise in parison design to achieve higher bottle production. Each and every bottle design, for optimum performance, has a unique parison design and temperature conditioning requirement which may or may not fit, for optimum productivity, the assumptions used in the design of the one-step equipment.

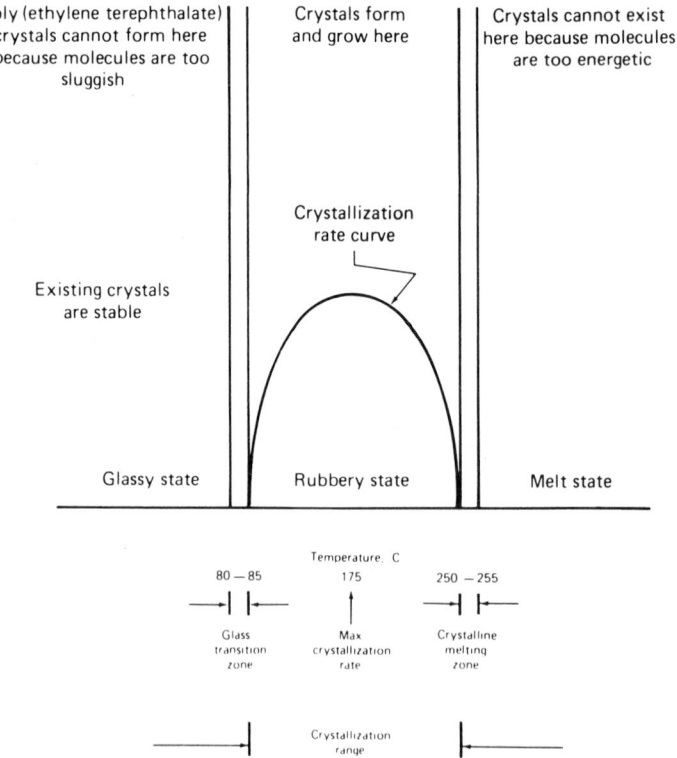

Figure 82.16.   Molders diagram of crystallization behavior of PET.

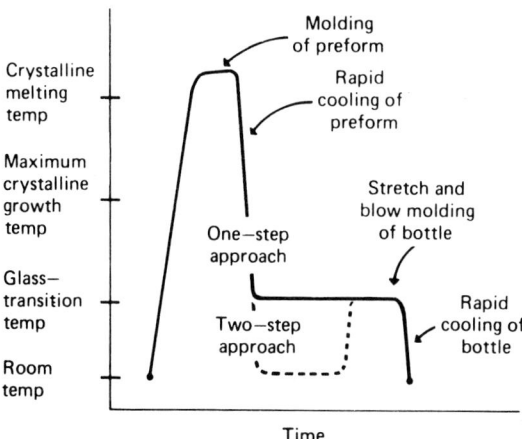

Figure 82.17. Basic stretch blow process. Courtesy of Jerome S. Schaul.

With injection stretch blow molding, the parison is virtually the same in principle as that used in injection blow molding. Both one-step and two-step process approaches are used. With the two-step process, the parison is injection molded in a separate machine, sorted, and later placed in an oven for temperature conditioning and blow molding. A rod is most often used inside the parison, in combination with high air pressure, to complete the stretch (Fig. 82.18). Injection stretch blow molding is most often used for PET resin. Several firms supply injection stretch blow-molding equipment in one-step and/or two-step configurations. Major suppliers include Nissei, Cincinnati-Milacron, Sidel, and Krupp Corpoplas. Equipment costs can range from $200,000 to more than $1.5 million or more.

Although it is related, the extruded parison involves a slightly different concept than the extrusion blow-molded parison. As before, both one-step and two-step approaches can be used. In the one-step approach, a tube is extruded and fed directly into an oven for conditioning. After conditioning, the tube is cut into parison lengths. Mechanical fingers grab both ends, stretching the parison. Two mold halves close where air pressure expands the stretched parison against the mold cavity. The two-step approach differs in that the extruded

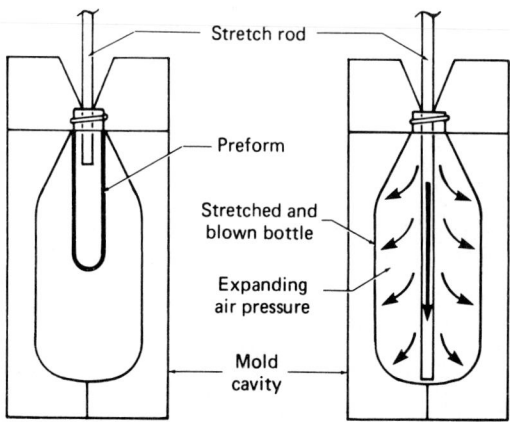

Figure 82.18. A temperature conditioned preform is inserted into the blow-mold cavity, then is rapidly stretched. Often a rod is used to stretch the preform in the axial direction with air pressure to stretch the preform in the radial direction.

tube is cooled and cut to length. Later, the cut tubes are placed in an oven for conditioning. Polypropylene is most often used with this stretch blow-molding method, although some PVC is used as well. This is no longer a common process approach. Commercial equipment was supplied by Deacon (two-step) and Bekum (one step).

With the extrusion blow-molded concept, the parison is shaped and temperature conditioned in a preform cavity in the same way a bottle is extrusion blow-molded. From this preform cavity, the parison is transferred to the bottle cavity, where a rod and air pressure combine to stretch and expand the resin. Although the one-step approach is the most common, a two-step approach, in the same fashion as the others, is feasible. PVC is most often stretch blow-molded using this method. Figure 82.19 shows a one-step extrusion blow stretch blow-molder. Suppliers of this equipment include Bekum and Battenfeld/Fischer.

82.7.4   Multilayer Blow Molding

All materials, whether metal, glass, or paper, have certain strengths and weaknesses, advantages and disadvantages. Many times two or more materials can be layered and combined to overcome weaknesses economically. Examples are chromium plated steel, laminated automobile winshields, and wax or polyethylene-covered paperboard. Multilayer blow molding is a process in which the strengths of two or more resins are combined to package a product far better than any of the resins could individually.

A few important characteristics or requirements of many bottles are cost, strength, clarity, product compatibility, and gas barrier. Polyethylene or polypropylene, for example, is relatively low in cost, approved for food contact, and an excellent barrier of water vapor. It is also a poor barrier of oxygen. As such, the material is not suited for packaging many oxygen sensitive foods requiring long shelf life. Poly(ethylene vinyl alcohol), on the other hand, is a relatively high cost material that provides an excellent barrier, but it is sensitive to water, which can deteriorate its properties. A thin layer of it sandwiched between two layers of polyethylene or polypropylene can solve the problem.

All of the basic blow-molding process methods have been used with multilayer blow molding. In each case, additional plasticizers or extruders are needed for each resin. The

Figure 82.19. Bekum BM04D continuous extrusion blow-molder for the manufacture of biaxially oriented PVC bottles. Production rate for 1-L bottle is 2000 per hour. Maximum bottle size is 2-L. Courtesy of Bekum.

continuous coextrusion process has been used for bottles up to 5 L in size. The accumulator head process has been used for drums and tanks up to 500 L in size. The coinjection process has been used for small bottles 500 mL in size. Suppliers of multilayer blow molding equipment include: Johnson Controls Inc., Bekum, Battenfeld/Fischer, Wilmington, and Kautex for continuous extrusion based machines; Nissei for injection based machines; and IHI for accumulator head based machines.

Related to the coinjection process is a thermoform insert process.[7] With this method, a coextruded sheet is thermoformed into an insert, which is placed on the core rod of an injection blow-molder just prior to parison/preform molding. The hot resin from the injection step softens the material in the insert, permitting it to be blow-molded in the next or blow station.

A common problem of multilayer materials is that the different layers will not stick to each other. An adhesive layer is often required to create the bond. As a result, three or more extruders are required. With the polyethylene/poly(ethylene vinyl alcohol) example above, five layers are actually required: polyethylene, adhesive, poly(ethylene vinyl alcohol), adhesive, and polyethylene.

E.I. du Pont de Nemours & Co. has developed a novel approach to gain some of the advantages of a coextruded blow-molded bottle without the cost and complexity of multiple extruders and expensive scrap reclaim systems. Only minor modification of the basic equipment is required. A barrier resin is tailored for physical blending with a base resin in much the same way that colorant is blended with a resin. However, instead of producing a homogenized mix, the barrier resin is permitted to laminate in the base resin (Fig. 82.20). Several random multiple layers of barrier resin of various lengths and sizes are created. Although no single layer completely covers the entire shape and surface of the bottle, barrier is created by the "tortuous path" the permeating gas must follow. A modified nylon material is the first to be used for this approach. Applications include fuel tanks and bottles for agricultural chemicals, thinners, and solvents.

82.7.5 Other Blow Molding Operations

Many related operations have been used to improve blow-molding production. In-mold labeling, fluorination surface treatment, and internal cooling systems are significant.

In-mold labeling is a process developed by Procter & Gamble Corp. while it was working with a few of the major custom blow-molded bottle manufacturers. A label with a heat-

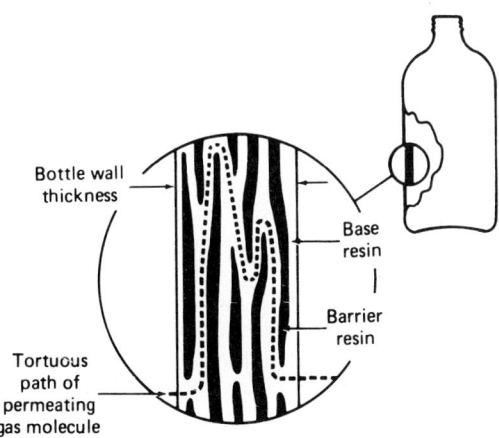

Figure 82.20. DuPont's blended resin method.

activated adhesive is automatically placed into the mold cavity and held by a vacuum. The expanding hot parison activates the adhesive, which creates a 100% bond.

Some of the advantages of the process include bottle weight reduction, improved label appearance, elimination of other high-speed, complex labeling equipment, and new package opportunities. A stiffer, stronger structure is created with 100% bond. Weight can often be removed from the bottle and it will still perform adequately. The 100% bond also improves label appearance by eliminating blisters and wrinkles. Although it is somewhat complex, the "pick and place" mechanism used to place the label into the mold cavity is often simpler than the high-speed equipment used to label on the filling line. New package design opportunities exist with the ability to place the label closer to or around bottle edges, which could further improve the strength.

The in-mold labeling process also has disadvantages. Production efficiency can drop from the slower cycle and complication added to the blow-molding process. The value of scrap is higher and more costly to reclaim. The investment in equipment and tooling limits product opportunities to those with very high production runs.

Fluorination surface treatment is a process to improve the gas barrier of polyethylene to nonpolar solvents.[8] A barrier is created by the chemical reaction of the fluorine and the polyethylene, which form a thin (20–40 nm) fluorocarbon layer on the bottle surface. Two systems are available for creating the layer. The "in-process' system uses fluorine as a part of the parison expand gas in the blowing operation. With it, a barrier layer is created only on the inside. The "post-treatment" system requires bottles to be placed in an enclosed chamber filled with fluorine gas. The method forms a barrier layer on both the inside and outside surfaces.

This surface treatment allows low-cost, blow-molded polyethylene bottles to be used for paint, paint thinner, lighter fluid, polishes, cleaning solvents, cosmetics, and toiletries. The approach can replace other higher cost resins or coextrusion processes.

Normally, a blow-molded part is cooled externally by the mold cavity, forcing heat to travel through the entire wall thickness. With the poor thermal conductivity of plastic resins, molding cycle times of heavy parts can be considerable. Internal cooling systems are designed to speed the mold cooling time, thus reducing costs by removing some of this heat from the inside. Several systems have been developed that incorporate three basic approaches: liquified gas, super-cold air with water vapor, and air exchange methods.

With the liquified gas system, immediately after the parison has been expanded, liquid carbon dioxide or nitrogen is atomized through a nozzle in the blow pin into the interior of the bottle.[9] The liquid quickly vaporizes, removing heat, and exhausts at the end of the cycle. In practice, this method has improved production rates by 25 to 35%. A disadvantage is the cost of the liquified gas. If its consumption is not precisely controlled, the total cost savings becomes minimal.

The super-cold air system with water vapor works much the same way. Here, very dry, subzero blowing air expands the parison. The expanded air is allowed to circulate through the bottle and exhaust. Immediately after the parison has expanded, a fine mist of water is injected into the cold air stream. As it flows, the water mist turns into snow. As the snow circulates through the container, it melts and then vaporizes. At the end of the molding cycle, the water mist is stopped, permitting the circulating air to purge and dry the interior before the mold is opened and the part is removed. Production rates can be improved as much as 50%.[10]

The air-exchange system is a far simpler method. Here, after the parison has been expanded, plant air is allowed to circulate through the bottle and exhaust.[11] Differential pressure inside the bottle is maintained at 80 psi (550 kPa) to ensure that the parison stays in contact with the mold cavity. Production rate improvements are modest; only 10–15% improvement can be expected.

The better internal cooling systems are often not justifiable with today's equipment because most blow-molding machines do not have the additional extruder plasticizing capacity to support the faster production rate. This is particularly true for the heavier bottles that would benefit most. As a result, only the low-cost air exchange systems have reached any degree of popularity.

82.8 TOOLING

Blow-mold tooling involves more than molds. Often, a series of interchangeable components are considered; for example, the extruder screw profile, neckring tooling, core rod tooling, extrusion heads and head tooling, blow pin tooling, and trim tooling.

82.8.1 Extruder and Screw Design

All blow-molding processes require the melting of plastic resin. The quality of this melt and the productivity of the blow-molding equipment depend heavily on extruder screw design. The designs used are the same as are used with profile extrusion and extrusion blown film technologies.[12] Two basic approaches are used: the single metering or compression screw (some of which are fitted with high-shear mixing tips); and the barrier flight screw.[13] and continuous extrusion blow-molders. A few continuous extrusion systems are also fitted with a melt screenpack or breaker plate. Some of the large industrial blow-molders and continuous extrusion blow-molders are fitted with a grooved barrel feed zone. The tapered grooves improve the processing of powdered ultrahigh molecular weight polyethylene.[14]

82.8.2 Injection Blow-mold Tooling Design

The injection blow-molding process requires two molds: one for molding the preform or parison, the other for molding the bottle. The preform mold consists of four major components: preform cavity, injection nozzle, neckring insert, and core rod assembly. The blow mold typically consists of three major components: bottle cavity, neckring insert, and bottom plug insert[15] (Figs. 82.21–82.25).

It is important to observe certain basic rules or constraints in the preform cavity design. First, the ideal core rod or cavity length-to-diameter (L/D) ratio is approximately 10:1 or

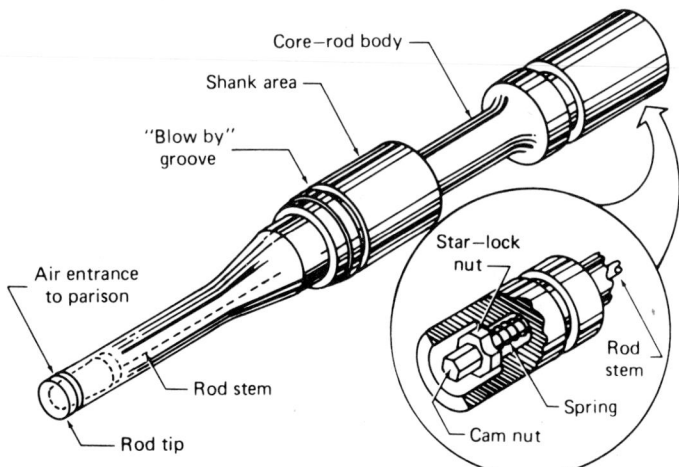

Figure 82.21. Typical bottom blow core rod showing its principal elements. The core-rod tip mechanism, which closes the air passage during the parison injection cycle, is shown in the enlarged view at left.

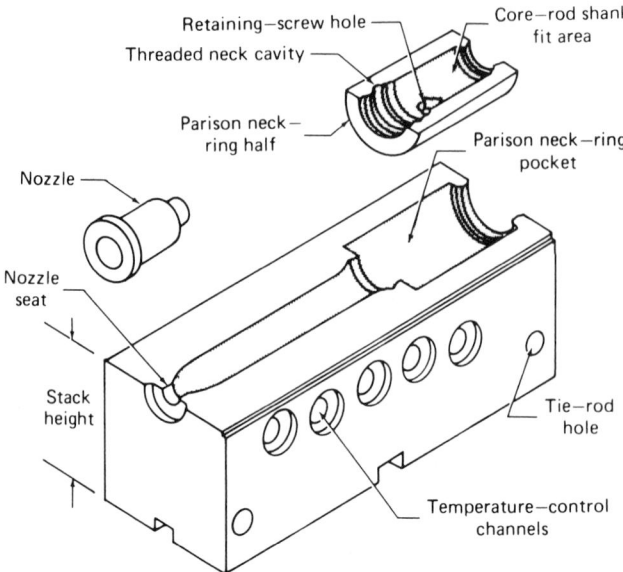

Figure 82.22.   Exploded view of one half of a parison mold cavity, with nozzle and neckring details.

less. Most often, this is a comparison of the overall height and the finish "E" diameter (see Fig. 82.3) of the bottle. This ratio ensures a minimum of core-rod deflection from injection pressures, which, in turn, will provide uniform material wall distribution and heat uniformity. Greater ratios have been used, but, in these cases, sliding pins are required to momentarily hold the end of the core rod on center during the material injection phase.

The ideal preform size to maximum bottle size (blow-up) ratio is 3:1 or less. Most often, this is a comparison of the maximum bottle diameter or width or depth and the finish "E" diameter. It is important to this ratio to provide uniform and consistent bottle cross-sectional wall distribution. If a greater ratio is used, the parison will tend to "float around" during expansion, which will increase the chance of an eccentric wall distribution.

The ideal parison wall thickness is between 0.08 and 0.20 in. (2 and 5 mm). A wall thickness greater than 0.24 in. (6 mm) is difficult to obtain the properly temperature condition

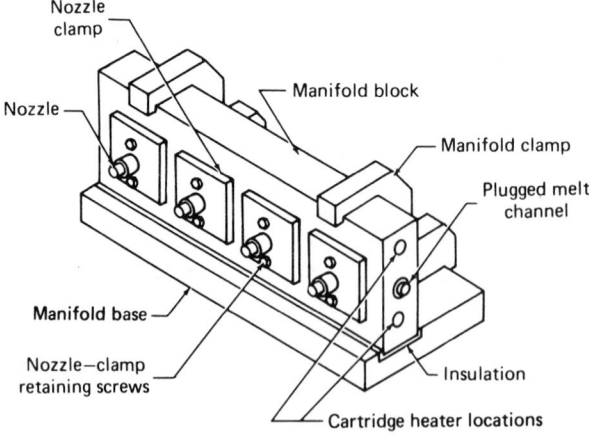

Figure 82.23.   Injection manifold for injection molding of parisons. Individual nozzles are clamped to the manifold block, which houses a hot runner for the melt.

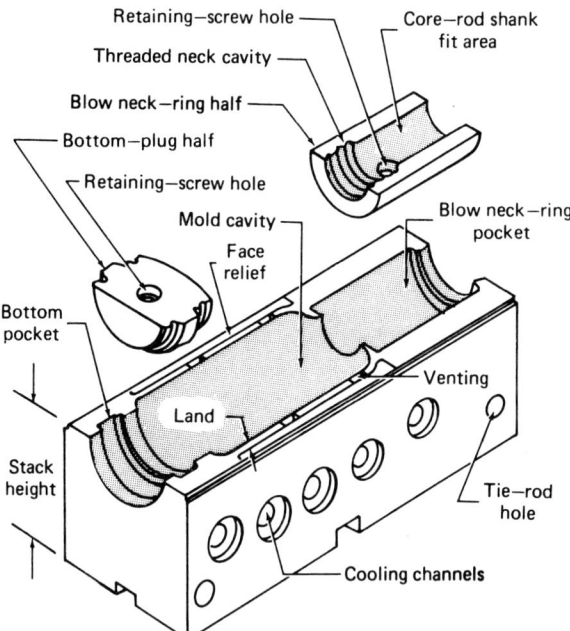

Figure 82.24. Exploded view of one half of an injection blow mold, with details of the bottom plug and neckring.

and can be unpredictable during expansion. A parison wall thickness of less than 0.08 in. (2 mm) can also be unpredictable. For a given weight parison, the thin wall approach would increase the projected area and thus possibly exceed the capacity of the press.

One important advantage of injection blow molding is the diametrical and longitudinal programming of the parison by shaping the parison mold cavity, core rod, or both. This is particularly important with oval bottles. This leads to another rule: in annular cross section, the heaviest area should not be more than 30 percent thicker than the lightest area. Generally, the shaping is done in the cavity; the core rod is kept round. If a greater ratio is used, then the selective fill of material during the injection phase will cause a vertical material weld

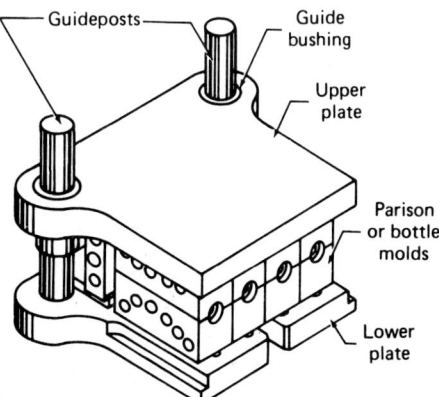

Figure 82.25. Die set for maintaining position and alignment of injection blow-mold cavities.

line to form in the bottle. This condition restricts the bottle ovality to 2:1; that is, the width should not exceed two times the depth.

In multiple cavity setups, each parison cavity is fitted with an injection nozzle of decreasing sizes. Flow of material through the injection manifold is balanced, thereby allowing each cavity to be filled at an equal rate.

The neckring insert has four functions or features:

- It forms the finish or threaded neck section of the bottle.
- Because it is an insert, it provides a relatively low-cost, easy method for changing the size of style of the finish.
- It firmly centers and locates the core rod in the parison cavity.
- It provides thermal isolation.

During the process the neck finish area of the parison must be cooled to retain its shape, but the remainder of the parison is kept hot for later expansion in the bottle cavity. Depending on the plastic material being molded, the temperature of the parison ranges between 149 and 275°F (65 and 135°C). The neckring insert is at times cooled as low as 41°F (5°C). The water lines for both the cavity and the neckring are usually drilled as closely as possible to each other, perpendicular to the cavity axis. Water flow is from one cavity to the next.

The core rod assembly also has four functions or features:

- It forms the interior of the preform.
- It supports the parison or the bottle during transfer.
- It is the valve where air enters to expand the parison. The valve is located either in the shoulder area or in the tip, the location being governed by the bottle or core rod length-to-diameter ratio. Wide mouth bottles (that is, core rods with low length-to-diameter ratios) will usually have a shoulder valve.
- The core rod has a "blow by" groove. This annular groove, located near the seating shank, 0.003 to 0.009 in (0.1 to 0.25 mm) deep, is needed to seal the parison and prevent excessive air loss during bottle blowing and to eliminate elastic retraction of the parison during the transfer between cavities.

A variety of materials are used to construct the parison cavity and core rods. For nonrigid polyolefin resins, the parison cavity is made of prehardened P-20 tool steel with a hardness between 31 and 35 Rc. For rigid resins, the parison cavity is made of A-2 tool steel air hardened to between 52 to 54 Rc. The parison neckring inserts for most resins are made of A-2 tool steel. The core rod, for greater strength, is made of L-6 tool steel, hardened to between 52 and 54 Rc. In all cases, the cavity surfaces are highly polished and chromium plated. The only exception is the neckring insert for polyolefin resins, which is occasionally sand blasted with a 120 grit.

The blow-mold bottle cavity defines the final shape of the part. Its ovality ratio is the only design constraint; that is, the cavity width should not exceed two times the depth. To compensate for resin shrinkage after molding, the cavity dimensions are enlarged slightly. Specific shrinkage rates vary with the types of resin and with process conditions. For nonrigid polyolefin resins, it is between 1.6 and 2.0%. For rigid resins, it is 0.5%. Slightly higher rates are usually applied to the heavier neck finish dimensions than to the body.

Vents are placed along the mold-parting surface to allow the escape of trapped air between the expanding parison and the cavity. If they are too deep, an objectional mark will be left on the bottle. Because 145 psi (1000 kPa) of parison expand air pressure is used in injection blow molding, these vents are kept to less than 0.002 in. (0.05 mm) deep.

The neckring insert is used in the bottle cavity in a similar manner to the way it is used in the parison cavity; however, the two are not identical. The thread diameter dimensions are between 0.002 and 0.010 in. (0.05 and 0.25 mm) larger in the bottle cavity than in the parison cavity. Unlike the parison neckring, the bottle neckring does not form the finish detail. Instead, it only secures the already formed neck. The additional size provides clearance, reducing the chance of distortion.

The bottom plug insert forms the bottom or push-up area of the container. In some molds this insert must be retractable. Generally the push-up of polyolefin bottles can be stripped without side action if the height is less than 0.2 in. (5 mm). With rigid resins, this height is reduced to 0.03 in. (0.8 mm). When side action is required, an air cylinder or cam or spring mechanism is used.

Either aluminum, steel, or beryllium copper is used for the bottle cavity and neckring. For polyolefin resins, #7075 aluminum is used. Surface finish is usually #120 grit sandblast which helps improve the venting of trapped air. For rigid resins, A-2 tool steel air hardened to 52 to 54 Rc is used. Surface finish is highly polished with chromium plating. Cast beryllium copper is often used for either application when minute detail is required. As with the parison cavity, water lines are usually drilled as close as possible to each other, perpendicular to the cavity axis.

Both the parison and bottle molds are mounted onto a die set, which is mounted to the platens of the injection blow-molder. Keyways in two direction, on each of the upper and lower platens, are used to precisely position the cavities. Guide posts and bushings are then incorporated to maintain precise alignment between the platens. To speed blow-molder set-up, the entire die set/mold assembly is exchanged during a job change. It is considered false economy to reuse the die set with another mold set.

Injection blow-mold tooling must be held to very precise tolerances. Dimensions often must be held to $+/-$ 0.0006 in. (0.015 mm). If it is poorly made, processing will suffer and bottle quality will be inconsistent and poor. Consider for a moment how closely the several core rods must be located fore and aft, and left and right of centerline with the parison and bottle cavities. If too tight, the mold could be damaged or the assembly could bind. If too loose, resin could flash around the shank area of the core rod or the core rod could shift sideways, causing uneven parison wall distribution. Consider further the many parts and sections of the mold set-up that must fit together and interchangeably with each other. Several core rods must fit the pocket of the parison or bottle cavities, which are stacked side by side next to each other on a die set. Clearly this need for precision is the most important factor in the high cost of injection blow-mold tooling; however, once properly made and set up, the injection blow-mold process can be extremely productive and trouble free.

82.8.2 Extrusion Blow-mold Tooling Design

Extrusion blow molding requires only one mold; however, since the parison is formed from an extruded tube and flash is a by-product of the molding process, several tooling elements combine to produce the finished bottle. These elements include the design of the extrusion head, head tooling, blow pin tooling, mold-cavity tooling, and trim tooling.

The extrusion head used in blow molding is one of three basic styles: the spider or axial flow head, the side feed or radial flow head, and the accumulator head.

In the simplest form of the spider or axial flow head (Fig. 82.26), a central torpedo is positioned in the melt flow path supported generally by two spider legs. The smooth and direct path with little or no place for material to hang up and degrade makes this head ideal for PVC resin. Cross-sectional areas of the melt flow path from the extruder, over the torpedo, and over the spider legs are carefully balanced to ensure an even and consistent pressure and flow. Occasionally the spider head style is used for polyethylene, in which it offers the advantage of very rapid color changes. However, off-set spider legs are required

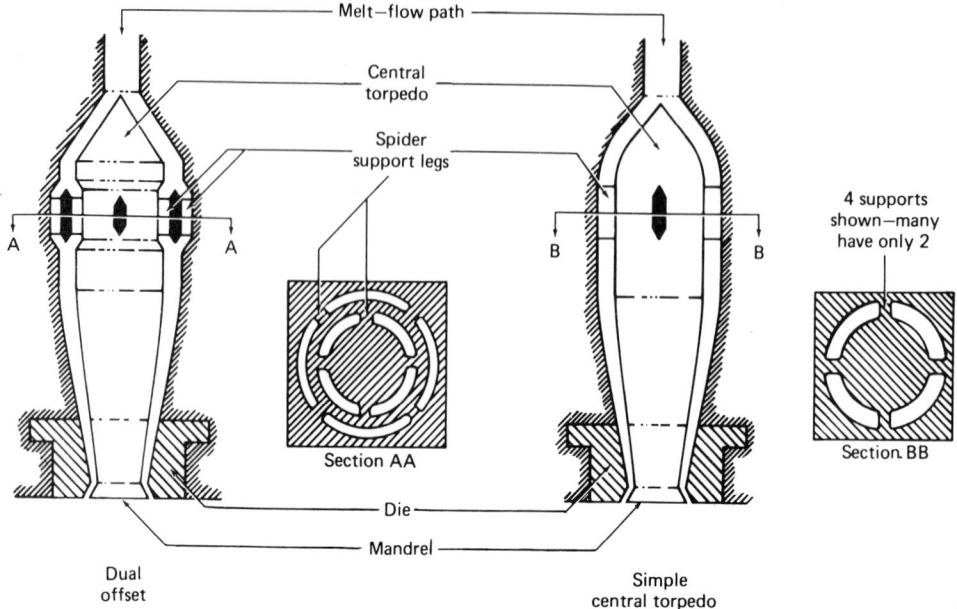

Figure 82.26.   Flow path of spider or axial flow head.

to mask the elastic memory retained by polyethylene of the legs. This requirement raises the manufacturing cost of the head considerably.

The side feed or radial flow head (Fig. 82.27) is a design most often used for polyethylene. Melt enters the head from the side, divides around the mandrel, and rewelds. As the melt moves downward, it enters the pressure ring area, creating a tremendous back pressure that helps ensure the resin reweld. The head is used for its simplicity and ruggedness, but has

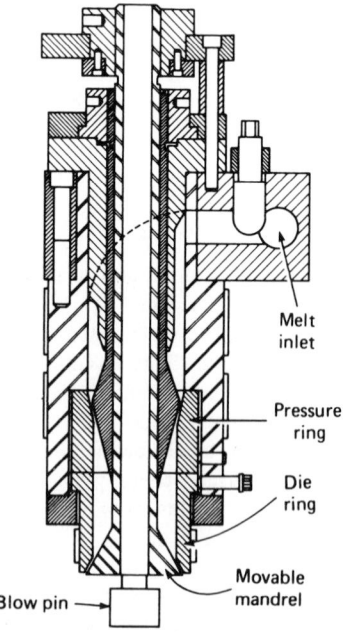

Figure 82.27.   Flow path side feed or radial flow head.

the disadvantage of relatively long color changes. Often during a color change, a fine trail of the previous color, which comes from reweld areas of the head, is noticeable on the parts for several hours of production. The problem has been reduced in recent designs by careful attention to streamlining and polishing of the flow path and, in some cases, the addition of a bleed screw to "pull off" the trail of bad color.

The accumulator head (Fig. 82.14) is considered a subcategory of intermittent extrusion blow molding. It is the combination of an extrusion head with a first in/first out tubular ram melt accumulator.

Regardless of the style, all heads can be fitted with parison programming. A parison programmer is a device that can change the gap or relationship between the head tooling die and mandrel while the parison is extruded. Thus, the wall thickness of the parison is changed and becomes ringed with sections of thinner and thicker material. These "rings" are located to correspond to specific sections of the bottle or blow-molded part where the part thickness may be too thin or too thick (Fig. 82.28). For many blow-molded parts, parison programming has the potential of reducing part weight and cost while improving performance and strength.

There are two styles of head tooling: convergent and divergent. As the name implies, convergent tooling has a land angle that converges toward a point. With divergent tooling the land angle flairs outward. Figure 82.29 illustrates basic nomenclature.

The ultimate goal of all head tooling design is a parison diameter. For proper blow molding, each bottle or part requires a parison of a specific diameter. The designer must first consider the basic shape of the part. For example, a bottle with a handle must have a parison large enough to permit the handle portion of the mold to capture part of it. Next, the designer must consider the parison blow-up ratio; that is, the change in size from the parison to the part. For extrusion blow molding the ratio is generally 4:1 or less.

Rarely is the parison diameter equal to the tooling diameter. The designer must estimate the degree of parison swell. Several resin, machinery, and processing factors will cause various amounts or changes in swell, among them:

- Resin molecular weight
- Resin molecular weight distribution

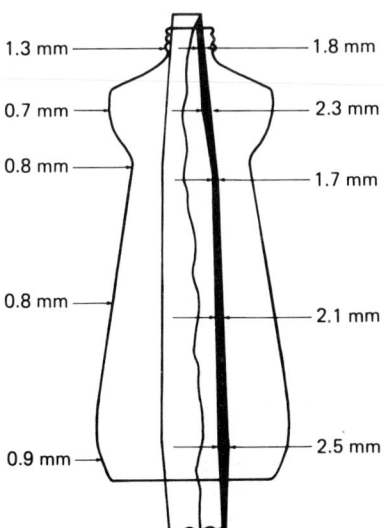

Figure 82.28. Example of improved material distribution in a blow-molded part from parison programming.

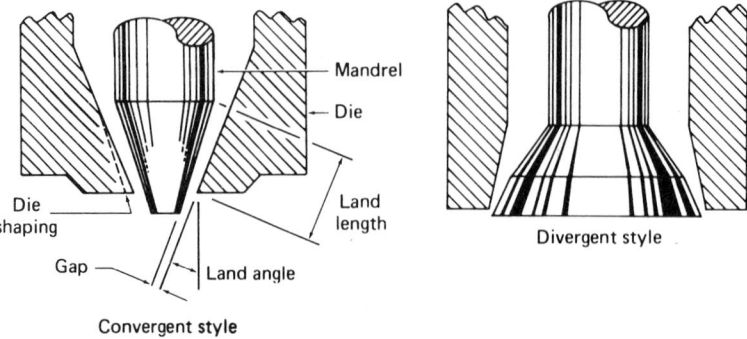

Mandrel

Die

Die shaping

Land length

Gap

Land angle

Convergent style

Divergent style

Figure 82.29. Head-tooling nomenclature.

- Tooling land length to gap ratio
- Tooling diameter to head diameter ratio
- Melt extrusion pressure
- Melt temperature
- Parison length

Table 82.2 gives approximate factors for parison die swell of common blow-mold resins.

As with parison programming, special shaping of the head tooling can also help redistribute material in the parison. As the parison is extruded, the shaped tool produces axial stripes of thinner or thicker material that correspond to specific sections of the blow-molded part. Usually the die is shaped or "ovalized" while the mandrel is kept round.

In most processes, the air that expands the parison against the mold cavity must enter through blow pin tooling, which is available in four basic styles: simple tube, needle, ram down or calibrated prefinish, and pull-up prefinish.

With the tube style, the mold is closed around the parison, which is closed and sealed around the tube. The tube usually enters the cavity through the center of the extrusion mandrel.

With the needle style, a hypodermic-shaped needle is pierced through the parison from some remote position in the mold after the mold has closed. The location of the hole on the part is usually inconspicuous or hidden from view.

Both the tube and the needle approaches are mainly used for industrial parts. When utilized for bottles, a post-machining or facing of the neck finish is required. Plastic chips falling into the bottle are a disadvantage, which led to the development of bottle prefinishing systems. In these systems the mold neckring and blow pin work together to form and size the neck finish to specification before the bottle is removed from the mold cavity.

With the ram down or calibrated prefinishing system, the critical top sealing surface of

TABLE 82.2. Approximate Factors for Parison Die Swell

	Swell, %
High density polyethylene	
Phillips type	15–40
Ziegler type	25–65
Low density polyethylene	35–65
Poly(vinyl chloride)	30–35
Polypropylene	35–55

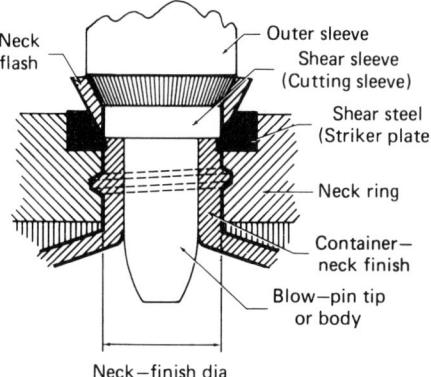

Figure 82.30. Calibrated neck finish.

the neck finish is held flat. Immediately after mold closing, the blow pin moves into the mold cavity, gathering a small amount of additional material to fill in and pack the neck finish area. The blow pin moves until the cutting or shear sleeve cuts through the parison flash, contacting the striker plate or shear steel mounted in the neck ring of the mold (Fig. 82.30).

Usually, the stroke of the blow-pin cylinder is adjusted to "bottom out" at the same time the cutter contacts the striker. The flash above the neck is easily separated, leaving the finish clean and flat. The blow pin also has a tip that forms the inside diameter; however, a precise dimension can not always be held for the entire length. Because the neck finish portion of the bottle is usually the heaviest, the blow pin is often cooled with circulating water. This helps to ensure a faster production cycle with accurate finish dimensions.

A very precise inside diameter can be formed with the pull-up prefinishing system, patented by Hoover Universal, Inc. As in the simple tube system, the mold closes on the parison, which then closes and seals around the pull-up blow pin. At the end of the molding cycle, just before mold opening, the blow pin "pulls up," shearing the inside diameter (Fig. 82.31). This system is used almost exclusively for plastic milk bottles.

As mentioned earlier, flash is a by-product of extrusion blow molding and it must be provided for in the design of the mold cavity tooling. Where the flash will exit from the cavity, clearance or a flash pocket is created. The depth of this flash pocket is critical,

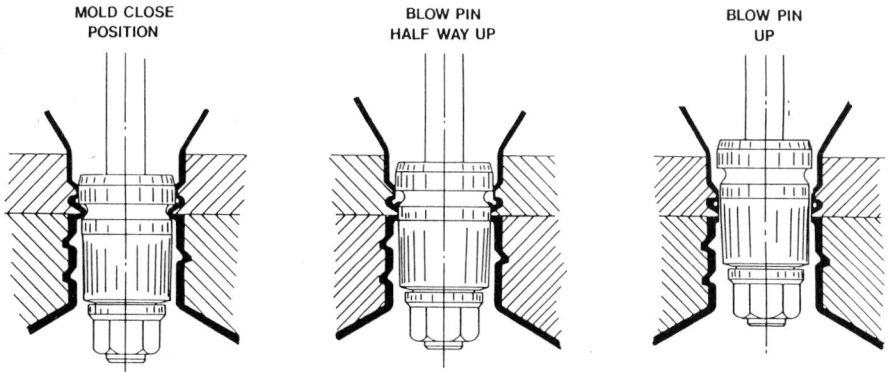

Figure 82.31. Pull-up prefinish system for center neck bottles. Inside diameter is sheared by action of the blow pin.

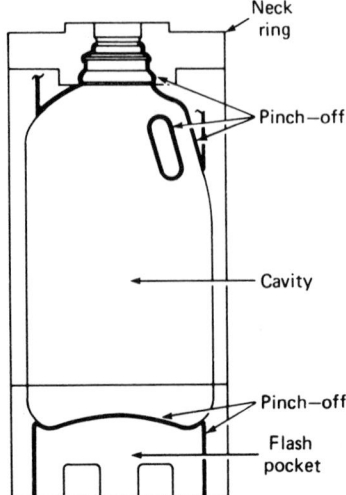

Figure 82.32. Typical extrusion blow-mold cavity.

particularly with large-diameter, thin-walled parisons. If too deep, the flash will not be adequately cooled and this may cause trimming problems. If too shallow, the flash will hold open the mold, creating a thick pinch-off that may also cause trimming problems. Dividing the cavity from the flash pocket is the pinch-off. The pinch-off has a dual role: it must seal the open ends of the parison and, at the same time, cut the flash from the bottle. Compression pinch-off is often used where some plastic material must be squeezed into the cavity to ensure a good weld (Figs. 82.32–82.34).

The cavity defines the shape of the part. As with other molding processes, cavity dimensions are enlarged slightly to compensate for resin shrinkage. Slightly higher shrinkage rates are used for polyolefins, particularly in the neck finish, for extrusion blow molding than for injection blow molding. The rate for the body is 2.0%; for the neck finish, as high as 3.5%. Rigid resins, however, remain about the same. Vents in an extrusion blow-mold

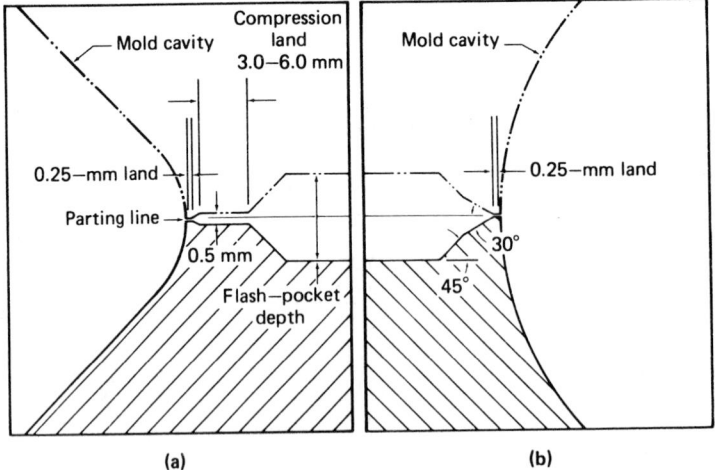

Figure 82.33. Pinch-off and flash-pocket detail. **(a)** Compression pinch-off **(b)** standard double-angle pinch-off.

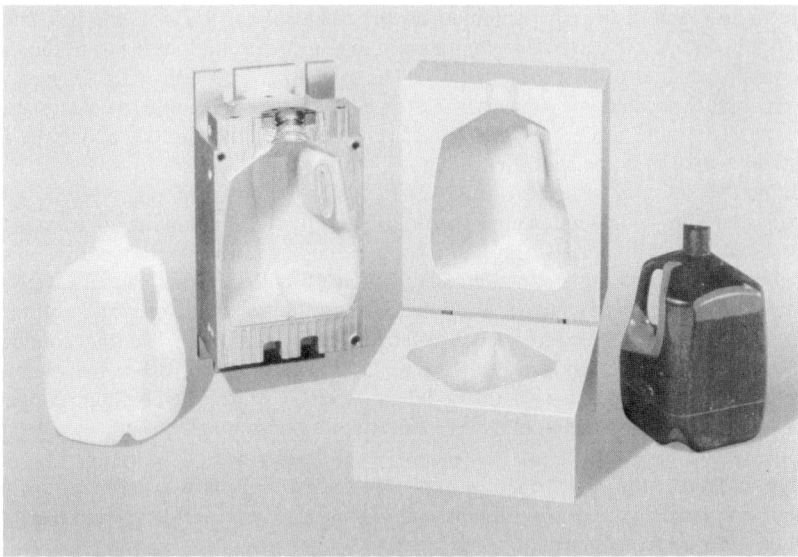

Figure 82.34. Many blow-molded cavities begin with a precision wood model from which a "stone" cavity pattern is created. The mold is cut using a tracer mill to duplicate the style.

cavity are often deeper as well: for some polyolefin molds, as deep as 0.003 in. (0.08 mm). The cavity surface is also rougher, with #60–#90 grit sandblast used for polyethylene. Polypropylene has a #240 grit surface. Only PVC molds have a high polish cavity surface. The primary reasons for these cavity differences are that the heavier neck finish area is often not cooled as well, which promotes additional shrinkage, and that lower parison expand air pressures are used [typically 90 psi (600 kPa)]. As a result, cavity surface imperfections are not as noticeable.

Unlike typical injection blow-mold cavities, extrusion blow-mold cavities are cooled in parallel from a manifold system. When possible, the neckring insert, bottom insert, and mold body are cooled with an individual circuit that permits each section to be set differently, which promotes a more even and uniform cooling of the part. Water lines are straight drilled holes parallel to the cavity axis with several right angle turns for greater turbulent blow. The lines are between 0.4–0.6 in. (10 to 15 mm) in diameter located approximately 1.5 diameters below the cavity surface. Spacing between the lines is between 3 to 5 diameters. Large cast aluminum industrial molds follow the same rules, but often the water lines are cast-in-place stainless steel tubing.

Trim tooling is needed to remove the flash from the blow-molded part after it is molded. The simplest way to remove the flash is manually, using a knife. The method obviously is suitable for very large industrial parts running on a relatively long molding cycle. Automatic flash trimming is used for high speed bottle production. Two basic approaches are utilized: in the blow-molder and downstream.

Although all types of parts have been trimmed in the blow-molder, the approach is most often used to remove the tail or bottom flash from small, nonhandled bottles. A sliding or pivoting plate is fitted into the flash pocket of the mold. A cylinder moving the plate pulls the tail flash free from the bottle prior to mold opening and part removal.

On bottles with handles, the flash is most often removed in a downstream trim press. In this press, the bottle is held in a contoured nest where a punch, also countered, knocks off the tail, handle, and top flash. The flash cut from the bottle by the mold pinch-off is held in place by a very thin membrane.

Wide mouth bottles are also trimmed down stream of the blow molder. A common approach is to use a "fly-cutter" to remove a dome and trim the finish inside dimension. Another approach is to mold a special "pulley" shaped dome above the part. Using a "spin-off" trimmer, the special dome engages a belt drive that rotates the bottle against a stationary knife blade. The knife blade, which is fitted into a sharp groove directly above the finish, progressively cuts the dome from the bottle as it rotates.

With the exception of large industrial blow-molds cast from aluminum, typically #A356, most extrusion blow-molds today are cut from #7075 or #6061 aluminum or from #165 or #25 beryllium copper materials. Beryllium copper offers the advantage of corrosion resistance and considerable hardness, making it the choice for PVC. It has the disadvantage of weighing about three times as much, costing about six times as much per cubic centimeter, and requiring about one third more time to machine as the aluminum. Thermal conductivity of the beryllium copper alloy is slightly poorer as well (Table 82.3). For polyolefin blow molding, some mold makers have combined the materials by inserting beryllium copper into the pinch-off area of an aluminum cavity, thereby obtaining a lightweight, easy to manufacture mold with excellent thermal conductivity and hard pinch-off areas.

Unlike injection blow molds that are mounted onto a die set, all extrusion blow molds are fitted with hardened steel guide pins and bushings. These guide pins ensure the two mold halves are perfectly matched.

The remaining hardware in the process, dies, mandrels, blow pin cutting sleeves, and neckring striker plates, are all made from tool steel hardened to 56-58 Rc.

82.9 AUXILIARY EQUIPMENT

Auxiliary equipment used for molding support functions is not different from the equipment used by other processes. The nature of the process requires the supply of clean compressed air. Injection blow molding usually requires a pressure of 145 psi (1000 kPa). Extrusion blow molding requires only 90 psi (600 kPa). However, stretch blow molding often requires a high 580 psi (4000 kPa) of pressure.

TABLE 82.3. Blow-mold Tool Materials

	Hardness Rockwell Brinell	Tensile Strength		Thermal Conductivity	
		psi	(MPa)	(W/M·K)	$\left(\dfrac{\text{Btu·ft}}{\text{h ft}^2\text{·°F}}\right)$
Aluminum					
A356	BHN-80	37,000	(255)	87.2	(151)
6061	BHN-95	40,000	(275)	97.1	(168)
7075	BHN-150	70,000	(460)	75.1	(130)
Beryllium copper					
25 and 165	RC-30 (BHN-285)	135,000	(930)	60.7	(105)
Steel					
0-1 and A-2	RC-52-60 (BHN-530-650)	290,000	(2000)	(20.2)	35
P-20	RC-32 (BHN-298)	145,000	(1000)	21.4	37

Water cooling is required to maintain hydraulic system and mold temperatures. Chilled water is often substituted to increase cooling efficiency, but, as in all cooling systems, turbulent flow is far more important for best efficiency than actual temperature.

Injection blow molding required as many as four mold water temperature controllers to fine tune the temperature of individual zones or areas of the preform cavity. Occasionally, water flow is also required in the core rod. In high temperature situations, oil circulating units are substituted.

To reduce the risk of the machine running out of material, virtually all systems use an automatic resin hopper loader.

Extrusion blow molding systems incorporate the automatic recovery of scrap resin trimmed from the bottle. A custom designed device to collect and route the trimmings to a granulator is usually involved. The chopped resin is then returned to the blow molder via a proportioning hopper loader where it is mixed with virgin resin.

Blow-molded polyethylene bottles that require decorating are usually flame treated. The quick exposure of the bottle to an oxygen rich flame will molecularly change the surface, permitting better adhesion.

The PET resin often used in stretch blow-molding is extremely sensitive to moisture and must be dried prior to molding into the preform. The moisture content should be less than 0.005%. To accomplish this, a desiccant, high temperature dryer is mounted directly to the extruder.

Lastly, many extrusion blow-molded bottles are leak tested for holes. The device quickly pressurizes the bottle and then measures the time needed for decay of the pressure. If decay occurs too quickly, the bottle is automatically rejected.

82.10 FINISHING/82.11 DECORATING

Clever decorating can make a relatively simple bottle design stand out. All major decorating processes can be used on blow-molded bottles and parts. These include:

- Engraving in the mold cavity with artwork or copy.
- Heat-transfer labeling where a preprinted, multicolor label on web stock is applied in one pass.
- Silk screening where as many as four or five colors are applied, one layer at a time.
- Hot sampling, a great way to apply bright foils.
- Pad printing, used for recessed areas not easily silk screened or hot stamped.
- Preprinted paper label with adhesive backing.
- Plastic sleeve label stretched over the bottle.
- Shrink bands, also used as a tamper evidence around the closure.
- In-mold labeling, in which the label is inserted into the mold prior to blow molding.
- For industrial nonpackaging applications, mask and spray paint systems.

In many cases, air pressure is applied inside the bottle to give it stiffness during the decorating. Often, more than one approach is used together to give a special effect.

BIBLIOGRAPHY

1. G.P. Kovach, "Forming of Hollow Articles," in E.C. Bernhardt, *Processing of Thermoplastic Materials,* Robert E. Krieger Publishing Co., 1959; reprinted 1974, pp. 511–522.
2. R. Holzman, "The Development of Blow Moulding from the Beginnings to the Present Day," *Kunststoffe* **69**(10), 704–711 (1979).

3. K. Stoeckhert, "Stretch Moulding," *Industrial & Production Engineering*, 62–70 (April 1980).

4. J. Presswood, *Oriented PVC Bottles: Process Description and Influence of Biaxial Orientation on Selected Properties,* 39th Annual Technical Conference, Society of Plastics Engineers, Inc., May 1981, pp. 718–721.

5. R.J. Abramo, *Fundamentals of Injection Blow Molding,* 37th Annual Technical Conference, Society of Plastics Engineers, Inc., May 1979, pp. 264–267.

6. H.G. Fritz, "Stretch and Injection Blow Moulding," *Kunststoffe* **71**(10), 687–699 (1981).

7. S. Date, *Co-Pak Multilayer Plastic Containers,* 5th Annual International Conference on Oriented Plastic Containers, Ryder Associates, Inc., March 1981, pp. 37–48.

8. "Surface Treatment Improves Polyethylene Barrier Properties," *Package Eng.,* 64–66 (Nov. 1981).

9. E. Jummrich, "Improving the Efficiency of CO_2 Interior Cooling in the Extrusion Blow Molding," *Industrial and Production Engineering,* 180, 184, 185 (Feb. 1981).

10. L.B. Ryder, "You Can Cool Blown Parts Faster—But it Takes a Miniblizzard," *Plastics Eng.* 22–27 (Jan. 1980).

11. L.B. Ryder, "Faster Cooling for Blow Molding," *Plastics Eng.,* 32–40 (May 1975).

12. S. Collins, "Screw Design for High-Performance Extrusion," *Plastics Machinery & Equipment,* 15–19 (May 1983).

13. R.A. Barr, "Screw Design for Blow Molding," 39th Annual Technical Conference, Society of Plastics Engineers, Inc., May 1981, pp. 734–735.

14. J. Sneller, "What Could Be New in Grooved Feed Extruders?" *Modern Plastics International,* 48–50 (March 1982).

15. J.R. Dreps, "Design Blow Molds With an Eye to Economy—Part I." *Plastics Eng.,* 34–39 (Jan. 1975); "Design Blow Molds With an Eye to Economy—Part II," *Plastics Eng.* 32–35 (Feb. 1975).

General References

W.W. Bainbridge, and B. Heise, *Design and Construction of Extrusion Blow Molds,* SPE Palisades Section, National Symposium—Plastic Molds/Dies: Design RETEC Oct. 1977, pp. 21-1–21-7.

D. Boes, "Parison Die for Blow Moulding—Design Criteria for the Spider Head," *Kunststoffe* **72**(1) 7–11 (1982).

C.C. Davis, Jr., *Materials for Plastics Molds and Dies.* SPE Palisades Section, National Symposium—Plastic Molds/Dies: Design RETEC, Oct. 1977, pp. 1-1–1-17.

R.D. DeLong, *Injection Blow Mold, Design and Construction,* SPE Palisades Section. National Symposium—Plastic Molds/Dies: Design RETEC, Oct. 1977, pp. 22-1–22-8.

Glossary of Plastic Terminology, Plastic Bottle Institute of the Society of the Plastics Industry, 1980.

M. Hoffman, "What You Should Know about Mold Steels," *Plast. Technol.,* 67–72 (April 1982).

C. Irwin, "High-Quality Molds for Extrusion Blow Molding," *Plastics Machinery & Equipment,* 57–59 (Sept. 1980).

C. Irwin, "Blow Molding," *Encyclopedia of Polymer Science and Engineering,* Vol. 2, 2nd ed., John Wiley & Sons, Inc., New York, 1985, pp. 447–478.

W. Kuelling and L. Monaco, "Injection Blow Molds: Here's How to Build Them," *Plast. Technol.,* 40–44 (June 1975).

B. Miller, "Engineering Plastics: Next Step for Blow Molding," *Plastics World,* 30–33, 87 (July 1983).

Operator's Guide—Controlling Shrinkage of HDPE Bottles, The Dow Chemical Co., Midland, Mich., 1979.

D.L. Peters, "Design Molds With Moving Sections to Blow Irregularly Shaped Parts," *Plast. Eng.,* 21–24 (Oct. 1982).

R.W. Saumsiegle, *The Three "E" System of Blow Molding Displacement Blow Molding,* 39th Annual Technical Conference, Society of Plastics Engineers, Inc., May 1981, pp. 727–728.

J. Szajna, "Designing Effective Plastic Bottles," *Food Drug Packaging,* 14–18 (May 1983).

83

CALENDERING

A. W. Coaker

6726 Chadbourne Dr., No. Olmsted, OH 44070

83.1 DESCRIPTION AND HISTORY OF PROCESS

Calendering is the extrusion of a mass of material between successive pairs of corotating, parallel rolls to form a film or sheet. The process is widely used in the thermoplastics and rubber industries.

There are two types of calendering: rolling sheet and viscous bank.

In rolling sheet calendering, which is used for textiles, paper, and nonwoven fabrics, a preformed web is passed through a calender to improve its appearance and properties (Fig. 83.1). In the viscous bank calendering, which is used for thermoplastics and rubber, the sheet is usually pressure-formed and surfaced in the calender, although sometimes a preformed plastic sheet is patterned or surfaced in the calender (Fig. 83.2). When plastic or rubber is laminated onto a substrate in a calender, the process is called lamination. An example is floor tile, which is made in specialized calenders.

83.2 MATERIALS

Some thermoplastics, thermoplastic elastomers, and elastomers are formed into supported and unsupported webs by calendering. These include poly(vinyl chloride) (PVC), acrylonitrile–butadiene–styrene (ABS), polyurethane, styrene–butadiene rubber (SBR), natural rubber, and many elastomeric compositions.

PVC and its copolymers—sometimes plasticized, sometimes blended with ABS or other polymers, and often filled and pigmented—comprise the largest volume of calendered thermoplastics. Analogous equipment and processes are used for other polymers. The calenders used for paper, textiles, and nonwoven fabrics are substantially different, however.

83.3 PROCESSING CHARACTERISTICS

When a polymeric system is developed for calendering, it is important to take into account the formulation of the thermoplastic or elastomeric stock; the design of and conditions on the calender; and the take-off, thickness measurement and control, and windup. Rheologically, the stock on the calender must be calenderable, which for some polymers happens only within a narrow range of temperature. Other considerations include whether the polymer is to be laminated to a fabric on the calender; whether the web is embossed or slit in-line; what finish is required (glossy, semimatte, or matte); and whether the product needs special properties such as optical clarity, biaxial orientation, or very low strain recovery.

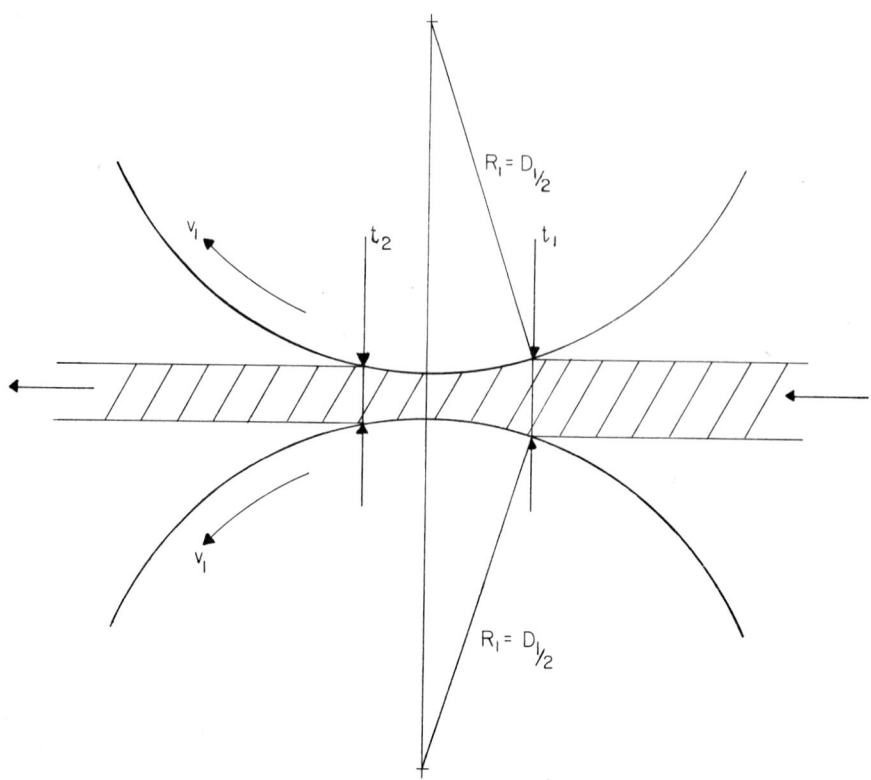

Figure 83.1. Simplest type of rolling sheet calendering.

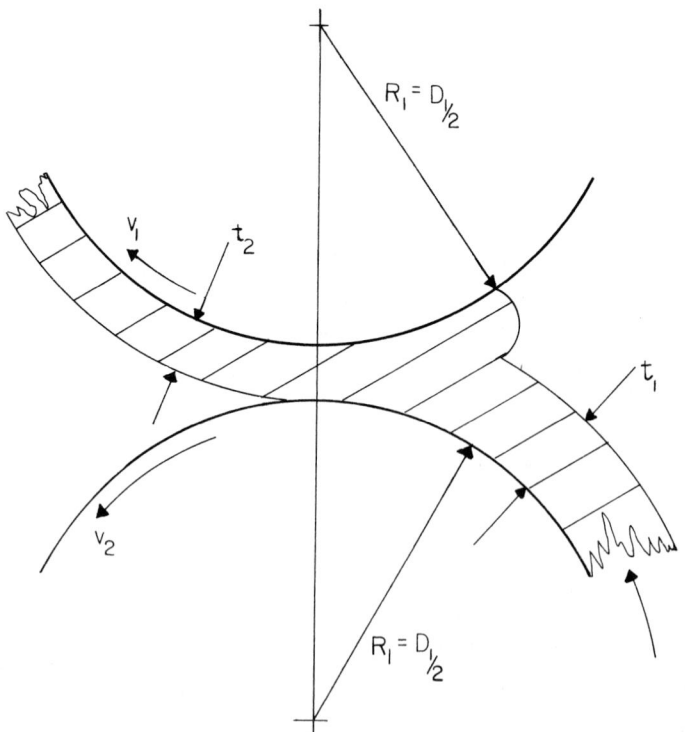

Figure 83.2. Viscous bank calendering.

MASTER PROCESS OUTLINE   **PROCESS** Calendering

Instructions:

For "ALL" categories use   Y = yes

For "EXCEPT FOR" categories use   N = no

                                    D = difficult

	Y Styrene-butadiene rubber (SBR)	□ **ALL Thermosets**
	Y Thermoplastic elastomers (TPEs)	**Except for:**
□ **ALL Thermoplastics**		
Except for:		Allyl Resins
		Phenolics
Acetal	Poly(methyl methacrylate)	Polyurethane Thermoset
Y Acrylonitrile–Butadiene– Styrene	Poly(methyl pentene) (TPX)	RP-Mat. Alkyd Polyesters
	Poly(phenyl sulfone)	RP-Mat. Epoxy
Cellulosics	Poly(phenylene ether)	RP-Mat. Vinyl Esters
Chlorinated Polyethylene	Poly(phenylene oxide) (PPO)	Silicone
Ethylene Vinyl Acetate	Poly(phenylene sulfide)	Urea Melamine
Ionomers	Polypropylene	
Liquid Crystal Polymers	Polystyrene	
Nylons	Polysulfone	
Poly(aryl sulfone)	Y Polyurethane Thermoplastic	□ **ALL Fluorocarbons**
Polyallomers	Y Poly(vinyl chloride)-PVC	**Except for:**
Poly(amid-imid)	Poly(vinylidine chloride)-Saran	
Polyarylate	Rubbery Styrenic Block	Fluorocarbon Polymers-ETFE
Polybutylene	Polymers	Fluorocarbon Polymers-FEP
Polycarbonate	Styrene Maleic Anhydride	Fluorocarbon Polymers-PCTFE
Polyetherimide	Styrene–Acrylonitrile (SAN)	Fluorocarbon Polymers-PFA
Polyetherketone (PEEK)	Styrene–Butadiene (K Resin)	Fluorocarbon Polymers-PTFE
Polyethersulfone	Thermoplastic Polyesters	Fluorocarbon Polymers-PVDF
	XT Polymer	

To manufacture specific products, appropriate formulations have to be developed first. Operating conditions for calendering each product are then optimized by iterative procedures from starting-point conditions suggested by theory or derived from analogous products. Calendering houses consider these conditions confidential, but they are described occasionally.[1,2]

83.3.1 Control of Fluxing Temperature

Fluxing units popular for calendering lines for elastomers and thermoplastics include batch-type Banbury mixers and continuous Farrel Continuous Mixers (FCMs), Buss Ko-Kneaders, and Planetary Gear Extruders. It is important that the stock delivered to the first calender nip be well fused, homogeneous in composition, and relatively uniform in temperature. The optimum average temperature for good fusion depends on the formulating ingredients. Typically, a rigid PVC formula based on medium molecular weight resin (intrinsic viscosity 0.90 to 1.15) has an optimum stock temperature of 356–374°F (180–190°C) at the first calender nip. For best calendering, there should be no cold volume elements below 347°F (175°C) and no hot spikes above 383°F (195°C). The necessity for close control of fusion and mixing conditions was predicted by work on the effects of flow/volume/element size, which varies with stock temperature, on the performance of PVC melts.[3]

Flexible PVC is normally calendered at temperatures 50–68°F (10–20°C) lower than rigid PVC. Typical calendering conditions for unsupported flexible PVC film and sheet are shown in Table 83.1.

TABLE 83.1. Typical Calendering Conditions for Flexible PVC

Roll No.	Heavy-gauge Product				Light-gauge Product			
	Roll temp.,		Roll speed,		Roll temp.,		Roll speed,	
	°F	°C	ft/min	m/min	°F	°C	ft/min	m/min
1	347	175	125	38	338	170	263	80
2	352	178	128	39	343	173	269	82
3	353	181	138	42	349	176	279	85
4	363	184	148	45	354	179	289	88

83.3.2 Calender Speed

Economics dictate that products be calendered under conditions that yield the highest wind-up speed consistent with good quality. However, there are numerous factors which limit calender speed. At a given calendering speed, more frictional heat is generated by the more highly filled and rigid stocks and by reduced web thickness. Excessive speed results in degradation or sticking of many products on which there is a stock temperature limitation. Optimal speeds of the final calender roll for double-polished, clear, rigid PVC products are substantially lower than for their flexible counterparts, which have lower melt viscosities at a given temperature and which degrade at about the same maximum temperatures as the rigid stocks. Air occlusion limits calender speed for some heavy-gauge products, whereas matteness, or microroughness of the film, or melt fracture in the last calender nip, limits speed on others.[4] Bank marks are minimized by optimizing formulas, calendering speeds, and roll temperatures so as to obtain the most orderly behavior of the rolling banks of stock at the calender-nip entrances.

Appropriate use of drawdown permits the wind-ups to be run substantially faster than the final calender roll on many thin, unsupported film products. On heavy gauge products, calenders and takeoffs are run almost synchronously.

Films and sheets with a high gloss taken off a highly polished final calender roll tend to stick more to the roll than their matte counterparts. Very soft webs also tend to stick to the final calender roll.

Generally, the fastest calender speeds are obtained using medium stiffness, moderately filled, matte products in a median thickness range.

83.3.3 Finishing

See Section 83.10.

83.3.4 Gauge

Heavy gauge products [those targeted for 0.02 in. (0.5 mm)] are drawn down at takeoff as little as is consistent with good stripping from the final calender roll [that is, from 0.021 to 0.020 in. (0.533 to 0.508 mm)]. Lighter gauge film coming off the calender [0.007 in. (0.18 mm) in thickness] typically is drawn down at a ratio of 1.2:1 and is wound up 0.006 (0.15 mm) thick.

83.3.5 Orientation

The temperatures affecting molecular orientation depend greatly on the formulation. Random copolymers of vinyl chloride, such as vinyl chloride–vinyl acetate copolymers, have significantly lower processing and orientation temperatures than PVC homopolymers. For medium molecular weight homopolymers, the thermoplastic or forming temperature range in which the polymer may be stretched without molecular orientation is above 374°F (190°C). If webs made from this type of PVC homopolymer are stretched at an orientation temperature optimized for their molecular weight and then rapidly cooled, the molecular orientation is frozen into the film and enhances the physical properties of the film in the direction of stretch. Biaxial orientation carried out in the orientation temperature range makes the film stronger both along and across the web.

Stretching carried out at still lower temperatures gives rise to strains that tend to recover slowly at room temperature, making products dimensionally unstable.

The temperature range in which desirable molecular orientations may be introduced depends on the molecular weight of the PVC and the formulation. It occurs generally in the range of 320 to 350°F (160 to 180°C) for medium molecular weight resins. Orientation introduced below 302°F (150°C) usually results in recoverable strains. For high molecular weight PVC, the corresponding temperatures are higher.

Film made from medium molecular weight PVC to be used in shrink-wrapping is uniaxially or biaxially stretched at a suitable temperature and rapidly cooled to freeze in the molecular orientation. After the wrapping operation, a heat tunnel may be used to activate the desired shrinkage.

Some films products used as tapes or candy wrappers are biaxially oriented to enhance their physical properties.

83.3.6 Embossing

See Section 82.11.

83.6 PROCESSING CRITERIA

83.6.1 Viscous Bank Process

Calendering of elastomers and thermoplastics is carried out in-line with numerous other operations, all of which must be carefully coordinated to achieve technical and commercial success. Two-roll milling is a mixing (qv) and fusion process frequently used in a calendering line that homogenizes fluxed stock prior to subsequent operations such as calendering. A typical sequence of operations for the manufacture of unsupported PVC film is shown in Figure 83.3; calendering is the crucial operation. Figure 83.4 shows the calendering operations schematically.

Recipes are developed for specific items made by particular calendering processes. A familiar example is the PVC swimming-pool liner, which normally contains high molecular weight PVC,[5] dialkyl phthalate plasticizers,[6] stabilizer, pigment, filler, lubricant, and fungicide.

The ingredients are each weighed to about +/−1% accuracy and charged to an intensive or ribbon blender in the following order: resins, plasticizer plus stabilizer, filler, lubricant, fungicide, and pigment. The main resin is a powder mixing-grade PVC that absorbs the plasticizer and liquid stabilizer to give a powder mix.

After intensive mixing or ribbon blending, the powder is cooled and passed through a metals diverter into a surge hopper, from which it is fed to the fluxing unit at the exact rate needed to feed the calender. Metal must be excluded because it would damage the calender rolls. Most fluxing units discharge stock with undesirably high temperature variations from point to point in the order of +/−36°F (20°C). The stock is therefore milled to assimilate loose blend and bring the viscous, fused mass to a more uniform temperature with excursions no greater than +/−9°F (+/−5°C). It is then fed, usually through a metals' alarm, into the first nip of the calender, typically a four-roll, inverted-L unit which has three nips, set at about 0.06, 0.03, and 0.015 in. (1.52 mm, 0.76 mm, and 0.38 mm), respectively, to

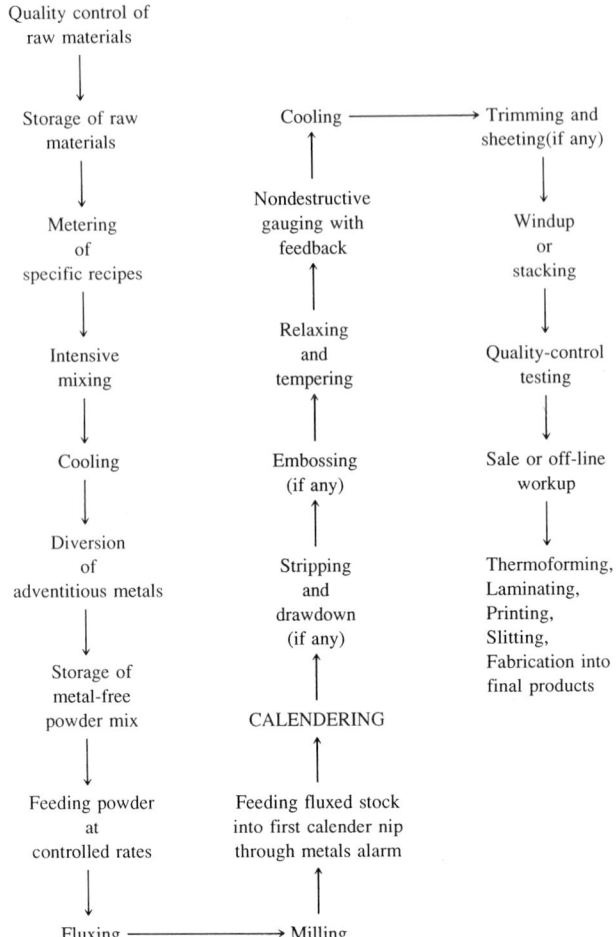

Figure 83.3. Sequence of operations for the manufacture of unsupported PVC film.

produce material in the 16–20 mil (0.4–0.5 mm) range. If there is no drawdown on stripping, the sheet comes off approximately 1.33 times thicker than the opening of the last nip, owing to the velocity profile in nip extrusion between corotating rolls.

The range of processing properties of elastomers and thermoplastics—from rigid to soft, flexible calendered film—is so wide that great differences in throughput are experienced from product to product, even in one line running different formulas to the same width and thickness.

A specialized vernacular exists for different segments of the calendering industry. For flexible PVC, film means flat goods greater than 6 mils (0.152 mm) thick; sheet means material thicker than this.[7]

When the tensile modulus of PVC exceeds 100,000 psi (689 MPa), it is considered rigid because it is not readily extensible. However, if it is in the form of a thin foil, it folds easily (just as metal foils do) even though the metal is inherently rigid. For rigid PVC, sheet refers to flat goods cut into individual rectangles and stacked prior to shipment; film means a continuous web rolled up on a core and shipped as roll goods.

Compared to other sheet-forming operations, the linear speed in calendering is generally high. Formulas and conditions must be well controlled to limit production of offgrade. Many

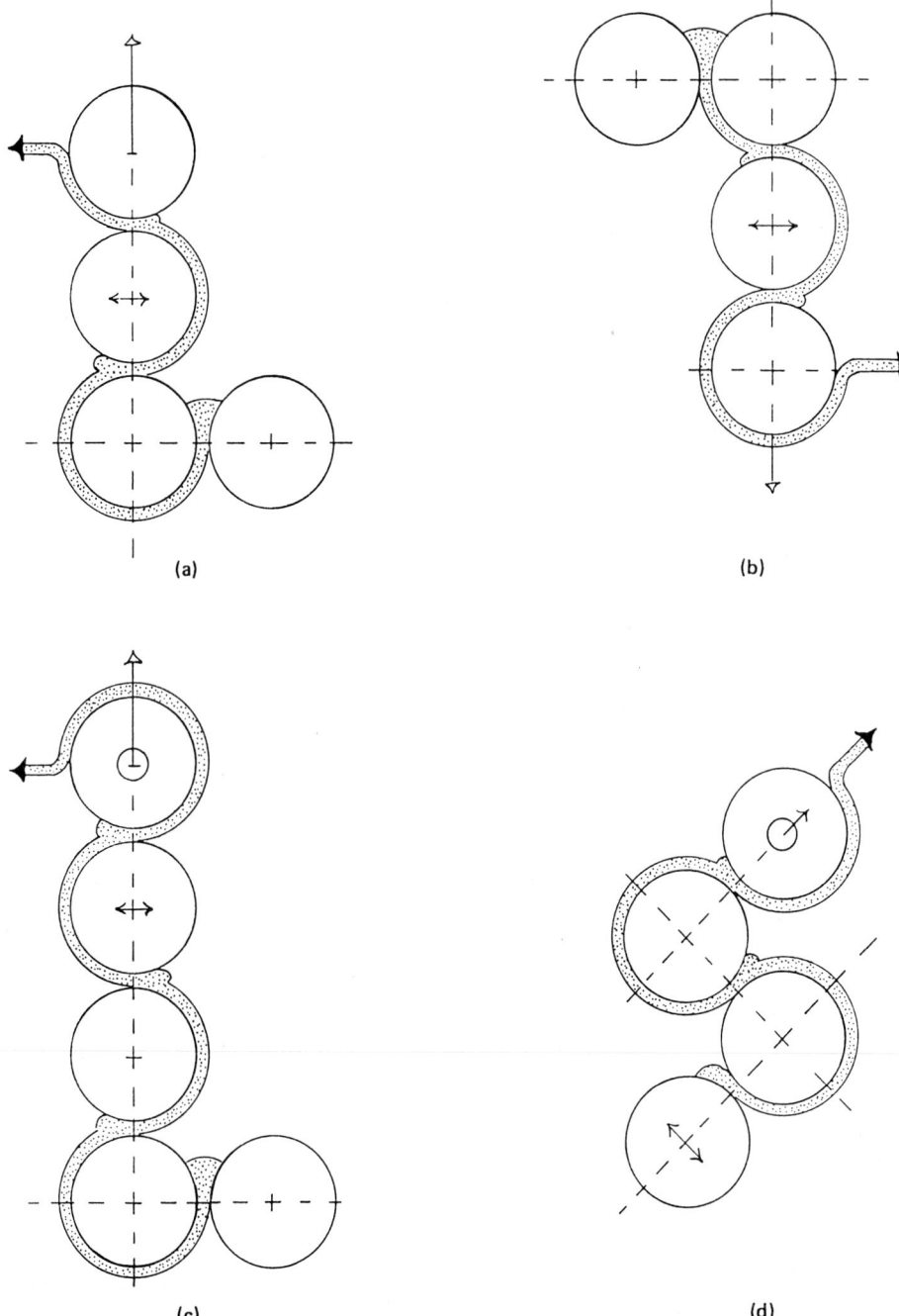

Figure 83.4. Calendering: **(a)** four-roll L calender; **(b)** four-roll inverted-L calender; **(c)** five-roll L calender; **(d)** four-roll Z calender. Courtesy of Marcel Dekker Inc.

operations must be maintained constantly is precise synchronization. When it is running at full speed and width, a modern calendering line is the most impressive sheet-forming operation.

83.6.2 Web Fed Process

A web is preformed, generally as a wet lap, for paper and nonwoven fabrics. The web then goes through drying and calendering or supercalendering to develop its desired properties.

Calenders for paper are generally wider and are designed with more nips than calenders for plastics; they also are run with lower nip pressures and at higher speeds. Typical nip-separating forces for web fed calendering processes are 100–3000 lb/in. (17.5–525.6 kN/m) of roll face as compared to 900–6000 lb/in. (157.7–1051.1 kN/m) for plastics calendering. Widths and thicknesses of the webs also differ considerably.

83.6.3 Theory

A theoretical understanding of the phenomena governing calendering is necessary to address the demands on a modern calender: high speed; small thickness tolerances of the finished sheet or film, both in the calendering direction and perpendicular to it; uniform structure and surface of the film; and economical operation.

Control of Thickness Across the Web. When a calendered web is wound up onto a core, close thickness control across the web is necessary. Calculations and experience in subsequent web-handling operations provide guidelines. For a web wound tightly on a core,

$$L = \frac{\pi}{1000} \left[t \left(\frac{D - d}{2t} \right)^2 + (d + t) \left(\frac{D - d}{2t} \right) \right] \tag{83.1}$$

where L, m = length of web on the roll; t = thickness in mm of web on the roll; D = outside diameter (OD) in mm of the roll; and d = inside diameter (ID) in mm of the roll.

If the web is thicker at the middle than near the outside edges and is wound up under controlled tension, the center of the roll is tight (hard) and the outside edges are loose (soft). Most calendered thermoplastic films shrink longitudinally, owing to recoverable strains introduced in the cooling and wind-up train. Where films are not constrained by tightness of the wind, shrinkage takes place within a few days of calendering and the web is then longest where originally it was thickest (usually at the center). For example, a thermoplastic web with an average thickness of 0.00502 in. (0.1275 mm) at its center, average thickness of 0.00492 in. (0.125 mm) near both edges at the time of windup onto a core 6.3 in. (16 cm) in diameter, and cut off on reaching 3.5 ft (1.05 m) OD has a length of 7526 yd (6635 m) at its center where it does not shrink because it is tight. Calculation shows that it will shrink to a length of approximately 7196 yd (6580 m) near the edges and become tight all the way across; the decrease in length is 0.84% at both edges of the roll.

If the roll has a commonly used width, such as 5.3 ft (1.6 m), and the web is to be printed or laminated onto a flat substrate, it will have enough "belly" in the center after shrinkage of the two edges to cause severe difficulties in some web-handling units. Also, if a roll width was slit down the middle and rewound as two 2.7 ft (0.8 m) rolls, each 2.7 ft (0.8 m) web would exhibit severe "race-tracking" or curving by virtue of having one side longer than the other. The magnitude of this effect can be calculated and expressed as a radius of curvature,[8] which, in the preceding example, would be about 317 ft (95 m). This is too small for satisfactory handling in some critical web-handling equipment. For satisfactory web handling without frequent adjustments to tension, relaxation, and temperature of the web, the race-tracking radius of curvature should be kept larger than 1333 ft (400 m).

Causes of Thickness Deviations. On a multiroll calender with a vertical stack of rolls, the separating forces exerted by the stock in the nips may be adjusted by operating conditions so that the middle rolls deflect very little. As the first approximation, it may be assumed that the #3 roll in a four-roll L-type or inverted-L calender does not deflect in balanced operation. Deflection of the #4 roll primarily determines gauge variation across the width of the web. For thin polymeric webs, roll-separating forces in the final nip may be as high as 6000 lb/in. (1051.1 kN/m), which may cause considerable deflection of the final calender roll. Unless this is compensated for, gauge variation will occur across the web.

Making the simplifying assumption that loading applied by the stock to the rolls is constant across the width of the web, tractable deflection expressions for rolls of simple geometry have been given for a final calender roll,[9] as illustrated in Figure 83.5. The deflections

$$y_{max} = \frac{5ql^4}{384EI}\left[1 + \frac{24}{5}\left(\frac{h}{l}\right) + 2\left(\frac{D}{l}\right)^2\right] \qquad (83.2)$$

and

$$y = y_{max} - \frac{5ql^4}{48EI}\left\{\left[\frac{3}{5} + \frac{12}{5}\left(\frac{h}{l}\right) + \left(\frac{D}{l}\right)^2\right]\left(\frac{x}{l}\right)^2 - \frac{2}{5}\left(\frac{x}{l}\right)^4\right\} \qquad (83.3)$$

where q = total nip loading on the roll adjusted for the distributed weight of the roll (in our example, 381.1 kN/m); l = width of the web (1.98 m); E = effective modulus of elasticity of the roll material (15.2 × 10^4 MPa); I = moment of inertia of the roll body (0.01251 m^4); D = diameter of the roll body (0.7112 m); d = diameter of the roll cavity (0.1778 m); h = roll neck overhang (0.381 m); y = roll deflection at distance x in mm from the roll center; and Y_{max}, maximum deflection in mm (at the roll center).

Other potential causes of thickness changes across the calendered web include nonhomogeneous rheology of the stock, problems with control of stock temperature and lubricity, equipment or control system malfunction, range of roll-separating force along the length of the roll face, and temperature gradients in the rolls or use of damaged calender rolls.

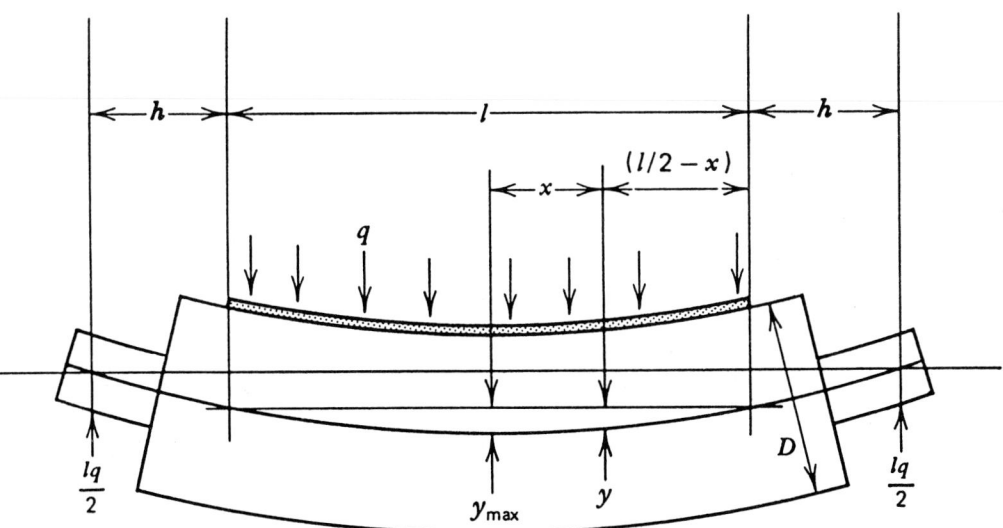

Figure 83.5. Deflection of a calender roll.[8] Courtesy of Marcel Dekker Inc.

Compensation for Calender-Roll Deflections. In the calendering of plastics, compensation for deflection of calender rolls is normally done by combining three methods: axis crossing or skewing, roll-straightening (often called roll counterbending), and roll crown (sometimes called roll profile). Roll crown may be ground or blocked on.

Axis crossing increases the rolls' separation at the ends but does not change their separation in the middle. The parameters for axis crossing are illustrated in Figure 83.6. Compensation is related to the amount of crossing[10,11] as follows:

$$z = 1,000 \left[C_0^2 \left(\frac{2x}{L} \right)^2 + D^2 \right]^{1/2} - D \qquad (83.4)$$

where z = rolls' gap compensation at point x owing to axis crossing in mm; C_0 = crossing at roll end-face in m; x = distance from transverse centerline of roll in m; L = length of roll in m; and D = diameter of rolls in m.

Roll straightening (or roll counterbending) is illustrated in Figure 83.7 and has been described and treated quantitatively.[11-14] This method involves the application of external bending moments to both ends of the shaft of a calender roll in order to compensate for deflection ($+$) or excessive roll crown ($-$), using relationships

$$\mu = \frac{Fa}{2EI} \left(x^2 - \frac{l^2}{4} \right) \qquad (83.5)$$

and

$$\mu_{max} = -\frac{Fal^2}{8EI} \qquad (83.6)$$

where F = force applied by the roll-straightening cylinders to the roll shaft in kN; a = moment arm between the main roll bearings and the point where the cylinders apply the straightening force in m; l = active length of the roll face in m; x = a variable distance from the transverse centerline of the calender roll towards the outside edge of the roll in m; E = effective modulus of elasticity of the material of the calender roll in MPa; I = moment of inertia of the roll body in m^4; μ = compensation owing to roll-straightening at point x in mm from the transverse centerline in m; and μ_{max} = compensation owing to roll-straightening at the transverse centerline in mm.

By convention, positive straightening is used to compensate for deflection caused by roll-separating forces in the nip of a calender; negative straightening is used to reduce excessive fixed crown on a roll, which has been ground on.

Axis crossing and roll-straightening quantitatively compensate for the results of the second-order term in equation 83.3, but ignore those of the fourth-order term. The so-called oxbow effect arises from these methods, which do not compensate exactly for roll deflection because they ignore the fourth-order term in equation 83.3 and because of inaccuracies of the isobaric and isothermal assumptions relating to treatment of the calender nips. An oxbow curve plots the difference between a correction and deflection curve.

Normal practice is to grind a crown onto the final roll of a calender. This compensates for a typical compromise deflection, which contains the exact reverse oxbow component to eliminate the oxbow effect when the roll deflects by the selected typical amount. A tractable equation to guide a roll grinder putting on crown with reverse oxbow has been given.[13]

Figure 83.8 shows the amount whereby axis crossing and roll-straightening overcompensate for deflection across a web when adjusted to give the nominal thickness at the center and edges. The maximum offset is at approximately 0.7 of the distance from the center to the edge of the web, and, in the example, is about 2 μm. This offset is large enough to produce web-handling problems on thin films if it is not compensated for, which is often

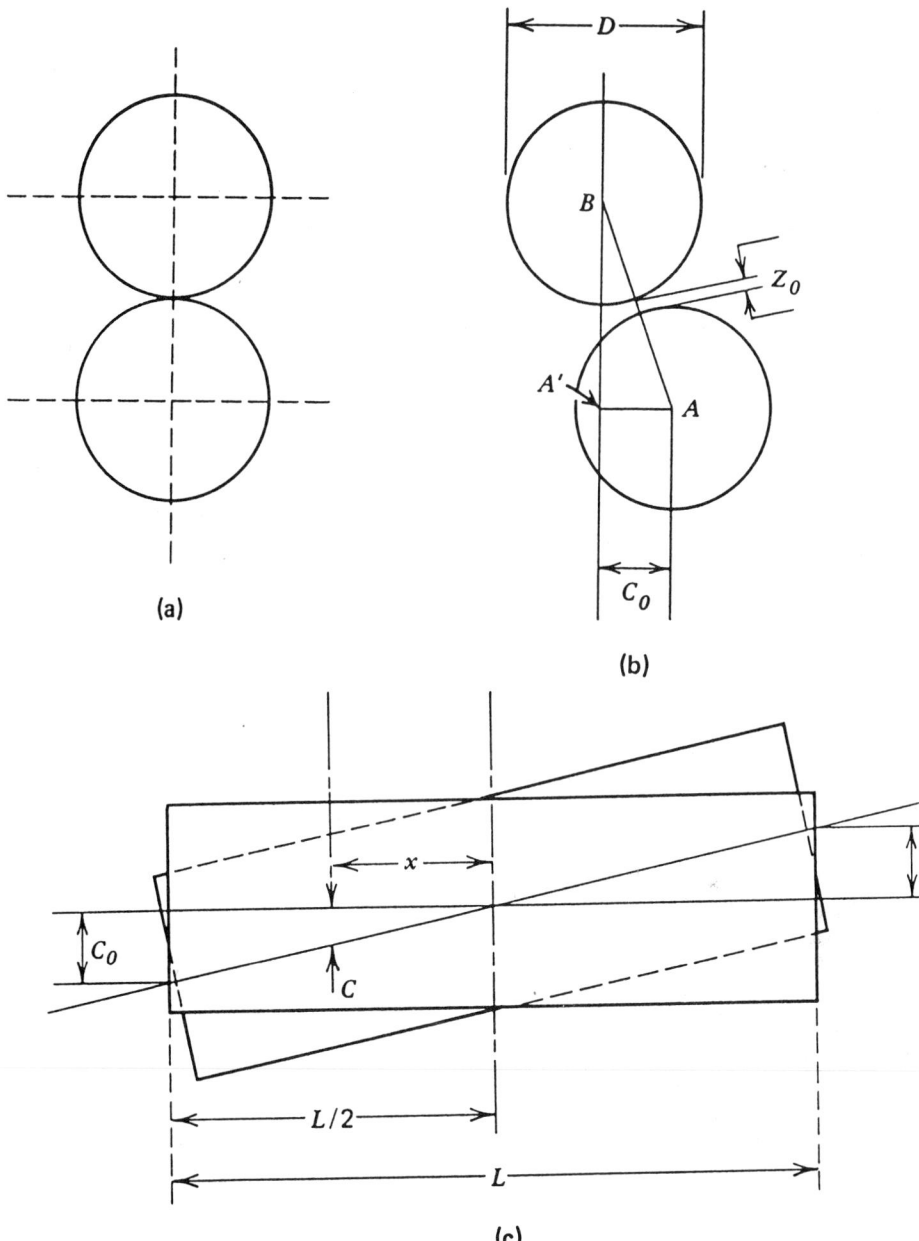

Figure 83.6. Axis crossing[8]: **(a)** end view of rolls prior to crossing; **(b)** near end of rolls crossed by movement C^0 of each end of lower roll, causing gap Z^0 at each end; **(c)** top view of crossed rolls, whose dimensions correspond to equation 4 in the text. Courtesy of Marcel Dekker Inc.

done by trial and error blocking of the calender rolls and then honing to the desired finish. Both of these require the services of a skilled craftsman and are expensive and time-consuming.

Estimating Roll-Separating Forces. Although it is limited by a number of simplifying assumptions, the following expression for calculating roll-separating forces in milling and calendering plastics has been used for years[15]:

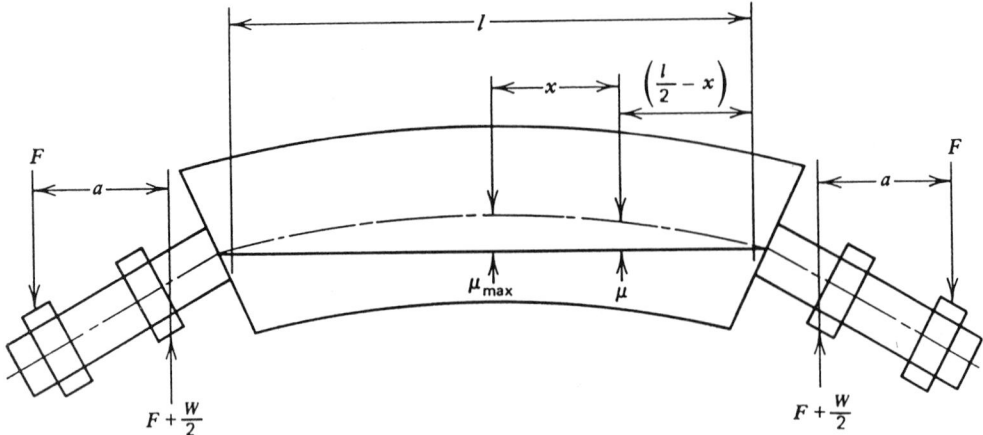

Figure 83.7. Roll-straightening[8]: W is the weight in kN of the roll, whose dimensions correspond to equations 5 and 6 in the text. Courtesy of Marcel Dekker Inc.

$$F = 2\mu Vrw \left(\frac{1}{h_0} - \frac{1}{H} \right) \tag{83.7}$$

where F = total separating force between rolls in kN, μ = viscosity of stock in the calender nip in Pa·s; V = peripheral speed of rolls in m/s; r = radius of rolls in m; w = width of web being calendered in m; h_0 = separation between rolls at the nip in mm; and H = height of stock at the nip entrance in mm.

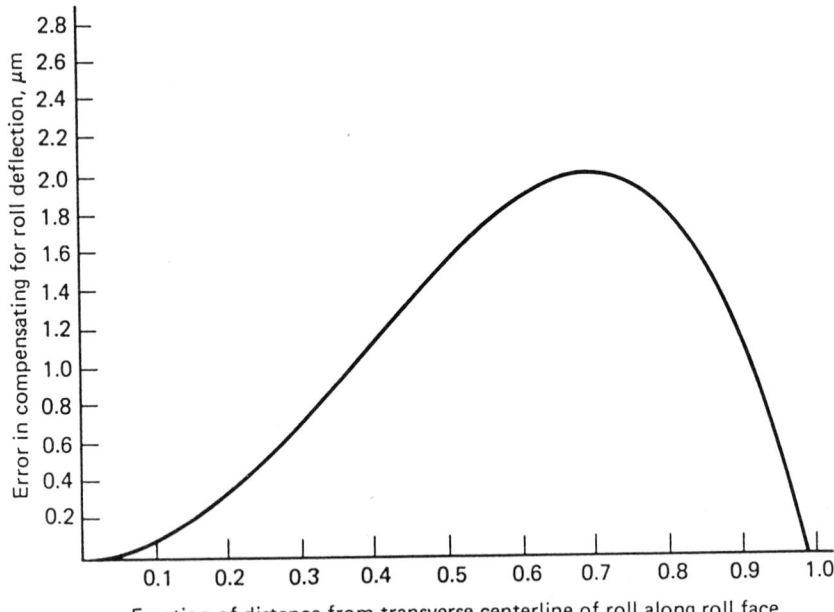

Figure 83.8. Overcorrection for deflection by axis crossing and roll-straightening when nominal gauge is achieved at the edge and center of the web.

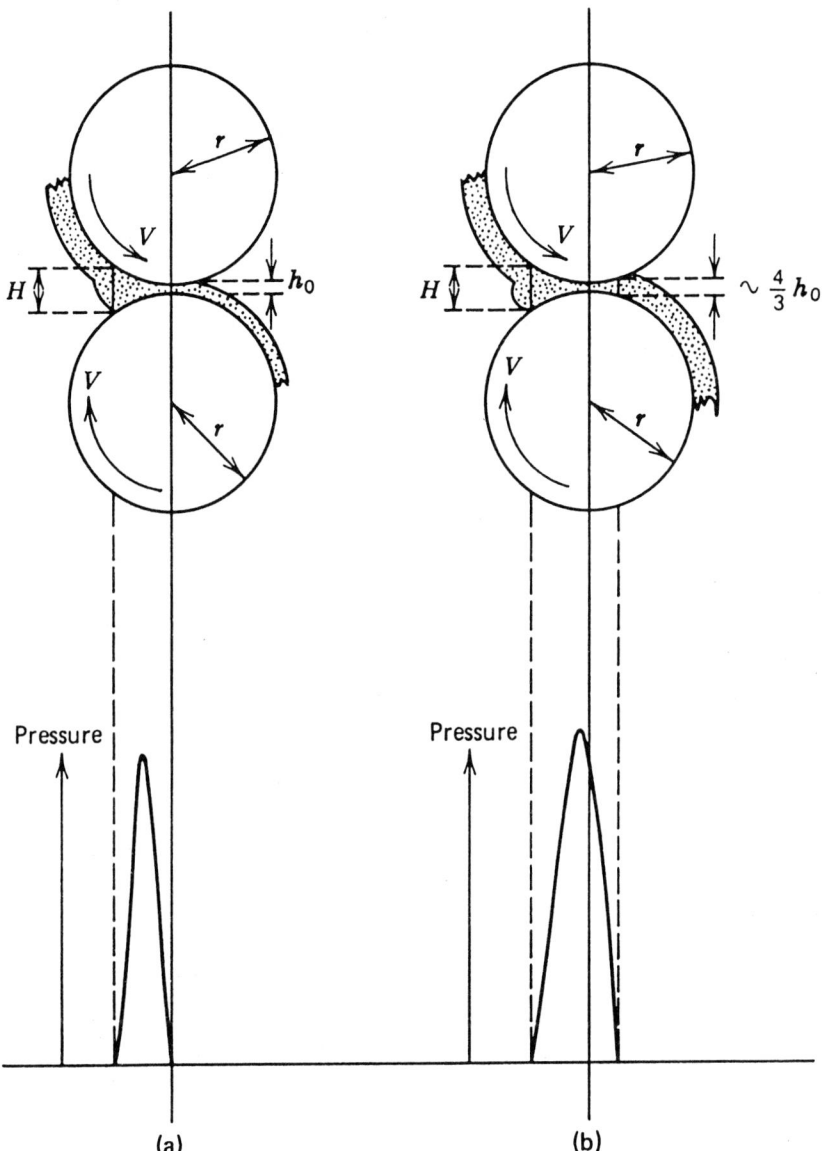

Figure 83.9. Roll-separating forces in a functioning calender nip[8]: **(a)** definitions used in reference 15; **(b)** behavior of plastics stocks from measurements in references 16, 17, 24, 25. Courtesy of Marcel Dekker Inc.

A cross section of a functioning calender nip is shown in Figure 83.9. Reasonably accurate results can be obtained from equation 83.7, provided that correct viscosity values are inserted.[16] A practical approach for determining appropriate viscosity values whereby performance on large calenders at commercial operating rates can be related to parameters measured with a capillary extrusion rheometer may be used.[17]

It is important to recall that the viscosity of most calenderable polymers decreases with increasing temperature. Passage of the stock through a calender nip generates considerable frictional heat. Temperature effects in calender nips may be observed with infrared radiation thermometers or other specialized equipment. Rate and thermal conductivity considerations[18] dictate that the frictional heat is only partially conducted into adjacent calender rolls under typical operating conditions. Therefore, successive nips contain progressively hotter stock, whose viscosity, therefore, differs from nip to nip.

In cases where rolls separating force F is measured and equation 83.7 is used to calculate what viscosity of stock gives that separating force, the calculated value of μ is referred to as the "adjusted viscosity" of the stock. This compares well with values measured at the average temperature of the stock in the nip being studied.

There is another simple equation that gives results within about 20% of those from equation 83.7.[19] Other expressions have been derived in early analyses of calender nips.[20-21]

A crucial advance in applying theory to processing situations was made when it was proven that interfacial slip occurs between rigid PVC stocks and metal processing equipment at processing temperatures.[22,23] A slip correction for predicting nip pressures was developed, without which predicted results were 75% higher than measured values.[24] Another method, which needs verification on a wider range of formulas, apparently permits calculation of roll torques and separating forces.[25]

Most practical calendering problems are solved by a combination of theoretical analysis and practical experiments in which calendering conditions and formulations are optimized simultaneously. Computer programs employ the theoretical equations to determine optimum crowns and other web profile compensation for given sets of conditions.[26]

83.7   MACHINERY

Many features now routinely incorporated in new calenders provide versatility, better quality, and higher operating rates. Complete control of friction ratios gives good tracking of stock via individual drives on all calender rolls. Close tolerances on film profiles are obtained from axis crossing, roll-straightening, and gap profiling of calender rolls, and greater profiling accuracy is attained by hot grinding of calender rolls at operating temperatures. Better stock-temperature control is achieved using drilled rolls with heat-transfer fluid circulated through them at high rates and using accurate ($+/- 3/4°F$ [1°C]) fluid-temperature control systems.

Less deflection at high separating force operating conditions can be achieved by the use of stiffer rolls, based on higher modulus steels or dual metal construction. Modern roller bearings have much less play in them than old-style journal bearings after some years of use. In older calenders severe cases of "roll float" caused by changes in the balance of forces on a roll were sometimes encountered. Today this can still occur, but to a lesser extent, on modern calenders. Preloading devices reduce roll float and thus provide better gap control between pairs of calender rolls, especially during start-up. Automated film-thickness measurements with feedback control to the calender and drawdown system also achieve better film-thickness control. Programming temperature changes at rates of 4.5°F/min (2.5°C/min) or less avoid roll distortion caused by to thermal shock.

Good drawdown and pick-off control is achieved by using tightly spaced multiroll pickoffs with grouped roll speed controls. Helical cavities in each pick-off roll allow high heat-transfer fluid-circulating rates. Quick opening devices on all nips reduce damage to rolls from running together in the event of power failure or stock runouts. Mechanical, square-cut devices with automated tension control and roll change as the roll reaches a preset diameter ensure better windup.

Quick roll-change capability has been developed for the last two calender rolls. Changes from high gloss to matte or to different profile rolls can be done in 10 hours rather than 80 hours, as formerly. Embosser rolls can be changed in minutes instead of hours.

Information on calendering equipment and auxiliary units is updated frequently by manufacturers such as Farrel Co., Adamson, Berstorff Corp., and EKK GmbH[27-30], and is periodically reviewed.[31] Equipment for the Luvitherm Process for making biaxially oriented, thin, rigid PVC films from emulsion PVC has been described.[28,32]

Modern radiation film-thickness gauges are furnished with computer systems that process the raw data from the measuring head and print out detailed production reports. They also control the gap in the last calender nip by varying rolls opening, axis crossing, and roll-straightening on the final calender roll.[33-35]

In order for radiation thickness gauges (also called beta gauges) to work effectively, the web composition must be exactly known and accurately controlled. The gauges measure attenuation of radiation, which is proportional to the quantity of matter in the measured area. For known formulas, this can be translated into web thickness. Good results have been reported.[36] However, beta gauges cannot compensate for deficiencies in older calenders or for improper control of calendering formulations.

Edge cuts in excess of 3 in. (76 mm) are normal today to avoid undesirable edge effects. On critical products it is now realized that close control of the rework ratio is necessary for uniformity of output. Excess rework from critical products is used in ones whose specifications permit this practice.

Typical calendering conditions for unsupported flexible PVC film and sheet are shown in Table 83.1.

83.10 FINISHING

For styling purposes, surface calendered goods require depths of matte (even roughness), which can be measured with a profilometer. Specific market segments use such terms as "satin," "Sheffield," and light matte to describe these finishes.

A one-sided matte finish may be applied with an embossing roll. However, a sharper, deeper, and more precise matte finish can be applied to film or sheet on the calender. To make two-sided matte products, such as two-sided credit-card stock, the last two calender rolls are matted. Only the last calender roll is matted for one-sided matte goods meeting critical specifications, such as print specialty film and one-sided credit card stock. For these applications the final calender roll or the last two rolls are sandblasted with aluminum oxide grit of controlled particle size, such as nominal 120 to 180 mesh.[37] For heavier mattes, coarser grit and more passes of the sandblasting unit on the rolls are required. Proper selection of metal for the calender rolls is also necessary for manufacture of these products. Chilled cast iron rolls take a sharp matte, which wears off slowly. Balancing the dwell of the web on the final two rolls is essential to obtain an even matte on both sides of two-sided matte goods.

83.11 DECORATING

83.11.1 Embossing

For PVC formulas in general, fusion, quenching and annealing phenomena are not yet fully understood and are the subject of continuing studies in physically simple situations.[38] The principle used in embossing PVC film and sheet is to deliver a relatively hot web to a relatively cold embossing roll. This imparts to the web and freezes in by rapid cooling the desired embossment pattern.

Expertise for embossing a wide range of rigid, semirigid and flexible PVC formulas has been developed. Film temperature, roll temperature, and pressure are optimized empirically for each formulation and pattern to be embossed. Both the steel embossing roll and the rubber back-up roll are internally cooled by a cooling medium. The latter, however, tends to stick to a hot PVC film unless it is also externally cooled in a water trough. This is because of the poor heat transfer of most rubbers, which allows a nonexternally cooled back-up roll surface to rise almost to the processing temperature of the PVC film.

TABLE 83.2.   Annual Calendering of PVC in the United States by Market Area, 10^3 tons[a]

Market	1983	1984	1985	1986	1987
Building and construction	109	114	105	107	120
Transportation	41	41	38	37	34
Furniture and furnishings	60	59	53	55	66
Consumer and institutional	116	93	96	105	110
Packaging	35	36	35	37	39
Electrical	5	5	4	5	5
Other	5	16	9	12	7
Totals	*371*	*364*	*340*	*358*	*381*

83.12   COSTING

The bulk of U.S. calendering output in plastics is PVC, with flexible sheet and film accounting for the greatest volume.[39] Interpretation of published statistics is difficult because of inaccuracies in distinguishing between calendered and extruded film and sheet goods. The figures reported in Table 83.2 reflect PVC resin content, when, in fact, all formulations contain additives and may include other polymers. The U.S. market for such products is much larger than that shown for domestic manufacture alone because of imports of roll goods and fully fabricated finished products. U.S. PVC sales for calendering are reported by market segment monthly[40] and annually.[41-43]

The capital cost of a new calendering line is 5×10^6 or more, exclusive of building costs.[31] In the United States, the fixed plus variable costs for operating a new line amount to \$450–550/h. Throughput for many rigid PVC calendered products ranges from 900–2000 lb/h (400 to 900 kg/h). The average range of conversion costs is \$1.08–2.47/lb (\$0.49–1.12/kg) for calendering rigid PVC.

Apart from material costs, the expense of running a large calender line amounts to \$350–560/h, depending on the degree of automation, investment, labor, power costs, and overhead. For a sheet 7.1 ft (2.13 m) wide, 0.02 in. (0.51 mm) thick, taken off at 91 ft/min (27.4 m/min) and wound up at an instantaneous rate of about 1076 lb/h (2370 kg/h), conversion cost is typically \$0.46/lb (\$0.21/kg). For example, for a PVC swimming-pool liner, the average raw materials adjusted for yield cost \$2.60/lb (\$1.18/kg), giving an inventory standard cost of roughly \$3.22/lb (\$1.46/kg). These costs are based on PVC resin at \$1.70/lb (\$0.77/kg) and plasticizer at \$2.42/lb (\$1.10/kg).

83.13   INDUSTRY PRACTICES

83.13.1   Specifications and Standards

Principal suppliers and customers issue product specifications on calendered plastics and rubber goods. Standards for calendered goods are updated less often. In addition to Federal and Military Standards, those issued by ASTM,[44] SPI,[45] and ANSI[46] may be consulted.

83.13.2   Health and Safety

Most of the hazards in calendering facilities are associated with the high speeds at which webs normally travel, the large forces involved in calendering, and the high temperatures. Most calendering plants have safety programs and carry out preventive maintenance. Calender safety regulations, applicable to the calender itself, are covered by Federal OSHA Section 1910.216 and by ANSI B28.1-1967 (R1972), as well as by some state codes.

BIBLIOGRAPHY

1. Eur. Pat. Appl. EP 73046 A2, (Mar. 2, 1983) E. Zentner and W. Bingemann (to Hoechst AG).
2. G. Hatzmann and M. Herner, *Kunstoffe* **68**(19), 561 (1978).
3. A.R. Berens and V.L. Folt, *Trans. Soc. Rheol.* **11**(1), 95 (1967); *Polym. Eng. Sci.* **8**(1), 5 (1968).
4. J.L. Bourgeois and J.F. Agassant, *J. Macromol. Sci. Phys.* **14**(3), 367 (1977).
5. *Thermoplastic Materials,* Bulletin PM-2, B.F. Goodrich Co., Cleveland, Ohio, 1983.
6. *Santicizer 711,* Bulletin PL-365, Monsanto Co., St. Louis, Mo., 1983.
7. *Whittington's Dictionary of Plastics,* Technomic Publishing Co., Inc., Stamford, Conn., 1983.
8. A.W.M. Coaker in I.L. Gomez, ed., in *Engineering With Rigid PVC: Processability and Applications,* Marcel Dekker, Inc., New York, 1984, Chapt. 8.
9. M.D. Stone and A.T. Liebert, *Tappi* **44**(5), 308 (1961).
10. G.F. Carrier, *J. Appl. Mech.* **72,** 446 (Dec. 1950).
11. K.J. Gooch, *Mod. Plast.* **34**(7), 165 (1957).
12. M. Jukich, *Mod. Plast.* **33**(8), 138 (1956).
13. R.C. Seanor, *SPE Tech. Pap. 12th ANTEC II,* 1956, p. 298.
14. R.C. Seanor, *Mech. Eng.* **79,** 293 (1957).
15. G. Ardichvili, *Kautschu.* **14,** 23 (1938).
16. J.T. Bergen and G.W. Scott, *J. Appl. Mech.* **73,** 101 (Mar. 1951).
17. F.D. Dexter and D.I. Marshall, *SPE J.* **12**(4), 17 (1956).
18. R.P. Sheldon and K. Lane, *Polymer* **6,** 77 (1965).
19. F.H. Ancker, *Plast. Technol.* **14**(12), 50 (1968).
20. D.D. Eley, *J. Polym. Res.* **1**(6) 529, 535 (1946).
21. R.E. Gaskell, *J. Appl. Mech.* **72,** 334 (Sept. 1950).
22. H. Munstedt, *J. Macromol. Sci. Phys.* **14**(2), 195 (1977).
23. J.C. Chauffoureaux, C. Dehennau, and J. Van Rijckevorsel, *J. Rheol.* **23**(1), 1 (1979).
24. J. Vlachopoulos and A.N. Heymak, *Polym. Eng. Sci.* **20**(11), 725 (1980).
25. J.F. Agassant and P. Avenas, *J. Macromol. Sci. Phys.* **14**(2), 213 (1977).
26. *Brownhill, Mod. Plast.* **61**(12), 52 (1984).
27. *Plastics Calender Lines,* Bulletin 233, Farrel Connecticut Div., Ansonia, Conn., 1982.
28. *Calender Plants for Processing Plastics,* Berstorff Corp., Charlotte, N.C., 1982.
29. *Adamson Rubber Calenders,* Adamson Div., Wean United, Inc., Pittsburgh, Penn., 1983.
30. Calandrette-Anlagen, EKK GmbH, Bochum, FRG, 1982.
31. D.V. Rosato, *Plast. World* **37**(7), 64 (1979).
32. *Technical Development of Plastics Calender Plants,* H. Berstorff Maschinenbau GmbH; reprint of W. Wockener, 1974.
33. *Profitmaster 5001 Process Management Systems,* LFE Corp., Waltham, Mass., 1980.
34. *Measurex System 2002,* Measurex Corp., Cupertino, Calif., 1981.
35. *Ohmart Webart Micro 2000,* Ohmart Corp., Cincinnati, Ohio, 1982.
36. *Plast. Technol.* **25**(5), 24 (1979).
37. *Linonblast Aluminum Oxide,* General Abrasive Div. of Dresser Industries, Inc., Niagara Falls, N.Y., 1983.
38. M. Gilbert and K.E. Kansari, *J. Appl. Polym. Sci.* **27,** 2553 (1982).
39. G.W. Eighmy, Jr., in *Modern Plastics Encyclopedia,* 1984–1985, McGraw-Hill, Inc., N.Y., 1984, p. 195.
40. *Monthly Statistical Report,* Compiled by Ernst and Whitney for the Society of the Plastics Industry's Committee on Resin Statistics, Washington, D.C.
41. *Mod. Plast.* **58**(1), 67 (1981).
42. *Mod. Plast.* **65**(1), 100 (1988).

43. *Mod. Plast.* **46**(1), 60 (1987).

44. *Index, Vol. 00.01.* American Society for Testing and Materials (ASTM), Philadelphia, Penn., 1984.

45. *Plastics Engineering Handbook,* 4th ed., Society of the Plastics Industry Inc. (SPI), Washington, D.C. (1976).

46. *ANSI Catalogue,* American National Standards Institute (ANSI), New York, 1984.

General References

R.H. Crotogino, *Temperature-Gradient Calendering* and R.H. Crotogino, S.M. Hussain, and J.D. McDonald, *Mill Application of the Calendering Equation,* Pulp and Paper Research Institute of Canada, Ponte Claire, Quebec, 1982.

Nipco and Tri-Pass Rolls, Bulletin 134; *Non-Woven Calenders—Calenders for Industrial Sheeting, Elastomers, Tapes, Coated Fabrics,* Bulletin 223, Farrel Machinery Group, Ansonia, Conn., 1983.

Super-Calenders and Action of the Filler, Appleton Machinery Co., Appleton, Wisc., 1983.

84

CASTING, POTTING AND ENCAPSULATION

John Hull

Hull Corporation, Hatboro, PA 19040

84.1 DESCRIPTION AND HISTORY OF PROCESS

Liquid reactive resins are often used to form solid shapes. Such resin systems harden or "cure" at room temperature or at elevated temperatures because of an irreversible cross-linking of complex molecules. This hardening process is distinctly different from that of resins in solution, which harden when the solvent is evaporated. The hardening of liquid reactive resins produces no by-products, such as gases or water or solvents. When such reactive resins are used as impregnants, they are sometimes referred to as "solventless systems."

The reaction of such systems is initiated when two or more components, usually a resin and a catalyst, are combined by mixing. Separately, the components are reasonably stable at room temperatures. Once mixed, they start to form long polymer chains, ultimately "gelling" or losing all fluidity, and subsequently become rigid. The reaction is exothermic (gives off heat) until cross-linking is complete. However, the reaction may be accelerated by external heat. The simplest processing with such resin systems is casting, in which the liquid resin mix is poured into a form or mold where it flows to fill the cavity and then hardens into a permanent configuration of the cavity (Fig. 84.1).

A major application of liquid resin casting is the production of simulated wood trim for furniture, ornate frames for wall clocks or mirrors, figurines, or other complex shapes. The ability of reactive resins to flow into small interstices of highly detailed mold surfaces makes it ideal for producing large numbers of identical shapes.

Another application is production of partially resilient components, such as industrial lift truck tires, principally from polyurethane resin systems.

A quite different application for liquid resin casting is to enclose and thereby protect an object, often an electrical device or circuit. Liquid reactive resins are good electrical insulators, physically strong, and form good barriers to moisture and gases. When the electrical device is placed in a form, such as a metal or plastic box or shell, that is intended to become an integral part of the finished product, the process is generally referred to as potting. If the form or shell is to be removed after the resin hardens, the process is termed encapsulation or embedment.

Some liquid reactive resin systems can be formulated such that, after mixing, they are chemically stable for many hours or even days at room temperature, but cross-link fairly rapidly at elevated temperatures. Resin systems of this type are often used to conformally coat an electrical device, such as a resistor or capacitors. In such a thixotropic dip process,

1117

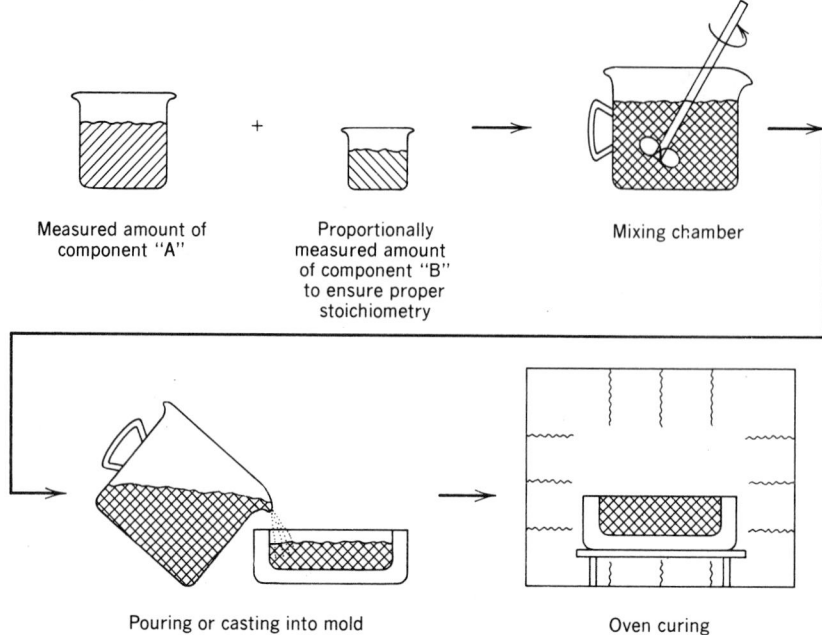

Figure 84.1. Hand proportioning, mixing, casting, and curing with a two-component liquid reactive resin system.

the resin mix is kept in an open container and the devices are dipped into the viscous liquid, then removed and passed through an oven. Heat speeds the reaction, causing the resin to harden in a conformal film around the device. The film thickness can be varied by controlling the viscosity of the mix or by repeated dipping and curing cycles.

In many electrical applications operated at high voltages or high frequencies, bubbles or voids in the liquid resins can cause electrical failures. To minimize the possibility of voids, part or all of the process may be performed under vacuum. The individual resin components are placed in a vacuum environment prior to mixing or immediately following mixing to "degas" the liquid and to remove any volatiles. Often, prior to the casting step, the devices to be potted and the forms or molds are exposed to vacuum, frequently with heat, to drive off any volatiles.

Sometimes processing must be done completely under vacuum: the resin is degassed, the devices are "baked out," mixing is performed, and casting is done—all under vacuum. Such processing is termed "vacuum potting," even though the product is removed from the mold. If the reactive resin must penetrate into many small spaces inside an object, such as the spaces between individual wires of a solenoid coil or transformer or small holes in a metal casting, the process is referred to as impregnation. In the impregnation process, the object to be impregnated is often "soaked" in vacuum, generally while it is being heated to remove all air or gases from the openings to be filled (just before the object is surrounded with the liquid resin). After the object is flooded with liquid catalyzed resin, vacuum is broken and the resin, usually a low viscosity system, penetrates the openings by capillary action that is aided by the atmospheric pressure. Such a process is termed vacuum impregnation. Its main purpose is to effect a solid, void-free object.

A related process that also uses low-viscosity liquid reactive resins is trickle impregnation, in which the catalyzed resin is allowed to drip on a transformer coil or other object with small openings. Capillary action draws the liquid into the openings at a rate slow enough to enable the air to escape as it is displaced by the liquid. When it is fully impregnated, the object is exposed to heat to cause the resin system to cure.

84.2 MATERIALS

Liquid reactive resin systems are thermosetting plastics, generally epoxies, polyesters, silicones, or polyurethanes. Occasionally elastomers, such as room-temperature vulcanizing (RTV) rubbers, flexible urethanes, polysulfides, and flexible silicones are used.

In applications where voids in the protective resin are not objectionable, thermosetting foams may be used. Foams offer the advantage of light weight (important for airborne devices) with good mechanical and moisture protection and improved thermal insulation.

Elastomers and foam systems are not used for impregnation.

A "syntactic" foam uses a conventional resin system with a "filler" of tiny hollow spheres, of perhaps 0.010 in. (0.3 mm) or less in diameter. The spheres are made from glass or from phenolic or other thermosetting resins. Such foam systems are rigid, generally of a specific gravity less than 1.0, and strong in compression and flexure.

84.3 PROCESSING CHARACTERISTICS

The simplest method of using liquid reactive resin systems involves hand weighing of resin components (for correct proportions), "dixie cup and popsicle stick" mixing, and atmospheric pouring into open molds. Such a method is obviously labor intensive and subject to human variables, including inaccurate weighing, inadequate mixing, and waste from careless pouring. On the other hand, the process requires no capital equipment, and affords a quick and easy evaluation for a given application or as an inexpensive means for prototype production.

MASTER PROCESS OUTLINE **PROCESS** Casting, potting, encapsulation

Instructions:
For "ALL" categories use Y = yes
For "EXCEPT FOR" categories use N = no
D = difficult

ALL Thermoplastics Except for:

Acetal — Y Poly(methyl methacrylate)
Acrylonitrile–Butadiene–Styrene — Poly(methyl pentene) (TPX)
Cellulosics — Poly(phenyl sulfone)
Chlorinated Polyethylene — Poly(phenylene ether)
Ethylene Vinyl Acetate — Poly(phenylene oxide) (PPO)
Ionomers — Poly(phenylene sulfide)
Liquid Crystal Polymers — Polypropylene
Nylons — Polystyrene
Poly(aryl sulfone) — Polysulfone
Polyallomers — Polyurethane Thermoplastic
Poly(amid-imid) — Poly(vinyl chloride)-PVC
Polyarylate — Poly(vinylidine chloride)-Saran
Polybutylene — Rubbery Styrenic Block Polymers
Polycarbonate — Styrene Maleic Anhydride
Polyetherimide — Styrene–Acrylonitrile (SAN)
Polyetherketone (PEEK) — Styrene–Butadiene (K Resin)
Polyethersulfone — Thermoplastic Polyesters
XT Polymer

ALL Thermosets Except for:

Allyl Resins
Phenolics
Y Polyurethane Thermoset
Y RP-Mat. Alkyd Polyesters
Y RP-Mat. Epoxy
RP-Mat. Vinyl Esters
Y Silicone
Urea Melamine

ALL Fluorocarbons Except for:

Fluorocarbon Polymers-ETFE
Fluorocarbon Polymers-FEP
Fluorocarbon Polymers-PCTFE
Fluorocarbon Polymers-PFA
Fluorocarbon Polymers-PTFE
Fluorocarbon Polymers-PVDF

At the other extreme, the process may utilize sophisticated proportioning, mixing, and dispensing machinery (Fig. 84.2), often with vacuum degassing of each resin system component under heat and agitation prior to the mixing and occasionally with casting performed in a vacuum chamber where the molds have been baked out under vacuum. This kind of process requires significant capital investment and is appropriate only when production demands are reasonably high and the economics of the process can be justified.

In between these two extremes are many variations, with the actual process selection being one that suitably satisfies the requirements for product quality, process cost, and production demand.

In handling liquid reactive resins, "cleanliness is next to godliness." When resins are cured, they adhere to whatever they contact and solvents cannot dislodge them. It is well worth the effort, therefore, to utilize disposable containers, to clean up any spilled resin before it cures, and to practice good housekeeping where the process is carried out.

Of vital importance in handling such resin systems is operator safety. Most of the materials are toxic. Some produce painful skin irritation and watering and burning of the eyes; others

Figure 4.2. Machine for automatic proportioning (volumetric), mixing (dynamic mixing head), and dispensing (foot pedal or timer controlled continuous or intermittent shot sizes), with manual solvent-type purge system to flush out catalyzed resin from mixing head. Partial disassembly and more thorough cleaning of mixing head is recommended after longer periods of nonuse. Courtesy of the Hull Corp.

cause respiratory problems for the operator. Skin salves, aspirators, and goggles are often used, and controlled ventilation systems in the casting room may be required. In general, the more automated the process becomes, the less human exposure there is to the toxic conditions.

Also on the subject of safety, one must realize that most solvents used to clean dispensing equipment are also toxic. Common solvents include trichloroethylene (TCE), a suspected carcinogen, and acetone and methyl ethyl ketone (MEK), which are definitely hazardous. Various commercial solvents are also used, but generally their safety is inversely proportional to their effectiveness. If cleanliness is maintained by careful handling of the resin system, the need for extensive use of cleaning solvents is minimized and, therefore, the process is made safer.

Aside from possible toxic problems with the resin system and the cleaning solvents, material waste is a major consideration in any production operation. Common resin systems are priced at from one to several dollars per pound, and waste can be costly. Yet, most operators feel it is wise to mix an excess of material to ensure they do not run out of it before the casting run is completed. The penalty is that partly filled containers of catalyzed resin must be thrown away at the end of the run. Once it is mixed, the resin will cross-link inexorably. Here again, the use of sophisticated (and often costly) process machinery to proportion, mix, and dispense the resin may be justified by the savings from minimizing waste.

Parts commonly potted with liquid resin systems are often electrical or electronic products such as transformers, coils, circuit packs made up of many components, cordwood modules, and so on. Many opti-electronic devices, light emitting diodes (LEDs), and alphanumeric (A-N) displays are potted in clear or translucent liquid resin systems to provide light transmittance as well as electrical insulation, mechanical protection, and a barrier to moisture and other environmental gases. Printed wiring boards and hybrid circuits are often potted in silicone rubbers or other elastomeric materials to ensure adequate protection of the circuits yet to permit access to components found defective after potting. The cured encapsulant may be cut away, the defective component replaced with a good one, and new liquid encapsulant cast into the opening to reseal the circuit.

Nonelectrical items formed by casting exhibit the mechanical properties pertinent to the resin system used. Their shapes and surface texture or details will faithfully reflect the cavity configurations and finish. Decorative objects often are spray painted to suggest the appearance of natural wood or to blend with any other decor.

Depending on the level of automation used in the potting process, costs may be fairly high. Costs are highest when little or no sophisticated machinery is used, but the improved results and the value added are often sufficiently necessary or desirable to warrant the expense of more labor-intensive potting.

84.4 ADVANTAGES/DISADVANTAGES

The principal advantage of the process is that it generally ensures the electrical and physical integrity of the part, or collection of parts, that are potted. It also lends itself to very low-volume or prototype production because of the basic simplicity of the process and because molds not subjected to high pressures are significantly cheaper then molds used for compression, transfer, and injection molding.

The principal disadvantages are the relatively high cost for high volume production (as compared to fluidized bed coatings, thixotropic dip, and encapsulation by transfer molding), the high material waste, and often, the messiness.

Another disadvantage that is not always a problem arises from the shrinkage of the resin system during cure. Shrinkage occurs progressively, starting where the heat of exotherm first generates and ending where the last area or volume segment cures, which may be minutes or even an hour later. Such high and nonuniform shrinkage puts physical stress on delicate inserts, which may damage the product. On nonelectrical castings, the shrinkage causes warpage, sink marks, and internal stresses. The shrinkage also makes it difficult to hold close dimensional tolerances on the finished product.

84.5   DESIGN

Because one of the principal applications of casting and potting is for electrical devices and circuits, the materials considered must ensure adequate dielectric strength, high dissipation factor (especially for frequencies of about 10 kHz and up), and arcing and tracking resistance (especially in high humidity or vacuum environments and in high voltage applications where corona discharge may tend to occur).

It is also important to recognize that shrinkage of the plastic during cure may tend to dislodge components being potted or put stress on wire bonds or solder joints.

If a minimum thickness is critical, particularly for the outer surfaces of the package, there are two possible approaches:

- Use a molded case or shell (with wall thickness adequate for the insulation required) in which the object may be potted.
- Make certain that the object is so constructed or supported in the mold cavity that no critical current-carrying connections or bonds or wires will touch the walls of the mold and be exposed after part removal.

In this regard, any parts, such terminals or contacts, that need to be exposed after cure should be so positioned in the mold (or sealed from the resin) that there is no need to remove cured resin from such areas.

If voids could cause problems, such as with high voltages or high frequencies, casting under vacuum is advisable. The resin system should undergo degassing prior to molding. If the item to be potted has any oil, grease, wax, solder flux, moisture, or other materials that volatilize under heat (either the heat of the cure or the exotherm), they should be removed prior to casting. A vacuum bake for a suitable length of time is often the best way to make certain that such volatiles are completely removed.

For nonelectrical applications, such as cast decorative simulated wood picture frames or cabinet trim, the part and mold must be designed so that air will not be trapped during pouring and leave a surface void. Furthermore, wide variation in section thickness may lead to unacceptable warpage and should be avoided, as should very thin sections. Because of the brittle nature of most of the compounds, they are easily broken in routine handling or use.

On the other hand, excessively thick sections may show sink marks after cure or even surface cracks caused by high localized exotherm temperatures. Except where they are needed for strength, thick sections represent material waste and can often be cored out on the back side to avoid such problems.

84.6   PROCESSING CRITERIA

Properties of cured resin systems are critically affected by the stoichiometry (that is, the ratio of the resin and catalyst). Correct ratios indicated by the supplier of the resin system ensure that cross-linking of the system is complete after cure.

Closely related to stoichiometry is proper mixing of resin components, so that the proportions are maintained throughout the mix. Improper mixing results in resin-rich and catalyst-rich areas, which manifest themselves as sticky places where no cures results or dry areas where the finished part has weak or powdery sections. The mixing part of the process is, of course, related to the miscibility of the resin components, and often to the viscosity ratios. It is difficult to mix very low viscosity catalysts into very high viscosity resins, for example, Raising the temperature of highly viscous resin system components generally lowers the viscosity, making mixing easier, but it may also accelerate the cure to the point where the material gels before it has finished flowing.

Proper curing cycles for the resin system are also recommended by the material supplier. Some resin systems cure at room temperature, requiring time (often several hours) to achieve maximum strength. Others must be heated to a specified temperature to initiate the reaction and then must be held at that temperature for a specified time period. More critical resin systems may necessitate the use of programmed heating, followed by a post-cure cycle lasting many hours. Ablative materials on re-entry nose cones of rockets, space vehicles, and rocket nozzles fall into this category.

All reactive resin systems exotherm: they generate heat during cure. Because the exotherm is internal, cast parts with heavy cross sections (large masses of material) should be made with material of lower exotherm to prevent overheating in those sections and subsequent resin degradation or breakdown.

Besides creating problems of excessive heat, exotherm may cause air or moisture in solution to expand into bubbles, which can lead to voids. This can happen even under properly controlled conditions (as when the air or volatiles have been fully absorbed at room temperature). Because such bubbles are undesirable, vacuum degassing of resin and catalyst prior to mixing is advisable. Similarly, most objects to be potted, as well as mold cavity surfaces, will have a monomolecular layer of water that can turn to vapor under the heat of reaction, causing voids. Vacuum bake out of both mold and part minimizes such voids.

Inasmuch as most reactive resin systems are adhesive, a mold release is necessary for parts cast in removable molds. Typical mold releases include carnauba wax, zinc stearate powder, or silicone or Teflon sprays. The amount of release necessary will depend on the smoothness of the cavity surface: more release is required on rougher surfaces. Too much mold release may cause rough or porous part surfaces. For optimum cosmetic results, mold cavity surfaces should be highly polished and nonporous. Such surfaces need only a very thin film of release, which may suffice for several castings before another coating is required.

The use of vacuum as a process parameter has been addressed earlier in this section. Pressure after casting is sometimes used to obtain faster penetration of the liquid resin into tight windings of coils and transformers or into pores or voids of metal castings that must be impermeable to gases or liquids after cure. Most impregnating equipment and many vacuum potters have provisions for applying positive pressure on the cast resin prior to cure, generally to 50 psi (0.3 MPa) and or occasionally to 100 psi (0.7 MPa). It is well to recognize that after casting under vacuum, the mere breaking of vacuum imposes an atmospheric pressure of about 15 psi (0.1 MPa) on the resin, assisting the penetration.

A word of caution regarding the use of vacuum for degassing and/or casting: All resin system components have a unique vapor-pressure curve with respect to temperature (Fig. 84.3). At a given temperature, there is a critical pressure (vacuum) where the resin system component will vaporize. If vacuum degassing or casting takes place at too low a pressure (too high a vacuum), one or more constituents of the resin system (usually of the catalyst) will boil off, leaving a resin system with properties different from those intended. The resin supplier may be able to suggest an appropriate vacuum level for its own system or this may need to be determined by experiment. In general, 3 to 5 mm Hg pressures are appropriate for drying room-temperature systems; higher pressures are necessary if higher temperatures are used.

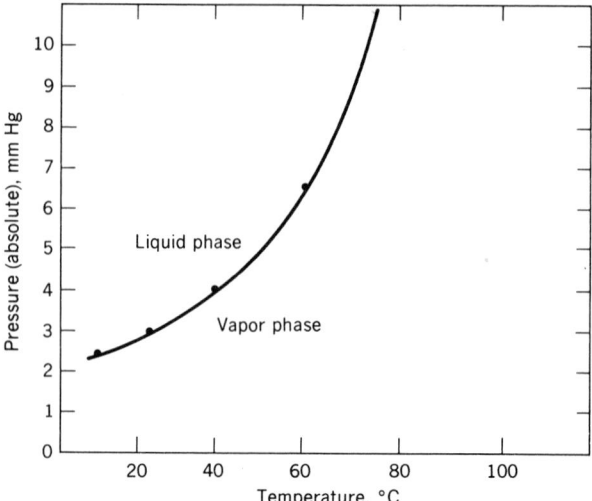

Figure 84.3. Vapor-pressure curve for catalyst component of a liquid reactive resin system. Degassing and casting pressures (vacuums) must be controlled so that resin system components stay in the liquid phase.

84.7 MACHINERY

A wide variety of equipment is available for casting, potting, and impregnating. Its selection must address the economics of the product being produced, the degree of precision necessary, and the production rate required. Proportioning can be accomplished manually with simple balances or scales. It can also be achieved with proportioning pumps ratioed appropriately or with positive displacement cylinders, one for the resin and one for the catalyst. These may be sized to give proper ratios with identical piston strokes (Fig. 84.4). Alternatively, identically sized piston and cylinders may be linked mechanically, allotting a full stroke to one piston and a proportional stroke to the other. Diaphragm pumps are also used with adjustable displacement to draw the resin components from reservoirs and to meter proper proportions into the mixing chamber. Gear pumps with drives ratioed to each other can be employed, but only when the resin and catalyst are relatively free of abrasive fillers.

After proportioning, resin and catalyst can be mixed with a dixie cup and a popsicle stick (as noted above), with a hand-held electrically driven mixer, or with heavy commercial agitators or stirrers in large vessels. These are all dynamic mixing devices. Mixing can be of an agitator type or a mechanical shear type, but the mechanical shear type is generally more effective and therefore faster.

In addition to dynamic mixing, static mixing may be used with many resin systems, particularly where the viscosities of resin and catalyst are about the same. Static mixing does away with moving parts in the mixing chamber and, thus, with seals and blades that need periodic cleaning and replacing. Static mixing is generally effected by use of a number of simple flow splitters positioned in a tube into which the two resin components are fed continuously in the proper proportions (Fig. 84.5). As the two streams enter the tube and meet each successive flow splitting element, the streams are divided and rotated 90°. Five or more flow splitting elements are used, depending on the degree of mixing required, to achieve homogeneous blending. The catalyzed resin exiting from the tube is usually cast directly into the mold or pot (Fig. 84.6).

Static mixing heads or tubes require a pressure differential to drive the mix through the tube. If the mix is highly viscous, the pressure differential may have to be fairly high, perhaps

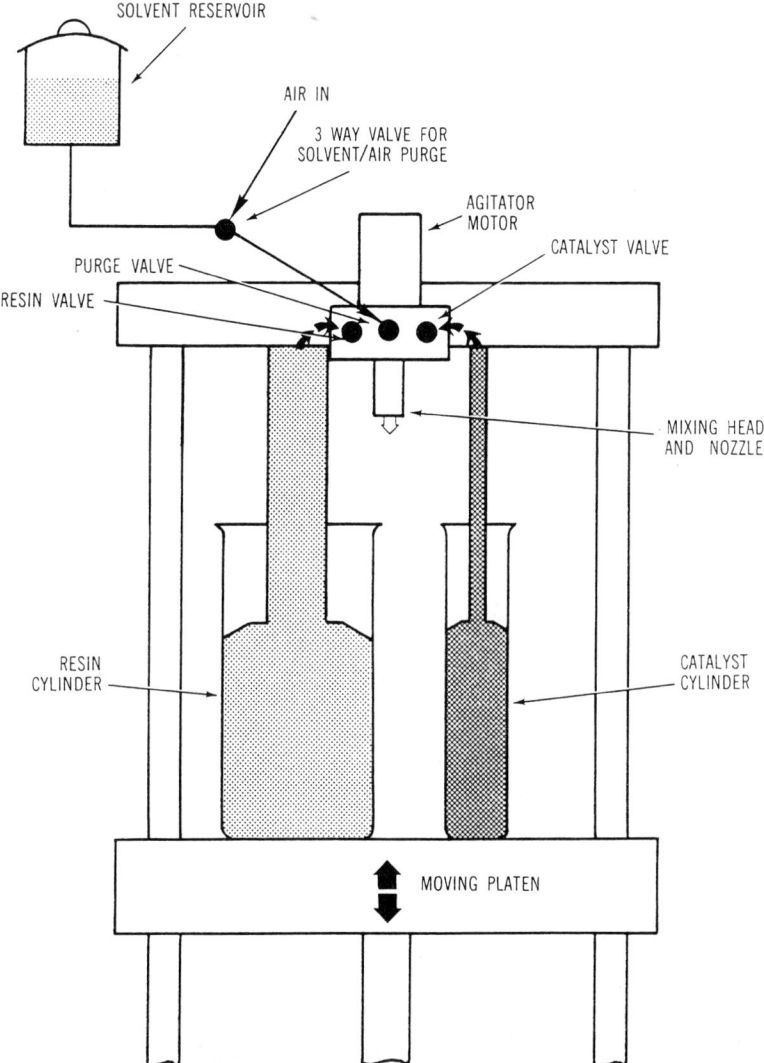

Figure 84.4. Automatic proportioning (volumetric), mixing (dynamic), and dispensing (foot pedal or timer controlled continuous or intermittent shot size) machine with manual solvent-type purge system.

100 psi (0.7 MPa), to achieve a reasonable rate of throughput. The size of diameter of the static mixer and the number of elements depends on the viscosity and desired throughput. Some static heads are disposable; others, generally of stainless steel, must be disassembled and cleaned at the end of each run.

If the mixing process delivers catalyzed resin into a holding tank or other container, the casting may be by manual pouring or by mechanical pumping. If the volume to be cast must be precisely controlled, such mechanical pumps can be positive displacement types with a timer or other means used to ensure correct charge size. In systems that proportion with position displacement pumps, the catalyzed resin is often delivered by continuous flow in a closed system from the proportioning pumps through the mixing head to the dispensing nozzle.

If the mix is to be cast under vacuum and is in a container at atmospheric pressure, it can be transferred simply by running a tube from the bottom of the mix container into the

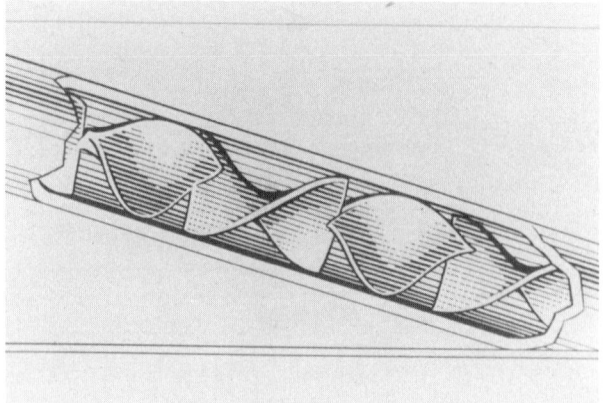

Figure 84.5. Typical flow-splitter elements in a static mixing head.

vacuum chamber, letting atmospheric pressure provide the moving force (Fig. 84.7). A simple adjustable tube clamp can be used to control resin flow or stop it entirely. Highly viscous mixes that do not flow may have to be physically transferred from the mixing chamber to the mold or work piece via a scoop or spatula.

When vacuum potting or casting is required, rotary oil-sealed vacuum pumps are generally used to evacuate, before mixing, the bell jar or chamber for resin and catalyst degassing and, before casting, the casting chamber. Often a solvent trap is required between the vacuum chamber and the vacuum pump to prevent volatiles from contaminating the vacuum pump oil. If very few volatiles are likely to be encountered, a vacuum pump with gas ballast provisions can eliminate the need for a solvent trap.

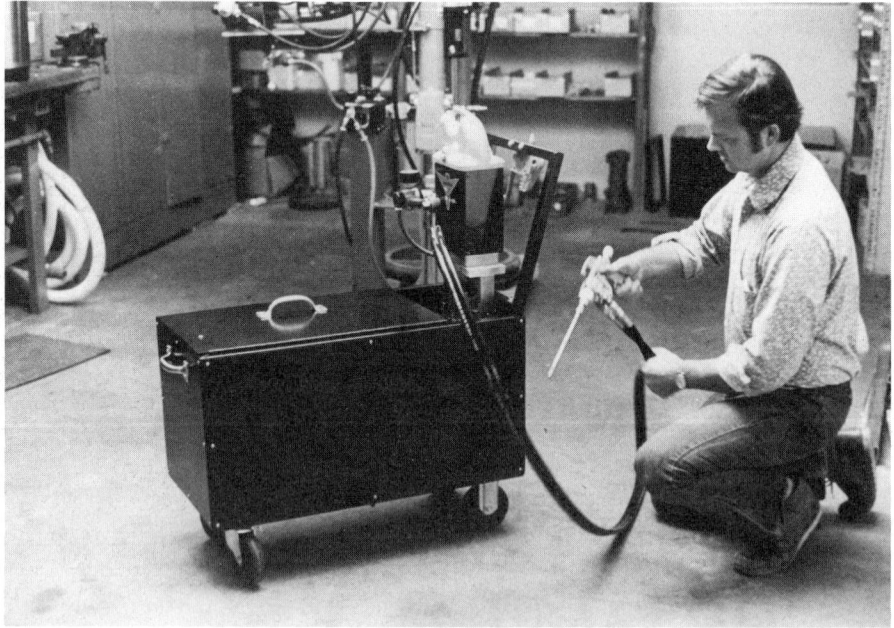

Figure 84.6. Dispensing machine with static mixing head and nozzle fed by flexible resin and catalyst lines for dispensing remotely from resin and catalyst reservoirs. Courtesy of Packaging Systems, Inc.

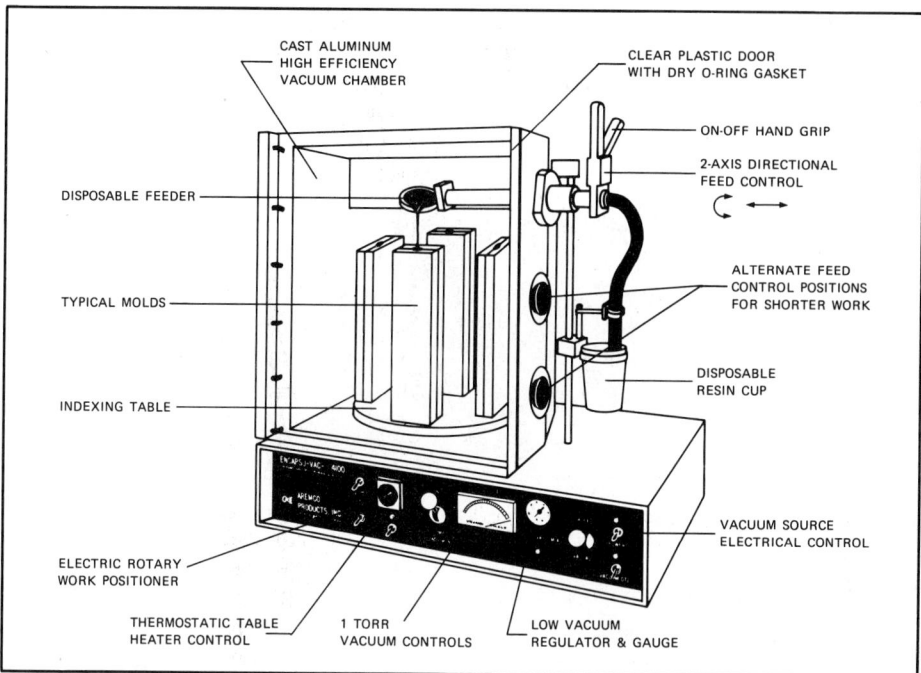

Figure 84.7. Simple bench-type vacuum potter for casting catalyzed resin under vacuum into evacuated mold. Degassing occurs as the mix flows over a heated plate prior to flowing into the open mold. Courtesy of Aremco Products, Inc.

Vacuum potters are also available as integrated systems that provide for degassing of resin or catalyst (with optional heat and/or agitation), mix under vacuum, bake out mold and parts under vacuum prior to casting, and cast under vacuum. The entire process keeps the resin system, molds, and parts to be potted under vacuum from start to finish (Fig. 84.8). Such potters may also be supplied with provisions for positive pressure to be applied after casting.

For continuously degassing resin or catalyst, stream degassers are available through which the liquid flows in a thin stream surrounded by vacuum. This enables air and volatiles to escape from a relatively fine stream without impeding the flow.

Temperature control in potting operations will be required if the resin or catalyst must be heated before mixing (generally to reduce viscosity, thereby simplifying and speeding up mixing). Heat may also be required for the mixing chamber, and, if vacuum bake out is required, for the molds and parts in the casting chamber. Thermocouple controlled electric heating is generally used, with electric "blankets" or bands around containers and electric cartridges in molds. For heating items inside the vacuum chamber, radiant heating may be used with heated thermostatically controlled chamber walls.

The switches, contactors, and motors of electric heaters may need to be explosion proof for safety reasons. In place of explosion proof electric motors, air motors are often used for agitators in degassing or mixing chambers.

Cured resin is a thermoset practically impervious to solvents, so care is required to keep catalyzed resins away from equipment and surroundings. Dynamic mixing heads, dispensing nozzles, molds, and surfaces on which molds rest must all be cleaned promptly at the end of a run, before the resin hardens. Purge systems for the mixing heads of dispensing machines and for mixing chambers are vital. They include solvent flow and washing steps for all surfaces in contact with catalyzed resin.

(a)

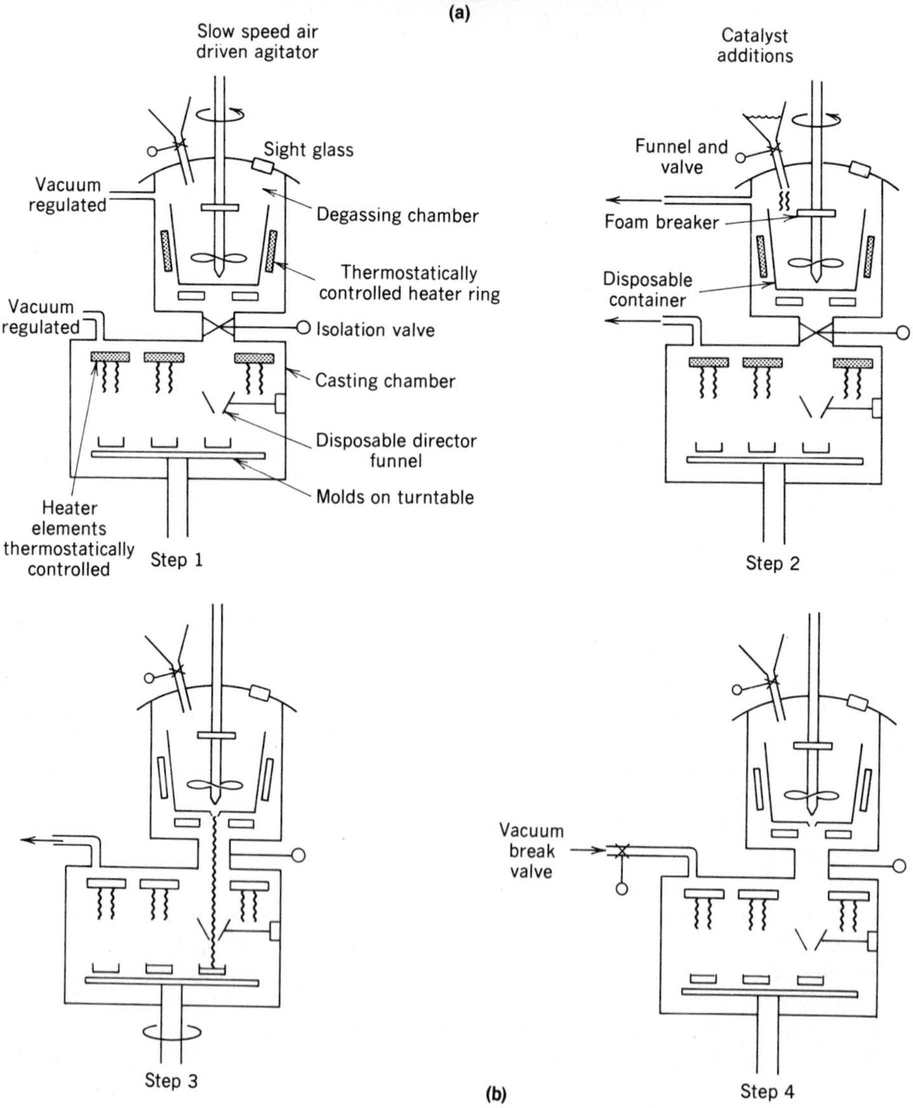

Slow speed air
driven agitator

Sight glass

Vacuum
regulated

Degassing chamber

Thermostatically
controlled heater ring

Vacuum
regulated

Isolation valve

Casting chamber

Disposable director
funnel

Molds on turntable

Heater
elements
thermostatically
controlled

Step 1

Catalyst
additions

Funnel and
valve

Foam breaker

Disposable
container

Step 2

Step 3

Vacuum
break
valve

Step 4

(b)

(c)

Figure 84.8. **(a)** Vacuum potter for resin and catalyst mixing and degassing under vacuum, mold bake out under vacuum, and casting under vacuum. **(b)** Vacuum potting process Step 1. Molds are heated under vacuum to remove all volatiles, while uncatalyzed resin is degassed with controlled vacuum and heat and gentle stirring. Vacuums in degassing and casting chambers can be different because of isolating valve. Step 2. After resin is degassed, catalyst is slowly added and mixed into the resin while maintaining vacuum in the degassing chamber. Step 3. When resin system is fully mixed, the isolation valve between the chambers is opened, the agitator shaft moves downward and pierces the bottom of the disposable degassing container. It retracts acting as a needle valve to control the flow from container into director funnel. Turntable is rotated until each mold is visually filled. Step 4. Vacuum is broken. Molds may be left in chamber under heat until resin system is cured. **(c)** Arrangement of casting molds on a rotary turntable in the casting chamber of a vacuum potter.

Many automatic dispensing machines, especially those arranged for intermittent dispensing, are furnished with audible purge alarms. Such alarms sound an alert for the operator after a given period of time (less than the gel time of the mix) following the previous dispensing shot. An even safer arrangement is the automatic purge system, which triggers the machine to spew out the contents of the mixing head if too much time elapses before the next shot.

Ideally, such units have a container into which the catalyzed mix flows. If not, the floor or work table will be messy, but even after resin cure it is easier to clean them than the mixing head and dispensing nozzle.

Automatic dispensing machines may be provided with adjustable timers or limit switch arrangements to enable consistent size shots, required for assembly line casting into uniform size molds.

84.8 TOOLING

Molds and cases for potting are generally of metal or plastic. Metal molds may be machined from aluminum or steel or may be constructed of welded sheet metal. Molds need provisions for part removal after cure; they often consist of two halves bolted together so they can be disassembled for part removal and periodic cleaning. A simple mold may be made by dipping a cold metal blank or form into molten pot metal and extracting the blank quickly. A layer of pot metal hardens on the outside of the blank, which is then removed and used for casting.

Such a mold may be used repeatedly until it is worn or damaged, after which it may be remelted for future use.

Plastic molds or potting cases and shells may be of thermosetting plastics (epoxies or phenolics) or thermoplastics (vinyl, polycarbonate, or polystyrene). If thermoplastic cases are used, the casting process must be controlled so that the plastic case never reaches a temperature that could melt it. Such thermoplastic molds or forms may be injection molded or thermoformed.

Detail molds for decorative furniture and appliance parts cast to look like wood grain are often made from flexible silicone rubber that can be "peeled off" the hardened part without damaging the mold. Such molds are generally made by casting around a carefully executed pattern of the final part and then removing the pattern.

84.9 AUXILIARY EQUIPMENT

In addition to the process equipment described above, batch or conveyorized ovens may be required for curing parts after casting. Sometimes vibrating tables are used on which molds are placed after atmospheric casting to enhance floating of the bubbles to the surface. The bubbles may simply burst when they reach the surface (especially if the vibrating mold is placed under a vacuum bell jar), but more often an operator must prick each bubble with a pointed tool as it appears at the surface so that the air or volatile can escape through the skin of the cast material.

84.10 FINISHING

When the cast or potted products have cured, further steps may be needed to complete the part.

If removable molds have been used, the gate or sprue often needs to be removed by sawing, filing, grinding, or sanding to present a smooth surface. Surface voids may need to be filled, usually by hand, using a spatula. Terminals or contacts of electrical devices may have a thin coating of cured resin, despite efforts in mold design or casting to prevent such coating. The coating may be removed by scraping, filing, or sanding, or it may be removed by a deflasher.

Inspection of finished parts may be purely visual for nonelectrical applications. Electrical devices will need to be inspected by electrical tests, possibly conducted under harsh environments.

Many cast or potted devices may be marked for identification. Hot stamping with or without decals is common, as is silk screening or tampon printing.

If the parts are furniture trim or appliance trim, they are often spray painted to simulate a wood appearance.

84.12 COSTING

Because of wide variations in processing techniques that are often labor intensive, with or without expensive processing equipment, parts costing varies with the specific part being molded.

The costs of cups, cases, shells, and resin systems (including an allowance for inevitable wastage) are straightforward. Disposable containers, static mixing heads, cleaning rags, and solvents must also be costed and allocated to each unit produced. Labor, floor space, amortization, and maintenance of equipment and molds, can all be costed using standard factory costing procedures.

With automatic dispensing systems and sophisticated vacuum potters, set up time is often extensive and costly. These costs should be allocated to each part produced during a run.

Shop overhead, general sales and administrative overhead, and profit can be applied to arrive at an appropriate selling price for the final product.

84.14   INDUSTRY ASSOCIATION

The following associations issue standards, regulations, and other information pertinent to casting and potting with liquid reactive resin systems: American Chemical Society, American Society of Testing Materials, Institute of Electrical and Electronics Engineers, National Electrical Machinery Association, Society of Plastics Engineers, and Society of the Plastics Industry.

84.15   INFORMATION SOURCES

A large number of periodicals occasionally publish information on the technology of casting and potting with liquid reactive resin systems. These include: *Design Engineering, Electrionics, Electronic Packaging and Production, Electronics, Modern Plastics, Plastics Compounding, Plastics Design Forum, Plastics Engineering, Plastics Machinery & Equipment, Plastics Technology,* and *Plastics World.*

In addition, several books published in recent years contain chapters on various aspects of the topic (see the General References).

Inasmuch as the technology of liquid reactive resins and their processing continues to develop, technical seminars are often presented on the subject. These seminars are sponsored by such groups as the Center for Professional Advancement, the Institute of Electrical and Electronic Engineers, the Society of Plastics Engineers, and others.

BIBLIOGRAPHY

General References

Harper and co-workers, *Application of Transfer Molding Techniques to Encapsulation of Complex Electronic Modules*, U.S. Army Missile Command, Redstone Arsenal, Ala., 1970. Contract No. DAA H01-68-C-2020 to Westinghouse Defense and Space Center, Aerospace Div., Baltimore, Md.

Proceedings of the Symposium on Plastic Encapsulation/Polymer Sealed Semiconductor Devices for Army Equipment, U.S. Army Electronics Research and Development Command, Ft. Monmouth, N.J., 1978.

Review of Manufacture and Application of Semiconductor Integrated Circuits to Army Items, U.S. Army Production Equipment Agency, Rock Island, Ill.

Volk and co-workers, *Electrical Encapsulation*, Reinhold Publishing Corporation, New York, 1962.

85

COMPRESSION AND TRANSFER MOLDING

John Hull

Hull Corporation, Hatboro, PA 19040

85.1 DESCRIPTION AND HISTORY OF PROCESS

Compression and transfer molding are the two main methods used to produce molded parts from thermosetting plastics—defined as plastics that polymerize under heat and pressure in an irreversible chemical reaction. The founding date of the thermosetting plastics industry is probably 1909, when Leo Baekland first produced phenol–formaldehyde resins. The trade name "Bakelite" of the Union Carbide Company is derived from Baeklands's name. The emphasis in the early years of thermosetting plastics was on their chemistry. Equipment for molding the materials was very crude, and cycle times were long.

In 1916, Novotny developed and patented a method for molding cylindrical printing plates. His process was the forerunner of transfer molding. In 1926, Frank Shaw obtained patents on the process called "transfer molding," and one of the first major commercial applications was the long body of an "automatic" pencil with the molded spiral grooves on the inside wall. An early patent on a fully automatic molding press was issued in the mid-1930s to Victor Zelov. From that date forward, semiautomatic and fully automatic molding of thermosets using both the compression and transfer processes has increased dramatically.

Natural and synthetic rubber falls into the category of thermosetting plastics, and the processes of compression and transfer molding are used extensively in the rubber industry.

The basic process of compression and transfer molding is as follows: A thermosetting molding compound is exposed to sufficient heat, approximately 300°F (149°C), to soften or plasticize it. The fluid plastic is held at the molding temperature under pressures as high as 2,000–4,000 psi (13.8–27.6 MPa) for a sufficient length of time for the material to undergo polymerization or cross-linking, which renders it hard and rigid. The part is then removed from the mold cavity.

85.1.1 Compression Molding (Fig. 85.1)

In its simplest form, compression molding consists of placing a quantity of thermosetting molding compound in the bottom half of an open heated mold. The upper and lower halves of the mold are bolted appropriately to the upper and lower press platens. The closing of the press brings the two mold halves together with high pressure; the opening of the press separates them. After the material is added, the press is closed, bringing the bottom half of the mold against the top half under pressure. As the mold halves come together, the plastic material begins to soften under heat and is formed by the pressure of the two mold

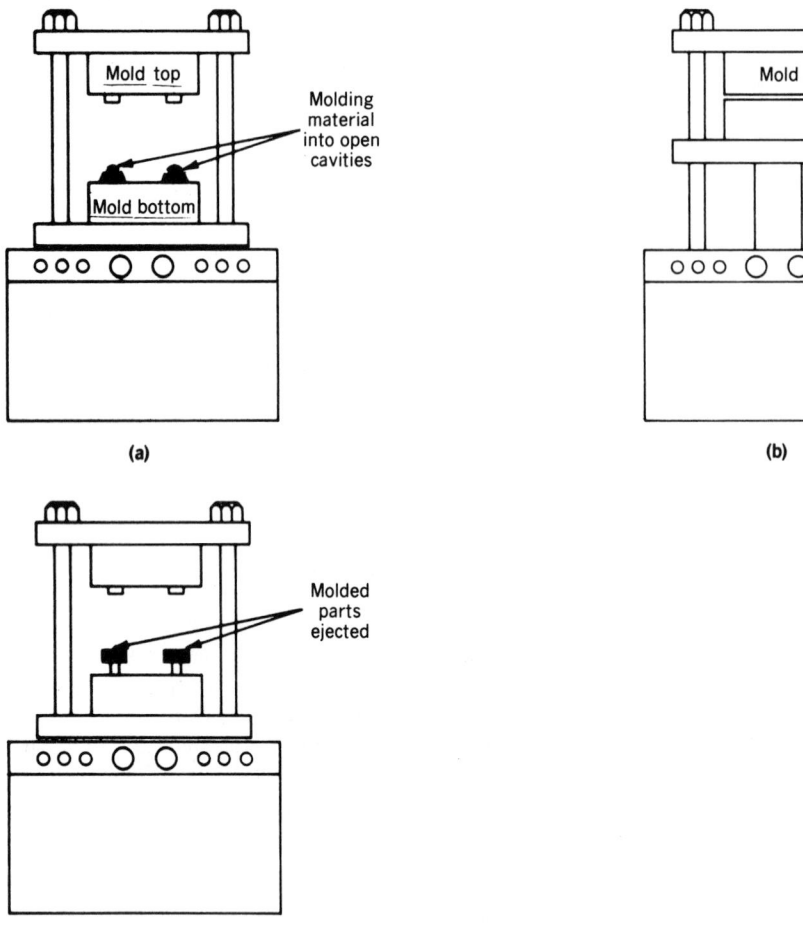

Figure 85.1. Compression molding sequence. (**a**) Molding material is placed into open cavities; (**b**) the press closes the mold, compressing material in the hot mold for cure; (**c**) the press opens and molded parts are ejected from the cavities.

halves. After an interval of time (which may be a minute or more, depending on the material and thickness of part), the press is opened and the plastic will have set up in the shape of the mold cavity.

Much compression molding is done using the semiautomatic technique. With this technique, the operator is required to introduce the plastic into the mold cavities and remove the molded parts from the cavities by hand. Once the cavities are loaded, the operator actuates appropriate controls on the press so that the press closes (possibly via a slow close action), effects the cure, and then opens automatically. In short, the press carries out a single cycle automatically but must be reactivated for each subsequent cycle.

In fully automatic compression molding (Fig. 85.2), the material is automatically fed into the cavities each cycle, the molded parts are removed automatically from the mold following each cycle, and the press recycles automatically.

85.1.2 Transfer Molding (Fig. 85.3)

For some applications, it is desirable to close the mold first and then introduce the molding compound in its fluid state through a small opening or gate leading to the mold cavity or

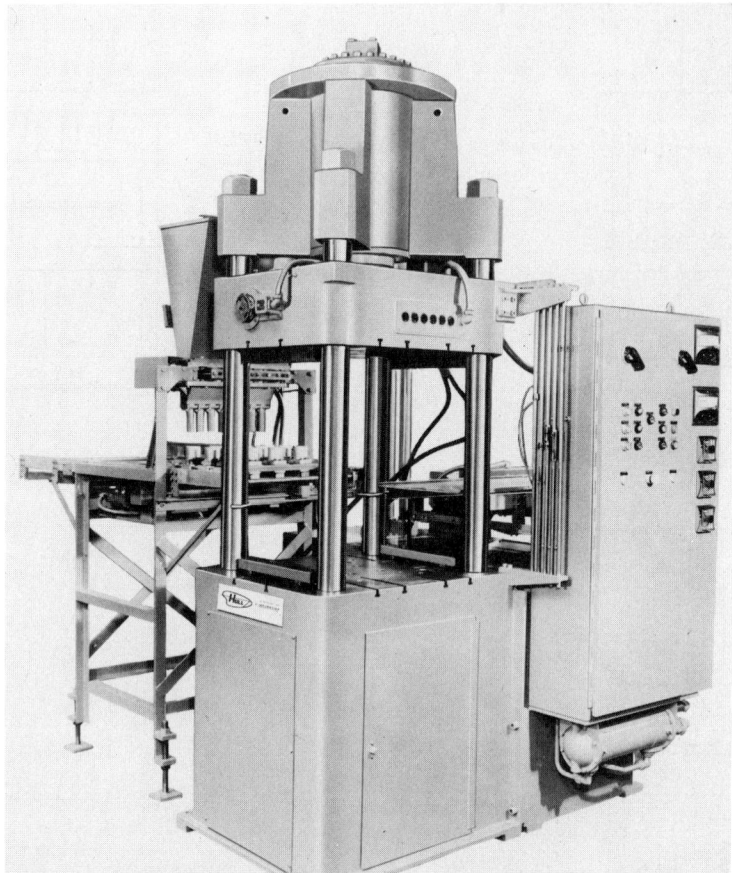

Figure 85.2. Fully automatic compression molding press.

cavities. This technique is called transfer or plunger molding. It is used frequently when mold sections are very delicate, when the molded part has thick sections [1/8 in. (3.2 mm) or more], or when an insert is retained in the cavity for molding in place. In such applications, closing the mold containing a molding compound that is not yet fully liquid (as is the case in transfer molding), flow speed and pressure can be controlled to minimize the possibility of any such damage.

In true transfer or pot-type transfer molding, the mold is closed and then placed in an open press. The charge of molding compound is introduced into an open pot at the top of the mold. The plunger is then placed into the pot and the press is closed. As the press closes, it pushes against the plunger, which in turn exerts pressure on the molding compound, forcing it down through a vertical passage called a sprue and through runners and gates into the cavities. After curing, the mold is removed from the press, the plunger is withdrawn, the mold is opened, and the parts are ejected.

Pot-type transfer molding may also be done with the bottom half of the mold bolted to the lower press platen and the plunger bolted to the upper press platen. The upper mold half, containing the pot, may then be manually placed over the lower mold half. Or, it may be suitably supported and guided so that the opening of the press separates the lower mold

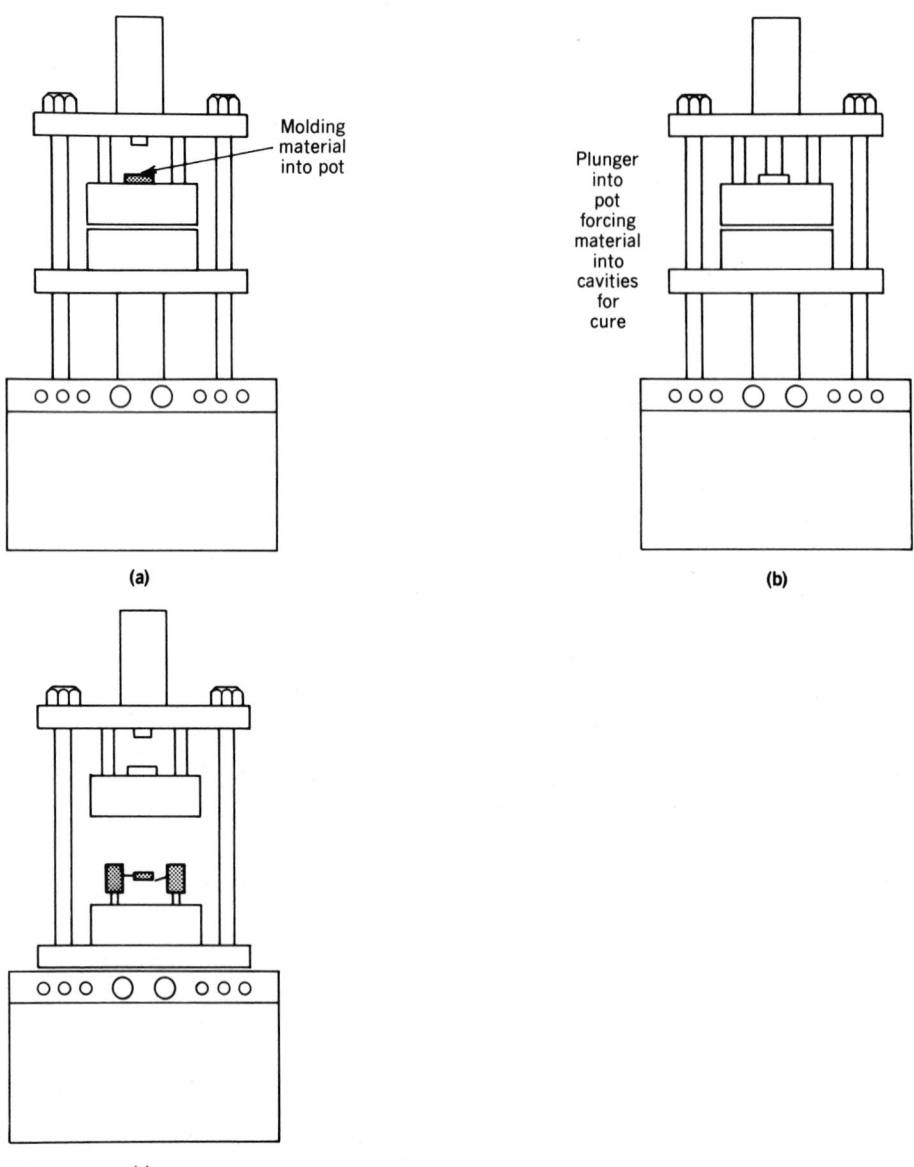

Figure 85.3. Transfer molding sequence. (**a**) the mold is closed and material is placed in the pot; (**b**) the plunger descends into the pot, causing material to melt and flow through runners into cavities; (**c**) after cure, the press opens, the plunger retracts, and parts are ejected with cull and runners.

half from the upper mold half and, in the same motion, pulls the plunger out of the pot. The procedure is reversed when the press closes.

Because much material is wasted in the large pot, it is generally more economical to use plunger transfer molding instead. In plunger molding, the plunger is essentially a part of the press rather than part of the mold. It is usually driven by a hydraulic circuit and a cylinder attached to the head of the press and can, therefore, be considerably smaller in diameter than the pot-type plunger. The mold is held closed by the clamping action of the press, independent of the plunger movement or force. The behavior of the molding compound is identical, however.

The decision on whether to use transfer or plunger molding depends on the equipment available, the type of mold desired, the economics of material wastage and press and mold costs, and so on. Inasmuch as most modern-day transfer molding utilizes hydraulically actuated plungers, the term transfer molding will be used throughout this section to refer to the plunger molding process, not the pot-type molding process.

Semiautomatic Transfer Molding (Fig. 85.4). In semiautomatic transfer molding, the operator actuates the press each cycle to close the mold and then manually introduces the molding compound into the transfer pot, which is in the top half of the mold. Next, the operator actuates appropriate controls to cause the plunger to descend into the pot. The

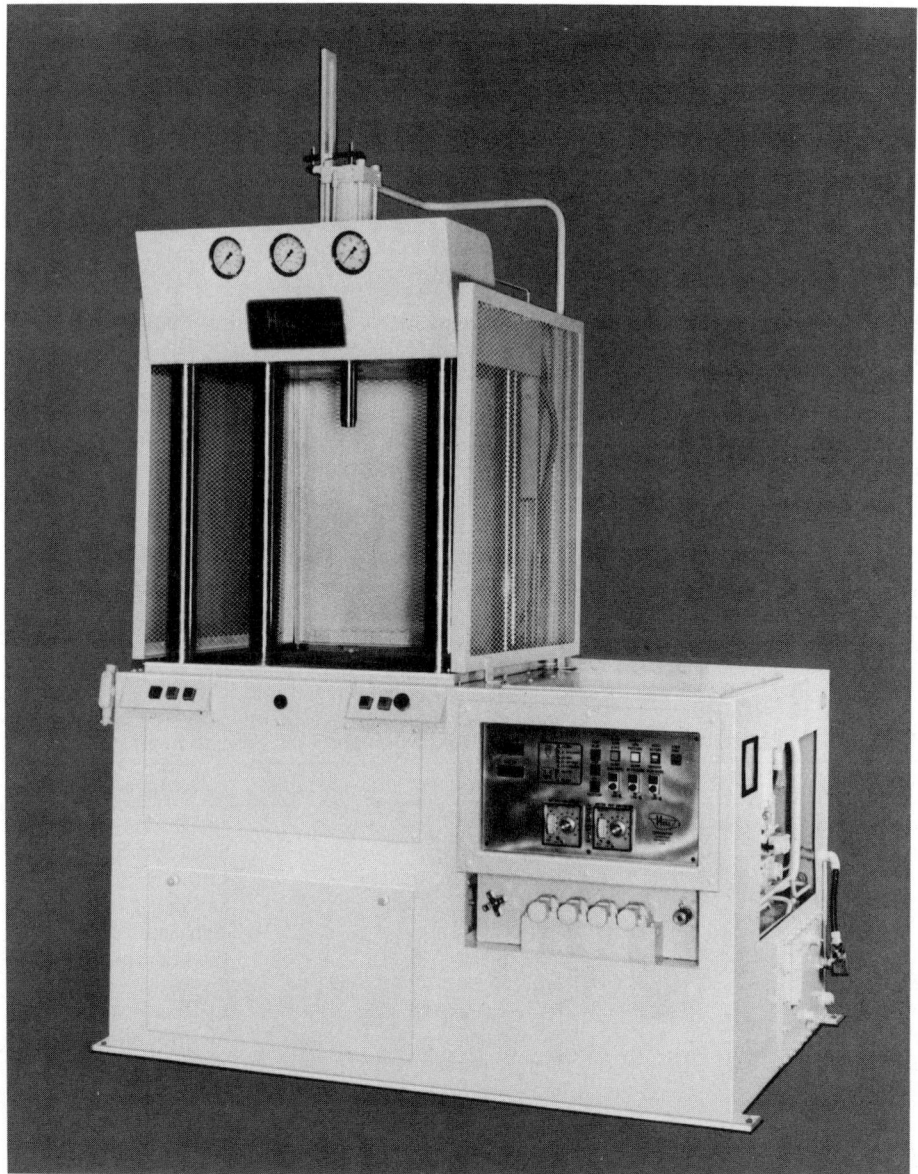

Figure 85.4. Semiautomatic transfer molding press.

press controls take over to time the cure cycle, open the mold, and eject the parts from the cavities using ejector pins in the mold. The operator lifts the parts from the ejector pins and initiates another cycle.

Fully Automatic Transfer Molding. In fully automatic transfer molding, the press recycles automatically, thereby feeding molding compound into the transfer pot and removing parts at the end of each cycle. Automatic transfer presses are often configured horizontally so that when the molded parts are ejected from the cavities, they will readily fall into a container or conveyor belt below the open mold.

85.2 MATERIALS

85.2.1 Common Thermosetting Plastics

The thermoplastics most commonly used for transfer molding are listed in Table 85.1. All of these materials will undergo cross-linking or polymerization at sustained heat and pressure in an irreversible chemical reaction. Material selection depends on the particular properties required of the final molded part, the dimensions of the part, and such characteristics as color (some materials are not available in a variety of colors), surface hardness, and so on.

85.2.2 Special Thermosetting Plastics

Several less common thermoset materials are available for applications requiring extreme properties.

Flock-filled Material. Flock-filled material includes small pieces of woven fabric interspersed with the basic resin formulation; for example, a flock-filled phenolic. Such a material exhibits excellent impact, tensile, compression, and flexural strength but may present a somewhat rough surface from a cosmetic point of view.

Glass Roving Filler. Glass roving consists of a yarnlike rope [approximately 1/4 in. (6.4 mm) in diameter] of glass fibers held together by uncured thermoset resin in such a manner that the rope is sufficiently flexible for easy insertion into a mold. Parts molded of glass roving filler are especially strong in impact and tensile strength. An outstanding example is

TABLE 85.1. Common Thermosetting Molding Compounds and Their Principal Applications

Material	Advantage	Applications
Phenol–formaldehyde	Low cost	
General-purpose	Durable	Small housings
Electrical grade	High dielectric strength	Circuit breakers
Heat resistant	Low heat distortion	Stove knobs
Impact resistant	Strong	Appliance handles, legs
Urea formaldehyde	Color stable	Kitchen appliances
Melamine formaldehyde	Hard surface	Plastic dinnerware
Alkyd	Arc resistant	Electrical switchgear
Polyester	Arc resistant	Electrical switchgear
Diallyl phthalate	High dielectric strength	Multipin connectors
Epoxy	Soft flowing	Encapsulate electronic components
Silicone	Withstands high temperature	Encapsulate high power electronic components

the reverse ring gear in some automatic transmissions for automobiles. In this application, the glass roving filled phenolic replaced an aluminum die-cast gear and was found to be quieter, more durable, and lighter in weight.

Another type of glass-reinforced thermosetting material is an extruded rope of larger diameter [3/4 to 1 1/2 in. (19–38 mm) in diameter], made of long glass fibers [1/2 in. (13 mm) or longer] and alkyd or long glass fibers and thermosetting polyester. In some instances, the "rope" is 2–3 in. (51–76 mm) in diameter and is supplied in "logs" with the length appropriate to the cavity for ease in manually loading uniform charges of material.

Bulk (or Dough) Molding Compound. Bulk (or dough) molding compound, referred to as BMC in the United States (DMC in the UK), is formulated from thermoset polyesters filled with glass fibers of lengths up to 1/2 in. (13 mm). These materials flow easily and prove exceptionally strong. Polyester also has good arc resistance, making this material ideal for heavy electrical switchgear.

Vinyl. Vinyl, which is a thermoset inasmuch as it cross-links but the cross-linking reaction is not irreversible, is often selected for phonograph records using the compression molding process. The material flows easily and reproduces the extremely fine details in the grooves of the records.

Encapsulating Grades. Very soft flow thermoset materials are required for molding around very delicate inserts. Vast quantities of electronic components, such as resistors, capacitors, diodes, transistors, integrated circuits, etc, are encapsulated with such soft flowing thermoset compounds principally epoxies, by transfer molding. Silicone molding compounds are used occasionally where higher environmental temperatures are required of the encapsulated part [up to 500°F (260°C) or more]. Polyester compounds, which are somewhat less expensive than epoxies or silicones, are also sometimes selected.

85.3 PROCESSING CHARACTERISTICS

85.3.1 Process

Compared to other techniques for shaping plastics, compression and transfer molding are fairly labor intensive, especially if they are semi-automated. The process requires lower capital investment than injection molding, particularly for semiautomatic compression and transfer molding. In general, molding cycles in compression molding are longer than those of injection molding. If the material used in transfer molding is preheated or preplasticized before it is introduced into the transfer pot, transfer molding cycles may be comparable to those of injection molding.

1. Recommended wall thickness/length of flow
   Minimum _____0.015 in._____ for _____NA_____ distance
   Maximum _____1½ in._____ for _____NA_____ distance
   Ideal _____NA_____ for _____NA_____ distance

2. Allowable wall thickness variations _____5_____ percent of nominal wall

3. Nominal wall-thickness tolerance
   Paralle to parting line _____± 0.005 in._____
   Perpendicular to parting line _____± 0.003_____

4. Radius requirements (See Figure 1)
   Outside: Minimum _____110_____ Maximum _____NA_____
   Inside: Minimum _____10_____ Maximum _____NA_____

5. Reinforcing ribs (See Figures 2a and 2b)
   Maximum thickness _____100_____
   Maximum height _____200_____
   Sink marks: Yes _Minor_ No _____

6. Solid pegs and bosses
   Maximum thickness _____100% W_____
   Maximum height _____4:1 L/D_____
   Sink marks: Yes _Minor_ No _____

7. Strippable undercuts
   Outside: Yes ___X___ No _____
   Inside: Yes ___X___ No _____

8. Molded-in threads
   Outside: Yes ___X___ No _____
   Inside: Yes ___X___ No _____

9. Hollow bosses (See Figure 3)
   Yes ___X___ No _____

10. Holes
    Perpendicular to parting line: Yes ___X___ No _____
    Parallel to parting line   Yes ___X___ No _____
    Through holes   Yes ___X___ No _____
    Blind holes   Yes ___X___ No _____

11. Draft angles
    Outside: Minimum _____1%_____ Ideal _____3–5%_____
    Inside: Minimum _____1%_____ Ideal _____3–5%_____

12. Trimming required
    Yes ___X___ No _____

13. Finished both sides
    Yes ___X___ No _____

14. Finishing Requirements
    Depends on application
15. Molded-in inserts
    Yes ___X___ No _____

16. Size limitations
    Length: Minimum _____NA_____ Maximum _____NA_____
    Weight: Minimum _____NA_____ Maximum _____NA_____

17. Tolerances
    Note: Similar to SPI Tolerance Charts attached.

18. Special process- and material-related design details
    (A) Core L/D
        1. 4:1 Injection and transfer molding
        2. 2:1 Compression molding
    (B) Gate location (transfer molding)
    (C) Ejector Pin Location, Top and/or Bottom
    (D) Part remain on force or part remain on cavity, on mold opening, vital for consistent automatic
        part removal

85.3.2 Parts

Thermoset molded parts are rigid and dimensionally stable, having properties relatively unaffected by exposure to elevated temperatures. They rarely deform under long-term stress (a deformation sometimes referred to as "cold flow"). In this respect, thermoset materials are considered far superior to thermoplastic materials.

Because thermoset materials withstand higher heat, they are used in many electrical applications, such as toaster legs and handles, lamp sockets, knobs on electric hot plates, etc.

Because thermoset materials offer good chemical resistance, they are often used as screw caps or closures on containers holding volatile solvents.

When they are reinforced with glass fiber or flock fillers, thermoset materials demonstrate very high impact strength. They are often used as garbage disposal housings, hammer handles, boat hulls, etc.

Many thermoset formulations are not suitable where color stability is important. Phenolics, as an example, are available in browns or blacks, but even the browns darken with age. The ureas and melamines and alkyds are often used where color stability is important; they are available in a variety of colors. Melamine dinnerware and urea housings for electric appliances are some common examples.

Thermoset materials can be molded with ribs, bosses, through holes, and blind holes into highly complex shapes. Thus they lend themselves to many structural components, such as wall plates, toggle switches in household wiring systems, circuit breakers, connectors, distributor caps, and under-the-hood automative applications.

With proper design, thermoset materials can be molded to quite precise tolerances that are stable under stress or temperature excursions. One example is the use of glass-filled thermoset phenolics in automotive disk brake pistons, an application requiring high strength, high heat resistance, dimensional stability, and close tolerances. In this particular application, although the parts are held to very close tolerances in molding, a final machining step is sometimes included to increase the diametral precision. In molded thermosets, the surface of the part will be identical to the cavity surface of the mold. As a result, the parts may be molded with an almost mirror finish, matte finish, or other texturing, depending on the mold cavity surface.

Threaded caps are often molded with thermoset materials. They require an unscrewing mechanism to remove the cap from the "force" or the male mold cavity half.

Molded-in inserts are frequently used the thermoset materials; for example, automotive distributor caps (molded with phenolics, alkyds, or polyesters) and multipin electrical connectors as used extensively in aircraft and computers.

Depending on the mold design, it is possible to mold through or blind holes in the vertical direction of the mold opening and also in other directions if hydraulic or mechanical side cores are used. Washing machine pump housings, for example, have holes running in two directions. The use of side cores is also common in molding thermoset bobbins or coil forms for solenoid coils and other electrical applications.

85.4 ADVANTAGES

85.4.1 Economics

Either compression or transfer molding may prove an ideal process for low volume production and for development or prototype work because the equipment required is less expensive than that for injection molding. Parts up to 10–12 in. (254–305 mm) in any dimension are suited to compression and transfer molding, but larger parts warrant consideration of other processes as well. For very large parts, machinery and mold costs shoot up rapidly. Extremely tiny parts, on the other hand, can often be made in multicavity molds at relatively low cost per part. Molds of several hundred cavities are not uncommon in some applications.

Using the semiautomatic compression or transfer process, parts requiring metal or other molded-in inserts are relatively simple to produce as compared to molding in inserts in fully automatic injection molding.

The low cost of many of the thermoset materials, principally phenolics, permits relatively inexpensive molded parts. The more exotic materials, such as the heavily glass-filled materials or the extremely soft flowing materials, are more costly (up to several dollars per pound), making them less competitive as regards material costs.

When applications call for maximum impact strength and dimensional stability, optimum materials are the fiber-reinforced phenolics or polyesters processed by semiautomatic compression molding. Thermoplastic materials can often withstand greater distortion without failure, but they lack acceptable dimensional stability under stress in many applications. The molding of fiber-reinforced materials with the injection molding process leads to high impact strength; however, compression molding results in less fiber breakage as well as more random fiber distribution and therefore even greater impact strength.

The structural design of some parts may result in some thick sections and some thinner sections. In general, the thickness of the section determines the curing time: Thicker sections require a longer time for cross-linking than thinner sections because of the slower heat transfer into the thick sections. This difference in cross-linking rate may lead to internal stresses in a molded part that has varying thicknesses. Therefore, an ideal design is one that has all wall or section thicknesses approximately the same.

Because the thermoset materials flow at relatively low viscosity, there is often a flash line or gate scar on the molded parts that must be removed for cosmetic purposes. Such finishing of the molded parts may be accomplished by simply tumbling the parts against each other in a rotating tumbling barrel or passing them through an air blast deflasher. This latter device blows a grit material that is softer than the molded plastic and can be directed against the flash and gate areas for flash removal without damage to the molded part. Occasionally, with heavily reinforced plastics, such flash must be removed with a trimming die or by manual filing or breaking.

Finishing costs are often higher with compression molding than with transfer or injection molding, principally because, with the latter two processes, the mold is fully closed before the material is introduced into the cavity. Other finishing costs may include tapping threads in molded holes or possibly drilling and tapping where the holes are not molded in. Occasionally, inserts are pressed into molded-in holes after molding as a secondary operation. Inserting such parts after automatic molding may prove more economical than using molded-in inserts with semiautomatic molding.

85.5 DESIGN

85.5.1 Economics

In selecting compression or transfer molding for a given part, a number of factors must be analyzed. Perhaps most important, after it has been determined that the part lends itself to compression or transfer molding, is the total production volume required over a given period of time. Such information can be reduced to a production rate; for example, a thousand parts per hour, a million parts per year, and so on. Estimating the cure and loading times of the process leads to a calculation of the number of cavities required to achieve the desired production rate. The number of cavities will indicate the size of the mold and the size of the press.

The actual characteristics of the part, as well as the production rate, will permit analysis of semiautomatic molding versus fully automatic molding. Generally speaking, the greatest economies are achieved with fully automatic molding when the volume requirements are high. Whether transfer molding is better than compression molding for a given application is usually not so much an economical determination (although production rates may be higher in a semiautomatic transfer operation) but rather is based on the requirements of the part design.

88.5.2 Part Specifications Affecting Material Selection

Some materials have lower coefficients of thermal expansion than others, with the highly filled materials showing greater stability. Thus, if parts to be molded require extremely close tolerances, glass- or mineral-filled formulations should be used. Tolerances in directions perpendicular to the parting line are better with transfer molding than with compression molding. This is because in transfer molding the mold is closed before the material is introduced. In compression molding, the mold is closing against the material, and a slightly oversize charge may flow between the mold surfaces before they are fully closed, causing wider variations in tolerances perpendicular to the parting line. On the other hand, with transfer molding, the material shrinks after cavity fill, more in the transverse direction of material flow as the cavity is being filled than in the longitudinal direction of flow. For that reason, in directions parallel to the parting line, tolerances are often easier to achieve in compression molding than in transfer molding.

The strength characteristics of the materials—flexural, compression, tensile, and impact—are most affected by the fillers used. For maximum impact strength, compression molding with high fiber filling is utilized. While such materials may also be transfer molded, the flow from transfer pot through runners and gates may cause the fibers to break or segregate and become oriented, resulting in lower mechanical strength.

In selecting materials for compression or transfer molding of a given part, the environment to which the part will be exposed during its life becomes critical. In electrical applications, for example, arc resistance and tracking resistance as well as dielectric strength are all important specifications. Some thermoset materials are better with respect to these characteristics than others: Many phenolics are designed as "electrical grade phenolics." If the part is subject to extreme moisture exposure, materials should be selected that have less tendency to absorb moisture. Higher temperature environments may eliminate some materials, particularly above 250°F (121°C). The silicone molding compounds can withstand temperatures as high as 700°F (371°C) in certain applications.

Another environmental consideration is the loading stress on the part: whether or not it will be subject to continuous bending or compression stress and whether such stress is cyclical or constant. In many instances, this one factor will mandate selection of a thermosetting material and therefore will require the part to be produced by compression or transfer or thermoset injection molding.

85.5.3 Molding Process

Once the part has been designed, the molding process can be considered in some detail. It is important, for example, to ensure that the material will flow in the cavity sufficiently to fill out all of the sections of the part, including bosses, ribs, and other protuberances. Part thickness and shape will determine the approximate curing time. Variations in part thickness lead to uneven curing times and may result in internal stresses. If reinforcing fibers are used, it is important to know that they will be appropriately dispersed throughout all sections of the part. Sections that are too thin may permit resin, but insufficient reinforcing fibers, to flow into the section.

85.6 PROCESSING CRITERIA

85.6.1 Temperature

Temperature is necessary to achieve cross-linking of the plastic and the appropriate viscosity for satisfactory flow of the material into all sections of the cavity. Optimally, the molding compound is heated frequently before it is introduced into the cavity or transfer pot. Such preheating is done either by a high frequency electronic method similar to microwave heating or a mechanical preplasticizing technique similar to extruding with a screw and barrel. Before the charge of molding compound is introduced into the mold, the temperature should be raised to ensure faster flow and curing. The mold temperature should be high enough to effect rapid curing of the part, but where lengthy flows are required (as in multicavity transfer molds with long runners and cavities distant from the transfer pot), too high a mold temperature may result in "precure," or solidification of the molding compound before it reaches the distant cavities. On the other hand, lower mold temperatures will slow down the polymerization and allow adequate flow time for the material (Fig. 85.5).

For optimum properties in a thermoset molded part, a post-curing operation is beneficial. Post-cure may involve holding the molded part in an oven at a temperature somewhat less than mold temperature but well above room temperature for several hours. Several steps with different temperature levels and times may be required.

85.6.2 Pressure

Pressure is necessary to ensure that the material flows into all parts of the cavity and that the mold stays firmly closed during the polymerization or cure. In a compression press, the tonnage force needed to close the mold and hold it closed can be calculated from the projected area of the cavities and the characteristics of the molding compound selected. In transfer molding, the mold must still be held closed against the force of material entering the cavities, but the more critical pressure is the pressure on the transfer plunger. It must be adequate to force the material through the runners and gates into the cavity and hold the material under sufficient pressure during polymerization. Some secondary effects of pressure include prevention of sink areas in the molded part and possible absorption of minimal quantities of air into the molded part in areas where cavity venting is inadequate or difficult.

85.6.3 Speed

In a compression press, speed control is required on the closing stroke. In general, high speeds are desirable to shorten overall cycles. It is important, however, to have a controlled

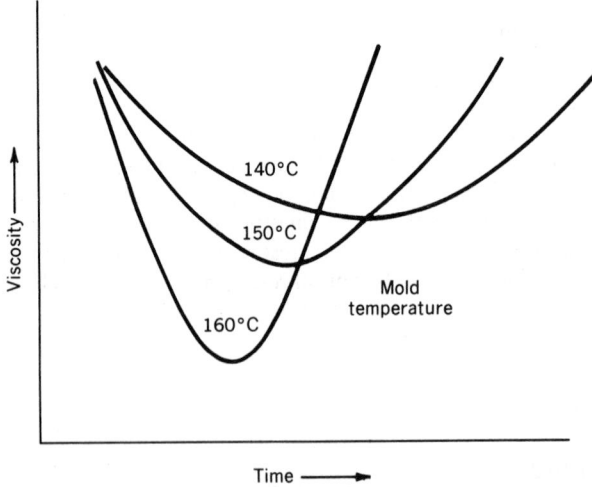

Figure 85.5. Higher mold temperatures shorten the flow time in the mold and accelerate cure.

slow down of the press closing stroke just before metal-to-metal contact of the mold so as to minimize damage to the mold and allow the material to flow to all parts of the cavity.

This is also true of transfer molding, but equally important is the speed of the transfer plunger as it descends into the transfer pot and forces the material from the pot to the cavities. In transfer presses, the plunger advances rapidly until it contacts the molding compound and then it is slowed automatically to a controlled rate to ensure optimum flow. If the flow is too rapid, excessive frictional heat may be generated in the molding compound as it flows through runners and gates, leading to precure. If the flow is too slow, the material may not absorb enough heat from the mold to reach the most distant part of the cavities; in this case, it cures while flowing. Modern controls enable plunger movement to be programmed with respect to velocity and pressure and also serve to minimize flash.

85.6.4 Material Flow Rating

Materials used in compression or transfer molding are available in several rates of flow. While the classification of flow is somewhat arbitrary, higher flow numbers indicate softer and longer flowing materials; lower flow numbers indicate stiffer materials. Softer flow materials usually require slightly longer cure times than stiff materials. After the mold is constructed, it may be necessary to try several different flows of the same formulation to determine the optimum flow rating.

85.6.5 Ejection Speed and Force

When the mold is opened, parts are ejected from the cavities or stripped off the male part of the cavity by ejector pins that push against the molded plastic. Because at the moment of mold opening thermoset plastics are not necessarily as rigid as they will be after cooling, too much force on an ejector pin may distort the molded plastic. Most modern presses have provisions for controlling both the speed of ejector pin travel and its force to minimize the possibility of distortion or breakage.

85.6.6 Vacuum Venting (Fig. 85.6)

As material flows into a cavity, particularly in transfer molding, air already in the cavity must be forced out through small vents machined in the cavity at the parting line. These small grooves must be small enough to prevent molding compound from flowing out of the cavity profusely. With some cavity configurations, however, venting is difficult through conventional means. Vacuum venting is a modification of the molding cycle to enable a vacuum pump to draw air out of the mold cavities while the molding compound is flowing down the runners and through the gates into the cavities. Vacuum venting requires certain construction techniques in the mold and necessary auxiliary vacuum equipment and controls, but the process of vacuum venting has been sufficiently developed that it can be adapted to most tranfer molding (and injection molding) applications.

85.6.7 Breathing Cycle

When some thermoset compounds experience heat before and during cure, they release volatiles that may be merely moisture and air in the material but also, in the case of many phenolics, may be condensation products, particularly water. As these volatiles develop, they cause increased pressure in the cavity. For proper cure, they must be allowed to escape. With compression molding of such materials, a breathing cycling may be necessary for releasing the volatiles. In operation, the mold is closed, a few seconds are allowed for heat transfer to the material, and the mold is opened [about 1/8 in. (3.2 mm)] for one or two seconds and then closed for the balance of the cure cycle.

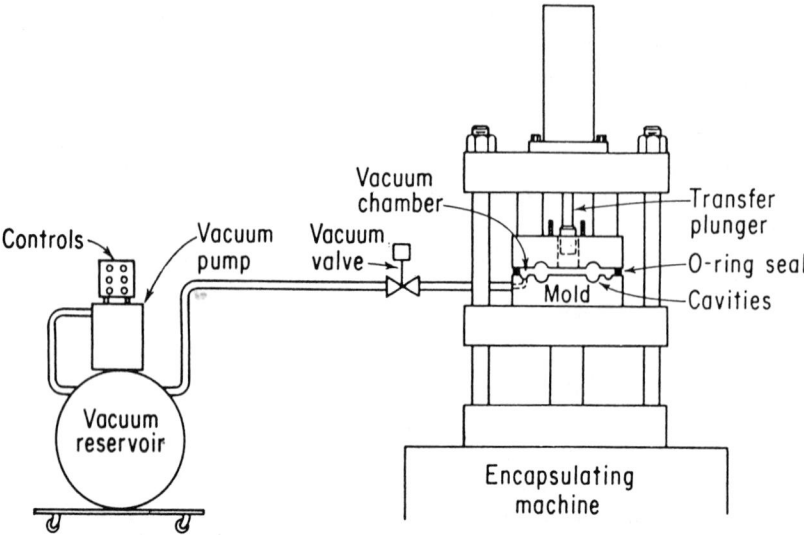

Figure 85.6. Vacuum venting. The vacuum line is connected to the manifold reaching all cavity vents. After the mold is closed and while the transfer plunger is forcing material into runners and cavities, the solenoid valve opens, drawing cavity air into the vacuum reservoir. The valve closes when the cavity is full.

85.7 MACHINERY

Compression and transfer molding machines have evolved into sophisticated equipment with complex hydraulics and mechanical features, electrical or electronic control systems, heating and cooling systems with precision controls, and other features designed to improve the molding process.

85.7.1 Specification (Fig. 85.7)

Probably the most common specification of compression and transfer molding machines is the tonnage of clamping force. The tonnage basically determines the projected area of cavity

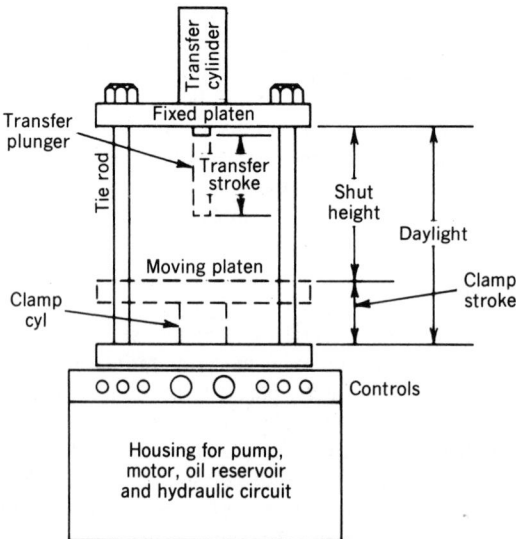

Figure 85.7. Typical specifications and terminology of compression and transfer presses.

or cavities that can be successfully molded with a particular compound requiring certain pressures for molding. Small presses are often in the 5–10 ton (41–98 kN) clamp capacity; large presses, up to 1000 or more tons (9800 kN).

The die space is another critical specification. It establishes the maximum width and length of a mold that will fit into a given press. Molds are generally rectangular in configuration; as the clamp tonnage increases, the die space increases proportionally.

Most compression and transfer presses are upward pressing; that is, the bottom half of the mold moves upward to a fixed top half of the mold for closing. However, some automatic presses are downward pressing; for example, many of the very large compression and transfer presses (over 500 tons (4900 kN) clamping capacity). This is desirable so that the operator will have easy access to the bottom half of the mold. From a molding standpoint, the success of the molding operation is rarely dependent on whether a press is upward or downward pressing.

Most presses have a single opening and accommodate a single mold. Occasionally, a press will have multiple openings for multiple molds. In melamine dinnerware molding, particularly for flat plates and trays, multiple opening presses are commonly used.

The term "daylight" refers to the open height of the press: the vertical distance between the bottom press platen and the top press platen when the press is open. The platen is the flat surface against which a mold half is mounted. If particularly deep parts are to be molded having a long dimension perpendicular to the parting line of the mold, the mold itself will be deep and considerable mold opening is required to ensure that the ejected parts will not come up against the other half of the mold. Such molds obviously require greater daylight than molds for thinner parts.

The shut height dimension for a molding machine is the vertical distance between platens when the press is fully closed. Even when it is fully closed, a press without a mold will have a shut height of perhaps 1 ft (30 cm) to several feet, depending on the tonnage.

The clamp stroke is the full travel of the moving platen during the press closing stroke. The amount of travel should be sufficient so that when the mold is open, molded parts can be easily removed from the cavities and an operator can conveniently inspect the surfaces and cavities of the open mold prior to the next cycle.

The transfer stroke is the distance traveled by the transfer plunger from full up to full down. In its downward position, the transfer plunger should protrude slightly below the surface of the top half of the mold so that the cull or residual material is fully pushed out of the transfer pot after each cycle. It should travel far enough upward that an operator has ample room to deposit the molding charge in the transfer pot at the beginning of each cycle. Pot diameters and pot depths will vary from mold to mold depending on the total size of charge needed for the particular mold. The transfer becomes particularly important when large charges are required.

Most compression and transfer presses now used are of the four post configuration; that is, having four strong vertical tie rods, one in each corner of the platen. These tie rods (or strain rods) are capable of tensile strength corresponding to the tonnage of the press. Occasionally, smaller compression or transfer presses use the two post configuration.

In "slab side" presses, the post or posts on the left and the post or posts on the right are replaced by a steel plate that takes the stress during the clamp cycle. One obvious disadvantage of the slab side press is inaccessibility to the mold from the left or the right side.

A C-frame press has, in effect, only one post and the stress of mold clamping is accommodated in a cantilever manner by the configuration of the press. C-frame presses are used where considerable access is needed to the mold area, possibly for automatically conveying inserts into the mold area and removing molded parts after molding.

The rotary press is another type of press used for certain types of automatic molding. It accommodates a number of smaller molds on a large platen and rotates the molds successively past a loading and unloading area. The rotary action of the table also effects the opening and closing of the molds, often by mechanical cam action or individual hydraulic cylinders.

85.7.2 Mold Closing Mechanism

Several techniques are used to move the platens together and apart. Most common is the straight ram, whereby a piston or occasionally several pistons are actuated with hydraulic fluid or air pressure to provide travel and force for the closing. Toggle presses (Fig. 85.8) utilize a much smaller cylinder and a mechanical linkage, or toggle system, to give a mechanical advantage to the force and travel of the piston in the cylinder. When the mold is fully closed, a considerably greater force is exerted on the mold halves than the smaller hydraulic or pneumatic cylinder could achieve by itself. In compression molding, maximum tonnage is not achieved with a toggle press until the mold is fully closed and may not be adequate when the mold is closing against the material at some distance from the fully closed position. On the plus side, the toggle press is extremely fast during the closing stroke and is particularly suited to transfer molding, where no force is needed until after the mold is fully closed.

On rotary presses, the closing and opening strokes may be affected by mechanical cams with extremely stiff springs that prevent damage to the mold when it is closed against some obstacle in the mold. Another type of press uses a lead screw for a mechanical closing and opening stroke. Screw presses are not in common usage except in unusual applications where this technique is optimum, as, for example, when particularly long cycles are required. In such presses, once the lead screw has been advanced to close the mold, power can be shut off and the clamp force will be held during cure without further use of power.

85.7.3 Power Unit

Hydraulic fluid is used to actuate the pistons most commonly used. Therefore, a press power unit consists of an electric motor, one or more hydraulic pumps, and appropriate valves and controls to ensure speedy and smooth operation. In older plants with many presses, a central accumulator is utilized to hold large amounts of hydraulic fluid under pressure. Valving

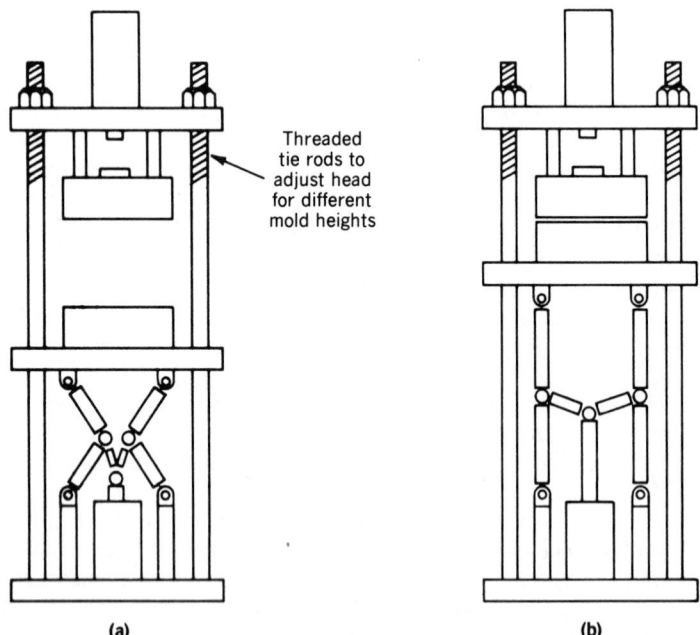

Threaded tie rods to adjust head for different mold heights

(a)   (b)

Figure 85.8. Toggle press: (**a**) open and (**b**) closed. Threaded tie rods are used to adjust the head for different mold heights.

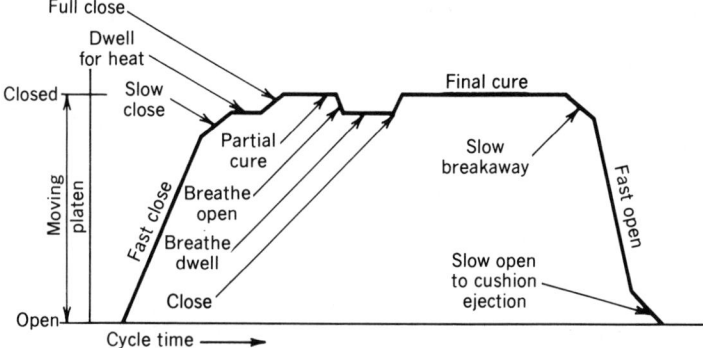

Figure 85.9. Typical compression cycle.

leads to each press to provide the necessary flow to the press cylinders. Such a central accumulator arrangement may provide economies in the initial installation for a large molding plant, but it also may introduce inaccuracies in the cycling times of high-speed presses, causing lower yields of satisfactory molded parts. Very small laboratory presses may be manually actuated much like a small hydraulic car jack.

85.7.4 Ejection

At the end of the molding cycle, parts must be ejected from the mold. The mold itself incorporates ejector pins and an appropriate ejector plate. The press actuates the ejector plate or plates. Frequently, the actuation is mechanical; it occurs during the final 0.5–1 in. (13–24 mm) of press opening stroke, when the rods actuating the ejector plates stop traveling and the mold continues its opening travel for that final fraction of an in (mm). If the press opening stroke has adjustable speed, then the speed of ejection can be controlled. For greater flexibility and greater control of both speed and ejection force, hydraulic ejection systems are utilized that are merely small cylinders and pistons built into press platens, actuated after mold opening. Press ejection systems may be top, bottom, or both top and bottom.

85.7.5 Motion and Cycle Controls (Figs. 85.9–85.10)

Early compression and transfer presses were usually opened and closed by the operator moving a lever on a manual hydraulic valve. Most present-day machines are equipped with

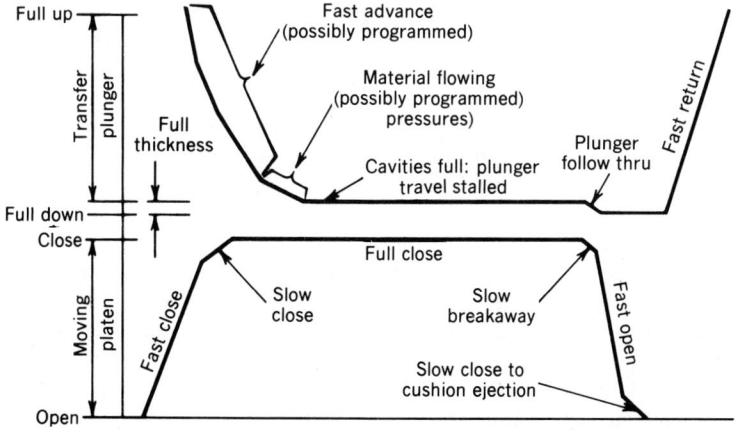

Figure 85.10. Typical transfer cycle.

solenoid-operated valves that are actuated either by a pushbutton or by the automatic control system of the press. Sequencing of such controls may be through electromechanical timers and relays, or, as on most modern presses, through solid state logic circuits and microprocessor controls. Such sophisticated control systems still send the electrical signals to the solenoid valves for closing and opening the press, adjusting pressures and speeds, and so on, according to the program.

During the clamp closing stroke, a fast approach is desirable followed by a slowing down to minimize mold damage during metal-to-metal contact. The fast and slow close is accomplished at a low pressure. High pressure is not brought to bear on the clamp platen until the machine senses that there is no unwanted object between the mold surfaces. "Soft close" is the terminology used to indicate that the mold will close under very low pressure until it has reached a point where high pressure can safely be applied without damaging the mold. After cure, the initial opening of the mold, referred to as "break away," is at a slow speed to avoid damage to the less than rigid part during this critical stage. Once the cavity is opened, the moving platen can proceed at a fast rate until the mold is fully opened. A press should have a pressure regulator to enable the maximum clamp pressure to be lowered from its rated value to lower values, particularly when small molds are used on a large press.

Movement of the transfer ram involves similar considerations. When the transfer plunger is beginning its descent, and before it encounters the molding compound, it is generally moving rapidly for a fast approach. At the time it contacts the material, a speed control is imposed to ensure that the ram will not cause the fluid material to flow too rapidly through runners and gates into the cavities. An accurate pressure control is also advisable at this stage of the transfer plunger movement so that adequate but not excessive pressure is exerted on the material. Sophisticated programmable controls enable stepped velocities and pressures during plunger travel. After cure, many presses allow the operator to preset whether the transfer ram will continue to descend slightly during the mold break away so that the excess material (or cull) is pushed out of the transfer pot before the plunger returns. Alternatively, the operator can preset whether the transfer plunger will move out of the pot prior to the break away. The actual return stroke of the transfer ram is at high speed.

For breathing a transfer mold, the transfer ram must be programmed to relieve all pressure before the mold breathe opening, to prevent the uncured material from flowing out as flash between the opened mold halves. Pressure to the transfer plunger is then reapplied after mold closing at end of the breathe.

85.7.6 Mold Heating

When molding thermoset compounds by compression or transfer methods, the mold is maintained at a constant temperature set for optimum polymerization of the material each cycle. Such temperatures range from 300–400°F (149–204°C). For optimum molding operations, temperature must be uniform across the surface of the mold and in the cavity areas, ideally to +/− 2°F (1.1°C). At present, electric heating is the most prevalent technique, utilizing multiple electric heating cartridges inserted in both the mold top half and bottom half, positioned to supply heat to all cavity areas. On larger molds, temperature controllers and sensing elements are often utilized in several zones.

An earlier method for mold heating involved the use of saturated steam channeled around the cavities. Steam has the advantage of rapid temperature recovery because of its tendency to condense in the steam channels when any lower temperature occurs, rapidly releasing the heat of condensation. But steam is hard to contain in a mold and in the vapor lines leading to and from the mold. It requires excessively high pressures when higher mold temperatures are required.

By contrast, circulating hydraulic oil can bring mold temperatures easily to 400°F (204°C) or higher. Self-contained oil heating and cooling systems are available that make this type of mold heating practical, although the cost of fluid-heated molds is generally higher than cost of electric cartridge-heated molds. Some molding compounds and molding cycles may require one temperature for cure but a lower temperature for optimum ejection. Such molds have provisions for both heating and chilling, and require longer molding cycles because of the time needed for heat transfer. They may have electric heaters for heating and water channels for cooling, or may utilize the circulating hydraulic oil system with programmed heating and chilling.

85.7.7 Fully Automatic Features

Fully automatic presses are capable of unattended operation, once they are appropriately set for the desired cycle. In addition to the normal opening and closing and ejecting actions of the press, fully automatic machines load the material and remove the molded parts each cycle.

For material loading, the molder often has the choice of loading preforms (compact tablets containing the exact amoung of molding material needed for a given cavity or charge) or granular powder. Granular powder loading, volumetrically metered, is most common in automatic compression machines. The machines include a hopper for holding a large amount of bulk material and a means of releasing from the hopper the correct amount of powder. The powder is either fed by gravity directly into mold cavities or into small loading cups that are mechanically moved into the mold area where they drop their charges of material.

To speed up the cycle and minimize mold wear, automatic molding machines often preheat the charges before they are delivered to the cavities. Preheating can be done by simply moving the charges into an area of infrared lamps for a programmed length of time before the charges are carried further into the mold area or they may be exposed to high-frequency preheating for more rapid and more uniform heating before being placed in the cavities.

Preheating may also be accomplished by a plasticizing screw and barrel system that delivers a slug of preplasticized material, which must then be mechanically transported into the cavity area.

Molded parts are frequently unloaded with a stripper comb that is brought in underneath the molded parts (which at this point in the cycle are resting on ejector pins that hold the parts an inch or more above the surface of the bottom mold half). The stripper comb is positioned completely beneath all the molded parts, the ejector pins are retracted, and the molded parts are left on the stripper plate. The plate then retracts out of the mold area and tilts so that the molded parts slide off the stripper comb onto a conveyor belt or into a barrel. The unloading mechanism usually includes a blow-off provision directing compressed air by nozzles against the entire top and bottom surfaces of the mold to remove any flash, dust, or contaminant before material is loaded for the next shot.

85.7.8 Costs

Table 85.2 gives approximate prices of molding machines and molds.

Operating costs of molding presses include electric power for the motor and the mold heaters. Both of these require their rated supply, but the actual current consumption is less than continuous because the mold heaters are turned on and off as the mold temperature controllers demand, and the motor for moving the clamp and transfer circuits is drawing less than rated power for part of each cycle. Floor space, insurance, and the cost of an operator or setup man are other costs that can be readily calculated. Machines are often depreciated over a two-year period. Maintenance costs of machines and molds are often estimated at 5% of the initial cost per year.

In a new installation, transportation costs as well as costs for rigging the machinery into place and tying in the electric power, cooling water, and air supply must be reckoned.

TABLE 85.2.   Approximate Prices of Typical Thermosetting Molding Equipment

Characteristics	Price, $
Molds	
Hand, single, small-cavity (dia 1 in.) (25 mm)	3,000
Hand, single, large-cavity (dia 4 in.) (102 mm)	4,500
Production, four-cavity (dia 3 in.) part (76 mm)	20,000
Production, fifty-cavity (dia 1 in.) part (25 mm)	50,000
Automatic preformers	
20 ton (2 in.) (51 mm)	20,000
35 ton (3 in.) (76 mm)	40,000
75 ton (4 in.) (102 mm)	50,000
Electronic preheaters	
1 kW	4,000
2 kW	5,000
5 kW	7,000
10 kW	10,000

Molding presses[b,c]

	10–20 tons	25–30 tons	30–75 tons	100–150 tons	200–300 tons	300–600 tons
Manual compression	2,000	6,000	10,000	20,000	30,000	50,000
Manual transfer	2,500	8,000	15,000	30,000	40,000	60,000
Semiautomatic compression	15,000	20,000	35,000	45,000	65,000	90,000
Semiautomatic transfer	20,000	25,000	40,000	55,000	75,000	100,000
Fully automatic compression	20,000	40,000	75,000	100,000	125,000	160,000
Fully automatic injection	20,000	40,000	60,000	100,000	150,000	175,000

[a] Prices as of 1984.

[b] Molds are either transfer or compression types, of hardened steel, highly polished, and chrome plated. The complexity of the part, the tolerances required, etc., affect the price of the mold. These prices assume relatively simple parts with standard right-angle corners, spherical surfaces, etc.

[c] Conversion factor, 1 ton-force = 0.81 kN.

85.8   TOOLING

Molds for compression and transfer molding are precision machined and may cost as much as the molding machine itself. This is particularly true for production molds.

85.8.1   Compression Molds

In their design and operation, compression molds fall into several categories. The simplest is a flash-type mold (Fig. 85.11) that allows any excess material in the cavity to flow out across the parting line. This type requires that an operator clean all flash from the mold surfaces before each shot. A fully positive compression mold (Fig. 85.12) requires great accuracy in charge size because when the mold cavity closes, there is no path for any excess material to escape. As a result, the full force of the clamp is imposed on the cavities to ensure maximum density of the molded parts. A semipositive compression mold permits excess material to be forced out through overflow vents in the cavity, restricting inflow to optimize back pressure and therefore density of the molded parts. In semipositive molds, the excess material cannot flow across the parting line as in the flash-type mold.

Hand molds are not bolted into the press; they are removed from the press each cycle to a work table where parts are removed. They are then closed and replaced in the press

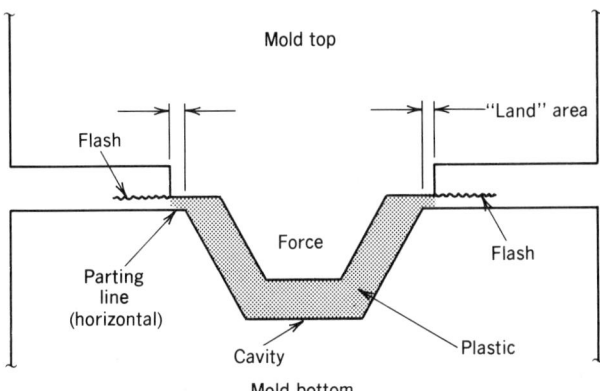

Figure 85.11. Flash-type compression mold.

after the next charge of material has been put in the cavities. Hand molds are best used for prototype or limited production work. Production molds are mounted in the press and left there for relatively long production runs (weeks or longer). They are self-contained with heating cartridges. Hand molds rest on heating platens mounted in the press.

85.8.2 Transfer Molds

Most transfer molds in use today are the plunger type with the plunger actuated by the hydraulic system of the press. Transfer molds also include both hand molds and production molds. Hand molds (Fig. 85.13) are removed from the press each cycle for unloading the molded parts and possibly loading inserts, after which the mold is closed and replaced on heating platens in the press for the next cycle. The top plate of the hand mold includes a sprue hold through which material from the transfer pot feeds into the mold cavities. Production molds (Fig. 85.14) are mounted in the press and are completely self-contained. If inserts are to be loaded, an operator places them directly in the open mold cavities or transports them to the mold using a loading frame that may hold a number of inserts for multicavity molds. When loading frames are utilized, they frequently are designed to remain in the mold during the molding cycle and yet not experience any pressure or force from the molding operation itself.

85.8.3 Mold Construction

The critical parts of a mold are the cavities, which must be precisely designed and manufactured so that the molded part will meet all dimensional specifications. For multicavity molds, hobbed cavities are frequently used. In the hobbing process, an impression is made from a master form into steel under very high pressure. Hobbed cavities are identical to each other in shape because of the method of production, and the cavities themselves have a hard surface resulting from the working of the steel during the cavity forming or hobbing process. In order to further harden a soft low-alloy steel after hobbing, carburizing or nitriding to a depth of 0.030 in. (0.8 mm) or more is done, followed by quenching.

Cavities may also be made by a process referred to as spark erosion machining or electric discharge machining (EDM). This process might be described as a "reverse welding" process. In effect, an electrode is brought into the vicinity of the cavity block. Under controlled voltage, particles of the tool steel leave the cavity block and flow toward the electrode. The process is performed in a constantly circulating oil bath to carry away the fine particles of steel and prevent their shorting the circuit between the electrode and the steel being eroded.

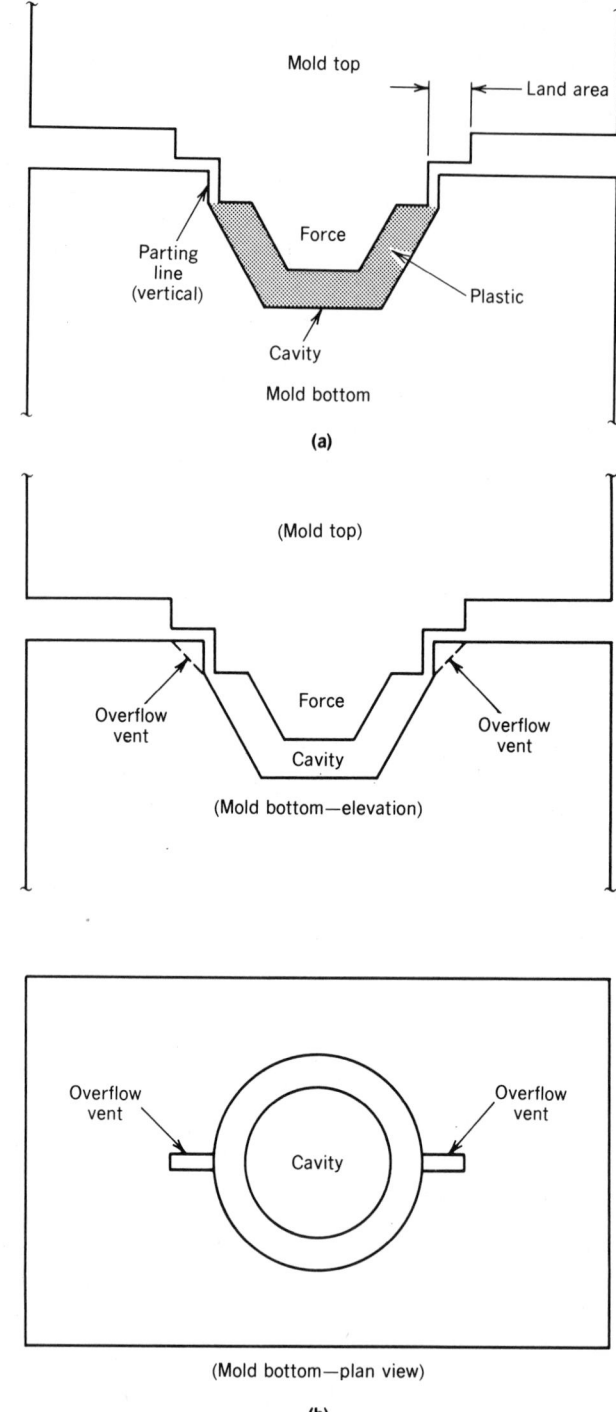

Figure 85.12.   (a) Fully positive compression mold. (b) Semipositive compression mold.

Figure 85.13. Hand mold.

If the electrode is shaped like the molded part, the process will yield a cavity with essentially the same shape. EDM has become a very common process for mold making. A third common process is machining of cavities, which often requires machining several precise cavity components that are subsequently assembled with each part being locked into place to produce the cavity. Cavities may need polishing after machining operations to achieve a smooth surface and ready release of the molded part. If ejector pins will be in the cavity, ejector pinholes must be machined after the hobbing or spark erosion or machining.

Production molds usually incorporate their own heating provisions, often electric cartridge heaters but sometimes channels for circulating oil or steam. Hand molds usually do not have self-contained heating provisions, but are heated by being placed on heating platens are mounted in the press. Heat lost when hand molds are removed from the press for unloading and loading before the next cycle is minimal. If the mold must be out of the press for any length of time between cycles, it is placed on a temperature-controlled heating plate outside the mold.

Most steels used in mold manufacturing are called tool steels and are capable of being hardened to Rockwell C56 or higher. The heat treatment may be case hardening, affecting only the surface, or it may be through hardening, which is generally preferable, Hobbing steels are not capable of heat treatment to those levels. Some tool steels contain up to 18% chromium and exhibit less tendency for corrosion. They have proven ideal for transfer

Figure 85.14. Production mold.

molding, particularly with softer flowing encapsulation compounds. Steels may be heat treated prior to machining if spark erosion or grinding is used. For final surface protection, flash chrome plating to a maximum thickness of 0.0005 in (0.013 mm) is often utilized. Chrome plating is done after machining and polishing of the cavity. The chrome plating gives an extremely hard surface to withstand the erosive effect of most thermosetting compounds.

In molds with only a few cavities and where molded-in inserts are required, the inserts are generally put in place by the operator, cavity by cavity. But in multicavity production molds, where there may be hundreds of cavities, loading inserts one by one becomes tedious, possibly dangerous to the operator, and limits productivity. For such multicavity molds with inserts, insert loading frames or carriers enable the inserts to be loaded into one frame on a work table during the cure of the previous cycle and then transported for a given shot with one movement by the operator. Two such loading frames are used: one stays in the mold during cure while the other one is unloaded and reloaded. Many ingenius designs can be found for appropriately handling inserts in production operations.

Occasionally a given part requires special features to be constructed in the molds, such as side cores. Side cores are used to mold holes, grooves, or undercuts in the molded part in a direction other than perpendicular to the parting line. In general, the mold is closed and a section of the cavity is moved into the cavity before the material is added. After molding, the side core must be retracted before the mold can be opened. When the molded part is finally ejected, no material has been molded in the area where the side core was positioned. Side cores may be cam actuated by the opening and closing stroke of the mold or hydraulically actuated by small hydraulic cylinders mounted on the mold and cycled in or out at the appropriate times during the molding cycle. When cam actuated or hydraulic actuated side cores are used, it is wise to have safety interlocks to prevent molds from being closed or opened when the cams or side cores are not in the correct position. Interlocks protect the mold and the side core.

Another special feature is vacuum venting of the cavities. This is not normally required for compression molding, where venting is usually adequate, but may be needed in transfer molding where the mold is closed before the material is introduced. In brief, vacuum-vented molds are comprised of vents in the cavities leading to a vacuum chamber. In addition to the vent area itself, a vacuum seal consisting of a high-temperature silicone rubber O-ring is placed around the entire parting surface of the mold.

If inserts are molded in, they must be adequately supported in the cavity when the mold is closed so that they are not displaced by the influx of molding material. Inserts are pushed into holes or slots in the cavity that satisfactorily support them. In some instances, inserts are flexible or are supported by weak extensions such as soft lead wires on a resistor (which are not rigid enough to prevent the part from shifting in the cavity during the molding operation). To make certain the inserts are positioned correctly, retractable positioning pins may be incorporated in a mold (Fig. 85.15). They are the ejector pins and are actuated so

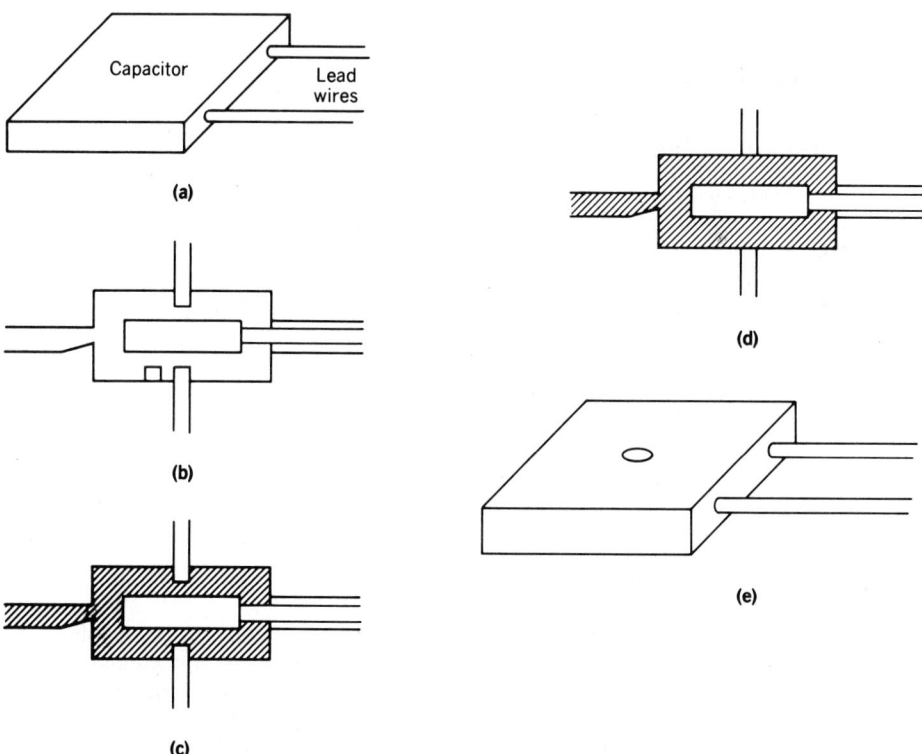

Figure 85.15. Retractable positioning pins. (**a**) Insert and lead wires (for example, a ceramic capacitor) before plastic is molded around it. (**b**) Ejector pins extended before cavity fill. (**c**) Pins restraining insert from moving up or down while material fills the cavity. (**d**) Cavity after filling but before the material hardens—the pins retract, allowing more material to fill the void left by the retracted pins. (**e**) Insert after molding, showing the flush-ejector pin scar.

that, during the initial introduction of material into the mold cavity, the pins project into the cavity to restrain any shifting of the inserts. Once the cavity is essentially filled with the molding material, but prior to polymerization, the pins are automatically retracted until their ends are flush with the surface of the cavity. Because there is still residual fluid material in the transfer pot and the transfer plunger is still under pressure, further flow of material into the cavity fills the voids left by retraction of the positioning pins. Inasmuch as the insert is well surrounded by molding compound at this stage, there is no further displacement. The molded part is cured with the insert in the desired position. Retractable positioning pins can be designed into a mold without adding appreciably to the cost. They are moved by small hydraulic cylinders mounted on the outside of the mold and actuated by a timer.

85.8.4 Mold Costs

Mold costs are hard to estimate. The cost of a production mold may well exceed the cost of the production press into which it is mounted. For example, a 299 ton (2900 kN) transfer machine for transfer molding may cost $70,000; a multicavity production mold for molded-in inserts (possibly a 240 cavity integrated circuit mold), $80,000. By contrast, small hand molds may be made very inexpensively from cold-rolled steel or even aluminum for less that $1,000. However, a multicavity hand mold made of through-hardened tool steel, highly polished and hard chrome plated, can cost $10,000 or more. In short, a mold is definitely capital equipment whose cost must be scrutinized in evaluating the economics of any molding program.

85.9 AUXILIARY EQUIPMENT

85.9.1 Preformer (Fig. 85.16)

Molding compounds usually come from the supplier in granular form at room temperature. A peformer is used to prepare compact tablets of molding compound of the desired weight for each cycle. Preforms are particularly useful in production compression molding or transfer molding with high frequency preheating. Preforms are easily transferred by an operator from the preformer to the preheater and then to the cavities or transfer pot.

The metering and compacting operation is done automatically at room temperature using a simple punch and die system to produce a cylindrical shape. Weight adjustments are easily made on the preformer. Some are designed for beside-the-press operation and may deliver preforms at the rate of 8 or 10 per minute. More expensive preformers will make large batches of preforms at a rate of 25 or 30 preforms per minute. High production preformers may serve a number of molded machines; a beside-the-press preformer can serve only one or two machines.

85.9.2 Preheater/Preplasticizer

Heating the charge of material before placing it into the cavities or transfer pot shortens cure, reduces erosion of cavity surfaces, and often improves the quality of the molded part. The most common preheating for the past several decades has been high-frequency preheating (Fig. 85.17). This device incorporates a machine containing electrodes; the preform is placed by an operator between the electrodes. When the preheater is actuated, polarity on the electrodes changes at 100 MHz, causing, in effect, each molecule in the preform to reposition itself at that frequency. Frictional heat raises the temperature of the preform from room temperature to perhaps 180°F (82°C) in 10 to 20 seconds, quite uniformly. The preform is then ready for transport to the cavities or transfer pot. At that temperature, with that rapid heating, the preform still holds its shape although it is soft and pliable if squeezed.

Figure 85.16. Preformer.

In many high-frequency preheaters, the cylindrical preform is rested on two horizontal rollers that rotate during the preheating. The rollers serve as the bottom electrodes, and the top electrode is a plate above the preform. Rotation during preheating ensures even greater uniformity of heat in the preform and therefore more consistent cycling in the press.

In recent years, screw preplasticizing of thermoset materials has become popular, particularly with phenolic molding compounds (Fig. 85.18). Preplasticizing is done beside the press; the preplasticized material is delivered from the extruder each cycle. Granular material is loaded into the hopper and the extruder is actuated by the operator. A screw augers the material from the hopper forward into the barrel. The frictional heat of screw rotation plus the heat in the jacket of the barrel brings the plastic up to a temperature of about 180°F (82°C) in less than 30 seconds. As the screw rotates and augers the material forward, it is reciprocated backward away from the nozzle of the extruder until the appropriate charge size has accumulated in front of it. At this point, screw rotation stops. When the operator is ready for the shot, he or she actuates the appropriate pushbutton and the screw advances forward, pushing the charge of plasticized material out the nozzle. The operator takes this charge and transports it to the cavities or the transfer pot. Obviously, with the screw preplasticizer, no preformer is needed.

A third method of preheating, sometimes used on automatic compression presses, is infrared preheating. In this arrangement, the molding compound is automatically distributed into a number of cups corresponding to the number of cavities in the mold. The plate holding these cups is moved under infrared heating lamps where the material is allowed to remain

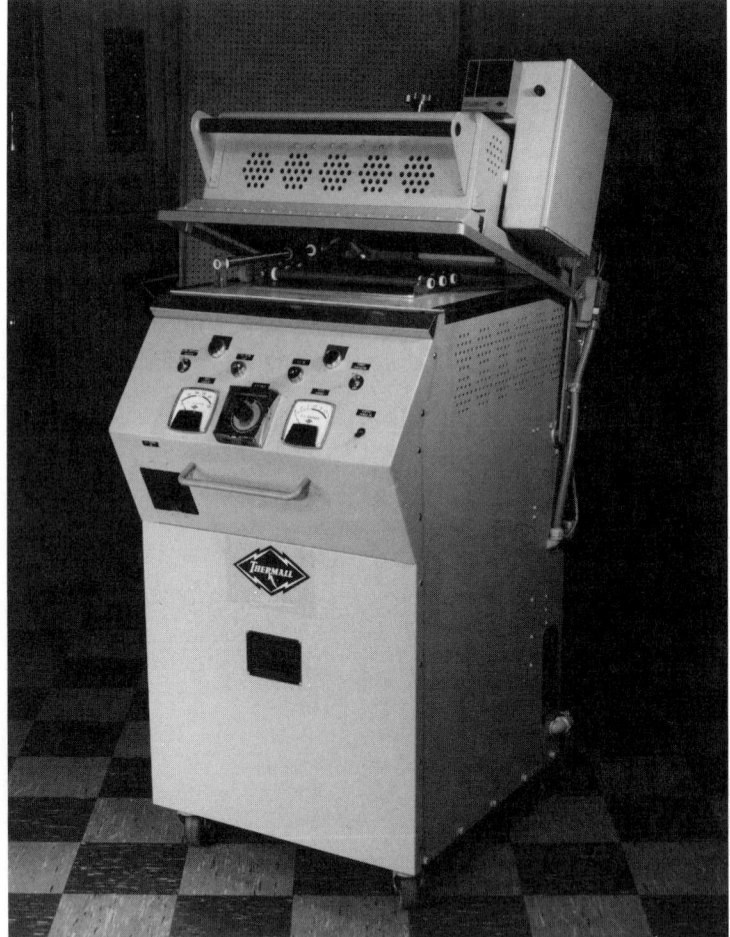

Figure 85.17.   High-frequency preheater (Courtesy of W.T. LaRose & Associates).

for the desired length of time before being transported into the mold. Infrared heating is slower than high frequency preheating and is not uniform throughout the charge, but, nevertheless, considerably speeds up the cycle and lessens mold wear.

85.9.3   Shrink Fixtures

After a molded part is removed from the cavity, it is still hot and the material is not fully rigid. Any internal stresses in the material may therefore cause the shape of the part to change while it is cooling. Where close tolerances are required, and especially with thin-sectioned of parts, dimensional accuracy is achieved by placing the hot molded part on a fixture near the press that will hold it until it has cooled. The fixture may be a spindle that is placed in a molded hole so that the hole dimension remains fixed; it may be merely a clamping mechanism to hold the molded parts flat against a cold, flat surface; or it may be any other restraining arrangement to enable the part to cool in the desired shape. Once cool, the part can be removed from the fixture with no further shrinkage or tolerance change.

Figure 85.18. Reciprocating screw preplasticizer for thermosetting plastics.

85.9.4 Post-Cure Oven

To obtain optimum physical properties, a more gradual cooling of the part is necessary. Post-curing cycles may involve one or more temperature levels to which the parts are cooled and held for a period of time before further cooling. Conveyorized ovens or sometimes batch ovens are located near the press for such cooling. The operator checks that the part advances to the desired temperatures during the slow cool.

85.9.5 Deflashing/Tumbling (Fig. 85.19)

After parts are removed from the mold, some flash may be present at the parting line or in other areas. Once the part is cooled, the flash becomes quite brittle and is removed as a secondary operation. The simplest method of flash removal is to place the molded parts in a slowly rotating tumbling barrel and allow them to be tumbled against each other for five or ten minutes. For many parts, such tumbling removes the unwanted flash and incurs minimal cost. On other parts, particularly those with complex configurations and possibly with inserts molded in, flash removal may require a precisely directed air blast of a grit material against certain sections of the molded part. Such air blast deflashers may be conveyorized automatic systems that transport the part past fixed nozzles, or they may be simpler arrangements where an operator wearing heavy rubber gloves holds a part under one or more fixed nozzles and moves the part so that the air blast is directed to appropriate areas. The grit used in an air blast system is often an organic media, such as ground walnut shells or apricot pits, or may be a polymeric material, such as small pellets of nylon, polycarbonate, and so on. Modern deflashers recycle the media and prove highly effective at efficiently removing unwanted flash.

Figure 85.19. Air pressure blast deflasher. Courtesy of Finmac Inc.

85.9.6 Work Loading Stations (Fig. 85.20)

In a production operation, using self-contained production molds or hand molds, the efficiency of the process is enhanced if appropriate provisions are made for carrying out each
step in the operation. Work loading stations are sometimes mounted right on the press so
that the operator can have at his fingertips the necessary tools for cleaning molds, removing
hand molds, ejecting parts, sliding the hand molds back into the press, etc. Considerable
thought should be given to practical means of shortening the cycle and reducing operator
effort.

85.9.7 Safety Provisions

All machinery and equipment involved in the molding process must be designed, manufactured, and operated for safety. The principal safety consideration is, of course, for the
operator and any persons who may be near the machine. A secondary safety consideration
is mold protection.

Operators should never place parts of their body between "pinch" areas. If it is necessary
to load inserts into cavities, long handled tongs or loading frams should be used (Fig. 85.18).
When the operator is blowing flash from a mold, the air nozzle should have a sufficiently
long enough extension that it is not necessary to put his/her hand between the mold surfaces.
Hot molds should have some thermal insulation in areas where an operator might frequently
touch the mold. If flash is blown from a mold, the particles may be sharp and should be
directed or collected in such a way that they will not enter the eyes or nostrils of anyone in
the vicinity.

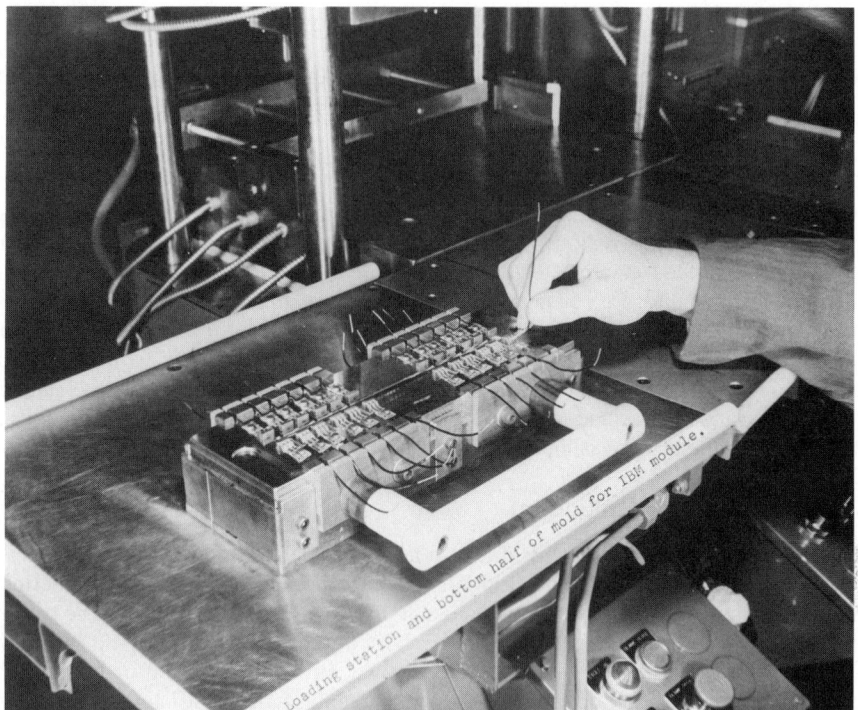

Figure 85.20. Work loading station.

Modern semiautomatic presses are equipped with "coincidental" pushbuttons. These require that operators use two hands to actuate the press close; that their hands push the appropriate control buttons simultaneously; and that pressure on the control buttons be maintained until the "dangerous" time is over. For example, in mold closing, an operator must push two buttons (more than 8 in. (20 cm) apart) simultaneously and hold the buttons until the press is completely closed. Should either button be released before the press is completely closed, the press motion will cease until the two buttons are pressed again. The same arrangement applies to the transfer plunger travel.

Mold protection involves some means of preventing high pressure from being applied to the clamp circuit until the mold has reached a metal-to-metal position under low pressure. This arrangement is particularly common with transfer molding machines.

Much has been written about safety in the workplace, and because molding presses and molds can be hazardous, full attention to safety is mandatory in any operation.

85.9.8 Robots

Whether the molding installation is for fully automatic or semiautomatic molding, robots may be incorporated to minimize human labor, expedite production, or ensure consistency of operations. Simple robots may be used to extract molded parts from the cavities, dropping them onto a conveyor belt or possibly onto a table for an operator to position or shrink fixtures. Robots may also be used to pour material from drums into hoppers or move loading frames in and out of the machine after an operator has prepared them. In short, robots are highly suited to molding operations but molding plants are only beginning to use them.

85.10 FINISHING

After molding, further operations may be necessary on the molded part in an area separate from the press.

85.10.1 Gate Scar Removal After Transfer Molding

In transfer molding, the material enters the cavity through a small opening called the gate. When the molded part is removed from the cavity, it may still be attached to the runner system utilized to feed material to the cavities. The part is easily broken away from the runner system, but the gate scar may include part of the runner protruding from the part or it may just be a rough spot on the molded part. Separation of the runner from the part is usually done by hand. Excessive material may be removed by a saw, particularly if it is glass-fiber-reinforced material. It may also be hand filed, sandpapered, or ground. Sometimes an automatic sanding machine is located next to the press and the operator merely brings the part against the moving sandpaper.

85.10.2 Holes (Not Molded In)

It may be necessary to drill holes or drill and tap provisions for mounting screws. Because thermoset plastics are quite brittle, drill bits must be selected that are suited to thermoset drilling, and the part must be held rigidly in a fixture when the drilling starts. The operator should be protected from chips flying in the vicinity of the drill bit. In deep holes and with dull drill bits, considerable heat may be generated during the drilling operation. With excessive heat, some thermoset materials become gummy, futher restricting the rotation of the drill. For highest quality, tools must be sharp and utilized properly.

85.10.3 Flash at Parting Lines (Compression Molding)

Because compression molds are closed with material already in the cavity, material finds its way between the mold surfaces and generates flash at the parting line. On reinforced materials, flash may be fairly strong and may require heavy tumbling or air blast or possibly beside-the-press filing, sandpapering, or grinding facilities for the operator to use. On items such as melamine dinnerware, where cosmetic appearance is critical, flash removal at the parting line may consist of gentle sanding followed by buffing and polishing with jewelers rouge. High-production melamine dinnerware plants often have such flash removal fully automated. The operator merely places the molded dishware into a fixture, and the automatic parting line polishing system takes over.

85.11 DECORATING

85.11.1 Metal Plating

Occasionally, molded parts need further decorations. Metal plating is performed on certain plastics for cosmetic or other purposes. Metal can be applied through electroless plating techniques or with vacuum metallizing processes. Because phenolics often outgas long after curing, plating over phenolics may eventually show small pin holes or blisters caused by the outgassing. Other plastics that do not outgas lend themselves to successful plating for long-term use.

85.11.2 Painting

Plastics may be painted by a simple dipping or spraying, including electrostatic spraying. Sometimes, paint is applied and then wiped off, leaving paint in molded-in lines, numbers, or letters. It is generally necessary to chemically clean the surface of thermoset molded parts before painting to remove the thin film of release agent that frequently is found on the surface of the molded part. Usually, a dip or wash into an appropriate solvent will suffice. In general, a matte finish on the molded part will accept paint more readily than a mirror finish.

85.11.3 Hot Stamping With Color Transfer

Even though thermosets are dimensionally stable, they respond well to marking by hot stamping. Shallow but adequate indentations on control knobs can be achieved by a hot stamping process. Color decals are used with hot stamping to put a more or less permanent color in the indentations.

85.11.4 Blasting (For Nongloss Finish)

If a molded part is going to be painted or if it will be stuck to another surface by an adhesive, simple sand blasting of the surface before painting or applying the adhesive will promote better bonding.

85.11.5 Foil Overlay

A very common method of decorating, called foil overlay, is used in compression molding to produce patterns on melamine dinnerware, knife handles, trays, and so on. With foil overlay, the mold is opened after partial cure and a melamine-impregnated paper pattern is laid on the part. Then the mold is closed for final cure. The impregnated paper pattern fuses to the part, producing, in effect, a molded-in pattern.

85.12 COSTING

In estimating the cost of molded parts, all aspects of the molding process must be considered.

85.12.1 Material

In any thermoset molding operation, one must calculate not only the material used in each part but also the scrap material that is not easily recovered. The part weight times the material cost per pound is easy to calculate. The calculation must include weight of flash, roughly at 5% for compression molding, and the weight of cull and runners, which can be more than the weight of the molded parts in a multicavity transfer mold.

85.12.2 Recycled Material

Over the years, efforts have been made to recover some value from thermoset scrap. In some plants, thermoset recyclers grind scrap into very fine particle size (160–200 mesh) and blend such reground material into virgin material at percentages up to 20%. The recycling process effectively reduces the cost of the molding compound utilized and also the cost of disposing of an otherwise hazardous waste.

85.12.3 Machine Operating Costs

In estimating machine operating costs, it is necessary to include not only the molding press but also the preformer, preheater, preplasticizer, and other capital equipment used with the press. Also pertinent are the amortization cost for the capital equipment plus the electric power cost, the cooling water cost (for a power unit), and the cost of compressed air for mold blow-off. Other costs are the manpower costs per hour (the operator and part-time supervisor); floor space costs (heating, lighting, and air conditioning); and costs of on-going maintenance of the equipment and the work area.

85.12.4 Set-Up Costs

Each time a mold is put into a press, additional costs are incurred including the time spent by individuals to install the mold and program in the cycle parameters and the time that the machine is not operating (but still depreciating). Most mold shops make a basic calculation of the average set-up costs and use that flat figure rather than maintaining costs on each individual set-up.

85.12.5 Mold Costs

Mold costs are not applicable if the mold is owned and maintained by a customer buying parts from the molder. Otherwise, mold costs include amortization costs (generally based on a two-year life), heating costs (based on the total wattage of the cartridge heaters used in the mold), and the average maintenance costs (sometimes estimated at 5% of the mold cost per year. Maintenance may involve periodic stripping and rechroming of the mold, replacement of broken ejector pins and burnt-out electric heating cartridges, and cavity rework after damage or wear.

85.12.6 Other Costs

In determining the real costs of molded parts, shop overhead, sales and administrative overhead, and desired profit should also be considered. In short, costing a mold operation requires consideration of the standard principles for costing any manufacturing process.

85.13 SHRINKAGE

See Standard Tolerance chart.

Standard Tolerance Chart

PROCESS Compression and Transfer Molding	MATERIAL Thermoset polyester, Phenolics, Melamines, Ureas, DAP, DAIP, Silicones, Urethane, Epoxy, Alkyd.

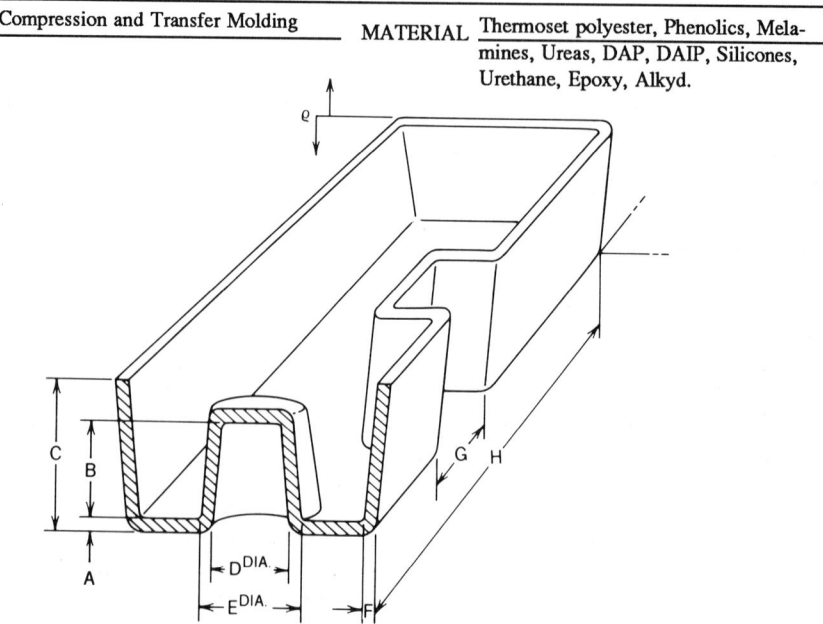

Values are based on a nominal 1/8 in. (3.2 mm) wall thickness and given in plus or minus in./in. and mm/mm.

FINE tolerance – Special care and added cost.
COMMERCIAL tolerance – Normal care required.
COARSE tolerance – Minimal care required, lowest cost.

Tolerance	A	B	C	D	E	F	G	H	
FINE	0.002	use for all dimensions							in./in.
									mm/mm
COMMERCIAL	0.003	use for all dimensions							in./in.
									mm/mm
COARSE	0.005	use for all dimensions							in./in.
									mm/mm

85.14 INDUSTRY ASSOCIATIONS

A number of industry associations provide support to a plastics mold.

The Society of Plastics Industries (SPI) is made up of companies producing plastics materials, plastics machinery, and plastics products. Activities are designed to further the basic interests and welfare of the plastics industry as a whole. SPI establishes safety standards, lobbies state and federal legislatures on legislation pertaining to plastics usage or disposal, and issues many publications beneficial to a better understanding of plastics.

The Society of Plastics Engineers (SPE) is an organization of approximately 25,000 technical persons engaged in plastics industries. SPE presents technical seminars and conferences and issues periodicals and other publications.

The I.T. Quarnstrom Foundation (ITQ), a division of SPE's Moldmaking Division, has publications about moldmaking and schools and universities that train moldmakers and mold designers.

Many plastics professionals are members of the American Chemical Society (ACS) and the American Institute of Chemical Engineers (AICHE). Their focus in on general chemistry as well as the chemistry of plastics and polymers and their application. Both groups conduct seminars, technical programs, and other forums.

The Plastics Institute of America (PIA) is specifically established to encourage and support research in areas of importance to the plastics industry. It solicits its members from companies in the industry who are supportive of such research.

85.15 INFORMATION SOURCES

Because plastics processing is continually evolving and new machines and new techniques are constantly appearing on the scene, the best sources of information at any point in time are recent periodicals serving the plastics industry. Some of the monthly publications include *Journal of Vinyl Technology, Modern Plastics, Plastics Engineering, Plastics Machinery & Equipment, Plastics World,* and *Polymer Composites.*

SPE publishes all the papers presented at its Annual Technical Conference (ANTEC), providing a valuable reference source. Several annual directories are published, often by the monthly periodicals, that contain excellent reference material on the basics of various processes and on the currently available materials and equipment.

86

EXPANDABLE POLYSTYRENE

John V. Wiman

BASF Company, South River Road and Cranberry, Jamesburg, NJ 08831

86.1 DESCRIPTION AND HISTORY OF PROCESS

Although polystyrene has been commercially available for more than 50 years, the methods of manufacturing and processing expandable polystyrene (EPS) were not patented until 1951. The 1951 patent holder is Badische, Anilin- & Soda Fabrik AG of the Federal Republic of Germany.

EPS, which is usually described as a cellular plastic, is most frequently made by polymerizing styrene, with or without the addition of small quantities of other polymers, in suspension. In the two most common methods of production, a volatile hydrocarbon is incorporated in the polymer either during the polymerization or by post-impregnation after polymerization. The resultant product is in the form of fine, discreet beads that soften and expand when they are heated. After expansion, the pre-expanded beads have a fine cell structure consisting of cells that are not interconnecting. In the normal pre-expanding process the beads are not restrained when they are heated and do not fuse together. They become suitable for molding after a short aging period.

Extruded or expanded cellular polystyrene foams, made from either mechanical or chemical frothing processes, are not to be considered in this section as they share few, if any, processing techniques with expandable polystyrene. They are generally not moldable except by vacuum forming or thermoforming in thin-wall sections.

The common energy source for EPS processing is live steam in both the pre-expander and in the mold. On rare occasion and for special applications hot air, radio frequency (RF) energy, or hot water can be used.

Small to medium articles are usually shape molded directly into the finished form in shape molding machines. Insulation and flotation material is usually block molded in large billets that are reduced to size using hot wire cutters or, occasionally, saws. Foam cups are cup molded. All the process forms use appropriately sized beads, in closed molds. These beads are aged after pre-expansion to permit the diffusion of air into the cells.

In the molding process, the pre-expanded beads are introduced into the closed mold and then heated. The beads soften and re-expand, filling up the interstitial volume and fusing one to another without the need for any external pressure. However, it is possible to apply external pressure for special purposes, such as to densify portions of the molding or to create hingelike areas in parts intended to be folded.

In normal molding, the density of the finished part will closely approximate the bulk density of the pre-expanded beads.

Expandable polystyrene offers, at moderate cost, a material with light weight, low thermal conductivity, energy and shock absorption capability, resistance to moisture, and a smooth finish making it attractive to the packaging, drink cup, and building product industries.

86.2 MATERIALS

Regular and modified EPS raw materials represent more than 98% of the market. Modified beads are those whose fire-retardant characteristic has been chemically modified through the addition of brominated or similar compounds. The benefit is that, when they are exposed to a fire or a source of ignition, the cellular materials molded from the modified beads either do not ignite as readily or do not burn as rapidly as those materials not so modified.

Modified materials are most often used for construction applications such as board stock, or for shape molding components or appliances requiring UL approval such as condensate drip trays in air conditioners or separator/insulation in refrigerators.

Since suspension polymerization generates a continuum of bead sizes from large to small, the sizing of EPS is by screen classification. The large bead cut is used for block molding and molding thick-wall sectioned items. The intermediate bead cuts are used for shape molding and some block molding. The small bead cut is used for cup molding and for thinner wall section shape molding.

Extra fine cuts may be used for molding special items such as very thin wall cups, parts requiring extra smooth surface finishes, and so on.

Extra large cuts may be used for such nonmolding applications as ceiling spray aggregates, EPS loose-fill extruder feedstock, and drainage board molding.

It is possible to color or tint EPS. Some raw materials are offered in basic colors. However, the production of colored materials is fraught with enough contamination, matching, and small-volume problems that molders are often encouraged to color their own raw materials. This can be accomplished relatively simply and inexpensively through the use of dry color or special wet dyestuffs. The benefit is that the molder need not inventory colored raw materials and can generate an almost infinite array of pastel tinted parts as needed.

The same equipment used for coloring can be employed by the molder to apply special antistat compounds to the beads so that molded parts can be produced having antistatic properties. These are important in packing designed for computer parts and other solid-state chip containing devices that can be damaged by surface electrostatic charges.

During polymerization, if larger quantities of nonstyrenic polymers are included wth the styrene, it is possible to produce expandable copolymers which, while not strictly EPS, can have special characteristics such as greater resistance to heat or more resiliency. These copolymers, together with such homopolymers as expandable polyethylene, may require special processing equipment or specifically modified versions of normal EPS equipment. Because of the shared processing technology and the end-use marketability of the molded product, such materials and equipment would be most logically found in the EPS shop.

The volatile hydrocarbon blowing agent almost universally incorporated into EPS is pentane, either in normal or isopentane form.

Alternatively, such hydrocarbons as butane and propane can be used, but the permeability of the polystyrene to them is so high as to make the shelf-life of the EPS extremely inconveniently short (1 to 3 days). Descriptions of the use of these so-called "wet gases" are found in early patents. The pentane served as as a plasticizer as well as a nucleating agent for the cell structure of the expanded bead. The wet gases and/or higher hydrocarbons have a "swelling" effect on the EPS. This can be a detriment or, under special circumstances, it can be used to advantage to produce beads for moldings with unique mechanical properties.

Materials have been produced that incorporate halogenated hydrocarbons, ostensibly to improve thermal conductivity and reduce the flammability of the EPS foam. Again, the tendency of the gases to migrate from the foam has hampered their commercialization.

Very little success has been achieved to date with practical attempts to make truly biodegradable EPS or perfumed or long-term odorized EPS. At various times, solvent-resistant copolymers have been produced but these have incorporated significant quantities of ACN, which makes them both flammable and their production toxicologically unattractive.

86.3 PROCESSING CHARACTERISTICS

The principal equipment required for EPS processing is unique to the material and its variants, such as expandable polyethylene. With the exception of such auxiliary/ancillary equipment as hot wire cutters for block foam and material granulators and grinders for raw material and scrap transport, all of the equipment for the EPS process is purpose-built and is nowhere else encountered. EPS shops deal with large volumes of low-density materials and dedicate significant portions of the shop to storage of pre-expanded beads in process and/or finished goods after molding. Block molding shops, in particular, require generously sized facilities with sufficient overhead space to store block [which may be 25 ft (8 m) in height] on end. In fact, EPS has been referred to as "encapsulated air."

The pre-expander is common to nearly all forms of molding. It differs principally in relationship to the size and density of the finished articles being molded. Block molding plants require the largest pre-expanders—those capable of expanding up to 3000 ft^3/h (85 m^3/h) of beads at densities in the nominal range of 1 lb/ft^3 (0.2 g/cm^3) but also those as low as 0.85 lb/ft^3 (0.01 g/cm^3).

In shape-molding shops one normally encounters expanders capable of processing up to 1000 parts per hour of beads at densities ranging from 1.2 to 2.5 lb/ft^3 (0.02 to 0.04 g/cm^3) or higher. Cup molding shops use expanders similar to shape molding shops; they may be slightly smaller and are especially well suited to expanding beads in the range of 3 to 4 lb/ft^3 (0.05–0.06 g/cm^3).

Special end-use applications of EPS, such as loose-fill packaging, require purpose-built combination expander/extruder systems for generating finished materials, through several steps, in the range of 0.25 to 0.5 lb/ft^3 (0.004 to 0.008 g/cm^3).

The most commonly encountered type of pre-expander (see Fig. 86.1) is a right vertical cylindrical vessel having an agitator through its vertical axis and stator bars horizontally spaced diametrically. At the bottom of the vessel is an inlet port for raw beads through which a variable speed auger screw 2 to 4 in. (51 to 102 mm) in diameter controls the feed rate of raw beads. At a point slightly below and upstream from the bead feed in the direction of the agitator rotation is a steam port 1 to 2 in (25 to 51 mm) in diameter through which controlled quantities of live steam are admitted as the heat/energy source to cause the beads to expand. The optimum steam pressure of 1–5 psig (7–35 kPa) is achieved and maintained by a system of multiple-stage simple regulators including steam separators and traps reducing steam from the primary source [which is commonly 125 psig (0.9 MPa)].

At the top of the pre-expander is a discharge chute, which may or may not be fixed in height, through which the pre-expanded beads discharge. The optimum L/D ratio of the pre-expander is 2:1 or greater. Modern pre-expanders designed for high-volume throughput frequently have L/D ratios of 3:1 or 4:1.

The pre-expanded beads discharging from the chute empty into a take-away system that may include a fluidized bed dryer for cooling the beads without mechanically damaging them and for removing as much of the moisture of pre-expansion as possible before the beads are transported to the aging bags or silos. Pneumatic transport is used almost exclusively in EPS plants; the transport system may consist of rotary air lock systems, venturi systems, or direct blower systems. However, the low density beads can suffer if they are put through high velocity systems before being sufficiently aged to withstand the abuse of transport.

The aging or conditioning of the beads takes place in storage silos or bags from 500 to 2000 ft^3 (14–56 m^3) in volume. The containers may be wire mesh rigid bins but most commonly are bags made from polypropylene woven monofilament hanging in frames arranged under duct work systems for filling the bags and over duct work systems for removing the raw materials and transporting them to the molding equipment. The time required for aging/conditioning is a function of the density of the beads and the type of the molding process. Aging times range from hours to 36 hours, longer aging times result in shorter cycles and more dimensionally stable finished parts.

Figure 86.1. Pre-expander for EPS. Courtesy Dingeldein & Herbert.

86.6 PROCESSING CRITERIA

86.6.1 Block Molding

A typical block mold (see Fig. 86.2) is a horizontal mold from 25 to 33 in. (63 to 84 cm) in width by 49 in. (124 cm) in height by up to 25 ft (8 m) in length. The length of the mold is usually determined by some multiple of the building product module of 8 ft (2.4 m).

Machines 18–20 ft (5.5–6.1 m) in length are supplied for use with billet spacers. The block mold consists of side walls in the form of steam chests, the inside of which is either drilled or perforated to permit steam to enter the pre-expanded beads after the mold has been filled and the lid has been closed. The lid, floor, and ends of the mold may or may not be steamed; this depends on the design of the individual machine manufacturer. In addition to the lid of the mold, one side wall of the machine and one end of the machine must be movable to permit the block to be removed after molding. A fill hopper is an integral part of the block mold; it may move horizontally over the mold when the lid is raised or may be lowered vertically down into fill position when the lid is open. Beads are gravity fed into the mold from the hopper, beneath which is a multiple port slide gate.

Figure 86.2. Typical block mold. Courtesy Wieser Maschinenbau Gesellschaft M.B.H.

After the block mold is filled, the lid is closed and latched either pneumatically or hydraulically. Next, low-pressure, high-volume steam at 20–45 psi (0.1–0.3 MPa) is admitted to the steam chests while the drains at the bottom of the chests and the vents in the top of the mold are still open. After the drains and vents are closed, the steam pressure in the chests registers resistance to flow because of the fusion of the beads. This is recorded as back pressure. When 5–10 psig (35–70 kPa) is reached, the steam is shut off. Alternatively, the steam pressure may be shut off by measuring the foam pressure within the mold through a pressure sensor in the lid.

At the end of the steam or fusion cycle, the positive pressure within the mold must be dissipated before it can be opened. This pressure reduction time (PRT) is achieved by cooling the mold by means of air movement through the steam chest or by vacuum or under pressure from a vacuum system. When the internal or cavity pressure is sufficiently low [0.5–1.5 psig (3.5–10 kPa)], the mold can be opened and the block ejected safely. The block mold can produce block at the rate of 6 to 18 parts per hour, depending on the technology used and the size and density of the block produced. High density materials require more molding time than low density materials.

On being discharged from the mold, the billets are moved to a storage area for one to three days (40 days for very special applications) and then are transported to the cutting area where hot wire cutters are used to generate board stock from the billets. The hot wire cutters can slab the billet and also cut off the excess length and width (see Fig. 86.3). Because of its volume, the material is normally shipped from the cutting area as rapidly as possible.

Today, specially shaped block molded material are generated via electric eye cutters, which follow a template and fabricate shapes useful in the packaging industry.

86.6.2 Shape Molding

Thermodynamically, all EPS molding processes are similar. However, the shape molding process differs from the block molding process in areas more profound than simply the parts generated and their size. This is evident in comparing the size of the parts molded by these processes versus the amount of metal required in the tools to constrain them during molding (that is, the metal in the tool that is thermodynamically active during molding). The ratio of metal to polymer in the shape molding process may be as high as 500 to 1000:1; in the block molding process, the ratio may only be 10:1. Another thermodynamic disadvantage for the shape molder is the size of the tool itself and the attendant difficulties of bringing in steam piping of sufficient size and placing drains or vents of sufficient size on the tool. The block mold, by definition, has relatively large-plane areas to which can be attached numerous large steam inlets; the shape mold has relatively small such areas that must be shared with stand-offs, filling devices, cooling systems, and other mechanical accoutrements.

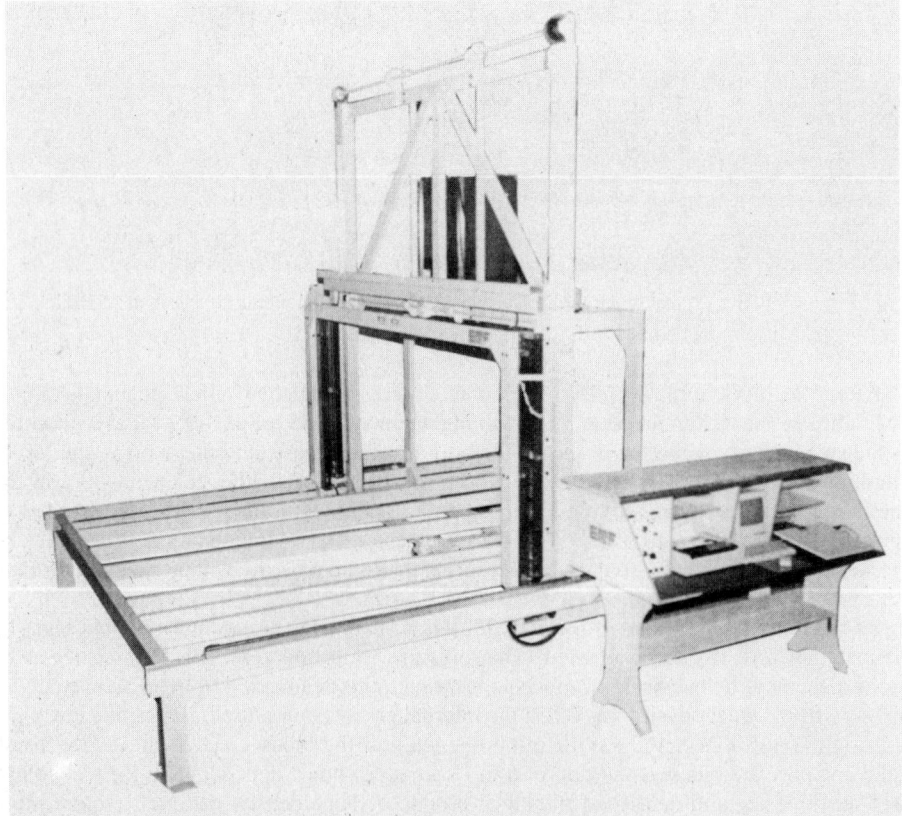

Figure 86.3. Computer-controlled hot wire cutter. Courtesy Engineering, Machines & Fabrication.

Figure 86.4. EPS molding machine for shapes. Courtesy Polipor Carcano.

The shape mold (see Fig. 86.4) consists of movable and stationary platens on which the cast aluminum mold is mounted suspended on a horizontal press frame containing the mold actuation system, the control console, the fill hopper, the major piping utility distribution, and the control elements. The actuation is usually direct hydraulic but toggles and/or air-over-oil systems are becoming increasingly popular because of their speed and low energy consumption. The machine can be fully or semi-automated.

The shape molding process beings when the mold is closed, having been filled with beads via pneumatic fill guns or (cold) slide runners. During the fill cycle some residual heat energy from the previous cycle is inescapably blown away.

Steaming is optimally accomplished by bringing low pressure steam into the top of the mold through a number of generous ports to create a steam curtain effect while the drains/vents at the bottom of the mold are held open. At a certain point, the drains are closed and the steam continues to be admitted until the beads begin to fuse, resisting the flow of steam and creating back pressure in the steam chest. This may be used as a cycle cut-off point. As in the block system, the internal pressure of the fused beads may be read to determine the end of the steam or fusion phase.

If the steam or fusion phase is the thermodynamic "upside," the cooling or PRT is the beginning of the "downside" of the thermodynamic cycle. Cooling is continued until the internal foam pressure is sufficiently reduced so that, when the mold is opened, the parts will not continue to expand. The parts are not necessarily "cool" at this point but are sufficiently stable to be removed from the machine.

Vacuum may be used as an assist during any of the cycle elements, including filling, steaming as air removal, or cooling as mechanical exfiltration of pressure plus part drying. When vacuum is used, the cycles are expected to be shorter and the parts somewhat drier (lighter) at the time of ejection.

Cycle times for shape molding have dropped steadily over the years. Frequently, machines can run at 120 shots per hour (compared to 2–3 min cycles in the early years). Part size and density still influence cycle times. The use of feedback control and vacuum molding have significantly reduced both cycle time and energy consumption parameters.

The shape molded parts thus ejected from the machine in modern, state-of-the-art molding may be dry enough to pack directly at the machine. It is not uncommon for parts to be dried either by standing or in heat tunnels prior to packing.

86.6.3   Cup Molding

The cup molding or thin-wall molding industry has historically been highly proprietary. The machinery is quite similar, all having been derived from the original developments of the Thompson Brothers in Phoenix, Ariz., in the late 1950s. The starting raw material for the cup molder is usually the smallest screening of EPS commercially available, which may or may not contain the usual surface lubricants and antistats found on cup or block molding beads.

The cup-molding process is thermodynamically similar to the block and shape processes, but, because of the nature of the cup and the absolute necessity for accurate registration of the male and female halves of the mold, the machines are normally limited to four cavities to minimize platen size. The thermodynamics of the process are further modified in recognition of the metal to polymer ratio of as much as 2000 to 3000:1.

Cup-molding machines are usually vertically actuated presses, arranged in banks, that automatically discharge the molded cups onto transport systems running down the machine rows. At the end of the rows the cups are oriented, tested, and packed nearly fully automatically, as befits a high-volume, low-cost item.

Over the years the density of the foam cup has dropped from 4 lb/ft^3 (0.06 g/cm^3) to approximately 2 lb/ft^3 (0.03 g/cm^3). As a result, it is possible in the 1980s to market a cup having the same or superior function as one made in the 1960s at only a slightly higher cost—despite doubling or quadrupling of raw material costs, energy costs, and machinery costs. The EPS cup has become truly ubiquitous.

86.8   TOOLING

Block molding, by definition, involves the use of a self-contained mold for the purpose of producing billets. Other than the incorporation of spacers in the end of the block mold to alter the length for special building product sizes, the mold is essentially unalterable. It is a fixed capital device having a fixed constant contribution to the cost of the board produced; there are no mold tooling costs involved for the board stock customer. The board stock customer, however, can incur special tooling costs if special hot wire set-ups are required that fall out of the normal range of adjustments in a standard hot wire cutter. Such special wire devices might include profiling shapes, jigs, or fixtures typically used to produce notches in the side of boards.

Since the shape-molding industry is overwhelmingly a custom-molding industry, the cost of tooling in shape molding is a significant factor. Tooling in the form of molds are either sand cast or die-cast aluminum matched mold halves that are self-contained or face-plates.

Self-contained tooling is intended to be attached directly to the machine platen and contains within it all or most of the plumbing, piping, and fittings required to mold including the steam inlets, the vents or drains, the fill ports, and the internal water spray cooling system.

Face-plate tools are intended to be mounted on semipermanent steam chests that are an integral part of the molding machine and need only contain the fill ports and ejector system (since the basic plumbing is part of the steam chest).

Although mold changing can be quicker with face-plate and steam chest combination tools, self-contained tools may be desirable because there are fewer restrictions in the layout of the cavities, more flexibility in the placing of the inlet fill ports, and the added feature of being able make a lighter tool.

Externally actuated spring-loaded ejector pins may or may not be required in the tooling. Air ejection may be used in place of mechanical ejection, depending on the part design. Self-contained tools may have fill guns inserted from the top, sides, or back, as opposed to steam chest tools, which can only use the top or sides. Slide runners are frequently used on steam chest tools. In EPS molding these runners are cold runners consisting of a simple ported face-plate covering a vertical channel in the tool through which beads are blown during the fill cycle but which must be evacuated before the steam cycle.

Modern tools are becoming lighter, in recognition of the lower internal pressures generated by state-of-the-art molding technology, which also serves to improve them thermodynamically.

The cost of tooling involves production of a pattern, production of the mold castings, finishing or machining the castings, coating the cavities with Teflon, if necessary, and inclusion of the mechanical ejector pins, if necessary.

Customarily, EPS tools are purchased by the custom molding customer through the molder from a mold maker in either a finished or semifinished form. The molder may finish the castings in-house.

Tooling costs are impossible to approximate because of the infinite variety of sizes and types of machines and the complexity of parts to be molded. They can range from $5,000–6,000 for a small tool to $10,000–15,000 for a large tool, which could be significantly more if special complexity or removable inserts are required.

Lead or production time for EPS tooling is commonly 8 to 12 weeks from the time of order; this is highly dependent on the particular tool. Tool life can be expected to range from 500,000 to 1 million cycles before major renovation is required, depending on the part being molded. A number of tools have been in operation for many millions of cycles and may, indeed, outlive their technological life. However, it may be desirable to buy a new tool to enjoy the benefits of new technology that cannot possibly be retrofitted into an old tool, so as to reduce costs through better cycles.

Self-contained tooling can generally be moved from shop to shop without great difficulty, but steam chest or face-plate tools require that a subsequent location have machines with appropriate steam chest types and sizes.

"Family molds" are generally avoided in the EPS industry because of the difficulties in blanking out a cavity in the event quantities of parts to be run on a particular tool are out of phase. Unless it is completely filled, an EPS tool cannot be properly controlled. However, it is not unusual on a mold to produce both the top and bottom of a part simultaneously if they are intended to be used as a pair. On multiple-part moldings involving typically a lid and a body, it is not uncommon for separate molds to be used in order to optimize the cycle time of each part and to maintain the flexibility of an inventory of sets.

86.10 FINISHING/86.11 DECORATING

EPS parts are rarely ejected from the machine requiring any additional shaping or sprue cutting. For special applications, it may be necessary to dry them in special drying tunnels. The customer and the molder must clearly understand the desired quality level of the part prior to contracting and molding.

After molding, it is possible to decorate the parts through the use of silk screens or self-adhesive labels. Occasionally, a job will require the insertion of a part prior to packing. These operations entail additional costs.

The usual method of packaging EPS is in lightweight, single-wall corrugated cartons; most molders have a standard high-volume RSC (regular slotted carton) designed to hold the most foam and fit most efficiently in commercial semi-trailers. It is also possible to ship large parts shipped strapped on flats or in light-gauge polyethylene bags. One disadvantage of the bag is its impermeability to moisture, which requires that the parts be as dry as possible prior to packing.

Block molded cut board intended for the construction industry is usually shipped stacked in a semi-trailer (for full truckloads) but may also be banded or strapped or may include a paper overwrap. material intended to be stored outside may be shrink-wrapped in film, which protects the material from weather, pilferage, and other forms of damage.

Foam cups are normally nested, packaged in polyethylene sleeves, and placed into master cartons as befits commodity items.

86.12 COSTING

The commercial cost to a volume construction consumer of modified commodity construction-grade EPS board stock, at a nominal density of 1 lb/ft^3 (0.02 g/cm^3), ranges from 8 1/2 to 10 cents per board-foot, depending on volume, adherence to "stock sizes," etc. Requirements for higher density, special finishing such as shiplapping, odd cuts, and so on will raise the price.

The cost of a shape molded part is highly dependent on such variables as density, thickness, and intricacy of molding, all of which increase cycle time. Obviously, part size and shape, which influence machine platen utilization, affect part cost. Although highly unreliable as a pricing criterion, the part cost on a per-pound basis can be estimated at from $2.50 to $4.00. It is only necessary, however, to consider the costing of a 1.25 lb/ft^3 (0.02 g/cm^3) EPS picture frame versus a 2.0 lb/ft^3 (0.03 g/cm^3) molded ball to understand the danger of a cost per pound estimate or the irrelevance in EPS of a cost per cubic inch.

87

EXTRUSION

Christiaan Rauwendaal

Manager, Process Research, R & D, RAYCHEM Corporation, 300 Constitution, Menlo Park, CA 94025

87.1 DESCRIPTION AND HISTORY OF PROCESS

Extrusion is the process of forcing material through an opening. Many different materials are processed by extrusion: metals, foodstuffs, ceramics, and plastics. The machine on which the process is performed is an extruder. The extruder is indisputably the most important piece of machinery in the polymer processing industry. Many different types of extruders are used to extrude plastics, the main ones being screw extruders, disk extruders, and reciprocating extruders. Screw extruders consist of at least one Archimedean screw that rotates in a stationary extruder barrel. Single-screw extruders have only one screw; multiscrew extruders have two or more screws. Single-screw extruders are the most common type of extruder used in the plastics industry, and, accordingly, are the focus of this section.

In standard single-screw extruders, the screw rotates but does not move axially. However, there are screw extruders designed to allow axial movement of the extruder screw in addition to rotational movement. These machines are called reciprocating screw extruders and are widely used in injection-molding machines. The ability to move the screw axially allows the reciprocating screw extruder to operate in a cyclic fashion, which is necessary in the injection molding process. The standard single-screw extruder operates in a continuous fashion, which is beneficial in the production of long lengths of product with constant cross sectional shape, such as tubing, film, and profiles.

The first machines for extrusion of plastic were built around 1935. They were primarily used to extrude rubber. Shortly after the introduction of the single-screw extruder, the twin-screw extruder was developed in Italy in the late 1930s. The early screw extruders were quite short with a length to diameter (L/D) ratio of about 5. They were generally heated with steam. Later, the machines became longer and electric heating began to displace steam heating. Modern single-screw extruders range in length from 20 L/D to about 30 L/D, with some specialty machines being considerably longer than 30 L/D. Most modern extruders now use electric heating.

87.1.1 Functional Process Description

There are six major functions performed in screw extruders: solids conveying, melting or plasticating, melt conveying or pumping, mixing, devolatilization, and forming. Devolatilization is not performed on all extruders, only on those specially designed to extract volatiles from the polymer. These devolatilization extruders are usually referred to as vented extruders. Two other functions that are not performed on all extruders are solids conveying and melting. Some extruders, called melt-fed extruders, are charged with molten polymer; their only function is to mix, pump, and force the material through die. In most extruders, called plasticating extruders, the machine is fed with solid material and, therefore, solids conveying and melting are important functions to be performed.

Solids Conveying. Solids are conveyed in two regions: the feed hopper and the screw extruder. The solids conveying in the feed hopper is generally a gravity flow of particulate polymer; the material moves down the feed hopper by its own weight. Unfortunately, this does not always work without problems. Some materials have poor bulk flow characteristics and may get stuck in the feed hopper. This problem is called arching or bridging. It is likely to occur when the particle shape is irregular, when the particle size is small, and when the internal friction of the bulk material is high. Highly compressible bulk materials are also quite susceptible to solids conveying instabilities. Feed hopper can be designed to avoid bridging if the appropriate bulk flow properties are known. The most important property is the internal shear strength of the bulk material as a function of normal stress: this functional relationship is referred to as yield locus (YL). For noncohesive or free-flowing materials the shear strength is a unique function of normal stress and also of consolidation time. These concepts and their application to feed hopper design have been described in detail by Jenike.[1,2] Theoretical and experimental work on gravity solids conveying in feed hoppers and the design criteria for feed hoppers are reviewed in a recent book on polymer extrusion.[3]

If a solid conveying problem does occur in the feed hopper there are several measures that can be taken to eliminate the problem. In many cases, a vibrating pad will be attached to the hopper to dislodge any bridges that may form. In some cases, stirrers are incorporated into the hopper to prevent the material from settling and consolidating. In other cases, a crammer feed is employed to force to bulk material from the feed hopper into the extruder. Figure 87.1 illustrates these devices.

Drag Induced Solids Conveying. Once the bulk polymer falls down into the screw channel the transport mechanism changes from gravity induced solids conveying to drag induced solids conveying. The polymer moves forward as a result of the rotation of screw in the stationary barrel. The frictional forces acting on the polymer particles are responsible for the motion of the polymer. The loose particles are compacted rather quickly into a solid bed. The compacting occurs because the polymeric particles are conveyed against a certain pressure. The pressure increase along the solids conveying zone causes the compaction of the solid bed. Once the solid bed is sufficiently compacted, it moves in plug flow. Thus, all elements of the solids bed at any cross section move at the same velocity; there is no internal deformation in the solid bed.

There are two main frictional forces acting on the solid bed; one at the barrel surface and one at the screw surface. It is important to realize that the frictional force at the barrel surface is the driving force for the solid bed and the frictional force at the screw surface is the retarding force on the solid bed. The magnitude of the frictional force is determined by the local coefficient of friction and normal stress between the two surfaces in sliding contact. The frictional stress τ can be expressed as:

$$\tau_f = f \cdot \sigma_n \qquad (87.1)$$

where f is the coefficient of friction and σ_n the normal stress. Equation 87.1 is generally known as Coulomb's law. Thus, the frictional force is large when the coefficient of friction and the normal stress are large. The fact that the frictional force at the barrel constitutes the driving force can be appreciated if an extreme situation is considered. If the frictional force on the barrel were zero, the solid bed would just rotate with the screw and never move forward. This situation is similar to a nut rotating on a bolt when it is free to rotate. However, when the nut is kept from rotating it will start moving foward along the bolt. In

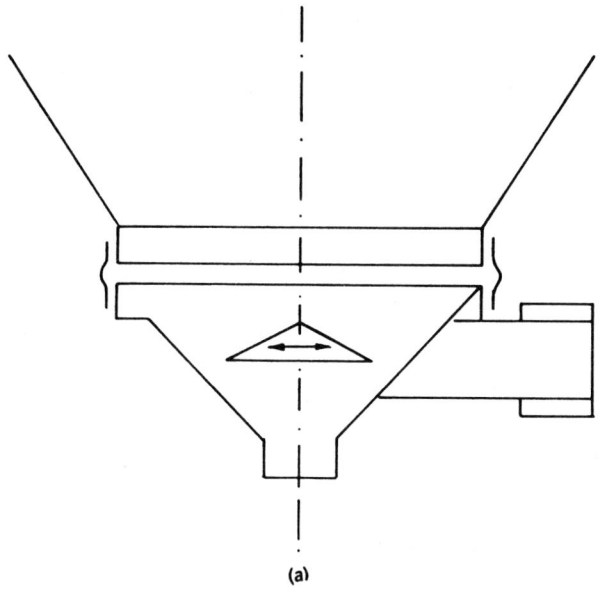

(a)

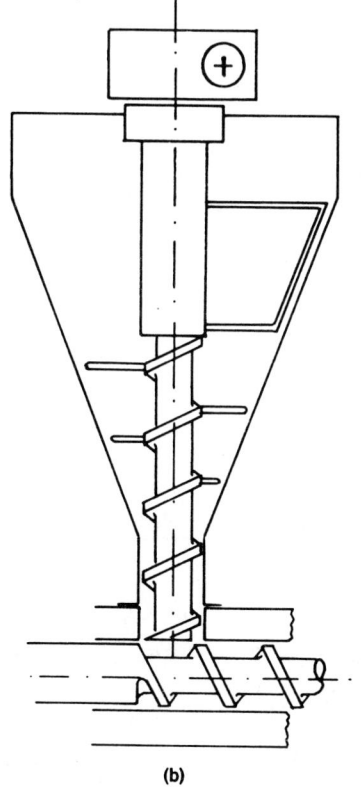

(b)

Figure 87.1. (a) Bin activator. (b) Crammer feeder with stripper arm.

a similar fashion, there will be forward movement of the solid bed in the extruder if there is a sufficiently large frictional force on the barrel.

From these considerations it becomes clear that for good solids conveying a large coefficient of friction (COF) on the barrel and a low COF on the screw are desirable. Thus, the screw surface should be smooth and slippery, while the barrel surface should be rough. A common way to increase the barrel surface roughness is by grooving the internal barrel surface. Development work on grooved barrel sections started in the Federal Republic of Germany in the 1960s. Nowadays, extruders with grooved barrel sections are very common in Europe and becoming more common in the United States. By considering the forces acting on an element of the solid bed and by taking a force balance in two perpendicular directions, an expression of the solids conveying rate can be derived. Darnell and Mol[4] developed the solids conveying theory of screw extruders about thirty years ago. The solids conveying rate is determined by the bulk density of the solids, screw and barrel geometry, screw speed, COF on barrel and screw, and the pressure gradient in the solids conveying zone. By using the theory, the benefits of a grooved barrel surface can be evaluated. This is shown in Figure 87.2, where the predicted solids conveying rate is plotted as a function of the pressure development in the solids conveying zone. Curve A shows the characteristic for a smooth barrel where the COF on the barrel is about the same as the COF on the screw. The solids conveying rate is rather sensitive to pressure; small changes in pressure cause large changes in solids conveying rate. Thus, the solids conveying process under these conditions is not very stable.

Curve B shows the characteristic for a grooved barrel section where the COF in the barrel is about two to three times larger than the COF on the screw. Here, the solids conveying rate is considerably higher compared to the smooth barrel. However, more important is the fact that the solids conveying rate has become rather insensitive to pressure. Small pressure fluctuations will not affect the solids conveying rate much at all, and the solids conveying process will be quite stable.

Melting. The solids conveying zone ends when the temperature in the solid bed reaches the polymer melting point and the thin film of molten polymer starts to form. Melting will start because of the heat conducted from the barrel heaters and the heat generated by friction along the screw and barrel surface. The frictional heat generation is generally quite sub-

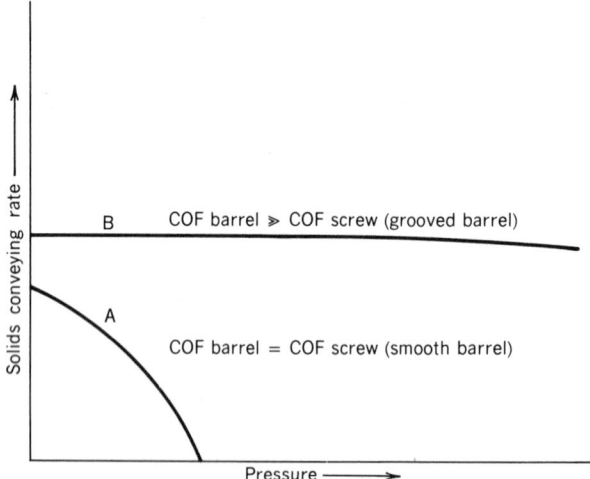

Figure 87.2.   Rate/pressure relationship for smooth and grooved barrel.

stantial. It is often possible to initiate melting without applying any external heat from the barrel heaters, using only the frictional heat generation.

Melting in single screw extruders was first studied qualitatively by Maddock[5] in the late 1950s. He performed screw extraction experiments to observe visually the progress of melting along the extruder screw. In the mid-1960s, Tadmor[6,7] performed similar extrusion experiments and proposed a general melting model. Based on the general melting model Tadmor developed a melting theory to describe the entire melting process in a single screw extruder quantitatively. This constituted a milestone in extrusion theory, because it permitted, for the first time, a complete theoretical description of the extrusion process from the feed hopper all the way to the die. The general melting model proposed by Tadmor is shown in Figure 87.3.

The major difference between the melting model shown in Figure 87.3 and the original model proposed by Tadmor is the varying upper melt film thickness. Tadmor originally assumed the melt-film thickness to be constant across the width of the solid bed. Shapiro[8] and Vermeulen and co-workers[9,10] checked the validity of this assumption and came to the conclusion that a constant melt-film thickness is physically not possible. The melt-film thickness has to grow along the width of the solid bed in order to accommodate the increased amount of molten material entering the melt film as a result of melting at the solid/melt interface.

The melting process occurs primarily at the interface of the upper melt film and the solid bed. This is a result of the generally large velocity difference between the solid bed and the barrel surface. The large relative velocity, combined with the small thickness of the melt film, causes a substantial amount of viscous heat generation in the upper melt film. There may or may not be a melt film between the screw surface and the solid bed. Even if there is a melt film, the melting at the lower interface will be quite small compared to the melting at the upper interface. This is because the relative velocity between the solid bed and the screw surface is generally small. Thus, there will be a little viscous heat generation in the lower melt film and, therefore, little melting.

The two major sources of heat for melting are the viscous heat generation in the melt film and the heat conducted from the barrel heaters. Under normal operating conditions, the contribution of the viscous heat generation is generally larger than that of the barrel heaters. The contribution of the viscous heat generation will increase with screw speed and polymer melt viscosity. It is important to realize that an increase in the barrel temperature does not always improve melting. Higher barrel temperatures will increase the heat conduction from the barrel heaters but will reduce the viscous heat generation in the upper melt film. The latter occurs because the temperature in the melt film will rise, prompting the viscosity of the material in the melt film to drop. As a result, the viscous heat generation in the melt film will decline. Thus, polymers whose melt viscosity is very sensitive to temperature are likely to exhibit reduced melting performance with increased barrel temperatures. Generally speaking, amorphous polymers have a rather strong temperature sensitivity

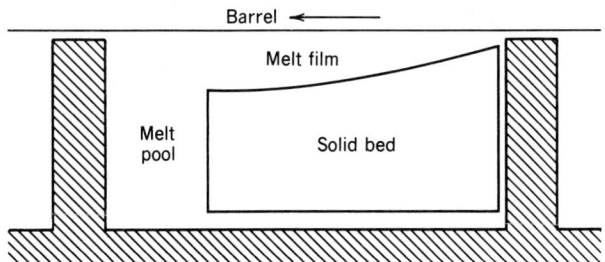

Figure 87.3. General melting model.

to viscosity, particularly materials such as PVC, ABS, PMMA, and PS. The melt viscosity of semicrystalline polymers tends to be rather insensitive to temperature.[3]

The melting rate depends strongly on the thickness of the melt film. The melt film thickness at the trailing flight flank is primarily determined by the radial clearance between the flight and barrel. The smaller the clearance, the thinner the overall melt film and the higher the melting rate. This is shown in Figure 87.4.

Thus, in order to maintain good melting performance it is important to maintain a small flight clearance in the melting region of the extruder. Unfortunately, screw and barrel wear often occurs right in the melting region, ie, in the latter part of the compression section of the extruder screw. Wear at this location is detrimental to the melting performance of an extruder and will eventually result in insufficient melting, which, in turn, will lead to melt temperature nonuniformities and pressure fluctuations. It is important, therefore, to inspect screw and barrel regularly to make sure there is no excessive wear in the melting region of the extruder. Wear in other sections of the extruder, such as the feed section or metering section, generally is less detrimental to the extruder performance than wear in the mid-section of the extruder. The single-screw extruder is quite an effective melting device. Effective melting is achieved by removing newly molten material as quickly as possible from the melt film, keeping the melt film relatively thin. This is a function of the drag flow in the upper melt film, which transports newly molten material to the melt pool. If newly molten material were to remain in the melt film, it would grow in thickness, and the melting rate would drop dramatically. The melting mechanism in the single screw extruder is described as conduction melting with drag induced melt removal.[12] By comparison, a ram extruder is an ineffective melting device because there is very little viscous heat generation in the polymer melt and the melt film thickness grows as melting proceeds, reducing the heat transfer to the solid/melt interface.

Melt Conveying. The melt conveying zone of the extruder starts where the melting zone ends: at the point where all the solid polymeric particles have melted. The melt conveying portion of the extruder acts as a simple pump. Forward motion occurs as a result of the rotation of the screw and the helical configuration of the screw flight. If the pressure is constant along the melt conveying zone, the output of the extruder equals the drag flow through the screw channel. The drag flow rate is given by:

$$M_d = \tfrac{1}{2} \rho_m^{v_{bz}} HW \qquad (87.2)$$

Figure 87.4. Effect of flight clearance on melting performance.

Where ρ_m is the melt density, v_{bz} is the down channel barrel velocity relative to the screw surface, H is the channel depth, and W is the perpendicular channel width.

The down channel barrel velocity is related to the screw diameter D, the screw speed N, and the flight helix angle θ by the following expression:

$$V_{bz} = \pi DN \cos \theta \tag{87.3}$$

The perpendicular channel width is related to the diameter, helix angle, and the flight width:

$$W = \pi D \sin \theta - w \tag{87.4}$$

where the w is the perpendicular flight width, which is often quite small compared to the channel width. A typical screw geometry is shown in Figure 87.5.

It is interesting to note that the drag flow rate is independent of the polymer melt viscosity. Thus, if the density is the same, the drag flow rate for olive oil will be the same as for polyethylene melt, polypropylene melt, etc. In many cases, the actual extruder output will be reasonably close to the drag flow rate. Accordingly, the drag flow rate can be used as an approximate prediction of the actual extruder output. In most actual extrusion operations, the pressure will not be constant along the melt conveying zone; there will be a pressure gradient, either positive or negative. If the extruder has to pump against a high diehead pressure, the pressure gradient will generally be positive. In this case, the actual extruder output will be less than the drag flow rate. If there is significant pressure development in

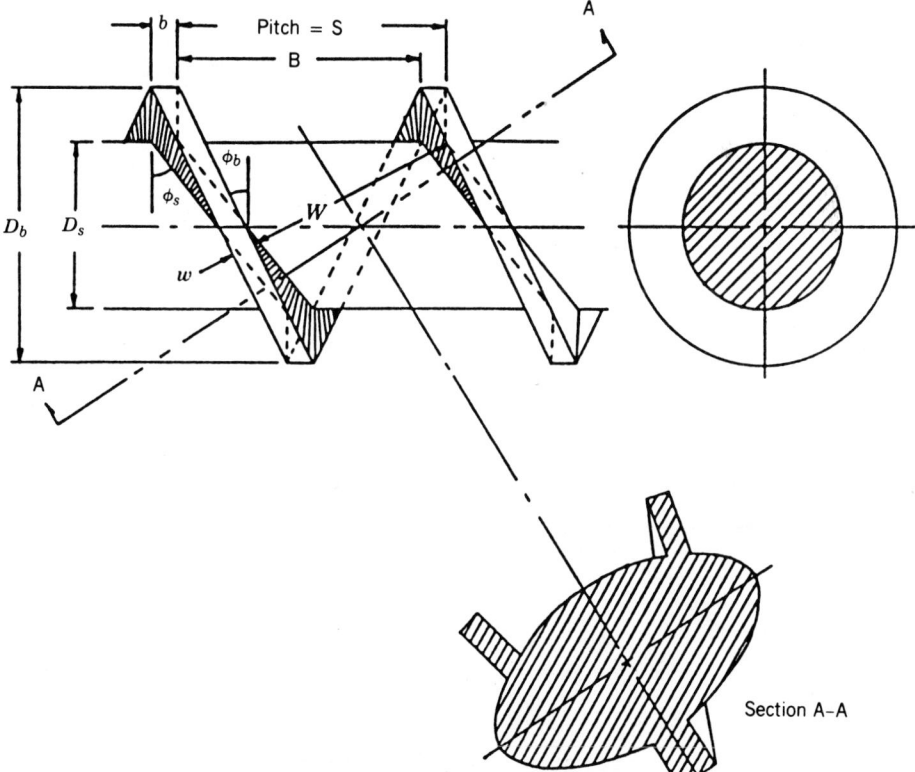

Figure 87.5. Typical screw geometry.

the solids conveying or melting zone of the extruder, the pressure gradient in the melt conveying zone will be negative. In such a situation the actual extruder output will be greater than the drag flow rate. This is shown in Figure 87.6, where the down channel velocity is plotted as a function of normal distance from the screw surface. When the pressure gradient is positive, the down channel velocities decrease. If the pressure gradient continues to increase it will reach a point where there will be no net output from the extruder. If the pressure gradient is negative, the down channel velocities increase. The effect of the pressure gradient on output can be evaluated from the following expression for the pressure flow rate:

$$\dot{M}_p = \frac{\rho_m WH^3 g_z}{12\mu} \tag{87.5}$$

where g_z is the pressure gradient in down channel direction and μ is the Newtonian viscosity of the polyer melt. Equation 87.5 is valid for Newtonian fluids. The total output is obtained by subtracting the pressure flow from the drag flow:

$$\dot{M} = \dot{M}_d - \dot{M}_p \tag{87.6}$$

Equation 87.6 can be used to predict the pumping rate for Newtonian fluids for any value of the pressure gradient assuming leakage flow over the screw flights can be neglected. Unfortunately, there is another complication in that polymer melts are not Newtonian. They behave as pseudoplastic fluids; that is, the viscosity decreases with increasing shear rate.

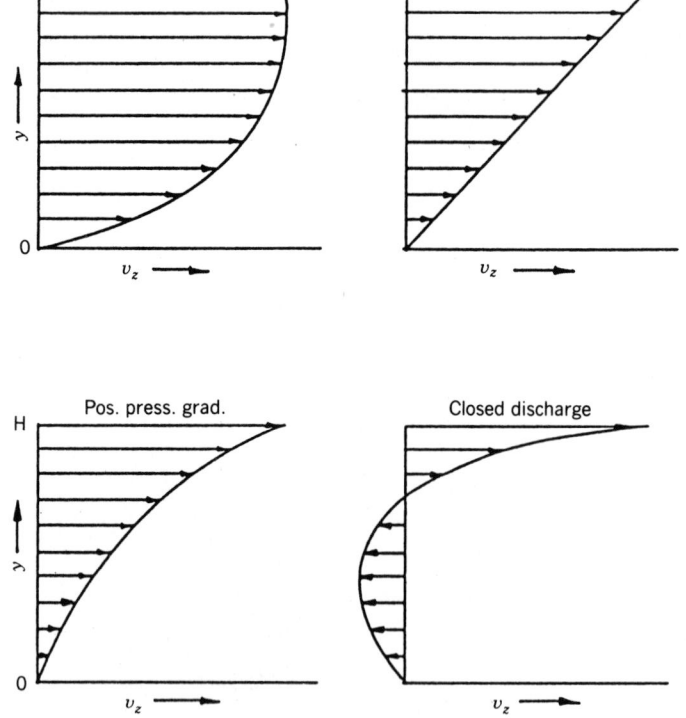

Figure 87.6. Velocity profiles at various values of the pressure gradient.

Polymer melts are often described as power law fluids; these are fluids whose dependence of viscosity on shear rate can be described by a power law equation:

$$\eta = m\dot{\gamma}^{1-n} \tag{87.7}$$

where η is viscosity, m the consistency index, $\dot{\gamma}$ the shear rate, and n the power law index.

The power law index indicates to what extent a fluid deviates from Newtonian flow behavior. If the power law index is close to unity the fluid is rather Newtonian. If the power law index is less than one half, the fluid is strongly non-Newtonian. Most of the large-volume commodity polymers fall into this latter category, such as PVC, HDPE, LDPE, and PP. If the polymer is strongly non-Newtonian the use of equations 87.4–87.6 to predict the extruder output will result in large errors. If the power law index n is known, the extruder output can be predicted with the following approximate relationship:[3]

$$M = \frac{4 + n}{10} \rho_m W H v_{b_z} - \frac{\rho_m W H^3 g_z}{(1 + 2n)4\eta} \tag{87.8}$$

where η is the local viscosity in the screw channel. The viscosity is generally evaluated at the Couette shear rate $\pi DN/H$. In actuality, the viscosity will not be uniform in the screw channel but will vary with shear rate and temperature.

If the screw channel is unrolled onto a flat plane, and the barrel is assumed to be moving with respect to a stationary barrel, the flow profiles in the screw channel can be depicted as shown in Figure 87.7, where r is ratio of pressure flow to drag flow.

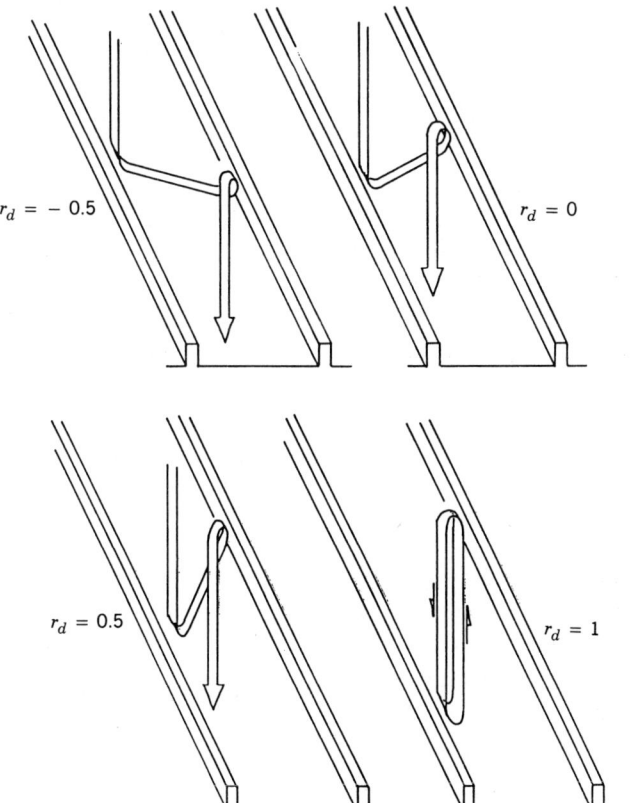

Figure 87.7. Flow fluid element if screw is unrolled onto a flat plate.

Die Forming. In this functional zone, the polymer is shaped and for this reason the die forming zone can be considered the most important functional zone. The die-forming zone is always a pressure consuming zone. The pressure built up in the preceding functional zones is used up in the die forming zone. The diehead pressure is the pressure required to force the polymer melt through the die. The pressure is not determined by the extruder but by the extruder die. The variables that affect the diehead pressure are

- The geometry of the flow channel in the die.
- The flow properties of the polymer melt.
- The temperature distribution in the polymer melt.
- The flow rate through the die.

When these variables remain the same, the diehead pressure will be the same whether a single-screw extruder or a twin-screw extruder is in use. The main function of the extruder is to supply homogeneous polymer melt to the die at the required rate and diehead pressure. The rate and diehead pressure should be steady, and the polymer melt should be homogeneous in terms of temperature and consistency. Die design is the one aspect of extrusion engineering that has remained more an art than a science. The obvious reason is that it is quite difficult to determine the optimum flow channel geometry from engineering calculations.

Description of the flow of the polymer melt through the die requires knowledge of the viscoelastic behavior of the polymer melt. The polymer melt can no longer be considered as purely viscous fluid because elastic effects in the die region are very important. Unfortunately, there are no simple constitutive equations that adequately describe the flow behavior of polymer melt over a wide range of flow conditions. Thus, a simple die flow analysis is generally very approximate; more accurate die flow analyses tend to be quite complicated. Many of the more accurate die flow analyses make use of the finite element method (FEM).

The objective of an extrusion die is to distribute the polymer melt in the flow channel in such a manner that the material exits from the die with a uniform velocity. The actual distribution will be determined by the flow properties of the polymer, the flow channel geometry, the flow rate through the die, and the temperature field in the die. If the die-flow channel geometry is optimized for one polymer for one set of conditions, a simple change in flow rate or in temperature can make the geometry nonoptimum. Except for circular dies, it is impossible to obtain a flow channel geometry that can be used, as such, for a wide range of polymers and for a wide range of operating conditions. For this reason, one generally incorporates adjustment capabilities into the die by which the distribution can be changed externally while the extruder is running. The flow distribution is changed in two ways: by changing the flow channel geometry via choker bars, restriction bars, valves, and so on, or by changing the local die temperature. Such adjustment capabilities complicate the mechanical design of the die but enhance its flexibility and controlability. Some general rules that are useful in die design are

- There should be no dead spots in the flow channel.
- There should be steady increase in velocity along the flow channel.
- Assembly and disassembly should be easy.
- Land length should be about 10 times land clearance.
- Avoid abrupt changes in flow channel geometry.
- Use small approach angles.

In die design, problems often occur because the product designer has little or no appreciation for the implications of the product design details on the ease, or rather, the difficulty of extrusion. In many cases, small design changes can drastically improve the extrudability of the product. Some basic guidelines in profile design to minimize extrusion problems are

- Use generous internal and external radii on all corners; the smallest possible radius is about 0.02 in. (0.5 mm).
- Maintain uniform wall thickness.
- Avoid very thick walls.
- Make interior walls thinner than exterior walls for cooling.
- Minimize the use of hollow sections.

Figure 87.8 illustrates applications of these guidelines to a few different profiles.

Calibration. In extruding of large products or using polymers with relatively little melt strength, the extrudate emerging from the extruder die is often led into a sizing die. The actual dimensions of the final extrudate are, to a large extent, determined by the sizing device, which is generally referred to as a calibrator. Its use is required if the emerging extrudate has insufficient melt strength to maintain the required shape. The calibrator is in close contact with the polymer melt and cools the extrudate. When the extrudate leaves the calibrator it has sufficient strength to be pulled through a haul-off device, such as a catapuller.

In a vacuum calibrator, a vacuum is applied to ensure good contact between the calibrator and the extrudate and to prevent collapse of the extrudate. The use of a vacuum calibrator is generally easier to maintain at a constant level than the use of positive air pressure within the extrudate. The latter tends to vary with the length of the extrudate is difficult to maintain when the extrudate has to be cut into discrete lengths.

Calibrators are useful when good, accurate shape control is important. When the requirements for shape control are less stringent, the extrudate shape is often maintained by support brackets placed downstream of the extrusion die. In fact, the extrudate shape can be modified substantially by support brackets. These can be useful because they allow the shape of the extrudate to be modified without changing the die but only by changing the shape of the support brackets.

As discussed by Michaeli,[11] there are five types of calibrators:

- Slide calibrators.
- External calibrators with internal air pressure.

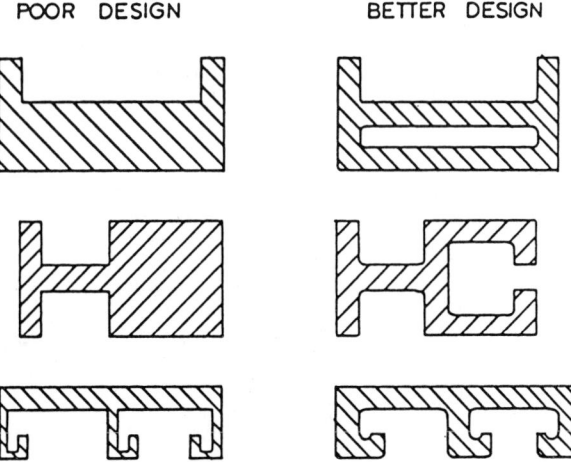

Figure 87.8. Profile design examples.

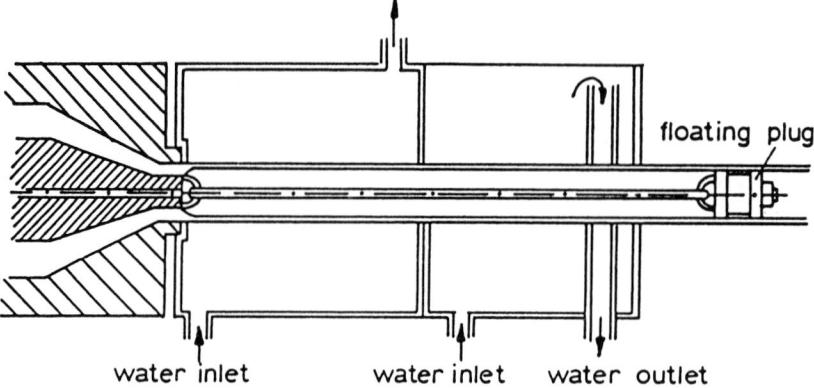

Figure 87.9. External calibrator with internal air pressure.

- External vacuum calibrators.
- Internal calibrators.
- Precision profile pultrusion (Technoform process).

Slide calibrators are used for simple and open profiles. The extrudate is more or less in contact with cooled plates and the profile is pulled through the calibrator, causing some amount of drawdown. An example of an external calibrator with internal air pressure is shown in Figure 87.9. The air pressure is maintained by a sealing mandrel located inside the extrudate downstream of the calibrator. The mandrel is attached to the tip by a cable to fix its position. This type of calibration is used in larger diameter pipe for PVC [$D>14$ in. (350 mm)] and PE [$D> 4$ in. (100 mm)]. An example of an external vacuum calibrator is shown in Figure 87.10. The profile is pulled through apertures with a reduced pressure applied to this section; the aperture action is followed by a simple water cooling section. External vacuum calibrators are used for hollow profiles and smaller-diameter pipe. An example of an internal calibrator is shown in Figure 87.11. This internal calibrator modifies the annular shape of the extrudate emerging from the die into a more or less triangular shape, which enables the shape to be modified just by the calibrator. However, internal calibration is not used very often. In precision profile pultrusion, a small amount of polymer is allowed to accumulate between the die and the calibrator. The accumulation is controlled by a sensor through adjustment of the haul off speed. The extrudate is pulled through a short, intensely cooled calibrator followed by a water bath. This type of calibration is referred to as the Technoform process; it was developed by Reifenhauser K.G.

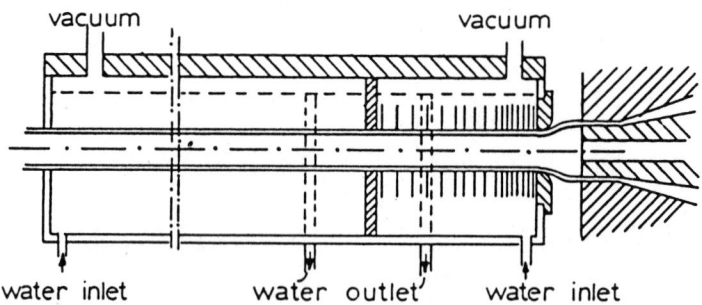

Figure 87.10. External vacuum calibrator.

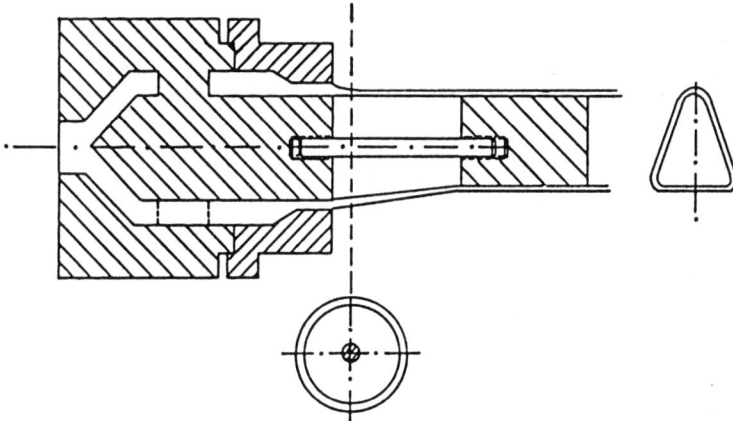

Figure 87.11. Internal calibrator.

87.2 MATERIALS

87.2.1 Thermoplastics

Thermoplastics are polymers that are solid at room temperature but soften or melt when the temperature is raised sufficiently high. When a thermoplastic is in the molten state it can flow and adopt the shape of an extrusion die or a mold. By cooling the shaped polymer melt, an extruded or molded product is made. If the product is not satisfactory it can be heated again to be reprocessed. Extrusion is the most common technique to process thermoplastics.

There are two types of thermoplastic materials: amorphous and semicrystalline. An amorphous thermoplastic does not contain highly ordered crystalline regions. When these materials are heated they do not melt because there are no crystallites to melt. Amorphous polymers simply soften when the temperature is raised. The softening starts when the temperature is raised above the glass-transition point of the polymer. Below the glass-transition point the material behaves as a relatively stiff, solid material. Examples of amorphous polymers are PS, PVC, PVA, PEO, PC, SAN, and ABS.

Semicrystalline polymers contain highly ordered crystalline regions called crystallites. The extent to which the material is crystalline is generally described by the percent crystallinity. One of the most crystalline polymers is HDPE, which has a percent crystallinity of around 90%. When a semicrystalline material is heated to its melting point, the crystallites will melt. Thus, the semicrystalline material undergoes a true first-order phase transition (unlike an amorphous material). Since the crystallites do not possess the same degree of crystalline perfection, the melting will not occur at one single temperature but over a melting range. The melting range can be as wide as 36–144°F (20–80°C). One polymer that is known for its narrow melting range is nylon; its melting range is just a few degrees. Examples of semicrystalline polymers are PE, PP, PVF2, PBT, PET, and EVA.

Most thermoplastics can be extruded reasonably well; however, there are some exceptions. When some materials are heated above their melting point, they do not transform in a flowable polymer melt. An example is ultrahigh molecular weight polyethylene (UHMWPE); this material behaves more like a solid than a melt above its melting point. As a result, such a material cannot be processed on a conventional extruder; it is normally processed by a sintering process. Another process that is used for such intractable materials is ram extrusion. When the ram is reciprocating, continuous profiles can be produced by this technique. Other such intractible materials are polyimide and polytetrafluoroethylene.

87.2.2 Thermosets

Thermosets are materials that undergo a chemical cross-linking reaction when the temperature is raised. As a result, the material hardens and becomes rigid, depending on the cross-link density. When cross-linking has taken place the chemical nature of the material is irreversibly changed; thermosets cannot be reprocessed as can thermoplastics. Examples of thermosets are phenolics, polyester, epoxies, and polyurethanes.

Thermosets can be processed on screw extruders before the cross-linking reaction takes place. In these applications, good temperature control is crucial. If the stock temperature in the extruder were to exceed the kick off temperature, cross-linking could start in the extruder and the entire machine could freeze up. Therefore, extruders used to extrude thermosets are generally quite short and have low compression ratios or no compression at all.

87.2.3 Elastomers

Elastomers are polymers with rubberlike properties; in particular, the ability to undergo large elastic deformations (several hundred percent). Elastomers can be divided into thermosetting elastomers (TSE) and thermoplastic elastomers (TPE). Thermosetting elastomers are formed and then cross-linked via the application of heat and pressure, a process called vulcanization, which is a relatively slow, time-consuming process. Examples of thermosetting elastomers are natural rubber, isoprene rubber, neoprene, and nitrile rubber. Thermoplastic elastomers were first introduced in the late 1960s. They have the advantage of being able to be processed on conventional polymer processing equipment. TPEs do not require the

---

MASTER PROCESS OUTLINE   **PROCESS**   Extrusion

Instructions:

For "ALL" categories use   Y = yes
For "EXCEPT FOR" categories use   N = no
                                       D = difficult

Y | **ALL Thermoplastics**
Except for:

Acetal	Poly(methyl methacrylate)
Acrylonitrile–Butadiene–	Poly(methyl pentene) (TPX)
Styrene	Poly(phenyl sulfone)
Cellulosics	Poly(phenylene ether)
Chlorinated Polyethylene	Poly(phenylene oxide) (PPO)
Ethylene Vinyl Acetate	Poly(phenylene sulfide)
Ionomers	Polypropylene
Liquid Crystal Polymers	Polystyrene
Nylons	Polysulfone
Poly(aryl sulfone)	Polyurethane Thermoplastic
Polyallomers	Poly(vinyl chloride)-PVC
Poly(amid-imid)	Poly(vinylidine chloride)-Saran
Polyarylate	Rubbery Styrenic Block
Polybutylene	Polymers
Polycarbonate	Styrene Maleic Anhydride
Polyetherimide	Styrene–Acrylonitrile (SAN)
Polyetherketone (PEEK)	Styrene–Butadiene (K Resin)
Polyethersulfone	Thermoplastic Polyesters
	XT Polymer

Y | **ALL Thermosets**
Except for:

Allyl Resins
Phenolics
Polyurethane Thermoset
RP-Mat. Alkyd Polyesters
RP-Mat. Epoxy
RP-Mat. Vinyl Esters
Silicone
Urea Melamine

Y | **ALL Fluorocarbons**
Except for:

Fluorocarbon Polymers-ETFE
Fluorocarbon Polymers-FEP
Fluorocarbon Polymers-PCTFE
Fluorocarbon Polymers-PFA
Fluorocarbon Polymers-PTFE
Fluorocarbon Polymers-PVDF

vulcanization cycle and, therefore, the mixing and compounding step can generally be eliminated. Examples of thermoplastic elastomers are styrene–butadiene copolymers, olefinic rubbers (EPR and EPDM), thermoplastic polyurethanes, and copolyesters.

Extruders used for warm feed extrusion are generally short and have deep-flighted screws to minimize temperature buildup in the material. Extruders used for cold feed extrusion are essentially the same as extruders used for thermoplastics.

87.3 EXTRUDED PRODUCTS

The number of products that can be made by the extrusion process is essentially unlimited. In its simplest form, the extruder will produce a product having a constant cross section such as pipe, sheet, or profile. These products can be flexible or rigid, depending on the polymer and the shape of the extrudate. However, by incorporating moving parts in the die or in downstream equipment, the cross section of the extruded product can be varied along the length of the extrudate. This is commonly done in extrusion blow molding where the spacing between tip and die can be varied to achieve a certain thickness profile. This technique is referred to as parison programming and is shown schematically in Figure 87.12. A similar technique is used in pipe extrusion to produce short sections having increased ID for use as fittings.

Another interesting technique is employed in the manufacture of corrugated tubing. A conveyor with the external shape of the corrugated tubing machined in its metal blocks is placed right up against the die. A positive internal air pressure forces the plastic up against the corrugated internal surface of the conveyor elements. The conveyor in this application acts as a calibrator with moving walls (Fig. 87.13).

Netting is another product that can be made by an extrusion process in which sections of the die can move. Here, the die has at least one movable section with a large number of grooves in it. The movable section rotates as it oscillates along another die section with grooves. As polymer melt flows out of the grooves, netting will be produced by the relative movement of the die sections.

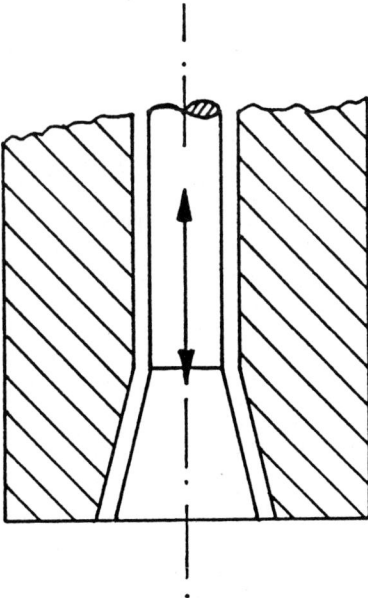

Figure 87.12. Parison programming by moving tapered tip.

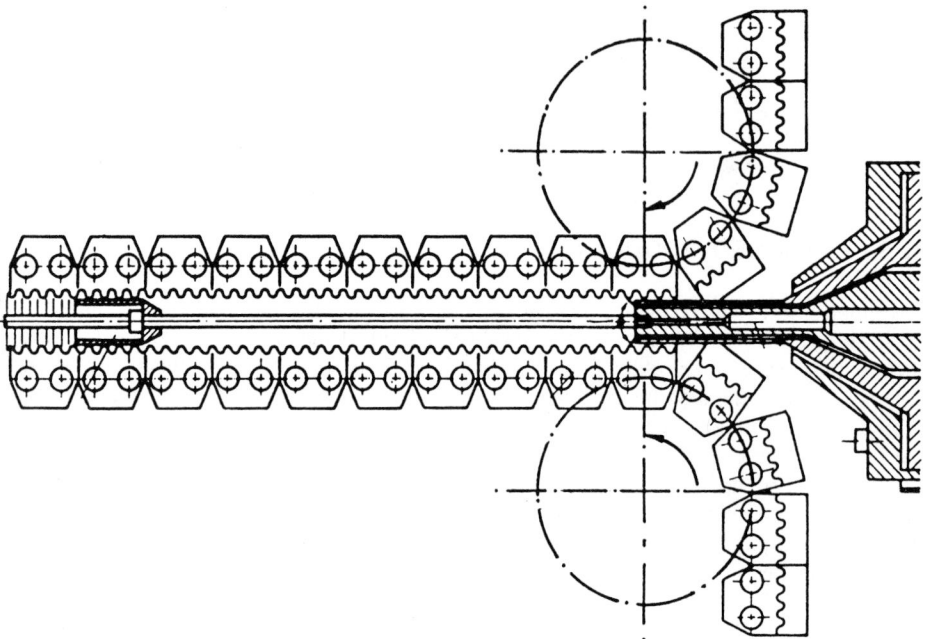

Figure 87.13. Setup to make corrugated tubing or pipe.

Multilayered products can be made by multiple extrusion, tandem extrusion, or coextrusion. In multiple extrusion the core of the product is extruded first; additional layers are then extruded onto the core in subsequent extrusion steps. In tandem extrusion the extruded core is fed into a second extruder that extrudes another layer onto the core. In coextrusion at least two extruders feed into one die where the different polymer streams are combined to form a coextruded product.

Coextrusion has become quite popular and is now a well-established technique in the extrusion industry. The two coextrusion systems in use are the feedblock system and the multimanifold system. In the feedblock system the polymers are combined at the die inlet and flow through the die together. This system is very simple but requires that the polymers to be extruded have very similar flow properties and process temperatures. An example of the feedblock sheet die setup is shown in Figure 87.14. In the multimanifold system the different polymer melt steams enter the die through separate inlet channels. The different melt streams are combined just before they exit the die. This system is more difficult to design and manufacture and, thus, more expensive. The advantage is that polymers with very different flow properties can be combined and that the individual layer thickness can be better controlled. An example of a multimanifold blown film die is shown in Figure 87.15.

87.4 ADVANTAGES/DISADVANTAGES

The extrusion process is ideally suited to make long lengths of a product having a constant cross section. Once the machine has reached stable operating conditions, the process requires little operator attention. The extrusion process lends itself well to automation. In some automated extrusion plants one operator handles a number of extrusion lines, keeping the labor cost quite low. A large number of plastics are available from which to obtain the

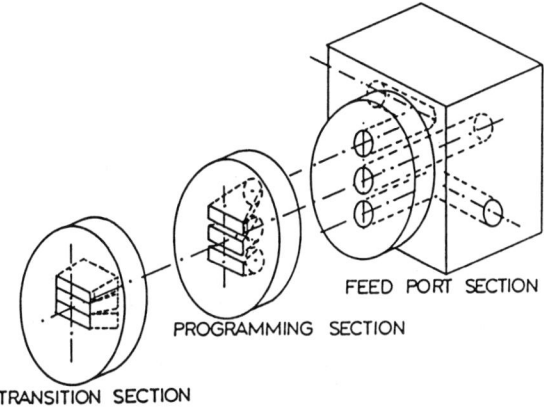

FEED PORT SECTION

PROGRAMMING SECTION

TRANSITION SECTION

Figure 87.14. Feedblock sheet die.

required product properties. With the large number of additives and fillers available today and the possibility of blending or alloying different polymers together, the number of extrusion compounds is essentially unlimited.

A disadvantage of extrusion is that it is, to some extent, a "black box" process. Polymer can be seen entering and exiting the extruder, but visual observation of the actual process is not possible. Therefore, when process instabilities occur, it is often difficult to determine the cause of the problem and to find the solution. Quick and accurate troubleshooting requires good instrumentation and a troubleshooter with thorough knowledge of the entire extrusion process. It is often difficult to obtain these two in combination, and this is why extrusion is still sometimes considered to be an art. However, as a result of a large number of scientific extrusion studies, the understanding of the theory and practice of extrusion has reached a level of reasonable maturity. Extrusion is now more of a science than an art, although there are still some aspects of extrusion that are not fully understood.

87.5 DESIGN

In the design of a product that is to be made by extrusion, it is good practice to involve a process engineer before the final product geometry is established. In many cases, small

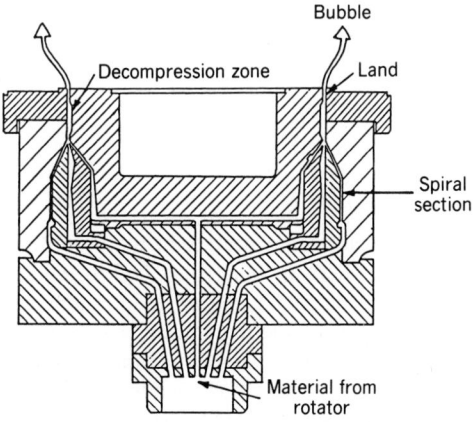

Bubble

Decompression zone   Land

Spiral
section

Material from
rotator

Figure 87.15. Multimanifold blown film die.

modifications can be made that greatly improve the extrudability of the product. Some guidelines for product design to minimize extrusion problems are given here in the subsection on die forming.

The ability to hold certain tolerances depends on the nature of the polymer. Rigid PVC allows a high degree of dimensional control because its swelling tendency at the exit of the die is quite small. This property, combined with the good melt strength and the relatively low cost of PVC, make it a very popular material in profile extrusion. Wall thickness can generally be maintained within ±8%. Overall product dimensions can usually be maintained within ±2%. Sharp radii should be avoided; the radius should normally be larger than 0.02 in. (0.5 mm).

Angles can be held to ±2° with rigid PVC or a similar material such as polystyrene. With flexible PVC and polyethylene angles can be held to ±5°.

In establishing specifications of the product tolerances, it is again good practice to consult with the extrusion engineer. If the dimensional tolerances are set closer than necessary, the result can be sustantial increases in both the tooling cost and actual extrusion cost. If the requirements for dimensional accuracy are very high, the extruder as such may be incapable of maintaining the required tolerances.

The extruder is not a positive displacement device, and the extruder output rate will vary with pressure, temperatures, and other process variables. The constancy of the extruder can be much improved by coupling the extruder with a gear pump. The main function of the extruder then becomes plasticating, and the pumping function is performed by the gear pump. Since the gear pump is a more positive displacement device, its output is quite constant. Extruders equipped with gear pumps are used in many applications where good dimensional control is very important; one example is the production of man-made fibers, where gear pump assisted extrusion has been standard practice for several decades.

87.6 PROCESSING CRITERIA

In order to control the process and to understand what is happening inside the extruder, a number of key process variables should be monitored on a continuous bases. Important process variables are

- Diehead pressure before screen pack.
- Diehead pressure after screen pack.
- Rotational speed of the screw.
- Temperature of polymer melt at the die.
- Power consumption of the extruder drive.
- Power consumption of barrel and die heaters.
- Temperature profile along the barrel.
- Cooling rate along length of extruder.
- Product take-up speed.

In addition to these, other variables may need to be monitored, such as the vacuum level in a vented extrusion operation where vacuum is applied to the vent port. Pressure measurement before and after the screen pack is useful to determine the pressure buildup as a result of contamination accumulating on the screens. Screen changes can be scheduled when the pressure differential reaches a certain value. In some automatic screen changers, the pressure differential is used to trigger a screen change without any operator intervention.

The measurement of diehead pressure is also important because it is an indirect indication of the extruder output rate. Fluctuation in diehead pressure in most cases can be correlated directly to variations of the extruded product. To assess the variation of pressure with time, it is convenient to use a chart recorder. Pressure fluctuations can be observed more easily from a chart than from a single instantaneous readout.

The rotational speed of the screw is important to monitor because it directly determines the extruder output. Fluctuations in screw speed will generally result in output fluctuations. As with diehead pressure, it is good practice to monitor screw speed with a chart recorder to provide a hard copy of the screw speed as a function of time.

The temperature of the polymer melt is another very important process variable. The melt temperature should be uniform and at the desired level. Melt temperature fluctuations will usually result in output fluctuations. Thus, it is important to monitor the melt temperature continuously, preferably by using a chart recorder.

The power consumption of the extruder drive is an indication of the resistance that the polymer exerts on the screw. This resistance against the rotation of the screw is primarily determined by the polymer melt viscosity and, to some extent, by the frictional characteristics of the particular polymer. Changes in drive power consumption at constant screw speed will affect the polymer stock temperatures and possibly the output. The power consumption of the barrel heaters is important in determining the overall power consumption and monitoring the local heat requirements along the extruder. Under steady conditions, the power to the heaters should be relatively constant. If the power consumption of a particular heat zone fluctuates widely on a persistent basis, there is a process instability that should be corrected. The instability could be caused by a faulty thermocouple, improper design or tuning of the controller, surging of the extruder, etc.

The temperature profile along the extruder is usually measured by thermocouples in the extruder barrel. Persistent deviations of the measured temperature from the set temperature are also indicative of a process control problem. For instance, if the measured temperature is far above the setpoint, even with the heater completely off and the cooling on full blast, there is too much internal heat generated by the frictional or viscous dissipation. This type of problem will result in overheating of the polymer and possibly degradation. Such a condition requires a drastic change in operating conditions or a different extruder screw.

The cooling rate along the extruder needs to be measured to determine how much cooling is required at the various locations. In a well-designed extrusion system, the need for cooling should be minimal. The energy extracted by cooling is lost and thus constitutes an unnecessary operating expense. Excessive cooling requirements along the extruder often indicate polymer overheating and improper screw design.

The product take-up speed should be monitored because sometimes variations in product dimensions are caused by the take-up and not by the extruder itself. In most such cases, the extruder is the first suspect, and troubleshooting efforts are directed towards the extruder. However, if the take-up is at fault, a readout of the take-up speed will be necessary to determine the source of the problem.

87.7 MACHINERY

The main elements of a single-screw extruder are the screw, barrel, feed hopper, die, heating and cooling system, control system, and the drive.

87.7.1 Extruder Screw

Figure 87.16 shows a typical screw for a single-screw extruder. This "standard extruder screw" consists of three geometrical sections: a feed section, compression section, and a metering section. The screw can also be described as a single flighted, constant pitch, single-stage extruder screw. Single flighted means that there is only one screw flight along the screw. Constant pitch refers to a constant helix angle of the screw flight along the length of

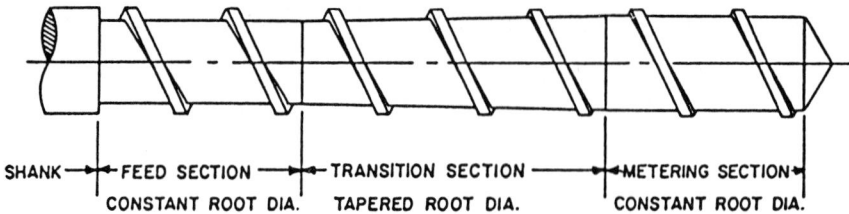

Figure 87.16. Typical screw for a single screw extruder.

the screw. Single stage indicates the presence of only one compression section along the screw.

The feed section is the first section of the screw. The channel depth is generally large to accommodate the relatively low density of the incoming bulk material. The length of the feed section is often about $5D$. In the feed section of the screw, the channel is mostly filled with solid polymer. The compression section of the screw (also called the transition section) has a gradually reducing channel depth. The metering section is the last section of the screw. Its channel depth is generally small to provide sufficient pressure-generating capability. The metering section is mostly filled with molten polymer.

In most standard temperature screws the OD of the screw is constant, with the radial clearance between the screw and barrel generally about one thousandth of the screw diameter. The helix angle or the pitch is also constant. Most screws have a helix angle that corresponds to a square pitch, which is an angle of 17.66°. This helix angle results in a flight pitch that equals the screw diameter.

Barrier Screws. Barrier-type extruder screws have enjoyed widespread popularity for a number of years. The barrier-type screw is different from the standard screw in the transition section. In the barrier screw the transition section has two flights, the main flight and a barrier flight. The barrier flight is generally narrower than the main flight and has a larger radial clearance. The clearance is chosen to allow the melt to flow across the barrier flight but not the solid polymer. Thus, the melt accumulates in the melt channel and the solid remains in the solids channel. The operating principle of the barrier screw is based on phase separation. The cross sectional area of the solids channel reduces to zero at the end of the barrier section. At the same time, the melt channel cross sectional area increases from zero to full channel cross section at the end of the barrier section. The barrier screw forces a certain melting profile on the polymer corresponding to the profile of the solids channel.

Only molten polymer can travel beyond the barrier section; thus, complete melting is almost assured. Another advantage is that the solid bed is more fully supported and less likely to break up, resulting in a more stable plasticating process. The benefits of barrier screws in terms of melting performance are marginal compared to nonbarrier screws.[3]

Mixing Elements. In order to improve the mixing capability of extruder screws, mixing elements are often incorporated towards the end of the screw. One can distinguish between dispersive and distributive mixing elements. A dispersive mixing element exposes the polymer melt to high shear stresses in order to break down gels or agglomerates. Examples of dispersive mixing elements are the Union Carbide mixing section and the shear blister. A distributive mixing element divides and combines the melt stream. Examples of distributive mixing elements are the Dulmage mixing section (**b**), the cavity transfer mixing section (**c**), and the pin mixing section (**a**) (Fig. 87.17).

Multistage Screws. Multistage screws are generally used for extraction of volatiles from the polymer. A two-stage extruder screw is essentially two single-stage screws in series along one shaft. Volatiles are removed through a vent port in the barrel at the extraction section of the screw (Fig. 87.18). Usually, a vacuum is applied at the vent port to improve the

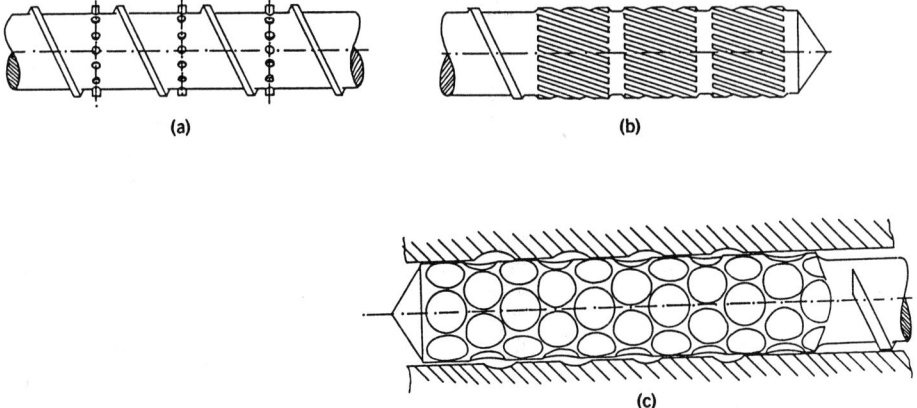

Figure 87.17. Distributive mixing elements (**a**) pin type (**b**) Dulmage type (**c**) cavity transfer type.

devolatilization effectiveness. For more complete devolatilization, two or more vent ports can be incorporated along the length of the extruder. In vented extrusion the screw design is very important to ensure proper operation of the extruder; the design of two-stage extruder screws is discussed in detail in the following publication.[13]

Multiscrew Extruders. Multiscrew extruders are machines with more than one extruder screw. The most important multiscrew extruder is the twin screw extruder. Intermeshing twin screw extruders have capabilities that often extend considerably beyond those of the single screw extruder. Some of the advantages of twin screw extruders are good mixing capability, good devolatilization capability, and good control over residence time (RT) and RT distribution.

Some disadvantages of twin screw extruders are that they are more expensive and their performance is difficult to predict.

A listing of the various types of intermeshing twin screw extruders is shown in Table 87.1.

There are very major differences among the various types of twin screw extruders. Therefore, it is hardly possible to make general statements about twin screw extruders. The various types of twin screw machines differ in geometry and applications. Low-speed corotating and counterrotating twin screw extruders are primarily used in profile extrusion. This application requires positive conveying characteristics, short residence time, and good control over stock temperature; counterrotating twin screw extruders are particularly suited for this application.

High-speed corotating twin-screw extruders are primarily employed in compounding applications, where use is made of controlled shear input and high throughout capability. Twin screw extruders are also starting to be used more and more as continuous chemical reactors.[14] Various polymers can now be polymerized in a twin screw extruder; examples include PETP, PBT, PA6, PA6/6, PMMA, etc.

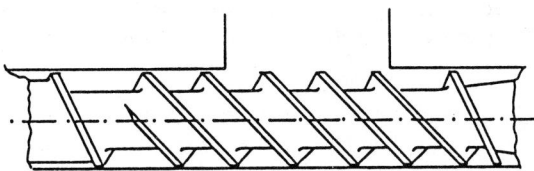

Figure 87.18. Two stage extruder screw-extraction section.

TABLE 87.1.   Classification of Twin Screw Extruders

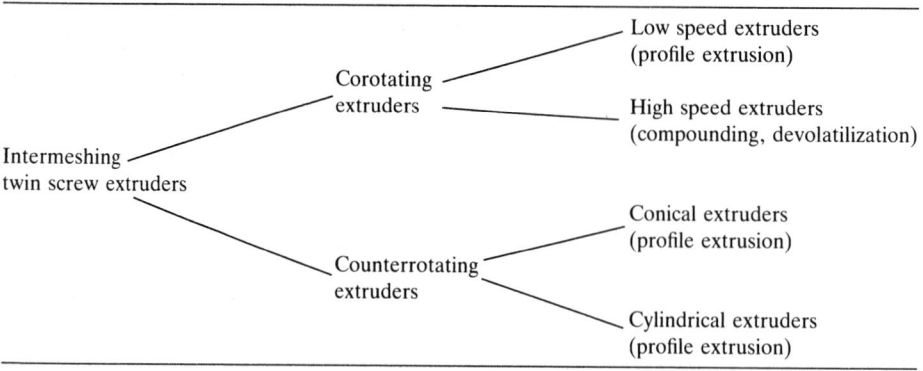

Intermeshing twin screw extruders
- Corotating extruders
  - Low speed extruders (profile extrusion)
  - High speed extruders (compounding, devolatilization)
- Counterrotating extruders
  - Conical extruders (profile extrusion)
  - Cylindrical extruders (profile extrusion)

87.7.2   Disk Extruders

Disk extruders utilize some type of disk to induce transport of the material. These machines are sometimes referred to as screwless extruders. A number of disk extruders have been designed over the last 30 years, among them the spiral disk extruder, drum extruder, stepped disk extruder, and elastic melt extruder. None of these machines has found widespread use in the polymer industry.

A relatively new disk extruder was conceived by Tadmor[15] and commercialized by Farrel. Material drops in the axial gap between two relatively thin disks mounted on a rotating shaft. The material will move with the disks, almost a full turn, until it meets a channel block. The channel block closes off the space between the disks and forces the polymer to flow to either an outlet channel or a transfer channel in the housing. The shape of the disks can be optimized for specific functions, such as solids conveying melting, devolatilization, melt conveying, and pumping. A detailed analysis of the diskpack machine can be found in Tadmor's book on polymer processing.[12]

87.7.3   Ram Extruders

Ram or plunger extruders are simple in design, rugged, and discontinuous in their mode of operation. Ram extruders are essentially positive displacement devices and are able to generate very high pressures. Because of the intermittent operation of ram extruders, they are ideally suited for cyclic processes, such as injection molding and blow molding. In fact, the early molding machines were almost exclusively equipped with ram extruders to supply the polymer melt to the mold. Certain limitations of the ram extruder have prompted a switch to reciprocating screw extruders. The two main limitations are limited melting capacity and poor temperature uniformity of the polymer melt.

Presently, ram extruders are used in relatively small shot-size machines and certain operations where their positive displacement characteristics and outstanding pressure generation capability are exploited. There are basically two types of ram extruders, single ram extruders and multiram extruders.

Ram extruders are used for extrusion of intractible polymers, such as UHMWPE and PTFE. They are also used in solid state extrusion, where the polymer is extruded just below its melting point. Solid state extrusion results in a highly oriented extrudate with very good mechanical properties.

87.7.4   Extruder Drive

Most extruder drives have a constant torque characteristic. As a result of the constant torque characteristic, the power is directly proportional to the motor speed. This means that the full motor power can only be utilized when the motor is running at its nominal speed.

Therefore, when power limitations are of concern, it is important to make sure that the motor operates close to its nominal speed. If it is running considerably below nominal speed, a gear change can be made to move the motor speed to the required level. Some gear boxes are equipped with a quick-change gear provision that allows a relatively quick change of the gear ratio. Such a feature enhances the flexibility of an extruder, particularly since a gear change on a regular gear box is a major job.

Motors. Most extruders use D-C motors conrolled by SCRs (silicon controlled rectifiers). The D-C motor is popular because it can be controlled over a wide speed range (about 30:1) with good speed regulation. It is relatively inexpensive and has good energy efficiency. The A-C motor, coupled with an eddy-current clutch, was used on many older extruders. The major drawback of the eddy-current clutch is that the energy efficiency is directly proportional to the speed difference between input and output shaft. As a result, the eddy-current clutch is very inefficient at low speeds. The new adjustable frequency A-C drive has a good speed range and speed regulation; however, its cost at this time is considerably higher than the D-C drive.

In the late 1980s there has been a strong move towards brushless D-C drives. They are more electrically efficient than conventional D-C drives, particularly at lower speed. Other significant advantages are reduced maintenance and wider speed range of up to 100:1.

Hydraulic drives are used very often on reciprocating extruders on molding machines. Hydraulic drives are not used very often on conventional extruders. Hydraulic drives have the advantage that the gear box can be eliminated by use of a low-speed, high-torque hydraulic motor. The overall energy efficiency of a hydraulic drive tends to be somewhat lower than a D-C drive and the price somewhat higher.[16] Another advantage of the hydraulic drive is that auxiliary equipment, such as the screen changer, can be operated from the same hydraulic power supply.

Reducer and Thrust Bearings. Extruders generally employ gear reducers to match the speed of the screw to the speed of the motor. A typical reduction ratio ranges from 15:1 to 20:1. Spur gears are often used in a two-step configuration; that is, two sets of intermeshing gears. A popular spur gear is the herringbone gear because the V-shaped tooth design practically eliminates axial loads on the gears. The efficiency of these gears is high, about 90% at full load and 96% at low load.

The thrust bearing assembly is usually located at the point where the screw shank connects with the output shaft of the drive. The thrust bearing capability is required because the extruder generally develops substantial diehead pressure in the polymer melt. This pressure is necessary to force the polymer melt through the die at the desired rate. However, since action equals reaction, this pressure will also act on the screw and force it towards the feed end of the extruder. The load on the thrust bearings is directly determined by the diehead pressure and the cross-sectional area of the screw.

Extruder manufacturers often give the rated life of the thrust bearing as a B-10 life. This is expressed in hours at a particular diehead pressure [usually 5000 psi (35 MPa)] and screw speed (usually 100 rpm). The B-10 life represents the life in hours at constant speed that 90% of an apparently identical group of bearings will complete or exceed before the first evidence of fatigue develops. Thus, 10 out of 100 bearings will fail before rated life!

The B-10 life at normal operating load should be at least 100,000 hours to obtain a useful life of more than 10 years from the thrust bearings. The predicted B-10 life at any diehead pressure and screw speed can be determined from the following relationship:

$$B\text{-}10\ (P,N)\ =\ B\text{-}10_{\text{st}} * \frac{100}{N} * \left(\frac{35}{P}\right)^k \tag{87.9}$$

where P is diehead pressure in psi (MPa), N is screw speed in rpm, $B\text{-}10_{\text{st}}$ is the $B\text{-}10$ at 100 rpm and 5000 psi (35 MPa), and k is a constant ($k = 3.00$ for ball bearings and $k = 10/3$ for roller bearings).

87.7.5 Heating and Cooling

The extruder barrel is equipped with heating and cooling systems to maintain a particular temperature profile along the barrel. Barrel heating consists usually of electrical resistance heaters clamped around the length of the barrel. They are usually grouped in three to six zones. Each zone has a separate thermocouple and temperature controller, thereby allowing independent temperature control.

Cooling can be accomplished by either air or water. Air cooling is usually done with blowers. The cooling system is divided into zones, such that each zone can be cooled independently. The cooling zones normally coincide with the heating zones. Water cooling is more effective than air cooling in terms of heat removal. However, smooth temperature control may be more difficult with water cooling.

Systems for heating and/or cooling of the extruder screw are not as common as they are for the extruder barrel. However, a screw heating or cooling capability can be added relatively easily. The heat transfer medium (water, oil, and so on) enters the hollow screw core and runs through a pipe to the end of the bore. The connection between screw and supply lines can be made with a rotary union to keep the system closed. It is also possible to stick a pipe into the screw core and drain back into the opening at the end of the screw. Obviously, the latter method provides a very simple method of screw temperature control. Provisions can be made to heat or cool only particular sections of the extruder screw. Electrical cartridge heaters have also been used for screw heating. In some cases, screw cooling is used to compensate for a screw cut too deeply. The effective channel depth is reduced by solidification of a layer of material on the screw surface. This mode of operation is not recommended since it wastes energy and probably means that an improper screw design is being utilized. It also widens the residence time distribution and can cause degradation problems in heat-sensitive materials. Screw cooling can also be used to reduce surging of the extruder.

A rather unusual method of screw cooling and heating involves the use of a heat pipe. This is a hermetically sealed pipe containing a heat transfer medium and generally a type of wick along the length of the pipe. If there is a sufficient temperature difference at the ends of the pipe, the heat transfer medium will flow and try to eliminate this temperature difference. The medium will flow in a circulatory fashion and create an efficient heat transfer. Heat pipes are used in many applications; mold cooling is one example.

A wick is often incorporated to improve the flow of the heat transfer medium through capillary action. The heat pipe is fitted closely in the screw bore with one end close to the metering end of the screw and the other end in the feed section end of the screw. Thus, excess heat from the metering section can be transferred to the feed section where heat is required. This can result in more even temperature profiles and more effective energy usage.

When considering the heating and cooling requirements of an extruder, it is important to take into account the size of the extruder. For small extruders the cooling requirements are less severe than for larger extruders. The surface area to channel volume ratio is large for small extruders; thus, cooling through conduction and radiation can be considerable. Temperature control, therefore, becomes more difficult as the extruder size increases.

87.7.6 Temperature Control

There are three types of temperature control: manual, off/off, and proportional control. Manual control is simple and inexpensive in terms of hardware. The amount of heat applied is varied with visual observation of the actual temperature. The applied heat can be changed with a potentiometer or variable transformer (variac, trafo) or by impulse control. The accuracy of this method is very much operator-dependent. Manual control should only be used for relatively short runs, such as for a short experiment; it is not recommended for continuous production.

On/off control was the first type of automatic temperature control. This type of controller compares the measured temperature with the temperature setpoint and simply switches off the power supply to the heater when the setpoint is exceeded. Under steady state conditions the temperature will cycle about the set point. The magnitude of the cycle will depend on the temperature lag between the thermocouple and heater. The location and depth of the thermocouple are important parameters affecting this thermal lag. In general, temperature excursions above and below the set point are large, making on/off control undesirable for situations where very close temperature control is required.

Proportional controllers apply an amount of power to the heater that is proportional to the measured temperature. The first systems were time proportioning control systems. In these systems, the "on" time is varied to control the power applied to the heater. This is basically a variable frequency on/off control. If the power switching cycle is considerably shorter than the natural cycle time, the temperature oscillation will die away and the temperature will stabilize at a steady value.

The time proportioning system was improved with the introduction of SCRs (silicon controlled rectifiers), which proportion a constant but smaller amount of power to the heaters. The proportional band is the temperature band in which power is proportioned to the heater. The set point is normally the upper limit of the band; at this point the heater input power is zero. The lower limit represents full power applied to the heater. If power is required to maintain the setpoint temperature, the actual temperature will be lower than the set point. This is the offset or droop error. The offset error can be reduced by decreasing the width of the proportional band. However, this will increase the initial temperature overshoot on start up. It will also adversely affect the damping of the oscillations after start up.

A disadvantage of the proportional-only controller is that it is based on steady state conditions. An upset in the thermal balance of the system will require a correction of the amount of power applied to the heater to maintain the same temperature. In an extruder, a change in throughput or a change in resin viscosity would cause a change in thermal balance. A proportional-only (P) controller is incapable of correcting for a change in the thermal balance. This can be a problem if frequent changes in operating conditions and/or polymer properties occur.

To circumvent this problem, an automatic reset can be incorporated in the control circuit (PI controller). This feature automatically eliminates the process droop. An electronic circuit picks up a possible change in steady-state conditions and adjusts the controller output to maintain setpoint conditions. The reset function includes adjustments that allow it to handle systems with widely varying characteristics.

Proportional control can be combined with rate or derivative control to eliminate large overshoots normally encountered during start up. These can be particularly severe when the time constant of the process is long, as it is in extrusion. An electronic circuit senses the rate of change in process temperature (the derivative of temperature with time) and produces a signal proportional to the rate of change. Thus, corrective action can be taken before the temperature variation becomes excessive. Proportional plus derivative (PD) control allows a narrow proportional band, reducing the offset error and improving control.

Proportional plus integral plus derivative (PID) control is used to eliminate the offset error automatically. This is achieved by integrating the error signal to provide a power control signal that reduces the error progressively to zero. Thus, the effect of the integrator is the same as manual trimming. The PID controller is capable of holding the setpoint without hysteresis; this is possible because of the sluggish response of the integrator.

Controlling the extrusion process temperature may require cooling instead of heating. Particularly at high screw speeds and with high viscosities, the internal viscous heat generation can be considerable. Thus, it is clear that dual controllers are required that are capable of controlling both heating and cooling. The requirement for cooling control is at least as stringent as that for heating control.

Oil or air-cooled systems can use stepless cooling systems with proportional valves and positioning motors. These systems are relatively expensive but they are reliable and require little maintenance. With water cooling, the cooling power is obtained by energizing a solenoid valve. For low temperatures (no flashing), a constant cycle rate is often used with variable pulse width. The pulse width varies in proportion to the cooling power required. At high temperatures, where water is flashed to steam, more intensive cooling is possible. In these cases a different cooling control can be used, one known as a constant pulse width system. When cooling is required, the solenoid is energized by a pulse signal of predetermined length. The frequency of the pulse is varied in proportion to the cooling power required.

87.7.7 Extruder Manufacturers

See Table 87.2.

87.8 DIE ASSEMBLY

The die is usually connected to an adaptor; in turn, the adaptor is connected to the extruder barrel. The purpose of the die is to shape the extrudate. This shape can be the final product shape or an intermediate shape that is later modified in a post-extrusion process step.

Quite often a breaker plate is incorporated between the barrel and the adaptor. The breaker plate is a thick circular disk drilled with many holes spaced closely together. The purpose of the breaker plate is to arrest the rotating motion of the extrudate as it is forced from the screw to the die exit. It can also be used to support screens to filter out contamination in the polymer. The screens are placed at the upstream side of the breaker plate. Screens can also be used to intentionally increase the diehead pressure and thereby obtain better mixing.

The die is one of the most important elements in extrusion. Unfortunately, die design is one of the most complicated aspects of polymer process engineering. It requires a thorough knowledge of the flow characteristics of the polymer and the product requirements. A description of the flow behavior of the polymer in the die usually entails highly complex mathematics unless very simple geometries are involved.

A good die designer, therefore, should be a good rheologist, mathematician, and a designer. A fundamental approach to die design can present great difficulties. For this reason, a largely empirical approach is used for many die designs, particularly for the more complicated geometries. This makes die design more of an art and places a high value on experience. However, empirical approaches rarely lead to fundamental improvements. One tends to use a design that used to work and make slight modifications to it. Design rules that work for one material may not hold for another material. Therefore, it is important to understand the complex interaction among material characteristics, die design, and operating conditions.

There is an extremely large number of different die designs. They can be categorized by the shape of the extrudate that is produced: annular geometries (pipe and tubing), circular geometries (rod, coated wire, filament, and so on), thin rectangular geometries (film and sheet), and, finally, the more complex geometries.

There is very little difference between a pipe and tubing die. The major distinction is one of size: small-diameter products [about 0.3 in. (1 cm) or less] are called tubing, larger diameter products are called pipe. Sometimes larger diameter products are called tubing if they are thin-walled and flexible. A typical pipe and tubing die is shown in Figure 87.19. As the polymer flows through the die body, it has to flow around the legs of the spider support of the core. After the spider support, the split streams recombine to a tubular stream. This can cause weld lines in the product and form a weak region that could result in premature failure of the product. A crosshead extrusion die is shown in Figure 87.20. The polymer flow makes a 90° turn and splits at the same time over the core tube. Thus, in the crosshead die, a weld line can form as well.

A crosshead die for a wire coating extruder is very similar to a crosshead tubing die; in fact, the same crosshead can be used for wire coating and tubing extrusion. Figure 87.20 shows the difference between pressure extrusion and tubing extrusion. Pressure extrusion is preferred when intimate contact between the wire and the polymer is required.

TABLE 87.2. List of Extruder Manufacturers

Manufacturer	Single-screw Extruders	Twin-screw Extruders
Akron Extruders, Inc. 1119 Milan St., Canal Fulton, OH 44614	1–2-$\frac{1}{2}$ in. (25–64 mm)	
A. Anger Derfflingerstr. 15, A-4017 Linz, Austria		Con. ctr. and 2$\frac{3}{8}$/5 in. 3$\frac{1}{2}$/6 in. (60/125 and 88/151 mm)
Axon S-26500 Åstorp, Sweden	$\frac{1}{2}$–4-$\frac{3}{4}$ in. (13–121 mm)	
Amut spa Via Cameri 16, I-28100 Novara, Italy	1$\frac{1}{8}$–6$\frac{1}{4}$ in. (30–160 mm)	Cyl. ctr 2$\frac{5}{8}$–6$\frac{1}{4}$ in. (67–160 mm)
Alpine American Corp. 5 Michigan Dr., Natick, MA 01760	1$\frac{3}{8}$–3 in. (35–75 mm)	
Baker Perkins (now APV Chemical Machinery) 1000 Hess St., Saginaw, MI 48601		2–8$\frac{1}{4}$ in. Corot(50–200 mm)
American Barmag P.O. Box 7046, 1101 Westinghouse Blvd., Charlotte, N.C. 28217	1$\frac{1}{8}$–12 in. (30–400 mm)	
Berstorff Corp. 8200-A Arrowridge Blvd., Charlotte, N.C. 28224	$\frac{3}{4}$–23$\frac{1}{2}$ in. (20–600 mm)	1$\frac{1}{2}$–10 in. Corot(40–258 mm)
Brampton Engineering 8031 Dixie Road South, Brampton, Ontario, Canada L6T3V1	1-$\frac{1}{2}$–3 in. (38–76 mm)	
C.W. Brabender Instruments, Inc. 50 East Wesley St., South Hackensack, NJ 07606	$\frac{3}{4}$–1-$\frac{1}{4}$ in. (19–31 mm)	1$\frac{3}{8}$ Cyl. ctr(35 mm)
John Brown Plastics Machinery, Ltd. Bath Road, Gloucestershire GL53TL UK	2–6 in. (51–52 mm)	
Buss, Inc. 2411 United Ln. Elk Grove Village, IL 60007	Ko-kneader 46–200 mm (1$\frac{7}{8}$–7$\frac{7}{8}$ mm)	
Cincinnati Milacron 4165 Halfacre Road, Batavia, OH 45103		2$\frac{1}{8}$–4$\frac{3}{8}$ in. Con. ctr.(55/110 mm) (−90/177 mm)
Dr. Collin GmbH Sportparkstrasse 2, D-8017 Ebersberg, Federal Republic of Germany	$\frac{7}{8}$–1$\frac{3}{4}$ in. (20–45 mm)	Cyl. co-and etr. 2 in. (50 mm)
Crown Products, Inc. 120 Factory Road, Addison, IL 60101	1-$\frac{1}{2}$–12 in. (76–305 mm)	
Crespi SpA Via Roncaglia 14, 20146 Milano, Italy	2–4$\frac{3}{4}$ in. (50–120 mm)	
Creusot-Loire 42 Rue d'Anjou, 75008 Paris, France		1$\frac{3}{4}$–9$\frac{7}{8}$ in. Carot(45–252 mm)
Custom Scientific Instruments, Inc. 13 Wing Dr., Cedar Knolls, NJ 07927		Disk extruder $\frac{3}{4}$ and 2 in. (19–51 mm)
Davis-Standard Div. U.S. Route 1, Pawcatuck, CT 02891	1.5–8 in. (38–203 mm)	
Egan Machinery, S. Adamsville Rd. Somerville, NJ 08876	1-$\frac{1}{2}$–6 in. (38–152 mm)	
Entwistle Co. Bigelow Street, Hudson, MA 01749	$\frac{3}{4}$–6 in. (19–152 mm)	

TABLE 87.2 (*Continued*)

Manufacturer	Single-screw Extruders	Twin-screw Extruders
Esde		$2\frac{1}{8}$–$3\frac{3}{4}$ in.
Mindener Strasze 39,		Co- and etr(55–95 mm)
4970 Bad Oeynhausen 1,		
Federal Republic of Germany		
Farrel Co.	Single screw up	
Ansonia, CT 06401	to 24 in.	
	(610 mm)	
Fairex	1–6 in.	
37 Rue Lavoisier, 77270 Villeparisis, France	(25–150 mm)	
Finnpack	$1\frac{1}{2}$–$4\frac{3}{4}$ in.	
37800 Toijala, Finland	(40–120 mm)	
Gatto Mach. Dev. Corp.	$1\frac{1}{4}$ in.	
45 Rabro Dr., Hauppauge, LI, NY 11788	(31 mm)	
Battenfeld Gloucester Engineering Co.	2–8 in.	
Blackburn Industrial Park, P.O. Box 900	(51–203 mm)	
Gloucester, MA 01930		
Goettfert	$\frac{3}{4}$–$2\frac{3}{8}$ in.	
Postfach 1220 - 6967 Buchen,	(20–60 mm)	
Federal Republic of Germany		
Guix	$2\frac{3}{8}$–$11\frac{7}{8}$ in.	
Apartado B, Cornella (Barcelona), Spain	60–300 mm	
Hartig Plastics Machinery Div.	$1\frac{1}{2}$–15 in.	
P.O. Box 791, New Brunswick, NJ 08903	(38–381 mm)	
HPM Corp.	2–6 in.	
820 Marion Rd., Mount Gilead, OH 43338	(51–150 mm)	
Ide America	$1\frac{1}{8}$–$3\frac{1}{2}$ in.	$2\frac{3}{4}$ in.
P.O. Box 515, Chagrin Falls, OH 44022	(30–90 mm)	Cyl ctr(70 mm)
Ital Spa	$1\frac{3}{4}$–$7\frac{7}{8}$	
21052 Busto Arsizio Va Italia	(45–200 mm)	
Killion Extruders, Inc.	$\frac{3}{4}$–$2\frac{1}{2}$ in.	
200 Commerce Rd., Cedar Grove, NJ 07009	(19–64 mm)	
Krauss-Maffei Corp.	$2\frac{3}{8}$–$4\frac{3}{4}$ in.	$3\frac{1}{2}$–$4\frac{3}{4}$ in.
7095 Industrial Rd., Florence, KY 41042	(60–120 mm)	Cyl. ctr(90–120 mm)
		$\frac{1}{2}$–$2\frac{3}{8}$/5 in.
		Con. ctr.(25/50–60/125 mm)
Kuhne Gmbh	1–6 in.	
Einstein Str., 5205 St. Augustin 3, Menden,	(25–150 mm)	
Federal Republic of Germany		
Chemiefaser Lenzing AG	$1\frac{1}{8}$–$4\frac{3}{4}$ in.	
A-4860 Lenzing, Austria	(30–120 mm)	
American Leistritz Extruder Corp.	2–6 in.	$1\frac{3}{8}$–$6\frac{5}{8}$ in.
169 Meister Ave., Somerville, NJ 08876	(50–150 mm)	Co and ctr(34–170 mm)
Macro Engineering Co. Ltd.	$2\frac{1}{2}$–6 in.	
1177 Invicta Dr., Oakville, ON	(64–152 mm)	
Ontario L6H 4M1, Canada		
Maillefer (now Nokia-Maillefer)	$1\frac{1}{8}$–$7\frac{7}{8}$ in.	
749 Ludlow Rd., South Hadley, MA 01075	(30–200 mm)	
Maris SpA		$1\frac{3}{8}$–$6\frac{5}{8}$ in.
Corso Monanisio 22, 10090 Rosta		Corot(35–170 mm)
(Torina), Italy		
Meaf B.V.	$2\frac{3}{8}$ in.	$1\frac{1}{8}$ in.
Postbus 98, 4400 AB Yerseke,	(60 mm)	Corot(30 mm)
the Netherlands		
MPM Polymer Systems, Inc.	1–6 in.	
P.O. Box 2066, Clifton, NJ 07015	(25–152 mm)	

TABLE 87.2 *(Continued)*

Manufacturer	Single-screw Extruders	Twin-screw Extruders
New Castle Ind., Inc.	$1\text{-}\frac{1}{2}\text{-}3\text{-}\frac{1}{2}$ in.	
1399 Countyline Rd., P.O. Box 7359	(38–89 mm)	
New Castle, PA 16101		
The Northampton Machinery Co. Ltd.	$1\text{-}\frac{1}{2}\text{-}6$ in.	
Balfour Rd., Northampton,	(38–152 mm)	
NN2 6JS, England		
NRM Corp	$2\text{-}\frac{1}{2}\text{-}6$ in.	
P.O. Box 25, Columbiana, OH 44408	(64–152 mm)	
Omipa SpA	$1\frac{1}{8}\text{-}6$ in.	$3\frac{3}{8}\text{-}4$ in.
via Maddalena 7-cas. post. 11,	(30–150 mm)	Cyl. ctr(85–100 mm)
21040 Morazzone (Varese), Italy		
Polysystems Machinery Mfg. Inc.	$1\text{-}\frac{1}{2}\text{-}6$ in.	
5191 Creekbank Rd., Mississauga,	(38–152 mm)	
Ontario L4W 1R3, Canada		
Prandi SpA	$3\frac{1}{2}\text{-}4\frac{3}{4}$ in.	
via Sempione-28040 Marano Ticino	(90–120 mm)	
(Novara), Italy		
Reeve UK Ltd.	$\frac{3}{4}\text{-}2$ in.	
Boyn Valley Rd., Maidenhead, Berkshire	(20–50 mm)	
SL6 4EJ, UK		
Reifenhauser-Van Dorn Co.	$1\frac{3}{8}\text{-}6$ in.	$2\frac{1}{8}\text{-}5\frac{1}{8}$ in.
35 Cherry Hill Dr., Danvers, MA 01923	(35–150 mm)	Cyl. ctr(55–130 mm)
Sterling Extruder Corp. (now APV)	$1\text{-}\frac{1}{2}\text{-}8$ in.	
901 Durham Ave., South Plainfield,	(38–203 mm)	
NJ 07080		
Wayne Machine & Die Co.	$\frac{3}{4}\text{-}2\text{-}\frac{1}{2}$ in.	
100 Furler St., Totowa, NJ 07512	(19–64 mm)	
Welding Engineers, Inc.	Non-	Ctr 0.8–10 in.
303 E. Church Rd., King of Prussia,	intermeshing	(20–254 mm)
PA 19406		
Welex Inc.	2–12 in.	
850 Jolly Road, Blue Bell, PA 19422	(51–305 mm)	
Werner & Pfleiderer Corp.		$1\frac{1}{8}\text{-}6\frac{3}{4}$ in.
663 E. Crescent Ave., Ramsey, NJ 07446		Corot(30–170 mm)
Windmoeller and Hoelscher Corp.	$1\text{-}\frac{3}{4}\text{-}6$ in.	
23 New England Way, Lincoln, RI 02865	(45–152 mm)	
Wilmington Machinery	$1\text{-}\frac{1}{2}\text{-}4\text{-}\frac{1}{2}$ in.	
3706 Boren Dr., P.O. Box 77419,	(38–114 mm)	
Greensboro, NC 27417		

Figure 87.21 identifies several types of sheet dies. The objective of all sheet dies is to deliver a sheet of uniform thickness. To achieve this, the opening of the die can be adjusted by, for instance, a flex lip adjustment. A choker bar in the center portion of the die provides another means of adjustment. Setting a particular temperature profile across the die is still another method of adjustment.

87.9 AUXILIARY EQUIPMENT

87.9.1 Conveying and Metering Equipment

In the area of materials conveying and metering, automation is starting to have a major impact. Conveying from the silo to the feed hopper often is accomplished via a pneumatic

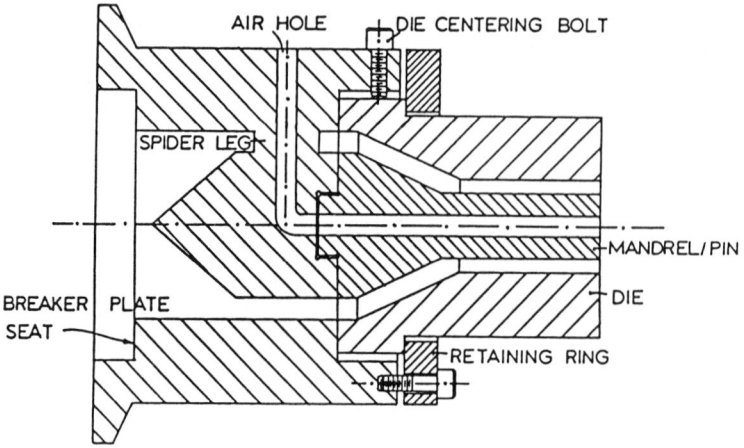

Figure 87.19. Typical pipe and tubing die.

materials handling system. The design of the pneumatic system depends on the bulk material characteristics, distance, and rate requirements. An alternative to a pneumatic conveying system is a system with flexible mechanical conveyors. Mechanical conveyors eliminate filters used in pneumatic systems and can handle difficult materials, such as scrap and powdered additives. Bridging in hoppers is counteracted by combinations of vibratory devices and aeration pads. Vertical conveying is often done by mechanical conveyors. A new device to do the same job is the spiral feeder. By subjecting the spiral to controlled vibrations, material can be made to travel up or down. This device can convey material as high as 22 ft (7 m) at a rate of about 350 ft^3 (10 m^3) an hour.

Bulk material is metered directly into an extruder by volumetric feeders and gravimetric feeders. Volumetric feeders are usually auger feeders in which the rotational speed of the auger is used to set the rate according to the transport characteristics and the density of the bulk material. Variations in bulk density will cause variations in the conveying rate. Gravimetric feeders continuously measure the weight of the material in the hopper and control the speed of the discharge device to obtain a constant rate of decline of the weight of the material in the hopper. In most systems, the total weight of both bulk material and hopper is measured.

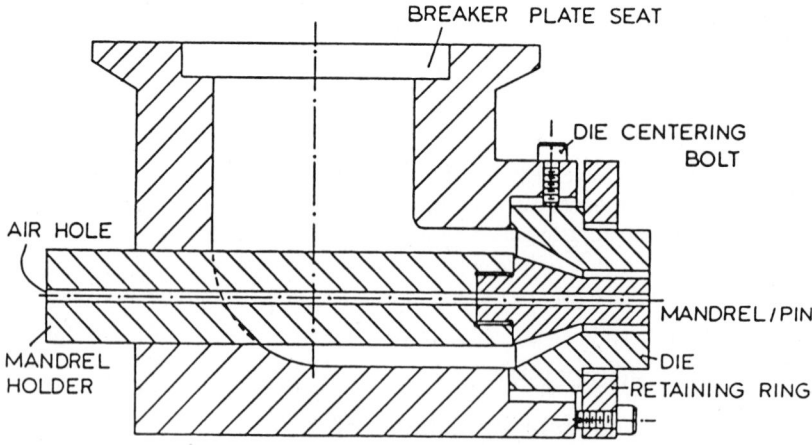

Figure 87.20. Crosshead tubing die.

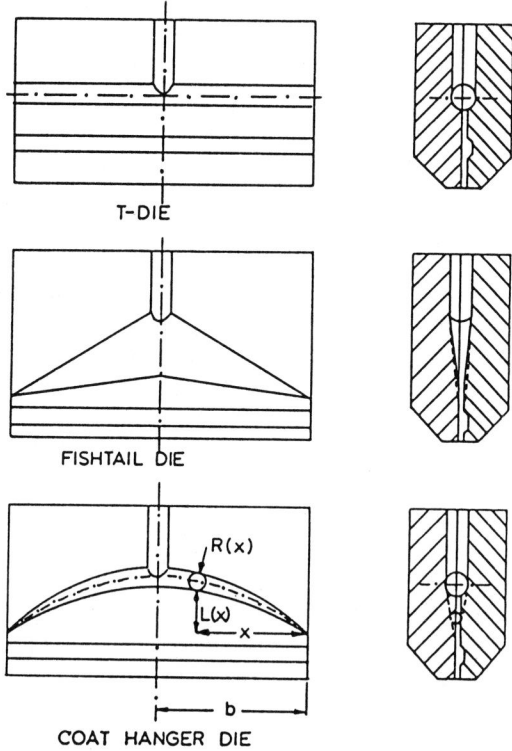

Figure 87.21. Three types of sheet dies.

Such metering devices are often linked together electronically to maintain the required ratio when several ingredients are fed simultaneously to an extruder. In this way, the output can be increased and decreased quickly without having to reset all the different feeders.

87.9.2 Dryers

Dryers are used for materials that contain a higher level of moisture than is acceptable in extrusion. For most materials, the moisture level has to be below 0.2% to yield acceptable extrudate quality; for some materials, this critical moisture level can be considerably lower (for example, about 0.02% for PC and 0.05% for bottle-grade PET). Excess moisture may be residual moisture from an underwater pelletizing operation, for instance, and is usually caused by the hygroscopic properties of the polymer. Hygroscopic polymers are polymers whose equilibrium moisture content under ambient conditions is above about 0.2%; examples include ANS, PMMA, PA, and etc.

The most common dryers are desiccant, hot air, and refrigerant type dryers. Desiccant dryers are now built with self-regenerating desiccant beds, which eliminate tedious manual regeneration. Small industrial desiccant dryers used on extruders and molding machines can reach dew points of $-22°F$ ($-30°C$). Closed loop cooldown of the desiccant beds is one method used to reduce energy consumption; dewpoint sensing is another. The regeneration heater is activated only when the measured dew point reaches the setpoint, instead of activation at regular time intervals. Some systems incorporate temperature measurement in the desiccant bed. When the temperature reaches setpoint, indicating complete regeneration, the heaters are shut off. This provides yet another means to reduce energy consumption. Refrigerant type hopper dryers can be used for materials that do not require a $-40°F$ dew point. Refrigerant dryers require about 30% less energy to operate than desiccant dryers. The dewpoint that can be obtained with refrigerant type dryers is about 32°F (0°C). Refrigerant dryers are easier to maintain and more compact than desiccant dryers but may not be suitable for all hygroscopic materials.

87.9.3 Size Reduction Equipment

In many extrusions and molding operations, the profitability of the process is dependent on the ability to recycle reground material back into the extruder efficiently. Granulators can be either centrally located or smaller dedicated granulators can be used locally. The knife design is very important to the proper functioning of a granulator. Mounting the rotor knives on an angle to the stationary knife results in a shear cut. This allows lower rotor speeds, reduces power consumption and enables handling of softer materials.

Two common types of pelletizers are the strand pelletizer and the die face pelletizer. Strand pelletizers produce more or less cylindrical pellets, while die face pelletizers produce more or less spherical pellets. This can result in significantly different solids conveying characteristics and air entrapment tendencies. Dicers handle material in web form and are capable of higher outputs than strand pelletizers. In film and fiber reclaim systems, in addition to size reduction, the material has to be densified so it can be recycled without conveying problems. Various systems are available to handle film and fiber scrap. Some systems first chop the scrap and then densify it with a frictional heater. Other systems use multistage cutting chambers. Another technique is direct repelletization without cutting or densification by feeding the scrap directly into a specially densified reclaim extruder.

Table 87.3 identifies manufacturers of auxiliary equipment and conveying and metering equipment.

87.10 FINISHING

A large number of finishing operations are used in the extrusion process. The finishing operation can be in-line, such as cutting, orientation, slitting, or stamping, or it can take place in a secondary off-line operation, such as thermoforming of an extrusion-formed sheet.

87.11 DECORATING

Various techniques are used to decorate extruded products, such as coloring, hot stamping, roll texturing, electroplating, painting, printing, and vacuum metallizing. For example, coloring can be accomplished using color concentrates mixed in with virgin pellets. Hot stamping is a process by which a decorative motif is applied to a product. A heated die is pressed against the product, transferring color or design from a coated film. Roll texturing is used to control the texture of the extruder product by passing it through a set of rolls with the required negative surface texture machined on each roll.

87.12 COSTING

The cost of an extruded part consist of several items: direct labor, materials, utility requirements, equipment amortization, building amorization, shipping, and packaging. If the extruded product requires a special die, the die generally does not work right away but has to be modified a few times before the correct extrudate shape and dimensions are obtained. This fine tuning of the tooling can become rather costly, and if the cost involved is not properly predicted, profitability can be adversely affected.

Consider an extrusion operation to produce film grocery bags. If the line is sized to a

TABLE 87.3. Auxiliary Equipment Manufacturers

Conveying and Metering Equipment Manufacturers

A.I.M.
Colortronics Systems, Inc.
Flexicon Corp.
Henderson Industries
K-Tron Corp.
Maguire Products
O.A. Newton & Sons Co.
Process Control Corp.
Thoreson-McCosh
Vibra-Screw Inc.

Acrison, Inc.
Conair Inc.
Foremost Machine Builders
IMS Co.
Littleford Bros.
National Bulk Equipment Inc.
Novatec, Inc.
Purnell International
Vac-U-Max
Whitlock, Inc.

Dryer Manufacturers

Advance Process Supply Co.
The Berlyn Corp.
Bry-Air, Inc.
Conair, Inc.
Dri-Air Div., Automated Assemblies Corp.
Novatec, Inc.
Process Control Corp.
Thoreson-McCosh Inc.
Vibra Screw Inc.

Bepex Corp.
Brown Plastics Engineering Co.
Colortronic Systems, Inc.
Day Mixing/Taylor Stiles
Foremost Machinery Builders
Plastic Process Equipment Inc.
Teledyne Readco
Universal Dynamics Corp.
Walton-Stout

Granulator and Pulverizer Manufacturers

Acme Plastics Machinery Co.
Alpine American Corp.
Bepex Corp.
C.W. Brabender Instruments, Inc.
Colortronic Systems
Corcoran Co.
Day Mixing/Taylor-Stiles
Entoleter, Alsteele Div.
Foremost Machine Builders
Gloucester Engineering Co.
HydReclaim Corp.
Miller Manufacturing Co.
Nelmor Co., Inc.
Polymer Machinery Corp.
Mikro Pul Corp.
Ramco Industries
Wor-Tex Corp.
Young Industries, Inc.

Air Products and Chemical Inc.
Ball & Jewell Div., Sterling Inc.
Berstorff Corp.
Buss-Condux, Inc.
Conair, Inc.
Cumberland Engineering
Donaldson Ind. Group, Majac Div.
Fitzpatrick Co.
Franklin Miller Inc.
HBM-Machinen GmbH
IMS Company
Mitts & Merrill Inc.
PallmannPulvisers Co. Inc.
Process Control Corp.
Pulverizing Machinery Division
Wedco, Inc.
Wussmont Co.

Pelletizer Manufacturers

Automatik Machinery Corp.
Berlyn Corp.
Black Clawson Co.
Bolton-Emerson, Inc.
Brown Plastics Eng. Co., Inc.
California Pellet Mill Co.
Cumberland Engineering, Inc.
Day Mixing/Taylor-Stiles
Farrel Co.
Battenfeld Gloucester Engineering Co.
Henion Bros., Inc.
Killion Extruders, Inc.
Rapid Granulator, Inc.
Sterling Extruder Corp.
Werner & Pfleiderer Corp.

Baker Perkins Inc.
Berstorff Corp.
Bolling Div., Intercole Bolling Corp.
C.W. Brabender Instruments
Buss-Condux, Inc.
Conair, Inc.
D & S Manufacturing Co., Inc.
Egan Machinery
Gala Industries, Inc.
Haake Buchler Instruments Inc.
Japan Steel works America
Plastics Equipment & Accessories Co., Ltd.
Reifenhauser-Nabco, Inc.
Wayne Machine & Die Co.

TABLE 87.4. Cost of Extruded Product

Cost Factor	$/h	$/lb
Material (resins + printing ink)	2000.00	1.00
Labor at $40/h[a]	160.00	0.080
Utility requirement at $0.07/kWh[b]	28.00	0.014
Equipment amortization[c]	28.20	0.014
Building amortization[d]	9.13	0.005
Packaging and shipping	20.00	0.010
Total cost	*2245.33*	*1.123*

[a]Assuming 4 operators required to run 10 extrusion lines.
[b]Assuming a power usage of 40 kW per extrusion line.
[c]Using 10 year straight line amortization.
[d]Using 25 year straight line amortization.

2-in. (50-mm) extruder, the total cost of the line can be assumed to be $350,000. If the plant contains 10 similar extrusion lines, the capital investment for processing equipment is $3,500,000. The capital investment for the building is estimated at $2,000,000. Assume that the line output is 200 lb/h. The various cost factors are shown in Table 87.4.

The total cost per pound is about $1.12. If 20 bags per pound is assumed, the direct cost per bag is $0.056. It should be noted that the material cost constitutes a major portion of the total direct cost. Thus, shopping around for the least expensive resin will be a worthwhile exercise.

It should be noted that this example is for a continuous 24-h/day operation. The makeup of the cost for a short profile extrusion run will be considerably different from the example of the film grocery bags. In short runs, the setup cost can be quite significant. The effect of setup cost on run length is qualitatively shown in Figure 87.22. In short extrusions, therefore, the customer is generally charged a considerable setup premium.

It should also be noted that cost procedures may differ quite substantially from one

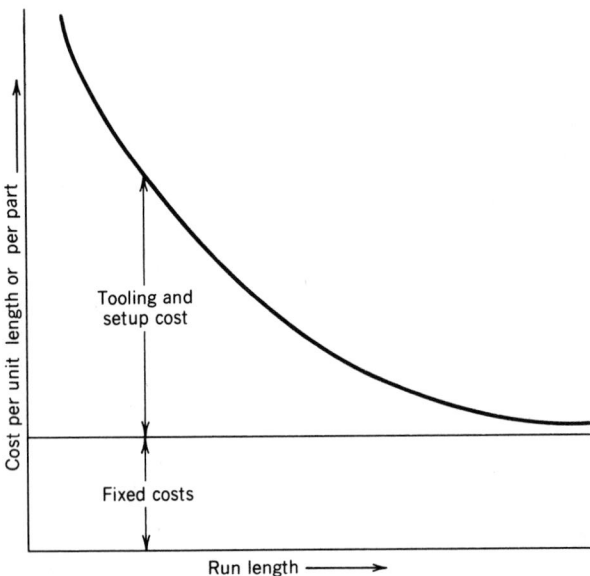

Figure 87.22. Effect of run length on part cost.

company to another. For instance, in some cases overhead is included in the labor cost; in other cases overhead is listed separately. In many cases a percentage gross markup is added to the cost, with the particular percentage varying widely. Overhead costs will depend to a large extent on the makeup of the company and its products. A company selling only a few standard products can operate at a relatively low level of overhead. However, a company with a large number of different products will have a higher level of overhead. Similarly, a company involved in research and development of new products will experience higher levels of overhead than a company selling only standard products.

87.13 INDUSTRY PRACTICES

A common industry practice for standard (stock) items is net thirty days. In practice, this often means that payment will be received about 45 to 60 days after receipt of invoice.

Nonstandard items that require development, custom designed tooling, and so on will usually be sold with a considerable down payment with the order. This is particularly true if the size of the order is large relative to the size of the company.

In selling extruded products, it is very important to draw up a complete set of product specifications agreed to by both supplier and customer. If the written specifications are not met, the responsibility lies with the supplier, if not, with the customer. In both cases, however, they will try to work out the problem together to minimize the damage.

87.14 INDUSTRY ASSOCIATIONS

87.14.1 United States

Extrusion Division of the Society of Plastics Engineers, Inc.
Plastics Institute of America
The Society of the Plastics Industry, Inc.
The Polymer Processing Society

87.14.2 United Kingdom

Plastics and Rubber Institute
Rubber and Plastics Research Association of Great Britain

BIBLIOGRAPHY

1. A.W. Jenike, "Gravity Flow of Bulk Solids," in *Bulletin No. 108,* Utah Engineering Experimental Station, University of Utah, Salt Lake City, 1961.
2. A.W. Jenike, "Storage and Flow of Solids," in *Bulletin No. 123,* Utah Engineering Experimental Station, University of Utah, Salt Lake City, 1964.
3. C.J. Rauwendaal, *Polymer Extrusion,* Carl Hanser Verlag, Munich, 1984.
4. W.H. Darnell and E.A.J. Mol., *SPE J.* **12,** 20 (1956).
5. B.H. Maddock, *SPE ANTEC, New York,* 303 (1959).
6. Z. Tadmor, *Polym. Eng. Sci.* **6,** 185 (1966).
7. Z. Tadmor and I. Klein, *Engineering Principles of Plasticating Extrusion,* Krieger, Malabar, Fla., 1970.
8. J. Shapiro, Ph.D. Thesis, Cambridge University, U.K. 1971.

9. J.R. Vermeulen, P.M. Gerson, and W.J. Beek, *Chem. Eng. Sci.* **26,** 1445–1455 (1971).

10. J.R. Vermeulen, P.G. Scargo, and W.J. Beek, *Chem. Eng. Sci.* **26,** 1457–1465 (1971).

11. W. Michaeli, *Extrusionswerkzeuge fuer Kunststoffe,* Carl Hanser Verlag, Munich, 1979.

12. Z. Tadmor and C. Gogos, *Principles of Polymer Processing,* Wiley-Interscience, New York, 1979.

13. C.J. Rauwendaal, *SPE ANTEC, Chicago,* 186 (1983).

14. L. Wielgolinski and J. Nangeroni, *Adv. Polym. Techn.* **3,** 99 (1983).

15. Z. Tadmor, P. Hold, and L. Valsamis, *SPE ANTEC, New Orleans,* 193 (1979).

16. Ref. 3, Chapter 3.

General References

Books on Extrusion

E.C. Bernhardt, ed., *Processing of Thermoplastic Materials,* Reinhold, New York, 1959.

N.M. Bikales, ed., *Extrusion and Other Plastics Operations,* Wiley, New York, 1971.

J.A. Brydson and D.G. Peacock, *Principles of Plastics Extrusion,* Applied Science Publishers, Ltd., London, 1973.

Der Extruder als Plastifizier-einheit, VDI-Verlag, Duesseldorf, 1977.

R.T. Fenner, *Extruder Screw Design,* Illiffe Books, Ltd., London, 1970.

E.G. Fisher, *Extrusion of Plastics,* Illiffe Books, Ltd., London, 1954.

A.L. Griff, *Plastics Extrusion Technology,* Reinhold, New York, 1968.

H. Hermann, *Schneckenmaschinen in der Verfarhrenstechnik,* Springer-Verlag, Berlin, 1972.

H.R. Jacobi, *Grundlagen der Extrudertechnik,* Carl Hanser Verlag, Munich, 1960.

L.B.P.N. Janssen, *Twin Screw Extrusion,* Elsevier, Amsterdam, 1978.

Kunststoff-Verarbeitung in Gespraech, 2 Extrusion, BASF, Ludwigshafen, 1971.

S. Levy, *Plastics Extrusion Technology Handbook,* Industrial Press Inc., New York, 1981.

F.G. Martelli, *Twin Screw Extrusion, A Basic Understanding,* Van Nostrand Reinhold, New York, 1983.

W. Mink, *Grundzuege der Extrudertechnik,* Rudolf Zechner Verlag, Speyer am Rhein, 1963.

H. Potente, *Auslegen von Schneckenmaschinen Baureihen, Modellgesetze und Ihre Anwendung,* Carl Hanser Verlag, Munich, 1981.

C. Rauwendaal, *Polymer Extrusion,* Hanser Publishers, Munich, 1986.

P.N. Richardson, *Introduction to Extrusion,* Society of Plastics Engineers, Inc., 1974.

G. Schenkel, *Kunststoff Extruder-Technik,* Carl Hanser Verlag, Munich, 1963.

G. Schenkel, *Plastics Extrusion Technology and Theory,* Illiffe Books, Ltd., London, 1966; published in the U.S. by American Elsevier, New York, 1966.

H.R. Simonds, A.J. Weith, and W. Schack, *Extrusion of Rubber, Plastics and Metals,* Reinhold, New York, 1952.

Z. Tadmor and I. Klein, *Engineering Principles of Plasticating Extrusion,* Van Nostrand Reinhold, New York, 1970.

Books on Polymer Processing

J. Frados, ed., *Plastics Engineering Handbook,* Van Nostrand Reinhold, New York, 1976.

J.M. McKelvey, *Polymer Processing,* Wiley, New York, 1962.

S. Middleman, *The Flow of High Polymers,* Interscience, New York, 1968.

R.M. Ogorkiewicz, *Thermoplastics: Effects of Processing,* Illiffe Books, Ltd., London, 1969.

J.R.A. Pearson, *Mechanical Principles of Polymer Melt Processing,* Pergamon, Oxford, 1966.

N.S. Rao, *Designing Machines and Dies for Polymer Processing With Computer Programs,* Carl Hanser Verlag, Munich, 1981.

S.S. Schwartz and S.H. Goodman, *Plastics Materials and Processes,* Van Nostrand Reinhold, New York, 1982.

Z. Tadmor and C. Gogos, *Principles of Polymer Processing,* Wiley, New York, 1979.

J.L. Throne, *Plastics Process Engineering,* Marcel Dekker, Inc., New York, 1979.

H.L. Williams, *Polymer Engineering,* Elsevier, Amsterdam, 1975.

Publications

Advances in Polymer Technology

Canadian Plastics

Chemical Engineering Science

European Plastic News

International Polymer Processing

Journal of Applied Polymer Science

Journal of the American Institute of Chemical Engineers

Kunststoffe

Modern Plastics

Newsletter of the Extrusion Division of the SPE

Plaste und Kautschuk

Plastics and Polymer

Plastics Engineering

Plastics Machinery and Equipment

Plastics Technology

Plastics World

Polymer

Polymer Engineering and Science

Plastverarbeiter

Preprints of SPE's Annual Technical Conference (ANTEC)

88

FLUIDIZED BED AND ELECTROSTATIC SPRAY TECHNIQUES

Thomas E. Jacques

Hysol Electronic Chemicals Div., The Dexter Corp., Olean, NY 14760

This chapter describes the application of plastics in the form of finely divided powder to create coatings on various substrates for a variety of purposes. These coatings may be decorative, essentially acting as paint, or functional, providing corrosion protection or electrical insulation. Currently, these processes are used for both thermoplastic and thermosetting materials.

88.1 FLUIDIZED BEDS

88.1.1 Description and History of Process

Modern fluidized bed techniques were first utilized by the chemical industry after World War II to maximize the surface contact area between a solid and a gas.[1] However, a gaseous dispersion of particles was used as a coating medium as early as 1940, by the Telegraph Construction and Maintenance Co. of England, which produced powdered polyethylene.[2] This powder became commercially available in 1946; many companies contributed to the subsequent development of the fluidized bed.[3] In 1957, it began to be used as an industrial coating process.[4]

A fluidized bed is a two-chambered vessel separated by a porous media that retains powder but allows free passage of gas (Fig. 88.1). The porous media can range from sintered metal to simple acoustical ceiling tile. Compressed gas, usually air, is pumped into the lower chamber and forced through the porous media into the powder in the upper chamber. The rising gas separates and suspends the powder, causing it to increase in volume to many times the height of the powder at rest until a steady state is reached. Then, the pressure drop of the rising gas will no longer support the weight of the powder. At this point, the gas being vented through the top of the bed forms an equilibrium with the gas being supplied.

The powder must be dry and free of agglomeration, so that the gas can completely surround individual particles. The movement of the gas suspends the powder in such a way that the fully raised bed acts very much like a liquid; hence the term, "fluidized bed."

The gas mixes the powder, resulting in uniformity throughout the bed. Most fluidized beds also have a mechanism to vibrate the bed. This improves the uniform distribution of the powder, especially in shallow beds, and assists the flow of powder during the dipping of parts to ensure that a uniform coating is achieved.

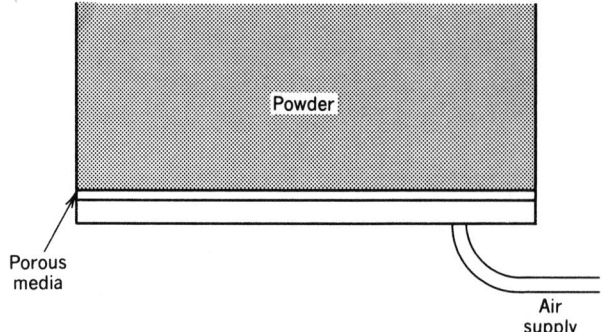

Figure 88.1.   Fluidized bed dipping process.

The powder particles range in size from less than 1 micron to about 200 microns. Particles larger than 200 microns are difficult to suspend. Particles smaller than 20 microns may create excessive dusting and release of particles from the top of the bed. Powders may also contain fluidizing aids to keep the powder free-flowing.

The actual coating process is uncomplicated; however, achievement of a uniform coating requires considerable skill. The part to be coated is heated to a temperature above the melt temperature of the powder and is then introduced into the fluidized bed. As the powder contacts the hot substrate, the particles adhere, melt, and flow together to form a continuously conforming coating. The part is removed from the bed when the desired coating thickness is obtained. On cooling, the plastic resumes its original characteristics. In the case of a thermosetting plastic, additional time at elevated temperature may be required to complete the cure of both the plastic and the coating.

88.1.2 Materials

Low melting plastics are most appropriate for fluidized beds. Higher melting plastics must have a sufficiently low melt viscosity that the particles can flow and fuse together to form a coating with good integrity. Thermoplastic materials in common use include nylon, poly(vinyl chloride), acrylics, polyethylene, and polypropylene. Other possible materials are thermoplastic urethane, silicones, EVA, polystyrene, or any other low-melt, low-viscosity plastic. Thermosetting plastics are limited largely to epoxy and epoxy/polyester hybrids since other thermosets, such as phenolic resin or DAP, give off by-products during cure. These volatiles can create voids in the coating, harming its integrity and function.

88.1.3 Processing Characteristics

The coatings produced by this process conform extremely well to the substrate, and with the correct process control, are of consistent thickness and appearance from part to part. The ability of the fluidized bed to continuously conform to and coat parts having unusual shapes and sizes permits a high degree flexibility in manufacturing. This is invaluable to the processer who wishes to coat one-of-a-kind products. In general, the coatings are smooth and glossy, with excellent appearance. They exhibit excellent adhesion to the substrate, providing the hermetic seal necessary for proper performance. Proper adhesion is critical for coatings designed to provide electrical insulation or corrosion protection.

Almost all the coatings applied by this process have a definite function, chiefly electrical insulation, but it can be used for applications that simply require a thick coating with powder. Thicknesses of to 0.1 in. (2.5 mm) or greater are easily attained. Fluidized bed coating is certainly the most efficient method for applying a thick coating in a single step, as repeated application and subsequent removal of volatiles is not necessary.

MASTER PROCESS OUTLINE   PROCESS   Fluidized Bed Coating

Instructions:
For "ALL" categories use   Y = yes
For "EXCEPT FOR" categories use   N = no
                                   D = difficult

☐ **ALL Thermoplastics**
  Except for:

	Y POLYETHYLENE
Acetal	Poly(methyl methacrylate)
Acrylonitrile–Butadiene–	Poly(methyl pentene) (TPX)
Styrene	Poly(phenyl sulfone)
Cellulosics	Poly(phenylene ether)
Chlorinated Polyethylene	Poly(phenylene oxide) (PPO)
Y Ethylene Vinyl Acetate	Poly(phenylene sulfide)
Ionomers	Y Polypropylene
Liquid Crystal Polymers	Y Polystyrene
Y Nylons	Polysulfone
Poly(aryl sulfone)	Polyurethane Thermoplastic
Polyallomers	Y Poly(vinyl chloride)-PVC
Poly(amid-imid)	Poly(vinylidine chloride)-Saran
Polyarylate	Rubbery Styrenic Block
Polybutylene	Polymers
Polycarbonate	Styrene Maleic Anhydride
Polyetherimide	Styrene–Acrylonitrile (SAN)
Polyetherketone (PEEK)	Styrene–Butadiene (K Resin)
Polyethersulfone	Thermoplastic Polyesters
	XT Polymer

☐ **ALL Thermosets**
  Except for:

    Allyl Resins
    Phenolics
    Polyurethane Thermoset
    RP-Mat. Alkyd Polyesters
  Y RP-Mat. Epoxy
    RP-Mat. Vinyl Esters
    Silicone
    Urea Melamine

☐ **ALL Fluorocarbons**
  Except for:

    Fluorocarbon Polymers-ETFE
    Fluorocarbon Polymers-FEP
    Fluorocarbon Polymers-PCTFE
    Fluorocarbon Polymers-PFA
    Fluorocarbon Polymers-PTFE
    Fluorocarbon Polymers-PVDF

Examples of electrical applications include small motor stators and rotors, electronic components (capacitors or resistors), and busbar. Other items coated using fluidized beds include transformer parts, valves, refinery equipment, and appliance and pump parts.

88.1.4 Advantages/Disadvantages

Advantages of powder coating by any method include:

- Process stability.
- Little or no waste.
- Environmental soundness (no solvents or by-products).
- Low energy requirements.
- Low capital investment.
- Manufacturing flexibility; ability to coat complex and varied shapes.
- High processing speed.
- Ability to obtain thick coatings in one step.

Probably the single biggest advantage of powder coating is the nearly 100% utilization of the material. With the proper recovery and recycling system, all product is recycled and used without the hazard or expense of solvents.

The stability of the fluidized bed process derives from the stability of the bed itself. The mixing of the gas keeps the powder well-distributed; the dipping process does not selectively remove particles in a manner that would change the character of the bed. Fluidized beds have been known to function for years with little attention other than replenishment of the powder supply.

Disadvantages include:

- High product cost
- Requirement that substrates withstand heating
- Thick coatings
- Dust as a health hazard
- For large parts, a large fluidized bed and much powder

Certain part configurations are not easily coated by this process. Parts with small, deep holes tend to plug off at the opening rather than coat uniformly. Tight U-shapes or parts that present large heat sinks in one area also coat unevenly. In general, the heat configuration of the product significantly influences the evenness of the coating.

88.1.5 Processing Criteria

The size and depth of the bed is determined by the size and quantity of the parts to be coated. Vibration of the bed aids powder mixing and flow around the substrate, and thereby ensures even coating. This is particularly important if the object to be coated has a large, flat surface where powder can come to rest and form a coating thicker than desired on the upper surface, or if the object has a highly irregular shape and intimate contact between the powder and the object is necessary to achieve a complete coating.

Heat transfer is very important. When a thick coating is to be applied, several dips may be required with a residence time between dips that permits heat to be released through the coating. This permits each additional dip to build on the prior coating. In the event that the part does not contain enough heat to achieve the desired thickness, the dipping cycle can be alternated with a reheating cycle. The velocity of the gas in the bed also affects the rate of cooling during the dip cycle.

Parts to be coated are usually immersed completely in the bed, suspended from wire, or held by cold fixtures so as to avoid powder build-up. Areas where coating is not desired are masked off with heat-resistant tape or mechanical masking fixtures. Mechanical fixtures must be cooled between dips to prevent heat build-up. Tapes must be removed while the coating is still in a molten state so that the coating does not peel back, creating a gap between it and the substrate. This adhesion loss can be very detrimental to the performance of the coating.

Most fluid bed coating is done with a "live" bed; that is, one through which gas is being circulated. However, coating of electronic components requires that the process be altered. Electronic components contain metal leads. If powder were to contact these leads, the plastic's inherent insulation characteristics could prevent electrical continuity and render the device useless.

To coat electronic components, the gas flow is stopped and the bed is allowed to settle for a few seconds just before the parts are dipped. Because the parts are small, they can easily penetrate the incompletely raised bed in which enough gas is trapped to keep the powder partially suspended. The bed is then vibrated to assist powder flow around the component and coating. This modification ensures that the leads are free of powder and the components are coated completely.

After application of the coating, reheat ovens may be used to reflow a thermoplastic material and form a perfectly smooth finish. For thermosetting plastics, such as epoxies, oven cure after coating is necessary.

88.2 ELECTROSTATIC SPRAY

88.2.1 Description and History of Process

Electrostatic spraying of plastic powders dates back to the early 1960s. The technique utilizes the principle that oppositely charged particles attract, a principle that has been used for many years in spraying solvent-based paints.[5]

In electrostatic spraying, plastic powder is first fluidized in a bed to separate and suspend the particles. It is then transferred through a hose by air to a specially designed spray gun (Fig. 88.2).

Powder can also be introduced into the transfer hose using a reservoir system. As it passes through the gun, direct contact with the gun and ionized air applies an electrostatic charge to the particles of powder. The contact area may be a sleeve that extends the length of the gun or merely small pins that extend into the passageway of the powder. For safety, the gun is designed for high voltage but low amperage. The powder continues through the gun and exits past a specially designed tip that forms the spray pattern of the powder.

The part to be coated is electrically grounded, attracting the charged particles. This not only reduces overspray, but also produces a more even coating. Parts to be coated may be preheated, thereby forming the coating immediately, or they may be coated cold. The electrostatic charge will hold the particles in place until heat is applied, although care must be taken not to jar the particles loose. Once heat is introduced, the particles melt and flow together, forming a continuous protective coating.

Electrostatic spraying is widely used in the solvent-based spray industry to reduce overspray and develop a more uniform coating. These advantages apply to spraying pow-

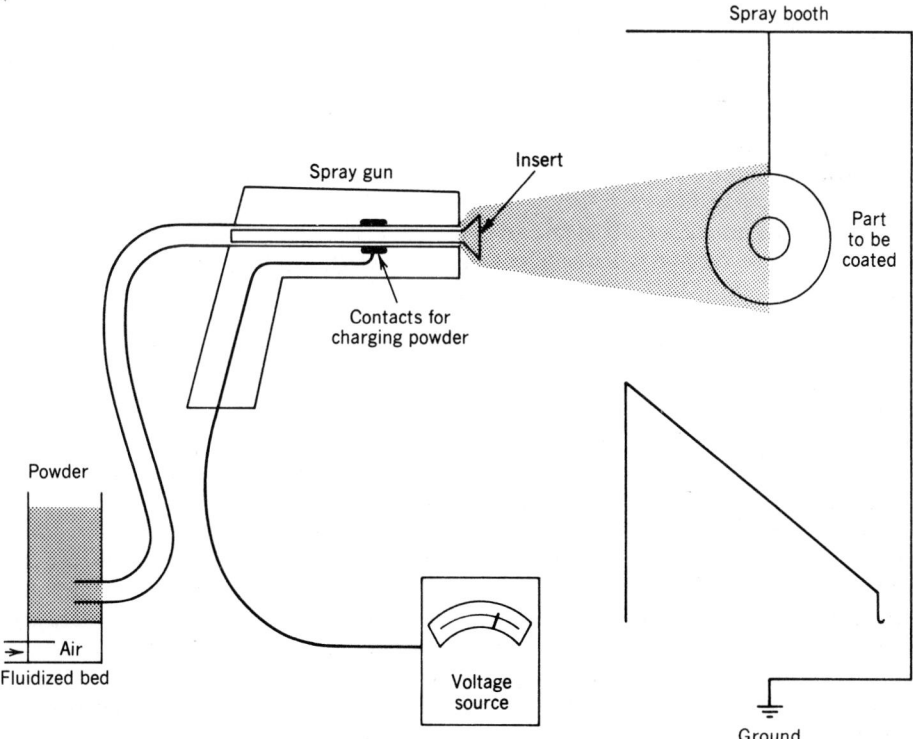

Figure 88.2. Electrostatic spraying process.

dered plastics, with the additional benefit of nearly 100% of the powder thanks to recovery and recycling systems.

88.2.2   Materials

The plastics used in spraying of powders are the same as those used in fluidized beds. The key characteristic for any plastic, thermoplastic or thermoset, applied as a powder is low melt viscosity, which enables the individual particles to flow together and form a continuous coating. The selection of the plastic depends on the end use of the coated part and its environmental conditions.

88.2.3   Processing Characteristics

The chief characteristic of electrostatically applied powdered plastic is the ability to produce a thin coating. It is the preferred method of producing a coating for [0.001–0.002 in. (0.025–0.050 mm)] thickness.

A continuous process, it is suited to automated assembly-line production. A well-designed recovery system, typically a fully enclosed spray booth, permits full utilization of the powder. The bottom of the booth is either a cone or an inverted pyramid. A downward draft of air in the booth, assisted by gravity, collects powder there from which it is transported to a storage area for recycling. Such as system minimizes dust in the environment and reduces worker exposure.

Much of what is termed decorative powder coating is very thinly applied by electrostatic spray. The primary function of the coating is that of paint. Appliances, laboratory instruments, transformer housings, engine parts, and chain link fences are among the many types of products so coated for decorative purposes.

Corrosion-resistant coatings are designed to prevent the corrosion of the underlying substrates. These include pipe coating, fencing, concrete reinforcing bars, valves, conduit, and pumps. Plastics used for corrosion protection are usually thermosetting, particularly epoxies, because their superior adhesion characteristics permit the formation of a tight bond between the coating and the substrate. The bond prevents water and other corrosive elements from penetrating the coating and creating a pathway for corrosion. Even when the coating is damaged, it can restrict corrosion to the damaged area.

Coatings that provide electrical insulation are also applied by electrostatic spray. Objects coated include electrical motor armatures and stators, electrical switchgear boxes, and magnet wire.

The particle size of sprayed powders is smaller than that of powders used in a fluidized bed. The average particle size is 30–60 microns; the largest, 100–120 microns. A smaller particle size is dictated by the mechanics of spraying and the need to apply a thin [0.001–0.002 in. (0.025–0.050 mm)] coating.

88.2.4   Advantages/Disadvantages

The advantages and disadvantages of powder coating are similar to that of fluidized bed coatings, but there are some restrictions on electrostatic spraying of powdered plastics.

Airborne dust from sprayed powder is a much greater hazard to worker health, and dust control systems are mandatory. An enclosed spray booth ensures both worker safety and powder recovery.

Airborne dust also constitutes a significant explosion hazard. Any finely divided organically based powder can explode when sufficient quantities are airborne and an ignition point is provided. Proper preventive measures must be followed strictly.

The thickness of a coating applied by electrostatic spray is limited. As the coating thickness builds during the process, the grounded object becomes insulated from the charged particles of powder as a result of interstitial air between the particles and the natural insulating properties of most plastics. The particles are no longer attracted to the object and effective coating is lost. Cold objects will completely cease coating; hot objects will only attract particles blown directly at them, altering the uniformity of the coating. The charge on the particle can be increased by increasing the voltage, which can slightly add to the achievable thickness. Coatings 50–75 microns thick can be applied electrostatically to cold objects; coatings up to 250 microns thick to hot objects.[5]

There are also some limitations on the configuration of the part to be coated. Depressions or recesses may not be deeper than their width. As a charged particle enters an enclosed area, it will be attracted to the closest area of grounding. On a deep recession, the particles will be attracted to the sides at the top of the opening, resulting in a thicker coating near the opening and little or no coverage at the bottom (Fig. 88.3). This is known as the Faraday Cage effect.

Parts must withstand heating to be coated by this process. They must also conduct electricity; that is, be able to be grounded. This is not a suitable method for coating insulating parts.

88.2.5 Processing Criteria

The electrostatic spray process is similar to the fluidized bed process. In most continuous coating processes, a conveyor-type line brings parts through the spray booth. The parts may come directly from the oven or proceed directly to an oven, depending on whether they are to be coated cold or hot. It is preferable to coat large parts cold and then apply heat because less energy is required to heat only the coating from the outside than to evenly heat the entire part.

Spraying hot parts can eliminate the need for electrical grounding, but at the expense of coating uniformity. Grounding is vital to coating uniformity.

The spray booth generally contains more than one spray gun at various angles and heights to ensure that all surfaces are coated evenly. Robotic control is possible, as is the flexibility to coat various configurations by repositioning the guns. The pattern of powder spray from the gun is directed to the needed areas. Part size, shape, and distance from the gun determine the spray pattern.

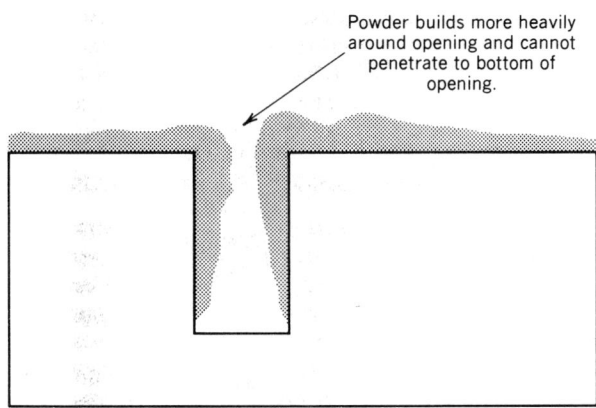

Figure 88.3. Faraday Cage effect.

BIBLIOGRAPHY

1. M. Elmas, *Fluidized Bed Powder Coating,* Powder Advisory Center, London, 1973.
2. Staff Report, *Paint and Varnish Production,* 45, (June 1962).
3. M. Elmas, *Paint and Varnish Production* (June 1962).
4. E. Gemmer, *Kuntstoffe* **47,** 510 (1957).
5. S. Kut, *Science and Technology of Surface Coatings,* Academic Press, New York, 1974, pp. 46–48.

89

INJECTION MOLDING (IM)

Irvin I. Rubin

President, Robinson Plastics Corp., 944 East 24th St., Brooklyn, NY 11210

89.1 DESCRIPTION AND HISTORY OF THE PROCESS

Injection molding is a major processing technique for converting thermoplastic and ther-mosetting materials into all types of products. In 1987, approximately 11×10^9 lb or 20% of the 54×10^9 lb of the thermoplastics sold in the United States were injection-molded. About 50% of the 9×10^9 lb of thermoset materials were injection-molded.

There are about 80,000 injection-molding machines in about 8,000 plants in the United States. They tend to cluster into groups of 10 to 12 machines, which is all a foreman and crew can efficiently handle. About 95% of the machines are either reciprocating screw or two-stage screws. The balance are plunger machines. About 50% are less than five years old, 35% are 5–10 years old and 15% more than 10 years old. In 1987, about 60% of the new machines purchased were imported.

The process is not new. John and Isiah Hyatt received a patent in 1872 for an injection-molding machine, which they used to mold camphor-plasticized cellulose nitrate (celluloid). In 1878, John Hyatt introduced the first multicavity mold. In 1909, Leo H. Baekeland introduced phenol–formaldehyde resins, which are now injection-moldable with the screw-molding machine.

The experimental and theoretical works of Wallace H. Carothers led to a general theory of condensation-polymerization that provided the impetus for the production of many poly-mers, including nylon. By the end of the 1930s, modern technology began to develop and great improvements in materials permitted injection molding to become economically viable.

Figure 89.1 shows a reciprocating screw injection-molding (IM) machine with a clamping capacity of 300 tons, using a 2 1/2-in. (64-mm) reciprocating screw that delivers a maximum of 20 oz (570 g) of polystyrene per shot. The machine is a tool for:

- Clamping the mold halves together.
- Raising the temperature of the plastic to a point where it will flow under pressure.
- Injecting the hot plastic into the mold.
- Cooling the plastic in the mold.
- Opening the mold and ejecting the hardened plastic piece.

The IM machine is an instrument for performing these functions automatically under controlled conditions of temperature, time, speed, and pressure. The nature of the process is such that the molecular structure, molecular weight, and molecular weight distribution (all of which affect melt viscosity), orientation, and crystallizability of the injected plastic play an important role and must be considered for their effects on end product's properties.

89.2 MATERIALS

See the Master Process Outline.

Figure 89.1.   2 1/2 in. (64 mm) reciprocating screw, 300-ton injection-molding machine. Courtesy of Reed-Prentice Division, Package Machinery Company.

MASTER PROCESS OUTLINE   **PROCESS**   Injection Molding

Instructions:
For "ALL" categories use   Y = yes
For "EXCEPT FOR" categories use   N = no
                                   D = difficult

Y	**ALL Thermoplastics** Except for:		Y	**ALL Thermosets** Except for:

Acetal	Poly(methyl methacrylate)
Acrylonitrile–Butadiene–	Poly(methyl pentene) (TPX)
Styrene	Poly(phenyl sulfone)
Cellulosics	Poly(phenylene ether)
Chlorinated Polyethylene	Poly(phenylene oxide) (PPO)
Ethylene Vinyl Acetate	Poly(phenylene sulfide)
Ionomers	Polypropylene
Liquid Crystal Polymers	Polystyrene
Nylons	Polysulfone
Poly(aryl sulfone)	Polyurethane Thermoplastic
Polyallomers	Poly(vinyl chloride)-PVC
Polyamid-Imid	Poly(vinylidine chloride)-Saran
Polyarylate	Rubbery Styrenic Block Polyms
Polybutylene	Styrene Maleic Anhydride
Polycarbonate	Styrene–Acrylonitrile (SAN)
Polyetherimide	Styrene–Butadiene (K Resin)
Polyetherketone (PEEK)	Thermoplastic Polyesters
Polyethersulfone	XT Polymer

	Allyl Resins
	Phenolics
	Polyurethane Thermoset
N	RP-Mat. Alkyd Polyesters
D	RP-Mat. Epoxy
N	RP-Mat. Vinyl Esters
N	Silicone
	Urea Melamine

Y	**ALL Fluorocarbons** Except for:

	Fluorocarbon Polymers-ETFE
	Fluorocarbon Polymers-FEP
	Fluorocarbon Polymers-PCTFE
	Fluorocarbon Polymers-PFA
N	Fluorocarbon Polymers-PTFE
	Fluorocarbon Polymers-PVDF

89.3 PROCESSING CHARACTERISTICS

Injection-molding produces parts in large volume at high production rates. Labor costs per unit are low and the process can be automated. The parts require little or no finishing. Many different surface finishes and colors are available. The same articles can be molded with different materials on the same equipment. Close tolerances can be maintained. Parts can be molded in a combination of plastic and fillers, such as glass, asbestos, talc, and carbon. Metallic and nonmetallic pieces can be inserted.

The process permits the manufacture of very small parts that are almost impossible to fabricate in quantity by other methods. Thermoplastic scrap losses are minimal, as runners, gates, and rejects can be reground and reused. Aside from die casting, it is the only commercial process in which the scrap can be reused immediately by regrinding and remolding at the machine. This process is the most economical way to fabricate many shapes.

The profit margins of the plastics industry, however, are very low. Molds, machinery, and auxiliary equipment are expensive, and three-shift operations are often necessary to compete. Process control may be poor and machinery may not operate consistently. Plastics vary from batch to batch. In addition, viscosity, temperature, and pressure in the mold are continually changing and cannot be measured. Quality is often difficult to determine immediately, and the long-term properties of the material are difficult to ascertain. Experience and good workmanship are essential.

89.3.1 Examples of Injection-molded Parts

IM parts are pervasive in our lives and our economy. A few typical places where one finds them are personal items (from combs to full-sized cribs), packaging (bottle caps), automotive (grilles), transportation (airplane trays), appliances (food processers), toys (chess sets), electrical (battery containers), electronic (circuit boards), medical (retractors), food (microwave dishes), disposables (cups), entertainment (CDs), and cosmetics (compacts).

If IM were stopped today, our manufacturing and economy would stop with it.

89.4 ADVANTAGES/DISADVANTAGES

89.4.1 Advantages of Injection Molding

- Parts can be produced at high production rates.
- Large-volume production is possible.
- Relatively low labor cost per unit is obtainable.
- The process is highly amenable to automation.
- Parts require little or no finishing.
- Many different surfaces, colors and finishes are available.
- Good decoration is possible.
- For many shapes, this process is the most economical way to fabricate.
- The same item can be molded in different materials without changing the machine or mold.
- Very small parts can be fabricated in quantity; this would be almost impossible by any other method.
- Minimal scrap losses result from regrinding and reusing of thermoplastic runners, gates, and rejects.
- Close dimensional tolerances can be maintained.
- Parts can be molded with metallic and nonmetallic inserts.

- Parts can be molded in a combination of plastic and fillers such as glass, asbestos, talc, and carbon.
- Aside from die casting, it is the only commercial process that can reuse its thermoplastic scrap immediately (regrind and remold at the machine).
- Compared to other materials (glass, aluminum, etc.), the amount of energy needed for manufacture is extremely low. From an energy point of view, the thermoplastic product can be reground after use and remolded or burned, both of which further reduce its energy cost.
- The inherent properties of the material give many advantages, such as high strength-to-weight ratios, corrosion resistance, strength, and clarity.

89.4.2 Disadvantages and Problems with Injection Molding

- The plastic industry has very low profit margins.
- Three-shift operations are often necessary to compete.
- Mold costs are high.
- Molding machinery and auxiliary equipment costs are high.
- Process control may be poor.
- Machinery is not consistent in operation, and controls do not directly measure what is supposed to be controlled.
- The possibility of poor workmanship is often present.
- Quality is often difficult to determine immediately.
- There is a lack of knowledge concerning the fundamentals of the process.
- There is a lack of knowledge about the long-term properties of the materials.
- Plastic cannot be made so that each pellet is the same. One must deal with averages of molecular weights and molecular configurations; these not only vary from pellet to pellet but, on a larger scale, from batch to batch. This causes an unsteady and varying operational state, which contrasts with low molecular weight chemicals, such as salt, sucrose, naphthalene, etc., that can be manufactured identically in structure and properties.
- To derive quantitative equations for flow and other properties needed in injection molding, one must know viscosity, temperature, and pressure. In a mold they are continually changing and not measurable. When applied in practice, the assumptions made may lead at best to some very questionable results. Experience and qualitative calculations have yielded far superior results.

89.4.3 Limitations of the Process

Theoretically, there are no limitations to the process other than those imposed by the physical properties of the material. One could build a two-stage machine as large as one wishes with molds to fit. The limitations, if any, are economic.

89.5 DESIGN

There are very few technical design limitations that derive from the process rather than from the materials.

As the thickness of the molded parts increase, the process approaches that of casting. The thicker the part, the longer the cycle and the higher the cost. For this reason, the overwhelming number of IM parts are less than 1/4 in. (6.4 mm) thick. One rarely sees a part as thick as 3/4 in. (19 mm).

Variations in wall thickness can create problems, particularly a surface depression called a sink mark.

An obvious limitation is that the part must be successfully ejected from the mold. Molds can be designed to overcome mechanical undercuts either automatically or by manipulation after the part is removed from the mold.

One problem that cannot always be overcome results from the movement of the cavities and cores during molding. This entrapment can cause frictional forces so great that the part will not be ejected. Then the mold has to be redesigned or the part changed, even to the extent of molding it in two parts with subsequent assembly.

Economic limitations result from the high costs of molds and equipment. For most items the production has to be sufficiently large that the amortized cost per piece will be acceptable. In a relatively very few instances, IM is the only way to manufacture the piece and the dollar cost has a different value.

Tolerances are material and cycle dependent. As in all processes, tolerance is a function of cost. Within the normal economic operating range, tolerances of $+/- 0.004$ to $0 0.010$ in./in. (mm/mm) are common. These are conditioned by the size of the piece, whether it is a parting line dimension and whether it is parallel or perpendicular to flow.

89.6 PROCESSING CRITERIA

89.6.1 Molecular Aspects of Injection Molding and Physical Properties of Molded Parts

The polymeric materials used in the molding process may be all amorphous or partly amorphous and partly crystalline. Varying amounts of frictional or shear energy and thermal energy from the heating units on the machine are required to provide the plastic flow conditions necessary for molding. The viscosity of the melt must be sufficiently low to allow the injection pressure to force the plastic into the relatively cold mold.

As the temperature increases, the energy absorbed by the polymer chains causes vibrational, rotational, and translational or segmental motion of the polymer molecules. This essentially Brownian motion results in a random molecular arrangement in the viscous liquid mass of molecules.

If a unidirectional force is applied to a polymer at rest (in a random configuration) above its glass-transition temperature, it begins to move away from the force. If the force is applied very slowly, in such a way that the Brownian movement can overcome the orienting forces caused by the flow, the mass of the polymer moves with a rate proportional to the applied stress. This is termed Newtonian flow.

As the molecules move more rapidly under the influence of higher pressure in the injection process, the chains untangle using varying amounts of the energy from the driving force. This does two things. First, compared to an untangled material that would not drain any energy away for untangling, it slows down the acceleration.

Secondly, the chains move so rapidly that the orientation cannot be removed by the Brownian movement. As the molecules separate they move further apart. The Van der Waal's forces holding them together decrease very rapidly with the sixth power of the distance, lowering the viscosity in addition to the effects of temperature. Therefore the increased shear rate (speed) is no longer proportional to the shear stress (force). This is non-Newtonian flow, characteristic of injection molding.

Since good molding requires constant viscosity, the speed of the injection plunger must be accurately controlled. This is best done with computer controlled machines having feedback systems.

As the flow rate increases, it reaches its final stage of maximum molecular orientation and no further reduction in entangling. Then, an increase in shear stress proportionally increases the shear rate, and the material again acts as a Newtonian fluid. Injection molding is not done in this shear range.

In thermoset IM, the viscosity does not substantially change with the flow rate. One controls velocity because too high shear rates create frictional heat, which sets the material prematurely in the runner system and prevents mold fill. Too rapid flow can also cause fillers to separate from the base resin.

89.6.2 Orientation

As the material hits the cold mold it freezes. Regardless of any orientation caused by the gate, the turbulence in the flow is enough to make random the outside molecular layer. This outside frozen layer is therefore relatively unoriented. Consider the next layers of polymer. Part of the polymer is frozen in the outside layers next to the cold wall. The flow of the material pulls the remainder of the molecules in the direction of flow. As would be expected, these layers are the most highly directional or oriented, and the most highly shear stressed. Nearer the center the molecules are less oriented. The closer the polymer is to the wall, the quicker it freezes. Because the outer layers thermally insulate the inner layers, the center remains warmer longer, allowing more time for Brownian movement, which disorients the material.

If a clear injection-molded part is placed between two polarizing filters, one of which is rotated, a characteristic series of colored bands appears primarily related to orientation stress in the part. This phenomenon is called birefringence.

The degree of birefringence (in the molded part) versus the distance from the mold wall is shown in Figure 89.2. The dashed line indicates the initial condition; the solid line, the final condition when the part cools. The reduction in orientation is caused by Brownian movement. Measurements of various moldings were taken with the aid of a polarizing microscope.[2,3] The peak orientation was between 0.025 and 0.030 in. (0.64 and 0.76 mm) from each side of a 0.100 in. (2.54 mm) thick slab. The two peaks were not identical in each specimen because of the difference in temperature of each side of the mold.

If our interpretations are correct, the milling off of about one-third of either side produces a part that has molecules randomly placed on one side and highly oriented on the other side. If the specimen is heated about its glass-transition temperature, T_g, the oriented molecules should assume a more random structure, thus causing a greater shrinkage in the oriented portion. The specimen should act like a bimetallic unit and bend in the direction of the oriented layer. This is what occurs Figure 89.3.[4]

Of course, orientation has an effect on the physical properties of the part. Regular poly-

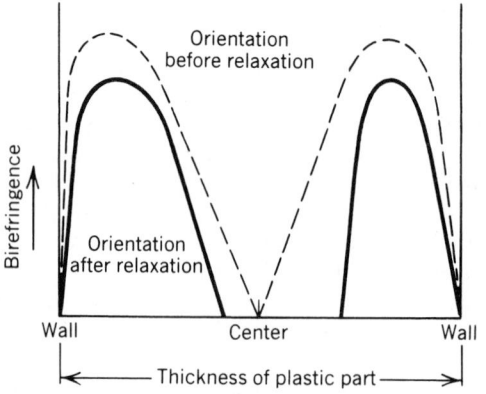

Figure 89.2.   Amount of birefringence vs distance from wall.[1]

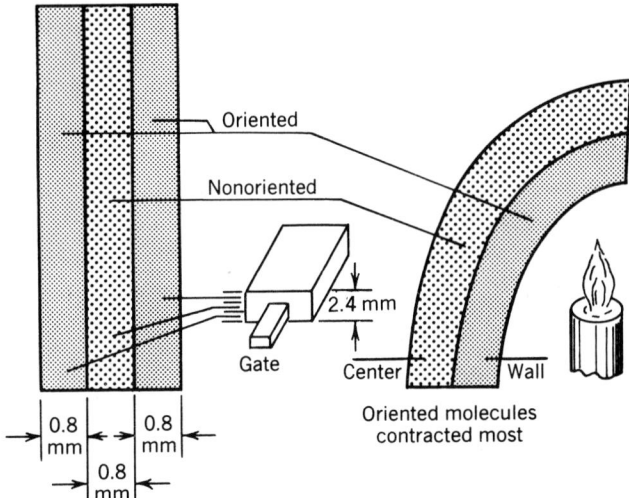

Figure 89.3. The effect of orientation/nonuniform shrinkage.[5]

styrene sheet, which has a tensile strength of 6000–7000 psi (41–48 MPa), is quite brittle. It is heated slightly above its T_g and stretched, which orients the molecules. While it is under tension, it is chilled to retain its new orientation. The tensile strength now is 9000–12,000 psi (62–83 MPa), depending on the elongation and processing temperature; the brittleness disappears. If the material is allowed to cool slowly its orientation disappears; the properties would be similar to those of the starting sheet.[6]

The effect of orientation (measured by birefringence) on the tensile strength, elongation at failure, and notched Izod impact test are shown in Figure 89.4.

The polymer is held together by two forces, C–C covalent linkages that have a disassociation energy of 83 kcal/mole (347 kJ/mol) and Van der Waal's forces that have a disassociation energy of 3 to 5 kcal/mole (12.5–20.9 kJ/mol) and decrease exponentially as the sixth power of distance. Van der Waal's forces are electrostatic in nature.

Therefore, tensile strength increases in the oriented flow direction because there are more C–C linkages in that direction than in the direction perpendicular to flow. These are much stronger than the van der Waal's forces, which are the major forces holding the polymer together perpendicular to flow. This also effects the percent elongation at failure and the notched Izod impact strength in the same way.

89.6.3 Effect of Molding Conditions on Orientation

Molding conditions, part thickness, and gate size affect orientation. The net orientation is the difference between that caused by flow and that lost by relaxation. Usually, a higher stock (or material) temperature produces less orientation because hotter material relaxes more. However, in thin sections the hotter material keeps the gate open longer, increasing the flow time and hence the birefringence.

Because of Brownian movement, the higher the mold temperature, the more time for relaxation and the lower the orientation. An increase in cavity thickness is one factor that has a dramatic effect on decreasing overall orientation. Because of the low thermal conductivity of plastic, the interior remains hotter longer, increasing relaxation.

Increasing the ram forward time materially increases orientation by permitting significantly longer flow time. This effect stops when the gate seals off. Higher injection pressures on the material increase orientation by inhibiting Brownian movement. The larger the gate size, the longer the seal-off time.

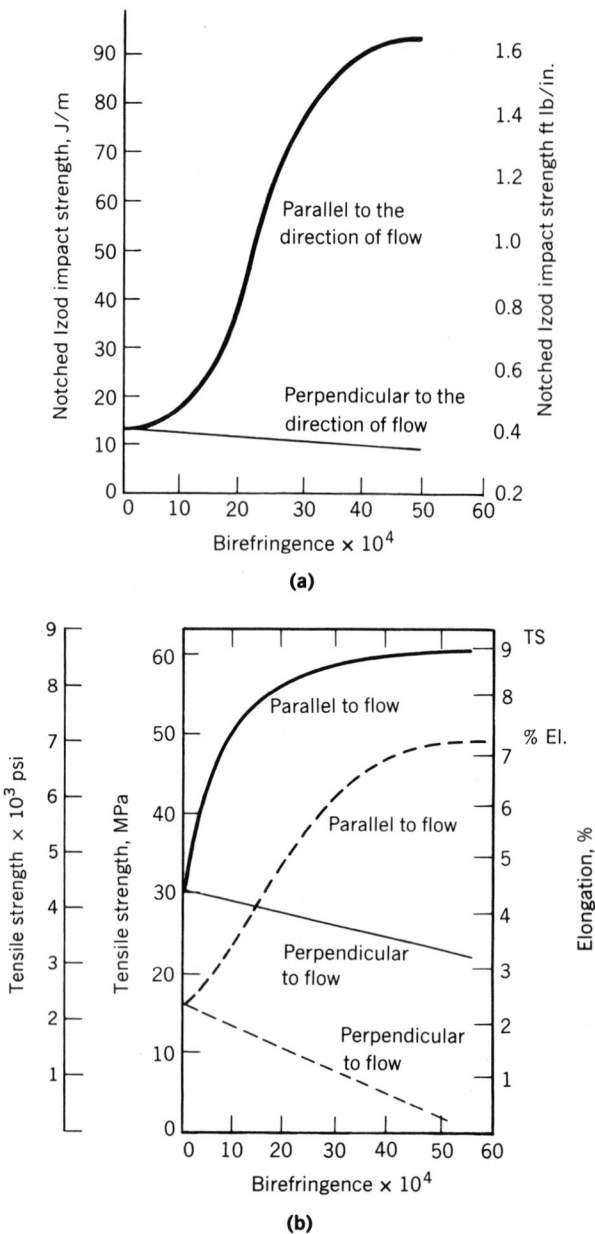

Figure 89.4. (**a**) Effect of orientation on impact strength; (**b**) Effect of direction of flow on physical properties of polystyrene.[7]

Some physical properties vary depending on the direction of flow, as evidenced by orientation. From the concepts of molecular structure, one would expect that there would be more shrinkage in the direction of flow because the C–C linkages are elastic and the Van Der Waal's forces are not.

In injection molding, the effect of different shrinkage parallel and perpendicular to flow is considerable. Consider the molding of center-gated 4-in. (102-mm) diameter, 0.060-in. (1.5-mm) thick polypropylene disk (Fig. 89.5). Consider a segment encompassing a 60°

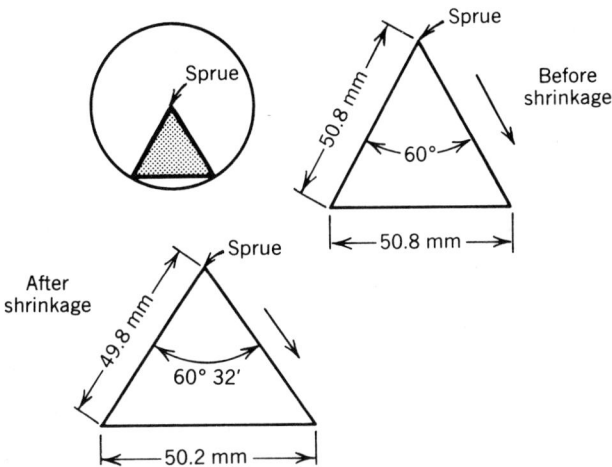

Figure 89.5. Warping of center-gated polyolefin part caused by different shrinkage parallel and perpendicular to flow.

angle. When the material flows in hot, it has a 2-in. (51-mm) dimension on each side of the equilateral triangle. The polypropylene typically shrinks 0.020 in./in. (mm/mm) in the direction of flow and 0.012 in./in. (mm/mm) perpendicular to flow. When the material cools, the radius is 1.960 in. (49.75 mm) and the chord 1.976 in. (50.15 mm).

The new angle is 60°, 32'. For the complete 360° circle, the increase is 3.2°. Unless the material is rigid enough to overcome this stress, the extra 3.2° of material will cause a warp. If this cover were thin-walled, it would mean that regardless of what was done during molding, a warp-free part could not be produced.

If a rectangular tray of the same material were molded Figure 89.6, a center gate (**a**)

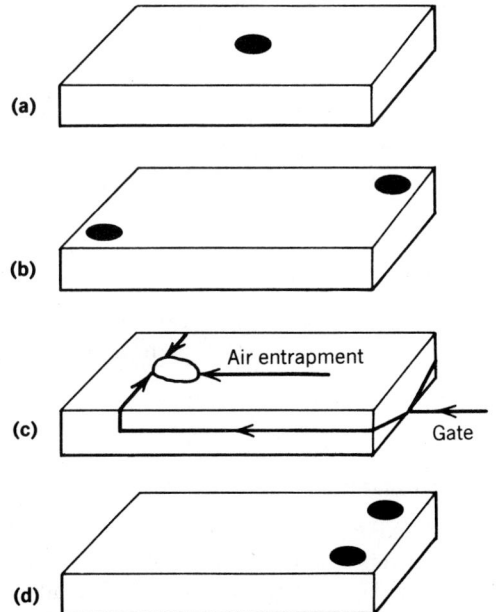

Figure 89.6. Effect of gate location on polyethylene tray.

would give a distorted tray unless the walls were heavy enough to overcome the stress. If it were gated as in (**b**), a severe radial twist would occur. If the part were edge-gated at one end as it is in (**c**), it would be warp-free but the material would flow around the rim on the parting line and trap air, giving either burns or poor welds. The best way to gate the piece would be to place two gates at one end. This would give maximum linear flow without air entrapment and produce a warp-free part.

One well-known cause for stress is packing at the gate. As the polymer cools, extra material is forced into the cavity by the injection ram to compensate for the thermal shrinkage as the part cools to room temperature. If too much material is forced in, it will compress the carbon–carbon linkages, overstressing the part at the gate area which can cause immediate or delayed failure. Special environmental conditions can also cause failure when a part is overpacked.

The main parameter is the ram forward time. Using an accelerated aging test, a center-gated polyethylene tumbler with a ram forward time of 10 s had no failures the first day and 7% 14 days later. Increasing the ram forward time to 25 s gave a one-day failure of 70% and a 14 day rate of 88%. Injection pressure and gate size also affect packing. Although packing is concentrated at the gate, it can extend throughout the parts. It is a prime cause for sticking in a mold. This is the result of better adhesion to the mold surface, deflecting the mold, and possibly distorting the core and cavity. Obviously, the amount of feed and the temperature affect the amount of plastic that flows into a cavity and hence the packing.

All materials shrink when they cool; however, crystalline materials shrink to a greater extent than amorphous materials because crystals take up less volume than an equal amount of amorphous materials. The degree of crystallization depends on the rate of quenching or temperature drop towards the mold temperature. Thus, temperature is a critical variable in influencing the properties of the finished part. Cool molds tend to freeze the polymer in the amorphous arrangement of the melt with lower levels of crystallinity. On the other hand, warmer molds permit the polymer to cool more slowly and thus permit more crystallization. The increased crystallinity results in greater shrinkage, high tensile strength, lower elongation, and greater hardness.

When a part is molded, a number of complex and opposing forces come into play. The hot material is injected into the cavity, where it shrinks as it cools. The material is compressed by the injection pressure, permitting more material to flow into the given volume to compensate for the shrinkage. The part cools during the injection process, continually reducing its volume. The amount of additional material forced in depends on the gate size, injection rate, temperature conditions, ram forward time, and pressure. This is why it is difficult to maintain really fine tolerances for injection-molded parts. The goal of proper injection molding is to so balance the machine and mold conditions so that acceptable parts are produced.

89.6.4 Molding Conditions

The chapter on Polypropylene has an excellent discussion on setting molding conditions and correcting molding faults that is applicable to most thermoplastic materials.

89.7 MACHINERY

89.7.1 Injection-molding Machine

The injection end of the first commercially successful injection-molding machine was a plunger. This type consists of a very heavy steel tube with a nozzle threaded into one end. The outside of the nozzle has a radius which fits into the reverse radius of the sprue bushing of the mold. It has a small hole (about 1/8-3/8 in. [3.2–9.5 mm]) through which the melted plastic material is forced by the injection plunger.

The inside of the tube is taken up by a steel spreader, or torpedo, which spreads the material around the inside wall of the tube. The tube, called a heating cylinder, is heated by electrical resistance heaters (heating bands) and controlled by thermocouples attached to pyrometers. The plunger is a round steel bar that pushes the cold material from the open end of the cylinder towards the hot nozzle end. It is volumetrically fed through a hole in the top rear of the cylinder.

Trillions of good parts were made by such plunger machines. The inherent problem with this system is that both melting and injection are done in one location, with each having different and opposing mechanical requirements.

The first revolution in machine design was separating the two functions of melting and shooting the material into the mold, by using two separate cylinders.

The first cylinder was the standard cylinder used for melting, mounted at 45 or 90° to the second, or shooting, cylinder or pot. The pot is a small-bore plunger cylinder without a torpedo. The two cylinders are connected by a valve. When the valve is open, the melting or plasticizing cylinder pushes the material into the nozzle end of the pot, forcing the plunger back. When the plunger moves back an experimentally predetermined distance, the filling stops and the valve is closed, because the backward motion of the plunger actuates a limit switch. At the appropriate time in the cycle the material is injected into the mold from the second cylinder, the so-called shooting cylinder or pot. This method of operation gives much better mixing of the plastic, more even temperature distribution, and better control of speed, pressure, and the amount of material.

This type machine is called a preplasticizing machine. Of the equipment in use today, 95% is preplasticizing; either with a reciprocating screw (Fig. 89.7) or a screw-pot (Fig. 89.8). In a reciprocating screw the melted material is collected in front of the screw, which continues to move backward against a controlled pressure as more material is melted. The area where the melted material is collected corresponds to the pot in a two-stage machine. When the screw is pushed forward by the hydraulic injection cylinder(s), a check valve (nonreturn valve) is closed. At this time the screw and nonreturn valve act like a plunger.

The plunger machine heats the material by conduction. Since plastic is an insulator, the material next to the outside wall is much hotter than that nearer the center. The laminar flow in the plunger-type cylinder does not reduce the temperature differential. This produces residual stresses in the molded parts, which can be very troublesome. The plunger machine

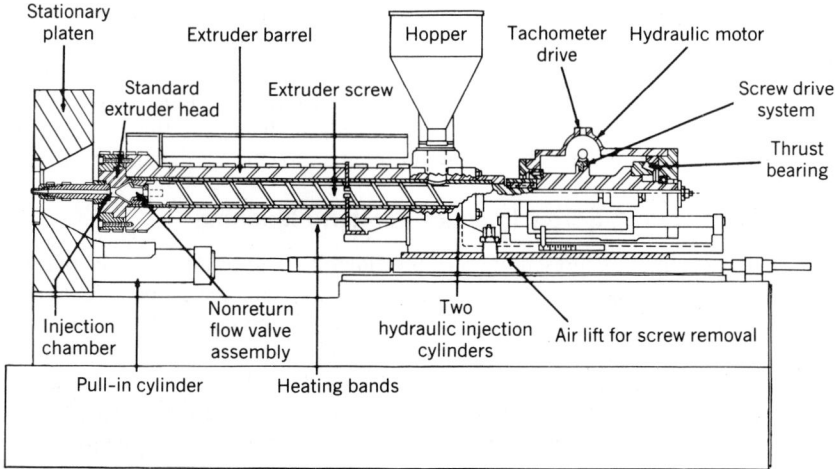

Figure 89.7. Schematic drawing of injection end of reciprocating screw machine. Courtesy of HPM Division of Koehring Co.

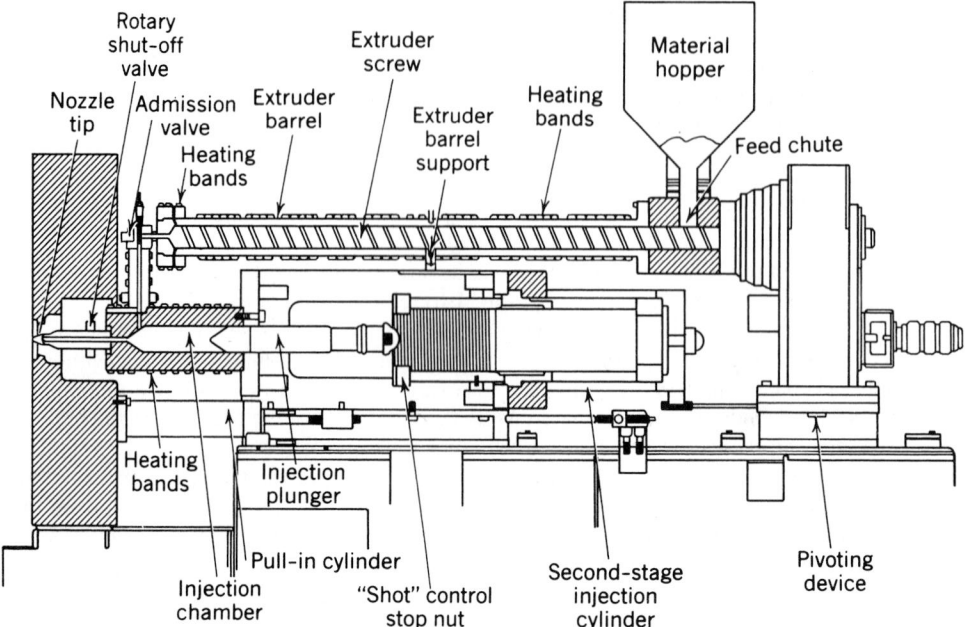

Figure 89.8. Injection end of two-stage screw-plunger machine. Courtesy of HPM Division of Koehring Co.

is now only used to produce special effects (mottling, for example); the two-stage plunger machine is obsolete. It is used in thermoset IM for glass-reinforced polyester because it reduces glass fiber breakage.

The second revolution in machine design was the reciprocating screw. The screw melts by friction. Heat is generated internally by the friction of the plastic molecules rubbing over each other as they are conveyed through the barrel. The screw is an excellent mixer. The net economic result of the superior melt and preplasticizing of the reciprocating screw was to rapidly make obsolete the plunger machine.

The third major change involves the use of the computer with feedback controls. The economies of the computer controlled machine with feedback increase productivity, quality, and operation so that it is cost effective to replace and/or convert existing machines.

Hydraulic molding machine circuits have not changed in the last 50 years. For a description of hydraulic components, circuits, and nonsolid-state electrical circuits, see ref. 8.

Solid-state controls and computers have caused radical changes in the electrical systems and circuits.[8] Such information is best found in the manufacturer's manuals.

89.7.2 Reciprocating Screw Injection-Molding Machine

Figure 89.9 is a schematic representation of the clamping end of an hydraulic machine. In order to minimize cost and power consumption, a small-diameter cylinder (not shown) is used to move the large clamp rapidly, using a prefill valve and gravity to get oil behind the large clamp ram. More than one cylinder can be used. The injection end of an in-line reciprocating screw plasticizing unit is shown in Figure 89.7. The ejection side of the mold is clamped to the moving platen. The mold has an empty space in the configuration of the part to be molded. This empty space if filled with melted plastic under high pressure.

The moving platen rides on four steel bars called tie rods or tie bars. The clamping force is generated by the hydraulic mechanism pushing against the moving platen and stretching tie rods.

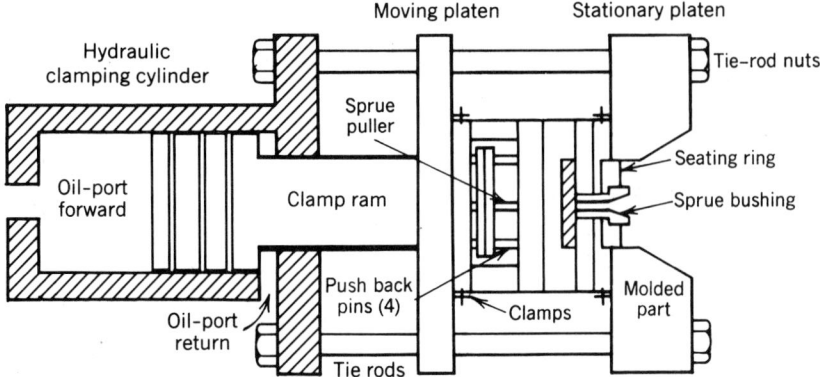

Figure 89.9. Hydraulic clamping system.

The molding process for reciprocating screw machines with hydraulic clamp include the following steps:

- Plastic material is put into the hopper. (The virgin powder is normally granulated to 1/8–3/16 in. (3.2–4.8 mm) spheres or cubes.
- Oil is sent behind the clamp ram, which moves the moving platen and closes the mold. The pressure behind the clamp ram builds up, developing enough force to keep the mold closed during the injection cycle. If the force of the injecting plastic material is greater than the clamp force, the mold opens, an unacceptable condition causing plastic to flow past the parting line on the surface of the mold. This condition produces flash, which must be removed or the piece has to be rejected and reground.
- The hydraulic injection cylinders are located on each side of the screw or there is a single hydraulic cylinder directly behind the screw. They bring the screw forward, injecting previously melted material into the mold cavity. The injection pressure is maintained for a perdetermined length of time. This is called the boost or hold pressure. Other than when molding PVC, a valve at the tip of the screw prevents material from leaking past the screw flights during injection. It opens when the screw is turning and melting material, permitting the plastic to flow in front of it to force the screw back.
- The oil velocity and pressure of the injection cylinder(s) are high enough to fill the mold as quickly as needed and maintain sufficient pressure to mold a part free from sink marks, flow marks, welds, and other defects.
- As the material cools, it becomes more viscous and solidifies to the point where it is no longer necessary to maintain injection pressure.
- The material is melted primarily by the turning of the screw, which converts mechanical energy into heat. It also absorbs some heat from the heating bands on the plasticizing cylinder (extruder barrel). As the material melts, it moves forward along the screw flights to the front end of the screw. The pressure generated by the screw on the material forces the screw, screw-drive system, and the hydraulic motor back, leaving a reservoir of plasticized material in front of the screw. The screw continues to turn until the rearward motion of the injection assembly hits a limit switch, which stops the rotation. This limit switch is adjustable, and its location determines the amount of material that remains in front of the screw (the size of the shot).

The pumping action of the screw also forces back the hydraulic injection cylinder(s). This return flow of oil from the hydraulic cylinder can be adjusted by the appropriate valve. This is called the "back pressure," which is adjustable from 0 to about 400 psi (2.8 MPa).

- Most machines retract the screw slightly at this point to decompress the material to prevent drooling out of the nozzle. This is called the suck back and is usually controlled by a timer.
- Heat is continually removed from the mold by circulating a cooling medium (usually water) through drilled holes in the mold. The amount of time needed for the part to solidify so that it can be ejected from the mold is controlled by the clamp timer. When it times out, the movable platen returns to its original position, opening the mold.
- An ejection mechanism separates the molded plastic part from the mold and the machine is ready for its next cycle.

In Figure 89.10 the back safety gate has been removed, showing the operator removing molded parts from a molding machine; the four tie bars on which the movable platen rides are visible. The rubber hoses circulate fluid for mold temperature control. Each cavity and core has its own set, so that the temperature on each can be controlled separately. The mold is held on the platen by clamps. A cutaway view of a 2 1/2 in. (64 mm) reciprocating screw, 300 ton machine is shown in Figure 89.11. The injection end is always to the right of the operator. This permits him or her to open the safety gate with the left hand and remove the molded pieces with the right hand, should the machine not be running automatically.

89.7.3 Controls

The location of the hydraulic and electrical controls varies widely and is obviously remote from the valves and solenoids. Some controls are mounted in separate units away from the molding machine. The controls can be as simple as buttons and dials on timers or as sophisticated as interfacing with a computer.

The oil reservoirs are mounted either on the floor or above the machines. Pressure may be read from gauges located anywhere, or from digital readouts on a panel. The hydraulic system generates a significant amount of heat that is removed by a water-cooled heat exchanger (water cooler). This water is usually from a water tower or well. Water is also used for cooling molds that usually come from a mechanical refrigeration unit.

The hydraulic and electrical systems should be designed for easy maintenance. Clear, easy to read, easy to understand, well-illustrated instruction manuals are needed. In purchasing a new machine, the instruction manual and quality of repair support should be a factor in selection.

89.7.4 Knockout (KO) Systems

Mechanical, hydraulic, or pneumatic knockout (KO) systems are always installed behind the moving platen and are used in conjunction with holes in the moving platen whose location and size have been standardized by the Society of the Plastics Industry (SPI). Occasionally, knockout systems are also installed on the stationary side.

In the mechanical system, a stationary KO plate on the machine has bars that pass through the platen and the back plate of the moveable half of the mold. KO pins or other devices are attached to the KO plate of the mold. The other end of the pin or device is in the molding area. After the part is molded, plastic is in direct contact with the top of the KO pins or devices. As the mold opens, the KO plate of the mold hits the bars that are attached to the stationary KO plate of the machine. This stops the motion of the mold KO plate. The rest of the mold continues to move back, leaving the molded part on the front end of the KO pins and thereby ejecting the piece.

There are four push-back pins attached to the four corners of the mold KO plate; they are flush with the mold surface when the mold is closed, and move forward with the KO

Figure 89.10.   Operator removing a molded part.

plate. When the mold closes, these pins contact the other side of the mold before the KO pins and force the mold KO plate back to its original position without damaging the tips of the KO pins. In mechanical KO systems, the operation is machine-dependent and the KO system only works when the machine is opened. To return the mold KO plate to its molding position, the machine has to be moved until the KO bars are below the top surface of the mold back up plate. This is an important disadvantage in many cam-acting molds.

A superior and more expensive method is to use two hydraulic or pneumatic cylinders to activate the machine KO plate. It can be moved backwards and forwards as often as desired (bumping) at various speeds and forces. It can be retracted or extended at any time in the cycle.

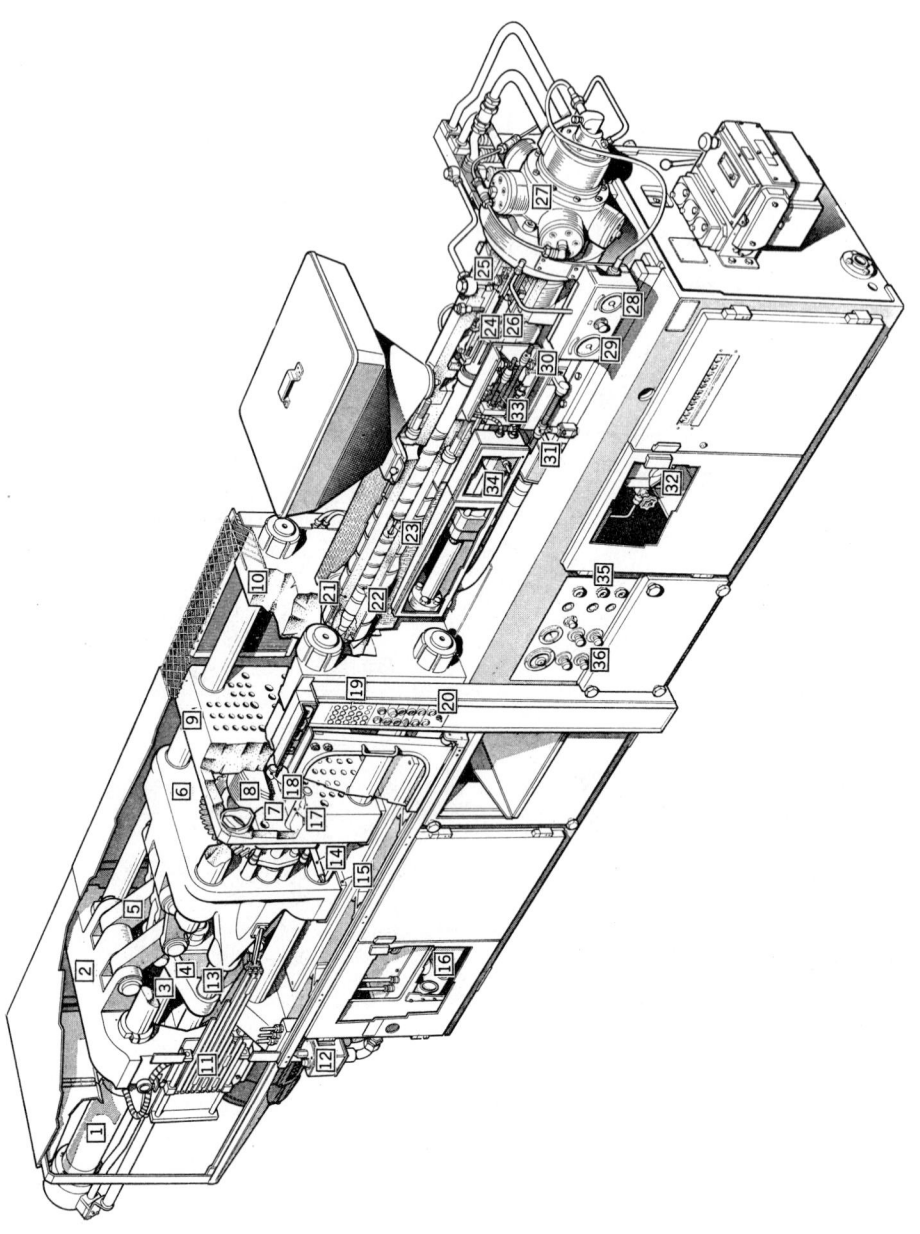

89.7.5 Clamping Systems

The machine is clamped either by a hydraulically operated toggle system or by a straight hydraulic ram. In both instances a hydraulic cylinder provides the force to stretch the tie bars, which causes the clamping action. In a straight hydraulic system, a large-diameter hydraulic cylinder is attached directly to the movable platen. The clamping force is rated in tons.

Force should not be confused with pressure. They are related by the equation:

$$F = P \times A \tag{89.1}$$

where F = force, lb (N), P = pressure (lb/in.^{2}, psi) (N/m^2), A = area (in.^{2}) (m^2).

For example, a press with a hydraulic clamp has a 20 in. (508 mm) diameter clamping cylinder. The maximum working line pressure is 2000 psi (13.8 MPa); the clamping force is

$$F = 2000 \text{ psi} \times 314.2 \text{ in.}^2 = 628,400 \text{ lb} = 314.2 \text{ tons}$$

$$F = 13.78 \text{ MN/m}^2 \times 0.203 \text{ m}^2 = 2.8 \text{ MN (314 US tons)}$$

This press would be called a 300-ton press. Clamping force is one of the main machine specifications. Machines available today range from 5.5 tons for a machine whose maximum is 10 g per shot to 5500 tons for a machine that can mold 54 lb (24.5 kg) of polystyrene per shot. A 1-ton clamp force corresponds to about 9 kN.

Hydraulic clamping[9] permits unlimited pressure selection, which can be continually monitored with a pressure gauge. As molds get bigger, longer strokes are required for the movable platen. The larger molds require larger clamp forces and larger-diameter cylinders. Cylinders of such size are very expensive. Use of a smaller-diameter cylinder for long rapid movement and a large-diameter, short-stroke cylinder for generating the clamp force reduces the cost.

There are many different systems of toggles.[10] Basically the hydraulic cylinder (Fig. 89.11, 1) is attached to a stationary plate (Fig. 89.11, 2). It moves forward, eventually spreading the toggle links (Fig. 89.11, 5) so that they are in a straight line and holds them there. The mechanical advantage is between about 20:1 to 30:1. The toggle system is usually less expensive to build than the hydraulic system. It requires good maintenance because wear reduces the clamping tonnage. Pressure adjustment is not as easily or accurately controlled as with an hydraulic system.

A clamping force is required to keep the mold closed to overcome the opening force generated by the injection of the plastic material. The amount of clamp force depends on the projected area of all the molding (parts and runner). The projected area is most conveniently estimated by looking from the heating cylinder to the clamping end. If the mold were transparent and everything molded (parts and runner) were black, the visible area of the black you would see is the projected area.

Figure 89.11. Cutaway view of a 2 1/2 in. (64 mm) reciprocating screw, 300-ton machine (Peco molding machine).

1, hydraulic cylinder; 2, tail stock plate; 3, hydraulic piston extension; 4, toggle crosshead; 5, toggle link; 6, moving back plate; 7, ejector plate; 8, mold height adjustment screw; 9, moving platen; 10, stationary platen; 11, limit switches and stops; 12, lubrication pump; 13, toggle cross head guide bar; 14, mold height adjustment mechanism; 15, moving plate support pad; 16, hydraulic tank; 17, ejector bar; 18, hydraulic ejector; 19, solenoid indicator lights; 20, manual control panel; 21, injection (heating) cylinder; 22, screw; 23, air tube and bore; 24, screw coupling; 25, bearing; 26, hydraulic motor drive shaft; 27, hydraulic motor; 28, screw speed indicator; 29, injection pressure gauge; 30, shot volume control mechanism; 31, retraction stroke limit switch; 32, screw speed control mechanism; 33, injection secondary pressure control; 34, hydraulic injection cylinder; 35, water valves; 36, hydraulic controls. Courtesy of *British Plastics*.

In other words, if a part is molded behind another part it would not be seen. Therefore, it would not be included in the projected area. It is also evident that the projected area of a 10 × 10 × 1/8 in. (25 × 25 × 0.32 cm) plaque and a container with maximum cross section of 10 × 10 in. (25 × 25 cm) that was 13 in (33 cm) deep would both have the same projected area, 100 in.^{2} (645 cm^2). Thus both would require the same clamp force. However the container requires much more steel around the sides of the cavity to contain the injection pressure.

For design purposes, a good approximation is 2 1/2 tons of clamp force for each square inch of projected area (0.4 ton/cm^2). In the example above, 100 in.^{2} × 2 1/2 t/in.^{2} (645 cm^2 × 0.4) = 250 ton. A machine with a minimum of 250-t clamp would be selected.

The clamp stroke is the maximum distance the clamp can move. The maximum daylight is the farthest distance the platens can separate from each other. The difference between maximum daylight and stroke is the minimum die thickness that can be put into the press and still maintain the clamping pressure. This minimum distance can always be decreased by adding a bolster plate in front of or behind the movable platen. These are important specifications. They tell how deep a piece may be molded and whether a mold of given depth will fit in the machine.

The clearance between the tie rods is the determining factor as to whether a mold of a given length or width fits. For example, a press has a 20 in. (51 cm) clearance vertically and an 18 in. (46 cm) clearance horizontally. Therefore, a mold less than 20 in. (51 cm) wide but over 20 in. (51 cm) long will fit vertically; a mold less than 18 in. (46 cm) high but over 18 in. (46 cm) long will fit horizontally. The length and width dimensions of a mold are often determined by the side parallel to the knockout plate. Molds often extend beyond the tie bars, and, with proper design, beyond the platen.

89.7.6   Injection End

The injection end has a hopper for holding the material. Hygroscopic materials such as nylons, polycarbonates, acrylics, acrylonitriles, and acetates require drying before molding.

There are three methods for drying plastics. In one method, a hopper dryer placed on top of the feed mechanism circulates hot, dry air (dehumidified in some instances) through the material. In the second method, the material is predried in shallow trays in an oven separate from the machine by hot, dry, dehumidified air. The third method employs a vented screw that melts the material and then decompresses it to atmospheric pressure, permitting the water and other volatiles to escape through an opening in the cylinder. The material is recompressed as it moves beyond the vent towards the nozzle and is then injected into the mold.

A magnet is placed in the hopper throat to catch any magnetic material accidentally introduced into the hopper or material.

89.7.7   The Injection Screw

A typical injection screw is shown in Figure 89.12. It is, in effect, a series of endless buckets; that is, a mechanical conveyor. The volumetric output equations of a screw are essentially a mathematical description of the screw's volume multiplied by its speed, less the plastics resistance to flow.

The material melts because molecules slide and rub over each other, a process called shearing. The resultant friction generates heat. Part of the polymer molecule adheres either to the wall of the barrel or to the screw; the conveying action of the system causes the molecules to move towards the nozzle end. Thus, the heat is directly generated into the material.

Heating bands are used to compensate for the black body radiation of the cylinder; otherwise it would be necessary to overheat the plastic. The bands are also utilized to adjust the final temperature of the plastic. About 70% of the energy input of the cylinder is used

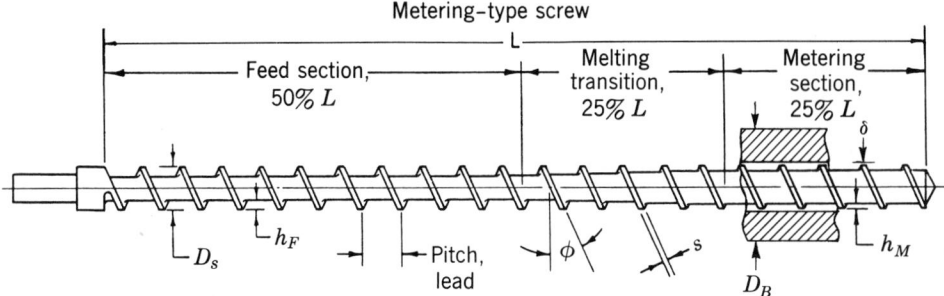

Figure 89.12. Injection screw.

for turning the screw (melting the material); about 27% is used for black body radiation losses; and 3% is used for bringing the material to the final temperature as set on the pyrometers.

The injection screw is divided into three parts. The first half of the screw compresses the material and removes the air. The flight depth (h_f) is constant. The second part (third quarter) of the screw has decreasing flight depths, thus reducing the volume. Reduction of the volume increases the friction of the polymer molecules as they rub over each other, increasing the temperature. At the end of this section all the material should be melted (plasticized). The third part (last quarter) of the screw is a pump. The flight depth (h_m) is constant but smaller than the first section because melted material takes up less volume than powder. All calculations about screw output relate only to this section.

On all injection screws one thread per screw diameter has been standardized. Thus, the ratio of the length over the diameter, L/D, indicates the number of turns (flights) there are on the screw. The reciprocating screw continually moves rearward under the hopper as it operates, reducing the number of available flights. An L/D of 20:1 is strongly recommended.

89.7.8 Torque

The work of melting a material is done by rotating a screw in a stationary barrel. The rotational force called torque is the product of the tangential force and the distance from the center of the rotating member. For example, if a 1-lb (4.45-N) weight were placed at the end of a 1-ft (0.305-m) bar attached to the center of the screw, the torque would be 1 ft × 1 lb or 1 ft-lb (1.36 Nm). Torque is related to horsepower:

$$\text{power, hp} = \frac{\text{torque (ft-lb)} \times \text{rpm}}{5252} \qquad (89.2)$$

$$\text{kW} = \frac{\text{torque (Nm)} \times \text{rpm}}{7124}$$

It is clear that the torque output of an electric motor of a given horsepower depends on its speed. A 30-hp (22-kW) motor has the following torque at various speeds:

Speed, rpm	Torque, ft-lb	(Nm)
1800	87.5	(119)
1200	133	(181)
900	175	(238)

The speed of a given horsepower motor is built into the motor. Changes in speed and torque can also be accomplished by changing the output speed of the motor by using a gear train. The change in torque varies inversely with the speed.

A-C motors develop a starting torque of almost twice the running torque. The screw has to be protected against overload to prevent screw breakage. This is not a problem with an hydraulic drive.

The drive must supply enough torque to plasticize at the lowest possible screw speed, but not enough to mechanically shear the screw. Changes in torque are needed because of the different processing characteristics of plastics. Much higher torque is required to plasticize polycarbonate than polystyrene. The speed of the screw regulates the quality and output rate of the polymer. A choice of two, and preferably three, torque speed ranges for handling the various materials is desirable. If a material is being molded with a minimum torque at a given speed, an increase in the speed increases the horsepower requirements.

The ability to easily control torque and speed is very important. The hydraulic motor has both infinitely adjustable control for the screw speed and for the constant torque output. Both are easily adjusted by valve settings. The system is designed to prevent torque high enough to break the screw. Because it is much lighter than an equivalent A-C motor and drive, it allows a lower melt pressure. These are some of the reasons why most machines use hydraulic motors to turn the screw.

The strength of the screw limits the input horsepower. As input horsepower is increased, a point is reached where the torque that can be provided is above the yield strength of the metal screw. The strength of the screw varies with the cube of the root diameter. For a typical injection screw with a 2 1/2 in. (64 mm) diameter at 200 rpm, the maximum permissible drive input is about 40 hp (30 kW). For a 3 1/2 in. (89 mm) diameter screw at 200 rpm, it is 75 hp (56 kW), and for a 4 1/2 in. (114 mm) screw at 150 rpm, it is 180 hp (134 kW). For a given horsepower, the slower the speed, the higher the torque (eq. 89.2). Too rapid a shear rate degrades the material. The shear rate is highest at the barrel wall. With present screw technology, the maximum surface speed is about 150 ft/min (45 m/min). Some of the more shear sensitive materials limit the surface speed to 100 ft/min (30 m/min).

89.7.9 Power Requirements

The work performed in screw plastication causes the material temperature to rise from room temperature to the molding temperature. Assuming that all the energy comes from turning the screw and that the mechanical efficiency is 100%, the work done would be the product of the average specific heat and the temperature rise. Neither of these assumptions is correct. A small amount of heat is supplied by the heating bands, and corrections for machine efficiency must be made. Since the screw acts as a pump, energy is also required when pressure is generated. This is relatively small and is disregarded in subsequent calculations. Therefore,

$$\text{Power} = C \times (T_p - T_f) \times Q + (\Delta P \times Q) \tag{89.3}$$

where C = average specific heat, btu/lb °F; T_p = temperature plasticized material, °F; T_f = temperature feed material, °F; Q = throughput lb/h, continuous rotation; ΔP = back pressure, psi.

In SI units, P = kW, C = J/kg °C, T_p = °C, T_f = °C, Q = kg/s, and P = N/m or Pa and converting into consistent units, equation 89.4 is obtained

$$HP = 0.00039 \times C \times (T_p - T_f) \times Q \tag{89.4}$$

and for 70% efficiency

$$HP = 0.00056 \times C \times (T_p - T_f) \times Q \tag{89.5}$$

For example, how much horsepower is required to plasticize high impact polystyrene [C = 0.42 btu/lb °F (1758 J/kg °C)] at the rate of 1 lb/h (7.26 × 10^{-4} kg/s) (Q = 1) when the molding temperature is 380°F (193°C) and the room temperature is 70°F (21°C)?

$$HP = 0.00056 \times 0.42 \times (380 - 70) \times 1 = 0.073 \text{ hp}$$

In SI units,

$$\text{Power} = 1758 \times 172 \times 0.000126 = 38.1 \ (J/s = \text{Watt})$$

$$= 0.0511 \text{ hp at } 100\% \text{ efficiency and } 0.073 \text{ hp at } 70\% \text{ efficiency.}$$

This is equivalent to 13.7 lb/h (6.2 kg/h) for each 1-hp (746-W) input. Molding materials range from 6 to 14 lb/h (2.7–6.4 kg/h) for each 1-hp (746 W) input.

Using a 30-hp (22.4-kW) motor, at a room temperature of 80°F (27°C), what is the maximum output of low density polyethylene [C = 0.8 btu/lb °F (3349 J/kg °C)] at 380°F (193°C)?

$$30 = 0.00056 \times 0.8 \times (380 - 80) \times Q$$

$$Q = 223 \text{ lb/h (101 kg/h)}$$

In SI units,

$$22,370 = 3349 \times 166 \times Q$$

$$Q = 0.04202 \text{ kg/s} = 144.9 \text{ kg/h at } 100\% \text{ efficiency and } 101 \text{ kg/h at } 70\% \text{ efficiency.}$$

Supposing the material temperature were raised to 450°F (232°C). Would the output increase?

$$30 = 0.00056 \times 0.8 \times (450 - 80) \times Q$$

$$Q = 181 \text{ lb/h (82 kg/h)}$$

In SI units,

$$\text{at } 70\% \text{ efficiency } 0.7 \times 22370 = 3349 \times Q = 0.0228 \text{ kg/s} = 82 \text{ kg/h}$$

Thus, a raise in the temperature lowers the maximum output.

The molder therefore molds at the lowest possible melt temperature. This gives maximum screw output and reduces the time needed for reversing the process; that is, polymer cooling in the mold.

Theoretically, the output of the machine (eq. 89.3) is independent of the screw diameter. If, for example, 2 1/2 and 3 1/2 in. (6.4 and 8.9 mm) screws, each having the same length/diameter (L/D) ratio and the same input drive power, are operated at their maximum capacity, they both deliver the same output [lb/h (kg/h)]. They deliver the same output (lb/h, kg/h). The plastic remains longer in the 3 1/2 in. (89 mm) screws. The reason for large screws is found in eq. 89.2. Screw speeds must be kept low to prevent polymer degradation. With constant horsepower, the slower speeds can develop a torque high enough to shear the screw. For a 2 1/2 in. (64 mm) screw at 200 rpm, the maximum permissible drive input is 40 hp (30 kW). Therefore, higher horsepowers (for higher output) require larger-diameter screws to prevent screw breakage. The torque a screw can safely carry is proportional to the cube of its root diameter. At 200 rpm, a 3 1/2 in. (89 mm) screw can handle 120 hp (90 kW); a 4 1/2 in. (114 mm) screw, 240 hp (179 kW).

It is obvious that in a screw machine the power rating available for screw rotation is a very important specification. Assuming similar efficiency for different screw designs, the maximum output, which is a primary concern of injection molders, is largely determined by the power rating of the screw.

89.7.10 Advantages of Screw Plasticizing

In a screw, the melting of the plastic is caused by the shearing action of the screw, which converts the mechanical energy of the screw drive into heat energy. The heat is applied directly to the material. This and the mixing action of the screw contribute to its major advantages as a plasticizing method:

- High shearing rates lower the viscosity, making the material flow easier.
- Good mixing results in a homogeneous melt.
- The flow is nonlaminar.
- The residence time in the cylinder is approximately three shots, compared to the 8 to 10 shots of a plunger machine.
- Most of the heat is supplied directly in the material so that the cycle can be delayed for a longer period before purging.
- The method can be used with heat-sensitive materials, such as PVC.
- The action of the screw reduces chances of material holdup and subsequent degradation.
- The screw is easier to purge and clean than the plunger.

89.7.11 Reciprocating Screw Systems

In a reciprocating (in-line) screw (Fig. 89.7), the material is fed from the hopper, plasticized in the screw, and forced past a one-way valve at the injection end of the screw. The material accumulates in front of the screw, forcing back the screw, its drive and motor, and the pistons of the two hydraulic injection cylinders. The return oil from these cylinders goes through a valve into the tank. This valve is called the back pressure control valve. It controls the mixing in the cylinder. The higher the pressure setting, the more complete the mixing.

When the screw reaches a certain position, it contacts a limit switch that stops the rotation. When the cycle is ready to inject, two hydraulic injection cylinders, one on each side of the carriage, move the screw assembly forward and use the screw as an injection ram. The one-way valve prevents the material from passing back over the flights. If this valve is not functioning correctly the screw rotates during injection, and the shot size cannot be controlled.

89.7.12 Screw-plunger System

In a screw-plunger (screw-pot or two-stage screw) system shown in Figure 89.8, a fixed screw is used for plasticizing. A reciprocating screw could be used, which would permit continual operation of the screw throughout the whole cycle.

The reciprocating screw and screw-pot are both preplasticizing systems. The difference is the location of the pot, which is in front of the reciprocating screw and is a separate cylinder in the two-stage machine. Most machines are of the reciprocating type, but screw-pot equipment offers significant advantages:

- Because the screw does not act as the injection ram, lighter bearings can be used. There is no need for the heavy thrust assemblies found on reciprocating screws. This reduces maintenance cost.
- The extruder barrel need only be strong enought to maintain the pressure of the material during plasticization, which is rarely greater than 5000 psi (35 MPa). In contrast, the barrel for the reciprocating screw must contain the 20,000 psi (138 MPa) pressure applied.
- There is less wear because the screw does not reciprocate.
- The connection between the two stages can be a ball check valve, which is trouble-free and easy to maintain. It presents minimum flow resistance.

- The nonreturn valve at the tip of a reciprocating screw wears rapidly, sometimes does not seat properly (preventing consistent molding), can cause wear in the barrel, retains and degrades material, and is much more expensive than a ball check valve.
- The small clearances between the plunger and barrel of the second stage help in degassing the material.
- The connection between the two stages results in better mixing of the melt.
- A very important advantage of a two-stage machine is that all the material goes over the full flights of the screw, receiving the same heat history. In a reciprocating screw, only the first material in passes over the full length of the screw.
- In a two-stage machine, the screw pumps only against the injection ram, which is floating in oil in the hydraulic cylinder. The reciprocating screw must push back the whole weight of the carriage and all the equipment on it. Therefore, the shot size control is considerably more accurate in the two-stage machine.
- Because part of the energy input is used to push back the heavy carriage, an in-line machine gives slightly less output per unit input.
- The two-stage machine has better injection speed control.
- Because of the preceding two factors a larger projected area can be molded.
- Extremely high injection pressures are available.
- The size of a pot in front a reciprocating screw is limited by the length of the feed section. If the screw goes too far back the material does not plasticize correctly. In a two-stage machine there is no theoretical limitation; thus, a 2 in. (51 mm) reciprocating screw normally has a maximum shooting capacity of 13 oz (319 g), whereas the same diameter screw can be readily used to shoot 100 oz (2.8 kg) in a two-stage machine.

There are a number of disadvantages of the two-stage machine:

- It requires two cylinders and two sets of heat controls.
- It is slightly more difficult to clean.
- It is slightly more difficult to set up.
- It does not process materials sensitive to high heat.
- Cylinders for molding thermosets and rubber are designed only for reciprocating screws.
- It takes up more space than a reciprocating screw.
- It costs more.

89.7.13 Intrusion Molding

When heavy sections are molded or when the shooting capacity of the machine is not adequate, intrusion molding can be used, where the screw turns continually and fills the cavity directly. When the cavity is filled, a cushion is extruded in front of the plunger, which then comes forward to supply the needed injection pressure.

89.7.14 Thermoset Molding

Thermosetting materials can be molded on reciprocating screw machines. The screw is designed differently and the barrel is usually heated with hot water. When thermosets are molded, the heating cylinder provides just enough heat to start exothermic cross-linking chemical reactions between the two monomers. The material must be transferred quickly into the mold and cannot be permitted to cool in the cylinder. If it does, the screw has to be removed and the thermosetting material chipped out.

89.7.15 Specifications—Injection End

Injection end specifications vary for the different methods of injection. They all include:

- The machine capacity, which is the maximum weight of material (in ounces) that can be injected in one shot and is based upon molding general-purpose (GP) polystyrene for thermoplastic material and GP phenolic for thermosetting material; for example, "16 oz (454 g) machine." The more accurate specification is the volume of material per shot [in.^{3} (cm^3)]. This is a measure of the geometry of the cylinder and feed system. The injection stroke is the maximum distance that the injection plunger can travel. The injection speed [in./min (cm/min)] is the speed of the injection plunger, usually with no material in the cylinder. The maximum injection rate [in./min or oz/min (cm/min or g/min)] is the rate at which the injection cylinder can eject fully plasticized material into the air. This is some what different from the speed achieved during molding.
- Standard injection pressure that can be placed directly on the materials is 20,000 psi (138 MPa). Material pressure greater than 20,000 psi (138 MPa) is required to mold some of the new materials. In a two-stage machine or in-line screw, this is easy to determine. It depends on the diameter of the screw or plunger, the piston diameter of the hydraulic injection cylinder, and the oil pressure. In straight plunger machines, the injection pressure on the material is sometimes given based on the same factors. There is at least a 30% loss in pressure exerted on the material from the feed end to the nozzle. The pressure loss is related to the granular condition of the cold pellets in the back of the cylinder.
- The plasticizing capacity is the most important injection end specification and the most difficult to verify. The SPI has adopted a specification developed by a technical committee of the Society of Plastics Engineers (SPE).
- The rate of recharging [oz/s (g/s)] tells how much material the screw can produce when it is running continuously. In most machines, the screw runs intermittently and the output is estimated based on the type of molding; a good average approximation is 50%. Some of the other injection end specifications are screw speed, screw diameter, screw torque, L/D, electrical heating specifications, and hopper capacity.

89.7.16   Specifications—Clamp End

The clamping end is hydraulic or toggle; its clamping force is rated in tons. Other specifications are

- The horizontal and vertical clearance between the tie bars gives the maximum size mold that will fit into the machine in each direction. Molding beyond the tie bars presents no difficulty. The size of the platens is self-explanatory. Molds can extend beyond the platens if the mold is strong enough or properly supported.
- The maximum daylight is the maximum distance between the two platens when the machine opened is fully opened. The stroke is the maximum distance the moveable platen can move. The difference between the two is the minimum mold thickness necessary to obtain full clamping force. From these data combined with the mold configuration and the piece part dimensions, it can be determined whether the part can be removed from the mold in the machine after molding.
- The knockout specifications indicate whether the system is mechanical, hydraulic or pneumatic. It gives the stroke of the knockout plate. The SPI has established a standard knockout hole pattern that is followed by all manufacturers.
- Speeds for the clamp and injection ends with no material are given. Low-pressure closing prevents the full clamp force from building up unless there is no obstruction on the mold surfaces. This is an excellent mold protection system; it is specified if it is available.

89.7.17   Other Specifications

Types of control (solid state, computer, computer with feedback), heating cylinder controls and specification, the number of motors and their sizes, the hydraulic pump capacities, water-cooling requirements, and the physical description of the machine are some of the other specifications.

89.7.18   How To Mold Quality Parts

Once the part and mold have been correctly designed and a quality mold built, it is relatively easy to make good parts.

Quality control specification are used as the guide. Different samples are molded under different conditions and records are kept. Each sample is sent through quality control and submitted to whatever tests are required and deemed advisable. This process is repeated until a set of conditions is found that will produce a fully acceptable part that is profitable to the molder.

It is essential for management to reproduce this part consistently on a 24 hour a day operation. The quality of the part is highly sensitive to the molding conditions. For example a plastic cup, such as used by the airlines, can be molded perfectly or in a way as to break in the carton after a few weeks or to crack when filled with a cold liquid or to crack when the mold opens—all by changing molding conditions.

There is no practical, nondestructive way to determine the quality of a molded parts. The only way to guarantee a quality part that meets specifications is to control the molding conditions within the experimentally determined parameters. This is why computer controlled machines with feedback are economically superior to noncontrolled, nonfeedback machines. They reproduce experimentally determined operating conditions with a much higher degree of accuracy and reduce, and may eliminate, manual inconsistencies. They will reproduce identical injection rate profiles that are needed to maintain identical viscosity from shot to shot.

89.7.19   Safety

Safety is everybody's business!!!

The molding machine is dangerous. It has caused fatalities, severe injuries, lost limbs, and fingers. The safety gate should be large enough and high enough to prevent contact with the platen area when the platens are closing. On smaller machines, a cover is essential. Both front and rear gates should be mechanically, electrically, and hydraulically interlocked so that the machine cannot operate unless the gate is fully closed.

At no time should repairs be made that require entry between the platens with the motor running. Moving parts such as toggles and cams should be guarded. The use of robots requires extensive guarding to prevent serious injury, particularly to the operator's eyes and head. The heating cylinder should be covered to prevent direct human contact with the heating bands and the electrical terminals. The nozzle end of the cylinder is guarded to prevent burning by hot or exploding material during purging. Experts and agencies should be consulted for detailed safety instructions.

The 1970 Occupational Safety and Health Act (OSHA) requires that the employee be provided with a safe workplace. This requirement must be observed.

89.8   TOOLING

89.8.1   Molds

A high-quality mold is absolutely essential for a profitable molding operation. An excellent mold is one that produces the designed part time after time without interruptions or deviations in cycle, with minimum downtime for repair, and with enough cooling to ensure minimum cycle time for the life of the mold. Customer, designer, molder, quality control, packaging, and moldmaker must agree on the design of the part, its parting line, ejection, surface finish, and so on. This must be done before the mold is built.

This information is ultimately put on a print that is initialled by all before production of the mold starts. It is then the responsibility of the molder and moldmaker to build a mold to produce a part to these specifications. Ultimately, the molder has the responsibility of producing an acceptable part.

The injection mold is the mechanism into which the hot plasticized material is injected and maintained under pressure and then cooled. When the plastic material has solidified sufficiently, the mold halves separate and the plastic pieces are ejected.

The quality of the part and its cost of manufacture are strongly influenced by mold design, construction, and excellence of workmanship.

As the machine capacity increases, the molds become larger and more expensive. Costs can range from as little as $2000 to $400,000 for a two-cavity automobile dashboard mold. This is only a small part of the investment. The original idea, market testing, samples, prototypes, advertising, selling, and the commitment for the initial order can be many times the cost of the tool.

Most critical in the production of a plastic part are the piece part and the mold design. Bad design does not necessarily result in piece part failure, though it may well do so. It does cause low productivity, high mold maintenance, and the probability of reduced part quality.

A poorly functioning mold requires excessive supervision in the plant and tool maintenance department. This can cause neglect of other operations. It is for this reason that the molder, moldmaker, and, on occasion, the purchaser should give maximum attention to the design. Unfortunately this is not always done.

The first step in designing a mold is to have an accurate, fully dimensioned drawing, one that notes tapers and where they start, tolerances, shrinkage specifications, surface finish specifications, the material in which the part is to be molded, part identification, and any other pertinent information. A model of the part to be molded or a CAD design is helpful, particularly if it or the drawing is complicated. Use of the model enables mold design to be significantly more reliable and molding problems to be more readily anticipated. At this time, the metal for the cavity and the core is selected. The cavity and core are those parts of the mold that, when held together by the closing of the mold, provide the air space into which the molten plastic is injected.

The following are decided next: the number of cavities; the parting line of the piece (where the faces of the cavities and core touch); the type and location of the gate (the entry point of the hot plastic into the cavity); the gating system (which brings the hot plastic into the cavity); the runner system (which brings the hot plastic from the plasticizing chamber to the gate); the method of ejection (which removes the molded plastic parts from the mold); the location of the ejecting devices; the location and size of the temperature control channels; the type and location of the venting system (which removes the air that is displaced by the incoming plastic); and the surface finish. The mold is designed and reviewed.

89.8.12   CAD/CAM in Moldmaking

There has been a steady decrease in the number of people building molds in the United States. Because of this and for its economic and technical value, computer-assisted design (CAD) and computer-assisted manufacture (CAM) are being rapidly adopted by the U.S. plastics industry.

CAD is extremely helpful the design of a part. Views, cross sections, projections, changes in sizes and colors, and mechanical and thermal analyses can be easily made by the computer.

CAD is also extremely useful as a drafting tool and can produce menu-driven drawings. Proper programming prevents common errors.

There are a number of software programs that deliver mold designs. However they are based on so many assumptions, some of which are incorrect, that they should be used with extreme caution. Some of the software is of help to very experienced mold designers.

The belief that if it comes from a computer it must be correct is rarely more harmful and unforgiving than in mold design.

In addition to its other functions, CAD/CAM transfers the mold design into metal cutting instructions. This common practice in the metal working industry is now being applied to mold making.

89.8.3 Mold Base

The steel parts that enclose the cavities and cores are called the mold base, mold frame, mold set, die base, die set, or shoe (Fig. 89.13). The sprue bushing is centered by the locating or seating ring on the stationary or injection platen, directly in line with the nozzle of the injection cylinder. The sprue has an opening in the center of a concave spherical surface whose counterpart is an equivalent convex surface with a hole on the nozzle of the injection heating cylinder. The opening of the sprue bushing must always be larger than the opening in the nozzle, so that when the plastic hardens it does not form an obstruction larger than the sprue opening and cause the sprue to stick. The sprue has a generous taper to facilitate easy removal of the plastic. The standard radii for sprue bushings and nozzles are 1/2 and 3/4 in. (13 and 19 mm).

The injection or top clamping or backup plate supports the cavity or A plate. The B plate, which is supported by its backup, plate has the cores attached to it; these match the cavities in the A plate.

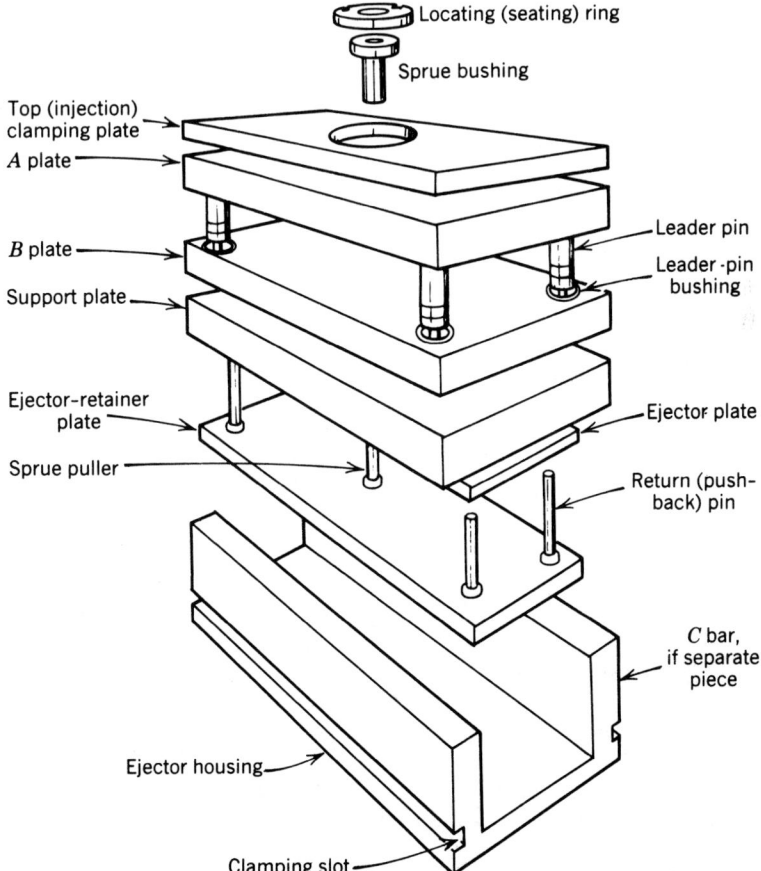

Figure 89.13. Standard mold base. Courtesy of D-M-E Corp.

Ejector or knockout (KO) pins and other KO devices are mechanically attached to the ejector plate assembly so that the molded parts can be knocked out and removed from the mold.

A number of companies manufacture standard mold bases and parts. Because of their high volume, they have equipment that usually makes their mold bases less expensive and superior to those manufactured by the moldmaker. In addition, replacement parts are standard and readily available to the molder at a low cost.

There are cavities and cores of such size or shape that a mold base is best built around them. Several types of steel are available for the mold base. It is strongly urged that the best quality be used for any mold that requires high-quality parts or a long production run.

89.8.4 Mold Types

The injection mold is identified descriptively by a combination of some of the following terms:

Parting Line
Regular
Irregular
Two-plate mold
Three-plate mold

Mold Material

Steel	Beryllium copper
Stainless steel	Aluminum
Prehardened steel	Brass
Hardened steel	Epoxy

Surface
Chrome plated
Electroless nickel
Etched
Sandblasted
EDM (electrical discharge machined)

Number of Cavities

Method of Manufacture

Machined	Pressure cast
Hobbed	Electroplated
Gravity cast	EDM (spark erosion)

Runner System
Hot runner
Insulated runner
Cold runners (for thermoset materials)

Gating

Edge	Diaphragm
Restricted (pin pointed)	Tab
Submarine	Flash
Sprue	Fan
Ring	Multiple

Ejection

Stripper ring   Removable insert
Stripper plate   Hydraulic core pull
Unscrewing
Cam

89.8.5 Two-plate Molds

The cross section of part of a regular two-plate injection mold is shown in Figure 89.14. The part being molded is a shallow dish. The cavities are gated on the edge. Temperature control channels are in both backup plates and in the cores and cavities. Because there is significant insulation between two pieces of metal, the use of channels in cores and cavities gives better and more efficient temperature control.

Support pillars are anchored between the back plate and the support plate, which is underneath the cores that are in the B plate. A machine knockout bar is shown. As mentioned, they remain stationary; as the moving platen returns, they stop the ejector plate. The mold opens on the parting line and an undercut in the sprue puller area pulls the molded sprue and runner with it. The part design and molding condition keep the plastic on the core. As the ejector mechanism works, the parts are pushed off the core, and the sprue puller moves forward out of its hole, freeing the sprue and allowing the parts to be removed.

In thermoset molding there is a tendency to flash around the KO pins so extractors, bumper bars, and combs are used for ejection.

89.8.6 Three-plate Molds

Suppose the dish were a deep cup and could be gated only in the top center section. The mold could be constructed as a one-cavity mold feeding directly from the sprue. If a multiple cavity mold were needed there are a number of alternatives, one of which is shown in Figure 89.15. Eight cavities, for example, could be located in two parallel rows of four. One cavity and core are shown with other significant parts of the mold.

The difference between this type of mold and the one illustrated in Figure 89.14 is that it separates between the A plate and the clamping plate as well as at the parting line. This is called a three-plate mold because plastic is molded between three plates. The plastic is injected through the sprue bushing into the runner, which is cut into the plate in a trapezoidal cross section. The flat back is on the pin plate. The plastic flows into the part through an auxiliary sprue bushing. Although this can be machined directly into the A plate and cavity, it is good practice to have a separate bushing so that it can be replaced or changed.

When the mold opens, the A and B plates move together. Sometimes this occurs normally. Other times latching or spring mechanisms are needed. The mold opens initially at the parting line (PL1). This breaks the gate and leaves the runner attached to the pin plate, because of the undercut pins (A) attached to the injection backup plate and extending into the runner. After the separation has occurred at PL1, the mold continues to open, separating at PL2. The molded pieces stay on the core. They are then ejected in a conventional manner, in this instance by a stripper plate and air (not shown). The pin plate is limited in its travel by stripper bolts B. When it is moved forward (by latches, chains, stripper bolts, ejector bars, or air cylinders) the runner is stripped off the undercut pins, and the plastic sprue is moved forward out of the bushing. The runner can be removed by hand, an air blast, or a mechanical system. The runner plate and A plate always stay on leader pins, The pins must be sufficiently long that the plates can separate far enough to remove the runner. It is good practice to support the pin plate on its own leader pins, attached to the backup plate. This prevents it from binding on the main leader pins.

This type of mold works very well, provided the workmanship is good quality and the

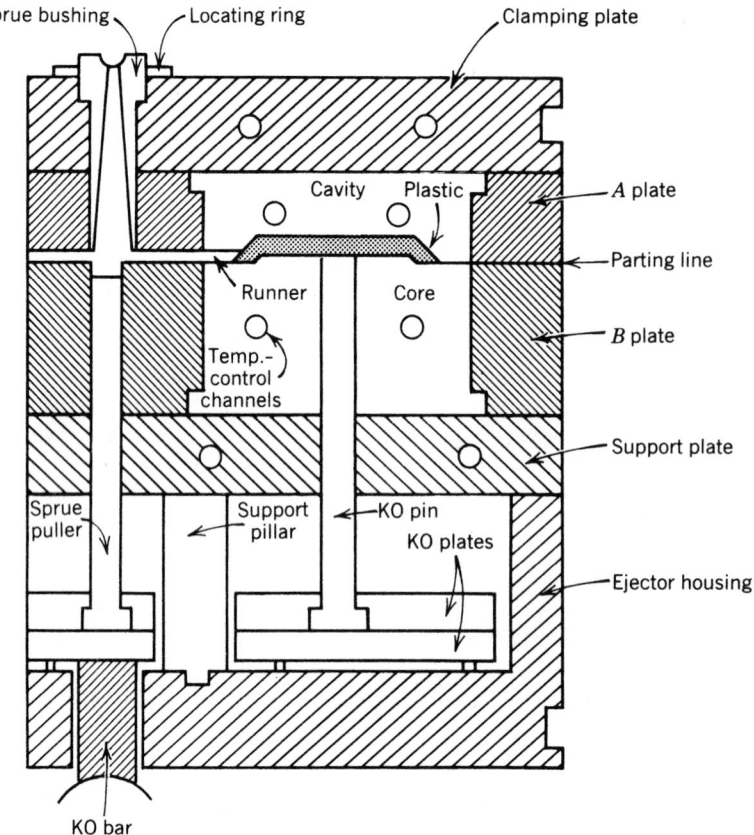

Figure 89.14.   Two-plate mold. Courtesy of Robinson Plastic Corp.

components fully sized and adequately designed. If not, cocking and binding occur relatively quickly on heavy molds. It is sometimes necessary to put an extra set of leader pins to support the A plate. These should not be used to line up the A and B plates. In other instances, small leader pins and bushings are put into the A and B plates to ensure good lineup and compensate for wear on the longer leader pins.

89.8.7   Hot Runner Molds

Thermoplastic runners are reground and reused. A logical extension of the three-plate mold overcomes this and is called a hot runner mold. This mold has a hot runner plate, which is a block of steel heated with electric cartridges, thermostatically controlled. This keeps the plastic fluid. The material is received from the injection cylinder and is forced through the hot runner blocks into the cavities. It is a more difficult mold to build and operate than a three-plate mold, but produces parts less expensively on longer-running jobs.[8]

89.8.8   Insulated Runner Molds

A combination between a hot runner mold and a three-plate mold is called an insulated runner mold.[11] The gating system is very similar to that of a three-plate mold except that the runners are very thick, at least 1 3/4 in. (45 mm) in diameter. There is no runner plate, and the backup plate and A plate are held together by latches. After the material is injected,

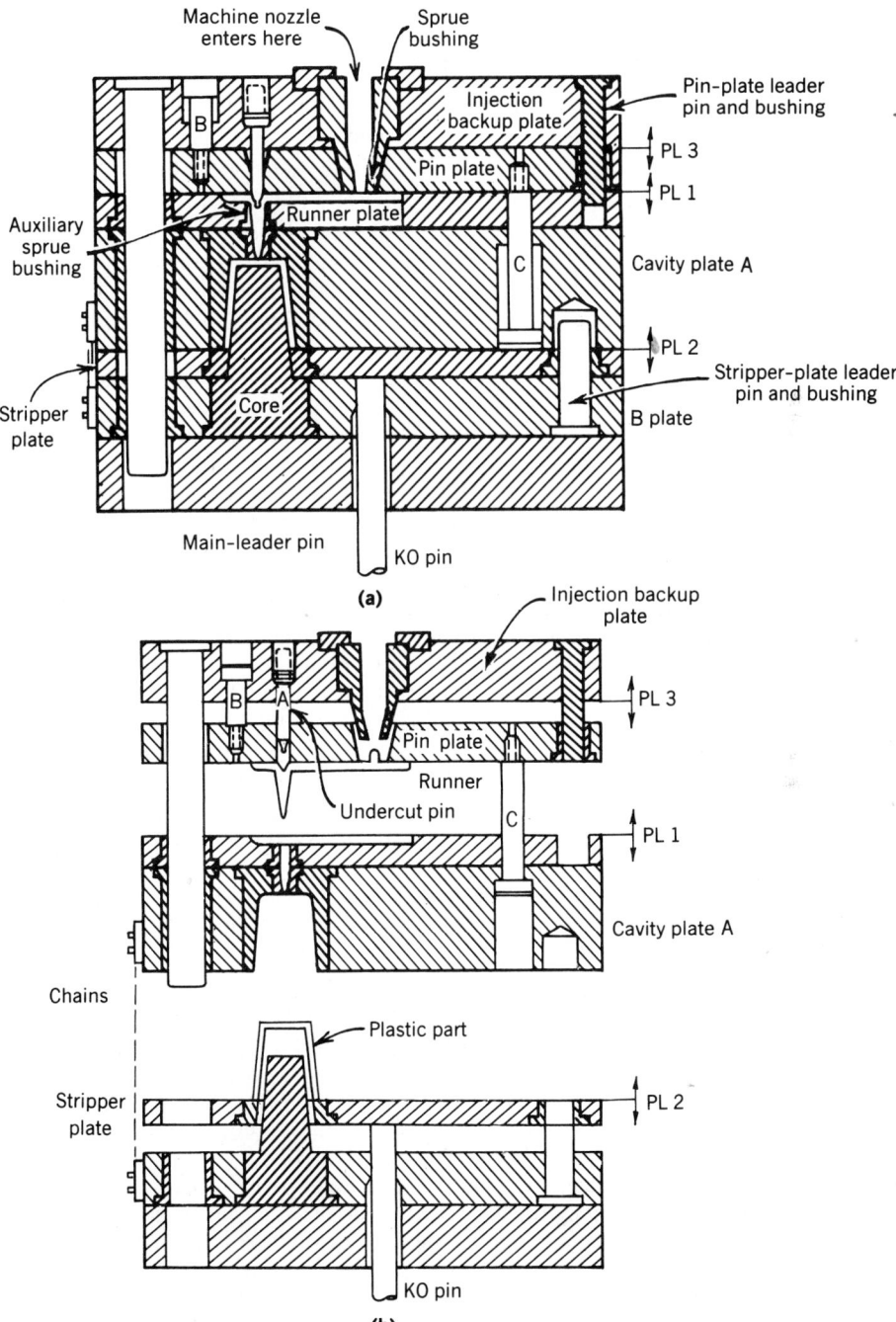

Figure 89.15. Three-plate mold in (**a**) closed position, and (**b**) in open position.

the outside of the runner freezes but insulates the center, permitting the core to remain fluid at molding temperatures and act as a hot runner. If the runner freezes during startup, the two plates are separated, the runner system is removed, and injection is started again. As soon as the runner reaches equilibrium, the mold is operated. These are more difficult to start and operate than a three-plate mold, but easier to run than a hot runner mold. The production characteristics are similar to those of a hot runner mold.

89.8.9   Materials for Cavities and Cores

The principal material of construction for molds is steel, followed by beryllium copper alloy. Brass, aluminum, and steel-filled epoxy are also used.

89.8.10   Steel

Steel is most often used for injection-mold sets, cavities, and cores. Information about steels and hardening is readily available in many books[8,12–15] and from steel manufacturers.

The type of steel selected depends on the end use, the size of the part, and the method of fabrication. It should be free from defects, have minimal distortion during heat treatment, be easily machinable, polish well, and weld readily.

To resist the stresses of injection molding and give reasonable protection against damage during production, steel has to have a minimum hardness. In the United States, it is usually designated on the Rockwell C scale. In other countries, it is the Brinell system. If the steel is too hard it becomes brittle. If it is too soft, it does not provide enough protection against damage and wear. A Rockwell C (R_c) of 50 to 55 gives good result. Steel this hard is difficult to machine even with carbides. It is easily worked by grinding and electrical or chemical removal equipment. The cavity or core is machined in the soft condition as it comes from the steel mill.

When carbon is added to iron, the alloy that is formed can be hardened and is called steel. When steel is heated to its critical temperature, it changes its structure. If this hot steel is quenched or quickly cooled, a hardened structure occurs. It is quenched in air, oil, or water (brine) and is correspondingly called air hardening, oil hardening, or water hardening steel. After these steels have been quenched they are hard and brittle and have to be tempered, or drawn. The steel is reheated to a lower predetermined temperature and cooled slowly under controlled conditions, which determine its final hardness.

To anneal or soften hardened steel, it is heated to a temperature just above its critical temperature and slowly cooled under controlled conditions. Hardening and annealing specifications are provided by steel manufacturers.

Iron is case-hardened by a process called carburizing. It is heated in contact with powdered carbon and absorbs the carbon on the surface, or case. The depth of the case depends on the time and temperature of the heating. The iron in the case combines with carbon and is now steel and susceptible to hardening. It has a hard outside and a soft inside. The parts are then annealed to produce the required surface hardness.

Mold parts are sometimes nitrided, which means they are subjected to ammonia gas at temperature up to 1200°F (635°C) for 50 to 90 hours. There is practically no distortion and a very hard, tough, thin case is produced.

The drawbacks to hardening are distortion and the possibility of cracking during heat treating. Corrections are difficult to make. To overcome this, a compromise between distortion, hardness, and machinability resulted in a series of prehardened steels. They range from R_c 28 to 44 and are readily machinable, though not as quickly or easily as soft steel. They are hard enough to give long satisfactory service. Most large-size molds are made in prehardened steel.

89.8.11   Beryllium–Copper

The second major material used in cavities is beryllium–copper. The material is an alloy of copper containing approximately 2 3/4 % beryllium and 1/2% cobalt.

An advantage of beryllium–copper is its high thermal conductivity.[16] It adds to or removes heat from a mold several times faster than steel. Since the time required for cooling a plastic in the mold is a function of heat removal, a beryllium cavity should give faster cycles. If cooling is the limiting factor in the molding cycle, this is true. The costs of beryllium and steel cavities are about the same, and the material selection depends on the mold properties desired. Beryllium–copper takes a very high polish, is not affected by water, and, when it is flash chrome-plated, gives an excellent, durable molding surface.

Other cavity and core materials used for sampling or very low production molds are brass, aluminum, and steel-filled epoxy. These are not materials of choice for production runs or quality parts.

89.8.12   Surface Finish

The surface finish of a mold affects appearance, ejectability, and mold cost. It is specified by comparing it with an injection-molded plaque with twelve different finishes. This is an SPI standard and can be purchased from them.

Molds are polished with abrasives, starting with coarse grits and finishing with grits as fine as 600 mesh. Materials for polishing are stones, emery cloth, carborundum in oil, and diamond compounds.

Steel is highly susceptible to rust, especially after polishing, and should be protected by a suitable rust preventive when it is not in use. Molds that are run cold should be brought to room temperature before being coated with a rust preventive. If they are not, water condenses on the mold and causes rust damage, particularly in humid air. Use of stainless steel or electroless nickel, or chrome-plating the mold prevents water damage. Maintenance of a proper polish is the molder's responsibility.

89.8.13   Equipment for Fabricating Cavities and Cores

Toolroom equipment is used for machining mold bases, cores, cavities, pins, blocks, and so on. Fabrication can be done electronically with punched tape and CAM systems.

A drill press is a tool that has a stationary table above which is a rotating, motor-driven shaft. The shaft contains a chuck to which the drill or other tool is attached. The drill moves up and down. It can be hand or automatically fed.

A miller is a drill press with a table that can be moved left or right, in or out, and up or down. The rotating shaft or spindle in the head, which can tilt, moves up and down. These movements can be automatically or manually controlled. A separate attachment has a stylus that moves to trace a three-dimensional replica of the part to be cut in steel. As the tracer (stylus) moves in one direction, the cutting tool moves in the same direction with a proportional movement. This is now called a duplicator.

A lathe has a rotating head to which the material to be cut is attached. The material rotates and the carriage, which contains the cutting tools, moves along the length of the bed or across it. The tail stock is equipped with a chuck for drilling and reaming.

A grinder has a rotating head to which is attached a grinding wheel. The table reciprocates left and right so that the circumference of the wheel does the grinding. It can move in and out at a predetermined distance per reciprocation of the table. The height of the grinding wheel above the work is accurately adjusted. The grinder is used to obtain accurate dimensions and a good surface finish. It readily grinds hardened steel.

Band saws consist of two wheels around which rotate an endless saw blade. Cut-off saws have straight blades that reciprocate.

89.8.14   Hobbing

Hobbing is the cold forming of metal. The term is used in plastics to designate the cold displacement of one material by another, caused by high pressures. For example, if a piece of plastic is left on a mold and the clamping pressure of the machine forces the plastic into the steel, the plastic is said to have hobbed itself into the mold. The term is used in mold making for the process that takes a hardened steel replica (hob) of the plastic part and, by means of high pressure, forces it usually into a soft iron block. Iron is very ductile. It flows around the hob, giving an identical but reversed impression. This is much the same as forcing a coin into a piece of clay.

Hobbing is a fast, economical way to produce multiple cavities. All the cavities are the same size compared to each other and the hob. A high polish on the hob is transferred to the cavity. Since the cavities are iron, they must be hardened by carburizing after they are machined to size. Figure 89.16 shows the hob (**a**) for a plastic column. The molded plastic part (**d**) is identical in shape but smaller than the hob because the plastic shrinks on cooling.[15] Point (**b**) is a hobbed cavity, and (**c**) is the completed, polished, carburized cavity ready for insertion into the mold.

89.8.15   Pressure Casting

Beryllium–copper cavities and cores are fabricated by pressure casting or, more accurately, hot hobbing. A hob is made proportional in size but larger than the finished plastic part. After hobbing, the beryllium shrinks as it cools just as plastic shrinks after molding. The hob is made of a good hot working die steel that will not deform under the temperatures and pressures of casting. For one or two cavities, a different alloy of beryllium can be used for the hob.

The hob is placed in the bottom of an insulated cylindrical container. The melted beryllium–copper is poured over it and a plunger exerts force on the beryllium and hob. When the beryllium is cool, the hob is separated and can be used again. The surface finish of the cavity depends on the quality of the hob. Because the beryllium is poured on in a liquid state, delicate fins and parts can be hobbed, something impossible with cold steel hobbing. There, high pressure would cause the hob to snap. The dimensions of the cavities are not as accurate as those from cold hobbing because of shrinkage factors. For most purposes, this is no problem.

When the parting line is uneven, hot hobbing can significantly reduce costs. The parting line can be cast so that a minimum of fitting is required. The cost of beryllium and steel cavities are similar, and the choice is based on engineering considerations.[7]

Beryllium can be readily cast using gravity alone and still give excellent surface reproductions. The casting is less dense than pressure casting and may be porous to water. This is readily solved.

89.8.16   Casting

Casting is readily adaptable for injection molds. Any metal can be successfully cast, particularly with the Shaw process. In this patented process, a sample (or a plaster reversal) of the part in plastic, wood, metal, or other material is cast against a patented ceramic slurry. The slurry is fired and gives a reverse ceramic reproduction with a micrograin structure filled with small air gaps. The gaps act as vents so that the molten metal can achieve a good reproduction of the surface. The resultant cavity is not as dense as those produced by other methods, and there may be very small pits. The appropriate shrinkage factors for the slurry, metal, and mold must be calculated. A new slurry casting must be made for each new cavity. The major advantages of casting are its speed and cost. A cavity can be made in less than a week. The economics depend on the size and nature of the part.[1]

89.8.17   Electroforming

Early methods of electroplating could not be used for mold cavities and failed primarily because the stress in the plating caused a deformation during molding. This has now been

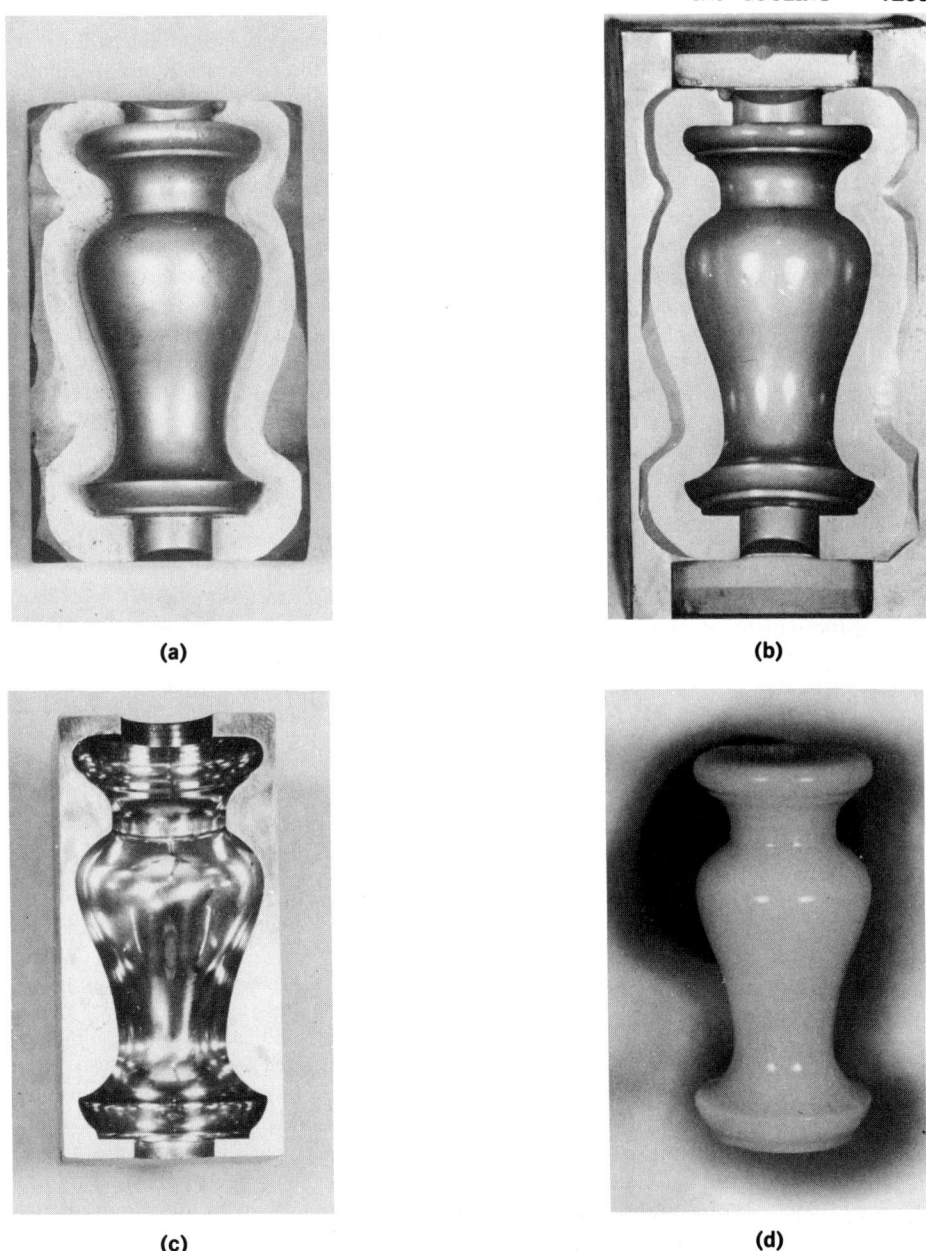

(a)

(b)

(c)

(d)

Figure 89.16. (a) Hob; (b) hobbed cavity; (c) finished cavity; (d) molded part. Courtesy of Robinson Plastics Corp.

overcome. A master, sometimes called a mandrel, is an exact reverse of the cavity. On it is plated approximately 0.15 in. (3.8 mm) of a nickel–cobalt compound at the rate of approximately 0.004 in. (0.1 mm) per 8 hours. Behind that is electroplated copper, which is harder than mold steel. The finished cavity has good dimensional stability, is rustproof, has high thermal conductivity, and is very precise—within 0.0001 in. (0.0025 mm).

Electroformed cavities are primarily used where accuracy is required, such as in gears, and where exact reproduction is needed. Irregular parting lines can be made with a near-perfect match. Irregular shapes that would be difficult to machine can be electroformed easily. Since the cavity is in the finished condition there is no distortion from hardening. The plating solution can be flushed into deep crevices, forming very narrow and thin slots.

89.8.18 Duplicating

This process is mechanical reproduction by means of cutting tools that are guided by a master, proportional in size to the desired finished parts. Duplicating is mostly used for large parts; hobbing and casting will usually reproduce a smaller one more economically. Large automatic duplicators, mechanically, tape, or computer controlled, are powerful horizontal millers with hydraulically controlled feeds. Maximum cutting speed is obtained with feedback and electronic techniques. Production of such effects as mirror images is easy. Small duplicators are often used in making hobs or engraving small designs, letters, and numerals on cavities and making carbon electrodes for spark erosion. A major disadvantage of duplicating is its poor surface finish.

89.8.19 Spark Erosion (EDM)

Steel is easily removed by an electrical discharge machine (EDM). An electrode, usually carbon or copper (though it can be made of any conducting material), is made in the reverse shape of the part to be produced. The steel and electrode are immersed in a circulating solution, which serves to flush away the eroded material which prevents a short circuit at the next spark and cools the work. When A-C power is rectified and charged into a capacitor system, the discharge between the electrode and the cavity creates a spark that erodes the steel. The electrode is eroded about one tenth as fast as the steel. Roughing electrodes shape the cavity; finishing electrodes bring it to size.

The process is accurate, produces good detail, can be used with hardened steel so that heat distortion from hardening and annealing is eliminated. It can also be used for cutting thin slots. By eroding on one plate rather than inserting cavities, the distance between cavities can be reduced and cooling can be improved. Cutting is relatively slow. The preparation of the electrodes and the operation of the equipment require good workmanship. Spark erosion is widely used in changing and correcting hardened steel cavities.

89.8.20 Tolerance

A mechanical tolerance is the total permissible variation of size, form, or location. It is indicated as a unilateral tolerance, in which the variation is from a given dimension in one direction only, or as a bilateral tolerance, in which the variation is permitted in both directions. Tolerances describe the part limits that should be controlled by function and aesthetics.

Suggested tolerances on molded parts in different plastics have been published by the SPI. Figure 89.17 shows tolerances for polystyrene. If finer tolerances are required, the rejection rate rises significantly and costs increase accordingly.

Nonmechanical tolerance such as performance specifications, aesthetic qualities, or color require care, clarity, and precision in their preparation. The important thing about all specifications is that they be clearly stated and the method of their measurement be documented before the part is committed to final production.

Piece-part tolerances are different from the functional tolerances required to make the mold. These are the sole responsibility of the molder and moldmaker.

89.8.21 Parting Line

When a mold closes the core and cavity meet, producing an air space into which the plastic is injected. This junction appears on the molded piece and is called the parting line. A piece

NOTE: The Commercial values shown below represent common production tolerances at the most economical level. The Fine values represent closer tolerances that can be held but at a greater cost.

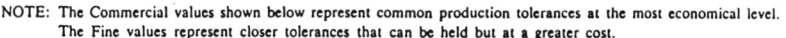

Drawing Code	Dimensions (Inches)	Plus or Minus in Thousands of an Inch
A = Diameter (see Note #1)	0.000 / 0.500 / 1.000 / 2.000	
B = Depth (see Note #3)	3.000 / 4.000	
C = Height (see Note #3)	5.000 / 6.000	

	6.000 to 12.000 for each additional inch add (inches)	Comm. ±	Fine ±
		.004	.002
D=Bottom Wall (see Note #3)		.0055	.003
E = Side Wall (see Note #4)		.007	.0035
F = Hole Size Diameter (see Note #1)	0.000 to 0.125	.002	.001
	0.125 to 0.250	.002	.001
	0.250 to 0.500	.002	.0015
	0.500 & Over	.0035	.002
G = Hole Size Depth (see Note#5)	0.000 to 0.250	.0035	.002
	0.250 to 0.500	.004	.002
	0.500 to 1.000	.005	.003
Draft Allowance per side (see Note #5)		1½°	½°
Flatness (see Note #4)	0.000 to 3.000	.007	.004
	3.000 to 6.000	.013	.005
Thread Size (class)	Internal	1	2
	External	1	2
Concentricity (see Note #4)	(T.I.R.)	.010	.008
Fillets, Ribs, Corners (see Note #6)		.015	.010
Surface Finish	(see Note #7)		
Color Stability	(see Note #7)		

REFERENCE NOTES

1 — These tolerances do not include allowance for aging characteristics of material.

2 — Tolerances based on ⅛" wall section.

3 — Parting line must be taken into consideration.

4 — Part design should maintain a wall thickness as nearly constant as possible. Complete uniformity in this dimension is impossible to achieve.

5 — Care must be taken that the ratio of the depth of a cored hole to its diameter does not reach a point that will result in excessive pin damage.

6 — These values should be increased whenever compatible with desired design and good molding technique.

7 — Customer-Molder understanding necessary prior to tooling.

Figure 89.17. Molding tolerances for polystyrene. Courtesy of the Society of the Plastics Industry Inc.

might have several parting lines if it has cam or side actions. The term parting line is usually restricted to the line that is related to the primary opening of the mold.

The selection of the parting line is largely influenced by the shape of the piece, method of fabrication, tapers, method of ejection, type of mold, aesthetic considerations, post-molding operations, venting, wall thickness, orientation, the number of cavities, and the location and type of gating.

89.8.22 Venting

When the hot plastic is injected into the mold it displaces air. In a well-built, properly clamped mold without vents, the injecting material may compress the air. The heat of compression might burn the material. The force of the compression might open the mold, causing flash. The resistance of the compressed air might prevent the mold from completely filling. These problems are alleviated by used of vents, which are usually ground on the parting line. Their size depends on the nature of the material and the size of the cavity.

A typical vent would be 0.002 in. (0.05 mm) deep and 1 in. (25 mm) wide. After a vent extended for 1/2 in. (13 mm), the depth would be increased to 0.03 in. (0.8 mm). Clearance between the knockout (KO) pins and their holes provides venting. Sometimes special pins are placed in the mold just for venting purposes.

The gate location has a lot to do with venting, and one is often restricted in gating because of the inability to vent the mold completely. Rapid injection is desirable in many moldings. Inadequate venting significantly slows down injection. The location and size of vents are still governed mainly by experience and trial and error. Vents are put in the obvious places before the mold is tested. Additional vents are added as required.

89.8.23 Ejection

After the part is molded it must be ejected from the mold. The molded part is ejected by KO pins, KO sleeves, stripper rings, stripper plates, or air, either singly or in combination. The quality of the molded piece is affected by the KO system.

Undercuts are pieces of mold material that obstruct the ejection of the molded part.

The geometry of the parts and the type of plastic material are the major factors in selecting the KO system. Most parts eject readily with a taper of 1° per side. Smaller tapers are admissible, if required. A high polish is not always necessary for easy ejection, but the direction of the polish is important. Draw polishing (stoning and polishing in the direction of ejection rather than randomly or perpendicular to ejection) is important in difficult cases. With some materials, such as olefins and nylons, fine sandblasting may help. Normally, a moderately polished surface does not present ejection problems.

The cross-sectional area of the KO pin or ring must be large enough to prevent damage by it. Aside from the obvious, when the KO pin goes through the molded part, serious stressing can be caused in the KO area. Birefringence studies of transparent molded parts show this clearly. Large-diameter KO pins are recommended.

Figure 89.18 is a schematic drawing of a stripper plate ejection system. The core pins are stationary. Around them are hardened replaceable stripper bushings that are mounted

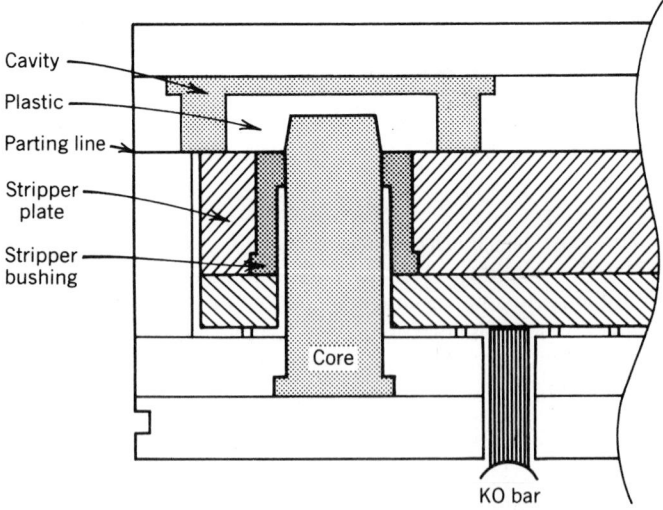

Figure 89.18. Stripper plate ejection system.

in the stripper plate. There is clearance in the lower part of the stripper bushings between the bushings and cores to minimize wear. The knockout bars cause the stripper plate to move in relation to the core pin, stripping the plastic off the core and leaving it either on the plate or free to fall off. Sometimes it is necessary to eject in two stages (double knockouts).

Figure 89.19 illustrates a condition called entrapped material. When the material flows in from the gate, it will deform the metal ring (B) outward, because of Hookean elasticity. When the injection pressure stops, the metal ring will snap back, squeezing and entrapping the ring of plastic between the metal ring (B) and the core (C). This generates huge frictional forces that are not relieved when the mold opens and the cavity moves away.

The solution is to remove the core pin before ejection, allowing room for the plastic to move and thus eliminating the extra friction caused by the deflection of the steel ring. In actual practice, the core would not be retracted but the core plate would be moved forward either by heavy springs or the KO system. A secondary KO system ejects the pieces.

Cam-acting molds are commonly used in injection molding when the parts being molded require undercuts and holes. The cams must be withdrawn before or while the mold is opening to permit ejection. They are also used for engineering considerations relating to ejection, venting, and gate locations.

Internal threads can be molded automatically by using a collapsible core.[5] Automatic unscrewing mechanisms include racks and pinions, gears, sprockets, electric motors, and hydraulic motors. Automatic unscrewing molds are considerably more expensive to build and maintain. An alternative is to use inserts in the mold that are removed with the piece and unscrewed on the bench; extra inserts save machine time. Other techniques include using threaded inserts in the mold, adding metal inserts in a postmolding operation, or tapping the hole in the plastic.

89.8.24 Runners

The runner is the connection between the sprue and the gate. It should be large enough to allow rapid filling and minimum pressure loss, but not so large as to increase the cooling cycle of the parts. The runner does not have to be completely hard at ejection. Most jobs

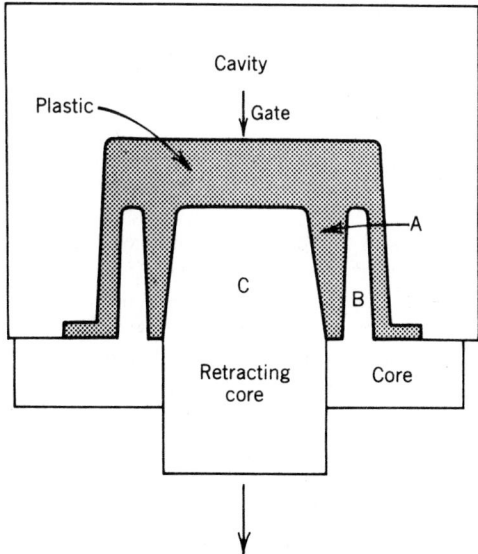

Figure 89.19. Use of retractable core to prevent ejection difficulties caused by entrapped material.

permit the runner to be reground and reused. Regrinding is expensive, wastes material, wastes energy, is a source of contamination, and is a place for entry of such foreign material as screw drivers and other metal parts.

The full, round runner is preferable because, for a given cross section, it permits the greatest flow. It has the highest ratio of cross-sectional area to circumference, minimizing the cooling effect. The material feeds from the center that has the hottest material. When a runner has to be on one side only, the best compromise is a trapezoidal shape. Half round and rectangular runners should not be used. The runner should be polished and the sharp corners broken. This results in less turbulence in the flow and slightly faster filling rates. The flow rate into the piece should be balanced by the gate, not by the runner.

89.8.25 Gates

The gate is the connection between the runner and the molded part. It must permit enough material to enter and fill the cavity, plus the extra amount required for thermal shrinkage. Gates are usually tapered toward the part to facilitate a clean break between the runner and the part.

Gates can be classified as large or restricted (pin pointed). Restricted gates are circular in cross section and, for most materials, do not exceed 0.060 in. (1.5 mm) in diameter. The more viscous and reinforced materials may have restricted gates as large as 0.115 in. (3 mm) in diameter. An example of a large gate, which is usually square or rectangular, is 1/4 in. (6.4 mm) wide by 3/16 in. (4.8 mm) high. Large gates are used for molding heavy sections and in circumstances where restricted gates create surface blemishes.

The restricted gate is successful because the viscosity of the plastic is sensitive to the shear rate; the plastic becomes less viscous with higher speed. As the material is forced through the small opening, its velocity increases. Once the gate is opened to the extent that it loses this shear rate/viscosity improvement, a much larger opening is required to obtain any acceptable flow. This is why there is a jump in size from a restricted to a large gate.

In thermoset molding a gate may be too restrictive. The filler may create an obstruction for itself but the resin can flow through it, straining out the filler from the resin.

Gates are also described by location, such as an edge gate, back gate, submarine gate, tab gate, and sprue gate. Figure 89.20 shows examples of various types of gates. A sprue gate feeds directly into the piece from the nozzle of the machine or a runner. It has the advantage of a short direct flow, with minimal pressure loss. Its disadvantages include the lack of cold slug, the possibilities of sinking around the gate, the high stress concentration around the gate, and the need for gate removal. Most single-cavity molds of any size are gated this way.

Edge gating is the most common. It can be of the large or the restricted type. If the edge gate is spread out, it is called a fan gate. If the gate is extended for a considerable length of the piece and connected by a thin section of plastic, it is called a flash gate. In thermoplastic molding, it is sometimes necessary for the gate to impinge upon a wall. This distributes the material move evenly and improves surface conditions. If walls are not available, a rectangular tab is milled into the piece and the gate is attached there. This is called a tab gate. Another advantage is the elimination of gate area blemishes when the tab gate is removed.

In gating into hollow tubes, flow considerations can dictate an even injection flow pattern. A single gate is not sufficient. Four gates 90° from each other produce four flow lines down the side of the piece, which may be objectionable. To overcome this, a diaphragm gate is used. The inside of the hole is filled with plastic directly from the sprue and acts as a gate. It must be machined out later. A ring gate accomplishes the same thing from the outside.

A submarine gate penetrates through the steel of the cavity. When the mold opens, the part sticks to the cavity and shears the piece at the gate. A properly located KO pin, using the flexibility of the plastic, ejects the runner and pulls out the gate. This type of gate is usually used in automatic molds.

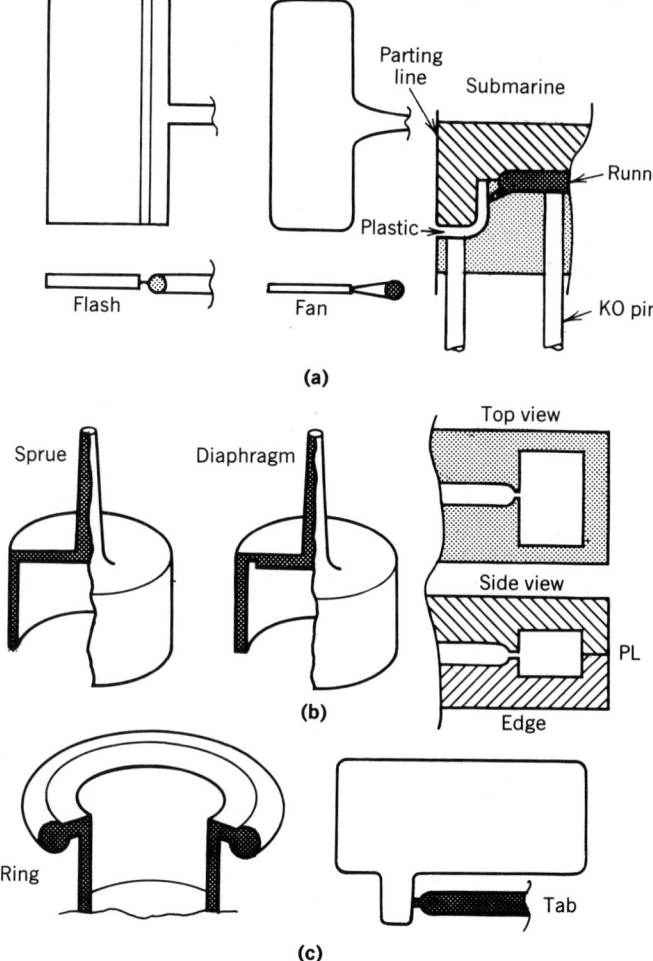

Figure 89.20. Various gating designs. Courtesy of Robinson Plastics Corp.

Restricted gates have the benefit of better mixing (because of the Reynolds effect). It is virtually impossible to mold a good variegated pattern (mottle) in thermoplastics without using a large gate.

In multicavity molds, the gate size must be adjusted so that the fill of all cavities ends at the same time. If not, severe molding and dimensional problems may develop. Such parts are subject to long-term failure. This is not true for thermosets.

89.8.26 Temperature Control

For consistent molding, accurate control of the mold temperature is required.[11] Refrigerated water and hot water to 200°F (93°C) are required at each machine. It is not always possible or necessary to predict the best temperature for a given mold and material. With thermostatically controlled temperatures, trial and error is not difficult. Different parts of the mold are often maintained at different temperatures. Equipment is available to provide refrigerated water either from one central unit or from smaller portable chillers. Separate circulating units with coils for cooling and electrical units for heaters are employed to control the temperature of the circulating fluid. Heat is supplied electrically to the mold with bands, cartridges, and strip heaters. They are best controlled with pyrometers, rather than non-feedback proportioning controls or auto transformers.

Even though the expression "heating a mold" is used, the purpose of the temperature control system is to remove heat from the plastic part at a controlled rate. Obviously, the lower the temperature of the controlling fluid, the quicker the heat is removed. The amount of heat removed depends on the material, the metal in which it is contained, the size of the cooling channels, their locations in relation to the molded parts, the cleanliness of the channels, the rate of flow of the heat exchange fluid, and its temperature. Air is an effective insulator; it is desirable to locate the cooling channels in the cavity and core. The minimum size of cooling channels should be 1/4 in. tapered pipe size, though 3/8 in. and 1/2 in. are preferable.

Mold temperature control is so important that molds are built at considerable extra cost to achieve greater cooling.

89.8.27 Automation

Today, all machines run automatically. What makes automatic molding automatic is the mold. There are a number of requirements for automatic molding. The machine must be capable of consistent, repetitive action. The mold must clear itself automatically. This means that all the parts have to be ejected using a runnerless mold, or that the runner and the parts have to be ejected in a conventional mold. There may be some method for assisting in the removal of the pieces, usually in the form of a wiper mechanism, air blast, or robot. All machines used automatically must have a low-pressure closing system to prevent the machine from closing under full pressure if there is any obstruction between the dies. The machine shuts itself off and/or an alarm is sounded.

Automatic molding does not necessarily eliminate the operator. Many times, an operator is present to pack the parts and perform secondary operations. Automatic molding gives a better quality piece and more rapid cycle. In automatic molding an experienced person usually attends several machines. Unless the powder feed and part removal are automated, this person takes care of those operations.

Automation is expensive to attain. It requires excellent machinery, controls, molds, trained employees, and managerial skill. When the quantity of a part permits, it is a very satisfactory and economical operation.

89.8.28 Mold Design Checklist

The mold design checklist (Table 89.1) is used as a final check of the mold drawing to make certain that no feature has been overlooked. It is also an excellent aid in mold design.

89.9 AUXILIARY EQUIPMENT

In IM, the major auxiliary equipment includes unmolded plastic-material handling systems, regrinders, hopper loaders, dryers, coloring, and molded-parts handling systems.

The plastic is colored before the material is shipped to the molder by compounding, adding small amounts of preprocessed material with a very high loading of the color, tumbling the colorant and the base resin together, or adding liquid colorant to the resin.

Molded parts handling can be as simple as an endless belt to remove the parts as they fall from the mold to a completely automated plant. Use of robots to remove parts from the mold automatically was common in the 1940s, though they lacked the sophistication and added features of today's robotics.

Robots are used to load (inserts) and unload machines, handle the molded parts, cut

TABLE 89.1. Check List For Molds

Piece Parts

1. Is this piece part drawing approved?
2. Have you read all the notes pertaining to the job?
3. Is the type of plastic material indicated?
4. Is the function, location, and use of the piece understood?
5. Can any changes be recommended to make a simpler or better piece?
6. Are the number of cavities correct?
7. Are tolerances indicated on all critical dimensions?
8. Can these tolerances be maintained?
9. Are the dimensions given including or excluding shrinkage?
10. What shrinkage factor is to be used?
11. Has adequate draft been specified?
12. Where does the draft start?
13. Have tapers been specified?
14. Has the parting line been approved?
15. Has the gate location been approved?
16. Is the gate location in the best possible place for maximum physical properties including orientation and packing?
17. Is the gate location in the best possible place for finishing?
18. In designing location of gate, will anticipated weld lines prove objectionable esthetically or mechanically?
19. Will the piece hang (stay) on the ejection side?
20. Has the ejector mechanism(s) been decided?
21. Have the location of the ejector mechanism(s) been approved?
22. Is the ejection mechanism(s) sufficient?
23. Has polish been specified, using SPE–SPI system?
24. Will the mold physically fit in the presses to be used?
25. Is the mold thicker than the minimum thickness required of the presses?
26. Is the stroke of the machine long enough to allow for part removal?
27. Is the ejection stroke of the machine long enough to allow for part removal?
28. Can the mold be clamped into the press?
29. Is the clamping capacity of the machine enough for the parts?
30. Is the injection capacity of the machine enough for the parts?
31. Do the ejector holes correspond with the ejection mechanism of the press?
32. Do we need knock out mechanism on injection side?
33. Are water lines located so that they will not be in the way of the operator?
34. Do water lines interfere with tie bars or other mechanisms?
35. Do water lines interfere with each other or are they too close to the platen?
36. In the event of requirements for heating the mold, are the heating elements and control units placed safely to be out of the operators way?
37. Have the dimensions of the locating ring been shown?

Mold Design

38. Are the mold plates and component parts strong enough for the piece?
39. Is there sufficient steel surrounding the cavities and cores?
40. Are there sufficient support pillars?
41. Is one leader pin and bushing unsymmetrical?
42. Will the leader pin enter before any other part of the mold?
43. Is there ample clearance for leader pins on the other side of the mold?
44. Is there sufficient travel for the ejector plate?
45. Is the ejector plate on its own leader pins and bushings?
46. Is the ejector plate strong enough?
47. If a stripper mold, is the stripper plate properly supported?
48. Have push back pins been provided?
49. Does the sprue bushing fit the machine?
50. Can the sprue bushing be made shorter? (especially in 3 plate molds)

TABLE 89.1. *(Continued)*

51. Have the dimensions of the sprue bushings been recorded?
52. Are there sufficient cooling channels in the mold and cavities?
53. Do the knockouts clear the water holes?
54. Are runners specified and VENTED?
55. Have gates been specified?
56. Have run-offs been provided when required?
57. Has venting for the runner system and cavities been specified?
58. In cam acting molds, have provisions been made for hardening moving parts?
59. In cam acting molds, have provisions been made for replacing worn parts and tightening cam?
60. In cam acting molds, can cam pins be replaced without removing the mold?
61. In cam acting molds, are the operators protected from moving plates and parts?
62. If there are electrical switches on the mold, have they been made safe?
63. If there are electrical switches on the mold, have they been made fail safe?
64. Can the electrical parts be replaced without removing the mold?
65. Have provisions been made for closing openings and depressions which might be filled up by flashed shots?
66. Are all steel and metal specifications shown?
67. Have the heat treating specifications been shown?
68. Have the surface specifications been shown (including Chrome Plating)?
69. If the mold has to be heated, have provisions for differences in expansion been made?
70. Have eye bolts been provided on both halves of the mold and on all heavy plates?
71. Where there are expendable parts such as springs, O-rings and switches, has a specification chart been provided?
72. Are bolt sizes specified? Stamp "M" for metric next to thread.
73. Are mold parts (sprues, etc) standard?
74. Have any spare parts to be furnished with the mold been designated?
75. Can mold and cavities be disassembled within a minimum of time?
76. Are all the component parts numbered so as to allow for proper reassembly?
77. Has the mold been properly marked for identification?
78. Are the dimensions on the prints the same as the dimensions on the mold?
79. Is there a schedule of completion dates for stages of the mold work?
80. Are there identifying marks on cavities and cores for proper re-assembly?
81. Are there identifying marks on all mold plates?
82. Is the type of steel marked on each piece?
83. Is there enough support on the "a" side under the 4-in. diameter of the seating ring?
84. Are there identifying marks on each molded piece?

gates, assemble, ultrasonically weld, finish, inspect, and package. Selection of a robotic system is an economic decision. Safety and environmental factors, engineering support systems, computer controls and interfaces, reaction of labor, the repetitiveness of the job, and human accuracy are other factors.

89.10 FINISHING

Plastic molded parts can be finished very easily with metal and woodworking equipment. Special cutting tools are available to improve productivity.

Molded parts can be ashed, buffed, tumbled, flame polished, and chemically polished. The surface can be sandblasted and etched chemically. The method of holding the work should be determined in the design of an IM part that will require finishing.

89.11 DECORATING

All commercially available decorating systems are used for IM parts. Again, the method of holding the work should be determined in the design of the part. Silicones, oils, and high stress can cause difficulties in decorating.

The cavity and core can also be decorated, imparting the reverse decoration to the molded parts. This is done by a photo-etching process. Sample injection-molded [8 × 11 in. (203 × 279 mm)] sheets are available with different wood grain, leather, and other patterns molded on them.

Characteristic finishes result from EDM finishes and sand blasting. Designs on molds are limited only by the customer's needs. When any design is molded on side walls, additional taper is needed for ejection.

Preprinted designs on plastic film can be inserted into the mold; after molding, they become an integral part of the molded piece.

89.12 COSTING

In IM, parts are not costed in the usual manner. For certain reasons the following system has evolved. While other systems might be better, in order to compete one is forced to quote the same way as one's competitor.

89.12.1 Material Costs

The cost of the material is related to its weight and cost per pound. The weight may be given, obtained by weighing a piece, or calculated from the drawing. For convenience, it is usually given in grams and used as pounds per thousand.

As an example, assume a part weighs 45.4 g. Multiplying by 2.2 will give 100 lb/1000 (45.4 kg/1000). The material is a special grade polystyrene and costs $0.88/lb. The material cost is $88 per 1000 pieces. There will be handling and waste charges. For materials below $1.00, most molders use between 7% and 10%; for materials between $1–2, 15%; and above that, up to 25%. Adding 10%, the cost per 1000 for material, the cost is now $96.80.

89.12.2 Molding Costs

The molding costs (an hourly rate) include everything except the cost of the material and those items that relate only to that part, such as packing, finishing, decorating, shipping, or special inspection.

The molding cost is based on the clamp size of the machine. It varies with the area: labor rates, utility rates, and local taxes strongly affect it. Plastics Technology magazine periodically conducts a survey on hourly rates and reports the results. In October 1988, the national average for hourly rates for machines less than 100 tons clamp was $27.09; 100–299 tons, $32.04; 300–499 tons, $42.03; 500–749 tons, $48.85; 750–990 tons, $57.78; and above 1,000 tons, $78.51.

Assume, for example, that a four-cavity mold and a machine produce 104 shots per hour. However, because of losses, the molder will only ship, bill, and be paid for 100 shots per hour. The production rate is 4 × 100 = 400 parts per hour. A 300-ton clamp machine is being used with an hourly rate of $42.03, so the cost is

$$\frac{\$42.03 \times 1000}{400}$$

or $105.08/1000 for molding. Packaging is $6.00/1000. Shipping is $5.00/1000. Hot stamping is $62.50/1000. There is a 4% loss in hot stamping, so 4% of the material plus the molding cost is added: $8.08/1000.

The customer is on the New York Stock Exchange, so he/she will pay any bills but he or she takes 90 days to do so. Financing costs add $4.50 per 1000.

The quotation to the customer is $96.80 + 105.08 + 6.00 + 5.00 + 62.50 + 8.08 + 4.50 = $287.96 per thousand parts.

89.13 INDUSTRY PRACTICES

89.13.1 Mold Ownership

The owner of the mold has the responsibility of insuring it for fire and theft. The molder is responsible for normal care and minor maintenance, such as polishing and replacing KO pins. Wear and tear are the responsibility of the owner.

A mold may be removed by the owner at any time unless a prior lien has been established. Most molders require that before such removal, all invoices for parts molded from the mold and the material purchased for orders from the mold have been paid for. Unless there is prior agreement, no charge is made for removal.

Most molders have prior agreements so that if the owner cannot be found or refuses to remove the mold, the molder may dispose of it.

In buying molds, deposits are required. The usual arragements are 1/3 with order, 1/3 upon receipt of molded parts, and 1/3 30 days after approval of parts, or 1/2 down and 1/2 30 days after approval of the molded parts.

89.14 INDUSTRY ASSOCIATIONS

The two major industry associations are the Society of Plastic Engineers (SPE), 14 Fairfield Dr., Brookfield Center CT 06805, and the Society of the Plastics Industry (SPI), 1275 K St. N.W., Suite 400, Washington, DC 20005.

Each has divisions devoted to molders and moldmakers. There is also the I.T. Quarnstrom Foundation, which is associated with the mold making division of the SPE.

89.15 INFORMATION SOURCES

A large number of monthly magazines, most of which are free, cater to the plastic industry and contain information pertaining to IM.

Many books have been and are being published. (References 8, 12–14, and 18 are books on injection molds with much valuable information.) The SPE publishes an annual list of books it has for sale. Similar information is available from major technical publishers.

There is a vast amount of free useful literature from the manufacturers of raw materials. Manufacturers catalogs and publications of mold steel suppliers are helpful.

Members of the technical divisions of the SPE will direct you to additional sources of information.

BIBLIOGRAPHY

1. I. Lubalin, *Mod. Plast.,* 147 (Oct. 1957).
2. R.L. Ballman and H.L. Tour, *Mod. Plast.,* 113 (Oct. 1960).

3. G.B. Jackson and R.L. Ballman, *SPE J.*, 1147 (Oct. 1960).

4. W. Woebcken, *Mod. Plast.*, 146 (Dec. 1962).

5. J. Andras, *SPE J.*, 35 (May 1967).

6. C.T. Hathaway, *SPE J.*, 567 (June 1961).

7. I. Thomas, *Mod. Plast.*, 101 (July 1961).

8. I.I. Rubin, *Injection Molding, Theory and Practice*, Wiley-Interscience, New York, 1972.

9. T. Erwin, *Plast. Technol.*, 35 (Jan. 1968).

10. T. Debrecceni, *Plast. Technol.*, 39 (Jan. 1968).

11. I.I. Rubin, *Plast. Eng.*, Part 1 (Feb. 1980); Part 2, 48 (Mar. 1980).

12. H. Gastrow, *Injection Molds: 102 Proven Designs*, Carl Hanser Verlag, Munich; Macmillan Publishing Co., Inc., New York, 1963.

13. L. Sors, Bardocz, and Radnoti, *Plastic Molds and Dies*, Van Nostrand Reinhold Co., New York, 1981, pp. 133–204.

14. G. Menges and P. Mohren, *Anleitung fur den Bau von Spritzgiesperkzeugen* (in German), Carl Hanser Verlag, Munich, 1974.

15. I. Thomas and E.W. Spitzig, *Mod. Plast.*, 115 (Feb. 1955).

16. W.J.B. Stokes, *SPE J.*, 417 (April 1960).

17. I.I. Rubin, *Adv. Plast. Technol.*, 65 (Jan. 1981).

18. R.C.W. Pye, *Injection Mold Design*, 2nd ed., George Goodwin, Ltd., London, 1978.

90

REACTION INJECTION MOLDING (RIM)

F. Melvin Sweeney
954 Helen Ave., Lancaster, PA

90.1 DESCRIPTION AND HISTORY OF PROCESS

Reaction injection molding (RIM) is a process for molding plastic parts from a two-part liquid chemical system. It uses equipment that meters reactants to an accuracy of 1%, mixes them by high pressure impingement, and dispenses them into a closed mold. In the mold the end groups react to form chemical linkages, producing block polymers.

The most common materials used in the RIM process are polyols, which contain two or more hydroxyl (—OH) groups, and isocyanates, containing two or more isocyanate (—N=C=O) groups. These reactive end-groups, so named because they occur at the ends of the chemical structure, react chemically to form a urethane linkage.

$$(-\overset{\overset{\displaystyle H}{|}}{N}-\overset{\overset{\displaystyle O}{\|}}{C}-O-)$$

The chemical system must be adjusted so that the number of hydroxyls and isocyanates balance and that all atoms in the reactive end-groups are used in the formation of urethane linkages. A simple linear polyurethane will have the block structure —ABABABAB—. Diagramatically, it will look somewhat like the structure shown in Figure 90.1.

The use of the letters A and B to represent diisocyanates and polyols is a convention that is almost universally accepted. Unfortunately, especially in earlier literature, this convention was not always followed. A better procedure is to use the terms ISO for that portion of the resin system containing the reactive isocyanate groups, and POLY for that portion of the resin system containing the hydroxyl groups.

In addition to polyurethane, the nylons, epoxies, and other reactive chemical systems such as polydicyclopentadiene have been processed by reaction injection molding. At this time, the polyurethanes are well-established as commercial systems applicable to RIM technology. Nylons are newer but are fully commercial at an earlier stage of development. Epoxies are more experimental at this time. Other systems are under investigation as candidates for reaction injection molding.

To produce a molded part by any of these systems requires precise, but quite realistic, process control. Figure 90.2 shows a simplified schematic of the RIM process.

The important elements of the process are conditioning, metering, mixing, and molding. All materials require precise temperature control. The flow properties of the liquid reactants

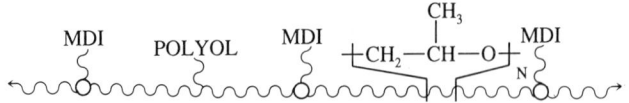

Figure 90.1.   RIM block polymer.

(viscosity) usually varies with the temperature, as does the density (weight/volume). For accurate metering the temperature must be controlled within very narrow limits. This is usually done by controlling the temperature of the reactants in the ISO and POLY holding or conditioning tanks. Temperature in the lines is normally controlled by recirculation between shots. For some chemical systems, such as nylon which is processed at a high temperature. the machines are designed with heated lines and temperature control devices for pumps, mixers, and other components.

The reactions characteristic of the RIM process are exothermic; that is, they give off heat. For this reason, the molds are also designed to control temperatures. There is a rule of thumb that organic chemical reaction rates double with each 18°F (10°C). While this rule provides guidance to the chemist doing organic chemical research, it is not recommended to RIM processors as a way to decrease cycle time. The type of linkage formed when end-groups react may change with temperature changes. Thus it is important to follow the system supplier's guidelines when processing material.

Polyurethane RIM systems have been commercial in the United States for about 30 years and a bit longer in Europe. It is still a rapidly changing field of technology. The automotive business in the United States accounts for most of the commercial RIM production.

General Motors made the first production RIM parts molded in the United States. The bumper cover for the Corvette was molded experimentally and soon became a production part on a limited production car. General Motors also developed production systems for molding very thin elastometric urethane parts that could be demolded in less than one

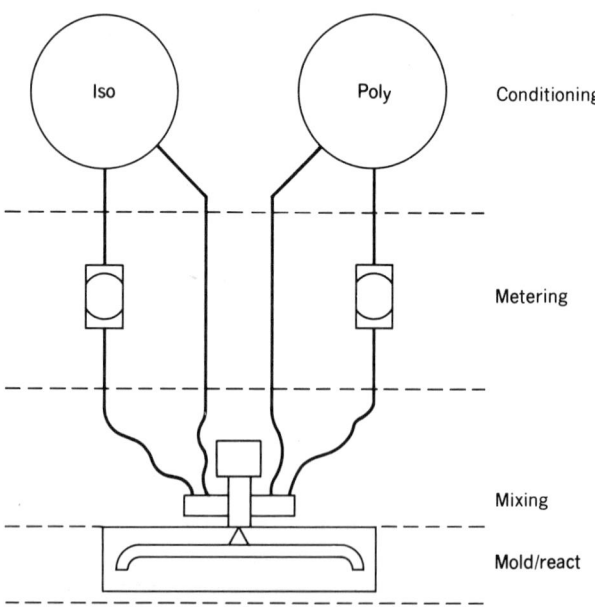

Figure 90.2.   RIM process schematic.

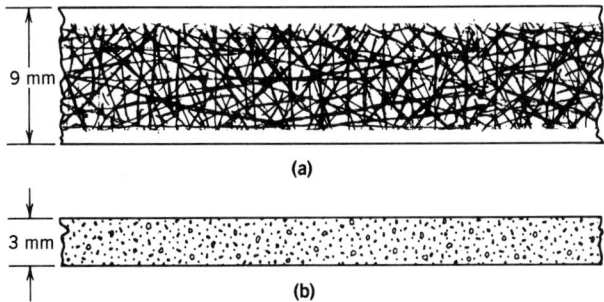

Figure 90.3. Structural RIM and elastomeric RIM in cross section: (**a**) structural; (**b**) elastomeric.

minute. The company also molded RIM nylon body panels that were used on selected production cars.

Ford used RIM urethane bumper covers for its high-production Fairlane and Zephyr models, and molded very large front end trim that incorporated the bumper cover, headlight openings, and grille.

Chrysler Omni and Horizon two-door models also used the multipurpose front end molding. Another interesting application of RIM was the simulated wood trim used on the Chrysler station wagons and sport convertibles.

A more recent development for RIM urethane elastomers, and to a lesser extent RIM nylons, is housings: computer housings, business machine housings, TV and radio cabinets, instrument cases, and similar electronic product enclosures. In addition to elastomeric RIM, some housings are molded from RIM structural foam.

Systems suppliers do not always clearly differentiate between elastomeric and structural RIM. This is not because they are deliberately misleading the customer, but because the boundary between the two is not always clear.

Elastomeric RIM is molded at high density in cavities having a thin cross section [usually 0.125 in. (3.2 mm) thick]. Structural foam has an interior foam structure, a density about one third that of elastomeric RIM, and is molded in thicker cross section [usually 0.375 in. (9.5 mm) thick]. Material usage is often comparable. However, the elastomeric process, requiring less expansion to fill the mold and having a thinner part cross section, can be molded in shorter cycle times. Cross sections of a typical foam and a typical elastomer are shown in Figure 90.3.

90.2 MATERIALS

The materials most commonly processed are polyurethanes and nylons. Work is also progressing on RIM epoxies and certain other polymers. The essential characteristics of materials for RIM processing is that they consist of two or more materials that react to form linkages. This produces a polymeric chain or network, depending on the starting materials.

90.2.1 Polyurethanes

Polyurethanes are formed through linkages that develop when a reactive hydroxyl end-group (—OH) reacts with a reactive isocyanate end-group (—N=C=O) to form a urethane linkage:

$$
\begin{array}{ccc}
\text{H} & \text{O} \\
| & \| \\
(-\text{N}-\text{C}-\text{O}-)
\end{array}
$$

MASTER PROCESS OUTLINE   **PROCESS**   RIM

Instructions:
For "ALL" categories use   Y = yes
For "EXCEPT FOR" categories use   N = no
                                                        D = difficult

☐ **ALL Thermoplastics**
   Except for:

Acetal	Poly(methyl methacrylate)
Acrylonitrile–Butadiene–	Poly(methyl pentene) (TPX)
Styrene	Poly(phenyl sulfone)
Cellulosics	Poly(phenylene ether)
Chlorinated Polyethylene	Poly(phenylene oxide) (PPO)
Ethylene Vinyl Acetate	Poly(phenylene sulfide)
Ionomers	Polypropylene
Liquid Crystal Polymers	Polystyrene
Y Nylons	Polysulfone
Poly(aryl sulfone)	Y Polyurethane Thermoplastic
Polyallomers	Poly(vinyl chloride)-PVC
Poly(amid-imid)	Poly(vinylidine chloride)-Saran
Polyarylate	Rubbery Styrenic Block
Polybutylene	Polymers
Polycarbonate	Styrene Maleic Anhydride
Polyetherimide	Styrene–Acrylonitrile (SAN)
Polyetherketone (PEEK)	Styrene–Butadiene (K Resin)
Polyethersulfone	Thermoplastic Polyesters
	XT Polymer

☐ **ALL Thermosets**
   Except for:

Allyl Resins
Phenolics
Polyurethane Thermoset
RP-Mat. Alkyd Polyesters
Y   RP-Mat. Epoxy
RP-Mat. Vinyl Esters
Silicone
Urea Melamine

☐ **ALL Fluorocarbons**
   Except for:

Fluorocarbon Polymers-ETFE
Fluorocarbon Polymers-FEP
Fluorocarbon Polymers-PCTFE
Fluorocarbon Polymers-PFA
Fluorocarbon Polymers-PTFE
Fluorocarbon Polymers-PVDF

Because the urethane linkage is the most common linkage, the polymer formed is called a polyurethane. The word "polymer" comes from the Greek "poly," meaning many, and "meros," meaning parts.

The number of polymer structures that can be formed using the urethane reaction is quite large. There are ways to build systems to produce polyurethanes having different physical properties.

If linear polyols are reacted with diisocyanates, as described in Figure 90.1, a flexible polyurethane will be formed. If a low boiling liquid, such as Refrigerant-11 (R-11), is incorporated into the system, the heat of reaction will produce a cellular structure. The resulting product will be a flexible urethane polymer or a flexible urethane foam.

If highly branched polyols are reacted with diisocyanates, a polymer network is formed as shown in Figure 90.4.

The physical properties of these materials can be varied by selecting polyols with shorter or longer polyol chains. The most common polyol chains are polyethers and polyesters. The thermoplastic chain composition itself plays a role in the physical properties of the end product. These segments are often referred to as "soft blocks."

In addition to changing the chain composition and length, the physical properties can be varied by blending up to approximately 10% of a triol into the basic system formulation. This produces branching in the "soft block" or "thermoplastic" segment of the block polymer. This is shown in Figure 90.5. Excessive triol modification may, in fact, diminish physical properties.

The use of short-chain extenders, such as 1,4 butane diol or ethylene glycol, blocks the

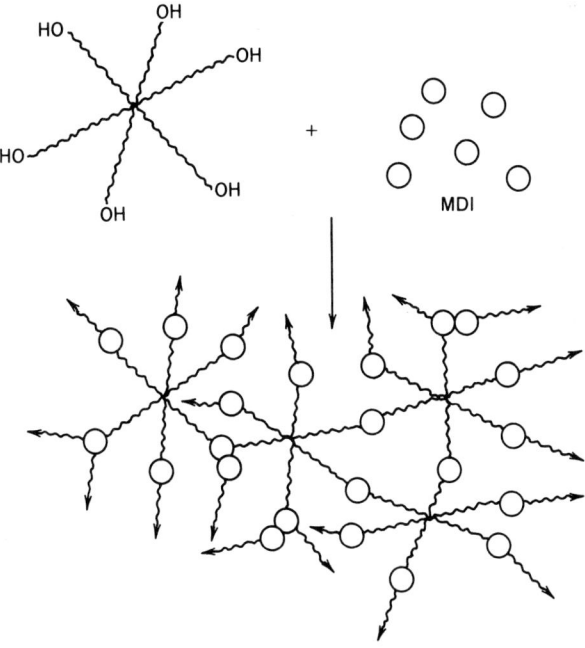

Figure 90.4. A highly branched polyol reacts with a diisocyanate to form a branched polymer, a network, or a matrix.

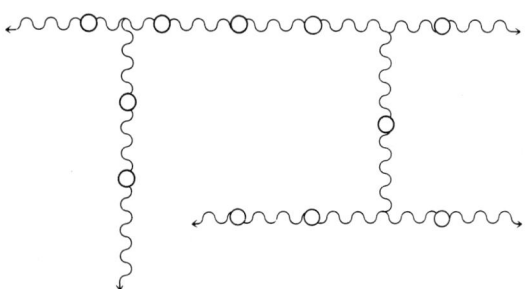

Figure 90.5. Cross-linking in the soft block using triols.

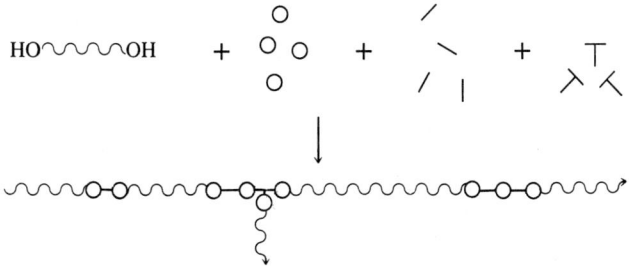

Figure 90.6. Use of short-chain diols and triols produce stiffer and stronger polymers by forming a larger "hard block."

isocyanate component into "hard blocks." Use of short-chain triols such as glycerine will produce cross-linking in the "hard" segments of the polymer chain as shown in Figure 90.6.

The use of blocking and cross-linking in the hard block tends to produce a stiffer, more rigid product. The hard blocks tend to be crystalline and reinforce the amorphous polymer, improving its strength.

Polymer chains and networks can be produced from a wide range of reactive raw materials, and a broad range of physical properties can be obtained.

The rest of this section explains how polyurethanes are formulated. To produce the desired molded end product, careful handling of the raw materials is essential in order to avoid moisture contamination, which changes the chemistry. Temperature control is necessary to keep the density, reactant viscosity, and reaction rate within process parameters.

90.2.2 Polyurethane Systems

Most urethanes are used as polyurethane systems. They may be formulated to include the following:

A	B
ISO	POLY
Diisocyanate	Polyol (diol)
	Polyol (triol modifier)
	Short chain diol
	Short chain triol
	Water (sometimes)

The above materials all react to form the polymer. One long-chain polyol is required. If linear, the urethane polymer will be flexible; if branched, the polymer will be rigid. Polyol selection and blended diol and triol additives modify the physical properties of a urethane block polymer.

$$R\text{-}11 \leftarrow \text{or} \rightarrow R\text{-}11$$
Surfactant
Catalyst 1
Catalyst 2
Additives

The function of the reactive portions of the system have been discussed in the previous section.

90.2.3 R-11

The heat of reaction boils this low-boiling liquid to produce, in the case of RIM elastomers, a packed part with good surface characteristics or, in the case of lower-density RIM structural foam, a cellular interior.

90.2.4 Surfactant

The surfactant is a material, which lowers liquid surface tension and forms films. It stabilizes the foam during the critical stage when the polymer chains are too short to produce a matrix with enough mechanical strength to support itself. It also helps produce a good part surface.

90.2.5 Catalyst

Catalysts do two things. They increase the reaction rate and often favor one reaction over another. This is important because the process is subject to competing reactions. Two catalysts are sometimes used to initiate the rapid formation of urethane linkages

$$
\begin{array}{cc}
\text{H} & \text{O} \\
| & \| \\
(-\text{N}-\text{C}-\text{O}-)
\end{array}
$$

in the initial part of the reaction period and to favor the production of stiffer urea linkages

$$
\begin{array}{ccc}
\text{H} & \text{O} & \text{H} \\
| & \| & | \\
(-\text{N}-\text{C}-\text{N}-)
\end{array}
$$

toward the end of the reaction when the part has fully expanded and the collapse of the incompletely polymerized molding is a possibility.

90.2.6 Additives

Additives are used to modify urethane products. Fillers provide mechanical reinforcement; pigments may be used to mask the nonuniform color of many polyurethane parts; and fire retardants are sometimes used to improve flammability ratings.

90.2.7 Nylons

Nylons are second to polyurethanes in their state of commercial development. Monsanto had two nylon chemical systems commercially available, using the trade name Nyrim. Allied Chemical had nylon RIM close to commercial introduction. At this time (1988), both companies have shelved their nylon RIM projects.

RIM nylons, like polyurethanes, form polymers very rapidly by the reaction of chemical end-groups. The linkages produced are as follows:

$$
\text{Polyestramide prepolymer} \; + \; (\text{CH}_2)_5 \quad \text{C}=\text{O}
$$
$$
\text{N}-\text{H} \longrightarrow \text{Nylon block}
$$

Caprolactam   copolymer

Equipment used to manufacture RIM polyurethane products must be extensively rebuilt to process RIM nylons. Figure 90.7 shows a typical process schematic for nylon RIM.

Because the viscosity of nylon RIM systems is low and the ingredients are quite reactive, leakage at the seals and the volumetric efficiency of the metering pumps may pose problems. Volumetric efficiency is the amount of material actually pumped divided by the volume displaced by the metering pumps. More leakage will occur between cylinder walls and pistons when the viscosity is low and when the pumps are adjusted to minimum displacement settings, or, in some cases, low pump revolutions per minute (rpms). As machine builders have become familiar with metering low viscosity reactants, they have avoided the metering pump problem by using lance or displacement pistons to meter hot low-viscosity fluids.

Conventional RIM machines are sometimes modified by enclosing the entire machine (except the pump drive motors) in a heated housing and providing heated lines. Two techniques for making heated lines are using hose brading to provide electrical resistance and

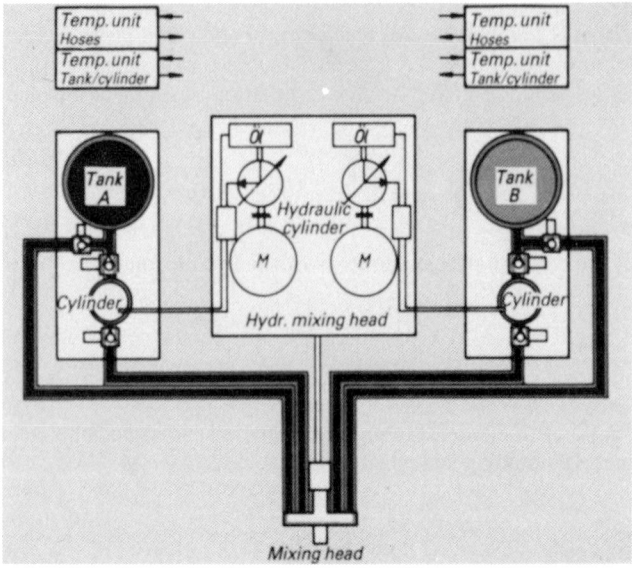

Figure 90.7.   Schematic of a RIM unit used to process nylon RIM. Courtesy of Krauss-Maffer.

circulating hot oil through the outer section of a concentric hose unit. The second is more commonly used.

Most manufacturers recommend having the machine designed specifically for nylon RIM systems.

The first commercial product made from nylon RIM was a front quarter panel (fender) for the Oldsmobile Omega Sport. Because of the excellent wear and self-lubricating properties of nylon, it is often used for mechanical components such as gears. Because of the excellent impact strength of nylon RIM, it has been used for bumper covers (Figure 90.8)

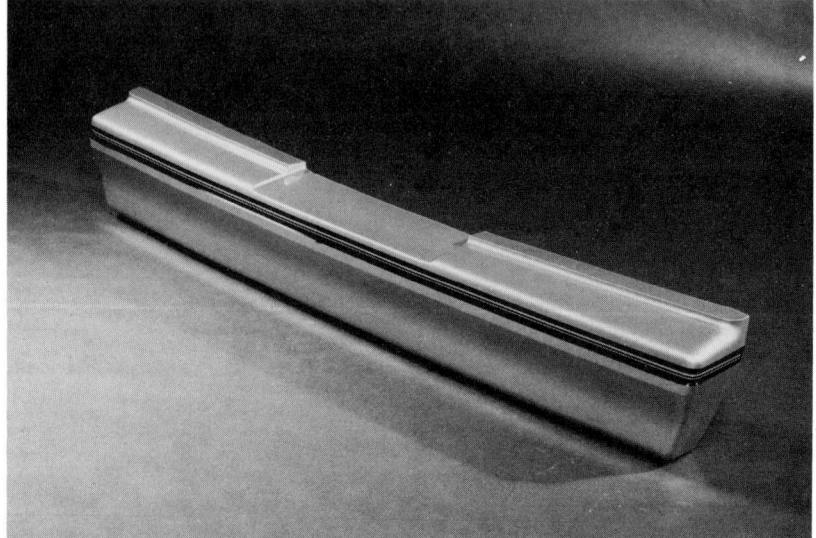

Figure 90.8.   Nyrim (nylon RIM) bumper cover. Courtesy of Monsanto Co.

and automobile fascia. It is also finding application in housings for business machines and electronics.

90.3 PROCESSING CHARACTERISTICS

90.3.1 Characteristics of the Process

The RIM process forms polymers from reactive liquid chemical systems by reacting end-groups to form linkages. If the polyols used to produce the polymer are linear, long chain flexible polymers are produced. If the polyols are highly branched, polymer networks are formed, and the product is rigid. Flexible polymers can be modified by tying chains together in the soft block using long chain triols or using short chain diols and triols to modify the hard block. By properly formulating the liquid reactants, a very wide range of systems can be developed, and a very wide range of products can be made from them.

90.3.2 Product Identification as a Guide to Selection

Normally the molder purchases polyurethane package systems. The ingredients are developed by the systems supplier. The molder determines the physical properties required by a customer or by the division in a large corporation using the item. He/she then selects a commercially available chemical "package system," processes it under the conditions specified by the supplier, and molds parts that will meet end-product specifications.

A typical physical property data sheet is shown in Table 90.1 for Union Carbide's RIM-2700 Urethane Elastomer. This example was chosen for illustrative purposes only and may not be typical of newer commercial products.

A good product information sheet provides the information listed in Table 90.1:
1. Products name: RIM-2700. 2. Product type: Urethane elastomer. 3. Manufacturing process: Reaction injection molding. 4. Product use: Automobile fascia. 5. Systems content: Specific isocyanate and specific polyol. 6. Reasons to consider this product: a. Low mold release required; b. Easy processing characteristics; c. Good physical properties.

The eight physical properties listed are typical for a product of this type. The test method notation identifies the ASTM number of the test procedure used to obtain the data tabulated.

TABLE 90.1. Typical Physical Properties

Property	Test Method	Value
Specific gravity		1.0
Shore D Hardness	ASTM D2240	58–62
Flexural modulus, Mpsi	ASTM D790	
at $-20°F$ ($-29°C$)		70–80
at 75°F (24°C)		25–29
at 158°F (70°C)		16–18
Modulus ratio,		
$-20°F/158°F$ ($-29°C/73°C$)		4.5
Ultimate tensile strength, psi	ASTM D412	3100–3500
(MPa)		(0.21–0.24)
Elongation at break, 0/0	ASTM D412	240–260
Die "C" tear, pli	ASTM D624	450–500
Heat sag, 1 h		
at 250°F (121°C) in.	ASTM D3769	0.15–0.30
(mm)		(3.9–7.6)

NIAX and Union Carbide are registered trademarks of Union Carbide Corporation, USA.

The control parameters for ASTM tests are quite specific, and products tested using the same test method should be directly comparable among different test laboratories and laboratory technicians.

A search of trade literature usually shows that several companies make polyurethane systems having competitive, though seldom identical, physical properties. Each product information sheet should identify the manufacturer and give addresses and phone numbers of sales offices.

90.3.3 Part Performance Specifications

The molded product must have certain characteristics to perform its function. An automobile-bumper-system cover must be strong, tough, and flexible. It must not sag during the high temperatures characteristic of paint-baking ovens. It must maintain good physical properties when it is subjected to a wide range of temperatures. The data provided suggest the product described is a good candidate for selection.

90.3.4 Part Types

The two major classifications for RIM parts are high-density, high-modulus, flexible elastomers and lower density structural foam parts. Automotive trim and fascia are usually elastomers. Furniture and equipment housings (especially when texture or sound-deadening are included in the product specifications) are frequently molded as structural foams.

90.4 ADVANTAGES/DISADVANTAGES

RIM is the logical process to select for molding very large parts. Capital requirements for RIM processing equipment are quite low when compared with the thermoplastic injection-molding equipment that would be necessary to mold parts of similar size. Energy requirements are much lower. Injection pressures for thermoplastic molding are measured in tons. The injection of liquid resins requires very little energy; however, the liquids are pressurized to 2000–3000 psi (13.8–20.7 MPa) to provide the energy required for impingement mixing. Molding pressures seldom reach 100 psi (0.7 MPa) maximum. There is a very wide range of reactive systems available, and a very wide range of products can be molded from them. Part sizes vary from shoe-sole size to very large parts, such as automobile body parts—bumper covers, quarter panels, and dashboard units. The largest part reported to have been manufactured was a motor boat hull weighing more than 600 lb (272 kg); however, the process had to be adjusted somewhat beyond the limits usually specified for RIM.

Although the RIM process is economically competitive for large parts, it is less competitive for smaller parts. For them, liquid casting systems are generally more expensive than equivalent thermoplastic molding compounds. It is important that the final cost analysis include both equipment and materials. RIM is a process which produces a part by a chemical reaction in the mold, so process control is necessary to ensure the production of good quality parts. Machine process technology has improved considerably over the past 20 years, and new developments are announced in each current trade journal.

The essential process parameters are conditioning, metering, mixing, and molding. Machine technology is sufficiently developed to control these parameters within processing specifications. In general, the liquid systems processed have been limited to a 3000 centipoise viscosity, and the systems normally processed have fallen well below this figure. Early literature suggests having the viscosities of the components roughly matched; but, by using adjustable impingement nozzles to adjust feed line back pressures, it is possible to mix materials with widely differing viscosities. Aftermixers built into the mold provide additional mixing.

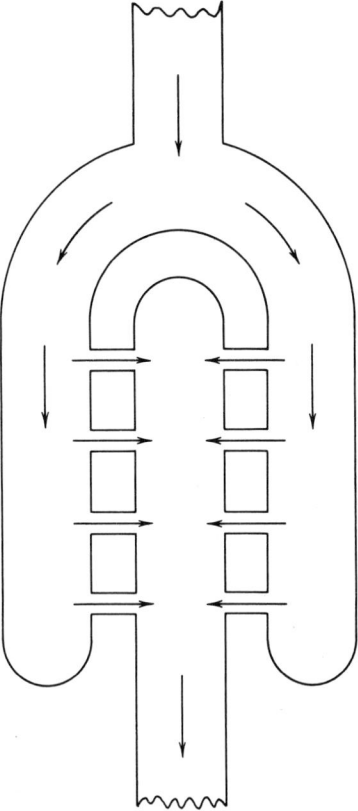

Figure 90.9. Typical aftermixer built into the RIM mold between the mixing head and the mold runner.

The aftermixer design shown in Figure 90.9 mixes ingredients a second time, again by impingement, and improves mixing in the head by increasing pressure in the mixing chamber itself. Newer mixing head designs are incorporating this feature into the mixing head, and several head modifications also incorporate this feature.

Temperature control throughout the process is essential, especially in the mold. The reaction is highly exothermic, and very fast chemical systems generate considerable heat within seconds. Process temperature control within the machine has not been a problem. Mold temperature control has been a problem if the part is poorly designed and part thickness is not uniform—a problem that is also common to thermoplastic injection molding. In thermoplastics, the part will have sink marks; in thermoplastic structural moldings and RIM, the part will bloat. The solutions to these problems are common to both processes: avoid thick sections, if possible; design a thick rib as two thin ribs; and use inserts to fill the centers of the thick section (or, in the case of RIM, process a slower chemical system formulated with less catalyst to minimize internal heat buildup). This technique is used in thick shoe soles and the large motor boat hull noted earlier.

90.5 DESIGN

The RIM process requires accurate metering of the two reactive chemicals to ensure that one reactive group A and one reactive group B are available to form an AB linkage. All reactive groups should be consumed in the process. The number of reactive groups combining to form a polymer chain must match, but the weight of material containing these groups may differ. The weight ratio of the two materials containing the same number of reactive groups may vary because the ratio of the weight of the reactive group to the whole polymer differs, but the number of reactive groups must be equal. Chemists use the term "equivalent weight" to define the weight of material containing one reactive group.

Pumps, which are used to accurately meter the liquid reactions, are described in Section 90.7.2. The two reactive groups are brought together to react by high-pressure impingement (described further in Section 90.7.3).

Conditioning and temperature control are accomplished by recirculating reactants from storage tanks designed to maintain raw material temperatures specified by the system supplier. These conditions are normally quite moderate for polyurethanes—85–100°F (30–38°C). The friction developed by metering pumps and by passing the reactants through impingement nozzles causes the reactant temperature to rise as the materials recirculate. Thus provision must be made to cool the raw materials down to the recommended conditioning temperature.

The mold-clamp pressures are not excessive because less than about 100 psi (0.7 MPa) is developed by the packing process (R-11 vapor pressure). However, the RIM parts are large, so the total force necessary to hold the mold closed is considerable (see Section 90.7.6).

The mold temperature controls the chemical reaction itself. Excessive heat develops double branching at linkage sites, which requires an equivalent weight correction. These double-branched linkages are unavoidable and do, in fact, improve part properties. However, excessive double branching is to be avoided. This is accomplished by controlling the temperature of the mold with cooling water. Because the molds are usually quite large, cooling water channels should be zoned in parallel to ensure uniform temperatures across the mold.

90.6 PROCESSING CRITERIA

Process details are described elsewhere in this section. Machine variables include: Temperature control (1. Viscosity, 2. Density, and 3. Reaction rate); Pressure control (1. Mixing); Ratio control (Matching equivalent weights); and Conditioning (Recirculation provides constant temperature and viscosity and keeps solids suspended in RIM systems).

For chemical system variables see sections 90.1, 90.2 and 90.3.

90.7 MACHINERY

The three primary pieces of equipment necessary to make RIM parts are the RIM machine, the mold clamp, and the mold. The two general types of metering machines are metering pump machines and displacement cylinder machines.

Several companies manufacture RIM machines that use variable displacement metering pumps to operate hydraulic cylinders, which actuate RIM system metering pistons.

RIM machines perform four functions:

- Recirculation to condition reactants
- Accurate metering
- Mixing by impingement.
- Dispersing mixed reactants directly into a closed metal mold.

90.7.1 Conditioning Tanks

Conditioning tanks are jacketed and/or contain tempering coils to maintain the process temperature required by the chemical system used. Liquid level control devices should be designed to keep the raw material level relatively constant. Adding 60 gallons (0.23 m^3) of raw material to an empty conditioning tank will upset processing temperatures and cause variations in viscosity, density, and reaction rates. Several systems that perform this function are available. Recirculated material re-enters the conditioning tank through a dip tube, thereby controlling the amount of the dissolved gas in the circulating reactants. Inert gas, such as dry nitrogen, is often introduced beween the metering pump and the mixing head to provide a gas bubble nucleus for the foaming reaction that finally fills and packs the mold. Agitators are usually recommended in both tanks.

90.7.2 Metering Pumps

Three types of positive displacement pumps are common to RIM machines. All can be adjusted to compensate for weight ratio differences characteristic of RIM chemical systems. The pumps are

- Bosch linear piston pumps used on smaller machines and for third-stream addition systems
- Rexroth axial piston pumps used on a range on machines from 8 to 160 L/min (four standard sizes are available.)
- Bosch radial piston pumps used on larger machines and where adjustment is seldom necessary

A newly developed gear pump is currently under test.

90.7.3 Mixing Heads

Mixing heads perform the following functions:

- Recycling reactants between shots
- Mixing by impingement
- Cleaning the mixing chamber at the end of the shot

Newer mixing heads often incorporate devices for raising pressure in the impingement zone. The most common use an angled mixing chamber which causes the mixed reactants to flow around a 90° bend in the head. Two clean-out pistons placed perpendicular to each other are required for cleaning out. In some cases, the second piston can be adjusted to baffle flow from the mixing zone. EMB machines use a single square piston and retractable side baffles.

Two common mixing heads are shown in Figures 90.10 and 90.11. The Krauss-Maffei head uses by-pass channels in the clean-out piston to recirculate raw materials. Reactants pass through the impingement nozzles, whether recirculating or mixing. The Battenfeld head uses by-pass channels cut in the mixer housing. Two pistons actuate the head. The manufacturer claims that keeping the channels well separated is a special advantage when processing low-viscosity reactants, such as nylon RIM systems.

90.7.4 Metering Pump Machines

Figure 90.12 shows a typical metering pump machine, showing the following elements:

- Conditioning tanks
- Tempering coils (or jacket)

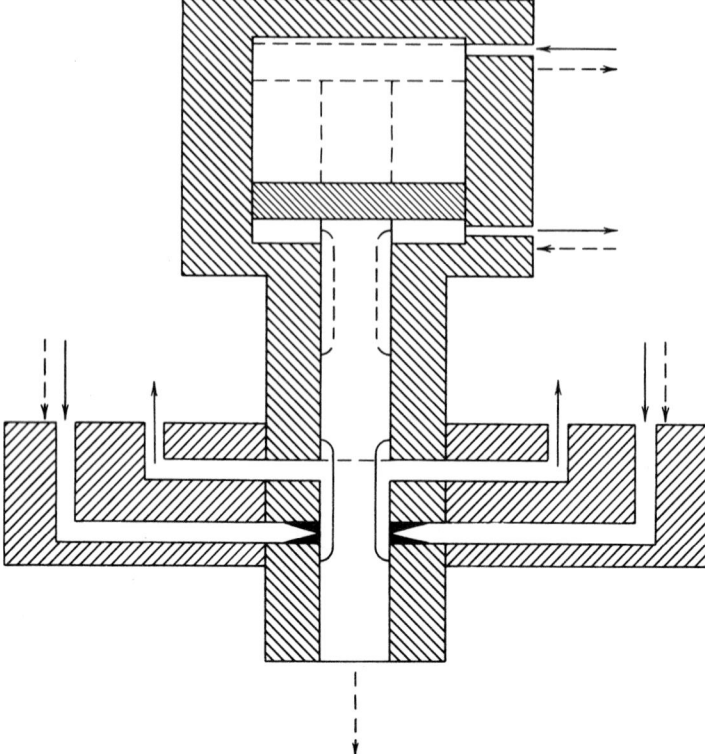

Figure 90.10. Krauss-Maffei mixing head with recirculating channels cut in the clean-out piston.

- Temperature control unit
- Automatic feed lines from bulk storage
- Recycle dip tube
- Pressure gauge
- In-line filters (paired)
- Metering pumps
- Drive motors
- Safety recycle line with electrical cut-off switches
- High-pressure line gauges
- Mixing head
- Hydraulic unit
- Automatic calibration unit.

Because the pressure drop in process lines is quite low compared to the very high pressure drop through impingement nozzles, the pressure differences along the lines is less than 1% of the pressure drop through the nozzles of the mixing head. For this reason one machine can meter reactants to several mixing heads along the process headers. Such a multistation processing unit is shown in Figure 90.13, complete with a bulk storage system.

Because the shot volume is determined by a timer, it is not necessary to make the same part at each station. The parts must use the same chemical system, however. An example

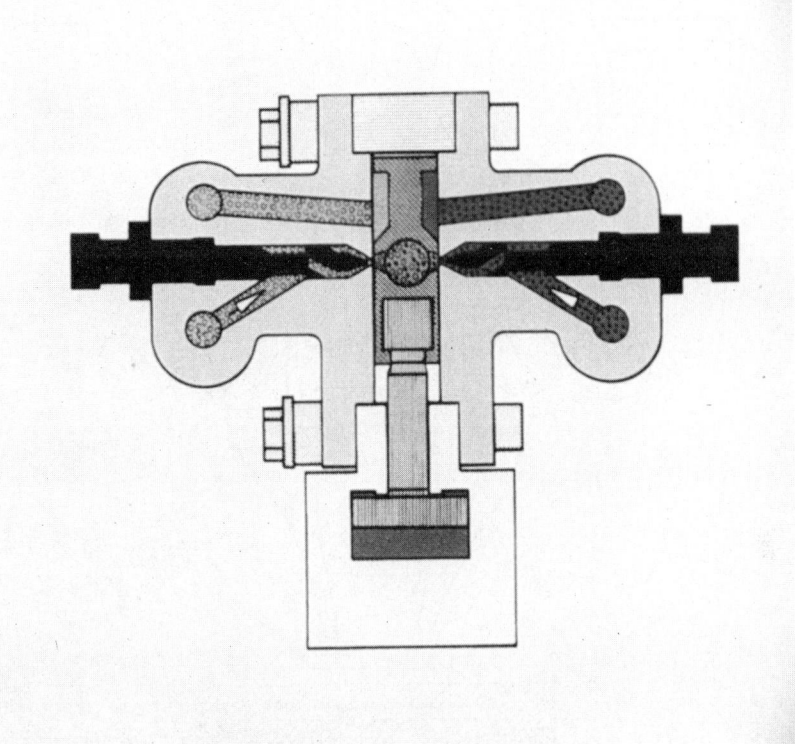

Figure 90.11.   Battenfeld mixing head with recirculating channels cut into the mixer body. The mixing chamber is shown 90° to the schematic, as it would appear looking through the end of the barrel.

of such parts are furniture components of different sizes that will eventually fit together. An interesting new development is to add color paste at individual mixing heads. By doing this, each clamp can mold differently colored products.

90.7.5 Displacement Piston Machines

These machines meter reactants using either displacement cylinders or lance pistons. The lance piston is more common because it allows for the circulation of filled materials, and the only leakage occurs at the lance packing where it can be easily detected. These designs are shown in Figure 90.14.

There are two basic types of displacement machines. One uses metering pumps to pump hydraulic oil to the pistons, which actuate the reactant displacement cylinders (Fig. 90.15). The other uses a single hydraulic pump to pressurize hydraulic oil and proportioning valves to determine flow rates to the hydraulic power pistons (Fig. 90.16). A potentiometer linked directly to the piston system provides continuous monitoring to the machine system.

In fact, through computer graphics, it is possible to provide a multicolor TV display of the machine in operation. The parts actually move, and flow patterns are shown as they occur. Changing conditions, such as temperature and pressure at key locations, are displayed simultaneously. The machine will automatically shut down if process variables are not within process specifications.

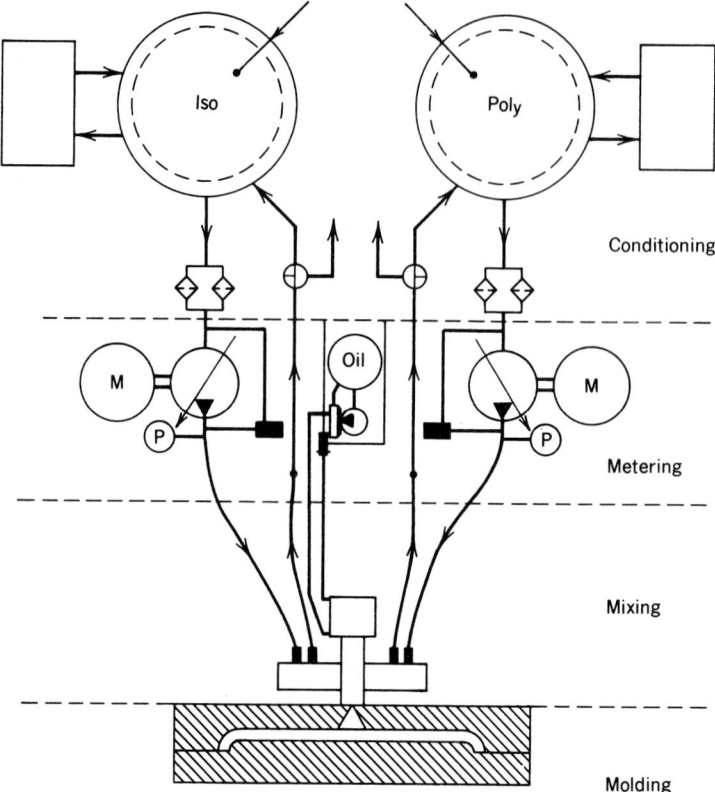

Figure 90.12. Detailed flow sheet for a metering pump RIM machine.

90.7.6 Clamps

Clamps perform the following functions:

- Support the mold
- Hold the halves of the mold together
- Open the mold
- Tilt to facilitate reactant flow as the chemical system expands.

Shoe-sole clamps are mounted on carousels. Some smaller molds are affixed to conveyor chains for line production. Automobile fascia molds open like a book to facilitate part removal.

90.7.7 Auxiliary Equipment

Single-unit displacement cylinders are designed to operate from one of the metering pumps connected to a hydraulic oil reservoir. For certain types of materials, it is necessary to increase pump motor horsepower and increase system design parameters to handle systems having higher viscosity or a high level of filler reinforcement.

Figure 90.17 shows a side-stream unit for adding catalysts, fire retardants, and special purpose additives that cannot be formulated into the basic chemical system.

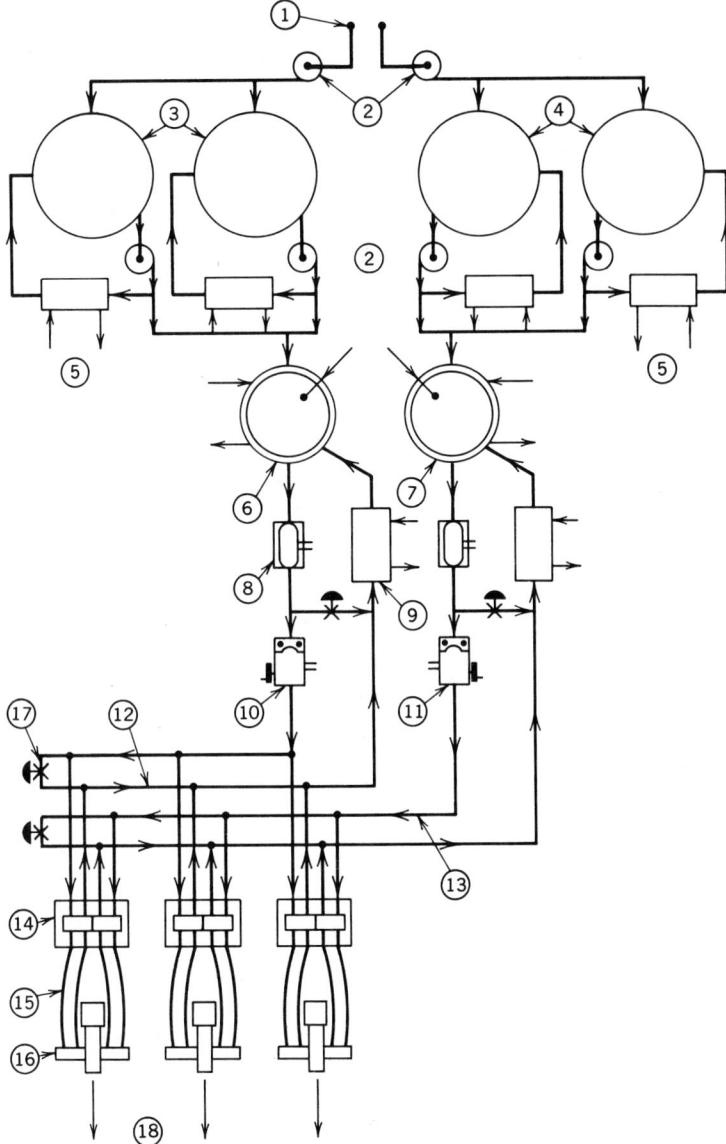

Figure 90.13. Complete multistation RIM processing system, showing bulk storage tanks and three mixing heads with switching station.

90.8 TOOLING

RIM parts are molded in metal molds. The mold performs the following functions:

- Conveys mixed reactants from the impingement mixing head to the cavity.
- Shapes the part by having a cavity cut to form a three-dimensional metal negative of the desired part shape.
- Controls the chemical reaction using cooling water to transfer the heat of reaction from

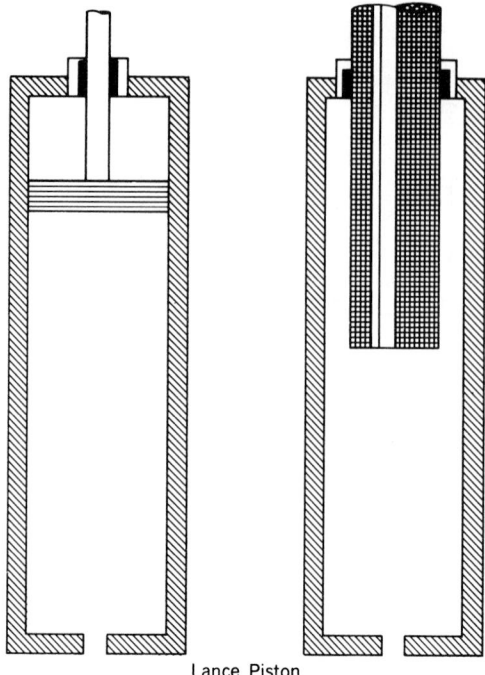

Lance   Piston

Figure 90.14.   Two types of metering cylinders.

the polymerizing urethane out of the system. For nylon, heat is used to initiate and control reaction rates.

The mold is, in fact, a chemical reactor. The reaction in reaction injection molding takes place in a completely filled mold cavity. The temperature of the mold plays a vital role in the polymerization of the reactants. For polyurethane, speeding the reaction rate by operating the mold at elevated temperatures is to be avoided, as this changes the type of linkages produced.

Pressure is developed in the mold primarily by the foaming and packing done by R-11 vaporization. Clamping units for holding the mold closed are lighter and less expensive than presses for injection molding. Indeed, the adhesive nature of polyurethanes sometimes causes difficulty in opening the clamp, especially when unseasoned molds and inadequate mold release agents are involved.

Most high-production RIM processes use machined steel molds. These molds have a reported working life of more than 1 million parts.

Some molders favor aluminum molds because aluminum has better heat transfer characteristics than steel and provides better temperature control in the mold cavity. Because it is softer, aluminum is easier to machine, and the lower machining costs may counter the higher cost of the base metal. Cast aluminum molds have been used for textured parts, most notably simulated-wood products such as cabinet doors. Aluminum molds have a reported working life of more than 100,000 parts. Mold damage by misplaced inserts and other foreign objects is more of a problem with aluminum than with steel.

Kirksite or zinc molds have been used, especially where good part texture is required. The heat transfer characteristics of Kirksite are not as good as those of steel. The metal is quite soft and dense. It is very prone to damage by foreign objects. Kirksite molds have been reported to have a mold life of 10,000 to 25,000 parts. They are mostly used by molders

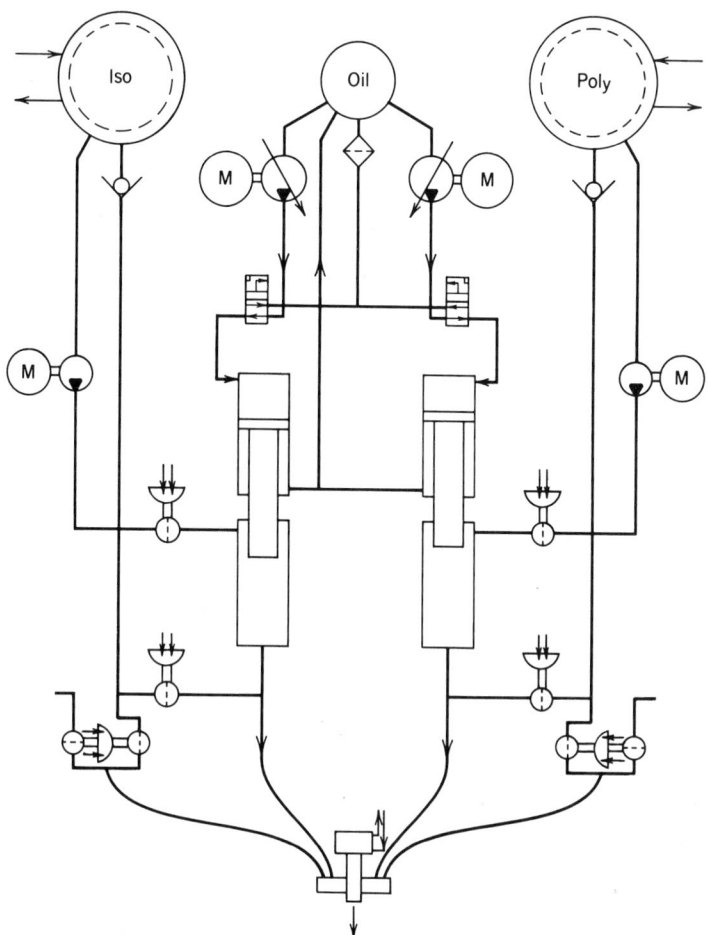

Figure 90.15.   Displacement cylinder machine using metering pumps to control the flow of hydraulic power.

having limited volume production runs and are usually remelted for making new molds. In general, the cavity is cast rather than machined.

Nickel shell molds, whether formed by vapor deposition or electroplating, provide a hard cavity for parts requiring finely detailed texture. The quality of a part made in a nickel shell mold will be excellent, especially when leather grain, wood grain, or other simulated natural textures are required. Nickel shell molds are most used for special products, such as automobile arm rests. Large nickel shell molds are often fastened into steel grid support units. Cooling coils are laid behind the nickel shell, and the grid box is filled with epoxy or, more likely, aluminum-filled epoxy, to provide structural integrity. However, with especially fast RIM systems, this construction system may fail.

Nickel has a rather high heat transfer rate; aluminum-filled epoxy has a comparatively low heat transfer rate. The nickel shell will heat up rapidly; the epoxy/aluminum composite will not. Not only will the temperatures differ, but the coefficients of expansion of the two materials do not match. Frequently, there is a shear failure at the nickel/epoxy interface. Although the discontinuity is not affected by the moderate molding pressures, adhesion by the newly molded polyurethane part frequently pulls the metal shell away from the epoxy

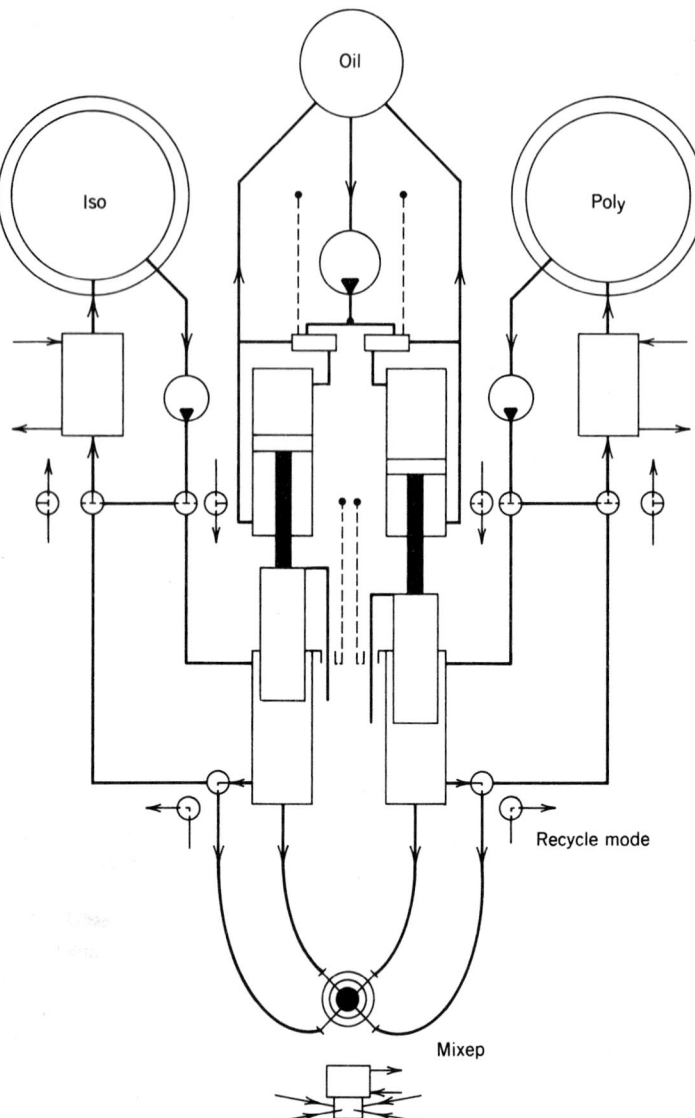

Figure 90.16. Displacement cylinder machine using proportional flow control valve to establish flow rates to the hydraulic power pistons.

composite backing when the mold is opened. This type of damage is difficult if not impossible to repair. Usually these failures are a problem when the mold shell is too thin and when other design parameters are ignored. The second most common error is the improper design of the tie-in system between the nickel shell and the steel grid. These design errors can be avoided, but failure to do so has given nickel shell RIM molds a bad reputation.

Epoxy molds have very poor heat-transfer characteristics, as do aluminum-filled epoxy and epoxy-bonded granual molds. However, these molds are used to molding RIM proto-types. To do this the RIM chemical system must be formulated with less catalyst to slow down the reaction rate, thereby increasing the required molding cycle time. Use of fast systems in epoxy molds may cause spalling on the mold cavity surface because of localized

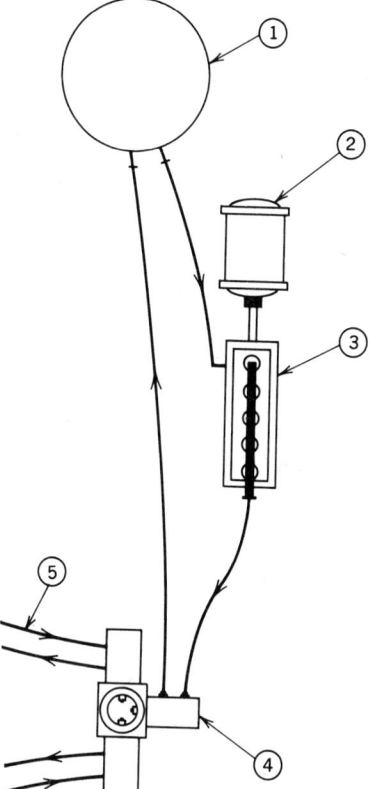

Figure 90.17. Third-stream addition unit for adding small amounts of special fluid additives.

overheating of the cavity surface. In addition, the part itself may overheat internally, causing failure in the polymer structure (cracking) that will markedly decrease the physical property characteristics.

The specially formulated RIM chemical system may be useful for molding a prototype to examine it for appearance, but it is not possible to obtain certain desirable engineering data, such as cycle time, or final part physical properties from such a prototype part. RIM molds are usually less expensive than thermoplastic injection molds of the same size.

90.9 AUXILIARY EQUIPMENT

Auxiliary equipment is added to the basic RIM machine for several purposes:

- To process filled polyol or reinforced RIM systems (RRIM)
- To add color pastes at the mixing head
- To add catalysts that cannot be made part of the RIM chemical system
- To add fire retardants that would not be stable in the RIM chemical system
- To add any additive that might not be stable in the RIM chemical system because they are reactive with the system reactants.

Because an auxiliary system must be designed as an integral part of the process (for example, a reactive fire retardant may react to form part of the polymer structure), the parameters of the chemical reaction and the mechanical characteristics of the auxiliary equipment must be related specifically to the basic processing equipment.

90.9.1 Bulk Storage

For high volume processors, bulk storage facilities are used to receive and condition the ISO and POLY chemical system raw materials. In transit, isocyanates tend to crystallize and polyols tend to gel, especially in cold weather. Receiving tanks are paired so conditioned material can be used in production while material in the second tank is being conditioned and brought up to temperature. It is imperative that a tank is never filled with the wrong reactant, even when it is empty. The reaction will pot the equipment, making a complete and very expensive rebuilding of the bulk storage unit mandatory.

Bulk handling equipment is available for blending chemical systems and systems containing fillers for RRIM processing. These blending units are usually custom built. RIM machine builders and chemical raw material suppliers can provide reliable guidelines for selecting a bulk equipment fabricator.

90.10 FINISHING

Most of the parts manufactured by the RIM process are quite large. The materials entering the mold are very fluid, having a viscosity usually below 3000 centipoise. An expanding agent is vaporized to fill the mold (up to 70% expansion for structured RIM) or pack the mold (possibly under 5% expansion for elastomeric RIM). The solid part is formed as the concurrent chemical reaction forms the polymer chain or network that gives the part its ultimate physical properties.

The expansion phase precedes the chain formation, so the mold is full while the reactants are relatively fluid. As the mold clamp pressures are low, flash is a characteristic problem with most RIM parts. This flash is usually manually removed using a knife or similar tool.

RIM parts (especially the polyurethanes) are usually removed from the mold before the chemical reactions that develop the physical properties are complete. The part is placed on a support jig that holds it in its final shape until it is fully cured. In some cases, this is done by simply setting the supported part aside for 12 to 24 h. More often the supported part is cured in an oven for several hours at temperatures of about 180°F (82°C). The final physical properties are thus developed after demolding.

Before finishing, mold release must be removed from the part (see Section 90.11).

Nylons, and some of the newer RIM raw materials, are completely reacted in the mold. Post-curing is not necessary.

90.11 DECORATING

RIM polyurethanes made with aromatic isocyanates (such as pure or polymeric MDI) have a tendency to darken as a result of the effect of ultraviolet light on the chemical ring structure of the MDI component. This does not noticeably affect the physical properties of the molded product itself. Soft white limestone or fine carbon black are often used as fillers to mask the effect of this color change for some products in which appearance is not a critical factor but merely a cosmetic requirement.

Polyurethane polymer systems manufactured with aliphatic isocyanates are light-stable, and products are molded in a wide range of bright colors. Especially interesting is the development of equipment to add color concentrate, usually dispersed in a polyol, directly into the mixing head attached to a given mold. The basic urethane formula is adjusted to compensate for the additional reactive polyol. Using this technique with a multiclamp RIM line, it is possible to mold different colors at each station using a single RIM machine.

In some cases, aromatic RIM systems are molded in color, then painted the same color. This technique eliminates the need to touch up every dent, nick, or scratch which would otherwise show up tan or white.

Usually, RIM parts are simply primed and painted. Before painting, however, the part is cleaned to remove mold-release agents. The most common mold releases are metal stearates (or soaps) that can be removed from the part by a water wash and paraffin waxes that are usually removed by solvent vapor degreasing. In some cases, a light particle (sand) blasting is used to prepare the surface. Silicone mold releases are to be avoided as they are very difficult to remove from the part, and paint will not stick to the silicone surface film. New internal mold releases are being developed for urethane RIM systems. They are specific to a given urethane system and are still in the development stage. Successful commercialization of internal mold-release compounds will greatly speed up part removal and decrease molding cycle times.

There are several systems for painting RIM parts. In some cases primers are applied and the finish coats are sprayed on. In general, coatings should be formulated specifically for application to RIM parts. Automobile fascia are often finished before shipment to the automobile assembly plant. Color matching is done to produce a uniform appearance between the metal body and the RIM fascia.

In some cases, the finish coats are applied to the entire car body assembly. In these applications it is necessary for the RIM part and any primers applied to it to withstand the very high baking temperatures characteristic of the finishing process; for example, a temperature of 350°F (177°C) for at least 20 minutes. These conditions vary at different auto body assembly plants.

90.12 COSTING

Because RIM products range in size from shoe soles to parts weighing several hundred pounds, and because equipment is usually designed and assembled for the manufacture of specific end products, information applicable to a given end product is beyond the scope of this chapter.

Elements to consider in making a cost study include the following:

- Normal raw material inventory.
- Normal finished part inventory.
- Cost of raw materials.
- Capital cost for equipment. When equipment is purchased, the initial cost is never the sole criterion. Some equipment comes fully equipped while others use a base price with optional extras. Before purchase, it is important to make certain that one machinery compared to another is comparably equipped.
- Equipment reliability. Some equipment is not designed for heavy-duty production operations.
- Cycle time and production schedule.
- Availability of spare parts. A long-term shutdown can be costly.

90.13 INDUSTRY PRACTICES

Because of the wide range of products made by the RIM process, industry practices will vary.

90.14 INDUSTRY ASSOCIATIONS

Polyurethane Manufacturers Association, Suite 203, 999 North Main Street, Glen Ellyn, IL 60137 (312) 858-2670.
Society of Plastics Engineers, 14 Fairfield Drive, Brookfield, CT 06805 (203) 775-0471,
Society of the Plastics Industry, 1275 K St. N.W., Suite 400, Washington, DC 20005 (202) 371-5200.

BIBLIOGRAPHY

General References

W.E. Becker, *Reaction Injection Molding,* Van Nostrand Company, New York.
E.N. Doyle *Development and Use of Polyurethane Products,* McGraw-Hill, New York. (This book provides a good general background on polyurethane technology. It is not specifically about RIM.)
F.M. Sweeney, *Introduction to Reaction Injection Molding.* Technomics Publishing Co., Lancaster, Penn.
F.M. Sweeney, *Reaction Injection Molding, Machinery and Processes,* Marcel Dekker, New York, NY.

Periodicals

Journal of Foamed Plastics (Technomic); *Journal of Polymer Science* (ACS); *Modern Plastics Plastics Engineering* (SPE); *Plastics Machinery & Equipment Plastics Technology Plastics World Polymer Composites* (SPE); *Polymer Engineering and Science* (SPE)

91

ROTATIONAL MOLDING

Philip T. Dodge

USI, Quantum Chemicals Div., 3100 Golf Rd., Rolling Meadows, IL 60008-4070

91.1 DESCRIPTION AND HISTORY OF PROCESS

Rotational molding, also referred to as rotomolding or rotational casting, is a plastics processing method for producing hollow parts. Most rotational molding incorporates the following steps:

- The molds or cavities are filled with a predetermined amount of powder or liquid. (This is called as charging the molds.)
- The mold halves are secured together by a series of bolts or clamps.
- The molds are placed in a heated oven and rotated biaxially.
- In the ensuing heating cycle the resin melts, fuses, and densifies into the shape of the mold being used to form the hollow object.
- The molds are placed in a cooling chamber in which a combination of air and water is used to cool the mold slowly, thereby maintaining the part's dimensional stability.
- The molds are removed from the cooling chamber, opened, and the finished parts are removed.
- The molds are recharged and these steps are repeated.

Biaxial rotation and heat have reportedly been used to form hollow objects as far back to the 1850s. The Dutch utilized this process successfully to mold hollow chocolate candies in the 1920s.

Markets for rotational molding have expanded as new resins have become available. At first, rotational molding was limited to vinyl plastisols. In the early 1960s, introduction of polyolefin and later low and high density polyethylenes gave rotational molders entry into new markets in which vinyl parts could not compete. In the early 1970s, cross-linkable and modified polyethylenes opened up additional market areas, especially the large tank market. The mid-1970s saw the development of linear-low polyethylenes. The 1980s brought a surge of nonpolyethylene resins into the rotational molding industry, among them nylon, polypropylene, and polycarbonate.

Most rotational molders will describe the process as an art, not a science.

91.2 MATERIALS

Rotational molding resins are used in powder form, ground to 35 mesh, and range in size from 74 to 2,000 microns.

Polyethylene remains the workhorse resin of the rotomolding industry, accounting for some 85% of all resins rotationally molded.

Resins with poor impact strength, low flowability (which makes them difficult to mold), or high cost are rarely selected for rotational molding.

MASTER PROCESS OUTLINE	PROCESS	Rotational Molding

Instructions:
For "ALL" categories use   Y = yes
For "EXCEPT FOR" categories use   N = no
                             D = difficult

[N] ALL Thermoplastics Except for:

	Y Polyethylene
	Polyethylene-UHMW
Acetal	Poly(methyl methacrylate)
Acrylonitrile–Butadiene–	Poly(methyl pentene) (TPX)
Styrene	Poly(phenyl sulfone)
Cellulosics	Poly(phenylene ether)
D Chlorinated Polyethylene	Poly(phenylene oxide) (PPO)
Ethylene Vinyl Acetate	Poly(phenylene sulfide)
Ionomers	Y Polypropylene
Y Liquid Crystal Polymers	Polystyrene
Nylons	Polysulfone
Poly(aryl sulfone)	Polyurethane thermoplastic
Polyallomers	Y Poly(vinyl chloride)-PVC
Poly(amid-imid)	Poly(vinylidine chloride)-Saran
Polyarylate	Rubbery Styrenic Block
Y Polybutylene	Polymers
Polycarbonate	Styrene Maleic Anhydride
Polyetherimide	Styrene–Acrylonitrile (SAN)
Polyetherketone (PEEK)	Styrene–Butadiene (K Resin)
Polyethersulfone	Thermoplastic Polyesters
	XT Polymer

[N] ALL Thermosets Except for:

Allyl Resins
Phenolics
Polyurethane Thermoset
RP-Mat. Alkyd Polyesters
RP-Mat. Epoxy
RP-Mat. Vinyl Esters
Silicone
Urea Melamine

[N] ALL Fluorocarbons Except for:

Fluorocarbon Polymers-ETFE
Fluorocarbon Polymers-FEP
Fluorocarbon Polymers-PCTFE
Fluorocarbon Polymers-PFA
Fluorocarbon Polymers-PTFE
Fluorocarbon Polymers-PVDF

91.3 PROCESSING CHARACTERISTICS

Rotationally molded parts impose fewer limitations than parts made by any other plastic processing method. All rotationally molded parts are hollow, with wall thicknesses that are even except at the corners, which are thicker.

Tanks ranging in size from 5 gallon to 22,000 gallons (0.003–83 m³) are a major application, as are containers for packaging and materials handling, portable outhouses, battery cases, light globes, vacuum cleaner and scrubber housings, garbage containers, surf boards, toys, traffic barricades, display cases, and ducting.

Standard design chart for _____ **Rotational molding** _____

1. Recommended wall thickness/length of flow, mm _____ Not applicable _____
   Minimum _____ for _____ distance
   Maximum _____ for _____ distance
   Ideal   _____ for _____ distance

2. Allowable wall-thickness variations
   _____ percent of nominal wall

3. Nominal wall-thickness tolerance
   Parallel to parting line _____ ± 20% _____
   Perpendicular to parting line _____ ± 20% _____

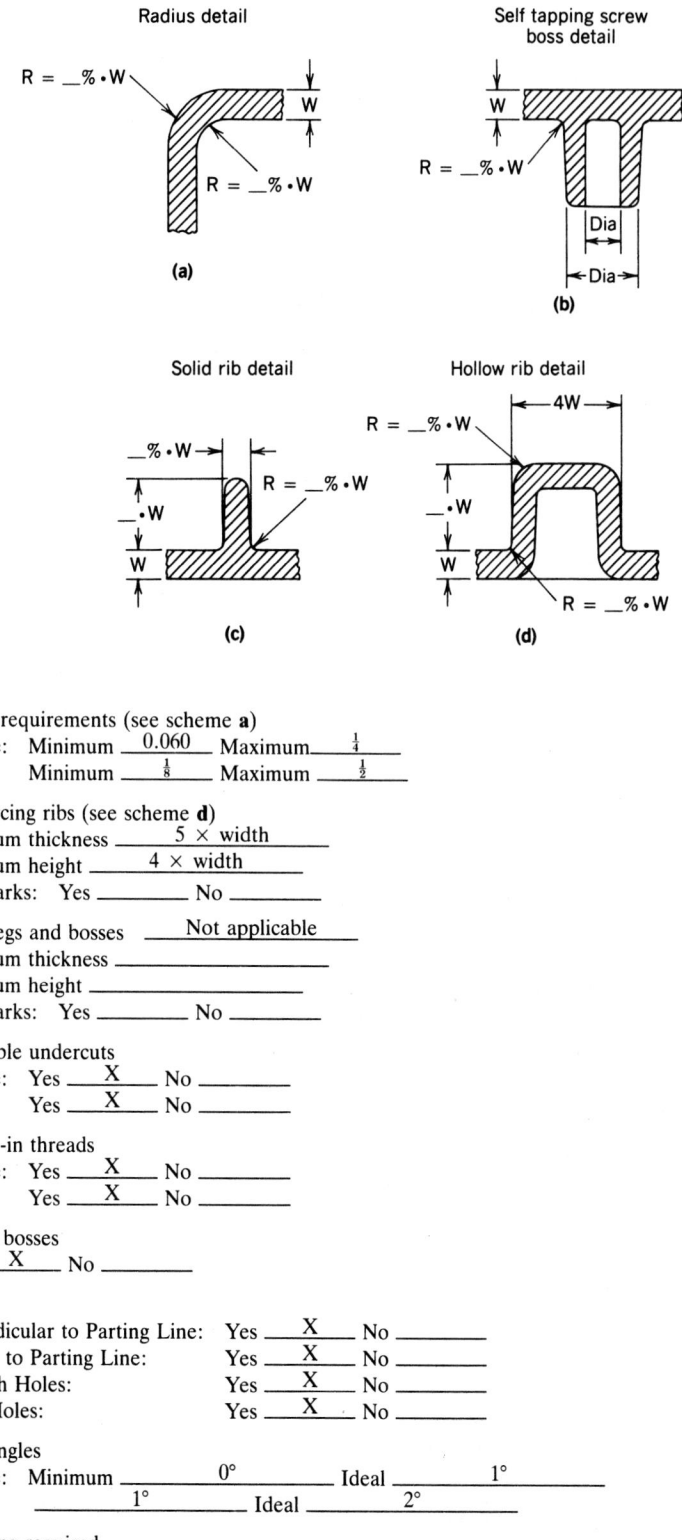

Radius detail

$R = \underline{\quad}\% \cdot W$

$R = \underline{\quad}\% \cdot W$

W

(a)

Self tapping screw
boss detail

W

$R = \underline{\quad}\% \cdot W$

Dia

Dia

(b)

Solid rib detail

$\underline{\quad}\% \cdot W$

$\underline{\quad} \cdot W$

W

$R = \underline{\quad}\% \cdot W$

(c)

Hollow rib detail

4W

$R = \underline{\quad}\% \cdot W$

$\underline{\quad} \cdot W$

W

$R = \underline{\quad}\% \cdot W$

(d)

4. Radius requirements (see scheme **a**)
   Outside: Minimum __0.060__ Maximum __$\frac{1}{4}$__
   Inside: Minimum __$\frac{1}{8}$__ Maximum __$\frac{1}{2}$__

5. Reinforcing ribs (see scheme **d**)
   Maximum thickness __5 × width__
   Maximum height __4 × width__
   Sink marks: Yes _____ No _____

6. Solid pegs and bosses __Not applicable__
   Maximum thickness _____
   Maximum height _____
   Sink marks: Yes _____ No _____

7. Strippable undercuts
   Outside: Yes __X__ No _____
   Inside: Yes __X__ No _____

8. Molded-in threads
   Outside: Yes __X__ No _____
   Inside: Yes __X__ No _____

9. Hollow bosses
   Yes __X__ No _____

10. Holes
    Perpendicular to Parting Line: Yes __X__ No _____
    Parallel to Parting Line: Yes __X__ No _____
    Through Holes: Yes __X__ No _____
    Blind Holes: Yes __X__ No _____

11. Draft angles
    Outside: Minimum __0°__ Ideal __1°__
    Inside: __1°__ Ideal __2°__

12. Trimming required
    Yes _____ No __X__

13. Finished both sides
     Yes _____ No __Not finished on inside__

14. Finishing Requirements
     _____None_____

15. Molded-in inserts
     Yes ___X___ No _____

16. Size Limitations _____None_____
     Length: Minimum _____ Maximum _____
     Weight: Minimum _____ Maximum _____

17. Tolerances
     Note: ___Chart attached___

18. Special process and material-related design details ____Not applicable____
     (A) Blow-up ratio _____ Blow molding
     (B) Core L/D _____ Injection & compression molding
     (C) Depth of draw _____ Thermoforming
     (D) Etc.

91.4   ADVANTAGES/DISADVANTAGES

91.4.1   Advantages

Design of parts for rotational molding is not limited by constraints common to conventional plastics design.

Mold and tooling costs are relatively inexpensive, compared to other processing methods, because a rotational mold does not require water channels to be drilled and does not need to withstand a clamping force, as injection and blow molds do. The low cost of a rotational mold makes part prototyping attractive.

Several different parts and colors may be molded on the same machine and even in the same cycle. Quick mold changes, which are important when short production runs are required, are possible with rotational molding. The process lends itself to the production of large hollow parts. [The largest part molded to date is a 22,500-gallon tank (84 m^3)].

Such secondary operations as trimming can be eliminated since rotationally molded parts have very little flash. They are relatively stress-free; the thicker corner sections add strength to the part.

Design criteria such as undercuts, intricate contours, molded-in inserts, and double-wall construction that can constrain design in other processes are commonplace in a rotationally molded part. The resultant part has a uniform wall thickness, which can be varied as the need dictates.

91.4.2   Disadvantages

In comparison to other plastics processing methods, rotational molding is slow. The kinds of resins available is limited. In the rotational molding field, the number of "experts" and the amount of professional publications are also limited.

91.5   DESIGN

Rotational molding gives the designer considerable freedom. Specific design guidelines are provided in the publication, *The Engineers' Guide to Designing Rotationally Molded Plastic Parts,* produced by the Association of Rotational Molders (ARM).

91.6   PROCESSING CRITERIA

Many rotationally molded parts are produced in the natural plastics color. If color is desired, suppliers can provide a precompounded colored resin, although some molders prefer to dry blend color into the resin. When this is done, proper addition and dispersion is critical. Such resins as nylon and polycarbonate may require a drying step before the parts can be molded.

Although scrap or regrind is not produced in large quantities, only a small percentage of regrind should be used with the virgin resin.

The grind of the material is extremely important. As noted, the industry standard is resin ground to 35 mesh powder. It is possible to utilize coarser or finer grinds, depending on the application.

The rotation ratio is important for proper wall distribution. For new parts, the correct ratio normally has to be established by trial and error. In generally, fewer rotations per minute will produce the most uniform wall thickness.

The molding cycle is also established by trial and error. The resin, the mold, and the final application must be taken into consideration to establish the most efficient cycle.

Although all resins process differently, most will share the following characteristic molding problems. Undercured parts will have a rough inner surface and many bubbles throughout the wall sections of the part. The impact strength of an undercured part will be poor. Overcured parts will be yellow in color as a result of resin degradation. Cold temperature impact strength may be poor.

It is advisable to contact the resin supplier's technical personnel about the characteristics of a properly cured rotationally molded part made from their resins.

91.6.1 Mold Release

Since most rotational molds are designed with little or no draft angle, it is important to condition the molds with a release agent. Molds are normally cleaned with a solvent and a lightly abrasive cloth to remove all foreign particles left on the surface during mold fabrication. A light coating of release is then applied and baked on. The resin used will dictate the amount of release required on a particular mold. After the initial application, several hundred parts can usually be molded before a stripdown of the mold is required and another baked-on is coating applied. During this time, some "touch up" of the mold may be required.

Many molders sandblast their molds to remove all buildup before reapplying the mold release. An especially effective release is needed at the parting line to aid in demolding.

91.6.2 Other Special Considerations

Parts that are not simple may require that the molds be custom machined. Special alignment pins may be needed to ensure proper fit. If inserts are called for, the mold may have to be machined to facilitate one or several of them.

91.7 MACHINERY

Equipment used in rotational molding is relatively simple. Models include the carousel, shuttle, and clamshell types.

The carousel machine (Fig. 91.1) is the one most commonly used. It consists of a heating station or oven, a cooling station, and, in many cases, an enclosed chamber and a loading and unloading station. The carousel will have three to four spindles or arms where the mold or molds will be mounted. Most carousels can rotate in a complete circle. The spindles are mounted on a central "hub" and are driven by variable motor drives.

New control systems allow each arm to operate independently, not only in movement but in control of oven temperature and molding time. Most new rotational molders incorporate microprocessors in the control system.

Natural gas fired ovens are preferred, with blowers to distribute the heat throughout the chamber. Oil or propane gas heated ovens are sometimes used.

The cooling station incorporates a forced air system (such as a fan) for initial cooling and a water system for cooling the molds and parts. Normally, a spray mist is preferred for even cooling. The cooling station may or may not be enclosed.

Standard Tolerance Chart

| PROCESS | Rotational Molding | MATERIAL | Polyethylene |

Values are based on a nominal 1/8 in. (3.2 mm) wall thickness and given in plus or minus in./in. and mm/mm.

FINE tolerance – Special care and added cost.
COMMERCIAL tolerance – Normal care required.
COARSE tolerance – Minimal care required, lowest cost.

Tolerance	A	B	C	D	E	F	G	H	
FINE	0.010	0.010	0.010	0.005	0.005	0.010	0.005	0.010	in./in.
									mm/mm
COMMERCIAL	0.015	0.020	0.015	0.010	0.010	0.015	0.010	0.015	in./in.
									mm/mm
COARSE	0.030	0.030	0.020	0.015	0.015	0.020	0.015	0.020	in./in.
									mm/mm

The shuttle type (Fig. 91.2) machine is appropriate for molding larger parts. A frame for holding one mold is mounted on a movable bed. Incorporated in the bed is the drive motors for turning the mold biaxially. The bed is on a track that permits the mold and bed to move into and out of the oven. After the heating cycle is complete, the mold is moved into the cooling station, which is not normally enclosed. A duplicate bed with mold is then sent into the oven from the opposite end.

The clamshell machine (Fig. 91.3) is the newest model on the market. It includes an enclosed oven that also serves as the cooling station. It utilizes only one arm; the heating, cooling, and loading/unloading stations are all in the same location.

When a long cycle time is not a consideration, an open-flame machine can be used to keep initial equipment costs low. In this case, the mold is rotated on a single axis. After the part is formed, excess resin is dumped out and the flame is turned off to allow for air cooling.

Machine maintenance is extremely important. A weekly maintenance inspection as well as a good inventory of spare parts can ensure fast and efficient repair work in case of emergency.

Standard Tolerance Chart

PROCESS _Rotational Molding_ _____   MATERIAL _Poly(vinyl chloride)_ _____

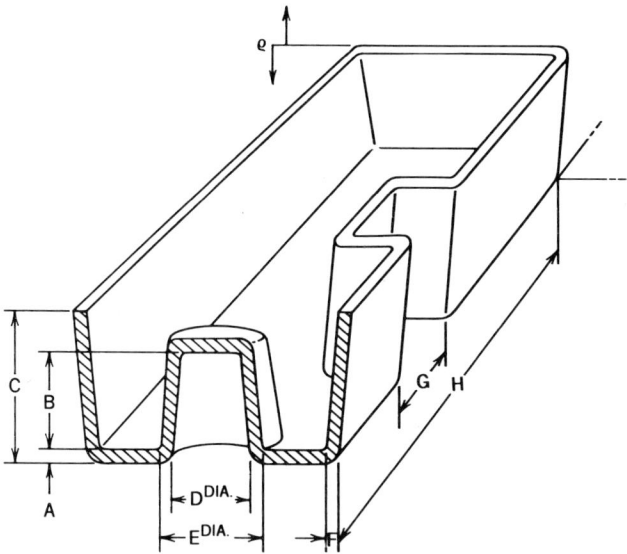

Values are based on a nominal 1/8 in. (3.2 mm) wall thickness and given in plus or minus in./in. and mm/mm.

FINE tolerance – Special care and added cost.
COMMERCIAL tolerance – Normal care required.
COARSE tolerance – Minimal care required, lowest cost.

Tolerance	A	B	C	D	E	F	G	H	
FINE	0.015	0.015	0.015	0.010	0.010	0.015	0.010	0.015	in./in.
									mm/mm
COMMERCIAL	0.020	0.025	0.020	0.015	0.015	0.020	0.015	0.030	in./in.
									mm/mm
COARSE	0.025	0.035	0.025	0.020	0.020	0.025	0.030	0.025	in./in.
									mm/mm

Several types of mold-filling devices are used. Although they can be automated, manual filling is more common. All that is needed is an accurate weighing device and a container for each of the resins to prevent cross-contamination. Some resins, such as nylon, require a nitrogen atmosphere in the mold. This is accomplished by routing the nitrogen through a channel in the spindle and connecting it to the mold with a rotary hose.

If the part has a large section where no resin buildup is desired, the mold can be shielded by an insulating material such as Teflon.

Rotational molds are very durable but do require a mold maintenance program. Components of a mold maintenance program are

• Proper cleaning with nonabrasive tools.
• Proper handling.
• Proper storage. Fabricated steel molds should be oiled to prevent rusting when they are not in use.

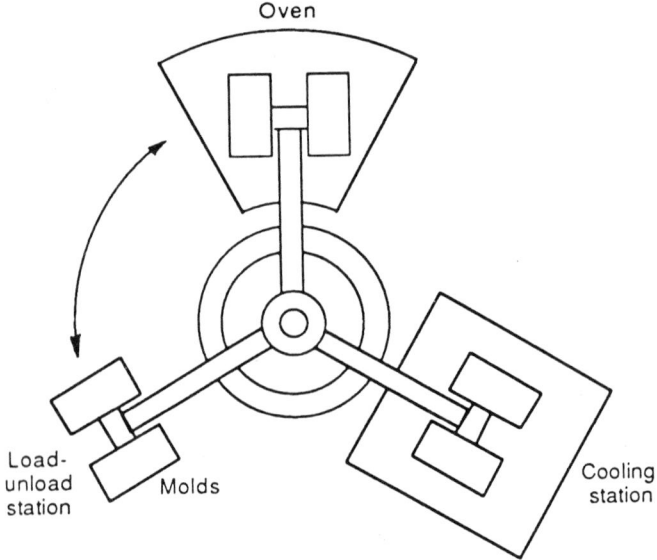

Figure 91.1. Example of a carousel machine.

91.8 TOOLING

Molds used in rotational molding are among the easiest and least expensive to fabricate. Two-piece molds are standard, but three-piece molds are sometimes required for proper finished part removal. Molds can be a simple, round object shape or a complex design with undercuts, ribs, and tapers.

A skilled person can fabricate sheet metal into a rotational mold as long as the part is not complex. In rotational mold design, it is important to consider:

- Heat transfer of the mold material.
- How is it to be mounted.
- What type of parting lines are needed.
- What type of clamping mechanisms are required.
- What type of venting is required.
- How well the material will stand up to the anticipated number of heating and cooling cycles.

Following are the chief mold types utilized.

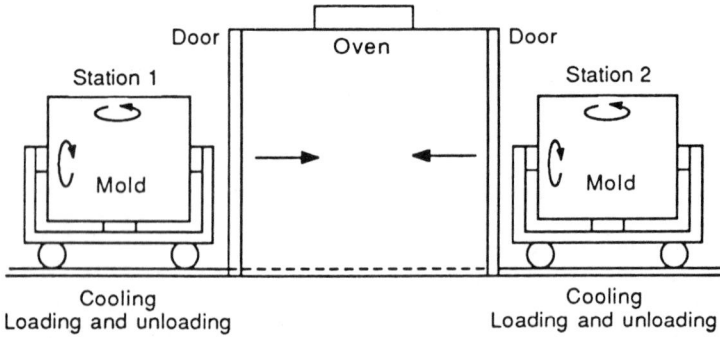

Figure 91.2. Example of a shuttle machine.

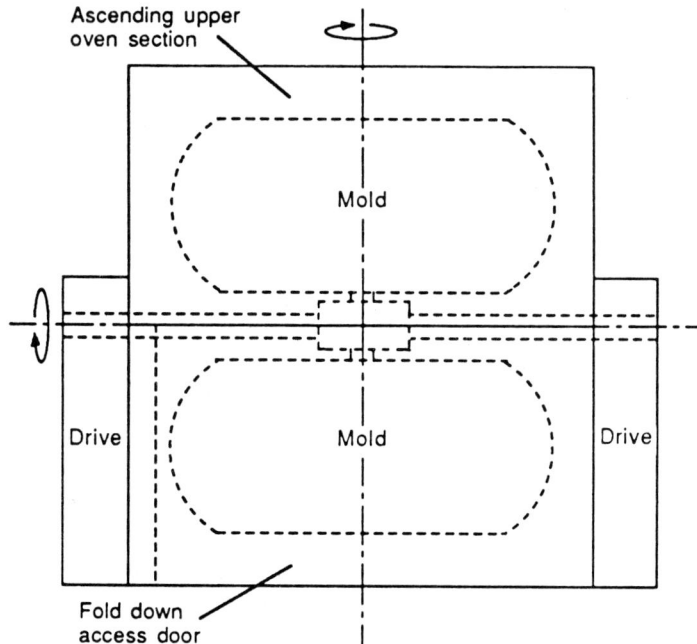

Figure 91.3. Example of a clamshell machine.

91.8.1 Cast Aluminum

Cast aluminum molds are by far the most popular and most frequently used mold type. Most small- to medium-sized parts are made with a cast mold. Cast aluminum has good heat-transfer characteristics and is cost effective when several molds of the same part are required. A drawback to cast molds is that they are porous and can be damaged easily.

91.8.2 Sheet Metal Molds

Sheet metal molds are used for larger parts. They are easy to fabricate; in many cases, mold sections simply have to be welded together. Sheet metal molds are cost effective when larger single molded parts are required.

91.8.3 Other Molds

Electroformed nickel molds create a part with very fine detail. Vapor-formed nickel molds will also give very fine detail but are more costly.

All molds used in rotational molding share the following characteristics:

Parting Lines. Because all molds have two or more sections, good parting lines are essential to achieving proper fit of the mold sections, minimize flash, and to provide proper formation of the finished part. Figure 91.4 shows parting lines commonly used for rotational molds.

Mold Mounting. Molds must be able to be mounted on the spindle or arm of the rotational molder. Large sheet metal molds are easily mounted by bolts or simple clamping systems. Cast aluminum molds utilize several small- to medium-sized molds mounted on the same spindle or arm.

A part called a "spider" is used to mount several cast aluminum molds. The spider consists of several arms or mounting legs to which each mold is attached, normally by bolts. In turn, the spider has one central mounting location that attaches to the machine spindle

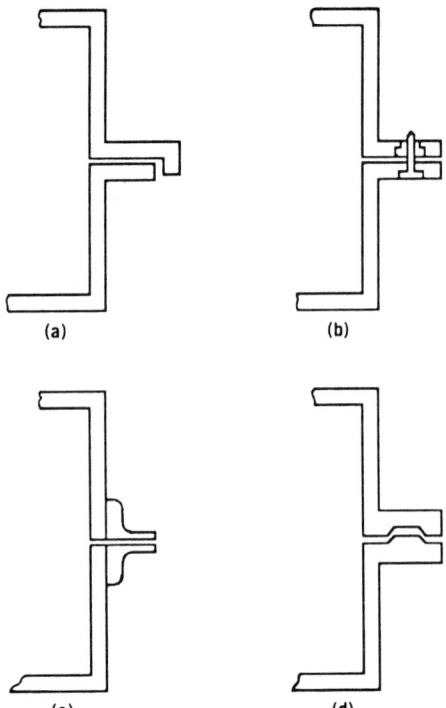

Figure 91.4.   Parting lines for rotational molds. **(a)** stepped, **(b)** pen and bushing, **(c)** fabricated, **(d)** tongue and groove.

by bolts. Use of the spider enables two or three dozen cast molds to be mounted on one central structure. The entire spider or just one or two large cast molds can be removed easily by a forklift or crane. This is important in typically short production runs involving a variety of parts. Figure 91.5 shows molds mounted on a straight arm and an offset arm.

Clamping Systems.   The best parting line cannot function properly without a good clamping system. The most common clamping system for small- to medium-sized parts is the "C" vise clamp, which can be purchased in any hardware store. Another popular clamping system is spring loaded clamps, which are welded onto the sections of a mold. On larger machines, nuts and threaded bolts are used; installation and removal is accomplished using an air gun (impact wrench).

Venting System.   Most rotational molds require a venting system to release the gas that builds up in the heating cycle. Vents range from 1/8 to 2 in. (3.1 to 51 mm) inner diameter, depending on the size of the mold. A good rule of thumb is 1/2-in. (13 mm) for each cubic yard (0.76 m^3) of volume in the part.

   Since vents will leave a hole in the molded parts, correct placement of them is essential. They should be located in an area that may be cut out of the finished part or where a patch will not detract from the appearance of the part. Vents must also be placed where water will not enter the part during the cooling cycle. If this happens, a "watertrack" mark will form on the hot resin, and this may be unacceptable in the finished product. Improper venting can also cause as blow holes in the parting line.

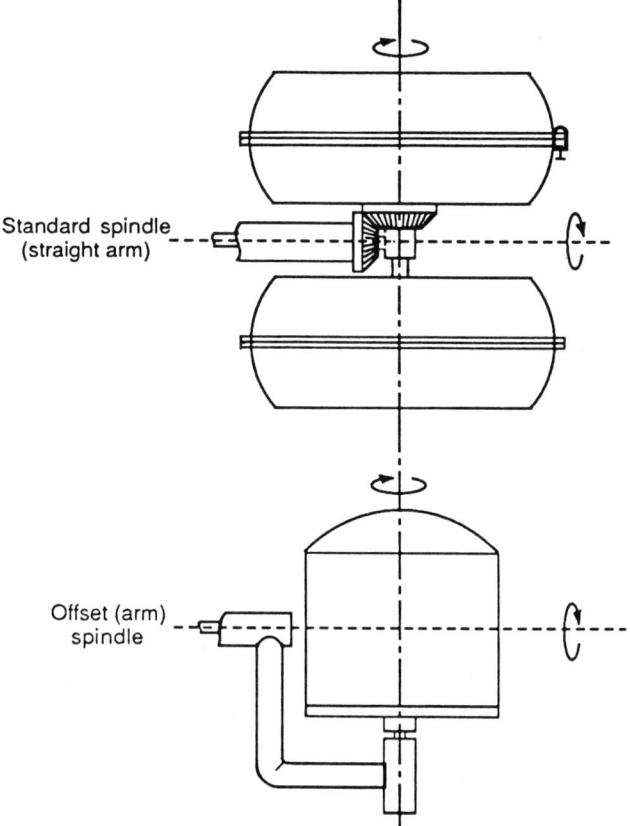

Figure 91.5. Molds mounted on a straight arm and an offset arm.

91.10 FINISHING

Most parts produced by rotational molding are complete products as they come out of the mold. However, in some cases, one or more secondary operation may be required.

Holes can be drilled automatically or manually. Additional fittings can be added by spin welding them onto the part. Sections of parts may be joined together by hot bar, hot gas, or ultrasonic welding. If a part is to be painted, the surface of most resins needs to be flame treated to ensure proper adhesion. Decals may be applied with a good adhesive.

As in all plastic processing, secondary operations are time consuming and should be kept to a minimum whenever possible.

91.12 COSTING AND INDUSTRY PRACTICES

Comparison of rotational molding parts on a piece part basis is difficult in that costs are related to such variables as resin type, application, and part quantity. In general, a rotationally molded part should be priced at five times the resin cost. For example, if 80 lb of resin at $0.50 per pound is used to produce a 250-gallon (0.95 m³) tank, the finished part should cost $40, plus the cost of any additional postmolding operations.

Resins costs for rotational molding cost are slightly high because of the need for additional stabilizers. Grinding costs about 5–7 cents per pound. Rotational molding machines range in cost from $40,000 for a small machine to $200,000 for large machines. If it is desirable to grind the resin in-house, the grinding system will cost $38,000–$70,000.

In comparison to blow molding, rotational molding offers the advantages of lower costs for equipment and molds and very broad design parameters. Blow molding offers a faster cycle time and many more resins to choose from.

In comparison to thermoforming, rotational molding offers lower mold costs and more design freedom. Thermoforming offers lower equipment costs, slightly shorter cycles, and have a better resin selection.

91.12.1 Safety

The rotational molding industry is a relatively safe industry.

Most of the resins used are nonhazardous. Suppliers providing resins that can cause some mild irritation will make available complete safety information on their handling and use.

Since most resins used for rotational molding are ground to 35 mesh, proper housekeeping procedures should be followed to reduce loose powder, thereby preventing possible dust explosion and fire.

A rotomolder is a relatively safe piece of processing equipment. Moving parts such as fans are enclosed with guards. Spindles and arms operate at low speed. The oven has a high limit switch to prevent oven temperature from exceeding the desired setpoint.

91.15 INDUSTRY ASSOCIATIONS AND INFORMATION SOURCES

The rotomolding industry is small. It is represented by the Association of Rotational Molders (ARM), 435 North Michigan Ave., Suite 1717, Chicago, IL 60611-4067; (312) 644-0828.

ARM is also the major information-disseminating organization in the rotational molding industry, having compiled a comprehensive library and the industry's first design manual. The Rotational Molding Development Center (RMDC) at the University of Akron was founded in 1986 to provide for the industry's future research needs.

92

STRUCTURAL FOAM

Stephen Ham

Cashiers Structural Foam Div., Consolidated Metco., Inc., P.O. Box 37, U.S. Hwy. 64, Cashiers, NC 28717

92.1 DESCRIPTION AND HISTORY OF PROCESS

Cellular or foam structures abound in nature. Animal bones and wood are examples demonstrating that a cellular structure provides an optimum strength-to-weight ratio. It was only logical that the plastics industry would eventually develop the technology to produce a foamed plastic with optimum strength. Pioneering efforts can be traced back to work in the 1930s with urethane foams.

Thermoplastic structural foam evolved from the injection-molding process. Its initial commercialization dates from the mid-1960s, to the efforts of Union Carbide Corporation and other material suppliers. Even at that time, foam technology was not new. It had been the practice to add small amounts of bicarbonate of soda or acetic acid to injection-molded materials to produce tiny bubbles in the center of thick sectioned parts. The expanding bubbles helped reduce surface depressions or sink marks.

The idea behind structural foam is to produce a thicker but less dense part with the goal of achieving more stiffness through increased wall sections. A standard beam deflection formula shows that doubling the wall thickness results in eight times the rigidity. In structural foam, the actual value is closer to three to four times the rigidity. This is the result of the lower density of material and the surface condition of the part, both of which are described later in this section.

It is important to understand that structural foam is a low-pressure [200–500 psi (1.4–3.4 MPa)] process, regardless of the material, equipment, wall thickness, or blowing agent used. The low-pressure pressure condition must occur inside the mold at some point to allow the foaming to occur. The basic process involves four steps:

- A premeasured "short shot" of extruded thermoplastic resin melt with a blended blowing agent is injected into a mold clamped at approximately 1/4 ton/in.^{2} (1400 kg/mm^2). The blowing agent is compressed during this stage (Fig. 92.1).
- Since this premeasured shot is less than the volume of the mold cavity, there is a pressure drop inside the mold that allows the blowing agent to expand (Fig. 92.2).
- The expansion pressure of the blowing agent [200–500 psi (1.4–3.4 MPa)] fills out the mold cavity (Fig. 92.3).
- The mold goes through a cooling cycle to allow the resin temperature to drop below

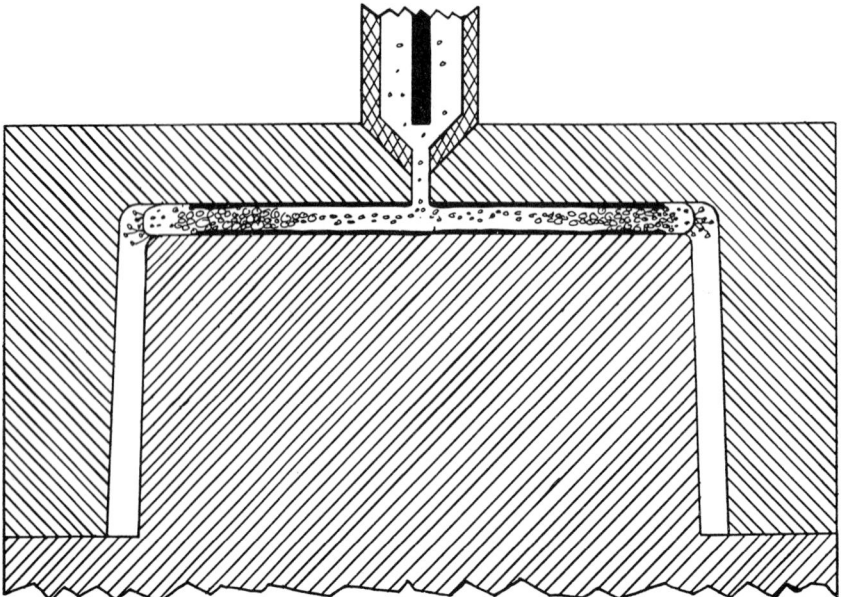

Figure 92.1.   Structural foam molding, step 1.

its distortion temperature. At this point, the mold is opened and the part is ejected from the mold. The process is then repeated. Low pressures inside the mold result in lower molded-in stresses. This is another benefit of the process.

92.2   MATERIALS

See the Master Process Outline.

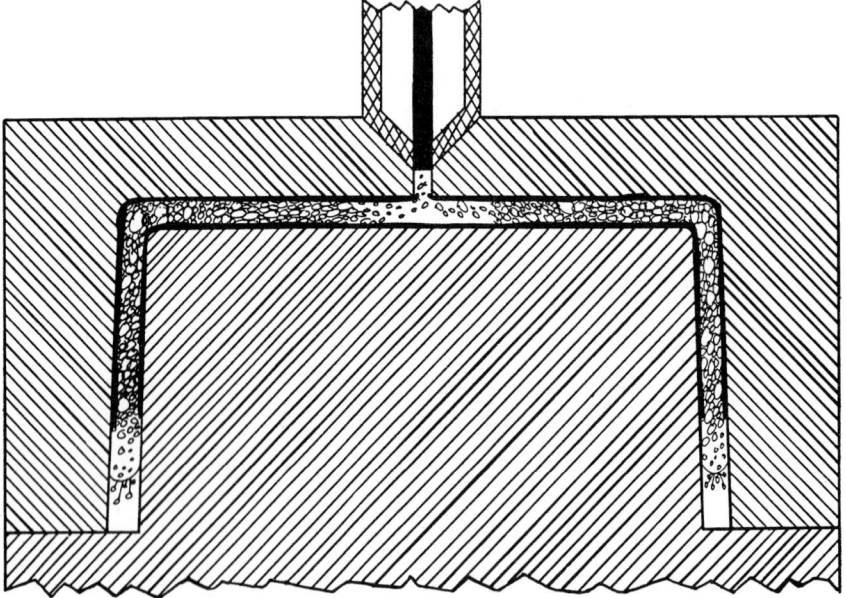

Figure 92.2.   Structural foam molding, step 2.

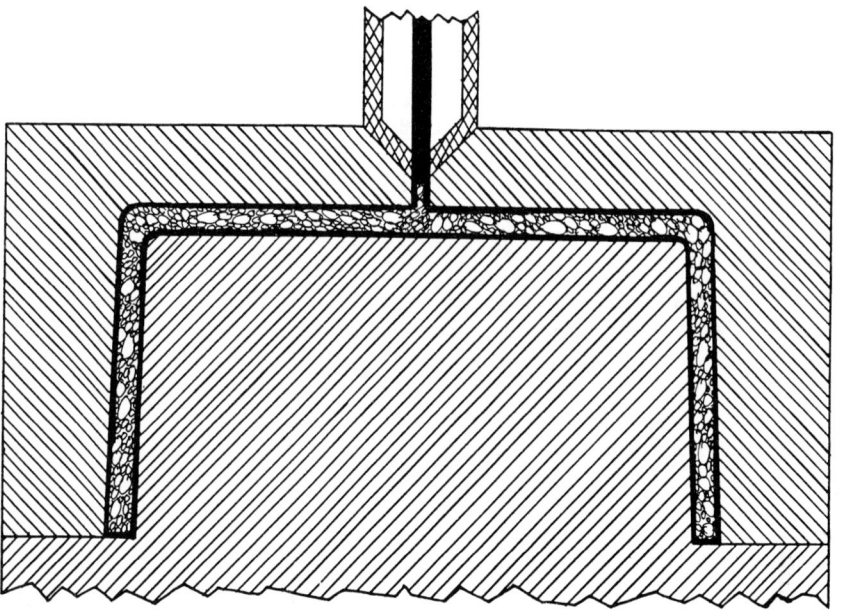

Figure 92.3. Structural foam molding, step 3.

MASTER PROCESS OUTLINE

PROCESS Structural Foam

Instructions:
For "ALL" categories use
For "EXCEPT FOR" categories use

Y = yes
N = no
D = difficult

☐ **ALL Thermoplastics**
Except for:

D	Acetal	Y	Polyethylene
Y	Acrylonitrile–Butadiene–		Poly(methyl methacrylate)
	Styrene	D	Poly(methyl pentene) (TPX)
	Cellulosics	D	Poly(phenyl sulfone)
D	Chlorinated Polyethylene		Poly(phenylene ether)
D	Ethylene Vinyl Acetate	Y	Poly(phenylene oxide) (PPO)
Y	Ionomers	D	Poly(phenylene sulfide)
	Liquid Crystal Polymers	Y	Polypropylene
D	Nylons		Polystyrene
D	Poly(aryl sulfone)	D	Polysulfone
	Polyallomers	D	Polyurethane Thermoplastic
D	Poly(amid-imid)	D	Poly(vinyl chloride) (PVC)
	Polyarylate	D	Poly(vinylidine chloride) (Saran)
D	Polybutylene		Rubbery Styrenic Block
Y	Polycarbonate		Polymers
	Polyetherimide		Styrene Maleic Anhydride
	Polyetherketone (PEEK)		Styrene–Acrylonitrile (SAN)
	Polyethersulfone	D	Styrene–Butadiene (K Resin)
		D	Thermoplastic Polyesters
			XT Polymer

N **ALL Thermosets**
Except for:

Allyl Resins
Phenolics
Polyurethane Thermoset
RP-Mat. Alkyd Polyesters
RP-Mat. Epoxy
RP-Mat. Vinyl Esters
Silicone
Urea Melamine

D **ALL Fluorocarbons**
Except for:

Fluorocarbon Polymers-ETFE
Fluorocarbon Polymers-FEP
Fluorocarbon Polymers-PCTFE
Fluorocarbon Polymers-PFA
Fluorocarbon Polymers-PTFE
Fluorocarbon Polymers-PVDF

92.3 PROCESSING CHARACTERISTICS

92.3.1 Blowing Agents

Almost every thermoplastic can be and probably has been "foamed," either in daily production or laboratory testing. The blowing agents used range from inert gas (called the physical process), to heat-activated concentrates (the chemical process). These are added to the molding resin during the melt stage.

The inert gas most commonly used is nitrogen. The early purpose-built, multiple-nozzle equipment utilized compressed nitrogen gas both for the blowing agent and for the injection pressure. As a blowing agent, it is injected into the barrel when the resin has reached the melt stage. Nitrogen gas is a very inexpensive blowing agent, but it is limited by rather poor gas dispersion and the need for a lengthy outgassing period of up to 14 days prior to painting.

Chemical blowing agents (CBA) were developed to allow the use of high-pressure injection machines for molding structural foam parts. These CBAs are heat activated concentrates that are added to the resin prior to the plastication step. They are available in several forms. Liquid dispersions are pumped directly into the feed throat of the barrel through a metering device. Powders and pelletized CBAs can be pure blended (tumbled) with the resin prior to plastication. Pelletized CBAs can also be metered in directly at the feed throat. Flake and bar stock forms of CBA requiring metering equipment are also available.

Generally, CBAs result in better surface finishes and improved physical properties because of better gas dispersion and thus improved cell structure. CBAs also require less outgassing time. These blowing agents are now used extensively on purpose-built, low-pressure machines as well as converted, high-pressure injection-molding machines.

All thermoplastic resins shrink to some degree as the resin temperature drops during the cooling sequence of the molding cycle. This shrinkage causes a buildup of stress. Part design is a critical factor in minimizing and controlling the effect of this material shrinkage. The initial packing pressure inside the mold contributes to these stresses. With structural foam, the cellular core offers additional stress absorption during the resin cooling sequence.

The benefits of low-stress parts include:

- Greater allowable wall thickness variations.
- Reduction of sink marks on appearance surfaces opposite ribs and bosses.
- Less warpage and improved dimensional tolerances.
- Improved chemical resistance.

These benefits of low-stress molding make parts consolidation a significant advantage of the structural foam process. Several parts can be combined into one molded part, thereby eliminating the assembly hardware and greatly reducing the required assembly labor. Parts consolidation has become almost an art form in the business machine industry in recent years where short product cycles have prompted continual exploration of structural foam's design flexibility.

In addition to low-stress parts, low-pressure molding offers other benefits. Less clamp tonnage is required on the molding equipment, which results in lower hourly machine rates and lower overall part cost. Also, lower in-mold pressures result in less costly molds. A general rule of thumb for determining the required clamping pressure is 1/4 ton/in.^{2} (1400 kg/mm^2) of projected surface area.

92.4 ADVANTAGES/DISADVANTAGES

The major drawback of structural foam is the surface finish of the molded part. While the blowing agent expansion provides low pressure molding and low stress parts, it also causes a surface condition referred to as "swirl." Swirl is the result of gas escaping from the leading edge of the material flow during the injection sequence. Figure 92.4 illustrates this condition.

As noted, structural foam offers many benefits as a processing technique of thermoplastics. High stiffness-to-weight is a natural benefit of a honeycomb structure. Very large parts can be produced using low clamp pressure by means of multiple injection nozzle molding equipment with oversized platens, made possible by the low pressure nature of structural foam.

The most important benefit, however, is the low level of residual stresses found in structural foam parts. Residual stresses are those forces in the internal molecular structure of the part that are present in the absence of external loading. While differences in thermal expansion constants, the inhomogeneity of shear, and elastic anisotropy in multiphase materials will result in minor amounts of stress, the major source of stress is nonuniform flow during processing and, to a lesser extent, nonuniform cooling. The effects of high levels of stress are nonuniform shrinkage, warpage, sink marks, loss of chemical resistance, and poor dimensional tolerances. The designer must be aware of the potential of residual stresses. For instance, a uniform wall thickness contributes to an even flow and therefore lower residual stresses. The designer must also adjust for differences in processes when specifying dimensional tolerances.

Structural foam exhibits much lower levels of residual stress than solid injection-molded parts for two reasons. First, the low filling pressure permits a more relaxed flow. Second, the inner layer of foam in the composite profile becomes a stress equalizer. The result of low residual stress is the ability to mold dimensionally accurate parts with less tendency to incur sink marks on the surface opposite ribs and bosses. The combination of these two factors makes structural foam a natural for combining several parts into one, reducing final assembly costs.

92.5 DESIGN

See the Standard Process Design Chart.

92.6 PROCESSING CRITERIA

92.6.1 Farrel/USM High-Pressure Structural Foam

United Show Machinery (USM) patented an expanding mold process that is a unique combination of structural foam and injection molding. The equipment utilized is a high-pressure reciprocating-screw injection-molding machine.

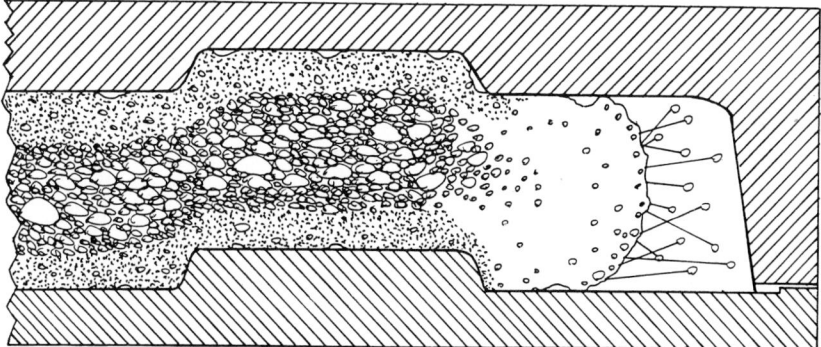

Figure 92.4. Causes of swirl.

STRUCTURAL FOAM (Low Pressure Injection)
STANDARD <u>PROCESS</u> DESIGN CHART
RECOMMENDED INSTRUCTIONS FOR WRITING PROCESS-RELATED DESIGN DETAILS

1. Recommended wall thickness/length of flow
   Minimum ___0.187___ for ___12 in.___ Distance
   Maximum ___0.312___ for ___28 in.___ Distance
   Idea ___0.250___ for ___20 in.___ Distance

2. Allowable wall-thickness variations ___100___ percent of nominal wall

3. Nominal wall-thickness tolerance
   Parallel to parting line ___0.002 at 1/4___ ton square inch
   Perpendicular to parting line ___0.004___

4. Radius requirements
   Outside: Minimum ___0.343___ Maximum ___0.406___
   Inside: Minimum ___0.093___ Maximum ___0.156___

5. Reinforcing ribs
   Maximum thickness ___0.75–1.00 wall thickness___
   Maximum height ___8 times wall___
   Sink marks: Yes ___ No __X__

6. Solid pegs and bosses
   Maximum thickness ___.75–1.00 × wall thickness___
   Maximum height ___6–8 × wall thickness___
   Sink marks: Yes ___ No __X__

7. Strippable undercuts
   Outside: Yes __X__ No ___
   Inside: Yes __X__ No ___

8. Molded-in threads
   Outside: Yes __X__ No ___ ultrasonic helicoils better
   Inside: Yes __X__ No ___ ultrasonic helicoils better

9. Hollow bosses
   Yes __X__ No ___

10. Holes
    Perpendicular to parting line: Yes __X__ No ___
    Parallel to parting line: Yes __X__ No ___
    Through holes: Yes __X__ No ___
    Blind: Yes __X__ No ___

11. Draft angles
    Outside: Minimum ___1°___ Ideal ___3°___
    Inside: Minimum ___1°___ Ideal ___3°___

12. Trimming required
    Yes ___ No __X__

13. Finished both sides
    Yes __X__ No ___

14. Finishing requirements
    _____

15. Molded-in inserts
    Yes __X__ No ___

16. Size limitations
    Length: Minimum ___ Maximum ___ Over 7 ft—limitation is the size of the machine
    Weight: Minimum ___ Maximum ___

17. Tolerances
    Note: Similar to SPI tolerance charts attached.

18. Special process- and material-related design details
    (A) Blow-up ratio _____ blow molding
    (B) Core L/D _____ injection and compression molding
    (C) Depth of draw _____ thermoforming
    (D) Etc.

The first step is to inject under very high pressure a molding material with preblended chemical blowing agent. In this step the mold is packed out under high pressure just like injection molding. The high in-mold pressure prevents blowing agent expansion, and, therefore, no swirl forms on the part surface (Fig. 92.5).

After an appropriate time has passed, the mold is expanded, increasing the volume of the mold cavity, which, in turn, causes a pressure drop. The pressure drop allows the blowing agent to expand within the integral skins that are already formed (Fig. 92.6).

There is a common misconception about the term "high-pressure structural foam." The term describes the injection pressure of the machine rather than the actual pressure inside the mold. The USM process comes the closest to being a true "high-pressure structural foam" process, but even it requires a low pressure stage to allow the foaming to occur.

In addition to the swirl-free surface finish, this process has the capability to greatly reduce part density, if required. The foaming can also be limited to just certain areas of the part by designating only those areas of the mold to expand. Some of the limitations include increased tooling costs and design limitations imposed by the mold expansion. Since higher clamp tonnage is required of the molding machine, somewhat higher unit costs can also be expected.

92.6.2 Coinjection Structural Foam

Coinjection structural foam, as the name would imply, is the nearly simultaneous injection of two dissimilar but compatible materials. The goal is to produce a part with the foam core encapsulated within a solid skin. Central to this technology is the molding equipment, which is a highly developed hybrid: half is a structural foam machine, half an injection-molding machine. Two separate barrels are utilized. One barrel produces the skin component, which is a material without a blowing agent. The other barrel provides the foam material with a chemical blowing agent. The blowing agent is heat activated but remains compressed under pressure. In some cases a low percent of blowing agent is utilized in the skin component as well, with only slight impact on surface finish. Injection pressure is provided by reciprocating screws that feed a special two-channel nozzle. The nozzle provides two concentric flow

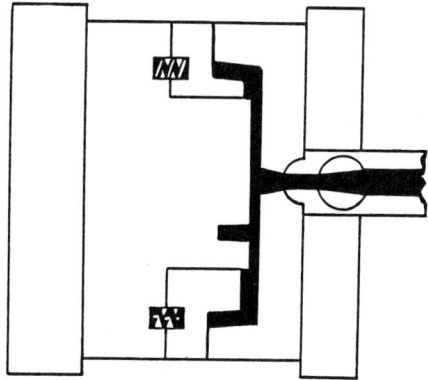

Figure 92.5 High-pressure stage before mold expansion.

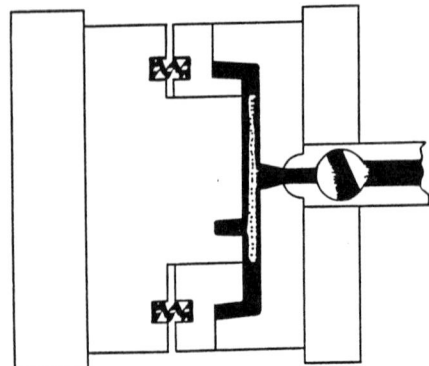

Figure 92.6 Low-pressure stage after mold expansion.

channels for the respective materials, thereby allowing the skin to encapsulate the core. The hydraulic nozzle is electronically controlled so as to allow for the variable injection velocity of each flow channel.

Figure 92.7 shows the first phase of coinjection structural foam, injection of solid skin to form the leading edge. Figure 92.8 shows the second phase, which follows almost simultaneously with injection of solid skin and a "short shot" of foamed core. The solid skin leading edge becomes smaller as it is stretched to the end of the flow. The final phase (Fig. 92.9) is the cut-off of the injection nozzle at a predetermined time when the mold cavity is less than completely full. The foamed core responds to the pressure drop by expanding to fill out the mold cavity. The leading edge of solid skin prevents the foam from breaking through, thereby effecting encapsulation.

It is noteworthy that while the materials are not mixed the core material is completely distributed. One of the unique features of the process is the ability to utilize different materials for the skin and core, an advantage when one material cannot provide all the required properties. It is also possible to use an inexpensive core material and a prime material for the skin, but this is not advisable for applications requiring Underwriters Laboratories approval. Table 92.1 identifies compatible materials for coinjection.

The major advantage of the process is the surface finish, which resembles that of injection molded parts when an injection grade material is used. The foamed core provides the advantages of low pressure structural foam.

92.6.3 Gas Counterpressure Structural Foam

The final swirl-free mold technique is gas counterpressure structural foam. The concept behind gas counterpressure is to utilize a gas-pressurized mold, which, through controlled

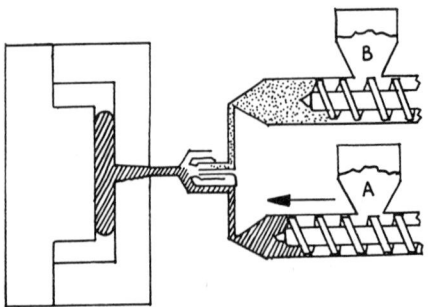

Figure 92.7. Injection of solid skin.

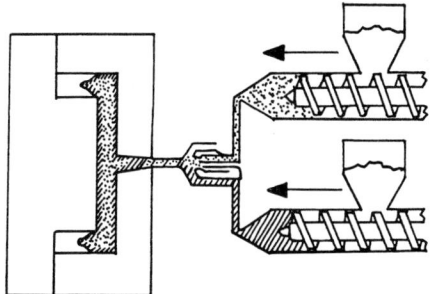

Figure 92.8. Coinjection of solid skin and foamed core.

venting, allows the foam expansion stages of the cycle to occur after a smooth surface has been formed. The purpose is to improve the molded surface finish by preventing the swirl formation. The counterpressure keeps the blowing agent compressed until solid skins are formed. Numerous development efforts, dating from the early 1970s, have yielded several different counterpressure technologies in Eurpoe, the United States, and Japan.

The basic concept requires a pressure-tight mold to be charged with gas counterpressure prior to the injection of plastic. In order to maintain pressure, the mold parting line, ejectors, moving slides, and nozzle seat must be sealed by some means. The sealing techniques vary considerably.

When the proper in-mold counterpressure is obtained, a predetermined short shot containing a dispersed and compressed blowing agent is injected into the mold. During the injection sequence, controlled venting should occur to allow consistent counterpressure as the unfilled mold volume is being reduced. The counterpressure prevents the blowing agent from expanding, thus eliminating swirl, which is normally caused by the gas bubbles being trapped between the molded surface and the skin of the part. Many techniques have been used successfully.

At a predetemined time, the counterpressure in the mold is released through exhaust vents. This venting process causes a pressure drop inside the mold cavity, thereby allowing an instantaneous expansion of the blowing agent, which is now encapsulated by solid skin. Normal mold cooling and part ejection cycles follow, and the sequence is repeated again.

The parts produced by gas counterpressure typically show 5–10% less density reduction than parts produced by conventional, low-pressure structural foam. This results from the formation of thicker skins. Gas counterpressure structural foam parts, which more dense, do display more uniform cell structure throughout the part, which improves physical properties.

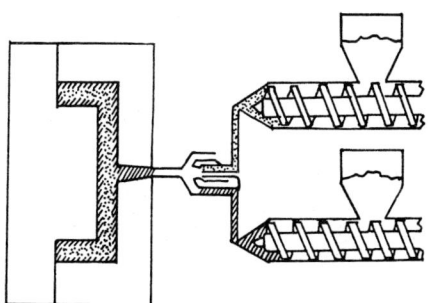

Figure 92.9. Blowing agent packing the mold.

TABLE 92.1. Compatibility of Materials for Coinjection[a]

Materials	ABS	Acrylic Ester Acrylonitrile	Cellulose Acetate	Ethyl Vinyl Acetate	Nylon 6	Nylon 6,6	Polycarbonate	HDPE	LDPE	Poly(methyl methacrylate)	Polyoxymethylene	PP	PPO	General-purpose PS	High-impact PS	Poly(tetramethylene terephthalate)	Rigid PVC	Soft PVC	Styrene Acrylonitrile
ABS		+	+	+			+	−	+	+		−		−	−	+	+	0	+
Acrylic ester acrylonitrile	+		+	−											0				+
Cellulose acetate	+	+		−										+				0	
Ethyl vinyl acetate	+	−	−		+	+		+	+			+		−			+		
Nylon 6				+				−	−			−							+
Nylon 6,6				+				−	−			−							
Polycarbonate	+													−	0				+
HDPE	−			+	−	−			+	−	0	−							
LDPE	+			+	−	−		+		+	−	+							
Poly(methyl methacrylate)	+							−	+		+	−		0	0		+	+	+
Polyoxymethylene								0	−	+									
PP	−			+	−	−		−	+	−			+	+	+		−	−	−
PPO												+		+	+				−
General-purpose PS	−		+	−			−			0		+	+						−
High-impact PS	−	0					0			0		+	+						−
Poly(tetramethylene terephthalate)	+																		+
Rigid PVC	+			+						+		−		−	−			+	+
Soft PVC	0		0					−		+		−		−	−		+		+
Styrene acrylonitrile	+	+			+		+		+	+	−	−	−	−	+		+	+	

[a] + = good adhesion. − = poor adhesion. 0 = no adhesion. Blank indicates no recommendation (combination not yet tested).

Source: Battenfeld

TABLE 92.2. Gracek Data on Percent Improvement of Counterpressure Over Conventional Structural Foam

Test	Copolymer Polypropylene, %	F.R.- Polystyrene, %	Mod-PPO, %	Polycarbonate 5% Glass
Instrumented Impact	40	20	23	59
Tensile strength	15	12	16	19
Tensile modulus	18	8	17	24
Flexural strength	12	11	10	28
Flexural modulus	11	11	11	21

Two independent tests have been published comparing the physical properties of counterpressure structural foam and conventional structural foam. In the first, Walter Gracek performed tests on specimens cut from actual molded parts in conjunction with his masters thesis at the University of Lowell. His data were presented by Greg Koski at the 1984 Structural Foam Conference sponsored by the Society of the Plastics Industry (SPI). The other testing was performed by General Electric Plastics and presented by Robert Johnson at the 1985 SPI Structural Foam Conference. Johnson's tests were performed on molded test plaques in accordance with ASTM procedures.

The primary differences between the data of these two studies relates to the differences between the samples—molded plaques and actual molded parts cut into plaques. All specimens are based on 0.250 in. (6.4 mm) wall thickness.

The Gracek data is summarized as follows in Table 92.2. It identified no difference in heat-deflection temperatures.

The Johnson data is summarized in Table 92.3. Here, the impact data represents the worse case situation as the affected area was the end opposite the sprue. Johnson found no effect on chemical resistance or Underwriters Laboratories flammability ratings. Johnson pointed out the design implications of the thicker skins of the counterpressure technique, their effect on moment of inertia calculations, and, hence, the greater load bearing capability of counterpressure parts at the same wall thickness as conventional structural foam parts.

TABLE 92.3. Summary of the Johnson Data

Material	S.F. Process:	Flexural Modulus psi (MPa)	Elongation, %	Notched Izod Impact ft-lb/in.	(J/m)
Polycarbonate 5% glass	Conventional	309,000 (2130)	6.7	28	(1490)
Polycarbonate 5% glass	Counterpressure	337,000 (2320)	8.3	74	(3950)
Polycarbonate no glass	Conventional	275,000 (1900)	11.5	18	(960)
Polycarbonate no glass	Counterpressure	299,000 (2060)	28.5	89	(4750)
Modified PPO	Conventional	289,000 (1990)	8.6	20	(1070)
Modified PPO	Counterpressure	313,000 (2160)	15.3	22	(1170)

The data presented are based on the densities normally expected in production situations. Generally, the counterpressure test specimens were 5% more dense than the corresponding conventional structural foam specimens. Improvements in physical properties exceed the 5% that one might expect.

The gas counterpressure technique requires machines with greater clamp tonnage because of the increased in-mold pressure caused by the gas counterpressure. The gas most commonly used for the counterpressure is nitrogen, but any inert gas will produce the desired results.

The amount of additional clamp pressure required ranges from 25 to 100%, depending on the venting technique used during the injection of plastic into the mold. This also explains why steel molds are generally required for this process. The result of this technology is the ability to produce parts with improved surface finishes in conventional structural foam machines. All that is required is a specially designed mold and a sequentially controlled pressure source. The surface finish is improved to the point where secondary painting can be eliminated in many applications. In highly cosmetic applications, the cost of painting is significantly reduced from one-half to one-third of the finishing cost of conventional structural foam.

The apparent increase in the density of molded parts is offset by improvement in physical properties.

92.7 MACHINERY

Structural foam was initially conceived as a method to produce large thermoplastic parts with the same advantages as the smaller parts produced by injection molding. Much of the equipment technology is the same as injection molding.

In the beginning, structural foam processors used purpose-built, two-stage injection, multinozzle machines characterized by oversized platens, material accumulators feeding built-in multiple-nozzle, hot runner manifolds, low clamp tonnage, and low injection pressure. The advantages were the ability to mold very large parts or several smaller parts simultaneously on the large molding platens.

Over the years, this family of equipment has evolved in response to the latest in structural foam theory. Today's refinements include two-stage injection through multiple gating nozzles, with sequentially controlled nozzles. Much higher injection pressures are available to ensure quick injection of material. Computer-aided design has substantially improved screw efficiency and through put. Digital and programmable controllers manage temperatures, injection and clamp pressures, timing, platen travel, and barrel operation. Figure 92.10 illustrates a multiple nozzle structural foam machine in operation.

Later refinements of structural foam-molding equipment include faster injection speeds to produce improved surface finishes (since less swirl can develop if the injection cycle is shortened). Increased injection speed is also helpful for parts with reduced wall sections where more packing pressure is required. Faster injection speeds, however, tend to produce higher cavity pressures and, therefore, higher molding-in stress.

The second major family of structural foam equipment is that of converted high-pressure injection molding machines. These are single-nozzle, reciprocating machines that were designed and built for injection molding. Two major modifications are necessary for this type of equipment to adequately manufacture structural foam parts. The first modification is a positive shut-off nozzle to allow for the short shot and the immediate pressure drop for foaming to occur. The second modification is installation of a "fast shot" package to the hydraulics of the reciprocating screw injection system. These machines do an adequate job of producing structural foam parts, but a certain inefficiency results from the high clamp tonnage. Structural foam requires about one-eighth of the clamp pressure of injection molding.

The third major category of equipment is a hybrid of the first two: purpose-built, single-

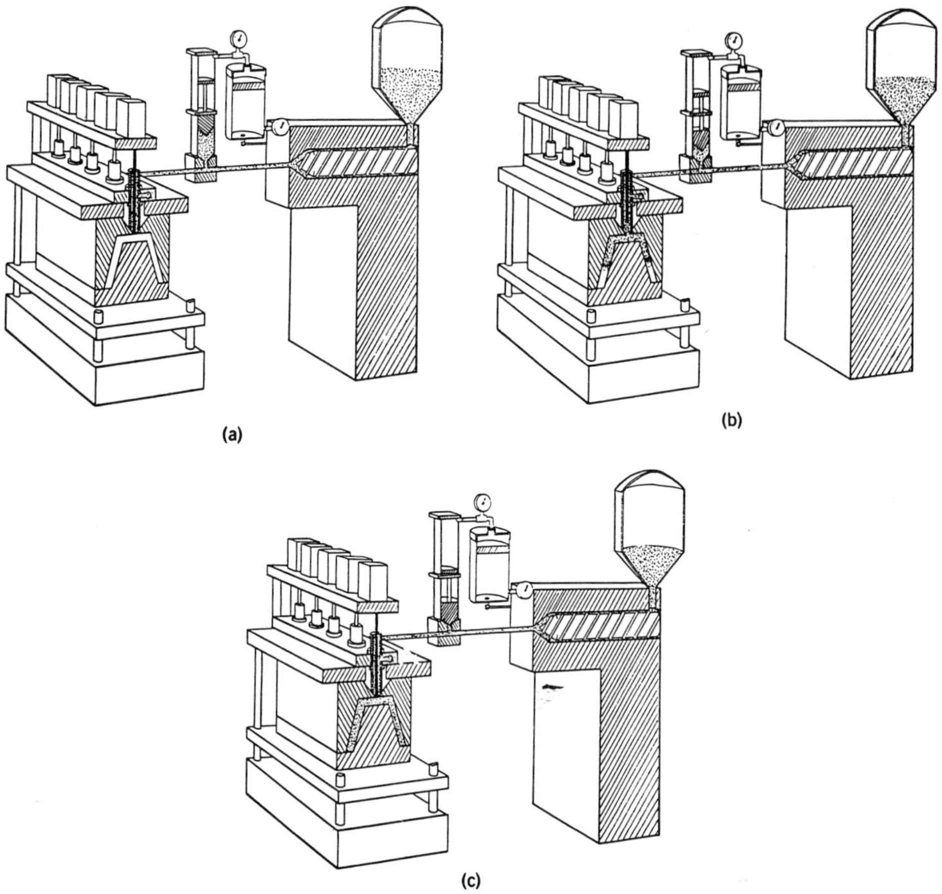

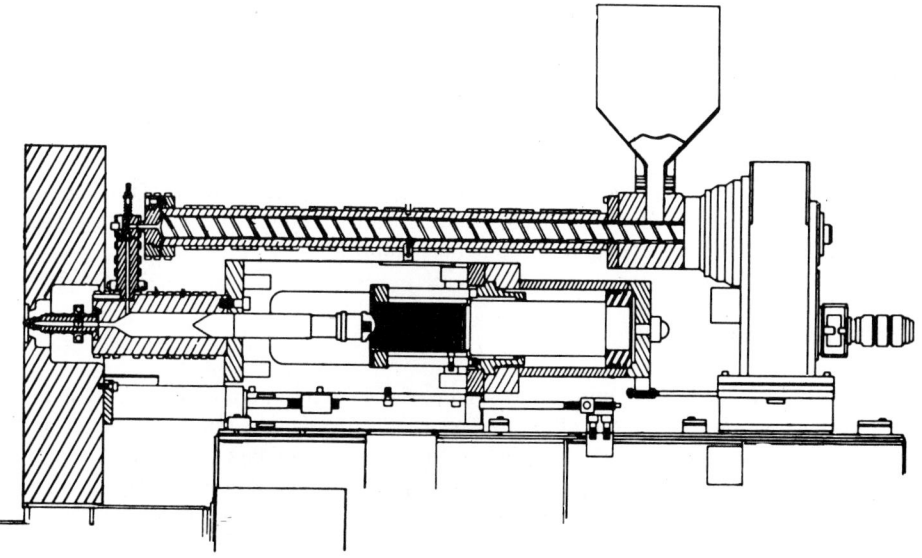

Figure 92.10. (**a**) Preinjection. (**b**) Injection of short shot. (**c**) Blowing agent pressure packs the mold.

Figure 92.11. Two-stage, single-nozzle injection machine.

nozzle, structural foam machines (Fig. 92.11). Generally these are two-stage machines with a material accumulator feeding a single nozzle that has shut-off capability. Fast injection speeds are obtained and oversize platens are utilized.

There are two other process types in very limited usage in the United States. The first is "foundry system" developed by Borg-Warner, in which unheated resin containing blowing agent is poured into a mold. The mold is sealed and then heated to activate the blowing agent. A water shower cooling cycle follows. The second is the TCM process, which involves pre-expanded thermoplastic resin injected into a clamped mold under low pressure. Further information on these is available from SPI.

92.8 TOOLING

The Tooling Checklist in the section on injection molding contains pertinent information for structural foam molding.

92.8.1 Structural Foam Molds

The structural foam mold has a cavity half, which defines the exterior of the part. The plane where these two mold halves come together is the parting line. In addition to defining the part, the mold also directs the material flow and controls the cooling. The mold has a built-in system of ejection that is activated by the moving platen of the molding machine. The part designer must take into consideration such tooling aspects as draft, ejection surfaces, and potential gate locations.

Being a low-pressure process, structural foam can utilize tooling with lighter construction. In the early days of the industry it was very common to use cast aluminum alloy molds cast to shape from a pattern and mounted to a steel mold base. Cooling jackets were either machined or cast. However, cast molds are less desirable because the shrinkage in the casting process combined with the molding shrinkage makes close tolerance parts almost impossible to mold. In addition, the casting process has an inherent porosity, creating imperfections in the surface and limiting the marketability of the parts molded.

Machined aluminum molds have replaced cast aluminum molds in the structural foam industry. The molds have steel bases, guided ejection, and steel inserts in wear areas. Tight tolerances can be machined directly into the mold. The part-to-part repeatability is excellent. Mold surfaces can be polished to achieve smooth ejection and better surface finish on the molded part. These tools are generally considered useful up to 50,000 parts, at which time parting line refurbishment is generally necessary. The swirl marks caused by the blowing agent and low pressure molding do limit the type of molded-in texture that can be defined on the part surface.

High volume applications or molded-in textures generally require more elaborate mold materials such as 4140, P-20, or HR 13 materials. The new swirl-free structural foam molding systems also generally require steel molds to obtain the best possible part surface and to cope with the higher pressures that are sometimes utilized.

92.10 FINISHING

Most structural foam parts require some degree of secondary operations, defined as those steps in manufacturing that occur after molding to enhance the function or cosmetics of the molded part. They generally include such areas as gate (molding appendage) removal, ultrasonic inserting, EMI/RFI shielding, paint preparations, painting, assembly, and packaging. To the extent of gate removal and packaging, virtually every part requires some secondary operation.

Standard Tolerance Chart

PROCESS Structural Foam _____   MATERIAL _____

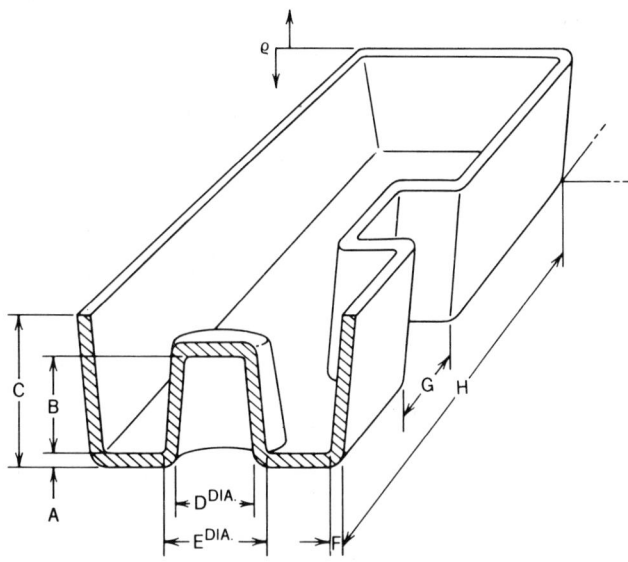

Values are based on a nominal 1/4 in (6.4 mm) wall thickness and given in plus or minus in/in and mm/mm.

FINE tolerance – Special care and added cost.

COMMERCIAL tolerance – Normal care required.

COARSE tolerance – Minimal care required, lowest cost.

Tolerance	A[a]					F[a]			
FINE	0.002	0.001	0.001	0.002	0.003	0.002	0.002	0.001	in./in.
									mm/mm
COMMERCIAL	0.005	0.005	0.005	0.004	0.005	0.005	0.004	0.0015	in./in.
									mm/mm
COARSE	0.020	0.010	0.010	0.004	0.010	0.020	0.004	0.004	in./in.
									mm/mm

[a]Actual tolerance (not in./in.)

The cosmetic requirements will dictate the degree of finish required. Of course, the type and quality of the mold and the actual molding operation itself has a tremendous impact on how the part is finished. Each molding material has its own finishing quirks. Part density determines the skin thickness, which affects the solvent reaction. Chemical blowing agents produce finer cell structure than physical blowing agents (nitrogen gas), thereby greatly reducing post blow problems. Proper finishing of structural foam can be a complicated business. The finisher certainly has to be aware of the molding conditions that produced the part.

Many structural foam parts receive no finish at all. Because as-molded structural foam has a swirled surface, unfinished parts are either not visible or their use does not warrant the expense of a swirl-free surface. Examples of these applications are abundant in internal part assemblies and in such industries as materials handling or public utility enclosures.

The following sequence of events applies to a highly cosmetic application such as a computer housing. Parts with a lesser degree of cosmetic requirement will require fewer of these steps.

In most structural foam applications the gate (molding vestige) must be removed so as not to interfere with the intended purpose of the part. This removal process might be clipping, routing, sawing, or simply breaking by hand. It is common practice to design a mold that produces parts with no gates. An example would be subgating via special ejectors that break off during ejection. Every effort should be made to locate the gates away from the appearance surfaces. This is determined by the mold design.

Mechanical secondary operations, such as ultrasonic inserting or bonding, are performed prior to painting. Structural foam reacts to ultrasonic and solvent bonding similarly to other thermoplastics parts. Ultrasonic bonding is commonly used to install threaded inserts. Structural foam has the ability to produce exceptionally strong bosses. The "hoop" stress normally associated with metal hardware in plastic bosses is absorbed in the foam core. This is also true of structural foam bosses designed for self-tapping (thread forming) screws and other hardware.

Many thermoplastic materials react to a vapor polish operation. This involves suspending the part in a solvent vapor (such as methylene chloride) for a time. Specialized equipment is required to safely handle production. The vapor attacks the thermoplastic causing an increase in the gloss.

Structural foam parts contain a cellular core consisting of expanded blowing agent pockets. These cells are under pressure and continue to outgas through the solid skin. This outgas period ranges from several minutes to a few weeks, depending on the permeability rate of the blowing agent, the molding resin selected, and the part configuration. A normal outgas period is 24 to 96 hours.

Traditionally, structural foam has been finished with a texture of one type or another because of the swirl pattern of structural foam. Improved molding technology has resulted in the ability to have texture-free and higher-gloss finishes. Commonly used paint systems include lacquers and acrylics as well as cross-linking (chemically cured) enamels such as polyurethanes, epoxies, and polyesters. Certain molding materials and finishing materials are especially compatible. In other cases, a barrier coat is applied to ensure paint adhesion. The appearance of the finish is judged by three elements: color, texture, and gloss. Each of these affects the appearance of the other. It is therefore necessary to set standards that include all three of these elements.

There are two approaches to a textured finish. The first is to have a molded-in texture on the part and a color-matched material. In some of the advanced molding processes, the parts may be acceptable as molded, but if cosmetic requirements dictate, a smooth color coat is applied over the texture. Again, in the case of some materials, a barrier coat is necessary prior to the color coat to ensure good adhesion.

The second approach is to apply a textured paint over a smooth molded part. The swirl pattern of low-pressure structural foam is normally finished with this approach. The amount of swirl will dictate the amount of surface preparation. A common approach is to apply a heavy filling-type primer and follow this with a light sanding operation. Where a barrier primer coat is necessary because of material/paint incompatibility, the sanding will be done first. Once the part surface is properly prepared, the textured paint is applied. Generally, this textured paint is applied in two coats with the first coat being a smooth coat.

The painted finish adds considerably to the part cost, often accounting for 25–40% of the price, but it does result in an appearance that is otherwise unobtainable. The following factors should be considered when deciding whether or not to specify painting:

- The effect of ultraviolet light from sunlight and fluorescent lights can darken or fade colors of unpainted plastic and, in some cases, affect physical properties.

- Matching of other components in the system may dictate a painted texture or a molded texture. A painted texture has a "warmer" look and is often specified to ensure good matching of components produced by a number of methods (such as sheet metal, injection, and foam).
- Cross-linked or chemically cured (polyurethane) finishes provide added chemical resistance to the basic plastic.
- As with injection molding, some part configurations result in knit lines and flow lines that are cosmetically objectionable and require a coat of paint to hide the defect.

92.12 COSTING

The cost of a molded part is determined by the raw materials, molding machine costs, secondary operations, operating overhead, and mold amortization. Although each processor will develop an individual pricing philosophy, certain fundamentals pertain throughout the industry.

The first step is to estimate the weight of the part. This can be accomplished by calculating the cubic volume and multiplying by the density of the material. Table 92.4 shows the weight conversion factors for common structural foam materials.

Once the weight is estimated, the raw material portion of the unit price can be figured by multiplying the cost per pound of the resin plus a handling and scrap factor. Generally, this handling factor is 10%.

Next, the cost of molding the part is calculated. The size and geometry of the part and the material selected will determine the size and therefore the cost of the molding machine to be used. Typically, clamp tonnage (1/4 ton/in.² (1400 kg/mm²) of exposed surface of the part) and shot capacity are the primary considerations. It is necessary to estimate the overall cycle time to produce each part. Generally, the thickest section of the part, part geometry, and material selection will determine how long the part must cool inside the mold before it is able to withstand the force of ejection. Once these two factors are known, it is then possible to figure the cost of molding. For example, if the machine rate is $60 per hour and the cycle is 120 seconds, the molding cost is $2 per part ($60 per hour divided by 30 parts per hour = $2 per part).

It is also necessary to figure the cost of setting up the machine and factor this into the unit cost. For instance, five hours of downtime of the machine for a run of 1000 pieces adds $0.30 to the unit cost ($60 per hour × 5 hours = $300 divided by 1000 pieces = $0.30 per piece).

The cost of secondary operations—ultrasonic inserting, prefinishing (gate removal and sanding), finishing or painting (if required), assembly, and packaging—must be figured

TABLE 92.4. Weight Conversion for Common Structural Foam Materials

	lb/in.³	Typically Overall Foamed Density
High density polyethylene	0.0280	0.77
High impact polystyrene	0.0288	0.80
PPE/PPO	0.0307	0.85
Polycarbonate, 5% glass	0.0325	0.90
FR polystyrene	0.0325	0.90
Polypropylene	0.0270	0.75
ABS, non FR	0.0307	0.85
Reference: water	0.03611	1.00

separately and added to the unit costs. It a common practice for the machine operator to do some of the secondary operations at the machine since structural foam parts require longer cycle times because of their oversized wall thickness.

The final unit cost factor is amortizing the tooling or mold cost over a projected number of pieces. This step determines the overall economics of the process. Parts and molds are quoted separately. The industry standard practice is to charge for the tooling prior to production deliveries.

Further information on industry practices is available from the SPI.

92.15 INFORMATION SOURCES

A complete history and background on the structural foam industry is available from Structural Foam Div. Society of the Plastics Industry, Inc., 1275 K St. NW, Washington, DC 20005 (202) 371-5200.

The SPI Structural Foam Plastics Div. conducts an annual conference and parts competition. Yearly proceedings are available through the SPI dating back to 1974.

In 1979 a business opportunity study was made by Marilyn Bakker of Business Communications Company in Stamford, Conn. This report and its 1982 update are available from: Business Communications Co., P.O. Box 2070, Stamford, CT 06906 Re: Business Report P-006

Plastics Design Forum and other trade magazines have given considerable coverage to structural foam over the years.

The Modern Plastics Encyclopedia offers information on material selection, design basics, and other topics. *Machine Design* magazine's annual "Materials Reference Issue" has an excellent section on polymer chemistry, which explains the basics of this subject structural foam.

BIBLIOGRAPHY

Following are sources on specialized topics:

General References

E.J. Baumrucker, *Coinjection of Structural Foam: How it Works, How it Affects Tool Design, Secondary Operations, Finishing, and Overall Part Cost* presented at the 1980 SPI Structural Foam Conference.

J.R. Thomas, *The Farrel High Pressure Structural Foam Process: Advances in Large Part Technology,* presented at the 1980 SPI Structural Foam Conference.

J.J. McRosky, *Coinjection Technology* presented at the 1984 Structural Foam Conference.

W. Gracek, *Counterpressure vs, Conventional Structural Foam: A Complete Physical Testing Report* presented at the 1984 SPI Structural Foam Conference.

D. Dreger, "Better Surface Finishes for Structural Foam Plastics" *Machine Design* (July 7, 1983).

R.B. Johnson, *Designing in Counterpressure Structural Foam* presented at the 1983 SPI Structural Foam Conference; published in Plastics Design Forum, July/Aug. 1985.

H. Echardt, *How To Develop a Successful Coinjection Application,* presented at 1986 SPI Structural Foam Conference.

M. Caropresso, *Technology Advancements in Counterpressure Structural Foam Processing,* presented at 1986 SPI Structural Foam Conference.

E. Ruhl, *Production of High Pressure Thermoplastic Structural Foam Parts Utilizing Expanding Mold and Gas Counterpressure on a Standard Injection Molding Machine,* presented at 1986 SPI Structural Foam Conference.

93

THERMOFORMING

William K. McConnell Jr.

President, McConnell Company, Inc., P.O. Box 11512, Fort Worth, TX 76110

93.1 DEFINITION AND HISTORY OF PROCESS

Thermoforming is a method of processing plastic resin into finished parts from sheet [0.010 in. (0.25 mm) thick and greater] or film (less than 0.010 in. (0.25 mm) thick]. In this chapter, sheet and film are considered one and the same. The plastic sheet is heated to its particular thermoforming temperature and immediately shaped to the desired configuration. At processing temperatures the sheet is very pliable and flexible, enabling it to be formed rapidly into great detail with a minimum of force. Pressure is maintained until the part has cooled. It is then trimmed on the selvage used to hold the sheet during processing.

The ancient Romans imported tortoise shell (Keratin) from the Orient and used hot oil to shaped this thermoplastic material into food utensils. During World War II, aircraft canopies, turrets, domes, relief maps, and many other items were vacuum formed. After this period, thermoforming rapidly expanded into a viable and profitable method of processing plastics. Today, thermoforming is one of the fastest, if not the fastest, growing methods of processing plastics.

93.2 MATERIALS, SHEET AND FILM

See the Master Process Outline. More information on each material is available in the specific section covering it in this Handbook.

93.2.1 Extrusion

Extrusion is the most common method of producing sheet and film for thermoforming. In this process, thermoplastic resin in the form of pellets, flakes, or powder is fed by gravity into a heated cylinder. The material is conveyed through this heated cylinder with a screw. It is brought to its particular processing temperature by conduction from the carefully controlled heated barrel and from the shear produced by conveying the material with the screw. A variable speed motor is utilized to rotate the screw.

It is extremely important that all resin and additives are thoroughly blended into a homogeneous mixture before being introduced into the hopper. The single biggest problem encountered in thermoforming is nonuniformly mixed materials. When regrind is conveyed from single or multiple containers to an efficient blender, the blend is commonly considered to be good. This is not necessarily so! In fact, more often than not, the extruder is not receiving a thoroughly homogenous mixture throughout the run. The reason is that a great variation can occur in the regrind and frequently does. Heat histories and degradation can

MASTER PROCESS OUTLINE | **PROCESS** Thermoforming

Instructions:
For "ALL" categories use   Y = yes
For "EXCEPT FOR" categories use   N = no
                                D = difficult

	ALL Thermoplastics Except for:	
Y	Polyester PETG CPET	
Y	Polyethylene	
Y	Polyethylene-UHMA	
Y	Polyethylene-FOAM	
D	Poly(methyl methacrylate)	
	Poly(methyl pentene) (TPX)	
D	Poly(phenyl sulfone)	
	Poly(phenyl ether)	
	Poly(phenylene oxide) (PPO)	
D	Poly(phenylene sulfide)	
Y	Polypropylene	
Y	Polystyrene	
D	Polysulfone	
Y	Polyurethane Thermoplastic	
Y	Poly(vinyl chloride)-PVC	
Y	Poly(vinyl fluoride)	
Y	Poly(vinylidine chloride)-Saran	
Y	Rubbery Styrenic Block Polyms	
	Styrene Maleic Anhydride	
Y	Styrene–Acrylonitrile (SAN)	
Y	Styrene–Butadiene (K Resin)	
	Thermoplastic Polyesters	
Y	XT Polymer	

ALL Thermoplastics Except for:

D	Acetal
Y	Acrylonitrile–Butadiene– Styrene
Y	Cellulosics
Y	Chlorinated Polyethylene
Y	Ethylene Vinyl Acetate
Y	Ionomers
	Liquid Crystal Polymers
D	Nylons
D	Poly(aryl sulfone)
Y	Polyallomers
D	Poly(amid-imid)
	Polyarylate
Y	Polybutylene
D	Polycarbonate
	Polyetherimide
	Polyetherketone (PEEK)
D	Polyethersulfone

N	**ALL Thermosets Except for:**
	Allyl Resins
	Phenolics
	Polyurethane Thermoset
	RP-Mat. Alkyd Polyesters
	RP-Mat. Epoxy
	RP-Mat. Vinyl Esters
	Silicone
	Urea Melamine

	ALL Fluorocarbons Except for:
D	Fluorocarbon Polymers-ETFE
	Fluorocarbon Polymers-FEP
D	Fluorocarbon Polymers-PCTFE
	Fluorocarbon Polymers-PFA
Y	Fluorocarbon Polymers-PTFE
	Fluorocarbon Polymers-PVDF

change drastically within small quantities. By the nature of thermoforming, from 10 to 70% of trim is generated continuously, depending on the particular part.

It has been demonstrated that, compared to virgin stock, reprocessed trim will have been partially degraded by virtue of its previous processing heat histories. Typically, the "number-average" molecular weight of the regrind is lower than that of the virgin resin, which results in a bimodal molecular weight feed to the extruder.

The success of the extruder in properly mixing the melt before extrusion is an important criterion for production of high-quality thermoformable plastic sheet. The bimodal weight feed is an example of a dual-viscosity system tht can cause sheet problems.

To overcome this very common problem, it is necessary to mix as large a batch of regrind as possible before feeding it to the blender. The plastic feed must be acceptably uniform in every way. The bulk density of the regrind must be controlled, frequently with 15,000 or 20,000 pound (6,800–9,000 kg) cyclone mixers. In order to control regrind feed quality after mixing in a cyclone-type blower, it may be necessary to pass the regrind a second time through a finer screen grinder. After the cyclone treatment, repelletizing through a vented reclaim process may be required. When it is precisely blended with virgin resin, "sweeteners," or other additives, this very consistent mix will extrude more easily and thermoform better. It is also important that this material be extruded with exactly the same specifications as to regrind mix, melt index, and so on each time. Where possible, regrind should be less than 50% of the mix. Under conditions where there is always more virgin resin, it is possible to keep to the sheet specifications more readily.

Once the thermoplastic resin is heated to its processing temperature, it is forced through a sheeting die to obtain the proper thickness and width. The sheet gauge for thermoforming should always be controlled with the die lips only—not with the cooling rollers. The hot sheet is fed from the die lips into the cooling roller stack. In most operations, a three-stack set of large-diameter temperature-controlled rollers is used. When embossing is required on one side, an etched center roller is used. Low pressure (to avoid excessive orientation) is applied with the other rollers. Common cooling roll temperatures are close to the heat-distortion temperature of the particular resin being extruded. When using high specific heat materials such as polyethylene or materials that require clarity, additional cooling rolls are frequently installed.

All sheet material has some orientation. Orientation is the stress added to the sheet during the extrusion process. For most amorphous materials, orientation in the finish sheet will be approximately 6–12% shrinkage in the extrusion direction and 1–3% growth in the transverse direction when extruded with "up-stack" equipment and several percent higher with "down-stack" equipment. Crystalline sheet will have 20–55% orientation in the extrusion direction. Polyolefin materials (in particular, high molecular weight polyethylene and high density polyethylene) will experience orientation shrinkage in the extrusion direction of 35–55%. Because of these sheet extrusion stresses, the thermoformed part will always shrink more in the direction of the extrusion than in the transverse direction. This tendency is aggravated in crystalline parts because of the higher shrinkage rates.

Sheet extrusion is a very complicated process because of the many variables encountered. Thermoforming also involves many variables; however, of all the variables and required technical expertise encountered during a total thermoforming operation, an average of 72% are attributed to the extrusion process. It is vital that quality control start with incoming resin at the sheet extrusion process and be carefully monitored throughout the complete extrusion operation. The ability to easily and quickly thermoform sheet samples as the sheet is being extruded is very important. The more the process is computerized, the more easily and accurately these variables can be controlled.

Biaxially Orienting Sheet Material. Orienting is a technique used to obtain higher physical properties than those present in the sheet under normal production. During the extrusion process, the sheet is extruded and cooled to the orienting temperatures. At these temperatures the sheet is biaxially stretched approximately 300%. While it is held in this oriented condition, the sheet is cooled to room temperature. Normal reheating for thermoforming and forming of parts does not anneal or stress relieve the sheet, even though the forming temperature is higher than the temperature used for orienting. An example is cell-cast acrylic sheet, which is biaxially oriented and thermoformed for all commercial airliner exterior passenger windows and many military fighter canopies.

Sheet Orientation Test With Radiant Heaters. This test is helpful in determining the percent of orientation in thermoplastic sheet.

1. Examine the sheet material to be tested and determine the direction of extrusion. Cut and identify test blanks exactly 10 in. (25 cm) square.
2. Where accuracy is needed, check and record thickness of each sample in 1–2 in. (25–51 mm) increments in both extrusion and transverse directions.
3. Powder both sides of sample sheet to prevent it from sticking to the support surface.
4. Ovens used for test should have approximately the same watt density and sandwich radiant heat as is used in production.
5. Fabricate a clamping frame using chicken wire or hardware cloth for support.
6. Place the test sample on the unrestrained support surface, put it in the oven, and heat it to same temperature used in production.

If there is considerable orientation, the test sample may have a tendency to curl onto itself. Thus it may have to be partially restrained by placing chicken wire or hardware cloth on top of it. In extreme cases, the test sample may have to be sandwiched between two aluminum sheets.

7. Remove the test sample and cool to room temperature for at least 24 hours.
8. Measure the dimensions along the extrusion direction. The change in dimensions is the orientation and is expressed in percent. For example, if the sample has shrunk in the extrusion direction from 10 in. (25 cm) to 8.5 in. (22 cm), the orientation is 15%.
9. Measurement along the trasverse direction will give the normal growth of the material caused by "necking" during extrusion. If the sheet grows in the transverse direction from 10 to 10-1/8 in. (25 to 26 cm), the material has a 1.5% growth factor. It is normal for the material to grow slightly in the transverse direction. Biaxially oriented materials will have large orientation of around 275–300%.

An unusually large amount of orientation requires corrective action. It is very important that the orientation is uniform and consistent throughout all of the material.

Register Stock. Anytime a distorted pattern is printed on the sheet and then thermoformed, it is essential tht the sheet be extruded with a minimum, very consistent orientation throughout the extrusion or calendering run.

For example, specifications for extruding high-impact styrene "register" stock are

• Stock temperature of 380–410°F (193–210°C).
• Grease spots from air nozzles eliminated.
• Automatic shears—Hand cutting of sheet creates styrene dust, and when several are cut at once, bevel the edges, requiring copper brush, Corona discharge treatment, or other method of removing dust and static.
• Use of white gloves because oil from fingers spot the sheet.
• Register stock can destabilize with some antistat sprays. For resin lubrication, use a proper stearate so that it burns out in barrel as extruded. Do not use external lubricants.
• Orientation should be checked by sandwich radiant heaters, 10-by-10 in. (25-by-25 cm) pieces across the web.
• The dies should be dammed with 3 in (76 mm) trim left on each side sheet.
• Aluminum shims should be used under sheet slitters so that the knife protrudes through the plastic and prevents edge buildup and edge orientation.
• Die lips should be set 0.005 in. (0.13 mm) under the sheet thickness.
• Take-off rolls should be the same speed as chrome rolls.
• Remove chatter lines if the center roll is too hot.
• Run all matte/matte surface materials with no beads.

Calendered vinyl and ABS are also frequently used as register stock. In fact, any thermoplastic material can be produced as register stock with special care.

Coextrusion. Coextrusion is a relatively recent development. By using a special adjustable feeder-block, which is installed just behind the sheeting die, multiple layers of various thermoplastics can be economically produced. Compatible materials in the various layers permit the trim to be readily reused.

Coextrusion has been a real boon to the thermoforming industry. It offers such features as:

• The ability to extrude a thin virgin cap sheet for appearance, including surface finish and match color.

- The ability to have a single sheet consisting of multiple layers of barrier materials for packaging food, medical, and industrial products.
- Extrusion of cap sheet consisting of regrind mixed with a virgin resin base for weathering properties.
- Extrusion of rigid or flexible skin and a rigid or flexible foam core.
- Production of flexible skins and a rigid core.
- Ability to extrude thermoplastic adhesive "tie" layers for different materials that do not normally bond together when coextruded.

93.2.2 Calendering

Calendering is a very fast and efficient method of manufacturing flexible and rigid vinyl sheet up to 45 mils (1.14 mm) thick. ABS and PP can also be calendered.

In this process, the sheet thickness is strictly controlled by four L-shaped gaping rolls. These rolls are run at the particular processing temperature of the material so that no stresses are set up in the sheet., After the resin is squeezed to the exact gauge, it immediately travels to the cooling rollers and on out for either sheeting or roll put-up.

93.2.3 Sheet and Film Purchasing Specifications Checklist

All material purchased should have written specifications. Items that should be considered include:

- Amount of orientation required, as tested by heating a 10-by-10 in. (25-by-25 cm) sample to thermoforming temperature using sandwich radiant heat and allowing it to cool and shrink unrestrained.
- Use of regrind (trim), according to what is acceptable for the particular application.
- Dimensional tolerances:
     Gauge tolerances;
     Width and length tolerances;
     Sheet flatness tolerance (polyolefins in particular);
     Out of square tolerances.
- Impact strength as determined by drop ball (dart) and Izod testing.
- Absence of moisture.
- Count of foreign matter, agglomeration, contamination.
- Finish of surface required—embossing, smoothness, gloss, pits, dimples.
- Tensile strength.
- Melt index.
- Density.
- Elongation.
- Molecular weight.
- Modulus of elasticity.
- Tests to be run either by the purchaser or the vendor.
- Packaging.
- Color.
- Fillers.
- Barrier protection, moisture vapor transfer rate (MVTR), oxygen permeability.
- Odor.

93.3   PROCESSING CHARACTERISTICS

The following are essential to the production of consistently top-quality thermoformed parts:

- Resin that is high in quality, homogenous, and consistently meets specifications.
- Sheet that is high in quality, homogenous, and consistently meets specifications.
- Consistent heating of the sheet consistently to the same proper, uniform thermoforming temperature.
- Rapid, superb vacuum with a minimum starting pressure of 28.5–29.5 in. (97–100 kPa) of mercury at sea level; compressed air with instant pressure available.
- Consistent, proper, uniform mold temperature.
- Consistent, efficient cooling of the part.
- Consistent trimming operations.

93.3.1   Heat

Assuming the sheet material meets specifications, temperature and vacuum and/or compressed air are the critical factors in the forming process. Any variation in the temperature of the hot sheet dramatically affects the hot (tensile) strength or elasticity of the plastic.

It is essential that the sheet material be heated very uniformly. With uniform heating, the faster the vacuum the better the material distribution because the material does not have a chance to cool off as it is being formed. The resulting part exhibits a minimum of internal stress and the best possible physical properties.

Some materials, such as cast acrylic in very deep draws, do not process well with fast vacuum forming. This material has a great deal of hot (tensile) strength, which permits the processor to use slower vacuum. However, a very hot mold must be employed.

When pressure forming is used, the material is moved even faster than it is with vacuum forming. Under optimum conditions the material distribution will improve and the parts will be even more stress free.

All thermoplastic materials have specific processing temperatures. Table 93.1 identifies the processing temperature ranges for some popular thermoforming materials.

Mold and Set Temperature.   The set temperature is the temperature at which the thermoplastic sheet hardens and can be safely taken from the mold. This is generally defined as the heat-distortion temperature at 66 psi (0.46 MPa). The closer the mold temperature is to the set temperature without exceeding it, the less will be the chance of internal stresses in the part. For a more rapid cycle time when post-mold shrinkage is a problem, post-cooling fixtures can be used so that parts may be pulled early.

Lower Processing Limit.   The lower processing limit is the lowest possible temperature for the sheet before it is completely formed. Material formed at or below this limit will have severely increased internal stress that later can cause warpage, lower impact strength, and poorer physical properties, another reason for rapid vacuum or forming pressure.

The least amount of internal stress is obtained by a hot mold, hot sheet, and very rapid vacuum and/or compressed air.

Orienting Temperatures.   Biaxially orienting the molecular structure of thermoplastic sheet approximately 275 to 300% at these temperatures and then cooling greatly enhances such properties as impact and tensile strength. Careful matching of heating, rate of stretch, mechanical stresses, and so on is required to achieve optimum results.

When oriented material is thermoformed, good clamping of the sheet is essential. The sheet is heated as usual to its proper forming temperature and thermoformed. The hot

TABLE 93.1. Thermoforming Processing Temperature Ranges

Material	1. Mold and Set Temperature,		2. Lower Processing Limit,		3. Orienting Temperature,		4. Normal Forming (Core) Temperature,		5. Upper Limit,	
	°F	(°C)	°F	(°C)	°F	(°C)	°F	(°C)	°F	(°C)
ABS	185	(85)	260	(127)	280	(138)	300	(149)	360	(182)
Acetate	160	(71)	260	(127)	280	(138)	310	(154)	360	(182)
Acrylic	185	(85)	300	(149)	325	(163)	350	(177)	380	(193)
Acrylic/PVC (DKE-450)[a]	175	(79)	290	(143)	310	(154)	340	(171)	360	(182)
Butyrate	175	(79)	260	(127)	275	(135)	295	(146)	360	(182)
Polycarbonate	280	(138)	335	(168)	350	(177)	375	(191)	400	(204)
Polyester, thermoplastic (PETG)[b]	170	(77)	250	(121)	275	(135)	300	(149)	330	(166)
Polyethersulfone	400	(204)	525	(274)	560	(293)	600	(316)	700	(371)
Polyethersulfone, glass-filled	410	(210)	535	(279)	560	(293)	650	(343)	720	(382)
Polyethylene, high density	180	(82)	260	(127)	270	(132)	295	(146)	360	(182)
Propionate	190	(88)	260	(127)	270	(132)	295	(146)	360	(182)
Polypropylene	190	(88)	265	(129)	280	(138)	310–330	(154–166)	331	(166)
Polypropylene, glass-filled	195	(91)	265	(129)	280	(138)	400+	(204+)	450	(232)
Polysulfone	325	(163)	374	(190)	415	(213)	475	(246)	575	(302)
Styrene	185	(85)	260	(127)	275	(135)	300	(149)	360	(182)
Teflon (FEP)[c]	300	(149)	450	(232)	490	(254)	550	(288)	620	(327)
Vinyl, rigid	150	(66)	220	(104)	245	(118)	280–285	(138–141)	310	(154)
Vinyl, rigid foam	162	(72)	240	(116)	260	(127)	300	(149)	350	(177)

[a]Trade mark of Polyeast Cork.
[b]Trade mark of Eastman.
[c]Trade mark of E.I. du Pont de Nemours & Co., Inc.

forming temperatures do not realign the molecular structure; therefore, the better properties of the oriented sheet are carried into the finished part.

Normal Forming Temperature. This is the temperature the core of the sheet should reach for proper forming conditions under normal circumstances. The normal forming temperature is determined by heating the sheet to the highest temperature at which it still has enough hot strength or elasticity to be handled, yet below the degrading temperature.

Upper Limit. The upper limit is the temperature at which the thermoplastic sheet begins to degrade or decompose. It is crucial to make sure the sheet temperature stays below the upper limit. When using radiant heat, the sheet surface temperature should be carefully monitored so that it does not degrade while the core is still heating to the forming temperature. These limits can be exceeded, for short a time only, with a minimum of impairment to sheet properties.

93.3.2 Vacuum and Compressed Air

As noted, under most conditions, a better part is obtained with faster vacuum and/or compressed air, which will also improve cycle time by creating a more intimate contact with the mold and giving more efficient cooling, better details, and close tolerances.

During the thermoforming process, the vacuum gauge should never drop below 25 in. (84 kPa) of Hg. (However, if the plant site is at an elevation of more than 2,000 ft (667 m), two large surge tanks should be used with a pressure valve between them. When the first tank drops below a set pressure, the valve kicks in the second tank and shuts off the first one.)

If the pressure drops as low as 20 in. (68 kPa) of Hg the machine should be shut down and the problem corrected. At this pressure it is possible to move the hot plastic sheet but not fast enough to obtain optimum physical properties. Lack of detail and warpage are typical results of insufficient pressure.

To optimize the vacuum system, it is important to eliminate all 90° elbows between the vacuum surge tank and the mold because each 90° angle slows the vacuum by 30%. Where necessary, flexible vacuum hose can be used with straight-in connections where this possible.

When a slower vacuum is unavoidable, it is advisable to use a full opening type ball or globe valve.

In order to achieve the rapid vacuum and necessary pressure to properly thermoform, it is essential to have a vacuum pump with a rated capacity of a minimum of 28.5 in. (97 kPa) Hg and a vacuum storage or surge tank. The surge tank or tanks should have a total volume of at least six times the cubic displacement that has to be evacuated. With long forming cycles, a surge tank permits use of a smaller vacuum pump. Short cycle pulsations are leveled out with a surge tank reservoir.

Vacuum Pressure Definitions. Psig Gauge pressure (in pounds per square inch) is the amount by which pressure exceeds the atmospheric pressure (negative in case of a vacuum)

Psia Absolute pressure (in pounds per square inch) is measured with respect to zero-absolute vacuum [29.92 in. (101 kPa) Hg]. In a vacuum system it is equal to the negative gage pressure subtracted from atmospheric pressure.

$$\text{Gauge Pressure} + \text{Atmospheric Pressure} = \text{Absolute Pressure}$$

$$1 \text{ in. of Hg} = 0.4912 \text{ psi of atmosphere on the part}$$

$$1 \text{ psi} = 2.036 \text{ in. of Hg}$$

Table 93.2 provides information about vacuum line and valve sizing.

93.3.3   Compressed Air Supply

A compressed air supply is subject to the same restrictions as vacuum. An adequate accumulator tank positioned as close as practical with properly sized lines, valves, and compressor is essential. Air should be very dry [dew point −40°F (−40°C)] and absolutely oil-free, particularly if it is used as an instrument or for prestretching, pressure forming, or part ejection. Caution should be exercised when air is exhausted from the pressure box before opening. All safety precautions applicable to pressure vessels must be obeyed. There must be appropriate safety ratings and over-pressure relief diaphragms. Pressure forming stations

TABLE 93.2.   Vacuum Line and Valve Sizing

Thermoforming Machine Forming Area	Vacuum Line and Control Valve Sizes
Up to 30 × 36 in. (76 × 91 cm)	1 in. diameter (25 mm)
36 × 48 in. up to 84 × 108 in. (91 × 222 × 213 × 274 cm)	1 1/2 in. diameter (38 mm)
96 × 120 in. and up (244 × 305 cm)	2 in. diameter (51 mm)

need to have pressure interlocks that prevent their being opened when internal air pressure is above a fixed (relatively low) level.

93.3.4   Heating Thermoplastic Sheet

Provided the sheet material is received completely within required specifications, the next most important factor in thermoforming is proper heating. The sheet must be heated uniformly throughout from the core to the surface and from the center area to the edge of the clamping frame. The only exception is where use of a special thermoforming technique dictates profile or area heating.

To control costs, the most efficient source of energy must be used. The theoretical amount of energy needed to heat a unit mass of sheet from room temperature to the forming condition can be calculated mathematically, using standard sources.

93.3.5   Heat Transfer Modes

There are three methods of heating: convection, conduction, and radiation (electromagnetic).

Convection. Convection is the slowest heating process. Convection heat transfer takes place when a material is exposed to a moving fluid (in thermoforming, recirculating hot air) at a different temperature. It is governed by Newton's Law of Cooling: $q = hA \, \Delta T$ (where hA is the convective heat transfer coefficient and ΔT is the temperature gradient between the sheet surface and the hot air).

Convection heating for thermoforming is accomplished utilizing a recirculating hot air oven. The oven temperature is carefully maintained at the thermoforming temperature of the particular material to be formed. Air is a great insulator and plastic materials absorb heat slowly, which is why this method is relatively slow. The specific heat of the particular material governs the heating cycle.

For example, a 0.125 in. (3.1 mm) thick acrylic sheet has a specific heat of 0.35. In a well-baffled hot air recirculating oven running at the forming temperature of 360°F (182°C), acrylic requires approximately 12.5 minutes of soak time to uniformly bring this mass up to the forming temperature. (This is about one minute for every 10 mils [0.01 in. (0.25 mm)] of thickness of the sheet. With radiant heat and using the proper wave length, 0.125 in. (3.1 mm) acrylic sheet can be brought to a core temperature of approximately 350–360°F (177–182°C) in 2.1 minutes.

The big advantage of convection heat is its superb uniformity of heating and its ability to keep the sheet surfaces from becoming hotter than the oven temperature. It is used especially for heating heavy gauge foam sheet, very thick solid sheet, sheet stock where the gauge is difficult to control accurately (off-gauge), sheet where surfaces might degrade easily if they are overheated. To speed up the cycle time, the oven can be run at a higher temperature than the normal forming range of the particular sheet; however, care must be exercised so that the surfaces of the sheet are not degraded. When the sheet is removed from the oven, it is then allowed to reach temperature equipilibrium before it is formed.

The most accurate hot air recirculating ovens are powered by electricity.

Conduction. Conduction is faster than convection heating but slower than radiant heating. Heat transfer by conduction takes place when a temperature gradient exists within a material; it is governed by Fournier's Law of Conduction:

$$q = -kA \, \frac{aT}{ax}$$

where q is the rate of heat transfer, aT/ax is the temperature gradient normal to the surfaces touching, A is area, and k is a material property independent of its shape called thermal conductivity or the "k" factor. The minus sign is inserted in accordance with the second law of thermodynamics.

Most conduction heater plates used in thermoforming are electrically heated PTFE-coated aluminum plates. With electricity, very good uniform heat can be maintained. The surface of the hot plates should have very even heating and the same heat sink distribution throughout. Sandwich heater panels are used, except when thin-gauge material is being processed. The most common use of hot plates is in roll-fed, in-line thermoformers, mainly in the packaging field.

As the thickness of the film or sheet increases, additional heating stations can be added to the machine so that the material can always be heated to the proper thermoforming temperature uniformly and more quickly than it can be cooled. As in convection heat, the contact plates are usually run at the same temperature as the forming temperature of the sheet to prevent degradation of the surface and to achieve extremely uniform heat, even when gauge varies within the sheet.

In sheet-fed machines, sandwich contact heaters are frequently used as a preheating station before the sheet enters the radiant ovens of single-station rotary machines in order to speed up the heating cycle. Contact heat provides a cooler environment around the machine than radiant heat; for most materials the heating plates are run at 300°F (149°C), compared to 900–1200°F (482–649°C) for radiant heating.

In some plants that extrude their own sheet and do not have good gauge control, contact heat is sometimes used so that uniform sheet temperature can be obtained. A change in sheet temperatures drastically affects the hot tensile strength of thermoplastic materials. When the sheet is close to its best forming temperature, a variation up or down by just a few degrees has a profound result. The hotter portion of the sheet will stretch much more than the cooler part.

However, if a nonuniform gauge sheet is heated so that the thick and the thin parts are the same temperature, the entire sheet ultimately possesses the same hot tensile strength. As a result, when the thermoforming technique used stretches the sheet, it stretches uniformly. This does not correct the original gauge variation, but, because the sheet is the same overall temperature, at least it expands proportionately.

Radiation. Radiation is the energy transferred between two separated bodies (in thermoforming, the sheet and the radiant heater surface) at different temperatures by means of electromagnetic waves. It is the most energy efficient way of heating sheet material.

Temperature is related to the intensity of heat or the electromagnetic energy possessed by any given body, which, having a measurable temperature, emits energy. One of the most important types of general radiation is that produced by a "black body." The radiation of a heated black body theoretically includes the whole spectrum, but the chief portions of measurable intensity important to the thermoformer lie in the visible and predominantly infrared range.

Infrared is usually correctly associated with heat. However, the specific wave length pattern related to a given temperature of a specific radiant heater is significant. This area of heat is the most efficient for thermoforming.

The temperature of the emitting surface of the most flexible radiant heater should be adjustable from 300°F (149°C) to 1300°F (704°C) where the radiated infrared waves vary from 6.8 to 2.97 microns. Most plastic sheet that is thermoformed requires an emitter temperature of 950–1190°F (510–643°C). Various wave lengths are required by different plastics, and several types of heater elements are used. In heating thicker foams, for example, a closed oven section may be desirable. The heaters would be run at around 6.8 microns or 300°F (149°C), which happens to be a very good absorption temperature for radiant heating of PE foam. With a closed oven having recirculating air, the foam can also be heated by convection heat, thereby achieving the best of both methods.

93.3.6 Heating Notes

$$P \text{ (power/wattage)} = V^2 \text{ (voltage)} \times R \text{ (resistance)}$$

- Reflection refers to angle of incidence = angle of reflection.
- When sandwich ceramic heaters are used at approximately 1100°F (593°C), 80–90% is converted to infrared radiant energy and 10–20% is converted to convection energy.
- Glazed ceramic heater panels are manufactured of 64% alumins and 36% clay.
- Efficiency of radiant heaters does not vary by the square of the distance; however, when a bank of heaters is moved from around 18 in. (46 cm) to within 6 in. (15 cm) of the sheet, heating efficiency picks up about 20–22%.

93.4 ADVANTAGES

Thermoforming offers several processing advantages:

- Relatively low tooling costs.
- Lower cost equipment.
- Pressure forming without expensive equipment.
- Pressure forming using compressed air up to 300 psi (2.0 MPa).
- Improved physical properties in the finished part.
- Processing of multiple-layer sheet material.
- Processing of parts with very high-strength reinforced composites.
- Processing of foam sheet.
- Production of very large parts.
- Production of thin parts with very small thickness-to-area ratio.
- Predecorating of sheet.

93.4.1 There Are Some Drawbacks to Thermoforming

- Amortizing the cost of producing sheet/film.
- Post-trimming.
- Reclamation of the selvage or trim material.
- Detail possible on mold side only.

93.5 DESIGN

This chapter reviews design considerations peculiar to thermoforming. More complete design information is given in the sections devoted to design.

93.5.1 Preliminary Considerations

Part Size. Extremely small parts may not be economical to thermoform, especially in heavier-gauge materials, because the parts may be difficult to handle and trim. One exception is thermoforming of thin-film packaging products.

On the other hand, large parts may be readily thermoformed, but it is important to consider the number of persons needed to load the machine and handle the part, possible shipping problems, and whether machinery is available to make the part. Sometimes, several smaller parts can function just as well as one large thermoformed part.

Manufacturing Capability. Before the design is finalized, it is important to know the capability of the fabricator, including such criteria as:

- Machine maximum platen size.
- Machine minimum depth of draw.
- Familiarity with required forming techniques.
- Trim methods.
- Experience in molding the material specified.
- Ability to hold the required tolerances.
- Experience and ability in the volume and cost range of the part being designed.

Tooling. The design can be influenced a great deal by the type of tooling to be used, and vice versa. For example:

- If prototype or wood tools are being used, the designer is very restricted as to materials and details that can be tooled and formed.
- When close tolerances are needed, temperature controlled metal tooling is usually required.
- Tooling includes trim and drill fixtures and jigs. Where close tolerances are required, tools for these secondary operations can cost nearly as much as the molds.
- Some more exotic engineering-grade materials cannot be molded on wood or cold molds.

The designer is responsible for selecting the material and should determine the forming techniques and the type of molds that will be required.

Styling. When the part is styled, care must be taken to make sure that shape and details are within acceptable limits for thermoforming. Great styling can be done with thermoforming parts, within these limits:

- All lines and details designed into the exterior of the part will appear also on the interior.
- All functional details designed on the nonappearance side must be acceptable where they appear on the exterior or appearance side.
- Part should be styled with generous radii, fillets, and draft. The sharp crisp edges loved by designers are difficult and costly to achieve in thermoformed parts. Outside radii should be at least one material thickness plus 0.030–0.060 in. (0.76–1.5 mm).
- Avoid having more than one texture or color on any one part.

93.5.2 Designing Aesthetics in Functional Detail

Many details formed in a thermoformed part are functional. Whatever functional detail is designed on one side of the part must be aesthetic on one or both sides. Some suggestions include:

- Ribs or crowning added to stiffen a surface should have balance and form.
- An offset added for stiffening can be sloped, tapered, and/or curved to improve appearance.
- A part molded from transparent material requiring partially clear areas can be heavily textured in the area not requiring clarity. The greater contrast makes the clear area look better.

- When an area is likely to cause webbing in forming, design in a rib or gusset to control the web.

93.5.3 Designing a Part to be Trimmed

A thermoformed part must be designed to be trimmed. Things to keep in mind about trimming include:

- Almost without exception, a thermoformed part must be trimimed from a blank.
- The simpler the trim, the less expensive the part. Simple trim reduces labor and permits more prospective vendors.
- Discuss preferred trim methods with prospective vendors.
- Know how the part will be trimmed and design in sufficient access to trim areas for the equipment to be used. Do not place a hole where a drill cannot reach.
- Design in molded reference points and surfaces.
- When possible, design a finish edge.
- When possible, design the part to be trimmed with one operation or at least at one trim station.
- Be familiar with the type of finish the different trim methods leave.

Common trimming methods are covered elsewhere in this chapter.

93.5.4 Assembly and Hardware

Many assembly methods work well with thermoformed parts, but each job is a special case and the assembly method should be given considerable thought. The more widely used methods of assembly include adhesives (single-component, two-component, pressure sensitive, hot melts, and so on), solvent bonding, ultrasonic welding, spin welding, and riveting (including pop rivets).

Some precautions to be remembered are

- When using adhesives, make certain they are compatible with the sheet material in which the part is molded.
- The surface of the part should mate intimately with parts to which it is attached.
- When using an adhesive that generates heat from exotherm, make sure that the maximum temperature reached does not exceed the deflection temperature of the sheet.
- When using solvent cement, use as little as possible and provide a path for evaporating solvent fumes to vent. These unvented fumes can attack and destroy the parts.
- When using ultrasonic or spin welding, be familiar with the process selected, discuss it with the potential vendor, and verify that there are proper provisions for the areas to be joined together.
- When joining parts with mechanical fasteners such as rivets, pop rivets, bolts, screws, and so on, make sure that all holes are oversize by 0.015–0.030 in. (0.4–0.8 mm) to preclude stress. Try to end up with the parts in compression using adequate-size washers or load spreading devices.
- When parts will experience a wide range of temperatures, make sure the type of adhesives or size of holes for fasteners allow for variations caused by thermal expansion.

93.5.5 Moldability

A moldable part will have a lower part price, more potential vendors, lower tooling costs, less development time and costs, and will leave the designer in a much more favorable position with everyone involved.

Making a part moldable by thermoforming means:

- Make the part no taller than is absolutely necessary.
- Make the ratio of part height to part minimum width as small as possible.
- Make the draw ratio or area ratio as small as possible.
- Make all outside radii and inside fillets as large as possible.
- Allow as much draft as possible on all visible areas of the part.
- If one small portion of a part will be difficult to mold, try to make that a separate part rather than making the job doubly difficult.
- If the design of a part requires a difficult molding technique, make sure that this part does not require the use of more than one difficult technique at the same time.
- Always design in a reference for trimming or hole drilling.
- Avoid joining surfaces at an outside or inside corner less than 90°.
- Avoid having more than one tall projection adjacent to another.
- When the design incorporates any of these criteria, look for a simpler way. There usually is one or more.

93.5.6 Tolerances

In thermoforming, the ability to hold tolerances is dependent on many factors, mostly the skill of several craftsmen: a patternmaker making a master pattern of wood, a foundryman pouring aluminum, a toolmaker, the lead person on a sheet extrusion line, the forming machine operator, the person trimming and drilling the part, and the person doing detail and deburring.

For a typical industrial thermoforming operation, normal limits would be:

- Formed details $+/-$ 0.030 in. (0.76 mm) up to 12 in. (30 cm), plus 0.002 (0.05 mm) per inch above 12 in. (30 cm).
- Drilled hole centers $+/-$ 0.015 (0.4 mm) up to 12 in. (30 cm), plus 0.001 in. (0.03 mm) per inch (25 mm) above 12 in. (30 cm).
- Drilled hole diameter $+/-$ 0.005 in. (0.1 mm) up to 3/8 in. (9.6 mm) diameter, plus 0.001 in. (0.03 mm) per 1/16 in. (1.6 mm) above 3/8 (9.6 mm) diameter.
- Trim dimensions $+/-$ 0.030 in. (0.76 mm) up to 12 in. (30 cm), plus 0.002 in. (0.05 mm) per inch (25 mm) above 12 in. (30 cm).

For a typical packaging thermoforming operation:

- Formed details $+/-$ 0.010 in. (0.3 mm) up to 6 in. (15 cm), plus 0.001 in. (0.03 mm) per inch (25 mm) above 6 in. (15 cm).
- Punched hole centers $+/-$ 0.10 in. (2.5 mm) up to 6 in. (15 mm), plus 0.001 in. (0.03 mm) per inch above 6 in. (15 mm).
- Punched hole diameter $+/-$ minus 0.002 in. (0.05 mm) up to 3/8 in. (9.5 mm) diameter, plus 0.001 in. (0.03 mm) per 1/16 in. (1.6 mm) above 3/8 in. (9.5 mm) diameter.
- Trim dimensions $+/-$ 0.010 in. (0.3 mm) up to 6 in. (15 mm), plus 0.001 in. (0.03 mm) per in. (25 mm) above 6 in. (15 mm).

The real key to establishing tolerances is to stay within the normal limits of the process and then loosen them wherever possible.

In thermoforming, close tolerances can only be held on the mold side. With a single mold (as opposed to matched molds), the wall thickness of the finished part can vary because of:

- Variation in sheet gauge. In the sheet extrusion process, the thickness constantly drifts from the high side to the low side of the extrusion tolerance.
- As the thermoforming sheet temperature changes during the forming process, the hot tensile strength changes greatly. This will, of course, vary wall thicknesses by a wide margin at times.
- Because of this big swing is hot tensile strength (stretch), it is not easy to maintain a uniform stretch in the sheet during thermoforming. When the hot sheet touches the mold, that part of the sheet is cooled below the temperature of the surrounding sheet. The cooler section does will not stretch uniformly, causing variation in wall thicknesses.

The most exacting tolerances are held on the mold side of a male mold or the male portion of a female mold. In the female cavities the surfaces have a tendency to shrink away from the mold as the part cools. Very good and fast vacuum along with pressure (compressed air) forming help reduce shrinkage.

93.6 PROCESSING CRITERIA

There are three basic methods of thermoforming: vacuum, pressure (compressed air), and mechanical.

93.6.1 Vacuum Forming

Vacuum is the most popular thermoforming method.

Drape Vacuum Forming. In drape vacuum forming (Fig. 93.1), a male or female mold is closed into the hot sheet. A vacuum is then applied to quickly remove the air between them: atmospheric pressure [14.7 psi (0.10 MPa)] moves the hot, pliable sheet against the mold, holding it there until it cools to the heat-distortion temperature.

In all thermoforming, it is very important that the vacuum and/or compressed air remain on the part until it has cooled to the heat distortion temperature!

Material distribution depends greatly on such variables as:

- Uniform heating of the sheet.
- Mold temperature and type of material.
- Rate of air evacuation (quality and speed of the vacuum and/or compressed air).

The hotter the mold and the faster the vacuum/air pressure, the better will be the material distribution. However, it is important not to exceed the required mold temperature.

Free Draw Vacuum Drawing. The sheet can also be freely drawn into a vacuum box and cooled while it only partially touches the mold or the clamping frame area. This technique, called free draw vacuum forming (Fig. 93.2), is used when the best available optics or surface is required. The portion of the sheet not touching the mold will have optics equal to that the sheet had originally.

93.6.2 Pressure Forming (Compressed Air)

When pressures greater than atmospheric [14.7 psi (0.10 MPa)] are required, pressure forming is used. Greater pressure creates such features as better detail, closer tolerances, faster cooling cycles, more strain-free parts, and better distribution of sheet material.

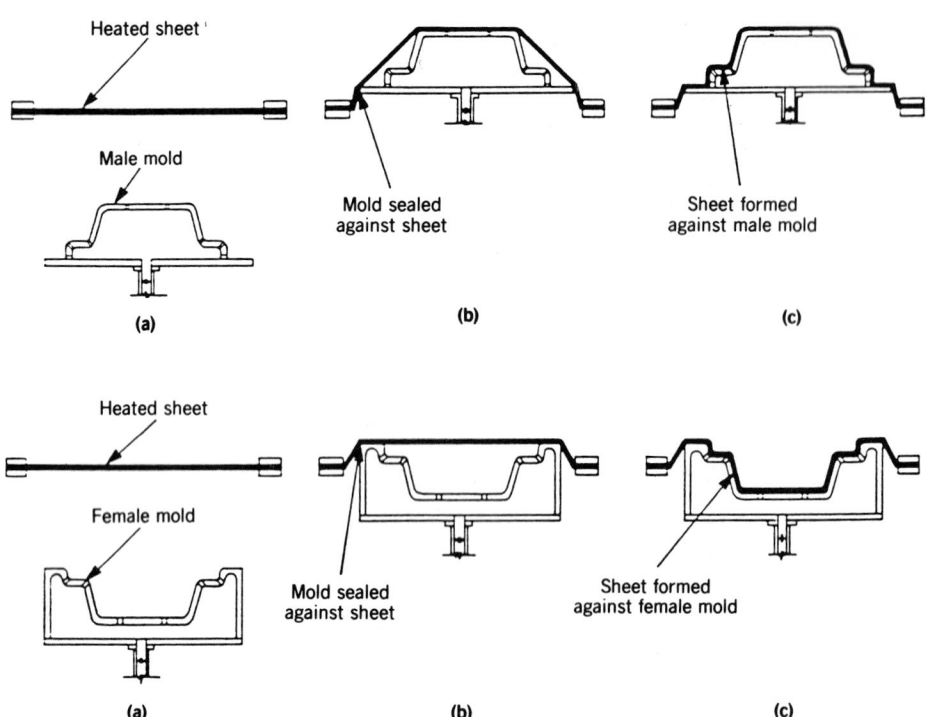

Figure 93.1. Drape vacuum forming.

A useful comparison of the per-part costs of pressure forming and injection molding has been prepared by Ken Maurus of GE Plastics for a 1000-part run:

• For injection molding, the tooling cost is $134,000; on a per-part basis the amortized cost is $134 per M. Materials and processing costs total $13.84 per M. The total cost per part is thus $147.84 per M.

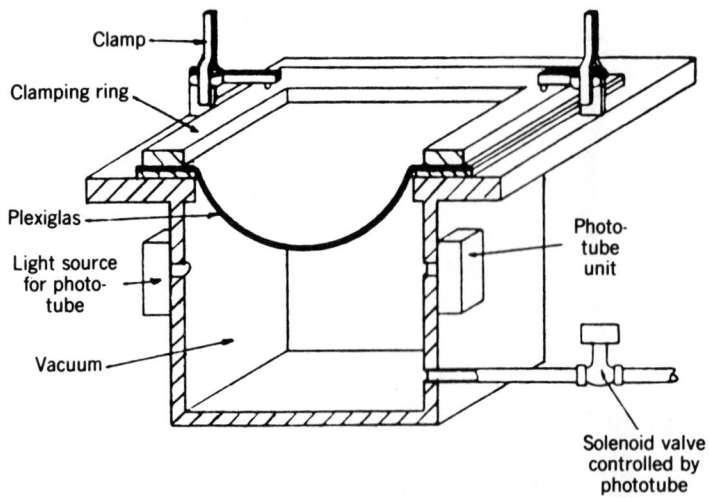

Figure 93.2. Vacuum free draw.

• For thermoforming, the tooling cost is $24,000; on a per-part basis the amortized cost is $24 per M. Materials and processing costs total $22.29 per M. The total cost per part is thus $46.29 per M.

Pressure Box or Plate Forming. In pressure box or plate forming (Figs. 93.3 and 93.4) 50 psi (0.34 MPa) pressure is common, but for some of the higher hot-strength materials, such as fiber-reinforced poly(phenylene sulfide), as much as 125 psi (0.86 MPa) is required.

To take full advantage of this technique, either the thermoforming machine mold platens must lock together or the mold and pressure box or plate must lock together. Hydraulically operated platens or mechanically locked platens (with pressure air bags or hoses to complete the seal) are recommended. Electrically driven platens usually will not hold the pressures; if they separate, the advantage of full pressure and vacuum until the part cools is lost.

Most roll fed in-line thermoformers and sheet fed pressure machines have 50 psi (0.34 MPa) compressed air and 29 in (98 kPa) of Hg vacuum as standard.

In pressure forming colorless polypropylene, the use of 100 psi (0.69 MPa) instead of 50 psi (0.34 MPa) and good vacuum will improve the clarity and the physical properties of the finished part. The addition of a nucleating clarifier, such as Millad 3905 by Milliken Chemicals, will greatly clarify extruded polypropylene sheet or film.

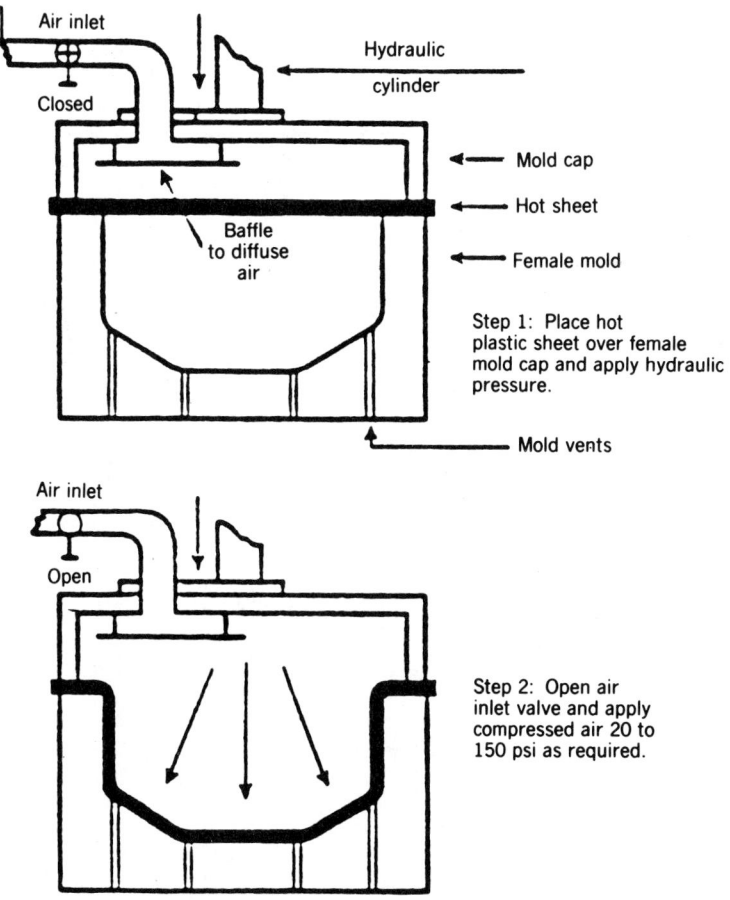

Figure 93.3. Pressure plate forming.

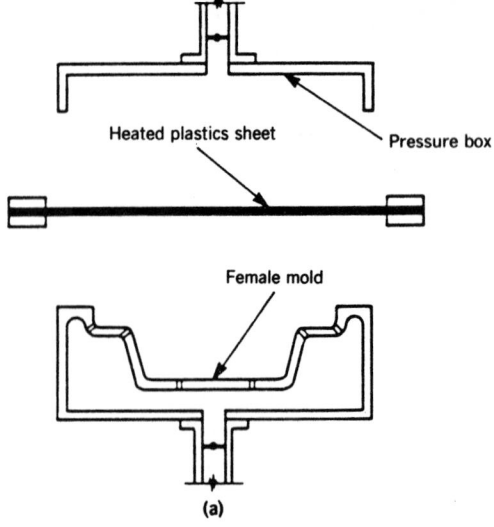

(a)

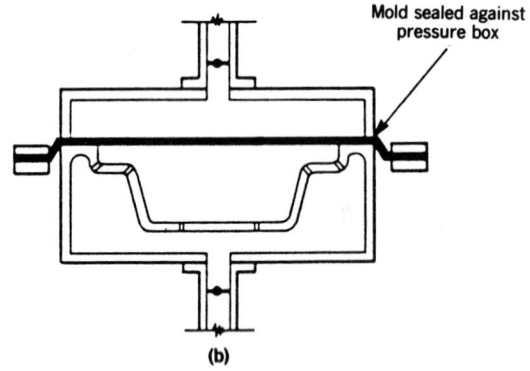

(b)

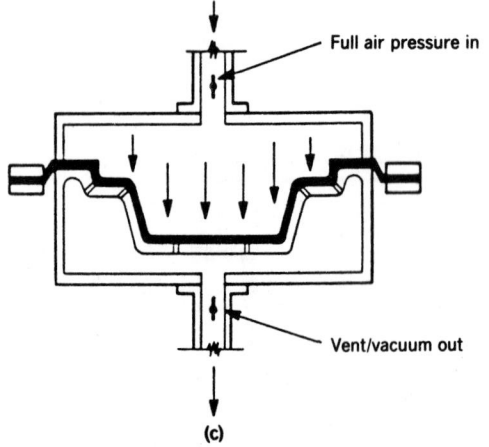

(c)

Figure 93.4. Pressure box forming.

Free Pressure Forming. With free pressure forming (Fig. 93.5), the hot sheet is sealed over a blow box so that only the periphery of the sheet is in contact with any tooling. Compressed air is injected into the box, pressurizing the sheet into the desired configuration. Bubble height can be controlled with a photo cell or microswitch, timed or estimated. Normal blowing pressures range from 20 to 120 psi (0.14 to 0.82 MPa). A typical application is skylights.

A variety of different shapes can be produced by varying the box contour or overpress on the top platen. Careful preferential heating of the sheet (including silicone blankets) can be effective for eccentric part contours.

Transparent bubbles with good optics are blown by gently placing the hot sheet on a blow table covered with a very low heat conducting material, such as 32-oz pool table felt, flocked suede rubber, or cotton flannel. To minimize or prevent surface "mark-off," the side of the sheet that goes against the table is underheated; the core and the opposite side are heated to forming temperature. Entrance of the compressed air is diffused to prevent spot cooling of the part. It is not necessary to heat the air as this forming technique only pressurizes the space between the hot sheet and the blow table. High pressures [such as 100 psi (0.69 MPa)], adequate air volume, and fast movement of the compressed air will improve clarity. A specific sheet temperature heated very uniformly and good vacuum is essential.

93.6.3 Mechanical Forming

True mechanical forming would employ no vacuum or compressed air to move the sheet. The forces necessary to move the sheet would be applied by mechanical or manual stretching, bending, compressing, stamping, pressure blanket, etc. In practice, plus assists, web catchers, side pulls, mechanically activated undercuts, and other helpers are necessary.

Stretch Forming. With stretch forming (Fig. 93.6), the hot sheet is stretched mechanically or by hand over or around a mold and clamped in place for cooling. This technique is frequently used in forming acrylic transparencies. It also works well where a part is in U-shaped or wrap-around shaped.

Ridge Forming. With ridge forming (Fig. 93.7), a frame structure is positioned so that only a small part of the mold touches the plastic sheet, thereby creating optics in the finished piece equal to those in the original sheet (except for the frame area). This technique only works with materials with high hot strength such as cast acrylic and butyrate. Display cases and boxes are typical applications.

Strip Heating. The easiest way of forming along a straight line is to use a strip heater (Fig. 93.8). The sheet is heated to just above the lower limit forming temperature on a straight

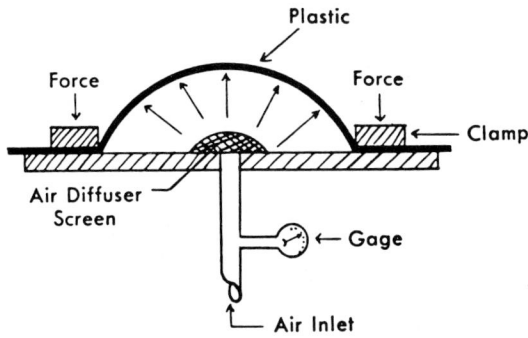

Figure 93.5. Free pressure forming.

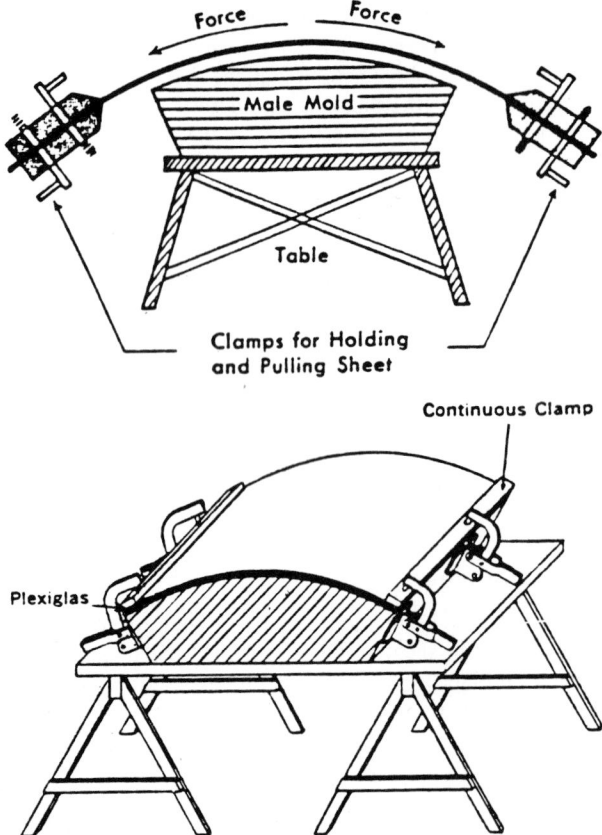

Figure 93.6.   Stretch forming.

line using tubular heaters or coiled nichrome resistance wire [0.026 in. (0.660 mm) diameter]. Instead of top and bottom (sandwich) strip heaters, it is usually preferable to heat a single side at a time. One side is heated for 40% of the total cycle; the sheet is flipped and the opposite side is heated for 50% of the cycle; and then the initial side is heated again for the last 10% of the cycle. In this way, the sheet is heated uniformly without blistering the surface.

After the narrow strip area is heated, the part is bent and placed in a jig to hold its shape while cooling. Several strip heaters and bending fixtures can be used at once to speed up production. Jigs should be designed so the heated sides will cool at the same rate on both sides to avoid spring back or bowing. Sharper, straighter bends can be achieved if a 90° V-groove for right angle bends is machined along the bend line on what will be the inside of the formed part.

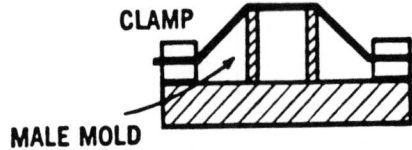

Figure 93.7.   Ridge forming.

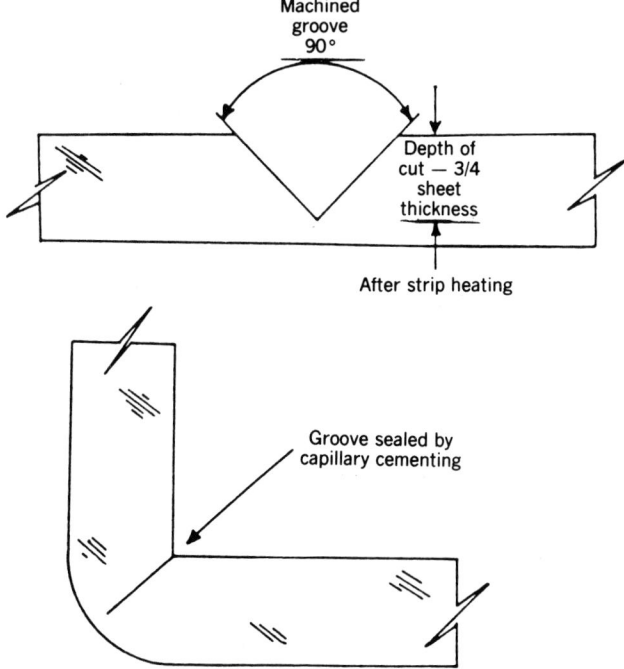

Figure 93.8. Strip heating.

It is possible to thermoform parts and then, as a post operation, strip form sections to obtain a more uniform wall thickness and smaller beginning blank size. This process lends itself to fast production of simple containers, store display fixtures, furniture, and many industrial items.

Matched Mold. The heated sheet is compression molded between two matching molds (Fig. 93.9). Mold cost is therefore greater. Platen press forces usually range from 5 to 50 psi (0.03 to 0.34 MPa); however, some materials and configurations may require higher pressures. This, of course, requires tooling to withstand the increased forces, which may result in higher costs and longer lead times. Foams, filled, and fiber-reinforced materials are popularly processed this way. It is important to note that both surfaces of the molds affect the finished part.

93.6.4 Prestretching

When more uniformity or a precise variation in wall thickness is desired, it may be necessary to prestretch the hot sheet before it touches the mold. As the hot sheet comes in contact with the mold it cools that part of the sheet, which increases its hot strength. As a result, the sheet stretches most in sections that have not yet touched the mold (Fig. 93.10)
Following are various techniques of prestretching.

Web Catchers. When boxlike male molds are used, the sheet tends to bridge or web at the corners and between multiple molds. This phenomenon is caused by the sheet draping across one or more molds and folding upon itself. To eliminate webbing on a single male mold, blocks can be placed at the corners to prevent the folding action.

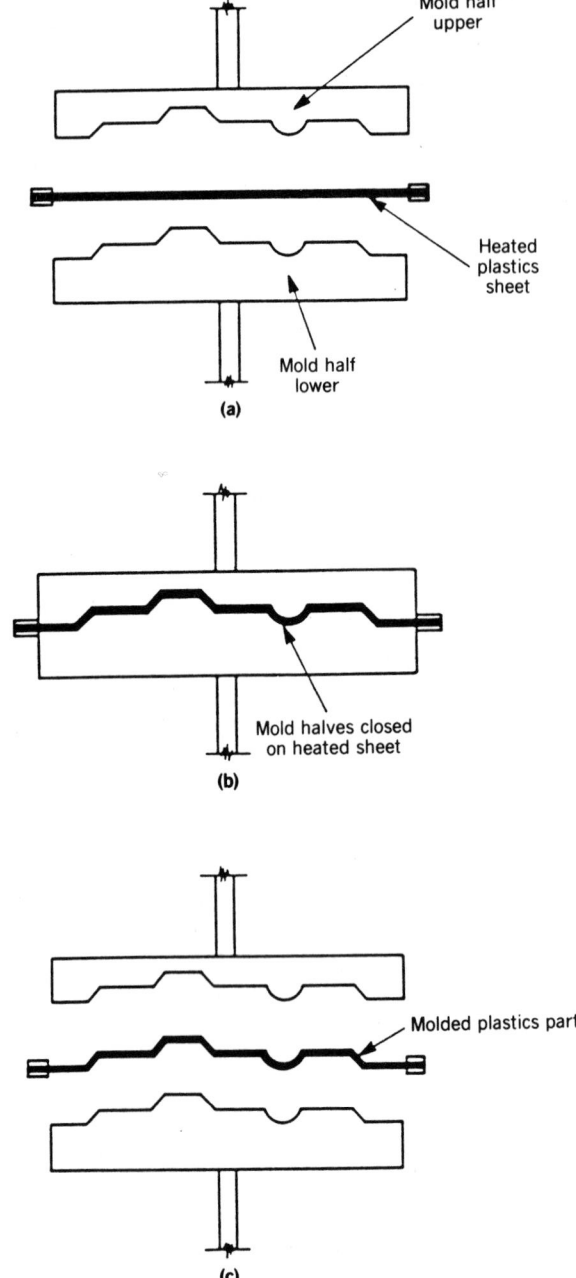

Figure 93.9. Mechanical forming with matched molds.

Snap Back (Male Mold). When prestretching is needed, snap back is nearly always a candidate because the tooling is the least expensive.

The prestretch box is pushed into the hot plastic sheet, causing a seal to form around its periphery. The hot material, which is still at forming temperature, is moved by vacuum or compressed air into a bowl shape. The sheet material is prestretched to about two thirds of the depth of the part. At this moment, the mold is forced into the bubble with the perimeter

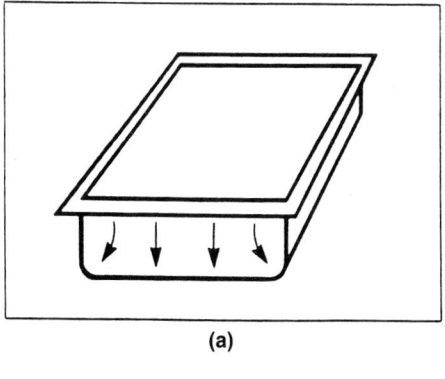

(a)

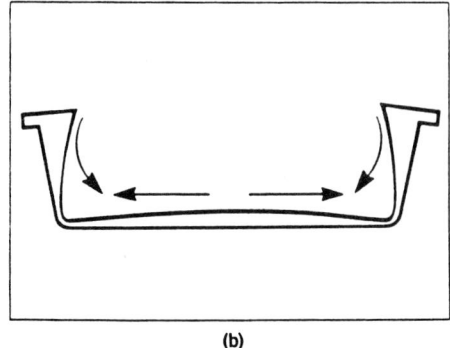

(b)

Figure 93.10. Variation of wall thickness. (**a**) End view of the direction of flow of heated film under vacuum during the thermoforming process. (**b**) Typical thermoforming profile showing downward and upward flow of heated film.

of the mold mating with the box for a vacuum seal and the flexible, hot plastic wrapping around the mold. At the same time the sheet is rapidly "snapped" to the detail of the mold surface by vacuum and the box is vented to atmosphere. After the sheet is pushed completely against the mold, the prestretch box is moved away from the part so cooling fans can speed up cooling time (Fig. 93.11).

Billow Snap Back (Male Mold). Instead of drawing the hot plastic into the prestretch box, it is billowed into a bubble shape. The height of the billow is usually about 65 to 70% the depth of the part the first time and is then adjusted as required. After the bubble has taken the proper shape, the mold is moved into it until the mold flange area seals against the box. Vacuum is applied and the sheet is snapped to the mold surface. This method normally produces a slightly more uniform wall thickness, but is a little more difficult to set up. Pressure in the box needs to be controlled by relief valves as the mold is forced into it.

Billow Snap Back (With Pressure Chamber). In Europe this method is called air slip and is very popular.

The bottom mold platen and a bottom moving heating panel are enclosed in a pressure chamber that is designed and constructed to withstand a minimum pressure of 10 psi (0.07 MPa). An outside top moving heater panel provides sandwich heat. With a top moving platen, great versatility in forming techniques is possible (Fig. 93.12).

The sheet is clamped into the top of the pressure chamber over the mold. After the sheet has been heated to forming temperature, the chamber is pressurized enough to prestretch the sheet as desired. Normally, very low pressures are used. After the bubble is shaped, the mold is moved into it and vacuum is applied. At the same time the chamber is vented to atmosphere. If a plug assist is needed it can be actuated with the top platen.

Billow Snap (Female Mold, No Plug). When it is not practical to use a plug assist with a female mold, some prestretching can be done by blowing a bubble and applying fast vacuum. For best results, the female mold should be mounted on the top platen. The mold is lowered into the hot sheet. After the mold seals off in the sheet the cavity is pressurized, blowing the sheet into a uniform bubble about 60–80% of the depth of the cavity. A very rapid vacuum is then drawn, which will add thickness to the bottom of the part.

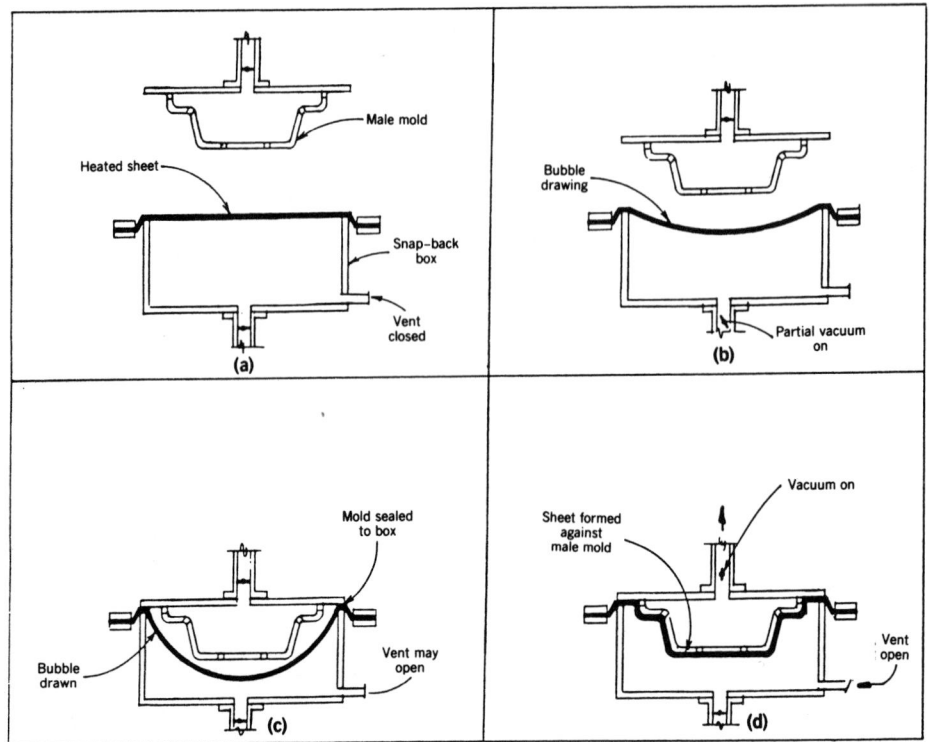

Figure 93.11. Vacuum snap-back forming.

Plug Assist (Female Mold). When it is necessary to use a female mold and prestretching is desired, a plug assist can be used (Fig. 93.13).

The mold is pushed into the hot sheet sealing around the perimeter. As the plug assist enters the hot sheet the air between the sheet and the mold is compressed, causing the sheet to billow around the plug. This action prevents the hot sheet from contacting the relatively cold mold as it is stretched into the cavity. The plug stops within about 10 percent of the bottom of the mold when vacuum and/or compressed air is rapidly applied, transferring the sheet from the plug to the mold (Fig. 93.13).

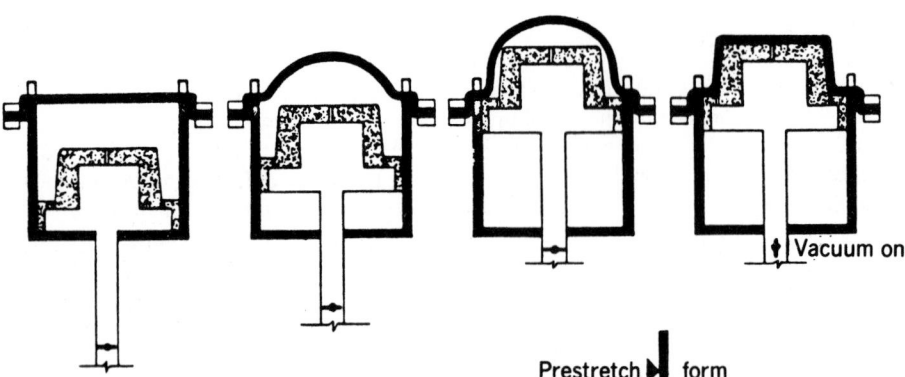

Prestretch ► form

Figure 93.12. Chamber billow up snap-back.

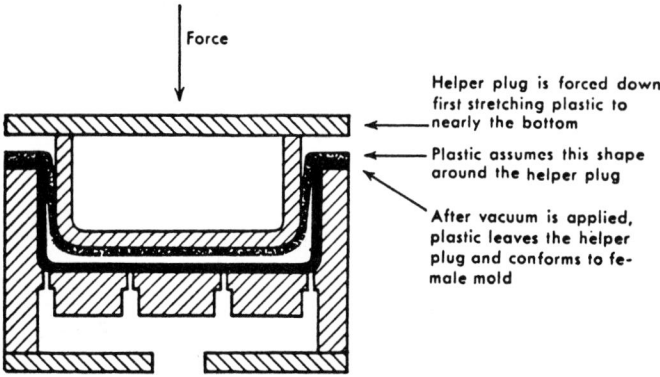

Figure 93.13. Plug assist (female mold).

Billow Plug Assist (Female Mold). When sheet fed thermoformers are used, this is a common prestretching technique. For best wall thickness distribution, the mold is mounted on the top movable platen with the plug on the bottom platen (Fig. 93.14).

The female mold is moved down into the hot sheet sufficiently to form an effective air seal. To ensure positive repeatable action, an adjustable microswitch is engaged as the mold bottoms out that pressurizes the mold cavity, thus blowing a bubble. An adjustable photoelectric eye stops the prestretching of the bubble at a predetermined height, usually about two thirds of the depth of the mold cavity.

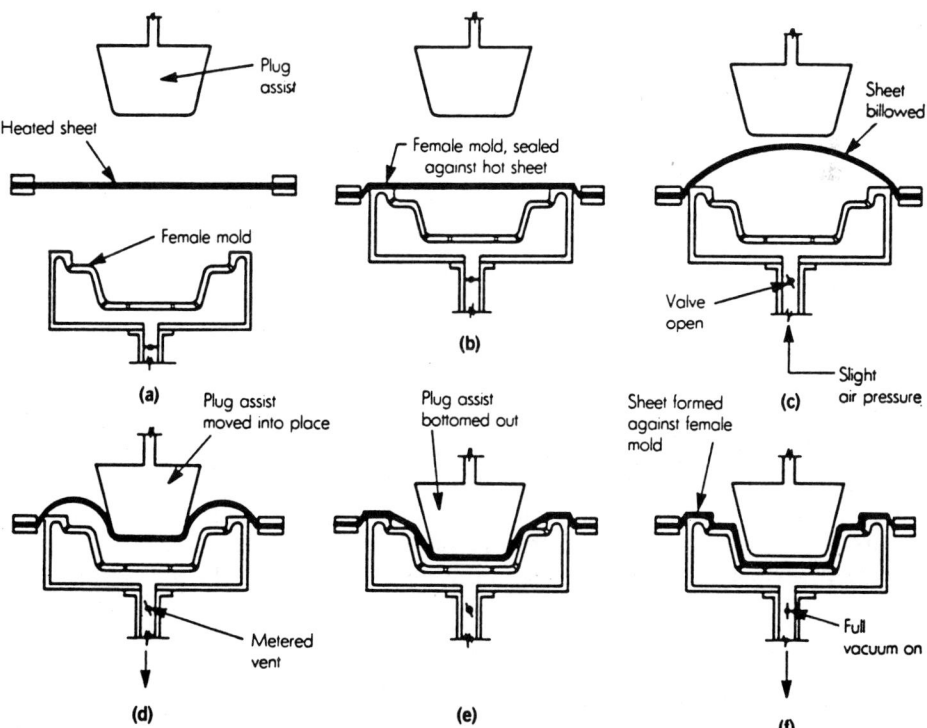

Figure 93.14. Billow molding with plug assist (top platen).

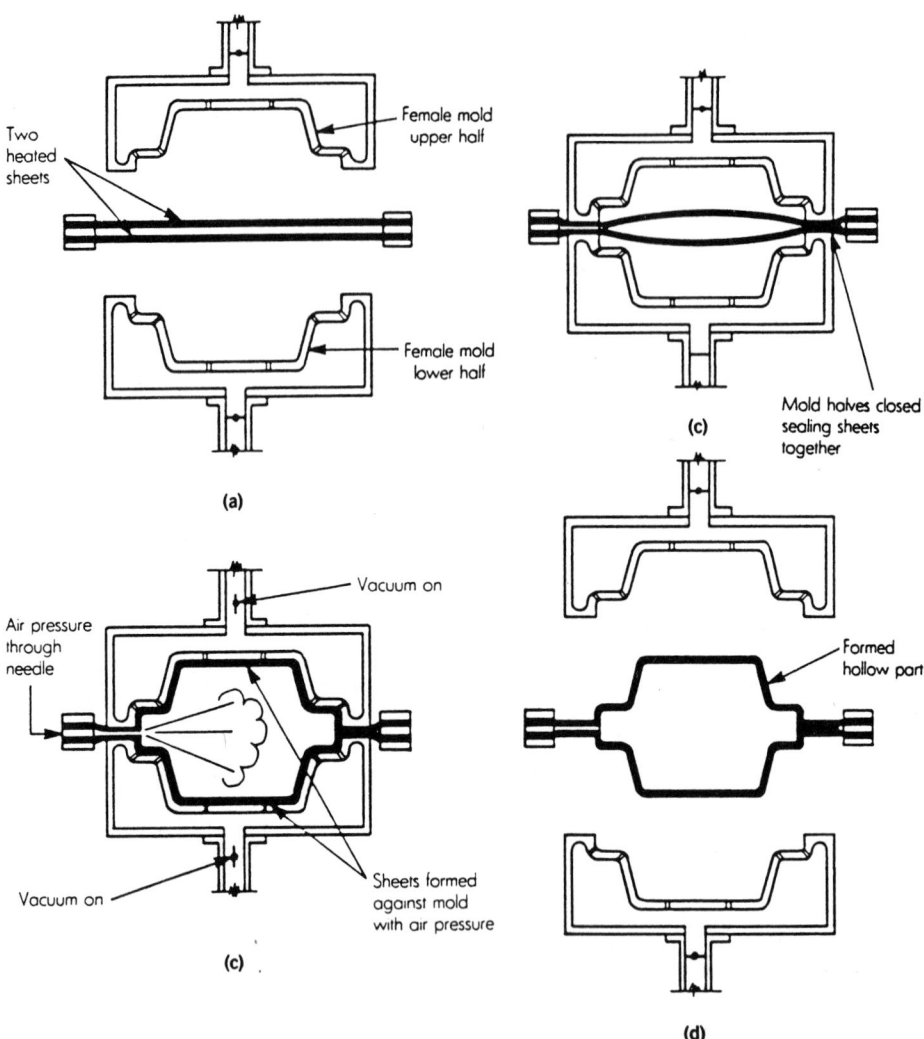

Figure 93.15. Twin sheet forming.

After the bubble has been properly shaped, the plug assist is moved into it, carrying the hot material into the mold cavity to within 5 to 10% of the bottom of the mold. At this point, the vacuum and/or pressure must be applied instantly. The plug assist should travel into the hot bubble at a rate fast enough that the compressed air in the mold cavity will keep the material against the plug until it has completed its stroke. If the part has much depth of draw, it will be necessary to partially bleed the compressed air from the compartment to prevent blowouts.

93.6.5 Twin-Sheet Forming

Great variety of shapes can be produced by traditional thermoforming techniques. However, hollow parts with small openings are a major exception. Such products can be made by bonding two thermoformed shells together, by blow molding, or by twin-sheet forming (Fig. 93.15).

Twin-sheet forming is also known as twin-shell forming and clamshell molding and employs two sheets to form a hollow part. Using a new process patented by W.J. Williams of Omico Plastics Inc. (Owensboro, Ky.), "foam-in-place" can be accomplished as the twin-sheet part is cooling in the mold.

In twin-sheet forming, two sheets of plastic are heated to forming temperature in separate clamping frames so sandwich heaters can be used. Two opposing molds are mounted on movable platens with vacuum and compressed air available to each. The two heated sheets are moved into the forming station between the opposite molds. The molds are pressed into the hot sheets with the matching perimeters of the molds engaging the sheets, which are at forming temperature. The pressure of the two molds squeezing the sheets together achieves an excellent bond very similar to thermowelding with heat and pressure. It is essential that the surfaces of the sheet be at or above forming temperature. Any other portion of the two halves can be joined together by the molds mating at that point. Vacuum is applied to both halves of the part, and compressed air is introduced between the sheets to complete the shape as the molds come together.

Two techniques of twin-sheet forming are use of a four-station rotary and use of a roll stock.

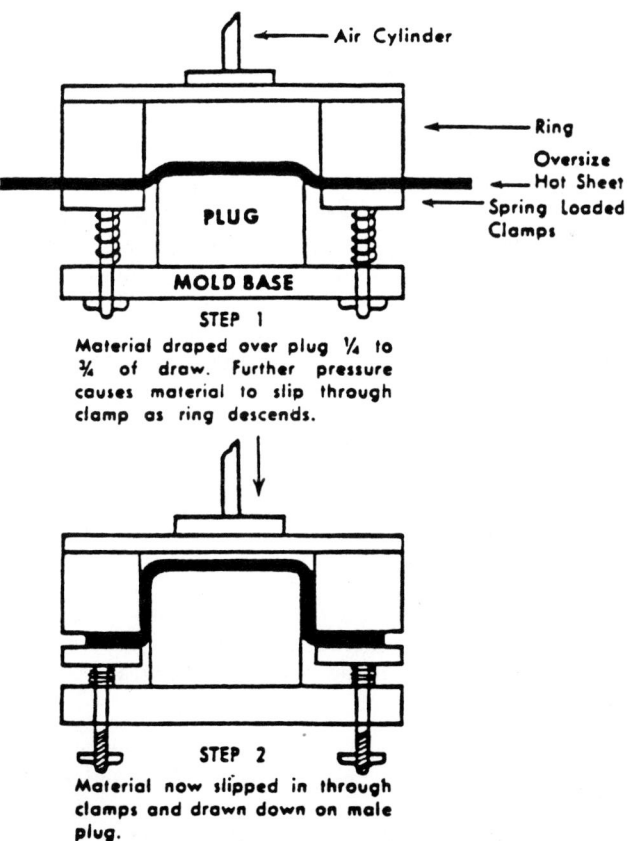

Figure 93.16. Slip forming using loose clamping frame.

Use Of A Four-station Rotary. Use of a four-station rotary is the most common technique of twin sheet forming. The four stations are

- Station #1: loading and unloading.
- Station #2: preheating the oven with top and bottom heater banks.
- Station #3: final heating the oven, also with top and bottom heater banks.
- Station #4: forming area with top and bottom movable platens having both compressed air and vacuum available to each platen.

Mold halves are fastened to opposing movable platens in the forming station. The first sheet of plastic is placed in a clamping frame and rotated to the first oven. While it is being heated, the second sheet of plastic (which can have a different color and/or finish) is placed in the second frame. When the first sheet reaches the forming station, the hot sheet can either be laid on top of the bottom mold with no vacuum or air pressure applied or held against the mold by vacuum and lowered out of the way. At this stage, orphan inserts for additional stiffness can be introduced.

The second sheet is usually timed to rotate into the forming station 15 seconds later. At this time both molds are moved into final position. Vacuum is applied to the molds and compressed air is blown between the hot sheets via a blow needle that is placed between the two hot sheets before the platens are finally clamped together. As with other types of thermoforming, carbon dioxide can be used as the positive pressure fluid to speed the cooling cycle.

At the same time, rigid urethane foam or other materials can be inserted according to the aforementioned patented process, which avoids the need for post-foaming tooling and produces a better part.

Some thermoforming machines are designed to permit plug assist techniques in either or both halves of the mold.

Twin Sheet Forming With Roll Stock. Two rolls of sheet stock are fed one over the other into a series of double stacked sandwich heater banks (two opposing heater panels installed directly over another series of heater panels). As the twin roll stock comes out of the heating area at forming temperature, it continues into the forming station. In this station the sheets are still kept about an inch or two (25–51 mm) apart, depending on the sheet gauge and the part configuration. The opposing molds are moved against the hot sheets, compressing the perimeters (and any other areas to be joined) together. Because the touching surfaces are at thermoforming temperature or above, they mate. The joint strength will be approximately 75 to 90% of the material strength. Vacuum is applied as soon as possible after the molds come together or just as they come together. Compressed air is usually injected between the two sheets at this time to complete the process. Exact timing of the vacuum and compressed air is of the utmost importance and will vary with different parts.

93.6.6 Slip Forming

In slip forming, the sheet is not gripped tightly as it is being formed. Not only does this allow the use of materials that have restricted stretch, such as carpet and fabrics, but it also enables more material to be taken out of the clamping frame area into the part when regular solid sheet is formed.

Loose Clamping Frame. The sheet is loosely held in the clamping frame, allowing the mold to drag the hot sheet through the clamp frame as it is moved into position. When the mold bottoms out the clamps close tightly or a clamping box is conveyed against the perimeter of the mold, thereby achieving a pressure seal. Vacuum and/or compressed air is then applied. Nonstretchable materials such as carpet, fabrics, and material with reinforcing nonstretchable fibers use matched molds with a loose clamping frame (Fig. 93.16).

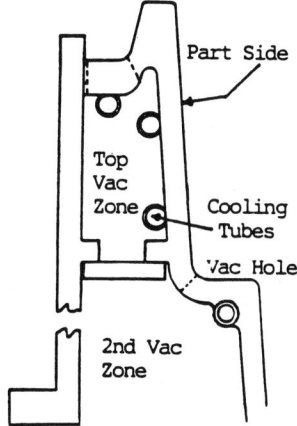

Figure 93.17. Slip forming with multiple vacuum zones.

Clamping Two Sides of Sheet. Using an automatically adjusting clamping frame with opposing sides moving in and out, the sheet can be held flat while it is heated. After the sheet has reached forming temperature, it is moved into the forming station. The opposing clamp sides gripping the sheet are activated toward each other, causing the plastic material to form a U-shape. As the mold or molds contact the sheet, it drapes across the mold. At this time a clamping ring overpress or a clamping box is positioned against the loose part of the sheet and the mold to form a pressure seal. Vacuum and/or compressed air is applied for the final shape. Matched molds can also be used at this time.

This technique is also used to wrap the sheet material partially around a cylindrical mold.

Use of Multiple Vacuum Zones. Where there are large flat areas in a mold that tend to freeze the material before that particular part has stretched as much as desired, multiple vacuum zones are employed (Fig. 93.17).

For example, in large containers with wide flat flanges, separate vacuum zones for the top portion (that includes the flat area and the rest of the mold) will greatly help in achieving a more uniform wall thickness. When utilizing a female mold, better distribution of the material will nearly always occur with the mold on the top platen. When a female mold is on the top platen and is pushed down into the sagging sheet, only the periphery of the mold touches the sheet. This action chills a cery minimum of the sheet. Vacuum is then applied only to the cavity part of the mold. As the sheet is swiftly drawn into the bottom of the mold, it stretches over the flange, which contains the second vacuum zone. Because the cavity (first vacuum zone) almost pulls the hot sheet completely against the mold, the second vacuum zone is quickly activated. This technique will move a lot of the extra thickness out of the flange area into the body of the part.

93.7 MACHINERY

93.7.1 Machine Safety

Safety is of utmost importance and should always be considered first.

93.7.2 Basic Machinery Requirements

Uniform Heat. Achieving uniform heat is the single most important function of thermo-forming machinery. Unless sheet to be formed is less than 0.030 in. (0.76 mm) in thickness, the oven section should consist of top and bottom sandwich heaters whose element temperature can be adjusted.

Radiation is by far the most commonly used method of heating.

Vacuum/Compressed Air Systems. Adequate surge and accumulator capacity should be located as close as is physically possible to the machine. No 90° elbows can be permitted in a vacuum system; instead, 45° elbows or curved "bends" with straight-in connections (if at all possible) should be used with a properly sized line.

Sheet Film Clamps. Clamps should provide enough uniform force to securely hold the film of the particular material to be formed.

Forming Platens. If moving platens are required, they must operate smoothly and parallel to each other and have enough force to move the mold in and out of the hot sheet. For pressure forming, they must be structured for pressures in use.

Control System. Microprocessors are desirable since, in thermoforming, repeatability is of great importance. Accurate control of all operations is essential.

93.7.3 Sheet-fed Single-station Machines

These machines are very popular. The most versatile will have a forming station with top and bottom movable platens. When air and vacuum are available to both platens with individual controls, all known thermoforming techniques can be used. Top and bottom sandwich-type heat with proper controls permit the use of all thermoplastic sheet and foam materials. Single-side heat can be used on thin gauge only.

93.7.4 Roll-fed Single-station Machines

These machines are usually used for packaging, blisters, and other thin-gauge work. As a rule, these are hand fed from a roll and cut with a guillotine each time parts are formed. They are popular among companies with lower volumes of in-plant skin packaging.

93.8 TOOLING

Well-designed and well-built tooling (molds, cooling fixtures and trimming fixtures) is a large part of achieving good, consistent thermoformed products. Molds with excellent temperature control and cooling efficiency are necessary for the production of economical, quality parts.

Following is a checklist of design considerations for any thermoforming tool:

- Male or female mold.
- Type of production.
    Prototype
    Complicated prototype
    Close tolerance prototype
    Part quality—low, medium or high production
    Delivery speed
    Pressure-formed part
    Break-away tool
    Post-cooling fixtures
    Trim fixtures

Special heat
Any other special requirements.
• Mold construction and material.

93.8.1 Male or Female Mold

In choosing whether to use a male or female mold, the following points should be taken into account:

• Which side of part needs the detail?
• Where on the part are tolerances needed, or critical?
• Male molds are cheaper to make than female molds.
• Are different parts required to fit tightly to each other or to be bonded together? If so, mating surfaces should be formed against the mold so tolerances can be controlled. This ensures that parts will go together easily and precisely.
• In female molds, multiple cavitites can be spaced closer together than male molds, thus utilizing a smaller blank.
• Closer tolerances can be held on male molds or male portions of a mold. As the part cools it has a tendency to shrink away from the mold in female sections, making very close tolerance more difficult.
• With female molds, the flange area wall thicknesses are the greatest and the bottoms of the cavities are the thinnest. With a male mold, this thickness variation is just the opposite.

During the forming process, the part of the hot sheet that touches any segment of the mold first will start to "freeze," causing a greater wall thickness in this area.

• Prestretching and profile heating techniques can be used with male or female molds to achieve a more uniform wall thickness or to add thickness variation where this is desirable.
• Are there undercuts in the part that will require break-away sections?
• Are matched molds required?

93.8.2 Mold Construction

Table 93.3 identifies the thermal conductivity properties of various tooling materials.

For prototypes and very short runs, the least expensive tool is generally wood. Wood (and epoxy and polyester also) enable the thermoformer to make his or her own tool. However, wood, epoxy, and polyester are such heat sinks it is almost impossible to achieve any consistency in temperature or a competitive cycle when they are used. Their heat-transfer rates (K factors) are below 1.

In higher-production or precision thermoforming, aluminum must be used for reasons of efficiency and competitive cost. Ease of manufacture and an excellent heat transfer rate characterizes the use of aluminum. Aluminum has a K factor of 115, approximately several hundred times greater heat conductivity than wood, epoxy, and polyester.

To produce consistent, quality parts at a competitive price, it is essential to control accurately the mold temperature, pressure on the part (vacuum and/or compressed air or matched molds), cooling rate from both sides of the part, length of time on the mold, and post-cooling if needed. Smaller molds are usually machined, as it is difficult to bend the stainless steel cooling tubes into sharp radii.

Cooling and vacuum channels are machined or gun-drilled. All seams and joints have to

TABLE 93.3. Tooling Materials Thermal
Conduction Properties

Material	"K" Factor[a]	Heat-Transfer Rate Factor Compared with Felt
Air (reference)	0.016	0.76
Wool felt	0.021	1.0
Spruce	0.052	2.5
Syntac 350[b]	0.070	3.3
Maple	0.094	4.5
Epoxy	0.131	6.2
Plaster of Paris	0.174	8.3
Alum-filled epoxy	0.50–0.99	24–47
Acrylic	1.4	67
Stainless steel 316	9.4	448
Bronze	20.5	976
Steel	26.0	1238
Kirksite	60.4	2876
Aluminum	115	5476

[a]K = Btu/(h)(ft^2)(°F).
[b]Syntactic foam by W.R. Grace.

be sealed tightly to avoid any leakage. The more uniform the wall thickness, the easier it is to uniformly cool the part and manage any warpage problems. Cooling plates are quite frequently used for shallow draw parts. When deep draw molds are used, cooling pins (heat pipes or thermal pins) can be used to achieve even cooling. Pencil sized holes are drilled in the vertical walls, about 1-1/2 in. (38 mm) apart, and slim, ultra-conductive pins are wedged in for a ream fit. They are relatively expensive, but very efficient.

Standard cooling plates, up to a 50-by-50 in. (127-by-127 cm) mold size, are available from Edward D. Segen & Co., Devon, CT 06460. Plates are in stock (only holes and plumbing to be added), so 7 to 10 day delivery is common. Addition of manifolds greatly enhances cooling.

For larger molds, stainless cooling tubes are bent and sand cast into the finished mold, enabling very consistent walls to be cast. Large molds demand a well-equipped, knowledgeable aluminum foundry, as a 1200 pound (550 kg) casting requires about 60,000 pounds (27,000 kg) of sand.

For the economical short run and prototype molds, it is best to find a good machine shop and aluminum foundry. To hold down costs on aluminum prototype and short run molds, small molds can be machined with no cooling channels; larger ones can be cast with no cooling tubes.

It is important to maintain the same wall thickness throughout the mold for uniform cooling. Ideal nominal wall thickness of molds up to a size of about 36-by-48 in. (91-by-122 cm) is 3/8 in. (0.953 mm). Above that size, a wall thickness of approximately 7/16 in. (11 mm) is used. Support posts or partitions should be placed on the back side of the mold shell every 6–8 in. (15–20 cm) so the vacuum pressure will not bow the mold during forming operations. Virgin Alcoa 356 or comparable grade aluminum should be specified.

These prototype molds are mounted on a 0.75 in. (19 mm) thick pattern grade birch plywood base. The aluminum can be preheated with a hot air gun or a propane torch. Mold temperature can be maintained between shots with the same type of heat or, if cooling is needed, a wet sponge or rag can be used with a bucket of water. A fine fog or mist water spray can also be helpful.

When making wood molds, it is important to use hard woods and avoid seams as much as possible. Heat tends to open seams. Epoxy and polyesters molds should be oven cured. Glass cloth reinforcement is also helpful. For one-shot prototypes with undercuts and negative draft, water soluble plaster can be used and then washed out after forming.

A fiber-glass part can be used as a prototype mold. The back side has to be supported, to avoid bowing or collapsing, and any needed vacuum holes must be drilled. The thermoformed part will be smaller than the original fiber-glass part because of to the shrinkage of the particular material used. If the finished part must be the same size as the fiber-glass part, the fiber-glass part should be sawed into as many pieces as necessary and separated according to shrinkage. Epoxy, glass cloth, or plywood can be used to finish the mold. A thermoformed, thermoplastic part can be used the same way. Caution is necessary to avoid warping the mold.

Metal spray has been used to manufacture a heavy mold skin. The spray is usually a zinc/aluminum alloy. A stainless steel-nickel alloy with a blowing agent can also be used to brush or pour a skin surface. The blowing agent is activated in an oven and thousands of microscopic vacuum holes are formed. The surfaces are usually built to at least 0.060 in. (1.5 mm) thick, with the better ones being much thicker. For both types, the finished mold shell is then backed up with a porous mixture for strength and vacuum. A common filler is epoxy coated aluminum slivers or beads. Copper cooling tubes can be inserted next to the surface shell before this backing is applied.

All thermoplastic materials shrink, the amount varying with the amount and type of orientation in the sheet when formed. The final part shrinkage can also vary with mold temperature (a hotter mold causes more shrinkage), time on the mold, whether it is formed on a male or female mold, and variations in the melt index of the sheet material. Various recessed areas, ribs, and projections will alter shrinkage from normal.

With these many variables, it is important to form prototype parts with "production-spec" sheet on easily modified prototype molds to determine the exact shrinkage. (Typical shrinkage values are given in each material chapter of this Handbook.)

Approximately 75% of a formed part's shrinkage occurs as the material cools from the processing temperature to the set temperature [assuming a material heat-distortion temperature at 66 psi (0.46 MPa).] On a male mold, additional draft can mitigate shrinkage.

93.8.3 Vacuum Holes and Slots

Fast, complete and continuous vacuum is a key ingredient in successful thermoforming. After the sheet material has been uniformly heated to its proper forming temperature, most of the internal stresses (orientation) have been relieved (annealed). If the hot sheet can be stretched into the exact shape of the part to be made while it is still at the forming temperature, without chilling any portion of it, the finished piece will have virtually no stresses. Consequently, the part will have the top physical and chemical properties of that particular plastic material.

The technique of free forming is easy but it is limited in application and relatively slow in cycle time. For most thermoformed parts, requirements dictate the use of molds to form against for both detail and cycles. In order to achieve the best all around results, an extremely fast vacuum and/or compressed air must be exerted.

Naturally, the larger and more frequent the vacuum openings are, the better. Where practical, slots should always be used for vacuum. This is most easily done with male molds, which can be mounted on a tooling plate with spacers, shims, or washers between the mold wall and the plate. This space should be as great as the sheet material and forming techniques will allow. With sheet materials like HDPE and PP, for example, that are in a soft, fluidlike condition at forming temperature, the slot will be from 3 mils [0.003 in. (0.076 mm)] to 10 mils [0.01 in. (0.25 mm)] thick. With 3/16–1/4 in. (4.8–6.4 mm) thick ABS, 20 to 40 mils [0.02 in. (0.50 mm) to 0.04 in (1.02 mm)] can be used. One or more large vacuum holes are made underneath the mold in the tooling plate for the main source to vacuum.

When the mold is solid, vacuum grooves will have to be fabricated on the reverse side. They should be 3/8 in. (9.5 mm) deep and a minimum of 3/8 in. (9.5 mm) wide. The grooves all over the mold bottom should be criss-crossed. Solid molds that are to be used on cooling plates need 30–50% of the mold bottom designed for direct, close-tolerance mating with the cooling plate.

Holes for vacuum should be drilled as large as possible without causing poor appearance on the finished piece of plastic or without allowing the vacuum to pull a hole in the finished part. All vacuum holes must be relieved with a 1/4 or 3/8 in. (6.4 or 9.5 mm) drill bit or a cone, to within 80 to 100 mils [0.08 in. (2.0 mm) to 0.1 in. (2.5 mm)] of the surface. A rough mold surface forms better than a polished one. For an attractive matte surface on the formed part, a #30 grit sand blast should be used. Polished surfaces of a mold have a tendency to trap air, thus giving a final part the look of having the chicken pox. Matte and embossed surfaces are easier to form and do not show small vacuum holes or scratches. Hole spacing should be 1 in. (25 mm) in recesses and corners and 1-1/2–3 in. (38–76 mm) on flat surfaces.

Core vents can also be used for help in faster air evacuation. Expanded polystyrene molders rely on core vents to rapidly inject steam. They are a small-diameter [approximately 3/8 in. (9.5 mm)] brass cylinder with one end closed. Narrow slots, commonly 2 to 8 mils [0.002 in. (0.50 mm) to 0.008 in. (0.20 mm)] are machined in the closed end. These are then swaged into the mold at desired locations.

Tubing can be cast into the shell of an aluminum mold for use with vacuum and plumbed accordingly. This greatly reduces the amount of air that needs to be evacuated.

Parts with a wide-flange being formed in a female mold can gain much better material distribution by using multiple cavity vacuum zones, which result in thinner flanges and thicker sides and bottom. In practice, the mold is built so that the back side of the mold flange area is sealed from the rest of the mold and plumbed separately. The mold is mounted on the top platen of the thermoformer and lowered into a hot sheet. At first, vacuum is only applied to the bottom section of the mold, thereby allowing the sheet to do more stretching over the flange area. Once the material is pulled into the bottom, the second vacuum zone around the flange area is induced.

A moat can be used anytime high mold shrinkage materials, such as polyolefins, are formed. Because of the high degree of shrinkage as the hot part cools, the thinner areas shrink sooner than the thicker sections. Stress lines are set up between the two, causing warpage in the outside areas. This also will occur in lower shrink sheet such as ABS where there are large flat flanges. Such warpage can be cured by placing a moat (trough or trench) just outside the trim line on male or female molds. The moat should be female, with the following dimensions:

Width = 3 times the starting thickness of the sheet
Depth = 2 times the starting thickness of the sheet

Use a raised, or male, ridge only as a last resort.

Porous molds give very, very fast vacuum and superb detail. As an example, automobile-door panels with fine stitching and embossing will show very realistic features using just good vacuum. This process of mold making is patented by Advanced Thermoforming Inc. in Detroit, Mich.

93.8.4 Temperature Control

In order to produce consistent, quality thermoformed parts efficiently, the mold temperature must be controlled. A hotter mold increases the final shrinkage. A 144 in. (366 cm) long HDPE part can be changed as much as 3 in. (7.6 cm) with variation in mold temperature.

The most effective and feasible way of accurately controlling the mold (and part) temperature is through the use of cooling tubes or channels in aluminum molds. Normally, cooling channels are spaced from 2–3 in. (50–76 mm) apart. The biggest single mistake made in designing and manufacturing temperature controlled tooling is lack of adequate cooling in the flange and/or thin partition areas. In most cases, the heaviest wall thickness of the formed parts is around the flange position close to the sheet clamping mechanism. This is the area that controls the cycle.

Tubing or channel diameter should be from 0.300–0.4375 in. (7.6–11.1 mm). Cooling manifolds should always be used in lieu of looping the ends of the cooling tubes together (although small molds can sometimes be efficiently cooled by just looping the connections). In order to take full advantage of cooling channels, a high turbulence of the cooling fluid must be maintained in the channels.

An increase in the flow rate from 2,000 Reynolds to 10,000 Reynolds increases the heat transfer coefficient by about nine times. In other words, the more turbulence, the better the cooling rate.

The in-and-out temperatures of the cooling fluid should never vary more than a total of 5°F (3°C) when parts are being thermoformed. For the best cooling, variance should be no more than 2 to 3°. If this is difficult to achieve, more zones of cooling need to be added. The direction of flow of coolant in adjacent channels should be alternated for more uniformity.

Aluminum molds that have no cooling tubes or channels can be made more efficient by mounting copper tubing on the back side. The tubing should be bent to fit, and the copper slightly tamped with a mallet to obtain a partial oval shape so better contact is made with the aluminum surface. It is important to either mechanically fasten the tubing to the mold or adhere it with a high-heat conductivity epoxy, such as Sycast 2850 KT 30 K factor compound from W.R. Grace or Zeston 17.5 K factor Z-10 material from Manville.

The ultimate cooling efficiency is achieved by incorporating manifolds in the plumbing of the coolant flow channels. On small molds, a 3/4–1 in. (19–25 mm) diameter copper tubing is used; on large molds, 1-1/2 in. (38 mm) diameter copper tubing.

93.8.5 Break-Aways and Undercuts

Parts to be thermoformed with undercuts and negative draft are quite commonly produced. A popular method of forming a part with an undercut flange area is to use a hinged undercut section. This can be done on four sides of a part, if required, with the corners mitered for a very close tolerance fit. They can be actuated by gravity, by the operator with a pole or a stick, by spring loading, or by mechanical means.

Removable sections (that come out with the part) are often used. The operator removes the section from the formed part and returns it to the mold before the next part is formed. The undercut section can also be mechanically actuated. In continuous operating machines, removable segments are brought into action with springs or air cylinder linkage. Threaded sections can be mechanically unscrewed.

When a slight undercut or reverse draft is necessary, stripper plates are often used quite effectively. They are either spring loaded or mechanically operated.

During a roll-fed, continuous machine thermoforming operation, a spring loaded or air-operated mechanical break-away section is usually used. Threaded sections can be unscrewed mechanically. Hard wood prototype tooling is very helpful in determining the exact configuration.

93.8.6 Orphaned Inserts

To increase the stiffness and strength of finished parts, molds can be designed with a recessed section to hold an insert such as a wooden rod or profile shape, metal rod, angle, or tubing. These are placed either by hand or mechanically onto the mold. During the thermoforming process the hot sheet wraps around the insert tightly, thus capturing it. It is important to leave room for the plastic to expand or contract longitudinally if the stiffener does not have the same coefficient of expansion and contraction.

To achieve rigidity in a flat part, high-density rigid urethane foam sheet, plywood, and other flat structural pieces can be used. The encapsulation is usually done on two opposing sides with the insert being small in relation to the size of the formed part.

A two-step forming for undercut parts is also practiced. The undercut portion is formed straight with no negative draft. After trim, the undercut area is strip heated and bent to shape.

93.9   AUXILIARY EQUIPMENT, 93.10   FINISHING, 93.11   DECORATING

93.10.1   Trimming Options

There are many efficient ways of trimming thermoformed parts. The process requires that a sheet or roll of plastic be clamped in some manner. As a result, it is necessary to select a safe and economical way of separating the portion of the finished part that is clamped and not desired from the finished part itself.

There are eight major trimming options:

- Routing and drilling
- Sawing
- Punching, compression, and shearing
- High-pressure water jet
- Thermal lasers, hot wire/knives
- Robotics
- Deburring
- Lip rolling

93.10.2   Routing and Drilling

The slowest method of trimming is by router; of hole making, by drilling. However, if compound trim shapes are required, these may be only options.

Routers. Routers are stationary mounted under a table surface, overhead in a drill press or in a C-structure frame platen, or on a vertical table board tilted back approximately 20°. They may also be hand operated.

When table mounted (horizontally or vertically), a sleeve or tube is frequently inserted around the cutting router bit and raised above the surface of the table just enough for the trim fixture to ride against. This sets a precise distance for the router to cut a flangelike area. A floor-type flange that the template can ride against can also be mounted on the table surface. If recessed pockets are encountered, a long shaft router bit can be mounted on a table with a cam-follower (ball bearing) fastened close to the top. The trim fixture can thus ride against the ball bearing, setting the distance that the router bit cuts into the flange surface. To trim a flush inside pocket, a veneer saw can be inserted in place of the router bit with the cam-follower just below the veneer saw.

Table mounted routers can also use a bolt or rod that is mounted above the part on a shelf or in an upper frame C-section for the template or the part itself to ride against.

Both electric and air routers are used. Air routers are preferable because they provide a higher rpm (around 35,000) giving a better and faster cut. Electric routers will range in speed from 18,000 to 25,000 rpm.

A developmental turbine air router operates at around 60,000 rpm.

Trim Fixtures. The key to accurate, quality trimming is the fixtures used. Well-designed, substantial trim fixtures always pay off. If they can be justified, all-metal trim fixtures should be made. In many cases, a skeleton framework with proper clamping mechanism (to hold the thermoformed part tightly while it is being trimmed) will work very well. Round, rectangular, or angle steel or aluminum tubing make very practical fixtures. When using plywood, a top quality birch, pattern grade or plywood should be selected. The plywood edges should be reinforced with metal to increase the template's life and provide more accuracy in the trimmed part.

For eccentric, compound shapes, fiber-glass trim fixtures with metal edges can be used. Combinations of wood and epoxy with metal-lined edges are also common.

Drilling templates should employ drill-jig bushings epoxied into precise position. It is essential for them to be positioned at a 90° angle to the drill.

If one operator will be hand routing small and large parts, C-type structures or gap-frame presses are often built with moving air-cylinder platens top and bottom. These will support the mating trim fixtures to hold the part securely. They are also built with a fixed bottom platen or table. This way, the operator can release or secure the part with a foot treadle and walk around the fixture or have the fixture rotate.

Another approach is to hang the top platen structure from the ceiling. A work table the size of the part or the trim fixture is moved under the hanging platen. The top part of the sandwich trim fixture is attached to the upper movable platen. Thus, after the sandwich trim fixture is closed securely, the operator can walk completely around the part during trimming. This same top fixture can also be held tightly to the formed part by vacuum without the hanging platen.

On very short runs a second operator goes ahead of the trimmer removing clamps and replacing them as the part is trimmed.

Sandwich-type metal, fiber glass, or wood and epoxy fixtures work quite well for router trim or drilling.

In deep drawn round, oval, or rectangular shaped parts, a metal frame can be bent and welded to exactly fit the outside of the finished part. An insert can then be made as a male wooden frame-plug to fit tightly inside the part, holding it securely while it is being trimmed. It is essential there be no chatter or vibration in the part as the cutting tool moves around it or as a hole is drilled.

Power or manually operated clam shells are also used, as well as two- and four-post air presses.

Router Bits and Drills. Router bits and drills should be carbide tipped and kept sharp. Most plastics are better cut with a drill that has zero rake on the cutting edge; in some cases, even a little negative rake is helpful. Resin suppliers are excellent sources for information on proper machining, cutting, and drilling tools.

In parts that do not require great accuracy for hole locations, a dimple can be formed in the part as a hole-locator and then drilled freehand. Many designers have a dimple formed wherever there will be an opening or hole cut as a quality control check. If a finished part is found with a dimple rather than an opening, it is sent back for rework.

Whenever practical, the designer should recess any holes or rectangular openings slightly past the part thickness. The back side can then be routed or sawed off, leaving an extremely smooth, sharp looking opening that follows the exact contours of the mold. This is especially popular in pressure forming.

93.10.3 Sawing

Sawing is a faster method of trimming than routing. It is necessary to select special saw blades for different thermoplastic materials. Again, one of the best sources of information about the type of blades to use is the resin supplier.

There are many sawing methods:

- Circular table saw
- Circular panel saws
- Radial saws
- Swing saws
- Veneer saws
- Band saws
- Horizontal band saws
- Jig saws
- Sabre saws
- Hole saws

Circular Saws. Saw blades with carbide tipped, square, and advance teeth (sometimes called triple chip teeth) are recommended for circular saws because they give a superior quality of cut with fewer blade changes and much faster cutting. To obtain the most satisfactory cuts possible with these blades, the saw and face plates must fit the arbor well [approximately 0.001 in. (0.025 mm) clearance] and run true. As soon as they start to dull, they should immediately be sent back for sharpening.

Table Saws. Table saws come in a variety of sizes from small, light-duty home workshop models to large, heavy production models. A medium-duty well constructed model with a 3/4 or 1 hp (560–746 W) motor is very popular. Generally, equipment of this type is used for cutting small parts to close outside dimensions. Special jigs are useful for holding the work and ensuring the accuracy of the cut. To prevent chipping, the height of a circular saw blade above the table should be just a little greater than the thickness of the section to be trimmed. The part must be held firmly so that the saw moves through the sheet in a straight line parallel to the saw blade without binding or chattering.

Circular saws should run at 8,000–12,000 rpm [a 12 in. (30 cm) diameter circular blade should run approximately 3000–4000 rpm for most materials].

Radial and Swing Saws. Radial and swing saws do the moving while the work is held stationary. They are used to make angle cuts and cross cuts in narrow material.

Panel Saws. Panel saws are used for large parts where there are straight flanged cuts to be trimmed. The saw and motor can either be mounted above or below the material to be cut. The work is clamped into position and the saw is fed through by hand or power. Vertical panel saws have the advantage of using less floor space, making it easier to load and unload large parts. Panel saws are not easily guarded and thus require careful thought for proper safety. A vacuum sawdust removal system is definitely recommended.

Band Saws. Band saws are normally used for curved sections but can also be used for shorter straight areas. Most newer band saws are built so that the speed of the blade can be changed to suit the work. Normally, the thicker the material the slower the speed to prevent overheating. Speed, feed, and thickness of stock should be such that each tooth cuts a clean chip. Depending on the material, thickness, and rate of feed, the blade should run at speeds of 2300–5000 rpm.

Metal-cutting blades are best for band saws. They stay sharp longer than the softer woodworking blades, but when dull, they should be thrown away. (This is cheaper and better in the long run than sharpening the wood-cutting blades.)

Special band-saw blades known as skip tooth or buttress saws have been developed especially for plastics. They are designed with an extra gullet capacity for large chips and maintain maximum strength, even in the narrower widths. They will keep sharp for longer periods of time, especially with the softer sheet materials.

This type of blade works well in 1/8–1/4 in. (3.1–6.4 mm) thick material with a 5 to 6 pitch teeth per inch (25 mm) and a bank speed of around 1500 fpm (460 mpm). On material that is greater than 1/4 in. (6.4 mm), the teeth should be 3 to 4 pitch teeth per inch and the speed around 1000 fpm (460 mpm). On thicknesses less than 1/8 in. (3.1 mm), metal cutting saw blades with 8 to 12 pitch teeth per inch (25 mm) and a speed of around 2000 fpm (6000 mpm) should be used.

Horizontal Band Saws. The horizontal band saw, a deep throated band saw laid on its side, is becoming popular. A table or other flat working surface with adjustable height is placed under the saw blade. Hat-shaped parts to be cut straight off under the flange are easily trimmed with this system. Such items aas luggage shells, ice-chest liners, and housings with a single horizontal cut are trimmed this way. The parts can powered automatically through the saw blade or moved manually.

For safety, guards can be built on the trim fixture itself.

Jig Saws. Jig saws can be used for cutting closed holes and openings in thin material. They do not cut well through thick sections because the stroke is short and the blade does not have a chance to clear the chips. When a jig saw is used, the feed must be light and the teeth cleaned off. Welding of the kerf (wet sawdust) behind the blade may be overcome by using two blades mounted side by side, placing a piece of corrugated fiber board beneath the plastic, or using a coolant. It is very important that the part be held securely to prevent vibration.

Sabre Saws. Sabre saws are chisel saws (sometimes called portable jig saws). As in all trimming, the part should be tightly supported to prevent vibration or chatter created by the reciprocating saw blade, which can tend to chip or crack the part. Chisel-type portable saws should be adjusted so that the cutting chisel stroke is about 3/16 in. (4.8 mm) greater than the thickness of the work to be cut. Two thicknesses of corrugated fiber board should be placed on the working surface under the part to be cut.

Veneer Saws. Veneer saws are small saw blades made approximately 4 in. (10 cm) in diameter and about 0.045 in. (1.1 mm) thick with 11 points and considerable set [0.025 in. (0.64 mm) both sides]. They are mounted on an arbor and are powered by high-speed electric or air router motors, which are quite commonly mounted underneath the trim table or in an overhead drill fixture to trim a deep part horizontally in the same plane all the way around. Because a saw blade cuts faster than router bit, it is preferable in these circumstances.

Like routers, veneer saws are difficult to guard and must be used with care. A veneer saw must not be used if there is vibration when the motor is running at full speed.

Hole Saws. Hole saws are a tubular tool with saw teeth filed on the lower end of the tube. The teeth have a set to cut a groove wider than the thickness of the tube wall. A shaft is fastened to the top of the tube so that it can be mounted in the drill press that is used to drive the saw. Usually a pilot drill and guide are provided in the center to locate and center the hole saw.

Hole saws can also be used in a free hand drill motor, using dimples in the formed part or other markings for location. In heavy gauge material the hole is cut half way through, then completed from the other side.

93.10.4 Punching, Compression, and Shearing

Punching, compression, and shearing are much faster methods of trimming by router or saw. However, to reduce costs of trimming fixtures and tools, the trim cut should be designed to be made in the horizontal plane. There are four main methods:

- Steel rule die punching (stamping) and compression
- Matched steel-rule die shearing
- Matched tool steel shearing
- Shearing and guillotining.

Steel Rule Dies. Steel rule dies are the most economical in this category and are normally used on materials 1/8 in. (3.2 mm) thick or less. They consist of highly hardened steel knife blades precisely bent and inserted into a piece of top grade pattern, birch plywood. A metal backing such as a sheet of aluminum is commonly mounted on the reverse side. Pieces of neoprene sponge rubber are bonded close to the knife blade to be used as ejectors or spring loaded pins or metal strips. The knife blades can also be welded to shape and to a metal backup plate. To speed tool changeover, it is important that all the dies have the same overall height and mounting locations. This is commonly standardized at 2-3/4 in. (70 mm). An obvious exception is for deep draw parts where the cutting is done from the deep side of the thermoformed area.

When steel rule dies are used, the backup plate that the knife bladese will cut against should always be harder than the material being cut. Stainless steel makes a good backup plate. However, unless the press has microprocessor controls such that the descending cutter is stopped precisely after it has cut through the material and goes no further, very precise tooling is required. Without these controls, stainless steel requires constant checking for proper adjustment.

Quite frequently a backup material of a hard plastic such as 3/8 in. (9.6 mm) polypropylene or nylon sheet is used. These are very resilient and long lasting, and can usually be rejuvenated after long use by heating in an oven. Better rejuvenation is achieved when a radiant heater is used for softening the surface enough to bring it back to its original condition.

In cutting materials that are brittle or heavier-gauge, it may be necessary to use a piece of hard rubber, polypropylene, or nylon as a backup material to prevent excessive cracking during the cutting operation.

The life of a steel rule die blade can be from 50,000 to 200,000 cuts before sharpening is necessary.

Matched Steel-rule Dies. Matched steel-rule dies are also used for longer life and better finish on the cutting edges. Forged knife blades are bent and welded into shape for both the top and bottom shearing surfaces. For example, on a HMWPE snow sled with compound curves in the trimmed flange, over a million and half pieces were trimmed with no signs of wear on the forged knife blades. These tools cost more than the steel rule die; nevertheless, they are a great deal cheaper than matched tool steel.

Matched Tool Steel Shearing Dies. Matched tool steel shearing dies are very common for long run products that are trimmed in line as well as post trimmed. Such products as disposable cups, creamers, blisters, and picnic ice-chest bodies are trimmed this way. Matched dies make a very good cut and give long wear. However, they are expensive and take a long time to produce. Stripper plates are normally used on trim fixtures and tooling that are spring loaded or mechanically actuated. These are appropriate for slight undercuts or the more brittle and/or notch-sensitive materials.

Shearing or Guillotining. Shearing or guillotining is frequently done when the thermo-formed part has a rectangular or square trim on the periphery. Parts are cut one side at a time.

Shearing or guillotining technology comes from metal and paper cutting shears. It is quite often used to precisely square blanks of sheet material or to rough trim parts before putting them into a final trim fixture.

Trim Presses. Included are air-operated arbor presses or four-post line pressure air and air-operated two- and four-post presses.

Air-operated presses are routinely used when there is only a small amount of trimming to be done and not much pressure is required.

Arbor presses, with 8 and 10 in. (20 and 25 cm) diameter short stroke cylinders and a 125 psi (0.9 MPa) compressed air system, have 6283 lb (28 kN) and 9817 lb (44 kN) of working force. They are practical for holes and special-size openings. Two- and four-post air operated presses do well on thin gauge.

Air Over Hydraulic Presses. Air over hydraulic pressees are very popular, especially with the microprocessor controls available. Air is used to speed up the cycle by retracting the dies rapidly, then moving them into place for the cut. Just before they touch the material, the hydraulic takes over. With the latest microprocessors, the pressure can be programmed such that the cut can be made under hydraulic pressure or a slow compression force. This gives a wider choice of material and thickness and saves tremendously on knife blades because of the accuracy of stop. New high-speed hydraulic pumps makes these quite effective units.

Hydraulically Operated Presses. Hydraulic presses usually have four adjustable mechanical stops built into the moving platen to halt the up or down platen action exactly and offer the most pressure. They operate very smoothly.

Straight hydraulic presses with microprocessor controls and high speed pumps are generally used for higher production. These are the most expensive for trimming but the most precise, heavy duty, and consistent. A 100 ton (9.8 kN) hydraulic platen press will cut up to 1000 linear in. (28 m) of steel rule die on PVC.

Fly Wheel Trimmers from the Metal Industry. Portable C-frame or insert style hydraulic cylinders can be mounted on a fixture in any position. Thus, difficult-to-trim holes and odd-shaped openings can be conveniently trimmed.

Clicker Die Cutting Press. The clicker die cutting press is a swing beam hydraulic post press developed in the leather industry in which the head swings away. The stroke is short, usually only 1 in. (25 mm). The precision of cut is more difficult to control than a two- or four-post press because there is only a single post. If a cut is made far from the post, it may have to be shimmed up as the job progresses.

Clamshell Presses. Clamshell-type trimming can be manually done or power operated. A hinged platen holds the cutting die.

Roller-Die Cutting Machines. Roller-die cutting machines are increasing in popularity because of their lower cost and ability to trim larger parts. This method originated in the printing industry, and consists of two large-diameter rubber-covered rollers that are spring-loaded and adjustable. The part to be trimmed is placed in its fixture with the steel rule die on the bottom or on top and then pushed into and between the rollers. The rollers are powered and draw the part through, applying pressure in just one small area across the transverse direction of the rollers. Thus much lower pressures are required.

Steel-Rulle Die Cutting Pressures. Typical cutting pressures for a few materials are shown in Table 93.4.

Hand Scissors. Hand scissors or even a paper cutter can be used on certain products.

Heating the Cutting Dies. Cutting dies are heated in instances where the finished part (or flange area) to be trimmed or the plate that the die will cut against is of brittle material and the finished trim needs to be of better quality. With brittle high-impact styrene and some other materials there may be feathered edges, strings, or cracks. One method of removing these flaws is to heat the steel-rule die-knife blade by fastening a thin tubing tightly around the outline of the die. Hot oil is run through the tubing to heat the cutting blade to a predetermined temperature (generally, close to or above the thermoforming temperature). The die can also be heated electrically by fastening flexible heating tape to the upper portion of the cutting die or using a metal backup plate with strip heaters.

For the most efficient cutting, a press should be used that, when activated, positions the cutting knife on top of the plastic and allows it to rest there for a short period of time. The heat softens the material so that the blade will shear as it continues to move through. With materials approximately 0.060 in. (1.5 mm), thick two to three seconds is usually ample; longer times may be necessary for heavier materials. The advantage of this method is that, besides improving the edge on the part, less pressure is required to perform the trim.

This type of die costs more, but its advantages, longer operating life without having to sharpen the die and low pressure requirements, usually can greatly offset the increased cost.

On many roll-fed in-line continuous thermoforming machines, the mold can be designed so that the part can be trimmed immediately after the vacuum has been drawn and while it is still on the mold. The material will still be very soft and require little pressure. Because it is always trimmed at the same temperature, the size will be consistent. With some larger sheet-fed parts, where the volume justifies the expense, the thermoformed part can be partially or completely trimmed on the same mold in the forming station. Volume normally dictates whether this is practical or not.

There are other methods of cutting the material involving heat to give better edge finish, tolerance, labor saving, and the ability to use lower pressure. The part can also be reheated to just below the heat-distortion point and then trimmed. Or, the metal plate against which the die cuts can be heated. The part to be trimmed is pushed onto this heated plate for 2–4 seconds, thereby heating the surface where the knife blade breaks out. This can minimize fracturing when the material is brittle. Of course, too much heat on the plate or too long a dwell time can cause warpage; consequently, timing is critical.

With continuous roll-fed in-line systems, the formed parts in the web are carried from the forming station onto a heated metal plate and allowed to dwell 2 to 8 seconds. A compression cut rather than a stamping cut leaves a better edge finish. Plate temperature can be adjusted for best results and, if necessary, a second hot plate station can be used.

TABLE 93.4. Force Required in lb per Linear in. (N/cm) of Cut in Following Materials

Gauge of Material, in.	CAB	Heated Acrylic—320°F 160°C	Acrylic/vinyl	OPS
0.040 (1.0 mm)	280 (490)		250 (440)	2000 (3500)
0.093 (2.4 mm)	650 (1140)		500 (880)	4650 (8130)
0.125 (3.2 mm)	875 (1530)	125 (220)	700 (1200)	
0.187 (4.7 mm)	1310 (2300)		1000 (1800)	

93.10.5 High-pressure Water Jet Trimming

Approximately 1100 water jet systems are currently being used to cut plastic parts. Cutting speeds of 1300 fpm (396 mpm) can be achieved in some materials in a straight cut. Average speed is determined by material thickness, type, and the contour to be cut.

Water jet cutting is the severing of a material by a very straight, very fine stream of water moving at high velocity and under extremely high pressure. Water is emitted at speeds up to 3000 feet (900 m) per second (approximately mach 3) and pressures up to 55,000 psi (379 MPa).

Noise levels can be high. As the water jet exits the material its speed reduces from supersonic mach 3 to subsonic and generates noise up to 105 decibels. These high levels can be brought within the OSHA standard with relative ease. A properly designed catcher needs to be installed on each job. Besides reducing the noise level it also carries off what little water is developed along with the kerf.

For tough materials, abrasives can be injected into the water stream for more efficient cutting. The closer the cutting head is to the surface, the better the finish of the cut. Thermoformed auto carpet, panels, tubs and showers from acrylic sheet that has been rigidized with fiberglass, HMWPE housings, fiber-reinforced composite parts, and many others are being very successfully trimmed by this method.

Water Jet Cutting Advantages. The advantages are

- Fewer frayed or jagged edges.
- Clean cutting separations.
- Cutting speeds can be increased.
- Minimum attention necessary.
- Same device cuts variety of different materials.
- No tool wear.
- No tool sharpening.
- No heat developed.
- Does not fracture and leave chips or foreign matter in the product.
- Provides dustless cutting.
- Minimal lateral forces are developed.
- Virtually no vertical tool loading.
- Allows use of lightweight, low-cost trim fixtures.
- Omnidirectional cutting is possible.
- Lower operating costs.
- Lower maintenance costs.

93.10.6 Thermal—Lasers, Hot Wire/Knives

Laser Cutting Equipment. Laser cutting equipment normally requires a larger initial investment than water jets. It does, however, provide about the thinnest possible cut, which can be very important in extremely tight-tolerance applications. It is much quieter and more reliable than water jets and requires much less maintenance. Lasers can easily vary the depth of cut, which can be a big advantage in some applications.

One disadvantages is that it does create more heat than water jets, which must be considered with heat-sensitive materials like some advanced composites. Heat also has a tendency to set up more internal stresses at the edge of the part, so that the trimmed part may have to be annealed afterwards.

Cutting rates vary according to material composition and thickness and the wattage of the unit. For example, an 1800 watt unit will cut 1/2 in. (13 mm) acrylic at a rate of about 120 in. (305 cm) per minute. It will also leave a nice polished edge on the acrylic.

Hot Wires. Hot wires can be used to cut certain materials that might have eccentric shapes or that are thin and have a compound curve. Some thermoformed PE foam parts also lend themselves to a resistance wire cut. An example of this would be a Texas-type ten gallon hat using a single strand hot resistance wire. Sometimes heated blades are used such that a resistance tape is placed over a steel saw blade and pushed through the material. Hot gas jets are not very popular because it is so difficult to keep them from scorching or degrading and discoloring the surface.

93.10.7   Abrasive Wheel Cutting

Thermoformed parts rigidized with fiber glass such that that straight edges are required are frequently cut with abrasive wheels. The use of 30 to 200 grit surfaces will produce a relatively smooth cut at high cut-off rates with about one half to one third the heat generated by toothed saws and 5 to 10% of the operating costs of toothed wheels. Only straight cuts are feasible with this method. Besides fiber-glass reinforced sheet, many types of thermoplastic materials can be so trimmed.

93.10.8   Robotics

Three- and five-axis robotic trimmers are becoming quite common for parts that cannot be trimmed on the thermoforming machine. Robotic trimmers offer three big advantages— accuracy, repeatability and labor savings. Once the part is fixtured, it can be trimmed automatically. Normal standard tolerances are $+/- 0.010$ in. (0.25 mm) on parts as long as 8 ft (2.4 m). Closer tolerances are available with special machines. Drills, routers, lasers, and water are all used economically and consistently.

The demand for quality and consistency in finished parts is causing these machines to become very useful to the thermoformer.

93.10.9   Deburring

When they are trimmed by punching or shearing, most materials can avoid deburring. However, tools must be kept sharp with an accurate and precise trim fixture setup.

Some materials cut more cleanly than others. A change in the formulation (such as adding more rubber to high impact styrene) can minimize or eliminate feathered or stringy edges. Because of the notch sensitivity of most materials, it is very important to have a smooth cut to obtain the maximum impact and physical properties in the finished part.

Deburring can be accomplished by scraping, filing, sanding, buffing, and flame polishing.

During the deburring operation, the operator must take care that the part is not deburred out of tolerance by removing too much material or nonuniformly removing material along the edges.

Scraping requires a sharp 90° angle in the scraping tool via either a rectangular or V-notch metal bar. The harder the metal, the longer it will last before requiring sharpening. Many types of files can be used, from the very fine to coarse. For sanding, hand or table operated power sanders are frequently used.

Buffing can smooth the edges as well as remove imperfections from the surface, particularly for clear materials. To avoid burning the surface, buffing wheels need to be lubricated with wax. Mild abrasives can be used to eliminate scratches. The final buffing can be done with a jeweler's rouge or toothpaste and then a lubricated buffing wheel.

Flame polishing is a fast method of barely melting the edge surface so a smooth appearance is achieved. An open flame or a hot air gun with a fine nozzle is passed across the surface. Heat polishing does set up more stresses in the edges, and on very brittle materials this may not be advisable.

93.10.10 Lip Rolling

Special devices are available from thermoforming machinery suppliers for post lip rolling of drinking cups.

93.10.11 Trim Summary

Exhaust and chip removal systems are highly recommended. With many materials, these are the only satisfactory way to ensure a well-trimmed part and a clean work area. The use of an antistat (or an antistat material) or ionized air also is extremely helpful in eliminating dust and dirt problems. Many trimming operations can be improved by applying an air–water mist or a soapy mist on the cutting tool and material as a lubricant.

A washing detergent mixed in small quantities with water for dipping or spraying the parts will reduce static for a number of hours or even days. Sheet material can be ordered that is electrostatically treated as it is extruded, producing a surface that will minimize dust attraction and improve bonding or foaming-in-place.

When using a heated die or heated table, the dwell time of the part and the die must be precisely controlled.

Hot oil for heating dies is normally heated to 600°F (316°C). For higher temperatures, electric strip heaters or flexible heat tapes should be used.

If parts are to be painted or used in a difficult bonding operation, it may be desirable to use electrically operated trim fixtures rather than compressed air. It is very difficult to keep compressed air completely oil-free. All compressed air systems should have very efficient dryers.

Reinforced fiber parts and filled material require a different approach to the trimming. Slow cutting speeds are important. To drill holes in glass-reinforced parts, drill speeds should not exceed 350 rpm though 1/4 in. (6.4 mm) diameter holes. For larger holes a slower speed must be used—no faster than 175 rpm. As with any trimming, the part has to be securely clamped to ensure stability while drilling or trimming. A band saw is usually the best means for cutting a reinforced part. The recommended blade speed is 1000 fpm (305 mpm). The slowest speed on the machine can be used to reduce blade wear; an alloy steel blade can be operated at higher speeds.

Diamond cutting wheels are used to make straight trims on reinforced plastics. When mineral-preinforced materials are trimmed, a coolant should be used to keep the temperature down. It is important not to bear down too heavily when trimming or sanding these materials.

93.10.12 Rigidizing

It is very difficult to thermoform parts with ribs or stiffeners, so parts with these requirements have to be rigidized some other way. There are three common methods of rigidizing thermoformed parts: bonded doublers, fiber-glass sprayup, and foaming-in-place.

93.10.13 Bonded Doublers

Two thermoformed parts are bonded together to add stiffness to the composite part. To be effective at least one portion of the part that is bonded should be as close to 90° angle as possible. The idea is to achieve an I-beam.

This is similar to spot welding corrugated sections of metal onto a plastic sheet for stiffness. In thermoforming, corrugated sections and thin gauge can also be laminated to a flat sheet or contoured shape. To be effective, the joint must have an excellent bond.

Orphan inserts can also be bonded in between the two parts to supply greater stiffness. Thermoformed inserts that match the contours of a particular product can be bonded in a shell-like case for both stiffness and practical packaging.

93.10.14 Fiber-glass Spray-up

A superb structural part can be obtained by spraying up chopped fiber glass and a compatible polyester resin with acrylic sheet or other type of thermoplastic that adheres well.

The process consists of thermoforming a thin skin [average wall thickness, 0.035 in. (0.9 mm)] a nominal finished wall thickness, and then spraying the reverse side with polyester resin reinforced with chopped fiber glass. To obtain better adhesion, the surface of the thermoformed part may be wiped with a chemical that creates an etched effect; a thin layer of fiber glass and resin is then sprayed on. The fiber glass must be quite tacky before the final build-up is applied.

As in bonding doublers, it is important that the thermoformed part and the polyester resin have a good chemical bond. If additional stiffness is needed, orphan inserts can also be added.

93.10.15 Foaming-in-Place

Inside and outside parts are first thermoformed and trimmed to size. The outside part is then placed into a female fixture that matches the contours of the outside surface. When it is in place a measured amount of polyurethane foam in a frothing state is added. The inside thermoformed part is placed in position and a mold matching the opposite side is clamped on. As the foam expands it chemically bonds to the sides of the parts, eventually filling up the cavities between the two.

ABS sheet material is popular in this application and is highly compatible with the proper grades of urethane foam. Again, an excellent bond between the foam and the skins must be achieved for full structural strength. In some cases, the surfaces of the thermoformed parts may need to be oxidized with an open natural gas flame, a Corona discharge treatment, or an adhesive spray. The thermoformed part to be foamed should have a wall thickness of 0.035 in. (0.9 mm).

Foaming fixtures must match the contours of the inside and outside of the part for good quality control.

93.10.16 Checklist for Troubleshooting

Start troubleshooting by stopping the machine. Make certain that the sheet is not left in the oven (if the oven retracts, be sure it is clear). Look for any obvious problems, then review the following list of possible causes.

- Pilot error
- Mechanical problems
- Sheet or film problems
- Heating
- Vacuum and/or compressed air
- Controls and electrical systems.

During troubleshooting and machine adjustments, change only one thing at a time.

If the initial investigation does not uncover the problem, review the thermoforming troubleshooting guide (Table 93.5) and the sheet extrusion troubleshooting guide (section 93.12.2) for possible causes of the problem.

TABLE 93.5. Thermoforming Troubleshooting Guide

Problem	Probable Cause	Suggested Course of Action
1. Blisters or bubbles	A. Heating too rapidly	1. Lower heater temperature 2. Use slower heating 3. Increase distance between heater(s) and sheet
	B. Excess moisture	1. Predry 2. Preheat 3. Heat from both sides 4. Lower heater(s) temperatures 5. Do not remove material from moisture-proof wrap until ready to use 6. Obtain dry material from supplier
	C. Uneven heating	1. Screen for uniform heat by attaching baffles, masks or screen 2. Check for heaters or screens out 3. Adjust individual heater temperatures for uniformity
	D. Wrong sheet type or formulation	1. Obtain correct formulation
2. Incomplete forming, poor detail	A. Sheet too cold	1. Heat sheet longer 2. Raise temperature of heaters 3. Use more heaters 4. If problem occurs repeatedly in same area, check for lack of uniformity of heat
	B. Clamping frame not hot before inserting sheet	1. Preheat clamping frame before inserting sheet
	C. Insufficient vacuum and/or compressed air	1. Check vacuum holes for clogging 2. Increase number of vacuum holes 3. Increase size of vacuum holes 4. Check for vacuum or air leak 5. Remove any 90° angles in vacuum system
	D. Vacuum not drawn fast enough	1. Use vacuum slots instead of holes where possible 2. Add vacuum surge and/or pump capacity 3. Enlarge vacuum line and valves avoiding sharp bends at tee and elbow connections 4. Check for vacuum leaks 5. Check vacuum systems for minimum 25 in. of Hg pressure
	E. Additional pressure needed	1. Use 20–50 psi air pressure on part opposite mold surface if mold will withstand this pressure 2. Use frame assist 3. Use plug, silicone slab rubber, or other pressure assist
3. Sheet scorched	A. Outer surface of sheet too hot	1. Shorten heat cycle 2. Use slower, soaking heat (lower temperature) 3. Move heater bands further from sheet

TABLE 93.5. (*Continued*)

Problem	Probable Cause	Suggested Course of Action
4. Blushing or change in color intensity	A. Insufficient heating	1. Lengthen heating cycle 2. Raise temperature of heaters
	B. Excess heating	1. Reduce heater temperature 2. Shorten heater cycle 3. If in some spot in sheet, check heaters
	C. Mold is too cold or hot	1. Heat mold or lower temperature
	D. Assist is too cold	1. Warm assist or use syntactic foam or felt covered plug
	E. Sheet being stretched too far	1. Use heavier gauge sheet or more elastic, deep draw formulation 2. Change mold design 3. Change forming technique
	F. Sheet cools before it is completely formed	1. Move mold into sheet faster 2. Increase rate of vacuum withdrawal 3. Be sure molds and plugs are hot
	G. Poor mold design	1. Reduce depth of draw 2. Increase draft (taper) of mold 3. Enlarge radii
	H. Sheet material not suitable for job	1. Try different sheet formulation or a different plastic material
	I. Uncontrolled use of regrind	1. Control percentage and quality of regrind
5. Whitening of sheet	A. Cold sheet stretching beyond its temperature yield point	1. Increase heat of sheet; increase speed of drape and vacuum
	B. Sheet material dry colored	1. If above action will not correct, check with sheet supplier for availability of other types of coloring. Some colors do not lend themselves to dry or concentrate coloring 2. A hot air gun can be used to diminish or eliminate whitened surfaces on formed part
6. Webbing, bridging or wrinkling	A. Sheet too hot causing too much material in forming area	1. Shorten heating cycle 2. Increase heater distance 3. Lower heater temperature
	B. Melt strength of resin too low (sheet sag too great)	1. Change to lower melt index resin 2. Ask sheet supplier for more orientation in sheet 3. Use minimum sheet temperature possible 4. Profile temperature of sheet
	C. Too much or too little sheet orientation	1. Have sheet supplier reduce or increase orientation
	D. Insufficient vacuum	1. Check vacuum system 2. Add more vacuum holes or slots
	E. Extrusion direction of sheet parallel to space between molds	1. Move sheet 90° in relation to space between molds
	F. Draw ratio too great in area of mold or poor mold design or layout	1. Redesign mold 2. Use plug or ring mechanical assist 3. Use female mold instead of male

TABLE 93.5. *(Continued)*

Problem	Probable Cause	Suggested Course of Action
		4. Add take-up blocks to pull out wrinkles
		5. Increase draft and radii where possible
		6. If more than one article being formed, move them farther apart
		7. Speed up assist and/or mold travel
		8. Redesign grid, plug or ring assists
		9. Use recessed pocket in web area
7. Nipples on mold side of formed part	A. Sheet too hot	1. Reduce heating cycle
		2. Reduce heater temperature
		3. Reduce temperature of sheet surface that contacts mold
	B. Vacuum holes too large	1. Plug holes and redrill with smaller bit
		2. Use slot vacuum
8. Too much sag	A. Sheet too hot	1. Reduce heating cycle
		2. Reduce heater temperature
	B. Melt index too high	1. Use lower melt index resin or different resin
		2. Have sheet supplier put more orientation in sheet
	C. Sheet area too large	1. Profile heat the sheet; use screening or other means of shading or giving preferential heat to sheet, thus reducing relative temperature of center of sheet
9. Sag variation between sheet blanks	A. Variation in sheet temperature	1. Check for air drafts through oven using solid screens around heater section to eliminate
	B. Wide sheet gauge variation	1. Replace sheet with proper gauge tolerance
	C. Sheet made from different resins; not a homogeneous mixture	1. Control regrind percentage and quality
		2. Avoid resin mix-ups
10. Chill marks or "mark-off" lines	A. Plug assist temperature too low	1. Increase plug assist temperature
		2. Use syntactic foam plug assist
		3. Cover plug with cotton flannel or felt
	B. Mold temperature too low—stretching stops when sheet meets cold mold (or plug)	1. Increase mold temperature not exceeding "set temperature" for particular resin
		2. Relieve molds in critical areas
	C. Inadequate mold temperature control	1. Increase number of water cooling tubes or channels
		2. Check for plugged water flow
	D. Sheet too hot	1. Reduce heat
		2. Heat more slowly
		3. Lower surface temperature of sheet
		4. Slightly chill surface of hot sheet contacting mold with forced air before forming

TABLE 93.5. *(Continued)*

Problem	Probable Cause	Suggested Course of Action
11. Bad surface markings	A. Pock marks due to air entrapment over smooth mold surface	1. Grit blast mold surface
	B. Poor vacuum	1. Add vacuum holes
		2. If pock marks are in isolated area, add vacuum holes to this area or check for plugged vacuum holes or vacuum leak
		3. Check entire vacuum system
	C. Mark-off due to accumulation of plasticizer on mold when using sheet with plasticizers	1. Use temperature controlled mold
		2. Have mold as far away from sheet as possible during heating cycle
		3. If too long, shorten heating cycle
		4. Wipe mold
	D. Mold is too hot	1. Reduce mold temperature
	E. Mold is too cold	1. Increase mold temperature
	F. Improper mold composition	1. Avoid phenolic or other "heat sink" glossy molds with clear transparent sheet
		2. Use aluminum molds where possible
	G. Mold surface too rough	1. Smooth surface
		2. Change mold material
		3. Sand blast mold surface with #30 shot grit
	H. Dirt on sheet	1. Clean sheet
		2. Use ionized air blow
	I. Dirt on mold	1. Clean mold
	J. Dust in atmosphere	1. Clean thermoforming area; isolate area if necessary and supply filtered air
		2. Use ionized air
	K. Contaminated sheet materials	1. If regrind is used be sure to *keep* clean and different materials stored separately
		2. Check supplier of sheet
		3. Use coex sheet with virgin
	L. Scratched sheet	1. Separate sheets with paper in storage
		2. Polish sheet
		3. Replace sheet
12. Shiny streaks on part	A. Sheet overheated in this area	1. Lower heater temperature in scorched area
		2. Shield heater with screen wire to reduce overheating
		3. Slow heating cycle
		4. Increase heater to sheet distance
	B. Bad sheet	1. Check with sheet supplier
13. Excessive shrinkage or distortion of part after removing from mold	A. Removed part from mold too soon	1. Increase cooling cycle
		2. Use cooling fixtures
		3. Use fan or vapor spray mist to cool part faster on mold
	B. Too much sheet orientation or nonuniform orientation	1. Replace sheet

TABLE 93.5. *(Continued)*

Problem	Probable Cause	Suggested Course of Action
14. Part warpage	A. Uneven part cooling	1. Add more water channels or tubing to mold 2. Check for plugged water flow 3. Cool part at same rate on both sides
	B. Poor wall distribution	1. Improve prestretching or plugging techniques 2. Use plug assist 3. Check for non-uniformity of sheet heating 4. Check sheet gauge
	C. Poor mold design	1. Add moat to mold at trim line 2. Add vacuum holes 3. Check for plugged vacuum holes
	D. Poor part design	1. Break up large flat surfaces with ribs where practical or make concave or convex
	E. Mold temperature too low	1. Raise mold temperature to just below "set-temperture" of sheet material
	F. Too much or nonuniform orientation in sheet	1. Check supplier
15. Poor wall thickness distribution and excessive thinning or holes in some areas when sheet stretched	A. Improper sheet sag	1. Use different forming technique such as mounting mold on top platen 2. Use vacuum snap-back technique 3. Use billow vacuum snap back 4. Use billow-up plug assist or vacuum snap-back into female mold 5. Use different melt index resin 6. Try more orientation in sheet
	B. Variations in sheet gauge	1. Consult supplier regarding commercial tolerances and improve quality of sheet
	C. Bad sheet	1. Check supplier for calendering of extruded sheet causing thin spots 2. Lack of complete, homogenous sheet
	D. Hot or cold spots in sheet	1. Improve heating technique to achieve uniform heat distribution; screen or shade as necessary 2. Check to see if all heating elements are functioning
	E. Stray drafts and air currents around machine	1. Enclose heating and forming areas
	F. Too much sag	1. Use screening or other temperature control of center areas of heater banks 2. Use lower melt index resin 3. Use more orientation in sheet
	G. Mold too cold	1. Provide uniform heating of mold to bring to proper temperature 2. Check temperature control system for scale of other stoppage

TABLE 93.5. *(Continued)*

Problem	Probable Cause	Suggested Course of Action
	H. Sheet slipping out of frame	1. Adjust clamping frame to provide uniform pressure 2. Check for variation in sheet gauge 3. Heat frames to proper temperature before inserting sheet 4. Check for nonuniformity of heat giving cold areas around clamp frame
16. Nonuniform prestretch bubble	A. Uneven sheet gauge	1. Consult sheet supplier 2. Heat sheet slowly in a "soak" type heat
	B. Uneven heating of sheet	1. Check heater section for heaters out 2. Check heater section for missing screens 3. Screen heater section as necessary
	C. Stray air drafts	1. Enclose or otherwise shield or screen machine 2. Check clamping frame air cylinders for leaks
	D. Non-uniform air blow	1. Baffle air inlet in prestretch box
17. Shrink marks on part, especially in corner areas (inside radius of molds)	A. Inadequate vacuum	1. Check for vacuum leaks 2. Add vacuum surge and/or pump capacity 3. Check for plugged vacuum holes 4. Add vacuum holes
	B. Mold surface to smooth	1. Grit blast mold surface with #30 grit
	C. Part shrinking away	1. May be impossible to eliminate on thick sheet with vacuum only; use 20–30 psi air pressure on part opposite mold surface if mold will withstand this pressure 2. Add moat to mold just outside trim line
18. Too thin corners in deep drawers	A. Improper forming technique	1. Check other techniques such as billow up plug assist, etc.
	B. Sheet too thin	1. Use heavier gauge
	C. Variation in sheet temperature	1. Profile sheet heating: adjust heating as needed by adding screens to portion of sheet going into corners or with panel heat lower temperature 2. Cross hatch sheet with markings prior to forming so movement of material can be accurately checked
	D. Variation in mold temperature	1. Adjust temperature control system for uniformity
	E. Improper material section or poor sheet	1. Consult sheet supplier or raw material supplier to be sure proper material is correctly extruded
19. Part sticking to mold	A. Mold or sheet temperature too high	1. Increase cooling cycle 2. Slightly lower mold temperature, not much less than recommended by resin manufacturer 3. Lower surface temperature on sheet side that contacts mold

TABLE 93.5. (*Continued*)

Problem	Probable Cause	Suggested Course of Action
	B. Not enough draft in mold	1. Increase taper 2. Use female mold 3. Remove part from mold as early as possible; if above "set temperature", use cooling jigs
	C. Mold undercuts	1. Use stripping frame 2. Increase air-eject air pressure 3. Remove part from mold as early as possible; if above "set temperature", use cooling jigs 4. Change mold design for undercut to break away
	D. Wooden mold	1. Grease with Vaseline 2. Use Teflon spray or zinc stearate 3. Lower surface temperature on sheet side that contacts mold
	E. Rough mold surface	1. Polish corners or all of mold 2. Use mold release 3. Use Teflon spray or zinc stearate
	F. Different melt index resin used in color concentrate	1. Have supplier use same grade resin in color concentrate
20. Sheet sticking to plug assist	A. Improper metal plug assist temperature	1. Reduce plug temperature 2. Use mold release 3. Teflon coat plug 4. Cover plug with felt cloth or cotton flannel 5. Use syntactic foam plug
	B. Wooden plug assist	1. Cover plug with felt cloth or cotton flannel 2. Grease with vaseline 3. Use mold release compound 4. Use Teflon spray or zinc stearate 5. Laminate wood plug surface with syntactic foam
21. Tearing of sheet when forming	A. Mold design	1. Increase radius of corner
	B. Sheet too hot	1. Decrease heating time or temperature 2. Check for uniform heat 3. Preheat sheet
	C. Sheet too cold (usually thinner gauges)	1. Increase heating time or temperature 2. Check for uniform heat 3. Preheat sheet
	D. Closing speed between mold and sheet	1. Reduce rate of closure
	E. Bad sheet	1. Check with supplier
22. Cracking in corners during service	A. Stress concentration	1. Increase fillets 2. In transparencies check with polarized light 3. Increase temperature of sheet

TABLE 93.5.   (*Continued*)

Problem	Probable Cause	Suggested Course of Action
		4. Be sure part is completely formed before some sections are too cool for proper forming, thus setting up undue stresses in these areas
		5. Change to a stress crack resistant resin
	B. "Under-designed"	1. Reevaluate design
	C. "Cold" formed	1. Vacuum *too* slow-speed up
		2. Be sure "core" of sheet is at forming temperature and then rapidly drawn to mold with excellent vacuum and/or compressed air
	D. Wrong resin	1. Change to proper material
	E. Bad sheet	1. Check with supplier

93.10.17 Sheet Extrusion Troubleshooting Guide

If there seems to be a problem with the material, call the sheet and/or resin supplier. Possible causes of sheet extrusion problems are listed below.

Surging. Causes are

- Starving screw—large regrind size—improperly mixed or blended—fused or clogged hopper.
- Insufficient back pressure.
- Nonuniform feedstock temperature.
- Too high feedstock temperature.
- Poor screw design (shows up as rpm increased).
- Moisture or volatiles.
- Excessive screw rpm.
- Extrusion heats not balanced.
- Air or gas entrapment.
- Improper barrel heats.
- Instrumentation lag.
- Voltage variations.
- Malfunction of temperature controllers, thermocouples, or heaters.

Orientation. Excessive or "Spot" causes are

- Too great a distance from die lips to roll nip.
- Excessive bead (nip) buildup.
- Variations in bead size.
- Large die lip opening [about 0.010–to 0.012 in. (0.25–0.30 mm) more than gauge of 0.120 in. (3.0 mm)]
- Low stock temperature.

- Excessive sheet tension from pull rolls.
- Polishing rolls too cold.
- Variations in regrind mix.
- Calendering sheet.

Lines (longitudinal), Herringbone Patterns, or Chicken Tracks. These problems can be caused by

- Moisture.
- Contamination.
- Insufficient polishing roll pressure.
- Excesesive bead buildup.
- Die damage.
- Damaged or dirty rolls.
- High stock temperature.
- Cold or blocked die.

Pits, Dimples. Causes are

- Air entrapment.
- Volatilization due to decomposition.
- Moisture.
- Contamination (material or machinery).
- Dust.
- Volatiles, entrapments between cooling rolls (sagging sheet, die lips-to-roll touching roll too soon).
- Material mixing incompatibility.
- Thickness variation.

Parabolic Lines. Causes are

- Nip too large.
- Surging.
- Starved feeding.
- Poor material mixing (causing pulsation).
- Excessive screw rpm with wrong screw (air or gas entrapment).
- Improper screw design.
- Localized heat patterns (excessive heat at die center).

93.12 COST ESTIMATING

Accurate cost estimating is vital. Important considerations steps in cost estimating are given below.

Materials. Cost estimating materials include:

- Type of material required and its forming characteristics.
- Availability of sheet or film.

- Source and freight.
- Cost.
- Size of sheet or width of web.
- Thickness required.
- Scrap value.
- Color and added cost.
- Paints, coatings, adhesives.
- Laminates and additives and their cost.
- Hardware of all kinds.
- Packaging materials.
- Preforming decoration such as screen printing.

Molding. Molding considerations include:

- Molding cycle time (large parts may require two operators).
- Assists required.
- Estimated molding spoilage.
- Machine setup time.
- Machine model required for job.
- Operations done during molding cycle.
- Multicavity forming of high-volume parts requiring special handling.
- Cleanliness required.

Trim. Trim considerations include:

- Each trim, drilling, and die cutting step required.
- Each trim setup.
- Spoilage from trimming.
- Handling and storage going through trim.

Detailing. To be considered:

- Scraping and deburring.
- Sanding.
- Filing.

Assembly. For cost estimating considerations:

- Ultrasonic welding, heat seal, spin weld.
- Riveting and bolting.
- Bonding, adhesives, and curing.
- Fiber glassing, including any secondary trim and detailing.
- Foam rigidizing including secondary trim and detailing.
- Staking of extrusions.
- Spoilage caused by assembly or rigidizing.

Finishing. Includes:

- Decorative painting.
- Protective coating.

- RFI/EMI conductive coating.
- Silk screening, printing.
- Flocking.
- Hot stamping.
- Antistatic coating.
- Spoilage caused by finishing.

Special Requirements. Includes:

- Annealing.
- Surface treatment.
- Flame treating.
- Antistatic.
- Spoilage.

Quality Control. Inspection for conformance to specs.

Testing. Testing includes falling dart; Flammability or flame spread; Smoke generation; Stress level; Peal; Static buildup; Adhesion; Permeability; Conductivity; RFI/EMI shielding.

Special Handling. Include the following:

- Expensive items requiring special care.
- Strippable coatings.
- Cleanliness required.

Clean and Pack. The amount needed is usually directly proportional to the price of the part.

Tooling. Tooling includes:

- Pattern work.
- Castings.
- Temperature control lines.
- Pressure quality tools.
- Machining or textured surface.
- Polished surface.
- Jigs and fixtures: cooling; each trim setup; bonding and ultrasonic welding; drilling; assembly; hot stamp or silk screen; painting.
- Forming assists: plugs; any required heaters or coatings; prestretch boxes.
- Testing tools.

93.14 INDUSTRY ASSOCIATIONS

Key thermoforming groups are the Thermoforming Division of the Society of Plastics Engineers and the Thermoforming Institute of the Society of the Plastics Industry, Inc.

94

VINYL PLASTISOL PROCESSING

Robert C. Rock

Vice President R&D, Plast-O-Meric Inc., 21300 Doral Road, Waukesha, WI 53187

94.1 DESCRIPTION AND HISTORY OF PROCESS

A vinyl plastisol is a dispersion of vinyl resin in plasticizer, usually incorporating heat and light stabilizers and color pigments. It may also contain mold-release aids or adhesion promoters, viscosity suppressants or viscosity builders, chemical blowing agents or mechanical air entrapment chemicals (for foams), and dehydrating agents or other resins or fillers to develop special characteristics.

Poly(vinyl chloride) resin was first commercially produced in Germany in the years 1912–1915. Plasticizing of PVC was advanced significantly in 1926, when Waldo Semon of B.F. Goodrich began to cast PVC from solvent solutions using solvents and plasticizers. This led to the first commercial plasticized PVC, Koroseal.

The first applications for vinyl plastisols were developed in the early 1930s in Germany. considerable use was made of those first crude plastisols for rubber replacement; for example, in rubberless rain coats. Early plastisols employed resins that were emulsion polymerized and dried. The drying step agglomerated the resin into coarse particles that had to be ground down in the making of the final plastisol. In the 1950s, B.F. Goodrich marketed the first true stir-in grade of emulsion-polymerized resin, which greatly simplified the making of a plastisol.

94.2 MATERIALS

Plastisols are flexible vinyl compounds that are compounded in a special way. They may contain from 25 to more than 500 parts of plasticizer for each 100 parts of vinyl resin. This corresponds to a Shore A hardness of about 98 down to an 8 durometer. Even harder compounds have been made in which part of the plasticizer is replaced by a reactive monomer that is cured when the vinyl is fused. The great range in possible hardnesses has enabled plastisols find use in widely varying applications.

When a plastisol is heated slowly, the first change that takes place is a slight lowering of viscosity. At a temperature betwen 120 and 200°F (49 and 93°C), the viscosity increases rapidly. This point (or range) is called the gel point (or gel range) of a given plastisol. At this point, the resin absorbs the plasticizer; in a very soft compound, the resin dissolves in the plasticizer. Because each resin particle remains a separate particle, the resultant gel has no useful physical properties. However, on further heating, the plasticized resin partially melts and flows into the plasticizer, a state called the fusion point (or range). On cooling, the material comprises the tough rubbery compound known as flexible vinyl.

This phenomenon is illustrated in Figure 94.1. An understanding of the mechanisms of gel point and fusion point, and their effect on molding and coating, is essential for satisfactory utilization of vinyl plastisols.

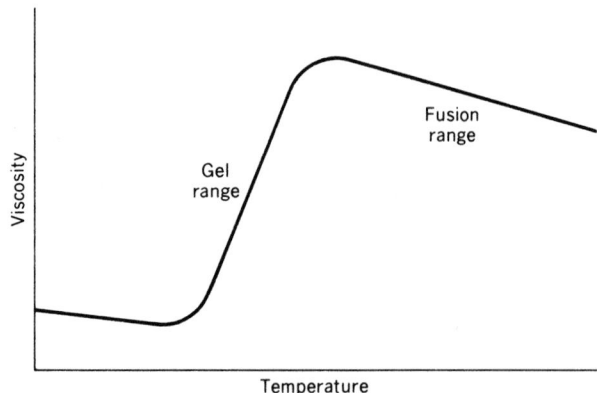

Figure 94.1.   Viscosity vs temperatures showing gel range and fusion range.

Because heat is the only essential element of vinyl plastisol processing, many different processing methods have evolved: dip molding (or coating), slush molding, rotational casting, open molding, closed molding, spray coating, continuous coating, and hot melt molding. Plastisols are essentially 100% solid materials; thus, there is no significant shrinkage or increase in density when they are fused. A small amount of shrinkage occurs because the vinyl compound usually shrinks more on cooling than the mold material does. This shrinkage is usually between 1 and 5%, with most compounds in the 2 to 3% range. Usually the softer a compound is, the more it will shrink.

MASTER PROCESS OUTLINE   **PROCESS**   Vinyl Plastisol
_____

Instructions:
For "ALL" categories use   Y = yes
For "EXCEPT FOR" categories use   N = no
                                    D = difficult

N	**ALL Thermoplastics** **Except for:**		

	Y	Vinyl plastisol
		Poly(methyl methacrylate)
Acetal		Poly(methyl pentene) (TPX)
Acrylonitrile–Butadiene–		Poly(phenyl sulfone)
Styrene		Poly(phenylene ether)
Cellulosics		Poly(phenylene oxide) (PPO)
Chlorinated Polyethylene		Poly(phenylene sulfide)
Ethylene Vinyl Acetate		Polypropylene
Ionomers		Polystyrene
Liquid Crystal Polymers		Polysulfone
Nylons		Polyurethane Thermoplastic
Poly(aryl sulfone)		Poly(vinyl chloride)-PVC
Polyallomers		Poly(vinyl fluoride)
Polyamid-imid		Poly(vinylidine chloride)-(Saran)
Polyarylate		Rubbery Styrenic Block Polyms
Polybutylene		Styrene Maleic Anhydride
Polycarbonate		Styrene–Acrylonitrile (SAN)
Polyetherimide		Styrene–Butadiene (K Resin)
Polyetherketone (PEEK)		Thermoplastic Polyesters
Polyethersulfone		XT Polymer

N	**ALL Thermosets** **Except for:**
	Allyl Resins
	Phenolics
	Polyurethane Thermoset
	RP-Mat. Alkyd Polyesters
	RP-Mat. Epoxy
	RP-Mat. Vinyl Esters
	Silicone
	Urea Melamine

N	**ALL Fluorocarbons** **Except for:**
	Fluorocarbon Polymers-ETFE
	Fluorocarbon Polymers-FEP
	Fluorocarbon Polymers-PCTFE
	Fluorocarbon Polymers-PFA
	Fluorocarbon Polymers-PTFE
	Fluorocarbon Polymers-PVDF

94.3 PROCESSING CRITERIA

94.3.1 Dip Molding

Dip molding and dip coating are really one and the same. In dip molding, the vinyl part is stripped off the mandrel or mold; in dip coating, the vinyl part becomes part of the finished product. The dip molding process consists of the following steps:

- Preheat the mold.
- Dip the mold in the plastisol and hold it in the plastisol for a specific period of time.
- Withdraw the mold and allow the excess plastisol to drain off.
- Return the mold to the oven and heat until the plastisol and the mold adjacent to the plastisol reaches 350°F (177°C).
- Cool the mold and part to 140°F (60°C).
- Strip off the part.

The purpose of mold preheating is to saturate the mold with enough heat to gel the plastisol. Because vinyl is a good insulator, heat transmits slowly through it. The thickness of the finished part is determined by the amount of heat transmitted to the plastisol. Thus, preheating temperatures are high for thin molds and lower for thick molds. Finished part thickness is also influenced by the length of time that the mold is in the plastisol.

The mold should be withdrawn from the plastisol slowly and smoothly. If it is not, lines will form on the part; if excessive draining occurs, there will be runs and streaks on the part. Ideally, the mold should contain sufficient residual heat and should be withdrawn slowly enough that the fluid plastisol runs off and the remainder gels immediately without running or dripping. When this cannot be accomplished, it may be possible to rack the parts so that runs drip from one corner; then, by inverting the part, the last drip can flow back.

In fusing the plastisol, it is essential that the plastisol and the mold reach 350°F (177°C). At this temperature, the plastisol possesses virtually 95% of its ultimate physical properties. Again, because the vinyl is a good heat insulator, it takes time for the heat to penetrate completely. It is often advantageous to set the oven temperature at 375 or 400°F (191 to 204°C) to shorten the fusion time. Care must be taken at higher temperatures to avoid exceeding the heat stability of the compound.

Cooling can be done by hanging in cool air, by water spray, or by actual dipping into water. With the latter technique, care must be taken so that water marks are not left on the part.

Parts are generally stripped from the mold at 130–140°F (54–60°C). At this temperature, the parts are still soft enough to stretch and pull over undercuts but cool enough not to be distorted by stretching.

Dip coating is done in the same manner except that the mold is part of the finished item and, therefore, the molded vinyl remains part of the finished item. Usually, vinyl plastisols do not adhere to metals or other mold materials. If the coating requires adhesion, a primer adhesive should be used. These are often lacquers that may be dipped, sprayed, or brushed on the part before preheating. They are available today as solvent-based or water-based adhesives.

Dip molding is used to produce automotive gear shift boots, slip-on tool grips, bird guards for the leads on large transformers, stethoscope tubes, electrical bus bar insulation boots, and medical inspection gloves. Dip coating is used for coating tool handles, insulating against heat or cold, insulating field coils in automotive starters, cushioning kitchen implements, covering sharp metal, and for electric insulation.

94.3.2 Slush Molding

One drawback of dip molded parts is that the air surface is the surface that visible. It can only be glossy or flat or grainy in texture. With slush molding, the mold surface becomes the exposed surface and any grain, special texture, or engraving on the mold is reproduced on the finished part.

Slush molding may be considered the reverse of dip molding. A female mold is used; it is filled with the plastisol.

Slush molding is done in the following sequence:

- Preheat the mold.
- Fill the mold with plastisol and hold to gel.
- Drain excess plastisol out of the mold.
- Heat to fuse the plastisol.
- Cool the mold and part.
- Strip the part from the mold.

As in dip molding, the mold is preheated sufficiently to gel the required thickness of plastisol. The mold is filled and held for several seconds before it is inverted and drained.

Sometimes the mold is filled cold and then heated to gel a skin on the mold. This can improve the reproducibility of the mold texture but can also cause the plastisol to gel on the air surface, resulting in poor draining and in lumps in the plastisol. The lumps must be screened out or they will redeposit on later castings, giving uneven thicknesses. The rest of the process is similar to dip molding, with an added trimming step.

Slush molding is used for automotive gear shift boots, arm rests and head rests, road safety cones, anatomical models, and doll parts and other toys.

94.3.3 Rotational Castings

Rotational casting of vinyl plastisols is similar to the rotational casting of powder resins, in that a measured amount of plastisol is placed in the mold and rotationally heated. This process is used to produce play balls, volleyballs, basketballs, doll heads and bodies, and the kinds of automotive parts produced by slush molding.

94.3.4 Open Molding

Open molding accounts for considerable consumption of vinyl plastisols. The process is very simple. The plastisol is poured into a mold. The mold and plastisol are heated to gel and fuse the plastisol. The mold is then cooled and the part is stripped from the mold. Inserts may be placed in the liquid plastisol before it is fused, and two or more colors may be placed in different parts of the mold.

The most important open molding application is automotive air filters. Open molding is also used to produce tablecloths, coin mats, truck flaps, and similar flat or relatively flat items. An exception is automotive oil filters, an application in which additives are incorporated in the plastisol to cause the plastisol to bond to the filter media and to the metal end cap. The plastisol then becomes both an adhesive and an end seal for the oil filter.

94.3.5 Closed Molding

Closed molding is a technique occasionally used with plastisols. The plastisol is poured or pumped into the mold, which is heated to fuse the plastisol and then cooled. Subsequently, the mold is opened and the part stripped out. Often a closed mold must be vented to accommodate the different expansion rates of the vinyl and the mold. Switch mats for automatic door openers are often made this way.

94.3.6 Spray Coating

Because they are liquid, plastisols may be sprayed onto molds or parts. The viscosity of such a special formulation is nonflowing after it is sprayed. Thicknesses of up to 50 mils (1.3 mm) can be achieved in a single pass on a vertical panel. The parts are heated and cooled, and then stripped off the mold or left on as a coating. Many tank linings are manufactured this way.

94.3.7 Continuous Coating

Platisols can be spread-coated on many substrates for many uses via a doctor blade, direct roll, or reverse roll. For example, plastisols are roll coated on primed metal for house siding. Cloth fabrics are saturated and coated in the manufacture of conveyor belts. Foamed vinyl fabrics are made by coating a thin layer of solid plastisol on embossed release paper, then coating a thicker coating of foam plastisol, and finally layering on a cloth scrim. The composite is fused and peeled from the release paper.

The wear layer of vinyl flooring is usually a coated clear plastisol, thereby making possible "no wax" flooring.

94.3.8 Silk-screened Inks

Plastisols can be silk screened on tee shirts or athletic uniforms, thus becoming a heat-fused ink. These are highly pigmented systems; application thicknesses of only a few mils (0.025 mm) are possible.

94.3.9 Hot Melt Plastisols

Soft plastisols (35 Durometer Shore A and softer) will melt and flow when they are heated to fusion temperatures. This hot, fluid vinyl can be poured or pumped into molds and cooled. Fishing worm, other baits, display items, and various novelties are made this way.

94.3.10 Organosols

Organosols are plastisols in which part of the plasticizer is replaced with a solvent. This makes it possible to spray or cast a harder film than is feasible with a pure plastisol. Because of the solvent, films are usually limited to 10 mils or less in thickness. Such films are used to replace paint films in applications needing the chemical resistance of vinyl.

94.4 ADVANTAGES/DISADVANTAGES

Plastisols have many advantages, most important is that they only require heat for processing; no pressure or mixing is necessary. This means that mold costs are very low and the overall processing equipment costs are, likewise, low. They are very versatile materials in that almost any additive can be incorporated for special effects as long as it is soluble in plasticizer or can be ground to a powder sufficiently fine to be suspended in the plastisol.

There are three main disadvantages to the use of vinyl plastisols. The first is that the cost of the dispersion grade of vinyl resin, which is necessary to form the suspension, is somewhat more expensive than the more common general-purpose type resins. Second, process times are slow, usually 4 to 20 minutes in length. Processing time can be minimized by use of several low-cost molds at once; as is done in the open molding of automotive air filter ends. Third, as manufactured, plastisol resins contain wetting agents and soaps from the original polymerization. This soap can cause clarity problems and somewhat limits electrical resistance.

94.8 TOOLING

Molds for plastisol molding can be made of almost any nonporous material that can tolerate the heat of processing. The most popular are cast aluminum and electroformed nickel; fabricated steel molds are also used extensively. Zinc is inappropriate; copper should be used cautiously. Zinc can catalyze early breakdown of vinyl; copper is only satisfactory when the plastisol is not overfused. If overfusion occurs, the part will be stained with copper chloride. Similarly, brass and bronze should only be used where the process is well controlled so that overfusion does not occur. It is possible to use glass and glazed ceramics despite their fragility; most medical inspection gloves are hot dipped on ceramic molds.

94.8.1 Coating Methods

Oven heating is the method most commonly used for preheating and fusing plastisols. Ovens may be gas fired or electric, convection or infrared, but, in all cases, they should maintain uniform heat and provide sufficient exhaust to vent the smoke produced by the hot plastisol properly.

There is a drawback to use of ovens: they are inefficient. Air is a poor medium for transmitting heat, and infrared heats only what it sees. Thus, various other methods of heat transfers are used, including hot plates for open molding, resistance heating for preheating rods, and hot molten salt baths for slush molding.

Cording molds (attaching a coil to a mold and using hot oil for heat and cold oil for cooling) is a technique occasionally used for slush molding, despite its very short cycle times.

As the foregoing examples demonstrate, any method of getting heat to the part can be used successfully.

Today, approximately 10% of all vinyl consumed in the United States for manufacturing is processed via plastisols. The many processes and systems that have evolved for converting vinyl plastisols have made this possible.

94.15 INFORMATION SOURCES

94.15.1 Processing Equipment Suppliers

E. B. Blue, 651 Connecticut Ave., S. Norwalk, CT 06854 (203) 838-8485.

Global Process Equipment, Inc., Sub. Chas. Ross & Son Co., 710 Old Willet Path, Box 12308, Hauppauge, NY 11788 (516) 234-0503.

Dempsey Industries, 802 North 4th St., Miamisburg, OH 45342 (513) 866-2345.

McNeil Akron, Inc., 96 East Crosier St., Akron, OH 44311 (216) 253-2525.

Rotational Engineering, Inc., 795 2nd St., Box 427, Berthoud, CO 80513 (303) 532-2194.

Rotatron Corp., 44 Da Vinci Dr., Bohemia, NY 11716 (516) 293-3176.

W. S. Rockwell Co., 200 Eliot St., Fairfield, CT 06430 (203) 259-1621.

Porbeck, Inc., St. Louis, MO 63100 (314) 231-0372.

94.15.2 Dispensing Equipment Suppliers

Lincoln St. Louis, St. Louis, MO (314) 383-5900.

Liquid Control Corp., North Canton, OH 44720 (216) 494-1313.

De Vilbiss Co., Toledo, OH (419) 470-2169.

Graco, Inc., Minneapolis, MN (612) 378-6000.

Alemite Chicago, IL (312) 589-1550.

Dicker International, Inc., 225 Broadway, New York, NY 10007 (212) 962-3232.

94.15.3 Mold Suppliers

Electroform. The following are suppliers of electroform molds.

Conforming Matrix, 749 New York Ave., Toledo, OH 43611 (419) 729-3777, Ray Sabetto.

T.V. Jay Co., 1767 Sunnyside Ave., Chicago, IL 60640 (312) 561-6886, Vic Krambo.

Bowers, 482 Seneca, Ridgewood, NY 01385.

Sheller Globe Iowa City Div., 2500 Hwy. 6, East Iowa City, IA 52240 (319) 338-9281.

Perfect Doll Moulds Corp., 103 Fifth Ave., New York, NY 10003 (2122) 255-7666, Alan Burnstein.

EMF, Colorado Springs, CO (303) 576-7733, Don Sheldon.

Plating Engineering, 1928 South 62nd St., Milwaukee, WI 53212, Jim Ritzenthaler.

Cast Aluminum. The following are suppliers of cast aluminum molds.

Wendel Pattern & Mfg. Co., 213 East Frank St., Kalamazoo, MI 49007 (616) 342-2432, Bill Wendel.

Portage Casting, Portage, WI.

Plasti-Clad Mold & Products, 1430 East Archwood Ave., Akron, OH 44306 (216) 773-3377, Bud Lamont.

Kelch Corp., Mequon, WI (414)377-5170.

95

MOLD MATERIALS—ALUMINUM

William Lewi

Valiant Plastics, 106-10 Dunkirk St., St. Albans, NY 11412

95.1 DESCRIPTION AND HISTORY

Aluminum is one of the most abundant metals. Although it makes up about 8% of the earth's crust, aluminum does not occur naturally in a metallic state. Aluminum is present in all soils and clay; in feldspar, mica, and other silicates; and in rocks and minerals. Aluminum ore may be extracted easily only from bauxite. In the United States, the largest deposits of bauxite are found in Arkansas. Bauxite deposits are also present in Jamaica, Dutch and British Guinea, France, Hungary, Italy, and the USSR.

Production of aluminum in large quantities was made possible by the discovery in the late 1880s of an electrolytic process by American and French scientists.

The bauxite ore is crushed, washed, and dried and then mixed with sodium hydroxide to form sodium aluminate. After being processed through a pressure filter, this mix becomes alumina and sodium hydroxide. The alumina is dried; subsequently, in an electrolytic cell, it is dissolved and forms pure aluminum on the negative electrode of the tank. The sodium hydroxide is returned for reuse.

The resultant silvery white metal has a density of only one third that of iron. However, after it is alloyed with magnesium, copper, silicon, zinc, nickel, chromium, or manganese, and then rolled, extruded, or forged, its strength can be greatly increased. Additional strength and hardness can be gained by heat treating the alloyed aluminum and stretching the rolled plates.

The use of aluminum for molds used in plastics processing has increased greatly in the last few years, thanks to such advantages as its heat conduction (only silver, gold and copper are better conductors); its ability to be cast either in sand or ceramic molds, making it an optimum material for blow molds; and its ease of machining, increasing its suitability for foam molding. The use of aluminum for compression molding is not recommended. The process of compressing and melting powder in an aluminum mold adds extra stress and wear on the parting line and will greatly reduce the life of the mold for this application.

Recent improvements in the grain enable aluminum molds to have high polish. Because its hardness compares favorably with the more expensive beryllium copper, bronze, brass, or cold rolled steel, injection molders have begun to consider and use aluminum for injection molds. Aluminum is not as permanent as hardened steel molds but, with proper design and care in molding, it can be just as good for many different molding materials and many shots.

95.2 DESIGN

Aluminum molds can be constructed either by duplicating a standard mold base completely in aluminum (mini-mold) or by mounting an aluminum "A" and "B" plate on an abbreviated standard mold base.

Since aluminum is not as hard as prehardened tool steel or hardened tool steel and it usually is not pocketed in the mold base, certain precautions must be taken in its design.

There must be more land around the cavity. A minimum of 1.5 to 2 times the depth is considered essential to contain the molding pressure. Draft on the sides shoulde be 2° per side or all ejector pins should be located so that the part will eject without the operator having to use any tool to remove the sticking part. Removal of a part with brass, aluminum, or even a plastic pick could easily damage the mold (Fig. 95.1). A taper of 5° is recommended to avoid galling of the mating aluminum surfaces. For core pulling, the use of hydraulic cylinders (Fig. 95.2) is recommended over the use of cam pins (Fig. 95.3). A combination of side coring through the use of cylinders and multiseal-offs present no problem (Fig. 95.4). The close grain of the aluminum permits a good polish, in many cases very close to that of a steel mold (Fig. 95.5 and 95.6). The good heat conductivity of aluminum permitted simple location of the waterlines to produce 25,000 trays with 40 pockets each. A steel mold was

(a)

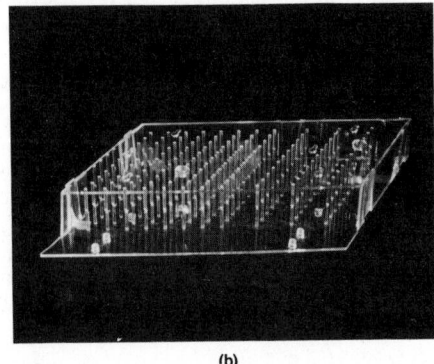

(b)

Figure 95.1. (**a**) Aluminum mold. (**b**) Molded part.

Figure 95.2. Core pulling using hydraulic cylinders.

Figure 95.3. Core pulling using cam pins.

Figure 95.4.   Cylinders and multiseal-offs.

then built to produce an additional 500,000 trays. A much more costly and complicated system had to be designed for the steel mold.

95.3   MACHINING

Since the aluminum plate should be a nonclad aircraft-quality grade 7075 T 6 (this being the best grade for molds), no heat treatment before or after machining is required. Grades such as 6061 T 6 or the Mohawk Alum grade Alpha Stock 79 (T 65 Temper) having a maximum thickness of 12 in. (30 cm) can also be used. In all instances, care should be taken to rough machine first to relieve the metal of any stress or strain produced by machining. Then, the outside should be machined to ensure that the mold is flat. A final machining of

Figure 95.5.   Polishing.

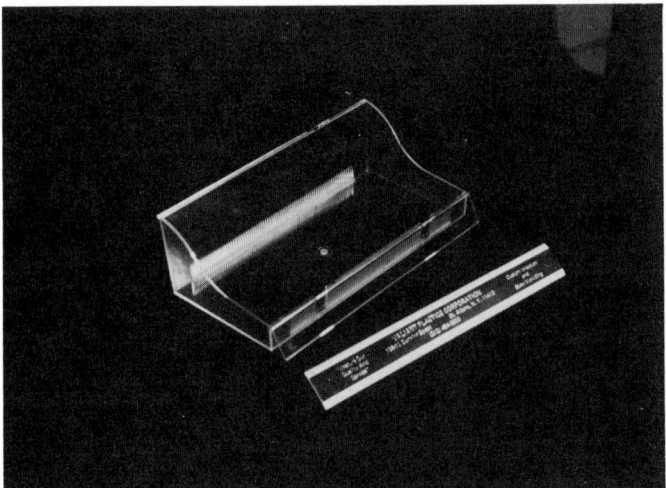

Figure 95.6. Tray molded with aluminum mold.

the cavity and core of the mold should be done with freshly sharpened cutters. All cutting edges should be perfectly smooth to avoid excess cutter marks. It is important to use higher cutter speeds and slower feeds for the final cut.

Because of the good heat conductivity of aluminum, location of waterlines is less critical. In most applications, they can be located with standard drilling equipment.

Should welding be required, Heli Arc Welding is recommended. A good mechanical or chemical precleaning is vital to ensure a sound and clean weld. Preheating to welding is advisable, particularly in large molds to avoid cold cracking. Preheating should not exceed 250°F (121°C) with a finish temperature always below 390°F (199°C) to avoid hot cracking. It is not beneficial to heat treat the mold after welding since it is very difficult to attain the original hardness in the weld.

The finished machined molds should never be polished with rotating polishing tools. The reason is that, on aluminum, they cannot be controlled sufficiently to prevent uneven surfaces or "orange peels," depending on the quality of the final machined surface. Tool marks should be removed with hand-held aluminum oxide stones, if necessary, as low as 120 grit and then using progressively finer grit (as high as 800 grit). White aluminum oxide soft stones from Gesswein will permit fast cutting with a good finish. The use of white polishing stones, followed with a final polish with Simichrome on a felt pad will give a very good finish.

For clear parts, the use of a 14-micron diamond compound, followed by a 6-micron compound and if necessary finished with a 3-micron compound on a felt stick, will give an excellent surface finish.

Where required, molds can be etched by most of the commercial mold decorators, but the patterns will be more pronounced than in steel. To a limited extent, molds can also be patterned by aluminum oxide sand blasting.

Aluminum molds can be plated for wear resistance and much greater surface hardness, but with an attendant loss of polish. Plating can be accomplished by hardcoating with an oxide film, such as Martin Hardcoat.

Nickel or chrome plating is possible but very difficult. It is highly recommended that a test bar machined from the same mold and with the same required finish be plated prior to submitting the entire mold to the plater.

95.4 IDENTIFYING ALUMINUM

95.4.1 Color Code

Tags or ends of bars (These colors do not apply to the ink used for identifying printing.)

Alum 6061 Blue
Alum 7075 Black
Alum Alpha Stock 79 N/A

95.4.2 6061 T 651 Description

6—Main alloying element, magnesium and silicon

0—No modification of original alloy

61—Specific alloy

T—Solution heat treated

6—Temper method, artificially aged

51—Additional processing for properties, stress relieved by stretching (the effect of alloying is to increase strength and corrosion resistance).

95.4.3 7075 T 651 Description

7—Main alloying element, zinc

0—No modification of original alloy

75—Specific alloy

T—Solution heat treated

6—Temper method, artificially aged

51—Additional processing for properties, stress relieved by stretching (the effect of alloying is to increase strength).

95.4.4 Alpha 79 T 65

No data available.
Weight per lb/in.^{3} (g/cm^3)
6061 T 6, 0.098 (2.7)
7075 T 6, 0.101 (2.8)
Alpha 79, 0.099 (2.7)

Chemical composition percent maximum unless shown as range:

6061 Silicon
0.40–0.8 iron, 0.7 copper, 0.15–0.40 manganese, 0.15
magnesium, 0.8–1.2 chromium, 0.04–0.35 zinc, 0.25
titanium, 0.15 others, 0.05–0.15 aluminum remainder

7075 Silicon
0.40 iron, 0.50 copper, 1.2–2.0 manganese, 0.30
magnesium, 2.1–2.9 chromium, 0.18–0.35 zinc, 5.1–6.1
titanium, 0.20 others, 0.05–0.15 aluminum remainder

TABLE 95.1. Mechanical Properties of Various Aluminum Alloys

Property	6061 T 6	7075 T 6	Alpha Stock 79 T 6[a]
Tensile strength, psi (MPa)	45,000 (310)	83,000 (570)	
Yield strength, psi (MPa)	40,000 (276)	73,000 (500)	67.1 (0.5)
Hardness, Brinell	95	150	
500-kg load, 10-mm ball			
Vickers			140–180
10 kg			
Ultimate shear strength	30,000 (207)	48,000 (330)	73.5 (0.5)
Modulus of elasticity	10×10^6 (69,000)	10.4×10^6 (72,000)	

Comparison of thermal conductivity, Btu/°F ft-h (J/ms · K)			
Aluminum	95.0 (164)		
Beryllium copper	64.0 (111)		
Prehardened steel P20	21.0 (36.3)		
Heat transmission			
Btu through 1-in. (3-mm) metal/in.² of surface for 1°F temperature difference/min			
Aluminum	0.1218		
Prehardened steel	0.0372		
Brass	0.0942		

[a]7.5 in. (191 mm) thick stock.

Alpha Stock 79
Chemical data not available

95.5 MECHANICAL PROPERTIES

See Table 95.1

95.6 INFORMATION SOURCES

H. DuBois and W. Pribble, *Plastics Molding Engineering Handbook,* D.V. Rosato and D. Rosato, *Injection Molding Handbook,* Reynolds Aluminum, Joseph T. Ryerson & Son, New York, Mohawk Aluminum Corp., Wallingford, Conn., Paul H. Gesswein & Co., Inc., Bridgeport, Conn.

96

MOLD MATERIALS, BERYLLIUM COPPER ALLOY

Robert E. Saxtan

Director of Product Engineering, NGK Metals Corp., PO Box 13367, Reading, PA 19612-3367

96.1 DESCRIPTION

Cast beryllium copper for plastics molds is available in five alloys. Table 96.1 shows the various alloy designations, chemical composition, and typical applications. Table 96.2 provides this information for wrought products.

96.2 MANUFACTURING

Beryllium copper alloys originate from the chemical processing of a beryllium-bearing ore for production of beryllium oxide. The oxide is charged into a direct arc furnace, along with carbon and copper, to produce a master alloy of approximately 4.0% Be (the balance being copper). The master alloy is cast into 5 lb (2.3 kg) pigs for sale to foundries and into larger pigs for use in producing the family of beryllium copper alloys. The final alloys are produced by adding master alloy to copper or beryllium copper mill scrap and appropriate additive elements, such as copper, nickel, and lead. This charge is usually melted in an induction furnace.

Products for the foundry industry are available as shot and 1 lb, 5 lb, or 15 lb (0.5, 2.3 or 6.9 kg) ingot. Material for wrought (that is, hot worked rather than cast) products is poured as semicontinuous cast ingot in various rectangular or round sizes for hot rolling, forging, or extrusion. Mill heats (metal or one analysis poured at the same time) generally run 4,000 to 8,000 lb (1,800–3,600 kg).

96.3 MOLD FABRICATION

The actual beryllium copper mold used in producing plastic parts can be made by one of several casting methods or by machining the cavity detail into a solid cast or wrought block. When casting, the excellent fluidity of the molten alloy allows for close reproduction of any detail in the pattern used. Pressure casting or hot hobbing and ceramic casting are the two most common casting processes.

Pressure casting consists essentially of pouring the mold alloy around a steel master hob or pattern and immediately applying pressure while the metal is molten. The metal then flows to conform precisely to the shape and surface finish of the hob. The hob must be

TABLE 96.1.   Beryllium Copper Casting Alloys

UNS Alloy No.	BERYLCO Alloy	Composition in % (Bal. Cu.)	Pouring Temp. Range, °F	Ingot Availability	Characteristics and Applications
C82400	165C 165CT .	Be 1.65–1.75 Co 0.20–0.30	1900–2050	Shot 1 lb 5 lb	165C-High strength, hardness, with good resistance to corrosion. Suitable for salt water immersion applications.
				15-lb notch bar	165CT-Properties same as 165C, but fine grain structure.
C82500	20C 20CT .	Be 1.90–2.15 Co 0.35–0.65 Si 0.20–0.35	1850–2050	Shot 1 lb 5 lb 15-lb notch bar	20C-High strength, hardness, wear resistance and excellent fluidity. Ideal for investment, shell, sand, ceramic castings. Meets requirements for Class 4 RWMA resistance welding.
					20CT-Properties same as 20C, but develops a fine grain structure particularly in slow cooling ceramic molds castings.
C82510	21C .	Be 1.90–2.15 Co 1.00–1.20 Si 0.20–0.40	1850–2350	Shot 1 lb 5 lb 15-lb notch bar	New beryllium copper alloy designed to produce an extremely fine grain structure, particularly where superheating is necessary to flow metal in long thin sections. Produces a superior as-cast finish when exceeding normal pouring temperatures of other beryllium copper casting alloys. Properties are similar to 20C or 20CT. Meets RWMA Class 4 requirements. May be cast in same types of molds as 20C.
C82600	245C 245CT .	Be 2.25–2.45 Co 0.35–0.65 Si 0.20–0.35	1850–2000	Shot 1 lb 5 lb 15-lb notch bar	245C-High strength, hardness, wear resistance and excellent fluidity. Low pouring temperature is ideal for pressure and permanent mold applications.
					245CT-Same as 245C, but develops fine grain structure when casting into slow cooling type molds.

TABLE 96.1. (*Continued*)

UNS Alloy No.	BERYLCO Alloy	Composition in % (Bal. Cu.)	Pouring Temp. Range, °F	Ingot Availability	Characteristics and Applications
C82800	275C 275CT	Be 2.50–2.75 Co 0.35–0.65 Si 0.20–0.35	1825–1925	Shot 1 lb 5 lb 15-lb notch bar	275C-Highest strength and hardness of BeCu alloys; excellent wear resistance and fluidity. For extremely fine detail pick-up in ceramic or pressure cast processes. Lowest pouring temperature of BeCu alloys allows most any type of molding process to be used. 275CT-Same as 275C, but develops fine grain structure when casting into slow cooling type molds.

designed to include fillets, wherever possible, so as to avoid sharp corners. A draft angle of at least 1.5° is also required to allow release of the hob from the solidified casting. Machining of the hob must be followed by polishing. The dimensions of the hob must allow for the shrinkage of beryllium copper from the pouring temperature to the room temperature, as well as the plastic shrinkage. The extra expense of producing the hob is offset when several molds of the same configuration are required.

A typical ceramic casting sequence involves pouring a quick setting liquid rubber around a pattern for the part. After setting, the rubber is stripped from the pattern. It is flexible enough that deep contours and re-entrant angles will be intact. A ceramic slurry in a gel form is then poured into the rubber mold. After this slurry sets by the hardening of the gel, the rubber is stripped off and the "green" ceramic reproduction of the part is fired in an oven for final strengthening. Appropriate gates and risers also are added prior to firing. At this point, the mold is ready to receive the molten beryllium copper.

Conventional sand castings can also be used for molds, depending on the requirements for surface finish.

TABLE 96.2. Beryllium Copper Wrought Alloys

Alloy	Composition		Features
Berylco 25 C17200	Beryllium Cobalt plus nickel Cobalt plus nickel plus iron Copper plus additive elements	1.80–2.00% 0.20% min 0.6% max 99.5% min	High strength Good conductivity Fatigue strength Wear resistance Nonmagnetic RWMA Class 4 material
Berylco 165 C17000	Beryllium Cobalt plus nickel Cobalt plus nickel plus iron Copper plus additive elements	1.60–1.79% 0.20% min 0.60% max 99.5% min	High strength Good conductivity Fatigue strength Corrosion resistance Nonmagnetic RWMA Class 4 material

After trimming the gates and risers, all castings are given an appropriate thermal treatment to establish the desired properties and then finish machining is performed to meet final dimensions.

96.3.1   Heat Treatment

Beryllium copper alloys are precipitation hardenable materials. Figure 96.1 shows the phase diagram for beryllium copper at the low end of the beryllium range. The total thermal treatment consists of two steps, solution annealing and age hardening.

Solution annealing is performed by heating the part into the range shown in the upper shaded area of the phase diagram. This step causes the beryllium to dissolve into the copper matrix and form a "solid solution." The recommended temperature range is 1450–1475°F (788–801°C); the holding time at temperature is based on 1 hour/in. (25 mm) of thickness, with a minimum of one hour. The parts are then rapidly cooled by quenching in cold water, after which they will be soft; about Rockwell B45-85. Temperature control is critical.

Beryllium copper will begin incipient melting at approximately 1575°F (856°C). At this point the grain boundaries will begin to melt, hot shortness will result, and the parts will crack either in quenching or later during the age hardening treatment. Temperatures below 1450°F (788°C) allow some of the beryllium to remain out of solid solution; thus it is not in the proper form for the subsequent age hardening step.

Electric or gas fired furnaces may be used for annealing. It is important to note that gas flames should not impinge on the parts. Salt bath furnaces should not be used for solution annealing, as the chlorides used can react with the beryllium copper and destroy any surface detail.

Age hardening consists of reheating the solution annealed parts into the range of the lower shaded area of Figure 96.1. The temperature used depends on the desired properties. Figures 96.2 and 96.3 are aging curves for solution annealed cast material and show the

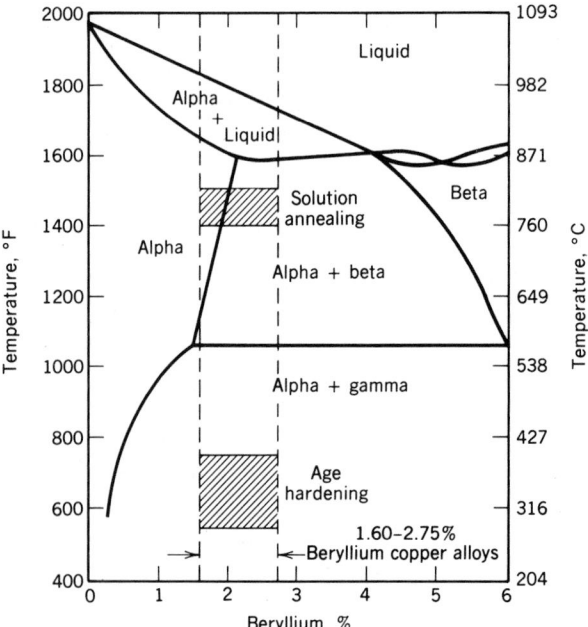

Figure 96.1. Copper-rich end of the beryllium copper equilibrium diagram, showing the solution annealing and age hardening temperature ranges.

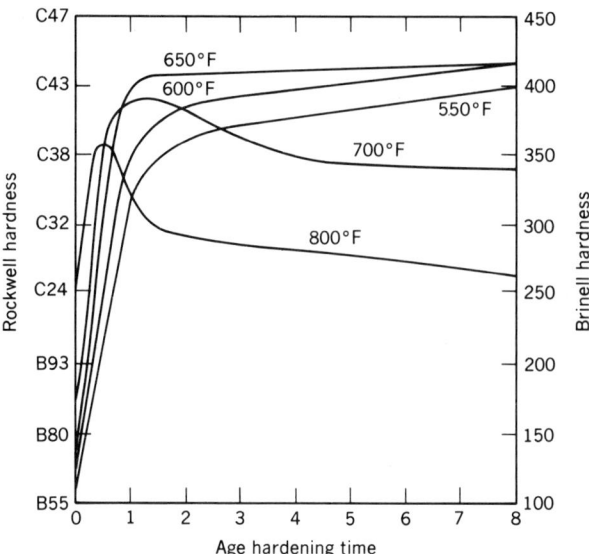

Figure 96.2. Typical cast beryllium copper aging curves for alloy C82500 and C82510.

various hardness levels that result from different combinations of temperature and time with two levels of beryllium content. Wrought products have slightly different aging curves.

The metallographic structure that existed at the solution annealing temperature is retained at room temperature because of the rapid cooling. This structure is unstable. When it is reheated to the aging temperature, it will tend to revert to that which is normal for the lower temperature. This takes place by the precipitation of a beryllium-rich phase (gamma) in the grain boundaries and within the grains. The controlled formation of this precipitate causes the increase in hardness and strength. The usual aging temperature range is 550–800°F (288–427°C). The range of 600–650°F (316–343°C) gives the highest hardness and is

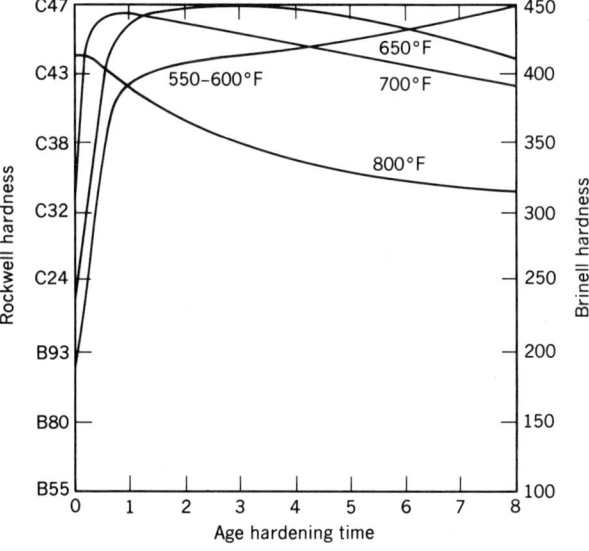

Figure 96.3. Typical cast beryllium copper aging curves for alloy C82800.

recommended for best wear life. The range of 750–800°F (391–421°C) improves ductility and toughness with some sacrifice in strength. Where high hardness is not required for service life, the higher aging temperatures are sometimes used to avoid the need for carbide tooling in machining.

Beryllium copper alloys undergo a volume shrinkage during age hardening because the precipitating phase is more dense than the matrix from which it forms. This shrinkage converts to approximately 0.003 in./in. (mm/mm) on a linear basis. It is predictable and must be compensated for in the dimensioning of the mold.

Circulating atmosphere electric furnaces are usually used for the aging treatment. Temperature control should be within +/− 10°F (6°C). Time control is critical when the higher temperatures are used because hardness drops off rapidly with time. Salt bath furnaces can be used for age hardening with no adverse effects to the material.

96.4 PROPERTIES

Table 96.3 shows typical properties for the beryllium copper cast alloys normally used for mold applications. Only C82400 and C82500 have approximate wrought alloy counterparts: C17000 and C17200, respectively. Their chemistries are somewhat different and wrought alloys are cold worked significantly, so that when tested in either the solution annealed or age hardened condition their mechanical properties are similar, but not identical. Some of the cast alloys, when peak-aged, can reach a hardness of Rockwell C45, but wrought alloys, due to lower beryllium content, will only reach Rockwell C41.

96.5 MACHINABILITY

It is possible to machine in the peak-aged condition (35 to 45 Rockwell C). Working in this final hardness state has the advantage that no dimensional change caused by shrinkage will take place after machining, as would be the case if machining were performed to final dimension and then aged.

Solution-treated beryllium copper (45 to 85 Rockwell B) can be machined like stronger manganese and aluminum bronzes. Material heat treated to 20 to 32 Rockwell C is less gummy and tends to form chips. At peak hardness (35 to 45 Rockwell C), lighter cuts and slower material removal rates are required. High speed tools are adequate for machining solution annealed beryllium copper, and solid carbide tools are preferred for the higher hardness age hardened varieties, particularly above Rockwell C 40.

Water-soluble cutting fluids provide the high cooling rates desired during machining. For heavy operations, such as tapping and deep drilling, lubrication of the tool and work piece with a mineral lard oil of 3–7% lard-oil content or a sulfurized mineral lard oil of 7% lard-oil content and a maximum sulfur content of 3% is recommended. Sulfur can cause discoloration on copper.

Machining by EDM can be completed at approximately the same rate as for brass. Reversed castings have been used as EDM electrodes. High conductivity copper electrodes, such as cadmium copper, appear to be the best electrodes for beryllium copper.

Alloys can be repaired or joined by tungsten inert-gas or metal inert-gas welding techniques employing beryllium copper rod. Silver brazing or soft solder are also useful for repairing and joining. Pinning and plugging techniques are also convenient for repairing and blocking off sections when cooling passages are drilled.

Wrought material can be obtained from the mill in the solution annealed condition or in the age hardened condition at whatever hardness level that is needed. In cases where extensive machining and/or maximum hardness is required, such as when starting with a solid block of wrought material, partial or underaging may be done prior to machining and

TABLE 96.3. Typical Properties of Beryllium Copper Cast Alloys Used for Mold Applications

UNS Alloy No.	BERYLCO Alloy	Composition	Density lb/in.³ (g/cm³)	Specific Gravity	Melting Range, °F (°C)	Modulus of Elasticity 10⁶ psi (GPa)	Thermal Conductivity Btu/ft.²/in./h/°F (at 68°F) × (w/m · K)	Thermal Expansion in./in./°F (68–392°F) (mm/mm/°C)	Condition	Heat Treat Conditions, °F[a]	Ultimate Tensile Strength, ksi (MPa)	Yield Strength 0.2% Offset, ksi (MPa)	Elongation, % in 2 in. (in 51 mm)	Rockwell Hardness
C82400	165C	Be 1.65–1.75 Co 0.20–0.30	0.298 (8.24)	8.26	1650–1825 (899–996)	18.5	825	9.4	As Cast		70 (483)	40 (276)	15	B 78
	165CT								Cast and Aged	3 h at 650	100 (640)	80 (5521)	3	C 21
									Solution Heat Treated	1475–1500	60 (414)	20 (1381)	40	B 59
									Solution H.T. and Aged	3 h at 650	155 (1069)	145 (1000)	1	C 38
C82500	20C	Be 1.90–2.15 Co 0.35–0.65 Si 0.20–0.35	0.292 (8.07)	8.09	1575–1800 (857–982)	18.5	725	9.4	As Cast		75 (517)	40 (216)	15	B 81
	20CT								Cast and Aged	3 h at 650	120 (827)	105 (724)	2	C 30
									Solution Heat Treated	1475–1500	60 (414)	25 (172)	35	B 63
									Solution H.T. and Aged	3 h at 650	160 (1103)	150 (1034)	1	C 43
C82510	21C	Be 1.90–2.15 Co 1.00–1.20 Si 0.20–0.40	0.298 (8.24)	8.26	1575–1800 (857–982)	18.5	725	8.6	As Cast		75 (517)	40 (276)	25	B 75
									Cast and Aged	3 h at 650	120 (828)	105 (724)	5	C 30
									Solution Heat Treated	1475–1500	60 (414)	25 (172)	40	B 63
									Solution H.T. and Aged	3 h at 650	160 (1103)	150 (1034)	1	C 42
C82600	245C	Be 225–2.45 Co 0.35–0.65 Si 0.20–0.35	0.292 (8.07)	8.09	1575–1750 (857–954)	19	675	9.4	As Cast		80 (552)	50 (345)	10	B 86
	245CT								Cast and Aged	3 h at 650	120 (828)	105 (724)	2	C 31
									Solution Heat Treated	1475–1500	70 (483)	30 (207)	12	B 75
									Solution H.T. and Aged	3 h at 650	165 (1138)	155 (1069)	1	C 45
C82800	275C	Be 2.50–2.75 Co 0.35–0.65 Si 0.20–0.35	0.292 (8.07)	8.09	1575–1700 (857–927)	19	675	9.4	As Cast		80 (552)	50 (345)	10	B 88
	275CT								Cast and Aged	3 h at 650	125 (862)	110 (759)	2	C 31
									Solution Heat Treated	1475–1500	80 (552)	35 (241)	10	B 85
									Solution H.T. and Aged	3 h at 650	165 (1138)	155 (1069)	1	C 46

[a]Metric units—cast and aged, 3 h at 343°C; solution treated at 802–815°C; Solution heat treated and aged for 3 h at 343°C.

1407

the final aging done after machining. This is accomplished by initially aging at 500–550°F (260–288°C) to bring the hardness up to Rockwell C 20 to 25. Time is extremely critical in this operation—one half to one hour maximum is preferred. To avoid nonuniform hardness caused by variation in the heat up time, this approach should be limited to sections with no more than a 10 sq in. (6,500 mm^2) cross section. The resulting hardness is readily machinable with high-speed tools and avoids the "gummy" solution annealed condition. In addition, most of the aging volume change will have taken place before machining. On completion of machining, the part is again age hardened to bring the hardness up to the level desired. The second treatment is usually done in the 600–650°F (316–343°C) range.

96.6   SURFACE TREATMENT

The cast surface of beryllium copper molds is usually of sufficient quality to require only minor finishing operations such as vapor blasting or liquid honing with glass beads in water. This will provide a matte finish, generally desired because it facilitates release of most plastics from the mold. Machined portions of the mold can be handled as other mold materials would be. Beryllium copper can be plated with copper, nickel, chromium, and electroless nickel to achieve coatings for special requirements. A hard beryllium copper substrate reinforces the harder chrome and electroless nickel plates, resulting in a highly durable surface suited to a wide range of molding conditions.

Mechanical or chemical cleaning may be employed to remove the scale that forms during heat treatment of molds.

96.6.1   Chemical Resistance

Beryllium copper will not rust, produce loose scale, or be attacked by most chemical agents encountered in injection molding, including lubricants, parting compounds, and the plastics resins themselves.

The exception is resins containing halogens, which emit hydrogen chloride or hydrogen fluoride fumes, and fire retardant plastics. If these are used, the tooling can be electroless nickel plated or chromium electroplated to protect it from corrosive attack.

97

MOLD MATERIALS–STEEL

James T. Christensen
Crucible Steel, P.O. Box 977, Syracuse, NY 13201

97.1 DESCRIPTION

A wide range of steels are used for plastic molds for a variety of reasons. Many mold shops and mold designers have considerable experience with a particular grade of steel that works in a specific application, but there is not any one, all-purpose tool steel with the ideal combination of properties for every mold use. Necessarily, trade-offs must be made in choosing the mold steel because each grade offers advantages and disadvantages.

Table 97.1 identifies the four basic categories of tool steel that are most popular for plastic molds: prehardened grades, including a stainless prehardened grade; air hardening grades, including two particle metallurgy grades; stainless steels, with one particle metallurgy grade; and oil hardening grade. There are some other grades not listed in this table in limited use in the plastic industry; they also fall into these four basic categories.

Table 97.1 also ranks the different grades of tool steel by machinability, grindability, and polishability, based on the typical application hardness most often used. The ranking is numeric, with the higher numbers indicating superiority in that property.

All of the tool steels listed in Table 97.1, except the three CPM grades, are melted in an electric arc furnace and refined in an AOD vessel. This is done to control the heat-to-heat uniformity of chemistry and the cleanliness levels in the steel. After melting, ingots are cast, upset pressed, and rolled to produce blocks and bars that are alike in structure throughout the cross section. The blocks are machined and supplied either in the spherodized annealed or the heat-treated (prehard) condition.

The CPM process used to produce the three high-vanadium tool steels listed in Table 97.1 is discussed later in this section.

Table 97.2 shows physical and mechanical properties for selected mold steels. These properties are listed by grade and a typical hardness for each.

97.2 IDENTIFYING STEELS

97.2.1 Prehardened Tool Steels

These steels are supplied already heat treated in the range of 260–350 BHN to eliminate size changes and extra finishing operations associated with hardening molds. For cores and cavities, these steels machine in the medium range; they can be polished to lens mold applications. The resulfurized or sulfur-added modificiations to AISI 4140 or 4150 enhance machinability to the high area; however, these modified steels cannot take on a high polish because of the sulfur additions. They are usually used for shoes, holders, and mold bases.

TABLE 97.1. Tool Steel Nominal Chemistries for Plastic Mold Steels

Type	AISI Symbol	C	M_N	S_I	C_R	N_I	M_O	V	W	Machinability[a]	Grindability[a]	Polishability[a]	Typical Hardness, R_c
Prehardened	P20	0.30	0.75	0.50	1.65		0.40			8.5	10	10	31
	4140	0.40	0.90	0.30	1.00		0.20			8.5	10	8	28
	T414	0.02	0.50	0.50	11.75	2.90				6	8	10	30
Air Hardening	S7	0.50	0.70	0.35	3.25		1.40	0.25		7.5	8	10	55
	H13	0.40	0.35	1.05	5.00		1.35	1.05		7	8.5	10	47
	A-2	1.00	0.85	0.30	5.25		1.10	0.25		5	7	8	61
	A-6	0.70	2.00	0.30	1.00			1.35		6	8	9	56
	D-2	1.55	0.30	0.45	11.50		0.80	0.90		3.5	3.5	6	61
	A-11 CPM10V[b]	2.45	0.50	0.90	5.25		1.30	9.75		4	3.5	7	61
	CPM9V[b]	1.78	0.50	0.90	5.25		1.30	9.00		4.5	4.5	7	55
Stainless	T420	0.35	0.45	0.50	13.00					6.5	8	10	50
	T440C	1.05	1.00	1.00	17.00		0.75			3.5	4	8	58
	CPM T440V[b]	2.20	0.50	0.50	17.50		0.50	5.75		3.5	4	7.5	59
Oil Hardening	01	0.90	1.25	0.30	0.50				0.50	7	9	9	61

[a]Low number means property is poor. High number means property is good.
[b]Crucible trade name.

TABLE 97.2. Physical and Mechanical Properties for Selected Plastic Mold Steels

Steel	Tensile Strength, ksi (MPa)	Tensile Modulus × 10⁶ psi (GPa)	Yield Strength 2% Offset ksi (MPa)	Thermal Conductivity Btu-ft/(h-ft.²-°F) °F		Thermal Conductivity W/m-k		Density, lb/in.³ (g/cm³)	Coefficient of Thermal Expansion in./in./°F × 10⁻⁶	°F	°C	mm/mm/°C × 10⁻⁶
P20 (311 BHN)	157 (1080)	30 (207)	136 (938)	200	24.7	93	42.7	0.284 (7.87)	5.84	100–500	38–260	10.5
				400	27.5	204	47.6					
4140 (302 BHN)	148 (1020)	30 (207)	95 (655)	200	24.4	93	42.2	0.282 (7.81)	7.10	68–800	20–427	12.8
				400	26.5	204	45.8		6.84	68–212	20–100	12.3
									7.62	68–752	20–400	13.7
T414 (300 BHN)	140 (966)	29 (200)	120 (828)	200	12.1	93	20.9	0.277 (7.67)	5.8	32–212	0–100	10.4
									6.1	32–600	0–316	11.0
S7 (207 BHN)	93 (641)	30 (207)	55 (379)					0.283 (7.84)	7.3	75–750	24–400	13.1
									7.6	75–1000	24–538	13.7
H13 (45HRc)	225 (1550)	30 (207)	200 (1379)	68	14.2	20	24.6	0.280 (7.76)	6.88	100–800	38–427	12.4
						199	25.1		7.00	100–1000	38–538	12.6
T420 (200 BHN)	95 (655)	29 (200)	50 (345)	200	14.4	93	24.9	0.276 (7.65)	5.7	32–212	0–100	10.3
									6.0	32–600	0–316	10.8
T440C (230 BHN)	110 (759)	29 (200)	65 (448)	200	14.0	93	24.2	0.276 (7.65)	5.7	32–212	0–100	10.3
									6.0	32–600	0–316	10.8

TABLE 97.3.   Attainable Hardness of Carburized CSM 2[a]

Tempering Temperature °F	(°C)	Case Hardness, R_c	Core Hardness, R_c
600	(316)	57–58	47–48
650	(343)	57–58	46–47
700	(371)	55–56	45–46
750	(399)	54–55	44–45
800	(427)	53–55	43–44
900	(482)	52–53	39–40

[a]Gas carburized—1600°F (871°C).
Furnace cooled to 1475°F (802°C).
Oil quenched and tempered 4 + 4 hours.
Section size—4 in. (101.6 mm) diameter bar.
Note: Larger sections than those tested will show lower hardness values; smaller sections will show higher hardness values.

P20. P20 is the mold steel most widely used in the plastic industry for all types of machine-cut molds. P20 is most commonly supplied in the prehardened condition (around 300 BHN). High hard material is available in hardnesses up to 350 BHN for added strength and durability.

P20 is also supplied to the annealed condition at around 200 BHN. It can be heat treated, carburized, or nitrided to high hardness if mold wear is a problem or if it is easier at 200 BHN to acquire a high optical finish than at the standard hardness of 300 BHN.

Table 97.3 shows the attainable hardness of both the case and core of gas carburized CSM2R (P20) for different tempering temperatures.

Figure 97.1 shows the case depth vs the carburizing time. Figure 97.2 shows case depth vs the nitriding time for both 200 BHN and 300 BHN prehardened blocks of P20, nitrided at 975°F for 24, 36, 48, and 60 hour cycles.

P20 can be plated with chrome or nickel to provide added wear and corrosion resistance for molding PVC or other engineering plastics. It is available in blocks up to 75,000 lb (3400 kg) for use in large automotive or other applications. Prehardened P20 is used in injection

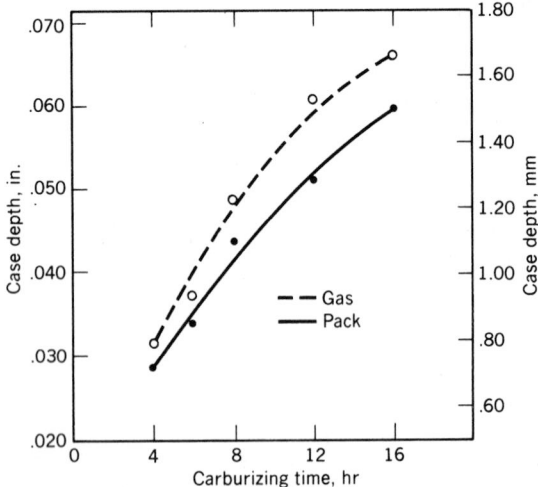

Figure 97.1.   Carburizing CSM 2: depth of case vs time.

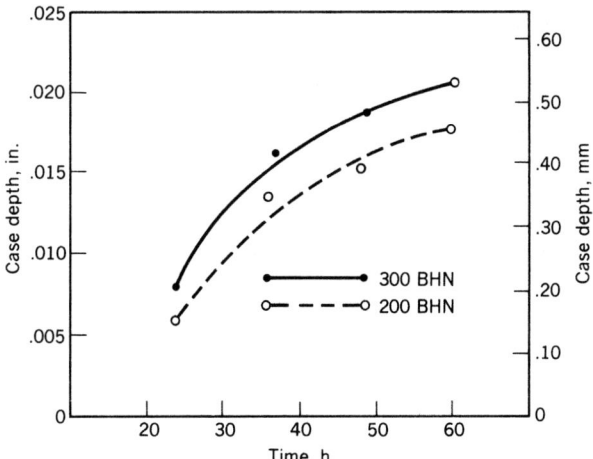

Figure 97.2. Nitriding 200 BHN and 300 BHN CSM2: depth of case vs time.

and transfer molding; carburized P20, in compression molding for greater surface strength and wear resistance.

4140. 4140 is an alloy steel supplied in the prehardened condition (around 300 BHN) and used for holders, shoes, and mold bases. It usually holds the plastic mold, which is inserted. Blocks of 4140 used for cores and cavities, especially the resulfurized modifications, should only be employed to produce nondecorative parts. It has good fatigue, abrasion, and impact resistance as well as resistance to softening at elevated temperatures.

T414. T414 prehardened tool steel is supplied at 300 BHN. T414 is designed primarily for those applications where a tool steel more corrosion-resistant than P20 is needed for molding all plastics, including vinyl-base and other corrosive plastics. T414 will not rust when molds begin to sweat or during storage. Polishability is excellent for this prehardened grade, making it ideal for optical finishes. Its general corrosion resistance is excellent in a variety of corrosive environments and is superior to that of T410 stainless. T414 eliminates the need for chromium plating and the attendant problems of stripping for replating and pitting in the replating operation.

T414 can be texturized best by using a ferric chloride and water mixture; this produces depths acceptable for most texturing finishes in the same time that it takes to texturize P20.

97.2.2 Air Hardening Tool Steels

Air hardening tool steels are distinguished by the method of quenching (attaining the hardness) after the austenitizing temperature. In the hardening phase, this group of tool steels is cooled more slowly than the oil or water hardening grades. The result is less intense strains and less distortion.

The most important elements for making steels capable of air hardening are chromium and molybdenum, which, when combined with carbon, produce hard carbides with exceptional wear resistance.

To obtain the required hardness levels in large blocks, some of these steels are quenched in oil. To avoid excess scaling, air hardening tools are sometimes flash quenched in oil to about 1000–1200°F (538–649°C) and then cooled in air. Adequate space must be allowed for uniform movement of air around the parts being cooled. Although it is not good practice to use forced air, a mild uniform flow from electric fans is often used to speed up the cooling action.

S7. S7 is a chrome, molybdenum tool steel supplied in the annealed condition at about 200 BHN. It is usually used in the 52–56 Rockwell C hardness range. Larger sections more than 6 in. (152 mm) thick should be given an interrupted oil quench and tempered immediately to the temperature producing the desired hardness. This should be done when the block becomes warm to the touch. S7 offers the combination of high shock resistance and toughness usually required in compression and automatic molding operations.

H13. H13 is a popular mold steel furnished in the annealed condition at about 210 BHN that must be heat treated after the mold is rough machined. This steel offers heat treatable hardnesses up to 53 HRc. It is often substituted for P20 in applications requiring higher hardness and better strength, such as compression molding. H13 resists softening at higher temperatures; its good heat check resistance is often useful in thermoset molding. To retard washing, high wear resistance applications can be handled by gas nitriding after finish grinding and polishing. Gas nitriding for 10 to 12 hours at 950°F (510°C) results in a case depth between 0.004 and 0.005 in. (0.10 to 0.13 mm) with a surface hardness of 65–70 HRc.

H13 is used for injection, compression, and transfer molds of intermediate hardness requiring good dimensional stability during heat treatment. H13 affords excellent polishability, which makes it a good candidate for lens mold applications.

A2. A2 is another air-hardening tool steel supplied in the annealed condition at 220 BHN. It can be heat treated to 60 HRc, making it suitable for compression molding and molding abrasive engineering resins. It is usually not available in large sections and is used for smaller molds. A2 has good dimensional stability during heat treatment, good toughness, fairly high abrasion resistance, and can be polished to an excellent finish. This makes A2 a good choice for smaller molds having complicated design.

A6. A6 is selected where high strength, good wear resistance, and good toughness are essential. It is supplied annealed at 220 BHN and requires a lower hardening temperature than A2, resulting in lower residual stresses and less chance of distortion in the furnished die. It can be hardened to 60 HRc and has excellent polishability at this maximum hardness range. It is used in injection, compression, and transfer molds.

D2. D2 is supplied annealed at around 235 BHN. It can be heat treated to a maximum of 62 HRc, at which it has low distortion and a high degree of safety. Its good abrasion resistance is important for molding reinforced plastics, although it is not usually available in large sections. D2 is used in small injection and compression molds and can be used to insert molds in abrasive applications. In its heat-treated condition, care in grinding is necessary to prevent abuse and grinding cracks.

Particle Metallurgy High-vanadium Tool Steels. Included among the air-hardening tool steels are two particle metallurgy high-vanadium tool steels (CPM 10V and CPM 9V). In the stainless steels is a third particle metallurgy grade, CPM T440V. These three grades are finding increased usage in molds, injection screws, barrel liners, and molding parts in such severe wear applications as glass-filled engineering resins.

The wear resistant properties in tool steels derive from both the heat treated hardness and the combination of hard, abrasion-resistant carbides in the microstructure. Of the carbides that are formed in most tool steels, vanadium carbides are the hardest and most wear-resistant. The others (in decreasing order) are tungsten, molybdenum, chromium, and iron carbides. Steels made with high vanadium levels do not represent a new developoment, but, in the past, they have had little success because of the size of the vanadium carbides produced in traditional air melted heats. Vanadium carbides produced in air melted heats are large, creating machining and grinding difficulties in both the annealed and heat treated condition as well as toughness limitations in service. The particle metallurgy process has overcome these limitations by making very small, uniformly distributed carbides in the microstructural matrix.

Another advantage of the particle metallurgy process over the traditional slow ingot solidification process is the production of steels that evidence no alloy segregation macroscopically and exhibit small, uniformly distributed carbides and fine grain sizes microscopically. Since the sulfides also remain fine and uniformly distributed, the steels can be resulfurized without loosing toughness.

The particle metallurgy process can also produce higher alloyed tool steel compositions with much higher levels of vanadium. The small, uniformly distributed vanadium carbides permit conventional machining with carbide and ceramic inserts, even when heat treated to the low 60s HRc. These steels are used in applications requiring wear resistance superior to those available in conventionally produced steels.

A11 (CPM 10V). A11 (CPM 10V), which was commercially introduced in 1978, is designed for the exceptional combination of wear resistance and toughness. It is supplied in the annealed condition at about 260 BHN and can be heat treated to a maximum of 65 HRc. Normal application hardness for A11 is 60–62 HRc, where the combination of toughness and wear resistance are optimum. In the 63–65 HRc range, maximum wear resistance and compressive strength are achieved.

Figure 97.3 compares the wear resistance of A11 with other tool steels in laboratory crossed-cylinder wear tests. Figure 97.4 illustrates the impact toughness of A11 vs D2 and M2, showing they are comparable at similar hardnesses.

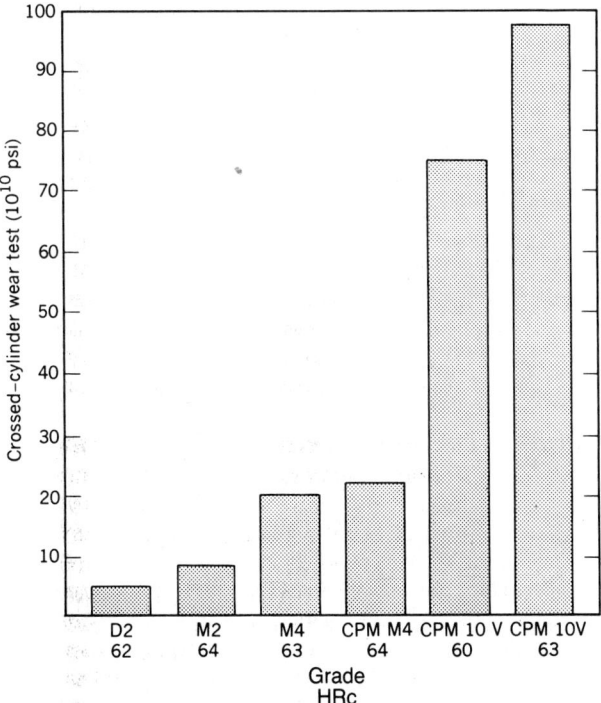

Figure 97.3. Wear resistance: A11 vs other tool steels.

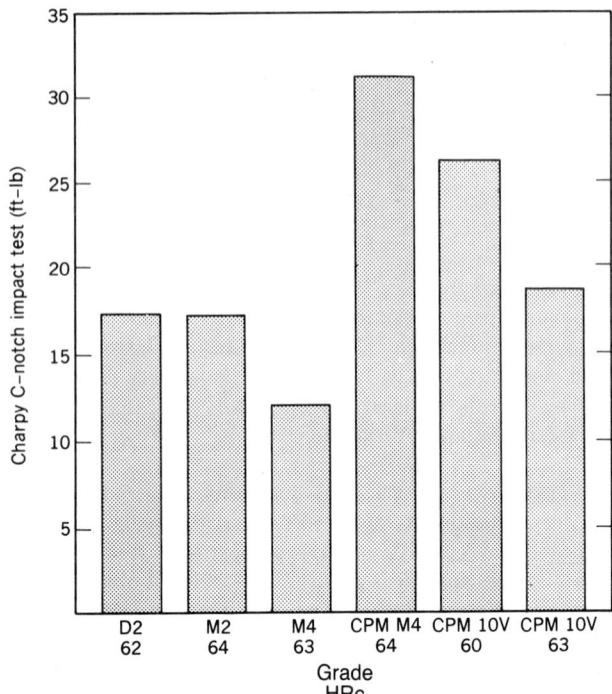

Figure 97.4.   Toughness of All vs D2 and M2.

A11 is used for inserts in molds running glass-filled resins and other engineering plastics, where mold wear is a problem. It is employed in injection-molding equipment in screws, barrel liners, nozzles, nonreturnable check, rings, granulator/pelletizer blades, and other high-wear parts.

CPM 9V.   CPM 9V is a high-vanadium tool steel whose chemistry is patterned after CPM 10V, although with less vanadium (9%) and carbon, to produce a high wearing tough tool. It is supplied in the annealed condition at about 240 BHN and can be heat treated to a maximum of 57 HRc. Normal application hardness range is 53–55 HRc. At this hardness, its wear properties are roughly half those of CPM 10V at 60 HRc but almost 10 times that of D2 at 62 HRc (see Figure 97.5).
   Figure 97.6 displays another advantage of CPM 9V, its superior toughness as compared to A2, D2, and CPM 10V.
   Although CPM 9V is fairly new grade to the plastics industry, it is expected to have a significant impact in tooling applications where heat check resistance, good toughness, and wear resistance are required.

97.2.3 Stainless Mold Steels

Stainless steels are selected for molds because of their corrosion resistance to PVC and other corrosive resins. Stainless molds will not corrode in the water lines or pit in storage between production runs. They eliminate the need to chrome plate and all the problems associated with stripping and rechrome plating. As a group, their polishability is excellent, especially at higher hardnesses, which makes them ideally suited for lens mold applications.

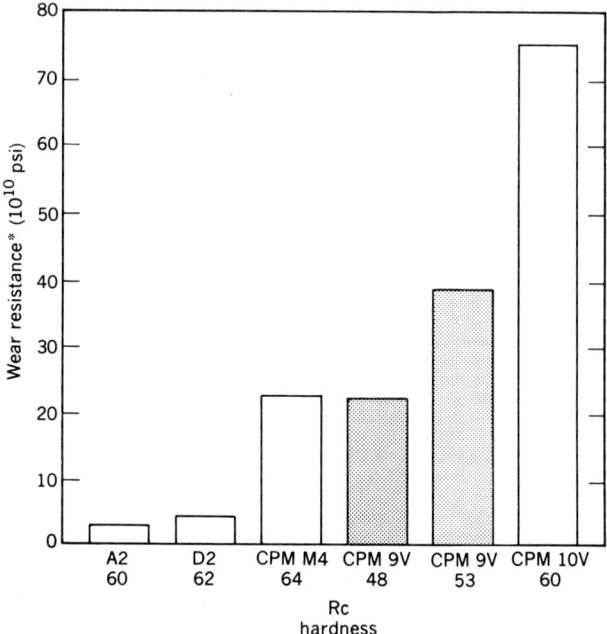

Figure 97.5.   Wear resistance: CPM 9V.

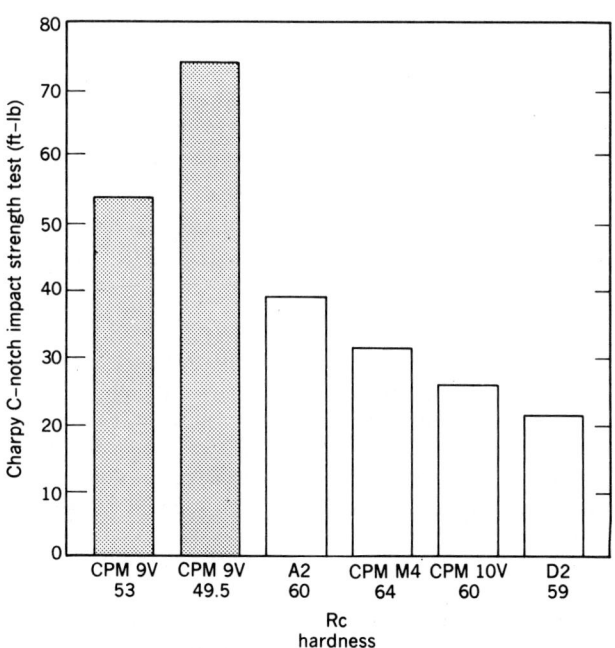

Figure 97.6.   Wear resistance: CPM 9V.

T420. T420 is a popular stainless mold steel furnished in the annealed condition at about 215 BHN for molds for all plastics. It can be readily machined in the annealed condition and has good dimensional stability in heat treatment. T420 can be heat treated to a maximum of 53 HRc. It is used frequently to eliminate corrosion problems associated with rusting water lines and mold surface pitting caused by condensation in operation or storage. T420 can be selected for injection, compression, and transfer molds used for PVC or other corrosive resins. It is also chosen for lens and glass molds because of its polishability.

T440C. T440C is another stainless steel selected for plastic mold applications and supplied in the annealed condition at 230 BHN. It can be hardened to the 60 HRc range, which results in higher strength and abrasion resistance than T420 but slightly lower corrosion resistance. This relates to its higher carbon content, forming more chromium carbides, which reduces the chromium remaining in solution. The result is slightly lower corrosion resistance than T420. T440C is used for smaller compression and injection molds.

CPM T440V. CPM T440V is the stainless version of the CPM high-vanadium tool steels. Its chemistry is patterned after T440C with almost 6% vanadium added for wear resistance. It is supplied in the annealed condition at about 260 BHN. Small molds and machinery parts that process corrosive, abrasive resins are excellent applications for CPM T440V. This grade can be heat treated to a maximum of 62 HRc, with recommended usage in the 58–60 HRc range. Typical applications include injection screws, bushings, nonreturn check ring valves, pelletizer blades, and mold inserts.

97.2.4 Oil Hardening Tool Steels

Oil hardening tool steels have substantial amounts of manganese and other alloys, which permit these steels to be hardened in oil. There is no case condition on these steels as there is in water hardening grades, and they will harden all the way through, even in relatively large sections (up to 2.5 in. (64 mm) round].

O1. O1 is supplied in the annealed condition at about 200 BHN. It can be heat treated to high hardnesses (61–63 HRc) and is used primarily for smaller molds or mold inserts where strength is a consideration over toughness. This steel may be hardened from fairly low temperatures with minimal size changes. O1 is considered a general-purpose tool steel used for injection, compression, and transfer molds.

97.3 HEAT TREATMENT

Molds are heat treated to produce combinations of strength, wear resistance, and toughness in the particular grade of steel selected so that it will perform in a specific way. (This applies to nitrided and carburized cases as well as the deep and through hardening grades.) Table 97.4 shows the annealing, hardening, and tempering temperatures with the expected results of heat treatment. Although heat treating represents a relatively small cost to a mold tooling project, if it is improperly done catastrophic failures can result. Before reviewing the actual steps in heat treating, two areas of concern must be identified—stress relieving and the furnace atmosphere.

97.3.1 Stress Relieving

Most mold steels are supplied in the annealed condition and are relatively low in residual stresses. Severe machining operations, such as hogging out a cavity or cutting, cause these residual stresses to increase and may result in unplanned distortion during heat treatment.

TABLE 97.4. Tool Steel Heat Treatment for Plastic Mold Steels

Steel	Annealing Temperature, °F (°C)	Hardness Annealed, BHN	Hardening Temperature, °F (°C)	Quenching Medium	Tempering Temperatures, °F (°C)	Hardness Heat-Treated, R_c
P20	1425–1450 (774–788)	180–210	1500–1550 (815–843)	Oil	1075–1150 (579–621)	30–36
4140	1550–1600 (843–871)	180–210	1550–1600 (843–871)	Oil	1100–1200 (593–649)	29–34
T414	1200–1300 (649–704)	210–230	1475–1550 (802–843)	Air	700–750 (371–399)	28–30
S-7	1500–1550 (815–843)	182–223	1725–1750 (940–954)	Air or air/oil	500–700 (260–371)	52–56
H-13	1575–1625 (857–885)	192–235	1800–1850 (982–1010)	Air	1000–1100 (538–593)	46–52
A-2	1550–1600 (843–871)	207–235	1750–1800 (954–982)	Air or air/oil	400–650 (204–343)	58–62
A-6	1400–1425 (760–774)	210–230	1500–1625 (815–885)	Air	400–600 (204–316)	55–59
D-2	1600–1650 (871–899)	217–255	1825–1875 (996–1024)	Air or air/oil	900–1000 (482–538)	54–60
CPM 10V	1600–1650 (871–899)	248–269	1950–2150 (1006–1177)	Air, salt or air/oil	1000–1025 (532–552)	58–65
CPM 9V	1625–1650 (885–899)	223–255	1850–2100 (1010–1149)	Air, salt or air/oil	1000–1100 (538–593)	46–57
T420	1550–1650 (843–899)	192–241	1800–1900 (982–1038)	Air or air/oil	400–750 (204–399)	48–54
T440C	1550–1650 (843–899)	190–215	1850–1900 (1010–1038)	Air or air/oil	300–500 (149–260)	55–58
CPM T440V	1625–1650 (885–899)	240–260	1850–2050 (1010–1121)	Air, salt or air/oil	300–500 (149–260)	52–62
01	1400–1440 (760–782)	183–212	1450–1500 (788–815)	Oil	350–450 (177–232)	58–63

After such severe machining operations, stress relieving is recommended to reduce the residual stresses and thereby keep the distortion during heat treatment to the level planned. If a deep cavity is being formed in a piece of mold steel, stress relieving is recommended after 80–90% of the material has been removed.

A mold should also be stress relieved after finishing, especially if this has involved much of grinding and polishing, to reduce the residual stress level of the mold going into service. Intermittent stress relieving can prevent premature cracking of a mold after a certain number of parts have been produced. For example, it can be done after a mold is taken out of service and is waiting for the next customer order. Stress relieving for annealed material should be done at 1200–1300°F (649–704°C) and held at this temperature for as many hours as the largest inch-thickness (25 mm) dimension. For example, a mold 5 in. (127 mm) thick should be held at this temperature for five hours and then air cooled. Prehardened or hardened material should be heated to 50–100°F (10–38°C) below the last tempering temperature and held at this temperature for an appropriate time before air cooling. If the stress relieving is done above the tempering temperatures, there will be a reduction in the hardness of the mold and some dimensional changes.

97.3.2 Furnace Atmospheres

For all tool steels, heat treating involves three stages: austenitizing, quenching, and tempering. Frequently, austenitizing is performed in two steps: a preheat and a high heat. In these stages, it is important to consider design and mass; these two factors determine the important relationship of time and temperature. Furnace atmospheres used in the heat treatment process must also be considered because of decarburization.

Decarburization occurs on steel surfaces heated above 1300°F (704°C) in oxidizing atmospheres and results in the loss of carbon on these surfaces. Carbon loss actually results in a two-toned steel, with a lower carbon outside area covering the desired inner composition. When this happens, a softer working surface is produced that usually leads to poor tool performance. Some tool steels have a greater tendency to decarburize than others, but those with higher carbon contents are more susceptible to this phenomenon.

To prevent decarburization, the furnace atmosphere can be an inert gas, a high-temperature salt bath, or no atmosphere (vacuum).

97.3.3 Preheating

Preheating is the first step in the heat-treatment process. It serves a number of purposes including minimizing decarburization, avoiding heavy scaling, and relieving stress caused by cold working. Preheating allows more uniform heating between the surface and the center of the mold because it equalizes the mold temperature before the mold is exposed to the high heat furnace. The preheat temperature varies with each grade but is generally above 1000°F (538°C) and seldom over 1600°F (871°C).

In large or complicated molds two preheats are recommended to reduce temperature differentials and thermal stress gradients.

97.3.4 High Heat

The high heat cycle is the most important step because during it the final conditioning of the steel is accomplished. Both the temperature and the time must be carefully monitored. Excessive temperature or time will cause grain coarsening and decarburization. Low temperatures or short times bring about inadequate solution of carbides or nonuniform structural conditions.

The high-heat time period is called the soak. The proper soak ensures that the entire section is thoroughly heated and that the steel will respond to the heat treatment in a uniform manner. Soak times depend on the grade of steel and the overall size of the piece being heat treated. Good operators know their furnace capabilities and make adjustments to make certain that the actual steel temperature is proper for the heat treatment of that grade.

97.3.5 Quenching

After austenitizing, the mold is quenched. During the quench, the actual hardening of the steel takes place. Different compositions of steels require different cooling rates to achieve full hardness. The decision to use a particular quenching medium is based on its ability to remove heat at the rate required by the particular steel and the mass of the section. The ideal medium will reduce the temperature quickly at first and then more slowly to accommodate the stresses, which develop in the lower temperature range, from hardening.

97.3.6 Tempering

Tempering is performed after the quench and involves a precisely controlled time–temperature relationship. Furthermore, control of the time between the quench and tempering is most vital. Since hardening takes place gradually at the end of the quench cycle, tempering too soon interrupts proper cooling and puts a stop to further hardening. In large sections with a significant temperature gradient, cracking may occur.

On the other hand, if the mold is permitted to stand for a time at room temperature, the stresses from hardening continue to build up to a point where rupture may occur. The general rule is to begin tempering as soon as the work can be held with the bare hands [between 125 and 150°F (52 and 66°C)].

The usual temperature range for tempering extends from 300 to 1200°F (149–649°C), depending on the grade and the hardness desired. At a minimum, the time period should be two hours; usually, an hour per inch (25 mm) at the smallest cross section.

97.4 SURFACE TREATMENT

There are several surface treatments used to prepare the mold surface for the final ordered part. These include polishing, electrical discharge machining (EDM), texturing (photoetching), and plating/coating.

97.4.1 Polishing

Polishing is the oldest and still most widely used technique for finishing the required surfaces. In most molds, a certain amount of polishing is required, even if other finishes are applied through texturing and plating. In general, the harder the steel surface is, the easier it is to bring it up to the required polished surface. The prehardened steel P20 and stainless steel T414 have excellent polishability and are used for lens mold applications. Occasionally orange peel or pitting problems are associated with these softer steels, but care in polishing can prevent this phenomenon. With certain grades, experience is the best guide in attaining the desired surface finish. In general, it is important to go slow; not use mechanical polishers; not skip any steps—through the paper grits, stones, and diamond polish; and avoid extreme pressures.

Orange peel and/or pitting can occur on any steel and usually appears as the surface is brought up with the diamond compounds to a mirror finish. These problems occur when the yield/tensile strength of the surface is exceeded. When this happens, the surface moves and actual pieces of the matrix are pulled out, causing pits. To correct this condition, the affected area must be stoned off to remove the pits, with the last stone before diamond polishing, stress relieving, and restoning making sure not to skip any steps. Hand polishing is recommended at light pressures.

97.4.2 Electrical Discharge Machining (EDM)

EDM can produce surface quality equal or better to that obtained with conventional machining. EDM often eliminates the need for secondary finishing operations. In the EDM operation, the EDM pulse actually melts away the surface of the steel being contacted by the spark. The recast "white" layer, which remains on the surface, contains brittle, untempered martensite. The newer generation of EDM pulse-generated power supplies reduces this white layer to less than 0.0005 in. (0.0127 mm). It is always a good idea to remove this white layer with subsequent stoning and polishing. If this layer is not removed, the work piece should be at least stress relieved.

97.4.3 Texturing

Texturing or photoetching is performed to give the surface of the molded part a different look. These different surfaces are made to look like leather, wood grain, and other common materials simulated by plastics. Most steels used for molds can be textured. Steels high in nickel are more difficult to etch because nickel resists acids used in the etching process. The stainless steels, which are high in chrome (12–15%), are etched successfully using different concentrations of ferric chloride and water mixtures.

The following simple precautions should be used when selecting a steel for molds to be textured:

- Use a good mold steel.
- Use the same steel for all components of the mold for uniform chemistry.
- Avoid nonuniform machining stresses.
- When welding, use welding rods of the same chemistry as the mold.
- Retemper thoroughly after welding heat treated molds.

The key word in successful texturing of molds is uniformity. The steel selected should be as alike as possible in all characteristics. Ideally, the steel for all molds in a family should be made from the same steel bar with the steel cut so that the grain runs in the same direction. The same processing methods and tools should be used on all molds when matching texture is desired.

97.4.4 Plating/Coating

Molds are given many plating or coating treatments to increase wear and corrosion resistance and aid in reducing release problems. There is no one plating or coating treatment that solves all these problems. The molder must decide which problem is causing the greatest loss in production and choose the plating or coating treatment designed to help solve that problem.

The following chart lists the plating or coating treatments for steels used for plastic molds and the reasons for choosing this treatment.

Plating or Coating Treatment	Purpose
Hard chrome	Reduce wear and corrosion.
Nickel (electroless and electrolyte)	Resist corrosion, improve wear; used to build up worn or undersized molds.
Nitriding	Increases surface hardness for increased wear, adds some corrosion resistance, and aids in polishing.
TiN (titanium nitride)	Increases surface hardness for increased wear and lubricity for faster part removal.
Ion implantation	Increases surface hardness for increased wear, adds some corrosion resistance, and improves mold-release characteristics.

97.5 INDUSTRY PRACTICES

When tool steels are selected for plastic molding, the mold designer must analyze all the requirements associated with each mold before choosing the steel with the best combination of properties.

Glass-filled or fiber-reinforced resins cause abrasion of the mold surface and wear feed screws and nozzles in plastic molding equipment. High wear resistance steels are selected in these applications. PVC materials generate corrosion of molds and molding equipment. Stainless grades are chosen in these molding applications to resist corrosion.

Certain pressures in the molding process, as in injection and compression molding, require higher strength tool steels. Operating temperatures and designed cycle times must be considered by the tool designer to evaluate the steel's heat check resistance and thermal conductivity. Heat treatment and surface treatment are other important decisions made by the mold designer to ensure that the tool lasts for the entire anticipated production schedule.

All these factors influence the life of a mold tool. A successful tool design achieves an intricate balance of the desired part design and material, the mold steel capabilities, and die making techniques.

BIBLIOGRAPHY

General References

Steel Products Manual–Tool Steel, American Iron and Steel Institute, Washington, D.C., Sept. 1981.

R.B. Dixon, *Advances in the Development of Wear Resistance: High-Vanadium Tool Steels for Both Tooling and Nontooling Applications,* American Society for Metals, Metals Park, Ohio, 1982.

B.S. Lement, *Plastic Mold Steels,* Climax Molybdenum Co. *Forming,* Vol. 2 of *Tool and Manufacturing Engineering Handbook,* 4th ed., June 1984.

Tool Steel for the Non-Metallurgist, Crucible Materials Corp., 1985.

W. Young, *Getting the Best Performance from Your Molds,* Tech. Literature 341/10, Hooker Chemicals & Plastics Corp., 1975.

Mold Finishing and Polishing Manual, I.T. Quarnstrom Foundation (available through the Society of Plastics Engineers), 1989.

98

SPRAYED METAL MOLDS

Leon Grant

TAFA Incorporated, Dow Road, P.O. Box 1157,
Bow (Concord), NH 03301-1157

98.1 DESCRIPTION AND HISTORY

Metal sprayed tooling is a means of reducing tooling costs for many plastic molding processes. In recent years, soaring tooling costs and the need for shorter production runs in the plastic molding field have prompted many companies to look at lower cost tooling methods. Processes such as structural foam, RIM, prototype injection molding, compression molding, and vacuum forming are highly adaptable to metal-sprayed tooling (Figs. 98.1–98.5).

In this type of toolmaking, an arc-wire torch is used to spray a Kirksite-type alloy for the tool surface, which is supported by special castable aluminum-filled resin or a cast metal alloy. The choice of backup material is a function of the molding process.

This toolmaking process can produce irregular-shaped plastic parts giving not only an exact copy of every surface detail but also of the precise form and dimension. Starting with a model made of almost any material, a Kirksite-type shell can be sprayed onto the model to the desired thickness.

This is accomplished without distortion or overheating of the model or pattern. The process is quick and less costly than conventional tooling methods, requiring only 15 minutes to spray 1 ft^2 (0.09 m^2) of model to a thickness of 1/16 in. (1.6 mm). There is no size limitation to the process; it has been used to make injection-molding tools the size of a dime as well as molds requiring 750 g of glass-filled nylon. Large autoclave tools, 70 ft^2 (6.5 m^2) in surface area and large RIM models requiring 40 lb (18 kg) of rigid RIM material have also been fabricated.

The process of arc spraying was developed in Europe some 35 years ago, but it has only been applied effectively as an alternative tooling medium for the plastics field in the last 10 years. The process was originally developed for spraying hard wire materials such as steel, nickel, copper, aluminum/bronze, and others to build up shafts and anticorrosion coatings.

The arc spray process was brought to the United States in the 1960s when significant progress was made in the development of lighter, more versatile, streamlined guns having the ability to spray lower melt alloys [900°F (482°C) and below] and thereby produce exceptionally strong, high-density coatings. In the early 1970s the first metal sprayed molds were made from the process, when it was discovered that thin shells of zinc and tin–zinc could be sprayed onto models of wood, plastic, and other materials. The shells backed with epoxy resins proved to be rugged enough to run in prototoype and short-run production environments. The shoe industry pioneered sprayed metal tooling.

More recently, a Kirksite-type alloy was developed to meet the need for a harder, more durable surface. Three times harder than pure zinc, with a melting temperature of 800°F

Figure 98.1. Vacuum form mold.

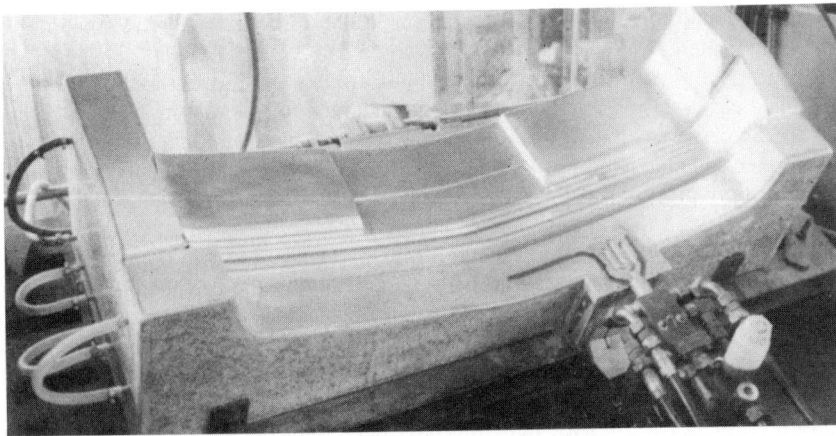

Figure 98.2. RIM automotive spoiler mold.

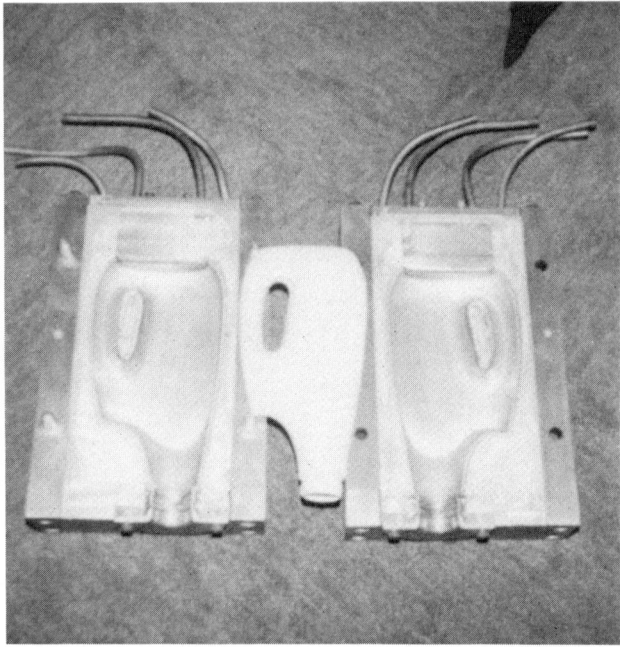

Figure 98.3. Blow mold.

(427°C), it has been a well accepted material for producing arc sprayed plastic molds. Today, many sprayed metal molds are used as short-run production tools whose tool life is dependent on severity of the application. In the past 10 years, deterrent factors have been overcome with the aid of new metallizing equipment, harder metals, improved procedures, and support materials.

At the present, most plastic molding processes have been tested for sprayed metal molds. Tables 98.1 and 98.2 indicate the expected mold life. Applications with lower pressures and temperatures are best suited for metal sprayed molds. Injection molding and compression

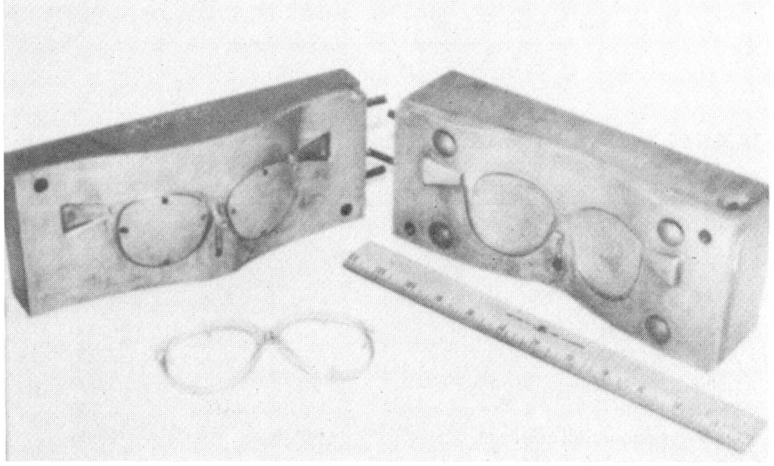

Figure 98.4. Prototype injection mold.

Figure 98.5. SMC mold.

TABLE 98.1. Processes Giving Longer Life Tools

Molding Process	Number of Parts
Cast urethane	100,000
Vacuum forming (see Fig. 98.1)	40,000
Metal spray foundry patterns	150,000
Investment casting	20,000
Structural foam	5,000
RIM (see Fig. 98.2)	10,000
Blow molding (see Fig. 98.3)	30,000

TABLE 98.2. More Severe Tool Applications

Molding Process	Number of Parts
Resin transfer molding	1,000
Injection molding of:	
Polyethylene	3,000
Styrene	3,000
Low-pressure styrene	30,000
ABS	3,000
Cellulose proprionate (see Fig. 98.4)	3,000
Polycarbonate	3,000
Glass- or mineral-filled resins	1,000
Compression molding (SMC) (see Fig. 98.5)	200
Autoclave tooling	2,000

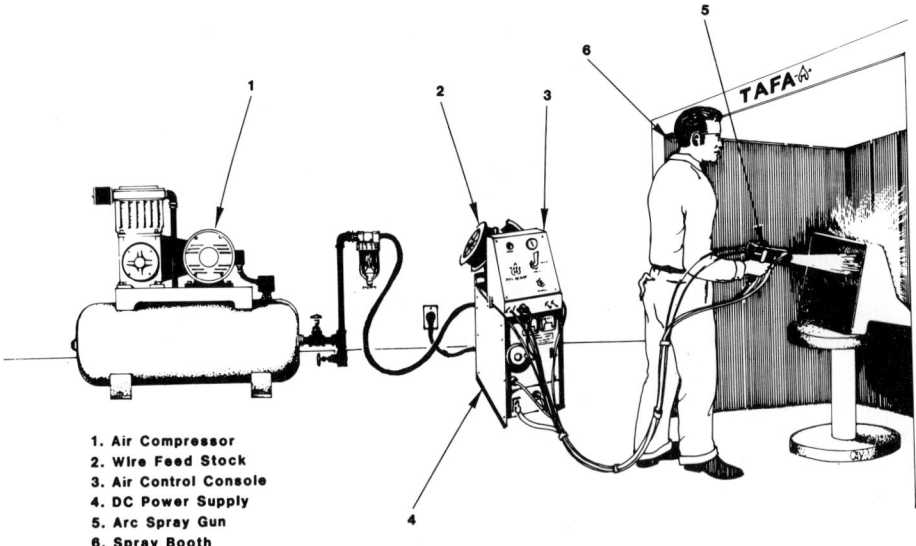

1. Air Compressor
2. Wire Feed Stock
3. Air Control Console
4. DC Power Supply
5. Arc Spray Gun
6. Spray Booth

Figure 98.6. Complete sprayed metal tooling system.

molding are higher pressure applications; here, metal-sprayed molds are best used as prototype tools.

98.2 PROCESSING CHARACTERISTICS

The arc spray system is simple. It consists of a low-voltage D-C power supply (much like that of an arc welder), a console for regulating the compressed air, the arc spray gun, and the wire feedstock. The system requires 50 cfm (1.4 m³/min) of compressed air and 5 kva of three-phase electrical power. Figure 98.6 illustrates a complete metallizing system.

Two electrically isolated wires are fed into the gun via an air motor. They intersect, forming an arc at the front of the gun. The arc melts the wire and the molten material is blown off by an atomizing jet as fine particles to form a dense metallic coating with high bond and cohesive strengths between particles. Figure 98.7 illustrates buildup of a sprayed coating.

One major advantage of the arc spray system is that it permits low-melt [800°F (427°C)] alloys to be sprayed onto wood and plastic models without overheating. This occurs because the molten particles are surrounded by the high-velocity air stream, which propels them

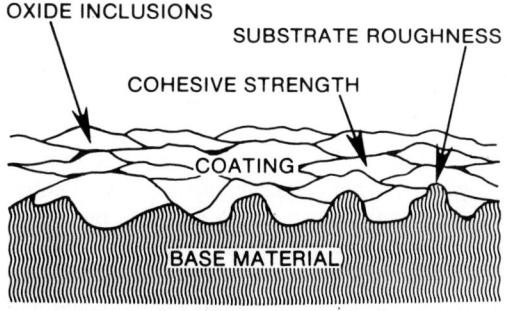

Figure 98.7. Cross section of Kirksite sprayed coating.

onto and cools the substrate. The particles solidify on impact after splatting and interlocking. When properly sprayed, the sprayed shell never reaches temperatures above 130°F (54°C).

The spray gun is about the size of a paint sprayer and is easy to use. Application is similar to that of a good paint sprayer. A flowing motion back and forth over the surface is desired at a distance of 6–8 in. (152–203 mm). A revolving workpiece helps to achieve a consistent coating thickness on the model. The spray pattern is approximately 3 in. (76 mm) in diameter; the spray rate is controlled by a valve on the gun. A practical spray rate range for pattern and toolmaking is 12–24 lb/h (5.5–11 kg/h), with exact rates depending on the size of the tool. The initial coat on all tooling surfaces is sprayed at low rates. Once the coating thickness is built up, the spray rate can be increased.

The wire feedstock comes in 25-lb (11.4-kg) spools; two spools are required to operate the equipment. The wire spools mount on the rear of the console, and the wire is fed to the gun via two flexible conduits sleeved in rubber.

The most common moldmaking wires are tin/zinc, zinc, and a Kirksite-type alloy. The equipment also has the ability to spray a variety of alloys, among them steel, nickel, copper, aluminum, and many others for many types of applications. These less commonly used wires are difficult to apply onto models because of high melt temperatures and high shrinkage rates.

98.3 PROCESSING CRITERIA

Metal-spray tooling requires five basic procedures. Although all are equally important, preparation is the most time consuming. However, it is time well-spent; poor preparation is often the source of problems that occur later in the manufacturing process.

Spray metal moldmaking procedures include:

- Model preparation
- Metal spraying the model
- Framing the mold
- Backup (or reinforcement) of metal spray shell
- Removal of the pattern

98.3.1 Model Preparation

Materials that can be used for models include wood, plastic, metal, clay, plaster, urethane, rubber, ceramic, and others. Depending on the model material, the surface should be sanded smooth and sealed with standard lacquer sealers. This is important because all imperfections become apparent in the sprayed surface when the metal shell is finally removed. The model should be mounted on a thick wood or metal surface that establishes the parting line. As noted, the preparation stage is the most time consuming procedure and must not be rushed. Good preparation is the key to successful metal sprayed tools.

When the model is prepared, a high-temperature barrier coat is applied to seal the model. It provides sealing for models of wax, clay, aluminum, silicone, and other materials with which the parting agent, PVA, is not compatible. The barrier coat is sprayed on with aerosol cans.

Lastly, the parting agent is applied as the final step in the preparation stage. The most popular parting agent used in sprayed metal tooling is a special poly(vinyl alcohol) (PVA). In the fabrication of sprayed metal tools, the PVA performs two functions. First, it provides good metal-spray adhesion to the pattern at the beginning stage of spraying the shell. Without PVA, the metal particles will not stick and are easily blown off. Second, it promotes release of the model when the tool is finished. The PVA is carefully and uniformly applied with a paint spray gun to a clean master in a dust-free atmosphere and allowed to dry.

98.3.2 Metal Spraying the Model

The parting agent should be thoroughly dry before the model is sprayed. The spray gun should be checked so that it is set at recommended voltage, amperage, and air-pressure levels. Metal spraying is done much the same as paint spraying. The ideal is to achieve a flowing motion over a large area, overlapping coats and building up a uniform coating. Patience is the key since the gun is capable of spraying at much higher rates than a model can accept. The initial coverage of the model is critical. The spray coating should be locked on the model by wrapping it over the edges of the mounting board or frame. For most applications, a coating of 60–80 mils (1.5–2 mm) is desirable.

In some molds, certain areas of the tool are difficult to spray because of their location or design. Here, metal inserts can be used. The metal inserts are sprayed in place and become part of the mold surface when the model is removed. In general, if a mold has deep recesses, projections, or ribs that are deeper than they are wide, inserts should be used. Figure 98.8 shows the proper use of an insert.

98.3.3 Framing the Mold

The mold is framed where internal pressures will be exerted on it when it is mounted in the molding machines. Prior to backing the metal shell, a frame should be constructed around it to contain the backup. The frame can be mounted in place before or after spraying. If the frame is mounted in place prior to backup, it is placed flush onto the pattern board. The choice of material is generally aluminum channel or bar stock to match the expansion characteristics of the sprayed shell and backup material. It is recommended that the frame be grit blasted with 24-mesh aluminum oxide before the model is sprayed to ensure that the sprayed shell bonds well to the aluminum.

If the frame is mounted after spraying, the procedure is to machine off a band of metal the exact width of the frame around the model. The frame is grit-blasted and set in place around the model.

If a deep frame must be used, it is sometimes advantageous to fabricate a subframe with 1/4–1/2 in. (6–13 mm) thick bar stock. The subframe is placed around the model prior to spraying and sprayed in place, making it easier to spray the model. An upper section of frame is then bolted to the subframe to the required height.

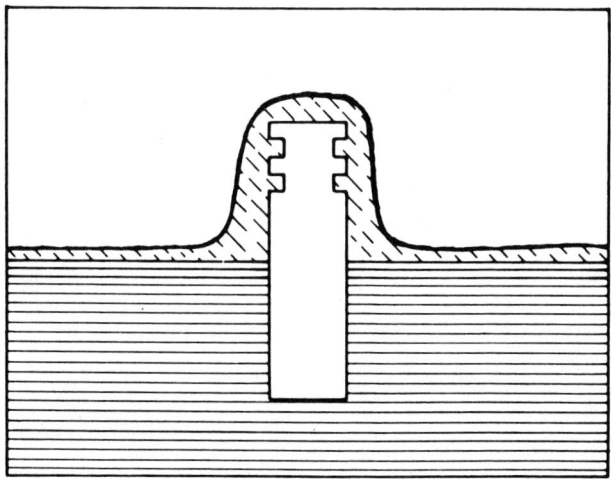

Figure 98.8. Proper use of metal insert.

TABLE 98.3. Physical Properties of Tafite 4321 Backup Resin

Property	Value
Compressive strength, psi (MPa)	10,500 (72)
Thermal conductivity, 1 Btu/h/ft/°F	
Tafite 4321 (60/40 formulation)	1.4
Tafite 4321 (85/15 formulation)	5.5
Epoxy	1.0
Dynamic viscosity, cP	1500–3000
Maximum exotherm temperature, °F (°C)	125 (52)
Linear shrinkage, %	1
Specimen Size: 10 × 2 × 3/4 in.	
(250 × 51 × 19 mm) = 0.1–0.2%	
Long-term heat resistance, °F (°C)	300 (149)
Shelf life, when stored at 68°F (20°C)	12 months

98.3.4 Backup (or Reinforcement) of the Sprayed Metal Shell

The most common backup material is a special methyl methacrylate resin highly filled with aluminum powder and aluminum peas (Tafite 4321 Backup Resin; see Table 98.3). The most widely used formulation is an 85% aluminum, 15% resin mixture. The 85% aluminum filler provides high thermal conductivity, very low shrinkage, and low exothermic reaction temperatures. It can be cast in large masses and is recommended for backup of RIM molds, PU foam molds, vacuum forming blow molds, structural foam molds, and large injection molds. Another advantage is its continuous heat resistance of 300°F (149°C) and 30 minute cure time.

A second type of backup material is a low melting bismuth-based alloy (Tafaloy 4328; see Table 98.4), which is recommended for backup of injection molds. Because of heat distortion and shrinkage problems, it is difficult to use a cast backup of zinc or Kirksite. Consequently, after a series of tests, this low melting point alloy was developed. It has a melt temperature of approximately 300°F (149°C) and actually expands when it is fully aged. Expansion is 0.0005 in./in. (mm/mm).

The thickness and complexity of the backup depends upon the pressure of the molding process in which the tool is used. Furthermore, this material can be recycled, allowing mold frames and bases to be used many times. This lowers the cost of a prototype mold program by 75%. The backup material flows much like a solder and can be cast void-free in large volume.

TABLE 98.4. Physical Properties of Tafaloy 4328 Backup Resin

Property	Value
Melting point, °F (°C)	300 (149)
Density, lb/in.³ (g/cm³)	10 (8.580)
Coefficient of thermal expansion, × 10⁻⁶ °C	15
Thermal conductivity (cal/cm²/cm/°C/s)	0.0500
Unfilled epoxy	0.0005
Aluminum filled epoxy	0.0020
Electrical conductivity (% of pure copper)	5.0
Brinell hardness	22
Tensile strength, psi (MPa)	8,000 (55)
Maximum sustained loads, psi (MPa)	
30 s	15,000 (183)[a]
300 s	

[a]The sprayed metal shell strength is higher, which permits higher mold pressures.

Injection molds as large as 400 lb (180 kg) have been made with backup casting. The backup is melted, cast, and reclaimed in alloy melting pots, whose use is recommended for close control of casting temperatures.

98.3.5 Removal of the Pattern

In general, if the pattern is properly prepared, it will release with little difficulty. Since the PVA is water soluble, tools can be submerged in warm water. Most often, the models are jacked or pulled out with the proper stripping setup. Many times, when the mold frame is held stationary and direct tension is applied, the model will pop loose from the mold. Any negative drafts will impede the stripping process, however.

98.3.6 Core and Two-Part Molds

In applications that require a matched metal mold, the process varies slightly in that the two halves are made separately. The model is mounted, establishing the parting line. On completion of the metal spraying and backup, the mounting board is removed, exposing side two of the model. The first half should be thoroughly cleaned before any metal framework is attached. The framework should be located with leader pins to eliminate chance movement during fabrication of the second half. The remaining procedure is the same. The final procedure is to separate the two halves. Once separated, the model is removed from the tool, resulting in a male and female cavity ready to be mounted in the molding machine.

Optional heating and cooling lines (usually copper) can be readily added to the mold sets. The cooling lines are fastened to the mold frames and positioned relative to the sprayed surface, as required. These are incorporated into the mold prior to backing up.

Ejection pins can also be utilized in metal sprayed tools. The design of the ejector pin system is dependent on the number of parts desired and complexity of the part molded.

98.4 ECONOMICS

When various tooling options for plastic molds are evaluated, the deciding factor is usually economics.

The investment for arc spray equipment can be considered low to moderate, depending on how elaborate the metal spray system becomes. The average capital investment for this process is $10,000 to $15,000.

It is difficult to evaluate the cost savings of sprayed metal tools versus other processes; however, a general rule of thumb is that as the amount of machining time required on a mold increases, the cost advantage increases for the sprayed metal process. When compound contours, complex angles, and irregular parting lines are inherent to the design of the plastic part, sprayed metal tools become very cost effective. Savings can range from 2 : 1 up to 20 : 1; the average cost savings is 70% versus machine molds.

99

PROCESS CONTROL, INTRODUCTION

Donald C. Paulson

Paulson Training Programs Inc., 160 West Street, Cromwell, CT 06416
The sections on Process Control were prepared with the assistance of Donald C. Paulson

Process control systems in their most elementary forms, have been used to control plastic processing machinery since the earliest days of the plastics industry. The first process control systems as basic as a light switch: were simple "on" and "off" temperature-control units. Next, sequence and timer controls were developed and along with them, adjustable hydraulic pressure controls. These are still the basic machine-control adjustments, no matter how sophisticated modern trappings may be.

The term used to define a system for controlling the sequence of machine events is called logic control. The sequence of each event in machine operation is governed by logic control circuitry, either electromechanical or electronic solid state. But logic control systems encompass only the machine *sequence* and safety interlocks. The *energy* that is given to the machine and the plastic being processed is controlled by open-loop or closed-loop control systems. In open-loop control, the operator makes the adjustments. For example, hydraulic pressure is often set manually and adjusted up or down by the machine operator.

In closed-loop control, a feedback control circuit automatically adjusts energy input to the machine based on one or more measurements. For example, barrel temperatures are often set on a temperature controller. A thermocouple measures barrel temperature and the controller power section supplies electricity to heater bands as required to maintain this set temperature. (However, the operator can still override the automatic control system by manually setting a new temperature on the controller).

Computer control has greatly expanded the functions and reliable repeat accuracy of computer of the machine control system. A computer can control a sequence of machine events (logic control) as well as heating, timing, pressure, and motion circuits, either in the open or closed-loop mode. Advent of the computer represents a major advance toward the ultimate goal of process control: reducing manufactured product variations to the zero-defect level by adjusting and controlling processing conditions so that all manufactured parts are identical and free of imperfections. This is, of course, not yet possible for any process. Variations in plastic parts occur because of variations in raw material (batch-to-batch and supplier-to-supplier, use of regrind plastic), variations in manufacturing environment, variations in mold cavities, wear of the machines, and molds, variations in electrical power, and many other factors.

Process control technology offers two alternative approaches to the reduction of product and process variations. The first is to *eliminate* the causes of variations. For plastic processes these variations are represented by differences in molecular length, additives, polymer flow behavior, ambient temperature, machine controls, power, machine wear, and other factors. Eliminating as many of these influences as possible *can* reduce part variability, but this approach is not practical. The primary reason is the high expense of such an optimum manufacturing facility and the non-Newtonian flow behavior of plastics.

The second and more pragmatic strategy is to compensate for process variations by adjusting machine-control settings. For example in thermoplastics molding, the viscosity of regrind plastic often is different from that of virgin resin due to slight alterations in molecular length caused by processing. It is not possible to change these molecular lengths back to the original. But it *is* possible to adjust the machine conditions to compensate for the change in viscosity. Closed-loop process control systems invariably use the compensation control technique as their primary control strategy rather than try to eliminate all process variations.

Any control system, no matter what type, has three basic elements: measurement, decision, and action.

A measurement used for process control must involve a property of the process or material that affects product quality. For example, the machine operating conditions are an important process condition affecting part quality and production rate. Therefore, a temperature measurement of the machine is often used in process-control systems. If direct temperature measurement of the plastic is feasible, than it can be used as a control measurement. A decision system determines if a control adjustment must be made. If, for example, a temperature measurement different from the preset temperature is called for, a change in energy input is initiated. In open-loop control, the decision step is a human one; in closed-loop control, a decision circuit calculates the amount of the change.

An action system is any device (in closed-loop mode) or person (open-loop) that makes the adjustment to any control setting.

Because of their fast response and pinpoint accuracy, closed-loop process control systems are finding increased applications in all types of plastics processing methods. Their earliest commercial use was in controlling parison wall thickness of extrusion blow-molded bottles. Programming of these parisons began to be used in the mid 1960s, when it became evident that profits and productivity were strongly dependent on high-volume production of defect free containers. This meant increasing parison thickness only in those areas where the most plastics was needed (ie, shoulders, neck finish, and bottom edges).

Closed-loop process control systems for injection-molding machines were developed shortly thereafter. The primary purpose was to minimize molded part dimensional variations, eliminate flash and short shots, and reduce part weight. In extrusion, closed-loop controls reduce sheet thickness and wire-coating variations, and improve pipe wall consistency. Process-control systems for thermoset processes are more recent developments. They measure exotherm of the thermoset, mold closure rate, and plastic pressure.

Control systems are often retrofitted to existing process machines. But increasingly, they are being offered by primary processing machine manufacturers as optional equipment to improve manufacturing control and machine capability.

100

PROCESS CONTROL, BLOW MOLDING

Denes B. Hunkar

President, Hunkar Laboratories Inc., 7007 Valley Ave. Cincinnati, OH 45244

100.1 INTRODUCTION

Parison programmers accordingly have been standard on new machines since the late 1970s. However, they do not all perform with equal effectiveness. Differences in technologies could mean significant material wasted and the production of unacceptable parts. What follows is a description of the state-of-the-art in controls for optimum productivity in extrusion blow molding.

The basic parison programmer has two fundamental functions: generation of the parison profile program and control of wall thickness (weight control). The program is triggered at the beginning of the cycle, and the program advance is synchronized with the machine cycle time or, in the case of accumulator/reciprocating screw machines, with the movement of the accumulator/screw. A block diagram of a typical closed-loop parison programmer is shown in Figure 100.1.

Essentially, control is focused on four areas: interpolation, the number of program points, machine synchronization, and the application of new parison control parameters.

100.2 INTERPOLATION

Even today there are common misconceptions regarding the importance of proper interpolation in parison programming. RC time-constant interpolation has been replaced by digital interpolation, which can follow a programmed contour in a far better way. RC interpolation has always been a compromise because the interpolation slope has to be preset to a program step of a given size, even though it might be larger or smaller than the one for which interpolation parameters had been calibrated. A typical program execution as a function of interpolation method employed is shown in Figure 100.2. The RC curve is obviously too slow in the A intervals and too fast in the B intervals. The tooling is moving either too slowly or too fast, and wall thicknesses are either too thin or too thick, regardless of the profile setting of the control. Digital interpolation follows the programmed profile much more closely.

An important advantage of digital interpolation occurs on multiple-head installations. Uneven tooling movements in multiple-parison installations means that some of the mandrels move faster than others. But the output of digital interpolation is always a finite slope. Not confronted by step functions, but dealing rather with a slope slewing rate that does not

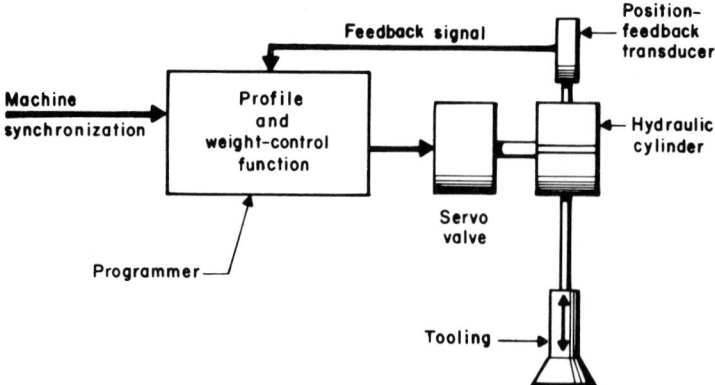

Figure 100.1.   A block diagram of a typical closed-loop parison programmer.

exceed the response of the system, the servo valve can stay in control at all times. Therefore, multiple-mandrel installations are now controlled so that the mandrels move with identical speeds.

100.3   PROGRAM POINTS

The ability of a parison programmer to follow even the simplest shapes is greatly dependent on the number of program points of the system. The difference in the control capability between a 10-point and a 32-point programmer can be seen in Figure 100.3. The desired curve or wall contour is compared with the curves executed by the two-programmer systems. It can be readily seen at the outset (interval A—the threaded-neck segment of the bottle) that more material must be used by the 10-point programmer than by the 32-point programmer to conform to parison wall thickness. The same applies to the wall-thickness requirement in the shoulder segment (interval B) and to the neck (interval C) and waste area (interval D).

Probably the most important point to be made is the considerable saving of material in the pinch-off area, where the drastic difference in the performance of the two programming systems becomes significant. Production results indicate that 2.5 g can be shaved off a 39-g

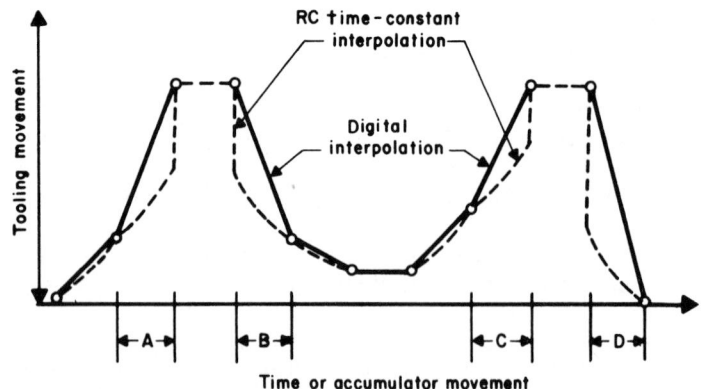

Figure 100.2.   Different results between RC and digital interpolations.

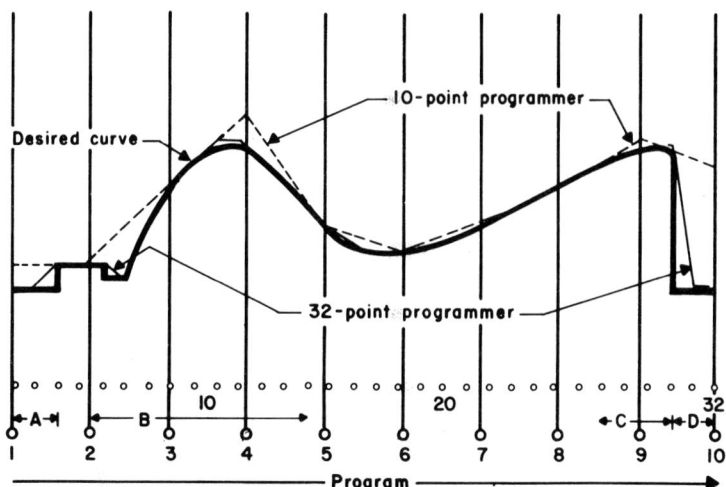

Figure 100.3. The difference in the control capabilities of a 10 and 32 point controller.

high-density polyethylene container at the pinch-off area when the programmer system is switched from a 20-point system to a 32-point system. The net material savings is 6.4%. And because other areas of the container were more precisely programmed, the actual top load strength of the container was improved.

100.4 MACHINE SYNCHRONIZATION

Early parison programmers and some of today's lower-cost models use a timer to advance the parison program. The timer is preset to the parison drop-time of the machine.

The problem with fixed-timer systems is that the program will no longer be in phase with the mold if the machine cycle changes and the program timer is not adjusted. Tests show that a change in machine cycle time of as little as 1.5% will affect the parison profile significantly enough to produce unacceptable changes in the wall and the shoulder areas. Timer-based parison programmers cannot be used on accumulator or reciprocating screw-type machines, since parison drop-rate is not a function of time, the program advance must be interlocked with the movement of the accumulator/screw.

For continuous extrusion machines, a synchronization feature automatically measures machine cycle time and distributes the program points over it equally. The parison is thus always programmed in phase with the mold, and the operator is spared another critical judgment.

Programs for accumulator and reciprocating-screw machines are based on position. A position transducer attached to the accumulator head or the reciprocating screw measures movement of the stroke. The program is automatically distributed over the stroke and advanced as a function of the incremental movement of the accumulator-screw. Program advancement is completely interlocked to the velocity of the moving member.

As a result of this development, the parison programmer can control the accumulator shot limit, the precompression of the shot can be controlled prior to the beginning of the programming phase, and the shot can be delivered to a cushion limit that permits first-in/first-out heads to retain a certain segment of the shot for two machine cycles. This provides a higher-temperature parison and improved pinch-off condition in molding large containers.

Synchronization can provide outputs to the machine as a function of accumulator movement. One control developed for this method is the "stop blow air" function; it permits the termination of the preblow as a function of parison growth.

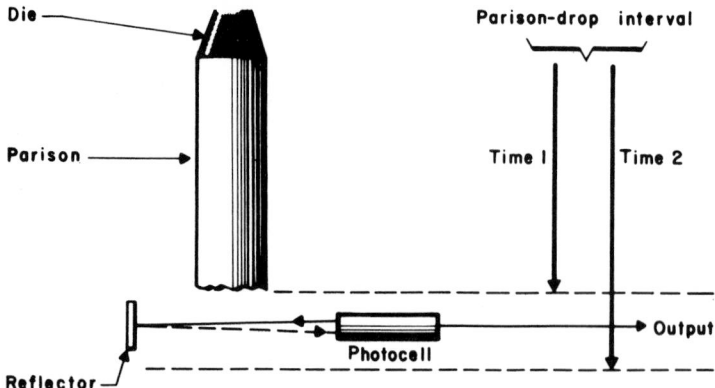

Figure 100.4.   Simple screw-speed parison-length control.

100.5   NEW CONTROL PARAMETERS

Parison-length control (screw control) is used to compensate for regrind variations, screw instability, and program phase lag. Of the several different methods of parison-length control currently in use, the most common is the simple screw-speed control shown in Figure 100.4.

Two timers and a photocell combine to decide on the suitability of a parison length and respond by changing screw speed to bring the parison length within tolerance. A third timer allows the system to check for proper mounting of the photocell and reflector.

Screw control of parison length is used extensively on continuous-extrusion machines. Recently it has been adopted for accumulator-type machines as well. The usual way to control parison length on accumulator equipment is to key it to the accumulator plasticate limit. The intent is to compensate for variations in melt density, which naturally affect parison flare and length.

Another method for controlling parison length in accumulator machines, proving interesting for manufacturers of gasoline tanks and other large components, modulates the basic gap opening of the die-height control as a function of variations in melt density while keeping accumulator stroke constant.

All parison programmers have a program offset control commonly called weight control, by which the die gap (overall parison wall thickness) can be manually varied without adjusting the parison profile. Figure 100.5 shows how the program profile offset operates.

Another device for controlling finished container weight is an electronic system, in which a weigh cell receives and weighs the container as it is ejected from the machine. A photocell then tracks its presence, an electronic scale with a high and low limit sets tolerances, and

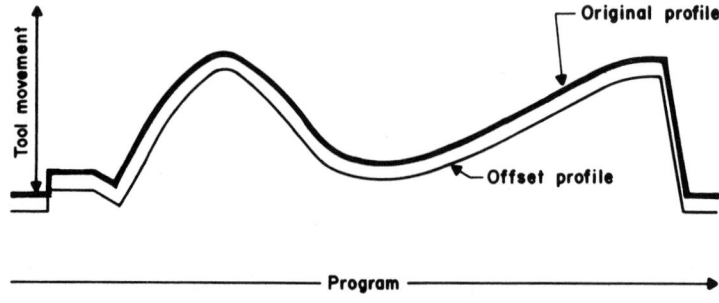

Figure 100.5.   Program profile offset.

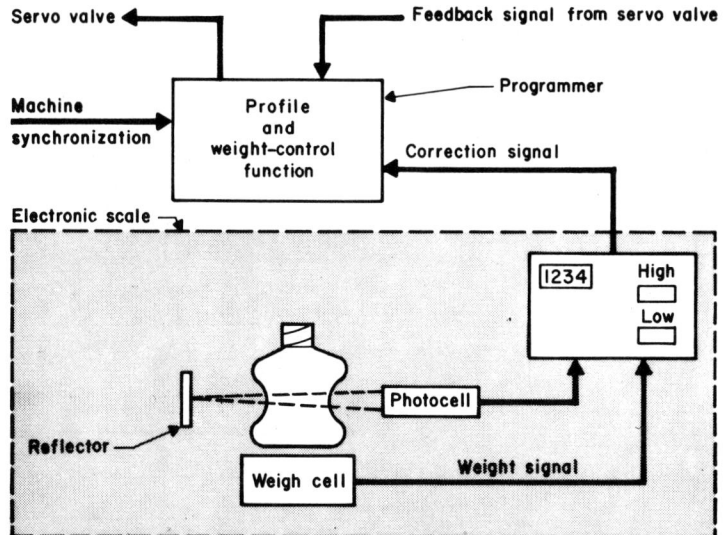

Figure 100.6. Closed loop blow-molding control.

a remotely controllable weight control in the parison programmer adjusts weight control to the parison wall thickness, thereby closing the loop around the entire machine (Fig. 100.6).

These systems were originally used only for containers that had to meet military specifications, or for packaging hazardous chemicals, where total container weight has to be guaranteed. Now, with cost pressures pushing molders to make containers that are very close to minimum-weight specifications, such systems may become standard for all types of containers.

100.6 RADIAL-DIE PROGRAMMING

A recent development for accumulator machines producing square or oval containers, or products with irregular-shaped cross sections is called radial-die programming. As the name suggests, it effects a change in the shape of the die from round to oval or ovalized square during the parison drop, achieving the same results as fixed ovalized tooling, but synchronously shaping the parison as it descends.

One method of radial-die programming is based on two semicircular tooling members that slide on each other to ovalize a round shape. Another uses a ring to restrict the formation of the parison wall to the sides of a square (Fig. 100.7). A movable restrictor ring installed in the die head is actuated by a hydraulic cylinder controlled by a servo valve and an additional parison programmer. Because the parison programmer is synchronized with parison drop, the hydraulic cylinder and restrictor ring can be programmed to laterally vary the parison wall on two or four of its sides.

100.6.1 Servo Valves

Hysteresis in servo valves can cause tooling to drift as much as $+0.05$ in. (1.3 mm), with a consequent drift in parison thickness. Temperature drift can have similar effects on wall thickness by causing an overall change in the die gap opening.

Valves are now available, especially those used at high temperatures (on top of the die head), have effectively eliminated valve hysteresis and drift, reducing the drift to $+0.0002$ in. (0.005 mm).

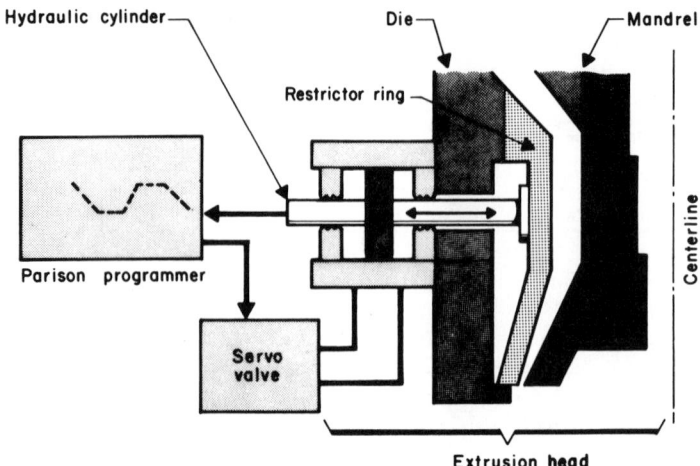

Figure 100.7. Radial die programming.

100.7 ROLE OF THE MICROPROCESSOR

Machine sequencing by microprocessor-based programmable controls has already made a major impact on extrusion blow-molding process control.

Microprocessor-based sequence controls do not produce better containers per se. Nor do they speed up machine cycles. Their reliability is no better than that of solid-state controls. But there are some important basic advantages to them.

Machinery builders can use the same controls on all of their machines, varying programs only. Manufacturing processes are simplified, volume is boosted, costs are reduced. Microprocessor-based programmable sequence controls should therefore reduce the cost of machine control.

Microprocessor-based sequence controls also can troubleshoot themselves, and such machine components as safety features (including those specified by customers) can be added at low cost. Thus, industry and government standards can be readily implemented.

The memory of most microprocessor-based sequence controls is sufficient to accept the sequencing of auxiliary equipment. Robots, trimmers and material feeders, for example, can be brought into the machine cycle making downstream equipment synchronous with the primary process and less costly to control.

Reprogramming machine sequences to suit special conditions is easier to do with microprocessor control than it is with solid-state controls. Machine sequence can be altered by a field programming device to incorporate special machine movements or sequences. Users can control programs to adapt to modifications and even to new processes.

100.8 MANAGEMENT INFORMATION SYSTEMS

The rising costs of material, machinery, labor, and other manufacturing elements encourage greater use of management information systems. A study of extrusion blow molders using such systems suggests that productivity can be improved by 12 to 20% on existing machinery.

Management information systems consist of a monitor/display cabinet interfaced to each machine and in turn connected to a central system that can range from a simple CRT monitor to a minicomputer complete with management software. The monitor cabinets are actually "intelligent" terminals that contain job and machine-related data and compare it to actual machine operation on a real-time basis. The monitor overseas several key machine parameters, signals and describes parameter deviation, evaluates part quality in terms of process parameter performance, accounts for downtime in several finite categories, constructs a production-control report, and communicates with the central computer site.

The central computer site periodically interrogates the files located in the monitor cabinets at each machine to facilitate plantwide production monitoring on a real-time basis. The central computer site also generates reports on demand or shift by shift, enabling management to see the production picture around the clock. The application of the management information systems is well accepted by the floor personnel. Operators benefit psychologically when they can measure their own performance against an established standard. Furthermore, since "intelligent" terminals are self-instructing and largely automatic, training problems normally associated with the installation of any new manufacturing system are minimized.

Management information systems are not costly luxuries. Plants machines having 6 to 10 machines can achieve a return on investment in less than one year. Obviously, in larger shops, where the central system may be shared by more machines, it is even sooner.

101

PROCESS CONTROL, ELEMENTS

Rodney J. Groleau

R.J. Groleau Associates, 401 E. Front St., Traverse City, MI 49684

There is a tendency among some processors (particularly smaller ones) to be put off by what is perceived to be the mystique of such automated manufacturing disciplines as process control. This attitude (which indeed may be fostered and perpetuated by the sometimes impenetrable linguistics of computer technology) deprives too many companies of the real bottom-line benefits that can be achieved with new technology.

However, there is a simple logic to the elements of process control. Once comprehended, it can go far toward dispelling the unfounded fears of processors unwilling to take the plunge. The basic components of process control incorporate:

- Measurement instruments (temperature, pressure, position, and time)
- Decision systems (manual, automatic on/off, proportional, adaptive, and programmed)
- Actuating systems (electrical, hydraulic, and servo valves).

These basic components are the foundation of all process control strategies. And process control is nothing more or less than the intentional adjustment of machine conditions in order to achieve consistency in plastics behavior and, by extrapolation, consistency in finished-part properties.

101.1 MEASURING INSTRUMENTS

101.1.1 Temperature

Two types of temperature measuring transducers are commonly used in plastics processing. They are the thermocouple and the resistance temperature detector (RTD). Thermocouples are more widely used than the RTDs and have been used for a longer time. Consequently, thermocouples are generally better understood.

Thermocouples do not measure temperature directly but respond to a temperature difference between two thermocouple junctions (Fig. 101.1). Standard thermocouples are of the iron–constantan type, made up of two dissimilar metals joined together in the thermocouple junction. Two thermocouple junctions make up any thermocouple measuring circuit. Generally, one junction is at ambient temperature and the other is exposed to the temperature to be monitored. In its simplest form, the thermocouple measures the difference between ambient temperature and the temperature to be measured. This is unsatisfactory for process control because ambient temperatures vary. Thus, if the uncompensated signal from a pair of thermocouple junctions were used for process control, the process temperatures would vary as ambient or room temperature varies.

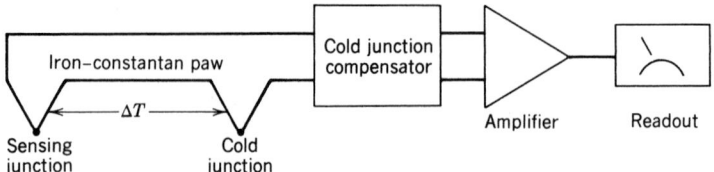

Figure 101.1.   Basic thermocouple schematic.

In laboratory situations the reference junction of the thermocouple set is placed in an ice bath maintained at 32°F (0°F). However, it is obviously inconvenient to use melting ice as a reference source. Consequently, the cold junction compensating circuit is generally employed in the process-control instrument, using thermocouples. This circuit measures the ambient temperature and electrically compensates for the reference junction thus deriving a true temperature output.

Thermocouples also are available in chromel–alumel types. They are used where temperatures above 700°F (371°C) are encountered. Because thermocouples provide a uniform output that does not vary sensor to sensor, they are easy to apply, and they provide a stable and reliable signal to process devices.

The resistance temperature detector, as its name implies, changes resistance with respect to temperature. The RTD is generally incorporated a Wheatstone bridge input circuit, with the RTD being one of the active arms of the circuit (Fig. 101.2). RTD's generally give higher outputs than thermocouples and measure temperature directly (rather than temperature difference as was the case with thermocouples). Consequently, they do not need cold junction compensation. Generally, RTDs provide sufficient output so that minimal amplification, if any, is necessary. The major problem with RTDs is that calibration changes with time, opening the potential for errors. In addition, RTDs are large with respect to thermocouples and have some limitations when used in small environments or where rapid response is necessary.

Another method of detecting temperature where electrical heating elements are employed is by detecting change in the resistance of the heating element and using it as a measure of the temperature being controlled. This technique (not widely employed due to the complexity of monitoring a circuit resistance under load) could become more widespread with the advent of microprocessors.

One of the problems with this technique is that the resistance from one heater element to another varies due to manufacturing techniques used. Consequently, it is costly and difficult to maintain constant heater resistance so that direct interchangeability without recalibration can be achieved. The major advantage of this technique is the absence of separate thermocouple or sensing wires for detecting the control temperature. Only the

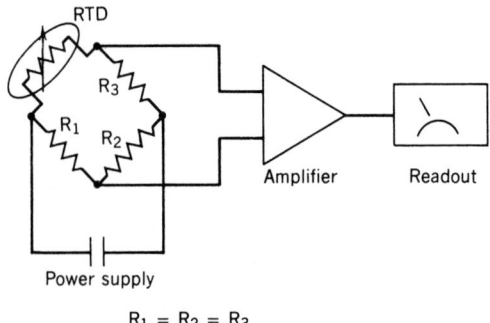

$$R_1 = R_2 = R_3$$

Figure 101.2.   Basic RTD schematic.

wires to the heating element itself are necessary. Thus, the number of wires can be cut virtually in half. This could be especially advantageous in runnerless molds where a large number of circuits are controlled by electrical means.

101.1.2 Pressure

Pressures to be measured in plastics processes are generally divided into three categories: hydraulic, and pneumatic. Pressure measurement can be further classified into two categories: melt pressure and mold pressure. Melt pressure is measured where the material is above the glass-transition temperature (T_g). This is most commonly measured in extrusion processes both before and after the screen pack or in the head of the extruder. To some limited degree it is also useful in injection molding as a nozzle pressure measurement. Mold pressure measured in the mold cavity, is the most common measurement in injection, compression, and transfer molding. All commercial pressure measurement devices fall into two categories: strain gauge transducers and piezoelectric transducers.

Strain gauge transducers utilize a phenomenon that any electrical conductor will change resistance to the flow of electrical current directly as the cross-section of that conductor changes. Thus, when a very fine wire is stretched, its resistance will go up and will in proportion to the amount of stretch or strain. Strain gauge transducers use strain gauge grids that are nothing more than very fine conductors generally etched in metal foil in a zigzag pattern, which increases their effective length in a very compact space. These strain gauges are attached to a transducer in such a way that their resistance changes as a result of forces applied to the sensing element of the transducer. Figure 101.3 shows a strain gauge element with strain gages attached along with the associated circuit configuration.

Piezoelectric transducers use a quartz crystal as a sensing element. When the crystal is subjected to a change in force, voltage is generated across the surfaces of the crystal; this voltage can be outputted to a signal conditioning device where it is integrated to create a pressure signal. Piezoelectric transducers are small, reliable devices which are very good for measuring pressures in processes where rapid transient signals are present. However, they are not generally suitable for measuring pressures that change slowly, such as in extrusion dies, because the integrating techniques causes the pressure signal to drift with time.

When piezoelectric transducers are used in injection molding, they must be rezeroed on every shot. They are also subject to stray signals due to capacitive changes around their lead wires. Consequently, care must be taken in routing these wires.

101.1.3 Position

Position measuring in plastics processing generally involves monitoring the stroke of an injection or transfer cylinder and the motions of the clamp. Position measuring devices

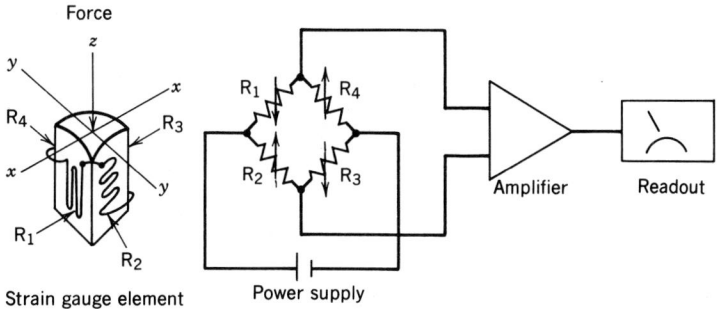

Figure 101.3.   Basic strain gauge transducer schematic.

generally fall into three categories: the linear potentiometer, the linear variable differential transformer (LVDT) and sonic position sensing.

The linear potentiometer is a linear variable resistor with a resistance element, generally made of conductive plastics, placed as a strip inside the body of the transducer. A slider is attached to a rod protruding from the body of the linear potentiometer. As the rod is moved, the slider resistance changes with respect to the ends of the potentiometer. This change in resistance can be measured by a signal conditioner and calibrated in the appropriate units of linear measurement. Linear potentiometers are most commonly used in injection, transfer and compression molding. They are simple and low cost and will perform for millions of cycles without problems.

Linear variable differential transformers (Fig. 101.4) consists of windings A, B and C. The transformer secondary windings are wound to produce opposing voltages and are connected in series. When the core is in the neutral or zero position, voltage induced into the secondary windings from winding C are equal and opposite, and the net output to the signal conditioner is zero. Displacement of the core by moving the input shaft of the LVDT decreases the coupling between the primary coil C and one of the secondary coils, while decreasing the coupling between the primary coil and the other secondary coil. Net voltage increases as the core is displaced from center, and the phase angle increases or decreases as a function of the direction in which the core is moved. A demodulator circuit can be used to produce a D-C output from this winding configuration; this D-C output can then be amplified and used for a readout or process control signal.

Sonic position sensing sold under the brand name Temposonic utilizes a system much like sonar to detect the position of a donutlike target as it slides back and forth along the detecting rod. Electronic signal conditioning converts this signal into a high level position signal directly proportional to position. Sonic devices are easy to install, are accurate to 0.001 inches and are highly reliable because there are no sliding parts. Sonic sensors are moderate to high priced relative to LVDTs and linear potentiometers.

101.1.4   Time

The measurement of time in plastics processes generally involves the use of plug-in electromechanical timers or solid-state timers generally embodied in microprocesors. The plug-in timer employs a synchronous motor that runs during the timing cycle preset on the face of the timer. When the timer runs out, a set of contacts are actuated to open or close appropriate circuits. This provides an output to the machine sequence logic to perform a function or change a condition in the process.

With the advent of microprocessors, electromechanical timers have been less widely used, since the microprocessor generally has a crystal controlled clock that provides a very accurate time base. Solid-state devices are generally more accurate than electromechanical timers, and thus provide more process repeatability.

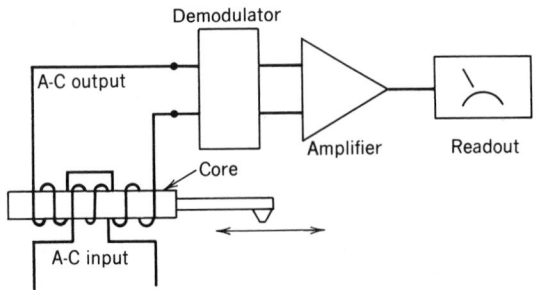

Figure 101.4.   LVDT basic schematic.

101.2 DECISION SYSTEMS

The output from transducers can be used to detect process conditions. With this information, the process decisions can be made, by a person in the control loop or by automatic means.

101.2.1 Manual

In a sense, all processes are closed loop, because in every process someone, or something, monitors the process, makes a decision, and acts on it. In an injection molding or an extrusion process, the operator decides, simply by examining the parts, what must be done to correct the situation. In automatic process control, key variables that predict the output of the process are monitored. Monitoring these variables allows corrections to be made before bad parts are made. In monitoring cavity pressures in a molding press, the quality of the finished part can be determined before the mold is opened. Using a readout, the operator can manually adjust the process to control fill rates and pressures, and thus achieve consistent parts. An operator-controlled mode is satisfactory if the operator can respond consistently to process variations. In automatic processes, however, it is generally not a suitable approach, unless alarms are used to alert nearby personnel to the need for process adjustments.

101.2.2 Automatic On/Off

In an automatic on/off control system the output of the process controller is not proportional or modulated as are more sophisticated controls. However, on/off control can be very effective if properly employed. For example, when the process temperature is below set point, output to the process heaters is switched on; once the set temperature has been reached, the heater is turned completely off.

101.2.3 Proportional

Proportional control is a refinement of automatic on/off control. When the error between the setpoint and the actual process condition is very large, the proportional controller reacts the same way as an on/off controller. However, as the process condition approaches the ideal, output of the controller begins to change in proportion to the amount of error between setpoint and actual condition.

This type of controller provides greater process stability and more accurate control. There are many refinements to the proportional controller. For example, integral and derivative functions can be added to respond to the rate of change of process conditions and to the error in the temperature once steady state conditions have been reached. These proportional-integral-derivative (PID) controllers provide an effective and sophisticated approach to many plastics processes.

101.2.4 Adaptive Control

Adaptive control is a technique that involves sensing conditions on one cycle of a cyclical process and adjusting the process for the next process cycle.

Although this technique has been employed over the past several years, it is now going out of vogue as statistical process control techniques are more widely employed. The reason for this is that adaptive control techniques only effectively correct for process disturbances which trend. These, by SPC definition, are abnormal special assignable causes that must be "fixed", not corrected for. In a process that has an absence of special causes, adaptive control will automatically "tamper" with the process in response to normal random variations. This "automated tampering" always makes the process less stable.

Adaptive algorithms based on SPC tests, which do not correct on each cycle but only in response to "allowable trends," may allow automatic centering of processes without over-correction when normal process variations are present. At this time, no adaptive control applications using these techniques are known.

101.2.5 Programmed Systems

Programmed systems are used to determine output with respect to time or position in a process. An example is programmed injection velocity for molding machines based on ram position. The output of the controller is generally a pre-set condition which changes with respect to time or position. Programmed devices can be coupled with closed loop feedback devices to give velocity profiles and fill-time control in an injection molding process.

101.3 ACTUATING SYSTEMS

Once a decision is made that something has to be adjusted, the actuating systems provide the means to affect the action. Actuating systems are divided into three categories: electrical, hydraulic, and pneumatic. Pneumatic systems are mainly used in a simple cylinder operated by solenoid valves for part diverters or simple cam actions.

101.3.1 Electrical Actuators

Relays consist of solenoids which activate one or more sets of contacts to provide outputs to a number of other devices, such as other relays or solenoids, to affect the process. Relays are electromechanical and consequently are subject to mechanical wear. Although relays have a relatively short life, they provide superior electrical isolation and when they fail, they generally fail in an "off" or safe condition. Relays also have high current-carrying capacities and are generally more tolerant of overloads than are solid-state devices.

Triac actuators represent the alternating current equivalent of a transistor. They respond to an input, providing the ability to switch high level A-C currents to a process heater solenoid or other electrical device. Triacs generate substantial amounts of electrical interference if not properly used.

Solid state relay actuators incorporate triacs as output devices. In addition to simple triacs, they have embodied in their circuitry optical coupling of inputs and zero crossing networks that cause the output current to be switched in such a way as to provide minimal electrical interference. Although are superior to triacs in performance, solid-state relays are generally more costly.

Hydraulic actuators are generally classified as solenoid valves, servo valves, and electro-hydraulic proportional valves. Solenoid valves are widely used to sequence process machinery. They use electrical solenoids connected to spool valves that route oil or air to various process circuits. Solenoid valves are on/off devices activated by triacs, relays, or solid state relays.

Servo Valves.   Servo valves are proportional devices that respond to proportional signals from process controllers. They modulate hydraulic pressure or flow to provide proportional output. Servo valves provide fast response and direct control of the process. They are sensitive to dirt and contamination and are much more delicate than other types of actuators. In addition, they are more costly than other types of control devices.

Electrohydraulic proportional valves are a compromise between servo valves and solenoid valves. These devices modulate pressure or flow in much the same way as do servo valves, but are slower in response. They are, however, far more tolerant to dirt and contamination and are less costly. In most plastic processes, electrohydraulic proportional valves are suitable for process control applications.

In addition to these actuation devices, manual pressure and flow control valves, provide a good degree of process repeatability if properly employed. Generally, the high spring forces built into manual valves give them excellent repeatability. In hydraulic flow control, manual-pressure-compensated flow control valves provide ample process repeatability with respect to flow rates or ram velocities.

102

PROCESS CONTROL, EXTRUSION

Sidney Levy

P.E., PO Box 7355, 4433 Dawn Avenue, LaVerne, CA 91750

There have been a number of advances in extrusion control systems, with most of the effort concentrated on single-screw equipment. Therefore, this analysis will concentrate on the single-screw systems, with some reference to twin-screws.

Two major control systems are involved in extrusion: control of primary machine output, and control of takeoff equipment. In all extruded products, dimensions are controlled by the relative takeoff rate to plastics delivery rate. Control of both output and takeoff rates is essential to maintaining product dimensions.

Extruder output control is a function of the melting conditions in the machine and the screw drive. In order to operate properly, the extruder barrel sections and dies must be maintained at the appropriate temperature. This can be done by discrete proportional control instruments. However, current technology involves combination controllers that read around the temperature sensors. This information, in conjunction with the reading from a melt thermocouple, is used to adjust barrel and die temperatures to values appropriate for the material. In just about all cases the sensors are thermocouples, and they are embedded in the barrel to a depth based on experience. In some cases, dual depth thermocouples are used to effect better control.

In single-screw extruders, energy supplied by mechanical work of the plastics material as it is conveyed through the machine is the major source of heat input. In some cases zone cooling is the major control condition. As a result, the control of the melt temperature is affected by machine operating conditions as well as by the heat/cool zones on the extruder. This affects the methods of control of the output of the extruder.

102.1 CONTROL OPTIONS

There are two ways to control output of a single-screw machine. Since most extruders are equipped with SCR drive systems, screw speed can be controlled electronically. An output sensor system will adjust output by feeding back a signal to the SCR motor drive control; this will adjust motor speed to compensate for any change in throughput. The other method uses a valve or other controllable flow restrictor at the output the machine before melt enters the die.

There are differences in effect between the two approaches. In screw-speed control, motor rpm will impart additional mechanical work to the polymer, causing a rearrangement of the melt bed and a consequent change in the temperature-control settings. In some cases this can lead to instability and surging in the extruder. In the controllable restriction mode, there is some additional heating at the expense of pressure as the die orifice is changed, but there is little effect on the melt bed or rearrangement of the temperature profile. Adjustable restriction can be with a melt valve driven by an electrohydraulically driven cylinder.

The result of a properly designed operating system for the single-screw extruder is to deliver melt to the die with consistent properties and at a constant predetermined rate. The extruder will then deliver a constant output shape through the die, minimizing the need for downstream pull rate control.

Constant extrudate shape and dimensions can also be maintained by varying the takeoff rate. Most pulling devices in extrusion lines are driven with variable-speed motors having SCR drives or variable-frequency A-C drives. Drive system speeds are electronically controlled so that pulling speed rates can be adjusted automatically to compensate for variations in extrudate dimensions. Error signals are generated from devices that include ultrasonic, nuclear, and optical intercept sensors.

To prevent system control loss, feedback signals generally control only a limited speed range. This is sufficient in most cases. But when dimensions exceed the control range an alarm is sounded so the operator can make manual adjustments.

Physical positioning of the sensors is important to good control. They should be located as close as possible to the die, since any time lag in reading will tend to destabilize the system. Some signal conditioning systems will anticipate the course of the change in machine output so that adjustments in pull speed can be made as fast as changes occur in extruder output. This capability avoids many of the problems that can result from downstream location of dimensional sensors.

102.2 HOW CONTROL SYSTEMS WORK

102.2.1 Sheet

In an automatically controlled line, the sheet is fed from the die to a three-roll stack which cools the sheet and does the final sizing between nips of the rolls. Rate of pull by the roll stack is controlled by feedback information from a thickness sensor that scans the sheet. The usual sensor is a nuclear gauge and the usual location is just downstream of the roll stack. One new control element adjusts the die lip opening across the die width in response to data from the scanning sensors and thus maintains constant plastics flow across the width of the sheet. The combination of roll stack speed control and die lip control assures optimum gauge control.

Location of the sensors past the roll stack is not necessarily optimum. The best location appears to be between the die and the roll stack. The roll stack tends to iron out cross-die variations but it does so with some densification of the material. Since the nuclear gauges read mass rather than thickness they can still read the cross-die variations in a muted form. Reflex nuclear gauges can be located between die and roll stack, but they need ruggedization to take the heat and corrosive exposure.

102.2.2 Pipe and Tubing

In the most commercial lines, the extrudate is run from the die into a vacuum sizer that sets the outside diameter. Varying pull speed within the limits of the range will vary wall thickness. Most systems for controlling pipe/tubing extrusion use an ultrasonic wall thickness gauge, normally located in the vacuum tank just past the sizing rings. In a closed-loop system, the output signal from the ultrasonic gauge is conditioned and fed back to the puller's SCR motor control to regulate speed and correct dimensional variations. In cases where the tubing is free drawn (as is the case with flexible vinyl compounds), a diameter measuring element is also needed. This is frequently an optical intercept gauge.

Additional control of wall thickness can be achieved with a variable-orifice die or by the controlling air inlet into the die. The latter procedure will expand or collapse the tubing slightly so that drawdown will generate the correct wall with thickness while puller control sets the outside diameter.

102.2.3 Profiles

In the automatic control of profiles it is usually necessary to monitor two dimensions: width and thickness. Although it is not uncommon to use two optical intercept units for sensing, the common practice is to use an optical intercept unit and either an ultrasonic or a nuclear thickness gauge. Where the extrudate can tolerate the contact, air gauges and roller contact gauges also can be used.

The constraints sometimes used on profiles, such as one-sided vacuum sizers and other fixtures, do not change the relative dimensions of the part. As a result, adaptation of automatic control to profiles is more difficult.

There are other characteristics of the extrudate monitored by control systems. For example, color of the extrudate can be monitored using a spectrophotometric device. This can be wide wavelength or limited to specific color bands and is used when the color of the product is critical. The output can be connected to a recorder or to an alarm, or to a unit that adjusts the color additive device.

An essential extrusion-control function is to monitor the number and location of flaws and other defects in the extrudate. One type of sheet-flaw detector is a laser reflection device that scans the sheet to detect anomalous reflections. When such a flaw appears, the device activates an event recorder that will mark the defect on the sheet. Another system is based on producing a static field on the extrudate. Anomalies in the static field are detected with an electric sensing device. The output is fed to an event recorder that triggers a marker to locate the defect. The technique will find gels, fisheyes, holes, and severely contaminated parts in the extrudate.

Other control determine the quality of the material. For example, ultrasonic units detect voids in the extrudate. Density of the extrudate can also be checked by sonic velocity through the extrudate. And optical, electrical and other properties are easily monitored with on-line sensors.

BIBLIOGRAPHY

J. Parnaby, A.K. Kochbar, and B. Wood, "Development of Computer Control Strategy for Plastic Extruders," *Polymer Engineering & Science*, **15**(8), (Aug. 1975).

S. Levy, "Gaging and Control of Extrusion Processes," *Plastics Machinery & Equipment*, **8**(6), 31 (June 1979).

S. Levy, "On-Line Gaging of Extrusion Dimensions," *Plastics Machinery & Equipment*, **7**(1) (Jan. 1978).

C.J.S. Petrie, "Mathematical Modeling and Systems Approach in Plastics Processing: The Blown Film Process," *Polymer Engineering & Science*, **15**(10), (Oct. 1975).

A.D. Schiller, "Omnyson Correlations with Process Conditions," Haake, Inc. Technical Bulletins TB 2002, 2003, 2004, and 2005.

103

PROCESS CONTROL, INJECTION MOLDING

Rodney T. Groleau

R.J. Groleau Associates, 401 E. Front St., Traverse City, MI 49684

Note: The end of this section has information on process control of compression and transfer molding.

103.1 INTRODUCTION

The key to implementing a successful injection molding process control strategy is to understand the molding process from the plastics point of view. What happens to the molten plastic as it flows through the machine into the mold and is cooled determines the characteristics of finished parts.

The reason that plastics variables are so important in injection molding is that long-chain molecules become aligned in the direction of flow. The rate of flow affects the degree of alignment of the molecules. As more molecules become aligned (faster flow rates), plastics viscosity decreases. This behavior is called non-Newtonian and is unique to plastics. It also makes the injection-molding process difficult to control. Correlations between finished part properties and machine conditions have statistically been shown to be extremely poor, however, almost all variations in finished part properties can be directly correlated to plastics variables. This is why it is important to measure plastics variables to achieve the maximum degree of process control.

There are essentially only four process variables in injection molding: (*1*) plastics pressure in the mold, (*2*) plastics flow rate, (*3*) plastics melt temperature, and (*4*) plastics cooling rate. All machine settings and environmental effects can be translated directly or indirectly into effects on these variables.

However, molding machines do not have controls marked "mold pressure," "plastics pressure," "plastics temperature," "flow rate," and "cooling rate." Instead, there are a number of machine controls that only *indirectly* influence these variables. To compound the problem, machine variables are often interrelated to the plastics variables in that one machine variable can cause several plastics variables to change. This clearly compounds the problem of injection molding control.

Because of the interrelationship of these variables, the standard molding process is extremely capricious. Any attempt to control must seek to isolate the plastics variables from each other as much as possible. Then, changing one control will not cause multiple changes in key parameters.

Injection pressure is the most important process variable because it relates directly to the degree of mold filling. Injection pressure relates to short shots and dimensional variations in parts. It also relates directly to the number of molecules in parts and, thus, directly influences part weight. Short shots, dimensional variations and weight variations account for over 80% of the causes of unacceptable parts. Controlling mold pressure can be a simple, cost effective way to get the most benefit at the least cost.

One of the problems in monitoring and controlling plastics pressure is that pressure in the mold cavity varies dramatically from hydraulic pressure. It also varies with distance of flow. In reality, the cavity pressure in a mold is actually a profile; its control is the key to complete control of the process. Change of cavity pressure is determined by other plastics variables, such as plastics temperature, flow rate, cooling rate, and mold temperature. If these variables are kept constant, only one cavity pressure sensing location is necessary to ensure proper control.

In some applications a hybrid approach to monitoring and controlling the process using cavity pressure is used. This entails using one sensor near the gate for control of cavity pressure with one or more sensors near the end-of-fill. By controlling the upstream in-cavity pressure and monitoring the last point to fill, cavity pressure profile can be maintained (See Fig. 103.1).

Flow rate is the second most important plastics variable for several reasons. It is not well controlled on a conventional molding machine. It influences viscosity and thus the plastics pressure in the mold on a machine not equipped with a mold pressure controller. Lastly, flow-rate variation changes the amount of orientation or molecular alignment during the filling process. This, along with cooling rate, determines the amount of orientation in the finished part.

The best measure of flow rate from the plastics point of view is to monitor actual time to flow plastics into the mold. This fill time is a pure variable for a given mold.

Programmed injection, another control element, is ram speed modification to provide nonuniform plastics flow rates. Programmed injection has many applications particularly in thick-section parts, optical parts and a variety of special applications. But if a mold can be used to make good parts in a conventional molding machine, programmed injection is usually

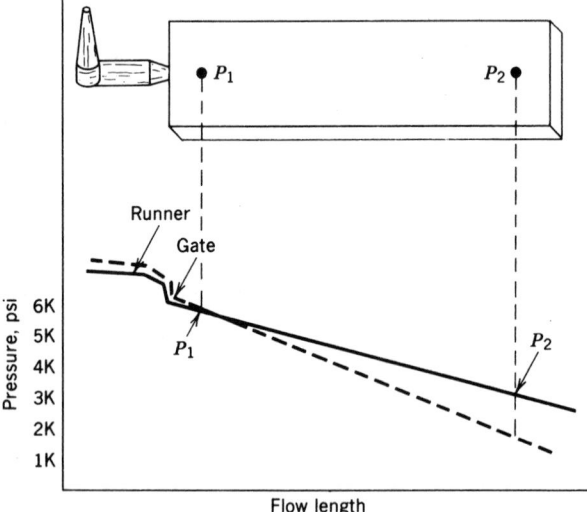

Figure 103.1. Peak cavity pressure profile. ———, Typical pressure profile; -----, profile shift due to a cold mold, cold material, fill rate change, or material viscosity change. $\Delta P = P_1 - P_2 =$ pressure loss across cavity.

not necessary. Another approach is to keep the injection rate consistent, shot after shot. On a machine equipped with a mold-pressure control system, increased first-stage injection pressure can ensure that sufficient energy is available to overcome viscosity variations. A simple but effective pressure compensated flow control can then be used to control fill speed. The flow control can be set by observing the plastics fill time reading on the control system and adjusting the valve to achieve the correct reading to make good parts. Fill time can be held to $+/-1\%$ using this technique.

Servo valve-based systems that allow the molder the latitude of using programmed injection will greatly reduce flow rate variations. However, increased cost and complexity are tradeoffs.

Mold pressure variations are the biggest single cause of part-quality variations. But this problem can be minimized by two-stage or boost cutoff control. The pressure-variation problem is illustrated in Figure 103.2.

Cycle 1 shows a typical cavity-pressure reading during the molding cycle. Cycle 2 shows a variation in plastics cavity conditions *even though machine conditions are constant*. It is this type of process variation that the boost cutoff eliminates.

Even with a noncompensated manual-flow control valve on the machine, improved uniformity in fill rates can be achieved utilizing the boost cutoff approach. However, with a manual pressure compensated valve, variations can be cut to as small as 1% fill rate control.

Peak mold pressure is determined by the boost cutoff setpoint and the speed of injection. With enough excess energy for cavity fill, the correct use of the flow control is to ensure the proper packing of the part. The point where the energy is turned off is called the boost cutoff or transfer setpoint. It is the mold pressure setting that switches from the first to second stage injection pressure. Mold pressure continues to rise after this until peak pressure is reached. This rise is called the "overshoot" of the system. It is caused by the response time of the molding system (generally small and constant) and the speed of the screw or plunger. The faster their speed, the longer it will take to decelerate and the higher the resultant cavity pressure. This part of the "overshoot" is variable depending primarily on fill rate.

To ensure that peak pressures are held constant, using the boost cutoff approach, it is important that fill rate be constant from cycle to cycle. This is accomplished by the correct utilization of flow control. To obtain the maximum benefit of a boost cutoff control system, proper flow control valves, properly sized for the application, must be utilized. If they are, a very cost-effective degree of control can be obtained.

103.2 THREE-STAGE CONTROL

A recent process-control development provides the simplicity of a mold-pressure controller with most of the capabilities of the servo-based controllers to achieve flow-rate modification. Three-stage controllers may incorporate ram position and/or hydraulic pressure to stage the machine between the various stages of the molding cycle when cavity pressure sensors are unavailable in a given mold. These controllers also can be used with programmed injection to provide optimum flexibility and control.

As its name implies, a three-stage controller divides the injection cycle into three different operations: fill, pack, and hold (Fig. 103.3). This allows mold fill rates to be completely independent of mold pack rates. The degree of mold packing (peak-mold pressure) is controlled independent of fill. The third stage, hold pressure, maintains the correct plastics pressure in the mold during solidification to prevent sinks. This type of controller is equipped with timing devices to allow monitoring of fill and pack time. The three-stage controller also controls peak mold pressure accurately because rate of pack is reduced, eliminating overshoot.

A typical three-stage control includes a valve assembly for the third stage control, a

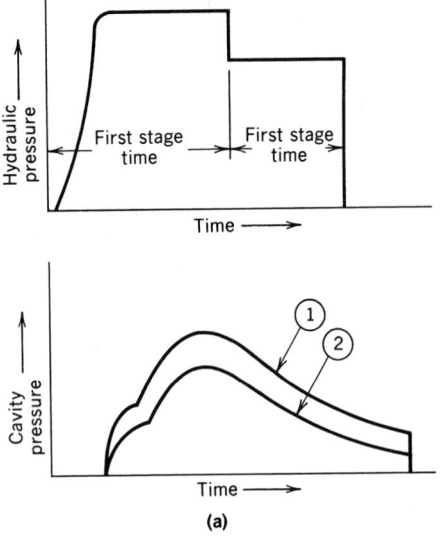

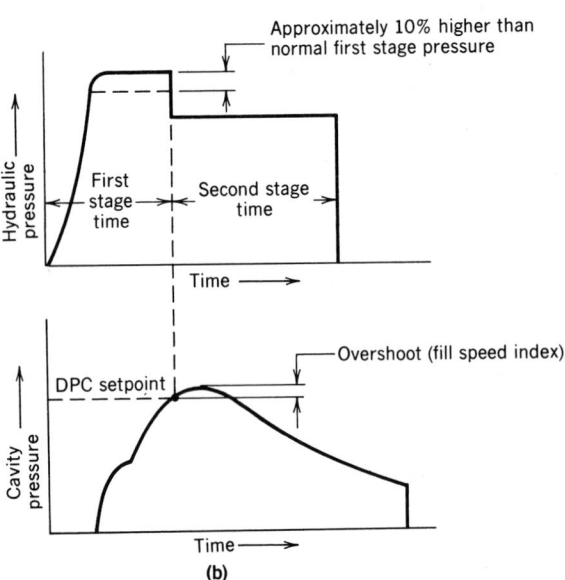

Figure 103.2. **(a)** Molding without boost cutoff (DPC). **(b)** Molding with DPC. Courtesy of D-M-E Company.

linear screw travel potentiometer for stroke control, the controller, a power interface box to interface the controller to the machine, a sensor interface box to interface the cavity pressure sensor (if used) to the controller, and a mold pressure sensor. Hydraulic pressure sensors can also be used with this type of controller. A diagram of typical three-stage control is shown in Figure 103.3. In this case, we are illustrating a three-stage control controlling the fill portion with screw position (stroke) and controlling the pack with the plastics pressure in the mold.

The stroke setpoint, which triggers the transition from fill to pack, is set to stop rapid

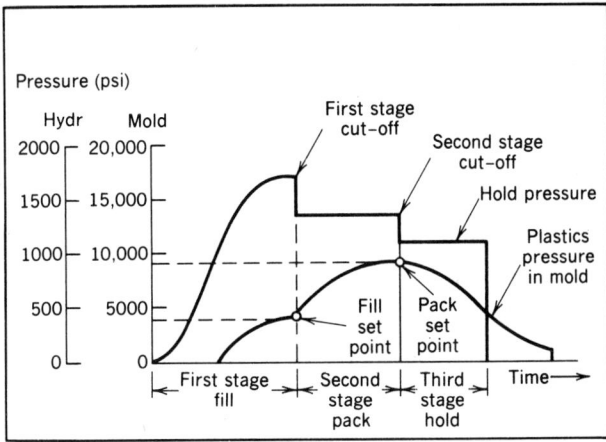

Figure 103.3. Three-stage closed loop control.

flow when the cavity is just filled, switching the machine from a high-pressure/high-volume fill to high-pressure/low-volume pack.

The pack rate is determined by the amount of hydraulic pressure used during the pack phase, the slower the rate, of the lower the pressure. The pack rate is adjusted by the molder. The amount of pack is controlled by the second stage or pack setpoint. When this setpoint is reached it switches the machine to the third stage low-volume holding pressure which maintains pressure while the material solidifies.

Other approaches to the use of three-stage allows the molder to fill with plastics pressure, pack with stroke; fill with stroke, pack with hydraulic pressure. Fill and pack transitions can be accomplished using only cavity pressure. Three-stage control offers optimum cost effectiveness and flexibility for controlling the variables of cavity pressure and fill rate.

103.3 OTHER MOLDING PROCESSES

Compression and transfer molding present their own pecularities, which must be recognized in the development and implementation of process control strategies.

103.3.1 Compression Molding

This is the simplest of the mold processes. There are three primary variables: the amount of material put into the mold, the temperature in each half of the mold, and the clamp tonnage generated.

The amount of material placed in the mold determines whether or not the mold cavity will be filled when the clamp is completely closed. Thus, charge weight is an important variable, which is normally controlled by pre-weighing the charge. Mold temperature determines the rate of cure, and is an important variable determining cycle time and part quality. Clamp tonnage determines whether or not the compression mold closes completely prior to part curing. Too little pressure will cause the mold to stay partially open resulting in a thick or possibly non-filled part. Too much clamp tonnage can damage the mold. Proper clamp tonnage is generally determined by trial and error.

A control technique called reaction detection takes place in the material in the mold while it is solidifying, as in sheet molding of polyester. Control is initiated by placing a thermocouple in the surface of the mold in such a way that it protrudes as much as 6 diameters into the part. When the mold is closed, the thermocouple monitors material temperature as the cycle progresses. The temperature in the mold will rise slowly until the exothermic reaction takes place. When this occurs, a rapid rise in temperature signals that cure is taking place.

Using this technique, compression molding cycles can be reduced by as much as 20%. The cycle will vary as necessary to ensure that each charge of material is sufficiently cured without adding safety cure time to the cycle. Obviously, on large compression molding machines this added productivity can enhance the profitability of the operation.

103.3.2 Transfer Molding

Process control in transfer molding is similar to that of an injection-molding machine, since the transfer machine is simply an injection-molding machine, in which material viscosity is lowered outside of the press, and prior to injection. In transfer molding of thermosets, the condition of the preform as it is placed in the transfer pot determines the viscosity of the material as it flows into the mold.

Large changes in viscosity can occur due to batch-to-batch variations of material and the pre-processing history of the preforms. Wide viscosity variations from shot to shot are the norm in thermoset molding. Consequently, pressure loss between the transfer pot and the mold can be large and erratic. This makes the use of process control centered around cavity pressure an even more effective approach than it is in thermoplastic injection molding.

The use of cavity pressure control coupled with pressure compensated flow controls on thermoset machines can make the process highly repeatable. Molding temperature of both halves of the mold is also important because mold temperature determines cure rate of the material and cure rate determines the degree of cross-linking and consequent part quality. Cure rate also determines cycle time. Single setpoint or boost cutoff control systems are normally used on thermoset transfer machines.

BIBLIOGRAPHY

General References

C.E. Beyer, "Pressure Control for Injection Molding Machines," *Technical Papers,* Vol III, ANTEC, Society of Plastics Engineers.

C.E. Beyer and R.B. McKee, "Temperature and Pressure Measurement in the Injection Machine Heating Cylinder," *Modern Plastics* (April, May and June, 1955).

D.C. Paulson, "Pressure Loss in the Injection Mold," *1967 Annual Technical Conference,* Society of Plastics Engineers.

R.J. Groleau, "Injection Molding Controls and Instrumentation," *Modern Plastics Encyclopedia,* Vol. 59, No. 10 A, McGraw-Hill, New York, 1982–1983.

R.R. White, "Injection Molding Controls and Instrumentation," *Modern Plastics Encyclopedia,* Vol 58, No 10 A, McGraw-Hill, New York, 1981–1982.

D.C. Paulson, "Guide to Injection Machine Control," *SPE J.* **27**(1) (Jan. 1971).

D.B. Hunkar, "How to Compensate for Flow Behavior in Injection Molds," *Plastics Engineering,* **33**(7) (July 1977).

R.J. Groleau, "Computer Aided Engineering for the Injection Molding Industry," in *Process Controls,* Hanser, Chapt. 3.2.

T.G. Beckwith and N.L. Buck, *Mechanical Measurements,* Addison-Wesley Publishing Company, Inc., London, England, 1961, pp. 120–300.

G.R. Partridge, *Principles of Electronic Instruments,* Prentice-Hall, Inc., Englewood Cliffs, N.J. , 1959, pp. 237–255.

104

PRODUCT DESIGN, INTRODUCTION

Glenn Beall

Consultant, 887 So. Riverside Drive, Gurnee, IL 60031
The sections on design were prepared with the assistance of Glenn Beall.

Plastics are materials of design, just as metals and glass are. Indeed plastics continue to replace these older materials in many major markets. Automotive components and beverage bottles are standout examples. But plastics are available in an almost infinite range of properties, freeing them from the structural-design limitations imposed on metal and glass. This very freedom, however, can also pose problems when developing a product in plastics and shepherding it through to the marketing stage. Any variation, even the slightest one, from the exact formulation specified at the outset of the program can spell trouble on the production line, and perhaps even disaster at the point of use. It is a prime reason why structural designers should be involved at every step of design evolution.

Designing with plastics is otherwise no different from the disciplines imposed by other materials. All new products should be subjected to a preliminary design review to make certain that the new structure satisfies established functional, economic and manufacturing requirements. These specifications must be clearly defined in the original design checklist (Table 105.1, Chapt. 105). Upon receiving this preliminary approval, the designer must then turn his/her attention to finalizing the detailed design and drawing of each component of the final product.

Many new products begin to fail at the stage of detailed part design, after the personally gratifying creative process has been completed. At this point, many creative designers have begun to lose interest in the new product, or worse, control over its development.

This is unfortunate because the designer is usually the best-informed individual with regard to the end-use requirements of the new product. It is especially unfortunate when a plastic is the material of design, because there will still be many critical design-related decisions to be made after the preliminary design has been finalized into a detailed engineering drawing. Still other equally important decisions will be required as the part is tooled, and eventually processed by one of the many plastics manufacturing techniques. A simple revision, such as a deeper draw or a sharper undercut, which might pose no problem with rubber, could make the specified plastic totally unsuitable for the intended end use.

The word "plastic" encompasses a family of more than 11,000 materials with widely varying physical properties. Plasticized PVC can be as soft and flexible as rubber; engineering polymers or reinforced epoxies can be stronger than metal. The part design requirements for metal parts are totally different from those for a similar sized and shaped plastic part. Reinforced epoxy and PVC parts are also designed according to totally different sets of design guidelines.

In addition to physical variations, plastics differ in their chemical, electrical, and other critical properties. The incorporation of additives and reinforcements adds yet another complication. So does the relatively recent advent of blends and alloys, making the materials

1463

choice even more bewildering, and putting a further premium on performance monitoring at every stage of product development.

The selection of methods for producing a finished plastic part is less complex, but only numerically so. There are more than 100 processing and finishing techniques available to produce plastic products. Specialized processes have been developed to produce special shapes in special materials. Of the major processes, injection molding is noted for producinig large quantities of intricate, precision thermoplastic parts. Blow molding is the technique for producing large quantities of hollow parts with relatively simple shapes. Details on these and other plastics processing techniques appear in other sections of this Handbook.

104.1 DESIGNING FOR THE PROCESS

Many plastic materials can be processed by more than one manufacturing technique. It is in this area that the part designer meets his/her biggest challenge. Since there are more than 11,000 plastic materials and over 100 primary and secondary processing techniques from which to choose, selecting the right combination is an essential design consideration.

Still, successfully designing polyethylene blow molded parts does not qualify a designer to know how to design a polyethylene part for production by rotational molding. Likewise, part design requirements for injection molded phenolic part are different from the ideal design requirements for a compression molded phenolic part. For example, the proper injection moldable proportions for a design detail as simple as a stiffening rib are totally different for nylon (a crystalline material with high mold shrinkage) and ABS, an amorphous one, which has low mold shrinkage. Furthermore, design guidelines would change if the same plastic material were processed by structural foam molding instead of by solid injection molding. The design guidelines would change yet again if glass-reinforcing fibers were added to the ABS or nylon, even though the type of processing or molding remained the same.

Designers must know precisely what dimensional tolerances are possible or practical for a given combination of plastic material and processing technique. Transfer-molded phenolic parts are well known for their dimensional stability. Thermoformed polyethylene parts, on the other hand, are difficult to hold to precision tolerances. The recommended tolerances for each family of plastic materials and the different processing techniques are covered in detail in this Handbook.

104.2 FINALIZING PART DESIGN

Faced with the problem of finalizing the design of a new plastic part for a material or process with which he/she is not familiar, the designer can enhance the part's chances of success by relying on the data presented in this handbook. First, one should check the processing sections to determine (1) whether the chosen plastic material can be processed by that technique, (2) what tolerances are possible with that process, and (3) which design guidelines are appropriate for the process, required draft angles, wall thickness versus part size, etc.

The designer should then turn to the materials sections and compare the tolerances and design guidelines established for specific materials. If the material and process are compatible, the designer then has a reliable starting point for his/her work.

At this point, the designer should go back to the design checklist to review and compare the original product specifications and the material and process combination's capability to provide those requirements. If all of the product's requirements can be met, it is then safe to go ahead and finalize part design.

In instances where the design guidelines or tolerances are not included in this Handbook, the designer would be well-advised to contact the plastic material supplier and request design guidelines. In the same context, the designer may want to consider submitting a copy of the preliminary part design to the material supplier for review by the firm's technical support staff. Their input can be very helpful in optimizing part design and processing technique.

105

PRODUCT DESIGN, BASIC PARAMETERS, BASIC REQUIREMENTS

Stuart Caren

Product Director, Unisys CAD/CAM, Inc., 2970 Wilderness Place, Boulder, CO 80301

Plastics permit a greater amount of structural design freedom than any other material. Plastic parts can be large or small, simple or complex, rigid or flexible, solid or hollow, tough or brittle, transparent, opaque or virtually any color, chemical resistant or biodegradable, and materials can be blended to achieve any *desired* property or combination of properties. In addition, the final part is affected by the part design and processing. The designer's knowledge of all of these variables can profoundly affect the ultimate success or failure of a consumer or industrial product.

Successful design of a plastic part is a complex undertaking. Although all of the sections in this Handbook contain information of value to designers, the five articles on design deal specifically with plastic product design.

Product design is accomplished in much the same way with plastic as with any other material, but there are some notable differences. The difficulty of the plastics product designer's job is complicated by three factors: the diversity of the materials from which to choose, the variety of processes that can be used, and the tradeoffs that can be made among design, material, and process.

As with any material, product design in plastics must also satisfy three basic requirements. The part must provide reliable end-use functionality. When designing the part, the design engineer must also take into account its manufacturability in a specified material and specified process. And the designer must be sure that the product will be economically feasible.

105.1 GETTING STARTED

Overall design or product conception can be initiated from many sources. The most obvious is the completely new product. Although such products are rare, they offer designers the opportunity to utilize their abilities fully. Depending on its complexity, a new product requires several months to several years before commercial introduction. More commonly, overall design is a modification of an existing part or product. This may be initiated by a company's need to make the product more attractive or easier for consumers to use. Manufacturing may request a new design to simplify assembly or minimize breakage; or management may demand that costs be reduced.

In some cases, achieving these goals requires only a material substitution or a minor design change. But often it mandates that the product be redesigned, different materials used, and the components made and assembled differently. The way in which a part is manufactured has a profound influence on its design. There are a number of processing techniques from which to choose, each of which produces a different type of part.

Closed molding (injection or compression) provides fine detail on all surfaces. Open molding (thermoforming or spray-up) provides detail only on the one side in contact with the mold, leaving the second side free-formed. Continuous production (extrusion and pultrusion) yields parts of continuous length. Hollow (rotational or blow) produces hollow parts. These processes can be used creatively to make other types of parts. For example, two molded or thermoformed components can be bonded together to form a hollow product, or many molded parts can be bonded to make a continuous part.

The designer must select the materials and processes which can produce the required part and still meet functional and cost requirements. Figure 105.1 shows the steps involved in defining, designing, and eventually approving a new part or product. The first step is a general product description: what it is to do, how it is to be used, where it is to fit, etc.

After the product is defined, the functional requirements and the cost value are established, which is then followed by a preliminary design.

After a preliminary design is completed and approved, it should be reviewed by the manufacturer's Engineering, Marketing, Manufacturing and Quality Assurance departments. Inevitably, some changes will be required. If they are found to be practical (capable of achievement without compromising product cost or functionality in the intended use environment), the design project can proceed to its next stage.

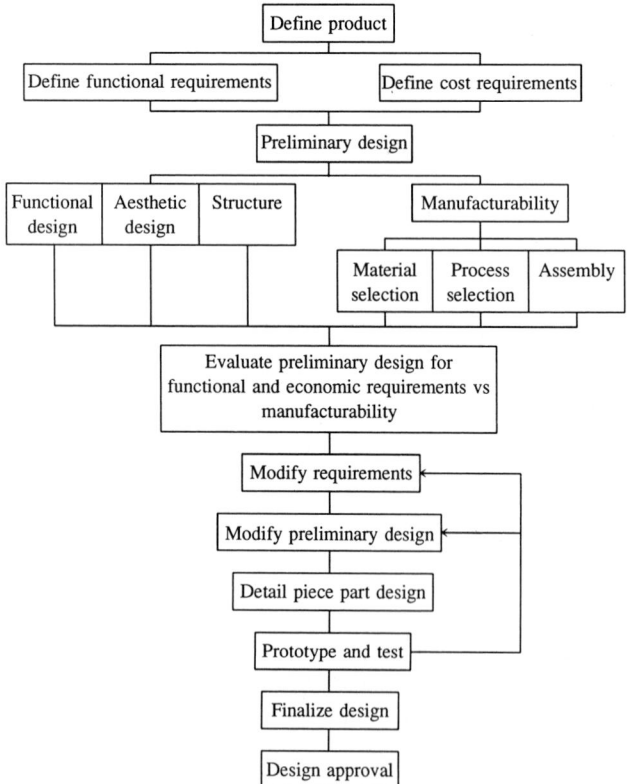

Figure 105.1. Product design flow chart. Copyright 1983 by Glenn Beall/Engineering Inc.

The next step is to prepare a detailed piece part design and drawing. Once the drawing is available, prototyping and testing can be initiated. Methods of prototyping vary greatly. In some cases, a painted model cut from polystyrene foam blocks will suffice. In other cases, a prototype must be made using the specified material and manufacturing process. Prototyping is essential, regardless of how it is done, in order to ensure that a product will perform as it is intended to.

The aim of product design or redesign is to achieve the best possible product at the least practicable cost. It is a dynamic procedure, with the key being *communication*. As the design project progresses and as more is learned, modifications may need to be made. Compromise may also be essential. For example, a superior design may cost more to produce than was originally estimated, but after an objective evaluation it may be determined that it is worth more. Thus, product cost can be increased without jeopardizing its chances for success. Similarly, prototypes usually show where additional strength is required or where a part is over designed.

A final note regarding overall product design procedure is that any design, no matter how good, can be improved. However, there comes a time when the design must be frozen and prototyping or production must begin. If this is not done, the new design will remain on the board until competition beats you to the marketplace.

105.2 FUNCTIONAL DESIGN

The main objective of prototyping and testing is to prove the acceptability of the design. Every part has its own requirements and all must be met to achieve a successful product. It is, therefore, wise to develop a written check list describing each requirement.

It is impossible to prepare a universal check list that can be used for any application. However, requirements that should be considered are assembled in Table 105.1. This check list can serve as a guide for any product line. It encompasses physical requirements, general requirements, appearance requirements, structural requirements, mechanical requirements, electrical requirements, environmental requirements, agency approvals required, assembly requirements, and special design features.

Physical requirements include size, weight, and capacity. Many parts must be a specific size to fit other components or assemblies. In these cases, critical fit dimensions must be noted. Packaging applications almost always have required capacity. Some special applications have specific weight or density requirements.

General requirements involve various factors that affect product design. For example, a reusable product must be designed differently from a single-use disposable. Low-volume parts require tooling different from that for large-volume parts. Tight dimensional tolerances are easier to hold in some materials and by some processes than others. A targeted factory cost will influence the designer's choice of material and process. If for some reason a specific material or process must be used it must be noted on the check list. Flammability or other requirements often must be met. Thus, the specific tests that the product must pass (such as Underwriters Laboratories UL-94) must be listed. It is also helpful to identify any labels or instructions that may be needed, whether they are to be molded into the part or attached to it as part of the assembly. Similarly, misuse and hazards should be considered early in the design stages of a product.

Appearance requirements involve both aesthetics and human engineering. The check list should include not only product color and texture, but also gate, parting line, and ejector locations and secondary plating, painting or other decoration. Therefore, it is essential to have the industrial designer involved early in the product design, before the design is finalized.

Structural design requirements must be known before any sensible design product can be launched. For instance, the type, amount and duration of load the product is expected

TABLE 105.1. Design Check List[a]

Part name _____

Description of application _____

_____

_____

Physical Requirements:

Size: Length: _____ Width: _____ Height: _____

Volumetric capacity (oz, lb, etc) _____

Must fit with _____

   Critical dimensions _____

Maximum (minimum weight) _____

Density (maximum or minimum) _____

Tight tolerances _____

General Requirements:

Durable (life expectancy) _____

Disposable _____

Quantity required _____

Target factory cost _____

Material _____

Process _____

Patents _____

Flammability _____

Labels, instructions, guarantees, etc (description and location) _____

Anticipated misuse, safety hazards and warnings _____

Product for export _____

Appearance Requirements:

Industrial design _____

   Aesthetics _____

   Human engineering _____

Match existing product line _____

Transparent _____

Color _____

Inside surface finish (SPE-SPI # _____) Other _____

Outside surface finish (SPE-SPI # _____) Other _____

Texture _____

   Texture depth _____ Pattern No. _____

Plated or metallized _____

Painting _____

Hot stamping _____

1468

TABLE 105.1. *(Continued)*

Parting line, pinch off or trim line location _____

Gate location _____

Ejector location _____

Mechanical Requirements:

Tensile loading

Load _____ Type _____

Duration _____

Flexural loading

Load _____ Type _____

Duration _____

Stiffness (flexural modulus) _____

Compressive loading _____

Creep, maximum allowable _____

Deflection, maximum allowable _____

Impact strength

At room temperature _____

High temperature _____

Low temperature _____

Shear strength _____

Hardness _____

Abrasion resistance _____

Environmental Requirements:

Operating temperature _____

Maximum _____

Minimum _____

Duration _____

Thermal expansion _____

Chemical resistance _____

Continuous contact with _____

Intermittent contact with _____

Occasional contact with _____

Outdoor exposure _____

Painting system

Solvent attach _____

Oven temperature _____

Vapor permeability _____

Electrical Requirements:

Volume resistivity _____

TABLE 105.1. *(Continued)*

Surface resistivity —————————————————————————————

Dielectric constant —————————————————————————————

Dissipation factor —————————————————————————————

Dielectric loss —————————————————————————————

Arc resistance —————————————————————————————

Electrically conductive —————————————————————————————

Transparent to microwaves —————————————————————————————

EMI/RFI shielding —————————————————————————————

Regulatory Agency Approvals Required:

Food and Drug Administration (FDA) —————————————————————

National Sanitation Foundation (NSF) —————————————————————

Underwriters Laboratories (UL) —————————————————————————

    Flammability rating (UL-94) —————————————————————————

    Temperature index (UL-746) —————————————————————————

    Other —————————————————————————————

American Society of Testing and Materials (ASTM) —————————————————

U.S. Pharmacopeia (USP) —————————————————————————————

Occupational Safety and Health Administration (OSHA) —————————————

American National Standards Institute (ANSI) —————————————————————

Department of Transportation (DOT) —————————————————————————

Society of Automotive Engineers (SAE) —————————————————————

National Electrical Manufacturers Association (NEMA) —————————————

State or local building codes —————————————————————————————

Federal Communication Commission (FCC) —————————————————————

Military specifications —————————————————————————————

Others —————————————————————————————

Export or foreign

    Canadian Standards Association (CSA) —————————————————————

    International Standards Organization (ISO) —————————————————————

    Others —————————————————————————————

Assembly Requirements:

To be assembled to —————————————————————————————

Assembly to be:

    Permanent —————————————————————————————

    Serviceable —————————————————————————————

    Leakproof —————————————————————————————

Type of assembly required:

    Mechanical—screws or inserts —————————————————————————

TABLE 105.1. *(Continued)*

Mechanical—press or snap fit _____

Adhesive _____

Solvent _____

Heat sealing _____

Electromagnetic bonding _____

Molded threads or fasteners _____

Flexibility in design of mating part _____

Special fixtures required _____

Machining required _____

Special tests _____

Special Design Features:

Bearing (low friction) surface _____

Sterilizable (ETO, γ radiation, steam) _____

Integral hinge _____

Threads _____

Undercuts _____

Insert molding _____

Two or more colors _____

Dual or co-extrusion _____

Other _____

to bear should be specified on the check list. Limits for allowable creep or deflection must be determined for the part in order for it to be properly designed. Other required mechanical properties, such as impact strength, hardness, or abrasion resistance must also be specified.

Mechanical properties of a product are influenced by the environment in which it is to be used. Specifically, the temperature at which the product is used will often affect its strength or stability. Chemicals can also affect plastics properties; the chemicals present in use environments must be known and their effects recognized. Likewise, ultraviolet light or paint solvents can affect the performance stability of a plastic product. Other environmental influences, such as vapor permeability or thermal expansion must be factored into the design decision.

Electrical requirements range anywhere from high resistance to conductive. All electronic enclosures must now provide shielding for Electromagnetic (EMI) and Radio Frequency Interference (RFI). The degree of shielding on these parts as well as the process must be specified. Other plastics applications, such as microwave cookware, have special requirements such as transparency to microwaves.

Every product must meet industry standards or government regulations. The most frequently needed approval is from the U.S. Food & Drug Administration, which has standards for everything from food packaging to medical implants. The second most common is probably Underwriters Laboratories, specifically its Subject 94 flammability rating. The American Society for Testing & Materials (ASTM) has standards for materials and performance tests. The U.S. Occupational and Safety Health Administration (OSHA) has requirements affecting almost anything in the workplace. There are many other regulations from industry and governmental agencies. The product manufacturer and the designer must know what standards have to be met.

Another factor that can significantly affect part design is the manner in which the product is assembled. Methods of assembly depend on whether the part is meant to be permanent, serviceable or leakproof, etc. Solvents or adhesives and welding techniques can be used for permanent assemblies, that cannot be disassembled for servicing. If the product need be disassembled a very limited number of times, self-tapping screws may suffice. But if regular disassembly is required, screws and molded-in or ultrasonic inserts should be used. Each type of assembly will affect the material selected, the manufacturing process and the overall product cost.

Many applications have special design requirements that do not fit into other check list headings. For example, bearing surfaces must combine low-friction and low-wear. Some medical applications require sterilization. The specific sterilization process (autoclaving, ethylene oxides, radiation, etc) will limit the variety of acceptable materials. Similarly, a good integral hinge can be made from polypropylene and only a limited number of other materials. Some electrical circuit boards are wave soldered, which requires a higher heat-resistant material than the application may otherwise require.

All these requirements must be considered before the designer starts work. Even minor changes or added requirements can drastically alter cost and performance calculations. It is always more efficient to design *right* the first time than to change a design when new requirements are added.

105.3 MANUFACTURING CONSIDERATIONS

There are usually several materials, design, and process combinations that can fulfill functional requirements. The job of the designer is to find the optimum combination.

Plastics are generally separated into thermoplastics or thermosets; the major difference being that thermoplastics can be remelted after they solidify; thermosets cannot. Therefore, the type of material will to a great extent dictate which processing techniques can be considered.

Because of the differences in materials, the plastic manufacturing methods are also generally known as thermoset or thermoplastic processes. This distinction is becoming less clearly defined as thermosets are developed for such traditionally thermoplastic processes as extrusion and injection molding, and as traditional thermoplastics are developed for such processes as reaction injection molding. Nonetheless, for a part designed for a particular process, only materials processable by that technique can be considered. A subsequent section discusses the steps involved in selecting the optimum materials and processes combination.

105.4 ECONOMIC DESIGN

Every product has a maximum value. This may be the price a consumer will pay or the cost level at which there is an acceptable return. To avoid excessive part cost, the designer must contend with vastly different material prices and manufacturing costs—including fabrication (molding, extrusion, thermoforming, etc), secondary trimming, painting, and assembly. The cost of tooling is also a significant contributor to product cost.

Materials cost is usually a major factor in the economics of a part or product. In addition to their wide range of properties, plastics vary greatly in cost. Some "commodity" materials, such as polyethylene, polypropylene and polystyrene, are priced in bulk at under $0.50 per pound. High-temperature engineering materials (such as polysulfone, silicone, and fluoro-

TABLE 105.2. Market (Not List) Prices[a]

Plastic	Cents/lb	Cents/in.^{3}
Thermoplastics		
ABS	82–120	3.1–4.8
Acetal	156	7.9
Acrylic	86–88	3.7
Cellulosics	129–131	5.9–6.0
EVA	69–95	2.3–3.2
Fluoropolymers	430–3000	47–238
Nylon (types 6, 66)	137–197	5.7–8.7
Nylon (others)	247–361	10.3–18.5
Polyarylate	200–280	8.8–12.3
Polybutylene	84–88	2.8–2.9
Polycarbonate	160–163	7.1–7.2
Polyester (PBT)	130–190	7.7–9.0
Polyethylene (LDPE)	57–65	1.7–2.2
Polyethylene (LLDPE)	52–70	1.7–2.4
Polyethylene (HDPE)	55–57	1.8–1.9
Poly(phenylene oxide)-based	110–181	4.3–7.0
Polypropylene (homopolymer)	48–53	1.6–1.7
Polypropylene (copolymer)	49–55	1.6–1.8
Polystyrene (GP crystal)	58–61	2.3–2.4
Polystyrene (high impact)	61–63	2.6
Polysulfone	382–430	17.9–20.2
Polyurethane	160–250	7.0–10.3
PVC	44–45	
SAN	83–87	3.2–3.3
Thermosets		
Alkyd	63–74	4.7–5.5
DAP	251–497	16.3–32.2
Epoxy	116–126	
Melamine	62–83	3.3–4.5
Phenolic	41–84	2.3–3.0
Polyester	59–69	
Polyurethane	86–93	
Silicones	581–640	37.6–41.5
Urea	50–72	2.7–3.9

[a]Beginning of 1989 from *Plastics Technology* magazine. Based on the formula: cost/in.^{3}-cents/lb × specific gravity (g/cc) × .0361.

polymers) can cost up to ten times as much. Table 105.2 lists the bulk material cost in 1989 for some of the more widely used plastics. Current prices are published regularly in the plastics magazines. Pricing information will make it easier to achieve a balance of function and economy. It must be noted, however, that material prices fluctuate. For current prices, the designer should consult the material's supplier and specify the grade, order quantity, and color that is planned for use.

Although most materials, including plastics, are sold by weight, they are used by volume. Plastics vary widely in density and specific gravity, which have a significant effect on cost per cubic inch. As an example, an engineering plastic that sells for $1.48 per lb. and has a specific gravity of 1.42 g/cc, would have a cost per cubic inch of 7.6 cents. A slightly more expensive competing material ($1.55 per pound) with a specific gravity of 1.10 g/cc would cost only 6.2 cents per cubic inch.

For determining the cost per cubic inch for the cost per pound the following formula can be used.

$$\text{Cost/in.}^3 = \text{cents/lb} \times \text{specific gravity (g/cc)} \times 0.0361$$

There is still another economic factor that must be considered. This is the minimum part thickness achievable with a given material without sacrificing performance.

Fabrication costs vary considerably, not only because of differences in the process, but also because of differences in part complexity, size and geographic location. Machine-hour rates differ among processes, and larger equipment demands higher rates. Also, some parts can be automated while others require manual handling and secondary operations. The Processing Sections provide information on costing for each process.

106

PRODUCT DESIGN, CONSIDERATIONS

Mort Blumenfeld

FIDSA, Industrial Design Consultant, 16566 San Tomas Drive, San Diego, CA 92128

Methods used to design parts in metals or other traditional materials seldom apply when designing with plastics. In spite of the wealth of technical information now available, designing with plastics is often as much art as science, and experience is the best teacher.

The use of the computer to aid in design and manufacturing coupled with the growing use of robots will have a significant impact on design.

Plastic materials perform in ways that differ distinctly from the traditional materials they often replace. Until fairly recently, there were a limited number of polymer types available, and then only with few variations in formulae. That is changing rapidly. An almost infinite number of variables are the result of custom compounding of polymers, and combining some into copolymers, terpolymers, multipolymers, interconnecting polymer networks, blends, alloys, and liquid crystal polymers.

Adding to this complexity is the increasing use of higher percentages of inorganic fillers, and organic and inorganic fibers. Fiber variations include milled, chopped, directional, oriented, short strand, long strand, mats, weaves, knits and hybridization by combining (for property optimization) such types as glass and carbon or aramid in cost-effective mixtures.

Besides fillers and fibers, a broad range of additives are employed to modify the basic resin. Among other things, they aid flow in molding, help meet fire safety codes, and toxicity rules, counter the destructive effects of ultraviolet light (UV), bacteria and mold, or simply to enhance appearance with dyes and pigments. It is easy to see that today's designers must be far more educated, experienced and sophisticated than their ancient forebears, the designer/craftsman.

106.1 MATERIAL SELECTION: THE BASICS

Material selection should begin with an open mind and should not narrowly limit itself to a preconception that plastic material is best.

A product or its components must meet carefully defined criteria, which are often weighted differently by management, sales, marketing, research and development, engineering, production, purchasing, technical service, and most important of all, the end-user. Material selection must be based upon an acceptable consensus.

There are more than 40 families of plastics to choose from, and they differ in more ways than their chemical composition, as previous design discussions in this volume make clear.

They also vary physically in the ways that their polymer chains are linked together, and this must be considered in making a selection for a specific use.

The lengths of the chains affect how plastics process. The presence or absence of branches on these chains affects how closely they can pack. Chains that pack loosely may produce weaker parts because of the lack of order; these are called "amorphous" plastics. Chains with little or no branches may be packed more closely; these orderly arrangements are known as "crystalline" plastics. Generally they are much stronger, more chemically resistant and often have slippery surfaces. They work well for mechanical parts, but are sometimes difficult to paint or cement together. Crystalline thermoplastics cannot be transparent. In processing they melt suddenly, flow easily and solidify rapidly. Examples include polyethylene, nylon, polypropylene and acetal. Amorphous thermoplastics have a more gradual melt transition when heated, take longer to become rigid, often have nonslip surfaces, in some cases are transparent, and are more easily painted, cemented and thermoformed than crystalline plastics. Some examples are polystyrene, vinyl, acrylic, cellulosic, polycarbonate, and polysulfone.

Another important factor affecting material selection is the way that amorphous and crystalline plastics contract in the molds as they cool. Crystalline thermoplastics shrink to a greater degree, and nonuniformly. Amorphous thermoplastics do not shrink as much during cooling in the mold, but they are more likely to have higher levels of internal stress if improperly molded. This will cause earlier part failures due to weathering, solvent exposure, overheating, or just mechanical stressing at less than expected limits.

Thermoplastics can be melted and reused. Thermosets chemically cross-link when polymerized and cannot be remelted.

Combinations of polymers can be mixed together. When the mix occurs within the polymer chain there is a chemical link-up and we refer to these polymers as "copolymers," a type of acetal being an example. When three are linked together, they are "terpolymers" such as acrylonitrile–butadiene–styrene (ABS). If different polymer chains are mixed together but with the chains discreet, they are called "alloys" (PC/PVC).

There are, of course, many other performance-related considerations that must go into the materials-selection decision. Many of them are detailed in other reports in the design section, and elsewhere in this Handbook. But the bottom line is this: When designing with plastics the "best" choice is the one that produces an acceptable part at the lowest cost. This is not always the lowest priced resin. Labor costs are the most critical factor. Material cost estimates should be based on cost per cubic volume, not cost per pound or metric equivalent.

It is a good idea to begin the selection process by first listing the important requirements, such as heat resistance, strength, weatherability, toxicity, flammability, code compliance with regulatory authorities and weight. Reject all candidates that fall far short of requirements. Set aside temporarily those that fall close by or miss by only a little, because ASTM values do not always bear relevance to unique part designs or end-uses, especially when close to limit figures.

Next, consider tooling cost, molding cost, decorating, joining and any other factors that may bear on the life expectancy of the product. The following comments briefly describe characteristics of the more common thermosetting and thermoplastics materials. These characteristics can be used to screen materials in the selection process. Each material is fully described in other sections of this Handbook.

106.2 THERMOSETS

These economical materials have outstanding electrical properties, flame resistance and chemical resistance. Until the mid-1960s, most thermosets could be processed only by compression molding, and were limited by the labor-intensive requirements of that process.

But now, like thermoplastics, thermosets can be injection-molded. This, together with improved impact strengths, makes thermosets attractive for applications requiring high performance characteristics. Common materials of this type are listed in the following.

Alkyds. Alkyds are easy to mold, have high heat resistance, and excellent electrical performance, and may be light-colored.

Allyls. Allyls have high heat and moisture resistance, good electrical performance in automotive and aerospace uses, good chemical resistance, dimensional stability, low creep.

Aminos. Melamine has excellent electrical properties, heat and moisture resistance, abrasion resistance (good for dinnerware and buttons); in high-pressure laminates melamine is resistant to alkalies and detergents. They are used for counter tops (FORMICA). Urea has properties similar to melamine and is used for wall switch plates, light-colored appliance hardware, buttons, toilet seats, and cosmetics containers.

Epoxy. Epoxy has perhaps the highest performance of all the thermosets. Very high strength in tension, compression, flexural loadings, very low shrinkages, hard, superior adhesion to other materials. Used with glass cloth to make circuit boards, tooling surfaces for metals and RP castings. Can be cured chemically without heat.

Phenolics. Phenolics are the low-cost workhorse of the electrical industry; low creep, excellent dimensional stability, good chemical resistance, good weatherability. Molded black or brown opaque handles for cookware are familiar applications. Also used as a caramel-colored impregnating resin for wood or cloth laminates, and (with reinforcement) for brake linings and many under-the-hood automotive electricals. One of the first thermosets to be injection-molded.

Polyester. Polyesters have an excellent balance of properties, a room-temperature cure and is a major resin used to make glass-fiber reinforced parts for automobiles, boats, and aircraft parts. Commodity types have moderate weatherability, high molding shrinkages with wavy surfaces and warpage. "Low-profile" additives reduce shrinkage and surface waviness to almost nil. This has led to major growth in such applications as automobile exterior body panels, instrument housings and microwave dinnerware in the form of bulk molding compounds (BMC) and sheet molding compounds (SMC). Recent improvements include the use of premium fibers (carbon, aramid, etc) and the addition of elastomerics to improve resistance to impact and torsional loading damages.

Polyurethane. Polyurethane durometers range from cushiony soft to glass hard with superior wear resistance as skateboard wheels, solid tires, floor coatings, marine finishes, etc. A major use for soft-foam is automotive bumpers; another is upholstery. Recent improvements include addition of fibers and fillers to reaction molded parts to improve cut strength resistance, and stiffer moduli to reduce waviness caused by heat and weathering.

Silicone. Silicone has excellent heat resistance, chemical resistance, good electricals, compatability with human body tissues, and a high cost. It cures and cross-links at ambient temperatures, catalyzed by moisture in the air. It is a good sealant and excellent for making flexible molds for casting. It is widely used for human implants.

106.3 THERMOPLASTICS

Thermoplastics come in greater variety than thermosets. They also lend themselves more readily to specialty compounding as copolymers, multipolymers, alloys and blends, often customized for cost-effective adaptation to specific application requirements. Unlike thermosets, they are in most cases reprocessable without serious losses of properties.

But they do have limitations of heat-distortion temperatures, cold flow and creep, and are more likely to be damaged by chemical solvent attack from paints, glues, and cleaners. When injection-molded, dimensional integrity and ultimate strength are more dependent on sound tool and part design and molding parameters than is generally the case with thermosets (where cross-linking of the polymer chains tends to offset such problems). Common thermoplastics include the following:

Acrylonitrile–Butadiene–Styrene. (ABS) is a terpolymer that provides a tough, hard, rigid plastic with adequate chemical, electrical and weathering characteristics, low water absorption, and resistance to hot-and-cold water cycles. Used for telephones, sports gear, automotive grilles, electronic instrument housings and furniture. It is electro-platable, good as a structural foam, and available as a tinted transparent.

Acetal. This crystalline polymer (and copolymer) is strong, stiff, and has exceptional resistance to abrasion, heat, chemicals, creep and fatigue. With a low coefficient of surface friction, it is especially useful for mechanical parts such as gears, pawls, latches, cams, cranks, plumbing parts, etc. It is chrome platable.

Acrylic. High optical clarity, the best weatherability, broadest color range and hardest surface of any untreated thermoplastic. Chemical, thermal and impact properties are good to fair. Normally an exterior material, used as optical lenses, automotive taillights, decorative nameplates, aircraft glazing, illuminated signs, medical devices, etc. A new use is as opaque colored sheeting thermoformed to produce an outer coating behind which glass-fiber-reinforced polyester resins are sprayed to produce camper tops, swimming-pool steps, plumbing fixtures with weatherability and repairability reported superior to polyester gel coats.

Cellulosics. Cellulosics are tough, transparent, hard or flexible polymers made from plant cellulose feedstock. With exposure to light, heat, weather and aging, they tend to dry out, deform, embrittle and lose gloss. Molding applications include: tool handles, control knobs, eyeglass frames. Extrusion uses: blister packaging, toys, holiday decorations, etc.

Fluoroplastics. Fluoroplastics have superior heat and chemical resistance, excellent electrical properties, but only moderate strength. Variations include PTFE, FEP, PFA, CTFE, ECTFE, ETFE, and PVDF. Used for bearings, valves, pumps handling concentrated corrosive chemicals, skillet linings, and as a film over textile webs for inflatables such as blimps and pneumatic sheds. Excellent human-tissue compatibility allows its use for implants.

Nylon (Polyamide). Nylon is a crystalline plastic and the first engineering-grade thermoplastic. Tough, slippery, with good electrical properties, but hygroscopic and with dimensional stability lower than most other engineering types. Also offered in reinforced and filled grades as a moderately priced metal replacement.

Phenylene Oxide-Based (PPO) Resin. This is one of the top choices for electrical applications, housings for computers and appliances, both neat and in structural foam form. It has superior dimensional stability, moisture resistance due to styrene components, which, however, cause some sacrifice of weather and chemical resistance. Used for automobile wheel covers, pool plumbing, consumer electronic external and internal components.

Polycarbonate. Polycarbonate is a tough, transparent plastic that offers resistance to bullets and thrown projectiles in glazing for vehicles, buildings, and security installations. It withstands boiling water, but is less resistant to weather and scratching than acrylics. It is notch-sensitive and has poor solvent resistance in stressed, molded parts and is used in coffee-makers, food blenders, automobile lenses, safety helmets, lenses, and many nonburning electrical applications.

Polyesters, Thermoplastics. Poly(butylene terephthalate) (PBT) is a crystalline polymer and an excellent engineering material. It has marginal chemical resistance but resists moisture, creep, fire, fats, and oils. Molded items are hard, bright colored, and retain impact strength at temperatures as low as $-40°F$ ($-40°C$). Uses include auto louvres, under-the-hood electricals, and mechanical parts. Poly(ethylene terephthalate) (PET) an amorphous polymer is available in an engineering grade, but is most widely used in beverage bottles.

Polyarylate. Polyarylate is a form of aromatic polyester (amorphous) exhibiting an excellent balance of properties, stiffness, UV resistance, combustion resistance, high heat-distortion temperature, low notch sensitivity, and good electrical insulating values. It is used for solar glazing, safety equipment, electrical hardware, transportation components and in the construction industry.

Polyethylene. This is the leading plastics family in total volume sold. Materials are inexpensive, easy to process and so versatile that they dominate the packaging and disposables fields. Crystalline in structure, they are varied by chain length, or molecular weight into low density (LDPE), linear low density (LLDPE), medium density (MDPE), high density (HDPE) and ultra high density (UHMWPE). Strong and flexible, though not transparent, they are very highly chemical resistant and difficult to cement or paint. They are blow molded into containers and bottles and molded into boxes, buckets, etc, which may be susceptible to stress cracking near the gates if care is not exercised. They are extruded for films, trash bags, and laminated coatings.

Polybutylene. Polybutylene is a polyolefin used for cold and hot water piping. As a blown film it is used for food packaging.

Polyetherimide. This is an engineering-grade amorphous thermoplastic polymer. It has superior strength, heat resistance, flame resistance, UV resistance and is transparent, although of amber brown color. Solvent resistance is especially good against aircraft grade fuels and lubricants, but it is attacked by methylene chloride and trichloroethane. Resistance to creep at lower stress loadings and good retention of strength at sustained high levels of heat are claimed to exceed those of other high performance engineering thermoplastics. Applications include printed circuit boards, heater housings, electrical components, steam sterilizable disposable and reusable parts.

Polyetheretherketone (PEEK). This is a high-temperature, crystalline thermoplastic used for high performance applications such as wire and cable for aerospace applications, military hardware, oil wells and nuclear plants. It holds up well under continuous 450°F (323°C) temperatures with excursions up to 600°F (316°C). Fire resistance rating is UL 94 V-0; it resists abrasion and long-term mechanical loads.

Polyimide. Polyimide is a high-cost heat and fire resistant polymer, capable of withstanding 500°F (260°C) for long periods and up to 900°F (482°C) for limited periods, without oxidation. It is highly creep resistant with good low friction properties. It has a low coefficient of expansion and is difficult to process by conventional means. It is used for critical engineering parts in aerospace, automotive and electronics components subject to high heat, and in corrosive environments. A racing engine with mostly polyimide parts has exceeded 15000 RPM without failure!

Polyphenylene Sulfide (PPS). PPS is able to resist 450°F (232°C), and has good low-temperature strength as well. It has low warpage, good dimensional stability, low mold shrinkage. Used for hair dryers, cooking appliances, and critical under-the-hood automotive and military parts.

Polypropylene. One of the high volume plastics has superior resistance to flexural fatigue stress cracking, with excellent electrical and chemical properties. This versatile polyolefin overcomes poor low temperature performance and other shortcomings through copolymer, filler, and fiber additions. It is widely used in packaging (film and rigid), and in automobile interiors, under-the-hood and underbody applications, dishwashers, pumps, agitators, tubs, filters for laundry appliances and sterilizable medical components.

Polystyrene. One of the high volume plastics, is low in cost, easy to process, has sparkling clarity, and low water absorption. But basic form (crystal PS) is brittle, with low heat and chemical resistance, poor weather resistance. High impact polystyrene is made with butadiene modifiers: provides improvements in impact strength and elongation over crystal polystyrene, accompanied by a loss of transparency and little other property improvement. A styrene–acrylonitrile (SAN) copolymer gains somewhat in strength and chemical and heat resistance. SAN is used for tinted drinking glasses, low-cost blender jars and water pitchers, and other consumer goods with longer life expectancies than ordinary PS.

Styrene Maleic Anhydride (SMA). SMA is a copolymer made with or without rubber modifiers. They are sometimes alloyed with ABS and offer good heat resistance, high impact strength and gloss but with little appreciable improvement in weatherability or chemical resistance over other styrene based plastics.

Expandable Polystyrene Beads (EPS). This is a modified PS prepared as small beads which, when steamed, expand to form lightweight, cohesive masses for forms used to pack fragile products for shipment. Similar dimensionally stable forms molded from EPS are used as cores for such products as automobile sun visors with surface overlays.

Polysulfone. Polysulfone is a high performance amorphous resin that is tough, highly heat resistant, strong and stiff. Parts are transparent and slightly clouded amber in color. Material exhibits notch sensitivity and is attacked by ketones, esters, and aromatic hydrocarbons. Other similar types in this group include polyethersulfone, polyphenylsulfone, and polyarylsulfone. They are used for medical equipment, solar-heating applications and other performance applications where flame retardance, autoclavability and transparency are needed.

Polyurethane, Thermoplastic (TPU). TPU has excellent properties except for heat resistance (only up to 250°F 121°C). TPU is used in alloys with ABS or PVC for property enhancement. Typical uses are in automobile fascias and exterior body parts, tubing, cord, shoe soles, ski boots and other oil and wear resistant products.

Poly(vinyl chloride) (PVC). PVC is a high-volume plastic that is low in cost, with moderate heat resistance and good chemical, weather and flame resistance. It qualifies for packaging, pipe and outdoor construction products (siding, window profiles, etc), and a host of low-cost disposable products (including FDA-grade medical uses in blood transfusion and storage). PVC comes in a variety of grades, flexible to rigid. They are tough, can be transparent (as in blow-molded bottles and jugs), and are also a good alloying resin to improve properties and reduce costs (ABS/PVC).

106.4 POLYMER MODIFICATION

If plastics material selection required merely the choice of polymer type, blends, alloys and multipolymers, the task of wise selection would be challenging enough, but modifiers must also be considered. Modifiers consist of a broad spectrum of additives, fillers and reinforcers. The combinations that can be specified and the percentage of each addition must be carefully controlled to avoid processing problems and the loss of important physical, chemical or thermal properties. All of the possible combinations result quite literally in a quantum leap in variables.

To give some idea of the complexity designers face, additives can be such things as antioxidants, antistatic agents, biocides, colorants, coupling agents, emulsifiers, flame retardants, foaming agents, fungicides, heat and light stabilizers, lubricants, mold-release agents, organic peroxides, plasticizers, preservatives, processing aids, slip agents smoke suppressants, viscosity depressants, and more.

Fillers include calcium carbonate, talc, kaolinite, alumina trihydrate, feldspar, silica, solid glass microspheres, hollow glass microspheres, agriculture by-products (such as rice hulls, corn starch, sawdust, and corn cobs), mica flake, carbon, metal powders, metal flake and wollastonite (a high aspect ratio mineral particulate). Fiber reinforcers include aramid, carbon, boron, glass (E type, C type, S type), hybrids (aramid/carbon, aramid/glass, carbon/glass), synthetic fibers, and plant fibers.

106.5 PROCESS SELECTION

Selection of the processing method should take place in the early stages of the design process, not after the design is nearly complete. Such an obvious axiom may seem superfluous, but unfortunately process selection is frequently an afterthought. It is as though most designs might be produced by nearly any process, which is not so.

The following brief descriptions of the capabilities and limitations the primary plastics processing techniques will be helpful in selecting the optimum manufacturing technique. See the sections in this Handbook on Processing, for more information.

Compression Molding. For thermoset plastics, partially polymerized preforms, pellets or liquids are placed in heated molds usually operated in vertical presses. The application of heat and pressure, as the mating steel molds close, melts and then cures the resin. The volume of the charges must be carefully measured to avoid short shots that will result in voids, or excess thickness with unacceptable flash around the parting line (like carelessly-poured waffles). Some thermoplastics are compression molded; for example, vinyl phonograph records. Transfer molding is a variant of this process. When fibrous glass preforms are inserted prior to application of the resin, the process may be called matched metal die molding.

Extrusion. Thermoplastic pellets are gravity-fed from a hopper into a heated barrel and propelled forward by a rotating screw, where melt occurs. The molten extrudate is forced through a steel die, cooled rapidly and rolled (flexible tubing, film), or cut to lengths and stacked (profiles, pipe, sheet). The process is continuous. If optically specular surfaces are called for, surface striation lines can be press polished with rollers. If textures like woodgrain or prismatic lens patterns are specified, these can be achieved with engraved rolls as the hot extrudate leaves the die. Die costs are relatively low, but long production runs are usually necessary to justify special shapes, colors or materials. Two or more colors or materials can be abutted, overlaid or concentrically layered by feeding from several extruders into one complex common die (coextrusion). Seven layer bottle parisons are currently being extruded for study.

Injection Molding. Thermoplastic polymers are gravity-fed from a hopper into a barrel, melted by a reciprocating screw and/or electric heat and are propelled forward by a ram (piston, plunger) or the screw (used as a plunger) into mating steel molds, which are cooled to below the heat-distortion temperature of the resin. The injected plastic material contracts as it cools (mold shrinkage) and shrinks. When cool enough to retain its shape, the plastic part is ejected from the mold.

Good part design requires adequate taper (draft) of side walls, radii at inside corners, minimal variations in wall cross-sections, use of ribs 60% or less of outer wall thickness for stiffness, strength, and minimal sink marks.

Thermosetting polymers can also be injection-molded. For these materials, the barrels on the injection-molding machine are heated by hot water to a point safely below cross-linking temperature; the polymer is then propelled by ram or screw feed into heated molds. After they cure the parts can be ejected while still hot, because they have already thermally set (or cross-linked).

Structural Foam Molding. This is an injection-molding process in which small to moderate amounts of gas bubbles are introduced into the molten polymer, with resulting foaming occurring within the part walls. Nitrogen gas or chemical blowing agents are introduced into the resin and held under pressure to prevent foaming until the melt is rapidly injected into the cavity. Because the foam cushions the pressure (like air in hydraulic brake lines), much lower clamp pressures are needed than for conventional injection molding. Unless special measures are used, molded foam parts will show surface swirl markings, necessitating painting for a quality appearance. Because thicker parts result from a given weight of material the sectional modulus increase provides more strength to structural foam parts. Thicker parts with insulating bubbles take longer to cool, increasing cycle time, but virtually eliminate sink marks.

Rotational Molding. This process is similar to slush casting of metals. Thin metal molds are charged with liquid or powdered polymer, premeasured to provide a coating on the inside of the mold to the specified thickness during melt cycle and subsequent chilling. Molds are mounted on "spiders," frame structures suspended on rotating arms projecting from a central hub. Charged molds are rotated biaxially and traversed through a heating chamber (which allows the polymer to melt and flow coat the inside of the molds) and into a chilled chamber. While molds are still rotating, the polymer cools and becomes firm, resulting in a part with exterior surfaces mirroring the mold's configuration and with interiors of soft and flowing surfaces occurring naturally.

Thermoforming. Thermoplastic sheet or film is first heated to above its softening range. Subsequent application of vacuum, air pressure or contoured molds shapes the material to its desired configuration. Deeply drawn parts may exhibit excessive thinning at stress points, or tearing of the softened sheet. Contact of the hot sheet with the mold surfaces will transfer surface textures in the form of mark-off. A great many products are thermoformed, including egg cartons, blister packages for foods and nonfoods, take-out food trays, skylights, signs, tote bins, carrying cases, and aircraft glazing.

Strip heaters (linear calrods) are used to soften localized areas to produce bends similar to brake forming of sheet metals. Thick sheets [0.1 in. (2.5 mm) or more] will take longer to heat to forming temperature, but will also provide more time to transfer from the oven to the mold setup. Thinner sheets heat quickly, but cool too fast to move very far, so heating is usually done with infrared sources mounted directly on the forming press near the tenter frame, which grips the periphery of the thin sheet and retains it during the entire process.

Crystalline polymers have such a sudden melt transition that more sophisticated automatic equipment is required to process them. Many disposable products are made this way, such as food packaging and medical products.

Thermoformed products often require secondary machining, cementing and finishing, and there is scrap loss during trimming.

Blow Molding. Both injection blow molding and extrusion blow molding require an intermediate preform called a parison. For extrusion blow molding a tubular parison descends vertically from the die of an extruder. Split molds (most often aluminum) close with low pressures on the parison, pinching off the top and bottom ends. Air is then blown in, and the parison stretches to fill the mold cavity, where it is chilled on contact with the mold walls, or with auxiliary means, freezing off the desired shape. Top and bottom pinch-offs can be removed automatically or manually depending on size, part complexity and tooling sophistication. Large jugs, bottles, trash containers, and double-walled typewriter and tool sets cases are some products that are made with this method.

Injection Blow Molding. Injection blow molding produces parisons on an injection-molding machine. These parisons unlike extruded ones can have varying wall thicknesses to control bottle dimensions, and often mold the complete complex closure design of the neck area before the final body-blowing stage. By means of a turret mechanism on the machine with the cores, still-hot parisons are rotated to a second set of cavities of the final blown configuration. Air is injected inside the parison and the parts are blown; however, the molded parisons can also be cooled, ejected, and shipped to a second area for later reheating and blowing. Containers are the main end-use for injection blow molding, but toys and specialty products are also made with this process.

Reaction Injection Molding (RIM). In this process, the two liquid components of polyurethane elastomer are brought together briefly inside an impingement-type mixing head, creating a homogeneous cream that is then injected at fairly low pressure into a closed mold where it cross-links to the desired shape. Parts produced can be hard or soft in durometer, dense or foamed, and they usually have a smooth exterior. Nylons, polyethylene, epoxies and polyesters all can be processed by RIM as well. For good rapid curing of parts, free of pinholes and surface blemishes, the two-component resins must be controlled accurately for temperature, pressure and viscosity. This is achieved by circulating the resins in the feed lines back to their supply tanks continuously until the next shot is initiated. Equally important is complete purging of the mixing head after each fill shot so that no waste material cross-links, blocking this sophisticated valve mixer head.

To improve cut strength resistance and provide stiffer moduli in RIM parts (especially for automotive exterior body components) chopped glass fibers and inorganic fillers have been added to the resins. This changes the process name to RRIM (*reinforced* reaction injection molding); precision components inside the mixing head must be made from a carbide material, with frequent replacement necessary due to wear. High-modulus RRIM and low-profile polyesters modified with elastomers are in a close competitive race for the large and lucrative automobile body panel market, replacing metals.

Composites Processing. Fiber reinforced plastics is a term that once referred only to thermosetting polyesters. This is changing rapidly; urethanes, epoxies and many thermoplastics are becoming increasingly important. Consequently, there are a number of craft-related as well as automated processes used within this industry.

Hand Lay-up. Hand lay-up, for example, is a labor-intensive process that calls for moderate worker skills. It requires only a one-sided mold, usually the female or cavity section. Featuring a smooth surface cast from a carefully crafted pattern, the mold is coated with a release agent prior to use. The part is produced in inverse order, beginning with the surfacing layer; sometimes a pigmented, UV-stabilized resin called a gel coat. Working inward, resins and fiber-reinforcing layers are added, consisting of mat or woven glass-fiber cloth. A roller is used to compact the fiber layers into the resin matrix and to work out trapped air. Structural members such as honeycomb, balsa or rigid foam blocks are sometimes added and encapsulated between the plies. The inner surface has an irregular, troweled look. Mold configurations can be altered at nominal costs. Part size and complexity thus are virtually unlimited. Boat hulls, corrosion-resistant tanks and ductwork, railcar fascias and large medical units such as CAT scanners are made this way.

Sprayup. Sprayup requires more skill than lay-up, and is a faster, sometimes automated variation on that process. The molds are similar to those used for hand lay-up. In front of a dual nozzle spray gun, polyester resin and catalyst meet in mid-air enroute to the mold surface. Glass fibers fed from a remote roll of roving, which looks like twine, pass through a rotary chopper atop the spray gun, and are integrated with the resin mixture as it is blown onto the mold surfaces. This process permits shorter cure cycles, and lends itself to more intricate contouring. However, resultant parts may not always attain the strength of woven or unidirectional fibering that is possible with other methods.

An important variation on sprayup uses a thin layer of thermoformed high impact acrylic sheet in lieu of a gel coat. Reinforcing sprayup is then applied on the rear side. The resulting product is more durable, chip and crack resistant, and more repairable than cross-linked gel coats. Bathtubs, camper tops, cycle fairings and pool accessories are examples of this process.

Pultrusion. This process utilizes the ultra high strength, continuous fibers oriented in tight parallel bundles in one long direction of parts with a constant cross-section design. The results look like extrusions but the process pulls rather than pushes the matrix through the profile die. Parallel strands of fibers are passed through a pot of catalyzed resin and saturated for good fiber wet-out. They are then pulled through a flexible orifice to remove excess resin before passing through the heated pultrusion die. The pultruded matrix is then drawn through an oven chamber for cure; and may be cut to desired lengths. Vaulting poles, pipe, tubing, structural beams and specialty shapes are examples of products made with this process. To provide radial strength when needed, pultrusions may be subsequently filament wound around the outside.

Filament Winding. This process begins like pultrusion, with roving fiber pulled through a resin pot and excess removed. However, instead of going through a die, the wet fibers are wound around a mandrel. Like fishing line being shuttled directly onto a reel, the wet fibers are directed around the core form to produce tubing. To produce tanks or other closed vessels, the shuttle proceeds around the ends of the forms as well. Mandrels can be rigid or made from inflatable bags that collapse for removal through a small opening after cure.

Centrifugal Casting. Centrifugal casting uses a metal cylinder rotating inside an oven. A long hollow lance passes down through the cylinder and a glass/resin mixture is sprayed inside as the lance is retracted. Centrifugal force holds the resin against the inside wall of the form until heat effects cure. Normal shrinkage during cure will permit removal of the casting without provision for taper or draft. High strength tanks, tubing, drive shafts telephone poles and other innovative products may be made with this process.

Matched Die Molding. This is basically the same as compression molding (the two names are used interchangeably). The only distinction is in the higher loadings and orientation of reinforcing fibers in matched die molding. Mating steel dies are mounted in a vertical press. Polyester premix in the form of preweighed chunks of dough or special low shrink polyesters like BMC and SMC are positioned and squeezed to shape with heat and high pressure; alternatively, a cloth or mat preform can be placed over the male die and liquid resin poured over it before molding to shape. The former method is used to make automobile body parts, aircraft sections, computer desktops, and the like. The latter method can be used to make luggage, outboard motors, etc. Inadequate placement of preforms or resins may result in voids (no resin in an area) or resin rich areas (no fibers in an area). Unless costs permit repairs, the entire part may have to be scrapped.

Resin Transfer Molding. This process is a relative newcomer to RP processing. It is a low-pressure, long-cycle answer to the problem of producing larger parts with good surfaces on both sides. Molds are made for each side in a manner similar to the ones used for hand lay-up or sprayup. Positioned for vertical opening, the upper mold has an injection port fitting located near the top center. With the molds in closed position, there is a tubular gasket seal around the parting line. This gasket is breached at several points by vent ports through which trapped air may escape. After closing the molds and clamping by mechanical, hydraulic or air actuated devices of nominal pressure, a mixing head is inserted into the top charging port. The catalyzed resin is slowly pumped in, infusing the reinforcing material and gently propelling trapped air ahead of it to the peripheral vent ports, which are then sealed off manually or by automated devices to prevent leakage of resin. Good fit and surface quality are reported. Bathtubs are among products made with RTM.

106.6 MATCHING PROCESS AND MATERIAL

A basic truism of structural design is that the material and the process selected profoundly affect the quality and the appearance of the product. For this reason, it is unwise to create a design first, and then decide on material and process. This seems obvious and logical, but it is frequently ignored in practice, especially when converting a metal design to a plastic material.

There is much to consider: Whether the material is heated first or worked at ambient temperature it is fluid or viscous during processing. Cooling as the part sets up results in different shrinkage rates for thicker versus thinner sections. This results in either external waviness or sink marks, or warpage and internal voids, as the part contracts. Flat surfaces are difficult to maintain. High speed of flow to fill the cavity of the mold is impeded going around square corners, so provision for radii and fillets is important. Attempting to flow past thin sections to fill wider sections beyond is difficult or impossible, because the flow thickens enroute, like plaque forming along walls of an artery. Even if both of these ills are avoided, the final result may still contain areas of high shear stresses invisible to the eye but waiting in ambush to cause failure later under extreme conditions previously thought to be well within the material's specifications.

However, when beginning to design a product or system, it is often wiser to permit the creative mind to freewheel, especially in the initial concept phases. Undue concern for technical aspects may inhibit creativity. After this initial conceptual stage, though, when ideas have been set down graphically and as the design begins to crystallize, it is wise to look carefully and explore the total aspects of the project with an eye to the most felicitous marriage of material and process.

Preliminary consideration of candidate materials, processes and tooling factors, configuration, thicknesses in section, ribs, bosses, holes, surface characteristics, color, graphics, decoration, and assembly methods will begin to impose some discipline on the part design as it evolves. In the middle and latter phases of the design cycle, two or three concepts should make their validity apparent to all involved. With luck, one will then be so obviously right that it will stand out without question.

When seeking the best choice among candidate materials the possibilities are numerous, and as time goes by they will become even more so. In narrowing the options, it is best to seek advice from engineering technical service or laboratory personnel of resin companies at the main or well-staffed regional offices. When choosing a molder, seek out those with experience close to the materials, end-use categories and sizes of your project. When prototype products are made, test them under the worst possible conditions, and if possible, simulating a longer time frame than the product is designed to withstand in normal service.

107

PRODUCT DESIGN, INDUSTRIAL DESIGN ASPECTS

Jordan I. Rotheiser

Jordan Rotheiser Design, 1725 McGovern Street, Highland Park, IL 60035

The industrial design aspects of plastic product design involve two distinctly different points of view. One concerns what the industrial designer needs to know about designing with plastics. The other involves what the design engineer needs to know about the industrial design function and other appearance elements in regard to plastics product design.

107.1 PLASTICS DESIGN FOR THE INDUSTRIAL DESIGNER

The industrual design profession has embraced plastics with enthusiasm for several reasons. First, plastics provide enormous freedom of form compared with traditional materials of design. They also permit part production that is faster and more consistent, and they do it all at a fraction of the cost for making nonplastic parts. This low product cost does not stem from the fact that plastics materials are low in cost; on a per-pound basis, they are actually more costly than many competing materials. But the processability and relatively low density of plastics (which translates into lower costs per cubic inch) gives them a big economic advantage. The net result is that the industrial designer can now achieve quality products at "disposable" price levels.

Colorability is another reason industrial designers select plastics for many products. Molding color into a part eliminates finishing and painting operations, thus reducing costs. Beyond cost, integral color also masks the nicks, chips and scratches that impair appearance during the life of the product. Color effects are almost limitless. Transparent, translucent, pearlescent, fluorescent, or marbleized colors are readily available for use in plastics.

Another design appeal of plastics is their ability to accept topical decoration. A permanently affixed multicolor label can be provided by means of heat transfer or hot stamping, for example. When a more secure surface is required, ie, for a keycap, the label can be placed directly in the mold and subsequently molded into the part as it is formed.

Two-color molding is another option. This is a process in which a part is first molded in one color and then (without demolding) a second cavity is placed over the part permitting a second color to be molded over a predetermined portion of it.

Other in-mold decoration processes are available, including a selection of patterns that can be etched into the mold surface. There are also twelve standard mold finishes, ranging from a very high polish usually reserved for lenses to a medium matte finish adequate to mask minor sink marks. A molded sample of these finishes is available from The Society of the Plastics Industry.

107.1.1 Choosing the Material

Nine thousand or so compounds exist in the plastics spectrum. Most of them however, differ only slightly from one another and are of little specific concern to the industrial designer. It is best to leave the exact material selection to the design engineer, who is far more knowledgeable in such areas as structure, strength, chemical resistance, and other essentials.

Some basic knowledge of plastics materials is important, however, if the industrial designer is to avoid making embarrassing proposals. Basically, there are two types of plastics: thermosets and thermoplastics. The difference between them is that thermosets undergo a chemical change when heated and cannot be melted for return to flowable form. Thermoplastics can be remelted and reused. Defective or used thermoplastic parts can be reground, mixed in with virgin material, and reused.

However, thermoplastics do not return completely to their original form. Each time they are remelted, they become weaker and more brittle. Therefore, there are limits on the percentage of regrind that is acceptable. It is important for industrial designers to know that such degradation also affects a plastics color. Darkening or yellowing results and that bright red you approved may become a maroon in production. Similar discoloration can take place even with virgin material if too much heat is applied during molding.

Recent developments in the field of injection molding of thermosets have refocused attention on these materials. For years this highly efficient manufacturing process was closed to thermosets because would they solidify in the molding machine. Because thermosets cannot be remelted, the injection barrel would have to be disassembled to remove the residue. Special techniques and devices have been perfected that reduce this problem to the level of insignificance.

The list of plastics that fall into the thermoset category is fairly short. One can be reasonably safe in assuming that anything not listed below is a thermoplastic. (Be aware, however, that there are also thermoplastic forms of polyester and polyurethane). Thermosets include alkyd, allyl (also known as diallyl phthalate), amino (urea and melamine), epoxy, phenolic, polyester, polyurethane, and silicone.

Alkyds and allyls are usually selected for demanding electrical and chemically resistant engineering applications. Epoxies, whose major use is in high-performance reinforced plastic parts, are often used in adhesives and coatings, although they are occasionally also used for electrical parts and prototype molds. Silicones have the unusual characteristic of being very slippery and are also used as a synthetic rubber. Phenolic, the largest-volume thermoset, is of limited interest because of poor colorability—it is principally available in black or near-black. That, for all practical purposes, leaves the aminos, polyesters, and polyurethanes as materials of design.

Aminos have the highest colorability of all of the thermosets. They are well suited to products like ashtrays or dinnerware, both of which must withstand elevated temperatures.

Polyesters are the workhorses of the reinforced plastics industry. They tend to be used mainly for large parts which would be impossible, or at least prohibitively expensive, to make by some other means.

Polyurethane, largely used as rigid or flexible foam, is also used for large parts, principally because it is the primary RIM material. RIM, short for reaction injection molding is a process well suited to such parts.

The world of thermoplastics is far broader, covering an enormous spectrum of property options. Some combinations of these properties are available only through plastics. For example, vinyl can provide transparency with flexibility; butyrate provides transparency with ductility. All the thermoplastics can be integrally colored and molded in highly complex one-piece shapes. Acrylic and polycarbonate offer high transparency rates, surpassing glass by two or three percent, and are shatter-resistant as well.

The designer's problem is to select from this bewildering array the plastic most suited to the design project at hand.

Some thermoplastics are priced too high to be considered for anything but specialized high-performance engineering applications. This group includes polyamide–imide, polyimide, polyphenylene sulfide, polysulfone, polyethersulfone, polyetheretherketone, and the fluoroplastics.

However, price per pound is not the sole determinant (or even the most important) in materials selection. Sometimes a plastic priced at 50 cents per pound can perform as adequately for a given application as one selling for twice as much. In such cases, selection of the higher-priced material would constitute over-engineering.

There are many ways to arrive at an intelligent and informed compromise that provides an optimum balance of cost and performance. One that designers might consider is the "cost-ladder" approach. It is based on the premise that, generally speaking, plastics generally increase in cost as they improve in physical properties. Therefore, the process of selection begins with the evaluation of the lowest cost material in the "ladder." If that material does not meet requirements, one goes up the ladder until the requirements are met or the product is priced out of the marketplace.

In the area of transparent rigid materials, the conventional "rungs" on the ladder include polystyrene, vinyl, PET, cellulosics, (acetate, butyrate and propionate), acrylic and polycarbonate. Polystyrene would be the least costly, and would therefore form the bottom rung of the ladder. Compared to higher-end clear materials, it would be slightly less transparent, more brittle, less scratch resistant and have poorer weathering resistance.

If any of these properties eliminates polystyrene, the next step up the ladder might be a copolymer of polystyrene and acrylonitrile known as SAN. If this also does not suffice, the next-highest material on the price-performance scale is evaluated.

In scaling the cost ladder, there are many factors to consider other than price and in-use performance reliability. For example, appliance housings (such as TV cabinets) often require a higher gloss surface and more rigidity than can be provided by a polyolefin material. Therefore, polystyrene may come to the fore. However, the same brittleness that detracts from its use as a transparent material eliminates it as a quality housing material. For this reason, polystyrene is often combined with acrylonitrile or with modified rubber to modify brittleness. Higher up on the ladder is the terpolymer ABS (acrylonitrile–butadiene–styrene). If greater strength is called for, one can move up further to poly(phenylene oxide) or poly(phenylene ether). They offer outstanding strength, toughness and dimensional stability, but suffer from some of the same chemical resistance problems (oils, greases, gasoline) as polycarbonate, which provides the practical ultimate in strength for housing applications.

Other materials factors will affect the designer's choice of materials include fiber reinforcement and structural foam.

Glass or other high strength fibers (and some minerals) are often added to polymeric materials to improve strength. While they may be necessary from an engineering standpoint, they have a negative effect on surface appearance. Parts made from reinforced plastics cannot easily be integrally colored, and generally must be finished and painted.

The result is the same when structural foam is used, but the reason is different. In the case of structural foam, a swirl pattern appears on the surface, giving the part a marbleized look. This effect can be minimized and in some cases eliminated with such newer processing procedures as counter pressure molding.

The use of structural foam extends the base material, thus providing a lower initial cost. An improvement in stiffness to weight ratio also is obtained, but physical properties of the material are reduced. Also, the cost savings are somewhat reduced by the longer molding cycle (the mold cannot be opened until the foaming action is completed). Foam parts require thicker walls. Traditionally, the optimum wall thickness was around 0.250 (6.4 mm) inches. The recent technique of thinwalling reduces wall thickness to 0.150 to 0.185 (0.4–0.5 mm) in., without sacrifice of performance.

107.1.2 Choosing the Manufacturing Process

The processes which industrial designers encounter most frequently are injection molding, thermoforming, blow molding or one of the reinforced plastics methods.

Most industrial designers are well aware of the capabilities of injection molding. Thermoforming, however, is an often overlooked alternative. This is due principally to the obsolete image of thermoforming as a technique limited to very simple shapes with large radiused corners, mainly for the packaging industry. However, recent improvements have greatly increased thermoforming's performance capabilities. It can now be considered for numerous lower value or large-part applications in which the cost of injection molds would be difficult to justify.

Blow molding is, of course, primarily used for bottles, although fuel tanks, kayaks and other large parts can also be so manufactured. An alternative is rotational molding, which can also be used for large parts.

All of these processes require something called draft. This is a detail of design in which any wall perpendicular to a parting line is tapered slightly (1 degree per side in most cases) in order to permit the molded part to be ejected easily—either out of the cavity or off the core. The draft angle is so small, that industrial designers often disregard it. However, 1 degree per side results in change of 0.035 in. per inch of depth or nearly 3/8 of an in. (9.5 mm) over a 10-in. (254 mm) length—and that can create severe problems. This is particularly true of the injection molding, for which a mold for a 10-in. (254 mm) part would be very costly.

Another problem associated with injection molding stems from the very freedom that makes it so appealing to begin with: its ability to make complex parts with recesses, openings and undercuts molded in—all in a single shot. The fact that these elements are possible does not mean they should be used with abandon. Often they require elaborate mechanisms to pull core or cavity sections to the side before the mold can be opened. These are costly, and they usually increase the cost of the molded product.

Parting lines are another element a competent designer should understand and prepare for. These lines occur wherever two separate steel sections come together. When a core meets a cavity, the parting line will result in a sharp edge. In order to eliminate that sharp edge, the parting line should be moved up into the cavity, at a sharp increase in mold cost. (It could still leave a seam line where the two pieces of steel meet).

The least expensive parting line is one that is flat. That is because it is easiest to make two flat pieces of steel match perfectly to avoid the possibility of molten plastic working its way into the gap (known as flash). If there is a step in the parting line, it becomes more difficult to match the steel pieces and therefore more expensive to make. If the parting line must be curved or irregular, it is obviously still more expensive.

Another form of parting line occurs when split cavity is required. A seam line will occur where the split takes place. A similar line will occur around the edge of a side action unless it is simply one where a hole is created. In that case, a sharp edge will be formed on the outside corner of the hole opening.

Gates are required to allow the molten material to flow into the mold. These cause considerable consternation among industrial designers because they are inevitably unsightly, more so with more complex shapes and harder-flow plastics that require larger gates. Be particularly aware of tunnel or submarine gates. These are popular because they require no further work in trimming the gate. However they can leave an unsightly mark on the side of the part, particularly on a polished surface. Gate location should be discussed with the engineer so that he/she can make an effort to design the part in such a fashion that the gate can be placed in the least obtrusive location consistent with good manufacturing practice.

One of the phenomena least understood by industrial designers and engineers is that of sink marks on moldment surfaces. These are caused by nonuniform wall thicknesses and internal ribs. Industrial designers can alleviate this problem by not creating contours that call for wall thickness variations exceeding 25%.

Certain contours can be of help to the engineer. For example, placing a gentle arc on a large surface can be used to hide warpage. Be careful, however, to avoid costly varying radii arcs when they are not truly necessary. Frequently they are not, since a single radius

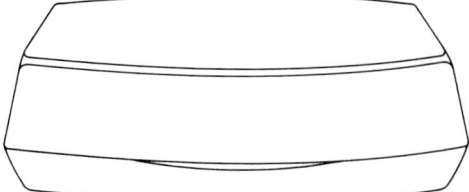

Figure 107.1. Distortion when two mating parts are in contact.

curve placed over a varying radius often reveals only an infinitesimal difference not visible to the average eye. Of particular concern are arcs in two directions. These can be a nightmare for both engineer and moldmaker.

107.1.3 Appearance Details

One of the more noticeable appearance problems associated with plastic parts has its roots in the difficulties related to molding distortion-free parts. When two mating parts have surfaces in direct contact (Fig. 107.1), the slightest distortion in either component becomes immediately obvious, giving the appearance of poor quality. This can be substantially controlled through the use of the semi-dovetail fitment shown in Figure 107.2, since each part will reinforce the other.

A generally more successful approach is to provide a visual separation of the mating elements. This can be accomplished with a slot as shown in Fig. 107.3. Variations of the slot concept include the v-groove, half v-groove or bead. They are effective, although generally less pleasing to the eye.

One other method of dealing with the problem is the skirt concept shown in Fig. 107.4. This device provides the additional advantage of permitting an inside surface of contact between the two parts while the outer edge or skirt is free to follow whatever contour is deemed desirable.

Distortion is not the only problem associated with the appearance of mating parts. Color matching is another. A reasonable color match is possible provided the same material is used for both parts. Still, potential discoloration problems exist in pigment or dye quality control, pigment mixing or dispersion, interaction with polymer additives, and changes in molding conditions.

If the parts are molded of different materials, the color-match problem becomes nearly insoluble. The identical pigment used in precisely the same proportions will result in different

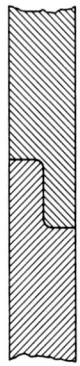

Figure 107.2. Semi-dovetail fitment.

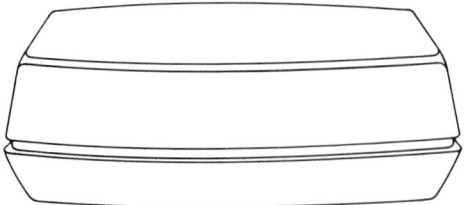

Figure 107.3. Use of a slot for visual separation of mating elements.

colors in different materials. One method of dealing with this phenomenon is to alter the surface of one of the parts. Normally, this is accomplished by etching the surface of the mold. The difference in texture will make it more difficult for the eye to distinguish the difference.

Another device is to separate the parts with a band of another color. This is usually accomplished with a finishing treatment.

107.1.4 Finishing and Decorating

The various processes used for achieving desired surface effects are described in detail in the chapter on Decorating. The industrial designer must have a thorough knowledge of the capabilities of each process, and what decorative effects can be achieved with each. Some thumbnail notes follow:

Hot Stamping. Much finer detail is now available. Serif type faces and type as small as 6 point (sans-serif) have been successfully applied in this manner. Another effective technique is to mold a raised surface into the part and hot stamp the top surface. Sharp, crisp lettering can be achieved with this method.

Heat Transfer. Multicolor effects are readily achieved with this process, since printing is done separately on film, and then transferred to the part. One should not attempt this in recesses unless the recess is shallow and there is plenty of clearance for the heated die.

Vacuum Metallizing. Generally regarded as a low-cost method of obtaining a metallized finish, this process does suffer from abrasion resistance and longevity problems. This process has been replaced by electroplating for applications requiring durability. But it is still the primary method of achieving gold or chromelike finishes for toys and cosmetics containers. It is also popular for second-surface (applied to the reverse side of the protected surface of a clear part). Available finishes include chrome, silver, gold, copper, and zinc. When the part's surface is left rough, a matte finish will result. It is important to remember that other metallic colors besides gold and chrome are available. Among them are bright and satin

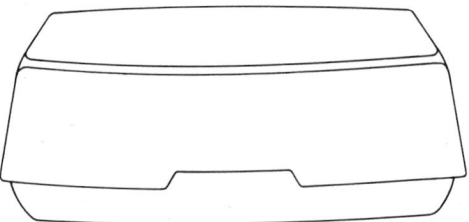

Figure 107.4. Skirt concept.

finish nickel, satin chrome, bright or oxidized copper and brass, black chrome, nickel and silver, aluminum, and nickel and gold.

Painting. The desire to avoid this costly process with its associated health and safety problems has been one of the motivating forces behind the switch from metal to plastic parts. Nonetheless, paint is necessary in many cases, particularly when heavily glass filled parts, structural foam parts or tose of different materials are involved. Spray painting is the most common form used on industrial parts. Because this process is highly skill-dependent, quality control has long been a problem. However, the introduction of programmable robotic sprayers has done much to alleviate this situation. Still, integral coloring is preferred whenever feasible. To avoid the necessity of hand masking, provision must be made for metal masks. On vertical walls, a draft of five to seven degrees is required (depending on the type of mask used) and the mask should extend 1/16th of an inch (1.6 mm) beyond the edge of the radius. Inside and outside radii less than 1/16th (1.6 mm) of an in. (more is better) will thin out in painting.

Fill-in painting is a method in which a recess is left in the surface of the part, the paint is placed in that recess, and the excess is wiped off. Care must be taken in the design of parts which use this process to avoid creating areas which are impossible to wipe.

Printing. Outside the packaging industry, screen printing is the usual technique used on plastic parts; although pad printing is a strong contender. Printing still seems to provide the highest quality of decoration.

In-mold Decoration. This term is used both to describe designs that are etched or engraved in the mold surface and the process of inserting a printed film into the mold, to be produced as an integral component of the finished part.

Etched surfaces can be drawn both parallel and perpendicular to the parting line. However, be alert to the fact that parallel to the parting line additional draft is required. A wide selection of patterns is available and new ones can be readily created.

Engraved designs and lettering normally have greater depth and fine detail. Parallel to the parting line, a side action (to clear the engraving) will be required in most cases. Therefore, they should be used only when absolutely necessary. Recessed letters and designs are to be avoided whenever possible. They collect dirt and they are costly to put into the tooling. Raised letters are less expensive to make and to maintain. In either case, sharp points, such as those found in the letters N, M, and W, are prone to breaking out when subjected to molding pressures over a period of time. Hence, it is wise to place designs and lettering in an insert in the mold. This will create an outline around the lettering (which can be incorporated into the design) and it will also make repairs and revisions far less costly. Generally, one should avoid the use of serif typefaces unless the letters are very large indeed. Artwork is prepared for engraving in the same manner as for printing. Regardless of which finishing method is used, it is important to consider its design requirements from the beginning. Many an engineer, preoccupied with the mechanical requirements of the part, postpones consideration of this aspect until the very end of the project, only to discover that major design revisions are necessary in order to meet appearance requirements.

107.2 INDUSTRIAL DESIGN EFFECTS ON THE DESIGN ENGINEER

The industrial design function involves a great deal more than appearance design. The designer is often called on to create the very concept of the product. In doing so, he or she will consider the utility, cost, innovation and human engineering aspects of the proposed product—in short, its basic appeal to the end-user.

It is beyond the scope of this chapter to examine these elements. Therefore the appearance aspects of industrial design plus a brief discussion of some of the problems relating to human engineering are presented here.

The appearance of the product cannot be treated lightly. Industry has learned that eye appeal translates quickly into buy appeal. Since the average purchaser rarely spends more than a few moments considering his or her purchase, the initial impact of the product is critical.

Appearance means something different to each discipline involved in the development of a new product. Industrial designers, design engineers, tool builders, and processors are each affected in a different way. Yet the cooperation of all is necessary in order to achieve the best possible appearance.

Concern for appearance generally translates to more work for the design engineer. It would certainly be far easier to construct a rectangular box or a drafted cylinder with a few appropriately placed screws. But a world full of rectangular boxes and drafted cylinders would be dreary indeed.

The principal problems associated with appearance factors are the development of contoured housings, space limitations, and assembly devices. Contoured housings are far more difficult to calculate than those with regular dimensions. In some cases, it is practically impossible to achieve complex shapes without creating wall thickness variations that cause sink marks and warpage.

The design engineer does have some allies in this struggle. One of them is the pattern-maker, who can often capture the contours based on an appearance model and a drawing—with many of the more difficult center point locations left unresolved. The pattern he or she creates (normally two to four times scale) is usually traced with a milling machine equipped for this purpose; the reduction in size eliminates any rough blends. For small parts, electrical discharge machining or hobbing may be used.

In some cases, a cast core and cavity may be called for. This process which a pattern of the same size as the finished product, with allowances made for both plastic and metal shrinkages. For products without tight tolerances, the appearance model can be used without going through the pattern phase. In this case, an additional coating is added to the model in order to account for the shrinkages. A slurry composed of refractory, binder and gel is poured over the pattern to create a mold from which the casting will be made. A coating equivalent to the wall thickness is then added so that the reverse side of the mold can be cast.

Although the computations and mold costs involved in development of the product's contours may be considerable, these problems are normally not as demanding of the design engineer's talents as those of space limitations and assembly methods.

107.2.1   Human Engineering

While the design engineer usually regards the problems of space limitations as being appearance related, they are most often the outcome of the industrial designer's concern for human engineering, or the proper relationship of the product to the human body. For example, coffee-cup handles must be comfortable to the hand; portable typewriters must be small enough to be carried by most people, etc.

The term "most people" is the point of dispute in many cases. Studies have been prepared that profile the physical characteristics of the "general" population. Although these studies are sometimes questioned, they do provide the basic anthropometric data upon which most designs are based.

Space limitation problems often derive from the need to accommodate those people who are physically in the extreme upper or lower percentiles. However, the costs of serving the extreme percentiles may be great enough to price a product out of the market. Thus, some people will forever be forced to drive automobiles equipped with blocks on the pedals while others will have to drive in a slouched position.

Clearly, human engineering requirements often dictate the size, weight, and form of a product. With the exception of the electronics industry (where miniaturization is highly advanced), this translates to smaller, lighter and contoured products as the industrial designer works from the "outside in" in the execution of his profession. Often this results in conflict with the customer company's designer engineer, who works from the "inside out" in the development of a product. Compromise becomes an important factor as he or she and others bring their requirements to bear on the appearance of the product.

107.2.2 The Molder's Role

Concern with appearance factors are often translated directly into problems for the tool builder and the part molder. In additioin to the contour difficulties previously discussed, the tool builder must be concerned with such critical factors as gate locations, mold cooling, venting and ejection devices. Furthermore, he or she may have to add side actions to achieve some part details. It is important to note that the tool builder is no mind reader. Appearance surfaces that must be free of gate, ejector or other surface marks must be so specified on the design drawing.

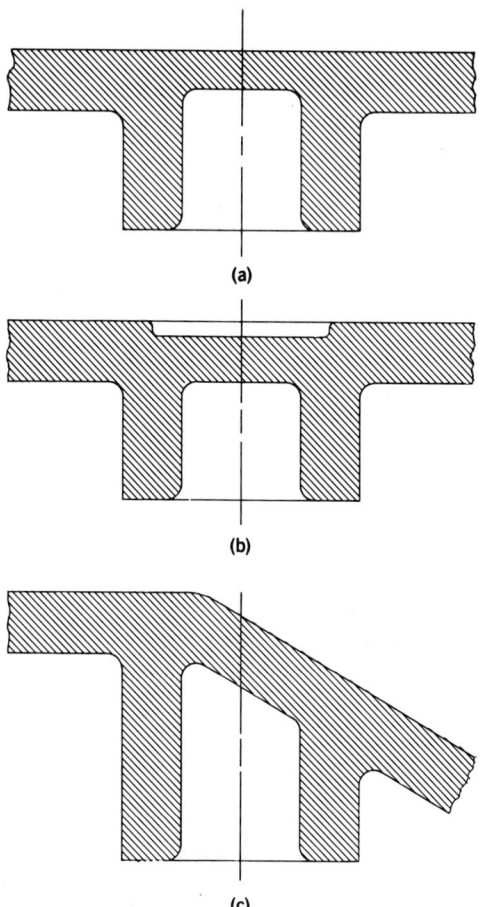

(a)

(b)

(c)

Figure 107.5. Reducing sink by recessing the core, recessing the surface or locating the boss point in the contour where a sharp corner occurs.

Normally, the molder is also involved with these decisions, since it is the molder who must use the mold. The molder must also deal with other appearance factors, such as, flash, sinks and distortion, that are associated with the temperature of the melt and the rate of cooling. If the melt temperature is lowered in order to control these conditions, the cavity may not be filled completely, an equally undesirable result. Once ejected, parts must be handled carefully to avoid scuffing or other surface damage.

Among the most persistent appearance problems a molder must deal with is sink marks which result from internal ribs, nonuniform wall thicknesses and internal bosses. Proper part design can usually eliminate the first two potentials for sink. But internal bosses are much more difficult to deal with. Since the boss must be strong enough to perform its function, one must live with the resultant sink or find another means of assembly.

There are, however, methods of reducing the amount of sink. First, the outer wall thickness can be reduced at the base of the boss hole. This can be accomplished by recessing the core or by recessing the surface (Fig. 107.5). Another method of reducing the amount of sink resulting from an internal boss is to strategically locate the boss at a point in the contour where a sharp corner occurs, also shown in Figure 107.5.

If the boss cannot be adequately reduced with these methods, there is the possibility of hiding it. This can be accomplished by using a matte or etched part surface above the boss. If a label is to be used, it may be possible to locate the boss under it.

Finally, there is the possibility of using structural foam. This process demonstrates much less tendency to sink. In some cases, the boss wall may be 90% of the outer wall without the occurrence of sink.

108

PRODUCT DESIGN, STRUCTURAL

John A. Jones

*Borg Warner Chemicals Co., Technical Center, P.O. Box 68,
Washington, WV 26181*

There are many structural properties that may govern the use of a particular material in a particular part. These properties include, among other things, density, refractive index, coefficient of friction and abrasion resistance, hardness, dielectric constant, strength and dissipation factor, electrical and thermal conductivity; and permeability to gases.

It is well known that mechanical loads on a structure induce stresses within the material comprising it. It is also well known that the magnitudes of these stresses depends on many factors, including forces, angle of loads, the rate and point of application of each load, the geometry of the structure, and the manner in which that structure is supported. The behavior of the material in response to these induced stresses determines the performance of the structure. If the material comprising a structure fails, then more often than not the structure will also fail to perform its intended function.

This discussion describes the following types of material behavior that must be examined and evaluated in any structural design project involving plastics:

- Short-term stress–strain behavior.
- Long-term viscoelastic behavior.
- Fatigue.
- Thermal expansion and contraction.
- Impact resistance.

Obviously, several of these are important to designers of structures of almost any material. But others are of special interest of those who design plastic products.

108.1 SHORT-TERM STRESS–STRAIN BEHAVIOR

Stress is defined as the force on a material divided by the area over which it acts. Consequently, the typical units of stress quantification are megapascals (MPa) or pounds per square inch (psi). If the area over which the force acts changes significantly because the material deforms, one can use that area to calculate stress (engineering stress) or use the prevailing area. The engineering stress concept used here is the more often-used.

Strain is defined as the deformation of a material divided by a corresponding undeformed dimension. The units of strain are meters per meter or inches per inch. Since strain is often regarded as dimensionless, strain measurements are typically expressed either as a percentage deformation or in microstrain units. One microstrain is defined as 10^{-6} meters per meter or inches per inch.

The relationship between stress and strain for a material constitutes its stress–strain behavior. For most materials this behavior is temperature-dependent, but for thermoplastics it is also time- and strain rate-dependent.

Thermoplastics are viscoelastic materials; they respond to induced stress by two mechanisms: viscous flow and elastic deformation. Viscous flow ultimately dissipates the applied mechanical energy as frictional heat and results in permanent material deformation. Elastic deformation stores the applied mechanical energy as completely recoverable material deformation. The extent to which one or the other of these mechanisms dominates the overall response of the material is determined by the temperature and by the duration and magnitude of the stress or strain. The higher the temperature, the most freedom of movement of the individual polymer molecules that comprise the thermoplastic, and the more easily viscous flow can occur.

Likewise, the longer the duration of material stress or strain, the more time for viscous flow to occur. Finally, the greater the material stress or strain, the greater the likelihood of viscous flow and significant permanent deformation. For example, when a thermoplastic part is loaded or deformed beyond a certain point, the material comprising it yields. Conversely, as the temperature or the duration or magnitude of material stress or strain decreases, viscous flow becomes less likely and less significant as a contributor to the overall response of the material; and the essentially instantaneous elastic deformation mechanism becomes predominant.

Consequently, changing the temperature or the strain rate of a thermoplastic may have a considerable effect on its observed stress–strain behavior. At lower temperatures or higher strain rates, the stress–strain curve of a thermoplastic may exhibit a steeper initial slope and a higher yield stress. In the extreme, the stress–strain curve may show the minor deviation from initial linearity and the lower failure strain characteristic of a brittle material. At higher temperatures or lower strain rates, the stress–strain curve of the same material may exhibit a more gradual initial slope and a lower yield stress, as well as the drastic deviation from initial linearity and the higher failure stain characteristic of a ductile material.

There are a number of different modes of stress–strain that must be taken into account by the structural designer, as follows:

Tensile Stress–Strain. Data are usually generated by clamping the two ends of a test specimen into the grips of a testing machine and then moving these grips apart at a specified rate. A displacement transducer called an extensometer measures the elongation of the reduced cross section gauge area of the specimen, while a load cell measures the tensile force exerted by the specimen on the testing machine. Stress and strain are computed from the measured load and elongation and are then plotted as a tensile stress-versus-strain curve for the material at the temperature and strain rate employed for the test.

A tensile stress–strain curve typical of many ductile plastics is shown in Fig. 108.1. As strain increases, stress initially increases approximately proportionately (from point 0 to point A). For this reason, point A is called the proportional limit of the material. From point 0 to point B, the behavior of the material is purely elastic; but beyond point B, the material exhibits an increasing degree of permanent deformation. Consequently, point B is called the elastic limit of the material. The first point of zero slope on the curve (point C) is identified with material yielding and so its coordinates are called the yield strain and stress (strength) of the material. The yield strain and stress usually decrease as temperature increases or as strain rate decreases. The final point on the curve (point D) corresponds to specimen fracture. This represents the maximum elongation of the material specimen; its coordinates are called the ultimate, or failure strain and stress. Ultimate elongation usually decreases as temperature decreases or as strain rate increases.

Brittle materials exhibit tensile stress–strain behavior different from that illustrated in Figure 108.1. Specimens of such materials fracture without appreciable material yielding. Thus, the tensile stress–strain curves of brittle materials often show relatively little deviation from the initial linearity, relatively low strain at failure, and no point of zero slope.

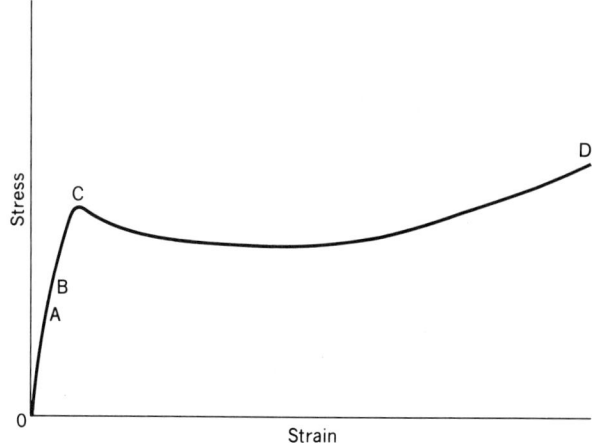

Figure 108.1. Tensile stress–strain curve typical of many ductile plastics.

Different materials may exhibit significantly different tensile stress–strain behavior at the same temperature and strain rate. Figure 108.2 illustrates this fact by presenting the room-temperature tensile stress–strain curves for a number of high-performance materials, including ABS, acetal, nylon, and polycarbonate. Figure 108.3 illustrates the temperature effect by presenting the tensile stress–strain curves of each material, determined at an elongation rate of 5 mm (0.2 in) per minute. Figure 108.4 illustrates the strain rate effect by presenting the tensile stress–strain curves of a typical ABS at elongation rates of 0.5, 5, and 51 mm (0.02, 0.2, and 2.0 in.) per minute, each determined at 23°C (73°F).

Tensile stress–strain data obtained per ASTM D638 for several plastics at room temperature and a 5 mm (0.2 inch) per minute elongation rate are shown in Table 108.1.

Flexural Stress–Strain. Data are generated by supporting a test specimen of uniform cross-section across a known span and deflecting the center of the specimen at a specified rate. A load cell measures the force required to bend the specimen as a function of total deflection.

Real differences between the tensile and the compressive yield stresses of a material may cause the stress distribution within the test specimen to become very asymmetric at high

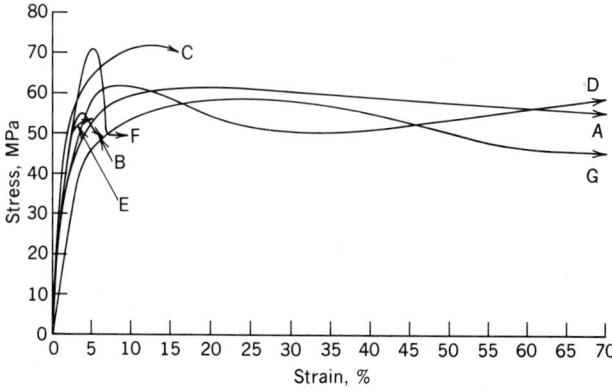

Figure 108.2. Room-temperature tensile stress–strain curves for seven thermoplastics at an elongation rate of 5 mm (0.2 in.) per minute. A, Celcon M25 acetal copolymer; B, Cycolac DH ABS; C, Delrin acetal homopolymer; D, Lexan 141 polycarbonate; E, Prevex PQA phenylene ether copolymer; F, Udel polysulfone; and G, Zytel 101 nylon (2.5% moisture content)

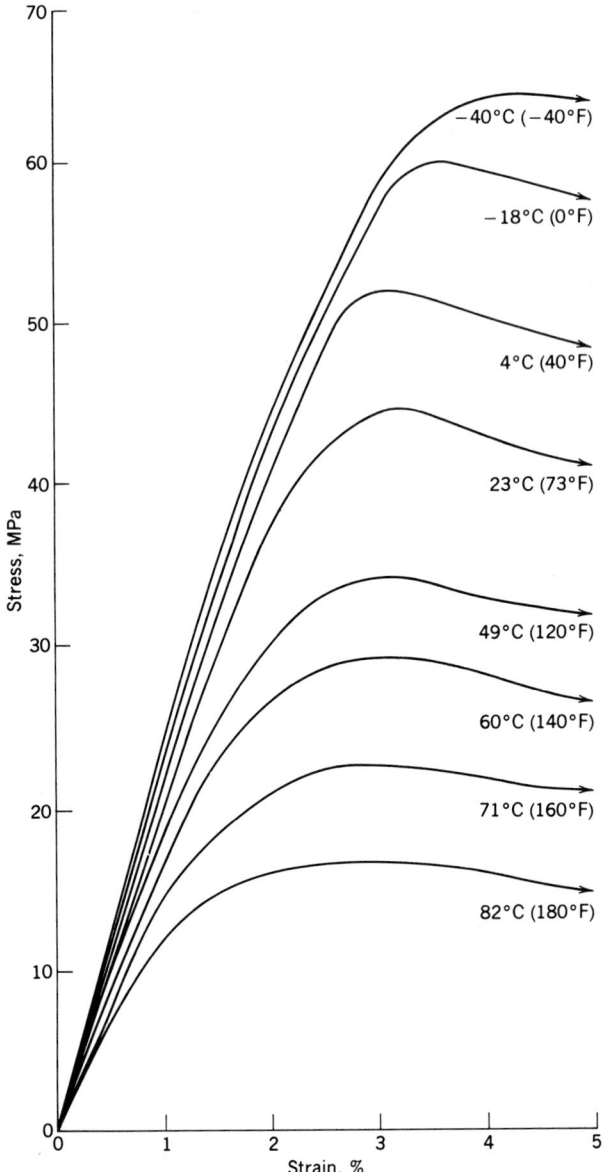

Figure 108.3.   Tensile stress–strain curves for a typical ABS at eight temperatures, at an elongation rate of 5 mm (0.2 in.) per minute.

strain levels. This causes the neutral axis to move from the center of the specimen toward the surface which is in compression. This effect, along with specimen anisotropy due to processing, may cause the shape of the stress–strain curve obtained in flexure to differ significantly from that of the tensile stress–strain curve.

Flexural stress–strain data obtained per ASTM D-790 for several plastics at room temperature and a 1.3 mm (0.05 in.) per minute deflection rate are shown in Table 108.2.

Compressive Stress–Strain.   Data are generated by placing a test specimen between the two flat, parallel faces of a testing machine and then moving these faces together at a specified

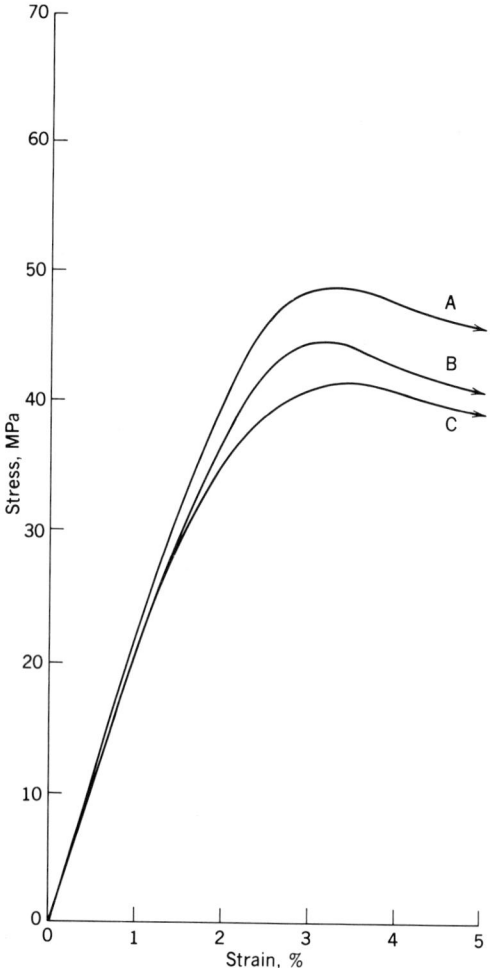

Figure 108.4.   Tensile stress–strain curves for ABS at three elongation rates at 23°C (73°F). A, 2.0 in./min (51 mm); B, 0.2 in./min (0.5 mm); 0.2 in./min (0.5 mm).

rate. A displacement transducer may be used to measure the compression of the specimen, while a load cell measures the compressive force exerted by the specimen on the testing machine. Stress and strain are computed from the measured load and compression, and these are plotted as a compressive stress-versus-strain curve for the material at the temperature and strain rate employed for the test.

In general, the compressive strength of a nonreinforced plastic or a mat-based plastic laminate is usually greater than its tensile strength. The compressive strength of a unidirectional fiber-reinforced plastic laminate is usually slightly lower than its tensile strength. Room-temperature compressive stress–strain data obtained per ASTM D695 for several plastics are shown in Table 108.2.

Shear Stress–Strain.   The shear mode involves the application of a load to a material specimen in such a way that cubic volume elements of the material comprising the specimen become distorted, their volume remaining constant, but with opposite faces sliding sideways with respect to each other. Shear deformation occurs in structural elements subjected to torsional loads and in short beams subjected to transverse loads.

TABLE 108.1. Room Temperature Tensile Stress–Strain Data for Several Plastics and Some Other Materials of Construction

Generic Material Type	Trademark	Grade	Modulus, MPa	Yield Stress, MPa	Elongation at Yield, %	Elongation at Break, %
ABS	Cycolac	DH	2,600	52	2.5	25–75
		GSM	2,200	43	2.5	25–75
		KJB	2,200	40	2.5	25–75
		L	1,800	34	3.3	25–75
Acetal copolymer	Celcon	M25	2,800	61	12	75
		M90	2,800	61	12	60
Acetal homopolymer	Delrin	100	3,100	68.9	12	75
		500	3,100	68.9	12	40
		900	3,100	68.9	12	25
Acrylic	Plexiglas	V052/045	2,960	72		5.4
		MI-7	2,239	48		
		DR	1,720	38	5.0	35
Nylon (DAM) (0.2% moisture)	Zytel	101		82.7	5	60
		ST 801				60
		158 L		60.7	7	150
		211		51.0	20	290
Nylon (50% RH) (2.5% moisture)	Zytel	101		58.6	25	>300
		ST 801				210
		158 L		51.0	40	>300
		211		40.7	30	285
Phenolic	Durez	29053[a]	19,310	55.2[b]		0.29
		152[a]	10,340	58.6[b]		0.57
		18441[a]	7,590	48[b]		0.63
Polycarbonate	Lexan	141	2,380	62	6–8	110
		940	2,240	62	6–8	90
Polyethylene	Dow	08064N		30		900
		10062N		29		900
		04052N		24		900
		08035N		14		500
Phenylene ether copolymer	Prevex	PQA	2,500	58	4–6	50–100
		VKA	2,500	55	4–6	50–100
Polypropylene	Pro-fax	6523	1,400	35.5	12	
		7523	1,200	27.3	13	<300
		8523	830	20.0	6.3	
Polystyrene	Fostarene	50	3,100	52		2.5
	Hostyren	360	2,070	31		30
		760	2,070	25		60
		840	1,930	25		50
Polysulfone	Udel	P-1700	2,482	70.3	5–6	50–100
		P-1710	2,482	70.3	5–6	50–100
		P-1720	2,482	68.9	5–6	50–100
Steel, structural ASTM A7-61T			200,000	230		
Brass, naval			100,000	170–340		
Aluminum, wrought 2014-T6			73,000	410		
Pine (southern long-leaf)			13,700			
Oak (white)			11,200			

[a]Injection molded specimens.

[b]At break.

TABLE 108.2. Room Temperature Flexural and Compressive Stress–Strain Data for Several Plastics and Some Other Materials of Construction

Generic Material Type	Trademark	Grade	Flexural Modulus, MPa	Flexural Yield Stress, MPa	Compressive Modulus, MPa	Compressive Stress, MPa	At
ABS	Cycolac	DH	2800	90	2600		
		GSM	2300	74	2200		
		KJB	2300	69	2200	42.4/45.1	10% yield
		L	1900	59	1800		
Acetal copolymer	Celcon	M25	2590	89.6		31/110	1%/10%
		M90	2590	89.6		31/110	1%/10%
Acetal homopolymer	Delrin	100	2620	98.6	4600	35.9/124	1%/10%
		500	2830	97.2	4600	35.9/124	1%/10%
		900	2960	96.5	4600	34.5/121	1%/10%
Acrylic	Plexiglas	V052/045	3170	110		117	Maximum
		MI-7	2239	72		72	Maximum
		DR	1720	62		41	Yield
Nylon (DAM) (0.2% moisture)	Zytel	101	2827			33.8	1%
		ST 801	1689			13.1	1%
		158 L	2034			16.6	1%
		211	1034				
Nylon (50% RH) (2.5% moisture)	Zytel	101	1207				
		ST 801	862				
		158 L	1241				
		211	745				
Phenolic	Durez	29053[a]		96.5[b]		193.1	Ultimate
		152[a]		89.6[b]		206.9	Ultimate
		18441[a]		82[b]		193.1	Ultimate
Polycarbonate	Lexan	141	2340	93.0	2380	86.1	Yield
		940	2240	90.9	2240	86.1	Yield
Polyethylene	Dow	08064N	1100				
		10062N	1100				
		04052N	861				
		08035N	410				
Phenylene ether copolymer	Prevex	PQA	2500	86	2500		
		VKA	2500	94	2500		
Polypropylene	Pro-fax	6523	1750	54.4			
		7523	1295	48.2			
		8523	1065	34.5			
Polystyrene	Fostarene	50		86.2			
Polysulfone	Udel	P-1700	2689	106.2	2579	96/276	Yield/break
		P-1710					
		P-1720					
Steel, structural ASTM A7-61T						230	Yield
Aluminum, wrought 2014-T6						430	Yield
Pine (southern long-leaf) (with grain)				101[b]		58.2	Ultimate
Oak (white) (with grain)				95.8[b]		48.5	Ultimate

[a] Injection-molded specimens.
[b] At break.

Shear stress–strain data can be generated by twisting a material specimen at a specified rate while measuring the angle of twist between the ends of the specimen and the torque exerted by the specimen on the testing machine. Maximum shear stress at the surface of the specimen can be computed from the measured torque, the maximum shear strain from the measured angle of twist.

The shear modulus of a material can be determined by a static torsion test or by a dynamic test employing a torsional pendulum or an oscillatory rheometer. The maximum short-term shear stress (strength) of a material can be determined from a punch shear test. Room-temperature shear stress–strain data obtained per ASTM D732 for several plastics are shown in Table 108.3.

Stress–strain data may guide the designer in the initial selection of a material. Such data also permit a designer to specify design stresses or strains either safely within the proportional/elastic limit of the material. On the other hand, if a vessel is being designed to fail at a specified internal pressure, the designer may choose to use the tensile yield stress of the material in the design calculations.

Designers of most structures specify material stresses and strains well within the proportional/elastic limit. This practice builds in a margin of safety to accommodate the effects of improper material processing conditions and/or unforeseen loads and environmental factors. This practice also allows the designer to use design equations based on the as-

TABLE 108.3.   Room Temperature Shear Stress–Strain Data and Poisson's Ratio for Several Plastics and Some Other Materials of Construction

Generic Material Type	Trademark	Grade	Shear Modulus, MPa	Shear Stress, MPa	At	Poisson's Ratio
ABS	Cycolac	DH	960	51.0	Ultimate	0.35
		GSM	810	37.9	Ultimate	0.35
		KJB	810	32.9	Ultimate	0.35
		L	660	30.0	Ultimate	0.36
Acetal copolymer	Celcon	M25/M90	1000	53	Ultimate	0.35
Acetal	Delrin	100/500	1330	65.5	Ultimate	0.35
homopolymer		900	1330	68.9	Ultimate	0.35
Acrylic	Plexiglas	DR		44.6		
Nylon (DAM)	Zytel	101		66.2	Ultimate	0.34–0.43
(0.2% moisture)		ST 801		57.9	Ultimate	0.34–0.43
		158 L		59.3	Ultimate	0.34–0.43
		211		62.7	Ultimate	0.34–0.43
Nylon (50% RH)	Zytel	101				0.35–0.50
(2.5% moisture)		ST 801				0.35–0.50
		158 L		55.8	Ultimate	0.35–0.50
		211				0.35–0.50
Phenolic	Durez	29053[a]		82.7	Ultimate	
Polycarbonate	Lexan	141	785	41.3	Yield	0.37
				68.9	Ultimate	
Phenylene ether	Prevex	PQA		62.6	Ultimate	
copolymer		VKA		66.2	Ultimate	
Polysulfone	Udel	P-1700	917	41.4	Yield	0.37
		P-1710		62.1	Ultimate	
		P-1720				
Steel, structural ASTM A7-61T			79200	120	Yield	0.27
Brass, naval			38000	280–310	Ultimate	
Aluminum,			30000	240	Yield	0.33
wrought 2014-T6				270	Ultimate	
Pine (southern long-leaf) (with grain)				10		
Oak (white) (with grain)				13.0		

[a]Injection molded specimens.

sumptions of small deformation and purely elastic material behavior. Other properties derived from stress–strain data include modulus of elasticity and tensile strength.

Modulus of elasticity is defined as the change in material stress corresponding to a unit change in material strain, over the range in which stress is proportional to strain. The typical units of modulus of elasticity are megapascals (MPa) or pounds per square inch (psi). A given material may exhibit several different moduli of elasticity. That derived from the tensile stress–strain curve is called Young's modulus (E). The modulus of elasticity derived from the shear stress–strain behavior of a material is called the modulus of rigidity and is usually represented by the symbol G.

Modulus of elasticity is one of the two factors that determine the stiffness or rigidity of structures comprised of a material. The other is the moment of inertia of the appropriate cross section, a purely geometric property of the structure. In identical parts, the higher the modulus of elasticity of the material, the greater the rigidity; doubling the modulus of elasticity doubles the rigidity of the part. The greater the rigidity of a structure, the more force must be applied to produce a given deformation.

It is appropriate to use Young's modulus to determine the short-term rigidity of structures subjected to elongation, beinding or compression. It may be more appropriate to use the flexural modulus to determine the short-term rigidity of structures subjected to bending, particularly if the material comprising the structure is nonhomogeneous, as foamed or fiber-reinforced materials tend to be. Finally, if a reliable compressive modulus of elasticity is available, it can be used to determine short-term compressive rigidity, particularly if the material comprising a structure is fiber-reinforced. To determine the short-term torsional rigidity of a structure, one must use the shear modulus of the material from which it is made. The room-temperature moduli of elasticity for several plastics and some other materials of construction are presented in Tables 108.1–108.3.

Poisson's ratio is the ratio of lateral strain to longitudinal strain under the condition of uniform, longitudinal stress within the proportional limit of a material. An example is stretching a cylinder of material. It will elongate and reduce its diameter.

Poisson's ratio can be determined directly during a uniaxial tensile test, using mechanical transducers (extensometers, strain guages, etc) or optical techniques to measure the simultaneous longitudinal extension and lateral contraction of the gauge section of a test specimen. The ratio can have a value from 0 to 0.5. A zero value implies no lateral contraction as the material elongates; a value of 0.5 implies that the material contracts sufficiently laterally to maintain constant volume and density as it elongates. Values less than 0.5 imply that density reduction due to void formation or crystallization is occurring during elongation. As Table 108.3 shows, Poisson's ratio for many thermoplastics lies in the range 0.3–0.5.

The application of Poisson's ratio is frequently required in the design of structures that are markedly two- or three-dimensional, rather than one-dimensional like a beam. For example, it is needed to calculate the so-called plate constant for flat plates that will be subjected to bending loads in use. The higher Poisson's ratio, the greater the plate constant and the more rigid the plate.

108.2   LONG-TERM VISCOELASTIC BEHAVIOR

Two of the most important types of long-term material behavior are viscoelastic creep and stress relaxation. Whereas stress–strain behavior usually occurs in less than an hour, creep and stress relaxation may continue over the entire life of the structure: 100,000 hours or more.

Viscoelastic Creep.   When a viscoelastic material is subjected to a constant stress, it undergoes a time-dependent increase in strain. This behavior is called creep. The viscoelastic creep behavior typical of many thermoplastics is illustrated in Figs. 108.5 and 108.6. At time

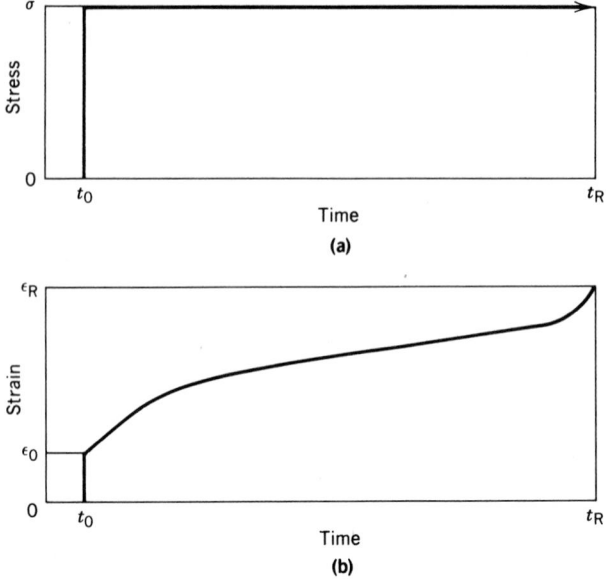

Figure 108.5. Viscoelastic creep behavior typical of many thermoplastics under long-term stress to rupture. **(a)** Input stress vs time profile. **(b)** Output strain vs time profile.

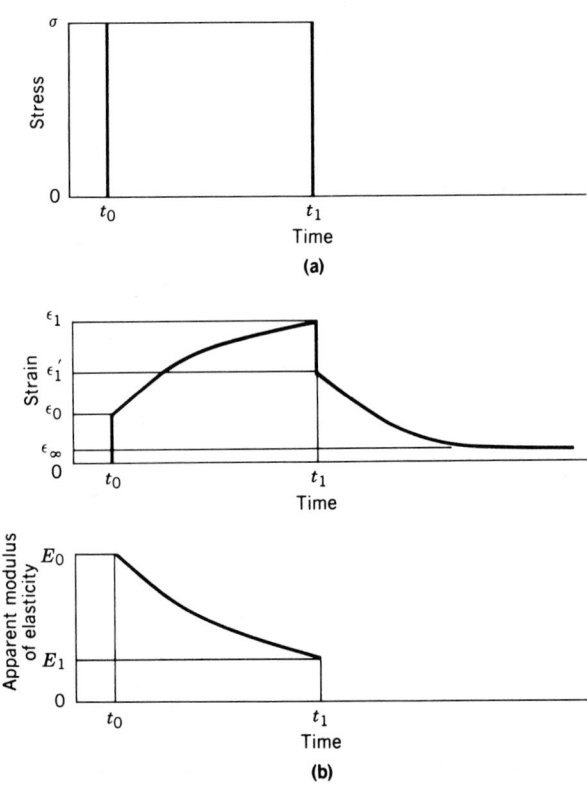

Figure 108.6. Viscoelastic creep behavior typical of many thermoplastics under shorter-term stress. **(a)** Input stress vs time profile. **(b)** Output strain vs time profile.

1506

t_0 the material is suddenly subjected to a constant stress which is maintained for a long period of time as shown in Fig. 108.5. The material responds by undergoing an immediate initial strain which increases to time t_R when it fails. In Fig. 108.6 the constant stress is maintained for a shorter time. The material undergoes an immediate initial strain at t_0 which increases to t_1 at time t_1. When the stress is removed, the material immedijately decreases in strain from ϵ_1 to ϵ_1 followed by a gradual decrease from ϵ_1, to a permanent residual strain. Although the creep behavior of a material could be measured in any mode, such experiments are most often run in tension or flexure. In the first, a test specimen is subjected to a constant tensile load and its elongation is measured as a function of time. After a sufficiently long period of time, the specimen will fracture—a phenomenon called tensile creep rupture (Fig. 10.8.7). In general, the higher the applied tensile stress, the shorter the time and the greater the total strain to specimen failure. Furthermore, as the stress level decreases, the fracture mode changes from ductile to brittle. In the second, a test specimen is subjected to a constant bending load and its deflection is measured as a function of time.

Viscoelastic creep data (Fig. 108.8) are usually presented in one of two ways. In the first, the total strain experienced by the material under the applied stress is plotted as a function of time. Families of such curves may be presented at each temperture of interest, each curve representing the creep behavior of the material at a different level of applied stress. Below a critical stress, viscoelastic materials may exhibit linear viscoelasticity; that is, the total strain at a given time is proportional to the applied stress. Above this critical stress, the creep rate becomes disproportionately faster. In the second, the apparent creep modulus is plotted as a function of time.

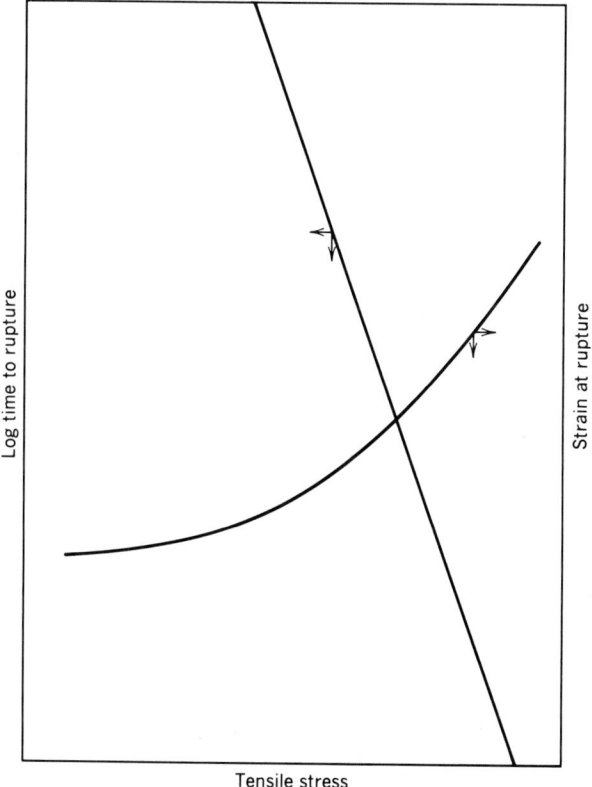

Figure 108.7.   Tensile creep rupture behavior typical of many thermoplastics.

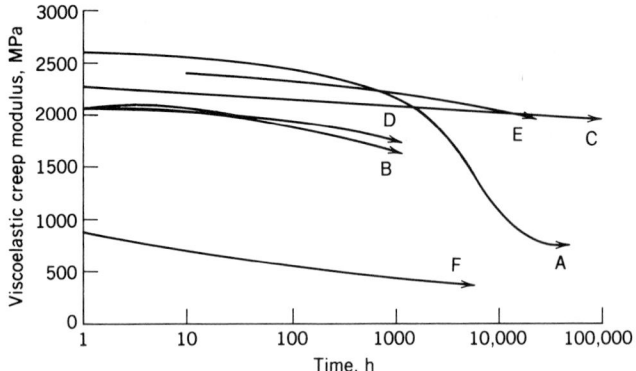

Figure 108.8. Room-temperature viscoelastic creep modulus versus time curves for six thermoplastics at constant stress levels. A, Celcon M90 acetal copolymer [3.4 MPa (500 psi)]. B, Cyclolac KJB ABS [13.8 MPa (2000 psi)]. C, Lexan 141 polycarbonate [20.7 MPa (3000 psi)]. D, Prevex PQA phenylene ether copolymer [13.1 MPa (1900 psi)]. E, Udel polysulfone [27.6 MPa (4000 psi)]. F, Zytel 101 nylon [2.5% moisture content, 13.8 MPa (2000 psi)].

The viscoelastic creep modulus may be determined at a given temperature by dividing the constant applied stress by the total strain prevailing at a particular time. Since the creep strain increases with time, the viscoelastic creep modulus must decrease with time (Fig. 108.6). Below its critical stress for linear viscoelasticity, the viscoelastic creep modulus versus time curve for a material is independent of the applied stress. In other words, the family of strain versus time curves for a material at a given temperature and several levels of applied stress may be collapsed to a single viscoelastic creep-modulus–time-curve if the highest applied stress is less than the critical value.

Figure 108.8 illustrates that different viscoelastic materials may have considerably different creep behavior at the same temperature. This figure shows the room-temperature apparent or viscoelastic creep modulus versus time curves for a number of performance resins. Figure 108.9 illustrates that a given viscoelastic material may have considerably different creep behavior at different temperatures.

Viscoelastic creep data are necessary in designing structures that must bear long-term loads. It is inappropriate to use an instantaneous modulus of elasticity to design such structures because they do not reflect the effects of creep. Viscoelastic creep modulus, on the

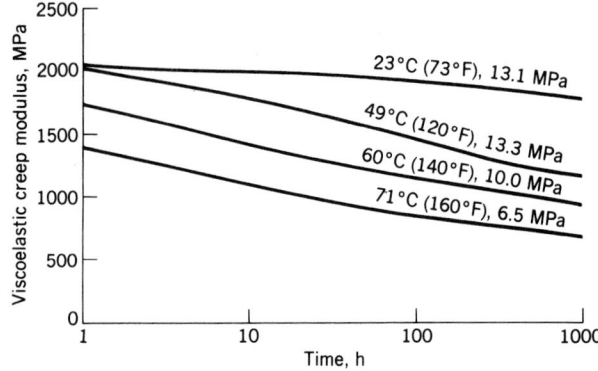

Figure 108.9. Viscoelastic creep modulus versus time curves for an engineering thermoplastic (polyphenylene ether copolymer) at four temperatures at indicated constant stress levels.

other hand, allows one to estimate the total material strain that will result from a given applied stress acting for a given time at the anticipated use temperature of the structure. The viscoelastic creep modulus is particularly useful to the designer because it may be substituted for Young's modulus to predict the long-term rigidity of loadbearing structures. Thus, creep data allow one to design a structure so that the stress within the material comprising it will remain at or below the desired level.

Viscoelastic Stress Relaxation. When a viscoelastic material is subjected to a constant strain, the stress initially induced within it decays in a time-dependent manner. This behavior is called stress relaxation. The viscoelastic stress relaxation behavior typical of many thermoplastics is illustrated in Figs. 108.10 and 108.11. The material specimen is a system to which a strain-versus-time profile is applied as an input and from which a stress-versus-time profile is obtained as an output. At time t_0 the material is subjected to a constant strain which is maintained for a long period of time, Figure 108.10. An immediate initial stress t_0 gradually approaches zero as time passes. In Figure 108.11 the constant strain is maintained for a shorter time, until t_1, when it is *suddenly* reduced to zero, this forces the specimen back to its original shape. The material responds by with an immediate initial stress which decreases at time t_1. When the applied strain is removed, the material responds with an immediate decrease in stress which may result in a change from tensile to compressive stress. The residual then gradually approaches zero.

The stress-relaxation behavior of a material is normally determined in either the tensile or the compressive mode. In these experiments, a material specimen is rapidly elongated or compressed to produce a specified strain level and the load exerted by the specimen on the test apparatus is measured as a function of time. Specimens of certain plastics may fail during tensile or flexural stress-relaxation experiments.

Viscoelastic stress-relaxation data are usually presented in one of two ways. In the first, the stress manifested as a function of time. Families of such curves may be presented at each temperature of interest, each curve representing the stress-relaxation behavior of the material at a different level of applied strain. Below a critical strain, viscoelastic materials

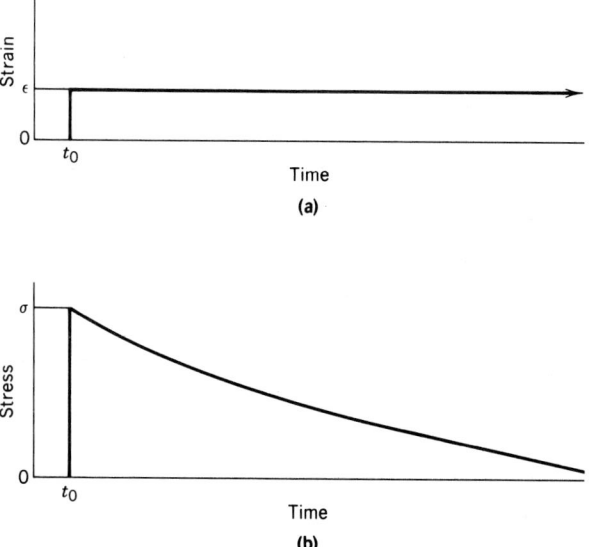

Figure 108.10. Viscoelastic stress relaxation behavior typical of many thermoplastics under long-term strain. **(a)** Input strain vs time profile. **(b)** Output stress vs time profile.

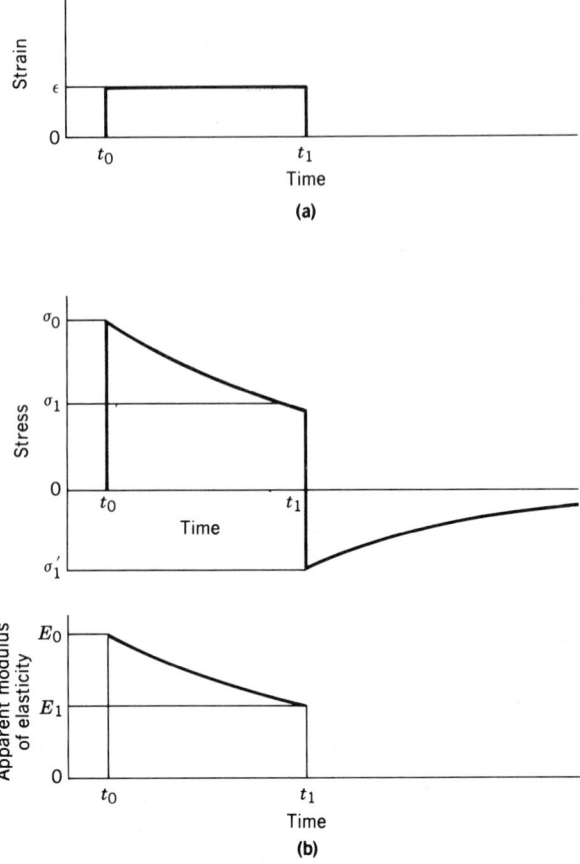

Figure 108.11. Viscoelastic stress relaxation behavior typical of many thermoplastics under shorter term strain. **(a)** Input strain vs time profile. **(b)** Output stress vs time profile.

may exhibit linear viscoelasticity; that is, the stress at a given time is proportional to the applied strain. Above this critical strain, the stress relaxation rate becomes disproportionately faster. In the second, the apparent stress relaxation modulus is plotted as a function of time. Apparent or viscoelastic stress relaxation modulus is a time- and temperature-dependent parameter which reflects the stress relaxation behavior of the material.

Although all viscoelastic materials undergo stress relaxation, the rate at a give temperature may differ considerably from material to material. Figure 108.12 shows the tensile stress versus time curves for acetal copolymers at 20°C (68°F) and 65% relative humidity at constant applied tensile strain levels of 1.0, 1.5, 2.0, 2.5, and 3.0%. Figure 108.13 illustrates that a given viscoelastic material may have considerably different stress relaxation behavior at different temperatures. Such data are necessary in designing structures that will be subjected to long-term deformation, including gaskets, springs, and force-fit components. The viscoelastic stress relaxation modulus allows one to estimate the material stress that will result from a given applied strain after a given time at the anticipated use temperature of the structure. It is particularly useful to the designer because it may be substituted for Young's modulus (E) in the appropriate elastic design equations to predict the long-term resiliency of such structures.

Stress-relaxation data enables the design of a structure so that the strain of the material comprising it will remain at the desired level. Too high a strain level and the material may cease to be linearly viscoelastic, which may lead to a significantly higher rate of stress

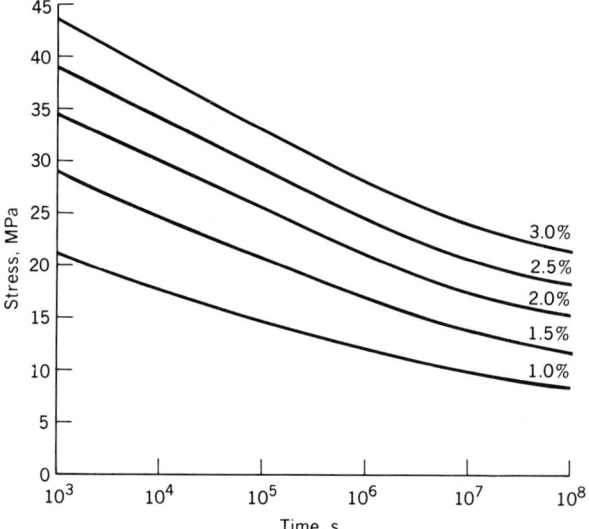

Figure 108.12. Tensile stress relaxation behavior of acetal copolymers at 20°C (68°F) and 65% relative humidity at five levels of constant applied strain.

relaxation. Too low a level, and the material stress may not be high enough to generate the required spring force.

Long-term Viscoelastic Behavior. The rate of creep and stress relaxation of thermoplastics increases considerably with temperature; these of the thermosets remain relatively unimportant up to fairly high temperatures. The rate of viscoelastic creep and stress relaxation at a given temperature may also vary significantly from one thermoplastic to another because of differences in the chemical structure and shape of the constituent polymer molecules. These differences affect the way the polymer molecules interact with each other, and hence their relative freedom of movement.

For the sake of practicality, viscoelastic creep and stress relaxation experiments are normally terminated at 1000 hours. Time–temperature superpositioning is often used to extrapolate this 1000-hour data to approximately 100,000 hours.

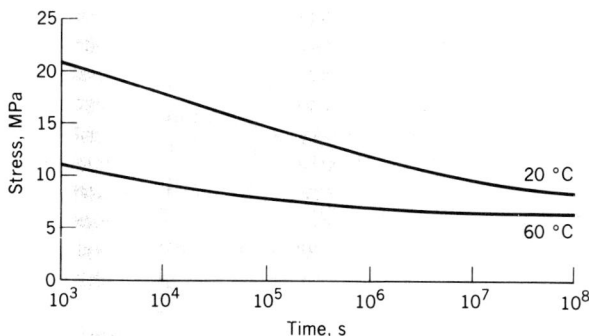

Figure 108.13. Tensile stress relaxation behavior of acetal copolymers at two temperatures at 1% constant applied strain.

108.3 FATIGUE

Material fatigue failure is the result of damage caused by repeated loading or deformation of a structure. The magnitudes of the stresses and strains induced by this repeated loading or deformation are typically so low that they would not be expected to cause failure if they were applied only once.

The fatigue behavior of a material is normally measured in either the flexural or the tensile mode. Testing may be carried out with either a constant amplitude of deformation or a constant amplitude of load. Specimens may be cracked or notched prior to testing, to localize fatigue damage and to permit the measurement of crack propagation rates.

In constant deflection amplitude fatigue testing, a material specimen is repeatedly bent to a specified maximum outer-fiber strain level; the number of cycles to specimen failure is observed. In constant flexural load amplitude fatigue testing, a bending load is repeatedly applied to a material specimen to produce a specified maximum outer-fiber stress level; the number of cycles to failure is observed. In both types of testing, the shape of the specimen causes failure to occur not at the support point but within a gauge section of reduced cross-section. Furthermore, in both types of testing, any mean stress–strain is possible. If the bending is symmetric with respect to the neutral axis, the material at the surfaces of the specimen is alternately subjected to equal magnitudes of tensile and compressive stress and strain; and the mean stress and strain are zero. If bending is asymmetric, a positive (tensile) mean stress and strain can be maintained in the material on one surface of the specimen while a negative (compressive) mean stress and strain or equal magnitude can be maintained on the opposite surface of the specimen.

Constant deflection amplitude fatigue testing is probably the less demanding of the two techniques, because any decay in the modulus of elasticity of the material due to hysteretic heating would lead to lower material stress at the fixed maximum specimen deflection. In the constant flexural load amplitude tests, maximum material stress is fixed, regardless of any decay in the modulus of elasticity of the material.

The test frequency used during fatigue evaluation of plastics is typically 30 Hz, and test temperature is typically 23°C (73°F). The behavior of viscoelastic materials is very temperature-and-strain rate-dependent. Consequently, both test frequency and test temperature to have a significant effect upon the observed fatigue behavior. Material fatigue data are normally presented in one of two ways: constant stress amplitude or constant strain amplitude, plotted versus the number of cycles to specimen failure to produce a fatigue endurance curve for the material. The fatigue testing of thermoplastics is normally terminated at 10^7 cycles.

Figure 108.14 shows a schematic fatigue endurance curve for a thermoplastic material.

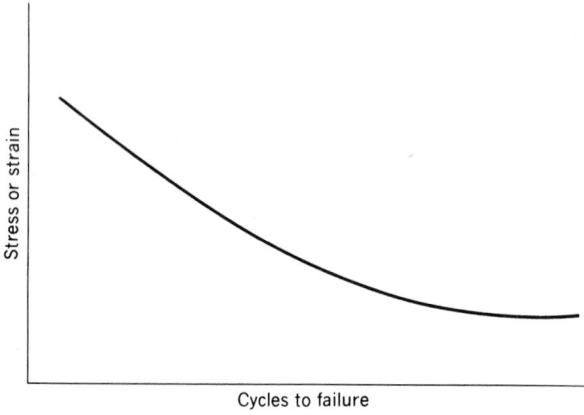

Figure 108.14.   Fatigue endurance curve typical of many thermoplastics.

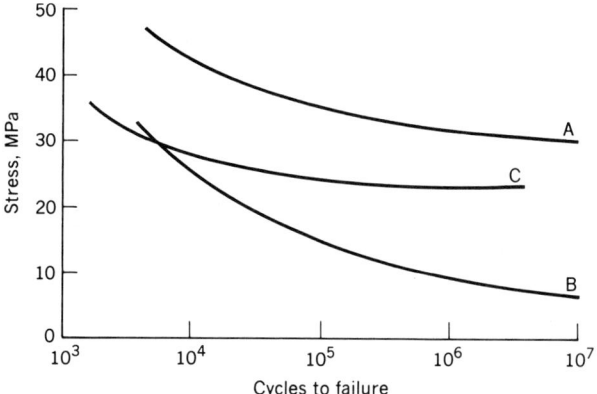

Figure 108.15. Room-temperature fatigue endurance curves for three thermoplastics at 30 Hz. A, Delrin 100 acetal homopolymer (50% RH); B, Udel polysulfone; C, Zytel 101 nylon (50% RH).

Two conclusions can be drawn from an inspection of this figure: (*1*) the higher the applied material stress or strain, the fewer cycles the specimen can survive; (*2*) the curve gradually approaches a stress or strain level called the fatigue endurance limit below which the material is much less susceptible to fatigue failure. Figure 108.15 illustrates that different materials may show different fatigue behavior at the same test temperature and test frequency.

108.4 THERMAL EXPANSION AND CONTRACTION

Unconstrained specimens of almost all materials respond to temperature increases by expanding and to temperature decreases by contracting. The coefficient of linear thermal expansion of a material is determined by varying the temperature of a representative test specimen, measuring its length as a function of temperature over the desired range, computing the total change in specimen length over that range, and then dividing that change in length by both the specimen length at the reference temperature and the total temperature excursion. In determining the coefficient of linear thermal expansion of plastics per ASTM D696, the temperature range is -30 to $+30°C$ (-22 to $86°F$); and the reference temperature is $23°C$ ($73°F$).

Table 108.4 illustrates that different materials may have widely different coefficients of linear thermal expansion. Plastics typically have coefficients that are considerably higher than those of other materials of construction, such as metals, glass, or wood. As the table shows, this difference may amount to a factor of 30.

Obviously, thermal expansion and contraction must be taken into account by the designer if critical part dimensions and clearances are to be maintained during use. Less obvious is the fact that parts may develop high stresses when they are constrained from freely expanding or contracting in response to temperature changes. These temperature-induced stresses can cause material failure directly or can produce forces that may cause the part to warp or buckle.

Plastics parts are often constraineg from freely expanding or contracting by rigidly attaching them to another structure made of a material with a lower coefficient of linear thermal expansion. When such composite structures are heated, the plastic component is placed in a state of compression and may buckle. When such composites structures are cooled, the plastic component is placed in a state of tension, which may cause the material to yield or crack. The precise level of stress in the plastic depends on the relative compliance of the component to which it is attached, and on assembly stress. For initially unstressed plastic components the maximum material stress in the plastic (ρ_p max) due to differential

TABLE 108.4.   Coefficients of Linear Thermal Expansion for Several Plastics and Some Other Materials of Construction

Generic Material Type	Trademark	Grade	Coefficient of Linear Thermal Expansion	
			$(10^{-5}\text{m/m-}°\text{C})$	$(10^{-5}\text{in./in.-}°\text{F})$
ABS	Cycolac	DH	7.0	3.9
		GSM	9.5	5.3
		KJB	9.9	5.5
		L	11.0	6.1
Acetal copolymer	Celcon	M25/M90	8.5	4.7
Acetal homopolymer	Delrin	100	10.4	5.8
		500		
		900		
Acrylic	Plexiglas	DR	10.1	5.6
Nylon (DAM) (0.2% moisture)	Zytel	101	8.1	4.5
		158 L	9	5
		211	7	4
Phenolic	Durez	29053[a]	1.48	0.82
		152[a]	3.6	2.0
		18441[a]	5.5	3.1
Polycarbonate	Lexan	141/940	6.75	3.75
Phenylene ether copolymer	Prevex	PQA	6.7	3.7
		VKA	6.8	3.8
Polypropylene	Pro-fax	6523	9.7	5.4
		7523	9.8	5.4
		8523		
Polysulfone	Udel	P-1700	5.6	3.1
		P-1710		
		P-1720		
Steel, structural ASTM A7-61T			1.17	0.65
Brass, naval			2.12	1.18
Aluminum, wrought 2014-T6			2.30	1.28
Pine (southern long-leaf)				
with grain			0.5	0.3
across grain			3.4	1.9
Oak (white)				
with grain			0.49	0.27
across grain			5	3
Glass		plate/crown	0.90	0.50
	Pyrex		0.32	0.18

[a]Injection molded specimens.

thermal expansion or contraction is given by the following equation:

$$\rho_{p\ max} \cong E_p(\alpha_m - \alpha_p)(\Delta T)$$

where E_p is the modulus of elasticity of the plastic at the temperature in question; α_m and α_p are the coefficients of expansion of the metal and plastic and ΔT is the temperature increase. This maximum stress in the plastic is approached as the relative compliance of the other component of the composite structure approaches zero.

To minimize the stresses induced by differential thermal expansion/contraction one must: (*1*) employ fastening techniques that allow relative movement between the component parts of the composite structure; (*2*) minimize the difference in coefficient of linear thermal expansion between the materials comprising the structure; or (*3*) minimize the temperature differences the structure will experience during use or shipment. Examples of proper fastening methods include the use of screws, bolts, spring clips, etc. with oversize holes, slots, or compliant bushings.

108.5 IMPACT RESISTANCE

The overall impact resistance of a structure is defined as its ability to absorb and dissipate the energy delivered to it during relatively high speed collisions with other objects without sustaining damage that would jeopardize its intended function. Several design features affect impact resistance. For example, rigidizing elements such as ribs may decrease a part's impact resistance, while less-rigid sections may absorb more impact energy without damage by deflecting elastically.

Likewise, dead sharp corners or notches subjected to tensile loads during impact may decrease the impact resistance of a part by acting as stress concentrators, whereas generous radii in these areas may distribute the tensile load and enhance the impact resistance. This point is particularly important for parts comprised of materials whose intrinsic impact resistance is a strong function of notch radius. Such "notch sensitive" materials are characterized by an impact resistance that decreases drastically with notch radius. Wall thickness may also affect impact resistance. Some materials have a critical thickness above which the intrinsic impact resistance decreases dramatically.

Methods employed to determine the impact resistance of plastics include: pendulum methods (Izod impact, tensile impact, and falling dart/Gardner impact) and instrumented techniques.

Izod Impact. This is the most widely used impact-resistance test for plastics. It requires specimens 0.318 to 1.270 cm (0.125 to 0.5 in.) thick, molded or cut from extruded sheet. A notch of 0.25 mm (0.01 in.) apex radius is then machined across the edge of the specimen to a depth of 0.254 cm (0.1 in.). These specimens are clamped so that the free end may be struck edge-on by a swinging pendulum impactor. Thus, during impact the specimen bends with the notched edge in tension. Energy lost by the pendulum and presumably absorbed by the test specimen during impact is divided by specimen thickness to arrive at the Izod impact energy, expressed as joules per meter of notch (J/m) or foot-pounds per inch of notch (ft-lb/in.). Test specimens of differing thickness may yield different results.

The Izod impact test is very sensitive to relatively small changes in material composition, morphology, and orientation. Therefore it is widely used by material manufacturers for quality control and by material suppliers and processors to detect contamination with other plastic mateials, thermally degraded material, incompatible regrind, dirt, etc.

Tensile Impact. Tensile impact testing employs unnotched specimens with gripping tabs at each end and an area of reduced cross section in the center. A uniform specimen thickness of 0.318 cm (0.125 in.) is standard. These standardized specimens are of two types: short and long. In general, the short specimen promotes a greater occurrence of brittle fracture, giving greater reproducibility but less differentiation among materials.

Test specimens are clamped into a pendulum impactor so that as its swings they will be subjected to a sudden tensile load. Energy lost by the pendulum during impact is divided by the minimum cross-sectional area of the specimen to arrive at the tensile impact energy, expressed as kilojoules per square meter (kJ/m^2) or foot-pounds per square inch (ft-lb/in.^{2}).

In addition to determining the general impact resistance of freshly molded plastic specimens, the tensile impact test is often used to measure the effect of long-term air exposure at elevated temperatures (heat aging).

The designer may use Izod and tensile-impact energy values to rank different materials during an initial screening of materials. However, impact data obtained from these tests do not always correlate well with final part performance. Room temperature Izod and tensile impact data for several plastics are shown in Table 108.5.

TABLE 108.5. Room Temperature Impact Resistance Data for Several Plastics

Generic Material Type	Trademark	Grade	Izod Impact Energy for 0.318 cm (0.125 in.) thick, Notched Specimens, J/m	Tensile-impact Short Specimen, kJ/m²	Energy Long Specimen, kJ/m²
ABS	Cycolac	DH	235	99–131[a]	
		GSM	374	102–115[a]	
		KJB	214	95	
		L	400	100–120[a]	
Acetal copolymer	Celcon	M25	85		190
		M90	75		150
Acetal homopolymer	Delrin	100	123		350
		500	74.7		200
		900	69.4		150
Acrylic	Plexiglas	V052/045	21[b]		
		MI-7	32[c]		
		DR	64[b]		
Nylon (DAM) (0.2% moisture)	Zytel	101	53	157	504
		ST 801	907		588
		158 L	53	153	611
		211	80		525
Nylon (50% RH) (2.5% moisture)	Zytel	101	112	231	1470
		ST 801	1068		1155
		158 L	75	218	945
Phenolic	Durez	29053[d]	19		
		152[d]	17		
		18441[d]	14		
Polycarbonate	Lexan	141	640–850	473–631	
		940	640	526	
Polyethylene	Dow	08064N	53		88
		10062N	48		105
		04052N	80		140
		08035N	130		81
Phenylene ether copolymer	Prevex	PQA	267	113[a]	
		VKA	293	86[a]	
Polypropylene	Pro-fax	6523	42.7		
		7523	133.5		
		8523	379		
Polystyrene	Fostarene	50	21		
	Hostyren	360	54		
		760	97		
		840	161		
Polysulfone	Udel	P-1700	69	341	
		P-1710	69	421	
		P-1270	69	336	

[a]0.159 cm (1/16 in.) thick specimens.
[b]Molded notch.
[c]0.635 cm (0.250 in.) thick specimen with molded notch.
[d]Injection-molded specimens.

Falling Dart–Gardner Impact. As the name implies, falling dart impact testing utilizes the kinetic energy of a free-falling projectile to assess the impact resistance of a plastic. A flat or contoured test specimen is rigidly clamped to a supporting metal ring and a metal dart of known weight is dropped onto the specimen from a known height. The end of the dart which impacts the specimen (the tup) is hemispherical, with a known diameter.

The Gardner falling weight test method is a variation of the falling-dart method. Test

2

specimens are supported by a metal ring but are not rigidly clamped in place. Moreover, the hemispherical tup is separate from the falling weight. The tup rests on the test specimen, serving as an anvil for the falling weight during impact.

Standard test specimens for falling dart/Gardner impact are normally flat disks of uniform thickness. Specimens of solid plastics are typically 0.318 cm (0.125 in.) thick; specimens of foamed plastics are typically 0.635 cm (0.250 in.) thick. For both cases, specimens are molded or cut from extruded sheet. The inside diameter of the supporting ring and the outside diameter of the tup may vary. In general, the larger the ring diameter, the more the specimen deflects during impact, and the higher the observed impact resistance of the specimen. The standard tup/support ring diameters for the falling dart impact test are 1.27/3.81, 1.586/7.62, and 3.81/12.7 cm (0.5/1.5, 0.625/3.0, and 1.5/5.0 in.). For the Gardner impact test, the standard tup diameter is 1.586 cm (0.625 in.); and the standard support ring inside diameters are 1.626, 3.175, and 7.62 cm (0.64, 1.25, and 3.0 in.).

To determine the impact resistance of a particular material, 20 to 26 identical test specimens are each subjected to a single impact of known energy. This energy is varied incrementally by changing dart weight or drop height. The Bruceton Staircase method is used to determine the impact energy at which 50% of the specimens tested would fail. Specimen failure is defined as a visible crack on the underside of the specimen. The typical units of falling dart impact resistance are joules (J) or foot-pounds (ft-lb.).

Falling dart and Gardner impact data are often used by a designer to select candidate materials for a particular application. To make proper comparisons of different plastics, however, the designer must know all the relevant background information, such as specimen thickness, support ring diameter, tup diameter, and mode of specimen preparation. Any of these test variables can have a significant effect on observed impact resistance. The falling dart and Gardner impact tests are also useful in determining the effects of surface treatments on the impact resistance of various plastics.

A shared advantage of falling dart and Gardner impact tests is that they subject plastic specimens to the same type of biaxial/multiaxial loads and stresses that an actual part would experience during impact.

Instrumented Impact Techniques. Without instrumentation, the pendulum impact tests and the falling weight impact tests described above yield only a single data point: the total impact energy necessary to cause specimen failure. With instrumentation, however, these tests can be used to gather a wealth of material behavior data during the impact event.

Instrumented impact techniques employ a load cell to measure the force exerted on the test specimen as a function of time. The only other measured quantities are the mass of the impactor or dart and its velocity just prior to impact. A computer stores load-versus-time data acquired during impact and calculates the bending load on the specimen versus its deflection and the specimen deflection as function of time. Electronic integration of the load-versus-deflection curve provides a plot of impact energy-versus-specimen deflection.

Such data allow one to determine: the time and the deflection (and hence strain) at which the material begins to yield and/or fracture; the initial slope of the load–deflection curve and hence the elastic modulus of the material at the temperature and strain rate employed for the test; the load (and hence stress) at yield and at maximum; and the energy necessary to cause the material to yield, to initiate specimen fracture, and to propagate the crack, as well as the total impact energy.

The instrumented falling dart impact test is superior to the non-instrumented test in the following ways: (1) It yields quantitative data for each specimen rather than a statistical statement about a number of specimens. (2) It does not require a subjective judgment by the operator of what constitutes a specimen failure. (3) It yields more consistent results with fewer specimens (typically five) required.

In the final analysis, the impact performance of a part is governed by many complicated and often interacting factors. None of the impact tests performed on standard test specimens should be regarded by the designer as an infallible predictor of actual performance. There is no substitute for testing the finished product. Tests that simulate as closely as possible real-life impact situations are the most reliable indication of impact resistance.

108.6   CONCLUSION

This discussion has presented information in five broad categories of material behavior that directly affect the structural performance of plastic parts. There are many other factors that also may dramatically affect plastic material behavior. Processing parameters such as melt temperature and fill rate, mold temperature, and packing pressure in injection molding, draw-down ratio in extrusion, and draw ratio in thermoforming all can significantly influence the observed properties of plastics.

Excessively high processing temperatures can increase the rate of thermal decomposition of plastics with generally deleterious effects. Too low a processing temperature can cause melt viscosity of these materials to be too high, thereby requiring excessive pressure to fill the mold or extrude the sheet or profile. High melt viscosity and high processing pressure can lead to high shearing stresses in the material and may impart increased molecular orientation to the final product. Thermoforming at sheet temperatures that are too low can lead to excessive tensile stresses. Likewise, low fill rate, high packing pressure, and low mold temperature in injection molding, high draw-down ratio in extrusion, or high draw ratio in thermoforming may cause increased bulk orientation in the resulting part. Molecular orientation makes the part anisotropic, ie, its properties are not the same when measured in different directions.

Several treatments can embrittle the surface of a plastic part, including contamination with certain chemical agents, oxidative degradation or cross-linking of the polymer molecules by heat aging or exposure to ultraviolet light; coating with brittle paints; and plating. Under certain circumstances cracks starting in this brittle surface layer can propagate into the underlying ductile plastic at sufficiently high velocity to induce a brittle failure of the entire

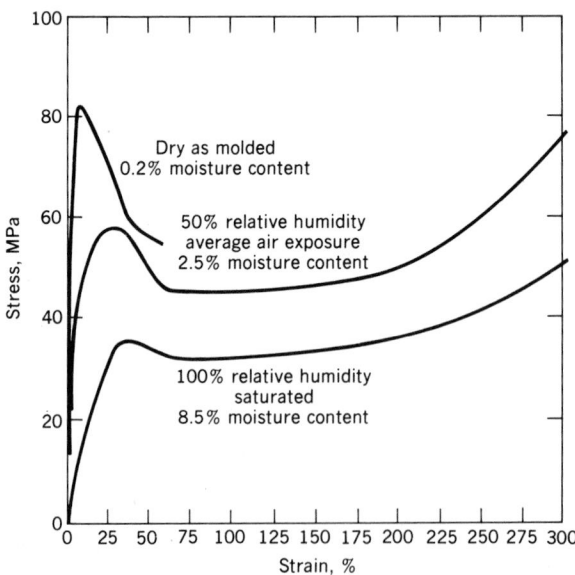

Figure 108.16.   Tensile stress–strain curves for nylon of three different moisture contents at 23°C (73°F). Courtesy of DuPont Engineering Polymer.

part. Surface embrittlement can have a significant effect on the observed stress/strain, fatigue, and impact behavior of the plastic.

Exposure to certain plasticizing agents (including water in some instances) can make some plastics more ductile. This phenomenon can also cause a significant increase in the observed rate of viscoelastic creep and stress relaxation at a given temperature. Because of its affinity for moisture and the great effect absorbed moisture has on its properties, nylon is widely cited to illustrate this effect. As shown in Tables 108.1 and 108.3 and in Figure 108.16, specimens of nylon that are dry as molded exhibit much less ductility and toughness than do specimens equilibrated with air at 50% or 100% relative humidity.

In light of the many types of behavior plastics can manifest and the considerable effect this behavior can have on the performance of the finished part, it behooves designers to become familiar with specific behavior characteristics of each plastic considered for an application. Sources of this information include the plastics supplier, compendia of plastics property data, computer-accessible plastic property databases, and this Handbook.

There are, of course, many other sources of information available in the literature (some of which were consulted for this discussion) that will be of value to structural designers working with plastics. A prime source for design information is the publication of the material manufacturers. See the Bibliography for some other recommended sources.

BIBLIOGRAPHY

General References

J.J. Aklonis and W.J. MacKnight, *Introduction to Polymer Viscoelasticity*, John Wiley & Sons, New York, 1984.

R.E. Chambers, "Behavior of Structural Plastics," in *Structural Plastics Design Manual*, Superintendent of Documents, U.S. Government Printing Office, Washington, D.C., 1979.

J.H. Faupel and F.E. Fisher, *Engineering Design*, 2nd. ed., John Wiley & Sons, New York, 1980.

J.D. Ferry, *Viscoelastic Properties of Polymers*, 3rd. ed., John Wiley & Sons, New York, 1980.

L.W. Fritch, "ABS Molding Variable; Property Responses," in D. Rosato, ed., *Injection Molding Handbook*, Van Nostrand Reinhold Co., New York, 1985.

F.J. Heger, "Considerations in Structural Design with Plastics," in *Structural Plastics Design Manual*, Superintendent of Documents, U.S. Government Printing Office, Washington, D.C., 1979.

R.W. Hertzberg and J.A. Manson, *Fatigue of Engineering Plastics*, Academic Press, New York, 1980.

M.J. Howard, managing ed., *Desktop Data Bank*, 6th ed., The International Plastics Selector, Inc., San Diego, Calif., 1983.

D.R. Ireland, "Instrumented Impact Testing for Evaluating End-use Performance," in R.E. Evans, ed., *Physical Testing of Plastics*, American Society of Testing & Materials, Philadelphia, Penn., 1981.

P. McGee, ed., "Section 8 (Plastics," *Annual Book of ASTM Standards,* American Society of Testing & Materials, Philadelphia, Penn., 1988.

"Design" section, *Modern Plastics Encyclopedia 1989*, McGraw-Hill Information Services Co., New York, pp. 398–430.

R.J. Roark and W.C. Young, *Formulas for Stress and Strain*, McGraw-Hill Book Co., New York, 1975.

M.D. Wolkowicz and S.K. Gaggar, "Effect of Thermal Aging on Impact Strength of ABS," *Polymer Engineering & Science,* **21**(9), 571–575 (June 1981).

109

DECORATING

Robert F. Cervenka
Philips Plastics Corp., Box 168, Phillips, WI 54555

Commercial techniques for decorating plastics are almost as varied and versatile as plastics themselves. Depending on end-use applications or marketing requirements, virtually any desired effect or combination of effects, degree of brightness, and shading of tone can be achieved on flexible or rigid plastic products. Solid colors (full strength or pastel) can be specified. So can transparents, pearlescents, fluorescents or any other permutation. In some cases decoration can be used to impart important secondary properties, such as electrical conductivity or static dissipation, for example.

The primary decorating technique is innate coloration achieved at the compounding stage. Although most thermoplastics (but not thermosets) are produced in natural white or colorless transparent form, color is usually added by direct feed or blending with base resin prior to the processing stage.

These colorants (or color concentrates) are available in a wide range of stock shades that have precise tinctorial values. Colors also can be matched to the exact spedifications of an individual customer; as with stock colors, these specifications are kept in computer memory to ensure batch-to-batch and order-to-order consistency. This is of particular importance when package or product color is an intrinsic component of a company's overall marketing strategy (such for example, as Kodak yellow or Tide orange). Decorating via color blending can also be utilitarian, as in color-coded wire-and-cable sheathing.

Beyond basic raw-materials coloring, designers have a large palette of decorating media at their disposal. Included are painting (or coating) processes, direct printing, transfer decoration, in-mold decoration, embossing, vacuum metallizing, sputtering, electroplating, and barrel plating.

109.1 PAINTING

Plastic processing has not only adapted existing painting (or coating) technology, but has expanded it—in methods of coating application and also in developing polymers to create new coatings. The technology of painting is so vast and offers so many options in methods of application and in the variety of coatings that the enhancements it can provide to plastic materials in both finish and function are truly unlimited.

Virtually all plastics, both thermoplastic and thermosetting, can be painted, with or without priming or other preliminary preparation procedures. It is usually possible to match material to coating to application method and achieve relatively easy processing that will eliminate preparation steps to accomplish desired surface finishes.

Plastic parts or materials can be coated by brushing, dipping, hand-spray painting, flow coating or roller coating; they can be automatically spray painted with rotating or recip-rocating spray guns; they can be electrostatically painted using either a conductive precoating procedure or conductive materials. It is also possible to paint areas selectively, through a variety of masking options. Recessed areas can be kept free of paint via the use of plug masks that will allow surrounding surfaces to be coated. The opposite of this situation is to paint recessed areas, keeping surrounding background areas free of paint through the use of lip masks.

Total surface coverages as well as masked operations can be adapted to almost unlimited geometries. Parts as small as knobs, pushbuttons, delicate nameplates and medallions are readily decorated via the painting process and done efficiently either by manual options (for low-volume requirements) or automatic systems (for higher volume requirements). Painting is routinely done on huge parts such as TV cabinets and exterior automotive components such as hoods and deck lids. Technical improvements in materials and coatings that enable plastics to survive automotive bake-oven cycles are being developed; they could hasten the advent of production cars with all-plastic bodies.

Painting operations have the advantage of being as simple or as sophisticated as the application may dictate. They also have the advantages of offering almost unlimited color options as well as great variety of surface finishes and final surface properties to meet such needs as ultraviolet light (UV) resistance, abrasion resistance, gloss, chemical resistance, etc. Limitations of painting on plastics are usually dictated by limitations of the base materials themselves. For example, to achieve extremely high abrasion resistance requires coatings that are cured at elevated temperatures that may exceed the temperature resistance of the base plastics. Due to the variety of plastics materials and coatings available. It is important to evaluate desired end-use properties to minimize the limitations of a painting application.

It is also important to design into the plastic part considerations for subsequent painting operations. Painting buildup may affect tolerances, designs may have to accommodate fix-turing to hold parts during the painting operation, and paint-break areas should be properly designed to accommodate plug or lip masks for selective coating.

It is extremely important that specifications for the base materials take into consideration the availability of the most workable paint formulations. Painting demands can vary from elementary hand spraying for noncritical functional coatings to extremely demanding high-quality cosmetic applicatioins that may require very sophisticated application equipment as well as very sophisticated coating formulations. Fortunately, there is a great variety of application equipment available, as well as a great variety in coatings. Thus, it is possible to match the end-use requirement with the proper coating and application system. End-use situations that can tolerate variations in coating thickness, drifts in color matching and wide latitude in plastic substrate and coating formulations will be able to tolerate the typical operator variations in a hand spraying procedure. But where the application requires close-tolerance coating thicknesses, and critical color matching, it is usually necessary to employ automatic flow coating or spraying systems. A very broad range of equipment is available for painting operations. It can be as simple as a self-contained manual spray gun with integral paint container, requiring only a compressed-air source to initiate painting. This is perfectly adequate for sample work, applications evaluations and low-volume production work. A step up in equipment is a centralized paint source that pressure-feeds paint to a hand-held spray gun; this permits painting on a larger volume scale.

Progressing beyond hand-held painting equipment, specific part geometry and part re-quirements dictate the type of machinery to be used. Automatic individual spray painting pieces of equipment use a rotating fixture that allows the part to be rotated under auto-matically operated and fixed spray guns. Equipment is also available that provides for stationary positioning of the part, with the spray guns revolving or reciprocating over a work surface. This is particularly useful to accommodate delicately masked parts.

For high volume production, painting equipment can become very sophisticated, even combining operations in one process. Typical is the paint fill-and-wipe operation used to fill depressed graphics to provide contrast to the background areas. Push-button or touch-tone telephone dialing systems are typical of this application. Parts to be decorated are automatically loaded into a fixture system that passes through a painting station where paint is applied, then wiped off, leaving the depressed engraving filled with paint.

Production painting of larger parts often is a completely automated operation involving robotized spraying arms that have been programmed to apply paint in a carefully orchestrated pattern that provides maximized speed and efficiency in part coating.

Tooling required for paint processing usually consists of holding devices to position or carry the part to be painted through the painting system and various types of masks for applications requiring controlled air coatings. As the painting operation becomes more sophisticated, it may become necessary to have very accurate holding devices machined specifically for individual articles. This may be necessary in order to fit the holding device to the part or fit the holding device to the masking. It is not uncommon for fits between holding fixture and part and holding fixture and masking to be held as close as plus or minus 0.002–0.003 in. (0.051–0.076 mm). These systems can cost as much as several thousand dollars.

Auxiliary equipment for painting ranges from a pressure pot feeding single hand-spray guns to totally automatic central systems incorporating mixing, agitating and metering of multiple components. These can be adapted for single component coatings or two-part coatings such as catalyst/resin type epoxy systems. Another important piece of auxiliary equipment is the air-makeup system which replenishes air exhausted out of the paint systems. Most painting systems are sensitive to changes in temperature and humidity; therefore makeup air systems are critical.

Contamination in the painting atmosphere is also an item of concern. Thus, filtering systems for air coming into the paint atmosphere and for removing paint from the atmosphere before exhausting are of extreme importance. A variety of filtering media are available to meet the specific needs of the volumes of air being exhausted and the particular coatings being utilized. They range from fibrous cloth to waterfalls and electrostatic devices.

Drying equipment to hasten part production also is an important auxiliary. It can be as simple as a storage conveyor system that allows parts to air dry over a period of time. However, some coatings require force drying by heated oven systems. Some parts require ultraviolet curing systems as the final operation.

Parts are usually market-ready after painting. There are instances, however, where parts require added secondary operations to correct overspray and/or handling defects. These situations are usually controllable through proper tooling where volumes allow, but may be a necessity for low-volume applications and geometries that do not lend themselves to total control during the painting process.

Many custom or job shops offer decorating as a service. Costing of painted articles is typically a time, material, reject factor computation.

Many manufacturers supply equipment for application of paint and equipment for control of painting atmospheres. A partial listing includes:

Binks Inc., Franklin Part, Ill.

Conforming Matrix Corp., Toledo, Ohio

Deco Tools, Inc., Toledo, Ohio

Dispatch Industries, Inc., Minneapolis, Minn.

Devilbiss Co., Toledo, Ohio

Graco Inc., Minneapolis, Minn.

Paasche Airbrush Co., Chicago, Ill.

Ronsbury Electrosatic, Indianapolis, Ind.

Coating sources are also invaluable for evaluations of painting application. They include:

Bee Chemical Co., Lansing, Ill.
Red Spot Paint & Varnish Co., Evansville, Ill.

Technical assistance or consulting recommendations can be directed to:

Federal Society of Coating Technology
1315 Walnut St.
Philadelphia, PA 19107
(215) 545-1506

Society of Manufacturing Engineers
Association for Finish Processes
One SME Drive
P.O. 903
Dearborn, MI 40121
(313) 271-1500

Society of Plastics Engineers
14 Fairfield Drive
Brookfield Center, CT 06805
Decorating Division

109.2   PRINTING

The primary printing processes used in plastics are rotogravure, flexography, silk screening, and pad printing (which uses techniques developed from letterpress printing).

Rotogravure is an appropriate title for a printing process that requires the use of an engraved metal cylinder or roller. The engraving or etching process on the surfaces of the metal cylinder result in recessed areas that pick up liquid coatings from a reservoir, so situated that the revolving cylinder will flood its surface with the liquid. A doctor blade or a squeegee action is used to wipe excess ink or coating from the foreground areas of the cylinder resulting in the recessed areas retaining the appropriate amount of printing ink. The ink-laden cylinder is then rolled over the medium to be printed. This process is suited to continuous printing of flat stock, or continuous strip or in some cases, individual sheets.

Attention paid to the proper formulation of printing ink will make the gravure process applicable to a great variety of plastic substrates. Virtually all thermoplastic film or sheet applications are printable by this process.

A good example of the capabilities of rotogravure is the printing of woodgrained patterns on carrier foil that will ultimately be utilized for hot-stamping applications. These patterns are very demanding in the quality of the printing process in order to achieve a look of real wood.

Woodgrain patterns may require the deposition of several coatings to achieve the proper effect. The etching process in producing the gravure cylinders is such that very fine tones and shading results can be achieved. Several engraved cylinders can be used in sequence for continuous printing.

Gravure Printing. Gravure printing is efficient for high volume production of continuous film. It produces very delicate and controllable pattern and color match definitions. The expense of making engraved or etched cylinder generally limits this printing process to higher volume applications and long or repetitive production runs.

Rotogravure printing of plastics usually requires that the substrate be in standard thickness ranges and widths. Material fed accessories, the printing press itself and take-off equipment are all fairly standardized.

Flexography. This can best be described as a marriage of rotary printing and letterpress printing. It uses a flexible printing plate, typically a metal–silicone rubber-bonded combination with the rubber surface processed to leave the printing surface raised over the background area. The raised and recessed areas on the printing plate can be fabricated through photographic etching and/or engraving. In the printing process, ink is transferred from a reservoir through a roller–doctor blade system onto the curved flexible printing plate; ink off the raised portions of the flexible printing plate is then transferred to the material to be printed; the printing material being conveyed to imprint position via the interpress positioning process.

Flexography is adaptable to shorter volume production requirements in that the material to be printed can be fed one at a time. Also, fabricating the silicone rubber–metal printing plate is less expensive than engraving or etching metal roller dies. The process is suitable for variety of applications, ranging from simple label film to decoration on molded parts, such as plaques, medallions or wall tile.

The flexible printing plates used in flexograph do not permit the very fine detail that can be achieved on etched metal surfaces such as utilized in gravure printing. There also are limits to the size and shape of articles that can be printed. Flexographic equipment is primarily designed for flat stock; deviations from that concept are not readily possible.

Silk Screening. As the name implies, silk screening involves the use of silk cloth or silk screen in the transfer of printing ink to articles to be printed. Integral to this process is the use of a suitable open-weave cloth (silk is still commonly but not exclusively used) stretched over a framework. The stretched cloth is selectively coated through the use of a stencil; this coated area resists the passage of printing ink, which can only penetrate through the uncoated open-weave surface. There are various ways to prepare the screen for printing, other than stenciling.

The silk-screen printing process involves the use of a rubber squeegee or metal flood bar to carry the ink over the open mesh areas of the screen. The squeegee action also forces ink through the open areas by pressure onto the surface to be printed.

Silk screening is a versatile process, with a lot of flexibility in inks that can be accommodated, therefore allowing it to be used for a wide variety of plastic substrates. However, it is advisable to determine the compatibility of inks or coatings and the intended substrate material.

The very nature of the process (drawing a squeegee over a fabric to force ink through open mesh areas) suggests some geometry limitations. It is not adaptable to complex shapes, but is effective on flat surfaces or the peripheral areas of cylinders.

Silk screening's cost is somewhat offset by the relatively low costs of producing screens, and its adaptability to manual screening for low-volume needs and to full automation where volumes demand is anticipated. Silk screening can also generate good design detail.

Coating materials may be used to put on protective coatings as well as for graphics.

In designing products for silk screening there must be enough room for the screen to fit. Projections and obstructions on nonflat surfaces may be overcome. Likewise, tolerances of plus or minus several thousandths of an inch, although obtainable and expensive, require planning in the design phase.

The materials used in silk screening can be altered to meet specific needs. For example, high volume applications may dictate that screens be made of woven stainless-steel threads rather than the less expensive (but less durable) silk, nylon, or polyester. Also, the detail of printing obtained from silk screening can be adjusted through proper selection of the screenmedia. The finer the mesh weave the greater detail that will result, which is a consideration when delicate graphics or patterns are required. Another way to achieve fine detail is to photographically remove emulsion or coating. There is also a photographically etched process utilizing an overlay material that does not migrate into the mesh; this results in clear, sharp lines without the interruptions or "peak/valley" effect that impregnated mesh and subsequent emulsion removal can produce.

It is important to control the squeegee action during silk screening. For proper control of ink deposition, it is necessary for the rubber squeegee to produce a smooth even stroke, beginning before contacting the imprint area and continuing over and past the imprint area into a reservoir or transfer stage. The squeegee also acts as a screen flooding devices, in some cases bringing the ink back over the imprint area in preparation for the next sequence and next article to be decorated. The screen must be supported prior to and after imprinting, to ensure that the screen is not stretched or deformed by the article being decorated.

Whether the equipment is manual or automatic, the same principles apply; the screen has to be raised to allow for the work piece to be positioned under the imprint area. The screen is then lowered to the proper spacing above the piece and the rubber squeegee is drawn across the screen surface forcing the ink through the open mesh imprint area onto the piece and then past the imprint area. The squeegee then reverses its direction and entraps the ink, again in front of it, flooding the imprint area to keep it from drying out.

Tooling for silk screening usually involves only the feed or holding devices required to position the substrate. Silk screening does not require a great deal of auxiliary equipment. It is important to note, however, that silk screening may require atmospheric condition control. For example, as humidity and temperature change the screen tension may change, resulting in different imprint dimensions. Also, those same temperature and humidity changes will affect quality and viscosity of the printing inks, which in turn will affect the quality of the imprint. Ultimate quality will require proper ventilation and controlled temperature and humidity of make-up air.

Pad Printing. Pad printing is a relatively new process. It uses printing principles and techniques from letterpress and flexography. The process involves the use of a smooth silicone pad to pick up ink from an engraved or etched plate and transfer the impression to the substrate. The engraved plate, known as a *cliche,* is produced in a manner similar to that of printing plates for offset or gravure roller printing. The uniqueness of pad printing has to do with the use of the silicone pad that picks the impression off the *cliche* and transfers it to the product to be decorated. The silicone pickup pad can be designed to meet virtually unlimited production part shapes and configurations. This ability has prompted tremendous growth in pad printing.

Pad printing process is extremely valuable for transferring imprints to grossly irregular surfaces incorporating concave and convex features. The silicone pad conforms to the surface of the individual item being printed in one cycle, and has the ability to adapt to part changes cycle to cycle. This is a very important attribute when production runs produce changing tolerances that result in slightly inconsistent imprint area surfaces. An additional capability of the process is that it can print several colors and impressions in one operation. Coatings can be layered when wet to accomplish multicolor designs with very accurate registration and impression quality. This same feature allows flexibility in imprint design to incorporate not only graphics or patterns in conjunction with each other but also background colors and protective coatings utilizing the same individual piece of equipment, and all within the same cycling. An example demonstrating pad printing capabilities is the imprinting of graphics onto the contoured surfaces of key caps such as are used in business machine keyboards. This application demonstrates the adaptation to contoured surfaces of the process, and the durability of impression that is achieved with some of the new printing inks.

Pad printing is also efficient for imprinting very delicate designs and/or graphics. It is an adaptable process, too; efficient to set up for short runs with manual loading and unloading, as well as readily adaptable to volume applications involving multiple colors and automatic part feeding. The silicone pads utilized in the process are somewhat expensive because they do wear out. Dependent upon the contours and the impression produced, the pad may only last for 20,000–40,000 cycles.

The ability of pad printing to adapt to varying contours and surface irregularities opens up many design opportunities for decorative applications. But for graphics or designs to be imprinted on irregular surfaces, it is important to understand the relationship between picking an imprint off a flat surface and transferring it onto a contoured surface. Depending on the degree of surface geometry change, it may be necessary to extrapolate the contoured dimensions to the flat surface; that is, the imprint on the flat surface will not look or dimension the same as the final result as viewed and measured on the contoured product.

Machinery for pad printing incorporates many principles of letterpress. The cycle starts with a doctor blade acting as a vehicle to bring the printing ink out of a reservoir, usually positioned to the rear of the printing plate, and flooding the printing plate with ink in a forward stroke. The doctor blade is then reversed to wipe off the printing plate, leaving ink only in the recessed imprint areas. The printing plate now being clean of ink (with the exception of the imprint area), the silicone pad is brought over the imprint and down into contact with the printing plate. Contact progresses from a central small area and spreads out as downward action forces the silicone pad to cover a greater area of the printing plate. The action then reverses itself with the pad rising, coming forward over the work piece, again with the contoured surfaces touching a small area first and processing over the entire pad area in contact with the work piece.

There are options in the design of the work station. It can be a single-station setup with the operator hand loading parts one at a time to the fixture. The work station also can be fully automated via the use of rotary tables where work stations will rotate into position, taking the printed part away and bringing in a fresh undecorated part. Vibratory feeders can fill the rotary tables. The parts removal is readily subjected to automation.

Pad printing equipment is typically very compact, quiet, and smooth operating, with great accuracy of repeatability. Typically electrically and/or air operated, the equipment is relatively expensive; a single imprint machine for even small parts costs in excess of eight or nine thousand dollars.

Pad printing equipment is a self-contained, stand alone system not requiring auxiliary equipment. It operates very satisfactorily in a variety of atmospheres. Even though temperature and humidity is not that critical for this equipment or process, the inks may be necessary to control the atmosphere for applications that are extremely critical in quality or automation.

Articles processed via pad printing require no secondary or finish operations. Should protective coatings be required they are normally incorporated into the printing process as an additional color or coating system.

Manufacturers of equipment from the various printing processes are as follows:

Rotogravure
Roto-Die Micrometrics, St. Louis, Mo.
Lumite Products Group, Slamanca, N.Y.
Kaumagraph Corp., Wilmington, Del.
Flexography
Anderson and Vreeland Inc., Fairfield, N.J.
Carraro/Amplas Inc., Green Bay, Wisc.
Wolverine Flexographic Mfg., Farmington, Mich.
Silk Screening
Advance Process Supply, Chicago, Ill.
Conforming Matrix, Toledo, Ohio

Pad Printing
Tampo Print America, Schaumburg, Ill.
United Silicone, Lancaster, N.Y.

Technical assistance and/or consulting recommendations can be obtained from:

Rotogravure
Gravure Technical Association
60 E. 42nd St.
New York, NY 10017

International Tape/Disc Association
10 Columbus Circle
New York, NY 10019
(212) 956-7110

Flexography
Flexographic Technical Association
95 West 19th St.
Huntington Station, NY 11746
(516)/271-4224

Silk Screening
Screen Printing Association
10015 Main St.
Fairfax, VA 22031

Pad Printing
Society of Plastic Engineers
14 Fairfield Drive
Brookfield Center, CT 06805
Decorating Division

109.3 TRANSFER DECORATION (HOT STAMPING)

This decoration technique is familiarly known as hot stamping, although the terminology "coated foil transferring" might be more appropriate. In this process, the printed coating on a carrier film is transferred onto a plastic surface. Secure adhesion is accomplished with the use of heat, pressure, and time under controlled conditions. The key to this process is the hot-stamping medium, which consists of a carrier film (usually a polycarbonate, polyester or cellulosic material) upon which various coatings provide the desired decorative effect and offer assistance to the processing procedure.

In hot stamping, the coated foil is placed over the plastic to be decorated, and a heated die forces the foil onto the plastic. The proper control of heat, pressure, and time transfers the coating off the carrier foil onto the plastic.

Hot-stamping process is a versatile decorating tool. A wide variety of coatings can be deposited on the carrier film, which accommodates the process for use on almost any thermoplastic material and many thermosets.

Coatings on the carrier medium can be formulated specifically for different plastic materials to enhance adhesion requirements, release off the carrier foil is controllable, and coatings can be formulated to react to proper heat exposure as dictated by the plastic substrate. The process itself is very flexible and can react to the condition of the plastic substrate, the temperature of the stamping die, the pressure being applied to the plastic substrate, and the timing of the stamping contact and over-all cycle. With so many variables available, it is not surprising that a great variety of materials can be used, all of them producing effective results.

Because producing the hot-stamp medium is really a coating process, it lends itself to a wide spectrum of effects. Coatings can be virtually of any color; metallic effects can be achieved with the deposition of micro-thin coatings of gold or silver or chrome; multiple coatings can be applied to achieve such effects as woodgraining, marbelizing, or multicolored designs. By embossing the surface of the carrier medium coating, it is also possible to achieve three-dimensional decorative effects.

Many end-use applications benefit from the hot-stamping process. These range from selective imprinting, such as on the windows of microwave-oven doors or covering large areas, such as the woodgrained surfaces of television cabinets. The process also is effective in highlighting raised areas such as borders and graphics on control panels used on consumer appliances, radios, television sets, etc.

Simplicity of operation is a key advantage of hot stamping. The process does not require any special atmosphere in which to operate, does not require air exhaust systems, and can be adapted to a range of part sizes. The process also lends itself to low volume production requirements as well as to fully automatic high volume applications.

There are some limitations, however. To control the decorative finish, it is necessary to ensure even contact of the die through the carrier medium onto the plastic surface. Even contact is necessary in order to control the heat, time and pressure variables on all surfaces of the coating to achieve the proper finish quality. If you can visualize trying to hot stamp into the inside of a sphere, for example, it would be impossible to have a flat transfer conform to the sphere's concave surfaces without wrinkling and/or creasing the coating.

Geometry of the surfaces to be decorated must, therefore, be considered when evaluating the use of hot stamping. Even so, with proper tooling it is possible to decorate surfaces other than flat areas. For example, it is possible to stamp around radiused corners, as on television cabinets, through the use of special holding fixtures, hot-stamping heads and part-rotating features. There is considerable flexibility of processing with hot stamping. Changing the temperature of the die surfaces, the degree of pressure being applied to the die through the carrier medium and onto the plastic, the amount of time that the die is in contact with the plastic, and coating additions and/or release agents all provide a lot of latitude for controlling the quality of the decorative surface.

A wide variety of hot-stamping equipment is available. This ranges from hand-held heated roller dies utilized in sample or prototype work to huge roller presses for laying finishes on continuous four-foot wide rolls of ABS thermoforming stock. The basic hot-stamp press consists of a table or bed area that holds a nest or fixture that in turn will hold in proper position the plastic part to be stamped. Immediately above the plastic part will be the area for the transfer medium to be held, usually in roll form. Above the imprint area is a die holder, which is a heated surface designed to receive the die that will transfer the coating onto the plastic part.

Hot-stamping presses normally utilize the downward action of the heated surface to the plastic part bed to also actuate a device that automatically advances the transfer medium as each part is stamped. For lower volumes and very large parts, it may be advantageous to manually load the parts one at a time. As volumes and size allows, it is possible to go to the other extreme, with parts automatically fed either via rotary turntables and vibratory feeders and automatic take-off devices to achieve very high rates of production.

Hot-stamping equipment has also been developed to meet a variety of special purpose applications. Equipment is available to do peripheral stamping such as decorating perimeters of rollers or dials, such as odometer wheels. Roller presses are available to accommodate automatic feeding on a conveyor. Multiple-head presses are available to provide multicolor capability through the use of rotary tables or shuttle tables feeding parts under the various head locations.

Tooling for hot stamping typically involves only the holding fixture for the part to be decorated and the heated die for transferring the coating to the part. It is important for the holding fixtures to have strong support under the area to be stamped. The heated die usually is a flat surface of metal or silicon rubber in applications where decorating is only on raised areas. The die is contoured for raised graphics on a contoured surface. Other than that, hot-stamping equipment is self-contained and needing no auxiliary equipment other than the compressed air source and electrical service. Virtually no parts finishing is required.

The cost of hot stamping is relative to the size of the surface to be decorated, the type of coating to be applied, the geometry of the surfaces involved, the size of the part itself, and volume considerations. Decorating media are usually sold at a cost per square inch. There may be different prices for different colors due to the different cost of pigmenting. Premiums are paid for metallic effects and multiple coating effects such as woodgraining or marbleizing. It is typical to hot stamp small items in large volumes (such as disposable drink stirrers) at costs of fractions of a cent per unit. It is also typical for parts to incur several dollars for the hot stamping (such as television cabinets, utilizing extremely large areas of decorative media).

Manufacturers that supply hot-stamp equipment and auxiliary devices such as foil-feed attachments, automatic part feeding systems and foil cutting devices are

Acromark Co. Inc., Berkely Heights, N.J.

Franklin Mfg. Co., Norwood, Mass.

Kensol-Olsenmark Inc., Melville, N.Y.

Peerless Hot Decorating Equip., Elmwood Park, N.J.

Transfer Print Foils Inc., East Brunswick, N.J.

United Silicone, Inc., Lancaster, N.Y.

Technical advise or consulting recommendations can be obtained from: Society of Plastics Engineers, 14 Fairfield Drive, Brookfield Center, Connecticut 06805. Ask to be referred to the current Decorating Division contact.

109.4 IN-MOLD DECORATING

As its name implies, in-mold decorating is a process in which a predecorated overlay or preform is placed in the mold, where the decorated element is fused to the molded part during the heating/cooling cycles of the molding operation. This can be done with either injection molding of thermoplastics or compression molding of thermosets.

In the case of injection molding, the procedure involves placement of the decorative component into the cavity of the injection mold prior to injection of the thermoplastic material. The decorative element can be loaded by hand or automatically. The manual operation usually involves use of a static wand to electrically charge the mold surface, causing the label to adhere to the mold surface. Automatic loading is readily accomplished with custom designed feed and take-off systems. These systems usually utilize a roll-feed concept in which labels are accurately positioned on a continuous strip to allow accurate indexing between the open cavity and core-mold components, and controlling registration before the mold closes for the injection cycle.

A variation of this process utilizes thermoformed sheet stock rather than foil or decal labels. This process requires the use of heavier sheet stock, formed to the exact shape of the surface to be decorated. The formed shape is die cut and the individual piece is then loaded into the injection-mold cavity. When resin is injected into the cavity, the two components are bonded into a one-piece molding with the decorated surface of the thermoform becoming the outside surface of the part.

In-mold decorating of thermoset materials requires a somewhat different procedure. In compression molding, where the thermoset raw material is loaded into the mold cavities in preform or powder state, it is not possible to preload the overlay or decal. This is because the thermoset requires high heat that would destroy the label element. The process therefore requires feeding the compression molding cavity with thermoset, then closing the mold and proceeding to mold up to a point just prior to final curing of the resin. At this point the mold is opened, the decal or overlay is placed over the plastic material, and the mold is again shut for completion of the in-mold decorating process.

A variety of thermoplastics are suited to in-mold decorating. Commonly used materials include polyethylene, polypropylene, polystyrene, ABS, polycarbonate, acrylic and the cellulosics. The in-molding principle also will accommodate functional engineering materials. In thermosetting, melamine is the most common material for in-mold decorating, chiefly because of its color capabilities.

In-mold decorating is a relatively predictable and controllable process. Its successful application requires the proper marriage of labeling and resinous materials. Decals can incorporate a variety of base films such as polyester, polyethylene, polypropylene, and polycarbonate. And they can be decorated by a variety of methods, from simple one-color silk screening to multiple coats of contrasting colors, tones, shadings and densities applied by lithography.

Producing foil for in-mold decorating is very similar to decal production, except that foil is supplied in continuous strips or roll form. Both foil and decals are supplied as thin materials, usually less than 10 mils (0.25 mm). The production of individual thermoformed inserts for in-mold decorating begins with the use of heavier stock, usually 0.02 in. (0.5 mm) thick, in sheet or roll form. The material can be decorated similar to decals or foil by using silk screen, lithography, or gravure printing. The foils utilized in thermosetting applications can be decorated in the same manner as those for the thermoplastic materials. The difference is the use of paper or fabric as carriers for thermosets as compared to thermoplastics materials for injection molding.

When in-mold decorating using decal or foil materials, it is important to load the film into the mold cavity so that the decorative coatings are bonded between the plastic part and the film (which will be the exterior surface). This results in a very durable decoration.

The molded-in decorating process is especially, useful when a number of colors are required to achieve the decorative effect. The decals, foils and thermoform stock can be decorated very economically and transferred in one step onto the plastic product. It has the advantage of providing a very durable decorative surface protected by the carrier film and coatings so much so that it is used in melamine dinnerware.

Part geometry is a limitation of this process. The ideal situation is to work on flat surfaces that are readily accessible when the mold is open. This allows for easy placement of the decals or thermoformed insert as well as for the controllable positioning during the molding cycle. It is possible to deviate from flat surfaces with certain areas and designs, and by using feeding and decal holding devices. This is especially true of thermoformed inserts. With proper attention to gating and material flow during injection, it is possible to produce complex shapes with the decorative surface conforming to those shapes.

When designing for in-mold decorating, part geometry, mold design, gating, knockout pin locations, leader pins and any other projections or interferences to feeding of the foil or inserts should be recognized.

In the case of thermoplastic in-mold decorative products, there are no secondary finishing requirements. But thermoset decal decorated products will require their typical deflashing and sanding operations. In-mold decoration involving thermoformed inserts requires very extensive "up front" equipment. In other words, total thermoform capability is needed, beginning with roll stock feeding into preheat stations then into a vacuum forming chamber to achieve close reproduction of thermform cavity shapes. Because these shapes must exactly match injection mold cavities, "rough" formed shapes require accurate die cutting equipment to trim them to proper size.

Resources from manufacturers of in-mold decorating equipment are rather limited. The majority of activity is involved with the manufacture of the actual decorating foil or decals, but there are some firms that specialize in the decal or foil-feed equipment such as:

Husky Injection Molding Systems Ltd.
530 Queen St. South
Bolton, Ontario, Canada
416/857-3240

A valuable source for processing information is manufacturers of the decorating media such as decals, foils or inserts.

Transfer Print Foils Inc., East Brunswick, N.J.

Gladen Corporation, Windsor, Conn.

Kurz-Hastings, Philadelphia, Penn.

Dri-Print Foils, Rahway, N.J.

For technical or consulting recommendations refer to:

Society of Plastics Engineers
14 Fairfield Drive
Brookfield Center, Conn. 06805
Decorating Division

109.5   EMBOSSING

Embossing is accomplished using a variety of procedures, but all produce the same effect: forming a tactile texture or pattern on sheet or film. The nature of the process (the use of heat and pressure to texture a semifinished substrate) limits embossing largely to thermoplastic materials. However, it can be adapted to thermoset composites, such as melamine-impregnated sheet stock.

Most commercial embossing is done with a two-roller system, in which one roller carries the embossing pattern and the other provides the essential pressure backup and feeding actions. The embossing roller has its texture (or pattern) applied through a variety of processes, including conventional engraving, chemical engraving, etching, and (more recently) laser cutting. The backup roller or forming roller can be made of various materials, with the choice dependent on the degree of texturing or patterning required. Typical rollers, are smooth steel, formed steel, or composites (rubber-coated steel). Matched steel rolls also are used for deep or very detailed embossing; this involvels the use of two engraved rollers. Such engraving is of male–female configuration, with the forming or backup roller having projections to match the depressions in the embossing or surface roller. The matched roller process provides maximum pattern definition at high roller speeds and reduces the need for roller maintenance.

A chemical-embossing process is available, in which the subject material is gravure printed using a combination of decorative inks to develop the pattern. One of the ink coatings contains an inhibiting agent that alters the material surface when it is heated.

Embossing can also be accomplished without rollers. A compression press or laminating press is used to texture sheet stock or laminates. In this process, textured aluminum foil is placed over the substrate and subjected to heat and pressure. This combined action transfers the pattern from the aluminum foil to the laminate; the foil is subsequently removed and discarded. An alternative to aluminum foil pressing offers multiple impression economies. Instead of using foil, textures are engraved on stainless steel plates that can be used in the press cycle time and time again.

Each of these embossing processes has advantages and disadvantages that must be considered. Roller systems are quite competitive for handling continuous film or sheet materials, for example. The foil or engraved steel laminating press approach is desirable for individual sheet stock preparation. Although each process must be judged individually, the number of embossing processes in conjunction with the variety of ways to produce textured surfaces offers great flexibility in meeting a variety of volume requirements and/or stock size requirements. Embossing is most frequently used as a method of decorating non-slip packaging materials, furniture laminates, building-panel laminates, vinyl wall coverings, textured foil for hot stamping, and other applications where the innate quality of three-dimensional printing can be a marketing advantage.

Manufacturers of equipment involved in printing and embossing are as follows:

Roto-Die Micrometrics, St. Louis, Mo.

Kaumagraph Corporation, Bloomington, Del.

Anderson & Vreeland, Inc., Fairfield, N.J.

Technical assistance and/or consulting recommendations can be obtained from:

Gravure Technical Association, 60 E. 42nd Street, New York, N.Y. 10017
Society of Plastics Engineers, 14 Fairfield Drive,
Brookfield Center, CT 06805
Decorating Division

109.6 VACUUM METALLIZING

Vacuum metallizing is a process of depositing metal onto plastic objects in an extremely low pressure atmosphere and at extremely thin deposition layers—measurable in fractions of microns. Central to this process is the use of a high vacuum chamber, which can range from very small 2 by 2 ft (0.6 × 0.6 m) laboratory units to high volume production units measuring 6 by 6 ft (1.8 × 1.8 m) or more. The function of the chamber is to hold and rotate the plastic parts to be decorated, to fixture the heat sources necessary to evaporate the thin metal coating, and to provide vacuum pumps for reducing internal pressures to less than one-half micron inch of mercury—required for achieving a clean, nondiscolored transfer to the plastic surfaces.

The metallizing process can be used on virtually all properly prepared thermoplastic and thermosetting materials. Different techniques are required for different materials in order to adapt the process most effectively.

Vacuum metallizing begins with preparation of the plastic part to be decorated. One constant essential is cleanliness, because the metallic coat will not adhere, and metallic finishes of all colors and shadings tend to magnify any surface imperfections. Attention to cleanliness starts right out of the injection mold, off the extruder or off the blow molder. If operators are involved, they should wear nylon gloves and the parts should immediately receive antistatic treatment, then be placed in contamination-free packaging.

Before parts are metallized, they usually require surface preparation in the form of a base coat. There are, however, situations where a base coat is not necessary—such as in transparent acrylic products where the aluminum is vaporized directly onto the surface and only requires back-up coating protection. An example would be steering-wheel hub medallions having three-dimensional second-surface (rear surface) effects. Almost always though, a base coat will be necessary to achieve a uniform high-gloss surface from the metallic deposit. The base coat may also be necessary to improve adhesion or to provide a barrier to the outgassing of certain materials when subjected to vacuum. There are many base coat materials on the market formulated to meet the requirements of specific plastic substrates. By working closely with suppliers of these coatings, it is possible to maximize quality in both appearance and function. This is a point of procedure whose importance cannot be emphasized.

Application of the base coat to the plastic object is by spraying or dipping. Parts can be manually sprayed (either one at a time or by racking small pieces on a holding rod or fixture). Parts can also be sprayed automatically, either singly or in a multiple, where groups of parts are racked. This is a common practice that uses machines, in which an operator loads a rack on a transverse-action mechanism that will pass the entire rack (full of parts) through a chamber housing rotating spray guns. In this method, parts are coated very uniformly cycle after cycle, without having to remove them from the rack as it travels through a baking atmosphere, into the vacuum chamber, out of the vacuum chamber, into a top coating operation, out of top coating, into a second baking operation, and then unloaded as a finished product.

Another coating process, but one that is very dependent on part geometry, is flow coating. Parts are mounted on racks or individual fixtures and transported through a spray environment or a dip into lacquer tanks that completely coat the parts and then spins off the excess lacquer. This assures a uniform, smooth finish. However, it is not advisable to apply this process to parts with narrow grooves or projections that would interfere with an even flow-pattern.

After base coating, the parts are baked to set up the coating, remove volatiles and enhance adhesion and appearance properties at temperatures that vary dependent on substrate and coating system used. Baking is usually done in electric or gas-heated forced-air ovens, or less often by UV systems, with the choice again dependent on economics, substrate and coating.

At this point, the parts are ready for loading into the vacuum chamber. Production applications entail a planetary carriage, a portable trolley that will hold and transport a large number of parts through a variety of fixturing devices to suit specific part geometries and sizes. In the planetary device, during the metal deposition, the parts are rotated to expose all surfaces of the objects to the metal source because the metal particles evaporate and travel in a straight line from the metal source. The entire loaded planetary trolley is placed within the vacuum chamber before the vacuum pumping cycle begins.

The pumping cycle usually involves two stages. The first stage is a roughing cycle that utilizes conventional piston-type vacuum or rotary pumps to evacuate the large initial volume of dense air out of the chamber. The second stage of vacuum pumping utilizes oil diffusion pumps to achieve the low vacuums required for proper metal evaporation. Oil diffusion pumps are stationary devices in which high-purity oil, formulated specifically for vacuum pumping purposes, is electrically heated. At an elevated temperature, the oil vaporizes and will rise within the diffusion pump chamber; as it rises it is impacted against down-slanting vanes. A cooling action, resulting from the oil rising away from the heat source and impinging on the downward acting vanes, traps air molecules in the chamber and forces them downward, densifying the air and moving it out of the diffusion pump. Pumping cycles can proceed quickly in newer high-capacity installations. Production cycles for even the largest chambers can be less than 15 minutes. When the low vacuum is achieved (approximately 0.5 micron in. of mercury), the aluminum vaporization action can take place. Electrically heated tungsten filaments melt the aluminum and subsequently vaporize it.

The vaporization cycle can be controlled manually or automatically. Heating of the filaments first results in the aluminum melting and, through capillary action, coating the surface of the tungsten filaments. At this point an additional electrical surge heats the filaments to a higher degree, causing the aluminum to evaporate and migrate from the heat source to cover the entire interior surfaces of the vacuum chamber. This "firing" of the aluminum takes only a few seconds; upon completion the vacuum of the chamber is released and the planetary fixtures are removed from the chamber. Parts are now ready for unloading.

The top coating procedure is typically a repeat of base coating, often using the same type of equipment. Top coating can also fulfill needs other than that of a protective coating. For example, it can incorporate the addition of texture or color to the metallic surface. Some interesting effects can be achieved with top coatings utilizing various colors and pigment levels. It is possible to achieve effects ranging from the very bright pure metal look (utilizing a clear top coating) to a very subdued, almost totally pigmented look with only a small degree of metalliic effect. Organic dyes used for metallic colors may not be light fast. Upon completion of the top coating application, parts are given a final baking to maximize their durability, adhesion, and environmental resistance (to perspiration or salt spray exposure, for example).

Metallizing is a very versatile process used in a great variety of applications. Examples range from highly decorative cosmetic closures to automotive grilles and instrument clusters. An important advantage is that it can replace metal parts with large savings in manufacturing costs and weight. The process can also serve functional needs, such as lamp reflectors or diffusion grids for overhead fluorescent lighting. The aluminum color looks like chrome. Since it will not corrode like chrome plating on metal, it has replaced such parts as automobile grilles and boat parts. Some applications require vacuum metallizing on interior surfaces of computer or communication equipment to achieve a degree of radio frequency interference shielding.

Some former limitations of metallizing are yielding to advanced technology. For instance, there have been tremendous improvements in both the baking temperatures that plastics can withstand and in the coating properties of formulations. Therefore, applications once closed to metallizing because of poor abrasion resistance, such as automotive grilles and instrument panel clusters, have become commonplace. It is also possible to metallize selectively by the use of masking devices. The decorative second-surface or three-dimensional see-through acrylic steering wheel buttons demonstrate this ability.

The evaporation process allows considerable design latitude for parts to be decorated by metallizing, because metallic finishes can be applied to rather confined areas as well as uniformly coating irregular contours.

It is necessary to allow for part tolerance changes caused by coating thicknesses that result from the base and top coat operations. These coatings can be as little as 0.001 in. (0.025 mm). Also, the baking operations used in metallizing dictate that care be taken in the design phase to minimize molded-in stress.

Vacuum metallizing equipment is expensive. A production-scale vacuum chamber and pumping system of 6 by 6 ft (1.8 by 1.8 m) can cost around $150,000. Coating equipment to support the process can be as inexpensive as a manual spray gun or as elaborate as automatic flow-coating or "ride-the-rod" systems which can cost more than $100,000.

A prime consideration is the need for atmosphere control, for health, processing parameters and assured cleanliness. Humidity control is extremely important, both for the application of coatings and to enhance pumpdown cycles (which are greatly extended in humid atmospheres). The cost of floor space, makeup air, exhaust systems and atmosphere-control investments can very easily exceed those of the vacuum system and coating equipment.

There is usually very little tooling required for metallizing individual parts, once the initial planetary fixturing is acquired. Holding devices or nesting fixtures are relatively simple; often a universal device can be used to hold a variety of parts. Tooling expense is greatest when selective plating is required. Masking may be necessary for both the coating operations and the metal deposition process. These masks can be made of metal stampings, vacuum formings, electroforms or pressure forms. Part geometry dictates the most feasible and/or economical choices.

Despite the relatively high cost of equipment and atmosphere controls, parts can be metallized very economically. The process is ideally suited to high production (and low unit cost), especially when using batch type processing. High volume applications, such as cosmetic container closures, can be decorated for pennies. Metallized large parts, such as automotive grilles are much less expensive than chrome plating. This is due in part to the low cost of materials used in the metallizing process.

There is some labor intensity in vacuum metallizing, even with the advance of automation. The labor factor is greatest in handling parts for racking and movement. Vacuum metallizing is typically an in-house capability, performed in line with the molding and other secondary operations. There are, however, many custom processors who specialize in vacuum metallizing.

Manufacturers that supply high vacuum equipment are

Stokes Division, Pennwalt Corp., Philadelphia, Penn.
Vacuum Technology Div. of Atlan-Tol Industries Inc., Cranston, R.I.

Manufacturers that supply equipment for the coating applications required in vacuum metallizing are

Binks Mfg., Franklin Park, Ill.
Conforming Matrix Corp., Toledo, Ohio
Devilbiss Co., Toledo, Ohio

Suppliers of base coatings and top coatings developed specifically for the metallizing process are

Bee Chemical Co., Lansing, Ill.
Red Spot Paint & Varnish Co., Evansville, Ind.

A source for technical information and/or consulting recommendations would be:

Society of Vacuum Coaters
1133 Fifteenth St. N.W.
Washington, D.C. (202) 429-9440

109.7 SPUTTERING

Like vacuum metallizing, sputtering is a process for depositing permanent metal coatings onto plastic substrates. It satisfies some decorative and functional demands, (ie, chrome effects and resistance to oxidation or corrosion) that conventional metallizing with aluminum cannot.

The term sputtering describes the removal of metal atoms from a base (metal) material utilizing a bombardment action. The bombardment is made of ions from a gaseous source such as argon and controlled within a magnetic field. The energy of the argon ions bombarded onto the target metal results in metal atoms being dispersed throughout the vacuum atmosphere and impinging or coating all surfaces exposed to the source material.

Sputtering can be used to coat virtually all thermoplastic and thermosetting materials. The process is virtually identical to vacuum metallizing. Its chief advantage is the ability to deposit a variety of metal coatings on plastic substrates. Metals and alloys such as chromium, stainless steel, copper, brass, titanium, tungsten, and aluminum can be utilized. This variety makes the sputtering process very useful for achieving desired decorative effects at reasonable cost. For example, the automotive industry accelerated the use of sputtering to meet its needs for economical chromelike coatings on front grilles, wheel covers and large interior instrument cluster trim pieces. The cosmetic, small appliance, furniture, and toy markets likewise have taken advantage of the decorative capabilities of sputtering. The ability to deposit metals that resist corrosion and oxidation, such as stainless steel alloys, has prompted the use of sputtering in the electronic industry for use on semiconductor integrated circuitry.

Metal deposition by sputtering is much thicker than that of metallizing, although the two processes are similar. Aluminum films deposited via vacuum metallizing will be in the micron inch range—a thickness of 1.5 microns (1.5×10^{-6} mm) is typical. Attempting to deposit heavier thicknesses of aluminum in vacuum metallizing results in loss of the bright aluminum finish due to excessive heat and oxidation during the evaporation process. The sputtering process can deposit heavier thicknesses, up to 10 micron in. (10^{-5} mm), without such impairment.

Product design considerations for sputtering and vacuum metallizing are identical. Allowances must be made for the coating buildups, positioning of the parts during the evaporation process, and all the handling considerations involved in achieving a quality decorative effect. Preparing plastic articles for sputtering requires all the care and consideration involved in vacuum metallizing. Cleanliness must be monitored and maintained at all processing stages, from part production through the final top coating operation.

Production of large articles and volume quantities of products can be achieved with vacuum systems adapted for sputtering. A typical installation will incorporate a large-diameter vacuum chamber five feet (1.5 m) and over in diameter and five or six feet long (1.5 or 1.8 m). These chambers are designed to be hoisted up and off the work load, which usually consists of several revolving planetary fixtures that hold items firmly in place via a variety of attachments and/or masking devices, all related to product geometry.

The heart of the sputtering process is its target or cathode component. The target is made of the metal to be sputtered. It is also possible to have this component plated with the metal to be sputtered, thereby allowing the target shape to be retained for future use. Once the plated coating is consumed, replating can make it reusable. The cathode is centrally located within the vacuum chamber, and is positioned between outer anode locations. An internal magnetic assembly generates a magnetic field to control the path of orbiting electrons produced by the ionizing material, argon gas.

With the sputtering arrangement centered in the vacuum chamber, the production parts are located around the sputtering source, usually fixtured or racked on rotating carriages. The entire chamber is then put through the pump-down sequence. This starts with the roughing vacuum pumps (rotary or piston types) and ends with diffusion pumping to lower pressures to the micron inch of mercury range. Sputtering starts at this level and usually lasts for several minutes, with sputtering action initiated by introducing argon gas into the vacuum chamber. At this stage, an electric field is activated between the anode and cathode arrangement; within this magnetic field the argon ions impinge on the cathode, causing metal atoms to dislodged from the target, radiating outward within the chamber and landing on production objects. The entire operation takes several minutes, after which the vacuum chamber is returned to atmospheric pressure. Production parts are then unloaded and the system is ready for the next cycle.

Tooling for the sputtering process includes racking and/or masking devices to hold production parts. The racked parts are mounted on carriages or rotating planetary devices to facilitate alignment of surfaces to be sputter coated. The geometry of the production items will dictate the complexity of the racking or fixturing required. It is not unusual for fixturing to consist of very simple standard spring steel prongs or hoods. It can also be as elaborate as individually electroformed masks, pressure formed masks and holding fixtures for articles requiring selective coating. Auxiliary equipment also is necessary for base coating and top coating operations, as described in the foregoing section on vacuum metallizing.

A sputtering process setup will usually cost in excess of several hundred thousand dollars.

Most systems are custom designed, and therefore relatively more expensive. Compared to alternative methods such as electroplating, however, it can be a very competitive process.

Manufacturers specializing in sputtering equipment:

Central Engineering, USA Limited located in South Windsor, Conn.
Leybold-Heraeus in Enfield, Conn.

A source for technical and/or consulting recommendations:

The Chemical Coaters Association, Box 241, Wheaton, IL 60187; (414) 353-4200.

109.8   ELECTROPLATING

Electroplating is a chemical process for depositing heavy metals on plastic substrates to achieve decorative effects and/or upgraded functionality. The preplating process, developed in the late 1950s has resulted in the ability to prepare a plastic surface with the characteristics necessary to accommodate a normal electroplating process. The relationship between pre-plating on plastic and subsequent electroplating on plastic is a critical one, and has resulted in a continuing improvement of the decorative and functional properties that can be achieved.

The key to electroplating on plastics is preparation of the plastic surface to receive metallic coatings that will be made electrically conductive. The principle of electroplating is to electrically conduct metal atoms such as copper, nickel and chrome off anodes placed within the plating baths through the plating solutions and onto the plastic production target. The target (or production part) acts as a cathode via connection to conductive plating racks. Conductivity of the plastic part is accomplished by attaching the part to the plating rack with metal holding devices, spring-loaded contacts or prongs. This arrangement securely attaches the plastic production article to the rack with its own contact point forming the continuity of the current flow from anode through the solution onto the plastic part.

However, electroplating on plastic is limited to a relatively few plastic materials. ABS is by far the most dominant material in this field. But because many end-use requirements cannot be met by ABS (such as automotive window cranks) other materials has been found that provide a match of physical properties and decorative capabilities. Commercially elec-troplatable materials now include filled nylons, filled polysulfones, modified poly(phenylene oxide), and modified or blended polycarbonates.

Electroplating on plastic is a complex procedure that requires total cooperation beginning with raw material formulation through molding and into the actual plating system. Materials to be electroplated are formulated specifically to meet molding and plating criteria. The molding process also is a very critical phase in proper plating procedures. Therefore it is extremely important that there be close communication between the molder or processor and the plater. Molding and plating under one roof offers the best insurance for quality and cost control.

Proper plating requires that molded parts be as free of stress as possible, and with handling procedures in place that will minimize poor plating adhesion (caused by overall high stress in the molded part, local stress points such as improper gate areas, contaminants that may result from handling, raw material contamination, and other factors). Many of these potential defect causes are not visible to inspection procedures and will only show up after plating—such as loss of adhesion due to molded in stress. One way to control accumulation of defective parts is to coordinate molding and plating so that results can be checked as quickly as possible.

The plating process begins by racking of molded parts. There are two options in handling parts through the entire plating process. One is to carry the parts on the same racks through preplating and electroplating. The other is to rack parts twice; once for the preplate process and again for the electroplating process. The choice is dependent on a variety of inputs. But basically it is a decision as to the amount of labor necessary to rack twice versus the single racking procedure. Experience and differing part geometries and size will dictate.

There is in fact a great degree of trial and error involved in achieving optimum quality in plating on plastic. The process begins with racked parts introduced into the preplate section. This involves a part-cleaning operation, a chemical etching action to remove the rubber and provide minute undercuts for good adhesion through various baths to neutralize the surface, and catalyzation to prepare the surface for a deposit of electroless nickel or copper. This preplate procedure is designed to create a surface on the plastic parts that will develop a bond between the plastic and the first nickel or copper deposit. These initial deposits are extremely thin, in the micron in. (10^{-6} mm) area.

When preplating is completed (and the plastic articles have a conductive coating), it is possible to proceed to the electroplating operation, which is very similar to conventional electroplating on metal. However, the idiosyncrasies of plastic properties demand close control of plating thicknesses in order to achieve desired final quality of adhesion, appearance, thermocycling, chemical resistance, etc.

The first step in electroplating is to move racked parts through a cleaning solution and into the first electrical deposit of nickel or copper in the micron in. (10^{-6} mm) range. This first deposit is designed to increase conductivity uniformly over the plastic surfaces without building up stress or concentrated conductive areas. The parts then proceed through conventional heavy metal plating baths, interspersed with rinse and cleaning baths. A 0.5 to 1 mil (0.01–0.25 mm) deposit of ductile copper comes first, followed by a somewhat thinner nickel coating, followed in turn by a final flash plating of chrome, applied in fractions of a mil (0.025 mm). The parts are then again put through a rinsing and cleaning operation to complete the electroplating process. The copper–nickel–chrome process is designed to meet most commercial plating demands. Precise control of each plating thickness is necessary to achieve the essential performance properties, such as thermocycling, adhesion, perspiration resistance, and gloss. There are, however, some applications that require an extra step in order to meet specific end-use requirements. For example automotive exterior salt-spray resistance requirements are achieved by adding another metal layer to the plating. This additional metal would be a "dura-nickel" or dull nickel plate, deposited on the bright copper before the bright nickel deposit.

Electroplated plastic parts have all the advantages of conventional chrome plating on metals. Obviously, the decorative results are important, but of equal importance are such functional advantages as durability, wear resistance, corrosion, conductivity, impact improvement over chrome plated metals, and design flexibility. Plated plastics are usually cost competitive with metal counterparts. Decorative variations of the plastic plating processes are available, including satin finishes, pure gold, black chrome, brasses, etc. Limitations of the process are typically related to limitations of the plastic substrate being used. Some base materials may not have adequate tensile strength, compressive strength, flex strength, temperature resistance or electrical properties. In addition, plating can alter plastics properties, so it is important to analyze each application to ensure that the final plated product will meet specifications.

It is extremely important that parts be designed to accommodate the plating process. For example, blind holes can cause plating problems by carrying solutions from bath to bath, possibly resulting in surface and/or physical property defects. Sharp corners, sharp parting lines or sharp edges will concentrate electrical current, resulting in plating build-up that may be not only unattractive but hazardous as well. Flat surfaces tend to result in poor appearance quality, but can be improved with slight crowning of the surfaces. Any flash will plate as a razor sharp edge.

Many of the part design considerations that go into proper molding to relieve stress are important. High stress areas will result in adhesion and/or appearance problems. Just plain good design principles, such as uniform wall sections, will greatly enhance plating quality.

Consideration must also be given to contact points for the electroplating portion of the process to ensure good control of plating continuity and plating thicknesses as well as hiding the contact area (which is almost always an undesirable blemish area). Gating location and gating design are important to control molded-in stress as well as improve surface finish. The overall size and geometry of parts to be plated must be evaluated, since they will be going through heated plating baths that tend to stress-relieve and change geometry if consideration is not given to thermal conditions. It is essential to determine allowable tolerances at the design stage to avoid excess plating build-ups that can result in tolerance control problems.

The plating process itself can result in some variables. Even when achieving desired quality levels of the plated surfaces, there may be variations in plating thicknesses and changes in tolerances of the finished products part-to-part, batch-to-batch and day-to-day. However, if the problem is addressed in the initial design phase, it can be avoided during production. Parts rejected for plating defects usually can be stripped and replated.

Electroplating equipment is typically specified to generate a certain amount of plated surface per hour. The principles of plating and the process will be very similar, regardless of the degree of production required. Once production objectives are determined, it is possible to design the plating process to meet that level of output.

Although hand-plating lines are available for electroplating, automation for high volumes is more usual. Basically, there are two automation options. One is the "return" system, with plating tanks placed in an oval or rectangular configuration with automatic rack-carrying devices to move the racks from bath to bath and tank to tank. Operators are needed only to load parts on a rack. The machine takes over at that point, going completely through the preplate, electroplate, unloading, rack stripping, and cleaning cycles, then back to the starting point. This system has the advantage of great automation but is somewhat limited in that it is continuous and uninterruptable. Time in the baths is controlled by the size of the baths rather than discretion of the operator. Return type systems usually have small individual tanks and use small individual racks. This can be cost-inefficient for plating large parts in volume, but it provides advantages in the close controls that are achievable.

Another automation option is the hoist system. Progression of racks through the system is usually preprogrammed to control the amount of rack dwell time within individual baths and the sequence rack travel to baths. There is thus a lot of flexibility to a hoist system; preprogramming offers the opportunity to skip baths and alter cycles should specific applications so require. The hoist system also allows for the use of much larger racks and plating tanks. One disadvantage is less control within the individual baths because of the large size of racks and baths.

A plating installation is an expensive investment. Even hand lines require a considerable capital outlay. An automatic return or hoist type plating system for producing approximately 250 square feet of plated surface per hour will easily cost in excess of $500,000 for the equipment only. Beyond this investment is the expense of chemicals, metal charging, initial racking, and floor preparation. After that, there still remains the cost of a waste-treatment system necessary to make the plating process legal. Its cost is at least equal to the basic equipment investment. Once the discharge waters are under control, attention must be paid to the disposability of accumulating heavy metal sludges or cake. State, City and EPA monitor electroplating plants very carefully.

Tooling for plating on plastics involves the racking components. Typically each change in part configuration to be plated will require new plating racks. Racks are not very expensive, but labor intensive. A typical rack to hold approximately 20 or 30 small electric razor housings may cost as little as $200. However a plating tank requires many racks. To maintain automation the cost of the large number of racks becomes a factor. Additionally, the limited life of racks require figuring replacement costs into the part cost.

A considerable amount of auxiliary equipment is required to operate, control and monitor plating on plastics. Part of this investment is necessary to control plating quality; equipment includes hot/cold thermo cycle chambers, and fixtures to check plating thicknesses, plating adhesion and chemical resistance. Laboratory equipment is needed for constant surveillance of plating machine baths relative to their chemical make-up, consistency and quality. There is also the need for equipment to monitor waste-treatment performance. Typically, this will require an atomic absorption analyzation to control water quality relative to the parts per million of copper, nickel, and chrome it will contain. Maintenance and control of the plating baths will also require a heat source, normally steam boilers, to generate and control bath temperatures. Very extensive and elaborate make-up air and air exhaust systems are necessary to control toxic and/or corrosive fumes given off by various baths in the process.

Plating on plastics can be an in-house capability, primarily due to the close coordination required between molding and plating. There are, however, job shop electroplaters.

The American Society of Electroplaters is an organization involved in the plating process and related activities in markets, processing and environment, and environmental involvement.

Manufacturers supplying equipment for the electroplating industry, which include plating machinery, plating chemicals and/or the waste treatment equipment, are

Harshaw Chemical Co., Cleveland, Ohio
MacDermid Inc., St. Louis, Mo.
Napco, Inc., Terryville, Conn.

Sources for technical and/or consulting recommendations are

American Society of Electroplated Plastics Inc.
1133 15th St. N.W.
Washington, D.C. 20005

American Electroplating Society, Inc.
1201 Louisiana Avenue
Winter Park, FL 32789

Barrel plating is a variation of electroplating, involving an alternative part-handling procedure. The principles are the same. However, barrel plating deviates from the contact-and-attachment rack design of electroplating to what could be described as a bulk plating concept. This concept also involves the use of metal anodes within the plating solution, but replacing the racking cathode design with a barrel design. (The barrel replaces the rack and becomes a cathode in the plating process).

With the barrel being the cathode and conductive to the plating machine electrical source, parts to be plated are loaded within the confines of the barrel. Barrel size [5 to 50 gal (0.02–0.2 m³)] and geometry will vary depending on capacities and geometries of production required.

In the production process, plastic parts are loaded into the barrel, not completely filling it. The loaded barrels are then submerged into the appropriate plating tanks and fully exposed, by rotation, to the current field of anode to cathode. The rotating action moves the plastic parts and mixes them to the point that, theoretically, each part will have some of its surface in contact with parts having contact with the barrel. Thus, all of the parts will receive some conductivity, making them receptive to metal deposition.

Barrel plating is generally used for functional, rather than decorative, coatings. It also is usually restricted to noncritical, nondecorative applications and plating of small items.

Reference material for barrel plating is the same as for electroplating.

110

RHEOLOGY

Pravin L. Shah

The Polymer Corp., 2120 Fairmont Ave., P.O. Box 422, Reading, PA 19603

Rheology is the science of deformation and flow of matter. The word rheology is derived from the Greek *rheos*, the study of flow. A body deforms when application of an appropriate force system alters its shape or size. A body flows when the degree of deformation changes with time.

Essentially, all thermoplastic resins (and many thermosetting resins) are required to undergo flow in the molten state during the course of product manufacture. Such important processing operations as extrusion, molding and calendering all involve the flow of molten polymers.

From the standpoint of the chemistry of macromolecules, the term "melt" should be applied only to semicrystalline polymers. In such systems, the crystalline regions undergo melting in a fixed temperature range; above melt temperature, the polymer behaves as a fluid rather than a solid. Amorphous polymers soften above their glass-transition temperature (T_g) and exhibit flow behavior similar to semicrystalline polymers. The flow behavior of polymeric melts cannot be considered to be purely viscous in character. The response of such materials is more complex, involving characteristics that are both viscous and elastic.

Understanding of rheology is important for a polymer manufacturer, who must tailor the company's products with respect to flow properties for a given application. Similarly, a process engineer finds viscous and elastic flow properties very useful in choosing the proper material or designing the proper fabricating device for his or her purpose. Such knowledge may relate either to the inherent behavior of the polymer melt or to its molecular origin, or both. In polymer processing, it is important to understand the dependence of melt viscosity on such variables as temperature, pressure, rate of shear, molecular weight, and structure. It is equally important to have available reliable experimental means of measuring viscous properties of a material.

Rheological behavior influences the mechanical behavior of a finished product. For example, orientation has dramatic effects on films, fibers, and blow-molded products. Of equal significance is the phenomenon of elasticity in the study of the rheology of high polymers. Elasticity manifests itself in elastic shear moduli, normal stresses, and die swell in the extrusion process. The basic cause of elasticity is the orientation of molecular segments during flow.

110.1 FUNDAMENTALS

A comprehensive discussion of the fundamentals of rheology is presented by Frederikson.[1] This chapter, will review the Newtonian and non-Newtonian flow behavior of various materials and present the flow models to describe types of fluids and their relationship of viscosity, shear stress, and shear rate. It will also cover the effects of temperature, pressure, shear rate, and molecular weight on polymer viscosity.

Classification of Fluid Behavior. An ideal fluid is one that is incompressible and has zero viscosity. It offers no resistance to shearing forces; hence, all shear forces are zero during the flow and deformation of an ideal fluid. Mathematical relationships have been developed for ideal fluids to describe their behavior for different physical situations, but their practical application was limited until the introduction of the boundary layer theory by Prandtl.

Prandtl showed that during the flow of liquid in a tube, frictional effects are confined to a thin fluid layer (called the boundary layer) adjacent to the solid surface. Flow outside the boundary layer is considered frictionless; therefore the mathematical models for ideal fluids can be applied in this region of flow. During flow of a *nonviscous* fluid past a solid boundary, the fluid adjacent to the boundary has a finite velocity determined by flow conditions and boundary geometry. For the flow of *viscous* fluids, no slip between the fluid and solid boundary is assumed. Due to limitations in the application of frictionless flow models to actual physical situations, classical theory of fluid dynamics was developed for Newtonian fluids, named after Sir Isaac Newton. Newtonian fluids are those for which viscosity is independent of the rate of shear. A typical shear-stress–shear-rate plot for such a fluid is shown in Figure 110.1 (Materials that cannot be characterized by Newtonian relationship are called non-Newtonian fluids, for which the viscosity at a given temperature is a function of the viscosity gradient.)

Non-Newtonian fluids can be further classified according to the manner in which viscosity varies with rate of shear.

Pseudoplastic materials are those in which viscosity decreases with rate of shear, but the material deforms as soon as shearing stress is applied. This class of materials has probably the greatest industrial application, since all high polymers and polymer melt follow this behavior. Figure 110.2 shows shear-stress shear-rate behavior of common non-Newtonian fluids, including Bingham (those that deform when they reach yield point) and Dilatant (those that show viscosity increase when shear rate rises). Comprehensive details of the classification of non-Newtonian fluids is covered by Skelland[2] and Frederikson[1].

Flow Models. Many mathematical models exist for describing the flow behavior of non-Newtonian fluids. The starting point for all non-Newtonian flow models is the basic isothermal Newtonian concept with modifications to fit the physical situation. Comprehensive details of the development of flow models have been published.[2-5]

A Newtonian flow model is derived as follows. For a fluid located between two parallel plates (*a*) moving and (*b*) stationary, separated by a distance (*dy*), the application of force

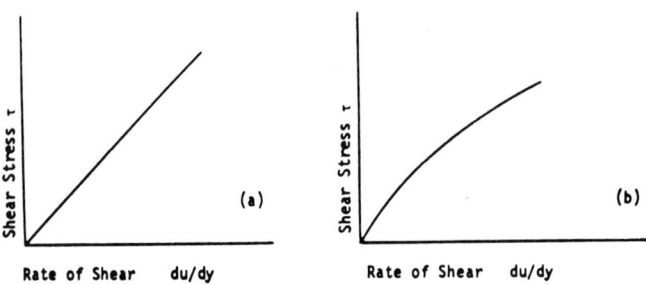

Figure 110.1.   Shear-stress and shear-rate relation for **(a)** Newtonian and **(b)** non-Newtonian fluids.

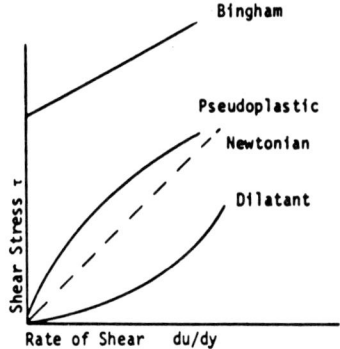

Figure 110.2. Shear-stress and shear-rate behavior of non-Newtonian fluids.

P to plate (a) results in a constant velocity (du) relative to plate (b). Assuming laminar flow and steady state, shear stress τ_{yx} can be defined as

$$\tau_{yx} = \frac{P}{A} \tag{110.1}$$

where A is the surface area of plate (a) (Fig. 110.3), subscript y shows the orientation of the surface, and x shows the direction in which stress acts. In time, t, a fluid particle will travel distance dut and the flow deformation will be

$$\gamma = \frac{du\,t}{dy} \tag{110.2}$$

where γ is defined as a strain.

Differentiating the above equation, one obtains rate of deformation:

$$\frac{d\gamma}{dt} = \frac{du}{dt} \tag{110.3}$$

where d/dy is rate of shear or velocity gradient.

According to the law for Newtonian fluids, shear stress is directly proportioned to rate of shear in laminar motion. Therefore:

$$\tau_{yx} = \mu\,\frac{du}{} \tag{110.4}$$

where γ is the Newtonian viscosity coefficient.

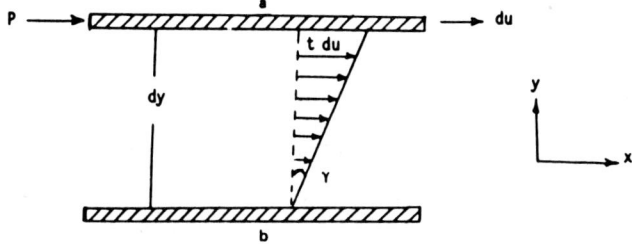

Figure 110.3. Laminar shearing motion between two parallel plates.

Equation 110.4 is the basic flow equation for Newtonian fluid and the starting point for non-Newtonian flow models. For laminar flow on non-Newtonian fluids, commonly used models are shown below.

Power law model:

$$\tau_{yx} = \mu \left(\frac{du}{dy}\right)^n \tag{110.5}$$

where n is defined as the flow index. For high polymers, n varies between 0 and 1, and it determines the degree of non-Newtonian behavior. For $n = 1$, equation 110.5 becomes the same as equation 110.4.

Prandtl Eyring model:

$$\tau_{yx} = A \sin^{-h} \left[\frac{1}{B} \left(\frac{du}{dx}\right)\right] \tag{110.6}$$

where A and B are constants.

Sisko model:

$$\tau_{yx} = A \left[\frac{du}{dy}\right] + B \left[\frac{du}{dy}\right]^n \tag{110.7}$$

Effect of Temperature on Polymer Viscosity. The viscosity of most polymers changes with temperature. The temperature dependence of viscosity for Newtonian fluids and for most polymer fluids at temperatures above the glass-transition temperature, follows the Arrhenius equation to a good approximation:

$$\eta = K e^{E/RT} \tag{110.8}$$

where η is the viscosity of the polymer, K is a constant characteristic of the polymer and its molecular weight, E is the activation energy, R is the gas constant and T is the temperature in degrees Kelvin. Constants in the Arrhenius equation can be evaluated by plotting the logarithm of viscosity against the reciprocal of absolute temperature, using shear stress or shear rate as a parameter. The data for most materials given straight lines over reasonably large range of temperature. Table 110.1 shows energy of activation for flow of certain polymers.

For amorphous polymers at temperatures less than 212°F (100°C) above glass-transition temperature, the Arrhenius equation does not fit the data well. In particular, one must be careful when extrapolating to lower temperatures in the region of T_g, where the activation energy is changing rapidly. In this region, the Williams-Landel-Ferry equation fits:

$$\eta = K \exp\left[1 - \frac{17.44(T - T_g)}{51.6 + T - T_g}\right] \tag{110.9}$$

Effect of Pressure on Viscosity. Viscosity is a function of the intermolecular forces that restrict molecular motion. These forces depend on the intermolecular distances, or the free volume. Therefore, changes in the parameters that affect free volume will also affect viscosity. Like temperature, pressure is another variable that causes changes in viscosity.

The free volume of a liquid is defined in various ways, but a common definition is the difference between actual volume and a volume in which such close packing of the molecules occurs that no motion can take place. The greater the free volume, the easier it is for flow

TABLE 110.1.   Energy of Activation for Flow of Polymers

Polymer	Energy of activation E,	
	kcal/g · mol	(kJ/g · mol)
Dimethyl silicone	4	(16.7)
Polyethylene (high density)	6.3–7.0	(26.3–29.2)
Polyethylene (low density)	11.7	(48.8)
Polypropylene	9.0–10.0	(37.5–41.7)
Polybutadiene (*cis*)	4.7–8	(19.6–33.3)
Polyisobutylene	12.0–15.0	(50–62.5)
Poly(ethylene terephthalate)	19	(79.2)
Polystyrene	25	(104.2)
Poly(α-methyl styrene)	32	(133.3)
Polycarbonate	26–30	(108.3–125)
Poly(1-butene)	11.9	(49.6)
Poly(vinyl butyral)	26	(108.3)
SAN (styrene–acrylonitrile copolymer)	25–30	(104.2–125)
ABS (20% rubber) (acrylonitrile–butadiene–styrene copolymer)	26	(108.3)
ABS (30% rubber)	24	(100)
ABS (40% rubber)	21	(87.5)

to take place. While free volume increases with temperature due to thermal expansion, an increase in hydrostatic pressure decreases free volume and increases the viscosity of a liquid.

Pressure can be quite high in capillary rheometers, injection-molding machines or extruders. In such operations, higher pressure can cause viscosity increase. However, viscous heating due to high shear in such operations results in lower viscosity than that due to pressure without viscous heat generation.

Effect of Shear Rate on Viscosity. Most polymer melts exhibit non-Newtonian behavior (apparent viscosity decreases as the rate of shear increases). Viscosity of polymer at high shear rates may be several orders of magnitude smaller than the viscosity of low shear rates. A typical shear rate vs viscosity curve is shown in Fig. 110.4. The melt has a Newtonian viscosity which is high at very low shear rates. Viscosity decreases nearly linearly with shear rate when plotted on a log-log scale (Fig. 110.5). In this linear range, the power law equation is applicable:

$$\tau = K\dot{\gamma}^n \qquad (110.10)$$

where K and n are constants, $\dot{\gamma}$ is shear stress and is rate of shear. For Newtonian liquids, $n = 1$: the value of n is less than 1 for non-Newtonian polymer melts.

The basic cause of non-Newtonian behavior of polymer melts is the orientation of molecular segments by the flow field. The phenomenon of elasticity is manifested from the orientation behavior of polymer melts. Entanglements increase the chances of orienting molecular segments.

Effect of shear rate on viscosity is explained by several theories. Grassley[6] and Bueche[7] assume that these are molecular entanglements which decrease as the rate of shear increases. These theories approximate the power law equation at the higher shear rates. For example, Bueche theory predicts that $n = 0.5$ in the power law equation for monodisperse polymers.

Effect of Molecular Weight on Viscosity. The molecular weight of a polymer is the most important factor affecting rheology. Polymer viscosity is approximately proportional to the

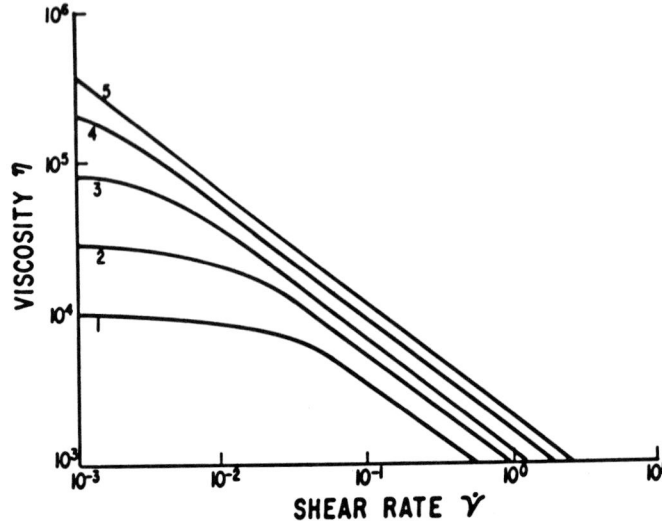

Figure 110.4. Typical log viscosity-log shear rate curves at five different temperatures. Curve 1 is for the highest temperature, and curve 5 is for the lowest temperature. For a typical polymer, the temperature difference between each curve is approximately 10°C.

weight average molecular weight (M_w) below its critical molecular weight (M_c). This relationship is expressed as follows:

$$\eta = K_1 M_w \text{ for } M_w < M_c \tag{110.11}$$

The viscosity of a polymer depends on M_w to a power equal to 3.5 at molecular weights above M_c.

$$\eta = K_2 M_w^{3.5} \text{ for } M_w > M_c \tag{110.12}$$

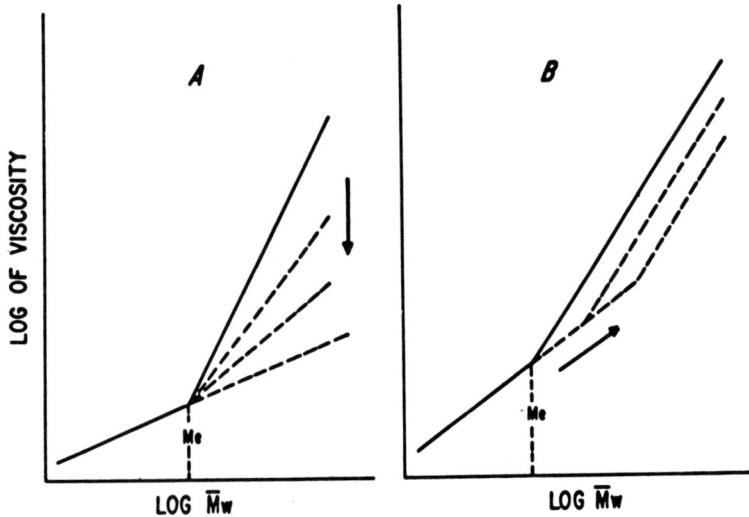

Figure 110.5. Two ways in which melt viscosity varies with molecular weight. Solid lines denote n_0. Dashed lines are for higher shear stresses, which increase in the direction of the arrows.

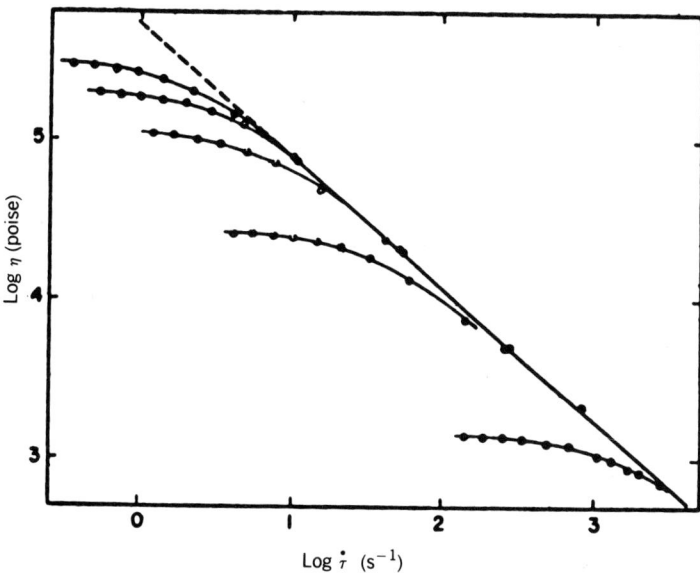

Figure 110.6. Viscosity as a function of shear rate for polystyrenes of different molecular weights at 183°C. In going from left to right the curves are for molecular weights of 242,000, 217,000, 179,000, 117,000, 48,500. From Ref. 8. Courtesy of Academic Press.

For most polymers the critical molecular weight is between 5,000 and 40,000. The relationship between molecular weight and viscosity is shown in Figure 110.6, where typical viscosity whear rate curves for polystyrenes are shown.

The distribution of molecular weights in a polymer also influences its rheology. For polymers with narrow molecular weight distribution, the weight average molecular weight is the most important. Polymers with broad molecular weight distribution are easier to extrude than those with narrow distribution.

110.2 INSTRUMENTS TO MEASURE FLOW PROPERTIES

In order to carry out meaningful engineering calculations for non-Newtonian fluid systems, it is essential to know the appropriate rheological properties. Many special purpose viscometers have been developed over the years; details of the commercially important ones have been published.[9-11]

Capillary Rheometers. These rheometers are widely used to study the rheological behavior of molten polymers. They are devices in which the relationship between shear rate and the shear stress is inferred indirectly from measurements of the pressure gradient and volumetric flow rate in a cylindrical tube of precisely known dimensions. As shown in Figure 110.7, a liquid polymer is forced by a piston or by pressure from a reservoir through a capillary and maintained at isothermal conditions by electrical temperature control methods. Extrusion pressure or volumetric flow rate can be controlled as the independent variable, with the other being the measured dependent variable. Under steady flow and isothermal conditions for an incompressible fluid, the viscous force resisting the motion of a column of fluid in the capillary is equal to the applied force tending to move the column in the direction of flow. Thus,

$$\tau = \frac{D\Delta P}{4L} \tag{110.13}$$

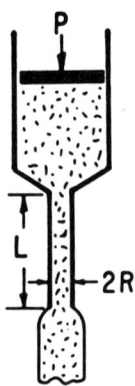

CAPILLARY

Figure 110.7. Capillary rheometer.

where R and L are radius and length of the column and P is pressure drop across the capillary. Shear stress τ is zero at the center of the capillary and increases to a maximum value at the capillary wall. In normal capillary rheometry for polymer melts, the flowing stream exits into the atmosphere, and the driving static pressure in the reservoir is taken to be ΔP. In such cases, "end effects" involving viscous and elastic deformation at the entrance and exit of the capillary should be taken into account when calculating the true shear rate at the capillary wall.

Shear rate in the capillary rheometer is calculated as:

$$\dot{\gamma} = \frac{4Q}{\pi R^3} \tag{110.14}$$

when Q is the volumetric flow rate through the capillary under a pressure drop P and R is capillary radius, and $\dot{\gamma}$ is the shear rate. Melt viscosity is expressed as:

$$\eta = \frac{\pi R^4 P}{8LQ} \tag{110.15}$$

or

$$\eta = \frac{\tau}{\dot{\gamma}} \tag{110.16}$$

There are three chief reasons why the capillary rheometer is widely used in the plastics industry:

(1) Shear rate and flow geometry in a capillary rheometer are very similar to conditions actually encountered in extrusion and injection molding. Therefore, melt viscosity data are more meaningful to correlate from a capillary rheometer to extrusion or injection molding of thermoplastics.

(2) A capillary rheometer typically covers the widest shear rate ranges; therefore, a process engineer finds it most practical to use and apply the data from a capillary rheometer to extrusion or molding processes.

(3) A capillary rheometer provides good practical data and information on the die swell, melt instability, and extrudate defects.

The main disadvantage of a capillary rheometer is that the rate of shear is not constant but varies across the capillary. Another disadvantage is the need to make a number of corrections in order to obtain accurate values of viscosity.

The measured values of polymer flow taken by capillary rheometers are often presented as plots of shear stress versus shear rate at certain temperatures. These values are called apparent shear stress and apparent shear rate at the tube wall. Typical values of apparent shear stress and rate obtained from a capillary rheometer should be corrected in order to obtain true values, as follows:

Entry and exit corrections should be made by subtracting the pressure drop of orifice from the pressure drop of capillary. The corrected value of shear stress is determined by the Bagley correction,[12] where ND is the length correction expressed as a function of radius

$$\tau = \frac{D\Delta P}{4(L + ND)} \tag{110.17}$$

A comprehensive study of the effect of pressure drop in capillary flow is reported in Ref. 13.

Correction for non-Newtonian behavior. For isothermal Newtonian flow, velocity profile is parabolic; however, non-parabolic velocity profile develops in non-Newtonian flow. The Rabinowitsch correction[14] is applied to eliminate this error as follows:

$$\dot{\gamma}_{wc} = \frac{3n + 1}{4n}\, \dot{\gamma}_{wa} \tag{110.18}$$

where subscript c stands for corrected value and a stands for apparent value of shear rate at tube wall.

Reservoir Losses. The rheometer barrel can be considered as a large capillary. Effect of reservoir on pressure drop in capillary can be avoided by using large capillaries (higher length-to-diameter ratio). However, there is a practical limit as to how long a capillary can be used for desired flow study. L/D greater than 80 is often recommended to correct this effect.

Effective Slip Near the Tube Wall. During the flow of polymer melt in a tube, the velocity gradient near the tube wall may induce some orientation of macromolecules, resulting in a phenomenon known as effective slip near the tube wall. This error can be corrected using the following equation according to Ref. 1:

$$Q = \pi U r^2 / r = R + \frac{1}{\tau^3} \int_0^{\tau_w} \tau_{rx}^2\, f(\tau_{rx})\, d\tau_{rx} \tag{110.19}$$

or

$$\tau_w = \frac{4U_s}{R} + \frac{4}{\pi R^3 \tau_w} \times \int_0^{\tau_w} \tau_r^2\, f(\tau_{rx})\, d\tau_{rx} \tag{110.20}$$

where U_s = effective slip velocity at τ_w.

By making a plot of $\dot{\gamma}$ versus $1/R$ at constant τ_w the slope of the line equals $4U_s$. This method is easy to work with, as discussed in Ref. 15.

Thermal Effects. During the laminar flow of a non-Newtonian polymer melt, viscous heat is generated in the fluid; it is maximum at the tube wall. Viscous heating increases when shear rate is increased, which results in reduction of fluid viscosity and introduces error not accounted for in the flow data. To date no appropriate method has been developed to take care of the thermal correction error. Recent literature[15,16] discusses the error contribution.

It should be emphasized that even though the above-mentioned corrections are significant for obtaining absolute values from flow data, comparison of flow properties can be made without applying corrections.[17]

Cone and Plate Rheometers. One advantage of the cone and plate rheometer (Fig. 110.8) is that it does not suffer from the complexities of stress and shear rate distribution that is common to capillary and coaxial-cylinder rheometers. Shear rate is the constant throughout the measurement. A disadvantage of this instrument is that it is limited to comparatively low shear rate ranges.

Among the more versatile cone and plate rheometers is the Weissenberg rheogoniometer. Not only can it be used to measure viscosities in simple shear, but it can also be used to determine the dynamic properties of viscoelastic materials. The unit is also set up to measure the normal stresses exhibited by viscoelastics, ie, those perpendicular to the plane of shear. Another verstaile instrument is a general purpose rotational rheometer based on a transducer system capable of measuring torque, normal force, and the forces in two directions orthogonal to the axis of rotation. This device is manufactured and sold by Rheometrics Inc., and as the Mechanical Spectrometer. It includes the widest possible range of rotational test geometrics.[11]

Parallel-plate Viscometers. As the name implies, a disk of viscous liquid is compressed between two parallel plates. Such a viscometer is best suited for fluids of very high viscosity at low shear rates. It cannot determine shear rates as easily as a capillary rheometer.

Coaxial Cylinder Viscometers. Also called concentric cylinder, this type of viscometer measures low-viscosity liquids. Typical is the Haake-Rotovisco. In this device, the cup is stationary, and the bob is driven through a torsion spring. A major advantage of the coaxial cylinder viscometer is the nearly constant shear rate throughout the entire volume of fluid

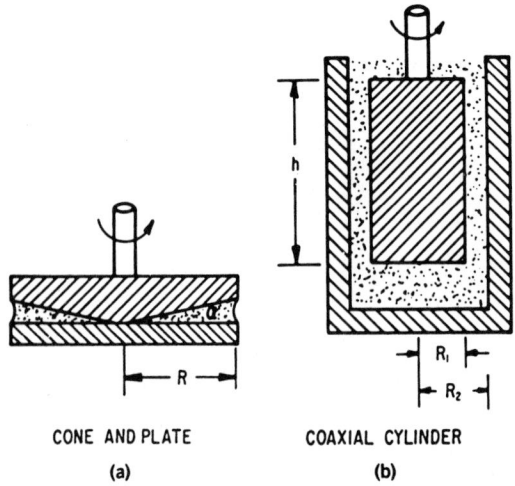

CONE AND PLATE   COAXIAL CYLINDER

(a)   (b)

Figure 110.8. Cone and plate and coaxial rheometers.

being measured. A disadvantage is the difficulty in filling with very viscous polymer melt. An extensive treatment of this class of viscometers is covered in Ref. 11.

Extensional Viscometers. These devices are of special value in such commercial processes such as biaxial orientation of film, which requires good understanding of the tensile or extensional viscosity of polymer melts. References 15 and 16 describe an instrument for this purpose.

Dynamic or Oscillatory Rheometers. Such instruments measure both viscous and elastic modulus of polymer melts (Fig. 110.9). In many polymer processing applications, both viscosity and elasticity are critical to a designer or a process engineer.

A discussion of many of the instruments used to measure complex viscous and elastic behavior is covered in Ref. 20. Simply explained, the polymer sample is deformed in shear or tension in a dynamic rheometer by an oscillating driver. The amplitude of the sinusoidal deformation is measured by a strain transducer. Because of the energy dissipated by the viscous fluid, a phase difference develops between the stress and the strain, and viscosity behavior is determined from the amplitude of the stress and the strain and the phase angle between them.

110.3 SIGNIFICANCE OF FLOW PROPERTIES

The rheological properties that are most critical in determining the processing characteristics of thermoplastics, as well as ultimate product quality are viscosity, elasticity, extrudate defects caused by melt fracture, land fracture, or post-extrusion swell.

Melt Viscosity. Perhaps the most important factor to a process engineer in predicting extrusion or molding behavior is melt viscosity. Factors involved in predicting processing melt temperature from melt viscosity are presented in Refs. 21 and 22. Higher melt viscosity resin usually results in an increased melt temperature at higher processing rates. The increase in melt temperature may cause such processing problems as material degradation, higher

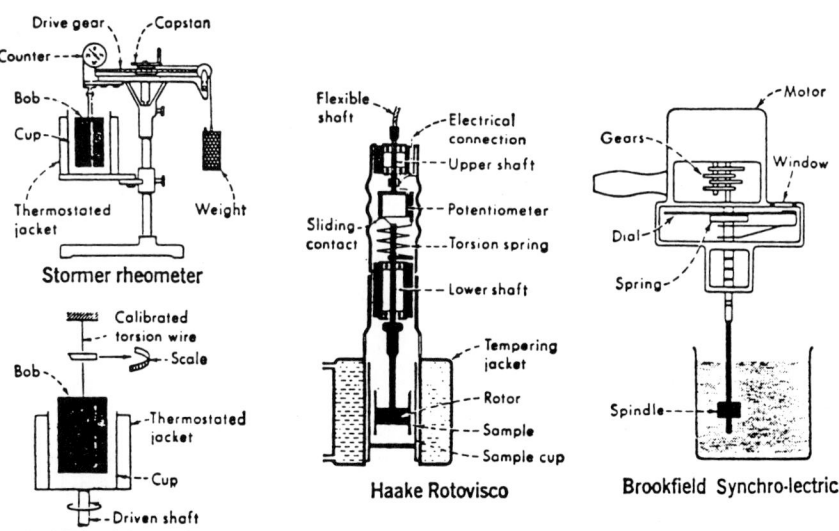

Figure 110.9.   Oscillatory rheometers.

TABLE 110.2. Shear Sensitivity of Polymers

Polymer	Temperature, °C	(°F)	μ_A at 100 s^{-1} / μ_A at 1000 s^{-1}
Acetal copolymer	180	(356)	2.1
Acrylic	200	(392)	2.3
Nylon-6	280	(536)	1.4
Nylon-6,6	310	(590)	1.2
Nylon-11	250	(482)	1.8
Nylon-6,10	280	(536)	1.6
Polyethylene	190	(374)	>2.6
Polypropylene	230	(446)	3.6–4.7
Polystyrene	240	(464)	3.9
Polycarbonate	270	(518)	2.7
Poly (vinyl chloride)	190	(374)	5.0–6.0

scrap rates, and increased machine downtime. To avoid such problems, a process engineer would benefit greatly by determining the shear and temperature sensitivity of a polymer prior to machine startup. The shear sensitivity of many conventional polymers is shown in Table 110.2.

Shear sensitivity is obtained by measuring the melt viscosity in a capillary rheometer at constant desired temperature at shear rate of 100 and 1000 s^{-1}. The ratio of viscosities at 100 and 1000 s^{-1} gives the shear sensitivity of a polymer. For example, Table 110.2 shows that nylon 6,6 has low shear sensitivity of 1.2, but PVC is highly shear sensitive (5.5).

Temperature sensitivity of many polymers is shown in Table 110.3. The data were obtained by taking the ratio of melt viscosity measured at two different temperatures at 1000 s^{-1} shear rate. As shown in the table, acetal was found to be least temperature sensitive (1.35): acrylic was found to be most (4.1). In other words, small change in processing temperature causes a major change in the melt viscosity of acrylic.

Elasticity Effect. The elasticity is most commonly observed in extrusion of thermoplastic materials. The phenomenon of elasticity is usually manifested in three ways during extrusion: (*1*) excessively large pressure drops in the capillary, (*2*) post-extrusion swell, and (*3*) extrudate defects.

Because the effects of elasticity are associated with the die-exit and take-off areas of the process, which in turn control product quality, the importance of these effects cannot be ignored.

TABLE 110.3. Temperature Sensitivity of Polymers

Polymer	T_1, °C	(°F)	T_2, °C	(°F)	μ_A, 1000 s^{-1}, T_1 / μ_A, 1000 s^{-1}, T_2
Acetal copolymer	180	(356)	220	(428)	1.35
Acrylic	200	(392)	240	(464)	4.1
Nylon-6	240	(464)	280	(536)	2.2
Nylon-6,6	270	(518)	310	(590)	3.5
Nylon-11	210	(410)	250	(482)	2.4
Nylon-6,10	240	(464)	280	(536)	2.0
Polyethylene	150	(302)	190	(374)	Usually <2.0
Polypropylene	190	(374)	230	(446)	1.3–1.5
Polystyrene	200	(392)	240	(464)	1.6
Polycarbonate	230	(446)	270	(518)	3.0
Poly(vinyl chloride)	150	(302)	190	(374)	1.45–2.0 (flexible-rigid)

Elastic effects are not always associated with high viscosity. Two liquids of the same viscosity have been shown to have different elastic effects.[23] Due to the elastic memory effect, polymer melts can store energy while flowing. This effect is commonly observed as post-extrusion swell when stored energy is released. Under high shear conditions, molten strand tends to swell to a cross-sectional area several times larger than the die. Reference 24 offers various theories to explain post-extrusion swell. These are polymer memory, normal forces, shear strain recovery, and rearrangement of the velocity profile at the die exit. In general, the degree of swelling increases in polymer melts with increase in shear rates.

In order to determine elastic effects in PVC, Sieglaff[25] reanalyzed the theoretical aspects of elastic responses proposed by Lanieve.[26] Sieglaff has shown that direct calculation of elastic energy of PVC melts (and other thermoplastics) from capillary rheometer measurement is possible. He concluded that the addition of low molecular weight nonelastic materials to PVC does not cause large changes in the elastic properties of the system. He also established that the recoverable elastic energy for PVC is critically dependent on temperature.

A comprehensive study of the elastic effects of certain thermoplastics over a wide range of shear rate, shear stress, die geometry, and temperature was made.[27] He expressed elasticity as the ratio of the cross-sectional area of the swollen extrudate to the area of the die. His conclusion: the elasticity increases linearly with shear stress. Swell therefore, becomes more severe as extrusion speed is increased. At constant shear stress, Kowalski established that swell decreases as L/D increases. Thus, when the die land length increases, the level of post-extrusion swell decreases.

Sieglaff[28] and Paradis[29] discuss the swelling effects of PVC compounds. Sieglaff showed that degree of swell increases with decreasing molecular weight of PVC at low shear rates. Paradis found no molecular weight dependence on swelling but indicated that this might be due to an insufficiently broad molecular weight range of samples. Carley[30] has discussed the importance of viscous, as well as elastic flow properties, for die design. He maintains that length-to-diameter ratio has a very important effect on swelling: the shorter the tube for a given flow rate and diameter, the less time there is for the stresses to set up as fluid enters the orifice to relax, so the more swelling there is as the melt emerges. He recommends two ways to compensate for swelling: The die opening can be reduced, or the extrudate can be pulled away from the die at a velocity at least as great as average velocity in the die. A combination of both approaches is suggested for better results. For comprehensive details on elasticity effects in polymer extrusion, reader should refer to Ref. 31.

Extrudate Defects. It is extremely important that the quality of thermoplastic extrudate have a smooth, glossy surface, excellent clarity, and constant cross-sectional dimensions. A great deal of experimental work has been done to understand rheological parameters related to product quality. Of all plastics processes, extrusion is the most readily amenable to flow analyses, due to steady-state conditions and well-defined boundaries. Extrudate defects are usually encountered when polymer melt is forced through a die or a capillary: They include melt fracture, land fracture, and post-extrusion swell (which has already been discussed).

Melt Fracture. The term "melt fracture" was first suggested by Tordella;[32] it refers to extrudates, produced at critical shear stress, whose shapes are distorted. Various theories have been reviewed and summarized to explain this defect.[33,34] This effect we thought to be caused by turbulent flow,[35,36] a theory contradicted by Tordella.[37]

Using photographic techniques, Tordella[38] and Bagley and Birks[39] shows that melt fracture can be associated with flow disturbances at die inlet (Figs. 110.10 and 110.11). They concluded that abnormal stresses in the entrance region of a capillary, cause elastic failure or shear fracture of viscoelastic fluids, resulting in distorted extrudates. An excellent discussion of unstable flow of molten polymers is given in Ref. 40. Reference 41 is a comprehensive study of the effect of material, geometric, and operating variables on the onset of melt flow

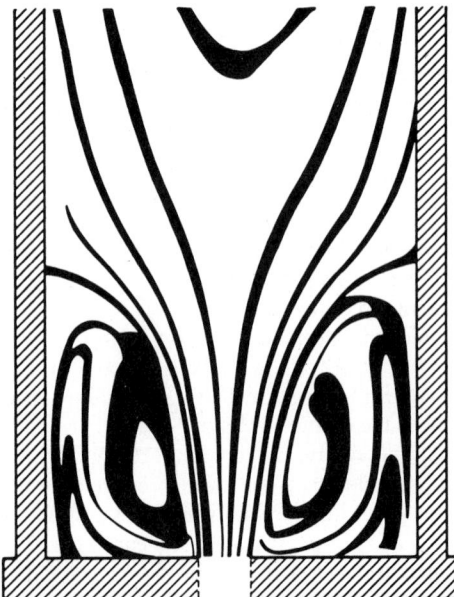

Figure 110.10.   Flow visualization. Steady-state flow pattern of low density polyethylene in inlet region of capillary, redrawn from photograph of Bagley and Birks. Black steaks are flow paths of black polyethylene filaments threaded into specimen of clear resin prior to inserting specimen into thermometer.

instabilities in extrusion. The results indicate that flow instability is strongly associated with the elastic properties of materials.

Land Fracture.   Land fracture is observed during extrusion of polymer melts through a die, and it is visually characterized as fine surface roughness. The main difference between melt and land fracture is that the former almost always is associated with a change in the slope of shear stress-rate curve; the latter does not alter the slope.

   The presence of surface roughness in extruded polyethylene samples has been demonstrated,[40] and it was concluded from birefringence studies that the wall of the capillary was

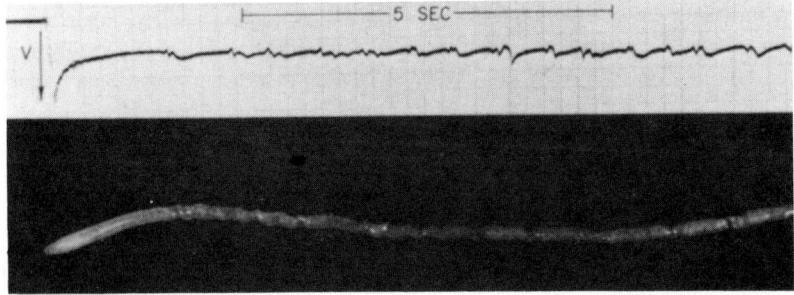

Figure 110.11.   Piston velocity or flow rate versus time record and corresponding low-density polyethylene specimen (melt index 2.0).[5] Piston velocity is zero at left of record and increases in downward direction.

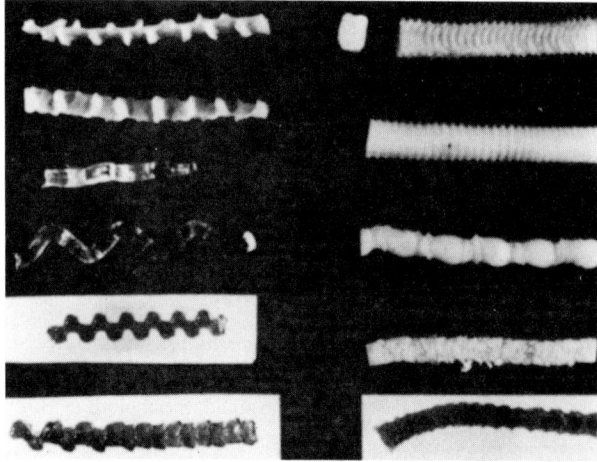

Figure 110.12. Unusual specimens extruded through circular capillaries. On left side of photographs are specimens having a "plane of symmetry," a/ig/ag configuration. Top and edge views of each specimen are shown. The first pair of specimens is linear polyethylene, the next are poly(methyl methacrylate). The top right specimen appears to show intense ripple and has a rectangular cross section edge, and end views of this specimen are shown. The next lower specimen has a ringed versus helical configuration, the middle specimen is similar except that the relative spacing between adjacent rings alternates between about 1 and 3. The second last specimen has a surface similar to scales on a fish. The left end of the lower right specimen exhibits a sevenfold rotational axis of symmetry, the rest of the specimen, fourfold. The left end was that which first emerged as the shear stress was applied by dropping away an abstruction from the exit of the capillary.

the site of this flow defect. A severe case of land fracture in extruded samples of various polymers has been reported[42] (Figs. 110.12–110.13).

This type of land fracture phenomenon in PVC samples extruded from a capillary rheometer has been reported[42] and the presence of land fracture in PVC samples from a capillary rheometer has documented.[44] These findings further indicate that surface roughness can be reduced and eliminated by addition of proper internal lubricants. Further studies are in progress.

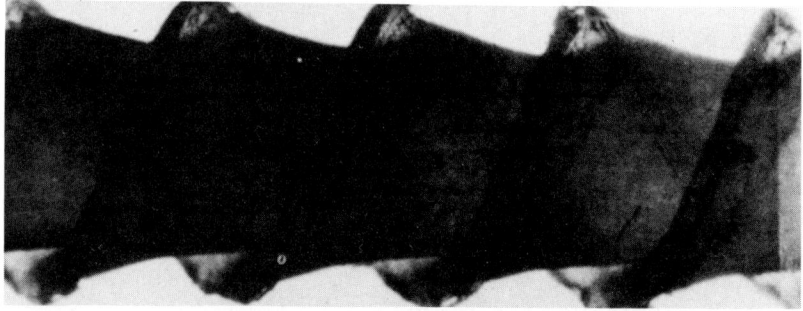

Figure 110.13. Double-entwined helix of poly(vinylidene fluoride) extruded at 240°C. The faint back shadow near the middle of the specimen shows that each convolution connects to its second rather than its first neighbor.

BIBLIOGRAPHY

1. A.C. Fredrikson, *Principles and Applications of Rheology,* Prentice Hall, New York, 1969.
2. A.H.P. Skelland, *Non-Newtonian Flow and Heat Transfer,* John Wiley & Sons, Inc., New York, 1967.
3. S.D. Cramer and J.M. Marchello, *AIChE Journal,* **14**(6), 11 (1968).
4. F.R. Eirich, *Rheology,* Vol. 3, Academic Press, New York, 1960.
5. J.M. NcKelvey, *Polymer Processing,* John Wiley & Sons, Inc., New York, 1962.
6. F. Bueche, *J. Chem. Phys.* **47,** 1942 (1962).
7. W.W. Graessley, *J. Chem. Phys.* **47,** 1942 (1967).
8. J. Stratton, *Colloid Interi. Sci.* **22,** 517 (1966).
9. Van Wazer and co-workers, *Viscosity and Flow Measurements,* Interscience, New York, 1962.
10. C.E. Nielsen, *Polymer Rheology,* Marcel Dekker, New York, 1977.
11. J.M. Dealy, *Rheology of Molten Polymers,* Van Nostrand, Reinhold Co., New York, 1982.
12. E.B. Bagley, *J. Appl. Phys.* **28,** 624 (1957).
13. R.L. Boles, H.L. David, and D.C. Bogue, *Polym. Eng. Sci.* **10,** 1 (1970).
14. B. Rabinowitsch, *J. Phys. Chem. A,* 145 (1929).
15. D.R. Hinrichs, *ASME HDT 5* **29,** 11 (1971).
16. M.R. Kamal and H. Nyun, *Polym. Eng. Sci.* **20,** 109–119 (1980).
17. P.L. Shah, *SPE Tech. Papers* **17,** 321 (1971).
18. C.D. Censon and R.J. Gallo, *Polym. Eng. Sci.* **11,** 174 (1977).
19. J. Meissner, *Rheol Acta* **8,** 78 (1969).
20. J.D. Ferry, *Viscoelastic Properties of Polymers,* John Wiley & Sons, Inc., New York, 1970.
21. P.L. Shah, *Polym. Eng. Sci.* **14,** 11, 773 (1974).
22. P.L. Shah, "Processing Melt Rheology," in *PVC Encyclopedia,* Marcel Dekker, New York, 1976, Chapt. 21.
23. A.S. Lodge, *Elastic Liquids,* Academic Press, New York, 1964.
24. J. Benbow and P. Lamb, *SPE Trans.* **3,** 1 (1963).
25. C.L. Sieglaff, *Polym. Eng. Sci.* **9,** 1 (1969).
26. H.L. Lanieve, Ph.D. Thesis, University of Tennessee, 1966.
27. R.C. Kowalski, Ph.D. Thesis, Brooklyn Polytechnic Institute, Brooklyn, N.Y., 1963.
28. C.L. Sieglaff, *SPE Trans.* **4,** 2 (1964).
29. R.A. Paradis, *Chem. Eng. Progr.* **62,** 12 (1966).
30. J.F. Carley, *SPE J.* **19,** 9 (1963).
31. E.B. Bagley and H.P. Schrieber, *Rheology,* Vol. 5, Academic Press, New York, Chapt. 3.
32. J.P. Tordella, *J. Appl. Phys.* **2,** 5, 6 (1956).
33. A.B. Metzner, *Ind. Eng. Chem.* **50,** 10 (1959).
34. J.J. Benbow and P. Lamb, *SPE Trans.* **3,** (1983).
35. H.K. Nason, *J. Appl. Phys.* **16,** 6 (1945).
36. R.F. Westover and B. Maxwell, *SPE J.* **13,** 8 (1957).
37. J.P. Tordella, *Trans. Soc. Rheo.* **1,** 1 (1957).
38. J.P. Tordella, *Rheol. Acta.* **1,** 2, 3 (1958).
39. E.B. Bagley and A.M. Birks, *J. Appl. Phys.* **31,** 3 (1970).
40. J.P. Tordella, *J. Appl. Poly. Sci.* **7,** 2 (1963).
41. T.F. Ballenger and co-workers, *Trans. Soc. Rheol.* **15,** 2 (1971).
42. E.B. Bagley, *J. Appl. Polym. Sci.* **7,** 1 (1963).
43. C.L. Sieglaff, *Polym. Eng. Sci.* **9,** 1 (1969).
44. P.L. Shah, *SPE J.* **27,** 1, 49 (1971).

General References

K. Weissenberg, *Proc. 1st International Congress of Rheology,* North Holland Pub., 1949.

C. Macasko and J.M. Starita, *SPE J.* **27,** 11, 38, (1971).

B. Maxwell and R.P. Chartoff, *Trans. Soc. Rheo.* **9,** 41 (1965).

C. Macosko and J.M. Starita, *SPE J.* **27,** 38 (Nov. 1971).

111

TESTING, QUALITY CONTROL

Len Buchoff

Elastomeric Technologies, Inc., 2940 Turnpike Drive, Hatboro, PA 19040

111.1 INTRODUCTION

Testing yields basic information about a plastic, its properties relative to another material, and its quality in reference to a standard. Most of all, it is essential to determination of the performance of a finished product.

Test methods generally require specimens produced by a standard method into a standard shape and conditioned in a standard way. Even so, it is difficult to predict with complete accuracy how a product will perform in actual use conditions, because real-life environment will usually differ substantially from the circumstances under which the sample is tested. Thus, test methods are usually a compromise—although they are essential to the development of new plastic materials—to choosing the correct plastic for a particular application, and to maintaining the quality of a plastic material and finished products made from it.

In order for a test to be useful, the tested values must be quantitative or at least rankable; the test results must be reproducible by different people using equivalent equipment at different locations; the tests must show differences in the properties under investigation but must not be sensitive to minor random variation; and test results should correlate with the in-use performance of the parts. Therefore all of the mechanical, environmental and other stresses that are expected must be imposed during testing for appropriate periods of time. Usually this procedure is done with a limited number of parts because of the time and expense involved. Tests can then be performed and/or developed whose results correlate with real-life experience. Improved materials and part designs can then be tested and compared against standards.

Some of the most useful tests do not yield results that can be used directly for designing a part. For example, the conditions under which testing for heat-distortion temperature (ASTM D648) are run are totally arbitrary. The end point, deflection of the sample by 0.010 in. (0.25 mm), has no theoretical basis. But heat-deflection temperature has been found, in many instances, to be related to the maximum practical operating temperature of a plastic fabricated in a particular way. Generally it is found that in a finished part loss of mechanical strength occurs at a few degress above or below the nominal heat-deflection temperature (HDT) of the plastic from which it is made. The exact number of degrees must be determined experimentally.

111.2 MECHANICAL TESTING

In plastics as in any other material of design, successful mass production depends upon inspection and control of sampling and testing. Engineering research and development depend on carefully planned, well-devised tests. Mechanical testing is basic and essential.

In a broad sense, mechanical strength refers to the ability of a structure to resist loads without failure, which may occur by rupture due to excessive stress or may take place owing to excessive deformation. In order to approximate the conditions under which a material must perform in service, a number of test procedures are necessary.

The method of loading is the most common basis for designating or classifying mechanical tests. There are five primary types of loading, governed by the stress condition to be induced: tension, compression, direct shear, torsion and flexure.

With respect to the rate at which load is applied, tests can be classified into three groups. If the load is applied over a relatively short time but slowly enough so that the speed of testing has a negligible effect on end results, the test is called static test. Such tests may be conducted over periods ranging from several minutes to several hours. If the load is applied very rapidly so that the effect of inertia and the time element are involved, the tests are called dynamic. If the load is applied suddenly as by striking a blow, the test is called an impact test.

If the load is sustained over a long period (months or even years), the test is called long-term. Test loads may be repeated many times, millions if necessary. The most important category of test in this group is the endurance or fatigue test, whose purpose is to determine the longevity of a component.

The environmental conditions to which plastic specimens are subjected before and during the tests have a major bearing on the results. These conditions include temperature, humidity, radiation (ultraviolet, atomic and other high-energy fields), and chemical exposure in the liquid or gas phase. The time of exposure is of vital importance.

Standard test methods and testing equipment are available to determine the performance properties of plastics materials. In stress testing (ASTM Designation D638), specimens are rectangular and molded or machined from fabricated plaques. Typically they are 1/8 in. (3.2 mm) thick, but the size can vary. Generally, the center section of the specimen is narrower than the ends so that failure will occur away from the section clamped in the tester jaws (Fig. 111.1). In conducting the test, the jaws may move apart at rates of 0.2 to 20 in./min. (5 to 508 mm/min.). Stress is automatically plotted against strain (elongation) as the test proceeds.

Stress–strain curves provide several types of information about the properties of the plastic under tensile stress. These include the following:

- Stiffness, or the ability to carry stress without changing dimension; given by Young's modulus of elasticity "E".

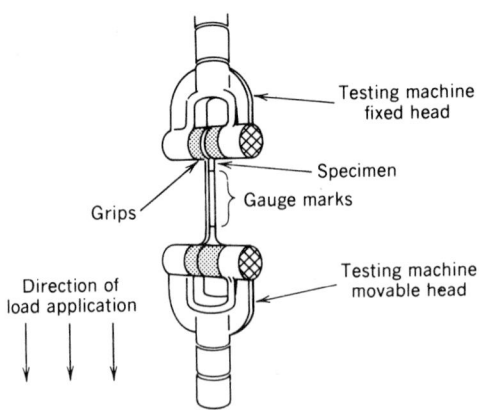

Figure 111.1.   Tensile stress testing, ASTM D638.

- Elasticity, the ability to carry stress without suffering permanent set indicated by the yield point and elastic limit.
- Resilience, the ability to absorb energy without suffering permanent set; given by the area under the elastic portion of the stress–strain curve.
- Strength, the ability to carry dead load; given by ultimate strength.
- Toughness, the ability to absorb energy and undergo large permanent set without rupturing; given by the total area under the stress-strain curve.

In flexural testing (ASTM Designation D790), specimens are bars of rectangular cross-section cut from fabricated shapes or molded to the desired finished dimensions. The specimen is placed on two supports spaced 4 in. (102 mm) apart. The load is applied at the center of the specimen to produce a specified rate of straining of the outer fiber and load-deflection data is taken (Fig. 111.2). Flexural strength is equal to the maximum stress in the outer fiber at the moment of crack or break. If no break has occurred in the specimen by the time the maximum strain in the outer fiber has reached 0.05 in./in.(mm/mm), the test is discontinued.

In testing for impact resistance (ASTM D256), the specimen is carefully notched in a precise manner and mounted in a clamp. A cantilevered beam traveling at 11 ft./s (3.4 m/s), strikes the specimen at the notch, breaking it (Fig. 111.3). The difference between the energy in the beam before and after breaking the specimen is reported as the Izod impact strength in ft-lb/in. of notch (J/m).

Tensile impact energy (ASTM D1822) is determined by affixing one end of the specimen to the cantilevered beam, while the other end is gripped by a crosshead (Fig. 111.4). The beam is swung in an arc, carrying the specimen between the jaws of the anvil. The crosshead hits the anvil, causing the specimen to fracture in tensile loading. Energy lost by the swinging beam is a measure of the impact strength of the specimen.

111.3 TEMPERATURE TESTING

The choice of a plastic for a particular application is strongly influenced by the effect of temperature on the material. Reversible changes of volume, hardness, strength, and electrical properties, and permanent changes in these properties must be taken into account.

An indication of the ability of a plastic to withstand load at elevated temperature is given by the "glass-transition temperature," or T_g generally determined from a plot of linear expansion vs temperature. At T_g the slope of the line, which is the thermal coefficient of expansion, abruptly changes. The upper use temperature of a plastic is often assumed to be a fixed number of degrees below the T_g. This can vary from 20 to 50°F (11–28°C) depending on the plastic and the criterion of performance.

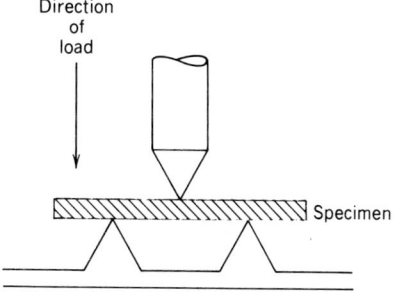

Figure 111.2. Testing for flexural properties of plastics, ASTM D790.

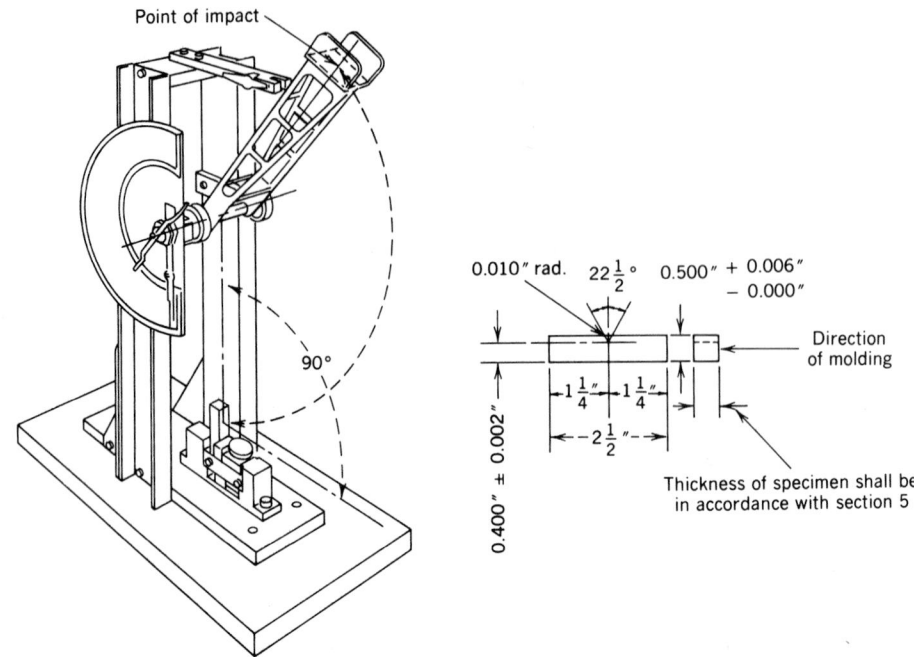

Figure 111.3.   Izod impact test, ASTM D256. Courtesy of ASTM.

One widely used measure of the relative behavior of plastics at elevated temperature is deflection temperature. As described in ASTM D648, a specimen $5 \times 1/2 \times 1/8$ to $1/2$ in. ($127 \times 13 \times 3.2–13$ mm) is used. It is placed on supports 4 in. (102 mm) apart and a load of 66 or 264 psi (0.46–1.8 MPa) outer fiber stress is applied on the center (Fig. 111.5). The temperature in the chamber is raised at a rate of 3.6°F (2°C) per minute; the temperature at which the bar has deflects 0.01 in. (0.254 mm) is reported as "deflection temperature at 66 (or 264) psi (0.46 or 1.8 MPa) fiber stress." Although this test is empirical and conditions must be maintained within narrow limits, deflection temperature at 264 psi (1.8 MPa) generally corresponds closely with T_g.

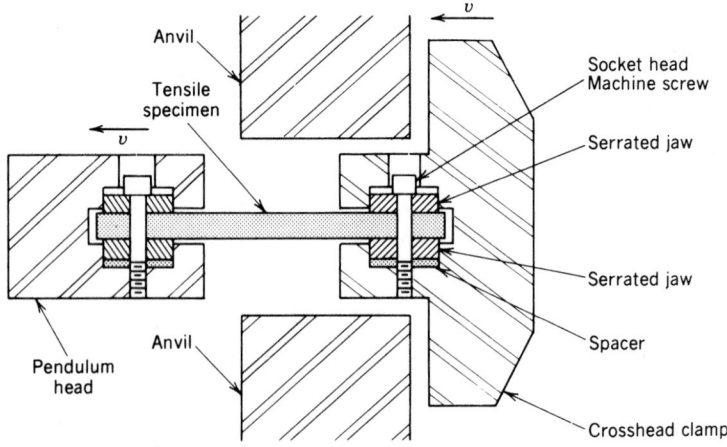

Figure 111.4.   Tensile impact energy testing, ASTM D1822. Courtesy of ASTM.

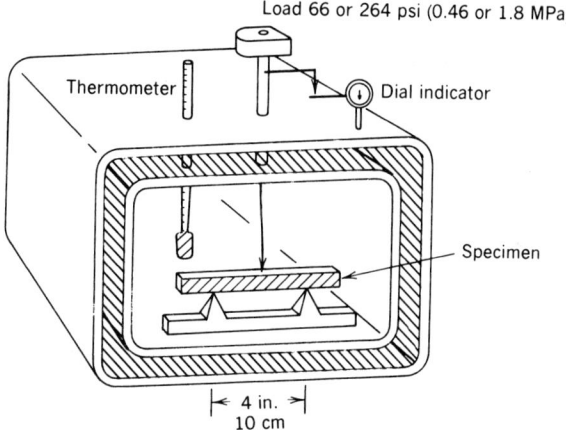

Figure 111.5. Deflection temperature test, ASTM D648.

Thermal analysis is often used to characterize plastics over wide temperature ranges. Various types of equipment automatically measure changes in properties of a few milligrams of material while it is heated. In differential thermal analysis and differential scanning calorimetry, heat given off or absorbed is recorded vs temperature. Information on phase transitions, crystallinity, chemical changes, and other effects can be easily and quickly determined.

Weight loss as a function of temperature is determined by thermal gravimetric analysis. This test can be carried out in a vacuum or in gases. Thermal mechanical analysis generates a plot of expansion vs temperature and is widely used to determine T_g and physical changes in the sample. Thermal analysis is useful in obtaining a "fingerprint" of the plastic sample to determine differences in various batches of the same material.

111.4 FLAMMABILITY TESTING

The ability of plastics to withstand heat and flame without burning is a critical factor in determining what material to use in products as diverse as electrical equipment, house furnishings, children's pajamas, and astronauts' uniforms.

Flammability testing is complicated by the effect of many factors on FR performance. Among these are thickness of the test sample, location of the flame contact, whether the sample is horizontal, vertical or some other angle, oxygen concentration and temperature of the surrounding environment, and air velocity.

Because of the difficulty and expense in running meaningful flammability tests, the Underwriter's Laboratory has been set up to perform this service. UL ratings (numerical and alphabetical) are universally accepted and used as reliable indicators of relative flame retardance. No claims of qualitative flame retardance can be made for any material; only the UL classification can be stated.

Limiting oxygen index (LOI) is one of the few flammability tests that yield reproducible numerical results. In this test a plastic specimen is mounted vertically and ignited at its top in an atmosphere of mixtures of nitrogen and oxygen. The LOI is the lowest percent of oxygen in the mixture that will support combustion. This test can measure the performance of plastics that do not burn in normal air (where oxygen concentration is 21%).

Because of the public's concept of "flammability" it cannot be overemphasized that "technical" flammability standards cannot be used alone to connote safety.

111.5 ELECTRICAL PROPERTIES TESTING

Plastics are often used in products because of their ability to insulate against the effects of electricity. The electrical properties of a material are strongly dependent on temperature, mechanical stress, contamination, and conditioning.

Resistivity of any plastic material is dependent on the geometry of the tested sample and the paths through which current flows during measurement. Resistance (ASTM D257) is measured by imposing a direct voltage between two electrodes attached to the sample and reading the resultant current. Resistance is calculated by the equation:

$$R = E/I$$

where R is resistance in ohms, E is the voltage, and I is the current in amperes.

Because the resistance of many plastics is very high, the current is very small and requires a sensitive instrument to measure it accurately. The resistivity of most plastics can vary over several orders of magnitude with variations in termperature and/or humidity. Therefore, the preparation of the samples and conditions under which they are tested must be very carefully controlled.

Surface resistivity is determined by measuring the surface resistance between two parallel electrodes attached to the sample surface. To convert this resistance to resistivity:

$$R_s = r_s \times L/D$$

where R_s is the surface resistivity in ohms, r_s = surface resistance in ohms, D is the distance between the electrodes and L is the length of the electrodes. D and L must be in the same units of length.

The volume resistivity of a plastic in the metric system is given in units of ohm-cm and is numerically equal to the volume resistance between opposite faces of a centimeter cube of the material. The volume resistance is measured between two electrodes attached to parallel sides of a sample of the plastic

$$R_v = r_v \times A/T$$

where R_v and r_v are volume resistivity and resistance, respectively. A is the area of the electrodes (cm^2), and T is the thickness of the sample (cm).

The dielectric strength (ASTM D149) of a dielectric material is the ratio of dielectric breakdown voltage to the thickness. It is given as volts per mil. In this test the voltage is applied between two electrodes on opposite sides of the sample. The voltage is increased at a fixed rate until dielectric breakdown occurs. The more rapid the increase, the higher the breakdown voltage. The electrode size is also important, the larger its size, the lower the breakdown voltage. The thickness of the sample must be specified because the dielectric strength (volts/mil) decreases with thickness although the dielectric breakdown voltage increases.

The dielectric constant and loss characteristics (ASTM D150) of an insulator can be critical in its effect on an alternating current system especially at high frequencies. In general it is desirable to have low dielectric constant when the plastic will be used to insulate components of the circuit. On the other hand, if the plastic will serve as dielectric in a capacitor, a high dielectric constant is needed. A low loss is generally required especially at high frequency as the power loss increases directly with frequency.

111.6 ACCELERATED OUTDOOR WEATHERING TESTING

Plastic products used in outdoor applications must retain their appearance and strength over long periods of time. Exposure to sunlight, temperature, oxygen, rain and humidity, dust and microorganisms produces change in color and degradation of physical properties. Because conditions vary markedly with location, testing must be done in multiple places. A further complication is the large variations of conditions over time in a single location. Because of these variables a large number of samples must be exposed to many types of environments for extended periods of time.

Accelerated tests and equipment have been developed to reduce the time and expense of evaluating the resistance of plastics to outdoor weathering and to evaluate the effectiveness of stabilizers.

The major factors causing deterioration are UV radiation, heat and water. The effects of UV radiation are simulated by arc lamps having spectra similar to sunlight. Other equipment employs cycles of UV, heat and water spray to simulate outdoor conditions.

There is some correlation between natural exposures and machine testing but these tests are at best indications of the effect of modifications of a plastic have on its ability to withstand weathering and should be viewed semi-quantitatively.

111.7 POST-MOLD STRESS TESTING

Stresses present in a plastic part are influenced by the pre-molding conditioning of the material, mold design, rate and pressure of molding, rate of cooling and other factors. These stresses can limit the operating temperature of the part, reduce its load-carrying capability, alter electrical properties and can cause distortion or destruction of the part in use environments.

If the part is transparent or translucent, stress can be determined with crossed polarizers. In this test the number of birefringences are counted. Stress is calculated from this number and the modulus of the plastic.

Internal stress can be determined by subjecting the molded part to a temperature slightly lower than the material T_g. The part will "unmold," or change shape to relieve the stresses. Observing the areas and amount of unmolding will qualitatively indicate the location and intensity of the stresses.

Certain solvents will not affect some unstressed plastics, but will cause them to crack when a stress level is exceeded. Examples of solvent-plastic pairs that have this property are kerosene–polystyrene, alcohol–poly(methyl methacrylate), and acetone–poly(vinyl chloride).

Stresses in some thermoset sheets can be found by machining material from one side of the part. The amount and direction of the resultant bending is a measure of stress in the part.

111.8 PROCESSABILITY TESTING

In order to ensure reproducible parts and consistent mold cycles, the flow rate of plastics must be kept within specified limits. A number of tests are useful in comparing the ability of different materials to fill a specific mold; they will also aid in controlling batch-to-batch uniformity.

The melt index (MI) tests the flow rates of extrudable thermoplastics, ASTM D1238. A specified amount of plastic is heated to a fixed temperature in a vertical cylinder. A piston loaded with a prescribed weight extrudes the plastics. Extruded samples are taken at intervals and weighed. The melt index (MI) of the plastic is reported as grams of material extruded in 10 minutes. The dimensions and surface finish of the cylinder, piston and orifice must be held within narrow limits to produce meaningful results.

Molding index of thermosetting powders (ASTM D-731) involves molding a cup in a compression mold of the flash type, having precisely defined dimensions. Molding index of the compound is the minimum force in pounds required to close the mold.

111.8.1   Spiral Mold Evaluation of Transfer Molding Compounds

In this test the material being evaluated is transfer-molded into a precisely characterized spiral groove. The temperature and pressure must be maintained at the specified values. The flow of the compound is reported as the number of inches that the material traveled before solidifying. This test is useful in determining how a material will fill a given mold and the change of flow as the material ages.

111.9   DENSITY

Density (weight per unit volume) is perhaps the most widely reported property of plastics. It is important in determining the weight of material in a part and therefore the cost of the material. Density is affected by the amount and type of filler used in a compound, the degree of crystallinity, the molding parameters and the environmental exposure of the part.

   ASTM D1505 is the test used to determine density of a material by the density-gradient technique. A gradient column is formed by mixing two liquids of differing densities in a procedure that produces a continually varying density along the height of column. A sample of the plastic is put into the column. The sample will fall to the point at which its density matches that of the liquid. A series of floats of different densities provide the calibration of the column.

111.10   QUALITY CONTROL TESTING

Zero-defect part production is a primary demand among today's quality-conscious manufacturers. It puts a premium on elimination of random variations that can cause rejection of a fabricated part.

   The purpose of statistical quality control is to determine when a process is going out of control and allow corrective action to be taken before rejected parts are produced.

   There are a number of ways for processors to approach zero-defect. One is to begin by identifying the properties which affect your part and you wish to monitor and by collecting data on them over an extended time. You then plot the average value of each property on a control chart vs time. These values should vary around a central line in a random way, (ie, the numbers should not show a trend). Control limit lines are drawn on the chart at the acceptable property extremes. If the values show a trend that tends toward a control limit, it is time to search for the factor that is causing this change. Various aspects of the process are critically examined to determine the problem and changes are made to correct the situation. The correction has been made when the measured property values again cluster around the control line.

BIBLIOGRAPHY

General References

ASTM Standards on Plastics, American Society for the Testing of Materials, Philadelphia, Penn.

Standard Tests on Plastics, Bulletin G1C, 9th ed., Celanese Corp., Chatham, N.J.

Underwriters Laboratories Inc. (UL), 207 East Ohio Street, Chicago, IL 60611, various publications

J.V. Schmitz, ed., *Testing of Polymers*, Vol. 1, Wiley-Interscience, New York, 1965.

R.W. Hertzberg and J.A. Mason, *Fatigue of Engineering Plastics*, Academic Press, New York, 1980.

L.E. Nielson, *Mechanical Properties of Polymers and Composites*, Marcel Dekker Inc., New York, 1974.

I.M. Ward, *Mechanical Properties of Solid Polymers,* Wiley-Interscience, New York, 1971.

J.A. Brydson, *Plastic Materials,* 4th ed., Butterworth Publishers Ltd., London, 1982.

M. Chanda and S.K. Roy, *Plastics Technology Handbook,* Marcel Dekker Inc., New York, 1987.

C.P. MacDermott, *Selecting Thermoplastics for Engineering Applications,* Marcel Dekker Inc., New York, 1984.

J.M. Margolis, *Engineering Thermoplastics: Properties and Applications,* Marcel Dekker Inc., New York, 1985.

112

LIQUID HEAT TRANSFER IN PLASTIC PROCESSING

Dewey Rainville

C.E.O., Universal Dynamics, Inc., 176 Cedar Street at Route 22, North Plainfield, NJ 07060*

112.1 INTRODUCTION

All plastic processing involves heat transfer. The heat added to raise the plastic to processing temperature must be removed so the product can be used. Heat is often added as frictional heat by using an extruder screw, calendar rolls, or mechanically. It is also added by electrical resistance using heater bands on barrels, tubular heaters, or radiant heat such as in heating plastic sheets for vacuum forming.

The rate that heat is applied by friction or electric resistance is controlled electrically or electronically, and this very important input control is covered elsewhere in the chapters on process control.

Controlling the rate of heat removal is important because of its major effect on the final product. Improper control leads to defects such as:

- Uncontrolled shrinkage.
- Warping of molded parts.
- Poor or non-uniform finish.
- Stress or strains in the parts.
- Failure to run at optimum production rates.

112.2 SYSTEMS

Liquid is used extensively for heat transfer and many types of accessories are available for its application and control. Efficient production and quality parts require the use of this equipment.

Water has the best ability to transfer heat and is used as a basis of comparison for other liquids. Its coefficient with heat is 1. It is circulated under pressure through the metals or other materials surrounding the plastic mold or device.

The majority of control systems involve circulating water or oil through channels in molds and other devices, such as chilled rolls on sheet lines, or in water baths.

Cool air is used on parts cooling conveyors and to cool the interior of extruder screws and barrels. The main use of hot air in processing is to pre-heat plastic in the feed hopper.

Electric heat is used in molds for manifolds, nozzles and cavity temperature control either directly or in conjunction with circulating liquid. Radiant heat is used to raise vacuum forming sheets to forming temperature. Extrusion dies are normally controlled by cartridge heaters.

Equipment consists of complete package units with a pump to circulate the fluid, electrically controlled heat, and cooling by replacing hot water with cold water or by circulating the process fluid through a manifold that can be cooled by separate circulation of tap, tower or refrigerated water. When tap or tower water is not cool enough, chillers are needed. Refrigerated water or ethylene glycol systems are used. These are available with water or air-cooled condensers.

Tap water or cooling tower water is sometimes used by pumping it directly to the process and then controlling the flow. To do this successfully, ideal temperatures for the process need to be approximately 55–75°F (13–24°C). This is never as profitable as an automatic temperature control unit using refrigerated water. Either individual chillers can be used at each processing unit or central chillers which cool large volumes of water and circulate it to many machines through insulated pipes.

Temperature control applications usually employ a unit for each process. Often more than one zone or separate temperature control systems are used to get the control where it is needed.

Water can be used up to 212°F (100°C) at atmospheric pressure and is used up to 250°F (121°C) under pressure sufficient enough to prevent steam. Above that temperature, oil or high temperature heat-transfer fluids that are fire retardant may be used up to 600°F (316°C). Temperatures up to 500°F (260°C) are common in molding phenolics and for some high-temperature thermoplastics.

Steam is a good heat-transfer fluid, but cannot be controlled accurately enough so it is rarely used. Water is the best transfer fluid and is the cleanest, and by far, the preferred choice.

Injection molding is the most common user of liquid heat transfer. Much of what we cover here also applies to other operations.

112.3 SAFETY

Above 150°F (66°C) a broken line can cause serious injury. Steam expands rapidly and represents a great danger. Heat transfer liquids above the boiling point of water under pressure are exceptionally hazardous. Caution should be employed in using the right fittings, pipes or hoses and procedures to avoid accidents. Consultation with a safety expert or your insurance company is desirable. Above 150°F (66°C), it is advisable to use electrical heating whenever possible.

112.4 BASIC WATER COOLING THEORY

The standard unit for measuring heat is the British Thermal Unit (Btu) (1055 J); 1 Btu is the amount of heat necessary to change the temperature of 1 lb (0.45 kg) of water 1°F (0.56°C). The unit of refrigeration is a commercial ton of refrigeration which is defined as the removal of heat at the rate of 200 Btu/min (211 kJ/min), or 12,000 Btu/h (13 MJ/h). The standard ton of refrigeration is 288,000 Btu (304 MJ), or the amount of Btus removed by a commercial ton of refrigeration in 1 day. It should be noted that the standard ton has the dimensions of heat while a commercial ton has the dimensions of heat divided by time. Table 112.1 gives some useful data for cooling calculations.

The branch of physics devoted to measuring the thermodynamic properties of moist air is known as psychrometry. Cooing towers, which are used in most plants, operate on psychrometric principles.

When a liquid changes its state to a gas (evaporation), it requires energy to loosen the

TABLE 112.1. Useful Data for Cooling Calculations

One standard ton = 288,000 Btu (304 MJ)
One commercial ton:
  removes 200 Btu/min (211 kJ/min)
          12,000 Btu/h (12.7 MJ/h)
          288,000 Btu/day (304 MJ/day)
    cools water

  20 lb (9 kg) 10°F (5.6°C)
200 lb (91 kg) 1°F (0.56°C)
   6 gal (23 L) 4°F (2.2°C)
 12 gal (46 L) 2°F (1.1°C)
 24 gal (91 L) 1°F (0.56°C)

Cooling 1000 gallons (3.9 m³) of water 25°F (13.9°C) requires
the removal of 208,000 Btu (219 MJ)

1 gal	= 231 in.³	= 3.785 L
1 gal	= 0.1337 ft³	= 0.003785 m³
1 gal/min	= 8.0208 ft³/h	= 0.2271 m³/h
1 ft³	= 7.42 gal	= 0.02832 m³
1 ft³/min	= 0.1247 gal/s	= 4.83 × 10⁻⁴ m³/h
1 ft³ of water	= 62.43 lb	= 28.3 kg
1 lb of water	= 0.016 ft³	= 4.6 × 10⁻⁴ m³
1 lb of water	= 0.1198 gal	
1 gal of water	= 8.345 lb	= 3.79 kg
1 Btu	= 778.2 ft lb	= 1055 J
1 hp	= 0.7068 Btu/s	= 746 W
1 kW	= 0.9478 Btu/s	= 1 kJ/s
1 Btu/s	= 1.055 kW	

Thermal conductivity:
1 Btu-min/h ft² °F = 0.1445 W/m · K

molecular bonds. It receives this energy in the form of heat, taking the heat from the surrounding substances, thus lowering the temperature. This heat is called the latent heat of vaporization or the latent heat. Thus, when water evaporates, it will remove heat from the surrounding water and air, lowering their temperature. This is the theoretical basis for cooling tower action. Conversely when gas is condensed, as in refrigerating systems, energy in the form of heat is liberated and the surrounding substance (the condenser) heats up.

Sensible heat is the heat required to change the temperature of the air or water without changing its state. It is a measure of the internal kinetic energy and changes with the absolute temperature of the body. The latent heat is potential energy and shows itself in changes of the physical state of the body (evaporation, condensation) and is not accompanied by any changes of temperature. The latent heat of evaporation of water (Btu/lb) at 60°F (15.6°C) is 1059.3 (2.5 MJ/kg), at 80°F (26.7°C) is 1048.1 (2.44 MJ/kg) and at 100°F (37.8°C) is 1036.7 (2.44 MJ/kg).

The sources of cooling water are

1. Rivers and lakes
2. Cooling ponds and wells
3. Spray ponds
4. Government water supplies
5. Cooling or evaporative towers
6. Mechanical refrigeration

Rivers and Lakes. These sources are rarely available to molding plants. They probably will require filtering, settling, or chemical treatment. A constant check is required because of the possibility of upstream pollution and other changes. The use of this water usually requires the permission of local authorities.

Cooling Ponds and Wells. Aside from the sources just mentioned the cheapest method of cooling water is by means of a cooling pond. This is a pond of water where hot water enters on one side and cold water is removed on the other side. While it is inexpensive it is also inefficient and often unsatisfactory for molding plants. It has a low heat-transfer rate and needs a large size. The cooling depends on air temperature, relative humidity, wind speed, and heat gain from the sun. During summer in middle northern latitudes the minimum water temperature expected would be about 86°F (30°C).

If the pond is used for cooling and hydraulic system of machines and refrigerating equipment, the size of a pond to handle 1000 gal/min (3.85 m³/min) would be approximately 60,000 ft² (5,600 m²). The pond has to be protected from children, algea, bacteria, and such other contaminators that may fall on an open body of water. Needless to say the water must be chemically treated and filtered. Well water can be used if the supply is adequate and the temperature low enough.

Spray Ponds. An improvement over cooling ponds is a spray pond, which is a body of water over which a spray system is installed. The nozzles are approximately 8 ft (2.4 m) above the water level. By presenting a much larger area for evaporization more cooling will occur. The disadvantages are those of an open water system, dependence on wind velocity, plumbing the spray, relatively high water losses, and possibility of the spray being a public nuisance.

Cooling Towers. An important source of water cooling in the molding plant is the cooling tower. There are two types, one that depends on prevailing winds (atmospheric) and the other that depends on a forced air feed by fans (mechanical) which is used in molding plants. When the fan is placed on the bottom of the tower, it is called a forced draft tower and when on top it is called an induced draft tower.

This type of tower has the advantage of a small ground area per unit cooling. It can be located anywhere including inside loft buildings; it requires low pumping head; it can control the temperature of the water more closely than of any of the previously mentioned systems; and it is economical in terms of water consumption. Its main disadvantages are that it has a high operating cost primarily for the air circulating fan. Its maintenance costs are comparatively high; it is subject to mechanical failure, and can present problems in removing the hot, moisture-filled exhaust air.

The hot water is pumped into the top of the cooling tower from where it falls by gravity over a grid, cooling itself during its fall. It is collected in the bottom or basin from which it is pumped out as the cooled water for the processing system. The cooling is done primarily by vaporization. It is estimated that between 10 and 20% of the cooling is done by convection heat transfer between the cool air and warm water. The water drops on slats to break its fall. The amount of cooling depends on the length of time the drop of water is exposed to vaporization. Since falling bodies are accelerated by gravity the slats effectively decelerate the droplets. Additionally it breaks the "Thomson effect," which hinders vaporization because of a difference in electrical potential at different points on the sphere of water, caused by surface tension effects. The slats act to break up the flow of water and form new drops, thus giving a larger surface area and breaking this thermal barrier at the surface. They are made either of redwood or polyethylene.

The cooling range of a tower is the difference in temperature between the water intake and outlet. The heat load of a tower is the number of Btu per minute (1055 J/min) removed by the tower. The circulation rate is the amount of water going through the tower per unit time. The heat load is the product of the circulation and the range. The amount of heat removed by the tower can be increased by the increasing of the area over which the water flows, increasing the amount of air flowing per unit time (velocity), raising the inlet water temperature and reducing the humidity of the air.

The principle involved in a cooling tower is the removal of sensible heat because of the difference in air and water temperature and the removal of latent heat by the change of state from water to water vapor of a small amount of the fluid. It takes approximately 1000 Btu (1 MJ) to evaporate 1 lb (0.454 kg) of water. This is the amount of heat required to cool 100 lb (45.4 kg) of water 10°F (5.55°C). Therefore for each 10° of cooling approximately 1% of the water circulated must be evaporated. This water plus the spray loss (tenths of a per cent) has to be replaced and is called the makeup rate.

All water contains dissolved chemicals. These are brought into the tower during makeup while pure distilled water departs during vaporization. Therefore in time the chemical composition of the water will cause scale and other problems in the molds and coolers. Water must be treated and filtered. The large velocity of air will cause dust and other particles to collect in the tower requiring cleaning. In northern climates during winter there is a possibility of the tower icing. This is overcome by simply bypassing some of the hot inlet water into the basin.

112.5 MECHANICAL REFRIGERATION

The mechanical refrigerator, like the cooling tower, removes heat from the system. The practical difference is that the heat removed by the cooling tower cools the water to temperatures which are controlled by atmospheric conditions. The heat removed in a mechanical refrigerator is removed at a temperature based on the design of the machine. For example, heat can be removed to cool water to 40°F (4.4°C) while having the removed heat dissipated at temperatures of 85 to 100°F (29.4–37.8°C). Mold temperature control is a basic requirement for accuracy and economy in molding. The temperature of the mold should be determined by the optimum molding conditions, not by the available temperature of the cooling water. For this reason mechanical refrigeration is required in a molding plant.

The simplest mechanical refrigeration system would be a closed box containing an open dish of a low boiling chemical (refrigerant), a fan, and a vent. If the refrigerant were liquid ammonia it would evaporate at minus 28°F (-33.3°C) at atmospheric pressure. One pound would absorb 589.3 Btu (622 kJ) in evaporating (latent heat of evaporation). If the temperature surrounding the box is above minus 28°F (-33.3°C), the heat absorbed by the ammonia in evaporating would come from the surrounding media. This is the same theory as water evaporating in a water tower. This method is not practical for many reasons. If the refrigerant were ammonia the odor and toxicity would make it unusable. The cost of the refrigerant would make it uneconomical. The technique would not allow for adequate temperature control. To overcome this the refrigerant is evaporated and mechanically condensed in a closed system and continually reused.

If the pressure is increased, the temperature at which the ammonia (refrigerant) will evaporate and condense is raised. At 47.6 psig (0.33 MPa), the temperature of vaporization is 32°F (0°C), at 92.9 psig (0.64 MPa) it is 60°F (15.6°C), and at 197.2 psig (1.36 MPa) is 100°F (37.8°C). Thus by changing the system's pressure the temperatures at which the change from liquid to vapor or vapor to liquid can be controlled. This means, in effect, that heat can be removed from the system at any convenient cooling temperature without depending on the atmospheric temperature as is required in a cooling tower. Machines designed for air-cooled operation are about 15% less efficient. For units up to 5 tons the convenience is worth the extra cost. These are mainly on portable units. For larger sizes water cooling is preferred.

In mechanical refrigerators the liquid refrigerant is charged into the receiver. The compressor is started. When the system operates, the gas is compressed. The condenser section removes heat from the vapor (either by air or water cooling) causing it to condense into liquid which is stored in the receiver, still under pressure. In large units the heat from this section can be used to help heat the plant. The gas is now expanded by an automatically controlled throttle valve which reduces the pressure so that the liquid refrigerant will evaporate. It does so in the evaporator absorbing heat from the surrounding environment. This causes the cooling. It then goes to the compressor where it is compressed again and the cycle repeated. In water cooled mechanical refrigerators, 2.4 gal/min (9.1 L/min) per ton of refrigeration of cooling water are required for each 10°F (5.55°C) of cooling. This is a good approximation for preliminary estimates, as many cooling towers used for injection molding operate at about that range during the summer months. The most common refrigerant is Freon F-12 (dichlorodifluromethane).

A mechanical refrigerator or a chiller is designed for outgoing water of a specific temperature. The most common is 50°F (10°C). Any deviations from this will change the efficiency of the unit. For example, water leaving at 60°F (15.6°C) would raise the efficiency to 120%, 40°F (4.4°C) would reduce it to 80%, and 30°F (−1.1°C) to 60%. Therefore a 10-ton chiller designed for 50°F (10°C) would deliver 12 tons at 60°F (15°C) and 6 tons of refrigeration at 30°F (−1.1°C).

The refrigerated water can be supplied either from a central system or portable coolers for each machine. The main advantages of the central system is low initial cost and freeing floor space around the molding machine. It has a number of disadvantages. It provides water at one temperature requiring elaborate mixing systems for mold temperature control. It is relatively inflexible in terms of capacity. At the initial installation one has to guess the cooling requirement for the future. Individual chillers can be bought as required. Molding can be scheduled for their maximum utilization.

112.6   COOLING REQUIREMENTS

Two convenient equations for determining cooling loads follow:

$$\text{ton} = \frac{\text{Btu}}{(12,000)} \text{ (h)} \tag{112.1}$$

$$\text{gal/min} = \frac{(\text{tons})(12,000)}{(\Delta t)(60)(8.3)} = \frac{(24)(\text{tons})}{\Delta t}$$

$$\Delta t = T_{\text{outlet}} - T_{\text{inlet}} \text{ °F} \tag{112.2}$$

Molding machines are usually cooled with tower water. If the temperature–humidity conditions are too severe mechanical refrigeration is added as required. Tower water is much more economical and should be used when possible.

The heat exchanger of a 16-oz (0.45 kg) 400-ton (4 MN) hydraulic clamp molding machine with 45 connected horsepower was instrumented to determine the heat removed. This averaged 25,000 Btu/h (26 MJ/h). It was relatively independent of the cycle time and ambient temperature. This is not indicative of the total heat loss as the machine radiates a considerable amount of heat energy. The cooler would require approximately two tons of refrigeration (eq. 112.1). A convenient approximation of machine cooling requirement is 1 ton/20 connected hp. If a water tower with a 6°F (3.3°C) approach or cooling range was used the machine would require approximately 8 gal/min of water (eq. 112.2).

Mold cooling requirements are relatively easy to estimate. The enthalpy of plastics is a measure of their heat content and given in Btu per pound. Graphs are available of enthalpy versus temperature. In crystalline material they include the heat of fusion. By subtracting the enthalpy at room temperature from the enthalpy of the material at the cylinder temperature the number of Btu to be removed is obtained. Table 112.2 shows this for some thermoplastics.

TABLE 112.2. Enthalpy Difference or Heat Content of Some Thermoplastics between Approximate Molding Temperature and Room Temperature

Thermoplastic	Btu/lb	kJ/kg
Polystyrene	155	360
Acetate	180	419
Acetal	180	419
Polypropylene	210	489
Nylon, 6	270	628
HD polyethylene	310	721
Nylon 6,6	340	791

These figures do not actually describe what occurs. When the molded part is removed from the mold, a considerable amount of heat is still in the part, which cools in the air. There is a significant radiation loss from the mold itself. The author molded a plaque of general purpose styrene 7 in. × 3 in. × 0.150 (178 × 76 × 4 mm). The heat loss through the mold water cooling was measured. The molded part was put in a calorimeter and the residual heat measured. The enthalpy graph of this particular material showed a heat content of 140 Btu/lb (325 kJ/kg) between molding and room temperature. There was 38 Btu/lb (88 kJ/kg) removed by the cooling water, 57 Btu/lb remained in the molded part and the balance of 45 Btu/lb (105 kJ/kg) was radiated from the mold. About 80 lb/h (36 kg/h) were molded. The amount of refrigeration required is 38 × 80 or 3040 Btu/h (325 kJ/kg). This is approximately one fourth of a ton. Using the enthalpy from the graph, 140 Btu/lb, one would expect that a ton of refrigeration would be needed. Practically the amount would vary with the geometry and thickness of the part and the size of the mold. Using 50 to 75% of the figures in Table 112.2 will give a good approximation of the required cooling.[1]

112.7   MOLD "HEATING" UNITS

The function of the fluid circulation through the mold is to control the rate of heat transfer, hence the cooling rate at the plastic. Elevated temperatures are used when slow cooling is required.

The temperature of the cooling medium will depend on the molding requirements. When the cooling medium is above room temperature, requiring the addition of heat, it is commonly called a mold heater, even though it is in effect cooling the mold. A mold heater is, in essence, a tank with a motor driven centrifugal pump recirculating a fixed amount of fluid from the tank through the mold. Adding heat to the fluid is done by electrical resistance heaters. When the molding conditions are on the border line of adding or removing heat from the circulating fluid, a coil attached to a cooling medium is inserted in the tank. A temperature sensing element activates the heating or cooling circuit for the temperature at which it is set. For temperatures above the boiling point of water nonaqueous fluids are used. It is essential to keep the fluid clean as rust scale and other contaminants seriously reduce the efficiency of the heat removal. When operated at high temperatures extreme care must be used in the selection and maintenance of the connecting hoses. A ruptured connector may result in serious burns.[2]

Mold temperature control units will accurately control mold temperature. A unit attached to a mold running a half pound shot of general purpose styrene at 83 cycles/h was instrumented. The inlet and outlet mold temperatures were read every 6 s. They were charted with the heat on-off and water on-off controls of the unit, the cycle time of the machine and the mold temperature. In a typical case with the cooling water and heating elements each cycling alternately every three shots, the inlet water temperature varied from 88.5 to 90.3°F (31.38–32.37°C), and the outlet water temperature from 89.7 to 91.5°F (32.09–33.08°C). The temperature difference between the outlet and inlet water was plotted for each 6-s reading. It varied from 2.0 to 2.9°F (1.11–1.66°C) with a mean of 0.9°F (0.5°C). The cycle of the curve followed that of the units heating cooling cycle. The mold temperature as read by a dial thermometer and pyrometer showed no readable change. By changing the molding conditions slightly so that cooling water was used all the time in the mold temperature unit the outlet temperature was 77.5°F (25.2°C) and the inlet temperature varied between 73.4 and 73.6°F (23.02–23.14°C). The variation between the difference of the inlet and outlet water never exceeded 0.2°F (0.36°C). The dial thermometer and pyrometer in the mold showed no change. In the first instance 4590 Btu/h (4.84 MJ/h) were removed and in the latter 4740 (5.0 MJ/h). These figures show that commercial units can produce accurate and consistent mold temperature control which is required for proper molding.

112.8 HEAT TRANSFER

The three methods for exchanging heat are radiation, convection, and conduction. Convection and conduction are the primary concerns. In the coolers for the molding machines the heat from the hot oil is exchanged into the tube walls, and from the tube walls into the circulating water. In the mold the heat from the plastic is transferred to the cavities and cores, which in turn transfers the heat to the mold temperature circulating medium. Some of the factors which affect the rate and amount of heat transfer are material of the container, size and shape of the container, rate of flow of both materials, temperature, viscosity, specific heat, thermal conductivity, density, and surface conditions of both sides of the container. The mathematics of these processes have not been quantitatively completed. Notwithstanding, a qualitative discussion of some of the factors affecting heat transfer is valuable.

The rate of heat removal equals the overall heat transfer coefficient times the area of exposed surface, times the difference in temperature between the two fluids (plastic and water).

$$Q = UA\Delta t \tag{112.3}$$

Q = rate of heat removal (Btu/h)

U = overall heat transfer coefficient in Btu/(h) (ft^2) (°F)

A = area (ft^2)

Δt = difference in temperature of the two fluids (°F)

This equation shows, as one would expect, that the lower the temperature of the cooling medium, the faster the heat removal. The rate could also be incresed by increasing the material temperature. This would be self-defeating because the higher removal rate would not compensate for the additional amount of heat to be removed, thus lengthening the cycle. The lower limit of the cooling temperature is the molding condition. Molds that are too cold may not fill, may develop surface blemishes and lower some physical properties.

The area of the cooling surface is limited by the geometry of the mold. Table 112.3 shows the effect of different size cooling channels. Using a 3/8-in. pipe instead of a 1/8-in. one will increase the cooling rate by a factor of 1.8. Large cooling channels are one of the easiest ways to reduce cycle time. Unfortunately this is often overlooked in mold design.

By use of electrical analogies the overall heat transfer coefficient is described as:

$$\frac{1}{U} = \frac{1}{h_1} + \frac{1}{h_2} + \frac{1}{h_3} \cdots + \frac{X}{k} \tag{112.4}$$

TABLE 112.3. Physical Characteristics of Drilled Mold Cooling Holes Related to Mold Temperature Control

Nominal Pipe Size	Tap Drill Used In Mold	ID		ID Area		ID Circumference,		Surface Area ft²/ft (m²/m) ft (m) of length		Ratio of Cooling Area to 1/8 in. Pipe	Capacity at 1 ft/s,		Capacity at 0.3 m/s,	
		in.	(mm)	in.²	(mm²)	in.	(mm)				gal/min	(lb/h)	L/min	(kg/h)
1/8	5/16	0.3125	(7.9)	0.0767	(50)	0.982	(24.9)	0.0818	(0.025)	1.0	0.24	(120)	0.91	(55)
1/4	7/16	0.4375	(11)	0.1503	(97)	1.374	(34.9)	0.115	(0.035)	1.4	0.47	(234)	1.8	(106)
3/8	9/16	0.5625	(14)	0.2485	(160)	1.767	(44.9)	0.147	(0.045)	1.8	0.77	(387)	2.9	(176)
1/2	11/16	0.6875	(17)	0.3712	(240)	2.160	(54.8)	0.180	(0.055)	2.2	1.15	(580)	4.4	(263)

where X = thickness of wall, ft; k = thermal conductivity of wall, Btu/(h)(ft^2) ÷ °F/ft = Btu/(h)(°F)(ft); h = individual heat transfer coefficients, Btu/(h)(ft^2)(°F).

This equation leads to some very interesting conclusions. It is important to notice that the heat transfer rate is controlled by the coefficient at the point of maximum resistance. For example ignoring X/k, if there are only two coefficients, h_1 = 20 and h_2 = 1000, U would equal 19.6. Suppose h_2 were changed from 1000 to 500, then U would equal 19.23. Therefore, even though one coefficient were changed by 50% it would only change the total coefficient by 2%. The film coefficient for water in the cooling system is approximately 1500. However, if scales, sludge and dirt enter the system this can drop to as low as 200 introducing serious resistance to heat transfer and probable increase in mold cycles. Therefore, clean circulating water and cooling channels are very important.[3]

For molds the X/K factor is important. The rate of heat removal will vary directly with the thermal conductivity of the mold material. Therefore if the K for beryllium is 70 and steel 24 the beryllium will remove or add heat to the plastic approximately three times as fast as steel. This is an important factor in mold material selection. It is also obvious that the closer the cooling channel is to the plastic (a minimum X) the higher the rate of heat removal. Cooling channel location should be designed so that there is even cooling of the mold surface. Since heat removal varies directly with the distance between the cooling channel and the mold, equally spaced circles from the cooling channel will be roughly the same temperature. They can be drawn on a mold layout and a good indication of the temperature profile of the cavity or core obtained.[4]

It is also evident that the highest heat transfer coefficient will occur when the cooling channels are directly in the cavity or core. If put in the surrounding mold base the controlling coefficient will be between the cavity and the mold base. Interface losses are very significant and should be avoided if possible. This is one of the advantages of EDMing cavities in one block.

The ability of the plastic to change temperatures is a factor in cooling the part. This is called the thermal diffusivity and is defined as the thermal conductivity divided by the product of the specific heat and the density. There is nothing the molder can do to change this as it is an inherent property of the material. It will explain why some materials cool more readily than others.[5]

It stands to reason that the velocity of the cooling media would affect the heat transfer rate. From dimensional analysis and experimental work the heat transfer coefficient is affected by, among other things, the Reynolds number.

There is a velocity factor in the Reynolds number. When the number is below 2100 there is laminar flow and the heat transfer coefficient (inside a tube) varies as the 1/3 power of the velocity. Above 2100, turbulent flow, it varies as the 0.8 power of the velocity. The probable reason for this is that the turbulent flow provides better mixing and the metal-water interface is broken more often. In turbulent flow, for example, if water flowing through the tube had a film coefficient of heat transfer of 300 Btu/(h)(ft^2)(°F) at a given velocity, and its velocity were doubled, the new film coefficient would be 300 ($2^{0.8}$) or 522. This means that increasing the velocity of the cooling fluid will increase the rate of heat transfer. This should not be overlooked. It may be necessary to increase the pumping capacity of the mold temperature control unit. The amount of fluid circulating can be easily determined with a water meter. With this information and the cooling channel dimensions the Reynolds number can be calculated.

When mold cooling seems inadequate the first things to be done are to clean the cooling system and mold channels, increase the velocity of the cooling medium and lower its temperature. If these methods do not work, consideration must be given to enlarging or adding to the cooling channels. Similarly, the heat exchanger for cooling the oil in the molding machine shoud be kept clean and periodically examined. If the machine overheats and the water temperature is normal, the cooler should be cleaned. If that does not help there probably is a malfunction in the hydraulic system permitting oil to bypass and generate heat.

The heat exchangers for cooling oil consists of tubes through which the cooling water flows, and a shell through which the hot oil, to be cooled, flows. They are mainly single pass exchanges; that is, the liquids flow in one direction. They can be connected in two ways. In parallel, the hot oil and the cold water enter at the same end so that the cooler oil and heated water will emerge at the other end. In counter flow the hot oil will enter at one end and the cold water will enter at the opposite end. Molding machines are connected counter flow since it will remove approximately 10% more heat from the oil.

This has relevance in mold cooling, as a mold is a heat exchanger with the hot plastic as a heat source and water as the cooling medium. Most molds have the hottest section at the sprue, primarily because of the radiation effect of the outside of the mold base. Attaching cold water to the outside of the mold (analogous to counter flow) will remove more heat than putting the cooling water directly into the sprue section first. Properly designed molds will permit the plastics engineer to adjust mold temperature accordingly.

112.9   MOLD TEMPERATURE CONTROL

The two reasons for providing for good mold temperature control are (a) economical and (b) part quality. The temperature control system includes the cooling fluid, means for its circulation, method of temperature control, and cooling channels in the mold. Its purpose is to remove heat from the plastic part at a controlled rate. The goal is the removal of heat as rapidly as possible so that the part can be removed from the mold in a condition which will result in acceptable pieces. This cannot be done without a good temperature control system which will permit the molding conditions to establish the mold temperature rather than the adequacy of the equipment.

An incorrect and inconsistent mold temperature will create serious difficulties. We shall assume an adequate system and discuss some of the problems caused by incorrect mold temperature. It is not always possible or necessary to predict the best temperature for a given mold and material. With thermostatically controlled temperatures, trial and error is not difficult. In many instances there will be several different temperatures maintained for different parts of the mold.

To cool any given part a specific number of Btu will have to be removed. Equation 112.3 shows that the greater the temperature difference between the plastic and the cooling fluid the higher the heat removal rate. Therefore, a lower mold temperature will permit the part to be removed more quickly.

112.10   SUMMARY

In the early days of plastic processing, heat removal was poorly controlled resulting in many failures and a poor public image. The addition of scientific temperature control was basic in raising the performance of plastic products which culminated in the impressive percentage of plastic material used today in the United States.

BIBLIOGRAPHY

1. C.E. Waters, "Sizing Chiller to Mold," *Modern Plastics,* 12 (April 1969).
2. J.D. Robertson, "A Practical Improvement in Mold Temperature Control," *SPE-J,* 72 (April 1969).
3. C.E. Waters, "Water Treatment Pays Off," *Modern Plastics,* 114 (March 1968).
4. H.A. Meyrick, "What You Should Know About Mold Cooling," *Modern Plastics,* 219 (Oct. 1963).

5. R.L. Ballman and T. Shushman, "Easy Way to Calculate Injection Molding Set-up Time," *Modern Plastics*, 126 (Nov. 1969).

General Reference

I.I. Rubin, "Mold Cooling," *Advances in Plastics Technology*, 65–93 (Jan. 1981).

*This article is taken in part from I.I. Rubin, Injection Molding, Theory & Practice, John Wiley & Sons, Inc., New York, 1972.

113

MATERIALS HANDLING

Dewey Rainville

*C.E.O., Universal Dynamics, Inc., Material Handling Division, 176 Cedar St.
at Route 22 North Plainfield, NJ 07060*

Materials handling is the business of transferring a resin from railcar or bulk storage silo to the infeed hopper of a processing machine, making necessary modifications along the way, then collecting and treating scrap generated during processing. On the surface, it appears to be a simple, low-technology operation, like stoking the furnace of a coal-burning steamship, then raking out the pit. In fact, though, it is a critical and complex process involving a number of essential and interrelated functions. A well-conceived and effective materials-handling operation is a prerequisite for achieving maximum productivity. It incorporates basic disciplines, none of which is more (or less) important than the others. They include:

1. Primary handling of incoming materials.
2. Drying and preheating of plastics.
3. Blending and feeding.
4. Loading and conveying.
5. Scrap reduction and collection.
6. Automated systems control.

113.1 PRIMARY HANDLING OF INCOMING MATERIALS

Bulk-material handling was in use for delivering and storing large quantities of products such as grain, flour, chemicals, and coal for many years before the arrival of plastics. These techniques were then applied to plastics, with a few modifications needed for the special needs of this new material.

As in any other business, the price-per-pound one pays for plastics raw materials declines with increasing volume. Thus hoppercar loads [200,000 lb (91,000 kg)] represent the most economical form of shipment. Truckloads [40,000 lb (18,000 kg)] are next highest on the ascending scale of resin pricing, followed by LTL (less than truckload), Gaylord bulk boxes [1000 lb (450 kg)], drums [200 to 300 lb (90–135 kg)], and bags [generally 50 lb (23 kg)].

It is uneconomical to keep a filled railcar at the plant siding for any length of time. Carriers impose a demurrage charge as a way to keep their rolling stock fluid. Thus, the bulk storage silo has been adopted by the plastics industry as the most economical and practical method of storing large quantities of resin. There are two basic methods of construction, bolted and welded.

Welded silos are available in aluminum, stainless steel and mild steel construction, and are fabricated at the manufacturer's plant. Interior epoxy coating and exterior finish are applied before shipment. The epoxy coating provides a smooth, corrosion- and abrasion-resistant lining for the tank interior surfaces. Brackets are welded to the tank prior to finishing to allow attachment of various accessory items such as ladders, guardrails, fill lines, etc. Welded silos are fabricated in both fully skirted and structural leg-types. In the fully skirted type, the tank side-wall extends down to the foundation and provides support for the structure. This enclosed area is extremely useful for storing such items as railcar unloader pumps, dehumidifiers and control panels, car unloading accessories and flex hose (Fig. 113.1). In the structural leg-type, the side wall terminates at the hopper, with the tank being supported by four "I"-beam legs welded to the tank (Fig. 113.2). Figure 113.3 shows a bolted silo installation with a railcar.

Bolted steel silos are made up of a series of steel panels which are assembled in the field. Each panel is formed at the manufacturers plant, at which time both the interior epoxy coating and exterior finish are applied.

The tank panel (staves) are approximately 8 ft (2.4 m) high, so that bolted tanks are available in 8-ft (2.4 m) increments, starting at 16 ft (4.8 m).

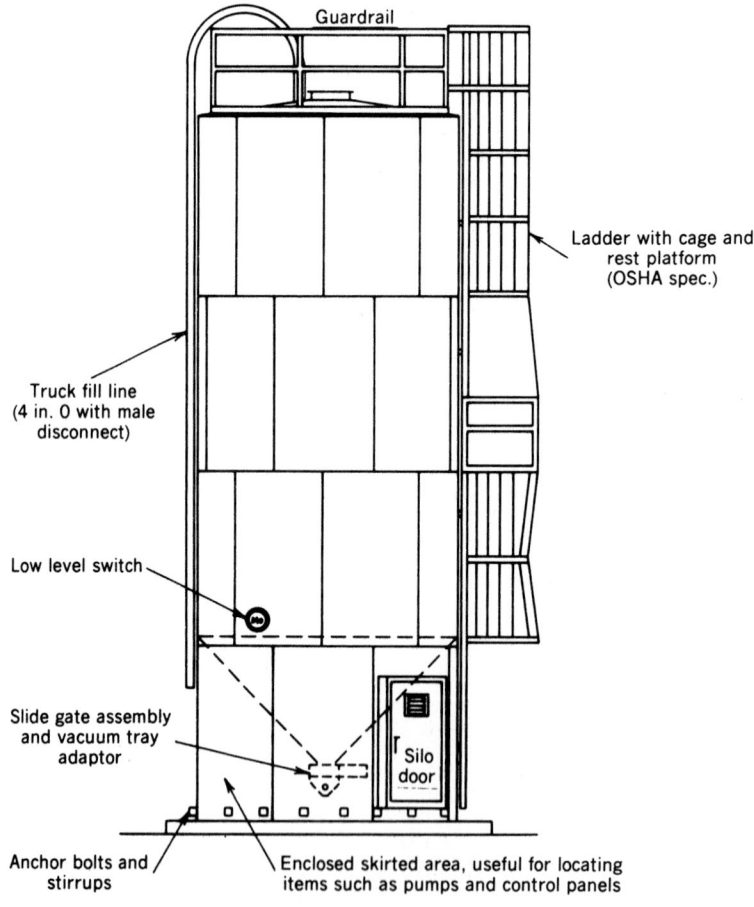

Figure 113.1.   Fully skirted bolted steel silo. Courtesy of Universal Dynamics, Inc.

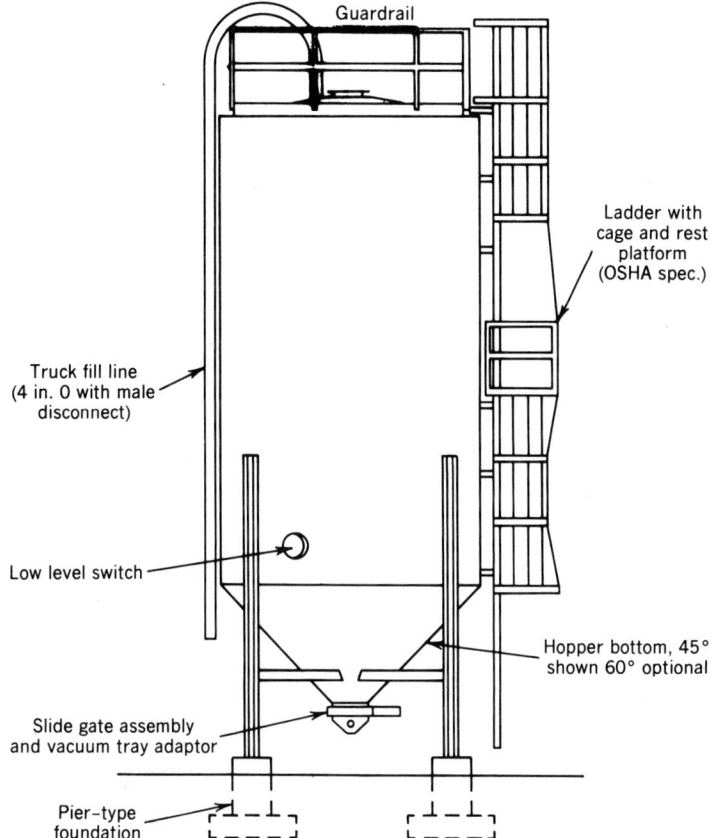

Figure 113.2.   Structural leg-type welded steel silo. Courtesy of Universal Dynamics.

113.2 METHODS OF SHIPMENT

Welded steel silos are shipped by the manufacturers on specially designed trucks of their own. Bolted silos are shipped from the manufacturers plant as a series of prefabricated steel panels. Silos usually have their final exterior finish before shipping.

113.3 FOUNDATION

Because of the weight and wind, loading of storage silos requires heavy, concrete slab foundations. Approval by a local professional engineer is desirable to assure that local codes are met.

113.4 SELECTION

Both bolted and welded silos are available in a number of sizes. Standard bolted tanks are offered in 9-ft, 12-ft, and 18-ft diameters (2.7, 3.6 and 5.5 m), with heights adjustable in 8-ft (2.4-m) increments, beginning with 24 ft (7.3 m). Welded silos are available in 9-ft,

Figure 113.3.   Bolted silo installation with railcar. Courtesy of Universal Dynamics, Inc.

10-ft and 12-ft (2.7, 3. and 3.6 m) diameters with heights in roughly 6-ft (1.8 m) increments. One limitation on welded tanks is the size which can be shipped over the road.

The size a customer selects will depend on whether he or she will receive the material by truck or in railcar. The most popular size tanks for truck-fill applications are

9 × 32 (2.7 × 9.8 m) bolted or welded 65,000 lb (30,000 kg)
12 × 24 (3.7 × 7.3 m) bolted or welded 75,000 lb (34,000 kg)
12 × 32 (3.7 × 9.8 m) bolted or welded 108,000 lb (49,000 kg)
10 × 32 (3 × 9.8 m) welded 77,000 lb (35,000 kg).

Popular railcar silos are

12 × 56 (3.7 × 17 m) welded [208,000 lb (94,000 kg)]
12 × 64 (3.7 × 19.5 m) welded [242,000 lb (110,000 kg)]
15 × 40 (4.6 × 12.2 m) bolted [215,000 lb (98,000 kg)]
15 × 48 (4.6 × 14.6 m) bolted [256,000 lb (116,000 kg)]

All volumes shown above take into account a 30-degree angle of repose, which is common to most plastic material. Capacity is based on 38 lb/ft^3 (1,610 kg/m^3) bulk density. In making a silo size selection, it is important to allow for a tank capacity that is equivalent to the volume of the carrier plus as large a safety margin as practical. Another factor that must be taken into consideration when sizing tanks is local law. Some zoning ordinances may restrict the overall height of industrial structures. If such a limitation exists, it may require going to a shorter tank of the next larger diameter. Delivery of plastic resin by bulk railcar has become increasingly popular because of significant savings which can reach $0.035/lb.

Material flow characteristics are important to silo selection. If the resin is a free-flowing pellet, a standard 45° hopper bottom is adequate. But if the resin has poorer flow characteristics, a 60° hopper bottom may be required. And if the material is extremely difficult, a specially engineered hopper with air pads or other flow-inducing devices is indicated. On simple, clean pellet storage applications, a plain clam-shell-type vent is adequate to allow air in and out of the tank. On very dusty or powder applications, bag-type or continuous self-cleaning filters may be required.

113.5 ACCESSORIES

A number of accessory items are required to make a tank a functional piece of equipment. These are listed as follows:

- Ladder, cage and rest platform that provide access to the tank deck from ground level. OSHA requirements dictate the design of this hardware. One requirement is that a rest platform must be provided for every 30 feet (9.1 m) of tank height.
- Guardrails, toeboards, and crossovers provide protection to individuals on top of the tank.
- Manholes enable inspection and access to the upper area of the silo.
- Truck fill lines (aluminum tubing 4-in. (102-mm) diameter with male disconnect fittings) allow transfer of material from truck to silo.
- Level switches provide high or low level signals. Rotating paddle-type are most commonly used, but others can be supplied at customer's request.
- Vacuum tray adaptors permit withdrawal of material by vacuum conveying systems.
- Slide-gate shutoffs ease shutoff of material flow in the event the take-off box requires cleaning or maintenance.
- Continuous level indicator allows remote readout of silo level. It is calibrated in tenths of a foot (30 mm).
- In addition, a variety of special items, such as silo dehumidification for hygroscopic materials is available.

113.6 BULK RESIN CONVEYING

The basic types of pneumatic conveying (vacuum system and combination negative/positive pressure system) have been applied to transfer resin from the railcar to the storage silo, and from the silo to the processing plant. Each of them possesses certain advantages.

Figure 113.4 shows a straight vacuum-type unloader for pelletized material. In this type of system, a vacuum pump is generally located in the skirt of the silo. Air return lines extend up to the vacuum hopper. On multiple-silo installations, each vacuum hopper has a vacuum line extending to a central area near the pump. A manual flex hose switching station is used to selectively draw vacuum on any silo loader.

Fill lines extend from each vacuum chamber to a central area. The silo fill lines are always equipped with male disconnect fittings; the air return lines are fitted with female disconnects. By utilizing stainless steel flex hose connections, unloading manifolds and accessories, the hookup to the railcar discharge is accomplished. When the pump is started, the unit functions identically to smaller vacuum loaders in that it runs for a period of time until the chamber is full. It then allows the material to dump into the silo. This process is repeated until either the silo is full or the railcar compartment is empty.

Figure 113.4.   Vacuum type unloader for pelletized materials. Courtesy of Universal Dynamics, Inc.

A high level switch must be used with this unloader. Most often, the rotating paddle-type is used. When material contacts the paddle, the unit is shut off. The pumping system itself incorporates a positive-displacement motor/blower assembly. The inlet of the pump is fitted with a two-stage secondary filter to prevent fines, which may have passed through the pellet screen, from entering the pump. The pump inlet also has a manual vacuum relief; in the event of a material line blockage, it will allow air to enter the system and prevent damage to the system.

The loader control panel, which can be mounted on the unit or at a remote point, contains motor starters, timers, high and low level switch lights, and (on multiple-silo systems) a selector switch to energize the proper high level switch for automatic operation. To change from loading one silo to another, one must change the flex hose connections to the material and vacuum line on the new tank and position the tank selector switch on the control panel to the new silo.

Figure 113.5 shows a typical combination negative/positive pressure vacuum system. Vacuum from the pump draws material into either a cyclone separator or filter receiver. The pellets pass through a rotary airlock and enter the blower discharge air stream, which is at positive pressure. The air/material mixture is transferred to the silos via stainless steel flex hose and fill lines. Combination units tend to be high-capacity systems used primarily with multiple-silo systems. One advantage is the fact that they require no equipment whatever on top of the silo, only a simple fill line. All maintenance is performed at ground level.

As with any type of pneumatic conveying system, transfer rate depends on material characteristics and distance. Railcar unloading has the added complication of always being a high-lift situation, anywhere from 40 to 70 feet (12.2 to 21.3 m) vertical. A 25-horsepower (19 kW) railcar unloader moving polyethylene pellets approximately 100 feet (30 m) horizontally and 60 feet (18 m) vertically will maintain an approximate throughput of 10,000 pounds per hour (4500 kg/h).

113.7 RAILCAR CONNECTIONS AND ACCESSORIES

The basic accessories required for unloading are railcar adaptor, air inlet filter, and hatch filter.

The railcar adaptor slips over the end of the discharge tube and is held in place by set screws. The railcar discharge is adjustable for air/material ratio. Adjustments must be made to establish the proper ratio for optimal material conveying. The introduction of air on the opposite side of the car provides for a much smoother material pickup.

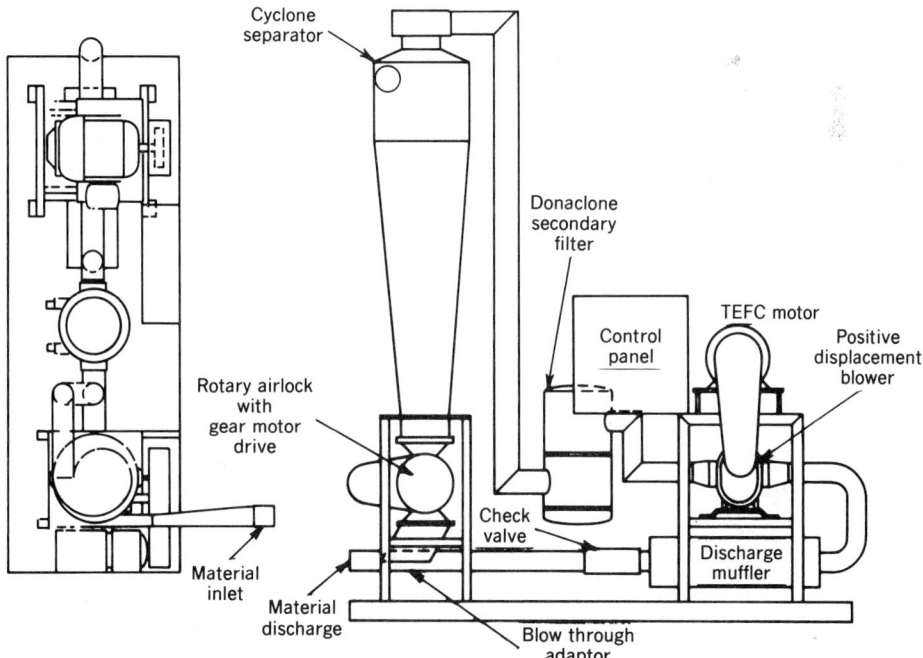

Figure 113.5. Typical combination negative/positive pressure vacuum system. Courtesy of Universal Dynamics, Inc.

The air inlet filter fits on the far side of the discharge and prevents contaminants from entering the conveying air stream from that point.

The hatch on the compartment being emptied must be open to allow air to take the place of the material being withdrawn. Contamination is prevented by placing the hatch filter over the opening.

The vast majority of resin is moved from storage silo to the interior of the plant by a simple, vacuum-type system as shown in Figure 113.6. Air is caused to move through a tube by means of a pump or blower. The back of the take-off box, where air enters the system, is fitted with a screen to prevent foreign matter from entering the system. The material to be conveyed is picked up by the air stream through a slot in the tube (the slot is adjustable so that the air/material ratio can be optimized for maximum conveying efficiency). The air and material mixture are conveyed to the vacuum chamber, where the pellets accumulate. When the chamber is full, a discharge flapper opens, allowing the material to dump. If a level switch extending into the receiving hopper is not satisfied, the cycle is repeated.

A simple and inexpensive way for a customer to enjoy the cost savings of bulk delivery is to use a basic silo-surge bin system as shown in Figure 113.7. These are silos sized to hold 60,000 to 80,000 lb (27,000–36,000 kg) of resin. Inside the plant, all that is required is a surge bin, usually of 2,000-lb (909-kg) capacity and a vacuum loader to transfer the material from the silo to a bin fitted with elevated legs and a manual shutoff device on the discharge. In this manner, Gaylords or drums can be filled by gravity and the material distributed to various processing or preprocessing machines by conventional means.

A principal point to remember about material distribution systems is the fact that material lines cannot be run the way ordinary plumbing, compressed-air and electrical conduits are installed. Because you are moving particles of material in an air stream, every attempt must be made to keep the runs as straight as possible with a minimum of turns. Long lengths of flex hose also should be avoided because of their negative effect on conveying rates.

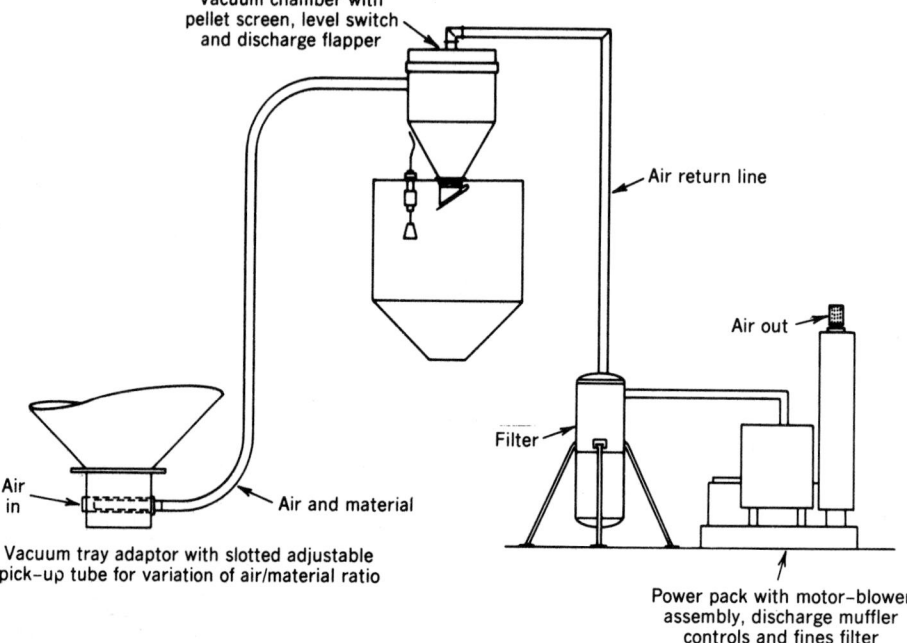

Figure 113.6. Simple vacuum-type system for conveying from storage silo to the interior of the plant. Courtesy of Universal Dynamics, Inc.

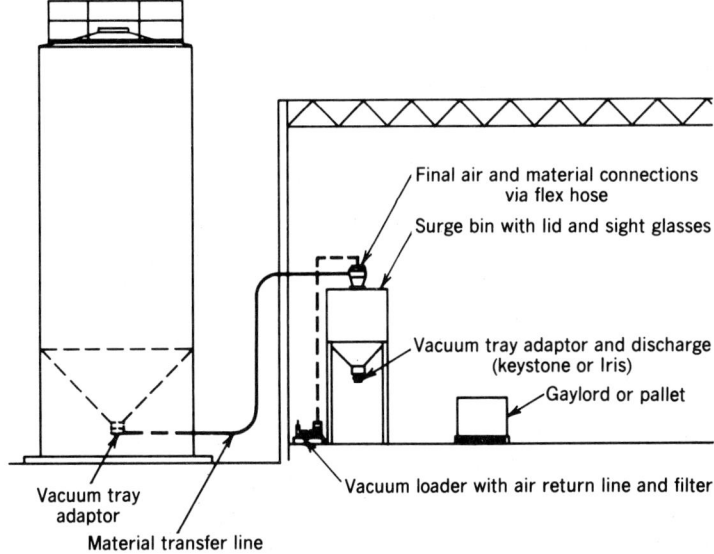

Figure 113.7. Simple silo-surge bin system. Courtesy of Universal Dynamics, Inc.

There are various ways of supporting the tubing. The most common method involves the use of brackets and clamps. These brackets, along with their accessory items, can be used to fabricate crossovers to span tracks or roadways and to suspend tubing from roof trusses, columns or other supports in the plant. With the material and air return lines thus run to the area of the vacuum hoppers, the final connections generally are made with flex hose to allow easy movement of the hopper should cleaning or servicing be necessary.

113.8 DRYING AND PREHEATING

For drying purposes, all plastics can be divided into two groups: hygroscopic (those that absorb water), and nonhygroscopic (those that do not). But plastics that do not absorb moisture may still be subject to surface moisture caused by condensation. This condition can result from bringing cold material into a warm room. Surface moisture can be removed by a hot air dryer. This is an inexpensive device that simply heats air and blows it up through the dry resin in a hopper to evaporate the moisture. There are several advantages to hot air drying:

1. The drying energy is not lost entirely; preheating of the material reduces the amount of frictional heat that must be later produced in the extruder.

2. Normally the temperature of material fed to the throat of a processing machine can vary from 30 to 60°F (-17 to 34°C), based on changes in room temperature, or other conditions related to ambient handling conditions. But a hot air dryer keeps the temperature at the throat constant. With so many variables in the plastic process, getting control over one more is very important.

3. Preheating reduces frictional wear on the extruder barrel and liner. It also reduces the thrust and wear on the screw bearing system. The load on the drive train is reduced, as is power consumption.

4. On an extruder that is running at maximum plasticating capacity, an output increase of 10 to 15% can be accomplished when resin is preheated.

Principles of Drying. When plastics absorb moisture, they can be dried by hot air. But room air has moisture in it depending on the humidity that day. Using the maximum drying temperature allowed for the particular plastic, the process arrives at an equilibrium. The only way the plastic can be dried to a lower moisture level is to lower the amount of water in the air—or lower the dewpoint in the system. Figure 113.8 shows a computerized dryer with an insulated drying chamber.

The main nonhygroscopic materials are polystyrene, polyethylene and polypropylene. Almost all engineering resins absorb moisture. Nylon, polycarbonate, PET, and ABS, are among the more common hygroscopic performance materials.

Although plastics generally absorb minute amounts of water when plastics absorb moisture, the amount is small. (saturation is normally below 0.2%.) There are exceptions. A major one is nylon, which can absorb up to 11% by weight. Thus, nylon scrap must be recycled promptly, and virgin resin is usually delivered in sealed bags, pre-dried.

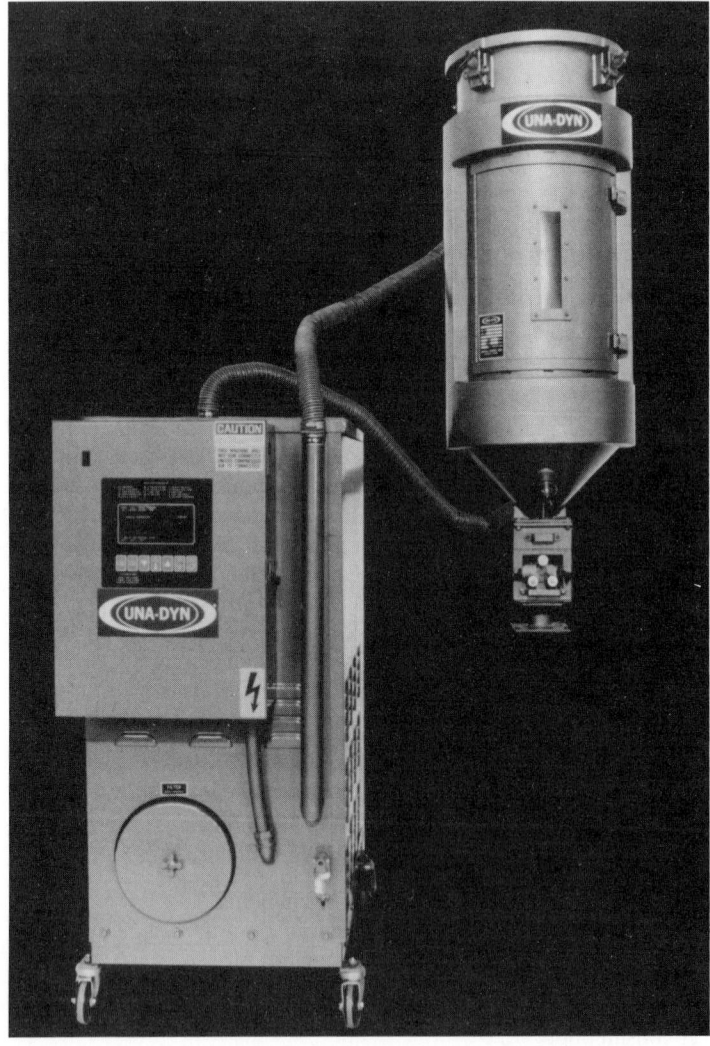

Figure 113.8.   Computerized dryer with an insulated drying chamber. Courtesy of Universal Dynamics, Inc.

Some special applications require drying plastics to below 0.002% moisture. These include PET for soda-bottle preform molds, polycarbonate for some requirements, and plateable-grade ABS (such as that for automotive applications).

Dehumidifying drying is accomplished by using a desiccant to dry the air before it is preheated and blown through the plastic. The most common desiccant is molecular sieve, but silica gel is also used. Normally, two or more desiccant beds are used. While one is drying the air, the other is being dried out at temperatures above 600°F (316°C). This is done by rerouting the air through valves or by rotating the beds.

Other drying methods, used to a lesser degree, include: steam jacketed high-intensity mixers, vacuum dryers, radiant heat, and radio frequency heat.

Dryer selection is based principally on the volume throughput required. Tables of exposure times enables one to determine the correct size hopper, for a given operation. Thus, throughput of 300 lb/h (136 kg/h) multiplied by 4-h exposure time equals a 1,200-pound (545 kg) hopper. The dehumidifier and blower are recommended by the manufacturer for each application, but are normally sized to agree with the hopper capacity.

Dryer maintenance must be diligent. Even though desiccant does not wear out through thousands of cycles, it will have to be removed and replaced if air going through is contaminated or contains plasticizers. Also, filters must be changed based on the material being run.

113.9 BLENDING

Processing of plastics would be much easier if only virgin resin had to be run. But, since scrap is always produced, an efficient way of getting it back into the system is needed. Fillers, reinforcements, chemicals, and additives are often compounded into the plastic at the point of product manufacture, where economics and flexibility require that they be applied in the final processing machine.

The expanding use of additives has in fact spawned many blending and feeding equipment innovations. Extrusion equipment itself has evolved, with better screw designs to get better final mixing and compounding action. This permits adding ingredients at the process level, eliminating the long-term planning and higher cost needed to have special compounds made.

Materials to be blended often have different particle sizes or bulk densities. If they are not properly mixed before going into the processing machine, the final product may not be made of a uniform mixture, and failure results. Fortunately, a wide variety of efficient equipment exists, to meet every blending and feeding requirement.

Central Blending. When a large machine requiring high throughput above 600 pounds per hour, or when several machines are running the same product and can receive their material from one blender, a central blender is used. It may be mounted on the floor above. Various types of feeders that are available to bring offline-produced blends to the hopper of feed throat of a processing machine. Vibrator feeders are the least expensive and also the least accurate.

If the material to be used is natural or is delivered precolored, then the only consideration is reintroducing regrind into the resin stream. The simplest method of doing this is by means of a dual-ratio valve system (Fig. 113.9). This generally is used in conjunction with a scrap grinder located beside the press and fitted with a vacuum pickup tube on the collection bin. An internally mounted shuttle valve assembly is incorporated within a standard vacuum receiver. The valve is positioned between two inlets with one connected to the virgin material source and the other to the grinder or other pickup point. The valve is operated by an air cylinder and controlled by a timer so that the system, during each loading cycle, draws first virgin and then regrind in whatever proportions are established on the controller. The setting on the dual-ratio valve timer must be set in proportion to the vacuum loading time cycle.

This system lends itself best to injection or blow molding operations where there are a number of machines, each using different colors or materials. Figure 113.10 shows a typical blender for scrap and virgin material.

If color is to be introduced into this type of system, an on-the-press feeder is required. The color injector is the simplest method used for accomplishing this type of application. The unit sits below the machine hopper as shown in Figure 113.11. As material passes through the feed throat, color is injected into the stream in a predetermined ratio.

If the machine/product mix of the plant lend themselves to it, great economy can be achieved by going to a centralized material blending system. Figure 113.12 shows schematically how this may be achieved. Blow molding plants generally are well-suited to this type of system in that they usually are limited to one material and two or three colors. Instead of having a grinder/blender system for each machine, we have one for each color located at some point remote from the machine area. Start-up scrap and reject bottles are brought back to the grinder. The virgin-regrind and color concentrate are metered and mixed in the blender. Individuals or common material lines extend from the machines to the blender are connected by flex hose and disconnects to blender collection bins. When a color change is desired on a particular machine, the line is disconnected from one blender; the vacuum hopper draws out whatever material is in the line; the machine hopper and vacuum chamber are wiped clean, and the line is reconnected to the blender with the desired color. Central dryers can be incorporated in systems of this type to remove moisture from all incoming material.

There are several main advantages to using such a central system.

- More economical; only one grinder-blender per color, rather than one per machine.
- Color matching problems are eliminated, since all of the same color material comes from the same blender rather than individual machine-mounted units.

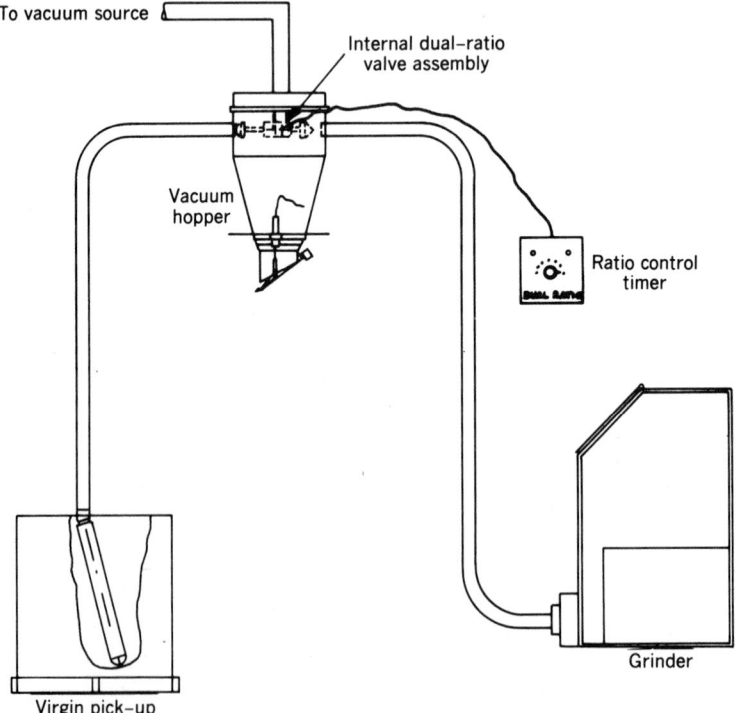

Figure 113.9.   Dual ratio loader. Courtesy of Universal Dynamics, Inc.

Figure 113.10. Blender for scrap and virgin materials. Courtesy of Universal Dynamics, Inc.

- Almost all material handling adjustments and maintenance are performed at floor level in one, centralized location.
- Grinder noise and dust are eliminated from the production area.

Continuous central blenders are composed of two to five feeders, each with its own hopper. Hopper sizes are selected to hold enough material to allow a warning time when the filling system fails and must be corrected before the material runs out and causes a shutdown of production. In the case of a large sheet line, shutdown and restart can cost several thousands of dollars.

For high production in plastic producing plants, older designs such as ribbon blenders, inclined rotary tube blenders, and high intensity mixers are often used for their high capacity. Other options include batch blenders, where ingredients are added and mixed as a batch. One such unit has a vertical screw that picks up the material at the bottom of the chamber and sprays it out of the top. Another common design is a drum tumbler. The elements are

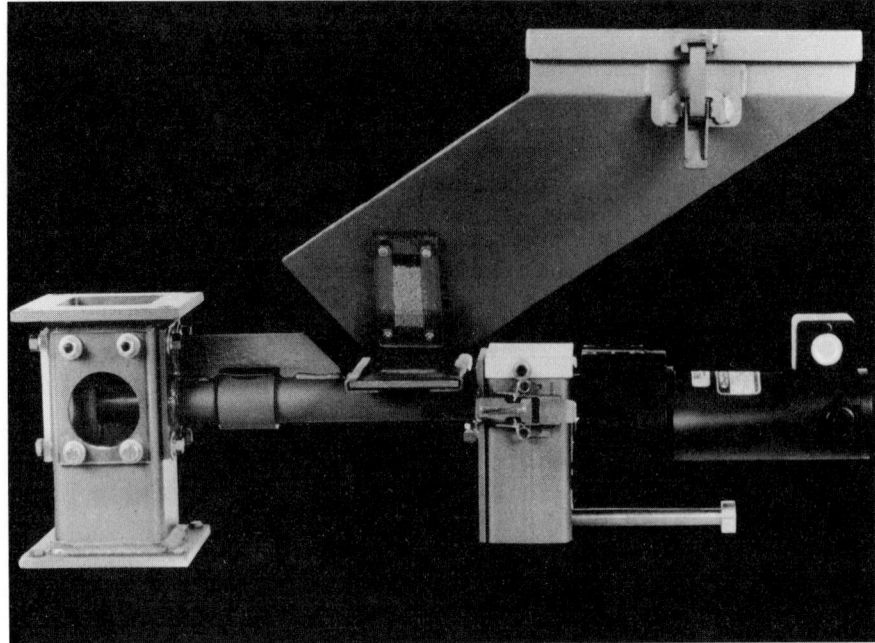

Figure 113.11.   Machine mounted color feeder. Courtesy of Universal Dynamics, Inc.

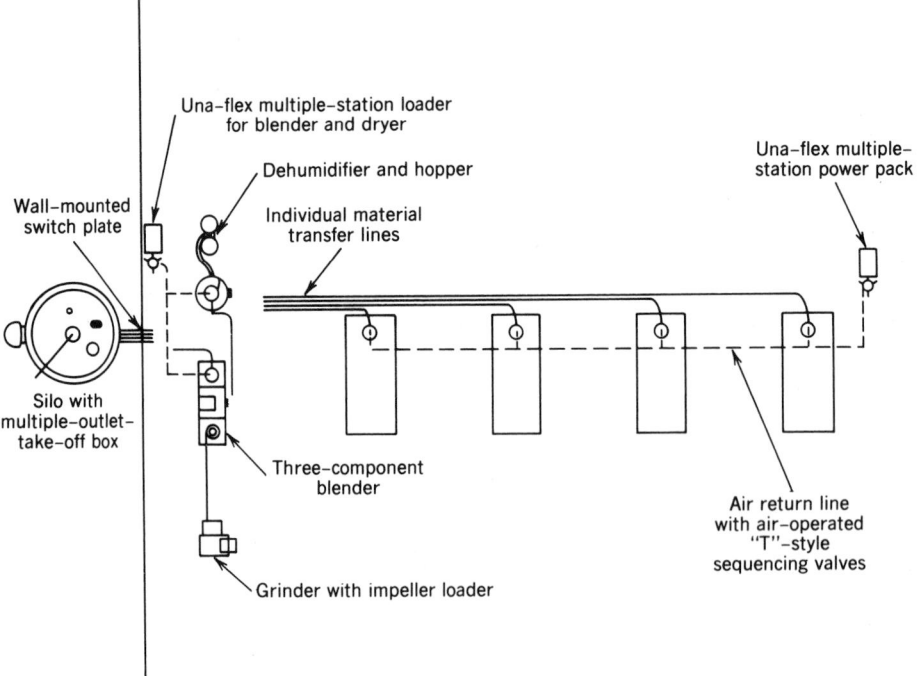

Figure 113.12.   Schematic of central dryer and blender. Courtesy of Universal Dynamics, Inc.

put into a drum and mounted into a machine that rotates it at an angle at about 15 rpm. This method is ideal for small batches.

On-machine blenders offer effective performance. In dual-tube loading, vacuum chamber picks up virgin resin in one pipe, and then switches to loading scrap. The proportions are determined by timing. Some mixing occurs when the chamber dumps into the machine hopper. Accuracy is low, but this is a simple and effective blending technique for noncritical applications.

113.10 ADDITIVE AUGER FEEDERS

Additives auger feeders are common on machine elements. For extrusion, D-C drive can be slaved to follow the extruder speed. For injection, the feeder turns only during recovery time. A common collection chamber permits up to four feeders to be mounted on one machine, but more than two is rare.

Powder feeders, with agitators in the hopper to prevent bridging are available. So are liquid pumps. The most common design uses a peristaltic principle, ie, gradually squeezing a tube filled with liquid. To clean, simply replace the inexpensive tube.

Continuous mixing also is accomplished in the screw of extrusion or molding equipment. Additives are fed into the stream of the base material, arriving at the processing screw as a very rough mixture. Complete mixing is accomplished while melting the plastic. Much progress has been done in improving the mixing action of the screw. (Pre-mixing may be advisable, however, to lessen the load on the screw.)

Various types of feeders are available to bring offline-produced to the hopper or feed throat of a processing machine. Vibratory feeders are the least expensive and also the least accurate. Auger feeders are most commonly used and are accurate to about 1 to 1.5 wt%. Weigh feeders are accurate and expensive, but they can be justified because close control allows precise measurement of color. What is saved by eliminating overuse can pay for the equipment. (Bulk feeders cannot compensate for differences in specific gravity.) Liquid pumps are effective and reduce costs; they are excellent where the proportion of color (or other additive) is small (less than 1.5 wt%). Rotary pocket disks for dosing are accurate for powders or for poor-flowing additives. They are often used way to add minute quantities.

113.11 LOADING AND CONVEYING

Loading of plastics originally meant opening a paper bag and pouring its contents into the processing machine hopper, in much the same way as coffee beans are prepared for grinding. The first electric-power hopper loader was made about 1960 by Erie Plastics, Erie, Penna. It consisted of a vacuum cleaner mounted over the hopper, with the bottom cut off and replaced by a mechanical flapper.

Today, most hopper loaders are electric units with a brush-type motor. The typical loader incorporates a device that activates a mercury or micro switch when the level of plastic in the hopper drops; this starts the motor of the loader, which works like a vacuum cleaner. A filter keeps dust and particles from entering the blower fan and the atmosphere.

Compressed-air Venturi-type loaders are now commonly used for small machines because less air is needed for low resin throughput. In operation, compressed air is drawn into a tube picking up the plastic pellets from a drum and conveying them to the machine infeed hopper. Unless a small cyclone is used on the delivery end, dusting may occur, however.

Multiple-station loaders represent an efficient advancement of the technology. This design connects two to 24 vacuum chambers to a central vacuum pump, usually a positive-displacement pump. A three-phase motor eliminates the need for brush replacement. The chambers can have individual filters, or pellet screens, but a separate filter is needed just forward of the pump to protect it.

If the machines are all running the same material, a common material line can be used. However, the need for a separate material line for each consuming machine is more common. Level switches in each hopper open "T" valves that direct vacuum to a specific chamber for material loading.

Loaders are available to suit every type of resin and every processing situation. They include dual-tube loaders (discussed in the previous section of this chapter), auger loaders, and impeller loaders.

Auger loaders, with both rigid and flexible augers, are used when excessive fines make air conveying difficult. They are not practical, however, where cleaning is needed for regular color or material changes. Impeller loaders are often mounted on granulators to blow the scrap to a cyclone over a blender or scrap bin.

Bulk loaders are really conveying devices. Applying as high as 50 hp (37 kW), they are used to move material several hundred feet at rates of 40,000 lb/h (18,000 kg/h). They find utility in the plastics industry in unloading railcars, loading silos or inplant storage bins for distances up to 300 feet (91 m).

Conveying technology itself is a highly developed science that deals with optimum line speeds, wear of components, horsepower requirements, and system controls. Much of this is covered in the opening section of this chapter.

Figure 113.12 shows material being transferred to machines by individual material lines. An alternative method is to utilize common material lines, which have one line per material source, rather than one with "Y" laterals at each use point. One advantage of a common material line system is that each line is always dedicated to the same material, thereby eliminating the possibility of cross contamination.

The common vacuum line has been almost universally adopted because of the savings in material and installation labor it permits.

It is also possible to distribute vacuum in a multiple-station system using either common or individual lines. Common vacuum lines are fitted with "T"'s at each station on which vacuum sequencing valves are located.

113.12 SCRAP REDUCTION AND COLLECTION

To put it as simply as possible, an efficient inplant scrap reclamation and reuse operation is nothing less than vital to a product manufacturer's profit picture. The price increases that have taken hold for themoplastics, both so-called commodity types and the more expensive engineering grades, make it indefensible to discard or otherwise waste the scrap material generated during product manufacture. Recovery of scrap from thermoplastic processes is a much more acute need than it is for thermosets. That is because of the different natures of these materials. Thermoplastics can be remelted; thermosets cannot; thus thermoplastic scrap recovery is relatively easy and economical; thermoset scrap recovery is just the opposite.

What makes recovery and reuse even more essential is the fact that ALL thermoplastic processes generate scrap. Some of it occurs at start-up, with material drooling out of the barrel while the processing machine is being brought up to satisfactory production level. Some of it is caused by human error or equipment malfunction, resulting in rejected parts. Some of it is intrinsic to the very nature of plastics processing. Sprues, runners and flash are produced as part of the product-molding operation; they must be removed and ground up for reuse.

Therefore, much technological effort has been focused on the development of scrap-reduction and recovery systems. It has yielded equipment to accommodate virtually every need.

There is no single "best" way to address the problem of reduction and handling of inplant-generated scrap. Continuous recycling is desirable, but cannot always be done. Scrap should go right back into the process, automatically. But collecting and immediately reusing it semiautomatically is better than collecting and storing it for later use.

Delayed recycling has problems that are difficult to overcome. Not only does it require supervision to keep track of scrap and reschedule it, but it takes floor space and collects dirt and contamination. The value of scrap drops sharply if not handled under careful control.

113.13 SCRAP-REDUCTION OPTIONS

Thermoplastic scrap grinders all operate on basically the same principle. They are composed of a rotor that holds blades (cutting-action and stationary) in an enclosure with an opening at the top or side for scrap entry and a perforated screen in the bottom of the chamber. The screen keeps oversize scrap particles in the cutting zone until they reach the size of the perforation diameter in the screen. They then drop out of the cutting chamber, into collection bins or moving conveyor belts for subsequent handling and return to the processing machine infeeds. Beyond that basic operating principle, there is a wide range of choice in types of grinder.

Beside-the-press. As the name implies, these scrap grinders are located beside injection-molding machines, which produce some scrap each cycle, Figure 113.13. The scrap is fed by the operator or automatically ground up and is either transferred to temporary storage or dropped into a chamber where it is blown (with a blower or impeller loader) to a cyclone in a storage area or to a blender. The grinders are designed so that the operator cannot reach into the chamber while operating or when cleaning without automatic granulator shutdown.

Auger Grinders. Instead of having an operator open the injection machine to take out the parts when the mold opens, the part falls automatically onto a conveyor underneath the machine Figure 113.14. The conveyor is designed to separate scrap from finished parts. As the parts come out of the mold, they are separated and go into one box while the scrap is diverted into an open auger feeder. This auger transports scrap into a grinder chamber. The big advantage is that the entire operation can be carried out without an operator.

Lump Grinders. Some scrap produced by the various processes is heavy and thick. The standard scrap grinder would jam easily on these chunks of heavy plastic. Therefore, machines have been designed specifically for "chewing up" these large pieces. This is done either by a large unit with a very heavy rotor, or by a grinder with staggered knives that nibble at the large lump and reduce it gradually. Lump grinders normally are used in a central grinding department rather than beside the operating machine.

Film Grinders. Plastic film is extremely difficult to granulate because it is very thin. Therefore, grinders are designed that will cut the film with a scissorslike rather than a chopping action. This is done by slanting the knives in the machine. The angle of the knife is such that a shearing action is accomplished. It is possible to cut plastic film in gauges as low as 0.001 in. (0.0254 mm) without tearing or fluffing the material. It is very important to avoid fluffing, because this decreases bulk density and increase the difficulties in melting, feeding and reprocessing the material. It also makes it more difficult to blend the material with virgin resin.

Because film scrap does not flow easily, film grinders always incorporate a blower discharge to keep air moving through to deliver scrap to a cyclone separator. Whenever possible, film scrap should be recycled automatically. One of the problems with film is that it develops an electrostatic characteristic that draws dust and dirt. Thus, the sooner it is reused, the less chance there is for contamination. If film scrap is on a roll, a grinder is used with feed rolls to pull the web off an unroll stand and feed it into the grinder. Trim from the edges

Figure 113.13. Beside-the-press grinder. Courtesy of Sterling, Inc.

of the film is either blown directly to the grinder as a ribbon or is chopped up first by a cutter-blower or an edge trim grinder Figure 113.15.

Large Part Grinders. Many injection-molded parts are several square feet in size. Such parts as automobile dashboards, surfboards, radio and television cabinets and industrial parts require grinders with feed openings of 3 by 6 ft (1 by 2 m) or more. These parts are always ground in a separate central grinding department because of the noise they generate. One of the problems with all scrap grinding is the production of noise. Grinders located in the manufacturing area must be soundproofed below 85 decibels; otherwise operating personnel would have to wear ear protection.

Figure 113.14.   Auger grinder. Courtesy of Sterling, Inc.

Pipe, Tube and Rod Grinders.   Extruded shapes that are in the form of long continuous
pieces represent a special scrap grinding challenge. A number of techniques have been
developed for this purpose. Some machines have staggered knives that initially break the
rod and tube into smaller pieces, which then go into a normal scrap grinder chamber. Others
actually chop off chunks with a shear action; these parts fall into a scrap grinding chamber.
The most common units for large-diameter pipe and rod are large, heavy grinders with very
strong rotors and knives. Horsepowers range from 75 to 150 (56–112 kw).

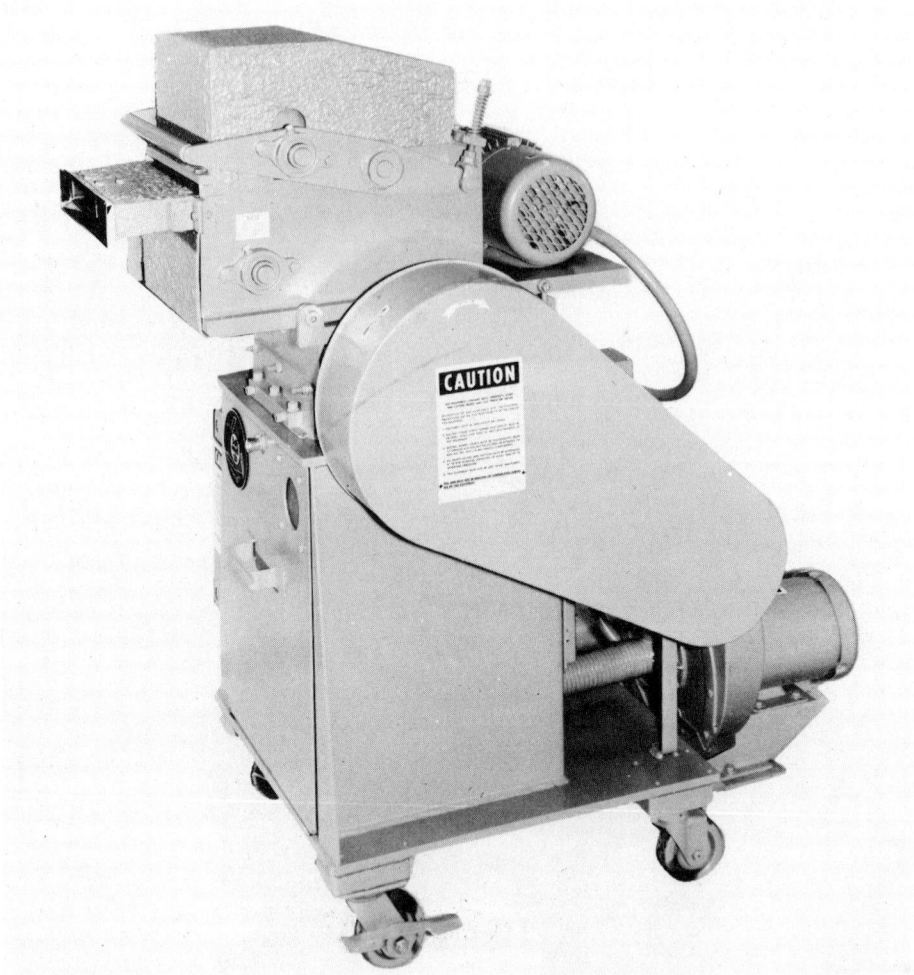

Figure 113.15. Feed roll grinder for sheet extrusion trim scrap. Courtesy of Sterling, Inc.

Sheet Grinders. Most of these units are designed to feed scrap directly back to a blender to be mixed with virgin material for immediate re-extrusion. Grinders are built so that the sheet can be fed without jamming. In most cases, the sheet is cut into smaller pieces by a shear at the infeed of the grinder and then fed in manually. There are other versions; among them a very large [150 horsepower (112 kW)] grinder with a throat opening as wide as 6 ft (3 m) that receives the complete sheet and grinds it automatically. These grinders are often buried in a pit under the floor so that the plastic can be diverted down into the pit and ground at the end of the line.

Strand and Sheet Pelletizers. A variation of the scrap grinder is a machine that is designed to make pellets for feeding to processor machines. These fall into two classifications:

Strand Pelletizers. This is a machine that receives a series of extruded strands about 1/8 in. (3.2 mm) in diameter and guides them through feed rolls into a rotating blade. The rotating blade and the feed rolls are timed together so that the length of cut produces uniform pellets.

The other type of pelletizer takes the entire sheet and produces pellets by the use of serrated knives. These knives are very expensive and there is a high-cost factor in maintaining them. They are used extensively for flexible poly(vinyl chloride) (PVC).

Hot Melt Grinders. In the wire-and-cable industry, there is a process where molten plastic coming out of the extruder must be ground up immediately and sent back through the system. The problem, of course, is that the hot material resists cutting is apt to simply smear and jam the inside of the grinder. This tendency is overcome by a very high flow of air through the grinder to keep the machine and the plastic cool.

Central Grinding. Central grinders are large machines for general scrap reduction operations. They are usually kept in a separate grinding room and are hand or conveyor fed. The machines are designed to be easy to clean because there is often a need to grind different colored plastics. The machines must be cleaned thoroughly between batches, to avoid contamination. Even so, the first few pounds of scrap put through the grinder are normally discarded to rid it of leftover plastic from the previous run.

Fine Grinders. There are applications where the use of plastics requires that it be reduced to a fairly fine mesh. These include coatings, fluidized bed operations, rotational molding and others. Because thermoplastics are heat sensitive, it is very difficult to grind them to a fine mesh without causing heat and fusion. Therefore, fine grinding embodies a requirement for cooling or special grinder design, Figure 113.16. This can be done in a variety of ways, as follows:

1. Cooling the rotor and body of the grinder with circulating water. This technique presents the danger of leaks, but is effective in dissipating heat buildup in the cutting chamber.

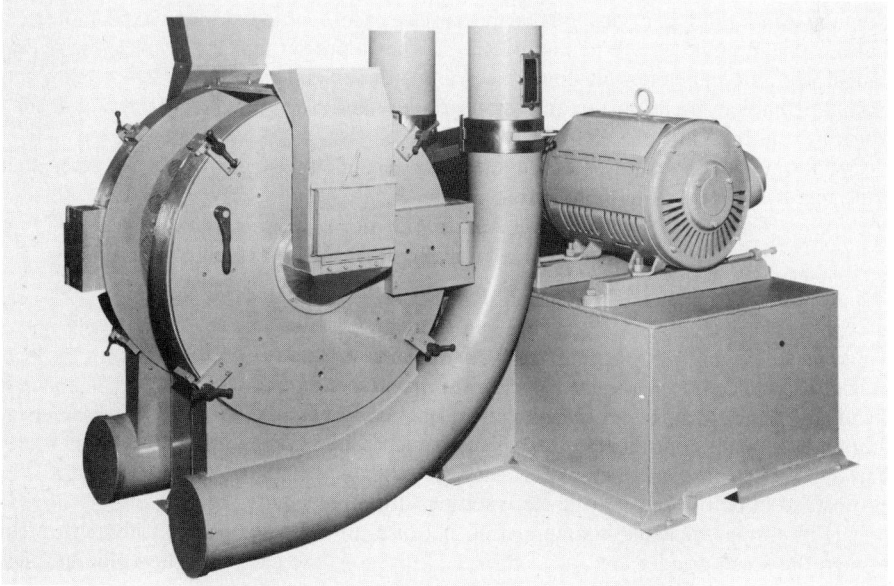

Figure 113.16. Fine grinding mill. Courtesy of Wedco, Inc.

2. The use of high-flow air is the most common cooling tactic used.

3. Solid dry ice or liquid nitrogen can be fed in with the plastic to be reduced. The additive evaporates as it cools, and it keeps the plastic rigid for fine grinding.

4. A specially designed machine pulverizes granular thermoplastic to a fine powder. It has counter-rotating rotors with filelike faces that reduce the granules rapidly and efficiently.

The production rate of a grinder is directly proportional to the number of knives, the length of the knives, and the size of the cutting chamber. The production is also affected by the size of the hole in the screen. If one is trying to grind fine and have holes as small as one-quarter of an inch, the plastic stays in the grinder longer in order to be reduced to that size. By simply going to a half-inch screen, you could almost triple the production from the same grinder. If a lot of cutting action is required, such as in some of the thinner film, the most important influence on production is the number of knives and the length of them. In other words, as one is rotating, the key factor is the number of inches of knives passing each other per hour.

Grinder Accessories. These incorporate a number of add-on equipment (some of it retrofittable) that extend the utility of the basic machine. Blower discharge to a cyclone is a common accessory. The scrap is simply picked up in either a blower or an impeller-loader, which uses mechanical force and air flow to move the scrap to a cyclone. The cyclone separates the air and the plastic scrap, which then falls into a collection bin, or directly into the hopper of a machine that will feed the scrap into further processing.

Feed rolls are accessories that continuously replenish the granulator. Sheet extrusion often requires either one set of very wide rolls for the whole sheet, or smaller grinders with feed rolls to take the edge trim for extrusion. Film feed rolls operate automatically from a rewind stand. And special feed rolls have been developed to accommodate unusual shapes.

Vacuum pickup is an accessory for removing granulated scrap from the bottom of a storage bin and conveying it to another location in the plant for further processing.

Overload control accessories can prevent serious damage to costly equipment, especially when scrap feed to the granulator is automatic. An electrical overload control is activated when the grinder motor begins to labor because of excess scrap load. The extra motor effort causes a rise in amperage. When amperage reaches a pre-set maximum level, the overload control will automatically shut down the granulator, or emit a warning signal. This action prevents complete meltdown in the scrap grinder, and also, protects the motor from overheating and burnout.

Cyclone feeding accessories are mounted on top of the grinder to receive scrap that is being removed by air from the process.

113.14 CARE, MAINTENANCE, AND SAFETY

The knives in a scrap grinder should be kept sharp and set to the manufacturer's recommended clearance between knives. While the machine is running, it gradually heats up and expands so that settings closer than 0.002 in. (0.05 mm) are not recommended. When heavy-duty work is being done, it is better to have a clearance between the knives of 0.005 (0.13 mm) and 0.008 (0.02 mm). However, the more flexible the material, the more apt it is to tear instead of cutting so that the close settings are more important.

It is important to avoid contamination because any metallic or rigid items that come through the scrap grinder are apt to chip the knives. These chips can then plug nozzles or damage the processing equipment; or, if they come back in the scrap, generate new chips in the knives.

The knives are often protected with permanent magnets in various parts of the system. In some cases, flat plate magnets are put in the throat of the grinder or in the feed chute where the scrap is coming into or out of the machine. In other cases, the discharge from the grinder has a grate magnet that collects any metallic particles to keep them from coming back through the system. Magnets are often installed on the inlet of the processing machine so that metal particles cannot enter and cause damage at that point.

Devices are available that detect metal in an air conveying stream and eject the contamination automatically.

Special care should be carried out to tighten all knife blades with a torque wrench rather than a normal wrench because it is important that the nuts do not loosen. If you do not use a torque wrench, you are leaving too much to the judgment of the maintenance personnel. The bolts that are used to hold in the knives are made of very special steel, but overtightening can lead to gradual elongation of the bolt, loosening of the knife and breaking of the bolt. When the head of the bolt breaks off, it is a very expensive accident.

113.15 SCREENS

Screens must be inspected regularly for deformation. They tend to bend out of shape because of the impact of the plastic on the screen. They should be removed regularly for cleaning when changing over from one material to another.

113.16 CONTROLS

Automated Systems Control. Most of the controls involved in plastic materials handling are electro-mechanical. However, as in injection molding and other processes, microprocessor-based controls are gaining favor rapidly because of the new capabilities they offer. Major advantages of microprocessor controls include:

1. Data accumulation and retention that can be used to simplify setup of future jobs or otherwise improve operating efficiency.
2. Ability to accomplish many functions simply, such as troubleshooting, actuating alarm signals, programming the time and duration of machine functions, etc.
3. Being able to communicate electronically with a central computer and to respond to orders by implementing instructions in its extensive data bank.
4. Remote positioning, for better use of plant space.
5. Ability to set up special programs in the field.

Microprocessor controls for the materials handling function encompass numerous important functions.

Plastic Dryer Controls. The plastic dryer controls can automatically troubleshoot up to 24 different potential problems, such as clogged filters, burned-out heaters, over-and-under temperature occurrences, etc. (Fig 113.17).

Central Loading System Controls. A central system will trigger an alarm if there is a loading failure on any of up to 24 lines, pinpointing the exact station where failure occurred and, even indicating the reason for malfunction. Many other features are incorporated, such as automatic setting of leading times per station and color graphic displays showing operation of valves, dump throats, and station being loaded.

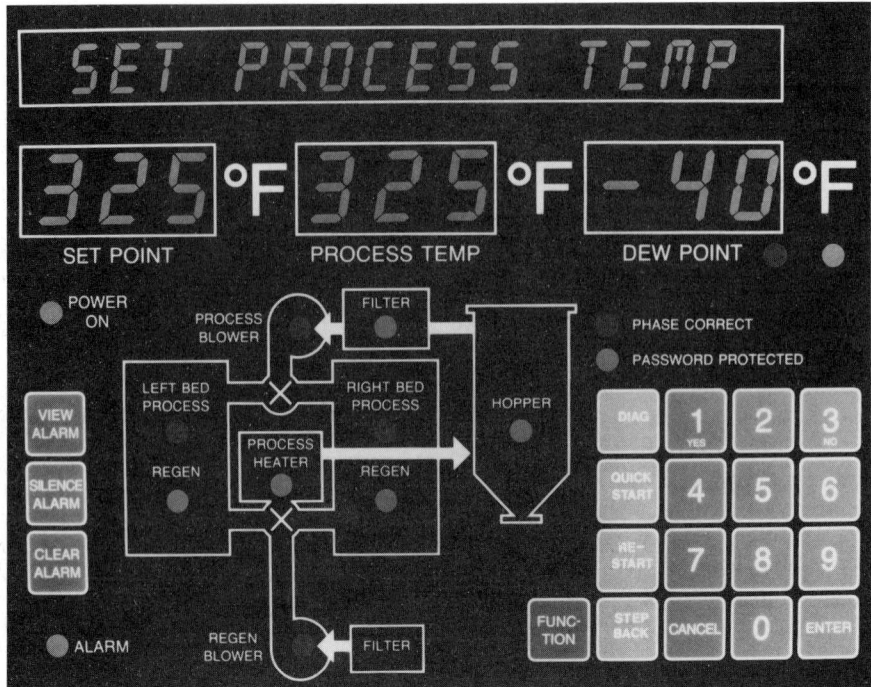

Figure 113.17. Micro processor dryer control panel with trouble shooting diagnostics. Courtesy of Universal Dynamics, Inc.

Additive Feeder Controls. Keep a running record of materials use while accurately feeding charges for injection to accuracies of 0.5%. They also can be slaved to extruder speed, with digital readouts in grams per shot or pounds per hour.

Silo System Controls. Will service up to 12 silos and two railcar unloaders. They store data on amount of material consumed as well as on remaining contents of any silo. This information can be called up on a continuous or as-need basis.

113.17 INDUSTRY COOPERATION

The Society of Plastics Industry is currently organizing cooperation of machinery manufacturers to arrive at a common "protocol" so that all equipment made by participating companies can communicate in the same plant with the same host computer. This is the trend of the future in all industries, and the plastics industry is enthusiastically embracing the challenge.

114

ROBOTICS IN INJECTION MOLDING

Fidel Ramos

107 Woodward Street, Newton Highlands, MA, 02161

Robots based automation for part and runner handling is neither new nor unproven in the plastics industry. Approximately 80,000 parts removal robots and sprue picker robots have been installed in plants around the world over the last 20 years. Robots are a combination of hard automation and general-purpose robot technology, and encompass two general types: parts-removal robots and sprue-picker robots. Both types are mounted on the top surface of the fixed platen of injection-molding machines, and are interlocked to machine controls to synchronize their motions with platen opening and closing. Some special-purpose versions mount on the side of the platens, or on the floor next to the machine. Regardless of their mounting location, their purpose is to remove parts and/or runners from between mold halves in the shortest amount of time without contacting the mold.

114.1 SPRUE PICKER ROBOTS

These units remove sprues or runners from subgated and three-plate molds. They often can remove lightweight parts attached to the runner system as well. Sprue pickers are differentiated from parts removal robots by their simple grippers and limited dexterity. They are designed to grab the sprue/runner with a simple pincers-type grip, remove it from the mold, extend beyond the safety gate of the machine, and release. Molded parts fall between the mold halves, and are collected in conventional ways. Neither mechanical separators nor operators are needed to separate parts from runners. A simple diverting chute or extended throat opening on a beside-the-press grinder, is usually required to direct runners after they are released by the sprue picker. Sprue pickers are available for molding machines sized from 20 to 1,000 tons (200–10,00 kN).

Top-entry robots used for runner removal are typically of the cylindrical coordinate type, with a linear vertical take-out motion used to lower and raise the gripper, and a linear horizontal strip stroke motion to extract the runner from the mold. A rotary axis pivots the take-out mechanism 90° to a horizontal position, so it may extend and release the runners beyond the safety gate of the molding machine. A typical sprue picker is shown in Figure 114.1.

114.2 PARTS-REMOVAL ROBOTS

Although more complex than sprue pickers, parts-removal robots provide greater benefits and capabilities by gripping and removing both parts and runners. These robots are available

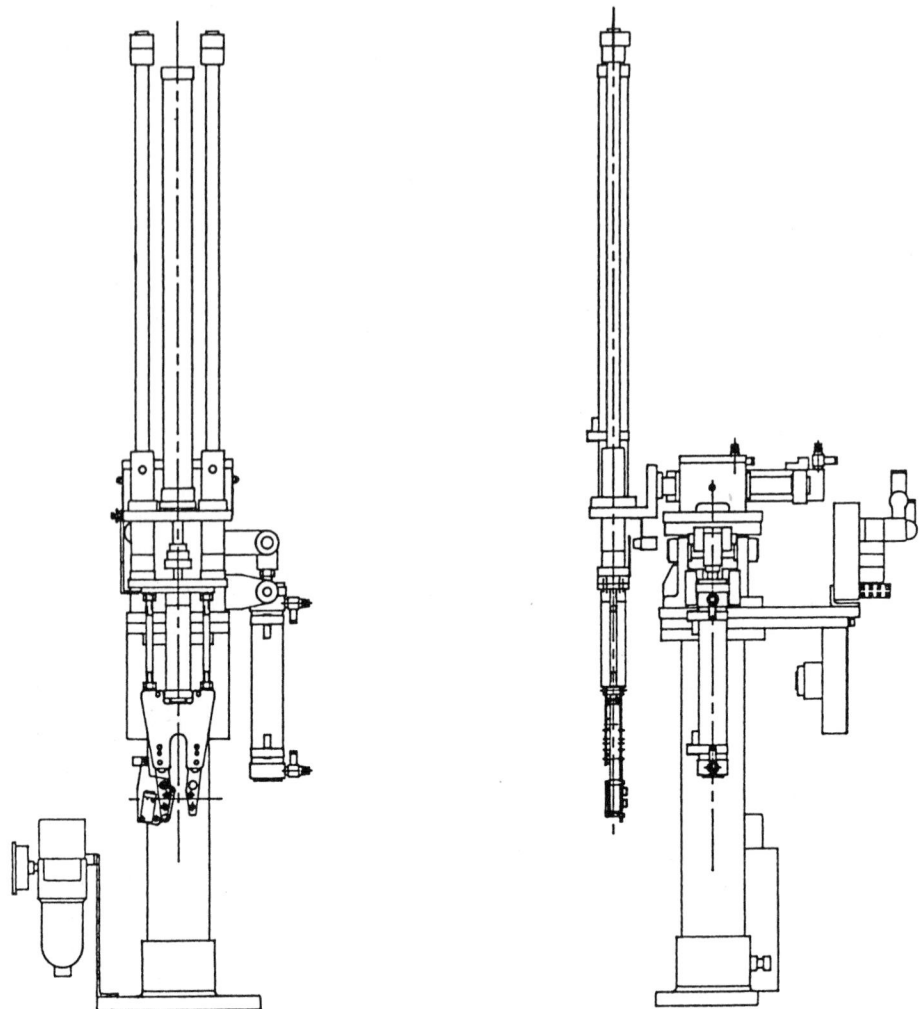

Figure 114.1. Typical sprue picker.

in many configurations, for applications including single-face molds, three-plate molds, and stack molds. Their capabilities are usually determined by the number and type of arms attached to the robot. Construction also is more robust than that of sprue pickers, allowing removal of large and heavy parts.

Most models are of top-entry design with three or more axes of motion. Due to their rectilinear motions, they generally provide greater control of the release point of the parts, and usually deposit parts beside the molding machine. Instead of releasing parts at a height near the top of the safety gate, the parts-removal robot lowers them to the centerline of the machine. This allows gentle release onto an indexing conveyor or other adjacent equipment for accumulation. Parts-removal robots are available for machines from 50 to 4,000 tons (500–40,000 kN). A typical double-arm parts-removal robot is shown in Figure 114.2.

Parts-removal robots are offered in a variety of configurations. Robots with double arms are available for use with three-plate molds. Double-arm robots enable the removal of parts with one arm and runners with the second arm. Single-arm models are used with hot runner or single face molds to grip parts and any runners using a single gripper.

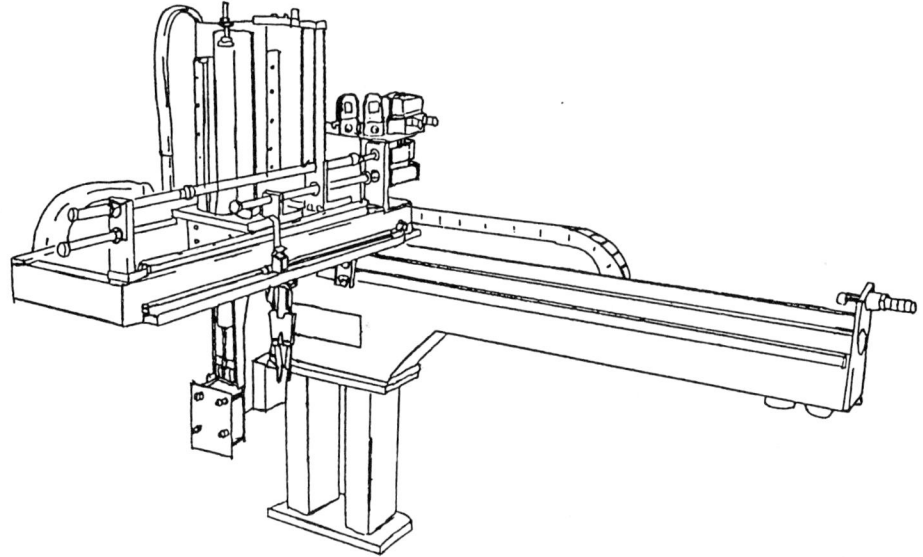

Figure 114.2. Pneumatic robot.

For new-generation stack molds, double-arm robots (each arm configured with its own parts gripper) allow removing parts from both sides simultaneously. Some molding applications require that inserts be placed in the mold. In these instances double-arm robots are used. One arm removes completed parts while the second arm places inserts. To accomplish insert loading efficiently, the mold should be designed to have the insert loaded to the stationary half and parts removed from the moving half.

Top entry robots are most often equipped with vacuum cup grippers (also known as end-of-arm tooling) for part removal. Vacuum cup tooling provides a flexible nonmarking means of removing multiple parts simultaneously at relatively low cost. Tooling components available from suppliers offer a high degree of flexibility enabling ease of reconfiguring end-of-arm tooling for a variety of different molds. Vacuum pick up of parts has the benefit of providing a simple, low cost means of verifying parts removal by means of a single vacuum sensor.

114.3 ROBOT VARIATIONS

Beginning in the late 1960s, general-purpose, multi-axis, floor mounted robots were applied to die-casting operations. They achieved widespread acceptance due to the very difficult operating environment of plants in that industry. Because of similarities in design of die casting and injection-molding machines of the time, applications of this type began to spread into the plastics industry as well. However, modifications were required to make the equipment practical for plastics processing.

Swing-arm robots, specifically designed for injection-molding machines, were developed and installed in limited numbers during the 1970s. These swing-arm robots were either powered mechanically by the movement of the molding press or by hydraulic or pneumatic cylinders. Mechanical driven swing-arm robots were fairly well accepted for runner removal applications and are still used today. It is estimated that 2500 swing-arm robots were installed in the U.S. Some early users experience poor reliability and significant down time on swing-arm robots since many designs were still developmental. For example, on mechanically driven robots, the velocity of movement of the arm was controlled by the opening and closing of the molding press; molders speeded up the molding press, they exerted extreme stress on the mechanism driving the arm, causing a high incidence of mechanical failure. Early swing-arm robots also were found to lack flexibility in parts removal; they were difficult to adjust and lacked the flexibility to place parts once removed from the molding press. Today, they are usually applied only in sprue picking or small-parts-removal applications on machines under 300 tons (3000 kN).

By the early 1980s, pneumatically driven top-entry robots began to appear in the United States. Most of the manufacturers were Japanese or European. These robots were special-purpose models, adapting general robot technology to injection molding.

With the advent of pneumatic robots, per-unit costs dropped, and integrating them to molding machines became easier and less costly. Their lower cost and dedicated design allowed molders to use one robot per molding machine, and gave them the flexibility to use the robot with any mold that would fit between the tie bars.

By the late 1980s, due in large part to the success of these models, foreign robot manufacturers had supplied more than 50% of the installed base of robots in use for part and runner removal. The functions of a typical top-entry parts-removal robot are shown in Figure 114.3.

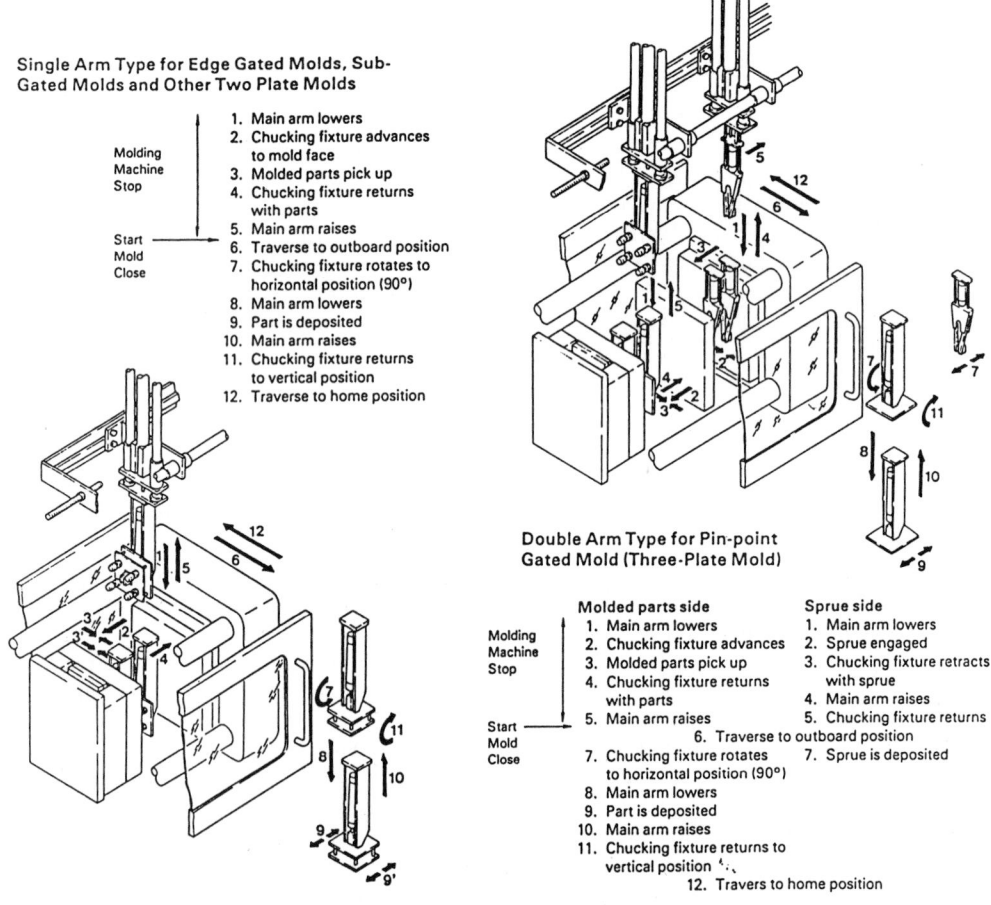

Single Arm Type for Edge Gated Molds, Sub-Gated Molds and Other Two Plate Molds

Molding Machine Stop

Start Mold Close

1. Main arm lowers
2. Chucking fixture advances to mold face
3. Molded parts pick up
4. Chucking fixture returns with parts
5. Main arm raises
6. Traverse to outboard position
7. Chucking fixture rotates to horizontal position (90°)
8. Main arm lowers
9. Part is deposited
10. Main arm raises
11. Chucking fixture returns to vertical position
12. Traverse to home position

Double Arm Type for Pin-point Gated Mold (Three-Plate Mold)

Molding Machine Stop

Start Mold Close

Molded parts side	Sprue side
1. Main arm lowers	1. Main arm lowers
2. Chucking fixture advances	2. Sprue engaged
3. Molded parts pick up	3. Chucking fixture retracts
4. Chucking fixture returns with parts	with sprue
5. Main arm raises	4. Main arm raises
	5. Chucking fixture returns
6. Traverse to outboard position	
7. Chucking fixture rotates	7. Sprue is deposited
to horizontal position (90°)	
8. Main arm lowers	
9. Part is deposited	
10. Main arm raises	
11. Chucking fixture returns to vertical position	
12. Travers to home position	

Figure 114.3. Injection molding machine automated by robot.

As more suppliers of parts-removal robots and sprue pickers competed for orders, new-generation models were developed. Improvements were made in the method of actuating robot motions. Over time, there was an accelerated transition from mechanical and hydraulic operation to all-pneumatic, pneumatic/electric, and all-electric servo drives. As the needs of injection molders were identified and better understood, robot suppliers began to offer special combinations of axis drives, depending on the relative importance of such factors as speed and flexibility. Japan and Europe were the leaders in developing these new-generation robots.

As the number robot designs stabilized, certain have become typical: On parts-removal robots for smaller molding machines [up to 500-ton (5,000 kN)], vertical motions are driven primarily by pneumatic cylinders because of their speed and the molders' desire to reduce mold open time. Traverse motions are generally by electric drive, to allow greater flexibility in stopping at multiple locations, although pneumatic drive is preferred when operating on short-molding-cycle applications. Larger robot models, over 500-ton (5,000 kN) are usually all-electric to allow maximum flexibility to set the stroke of each motion, and eliminate the need for manual robot adjustments.

Special molding applications such as compact disk (CD), require special-configuration robots for best performance. For example, CD manufacturing requires an extremely clean (class 100 or class 10) environment, often needing filters and air curtains over the platen area. Special-purpose side entry robots are ideal in CD production because they do not interfere with the air-curtain or generate particulates over the top of the mold that would contaminate the critical disk surface.

Side entry robots are also often used with molds with extremely fast cycle times (under 10 s). Their in-and-out motions require less travel time than top-mounted robots that must travel down, up, across, down, and back again. However, although side-entry robots are faster, they usually have less adjustment range for the part drop-off location.

114.4 AUTOMATION IN THE U.S. PLASTICS INDUSTRY

The U.S. plastics industry lagged behind the rest of the industrialized world in applying parts removal and sprue-picker-robot technology. Molders in Europe and Japan were the early users of this technology, allowing them to increase their productivity and product quality, while reducing their costs. In the U.S. many robot models are available from both domestic manufacturers and foreign suppliers.

In the application of robots to injection molding, Japan is significantly ahead of all other countries, and therefore, much can be learned from their experience. Some estimates are that over 80% of Japan's molding machines (60,000 machines) were equipped with robots for runner and/or part handling by 1988. In the U.S., less than 5,000 of the approximately 80,000 machines in operation are automated. On a cost basis alone, U.S. molders are at a clear competitive disadvantage. The Japanese are already recognized as a low-cost, high-quality producer. As Japanese and European molding companies locate new highly automated molding plants around the world, parts removal robots and sprue pickers become more necessary to remain competitive.

114.5 BENEFITS

By carefully analyzing the direct and indirect benefits of parts-removal robots and sprue pickers, molders will usually identify sufficient savings to return their investment in automation within one year. The benefits of automation include:

- Pacing the molding machine to maintain consistent molding cycle times
- Ability to remove and handle delicate or appearance parts without damage or surface marring

- Automatic mold protection by 100% verification that parts and runners have been removed from the mold
- Part orientation in the mold, maintained by the robot after ejection, facilitates packaging or secondary operations
- Cavity separation is maintained to facilitate quality control
- Safety and working conditions are improved; operators no longer must reach into the machine clamp area to demold parts.

It is important to understand the molder's operations, objectives and expected results before undertaking robotic automation of injection molding. The following examples identify some specific problems encountered by injection molders, and suggest how robots can minimize or eliminate them.

A common problem encountered in two-plate, subgated molds and three-plate pin-gated molds is effective part runner separation. There are three methods to accomplish this separation: 1. Run the molding machine on automatic cycle and accumulate parts and runners in a bin or box for manual separation. 2. Use an unscrambler to separate components. 3. Install a sprue-picker robot for runner removal directly from the mold as the machine cycles.

The problems associated with the first two methods are as follows:

Problem Potential	Result	How To Check
1. Safety; runners have sharp edges and often become entangled.	Possible operator injury from scratches or cuts.	Review incidence and causes of visits to nurse.
2. Operators or unscrambler lose parts that become entangled or carried over in runners, or because of improper adjustment of unscrambler.	Low yield; less than acceptable part and runner separation.	Monitor 24-h production yield by comparing cycle count with actual part count.
3. Batch loading of runners to grinders.	Possibility of loading runners to the wrong grinder, causing matial contamination.	Check records or set up a system to evaluate the incidence and value of loss.
	Grinders become overloaded, resulting in excessive downtime and maintenance. Grinders therefore are typically oversized, resulting in higher maintenance and operating costs.	Evaluate the incidence and cost of grinder maintenance. Evaluate power consumption of current grinder vs properly sized grinder.
	Direct load to hopper is inconsistent because supply of regrind material also is inconsistent, causing variations in ratio of virgin to regrind.	Evaluate the ratio of regrind to virgin at regular intervals during each shift. Evaluate loss yield attributed to poor material mixing.
4. Grinders located near product.	Blowback from the grinder can contaminate parts and causes rejects and/or jams up in downstream automation equipment.	Review causes of rejects and/or incidence of downstream equipment traceable to grinder dust or particles.

Problem Potential	Result	How To Check
5. Runners do not fall free of mold.	Machine will have to set with multiple K.O. (ejector) strokes resulting in increased cycle time.	Count. K.O. strokes and review mold maintenance records for ejectors and K.O. mechanism. Review machine maintenance record for K.O.-related repair. Put a counter on the machine, tied into low pressure safety to determine incidence rate. Check incidence of interruption of automatic cycle; incidence of mold repair.
6. Runners build up under mold in chute or conveyor.	Interruption of automatic cycle, causing operator intervention and adding an uncontrolled variable to the process.	Same as above.
7. Grease, oil, water fall onto conveyor or container located under or beside the press.	Some loss of product due to direct contamination, causing lower yields or rework.	Compare cycle count times, cavities to actual count for 24-h period. Watch for operators cleaning or wiping parts.
8. Contaminants from under the mold contact the runner.	Regrind is contaminated, in turn causing contamination of molded parts, resulting in questionable quality and lower yield.	Check the loss of material or yield caused by contaminated regrind; institute clean-up procedures.
9. Runner hangs up and must be removed by operator each cycle.	High labor cost. Uncontrollable process variable take-out time, causing product deviation and low yield.	Measure mold open time 4 times per shift and compare shortest to longest. Calculate theoretical yield and compare it to actual.

Any one of the preceding nine examples alone can provide justification for purchasing and installing a sprue picker. The use of a sprue picker can overcome all these problems.

For instance, the sprue picker grips and removes runners from the tip or side of the molding machine. This eliminates runners hanging-up in the machine, or operator handling and sorting of runners. On machines with automatic cycles, sprue pickers eliminate problems associated with contaimination of regrind caused by runners falling and contacting grease on molding machine surfaces.

Mechanical gripping of the runner system and verification of that removal is complete, eliminating the need for multiple ejector strokes and/or high-speed mold opening or ejector action to "blast" the runner free. Mold is protected by preventing clamp closing on runners. The gripper also always goes to the centerline of the mold to pick the runner. Since nearly all runner systems have, or can have, a center pickup point, it minimizes the need for

TABLE 114.1.   Robotics Cost Justification: A Checklist

Ideal Conditions for Sprue-picker Installations:

Manual sorting of runner/parts
Sprues periodically hang up in mold
Multiple knock-outs required to clear sprue
Contaminated regrind from sprues hitting tie bars

Ideal Conditions for Parts-removal Robot Installations:

Semi-automatic cycles with operators on the gates
Low yields from inconsistent cycles
Delicate or cosmetic parts requiring gentle handling
Multicavity molds requiring cavity separation
Maintenance of part orientation required to facilitate packaging and secondary operations
Mold damage from parts sticking in the mold
Multiple knock-out required to clear part

Key Elements to Consider in a Robot Cost Justification

1. Increased yield of good parts per hour from:

   Cycle consistency
   Reduced quality control rejects/damage
   Reduced scrap due to contaminants
   Increased plant capacity with no capital cost

2. Cycle-time differential from:

   Reductions in cycle times (more parts per hour)
   Reduced time to produce a given job
   Increase in available time for additional jobs
   Reduction in cost per part due to shorter cycle times

3. Reduction in labor cost and content

   Changes in number of machines serviced per operator
   Improved quality reduces need to inspect individual parts
   Elimination of manual parts sorting
   Possible integration of secondary automation at press
   Robots operate reliably three shifts, seven days per week

adjustment of the sprue picker between mold changes. This ensures that runners are consistently removed.

Because runners are removed each cycle (verified by means of a detector switch on the robot), 100% separation of parts and runners is achieved. Uniform flow of runners to the grinder assures consistent loading of the grinder, thus allowing for a smaller grinder size. By grinding the runners automatically at the press, regrind can be supplied continuously to the hopper for re-use.

Finally, unlike separators, the runner drop-off point to the grinders can be set to the rear of the press. This places the location of the grinder farther from the product and reduces the likelihood of contamination.

114.6 AUTOMATING WITH PARTS-REMOVAL ROBOTS

More complex than sprue pickers, parts-removal robots offer many additional capabilities and benefits, Figure 114.4. The most easily identified opportunity for parts-removal robots, with the greatest financial payback, is to automate molding machines which are currently operator controlled.

This is because operators cost money and introduce inconsistency into the molding process

Figure 114.4. Functions of a typical top-entry parts-removal robot.

by creating fluctuations in the molding cycle that affect part-to-part consistency, quality, and productivity. Since the molding cycle is set to compensate and tolerate, a wide range of operator caused variables, automation will cause significant improvement.

Parts-removal robots are likewise capable of dramatically improving productivity by eliminating variables that affect yield. The most obvious losses in yield result from: inconsistent cycles; availability of operators to run machines; operator fatigue; inability of operator to keep up with output; improper sorting of family molded parts; inability to separate a defective cavity; parts damaged from scratches or mechanical deviation caused by dropping parts on automatic cycle; contamination of parts from water, grease, oil or particulate from under the molding machine; inconsistent mix of regrind and virgin material; mold damage from parts or runners sticking in the mold and associated down time.

Below is a list of some of these problems, their results, and means of identifying or qualifying their impact on the molding operation.

As was the case with sprue pickers, any one of the preceding eight examples alone provides the justification to purchase and install a parts-removal robot.

Parts-removal robots improve operator safety by eliminating hazards posed by handling hot parts and hot mold surfaces.

Problem Potential	Result	How To Check
1. Operator removes parts	High labor cost. Inconsistent cycle time, causing part deviation and frequent adjustment of machine parameters. Danger to operator.	Compare counted yield to actual yield; check for short shots, flash, etc.
2. Inconsistent regrind mix due to irregular or batch regrinding of runners	Low yields. Inconsistent part quality.	Monitor rejects, evaluate ratio of regrind to virgin at intervals during 24-h period.

Problem Potential	Result	How To Check
3. Parts drop into box or conveyor	Surface scratches; part contamination from grease, oil, water, particulates; longer cooling time to prevent part damage from dropping; increased labor costs to clean parts.	Monitor loss in yield due to scratches, decrease cooling time and calculate increased yield achievable by automated part removal; monitor loss of yield due to particulate contamination or the need to wash parts.
4. Operator sorts family of molded parts	High labor cost; probability of sorting errors.	Calculate labor cost; monitor rejects.
5. Operator sorts cavities	High labor cost; probable missorts.	Same as above.
6. Multi-cavity mold with bad cavity	Mold run with blocked cavity is out of balance; manual sorting of cavities required, causing delay and high labor cost.	High potential for additional rejects; check incidence of part sort for bad cavity.
7. Part sticks in mold having side action, core pull, or delicate sections	Mold damage, production delays, change in production schedules	Evaluate incidence of mold repair and associated costs including downtime.
8. No operator available	Loss of production; loss of business; loss of profit	Monitor causes of downtime.

Because the robot paces the molding machine, cycle times are more consistent. This allows shorter cycle times, improved yield, and fine tuning of process parameters. Consistent cycles, for example, allow consistent mold surface temperatures, and uniform material residence time. Positive gripping, detection, and removal of parts and runners also protects molds from closing on trapped parts and runners, and allows smaller mold-open daylight and reduced ejector stroke and cycles—resulting in shorter mold-open times.

The quality of delicate or cosmetic parts is improved by positive gripping, handling and placement of molded parts. Contamination from below-platen water, grease and oils on dropping parts is eliminated, as well as hot runners sticking to parts. Reduced cooling time in the mold is possible since the part-removal robot can handle high-temperature parts, and provide additional cooling time while it transfers parts to beside-the-press conveyors.

Separation of parts by cavity in the robot grippers allows segregation for quality control reasons, traceability by cavity, and separation by cavity of family molds. Also, maintenance of part orientation in the gripper facilitates such secondary operations as layer packing, packaging, decorating processes, and assembly.

115

COMPUTER INTEGRATED MANUFACTURING (CIM)

Jeff Rainville

Universal Dynamics Corp., 13614 Dawson Beach Road, PO Drawer X, Woodbridge VA 22194-0396

The benefits that plastics products manufacturers can derive from computer integrated manufacturing (CIM) have been well documented. And although most processors have not yet taken full advantage of the technology, there is little doubt that the CIM concept is progressing rapidly, and will one day be commonplace manufacturing procedure.

The ability of CIM to network similar and dissimilar machinery in a processing plant into a smooth-working entity where changes in processing parameters, or even entire job change-overs, can be made swiftly and easily has enormous profits-enhancing potential in a business world where global competition is intensifying year by year.

115.1 ADVANTAGES

Although the disciplines of computer integrated manufacturing may seem complex, they are really extensions and refinements of now-familiar processing technology. And the benefits are substantial. They include improved manufacturing control; more efficient and effective planning; improved product quality; ease in tracking materials and parts; and improved productivity.

Improved Manufacturing Control. CIM offers the ability to put real-time status and historical job information at an operator's fingertips, literally. It obviates the former need for labor-intensive paper systems in order to track performance and resource utilization. It is even possible to supervise remote locations, and to obtain direct electronic interface with original equipment manufacturers for diagnostics assistance. CIM offers manufacturing control capabilities that have never before been economically viable.

More Efficient Planning. Combining historical data verification with job modeling is a powerful forecasting tool. It enables a manufacturer to use machine data, such as cycle time, reject rates, and down-time, to predict the future performance of similar jobs accurately. Such accurate predictions improve the ability to order schedules by better planning of machine utilization and deliveries. An important result is an improved assurance in pricing competititve jobs for customers. Equally important is that customer confidence will be enhanced.

Improved Product Quality. Real-time monitoring and alarm systems minimize or eliminate the danger that processing operations will vary from pre-set standards and thus endanger product quality. Control limits can be set tight enough to indicate a subtle variation *before* the quality of the finished part can be effected. Fast, corrective action can turn potential rejects into shippable parts. Such action may be taken at the computer without requiring manual intervention. A side benefit is a reduction in the cost to assure quality control by the reduction of quality control labor.

Parts and Material Tracking. This is accomplished by multiplying the number of cavities by the number of good cycles. A cycle is "good" if it falls within the standard processing parameters. Reject information, such as closed-off cavities, is supplied by the machine operator. In extrusion processes, output can be calculated by counting the cycles of cutoff devices or footage counters.

Improved Productivity. Any reduction in reject rates will automatically increase the production of quality parts. Additionally, sensitive jobs can be run faster because CIM provides greater awareness of process parameters. And the improved forecasting it makes possible reduces the number of tooling changes required, thus increasing machine uptime.

115.2   SYSTEM HARDWARE AND SOFTWARE

Their interrelated functions, as outlined below, are essential to a well-coordinated CIM system.

Microprocessor-based Controls. These controls have many advantages over electromechanical controls. They are programmable, which offers flexibility and they can communicate, which makes system interface possible. They are very fast, which permits the control of many signals.

Programming flexibility is a key feature in meeting customer requirements. Microprocessors can be programmed by the user to do any number of essential jobs, e.g., adjusting the bulk density factor on a blender or a batch recipe. Microprocessors have bidirectional communication capability that enables them to be interfaced directly to a machine controller, or to a CIM central processor (CPU). Transmission of data makes possible real-time monitoring of setpoints and machine operating status, and allows a machine controller or CPU to request action, such as a change in set points or standby.

One of the most important benefits of a microprocessor is the speed at which it operates. In the same amount of time it takes to pull in an electromechanical relay, a microprocessor may check all the inputs and outputs of a machine, do several calculations, and communicate to the machine controller. Such speed allows all the operating parameters and communications to be accomplished by a single chip.

Cell controllers are microprocessors that act as "traffic cops" for communications among several controllers. For example, a cell controller on an injection-molding machine may control communications to dryers, loaders, feeders, and hot-runner mold controls, all this while feeding processing information to the machine controller. A cell controller may act as a monitor, by centralizing a display representing all the connected machines. Or it may act as translator for a nonconforming device, such as a foreign-made machine with a nonconforming protocol. Translation software could be written for the cell controller more economically than modifying the system software or re-engineering the networking hardware.

Microprocessor-based Monitors. These employ sensors to detect changes in machine status. In some cases they have outputs that are programmable, like providing an output to activate a conveyer belt after a certain number of parts have been molded. Most monitors interface with both analogue and digital devices. Examples are switch contacts with timers for cycle analysis (cycle time, hold time, fill time), thermocouples for sensing temperature, and transducers for sensing melt pressure or hydraulic pressure.

Computer. This familiar high speed electronic machine performs mathematical or logic calculations. It is commonly referred to as a central processing unit (CPU). Its basic operating software is stored in RAM (random access memory), while system data are stored on either a hard disk (contained in the CPU itself) or on a removable floppy disk. The hard disk is most commonly used in CIM systems because of its large storage capacity and fast access speed.

Monitors. They assist operational personnel by screen display of text and graphics that provide dramatic and easy-to-comprehend real-time illustrations of processing events. Remote monitors can be distributed throughout the plant for easy viewing. With the advent of more sophisticated software, higher-resolution monitors provide even more detailed graphics.

Printers. Printers create hard-copy text reports of processing events for record-keeping or analysis. Remote printers may be used to print product labels or shift reports.

Bidirectional communication is the ability to transmit and receive data, and involves various disciplines, depending on user needs. These are cellular manufacturing, network buss, star or daisy-chain configuration, device addressing, and signal checking.

Cellular manufacturing is a system of structuring software that groups controllers into logical manufacturing cells. For example, an extruder line (incorporating dryer, loader, feeder, cutoff, puller, and process monitor) could be one cell. Alternatively, the extruder and process monitor could be one cell and the dryer and other auxiliaries could be another. Cellular system screens permit a single cell indicator to identify cell condition, reflecting all the controllers in the cell.

The network buss may be any of several configurations: parallel, simplex, serial half duplex, or serial full duplex.

Parallel operation is primarily used for short data runs, but it has the disadvantage of requiring many wires; this becomes expensive and cumbersome when long runs are required. Simplex operation is serial, and is limited to communicating in one direction (eg, to the printer or CRT). Serial half duplex operation is bidirectional, with the provision that only one station transmits at a time (limited band width). Serial full duplex operation is both serial and bidirectional, with simultaneous operations allowed (higher band width required).

In star configuration, the total number of devices on a network is limited to the fanout of the command console, usually about 30 devices. Repeaters can be employed to increase the fanout, but this increases the systems complexity. In daisy chain configuration, each device on the network is itself a repeater, therefore the total number of devices on the network is limited only by the limitations of the addressing scheme.

Device Addressing. Since the command console transmits to all devices simultaneously, it is necessary to implement a scheme for identifying which device is the "target" of the communication. This is done by assigning addresses to each device. As the command console sends out a poll request, for example, the signal will be preceded by a "header," which consists of the device type and address of the target device. Only if the receiving device recognizes its own device type and address will it attempt to transmit an acknowledgment back to the command console.

Signal Checking. In order to ensure message integrity, data checking is performed on each message sent and received, to and from the devices and the command console. This is done by means of a "checksum" or a more sophisticated cyclic redundancy check (CRC).

System software is stored in the random access memory of the central computer. Its logic causes the computer to continually poll each device on the network, control the graphic presentation of the information, supervise data storage on the hard disk, issue request for action commands, and generate reports. There are many languages available in which to write system software; each must be compatible with operating software and computer hardware.

115.3 MONITOR CAPABILITIES

The ability of electronic instrumentation to determine exactly what is happening during the plastics-processing operation, and exactly why and where it is happening, is at the very heart of a successful computer integrated manufacturing setup. In fact, without accurate, fast-response monitoring, there would be no CIM. Many factors are involved, as follow:

Real-time Machine Status. Real-time implies the "instantaneous" readout of a machine's status. It is important to understand the effect of scan time on a CIM system. The number of controllers, message transmission rate, and the speed of the CPU itself can greatly affect scan time, which typically ranges from a few seconds to a few minutes. Systems that transmit large amounts of data tend to use buffers to store data, and make fewer large transmissions. Newer systems switch into a "transparent" mode, which utilizes a binary code (instead of ASCII characters) to cut down the transmission time.

Set Points. Set points are operator-adjustable values such as machine temperatures, time cycles, pressures, and positions. Set points are targets by which the process is controlled. They are critical to process evaluation, and are an important consideration in any CIM system.

Variables. Variables are actual values that change inherently with the process or machine operation variables have a measurable value such as temperature, pressure, dew point, dimensions, weight or time.

Alarms. Alarms are variables that activate a response or special notation. They are tracked separately to stimulate action or facilitate machine or process diagnosis. Alarms are grouped into categories to prioritize the type of action required. For example, alarms that indicate safety hazards or critical processing variances may be coded red indicating the need for immediate action; alarms that indicate deferred action should be coded yellow. If the action categories of alarms are over-used, however, the reliability of response will decrease.

Production and Process Monitoring. Many levels of such monitoring are available. The simplest and least expensive is the cycle monitor, which takes an output from the machine and is utilized in a number of ways to give important management information. Sophisticated systems take process variable data from the machine and store data for statistical purposes. Units that monitor process variables also perform do all the same functions as simple cycle monitors.

Cycle Monitor with Alarms. This is the most basic of plant management tools, with the simplest installation and lowest cost. When tied into a central computer with good management software, this level of control provides an amazing amount of information to assist in managing an operation.

For example, it lets management know at a glance the status of each machine in the plant, since each job is assigned a standard cycle, used to calculate job and operator efficiency. Current machine status is displayed on the monitor in color, with green for "on standard", yellow for "below standard", and red for "machine down". Average cycle time, last cycle time and total number of cycles are normally displayed for each machine on the monitor.

An example of cycle time application would be a blow-molding machine (which should have a consistent cycle time, producing a specific number of bottles each time). A consistent cycle means that the machine is functioning correctly. By counting cycles, one can calculate a relatively accurate production quantity. When cavity and part standards are added to the software, production scheduling and job historical data can be easily accumulated. Job scheduling can be done automatically in the system software, with standards being transferred directly from the computer to the monitor panel. In noncyclical processes (such as an extrusion), extruder and accessory equipment cycles or screw speed could be monitored. At any given time, the monitor boxes give a visual color display of the machine status. The information is simultaneously transferred to the central PC and is included in the current display on the standard monitor.

Process Monitor with Alarms.   Perhaps the largest growth area in centralized monitoring and CIM systems is the ability to monitor process variables. Typically, this monitoring function signals the operator (with a flashing yellow display) when something is running out of control. Some of these monitored variables are

1. Injection time/pressure. An input to the monitoring device is connected to a machine output (such as the solenoid contacts). An additional input to the unit would be a hydraulic transducer, to measure oil pressure during the injection cycle of the machine. Variations in time or pressure would signal machine or material problems, that could seriously affect part quality.

2. Holding time/pressure. One input is connected to the machine electrical circuit, while the same hydraulic transducer that measures injection pressure is used to measure holding pressure and monitor variations in time or pressure.

3. Screw time. One input is connected to the machine, which will tell the monitor when the screw circuit is energized. Variations in time will relate directly to cycle time and are indications of screw wear, material variations, back pressure and hydraulic or temperature problems in the machine itself.

4. Back pressure is measured using the same hydraulic pressure transducer that is measures injection and holding pressure. The reading is taken as an average during screw recovery time. Variations indicate machine hydraulic problems that cause variations of plastic melt consistency and temperature, with the possibility of severe processing problems.

5. Temperatures. When temperature sensors are connected to the monitor, the data they collect can be used as an indicator of material consistency and machine performance; hence part quality.

6. Shot size/screw position. Monitoring is performed by a measuring device installed on the machine. Although expensive, monitoring of shot size and cushion can be an important indicator of process and material consistency.

115.4   OTHER PROCESSES AND VARIABLES

Processes such as extrusion or blow molding have their own special monitoring applications, with significant payback potentials.

Extrusion.   Monitoring head pressure, eg, gives a good indication of the condition of the machine and its drive, screw, and temperature control. Melt and die temperature monitoring are likewise essential; temperature variations can have significant negative effect on product quality. Melt temperature is a function of screw speed, back pressure and material in addition to the barrel temperature. It is also an indication of the effectiveness of barrel cooling and screw wear.

Extruder screw speed also has a direct relationship to material output, temperature and part quality. If a large variation is present, product quality will probably suffer. And since resin throughput is a function of temperature, screw speed and material, monitoring is required because variations could mean poor product quality.

Blow Molding. Companies that blow mold plastic bottles or other hollow structures can achieve significant benefits from a monitoring system. In one documented instance, a plant with 19 machines experience 94 down-time occurrences in 24 hours. Installation of a monitoring system pinpointed the problem quickly. The molder estimates that monitoring saves more than $80,000 a year in the costs of maintenance and idle equipment.

Monitoring in blow-molding operations is essential to container quality. For example, if a screw speed varies bottles will probably have wall thickness and weight variations along with possible material degradation. Both conditions adversely effect product performance.

Variations in manifold/melt temperature will create problems similar to screw speed variations. In some materials such as PVC, burning may occur. And if pressure is not consistent, bottle volume and quality may suffer.

115.5   DATA ACQUISITION/MANAGEMENT

In order to generate data that has CIM utility, three basic decisions must be made. First; what type of report is wanted? Second; on what machine or group of machines is the report to be based? Third; what is time window of data to be considered? Reports are the most powerful tool of any CIM system. Used properly they can aid machine and process diagnosis, predict future manufacturing efficiencies, and alert the operators to deteriorating conditions prior to down time or machine failure. Additionally, printed or stored reports enable the cause or product failures to be traced for an unlimited period of time.

Data acquisition permits material use to be tracked on different levels in travel through the plant. Each material is tracked separately by name and number. The first level is measured as it passes through silos. The second level may be central blenders feeding several machines. The third level may be at the machine itself, tracking virgin, regrind, and color. Other essential reporting;

Event History. The event history report identifies each event and the precise time and date on which it occurred. This may be used for process verification or as a "who-done-it?" drill. Either way it is a powerful supervisory tool. Just the fact that a record will be made encourages operators to think twice before acting.

Alarm History. This report indicates the nature of the alarm, when it took place, and (in some cases) what cycle the machine was in at the time. This report is generally used to aid in machine diagnosis.

Elapsed Time Report. This indicates the up-time in hours and minutes for each machine. This is a very effective way to schedule routine maintenance, such as oil changes and filter replacement.

Process and Production Reports. There are many types of reports available to the user of a central monitoring/CIM system. The most often utilized are the shift and job reports preprogrammed by the manufacturer. Some systems allow the user to program the report to fit specific needs. This is a valuable feature in that it allows the user to report only essential information. Reports can be printed out or transferred directly to the plant's mainframe computer. No longer does management have to wait 24 hours to find out what happened on the previous day.

When tied into a management system that calculates cost, a company can know just how much money was made or lost the day before. Sales personnel can also get accurate information on what is available to be shipped. In some cases, customers are even demanding that molders provide a disk on audio cassette containing statistical information and reports with each shipment.

Production Reports. A variety of processing information is contained in production reports and can include as much of this information as management would ever want. For example, cycle-related information can be calculated based on the monitor such as:

Last Cycle
shift average cycle
average last ten cycles
efficiency based on standard
Parts
parts per cycle (set)
estimated parts produced
number of reject parts (manually entered)
total parts this order
parts to go efficiency
Down Time
number of occurrences
   per shift
   this job
total amount of down time
   per shift
   this job down time by reason (up to 11 reasons)
   per shift
   this job
Quality Reports
number of rejects
   per shift
   this job
type of rejects (up to 11 reasons)
   per shift
   this job

115.6 MACHINE AND PROCESS VARIABLES

Unfortunately for the processor, no single process parameter is the key to good process monitoring. There is no magic universal formula, for all the information that a processor could apply to know the precise effects of parameter changes. Because of the relationship of part quality to so many intermeshing variables that interrelate, that monitoring a number of variables and charting them on a data base is nothing short of vital.

Control of machine and process variables have reached the level of sophistication that statistical quality control and statistical process control (much in demand by customers) are easily achievable. In reality, with a monitoring system in place, the plant is continuously doing statistical quality/process control. (Although the system is not in itself a control, the information it gathers is so used.) When properly integrated into the production system, statistic quality and process control pays for itself. In most cases a 5% to 20% increase in productivity is achieved by the addition of an automatic reporting system with this capability.

Statistical quality control software exists (called SQC pack) that performs multiple statistical functions in a user friendly manner. When integrated into monitoring system software, it becomes an extremely powerful tool. It also has the capability to interface with measuring devices such as a digital weighing scale, electronic micrometer, or can take manual entry of user variables.

115.6.1 Capabilities

Sampled process parameters may be graphed to show such things as their range, standard deviation and the number of samples out of the control limits.

115.6.2 Variable Interface with System Software

Many process variables can be automatically sampled in a way that the data may be easily recalled and statistical calculations performed on them.

115.6.3 Process Analysis and Trending

When multiple process parameters are displayed and printed in graphs, then compared to each other, it becomes relatively easy to spot trends and problems in the process.

115.6.4 Trouble Shooting

An example of using the data base and statistical software would be when a machine is going from one problem to another, such as flash in injection molding and then to short shots. By analysis of the graphs of such parameters as screw time, fill time, temperature, cycle time and holding pressure, it is possible to isolate a machine, material or labor problem. (If the cycle time shows a wide variation and all other variables are in tolerance, the operator probably is at fault.)

Part-analysis information can be entered into the system either manually or with an electronic measuring device. When statistical charts are generated of weights and dimensions, it is easy to spot a problem.

The typical customer of a custom processor requires a statistical sampling of parts for quality, with typical requirements of a 95% confidence level (normally considered to be "3 sigma" or standard deviation) that all of the product is within specifications based on statistical sampling. Quality control personnel use charts to sample product, and the monitor uses process parameters which are generated over time with experience. The usual number of samples required for statistical validity sample is 100. SQCpack requires a minimum of 100 samples.

115.7 CONTROL/REQUEST FOR ACTION

Although the word "control" is frequently used as part of system specification, a more accurate phrase is "request for action" (control implies definite and spontaneous action, which cannot be assured in a network environment). An action is not definite because there may be a transmission or receiving error. Although most programs use interrupts to request action, there may still be a delay before the action takes place. Remote control can be useful in an emergency situation, but it is not the same as hitting an emergency stop button.

For safety, therefore, if a machine is in standby condition, it should be clearly visible from the control panel. Normal safety procedures should prevail without relying on the CIM system to alarm or take corrective action. Password or key protection should also be provided to screen personnel who have access to request-for-action commands or system configuration software.

115.8 REMOTE SYSTEM INTERFACE

This type of interface utilizes a modem to connect a remote computer to the CIM system CPU via telephone line. A remote computer can have full access to the system software for monitoring, request-for-action commands, or report generation. State-of-the-art multiplant supervision can be achieved with a remote interface. This is especially attractive when combined with production or process monitoring. Data files can be downloaded on a job and run in a remote statistical quality control or a spread sheet program for further evaluation.

Original equipment manufacturers can gain direct access to machines for operational diagnostics. For example, a dryer manufacturer could review data reports and exercise the machine while monitoring the results. There is little doubt that any problem would be identified and corrective action recommended. Processing experts could review processing parameters and machine performance for quick advice on processing techniques.

116

FASTENING AND ASSEMBLY

George M. Loucas

Fasteners Div., Controls and Fasteners Group, TRW, Glen Road, Mountainside, NJ 07092

Today's designers have at their disposal a wide array of mechanical fastener systems for joining components to similar or dissimilar materials. This method of attachment employs familiar self-tapping screws, clips, and expansion rivets. Also, available are soft or heat-treated spring-steel products that conform to screw threads or make their own threads or push onto rivets, studs, wire, and shafts.

Whatever type of mechanical fastener is used, it must satisfy the following criteria:

- To provide reliable assembly clamping force in pressing two or more panels or units together.
- To prevent the assembly from coming undone.
- To resist back-off or loosening caused by fatigue and vibration as a result of wear from repeated oscillation and frictional changes.
- To resist external axial and transverse forces acting on a joint, causing tensile and bending stresses.
- To combat creep over an extended time period.
- To compensate for environmental temperature fluctuations.
- Permit removal of the part or panel for servicing or repair, and to be replaceable for reattachment.

116.1 APPLICATION CONSIDERATIONS

With these criteria in mind, it is then necessary to analyze the fitness of a particular fastener for the job at hand. For example, fasteners for threading applications must be able to fit complementary screw threads, or form them on round studs or shafts. Fasteners for push-on or push-in applications must be applied to rivets, studs, wires or shafts, or snapped into a mating panel opening.

Some key considerations in determining the type of fastener to use: clamping force needed; nature of the mating part, its hardness and finish; tightening torque required to produce the desired clamping force; maximum torque that can be applied before affecting the strength of the screw thread or stud materials before strip-out failure; and susceptibility of the plastic assembly to creep or failure phenomena.

The relationship of correct tightening torque to obtain desired clamping force must be always kept in mind. Torque may vary because it is dependent on the friction of the mating threads (for multi- or single-threaded nuts) and on interference of the self-threading fastener or the stud, wire, or shaft. In both cases the friction of the fastener (or its mating part) with the panel influences torque.

Another factor to be considered is whether the application is to be by hand or power tool. The amount of panel resistance to assembly pull-up contributes also to the tightening relationship.

The push-on or push-in fastener must be able to grip the stud firmly to prevent failure in use. Resilient arched or dished fasteners are among the best choices to guard against looseness in stud assemblies.

Structure of plastic walls is a paramount consideration when mechanical fastening is contemplated. Sufficient thickness and taper is necessary to sustain assembly stresses from screw or component sections, and molded-in holes must have correct interference for acceptable anchorage. Optimum wall thickness depends on a wide range of variable factors, so testing is of paramount importance in order to minimize product failure and guard against over compensation that can eat up profits.

TABLE 116.1. Type Designation of Tapping Screws and Metallic Drive Screws[a]

Type	A S A	Manuf.	Federal
	A	A	A
	B	B or Z	B
	BP	BP or ZP	BP
	C	C	C
	D	1	CS Alt. #1
	F	F	CF
	F	F	CF
	G	G	CS Alt. #2
	T	23	CG
	BF	FZ	BF
	BG	H	BG
	BT	25	BG

[a]Ref. 1.

TABLE 116.2. Diameters and Thread Pitches for Steel Tapping Screws

Nominal Size[a]	Thread Types[b] AB,B,BF,BT,BSD			Thread Types[b] F,T,SW,SF,TT,CSD		
	Dia × Pitch, mm	Dia, in.	Threads per in.	Dia × Pitch, mm	Dia, in.	Threads per in.
2	2.2 × 0.79	0.087	32	2.2 × 0.45	0.087	56
4	2.8 × 1.06	0.110	24	2.8 × 0.64	0.110	40
6	3.5 × 1.27	0.138	20	3.5 × 0.79	0.138	32
8	4.2 × 1.41	0.165	18	4.2 × 0.79	0.165	32
10	4.8 × 1.59	0.189	16	4.8 × 1.06	0.189	24
12	5.5 × 1.81	0.217	14	5.5 × 1.06	0.217	24
14	6.3 × 1.81	0.248	14	6.3 × 1.27	0.248	20
16	7.9 × 2.12	0.311	12	7.9 × 1.41	0.311	18
18	9.5 × 2.12	0.374	12	9.5 × 1.59	0.374	16
20				11.1 × 1.81	0.437	14
22				12.7 × 1.95	0.500	13

[a]Nominal sizes are designated as numbers without significance or direct relationship to thread major diameter, except that increasing numbers represent increasing screw sizes.
[b]Diameters and pitches are in millimeters.

116.2 MECHANICAL FASTENING SYSTEMS

The following are some of the more frequently utilized mechanical fastening systems.

Self-taping Screws. This is a simple and inexpensive fastening system requiring that a boss be molded into a panel or part at appropriate predetermined locations. When there is sufficient boss material and the screw is tightened correctly, the assembly has excellent structural integrity. However, splitting of the boss or strip-out of formed threads in the boss are potential failure modes that render the assembly nonfunctioning. The problem can be remedied with cap fasteners (discussed later in this chapter). Recommended plastic panel

TABLE 116.3. Approximate Hole Sizes for Steel Type B[a] Thread Forming Screws, In Plastics[b]

Screw Size	Phenol Formaldehyde		Cellulose Acetate Cellulose Nitrate Acrylic Resin Styrene Resin		
	Hole Required	Drill Size No.	Hole Required	Drill Size No.	Minimum Penetration in Blind Holes
2	0.078	47	0.078	47	$\frac{3}{16}$
4	0.099	39	0.093	42	$\frac{1}{4}$
6	0.128	30	0.120	31	$\frac{1}{4}$
7	0.136	29	0.128	30	$\frac{1}{4}$
8	0.149	25	0.144	27	$\frac{5}{16}$
10	0.177	16	0.169	18	$\frac{5}{16}$
12	0.199	8	0.191	11	$\frac{3}{8}$
$\frac{1}{4}$	0.234	$\frac{15}{64}$	0.221	2	$\frac{3}{8}$

[a]Otherwise designated as Type Z.
[b]From Ref. 2.
Note: Because conditions differ widely, it may be necessary to vary the hole size to suit a particular application. All dimensions are given in inches.

TABLE 116.4. Approximate Hole Sizes for Steel Types BF, BG and BT Thread Cutting Screws in Plastics[a]

Screw Size	Phenol Formaldehyde				Cellulose Acetate, Cellulose Nitrate, Acrylic Resins, Styrene Resins			
	Hole Required	Drill Size No.	Depth of Penetration		Hole Required	Drill Size No.	Depth of Penetration	
			Min.	Max.			Min.	Max.
2–32	0.0781	$\frac{5}{64}$	$\frac{3}{32}$	$\frac{1}{4}$	0.076	48	$\frac{3}{32}$	$\frac{1}{4}$
3–28	0.089	43	$\frac{1}{8}$	$\frac{5}{16}$	0.089	43	$\frac{1}{8}$	$\frac{5}{16}$
4–24	0.104	37	$\frac{1}{8}$	$\frac{5}{16}$	0.0995	39	$\frac{1}{8}$	$\frac{5}{16}$
5–20	0.116	32	$\frac{3}{16}$	$\frac{3}{8}$	0.113	33	$\frac{3}{16}$	$\frac{3}{8}$
6–20	0.125	$\frac{1}{8}$	$\frac{3}{16}$	$\frac{3}{8}$	0.120	31	$\frac{3}{16}$	$\frac{3}{8}$
8–18	0.147	26	$\frac{1}{4}$	$\frac{1}{2}$	0.144	27	$\frac{1}{4}$	$\frac{1}{2}$
10–16	0.1695	18	$\frac{5}{16}$	$\frac{5}{8}$	0.166	19	$\frac{5}{16}$	$\frac{5}{8}$
12–14	0.1935	10	$\frac{3}{8}$	$\frac{5}{8}$	0.189	12	$\frac{3}{8}$	$\frac{5}{8}$
$\frac{1}{4}$–14	0.228	1	$\frac{3}{8}$	$\frac{3}{4}$	0.221	2	$\frac{3}{8}$	$\frac{3}{4}$

Note: Because conditions may vary, it may be necessary to change the hole size to suit a particular application.
[a]From Ref. 3.

screw-hole sizes for different self-tapping screws are shown in Tables 116.1, 116.2, 116.3, and 116.4.

Multi-thread Inserts. There are machined metal components that can be affixed to plastic products by insert molding, pressed mechanically with or without heat, or by ultrasonic means. They must be attached carefully to avoid stress cracking. The inserts accommodate many screw thread sizes (Fig. 116.1).

Clip Fasteners. The heat-treated spring-steel "J" or "U" clip fastener is a low-cost fastening option. The product may have a single helical tooth form to fit a screw thread, or a barrel section that has already been tapped with a thread. The clip is attached to the panel by spring pressure or by a separate retaining tab which locates the fastener properly over the panel hole (Fig. 116.2).

Multithread "Tee" Nuts. These are manufactured from low carbon steel. They are distinguished by a tap threaded barrel, drawn out of a round flange which may have panel anchoring barbs and in some cases a convolution to prevent loosening when the screw or bolt is tightened (Fig. 116.3).

Locknuts. Many types of locknuts are available, including multithread nuts with fiber and nylon inserts or distorted threads to help prevent backoff and subsequent loosening or fasteners. Adhesives that mix and set when the screw or bolt is applied also are used to combat backoff.

Even so, plastic inserts, nuts, or adhesives can lose their original effectiveness when

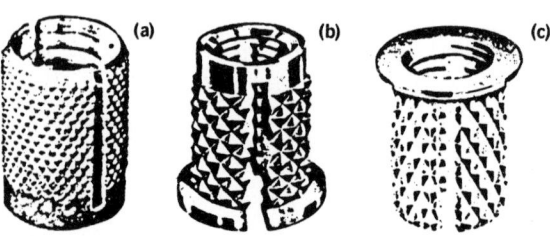

Figure 116.1. Multi-thread inserts.

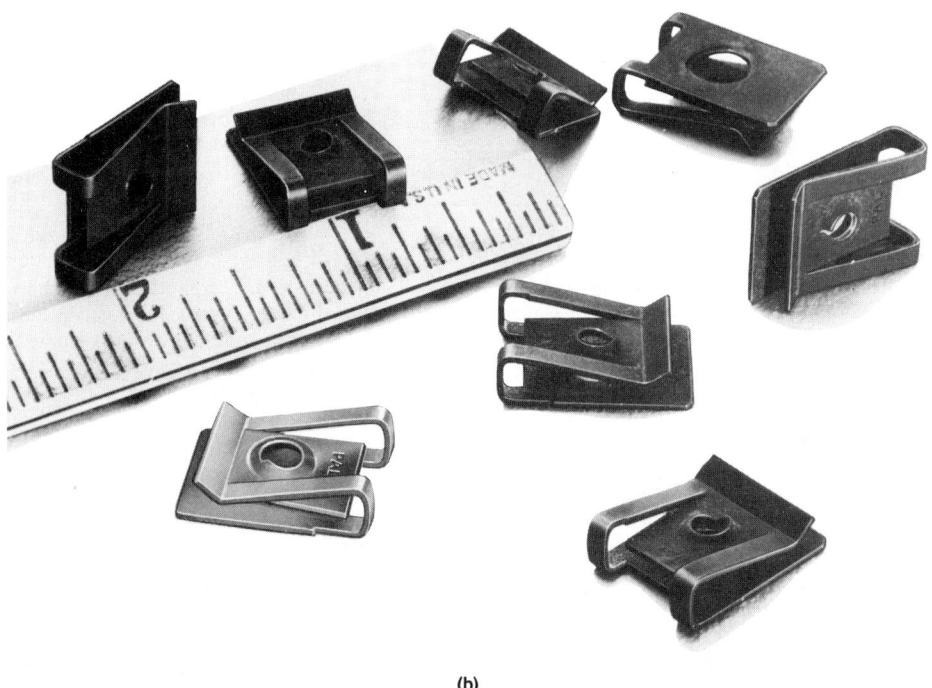

(b)

Figure 116.2. "J" and "U" clip single and multi-thread fasteners. (**a**) New applications for "J" and "U" types. (**b**) New single thread "U" nut.

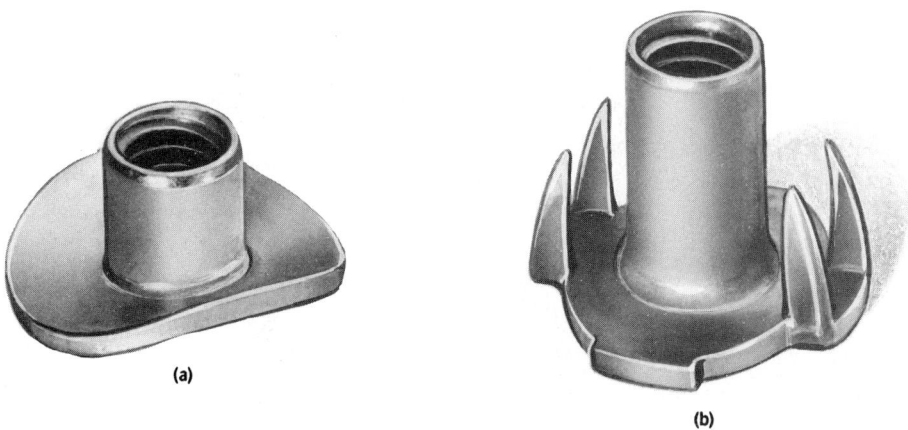

(a)

(b)

Figure 116.3. Multi-thread "Tee" nuts. (**a**) Round base plain. (**b**) Staple prongs. (Carr fasteners).

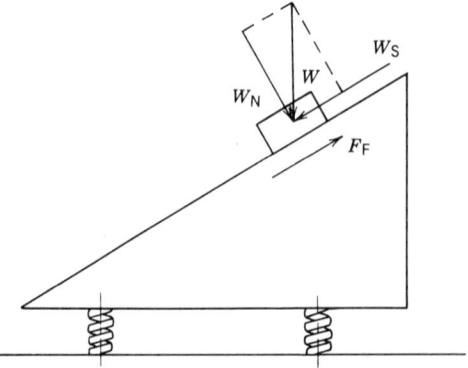

Figure 116.4. General accepted theory of vibration loosening of fastener joint. Vibration motion alleviates the frictional affect of a nut tightened on a bolt and therby causes loosening. For example: Consider a weight resting on an inclined plane: If static frictional force F_F exceeds the sliding force component W_S of the weight W, the body remains at rest. If the plane surface, however is vibrated, the effective coefficient of friction is reduced and as the vibratory motion becomes greater sliding occurs even though the W_N and W_S forces remain unchanged. The above analogy can be applied to a screw thread because it also is an incline plane.

subjected to time and temperature changes, or to frequent removal and retightening (Fig. 116.4). Positive resistance to vibration loosening can be achieved with heat-treated spring-steel fasteners.

Locknuts come in various configurations: regular hexagonal, open and closed top hexagon acorns, resilient arch, hexagon with washer flange, free-spinning hexagon washer, "J" and "U" clips, and wing types (Fig. 116.5). Many locknuts are of the single helical thread variety, which produce a double locking action when they are tightened (Fig. 116.6).

The advantages of these fasteners are light weight, vibration resistance, space saving, serviceability, and the ability to be removed easily and reused.

Cap Fasteners. These fasteners are available in various designs. Their function is to fit over the plastic boss to prevent it from splitting when self-tapping or regular machine screws

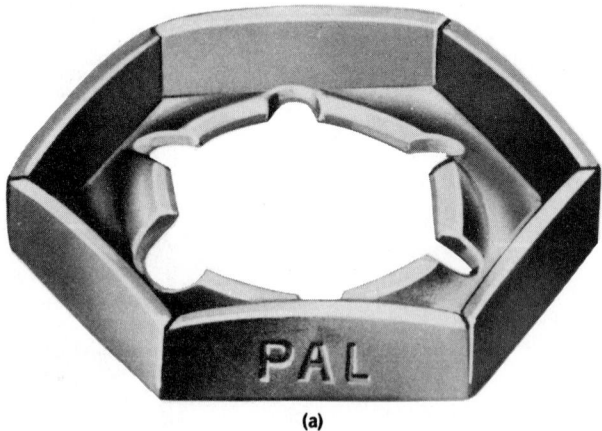

(a)

Figure 116.5. Multi and single thread locknuts. (**a**) Regular type; (**b**) acorn type, closed top; (**c**) acorn type, open top; (**d**) tension type; (**e**) wing type; (**f**) inverted type. (Palnut)

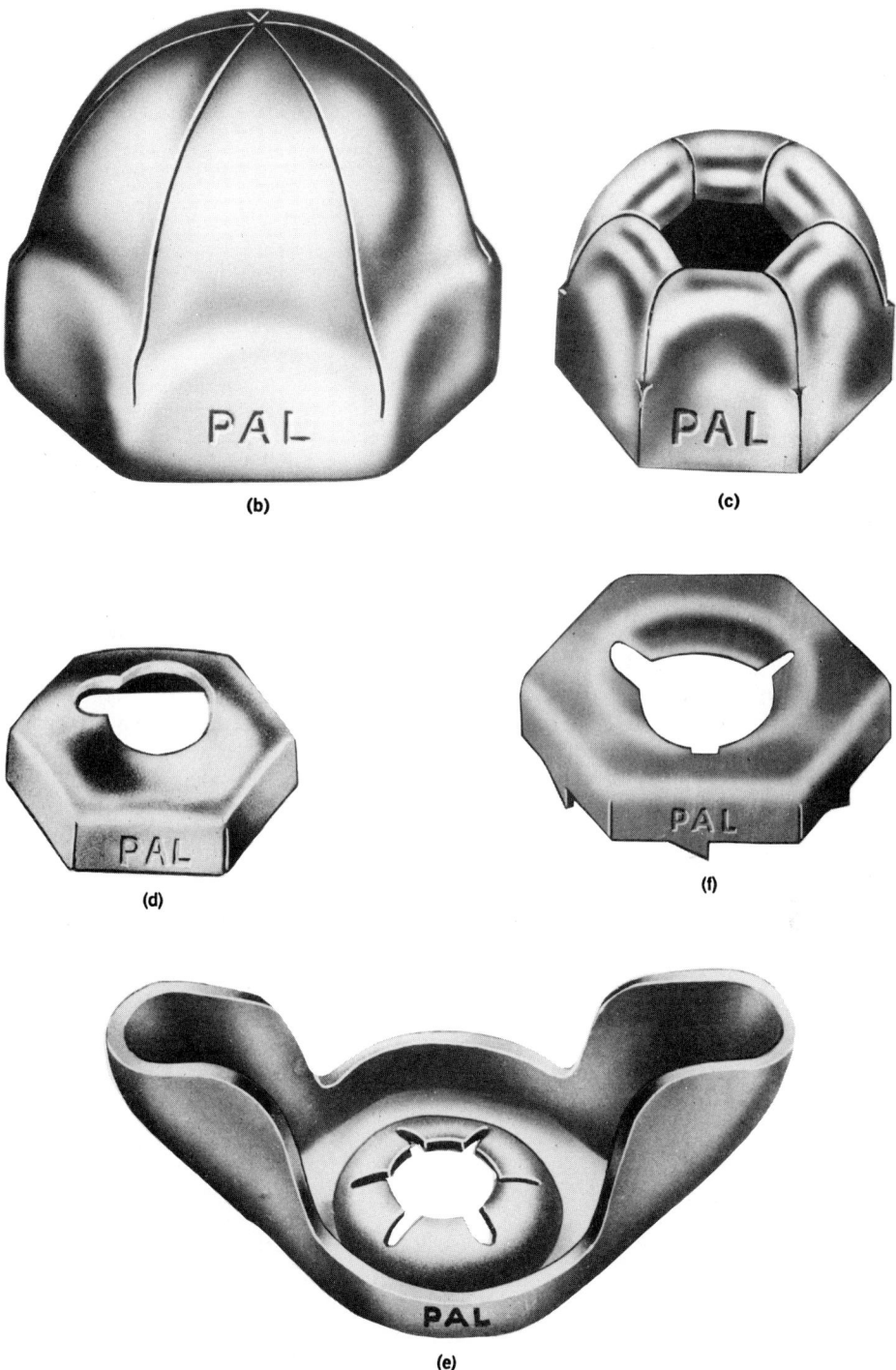

Figure 116.5. (*Continued*)

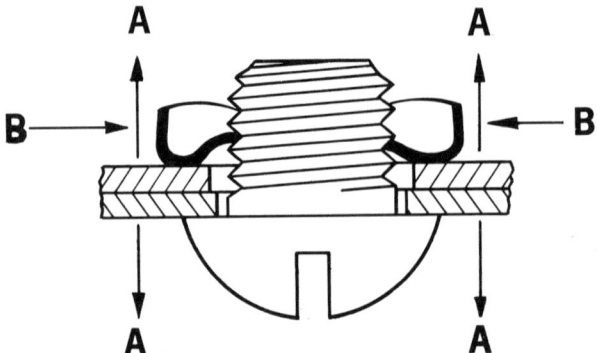

Figure 116.6. Double locking action. When locknuts are seated and tightened, powerful spring forces (A-A) are exerted upward against the bolt threads and downward against the assembly. At the same time, spring forces (B-B) are exerted inward, gripping the root of the bolt thread like a chuck. (Palnut).

are threaded through the helical tooth form and into the hole. (Recommended inside diameters are shown in Tables 116.3 and 116.4).

Major advantages are they clamp the sides of bosses firmly when the screw is tightened, and that they form helical threads if the original plastic hole threads are stripped. Note: side barbs on some fasteners prevent cap turning during tightening (Fig. 116.7).

Push-on Fasteners. The holding strength of a push-on mechanical fastener system depends on the teeth bite into the stud. Resilient parts produce a viselike grip when the arch is flattened, creating an axial stud tensile force that acts to clamp the assembly securely together. There are many flat-type push-on devices, light and heavy duty (Figs. 116.8 and 116.9).

Push-on, acorn or hat fasteners with closed tops are a little different. Their function is to serve in stopping rather than clamping.

Some of these fasteners are removable by squeeze relaxation of the tooth grip on the

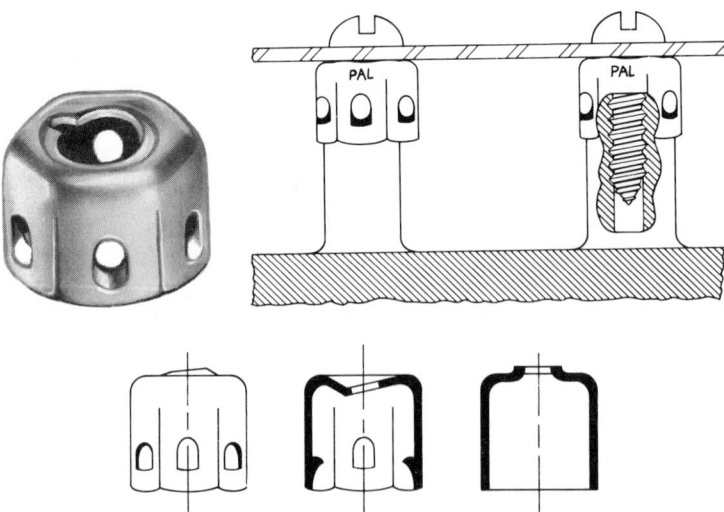

Figure 116.7. Single thread cap fasteners (Plastic World, Palnut).

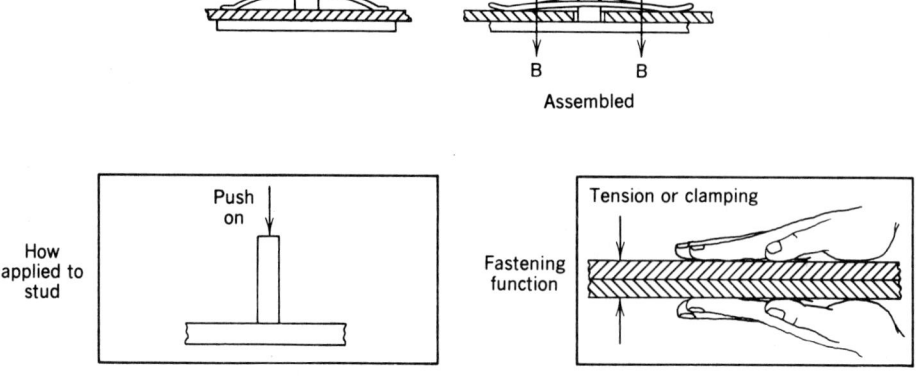

Figure 116.8. Push-on fasteners (Palnut).

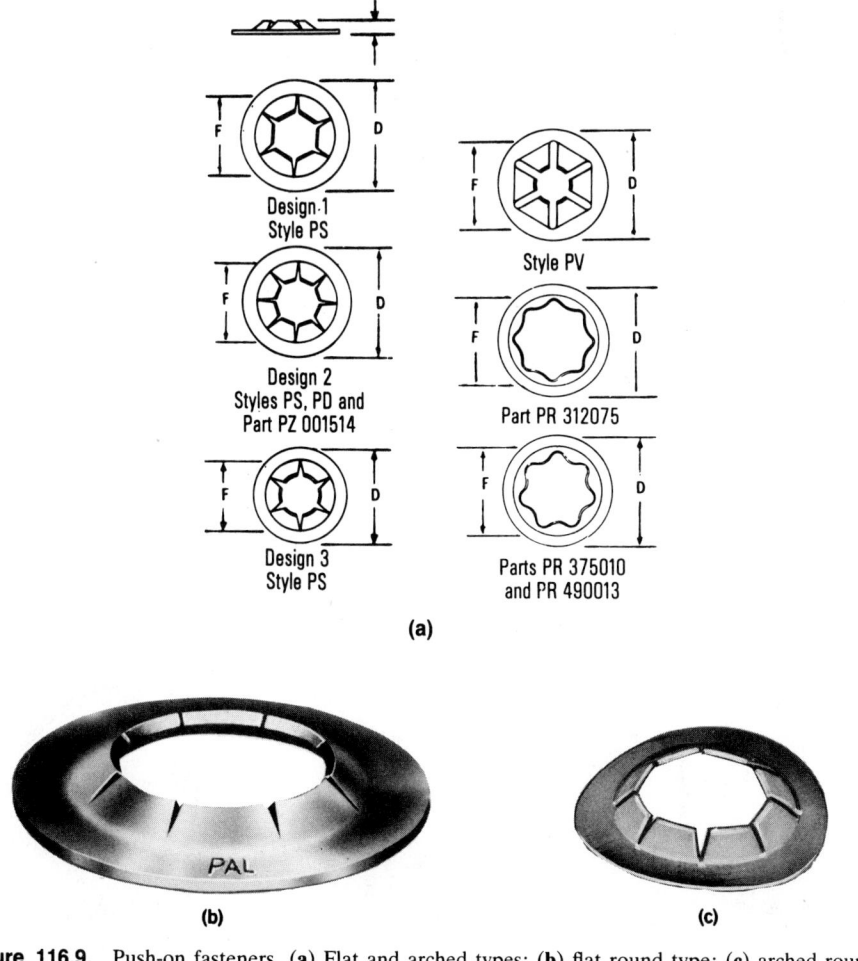

Figure 116.9. Push-on fasteners. (**a**) Flat and arched types; (**b**) flat round type; (**c**) arched round. (Palnut).

Figure 116.10.   Removable type. (Palnut).

stud (Fig. 116.10). Push-on self-threading fasteners are light-duty nuts that come in regular hexagonal, hex with washer flange, and wing configurations. They are applied to the stud by pushing and then turning one-third of a revolution. Care must be exercised not to overtighten because of the fragility of these fasteners.

Self-threading Fasteners. The many forms that are available are regular hexagonal, open or closed-top acorn, hex with washer flange, or wing. All single-thread designs function the same way. The tooth form plows a coarse thread for fast assembly. Nuts start on almost blunt studs and seat tighten reliably even when the stud is as much as 20° off the perpendicular to the panel surface (Figs. 116.11 and 116.12).

Advantages are special inserted threaded studs are unnecessary; molded-in studs on the unit to be assembled become part of the fastening system; prevailing torque backoff resistance is already part of the design; and use of a lock washer is unnecessary.

Panel Fasteners

Blind Push-on Fasteners. Are pre-assembled to the stud; barbs on fastener legs keep it secured to the stud. This type of fastener has three curved knee-shaped legs, equally spaced for optimum holding performance (Fig. 116.13**a**).

Figure 116.13**b** shows a blind push-on fastener that acts as an interference stop by the gripping action of similar leg barbs on the stud.

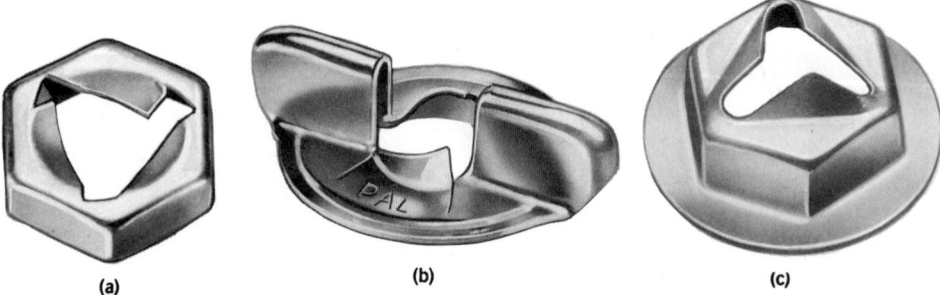

(a)   (b)   (c)

Figure 116.11.   Push-on self threading fasteners. (**a**) Regular; (**b**) zip twist; (**c**) washer. (Palnut).

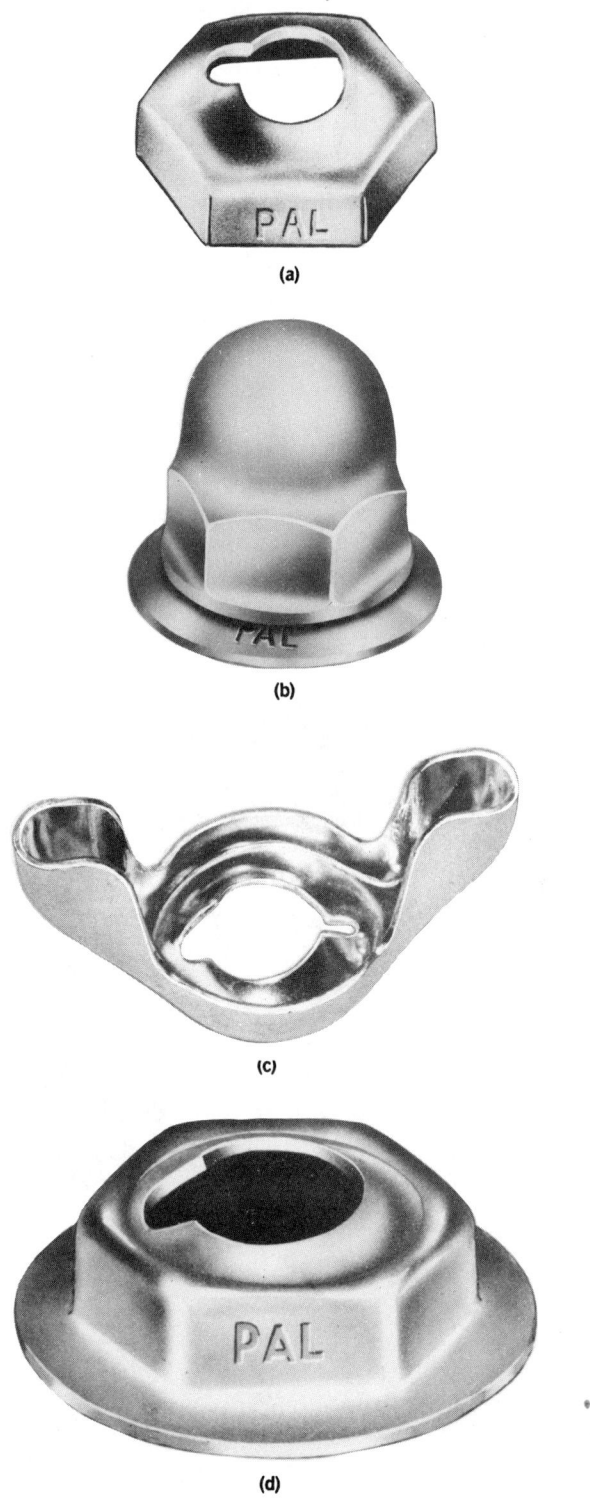

Figure 116.12. Self threading fasteners. (**a**) Regular; (**b**) capped washer; (**c**) self threading wing nut; (**d**) washer. (Palnut).

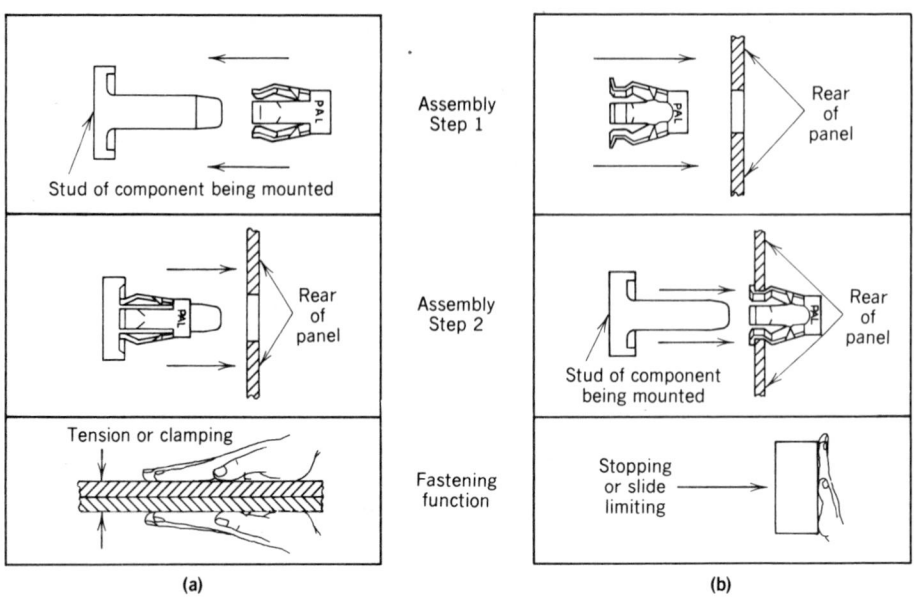

Figure 116.13. (**a**) Panel fasteners. (**b**) Blind push on fasteners. (Palnut).

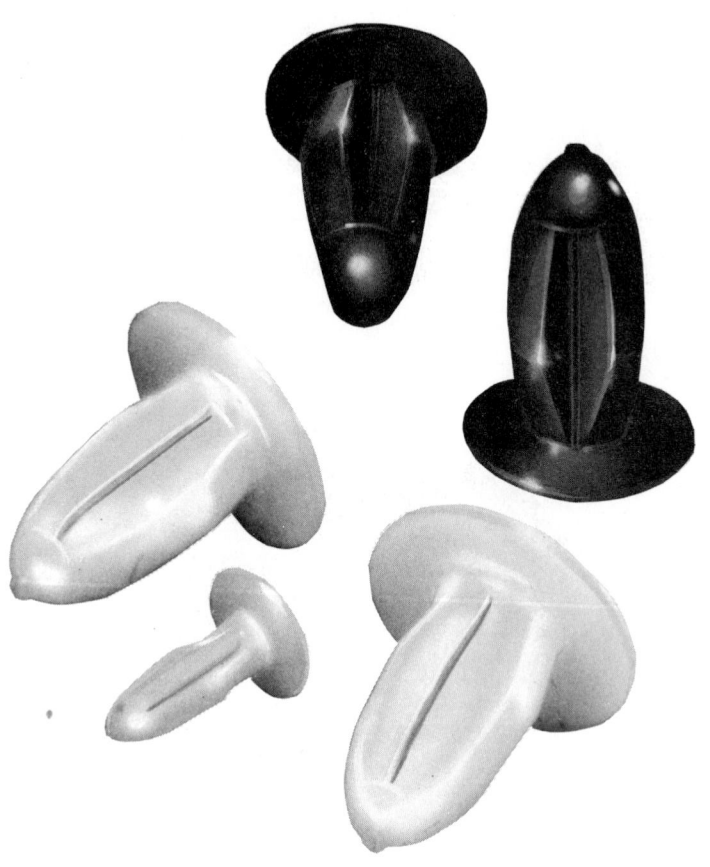

Figure 116.14. Plastic accordian rivets.

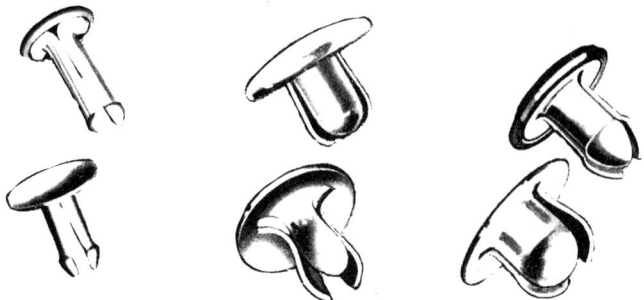

Figure 116.15. Shoulder panel fasteners.

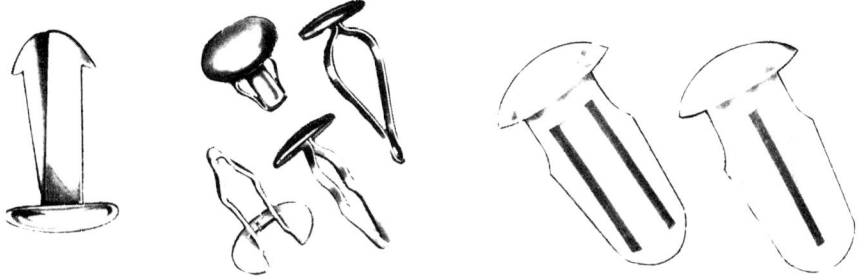

Figure 116.16. Dart fasteners.

Plastic Accordion Rivets. These fasteners can replace screws, blind rivets, and other forms. The barrel compresses, allowing passage through the panels to be assembled without scratching them (Fig. 116.14).

Shoulder Panel Fasteners. These fasteners perform the function of a removable rivet. They can hold two or more thickness of materials together. Panels are retained firmly between the shoulders of the stud and the underside of the flange or cap (Fig. 116.15).

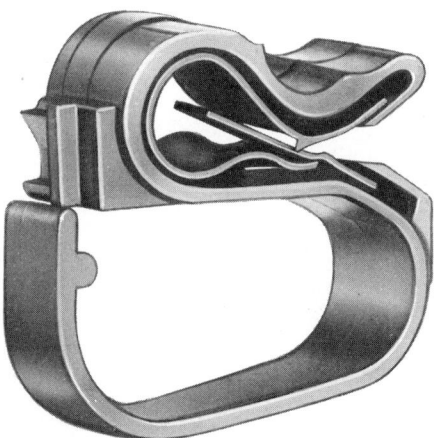

Figure 116.17. Tubing and wiring fasteners.

Dart Fasteners. Dart fasteners come in single or double loop studs. The resilient spring-steel loop enters the panel hole easily, and the gradual sholulder provides takeup for fastening noncompressible materials, thus assuring vibration-free assembly (Fig. 116.16).

Plastic Ratchet Types. These can be used as removable rivets or as studs to assemble components.

Tubing and Wiring Fasteners. A simple plastic snap-in loop clip holds tubing and wiring and snaps into panel hole to support the bundle. This type of fastener is widely used in automotive and aircraft assembly (Fig. 116.17).

BIBLIOGRAPHY

1. ASA B18.6.4-1958, Slotted and Recessed Head Tapping Screws, The American Society of Mechanical Engineers.
2. *Ibid.*, p. 139.
3. *Ibid.*, p. 143.

General References

V.K. Stokes, "Joining Methods for Plastics and Plastic Composites," *Polym. Eng. & Sci.* **29**(19), 1310–1324 (Oct. 1989).

J.R. Vinson, "Mechanical Fastening of Polymer Composites," *Polym. Eng. & Sci.* **29**(19), 1332–1339 (Oct. 1989).

117

ADHESION AND SOLVENT BONDING

S.C. Temin

20 Rainbow Pond Drive, Walpole, MA 02081

Industrial practice often requires that plastic parts be affixed to other plastics or to nonplastic materials. Of the methods available to engineers who must make judgements relative to the most suitable assembly technique, adhesive bonding must be seriously considered.

Adhesive bonding of plastic substrates is already a common practice; even greater utilization can be expected with the increasing availability of improved adhesives. Currently, adhesives compete with welding, fusion, and mechanical fastening. However, adhesive bonding offer special advantages. Not least among them is the opportunity of weight savings compared with metal fasteners. It is of special appeal in the automotive industry, where vehicle lightweighting is crucial.

117.1 ADVANTAGES

The use of adhesives may indeed be the only sensible route in some instances. The reasons are as follows:

1. The ability to attach materials that are impractical to bond in other ways. It is obvious that screws or bolts are not suitable for paper, glass or other brittle materials, thin films, fibers, and foams. Moreover, the shape of the substrate may render mechanical fastening difficult or impossible.

2. The ability to join dissimilar materials. Significant differences in thermal coefficient of expansion may disqualify mechanical fastening. Thermal fusion is also disqualified for unlike polymers. Adhesive bonding of dissimilar materials often yields composites with enhanced properties by utilizing the best features or physical properties of each component.

3. Uniform distribution of stress. In a simple bonded overlap, any load applied to the joint is distributed evenly over the entire bonded area as if the material were continuous. In terms of stress vectors in both tensile stress (away from the bond) and shear stress (across the bond), all of the adhesive contributes to opposing the stress. In contrast the holes required for rivets, nails, screws or bolts are stress concentrators or focal points that contribute to failure. Adhesive bonded assemblies generally can resist shock, vibrational and fatigue failure better than mechanically attached assemblies. This often means that adhesive-bonded members can be thinner in cross-section and lighter in weight.

4. Reduction in joint weight. This feature is partially related to the better performance possible with lighter weight parts because of the elimination of stress concentration points. Reduction in weight is possible because adhesives are generally of low specific gravity, and applied preferentially in thin layers. In many joints the forces of adhesion exceed the cohesive strength of the adherends.

5. Environmental protection. Although usually more important in metal-to-metal joining, the continuous contact between surfaces made by adhesives seals out corrosive or harmful agents. Both gases and liquids can be excluded by the adhesive.

6. Resistance to vibration. Elastomeric adhesives resist the fatigue and subsequent failure that is brought on by repeated deformations. In general the more flexible the adhesive, the greater the resistance to strain occurring in the adherends.

7. Smoother surfaces. Adhesive bonding, like thermal bonding, eliminates such imperfections as rivet-point dimpling, surface spots or projecting portions of devices that are common with mechanical fastening. This reduces finishing costs and makes possible more esthetic and more functional product design.

117.2 BASIC REQUISITES

All adhesives must be liquids, or at least capable of some degree of flow. A primary requirement for good adhesion is that the adhesive and adherend surfaces must be in close contact, since the attractive forces that promote adhesion vary as the inverse sixth power of the intermolecular distances. To get substrates close enough to effect a good bond, their surface roughnesses must be made to correspond. In essence, the adhesive must be a fluid because the solid adherend is never smooth in a microscopic sense. For solid polymers, this means that the adhesive must be melted or dissolved in a suitable solvent before application.

The adhesive must also be sufficiently mobile to quickly penetrate holes and depressions in the solid surface. If voids are left when the adhesive solidifies, the joint will be weak.

Finally, the adhesive must solidify to provide sufficient strength to resist unbonding stresses. (Pressure-sensitive adhesives, of course, do not solidify; their function is best described in terms of their viscoelastic properties). Solidification occurs either through evaporation of a solvent or through chemical reaction (polymerization), frequently called curing. Drying or curing converts the fluid material to a solid, which can be either rigid or elastomeric.

Wetting of the solid surface requires the selection of an adhesive that is suitable in terms of the surface energy or surface tension of the adherend. Any adhesive supplier can furnish this information. However, it is often necessary to alter the surface of the adherend to make it receptive to adhesives.

117.3 TYPES

There are five basic physical types of adhesives: curable liquids, solvent cements, hot melts, aqueous, and pressure-sensitive. The curable liquids undergo a chemical reaction or polymerization to attain a nonfluid condition. Often these consist of two parts, each of which is stable by itself but when mixed together undergo a chemical reaction. Solvent cements are simply dilute solutions of fully reacted adhesives that solidify by loss of solvent. Hot melts are 100% solids that become liquids on heating above their melting or softening points; they are applied hot and solidify on cooling.

Aqueous adhesives dissolved or dispersed in water may either simply dry to the proper physical form or undergo a reaction when the water is removed. Solvent-activated adhesives are deposited at the bond line in solid form and rewet with solvent just before making the assembly. As with all solvent-borne adhesives, better joints are formed with porous substrates that permit a faster evaporation of solvent.

Pressure-sensitive adhesives do not change form or harden but behave as high-viscosity fluids which, because of an increase modulus as the shear or deformation rate is increased, can be easily removed from the substrate. Pressure-sensitives are frequently used as temporary anchorage for components that must be repositioned.

Another way of looking at adhesives is in terms of their chemical composition. Table 117.1 lists in a simplistic sense the more common physical and chemical types (the actual chemistry of each is more complex and beyond the scope of this chapter). In addition to unlisted additional compositions that fall in each category, there are many mixtures or blends of two or more chemical types, called hybrids. Table 117.2 lists common adhesives and their typical properties.

117.4 LIMITATIONS

A key disadvantage of adhesive bonding is the need for fixtures or clamps in many such operations. Another is the slowness, relative to mechanical fastening or fusion welding, of adhesive bonding. Other disadvantages include the following:

Limited Shelf Life. Adhesives that depend on a curing reaction and are mixed beforehand, have a relatively short pot or storage mix. Some other adhesives must be carefully stored

TABLE 117.1. Adhesives Classified by Type

Types	Examples
Physical	
Curable liquids (one or two-component)	Epoxies, urethanes, acrylics, silicones
Solvent cements (includes water)	Acrylics, elastomers, vinyls, cellulosics, urethanes
Hot melts (100% solids)	Polyamides, EVA, polyesters, urethanes
Aqueous (includes dispersions)	PVA, epoxies, silicones, aminoplastics, rubbers, phenolics
Pressure sensitive	Double-faced tapes
Chemical	
Anaerobic	Methacrylates
Contact	Rubber
Cyanoacrylate	Methyl, ethyl esters
Emulsions	PVA, PVC, rubber, EVA, acrylic, chloroprene
Hot melt	Polyurethane, polyamide, EVA, polyester
Solvent-based (Cements)	Acrylic, chloroprene, nitrile, nitrocellulose phenolic, PU rubber, vinyl
High temperature use	Epoxy, PU, silicone, polyamides
Room temperature cure	Acrylic, cyanoacrylate, epoxy, PU

TABLE 117.2. Common Adhesives

Adhesive	Properties
Acrylic	Room temperature cure
Phenolic	Most require heat
Epoxy	Broad range of properties
Modified polyolefin	Hot melts
Model cement	Resin in solvent
Polyurethane	Broad range
Pressure-sensitive	Mostly in tape form
Emulsions (vinyl, acrylic, rubber)	No solvent
Sealants	Caulking (butyls, silicones)
White glues	PVA dispersion

in order to avoid premature reaction. Additionally, two-part adhesives have a working life that depends on their rate of reaction; the faster the cure, the short pot life. For this reason it may be desirable to use dispensing machines that blend the two components immediately prior to application. In general, if rapid cure under ambient conditions is required, care must be taken not to pre-mix large quantities of the two components.

Need for Surface Preparation. Surface contamination of plastics can prevent the necessary intimate contact between adhesive and adherend. For example, low molecular weight substances can bloom to the surface, as can certain additives, creating a "film" that defeats the purpose of the adhesive.

Solvent Retention. Where solvents are used, or where a product of the curing reaction is volatile, there is always the danger that failure to dry the adhesive adequately can cause bubbles or voids. These imperfections can seriously weaken the glue line, since they function as stress concentrators that prevent intimate and complete contact of the two surfaces.

Temperature Limitations. At elevated temperatures, polymeric adhesives become unstable and will degrade. Some adhesives also become brittle and ineffective at low temperatures. Thus, temperature considerations are important in adhesive selection, and could in fact affect the decision as to whether adhesive bonding is suitable for a particular application.

Material Limitations. A primary disadvantage of adhesive bonding (excepting pressure-sensitives) or welding is the inability to disassemble the parts for repair or inspection.

117.5 SELECTION FACTORS

Determination of the relative merits of a particular adhesive, involves a number of factors: need for a primer, amount of surface preparation, cure time, fixture time, need for heating, adhesive viscosity, cost, and properties of the cured adhesive.

Some of these considerations relate to the cost of the operation. Elimination of surface preparation and the need for a primer can be important factors in selecting an adhesive. If cure time or consequent fixture time is excessive, the resulting time delays may mitigate against an otherwise useful adhesive. If heating is necessary (either to effect complete cure or to reduce the time for reaction or solvent evaporation) an additional cost is involved. However, most adhesive operations require realtively low capital costs.

Other selection factors relate to the nature of properties of an adhesive. Viscosity must be suitable—either low enough to ensure good wetting of the adherend or high enough to prevent run-off on vertical surfaces. Further, the modulus of the cured or dried adhesive is a factor in terms of the flexibility or resistance to deformation required for the bond line. In turn, these considerations depend on the nature of the plastic and the anticipated use-life stresses or extent of strains.

A most critical factor in selection, of course, is adhesion to the plastic. Not all adhesives will function satisfactorily with all plastic surfaces; even within classes of adhesives or plastics, individual members differ in behavior.

Consideration of environmental resistance depends on the desired use life for the bonded assembly. For instance, water resistance obviously is required if the assembly is intended for outdoor exposure. Phenolics, epoxies, and acrylics are examples of adhesives that weather well. Similarly, the part may have to be exposed to oils or solvents and an adhesive should be chosen with that in mind. Also, certain rubber-based adhesives are contraindicated if long use life is contemplated, since polymers with unsaturated backbones undergo an oxidative degradation with time, which is accelerated by heat or light.

117.6 SURFACE PREPARATION

Improved adhesion can be obtained when the surface of a plastic part is prepared correctly. The adhesive must wet, spread and penetrate the rough surface of the adherend.

Cleanliness is a basic requirement in order for the adhesion to make intimate contact with the substrates to be bonded. Surface preparation entails both obtaining a clean surface and, if necessary, altering the surface tension of the substrates to correspond to the surface tension of the adhesive.

Roughening of the plastic surface provides an increased area for bonding and strengthens the joint against shear forces. Accordingly, instructions for surface preparation usually entail both cleaning and abrading the surface. The extent of surface preparation required for a particular bond depends on the physical requirements for the assembled part; more extensive preparation is indicated in instances where extreme in-use stress will be placed on the joint. Thus surface preparation may consist of one or more of the following: solvent cleaning; abrading and solvent cleaning; chemical treatment (to alter surface energetics).

Solvent Cleaning. For many operations, particularly where the joint will not be subjected to severe in-use stresses, cleaning the surface may require only a wipe with a suitable solvent or detergent solution. Different solvents are recommended for wiping various plastic surfaces. The choice of solvent is dictated by the solubility properties of the plastic. For example, most common solvents other than alcohols severely attack polystyrene, whereas, nylons are resistant to the same solvents. As a generalization, crystalline polymers are less likely to be attacked by common solvents.

Often, parts are solvent cleaned by more elaborate procedures, such as vapor degreasing, ultrasonic vapor degreasing, or immersion in a series of cleaning agents, which tend to give cleaner surfaces than wiping with cloth or paper. Commonly used solvents are halogenated to minimize flammability concerns. They include methylene chloride, perchloroethylene and similar materials Table 117.3.

Abrading and Solvent Cleaning. In the simplest case, plastic components can be wiped with a solvent, abraded with emery cloth and then wiped again to ensure removal of all debris. The plastic surface is not changed chemically but is altered physically; because some of the plastic may be removed the roughened surface now offers the possibility of mechanical interlocking Table 117.3.

TABLE 117.3. Plastic Surface Cleaning[a]

Plastic	Method
Acrylics	Wipe with methanol
Aminoplasts	Scrub (detergent), scour, rinse
Cellulosics	Wipe with alcohol, scour
Epoxy	Degrease (acetone, MEK)
Fluorocarbon	Sodium (solubilized), acetone wash
Nylon	Phenol (12% aqueous); 1:1 resorci-nol–ethanol
Polyamide	Solvent-clean (acetone, MEK), scrub
Polyolefin	Oxidize (chromate or radiation)
Polystyrene	Wipe with alcohol
Polyurethane	Wipe with acetone or MEK
Rubbers (NBR, SBR, NBR, Neoprene)	Scrape, scrub with toluene (immerse in acid)
Reinforced plastics	Wipe with MEK

[a]Most surfaces (except those chemically altered) are preferably abraded or roughened.

Chemical Treatment. Where optimum adhesion is required, chemical treatment of surfaces may be required. For low energy plastics, such as polyolefins and fluorocarbons, the chemical nature of the plastic surface must almost always be altered in order to give a more polar (higher-energy) surface.

Chemical alterations can be drastic in the case of the low energy plastics. For instance, chromic oxidation involves using a liquid used to remove organic contaminants from glassware. It is hazardous to use, and proper safeguards must be exercised. Similarly the defluorinating solution (utilizing a form of sodium) is dangerous and requires expertise in disposal.

To avoid the problem of hazardous chemicals, sheets of fluorocarbons can be purchased already chemically modified.

An alternative approach to activating plastic surfaces that does not require the use of hazardous fluids is radiation treatment. Such treatment renders polyolefins more amenable to adhesive bonding. Other techniques for enhancing adhesion of plastics include flaming and plasma treatments. Detailed instructions on surface preparation for plastics are available in the literature.

117.7 SOLVENT CEMENTING

A variation of adhesive bonding, solvent cementing is particularly useful for noncrystalline thermoplastics. Crystalline polymers generally do not dissolve in ordinary solvents at room temperature.

Solvent cementing depends on active solvents to soften and swell the plastic surfaces; after assembly and evaporation of the solvent, a monolithic clear joint is obtained. Individual solvents are seldom used; a combination of solvents is a more common way to attain the properties desired. Solvents frequently contain some dissolved polymer (of the same type as that to be bonded), to aid in gap filling and to speed up drying. With cements thickened with dissolved polymer, it may be necessary to mask the area around the joint to facilitate removal of excess cement squeezed from the bond line by clamping pressure.

Masking has general application in solvent cementing when the parts to be assembled are soaked in the solvent as a preliminary to adhesion. Cellophane tape rather than masking paper, is preferred. Parts can also be masked using commercially available masking compounds based on animal glues or gelatins. The masking compounds are applied hot and carefully stripped away for application of the solvent cement.

Another method used in joining is capillary action. Fine wires are used as shims, and the adhesive applied from a dropper or needle. After capillary action has adequately spread the cement, the wires are removed.

Table 117.4 gives some examples of bonding with solvents or solvent cements. The plastics listed are not necessarily always bonded this way, however. For example, ABS can be readily joined using epoxy or acrylic adhesives. This avoids the long drying times associated with solvents. Care must be taken in choosing solvents for polystyrene; caution must be exercised to avoid solvent crazing.

Solvent cementing is the method of choice for many plastics. Acrylics, for example, are readily bonded with such chlorinated solvents such as methylene chloride, ehtylene chloride 1,1,2-trichloroethane, or chloroform. If a more viscous cement is desired, a solution of acrylic chips (from 2 to 8% by weight) in the chlorinated solvent can be used. Cellulosics also are good candidates for solvent bonding. A typical cement for cellulose acetate is a 10% solution of the polymer in a mixture of acetone and methyl cellosolve. For acetate butyrate, a cement based on equal parts of acetone and ethyl acetate is useful.

Poly(vinyl chloride) is frequently bonded using solvents like acetone or methyl ethyl ketone. Sometimes more powerful solvents like cyclohexanone or tetrahydrofuran have to be used. Here the cements not only contain polymer (PVC) but plasticizer as well.

It is important to remember that solvent cementing is generally employed only when the parts to be joined are of the same plastic. Finding a mutually satisfactory solvent mixture for two different plastics is often difficult.

117.8 PREFERRED ADHESIVES

Choosing an adhesive for a particular plastic is not so easy as one might suppose. The adhesive providing the strongest bond to a given plastic may not necessarily be the best for a particular job. Primary consideration must be given to the form of the plastic, to its modulus, and to the conditions to which the bonded assembly will be subjected in actual use.

TABLE 117.4. Solvents for Cementing

Plastic[a]	Recommended Solvents[b]
ABS	Blends of acetone, MEK, MIBK, THF or methylene chloride
Acetal	Methylene or ethylene chloride
Cellulose acetate	Acetone or MEK with methyl cellosolve
Poly(methyl methacrylate)	Methylene chloride, chloroform, trichloroethylene
Phenylene oxide	Chlorinated solvents, xylene/MIBK (25/75)
Poly(butylene terephthalate)	Hexafluoroisopropanol
Polycarbonate	Methylene chloride
Polystyrene	Methylene chloride, ethyl, acetate, MEK, trichloroethylene
Polysulfone	Methylene chloride
Poly(vinyl chloride)	Mixed solvents (THF, MEK, MIBK, dioxane) with plasticizer

[a]Usually 1 to 7% of plastic is dissolved in solvent.
[b]MEK = methyl ethyl ketone, THF = tetrahydrofuran, MIBK = methyl isobutyl ketone.

Flexible parts should be bonded with flexible adhesives. When bonding dissimilar materials, the adhesive must be resilient enough to allow for differences in coefficients of thermal expansion.

Choice is complicated by the fact that an unfilled plastic behaves differently from the same material reinforced. It frequently happens that adhesives that give relatively low bond strengths with the unfilled plastic will give greatly improved performance with the same material filled with a polar inorganic substance, particularly after abrading. As earlier noted, organic polymers are low energy materials, in contrast to the higher-energy surface (like glass) to which most adhesive readily adhere.

Table 117.5 lists a number of plastics along with recommended adhesives. However, resin formulations often differ within the same family, and this can affect the performance of the adhesive. For example, structural acrylics are often formulated with different monomers and a particular grade may be more (or less) receptive to one adhesive than to another. However, adhesives can be formulated to vary considerably in polarity and in terms of the flexibility of the cured adhesive. Where flexibility is crucial, rubber-based adhesives may be

TABLE 117.5.   Recommended Adhesives for Bonding Plastics

Plastic	Adhesives
ABS	Epoxy, urethane, acrylic, nitrile–phenolic, cyanoacrylate
Acetal	Epoxy, phenolic, polyester, EVA, cyanoacrylate
Cellulose acetate	Urethane, resorcinol–formaldehyde, nitrile–phenolic, rubber-based
Elastomers	Pressure-sensitive based on similar elastomer, urethane
Epoxy	Epoxy (with primers), nitrile–phenolic, acrylic, polyester, resorcinol–formaldehyde
Fluorocarbon	Urethane
Nylon	Phenolic, epoxy, polyamide hot melt
Phenolic	Epoxy, hybrid, phenolic, poly(vinyl acetate), urea–formaldehyde, acrylic, urethane
Phenyiene oxide	Polysulfide epoxy, silicone, rubber-based, acrylic cyanoacrylate
Poly(methyl methacrylate)	Cyanoacrylate, nitrile phenolics, epoxy, urethane
Polycarbonate	Epoxy, acrylic, urethane, silicone, cyanoacrylate
Polyolefin (untreated)	Rubber-based, EVA, modified polyethylene
Polyester (linear)	Polyester, cyanoacrylate, nitrile rubber, urethane, acrylic
Polyester (unsaturated)	Acrylic, urethane, neoprene, nitrile–phenolic, epoxy
Polysulfone	Epoxy, vinyl–phenolic, rubber-based, urethane, polyester, cyanoacrylate
Polystyrene	Vinyl acetate–vinyl chloride emulsions, acrylic, polyamide, urethane, epoxy, polyester, hot melt, urea–formaldehyde
Polyurethane	Pressure-sensitive rubber emulsion, polyester, epoxy, phenolic, urethane, nitrile rubber
Poly(vinyl chloride)	
Flexible	Nitrile rubber, neoprene, urethane
Rigid	Epoxy, urethane, acrylic, nitrile rubber, silicone, nitrile–phenolic
Silicone	Silicone
Urea–formaldehyde	Epoxy, nitrile–phenolic, phenolic, polyester, neoprene, cyanoacrylate

preferred over others that have higher adhesion values. Cyanoacrylates, for instance, cure so quickly that jig or fixture time may be avoided, but they are usually quite inflexible.

Plastics and adhesives designed for very high temperature use are not covered in this discussion because there are some special problems with these systems. Manufacturers, particularly those in the aerospace industry, may wish to consult government sources for recommendations.

A comment is in order about a relative newcomer, acrylic structural adhesives. Their great advantage is hardening to handling strength in seconds without heat. This capability translates into markedly reduced process costs and greatly increased process speeds. Structural acrylics are much tougher and higher in impact resistance than cycanoacrylates, and are more resistant to environmental factors.

Offshoots of this adhesive category are UV-curable and visible light-curable types. Otherwise similar to chemically curing grades, they solidify in less than 10 seconds. Their utility for certain applications utilizing transparent plastics is promising.

117.9 JOINT DESIGN

There are a variety of ways to effect adhesive joining of plastic components (lap, scarf, strap, etc). Whichever is used, it is always best to design joints to minimize peel stresses. It is important that the joint be designed so that the adhesive is subjected to compressive and shear forces in use.

The effect of joint design can be summarized:

- Butt   Unsatisfactory
- Lap   Practical
- Scarf   Very good
- Joggle lap   Good
- Strap   Fair
- Double strap   Good
- Recessed double strap   Good, expensive

Other generalizations about adhesive bonding:

1. Width of the bonding area increases joint strength linearly; increasing the length of the bonded area, although beneficial, does not make as great a contribution to strength.

2. Thickness of the bond line should be controlled to about 4 to 6 mils (0.1–0.15 mm) of adhesive. A greater gap in the distance between adherends has a deleterious effect on peel and cleavage strength of the bond.

3. Stiff adherends (high-modulus plastic substrates) are less sensitive to joint geometry than flexible adherends.

117.10 APPLYING THE ADHESIVE

Techniques for applying liquid adhesives depend on the sophistication of the job as well as on production-volume needs. Thus, applicator equipment ranges all the way from conveyorized and automated dispensing systems to eyedroppers or brushes. Typical single type applications are

- Spray gun (air or airless)
- Brush

- Glue gun
- Dip vessel
- Pad
- Roller
- Spatula/trowel

Continuous dispensers include:

- Roll coater
- Knife coater
- Flow brushes
- Conveyor dip
- Hot melt
- Metering pumps
- Metering machines

Because many adhesives are two-part formulations, processors can take advantage of equipment that automatically meters controlled amounts of each component, mixes them, and dispenses them to the substrate. Continuous bead or intermittent drops can be programmed. Such equipment is especially useful when broad, flat surfaces are to be joined. For smaller jobs, hand-held guns are commonly used.

118

ULTRASONIC ASSEMBLY

Sylvio Mainolfi

Branson Ultrasonic Corp., 94 Prospect St., Watertown, Conn. 06795

Ultrasonic assembly is a fast, clean, efficient, and economical method of assembling or continuously processing molded thermoplastic parts or synthetic fibrics and films. Since its successful commercial introduction more than 25 years ago, ultrasonic assembly technology has been used extensively in all major industry segments, including automotive, appliance, electrical and electronic, medical packaging, and textiles, to join plastic to plastic and plastic to metal or other nonplastic materials. It replaces or precludes the use of adhesives, glues, solvents, mechanical fasteners, or other consumables.

118.1 BASIC THEORY AND PRINCIPLE OF OPERATION

Everything that makes a sound vibrates and everything that vibrates makes a sound; however, not all sounds are audible. Ultrasound literally means beyond sound–sound beyond the audible spectrum. Ultrasonics refers to sound (mechanical vibratory energy) above 18,000 Hertz (cycles/s, Hz), which is the appropriate level of human hearing.

In ultrasonic welding a thermoplastic material is subjected to high frequency mechanical vibrations (usually 20 to 40 kHz). Through a combination of surface and intermolecular friction, a rise in temperature is produced at the joint interface. If the rise in temperature is sufficient to "melt" the material, a molecular bond will result and a uniform weld is produced.

The essential components required to apply this ultrasonic energy are the power supply, converter, booster, horn, and actuator (Fig. 118.1). The power supply (ultrasonic generator) converts 50/60 Hz current to high-frequency electrical energy, which is supplied to the converter, a component that changes electrical energy into mechanical energy at ultrasonic frequencies. Attached to the converter is a booster (amplitude-modifying device), which can either increase or decrease the amplitude of vibration to the horn (amplitude referring to the peak-to-peak excursion of the horn at its workface).

The horn is usually a one-half wavelength long metal bar dimensioned to be resonant at the applied ultrasonic frequency that transfers vibratory energy to the workpiece. The mechanical assembly that holds the converter, booster, and horn is referred to as the actuator (stand, press). Its function is to bring the horn into contact with the workpiece, apply force, and retract the horn at the completion of the welding cycle.

118.2 POLYMERS

When a potential application for ultrasonic assembly is being considered, the polymer must be known because each polymer reacts differently to ultrasonic energy (see Table 118.1).

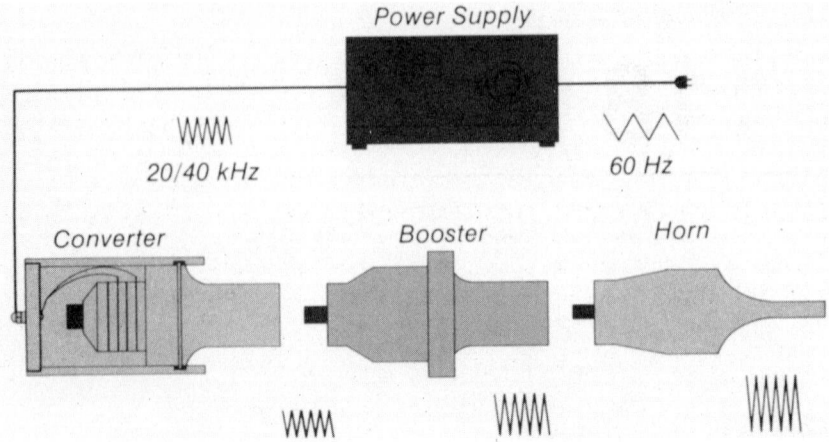

Figure 118.1. Components for supplying ultrasonic energy.

There are two basic polymer families; thermoplastics and thermosets. A thermoset polymer undergoes an irreversible chemical reaction when it is processed and cannot be remelted and reformed. A thermoplastic can be remelted and reformed with the introduction of heat and pressure; this makes it ideally suited for ultrasonic assembly.

In discussing thermoplastics, it should be noted again that each thermoplastic resin reacts somewhat differently to ultrasonic energy. To better understand why different thermoplastics behave as they do, it is necessary to examine the specific characteristics or properties that affect ultrasonic energy requirements. The most important of these will be discussed in detail.

TABLE 118.1. Ultrasonic Weldability[a] of Amorphous and Crystalline Polymers

Material	Excellent			Good			Fair	Poor	No
	10	9	8	7	6	5	4	3	2	1	0

Amorphous

Polystyrene

SAN-NAS-ASA

ABS

Phenylene-
 oxide based resins

Acrylic

Polycarbonate

Polysulfone

Crystalline

Nylon

Polyester

Acetal

Poly(phenylene
 sulfide)

Polypropylene

Polyethylene

Fluoropolymers

[a]Weldability is a function of joint design, energy requirements, amplitude, and fixturing.

118.2.1   Physical Structure

The physical structure of a polymer in the solid state (the way in which its molecules are arranged) helps determine a resin's physical properties, which include its melting and welding characteristics. The structure has a direct bearing on the energy transmission characteristics of the polymer. Polymer structure is classified as either amorphous or crystalline.

Amorphous polymers are characterized by a random molecular arrangement whose chains vary in length. Mechanical vibratory energy can generally be readily transmitted through this type of structure with little attenuation occurring.

Crystalline polymers have an orderly arrangement of molecules that repeat in precise patterns interspersed in an amorphous matrix. The crystals can be thought of as flat or coiled springs. This springlike structure has a dampening efect, energy is absorbed, making it more difficult to transmit vibratory energy through the part to the joint interface.

118.2.2   Melt Characteristics

In addition to different energy transmission characteristics, amorphous and crystalline polymers have different melt characteristics (Table 118.2).

With amorphous materials, one must consider their glass-transition temperatures (T_g). When heated, amorphous polymers gradually soften, passing from a rigid, glassy state through the glass-transition region. Once the temperature is above T_g, the polymer enters liquid flow. Resolidification is the reverse process. The energy requirements (specific heat) of amorphous polymers remain constant as the temperature increases (Fig. 118.2).

Crystalline materials have a sharp melting temperature (T_m) for their crystals. The polymer remains rigid until the crystals melt, after which it quickly reaches a liquid state, allowing welding to take place. A very high energy level is required to break down the springlike structure and achieve a true melt state. Once the application of ultrasonic energy into the material ceases, heat is dissipated instantaneously, producing quick resolidification.

Crystalline materials require more energy than amorphous materials. As the temperature approaches T_m, a much higher level of energy is required because of the high specific heat values (D_p) of these materials (Fig. 118.3).

118.2.3   Stiffness

In this case, stiffness refers to the dynamic shear modulus (G), or modulus of elasticity. In general, the higher the shear modulus (G), the more readily energy will be transmitted through the structure to the joint interface.

In most applications, the polymers to be assembled are the same. However, there are applications where parts made of different thermoplastic resins have to be welded together. Many factors affect the success of welding of dissimilar polymers, among them a similar melt temperature and like molecular structures. A similar melt temperature, no more tha a 40°F (22°C) difference, is a basic requirement. If there is a difference of greater than 40°F, one polymer can soften rapidly before the second polymer can begin to soften. With respect to like molecular structures, a close examination of compatible thermoplastics will reveal whether "like radicals" are present. If they are, it is more likely that the dissimilar polymers will be compatible. It is important to note, however, that these factors are general guidelines and do not necessarily ensure successful welding even if they are met.

Before discussing the weldability of the individual polymers, the difference between near-field and far-field welding must be understood. Near-field welding refers to a joint interface 1/4 in. (6 mm) or less from the area of horn contact, while far-field welding refers to a joint interface more than 1/4 in. (6 mm) from the horn contact area. The farther the distance from the area of horn contact to the joint, the more significance the energy transmission properties of the resin have on the results. In general, far-field welding is more successful

TABLE 118.2. Characteristics of Welding[a]

Material	Ease[b] Near Field	Ease[b] Far Field	Swaging and Staking	Inserting	Spot Welding	Vibration Welding
					Ultrasonic Welding	
Amorphous Resins						
ABS	E	G	E	E	E	E
ABS/Polycarbonate alloy (Cycoloy 800)	E–G	G	G	E–G	G	E
Acrylic[c]	G	G–F	F	G	G	E
Acrylic multipolymer XT-polymer)	G	F	G	G	G	E
Butadiene–styrene (K-Resin)	G	F	G	G	G	G
Cellulosics—CA, CAB, CAP	F–P	P	G	E	F–P	E
Phenylene-oxide based resins (Noryl)	G	G	G–E	E	G	E–F
Poly(amide-imide) (Torlon)	G	F				G
Polycarbonate[a]	G	G	G–F	G	G	E
Polystyrene (general purpose)	E	E	F	G–E	F	E
Rubber modified (high-impact)	G	G–F	E	E	E	E
Polysulfone[a]	G	F	G–F	G	F	E
PVC (rigid)	F–P	P	G	E	G–F	G
SAN-NAS-ASA	E	E	F	G	G–F	E
Crystalline Resins[d]						
Acetal	G	F	G–F	G	F	E
Fluoropolymers	P					F
Nylon[d]	G	F	G–F	G	F	E
Polyester (thermoplastic)	G	F	F	G	F	E
Polyethylene	F–P	P	G–F	G	G	G–F
Poly(methyl pentene) (TPX)	F	F–P	G–F	E	G	E
Poly(phenylene sulfide)	G	F	P	G	F	G
Polypropylene	F	P	E	G	E	E

[a]E = Excellent, G = Good, F = Fair, P = Poor.

[b]Ease of welding is a function of joint design, energy requirements, amplitude, and fixturing. Near field welding refers to joint ¼ in. (6.35 mm) or less from area of horn contact; far field welding to joint more than ¼ in. (6.35 mm) from contact area.

[c]Cast grades are more difficult to weld due to high molecular weight.

[d]Moisture will inhibit welds.

[e]Crystalline resins in general require higher amplitudes and higher energy levels because of higher melt temperatures and heat of fusion.

(Courtesy of Branson Ultrasonics Corp.)

with amorphous materials while near-field is generally recommended with crystalline materials.

118.2.4 Weldability of Amorphous and Crystalline Polymers

To gain a practical knowledge of weldability, it is necessary to examine the polymers individually. The accompanying chart and bar graph compare the relative weldability for the more common polymers. To simplify matters, the various resins have been categorized

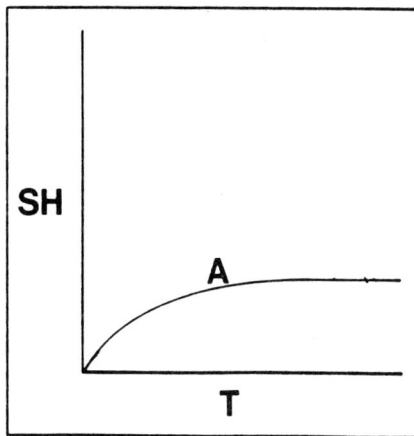

Figure 118.2. Energy requirements of amorphous polymers.

according to their physical structure, amorphous vs crystalline, and rated according to their near-field vs far-field characteristics.

118.2.5 Variables that Affect Weldability

External variables that influence weldability are hygroscopicity, mold-release agents, fillers, alloys, flame retardants, regrind, and pigments. These must be recognized and dealt with if repeatable production results are to be achieved.

118.2.6 Hygroscopicity

Hygroscopicity is the ability of a material to absorb moisture. Virtually all thermoplastics suffer to some extent from hygroscopicity, but nylon and, to a somewhat lesser degree, polycarbonate and polysulfone are thermoplastics most affected.

118.2.7 Mold-Release Agents

External release agents (such as zinc and aluminum stearates, silicones, and fluorocarbons) are commonly used to facilitate removal of parts from injection molds. These lubricants are

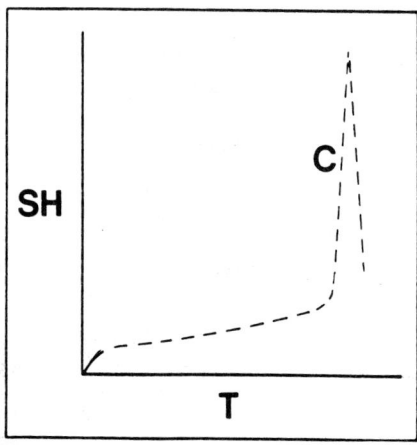

Figure 118.3. Energy requirements of crystalline polymers.

normally sprayed directly into the mold cavity and are transferred to the surface of the molded part. They act as a contaminant, impeding the buildup of heat caused by friction and thereby adversely influencing welding. In addition, the amount of mold release is not constant from piece to piece.

If mold release is present at the weld surface, it can sometimes be removed by immersing the part in soapy water or into a solvent cleaning bath (Freon being a popular solvent). If the use of a mold-release agent is absolutely necessary then a paintable grade of mold release agent is recommended.

118.2.8 Fillers

Physical properties of polymers can be altered through the use of such fillers as talc, glass, carbon, and calcium carbonate. The addition of fillers usually increases the stiffness of a polymer, which enhances the ability of the resin to transmit vibratory energy (weldability). However, it is important to remember that the relationship of the percentage of filler to improved weldability is favorable only within certain limits.

No special considerations are necessary when the filler content is below 10%.

Filler content in the range of 40% can leave insufficient resin at the joint surface, causing inconsistent results. Thus, joint design and molding conditions should be considered carefully.

118.2.9 Flame Retardants

The addition of a flame retardant inhibits ignition and/or modifies the melt characteristics of the polymer and can also adversely affect the weldability of the resin. The amount and type of retardant vary, depending on which test requirements are to be met. Flame retardants can be compensated for by modifying the joint design and/or altering the equipment set-up parameters.

118.2.10 Plasticizers and Impact Modifiers

The addition of plasticizers or impact modifiers increases the resiliency and flexibility of a polymer; by their very nature, these properties impede a material's ability to transmit mechanical vibratory energy.

When a highly plasticized or impact-modified grade is welded, the applied energy is absorbed in the material. This makes it necessary to increase the amount of energy supplied to the parts in order to achieve acceptable welding. Also, the greater the distance between the horn contact surface and the joint interface, the greater the potential for attenuation. Generally, near-field design is recommended to reduce energy absorption. It can be expected that impact grades of polymers will require higher energy and amplitude levels than non-impact grades.

118.2.11 Regrind

The addition of regrind to a thermoplastic usually does not adversely effect weldability. Since no foreign substances are introduced and assuming the material has not been degraded, few problems should be encountered.

118.2.12 Pigments

Liquid or dry pigments or colorants do not normally interfere with welding. Certain oil-base pigments can have a negative impact as will high concentrations of pigment. Occasionally, white pigment has caused problems. Titanium dioxide is the chief pigment used to obtain white parts. It is inorganic, and can act as a lubricant, thereby inhibiting welding.

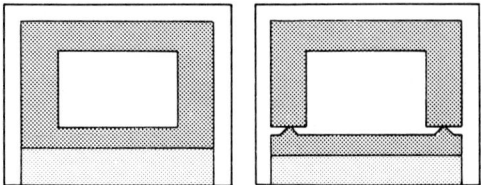

Figure 118.4. Welding.

118.3 ASSEMBLY TECHNIQUES

The ultrasonic process can be used to perform a number of different assembly techniques. These should be considered at a very early stage in the development of an application to ensure that acceptable results will be obtained.

118.3.1 Welding

Welding is the process of generating melt at the mating surfaces of two thermoplastic parts. When ultrasonic vibrations stop, the molten material solidifies and a weld is achieved. The resultant joint strength approaches that of the parent material; with proper part and joint design a clean cosmetic appearance is achieved. Ultrasonic welding allows fast, clean assembly without the use of any consumables (Fig. 118.4).

118.3.2 Swaging

Swaging is defined as mechanically capturing another component of an assembly by ultrasonically melting and reforming a ridge of plastic. Advantages of this method include speed of processing, less stress buildup, good appearance, and the ability to overcome material memory (Fig. 118.5).

118.3.3 Insertion

Insertion is the embedding of a metal component (such as a threaded insert) in a preformed hole in a thermoplastic part. High strength, reduced molding cycles, and rapid installation with no stress buildup are some of the advantages (Fig. 118.6).

118.3.4 Staking

Staking is the process of melting and reforming a thermoplastic stud to lock a dissimilar material in place mechanically. Short cycle times, tight assemblies, good appearance of the final assembly, and elimination of consumables are possible with this method (Fig. 118.7).

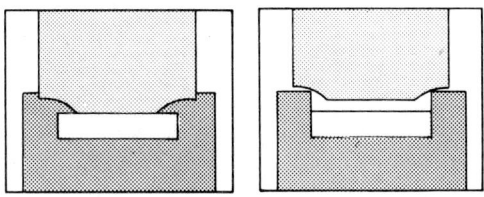

Figure 118.5. Swaging.

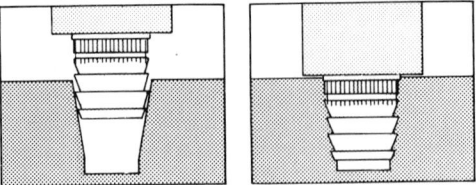

Figure 118.6.   Insertion.

118.3.5   Spot Welding

Ultrasonic spot welding is an assembly technique for joining two thermoplastic components at localized points without the necessity for preformed holes or an energy director. Spot welding produces a strong structural weld and is particularly suitable for large parts, sheets of extruded or cast thermoplastic, and parts with complicated geometry and hard-to-reach joining surfaces. A smooth surface appearance is produced (Fig. 118.8).

118.3.6   Textile/Film Sealing

Fabric and film sealing utilizes ultrasonic energy to join thin thermoplastic materials. Clear, pressure-tight seals in films and neat, localized welds in textiles may be accomplished. Simultaneous cutting and sealing is also possible. A variety of patterned anvils are available to provide both decorative and functional "stitch" patterns (Fig. 118.9).

118.3.7   Slitting

Slitting is the use of ultrasonic energy to separate and edge-seal knitted woven and nonwoven synthetic materials as well as thermoplastic films. Smooth, sealed edges that will not unravel are possible with this method. There is no "bead" or buildup of thickness on the slit edge to add bulk to rolled goods (Fig. 118.10).

118.4   ADVANTAGES/DISADVANTAGES

For ultrasonic assembly, the advantages far outweigh the limitations. The major advantages are

- Very fast cycle times: Actual welding cycles are usually less than 1 second with total manual cycles (loading, welding, and unloading) ranging from 3 to 5 seconds.
- Very cost competitive: When the cost of ultrasonic equipment is compared to the cost of equipment for other assembly processes (spin welding, hot plate welding, or dielectric sealing, for example), ultrasonics compares favorably.
- No consumables required: Adhesives, glues, solvents, and mechanical fasteners all cost

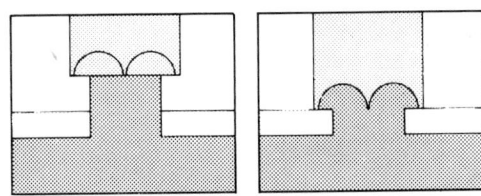

Figure 118.7.   Staking.

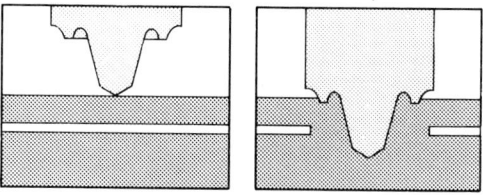

Figure 118.8. Spot welding.

something and some are labor intensive as well. These consumables are not required with ultrasonic assembly.

- Clean, safe, energy efficient: No messy consumables are required. The toxic dangers of solvents are avoided. Ultrasonic energy is supplied only on demand, keeping energy costs at a minimum.
- Flexibility: Ultrasonic equipment can be operated cyclically or continuously. The equipment can be used to perform many types of assembly operations, as noted above. Better equipment utilization is provided because a welder does not have to be dedicated to one operation.
- Design latitude: Some processes lock the designer into one assembly approach. Ultrasonic offers the designer much greater latitude to choose from a number of different techniques. One primary example is that of an instrument panel where staking, spot welding, or straight welding can all be used.

As with all techniques, certain limitations exist:

- Part size: The size of parts that can be successfully assembled ultrasonically are directly related to the size of horns available. The practical size limits are 6 in. (152 mm) in diameter and 5-by-5 in. (127-by-127 mm) square, although there are many exceptions [9-by-12 in. (229-by-305 mm)] rectangular, for example.
- Part geometry: The ability to make a horn to match a contoured or odd-shaped part successfully can also be a limiting factor. When a horn workface is relieved, uneven amplitude across the face can result. This may produce a nonuniform transfer of energy to the joint interface.
- Far-field welding: As the distance from the horn contact area to the joint interface increases, the ability to achieve acceptable results decreases. The polymer being welded also plays a large part in influencing the results. The designer should consult with someone knowledgable in ultrasonic assembly before designing a part for welding.

118.5 DESIGN

When a part is either too difficult, complex, or expensive to be molded as one piece, design engineers can often turn to ultrasonic assembly. Most components are joined together

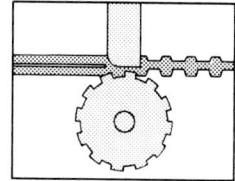

Figure 118.9. Textile/film sealing.

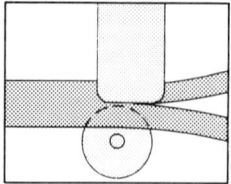

Figure 118.10.   Slitting.

through the welding process, but other ultrasonic assembly techniques can also be utilized. Once the piece part requirements are known, the ultrasonic technique is chosen, and the weld design is optimized.

118.5.1   Welding

If the application requires a hermetic seal or a strong structural weld, the designer should consider ultrasonic welding. Whatever joint design is chosen it should meet the three basic requirements for a sound joint design:

- A small initial contact area should be established between the mating parts. Less applied energy (and therefore time) is required to start and complete the "melt down," or collapse. A reduction in the time the vibrating horn remains in contact with the part reduces the potential for scuffing of the part. Overall, less material will be moved during welding, which will result in less flash.

- A means of alignment must be included so that the mating parts do not become mis-aligned during the welding operation. Alignment pins and sockets or tongues and grooves are often used. The horn and/or fixture should not be used to provide part alignment because some form of marking usually results if this is attempted.

- Uniform contact between the parts should be provided, which means that the mating surfaces should be in intimate contact along the entire joint. The joint area should also be in a single plane, if possible. This will ensure uniform energy transfer, which, in turn, will produce a uniform (and therefore controllable) melt down, reducing the potential for flash.

The designer must next consider the thermoplastic material to be assembled. Amorphous thermoplastics generally require an energy-director joint design; crystalline thermoplastics necessitate the use of a shear or interference joint design. Lastly, the functional requirements of the application should help determine the final part configuration.

Figure 118.11**a** illustrates a simple butt joint modified with an energy director showing the proportions desired before the weld and indicating the resultant flow of material during the weld cycle. Some flash is produce with this design; however, the amount is usually minimal. Parts should be dimensioned to allow for the dissipation of the energy director material throughout the joint area, as illustrated. Practical considerations suggest a minimum height of 0.010 in. (0.25 mm) for the energy director for easy-to-weld resins. Larger energy directors may be necessary for certain high-energy resins; that is, those with low stiffness or high melt temperatures amorphous resins such as cellulosics, polycarbonate, or polysulfone. For them, a minimum height of 0.020 in. (0.5 mm) is generally preferred.

Figure 118.11**b** shows the same basic energy director design with a skirt. This skirt acts to conceal flash on the appearance side of the finished assembly and provides a means of aligning the mating parts.

A typical mistake with the energy director design is beveling one joint face at a 45° angle. Figure 11**c** shows the result of this design. This is to be avoided!

Figure 118.11**d** illustrates a step joint with an energy director used for alignment for applications where excess melt or flash on the appearance side would be objectionable.

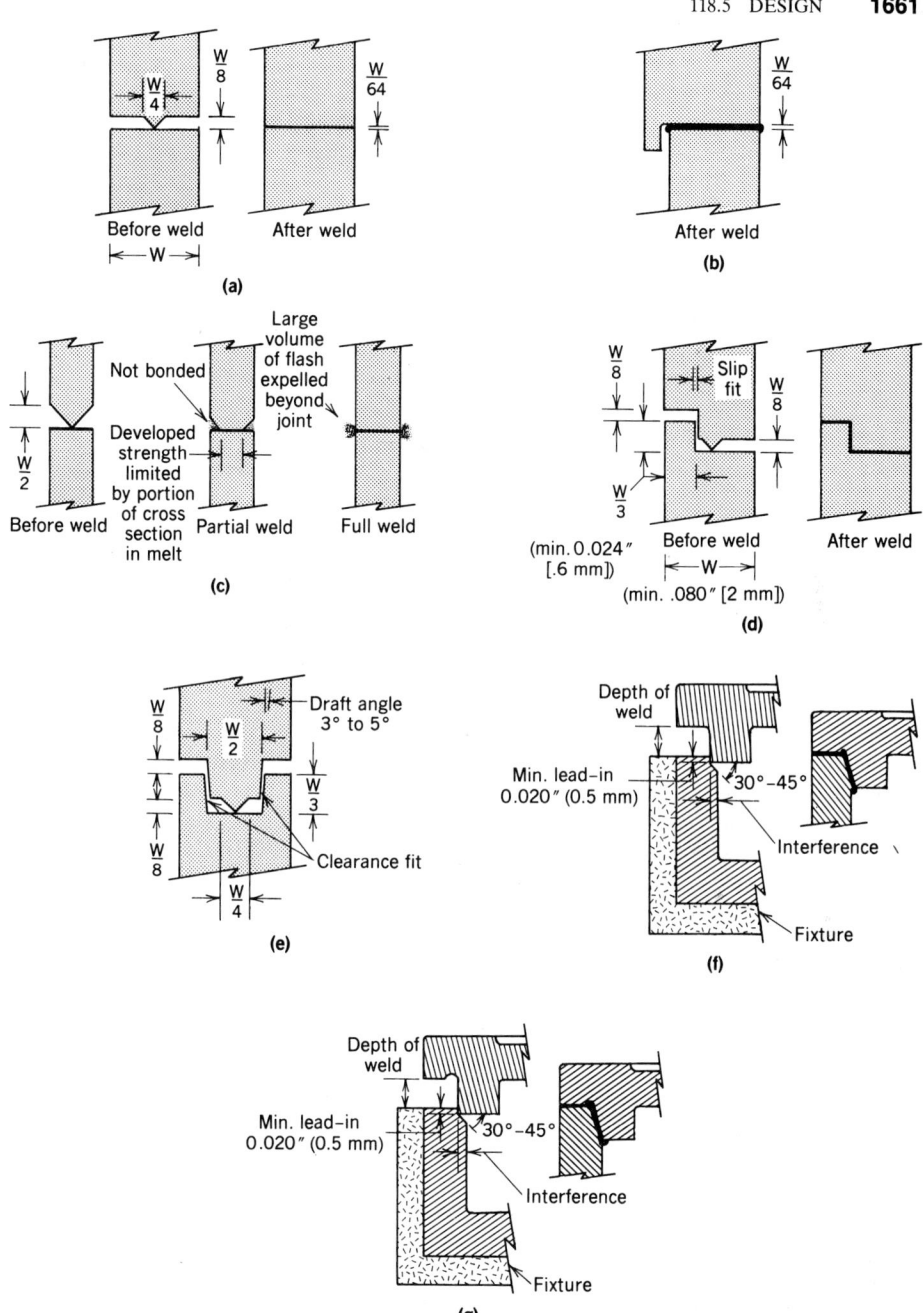

Figure 118.11. (**a**) Simple butt joint moidified with an energy director. (**b**) 11a design with skirt. (**c**) Typical mistake—beveling one joint face at a 45° angle. (**d**) Step joint with energy director. (**e**) Tongue-and-groove joint with energy director. (**f**) Shear joint for hermetic seal and crystalline materials. (**g**) Lead-in and flash trap for 11**f**.

A tongue-and-groove joint with an energy director (Fig. 11**e**) is used primarily to prevent flash on both sides of the welded surface. The need to maintain clearance on both sides of the tongue, however, makes this part more difficult to mold. Draft angles can be modified in accordance with good molding practices, but interference between the elements must be avoided.

Figure 11**f** shows the shear joint used when a strong hermetic seal is needed; it is especially recommended for crystalline resins such as nylon, acetal, thermoplastic polyester, polyethylene, and poly(phenylene sulfide). Since crystalline resins change rapidly from a solid to a molten state over a narrow temperature range, an energy director type of joint is not preferred because the molten resin from the director will rapidly solidify before it is able to fuse with the abutting surface. With the shear joint, welding is accomplished by first melting the small, initial contact area and then continuing the melt, with a controlled interference along the vertical walls as the parts telescope together. A lead-in is required for self-locating, and a flash trap can be incorporated if necessary (Fig. 11**g**).

The strength of the welded joint is a function of the vertical dimension of collapse of the joint (that is, the depth of the weld) and can be adjusted to meet the requirements of the application. If the joint strength is to exceed the wall strength, a depth of 1.25 times the wall thickness is recommended.

Table 118.3 gives a general guideline for interference and part tolerance in relation to maximum part dimension. It should be noted that a shear joint is not recommended for parts with a maximum dimension of 3.5 in. (89 mm) or greater. This is caused by the difficulty of holding the molding tolerances necessary to obtain consistent results. An energy director type joint is recommended for parts falling into this category.

One important consideration often overlooked when parts are designed for ultrasonic assembly is the horn contact surface on the part. To achieve consistent results, the horn must be in intimate contact with the part directly over the joint area. With large of odd-shaped parts this is sometimes difficult to achieve. One proven method of ensuring proper horn contact is to design the part with a raised surface (Fig. 118.12).

There are also a number of other design considerations that should be addressed if acceptable results are to be achieved.

As in any design, sharp corners are areas of localized stress (stress risers). When a molded part with stress risers is subjected to the high-frequency mechanical energy, damage such as fracturing or burning may occur in the high-stress areas. This can be remedied by having a generous radius on corners, edges, and junctions.

Holes or voids in the part being contacted by the horn create an interruption in the transmission of the applied ultrasonic energy. Little or no welding will occur directly beneath an opening. If no openings are located in the mating part, it is recommended that it be contacted by the horn.

Appendages, tabs, or other details molded on to the interior or exterior surfaces of the

TABLE 118.3. General Guideline for Interference and Part Tolerance in Relation to Maximum Part Dimension

Maximum Part Dimension	Interference per Side (Range)	Part Dimension Tolerance
Less than 0.75 in. (18 mm)	0.008 in to 0.012 in. (0.2 to 0.3 mm)	±0.001 in. (±0.025 mm)
0.75 in. to 1.50 in. (18–35 mm)	0.012 in. to 0.016 in. (0.3 to 0.4 mm)	±0.002 in. (±0.05 mm)
Greater than 1.50 in. (35 mm)	0.016 in. to 0.020 in. (0.4 to 0.5 mm)	±0.003 in. (±0.075 mm)

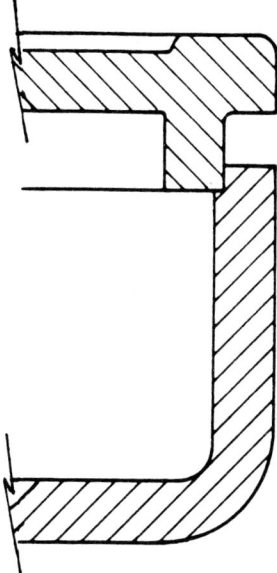

Figure 118.12. Raised horn contact surface.

molding can be adversely affected by the applied mechanical energy, resulting in fracturing. The following recommendation can minimize or eliminate this problem:

- The addition of a generous radius to the area where appendages intersect the main part
- Dampening of vibration with proper fixturing.

Diaphragming is an oil-canning effect with related burn-through that typically occurs in thin-walled sections of the part contacted by the horn. This can be corrected by one or a combination of the following design modification:

- Thicker wall section
- Internal support ribs
- Cold well or slug

118.5.2 Staking

Staking is the controlled melting and reforming of a plastic stud or boss to capture or lock another component, such as a printed circuit board or display window. The plastic stud (or boss) protrudes through a hole in the component to be staked in place. Ultrasonic energy is transferred to the head of the stud, which melts and flows, filling the volume of the cavity in the horn face to produce the finished head. Proper stud design produces optimum strength and appearance with minimum flash. The popularity of staking with ultrasonics may be attributed to its inherent advantages:

- Short cycle time
- Tight final assembly with virtually no tendency of springback in the finished head
- Ability to stake multiple studs simultaneously

- Elimination of consumables such as mechanical fasteners
- The stud is a nonconductor

Several configurations for stud/horn cavity design are available. Design selection depends primarily on the requirements of the end application and the physical size of the stud(s) being staked. The principle of staking is the same for most configurations: the area of initial contact between the horn and stud should be kept to a minimum, concentrating the energy to produce a rapid reforming.

The standard profile stake (Fig. 118.13**a**) is most commonly used for studs having a diameter between 1/16 and 1/8 in. (1.6 and 3.2 mm). The top of the stud is flat; the melt is initiated by a sharp point in the horn cavity. The head produced is clean, neat, and strong. The standard design is ideal when staking nonabrasive thermoplastics, as it is both rigid and resilient.

The dome stake (Fig. 118.13**b**) is recommended for a stud having a diameter of 1/8 in. (3.2 mm) or less or when multiple studs are to be staked. The top of the stud should be tapered (with a 90° included angle for best results), at the point at which material melt initiates. The head produced is clean, neat, and strong, satisfying the requirements of most staking applications. The alignment between the stud and horn is not as critical as it is with the standard profile. The dome configuration is also less susceptible to wear, which can sometimes be a problem with the standard profile, especially when abrasive materials are being staked.

The hollow stake (Fig. 118.13**c**) is highly recommended for studs with diameters of 5/32 in. (4.0 mm) or greater. The hollow stud reduces the volume of material to be staked, thus reducing the cycle time required to reform the stud. Hollow studs (or bosses) also offer certain molding advantages, such as the prevention of sinks and internal voids. The head produced with this method is strong and allows for disassembly for repair purposes in cases where the formed head can be easily removed, the assembly repaired, and a self-tapping screw used to reassemble the part.

The knurled stake (Fig. 118.13**d**) is designed for simplicity and a rapid rate of assembly. It is recommended for use when appearance and strength are not critical. This design does not need a dimensioned horn cavity and allows for multiple stakes without concern for precise horn alignment or stud diameter. Toy and packaging applications are prime areas of use.

The flush stake (Figure 118.13**e**) is recommended for applications that require that the formed head not protrude above the top surface of the part being contained. The tapered stud design used in dome staking is recommended, along with a flat-faced horn.

One very important design consideration is the use of a generous radius at the base of the stud. If this detail is overlooked, cracking can and will occur!

118.5.3 Insertion

Insertion is the process of encapsulating a metal component in a plastic part. With ultrasonics, this is done either by contacting the metal insert and driving it into the plastic part or by contacting the plastic part and driving it onto the insert (Fig. 118.14**a**). Typical applications include computer terminals, appliance housings, and electrical connectors.

A hole, slightly smaller in diameter than the insert it is to receive, is preformed either by being molded or drilled into the plastic part. This hold provides a certain amount of interference and also serves as a means of alignment, guiding the insert into place. The metal insert is usually designed with knurls, flutes, undercuts, or threads to resist loads imposed on the finished assembly. Ultrasonic energy travels through the driven component to the interface of the metal insert and plastic. Frictional heat, caused by the insert vibrating against the plastic, causes the plastic to momentarily melt and flow into the undercuts, and

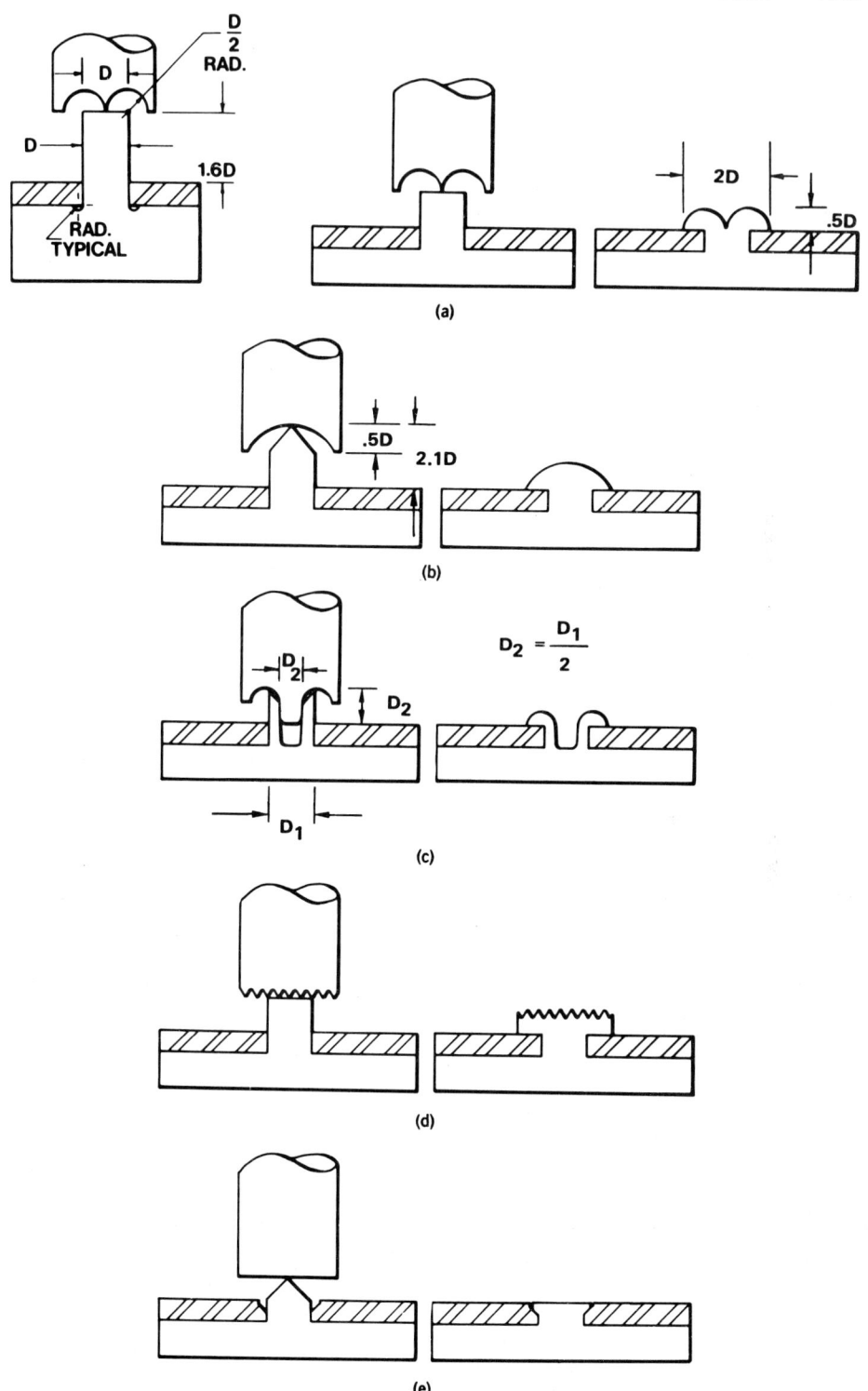

Figure 118.13. (**a**) Standard profile stake. (**b**) Dome stake. (**c**) Hollow stake. (**d**) Knurled stake. (**e**) Flush stake.

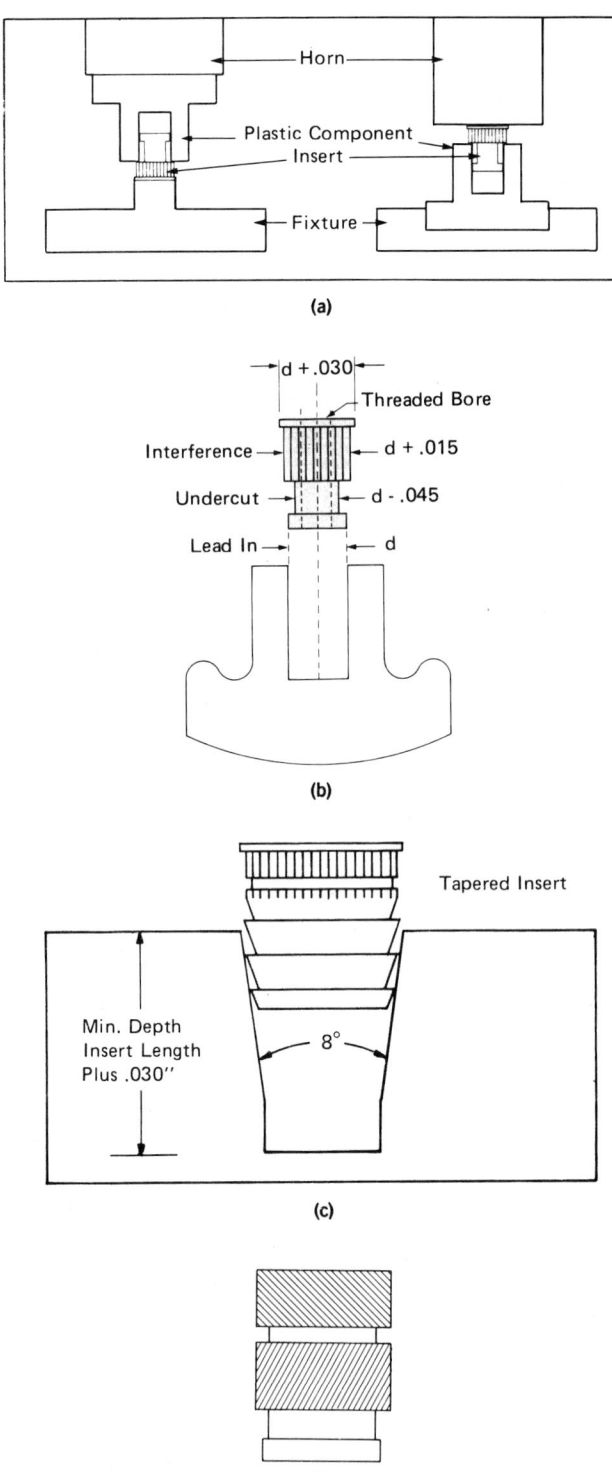

Figure 118.14. (**a**) Insertion of metal parts. (**b**) Multiple undercuts in metal inserts. (**c**) Long axial knurls on metal inserts. (**d**) Opposing diagonal knurls.

so on, permitting the insert to be driven and held in place. When ultrasonic energy ceases, the plastic resolidifies and the insert is permanently secured.

The popularity of inserting with ultrasonics may be attributed to its inherent advantages:

- Short cycle times
- Minimal induced stress around the insert
- Reduced molding cycles through the use of automated cycle
- Ability to drive in miltiple inserts simultaneously

The functional characteristics of an application usually determine the insert design. A sufficient volume of plastic must be displaced to fill the undercuts, flutes, knurls, threads, and/or contoured areas of the insert. This controlled flow of plastic locks the insert in place, producing the strength necessary for the application.

A typical insert designed for tensile strength should have multiple undercuts to provide maximum pullout resistance (Fig. 118.14**b**). The hole may also be tapered to accept a tapered insert. This permits accurate positioning of the insert and usually reduces installation time. For maximum torque strength, an insert should have long axial knurls to provide maximum rotational resistance (Fig. 118.14**c**). An insert designed to prevent jack-out should have opposing diagonal knurls and multiple undercuts to provide maximum jack-out torque resistance (Fig. 118.14**d**).

In all cases, the acceptance hole should be designed so that the insert does not bottom out. It should also be of sufficient depth so that threaded screw does not reach the bottom of the final assembly.

Other considerations often overlooked are

- Final position of the insert: After seating, the top of the insert should be flush or slightly above the surface of the plastic part (Fig. 118.15).
- Boss diameter: The wall thickness of a boss receiving an insert should be 0.5 to 1.0 times the insert diameter to prevent wall fracture.

Most insert applications involve the installation of standard threaded-bore inserts, but other metal components can be ultrasonically inserted. These include eyeglass hinges, machine screws, threaded rods, roll pins, metal shafts, electrical contacts, and terminal connectors.

118.5.4   Textile and Film Welding

Ultrasonic technology can be used to bond film, woven goods, and nonwoven goods in both wide-roll and individual products (Table 118.4). The success of this technology is also pro-

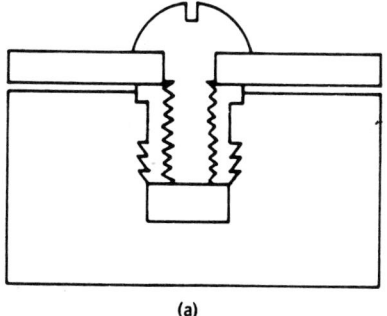
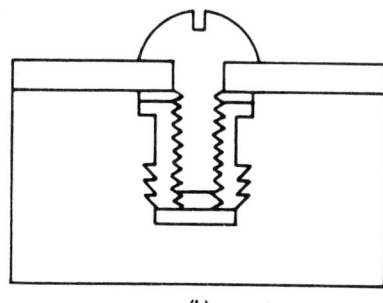

(a)   (b)

Figure 118.15.   Final position of inserts. (**a**) Preferred; (**b**) poor.

TABLE 118.4. Characteristics, Ease of Weldability[a]

Material	Characteristics	Uses	Woven	Nonwoven	Knitted	Coated Materials	Laminates	Film
Polyester	Resistant to most organic solvents and chemicals, strong, abrasion resistant	Sheets, filters, clothing, quilts, disposable garments, packaging materials, recording tapes, mattress pads, conveyor belts, sails, fiberfill, laminates	G	E	G	E	E	E
Nylon	Resistant to most organic chemical solvents, strong, abrasion resistant, elastic	Hook and loop material, carpet, lingerie, filters, cooking bags, meat bags, rainwear, camping gear, seat belts	E–G	E–G	G	G	G	G
Polypropylene	Good chemical resistance and wicking characteristics	Carpet backing, bagging, upholstery, tents, outdoor furniture, snack food packaging	G	E	G	G	G	G
Polyethylene	Tough, flexible, inexpensive, high density polyethylene welds best	Packaging films, laminates, disposable clothing		E		G	G–P	F–P
PVC	Resistant to water and many chemicals, good insulating characteristics; NOTE: the addition of plasticizers to impart flexibility in the material inhibits weldability	Films, shrink packaging, tarps, outdoor furniture	G–P			G–P	G–P	G–P
Acrylic	Unaffected by most detergent solutions, inorganic acids and alkalines. Attacked by aromatic hydrocarbons, esters, and ketones.	Knitting yarns, filters, awnings, blankets, sportswear	F–P	F–P	F–P	F–P		F–P
Urethane	Thermoplastic urethane is weldable	Rainwear, sponges, filters		G		E–G		

[a]E = Excellent, G = Good, F = Fair, and P = Poor.
(Courtesy of Branson Ultrasonic Corp.)

viding a foundation for using ultrasonic energy to laminate and even form nonwoven fabrics and products.

The potential that exists with this bonding process is a direct result of its inherent advantages:

- Energy is only expended at the precise location and area of the bond site, which has produced speeds approaching 500 ft/min (152 m/min) for certain applications.
- Consumables such as chemical binders, needles, and thread are not required.
- Excellent bond integrity is obtained because heat energy is generated and released within the fibers, thus minimizing material degradation. With thermal bonding, for example, heat energy has to be conducted through the fibers to be bonded, thereby requiring excessive heat energy, which can cause degradation.

Successful production applications include the Pinsonic process, where webs of nonwoven fabric, fiberfill, and woven shell fabric are laminated into bedspreads and mattress pads, and seals on sanitary napkins processed on a rotary drum, surgical face masks and caps, and various types of absorbent pads.

118.6 APPLICATIONS

Successful applications are the bottom line when it comes to applying the design guidelines that have been discussed. Some examples for each assembly technique follow.

118.6.1 Welding

- Thermal mugs, cups, and bowls
- Starbird and E.T. toys
- Automotive tail lamps and side markers
- Medical filters and reservoirs
- Home appliances, including coffee makers and vacuum cleaners.

118.6.2 Swagging/Forming

- Closed-end tubes
- Disposable batteries
- Digital thermometer shrouds.

118.6.3 Insertion

- Eyeglass frame hinges
- Battery cover of the Swatch disposable watch.

118.6.4 Staking

- Video tape reels
- Thermos jug lids
- PC boards to calculator and computer housings.

118.6.5 Spot Welding

- Automotive instrument panels and arm rests
- Recreational vehicles and golf carts.

118.6.6 Textile and Film Welding

- Pinsonic bedspreads, thermal drapes
- Drink and food pouches
- Ribbon cartridge splicing
- Diapers, sanitary products, incontinent devices.

116.6.7 Slitting

- Finished goods including tablecloths, handkerchiefs, bed sheets, and flags.

118.7 PROCESS VARIABLES

A basic understanding of weld energy is helpful because the success of ultrasonic assembly is based on its dissipation in the form of heat at the joint or bond area.

Weld energy is the product of the average power dissipated in the joint and the weld time ($E = P \times t$). Power is applied as a function of velocity and force. Velocity is defined as the product of amplitude and frequency (vel = amp $\times$ freq). Because the frequency is considered a constant, velocity becomes strictly a function of amplitude. Force, for the purpose of this discussion, refers to clamping force and is dependent on the air cylinder size, and gauge pressure.

It is apparent that amplitude, clamping force, and weld time are the main variables that require operator attention.

Amplitude, which is defined as peak-to-peak displacement of the horn at its workface, is determined by the choice of booster and the size and shape of the horn itself.

Clamping force is generated by the pneumatic system (actuator) and is easily adjusted and controlled by gauges and valving.

Weld time, the length of time ultrasonic energy is applied to the workpiece, is controlled by the programmer. Adjustment of weld time is accomplished via potentiometer adjustment or keypad entry.

In that energy applied to the worpiece is influenced by amplitude, clamping force, and weld time, a change to any or all three will result in a change to the amount of energy to the workpiece. Amplitude is the most important of these variables.

With this understanding of energy, one must next consider the actual mode of welding to be used for the process. There are five basic modes:

- Continuous: Ultrasonics are either on or off. This mode is utilized with a handgun or for constant sonics applications. External programming is provided by the end user.
- Time: Ultrasonics are turned on for a predetermined period of time. The amount of energy into the workpiece will vary from cycle to cycle, depending on part design, part tolerances, horn coupling, etc.
- Energy: Ultrasonics are applied to the workpiece until a predetermined energy level is reached. Weld time is automatically adjusted to ensure that the preset energy level is achieved.

- Time/energy compensation: A specific weld time along with energy limits are set. If the total energy into the workpiece reaches the upper energy limit before the specified weld time, ultrasonics are turned off. If energy into the workpiece does not reach the lower energy limit in the specified weld time, the time is extended until the lower energy limit is reached. Time will be extended up to 50 percent of the original weld time setting. If the lower energy isd still not reached, the cycle will short.
- Distance: Ultrasonics are turned on until a predetermined melt down is achieved. This mode can also be used to achieve a consistent, finished part height. Figure 118.16 shows an "intelligent" actuator used when this mode is required.

The mode of welding that is chosen is based on the end requirements of the application. Each mode has associated alarms, which can signal suspect parts—that is, parts that do not fall within preset mode parameters.

118.8 EQUIPMENT

Equipment selection for ultrasonic assembly is an important consideration both in terms of cost and end results. There are three steps involved in choosing the proper equipment: configuration, power level, and level of controls.

A basic ultrasonic system can be either integrated or modular in construction. The configuration chosen is usually based on the mode of operation—manual, semiautomated, or fully automated.

For manual operation, an integrated welder is recommended (Figure 118.17). An integrated welder is a self-contained unit with the power supply, actuator, and acoustic components packaged as a single entity. The benefits associated with an integrated welder include lower capital expenditure, ease of service, and reduced floor space requirements.

When part volumes begin to increase and a manual system cannot meet production requirements, some form of semiautomation is usually considered. Popular standards include rotary indexing tables and in-line conveyor systems. Modular systems are recommended for these situations (Fig. 118.18). An actuator is mounted on the table or conveyor with the power supply remotely located, usually at or near the system controls. The benefits of a modular scheme include quick changeover of individual components, flexibility in changing power levels, and a smaller envelope at the point of operation.

For high-volume production, totally automated syustems have become the norm. Basic ultrasonic components have gained in popularity for use in these systems. A converter, booster, and horn are mounted stationary with the workpiece moved to the horn, or these components are integrated with an end-user supplied slide, robot, or cam-driven mechanism. As with semiautomated systems, the power supply is remotely mounted. Individual components are chosen when space is at a premium, special mechanics are required, or system cost is critical. Packaging and textile OEMs as well as custom machine builders design their systems around ultrasonic components.

Once the proper configuration has been determined, the next step is to determine the appropriate power level. The power level is determined by the application; however, there are other considerations, such as cycle rates and material, that influence the power level. Power supplies are available from 150 watts to 3200 watts at 20 kHz and from 150 watts to 700 watts at 40 kHz.

The last step in the equipment selection process is choosing the level of controls required. Controls that are integrated into the power supply, range from the basic (end-user supplied) to analog control to microprocessor-controlled power supplies (Fig. 118.19). The level of controls is determined by the degree of process control required. Controls can be supplied by the end-user or an ultrasonic equipment manufacturer.

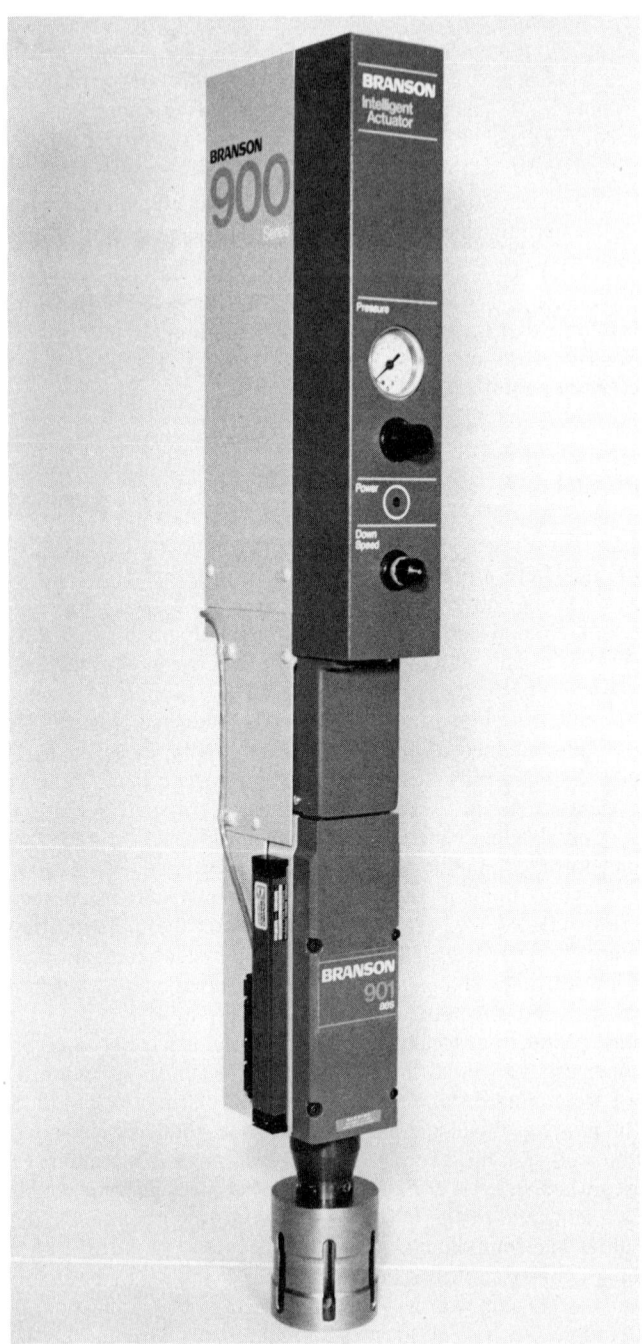

Figure 118.16.   Intelligent actuator.

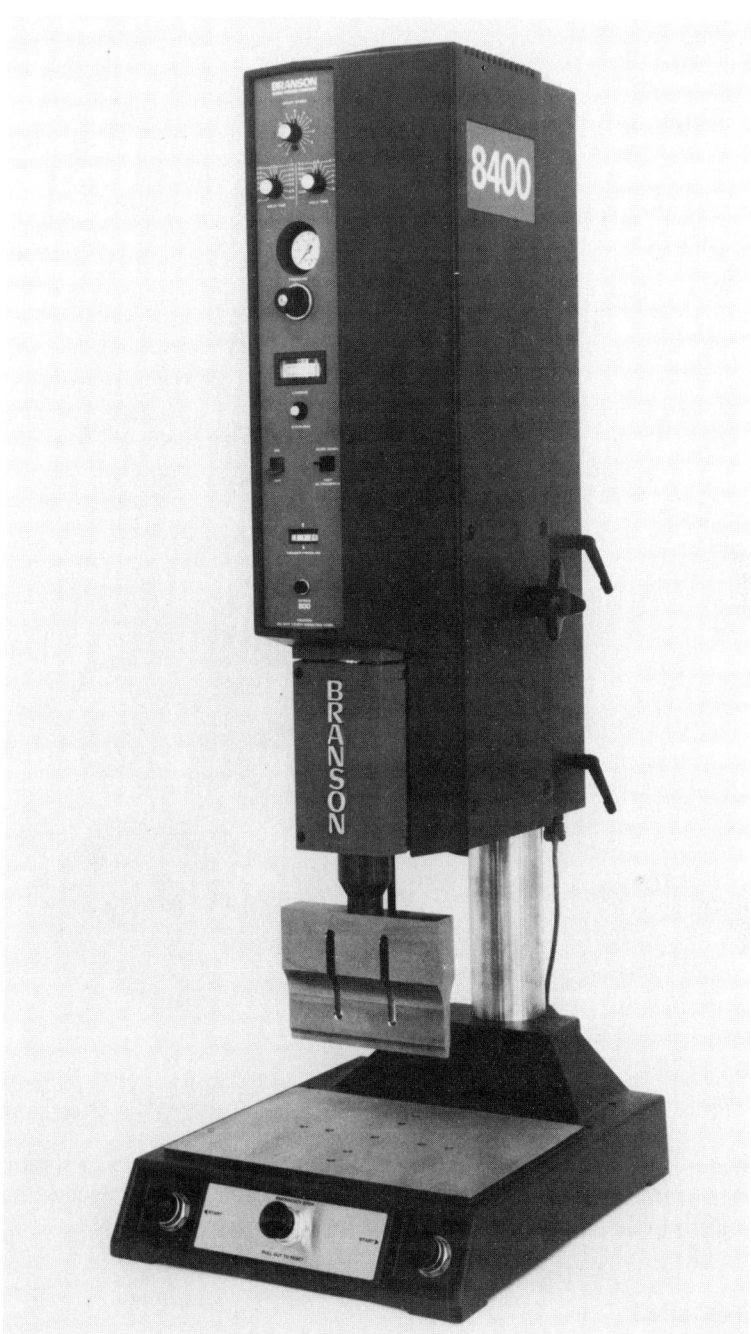

Figure 118.17. Integrated welder.

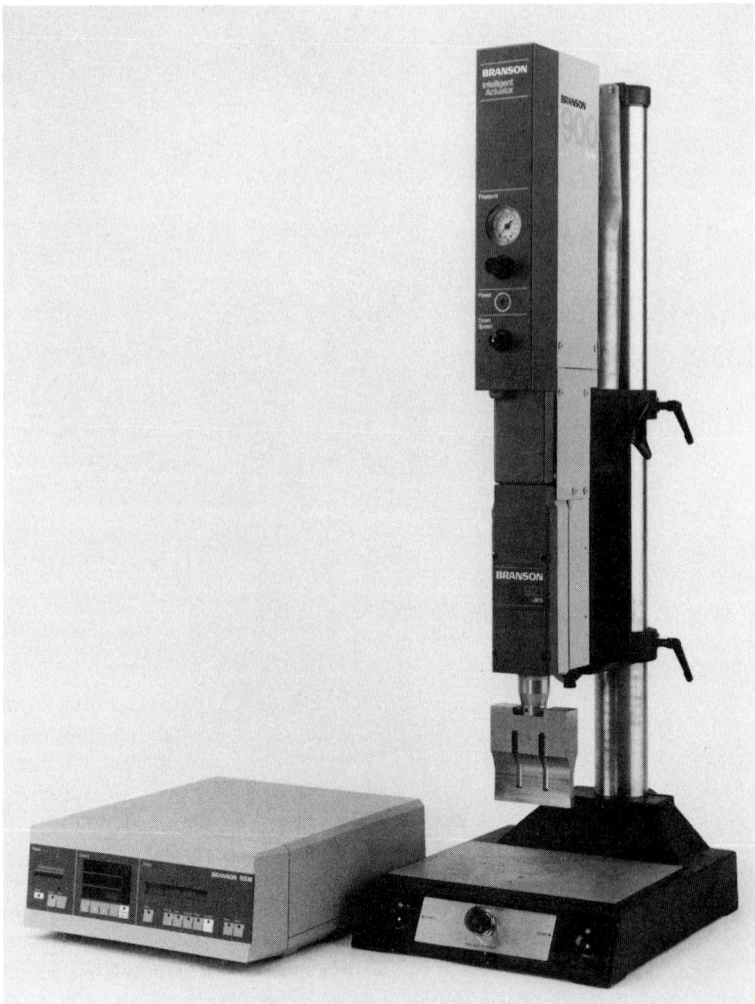

Figure 118.18. Semiautomated modular system.

Most major ultrasonic equipment suppliers offer engineering support to assist in equipment selection. Branson Ultrasonics Corporation, for example, offers engineering design and after-sales service support on a worldwide basis.

118.9 TOOLING

In many applications, tooling is the most critical aspect of the setup. Tooling consists of an ultrasonic horn and a fixture.

The horn is usually a one-half wavelength long metal bar dimensioned to be resonant at the applied ultrasonic frequency. When a horn is driven at its resonant frequency, the ends will expand and contract longitudinally about its center (nodal point), alternately lengthening and shortening the horn. The peak-to-peak movement at the horn face (end) is referred to as amplitude. As stated earlier, amplitude is the primary variable in determining output

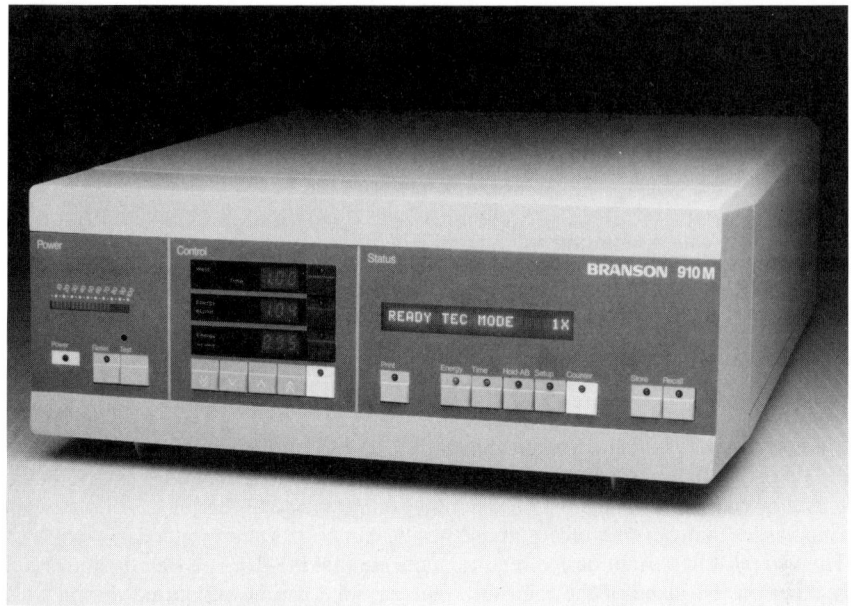

Figure 118.19. Microprocessor controls.

power to the workpiece and is an important consideration in the design of a horn. Gain, the ratio of output to input amplitude, is a function of the cross-sectional area between the input and output sections of the horn. As an example, if the cross-sectional area of the out end is less than the input end, the gain ratio will be greater than 1.0 and the corresponding amplitude will be increased. There are many different horn shapes, each designed to alter the gain to a specified ratio.

Materials used for horns have two main characteristics—low acoustical impedance (low losses at ultrasonic frequencies) and high fatigue strength. Aluminum, titanium, and certain steel alloys are commonly used in the manufacture of ultrasonic horns.

Aluminum is used for low-amplitude or large-part applications. Titanium is recommended for high-amplitude applications and exhibits better wear characteristics than aluminum. Heat-treated steel alloy horns offer high wear resistance but have high losses, limiting their use to low-amplitude, high wear applications such as insertion. Carbide-faced titanium horns are recommended for high amplitude, high wear applications.

Fixturing, an aspect of ultrasonics assembly that is often overlooked, has two main purposes—to provide proper alignment between the workpiece and horn and to provide the proper support to the joint area. Other factors, such as polymer, part geometry, and wall thickness, affect transmission of mechanical energy to the joint interface and should be considered when designing a fixture.

Fixturing materials can be resilient or rigid in nature. The type of assembly operation and the polymer being assembled influence the material choice. Rigid fixturing can be utilized for most applications. Resilient materials are used when marking or contoured parts present problems.

118.10 SUMMARY

Properly utilized, the ultrasonic assembly process is a powerful tool. When care is taken in polymer selection, part design, and equipment procurement, the process provides an efficient, consistent, and trouble-free method of asssembling plastic parts. This method has truly earned its place in manufacturing over the last 20 years.

118.11   VIBRATION WELDING

Large and irregularly shaped parts cannot be assembled using ultrasonics. A variation of friction welding called vibration welding may be used. Currently (1989) the largest size is about 16 × 22 in. (400 × 560 mm).

Frictional heat is generated by clamping two plastics parts together, holding one stationary, and vibrating the other through a reciprocating displacement of 0.040–0.160 in. (1–4 mm) at a frequency of 120–240 cycles per second. When the material is melted, the parts are aligned and pressure is maintained until the plastic solidifies giving a bond strength approaching that of the plastic itself.

Since vibration welding does not depend on the transmission properties of the plastic itself, almost all thermoplastics made by any process (including foaming) can be welded as long as they can vibrated relative to each other on the plane of the joint. All the dissimilar thermoplastics that can be welded ultrasonically, can be vibration welded.

The two most important design requirements are that the parts be free to vibrate against each other on the plane of the joint and that the joint can be supported during welding. The basic joint design is the simple butt joint. If excess oozing of melted plastic is unacceptable, than other joints can be used. The thinnest wall section practical is 0.32 in. (8 mm). Walls this thin need support during welding.

119

MACHINING OF PLASTICS

John Hull

Vice Chairman, Hull Corp., Hatboro, PA 19040

One of the many virtues of plastics is that they can often be processed directly into final form with little or no additional finishing. But there are situations when machining operations *are* required after processing. One example is the turning required on thermoset disk brake pistons and cylinders to ensure perfect roundness. Another is the insertion of threaded metal lugs in automatically molded connectors. Gate scar removal on injection and transfer molded parts, and cutoff of extruded pipe and profiles are other common machining requirements. Most thermoformed parts require post machining. Not to be overlooked are prototype requirements, where total machining of one or more parts is an economical way to evaluate material selection or part design prior to their going into production. Fabricating plastics, particularly clear acrylic, combining their decorative and functional properties is a business in itself.

When machining is required, the equipment and components used are very similar to those for working with wood or soft metals. The differences between tools used for plastics and those used for other structual materials arise from the intrinsic nature of plastics. Thermoplastics soften at elevated temperatures. Styrenic and vinyl materials, for example, become quite soft at 212°F (100°C) while high-performance engineering plastics soften at temperatures of 392°F (200°C) and higher. Even phenolics, and other common thermosets, noted for their resistance to deformation under heat, have upper temperature limits. But these are not reached in typical machining operations.

119.1 CONTROLLED HEAT IS ESSENTIAL

Most machining operations—sawing, turning, grinding, etc,—generate localized heating in the materials being machined. The concentrated heat leads to a flowing or deforming of the plastic, which presents the danger of part and tool damage through tearing, tool binding, galling, smearing, or other damage. After machining, the part may not conform to dimensions, may be cosmetically altered, and may even have lost some of the intended physical or electrical properties in those overheated localities. The low thermal conductivity of plastics increases the problem of heat dissipation. The basic principle to be observed in plastics machining, therefore, is to prevent generated heat from approaching the melt temperature of the material—a temperature that will be unique to each polymer, and may even be different within families of any one polymer, depending on additives used.

For that reason, the recommended feeds and speeds, cutting-edge angles and clearances, etc, discussed in this article must be considered only as guides or "starting points" for preliminary trials in a machining operation. The machinist will have to establish the precise process parameters that yield optimum results. Suppliers of plastics usually can recommend detailed machining instructions; they should be consulted when solutions to processing problems are not readily found.

To avoid overheating the plastic in the vicinity of the machining area, several precautions are basic:

1. Use only very sharp cutting tools, and keep them sharp.
2. Be sure the tool is cutting, not scraping.
3. Keep close control of the rates of feed and speed that have been selected for the process.
4. Use coolants if the machining operation is of relatively long duration (which could be as little as 10 seconds for a drilled hole, for example). Cooling can be done by air blast, cutting oil, or liquid ot vapor spray (possibly with a water-soap solution).
5. Provide for frequent removal of chips by retracting the saw, drill bit, or tap. Avoid deforming a part during the cutting operation by applying too heavy a force; take smaller "shavings" when turning, as opposed to deep cuts. Use a pillow block or support to prevent a long length of stock from bending or sagging during sawing, turning, or milling.

Many thermoplastics are resilient and therefore tough enough to withstand the impact of a fast-moving cutting heads. However, others, such as acrylics, polystyrenes and polycarbonates, and most thermoset materials, are brittle and require special attention to avoid or minimize chipping and cracking. Sharp tools are especially critical with such materials, and it is sometimes necessary to preheat the material slightly before cutting to reduce the possibility of fracturing.

A further requirement is to remove swarf (melt residue) promptly and frequently during cutting operations. Swarf from both thermoplastic and thermoset machining operations quickly becomes gummy in the presence of heat, causing tools to bind and surfaces to gall.

Other intrinsic characteristics of plastics must be recognized and dealt with when a specified machining requirement is presented. For example, the "memory" common to most thermoplastics makes machining to close tolerances difficult because of material recovery from deformation when the cutting forces stop. Coupled with the memory factor is the fact that the modulus of elasticity of most thermoplastics may be less than 10% that of metals or wood. Plastics' relatively high coefficient of thermal expansion leads to binding or seizing of the drill bit or saw blade. Clearances and good "sets" in such tools become important.

119.2 CHECKLIST FOR MACHINING

The following recommendations are general, not specific.

Drilling. Heat dissipation and swarf removal are particularly critical. Surfaces of drill bits contacting the material, including surfaces of flutes, should be highly polished and flash chrome plated. Drill bits manufactured specifically for plastics machining are commercially available, but they are not necessarily suited to all plastics. Speeds of rotation and advance are generally decreased with increased hole diameters and with increased hole depths. Rotational speeds may generally be faster with harder materials, both thermoplastics and thermosets (Table 119.1).

Reaming. After drilling, reaming may be done with conventional tools, providing they are well sharpened. Helically fluted reamers are preferred, particularly for blind holes. Cooling with water or light machine oils is recommended (Table 119.2).

TABLE 119.1. Drilling[a]

Plastic	Hardness	Condition	Speed, fpm (m/min)	Feed, in./rev (m/rev)[b]	Tool Material, Grade AISI C[c]
Acrylic, acetal, polycarbonate, polysulfone polystyrene	60–120 R_m	Cast, molded or extruded	100–200 (30–60)	A	D
ABS, polyarylether, polypropylene, polyethylene, cellulose acetate	50–120 R_r	Cast, molded or extruded	150–200 (46–76)	A	D
Fluorocarbons	74–95 R_r	Molded or extruded	100–200 (30–60)	A	D
Nylons (unfilled)	78–120 R_r	Molded or extruded	100–250 (30–76)	A	D
Nylons (35% glass) 6; 6,6	78–120 R_r	Molded	75–150 (23–46)	A	E
Nylons (35% glass, 6,10; 6,12	40–50 R_e	Molded	65–125 (20–38)	A	E
Epoxy, melamine, phenolic	100–128 R_m	Cast or molded	100–200 (30–60)	B	D
Furan, polybutadiene	40–100 R_r	Cast	100–200 (30–60)	B	D
Silicone	15–65 Shore A	Cast or molded	65–125 (20–38)	B	D
Silicone (glass-filled)	80–90 R_m	Molded	65–125 (20–38)	B	E
Polyimide	40–50 R_e	Molded or extruded	100–200 (30–60)	A	D
Polyimide (glass-filled)	109–115 R_m	Molded	65–125 (20–38)	B	E
Polyurethane	65–95 Shore A	Cast	100–200 (30–60)	B	D
Polyurethane	55–75 Shore D	Cast	150–250 (46–76)	A	D
Allyl (DAP)	95–100 R_m	Cast	100–200 (30–60)	A	D
Allyl (fiber filled)	106–115 R_m	Molded	75–150 (23–46)	C	E

[a]Reprinted Ref. 1. Courtesy of the Machinability Data Center and Metcut Research Associates Inc.

[b] Nominal Hole Diameter in. (mm)

	1/16 (1.5)	1/8 (3)	1/4 (6)	1/2 (12)	3/4 (18)	1 (25)	1 1/2 (35)	2 (50)
A	0.001 (0.025)	0.002 (0.050)	0.004 (0.10)	0.005 (0.13)	0.006 (0.15)	0.008 (0.20)	0.010 (0.25)	0.012 (0.30)
B	0.001 (0.025)	0.002 (0.050)	0.003 (0.075)	0.004 (0.10)	0.005 (0.13)	0.006 (0.15)	0.008 (0.20)	0.010 (0.25
C	0.001 (0.025)	0.002 (0.050)	0.004 (0.10)	0.006 (0.15)	0.008 (0.20)	0.010 (0.25)	0.012 (0.30)	0.015 (0.40)

[c]D = M 10, M7, M1 (S2, S3)

E = T 15 M 42 (S 9, S11) [Any premium HSS M 41–47 (S 9–12)].

Threading. Taps for threading should be slightly oversized, and threading dies undersized, especially for the softer thermoplastics, to allow for memory recovery of the plastic after cutting forces have been relaxed. Hole size and rod size should allow for only 75% of thread height, in order to minimize breakage or "stringing" of sharp threads. Class I threads are most practical, but Class II and III are possible. Parts should be annealed after tapping or threading. Taps and dies should have large, highly polished flutes, preferably flash chrome plated, and taps should have ample back clearance (Table 119.3).

Turning and Milling. Vital to precision turning of plastics is prevention of deflection. The part must be well supported as close to the cutting tool as practical. When such close support is not feasible (as, for example, in putting a taper on a long cylindrical rod), follow rests

TABLE 119.2. Reaming[a]

Plastic	Hardness	Condition	Speed, fpm (m/min)	Feed, in./rev (m/rev)[b]	Tool Material, Grade AISI C[c]
Acrylic, acetal, polycarbonate, polysulfone polystyrene	60–120 R_m	Cast molded or extruded	65–130 (20–40)	A	D
ABS, polyarylether, polypropylene, polyethylene, cellulose acetate	50–120 R_r	Cast molded or extruded	100–165 (30–50)	A	D
Fluorocarbons	74–95 R_r	Molded or extruded	65–130 (20–40)	A	D
Nylons (unfilled)	78–120 R_r	Molded or extruded	65–165 (20–50)	A	D
Nylons (35% glass) 6; 6,6	78–120 R_r	Molded	50–100 (15–30)	B	E
Nylons (35% glass) 6,10; 6,12	40–50 R_e	Molded	45–85 (14–26)	B	E
Epoxy, melamine, phenolic	100–128 R_m	Cast or molded	65–130 (20–40)	A	D
Furan, polybutadiene	40–100 R_r	Cast	65–130 (20–40)	A	D
Silicone	15–65 Shore A	Cast or molded	45–85 (14–26)	A	D
Silicone (glass-filled)	80–90 R_m	Molded	45–85 (14–26)	B	E
Polyimide	40–50 R_e	Molded or extruded	65–130 (20–40)	A	D
Polyimide (glass-filled)	109–115 R_m	Molded	45–85 (14–26)	B	E
Polyurethane	65–95 Shore A	Cast	65–130 (20–40)	B	D
Polyurethane	55–75 Shore D	Cast	100–165 (30–50)	A	D
Allyl (DAP)	95–100 R_m	Cast	65–130 (20–40)	A	D
Allyl (fiber filled)	106–115 R_m	Molded	45–85 (14–26)	B	E

[a]Reprinted Ref. 1. Courtesy of the Machinability Data Center and Metcut Research Associates Inc.

Reamer Diameter
in. (mm)

	1/8 (3)	1/4 (6)	1/2 (12)	1 (25)	1 1/2 (35)	2 (50)
A	0.002 (0.050)	0.002 (0.050)	0.004 (0.10)	0.006 (0.15)	0.008 (0.20)	0.010 (0.25)
B	0.002 (0.050)	0.002 (0.050)	0.003 (0.075)	0.003 (0.075)	0.010 (0.25)	0.010 (0.25)

[c]D = M1, M2, M7, C-2 (S3, SS4, S2, K10)
E - C2 (S9, S11, K10) [Any premium HSS - T 15, M33, M41-47 (S9-12)]
Based on 4 flutes (1/8 and 1/4), 6 flutes (1/2), and 8 flutes (1 +).

TABLE 119.3. Tapping[a]

All thread sizes	Speed,[b] fpm (m/min)
All cast, molded and extruded thermoplastics and thermosets	50 (15)
All filled thermoplastics and thermosets	25 (8)

[a]Reprinted from Ref. 1. Courtesy of the Machinability Data Center and Metcut Research Associates Inc.

[b]HSS tool steel AISI-M 10, M 7, M1 ISO-S2. S3. These speeds are for tapping 65% to 75% threads in shallow through holes. Reduce speed when tapping deep holes, blind holes of higher percentage threads.

and/or box tools are recommended. Box tools have two cutting heads, on opposite sides of the work piece, or more than two heads, spaced uniformly around the part, to insure minimum bending of the workpiece. Local deflection into the plastic due to the cutting force of each cutting head is, of course, still present with box tools. Keep feed pressure minimal to avoid overheating, and use surface speeds up to 600 ft/min (180 m/min) for the more resilient materials, higher for the harder materials. Do not expect smooth finishes on machined thermosets, especially filled thermosets. Procedures for machining thermosets are similar to those for brass, but tool wear is extreme. Lubricants are rarely used, but air jets in the cutting area aid in heat dissipation. Dust and shavings should be removed with a suction hose close to the cutting area (Tables 119.4 and 119.5).

Sawing. Circular saws and band saws are most commonly used with plastics. Cooling is highly recommended. Circular saw blades should be hollow ground. Set on band saw teeth should be greater than that used for steel. Narrow-faced abrasive wheels are often used for cutoff operations. ANSI and ISO C60-PB abrasive wheels are used for thermoplastics and thermosets and ANSI and ISO copper bound D60-N100M wheels for composites. Feed speeds will vary with materials, but should be unforced. Saw blades must be kept very sharp when cutting thermosets. Carbide cicular saw blades and abrasive cutoff wheels are recommended. Suction hoses should be used to remove dust (Tables 119.6 and 1J⁰ 7).

119.3 SHEARING, GUILLOTINING, BLANKING, AND PUNCHING

Sheet plastic is often cut to size by shearing rather than sawing. Thermoformed parts generally need to be trimmed after forming, frequently with trim dies. Simple arbor presses, or automated punch presses and kick presses, with conventional dies and guillotines are commonplace.

Guillotining. (Preheat acetal and acrylic to 174–194°F (80–90°C)

Material	Maximum Sheet Thickness	
	THERMOPLASTICS	
	in.	(mm)
Acetal	1/16	(1.6)
Acrylic, nylon	1/8	(3.2)
Polycarbonate, polyolefins,		
PTFE	1/4	(6.4)
	PHENOLICS	
Paper base laminate	0.04 in.	(1.0)
Slope of moving blade	3/64 in./ft	(0.4 mm/m)
Fabric base laminate	3/32 in.	(2.4)
Slope of moving blade	1/4 in./ft	(1.9 mm/m)

Punching Phenolic Laminate. If thickness less than 0.04 in. (1 mm), punch cold. Greater thickness, heat to 176–194°F (80–90°C). Punches and dies for apertures slightly oversize [about 0.001 in. per 1/64 in. thickness (0.0254 mm per 0.4 mm)]. Blanking tools undersize, same amount. Clearance: 0.0005–0.0015 in. (0.0127–0.0381 mm) increasing from low to high as thickness increases.

Laser Machining. This relatively new process is highly practical with many thermoplastics. Its principle is instant vaporization with minimum heat dissipation to areas of the plastic

TABLE 119.4. Turning[a]

Plastic	Hardness	Condition	Depth of cut, in. (mm)	High Speed Steel			Carbide[c]	
				Speed, fpm (m/min)	Feed, ipr (mm/r)	Tool[b] Material	Speed, fpm (m/min)	Feed, ipr (mm/r)
Acrylic, acetal, polycarbonate, polysulfone polystyrene	60–120 R_m	Cast, molded or extruded	0.040 (1)	400 (120)	0.005 (0.13)	A	600 (185)	0.005 (0.13)
			0.150 (4)	350 (105)	0.008 (0.20)		550 (170)	0.010 (0.25)
			0.300 (8)	300 (90)	0.010 (0.25)		500 (150)	0.012 (0.30)
ABS, polarylether, polypropylene, polyethylene, cellulose acetate	50–120 R_r	Cast, molded or extruded	0.040 (1)	450 (135)	0.005 (0.13)	A	700 (215)	0.005 (0.13)
			0.150 (4)	400 (120)	0.008 (0.20)		650 (200)	0.010 (0.25)
			0.300 (8)	350 (105)	0.010 (0.25)		600 (185)	0.010 (0.25)
Fluorocarbons	74–95 R_r	Molded or extruded	0.040 (1)	400 (120)	0.005 (0.13)	A	600 (185)	0.005 (0.13)
			0.150 (4)	350 (105)	0.008 (0.20)		550 (170)	0.010 (0.25)
			0.300 (8)	300 (90)	0.010 (0.25)		500 (150)	0.012 (0.30)
Nylons (unfilled)	78–120 R_r	Molded or extruded	0.040 (1)	500 (150)	0.005 (0.13)	A	800 (245)	0.005 (0.13)
			0.050 (4)	450 (135)	0.010 (0.25)		700 (215)	0.010 (0.25)
			0.300 (8)	400 (120)	0.012 (0.30)		650 (200)	0.012 (0.30)
Nylons (35% glass) 6; 6,6	78–120 R_r	Molded					600 (185)	0.005 (0.13)
							550 (170)	0.008 (0.20)
							500 (150)	0.010 (0.25)
Nylons (35% glass, 6,10; 6,12	40–50 R_c	Molded					500 (150)	0.005 (0.13)
							450 (135)	0.008 (0.20)
							400 (120)	0.010 (0.25)
Epoxy, melamine, phenolic	100–128 R_m	Cast or molded	0.040 (1)	500 (150)	0.005 (0.13)	A	800 (245)	0.005 (0.13)
			0.150 (4)	450 (135)	0.010 (0.25)		700 (215)	0.010 (0.25)
			0.300 (8)	400 (120)	0.015 (0.40)		650 (200)	0.015 (0.40)
Furan, polybutadiene	40–100 R_r	Cast	0.040 (1)	250 (76)	0.005 (0.13)	B	450 (135)	0.005 (0.13)
			0.150 (4)	200 (70)	0.010 (0.25)		400 (120)	0.008 (0.20)
			0.300 (8)	175 (53)	0.015 (0.40)		350 (105)	0.010 (0.25)

			Depth of Cut, in. (mm)	Speed, fpm (m/min)	Feed, in./rev (mm/rev)	Tool Material[b]	Speed, fpm (m/min)	Feed, in./rev (mm/rev)
Silicone	15–65 Shore A	Cast or molded	0.040 (1)	200 (60)	0.005 (0.13)	B	450 (135)	0.005 (0.13)
			0.150 (4)	175 (53)	0.010 (0.25)		400 (120)	0.008 (0.20)
			0.300 (8)	150 (46)	0.015 (0.40)		350 (105)	0.010 (0.25)
Silicone (glass-filled)	80–90 R_m	Molded	0.040 (1)				400 (120)	0.005 (0.13)
			0.150 (4)				350 (105)	0.008 (0.20)
			0.300 (8)				300 (90)	0.010 (0.25)
Polyimide	40–50 R_c	Molded or extruded	0.040 (1)	500 (150)	0.005 (0.13)	A	800 (245)	0.005 (0.13)
			0.150 (4)	450 (135)	0.010 (0.25)		700 (215)	0.010 (0.25)
			0.300 (8)	400 (120)	0.015 (0.40)		700 (215)	0.015 (0.40)
Polyimide (glass-filled)	109–115 R_m	Molded	0.040 (1)				500 (150)	0.005 (0.13)
			0.150 (4)				450 (135)	0.010 (0.25)
			0.300 (8)				400 (120)	0.012 (0.30)
Polyurethane	65–95 Shore A	Cast	0.040 (1)	250 (76)	0.005 (0.13)	B	450 (135)	0.005 (0.13)
			0.150 (4)	200 (60)	0.010 (0.25)		400 (120)	0.008 (0.20)
			0.300 (8)	175 (53)	0.015 (0.40)		350 (105)	0.010 (0.25)
Polyurethane	55–75 Shore D	Cast	0.040 (1)	300 (90)	0.005 (0.13)	A	500 (150)	0.005 (0.13)
			0.150 (4)	250 (76)	0.010 (0.25)		450 (135)	0.010 (0.25)
			0.300 (8)	200 (60)	0.015 (0.40)		400 (120)	0.015 (0.40)
Allyl (DAP)	95–100 R_m	Cast	0.040 (1)	400 (120)	0.005 (0.13)	A	600 (185)	0.005 (0.13)
			0.150 (4)	350 (105)	0.008 (0.20)		550 (170)	0.010 (0.25)
			0.300 (8)	300 (90)	0.012 (0.30)		500 (150)	0.015 (0.40)
Allyl (fiber filled)	106–115 R_m	Molded	0.040 (1)	250 (76)	0.005 (0.13)	B	500 (150)	0.005 (0.13)
			0.150 (4)	200 (60)	0.010 (0.25)		450 (135)	0.010 (0.25)
			0.300 (8)	175 (53)	0.015 (0.40)		400 (120)	0.012 (.030)

[a]Reprinted from Ref. 1. Courtesy of the Machinability Data Center and Metcut Research Associates Inc.

[b]Tool Material A = AISI-M2, M3 ISO-S4, S5

B = AISI-T 15, M2, M31, M41-47 ISO-S9-S12

[c]Carbide tool is uncoated and either brazed or indexable.

TABLE 119.5. Face Milling[a]

Plastic	Hardness	Condition	Depth of cut, in. (mm)	High Speed Steel Speed, fpm (m/min)	Feed, per tooth, in. (mm)	Tool[b] Material	Carbide[c] Speed, fpm (m/min)	Feed, per tooth, in. (mm)
Acrylic, acetal, polycarbonate, polysulfone polystyrene	60–120 R_m	Cast, molded or extruded	0.040 (1)	400 (120)	0.005 (0.13)	A	650 (200)	0.005 (0.13)
			0.150 (4)	350 (105)	0.008 (0.20)		600 (185)	0.007 (0.18)
			0.300 (8)	300 (90)	0.010 (0.25)		550 (170)	0.009 (0.23)
ABS, polarylether, polypropylene, polyethylene, cellulose acetate	50–120 R_r	Cast, molded or extruded	0.040 (1)	450 (135)	0.005 (0.13)	A	750 (230)	0.004 (0.10)
			0.150 (4)	400 (120)	0.008 (0.20)		700 (215)	0.007 (0.18)
			0.300 (8)	350 (105)	0.010 (0.25)		650 (200)	0.009 (0.23)
Fluorocarbons	74–95 R_r	Molded or extruded	0.040 (1)	400 (120)	0.005 (0.13)	A	650 (200)	0.004 (0.10)
			0.150 (4)	350 (105)	0.008 (0.20)		600 (185)	0.007 (0.18)
			0.300 (8)	300 (90)	0.010 (0.25)		550 (170)	0.009 (0.23)
Nylons (unfilled)	78–120 R_r	Molded or extruded	0.040 (1)	500 (150)	0.006 (0.15)	A	850 (260)	0.006 (0.15)
			0.050 (4)	450 (135)	0.010 (0.25)		750 (230)	0.008 (0.20)
			0.300 (8)	400 (120)	0.014 (0.36)		700 (215)	0.010 (0.25)
Nylons (35% glass) 6; 6,6	78–120 R_r	Molded					650 (200)	0.004 (0.10)
							600 (185)	0.006 (0.15)
							550 (170)	0.008 (0.20)
Nylons (35% glass, 6,10; 6,12)	40–50 R_c	Molded					550 (170)	0.004 (0.10)
							500 (150)	0.006 (0.15)
							450 (135)	0.008 (0.20)
Epoxy, melamine, phenolic	100–128 R_m	Cast or molded	0.040 (1)	500 (150)	0.006 (0.15)	B	850 (260)	0.004 (0.10)
			0.150 (4)	450 (135)	0.008 (0.20)		750 (230)	0.006 (0.15)
			0.300 (8)	350 (105)	0.010 (0.25)		700 (215)	0.008 (0.20)
Furan, polybutadiene	40–100 R_r	Cast	0.040 (1)	250 (76)	0.005 (0.13)	C	475 (145)	0.004 (0.10)
			0.150 (4)	200 (60)	0.008 (0.20)		425 (130)	0.006 (0.15)
			0.300 (8)	150 (46)	0.010 (0.25)		375 (115)	0.008 (0.20)

Material	Hardness	Condition	Depth of cut, in. (mm)	Speed, fpm (m/min)	Feed, ipr (mm/rev)	Tool material[c]	Speed, fpm (m/min)	Feed, ipr (mm/rev)
Silicone	15–65 Shore A	Cast or molded	0.040 (1)	200 (60)	0.005 (0.13)	C	475 (145)	0.004 (0.10)
			0.150 (4)	175 (53)	0.008 (0.20)		425 (130)	0.006 (0.15)
			0.300 (8)	125 (38)	0.010 (0.25)		375 (115)	0.008 (0.20)
Silicone (glass-filled)	80–90 R_m	Molded					425 (130)	0.004 (0.10)
							375 (115)	0.006 (0.15)
							325 (110)	0.008 (0.20)
Polyimide	40–50 R_c	Molded or extruded	0.040 (1)	500 (150)	0.005 (0.13)	B	850 (260)	0.005 (0.13)
			0.150 (4)	450 (135)	0.008 (0.20)		750 (230)	0.008 (0.20)
			0.300 (8)	350 (105)	0.010 (0.25)		650 (200)	0.010 (0.25)
Polyimide (glass-filled)	109–115 R_m	Molded					525 (160)	0.005 (0.13)
							475 (145)	0.008 (0.20)
							425 (130)	0.010 (0.25)
Polyurethane	65–95 Shore A	Cast	0.040 (1)	250 (70)	0.005 (0.13)	C	475 (145)	0.004 (0.10)
			0.150 (4)	200 (60)	0.008 (0.20)		425 (130)	0.006 (0.15)
			0.300 (8)	175 (53)	0.010 (0.25)		375 (115)	0.008 (0.20)
Polyurethane	55–75 Shore D	Cast	0.040 (1)	300 (90)	0.005 (0.13)	B	525 (160)	0.004 (0.10)
			0.150 (4)	250 (76)	0.008 (0.20)		475 (145)	0.006 (0.15)
			0.300 (8)	200 (60)	0.010 (0.25)		425 (130)	0.008 (0.20)
Allyl (DAP)	95–100 R_m	Cast	0.040 (1)	400 (120)	0.005 (0.13)	A	650 (200)	0.006 (0.15)
			0.150 (4)	350 (105)	0.008 (0.20)		600 (185)	0.008 (0.20)
			0.300 (8)	300 (90)	0.010 (0.25)		550 (170)	0.010 (0.25)
Allyl (fiber filled)	106–115 R_m	Molded	0.040 (1)	250 (76)	0.005 (0.13)	C	525 (130)	0.006 (0.15)
			0.150 (4)	200 (60)	0.008 (0.20)		475 (115)	0.008 (0.20)
			0.300 (8)	175 (53)	0.010 (0.25)		425 (100)	0.010 (.025)

[a]Reprinted from Ref. 1. Courtesy of the Machinability Data Center and Metcut Research Associates Inc.

[b]A = AISI-M2. M3 ISO-S4, S5
B = AISI-M2. M7 ISO-S2. S4
C = AISI-T15. M42 ISO-S9-S11 (Any premium HSS (T15, M33, M41–47 or S9–12).

[c]Carbide tool is uncoated and either brazed or indexable, AISI-C2, ISO-K20, M20.

TABLE 119.6. Circular Sawing[a,b]

Plastic	Hardness	Condition	Solid Stock Diameter or Thickness, in. (mm)	Pitch in./tooth (mm/tooth)	Cutting Speed fpm (m/min)	Feed[c]
Acrylic, acetal, polycarbonate, polysulfone polystyrene	60–120 R_m	Cast, molded or extruded	1/4–3 (6–80)	0.12–0.50 (3–12)	300 (90)	A
			3–6 (80–160)	0.40–0.70 (10–18)	250 (76)	
			6–9 (160–250)	0.60–0.80 (15–20)	200 (60)	
			9–15 (250–400)	0.70–1.00 (18–25)	150 (46)	
ABS, polarylether, polypropylene, polyethylene, cellulose acetate	50–120 R_r	Cast, molded or extruded	1/4–3 (6–80)	0.12–0.50 (3–12)	350 (105)	A
			3–6 (80–160)	0.40–0.70 (10–18)	300 (90)	
			6–9 (160–250)	0.60–0.80 (15–20)	220 (67)	
			9–15 (250–400)	0.70–1.00 (18–25)	170 (52)	
Fluorocarbons	74–95 R_r	Molded or extruded	1/4–3 (6–80)	0.12–0.50 (3–12)	300 (90)	A
			3–6 (80–160)	0.40–0.70 (10–18)	250 (76)	
			6–9 (160–250)	0.60–0.80 (15–20)	200 (60)	
			9–15 (250–400)	0.70–1.00 (18–25)	150 (46)	
Nylons (unfilled)	78–120 R_r	Molded or extruded	1/4–3 (6–80)	0.12–0.50 (3–12)	350 (105)	A
			3–6 (80–160)	0.40–0.70 (10–18)	300 (90)	
			6–9 (160–250)	0.60–0.80 (15–20)	220 (67)	
			9–15 (250–400)	0.70–1.00 (18–25)	170 (52)	
Epoxy, melamine, phenolic	100–128 R_m	Cast or molded	1/4–3 (6–80)	0.12–0.50 (3–12)	350 (105)	A
			3–6 (80–160)	0.40–0.70 (10–18)	300 (90)	
			6–9 (160–250)	0.60–0.80 (15–20)	220 (67)	
			9–15 (250–400)	0.70–1.00 (18–25)	170 (52)	
Furan, polybutadiene	40–100 R_r	Cast	1/4–3 (6–80)	0.12–0.50 (3–12)	150 (46)	B
			3–6 (80–160)	0.40–0.70 (10–18)	120 (37)	
			6–9 (160–250)	0.60–0.80 (15–20)	100 (30)	
			9–15 (250–400)	0.70–1.00 (18–25)	80 (24)	
Silicone	15–65 Shore A	Cast or molded	1/4–3 (6–80)	0.12–0.50 (3–12)	100 (30)	B
			3–6 (80–160)	0.40–0.70 (10–18)	80 (24)	
			6–9 (160–250)	0.60–0.80 (15–20)	50 (15)	
			9–15 (250–400)	0.70–1.00 (18–25)	50 (15)	

Material	Hardness	Form	Thickness, in. (mm)	Feed range	Speed, fpm (m/min)	Tool[b],[c]
Polyimide	40–50 R_c	Molded or extruded	1/4–3 (6–80)	0.12–0.50 (3–12)	350 (105)	A
			3–6 (80–160)	0.40–0.70 (10–18)	300 (90)	
			6–9 (160–250)	0.60–0.80 (15–20)	220 (67)	
			9–15 (250–400)	0.70–1.00 (18–25)	170 (52)	
Polyurethane	65–90 Shore A	Cast	1/4–3 (6–80)	0.12–0.50 (3–12)	150 (46)	C
			3–6 (80–160)	0.40–0.70 (10–18)	120 (37)	
			6–9 (160–250)	0.60–0.80 (15–20)	100 (30)	
			9–15 (250–400)	0.70–1.00 (18–25)	80 (24)	
Polyurethane	55–75 Shore D	Cast	1/4–3 (6–80)	0.12–0.50 (3–12)	200 (60)	A
			3–6 (80–160)	0.40–0.70 (10–18)	150 (46)	
			6–9 (160–250)	0.60–0.80 (15–20)	120 (37)	
			9–15 (250–400)	0.70–1.00 (18–25)	100 (30)	
Allyl (DAP)	95–100 R_m	Cast	1/4–3 (6–80)	0.12–0.50 (3–12)	250 (76)	A
			3–6 (80–160)	0.40–0.70 (10–18)	200 (60)	
			6–9 (160–250)	0.60–0.80 (15–20)	150 (46)	
			9–15 (250–400)	0.70–1.00 (18–25)	120 (37)	
Allyl (fiber filled)	106–115 R_m	Molded	1/4–3 (6–80)	0.12–0.50 (3–12)	150 (46)	B
			3–6 (80–160)	0.40–0.70 (10–18)	120 (37)	
			6–9 (160–250)	0.60–0.80 (15–20)	100 (30)	
			9–15 (250–400)	0.70–1.00 (18–25)	80 (24)	

[a]Reprinted from Ref. 1. Courtesy of the Machinability Data Center and Metcut Research Associates.
[b]High speed steel blade, AISI M2, M7, ISO-S2, S4.
[c]Feed: in./tooth (mm/tooth).

	1/4–6 in. (6–160 mm)	6–15 in. (160–400)
A =	0.004 (0.10):	= 0.006 (0.15)
B =	0.003 (0.075)	= 0.005 (0.13)
C =	0.003 (0.075)	= 0.004 (0.10)

outside the precise area where the focused beam "sees" the work piece. Thermoset materials tend to char under such exposure and are far less suited to the process. Holes, slits, closely spaced perforations, and edge cuts are common with laser machinings, especially with thin sheets and film. Dimensional accuracy depends on technique, and rarely can be as precise as with metal-cutting tools.

Miscellaneous Machining Processes. Waterjet cutting, or hydrodynamic machining with high-velocity fluid jets, has proved successful on such widely diversified materials as plastic acoustic tile, laminated paperboard, asbestos-filled phenolic, composites, and urethane foam, and rubber. Ultrasonic machining is used generally in conjunction with mechanical drilling, on hard, brittle plastics. Electrical discharge machining, effective with metals, also can be used on plastics (providing they are sufficiently conductive).

119.4 FINISHING OPERATIONS

After the machining step, parts may require filing, grinding, sanding, polishing, and/or buffing to produce the desired surface or edge finish. Manual or mechanical files and burrs require very sharp teeth and large smooth rounded gullets to minimize buildup of swarf. A surface speed up to 1000 ft/min. (300 m/min.) is generally satisfactory, whether by moving the file past the work piece or by rotating or moving the work piece past a stationary file.

Filing of the softer thermoplastics is difficult, whereas the harder ones (styrenics, acrylics, polycarbonates) are easily filed, as are most thermosets. Fiber filled materials often show exposed fiber ends after filing, with some ends protruding from the surface. On such hard materials, file toward a solid portion of the piece to minimize chipping. For the softer thermoplastics, scraping or even carving with a hand tool is often best for edge finishing.

Sanding and grinding machines, either belt or disk, are very effective with most plastics. Speed control is important to ensure adaptability to the specific type of plastics being processed. Coarse wheels and belts may be run dry (use suction hoses to remove the dust), whereas fine grinding and sanding operations are often run wet to minimize heat and dust and to achieve a finer finish.

TABLE 119.7. Band Sawing

Material	Thickness, in.[a]	Teeth/in.	(Teeth/cm)	Speed, ft/min	(m/min)
Most thermosets {	Less than 1/4	10	(4)	6,000	(1,800)
	Greater than 1/4	5	(2)	4,000	(1,200)
PVC } {	Less than 1/4	10	(4)	1,000	(300)
Cellulose acetate	Greater than 1/4	6	(2)	750	(200)
Polystyrene }		12	(5)	1,000	(300)
PPO {	Less than 1/4	8	(3)	2,000	(600)
Polycarbonate	Greater than 1/4				
	Less than 1/4	14	(6)	10,000	(3,000)
Acrylic {	1/4–1	8	(3)	7,000	(2,100)
	Greater than 1	3	(1)	5,000	(1,500)
Nylon } {	Less than 1/4	8	(3)	4,000	(1,200)
Acetal	Greater than 1/4	5	(2)	3,000	(900)
Polyethylene }		12	(5)	5,000	(1,500)
PTFE {	Less than 1/4	8	(3)	4,000	(1,200)
Polypropylene	Greater than 1/4				
ABS	Less than 1/4	6	(2)	2,000	(600)

[a]1/4 in. = 6.4 mm.

Optimum grit sizes for wheels and belts will vary for various plastics. For acrylics, as an example, sandpapers from grades 320 to 600A prove practical; for polycarbonates, 180 is the finest grit size recommended for a high-luster surface. Belt speeds of 4000 surface feet/ min (1200 surface meters/min) are common for most thermoplastics and thermosets. Pressure of the part against the belt or wheel should be moderate, in order to avoid excessive heat.

More detailed and specific information can be provided by producers of plastic resins and compounds, as well as finishing-equipment suppliers.

BIBLIOGRAPHY

1. *Machining Data Handbook,* 3rd ed., Machinability Data Center and Metcut Research Associates, Inc. Cincinatti, Ohio, 1980.

General References

A. Kubayashi, *Machining of Plastics,* McGraw-Hill, Inc., New York, *Modern Plastics Encyclopedia, 1986–1987,* McGraw-Hill, Inc., New York, pp. 383–96.

J. Frados, *Plastics Engineering Handbook of the Society of the Plastics Industry, Inc.,* 4th ed., Van Nostrand Reinhold, New York, 1976.

APPENDICES

TABLE 1. Abbreviations Alphabetized by "Abbreviation"

Abbreviation	Meaning	Abbreviation	Meaning
AA	Acetic Aldehyde Acetaldehyde	COGSME	The Composite Group of the SME
ABS	Acrylonitrile–Butadiene–Styrene	CPE	Chlorinated Polyethylene
ACS	American Chemical Society Acrylonitrile Chlorinated PS	CPI	Condensation Reaction Polyimides
		CPSE	Consumer Product Safety Commission
		CPU	Central Processing Unit
AGVS	Automatic Guided Vehicle System	CPVC	Chlorinated Poly(vinyl chloride)
AI	Artificial Intelligence	CR	Controlled Rheology
AMC	Alkyd (Molding Compound)	CRC	Cyclic Redundancy Check
AMS	Alpha Methyl Styrene	CRT	Cathode Ray Tube
API	Addition Reaction, Polyimides	CTE	Coefficient of Thermal Expansion
AR	Alkaline Resistant (Glass)	CTFE	Polymonochlortrifluorethylene
ARM	Association of Rotational Molders	CVD	Chemical Vapor Deposition
		DAIP	Diallyl Isophthalate
ASA	Acrylic–Styrene–Acrylonitrile Acrylonitrile–Styrene–Rubber Copolymers	DAP	Diallyl Phthalate
		DGEBA	Diglycidyl Ether of Bisphenol A
		DIDP	Diisodecyl Phthalate
ASME	American Society of Mechanical Engineers	DIOP	Diisooctyl Phthalate
		DMC	Dough Molding Compound
ASTM	American Society of Testing Materials	DNC	Direct Numerical Control
ATH	Aluminum Trihydrate Hydrated Alumina	DNP	Dinonyl Phthalate
		DOP	Dioctyl Phthalate
ATPE	Aromatic Amine Terminated Polyether	DSMO	Dimethyl Sulfoxide
		DTUL	Distortion Temperature Under Load
Be	Beryllium		
BMC	Bulk Molding Compound	EB	Ethyl Benzene
CA	Cellulose Acetate	EC	Ethyl Cellulose
CAB	Cellulose Acetate Butyrate	ECPE	Spectra Extended Chain Polyethylene Fibers
CAD	Computer Aided Design		
CAM	Computer Aided Manufacturing	ECTFE	Ethylenechlorotrifluorethylene
		EDM	Electrical Discharge Machine
CAN	Cellulose Acetate Nitrate		
CAP	Cellulose Acetate Proprionate	EMA	Ethylene–Methyl Acrylate
CBA	Chemical Blowing Agents	EP	Epoxy
CD	Compact Disk	EPR	Ethylene–Propylene Rubber
CDB	Conjugated Diene Butyl		
CIM	Computer Integrated Manufacturing Computer Integrated Module	EPS	Expanded Polystyrene
		ESC	Environmental Stress Cracking
		ESCR	Environmental Stress Cracking Resistance
CMM	Coordinate Measuring Machines	ESD	Electrostatic Dissipation
CNC	Computer Numerical Control	ETFE	Ethylenetetrafluorethylene
		EVA	Ethylene Vinyl Acetate

TABLE 1. *(Continued)*

Abbreviation	Meaning	Abbreviation	Meaning
EVOH	Ethylene Vinyl Alcohol	LDPE	Low Density Polyethylene
FDA	Food & Drug Administration	LED	Light Emitting Diode
FEA	Finite Element Analysis	LIM	Liquid Injection Molding
FEM	Finite Element Modeling (Analysis)	LLDPE	Linear Low Density Polyethylene
FEP	Fluorethylene–Propylene Copolymer	LMC	Low Pressure Molding Compound
FF	Furan–Formaldehyde	LOI	Limiting Oxygen Index Loss on Ignition
FIDSA	Fellow, Industrial Designers' Society of America	LVDT	Linear Variable Differential Transformer
FPVC	Fluorinated Poly(vinyl chloride)	MA	Maleic Anhydride
	Flexible Poly(vinyl chloride)	MAP	Manufacturing Automation Protocol
FR	Fiber Reinforced Flame Retardant	MBS	Methacrylate–Butadiene–Styrene
FRP	Fiber Reinforced Plastic	MDI	Diphenylmethane 4, 4'-Diisocyanate
GMP	Good Manufacturing Procedure	MDPE	Medium Density Polyethylene
GPPS	General Purpose Polystyrene	MEK	Methyl Ethyl Ketone
HDI	Hexamethyl Diisocyanate	MEKP	Methyl Ethyl Ketone Peroxide
HDPE	High Density Polyethylene	MER	Chemical Repeating Unit
HDT	Heat-Deflection Temperature	MF	Melamine–Formaldehyde
		MI	Melt Index
HIPS	High Impact Polystyrene	MIPS	Medium Impact Polystyrene
HM	High Modulus		
IM	Injection Molding	MPDA	*m*-Phenylene Diamine
IMM	Injection Molding Machine	msi	psi $\times$ 10^6
		MVT	Moisture Vapor Transmission
IMR	Internal Mold Release		
IPDF	Isophoron Diisocyanate	MW	Molecular Weight
IPN	Interpenetrating Polymer Networks	MWD	Molecular Weight Distribution
ISCC	Intersociety Color Council (NBS)	NAPCOR	National Association for Plastic Container Recovery
ISO	Reactive Isocyanate in RIM	NBS	National Bureau of Standards (now called National Institute of Standards and Technology)
	International Standards Organization		
ITP	Interpenetrating Thickening Process		
ITQ	I.T. Quarnstrom Foundation	NEAT	Nothing Else Added To It
IV	Inherent Viscosity	NMA	Nadic Methyl Anhydride
JIT	Just in Time	OPP	Oriented Polypropylene
JND	Just Noticeable Differences	OSHA	Occupational Safety & Health Act
KO	Knock Out (Pin)	PA	Polyamide (Nylon)
ksi	psi $\times$ 10^3	PAI	Polyamid–Imide
L/D	Length/Diameter Ratio	PAN	Polyacrylonitrile
LCP	Liquid Crystal Polymer	PAS	Polyarylsulfone

1694

TABLE 1. *(Continued)*

Abbreviation	Meaning	Abbreviation	Meaning
PB	Polybutylene	PU	Polyurethane
PBI	Polybenzimidazol	PUE	Polyurethane Elastomer
PBT	Poly(butylene terephthalate)	PUF	Polyurethane Foam
		PUR	Polyurethane
PC	Polycarbonate	PVA	Poly(vinyl alcohol)
PC/A	Polycarbonate/Acrylic	PVDC	Poly(vinylidene chloride)
PCTFE	Polychlorotrifluoroethylene	PVDF	Poly(vinylidene fluoride)
PDCP	Polydicyclopentadiene	PVF	Poly(vinyl fluoride)
PDCPD	Polydicyclopentadiene	$R_\#$	Rockwell Hardness,
PE	Polyethylene		subscript $_\#$ indicates
PEC	Polyphenylene Ether Copolymer		scale
		RAM	Random Access Memory
PEEK	Polyetherether Ketone	RC	Resistance-Condenser
PEI	Polyetherimide	RF	Radio Frequency
PEKK	Polyetherketone Ketone	RIM	Reaction Injection Molding
PEO	Polyphenylene Oxide		
PES	Polyethersulfone	RLP	Reactive Liquid Polymer
PET	Poly(ethylene terephthalate)	RMPS	Rubber Modified Polystyrene
PETG	Poly(ethylene terephthalate) Glycol Modified	RP	Reinforced Plastics
		RPM	Revolutions Per Minute
		RPVC	Rigid Poly(vinyl chloride)
PF	Phenol–Formaldehyde	RRIM	Reinforced Reaction Molding
PFA	Perfluoroalkoxy (Resin)		
PHBA	*p*-Hydroxybenzoic Acid	RT	Residence Time
PHNA	*p*-Hydroxynaphtheic Acid	RTD	Resistance Temperature Detectors
PHR	Parts Per Hundred		
PI	Polyimide	RTM	Resin Transfer Molding
PIA	Plastic Institute of America	S/MMA	Styrene/Methyl Methacrylate
PIB	Polyisobutylene	SAMPE	Society for Advancement of Materials & Process Engineers
PID	Proportional Integral Derivative		
	Proportional Integral Derivative (Controlled)	SAN	Styrene Acrylonitrile
		SARA	Superfund Amendments & Reauthorization Act
PL	Parting Line		
PMDA	Pyromellitic Dianhydride	SBR	Synthetic Butyl Rubber
PMMA	Poly(methyl methacrylate) (Acrylic)	SBS	Styrene–Butadiene–Styrene
PMP	Poly(methyl pentene)	SCR	Silicone Controlled Rectifiers
PMS	Paramethyl Styrene		
POLY	Reactive Polyols in RIM	SEBS	Styrene/Butylene–Butylene/Styrene Elastomers
POM	Acetal (Polyoxymethylene)		
PP	Polypropylene	SI	Silicone
PPBP	*p,p'*-Biphenol	SIS	Styrene–Isoprene–Styrene
PPE	Poly(phenylene ether)	SMA	Styrene Maleic Anhydride
PPS	Poly(phenylene sulfide)		
PPSS	Poly(phenylene sulfide sulfone)	SMC	Sheet Molding Compounds
PRT	Pressure Reduction Time	SME	Society of Manufacturing Engineers
PS	Polystyrene		
PSI	Pound per Square Inch	SPC	Statistical Process Control
PTFE	Polytetrafluoroethylene		
PTMG	Poly(tetramethylene glycol)	SPE	Society of Plastic Engineers

TABLE 1. *(Continued)*

Abbreviation	Meaning	Abbreviation	Meaning
SPI	The Society of the Plastics Industry, Inc.	TPE	Thermoplastic Elastomer
		TPU	Thermoplastic Polyurethane
SPPF	Solid Phase Pressure Forming	TPX	Poly(methyl pentene)
SQC	Statistical Quality Control	UF	Urea–Formaldehyde
		UHMWPE	Ultra High Molecular Weight Polyethylene
SRIM	Structural Reaction Injection Molding	UL	Underwriters Laboratories
STC	Structural Thermoplastic Composite Sheet	UTL	Use Temperature Limit
T/L	Truckload	UV	Ultraviolet Light
T_g	Glass-transition Temperature	VA	Vinyl Acetate
		VAE	Vinyl Acetate–Ethylene
T_m	Melting Temperature	VC	Virtually Cross-Linked
TBPB	*Tert*-Butyl Perbenzoate	VCM	Vinyl Chloride Monomer
TBPU	*Tert*-Butyl Peroctate	VDC	Vinylidene Chloride
TCE	Trichloroethylene	VLDPE	Very Low Density Polyethylene
TDI	Toluene Diisocyanate		
TFE	Polytetrafluorethylene	VT	Vicat Temperature
THF	Tetrahydrofuran	WVT	Water Vapor Transmission
TPA	Terephthalic Acid		

TABLE 2. Abbreviations Alphabetized by "Meaning"

Abbreviation	Meaning	Abbreviation	Meaning
POM	Acetal (Polyoxymethylene)	CAM	Computer Aided Manufacturing
AA	Acetaldehyde	CIM	Computer Integrated Manufacturing
AA	Acetic Aldehyde		Computer Integrated
ASA	Acrylic–Styrene– Acrylonitrile		Module
ACS	Acrylonitrile Chlorinated PS	CNC	Computer Numerical Control
ABS	Acrylonitrile–Butadiene– Styrene	CPI	Condensation Reaction Polyimides
ASA	Acrylonitrile–Styrene– Rubber Copolymers	CDB	Conjugated Diene Butyl
API	Addition Reaction, Polyimides	CPSE	Consumer Product Safety Commission
AR	Alkaline Resistant (Glass)	CR	Controlled Rheology
AMC	Alkyd (Molding Compound)	CMM	Coordinate Measuring Machines
AMS	Alpha Methyl Styrene	CRC	Cyclic Redundancy Check
ATH	Aluminum Trihydrate	DAIP	Diallyl Isophthalate
ACS	American Chemical Society	DAP	Diallyl Phthalate
ASME	American Society of Mechanical Engineers	DGEBA	Diglycidyl Ether of Bisphenol A
ASTM	American Society of Testing Materials	DIDP	Diisodecyl Phthalate
		DIOP	Diisooctyl Phthalate
ATPE	Aromatic Amine Terminated Polyether	DSMO	Dimethyl Sulfoxide
		DNP	Dinonyl Phthalate
AI	Artificial Intelligence	DOP	Dioctyl Phthalate
ARM	Association of Rotational Molders	MDI	Diphenylmethane 4,4'-Diisocyanate
AGVS	Automatic Guided Vehicle System	DNC	Direct Numerical Control
		DTUL	Distortion Temperature Under Load
Be	Beryllium	DMC	Dough Molding Compound
BMC	Bulk Molding Compound		
CRT	Cathode Ray Tube	EDM	Electrical Discharge Machine
CA	Cellulose Acetate		
CAB	Cellulose Acetate Butyrate	ESD	Electrostatic Dissipation
CAN	Cellulose Acetate Nitrate	ESC	Environmental Stress Cracking
CAP	Cellulose Acetate Proprionate	ESCR	Environmental Stress Cracking Resistance
CPU	Central Processing Unit	EP	Epoxy
CBA	Chemical Blowing Agents	EVA	Ethylene Vinyl Acetate
MER	Chemical Repeating Unit	EB	Ethyl Benzene
CVD	Chemical Vapor Deposition	EC	Ethyl Cellulose
		EVOH	Ethylene Vinyl Alcohol
CPE	Chlorinated Polyethylene	EMA	Ethylene–Methyl Acrylate
CPVC	Chlorinated Poly(vinyl chloride)	EPR	Ethylene–Propylene Rubber
CTE	Coefficient of Thermal Expansion	ECTFE	Ethylenechlorotrifluorethylene
		ETFE	Ethylenetetrafluorethylene
CD	Compact Disk	EPS	Expanded Polystyrene
CAD	Computer Aided Design		Spectra

TABLE 2. *(Continued)*

Abbreviation	Meaning	Abbreviation	Meaning
ECPE	Extended Chain Polyethylene Fibers	LED	Light Emitting Diode
		LOI	Limiting Oxygen Index
FIDSA	Fellow, Industrial Designers' Society of America	LLDPE	Linear Low Density
		LVDT	Linear Variable Differential Transformer
FR	Fiber Reinforced		
FRP	Fiber Reinforced Plastic	LCP	Liquid Crystal Polymer
FEA	Finite Element Analysis	LIM	Liquid Injection Molding
FEM	Finite Element Modeling (Analysis)	LOI	Loss on Ignition
		LDPE	Low Density Polyethylene
FR	Flame Retardant		
FPVC	Flexible Poly(vinyl chloride)	LMC	Low Pressure Molding Compound
FPVC	Fluorinated Poly(vinyl chloride)	MPDA	*m*-Phenylene Diamine
		MA	Maleic Anhydride
FEP	Fluorethylene–Propylene Copolymer	MAP	Manufacturing Automation Protocol
FDA	Food & Drug Administration	MDPE	Medium Density Polyethylene
FF	Furan–Formaldehyde	MIPS	Medium Impact Polystyrene
GPPS	General Purpose Polystyrene		
		MF	Melamine–Formaldehyde
T_g	Glass-transition Temperature	MI	Melt Index
		T_m	Melting Temperature
GMP	Good Manufacturing Procedure	MBS	Methacrylate–Butadiene–Styrene
HDT	Heat-Deflection Temperature	MEK	Methyl Ethyl Ketone
		MEKP	Methyl Ethyl Ketone Peroxide
HDI	Hexamethyl Diisocyanate		
HDPE	High Density Polyethylene	MVT	Moisture Vapor Transmission
HIPS	High Impact Polystyrene	MW	Molecular Weight
HM	High Modulus	MWD	Molecular Weight Distribution
ATH	Hydrated Alumina		
ITQ	I.T. Quarnstrom Foundation	NMA	NADI Methyl Anhydride
		NAPCOR	National Association for Plastic Container Recovery
IV	Inherent Viscosity		
IM	Injection Molding		
IMM	Injection Molding Machine	NBS	National Bureau of Standards (now called the National Institute of Standards and Technology)
IMR	Internal Mold Release		
ISO	International Standards Organization		
IPN	Interpenetrating Polymer Networks	NEAT	Nothing Else Added To It
ITP	Interpenetrating Thickening Process	OSHA	Occupational Safety & Health Act
ISCC	Intersociety Color Council (NBS)	OPP	Oriented Polypropylene
		PPBP	*p,p'*-Biphenol
IPDF	Isophoron Diisocyanate	PHBA	*p*-Hydroxybenzoic Acid
JIT	Just in Time	PHNA	*p*-Hydroxynaphtheic Acid
JND	Just Noticeable Differences	PMS	Paramethyl Styrene
		PL	Parting Line
KO	Knock Out (Pin)	PHR	Parts Per Hundred
L/D	Length/Diameter Ratio	PFA	Perfluoroalkoxy (Resin)

TABLE 2. *(Continued)*

Abbreviation	Meaning	Abbreviation	Meaning
PF	Phenol–Formaldehyde	PID	Proportional Integral Derivative
PIA	Plastic Institute of America	PID	Proportional Integral Derivative (Controlled)
PAN	Polyacrylonitrile		
PAI	Polyamid–Imide	ksi	psi $\times$ 10^3
PA	Polyamide (Nylon)	msi	psi $\times$ 10^6
PAS	Polyarylsulfone	PMDA	Pyromellitic Dianhydride
PBI	Polybenzimidazol	RF	Radio Frequency
PB	Polybutylene	RAM	Random Access Memory
PBT	Poly(butylene terephthalate)	RIM	Reaction Injection Molding
PC	Polycarbonate	ISO	Reactive Isocyanate in RIM
PC/A	Polycarbonate/Acrylic		
PCTFE	Polychlorotrifluoroethylene	RLP	Reactive Liquid Polymer
PDCP	Polydicyclopentadiene	POLY	Reactive Polyols in RIM
PDCPD		RP	Reinforced Plastics
PEEK	Polyetherether Ketone	RRIM	Reinforced Reaction Molding
PEI	Polyetherimide		
PEKK	Polyetherketone Ketone	RT	Residence Time
PES	Polyethersulfone	RTM	Resin Transfer Molding
PE	Polyethylene	RTD	Resistance Temperature Detectors
PET	Polyethylene terephthalate		
		RC	Resistance-Condenser
PETG	Polyethylene terephthalate Glycol Modified	RPM	Revolutions Per Minute
		RPVC	Rigid Poly(vinyl chloride)
PI	Polyimide	R$_{\#}$	Rockwell Hardness, subscript$_\#$ indicates scale
PIB	Polyisobutylene		
PMMA	Poly(methyl methacrylate) (Acrylic)	RMPS	Rubber Modified Polystyrene
PMP	Poly(methyl pentene)	SMC	Sheet Molding Compounds
TPX			
CTFE	Polymonochlortrifluorethylene	SI	Silicone
PPE	Poly(phenylene ether)	SCR	Silicone Controlled Rectifiers
PEC	Polyphenylene Ether Copolymer	SAMPE	Society for Advancement of Materials & Process Engineers
PEO	Polyphenylene Oxide		
PPS	Poly(phenylene sulfide)		
PPSS	Poly(phenylene sulfide sulfone)	SME	Society of Manufacturing Engineers
PP	Polypropylene	SPE	Society of Plastic Engineers
PS	Polystyrene		
TFE	Polytetrafluoroethylene	SPPF	Solid Phase Pressure Forming
PTFE	Polytetrafluoroethylene		
PTMG	Poly(tetramethylene glycol)	ECPE	Spectra
		SPC	Statistical Process Control
PUR	Polyurethane		
PU		SQC	Statistical Quality Control
PUE	Polyurethane Elastomer		
PUF	Polyurethane Foam	SRIM	Structural Reaction Injection Molding
PVA	Poly(vinyl alcohol)		
PVF	Poly(vinyl fluoride)	STC	Structural Thermoplastic Composite Sheet
PVDC	Poly(vinylidene chloride)		
PVDF	Poly(vinylidene fluoride)	SAN	Styrene Acrylonitrile
PSI	Pound per Square Inch	SMA	Styrene Maleic Anhydride
PRT	Pressure Reduction Time		

TABLE 2. (*Continued*)

Abbreviation	Meaning	Abbreviation	Meaning
SBS	Styrene–Butadiene–Styrene	TDI	Toluene Diisocyanate
		TCE	Trichloroethylene
SIS	Styrene–Isoprene–Styrene	T/L	Truckload
SEBS	Styrene/Butylene–Butylene/Styrene Elastomers	UHMWPE	Ultra High Molecular Weight Polyethylene
		UV	Ultraviolet Light
S/MMA	Styrene/Methylmethacrylate	UL	Underwriters Laboratories
SARA	Superfund Amendments & Reauthorization Act	UF	Urea–Formaldehyde
		UTL	Use Temperature Limit
SBR	Synthetic Butyl Rubber	VLDPE	Very Low Density Polyethylene
TPA	Terephthalic Acid		
TBPB	*Tert*-Butyl Perbenzoate	VT	Vicat Temperature
TBPU	*Tert*-Butyl Peroctate	VA	Vinyl Acetate
THF	Tetrahydrofuran	VAE	Vinyl Acetate–Ethylene
COGSME	The Composite Group of The SME	VCM	Vinyl Chloride Monomer
		VDC	Vinylidene Chloride
SPI	The Society of the Plastics Industry, Inc.	VC	Virtually Cross-Linked
		WVT	Water Vapor Transmission
TPE	Thermoplastic Elastomer		
TPU	Thermoplastic Polyurethane		

TABLE 3. Cycle Time (in seconds) into Shots per Hour and Hours for Molding 1000 Shots; Grams into Pounds per 1000 Pieces; Grams into Ounces[a]

Shots per Hour	Hours per 1000 Shots	Grams or Cycle Time (s)	Pounds per 1000 Pieces	Grams to Ounces	Shots per Hour	Hours per 1000 Shots	Grams or Cycle time (s)	Pounds per 1000 Pieces	Grams to Ounces
3600	0.2778	1	2.205	0.035	71	14.17	51	112.4	1.804
1800	0.5556	2	4.409	0.071	69	14.44	52	114.6	1.839
1200	0.8333	3	6.614	0.106	68	14.72	53	116.9	1.875
900	1.111	4	8.819	0.141	67	15.00	54	119.1	1.910
720	1.389	5	11.02	0.176	65	15.28	55	121.3	1.946
600	1.667	6	13.23	0.212	64	15.56	56	123.5	1.980
514	1.944	7	15.43	0.247	63	15.83	57	125.7	2.016
450	2.222	8	17.64	0.282	62	16.11	58	127.9	2.051
400	2.500	9	19.84	0.318	61	16.39	59	130.1	2.086
360	2.778	10	22.05	0.353	60	16.67	60	132.3	2.117
327	3.056	11	24.25	0.388	59	16.94	61	134.5	2.152
300	3.333	12	26.46	0.423	58	17.22	62	136.7	2.187
277	3.611	13	28.66	0.459	57	17.50	63	138.9	2.223
257	3.889	14	30.87	0.494	56	17.78	64	141.1	2.258
240	4.167	15	33.07	0.530	55	18.06	65	143.3	2.293
225	4.444	16	35.27	0.564	55	18.33	66	145.5	2.328
212	4.722	17	37.48	0.600	54	18.61	67	147.7	2.363
200	5.000	18	39.68	0.635	53	18.89	68	149.9	2.399
189	5.278	19	41.89	0.670	52	19.17	69	152.1	2.434
180	5.556	20	44.09	0.701	51	19.44	70	154.3	2.469
171	5.833	21	46.30	0.741	51	19.72	71	156.5	2.505
164	6.111	22	48.50	0.776	50	20.00	72	158.7	2.540
157	6.389	23	50.71	0.811	49	20.28	73	160.9	2.575
150	6.667	24	52.91	0.847	48	20.56	74	163.1	2.610
144	6.944	25	55.12	0.882	48	20.83	75	165.4	2.646
138	7.222	26	57.32	0.917	47	21.11	76	167.6	2.681
133	7.500	27	59.53	0.952	46	21.39	77	169.8	2.716
129	7.778	28	61.73	0.988	46	21.67	78	172.0	2.751
124	8.056	29	63.93	1.023	45	21.94	79	174.2	2.787
120	8.333	30	66.14	1.058	45	22.22	80	176.4	2.822
116	8.611	31	68.34	1.094	44	22.50	81	178.6	2.857
112	8.889	32	70.55	1.129	44	22.78	82	180.8	2.893
109	9.167	33	72.75	1.164	43	23.06	83	183.0	2.928
106	9.444	34	74.96	1.199	43	23.33	84	185.2	2.963
103	9.722	35	77.16	1.235	42	23.61	85	187.4	3.000
100	10.00	36	79.37	1.270	42	23.89	86	189.6	3.034
97	10.28	37	81.57	1.305	42	24.17	87	191.8	3.069
95	10.56	38	83.78	1.340	41	24.44	88	194.0	3.104
92	10.83	39	85.98	1.378	41	24.72	89	196.2	3.140
90	11.11	40	88.19	1.411	40	25.00	90	198.4	3.175
88	11.39	41	90.39	1.446	40	25.28	91	200.6	3.210
86	11.67	42	92.59	1.481	39	25.56	92	202.8	3.245
83	11.94	43	94.80	1.517	39	25.83	93	205.0	3.281
82	12.22	44	97.00	1.552	38	26.11	94	207.2	3.316
80	12.50	45	99.21	1.588	38	26.39	95	209.4	3.352
78	12.78	46	101.4	1.623	37	26.67	96	211.6	3.386
77	13.06	47	103.6	1.658	37	26.94	97	213.9	3.422
75	13.33	48	105.8	1.693	37	27.22	98	216.1	3.457
73	13.61	49	108.0	1.729	36	27.50	99	218.3	3.492
72	13.89	50	110.2	1.764	36	27.78	100	220.5·	3.528

[a]Tables for converting:
A. Cycle time (seconds) into shots per hour.
B. Cycle time (seconds) into hours of molding for 1000 shots.
C. Weight (grams) into pounds per 1000 pieces.

D. Weight (grams) into ounces.
1 g = 0.0353 oz   1 lb = 454 g
1 oz = 28.3 g   1 g = 0.0022 lb

TABLE 4. Fractional Diameters, Decimal Diameters, Millimeter Diameters, Areas of Circles, Circumferences of Circles, Surfaces of Spheres, and Volumes of Spheres

Fractional Diameter	Decimal Diameter, in.	Millimeter Diameter	Area of Circle in.^{2}	Circum. of Circle, in.	Surface of Sphere, in.^{2}	Volume of Sphere, in.^{3}
1/64	0.015625	0.397	0.00019	0.04909	0.00076	
1/32	0.031250	0.794	0.00077	0.09818	0.00308	0.00002
3/64	0.046875	1.191	0.00173	0.14726	0.00692	0.00006
1/16	0.062500	1.588	0.00307	0.19635	0.01228	0.00013
5/64	0.078125	1.984	0.00479	0.24544	0.01916	0.00025
3/32	0.093750	2.381	0.00690	0.29452	0.02761	0.00043
7/64	0.109375	2.778	0.00939	0.34361	0.03756	0.00068
1/8	0.125000	3.175	0.01227	0.39270	0.04908	0.00102
9/64	0.140625	3.572	0.01553	0.44179	0.06212	0.00145
5/32	0.156250	3.969	0.01917	0.49087	0.07668	0.00200
11/64	0.171875	4.366	0.02320	0.53996	0.09280	0.00266
3/16	0.187500	4.763	0.02761	0.58905	0.11044	0.00345
13/64	0.203125	5.159	0.03240	0.63814	0.12960	0.00439
7/32	0.218750	5.556	0.03758	0.68722	0.15032	0.00548
15/64	0.234375	5.953	0.04314	0.73631	0.17256	0.00674
1/4	0.250000	6.350	0.04909	0.78540	0.19636	0.00818
17/64	0.265625	6.747	0.05541	0.83449	0.22164	0.00981
9/32	0.281250	7.144	0.06213	0.88357	0.24852	0.01165
19/64	0.296875	7.541	0.06922	0.93266	0.27688	0.01369
5/16	0.312500	7.938	0.07670	0.98175	0.30680	0.01598
21/64	0.328125	8.334	0.08456	1.0308	0.33824	0.01849
11/32	0.343750	8.731	0.09281	1.0799	0.37124	0.02127
23/64	0.359375	9.128	0.10143	1.1290	0.40572	0.02430
3/8	0.375000	9.525	0.11045	1.1781	0.44180	0.02761
25/64	0.390625	9.922	0.11984	1.2272	0.47936	0.03120
13/32	0.406250	10.319	0.12962	1.2763	0.51848	0.03511
27/64	0.421875	10.716	0.13978	1.3254	0.55912	0.03931
7/16	0.437500	11.113	0.15033	1.3744	0.60132	0.04385
29/64	0.453125	11.509	0.16125	1.4235	0.64500	0.04870
15/32	0.468750	11.906	0.17257	1.4726	0.69028	0.05393
31/64	0.484375	12.303	0.18426	1.5217	0.73704	0.05949
1/2	0.500000	12.700	0.19635	1.5708	0.78540	0.06545
33/64	0.515625	13.097	0.20881	1.6199	0.83524	0.07177
17/32	0.531250	13.494	0.22165	1.6690	0.88660	0.07848
35/64	0.546875	13.891	0.23489	1.7181	0.93956	0.08562
9/16	0.562500	14.288	0.24850	1.7671	0.99400	0.09318
37/64	0.578125	14.684	0.26250	1.8162	1.05000	0.10115
19/32	0.593750	15.081	0.27688	1.8653	1.10752	0.10958
39/64	0.609375	15.478	0.29164	1.9144	1.16656	0.11846
5/8	0.625000	15.875	0.30680	1.9635	1.22720	0.12783
41/64	0.640625	16.272	0.32232	2.0126	1.28928	0.13765
21/32	0.656250	16.669	0.33824	2.0617	1.35296	0.14798
43/64	0.671875	17.066	0.35454	2.1108	1.41816	0.15880
11/16	0.687500	17.463	0.37122	2.1598	1.48488	0.17013
45/64	0.703125	17.859	0.38829	2.2089	1.55316	0.18200
23/32	0.718750	18.256	0.40574	2.2580	1.62296	0.19442
47/64	0.734375	18.653	0.42357	2.3071	1.69428	0.20737

TABLE 4. (*Continued*)

Fractional Diameter	Decimal Diameter, in.	Millimeter Diameter	Area of Circle in.²	Circum. of Circle, in.	Surface of Sphere, in.²	Volume of Sphere, in.³
3/4	0.750000	19.050	0.44179	2.3562	1.76716	0.22089
49/64	0.765625	19.447	0.46038	2.4053	1.84152	0.23496
25/32	0.781250	19.844	0.47937	2.4544	1.91748	0.24967
51/64	0.796875	20.241	0.49873	2.5035	1.99492	0.26495
13/16	0.812500	20.638	0.51849	2.5525	2.07396	0.28084
53/64	0.828125	21.034	0.53862	2.6016	2.15448	0.29736
27/32	0.843750	21.431	0.55914	2.6507	2.23656	0.31451
55/64	0.859375	21.828	0.58003	2.6998	2.32012	0.33230
7/8	0.875000	22.225	0.60132	2.7489	2.40528	0.35077
57/64	0.890625	22.622	0.62298	2.7980	2.49192	0.36989
29/32	0.906250	23.019	0.64504	2.8471	2.58016	0.38971
59/64	0.921875	23.416	0.66747	2.8962	2.66988	0.41021
15/16	0.937500	23.813	0.69029	2.9452	2.76116	0.43143
61/64	0.953125	24.209	0.71349	2.9943	2.85396	0.45335
31/32	0.968750	24.606	0.73708	3.0434	2.94832	0.47603
63/64	0.984375	25.003	0.76104	3.0925	3.04416	0.49943
1	1.000000	25.400	0.78540	3.1416	3.14160	0.52360
1/32	1.03125	26.193	0.83525	3.2398	3.34100	0.57418
1/16	1.06250	26.988	0.88664	3.3379	3.54656	0.62804
3/32	1.09375	27.781	0.93956	3.4361	3.75824	0.68509
1/8	1.12500	28.575	0.99403	3.5343	3.97612	0.74551
5/32	1.15625	29.369	1.05001	3.6325	4.20004	0.80937
3/16	1.18750	30.163	1.10754	3.7306	4.43016	0.87681
7/32	1.21875	30.956	1.16659	3.8288	4.66636	0.94785
1/4	1.25000	31.750	1.22719	3.9270	4.90876	1.0227
9/32	1.28125	32.544	1.28931	4.0252	5.15724	1.1013
5/16	1.31250	33.338	1.35297	4.1233	5.41188	1.1839
11/32	1.34375	34.131	1.41817	4.2215	5.67268	1.2705
3/8	1.37500	34.925	1.48490	4.3197	5.93960	1.3611
13/32	1.40625	35.719	1.55316	4.4179	6.21264	1.4560
7/16	1.43750	36.513	1.62296	4.5160	6.49184	1.5553
15/32	1.46875	37.306	1.69429	4.6142	6.77716	1.6589
1/2	1.50000	38.100	1.76715	4.7124	7.06860	1.7671
17/32	1.53125	38.894	1.84155	4.8106	7.36620	1.8798
9/16	1.56250	39.688	1.91748	4.9087	7.66992	1.9974
19/32	1.59375	40.481	1.99495	5.0069	7.97980	2.1195
5/8	1.62500	41.275	2.07395	5.1051	8.29580	2.2468
21/32	1.65625	42.069	2.15448	5.2033	8.61792	2.3788
11/16	1.68750	42.863	2.23655	5.3014	8.94620	2.5161
23/32	1.71875	43.656	2.32015	5.3996	9.28060	2.6584
3/4	1.75000	44.450	2.40530	5.4978	9.62120	2.8062
25/32	1.78125	45.244	2.49196	5.5960	9.96784	2.9589
13/16	1.81250	46.038	2.58016	5.6941	10.32064	3.1177
27/32	1.84375	46.831	2.66990	5.7923	10.67960	3.2817
7/8	1.87500	47.625	2.76117	5.8905	11.04468	3.4514
29/32	1.90625	48.419	2.85398	5.9887	11.41592	3.6268
15/16	1.93750	49.213	2.94832	6.0868	11.79320	3.8083
31/32	1.96875	50.006	3.04419	6.1850	12.17676	3.9954

TABLE 4. *(Continued)*

Fractional Diameter	Decimal Diameter, in.	Millimeter Diameter	Area of Circle in.^{2}	Circum. of Circle, in.	Surface of Sphere, in.^{2}	Volume of Sphere, in.^{3}
2	2.00000	50.800	3.1416	6.2832	12.5664	4.1888
1/32	2.03125	51.594	3.2406	6.3814	12.9624	4.3879
1/16	2.06250	52.388	3.3410	6.4795	13.3640	4.5939
3/32	2.09375	53.181	3.4430	6.5777	13.7720	4.8054
1/8	2.12500	53.975	3.5466	6.6759	14.1864	5.0243
5/32	2.15625	54.769	3.6516	6.7741	14.6064	5.2486
3/16	2.18750	55.563	3.7583	6.8722	15.0332	5.4809
7/32	2.21875	56.356	3.8664	6.9704	15.4656	5.7185
1/4	2.25000	57.150	3.9761	7.0686	15.9044	5.9641
9/32	2.28125	57.944	4.0873	7.1668	16.3492	6.2155
5/16	2.31250	58.738	4.2000	7.2649	16.8000	6.4751
11/32	2.34375	59.531	4.3143	7.3631	17.2572	6.7410
3/8	2.37500	60.325	4.4301	7.4613	17.7204	7.0144
13/32	2.40625	61.119	4.5475	7.5595	18.1900	7.2942
7/16	2.43750	61.913	4.6664	7.6576	18.6656	7.5829
15/32	2.46875	62.706	4.7868	7.7558	19.1472	7.8775
1/2	2.50000	63.500	4.9087	7.8540	19.6348	8.1813
17/32	2.53125	64.294	5.0322	7.9522	20.1288	8.4910
9/16	2.56250	65.088	5.1572	8.0503	20.6288	8.8103
19/32	2.59375	65.881	5.2838	8.1485	21.1352	9.1357
5/8	2.62500	66.675	5.4119	8.2467	21.6476	9.4708
21/32	2.65625	67.469	5.5415	8.3449	22.1660	9.8121
11/16	2.68750	68.263	5.6727	8.4430	22.6908	10.164
23/32	2.71875	69.056	5.8054	8.5412	23.2216	10.521
3/4	2.75000	69.850	5.9396	8.6394	23.7584	10.889
25/32	2.78125	70.644	6.0754	8.7376	24.3016	11.264
13/16	2.81250	71.438	6.2126	8.8357	24.8504	11.649
27/32	2.84375	72.231	6.3515	8.9339	25.4060	12.040
7/8	2.87500	73.025	6.4918	9.0321	25.9672	12.443
29/32	2.90625	73.819	6.6337	9.1303	26.5348	12.852
15/16	2.93750	74.613	6.7771	9.2284	27.1084	13.272
31/32	2.96875	75.406	6.9221	9.3266	27.6884	13.699
3	3.00000	76.200	7.0686	9.4248	28.2744	14.137
1/16	3.06250	77.788	7.3662	9.6211	29.465	15.039
1/8	3.12500	79.375	7.6699	9.8175	30.680	15.979
3/16	3.18750	80.963	7.9798	10.014	31.919	16.957
1/4	3.25000	82.550	8.2958	10.210	33.183	17.974
5/16	3.31250	84.138	8.6179	10.407	34.472	19.031
3/8	3.37500	85.725	8.9462	10.603	35.784	20.129
7/16	3.43750	87.313	9.2806	10.799	37.122	21.268
1/2	3.50000	88.900	9.6211	10.996	38.484	22.449
9/16	3.56250	90.488	9.9678	11.192	39.872	23.674
5/8	3.62500	92.075	10.321	11.388	41.284	24.942
11/16	3.68750	93.663	10.680	11.585	42.720	26.254
3/4	3.75000	95.250	11.045	11.781	44.180	27.611
13/16	3.81250	96.838	11.416	11.977	45.664	29.016
7/8	3.87500	98.425	11.793	12.174	47.172	30.466
15/16	3.93750	100.013	12.177	12.370	48.708	31.965

TABLE 4. *(Continued)*

Fractional Diameter	Decimal Diameter, in.	Millimeter Diameter	Area of Circle in.²	Circum. of Circle, in.	Surface of Sphere, in.²	Volume of Sphere, in.³
4	4.00000	101.600	12.566	12.566	50.264	33.510
1/16	4.06250	103.188	12.962	12.763	51.848	35.102
1/8	4.12500	104.775	13.364	12.959	53.456	36.751
3/16	4.18750	106.363	13.772	13.155	55.088	38.443
1/4	4.25000	107.950	14.186	13.352	56.744	40.195
5/16	4.31250	109.538	14.607	13.548	58.428	41.991
3/8	4.37500	111.125	15.033	13.744	60.132	43.847
7/16	4.43750	112.713	15.466	13.941	61.864	45.749
1/2	4.50000	114.300	15.904	14.137	63.616	47.713
9/16	4.56250	115.888	16.349	14.334	65.396	49.723
5/8	4.62500	117.475	16.800	14.530	67.200	51.801
11/16	4.68750	119.063	17.257	14.726	69.028	53.923
3/4	4.75000	120.650	17.721	14.923	70.884	56.116
13/16	4.81250	122.238	18.190	15.119	72.760	58.354
7/8	4.87500	123.825	18.665	15.315	74.660	60.663
15/16	4.93750	125.413	19.147	15.512	76.588	63.019
5	5.00000	127.000	19.635	15.708	78.540	65.450
1/16	5.0625	128.588	20.129	15.904	80.516	67.929
1/8	5.1250	130.175	20.629	16.101	82.516	70.482
3/16	5.1875	131.763	21.135	16.297	84.540	73.085
1/4	5.2500	133.350	21.648	16.493	86.592	75.767
5/16	5.3125	134.938	22.166	16.690	88.664	78.497
3/8	5.3750	136.525	22.691	16.886	90.764	81.308
7/16	5.4375	138.113	23.221	17.082	92.884	84.168
1/2	5.5000	139.700	23.758	17.279	95.032	87.113
9/16	5.5625	141.288	24.301	17.475	97.204	90.107
5/8	5.6250	142.875	24.850	17.671	99.400	93.180
11/16	5.6875	144.463	25.406	17.868	101.624	96.322
3/4	5.7500	146.050	25.967	18.064	103.868	99.549
13/16	5.8125	147.638	26.535	18.261	106.140	102.813
7/8	5.8750	149.225	27.109	18.457	108.436	106.181
15/16	5.9375	150.813	27.688	18.653	110.752	109.588
6	6.0000	152.400	28.274	18.850	113.096	113.096
1/16	6.0625	153.988	28.867	19.046	115.468	116.659
1/8	6.1250	155.575	29.465	19.242	117.860	120.303
3/16	6.1875	157.163	30.069	19.439	120.276	124.022
1/4	6.2500	158.750	30.680	19.635	122.720	127.832
5/16	6.3125	160.338	31.297	19.831	125.188	131.695
3/8	6.3750	161.925	31.919	20.028	127.676	135.656
7/16	6.4375	163.513	32.548	20.224	130.192	139.671
1/2	6.5000	165.100	33.183	20.420	132.729	143.791
9/16	6.5625	166.688	33.824	20.617	135.296	147.965
5/8	6.6250	168.275	34.472	20.813	137.888	152.250
11/16	6.6875	169.863	35.125	21.009	140.500	156.583
3/4	6.7500	171.450	35.785	21.206	143.140	161.031
13/16	6.8125	173.038	36.450	21.402	145.800	165.527
7/8	6.8750	174.625	37.122	21.598	148.488	170.141
15/16	6.9375	176.213	37.801	21.795	151.204	174.812

TABLE 4. *(Continued)*

Fractional Diameter	Decimal Diameter, in.	Millimeter Diameter	Area of Circle in.²	Circum. of Circle, in.	Surface of Sphere, in.²	Volume of Sphere, in.³
7	7.0000	177.800	38.485	21.991		
1/8	7.1250	180.975	39.871	22.384		
1/4	7.2500	184.150	41.282	22.776		
3/8	7.3750	187.325	42.718	23.169		
1/2	7.5000	190.500	44.179	23.562		
5/8	7.6250	193.675	45.664	23.955		
3/4	7.7500	196.850	47.173	24.347		
7/8	7.8750	200.025	48.707	24.740		
8	8.0000	203.200	50.265	25.133		
1/8	8.1250	206.375	51.849	25.525		
1/4	8.2500	209.550	53.456	25.918		
3/8	8.3750	212.725	55.088	26.311		
1/2	8.5000	215.900	56.745	26.704		
5/8	8.6250	219.075	58.426	27.096		
3/4	8.7500	222.250	60.132	27.489		
7/8	8.8750	225.425	61.862	27.882		
9	9.0000	228.600	63.617	28.274		
1/8	9.125	231.775	65.397	28.667		
1/4	9.250	234.950	67.201	29.060		
3/8	9.375	238.125	69.029	29.452		
1/2	9.500	241.300	70.882	29.845		
5/8	9.625	244.475	72.760	30.238		
3/4	9.750	247.650	74.662	30.631		
7/8	9.875	250.825	76.589	31.023		
10	10.000	254.001	78.540	31.416		
1/8	10.125	257.176	80.516	31.809		
1/4	10.250	260.351	82.516	32.201		
3/8	10.375	263.526	84.541	32.594		
1/2	10.500	266.701	86.590	32.987		
5/8	10.625	269.876	88.664	33.379		
3/4	10.750	273.051	90.763	33.772		
7/8	10.875	276.226	92.886	34.165		
11	11.000	279.401	95.033	34.558		
1/8	11.125	282.576	97.205	34.950		
1/4	11.250	285.751	99.402	35.343		
3/8	11.375	288.926	101.62	35.736		
1/2	11.500	292.101	103.87	36.128		
5/8	11.625	295.276	106.14	36.521		
3/4	11.750	298.451	108.43	36.914		
7/8	11.875	301.626	110.75	37.306		
12	12.00	304.801	113.10	37.699		
1/4	12.25	311.150	117.86	38.485		
1/2	12.50	317.500	122.72	39.270		
3/4	12.75	323.850	127.68	40.055		

TABLE 4. *(Continued)*

Fractional Diameter	Decimal Diameter, in.	Millimeter Diameter	Area of Circle in.^{2}	Circum. of Circle, in.	Surface of Sphere, in.^{2}	Volume of Sphere, in.^{3}
13	13.00	330.201	132.73	40.841		
1/4	13.25	336.550	137.89	41.626		
1/2	13.50	342.900	143.14	42.412		
3/4	13.75	349.250	148.49	43.197		
14	14.00	355.601	153.94	43.982		
1/4	14.25	361.950	159.48	44.768		
1/2	14.50	374.650	165.13	45.553		
3/4	14.75	374.650	170.87	46.338		
15	15.00	381.001	176.71	47.124		
1/4	15.25	387.350	182.65	47.909		
1/2	15.50	393.700	188.69	48.695		
3/4	15.75	400.050	194.83	49.480		
16	16.00	406.401	201.06	50.265		
1/4	16.25	412.750	207.39	51.051		
1/2	16.50	419.100	213.82	51.836		
3/4	16.75	425.450	220.35	52.622		
17	17.00	431.801	226.98	53.407		
1/4	17.25	438.150	233.71	54.192		
1/2	17.50	444.500	240.53	54.978		
3/4	17.75	450.850	247.45	55.763		
18	18.00	457.201	254.47	56.549		
1/4	18.25	463.550	261.59	57.334		
1/2	18.50	469.900	268.80	58.119		
3/4	18.75	476.250	276.12	58.905		
19	19.00	482.601	283.53	59.690		
1/4	19.25	488.950	291.04	60.476		
1/2	19.50	495.300	298.65	61.261		
3/4	19.75	501.650	306.35	62.046		
20	20.00	508.001	314.16	62.832		
1/4	20.25	514.350	322.06	63.617		
1/2	20.50	520.700	330.06	64.403		
3/4	20.75	527.050	338.16	65.188		

TABLE 5. Draft Angles per Side[a]

Depth	1/4°	1/2°	1°	1 1/2°	2°	2 1/2°	3°	4°	5°	7°	10°	
1/32	0.0001	0.0003	0.0005	0.0008	0.0011	0.0014	0.0016	0.0022	0.0027	0.0038	0.0055	1/32
1/16	0.0003	0.0006	0.0011	0.0016	0.0022	0.0027	0.0033	0.0044	0.0055	0.0077	0.0110	1/16
3/32	0.0004	0.0008	0.0016	0.0025	0.0033	0.0041	0.0049	0.0066	0.0082	0.0115	0.0165	3/32
1/8	0.0005	0.0010	0.0022	0.0033	0.0044	0.0055	0.0066	0.0088	0.0109	0.0153	0.0220	1/8
3/16	0.0008	0.0016	0.0033	0.0049	0.0065	0.0082	0.0098	0.0130	0.0164	0.0263	0.0331	3/16
1/4	0.0011	0.0022	0.0044	0.0066	0.0087	0.0109	0.0131	0.0174	0.0219	0.0351	0.0441	1/4
5/16	0.0014	0.0027	0.0055	0.0082	0.0109	0.0137	0.0164	0.0218	0.0273	0.0384	0.0551	5/16
3/8	0.0016	0.0033	0.0065	0.0098	0.0131	0.0164	0.0197	0.0262	0.0328	0.0460	0.0661	3/8
7/16	0.0019	0.0038	0.0076	0.0115	0.0153	0.0191	0.0229	0.0306	0.0383	0.0537	0.0771	7/16
1/2	0.0022	0.0044	0.0087	0.0131	0.0175	0.0218	0.0262	0.0350	0.0438	0.0614	0.0882	1/2
9/16	0.0023	0.0050	0.0098	0.0147	0.0197	0.0245	0.0295	0.0394	0.0493	0.0691	0.0992	9/16
5/8	0.0027	0.0054	0.0109	0.0164	0.0218	0.0273	0.0328	0.0436	0.0547	0.0767	0.1102	5/8
11/16	0.0030	0.0060	0.0110	0.0180	0.0230	0.0300	0.0361	0.0480	0.0602	0.0844	0.1212	11/16
3/4	0.0033	0.0065	0.0131	0.0196	0.0262	0.0328	0.0393	0.0524	0.0656	0.0921	0.1322	3/4
13/16	0.0036	0.0071	0.0142	0.0212	0.0284	0.0355	0.0426	0.0568	0.0711	0.0998	0.1432	13/16
7/8	0.0038	0.0076	0.0153	0.0229	0.0306	0.0382	0.0459	0.0612	0.0766	0.1074	0.1543	7/8
15/16	0.0041	0.0082	0.0164	0.0245	0.0328	0.0409	0.0492	0.0656	0.0821	0.1151	0.1653	15/16
1	0.0044	0.0087	0.0175	0.0262	0.0349	0.0437	0.0524	0.0698	0.0875	0.1228	0.1763	1

[a]The table's use is illustrated in the following example. Find the difference in diameter from top to bottom of a round cavity 3/4 in. deep with a taper of 2 1/2° per side. From the table the taper per side is 0.0328 in., the difference being 2 × 0.0328 = 0.0656 in. Since the tables are additive, the same results could have been obtained by multiplying the 2 1/2° per side by 2 (one for each side) or 5° and read the results in the 5° column for 3/4 in. = 0.0656 in. Similarly, the taper for 1 1/4° can be found by adding the 1/4° and 1° columns.

TABLE 6. Conversion of Rockwell C and Brinell Hardness Scales

Rockwell C Scale 120° Cone 150 kg Load	Approximate Brinell Number	Rockwell C Scale 120° Cone 150 kg Load	Approximate Brinell Number
15	200	43	405
16	205	44	415
17	210	45	427
18	214	46	440
19	218	47	452
20	223	48	463
21	227	49	475
22	233	50	487
23	239	51	503
24	245	52	514
25	250	53	526
26	255	54	538
27	263	55	552
28	268	56	565
29	274	57	578
30	280	58	590
31	287	59	607
32	295	60	620
33	304	61	635
34	313	62	650
35	322	63	665
36	330	64	680
37	340	65	695
38	350	66	710
39	360	67	727
40	370	68	743
41	380	69	760
42	390	70	775

TABLE 7. Temperature Conversion

°C	*	°F	°C	*	°F	°C	*	°F	°C	*	°F
−168	−270	−454	−1.11	30	86.0	30.6	87	188.6	277	530	986
−162	−260	−436	−0.56	31	87.8	31.1	88	190.4	282	540	1004
−157	−250	−418	0	32	89.6	31.7	89	192.2	288	550	1022
−151	−240	−400	0.56	33	91.4	32.2	90	194.0	293	560	1040
−146	−230	−382	1.11	34	93.2	32.8	91	195.8	299	570	1058
−140	−220	−364	1.67	35	95.0	33.3	92	197.6	304	580	1076
−134	−210	−346	2.22	36	96.8	33.9	93	199.4	310	590	1094
−129	−200	−328	2.78	37	98.6	34.4	94	201.2	316	600	1112
−123	−190	−310	3.33	38	100.4	35.0	95	203.0	321	610	1130
−118	−180	−292	3.89	39	102.2	35.6	96	204.8	327	620	1148
−112	−170	−274	4.44	40	104.0	36.1	97	206.6	332	630	1166
−107	−160	−256	5.00	41	105.8	36.7	98	208.4	338	640	1184
−101	−150	−238	5.56	42	107.6	37.2	99	210.2	343	650	1202
−95.6	−140	−220	6.11	43	109.4	38	100	212	349	660	1220
−90.0	−130	−202	6.67	44	111.2	43	110	230	354	670	1238
−84.4	−120	−184	7.22	45	113.0	49	120	248	360	680	1256
−78.9	−110	−166	7.78	46	114.8	54	130	266	366	690	1274
−73.3	−100	−148	8.33	47	116.6	60	140	284	371	700	1292
−67.8	−90	−130	8.89	48	118.4	66	150	302	377	710	1310
−62.2	−80	−112	9.44	49	120.2	71	160	320	382	720	1328
−56.7	−70	−94	10.0	50	122.0	77	170	338	388	730	1346
−51.1	−60	−76	10.6	51	123.8	82	180	356	393	740	1364
−45.6	−50	−58	11.1	52	125.6	88	190	374	399	750	1382
−40.0	−40	−40	11.7	53	127.4	93	200	392	404	760	1400
−34.4	−30	−22	12.2	54	129.2	99	210	410	410	770	1418
−28.9	−20	−4	12.8	55	131.0	100	212	413.6	416	780	1436
−23.3	−10	14	13.3	56	132.8	104	220	428	421	790	1454
−17.8	0	32.0	13.9	57	134.6	110	230	446	427	800	1472
−17.2	1	33.8	14.4	58	136.4	116	240	464	432	810	1490
−16.7	2	35.6	15.0	59	138.2	121	250	482	438	820	1508
−16.1	3	37.4	15.6	60	140.0	127	260	500	443	830	1526
−15.6	4	39.2	16.1	61	141.8	132	270	518	449	840	1544
−15.0	5	41.0	16.7	62	143.6	138	280	536	454	850	1562
−14.4	6	42.8	17.2	63	145.4	143	290	554	460	860	1580
−13.9	7	44.6	17.8	64	147.2	149	300	572	466	870	1598
−13.3	8	46.4	18.3	65	149.0	154	310	590	471	880	1616
−12.8	9	48.2	18.9	66	150.8	160	320	608	477	890	1634
−12.2	10	50.0	19.4	67	152.6	166	330	626	482	900	1652
−11.7	11	51.8	20.0	68	154.4	171	340	644	488	910	1670
−11.1	12	53.6	20.6	69	156.2	177	350	662	493	920	1688
−10.6	13	55.4	21.1	70	158.0	182	360	680	499	930	1706
−10.0	14	57.2	21.7	71	159.8	188	370	698	504	940	1724
−9.44	15	59.0	22.2	72	161.6	193	380	716	510	950	1742
−8.89	16	60.8	22.8	73	163.4	199	390	734	516	960	1760
−8.33	17	62.6	23.3	74	165.2	204	400	752	521	970	1778
−7.78	18	64.4	23.9	75	167.0	210	410	770	527	980	1796
−7.22	19	66.2	24.4	76	168.8	216	420	788	532	990	1814
−6.67	20	68.0	25.0	77	170.6	221	430	806	538	1000	1832
−6.11	21	69.8	25.6	78	172.4	227	440	824	0.56	1	1.8
−5.56	22	71.6	26.1	79	174.2	232	450	842	1.1	2	3.6
−5.00	23	73.4	26.7	80	176.0	238	460	860	1.7	3	5.4
−4.44	24	75.2	27.2	81	177.8	243	470	878	2.2	4	7.2
−3.89	25	77.0	27.8	82	179.6	249	480	896	2.8	5	9.0
−3.33	26	78.8	28.3	83	181.4	254	490	914	3.4	6	10.8
−2.78	27	80.6	28.9	84	183.2	260	500	932	3.9	7	12.6
−2.22	28	82.4	29.4	85	185.0	266	510	950	4.5	8	14.4
−1.67	29	84.2	30.0	86	186.8	271	520	968	5.0	9	16.2

[a]°F = 9/5 °C + 32

°C = 5/9 (°F − 32)

°K (Kelvin) = °C + 273.16

°R (Rankine) = °F + 459.69

TABLE 8. Specific Gravity into Grams per Cubic Inch and Ounces per Cubic Inch[a]

Specific Gravity	g/in.^{3}	oz/in.^{3}	Specific Gravity	g/in.^{3}	oz/in.^{3}
0.90	14.75	0.522	1.20	19.66	0.693
0.91	14.91	0.526	1.21	19.83	0.699
0.92	15.08	0.532	1.22	19.99	0.705
0.93	15.24	0.537	1.23	20.16	0.711
0.94	15.40	0.543	1.24	20.32	0.717
0.95	15.57	0.549	1.25	20.48	0.722
0.96	15.73	0.545	1.26	20.65	0.728
0.97	15.90	0.561	1.27	20.81	0.734
0.98	16.06	0.566	1.28	20.98	0.740
0.99	16.22	0.572	1.29	21.14	0.745
1.00	16.39	0.578	1.30	21.30	0.751
1.01	16.55	0.584	1.35	22.12	0.780
1.02	16.72	0.589	1.40	22.94	0.809
1.03	16.88	0.595	1.45	23.76	0.838
1.04	17.04	0.601	1.50	24.58	0.867
1.05	17.21	0.607	1.55	25.40	0.896
1.06	17.37	0.613	1.60	26.22	0.925
1.07	17.53	0.618	1.65	27.04	0.953
1.08	17.70	0.624	1.70	27.86	0.982
1.09	17.86	0.630	1.75	28.68	1.011
1.10	18.03	0.636	1.80	29.50	1.040
1.11	18.19	0.641	1.85	30.32	1.069
1.12	18.35	0.647	1.90	31.14	1.098
1.13	18.52	0.653	1.95	31.96	1.127
1.14	18.68	0.659	2.00	32.77	1.156
1.15	18.85	0.665			
1.16	19.01	0.670			
1.17	19.17	0.676			
1.18	19.34	0.682			
1.19	19.50	0.688			

[a] 1 oz = 28.3 g
1 g = 0.0353 oz
Specific gravity $\times$ 16.387 = g/in.^{3}
Specific gravity $\times$ 0.5778 = oz/in.^{3}
Specific gravity $\times$ 0.0361 = lb/in.^{3}
Specific gravity $\times$ 62.4 = lb/ft^3
lb/ft^3 $\times$;0.01604 = specific gravity

TABLE 9. Conversion from Millimeters into Inches

Milli-meters	Inches	Milli-meters	Inches	Milli-meters	Inches	Milli-meters	Inches	Milli-meters	Inches
1	0.0394	51	2.0079	101	3.9764	151	5.9449	201	7.9134
2	0.0787	52	2.0472	102	4.0157	152	5.9842	202	7.9527
3	0.1181	53	2.0866	103	4.0551	153	6.0236	203	7.9921
4	0.1575	54	2.1260	104	4.0945	154	6.0630	204	8.0315
5	0.1968	55	2.1653	105	4.1338	155	6.1023	205	8.0708
6	0.2362	56	2.2047	106	4.1732	156	6.1417	206	8.1102
7	0.2756	57	2.2441	107	4.2126	157	6.1811	207	8.1496
8	0.3150	58	2.2835	108	4.2520	158	6.2205	208	8.1890
9	0.3543	59	2.3228	109	4.2913	159	6.2598	209	8.2283
10	0.3937	60	2.3622	110	4.3307	160	6.2992	210	8.2677
11	0.4331	61	2.4016	111	4.3701	161	6.3386	211	8.3071
12	0.4724	62	2.4409	112	4.4094	162	6.3779	212	8.3464
13	0.5118	63	2.4803	113	4.4488	163	6.4173	213	8.3858
14	0.5512	64	2.5197	114	4.4882	164	6.4567	214	8.4252
15	0.5905	65	2.5590	115	4.5275	165	6.4960	215	8.4645
16	0.6299	66	2.5984	116	4.5669	166	6.5354	216	8.5039
17	0.6693	67	2.6378	117	4.6063	167	6.5748	217	8.5433
18	0.7087	68	2.6772	118	4.6457	168	6.6142	218	8.5827
19	0.7480	69	2.7165	119	4.6850	169	6.6535	219	8.6220
20	0.7874	70	2.7559	120	4.7244	170	6.6929	220	8.6614
21	0.8268	71	2.7953	121	4.7638	171	6.7323	221	8.7008
22	0.8661	72	2.8346	122	4.8031	172	6.7716	222	8.7401
23	0.9055	73	2.8740	123	4.8425	173	6.8110	223	8.7795

n	value	n	value	n	value	n	value	n	value
24	0.9449	74	2.9134	124	4.8819	174	6.8504	224	8.8189
25	0.9842	75	2.9527	125	4.9212	175	6.8897	225	8.8582
26	1.0236	76	2.9921	126	4.9606	176	6.9291	226	8.8976
24	0.9449	74	2.9134	124	4.8819	174	6.8504	224	8.8189
25	0.9842	75	2.9527	125	4.9212	175	6.8897	225	8.8582
26	1.0236	76	2.9921	126	4.9606	176	6.9291	226	8.8976
27	1.0630	77	3.0315	127	5.0000	177	6.9685	227	8.9370
28	1.1024	78	3.0709	128	5.0394	178	7.0079	228	8.9764
29	1.1417	79	3.1102	129	5.0787	179	7.0472	229	9.0157
30	1.1811	80	3.1496	130	5.1181	180	7.0866	230	9.0551
31	1.2205	81	3.1890	131	5.1575	181	7.1260	231	9.0945
32	1.2598	82	3.2283	132	5.1968	182	7.1653	232	9.1338
33	1.2992	83	3.2677	133	5.2362	183	7.2047	233	9.1732
34	1.3386	84	3.3071	134	5.2756	184	7.2441	234	9.2126
35	1.3779	85	3.3464	135	5.3149	185	7.2834	235	9.2519
36	1.4173	86	3.3858	136	5.3543	186	7.3228	236	9.2913
37	1.4567	87	3.4252	137	5.3937	187	7.3622	237	9.3307
38	1.4961	88	3.4646	138	5.4331	188	7.4016	238	9.3701
39	1.5354	89	3.5039	139	5.4724	189	7.4409	239	9.4094
40	1.5748	90	3.5433	140	5.5118	190	7.4803	240	9.4488
41	1.6142	91	3.5827	141	5.5512	191	7.5197	241	9.4882
42	1.6535	92	3.6220	142	5.5905	192	7.5590	242	9.5275
43	1.6929	93	3.6614	143	5.6299	193	7.5984	243	9.5669
44	1.7323	94	3.7008	144	5.6693	194	7.6378	244	9.6063
45	1.7716	95	3.7401	145	5.7086	195	7.6771	245	9.6456
46	1.8110	96	3.7795	146	5.7480	196	7.7165	246	9.6850
47	1.8504	97	3.8189	147	5.7874	197	7.7559	247	9.7244
48	1.8898	98	3.8583	148	5.8268	198	7.7953	248	9.7638
49	1.9291	99	3.8976	149	5.8661	199	7.8346	249	9.8031
50	1.9685	100	3.9370	150	5.9055	200	7.8740	250	9.8425

TABLE 10. Conversion from Inches into Millimeters[a]

in.	mm	in.	mm	in.		mm	in.		mm
1	25.4	34	863.6	1/64	0.015625	0.396875	33/64	0.515625	13.096875
2	50.8	35	889.0	1/32	0.031250	0.793750	17/32	0.531250	13.493750
3	76.2	36	914.4	3/64	0.046875	1.190625	35/64	0.546875	13.890625
4	101.6	37	939.8	1/16	0.062500	1.587500	9/16	0.562500	14.287500
5	127.0	38	965.2	5/64	0.078125	1.984375	37/64	0.578125	14.684375
6	152.4	39	990.6	3/32	0.093750	2.381250	19/32	0.593750	15.081250
7	177.8	40	1016.0	7/64	0.109375	2.778125	39/64	0.609375	15.478125
8	203.2	41	1041.4	1/8	0.125000	3.175000	5/8	0.625000	15.875000
9	228.6	42	1066.8	9/64	0.140625	3.571875	41/64	0.640625	16.271875
10	254.0	43	1092.2	5/32	0.156250	3.968750	21/32	0.656250	16.668750
11	279.4	44	1117.6	11/64	0.171875	4.365625	43/64	0.671875	17.065625
12	304.8	45	1143.0	3/16	0.187500	4.762500	11/16	0.687500	17.462500
13	330.2	46	1168.4	13/64	0.203125	5.159375	45/64	0.703125	17.859375
14	355.6	47	1193.8	7/32	0.218750	5.556250	23/32	0.718750	18.256250
15	381.0	48	1219.2	15/64	0.234375	5.953125	47/64	0.734375	18.653125
16	406.4	49	1244.6	1/4	0.250000	6.350000	3/4	0.750000	19.050000
17	431.8	50	1270.0	17/64	0.265625	6.746875	49/64	0.765625	19.446875
18	457.2	51	1295.4	9/32	0.281250	7.143750	25/32	0.781250	19.843750
19	482.6	52	1320.8	19/64	0.296875	7.540625	51/64	0.796875	20.240625
20	508.0	53	1346.2	5/16	0.312500	7.937500	13/16	0.812500	20.637500
21	533.4	54	1371.6	21/64	0.328125	8.334375	53/64	0.828125	21.034375
22	558.8	55	1397.0	11/32	0.343750	8.731250	27/32	0.843750	21.431250
23	584.2	56	1422.4	23/64	0.359375	9.128125	55/64	0.859375	21.828125
24	609.6	57	1447.8	3/8	0.375000	9.525000	7/8	0.875000	22.225000
25	635.0	58	1473.2	25/64	0.390625	9.921875	57/64	0.890625	22.621875
26	660.4	59	1498.6	13/32	0.406250	10.318750	29/32	0.906250	23.018750
27	685.8	60	1524.0	27/64	0.421875	10.715625	59/64	0.921875	23.415625
28	711.2	61	1549.4	7/16	0.437500	11.112500	15/16	0.937500	23.812500
29	736.6	62	1574.8	29/64	0.453125	11.509375	61/64	0.953125	24.209375
30	762.0	63	1600.2	15/32	0.468750	11.906250	31/32	0.968750	24.606250
31	787.4	64	1625.6	31/64	0.484375	12.303125	63/64	0.984375	25.003125
32	812.8	65	1651.0	1/2	0.500000	12.700000	1	1.000000	25.400000
33	838.2	66	1676.4						

TABLE 10. (*Continued*)

in.	mm	in.	mm	in.	mm
0.001	0.0254	0.01	0.254	0.1	2.54
0.002	0.0508	0.02	0.508	0.2	5.08
0.003	0.0762	0.03	0.762	0.3	7.62
0.004	0.1016	0.04	1.016	0.4	10.16
0.005	0.1270	0.05	1.270	0.5	12.70
0.006	0.1524	0.06	1.524	0.6	15.24
0.007	0.1778	0.07	1.778	0.7	17.78
0.008	0.2032	0.08	2.032	0.8	20.32
0.009	0.2286	0.09	2.286	0.9	22.86

[a]Courtesy DME Co.

TABLE 11. Commonly Used Metric Conversions[a]

Area

1 inch^2	in.^{2}	= 645.6 mm^2
1 inch^2	in.^{2}	= 0.000 645 m^2
1 foot^2	ft^2	= 0.092 9 m^2
1 yard^2	yd^2	= 0.836 m^2

Density

1 gram per cc	g/cc	= 0.036 1 lb/in.^{3}
1 pound per cu ft	lb/ft^3	= 16.02 kg/m^3
1 pound per cu ft	lb/ft^3	= 62.37 g/cc
1 pound per cu in.	lb/in.^{3}	= 0.027 7 kg/m^3

Electrical

1 volt per mil	V/mil	= 25.4 kV/mm

Energy-Work

1 erg	dyn·cm	= 1 × 10^{-7} J
1 Newton-meter	N·m	= 1 J
1 watt-second	W·s	= 1 J
1 foot-pound	ft·lbf	= 1.356 J
1 horsepower-hour	hph	= 2.685 × 10^6 J
1 kilowatt-hour	kWh	= 3.6 × 10^6 J
1 calorie	cal	= 4.19 J
1 British thermal unit	Btu	= 1,055 J
1 British thermal unit	Btu	= 0.000 11 cal
1 inch-pound	in·lbf	= 0.133 J

Force

1 dyne	dyn	= 1 × 10^{-5} N
1 kilogram-force	kgf	= 9.81 N
1 pound-force	lbf	= 4.448 N
1 ton-force	tf	= 9,810 N

Impact Strength

1 foot pound per inch^2	ft·lb/in.^{2}	= 0.475 5 kJ/m^2
1 foot pound per inch of notch (Izod test)	ft·lb/in.	= 0.018 73 J/m

Length

1 kilometer	km	= 1,000 m
1 centimeter	cm	= 0.02 m
1 millimeter	mm	= 0.001 m
1 micron	μm	= 1 × 10^{-6} m
1 nanometer	nm	= 1 × 10^{-9} m
1 Angstrom unit	A	= 1 × 10^{-10} m
1 inch	in.	= 0.025 4 m
1 foot	ft	= 0.304 8 m
1 yard	yd	= 0.914 4 m
1 mile	mile	= 1,609 m
1 inch/inch per °F	in./in./°F	= 0.556 mm/mm/°C

Mass

1 gram	g	= 0.028 4 kg
1 ounce	oz	= 28.4 g
1 pound	lb	= 0.453 6 kg
1 ton	t	= 1,000 kg
1 ton US	tn	= 907 kg
1 ton UK	ton	= 1,016 kg

Power

1 kilowatt	kW	= 1,000 W
1 horsepower	hp	= 746 W
1 foot pound per second	ft·lbf/s	= 1.356 W

TABLE 11. *(Continued)*

Specific Energy

1 calorie per gram	cal/gm	= 4,190 J/kg
1 Btu per pound	Btu/lb	= 2,326 J/kg
1 hph per pound	hph/lb	= 5.92×10^6 J/kg
1 kWh per kg	kWh/kg	= 3.6×10^6 J/kg
1 kWh per kg	kWh/kg	= 1.644 hph/lb

Stress

1 dyne per cm^2	dyn/cm^2	= 0.1 Pa
1 Newton per meter^2	N/m^2	= 1 Pa
1 Joule per meter^3	J/m^3	= 1 Pa
1 atmosphere	atm	= 1.013×10^5 Pa
1 mm mercury	mm Hg	= 133.3 Pa
1 mm water	mm H$_2$O	= 9.81 Pa
1 bar	bar	= 1×10^5 Pa
1 pound per inch^2	psi	= 6,890 Pa
1 megapascal	MPa	= 145 psi
1 ton per inch^2	tn/in.^{2}	= 140.7 kg/cm^2

Temperature

1 degree Fahrenheit	°F	= 9/5 °C + 32
1 degree Celsius	°C	= 5/9 (°F + 32)
1 degree Kelvin	K	= °C + 273.16
1 degree Rankine	°R	= °F + 459.69

Thermal Conductivity

1 cal/cm·s·°C	= 419 J/ms·K
1 kcal/m·h·°C	= 1.163 J/ms·K
1 Btu/ft·hr·°F	= 1.73 J/ms·K
1 Btu/in.·ft^2·h·°F	= 0.144 J/ms·K
1 Btu/ft·s·°F	= 6,230 J/ms·K
1 W/m·K	= 1 J/ms·K

Viscosity

1 poise	poise	= 0.1 Pa·s
1 poise	poise	= 14.5×10^{-6} psi·s
1 poise	poise	= 1 dyn/cm^2
1 pound-second/in.^{2}	psi·s	= 6,897 Pa·s

Volume

1 cubic centimeter	cc	= 1×10^{-6} m^3
1 liter	L	= 0.001 m^3
1 milliliter	mL	= 1×10^{-6} m^3
1 cubic inch	in.^{3}	= 1.639×10^{-5} m^3
1 cubic foot	ft^3	= 0.028 32 m^3
1 gallon US	gal US	= 0.003 785 m^3
1 gallon UK	gal UK	= 0.004 546 m^3

A	=	Angstrom Unit	g	=	Gram	MPa	=	Megapascal
atm	=	Atmosphere	h	=	Hour	N	=	Newton
Btu	=	British Thermal Unit	Hg	=	Mercury	oz	=	Ounce
C	=	Degrees Celsius	Hp	=	Horsepower	Pa	=	Pascal
cal	=	Calorie	in.	=	Inch	psi	=	Pounds per Square Inch
cc	=	Cubic Centimeter	J	=	Joule	R	=	Degrees Rankine
cm	=	Centimeter	K	=	Degrees Kelvin	s	=	Second
cu	=	Cubic	kg	=	Kilogram	t	=	Ton
f	=	Force	L	=	Liter	tn	=	Ton US
dyn	=	Dyne	lb	=	Pound	ton	=	Ton UK
F	=	Degrees Farenheit Force	m	=	Meter	V	=	Volts
ft	=	Foot	mL	=	Milliliter	W	=	Watt
gal	=	Gallon	mm	=	Millimeter	yd	=	Yard

TABLE 12. Mensuration Formulas

Annulus	Area	= difference of the square of the diameters × 0.785
Circle	Area	= diameter^2 × 0.785
Circle	Circumference	= diameter × 3.14
Cone	Surface area	= area of base + 3.14 [(radius)(slant height)]
	Volume	= ⅓ (area of base)(vertical height)
Conic	Frustrum	= 0.2618 (height)($D^2 + Dd + d^2$)
Cube	Surface area	= 6 (length of side)^2
	Volume	= (length of side)^3
	Diagonal	= 1.732 (length of side)
Ellipse	Area	= 0.785 [axis (A) × axis (B)]
Hexagon	Area	= 0.866 (diameter inscribed circle)^2
Octagon	Area	= 0.828 (diameter inscribed circle)^2
Pyramid (regular)	Area of sides	= ½ (length of base)(vertical height)
Pyramid	Volume	= ⅓ (area of base)(vertical height)
Sphere	Area	= 3.14 (diameter)^2
	Volume	= 0.524 (diameter)^3

TABLE 13. Conversion Pounds into Kilograms

lb	0	1	2	3	4	5	6	7	8	9
0	0.00	0.45	0.91	1.36	1.81	2.27	2.72	3.18	3.63	4.08
10	4.54	4.99	5.44	5.90	6.35	6.80	7.26	7.71	8.16	8.62
20	9.07	9.53	9.98	10.43	10.89	11.34	11.79	12.25	12.70	13.15
30	13.61	14.06	14.51	14.97	15.42	15.88	16.33	16.78	17.24	17.69
40	18.14	18.60	19.05	19.50	19.96	20.41	20.87	21.32	21.77	22.23
50	22.68	23.13	23.59	24.04	24.49	24.95	25.40	25.85	26.31	26.76
60	27.22	27.67	28.12	28.58	29.03	29.48	29.94	30.39	30.84	31.30
70	31.75	32.21	32.66	33.11	33.57	34.02	34.47	34.93	35.38	35.83
80	36.29	36.74	37.19	37.65	38.10	38.56	39.01	39.46	39.92	40.37
90	40.82	41.28	41.73	42.18	42.64	43.09	43.54	44.00	44.45	44.91

TABLE 14. Conversion Kilograms into Pounds

kg	0	1	2	3	4	5	6	7	8	9
0	0.0	2.2	4.4	6.6	8.8	11.0	13.2	15.4	17.6	19.8
10	22.0	24.3	26.5	28.7	30.9	33.1	35.3	37.5	39.7	41.9
20	44.1	46.3	48.5	50.7	52.9	55.1	57.3	59.5	61.7	63.9
30	66.1	68.3	70.5	72.8	75.0	77.2	79.4	81.6	83.8	86.0
40	88.2	90.4	92.6	94.8	97.0	99.2	101.4	103.6	105.8	108.0
50	110.2	112.4	114.6	116.8	119.0	121.3	123.5	125.7	127.9	130.1
60	132.3	134.5	136.7	138.9	141.1	143.3	145.5	147.7	149.9	152.1
70	154.3	156.5	158.7	160.9	163.1	165.3	167.6	169.8	172.0	174.2
80	176.4	178.6	180.8	183.0	185.2	187.4	189.6	191.8	194.0	196.2
90	198.4	200.6	202.8	205.0	207.2	209.4	211.6	213.8	216.1	218.3

TABLE 15. Conversion from MPa to psi

MPa	psi	MPa	psi	MPa	psi	MPa	psi
1	145	26	3,770	51	7,395	76	11,020
2	290	27	3,915	52	7,540	77	11,165
3	435	28	4,060	53	7,685	78	11,310
4	580	29	4,205	54	7,830	79	11,455
5	725	30	4,350	55	7,975	80	11,600
6	870	31	4,495	56	8,120	81	11,745
7	1,015	32	4,640	57	8,265	82	11,890
8	1,160	33	4,785	58	8,410	83	12,035
9	1,305	34	4,930	59	8,555	84	12,180
10	1,450	35	5,075	60	8,700	85	12,325
11	1,595	36	5,220	61	8,845	86	12,470
12	1,740	37	5,365	62	8,990	87	12,615
13	1,885	38	5,510	63	9,135	88	12,760
14	2,030	39	5,655	64	9,280	89	12,905
15	2,175	40	5,800	65	9,425	90	13,050
16	2,320	41	5,945	66	9,570	91	13,195
17	2,465	42	6,090	67	9,715	92	13,340
18	2,610	43	6,235	68	9,860	93	13,485
19	2,755	44	6,380	69	10,005	94	13,630
20	2,900	45	6,525	70	10,150	95	13,775
21	3,045	46	6,670	71	10,295	96	13,920
22	3,190	47	6,815	72	10,440	97	14,065
23	3,335	48	6,960	73	10,585	98	14,210
24	3,480	49	7,105	74	10,730	99	14,355
25	3,625	50	7,250	75	10,875	100	14,500

TABLE 16. Conversion from psi to MPa

psi	MPa	psi	MPa	psi	MPa	psi	MPa
100	0.69	2,600	17.94	5,100	35.19	7,600	52.44
200	1.38	2,700	18.63	5,200	35.88	7,700	53.13
300	2.07	2,800	19.32	5,300	36.57	7,800	53.82
400	2.76	2,900	20.01	5,400	37.26	7,900	54.51
500	3.45	3,000	20.70	5,500	37.95	8,000	55.20
600	4.14	3,100	21.39	5,600	38.64	8,100	55.89
700	4.83	3,200	22.08	5,700	39.33	8,200	56.58
800	5.52	3,300	22.77	5,800	40.02	8,300	57.27
900	6.21	3,400	23.46	5,900	40.71	8,400	57.96
1,000	6.90	3,500	24.15	6,000	41.40	8,500	58.65
1,100	7.59	3,600	24.84	6,100	42.09	8,600	59.34
1,200	8.28	3,700	25.53	6,200	42.78	8,700	60.03
1,300	8.97	3,800	26.22	6,300	43.47	8,800	60.72
1,400	9.66	3,900	26.91	6,400	44.16	8,900	61.41
1,500	10.35	4,000	27.60	6,500	44.85	9,000	62.10
1,600	11.04	4,100	28.29	6,600	45.54	9,100	62.79
1,700	11.73	4,200	28.98	6,700	46.23	9,200	63.48
1,800	12.42	4,300	29.67	6,800	46.92	9,300	64.17
1,900	13.11	4,400	30.36	6,900	47.61	9,400	64.86
2,000	13.80	4,500	31.05	7,000	48.30	9,500	65.55
2,100	14.49	4,600	31.74	7,100	48.99	9,600	66.24
2,200	15.18	4,700	32.43	7,200	49.68	9,700	66.93
2,300	15.87	4,800	33.12	7,300	50.37	9,800	67.62
2,400	16.56	4,900	33.81	7,400	51.06	9,900	68.31
2,500	17.25	5,000	34.50	7,500	51.75	10,000	69.00

TABLE 17. SI Prefixes

Multiplication Factor	Prefix	Symbol
1 000 000 000 000 000 000 = 10^{18}	exa	E
1 000 000 000 000 000 = 10^{15}	peta	P
1 000 000 000 000 = 10^{12}	tera	T
1 000 000 000 = 10^{9}	giga	G
1 000 000 = 10^{6}	mega	M
1 000 = 10^{3}	kilo	k
100 = 10^{2}	hecto[a]	h
10 = 10^{1}	deka[a]	da
0.1 = 10^{-1}	deci[a]	d
0.01 = 10^{-2}	centi[a]	c
0.001 = 10^{-3}	milli	m
0.000 001 = 10^{-6}	micro	μ
0.000 000 001 = 10^{-9}	nano	n
0.000 000 000 001 = 10^{-12}	pico	p
0.000 000 000 000 001 = 10^{-15}	femto	f
0.000 000 000 000 000 001 = 10^{-18}	atto	a

[a]To be avoided where practical, except as noted in 3.2.2.

TABLE 18. Typical Ranked Properties of Unfilled Resin, Density[a]

Material	Density, lb/ft^3	Density, g/cc
Poly(methyl pentene) (TPX)	52.0	0.834
Polyallomers	56.0	0.898
Polypropylene	56.4	0.903
Polyethylene (LLDPE)	57.2	0.918
Polyethylene (LDPE)	57.7	0.925
Polyethylene (UHMW)	58.1	0.930
Polybutylene (pipe grade)	58.5	0.935
Ionomers	58.6	0.940
Ethylene vinyl acetate	59.0	0.944
Polyethylene (HDPE)	59.5	0.954
Styrene–butadiene	63.0	1.010
Rubbery styrenic block polymers	63.5	1.015
Acrylonitrile–butadiene–styrene	63.9	1.025
Polystyrene (GP)	65.4	1.050
Polystyrene (HI)		1.050
Poly(phenylene ether)	67.0	1.080
Styrene–acrylonitrile		1.080
Poly(phenylene oxide) (PPO)		1.080
Alkyd polyesters cast, flexible	69.0	1.100
Styrene maleic anhydride		1.100
XT polymer	69.8	1.120
Nylon 6	70.4	1.130
Nylon 6,6	71.1	1.140
Poly(methyl methacrylate) (IM)	73.5	1.180
Cellulose butyrate		1.180
Polyurethane (thermoplastic)	75.0	1.200
Polycarbonate		1.200
Polyarylate		1.200
Vinyl esters	76.0	1.220
Cellulose proprionate		1.220

TABLE 18. *(Continued)*

Material	Density, lb/ft^3	Density, g/cc
Chlorinated polyethylene	76.4	1.170
Polysulfone	77.3	1.240
Epoxy cast	78.0	1.250
Polyurethane (thermoset)	79.0	1.270
Polyetherimide		1.270
Cellulose Acetate	79.8	1.280
Alkyd polyesters cast rigid	80.0	1.290
Poly(phenyl sulfone)		1.290
Thermoplastic polyester (PBT)	81.7	1.310
Polyetherketone	82.4	1.320
Poly (aryl sulfone)	85.0	1.370
Polyethersulfone		1.370
Thermoplastic polyester (PET)	87.3	1.400
Polyamid–imid		1.400
Phenolics, GP, IM		1.400
Acetal, copolymer	88.0	1.410
Acetal, homopolymer	89.0	1.420
Poly(vinyl chloride), flexible	90.0	1.440
Poly(vinyl chloride), rigid	91.0	1.450
Urea, alpha cellulose-filled	93.5	1.500
Melamine, alpha cellulose-filled		1.500
Poly(phenylene sulfide)	98.0	1.570
Phenolics + mineral + cellulose (CM)	98.5	1.580
Liquid crystal polymers	106.0	1.700
Fluorocarbon polymers (ETFE)		1.700
Poly(vinylidine chloride) (Saran)		1.700
Fluorocarbon polymers (PVDF)	110.0	1.770
Diallyl phthalate	112.0	1.800
Fluorocarbon polymers (PCTFE)	133.0	2.130
Fluorocarbon polymers (PFA)		2.130
Fluorocarbon polymers (FEP)	134.0	2.150
Fluorocarbon polymers (PTFE)	137.0	2.200

[a]Fillers and reinforcements have a major effect. Significant variations exist in each material. Consult material manufacturers for further information.

TABLE 19. Typical Ranked Properties of Unfilled Resin, Tensile Strength[a]

Materials	Tensile Strength	
	psi	MPa
Rubbery styrenic block polymers	750	5
Alkyd polyesters cast, flexible	2,000	14
Polyethylene (LDPE)		14
Chlorinated polyethylene		14
Ethylene vinyl acetate	2,500	17
Poly(vinyl chloride) (flexible)		19
Polyethylene (LLDPE)		19
Fluorocarbon polymers (PTFE)	3,000	21
Fluorocarbon polymers (FEP)		21
Polyethylene (HDPE)	3,500	24
Polyallomers		24
Styrene–butadiene	3,800	26
Polystyrene (HI)	4,000	28

TABLE 19. *(Continued)*

Materials	Tensile Strength	
	psi	MPa
Polyethylene (UHMW)	4,500	31
Cellulose butyrate		31
Fluorocarbon polymers (PFA)		31
Ionomers	4,800	33
Polybutylene (pipe grade)	5,000	35
Poly(vinlylidine chloride) (Saran)		35
Polypropylene		35
Acrylonitrile–Butadiene–Styrene	5,500	38
Cellulose acetate	5,800	40
Fluorocarbon polymers (PCTFE)		40
Polyurethane (thermoplastic)	6,300	43
Urea, alpha cellulose-filled		43
Polystyrene GP	6,500	45
Fluorocarbon polymers (PVDF)	6,700	46
Poly(vinyl chloride) (Rigid)	6,800	47
Styrene maleic anhydride	7,000	48
Phenolics, GP, IM		48
Polyphenylene oxide (PPO)		48
Fluorocarbon polymers (ETFE)		48
Diallyl phthalate	8,000	55
XT polymer		55
Thermoplastic polyester (PBT)		55
Melamine, alpha cellulose-filled		55
Phenolics + minerals + cellulose, CM	8,500	59
Epoxy cast		59
Acetal copolymer	8,800	61
Polycarbonate	9,000	62
Poly(methyl methacrylate) IM		62
Alkyd polyesters cast rigid	9,500	66
Polyarylate		66
Styrene–acrylonitrile	10,000	69
Acetal homopolymer		69
Polysulfone		69
Poly(phenyl sulfone)	10,500	72
Vinyl esters	11,000	76
Poly(phenylene ether)		76
Nylon 6	11,600	80
Poly(aryl sulfone)	12,000	83
Polyethersulfone	12,200	84
Poly(phenylene sulfide)	12,500	86
Nylon 6,6	13,000	90
Polyurethane (thermoset)		90
Polyetherketone	13,300	92
Liquid crystal polymers	13,600	94
Polyetherimide	15,200	105
Thermoplastic polyester (PET)	18,000	124
Polyamid–imid	27,000	186
Poly(methyl pentene) (TPX)	28,000	193

[a]Significant variations exist in each material. Fillers and reinforcements may have a major effect. Consult material manufacturers for further information.

TABLE 20. Typical Ranked Properties of Unfilled Resin, Tensile Modulus[a]

Material	Tensile Modulus	
	psi	MPa
Rubbery styrenic block polymers	800	5
Polyurethane (thermoplastic)	900	6
Poly(vinyl chloride) (flexible)	1,000	7
Alkyd polyesters cast rigid	4,700	33
Polyethylene (LLDPE)	24,000	162
Polyethylene (LDPE)	32,500	224
Polyphenylene ether	37,000	755
Ionomers	50,000	345
Polybutylene (pipe grade)	55,000	380
Fluorocarbon polymers (PTFE)	60,000	410
Polyurethane thermoset		410
Poly(vinylidine chloride) (Saran)	75,000	520
Fluorocarbon polymers (FEP)	80,000	550
Fluorocarbon polymers (PFA)		550
Poly(methyl pentene) (TPX)	116,000	800
Fluorocarbon polymers (ETFE)	120,000	830
Cellulose butyrate	125,000	860
Polypropylene	200,000	1,380
Styrene–butadiene		1,380
Fluorocarbon polymers (PCTFE)	207,000	1,400
Fluorocarbon polymers (PVDF)	230,000	1,600
Acrylonitrile–butadiene–styrene	280,000	1,960
Polyarylate	290,000	2,000
Poly(phenyl sulfone)	310,000	2,100
Polystyrene HI		2,100
Polycarbonate	345,000	2,400
Styrene maleic anhydride	350,000	2,400
Polyethersulfone		2,400
Epoxy cast		2,400
Thermoplastic polyester (PBT)	360,000	2,500
Poly(phenylene oxide) (PPO)		2,500
Polysulfone		2,500
Poly(aryl sulfone)	385,000	2,700
Poly(methyl methacrylate) IM	398,000	2,700
Acetal copolymer	410,000	2,800
XT polymer	430,000	3,000
Nylon 6		3,000
Polyetherimide		3,000
Acetal homopolymer	450,000	3,100
Polystyrene GP		3,100
Thermoplastic polyester (PET)		3,100
Vinyl esters	460,000	3,200
Poly(phenylene sulfide)	480,000	3,300
Nylon 6,6	493,000	3,400
Polyetherketone	520,000	3,500
Styrene–acrylonitrile		3,500
Polyamid–imid	650,000	4,500
Phenolics, GP, IM	1,200,000	8,300
Urea, alpha cellulose-filled	1,300,000	9,000
Melamine, alpha cellulose-filled	1,350,000	9,300
Phenolics + minerals + cellulose, CM	1,500,000	10,300
Diallyl phthalate	1,600,000	11,000
Liquid crystal polymers	1,870,000	12,900

[a]Significant variations exist in each material. Fillers and reinforcements may have a major effect. Consult material manufacturer for further information.

TABLE 21. Typical Ranked Properties of Unfilled Resin, Flexural Modulus

Material	Flexural Modulus	
	psi	MPa
Rubbery styrenic block polymers	1,300	9
Ethylene vinyl acetate	8,500	59
Poly(vinyl chloride) (rigid)	13,000	90
Alkyd polyesters cast rigid	16,000	110
Ionomers	32,000	220
Poly(phenylene ether)	36,000	250
Polyethylene (LLDPE)	45,000	310
Polybutylene (pipe grade)	55,000	380
Polyurethane (thermoplastic)		380
Fluorocarbon polymers (PTFE)	60,000	413
Polyurethane (thermoset)		413
Fluorocarbon polymers (FEP)	95,000	655
Fluorocarbon polymers (PFA)	100,000	690
Poly(methyl pentene) (TPX)	109,000	750
Polyallomers		750
Polyethylene (HDPE)	170,000	1,170
Cellulose butyrate	175,000	1,200
Fluorocarbon polymers (PCTFE)	190,000	1,250
Fluorocarbon polymers (ETFE)	200,000	1,400
Fluorocarbon polymers (PVDF)	218,000	1,500
Styrene–butadiene	230,000	1,600
Polypropylene	245,000	1,700
Cellulose acetate	280,000	1,900
Polyarylate	300,000	2,100
Acrylonitrile–butadiene–styrene		2,100
Poly(phenylene oxide) (PPO)	325,000	2,200
Thermoplastic polyester (PBT)	330,000	2,300
Poly(phenyl sulfone)		2,300
Polystyrene, HI		2,300
Polycarbonate		2,300
Nylon 6	350,000	2,400
Acetal copolymer	375,000	2,600
Polyethersulfone		2,600
Polysulfone	390,000	2,700
Poly(methyl methacrylate) IM		2,700
XT polymer	400,000	2,750
Poly(aryl sulfone)		2,750
Styrene maleic anhydride	405,000	2,800
Acetal homopolymer		2,800
Nylon 6,6	420,000	2,900
Thermoplastic polyester (PET)	450,000	3,100
Polystyrene, GP		3,100
Polyetherimide	480,000	3,300
Styrene–acrylonitrile	490,000	3,380
Vinyl esters	500,000	3,450
Polyetherketone	530,000	3,650
Polyamid–imid	710,000	4,900
Melamine, alpha cellulose-filled	1,100,000	7,600
Poly(phenylene sulfide)	1,200,000	8,300
Liquid crystal polymers	1,320,000	9,100
Urea, alpha cellulose-filled	1,500,000	10,300
Diallyl phthalate	1,700,000	11,800
Polyethylene (UHMW)	16,000,000	110,000

*a*Significant variations exist in each material. Fillers and reinforcements may have a major effect. Consult material manufacturer for further information.

TABLE 22. Typical Ranked Properties of Unfilled Resin, Elongation to Failure[a]

Elongation to failure	%
Melamine, alpha cellulose-filled	1
Urea, alpha cellulose-filled	
Phenolics + minerals + cellulose, CM	
Polystyrene HI	2
Polyallomers	
Poly(phenylene sulfide)	
Liquid crystal polymers	
Polystyrene GP	3
Acrylonitrile–butadiene–styrene	
Styrene–acrylonitrile	
Polyamid–imid	
Poly(phenylene ether)	
Diallyl phthalate	4
XT polymer	
Epoxy cast	5
Fluorocarbon polymers (PCTFE)	
Alkyd polyesters cast rigid	
Poly(aryl sulfone)	6
Vinyl esters	
Thermoplastic polyester (PET)	7
Polyetherimide	8
Fluorocarbon polymers (PVDF)	15
Styrene maleic anhydride	
Nylon 6,6	20
Poly(vinylidine chloride) (Saran)	25
Fluorocarbon polymers (PFA)	30
Fluorocarbon polymers (FEP)	35
Cellulose acetate	38
Fluorocarbon polymers (PTFE)	40
Polyetherketone	50
Polyarylate	
Acetal homopolymer	
Poly(vinyl chloride) (rigid)	60
Acetal copolymer	
Poly(phenyl sulfone)	
Polyethersulfone	
Cellulose butyrate	64
Nylon 6	75
Polysulfone	
Styrene–butadiene	80
Polycarbonate	110
Thermoplastic polyester (PBT)	175
Alkyd polyesters cast, flexible	
Polybutylene (pipe grade)	280
Poly(vinyl chloride) flexible	300
Polyethylene (UHMW)	
Ionomers	400
Polypropylene	
Polyethylene (LDPE)	
Polyethylene (HDPE)	600
Polyethylene (LLDPE)	700
Rubbery styrenic block polymers	800
Ethylene vinyl acetate	
Chlorinated polyethylene	
Polyurethane (thermoplastic)	
Polyurethane (thermoset)	1000

[a]Significant variations exist in each material. Fillers and reinforcements may have a major effect. Consult material manufacturer for further information.

TABLE 23. Typical Ranked Properties of Unfilled Resin, Notched Izod Impact Strength[a]

Material	Notched Izod Impact Strength	
	ft.lb/in. notch	J/m
Alkyd polyesters cast rigid	0.30	16
Phenolics, GP, IM		16
Urea, alpha cellulose-filled	0.31	17
Styrene–acrylonitrile	0.35	19
Polystyrene GP		19
Vinyl esters	0.39	21
Styrene–butadiene		21
Poly(phenylene sulfide)	0.40	21
Epoxy cast	0.60	32
Diallyl phthalate		32
Poly(methyl methacrylate) IM		32
Poly(vinylidine chloride) (Saran)	0.70	37
Polypropylene		37
Nylon 6,6		37
Thermoplastic polyester (PBT)	1.00	53
Nylon 6		53
XT polymer		53
Melamine, alpha cellulose-filled		53
Polyetherimide		53
Polysulfone	1.30	70
Acetal copolymer		70
Liquid crystal polymers	1.60	85
Poly(aryl sulfone)		85
Polyetherketone		85
Polyethylene (HDPE)	1.70	91
Acetal homopolymer	1.80	96
Polyethersulfone	2.00	107
Polyallomers		107
Polycarbonate	2.30	123
Polystyrene HI	2.50	133
Cellulose butyrate	2.90	155
Polyamid–imid	3.00	175
Cellulose acetate	3.50	187
Polyarylate	4.20	224
Fluorocarbon polymers (PCTFE)	5.00	267
Poly(phenylene ether)	6.00	321
Fluorocarbon polymers (PVDF)	6.40	340
Styrene maleic anhydride	7.00	370
Alkyd polyesters cast, flexible		370
Poly(phenylene oxide) (PPO)		370
Acrylonitrile–butadiene–styrene	8.30	440
Ionomers	11.40	608
Poly(phenyl sulfone)	12.00	640
Poly(vinyl chloride) (rigid)		640
Polybutylene (pipe grade)	13.50	720

TABLE 24. Typical Ranked Properties of Unfilled Resin, Dielectric Strength[a]

Material	Dielectric Strength	
	V/mil	kV/mm
Phenolics, GP, IM	230	9
Polyurethane, thermoset		9
Styrene–butadiene	300	12
Melamine, alpha cellulose-filled		12
Urea, alpha cellulose-filled	350	14
Polycarbonate	380	15
Poly(aryl sulfone)		15
Poly(phenyl sulfone)		15
Phenolics + minerals + cellulose, CM		15
Polyethersulfone	400	16
Polyarylate		16
Alkyd polyesters cast, flexible		16
Poly(vinyl chloride) (flexible)		16
Styrene–acrylonitrile		16
Ionomers		16
Diallyl phthalate		16
Thermoplastic polyester (PBT)		16
Styrene maleic anhydride	420	17
Polysulfone		17
Polyetherketone	480	19
Epoxy cast	500	20
Polystyrene (HI)		20
Acetal homopolymer		20
Alkyd polyesters cast rigid		20
Polystyrene (GP)		20
Acetal copolymer		20
Polyethylene (UHMW)		20
Chlorinated polyethylene		20
Liquid crystal polymers		20
Polyamid–imid		20
Poly(methyl methacrylate) (IM)		20
Nylon 6		20
Nylon 6,6	600	24
Polyethylene (HDPE)		24
Poly(phenylene sulfide)		24
Acrylonitrile–butadiene–styrene		24
Poly(phenylene oxide) (PPO)	640	25
Fluorocarbon polymers (PTFE)		25
Thermoplastic polyester (PET)	650	26
Polypropylene	700	28
Poly(phenylene ether)		28
Polyetherimide	830	33
Rubbery styrenic block polymers	900	35
Polyallomers	950	37
Fluorocarbon polymers (FEP)	1300	51
Fluorocarbon polymers (PVDF)		51
Fluorocarbon polymers (ETFE)	1500	60
Polymethyl pentene (TPX)	1650	65
Fluorocarbon polymers (PCTFE)	2600	102
Fluorocarbon polymers (PFA)		102

[a]Significant variations exist in each material. Fillers and reinforcements may have a major effect. Consult material manufacturer for further information.

TABLE 25. Typical Ranked Properties of Unfilled Resin, Deflection Temperature[a]

Material	Deflection Temperature at 264 psi	
	°F	°C
Ionomers	95	35
Poly(methyl pentene) (TPX)	104	40
Polyethylene (LDPE)	108	42
Polyethylene (UHMW)	115	46
Polyallomers	120	49
Poly(vinylidine chloride) (Saran)		49
Fluorocarbon polymers (PFA)	122	50
Fluorocarbon polymers (FEP)	129	54
Thermoplastic polyester (PBT)	130	54
Polypropylene	131	55
Fluorocarbon polymers (PTFE)	132	56
Polybutylene (pipe grade)	135	58
Cellulose butyrate	136	58
Cellulose acetate	142	61
Nylon 6	149	65
Fluorocarbon polymers (PCTFE)	158	70
Poly(vinyl chloride) (rigid)	160	71
Fluorocarbon polymers (ETFE)	165	74
Cellulose proprionate	167	75
Styrene–butadiene	171	77
Polystyrene HI	180	82
Fluorocarbon polymers (PVDF)	185	85
Poly(methyl methacrylate) (IM)	188	87
XT polymer		87
Poly(phenylene oxide) (PPO)	190	88
Acrylonitrile–butadiene–styrene	210	99
Polystyrene (GP)	212	100
Styrene–acrylonitrile	220	104
Nylon 6,6		104
Acetal copolymer	230	110
Styrene maleic anhydride	232	111
Poly(phenylene ether)	240	116
Vinyl esters		116
Urea, alpha cellulose-filled	266	130
Polycarbonate	270	132
Acetal homopolymer	277	136
Phenolics, GP, IM	302	150
Polyetherketone	320	160
Diallyl phthalate	325	163
Epoxy cast	340	171
Polysulfone	345	174
Polyarylate		174
Phenolics + minerals + cellulose, CM	361	182
Melamine, alpha cellulose-filled		182
Polyetherimide	392	200
Polyethersulfone	397	203
Poly(phenylene sulfide)	400	204
Poly(aryl sulfone)		204
Poly(phenyl sulfone)		204
Polyamid–imid	500	260
Liquid crystal polymers	606	319

INDEX